MODERN TEXTBOOK OF ZOOLOGY

INVERTEBRATES

[ANIMAL DIVERSITY-1]

12TH EDITION

R.L. KOTPAL

FORMERLY – HEAD,
DEPARTMENT OF ZOOLOGY

MEERUT COLLEGE
MEERUT

RASTOGI
PUBLICATIONS

SHIVAJI ROAD, MEERUT- 250002 , INDIA

MODERN TEXTBOOK OF ZOOLOGY

INVERTEBRATES

[ANIMAL DIVERSITY-1]

A TEXTBOOK FOR UNIVERSITY STUDENTS

12TH EDITION

MODERN TEXTBOOK OF ZOOLOGY

INVERTEBRATES

[ANIMAL DIVERSITY-1]

12TH EDITION

ISBN : 978-81-938875-5-4

TITLE CODE NO. : **Z–1**

Note : All possible care has been taken in the editing, proof reading and printing of this book but in case of any ommission/mistake which might have crept in the book, neither the author nor the publisher shall be held responsible for the same. The publisher shall feel grateful for the suggestions received from the readers for further improvement of the book.

TWELFTH EDITION (REPRINT) : **2026–2027**

PRICE : **₹ 925.00** ONLY

PUBLISHED BY RAKESH KUMAR RASTOGI FOR RASTOGI PUBLICATIONS, 'GANGOTRI' SHIVAJI ROAD, MEERUT - 250002
PRINTED AT INDIAN PRESS, MEERUT. INDIA.
website: www.rastogipublications.com email: info@rastogipublications.com

MOBILE: 9412700277/9412708462
(11AM to 6PM, Sunday Closed)

(JMD/12E/0040019)

Preface to the Present Edition

It is a matter of great pleasure and satisfaction to present before the readers, the Twelfth Edition of the present book ***'Modern Text Book of Zoology: INVERTEBRATES'***. In the light of the modern technology & studies, many Universities have revised their undergraduate syllabi in Life Sciences. This has necessitated the revision of the book, and also to incorporate the latest information on the subject.

The first few chapters of the book have been fully revised, reset and updated, while rest of the chapters have also been revised accordingly. Large number of illustrations have been redrawn or re-labelled. In every chapter, sufficient number of variety of questions viz: long answer type, short answer type, very short answer type, multiple choice questions and fill in the blanks etc are included.

The present edition of the book is printed in double colour. We are sure, that this will further improve the clarity of many illustrations and students shall find it more useful while reading the text matter.

We are thankful to Dr. Reena Mathur (Formerly, Professor & Head, Department of Zoology, University of Rajasthan, Jaipur) for her valuable co-operation and suggestions in the revision of this book. She has provided latest information and material for updating several chapters in the book.

Our sincere thanks are also due towards the learned teachers, reviewers and editors who have sent their valuable suggestions, for the improvement of the book.

Suggestions for further improvement of this book are cordially invited from the learned readers.

—The Publisher

Preface to the First Edition

We are living in an age of Science where man has reached to moon and has made remarkable achievements in various scientific spheres. A number of highly complicated and technical instruments have been devised to enable the modern researcher to evolve new ideas as theories. The electron microscope, for instance, is an indispensable companion of a growing biologist. The old concepts are being declared obsolete day by day, more so in Biology. Many a concept in Cytology and Histology has undergone a remarkable change. The fast changing face of Zoology has created a gap and has given an impetus to the authors to present the ***"Modern Text Book of Zoology : INVERTEBRATES"*** for the young students preparing for the degree and honors courses of various universities in India and abroad.

The authors, with their long teaching and writing experience, have attempted to produce a complete book on Invertebrates. For a better study of Invertebrate Phyla, introductory chapters on the basic principles of Zoology, Taxonomy and Cytology have been incorporated. The various phyla have been treated in a simple and lucid style and are in accordance with the syllabi of all the Indian universities. The approach to the discussion of all phyla is very simple so as to impart to the students a clear and vivid understanding. According to the scheme of treatment the important animal types of each phylum have been dealt with first, and efforts have been made to present their elaborate and up-to-date account. This is followed by a chapter on characters and classification and brief description of other important types of the phylum. Further, a chapter on topics of significance and general interest pertaining to the phylum has also been added to make the treatment more elaborate. Relevant comparative accounts have also been given where considered necessary.

It has been the constant endeavour of the author to furnish maximum substance, keeping in view the limitations of size and bulk of the book. Efforts have been made to condense the matter as far as practicable. The size of print has been selected according to the needs of the text. Special attention has been paid to the preparation of illustrations, these have been designed with utmost care and accuracy and conform to the description in the text. They are quite simple and capable of being reproduced by the students without much difficulty. An exhaustive list of books suggested for further study will be of great help to the students.

Suggestions for the improvement of the book will be thankfully acknowledged and incorporated in further editions.

July, 1973

—The Authors

Contents

1. Animal Diversity and Taxonomy 1–21

Animal Diversity; 1
Species Estimates; 3
Need of Classification; 5
Taxonomy; 6
Significance of Classification; 7
Brief History of Classification; 7
Nomenclature of Organisms; 8
Groups in Classification; 9
Modern Classification of Animal Kingdom; 11
Animal Kingdom; 11
Modern Molecular Methods in Taxonomy: (Use of DNA and RNA Sequences to Infer Evolutionary Relationships); 14

2. The Invertebrates 22–37

Why we Study Invertebrates?; 22
What is an Invertebrate ? 23
Invertebrate Phyla; 23
Invertebrata *versus* Non-Chordata; 23
Diversity of Invertebrates; 24
Major and Minor Phyla; 26
Phylogeny of Invertebrates; 28
Outline Classification of Invertebrates; 29

SUBKINGDOM PROTOZOA

3. *Euglena viridis* 38–49

Systematic Position; 39
Habits and Habitat; 39
Culture of *Euglena*; 40
Structure; 40
Locomotion; 42
Nutrition; 44
Respiration; 44
Osmoregulation and Excretion; 44
Behaviour; 45
Reproduction; 46
Encystment; 47
Other Euglenoid Flagellates; 48
1. *Euglena spirogyra,* 2. *E. gracilis,*
3. *Astasia longa,*
4. *Peranema trichophorum.*

4. *Trypanosoma gambiense* 50–59

Polymorphic Forms of Trypanosomes; 51
Sleeping Sickness; 56
(Symptoms & Pathogenesis Diagnosis, Treatment Prevention or Prophylaxis).
Some Important Trypanosomes of Man and Domestic Animals; 57

5. *Leishmania donovani* 60–67

Species of *Leishmania* Infecting Man; 62
Kala-azar or Dumdum Fever; 65
(Occurrence, Symptoms, Diagnosis, Treatment and Prevention).

6. *Amoeba proteus* 68–84

Culture of *Amoeba*; 69
Biological Significance of *Amoeba*; 80
Other Species of *Amoeba*; 81
1. *Amoeba verrucosa,* 2. *A. Discoides,*
3. *A. radiosa,* 4. *A. dubia,* 5. *A terricola.*

7. *Entamoeba histolytica* 85–93

Amoebic Dysentery or Amoebiasis; 89
(Symptoms Modes of Transmission, Diagnosis, Therapy, Prevention);
Other Entamoebae; 91
1. *Entamoeba hartmanni,*
2. *Entamoeba coli,* 3 . *E. gingivalis.*

8. *Elphidium crispum (=Polystomella crispa)* 94–99

9. *Plasmodium vivax* : The Malarial Parasite 100–114

Human Malaria; 106
(Early History, Types of Malaria, Symptoms and Pathogenesis, Duration of Infection, Replapses and Control of Malaria),
Anti-Malaria Campaign in India; 111.

10. *Monocystis agilis* 115–123

Laboratory Study of *Monocystis*; 116
Comparison of *Monocystis* and *Plasmodium*; 120
Effects of Parasitism; 122

11. *Eimeria tenella* 124–128

Coccidiosis; 127
(Pathogenesis, Diagnosis, Prevention, Treatment).

12. *Paramecium caudatum* : The Slipper Animalcule 129–147

Culture of *Paramecium*; 130
Cytoplasmic Particles of *Paramecium*; 145

13. *Vorticella* : The Bell Animalcule 148–157

Comparison of Conjugation in *Paramecium* and *Vorticellia;* 155

14. Protozoa : Characters, Classification and Types 158–172

Definition; 158
General Characters; 158
Classifiaction; 159
Other Types of Protozoa; 163
1. *Chlamydomonas,* 2. *Volvox,* 3. *Noctiluca,* 4. *Ceratium,* 5. *Trichomonas,* 6. *Giardia,* 7. *Arcella,* 8. *Difflugia,* 9. *Actinophrys,* 10. *Radiolarian Ooze,* 11. *Gregarina,* 12. *Opalina,* 13. *Balantidium,* 14. *Nyctotherus,* 15. *Stentor,* 16. *Ephelota.*

15. Protozoa : General Account 173–196

Concept of Protista; 173
Status of Protozoa; 174
Protozoa Unicellular or Acellular; 175
Body Covering and Skeleton; 176
Nucleus; 177
Locomotor Organelles and Locomotion; 178
Nutrition in Protozoa; 182
Contractile Vacuoles and Osmoregulation; 184
Reproduction in Protozoa; 185
Economic Importance of Protozoa; 188
Parasitism in Protozoa; 190
Protozoa and Human Diseases; 192

SUBKINGDOM METAZOA

16. Organization of Metazoa 197–211

What are Metazoa ?; 197
Metazoa *versus* Metaphyta; 197
Lower and Higher Metazoa; 198
Metazoan Organization; 198
Symmetry and its Significance; 199
Cephalization and Polarity; 200
Metamerism; 200
Comparison of True Metamerism and Pseudometamerism; 205
Body Cavity or Coelom; 205
Levels or Grades of Organization; 209

17. Origin of Metazoa 212–220

Introduction; 212
Premetazoan Ancestors; 212
Theories of Origin of Metazoa; 213

18. Organization of Bilateria 221–224

Bilateral Phyla; 221
Theories of Origin of Bilateria; 222

LOWER INVERTEBRATES

19. *Leucosolenia* : An Asconoid Sponge 225–233

20. *Scypha (=Sycon)* : A Syconoid Sponge 234–246

21. Porifera : Characters, Classification and Types 247–254

General Characters; 247
Classification; 248
Other Types of Porifera; 249
1. *Olynthus* Stage; 2. *Euplectella,* 3. *Hyalonema,* 4. *Cliona,* 5. *Chalina,* 6. *Spongilla* and 7. *Euspongia.*

22. Porifera : General Account 255–270

Canal System ; 255
Skeleton; 259
Reproduction; 261
Regeneration; 263
Animal Nature; 264
Individuality of Sponges; 264
Origin of Sponges; 265
Affinities and Systematic Position of Sponges; 265
Economic Importance of Sponges; 267
Sponge Industry; 268.

23. *Hydra* 271–294

Collection and Preservation; 272
Regeneration in *Hydra*; 290
Immortality in *Hydra*; 291

24. *Obelia* : A Sea-Fur 295–310

Comparison of Polyp (Hydranth) and Medusa; 307
Derivation of Medusa from Polyp; 308

25. *Aurelia* : A Jelly-Fish 311–325

Contrast Between *Obelia* and *Aurelia*; 324

26. *Metridium* : A Sea Anemone 326–334

Comparison of Hydrozoan Polyp and Sea Anemone; 333

27. Coelenterata : Characters, Classification and Types 335–349

General Characters; 335
Classification; 336
Other Types of Coelenterata; 338
1. *Bougainvillae,* 2. *Hydractinia,* 3. *Tubularia,* 4. *Pennaria,* 5. *Sertularia,* 6. *Plumularia,* 7. *Campanularia,* 8. *Millepora,* 9. *Physalia,* 10. *Velella,* 11. *Porpita,* 12. *Tubipora;* 13. *Alcyonium,* 14. *Gorgonia,* 15. *Corallium,* 16. *Pennatula,* 17. *Fungia,* 18. *Madrepora (=Acropora),* 19. *Astraea,* 20. *Meandrina (=Meandra),* 21. *Favea.*

28. Coelenterata : General Account 350–359

Polymorphism; 350
Corals; 353
Coral Reefs; 356

29. Ctenophora 360–365

General Characters; 360
Classification; 361
Affinities; 361
Types of Ctenophora; 362
1. *Pleurobrachia* and *Hormiphora,* 2. *Velamen,* 3. *Ctenoplana,* 4. *Coeloplana.*

30. *Dugesia (Euplanaria)* : A Planarian 366–378

Regeneration; 377

31. *Fasciola hepatica* : The Sheep Liver Fluke 379–397

Features of Significance in Life Cycle of *F. hepatica*; 393
Parasitic Adaptations of *Fasciola*; 395
Liver Rot or Fascioliasis; 395 (Infection, Pathogenesis, Therapy, Prophylaxis).

32. *Taenia solium* : The Pork Tapeworm 398–416

Parasitic Adaptations of *Taenia*; 411
Comparative study of *Fasciola* and *Taenia*; 412
Cestodiasis; 414 (Infection, Pathogenesis, Therapy, Prophylaxis).

33. Platyhelminthes : Characters, Classification and Types 417–429

Definition; 417
Morphological Advances; 417
Phylogenetic Significance; 418
General Characters; 418
Classification; 418
Other Types of Platyhelminthes; 421
1. *Convoluta*, 2. *Bipalium*, 3. *Thysanozoon*, 4. *Diplozoon*, 5. Polystomum, 6. *Opisthorchis* (=*Clonorchis*), 7. *Schistosoma*, 8. *Paragonimus*, 9. *Moniezia*, 10. *Taenia saginata*, 11. *Echinococcus granulosus*, 12. *Dipylidium*, 13. *Hymenolepis*.

34. Superphylum Aschelminthes : The Pseudocoelomate Phyla 430–435

Derivation of Name; 430
General Characters; 431
Classification or Major Groups; 431
Inter-relationships; 434

35. *Ascaris lumbricoides* : The Common Roundworm 436–455

Parasitic Adaptations of *Ascaris*; 451
Ascariasis; 452
(Infection, Pathogenesis, Therapy, Prophylaxis).

36. *Ancylostoma duodenale* : The Common Hookworm 456–461

Ancylostomiasis; 460
Another hookworm; 460
(Necator americanus).

37. *Wuchereria bancrofti* : The Filarial Worm 462–466

Filariasis or Elephantiasis; 465
(Diagnosis, Pathogenesis, Control).

38. Phylum Nematoda : Characters, Classification and Types 467–475

General Characters; 467
Classification; 468
Common Nematode Parasites; 469
1. *Trichuris trichuru,*
2. *Trichinella spiralis,* 3. *Rhabditis,*
4. *Enterobius vermicularis,*
5. *Dracunculus medinensis,*
6. *Loa loa,* 7. *Tococara canus.*

39. Helminthes : General Account 476–483

Evolution of Parasitism in Platyhelminthes; 476
Parasitic Adaptation of Flatworms; 477
Helminths and Human Diseases; 478
Nematode Parasites of Man and Animals; 481.

HIGHER INVERTEBRATES

40. *Neanthes* or *Nereis virens* : A Clamworm 484–502

Typical Trochophore Larva; 500
Adaptations of *Nereis* ; 501

41. *Pheretima posthuma* : The Indian Earthworm 503–532

Regeneration and Grafting; 527
Adaptations of Earthworm; 527
Economic Importance; 527

42. *Hirudinaria granulosa* : The Indian Cattle Leech 533–557

Parasitic Adaptations of Leeches; 555

43. Annelida : Characters, Classification and Types 558–570

General Characters; 558
Classification; 559
Other Types of Annelida; 560
1. *Aphrodite,* 2. *Polynoe,* 3. *Eunice,*
4. *Phyllodoce,* 5. *Chaetopterus,*
6. *Arenicola,* 7. *Amphitrite,*
8. *Terebella,* 9. *Megascolex,*
10. *Lumbricus,* 11. *Tubifex,*
12. *Pontobdella,* 13. *Hirudo medicinalis,* 14. *Polygordius,*
15. *Sipunculus.*

44. Annelida : General Account 571–590

Comparative Accounts of *Nereis, Pheretima* and *Hirudinaria*; 571 (1. Digestive System, 2. Excretory System, 3. Reproductive System)
Origin of Coelom and Metamerism; 576
Adaptive Radiation in Annelida; 576
Segmental Organs in Annelida; 582 (Coelomoducts and Nephridia)
Regeneration; 585
Asexual Reproduction in Polychaeta; 586
Economic Importance; 588.

45. *Palaemon* or *Macrobrachium malcolmsonii* : The Indian Freshwater Prawn 591–628

Penaeus indicus; 620
Crustaceans as Source of Food; 621
Prawn Fishery in India; 622

46. *Palamnaeus* : The Indian Scorpion 629–644

47. *Periplaneta americana* : The Common Cockroach 645–673

Other Cockroaches; 668
1. *Blatta orientalis,*
2. *Blatella germanica,*
3. *Periplaneta australasiae.*

48. The Grasshopper or Locust 674–691

Economic Importance of Grasshopper; 687
Grasshopper Control; 689
Locust Migration; 689
Locust Phase Theory; 689.

49. *Musca domestica* : The Housefly 692–700

Economic Importance of Housefly; 697
Fly Control; 698.

50. *Culex* and *Anopheles* : The Mosquitoes 701–710

Comparison of Life Histories of *Culex* and *Anopheles*; 707
Mosquito-Borne Diseases; 707
Mosquito Control; 707.

51. *Apis* : The Honey Bee 711–724

Economic Importance of Honey Bee; 719
Apiculture; 719.

52. *Bombyx mori* : The Silkworm 725–731

Sericulture; 727
Sericulture in India; 729.

53. Arthropoda : Characters, Classification and Types 732–762

General Characters; 732
Classification; 733
Other Types of Arthropoda; 740
1. *Limulus,* 2. Spiders, 3. Mites,
4. Ticks, 5. *Daphnia*, 6. *Cyclops,*
7. *Cypris,* 8. *Lepas,* 9. *Balanus,*
10. *Sacculina,* 11. *Squilla,*
12. Lobster (*Palinurus),* 13. *Hippa,*
14. *Eupagurus,* 15. *Cancer,*
16. Millipeds (*Spirobolus*),
17. Centipedes (*Scolopendra),*
18. *Lepisma,* 19. *Locust,* 20. *Gryllus,*
21. Praying Mantis, 22. Stick Insect,
23. Termites, 24. Dragon Fly,
25. *Pediculus* (Human Louse)
26. *Cimex* (Bed Bug), 27. *Cicada,*
28. Earwig, 29. Butterfly, 30. Moth,
31. Beetle, 32. *Drosophila,*
33. Fleas *(Pulex)*, 34. Wasp,
35. Ants.

54. Arthropoda : General Account 763–802

Biological Success; 763
Larvae of Crustacea; 763
Mouth Parts of Insects; 767
Economic Importance of Insects; 769
Metamorphosis in Insects; 772
Hormonal Control of Metamorphosis; 773
Insect Control; 774
Methods of Insect Control; 774
Respiration in Insects; 777
Body Wall of Arthropods; 783
Respiratory Organs in Arthropoda; 786
Economic Importance of Arthropoda; 789
Social Life in Insects; 791
Crustacean Parasites; 795.

55. *Peripatus* 803–808

56. *Pila* : An Apple Snail 809–832

57. *Unio (Lamellidens)* : Freshwater Mussel 833–852

Comparative Study of Respiratory Systems of *Pila* and *Unio*; 850
Comparative Account of Nervous Systems of *Pila* and *Unio*; 851.

58. *Neopilina* 853–858

Discovery; 853
External Features; 854
Internal Structure; 855
Relationships of Monoplacophora and Origin of Mollusca; 856.

59. *Sepia* : The Cuttle-fish 859–870

60. Mollusca : Characters, Classification and Types 871–884

Definition; 871
Derivation of Name; 871
Evolutionary Characters; 872
General Characters; 872
Classification; 872
A Few Other Molluscs; 876
1. *Chiton,* 2. *Dentalium,* 3. *Patella,* 4. *Cypraea,* 5. *Aplysia,* 6. *Helix,* 7. *Doris,* 8. *Limax,* 9. *Mytilus,* 10. *Ostrea,* 11. *Pecten,* 12. *Peral oyster,* 13. *Solen,* 14. *Teredo,* 15. *Loligo,* 16. *Octopus,* 17. *Nautilus.*

61. Mollusca : General Account 885–906

Podium or Foot in Mollusca; 885
Respiration in Mollusca; 891
Torsion in Gastropoda; 895
Pearl Culture; 899
Chank Fisheries; 900
Lime Fisheries; 901
Economic Importance of Mollusca; 901
Molluscan Larvae; 903
Shell in Mollusca; 903.

62. *Asterias* : A Sea Star 907–925

63. Echinodermata : Characters, Classification and Types 926–936

Historical; 926
Derivation of Name; 926
Definition; 927
General Characters; 927
Classification; 927
A Few Other Echinoderms; 930
1. *Pentaceros,* 2. *Astropecten,* 3. *Ophiothrix,* 4. *Gorgonocephalus,* 5. *Echinus,* 6. *Clypeaster,* 7. *Cucumaria,* 8. *Thyone,* 9. *Antedon.*

64. Echinodermata : General Account 937–949

Larval Forms in Echinodermata; 937
Autotomy and Regeneration; 942
Water-Vascular System; 943
Symmetry in Echinodermata; 948.

65. *Balanoglossus* : A Tongue Worm 950–968

Affinities & Systematic Position of *Balanoglossus* (Hemichordata); 964.

66. Hemichordata : Characters, Classification and Types 969–974

Definition; 969
General Characters; 969
Classification; 970
Other Hemichordates; 971
1. *Saccoglossus,* 2. *Ptychodera,* 3. *Rhabdopleura,* 4. *Cephalodiscus,* 5. *Atubaria.*

❏ Glossary of Technical Terms 975–980

❏ Selected References 981–983

List of Tables

Table 2.1.	Contrast between Lower and Higher Invertebrates.
Table 2.2.	Comparison (Differences) between Invertebrates and Vertebrates.
Table 2.3.	Major Characteristics of Principal Animal Phyla.
Table 2.4.	Major Groups and Phyla in Animal Classification.
Table 4.1.	Some Important Trypanosomes of Man and Domestic Animals.
Table 5.1.	Species of *Leishmania* infecting Man and other Animals.
Table 9.1.	Comparative Account of 4 Species of Human Infecting *Plasmodium.*
Table 10.1.	Comparison of *Monocystis* and *Plasmodium.*
Table 15.1.	Comparison of Conjugation in *Vorticella* and *Paramecium.*
Table 16.1.	Differences between True Metamerism and Pseudometamerism.
Table 24.1.	Difference of Polyp (Hydranth) and Medusa of *Obelia.*
Table 25.1.	Contrast between *Obelia* and *Aurelia.*
Table 26.1.	Differences between a Hydrozoan Polyp and a Sea Anemone.
Table 32.1.	Comparative Study of *Fasciola* and *Taenia.*
Table 44.1.	Comparative Account of Digestive System of *Nereis, Pheretima* and *Hirudinaria.*
Table 44.2.	Comparative Account of Excretory System of *Nereis, Pheretima* and *Hirudinaria.*
Table 44.3.	Comparative Account of Reproductive System of *Nereis*, *Pheretima* and *Hirudinaria.*
Table 45.0.	Branchial Formula of *Palaemon.*
Table 45.1.	Some Important species of Prawns in the Order of their Abundance in Landings.
Table 45.2.	Statewise Landings of Prawns in India.
Table 48.1.	Differences between Solitary and Gregarious Phases of Locusts.
Table 50.1.	Comparison of Life Histories of *Culex* and *Anopheles.*
Table 54.1.	Diseases Transmitted by Insects.
Table 54.2.	Some Interesting Habits of Bees, Ants and Termites Regarding their Nest and Broods.
Table 57.1.	Comparative Study of Respiratory System of *Pila* and *Unio.*
Table 57.2.	Comparative Account of Nervous Systems of *Pila* and *Unio.*

1

Animal Diversity and Taxonomy

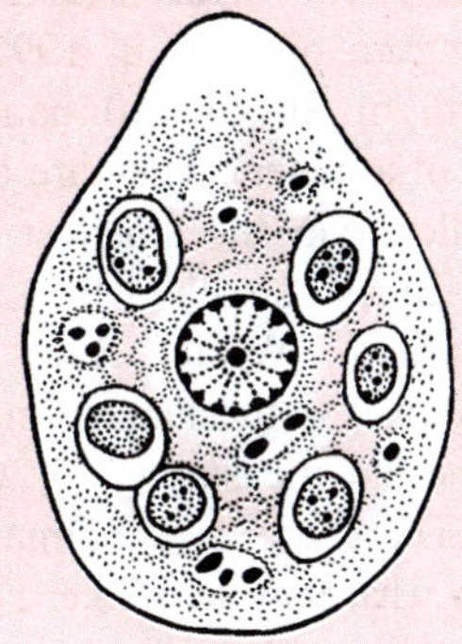

ANIMAL DIVERSITY

Study of animal life, function, reproduction, and interaction with their environment is exciting, fascinating, and inspiring. The estimated number of animals and plants on our blue planet described and classified by Biologists is about 1.9 million (2010) and this is less than one quarter of the total species estimated or yet to be discovered and described. Every year scientists record about 18,000 new species of plants and animals. Out of this approximately 13,500 are new invertebrates. So far 1,305,250 invertebrates have been discovered, and so many species are waiting to be described.

Between 1999 and 2009 more than 1,200 new species of plants and animals were discovered in 'Amazon'. Details of newly discovered species can be obtained by visiting following websites :

http://wwf.panda.org	— World Wildlife Foundation
zsi.gov.in	— Zoological Survey of India
http://news.mongabay.com	— Also gives information on newly discovered species
www.zsi.com	— Zoological Society of Land
www.wcs.org	— World Conservation Society
http://www.iucn.org	
http://species.asu.edu	— International Institute for Species Exploration
http://indianbiodiversitytalk. biogspot.in	
www.wii.gov.in	— Wildlife Institute of India.

As evident from above a large number of animals inhabit the earth today and many more have lived here in the past and equally more are waiting to be listed. These animals vary in their shape and size, habit and habitat, colour, form and weight as well as in their distribution. The nanobes are the smallest living organisms known on the earth. They were discovered in ocean by **Dr. Phillipa Uwins**, from Queensland University (Australia) during an oil drill in 1991. The nanobes have a length of 20 to 150 nanometers (the millionth part of a meter). They are almost a thousand times smaller than the smallest cells in our body and ten times smaller than the smallest known bacterium. While Blue whale (*Balaenoptera musculus*) is the largest animal weighing 100-200 tonns and measuring 25-30 meters in length.

There are few animals which are different from what they look like. For instance, a bat which looks like a bird is not really a bird but is a mammal like a man. Likewise a whale which generally known as a fish is not a fish. These things appear to be surprising to a common man. A considerable part of animal kingdom on one hand is so small in size that they can only be observed with the help of a microscope and many of them live in water. On the other hand giant sized animals like Elephant, Rhino, Hippo live on the earth. Animals of all kinds *i.e.,* round, flat, elongated, fat, colourful as well as scary in shape and size exist in nature. There are certain animals which are known only by fossil remains.

Diversity amongst animals does not exist in their shape and size only but also in their species, number and habitat. Starting from the indefinitely deep vents in oceans to snow covered mountain peaks and from the poles of the earth to the equator, the animals inhabit all the places. From the point of view of habits and habitat some animals live in **fresh water** ponds and rivers. The animals living in flowing water in a river are called **lotic** and those inhabiting the stagnant water of ponds, pools, lakes and swamps, are called **lentic**. Amongst the **marine animals**, a few living near the water surface are **pelagic**, those living in the depths of the oceanic water are **abyssal**. A few others living near the oceanic bottom are **littoral** and still other leading a sedentary life on the sea-bottom are **benthic**. In the same way some animals living under the surface of the earth are **subterranean**, others inhabiting the earth surface are **terrestrial**, some of them living on trees are **arboreal** and still others flying in air are **aerial**. The fast runners on the earth are **cursorial** while with digging habit are called **fossorial** animals. Therefore it is clear that with this great diversity of animals their study is not so easy. There are water areas where big rivers join the ocean, called **estuarine water** and have lesser salinity than sea water. A few fishes are unable to survive throughout their life in marine water as for example salmon and trouts. These fishes come for spawning to the estuaries and are called **anadromous** fishes in contrast to those which migrate from the estuaries to the sea for the purpose of spawning are called **catadromous** fishes like fresh water eels. Based on the salanity tolerance the fishes are **euryhaline**, migrating between fresh water estuaries and salt water and have wide salinity tolerance. On the other hand fishes having a narrow range of salinity tolerance remain confined either to fresh water or salt water are called **stenohaline**.

2010 was Declared as International Year of Biodiversity

The United Nations proclaimed year 2010 as the International Year of Biodiversity and people all over the world are working to safeguard this irreplaceable natural wealth and reduce biodiversity loss.

The International Year of Biodiversity offers a unique opportunity to increase understanding of the vital role that biodiversity plays in sustaining life on Earth.

The 1992 Convention on Biological Diversity (CBD) carries an internationally accepted definition of biodiversity as "... the variability among living organisms from all sources, inter alia, terrestrial, marine and other aquatic ecosystems and ecological complexes of which they are part, this includes diversity within species, between species and of ecosystems" (**Glowka** *et al.,* 1994).

According to the Millenium Ecosystem Assessment, the total number of species on Earth ranges from 5 to 30 million and out of these only 1.7 to 2 million species have been formally identified.

Group Number of Described Species

The following are the number of group members of the described species.

Bacteria and Blue-green Algae 4,760; Fungi 46,983; Algae 26,900; Bryophytes (mosses and liverworts) 17,000; Gymnosperms (conifers) 750; Angiosperms (flowering plants) 250,000; Protozoans 30,800; Sponges 5,000; Corals and Jellyfishes 9,000; Roundworms and Earthworms 24,000; Crustaceans 38,000; Insects 751,000; Other Arthropods and Minor invertebrates 132,461; Molluscs 50,000; Starfishes 6,100; Fishes (teleosts) 19,056; Amphibians 4,184; Reptiles 6,300; Birds 9,198 and Mammals 4,170. The total No. of species is **1,435,662**.

About 1.8 million have been described by scientists (**Hilton-Taylor** *et al.* 2008), but conservative estimates suggest that there are 5-15 million species alive today (May, 2000), since many groups of organisms remain poorly studied. Over 15,000 new species are described every year (**Dirzo** and **Raven** 2003; www. nature.com)

The 1.8 million species described by **Science** are incredibly diverse. They range from tiny, single-celled microbes like *Nanoarchaeum equitans,* 400 nm in diameter living as parasites on other microbes in thermal vents at temperatures of 70-98ºC (**Huber** *et al.*, 2002). Since 2000, 53 new species of primates have been described (IUCN 2008) including a new species of Brazilian monkey, Mura's saddleback Tamarin.

Invertebrates play a vital role in the survival of other living organisms. These are involved in keeping the environment active and provide us with rich soil, clean water and food. Herbivory is one of these services. Herbivores help the nature by releasing their faeces on the soil which is produced by eating plants etc., after decomposition of the faeces the fertility of the soil increases. Herbivores also help in pollination, prunning the old parts of the plants and flowers. This phenomenon is helpful in the growth of new organs and parts and also helps in seed dispersal and parasitism etc.

Invertebrates are widely present every where all around us on the earth. They have been recorded in the upper reaches of the atmosphere, in the driest deserts and in the cannopies of the wettest rainforests. They can even be found in the frozen Antarctic (some mites and springtails can withstand temperature of – 35°C) or on the ocean floor of the Abyssal Sea which reaches the depth up to 11,000 meters. Invertebrates are undisputed heavy weights of the planet. If we take the weight of the entire animal kingdom then arthropods alone will be of more than 85% of that weight.

It is estimated that five to eight million insect species have not been identified or discovered, while approximately 10,000 species of chordates awaiting the discovery and description. A good number of oceanic species diversity still remains unidentified.

SPECIES ESTIMATES

An estimate of the numbers of species groupwise in world and India is described in the table 1 of Zoological Survey of India, Kolkata, given on the next page.

The sheer number and mass of invertebrates reflects their enormous ecological impact. Admittedly, some have a negative impact on humans, either by harming us directly as disease agents, or other harmful effects. Invertebrates are the part of nearly every food chain, either directly as food for fishes, amphibians, reptiles, birds, mammals or indirectly, as agents in the endless recycling of nutrients in the soil. Insects, worms, and mites are extremely important in helping microbes to break down the dung and dead plant or animal matter. 99% of human and animal waste is decomposed by invertebrates.

Some invertebrates are "Keystone Species" (small in number, form small part of the ecosystem by measures of biomass or productivity but play important role in the maintenance of biotic communities). Coral reefs are perhaps the most dramatic example, providing a wide range of niches for diversity of plants and perhaps one-third of all fish species.

Faunal Resources of India

India is very rich in terms of biological diversity due to its unique biogeographical location,

Table 1. **Number of Animal Species known from India (Updated by Zoological Survey of India, December, 2016).**

Kingdom	Phylum	Number of species			
		World* (living + fossil)	World* (living)	India	%
Protista	Phylum Protozoa	36,400 (excluding fossil)	36,400	3,510	9.64
Animalia	Phylum Mesozoa	122	122	10	8.02
	Phylum Porifera	11,055	8,838	545	6.16
	Phylum Cnidaria	17,702	11,522	1,396	12.12
	Phylum Ctenophora	199	199	19	9.55
	Phylum Platyhelminthes	29,488	29,487	1,738	5.89
	Phylum Rotifera	2,049	2,049	466	2.24
	Phylum Gastrotricha	828	828	162	19.56
	Phylum Kinorhyncha	196	196	10	5.10
	Phylum Nematoda	25,043	25,033	2,914	11.63
	Phylum Acanthocephala	1,461	1,330	301	22.63
	Phylum Spiuncula	156	156	41	26.28
	Phylum Echiura	198	198	47	23.73
	Phylum Annelida	17,426	17,388	1,024	5.89
	Phylum Onychophora	187	183	1	0.53
	Phylum Arthropoda	13,02,809	12,57,040	75,528	6.00
	Subphylum Chelicerata	1,15,992	1,13,773	5,945	5.23
	Class Arachnida	1,14,275	1,12,442	5,907	5.25
	Class Merostomata	103	4	2	50.00
	Class Pycnogonida	1,346	1,335	36	2.69
	Subphylum Crustacea	73,141	67,735	3,796	5.61
	Subphylum Hexapoda	10,80,760	10,63,533	65,409	6.15
	Class Collembola	8,187	8,162	324	3.97
	Class Diplura	976	975	18	1.85
	Class Protura	816	816	20	2.45
	Class Insecta	10,70,781	10,53,578	65,047	6.17
	Subphylum Myriapoda	12,010	11,999	378	3.15
	Class Chilopoda	3,118	3,112	101	3.25
	Class Diplopoda	7,842	7,837	270	3.45
	Class Symphyla	204	204	7	3.43
	Phylum Phoronida	16	16	3	18.75
	Phylum Bryozoa (Ectoprota)	11,652	6,186	327	5.29
	Phylum Entoprocta	186	186	10	5.37
	Phylum Brachiopoda	7,390	392	8	2.04
	Phylum Chaetognatha	186	170	44	25.88
	Phylum Tardigrada	1,335	1,167	30	2.57
	Phylum Mollusca	1,18,062	84,978	5,189	6.11
	Phylum Nemertea	1,368	1,368	6	0.43
	Phylum Echinodermata	20,550	7,550	777	10.29
	Phylum Hemichordata	162	139	14	10.07
	Phylum Chordata	89,955	71,526	6,573	9.08
	Subphylum Cephalochordata	33	33	6	18.18
	Subphylum Urochordata	2,804	2,804	516	18.40
	Subphylum Vertebrata [= Craniata]	88,512	68,689	6,051	6.85
	Class Pisces	37,172	34,362	3,324	9.70
	Class Amphibia	8,007	7,667	388	5.06
	Class Reptilia	16,123	10,450	572	5.47
	Class Aves	11,241	10,357	1,340	12.93
	Class Mammalia	15,969	5,853	427	7.29
	Total (Animalia)	**1,664,289**	**15,29,953**	**97,183**	**6.35**
	Grand Total (Protista + Animalia)	**1,700,689**	**15,66,353**	**1,00,693**	**6.42**

diversified climatic conditions and enormous ecodiversity and geodiversity. India's immense biological diversity encompasses ecosystems, populations, species and their genetic make-up. This diversity can be attributed to the vast variety in physiography and climatic situations resulting in a diversity of ecological habitats ranging from tropical, sub-tropical, temperature, alpine to desert. According to world biogeographic classification, India represents two of the major realms (the Palaearctic and Indo-Malayan) and three biomes (Tropical Humid Forests, Tropical Dry/ Deciduous Forests and Warm Deserts/Semi-Deserts). Wildlife Institute of India has proposed a modified classification which divides the country into 10 biogeographic regions : (i) Trans-Himalayan, (ii) Himalayan, (iii) Indian Desert, (iv) Semi-Arid, (v) Western Ghats (vi) Deccan Peninsula, (vii) Gangetic Plain, (viii) North East India, (ix) Islands and (x) Coasts. In the light of Biodiversity Convention, India holds a unique position with the priority of conservation of natural resources and sustainable development. Infact, in 2% of world's total land surface, India is known to have over 7.50% of the animal species.

Biologists every year document about 13,500 new species of invertebrates. The species totals do not include domestic animals such as sheep, goats and camels. Nor do they include single-celled organisms such as bacteria.

(http://www.currentresults.com 2012).

Animal discoveries by zoological survey of India by 2016 are mentioned below :

Species of Animals new to science from India.

Phylum/Group	No. of species
Porifera	1
Cnidaria	4
Platyhelminthes	10
Nematoda	15
Annelida	1
Arachnida	27
Crustacea	9
Collembola	4
Insecta	197
Pisces	12
Amphibia	12
Reptilia	6

Species of Animals new records from India.

Phylum/Group	No. of species
Protozoa	2
Cnidaria	8
Platyhelminthes	1
Arachnida	19
Crustacea	1
Insecta	30
Bryozoa	8
Mollusca	1
Echinodermata	2
Ascidiacea	5
Pisces	4

NEED OF CLASSIFICATION

For keeping books of different subjects different almirahs are used in a good library so that when we need a particular book of History, Geography, Hindi or English literature, Chemistry, Physics, Botany or Zoology etc, we can atonce take out that book from that section or almirah. If these books are not kept in such systematic way, can we take out a required book out of the assemblage of thousands of books ? In the same way inspite of a great diversity amongst the animals, the animal kingdom is well arranged and is very systematic. For systematic study animals are divided into minor and major groups on the basis of similarities and differences. Each group has been given a particular name depending upon their characteristics. For example different kinds of birds are included in class **aves** based on their structural and physiological similarities. In the same way lizards, snakes and tortoises have been included in class **reptilia**.

Species description is seen as an old-fashioned way of doing research. It is very important to know the living organisms around us, and careful and accurate identification and classification are of vital importance. Without taxonomy, nobody would be sure of the identity of organisms they were interested in, or whether they belonged to the same or different species as the organisms studied by others. Phylogeny can be defined as the evolutionary history of a group or lineage, the origin and evolution of higher taxa, or the natural process or repeated irreversible splitting of populations (see **Lincoln** *et al.*, 1998; **Wagele**, 2005).

TAXONOMY

The study of identification and relationships among living organisms is called **systematics**. The term systematics was used by **Linnaeus** (1707-1778). The branch of biology dealing with identification, nomenclature and classification of organisms is called TAXONOMY. The term **Taxonomy** was coined by **A.P. de Candolle** (1813). It is derived from Greek words (**taxis**, arrangement and **nomos**, law). Structural similarity or homology forms the basis of classification. Systematics is the scientific study of diversity of organisms. The term systematics stems from the latinized Greek word '**systema**'. The **classical systematics** (old systematics) was first employed by **Plato** and **Aristotle** and was accepted by **Linnaeus**. In this system purely morphological definition of species was employed. Species are known by a single or few specimens only. It explains **typological concept** in which one or a few individuals were thought to give information about characters of the species. According to typological concept species is **static** and remains **fixed**. Variations were supposed to be caused by imperfect expressions of traits. Individuals were considered merely expressions of the same type. Taxonomy is the one of the oldest and most fundamental branch of biology. Though enough work has been done in other branches even then the importance of taxonomy can not be ignored. New species and types of animals and plants are being identified every day and their classification and characters are studied with the help of Taxonomy. In addition to this, the use of scanning, electron microscopy and the information generated by other fields of science such as, ecology, cytology, genetics, biochemistry, molecular biology specially DNA, animals are being restudied and replaced in phylogenetic trees.

The concept of **new systematics** (neosystematics or biosystematics) was given by **J. Huxley** (1940). It deals with subspecies and populations to explain **population systematics**. New systematics considers besides morphology, other aspects also like ecology, biochemistry, physiology, cytology and genetics.

New systematics has led to new branches of taxonomy like **morphotaxonomy** based on morphological characters, **karyotaxonomy** based on nucleus and chromosomes. **Cytotaxonomy** is based on cytoplasm (cytochrome C), **experimental taxonomy** based on experiments determining genetic interrelationships, **biochemical taxonomy** on the basis of biochemical studies, **chemotaxonomy** based on particular chemicals like secondary metabolites and **numerical taxonomy** or **phenetic** or **adansonian taxonomy** explained by **Adanson** (1963) based on statistical methods which are used in classification.

Latest in the taxonomy is the use of DNA and computers. DNA is more stable than protein in the environment; this raises the possibility of doing DNA sequencing on well preserved (in snow and amber) extinct organisms. Egyptian mummies over 2000 years old and human remains in Florida that are at least 7500 years old have yielded samples of DNA that were successfully sequenced. In 2006, a new record was reported — that the DNA was sequenced from 8,00,000 years old elephant fossils.

Protein sequencing provides a tool for establishing homologies from which genealogies can be constructed and phylogenetic trees are drawn. With such information, one can reconstruct an evolutionary history of the molecule and thus of their respective owners. For example : Cytochrome C is an ancient molecule, and it has evolved very slowly. Even after more than 2 billion years, one-third of its amino acids are unchanged. This conservatism is a great help in working out the evolutionary relationships between distantly-related creatures like fish and humans.

Taxonomists are using the **chromosome painting** method to compare the entire genome, in which a fluorescent label to the DNA of individual chromosomes of one species (*e.g.,* human) is attached. Then it is exposed to chromosomes of another species (*e.g.,* chimpanzee and orangutan). Regions of gene homology hybridizes taking up the fluorescent label and the "painted" chromosomes are examined under the microscope to see which species is closer to human ?

SIGNIFICANCE OF CLASSIFICATION

1. Convenience of study. Out of millions of types of organisms it is difficult to study each of them. The study of selected animals of a particular group gives an idea about the remaining animals of that group, for example study of rabbit gives the knowledge about all animals of class mammalia.

2. Knowledge of affinities. The knowledge about relationships or affinities of different animals with other animal species comes from classification.

3. Knowledge of sequence of evolution. Classification gives the sequence of evolution of animals, like evolution of sponges from single-celled protozoa and that of arthropoda from annelida, could only be known through classification.

4. Knowledge of connecting links. The transitory stage between two groups of animals is known as **connecting link**. It gives the sequence of evolution of animals. With the help of classification definite position of connecting links can be made clear.

5. Knowledge of adaptation. By classification we come to know about such characters of animals with which they adapt to an environment.

6. Knowledge of phylogeny. Animals of one group are evolved from a common ancestor, is known through classification because this is based on phylogeny. We know that all animals of class mammalia have evolved from one common ancestor.

Fig. 1. Carolus Linnaeus : Father of modern taxonomy.

BRIEF HISTORY OF CLASSIFICATION

The Greek philosopher **Aristotle** (384-322 B.C.), known as Father of Zoology in his book *Historia Animalium* he classified the animals and this was followed for about 2000 years.

Following was his system of classification :

Group I. Enaima. It included the vertebrate animals having red coloured blood.

Class 1. Viviparous : Animals which give birth to young ones *e.g.,* man, whale and other mammals.

Class 2. Oviparous. Egg laying animals *e.g.,* amphibians, reptiles, fishes and aves.

Group II. Anaima. It included the invertebrate animals without red coloured blood.

Class 1. Cephalopoda. It contained giant bodied molluscs.

Class 2. Crustacea. It included crabs and shrimps.

Class 3. Insects and spiders. It included different types of insects and spiders.

Class 4. Mollusca and echinodermata. It included other molluscs and star fishes.

Class 5. Sponges and coelenterata. It included sponges and sea animones.

Aristotle classified animals on the basis of their body size, habit and habitat.

Species is a Unit of Classification

The word **species** was coined by **John Ray** (1628-1705). The concept of biological species was put forward by **Mayer**. A Swedish botanist **Karl von Linne** who later on became popular by the name of **Carolus Linnaeus** (1707-1778) wrote the book *Systema Naturae* (1735). In this book he presented the system of nomenclature for plants and animals known as **binomial nomenclature**. This book is known as the "Dictionary of Classification" and Linnaeus is known as the **'Father of Modern Taxonomy.'**

There is a tendency among young and modern ecologists to view phylogeny, taxonomy and museums and herbaria as "dusty" places with old people who are old-fashioned postage-stamp collectors types. The results is that taxonomical experts retire and are not replaced, fundamental basic zoology and botany are disappearing from university curricula and new researchers in molecular biology, instrumentation, conservation, environmental studies and biodiversity are fast spreading.

NOMENCLATURE OF ORGANISMS

While classifying any animal its nomenclature is required first. There are two types of names of organisms : (i) **Common** or **Vernacular names** and (ii) **Scientific names**. The well known animals and plants are known by different names in different countries and language of the world. An organism may be popular by different local names in different parts of a country. This poses a great problem in the study of an organism because a man of other place or language can not understand about that organism. When **Seton** (1929) wanted to study a big member of cat family, he realised that in different parts of America it is known by different names, *e.g.,* painter, cougar, puma, panther, mountain lion, catamount, variment-red tiger, brown tiger, sneak cat, king cat, purple panther, mountain devil, mountain screamer and mountain demon. The easiest solution to all such problems was, only to assign one scientific name for this animal : *Felis concolor,* so that it can be studied easily throughout the world.

Mononomial nomenclature. To assign name of one word in classification is called mononomial nomenclature. While writing the names of **taxa** or **supraspecific groups**, mononomial nomenclature is used *e.g.,* name of family, order, class, or phylum, etc.

Binomial nomenclature. The system of writing scientific names of the organisms adopted by **Linnaeus** is called binomial nomenclature. He gave system of writing the name of organisms in two words. The first word of the name belongs to the **genus**, other word belongs to that specific plant or animal and is called species. According to the binomial nomenclature the very popular Indian bird, commonly called **Gauraiya** in India, is named as *Passer domesticus,* the dog as, *Canis familiaris,* the house fly as *Musca domestica,* and the lion as *Panthera leo.*

Even Linnaeus changed his real name **Karl von Linne** according to the binomial nomenclature to **Carolus Linnaeus**.

Trinomial nomenclature. There are certain organisms which have **subspecies**. The name of their subspecies is written after the name of their genus (generic name) and the species (specific name). These subspecies occur in different regions and have different characteristics. Therefore, the system of writing names in three words is known as **trinomial nomenclature**. The scientific name of common crow is *Corvus splendens.* It is mainly found in India, Myanmar and Sri Lanka but it is different in each country. It is therefore, divided into three subspecies. In India it is called *Corvus splendens splendens,* in Myanmar (Burma) it is *Corvus splendens insolens* and in Sri Lanka it is

Corvus splendens protegatus. In the same way the subspecific name of Indian lion is *Panthera leo persica* and the modern man is *Homo sapiens sapiens.*

Cladistic analysis. This is a method to classify organisms to mach their evolutionary history. Common phylogenetic, features are used to establish relationship between organisms with the help of sophisticated computer programmes the sort out the organisms according to their evolutionary structures.

Rules of scientific nomenclature. While writing the scientific names it is necessary to follow certain **international rules** which are :

(1) The scientific name (generic and specific both) should be written or printed in *Italics.*
(2) The generic name must start with capital letter and the specific name with small or lower letter.
(3) The generic name should be followed by the specific name.
(4) In case different scientists have named the same genus or species differently, the name first published be accepted.
(5) A family name is formed by suffix -IDAE to the names of the genus and name of a sub family by suffix-INAE.

There are three phases of taxonomy : alpha, beta and gamma :

Alpha taxonomy. A biologist who identifies a species of bird and places it in its scheme of classification, modifying the classification if necessary, is working at the level of alpha taxonomy. **William BertramTurrill** introduced the term "alpha taxonomy" in a series of papers published in 1935 and 1937. At the level of alpha taxonomy, the organisms are identified, named and characterised. The term "alpha taxonomy" is primarily used today to refer to the discipline of finding, describing, and naming taxa, particularly species.

Beta taxonomy. A biologist who examines, identifies, describes and names a species of birds is working at the level of beta taxonomy. **Ernst Mayer** in 1968 defined beta taxonomy as the classification of ranks higher than species. This stage is also known as macro taxonomy or synthesis taxonomy. At this level the species are arranged or classified into a natural system of higher and lower categories.

Gamma taxonomy. A biologist who studies a group of birds for the purpose of working out their Phylogenetic relationship is working at the level of gamma taxonomy. This phase of taxonomy is also called **biological taxonomy** and is concerned with the study of **speciation**. It is involved primarily with the analysis of intra specific variations which leads to the study of origin, evolution and determination of biological species and its subspecies.

In the 1950s **R.R. Sokal** and **P.H.A. Sneath** advocated the use of **numerical taxonomy** to remove the subjectivity of classification. In their method, all relevant characters (morphological features) of a group of organisms are numerically coded and the overall similarity is calculated by use of a mathematical **algorithm**. Initially, many traditional taxonomists rejected numerical taxonomy, since its results sometimes contradicted their own decade-long studies of comparative morphology. However, nearly all modern taxonomists currently use numerical methods in taxonomy.

GROUPS IN CLASSIFICATION

In order to classify the organisms they are divided into catagories, ranks or groups of different status. The supraspecific catagory assigned in classification are called **taxa** (sing. taxon). In classification these groups or taxa are arranged in a definite order which is known as hierarchy of classification. There are six main **taxa** : (i) phylum, (ii) class, (iii) order, (iv) family, (v) genus, and (vi) species.

The words species, genus and family were first used by **John Ray**, words class and order were used by **Linnaeus** and the term phylum was given by **George Cuvier**.

TAXA OR RANKS

KINGDOM
↑
PHYLUM
↑
CLASS
↑
ORDER
↑
FAMILY
↑
GENUS
↑
SPECIES
↓
SUBSPECIES
↓
CLINES
↓
DEMES

SUPRASPECIFIC GROUPS (Kingdom to Species)

INFRASPECIFIC GROUPS (Species to Demes)

Species

In the system of classification the species (plural is also species), constitutes the small and large unit of taxonomy. **Linnaeus** gave the traditional concept of species in *Sistema Naturae*. It has been variously defined by different scientists. According to **John Ray** all members produced from similar parents are kept in the same species. **Mayer** gave the concept of biological species as interbreeding group of organisms. **Linnaeus** described that animals with similar structural and morphological resemblance from the same species. It is also known as 'Morphological Concept'. Genetic species concept was given by **Lotsy** (1918). According to this a species is a group of genetically identical individuals.

By hybridization or cross-breeding between two almost similar species a new species resembling both the species can be produced *e.g.*, a **mule** is formed from a female horse and male ass, a **tiglion** from a tiger and a lioness, a **zabroid** from an ass and a zebra, and **panthopard** from a female panther and a male leopord etc. The modern scientists are of the opinion that a species is a population of such organisms which have a **gene pool**, they have similar adaptation to the environment and have the capacity to change into new species through **organic evolution**.

Modern Concept of Species

The modern concept of species is biological species or genetic species. **Mayer** (1942) has defined species as a "population of interbreeding individuals." There is a free gene flow in an interbreeding population. The genetic changes appearing in a part of population of the species can ultimately appear in the descendents of the species. The population thus tends to evolve as a whole.

An animal species may have following characteristics :

(1) Each species possesses a common gene pool with a free gene flow.
(2) Each species is in a process of continuous adjustment to its environment.
(3) Each species occurs in an ecological niche not necessarily occupied by other species.
(4) Each species possesses a constellation of isolating mechanisms that directly or indirectly prevent exchange of genes with related species.
(5) Each species has the capacity to give rise to new species.

Often the individuals comprising a species can be subdivided into smaller groups known as **subspecies**. These differ in minor but very distinct characters. Each subspecies occupies a separate range, and the members at the borderline of two adjacent subspecies have characteristics common to both.

Types of Species

Species can be of different types :

1. Agamo species. Species reproducing asexually are called agamo species.

2. Gamo species or biological species. Species reproducing sexually are called gamo species.

3. Sympatric species. Two or more species living in a common habitat are called sympatric species.

4. Allopatric species. Two or more species living in different habitat or different geographical conditions are called allopatric species.

5. Synchronic species. Two or more species found in the same time period are called synchronic species.

6. Allochronic species. Two or more species which are found in different time periods are called allochronic species.

7. Palaeospecies or Palaeontological species. Such species which are now available in the form of fossils are called palaeospecies. They are also known as **fossil species**.

8. Neontological species. Species which are found living at present are called neontological species.

9. Polytypic species. Species which are found in more than one subspecies are called polytypic species.

10. Monotypic. Species which are not subdivided in to subspecies.

11. Sibling species. These are allopatric species which are morphologically similar but are found in different yet adjacent habitats.

Law of Priority

If two or more than two names are given for a single animal species by different scientists at different time period, then the name first published in a scientific literature stands valid. This is the Law of Priority.

MODERN CLASSIFICATION OF ANIMAL KINGDOM

The entire living world was divided by **Linnaeus** into two kingdoms :

1. Animal kingdom. Animal world constitutes the animal kingdom. They do not possess chlorophyll, perform locomotion and they do not possess cell wall made up of cellulose.

2. Plant kingdom. The plant world constitutes the plant kingdom. They possess chlorophyll, cell wall made up of cellulose and they have no power of locomotion.

Ernst Haeckel, 1894 proposed two kingdoms but they could not get proper recognition :

1. Monera kingdom. Haeckel proposed it for enucleated or prokaryotic cell like blue green algae and bacteria.

2. Protista kingdom. Those living species which possess characters of both animals and plants were included by Haeckel in this kingdom. They include one celled eukaryotes.

Animals have been classified in various ways by various authors. The present classification is mainly based on the book, *General Zoology* by **Tracy. I. Storer** and **Robert L. Usinger**. They have classified the animal kingdom into 23 phyla (see Table 2).

ANIMAL KINGDOM

The animal kingdom has been divided into two subkingdoms :

Subkingdom I. Protozoa

These are unicellular animals which have been included in one phylum **Protozoa**.

Examples : *Euglena, Amoeba, Paramecium, Balantidium* and *E. coli* etc.

Subkingdom II. Metazoa

These are multicellular animals. Subkingdom metazoa has been divided into three Infra-kingdoms.

Infra-kingdom 1. Mesozoa

These are parasitic, worm like animals and lack tissues. They have been included in phylum **Mesozoa**. Example : *Dicyema.*

Infra-kingdom 2. Parazoa

They are marine as well as fresh water forms and have cellular grade of body organisation. They lack-tissues, choanocyte cells are found. They have

Table 2. **Outline classification of Animal Kingdom.**

ANIMAL KINGDOM

Sub Kingdom

- I. Protozoa
  - *Phylum I. Protozoa*
- II. Metazoa
  - I. Mesozoa
    - *Phylum 1 Mesozoa*
  - 2. Parazoa
    - *Phylum 1 Porifera*
  - Infrakingdoms 3. Enterozoa
    - A. Radiata
      - *1. Cnidaria*
      - *2. Ctenophora*
    - B. Bilateria
      - (A) Acoelomata
        - *1. Platyhelminthes*
        - *2. Nemertinea*
      - (B) Pseudocoelomata
        - *1. Aschelminthes*
        - *2. Entoprocta*
        - *3. Acanthocephala*
      - (C) Eucoelomata
        - *1. Ectoprocta*
        - *2. Brachiopoda*
        - *3. Phoronida*
        - *4. Mollusca*
        - *5. Annelida*
        - *6. Sipunculida*
        - *7. Echiuroidea*
        - *8. Arthropoda*
        - *9. Chaetognatha*
        - *10. Echinodermata*
        - *11. Pogonophora*
        - *12. Hemichordata*
        - *13. Chordata*

been included in phylum **Porifera**. *e.g.,* all sponges.

Infra-kingdom 3. Enterozoa

Tissues and organs are found. They include all the remaining metazoans. Infra-kingdom Enterozoa has been divided into two divisions :

Division A. Radiata. Radially symmetrical animals which do not possess organs *i.e.,* their body organization is of tissue grade. Two phyla are included in this division :

Phylum (a). Cnidaria. Nematoblast cells are found, tentacles present and comb plates are absent, *e.g., Obelia* and *Hydra.*

Phylum (b). Ctenophora. They possess comb plate which helps in locomotion. Nematoblasts and tentacles absent, *e.g., Beroe.*

Division B. Bilateria. Bilaterally symmetrical animals, triploblastic and have organ grade of body organisation. Bilateria has been divided into 3 sections :

Section *1. Acoelomata.* No body cavity or coelom. Space between body wall and visceral organs is occupied by mesoderm formed of parenchyma. It includes two phyla :

Phylum (a). Platyhelminthes. Parasitic and flat worms, *e.g.,* tapeworm *Taenia,* liver fluke *Fasciola.*

Phylum (b). Nemertinea (Rhynchocoela). Thin, elongated, ribbon like aquatic animals *e.g., Stichostemma.*

Section *2. Pseudocoelomata.* Pseudocoel found in the adult stage, parenchyma fills the spaces between body wall and visceral organs in young stage. This section includes three phyla :

Phylum (a). Aschelminthes. Slender, elongated and round body, mouth and anus at two different ends of the body. *e.g., Ascaris.*

Phylum (b). Entoprocta. Fixed and stalked animals, mouth and anus located at one end close to each other, *e.g., Loxosoma.*

Phylum (c). Acanthocephala. Flat bodied parasites, no alimentary canal, thorny proboscis at anterior end *e.g., Acanthocephala.*

Section *3. Eucoelomata.* They possess true coelom lined with mesoderm. Section Eucoelomata includes 13 phyla :

Phylum (a). Ectoprocta or Bryozoa. Aquatic, minute, colonial moss like animals. The digestive tract U-shaped, *e.g., Bugula.*

Phylum (b). Brachiopoda. Body covered with shell formed of calcium carbonate, interior with two spiral arms called **lophophore**. Commonly called **lampshells.** *e.g., Lingula.*

Phylum (c). Phoronida. Worm like cylindrical, unsegmented, live in self secreted membranous tube, lophophore is horse shoe-shaped and its digestive tract is U-shaped, *e.g., Phoronis.*

Phylum (d). Mollusca. Soft bodied animals having a membranous covering called **mantle** which secretes **calcarious shell**. Mostly aquatic forms respiring by gills and possess a muscular foot for locomotion *e.g., Unio,* and *Pila* etc.

Phylum (e). Annelida. Segmented worms underground or aquatic, having setae for locomotion. Body cavity is true coelom. *e.g.,* Earthworm, *Nereis.*

Phylum (f). Sipunculoidea. Unsegmented body without setae, *e.g., Sipunculus* or peanut worm.

Phylum (g). Echiuroidea. Anterior end of the body projected as proboscis infront of the mouth. Anus at the posterior end. A pair of ventral setae, *e.g., Echiurus.*

Phylum (h). Arthropoda. Body covered with exoskeleton formed of chitin, jointed and paired and appendages. Body divided into three parts : head, thorax and abdomen. Compound eyes *e.g.,* cockroach, scorpion, and prawn etc.

Phylum (i). Chaetognatha. Marine animals, slender and transparent body. Mouth surrounded by spicules, bristles or hooks, *e.g.,* Arrow worm or *Sagitta.*

Some Important Taxonomists

1.	**Aristotle**	:	Called as father of Zoology
2.	**Johnston**	:	Coined the name Mollusca.
3.	**Leeuwenhoek**	:	Observed Protozoa and discovered *Hydra* in 1703.
4.	**John Ray**	:	Gave the term species.
5.	**Jacob Klein**	:	Gave the name Echinodermata.
6.	**Carolus Linnaeus**	:	Father of Modern Taxonomy.
7.	**Lamark**	:	Coined the name Annelida.
8.	**Gold-fuss**	:	Coined the word Protozoa.
9.	**Robert Grant**	:	Coined the word Porifera.
10.	**Eschboltz**	:	Gave the word Ctenophora.
11.	**Von Siebold**	:	Gave the name Arthropoda.
12.	**Leuckart**	:	Coined the word Coelenterata.
13.	**Gegenbaur**	:	Coined the name Platyhelminthes.
14.	**Haeckel**	:	Formed the group Protista and Monera.
15.	**Sollas**	:	Gave the name Parazoa.
16.	**Bateson**	:	Used the word Hemichordata.
17.	**Mayer**	:	Gave the modern definition of biological species.
18.	**Grobben**	:	Gave the name Aschelminthes.

Phylum (j). Echinodermata. Marine animals with pentamerous body provided with spines in skin, water-vascular system and tube feet for locomotion are present, *e.g.,* Starfish and sea cucumber.

Phylum (k). Pogonophora. Slender body shaped like a long tube. No digestive tract, beard like ciliated tentacles, *e.g.,* Beard worm or *Spirobrachia.*

Phylum (l). Hemichordata. Unsegmented body, respiration by gills, rudimentary notochord *e.g.,* Tongue worm or *Balanoglossus.*

Phylum (m). Chordata. Notochord present at some stage of life. Hollow nerve cord on the dorsal side and heart on the ventral side of the body. They are aquatic, terrestrial or aerial, *e.g.,* fish, frog, snake, bird, dog and man.

In this way **Storer** and **Usinger** divided the animal kingdom into **23 phyla**. Out of these, **10** are **major phyla** and the remaining are **minor phyla**.

MODERN MOLECULAR METHODS IN TAXONOMY (Use of DNA and RNA Sequences to Infer Evolutionary Relationships)

Introduction

In earlier years, classifying millions of different species of living organism has been a Herculean task for taxonomists. Taxonomy was limited to anatomical and physiological comparisons. It was due to the lack of appropriate technology to analyze the bio-molecules *i.e.*, DNA, the molecule of heredity of living organisms. The theoretical frameworks for molecular **systematics** were described by **Emile Zuckerkandl** (1960s), **Emanuel Margoliash** (1970s), **Linus Pauling** (1980s), and **Walter M. Fitch** (1990s) in their works.

This part of the chapter would introduce you to the fast emerging latest concepts, technology and new terms in molecular taxonomy. This is important to emphasize here that this technology is still very expansive and out of reach of and individual researcher.

Modern taxonomists are classifying animals at the molecular level called **phylogenetic systematics** or **cladistic taxonomy**. Besides using age old conventional methods such as comparative and functional anatomy embryology, immunology, serology, biochemistry and population genetics, **Phylogenetics** is a science of classification that arranges organisms by their **degree of evolutionary relatedness** and not just by their morphological characteristics. This new approach is resulting in a complete metamorphosis of traditional taxonomic categories for every living thing. This field evolved after gaining knowledge of basic genetics, including DNA replication, protein synthesis, mutation and recombination. *Arabidopsis thaliana* was the first plant genome which was sequenced.

Phylogenetics is the science of estimating and analyzing evolutionary relationships. Phylogenetic relationships among micro-organisms are especially difficult to discern. Molecular biology often helps in determining genetic relationships between different organisms. Nucleic acids (DNA and RNA) and proteins are '**information molecules**' which retain a record of an organism's evolutionary history. In molecular phylogeny, the relationships among organisms or genes are studied by comparing homologues of DNA or protein sequences.

Phylogenetics is a method of classification based on the study of evolutionary relationships between groups of organisms, and the integration of proper names of groups of organisms into a hierarchical system which reflects their phylogeny. Without taxonomy, there would be no meaningful genome projects, and medical science, for example, whould be seriously compromised. Without taxonomy, we could not begin to understand biodiversity and the related issue of conservation (**Nature**, 2002).

Streams of Thoughts/Schools Dealing with Taxonomy and Phylogenetics

1. Evolutionary taxonomy. It is an early school of phylogenetic conclusions, which recognises that similarity among species could arise either because species were closely related or because of convergent or parallel evolution.

2. Numerical taxonomy or phenetics. In early 1960s, a group of statisticians and biologists introduced a new approach, known as **numerical taxonomy** or **phenetics**, initially developed by **Sokal** and **Sneath** (1963). This school considered that organisms should be grouped on the basis of overall similarity, independently of whether these groupings represent phylogeny. Numerical taxonomy was popular, however, its use has declined since the clusters do not necessarily reflect true phylogenetic relationships.

3. Phylogenetic systematics (**or Cladistics sensu lato**). In 1966, **W. Hennig** published the book *Phylogenetic Systematics*, giving rise to the highly influential school of systematics known as **Cladistics**.

Taxonomy is getting fashionable and topics like **Phylocode** and **Bar Coding** are among the most controversial and discussed subjects in taxonomy today.

Major Landmarks in DNA Sequencing

(1) **1953**–Discovery of the structure of the DNA–double helix.

(2) **1972**–Development of recombinant DNA technology, which permits isolation of defined fragments of DNA. Prior to this, the only accessible samples for sequencing were from bacteriophage or virus DNA.

(3) **1977**–The first complete DNA genome to be sequenced was that of bacteriophage ψ × 174.

(4) **1977–Allan Maxam** and **Walter Gilbert** published "DNA sequencing by Chemical Degradation". **Frederick Sanger**, independently, published "DNA Sequencing with Chain-terminating Inhibitors".

(5) **1984**–Medical Research Council scientists decipher the complete DNA sequence of the Epstein-Barr virus, 170 kb.

(6) **1986–Leroy E. Hood's** Laboratory at California Institute of Technology and **Smith** announce the first semi-automated DNA sequencing machine.

(7) **1987**–Applied Biosystems marketed first automated sequencing machine, the model ABI 370.

(8) **1990**–The U.S. National Institutes of Health (NIH) begins large-scale sequencing trials on *Mycoplasma capricolum*, *E. coli*, *Caenorhabditis elegans*, and *Saccharomyces cerevisiae*.

(9) **1991**–Sequencing of human expressed sequence tags begins in Craig Venter's Lab, an attempt to capture the coding fraction of the human genome.

(10) **1995–Craig Venter**, **Hamilton Smith**, and colleagues at The Institute for Genomic Research (TIGR) published the first complete genome of a free-living organism, the bacterium *Haemophilus influenzae*.

(11) **1996–Pal Nyren** and his student **Mostafa Ronaghi** at the Royal Institute of Technology in Stockholm published their method of pyrosequencing.

(12) **1998–Phil Green** and **Brent Ewing** of the University of Washington published '**phred**' for sequencer data analysis.

(13) **2000–Lynx Therapeutics** published and marketed 'MPSS' - a parallelized, adapter/ligation-mediated, bead-based sequencing technology, launching '**next-generation**' sequencing.

(14) **2001**–A draft sequence of the human genome was published.

Molecular biology and **molecular taxonomy** have provided a great deal of evidences for the evolutionary theory of the origin of species. **Molecular genetic** methods provide taxonomy with a powerful tool to re-identify or re-describe species as they apply to all organisms irrespective of developmental stage, sex, or body part.

Methods in Molecular Systematics

A decade ago, **molecular techniques** in taxonomy were confined to only a few particularly 'difficult' taxa. But now because of standardized and comprehensive molecular sampling of biological diversity, involving the existing taxonomic expertise, has created extensive reference databases called '**bio-bank**'. It is a type of **bio repository** that stores biological samples usually of human to be used in research. Because of these, it has become much easier to recognize species and their inter relationships.

Allozyme fingerprinting. Different alleles produce slightly different proteins which migrate differently on an electrically charged gel. Takes about 4 hours per gel, but up to about 30 samples can be run at a time.

DNA sequencing. Expensive but cost has reduced considerably. Most of the process has now been automated.

Examples

Recently, with the advent of modern molecular biology, the amount of **DNA** sequence data from diverse taxa has simply exploded. This advancement has opened new avenues to analyze the relationships among organisms. In contrast to anatomy, which is largely a qualitative analysis, **DNA** and **protein** comparisons permit the construction of quantitative comparisons among organisms. Furthermore, since DNA is the material of heredity and is, in a sense, a record of a animal's genetic ancestors, modern molecular biology allows the direct assessment of an **organism's genealogy**.

Skunks

A whole new field has been opened, for example, it was long assumed that **skunks** were part of family **Mustelidae**. The mustelids of the group mammals, includes all the weasels, otters, badgers, polecats and the wolverine and minks. Skunks possess anal scent glands that produce stinking liquid, which most of the mustelids also possess and produce. Skunks resemble morphologically with other mustelids.

When we obtain sequences of a section of a certain gene from the members of several taxa, the difference between the sequences of these genes allows us to establish the probable order of branching off of the relevant taxa from the common ancestor.

In **1997**, an **mtDNA analysis of skunk** was done. By traditional classification they have put them in wrong family, on the basis of morphological similarities. Skunks were reclassified and put into family called Mephitidae along with the African Zorilla that also resembles skunk species (*Spilogale*). Although direct phylogenetic relationships between skunks and the zorilla were not largely accepted in twentieth century. But DNA study revealed that two animals called '**stink badgers**' which are natives of Indonesia and the Philippines, are actually the only Old World representatives of the skunk lineage.

Giant Panda

All living things have DNA which provides a complete and accurate record of ancestry. In the strands of DNA, one can figure out how the evolution of a particular organism happened. For example, the taxonomic status of the **Giant Panda** was discussed for decades. Some authorities placed it in the family Procyonidae, while others placed it in the bear family Ursidae. **Red pandas**, which very much look-like raccoons, have a chromosome number that is similar to the **Giant Panda**.

Giant Pandas have 42 chromosomes, while **Red Pandas have 36**. Both have a 'wrist' that is made from the radial sesamoid bone. Both animals eat a lot of bamboo, and this 'wrist' helps them hold the bamboo. Traditionally, these animals were believe to be close relatives-that is why we call both of them **pandas**. Therefore, they were put in the same family.

The bears have 74 chromosomes. This argument was used to deny the giant panda to have a place in the bear family– until a reciprocal **chromosome painting analysis** revealed that pandas have more chromosomes. Chromosome is a technique for visualizing **chromosome aberrations** using fluorescently labelled DNA probes which are hybridized to chromosomal DNA. One chromosome of giant panda is often similar to 2 chromosomes of other bears. The difference is due to their two chromosomes which are fused together. In a way giant Panda has about 84 chromosomes, this brings Panda closer to bears who have 72 chromosomes, alongwith many other morphological similarities.

Further, **Mitochondrial DNA analysis,** has revealed that **Red Pandas** do not belong to family Procyonidae. They are currently listed in a different family of its own called family **Ailuridae**. This family comprises of the Red Panda (the sole living representative) and its extinct relatives.

Human

A laymen generally ask that who is our closest non-human relative ? By DNA evidence, **humans** and two species of **chimpanzee**, the common chimpanzee (*Pan troglodytes*) and the bonobo chimpanzee (*Pan paniscus*), are closely related. We share about 99% of our genomes with them; we had a common ancestor four to six million years ago. If we use the comparative morphology method, the results would be different. **Convergent** and **parallel** evolution are major factors that can make **morphology** more similar. The organisms are more closely related than they actually are. Considering all this the **molecular evidence** suggests that the two species of **orangutan** namely Bornean orangutan (*P. pygmaeus*) and the Sumatran orangutan (*P. abelii*) split off from the common ancestor of all the African apes between 15.7 and 19.3 million years ago. The orangutan genome is only 97% similar to that of a human. This makes chimpanzees our closest relatives then orangutans and gorillas. Other animals like mouse the difference in amino acids with human is only 16, with horse 18, bird 35, frog 62 and shark 79. This indicates that mouse is closest to human beings.

These examples give you an idea, that with the advent of molecular biology, the new techniques and instrumentation are throwing the light on the phylogeny and taxonomy of the animals.

Human Genome Project (HGP)

This project was started in 1990 and was completed in 2003. A project coordinated by the U.S. Department of Energy and the National Institutes of Health. During the early years of the HGP; additional contributions were from United States, the United Kingdom, Japan, France, Germany, Spain, China and others. Following were the projects :

- To identify all the approximately 20,000-25,000 genes in human DNA.
- To determine the sequences of the 3 billion chemical base pairs that make up human DNA.
- To store this information in databases.
- To improve tools for data analysis.
- To transfer the related technologies to the private sector.
- To address the ethical, legal, and social issues (ELSI) that may arise from the project.

New Terms and Methods

The general interest about biodiversity conservation, the advances of internet and web pages, the progress in molecular techniques, the development of statistics in phylogeny, and the global projects are giving some hope to new techniques like **Phylocode** and **Bar Coding**.

It is necessary for the students to get familiar with new terms like **International Code of Phylogenetic Nomenclature (ISPN)**, also known as the **PhyloCode**. Two zoologists from USA, **Kevin de Queiroz** and **Jacques Gauthier**, started the discussions in **1990's** and laid the **theoretical foundation** to a new nomenclatural code for all organisms, which was named as PhyloCode. In 1998 a meeting in Harvard drew out the lines for a **PhyloCode**, and a first draft was published on the web in 2000. The PhyloCode is controversial, and has led to a worldwide and very interesting debate.

New Terminology

It would be beneficial for a student to get familiar with following new terms :

1. Computational phylogenetics. It is the application of computational **algorithms**, methods and programmes to phylogenetic analysis. The goal is to assemble a **phylogenetic tree** representing a hypothesis about the evolutionary ancestry of a set of genes, species or other taxa. These techniques have been used to explore the **family tree** of species and the relationships between specific genes shared by many types of organisms. Traditional phylogenetic relies on morphological data obtained by **measuring** and **quantifying** the **phenotypic properties** of representative organisms, while the more recent field of **molecular phylogenetic** uses, **nucleotide sequences**, encoding genes or acid sequences encoding proteins as the basis for classification. Many forms of molecular Phylogenetics are closely related and make extensive use of **sequence alignment** in constructing and refining phylogenetic trees. These are used to classify the evolutionary relationships between **homologous genes**. These are represented in the **genomes of divergent species**. The phylogenetic trees constructed by **computational methods** are unlikely to reproduce perfectly the evolutionary tree. This tree that represents the historical relationships between the species being analyzed.

2. Molecular phylogenetics. It is the analysis of hereditary molecular differences, mainly in DNA sequences, to gain information on an organism's evolutionary relationships. The result of a molecular phylogenetic analysis is expressed in a phylogenetic tree. Molecular phylogenetics is one aspect of **molecular systematics**. A broader term that also includes the use of molecular data in taxonomy and biogeography.

Every living organism contains DNA, RNA, and proteins. In general, closely related organisms have a high degree of similarity in the molecular structure of these substances, while the molecules of organisms distantly related usually show a pattern of dissimilarity.

3. Molecular clock. A technique in molecular evolution that uses fossil constraints and rates of molecular change to deduce the time in geologic history when two species or other taxa diverged. Conserved sequences, such as mitochondrial DNA, are expected to accumulate mutations over time, and assuming a constant rate of mutation providing a **molecular clock**. Which is based on the **Molecular Clock Hypothesis (MCH)**. Molecular phylogeny uses such data to build a '**relationship tree**' that shows the probable evolution of various organisms.

Another application of molecular phylogeny is in DNA barcoding, wherein the species of an individual organism is identified using small sections of mitochondrial DNA.

4. GenBank. It is the genetic sequence data base of the **National Institute of Health (NIH)**. It includes collection of all available DNA sequences of published papers. Currently > 28,000 million bases from > 250,000 species (**Harris**, 2003) are rising daily. Most of the journals, before paper publication, require that sequences used in the paper must be registered in data bases such as GenBank, where they are publicly available and can be used by any one.

5. DNA barcoding. In February 2005, the **First International Barcoding of Life Conference** was held in London, convened by the **Consortium for the Barcode of Life (CBOL)**. DNA barcoding is now being proposed as a way to catalogue life. This new technology makes use of short but specific **DNA tags** or '**barcodes**' to distinguish one species from another. It uses a small part of the mitochondrial genome, 650 to 750 bases of the cytochrome **c oxidase I gene (COI)** to provide a unique fingerprint for each species. Most of the eukaryotes, COI variation appears to be lower among individuals within a species than among those from different species (**Marshall**, 2005). **Paul Hebert** (2008) is a molecular biologist and Director of the Biodiversity Institute of Ontario at the University of Guelph in Canada. He is best known for his discovery and proliferation of DNA barcoding, which uses a specific section of genetic code to identify any species. He is considered the '**Father of Bar Coding**'.

6. SNP (Single Nucleotide Polymorphisms). Molecular systematics has been made possible by the availability of techniques for DNA sequencing. Which allow the determination of the exact sequence of nucleotides or bases in either DNA or RNA. More and more sophisticated DNA analysis, including the use of **SNP (Single Nucleotide Polymorphisms) chips** to examine large parts of the genome, are revolutionizing the way of modern zoologists classifying the organisms.

A review by **Guerra-Garcia**, **Espinosa** and **Garcia-Gomez** (2008) on Trends in Taxonomy gives an overview of the subject. The difficulties it is facing are lack of funding for taxonomic study, availability of less number of young taxonomists, and no seperate recognition of the taxonomist in the society.

V. M. Sathish Kumar (2011) of Zoological Survey of India, has written a review Taxonomy in India in the 21st Century : Call for a Digital

Revival. The abstract reads "The biodiversity of the planet is under threat due to various reasons. For the conservation and management of the bio-resources we need to precisely identify a species, so that the species can be accessed and managed accordingly. This is the very first basic step for every taxonomist. Though traditional taxonomy has been used in systematics for over 250 years with the identification of more than 1.75 million species. In many countries like India, traditional taxonomy is yet to adopt the latest molecular techniques available for a better and accurate identification of taxa. Enhanced taxonomic infrastructure and research tools can certainly allow Indian taxonomists to speed up the process of species exploration, description and classification. It will enable the world scientific community to get access to India's flora and fauna dynamically without excuses of reduced funding or dearth of taxonomists for not mapping the biological diversity of India. This article overviews the latest digital advancements to revolutionize the taxonomical research in India in the current century."

Institutes of Molecular Taxonomy

It is not possible to list all the Universities, Departments and Institutes dedicated to Molecular Taxonomy, but a few leading have been named here :

1. **'Molecular Taxonomy Lab' (MTL)** of the N.H.M.C was established in 2001 and it is currently supervised by **Dr. Parmakelis Aris** and **Poulakakis Nikos**. It is a fully equipped molecular laboratory supporting the research projects of the N.H.M.C that involve phylogenetic approaches and study of the genetic structure of animal and plant populations.

2. **The Molecular Taxonomy Laboratory of the Hungarian Natural History Museum** was established in 2006. The research scope of the laboratory covers molecular taxonomy and phylogenetics, phylogeography, population biology and conservation biology.

   These studies are carried out in close cooperation with the Departments of Botany, Zoology and Anthropology. The laboratory has been equipped with latest and up-to-date set of instruments.

3. **Department of Evolution Genomics and Systematics, Evolutionary Biology Centre (EBC), Uppsala University, Norbyvagen 18D, Uppsala 75236, Sweden.**

4. **Department of Evolutionary Biology.** Zoological Institute, Technical University of Braunschweig, Spielmannstrasse 8,38106 Braunschweig, Germany.

5. **Department of Biodiversity and Evolutionary Biology,** Museo Nacional de Ciencias Naturales, CSIC, C/José Gutierréz Abascal 2, Madrid 28006, Spain.

6. **The Yale Institute for Biospheric Studies, Molecular Systematics and Conservation Genetics Center** (YIBS-MSCG) was established in 1998.

7. **Institute for Molecular Systematics,** School of Biological Sciences Victoria University, Wellington, New Zealand.

8. **The Laboratory of Molecular Taxonomy** (LMT) was established in 2000 as a joint project of the Institute of Parasitology and Faculty of Biological Sciences, University of South Bohemia, Czech Republic. The laboratory is designed to study molecular evolution of single celled eukaryotes.

This field of research still has to go a very long way involving interest of young scientists, their hard work, determination and expensive technologies to sequence the genomes of millions of organisms to come to a more modern way of finding out who is close to whom genetically and not just morphologically.

Important Questions

➤➤*Long answer type questions*

1. Write an essay on animal diversity.
2. Why is there a need of classification ? What is the significanse of classification.
3. Write a note on nomenclature of organisms. Write a note on taxa or groups in classification.
4. Explain modern classification of Animal Kingdom.
5. Write in detail on Molecular Taxonomy, giving examples now this technology is throwing a new light on classification?
6. Write briefly on PhyloCode and bar coding, biobanks, molecular clock, Genbank.
7. What are the Modern Molecular Methods in taxonomy ? Explain.

➤➤*Short answer type questions*

1. Define species giving one example.
2. Briefly describe significance of classification.
3. What is the main characteristic of phylum Echinodermata ? Write zoological names of four animals of this phylum.
4. Who gave binomial nomenclature ?

➤➤*Fill in the blanks*

1. A biologist who examines, identifies, describes and names a species of birds is working at the level of .... taxonomy.
2. A biologist who identifies a species of bird and places it in its scheme of classifcation, modifying the classification if necessary, is working at the level of .... taxonomy.
3. A biologist who studies a group of birds for the purpose of working out their phylogenetic relationship is working at the level of .... taxonomy.

➤➤*Multiple choice questions*

1. Linnaeus is associated with :
   (a) inheritance of acquired characters
   (b) binomial nomenclature
   (c) law of independent assortment
   (d) law of limiting factors
2. Who propounded binomial system of nomenclature :
   (a) Carolus Linnaeus (b) Gregor Mendel
   (c) Pasteur (d) Darwin
3. Radial symmetry is best seen in :
   (a) Mollusca (b) sponge
   (c) star fish (d) fishes
4. Inter-breeding animals are members of the same :
   (a) order (b) species
   (c) genus (d) family
5. One celled animals and plants have been kept in :
   (a) Monera (b) Protista
   (c) Protozoa (d) Parazoa
6. Animals living on the surface of sea water are called :
   (a) lotic (b) pelagic
   (c) benthic (d) lentic
7. Branch of biology that deals with classification :
   (a) histology (b) taxonomy
   (c) physiology (d) geology
8. Sometimes different scientists give different names of the same species, in such situation :
   (a) all these names are cancelled and a new name is put forward
   (b) people are free to choose any name
   (c) first name is considered right
   (d) last name is considered right
9. Which is a species :
   (a) carnivora (b) *Canis*
   (c) *Canis femiliaris*
   (d) Mammalia
10. According to the system of binomial nomenclature every organism has :
    (a) two names one for the place to which it belong and the other of the discoverer
    (b) one name established by two scientists
    (c) one scientific name of which first part is genus and other species
    (d) two names one scientific and one common
11. A new species is formed when :
    (a) because of new genic populations organism with new traits are born
    (b) exchange of parts between homologous chromosomes takes place

(c) exchanges in the genotype in population leads to sexual isolation
(d) the exchange of parts of chromosomes during gametogenesis leads to the formation of new genotype

12. A new species can be formed when :
(a) individual variations leads to sexual isolation
(b) there is no relationship between phenotype and genotype
(c) an individual is produced by the combined effects of phenotypic and environmental influences
(d) common genotype of similar species

13. Inter-breeding animals belong to the same :
(a) genus (b) species
(c) family (d) order

14. In which character vertebrates differ from invertebrates:
(a) coelom (b) tail
(c) dorsal tubular nerve (d) brain cord

15. What is the basic unit of classification :
(a) order (b) phylum (c) genus (d) species

16. Which is the matching set in classification :
(a) star fish, jelly fish, cuttle fish, *Octopus*
(b) leech, locust, sea urchin, lobster
(c) *Nereis,* planaria, round worm, earthworm
(d) millipede, crab, centipede, cockroach

17. Which animals are of similar genus :
(a) *histolytica* and *coli*
(b) *histolytica* and *Plasmodium*
(c) *histolytica* and *falciparum*
(d) *histolytica* and *bancrofti*

18. Animals active at day time are called :
(a) nocturnal (b) herbivore
(c) diurnal (d) insectivore

19. Artificial system of classification was first adopted by :
(a) Simpson (b) Linnaeus
(c) Dobzhansky (d) Mayr

20. Carolus Linnaeus system of nomenclature is :
(a) natural (b) artificial
(c) phylogenetic (d) binomial

21. Which scientist made significant contribution in the field of animal classification :
(a) Oparin (b) Linnaeus
(c) Lamarck (d) Pasteur

22. Branch of science concerned with classification of animals :
(a) parasitology (b) ecology
(c) genetics (d) taxonomy

23. Who wrote the book '*Systema Naturae*' :
(a) Lamarck (b) Darwin
(c) Buffon (d) Linnaeus

24. Protista obtain food as :
(a) chemosynthesis (b) photosynthesis
(c) holotrophic
(d) photosynthesis, holotrophic, symbiosis

25. Phylogenetic classification is based on :
(a) external similarity (b) common evolution
(c) habit and habitat (d) utilitarian system

26. Radial symmetry is generally exhibited by animals :
(a) which are attached with substratum
(b) which are aquatic
(c) which have ciliary feeding
(d) which have one opening of alimentary canal

27. Taxonomically a species is :
(a) a group of evolutionary related populations
(b) population having same evolutionary basis
(c) catagory to which most taxonomic information is attached
(d) fundamental unit in phylogenetic history of organisms

28. Linnaeus gave natural system of classification based on :
(a) morphology (b) anatomy
(c) evolutionary trends (d) all of these

29. Radial symmetry is usually associated with :
(a) aquatic life (b) creeping mode of life
(c) lower grade of organisation
(d) sedentary mode of life

30. A taxon with reference to classification of organisms can be explained as :
(a) group of organisms based on chromosome number
(b) group of similar species
(c) group of rank of organisms
(d) group of similar genera

Answers

➤➤*Multiple choice questions*

1. (b)	2. (a)	3. (c)	4. (b)	5. (b)	6. (b)	7. (b)	8. (c)	9. (c)	10. (c)
11. (c)	12. (a)	13. (b)	14. (c)	15. (d)	16. (d)	17. (a)	18. (c)	19. (b)	20. (d)
21. (b)	22. (d)	23. (d)	24. (d)	25. (b)	26. (b)	27. (b)	28. (d)	29. (d)	30. (c)

➤➤*Fill in the blanks*

1. beta, 2. alpha, 3. gamma.

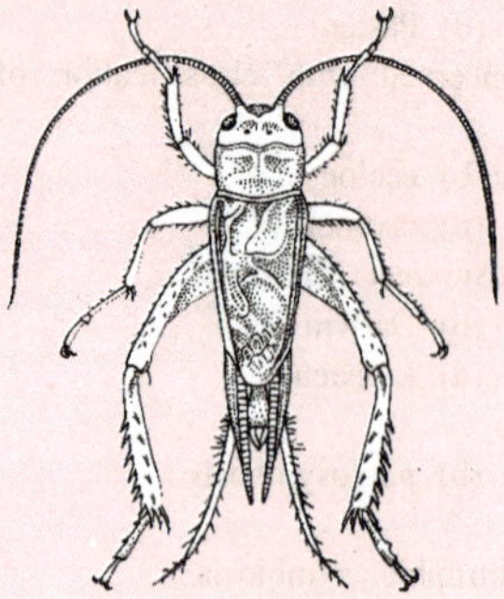

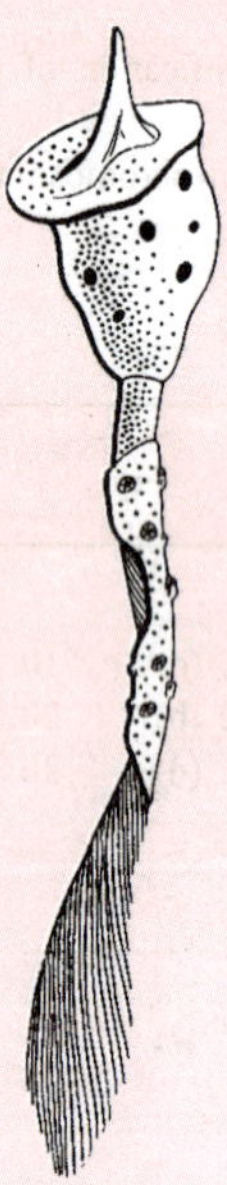

2

The Invertebrates

WHY WE STUDY INVERTEBRATES ?

Famous Biologist **E. O. Wilson** has called the invertebrates as "little things which run the world." It is because of their number, variety and influence on the larger organisms and even entire ecosystem. These different creatures like butterflies, beetles, bees, ants, dragonflies, snails, lobsters and starfishes etc., are primary builders of a healthy environment. They build amazingly colourful coral reefs of the oceans, and serve as food for countless other animals.

Invertebrates are a part of nearly every food chain, either directly, as food for fishes, amphibians, reptiles, birds, mammals or indirectly, as agents in the endless recycling of nutrients in the soil. Food webs are often dependent on invertebrate species performing essential services such as pollination or seed dispersal. We can not think to live on the earth without the presence of invertebrates.

The Animal Kingdom has categorized just over a million scientifically described species, in thirty five phyla. Phylum Arthropoda (insects, spiders, crustaceans, millipedes and centipedes among others) has an estimated 9,99,059 identified species, or 85% of the total identified animal species. A part from arthropods the invertebrates (excluding viruses and bacteria) number reaches to 12,38,000 or 94%. The phylum Chordata, which includes fish, birds and mammals, contains nearly 50,000 (5%) species, of which nearly 4600 (0.04%) are only mammals. It is estimated that five to eight million insect species have not been identified or discovered. Undescribed 5,000 to 10,000 species of Chordates are waiting for identification and description. Certain marine taxa, particularly small and deep sea organisms, are very poorly known, further telling that oceanic species diversity still remains underestimated.

The study of invertebrates have aquainted us with their involvement in the serious effects on the sustainability of several agricultural production. The invertebrates and micro-organisms play a vital role in maintaining and enhancing the soil fertility, detoxifying of pesticides and other pollutants, and also in biological control of agricultural pests.

WHAT IS AN INVERTEBRATE ?

The described species of animals are over one million. Of these, about 5% animals possess a backbone or vertebral column and are known as **Vertebrates**. The remaining animals which lack backbone or vertebral column are referred to as **Invertebrates**. All such animals are included in a group, '**Invertebrata**'. This word is a semantic blanket that covers a wide range of animals regardless of their size, shape, morphological characteristics and phylogenetic relationship. We know that worms, butterflies, sponges, corals, snails, crabs and strafishes are not alike in any specific structure, even then all have been included in 'Invertebrata'. This has been possible only due to the fact that these animals do not possess a vertebral column.

INVERTEBRATE PHYLA

Presently there are about 30 invertebrate phyla, which are characterised by a unity of basic structural pattern in each of them. This means that in each phylum, the anatomical features are constructed on the same ground plan in many respects though the members may differ in external features. The common anatomical ground plan implies a unique network of relationship among the groups of structural units which compose it. Other significant feature of inter-relationships among the members of the same phylum is functional. All animals of a group work as efficient machines with similar functional integration. Another important character, which binds members of the individual phylum with one cord, is the **common ancestry**. It has been confirmed by evolutionary studies that all the members of an individual phylum have been derived directly or indirectly from a common primitive ancestral type. Thus, the 30 phyla display 30 patterns, each manifesting a characteristic, anatomical and functional integrity and common ancestry.

INVERTEBRATA *VERSUS* NON-CHORDATA

Animals are often distinguished into two major categories, the '**Chordates**', and the '**Non-Chordates**'. This division is based on the presence or absence of a singular character, the **notochord**, a stiff rod-like structure which serves as an internal skeleton. **Phylum Chordata** includes those animals which possess notochord during any period of their life. **Subphylum Vertebrata** includes those animals in which the supporting action of notochord is partly or wholly taken over by a segmented vertebral column.

Animals without notochord are called **Non-Chordates**, and those with vertebral column are called **Chordates**. The terms **Non-Chordates** and **Invertebrates**, are often used for each other as synonyms, but strictly speaking they are not synonymous. For example, the Protochordates are all invertebrates (due to lack of vertebral column), but they are still chordates (due to presence of notochord). Thus, all the non-chordates are invertebrates whereas all the invertebrates are not non-chordates (*e.g.,* protochordates).

This means that the Invertebrates minus Protochordates are equal to Non-Chordates. The present book describes only those Invertebrates which are Non-chordates. It does not deal with the other Invertebrates, *i.e.,* Protochordates, which have been described in 2nd volume of the book **Vertebrates**.

It must be obvious by now that it is more scientific to use the term '**Non-Chordates**' for animals without a notochord. But in usage, they are still called the '**Invertebrates**' which is a broader and older term.

DIVERSITY OF INVERTEBRATES

As already stated, the Invertebrates represent a heterogeneous assemblage which includes such diverse forms as sponges, corals, worms and butterflies. Accordingly, they exhibit a great diversity of form, structure, physiology, habit and habitat, etc.

1. Numerical strength. About one and a quarter million species of living animals are known at present, of these only 5% *i.e.,* about 50,000 species belong to the chordates. The remaining 95% *i.e.,* about 1.2 million species belong to the invertebrates. Out of these nearly one million are Arthropods in which class insecta alone includes more than 8,67,000 species. Other major invertebrate groups include more than 66,000 molluscs, 31,000 protozoans, 4,500 sponges, 10,000 coelenterates, 17,500 platyhelminthes, 20,000 nematodes, 12,700 annelids and 6,200 echinoderms.

It has been calculated that the number of **extinct** species is around seven times the number of **living** species. There might be approximately 7 to 8 million invertebrate species in all.

2. Size. It is a general principle that every organism is distinguished by a characteristic size. The invertebrate animals range in size from the microscopic protozoans to the large-sized cephalopods. At the lowest extremity, the malarial parasite (*Plasmodium*) is so small that it occupies nearly one-fifth of a human red blood corpuscle. At the uppermost extremity, a species of the giant squids (*Architeuthis*) of North Atlantic has been reported as bigas 16.5 meters in length, including the tentacles.

3. Shape. Animals of nearly all shapes are included amongst the invertebrates. The irregular everchanging body shape of *Amoeba,* plant-like appearance of many sponges and coelenterates, leaf-like and ribbon-shaped flatworms, elongated and vermiform annelids, nemerteans and nematodes, star-shaped star-fishes, etc., display spectra of body shapes.

4. Symmetry. All symmetries are represented by the invertebrates. Protozoans show bilateral as well as radial symmetry and some are asymmetrical. Sponges are either asymmetrical or radially symmetrical. Coelenterates are radially symmetrical. Ctenophores show biradial symmetry. The members of remaining phyla are mostly bilaterally symmetrical. Spherical symmetry too is represented in invertebrates, principally in some spherical protozoans like *Heliozoa* and *Radiolaria.*

5. Habits and habitat (Ecology). Changes in morphology and function have enabled animals to live in different kinds of habitats. The invertebrates occupy a great variety of habitats and have adapted themselves to different modes of life. They are found in the seas, in fresh waters, in air and on all parts of land from snow-covered mountains to deserts. Protozoans are cosmopolitan in distribution, found in all habitats and are free living, parasitic or commensal. Sponges and coelenterates are aquatic animals, chiefly marine, but helminths again exhibit all forms of life. Annelids also occur in different habitats and some (*e.g.,* leeches) are sanguivorous in habit. The most diverse and biologically successful among all animals are the arthropods. They live in all sorts of habitats and consume the largest amount and variety of food. The insects are the only invertebrates capable of flight. Molluscs are mostly marine, but snails and clams live in freshwater while some species of snails and slugs are terrestrial. Echinoderms are exclusively marine.

6. Grades of organization. Invertebrates show all grades of organisation. **Protoplasmic**

grade is found in Protozoa, as all activities at this level are confined within the limits of a single plasma membrane (plasmalemma). **Cellular grade** is characteristic of sponges because in a sponge only cells exhibit division of labour for performing specialized functions. **Cell-tissue grade** is seen in coelenterates as their cells are not only specialized for different functions but also certain similar cells gather together to form tissues as well. A noteworthy example is the nerve net formed by nerve cells and their processes. **Tissue-organ grade** appears in flatworms with the arrangement of tissues to form organs. **Organ-system grade**, where organs join together in a system to perform some function, is typical of all higher invertebrate forms.

7. Types of coelom. In some many-celled invertebrates, like sponges and coelenterates, the body is a double-layered sac surrounding a single cavity, which opens to outside through a mouth. Such animals are **acoelomate** as they have no coelom (*e.g.,* from Acoelomate Phyla Platyhelminthes). Other invertebrates have a cavity in between the body wall and the gut. This cavity is **pseudocoelom** in nematodes as it is not lined by mesoderm (*e.g.,* from Pseudocoelomate-Phyla Acanthocephala, Aschelminthes). In higher invertebrates, the coelom is lined by the mesoderm and hence it is the **true coelom**.

8. Diversified respiratory surfaces. The protozoans, sponges, coelenterates and many worms have a direct diffusion of gases between the organism and the environment. In most of the annelids, the exchange of gases is through moist skin. **Gills** are common in most of the higher invertebrates. Echinoderms use **dermal branchiae** and **tube feet** for the above purpose. In insects the **tracheal system** is adapted for aerial respiration. Sea cucumbers have **respiratory trees** acting as respiratory organs.

9. Diversified excretory mechanisms. In protozoans, sponges and coelenterates excretion is by **direct diffusion** through cell membranes. Flatworms possess characteristic **flame cells**, while annelids and molluscs use **nephridia** for the purpose. The excretory devices in insects are **Malpighian tubules**. Echinoderms and some other invertebrates have amoeboid cells or **phagocytes** for storage and disposal of excretory products outside the body.

10. Varied modes of reproduction. Among invertebrates, mode of reproduction varies from simple **asexual binary fission** to most complicated **sexual reproduction**. In certain kinds, **parthenogenesis** (Gr., **parthenos**, virgin + **genesis**, origin) is also met with, in which an unfertilized egg develops into a complete individual. It occurs in rotifers, bees, some other insects and certains crustaceans. **Paedogenesis** (Gr., **pais**, a child + **genesis**, origin), in which eggs produced by immature individuals develop into larvae, is seen in some invertebrates like the gall fly, *Miastor*. In sexually reproducing invertebrates **herma-phrodites** or bisexual forms are met with, particularly in coelenterates, platyhelminthes, annelids and crustaceans. Fertilization is either internal or external. Development is direct or indirect. In the latter case the development includes both larval stages and metamorphosis.

Animal Diversity Web (ADW; https://animaldiversity.org/) is an online non-profit site created by **Philip Myers**, a former biology professor at the University of Michigan. This site offers database that has the natural history, classification, species characteristics, conservation biology, and distribution information on thousands of species of animals. It includes thousands of photographs, hundreds of sound clips, and a virtual museum.

The ADW acts as an online encyclopaedia, with each individual species account displaying basic information specific to that species. Each species account includes geographic range, habitat, physical description, development, ecosystem roles, reproduction, life span, communication and perception, behaviour, food habits and conservation status. The website also offers a cell phone app for handy reference.

MAJOR AND MINOR PHYLA

It is customary to divide the invertebrate phyla into **major** and **minor** phyla. The concept of major and minor phyla depends upon two factors: (i) **number of species and individuals**, and (ii) **their participation in ecological communities**.

On the basis of first factor, 11 phyla **(Protozoa, Porifera, Coelenterata, Platyhelminthes, Rotifera, Nematoda, Mollusca, Annelida, Arthropoda, Ectoprocta and Echinodermata)** seem to be clearly major, as is evident from the species number given in **Table 1** (page 23). Whereas, the minor phyla make up only a fraction of animal communities. On the basis of second factor, if the phyla are represented in great majority of ecological communities, they would be regarded as major phyla. On this basis, the two phyla, **Rotifera** and **Ectoprocta**, cannot be considered as major phyla. Although they are greater in species number, but they are included in minor phyla because of their limited participation in animal communities. Thus, utilizing a combination of the above two factors, we can regard only 9 as major phyla and the rest as minor phyla.

Lower and Higher Invertebrates

The invertebrate phyla are usually referred to as **lower** and **higher** invertebrates. Lower invertebrates are generally smaller in size and simple in body organisation. They are believed to have originated, in the main lines of evolution, near the base of the phylogenetic tree of Animal Kingdom. **Protozoa**, **Porifera**, **Coelenterata**, **Platyhelminthes** and **Nematoda** fall in the category of lower invertebrates. On the other hand, the higher invertebrates are generally larger in size and complex in body organization. These take up higher positions in the phylogenetic tree of the Animal Kingdom. **Annelida**, **Arthropoda**, **Mollusca** and **Echinodermata** are higher invertebrates.

Table 1. **Contrast between Lower and Higher Invertebrates.**

Lower Invertebrates	Higher Invertebrates
1. Generally smaller in size.	1. Generally larger in size.
2. Body organization simple.	2. Body organization complex.
3. Radial, biradial or no symmetry.	3. Bilateral symmetry.
4. Germ layers wanting or 2 or 3 germ layers.	4. Three germ layers present.
5. No coelom or a pseudo-coelom.	5. True coelom.
6. Generally no separate mouth and anus.	6. Mouth and anus separate.
7. No muscular gut.	7. A true muscular gut.
8. Blood vascular system not well developed.	8. A well developed blood vascular system present.

Contrast between Lower and Higher Invertebrates

On the basis of some important features, a contrast may be made between the lower and higher invertebrates as shown in the Table 1.

Invertebrates *versus* Vertebrates (Comparison)

As already mentioned earlier, in the traditional system of classification the entire Animal Kingdom has been divided into two major groups: **Invertebrata** (or Non-Chordata) and **Vertebrata** (or Chordata). The former includes those animals that do not possess a vertebral column or backbone, while the later is characterised by its presence. No doubt, the division of animals into invertebrates and vertebrates is purely artificial, but this has a practical applicability in their systematic study.

The higher invertebrates share many structural peculiarities with the vertebrates, such as axiate body plan, bilateral symmetry, triploblastic and coelomate condition, metameric segmentation and organ-system grade of organization. However, the two groups have several fundamental differences Table 2.

Table 2. **Comparison (Differences) between Invertebrates and Vertebrates.**

FEATURES	INVERTEBRATES (Non-Chordates)	VERTEBRATES (Chordates)
1. Symmetry	1. Radial, biradial or lacking.	1. Bilateral.
2. Metamerism	2. True or pseudometamerism or lacking.	2. True metamerism.
3. Post-anal tail	3. Lacking.	3. Usually present projecting beyond anus.
4. Grade of organization	4. Protoplasmic to organ-system.	4. Organ system.
5. Germ layers	5. 2 (diploblastic), 3 (triploblastic) or lacking.	5. 3 (triploblastic).
6. Coelom	6. Acoelomate, pseudocoelomate or truly coelomate.	6. Truly coelomate.
7. Limbs derivation	7. From same segment.	7. From several segments.
8. Notochord	8. Notochord or backbone lacking.	8. Present at some stage or replaced by a backbone made of ringlike vertebrae.
9. Gut position	9. Dorsal to nerve cord.	9. Ventral to nerve cord.
10. Pharyngeal gill-slits	10. Absent.	10. Present at some stage of life.
11. Anus	11. Opens on the last segment or absent.	11. Differentiated and opens before the last segment.
12. Blood vascular system	12. Open, closed or absent.	12. Closed and much developed.
13. Heart	13. Dorsal, lateral or absent.	13. Ventrally placed.
14. Dorsal blood vessel	14. Blood flows anteriorly.	14. Blood flows posteriorly.
15. Hepatic portal system	15. Absent.	15. Present.
16. Haemoglobin	16. In plasma or absent.	16. In red blood corpuscles.
17. Respiration	17. Through body surface, gills or tracheae.	17. Through gills or lungs.
18. Nervous system	18. Solid.	18. Hollow.
19. Brain	19. Above pharynx or absent.	19. Dorsal to pharynx in head.
20. Nerve cord	20. Double, ventral, usually bearing ganglia.	20. Single, dorsal, without ganglia.
21. Segmental nerve roots	21. Dorsal and ventral roots not separate.	21. Dorsal and ventral roots separate.
22. Reproduction	22. Asexual reproduction predominant.	22. Sexual reproduction predominant.
23. Regeneration power	23. Usually good.	23. Usually poor.
24. Body temperature	24. Cold-blooded.	24. Cold or warm-blooded.

Vertebrates can be pictured as 'upside-down' invertebrates oweing to the position of their nervous and circulatory systems (Figs. 1 and 2).

Classification of animals and the total number of species described varies from author to author, this can be seen on websites of important institution, societies such as Zoological Survey of India, IUCN, FAO, UNEP, and many others that have been mentioned in text at relevant places.

Though number of species discovered and described has been already mentioned still, a large number of species are either waiting to be described or discovered. Table 3 gives the estimate of the total number of invertebrate species on earth which is approximately 6.7 million.

Table 3.

Invertebrate Animals	
Insects	5,000,000
Arachnids	600,000
Molluscs	200,000
Crustaceans	150,000
Echinoderms	14,000
Others	791,830
Total Invertebrates	**6,755,830**

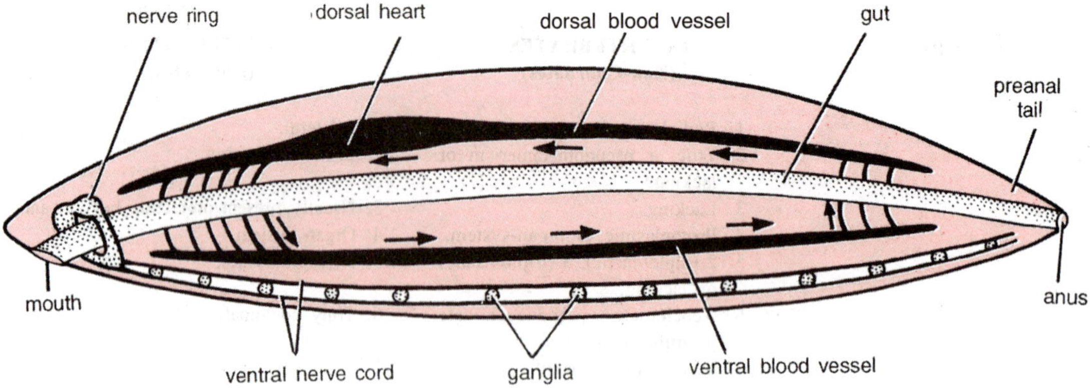

Fig. 1. Generalized invertebrate organization (diagrammatic).

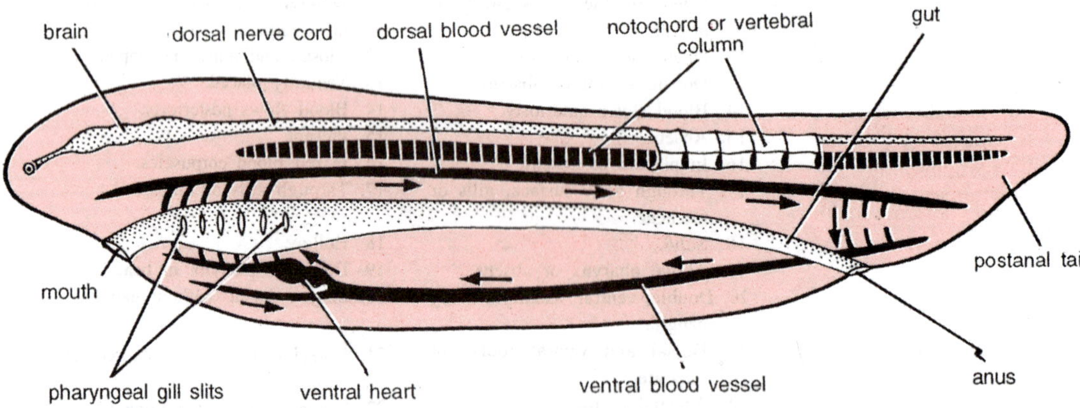

Fig. 2. Generalized vertebrate organization (diagrammatic).

Biologists every year document about 13,500 new species of invertebrates.

(http://www.currentresults.com 2012)

There is a vast list of animals identified and reported of all major and minor phyla of invertebrates.

PHYLOGENY OF INVERTEBRATES

Evolutionary history of invertebrate animals suggests that the diverse phyla of invertebrates, as we see them today, have evolved as the result of gradual changes spread over an immense period of geological time. In the absence of the fossil record of Archaeozoic era, it is rather impossible even to guess about the animal ancestry. Zoologists believe that the phylum Protozoa owes its origin to primitive algae from which first arose flagellate Protozoa or Mastigophora. In course of time they gave rise to other protozoans as well as the sponges (Porifera). The coelenterates are believed to have evolved from a primitive multinucleate ciliate protozoan. Firstly, it was syncytial in structure, but later on became cellular and achieved bilateral form. It resembled a planula larva. From planula-like structure which probably arose from the coelenterates, the ctenophores and the acoelomate flatworms. From primitive acoelomate flatworms arose other platyhelminths, aschelminths and nemerteans. Since a trochophore larva appears in the development of annelids and molluscs, it is believed that a trochophore-like animal was the ancestor of both these groups. The trochophore-like ancestor itself seems to have evolved from the planula-like metazoan ancestor. Further, the Onychophora, which are connecting links between Annelida and Arthropoda, have led us to presume that the arthropods have evolved from annelids.

In the development of the echinoderms, the dipleurula stage is of common occurrence. It establishes the fact that dipleurula-like creature must have been the ancestor of all the echinoderms. Like trochophore, the dipleurula-like ancestor is also considered to be derived from the planula-like ancestor.

The hemichordates and the chordates also have their ancestral links with Dipleurula.

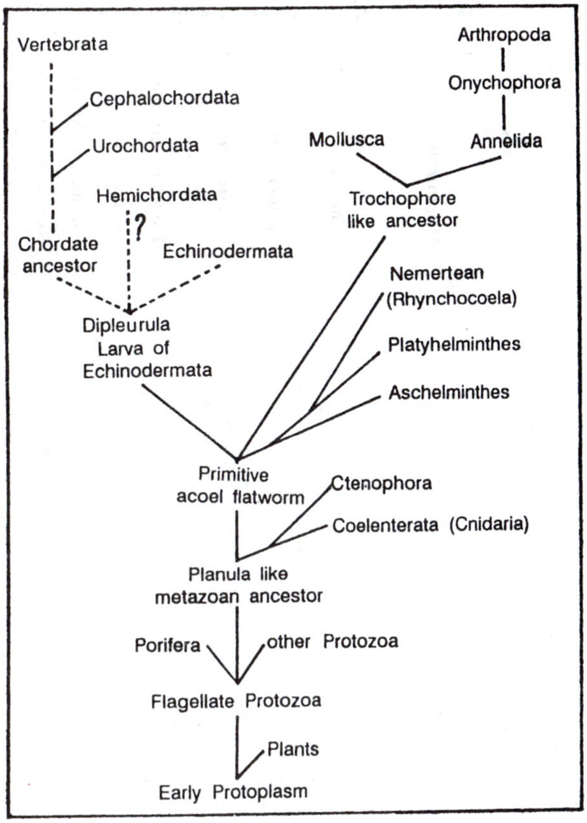

Fig. 3. Hypothetical phylogenetic tree of major animal groups.

Major Characteristics of Invertebrates

As mentioned earlier, '**Invertebrata**' is not a natural group and the animals included do not show a single common character except for the absence of **notochord**. However, the zoologists often find it convenient to combine phyla under a few larger taxonomic units or categories. These are based on major characteristics or patterns because of certain common embryological and anatomical features. Table 4 shows the major characteristics or patterns which are used in the classification of invertebrates into phyla. The table also shows distinctive features of principal animal phyla.

OUTLINE CLASSIFICATION OF INVERTEBRATES

Animals have been classified in various ways by different authors. It mainly depends upon what criteria and qualities the classifier has in mind. Number of phyla recognized by any particular author depends upon the extent, these are put in different classes or recognized as independent phyla. Broadly speaking the whole animal kingdom has been divided in to 30 phyla and 68 classes. These classes are further divided in to 350 orders. The following synoptic classification is based on the works of **Hyman**, **Barnes** and **Storer** with slight modification. Only the major taxonomic categories (subkingdoms, branches, grades, divisions, etc.) and phyla are listed with their diagnostic characters.

KINGDOM ANIMALIA

This is the largest group of the animal classification. It includes the entire fauna (animal population) of the world. It is divided into two

Table 4. **Major Characteristics of Principal Animal Phyla.**

Cells	Germ layers	Organization	Symmetry	Segmentation	Digestive tract	Excretory Organs	Coelom	Circulatory System	Repiratory System	PHYLA	Distinctive Features (exceptions omitted)
1	–	Proto-plasmic	–	–	–	–	–	–	–	PROTOZOA	Microscopic, Single or colonies of like cells.
Cells many. Arranged in layers or tisues (METAZOA).	2. Diploblastic	Cellular	Radial biradial	a	Incomplete without Anus	a	a	a	a	PORIFERA	Bodywall perforated by pores and canals.
		Tissue Grade		a		a	a	a	a	COELENTERATA	Nematocyts present. Digestive tract sac like.
		Organ Grade		a		a	a	a	a	CTENOPHORA	Biradial comb plates for locomotion.
	3. Triploblastic		Bilateral	a		a	fc	a	a	PLATYHELMINTHES	Flat, soft. Digestive tract much branched.
		Organ-system Grade		a	Comp-lete	ps	fc	a	a	ACANTHOCEPHALA	Worm-shaped. Proboscis hook-bearing.
				a	Complete with Anus. Protostomic	a	fc	+	a	NEMERTINEA	Slender, ciliated. Proboscis soft, eversible.
				a		ps	a	a	a	NEMATOMORPHA	Thread-like. Mouth may be absent.
				a		ps	fc	a	a	ROTIFERA	Microscopic. Cilia on oral disc.
				a		ps	fc	a	a	GASTROTRICHA	Microscopic. Surface spiny. Cilia ventral.
				a		ps	+	a	a	NEMATODA	Cylindrical. Cuticle tough. Cilia absent.
				a		lo	a	a	a	BRYOZOA	Moss-like colonies. Lophophore present.
				a		lo	+	+	a	BRACHIOPODA	With dorsal & ventral calcreous shell and lophopore.
				a		h sc	+	+	+	MOLLUSCA	Soft. External limy shell of 1,2 or 8 parts.
				+		sc	+	+	+ a	ANNELIDA	Metamerically segmented. Setae present.
				+		ps	+	a	a	TARDIGRADA	4 Pairs of unjointed legs. Bearing claws.
				a		ps	+	+	a	ONYCHOPHORA	Skin soft. Paired nephridia and tracheae present.
				+		h sc	+	+	+	ARTHROPODA	Chitinous exoskeleton. Jointed appendages.
				a	Complete Deuterostomic	en	+	a	a	CHAETOGNATHA	Arrow-shaped. Transparent. Lateral fins.
			Bilateral Radial	a		en	+	+	+	ECHINODERMATA	5-parted radial symmetry. Tube feet. Spiny endoskeleton.
				+		en	+	+	+	CHORDATA	Notochord. Dorsal tubular nerve cord. Gill slits. Limbs or fins for locomotion.

a = Absent; +, Present; en = Enterocoelom; fc = Flame cells; h = Haemocoel; lo = Lophophorate coelomata; ps = Pseudocoelom; r = Reduced; sc = Schizocoelom; v = Various or none.

Table 5. **Major Groups and Phyla in Animal Classification.**

Major Groups of Animals

- **KINGDOM ANIMALIA**
  - *SUBKINGDOM I Protozoa*................
  - *SUBKINGDOM II. Metazoa*
    - ***Branch 1.*** Mesozoa.............
    - ***Branch 2.*** Parazoa...............
    - ***Branch 3.*** Eumetazoa
      - *Grade A.* Radiata.....................
      - *Grade B.* Bilateria
        - *Division 1.* Protostomia
          - (i) Subdivision Acoelomata
          - (ii) Subdivision Pseudocoelomata
          - (iii) Subdivision Lophophorate Coelomata
          - (iv) Subdivision Schizocoelous Coelomata
        - *Division 2.* Deuterostomia
          - (v) Subdivision Enterocoelous Coelomata

	Phyla	Number of Species
●	1. Protozoa	31,250
+	2. Mesozoa	71
●	3. Porifera (Sponges)	4562
●	4. Coelenterata or Cnidaria	9923
+	5. Ctenophora	100
●	6. Platyhelminthes (Flatworms)	17511
+	7. Rhynchocoela /Nemerteanea (Ribbon Worm)	1200
+	8. Acanthocephala	800
+	9. Entoprocta	150
+	10. Rotifera	2500
+	11. Gastrotricha	3,000
+	12. Kinorhyncha	100
●	13. Nematoda (Roundworms)	30,028
+	14. Nematomorpha	320
+	15. Phoronida	11
+	16. Ectoprocta	5,000
+	17. Brachiopoda	300
+	18. Priapula/Priapulida	17
+	19. Sipunculida	145
●	20. Mollusca	66,535
+	21. Echiura/Echiurida	127
●	22. Annelida (Segmented Worms)	12,701
+	23. Tardigrada	514
+	24. Onychophora	100
●	25. Arthropoda	999059
+	26. Pentastomida	70
+	27. Chaetognatha	111
●	28. Echinodermata	6223
+	29. Pogonophora	145
+	30. *Chordata	46499
	*Hemichordata	120
	*Urochordata	1,250
	*Cephalochordata	25
+	31. Cyclinophora	3
+	32. Loricifera	21
+	33. Micrognathozoa	1
+	34. Myxozoa	1300
+	35. Orthonectida	20
+	36. Placozoa	1
+	37. Rhombozoa	75
+	38. Xenoturbellida	2
+	39. Onychophora (velvet worms)	100
+	40. Cyclinophora	3

● Shows major phyla discussed in detail in this volume. + Shows minor phyla. * Described in second volume on VERTEBRATES. [Note : Approximate numbers of described living species of major metazoan phyla from Brusca and Brusca (1990) and Nielsen (1995), Alfred, J. R. B. (1998) and wikipedia.org and updated by Animal Discovery 2011, compiled by : Director, Zoological Survey of India, 2012].

Contrasting features of Protozoa and Metazoa are listed in the following table :

SUBKINGDOM I. PROTOZOA	SUBKINGDOM II. METAZOA
1. Microscopic, unicellular animals.	1. Usually large, mulitcellular animals.
2. Structure simple with subcellular organelles.	2. Structure complex with strong cellular differentiation.
3. Centriole absent.	3. Metazoan cells contain centrioles.
4. Grade of organization protoplasmic.	4. Grades cellular, cell-tissue, and tissue-organ-system.
5. Little physiological division of labour.	5. Physiological division of labour well marked.
6. Life cycle with more than one generation universal.	6. Life cycle comprising more than one generation is rare.
7. Asexual reproduction universal.	7. Asexual reproduction occurs only in lower Metazoa.
8. Sexual reproduction rare.	8. All reproduce sexually.
9. Conjugation occurs between adults (hologamy).	9. Conjugation occurs between uninucleate sperm and ovum (syngamy).
10. When a cell divides, the daughter cells become separated as independent animals.	10. Fertilized egg repeatedly divides. Resulting daughter cells remain cohered to form an embryo.
11. Form of individuals may vary even in the same species.	11. Form of body is definite for all members of a species.
12. Natural death does not take place due to lack of a body; hence often termed immortal.	12. Natural death takes place, hence termed mortal.

SUBKINGDOM I. PROTOZOA

Unicellular, microscopic animals. No tissues.

Phylum 1. Protozoa (First animals). Protozoans. About 31,500 species. Solitary or colonial. Cell organelles specialized. Single to many nuclei. Nutrition holozoic, holophytic or saprozoic. Free living, commensal, symbiotic or parasitic. Freshwater, marine or moist terrestrial. Examples : *Euglena, Trypanosoma, Amoeba, Paramecium, Monocystis,* etc.

SUBKINGDOM II. METAZOA

Multicellular animals. Cells arranged in layers or tissues. Metazoans are subdivided into three branches: **Mesozoa**, **Parazoa** and **Eumetazoa**. Their contrasting features are given in the following table :

Characters	Branch 1. MESOZOA	Branch 2. PARAZOA	Branch 3. EUMETAZOA
1. Habit	1. Sessile, parasitic.	1. Sessile.	1. Mostly mobile.
2. Tissues	2. Absent. Few cells in 2 layers.	2. Poorly defined tissues.	2. Well defined tissues.
3. Organs	3. Not formed.	3. Not formed.	3. Well formed.
4. Digestive tract and mouth	4. Absent.	4. Absent.	4. Mostly with mouth and digestive tract.
5. Digestion	5. Intracellular.	5. Intracellular.	5. Mostly extracellular.
6. Body surface	6. Not porous .	6. Porous.	6. Not porous.
7. Body cavities	7. Absent.	7. One to many internal	7. Body cavities not lined by choanocytes.

BRANCH 1. MESOZOA

Digestive cells few, external, ciliated. No organs or tissues.

Phylum 2. Mesozoa (Middle animals). About 50 species. Worm-like, small, endoparasites of marine invertebrates. Body with an outer single layer of ciliated digestive cells enclosing one or several reproductive cells. Examples : *Dicyema, Rhopalura,* etc.

BRANCH 2. PARAZOA

Digestive cells many, internal, flagellated. No digestive cavity and mouth. Tissues poorly defined. Organs absent.

Phylum 3. Porifera (Pore bearers).Sponges. About 4,500 species. Body usually irregular with numerous pores and water canals, some lined by flagellated collar cells or choanocytes. Skeleton of minute calcareous spicules made of spongin fibres. Solitary or colonial. Sessile. Marine, a few freshwater. Examples : *Leucosolenia, Scypha, Sycon, Euspongia,* etc.

BRANCH 3. EUMETAZOA

Many-celled animals with organs, mouth and digestive cavity. Subdivided into two grades : **Radiata** and **Bilateria.** Their contrasting features are given in the following table :

GRADE A. RADIATA	GRADE B. BILATERIA
1. Body radially or biradially symmetrical. Sometimes bilateral symmetry which is a secondary adaptation.	1. Body bilaterally symmetrical. Sometimes radial symmetry which is a secondary adaptation.
2. Organ-systems are incipient, *i.e.,* not well marked.	2. Organ-systems well marked.
3. Mesoderm is not developed. Diploblastic.	3. Mesoderm is well developed. Triploblastic.
4. Coelomic cavity invariably absent.	4. Coelomic cavity a pseudocoelom or true coelom or absent.
5. Tentacles with nematocysts. Comb plates in some.	5. Tentacles, if present, without nematocysts. No comb plates.
6. Principal external opening of digestive cavity is mouth.	6. External openings of digestive cavity are mouth and anus.

Grade A. Radiata

Radially symmetrical, tentaculate, diploblastic animals with few organs. Digestive cavity opens externally through mouth.

Phylum 4. Coelenterata (With hollow intestine) or **Cnidaria** (Nettle-bearing). About 10,000 species. Symmetry radial or approaching bilateral. Two or three layers of cells. Mouth encircled by tentacles bearing nematocysts. Sac-like gastrovascular cavity. Sessile or free swimming. Solitary or colonial. Marine or freshwater. Examples : *Hydra, Obelia, Aurelia,* Corals, etc.

Phylum 5. Ctenophora (Comb-bearers). Comb-jellies. About 100 species. Symmetry biradial. 2 tentacles and 8 longitudinal rows of ciliated comb plates for locomotion. No nematocysts. No anus. Free swimming, marine. Examples : *Pleurobrachia, Cestum, Beroe, Ctenoplana, Coeloplana,* etc.

Grade B. Bilateria

Bilaterally symmetrical triploblastic animals with organ-systems. Digestive tract complete with anus. Mesoderm present. It is subdivided into two divisions : **Protostomia** and **Deuterostomia**. Their contrasting characters are listed in the following table :

DIVISION 1. PROTOSTOMIA	DIVISION 2. DEUTEROSTOMIA
1. Mouth arises from blastopore or from the anterior margin of blastopore.	1. Mouth arises anteriorly some distance away from blastopore.
2. Coelom absent, pseudocoelom persistent blastocoel or true coelom. Developed as schizocoel by a split of mesoderm.	2. Coelom developed as enterocoel by fusion of gut pouches.
3. Cleavage spiral and determinate.	3. Cleavage radial and indeterminate.
4. Larva trochophore.	4. Larva tornaria or bipinnaria.

Division 1. PROTOSTOMIA

Cleavage spiral and determinate. Mouth arises from or near blastopore. Protostomes are divided by Barnes into 4 subdivisions : (i) Acoelomata, (ii) Pseudocoelomata, (iii) Lophophorate coelomata and (iv) Schizocoelous coelomata.

SUBDIVISION (I) ACOELOMATA

No body cavity or coelom. Space between body wall and digestive cavity is occupied by mesenchyme parenchyma.

Phylum 6. Platyhelminthes (Flatworms). About 17,500 species. Body dorso-ventrally flattened. Digestive tract branched or absent. No anus and circulatory system. Free-living or parasitic. Marine, freshwater, a few terrestrial. Examples : *Planaria, Fasciola, Taenia,* etc.

Phylum 7. Rhynchocoela or **Nemertinea** (Ribbon worms). About 900 species. Body dorso-ventrally flattened and ciliated. With mouth and anus. Proboscis eversible. Mostly marine, few terrestrial and freshwater. Examples : *Cerebratulus, Lineus,* etc.

SUBDIVISION (II) PSEUDOCOELOMATA

Body cavity a pseudocoelom which is a persistent blastocoel, not lined by mesoderm. Anus present.

Phylum 8. Acanthocephala (Spiny-headed worms). About 800 species. Minute worm-like endoparasites. No digestive cavity. Protrusible proboscis with recurved spines. Examples : *Acanthocephalus, Gigantorhynchus,* etc.

Phylum 9. Entoprocta. About 60 species. Sessile. Body of calyx and slender stalk. Digestive tube U-shaped. Mouth and anus close together and surrounded by a tentacular crown. Mostly marine. Solitary or colonial. Examples : *Pedicellina, Loxosoma, Urnatella,* etc.

***Superphylum* Aschelminthes**

Sac worms. An assemblage of pseudocoelomates with an anterior mouth, posterior anus and straight digestive tube. Predominantly aquatic. Free living, epizoic or parasitic. Includes 5 classes (Rotifera, Gastrotricha, Kinorhyncha, Nematoda and Nematomorpha) which are also recognized as phyla.

Phylum 10. Rotifera (Wheel animalcules). About 2500 species. Microscopic. Anterior end with a ciliated crown. Pharynx with internal jaws. Mostly freshwater, some marine. Examples : *Philodina, Rotatoria, Hydatina,* etc.

Phylum 11. Gastrotricha (Hairy stomach worms). About 3,000 species. Microscopic. Ventral surface flattened and ciliated. Cuticle with spines, plates or scales. Freshwater and marine. Examples : *Chaetonotus, Macrodasys,* etc.

Phylum 12. Kinorhyncha (Jaw-moving worms). About 100 species. Small. Cuticle segmented and with recurved spines. Spiny anterior end or proboscis retractile. Marine. Examples : *Echinoderes, Pycnophyes,* etc.

Phylum 13. Nematoda (Round worms). About 30,000 species. Body slender and cylindrical. Cuticle tough, often ornamented. Radial or biradial arrangement of structures around mouth. Free-living or parasitic. Freshwater, marine or in soil. Examples: *Ascaris, Trichinella, Wuchereria, Ancylostoma, Enterobius,* etc.

Phylum 14. Nematomorpha or **Gordiacea** (Horsehair worms). About 250 species. Body long, thread-like. Larval stage parasitic in insects. Adult free-living in water or damp soil. Examples : *Nectonema, Gordius,* etc.

SUBDIVISION (III) LOPHOPHORATE COELOMATA

Coelom develops as schizocoel or enterocoel. With a crown of hollow tentacles (lophophore) surrounding mouth. Head indistinct. Digestive tract U-shaped.

Phylum 15. Phoronida. About 11 species. Worm-like unsegmented body enclosed in a chitinous tube. Lophophore horseshoe-shaped. Marine. Examples : *Phoronis, Phoronopsis.*

Phylum 16. Bryozoa or **Ectoprocta** (Moss animals). About 4,000 species. Sessile moss-like colonies. Body enclosed in a gelatinous, chitinous or calcareous covering. Lophophore V-shaped or circular. Mostly marine, few freshwater. Examples : *Plumatella, Bugula, Pectinatella,* etc.

Phylum 17. Brachiopoda (Lamp shells). About 300 species. Body enclosed in two unequal calcareous shell valves. Lophophore W-shaped. Marine. Examples : *Lingula, Crania, Terebratulina,* etc.

SUBDIVISION (IV) SCHIZOCOELOUS COELOMATA

Coelom is a schizocoel which originates as a space by the splitting of the embryonic mesoderm.

Phylum 18. Priapulida. 15 species. Sausage or cucumber-shaped marine animals with a swollen anterior introvert or proboscis. Body surface covered with spines and tubercles. Peritoneum of coelom greatly reduced. Examples : *Priapulus, Halicryptus* etc.

Phylum 19. Sipunculida (Peanut worms). About 145 species. Body elongated and cylindrical with retractile anterior end (introvert). Lobes or tentacles around mouth. Anus dorsal. Marine. Examples : *Spipunculus, Aspidosiphon, Phascolosoma,* etc.

Phylum 20. Mollusca (Soft-bodied animals). About 66,500 species. Body soft unsegmented, with ventral muscular foot. Mantle with shell glands. External limy shell of 1,2 or 8 parts. Terrestrial, freshwater and marine. Examples : Chitons, Snails *(Pila),* Mussels *(Unio),* Squids *(Loligo),* etc.

Phylum 21. Echiurida (Adder-tailed worms). About 127 species. Body cylindrical, unsegmented, with anterior retractile proboscis. One pair of large ventral setae below mouth. Marine. Examples : *Echiurus, Urechis,* etc.

Phylum 22. Annelida (Ringed worms). About 12,700 species. Body elongate, metamerically segmented. Setae for locomotion. Terrestrial, freshwater and marine. Examples : Earthworms *(Pheretima), Nereis,* Leech *(Hirudinaria),* etc.

Phylum 23. Tardigrada (Water bearers). About 514 species. Minute. Body segmented with 4 pairs of unsegmented legs terminating in claws. Freshwater, terrestrial and marine. Examples : *Echiniscus, Hypsibius,* etc.

Phylum 24. Onychophora (Claw-bearers). About 100 species. Worm-like unsegmented body covered by thin cuticle. A pair of anterior antennae. Many pairs of short stumpy legs ending in claws. Moist soil. Examples : *Peripatus, Peripatopsis,* etc.

Phylum 25. Arthropoda (Joint-footed animals). About 9,99,059 species. Body segmented with jointed appendages. Exoskeleton chitinous. Coelom vestigial. Body cavity haemocoel. Terrestrial, freshwater and marine. Examples : Prawns, Scorpions, Flies, Centipedes, etc.

Phylum 26. Pentastomida (Tongue worms). About 70 species. Worm-like unsegmented body with two anterior appendages germinating in claws. Blood-sucking endoparasites of vertebrates. Examples : *Cephalobaena, Porocephalus.*

Division 2. DEUTEROSTOMIA

Cleavage radial and indeterminate. Mouth arises some distance away from blastopore.

SUBDIVISION. ENTEROCOELOUS COELOMATA

Coelom is an enterocoele which originates as pouches of embryonic gut (archenteron).

Phylum 27. Chaetognatha (Arrow worms). About 111 species. Small elongated transparent body bearing postanal tail and lateral fins. Anterior end with grasping spines. Planktonic and marine. Examples : *Sagitta, Spadella,* etc.

Phylum 28. Echinodermata (Spiny-skinned animals). About 6223 species. Secondarily pentamerous radial symmetry. Calcareous endoskeleton of plates bearing external spines. A part of coelom as water vascular canals. Locomotion by tube feet. Examples : Starfish, Brittle stars, Sea urchins, Sea lilies, etc.

Phylum 29. Pogonophora (Beard worms). About 145 species. Body long, enclosed in a chitinous tube. Anterior end with one to many tentacles. No digestive tract. Deep water and marine. Examples : *Siboglinum, Spirobrachina, Polybrachia,* etc.

Phylum 30. Hemichordata (Acorn worms). About 120 species. Body worm-like divided into proboscis, collar and trunk. With gill slits. Embryo lacking a typical notochord. Marine. Examples : Acorn worms *(Balanoglossus), Cephalodiscus,* etc.

Important Questions

➤➤*Long answer type questions*

1. What is an invertebrate ? On what basis the animals have been classified as invertebrates and vertebrates ?
2. Give a historical account of assigning the invertebrate animals to various taxonomic phyla. Name the present invertebrate phyla mentioning the approximate number of species contained by them.
3. Elucidate the concept of major and minor phyla. Which phyla of invertebrates are treated as major phyla and why ?
4. Give the main points of differences between invertebrates and vertebrates.
5. Outline the phylogenetic account of the invertebrates.
6. Give an outline classification of animals enumerating characteristic features of different phyla.
7. Distinguish between — (i) Protozoa and Metazoa, (ii) Coelomata and Acoelomata, (iii) Radiata and Bilateria, (iv) Protostomes and Deuterostomes.

➤➤*Short answer type questions*

1. Give four important characters of phylum Protozoa. On which characters classification of Protozoa is based ?
2. Write names of two invertebrate animals found near your house. Classify then upto orders.
3. Classify these animals— (i) mosquito, (ii) liver fluke, (iii) leech.
4. Give natural habitat of— (i) *Entamoeba,* (ii) *Hydra,* (iii) *Ascaris* and (iv) *Plasmodium.*
5. Give characteristics of phylum Arthropoda. Give four examples of this group.
6. Classify any two animals upto class—(i) Tapeworm, (ii) Star fish, (iii) Locust, (iv) Earthworm, (v) Honey bee.
7. Give two important characteristics of each—(i) sponge, (ii) *Hydra*
8. Classify these animals upto class giving one character of each—(i) *Octopus,* (ii) grasshopper, (iii) *Nereis.*
9. Differentiate between jelly fish and starfish.

➤➤*Multiple choice questions*

1. Which group includes all marine animals :
   (a) Mollusca (b) Porifera
   (c) Echinodermata (d) Coelenterata
2. Platyhelminthes are generally called :
   (a) round worms (b) flat worms
   (c) tubiculous worms (d) blind worms
3. Myriapod has :
   (a) chitinous exoskeleton, ventral nerve cord, three pair of antennae
   (b) chitinous exoskeleton, ventral nerve cord, many body segments, one pair of antennae
   (c) soft body, ventral nerve cord, many body segments, two pair of antennae
   (d) chitinous exoskeleton, dorsal nerve cord, three body segments, one pair of antennae
4. Which is an annelid :
   (a) ant (b) crab
   (c)*Nereis* (d) *Octopus*
5. Chitinous exoskeleton is found in :
   (a) birds (b) turtle
   (c) insects (d) fishes
6. Corals are formed by :
   (a) Molluscs (b) Coelenterates
   (c) Protozoans (d) Echinoderms
7. Medusae are found in the life history of :
   (a)*Obelia* (b) *Hydra*
   (c) star fish (d) none of them
8. Conjugation in *Parameciun* is for :
   (a) binary division (b) sexual reproduction
   (c) asexual reproduction(d) metamorphosis
9. Alimentary canal is not found in :
   (a) Arachnida (b) Apoda
   (c) Gastropoda (d) Cestoda
10. Sessile (attached) animals are marine because :
    (a) for protection from terrestrial animals
    (b) protection against bacteria
    (c) they can produce water current to obtain food and oxygen
    (d) can enjoy water life
11. The light blue colour of blood in arthropods may be due to :
    (a) haemopoitin (b) haemoglobin
    (c) haemocyanin (d) none
12. Why *Amoeba* has been kept in protozoa :
    (a) due to contractile vacuole
    (b) because of nutrition being insectivorous
    (c) cell wall
    (d) acellular body
13. *Palaemon* belongs to which phylum :
    (a) Crustacea (b) Insecta
    (c) Arthropoda (d) Mollusca
14. Which is not an insect :
    (a) mosquito (b) spider
    (c) housefly (d) bedbug

15. Common character of mosquito, rat flea, bedbug and leech is :
(a) all have nephridia
(b) saliva contains anticoagulant
(c) lay eggs in stagnant water
(d) all are insects

16. Body is unsegmented in :
(a) earthworm (b) glow-worm and scorpion
(c) *Ascaris* and hook-worm
(d) tape-worm

17. Ecdysis is not found in :
(a) insect (b) polychaete
(c) snake (d) all

18. Turbellarians are :
(a) independent round-worms
(b) independent flat-worms
(c) parasitic tape worms
(d) parasitic-worms

19. *Sepia* and *Octopus* are two common marine animals. They show active movement in water by :
(a) movement of oral arms
(b) jet of water propelled through siphon
(c) attachment with other animals through its suckers
(d) undulating movement of lateral fins

20. Skin is covered in Arthropoda with :
(a) mucous (b) chitinous cuticle
(c) unstriated muscles
(d) shell formed of calcium

21. Mesogloea is found in which group of animals :
(a) Porifera (b) Coelenterata
(c) Hydrozoa (d) Annelida

22. Class in which the animals have 3 pair of jointed legs:
(a) Arthropoda (b) Crustacea
(c) Insecta (d) Gastropoda

23. Which phylum has highest number of species :
(a) Protozoa (b) Metazoa
(c) Arthropoda (d) Insecta

24. Respiratory pigment in insects :
(a) haemoglobin (b) cytochrome
(c) haemocyanin (d) none

25. All worms are :
(a) radially symmetrical (b) asymmetrical
(c) triploblastic (d) diploblastic

26. Wasp, gilamonster, centepede, scorpion and honey bee are :
(a) terrestrial (b) viviparous
(c) social (d) poisonous

27. Main characteristic of Arthropoda :
(a) chitinous exoskeleton, external segmentation, jointed legs
(b) external segmentation, hair, 3 pairs of jointed legs
(c) external skeletal scales, segmentation, one pair antennae
(d) external segmentation, one pair chelicerae, one pair pedipalpi.

28. Glow worm is a :
(a) Mollusc (b) Annelid
(c) insect (d) Helminth

29. Which is a coelenterate :
(a) sea cucumber (b) sea cow
(c) sea anemone (d) sea horse

30. Character of sporozoa :
(a) aquatic life (b) holophytic nutrition
(c) spore formation (d) holozoic nutrition

31. Metazoa includes :
(a) Sporozoa (b) Sarcodina
(c) Cnidaria (d) Protista

32. Pearl oyster belongs to :
(a) Gastropoda (b) Cephalopoda
(c) Scaphopoda (d) Pelecypoda

33. Match the species of the same phylum :
(a) scorpion, spider
(b) scorpion, *Obelia, Hydra*
(c) cockroach, earthworm, *Taenia solium*
(d) starfish, *Obelia, Hydra*

34. Classification of phylum Porifera is based on :
(a) canal system
(b) spicules
(c) shape of choanocytes
(d) ascocytes

35. What is common between earthworm, leech and centipede :
(a) they have malpighian tubules
(b) they are hermaphrodite
(c) they have ventral nerve cord
(d) they have no legs

Answers

➤➤*Multiple choice questions*

1. (c) 2. (b) 3. (b) 4. (c) 5. (c) 6. (b) 7. (a) 8. (b) 9. (d) 10. (c)
11. (c) 12. (d) 13. (c) 14. (b) 15. (b) 16. (c) 17. (b) 18. (b) 19. (b) 20. (b)
21. (b) 22. (c) 23. (c) 24. (d) 25. (c) 26. (d) 27. (a) 28. (c) 29. (c) 30. (c)
31. (c) 32. (d) 33. (a) 34. (b) 35. (c)

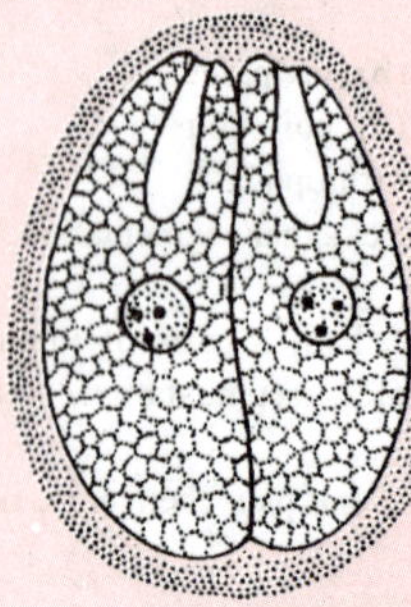

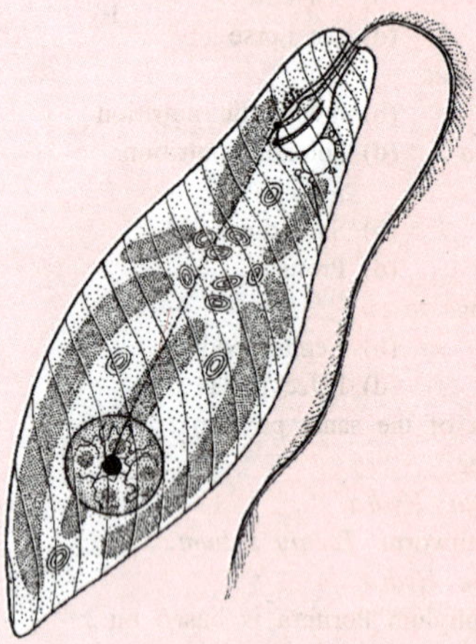

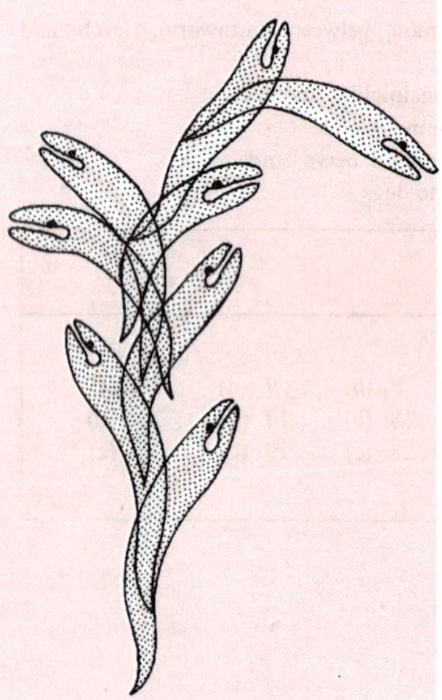

3

Euglena viridis

There are a large number of small protozoans which move with the help of one or more whip-like structures, called **flagella**. All these unicellular organisms are included in the superclass **Mastigophora** (Gr., **mastix**, whip + **pherein**, to bear) or **Flagellata** (L., **flagrum,** a whip). The flagellates are either **plant-like** typically having chlorophyll-bearing plastids, the **chloroplasts** or **chromatophores**; or **animal-like**, with no such plastids. The former are **autotrophic** due to their capacity to synthesize organic food by photosynthesis, whereas the later are **heterotrophic** as they utilize pre-synthesized food of their environments. These two groups of flagellates belong to two separate classes : **Phytoflagellata** or **Phytomastigophorea** and **Zooflagellata** or **Zoomastigophorea**, respectively.

Euglena is a typical example of Mastigophora. It is a phytoflagellate as it possesses both chloroplasts as well as flagella. It is autotrophic in sunlight, but becomes heterotrophic in dark. Because of its two-fold nutritional abilities, it is usually studied as a plant as well as an animal. This plant-animal organism is important as it serves as a key organism in research on photosynthesis, chloroplast structure, photoreception and flagellar activity. The species of *Euglena* which is commonly used for classroom study is *Euglena viridis*. There are 800 species of *Euglena* found in the world.

Euglena viridis

As the name implies, *Euglena viridis* is a green organism with an eye-like photoreceptive structure (Gr., **eu**, true + **glene**, eyeball or eye pupil + L., **viridis**, green).

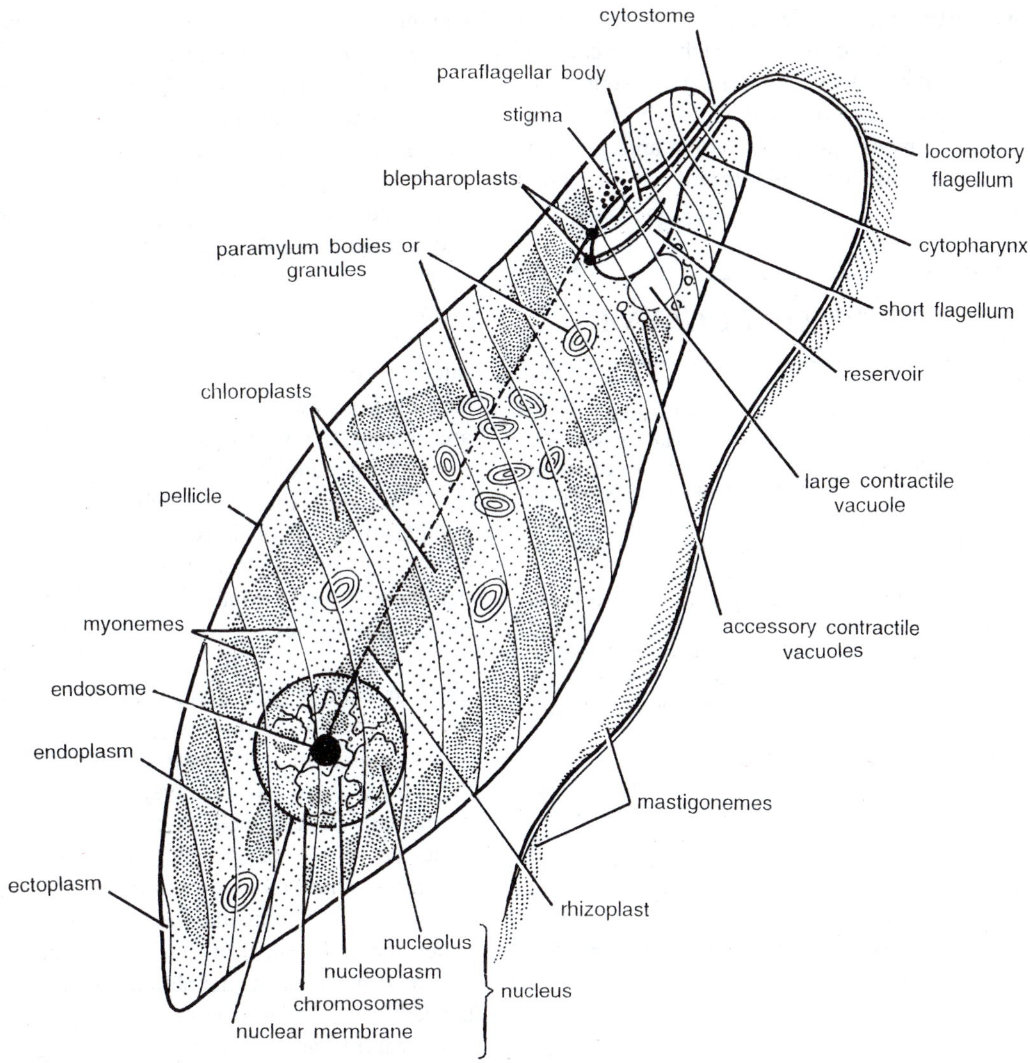

Fig. 1. *Euglena viridis.*

SYSTEMATIC POSITION

Phylum	Protozoa
Subphylum	Sarcomastigophora
Superclass	Mastigophora
Class	Phytomastigophorea
Order	Euglenida
Genus	*Euglena*
Species	*viridis*

HABITS AND HABITAT

Euglena viridis is a solitary and free-living freshwater flagellate. It occurs in freshwater ponds, pools, ditches and slowly-running streams, where there is a considerable amount of vegetation. It is fairly active and often found at various depths below the surface of water. Ponds in well-maintained gardens, containing decaying nitrogenous organic matter, such as faeces of animals, leaves, twigs, etc., are good sources of this organism. Sometimes it is so abundant that the water appears green in colour and it seems as if a film of scum is present on the pond surface.

CULTURE OF *EUGLENA*

Euglena can be cultured and studied easily in the laboratory. An easy way to prepare a culture is to fill clean and wide-mouthed large bottles with boiled tap or pond water and then add 20 boiled wheat grains to each bottle. Keep these bottles in a sunny place for a week. Wheat grains may be added monthly to maintain the culture.

Another method is to prepare a culture of *Euglena* in manure solutions. Horse or cow manure is boiled in pond or distilled water. These manure solutions are allowed to stand 36 to 48 hours and then inoculated with euglenae. In the laboratory, the euglenae grow and multiply well in a jar exposed to indirect sunlight.

If kept in dark or if grown in such a media which is rich in certain organic nutrients, some photosynthetic species of *Euglena* will lose their chlorophyll and will resemble like the genus *Astasia.* Irreversible loss of chlorophyll can be induced by the treatment of Streptomycin at a higher than normal temperature. Results of such experiments shows that members of the genus *Astasia* and perheps of some other colourless genera have been derived secondarily from photosynthetic species.

STRUCTURE

1. Shape and size. *E. viridis* is a small microscopic organism measuring about 60μ (microns) in length. The body is elongated and spindle-shaped. The anterior end is rounded or blunt, the middle part is somewhat wider and the posterior end is pointed. From the anterior end arises a whip-like flagellum which is seen moving when *Euglena* is progressing forward.

2. Pellicle. The body of *Euglena* is enveloped by a distinct, thin, elastic, tough **pellicle** which lies beneath the plasma membrane. It is flexible enough to permit movements. It is made of protein, and not cellulose, so that it is not homologous to plant cell wall. Electron microscopy has revealed that the pellicle consists of thin, firm, elastic and helically disposed **strips**. These pellicular strips fuse at both the ends of the cell body and each bears a groove along one edge and a ridge along the other. The edges of adjacent strips overlap and articulate along their entire length in such a way that the ridge of one fits into the groove of the other. Under the light microscope, these articulating edges appear as striations, called **myonemes**. Beneath the pellicle are present a row of mucus-secreting **muciferous bodies** and bundles of **microtubules**.

3. Cytoplasm. The pellicle encloses the cytoplasm which is divisible into two zones :

(a) Ectoplasm. It is the clear dense narrow peripheral zone containing microtubules and muciferous bodies.

(b) Endoplasm. It is the more fluid-like granular central zone containing the nucleus and several types of inclusions.

4. Reservoir. At the blunt anterior end of the body, there is an invagination forming a permanent flask-shaped cavity. It consists of a wide **reservoir** or **flagellar sac** which leads through a short tubular canal, the **cell gullet** or **cytopharynx**, to outside. Its external opening is called the **cell mouth** or **cytostome**. The reservoir is lined by plasma membrane without pellicle beneath it. The names 'cytostome' and 'cytopharynx' are misnomer because *Euglena* does not ingest solid food.

5. Flagellum. At the anterior end of the body, a single thread-like flagellum arises through the cytostome. Typically it is made of an axial elastic filament or **axoneme**, covered by a protoplasmic sheath.

Electron microscopy has revealed that the flagellum is not just one but paired, the other being smaller and confind within the reservoir. The two flagella originate from two tiny granules, the **blepharoplasts** (Gr., **blepharon**, eyelid + **plastos**, formed) or **kinetosomes** which lie embedded in the cytoplasm near the base of the reservoir. The long flagellum bears a lateral swelling, the **paraflagellar body**, near its base

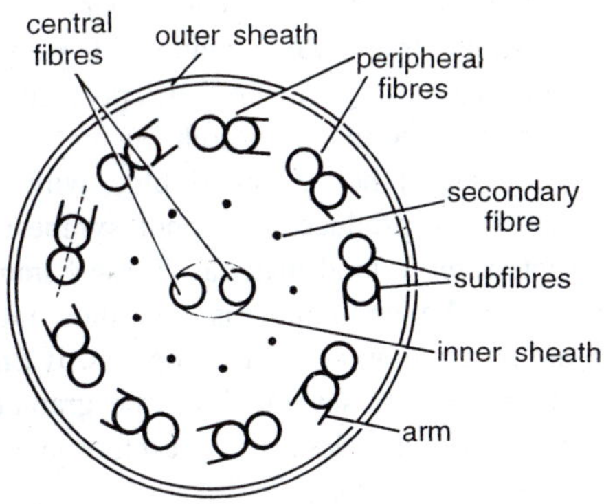

Fig. 2. *Euglena.* T.S. flagellum (diagrammatic).

within the reservoir. This body acts as a photoreceptor and probably contains **lactoflavin** as sensitizer.

Each flagellum consists of **2 central** and **9 peripheral fibres**. The central fibres are enclosed in an inner membranous sheath. Each central fibre is single, whereas the peripheral fibres are paired, that is, each made of two **sub-fibres**. Of each peripheral fibre, one of the two sub-fibres bears a double row of short projections, or **arms**, all pointing in the same direction. The whole flagellum is enveloped by an outer sheath which is continuous with the plasma membrane. In the space between the peripheral and central fibres lie **9 secondary fibres** which are somewhat inconspicuous. Along the side of the bundle of fibrils runs a rod-like structure. On the long axis of the flagellum is found a unilateral row of hair-like contractile processes, called **mastigonemes**. According to the investigation of **Manton** (1959), these processes arise laterally from two of the nine peripheral fibres. This type of flagellum is known as **stichonematic**.

6. Nucleus. A single, large, spherical or oval, vesicular nucleus lies near the centre of endoplasm, usually towards the posterior end of the body. It contains a large solid central body or nucleolus of chromatin, called **endosome** or **karyosome**. Its function remains controversial. The nuclear membrane is a double membrane which is perforated by pores. According to electron microscope, the nucleoplasm contains several

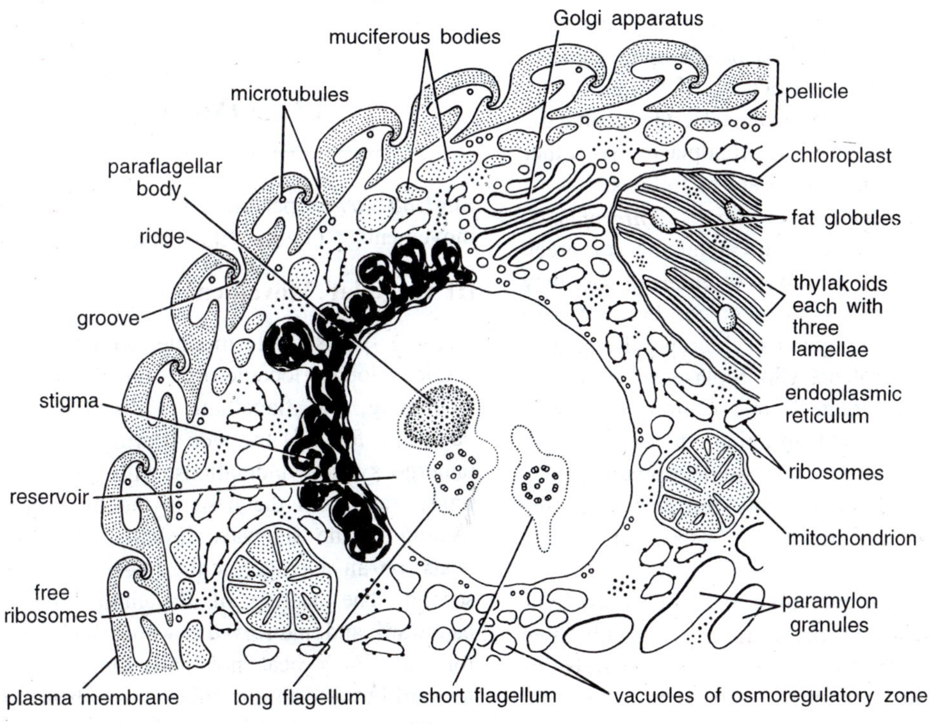

Fig. 3. *Euglena.* An electron micrograph (diagrammatic) of a portion of the T.S. body, passing through the reservoir.

nucleoli and a large number of granular and thread-like chromosomes.

7. Contractile apparatus. Associated with the reservoir occurs a dense osmoregulatory zone. It includes a large central contractile vacuole. It is surrounded by several smaller accessory vacuoles, which probably fuse together to form the larger vacuole. These vacuoles play a role in the discharge of water along with some waste products of metabolism to outside, via the reservoir, cytopharynx and cytostome.

8. Stigma. Lying near the reservoir, on side opposite to that of contractile vacuole, is a discoid or shallow cup-shaped red pigment spot, the **eye-spot** or **stigma**. According to **Leedale** (1966), the stigma is composed of a plate of lipid droplets, each charged with a red pigment, the **carotenoid**.

The stigma and the paraflagellar body together form a photoreceptor apparatus. When *Euglena* is moving toward light, the receptor is illuminated. The stigma evidently serves as a shield. When it changes direction, the shadow of pigment falls on the receptor. *Euglena* orients itself in such a way that the photoreceptor is continuously exposed. Thus, the animal, which depends upon sunlight for photosynthesis, can orient itself towards light.

9. Endoplasmic inclusions. Besides contractile apparatus and stigma, the endoplasm has other important inclusions which are as follows :

(a) Chromatophores or chloroplasts. Suspended in cytoplasm are a number of green bodies, the **chromatophores** or **chloroplasts**, they give the *Euglena* its green colour because they contain the green pigments, chlorophyll a and b, along with β-carotene. Chloroplasts of *E. viridis* are slender and radiate from a central point so as to form a star-shaped grouping. In the centre of each chloroplast is a single proteinaceous **pyrenoid**, which is probably the centre for the formation of a starch-like substance, called **paramylum**.

A chloroplast contains groups of chlorophyll-bearing lamellae or **thylakoids**, each with three lamellae. These are placed in the **matrix** or **stroma**, also containing ribosomes and fat globules. Each chloroplast is bounded by a triple membrane envelope (**Sleigh**, 1973).

(b) Paramylon. The endoplasm contains several small free oval granules of paramylon or paramylum which is a polysaccharide (β-1, 3 glucan) similar to starch but not identical with it, as it is not coloured blue with iodine solution. It is produced by photosynthesis and represents reserve food material. These Paramylum bodies may be so numerous that they interfere with the observation of the chloroplasts. If *Euglena* is kept in dark for several days, the granules for paramylum become less and the chloroplasts are more easily seen.

(c) Other cytoplasmic structures. The cytoplasm contains other subcellular organelles as well. The **Golgi bodies** are piles of large flattened sacs with minute vesicles pinching off from them. The **endoplasmic reticulum** is in the form of small interconnecting tubules and vesicles. The **mitochondria** are with tubular cristae and are more in number near the reservoir. The **ribosomes** occur scattered freely in the endoplasm and also on the endoplasmic reticulum and in the chloroplasts.

LOCOMOTION

Euglena viridis performs two different kinds of movements : (1) **flagellar**, and (2) **euglenoid**.

[I] Flagellar Movement

Euglena swims freely in water with the help of a single, long locomotory flagellum. During swimming, the flagellum is directed obliquely backward towards the side bearing the stigma. It undergoes spiral undulations, with waves, that are transmitted from the base to the tip, causing its beating or the sideways lashing. The flagellum beats, on an average, at the rate of 12 beats per second. The beating of flagellum drives water backward and induces the whole body to move forward. Each beat not only throws the body forward but also to one side. Thus, when the beats are repeated over and over, the *Euglena* revolves in circles or **gyrates**. As the flagellum is directed

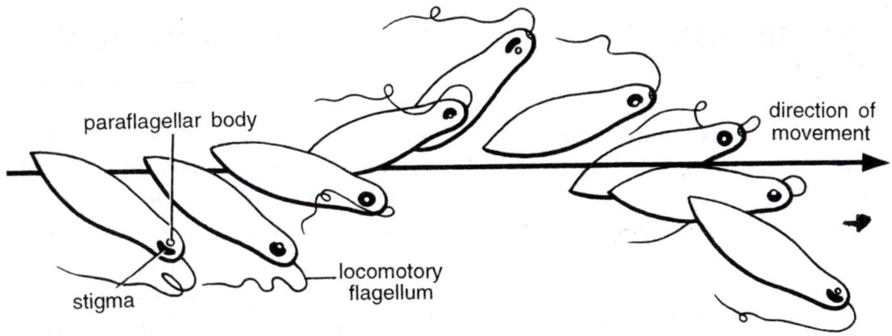

Fig. 4. *Euglena.* Successive stages in the flagellar movement.

obliquely backward to the long axis of the body, the organism also **rotates** on its axis. It has been calculated that *Euglena* rotates at the rate of one turn per second.

Movement of a flagellum involves contraction of its nine peripheral fibres (Fig. 2). Their position is ideal for the undulating actions as they could exert bending around the flagellar axis. The energy for the contractile action of fibres is supplied by ATP (adenosine triphosphate) formed in the mitochondria, which are included in the blepharoplasts. The function of mastigonemes, present on the flagellum of *Euglena,* is not yet known.

[II] Euglenoid Movement

The pellicle possesses considerable flexibility that enables *Euglena* to perform peristaltic activity. This activity brings about worm-like wrigglings or writhing movements while the animal creeps on the bottom. As the peristaltic waves pass, the body becomes shorter and wider first at the anterior end, then in the middle and later at the posterior end. This characteristic movement, found in *Euglena* or some other Protozoa, is commonly known as the **euglenoid** or **metabolic movement**.

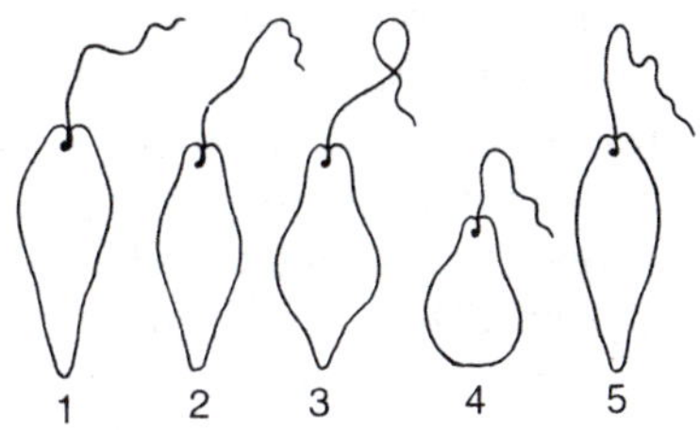

Fig. 5. *Euglena.* Successive stages in euglenoid movement.

During the euglenoid movement, the adjacent pellicular strips bend and move against one another, probably the ridge of one sliding in groove of the other. The sliding of the ridges in the grooves is lubricated by the secretion of underlying muciferous bodies.

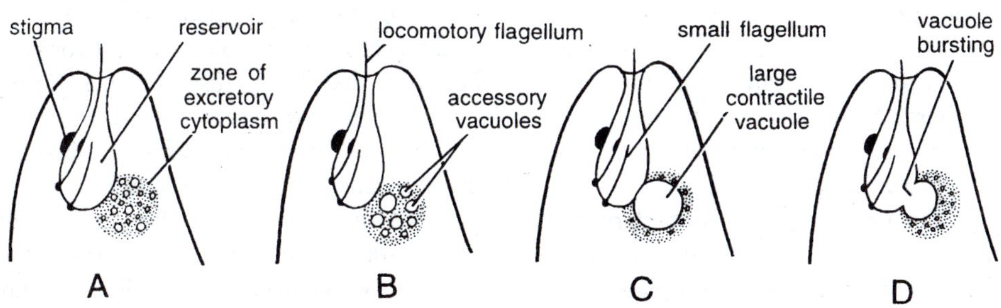

Fig. 6. *Euglena.* Successive stages of diastole and systole of contractile vacuole during excretion and osmoregulation.

NUTRITION

In *E. viridis,* nutrition is **autotrophic** or **holophytic** as well as **saprophytic**. Such a dual mode of nutrition is frequently referred to as **mixotrophic**. There is no evidence of animal-like or **holozoic** nutrition in *Euglena.*

[I] Holophytic or Autotrophic Nutrition

The chief mode of nutrition in *Euglena* is **holophytic** or **autotrophic**. That is, like green plants, it can manufacture its own food, in the presence of sunlight, by the process of **photosynthesis** with the aid of the chlorophyll present in the chloroplasts., The chlorophyll absorbs energy from sunlight. With this energy water reacts with CO_2 in a series of steps forming a hexose sugar. This is then transferred into a kind of polysaccharide, called **paramylum** or **paramylon**. It differs from the true starch in that it does not become blue with iodine solution. It is stored up for future use either scattered in the form of refractile granules in the endoplasm, or deposited around one or more proteinaceous bodies, the **pyrenoids**. Paramylon is found more abundantly in well-fed individuals.

[II] Saprophytic or Saprozoic Nutrition

In prolonged darkness, *Euglena* loses its chlorophyll and green colour. It becomes **etiolated**, that is, becomes pale or white, yet it continues to live and perform all the life activities. In the absence of sunlight, *Euglena* lives by the **saprophytic** or **saprozoic** method, which means the products of decaying organic substances dissolved in surrounding water are absorbed through its general body surface. It seems that *Euglena* secretes digestive enzymes that are typically animal-like in nature.

Generally, the chloroplasts lost in dark are regained in the light. But in *E. gracilis,* the change is permanent; the lost chloroplasts can never be recovered.

Pinocytosis has also been observed to take place at the base of the reservoir for the intake of proteins and other large molecules.

RESPIRATION

Euglena respires with the help of free oxygen dissolved in water, which diffuses in through the pellicle. This oxygen brings about oxidation reactions catalysed by enzymes present in the mitochondria. . The energy so liberated is entrapped in the high energy phosphate bonds of ATP, which supplies energy for metabolic activities. As a result of oxidation reactions, water and CO_2 are formed as by-products. In sunlight, it is likely that this CO_2 is utilized for photosynthesis, but in dark it is liberated to outside by diffusion through the general body surface.

OSMOREGULATION AND EXCRETION

The osmoregulatory function, that is, the removal of excessive water entering into the body by endosmosis, is performed by an anteriorly placed **contractile apparatus**. Outer pellicle is permeable to surrounding water, which continues to enter into euglena's body by endosmosis which is the result of differential concentrations of body protoplasm and surrounding freshwater. In *E. viridis,* the contractile apparatus consists of a large contractile vacuole surrounded by numerous small accessory vacuoles. The cytoplasm secretes the excessive water into these smaller vacuoles which, in their turn, drain into the larger vacuole. It is also likely that the accessory vacuoles are first formed which when fully extended, fuse together forming the larger vacuole. The larger vacuole finally empties into the reservoir. The process involves the **diastole** (increase in volume), and **systole** (decrease in volume) of the large contractile vacuole. In diastole, the contractile vacuole is filled with water, while in systole it is emptied to throw its watery content into the reservoir. It is believed that the postero-lateral wall of the contractile vacuole, adjacent to the reservoir, is very unstable and it bursts at the systole.

Ammonia, the nitrogenous waste product, resulting from catabolism, passes out by diffusion through the general surface of the body. Excretory substances may also be emptied by the contractile vacuole into the reservoir. It has been suggested that the dense zone of cytoplasm around the contractile apparatus is both **osmoregulatory** and **excretory** in function. It secretes water as well as excretory products into the lumen of the vacuole.

BEHAVIOUR

Euglena reacts to a variety of stimuli in the same manner as other Protozoa do.

[I] Reaction to Light

It avoids strong lights as well as shady areas, but reacts positively to a moderately intense light such as that from a window. Like most motile and free-living uni-cellular organisms containing chlorophyll, it orients itself parallel to a beam of ordinary light and swims towards the source of illumination. In a dish containing culture, most of the individuals are found to aggregate on the side towards the light. If the dish is placed in direct sunlight and one half of it is covered, the organisms avoid the region of direct sunlight as well as the shade but gather in between the two in a small band (Fig. 7).

[II] Shock Reaction

Investigations of Mast and his pupils have clearly revealed the part played by the photoreceptive stigma in the orientation of *Euglena.* Being positively phototactic, *Euglena* swims towards the source of light. It adopts a spiral course, rotating and gyrating around its body axis so that the paraflagellar body remains continuously exposed to light and uniformly illuminated. However, if the front source of light is screened and a new lateral beam of light is thrown, then *Euglena* produces a **shock reaction** and, by **trial** and **error**, reorients itself towards the new source of light. This shock reaction is explained as follows :

When the stigmal side of body faces a lateral beam of light, the paraflagellar body is shaded by the stigma once in each rotation. Thus, the base of flagellum is alternately darkened and illuminated in lateral light. Each darkening of the paraflagellar body excites its photoreceptor to produce a minor shock reaction. This effects flagellar action in such a way that the body bends at right angles, turning the flagellar end gradually towards the new light source. Light being essential for photosynthesis, this specialization brings the animal into light which is of distinct advantage in its nutrition.

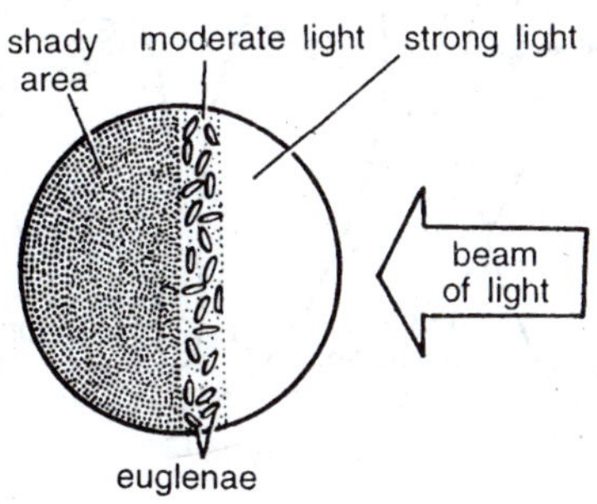

Fig. 7. *Euglena* showing reaction to light.

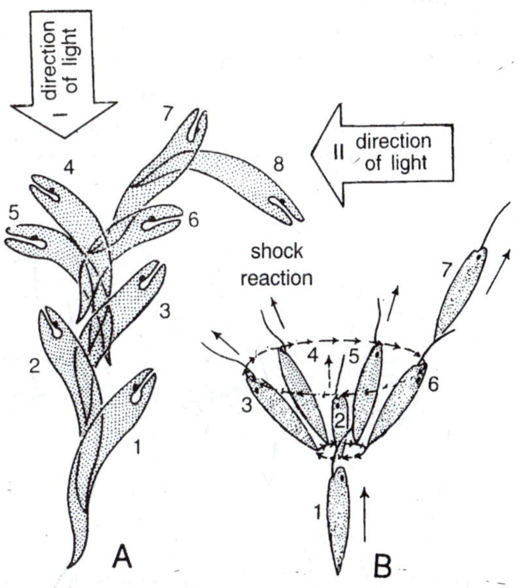

Fig. 8. *Euglena.* **A.** Positions 1-4, when moving towards I direction of light. Position 5, in shock reaction. Positions 6-8, when moving towards II direction of light. **B.** Avoiding reaction on a trial and error pattern.

[III] Avoiding Reaction

Euglena also responds to mechanical, thermal and chemical stimuli showing an **avoiding reaction** on a **trial** and **error** pattern (**phobotaxis**). In most cases the animal slows down, stops or even moves backwards. The posterior end of the body may act as a pivot, while the anterior end traces wider circles of gyrations and swims forward in a new spiral path.

REPRODUCTION

There is no evidence of sexual reproduction in *E. viridis*. It multiplies asexually by **binary** and **multiple** fissions and undergoes **encystment**.

[I] Binary Fission

Transverse binary fission is unknown in *Euglena.* Under favourable conditions of water, temperature and food availability, *Euglena* divides by a simple **longitudinal** binary fission. The longitudinal division is always **symmetrogenic**, that is, the parental *Euglena* divides into two daughter individuals, where one is the plane mirror image of the other. First the nucleus divides into two by **mitosis** which is followed by the division of cytoplasm (**cytokinesis**). The unusual feature which is observed in nuclear division is the persistence of nuclear membrane. In **prophase stage**, all the nucleoli (endosomes) fuse together into a single **nucleolar body** and each chromosome splits longitudinally into two daughter chromosomes or **chromatids**. In **metaphase stage**, the paired chromatids come to lie in a longitudinal plane. The microtubules are present in nucleus but they do not form a spindle. In **anaphase stage**, the paired chromatids separate and move towards their respective poles. It has been suggested that the movement of chromatids is autonomous, with mutual repulsion. The nuclear membrane begins to constrict longitudinally. In the **telophase stage**, constriction in nuclear membrane deepens and finally separates the nucleus into two daughter nuclei. The nucleolar body also splits into two halves, each taking its place in the daughter nucleus of its own side. Next follows the **cytokinesis**. A longitudinal furrow appears in the cytoplasm, beginning at the anterior end, which deepens and finally divides *Euglena* into two daughter euglenae.

Organelles of the anterior end, such as the reservoir, cytopharynx, cytostome, flagella, stigma and contractile vacuole duplicate by the formation of a new set of these structures. However, a new set of flagella arises from new basal bodies which appear in the vicinity of old basal bodies. Multiplication of basal bodies usually precedes cell division.

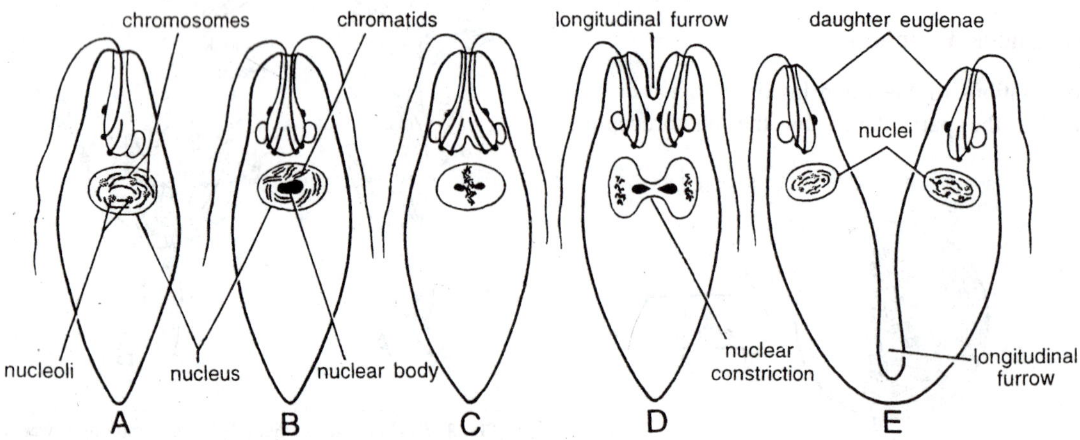

Fig. 9. *Euglena.* Stages of longitudinal binary fission. **A.** Interphase. **B.** Prophase. **C.** Metaphase. **D.** Anaphase. **E.** Telophase and Cytokinesis.

[II] Multiple Fission and Palmella Stage

Under inactive periods, *Euglena* undergoes multiple fission in an encysted condition. Movement ceases altogether, flagellum is thrown off and *Euglena* becomes rounded and embedded in an extensive, thick and mucilaginous coat or **cyst** which is secreted by the muciferous bodies. Encystment is usually followed by repeated longitudinal binary fissions with the formation of several daughter individuals (16 or 32), embedded within a mucilaginous mass. The daughter individuals within the mucilaginous mass secrete their own mucilaginous coats or cysts. This resembles the palmella stage of many algae, such as *Chlamydomonas*. Later these daughter individuals acquire flagella and escape to grow into adult euglenae.

ENCYSTMENT

Encystment takes place as a protective measure to tide over unfavourable conditions such as the lack of food and oxygen, draught, excessive heat, etc. A cyst-wall is secreted in the form of a thick, spherical, yellowish brown and gelatinous covering, composed of a special carbohydrate. In different species of *Euglena,* cyst may be thick (2-4 layered), stalked or operculated with the organism lying centrally or eccentrically in it. The encysted animal not only successfully withstands the adverse conditions of life, but also enjoys a far and wide dispersal. When the conditions become favourable, the animal becomes active and emerges from the cyst to resume its normal free-swimming life (Fig. 10).

As already stated above, the individual may undergo a single or several divisions, resulting in two or many new individuals (**palmella stage**).

Euglena as an Animal

Euglena is studied as an animal as well as a plant. It is more an animal than a plant because of the following reasons :

(1) Absence of a cellulose cell-wall overlying the plasma membrane.
(2) Presence of centrioles forming the blepharoplasts or kinetosomes.
(3) Reserve food is paramylon which is not a true starch.
(4) Presence of paraflagellar body, a sensory (photoreceptive) organelle.
(5) Moves from place to place like an animal.
(6) Responds to various stimuli like an animal.
(7) Pinocytosis and probably holozoic nutrition takes place.

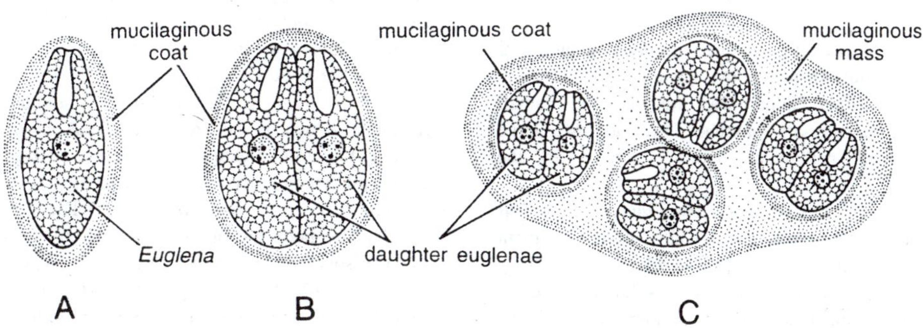

Fig. 10. *Euglena.* Multiple fission and encystation. **A.** Encysted individual. **B.** Fission in encysted condition. **C.** Palmella stage.

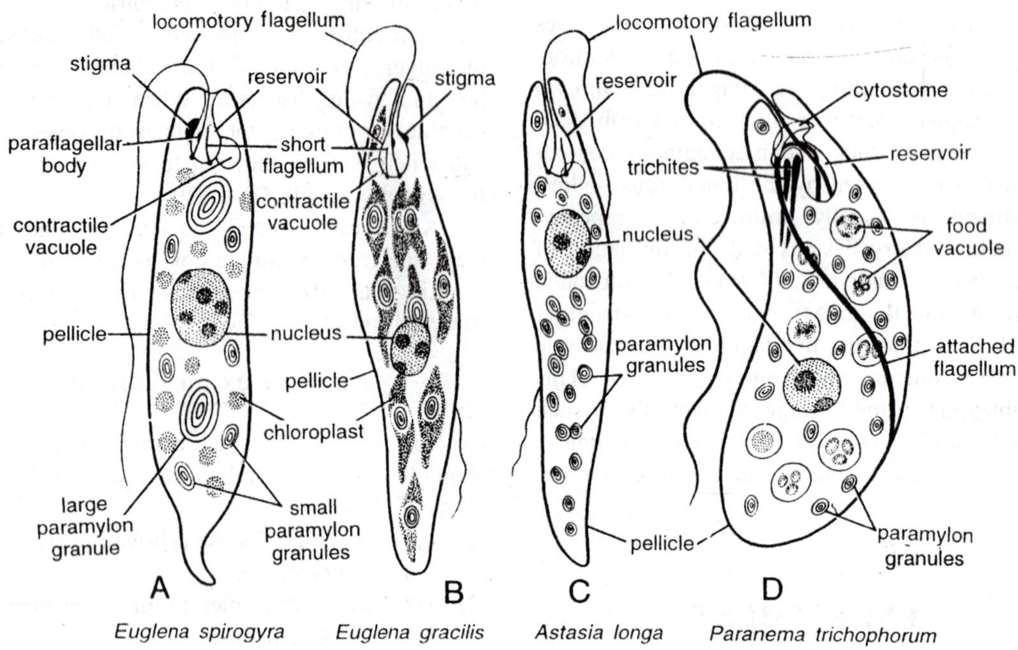

Fig. 11. Some euglenoid flagellates.

OTHER EUGLENOID FLAGELLATES

1. ***Euglena spirogyra.*** It is a large *Euglena,* about 100μ long, with a spindle-shaped body. Posterior end of body is somewhat tail-like. Chloroplasts are numerous, small and discoid. Pyrenoids are wanting. There are two large and many small paramylon bodies scattered in the cytoplasm.

2. ***Euglena gracilis.*** It is a small *Euglena* about 50μ long, with a spindle-shaped body. Chloroplasts are about 10 in number, large, flat and plate-like each with a proteinaceous pyrenoid. Once in darkness for a few days, the Chloroplasts are bleached off permanently and the chlorophyll does not appear again. Cytoplasm contains many paramylon bodies, each attached to a chloroplast.

3. ***Astasia longa.*** It is a euglenoid flagellate, which feeds by osmotrophy in the absence of chloroplasts. Stigma and the paraflagellar body are also wanting. Cytoplasm contains numerous small free paramylon bodies. Zoologists believe that *Astasia longa* is the bleached form of *E. gracilis.*

4. ***Paranema trichophorum.*** It is another euglenoid flagellate which is holozoic and feeds by phagotrophy upon quite large organisms. Of the two flagella, one is long and locomotory, while the other is trailing and adhered to the body surface. The stigma and paraflagellar body are absent. Cytopharynx or gullet bears an accessory rod apparatus, the **trichites**, for puncturing the body wall of a large prey. Cytoplasm contains food vacuoles and numerous small free paramylon bodies.

Important Questions

➤➤*Long answer type questions*

1. Describe the occurrence, structure and modes of locomotion, nutrition and reproduction of *Euglena.*
2. Draw a magnified view of the anterior end of *Euglena* and describe the structures present therein.
3. Give the detailed structure of *Euglena* as revealed by electron micrographs.
4. Describe the different modes of nutrition in *Euglena.*
5. Justify the statement that *Euglena* is more an animal than a plant.
6. Give arguments for treating *Euglena* as an animal, as a plant and as a protist.
7. Write short notes on : (i) Euglenoid movement, (ii) Encystment of *Euglena,* (iii) *Euglena gracilis,* (iv) Mixotrophic nutrition, (v) Palmella stage, (vi) *Peranema trichophorum,* (vii) Structure of flagellum, (viii) Stigma.

➤➤*Short answer type questions*

1. Describe the pyrenoid bodies in *Euglena.*
2. What is the type of nutrition in *Euglena* ?
3. What is the palmella stage in reproduction in *Euglena* ?
4. Encystment in *Euglena* is essential, why ?
5. Describe the flagellum of *Euglena.*
6. Give at least two reasons each for considering *Euglena* (i) as an animal, (ii) as a plant, (iii) as a protist.
7. What is the structure and function of stigma in *Euglena* ?
8. Write on the binary fission in *Euglena.*
9. Name an animal which is able to perform heterotrophic as well as autotrophic nutrition ?
10. Name the animal which show palmella stage ?

➤➤*Multiple choice questions*

1. Coprozoic nutrition is found in :
   (a) *Balantidium* (b) *Euglena*
   (c) *Noctiluca* (d) *Copromonas*
2. Holophytic (autotrophic) nutrition is found in :
   (a) *Paramecium* (b) *Balantidium*
   (c) *Euglena* (d) *Nyctotherus*
3. Saprophytic or saprozoic nutrition is found in :
   (a) *Euglena* (b) *Leishmania*
   (c) *Amoeba* (d) *Monocystis*
4. Photoreceptor organ of *Euglena* is :
   (a) stigma near reservoir
   (b) paraflagellar body at the base of flagellum
   (c) stigma and paraflagellar body
   (d) chromatophore
5. Lactoflavin (sensitizer) found in :
   (a) eye spot (b) paraflagellar body
   (c) Golgi bodies (d) mitochondria
6. *Euglena* is green in colour which is due to :
   (a) chromatophores (b) leucoplasts
   (c) carotene (d) pyrenoid and paramylum
7. *Euglena* belongs to class :
   (a) Phytomastigophorea
   (b) Zoomastigophorea
   (c) Rhizopodea
   (d) Actinopodea
8. Connecting link between plant and animal kingdom :
   (a) *Paramecium* (b) *Euglena*
   (c) bacteria (d) virus
9. Which is oldest one of the animals :
   (a) *Amoeba* (b) *Paramecium*
   (c) *Euglena* (d) *Opalina*
10. Osmoregulation in *Euglena* occur with the help of :
    (a) vacuole (b) food vacuole
    (c) contractile vacuole (d) none
11. Reserved food in *Euglena* is in form of :
    (a) starch (b) glycogen
    (c) fatty acids (d) paramylon
12. Encystment take place in condition of :
    (a) lack of food (b) drought
    (c) excessive heat (d) all of them

Answers

➤➤*Multiple choice questions*

1. (d) 2. (c) 3. (a) 4. (c) 5. (b) 6. (a) 7. (a) 8. (b) 9. (c) 10. (c)
11. (d) 12. (d)

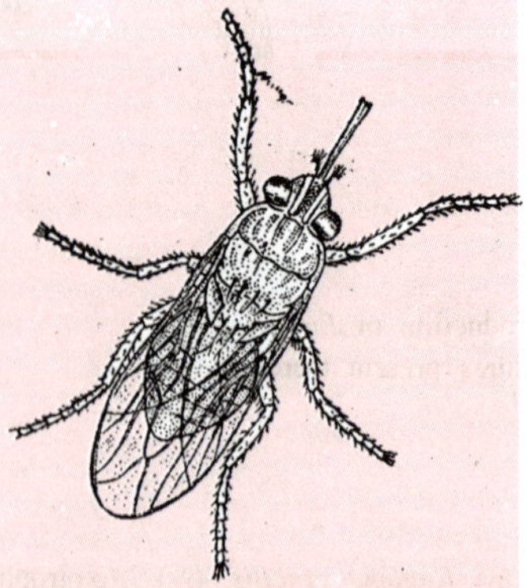

4

Trypanosoma gambiense

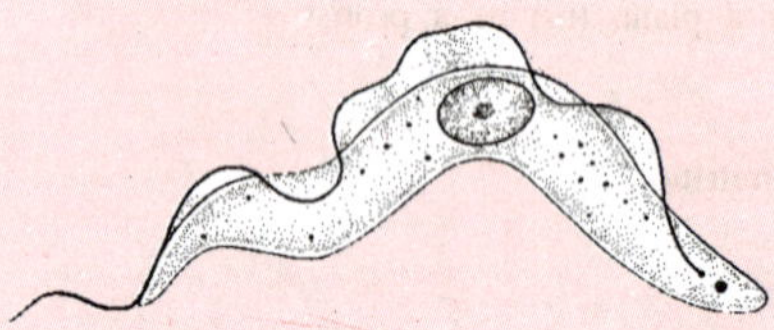

The animal-like flagellates, or **zooflagellates**, include great many types of protozoans, free-living as well as parasitic. They differ with **phytoflagellates** in the absence of chlorophyll-bearing chloroplasts and the paramylon granules. But they definitely show resemblance in having one or more flagella. The parasitic zooflagellates include a number of organisms, which inhabit the blood and tissues of man and other vertebrates. The most common ones are the trypanosomes, belonging to the genus *Trypanosoma* (Gr., **trypanon**, auger+**soma**, body). They parasitize mostly the blood of vertebrates. They are characterized by the possession of a **flagellum** which arises near the posterior end of the body, extends forward as the border of an **undulating membrane**, and emerges as a free filament at the anterior end. Some species are nonpathogenic doing little or no harm to the host. Others, such as *T. gambiense,* are highly pathogenic producing severe diseases in man and animals.

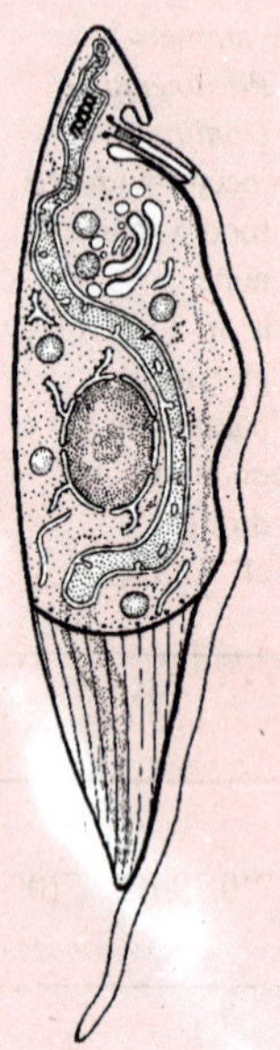

Trypanosoma gambiense

SYSTEMATIC POSITION

Phylum	Protozoa
Subphylum	Sarcomastigophora
Superclass	Mastigophora
Class	Zoomastigophorea
Order	Kinetoplastida
Genus	*Trypanosoma*
Species	*gambiense*

POLYMORPHIC FORMS OF TRYPANOSOMES

At least four forms of trypanosomes are recognized on the basis of the positions of kinetoplast and blepharoplast and the course taken by the flagellum. Two or more such forms occur either in one or both the hosts in the life cycles of various species of *Trypanosoma*. These forms are as follows :

1. **Leishmanial (amastigote).** Round or oval form with a nucleus, blepharoplast and kinetoplast. Flagellum reduced and fibril-like, embedded in cytoplasm.

2. **Leptomonad (promastigote).** Body elongate, nucleus large and anteriorly located blepharoplast and kinetoplast. Flagellum short and unattached.

3. **Crithidial (epimastigote).** Body elongate. Blepharoplast and kinetoplast placed immediately anterior to nucleus. Undulating membrane inconspicuous.

4. **Trypanosomid (trypomastigote).** Body elongate and slender. Blepharoplast and kinetoplast situated at or near posterior end. Undulating membrane conspicuous.

1. Ecology discovery. *T. gambiense* as a human parasite was first observed by **Forde** in 1901. **Sir David Bruce** (1895) discovered that sleeping sickness is transmitted by tsetse fly.

2. Distribution. *T. gambiense* is confined mainly to Central and West Africa particularly in

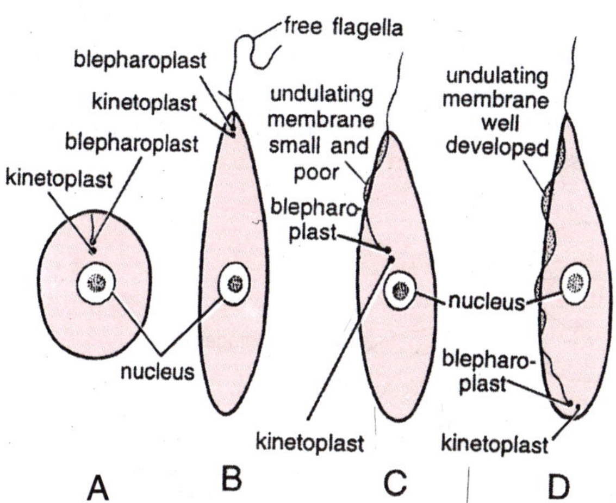

Fig. 1. Polymorphic forms of trypanosomes. **A.** Leishmanial. **B.** Leptomonad. **C.** Crithidial. **D.** Trypanosomid.

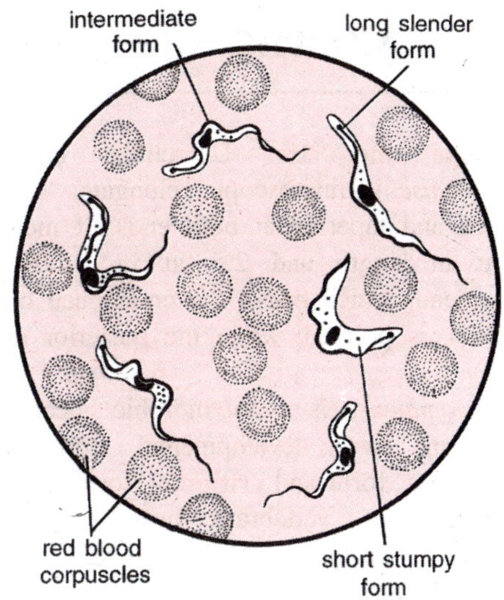

Fig. 2. *Trypanosoma gambiense.* Various forms in human blood plasma.

Nigeria and Congo. Areas near the rivers and lakes have the greatest incidence of infection, since the insect vector, the tsetse fly, lives there in low and thick vegetation.

3. Habits and habitat. *T. gambiense* causes the dreaded disease of African sleeping sickness in man. The parasite lives in the blood stream and in the lymph glands of infected persons in early stages of disease. Later, it invades the cerebrospinal fluid of the central nervous system causing fever, anaemia, lethargy and death. The chief vector which transmits the trypanosome from man to man is the tsetse fly, *Glossina palpalis,* but *G. tachinoides* is also known to be an important transmitter.

Occasionally, the trypanosome utilizes domestic and wild mammals, such as monkeys, dogs, pigs, antelopes, buffaloes and reed bucks, as **temporary** or **reservoir hosts**. In these hosts, the parasite does not undergo any developmental cycle but simply waits for its introduction into the human organism. Entry into and exit from reservoir hosts are always through the bite of tsetse fly, which feeds upon the blood of man as well as the reservoir hosts.

MORPHOLOGY

1. Shape and size. Unicellular body of *T. gambiense* is microscopic, elongate, leaf-like, flattened and tapering at both ends. It measures 10-40μ in length and 2.5-l0μ in width. The anterior end, which marks the emergence of free flagellum, is pointed, while the posterior end is blunt.

T. gambiense is a polymorphic species and occurs in two developmental forms, the **trypanosome** form and **crithidial** form. In blood stream of its vertebrate host (man), the **trypanosome** form may be long slender or short stumpy or intermediate between the two. In invertebrate host (**tsetse fly**), the **trypanosome** form is long slender in midgut, and short stumpy in salivary glands. **Crithidial** form is represented only in salivary glands of tsetse fly.

2. Pellicle. The whole body is surrounded externally by a thin, elastic and firm covering the **pellicle**. It is supported by the fine fibrils, called **microtubules**, which run along the whole length of body. It is believed that microtubules help to maintain the shape of the organism while it swims in the blood plasma.

3. Flagellum. Trypanosome is uniflagellate as it bears a single flagellum. It arises from a **basal body (kinetosome)**, situated near the posterior end of body. It runs forward, attached along the entire length of body, and becomes free at the anterior end. In a transverse section, seen under electron microscope, the flagellum shows the usual 9 + 2 internal fibril arrangement, as in case of *Euglena* (Fig. 2, page 40). It also bears unilateral hair as in *Euglena.* No special structures attaching the flagellum with pellicle have been observed. When the trypanosome moves in blood, the undulating waves pass from tip to base of flagellum, a condition unlike that of *Euglena* in which waves pass from base to tip.

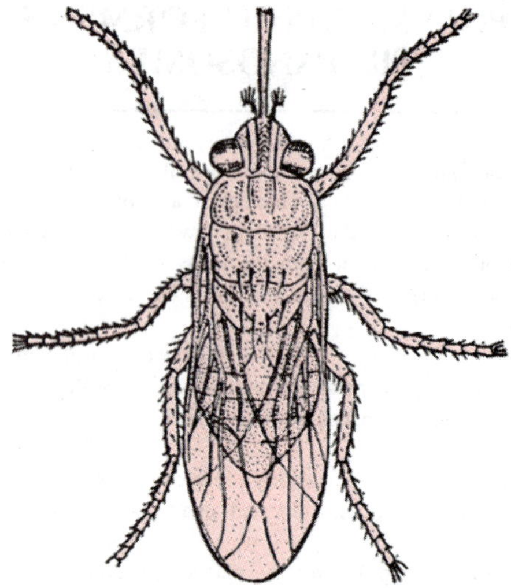

Fig. 3. Tsetse fly.

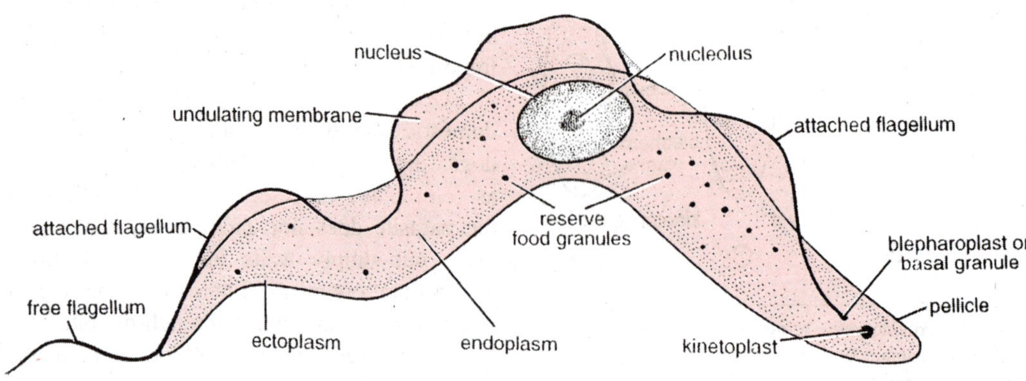

Fig. 4. *Typanosoma gambiense.* Structure under light microscope.

4. Undulating membrane. When the flagellum beats, the pellicle, to which it is attached, is pulled up into a membranous fold called **undulating membrane**. This is believed to be an adaptation for locomotion in the viscous fluid (blood).

5. Cytoplasm. The cytoplasm is enclosed within the pellicle. It has not been observed to be differentiated into ectoplasm and endoplasm.

Electron microscopy has revealed that there is present a reservoir corresponding to the **reservoir** of *Euglena,* but at the posterior end of body. The **basal body** or **blepharoplast** of flagellum is located at the base of this reservoir. An elongated and uneven **mitochondrion**, with tubular cristae, is present in the cytoplasm extending from one end of the body to the other. Just near blepharoplast, the mitochondrion forms a disk-like structure, the **parabasal body** or **kinetoplast**. It includes a double-stranded DNA helix, which lies antero-posteriorly within the disc. A **Golgi apparatus** is present in between the reservoir and nucleus. **Endoplasmic reticulum** is present throughout the cytoplasm. **Ribosomes** occur on the endoplasmic reticulum as well as scattered freely in cytoplasm. Cytoplasm also contains numerous greenish refractile granules, sometimes called **volutin granules**. These are supposed to be stored food particles, chiefly glycogen and phosphates. Various small **vacuoles** are also seen in the cytoplasm, some with hydrolytic enzymes (lysosomes).

6. Nucleus. Within the cytoplasm is present a large, oval and vesicular nucleus. The nuclear membrane is a double unit membrane and bears pores. The **nucleolus** or **endosome** is centrally placed, surrounded by a clear space. Numerous chromosomes remain scattered within the nucleoplasm.

METABOLISM

In the absence of mouth, the trypanosome obtains its nourishment by osmotrophy from blood plasma. Its long slender form, in the human blood stream, absorbs a large quantity of blood glucose that undergoes glycolysis (anaerobic) to release energy necessary for metabolism. At this stage, the mitochondrion is without cristae and thus inactive. Under conditions, when blood produces antibodies to prevent absorption of glucose, the mitochondrion develops cristae and becomes active to synthesize a series of enzymes. The enzymes in Krebs cycle oxidise pyruvic acid (a product of glycolysis) to CO_2 and water and yield energy (aerobic). At this stage, trypanosomes stop multiplying and take up first the intermediate forms and then short stumpy forms. In the body of tsetse fly, the trypanosome forms in mid-gut and crithidial forms in salivary glands have extensive network of cristae-containing mitochondrion as the parasite ceases to get blood glucose and thus depends wholly on pyruvic acid. The metacyclic trypanosomes, in salivary glands, do not feed and their mitochondrion assumes a cristae-less tube possibly indicating reversion to mitochondrial inactivity at this stage. The DNA of kinetoplast plays some yet unknown role in the mitochondrial activity.

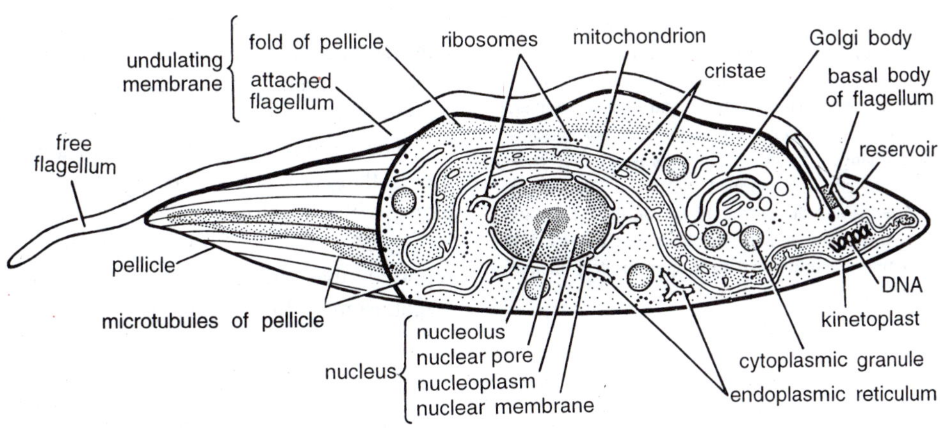

Fig. 5. *Trypanosoma.* Structure under electron microscope (diagrammatic).

The gaseous exchange in respiration and the elimination of excretory products take place by diffusion through the pellicle. The osmoregulatory mechanism is altogether absent.

REPRODUCTION

Sexual reproduction is unknown in *T. gambiense*. Asexual reproduction takes place by **longitudinal binary fission**. First, the kinetoplast and the blepharoplast divide to become paired. The old flagellum remains attached to one of the blepharoplasts, while a new one begins to grow out from the other blepharoplast. The nucleus divides by mitosis into two daughter nuclei. Finally, the body splits mid-longitudinally, from the anterior end backwards, forming two daughter trypanosomes.

LIFE CYCLE

Hosts. *T. gambiense* is **digenetic** that is, it completes its life cycle in two hosts. (Gr., **di**, double + **genos**, race). The **primary** or **principal** or **definitive host** is man and the **intermediate host** or **vector** is a blood-sucking insect called tsetse fly (*Glossina palpalis*). The mammals, like antelopes, pigs, buffaloes, etc., often act as **reservoir hosts** harbouring the parasite.

Principal, primary or definitive host. An animal in which the parasites feed and attain sexual maturity.

Intermediate host or vector. An animal which transmits the parasite from one individual of principal host to another. It is termed **mechanical** when the parasite does not undergo developmental changes in it. If the parasite undergoes developmental changes, it is termed **biological**. In such a host the parasite passes the sexual phase of its life cycle.

Reservoir host. An animal which is utilized by the parasite as a temporary refuge until the appropriate principal host is reached.

Life Cycle in Man

1. Infection. The infection to man by *T. gambiense* is always initiated by the bite of the tsetse fly, *Glossina palpalis,* which harbours the infective **metacyclic** forms in the lumen of its salivary glands. When the fly feeds upon mans blood, it releases the contained trypanosomes into his blood stream. While feeding, the fly grips the skin surface with its claws and punctures it with its proboscis. To prevent the clotting of blood, the fly first introduces some saliva containing an anticoagulant, along which the infective trypanosomes too are inoculated.

2. Multiplication. All stages of parasite in man are extracellular as they are present in the blood plasma and not inside blood cells. In human blood, the metacyclic forms, which are devoid of a free flagellum, become transformed into **long slender** forms with flagella. These swim freely by the beating of their free flagellum and the vibratile movements of the undulating membrane and multiply actively by longitudinal binary fission. The multiplying trypanosomes obtain energy by anaerobic glycolysis.

3. Metamorphosis. When the absorption of glucose ceases due to antibodies produced in blood, the glycolysis is hampered and the trypanosomes stop multiplying. They now shrink to short **stumpy** forms, which are devoid of a free flagellum. During transformation from long slender forms to short stumpy forms, the **intermediate** forms with, somewhat shortened body and a small free flagellum also appear. The intermediate as well as the short stumpy forms depend on their energy supply upon the aerobic oxidation of pyruvic acid.

The stumpy forms do not feed and ultimately die if they are not sucked up by the tsetse fly along with the blood meal.

4. Relapse of infection. It has also been reported that some of the long and slender trypanosomes do not undergo transformation, but

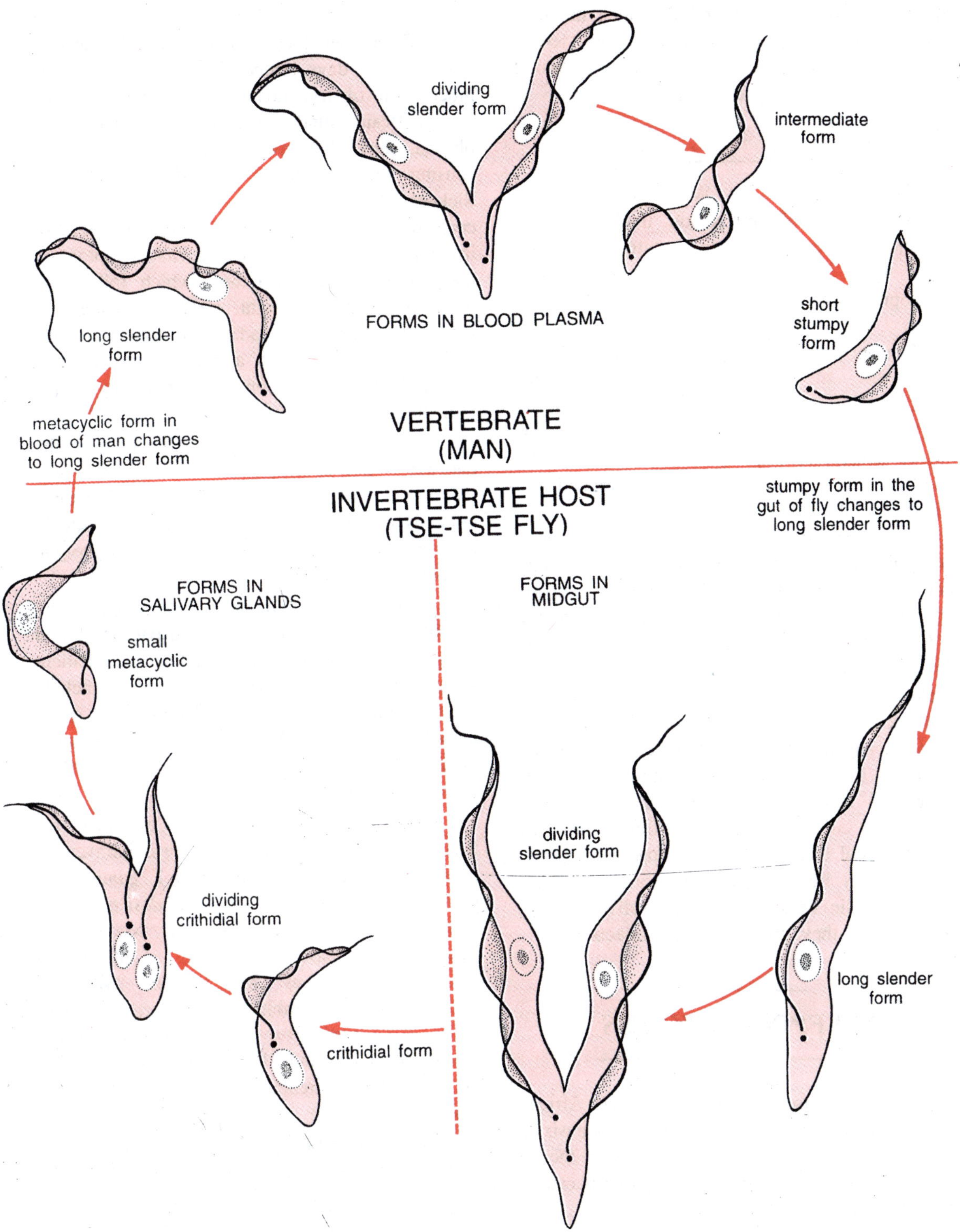

Fig. 6. *Trypanosoma gambiense.* Stages of life cycle in man and tsetse fly.

change the antigens of blood to which the host has produced antibodies. These unaltered slender forms survive and continue to multiply in blood leading to future relapses of the infection.

Life Cycle in Tsetse Fly

1. Transfer to tsetse fly. When tsetse fly sucks the blood of an infected person, it also takes in the short stumpy forms along with the sucked blood. It is the stumpy forms which continue development in the vector.

2. Development in mid-gut. Further development proceeds in the insects mid-gut within **peritrophic membrane**. Here the parasites transform into long slender forms and multiply by longitudinal binary fission. The kinetoplast moves farther from the posterior end of body. The energy-yielding process is related to mitochondrial oxidation of pyruvic acid.

3. Development in salivary glands. Later, the long slender forms make their way into salivary glands via the oesophagus and mouthparts of insect. Here they metamorphose into the **crithidial** forms with shortened body, reduced free flagellum and the kinetoplast in front of the nucleus. The mitochondrion develops an extensive network of cristae and the parasite respires more economically as the blood glucose gradually declines. The crithidial forms multiply in the lumen of salivary glands and transform into slender **metacyclic** forms. At this stage, the mitochondrial activity is again suppressed. When the tsetse fly bites a healthy person, it transfers the metacyclic forms, along with saliva into his blood where they initiate another infection.

SLEEPING SICKNESS

T. gambiense causes the disease of West African Sleeping Sickness or Gambian trypanosomiasis in man. It is different from Encephalitis or the so called American Sleeping Sickness which is caused by a filterable virus.

1. Symptoms and pathogenesis. Bite of tsetse fly causes local irritation which subsides after a few days. Infection is caused by the inoculation of trypanosomes into human blood stream. Besides undergoing development in blood plasma, they also invade the lymphatic glands causing their swelling. The disease Sleeping Sickness is caused when the parasites invade cerebrospinal fluid of central nervous system. An irregular recurrent fever is the first symptom of the disease. Other symptoms which follow are weakness, loss of weight, anaemia, increase in pulse rate and severe headache. In due course the patient falls asleep, first at regular intervals and then lies prostrate in coma. Death is always the ultimate fate.

The cause for the injury of the host by the parasite is not known, but **Laveran** (1913) and **Rondoni** (1913) attribute it to toxicity. The toxic effect is produced when the trypanosomes die in the cerebrospinal fluid and their autolytic decomposition starts.

2. Diagnosis. Diagnosis is done by identification of trypanosomes in a patient's blood sample by microscopic examination. Patient samples that can be used for diagnosis include chancre fluid, lymph node aspirates, blood, bone marrow. In Sleeping Sickness stage identification of trypanosomes is presence of cerebrospinal fluid (neurological phase).

3. Treatment. The standard treatment for first-stage (haemolymphatic) disease is intravenous or intramuscular **Pentamidine** (for *T. gambiense),* or intravenous **Suramin** (for *T. rhodesiense*). The modern standard treatment for second-stage (neurological phase) disease is : Intravenous **Melarsoprol** sold as "Mel B" and "Melarsen Oxide-BAL." Combination therapy with **Eflornithine** amd **Nifurtimox** is safer, easier and more effective and has been recommended as first-line treatment for second-stage (for *T. gambiense*).

No one should begin treatment for Sleeping Sickness until the diagnosis has been confirmed by laboratory after blood test.

Table 1. **Some Important Trypanosomes of Man and Domestic Animals.**

Species	Geographical distribution	Average length in μ width	Position of kinetoplast	Position of nucleus	Posterior end	Free flagellum	Primary hosts	Disease	Transmission by
1. *T. gambiense*	Central and W. Africa	10-40×2.5-10	Sub-terminal	Central	Blunt	Long	Man, antelope, monkeys, pigs, buffaloes, dogs	Gambian or W. African Sleeping sickness	Blood sucking tsetse fly *Glossina palpalis*
2. *T. rhodesiense*	Central and E. Africa	20 × 1.5	Sub-terminal	Central	Blunt	Long	Man, domestic animals	Rhodesian or E. African Sleeping Sickness	Biting of *Glossina* spp.
3. *T.cruzi*	S. and Central America	15-24 × 1.5	Sub-terminal	Central	Short pointed	Long	Man, cats, dogs, monkeys.	Chagas disease	Faeces of kissing bug *Triatoma*
4. *T. lewisi*	N. America	21-36 × 2	Not terminal	Slightly anterior	Long pointed	Long	Rats	Non-pathogenic	Faeces of infected rat-flea
5. *T. brucei*	Africa	30 × 1.5	Sub-terminal	Central	Blunt	Long	Horses, donkeys, mules, camels, cattle	Nagana	*Glossina morsitans*
6. *T. evansi*	N. Africa, Asia, S. China, N. and S. America, India	20-28 × 1.5	Sub-terminal	Central	Blunt	Long	Horses, camels, mules donkeys, cattle, dogs, elephants	Surra	Biting of Tabanid fly
7. *T. equiperdum*	S. America	20-28 × 1.5	Sub-terminal	Central	Blunt	Long	Horses, mules, donkeys, cattle	Dourine	During coitus
8. *T. equinum*	S. America	20-28 × 1.5	Sub-terminal	Central	Blunt	Long	Horses	Mal de caderas	Biting of Tabanid fly
9. *T. vivax*	W., Central and E. Africa	21-26 × 2	Sub-terminal	Central	Blunt rounded	Long	Ruminants Equines	Virulent	Biting of *Glossina* spp.
10. *T. congolense*	W. Central E. Africa	12-18 × 1.2	Sub-terminal	Central	Blunt	Absent	Cows, other domestic animals	Bovine Trypanosomiasis	Biting of *Glossina* spp.
11. *T. simiae*	Africa	17-18 × 1.2	Sub-terminal	Central	Blunt	Absent	Pigs, monkeys, sheep, goats	Virulent	Biting of *Glossina* spp.
12. *T. suis*	Tanzania, Congo	15 × 3.5	Sub-terminal	Central	Very pointed	Short	Pigs	Virulent	Biting of *Glossina* spp.
13. *T. theileri*	N. America	60-100 × 2.3	Sub-terminal	Central	Long pointed	Short	Cattle	Non-pathogenic	Faeces of tabanids.

4. Prevention (Prophylaxis). Prevention of African Sleeping Sickness depends upon the eradication of vector, *i.e.,* the tsetse fly. The flies can be eradicated by clearing out bushes and low trees along rivers in the endemic areas. Spraying of DDT over bushy areas in the vicinity of villages is effective in controlling the tsetse flies. Domestic and wild animals, which act as reservoir hosts, may be put to wholesale slaughter, but this method is not practicable to operate. Further protection is obtained by injecting one gm. of **Suramin** (Antrypol) every two or three months.

Important Questions

➤➤*Long answer type questions*

1. Describe the structure and life history of any trypanosome studied by you.
2. Give the detailed structure of a trypanosome as revealed by the electron microscope.
3. What is African Sleeping Sickness? How it is caused? Describe its transmission and pathogenesis. Suggest methods for its control.
4. Write an essay on *Trypanosoma* and diseases.
5. Give the classification, characteristics and life-history of the Protozoa causing Sleeping Sickness.

➤➤*Short answer type questions*

1. What is the zoological name of the vector of *Trypanosoma gambiensel* ?
2. Name the natural reservoirs of *Trypanosoma gambiense.*
3. Name the vertebrate which retains *Trypanosoma* without any harmful effects to its own self.
4. With reference to *Trypanosoma,* explain polymorphism.
5. With the help of labelled diagram explain the changes taking place in *Trypanosoma gambiense* in the body of tse-tse fly.
6. Give the name all forms of *Trypanosoma* which shows polymorphism.
7. Name the primary host and intermediate host of *T. gambiense.*

➤➤*Multiple choice questions*

1. *Trypanosoma gambiense* causes the disease :
   (a) beri beri (b) scurvy
   (c) sleeping sickness (d) malaria
2. Some times trypanosome passes into internal organs like spleen, lungs where they loose flagella and become oval shaped. What is its name ?
   (a) parabasal body (b) latent bodies
   (c) Rhizoplast (d) *Leishmania*
3. In the body of tse-tse fly, three morphological forms of *Trypanosoma gambiense* are observed; they are metacyclic form, long and slender form and :
   (a) megalospheric form (b) microspheric form
   (c) crithidial form (d) primary form
   (e) secondary form
4. Tse-tse fly is the vector of the disease :
   (a) typhoid (b) kala azar
   (c) gambian fever (d) oriental sores
5. *Trypanosoma* is transmitted by :
   (a) contamination (b) inoculation
   (c) contagious (d) touch
6. The reservoir host for *T. gambiense* is :
   (a) rat (b) pig
   (c) man (d) tse-tse fly
7. Chagas disease is caused by :
   (a) *T. gambiense* (b) *T. cruzi*
   (c) *T. evansi* (d) *T. rhodesiense*
8. *Trypanosoma gambiens* lives in the human body in :
   (a) lymph
   (b) blood
   (c) cerebrospinal fluid
   (d) blood and in cerebrospinal fluid
9. The main symptom for Gambian fever is :
   (a) headache (b) dysentry
   (c) diarrhoea (d) insomnia
   (e) pain at the joints
10. The infective stage of *Trypanosoma* is :
    (a) long and slender form
    (b) crithidial form
    (c) metacyclic form
    (d) intermediate form

11. *Trypanosoma* causing sleeping sickness belongs to class :
    (a) Rhizopoda (b) Mastigophora
    (c) Sporozoa (d) Ciliata
12. Who discovered the disease sleeping sickness :
    (a) Ronald Ross (b) Forde
    (c) David Bruce (d) Lamble
13. The vector for *Trypanosoma* :
    (a) *Glossina palpalis* (b) *G. tachinoides*
    (c) both (d) none
14. Nourishment in *Trypanosoma* :
    (a) autotrophy (b) chemotrophy
    (c) osmotrophy (d) none
15. Which is nonpathogenic :
    (a) *T. gambiense* (b) *T. lewisi*
    (c) *T. cruzi* (d) *T. evansi*
16. Chagas disease due to infection of :
    (a) *T. gambiense* (b) *T. rhodesiense*
    (c) *T. cruzi* (d) *T. brucei*
17. Osmoregulatory mechanism absent in :
    (a) *Paramecium* (b) *Amoeba*
    (c) *Euglena* (d) *Trypanosoma*
18. Caused of surra disease in cattle :
    (a) *T. evansi* (b) *T. brucei*
    (c) *T. lewisi* (d) *T. cruzi*
19. *Trypanosoma* finally invades :
    (a) cerebrospinal fluid (b) brain
    (c) liver (d) blood
20. *Trypanosoma* belongs to the class :
    (a) Sarcodina (b) Zooflagellata
    (c) Ciliata
    (d) Sporozoa
21. In *Trypanosoma gambiense* :
    (a) reproduction is by multiple fission
    (b) there are two nuclei, a micronucleus and a macronucleus
    (c) there are two locomotary organelles, a flagellum and an undulating membrane.
    (d) tsetse fly has no role to play in life cycle.
22. Which is true about *Trypanosoma*?
    (a) monogenetic
    (b) polymorphic
    (c) facultative parasite
    (d) non-pathogenic

Answers

➤➤*Multiple choice questions*

1. (c) 2. (d) 3. (c) 4. (c) 5. (b) 6. (b) 7. (b) 8. (b) 9. (a) 10. (c)
11. (b) 12. (c) 13. (c) 14. (c) 15. (b) 16. (c) 17. (d) 18. (a) 19. (a) 20. (b)
21. (c) 22. (b).

5

Leishmania donovani

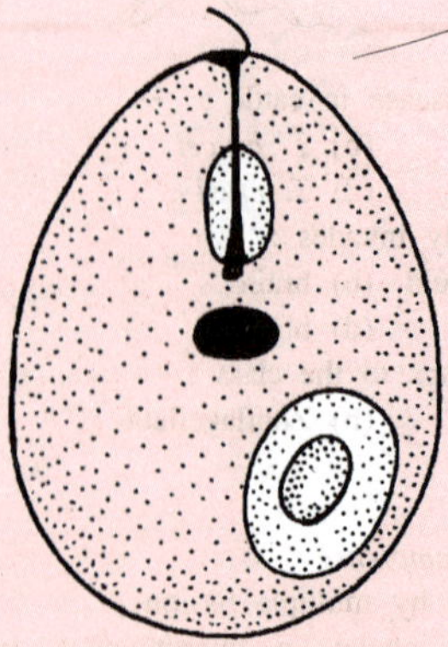

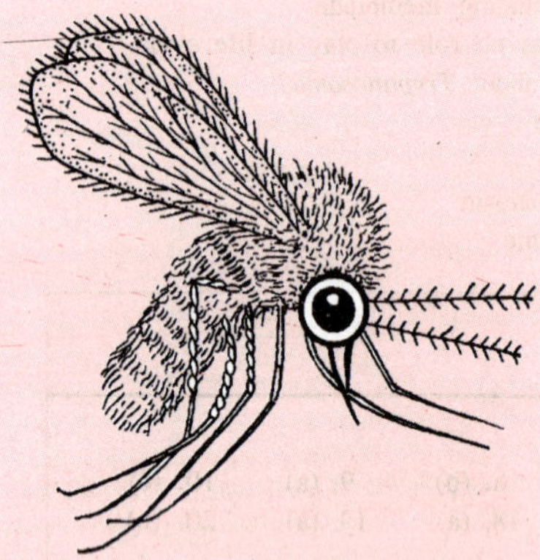

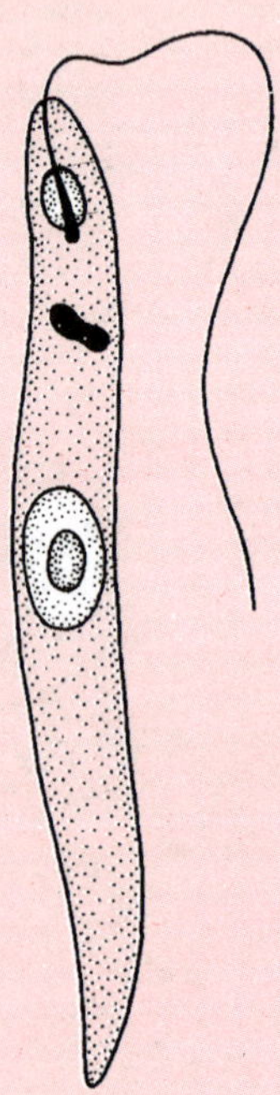

Leishmania is an important pathogenic zooflagellate genus closely related to *Trypanosoma.* Various species of *Leishmania* (***Table 1***) infect man, cattle, dog, sheep, horse, etc. and cause serious diseases collectively known as *Leishmaniasis.* All types are carried by the blood-sucking sandflies of the genus *Phlebotomus,* and all are intracellular parasites in leucocytes of blood or in cells of liver and spleen. *Leishmania donovani* causes a malaria-like oriental disease in man called **Kala-azar, Dumdum** fever or **Black fever**. In total 53 species are known so far.

Leishmania donovani

SYSTEMATIC POSITION

Phylum	Protozoa
Subphylum	Sarcomastigophora
Superclass	Mastigophora
Class	Zoomastigophora
Order	Kinetoplastida
Genus	*Leishmania*
Species	*donovani*

1. Ecology discovery. Genus *Leishmania* was created by **Ross** in 1903. The species *L. donovani* was reported simultaneously by **Leishman** from London (1903) and **Donovan** from Madras (1903), hence the name *Leishmania donovani.*

2. Distribution. *L. donovani* infects man in India, China, Russia, Mediterranean countries, and in parts of Africa and South America. In India, it is endemic in Assam, Bengal, Bihar, Orissa, Tamilnadu and eastern parts of Uttar Pradesh.

3. Habits and habitat. In man, *L. donovani* lives as an intracellular parasite in leucocytes or cells of liver, spleen, bone marrow, lymphatic glands, etc. It is the causative agent of the disease known as **Kala-azar** resulting in fever, enlargement of spleen and a reduction in the number of white corpuscles in blood. It is transmitted through the bite of sandflies.

MORPHOLOGY

[I] Shape and Size

The genus *Leishmania* occurs only in two forms or stages, **leishmanial** and **leptomonad**, which alternate between a vertebrate (man) and an invertebrate (sandfly) host. The two forms are recognized as follows on the basis of the positions of their kinetoplast and blepharoplast, and the course taken by the flagellum.

1. Leishmanial or amastigote form. It occurs intracellularly in blood cells or reticulo-endothelial cells of the vertebrate host or man. It is microscopic, rounded or oval, with a central or eccentric nucleus, blepharoplast and kinetoplast, but no free flagellum. It measures 2 μ to 4 μ in diameter.

2. Leptomonad or promastigote form. It is found in the midgut of the invertebrate host or sandfly. It is elongated, slender and spindle-shaped with a large centrally placed nucleus, blepharoplast, kinetoplast and a long free flagellum. A fully formed promastigote stage measures 15-20 μ in length and 1-2 μ in width.

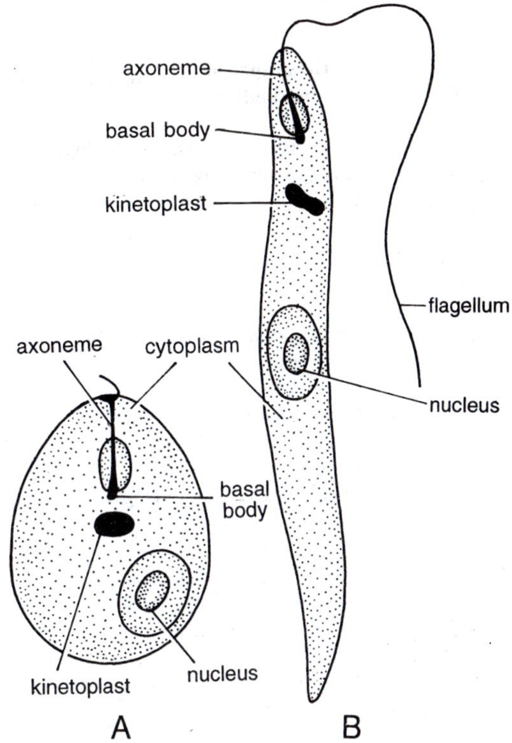

Fig. 1. Two forms of *Leishmania*.
A. Amastigote or leishmanial form in man.
B. Promastigote or leptomonad form in sandfly.

[II] Cell Membrane

The whole body is covered externally by a very thin, delicate, elastic and firm covering or **pellicle**. It gives definite shape to the body and does not form an undulating membrane.

[III] Flagellum

Leishmania is uniflagellate, bearing a single flagellum. In the leptomonad form of parasite in sandfly, the flagellum is long and free. It arises from a minute **basal body** or **blepharoplast** situated near the anterior end. Closely associated

Table 1. **Species of *Leishmania* infecting man.**

Species	Geographical distribution	Primary hosts	Position	Disease produced & symptoms	Transmission in host
1. *Leishmania donovani*	India, China, North Africa, South Europe, Russia, South America	Man, dogs, cats, etc.	Intracellular in leucocytes and cells of spleen, liver, bone marrow and lymph glands.	Kala-azar or Dumdum fever with fever, anaemia, enlargement of spleen and liver and darkening of skin. Also Dermal leishmanoid.	Cyclic, through bite of blood-sucking sandfly *Phlebotomus argentipes* in India.
2. *L. infantum*	Mediterranean areas. Considered identical to *L. donovani*	Children	Intracellular	Infantile Kala-azar with enlargement of spleen.	Bite of sandfly
3. *L. chagasi*	South America. Identical to *L. donovani*	Man	Intracellular	S. American Kala-azar	Bite of sandfly
4. *L. canis*	Mediterranean areas. Identical to *L. donovani*	Dogs	Intracellular	Similar to visceral leishmaniasis or kala-azar	Bite of sandfly
5. *Leishmania tropica*	Syria, Arabia, Iraq, Iran, Central Asia, Central and Western India	Man, dogs, cats, etc.	Intracellular in skin	Skin lesions called Oriental sores spps.	Bite of Sandfly *Phlebotomus*
6. *Leishmania brasiliensis*	Mexico, Central and South America	Man dogs	Intracellular in mucous membrane of nose & throat	Naso-pharyngeal leishmaniasis called Espundia in Brazil, Uta in Peru, etc.	Bite of *Phlebotomus intermedius*

with the blepharoplast lies a disc-shaped **parabasal body** or **kinetoplast**. There is, however, no cytostome. In the leishmanial form of parasite in man, there is no free flagellum, as it is greatly reduced, fibril-like and lies embedded in cytoplasm. A transverse section of flagellum under electron microscope shows the typical 9 + 2 internal fibril arrangement, as in case of *Euglena* (see Fig. 2. page 40).

[IV] Cytoplasm

Underneath pellicle, the body cytoplasm is colourless, homogeneous and not differentiated into ectoplasm and endoplasm. Electron microscopic studies show that cytoplasm is marked by longitudinal **striations** or **microtubules** which may be contractile. Other structures present in cytoplasm are the blepharoplast, kinetosome, rhizoplast, Golgi body, mitochondrion, vacuole and nucleus.

[V] Nucleus

A single large spherical nucleus lies eccentrically or in the middle of the body. It is vesicular and with a distinct central **karyosome** or **nucleolus**. It is covered by a double unit membrane with pores and measures about 1μ in diameter.

METABOLISM

As in *Trypanosoma,* a mouth or cytostome is also lacking in *Leishmania,* so that **nourishment** is obtained saprozoically through body surface from host cells. Gaseous exchange in **respiration** and **elimination** of excretory products also occur by diffusion through body surface. **Sexual reproduction** is unknown and **asexual**

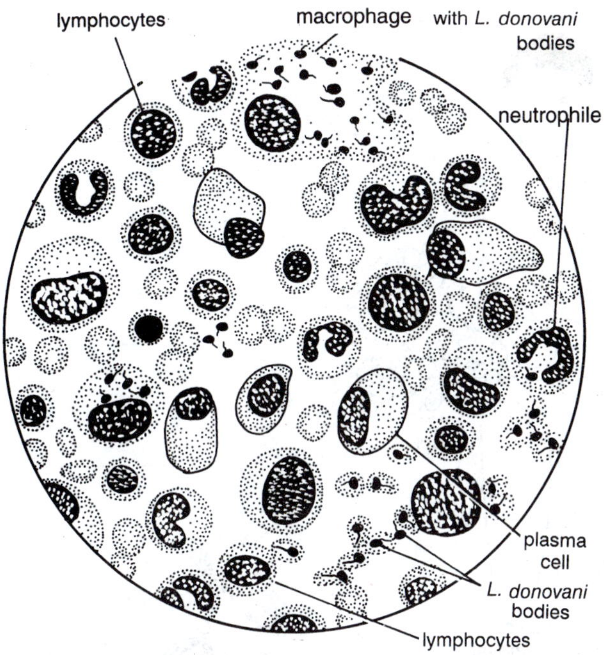

Fig. 2. Smear of bone marrow fluid showing free and phagocytosed amastigote forms *of L. donovani.*

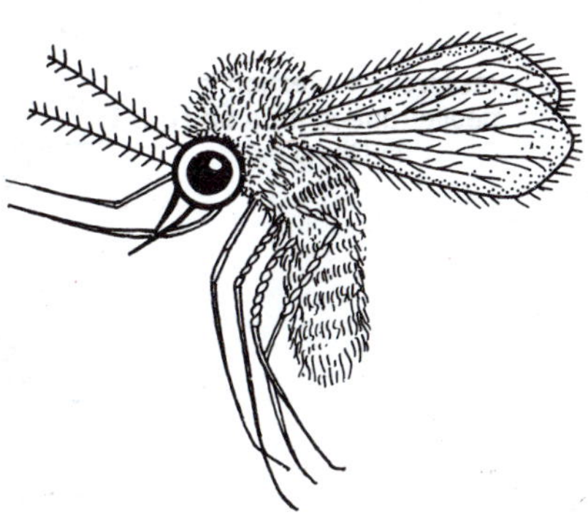

Fig. 3. Sandfly *(Phlebotomus).*

reproduction always takes place by longitudinal binary fission in the same manner as already described in case of *Trypanosoma.*

LIFE CYCLE

Hosts. Like *Trypanosoma, Leishmania* is also a **digenetic** parasite and requires two hosts for completion of its life cycle. **Primary** or **principal host** is a vertebrate or man, in which the parasite feeds and multiplies asexually. The **secondary** or **intermediate host** or **vector** is an invertebrate or blood-sucking insect or sandfly, belonging to the genus *Phlebotomus.* Some mammals like dogs, jackals (Russia) and gerbils and ground squirrels (Kenya) also serve as **reservoir hosts** in which the parasite does not undergo any change but simply waits for its introduction into the human host. In India, canine leishmaniasis does not exist so that man is the main and sole host. Under laboratory conditions small burrowing rodents or hamsters (*Cricetulus griseus*) are also susceptible to infection and utilized for testing the drugs for *Leishmania* infections.

[I] Life Cycle in Man

1. Infection. In India, *L. donovani* is transmitted to man by the sandfly *Phlebotomus argentipes.* The insect vector, which has fed on some suitable fruit or plant juice after infected human blood meal, shows an enormous number of parasites in its buccal cavity and pharynx. Salivary glands of the vector are not involved as in case of a tsetse fly transmitting the trypanosomes. When such a sandfly bites a man, it liberates the parasites in the skin wound caused by its proboscis. However, the actual mode of transmission is not clear in India. Perhaps the Indian vector (sandfly) does not bite but spreads infection by being crushed possibly by slapping.

Multiplication. The parasites introduced by sandfly into human body are in the promastigote or leptomonad form. Some of them entering the blood circulation directly become destroyed. While those entering the cells of reticulo-endotheleal system (liver, spleen, bone marrow and lymph nodes) change into amastigote or leishmanial forms. These undergo slow multiplication by binary fission so that the host cells become greatly enlarged.

Spread of infection. When the number of parasite reaches 50 to 200 or even more, the host

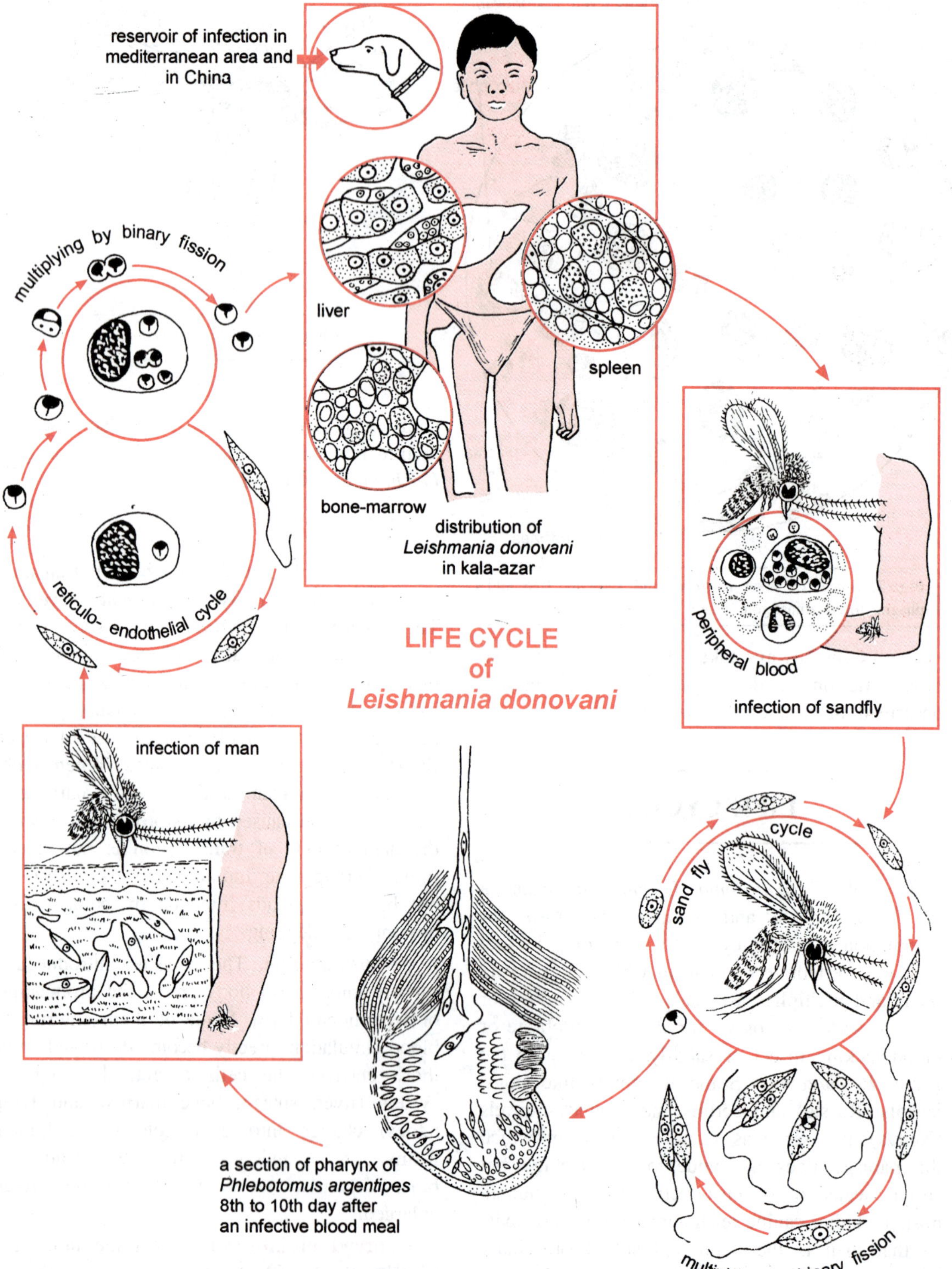

Fig. 4. *Leishmania donovani.* Complete life cycle.

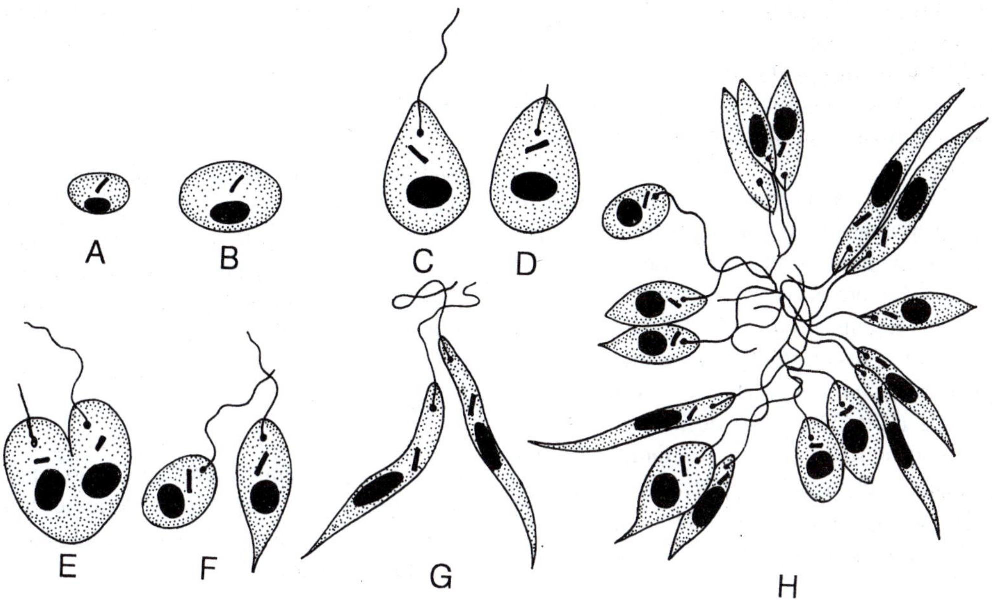

Fig. 5. *L. donovani.* Stages of development from amastigote to promastigote form and binary fission.

cell ruptures. The liberated parasites are taken up by new host cells and the multiplication cycle is repeated so that the reticulo-endothelial system becomes progressively infected . Some of the free amastigotes become phagocytosed by neutrophils and monocytes (macrophages). These heavily parasitised cells wander through the general blood circulation leading to a general infection.

[II] Life Cycle in Sandfly

1. Transfer to sandfly. When a sandfly sucks blood of an infected person, it obtains free amastigotes as well as the parasitised neutrophils and monocytes along with the blood-meal.

2. Development in sandfly. In the midgut of sandfly, the amastigote forms become elongated and acquire a free flagellum , thus changing into the promastigote forms. These multiply by longitudinal binary fission. In 6 to 9 days, the number of parasites becomes enormous and they spread into the pharynx and buccal cavity. The salivary glands are not infected. Transmission into a new host occurs when such a heavily infested sandfly bites the host.

KALA-AZAR OR DUMDUM FEVER

1. Occurrence. Kala-azar, also known as Black fever or Dumdum fever, is a serious oriental disease of man. It occurs in India, China, Mediterranean countries and parts of Africa and South America. Its causative agent is a pathogenic flagellate, called *Leishmania donovani,* which is transmitted by the bite of small blood-sucking sandflies called *Phlebotomus argentipes.*

2. Symptoms and pathogenesis. Incubation period is long, from 3 to 6 months, and symptoms may appear even after 2 years. Early symptoms of Kala-azar include swelling, high fever and enlargement of spleen and liver. It is followed by general weakness, emaciation, anaemia due to reduction in number of blood cells, and a peculiar darkening of skin. The word 'Kala-azar' has been derived from two Indian words, **kala** and **azar**, meaning 'Black sickness'. In advanced stage, skin becomes dry, rough and dark or pigmented. Hair becomes brittle and falls out. If not properly treated, the patient dies within 2 years. Death is generally due to secondary infections by bacteria or viruses. The defense mechanism of body becomes so weak that the patient is unable to resist them.

3. Diagnosis. Kala-azar can be diagnosed by microscopical examination of blood film or biopsy material taken from spleen or bone marrow of patient, for the presence of amastigote forms of *L. donovani.* Examination of W.B.C. count shows decrease of neutrophils but increase of lymphocytes and monocytes. Number of erythrocytes (R.B.C.) is also decreased.

4. Treatment (Therapy). In India, effective and affordable treatment for Visceral Leishmaniasis is sorely lacking and parasites or insect vectors are becoming increasingly resistant to existing anti-parasite drugs.

The traditional treatment is with **pentavalent antimonials** as **Sodium Stibo Gluconate inj.** and **meglumine antimonite.** Resistance is now common in India, and rates of resistance have been shown to be as high as 60% in parts of Bihar.

The treatment of choice for Visceral Leishmaniasis in India is now **Amphotericin B** in its various liposomal preparations like **AmBisome** and **Amphocil**. A newer oral drug having 95% cure rate **Miltefosine** is being evaluated.

On February 2012, the world's first clinical trial for Leishmaniasis vaccine was launched by the Non-profit Infectious Disease Research Institute. The vaccine was designed by fusing two *Leishmania* parasite proteins with an adjuvant. The first trial took place in Washington (state) and is followed by a trial in India.

5. Prevention (Prophylaxis). Preventive measures include :

(a) Eradication of the insect vector, i.e., sandfly. In endemic areas, low trees and bushes, etc. should be cleared out. Periodic fumigation and spray of insecticides of sleeping quarters should be done.

(b) Attack on the parasite. In areas where dogs act as reservoir hosts, all street dogs should be killed. Control measure in India should be proper treatment campaign.

(c) Personal defense. For avoiding bite of sandflies, use mosquito-nets or screens and avoid sleeping on ground floors.

Important Questions

➤➤*Long answer type questions*

1. What causes Kala-azar ? Give a detailed description of the causative agent and its preventive measures.
2. Give an account of the structure and life cycle of *Leishmania donovani.*
3. Provide an account of the different species of *Leishmania* and the diseases caused by them.

➤➤*Short answer type questions*

1. Who first described the Leishmania donovani?
2. How *Leishmania donovani* enters the host?
3. What are chrithidial forms of *Leishmania donovani?*

➤➤*Multiple choice questions*

1. Intermediate host of *Leishmania* is :
   (a) *Phlebotomus* (sandfly)
   (b) tse-tse fly
   (c) mosquito (d) house fly
2. Who discovered *Leishmania* :
   (a) David Bruce (b) Ross
   (c) Charles Laveran (d) Lancisi
3. *Leishmania* causes :
   (a) kala azar (b) malaria
   (c) dysentery (d) sleeping sickness
4. Black fever is caused by :
   (a) *Plasmodium* (b) *Trypanosoma*
   (c) Giardia
   (d) *Leishmania*

5. Leptomonad form of *Leishmania* is found in :
   (a) man (b) sandfly
   (c) tse-tse fly (d) bed bug
6. *Leishmania* is intracellular parasite of :
   (a) leucocytes
   (b) hepatic cells
   (c) cells of spleen
   (d) all
7. Reservoir host for *Leishmania* :
   (a) dog (b) jackal
   (c) squirrel
   (d) all of these
8. Due to infection of *Leishmania* occur :
   (a) Kala-azar (b) Oriental sore
   (c) Esphundia (d) all
9. Cause of Kala-azar :
   (a) *Leishmania donovani*
   (b) *L. infantum*
   (c) *L. chagasi* (d) *L. canis*
10. General body surface of *Leishmania* perform :
    (a) nourishment (b) respiration
    (c) excretion (d) all
11. Each of the following statements concerning kala-azar is correct except :
    (a) Kala-azar is caused by *Leishmania donovani*
    (b) Kala-azar is transmitted by the bite of sandflies
    (c) Kala-azar occurs primarily in rural latin America
    (d) Kala-azar can be diagnosed by finding amastigotes in bone marrow

Answers

➤➤*Multiple choice questions*

1. (a) 2. (b) 3. (a) 4. (d) 5. (b) 6. (d) 7. (d) 8. (d) 9. (a) 10. (d) 11. (c)

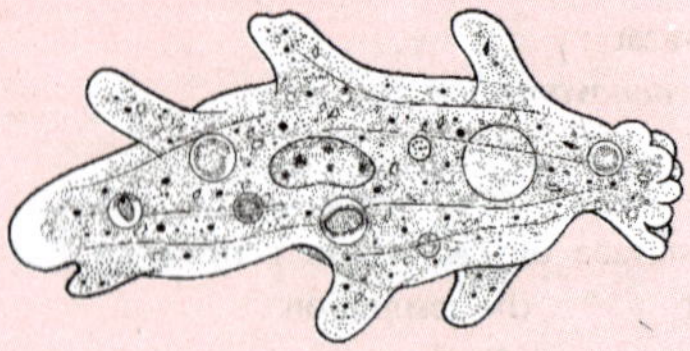

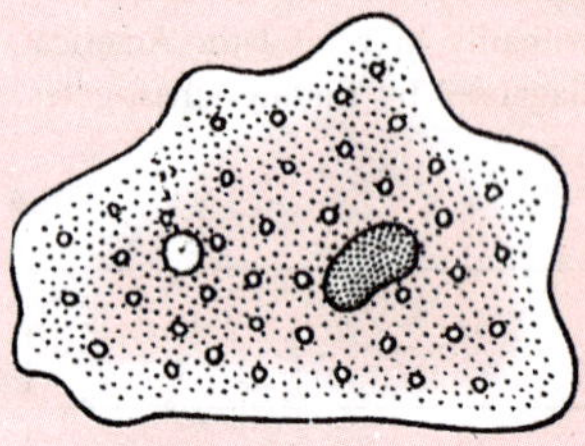

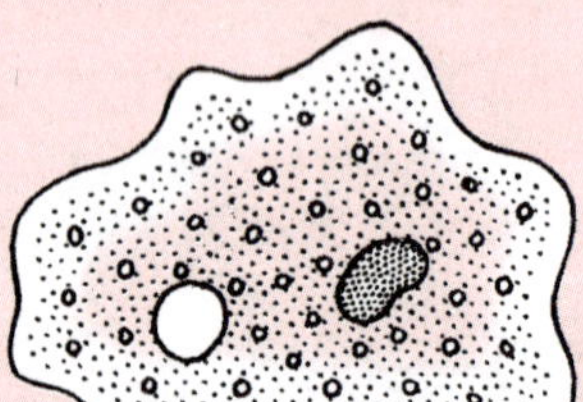

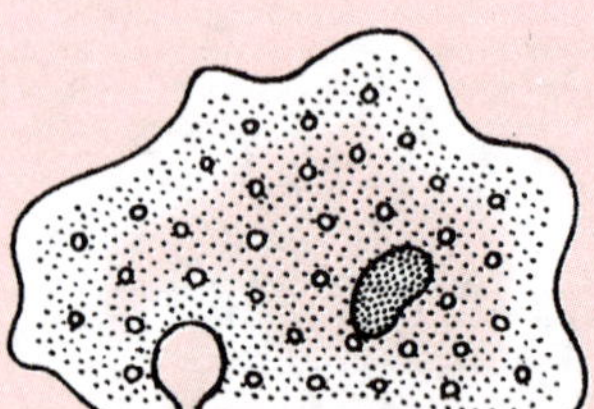

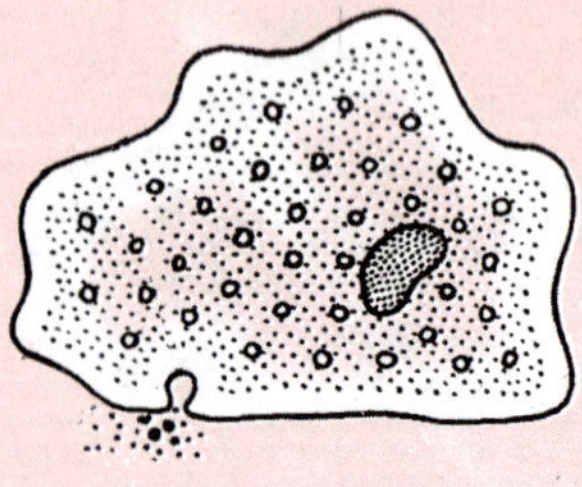

6

Amoeba proteus

Amoeba is the most popular, free-living available protozoan. It is regarded as the lowest form of animal, as its body consists of a mere speck of protoplasm. The body resembles a tiny blob of jelly, and yet it contains all the equipment required by the organism to perform all the vital functions of life, such as movement, nutrition, respiration, excretion, reproduction, etc. It serves as an interesting and suitable material for laboratory studies by the students because it is a large protozoan, very slow in locomotion and easy to obtain. *Amoeba* belongs to the superclass **Sarcodina**. There are several species of *Amoeba,* but most commcnly studied, is the large freshwater species *Amoeba proteus.* This is also a useful experimental animal.

Amoeba proteus

The name *Amoeba proteus* has been derived from two Greek words — *Amoeba* from **amoibe** which means change, and *proteus* after the name of the mythical Greek sea-god **Proteus**, who could change into any shape. Thus, the interesting feature associated with this animal is that it has no shape, or rather, its shape is constantly changing. *Amoeba* was first described by **Rosenhof** in 1755. **H.I. Hirschfield** (1962) has given a full and comprehensive account of the biology of *Amoeba.*

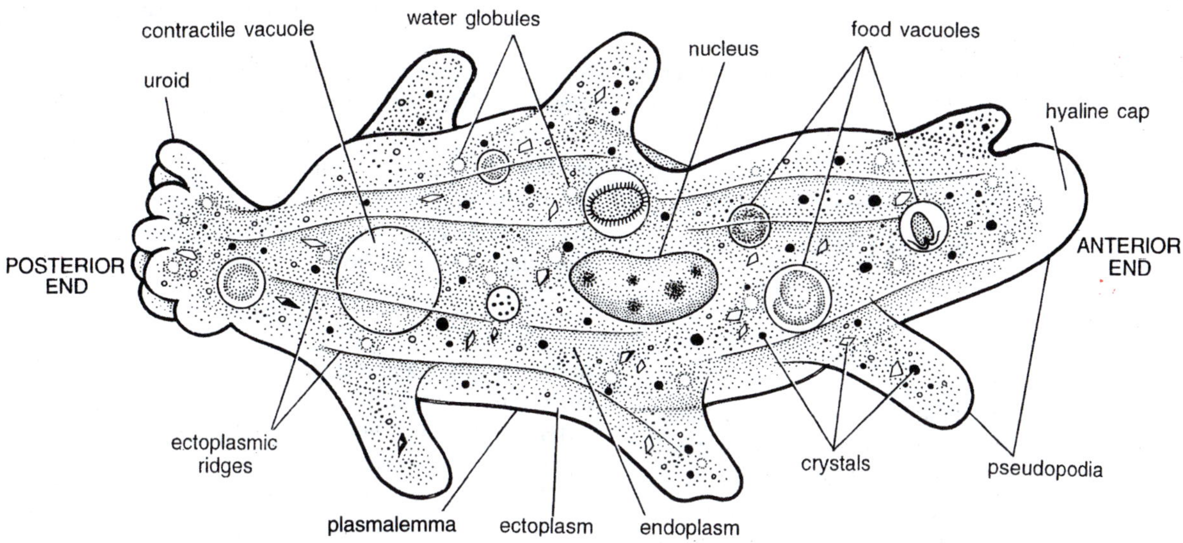

Fig. 1. *Amoeba proteus.*

SYSTEMATIC POSITION

Phylum	Protozoa
Subphylum	Sarcomastigophora
Superclass	Sarcodina
Class	Rhizopodea
Subclass	Lobosia
Order	Amoebida
Genus	*Amoeba*
Species	*proteus*

HABITS AND HABITAT

Amoeba proteus is widely distributed. It is commonly found on the bottom mud or on underside of aquatic vegetation in freshwater ponds, ditches, lakes, springs, pools and slow-running streams. It is rarely found in free water as it requires a substratum to glide on from place to place. It occurs in abundance in those waters which contain bacteria and organic substances such as aquatic vegetation, leaves and twigs. Sides of lotus ponds and water troughs are good places for their collection. It moves and feeds with the help of false feet or **pseudopodia**, formed as a result of streaming flow of cytoplasm. *Amoeba* has great power of **regeneration**. If an individual is cut into pieces, accidently or in a laboratory, every piece containing a part of nucleus grows into a complete *Amoeba*.

CULTURE *OF AMOEBA*

Amoeba may be obtained for class study by scraping decaying vegetation from bottom of a pond. When the scraping is allowed to settle in a wide-mouthed container, amoebae of different kinds may be found in the sediment and sorted with the help of a fine pipette under a binocular microscope.

A temporary culture of *Amoeba* can be prepared in laboratory by hay-infusion method. Decaying aquatic weeds or other organic substances such as hay, dry leaves, twigs, seeds, etc., are taken and boiled with sufficient amount of fresh water. To prepare culture medium, tap water should not be used because it is usually chlorinated. After boiling for about 15 minutes, water is filtered and the filtrate is allowed to cool. To this filtrate a few *Amoeba*-rich water drops are added and the amoebae are allowed to multiply for 2 or 3 days. A culture can also be maintained in laboratory by keeping amoebae in small covered petri dishes containing a few boiled wheat grains. The latter serve as food for bacterial growth on which other micro-organisms feed and which in turn serve as food for amoebae. New culture should be started from time to time by putting together fresh wheat grains and a few amoebae from older cultures.

STRUCTURE

[I] Shape and size

Amoeba is a unicellular, microscopic animalcule and measures about 250 to 600μ (microns) in maximum diameter. To the naked eye, the larger *Amoeba proteus* is just visible as a whitish blob. Under the microscope it appears as an irregular, colourless and translucent mass of protoplasm, continually changing its shape by sending out and withdrawing finger-like processes, called **pseudopodia**. When it withdraws all its pseudopodia, it becomes spherical in shape.

Far from being shapeless, *Amoeba* is considered to have a definite polarity, *i.e,* it has definite anterior and posterior ends. At the anterior end, the animal puts out pseudopodia, while the posterior end is marked by a wrinkled region, called **uroid**.

[II] Pseudopodia

Pseudopodia or **false feet** (Gr., **pseudo**, false + **podium**, foot) are irregular blunt processes of the cell body. These are of variable size and are capable of protruded or retracted, often with considerable speed. These are formed as a result of liquefaction and flowing forward of the cytoplasm. As many pseudopodia are formed simultaneously *A. proteus* is a 'polypodial' species. Its pseudopodia are large and broad with rounded tips. Such pseudopodia are called **lobopodia**. They move by 'pressure flow" mechanism and assist the animal in locomotion and food ingestion.

[III] Plasmalemma

Amoeba possesses no pellicle or cell wall. The body is covered by a very thin, delicate plasma membrane or unit membrane, called **plasmalemma**. This membrane is selectively permeable, *i.e.,* water and some small solute molecules can pass freely through it in both directions, but certain large molecules cannot pass. An unusual feature associated with plasmalemma is the presence of numerous fine, ridge-like extensions on its outer surface. Zoologists believe that these have adhesive properties and serve to bind the animalcule to the substratum.

[IV] Cytoplasm

Inside the plasmalemma, there is a dense mass of cytoplasm containing several organelles. It is differentiated into two fairly distinct zones, an outer **ectoplasm** and an inner **endoplasm**.

1. Ectoplasm. Lying immediately beneath the plasmalemma, ectoplasm is thin, clear and transparent. It is somewhat rigid, contractile and under tension. It is most clearly visible at the tip of a pseudopodium where it forms a **hyaline cap**.

2. Endoplasm. Completely surrounded by ectoplasm, the endoplasm forms the bulk of the animal. It is fluid-like, granular and semi-transparent. As the ectoplasm is under tension, the endoplasm must also be under pressure of the ectoplasm.

According to **Mast**, endoplasm occurs in two colloidal states. The peripheral viscid or **gel state is termed plasmagel** and the central flowing or **sol** state is termed **plasmasol. Plasmagel** forms a tube, through which flows the **plasmasol**. The two colloidal states of endoplasm are inter-changeable. However, the electron microscope has not revealed the occurrence of two colloidal phases in endoplasm. It is believed that it is the ectoplasm which is in **gel** state, while the endoplasm is in **sol** state. According to **De Bruyn**, protoplasm can be thought of as a "three-dimensional network" of protein chains, linked together by cross-linkages of side chains. **Gel** state is due to protein chains fully expanded and **sol** state is the result of contraction of such chains.

[V] Endoplasmic Organelles

A number of organelles, visible under light microscope, are found within the endoplasm. These organelles comprise of the nucleus, contractile vacuole, food vacuoles and water globules, etc.

1. Nucleus. In *A. proteus,* nucleus is a single large flattened, discoidal and slightly biconcave disk like a human erythrocyte. It may be lying

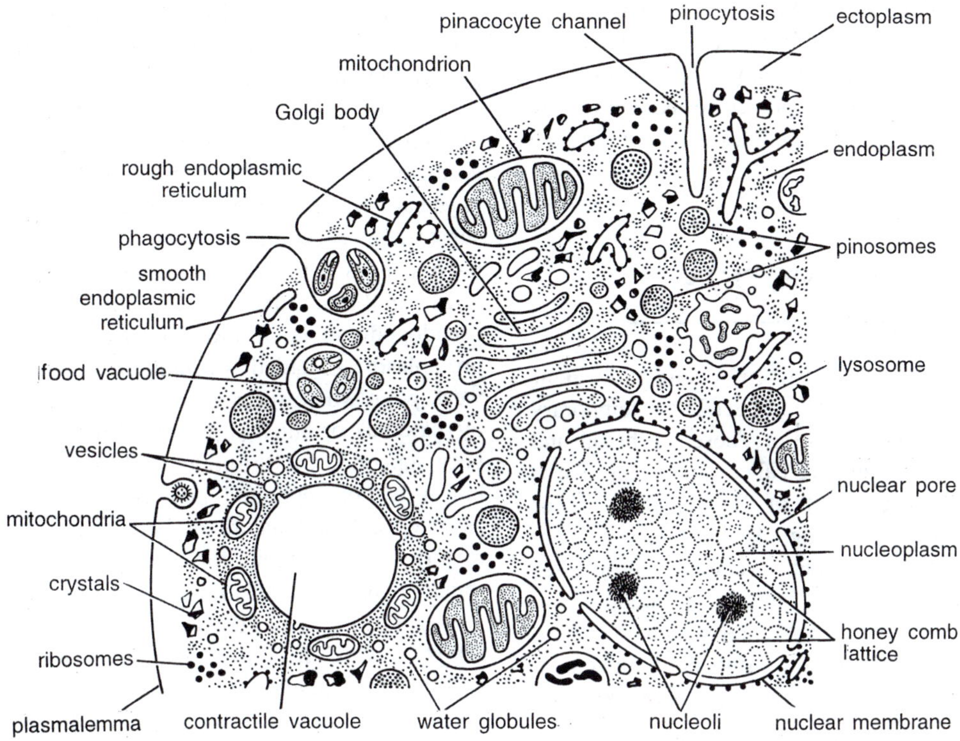

Fig. 2. *Amoeba.* A diagrammatic representation of the structure of a portion as revealed by electron microscopy.

anywhere in the endoplasm. It may be difficult to see the nucleus in a living animal, but it can be easily seen with the phase contrast microscope or after fixing and staining the *Amoeba* in a drop of **methyl green acetic acid**. The nucleus is granular and refractive to light. It is bounded by a thin **nuclear membrane** which is double and intercepted by pores. A honeycomb-like lattice is found below the inner **nuclear membrane**. It probably plays some part in maintaining the flattened form of the nucleus. The **nucleoplasm** is a clear achromatic substance with scattered refringent **chromatin granules**. A few nucleoli are seen in the nucleus of fixed and stained specimen, but they do not exist in the living animal.

2. Contractile vacuole. The endoplasm, at its posterior end, contains a single, clear rounded and pulsating **contractile vacuole**, filled with a watery fluid and enclosed by a unit membrane. Surrounding this membrane is a region containing many tiny feeder vacuoles and mitochondria. It helps in the osmoregulatory and excretory activities of the animal.

3. Food vacuoles. A number of spherical spaces, small and large, containing water and food in various phases of digestion, occur in the endoplasm. These are the **food vacuoles**. These are formed when *Amoeba* engulfs food with a drop of water. Digestion takes place inside these **food vacuoles**. They disappear with the egestion of non-digestible food from the body.

4. Water globules. These are several, small and spherical vacuoles filled with a colourless fluid.

5. Other organelles. In addition to these organelles, the endoplasm shows, in electron microscope, various inclusions characteristic of an animal cell. The **endoplasmic reticulum**, besides forming a network of tubules, occurs as vesicles. The **ribosomes** occur on some of the endoplasmic vesicles as well as scattered in the endoplasm. The **Golgi bodies** are present as several groups of sac-like tubes. The **mitochondria** are more or less oval and have **tubular cristae**. Plate-like or bipyramidal **crystals** and stored food abound in the endoplasm. The **lysosomes** are membrane-bound spherical bodies, scattered in the endoplasm. Thin filamentous microtubules abound in the endoplasm.

LOCOMOTION

Amoeba shows characteristic **amoeboid movement** by the formation of finger-like temporary processes, the **pseudopodia**. These are broad and with rounded tips and are called **lobopodia**. These are formed as a result of flowing forward of cytoplasm. When moving, *Amoeba* may put out several pseudopodia but eventually progresses through just one broad anterior pseudopodium. *Amoeba* moves with an average speed of 1μ/sec or one micron per second. Amoeboid movement is considered to be the most primitive kind of animal movement.

Theories of Amoeboid Movement

How the cytoplasmic flow is effected and the pseudopodia are formed is not clearly known. Since the first observation of amoeboid movement by **Rosel von Rosenhof** in 1755, several theories, or hypotheses, have been proposed to explain the mechanism of **amoeboid movement**, but the evidence supporting them is unfortunately insufficient. A brief description of these theories is as under.

1. Contraction-hydraulic theory. Schultze (1875) was of the view that plasmagel (ectoplasm) undergoes contraction at the posterior end and causes protoplasmic currents to flow forwards, pushing the more fluid-like plasmasol (endoplasm) forward. This results in the formation of pseudopodium and propelling the body ahead.

2. Surface-tension theory. Berthold (1886) explained that the amoeboid movement is due to a difference in surface tension between the physical characteristics of body surface and substratum. According to this view, amoeboid movement is comparable with the movement of a **fluid globule**, like a mercury droplet. A pseudopodium is formed by an out flow of protoplasm (so-called fountain streaming) from a weak point where surface tension becomes reduced by external or internal factors. This theory had been supported by **Rhumbler** and **Butschli** (1898), but it is not supported now-a-days. The theory assumes a liquid nature of body surface, but in majority of amoeboid forms outer body surface remains rigid and gelatinized.

3. Rolling movement theory. Jennings (1904), with his investigations on *Amoeba verrucosa,* explained that amoeboid movement takes place due to rolling movement of body surface comparable to rolling movement of a fluid-filled sac on a substratum. He observed in *A. verrucosa* that a carbon particle on amoeba's upper surface first passes forward and then turning downwards along the anterior tip, remains on the lower surface for a time as the body rolls forward, and then passes upward at the posterior end to repeat the cycle.

Jenning's finding may be correct for *A. verrucosa* which is devoid of pseudopodia, but it cannot be applied to *A. proteus* which moves with pseudopodia.

4. Walking movement theory. Dellinger (1906) studied locomotion in *A. proteus* and came to the conclusion that a contractile substance present in the cytoplasm is mainly responsible for

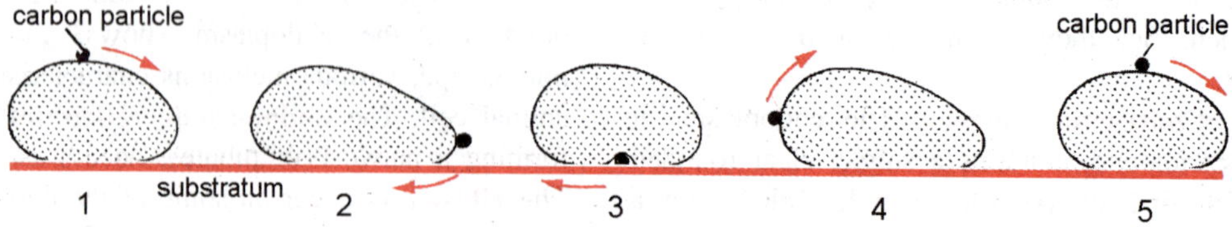

Fig. 3. *Amoeba verrucosa* showing rolling movement.

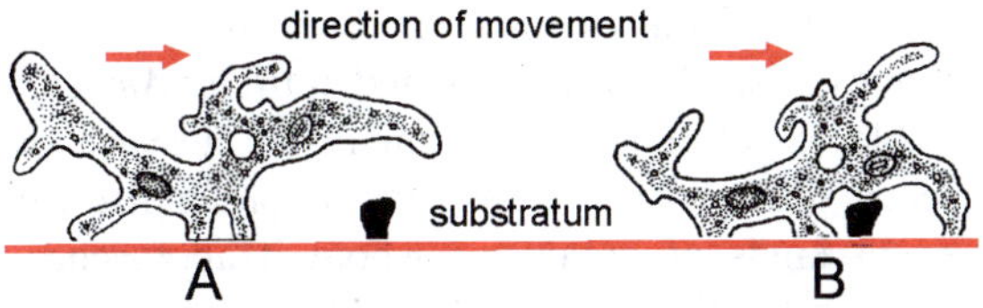

Fig. 4. *Amoeba* showing walking movement.

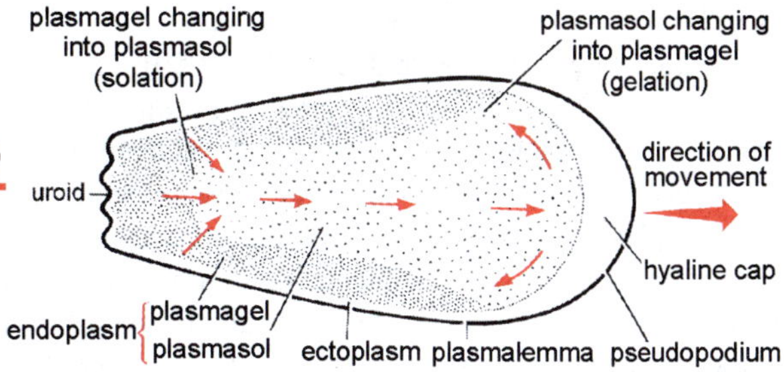

Fig. 5. Amoeboid movement after 'sol and gel' theory by **Mast**.

the amoeboid movement. According to this theory, the extended pseudopodia become attached to the substratum and then contract to pull the body forward. Seen in a profile, *Amoeba* appears to walk on its leg-like pseudopodia.

5. Sol-gel theory. This theory, first put forward by **Hyman** (1917) and later supported by **Pantin** (1923-26) and **Mast** (1925), is the most widely acceptable view, now-a-days. It attributes amoeboid motion to a change in the consistency of cytoplasm. Based on spontaneous sol-gel phenomenon of protoplasm, it offers the best explanation for amoeboid locomotion.

According to this theory, cytoplasm of *Amoeba* is differentiated into a clear outer **ectoplasm** and a granular inner **endoplasm**. The latter is further distinguished into an outer stiffer and jelly-like region, the **plasmagel**, and an inner fluid region, the **plasmasol** (**Mast**, 1925).

According to the **sol-gel** or **change of viscosity theory**, amoeboid movement involves four processes occurring simultaneously: (i) The outermost thin, elastic cell membrane or **plasmalemma** becomes attached to the substratum. (ii) There is a local partial liquefaction of the plasmagel at the anterior end. This causes the central plasmasol, under tension, to flow forward and force the plasmagel against this weakened area to produce a bulge, the beginning of the pseudopodium. As plasmasol enters the newly formed pseudopodium, it rapidly changes into plasmagel around the periphery, thus forming a gelatinized tube within which the plasmagel continues to flow forward. (iii) Posteriorly, inner surface of contractile plasmagel undergoes **solation**, so that a constant flow of plasmasol is maintained from behind forwards, in the direction of movement. (iv) The outer tube of elastic plasmagel contracts and moves from in front backwards, while the main bulk of body travels forwards. The plasmagel thus exerts a squeezing motion from the sides and rear of amoeba, forcing the plasmasol ahead. At the tip of pseudopodium the endoplasm is changed to ectoplasm.

6. Folding and unfolding theory. As already noted, protoplasm can be thought of as a three-dimensional network of protein chains linked

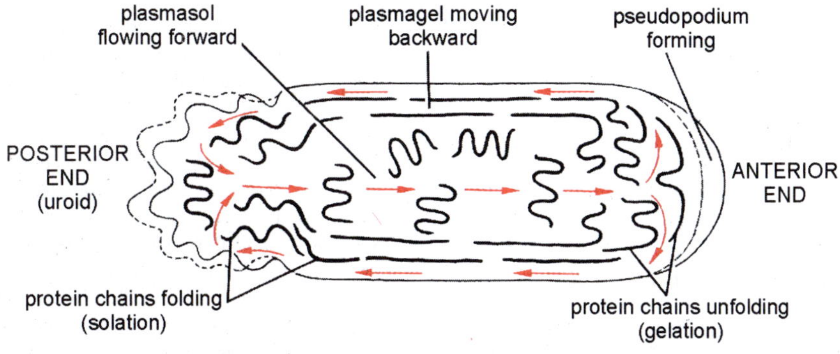

Fig. 6. Amoeboid movement after 'folding and unfolding' theory by **Goldacre** and **Lorch**.

together by cross linkages of side chains. According to **Goldacre** and **Lorch** (1950) and **Goldacre** (1952), the folding and unfolding of protein chains causes contraction and relaxation of protein molecules. They suggested that **sol** state of protoplasm is due to the folding of protein molecules and **gel** state is due to their unfolding. When *Amoeba* progresses, the cortical plasmagel at the posterior end contracts (**folding**). It is then liquefied to form **plasmasol** which is forced through the central endoplasm to flow forward. At the anterior end the plasmasol is converted back into plasmagel (**unfolding**), which forms the advancing pseudopodium. For the folding and unfolding processes, a considerable amount of energy is invested which comes from ATP (adenosine triphosphate).

7. Front or fountain-zone contraction theory. Allen (1961) suggests that, on the molecular level, amoeboid movement is a type of slow contraction, similar in many ways to muscle contraction. The endoplasm is believed to contain long protein chains, which undergo contraction at the anterior end of the body so that the *Amoeba* is pulled forward. It visualises the axial stream of endoplasm diverted in regular streamlines, thus creating a **fountain zone**; the endoplasm is constantly converted to ectoplasm anteriorly and ectoplasm to endoplasm posteriorly.

8. Reversible gel-sol transformation theory. Advocated by **Yagi** (1961) and **Marsland** (1964), this theory is the most accepted explanation of amoeboid movement. This theory suggests that solation at the anterior end occurs into which endoplasm flows under pressure generated by contraction of the cortical plasmagel at the posterior end. This results in propulsion of *Amoeba*.

NUTRITION

Food and Selection of Food

Amoeba is carnivorous and its mode of nutrition is **holozoic**, *i.e.*, it feeds by **phagocytosis**, a mechanism in which food is engulfed in a solid form. The food consists of bacteria, diatoms, desmids, flagellates, ciliates and rotifers. *Amoeba* is able to select its food even in the absence of special sense organs. It displays its preference for certain kinds of organisms whom it approaches, and avoids others. It is also able to distinguish organic food particles from inorganic sand particles.

1. Ingestion. *Amoeba* captures and engulfs its prey by means of pseudopodia. Pseudopodia are formed at the points, where the food comes in contact with the surface of the body.

According to **Rhumbler** (1930), *Amoeba* can ingest food in four possible ways, depending on the nature of food.

1. Import. This method involves the taking in of food, such as an algal filament, on contact. The food passively sinks into the body by rupturing the plasmalemma and ectoplasm at the point of contact. The ruptured site soon heals up.

2. Circumfluence. When *Amoeba* comes in contact with a less active or motionless organism, like bacteria, it extends its pseudopodia around the organism and envelops it completely in a **food vacuole** within cytoplasm. The enveloping pseudopodia always maintain intimate contact with the surface of prey.

3. Circumvallation. By this method, *Amoeba* engulfs an active prey like a flagellate or ciliate. *Amoeba* sends out pseudopodia around 'the prey forming a cup-like structure, called **food cup**. The pseudopodia are not in intimate contact with the prey during engulfment. The opening of food cup constricts, leaving a pore which soon closes. In this way, a non-contractile **food vacuole** is formed with the prey in a considerable amount of water.

4. Invagination. In. this method the food is adhered by the toxic and sticky secretion of the ectoplasm. The food organism is sucked in, upon contact with the ectoplasm, by the formation of an **ectoplasmic tube**. This tube, upon engulfment, takes the form of a food vacuole.

5. Pinocytosis. Pinocytosis is the 'cell drinking' phenomenon also round to occur in *A. proteus.* It assist in the ingestion of liquid food. It has been experimentally demonstrated that pinocytosis is induced by proteins, inorganic ions and certain dyes. When an *Amoeba*, which has been starved for two days, is placed in 1% sodium acetate solution, it becomes star-shaped with pseudopodia-like projections. Many of these projections show to contain **pinocytosis channels** running from the surface deep into the body. It is thought that plasmalemma bound with colloidal material flows into the pinocytosis channels. From the inner ends of these channels, **pinocytosis vesicles** or **pinosomes**, containing engulfed liquid material, become separated off. It has not yet been confirmed whether this phenomenon is one of the means of ingestion of food in *Amoeba* under normal conditions.

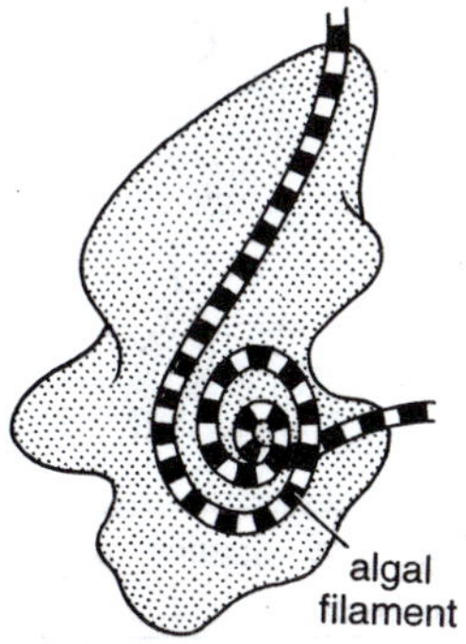

Fig. 7. *Amoeba* ingesting an alga by import.

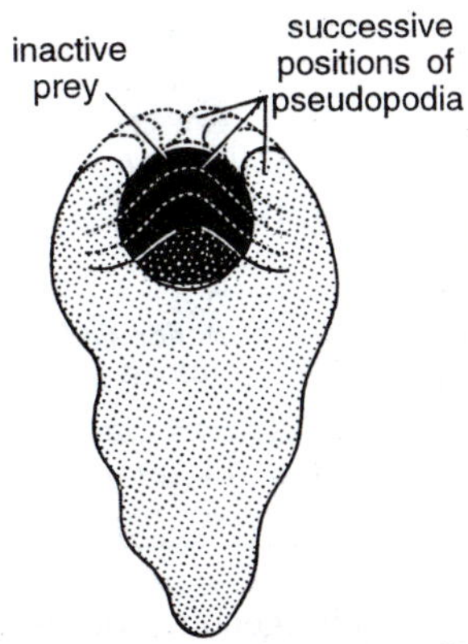

Fig. 8. *Amoeba* feeding by circumfluence.

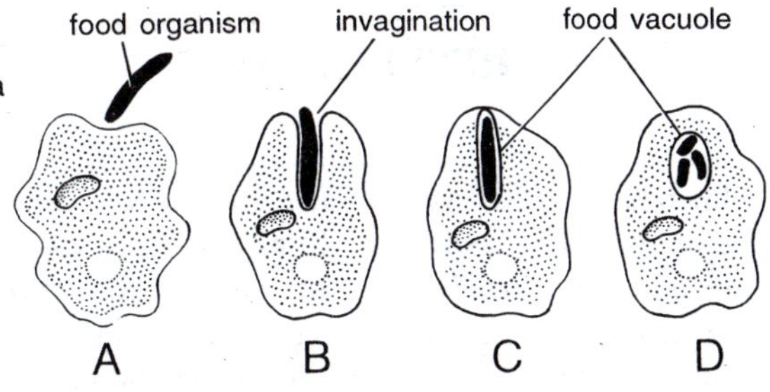

Fig. 10. *Amoeba* ingesting food by invagination.

2. Digestion. The **food vacuole** or the **gastric vacuole**, formed by the extension and joining of the pseudopodia that capture the prey, is comparable to the gut of higher animals. Its membrane wall resembles in all respects the plasmalemma. Several such vacuoles, small and large, may be seen in an actively feeding *Amoeba*.

Lysosomes, containing digestive enzymes, fuse with the food vacuole and digestion of contained food organism slowly proceeds. It has been demonstrated that the reaction in the food vacuole is at first acidic and later on becomes alkaline. In acidic medium, the organism is killed and in alkaline medium, digestion of starch, proteins and fats takes place by the enzymes **amylase, protease** and **lipase,** respectively.

3. Absorption and assimilation. As digestion goes on, the food vacuoles gradually shrink in size. The food vacuoles keep on moving in the endoplasm due to its streaming movement, called **cyclosis**. The digested food diffuses out into the endoplasm. Completion of digestion requires about 30 hours. The digested food is assimilated to form new protoplasm.

4. Egestion. Egestion of undigested residue takes place at any point on the surface of body. In *Amoeba*, there is no definite exit for this purpose. In the actively moving *Amoeba* the much reduced food vacuoles, containing undigestible residue, are shifted backwards and discharged at the posterior end, as the animal moves on.

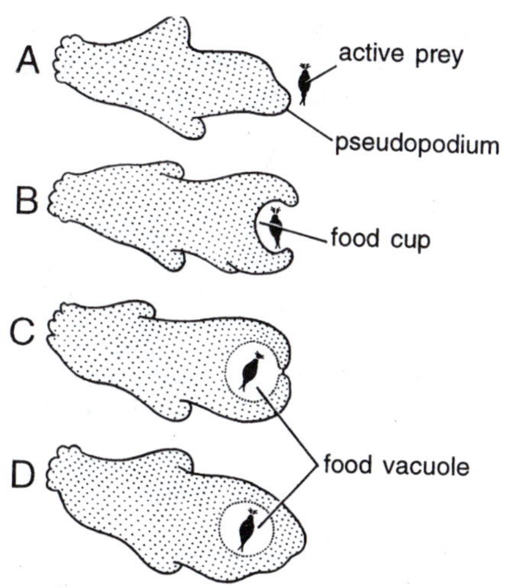

Fig. 9. *Amoeba* ingesting a rotifer by circumvallation.

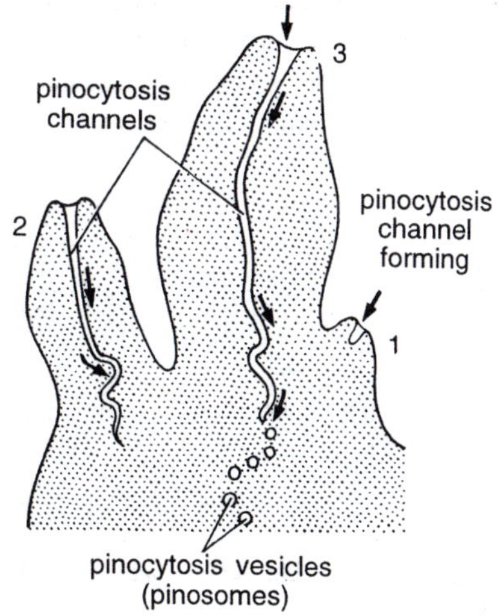

Fig. 11. *Amoeba.* Pinocytosis.

RESPIRATION

Interchange of oxygen coming inwards and carbon dioxide going outwards forms the process of respiration. *Amoeba* has no special respiratory organs and no respiratory pigments but there is a free exchange of gases by diffusion through the general body surface (plasmalemma) which is permeable to the gases dissolved in water. Oxygen constantly diffuses into the cytoplasm for its concentration in water is always higher than in the cytoplasm. The oxygen brings about enzymatically assisted oxidation of carbohydrates, fats and even proteins and breaks them into simpler compounds. The energy liberated in the oxidation reactions is stored in the high energy bonds of adenosine-triphosphate or ATP, like that of any other cell. The oxidation of carbohydrates and fats results in the formation of metabolic wastes (carbon dioxide and water). CO_2 diffuses out in the surrounding water as well as in the water discharged by the contractile vacuole.

EXCRETION

In the body, metabolism of fats and carbohydrates produces CO_2, whereas metabolism of proteins produces ammonia. Removal of nitrogenous wastes from body is called **excretion**. *Amoeba* is an **ammonotelic** animal because it excretes

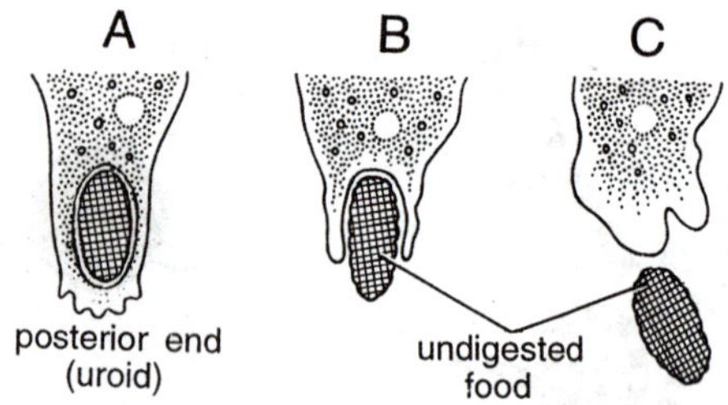

Fig. 12. *Amoeba* showing stages of egestion of undigested food.

nitrogenous wastes in the form of ammonia. As no special organelles are present for excretion, ammonia is excreted by diffusion from the general body surface in the surrounding water. Some amount of ammonia is dissolved in water which is discharged through the contractile vacuole.

OSMOREGULATION

The protoplasm of *A. proteus* is of higher concentration than the freshwater of its environment. This causes the entrance of water into the body of *Amoeba* by **endosmosis** through the semi-permeable plasmalemma. Not only does water enter in by endosmosis but also some water is formed within the cytoplasm as a result of metabolic activity and some gets in along with ingested food organisms. This necessitates to get rid of excess water to prevent swelling and rupture of amoeba's body. The mechanism which effects

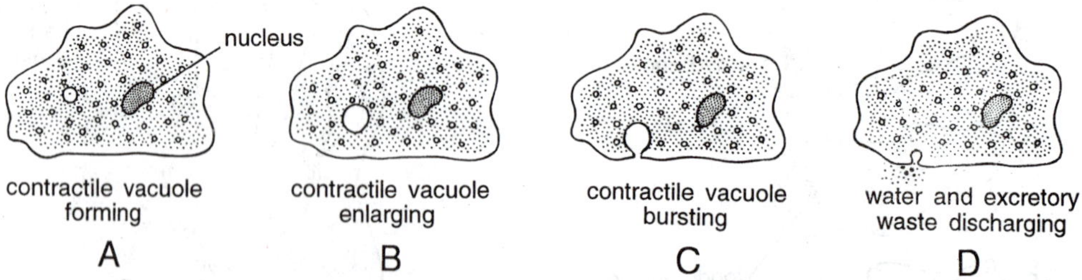

Fig. 13. *Amoeba* showing various stages of osmoregulation and excretion.

the water regulation is called **osmoregulation**. It takes place through the contractile vacuole. The main function of contractile vacuole is osmoregulation, although CO_2 and nitrogenous waste substances are also excreted through it.

The exact nature regarding the working of contractile vacuole is not known. *Amoeba* is **hypertonic** to the surrounding water, so that water is actively secreted into contractile vacuoles by endosmosis through vacuolar membrane. Tiny membrane-bound water-filled feeder vacuoles also get incorporated into contractile vacuole filling it with water. As water continues to fill the vacuole, it increases in size. When the vacuole becomes fully expanded, it comes to lie in the zone of gelated ectoplasm, where the ectoplasmic pressure results in its contraction and bursting. This contractile vacuole disappears and a new one begins to form in the endoplasm. Due to force of contraction, the contents of contractile vacuole, that is H_2O, CO_2, ammonia, etc. are discharged to outside.

BEHAVIOUR

Behaviour involves the manner in which an animal responds to the environmental conditions. The responses to stimuli are called **taxes** (singular, **taxis**). A taxis may be either **positive**, in which the organism moves towards the stimulus, or **negative**, in which the organism moves away from the stimulus. *Amoeba proteus* shows both types of taxes, positive as well as negative, specifically to different stimuli. There are two views as to the effectiveness of the responses. According to one view, the response of *Amoeba* to a given stimulus is automatic or directed (**topotaxis**), that is, *Amoeba* either moves towards the stimulus or avoids it, depending upon the nature of the stimulus. Another view holds that response to a stimulus is undirected (**phobotaxis**), that is, *Amoeba* always avoids a stimulus and moves here and there to get a favourable environment (Fig. 14).

Taxes are named according to the nature of stimulus. With respect to the kinds of stimuli, taxes are classified as follows :

1. **Thermotaxis** (response to heat). *Amoeba* responds negatively to both low and high temperatures. Optimum temperature lies between 20°C and 25°C. It ceases all activities at temperatures above 35°C.
2. **Phototaxis** (response to light). *Amoeba* avoids both direct sunlight and total darkness. It responds positively to normal or weaklight.
3. **Thigmotaxis** (response to contact). A floating *Amoeba* responds positively to those objects upon which it glides or rests. It will back away from contact with a foreign object or a probe while crawling or resting.
4. **Chemotaxis** (response to chemical substances). *Amoeba* is negatively chemotactic to strong solutions of alkalies, salts and sugars. It also avoids sand particles or some other obstacles in the way. It responds positively to the food organisms.
5. **Rheotaxis** (response to current of air or water). *Amoeba* prefers to be drifted along the flowing water.
6. **Galvanotaxis** (response to constant electric current). When an electric current is passed through water, containing *Amoeba,* it stops moving, withdraws pseudopodia and becomes globular. In a weak electric current it moves towards negative pole (cathode) and avoids positive pole (anode).
7. **Geotaxis** (response to gravity). The response of *Amoeba* to gravity is mostly positive as it drops to the bottom of container filled with water.

Amoeba does not have a nervous system and special sense organs for the condition and perception of stimuli. The sensitivity is solely the function of the protoplasm.

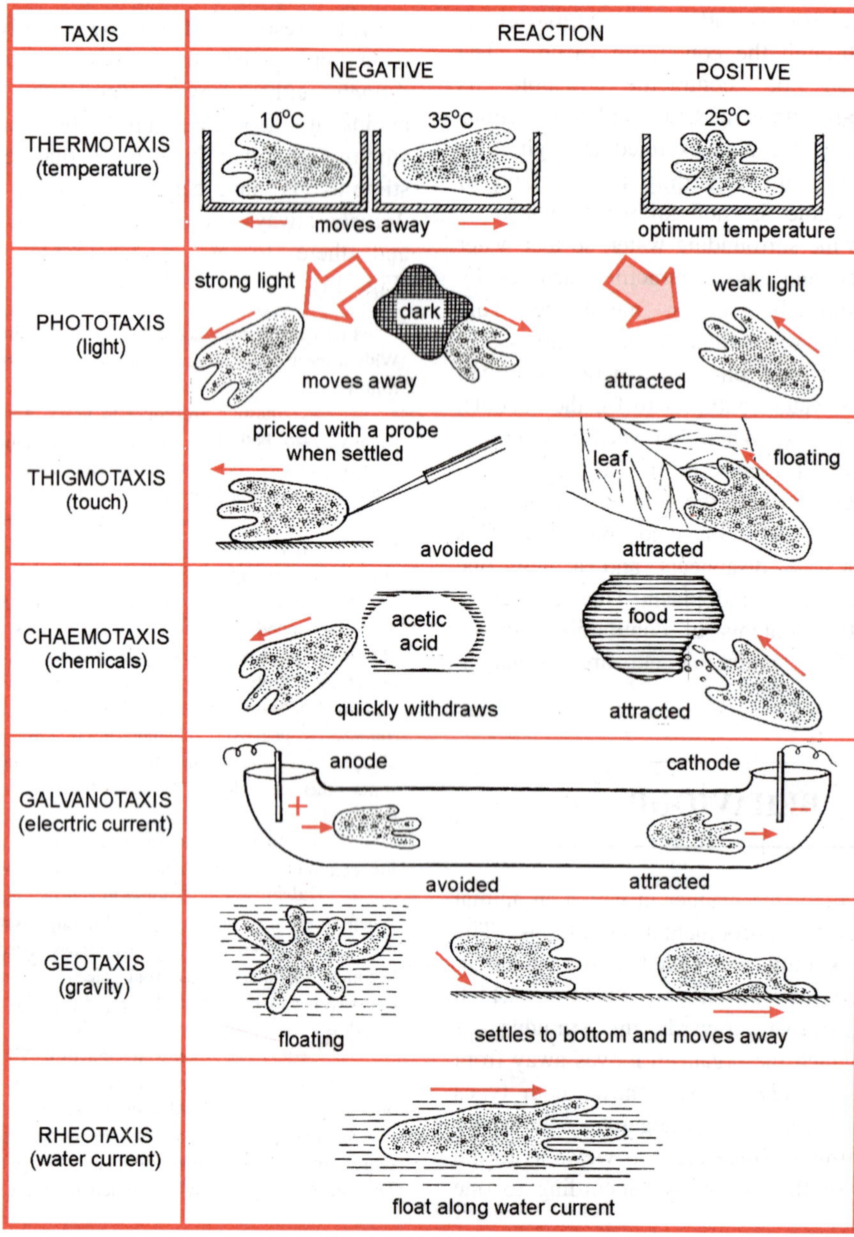

Fig. 14. *Amoeba.* Reactions to various stimuli.

REPRODUCTION

Amoeba proteus does not reproduce **sexually** by mating, that is, by the fusion of sex cells or gametes. Reproduction is essentially **asexual** and takes place by various methods such as binary fission, multiple fission and sporulation.

[I] Binary Fission

It is the most common mode of reproduction. It results in the division of parent *Amoeba* into two daughter amoebae. Division involves the nuclear

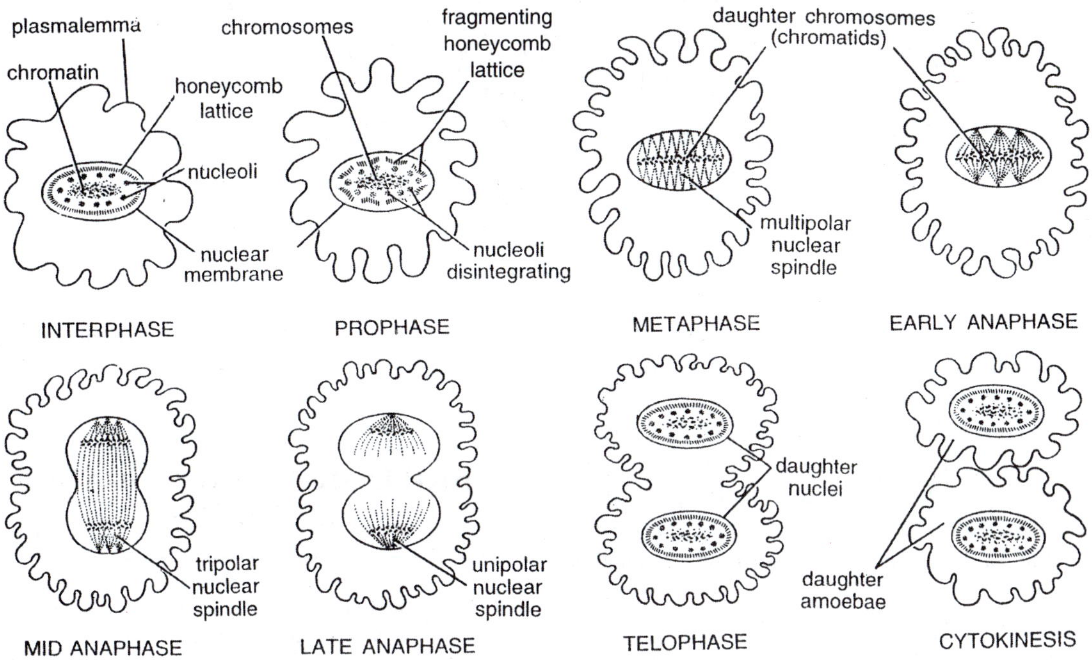

Fig. 15. *Amoeba.* Stages in binary fission.

division followed by cytoplasmic division. The animal divides by mitosis and involves the stages prophase, metaphase, anaphase and telophase (Fig. 15).

1. Prophase. In the prophase stage, *Amoeba* withdraws its pseudopodia and becomes somewhat rounded. Cytoplasm loses its transparency to a large degree and contractile vacuole disappears. The honeycomb-like lattice underneath nuclear membrane first fragments and then disappears. The nucleoli disintegrate. A very large number (500 to 600) of very small chromosomes emerge in the central nucleoplasm.

2. Metaphase. The metaphase stage is marked by the arrangement of chromosomes at the equator. Each chromosome splits longitudinally and becomes paired. Daughter chromosomes, on each side, become attached to the spindle fibres arising from multiple poles, situated within the nuclear membrane.

3. Anaphase. In the anaphase stage, daughter chromosomes move towards opposite poles and constriction of nuclear membrane begins in the middle. Nuclear spindle, which was **multipolar** at metaphase, becomes **unipolar** at the end of anaphase.

4. Telophase. During telophase stage, the constriction of nuclear membrane is completed and the nucleus is finally divided into two daughter nuclei. In each daughter nucleus the lattice is formed underneath the nuclear membrane and the nucleoli reappear.

5. Cytokinesis. Next follows cytokinesis. *Amoeba* stretches and constricts in the middle. Large pseudopodia are formed at opposite poles, drawing both the daughter amoebae in opposite directions. It ultimately divides parent *Amoeba* into two daughter amoebae which grow, and in their turn, repeat the same cycle of binary fission.

In *Amoeba*, reproducing by binary fission, the parent becomes wholly merged in the offspring. Thus, there exists a continuity of life, so that *Amoeba* is potentially immortal. However, death may come by starvation, accident or some other misfortune.

[II] Multiple Fission and Encystment

According to some earlier workers, such as **Scheel** (1899) and **Carter** (1915), *Amoeba* forms a cyst

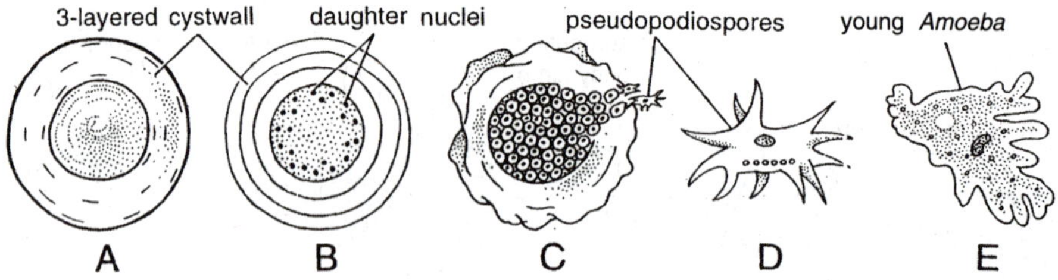

Fig. 16. *Amoeba.* Encystment and multiple fission.

and reproduces by **multiple fission**, during adverse environmental conditions. The animal secretes a three-layered, protective, chitinous **cyst** around it and becomes inactive. Inside the cyst, the nucleus repeatedly divides to form several daughter nuclei, which arrange themselves near the periphery. Each daughter nucleus becomes enveloped by a small amount of cytoplasm, thus forming a daughter *Amoeba*, called **amoebula** or **pseudopodiospore**. When favourable conditions arrive, the cyst breaks off liberating the young pseudopodiospores, each with fine pseudopodia. They feed and grow rapidly to become adults and lead an independent life.

Multiple fission, preceded by encystment is no longer believed to occur in *Amoeba* by modern workers, some of them maintained *A. proteus* for 28 years without ever forming a cyst.

[III] Sporulation

According to **Taylor**, during unfavourable conditions, *A. proteus* multiplies by sporulation without encystment. Nucleus breaks into several small fragments or **chromatin blocks**. Each block develops a nuclear membrane, becomes surrounded by a little cytoplasm and develops a **spore-case** around it. With the disintegration of parent body, about 200 such **spores** are liberated, each hatching into a small **amoebula** under favourable conditions. However, evidence has been lacking in support of sporulation in *Amoeba*.

[IV] Conjugation

Some observers have described a temporary fusion between two amoebae. After some time they become separated again. It is said that this temporary union enables the two amoebae to lead a more active and vigorous life. This phenomenon is termed **rejuvenation**. However, it has not been confirmed by other workers.

BIOLOGICAL SIGNIFICANCE OF *AMOEBA*

(1) *Amoeba* depicts organization of a protoplasmic mass or a single cell into a complete organism.
(2) Binary fission of *Amoeba* gives a clear-cut understanding of the mitotic division of a cell.
(3) The taxes or responses of *Amoeba* represent the early beginning of sensitivity in animals.
(4) Different organelles of *Amoeba* give the first indication of division of labour concerning vital activities.
(5) The great number of chromosomes (500 to 600) present in the nucleus of *Amoeba* suggests the occurrence of isolated genes, which in higher animals are located in chromosomes.
(6) *Amoeba* gives a faint idea regarding the anatomical structures of higher animals. For example, the food cup is comparable to the buccal cavity, food vacuole to gut, pseudopodia to legs, contractile vacuole to urinary bladder or kidney, and so on.

Amoeba Interesting Facts

Here are some interesting facts regarding *Amoeba*:

- As the *Amoeba* consumes food, its (prey) can stay alive for a while.
- To prevent cannibalism, the *Amoeba* secretes a chemical which keeps other amoebae away. This type of chemical secretion can also protect the *Amoeba* from predators.
- The *Pelomyxa palustris* is the largest of all *Amoeba* that are found in pond water.
- The amoebae are mostly known as pathogens. However, it is interesting to know that most of the amoebae recovered or obtained from the environment by scientists are non-pathogenic.

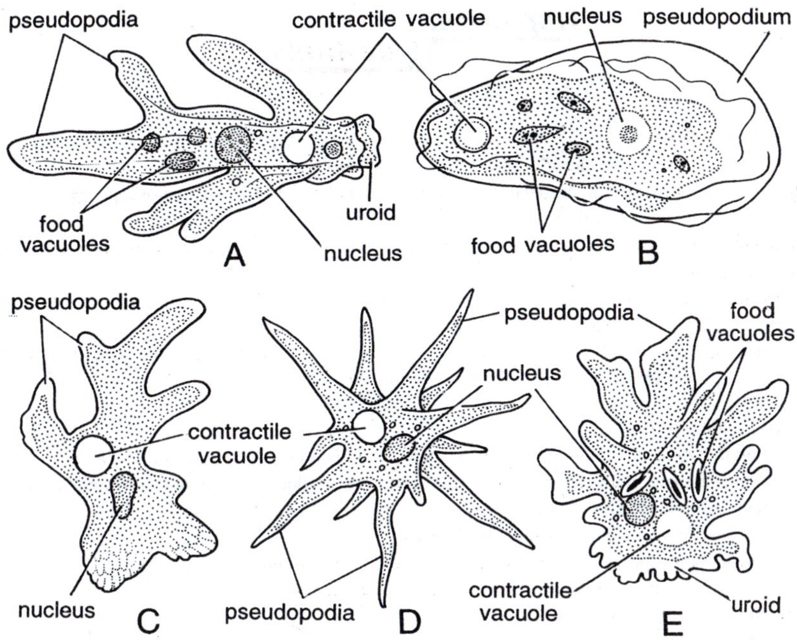

Fig. 17. Some common species *of Amoeba.* **A.** *A. proteus.* **B.** *A. verrucosa,* **C.** *A. discoides.* **D.** *A. radiosa.* **E.** *A. dubia.*

OTHER SPECIES OF *AMOEBA*

Numerous species, marine as well as freshwater, belong to the genus *Amoeba.* They greatly vary in their shape, size and in the characters of the nucleus and pseudopodia.

[I] *Amoeba proteus*

First described by Leidy in 1879, is perhaps the commonest and one of the largest freshwater species. This species is also used for teaching and demonstration of living protozoans in the class. The body is about 0.6 mm in diameter and provided with a few large and blunt pseudopodia. A characteristic feature is the presence of definite longitudinal ectoplasmic ridges. The nucleus is biconvex, lens-shaped or disc-like, and may become folded and convoluted in older forms. According to Taylor, this species forms endogenous spores in an unencysted condition. It does not normally feed upon diatoms.

[II] *Amoeba verrucosa*

It is another common freshwater species. The body, having a wrinkled surface, is nearly 0.2 mm. in diameter. The pellicle is tough and the pseudopodia are many, short and blunt. The nucleus is ovoid and contains a large endosome. The movements of this species are very sluggish. It covers the distance of about 0.5mm in a second.

[III] *Amoeba discoides*

It has a body about 0.4 mm in diameter and with only a few blunt pseudopodia. The biconcave, discoidal nucleus never becomes folded like that of *A. proteus* and the ectoplasmic ridges are also not known.

[IV] *Amoeba radiosa*

It is also a small freshwater species up to 0.04 mm in diameter. The body is oval but the animal looks star-shaped on account of the presence o.f slender, radiating pseudopodia, which may be stiff and straight or curved and tapering.

[V] *Amoeba dubia*

It is about 0.4 mm in diameter with a smooth body surface and forming several flat pseudopodia. The nucleus looks spherical in front view but oblong in profile. The endoplasm is laden with a few large particles of various shapes. Diatoms form an important item of their diet.

[VI] *Amoeba terricola*

This species occurs in damp earth. The atmosphere contains the spores or cysts of *Amoeba* in large numbers. Several species of allied genera are also commonly known as amoebase *e.g., Entamoeba histolytica, Pelomyxa palustris, etc.*

Important Questions

➤➤Long answer type questions

1. Describe the habit, structure, physiology and behaviour of *Amoeba.*
2. Give an account of the structure of *Amoeba proteus* as seen under the electron microscope. How would you distinguish *Amoeba* from a typical animal cell ?
3. Write detailed notes on : (i) Locomotion in *Amoeba,* (ii) Modes of reproduction in *Amoeba,* (iii) Nucleus of *Amoeba,* (iv) Osmoregulation in *Amoeba,*
4. Justify the statement that "binary fission in *Amoeba* is essentially an act of mitosis."
5. Discuss the biological significance of *Amoeba.*
6. Write short notes on : (i) Amoeboid movement, (ii) Circumvallation, (iii) Contractile vacuole, (iv) Holozoic nutrition, (v) Lobopodia, (vi) Pinocytosis, (vii) Plasmalemma, (viii) Phototaxis.

➤➤Short answer type questions

1. Is the interior of the *Amoeba* homogenous or is there an indication of various organelles?
2. Does the cytoplasm remain stationary as the *Amoeba* moves or does it shift as movement proceeds?
3. What is the mode of nutrition in *Amoeba* ?
4. What is the function of contractile vacuole in *Amoeba* ?
5. What are the pseudopodiospores in *Amoeba* and why they are essential in reproduction ?
6. Explain the formation of pseudopodia basing on the sol-gel theory.
7. How *Amoeba* feeds on flagellates and ciliates ?
8. Describe sporulation in *Amoeba.*
9. Who was describe *Amoeba* first ?
10. How are the culture of *Amoeba* in laboratory ?
11. What is a hyaline cap ?
12. Who was observed first to amoeboid movement ?

➤➤Multiple choice questions

1. The fresh water *Amoeba* and intestinal *Amoeba* are alike in :
   (a) the possession of single contractile vacuole
   (b) the absence of cilia
   (c) the structure of the cyst
   (d) their mechanism of dispersal
2. *Amoeba* is capable of regeneration. This is possible only from :
   (a) a nucleated bit of *Amoeba*
   (b) an anucleate bit of *Amoeba*
   (c) a young *Amoeba*
   (d) an old *Amoeba*
3. Name a digenic parasite from protozoa :
   (a) *Entamoeba histolytica*
   (b) *Polystomella*
   (c) *Plasmodium*
   (d) *Copromonas*
4. *Amoeba* was discovered by :
   (a) Ross (b) Rosenholf
   (c) Lamble (d) Losch
5. During extreme heat a fresh water *Amoeba* living in a pond :
   (a) will form a gamete
   (b) will form a cyst
   (c) change to parasitic phase
   (d) will eliminate large parts of DNA
   (e) elimination of excessive food
6. The food capturing orgenelle of *Amoeba* is :
   (a) food vacuole (b) contractile vacuole
   (c) pseudopodia (d) nucleus
7. Which is the most widely accepted theory of locomotion in *Amoeba* :
   (a) walking movement theory
   (b) rolling movement theory
   (c) sol-gel theory
   (d) surface tension theory
8. Which type of pseudopodium is found in *Amoeba* :
   (a) rhizopodium (b) actinopodium
   (c) reticulopodium
   (d) axopodium
9. Surface tension theory explains the theory of :
   (a) amoeboid movement
   (b) caterpillar movement of *Hydra*
   (c) ciliary beat
   (d) jerking movement of *Euglena*
   (e) tension development on the surface during the movement of *Paramecium* in water
10. Many genera of foraminifera are dimorphic. What is meant by dimorphism :
    (a) existence of an animal in many forms
    (b) presence of two types of pseudopodia in foraminifera
    (c) existence of an animal in two forms which differ in characters

(d) presence of different types of organelle in the same animal
(e) alternation of generation is known as dimorphism

11. Sol-gel theory was first given by :
(a) Pantin (b) Hyman
(c) Mast (d) Mast & Pantin

12. Pseudopodia are formed in an *Amoeba* :
(a) when it comes in contact with a food particle
(b) by the movement of the surrounding water
(c) by exchange of salts with the medium
(d) by sol-gel transformation of the cytoplasm

13. *Amoeba* is placed in phylum Protozoa, because of :
(a) presence of cell wall
(b) phagocytic mode of nutrition
(c) acellular body
(d) presence of contractile vacuole

14. Egestion of undigested food in *Amoeba* takes place through :
(a) circumvallation
(b) a temporary rupture of its surface membrane
(c) the hyaline cap formed at its advancing end
(d) pinocytosis

15. *Amoeba* reacts :
(a) negatively to both weak and strong light
(b) positively to strong light and negatively to weak light
(c) positively to both weak and strong light
(d) negatively to strong light and positively to weak light

16. '*Amoeba* is immortal', it can be explained by Germplasm theory of :
(a) Heitzman (b) Schultz
(c) Berthold (d) Weisman

17. *Amoeba* belongs to class Sarcodina, which is characterised by :
(a) acellular body
(b) pseudopodia for locomotion
(c) uninucleate body
(d) presence of contractile vacuole

18. Normally *Amoeba* is not found :
(a) crawling (b) feeding
(c) secreting a cyst (d) responding to stimuli

19. Body of *Amoeba* has permanent :
(a) shape (b) organelle for locomotion
(c) anterior end (d) food vacuole

20. A contractile vacuole is developed when :
(a) fresh water *Amoeba* is placed in sea water
(b) fresh water *Amoeba* is placed in salt free water
(c) marine *Amoeba* is transferred in hypotonic medium
(d) sea water enters in body of marine *Amoeba*

21. *Amoeba* shows positive response for :
(a) dim and intense light
(b) solid objects
(c) cathode
(d) chemicals

22. Minimum number of daughter amoebae are produced, when surrounding water has :
(a) no food (b) less food
(c) plenty of food (d) high temperature

23. Simplest mode of locomotion is :
(a) swimming (b) creeping
(c) floating (d) walking

24. Plasmalemma of *Amoeba* is :
(a) impermeable (b) least regenerative
(c) site for excretion (d) not for respiration

25. If an *Amoeba* is kept in distilled water its contractile vacuole will :
(a) disappear
(b) become swollen and full of water
(c) work faster (d) have no effect

26. Regarding locomotion in *Amoeba* Hymen (1917) postulated :
(a) theory of rolling movement
(b) surface tension theory
(c) walking movement theory
(d) sol-gel theory

27. If an *Amoeba* is put in medium A from fresh water its contractile vacuole disappeared but when it is transferred to medium B its contractile vacuole reappeared. What is the difference between medium A and B :
(a) medium A is hypotonic
(b) medium B is hypertonic
(c) medium A is hypertonic
(d) medium A is isotonic

28. Nitrogenous wastes in *Amoeba,* are excreted through :
(a) plasmalemma (b) food vacuoles
(c) contractile vacuole (d) none of these

29. When the prey is very active, *Amoeba* ingests it by :
(a) circumfluence (b) circumvallation
(c) invagination (d) none of these

30. When *Amoeba* is subjected to weak and steady electric current :
(a) it dies (b) it remains at anode
(c) it moves towards the cathode
(d) it remains in the middle

31. Sol-gel Theory was proposed by :
(a) Mast (b) Hymen
(c) Pantin
(d) Mast & Pantin

32. Mode of nutrition in *Amoeba* is :
(a) holozoic (b) mixotrophic
(c) saprophytic (d) holophytic

33. Medium inside food vacuole of *Amoeba* is :
(a) acidic (b) alkaline
(c) I acidic then alkaline
(d) I alkaline then acidic

34. *Amoeba* respires through its :
(a) plasmalemma
(b) general body surface
(c) cytoplasm (d) pellicle

35. Excretion in *Amoeba* occurs by :
(a) contractile vacuole (b) food vacuole
(c) plasmalemma (d) all of these

36. In *Amoeba* binary fission occurs when :
(a) food is abundant
(b) temperature is suitable
(c) both
(d) pond water dries up

37. Regarding locomotion in *Amoeba* Pantin supported :
(a) theory of rolling movement
(b) surface tension theory
(c) walking movement theory
(d) sol-gel theory

38. An *Amoeba* living in fresh water respires by means of :
(a) nucleus (b) food vacuole
(c) plasmalemma (d) pseudopodia

39. A full-grown *Amoeba* undergoes binary fission. The total surface area of one daughter *Amoeba* soon after division is likely to be :
(a) slightly less than that of parent *Amoeba*
(b) equal to the half of that of the parent *Amoeba*
(c) slightly more than half of that of the parent *Amoeba*
(d) less than half of that of the parent *Amoeba*

40. What is common between an *Amoeba* and a leucocyte?
(a) contractile vacuole (b) sporulation
(c) encystment (d) pseudopodia

41. Locomotory organelles in *Amoeba* are :
(a) lobopodia (b) cilia
(c) flagella (d) myonemes

42. *Amoeba* was discovered by :
(a) Lancisi (b) Leeuwenhoek
(c) Rosenhoff (d) Ross

43. During digestion in *Amoeba* the medium is first :
(a) acidic (b) highly alkaline
(c) neutral (d) slightly alkaline

44. The function of the contractile vacuole in *Amoeba* is :
(a) excretion (b) osmoregulation
(c) none (d) both

45. A 'lobopodia' is derived of :
(a) undulating membrane (b) pseudopodia
(c) flagella (d) cilia

46. Gel-sol transformation theory put founded by :
(a) Yogi and Marsland (b) Allen
(c) Hymen (d) Jenning

47. *Amoeba* is :
(a) ammonotelic (b) ureotelic
(c) urecotelic (d) all

48. *Amoeba* shows :
(a) phototaxis (b) chemotaxis
(c) thermotaxis (d) all

49. Destruction of nucleus in *Amoeba* result in :
(a) no change
(b) quick locomotion
(c) slowing down of metabolic activities
(d) immediate death

50. Nuclear DNA of *Amoeba* is :
(a) a double helix (b) single stranded
(c) circular (d) like a clover leaf

51. Pseudopodia of *Amoeba* is composed of :
(a) ectoplasm (b) endoplasm
(c) both (d) none

52. How is *proteus* pronounced?
(a) Pro TEE uhs (b) Pro TAY uhs
(c) Pro TAY oos

53. Which is the disease caused by *Amoeba*?
(a) Amoebic dysentery
(b) *Entamoeba hartmani*
(c) contractile vacuole

54. What is the inside of the *Amoeba* made up of?
(a) water
(b) membrane
(c) cytoplasm

55. What happens if you cut an *Amoeba* in half ?
(a) both halves will move away and live normally.
(b) one side will get the nucleus, and the other won't.The side without the nucleus will die.
(c) It will instantly die.

56. How big is an *Amoeba* on an average ?
(a) 40 microns
(b) 250-600 microns
(c) 740 microns

Answers

➤➤*Multiple choice questions*

1. (b)	2. (a)	3. (c)	4. (b)	5. (b)	6. (c)	7. (c)	8. (a)	9. (a)	10. (c)
11. (b)	12. (d)	13. (c)	14. (b)	15. (d)	16. (d)	17. (b)	18. (d)	19. (b)	20. (b)
21. (c)	22. (c)	23. (b)	24. (c)	25. (b)	26. (d)	27. (c)	28. '(a)	29. (b)	30. (c)
31. (b)	32. (a)	33. (c).	34. (b)	35. (c)	36. (c)	37. (d)	38. (c)	39. (b)	40. (d)
41. (a)	42. (c)	43. (a)	44. (d)	45. (b)	46. (a)	47. (a)	48. (d)	49. (c)	50. (a)
51. (c)	52. (a)	53. (a)	54. (c)	55. (b)	56. (b).				

7

Entamoeba histolytica

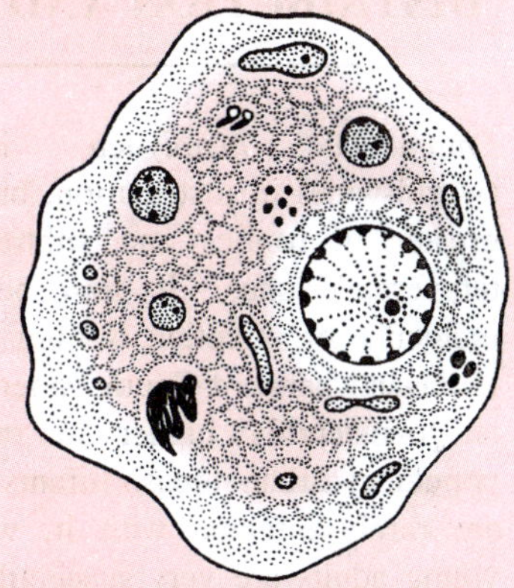

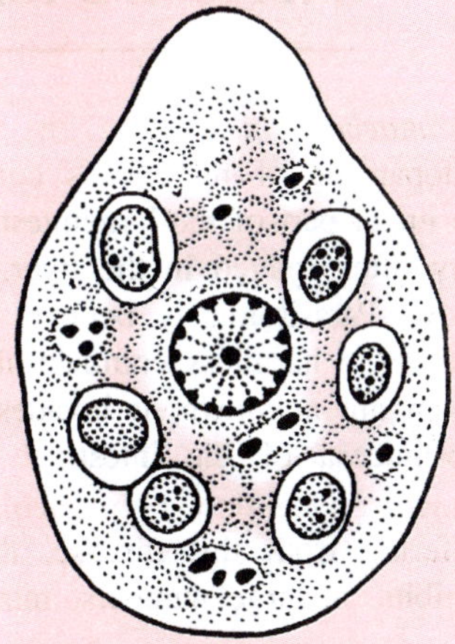

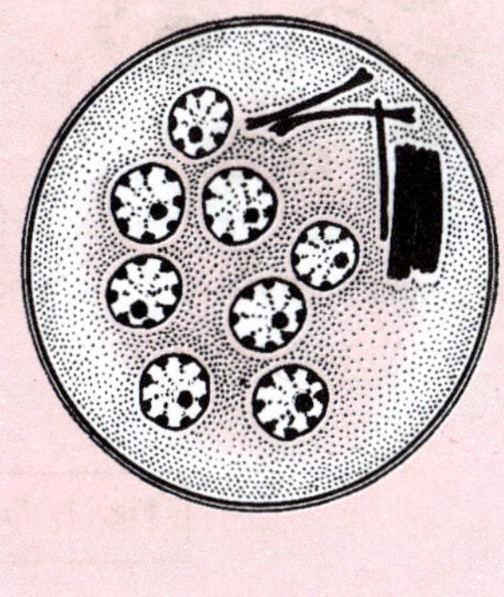

Apart from free-living forms, certain amoebae lead a commensal or parasitic life in the intestine of man and other animals. The common parasitic genera are *Entamoeba, Endamoeba, Endolimax, Iodamoeba* and *Dientamoeba.* Of the species belonging to these genera, *Entamoeba histolytica* (Gr., **entos**, within + **amoeba**, change + **histos**, tissue + **lysis**, dissolve), which is the causative organism of **amoebic dysentery** or **amoebiasis** in man, is undoubtedly the most studied of all parasitic amoebae.

Entamoeba histolytica

Entamoeba histolytica was first discovered by **Lambl** in 1859, and its pathological nature was described by **Friedrick Losch**, a Russian zoologist in 1875. He discovered this protozoan in the faeces and intestinal ulcers of a dysentery patient and succeeded in transferring it to puppies.

SYSTEMATIC POSITION

Phylum	Protozoa
Subphylum	Sarcomastigophora
Superclass	Sarcodina
Class	Rhizopodea
Subclass	Lobosia
Order	Amoebida
Genus	*Entamoeba*
Species	*histolytica*

DISTRIBUTION AND INCIDENCE

Entamoeba histolytica is cosmopolitan (world-wide) in distribution, but its prevalence is greater in tropics and subtropics than in the temperate zones. It has been reported that incidence of infection is high in Mexico, China, India and parts of South America. The incidence is considerably higher in rural and densely-populated urban areas. Infants under a year old are rarely infected with it, while children and young adults are very susceptible to the parasite. It is estimated that 10% of world population is suffering from amoebiasis.

HABITS AND HABITAT

Entamoeba histolytica is a microscopic endoparasite of man. It is commonly found in the upper part of the l:rgeintestine (colon) and is very often lodged in the liver, lungs, brain and testes. Parasitologists believe that this parasite lives as a harmless commensal but, for reasons which are unknown, it invades the mucosa and sub-mucosa of the intestinal wall and causes **amoebic dysentery** or **amoebiasis**. In a person suffering from this disease, the intestinal wall exhibits, in large numbers, minute **ulcers** which are formed as a result of the dissolution and destruction of the mucous lining of intestine. This is accomplished with the aid of a tissue-dissolving substance secreted by the parasite.

A successful culture of *E. histolytica* was first made by **Boeck** and **Drbohlav** (1925) using solidified blood agar or solidified egg slopes covered with Locke's solution. The growth of *E. histolytica* requires in culture the presence of starch or rice flour and some metabolic associates, such as enteric bacteria, organism *t* (a non-pathogenic bacterium) or the parasitic flagellate *T. cruzi*, living or dead. Micro cultures are made from a single washed amoebae in micro-tubes (4×50 mm) containing a medium of thioglycollate preparation, horse serum and an overlay of a rich culture of *T. cruzi* (Philip's medium). *E. histolytica* has also been grown in a serum medium containing thioglycollate and penicillin inhibited streptobacilli which are living but not multiplying (Shaffer and Frye's medium).

MORPHOLOGY

Entamoeba histolytica is a small microscopic parasitic *Amoeba*. In its life cycle, it occurs in three distinct forms : (i) **trophozoite** or **magna** form, (ii) **precystic** or **minuta** form, and (iii) **cystic** forms.

[I] Trophozoite or Magna Form

Trophozoite of *E. histolytica* is also known as the **trophic** or **magna** form. It is the most active, motile and feeding form which is pathogenic to

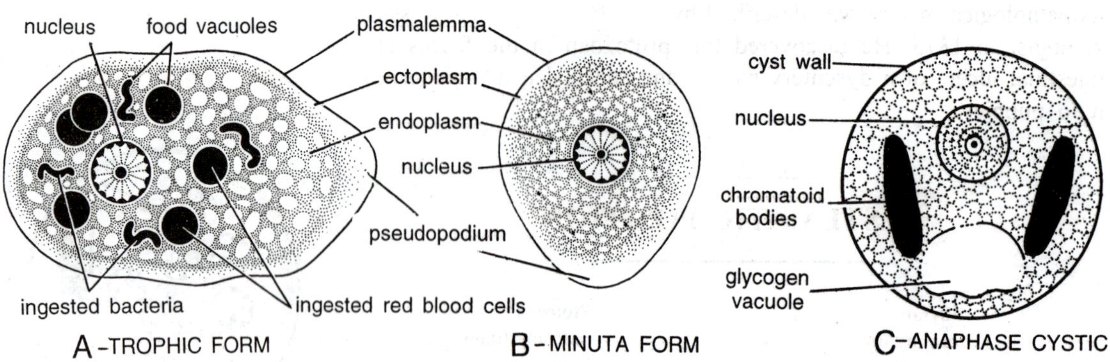

Fig. 1. *Entamoeba histolytica.* Structure of three stages of life cycle.

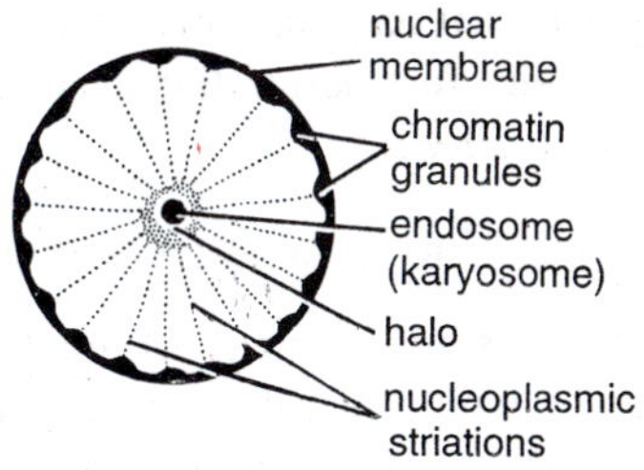

Fig. 2. *Entamoeba histolytica.* Vesicular nucleus.

man. It lives in the mucous and submucous layers of the large intestine of man. It usually measures 20 to 30μ in diameter and, more or less, resembles the common *Amoeba* in all structural details. The outermost body covering or **plasmalemma** is a thin, elastic and semi-permeable membrane. The cytoplasm is differentiated into the outer clear and non-granular **ectoplasm** and the central more fluid and granular **endoplasm**.

1. Nucleus. Endoplasm contains a nucleus which is seldom visible in a living organism. After being fixed and stained, it appears as a vesicular structure 4 to 6μ in size. It is bound by a thin and delicate **nuclear membrane** whose inner surface is encrusted with fine **chromatin granules** arranged in a peripheral ring. In the centre of nucleus is a small dot-like **endosome** or **karyosome** often surrounded by a clear area or 'halo'. The fluid-filled space between nuclear membrane and endosome is marked by spoke-like striations of **chromatin material**.

Other inclusions of the endoplasm are **food vacuoles** that enclose the ingested red blood corpuscles (erythrocytes), white blood corpuscles (leucocytes), fragments of epithelial cells and bacteria. **Contractile vacuoles** are wanting in *E. histolytica* since it inhabits an isotonic environment. The osmotic concentration of its body protoplasm equals to that of the intestinal fluid of the host and hence no water enters the organism by osmosis. Thus, there is no need for osmoregulation.

2. Pseudopodium. When the *E. histolytica* moves forward a large, broad and blunt pseudopodium is formed. Hence the parasite is typically **monopodial** (Gr., **monos** single +**podos** foot). It is only at this anterior end that the outer clear ectoplasm is sharply differentiated from the inner granular endoplasm. With a single pseudopodium the trophozoite moves with a forward-flowing movement. This resembles the crawling of a garden slug (*Limax* spp.) and hence the movement is sometimes called 'limax-type movement'.

Nutrition in trophozoite is **holozoic**. It feeds by phagocytosis. Food particles are engulfed at the posterior end, where plasmalemma is more sticky. Food particles adhere to it and plasmalemma is invaginated into cytoplasm carrying the food particles along with it. The invaginated plasmalemma now separates off from the main plasmalemma to become food vacuole.

[II] Precystic or Minuta Form

Precystic or minuta form of *E. histolytica* is small, spherical, non-motile and non-feeding form. It measures 12 to 15μ in diameter. In structural details it resembles the trophozoite except that it is smaller in size and the food vacuoles are absent. It lives in the lumen of large intestine and is non-pathogenic to man. But when resistance of host's body is low, it changes into magna form and invades the tissues of intestine.

[III] Cystic Form

Under normal condition, minuta forms undergo encystation. It becomes rounded and surrounded by a thin, highly resistant and refractile **cyst wall**. A mature cyst is a spherical body 10 to 12μ in diameter. Its cytoplasm is clear and contains one or two **glycogen masses** (reserve food) and one or more characteristic, refractile, bar-like **chromatoid bodies** or **chromidial bars** with rounded ends. Both glycogen masses and chromidial bars gradually disappear. Chromidial bars are made of ribonucleoprotein which disperses throughout cytoplasm with their disappearance. **Nucleus** retains the characters of the trophozoite. To start with, the cyst is **uninucleate** but its nucleus divides to form a **binucleate** and finally a **tetranucleate** or **quadrinucleate** cyst.

REPRODUCTION AND LIFE CYCLE

[I] Hosts

E. histolytica is **monogenetic** (Gr., **monos**, **one** + **genos**, race), *i.e.*, only one host is required for its complete life cycle. Pigs, dogs, rats and rabbits are supposed to be the **reservoir host**. It undergoes reproduction and completes its life cycle only in man.

[II] Binary Fission

Trophozoites multiply asexually by **binary fission** within the wall of large intestine. Its nucleus undergoes **mitosis** without the disappearance of the nuclear membrane. The mitosis is followed by **cytokinesis** or division of the cytoplasm. As a result, two daughter organisms are formed. They grow rapidly in size, feeding upon bacteria and host tissue elements, and in their turn, again multiply by binary fission. Some of them invade fresh host cells, while others become the precystic or minuta forms.

[III] Encystation

The **precystic** or **minuta** forms encyst only in the lumen of the intestine and not in the tissues. They round up and secrete a thin, refractile, tough and flexible **cyst wall** around them. The cysts at this stage are **uninucleate**. The single nucleus of the cyst soon undergoes a mitotic division to form two nuclei, leading to the formation of **binucleate cysts**. A second mitotic division divides the two nuclei into four and thus **tetranucleate cysts** are formed. The whole process of encystation is completed within a few hours.

Tetranucleate cysts constitute the transmissive or infective stage. They do not develop further but pass out of the host in the faeces. They can remain alive for 10 days in moist stool. They are very susceptible for desiccation and die at a temperature of 50°C.

[IV] Transmission and Infection

Infection depends upon intake of food or water contaminated with faecal matter containing tetranucleate cysts of *E. histolytica.* The food handlers, like cooks, meatsellers, hawkers, etc., who themselves are infected and are unhygienic by nature, act as cyst passers. Untreated human faeces voided by children and adults on open grounds or in crop and vegetable fields (which is the usual practice in rural areas in India) is a common source of infection. Houseflies and cockroaches, which are coprophagous (feeding upon faecal matter), carry viable cysts on their legs or in their intestine and transfer them to unprotected foodstuffs.

[V] Excystation

In the new host, the ingested cysts pass down the alimentary canal and reach the small intestine. The cyst wall protects them from the action of host's gastric juice during their passage through the stomach. After 5 or 6 hours, excystation takes place as the cyst wall is digested by trypsin in small intestine releasing the tetranucleate *Amoeba*, called the **excystic** *Amoeba* or **metacyst**.

[VI] Metacystic Development

Each metacyst immediately proceeds to divide by binary fission. Its nuclei divide in a specific pattern, accompanied by simultaneous cytoplasmic divisions, to produce 8 small, uninucleate amoebulae or **metacystic trophozoites**. These metacystic trophozoites pass into the large intestine, invade the mucous lining and grow into mature trophozoites.

Note: Both the processes of excystation and encystations may occur in one and the same host but after the formation of cysts a transference to another new host is required for the continuation of a species.

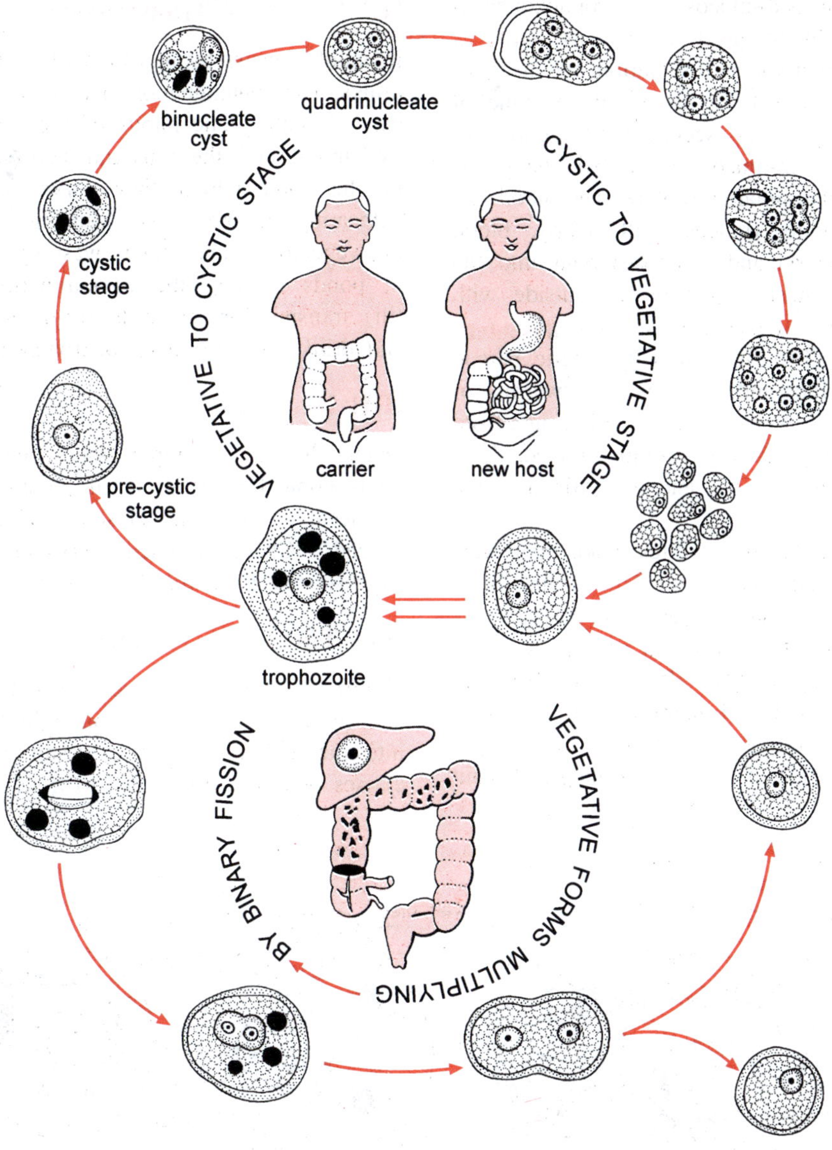

Fig. 3. *Entamoeba histolytica.* Reproduction and life history.

AMOEBIC DYSENTERY OR AMOEBIASIS

[I] Symptoms and Pathogenesis

As already noted, *E. histolytica* lives in the lumen of large intestine of man as harmless minuta forms. Whenever resistance of gut is lowered in infected people, these become pathogenic, change to magna forms and invade the intestinal wall. They make their way deep into sub-mucosa by eating through mucosa of the intestinal wall. Here they multiply by binary fission and spread radially outward to form flask-shaped **ulcers** containing cellular debris, lymphocytes, blood corpuscles and bacteria. This causes formation of **abscesses** in intestinal wall.

Penetration into sub-mucosa by trophozoites is made possible by **histolysis** as well as **cytolysis**. The mechanism involves dissolution and necrosis of tissues and cells by a proteolytic enzyme of the nature of **histolysin** secreted by trophozoites themselves. As sub-mucosa is eroded, the ulcers burst and blood capillaries rupture. The blood and the ulcer contents (mucus, cell debris, blood corpuscles, bacteria and amoebae) pour into the lumen of the intestine and passes outside with stool. This characterizes the **amoebic dysentery** or **amoebiasis**. The stool of a dysenteric person is usually acidic and consists of swarms of entamoebae as well. Person suffering from amoebic dysentery has repeated blood-mixed, slimy and foul-smelling motions and is confined to the lavatory.

Sometimes, the trophozoites make their way, through blood circulation, into the brain, liver, spleen, lungs and gonads. Here also they destroy the tissues and cause formation of **abscesses**, or cavities containing pus. The affected liver becomes enlarged, congested and painful to touch. This pathological condition is referred to as **amoebic hepatitis**. Formation of abscesses in lung and brain usually prove fatal.

[II] Modes of Transmission

It mainly occurs by ingestion of cysts in food or drinks. The contamination of food or drinks occurs by : (i) unhygienic habits of food handlers who by habit scratch the anus and then put the fingers in the food which they serve, (ii) habit of defecating in open fields causing contamination of vegetables and then washing the bottom in ponds causing the contamination of water, (iii) transmission of cysts from stools to food and drinks by flies and cockroaches.

[III] Diagnosis

Diagnosis, in a simple way, consists in the microscopical detection of trophozoites or .cysts in faecal smears. The presence of white, stone-shaped 'Charcot-Leyden' crystals in faeces suggests the *E. histolytica* infection.

[IV] Therapy (Treatment)

E. histolytica infections occur in both the intestine and (in people with symptoms) in tissue of the intestine and/or liver. As a result, two different classes or drugs are needed to treat the infection,

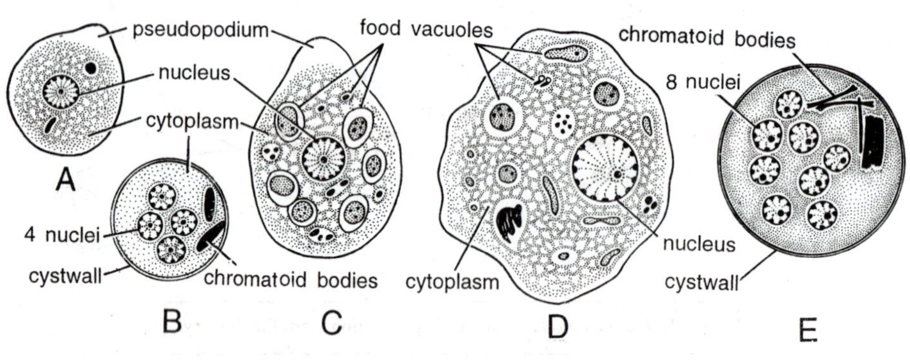

Fig. 4. Other entamoebae of man. **A.** Trophozoite of *E. hartmanni,* **B.** Cyst of *E. hartmanni,* **C.** Trophozoite of *E. gingivalis.* **D.** Trophozoite of *E. coli.* **E.** Cyst of *E. coli.*

one for each location. Such anti-amoebic drugs are known as amoebicides.

1. Tissue amebicides. Metronidazole, or a related drug such as tinidazole, secnidazole or ornidazole, is used to destroy amoebae that have invaded tissue.

2. Lumenal amebicides. Since most of the *Amoeba* remain in the intestine lumen when tissue invasion occurs, it is important to get rid of luminal *Amoeba* and their cysts to prevent re-infection. Paromomycin, Idoquinol and Diloxanide furoate (Amicline) have been shown to be effective.

[V] Prevention (Prophylaxis)

Prevention of infection is entirely a matter of hygiene, both personal as well as municipal. For personal hygiene it is suggested to adopt the following habits :

(1) Washing hands with soap and water after handling dirty articles, before taking meals and after using toilet.
(2) Cutting finger nails regularly.
(3) Avoiding use of unboiled water and improperly washed vegetables and raw salads.
(4) Protection of foods and drinks from contamination by houseflies, cockroaches, etc.
(5) Avoiding passing out of stool on open grounds, street sides or vegetable fields.

The municipal hygiene is the responsibility of the town areas, municipalities or other local bodies. They must take the following preventive measures :

(1) Proper sanitation of roads, streets, lanes and open drains etc.
(2) Purification of drinking water.
(3) Proper disposal of sewage.
(4) Covering of the food articles by the traders.
(5) Chemical treatment of human faeces to be used as fertilizer.
(6) Periodical examination of food handlers to find out whether they are infected with *E. histolytica.* On positive findings they should be treated properly.

OTHER ENTAMOEBAE

[I] *Entamoeba hartmanni*

This *Entamoeba* closely resembles the minuta form of *E. histolytica.* It also lives in the lumen of large intestine and invades intestinal tissue. Trophozoites measure 9 to 14μ in diameter. Nucleus is more dense. Mature cysts are tetranucleate and are less then 10μ in diameter (the size is smaller than the cysts of *E. histolytica). E. hartmanni* too causes amoebic dysentery but is supposed be less harmful.

[II] *Entamoeba coli*

This *Entamoeba* is said to be non-pathogenic as it does not invade the host tissue. Living in the lumen of intestine it may produce digestive complications. Trophozoite of this species measures between 20 to 40μ in diameter. It is less active and forms a small pseudopodium. Cytoplasm never contains red blood corpuscles. Central karyosome or endosome of nucleus is minute and is generally eccentric. Cysts measure between 15 and 20μ in diameter and contain eight nuclei. Chromatoid bars are thin with sharp ends.

E. coli occurs in the intestine of about 50% of human population. Infection takes place when cysts in stools are voided by infected persons to outside and these are then taken by other men with contaminated food and water.

[III] *Entamoeba gingivalis*

This *Entamoeba* dwells in mouth, usually between teeth and gums. Trophozoite measures about 12 to 20μ in diameter. Ectoplasm is crystal clear and central granular endoplasm is vacuolated. Endoplasm contains a vesicular nucleus with a large central endosome, and numerous food vacuoles containing bacteria and nuclei of leucocytes. Pseudopodia are usually many with broad and rounded ends.

E. gingivalis occurs in mouths of nearly 70% of human population. Its incidence is high in those persons who suffer from pyorrhoea. It has been the belief for a long time that *E. gingivalis* is the causative organism for pyorrhoea, but it has now been established that pyorrhoea is caused as a result of bacterial infection. *Entamoeba,* of course, aggravates pyorrhoea by destroying tissues in gums. The species does not form cysts, since it is directly transmitted from mouth to mouth by contact during kissing or in feeding.

Important Questions

➤➤Long answer type questions

1. Give an account of the bionomics, morphology, life history and economic importance of *Entamoeba histolytica.*
2. Describe the life history and pathogenicity of the organism causing amoebic dysentery. Suggest methods for the control of the disease.
3. Distinguish between *Amoeba* and *Entamoeba.* Justify the statement that *Entamoeba histolytica* is not pathogenic but behaves as a pathogenic organism.
4. Write short notes on :
   (i) Chromatoid bodies, (ii) Encystment, (iii) *Entamoeba coli,* (iv) *Entamoeba gingivalis,* (v) Metacysts, (vi) Minuta, (vii) Nucleus of *Entamoeba.*

➤➤Short answer type questions

1. Name the rhizopodan parasite causing dysentery.
2. What are two types of reserve food materials found in the cyst of *Entamoeba histolytica* ?
3. What is amoebiosis ? What are its pathological effects ?
4. Compare the structure of pseudopodia of *Entamoeba histolytica* and *Elphidium.*
5. What is the mode of transmission of *Entamoeba histolytica* ?
6. What are the diseases caused by *Entamoeba histolytica* ?
7. Which one of the following organism is closely related to *Entamoeba histolytica.* Is it *Escherichia coli* or *Entamoeba coli ?*
8. How are cysts adapted for the survival outside the host ?
9. What are the organs or glands infected by *Entamoeba histolytica* ?

➤➤Multiple choice questions

1. R.B.C. is present in the food vacuole of :
   (a) *Entamoeba histolytica*
   (b) *Amoeba proteus*
   (c) *Trypanosoma gambiense*
   (d) *Leishmania donovani*
2. What disease is caused by *Entamoeba histolytica :*
   (a) Amoebic dysentery (b) Gambian fever
   (c) Typhoid (d) Malaria
3. Pathological nature of *Entamoeba histolytica* was discovered by :
   (a) Lamble (b) Losch
   (c) Grassi (d) Lavern
4. The infective stage of *Entamoeba histolytica* is :
   (a) uninucleate cyst (b) binucleate cyst
   (c) tetranucleate cyst (d) minuta form
5. Besides the bead-like chromatin granules what is present inside the nucleus of *Entamoeba histolytica :*
   (a) parabasal body (b) contractile vacuole
   (c) micronucleus (d) karyosomes
6. *Entamoeba histolytica* is found in :
   (a) the rectum of man
   (b) the intestine of man
   (c) the oral cavity of man
   (d) the stomach of man
7. Amoebic dysentery is caused by :
   (a) *Amoeba proteus*
   (b) *Entamoeba histolytica*
   (c) *Plasmodium vivax*
   (d) *Taenia solium*
8. *Entamoeba histolytica* causes :
   (a) kala-azar (b) filariasis
   (c) dysentery (d) sleeping sickness
9. *E. histolytica* is found in :
   (a) colon (b) intestine
   (c) both (d) none
10. *Entamoeba histolytica* is without :
    (a) pseudopodium (b) contractile vacuole
    (c) food vacuole (d) nucleus
11. Cysts of amoebic dysentery spread by :
    (a) housefly (b) mosquito
    (c) men (d) none of these
12. Amoebic dysentery in man is caused by :
    (a) *Amoeba proteus* (b) *Entamoeba histolytica*
    (c) *Entamoeba coli* (d) *Entamoeba gingivalis*
13. In *Entamoeba histolytica* the normal trophozoites are called :
    (a) sporulets (b) minuta forms
    (c) monogenetic forms (d) magna forms
14. *Entamoeba histolytica* has no need for :
    (a) feeding (b) growth
    (c) osmoregulation (d) reproduction
15. Precystic stage of *Entamoeba* is called :
    (a) magna form (b) minuta form
    (c) feeding trophozoite (d) growing trophozoite
16. *Entamoeba histolytica* is a human parasite usually found in :
    (a) intestine (b) liver
    (c) blood (d) none of the above

17. The minuta form of *Entamoeba histolytica* differs from *Amoeba proteus* in having :
(a) one pseudopodium formed of ectoplasm
(b) one pseudopodium formed of endoplasm
(c) no food vacuoles
(d) small contractile vacuole

18. *Entamoeba histolytica* differs from *Amoeba proteus* in having :
(a) no pseudopodium (b) no food vacuoles
(c) reproduction only by encystment
(d) no contractile vacuole

19. Encysted stage is not known to occur in life of :
(a) *Entamoeba gingivalis*
(b) *Entamoeba coli*
(c) *Entamoeba histolytica*
(d) *Amoeba*

20. Monopodial locomotion does not occur in :
(a) *Entamoeba histolytica*
(b) *Entamoeba coli*
(c) *Entamoeba gingivalis*
(d) *Entamoeba*

21. Most active form of *Entamoeba histolytica* :
(a) trophozoite (b) pericyclic
(c) cystic (d) all

22. *Entamoeba histolytica* has :
(a) one contractile vacuole
(b) two contractile vacuoles
(c) three contractile vacuoles
(d) no contractile vacuole

23. Which of the following has only one host :
(a) *E. histolytica* (b) *T. gambiense*
(c) *T. solium* (d) *P. vivax*

24. The excretion in *Entamoeba histolytica* take place :
(a) food vacuole (b) contractile vacuole
(c) general body surface
(d) endosome

25. How many young amoebae hatch out from a cyst of *E. histolytica* :
(a) one (b) two
(c) four (d) six

26. Cyst of *E. histolytica* spread by :
(a) mosquito (b) bed bug
(c) mouse (d) house fly

27. Chromatoid body in *Entamoeba histolytica* are found in :
(a) cyst (b) trophozoite
(c) meta cyst (d) minuta

28. For prevention against amoebiosis we should :
(a) eat spicy food (b) use mosquito net
(c) drink boiled water
(d) eat plenty of food

29. The meta cyclic form has .... nucleus :
(a) one (b) two
(c) four (d) eight

Answers

➤➤*Multiple choice questions*

1. (a)	2. (a)	3. (a)	4. (c)	5. (d)	6. (a)	7. (b)	8. (c)	9. (a)	10. (b)
11. (a)	12. (b)	13. (d)	14. (c)	15. (b)	16. (a)	17. (c)	18. (d)	19. (a)	20. (c)
21. (a)	22. (d)	23. (a)	24. (c)	25. (a)	26. (d)	27. (a)	28. (c)	29. (c)	

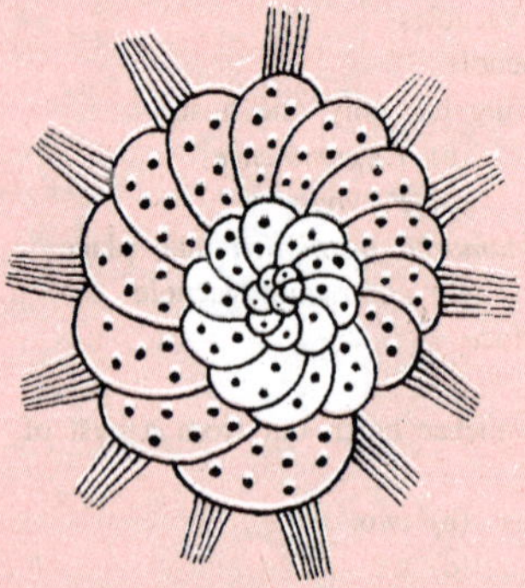

8

Elphidium crispum (=*Polystomella crispa*)

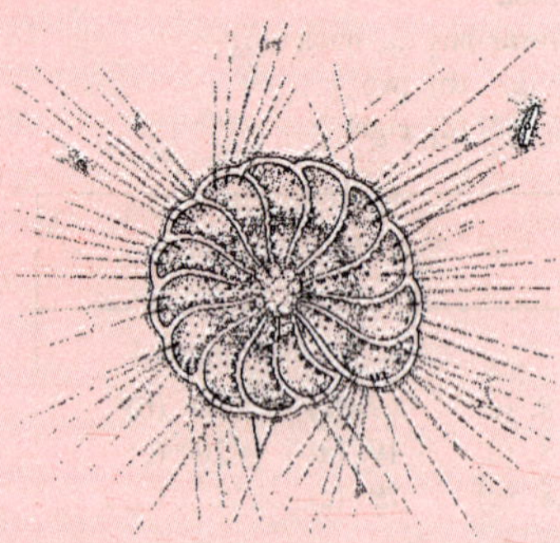

Not all Protozoa are naked, some are enclosed in shells. The most common shelled protozoans are **foraminiferans** belonging to the order **Foraminiferida** (L., **forare**, pores + **ferre**, to bear). Their shells are typically many-chambered and perforated all over with small pores, through which extend long and fine pseudopodia. Almost all the members of the order are marine. When they die, their shells fall in a steady rain to the ocean floor and contribute to the formation of bottom sediments, called **ooze**.

Elphidium crispum (*Polystomella crispa*)

Many foraminiferans exhibit the phenomenon of **dimorphism**, *i.e.*, individuals of a single species occur in two distinct forms, which differ in the structure of shell, number of nuclei and mode of reproduction. *Elphidium* (=*Polystomella*) is a typical representative of the dimorphic forms. The most common species is *Elphidium crispum*, whose biology is dealt here in detail.

SYSTEMATIC POSITION

Phylum	Protozoa
Subphylum	Sarcomastigophora
Superclass	Sarcodina
Class	Rhizopodea
Subclass	Granuloreticulosa
Order	Foraminiferida
Genus	*Elphidium*
Species	*crispum*

OCCURRENCE : HABIT AND HABITAT

E. crispum is a free-living marine foraminiferan and occurs in the littoral zone of the sea down to about 600 meters. It is commonly found creeping on the sea weeds.

STRUCTURE

[I] Dimorphic Forms

E. crispum is a dimorphic species, the two forms being the **microspheric** and **megalospheric**. The microspheric form is small, multinucleate, with a smaller initial chamber and asexual mode of reproduction. The megalospheric form is large, uninucleate, with a larger initial chamber and sexual mode of reproduction.

[II] Shell

Shell of *E. crispum* is small, translucent, pale yellow in colour and spherical or oval in shape. It measures 1 mm in diameter. The shell consists mainly of calcium carbonate with small amounts of other inorganic compounds such as silica and magnesium sulphate. The shell is **polythalamus** or **multilocular**, *i.e.*, it consists of many chambers. It is coiled in a **planospiral** manner, *i.e.*, all the coils lie in the same plane or spirally flat. All the chambers are filled with cytoplasm. The shell is **perforated**, that is, the walls of chambers bear minute pores through which cytoplasm becomes continuous from one chamber to another, and also extends to outside forming pseudopodia.

Each chamber of shell is somewhat elongated laterally. Its anterior surface is convex while the posterior surface is concave. The concave posterior surface bears **retral processes** filled with cytoplasm. Minute pores occur in the spaces between retral processes which act as open doors for the emergence of the pseudopodia to outside. Peripheral part of shell is rigid and consists of a continuous rim, called **keel**, whereas the central part is the rounded **umbo** which becomes prominent due to deposition of more calcium carbonate.

Formation of shell begins with an initial single chamber, the **proloculum.** As the animal grows in size, successive chambers are laid down in a spiral manner, forming whorls. Each chamber is larger than the preceding one and each new whorl partially overlaps and conceals the older one. Thus, only the last whorl, which is the largest and most recently formed, is visible from outside. The overlapping portions of the chambers are referred to as **alar processes** (Fig. 1).

[III] Cytoplasm

All the shell chambers of living *E. crispum* are filled with a mass of **inner cytoplasm**. Besides, a thin layer of **outer cytoplasm** invests the entire

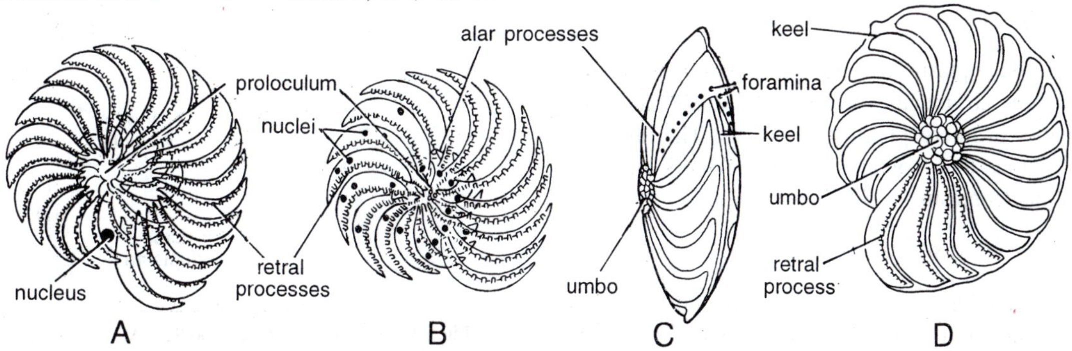

Fig. 1. *Elphidium cirspum.* Decalcified and stained specimens. **A.** Megalospheric individual. **B.** Microspheric individual. **C.** Endon view of the shell. **D.** Lateral view of the shell.

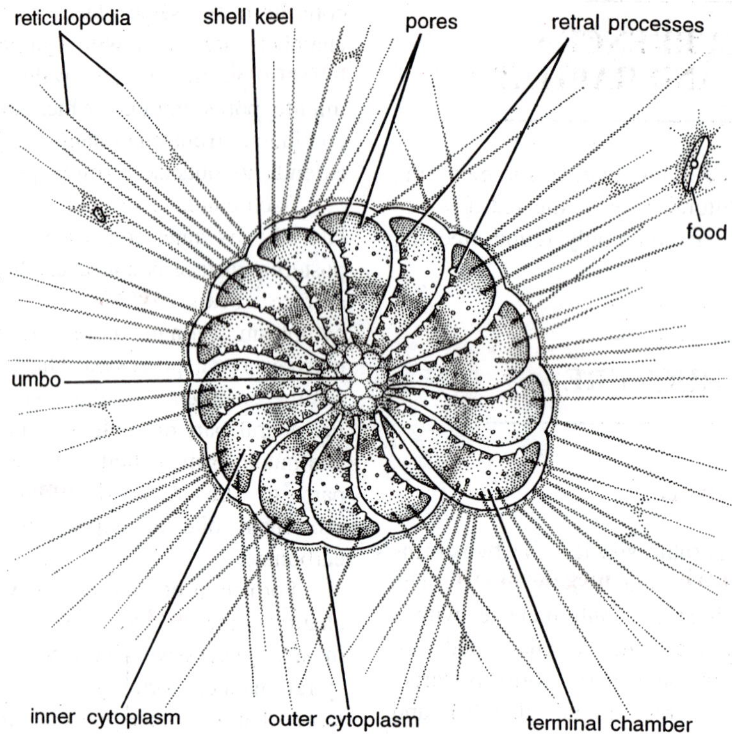

Fig. 2. *Elphidium crispum* (living).

shell, which makes the shell internal rather than external. Part of cytoplasm within the shell is also called **endoplasm** and that outside it, the **ectoplasm**. Endoplasm contains nucleus, (or nuclei), food particles, minute vacuoles, Golgi bodies, mitochondria, endoplasmic reticulum, ribosomes and colour granules or **xanthosomes**.

[IV] Nucleus

Inner cytoplasm or endoplasm contains a single nucleus in megalospheric individuals and many nuclei in microspheric individuals. The nucleus is vesicular and contains many nucleoli embedded in nucleoplasm.

[V] Reticulopodia

Pseudopodia of *E. crispum* are typically foraminiferan because of their thread-like branching and anastomosing nature. Such pseudopodia are known as **reticulopodia** or

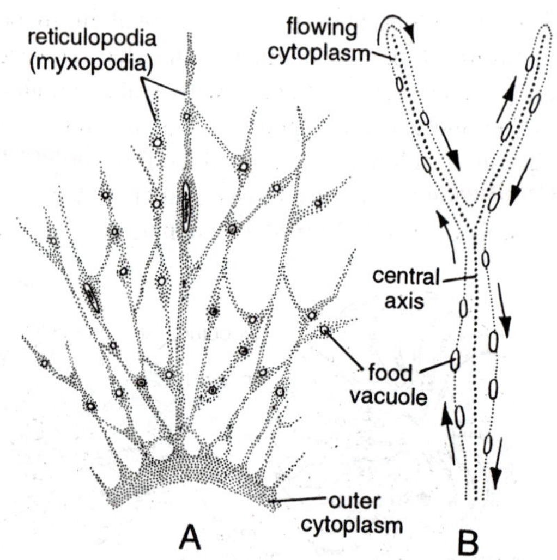

Fig. 3. *Elphidium.* Reticulopodia. **A.** A group of reticulopodia. **B.** A reticulopodium showing streaming circulation of cytoplasm.

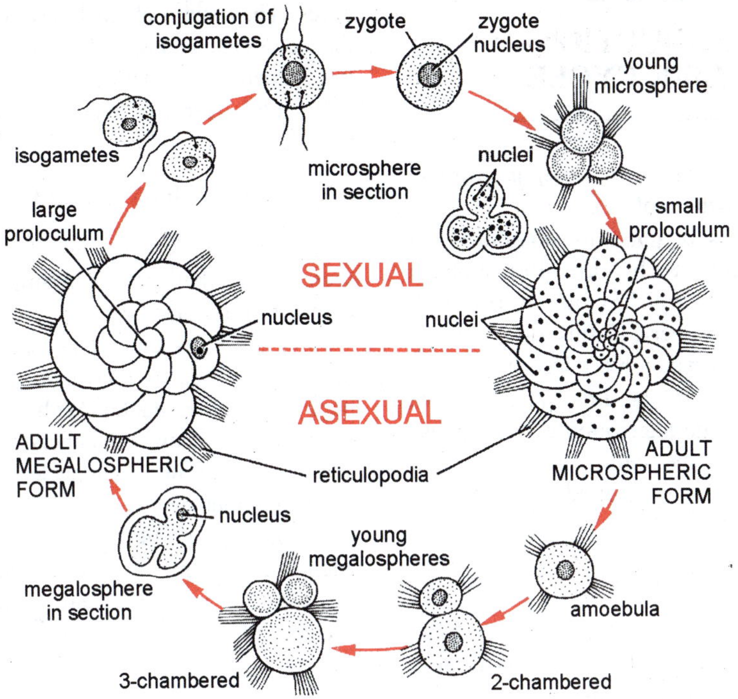

Fig. 4. *Elphidium.* Reproduction and life cycle.

myxopodia. Each consists of two regions—inner fibrillar **axis** and outer fluid-like **cortex**. It has been observed that in the network of reticulopodia cytoplasm shows streaming circulation, as is evident from the movement of granules in opposite directions along the two sides of axis. Pseudopodia may be withdrawn within the shell at lightning speed or may be extended out a short distance in water like little rockets and wave about bending, undulating, quivering and putting out side branches, which meet and fuse and so establish a reticulum. Reticulopodia help the organism in locomotion, nutrition and in the construction of shell and cyst walls.

LOCOMOTION

E. crispum creeps slowly over the substratum on the sea bottom with the help of its reticulopodia, which are arranged in bundles around the shell. With the contraction of distally placed bundles the body is pulled or dragged forward.

NUTRITION

E. crispum is typically **holozoic**. It feeds upon minute organisms like diatoms, other protozoans, crustacean larvae, etc. The net-like reticulopodia are thought to secrete an external mucous layer to trap the food organisms. The mucous layer contains proteolytic secretions which paralyze the prey and initiate its digestion even during capture. The captured food, enclosed in a food vacuole, is drawn into endoplasm by the withdrawal of reticulopodial filaments towards the interior of body. Digestion is normally completed outside the shell and the products of digestion pass into the endoplasm.

REPRODUCTION AND LIFE CYCLE

The two distinct forms of *E. crispum* actually display the phenomenon of **sexual dimorphism**. Megalospheric forms or **gamonts** are the sexual forms. They reproduce by the formation of gametes (gamogony), which unite in pairs and form zygotes. Microspheric forms or **agamonts** are the asexual forms which reproduce by multiple fission (agamogony). In the life cycle, the two forms display a definite alternation of generations. According to Jepps, the entire cycle of sexual and asexual phases is completed in about two years.

[I] Sexual Phase (Gamogony)

1. Megalospheric form or gamont. The sexual megalospheric form or gamont is somewhat larger in size and possesses a large embryonic chamber, the **proloculum**. It remains uninucleate throughout its life.

2. Isogamy. Each gamont produces numerous identical biflagellate gametes or **isogametes**. The formation of gametes take place inside the parent shell. The single nucleus divides repeatedly by mitosis which is accompanied by the fragmentation of entire cytoplasm. Each fragmented cytoplasmic portion, along with a daughter nucleus organizes into a gamete. Each gamete is a spherical bit of cytoplasm and possesses two motile flagella, a nucleus and a fairly large fat globule.

3. Fertilization. The gametes are released into the surrounding sea water through the shell pores. They swim about freely for some time and then elongate to become conical in shape. Later, the gametes, originating from different gamonts, unite and fuse in pairs, resulting in diploid **zygotes**.

[II] Asexual Phase (Agamogony)

1. Microspheric form or agamont. The zygote secretes the shell around a small initial chamber or proloculum and becomes a young **agamont** or **microsphere**. Its diploid nucleus divides first by meiosis and then by mitosis to give rise to several haploid daughter nuclei. With further growth and development, new chambers are successively added, resulting in a multinucleate and multilocular adult agamont or **schizont**. The nuclei are found in cytoplasm of all the chambers and they are encircled by numerous chromatidial granules. The microspheric individuals are scarce in occurrence because of the remote chances of the fusion of gametes.

2. Schizogony. Microsphere or agamont reproduces asexually by schizogony. Entire cytoplasm containing several nuclei creeps out of the shell and takes the form of a lump around it. This lump now organizes into a large number of small uninucleate *Amoeba* like cells, called **amoebulae** or **agametes**.

3. Development of agametes. The agametes now become detached from the parental shell. Each one secretes a shell around a large proloculum and becomes a young gamont or **megalosphere**. It repeatedly secretes additional chambers and transforms into a uninucleate and multilocular adult gamont or megalospheric individual.

Important Questions

➤➤Long answer type questions

1. Give a detailed account of the structure and life history of a foraminiferan studied by you.
2. Describe the phenomenon of dimorphism in *Elphidium crispum*. Give a brief account of its life history.
3. Give an account of the structure and formation of shell in *Elphidium (=Polystomella)*.
4. Describe the asexual phase of reproduction in *Polystomella*.

➤➤Short answer type questions

1. Mention the two forms of *Elphidium*.
2. Mention two differences between the nuclei of microspheric and megalospheric forms of *Elphidium*.
3. What is the type of shell in *Elphidium*.
4. What is metagenesis ? With the aid of diagrams illustrate the phenomenon with reference to *Elphidium*.
5. Explain alternation of generation in *Elphidium*. How it differs from *Amoeba* and *Paramecium* ?

➤➤Multiple choice questions

1. The mode of nutrition in *Elphidium* is :
   (a) saprozoic (b) holophytic
   (c) coprozoic (d) mixotrophic
   (e) holozoic
2. Locomotion in *Elphidium* is by :
   (a) filopodia (b) lobopodia
   (c) axopodia (d) reticulopodia
3. Dimorphic form occur :
   (a) *Paramecium* (b) *Euglena*
   (c) *Polystomella* (d) *Amoeba*
4. Which one of them have shell :
   (a) *Paramecium* (b) *Polystomella*
   (c) *Trypanosoma* (d) all
5. The free living marine protozoan is :
   (a) *Paramecium* (b) *Amoeba*
   (c) *Euglena* (d) *Polystomella*
6. *Polystomella* belong to order :
   (a) Foraminiferida (b) Euccosida
   (c) Amoebida (d) Eugregarinida

Answers

➤➤Multiple choice questions

1. (e) 2. (d) 3. (c) 4. (b) 5. (d) 6. (a)

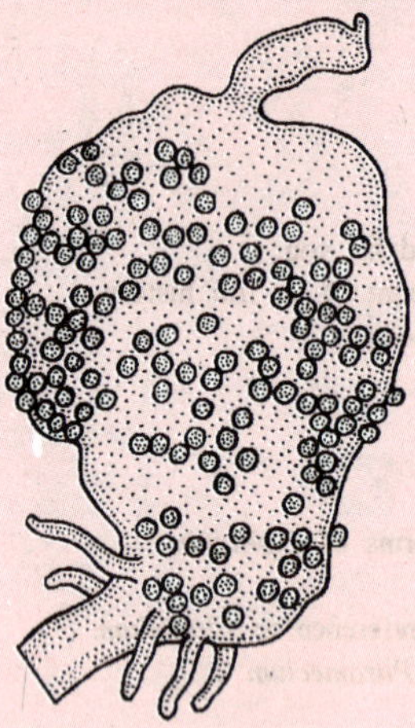

9

Plasmodium vivax : The Malarial Parasite

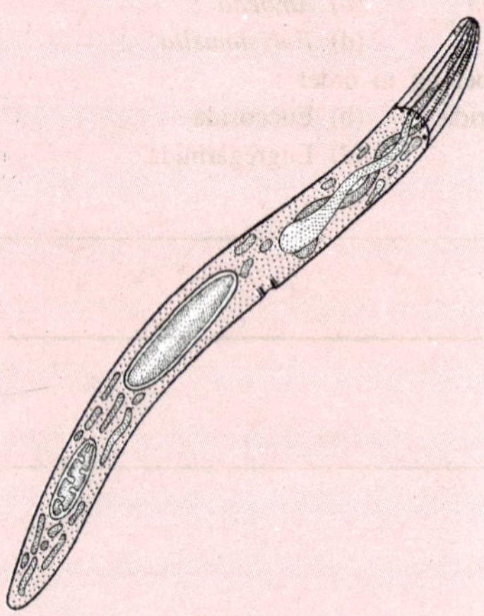

Members of subphylum Sporozoa are all parasites and without organelles of locomotion as adult. These are either intra or intercellular parasites of both invertebrates and vertebrates and many of them are causative organisms of dreadful diseases like malaria, various cattle fevers, coccidiosis in chickens, epidemic deaths in cultivated honeybees and silkworms, etc. The most interesting sporozoan genus is *Plasmodium* whose 60 known species cause malaria in man and other annuals. Because of their malaria-causing abilities, the species are commonly referred to as **malarial parasites***. All these reside in the red blood corpuscles, reproduce in them and destroy them. Mosquitoes are the transmitting agents or *vectors* of the malarial parasites. About 200 species have been described.

Plasmodium vivax

Four species of *Plasmodium are* known to cause different types of malaria fever in man. They are *P. vivax, P. ovale, P. malariae* and *P. falciparum.*

The following description belongs to *P. vivax* in particular, which is the most common type of malarial parasite.

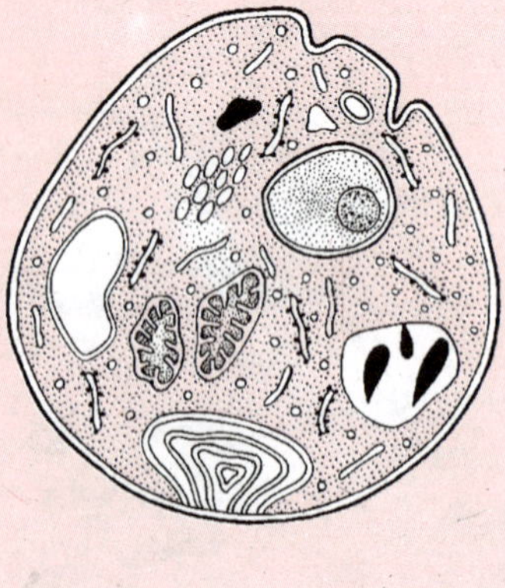

* According to some, the term 'malarial parasite' is incorrect, it should be 'malaria fever parasite'.

SYSTEMATIC POSITION

Phylum	Protozoa
Subphylum	Sporozoa
Class	Telosporea
Subclass	Coccidia
Order	Eucoccida
Suborder	Haemosporina
Genus	*Plasmodium*
Species	*vivax*

DISTRIBUTION OF *PLASMODIUM*

Geographical distribution of the species of *Plasmodium* is widespread in tropical and temperate countries. Where the migratory birds are hosts, the parasites are spread all over the world. Of the human infecting parasites, *P. vivax* is the most widely distributed. It prevails mainly in temperate regions of the world. *P. falciparum* is confined to the warmer parts only. *P. malariae* follows *P. vivax* in distribution. *P. ovale,* though widespread, is the rarest of the four.

LIFE CYCLE OF *PLASMODIUM VIVAX*

P. vivax is the most common of the human infecting malaria fever parasites. It is the causative organism of **benign tertian** or **vivax malaria**, which is characterized by a 48-hour cycle between the first malaria fever and subsequent recurrence of chills and fever. It is an intracellular parasite in man, living in the red blood corpuscles and liver cells, while extracellular in mosquito, living in its alimentary canal and salivary glands.

Hosts

The life cycle, of *P. vivax,* is very complicated. Being **digenetic**, it is completed in two hosts, man and mosquito.

1. Man. Asexual cycle is passed in man in two phases. First phase in liver cells is called **liver schizogony**. It involves multiple fission forming **merozoites** at the end. Second phase is completed in red blood corpuscles. It is known as **erythrocytic schizogony** which forms **gametocytes** at the end.

2. Mosquito. Sexual cycle is completed in female *Anopheles* mosquito. It also involves two phases, gametogony and sporogony. **Gametogony** is concerned with the production and fusion of **gametes**, whereas **sporogony** is postzygotic multiplication resulting in the formation of infective individuals, the **sporozoites**.

Asexual Cycle of *P. vivax* in Man

[I] Infection

A healthy person acquires infection when a female *Anopheles* mosquito, containing infective stages of parasite (sporozoites) in its salivary glands, bites him for sucking his blood. The mosquito punctures the host's skin by its proboscis and first introduces some saliva into blood stream. Along with saliva, thousands of sporozoites contained therein are also inoculated. Purpose of pouring saliva is to check clotting of blood, as it contains an anticoagulant.

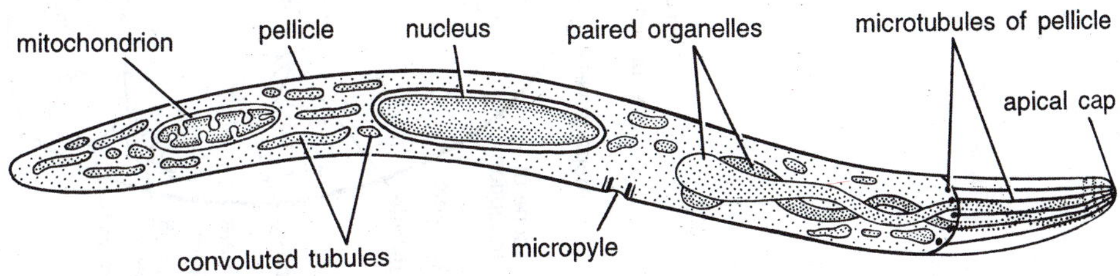

Fig. 1. *Plasmodium.* Structure of a sporozoite as revealed by electron microscope.

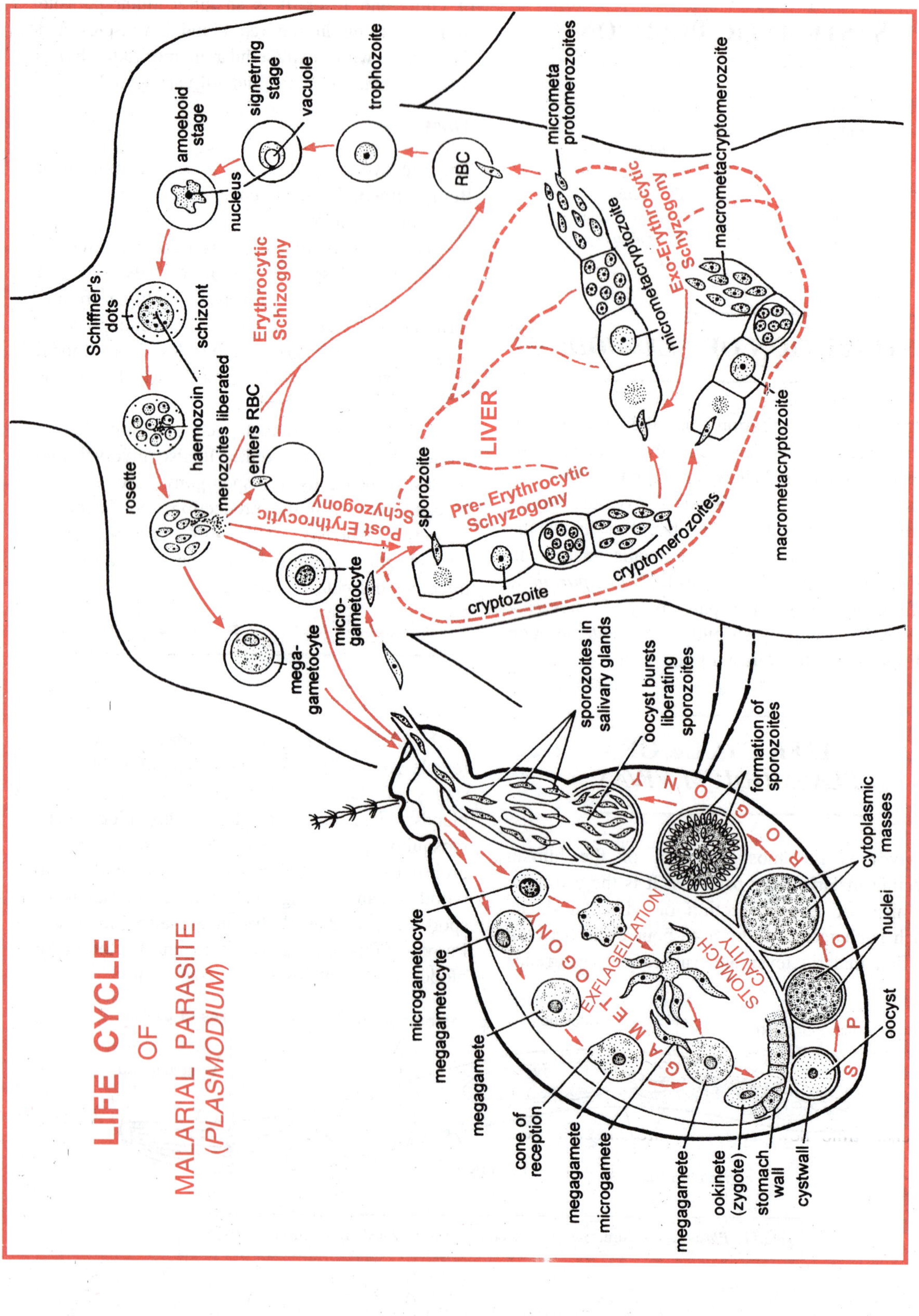

Fig. 2. Life cycle of *Plasmodium vivax.*

[II] Sporozoites

Sporozoites represent the infective forms of parasite. These are small spindle-shaped, slightly curved or sickle-shaped, and uninucleate organisms, measuring 11-12 μ in length and 0.5-1 μ in width.

Electron microscope reveals the complicated gross anatomy of sporozoite. External covering is the firm and elastic **pellicle**, which contains longitudinally arranged contractile **microtubules** for its wriggling movements. At the anterior end is the **apical cap**, made up of three or more concentric rings, into which open the secretory **paired organelles**. These are supposed to secrete lytic juice to facilitate penetration into liver cells. A **micropyle** represents the cytostome of other protozoans. Many **convoluted tubules** of unknown function are found along the whole body of sporozoite. There is a single vesicular **nucleus** with a centrally-located nucleolus and a **mitochondrion** with tubular cristae.

[III] Liver Schizogony

Once within the human blood, the sporozoites circulate for about half an hour and then get into liver to invade the hepatic cells. Here they multiply asexually by schizogony. Liver schizogony has two phases, **pre-erythrocytic** and **exo-erythrocytic**.

1. Pre-erythrocytic phase. After penetrating a hepatic cell each sporozoite becomes a **cryptozoite**. It grows for a number of days and becomes a spherical and non-pigmented **schizont**. It divides by schizogony (multiple fission) and forms a large number of uninucleate cells, the **cryptomerozoites**. Their reported number varies from one thousand to several thousands. They are liberated when the liver cell bursts. This is the end of pre-erythrocytic phase. During pre-erythrocytic schizogony, blood remains sterile and its inoculation does not produce infection.

2. Exo-erythrocytic phase. Cryptomerozoites enter fresh liver cells to become **metacryptozoites**. They undergo schizogony similar to the previous one producing enormous number of **metacrypto-merozoites**. This may be repeated several times and each time new liver cells are infected. All these succeeding schizogonic divisions are referred to as **exoerythrocytic** or **phanerozoic schizogony**.

It has been reported that metacrypto merozoites produced are of two types. Smaller and more numerous are **micro-metacrypto merozoites**. They enter the red blood corpuscles to start the erythrocytic stage. Larger and less numerous are **macro-metaciypto merozoites**. They invade fresh liver cells to continue the exo-erythrocytic schizogony. Both types of schizogony continue side by side.

Pre and exo-erythrocytic phases of parasite remain immune to the resistance of host and parasites are not susceptible to the action of any anti-malarial drug. Also little damage to the host is done during this stage.

3. Pre-patent and incubation periods. The duration between initial sporozoite infection and first appearance of parasites in blood is termed the **pre-patent period**. It varies from species to species, being of about 8 days in *P. vivax*. The period between infection and appearance of first malarial symptoms is the **incubation period** which is of about 10-17 days (average 14 days) in *P. vivax*.

[IV] Erythrocytic Schizogony

Micro metacryptomerozoites, after escaping into blood stream, invade the erythrocytes or red blood corpuscles. This starts the erythrocytic schizogony which includes the following stages :

1. Trophozoite stage. Inside R.B.C., the micro-metacryptomerozoite becomes rounded and modified into a young **trophozoite**.

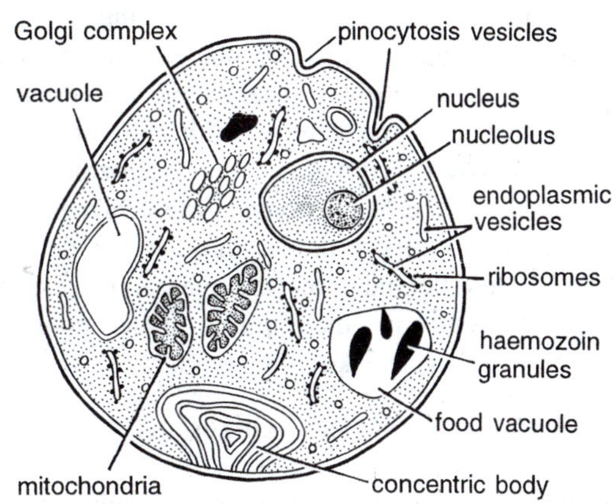

Fig. 3. *Plasmodium.* Electron micrograph of a trophozoite.

2. Signet ring stage. As the trophozoite grows in size, a central vacuole is developed so that the nucleus is pushed to one side into peripheral cytoplasm. This stage is clincially referred to as **signet ring stage** as it resembles the signet ring with the peripherally located nucleus looking like the gem of the ring.

Signet ring trophozoite ingests a large portion of cytoplasm of red blood corpuscle forming a food vacuole into which it secretes digestive enzymes. Besides, a number of pinocytotic vesicles are also formed around the periphery of parasite into which digestion takes place. The enzymes bring about proteolysis of blood haemoglobin, which breaks down into its protein component and **hematin**. Protein is used as food by the trophozoite, while the unused hematin forms the toxic malarial pigment, called **haemozoin**.

3. Amoeboid stage. Meanwhile, the signet ring trophozoite develops into an active, **amoeboid trophozoite**. It sends pseudopodial processes into cytoplasm of the blood corpuscle. At this time, small red eosinophilic granules appear in the cytoplasm of the host corpuscle which are known as **Schuffner's granules**.

4. Schizont. The amoeboid trophozoite, after active feeding, becomes rounded, grows in size and becomes **schizont**. It now undergoes schizogony. Its nucleus divides to form 12-24 nuclei which get arranged at the periphery and cytoplasmic masses surrounding them. Each cytoplasmic mass with one nucleus becomes an oval-shaped **merozoite**. The haemozoin granules are gathered at the centre. With the rupture of the red blood corpuscle, the merozoites are liberated into the blood plasma. These invade fresh corpuscles to repeat the erythrocytic cycle. The 'ghost', left behind after the merozoites escape, are destroyed in the spleen. One complete erythrocytic cycle takes 48 hours in *P. vivax.*

[V] Post-erythrocytic Schizogony

Sometimes some merozoites produced in erythrocytic schizogony reach the liver cells and undergo schizogonic development in liver cells. This is referred to as **post-erythrocytic schizogony**.

[VI] Formation of Gametocytes

When the impulse to multiply asexually by schizogony is exhausted, the merozoites do not proceed ahead with erythrocytic cycle, but, after entering the red blood corpuscles, increase in size to become rounded **gametocytes**. These show sexual dimorphism, being of two types. The male or **microgametocyte** is smaller and contains a large diffused nucleus. The female or **megagametocyte** is larger with a small compact peripheral nucleus. The gametocytes do not divide, but remain as intracellular parasites within their host's blood corpuscles, until they either die or are ingested by the vector, in which they continue their development.

Sexual Cycle of *P. vivax* in Mosquito

[I] Ingestion by Mosquito

When a female *Anopheles* mosquito sucks blood of the infected person, containing gametocytes in R.B.C., any other phases of the parasite and R.B.C.s are digested. The gametocytes however survive. These are liberated and become lodged in the cavity of gut.

[II] Gametogony

Development of gametes from gametocytes is known as **gametogony** or **gametogenesis**. Like gametocytes, the gametes are also of two **types microgametes** and **macro-gametes**. In other animals and most Protozoa, reduction division occurs during the formation of gametes. Thus only gametes are haploid while zygote and other stages are diploid. But in *Plasmodium,* according to some workers (**Bano**, 1959), meiosis or reduction division takes place in the first divisions of zygote. Consequently, only zygote is diploid, while gametes and all other stages of life cycle are haploid, as in higher plants.

1. Microgametes. The male or microgametocyte begins to undergo a process, called **exflagellation**, in the midgut of mosquito. The drop in temperature, due to transfer from warm-blooded human host to cold-blooded insect, provides the stimulation for the process. In each microgametocyte, nucleus divides by mitosis to produce 6-8 haploid daughter nuclei, which assemble at the periphery. The cytoplasm outgrows into long, thin and flagella-like projections and a daughter nucleus enters each one of them. These projections break away as mature male or **microgametes** (sperms). Each measures from 20-25μ in length.

2. Megagametes. Female megagametocyte undergoes some reorganization and becomes a female gamete, that is, **megagamete** or **macrogamete** (ovum) which is ready for fertilization.

[III] Fertilization

The megagamete gives out a small cytoplasmic projection, the **cone of reception** or **fertilization cone**. Nucleus of megagamete comes to lie near its receptive cone. When a lashing microgamete comes in contact with a megagamete, the former penetrates the latter through its receptive cone and **fertilization** or **syngamy** takes place. A complete fusion of nuclei and cytoplasm of two gametes occurs, resulting in the formation of **zygote** with a single diploid nucleus or **synkaryon**. Syngamy is **anisogamous** as the uniting male and female gametes are dissimilar.

[IV] Ookinete

The zygote remains rounded and motionless for some time, but soon it becomes elongated, vermiform and motile. It performs wriggling or gliding movement, and is known as the **vermicule** or **ookinete**. It measures about 15-22μ in length and 3μ in width. It moves exhibiting gregarine gliding-bending movements and peristaltic contractions and comes to lie against the peritrophic membrane of gut.

Electron microscope has revealed the presence of a central irregular **nucleus**, dense cytoplasm, brown pigment granules, many **mitochondria** and **ribosomes** in the ookinete. This suggests that a very rapid synthesis of protein takes place within this stage of the parasite. Motile nature of ookinete is due to the presence of ectoplasmic contractile fibrils, the **microtubules**.

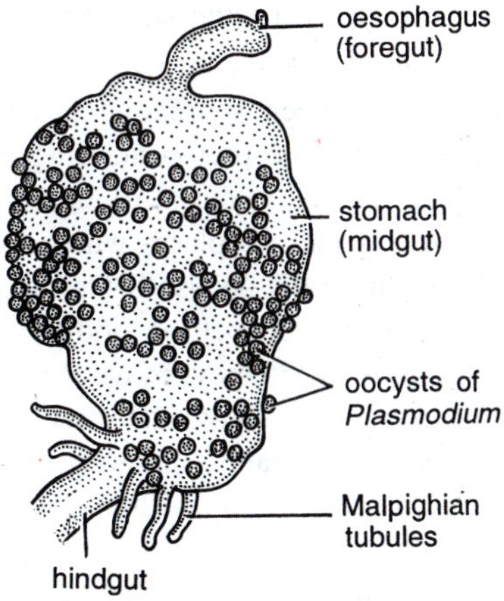

Fig. 4. Stomach or midgut of an infected female *Anopheles* mosquito with oocysts *of Plasmodium.*

[V] Encystment

Ookinete penetrates through the wall of midgut (stomach) to settle down just under the thin membrane that separates midgut from haemocoel. Here it becomes spherical and begins to encyst. The **cyst wall** is thin, membranous and elastic, secreted partly by ookinete and partly derived from mid-gut tissues of mosquito. The encysted zygote is called **oocyst** or **sporont**. The oocysts grow in size and they can now be seen on the outside of midgut or stomach as transparent rounded structures. These may be 50 or more in number.

[VI] Sporogony

Each oocyst now enters a phase of asexual multiplication known as **sporogony**. Its nucleus divides first by meiosis (post-zygotic) and subsequently by mitosis (**Bano**, 1959), forming an enormous number of small haploid nuclei. At the same time, the cytoplasm develops large vacuoles and takes up a sponge-like structure, in which numerous irregular cytoplasmic masses are formed connected by protoplasmic strands. The daughter nuclei arrange themselves along the free margins of cytoplasmic masses. Later, slender and finger-like processes are given out from these cytoplasmic bodies and a daughter nucleus migrates into each of

* **A sporoblast** stage, as found in *Monocystis,* is absent in *Plasmodium* in which the sporozoites are formed directly from the oocysts.

them. In this way, about 10,000 minute, slender and sickle-shaped bodies, called **sporozoites**, are formed from each oocyst*. Each sporozoite has tapering ends and a broad middle part containing a single nucleus. When mature oocysts rupture and are separated from residual cytoplasm which look like a tangled mass within the oocyst. These sporozoits are liberated into haemocoel or body cavity of mosquito. Being motile, they move to different organs in the body of mosquito, but many of them penetrate the salivary glands. In mosquito, whole sexual cycle is completed within 10-20 days depending upon temperature.

The mosquito now becomes infective. According to one estimation, salivary glands of a single infected-mosquito may contain as many as 200,000 sporozoites. When it bites a healthy person, thousands of sporozoites are injected in his blood along with saliva and start the cycle again.

Fig. 5. Ronald Ross (1857-1932). In 1897, he proved that malaria is caused by a protozoan blood parasite (*Plasmodium*) transmitted by female *Anophelese* mosquito. in 1902, he received the Nobel Prize in Medicine and Physiology.

HUMAN MALARIA

Of all the contagious diseases caused by **Protozoa**, malaria diseases are the most destructive for man. They are widespread in the tropics and subtropics and also in certain areas of the temperate zones.

Early History

In the past, malaria was thought to be caused by foul gases emanating from marshes, hence the disease was named **malaria** (Italian; **mala**=bad; **aria** = air) meaning "**bad air**". In 1716, Lancisi established a connection between abundance of mosquitoes and malaria. In 1880, a French doctor, **Charles Laveran**, discovered the malarial parasite *Plasmodium* in human blood. In 1885-86, Golgi studied the erythrocytic schizogony of malarial parasite in man. In 1897, **Ronald Ross** discovered the oocysts of *Plasmodium* on the stomach of female *Anopheles* mosquito. In 1898, Ross worked out the life cycle of avian malarial parasite (*P. paraecox)* in mosquito; while Grassi, Bignami *et. al.* demonstrated the life cycle of human malarial parasite *(P. vivax)*. In 1902, Ronald Ross was awarded the Nobel Prize in Medicine and Physiology. In 1948, **E. Short** worked out the pre-erythrocytic schizogony of *P. vivax* in liver cells. In 1966, **P.C.C. Garnham** wrote a detailed monograph of malarial parasites, while their fine structure was reviewed in 1969 by **M. Rudzinska**

Types of Malaria

The following four different types of human malaria are recognized on the basis of the period of recurrence of fever.

1. Tertian, benign tertian or vivax malaria. The causative agent of this type of Malaria is *P. vivax.* It is characterized by the recurrence of fever every third day, *i.e.,* after 48 hours. This type of malaria does not result in death of the patient. The incidence of the disease is worldwide, mainly in temperate regions.

2. Quartan malaria. *P. malariae* is the causative organism of quartan malaria, which is characterized by the recurrence of fever every fourth day, *i.e.,* at intervals of 72 hours. It is well known for its longevity, 40 years or more in untreated persons. Though it ordinarily does not prove fatal to the patient, the chronic infections sometimes give rise to lethal kidney conditions. The disease is of worldwide occurrence, but it is mainly confined to tropical and subtropical regions.

Table 1. **Comparative account of four species of human infecting *Plasmodium*.**

Characters	*P. vivax*	*P. malariae*	*P. ovale*	*P. falciparum*
1. Geographical distribution	Worldwide, in tropical, sub-tropical and warmer temperate regions.	Worldwide, in tropical and sub-tropical regions.	Tropical Africa.	Worldwide, in tropical, tropical, subtropical and warmer temperate regions.
2. Duration of pre-erythrocytic cycle (pre-patent period)	8 days	7-12 days	9 days	5-6 days
3. No. of metacryptozoites formed per schizont.	10,000	2,000	15,000	40,000
4. Duration of erythrocytic cycle	48 hours	72 hours	48 hours	36-48 hours
5. No. of merozoites formed per schizont.	12-24	6-10	6-12	18-24
6. Incubation period (average)	14 days	18-24 days	14 days	12 days
7. Signet ring form in R.B.C.	About 1/3 diameter of R. B. C. Large ring with vacuole and usually one chromatin dot.	About 1/3 diameter of R.B.C. Usually one ring and one chromatin dot.	About 1/3 diameter of R.B.C. Usually one ring and one chromatin dot.	Small ring situated at edge of R. B. C. Sometimes 2 chromatin dots.
8. Late trophozoite in R.B.C.	Amoeboid and large.	Compact and often band-shaped, small.	Compact and small slightly amoeboid.	Compact and medium-sized. Rarely seen in peripheral blood.
9. Schizont in R.B.C.	Large, 10μ diameter. With 2-24 merozoites.	Medium sized, 7μ diameter. With 6-12 merozoites.	Medium sized, 7μ diameter. With 6-12 merozoites.	Small, 5μ diameter. Not seen in peripheral blood.
10. Gametocytes in R.B.C.	Round. Fill R.B.C. Male 9μ, Female 10-11μ. Pigment granules evenly distributed.	Round or ovoid, Fills R.B.C., 7μ. Pigment granules at centre and periphery.	Round or ovoid. Fill 3/4 of R. B. C., 9μ. Pigment granules, evenly distributed.	Crescentric. Male 9-11μ Female 12-14μ. Pigment granules around nucleus.
11. Host R.B.C.	Enlarged, Red Schuffner's dots usually seen.	Not enlarged. Ziemann's dots seen.	Slightly enlarged, Schuffner's dots seen.	Not enlarged. Greenish Maurer's dots or clefts seen.
12. Pigment (haemozoin)	Yellowish-brown. Fine granules and rodlets.	Dark brown to black. Abundant as coarse granules.	Dark brown. Less abundant coarse granules	Dark brown or black. Coarse granules in a compact mass.
13. Microgametes formed	4 to 8	2 to 5	...	4 to 8
14. Duration of sexual cycle at 25°C.	10 days	25-28 days	16 days	10-12 days
15. Type and effects of Malaria fever	Benign tertian. Death rate low	Quartan. Severe.	Ovale or mild tertian. Severe	Malignant tertian. Death rate high.

3. Ovale or mild tertian malaria. This type of malaria resembles very much to the tertian malaria and is caused by *P. ovale.* The fever recurs every third day or at intervals of 48 hours. Ovale malaria is not greatly harmful and is mainly confined to tropical Africa.

4. Aestivo-autumnal, malignant tertian or pernicious malaria. The causative organism of this malaria is *P. falciparum.* In this also the fever cycle is of 48 hours. The fever is often fatal to the patient as it affects the brain. Its incidence follows that of *P. vivax.*

5. Quotidian malaria. When more than one species of *Plasmodium* infect the patient, or when 2 or 3 generations of parasites mature on successive days, the fever is repeated daily with an interval of 24 hours.

Symptoms and Pathogenesis

Symptoms of malaria first appear several days after the infection of the malaria parasite in man. This interval of time or the **incubation period** is utilized by the parasites to increase their progeny. To establish malarial symptoms, it is necessary that a large number of organisms must continue erythrocytic cycle at a time.

Malaria is quite a dramatic disease. Each attack of fever shows three successive stages :

1. Cold stage. At the onset of malaria fever, the patient suffers from a severe shaking chill. He feels so cold that his teeth chatter though he may be covered with a huge pile of blankets. Cold stage lasts 20 minutes to one hour.

2. Hot stage. As the chill subsides, the body temperature rises as high as 41°C or 106°F. The patient feels very hot with a terrible headache. It lasts one to four hours.

3. Sweating stage. As the temperature lowers down, the patient sweats profusely. Finally, the fever comes down, temperature becomes normal and patient becomes comfortable until the next attack which takes place at regular intervals of 48 hours in case of vivax malaria. The total

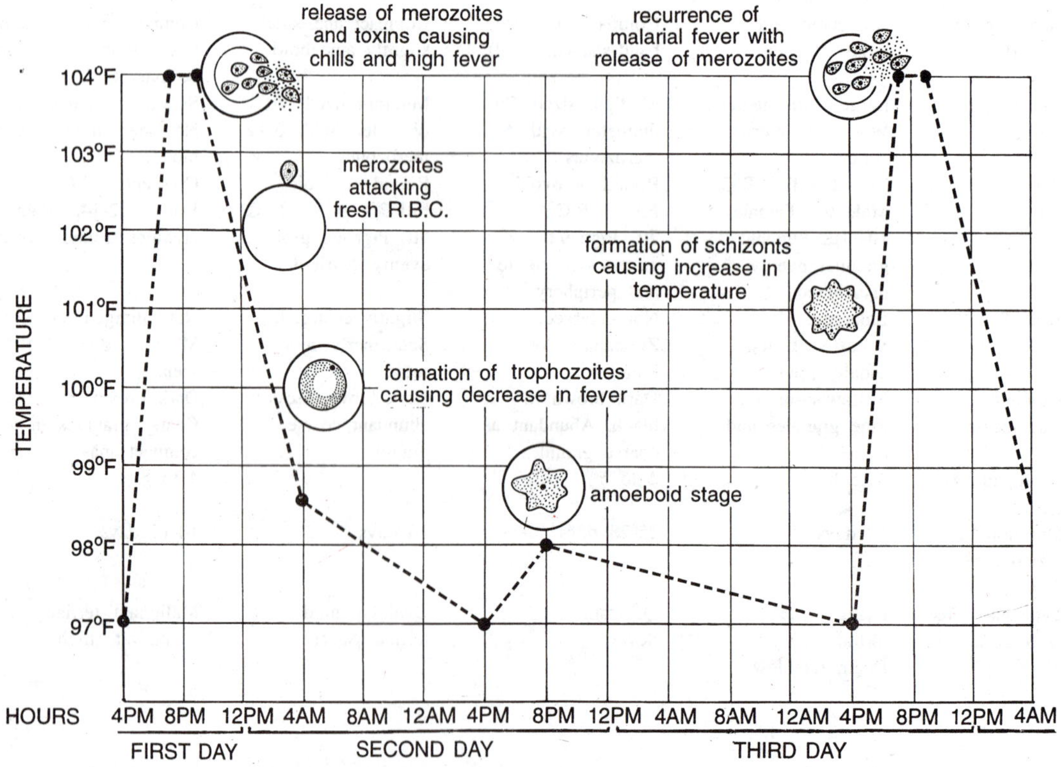

Fig. 6. *Plasmodium.* Fever cycle in vivax malaria.

duration of paroxysm is 6 to 10 hours depending on the species of *Plasmodium.*

Malaria fever occurs when schizonts in red blood corpuscles burst and set free their contained merozoites and malarial pigment **(haemozoin)** in the blood plasma. Bursting of schizonts tends to be synchronous as they all burst at the same time. Haemozoin is said to be toxic and so induces high fever and shivering (Fig. 6).

In infections with *Plasmodium* species, **anaemia** is inevitable. The reasons for the origin of **anaemia** are as follows :

(1) Destruction of erythrocytes on liberation of merozoites.
(2) Infected erythrocytes become more fragile, rupture easily and are destroyed.
(3) The enlarged spleen due to malarial infection releases a lytic substance, **lysolecithin**, which destroys erythrocytes.
(4) Malarial parasites produce **haemolysin** (an antibody), which brings about **haemolysis** of some normal erythrocytes.

In chronic cases, spleen becomes enlarged and fibrosed so much that it ruptures under the influence of trauma.

The falciparum infection often results in thrombosis of visceral capillaries. Death takes place when capillaries of brain become plugged with both the parasites and malarial pigment. Another very serious outcome of the falciparum infection is **black-water fever,** which is characterized by the wholesale destruction of patient's erythrocytes and excretion of liberated haemoglobin in urine.

Duration of Infection

If the patient is not given any treatment and if death does not result rapidly, after a variable number of erythrocytic schizogonic cycles, the host's antibodies will destroy all or most of the erythrocytic schizonts. This will result in the disappearance of the malarial symptoms and occurrence of latency. However, the prarasite persists either in the liver cells (*P. vivax, P. ovale, P. malariae*), or in the capillaries of various viscera (*P. falciparum*). In case of vivax malaria, the infection persists for atleast two years, in quartan malaria for five years and in malignant malaria for six months.

Relapses

Malarial relapses are fairly common in the human malarias except the ovale malaria. In tertian malaria relapses have been observed to occur after 6-12 months of primary infection. The short term relapses have been reported in malignant malaria. In quartan malaria relapses occur after long periods of latency.

Control of Malaria

Malaria has posed a whole series of biological problems. Symptoms of malaria were recognized atleast 2000 years ago. Even at that time people realized that the disease was more prevalent among people, who lived close to marshy grounds. No further progress was made until the 17th century when the bark of *Cinchona* tree, growing in the New World, was found to be excellent for the treatment of this disease. Its active ingredient was **quinine** and this was in use well upto present century. Now medical science has gone a long way to control this disease.

In India, prevention and control of malaria is treated as a national problem and the anti-malaria department of the government deals with it. The war waged on malarial parasite is fought by attacking it simultaneously on three major fronts. In other words, all the control measures fall under the following three categories :

(1) **Elimination** or destruction of vector, *i.e., Anopheles* mosquito.
(2) **Prophylaxis**, *i.e.,* prevention of infection.
(3) **Treatment** of the infected patient.

[I] Destruction *of Anopheles*

If Anopheline mosquitoes are completely destroyed, the infection from one person to another can be checked. It is the most effective and the surest way of controlling malaria. It is achieved by using effective insecticides and by draining swamps to destroy the breeding places of mosquitoes.

1. Destruction of adult mosquitoes. Adult mosquitoes can be most effectively combated in dwellings. The following methods can be employed to kill them :

(a) Killing by hands. The mosquitoes which approach or bite may be killed by hands to save oneself as well as others.

(b) Traps. These are small boxes made of wire gauze and internally lined by black paper or cloth. The mosquitoes tend to enter the boxes and then can be killed by closing them.

(c) Fumigation. Mosquitoes can be driven out of the houses or killed by fumigation. Sulphur, pyrethrum, cresol, tarcamphor or other derivatives of naphtha are burnt to produce poisonous fumes. Smoke of garlic is now considered to be effective in driving the mosquitoes out.

(d) Spraying. Mosquitoes can be killed inside human dwellings by using **mosquito-bombs** or by spraying DDT, flit, pyrethrum and other insecticides.

(e) Sterilization. Sterilization of mosquitoes is now being achieved in some parts of the world, particularly Japan. This method is very effective in controlling mosquito population.

2. Elimination of breeding places. Swampy areas, marshes and stagnant waters, such as pools, ponds, pits and ditches must be drained off as they are breeding places of mosquitoes. If possible, the breeding grounds may be filled up with earth or stones, etc. Domestic species can be largely controlled by eliminating receptacles that hold water, such as tin-cans, buckets, cisterns, barrels, etc. Bushes and shrubs should be cleared off. Open drains should be closed or made underground.

3. Destruction of larvae and pupae of mosquitoes. It is easier and more convenient to kill the aquatic larvae and pupae than the winged adults of mosquitoes.

(a) Proper drainage. The young stages of mosquitoes can be prevented from flourishing by not allowing water to stand in drains, streams, ditches, etc. A speedy flow of water can be maintained, so that the developing forms get away into larger streams and rivers, where they generally die.

(b) Oil screens. Surface of water may be covered by spreading petroleum, paraffin oil, crude oil or kerosene oil. The thin surface oil film prevents the larvae and pupae from carrying on their normal respiration and they die.

(c) Chemical larvicides. Oil solutions or emulsions of DDT, DDD and Benzene hexachloride are effective larvicides. Dusts containing Paris green, DDT or BHC are effective in the control of surface feeding *Anopheles* larvae. Larger bodies of water can be sprayed with these poisonous substances by aeroplanes. Extracts of aquatic weed *Ceratophyllum demersum* control mosquito larvae in ponds. When tulsi (*Oscimum sanctum)* seeds are introduced in water they swell and the mucilage coating attracts and traps the mosquito larvae which eventually die.

(d) Biological methods. Larvicidal fishes, like stickle-backs, minnows *(Gambusia)* and trouts, etc., ducks and aquatic nymphs and adult insects like dragon-flies, which are natural enemies of mosquito, may be introduced in ornamental fountains, ditches, ponds, lakes, canals, tanks, etc. They feed upon mosquito larvae and pupae.

[II] Prevention of Infection

Protection of healthy persons in malaria prone from being infected can be done through the use of new insect repellents, nets, gloves and by screening the bedroom windows.

1. Defence against mosquito bites. Mosquitoes can be prevented from biting by adopting various protective measures, like :

(1) Building houses on high grounds.
(2) Screening of doors, windows and ventilators.
(3) Use of mosquito nets.
(4) Use of anti-mosquito creams (repellents) on exposed parts of the body.

2. Use of prophylactic drugs. Certain anti-malaria drugs, such as **Quinine, Paludrine, Daraprim, Proguanil** and **Pyrimethamine,** if taken in small daily or weekly doses, are effective in keeping off malaria. They kill sporozoites before they become estabilished in liver.

3. Reduction of susceptibility to infection. Chances of infection can be greatly minimized by maintaining proper health. It can be achieved by getting proper nourishment, by avoiding exposure to bad conditions and by following regularity in life.

[III] Treatment of Malaria

The treatment of malaria depdends on the severity of the disease. **Uncomplicated malaria** is treated with oral medications.

Tretment of *P. vivax* cases. Chloroquine remains the treatment of choice for *vivax* malaria. *Vivax* malaria relapses due to the presence of hypnozoites in the liver. The relapse rate in *vivax* malaria in India is around 30%. For its prevention, **Primaquine** may be given under medical supervision.

Treatment of *P. falciparum* cases. Many areas in India have developed resistance to **Chloroquine**. According to WHO guidelines 2010, artemisinin-based combination therapies (ACTs) are recommended as anti-malarial treatments for uncomplicated malaria caused by *P. falciparum.*

What is ACT ? ACT consists of an artemisinin derivative combined with a long acting antimalarial (amodiaquine, lumefantrine, mefloquine or sulfadoxine-pyrimethamine). **The ACT used in the national programme in India is artesunate + sulfadoxine-pyrimethamine (SP)**. Presently, **Artemether + Lumefantrine** fixed dose combination and blister pack **artesunate + mefloquine** are also available in the country.

Treatment of complicated (Severe) malaria. In severe *falciparum* malaria, it is recommended that rapid clinical assessment and confirmation of the diagnosis is made (by OPtiMAL) followed by administration of **intravenous** (IV) or **intramuscular** (IM) **Artesunate Injection** for a minimum of 24 hours. Quinine is an acceptable alternative if artesunate injection is not available. Therefore, it is recommended to complete treatment under medical supervision.

Anti-malaria Campaign in India

About two decades ago malaria was the most widespread disease in India. With the assistance of World Health Organization (WHO), the Ministry of Health of the Government of India launched a National Malaria Control Programme in 1962 and almost controlled it by destroying the mosquitoes with D.D.T. and other insecticides. For the last few years the mosquitoes have appeared again in plenty and the malaria infection has begun. The D.D.T. seems to have become ineffective as mosquitoes have developed immunity for it. Also a change in the behaviour of the mosquitoes has been noted. They now bite and rest outdoors instead of biting humans indoors and resting on the walls. The **Vector Control Research Center**, Pondicherry, of the Indian Medical Research Board is now busy in finding out the alternative means for the eradication of mosquitoes. It has discovered a species of water-bug (fam. Notonectidae) which feeds on the eggs and larvae of mosquito. The **National Institute of Communicable Diseases**, Delhi, and the **Postgraduate Medical Institute**, Chandigarh, are busy in developing the **vaccines** for malaria. Two such vaccines are being testified.

With new strategies adopted by Government of India such as increased financial allocation to the anti-malarial operations and the treatment of every fever case with drugs readily made available, there has been decline in the incidence of malaria recently.

Important Questions

➤➤*Long answer type questions*

1. Give an account of the life history of the malarial parasite in man and suggest methods for controlling malaria.
2. What do you understand by 'Intermediate Host' ? Describe the life cycle of the malarial parasite in mosquito.
3. Give an illustrated account of the life cycle of *Plasmodium vivax.*
4. How does malaria spread from one man to another ? What steps would you suggest to protect your household from this disease ?
5. Describe briefly the life-history *of Plasmodium* and compare it with that of *Monocystis.*
6. Write an essay *on Plasmodium* and malaria.
7. Write short notes on: (i) Control of malaria, (ii) Exoerythrocytic cycle, (iii) Incubation period, (iv) Ookinete, (v) Prepatent period, (vi) Schizogony, (vii) Sporogony.

➤➤*Short answer type questions*

1. Mention the names of human body cells which are prove to the attack of *Plasmodium.*
2. Mention the four species of *Plasmodium* and the types of malaria caused by them.
3. What is meant by signet-ring stage ?
4. To which order does malarial parasite belong ?
5. Define schizogony.
6. How is malaria transmitted from one person to the other ? Give an illustrated description of the cycle of Ross.
7. What are the Tertian and Quartan agents ? What are their causative agents ?
8. Write a brief account of the contributions made by Ronald Ross giving an approximate date.
9. Distinguish between the schizont and cryptoschizont stages of *Plasmodium.*
10. Differentiate between pre-patient and incubation period.
11. Compare the micro- and macro-gametocytes of *Plasmodium.*
12. Distinguish between four species of human malarial parasites based on the type of malaria they produce.
13. Describe the sexual cycle of the malarial parasite.
14. Draw a labelled diagram to illustrate sporogony in *Plasmodium.*

➤➤*Multiple choice questions*

1. The shivering characteristic of malaria occurs when :
   (a) merozoites are liberated from RBC with toxin
   (b) when schizonts enter the RBC
   (c) when the schizonts are on the reticulo-endothelial cells
   (d) when the signet ring is formed
2. Besides R.B.C. *Plasmodium* attacks one more type of human body cells during its normal course of life cycle. Name that human body cells :
   (a) liver cells (b) muscle cells
   (c) columnar cells of intestine
   (d) nerve cells
3. What is cryptomerozoite of *Plasmodium* ?
   (a) product of sporogony in the gut of mosquito
   (b) product after sexual fusion of gametes in the gut of mosquito
   (c) product of schizogony
   (d) product of hologamy in RBC
   (e) product of gametogamy in RBC
4. The intermediate host of malarial parasite is :
   (a) man (b) female *Anopheles*
   (c) *Culex* (d) tse-tse fly
5. In the life history of *Plasmodium* exflagellation occur in the :
   (a) merozoite (b) microgametocyte
   (c) macrogametocyte (d) cryptozoite
6. Malaria is transmitted by the :
   (a) female *Culex* mosquito
   (b) male *Anopheles* mosquito
   (c) female *Anopheles* mosquito
   (d) male *Culex* mosquito
7. Vegetative phase of *Plasmodium* in R.B.C. of man is called :
   (a) sporozoite (b) trophozoite
   (c) merozoite (d) cryptozoite
8. One of the following species of *Plasmodium* is not found in India :
   (a) *falciparum* (b) *ovale*
   (c) *malariae* (d) *vivax*
9. *Plasmodium* is the causative agent of malaria was discovered by :
   (a) Lavern (b) Ross
   (c) Golgi (d) Lamble
10. Which species of *Plasmodium* show longest incubation period ?
    (a) *vivax* (b) *falciparum*
    (c) *malariae* (d) *ovale*
11. The malarial parasite is introduced into the blood of man as a :
    (a) metacryptozoite (b) crypotozoite
    (c) schizont (d) sporozoite
12. *Plasmodium* belongs to the class :
    (a) ciliophora (b) rhizopoda
    (c) mastigophora (d) sporozoa
13. Trophozoite of *Plasmodium* lives in :
    (a) red blood corpuscle (b) lymph
    (c) liver (d) blood plasma
    (e) lung cavity
14. Malaria is caused by :
    (a) *Ascaris* (b) foul air
    (c) *Plasmodium* (d) mosquito
15. Shivering in malaria occurs when :
    (a) liver cycle is completed
    (b) merozoites and toxins are released into blood
    (c) gamonts are formed
    (d) mosquito ingests infected human blood
16. The infective stage of *Plasmodium* in man is :
    (a) schizont (b) merozoite
    (c) sporozoite (d) trophozoite
17. Infective stage of *Plasmodium* is :
    (a) trophozoite (b) ookinete
    (c) sporozoite (d) schizont
18. The formation of gametocyte of malarial parasite takes place :
    (a) in the stomach of female *Anopheles*
    (b) in the stomach of male *Anopheles*
    (c) in the blood of man
    (d) in the salivary gland of *Anopheles*

19. Schizont is a stage in the life history of malarial parasite occurring in :
(a) red blood cells of man
(b) stomach of *Anopheles*
(c) salivary glands of *Anopheles*
(d) blood of *Anopheles*
20. Sporogony of malarial parasite occurs in the :
(a) liver of man (b) R.B.C. of man
(c) stomach wall of mosquito
(d) salivary gland of mosquito
21. Malignant malaria is due to infection of :
(a) *Plasmodium ovale* (b) *P. faliciparum*
(c) *P. malariae* (d) *Wuchereria bancrofti*
22. Prepatent period is :
(a) time required by sporozoite to become cryptomerozoite
(b) time required to complete I liver cycle
(c) 10 days (d) all of these
23. In the life-cycle of *Plasmodium vivax,* the microgametocyte undergoes exflagellation in the :
(a) stomach of the female *Anopheles*
(b) erythrocyte of man
(c) liver cells of man
(d) blood plasma of man
24. The erythrocytic phase in the life-cycle of *Plasmodium vivax* is completed in :
(a) 24 hours (b) 36 hours
(c) 48 hours (d) 72 hours
25. Asexual reproduction during schizogony of malarial parasite is :
(a) sporulation (b) binary fission
(c) multiple fission (d) amitosis
26. In life cycle of malarial parasite, amoeboid stage is found :
(a) inside the gut of mosquito
(b) inside the cyst attached to the wall of mosquito stomach
(c) inside the liver cells of man
(d) in RBCs of man
27. The zygote of *Plasmodium vivax* divides repeatedly to produce :
(a) sporozoites (b) cryptomerozoites
(c) micrometacryptomerozoites
(d) metacryptozoites
28. In life cycle of *Plasmodium vivax* after a number of erythrocytic cycles, some merozoites invading fresh RBCs grow to :
(a) macrometacryptozoites
(b) schizozoites
(c) gametocytes (d) signet ring stage
29. The incubation period of *Plasmodium vivax* is :
(a) 48 hours (b) 72 hours
(c) 120 hours (d) 240 hours
30. In life history of malaria the cells produced by Schizogony are :
(a) sporozoites (b) schizonts
(c) cryptozoites (d) merozoites
31. Life cycle of Malarial Parasite is :
(a) monogenetic (b) digenetic
(c) not in *Anopheles* (d) none of these
32. Infective stage of *Plasmodium* to man is :
(a) gamont (b) sporozoite
(c) trophozoite (d) merozoite
33. Schizont of *Plasmodium* is found in :
(a) liver (b) blood
(c) both (d) mosquito
34. The stage of *Plasmodium* that appears in human blood during night only :
(a) schizont (b) sporozoite
(c) gamonts (d) microgametocyte
35. Charles Laveran :
(a) confirmed discovery of malaria
(b) discovered malaria (c) first observed oocysts
(d) described life cycle of *Plasmodium*
36. Black water fever is caused by :
(a) *Plasmodium ovale* (b) *Plasmodium falciparum*
(c) *Plasmodium vivax* (d) *Plasmodium malariae*
37. *Plasmodium* is a parasite of human :
(a) blood (b) liver
(c) spleen (d) bone marrow
38. *Plasmodium* belongs to class sporozoa which is characterised by :
(a) pseudopodia (b) myonemes
(c) no distinct locomotory organelles
(d) flagella
39. Asexual reproduction called sporogony occurs in *Plasmodium :*
(a) in man only
(b) in mosquito
(c) both in man and female mosquito
(d) in female mosquito only
40. Sporozoite of *Plasmodium* is :
(a) amoeboid and uninucleate
(b) pelliculate, uninucleate and with numerous mitochondria
(c) capable to pass across a blood capillary
(d) multinucleate
41. Prepatent period in Malarial infection is :
(a) interval between inoculation and fever
(b) duration between inoculation and Start of blood cycle
(c) interval between two liver cycles
(d) duration between repetition of fever
42. What is cryptomerozoite of *Plasmodium* ?
(a) a product of gametogamy
(b) a product of schizogony
(c) a product of sporogony in mosquito
(d) none of these
43. An intracellular parasite is :
(a) *Entamoeba* (b) *Ascaris*
(c) *Trypanosoma* (d) *Plasmodium*
44. Most dangerous species of *Plasmodium* is :
(a) *malariae* (b) *falciparum*
(c) *ovale* (d) *vivax*

45. In India, urban malaria is transmitted by :
(a) *Culex quinquefasciatus*
(b) *Anopheles annularis*
(c) *Anopheles subpictus* (d) *Anopheles stephensi*

46. What is the duration of incubation period of *P. falciparum* :
(a) 12 days (b) 10 days
(c) 14 days (d) None of the above

47. Erythrocytic cycle of *P. vivax* completed in :
(a) 24 hours (b) 48 hours
(c) 72 hours (d) 96 hours

48. Active feeding stage of *Plasmodium* :
(a) trophozoite (b) sporozoite
(c) merozoite (d) metacryptozoite

49. Malaria parasite can best obtained from its patient :
(a) one hours before rise the temperature
(b) five hours after temperature
(c) when temperature rise with rigor
(d) any time

50. Schuffner's dots in a patient of malaria, are found :
(a) RBC (b) gametocyte of *Plasmodium*
(c) liver cells (d) signet ring trophozoite

51. Toxic material formed after proteolysis of Hb :
(a) haematin (b) hemozoin
(c) heam (d) Schuffner's dot

52. Sexual phase in the life cycle of *Plasmodium* occurs in :
(a) gut of mosquito (b) salivary gland of mosquito
(c) blood of man (d) body cavity of mosquito

53 The malarial parasite was discovered in the RBC's of human beings by :
(a) Haldane (b) Laveran
(c) Ronald Ross (d) Stephans

54. Sir Ronald Ross discovered that :
(a) malaria is caused by Protozoan
(b) malaria is caused by foul air
(c) four species of *Plasmodium* cause malaria
(d) malaria is transmitted by *Anopheles*

55. Life history of *Plasmodium* in *Anopheles* was first described by :
(a) Grassi (b) Ronald Ross
(c) Laveran (d) Lancisi

56. Which set of species belong to same genus ?
(a) *sapiens* and *cristatus*
(b) *proteus* and *gingivalis*
(c) *coli* and *vivax*
(d) *vivax* and *falciparum*

57. Which one caries malarial parasite to the host ?
(a) Male *culex* (b) Male *Anopheles*
(c) Male *Aedes* (d) None of the above

58. What is meant by intracellular parasite ?
(a) paraite on cell surface
(b) parasite between cells
(c) parasite around cell
(d) parasite inside cell

59. Which one of the following is the correct matching of the causative germ and the disease it causes ?
(a) *Wauchereria bancrofti*- Gambia fever
(b) *Anopheles maculipenis*- Malaria
(c) *Leishmania donovani*- Sleeping sickness
(d) *Glossina palpalis*- Kala azar

60. In India, urban malaria is transmitted by :
(a) *Culex quinquefasciatus*
(b) *Anopheles annularis*
(c) *Anopheles subpictus*
(d) *Anopheles stephensi*

Answers

➤➤*Multiple choice questions*

1. (a)	2. (a)	3. (c)	4. (b)	5. (b)	6. (c)	7. (b)	8. (b)	9. (a)	10. (c)
11. (a)	12. (d)	13. (a)	14. (c)	15. (b)	16. (c)	17. (a)	18. (b)	19. (a)	20. (c)
21. (b)	22. (a)	23. (a)	24. (c)	25. (c)	26. (d)	27. (a)	28. (c)	29. (a)	30. (c)
31. (b)	32. (c)	33. (a)	34. (a)	35. (b)	36. (a)	37. (a)	38. (c)	39. (d)	40. (c)
41. (a)	42. (b)	43. (a)	44. (b)	45. (d)	46. (a)	47. (b)	48. (a)	49. (c)	50. (a)
51. (b)	52. (a)	53. (b)	54. (d)	55. (a)	56. (d)	57. (d)	58. (d)	59. (b)	60. (d)

10

Monocystis agilis

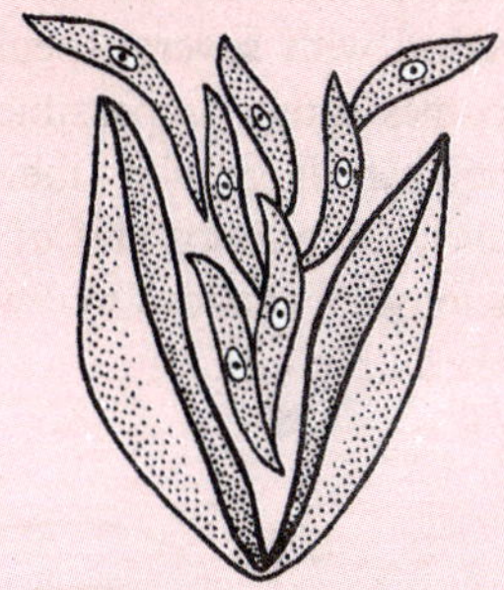

Unlike digenetic *Plasmodium,* there are some sporozoan types which are parasitic in one host only. The **gregarines**, parasitic in invertebrates and ascidians, fall into this category. These are most commonly found in the coelomic epithelial cells, gut and reproductive organs of their hosts. Among those gregarines that exhibit sexual reproduction only, two forms are recognized, **acephaline,** and **cephaline**. In **acephaline** forms, body is not divided into chambers and anterior end does not bear an organ for attachment. **Cephaline** forms possess the chambered body as well as the organ for attachment. *Monocystis* (Gr., *monos,* single + *kystis,* bladder) is an acephaline gregarine. The most common species are *M. agilis, M. lumbrici, M. magna, M. beddardi, M. pheretimi, M. lloidi* and *M. bengalensis.* A generalized account of the genus *Monocystis* has been given here, with particular reference to *M. agilis.*

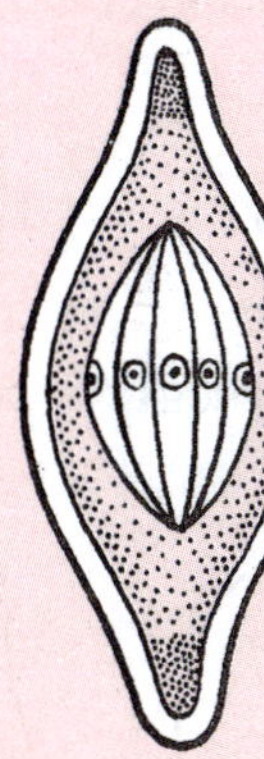

Monocystis agilis

SYSTEMATIC POSITION

Phylum	Protozoa
Subphylum	Sporozoa
Class	Telosporea
Subclass	Gregarinia
Order	Eugregarinida
Genus	*Monocystis*
Species	*agilis*

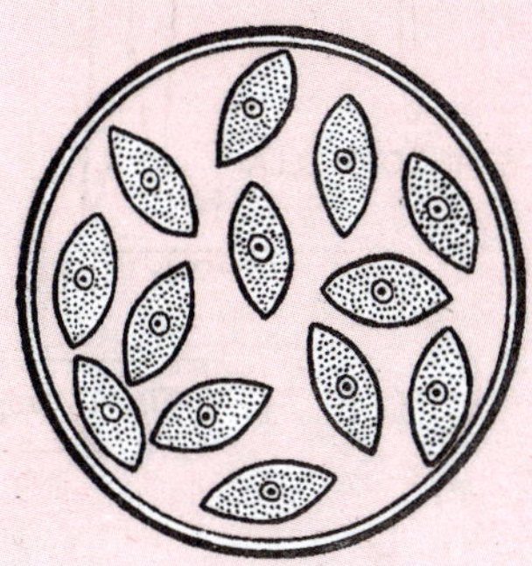

HABITS AND HABITAT

Species of *Monocystis* are typically endoparasites of earthworms and occur in their coelom and seminal vesicles. Besides *Monocystis,* there are also found the life cycle stages of other aliied genera, such as *Enterocystis, Rhabdocystis, Apolocystis, Nematocystis, Dirhynchocystis, Dendrocystis, Cephalocystis, Rhynchocystis, Zygocystis, Pleurocystis, Stomatophora, Choanocystella,* and so on. When an earthworm is infected with several species of these parasites, it becomes rather impossible to relate the stages of the parasites with their respective genera. However, the trophozoites of different genera have characteristic shapes by which they may be identified (Fig. 1).

STUDY OF *MONOCYSTIS* IN LABORATORY

Monocystis is easily obtained for study in the laboratory. Get a living, sexually mature earthworm and narcotise it by adding a few drops of **chloretone** in water containing the animal. Pin it down and then make a slit in the dorsal body wall from about X to XV segments. You will observe yellowish-white bodies extruding through the slit. These are the seminal vesicles. Pick up these bodies with forceps and keep in a watch-glass in 0.7% NaCl solution. Make a paste of the seminal vesicles with **corrosive sublimate** (mercuric chloride in water). Spread a drop of the paste on the cover-slip which should then be placed on the glass slide. Examine it under the microscope for trophozoites, cysts, spores and other developmental stages of the parasite. A permanent slide can also be prepared by the usual method of staining and dehydration, etc.

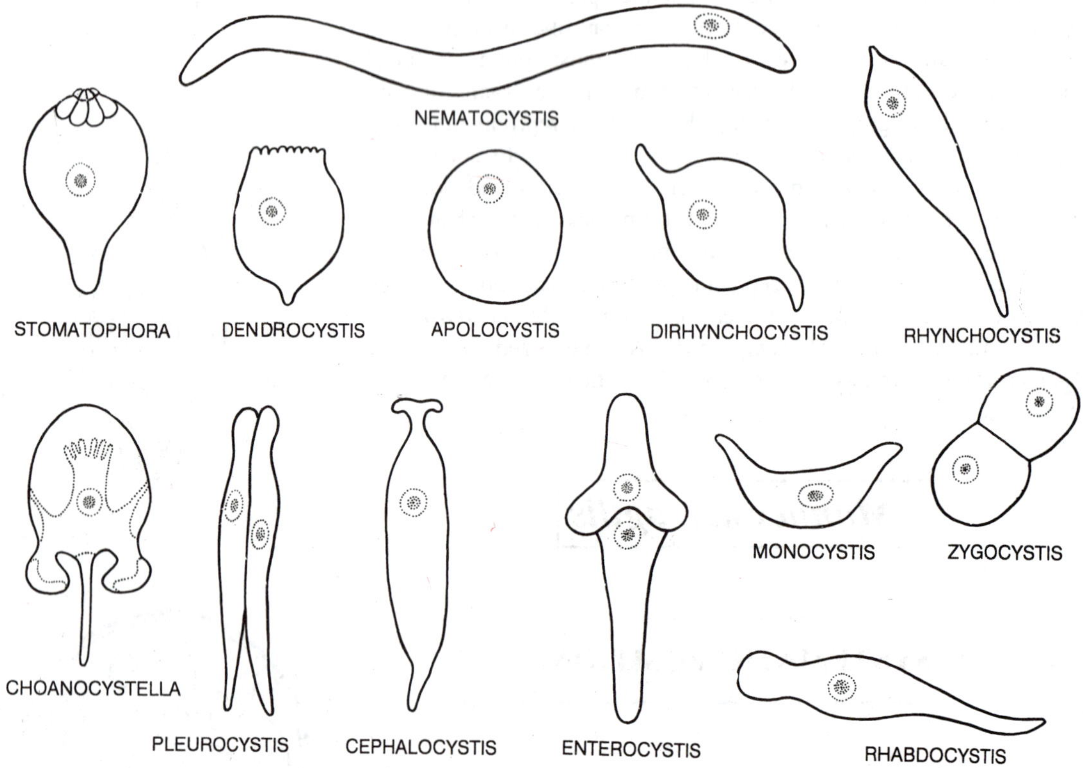

Fig. 1. Trophozoites of a few allied genera of *Monocystis* and their shapes.

STRUCTURE OF TROPHOZOITE

Trophozoite is the adult and feeding stage of parasite. It is found in seminal vesicles of earthworm. It develops within the sperm ball (**sperm morula**) which is a group of developing **sperms** or **spermatozoa** of the host. Sometimes, it is seen surrounded by a thin layer of degenerating sperm morula, to which the tails of dead spermatozoa are attached giving it a strongly ciliated appearance. When sperm tails are separated, the trophozoite becomes free.

[I] Shape and Size

Trophozoite is microscopic and unicellular. The actual shape and size differ according to the stage of development. The youngest stages are usually rounded or oval and about 5 μ long. A full grown mature trophozoite is elongated, spindle-shaped, flattened and worm-like organism with tapering body ends. It may reach a size of about 500μ long and 40μ broad and is visible to the naked eye.

[II] Pellicle

Body of trophozoite is bounded externally by a definite thick and firm **pellicle**. It may be modified in various ways, *i.e.,* it may be striated or thrown into ridges and furrows. It contains longitudinally arranged contractile **microtubules**.

[III] Cytoplasm

The cytoplasm has a sharp differentiation between outer clear **ectoplasm** and inner granular **endoplasm**. Ectoplasm is further divided into three layers : (i) outer epicyte, (ii) middle **sarcocyte** and (iii) inner **myocyte**. The innermost layer, or myocyte consists of longitudinal and transverse contractile fibrils or **myonemes**, which are supposed to enable the trophozoite to perform its characteristic metabolic movements.

The endoplasm contains in abundance the granules of **paraglycogen** (a special form of glycogen), fat globules and sometimes **volutin**, a protein rich in phosphorus and nucleic acid.

Electron micrograph of *Monocystis* has shown the presence of Golgi bodies and mitochondria as has already been noted in various protozoans. The mitochondria are more in number at the periphery.

[IV] Nucleus

Nucleus is single and vesicular with spherical or ellipsoidal form. It contains usually one **nucleolus** or **karyosome** (sometimes more), which stains deeply with iron haematoxylin. Nuclear membrane is delicate and bears pores. Nucleoplasm contains four chromosomes (haploid number).

PHYSIOLOGY

[I] Locomotion

Monocystis does not possess special organs for locomotion. It shows wriggling or gliding movements brought about by the rhythmic contractions and relaxations of myonemes. Such a type of movement, characteristic of a monocystid gregarine, is known as **gregarine movement**. This is similar to the **euglenoid movement** of *Euglena.*

[II] Nutrition

Nutrition in *Monocystis* is **saprozoic**. It absorbs its nourishment from the fluid of the seminal vesicle by osmotrophy through the general surface of the body. It is quite likely that it may be secreting digestive enzymes upon the outer fluid and rendering the protoplasmic particles soluble for their absorption. In some species, like *M. buccalis,* the anterior end of the body is modified for ingesting the nutrient from its vicinity. Reserve food material is stored in endoplasm in the form of **paraglycogen** granules.

[III] Respiration

The actual mechanism of respiration in *Monocystis* is not known. It is believed that the parasite obtains oxygen by diffusion through its pellicle from the cell contents of host's sperm morula, when the trophozoite is within it, or from the fluid of the seminal vesicle. The mitochondria synthesize respiratory enzymes for oxidation reactions involved in aerobic conversion of pyruvic acid to CO_2 and water.

[IV] Excretion

Carbon dioxide and nitrogenous waste products, produced as a result of metabolism, are eliminated by diffusion into the surrounding fluid.

REPRODUCTION AND LIFE CYCLE

Monocystis is **monogenetic**, *i.e.,* its life cycle is completed in a single host which is earthworm. Only zygotes are diploid while all other stages (gamonts, gametes, etc.) are haploid. Asexual multiplication by schizogony does not occur. The life cycle proceeds as follows :

[I] Gamontogamy

This is a method of sexual reproduction which involves the pairing of gamonts, formation of gametes and fertilization. It takes place as follows :

1. Syzygy. After a period of feeding, growing and wandering about, each trophozoite becomes an oval reproductive body, the **gamont** or **gametocyte**. The gametocytes become shortened and spherical and associate in pairs. The two then secrete a common resistant and protective cyst wall, the **gametocyst** or **gamontocyst**, around themselves. The cyst wall has two layers, an outer thick and rigid and an inner thin **endocyst**. Within the cyst wall, the gametocytes never fuse or conjugate. This type of pairing of the gametocytes is referred to as **syzygy**.

2. Gametogony. Within the gametocyst, each gametocyte undergoes nuclear multiplication by mitosis. The nuclei, thus formed, move to the periphery and each gets surrounded by a small amount of cytoplasm (cytomere). These uninucleate cytoplasmic bodies become free as **gametes**, leaving a certain amount of **residual cytoplasm**

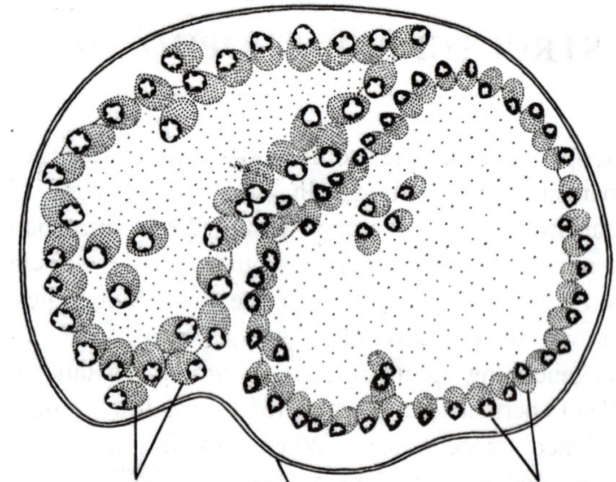

Fig. 2. *Monocystis.* Formation of anisogametes within a gametocyst.

in the centre of the gametocyst. The gametes are **anisogametes**, and contain four chromosomes (haploid number) each. All gametes from one gametocyte are of the same sex.

> According to Hahn, the gametes are of unequal size and shape *i.e.,* the gametes are **anisogamous**. The male or microgametes are small and pointed and the female or macrogametes are large and rounded (Fig. 3). According to **M. A. Sleigh** (1973), the gametes are **isogametes**.

3. Syngamy. The gametes mingle within the gamontocyst and fuse in pairs to produce diploid zygotes. It is presumed that the two gametes uniting in a pair come from different gametocytes of opposite sexes. Numerous diploid zygotes are formed within a single gametocyst.

[II] Sporogony

Each zygote, which is first spherical, transforms itself into a unicellular oval body, called **sporoblast**. It soon secretes a thick cyst wall, the **sporocyst** or **zygocyst**, around itself and becomes a **spore** which is biconical in shape. It possesses a mucoid plug at either end. The spore now acquires the characteristic form of a boat or canoe resembling a diatom of the genus *Navicella*. This stage is therefore frequently referred to as **pseudonavicella** (Gr., **pseudo**, false + **navum**, boat). Within the sporocyst the spore undergoes three successive nuclear divisions, producing eight daughter nuclei, the first division being the

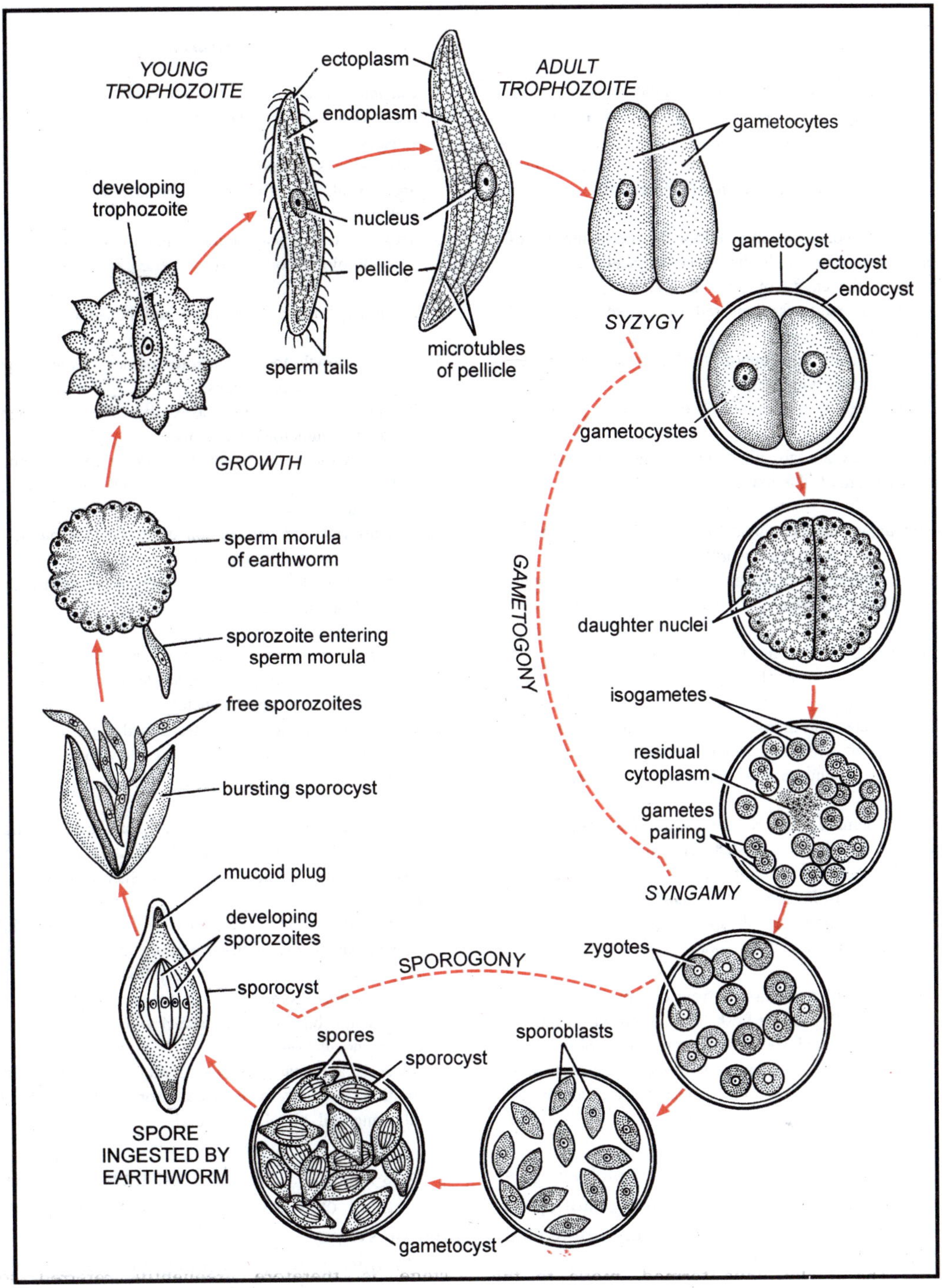

Fig. 3. *Monocystis.* Life cycle.

Table 1. **Comparison of *Monocystis* and *Plasmodium*.**

Monocystis	*Plasmodium*
1. Monocystis is a parasite in seminal vesicles or in coelomic cavity of posterior segments of earthworm.	1. *Plasmodium* is a blood parasite living in the liver and red blood cells of man or gut of female *Anopheles* mosquito.
2. Life cycle is simple, involving a single host (earthworm). No alternation of generations.	2. Life cycle is complicated involving two phases-asexual and sexual. Iternating in two hosts a vertebrate (man) and an invertebrate (mosquito).
3. Young trophozoite is intracellular in sperm morula of earthworm, but adult becomes extracellular, living freely in the cavity of seminal vesicles.	3. Trophozoite is always intracellular living either in liver cells or blood corpuscles of vertebrate host.
4. Adult trophozoite is elongated worm-like, active and motile.	4. Adult trophozoite is rounded and non-motile.
5. During growth, the trophozoite does not pass through ring and amoeboid stages.	5. During growth the trophozoite passes through signet ring and amoeboid stages.
6. Trophozoites do not multiply asexually by schizogony.	6. Trophozoites multiply asexually by schizogony producing numerous merozoites.
7. Gametocytes are all alike and formed directly by a modification of the mature trophozoites.	7. Gametocytes are differentiated into male and female and formed by a modification of some of the merozoites.
8. Gametogony and conjugation or syngamy take place in the same host (earthworm).	8. Gametocytes and conjugation occur when gametocytes are transmitted into the gut of mosquito.
9. Before gametogony, gametocytes arrange in pairs (syzygy) and secrete a common cystwall around each pair.	9. Gametocytes do not associate in pairs (syzygy) and do not undergo encystment.
10. Gametes formed by the two gametocytes of each pair are gamous, non-motile and many.	10. Gametes are anisogamous. A macrogametocyte forms a single motile macrogamete, while a microgametocyte produces 2 to 8 microgametes.
11. Zygotes are non-motile, each secreting a sporocyst around itself while still inside the original gametocyst, and thus turning into a sporoblast and later into a spore.	11. Zygotes are motile which bore into the gut wall of the host and undergo encystment to form oocysts. Sporoblasts are not formed.
12. A large number of spores are produced having a thick resistant covering to withstand unfavourable conditions of the outside world and the high mortality, during transference from one host to another.	12. A smaller number of oocysts are produced having a thin and non-resistant covering because these never pass to the outside world to face dangers and destruction.
13. Each spore produces 8 sporozoites which remain protected inside the spore membrane during transmission.	13. Each oocyst forms hundreds of sporozoites which remain naked, *i.e.*, without a spore membrane because they never live outside the host body.
14. Exact mode of transmission from one host to another in unknown. Probably a fresh 'host' worm ingests them with contaminated soil or droppings of a bird.	14. Transmission from one man to another takes place through the bite of the female *Anopheles* mosquito by direct inoculation.
15. There is no prepatent period as the sporozoites do not develop until they reach the seminal vesicles and enter the germ cells of earthworm	15. There is a prepatent period during which sporozoites multiply asexually in liver cells for a few days before attacking red blood corpuscles.

reduction division. Each daughter nucleus gets surrounded with a cytoplasmic mass, resulting in the formation of eight minute, elongated and sickle-shaped haploid **sporozoites** or **zoites**. These are bundled together within the sporocyst. The 8 sporozoites or zoites can be counted only in cross-sections. All these changes occur within the original cyst wall (gametocyst) formed at the time of syzygy.

At this stage the original cyst wall ruptures and spores are liberated into the cavity of seminal vesicle of earthworm. Rupture of original cyst wall

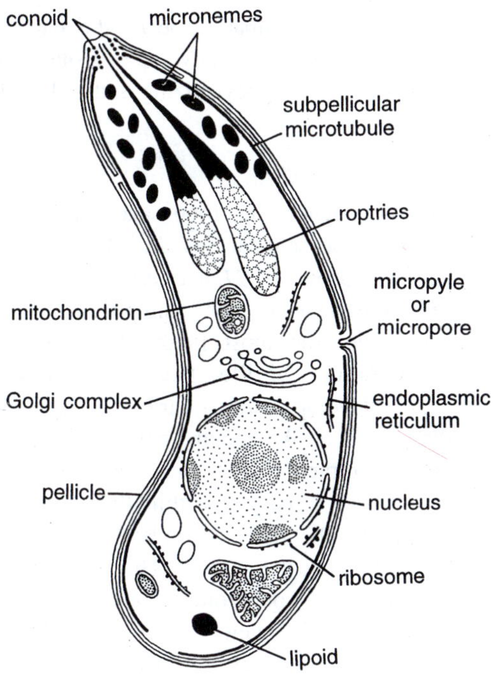

Fig. 4. *Monocystis.* Sporozoite as revealed by electron microscope (diagrammatic).

may be assisted by the swelling of residual cytoplasm. Liberated spores wait for their transmission to a new earthworm host. In those species of *Monocystis,* which undergo reproduction and development in coelomic cavity, the spores are found in posterior segments of body.

[III] Transference

The exact manner in which the spores of *Monocystis* are transmitted from one worm to another is not known with certainty. It may take place in any of the following ways :

1. During copulation. Occasional presence of spores in cocoons and sperm ducts of earthworms had led to the presumption that they get transferred from one host to another during copulation along with spermatozoa and seminal fluid. However, there seems to be no direct proof in support of this theory.

2. Death of host. When the host-worm dies and decays, the spores get scattered in soil. When infected soil is eaten by another worm, these also enter its alimentary canal.

3. By birds. When infected worm is devoured by a predator, such as a bird, the spores would pass out unaltered with its excrement or droppings and subsequently swallowed up by other worms. Spores have been actually found in the gut contents of various birds.

4. Autotomization. In certain species of *Monocystis,* spores are found in coelomic cavity of posterior segments of earthworm. When posterior body is autotomized, the spores are liberated in soil. With ingestion of contaminated soil, by a fresh earthworm, spores too are ingested. Within the alimentary canal of a new host, the sporocyst breaks or possibly dissolves by the action of the digestive enzymes liberating sporozoites into lumen of alimentary canal.

[IV] Sporozoites

The sporozoites are minute spindle-shaped protoplasmic bodies resembling in some detail the sporozoites of *Plasmodium.* Each contains a single nucleus, one or more mitochondria, Golgi complex and granules of reserve food, chiefly carbohydrate. Outer pellicle has longitudinally arranged sub-pellicular contractile microtubules. At the anterior region is a pair of secretory organelles or **roptries**. Their secretion helps in penetration through the tissues. Anterior end also bears **conoids** and **micronemes**, whose functions are not well understood.

[V] Invasion of Seminal Vesicles

With the help of contractile action of sub-pellicular micro-tubules, the sporozoites make active movements. They make their way into the epithelial cells of the gut mucosa, one sporozoite entering one cell. Here they grow in size and then escape into the seminal vesicles. In the seminal vesicles, each sporozoite enters a central cell of the sperm morula (a group of sperm mother cells).

[VI] Development of Trophozoite

In the sperm morula, the sporozoite feeds and grows into a young **trophozoite**. It exhibits hair-like sperm tails around its body, which are the remnants of degenerating sperms. With the degeneration of the sperm tails as well, the trophozoite becomes free in the cavity of the seminal vesicle as an adult trophozoite. Thus, the trophozoites are at first intracellular and later extracellular.

EFFECT OF PARASITISM

[I] On the Host

Although every worm is infected with *Monocystis,* its fertility is not greatly impaired, since most of the seminal vesicles are not involved. The trophozoites invade and grow inside sperm morulae, inhibiting the development of spermatogonia and spermatids. The worm is apparently able to combat some of the parasites by forming a resistant envelope around the trophozoites or by phagocytosizing and killing the resistant spores.

[II] On the Parasite

With an abundance of food and almost no competition inside seminal vesicles of earthworm, the adult trophozoite of *Monocystis* undergoes an extreme retrogression or simplification of structure. Under the most favourable, easy and uniform conditions of life, the parasite has no need to develop the various vegetative organelles, such as the mouth, gullet, locomotory structures, contractile vacuoles, etc. But the entire life cycle is passed within the same host and to continue the race, transmission to another host worm becomes essential. This is done by the production of sporozoites contained within spores having resistant coverings. During transmission the spores pass through the outside world, where danger of enemies and competition around, resulting in their large scale destruction. To compensate this high mortality, huge numbers of potential individuals are produced to ensure that a sufficient number may survive to continue the race.

Important Questions

➤➤*Long answer type questions*

1. Describe the structure, physiology and life history of *Monocystis.*
2. Compare the main events in the life cycle of *Monocystis* and *Plasmodium.*
3. Write short notes on : (i) Gametogony in *Monocystis.* (ii) Gregarine movement. (iii) Nutrition in *Monocystis.* (iv) Syzygy.

➤➤*Short answer type questions*

1. What is pseudonavicella ?
2. Which is the host for *Monocystis* ?
3. Give the structure of trophozoite in *Monocystis.*
4. Describe the syzygy in *Monocystis.*
5. Give the description of sporozoites in *Monocystis.*
6. Give an account of autotomization in *Monocystis.*
7. Distinguish between isogamy and anisogamy.
8. Distinguish between cephaline and acephaline forms.

➤➤*Multiple choice questions*

1. Life cycle of *Monocystis* includes :
   (a) gametogony and sporogony
   (b) schizogony and gametogony
   (c) syzygy and gametogony
   (d) sporogony only
2. *Monocystis* belongs to the order :
   (a) Coccidia (b) Gregarinida
   (c) Haemosporidia (d) Myxosporidia
3. *Monocystis* trophozoite is found in :
   (a) seminal vesicles of earthworm
   (b) seminal vesicles of cockroach
   (c) both (d) none
4. Monogenetic parasite is :
   (a) *Trypanosoma*
   (b) *Plasmodium*
   (c) *Monocystis*
   (d) *Leishmania*
5. *Monocystis* is a parasite of :
   (a) man (b) cattle
   (c) earthworm
   (d) mosquito
6. Trophozoite of *Monocystis* is a stage :
   (a) nutritive (b) genetic
   (c) infective (d) transform

Answers

➤➤*Multiple choice questions*

1. (a) 2. (b) 3. (a) 4. (c) 5. (c) 6. (a)

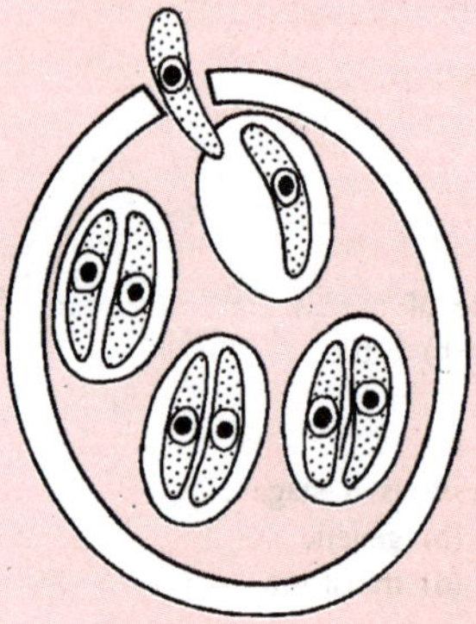

11

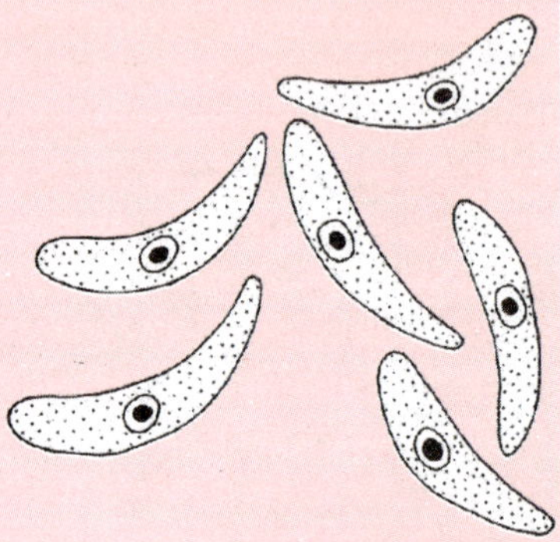

Eimeria tenella

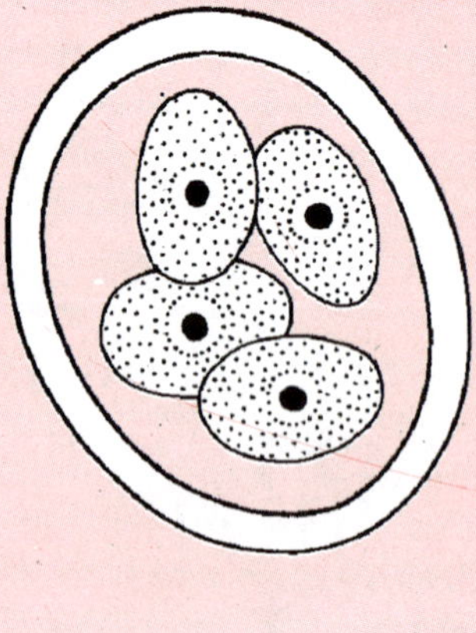

Eimeria, Isospora and related genera belong to subclass Coccidia of protozoan subphylum Sporozoa. These coccidians are parasites of various invertebrates and vertebrates. They are generally intracellular parasites of epithelial lining of host's alimentary canal. They cause a severe disease condition known as **coccidiosis** in their respective hosts. Two species of *Eimeria* cause heavy losses to the poultry industry in India. *E. tenella* affects the caeca of chicken and causes **caecal coccidiosis**. Another species, *E. mitis,* attacks the anterior part of ileum of adult birds. Some important mammalian species are *E. stiedae* (in bile duct of rabbit), *E. bovis* (in intestine of cattle), and *E. canis* (in small intestine of dog). The life cycle of *E. tenella* is dealt here in detail as a representative type of the group.

Eimeria tenella

SYSTEMATIC POSITION

Phylum	Protozoa
Subphylum	Sporozoa
Class	Telosporea
Subclass	Coccidia
Order	Eucoccida
Suborder	Eimeriina
Genus	*Eimeria*
Species	*tenella*

LIFE CYCLE

E. tenella is **monogenetic**, *i.e.,* its life cycle involves only one host (fowl). Various stages of complicated life cycle may conveniently be described under two phases, asexual cycle or **schizogony** and sexual cycle involving **gametogony** and **sporogony**. Much of the life cycle is intracellular.

Asexual Cycle (Schizogony)

There are 2 (sometimes 3) generations of schizonts and merozoites which are morphologically distinguishable.

[I] First Generation Schizogony

The beginning of complex life cycle of *E. tenella* may be taken to be initiated by the infection of epithelial cells of host caecum by sporozoites.

1. Infection by sporozoites. The infected fowl passes out oocysts or zygocysts with faecal matter. When these oocysts are swallowed by a new or healthy bird, its digestive juices dissolve the cystwall and release the sporozoites (or zoites) which enter the gut epithelial cells. Within the epithelial cells, the sporozoites grow in size and multiply by schizogony.

2. Sporozoite. It is the earliest intracellular stage of parasite. It is elongated, slightly curved, microscopic unicellular organism with one end

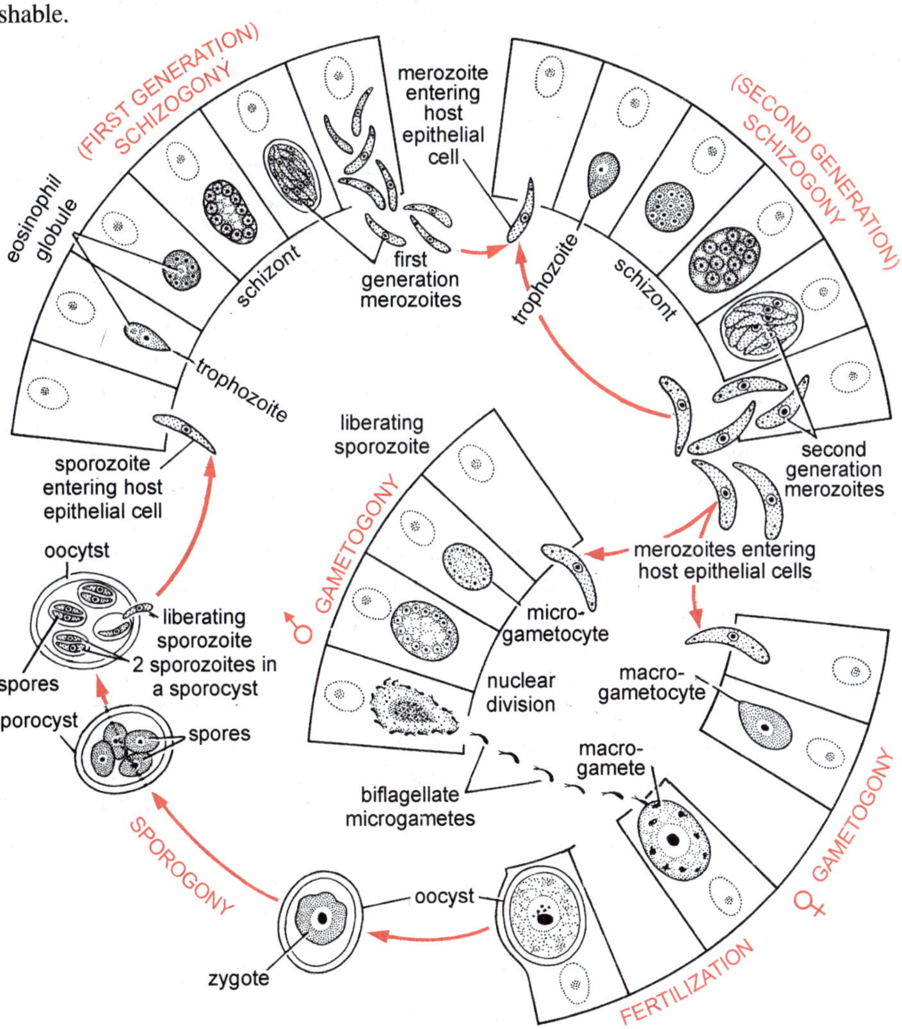

Fig. 1. *Eimeria tenella.* Life cycle.

pointed and the other end blunt. The pellicle, forming external envelope, contains longitudinally arranged contractile microtubules which help in wriggling movements of the organism. Cytoplasm includes a vesicular nucleus, a mitochondrion, Golgi bodies, endoplasmic reticulum, ribosomes, lysosomes and vacuoles containing reserve food, etc. Penetration of sporozoite into host epithelial cell is facilitated by the lytic secretion stored in the roptries of parasite.

3. Trophozoite. The sporozoite soon increases in size and becomes pear-shaped. This stage of the parasite is called trophozoite. The characteristic feature of trophozoite is the presence of an **eosinophil globule** at its blunt end. This globule stains brightly with eosin and black with haematoxylin.

Trophozite absorbs nourishment through its general surface from the surrounding cytoplasm of the host cell. It grows considerably so that the host cell and its nucleus show hypertrophy but are not destroyed.

4. Schizont. Eventually, the trophozoite undergoes multiple fission by schizogony. Repeated mitotic nuclear divisions are followed by the formation of cytoplasmic masses. This results in a multinucleate first generation **schizont**. It is somewhat oval in shape with vacuolated cytoplasm and numerous small peripherally arranged nuclei which are ill-defined. The eosinophil globule persists in its cytoplasm.

5. Merozoites. Within the schizont, the peripherally arranged nuclei with cytoplasmic masses become pyriform bodies, the **merozoites**. As many as 900 merozoites may be produced by a single first generation schizont. Each merozoite is a very small and inconspicuous body, measuring about 2-4μ in length and 1-1.5μ in width. Its one end is pointed while the other end is rounded. The pointed end contains a terminal granule and the rounded end bears a number of cytoplasmic granules. The vesicular nucleus takes up the central position. The host cells containing schizonts are hypertrophied and are finally extruded into the lumen of the caecum. The schizonts burst and the first generation merozoites escape to invade other epithelial cells.

[II] Second Generation Schizogony

The liberated merozoites set up new infections and schizogony may be repeated several times causing serious damage to epithelial cells.

1. Invasion of fresh host cells. A first generation merozoite invades a fresh epithelial cell and comes to lie within its cytoplasm. It soon grows into a trophozoite.

2. Trophozoite. The trophozoite resembles its counterpart of the first generation, except that it lacks an eosinophil globule. It feeds upon the host cytoplasm and its nucleus undergoes repeated mitotic divisions forming a multinucleate schizont.

3. Schizont. The second generation schizont differs very much from the first generation schizont. It is larger in size and the nuclei are scattered throughout its cytoplasm. Each nucleus has a distinct nucleolus or karyosome.

4. Merozoites. Within the schizont the cytoplasmic divisions result in the aggregation of cytoplasmic bits around each nucleus. These nucleated cytoplasmic bits metamorphose into the second generation merozoites.

A second generation merozoite is also elongated and pyriform with a rounded and a blunt end. They are larger than the first generation merozoites, measuring about 16 μ in length and 2 μ in width. Their number in a single schizont is about 250 which is less than the first generation. When the infected host cells become hypertrophied, the schizonts burst and the merozoites escape. They soon penetrate other epithelial cells lining the caecum.

Sexual Cycle (Gametogony and Sporogony)

A few second-generation merozoites may produce third-generation schizonts. These are not common, small and each produces 4 to 30 merozoites. These probably lead to the formation of infective oocysts discharged with the faeces.

However, most of the second generation merozoites do not repeat schizogony. Instead, for reasons not understood, they initiate the sexual cycle. They develop into **gametocytes** which produce **gametes**.

[I] Gametogony

The gametocytes or gamonts are sexually distinguished into two types : male or **microgametocytes** and female or **macrogametocytes**. Gametes also differ in size and number. Thus **anisogamy** is well-marked.

1. Microgametocytes. After invading the host epithelial cells, some of the merozoites become male gamonts or microgametocytes immediately. A microgametocyte is small, oval and measures about 5.5-18μ in length. It shows repeated nuclear divisions and gives rise to numerous small, comma-shaped biflagellate male cells or **microgametes**. Each of them bears a nucleus, a mitochondrion and two flagella.

2. Macrogametocytes. The female gamont or **macrogametocyte** is oval but comparatively larger, measuring 8-25μ in length. Their cytoplasm, especially near periphery, bears irregular masses of basophil reserves. Each macrogametocyte develops into a single female cell or **macrogamete**.

[II] Fertilization

Details of fertilization are not known. The microgametes escape from the host cell and invade the adjoining cells containing the macrogametes. A microgamete fuses with a macrogamete and thus fertilization takes place inside the host cell. The resulting fertilized cell or **zygote** is diploid.

[III] Oocyst

The zygote, while still within the host cell, secretes a thick double-layered cyst wall around itself. It has a thick outer and a delicate inner layer. According to **Monne** and **Honig** (1954), the outer thick layer is quinone-tanned protein, while the inner layer is a lipo-protein layer. The zygote now is referred to as the **oocyst** or **zygocyst**.

The mature oocyst is oval in shape and measures 20-26μ by 16.5-23μ. It does not undergo further development in the host cell. The completed oocysts fall in the cavity of the intestine by the rupture of the host cells. From here they are expelled outside the bird's body in faeces. The passing out of oocysts in enormous numbers may be seen on the seventh or eighth day of infection. The oocysts survive for a longer time under aerobic conditions.

[IV] Sporogony

Within 24 hours the zygote, within oocyst divides by meiosis into four haploid cells, called the **sporoblasts** or **zoitoblasts**. Each sporoblast then secretes a tough wall, the **sporocyst** or **zoitocyst** around itself and becomes a **spore**. Each spore divides mitotically to produce two sporozoites or **zoites**. Thus a mature oocyst includes eight sporozoites, two within each sporocyst. *E. tenella* is thus **tetrasporous** and **dizoic**. Division of encysted zygote into sporozoites generally takes place outside the host after the oocysts pass out with faecal matter. Under suitable temperature and moisture conditions, each oocyst sporulates in about 21 hours and becomes infective.

[V] Transference of Oocysts

The infective oocysts are easily passed on into other members of the flock through direct ingestion of droppings (faecal matter) or contaminated food.

[VI] Liberation of Sporozoites

Into the gut lumen of new host, the sporozoites liberate out through a pore or micropyle in the oocyst, under the influence of mechanical pressure and the enzymes pepsin and trypsin, and set up new infections.

COCCIDIOSIS

1. Pathogenesis. *E. tenella* causes a fatal disease of chickens called **caecal coccidiosis**. It is caused due to extensive destruction of the caecal epithelium, with sloughing off the walls and severe haemorrhage. The symptoms include bloody droppings, pale face and shanks, bloody vent and enlarged caeca distended with blood or with yellowish grey cheese-like ceres. In advanced stages, the infected birds show restlessness, drooping wings, suffled feathers and unsteady gait. Both the young and the old birds are susceptible to coccidiosis, but the infants are more severely injured as compared to the adults.

2. Diagnosis. Correct diagnosis of coccidiosis caused by *E. tenella* is possible by observing the symptoms and determining the presence of oocysts in the droppings. The microscopical examination of the faecal smear is done to confirm the presence of the oocysts.

3. Prevention. Infection of young chickens may be prevented through proper hygienic conditions and with good range management. Floors of poultry houses, yards and ranges must be cleaned periodically by scrubbing them through the hot lye solution. Equipment used for rearing birds should also be subjected to cleaning with the same solution.

As moist litter is favourable for keeping the oocysts viable, it is recommended that the litter should be heaped to develop its temperature to 125°F. NH_3 destroys the oocysts in dry litter.

Coccidiostatic drugs, which mostly include sulpha drugs, are added in prescribed doses in the poultry feed and water for a few days, mostly at the time when infants are being reared, or when coccidiosis is apprehended. These drugs help the birds to develop immunity from the coccidial organisms.

4. Treatment. The chickens who have caught coccidiosis are treated with sulpha drugs. **Sulphamezathene** and **Sulphaquinoxaline**, taken at the rate of 0.5% with drinking water for about a week, promise complete relief. Because of its low toxicity, sulphaquinoxaline is generally recommended.

Important Questions

➤➤*Long answer type questions*

1. Give an illustrated account of the life history *of Eimeria.*
2. How is Coccidiosis caused in poultry ? Suggest measures for its control.

➤➤*Short answer type questions*

1. Name the sporozoan parasite causing caecal coccidiosis in chickens.
2. What is the coccidiosis ?
3. What are gametogony and sporogony ?
4. Give the name of protozoan which is known as parasite of birds.

➤*Multiple choice questions*

1. *Eimeria* belongs to :
   (a) plasmodroma (b) ciliophora
   (c) cnidospora (d) sporozoa
2. *Eimeria* caused :
   (a) kala-azar (b) dysentery
   (c) coccidiosis (d) no any disease
3. Sexual cycle in *Eimeria* involving :
   (a) gametogony (b) sporogony
   (c) both (d) none the above
4. The symptoms of caecal coccidiosis chickens are
   (a) dropping wings
   (b) suffled feathers
   (c) unsteady gait
   (d) all
5. The best treatment for coccidiosis :
   (a) antiviral drugs (b) antibiotics
   (c) sulpha drugs
   (d) antipyretic

Answers

➤➤*Multiple choice questions*

1. (d) 2. (c) 3. (c) 4. (d) 5. (c)

12

Paramecium caudatum : The Slipper Animalcule

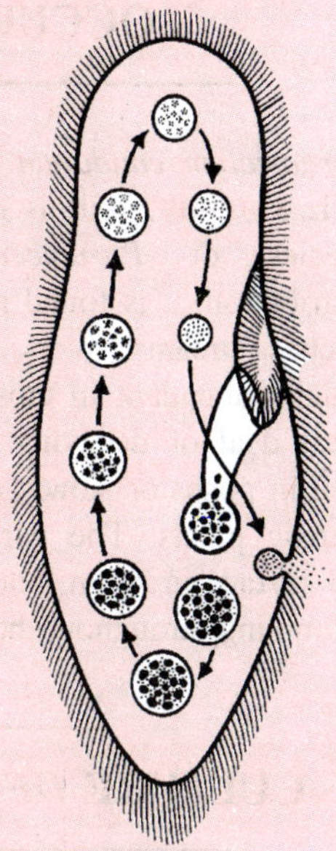

Protozoa which move with the help of **cilia** are called **ciliates** and are included in the subphylum *Ciliophora.* Besides the possession of these locomotor organelles, all ciliates possess two types of nuclei, **macronucleus** and **micronucleus** (nuclear dimorphism), and a unique form of sexual reproduction (**conjugation**). *Paramecium* is a typical ciliate genus containing about 10 known species differing in shape, size and structure. The largest species, *P. caudatum,* has a single comparatively large and compact micronucleus with chromatin material scattered throughout the nucleoplasm. It occurs widely and abundantly. *P. aurelia* has two micronuclei while *P. multimicronucleatum* has many micronuclei. *P. bursaria* is green due to presence of symbiotic alga, **zoochlorella**. In the following text, the biology of *P. caudatum* is treated in detail.

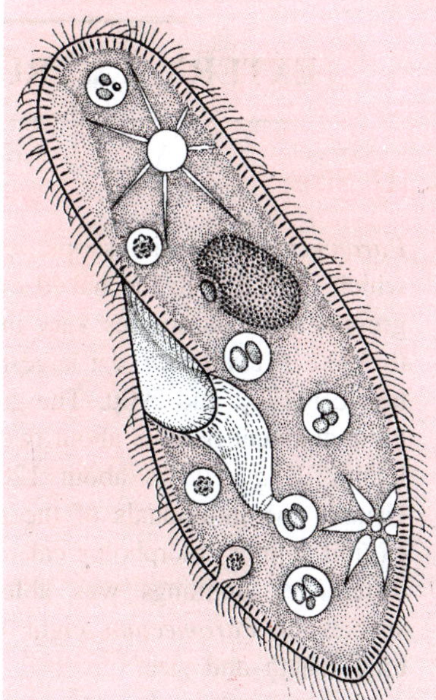

Paramecium caudatum

SYSTEMATIC POSITION

Phylum	Protozoa
Subphylum	Ciliophora
Class	Ciliata
Subclass	Holotricha
Order	Hymenostomatida
Suborder	Peniculina
Genus	*Paramecium*
Species	*caudatum*

OCCURRENCE

Paramecium caudatum (Gr., **paramekes**, oblong + L. **caudata**, tail) is one of the most common species of *Paramecium* having worldwide distribution. It is found in freshwater ponds, pools, ditches, streams, rivers, lakes, reservoirs, etc. It is usually abundant in those waters which contain a great deal of decaying organic matter. It thrives well in ponds or slowly running streams containing aquatic plants. The paramecia often gather near the surface in scum, but are usually seen actively swimming throughout the water in which they live.

CULTURE *OF PARAMECIUM*

Paramecium is easily grown in wide mouthed jars with glass covers, three-quarter filled with boiled pond water or Chalkey's medium (NaCl 80 mg, $NaHCO_3$ 4mg, KC1 4mg, $CaCl_2$ 4mg, CaH_4 $(PO_4)2H_2O$ 1.6 mg, dissolved in one litre of distilled water), and with 7-12 drops of skimed-milk added weekly. The jars are kept away from direct light to allow bacteria to flourish which serve as food for the multiplying paramecia.

EXTERNAL STRUCTURE

[I] Size

Paramecium is a microscopic, elongated organism which is visible to the naked eye as a whitish or greyish spot. Its species vary in length from 80μ to 350μ. *P. caudatum,* the largest species, measures between 170μ and 290μ. The greatest diameter of the cylindrical body is about two-third of its entire length. *P. aurelia* is about 120μ to 250μ long. Usually the individuals of the same species may exhibit minor morphological and physiological differences. Jennings was able to find in one species of *Paramecium* eight races differing in total length and size.

[II] Shape

Paramecium is often described as slipper shaped, cigar-shaped or spindle-shaped. Its shape is usually constant and in general asymmetrical. Because of its slipper-like shape, the *Paramecium* is sometimes called the **slipper animalcule**. Joblot assigned the name 'chausson' to *P. caudatum* which means slipper-shaped animalcule. The body is elongated, blunt and rounded at the anterior end and somewhat pointed at the posterior end. In cross section, it is circular with greatest diameter behind the centre of body. The anterior half of body is slightly twisted. The body is distinguished into an oral or ventral surface and an aboral or dorsal surface.

[III] Oral Groove

Ventral surface of body bears a prominent, oblique and shallow depression, called **oral groove**. It originates from the middle of body and extends to the left side of anterior end. Posteriorly, the oral groove leads into a deeper conical **vestibule** which in turn communicates with a **buccal cavity** having a basal mouth or **cytostome**.

[IV] Pellicle

External envelope of body is a living, clear, firm and elastic cuticular membrane, the **pellicle**. When stained specimens are observed under light microscope, the pellicle appears to be a regular series of polygonal (or hexagonal) depressions with their raised rims. A single cilium emerges out from the middle of each polygon or circumciliary space. Electron microscopic studies by **Ehret** and **Powers** (1959) have revealed that the polygons are defined by a corresponding regular series of cavities, the **alveoli**. In fact, it is the pit in the centre of each **alveolus** which forms a polygon. All the alveoli collectively form a continuous alveolar layer, which is delimited by an **outer alveolar** and **inner alveolar membrane**. The outer alveolar layer lies in close contact beneath the outer cell membrane (not shown in the diagram). Thus, the

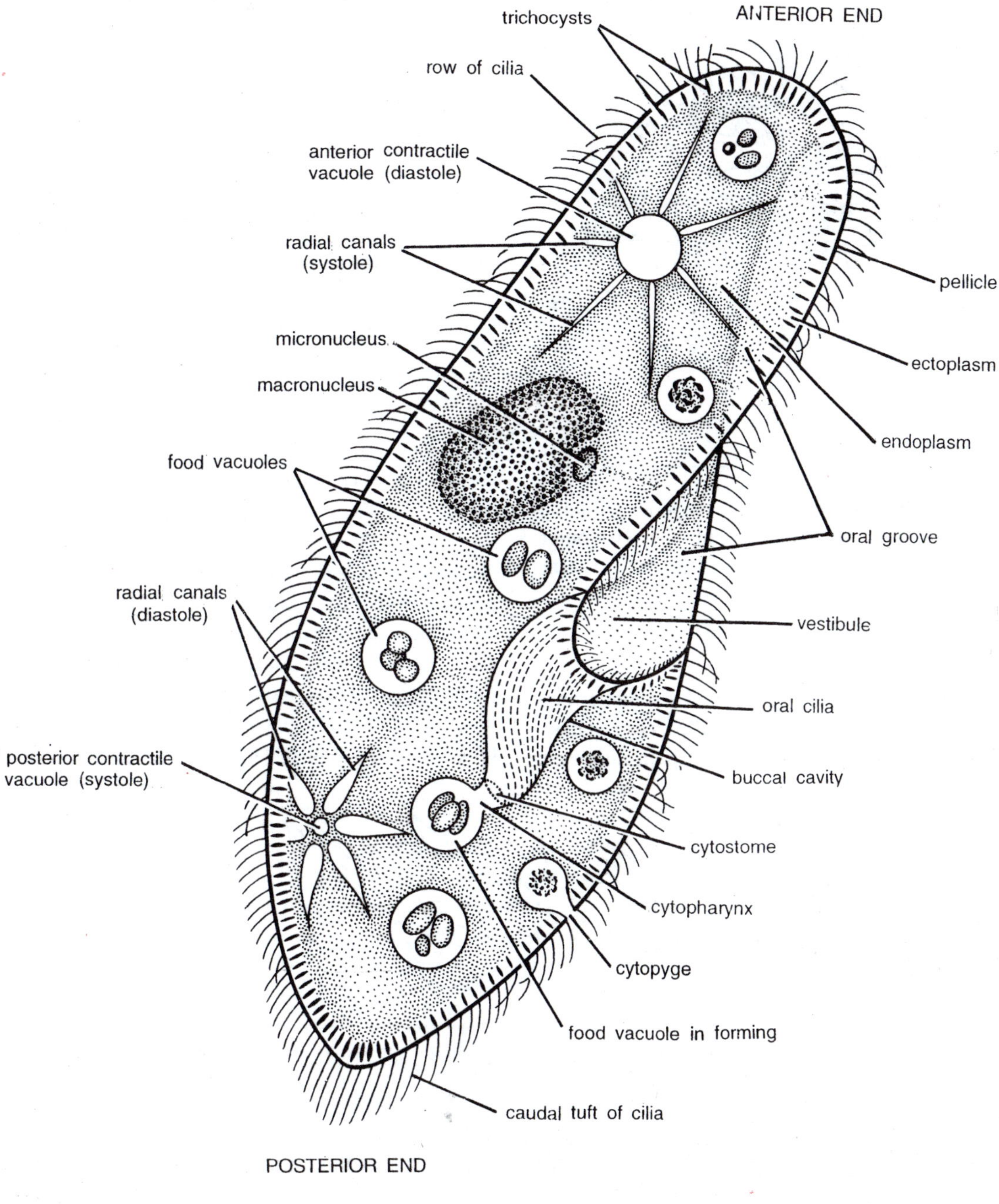

Fig. 1. *Paramecium caudatum.*

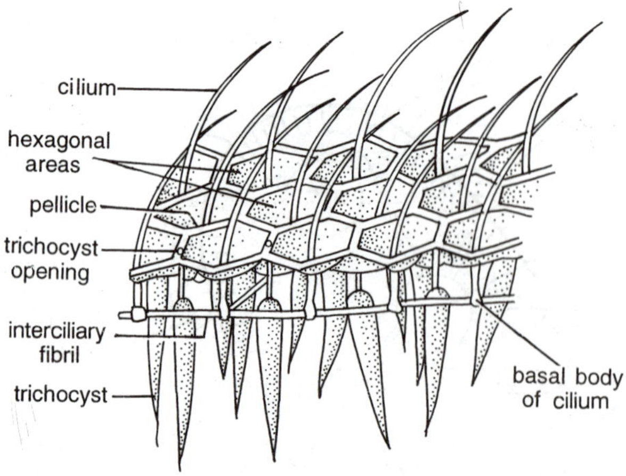

Fig. 2. *Paramecium.* Diagrammatic surface view of a small area of pellicle.

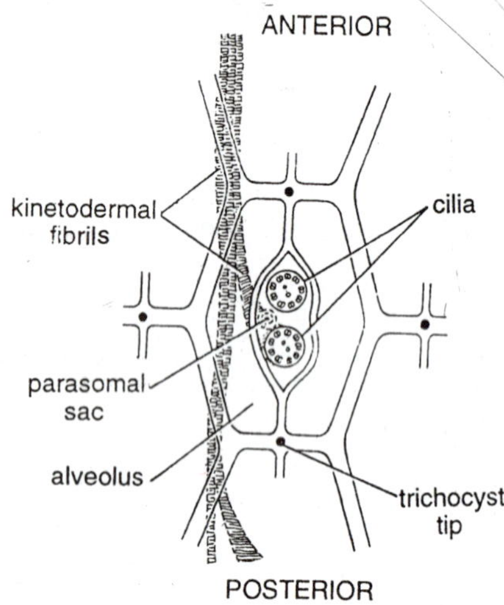

Fig. 3. *Paramecium.* Diagram of a hexagonal area or ciliary field.

pellicle of *Paramecium* includes a series of three membranes : (i) outer cell membrane, (ii) outer alveolar membrane, and (iii) inner alveolar membrane.

[V] Cilia

The entire body surface is covered by numerous, tiny, hair-like fine projections, called **cilia**. These measure 10-12μ in length and 0.27 μ in diameter. As already stated, one cilium (2 in *P. bursaria*) arises from the centre of each polygonal depression (circumciliary space) of pellicle. There are 10,000 to 14,000 cilia covering the whole body surface. These motile organelles are arranged in regular longitudinal rows. Their length is uniform throughout, except for a few longer cilia

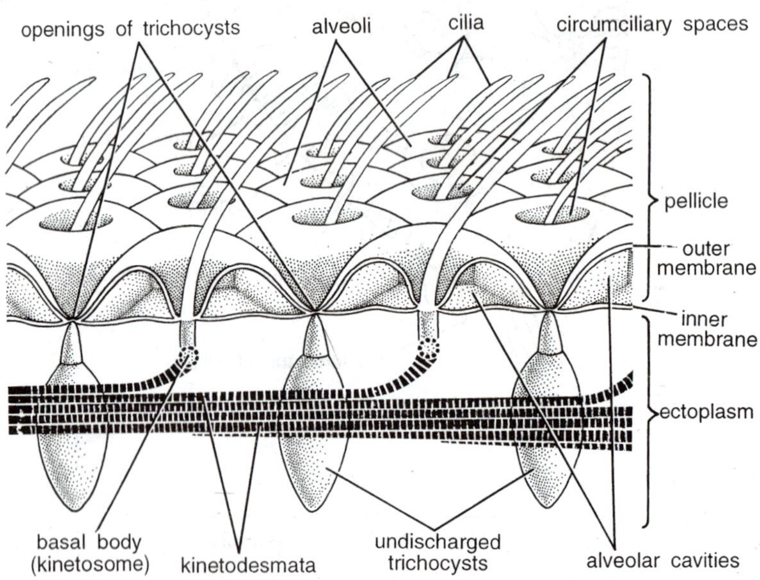

Fig. 4. *Paramecium.* A diagrammatic three-dimensional electron microscopic representation of a portion of pellicle and infraciliary system.

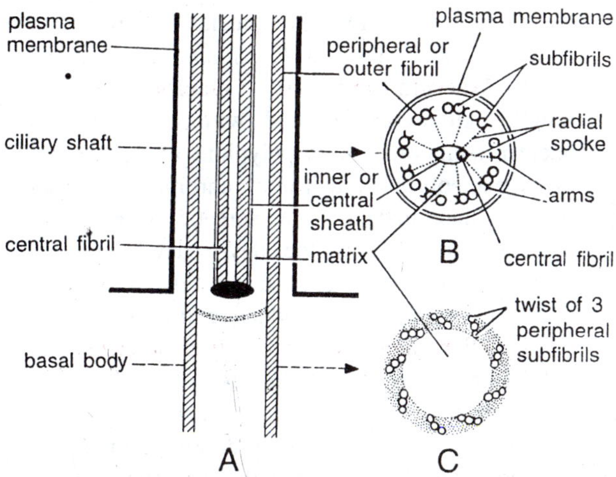

Fig. 5. Structure of a cilium and its basal body. **A.** Cilium in L.S. **B.** Free part of cilium in T.S. **C.** Basal body in T.S.

at the extreme posterior end of the body, forming a **caudal tuft**, hence the species name *caudatum.*

Electron microscopy has shown that a cilium has the same fundamental structure as has been seen in case of a flagellum (Fig. 5). It consists of a fluid matrix, surrounded by an outer membranous sheath, which is continuous with the outer cell membrane of body. Within matrix are 9 peripheral longitudinal fibres, which run along the whole length of cilium body. Each fibre is formed of two sub-fibres, one of which carries a double row of short arms or projections, all running in the same direction (clock-wise). In the centre of matrix are two single fibres, which are enclosed within an inner membranous sheath. In between the central and peripheral fibres are nine additional accessory fibres.

INTERNAL STRUCTURE

[I] Cytoplasm

Within pellicle, the cytoplasm of body is clearly differentiated into two regions.

1. Ectoplasm. The narrow, peripheral, clear and dense zone is called the **ectoplasm** or **cortex**. It includes the structure of the infraciliary system and the trichocysts.

2. Endoplasm. The large, central, granular and semi-fluid zone is the **endoplasm** or **medulla**. It includes the usual cell components like **mitochondria, Golgi bodies ribosomes, crystals, reserve food granules,** etc. In *P. bursaria,* the endoplasm is filled with symbiotic *Zoochlorella,* a unicellular chlorophyll-bearing alga. Prominent endoplasmic inclusions are **nuclei, contractile vacuoles** and **food vacuoles**.

[II] Infraciliary System

Immediately beneath the pellicular alveoli is located the infraciliary system constituted by the basal bodies and kinetodesmata.

1. Basal bodies. The base of each cilium is produced into a tube-like structure, called **basal body** or **kinetosome**. It is formed by the thickened basal ends of peripheral fibres of cilium. The central fibres do not enter into it. The wall of basal body consists of 9 triplets of sub-fibres. The basal bodies are self duplicating units and progenitors of new cilia. Each basal body is either a centriole or its derivative.

2. Kinetodesmata. Associated closely with basal bodies of cilia and lying in the ectoplasm is a system of specialized striated fibrils, called **kinetodesmal fibrils**. A single fibril or **kinetodesmos** arises from the kinetosome or basal body of each cilium and runs anteriorly somewhat tapering along the course. It joins its counterparts from the posterior kinetosomes, forming a bundle of overlapping longitudinal fibrils, called **kinetodesma** (pleural, **kinetodesmata**). The number of fibrils in each kinetodesmata constantly remains (5), because the individual fibrils do not run anteriorly farther than 5 basal bodies. It has been suggested that fibrils coordinate ciliary beat and movement, but the evidence is very conflicting.

The kinetosomes of a longitudinal row plus their kinetodesmata constitute a structural unit, called the **kinety**. A kinety system is apparently characteristic of all ciliates. It is said that the pattern of infraciliature plays an important role in the morphogenesis of Protozoa. For example, in *Paramecium,* one set of kinety, is solely responsible for the development of mouth structure. A new mouth fails to develop if this kinety is removed experimentally.

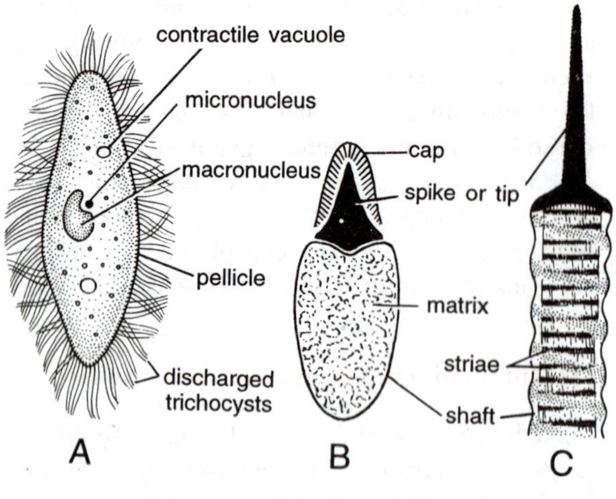

Fig. 6. *Paramecium.* **A.** Animal with discharged trichocysts. **B.** Undischarged trichocyst. **C.** Apical portion of discharged trichocyst.

[III] Trichocysts

Trichocysts are peculiar rod-like or oval organelles present throughout the ectoplasm alternating with basal bodies and oriented at right angles to the body surface. These were first seen in *Paramecium* by Elis. These are very small in size, measuring about 4μ in length. Each trichocyst consists of an elongated **shaft** and a terminal pointed **tip**, called the **spike** or **barb**, covered by a **cap**. The matrix of shaft consists of a dense mass of a fibrous protein, called **trichinin**. Its fibres remain condensed forming a cross-striated lattice work.

Function of trichocysts is not well known. It is believed that these discharge and anchor the animal to a firm substratum when it feeds upon bacteria. Others believe that these are organelles of defence.

Discharge of trichocysts is triggered by mechanical, chemical or electrical stimulation. It occurs in a span of a few milliseconds. When fully discharged the shaft becomes a long cross-striated rod and measures about 40μ in length. It is believed that the discharge process consists of an unfolding of the lattice of trichinin fibres.

[IV] Nucleus

Paramecium is **heterokaryotic** as it possesses two types of nuclei. In *P. caudatum,* there is a large **macronucleus** and a small **micronucleus**. Besides the macronucleus, two micronuclei are present in *P. aurelia* and many in *P. mutimicronucleatum.*

1. Macronucleus. The **macronucleus** is roughly kidney-shaped and with inconspicuous nuclear membrane. It is polyploid and possesses many nucleoli and much more chromatin material (DNA). Macronucleus is the somatic or vegetative nucleus and controls the day-to-day metabolic activities of the cell. It is derived from micronucleus during reproductive processes.

2. Micronucleus. The **micronucleus** is lodged in a depression on the surface of the macronucleus. It is usually spherical, with a nuclear membrane and with diploid number of chromosomes. It contains a definite nucleolus in *P. aurelia,* while in *P. caudatum* the nucleolus is absent. It controls the reproductive activities of the organism.

[V] Contractile Apparatus

In *Paramecium,* unlike *Amoeba* and *Euglena,* there are two **contractile vacuoles**, occupying somewhat fixed positions in endoplasm. One vacuole lies near each end of body, close to the dorsal surface. Each of them is surrounded by a circlet of 6 to 10 long, narrow, spindle-shaped **radial canals** (afferent pulsating canals) extending far into cytoplasm. Each contractile vacuole opens to

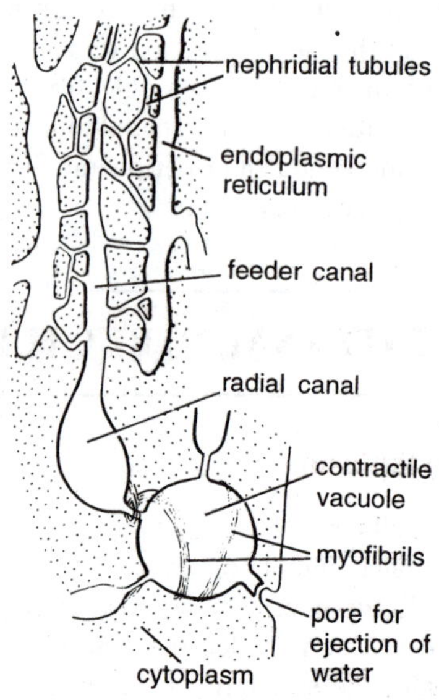

Fig. 7. *Paramecium.* Contractile apparatus.

outside through a permanent pore in pellicle of dorsal side of body. The two contractile vacuoles do not lose their identity when water is expelled.

Electron microscopy has revealed that each contractile apparatus includes some of the tubules of the endoplasmic reticulum, nephridial tubules, feeder canals, accessory vacuoles (radial canals) and the main contractile vacuole. The accessory vacuoles or radial canals are, in fact, the ampullae of feeder canals.

[VI] Food Vacuoles

Numerous non-contractile food vacuoles, recently termed **gastrioles** by **Vokovsky**, can be seen moving with the streaming endoplasm (cyclosis). They differ in shape and size according to the nature of ingested food particles, but mostly they are rounded in form.

[VII] Oral Apparatus

In *Paramecium,* oral groove leads ventrally and posteriorly as a tubular structure, called **vestibule**. It leads directly into a wide tubular passage, the **buccal cavity**. In its turn it opens into a narrow gullet or **cytopharynx** through a narrow aperture, the **cytostome**. The cytopharynx, at its proximal end, forms a **food vacuole**.

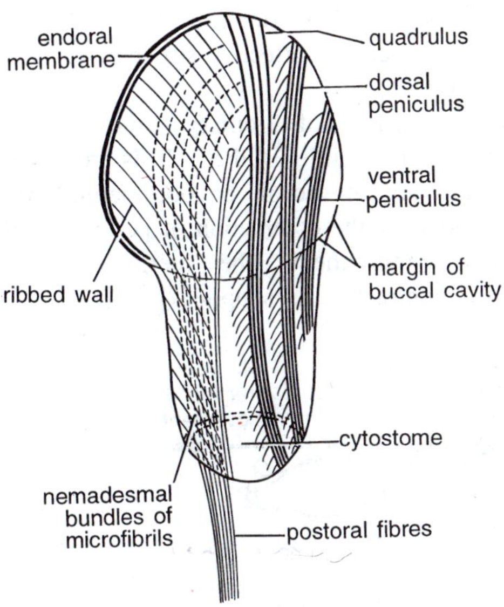

Fig. 8. *Paramecium.* Oral apparatus showing buccal ciliature.

Buccal cavity, at right side, is bordered by a row of cilia forming the **endoral membrane**. At left side are three groups of four rows of cilia, extending from the rim of the opening to the posterior end of buccal cavity. These are **ventral peniculus, dorsal peniculus** and **quadrulus**. These ciliary rows constitute the **membranelles**. From endoral membrane a ribbed pellicle extends upto cytostome. **Nemadesmal fibres** run dorsal to the ribbed pellicle and extend as **post-oral fibres** around cytopharynx. Rows of normal somatic cilia line the wall of vestibule.

[VIII] Cytopyge

Near posterior end of body, a little behind cytostome and a little to the right side, a small portion of ectoplasm and pellicle is somewhat weak. Here, at the time of **egestion**, a minute aperture called **cell anus**, **cytopyge** or **cytoproct**, is visible. It is, however, difficult to say whether it is a permanent opening with tightly-closed lips or a temporary opening formed at the time of egestion.

LOCOMOTION

Paramecium has a streamlined body which enables it to swim about in water with a minimum amount of friction. The rapid swimming is facilitated by the beating of fine and hair-like cellular organelles, called **cilia**, that cover the animal's entire cell-body. *Paramecium* moves with a speed of 1500μ or more per second.

[I] Ciliary Beats

During movement, a cilium oscillates like a pendulum. Each oscillation comprises a fast **effective stroke** and a slow **recovery stroke**. During the **effective stroke** or the strong backward lash, the cilium becomes slightly curved and rigid and strikes the water like an oar, so that body is propelled forward in opposite direction of stroke. The **recovery stroke** which follows immediately brings the cilium again into position for the next effective stroke (Fig. 10).

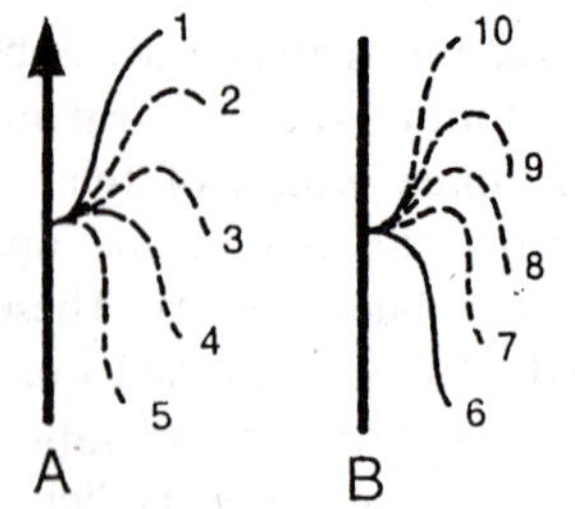

Fig. 9. Diagrams illustrating ciliary movements of a single cilium (after Gray). **A.** Effective stroke. **B.** Recovery stroke.

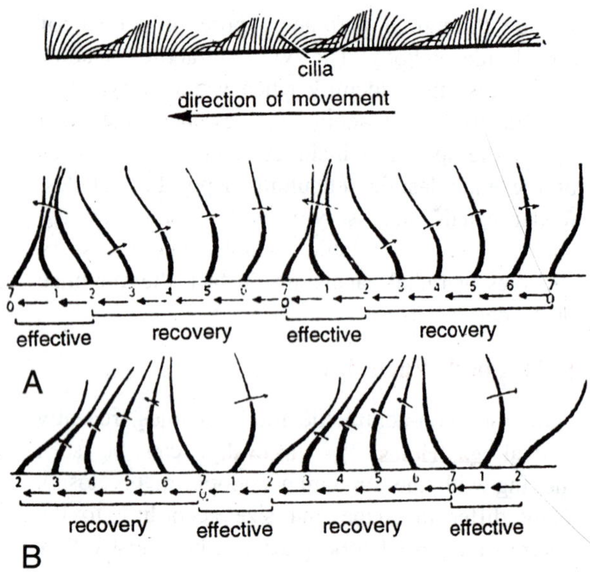

Fig. 10. A. A single row of cilia showing metachronal rhythm. **B.** Cilia 1-7 indicate recovery stroke. Cilia 8-12 indicate effective stroke.

All the cilia of body do not move simultaneously and independently but progressively in a characteristic wave-like manner, called **metachronal rhythm**. The cilia in a longitudinal row beat in a characteristic wave beginning at the anterior end and progressing backwards. Consequently, a cilium in a longitudinal row always moves in advance of the one behind it. All the cilia of a transverse row beat simultaneously or synchronously. During forward movement of *Paramecium* the metachronal waves pass from the posterior end forwards.

[II] Mode of Swimming

The animal does not follow a straight tract but rotates spirally like a rifle bullet along a left-handed helix. The reason for this is twofold. Firstly, the body cilia do not beat directly backwards but somewhat obliquely towards right, so that the animal rotates over to the left on its long axis. Secondly, the cilia of oral groove strike obliquely and more vigorously so as to turn the anterior end continually away from the oral side and move in circles. The combined effect causes the movement of animal along a fairly straight path, rotating about its axis in an anticlockwise direction.

In backward movement a *Paramecium* follows a straight course. In this case the metachronal wave passes from anterior end backwards. This is due to the fact that effective stroke is carried out anteriorly.

Mechanism of ciliary movement in *Paramecium* is little studied. It is now known that cilia are moved in a coordinating system. They move by the contraction of peripheral fibres located within them. The energy needed for fibrillar contraction is supplied by ATP.

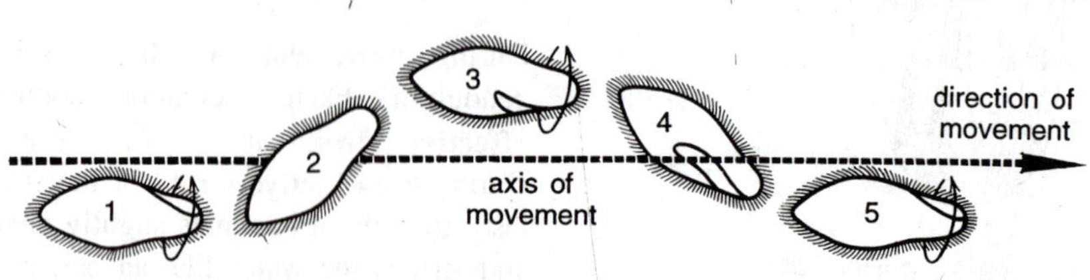

Fig. 11. Anticlockwise spiral path followed by a swimming *Paramecium.*

NUTRITION

[I] Food

Paramecium feeds in the **holozoic** manner, like *Amoeba.* The food consists chiefly of bacteria which float in water in which it lives. It has been estimated that 2 to 5.5 million individuals of *Bacillus coli* are devoured in 24 hours by a single *Paramecium.* In a sense, paramecia are also beneficial to bacteria, lest they might reproduce too rapidly as to endanger their own existence by overcrowding. It also feeds upon small Protozoa, unicellular plants (algae, diatoms, yeasts, etc.) and small bits of animals and vegetables. It will reject most of the non-digestible material and devour certain kinds of food. One species, *P. bursaria,* is interesting, being green in colour due to the presence of numerous unicellular alga, the symbiotic *Zoochlorella* in its endoplasm. It can thus live holophytically for long periods on food substances manufactured by *Zoochlorella.* During scarcity of food, it can digest even its own *Zoochlorella* and can live apparently indefinitely without them.

[II] Feeding Mechanism

Paramecium swims to places where it can get its food. Its food catching apparatus is much more specialized than that of *Amoeba* and *Euglena.*

Food is ingested by a definite cell mouth or **cytostome** lying at the bottom of buccal cavity. The constant lashing movements of cilia of oral groove drive a current of water with food particles towards the vestibule. Ciliary tracts of vestibule direct the food particles into buccal cavity. *Paramecium* is a selective feeder. According to **Mast** (1947), many kinds of particles may be carried with water current into vestibule, but only selected ones are passed on inside the buccal cavity. Rest of particles are rejected, that is, discharged to outside. Passage along which ciliary action drives selected food particles, is termed the **selection path**, whereas passage along which unwanted food particles are driven outside vestibule, is the **rejection path**.

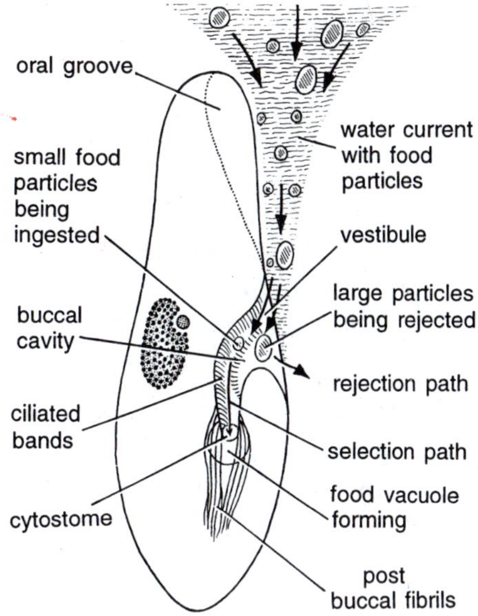

Fig. 12. *Paramecium* receiving food particles with water current drawn into buccal cavity by ciliary action.

Beating of cilia of membranelles of buccal cavity drives the selected food particles through cytostome into cell gullet or cytopharynx. The food now gradually collects at the bottom of cytopharynx into a membranous vesicle which is later nipped off as a **food vacuole**. Another food vacuole may be formed within 1 to 5 minutes depending upon the supply of food and the rate of feeding.

[III] Digestion

Each food vacuole consists of food particles surrounded by a thin film of water. Rapid and irregular movement of endoplasm does not occur in *Paramecium,* but the food vacuole is circulated around the body along a more or less definite path by a slow streaming movement of endoplasm, known as **cyclosis**. Several vacuoles may be seen thus circulating in a definite direction in the endoplasm of a well-fed *Paramecium.* The vacuoles are carried first posteriorly, then forward and aborally and again posteriorly and orally up to cytopyge. Digestion and assimilation of food take place during this journey. Digestive enzymes (proteases, carbohydrases, lipases) are secreted by

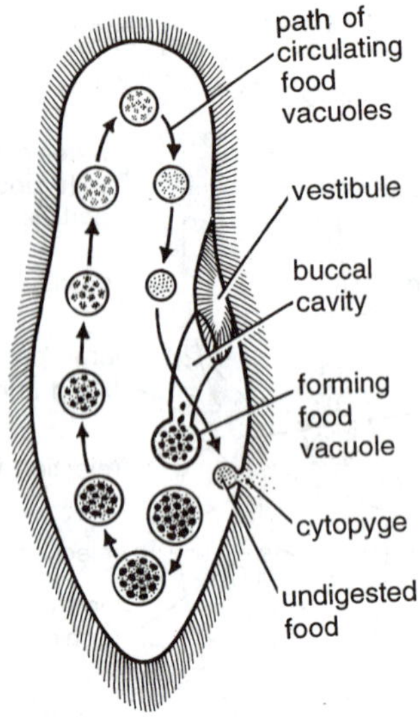

Fig. 13. *Paramecium* showing cyclosis and path of food vacuoles in endoplasm.

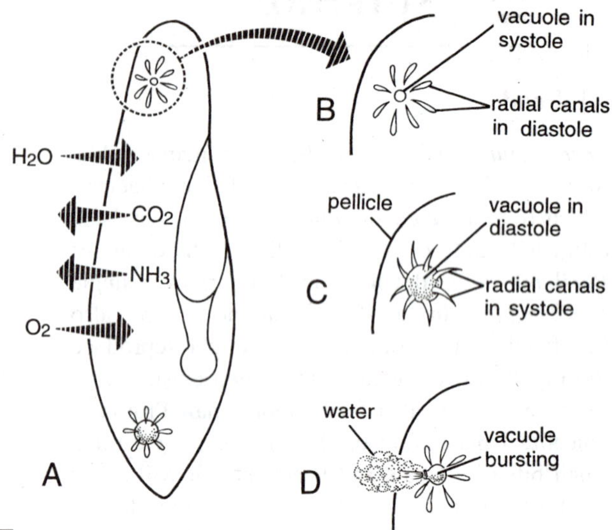

Fig. 14. *Paramecium.* Diagrammatic representation of respiration, excretion and osmoregulation.

the lysosomes into the food vacuoles. As in *Amoeba,* the contents of a vacuole first become increasingly acidic, but later gradually become alkaline. This can be demonstrated with the help of Congo Red and other indicator dyes. The alkaline phase results from the secretion of enzymes within an alkaline medium into the vacuole. Products of digestion (glycogen and fat globules) are diffused into the surrounding cytoplasm and either stored or used for vital activity and growth.

[VI] Egestion

The vacuole gradually becomes smaller as digestion and absorption proceed. Finally, the undigested residual matter is eliminated from body, through a definite **anal spot** or **cytopyge** on ventral surface, posterior to cytostome. The cytopyge is of the nature of a potential cell anus as the undigested matter is always discharged at this spot.

RESPIRATION AND EXCRETION

Respiration takes place, as in *Amoeba* and other freshwater Protozoa, by diffusion through the semi-permeable pellicle. Oxygen dissolved in water is diffused in and used for oxidation of protoplasmic molecules. Catabolic waste products such as CO_2 and nitrogenous matter (NH_3) simply diffuse out into external water because their concentration is always higher in body. Crystals present in cytoplasm are in fact excretory products, which get dissolved and eliminated with the fluid of contractile vacuoles.

OSMOREGULATION

The function of the contractile vacuoles in *Paramecium* is **osmoregulation**.

An excess of water accumulates in body because of continuous endosmosis, the concentration of body cytoplasm being higher than that of external medium. Small quantities of water are also taken in along the ingested food. This excess of water is got rid off by means of contractile vacuoles which contract (**systole**) and expand (**diastole**) at regular intervals, assisted by the contractility of **myofibrils** (see Fig. 7).

Water from cytoplasm is secreted into some of the tubules of endoplasmic reticulum from where it flows down the nephridial tubules into feeder canals to accumulate in latter's ampullae (radial canals). The ampullae converge and discharge into contractile vacuole. When vacuole has grown to its maximum size, it contracts and discharges to the exterior, through a pore in pellicle on dorsal side. Posterior contractile vacuole pulsates faster than anterior vacuole because of the large amount of water being delivered into posterior region by cytopharynx.

BEHAVIOUR

The way in which an organism establishes an active relationship to its environment is called **behaviour**. It is largely determined by the environmental influences or **stimuli** to which the organism is subjected. The responses of *Paramecium* to various kinds of stimuli, such as light intensity, temperature, concentration of O_2, CO_2 and different chemicals in water are interesting. These produce definite behavioural patterns or **reactive behaviour**. The response is **positive** if the animal moves towards a stimulus, and **negative** when it moves away.

Avoiding Reaction

It is perhaps the most important mode of behaviour exhibited by *Paramecium.* If a fast swimming individual strikes a solid object, it moves back for a short distance, turns on its side and swims forward again but at an angle to its original path. If it again collides with an obstacle, it shows the same negative reaction which is repeated until the animal passes the obstacle or becomes exhausted.

Trial and Error Reaction

Paramecium can also learn by **trial** and **error reaction**, involving a series of experiments on the part of animal. It constantly tests water just ahead by drawing it in its oral groove in the form of a cone. If water is too hot or too cold or if it contains an irritating chemical substance, the animal shows an avoiding reaction. It immediately backs up and pivots upon its posterior end, while the anterior end swings in a circle. Again it swings forward but in a different direction. If this avoiding reaction once more brings it into the region of the stimulus, it is repeated. These reactions help the animal to avoid undesirable environment without actually getting into it. Moreover, these bring the animal sooner or later, into the most favourable part of environment.

The responses of *Paramecium* to different stimuli may be summarised as under :

1. Temperature. Response to temperature is **thermotaxis.** Optimum temperature for *Paramecium* lies between 24°C and 28°C. An avoiding reaction is given to the temperatures higher or lower than this, until the animals escape or get killed.

2. Light. Response to light is **phototaxis**. Paramecia do not respond to ordinary changes of light, but a negative response is shown to strong light, darkness and ultra-violet rays.

3. Touch. Response to contact or **thigmotaxis** is variable. If the more sensitive anterior end is strongly touched with a solid object, the avoiding reaction is given. But a slow-moving *Paramecium* frequently comes to rest in contact with an object, such as an alga or a plant stem, which can provide rich supplies of food.

4. Chemicals. Chemotaxis or response to chemicals is negative in most of the cases. The animals show a definite avoiding reaction and do not enter a drop of weak salt solution. However, a positive reaction occurs with a drop of weak acid solution. The animals also find and select their food in this manner.

5. Water current. Paramecia show a positive **rheotaxis**, orienting themselves with their anterior ends upstream and swimming against the current.

6. Electric current. A positive **galvanotaxis** is shown to weak electric current, the animals moving towards the negative pole (cathode). A strong current, however, causes them to move backward towards the anode, finally to disintegrate and die.

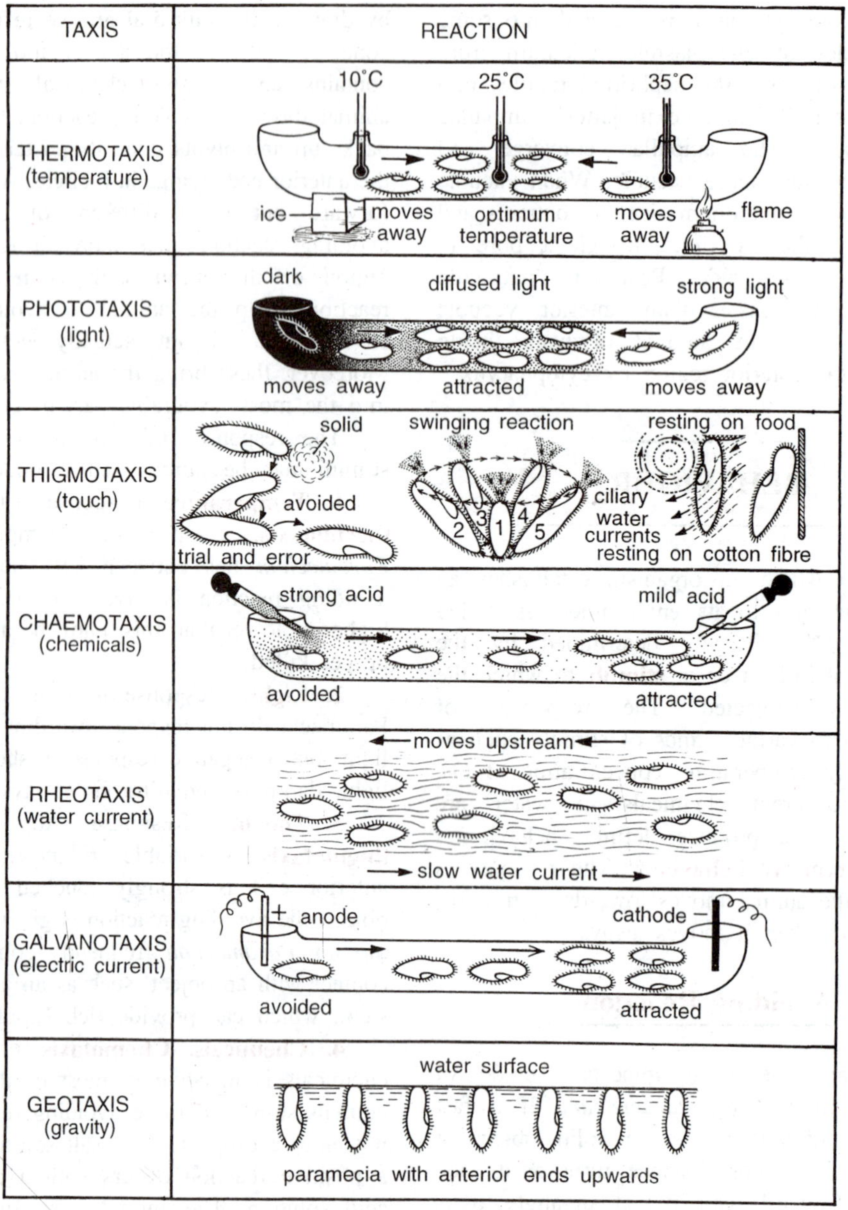

Fig. 15. *Paramecium.* Responses to different stimuli.

7. Gravity. Paramecia generally show a negative **geotaxis** or response to gravity as seen in a culture contained in a test tube, where they gather close to the surface film with their anterior ends pointed upwards. If paramecia are introduced in an inverted water-filled U-tube stoppered at both ends, they immediately move upward into the horizontal part of the tube. When, in moving across the tube, they find their path going downward, they reverse their direction of movement.

REPRODUCTION

Paramecium reproduces asexually by transverse binary fission and also undergoes several kinds of nuclear reorganisations, such as conjugation, endomixis, autogamy, etc. Under certain conditions of food and temperature, it undergoes encystment.

Transverse Binary Fission

During favourable conditions, *Paramecium* commonly reproduces by transverse or horizontal **binary fission**, which is at right angles to the longitudinal axis of body. *Paramecium* stops feeding and its oral groove and buccal structures begin to disappear. While this is happening, the **micronucleus** starts dividing by the complicated process of mitosis, the nuclear membrane remaining intact. Micronucleus first increases slightly in size and then chromosomes, numbering from 36 to 150, depending upon the race, begin to appear. Each chromosome splits longitudinally to form two chromatids (**prophase stage**). Paired chromatids now get arranged on the nuclear spindle at its equatorial plane (**metaphase stage**). This is followed by separation apart of chromatids and elongation of micronucleus (**anaphase stage**). By the last stage (**telophase stage**), micronucleus becomes very much elongated and its two ends become organised into two daughter micronuclei. The daughter micronuclei then separate. Simultaneously, the **macronucleus** divides amitotically by simply becoming elongated and constricted in the middle. Two oral grooves now begin to form, one in the anterior half and the other in the posterior half. Two original contractile vacuoles remain, one in each half of the dividing parent individual. Two new contractile vacuoles are later formed. Two new buccal structures also appear. In the meantime, a constriction furrow appears near middle of the body. It deepens and ultimately the cytoplasm is completely divided, resulting into two daughter paramecia. Of the two daughter paramecia, the anterior one is called **proter** and the posterior, **opisthe**. These grow to full size and divide again by fission.

P. caudatum divides 2-3 times in a day by binary fission. The process is completed in about 30 minutes, though separation of daughter paramecia takes about one hour or more. The term **clone** is used to refer to all the individuals that are produced asexually from one parent *Paramecium*. All the members of a clone are genetically alike.

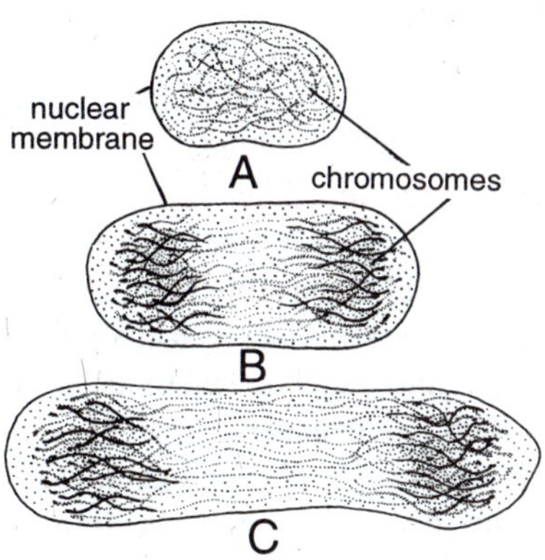

Fig. 16. *Paramecium.* Chromosomal movement in mitosis of micronucleus.

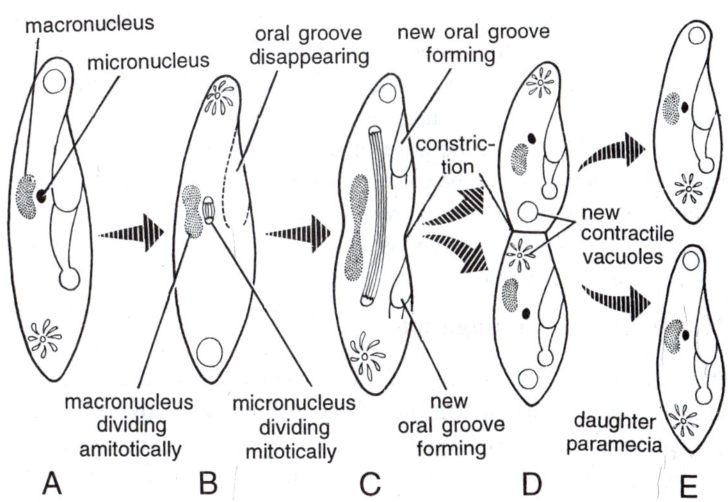

Fig. 17. *Paramecium.* Stages showing binary fission.

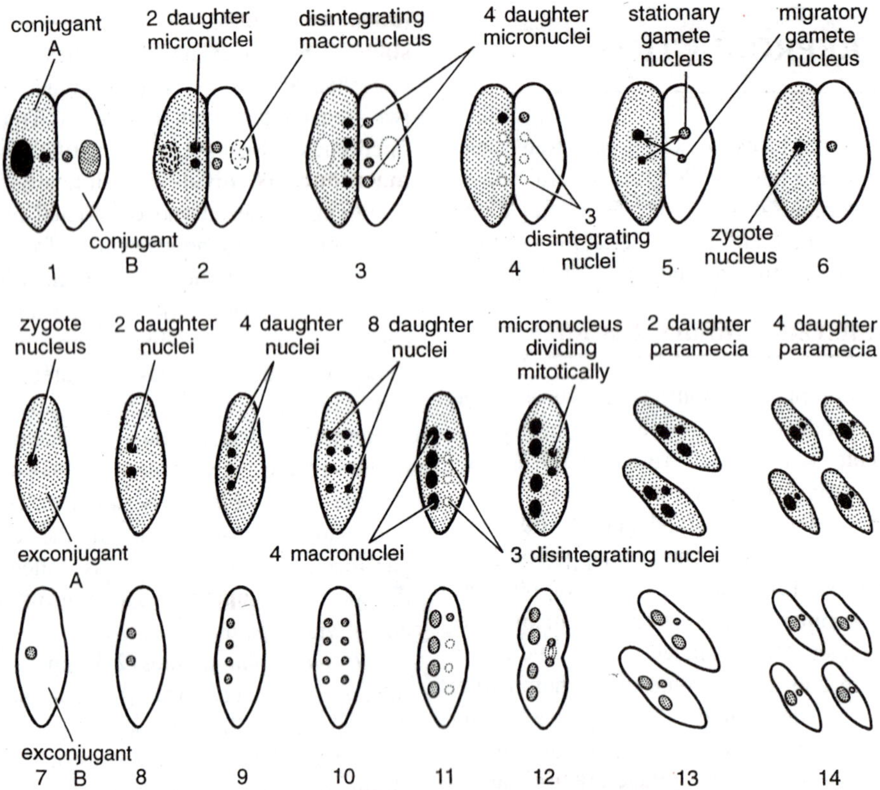

Fig. 18. *Paramecium.* Stages in conjugation.

Conjugation

Paramecium undergoes a sexual phenomenon, which is called **conjugation**. It is frequently referred to as sexual reproduction, but it is simply a temporary union of two individuals of one and the same species for the purpose of exchanging a part of their micronuclear material. This remarkable process in *Paramecium* occurs frequently between binary fissions and is necessary for the continued vitality of the species.

[I] Process of Conjugation

The details of this process differ slightly in different species of *Paramecium.* The following account refers to *P. caudatum.*

In conjugation, two individuals or **preconjugants**, from two different mating types, come in contact ventrally and unite by their oral grooves. They stop feeding and their buccal structures disappear. The pellicle and ectoplasm degenerate at the point of contact and a **protoplasmic bridge** is formed between the two individuals, which are now called the **conjugants**. While so united, like the 'Siamese twins', the conjugating pair continues to swim actively and a sequence of complicated nuclear changes takes place in each animal.

The vegetative **maeronucleus** simply breaks up into fragments, which are later absorbed by cytoplasm. The diploid **micronucleus** of each conjugant first grows in size and then divides by meiosis. Thus, 4 haploid daughter micronuclei are produced of which 3 degenerate or become pycnotic and disappear in each conjugant, while the remaining one divides by mitosis forming 2 unequal **pronuclei** or **gamete nuclei**. Smaller one is the active **migratory gamete nucleus** and the bigger one is the passive **stationary gamete nucleus**. The migratory nucleus of one conjugant

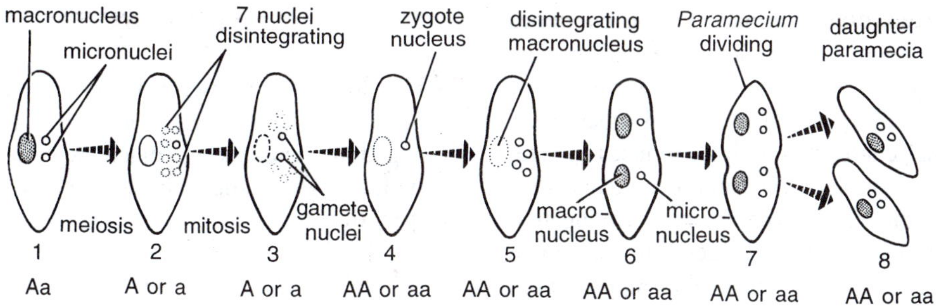

Fig. 19. *Paramecium aurelia.* Stages in autogamy.

then passes through the protoplasmic bridge into the other individual and fuses with its stationary nucleus, forming a single diploid **zygote nucleus** or **synkaryon**. The complete fusion of two nuclei from two different individuals forming a zygote nucleus is termed **amphimixis**.

The two pairing paramecia, after a union of about 12 to 48 hours, separate and are now called **exconjugants**. In each exconjugant, the zygote nucleus divides by mitosis three times in rapid succession producing 8 nuclei, of which 4 enlarge to become **macronuclei** and other 4 become **micronuclei**. Three micronuclei disintegrate and disappear, while the remaining micronucleus divides, with binary fission of exconjugant. Thus, from each exconjugant two daughter paramecia are obtained, each containing 2 macronuclei and one micronucleus. The micronucleus again divides with the division of each daughter *Paramecium*, forming two individuals each containing one macronucleus and one micronucleus. Thus, each conjugant produces four daughter individuals at the end of conjugation.

[II] Factors and Conditions of Conjugation

Conjugation is very complex physiologically. The factors and conditions governing conjugation are several and these may also vary with the species.

(1) Conjugation does not occur under favourable living conditions. Starvation or shortage of food and a particular bacterial diet or certain chemicals are said to induce conjugation in some species.

(2) A certain range of light and temperature, differing with species, is said to be essential for conjugation to occur.

(3) In *P. caudatum,* conjugation usually starts early in morning and is continued till afternoon.

(4) The conjugating individuals are usually smaller in size (210μ long) than the normal individuals (300-350μ long).

(5) A definite state of nutrition is indispensable since starved or overfed individuals generally will not conjugate.

(6) Maupas maintains that individuals must have passed through a desirable number of asexual generations (period of immaturity) before they become sexually mature and conjugate.

(7) The pairing conjugants are **isogamous** and there is no morphological sexual dimorphism into male and female conjugants.

(8) Conjugation never takes place among the members of a 'pure line', that is among the descendants of a single individual. It occurs only between individuals belonging to two different **mating types**. Thus, a sort of physiologically sexual differentiation exists in *Paramecium.*

(9) Agglutination favours conjugation. It is the interaction of mating type substances (proteins) which are localized in cilia.

[III] Significance of Conjugation

The significance of conjugation has been much discussed but it still remains uncertain. The following functions or effects are attributed to this process :

1. Rejuvenation. If binary fission continues repeatedly for several generations, the *Paramecium* loses its vigour and enters upon a period of depressed physiological efficiency and senescence. The individual ceases to multiply, reduces in size, degenerates in organization and eventually dies off. To avoid this senile decay of race, conjugation is resorted to and the process seems to rejuvenate and revive the lost vigour for asexual reproduction.

However, **Woodruff** and **Jennings** do not support the view that conjugation helps in rejuvenescence. Woodruff succeeded in maintaining a culture of paramecia for nearly 36 years, resulting in hundreds of thousands of generations without resort to conjugation.

2. Nuclear reorganization. During conjugation the nuclear apparatus is reorganized and a readjustment occurs between it and the cytoplasm. Probably the macronucleus loses its potentialities in performing its manifold metabolic activities. Its replacement by a new macronucleus brings renewed vigour and vitality to accelerate the metabolic activities.

3. Hereditary variation. During asexual reproduction by fission, the hereditary material of the parent passes unchanged on to the progeny, so that all the descendants of one *Paramecium* have the same inheritance. The periodic occurrence of conjugation, however, ensures **inherited variation**. It brings about the blending of two lines of ancestry just as bisexual reproduction does.

[IV] Genetic Consequences of Conjugation

If conjugation takes place between two paramecia, one homozygous for a dominant gene (AA) and the other homozygous for its recessive gene (aa), the first generation would be heterozygous (Aa). If the two conjugants are already heterozygous (Aa), then the resulting progeny would be either homozygous or heterozygous, depending upon which gene gets eliminated at the stage of disintegration of three micronuclei in each conjugation.

Autogamy

W.F. Diller (1936) described a process of nuclear reorganization in *P. aurelia,* resembling conjugation, but taking place within a single individual. He called it **autogamy** or **self-conjugation**.

[I] Process of Autogamy

During autogamy in P. *aurelia,* the 2 diploid micronuclei divide by meiosis to form eight haploid daughter nuclei. Seven of them disintegrate, whiie the remaining haploid micronucleus undergoes a mitotic division forming 2 gamete nuclei. Meanwhile, the macronucleus grows into an irregular skein-like mass, which breaks into pieces later to be absorbed in the cytoplasm. The two gamete nuclei enter a protoplasmic cone temporarily formed near cell mouth and then fuse together to form a completely homozygous diploid zygote nucleus or **synkaryon**. This divides twice to yield 4 nuclei, 2 of which become macronuclei and 2 micronuclei. The cell body and the micronuclei then divide to form two daughter individuals, each with a new macronucleus and 2 micronuclei. Autogamy rejuvenates *Paramecium.*

[II] Genetic Consequences of Autogamy

If autogamy takes place in a *Paramecium* heterozygous for a dominant gene (Aa), the resulting progeny will depend upon the survival of the gene A or a. If the gene A survives, it will lead to AA individuals or vice versa. Thus, autogamy always results in homozygosity.

Cytogamy

In 1940, **R. Wichterman** reported, in *P. caudatum,* a sexual process without nuclear exchange, termed **cytogamy**. The process resembles **conjugation** in that two small paramecia (200μ long) temporarily fuse by their oral surfaces. The early nuclear divisions are also similar to those of conjugation; but there is no nuclear exchange between the individuals (**cytogamonts**). But, two haploid **gamete nuclei** in each individual are said to fuse to form a **synkaryon**, as in **autogamy**. The process is completed in about 13 hours.

Endomixis

Endomixis (Gr., **endon**, within + **mixis**, mingling) is an interesting phenomenon involving a total internal nuclear reorganization within a single individual in a culture of a pedigreed race of *Paramecium*, taking place in the absence of conjugation. **Woodruff** and **Erdmann**, in 1914, first of all reported endomixis in the bimicronucleate species, *P. aurelia*, occurring periodically at regular intervals of about 30 days. The whole process may be summarized as follows :

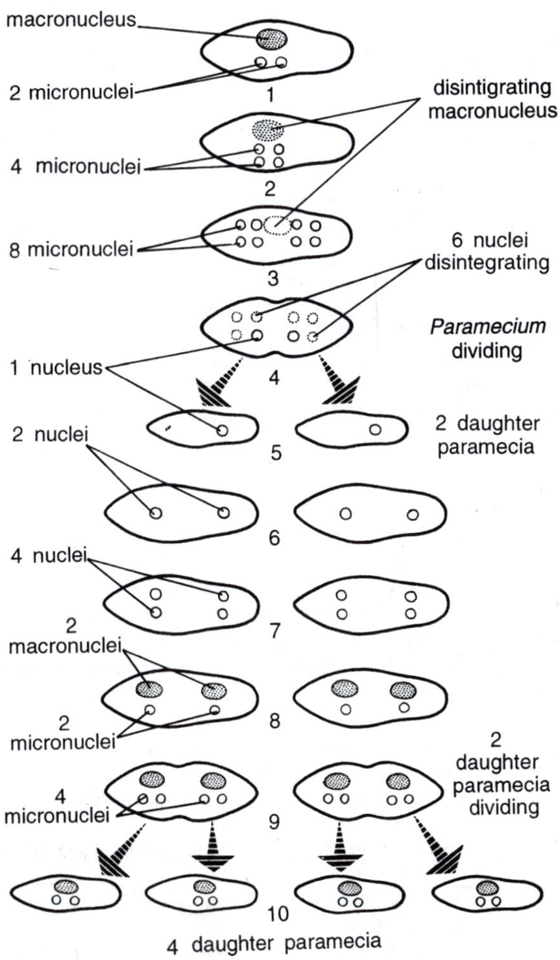

Fig. 20. *Paramecium.* Stages in endomixis.

The vegetative **macronucleus** degenerates and disappears, while the **micronuclei** divide twice by mitosis forming 8 **daughter nuclei** of which 6 degenerate. At this stage *Paramecium* also divides, and each daughter paramecia receive one micronucleus. This micronucleus divides twice forming 4 nuclei, 2 of which become macronuclei and 2 micronuclei, in each individual.

The micronuclei again divide with the binary fission of *Paramecium* into two daughters, each getting one macronucleus and 2 micronuclei. Thus, four daughters are produced from a single parent bringing about an intracellular nuclear reorganization and readjustment between the cytoplasm and the nuclear apparatus in each individual.

CYTOPLASMIC PARTICLES IN *PARAMECIUM*

1. *Kappa* particles. At the time of mixing of two races of Paramecia for conjugation, **T.H. Sanneborn** found that sometimes one race survives and the other dies out. These two races have been designated as **killer** and **sensitive**, respectively. It was found out that in the individuals which survive (**killers**) occur special self-replicating cytoplasmic bodies containing DNA, called **kappa particles**. These are associated with the production of a killing substance, **paramecin**. This substance diffuses out into the surrounding water and causes the death of the sensitive (kappa free) individuals. In the course of studies about the kappa particles, it was found that a dominant gene (K) in nucleus is necessary for kappa to exist, multiply and produce paramecin. Kappa particles provide an example of cytoplasmic inheritance. They are transmitted directly by cytoplasmic genes (plasmagenes) from cytoplasm of parent cell to the daughter cells, and not by nuclear genes as in ordinary heredity.

2. *Pi* particles. These are mutant forms of kappa particles. They do not release any toxic substance meant for killing those which are without such particles, that is, the sensitives.

3. *Mu* particles. These particles are also killers and kill the mate without such particles, during conjugation.

4. *Lambda* particles. These particles are borne by killer paramecia and cause the sensitive paramecia to lyse or disintegrate.

Important Questions

➤➤*Long answer type questions*

1. Describe the detailed structure of the pellicle and the infraciliary system of *Paramecium.*
2. Describe the modes of locomotion and nutrition in *Paramecium.*
3. Describe the feeding mechanism and the process of digestion in *Paramecium.*
4. Compare the modes of locomotion and nutrition of *Amoeba, Euglena* and *Paramecium.*
5. How is osmoregulation effected in *Paramecium* ?
6. Give an account of conjugation in *Paramecium* and discuss its significance.
7. Describe the processes of autogamy and endomixis in *Paramecium.*
8. Write short notes on : (i) Amphimixis, (ii) Autogamy, (iii) Binary fission in *Paramecium,* (iv) Buccal ciliature of *Paramecium,* (v) Cyclosis, (vi) Endomixis, (vii) Kappa particles, (viii) Synkaryon, (ix) Trichocysts, (x) Conjugation.

➤➤*Short answer type questions*

1. How do the *Paramecium* feed ?
2. Mention the organ of offence and defence in *Paramecium.*
3. What is endomixis ?
4. What are the two functions of contractile vacuoles of *Paramecium* ?
5. What is meant by undulating membrane ? What function does it serve ?
6. Give two differences between endomixis and autogamy.
7. Write the food procuring mechanism exhibited by *Paramecium.*
8. What is a trichocyst ? What are its functions ? Describe the mechanism of its discharge.
9. Distinguish between conjugation and autogamy.
10. What is nuclear dimorphism ? Explain the phenomenon with reference to *Paramecium.*
11. Differentiate between cytogamy and autogamy, and endomixis and hemixis.
12. Give an account of nuclear reorganization processes in *Paramecium.*
13. Distinguish between synchronous and metachronous beating of cilia in *Paramecium.*
14. Distinguish between macro and micro-nucleus of *Paramecium* on the basis of their functions.
15. Distinguish between effective stroke and recovery stroke of a flagellum.
16. With reference to plane of division distinguish binary fission in *Paramecium* and *Trypanosoma.*
17. Genetic information in *Paramecium* is contained in which organelle ?
18. Protozoa reproduce by several methods. Which protozoan reproduces both by binary fission and conjugation ?

➤➤*Multiple choice questions*

1. Movement of food vacuole in *Paramecium* along a definite path is known as :
   (a) cytokinesis (b) cyclosis
   (c) endomixis (d) metagenesis
2. In *Paramecium* the division of Macro-nucleus during binary fission is :
   (a) mitotic (b) amitotic
   (c) meiotic (d) prenuclear
3. The functions of the trichocysts are :
   (a) offence and defence
   (b) narcotising prey
   (c) to attain resting condition
   (d) all of the above mentioned
4. The main functions of the contractile vacuole is :
   (a) pumping out excess water
   (b) excretion
   (c) osmoregulation
   (d) respiration
5. What is a trichocyst ?
   (a) spindle shaped structure below pellicle
   (b) interlacing
   (c) fusion of cilia and flagella
   (d) modification of the contractile vacuole a structure concerned with photosynthesis
6. Mention a function of the neuro-motor system of *Paramecium :*
   (a) co-ordination of ciliary beat
   (b) co-ordination of various stimuli
   (c) control of digestion
   (d) co-ordination of respiratory movement
   (e) control of osmoregulation
7. Which of the following helps in anchorage and defence of *Paramecium* ?
   (a) nematocyst (b) oocyst
   (c) trichocyst
   (d) statocyst

8. Autogamy and conjugation are sexual processes, because :
(a) recombination of genes takes place
(b) the individual is rejuvenated
(c) two individuals are involved
(d) fusion of haploid nuclei occurs
9. The number of nucleus in *Paramecium* is :
(a) one (b) two
(c) three (d) four
10. Who discovered *Paramecium* :
(a) Hill (b) Ross
(c) Lavine (d) Grassi
11. 'Hay-infusion method' use for the culture :
(a) *Amoeba* (b) *Paramecium*
(c) *Euglena* (d) all
12. 'Caudal tuft' present in :
(a) *Amoeba* (b) *Paramecium*
(c) *Euglena* (d) *Trypanosoma*
13. Vegetative function is controled by :
(a) micronucleus (b) macronucleus
(c) both (d) none
14. How many food vacuole present in *Paramecium* :
(a) one (b) two
(c) four (d) numerous
15. Among them which one is filter feeder :
(a) *Paramecium* (b) *Amoeba*
(c) *Trypanosoma* (d) *Monocystis*
15. A protozoa feeds on protozoans :
(a) *Paramecium* (b) *Amoeba*
(c) *Trypanosoma* (d) *Plasmodium*
17. Protozoa which able to creep on a substratum :
(a) *Amoeba* (b) *Paramecium*
(c) *Euglena* (d) none
18. How many paramecia produce after the conjugation :
(a) two (b) four
(c) eight (d) sixteen
19. Autogamy occurs only in :
(a) *Paramecium caudatum*
(b) *P. aurelia*
(c) *Amoeba* (d) *Trypanosoma*
20. Cytogamy occurs in :
(a) *Paramecium caudatum*
(b) *P. aurelia*
(c) *Amoeba* (d) none
21. Nuclear dimorphism is shown by :
(a) *Amoeba* (b) *Paramecium*
(c) *Trypanosoma* (d) *Plasmodium*
22. In the cell body of *Paramecium,* the elimination of undigested residual matter takes place through the :
(a) cutosine (b) anus
(c) gullet (d) cytopyge
23. The cilia in *Paramecium* are :
(a) all equal
(b) all unequal
(c) longer at posterior end
(d) longer at anterior end
24. In *Paramecium* cyclosis is associated with :
(a) locomotion (b) respiration
(c) digestion (d) excretion
25. Protozoa reproduce by several methods. Which protozoan reproduces both by binary fission and conjugation :
(a) *Amoeba* (b) *Euglena*
(c) *Monocystis* (d) *Paramecium*
26. In *Paramecium*, conjugation takes place for :
(a) rejuvenation (b) sexual reproduction
(c) asexual reproduction
(d) fragmentation
27. In *Paramecium*, both conjugation and autogamy are sexual processes because of :
(a) involvement of two individuals
(b) fusion of two haploid nuclei
(c) gene recombination
(d) rejuvenation
28. Genetic information in *Paramecium* is contained in :
(a) micronucleus (b) macronucleus
(c) both micronucleus and macronucleus
(d) mitochondria
29. When faced with a foreign body in water, *Paramecium*:
(a) displays avoiding reaction
(b) continues undisturbed
(c) fights its way through
(d) rounds up and remains static
30. Which of the following statement is correct regarding *Paramecium* :
(a) it moves away from alkalies
(b) it moves towards strong acids
(c) it moves towards intense light
(d) none of the above

Answers

➤➤*Multiple choice questions*

1. (b)	2. (a)	3. (d)	4. (c)	5. (a)	6. (a)	7. (c)	8. (b)	9. (b)	10. (a)
11. (b)	12. (b)	13. (b)	14. (d)	15. (a)	16. (a)	17. (a)	18. (b)	19. (b)	20. (a)
21. (b)	22. (d)	23. (c)	24. (c)	25. (d)	26. (b)	27. (b)	28. (a)	29. (a)	30. (a)

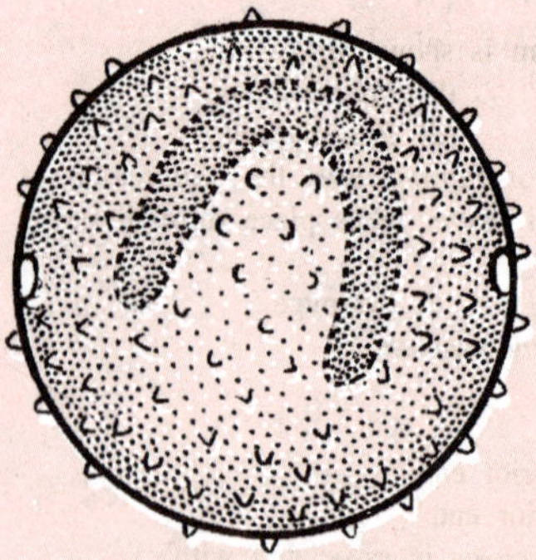

13

Vorticella : The Bell Animalcule

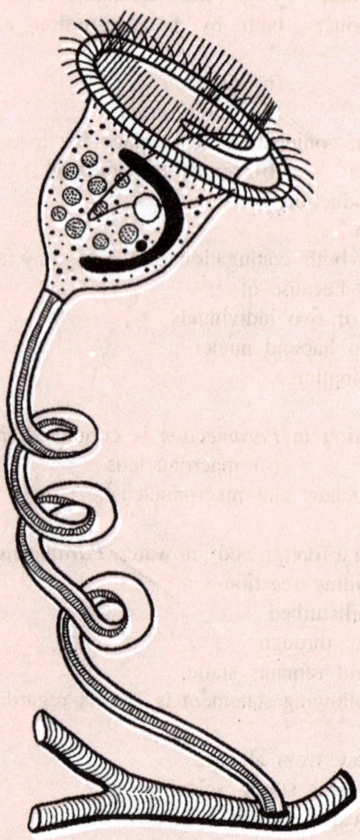

Unlike *Paramecium,* certain ciliates occur as sessile individuals, attached by a stalk to the substratum. *Vorticella* (L., **vortex,** a whirlpool) is such a ciliate. Its body is almost bell-shaped or campanulate with free anterior end surrounded by cilia. The following text is generalized and mostly relates to *V. microstoma, V. campanula* and *V. nebulifera.* Over 200 species have been described.

SYSTEMATIC POSITION

Phylum	Protozoa
Subphylum	Ciliophora
Subclass	Peritricha
Order	Peritrichida
Suborder	Sessilina
Family	Vorticellidae
Genus	*Vorticella*

HABITS AND HABITAT

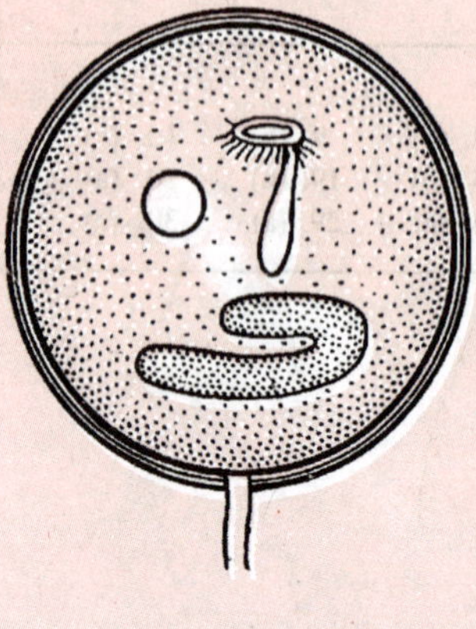

Vorticella is an extremely common, solitary and stalked ciliate, often found in freshwater ponds, lakes, rivers, streams, etc. Several individuals often occur socially in large groups anchored by their large contractile stalks to aquatic plants, animals, stones, twigs, etc. The contractile stalk permits the body to expand and contract frequently. Under certain conditions the body becomes detached from its stalk and swims freely. There are about 200 species, all of which are solitary. They occur in large groups but are never colonial. They found abundantly in stagnant water containing decaying organic matter,

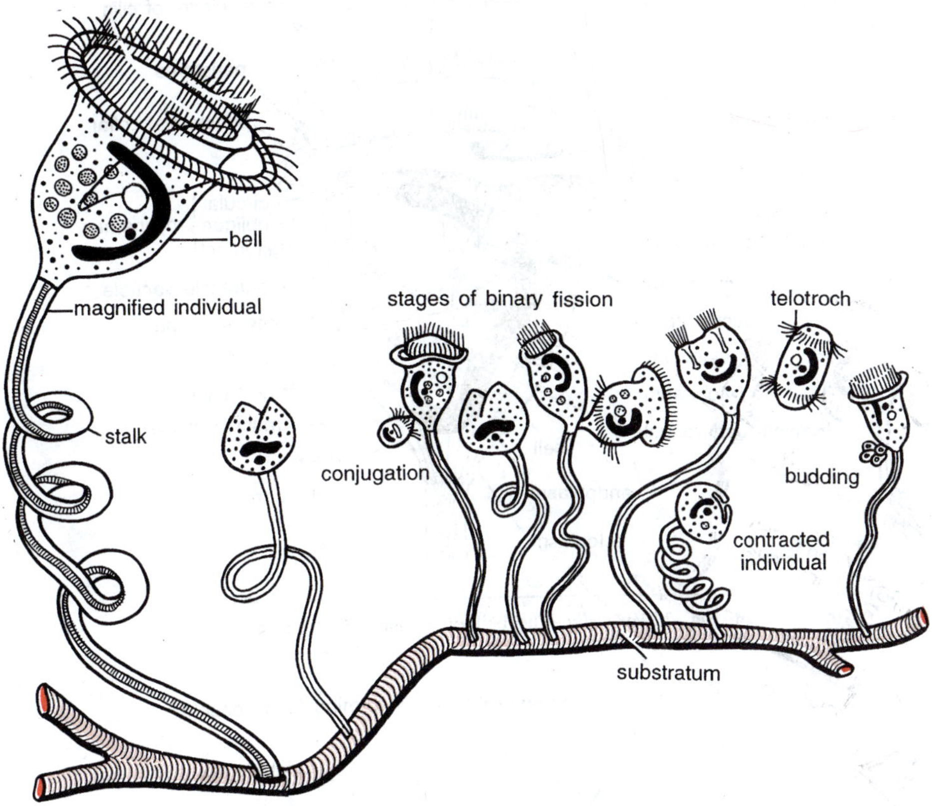

Fig. 1. *Vorticella.* A group of individuals in different stages.

feeding largely on bacteria. But some species, such as *V. campanula* and *V. nebulifera,* can live only in uncontaminated waters, where bacterial growth does not become too great. Some species are marine, some are epizoic and a few are parasitic. *V. microstoma* is the one which is more frequently seen in laboratory infusions.

STRUCTURE

Shape, Size and Colouration

Vorticella has a campanulate or bell-shaped asymmetrical body. To its base is attached a long contractile stalk. Both body and stalk are capable of great individual variations in size and form. The bell of largest species (*V. campanula*) varies in size up to 160μ long and 100μ broad, while the length of stalk varies from 55 to 4150μ. Body of the smaller species, *V. microstoma,* measures about 35μ by 55μ. *V. nebulifera* is greenish and *V. campanula* bluish in colour.

Bell

The body proper of each individual is shaped like a solid inverted bell. Its detailed structure is as follows :

1. Peristome. The broader free end of bell, with a slightly convex central area, is termed the **peristomial disc**. Its outer margin is thickened so as to form a prominent rim called **peristomial collar, border** or **lip**. It is separated from peristomial disc by a shallow, circular and marginal depression, known as **oral groove** or

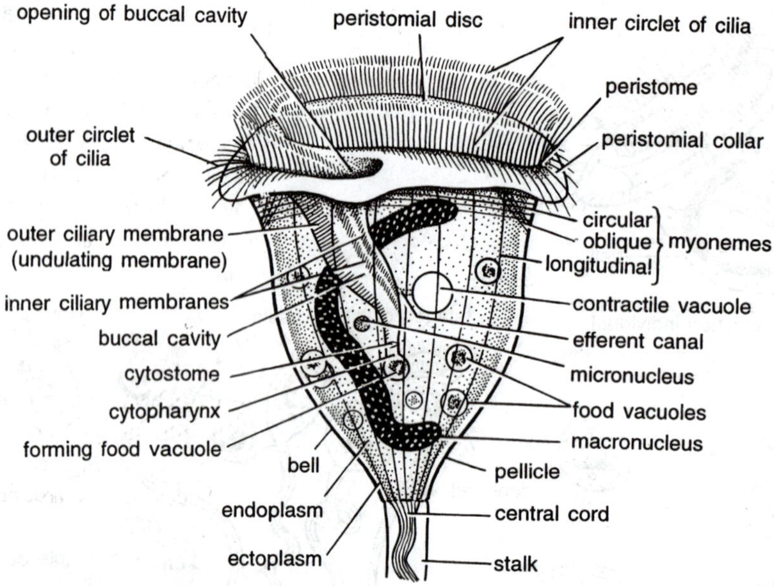

Fig. 2. *Vorticella.* The bell of an individual (magnified).

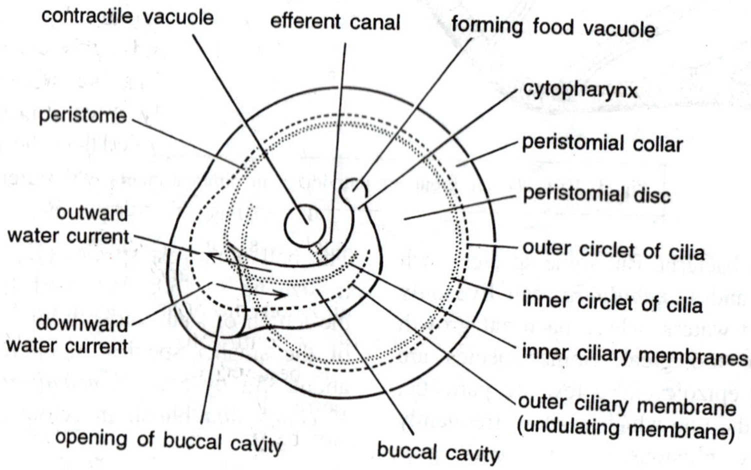

Fig. 3. *Vorticella.* Peristome as seen from above.

peristome. Cilia are inserted in this groove. The collar can close over the disc when the animal is retracted.

2. Food passage. On one side of the disc the peristomial groove is deeply invaginated forming a funnel-like **vestibule** or **buccal cavity**. At its inner end, the buccal cavity opens through a cell mouth or **cytostome** into cell gullet or **cytopharynx,** which ends in a food **vacuole** in endoplasm.

3. Oral ciliature. There are no somatic or body cilia in *Vorticella.* Only the peristomial groove bears the oral or aboral cilia arranged in a whirlpool manner. They form three concentric rows of ribbon-like membranes, which are arranged into an inner and an outer circlet. The inner circlet includes two membranes, while the outer circlet is a single membrane. Bases of all the three membranes are close together. The three ciliary membranes wind in a counter-clockwise direction around the margin of disc and then continue downward into buccal cavity. Inside buccal cavity the ciliary membranes separate. Two inner membranes run along the inner wall of buccal cavity, and the outer membrane follows its outer wall. Cilia of outer membrane become long and fused together forming an **undulating membrane**. Cytopharynx is devoid of cilia.

4. Ectoplasm. Body cytoplasm is differentiated into an outer layer of clear **ectoplasm** and an inner granular fluid of **endoplasm**. Ectoplasm has no trichocysts. It consists of an elastic outer **pellicle** and a thicker inner layer of contractile fibres or **myonemes**.

(a) Pellicle. Pellicle forms an outer envelope of the bell and also lines the buccal cavity. It is especially thick at the base of the bell. It is devoid of cilia but their basal granules are present in circles showing that their cilia are lost. Basal granules can be demonstrated by the **Klein silver-line* method**. Besides, on closer examination, the pellicle shows indistinct circular striations where the cilia might have been.

(b) Myonemes. Myonemes form a system of variously running fibres which function differently on contraction. Longitudinal myonemes shorten the body, oblique myonemes pull the disc inwards and circular myonemes contract and close the peristomial border over cilia and disc. Myonemes running parallel down the sides of bell are more visible at the base where they converge before entering the stalk.

5. Endoplasm. The granular endoplasm contains nuclei, contractile and food vacuoles, etc.

(a) Nuclei. Endoplasm contains a large, elongated, rod-like and horseshoe-shaped **macronucleus**. It is highly polyploid with a large amount of chromatin material scattered in the nucleoplasm. A tiny **micronucleus** occurs in close association with the macronucleus. It is rarely seen in living animals. To see the micronucleus it is very necessary to stain it with acetocarmine or any of the more permanent stains.

(b) Contractile vacuole. Located near buccal cavity is a clear and permanent pulsating space or contractile vacuole. It is about 7μ in diameter when fully distended. It shows diastole (expansion) and systole (contraction) in a rhythmic manner and pours its contents into an **efferent canal**, which leads into the buccal cavity. In some species there are two contractile vacuoles as in *V. picta* and *V. monilata.*

(c) Food vacuoles. In the endoplasm are present many large and small food vacuoles containing food particles. They are formed at the inner end of cytopharynx and constrict off as rounded spheres into the endoplasm.

(d) Cytopyge. The undigested food particles are passed into buccal cavity through a temporary or permanent opening, the **cytopyge** or **cytoproct**, formed at a weaker spot of the ectoplasm.

Stalk

The bell-shaped body of *Vorticella* remains attached to the substratum by a long, uniformly thin, unbranched and highly contractile **stalk**. It is also devoid of cilia. The older view is that the stalk is formed only by ectoplasm. But, electron microscopy has revealed that the stalk consists of two parts, central canal and outer sheath. Central canal consists of endoplasm, a contractile cord which is a specialized myoneme, called **spasmoneme**, and some granules, possibly mitochondria. Spasmoneme is a bundle of spiral fibrils continuous with the myonemes of the bell. It is believed that contractility of stalk is due to contractile ability of spasmoneme. On the stalk may be present commensal bacteria.

LOCOMOTION

Normally, the cilia remain very active and movements of stalk are frequent and rapid. When the animal is feeding, its stalk remains fully extended and the bell sways to and fro like a flower in the breeze. Individuals do not move in unison; each individual sways to its own rhythm.

* **Klein** (1926) used silver nitrate to demonstrate a complicated silver-line system, consisting of a network of fibrils, blackened.

But *Vorticella* is extremely sensitive to any mechanical stimulus. When irritated, all activities cease instantly. Stalk is retracted into a close and delicate spiral, cilia are stilled, disc is pulled in and closed over by peristomial lip and the rounded body is brought close to bottom, where it rests motionless until the danger has passed.

NUTRITION

Nutrition is holozoic as in *Paramecium.* A water current produced by adoral cilia of peristome brings bacteria and other small organic particles into the peristomial groove, from where they are further carried into buccal cavity. Within buccal cavity the water current flows downward between inner and outer ciliary membranes and outward between inner wall of buccal cavity and inner ciliary membranes. Food particles are driven towards cytostome along the outer membrane helped by undulations of two inner membranes. Terminal end of cytopharynx with food particles is pinched off along with some water into the endoplasm forming food vacuoles one after the other. Formation of food vacuoles may be compared with the formation of soap bubbles at the end of a pipe. In endoplasm, the food vacuoles move in an irregular cyclosis and not in a regular cyclosis like that of *Paramecium.* Digestion is similar to that of *Paramecium* and the medium of food vacuoles is first alkaline, then acidic and then it becomes alkaline till absorption. Excess of digested food forms refractile glycogen granules stored in endoplasm.

RESPIRATION, EXCRETION AND OSMOREGULATION

Respiration and excretion take place by diffusion as in *Paramecium.*

A single, large and pulsating contractile vacuole is present in endoplasm between the disc and buccal cavity. It opens into buccal cavity through a permanent opening and pulsates rhythmically showing diastole and systole phases. At diastole phase, the excess water from endoplasm is secreted into it and at systole phase, the water is expelled into buccal cavity.

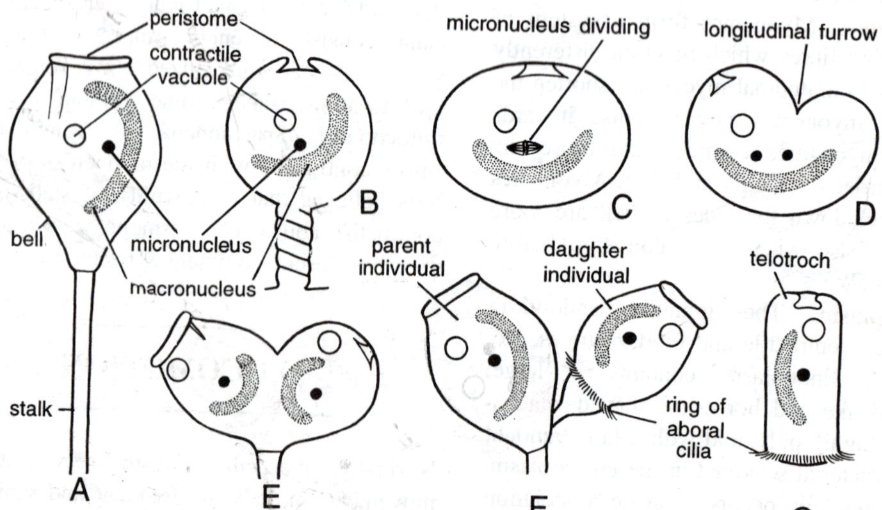

Fig. 4. *Vorticella.* Stages of binary fission.

CONTRACTILITY AND IRRITABILITY

The striking characteristic of *Vorticella* is its extreme irritability. It very readily responds to external stimuli. Even a flash of change in its surrounding water or the touch of a minute particle will bring about change in the relative position of its parts. First to respond is always the stalk, which becomes coiled into a close spiral to reduce its size. Then the disc is withdrawn and peristome closes over it. As a result the body form becomes somewhat globular.

REPRODUCTION

Reproduction takes place asexually by longitudinal binary fission and at intervals by conjugation. Encystment also occurs under unfavourable conditions.

[I] Longitudinal Binary Fission

The organism splits into two by longitudinal binary fission. During fission, *Vorticella* closes its peristome over the disc and the body becomes depressed and transversely elongated. The elongated, curved macronucleus shortens to become straight and comes to lie transversely in the middle of body. Macronucleus now divides into two daughter nuclei amitotically, whereas the tiny micronucleus divides by mitosis. Meanwhile, a vertical constriction develops at the free distal end, and then passes down the length of bell to one side of stalk. It divides the animal into two somewhat unequal parts or daughter individuals. Smaller daughter individual is without a stalk, while the larger daughter individual retains the parental stalk. Smaller individual acquires a ring of aboral cilia and develops a new contractile vacuole. It becomes cylindrical, gets detached basally and is now called a **telotroch**. It swims about rapidly by its posterior circlet of cilia keeping its posterior or aboral end foremost. Later, it settles down by its aboral end, which also has a short adhesive disc or **scopula**, which is a concavity bordered by a projecting rim and stiff cilia-like projections. Scopula secretes a stalk by which the telotroch gets fixed. Now the scopula is lost, the bell expands and a new peristomial disc is formed so that it metamorphoses soon into an adult. The whole process of binary fission is completed in about 20 to 30 minutes. The larger individual, retaining old disc, may be called the **parent**, while the smaller individual, or telotroch, the **offspring**.

[II] Conjugation

In *Vorticella,* sexual reproduction takes place by conjugation. It is known as **anisogamontogony** because the two conjugating individuals (**anisogametes**) are dissimilar in size. **Macroconjugant** is the larger, stalked and sessile individual (the female) looking like an ordinary

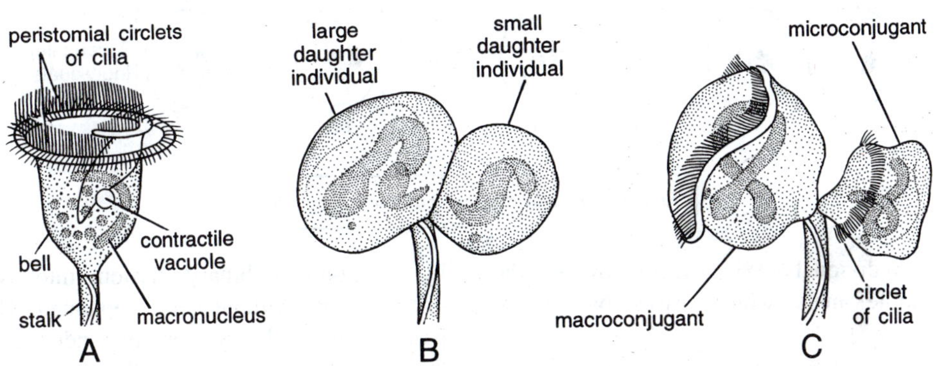

Fig. 5. *Vorticella.* Stages showing formation of micro and macroconjugants.

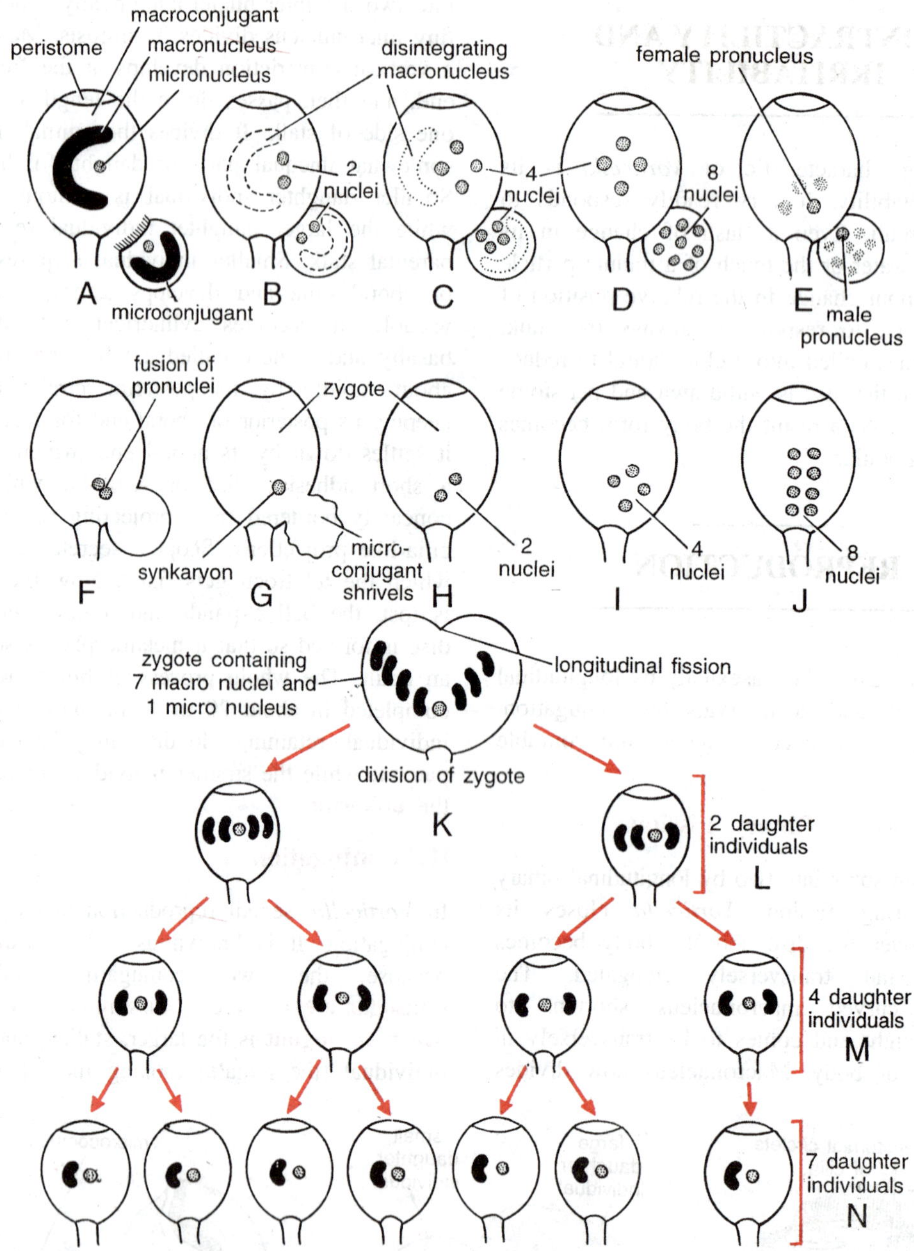

Fig. 6. *Vorticella.* Stages showing conjugation.

vegetative individual. **Microconjugant** is the smaller motile male which arises by special divisions.

1. Formation of micro and macroconjugants. *Vorticella* first divides asexually by longitudinal binary fission into two very unequal parts without being separated. The larger part is the ordinary somatic individual, while the smaller part is called the **microconjugant**. In some species (*V. nebulifera*), more than one

Table 1. **Comparison of Conjugation in *Paramecium* and *Vorticella.***

Characters	*Paramecium*	*Vorticella*
1. Conjugants	1. Conjugants are **isogamous**, *i.e.*, similar in size.	1. Conjugants are **anisogamous**, *i.e.*, dissimilar in size.
2. Dimorphism of preconjugants	2. No morphological sexual dimorphism in male and female preconjugants.	2. Dimorphism distinct. Male or **microconjugant** is smaller than female or **macroconjugant**.
3. Formation of conjugants (Parentage)	3. Conjugating individuals are smaller 1250μ in size than normal individuals (300μ) and come from different parents.	3. Produced by unequal longitudinal binary fission of a single parent.
4. Motility	4. Pairing conjugants continue to swim actively while attached together like Siamese twins.	4. Male or microconjugant swims actively before fixing upon a macroconjugant which is stalked and sessile.
5. Mating types	5. Conjugation occurs between two conjugants belonging to two different mating types which are physiologically different.	5. Both conjugants show morphological as well as physiological differences and come from two different parents.
6. Union	6. Two conjugants unite temporarily for exchange of nuclear material.	6. Union of two conjugants is permanent.
7. Exchange of nuclei	7. As in *Vorticella* male pronucleus of one conjugant enters the other conjugant and fuses with its female pronucleus (amphimixis) to form synkaryons.	7. As in *Paramecium,* mutual exchange of pronuclei takes place but synkaryon is formed only in macroconjugant.
8. Postconjugants	8. After amphimixis, both conjugants separate as exconjugants.	8. After amphimixis, macroconjugant absorbs the contents of microconjugant whose pellicle falls off.
9. Further divisions	9. Each exconjugant undergoes further nuclear and cytoplasmic divisions, independently.	9. Microconjugant degenerates. Only macroconjugant undergoes further divisions.
10. Offsprings	10. Each exconjugant produces 4 daughter paramecia on completion of conjugation.	10. Only macroconjugant gives rise to 7 daughter individuals or *Vorticella.*
11. Nature of conjugation	11. Conjugation is not a sexual reproduction. It is merely an exchange of nuclear material. There is no fusion of cytoplasm.	11. Conjugation involves fusion of nuclei and cytoplasm of both conjugants. It is intermediate between conjugation and syngamy.

microconjugant are produced by repeated fissions. Each microconjugant acquires a girdle of cilia at the posterior end, becomes detached from stalk and swims about actively in water. Swimming is an adaptation ensuring conjugation in sessile species. Microconjugant differs from telotroch in being much smaller in size, in never metamorphosing into adult and in never forming a stalk. Microconjugants never feed or encyst and survive only for about 24 hours after which they die. **Macroconjugants** are produced by the normal stalked individuals, in which certain nuclear modifications have taken place. They are morphologically similar to the normal trophic individuals, but are specialized physiologically and can attract microconjugants for about two hours. Macroconjugants are stationary and passive, while microconjugants are motile and active.

2. Fusion of conjugants. A microconjugant swims and fuses by its aboral end with a macroconjugant in the lower third of its body. After attachment, microconjugant loses cilia and throws off its pellicle. The two conjugants differ in their further development. Macronuclei of both conjugants disintegrate and finally disappear. In microconjugant, three nuclear divisions take place forming 8 nuclei, out of which 7 disintegrate and the remaining one becomes the **male pronucleus**.

In macroconjugant, micronucleus undergoes two divisions forming 4 nuclei, 3 of which disintegrate and the remaining one becomes the **female pronucleus**. In both cases, first division is reduction division and hence the male as well as female pronuclei are haploid.

Later, the male pronucleus of microconjugant migrates into cytoplasm of macroconjugant and fuses with its female pronucleus, resulting in a diploid **zygote nucleus** or **synkaryon**. After this the microconjugant shrivels and perishes. Fertilized macroconjugant is now termed as **zygote**.

In case of some species, like *V. monilata,* one more division of the remaining pronucleus in both the conjugants takes place, so that each conjugant contains two pronuclei, one stationary and the other migratory. The mutual exchange of migratory pronuclei takes place. However, the synkaryon is formed only in the macroconjugant, whereas the microconjugant along with its contained nuclei is perished and absorbed.

3. Post-zygotic development. Synkaryon of zygote undergoes three mitotic divisions to form 8 nuclei. Out of these, 7 nuclei become macronuclei and the remaining one, the micronucleus. Micronucleus now divides mitotically producing two daughter nuclei followed by cytokinesis. Thus, two daughter cells are formed, one with 4 macronuclei and one micronucleus and the other with 3 macronuclei and one micronucleus. Each daughter cell and its micronucleus divide twice. Daughter cell with 4 macronuclei forms 4 daughter individuals, and daughter cell having three macronuclei gives rise to only three individuals, each having one micronucleus and one macronucleus. The 7 daughter cells thus produced begin to grow, acquire stalks and finally become adults.

DISPERSAL AND ENCYSTATION

Under unfavourable conditions a normal individual of *Vorticella* may also grow a posterior ring of cilia to become a **telotroch**. The telotroch then breaks

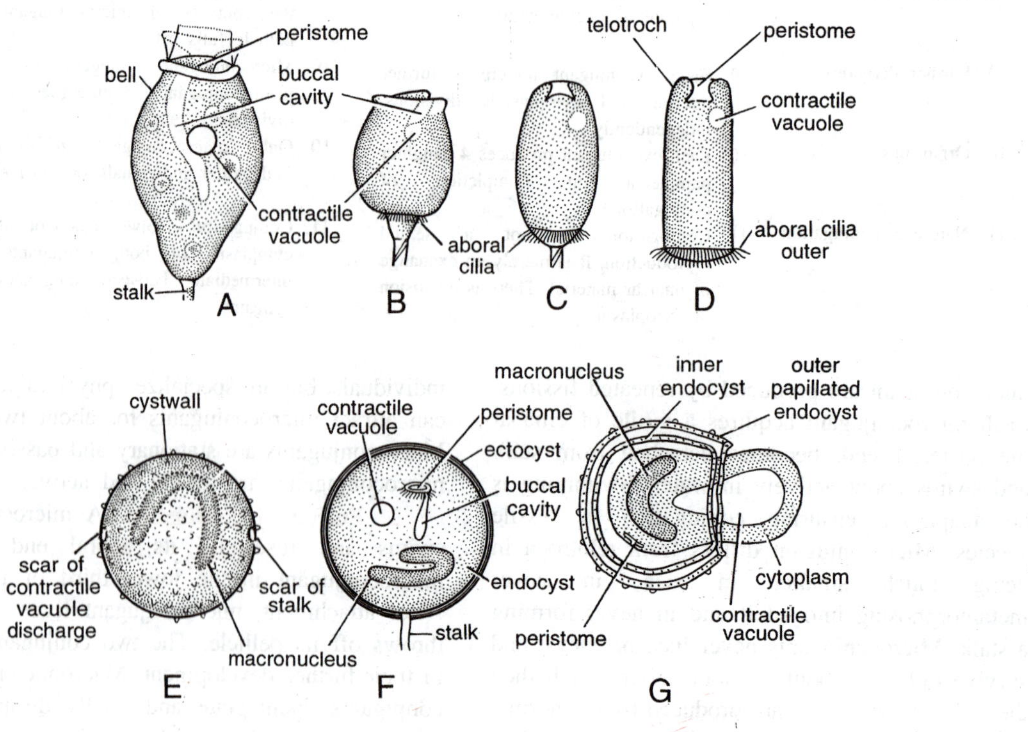

Fig. 7. *Vorticella.* **A-D.** Formation of telotroch. **E.** Complete cyst. **F.** Cyst in section. **G.** Excystation.

loose from its stalk, swims away to a more favourable spot and there becomes fixed again by growing a stalk. *Vorticella* is also known to tide over adverse conditions by undergoing **encystation**. The whole bell-body encysts on the stalk from which it finally breaks off. The stalk contracts to its maximum and the bell-body loses water, rounds off and secretes around it a gelatinous **ectocyst**. It soon becomes a wrinkled membrane and underneath it develops a double-layered **endocyst**. Myonemes become indistinct and pulsating rate of contractile vacuole goes very low, till it disappears. After some time, peristome is absorbed. Completed cyst of *V. microstoma* measures about 38μ in diameter and its poles are marked by two scars. One scar represents the point of attachment with stalk, while the other scar indicates the last point of escape for excreted water. On the return of favourable conditions excystation takes place. Contractile vacuole becomes enlarged and the organism grows an aboral circlet of cilia to become a telotroch. After some time free swimming, telotroch gets fixed to some time substratum and grows into an adult *Vorticella*.

Important Questions

➤➤*Long answer type questions*

1. Give an illustrated account of the structure of *Vorticella*.
2. Describe the process of conjugation in *Vorticella* and compare it with that of *Paramecium*.
3. Write short notes on : (i) Peristome, (ii) Scopula, (iii) Telotroch.

➤➤*Short answer type questions*

1. Name any one distinguishing feature of *Vorticella* organism.
2. Describe the telotroch of *Vorticella*.
3. Give the structure of bell in *Vorticella*.
4. Write on the conjugation in *Vorticella*.
5. Distinguish between micro- and macro-conjugants in *Vorticella*.
6. What is the size of the bell of largest species of *Vorticella* i.e. *V. campanula* ?
7. Give a brief account on contractility and irritability of *Vorticella* ?

➤➤*Multiple choice questions*

1. *Vorticella* belongs to the order :
   (a) spirotricha (b) oligotricha
   (c) chonotricha (d) peritrichida
2. Common name of *Vorticella* :
   (a) whorl animalcule (b) bell animalcule
   (c) ball animalcule (d) both
3. Contraction of body occurs by :
   (a) pellicle (b) myonemes
   (c) cytoplasm (d) none
4. Fresh water Protozoa :
   (a) *Euglena* (b) *Amoeba*
   (c) *Vorticella* (d) all
5. Telotroch present in :
   (a) *Paramecium*
   (b) *Vorticella*
   (c) *Euglena*
   (d) all
6. Nutrition in *Vorticella* is :
   (a) Holozoic (b) Saprozoic
   (c) Mixotropic (d) Autotrophic
7. Cytopyge is found in :
   (a) *Amoeba* (b) *Euglena*
   (c) Mixotropic (d) *Paramecium*

Answers

➤➤*Multiple choice questions*

1. (d) 2. (b) 3. (b) 4. (b) 5. (b) 6. (a) 7. (c).

14

Protozoa : Characters, Classification and Types

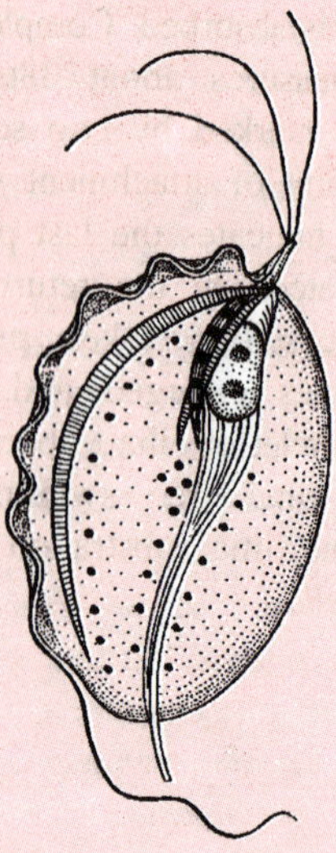

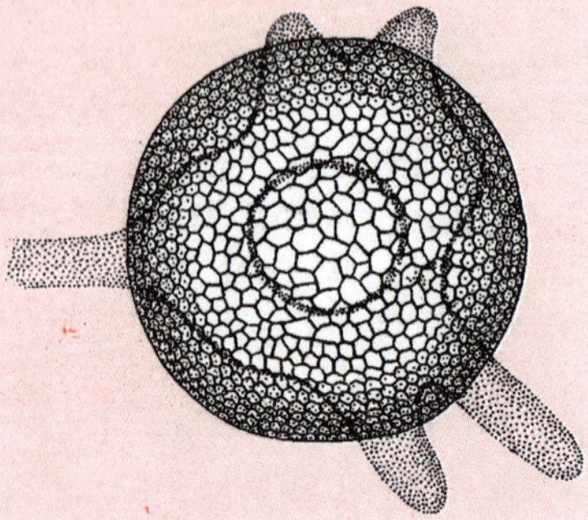

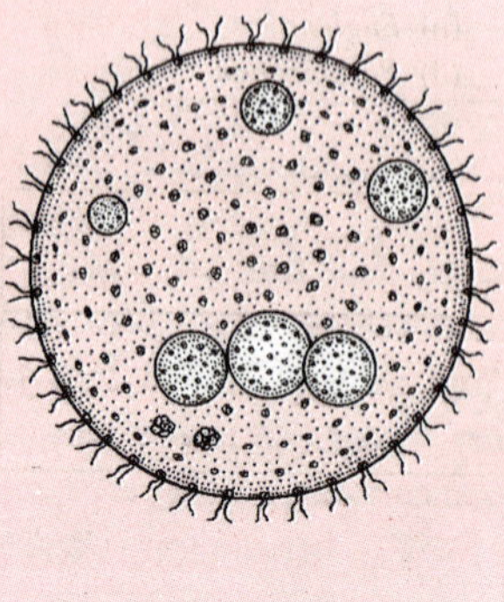

DEFINITION

Protozoa may be defined as microscopic and acellular animalcules, without tissues and organs, having one or more nuclei, but no nucleus ever incharge of a specialized part of cytoplasm. They exist either single or in colonies which differ from a metazoan in having all the individuals alike except when engaged in reproductive activities.

GENERAL CHARACTERS

1. Small, usually **microscopic** animalcules, ordinarily not visible without a microscope.
2. Simplest and most **primitive** of all animals, with protoplasmic grade of organization.
3. Body **unicellular**, containing one or more nuclei which are monomorphic or dimorphic.
4. **Solitary** or forming loose **colonies** in which individuals remain alike and independent.
5. Body symmetry none, bilateral, radial or spherical.
6. Body **naked** or bounded by a **pellicle** and often provided with simple to elaborate **shells** or **exoskeletons**.
7. Body form usually constant, varied in some, while changing with environment or age in many.

8. The single cell body performs all the essential and vital activities, which characterize the animal body; hence only subcellular **physiological division of labour**.
9. Locomotor organelles are finger-like **pseudopodia** or whip-like **flagella** or hairlike **cilia** or absent.
10. Nutrition **holozoic** (animal-like), **holophytic** (plant-like), **saprozoic** or **parasitic**. With or without definite oral and anal apertures. Digestion occurs intracellularly inside food vacuoles.
11. Respiration and excretion through general surface or through **contractile vacuoles**, which serve mainly for **osmoregulation**.
12. Reproduction **asexual** by binary or multiple fission and budding, and **sexual** by conjugation of adults (hologamy) or by fusion of gametes (syngamy).
13. Life history often complicated with alternation of asexual and sexual phases.
14. **Encystment** commonly occurs to help in dispersal as well as to resist unfavourable conditions of food, temperature and moisture.
15. **Free-living** Protozoa mostly aquatic, inhabiting fresh and sea waters and damp places. **Parasitic** and **commensal** Protozoa live over or inside the bodies of animals and plants. Sufficient moisture is essential in their environment.
16. The single-celled individual not differentiated into somatoplasm and germplasm; therefore, exempt from natural death which is the price paid for the body.
17. Protozoans evolved 3.8–4.5 billion years ago, presently has 36,400 living species.

CLASSIFICATION

Phylum Protozoa is a large and varied group and poses a number of problems in its classification. The conventional scheme followed by **Hyman** (1940), **Hickman** (1961) and **Storer** (1965) etc. recognizes 2 subphyla on the basis of organs of locomotion and 5 classes, briefly outlined as follows :

Subphylum A. Plasmodroma

Locomotory organelles are flagella, pseudopodia, or none. Nuclei of one kind.

Class 1. Mastigophora. Move by one to many flagella. Ex. *Euglena.*

Class 2. Sarcodina. Move and capture food by pseudopodia. Ex. *Amoeba.*

Class 3. Sporozoa. No locomotory organs. All parasitic. Spore-formation common. Ex. *Plasmodium.*

Subphylum B. Ciliophora

Cilia or sucking tentacles throughout or at certain stages. Nuclei of 2 kinds.

Class 4. Ciliata. Move by cilia throughout life. Ex. *Paramecium.*

Class 5. Suctoria. Move by cilia as young and by tentacles as adult. Ex. *Podophyra.*

The following classification of Protozoa is based on the scheme given by the Committee on Taxonomy and Taxonomic Problems of the Society of Protozoologists, and mainly proposed by **B.M. Honigberg** and others (1964). It divides Protozoa first into 4 subphyla : (1) **Sarcomastigophora**, (2) **Sporozoa**, (3) **Cnidospora**, and (4) **Ciliophora**. Only important orders have been mentioned here.

Subphylum I. Sarcomastigophora

Locomotor organelles pseudopodia or flagella or both. Nuclei of one kind (monomorphic).

Superclass A. Mastigophora (=Flagellata)

1. Simple, primitive, with firm pellicle.
2. Locomotor organelles flagella.
3. Nutrition autotrophic or heterotrophic, or both.

CLASS 1. Phytomastigophorea (= Phytoflagellata)

1. Chlorophyll-bearing chromatophores present.
2. Nutrition mainly holophytic by phototrophy.
3. Reserve food starch or paramylon.
4. Flagella 1 or 2, sometimes more.

Order 1. Chrysomonadida

1. Small, with thin pellicle, often amoeboid. Flagella 1 to 3.
2. Gullet absent. Stigma often present.
3. Chromatophores 1 or 2, yellow or brown, and discoidal.
4. Starch absent. Leucosin and fats may be present.
   Examples : *Chrysamoeba, Synura, Ochromonas, Dinobryon.*

Order 2. Cryptomonadida

1. Small, with a rigid pellicle. Flagella 2.
2. Anterior gullet reaches up to middle of body.
3. Chromatophores 2, yellow, brown or colourless.
4. Reserve foodstuff starch, sometimes oils.
   Examples : *Chilomonas, Cryptomonas.*

Order 3. Euglenida

1. Large, pellicle thick and firm. Flagella 1 or 2.
2. Anterior end with a gullet leading into a reservoir.
3. Chromatophores numerous, green or colourless.
4. Reserve foodstuff paramylon and oils.
   Examples : *Euglena, Peranema, Phacus, Copromonas.*

Order 4. Volvocida (= Phytomonadida)

1. Small, with rigid cellulose covering (theca).
2. No gullet. Flagella 2 to 4.
3. Chromatophore green, usually cup-shaped.
4. Reserve foodstuff starch and oils.
   Examples : *Chlamydomonas, Volvox.*

Order 5. Chloromonadida

1. Small, dorso-ventrally flat. Pellicle delicate.
2. Gullet present.
3. Chromatophores green and numerous.
4. Reserve foodstuff oils.
   Examples : *Vacularia, Coelomonas, Gonyostomum.*

Order 6. Dinoflagellida

1. Small, planktonic. Naked, amoeboid or with a thick cellulose theca.
2. Gullet present or absent. Flagella 2.
3. Chromatophores yellow or brown.
4. Reserve foodstuff starch or oils or both.
5. Some are bioluminescent.
   Examples : *Noctiluca, Ceratium.*

CLASS 2. Zoomastigopohrea (= Zooflagellata)

1. Chlorophyll or chromatophores absent. Mostly parasitic.
2. Nutrition holozoic or saprozoic.
3. Reserve food glycogen.
4. Flagella one to many.

Order 1. Rhizomastigida

1. Small, amoeboid, chiefly freshwater.
2. Locomotion by 1-4 flagella and pseudopodia.
   Examples : *Mastigamoeba, Dimorpha.*

Order 2. Kinetoplastida

1. No gullet. Kinetoplast present.
2. Flagella 1 to 4. No definite pellicle.
3. Mostly parasitic forms living in blood.
   Examples : *Bodo, Leishmania, Trypanosoma.*

Order 3. Choanoflagellida

1. A collar round the base of a single flagellum.
2. Free-living, solitary or colonial.
   Example : *Proterospongia.*

Order 4. Diplomonadida

1. Bilaterally symmetrical, binucleate, with delicate pellicle and often with a cytostome.
2. Flagella 3 to 8. One often trailing or forming border of an undulating membrane.
3. Mostly intestinal parasites.
   Examples : *Hexamita, Giardia.*

Order 5. Hypermastigida

1. Highly specialized, numerous flagella.
2. Kinetosomes arranged in a circle, plate or longitudinal or spiral rows.
3. Mouth absent. Food ingested by pseudopodia.
4. Gut parasites of termites and cockroaches.
   Examples : *Lophomonas, Trychonympha.*

Order 6. Trichomonadida

1. Flagella 4 to 6. One flagellum trailing.
2. Parasites of vertebrates.
   Example : *Trichomonas.*

Superclass B. Opalinata

1. Entire body covered by cilia-like flagella.
2. Nuclei 2 to many, monomorphic.
3. Reproduction by symmetrogenic binary fission or by syngamy of anisogametes.
4. Parasitic mainly in frogs and toads.
   Examples : *Opalina, Zelleriella.*

Superclass C. Sarcodina (=Rhizopoda)

1. Body mostly amoeboid without definite pellicle. Some with a skeleton of some kind.
2. Locomotion by pseudopodia.
3. Nutrition holozoic or saprozoic.

CLASS 1. Rhizopodea

Pseudopodia as lobopodia, filopodia or reticulopodia, without axial filaments.

Subclass (a) Lobosia

Pseudopodia as lobopodia.

Order 1. Amoebida

1. Body amoeboid, naked, without skeleton.
2. Nucleus with honeycomb lattice.
3. Largely freshwater and free-living. Many parasitic.
   Examples : *Amoeba, Entamoeba, Pelomyxa.*

Order 2. Arcellinida (= Testacida)

1. Body enclosed in one-chambered shell of pseudochitin, with a single opening through which lobopodia protrude.
2. Free-living, mostly freshwater.
   Examples : *Arcella, Difflugia, Euglypha.*

Subclass (b) Filosia

Pseudopodia as filopodia. Naked or with a shell with single aperture.

Examples : *Allogromia, Penardia* (naked).

Subclass (c) Granuloreticulosia

Pseudopodia delicate granular reticulopodia.

Order Foraminiferida

Large sized with uni- or multichambered calcareous shell with one or more openings through which reticulopodia emerge.

Examples : *Globigerina, Elphidium* (=*Polystomella*)

CLASS 2. Actinopodea

Pseudopodia mainly axopodia with axial filaments, radiating from a spherical body.

Subclass (a) Heliozoia

1. Spherical protozoans, called sun-animalcules.
2. Pseudopodia (axopodia) radiating.
3. Naked or skeleton of siliceous scales or spines.
   Examples : *Actinophrys, Actinosphaeriuri.*

Subclass (b) Radiolaria

1. Body naked or with perforated chitinoid central capsule separating ectoplasm from endoplasm.
2. Reticulopodia, axopodia or filopodia.
3. Skeleton mostly of siliceous spicules or of strontium sulphate.
   Examples : *Collozoum, Thalassicola.*

Subclass (c) Acantharia

1. Imperforate non-chitinoid central capsule without pores.
2. Skeleton of strontium sulphate.
3. Pseudopodia are axopodia.
   Example : *Acanthometra.*

Subclass (d) Proteomyxidia

1. Pseudopodia are filopodia.
2. Mostly parasites on algae.
   Examples : *Vampyrella, Pseudospora.*

CLASS 3. Piroplasmea

Small parasites in red blood cells of vertebrates.

Example : *Babesia* (formerly included with Sporozoa, but its species do not produce spores).

Subphylum II. Sporozoa

Locomotor organelles absent. Spores usually present. Exclusively endoparasites.

CLASS 1. Telosporea

Spores without polar capsules and filaments, naked or encysted.

Subclass (a) Gregarinia

1. Mature trophozoites large, extracellular in host's gut and body cavities.
2. Each spore produces 8 sporozoites.
3. Parasites in invertebrates.
   Examples : *Monocystis, Gregarina.*

Subclass (b) Coccidia / Apicomplexa

1. Mature trophozoites small and intracellular.
2. Each oocyst produces many sporozoites.
3. Blood or gut parasites of vertebrates.
   Examples : *Eimeria, Isospora, Plasmodium.*

CLASS 2. Toxoplasmea

Spores absent. Only asexual reproduction.
Example : *Toxoplasma.*

Class 3. Haplosporea

Spore cases present. Only asexual reproduction.
Example : *Ichthyosporidium.*

Subphylum III. Cnidospora

Spores with polar filaments present.

CLASS 1. Myxosporidea

1. Spores large, developed from several nuclei.
2. Spores with two or three valves.
3. Parasites mostly in fishes.
   Examples : *Myxidium, Myxobolus, Ceratomyxa.*

CLASS 2. Microsporidea

1. Spores small, developed from one nucleus.
2. Spores with a univalved membrane.
3. Intracellular parasites in arthropods and fishes.
   Example : *Nosema.*

Subphylum IV. Ciliophora

Presence of cilia as locomotor and feeding organelles at some stage in the life cycle. Nuclei of 2 kinds (dimorphic).

CLASS Ciliata (=Infusoria)

1. Locomotor organelles numerous hair-like cilia, present throughout life.
2. Definite mouth (cytostome) and gullet present except in a few parasitic forms. Anal aperture (cytopyge) permanent.
3. One or more contractile vacuoles present even in marine and parasitic types.
4. Mostly two kinds of nuclei, large *macronucleus* and smaller *micronucleus.*

Subclass (a) Holotricha

1. Body cilia simple and uniform.
2. Buccal cilia mostly absent.

Order 1. Gymnostomatida

Large ciliates without oral ciliature. Cytostome opens directly. No vestibule.
Examples : *Coleps, Didinium, Prorodon, Dileptus.*

Order 2. Trichostomatida

With vestibular but no buccal ciliature.
Examples : *Balantidium, Colpoda.*

Order 3. Chonotrichida

No body ciliature. A spirally coiled apical funnel contains vestibular cilia.
Examples : *Spirochona, Lobochona.*

Order 4. Apostomatida

Spirally arranged body cilia. Cytostome midventral.
Example : *Hyalophysa.*

Order 5. Astomatida

Body ciliation uniform. Cytostome absent.
Examples : *Anoplophyrya, Maupasella.*

Order 6. Hymenostomatida

Body ciliation uniform. Buccal cavity ventral with ciliary membranes.
Examples : *Colpidium, Paramecium.*

Subclass (b) Peritricha

1. Adult without body cilia.
2. Apical end with buccal cilia.

Order : Peritrichida

Examples : *Vorticella, Carchesium.*

Subclass (c) Suctoria

1. Sessile and stalked body.
2. Young with cilia, adult with suctorial tentacles.

Order : Suctorida

Examples : *Acineta, Ephelota, Podophyra.*

Subclass (d) Spirotrichia

1. Reduced body cilia.
2. Buccal cilia well marked.

Order 1. Heterotrichida

Body cilia short. Uniform or absent.

Examples : *Stentor, Bursaria, Spirostomum.*

Order 2. Oligotrichida

Body cilia reduced or absent. Buccal membranes conspicuous.

Examples : *Strombidium, Halteria.*

Order 3. Hypotrichida

Dorso-ventrally flattened. Fused cilia forming ventral cirri.

Examples : *Euplotes, Stylonchia.*

OTHER TYPES OF PROTOZOA

1. *Chlamydomonas.* *Chlamydomonas* is a typical genus of the order Phytomonadida. The microscopic, unicellular and solitary individuals are found in stagnant freshwater ponds. Ovoid or flattened body is surrounded by a thick and delicate cellulose cell wall or capsule. It bears two long flagella, each arising from a blepharoplast in cytoplasm. The cytoplasm includes a vesicular nucleus, a cup-shaped chloroplast, with a pyrenoid, a red stigma, a pair of small contractile vacuoles and reserve food granules as starch and oil droplets. Nutrition is mainly holophytic but osmotrophic tendency is also seen. Asexual reproduction is by binary fission, often in palmella stage. Sexual reproduction is by isogamy, and in some, by anisogamy.

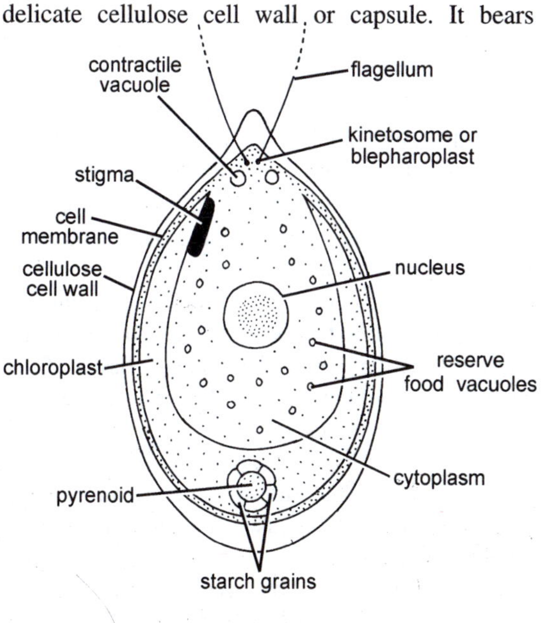

Fig. 1. *Chlamydomonas.*

2. *Volvox.* *Volvox* is a beautiful, colonial phytomonad commonly found in freshwater ponds and lakes. Colony looks like a small green, hollow sphere filled with a watery jelly. The sphere contains thousands of zooids (somatic cells), arranged in a single peripheral layer. Each zooid has 2 flagella, 2 or more contractile vacuoles, a large cup-shaped chloroplast, a single nucleus, a red stigma but no gullet. Fine protoplasmic strands connect adjacent zooids. Colony swims by rolling due to combined action of flagella. Nutrition is holophytic. Colony increases in size by binary fission of the surface somatic cells. Asexual reproduction takes place from reproductive zooids, mostly situated in the posterior part of colony. Some of these zooids, called **parthenogonidia**, repeatedly divide to form daughter colonies, which are liberated by the disintegration of parent colony. Sexual reproduction takes place by specialized reproductive zooids, called **antheridia** and **archegonia**, which form biflagellate **microgametes**

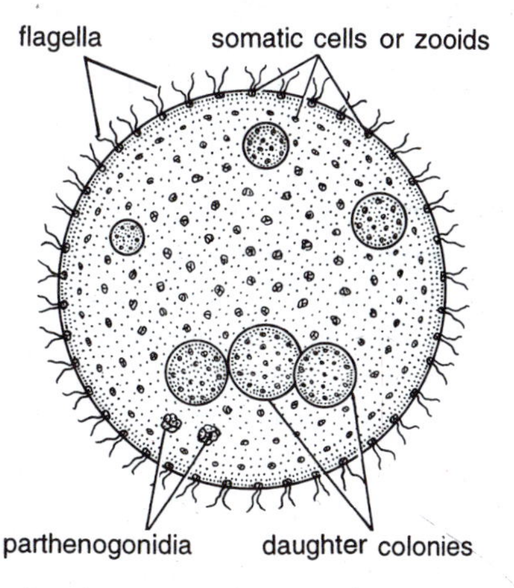

Fig. 2. *Volvox glabator.* Vegetative colony.

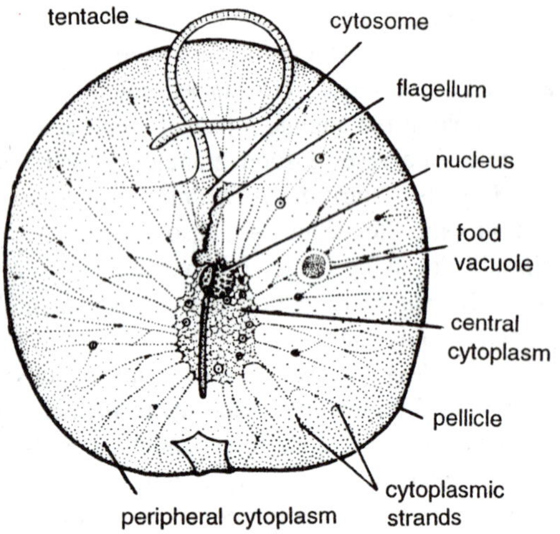

Fig. 3. *Noctiluca.*

and large non-flagellate **macrogametes**, respectively. Fertilization occurs in water. Zygote acquires a resistant cyst wall. Under favourable conditions, it germinates to form a new colony.

3. *Noctiluca.* *Noctiluca* is a marine and pelagic dinoflagellate, occurring in abundance on the water surface near shores. Spherical body is naked, jelly-like and slightly pink in colour. Mid-ventrally, there is a large pouch-like **cytostome** leading into an elongated cytopharynx, which bears a small delicate flagellum and a large motile **tentacle**. Cytoplasm is much vacuolated at the periphery, but is dense centrally which contains a single nucleus. Locomotion is affected by the locomotor tentacle. Nutrition is exclusively holozoic in the absence of chloroplasts. Reproduction takes place by binary fission and multiple fission (sporulation). During sporulation, cytoplasm forms a multinucleate surface disc, which by exogenous budding gives rise to numerous uninucleate and uniflagellate swarmers or spores. These spores behave as **luciferin** which under the influence of an enzyme, **luciferase**, emits light.

4. *Ceratium.* *Ceratium* includes marine as well as freshwater species. The microscopic flattened body is armoured by **thecal plates**, which are sculptured with spines and pits. It is produced into an anterior and generally two posterior spines.

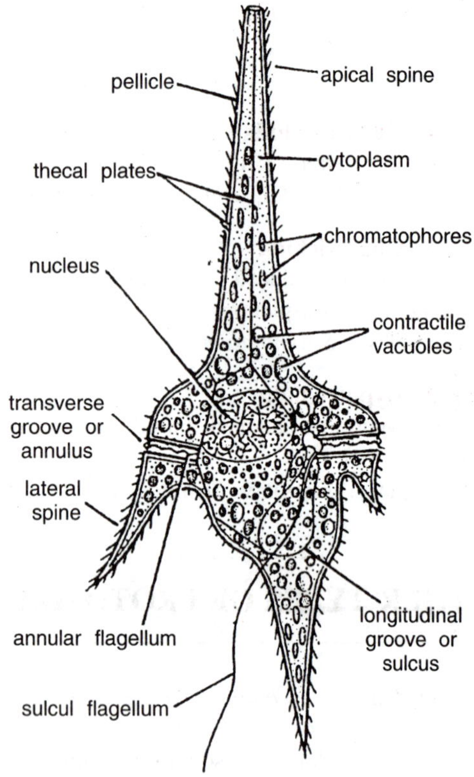

Fig. 4. *Ceratium.*

The shell is grooved by a transverse and a longitudinal furrow, each containing a flagellum. The deep transverse groove or **annulus** encircles the body and contains the transverse flagellum. The longitudinal groove or **sulcus** is provided with large locomotor sulcul flagellum. The cytoplasm contains a large oval nucleus, numerous chromatophores and a complex system of contractile vacuoles. The animal is mainly holophytic, but a few species may be holozoic. The reproduction is by oblique binary fission. Cyst formation sometimes occurs. Several species are known to be luminescent.

5. *Trichomonas*. *Trichomonas* is a multiflagellate and the most common endoparasite of invertebrates as well as vertebrates. Its body is roughly egg-shaped or pear-shaped, slightly tapering posteriorly. Four flagella arise at the anterior end from a group of basal bodies forming the **basal body complex**. The recurrent or trailing flagellum is directed posteriorly bordering an

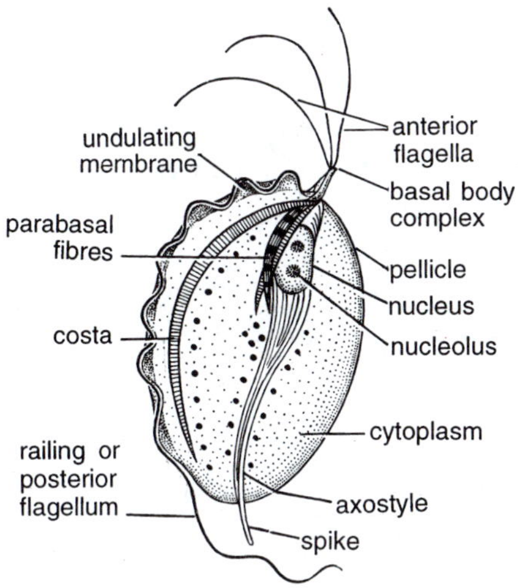

Fig. 5. *Trichomonas tenax.*

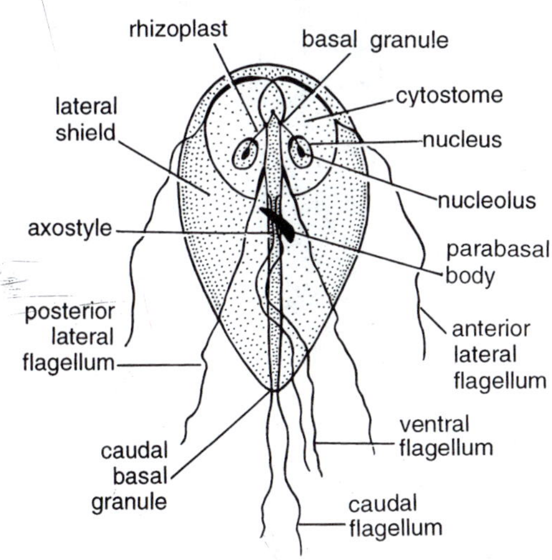

Fig. 6. *Giardia lamblia.*

undulating membrane. Two large striated **parabasal fibres** arise from the basal body complex. A contractile fibre **costa** runs along the line of attachment of undulating membrane. Body is supported by a central red or cylinder of microtubules, the **axostyle**. It frequently projects from the posterior end of body as **spike** and serves to anchor the animal while it feeds. Endoplasm contains an oval **nucleus** with few nucleoli. Broad anterior end of body bears ventrally a **cytostome** used for ingestion while feeding holozoically. Saprozoic mode of feeding is also met with. Reproduction takes place exclusively by longitudinal binary fission.

Parasitic human species of *Trichomonas* are *T. tenax* of mouth, *T. hominis* of colon and *T. vaginalis* of vagina. Of other animals, *T. foetus* of cattle and *T. gallinae* of doves and pigeons are important. The disease caused by species of *Trichomonas* is called **trichomoniasis**.

6. *Giardia.* *Giardia,* commonly nicknamed as the 'Grand Old Man of the Intestine', is a diplomonadid parasitic flagellate occurring in the intestine of man and other animals. *Giardia (lamblia) intestinalis* of small intestine (duodenum) of man possesses a bilateral symmetrical body and measures 10-18μ in length. Flagella are four on each side, all springing from the middle and hinder region of body. Basal granules of foremost flagella are interconnected and also connected with a pair of conspicuous vesicular nuclei lying just posterior to them. A single axostyle forms the longitudinal axis of body. A parabasal body is prominent. Reproduction in human intestine takes place by longitudinal binary fission. Infection is by uninucleate cysts which pass out in faeces and infect other persons through contaminated food and water. *Giardia* interferes with the normal digestive processes.

7. *Arcella.* *Arcella* commonly occurs in stagnant fresh water containing much vegetation and also in moist forest soils. It secretes a yellow to brown, thick, hard, transparent and hemispherical or disc-shaped **shell** or **test** around itself and carries it along as it moves. The shell is made of **pseudochitin**, a proteinous substance containing silica and iron. It is finely sculptured and arched above but the ventral surface shows an inverted funnel-like depression leading to a single central opening, the mouth or **pylome**. One to six small, hyaline, finger-like, simple or branched lobopodia are extended through the pylome and serve as locomotor and feeding organelles. Body is attached to inner wall of test by ectoplasmic strands, known as **epipodia**. Endoplasm contains various reserve food particles, food vacuoles and contractile vacuoles. One or more vacuoles are said to be non-contractile and filled with a gas, to which probably the organism

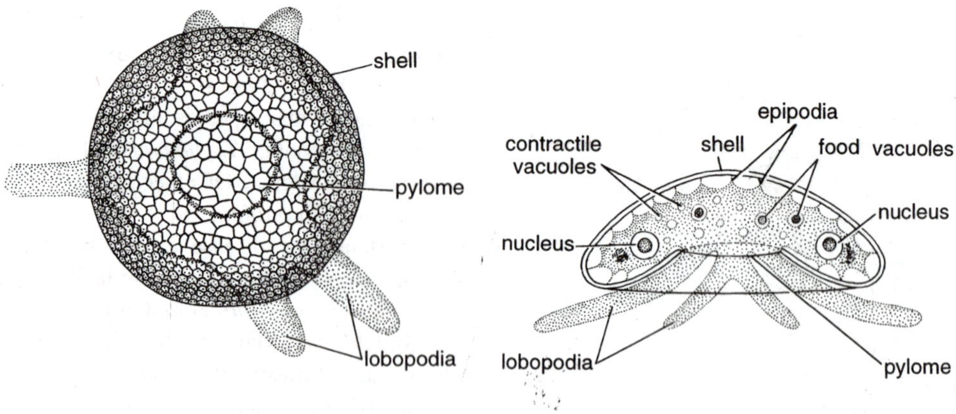

Fig. 7. *Arcella vulgaris.* **A.** Dorsal view. **B.** Side view.

owes it buoyancy. Two vesicular **nuclei** are present in *A. vulgaris.* Asexual reproduction takes place by binary fission. Multiple fission is also said to take place.

8. *Difflugia.* *Difflugia* is a free-living test-acid, occurring in small leaf-chocked freshwater puddles, ponds, ditches or even in moist soil. It builds around itself a beautifully symmetrical shell or test by cementing together sand grains, diatom shells, spicules of freshwater sponges and other foreign substances by a sticky chitinous secretion. Test may be pyriform, vase-shaped or flask-shaped according to the species and has a distinct neck with a terminal mouth opening or **pylome**. As many as half a dozen slender lobose pseudopodia (lobopodia) may protrude out from **pylome**. Numerous contractile and food vacuoles are scattered around the periphery of endoplasm. A single **nucleus** is centrally located in endoplasm occupying the bulbous part or **fundus** of shell. *Difflugia* feeds upon microorganisms and filamentous algae such as *Spirogyra.* It reproduces typically by binary fission, but multiple fission has also been reported. *Difflugia oblongata* remains active in spring and summer, but becomes encysted within its test in late fall.

9. *Actinophrys.* A common shell-less heliozoan is *Actinophrys sol,* which lives in stagnant fresh water, floating about among vegetation or rolling over bottom. Mechanism of its rising or sinking in water is not well understood. Its minute spherical body measures about .05 mm in diameter. Protoplasm is not distinctly divided into endoplasm and ectoplasm. Inner endoplasm is not vacuolated but contains a large central nucleus and food vacuoles. Outer ectoplasm is packed with numerous, blister-like vacuoles, filled with a clear fluid. Usually a single and large contractile vacuole is found near the surface of body. Numerous, slender pointed and stiff pseudopodia (axopodia) radiate from all sides of body. Their axial filaments run through cytoplasm and converge around the centrally

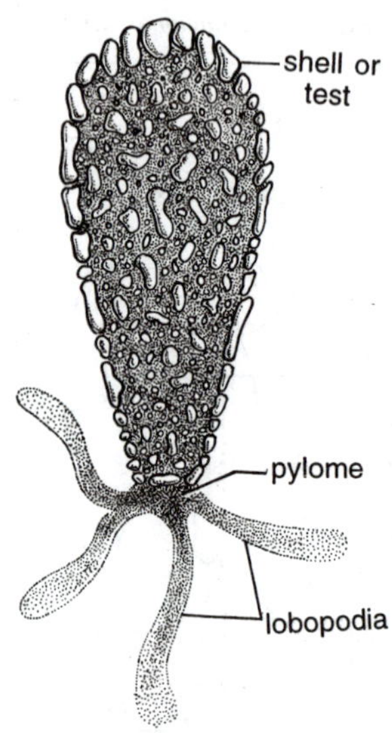

Fig. 8. *Difflugia.*

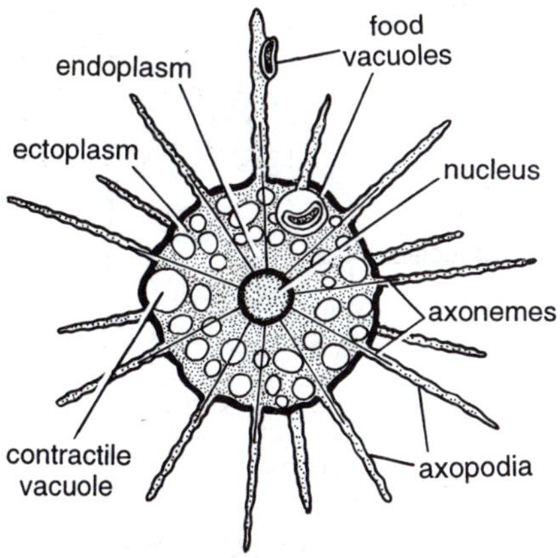

Fig. 9. *Actinophrys sol.*

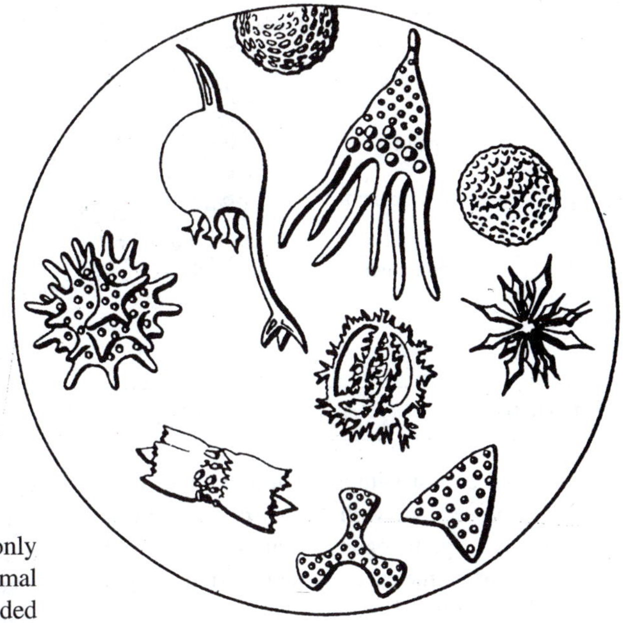

Fig. 10. Radiolarian ooze.

placed nucleus. *Actinophrys sol* is commonly known as the "sun animalcule" because the animal with radiating axopodia resembles sun surrounded by rays of light. Axopodia are used for gathering food and not for locomotion. Food organisms are paralysed when they come in contact with pseudopodia and devoured, as in *Amoeba.* Binary fission occurs with or without formation of a cyst wall. Self-fertilization (autogamy) or paedogamy takes place in encysted condition, when two daughter gametes become fused and the zygote, after resting for some time, divides into two daughter cells, which eventually emerge out of their cysts and develop into independent adults.

10. Radiolarian ooze. Radiolaria are exclusively marine and planktonic Sarcodina, found mostly in deep sea, particularly the tropical Pacific, floating or drifting at varying depths. They resemble heliozoans in their outer appearance, but their cytoplasm is divided by a perforated and chitinous membrane or **central capsule**, into an intracapsular or **inner medulla**, and an extracapsular or **outer cortex**. Medulla contains one or more nuclei and oil globules. Cortex is vacuolated, pigmented and often contains yellow, unicellular and symbiotic algae. Numerous vaucoles filled with a light fluid give a frothy appearance to cytoplasm and reduce specific gravity of organism. Besides central capsule, extracapsular cytoplasm may contain scattered, siliceous spicules. In majority, however, a hard, transparent and symmetrical skeleton of silica (SiO_2) or strontium sulphate is present, often of bizarre shape and beauty. It usually forms an intricate lattice-work, through the opening of which are extended fine, stiff and ray-like pseudopodia, usually without conspicuous axial filaments. When these animals die, their minute but hard skeletons sink to the sea-bottom and form the **Radiolarian ooze**, which covers nearly 50,000,000 SQ kilometers miles of the ocean floor, particularly of the Pacific Ocean. Their fossils are abundant in rocks such as flint and chert. They also constitute a part of the "Tripoli Stone", used in abrasive powders for polishing metals.

11. *Gregarina.* Species of *Gregarina* live as intestinal parasites of cockroaches, grasshoppers, crickets, meal-worms, crayfish, centipedes and other arthropods. The better known species are *G. blattarum* of oriental cockroach, *G. oviceps* of crickets and *G. polymorpha* of mealworm.

Body of adult trophozoite or sporont is divisible into two chambers, anterior small **protomerite**, and posterior large **deutomerite**, containing nucleus. Sporonts associate in pairs

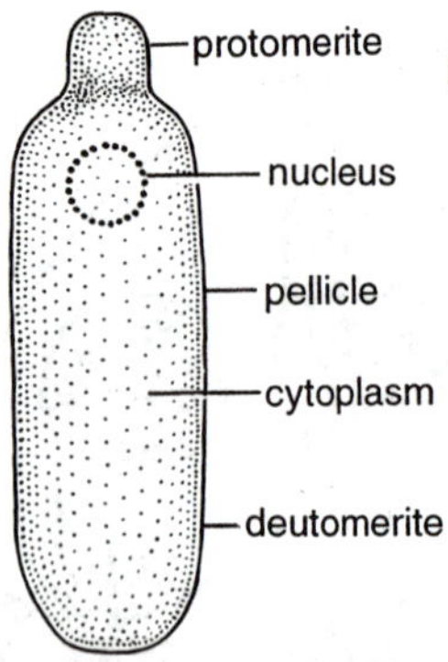

Fig. 11. *Gregarina blattarum.*

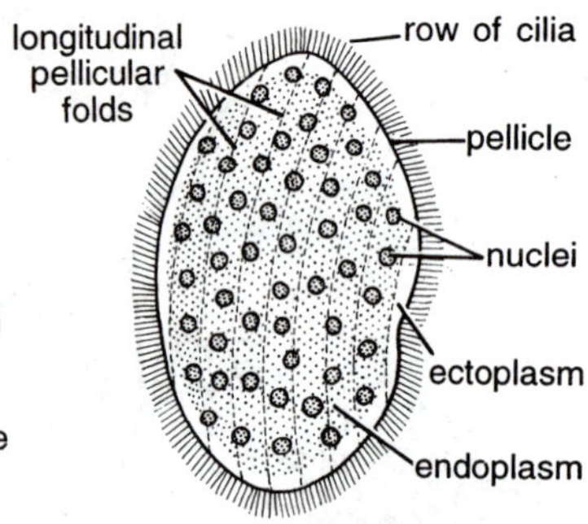

Fig. 12. *Opalina.*

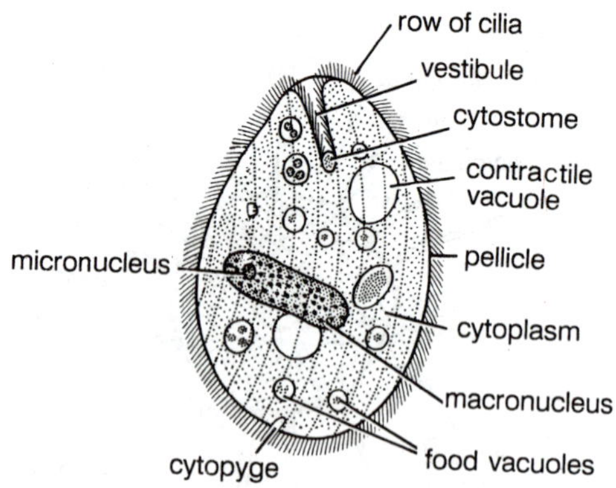

Fig. 13. *Balantidium.*

(syzygy) and encyst within a common gametocyst. Cysts are passed out with faeces of host. Within gametocysts, gametogony takes place by production of gametes and their fusion in pairs resulting in zygotes. These encyst within sporocysts and form spores. Three rapid divisions occur within spore, resulting in eight sporozoites. Gametocysts, containing spores, are discharged with host's faeces. These rupture to liberate spores which are ingested by another host. In the gut of new host spores hatch and each of the liberated sporozoites enters an epithelial cell and grows into an intracellular trophozoite. At this stage, trophozoite has an apical cap, the **epimerite**, for attachment to host cell. Older trophozoites lose their attachment to the host cells and become extracellular.

12. *Opalina.* The best known genus of superclass Opalinata (subphylum Sarcomastigophora) is *Opalina.* It lives as an endoparasite or endocommensal in large intestine of Anura. Commonest species is *O. ranarum* which is found in rectum of frogs and toads. Its body is flattened, leaf-like and oval in outline. It may reach a length of about 1 mm and may be seen with naked eyes. Thin but tough pellicle is uniformly covered by rows of innumerable and equal-sized cilia (flagella). Between rows of cilia, pellicle is thrown into narrow longitudinal folds, supported by microtubular fibrils. Mouth, gullet, cytopyge, food vacuoles and contractile vacuoles are absent. *Opalina* feeds by pinocytosis, ingesting liquid food contents of rectum of host through general body surface. A large number of small, spherical and similar nuclei are found evenly distributed in endoplasm, which also contains several minute spindle-like or ovoid granules of unknown character. Each nucleus has both trophochromatin and idiochromatin.

During breeding season of host, *Opalina* undergoes transverse as well as longitudinal binary fission's to produce smaller individuals, each with only three to six nuclei and few rows of cilia. These encyst and pass out in faeces of host. If cysts are swallowed by tadpoles, along with weeds, they hatch in rectum as gametocytes, which divide by a series of longitudinal fissions to form uninucleate, club-shaped gametes. These include small slender microgametes and large macrogametes. These conjugate in pairs to form zygotes which encyst for a while as **zygocysts**. With growth and rapid nuclear division, each zygote reaches the normal multinucleate adult stage.

13. *Balantidium.* *Balantidium* is a trichostomatid ciliate parasite, inhabiting the intestine of man, pigs, sheep, guinea pigs, camels opossums, ostriches, fish, cockroaches, etc. They are present in large numbers in rectal contents of frog. Of about 50 species of *Balantidium* only one species, *B. coli,* infects man. The organism is large and ovoid and measures about 60-70μ × 40-60μ. Entire outer surface is covered by longitudinal, slightly spiral rows of cilia. At the anterior end is a vestibule leading to the cell mouth or cytostome. It is lined by longer cilia.

Cytoplasm includes two contractile vacuoles, a macronucleus, and food vacuoles. At the hinder end is a permanent pore or cytopyge for egestion of undigested residue. Nutrition is holozoic and food is carried down into the vestibule forming food vacuole with a drop of water. Reproduction takes place by asexual binary fission. Infection takes place through cysts in contaminated food and water.

14. *Nyctotherus*. *Nyctotherus* is the intestinal, endo-commensal of amphibians and some invertebrates. Its important species are *N. cordiformis* in rectum of frogs and *N. ovalis* in intestine of cockroaches. Slightly compressed body is oval measuring about 60-120 μ in length. Outer surface bears longitudinal and slightly spiral rows of cilia. On lateral side is a vestibule leading to cell-mouth or cytostome. It bears somewhat longer cilia. Cytoplasm includes a contractile vacuole, a macronucleus, a micronucleus and food vacuoles. Cytopyge, a permanent pore for the egestion of food, is located at its hinder end. Nutrition is holozoic and reproduction takes place by binary fission and conjugation.

15. *Stentor*. *Stentor* is a free-swimming or sessile ciliate, mostly found in pond water contaminated with sewage. The sessile form takes a funnel or trumpet-shaped appearance. Oral surface bears a broad vestibule leading into buccal cavity, which is lined by a row of membranelles. Buccal cavity leads into cytopharynx through a cytostome. Macronucleus is greatly beaded or chained and several micronuclei lie close to it. A single contractile vacuole lies anteriorly and fed by a single and long radial canal. Nutrition is holozoic like *Paramecium* and reproduction takes place by oblique binary fission and conjugation.

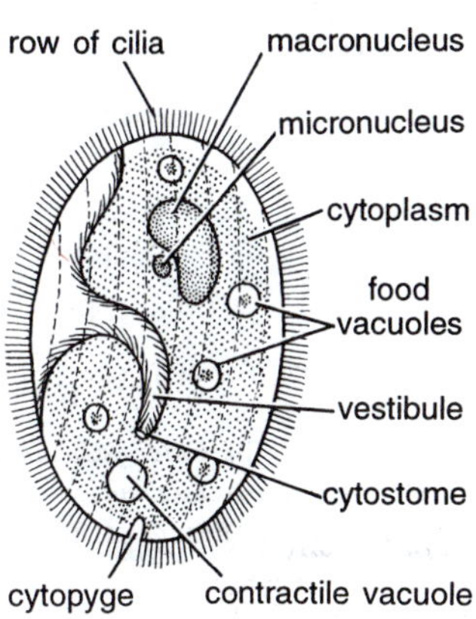

Fig. 14. *Nyctotherus.*

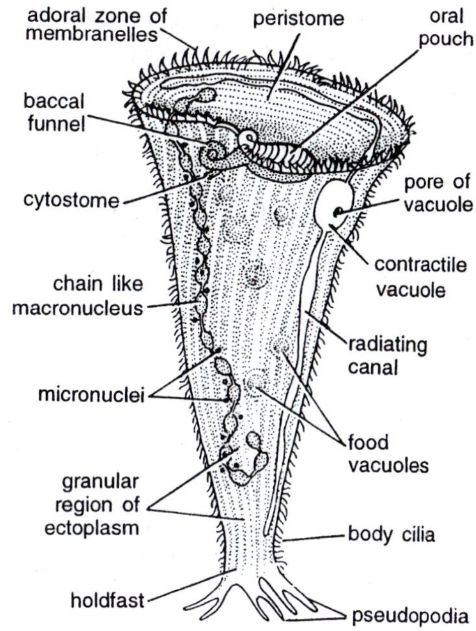

Fig. 15. *Stentor.*

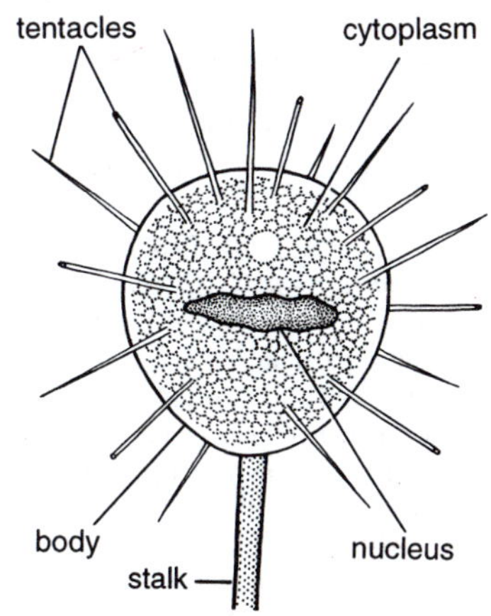

Fig. 16. *Ephelota.*

16. *Ephelota*. *Ephelota* is a marine attached form of subclass Suctoria (Class Ciliatea). Spherical, pedunculate body is not seated in a cup. The tentacles, borne distally, are of two kinds: (i) long, pointed and prehensile, used for piercing and (ii) short, cylindrical and knobbed, used for sucking. Reproduction occurs by a peculiar process of multiple exogenous budding. In *Ephelota gemmipara,* distal end of body projects into several buds, each receiving a branch of nucleus. Finally, the buds detach and acquire cilia on one surface. After a short, active and free-swimming existence, they lose cilia and grow into tentaculate adults.

Important Questions

➤➤Long answer type questions

1. Give the distinctive characters of Protozoa. Classify the phylum up to classes giving diagnostic characters and examples.
2. What is the basis of classification of phylum Protozoa into classes ?
3. Write short notes on : (i) *Ceratium,* (ii) *Gregarina,* (iii) *Leishmania,* (iv) *Noctiluca,* (v) Rectal ciliates.

➤➤Short answer type questions

1. Name a symbiotic protozoan present in digestive tract of termites.
2. What is the study of protozoans called as ?
3. Name the protozoan that lives in the intestine of wood eating termites.
4. Give an example for coprozoic protozoan.
5. Which order of protozoa shows division of the cytoplasm as central capsule ?
6. Give an example of a protozoan with
   (a) holophytic nutrition, (b) mixotrophic nutrition
7. Give any two reasons why *Opalina* is no longer included in ciliata ?
8. Classify *Opalina* upto order.
9. What is holozoic nutrition ? Give two examples from Protozoa.
10. State two important characters of the order Heliozoa. Name two examples.
11. Write note on *Opalina.*
12. Write note on Heliozoa.
13. Give the distinguishing character and one example of Haemosporidea.
14. Give the taxonomic position of the following animals :
    (a) *Monocystis* (b) *Polystomella*
    (c) *Amoeba* (d) *Euglypha*
    (e) *Paramecium* (f) *Giardia*
15. With reference to their habitat distinguish between coccidia and haemosporidia.
16. What is osmoregulation ? Give an account of the process of osmoregulation in Protozoa.
17. Mention three characters to show the affinities of suctoria with ciliata.
18. Differentiate between lobopodium and reticulopodium.
19. Distinguish between dinoflagellata and hypermastigina based on their flagellar condition.

➤➤Fill in the blanks

1. Conjugation is found in the class .....
2. *Balantidium coli* causes a disease called .... in man.
3. The plane of binary fission in ciliophora is .....

➤➤Multiple choice questions

1. Tentacles are found in :
   (a) *Stylonychna* (b) *Ephelota*
   (c) *Euglypha* (d) *Euplectella*
2. One of the important characteristic features of the order chrysomonadida is :
   (a) the presence of pseudopodium
   (b) the absence of gullet
   (c) the presence of gullet
   (d) the attachement of a long flagellum without gullet
   (e) absence of flagellum
3. The presence of a long flagellum and a short tentacle is the characteristic feature of :
   (a) *Copromonas* (b) *Chlamydomonas*
   (c) *Noctiluca* (d) *Euglena*
4. One of the characteristic feature of Telosporea is :
   (a) uninuclear trophozoites
   (b) multinuclear trophozoites
   (c) presence of shells
   (d) presence of an eye spot
   (e) elongate sporozoite

5. Class sporozoa includes parasitic protozoa in which locomotory organelles are :
(a) flagella (b) absent
(c) both cilia and pseudopodia
(d) of various types
6. Locomotion in young sporozoans is by :
(a) gliding (b) ciliary
(c) amoeboid (d) euglenoid
7. An example of sporozoan is :
(a) *Elphidium* (b) *Monocystis*
(c) *Difflugia* (d) *Vorticella*
8. Contractile vacuole is present :
(a) in marine protozoa (b) in parasitic protozoa
(c) in fresh water protozoa
(d) in all protozoa
9. Which of the following do not have locomotory organelles :
(a) Rhizopoda (b) Flagellate
(c) Ciliata (d) Sporozoa
10. Trichocysts are found in :
(a) ectoplasm of *Euglena*
(b) ectoplasm of *Paramecium*
(c) endoplasm of *Euglena*
(d) endoplasm of *Paramecium*
11. The mononucleate form is :
(a) *Balantidium* (b) *Chrysamoeba*
(c) *Stylonychia* (d) *Pelomyxa*
12. *Trichomonas* belongs to :
(a) protomonadina (b) dinoflagellata
(c) schistoflagellata (d) polymastigina
(e) trichomonadida
13. The characteristic feature of class Sporozoa is the presence of :
(a) flagella (b) parasitism
(c) chlorophyll (d) amoeboid movement
(e) cilia
14. The primary grouping of the protozoans is based on their :
(a) mode of feeding (b) locomotory organelles
(c) mode of reproduction
(d) size and shape
15. Class mastigophora includes the protozoan in which locomotory organelle is :
(a) lobopodium (b) flagella
(c) cilia (d) absent
(e) pseudopodia
16. The phylum protozoa is divided into sub-phyla on the basis of their :
(a) nuclear structure (b) locomotory structure
(c) mode of reproduction
(d) mode of nutrition (e) habitat
17. The presence of slender pseudopodia with interlacing branches to form a net work is the characteristic feature of the order :
(a) Radiolaria (b) Lobosoa
(c) Mycetozoa (d) Heliozoa
(e) Foraminifera
18. Order Lobosa includes :
(a) *Pelomyxa* (b) *Elphidium*
(c) *Globigerina* (d) *Euglypha*
19. *Chilomonas* belongs to the order :
(a) Chrysomonadina (b) Euglenodina
(c) Cryptomonadina (d) Phytomonadina
20. Order suctorida includes :
(a) *Trichonympha* (b) *Giardia*
(c) *Lophomonas* (d) *Ephelota*
21. *Proterospongia* is included in order :
(a) Euglenida (b) Rhizomastinoida
(c) Choanoflagellida (d) Rhizomastigida
22. *Volvox* is included in order :
(a) Phytomonadida (b) Euglenida
(c) Chrysomonadida (d) Dinoflagellata
23. *Ceratium* is included in order :
(a) Mastigophora (b) Dinoflagellata
(c) Euglenida (d) Ciliata
24. What are the locomotor organelles of Mastigophora :
(a) many pseudopodia
(b) many pseudopodia and cilia
(c) a flagellum and many pseudopodia
(d) reticulate pseudopodia and flagellum
(e) flagellum
25. Karyosome consists of two materials. What is the achromatic substance present in it ?
(a) plastin (b) chromosomes
(c) paramylum (d) eyespot
26. The shell of Polystomella is made up of :
(a) calcium carbonate (b) silicon
(c) chitin
27. Cytopyge is found in :
(a) *Amoeba* (b) *Paramecium*
(c) *Euglena* (d) *Trypanosoma*
28. The locomotory organelle of foraminifera are :
(a) lobopodia (b) filopodia
(c) axopodia (d) reticulopodia
29. Which of the following protozoans is unlikely to have a contractile vacuole ?
(a) *Paramecium* (b) *Plasmodium vivax*
(c) *Amoeba proteus* (d) *Euglena*
30. Losch discovered that :
(a) *Entamoeba histolytica* causes amoebic dysentery
(b) trophozoite of *Plasmodium* has brownish yellow pigment granules, called haemozoin granules
(c) *Trypanosoma* causes sleeping sickness
(d) digestion in *Amoeba* is similar to that of higher animals
31. Grand old man of the intestine is :
(a) *Amoeba* (b) *Entamoeba*
(c) *Trypanosoma* (d) *Giardia*
32. The disease caused by the infection of *Trichomonas* :
(a) amoebiosis (b) kala-azar
(c) sleeping sickness (d) trichomoniasis
33. 'Tripali stone' is characterise presence of :
(a) foramaniferian (b) radiolarian
(c) *Actinosphaerium* (d) all of them

34. Which show endocommensal form in large intestine of anura :
(a) *Opalina* (b) *Monocystis*
(c) *Trypanosoma* (d) none
35. *Balantidium* in housing the intestine of :
(a) man (b) pig
(c) sheep (d) fish
(e) all
36. In Protozoan classification is based on :
(a) Locomotary organelle
(b) Mode of Feeding
(c) Size and Shape
(d) Mode of reproduction
37. If certain Protozoan has two contractile vacuoles instead of one it suggests that :
(a) 'Drink more water'
(b) respiratory rate is extremely high
(c) excretes more water
(d) the temperature of environment is low
38. Protozoan with pseudopodia as locomotary organelles are :
(a) Ciliata (b) Mastigophora
(c) Sporozoa (d) Sarcodina
39. Which Protozoan resembles the ancestral form from which the plants and animals evolved :
(a) *Amoeba*
(b) *Paramecium*
(c) *Plasmodium*
(d) *Euglena*
40. Which of the following is a shelled Protozoan
(a) *Arcella* (b) *Elphidium*
(c) *Difflugia* (d) All
41. Presence of two types of nuclei, micronucleus and macronucleus is characteristics of :
(a) Sporozoan (b) Flagellata
(c) Ciliata
(d) Sarcodina
42. If all ponds and puddles are destroyed the organism likely to be destroyed is :
(a) *Leishmania* (b) *Plasmodium*
(c) *Trypanosoma* (d) *Ascaris*
43. Schuffner's dots in a patient of Malaria are found in :
(a) RBC's
(b) Gametocytes of *Plasmodium*
(c) Liver cells
(d) Signet- ring trophozoites

Answers

➤➤*Fill in the blanks*

1. ciliata, 2. chronic dysentery, 3. perkinetal.

➤➤*Multiple choice questions*

1. (b) 2. (b) 3. (c) 4. (a) 5. (b) 6. (a) 7. (b) 8. (c) 9. (d) 10. (b)
11. (b) 12. (d) 13. (b) 14. (b) 15. (b) 16. (b) 17. (e) 18. (d) 19. (c) 20. (a)
21. (c) 22. (a) 23. (b) 24. (e) 25. (a) 26. (b) 27. (d) 28. (b) 29. (a) 30. (a)
31. (d) 32. (d) 33. (b) 34. (a) 35. (e). 36. (a) 37. (d) 38. (d) 39. (d) 40. (d)
41. (c) 42. (b) 43. (a).

15

Protozoa : General Account

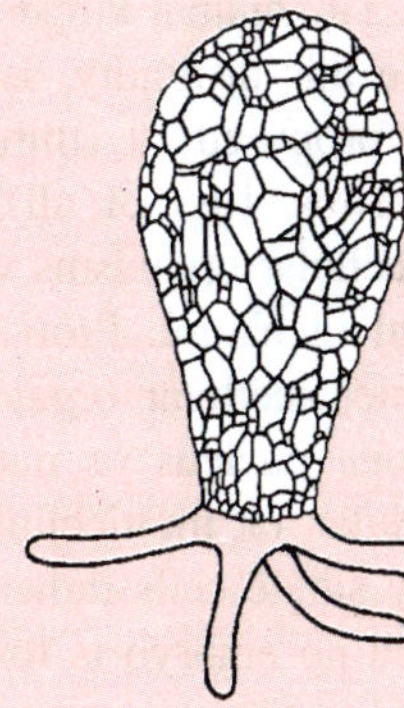

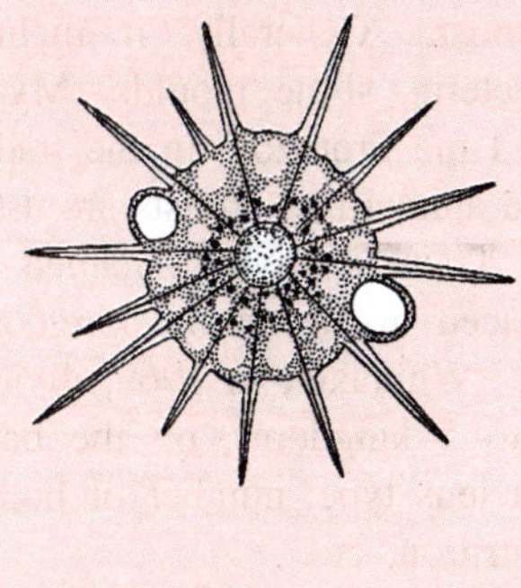

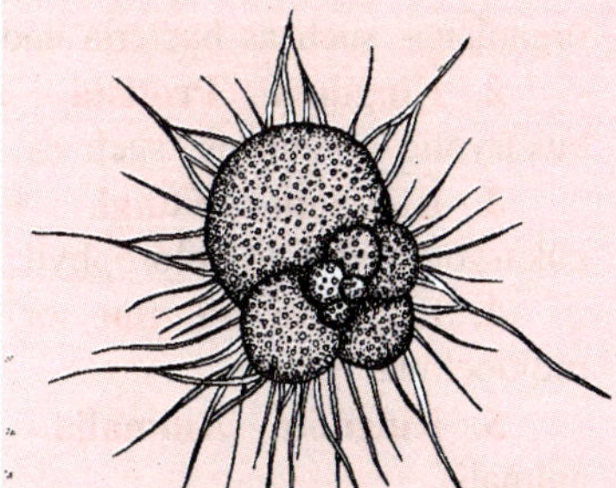

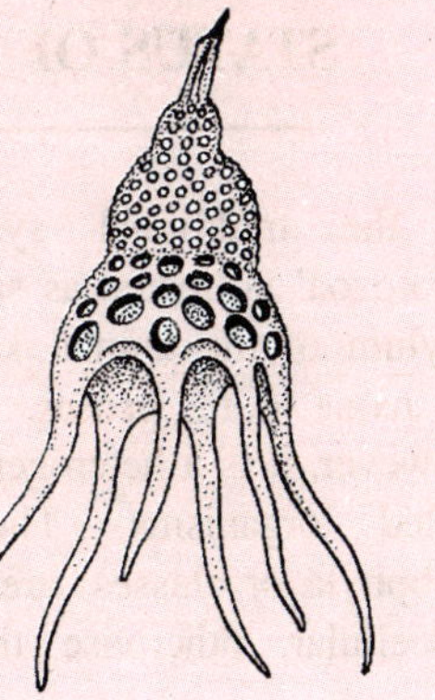

CONCEPT OF PROTISTA

There are certain basic differences between higher animals and higher plants. In general, plant cells are characterized by cellulose cell walls, but animal cells are devoid of cellulose cell walls. All plants synthesize complex organic foodstuffs by photosynthesis, with the help of chlorophyll present in cells, so nutrition is holophytic. But animals are unable to prepare their own food, due to lack of chlorophyll, and they have to depend on plant material, so their mode of nutrition is holozoic or saprophytic. All plants are stationary, but most animals are able to move.

On the basis of the above differences, all multicellular organisms can be clearly grouped either as animals or as plants. For example, roses, geraniums and oaks are distinctly plant-like organisms, whereas horses, lions and elephants are distinctly animal like organisms.

However, the above criteria cannot be applied to non-multicellular or non-cellular organisms (Protozoa and Protophyta) because of numerous difficulties. Sometimes it is difficult to tell whether non-cellular organisms are plants or animals. Some have characteristics of animals, some of plants, and many of both. Thus, botanists claim some, zoologists claim some, and many are claimed by both. For example. *Euglena* is treated as an animal in zoology textbooks, and as a plant in botany textbooks. Similarly, two genera of the protozoan order Cryptomonadina (*Cryptomonas* and *Cheilomonas*) are morphologically very similar, but chlorophyll is present in *Cryptomonas* and absent in *Cheilomonas.* Placing them in two different kingdoms on the basis of chlorophyll would mean widely separating these two closely related forms.

Considering these facts, it is difficult to classify all living organisms simply in two kingdoms, *i.e.*, animal kingdom and plant kingdom. To remove this difficulty, as early as 1866, **Ernst Haeckel** proposed a third kingdom *Protista*, (Gr., **protistos**, first of all), to include all those non-multicellular organisms which are neither plant nor animal in nature. Protista may be defined as microscopic, acellular organisms with no definite nucleus, one nucleus or many nuclei. They may be unicellular or multicellular, their gametes are formed by single cells rather than by multicellular gonads and no embryo is formed from the zygote.

However, there is no general agreement as to which organisms should be placed in the kingdom Protista. Generally, it includes such forms as bacteria, slime moulds (Mycetozoa, Myxophyta) and the Protozoa. In the traditional classification, the animal like protists are usually lumped together in a single phylum, named 'Protozoa', which is placed in the animal kingdom.

Whittaker, in 1969, divided all the organism; into 5 kingdoms, on the basis of cell structure, nucleus type, number of body cells and mode of nutrition, etc.

1. Kingdom Monera. For prokaryotic organisms, such as bacteria and blue-green algae.

2. Kingdom Protista. For single-celled eukaryotic organisms, such as Protozoa.

3. Kingdom Fungi. For multicellular, eukaryotes without chlorophyll, such as *Rhizopus*.

4. Kingdom Plantae or **Metaphyta.** For multicellular plants.

5. Kingdom Animalia. For multicellular animals.

STATUS OF PROTOZOA

In the traditional system of classification, 'Protozoa' is treated as subkingdom or as the first phylum of the animal kingdom, as indicated by its name (Gr., **protos**, first + **zoon**, animals). However, it is a heterogeneous grouping of single-celled organisms. The principal subgroups (subphyla or classes) are similar chiefly in being unicellular, otherwise they are distinctive and dissimilar in many other respects and display extreme diversity. They display all types of symmetry, a great range of structural complexity, and adaptation for all types of environmental conditions. Their evolution parallels that of multicellular animals and their origin is perhaps polyphyletic.

It seems improper to call protozoan organisms as 'animals' because sometimes it is difficult to tell whether the members are plants or animals, as they are intermediate forms having both plant and animal characteristics. Common examples are *Euglena* and *Volvox* having chlorophyll, which are considered animals by zoologists and plants (algae) by botanists. Most of the free-living flagellates are thus placed by phycologist under different phyla of algal organisms as they contain, chlorophyll and show plant-like autotrophic nutrition. They are much more closely related to multicellular green algae than to other flagellates. Similarly, of the two structurally similar cryptomonad genera, *Cryptomonas* has chlorophyll while *Cheilomonas* is without chlorophyll. On the basis of presence or absence of chlorophyll, these two genera would be placed separately in plant and animal kingdoms, respectively, which seems unrealistic. Further, some dinoflagellates have cellulose cell wall like plants, but do not have chlorophyll and are holozoic. It would not be proper to describe some dinoflagellates as plants and others as animals.

One solution to these difficulties is to remove the protozoans from animal kingdom and place them under a new kingdom, called *Protista,* along with some algal and fungal groups. This was done by **Haeckel** (1866) who created Protista, and by **Whittaker** (1969) who proposed 5 kingdoms including Protista. In the kingdom Protista, the different classes of Protozoa (Flagellata, Sarcodina, Sporozoa, and Ciliata) become independent phyla. This arrangement justifies to abandon Protozoa as a phylum in the absence of any unifying characters in different protozoan classes.

However, in the present textbook, we will consider Protozoan representing a single phylum of animal kingdom, according to the old, classical or traditional concept, which is still adopted by most protozoologists as well as textbooks.

PROTOZOA UNICELLULAR OR ACELLULAR

Study of invertebrates begins with Protozoa. They were originally defined as **unicellular** animals, but now referred to as **acellular** whether Protozoa should be considered unicellular or acellular, depends on the concept of the 'cell', itself.

1. Unicellular. According to the 'cell theory' of **Schleiden** and **Schwann** (1839), all plants and animals are made up of cells as units. If we define the cell as the unit of animal body, then Protozoa are **unicellular** animals, consisting of a single cell. According to this view, body of a free-living protozoan is morphologically equivalent to a single cell of a metazoan body. Supporters of this view claim that Metazoa evolved by the aggregation of separate unicellular protozoan individuals forming a multicellular colony (Flagellate or Colonial Theory).

2. Acellular or non-cellular. If we regard the animal body divisible into units, called cells,

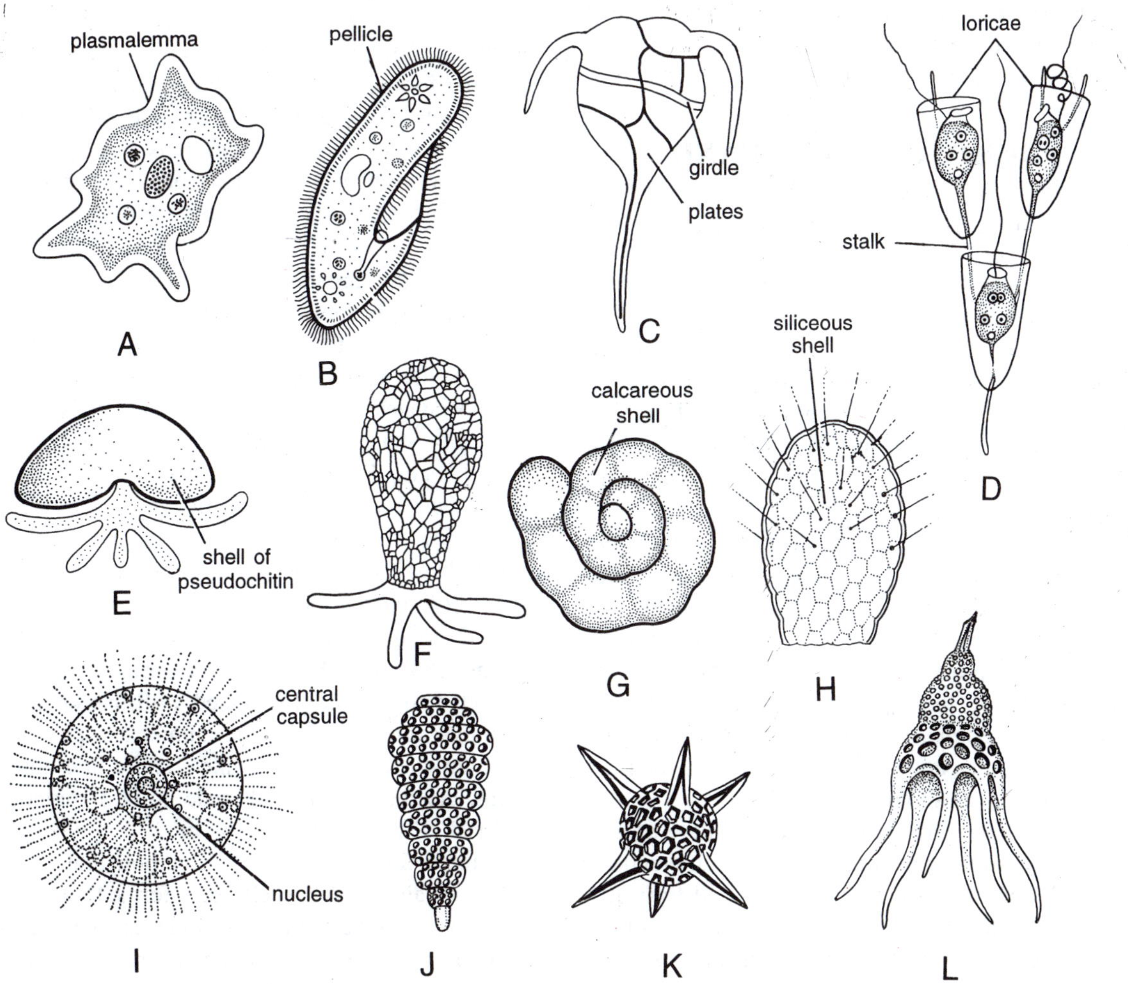

Fig. 1. Protozoan skeletons. **A.** Plasmalemma of *Amoeba,* **B.** Pellicle *of Paramecium.* **C.** Thecal plates of *Ceratium.* **D.** Lorica of *Poteriodendron.* **E.** Pseudochitinous shell of *Arcella.* **F.** Sand grain shell of *Difflugia.* **G.** Calcareous shell of *Discorbis.* **H.** Siliceous shell of *Euglypha.* **I.** *Thalassicola* showing central capsule. **J.-K.** Radiolarian skeletons showing lattice network. **L.** Helmet-shaped radiolarian skeleton.

then Protozoa are **acellular** or **non-cellular** animals whose body is not divisible in cells. Due to this view a single celled protozoan body is not merely a loose cell, moving about. It is a complete animal performing all the vital functions of life, such as nutrition, excretion, respiration, reproduction, etc. In structure, a protozoan cell may be compared with an individual cell of the metazoan body, but physiologically it is equivalent to the whole metazoan body. Dobell, therefore, defines Protozoa as cellular or non-cellular rather than unicellular animals. This view implies that metazoa arose by cellularisation of a multinucleate syncytial ciliate (Syncytial Theory).

BODY COVERING AND SKELETON

In Protozoa, cytoplasm remains separated from external environment by a cell envelope. This is of special importance as this covering protects the body from harmful influences of external environment, permits a controlled exchange of substances across it, perceives mechanical and chemical stimuli and establishes contact with other cells. It may take the following forms :

1. Plasmalemma. In some forms, like *Amoeba, Chrysamoeba* and *Sporozoa* body covering is a thin plasma membrane or **plasmalemma**. It possess the typical bilayered lipid and protein ultra structure in general. Flexibility of this membrane allows change in shape. In *Amoeba,* it is provided with longitudinal ridges of mucopolysaccharides which help in adhesion to substratum.

2. Pellicle. In other Protoza, like *Euglena,* the body covering is in the form of a differentiated pellicle, which is somewhat thicker and firm. It is underlined by the plasma membrane and is formed of proteins. The rigidity of pellicle gives definite shape to the body. The thickened pellicle in some of the more specialised Protozoa is variously ridged and sculptured, as in *Paramecium, Coleps,* etc.

3. Skeletons. In different groups of Protozoa, various kinds of permanent non-living external (or internal) layers are secreted. These may be composed of organic (gelatin, cellulose or pseudochitin) and inorganic (silica, calcium carbonate) materials, which are often impregnated with foreign bodies. These constitute the Protozoan skeleton and include **cyst**, **theca**, **lorica** and **test** or **shell**.

(a) Cyst. It is a temporary sheath, and is formed both in free-living and parasitic Protozoa when they pass into a dormant state. Exhaustion of food supply, drought and putrefaction favour encystment.

(b) Theca. Many dinoflagellates bear a coat of closely-fitted armour of cellulose, called **theca**. In some forms, theca is composed of two valves, while in a majority of dinoflagellates (*e.g., Ceratium, Glenodinium*) it is differentiated into a number of plates laid out in a definite pattern and variously sculptured or ornamented.

(c) Lorica. It is a covering which fits less closely to the organism than theca. It may be gelatinous or tectinous. It is usually a cup or vase-like structure with an opening, through which emerges the anterior part of organism's body or its appendages. Base of lorica in sessile species may be attached directly to the substratum (*e.g., Phacus, Salpingoeca*) or may end in a stalk (*e.g., Monosiga*). In colonial loricated individuals, one lorica may be attached to another directly (*e.g., Dinobryon*) or by means of a stalk (*e.g., Poteriodendron*).

(d) Shell or test. Shells or tests are widespread among Protozoa. These are coverings in loose contact with the body, provided with one or more openings, through which the animal can protrude itself. In shelled amoebae, like *Arcella* and allied forms, the shell is thin and made up of a chitinous material, called **tectin** or **pseudochitin** (a combination of proteins and carbohydrates). In *Difflugia* and others shells are formed of sand particles and other foreign substances like pieces of foraminiferan shells and sponge spicules, which are embedded in secreted matrix acting as cement. Foraminiferans secrete mostly a calcareous shell, made of calcium carbonate. It may be single chambered or multichambered and dimorphic in some cases (*e.g., Elphidium, Discorbis*). Siliceous shells, made up of silica, are found in some

Rhizopoda (e.g. *Euglypha*). Usually the silica forms polygonal, square or elliptical plates which are formed within cytoplasm then transferred to the test. Some heliozoans also secrete siliceous pieces, which are embedded in an outer gelatinous covering.

4. Internal skeleton. Continuous internal skeleton is found in radiolarians. It lies in between the ectoplasm and endoplasm, forming the so-called **central capsule**, (*e.g.*, *Thalassicola*). It consists of gelatin, pseudochitin, silica or strontium sulphate. It is perforated by one, a few or many pores. In many radiolarians the skeleton consists of a lattice network, which is variously sculptured and ornamented.

NUCLEUS IN PROTOZOA

All protozoans possess nuclei. Mastigophora, Sarcodina and Sporozoa have single or multiple nuclei, identical in structure and function. In Ciliata, there are generally two types of nuclei, **macronucleus** and **micronucleus**, which differ in size, structure and behaviour during reproduction. Only the micronucleus takes an active part in sexual reproduction and is regarded as **generative** nucleus. Whereas the macronucleus is associated with trophic or metabolic activities and is regarded as **somatic** nucleus.

Protozoan nuclei present varying appearances. This diversity is due to differential coiling of chromosomes in a resting nucleus and also to different proportions of nuclear inclusions. **Vesicular nucleus** is more common in distribution.

Vesicular nucleus is more common in distribution. It contains a distinct nuclear membrane, considerable amount of nucleoplasm and one or more prominent bodies, the **nucleoli** or **endosomes** or **karyosomes**. This type of nucleus is usually found in most species of Sarcodina and Mastigophora. Micronucleus of *Paramecium aurelia* is also of this type. **Compact**

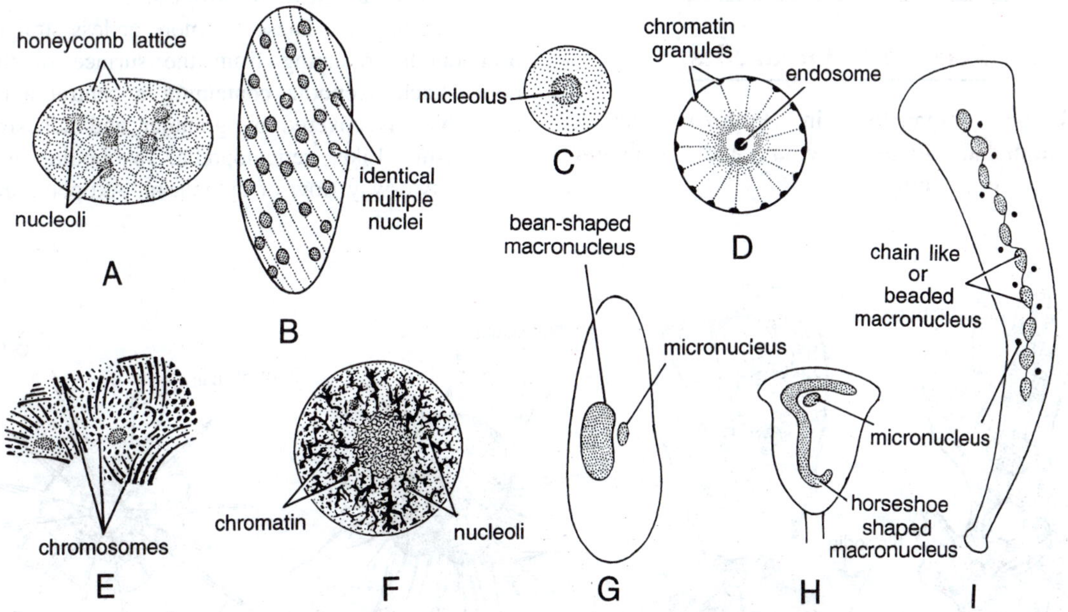

Fig. 2. Protozoan nuclei. **A.** Vesicular nucleus *of Amoeba proteus.* **B.** Multiple nuclei of *Opalina.* **C.** Vesicular micronucleus of *Paramecium aurelia.* **D.** Vesicular nucleus of *Entamoeba.* **E.** Chromosome nucleus of *Amoeba sphaeronucleus.* **F.** Polyploid nucleus of *Aulacantha scolymantha.* **G.** Compact macronucleus and vesicular micronucleus of *Paramecium caudatum.* **H.** Horseshoe shaped micronucleus of *Vorticella.* **I.** Beaded macronucleus of *Stentor.*

or **massive nucleus** contains an inconspicuous nuclear membrane, and small amount of nucleoplasm with evenly distributed minute granules of chromatin material. Macronucleus of ciliates and nuclei of Dinoflagellida are of this type. Ciliate macronucleus is generally rounded, oval or bean-shaped (*Paramecium*), but often becomes elongated in the form of C (*Euplotes*), or bent like a horse-shoe (*Vorticella*). It may be monoliform or shaped like a string of beads (*Stentor, Spirostomum*), or becomes branched (*Ephelota*).

In **nucleus** of Dinoflagellida and others, chromosomes are retained in interphase. In **polyploid nucleus**, chromatin material is distributed radially and nucleoli remain scattered in chromatin. This type of nucleus occurs in many radiolarians.

LOCOMOTOR ORGANELLES AND LOCOMOTION IN PROTOZOA

Locomotor Organelles

Locomotor organelles in Protozoa include **pseudopodia**, **flagella**, **cilia** and **pellicular contractile structures**.

[I] Pseudopodia

Pseudopodia or false feet are temporary structures formed by the streaming flow of cytoplasm. Sarcodina move with these structures. On the basis of form and structure, pseudopodia are of the following four types :

1. Lobopodia. These are lobe-like pseudopodia with broad and rounded ends, as in *Amoeba*. These are composed of both ectoplasm as well as endoplasm. Lobopodia move by pressure flow mechanism.

2. Filopodia. These are more or less filamentous pseudopodia, usually tapering from base to the pointed tip, as in *Euglypha*. Unlike lobopodia, the filopodia are composed of ectoplasm only. Sometimes they may branch and form simple or complex networks.

3. Reticulopodia. The reticulopodia (rhizopodia or myxopodia) are also filamentous. Filaments are branched and interconnected profusely to form a network. This type occurs in foraminiferans (*e.g., Globigerina*). Reticulopodia display two-way flow of cytoplasm.

4. Axopodia. These are more or less straight pseudopodia radiating from the surface of the body. Each axopodia containing a central axial rod which is covered by granular and adhesive cytoplasm. Like reticulopodia, axopodia also display two-way flow of cytoplasm. Axopodia are

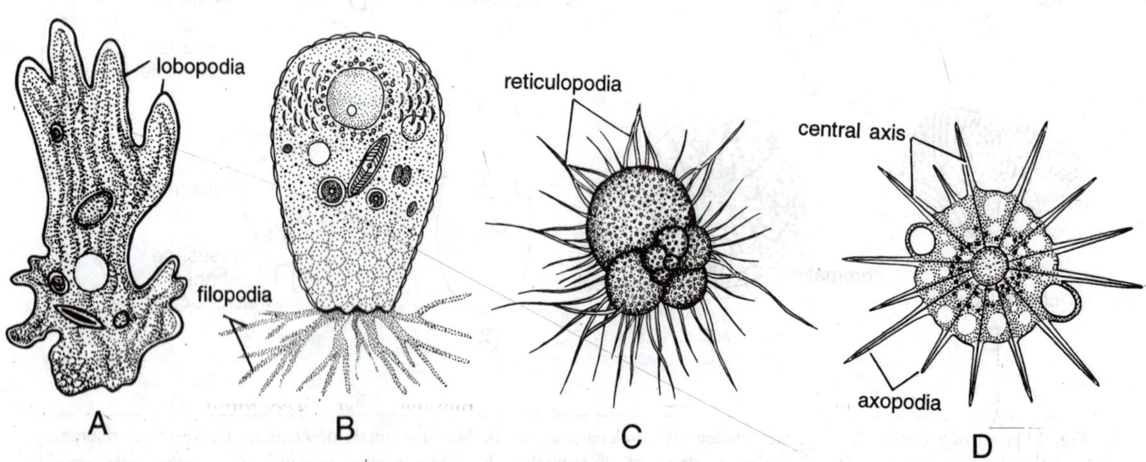

Fig. 3. Types of locomotor organelles. **A.** Lobopodia of *Amoeba proteus.* **B.** Filopodia of *Euglypha.* **C.** Reticulopodia of *Globigerina.* **D.** Axopodia of *Actinophrys sol.*

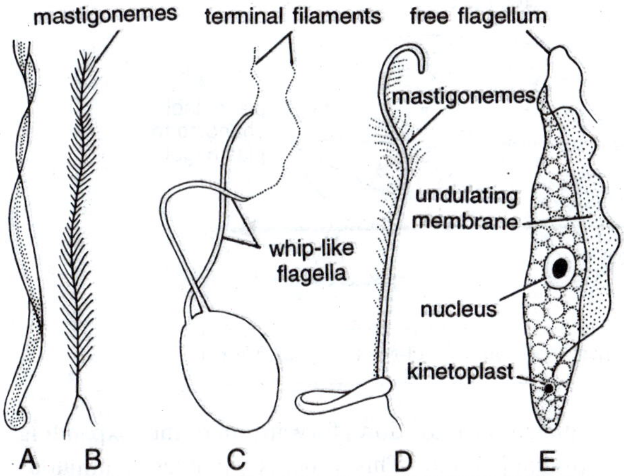

Fig. 4. Types of flagella in Protozoa. **A.** Flagellum of *Trachelo monas.* **B.** Flagellum of *Euglena* with mastigonemes. **C.** Whip-like flagella of *Polytoma.* **D.** Flagellum of *Urcoelus* with mastigonemes. **E.** Undulating membrane of *Trypanosoma* with a flagellum.

characteristic of heliozoans, such as *Actinosphaerium* and *Actinophrys.*

[II] Flagella

Flagella are the locomotor organelles of flagellate Protoza, like *Euglena, Trypanosoma,* etc. These are thread-like projections on the cell surface. A typical flagellum consists of an elongate, stiff axial filament, the **axoneme**, enclosed by an **outer sheath**. In axoneme, nine longitudinal peripheral paired fibres form a cylinder, which surrounds the two central longitudinal fibres, enclosed by a membranous inner sheath. Each of the peripheral pairs bears a double row of short arms. Axoneme arises from a basal granule, the **blepharoplast** or **kinentosome**. Mostly, it is a cylindrical body formed by the bases of peripheral fibres. Blepharoplasts are derived from centrioles, as the two structures are homologous.

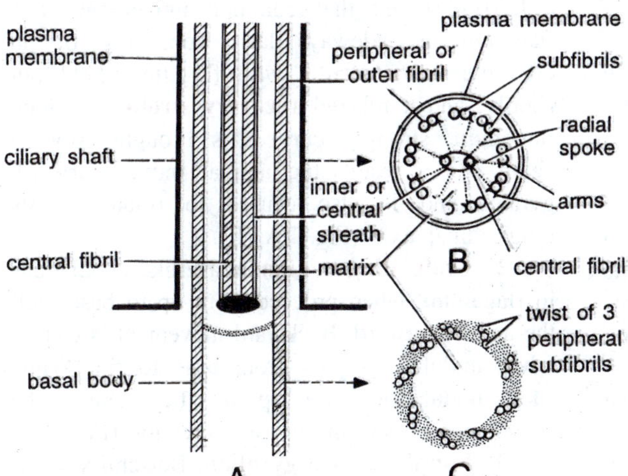

Fig. 5. Structure of a cilium. **A.** In L.S. **B.** In T.S. **C.** Basal body in T.S.

Fibres of axoneme remain embedded in a fluid matrix. In between the outer ring of peripheral fibres and inner ring of central fibres, mostly occur nine accessory fibres. In certain groups of Mastigophora are found flagellar appendages or **mastigonemes** extending laterally from the outer sheath.

Number and arrangement of flagella vary in Mastigophora from one to eight or more. Free-living species have usually one or two, while in parasitic species, the number ranges from one to many.

[III] Cilia

Cilia, characteristic of Ciliata, resemble flagella in their basic structure. These are highly vibratile small ectoplasmic processes. Electron microscope reveals the presence of an external membranous sheath, continuous with plasma membrane of cell surface and enclosing the fluid matrix. Running along the entire length of body of cilium are nine paired peripheral fibres and two central fibres, all embedded in a structureless matrix. Central fibres are enclosed within a delicate sheath. In between the outer and inner fibre rings are present nine spoke-like **radial lamellae**. In addition to these, one **sub-fibre** or **microfibre** of each peripheral pair bears a double row of short, projections, called **arms**, all pointing in the same direction.

Each cilium arises from a thickened structure, the basal granule, basal body or blepharoplast. According to **Lenhssek** and **Henneguy** (1898), basal granules are centrioles or their derivatives. Basal granules show nine peripheral subfibre triplets, each disposed in a twist-like fashion. In many

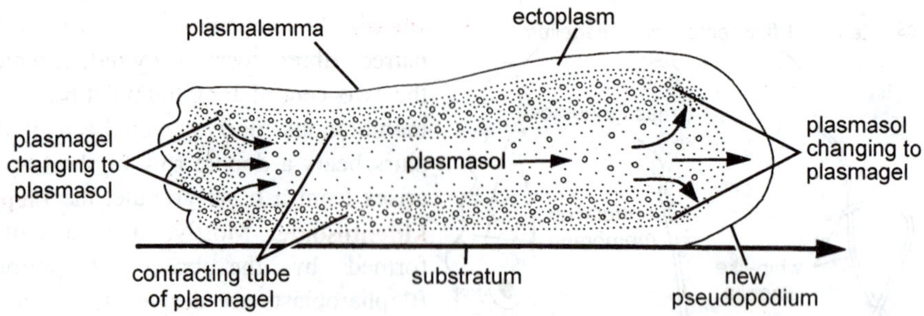

Fig. 6. *Amoeba* showing amoeboid movement according to sol-gel theory of Mast.

species, cilia become fused variously forming compound organelles, such as undulating membranes (*Pleuronema*), membranelle (*Vorticella*), and cirri (*Euplotes*).

[IV] Pellicular Contractile Structures

In many Protozoa are found contractile structures, in pellicle or ectoplasm, called **myonemes**. These may be in the form of ridges and grooves (*e.g.*, *Euglena*), or contractile myofibrils (*e.g.*, larger ciliates), or microtubules (*e.g.*, *Trypanosoma*).

Methods of Locomotion

Basically there are four known methods by which Protozoa move : (i) Amoeboid movement, (ii) Flagellar movement, (iii) Ciliary movement, and (iv) Metabolic movement. Speed of locomotion varies from 0.2μ to 3μ per second in amoeboid forms, I5μ to 300μ in flagellates, and 400μ to 2000μ in ciliates.

[I] Amoeboid Movement

It is characteristic of all Sarcodina and certain Mastigophora and Sporozoa. It consists in the formation of pseudopodia by the streaming flow of cytoplasm in the direction of movement. Locomotion by pseudopodia is possible only over a surface. We still do not know precisely about the mechanism involved in the formation of pseudopodia, but the most convincing theory at present is that it depends upon active contraction of the ectoplasmic tube (plasmagel) at the posterior end of the body. This leads the endoplasm (plasmasol) to flow forward into the expanding pseudopodium. This process involves continuous solation at the posterior end and gelation at the anterior end (Fig. 6). This is called **sol-gel** or **change of viscosity theory** by **Mast** and **Pantin** (1925). It was further developed by **Goldacre** and **Lorch** (1950) and by **Allan** and **Rosalansky** (1958). Other aspects and theories of amoeboid locomotion have been discussed at length in the chapter on *Amoeba*.

[II] Flagellar Movement

It is characteristic of Mastigophora which bear one or more flagella. The flagella need liquid medium for movement or locomotion. Three types of flagellar movements have been recognized :

1. Paddle stroke. Ulehla and **Krijsman** (1925) observed that common movement of a flagellum is sideways lash, consisting of an effective **down stroke** with flagellum held out rigidly, and a relaxed **recovery stroke** in which flagellum, strongly curved, is brought forward again. As a result, the animal moves forward, gyrates and is also caused to rotate on its longitudinal axis (Fig. 7A, B).

2. Undulating motion. Wave-like undulations in flagellum, when proceed from tip to base, pull the animal forward. Backward movement is caused when undulations pass from base to tip. When such undulations are spiral, they cause the organism to rotate in opposite direction (Fig. 7D).

3. Simple conical gyration. Butschli's screw theory postulates a spiral turning of flagellum like a screw. This exerts propelling action, pulling the animal forward through water with a **spiral**

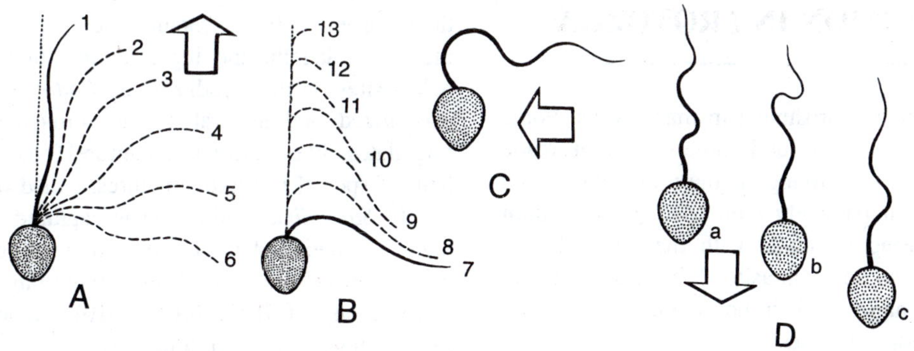

Fig. 7. Flagellar movement. **A.** Effective stroke. **B.** Recovery stroke. **C.** Lateral movement. **D.** Backward movement.

Fig. 8. Successive stages of gyrating flagellar movement.

rotation as well as **gyration** (revolving in circles) around the axis of movement (Fig. 8).

The mechanism producing flagellar beat is not exactly known. It is believed that some or all of the axonemal fibres are involved. According to the latest **sliding tubule**. Theory of flagellar (or ciliary) movement, adjacent doublets slide past each other, causing the entire flagellum or cilium to bend. Cross bridges are formed and energy utilized for the process is supplied by adenosine triphosphate (ATP).

[III] Ciliary Movement

Most ciliates appear to move in a spiral path, rotating on their axis as they go. Spiral movement is due to in opposite directions on the two sides of the pseudopodial filaments oblique strokes of all body cilia working together and striking in the same direction. Cordination of ciliary movement is due to fact that basal bodies of all cilia are linked by kinetodesmata. Cilia also need liquid medium for their movements. Large ciliates are the swiftest swimmers, and the champion of them may be named *Paramecium caudatum.*

Ciliary action resembles the swing of a pendulum except that it is more rapid in one direction. Backward and forward vibrations produce a paddle stroke effect. Backward effective stroke is more active during which movement is brought about, while forward recovery stroke produces no significant movement. While moving, the succession of beats are coordinated in the well-known pattern of metachronal rhythm, conventionally compared to the passage of wind over a field of wheat (Fig. 7).

Ciliary locomotion has been discussed at length in the chapter on *Paramecium.*

As regards the essential mechanism of ciliary movement, little definite can be said. The evidences strongly suggest that ciliary movement is based on contraction of peripheral fibres.

[IV] Metabolic Movements

This is typical of certain flagellates (*e.g. Euglena*) and most sporozoans at certain stages of their life cycles. Such organisms are seen to show gliding or wriggling or peristaltic movement. Contractile myonemes or microtubules, present in their pellicular walls, are responsible for this type of movement. Movements of this kind are usually also referred to as **gregarine movements** since they are characteristically exhibited by most gregarines.

NUTRITION IN PROTOZOA

Protozoa obtain nourishment in many ways. Some synthesize their own food, others have it made for them by algae living in their cytoplasm and still other wait passively until food comes within reach, capturing it almost with the outer sheath. Some Protozoa lead parasitic life, usually doing no or little harm to their hosts, but occasionally causing serious diseases.

All types of nutrition are found in Protozoa, such as holophytic, holozoic, saprozoic, mixotrophic and parasitic, etc.

Holophytic Nutrition

All those phytoflagellates possessing chloroplasts or chromatophores synthesize their food by photosynthesis. As energy is supplied by sunlight to carry on food-making activity, this method involving self-feeding is also referred to as **autotrophic phototrophy**. Carbon dioxide and water, acting as raw materials, enter into a complex cycle of chemical reactions and produce dextrose sugar. From dextrose paramylum may be formed which is especially characteristic of the euglenoid flagellates.

Holozoic Nutrition

Majority of free-living Protozoa derive nourishment by ingesting other organisms, both animals and plants. Such Protozoa are called **holozoic**, and mode of nutrition is said to be **holozoic nutrition**. All Sarcodina are strictly holozoic with the exception of some parasitic species. This mode of nutrition involves development of organelles for food capture, ingestion, digestion and egestion of indigestible residues.

[I] Food and Feeding

Food of holozoic Protozoa consists of microorganisms like other protozoans, bacteria, diatoms, rotifers, crustaceans, larvae, etc. The method involved in ingestion of these organisms is referred to as **phagotrophy** or **phagocytosis**.

Colourless **flagellates** and those who have lost their chromatophores capture their solid food by means of flagella and ingest them either through naked sites on their bodies (*e.g., Bodo*) or through specialized oral apparatus (*e.g., Euglena*). Some flagellates, like *Peranema,* capture food with the help of special rod-like structures, called **trichites**.

In *Sarcodina,* solid food is captured with the help of pseudopodia. This type of feeding is said to be **amoeboid nutrition**. According to the observations of **Rhumbler** (1910), amoebae are said to ingest food in four ways :

1. By import. This method involves taking in food into body upon contact with very little movement on the part of organism as there is

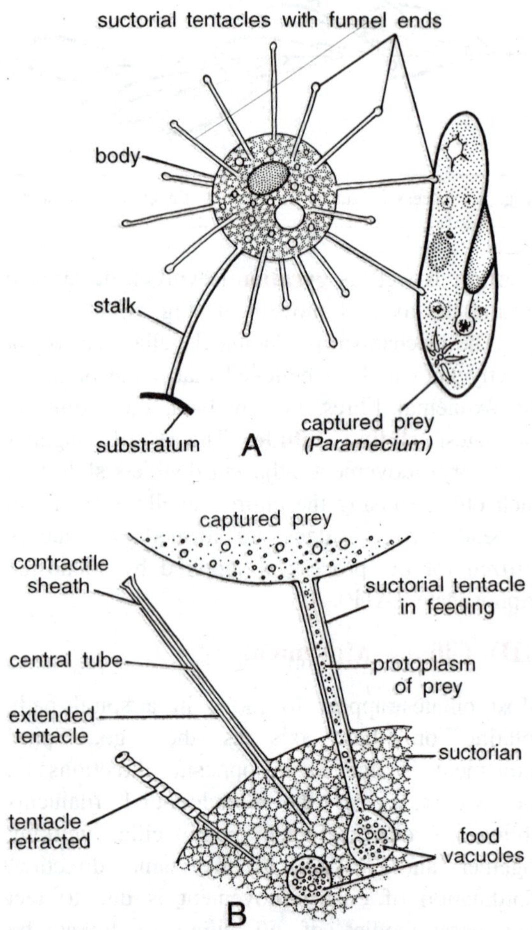

Fig. 9. Feeding in Suctoria. **A.** *Podophrya* sucking *Paramecium.* **B.** Prey's protoplasm flows into the suctorian's body through the tubular tentacle.

some chemical attraction between it *and Amoeba.* Passive organisms such as algae are imported like this (see Fig. 7 : Ch. 6).

2. By circumfluence. By this method *Amoeba* engulfs the food organism upon contact by rolling over it in order to arrest it completely. This happens when less active organisms are fed upon such as bacteria (see Fig. 8 : Ch. 6).

3. By circumvallation. This method is applied when *Amoeba* happens to feed upon an active prey. It sends out pseudopodia to engulf the desired prey, while it is still some distance away. The pseudopodia surround food without touching it. Later, these pseudopodia fuse at their tips and the so-called food vacuole is formed which is engulfed as a whole (see Fig. 9 : Ch. 6).

4. By invagination. It is a method of sucking in the food particle upon contact with ectoplasm. The food particle then sinks into endoplasm (see Fig. 10 : Ch. 6).

Heliozoans (sun animalcules) and **radiolarians** hold and pull the prey that comes within reach by their axopodia which act as traps. After capture the axopodia withdraw and the prey passes into deeper cytoplasm.

Foraminiferans produces delicate reticulopodia forming a net. A granular mucoid film is present on the reticulopods upon which stick the food particles on contact which are then dragged into inner cytoplasm (see Fig. 3 : Ch 8).

In **ciliates**, the oral apparatus meant for foodcapturing is well developed. By the beating of cilia of oral groove, food is taken into buccal cavity and then driven towards mouth or cytostome by ciliary bands (membranes, membranelles) into cytopharynx which strips off food vacuoles containing food particles into the endoplasm (see Fig. 12 : Ch. 12).

Suctorians feed on other ciliates with the help of their tentacles which are usually knobbed at the tip. Each tentacle consists of a rigid central tube surrounded by a contractile sheath. As soon as the prey is adhered to the tips of tentacles, it is paralysed by some toxin (hypnotoxin). It is then gradually sucked into the suctorian body through their central tubes (Fig. 9).

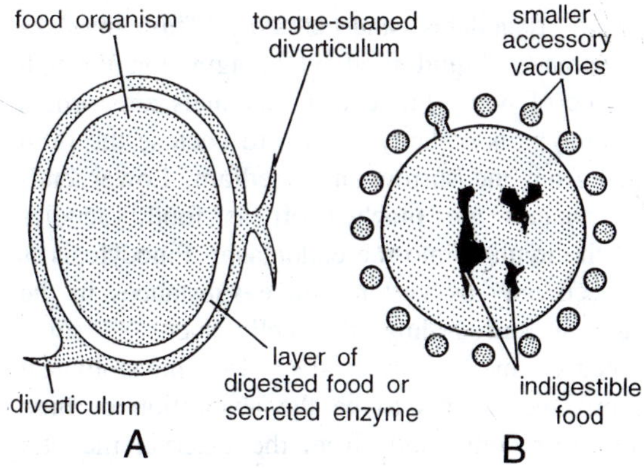

Fig. 10. Food vacuole. **A.** Recently formed food vacuole. **B.** An older food vacuole.

[II] Digestion

Food in Protozoa is digested within food vacuoles, which usually keep on circulating in the endoplasm. Within food vacuoles, the reaction is at first acidic which later on becomes alkaline. Proteolytic and carbohydrate splitting enzymes have been demonstrated in many protozoan organisms. In the acidic medium, proteins are converted into dipeptides and, in the alkaline medium, dipeptides are converted into amino acids. Hydrolysis of carbohydrates takes place in alkaline medium. Certain Protozoa are also said to digest fats. Digestive enzymes are furnished by lysosomes which fuse with the food vacuoles.

[III] Absorption

Digested food gets diffused into endoplasm, where it is assimilated into protoplasm. Excess food may be stored as glycogen, paramylum, lipids, chromatoid bodies, etc.

[IV] Egestion

Indigestible residue of food is expelled from the hinder part of body in case of moving *Amoeba* at any point. Ciliates often possess an anal opening, the **cytopyge** or **ectoproct**, for this purpose.

Pinocytosis

In addition to phagocytosis, **pinocytosis** or **cell-drinking** has been reported in *Amoeba* and certain flagellates and ciliates. This involves ingestion of liquid food by invagination through surface of body. Pinocytosis channels are formed at some parts of body surface to enclose the fluid food from the surrounding medium. Lower ends of channels are pinched off as food vacuoles which circulate into the endoplasm. Pinocytosis is induced only by certain active substances in the medium surrounding the cell, such as some proteins and many salts. Its physiological significance seems to be the absorption of high molecular compounds from the external medium (see Fig. 11 : Ch 6).

Saprozoic Nutrition

Saprozoic nutrition involves absorption of food by osmosis, *i.e.*, through general surface of body. This method of food-getting is referred to as **osmotrophy**. Food consists in the form of solution of dead organic matter, rendered so by the decomposing bacteria. This mode of nutrition is found in *Mastigamoeba,* and some colourless flagellates (*e.g.*, *Chilomonas, Astasia, Polytoma*). Dissolved food materials, upon which the saprozoic protozoans subsist, are proteins and carbohydrates.

Myxotrophic Nutrition

This is a combination .of more than one mode of nutrition. Many Protozoa using photosynthesis as a means of food-synthesis also take in some part of their diet in dissolved form by osmotrophy or solid form by phagotrophy. Flagellates like *Euglena* and *Peranema* nourish themselves by this method.

Nutrition of Parasites

The food-getting mechanisms used by parasitic Protozoa are generally the same as those of their non-parasitic relatives. Many intestine-inhabiting Zoomastigophorea (*Trichomonas*) have a distinct mouth or cytostome through which food particles are ingested by phagotrophy. Many parasitic ciliates, like *Nyctotherus* and *Balantidium,* do the same. Parasitic Sarcodina of the genus *Entamoeba* feed by phagotrophy, at least at certain stage of their life cycle. Zooflagellates inhabiting blood (*e.g.*, *Trypanosoma*) feed by **osmotrophy**. Osmotrophic forms may be either **coelozoic** or **histozoic**. *Opalina,* which is found in the rectum of frog, is coelozoic and absorbs all its food through the cell surface. The young trophozoite of *Monocystis* is histozoic within the sperm morula and it feeds upon the sperms substance by osmotrophy. Parasitic saprozoic forms may also use directly the serum of their host's blood.

CONTRACTILE VACUOLES AND OSMOREGULATION

Contractile vacuoles are pulsating organelles found in freshwater Sarcodina, Mastigophora and Ciliata. They also occur in marine ciliates. They are altogether absent in parasitic forms of the protozoan classes, except Ciliata, where they occur frequently.

Contractile Vacuoles in Different Protozoa

In *Amoeba,* the contractile vacuole is in its simplest form, being a spherical vesicle bounded by a limiting membrane of about 0.5μ in thickness. It is surrounded by a circlet of mitochondria which furnish energy for its pulsating activity. It is thought to be a temporary structure formed by the fusion of numerous small accessory or feeder vacuoles. In protozoans, with a more defined body

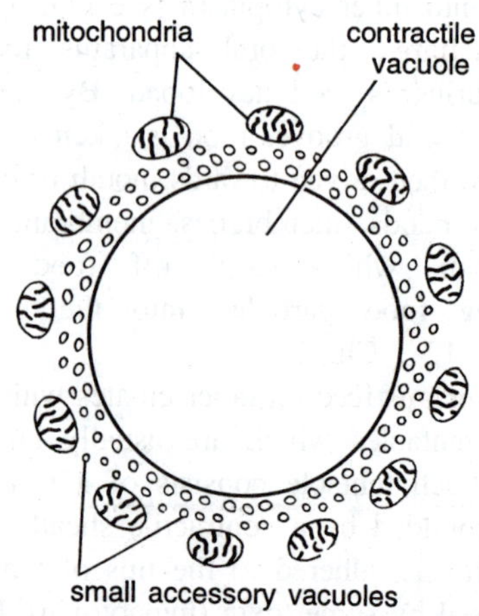

Fig. 11. Contractile vacuole of *Amoeba* surrounded by mitochondria.

form, like *Euglena,* the contractile vacuole has a fixed position. Formation of main vacuole is contributed by a few accessory vacuoles. Sporozoans have no contractile vacuoles. Ciliates, both marine and freshwater, have contractile vacuoles of more advanced type. *Paramecium* has two vacuoles, one located at the anterior end and other at the posterior end of body. The main larger vacuole is fed by a varying number of radial canals or afferent canals. It opens to outside through a small pore in the organism's body surface.

Electron microscopy has revealed that the bounding membrane of the contractile vacuole is a lipoprotein membrane, similar to the cell membrane.

[II] Mechanism of Osmoregulation

The function of contractile vacuole is **hydrostatic** or **osmoregulatory**. Water in freshwater Protozoa enters the organism by endosmosis and during feeding. The reason is that cytoplasm of a freshwater protozoan represents a medium denser than surrounding water and outermost plasmalemma or pellicle acts as a semi-permeable membrane. Thus water continuously flows inside to dilute the cytoplasm. If the organism does not possess a mechanism to get rid of this excess water, it will swell to the point of rupture and dissolution. The mechanism which is assumed to effect water regulation is the contractile vacuole. The vacuole periodically increases in volume (diastole) to get filled with water and contracts (systole) to discharge its water content to the surrounding environment. But the exact mechanism of its working is little understood. Three theories have been put forth to explain it. These are as follows :

1. Osmotic theory. According to this theory, water from cytoplasm enters the contractile vacuole by osmosis.

2. Filtration theory. According to this theory, hydrostatic pressure forces water through vacuolar membrane. But, Kitching is of the view that this theory cannot be applied to the working of contractile vacuole as it is only connected to outside medium at systole. Thus, internal hydrostatic pressure could not be relieved by filtration of fluid into vacuole, since the total volume within the limiting cell membrane would not be reduced.

3. Secretion theory. This theory holds that water is actively secreted into vacuole during diastole through vacuolar wall. This theory is most widely accepted. Systole and diastole are attributed to the contractile nature of the vacuole's wall.

REPRODUCTION IN PROTOZOA

Asexual Reproduction

1. Binary fission. This involves the division of one individual into two approximately equal parts. The division is not a mere fragmentation but a complicated process of mitosis, during which nuclear division or **karyokinesis** is always followed by the division of cytoplasm or **cytokinesis** (Fig. 12). Division or fission may be either in a transverse plane (*e.g., Paramecium*), or in a longitudinal plane (*e.g., Euglena*), or in an oblique plane (*e.g., Ceratium*) or in any plane (*e.g., Amoeba*). The two daughter organisms produced as a result of binary fission carry all the cytoplasmic organelles of the parent individual. Some organelles like mitochondria, divide at the time of division, while others, like oral apparatus, flagella, and contractile vacuoles, are formed afresh by one of the daughters. In shelled sarcodina (*e.g., Euglypha, Arcella*) a mass of protoplasm extrudes from the opening of shell, which secretes a new shell. This double-shelled organism now divides into two. In ciliates (*e.g., Paramecium*), during fission, mega or macronucleus divides amitotically and micronucleus by usual mitotic division. Some Protozoa divide only in the encysted stage *e.g., Colpoda, Tillina*).

[II] Plasmotomy

It is a special type of binary fission concerned with the division of multinucleate Protozoa into two or more smaller multinucleate daughter individuals. Plasmotomy takes place in *Pelomyxa,* opalinids and some other forms, etc. (Fig. 13).

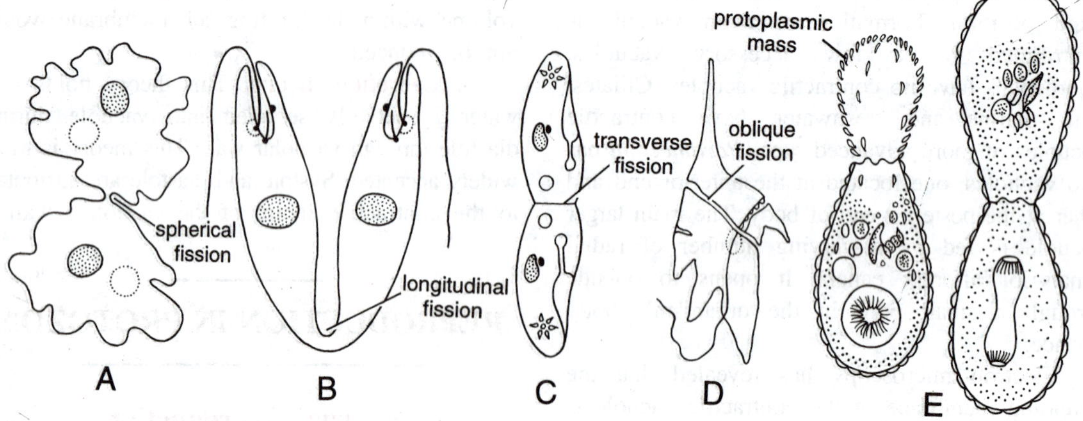

Fig. 12. Binary fission in Protozoa. **A.** *Amoeba* (irregular). **B.** *Euglena* (longitudinal). **C.** *Paramecium* (transverse) **D.** *Ceratium* (oblique). **E.** *Euglypha* (Two stages in fission).

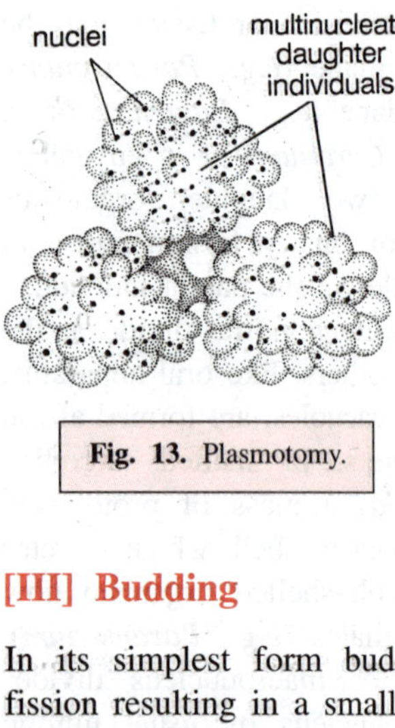

Fig. 13. Plasmotomy.

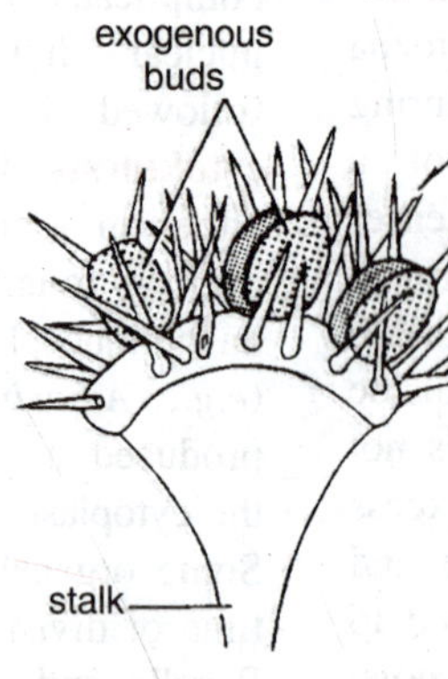

Fig. 14. Multiple budding in *Ephelota.*

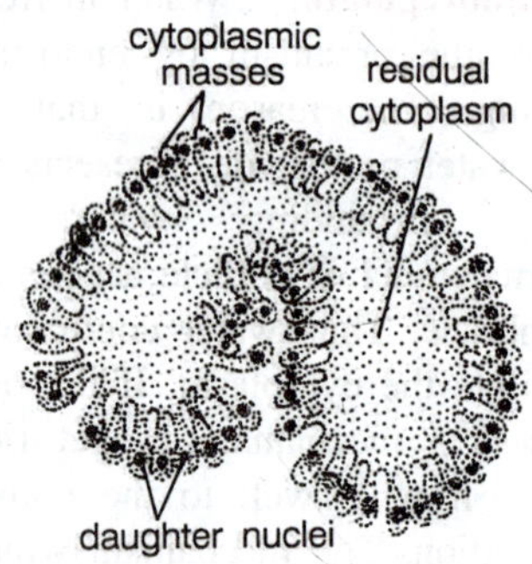

Fig. 15. Multiple fission.

[III] Budding

In its simplest form budding implies modified fission resulting in a small daughter individual in the form of a bud. When the bud breaks off, it grows to full size. When a parental body produces only one bud it is **monotonic** (e.g. *Vorticella*), while in **multiple budding**, several buds are formed simultaneously (*e.g.*, *Ephelota*).

[IV] Multiple Fission

During **multiple fission** or **sporulation**, nuclear division is not followed immediately by division of cytoplasm. First, nucleus undergoes a series of divisions either by repeated binary fissions as in *Plasmodium,* or by simultaneous multiple divisions, as in *Aggregata.* The body thus becomes multinucleate. Later, the body cytoplasm divides into as many parts as there are daughter nuclei which usually arrange themselves at the periphery, each getting surrounded by a fragment of cytoplasm. Thus, the parent body simultaneously divides into as many daughter individuals as there are nuclei. Parent cell usually leaves behind some residual cytoplasm which disintegrates afterwards. Number of offspring greatly varies among different and the same species and sometimes runs into thousands.

Multiple fission is quite common in Foraminifera, Radiolaria, Sporozoa and certain Mastigophora. The process receives different names according to the particular period in life cycle when it occurs.

Schizogony. In this process, a series of nuclear divisions results into numerous daughter nuclei. This is followed by the formation of cytoplasmic buds, each containing a nucleus. The buds are pinched off to grow directly into new organisms.

[V] Plasmogamy

In certain Rhizopoda and Mycetozoa, two or more individuals may fuse by their cytoplasm to form a plasmodium, in which the nuclei remain distinct, and they separate again unchanged afterwards. The process, which is thus non-sexual and not syngamy, is called **plasmogamy** and sometimes serves the purpose of digestion of large prey.

Sexual Reproduction

In Protozoa, sexual reproduction takes place mainly by two processes : (i) **syngamy** or fusion of two sex cells, and (ii) **conjugation** or temporary contact of two protozoans with nuclear exchange.

[I] Syngamy

Syngamy is the complete fusion of two sex cells or gametes, resulting in the formation of zygote. The fusion nucleus of zygote is called **synkaryon**. Depending upon the degree of differentiation displayed, by the fusing gametes, syngamy is of the following types :

1. Hologamy. The two ordinary mature protozoan individuals do not form gametes but themselves behave as gametes and fuse together to form zygote. Hologamy occurs in a few Sarcodina and Mastigophora (*e.g., Copromonas*).

2. Isogamy. When two fusing gametes are similar in size and shape but differ in behaviour, they are called **isogametes** and their union, **isogamy**. Isogametes are generally produced by multiple fission. Isogamy is common in Forminifera (*e.g., Elphidium*), Gregarinia (*e.g., Monocystis*) and Phytomonadida (*e.g., Chlamydomonas*).

3. Anisogamy. When two fusing gametes, differ morphologically as well as in behaviour, they are called **anisogametes**. Usually small and motile gametes are the male or **microgametes** and large non-motile ones are the female or **macrogametes**. Fusion of such dissimilar gametes is **anisogamy**. This mode of sexual reproduction is widely seen in Sporozoa (*e.g., Plasmodium*) and Phytomonadida (*e.g., Volvox*).

4. Autogamy. It is the fusion of gametes derived from the same parent cell, as in *Actinophrys* and *Actinosphaerium.* In *Actinophrys,* during sexual reproduction, pseudopodia are withdrawn and a cyst is formed. Now meiotic division takes place and two daughter nuclei with half number of chromosomes are formed. No cell division takes place. After sometime, gametic nuclei fuse to form a zygote nucleus.

[II] Conjugation (Amphimixis)

This involves temporary union of two individuals, called **conjugants**, usually at oral or buccal regions of their body. It is characteristic of Suctoria and holotrich ciliates.

Fusion of protoplasm occurs at the place of contact. Macronuclei break up and disappear. Micronuclei undergo meiotic division now all but one micronuclei degenerate. This remaining micronucleus again divides forming two gametic micronuclei. Out of these two, one is considered **male pronucleus** and other **female pronucleus**. Male pronucleus of one conjugant moves through fused protoplasm into the other conjugant. In each conjugant, these male and female pronuclei fuse together forming a **zygote nucleus**. Now two individuals separate and are called **exconjugants**. Each exconjugant undergoes further nuclear and cytoplasmic divisions forming four daughter individuals.

Association in conjugation is not at random but indicates a high level of specialization. Sonneborn has recognized different syngens in a species of *Paramecium* and each syngen includes two mating types. Conjugation can only take place between individuals of the same syngen but belonging to opposite mating types. The unique feature of conjugation is an exchange of hereditary material so that each conjugant benefits from a renewed hereditary constitution.

Conjugation in *Paramecium* has been discussed at length in Chapter 11.

Parthenogenesis

In *Actinophrys,* the gametes which fail at crossfertilization, develop parthenogenetically. It also occurs in *Chlamydomonas* and others when syngamy has been missed. Individuals of *Polytoma,* which are potential gametes, can grow and divide parthenogenetically.

Regeneration

Most Protozoa can regenerate their lost parts, as normally displayed at fission or encystment. Parasitic Protozoa usually have slight regenerative capacity. Nucleus plays an important role in the process. Relative quantities of nuclear and cytoplasmic materials and the size of the broken piece affect the rate and result of the process of regeneration.

ECONOMIC IMPORTANCE OF PROTOZOA

Protozoa are unicellular microscopic organisms found almost everywhere, in water, in moist soil, in air, or even within the bodies of other animals and plants. In this fast-moving **Age of Mammals**, dominated by man, it may appear at first glance that these minute animals would have little or no economic significane. However, such is not the case, and Protozoa exert far more influence in worldly affairs than is generally thought. They are **harmful** as well as **useful**, but the harmful species are relatively smaller in number as compared with the useful species.

Useful Protozoa

1. Helpful in sanitation. Numerous holozoic Protozoa feed on putrefying bacteria in various bodies of water and thus help indirectly in the purification of water. These Protozoa play an important part in the sanitary betterment and improvement of the modern civilized world in keeping water safe for drinking purpose.

2. Planktonic Protozoa as food. Protozoa floating in the plankton of sea provide directly or indirectly the source of food supplies to man, fish and other animals. They form one of the first links in the numerous and complicated food chains that exist in the oceans of the world. Clams and young fish feed extensively on aquatic insect larvae, small crustaceans, worms, etc., all of which take Protozoa as food. Thus, Protozoa indirectly form the food of fish, clams and other animals, which in their turn are consumed by man.

3. Symbiotic Protozoa. Some protozoans are found in symbiotic relationship with other organisms. This association is usually beneficial to both the partners. The two partners become so dependent on each other, that one cannot get along without the other, and their separation results in the death of both. Most outstanding examples of symbionts among the Protozoa are several intestinal flagellates (*Trichonympha, Colonympha* etc.) of termites and woodroaches. According to Cleveland, these flagellates are extremely vital for the very existence of their hosts. They digest cellulose converting it into soluble glycogen substances for their hosts as well as for themselves.

4. Oceanic ooze and fossil Protozoa. Tiny skeletons of dead pelagic Foraminifera, Radiolaria and Heliozoa sink to the sea bottom forming the soft mud or **oceanic ooze**. These tiny skeletons are made of silica or calcium carbonate. Over countless millions of years these skeletons, deposited on the floor of ocean, became solid and fossilized and converted into some important sedimentary rock strata found all over the world. These have been put to various commercial uses, such as filtering agents, abrasives, chalk, building stones, etc. The white **chalk** cliffs of Dover, England, and the **limestone** beds of Paris, Cairo and North America are composed almost exclusively of the fossil Foraminiferida. It is hard to believe that great city of Paris is indebted to these obscure animals. Most of the buildings of Paris are made of limestone composed almost exclusively of the shells of the foraminiferan genus, *Miliolina.* Similarly, the great pyramids of Egypt were carved from the limestone deposits made by tests of an early Tertian foraminiferan, *Nummulites.* Radiolarian fossils are abundant in the hard rocks of the nature of **flint** and **chert**. They also constitute a part of Tripoli Stone, which is used in abrasive powders of materials.

5. Protozoa in study. Protozoa are singlecelled organisms, possessing forms and functions like those of metazoan cells. They are studied in laboratories for the comprehension and application of biological principles. Due to their minute size and quick reproduction, they are studied by geneticists for heredity and variations. Knowledge of fossil Protozoa is essential for the students of geology and palaeontology. Protozoa are the progenitors of all metazoans alive today. Their study helps in understanding the probable beginning of organic matter and the origin and evolution of life on Earth. Study of physiology of Protozoa has contributed much to our knowledge about the physiology of animal cell.

Harmful Protozoa

1. Soil Protozoa. Several species of Protozoa, present in large numbers in soil, feed open the nitrifying bacteria, and thus decline their activity and consequently tend to decrease the amount of nitrogen given to soil by the nitrifying bacteria.

2. Water pollution. Whereas some Protozoa are helpful in water sanitation, others become responsible for water contamination or pollution. The Protozoa of faecal origin belong to this latter category. Some free-living Protozoa (*e.g., Uroglenopsis*) also pollute water by producing aromatic and oily secretions with objectionable odours, which render water unfit for human consumption.

Some bioluminescent dinoflagellates, such as *Noctiluca, Gymnodinium* and *Gonyaulax,* living in sea, sometimes multiply so extensively as to turn the water red with their bodies. The phenomenon is known as **blooming** and is the cause of 'red tides', often experienced in the sea. Outbreaks of this 'red water' often gives a foul and disagreeable odour to the ocean water. Large concentrations of these flagellate protozoans may even lead to destruction of fish and poisoning of edible molluscs, such as clams, oysters and mussels, etc., making them unfit for human consumption.

3. Pathogenic Protozoa. Some Protozoa cause diseases in man as well as animals and these are termed **pathogenic Protozoa**. They occur in all classes of Protozoa.

(a) Pathogenic sarcodines. There are two common genera of parasitic amoebae, *Entamoeba* and *Endamoeba,* which live in the intestine of man and of other animals. Only two species of *Entamoeba*, *E. histolytica* of man and other mammals and *E. invadens* of reptiles are known to be seriously pathogenic. *E. histolytica* is responsible for amoebic dysentery or amoebiasis in man, which occurs in about 60-70% Indian population. *E. invadens,* occurring in the colon of reptiles, causes reptilian amoebiasis.

(b) Pathogenic flagellates. Pathogenic species of parasitic flagellates are included in the genera *Leishmania, Trypanosoma, Histomonas, Trichomonas* and *Giardia.* Three pathogenic species of *Leishmania* have been known to cause severe diseases in man. *L. donovani* causes Kala-azar, a disease of the spleen and liver, *L. tropica* causes a peculiar type of skin lesion (cutaneous leishmaniasis) and *L. brasiliensis* causes infection of nasopharynx and skin lesion. These are transmitted by sandflies of the genus *Phlebotomus.* Parasitic species of *Trypanosoma* in mammals cause worst diseases. *T. gambiense,* is the causative agent of fatal African sleeping sickness.

T. Rhodesiense, T. cruzi, T. equiperdum, T. evansi, and *T. brucei* are other common pathogenic species.

Histomonas meleagridis is the parasitic mastigamoeba. Of the parasitic species of *Trichomonas, T. vaginalis* is the causative organism of vaginal trichomoniasis or vaginitis in human females. *T. foetus* causes trichomoniasis of cattle in U.S., and *T. gallinae* is pathogenic in doves, pigeons, turkeys and chickens. Of the numerous species of *Giardia, G. intestinalis* (= G. *lamblia*) of man causes enterocolitis.

(c) Pathogenic Sporozoans. Protozoan superclass Sporozoa is exclusively of parasitic forms. Though most of sporozoans are harmless, yet some genera like *Plasmodium, Eimeria, Isospora* and *Babesia* include pathogenic species. Species of *Plasmodium* are called malaria parasites as they cause the disease of malaria. Four species of *Plasmodium,* namely *P. vivax, P. malariae, P. ovale* and *P. falciparum* cause malaria in man. Malaria is caused by *P. cyanomolgi* in monkeys, by *P. verghei* in tree rats and by *P. gallinaceum* in jungle fowl of Asia. Pathogenic species of *Eimeria* cause coccidiosis in chickens and rabbits. *E. tenella* and *E. mitis* infect chickens, whereas *E. mana* and *E. steidae* infect rabbits. *E. canis* in dogs, *E. felina* in cats, *E. bovis* in cattle and *E. intricata* in sheep and goats are also common. *Isospora,* intestinal parasites of man and other animals, include one truly pathogenic species of man, *hominis, I. felis, I. bigemina* and *I. rivolta* infect cats and dogs and occur in mucous membranes of ileum. Their transmission is by cylindrical oocysts. Species of *Babesia* are intra-erythrocytic parasites of various vertebrates. *Babesia bigemina* of cattle causes the lethal haemoglobinuric fever, red-water fever or Taxas fever. *B. equi* in horses, *B. rohdani* in rodents, *B. felis* in cats, *B. motasi* in goats, etc., cause malignant jaundice, anaemia and fever in their respective hosts.

(d) Pathogenic ciliates. Balantidium coli is the only important ciliate pathogenic parasite. It is found in the intestine of man and often in frogs.

PARASITISM IN PROTOZOA

Parasite and Parasitism

The word **parasite** has been derived from two Greek words, **para** = beside, and **sitos** = **food**, which means eating beside one another. *Parasites* may be defined as "the species which exist at the expense of certain other species, called **hosts**, and are biologically and economically closely connected with them throughout their life-span." **Parasitism** is an association between the parasites and their hosts. It may be defined as "an association between two organisms of such kinds that one (parasite) lives and feeds, temporarily or permanently, either in or on the body of the other (host)".

Parasitic species occur in all groups of Protozoa, and one of them, Sporozoa, is exclusively parasitic. A brief account of parasitism in Protozoa may be given under the following headings :

Types of Parasites

1. Ectoparasites. Those Protozoa which inhabit the external surface of their hosts. *Hydramoeba hydroxena,* feeding on the ectodermal cells of *Hydra* and *Ichthyophthirius multifilis,* burying in the epidermis of freshwater fishes, are examples of ectoparasitic Protozoa.

2. Endoparasites. Those parasites which live inside the body of their hosts. These are divided into four categories :

(a) Parasites of digestive tract. Those endoparasites which dwell inside the lumen of alimentary canal of hosts. *Giardia lamblia,* a parasitic flagellate, *Entamoeba histolytica,* a parasitic *Amoeba, Isospora hominis,* a parasitic coccidian, *Balantidium coli,* a parasitic ciliate, are all intestinal endoparasites of man.

(b) Parasites of mouth. Those endoparasites which reside in mouth cavity of hosts. In man, *Entamoeba gingivalis* and *Trichomonas tenax* are found in pockets between the gums and teeth.

(c) Parasites of genital tract. Those endoparasites which inhabit the genital tract of hosts. In human female, *Trichomonas vaginalis* lives in vagina.

(d) Parasites of body tissues. Those parasites which live within tissues of hosts and may enter through skin or from digestive tract. Species of *Trypanosoma, Leishmania, Plasmodium* and *Babesia* are common blood parasites of vertebrates. Species of *Eimeria* and *Isospora* occur in the epithelial lining of gut of their respective hosts.

3. Hyperparasites. These are Protozoa parasitizing other species of parasitic Protozoa. For instance, the opalinid (*Zelleriella*) which lives in the frog's intestine, is parasitized by a certain *Amoeba*. *Nosema notabilis* parasitizes *Sphaerospora polymorpha* which is a parasite of urinary bladder of toad fish.

4. Pathogenic parasites. Most of the parasitic protozoans do not cause disease conditions in their hosts except producing minor symptoms. On the other hand, certain parasites act as disease-causing organisms in man and other animals. Such parasites are referred to as **pathogenic parasites**. Important pathogenic parasites of man are *Leishmania donovani, Trypanosoma gambiense, Plasmodium vivax, Entamoeba histolytica,* etc.

Host Specificity

Two general trends are exhibited regarding the development of host specificity in parasitic Protozoa. Firstly, some parasites can successfully parasitize a wide variety of hosts. *Trypanosoma, Entamoeba* and *Eimeria* belong to this group. Secondly, some parasites have become restricted to only a few specific host. *Coccidia* of mammals (*e.g., Plasmodium*) are such parasites.

Transmission

Protozoan parasites have different ways of infecting their hosts. *Entamoeba gingivalis* is transferred directly from one man to another through mechanical contact, like kissing (direct transfer). Some, such as *Entamoeba histolytica* and *Eimeria tenella,* are transferred by cysts in water or food (contaminative transfer). Species of *Trypanosoma, Plasmodium,* etc. are transmitted by certain invertebrate vectors (inoculative transfer). Transmission by invasion of ovary or egg takes place for species of *Babesia*. Placental transference has been reported for *Plasmodium* in man (congenital transfer).

Life Cycle

Many parasites, such as *Eimeria* and *Monocystis,* have only a single host during their life cycle, only a part spent outside the host. These are called **monogenetic** parasites. Other protozoan parasites (*e.g., Plasmodium, Trypanosoma*) have two or more hosts, belonging to widely separated animal groups. The two hosts are usually designated as the **primary host**, in which the parasite's ancestors evolved and the **secondary host** or **vector,** which acts as a transmitting agent for the parasite to the other host. These are called **digenetic** parasites. If the parasite undergoes part of its life cycle in vector, its transmission is called **cyclical**, if not, it is referred to as **mechanical** transmission. Besides the two hosts, some other animals may be infected by the parasites and serve as a source for infecting other animals. These animals constitute the **reservoir hosts**.

Effects of Parasites on Their Hosts

Parasitic Protozoa bring about change within their hosts to a lesser or greater degree. Some prove to be injurious for their hosts, while others produce almost no effect. *Entamoeba histolytica* destroys host's large intestine causing large ulcerations. *Eimeria stiedae* is known to cause hyperplasia of the hepatic cells of rabbits. The coccidian parasite of earthworm, *Polymnia nebulosa,* brings about hypertrophy of the sperm mother cells. The malarial parasite of birds, *Plasmodium gallinaceum,* clogs the fine blood capillaries. Similar examples are not few but numerous in the world of protozoan parasites.

PROTOZOA AND HUMAN DISEASES

Most of the diseases affecting human beings are caused by living organisms. Of the parasitic species of **Bacteria**, **Protozoa**, **Fungi**, **Viruses**, **Helminths** and **Arthropods**, some are pathogenic for man. Diseases occur as a result of interaction between pathogens (parasites) and human host under specific environmental conditions. This interaction is referred to as **infection** or **infectious process**, the manifestation of which is always some **infectious disease**. Disease is caused when pathogenic parasites alters the human body's normal vital activities as a result of morphological and physiological damage.

Pathogenic parasites, belonging to various groups of animal kingdom, share the following common characteristic features :

(1) All are **obligatory parasites**, *i.e.*, they depend for their existence upon their hosts.

(2) All show **host specificity**, *i.e.*, they harbour selective hosts.

(3) They secrete toxic substances which cause the particular disease.

(4) They complete their life cycle in one or more than one host.

Parasitic species occur in all classes of Protozoa. Approximately 15 different genera have been found living as parasites within the human body. While the majority of these have relatively little effect upon their hosts, certain pathogenic parasites cause some of the worst human diseases. Some of these are described below.

Amoebiasis

Amoebiasis, also known as **amoebic dysentery**, is caused by *Entamoeba histolytica* (Sarcodina). Its trophozites, that penetrate the wall of hosts intestine (colon), secrete histolytic enzymes and feed upon its cells causing the formation of ulcers. These ulcers rupture and discharge blood and mucus into the intestine that pass with stools. The trophozoites, under certain circumstances, reach liver, lung and brain where they cause abscess formation.

No intermediate host is involved in the life cycle of *E. histolytica*. Transmission of parasite from man to man takes place through tetranucleate cysts. Prior to cyst-formation the trophozoite changes into a smaller **minuta form**, which then encysts to form a tetranucleate cyst. These cysts are excreted with faecal matter and contaminate food and water and spread into new hosts. Houseflies help in its rapid spread.

Amoebic dysentery is endemic in warm countries. **Treatment** of amoebic dysentery or amoebiasis starts with a course of the antiamoebic drug **metronidazole** to finish off the parasite (trophozoites) and a further course of **diloxanide furoate, paromycin** or **idoquinol** to finish off the cysts.

Diarrhoea

Diarrhoea, which is characterized by loose bowels, is caused by a flagellate parasite, *Giardia* (=*Lamblia*) *intestinalis*. It harbours the small intestine of man. Its cell body is pear-shaped with dorsal side convex and ventral side flattened and deepened anteriorly to form a suction groove. It bears two nuclei and four pairs of flagella arranged symmetrically.

The parasite divides by binary fission to multiply rapidly and feeds upon amino acids and vitamins contained in food within the intestine. This causes intestinal disorders leading to epigastric pain, abdominal discomfort, loss of appetite stomach cramps, burping (often sulphurous), greasy stools, headache and diarrohea.

Transmission of parasite takes place through cysts which are excreted with faeces and enter new hosts in food or water. Infection of *Giardia* is more in children than in adults. **Treatment** is often not necessary as the infection is chronic and resolves by itself. However when it is acute medications like **Metronidazole** or **Albendazole** (a five to 10 day long course) are needed to treat it.

Trypanosomiasis

Trypanosomiasis (also called Trypanosomosis). These are two diseases caused by flagellate parasitic protozoan trypanosomes of the genus *Trypanosoma*.

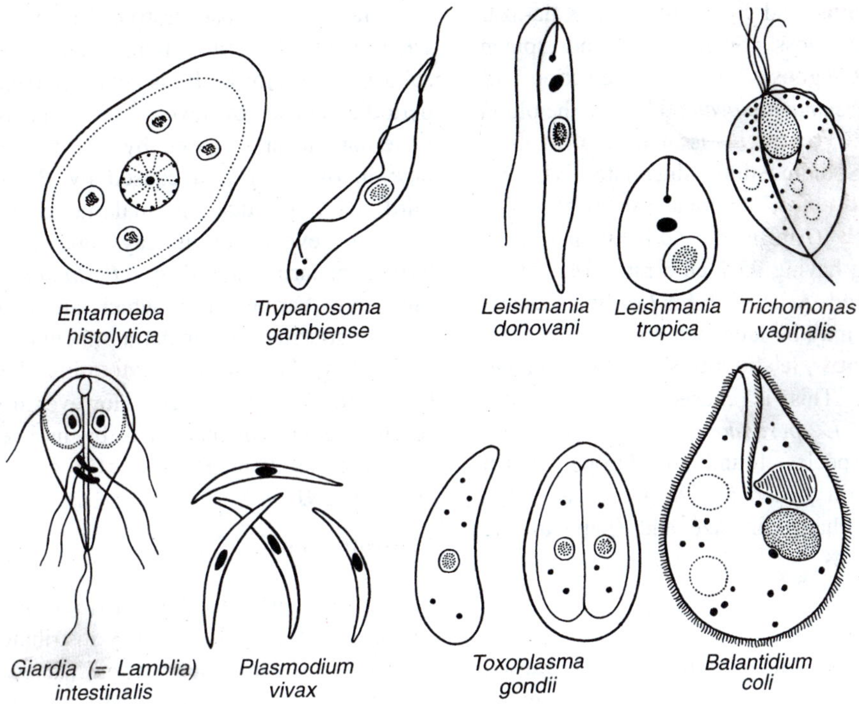

Fig. 16. Some pathogenic protozoan parasites of man.

1. African sleeping sickness. This disease is caused by protozoan parasite *Trypanosoma brucei gambiense* (90% cases) and *Trypanosoma brucei rhodesiense* (10% cases). **Spread** of disease of humans is via insect vector tsetse bites. **Signs and symptoms** of African sleeping sickness include skin sore, lymph node enlargement, malaise, seizures, and difficulty in talking and walking. These symptoms can develop over many years and if not treated, the person dies. **Diagnosis** is by detection of trypanosomes in a patients chancre fluid, lymph node aspirates, blood, bone marrow and cerebrospinal fluid. **Treatment** for haemo-lymphatic phase is intravenous or intramuscular **Pentamidine** (for *T. b. gambiense*) or intravenous **Suramin** (*T. b. rhodesiense*). The treatment for neurological phase is intravenous. **Melarsoprol** (Mel B and Melarsen Oxide- BAL) Combination therapy with **Eflornithine** and **Nifurtimox** is an alternative for neurological phase (for *T. b. gambiense*).

2. Chagas Disease (Also called **American Trypanosomiasis**). This disease is caused by protozoan parasite *Trypanosoma cruzi*. **Spread** of disease to humans is by an insect vector, the blood sucking 'kissing bugs' of genus *Tiratoma.* The disease is widespread in poor, rural areas of Mexico, Central America, and South America. **Signs** and **symptoms** of Chagas disease is a local swelling at the site of infection. **Treatment** is with **Benznidazole** and **Nifurtimox**

Leishmaniasis

Leishmaniasis is a disease caused by flagellate protozoan parasite of *Leishmania* genus. **Spread** of disease to humans is by the bites of an insect vector, female sand flies of genus *Phlebotomus.*

1. Visceral leishmaniasis (Also called Kala-azar). This disease is most severe and potentially fatal (only second to malaria) form, caused by *L. donovani* and *L. infantum* found in India and

East Africa. **Signs** and **Symptoms** of Kala-azar include weight loss, fever, liver and spleen enlargement. **Diagnosis** is by detection of **amastigotes** (*Leishmandonovan* bodies) in the blood by microscopy. **Treatment** is with **Pentavalent Antimonials** (Sodium stibo Gluconate INJ) and **Meglumine Antimonite**. Alternate treatment is now **Amphotericin B** (AmBisome and Amphocil). A newer oral drug having 95% cure rate **Miltefosine** is being evaluated. A **Vaccine** for Leishmaniasis is under trial and may be launched soon.

2. Cutaneous leishmaniasis (Also called Oriental Sore). This is caused by *L. tropica, L. brasiiensis, L. peruana, L. guyanesis,* and *L. mexicana* species found in Europe, North Africa, and Latin America. **Symptoms** are skin sores that can change in size and shape as the disease progresses.

Trichomoniasis

This disease is also called **vaginitis** and is caused by the species of *Trichomonas,* the flagellate protozoans. Its body is rounded with one nucleus, an axostyle, a parabasal body, 4-6 flagella and one backwardly directed flagellum. Most common pathogenic species is *Trichomonas vaginalis* that inhabits vagina of women and causes **vaginitis**. Disease is characterized by inflammation, burning sensation, itch and frothy vaginal discharge. Transmission is always through sexual intercourse by male members who act as intermediaries. Even in man, the infection of urethra and prostate is not uncommon. Treatment is by taking **Metronidazole**.

Malaria

Malaria is caused by the species of *Plasmodium,* the sporozoan parasites. It is transmitted through bite of female *Anophels* mosquito. In man, the parasite attacks liver cells and red blood cells. A toxic substance, **haemozoin**, released by parasite causes malaria.

Malaria is most destructive for man. It is widespread in the temperate, tropics and subtropics regions. It is characterized by the periodic attacks of fever. The fever is repeated in tertian malaria caused by *P. vivax* every third day; in ovale malaria caused by *P. ovale* every third day; in quartan malaria caused by *P. malariae* every fourth day, and in malignant tertian malaria caused by *P. falciparum* every third day. Various drug which are now used in the treatment of malaria include **Quinine, Atebrin, Chloroquine, Camoquine, Pamaquine, Paludrine, Daraprim,** etc. However treatment of severe *Falciparum* malaria is by intravenous (IV) or intremuscular (IM) **Artesunde INJ** or **Quinine** INJ.

Toxoplasmosis

This disease is caused by *Toxoplasm gondii,* a sporozoan parasite. It is widely distributed all over the world. The parasite occupies the cells of the reticulo-endothelial and central nervous systems. The parasites multiply by **endodyogeny** but under certain conditions large cysts are also produced. Usually the parasite remains in hosts body without causing any symptoms, but if an infected woman conceives, abortion takes place. If infection of *T. gondii* occurs during pregnancy, in third trimester then nervous system of the developing foetus is affected and the infant dies after birth. **Treatment** is with oral medications like **Pyrimethamine** (*Daraprim*) **Sulfadiazine** (*Microsulfon*), and **folinic acid**.

Balantidial Dysentery

This disease is caused by *Balantidium coli,* an intestinal ciliate. It is characterized by diarrhoea and ulceration of the large intestine. Transmission of parasite to a new host takes place through cysts in contaminated food or water. **Treatment** is by oral drugs like **Tetracycline**, **Carbasome Metronidazole** or **Diiodohydroxyquin**.

Important Questions

➤➤*Long answer type questions*

1. Mention the chief types of locomotor organelles of Protozoa. Explain briefly how they bring about locomotion.
2. Describe the organs of locomotion in *Amoeba, Euglena, Paramecium* and *Monocystis.*
3. Write an essay on : (i) Nutrition in Protozoa, (ii) Parasitism in Protozoa, (iii) Protozoa and Human Diseases, (iv) Body coverings and Skeletons in Protozoa, (v) Locomotion in Protozoa.
4. Give a brief account of the modes of reproduction in Protozoa.
5. Discuss the economic importance of Protozoa.
6. Distinguish between : (i) Amphimixis and Syngamy. (ii) Cytostome and cytopyge, (iii) Gametogony and sporogony, (iv) Macronucleus and micronucleus, (v) Saprozoic and saprophytic nutrition.
7. Name some important pathogenic species of Protozoa that cause diseases in man. Describe the methods of their control.
8. Write short notes on : (i) Amoebiasis, (ii) Contractile vacuole and osmoregulation, (iii) Euglenoid movement, (iv) Holozoic nutrition, (v) Hyperparasites, (vi) Leishmaniasis, (vii) Lobopodia, (viii) Mixotrophic nutrition, (ix) Multiple fission, (x) Nucleus in Protozoa, (xi) Parasitism, (xii) Pathogenic Protozoa, (xiii) Skeleton in Protozoa, (xiv) Symbiotic Protozoa, and (xv) Trypanosomiasis.

➤➤*Short answer type questions*

1. Which protozoan parasite exhibits schyzygy ?
2. What is incubation ?
3. What is exflagellation ?
4. Name a symbiotic protozoan present in digestive tract of termites.
5. What is the study of protozoans called as ?
6. Give an example for coprozoic protozoan.
7. Give an example of a protozoan with (a) holophytic nutrition, (b) mixotrophic nutrition
8. Name the different forms of flagella found in Protozoa.
9. With reference to Protozoa, define commensalism and give an example.
10. Mention the different types of flagella found in Mastigophora.
11. Define symbiosis with reference to protozoa and give an example of a type where both organisms are animals.
12. What is holozoic nutrition ? Give two examples from Protozoa.
13. With reference to Protozoa, what is meant by syngamy ?
14. Distinguish between sporonts and sporozoites.
15. Describe the structure and types of cilia in ciliophora.
16. With reference to their habitat distinguish between coccidia and haemosporidia
17. What is osmoregulation ? Give an account of the process of osmoregulation in Protozoa.
18. Write the functional value of chromatophores in phytomastigina. Do the chromatophores have any bearing on the phylogenetic implications of the flagellates ?
19. Differentiate between lobopodium and reticulopodium.
20. Distinguish between dinoflagellata and hypermastigina based on their flagellar condition.
21. Compare the organization of the Flagellata and Ciliophora and give reasons for believing that ciliophora are more advanced than the flagellata.
22. Distinguish between paedogamy and hologamy in Protozoa.

➤➤*Multiple choice questions*

1. Alternation of sexual and asexual generation is known as :
   (a) digenesis
   (b) metagenesis
   (c) metamorphosis
   (d) dimorphism
2. What is meant by intracellular parasite ?
   (a) parasite on the outer surface of cells
   (b) parasite between the cells
   (c) parasite inside the cells
   (d) parasite which interconnects neighbouring cells
3. What is a vector ?
   (a) disease transmitting host
   (b) natural reservoir of disease
   (c) non pathogenic protozoa
   (d) pathogenic protozoa
4. Down stroke and recovery stroke are the two phases of :
   (a) movement of contractile vacuole
   (b) flagellar movement
   (c) movement of food vacuole
   (d) amoeboid movement

5. The presence of slender pseudopodia with interlacing branches to form a net work is the characteristic feature of the order :
(a) radiolarian (b) lobosoa
(c) mycetozoa (d) heliozoa
(e) foraminifera
6. Contractile vacuole is noticed among *Amoeba* in :
(a) sea water (b) fresh water
(c) both (d) none
7. One-celled animals reproduce by :
(a) budding (b) splitting
(c) cutting (d) regeneration
8. The contractile vacuole in protozoans is chiefly concerned with the process of :
(a) digestion (b) osmoregulation
(c) excretion (d) assimilation
9. Food capture is the function of :
(a) food vacuole (b) reticulopodium
(c) eyespot (d) paramylum
10. Karyosome consists of two materials. What is the achromatic substance present in it ?
(a) plastin (b) chromosomes
(c) paramylum (d) eyespot
11. The shell of Polystomella is made up of :
(a) calcium carbonate (b) silicon
(c) chitin
12. Cytopyge is found in :
(a) *Amoeba* (b) *Paramecium*
(c) *Euglena* (d) *Trypanosoma*
13. The locomotory organelle of foraminifera are :
(a) lobopodia (b) filopodia
(c) axopodia (d) reticulopodia
14. The function of contractile vacuole is increased when :
(a) it is needed to remove waste products
(b) it has stopped feeding
(c) the cell membrane is permeable to water
(d) the concentration of water outside is higher than that inside
(e) the pond water contains more solute than the animal's body
15. Contractile vacuole in protozoans serves the function of :
(a) excretion (b) osmoregulation
(c) digestion (d) respiration
16. What would happen if some fresh water protozoans are placed in a medium of high osmotic potential ?
(a) the contractile vacuole will work faster
(b) the contractile vacuole grows larger
(c) the contractile vacuole shows no change
(d) the contractile vacuole does not form
17. The probable function of contractile vacuole is :
(a) to remove salts only
(b) to remove excess water
(c) to remove undigested food particles
(d) to transport water from one location to the other
18. When we transfer a protozoan living in sea to a fresh water medium :
(a) it bursts
(b) it enlarges but shrinks again
(c) it shrinks
(d) it remains static
19. If marine protozoan is placed in fresh water its body will :
(a) rupture
(b) shrink
(c) do not change
(d) first enlarge then become normal
20. Cilia in *Paramecium* arise from :
(a) basal granules (b) pellicle
(c) cytopharynx (d) trichocysts
21. Biolumniscence represented by :
(a) *Noctiluca* (b) *Giardia*
(c) *Volvox*
(d) *Trichomonas*
22. Subcellular physiological division of labour occur in :
(a) Protozoa (b) Porifera
(c) Cnidaria (d) all

Answers

➤➤*Multiple choice questions*

1. (b)	2. (c)	3. (a)	4. (b)	5. (e)	6. (b)	7. (b)	8. (b)	9. (b)	10. (c)
11. (a)	12. (b)	13. (d)	14. (c)	15. (b)	16. (d)	17. (b)	18. (b)	19. (b)	20. (a)
21. (a)	22. (a)								

16

Organization of Metazoa

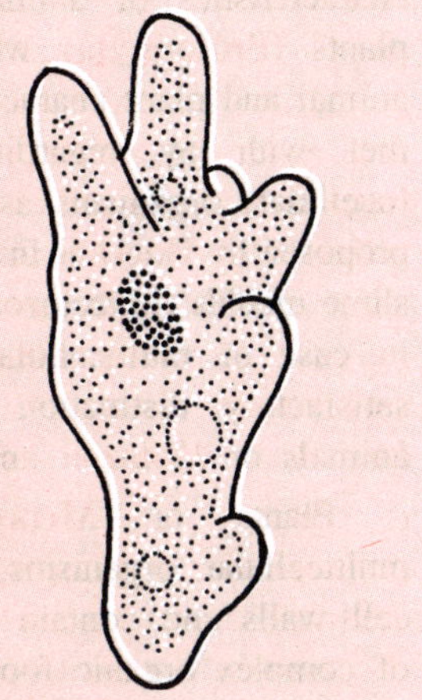

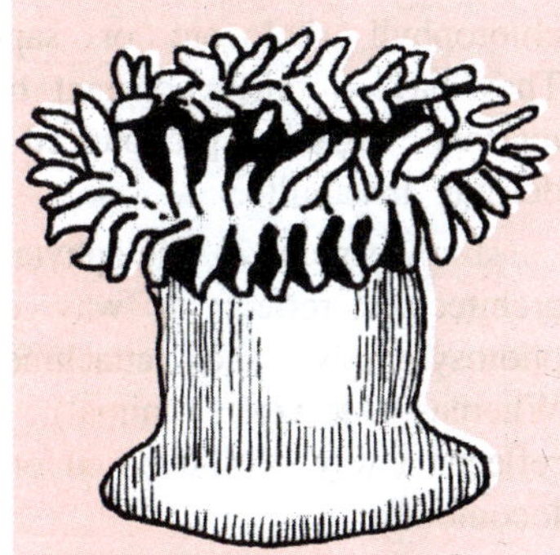

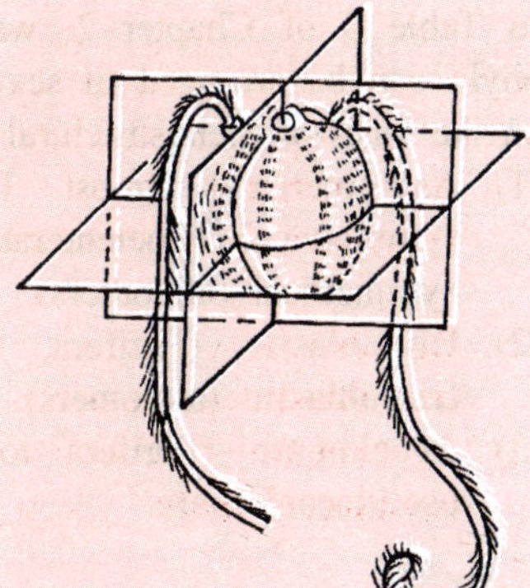

WHAT ARE METAZOA ?

On the basis of the number of cells forming the body, the Animal Kingdom is generally divided into two Sub-kingdoms, **Protozoa** and **Metazoa**. Members of the **Subkingdom Protozoa** are unicellular or acellular animals composed of single cells or colonies of like cells. On the other hand, members of the **Subkingdom Metazoa** are multicellular animals composed of many cells, usually arranged in distinct layers, tissues and organs.

Metazoans may be defined as holozoic multicellular organisms which develop from embryos. Their body cells are generally differentiated into tissues and organs* which are specialized for different functions. Their gametes are never formed within unicellular structures but are produced within multicellular sex organs or atleast within surrounding somatic cells. This definition holds good to differentiate the Sub-kingdom Metazoa from the Subkingdom Protozoa of the Animal kingdom.

In short, Metazoa can be defined as multicellular animals, or as cellular animals, or simply as animals.

METAZOA *VERSUS* METAPHYTA

The usual concept of plants and animals does not apply satisfactorily to non-multicellular or acellular organisms. Some of them have the

* Body of all Metazoans is not organized on tissue and organ level of organization. Poriferans are at cellular grade of organization and coelenterates are at tissue grade of organization.

characteristics of animals **(Protozoa)**, some of plants **(Protophyta)** while others possess both animal and plant characteristics. This difficulty is met with by regarding all non-multicellular (acellular) organisms as **Protista**. The term was proposed by **Haeckel** in 1866. It included bacteria, slime moulds (Mycetozoa) and Protozoa. However, in case of multicellular organisms, a fair and satisfactory distinction can be made between animals or **Metazoa** and plants or **Metaphyta**.

Plants or **Metaphyta** are autotrophic multicellular organisms. They possess cellulose cell walls and contain chlorophyll for synthesis of complex organic foodstuffs by photosynthesis (holophytic nutrition). They have inconstant body form, mostly external organs and limited movement (stationary).

On the other hand, animals or **Metazoa** are heterotrophic multicellular organisms. They do not have cellulose cell walls and lack chlorophyll (holozoic or saprozoic nutrition). They have a fairly constant body form, mostly internal organs, pronounced movement and definite irritability.

In general, Metaphyta (plant) body architecture reflects a way of life based on photosynthesis and attachment or sessility. Whereas Metazoa (animal) body architecture reflects a way of life based on alimentation and locomotion.

LOWER AND HIGHER METAZOA

In Table 3 of Chapter 2, we saw that animal phyla can be arranged in several different ways on the basis of their structural traits, as follows :

(1) **Asymmetrical** (most Porifera), **radially symmetrical** (Coelenterata) and **bilaterally** symmetrical (all others).

(2) **Diploblastic** (Porifera, Coelenterata) and **triploblastic** (all others).

(3) **Acoelomate** (Porifera to Platyhelminthes), **pseudocoelomate** (Entoprocta, Acanthocephala, Aschelminthes) and **coelomate** (all others).

(4) **Segmented** (Annelida, Arthropoda, Tardigrada, Chordata) and **unsegmented** (all others).

(5) **Chordates** (Protochordata, Vertebrata) and **non-chordates** (all others).

(6) **Vertebrates** (higher chordates) and **invertebrates** (all others).

1. Lower Metazoa. Porifera, Coelenterata, Platyhelminthes, Aschelminthes, Entoprocta and Acanthocephala are generally called **Lower Metazoa**. They are unsegmented, radially or bilaterally symmetrical, diploblastic or triploblastic, and acoelomate or pseudocoelomate.

2. Higher Metazoa. The remaining phyla (Mollusca, Annelida, Arthropoda, Echinodermata and Chordata) are known as **Higher Metazoa**. They are triploblastic and truly coelomate animals. Except chordata, all are non-chordates or invertebrates.

However, the Lower and Higher Metazoan Phyla, based on structural traits, must not be confused with the **Minor** and **Major Phyla** which are based on different criteria, such as the number of species and individuals and their participation in ecological communities (See chapter 2).

METAZOAN ORGANIZATION

Relationships between animals, or groups of animals, are best explained by their comparative morphology and embryology. Animals show various patterns in their morphology. The gross external morphology of animals falls under a limited number of patterns (criteria). These include form of animals (**symmetry**), arrangement of body parts in segments (**metamerism**), formation of a head (**cephalization**), and progressive sequence of specialization of structure (**levels or grades of organization**). Similarly, the criteria of internal morphology are differences in formation of body cavity (**coelom**) and reproduction (**embryology**), etc.

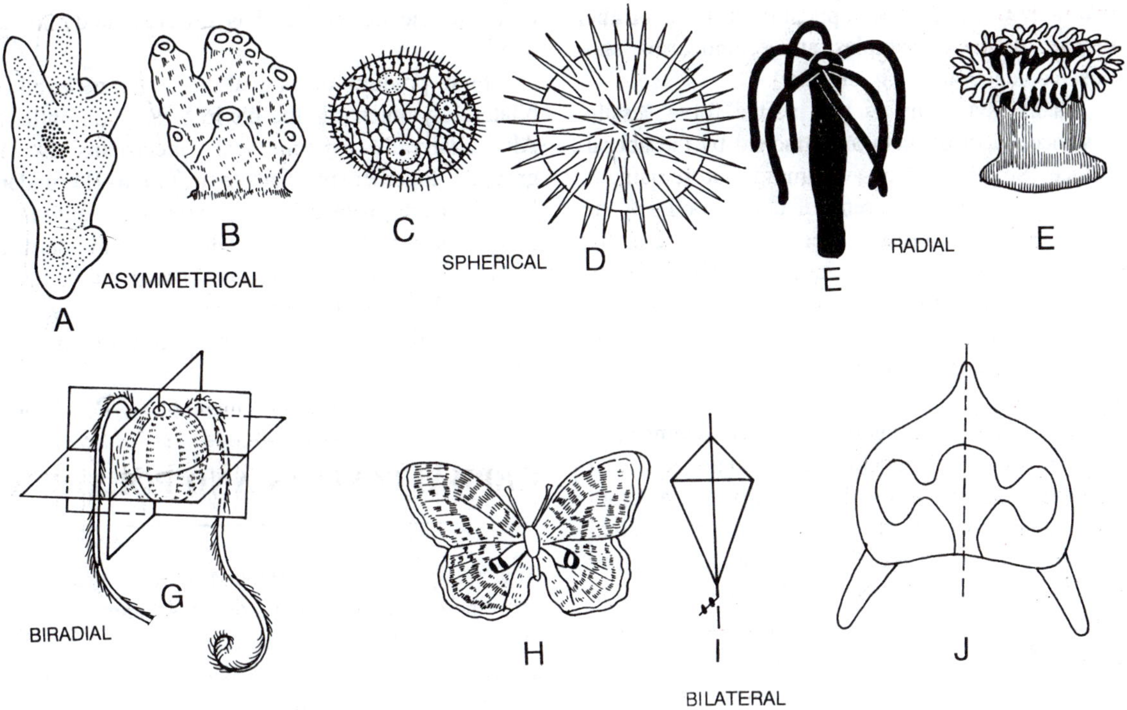

Fig. 1. Types of symmetry in animals.

SYMMETRY AND ITS SIGNIFICANCE

Symmetry means an arrangement of body parts into geometrical designs. It refers to the division of body into equal parts by lines or planes. Current zoological ideas about symmetry derive chiefly from Haeckel. An animal is called **symmetrical** when a plane passing through its centre will divide it into similar halves. When an animal cannot be divided into like parts by a plane, it is called **asymmetrical**. All animals are either asymmetrical or symmetrical. Examples of asymmetrical animals are most sponges, some Protozoa (*Amoeba*) and few others.

Certain terms are often used when explaining symmetry. An **axis** is an immaginary line passing through the centre of body, such as longitudinal axis and oral-aboral axis. Either end of an axis is termed **a pole**. Thus, each axis has two poles. A **plane** of symmetry is a straight line that divides organisms into corresponding halves.

Metazoa display only two types of symmetry, **radial** and **bilateral**. Two other types of symmetry are also recognized, **spherical** and **biradial**. Protozoa are not only asymmetrical but display all four types of symmetry in their diverse body forms.

1. Spherical symmetry. It is found in animals whose body has the shape of a sphere. All planes that pass through the centre will cut it into similar halves. Some protozoans (*e.g.*, *Volvox*, Heliozoa, Radiolaria) have spherical symmetry, and it is adapted for free-floating or rolling movements.

2. Radial symmetry. The body is in the form of a flat or tall cylinder. Many similar body parts, called **antimeres**, are arranged around one main, central or longitudinal axis in a circular or radiating manner like the spokes of a wheel. All the lines passing through this longitudinal axis, in any plane, will divide the body into equal halves or antimeres. The surface having mouth is the

oral surface, and the opposite surface is the **aboral surface**. Examples are echinoderms and most coelenterates (*e.g.*, *Hydra*).

Radial symmetry is best suited for a sessile existence. Most of them are attached by the aboral surface. Some are free-swimming but remain at the mercy of water currents. Due to similarity of antimeres, their sensory receptors are equally distributed all around the periphery. This enables them to receive stimuli and to meet the chalanges of the environment equally from all directions. They can obtain food or repel enemies from all sides.

In the animal kingdom, radially symmetrical phyla are Porifera, Coelenterata, Ctenophora and Echinodermata. Out of these, only Coelenterata and Ctenophora display a fundamental radial symmetry. Both the phyla were grouped together by **Hatschek** (1888-91) under the Division **Radiata**. Adult Porifera are mostly asymmetrical, but they start life from a radially symmetrical larva. On the other hand, larval stage of Echinodermata has bilateral symmetry, but the adults become radially or pentaradially symmetrical.

3. Biradial symmetry. It is a variant form of radial symmetry found in Ctenophora and most Anthozoa (*e.g.*, anemones), and is best fitted for a floating life. Such symmetry has only 2 pairs of symmetrical sides. There are only 2 planes of symmetry, one through the longitudinal and sagittal axises, and the other through the longitudinal and transverse axises, which will divide the animal into equal halves.

4. Bilateral symmetry. In most higher animals, the longitudinal axis of body runs from the anterior end (head) to the posterior end (tail). There is a single plane, the **median longitudinal** or **sagittal** plane, through which the body can be divided into two similar right and left halves. This is called **bilateral symmetry**. Besides right and left sides, an upper or **dorsal** surface and a lower or **ventral** surface are also recognizable, which are unlike because they are exposed to different conditions.

Bilateral symmetry is characteristic of the most successful and higher animals, including the remaining invertebrates and all vertebrates. In most of them, the anterior end is differentiated into a **head**.

First phylum of animal kingdom to exhibit bilateral symmetry is the phylum Platyhelminthes. All bilaterally symmetrical metazoans were grouped together by **Hatschek** (1988-91) under the Division **Bilateria**. As already mentioned earlier, some Bilateria, such as echinoderms, display a radial symmetry which has been secondarily derived from bilateral ancestors due to assumption of an attached mode of life by adults.

CEPHALIZATION AND POLARITY

Bilateral symmetry is correlated with the locomotor movements brought about by these animals. One end of their body, usually containing the mouth, always moves forward in a particular direction. It is the first to come in contact with the environment, so that there is great concentration of nervous tissue and sense organs at this anterior end called **head**. The posterior or rear end is usually equipped with some locomotory organ. This modification of anterior or oral end of the animal into a definite head is called **cephalization** which is characteristic of most bilateral animals.

Cephalization is always accompanied by a differentiation along an antero-posterior or oral aboral axis. This condition is known as **polarity**, and it usually involves gradients which refers to ascending or descending activities between anterior and posterior ends.

METAMERISM

Meaning of Metamerism

When the segmentation in bilateral animals, such as annelids, involves a longitudinal division of the body into a linear series of similar sections or parts, it is termed **metameric segmentation** or **metamerism**. Each section or part is called a

segment, somite or **metamere** (Gr., **meta**, after; **meros**, part). Each metamere typically repeats some or all of the various organ units. The term metamerism is applied only when organs of mesodermal origin are so arranged. The primary segmental divisions are the body wall musculature and sometimes the coelom. This in turn imposed a corresponding metamerism on the associated supply systems (nerves, blood vessels and excretory organs). Longitudinal structures such as gut, principal blood vessels and nerves extend the entire length of body, passing through successive segments. While other structures, such as gonads, are repeated in each or few segments only.

Metamerism is always limited to the trunk region of the body. The **head** (or acorn) represented by the prostomium and bearing the brain and sense organs, and the **pygidium**, represented by the terminal part of the body which carries the anus, are not metameres. New segments arise just in front of the pygidium; thus the oldest segments lie just behind the head.

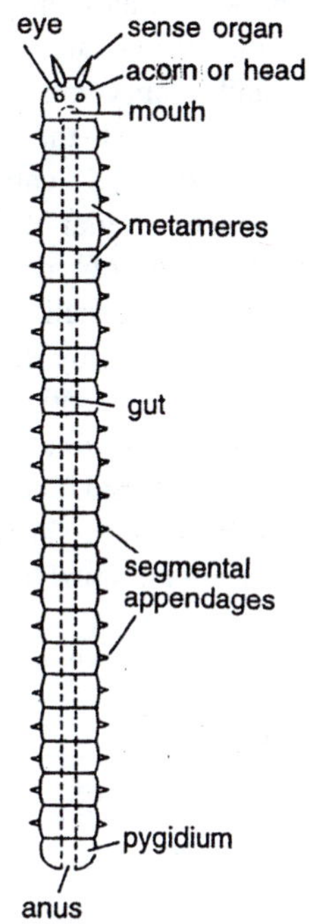

Fig. 2. A typical metameric animal, such as an annelid worm.

Metameric Animals

Metameric segmentation of the body, encountered for the first time in Annelida, is of considerable interest because the most successful groups of animal kingdom, *i.e.,* Arthropoda and Vertebrata, also have their parts metamerically repeated. At least one group of Mollusca (Monoplacophora) also exhibits metamerism: Metameric segmentation seems to have evolved three times independently in animal kingdom (i) in the annelids-arthropods, (ii) in the chordates and in (iii) cestodes.

External and Internal Metamerism

Metamerism is conspicuously visible in most annelids, both **externally** as well as **internally**. The common earthworm is a good illustration of both external and internal metamerism. Its body consists of a great number of similar segments and all the body organs, such as musculature, setae of locomotion, blood vessels, nerves, ganglia excretory organs and gonads, etc., are repeated segmentally. Even the coelom is divided into segmental compartments by the intersegmental transverse mesenteries, called **septa**. Only the digestive tract remains unaffected, but it also extends through every segment. In arthropods, metamerism is chiefly **external**, while man and other vertebrates show an **internal metamerism** of body muscles, nerves, certain blood vessels, vertebrae and ribs.

Complete and Incomplete Metamerism

In annelid worms, metamerism is **complete,** affecting practically all the systems. The metameres are essentially alike or **homonomous**, each having segmental blood vessels, nerves, nephridia and coelomoducts. This condition is called **homonomous metamerism**. On the other hand, higher animals such as arthropods and vertebrates, show **incomplete** metamerism. Because of division of labour, the segments or metameres of different regions of their body become greatly dissimilar. This is called **heteronomous metamerism**. However, incomplete metamerism should not be confused with the repetition of single organs such as shell plates or gills in certain unsegmented animals like molluscs.

In arthropods and vertebrates, metamerism is more complete and metameres are uniform and clear in the larval and embryonic stages. But, metamerism becomes obscure in the adut due to subsequent specialization or modification, so that the segments are no longer similar. It may result from simplification, by loss of metameres, by fusion of segments (cephalization), by differentiation between segments, by disappearance of organs, or by development of other structures, such as limbs. Heteronomous condition always appears first at the anterior end and progresses posteriorly. In segmented animals, varying degrees of specialization are met with some of which are extreme.

Origin and Evolution of Metamerism (Theories)

How metamerism has been brought about is still doubtful. No satisfactory reason can be given for the origin of metamerism. Various hypothesis have been proposed to explain the origin of metamerism, but none is acceptable in the absence of convincing evidence. The main theories concerning the origin of metamerism emphasize primarily either repetition of organs or mesodermal segmentation and correlate it with the origin of coelom.

1. Pseudometamerism theory. This theory stresses that metamerism developed secondarily as a result of repetition of body parts, such as muscles, nerves, nephridia, coelom, blood vessels, etc., in a single individual.

Such serial repetition of organs, such as testes, yolk glands and transverse connectives of two nerve cords, is seen in some elongated turbellarians and nemerteans. Later, a segmented condition was obtained by the formation of cross-partitions in between them, so that each segment received a repeated part of each system. This process is witnessed even today in the formation of somites in larval and adult stages of some Annelida, in which cross-partitions develop after the basic segmentation is already laid down. Such segmentation was probably an adaptation for an undulatory mode of swimming. However, all ribbon-like animals swim in this way, whether segmented or not. This theory is supported by **Hyman** (1951) and **Goodrich**.

2. Cyclomerism theory. Originally proposed by **Sedgwick** (1884) and greatly supported by **Remane** (1950, 1963), the cyclomerism theory is a corollary of the enterocoelous theory for the origin of mesoderm and coelom. This theory assumes that coelom originated in some ancestral radiate actinozoan coelenterate, through the separation of four gastric or enterocoelic pouches from the central digestive cavity or gut. Division of two pouches resulted into three pairs of coelomic cavities —the protocoel, mesocoel and metacoel, in the protocoelomate or ancestral coelomate. Loss of protocoel and mesocoel led to the unsegmented coelomates, such as molluscs and sipunculids. Later subdivision of metacoel produced primary segments, leading to the segmented annelids. The phylogenetic implication of this theory is that all bilateral metazoans were originally segmented and coelomate, and that the acoelomate unsegmented groups (flatworms, nemerteans) have lost these characters secondarily.

3. Corm or fission theory. According to this theory, metameric segmentation resulted when some non-segmented ancestor divided by transverse fission repeatedly or by asexual budding producing a chain of sub-individuals or zooids, united end to end due to their incomplete separation. This occurs in some Platyhelminthes and Annelida even today. Later, with the passage of time, these subindividuals or segments gradually became integrated morphologically as well as physiologically into one complex individuality. A segmented animal, according to this view, is a chain of completely coordinate subindividuals. This theory was supported largely during the 19th century and greatly elaborated by **Perrier** (1882). The chief objection to this theory is the lack of gradations of age in such a chain of zooids, which is true of segments in a metameric animal, such as an annelid. In platyhelminthes and scyphozoan strobilae the sequence of zooid formation is never serial, fission occurs always somewhere in the middle of the chain. In cestodes the proglottids are serially arranged but in a reversed order. Another objection is that reproduction by fission

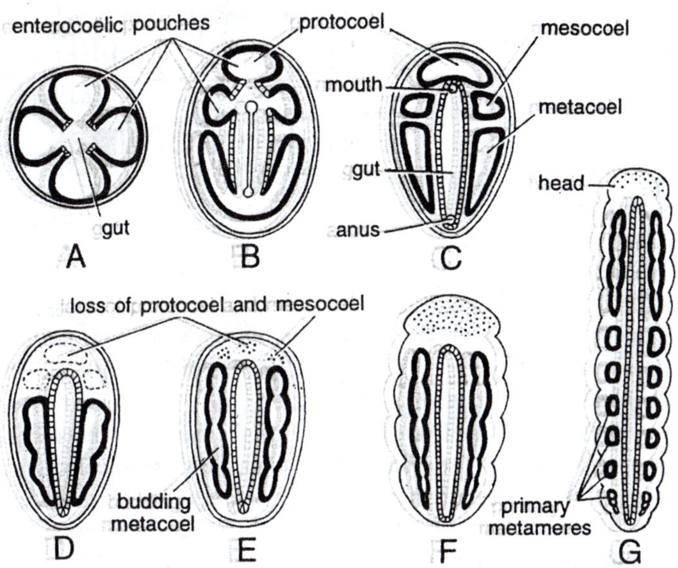

Fig. 3. Diagrams illustrating the Enterocoel-Cyclomerism theory of metamerism. **A-C.** Transformation of enterocoelic pouches of a radiate ancestor into coelomic pouches of a bilateral ancestor. **D-E.** Loss of protocoel and mesocoel. **F-G.** Subdivision of metacoel producing primary metameres.

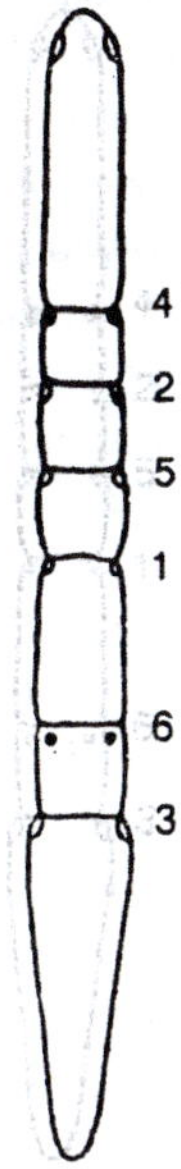

Fig. 4. Diagram showing development of a chain of zooids in rhabdocoel flatworm *Stenostomum*, according to Corm Theory. Numbers indicate sequences of fission planes.

is usually confined to sessile animals, whereas the ancestors were probably free-swimming. Moreover, fission is a more extensive and disruptive division of body than is metamerism.

4. Embryological theory. It explains original metameric segmentation mainly as an embryological accident. It suggests that mechanical stresses in the mesoderm of the elongating embryo or larva resulted in its fragmentation leading to segmental repetition of mesodermal derivatives in the adult.

5. Locomotory theory. It is an amalgamation of pseudometamerism and embryological theories. It postulates that metamerism evolved as an adaptation to locomotion of different kinds. Annelid metamerism probably evolved as an adaptation for burrowing and chordate metamerism as an adaptation for undulatory, serpentine swimming movements.

R.B. Clark (1964) suggests that coelom evolved initially as a hydraulic skeleton to facilitate locomotion in response to the increasing body size. According to Clark, metamerism also evolved as an adaptation to burrowing in annelids. The ancestors of Annelida were in all probability elongate coelomate animals which burrowed in marine sand and mud. The evolution of compartmented coelom, due to development of septa and metameric segmentation, localized the function of the hydraulic skeleton. This allowed only part of the body to contract while other parts in the longitudinal axis relaxed. The locomotory movements can be more continuous and better controlled if the action of the bodywall muscles is localized, *i.e.*, restricted to sections of the body. Metamerism permits such localization and accounts for the evolution of this condition in annelids. The localization enabled a strong peristaltic wave to propagate down the body, an efficient type of locomotion for worm-like burrowers. Thus, annelid metamerism is basically a modification of the coelom and the bodywall muscles. The initial segmentation of the coelom and bodywall muscles led to subsequent segmental organization for the nervous, circulatory and excretory systems.

In chordates, metamerism evolved as an adaptation for undulating swimming movements. According to **Berrill** (1955), notochord evolved in the early chordates to provide support for the body, followed by metameric segmentation of bodywall musculature, so that alternate waves of contraction sweeping down the body could enable strong swimming. The initial muscle segmentation caused segmentation of the nervous and circulatory systems.

In cestodes, metamerism evolved as a response to reproduction. A reproductive package was formed which, once it had performed its function, was expendable. Proglottids are not additional to the body but are the body and so have to carry a complete series of organs.

Conclusion. Metameric segmentation met in, one phylum is not necessarily similar to that in another. It probably arose independently in more than one line of evolution, each time in adaptation to a major advantage for the group in question. It probably evolved in response to burrowing in annelids, to swimming in chordates and to reproduction in cestodes. Therefore, we cannot device one explanation for all the cases of segmentation or think of a common ancestor of all sorts of segmented animals. Metameric segmentation here or on page 576.

Significance of Metamerism

It is not clear just how and why metamerism evolved or how the primitive ancestors were benefitted by it. Probably specialization of metameres for particular functions showed an advancement.

It is possible that segmentation was initiated in the musculature of an elongated swimming worm. This breaking up of the body into metameres would facilitate swimming movements. Metamerism helps in locomotion in several ways. The coordination of muscular action and fluid-filled coelomic compartments cause efficient swimming and creeping which is an advancement over the simple ciliary and creeping movement of lower invertebrates. Fluid-filled coelomic compartments also provide hydrostatic skeletons for burrowing. Precised movements can take place by differential turgor effected by flow of coelomic fluid from one part of the body to the other.

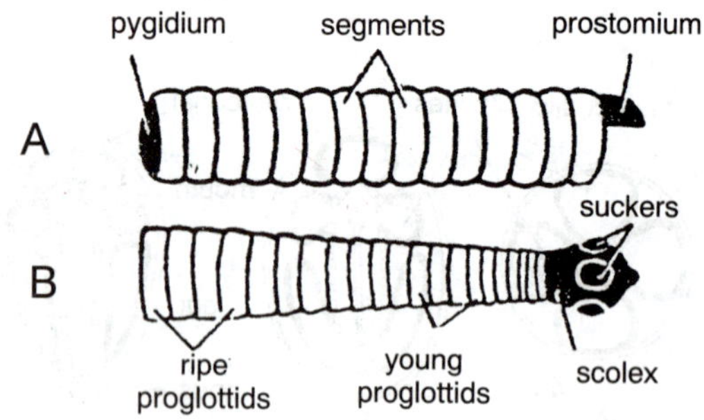

Fig. 5. Diagrams to illustrate differences between **A.** True metamerism of a generalized annelid. **B.** Strobilization of a generalized tapeworm.

Another advantage of segmentation or metamerism is the opportunity for different segments to specialize for different functions, thus leading to a rapid evolution of high grade of organization. It is not clearly marked in annelids, but is well developed in arthropods. Metamerism has, therefore, contributed towards the greater complexity of animal bodies and rapid evolution of high organization in animals. Thus, some indication of primitiveness of an animal can be determined by the degree of segmentation it displays.

The fact that cestodes, annelids, arthropods and chordates have metamerism does not necessarily indicate a close relationship among them, for the metameric condition may have arisen independently by convergent evolution.

Pseudometamerism

True metamerism, as shown by annelids, must not be confused with the **pseudometamerism** or **strobilization** of the tapeworms. It refers to superficial segmentation and could be termed body annulation. True segments of annelids are laid down in the embryonic stage. Whereas proglottids of tapeworm are not true metameres but rather complete reproductive individuals produced by strobilization, a type of budding, with the buds remaining attached. Table 1 shows the important differences between the two. However, the modern view now gaining favour is that cestodes are indeed metamerically segmented, although their

Table 1. **Differences between True Metamerism and Pseudometamerism.**

True metamerism	Pseudometamerism
1. Number of segments is generally constant for each species; *i.e.*, new segments are not added to the body after maturation except in asexual reproduction.	1. Number of segments or proglottids forming the body is not fixed as new segments are continually added throughout life.
2. Growth occurs due to simple elongation of pre-existing segments. The segments and ends of body have a fixed relationship to one another throughout life.	2. Growth occurs due to addition of new segments from a region of proliferation, just behind the scolex.
3. All segments are of the same age and at the same stage of development.	3. Proglottids differ from one another in age and in the degree of development.
4. Segments are functionally interdependent and integrated. Working in co-ordination, they preserve the individuality of body. For example, in a worm, during locomotion, muscles of each segment contract in a regular sequence so that rhythmical waves of contraction pass over the whole body which moves forward in an orderly manner.	4. Segments or **proglottids** are independent and self-contained units, each having a full set of sex organs and a portion of excretory and nervous systems. They are productive units developed for detachment. A tapeworm represents a sort of colony or strobila made of a linear rows of incomplete individuals.

BODY CAVITY OR COELOM

Definition

A body cavity can mean any internal space, or a series of spaces present inside body. Whereas coelom or true body cavity generally refers to a large fluid-filled space (cavity) lying between the outer bodywall and the inner digestive tube. It arises as a secondary cavity between two layers of embryonic mesoderm and contains most of the visceral organs.

Thus, a true coelom may be defined as a secondary body cavity, formed by the splitting of mesoderm during embryonic development and bounded on all sides by a definite coelomic epithelium or peritoneum. It contains a colourless coelomic fluid, the excretory organs open into it and the reproductive organs arise from its walls.

However, all animals do not possess a coelom, such as sponges, coelenterates, ctenophores, flatworms and proboscis worms. They are said to be **acoelomate**.

Types of Coelom

Two types of body cavities or coelom occur in Bilateria, **primary** and **secondary**.

1. Primary coelom. It is also called a false coelom or **pseudocoelom**. It is derived from the blastocoel of the embryo, rather it represents a persistent blastocoel ? Internal organs remain free in it since it is not bound by peritoneum or mesoderm. It is a space enclosed by ectoderm on the outside and endoderm (digestive tract) on the inside, and not by mesoderm on both sides. Such a pseudocoelom occurs in many worm-like animals including rotifers and roundworms, etc.

2. Secondary coelom. In more highly developed Bilateria, the blastocoel is gradually obliterated by the embryonic archenteron, without forming a primary coelom. Instead, a **secondary** or **true coelom** or **eucoelom** develops within the embryonic mesoderm and lined by a characteristic layer of flattened mesodermal epithelial cells, known as **peritoneum**, which also surrounds the internal organs of the body. A true coelom probably appeared for the first time in annelids.

Coelomic Division of Metazoa

On the basis of presence or absence of coelom, the Metazoa are divided into 3 major groups as follows :

1. Acoelomata. No body cavity or coelom is present. Embryonic mesoderm remains as a solid layer, space between endoderm (gutwall) and ectoderm (bodywall) is filled with mesenchyme and muscle fibres. Examples : Porifera, Coelenterata, Ctenophora, Platyhelminthes and Nemertinea.

2. Pseudocoelomata. Body space is a pseudocoelom or false coelom. It is a persistent blastocoel enclosed between outer ectoderm and

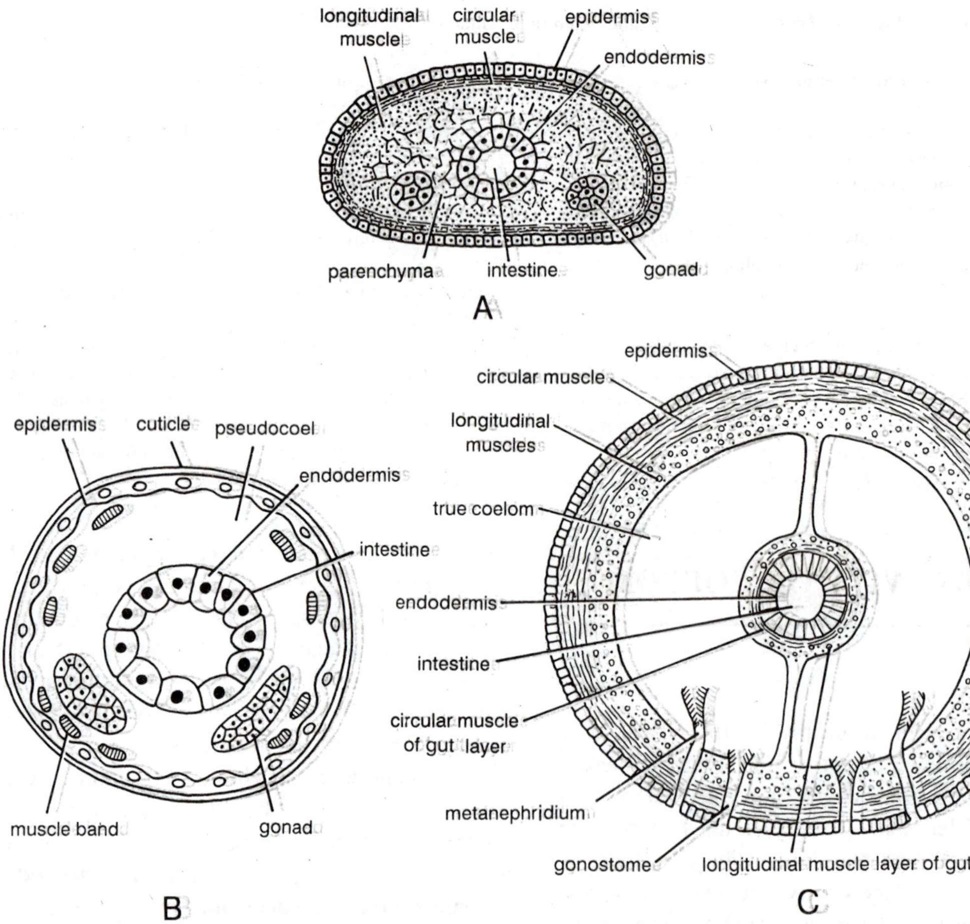

Fig. 6. Main types of body arrangements in Metazoa on the basis of coelom (In diagrammatic cross-sections). **A.** Acoelomate, with no body cavity. **B.** Pseudocoelomate, with body cavity not bounded by mesoderm. **C.** Coelomate or Eucoelomate with body cavity enclosed by mesoderm.

inner endoderm, and not lined by mesoderm. Examples : Acanthocephala, Ectoprocta and Aschelminthes (Rotifera, Gastrotricha, Kinorhyncha, Nematoda, Nematomorpha).

3. Coelomata or eucoelomata. Body space is a true coelom, enclosed by mesoderm on both sides. Remaining phyla of Bilateria, from Annelida to Arthropoda, belong to Coelomata.

There are 3 different ways in which entomesoderm and coelom can arise during embryological development. Accordingly, **Hyman** (1951) further divides coelomate Bilateria in 3 groups as follows :

(a) Schizocoelomata. Coelom arises by a splitting of endomesodermal bands which originate from blastoporal region of larva and extend between ectoderm and mesoderm. It is a true coelom called a **schizocoel**. Examples : Most of the Protostomia (Annelida, Arthropoda, Mollusca, etc.).

(b) Mesenchymal coelomata. It is seen only in Phoronida in which mesenchymal cells rearrange to enclose a space or coelom, which is regarded an aberrant schizocoel.

(c) Enterocoelomata. Coelom arises in the form of mesodermal pouches from larval archenteron. After separation from endoderm, the pouches fuse and expand until they touch the gut and bodywall. Since the coelom arises from larval enteron, it is called an **enterocoel**. Examples : Deuterostomia (Chaetognatha, Echinodermata, Hemichordata, and Chordata) and Brachiopoda.

Modifications of Coelom

There are modifications of coelom in different animals. As already described, in aschelminths, it forms a **pseudocoelom** not lined with mesoderm. In arthropods, it is formed by scattered spaces, collectively known as **haemocoel**, in which blood circulates. In annelids, coelom is divided by **septa** into chambers corresponding to the somites. In mammals, coelom is divided by a muscular **diaphragm** into thoracic and abdominal cavities, with a separate **pericardial cavity** around heart.

Significance of Coelom

Evolution of coelom is of great significance in animals. It plays an important role in the progressive development of complexity of structure. It permits greater size and contributes directly to the development of excretory, reproductive and muscular systems of body. In a triploblastic animal (*e.g.*, earthworm), the appearance of a perivisceral coelom between gut and bodywall leads to several advantages.

(1) It surrounds the internal organs like a water jacket and protects them from external shocks.

(2) It provides flexibility to the body. It provides space and gives the digestive tract and other internal organs freedom of movement and opportunity for enlargement, further differentiation and greater activity.

(3) The coelomic fluid often functions as a hydraulic skeleton and also serves as a circulatory medium for the transport and distribution of nutritive substances and gases.

(4) The excretory matter is collected into coelomic fluid and then passed out of the body through nephridia.

(5) The gonads arise from its coelomic epithelium and project into coelom. Ova and sperms are extruded through special gonoducts connecting the coelom with the exterior.

No doubt, the evolution of coelom made a major advance in the evolution of Metazoa. It is evident from the great increase in size and diversity in structure and ways of life met within coelomate phyla in comparison with acoelomate and pseudocoelomate phyla.

Evolution of Coelom (Theories of Origin)

Origin of coelom in Metazoa is of great evolutionary significance. It has recently been extensively discussed by **R.B. Clark** (1964). There is no direct evidence of evolution of coelom available from palaeontology. There is only indirect evidence which is based mainly on the embryology of present day metazoans. This has been differently interpreted by different workers resulting in many conflicting theories about the origin and evolution of coelom. Out of these, there are 4 principal theories as follows :

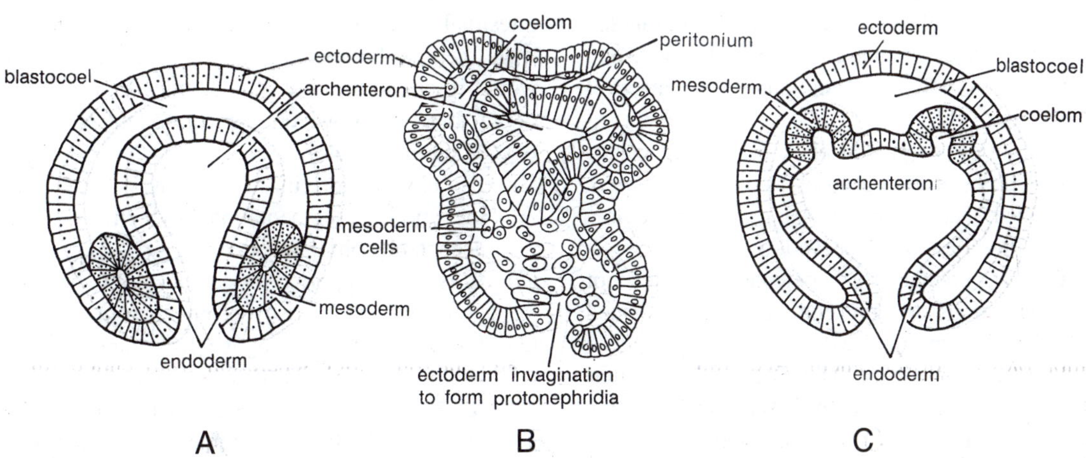

Fig. 7. Three types of mesoderm and coelom formation. **A.** Schizocoelous. **B.** Mesenchymal. **C.** Enterocoelous.

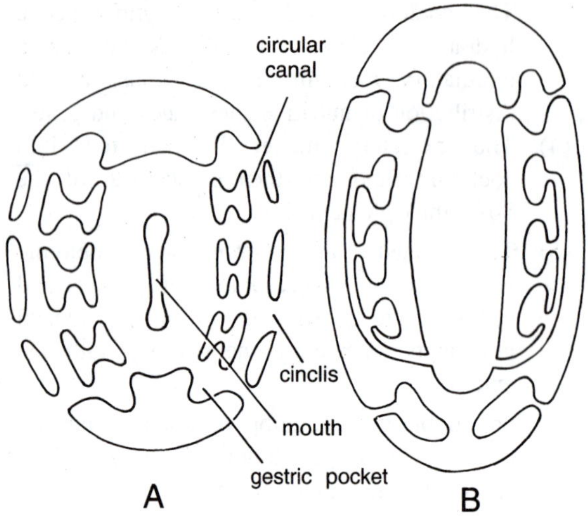

Fig. 8. Diagrammatic representation of coelom formation according to Enterccoel Theory. A. Formation of gastric pockets in an anthozoan. B. Gastric pockets become coelomic pouches.

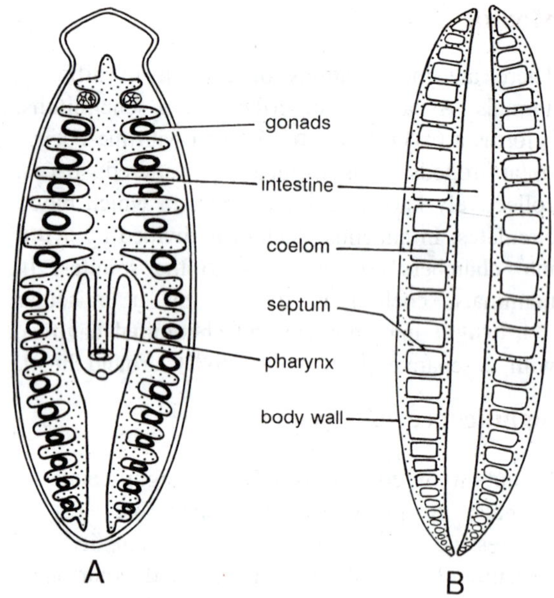

Fig. 9. Gonocoel theory of coelom formation. A. Gonads alternating with intestinal branches in a triclad flatworm. B. Coelomic sacs or compartments in an annelid, formed by fusion of adjacent expanded gonads.

1. Enterocoel theory. The theory holds that the primitive mode of coelom formation was enterocoelous. The idea was first proposed by **Lankester** (1875) and later modified by **Sedgwick** (1884) and till recently by **Hartman** (1963) and **Ramne** (1963). According to this theory, the bilateral metazoan ancestor of coelenterates had gastric pockets which become separated from the central digestive cavity to form coelomic pouches (Fig. 8). However, the theory is mainly objected on the following grounds :

(1) Gastric pouches occur in highly organized coelenterates, such as Scyphozoa and Anthozoa, which are not suitable for ancestral types.
(2) Sealing off of gastric pockets in the ancestor would defeat the purpose for which they were formed, that is, for increasing the surface area for digestion and absorption.
(3) Gastric pockets of coelenterates are not evaginations of gut endoderm, but formed differently by ingrowth of bodywall-septa.

2. Gonocoel theory. This is the most popular theory of origin of coelom. According to this theory, coelom represents a persistent expanded gonadial cavity or gonocoel. The theory was proposed in 1885 by **Bergh**, based on an idea earlier expressed in 1878 by **Hatschek**.

Meyer (1890) considered coelom initially unsegmented, arising from a single large pair of gonads. However **Bergh** and **Lang** (1903) believed that coelom initially arose in a segmented condition. In nemerteans and flatworms, especially in the triclad turbellarian *Procerodes lobata* (Fig. 9), they observed intestinal branches alternating with a linear series of gonads. If the intestinal branches withdraw and gonads enlarge with their walls meeting to become septa, they would resemble the row of coelomic compartments of an annelid. However, the chief defects of the Gonocoel theory are :

(1) It links the origin of coelom with that of metameric segmentation which is not acceptable phylogenetically.
(2) It is not supported by embryology because gonads do not arise before coelom.
(3) It does not account for unsegmented coelomate phyla, which are in majority, and there is no evidence that they have originated from segmented ancestors. **Goodrich** (1946) and **Hyman** (1951) have rejected the Gonocoel theory. They consider that metameric segmentation was preceded

by a pseudometameric distribution of organs (such as gonads) in an elongated worm-like body.

3. Nephrocoel theory. According to this theory, proposed by **Lankester** in 1874, coelom originated as the expanded inner end of a nephridium. The theory is not seriously held because, (i) Protonephridia have been described in coelomates, and (ii) Some coelomate groups, such as Echinodermata, do not have excretory organs.

4. Schizocoel theory. According to this theory, coelom has a mesenchymal origin and has no relation with nephridia and gonads of Lower Bilateria and endodermal pouches of Radiata.

Conclusion. None of the above theories satisfactorily explains the origin of coelom. Firstly, they do not explain the advantages of intermediate stages passed through during the course of evolution. Secondly, the connection between evolution of coelom and metameric segmentation together, has not been made clear. Thirdly, no exact nature of a coelom has been defined. Which cavities should be regarded coelomic and which not, has not been explained. Moreover, there is no evidence showing that secondary body cavity is homologous (with same origin) throughout the animal kingdom. Probably it is polyphyletic in origin. As **Clark** postulates, coelom might have arisen independently a number of times and in various ways in different animal groups — as a persistent blastocoel in pseudocoelomates, as an enterocoel in deuterostomes, as a schizocoel in protostomes, and as a gonocoel, etc.

LEVELS OR GRADES OF ORGANIZATION

All living plants and animals are made of cells. A cell is the unit of structure (histology) and function (physiology) of animals and plants, so that, differentiation of cells is always accompanied by physiological division of labour. Depending on the number and degree of complexity or specialization of cells present in individuals, we can divide animals into the following patterns, levels or grades of organization : (i) protoplasmic, (ii) cellular, (iii) cell-tissue, (iv) tissue-organ, (v) and organ-system. These are approximately in the order in which they have evolved. Most animals fit neatly into one of these groups or categories although some animals appear to be intermediate between certain categories. For example, sponges only show a hint of tissue (cellular level), but not true tissue as shown by coelenterates (cell-tissue level).

1. Protoplasmic or acellular level. This type of organization occurs in Protozoa or acellular protists. All the activities are performed by one-cell body. However, specialized cytoplasmic structures, or the organelles, carry on specific functions, thus illustrating division of labour.

Colonial protozoans, suth as *Volvox,* are differentiated into somatic and sex cells. But, the somatic cells do not show specialization and division of labour. This distinguishes colonial Protozoa from Metazoa.

2. Cellular level. Multicellular animals such as mesozoans and sponges, are made of loose association of cells. All the cells are in direct contact with the environment, that is water, so that respiration, excretion, etc. take place by the general body surface. However, some division of labour occurs in certain cells concerned with reproduction (germ cells) and nutrition (choanocytes). But, such specialized cells do not form definite tissues. All cells act more or less independently and show little coordination. Sponges and mesozoans do not attain a higher level than this and are considered to be the simplest metazoans. Certain colonial protozoans are considered to have reached the cellular level.

3. Cell-tissue level. Multicellular animals, such as *Hydra* and jelly fish, are supposed to make the beginning of the tissue plan. Most of their cells remain scattered, at the cellular level, and not organized into tissues. Nerve cells appear for the first time but nerve ganglia are absent. But the nerve cells and their processes remain interconnected forming a **nerve net** which is a good example of a tissue with the function of transmission of impulses and coordination. Thus, first truly metazoan phylum of tissue grade organization is Cnidaria or Coelenterata. Their body is generally radially symmetrical and made

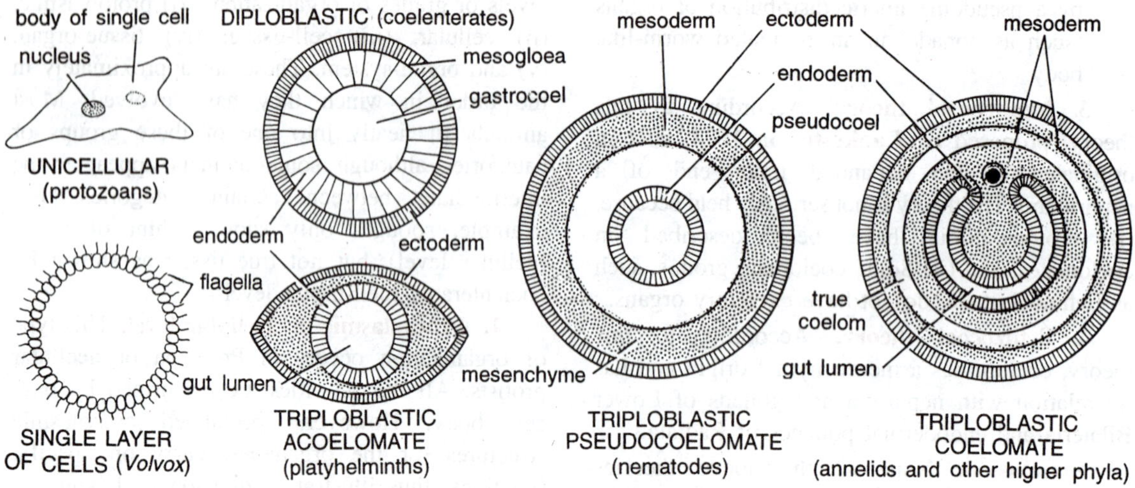

Fig. 10. Levels or grades of body organization in animal kingdom.

of only two germ layers, ectoderm and endoderm. Such animals are called **diploblastic**.

4. Tissue-organ level. Further step of advancement is the aggregation of tissues into organs which have a more specialized function than tissues. Platyhelminths are the first bilaterally symmetrical, acoelomate metazoans to have reached the tissue-organ level of organization. They have a number of well-defined organs such as eyespots, proboscis, gonads, etc. They are also **triploblastic** because they have a third or middle cellular germ layer, called **mesoderm**, which lies between outer ectoderm and inner endoderm.

5. Organ-system level. In higher animals, several organs are associated to form a distinct system concerned with a specific function, such as digestion, respiration, circulation, excretion and reproduction. This is the highest level of organization. This is first seen in a group of marine worms known as nemerteans. They have a complete digestive system separate and distinct from circulatory system.

Members of superphylum Aschelminthes, such as nematodes, are bilaterally symmetrical, triploblastic and pseudocoelomate metazoans. They have a pseudocoelom but lack a true secondary coelom. On the other hand, annelids, arthropods, molluscs, echinoderms and chordates are triploblastic and eucoelomate because they possess a true coelom.

Important Questions

➤➤*Long answer type questions*

1. What are Metazoa and how do they differ from Metaphyta ?
2. Define true metamerism and give the various theories of its origin and evolution in Metazoa.
3. What is coelom? Describe its condition throughout Metazoa.
4. Give an illustrated account of the various grades or levels of structure met with in the animals.
5. Write short notes on : (i) Cephalization and polarity, (ii) Enterocoel, (iii) Pseudocoelom, (iv) Pseudometamerism, (v) Schizocoel, (vi) Symmetry of animals.

➤➤*Short answer type questions*

1. What are metazoa. How you define ?
2. Define metaphyta ?
3. Give differences between diploblastic and triploblastic.
4. Differenciate between asymmetrical and symmetrical.

5. What is radial symmetry ?
6. Define pseudocoelomate ?
7. What is biradial symmetry ?
8. Give an account about metamerism.
9. How you define somite ?
10. Define coelom ?
11. Give the name of pseudocoelomate phyla ?
12. What is schizocoelome.
13. Define and differentiate between schizocoelome and enterocoelome.
14. Name the acoelomate group of animals.
15. Give two names of metamerically segmented animals.

➤➤*Fill in the blanks*

1. Haemocoelome present in........
2. In mammals coelome divided by ....
3. Enterocoel theory put farward by........
4. Protozoans are........grade of organism.
5. Hymen rejected the........theory of coelome origin.

➤➤*Multiple choice questions*

1. Animals which develop mouth away from blastopore are called :
   (a) protostomia (b) deuterostomia
   (c) parazoa (d) metazoa
2. Which phyla are radially symmetrical :
   (a) Cnidaria and Annelida
   (b) Echinodermata and Arthropoda
   (c) Platyhelminthes and Nematoda
   (d) Cnidaria and Protozoa
3. Pseudocoelom develops from :
   (a) Mesoderm (b) Blastocoel
   (c) Archenteron (d) Gut
4. Radial symmetry generally occurs in :
   (a) Free swimming organisms
   (b) Sessile organisms
   (c) Terrestrial organisms
   (d) Parasitic organisms
5. Which is characteristic of Protostome :
   (a) Spiral and determinate cleavage and schizocoel
   (b) Spiral and determinate cleavage and enterocoel
   (c) Spiral and indeterminate cleavage and enterocoel
   (d) Spiral, indeterminate cleavage and schizocoel
6. Syncytial theory explains :
   (a) Formation of coelom
   (b) Symmetry of animals
   (c) Origin of life (d) Origin of metazoa
7. Which animal transforms from bilateral to radial symmetry in its life history :
   (a) *Obelia* (b) Starfish
   (c) Sponge (d) *Hydra*
8. True coelom is the space between alimentary canal and body wall enclosed by layers of :
   (a) Mesoderm on one side and ectoderm on the other
   (b) Endoderm on one side and ectoderm on the other
   (c) Mesoderm on both sides
   (d) Ectoderm on both sides
9. Coelom produced from mesoderm cells is called :
   (a) Hydrocoel (b) Enterocoel
   (c) Schizocoel (d) Pseudocoel
10. On the basis of body organization evolution of animals are grouped as :
    (a) Parazoa and Metazoa
    (b) Protozoa and Metazoa
    (c) Protozoa and Parazoa
    (d) Metazoa and Eumatozoa

Answers

➤➤*Fill in the blanks*

1. Arthropods, 2. Amuscular diaphragm, 3. Lankester, 4. acellular, 5. gonocoeal.

➤➤*Multiple choice questions*

1. (b) 2. (d) 3. (b) 4. (b) 5. (a) 6. (d) 7. (b) 8. (c) 9. (c) 10. (b)

17

Origin of Metazoa

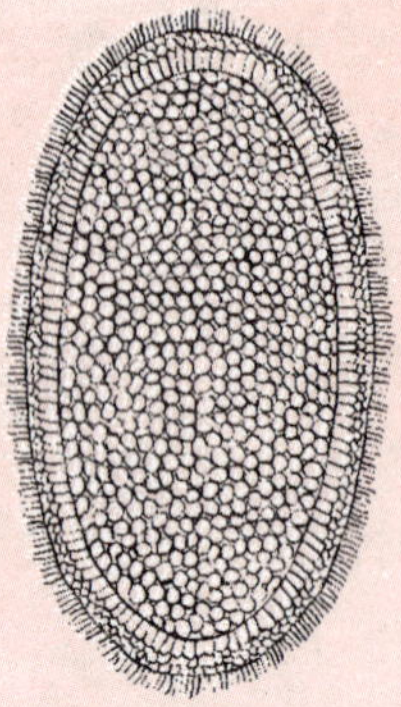

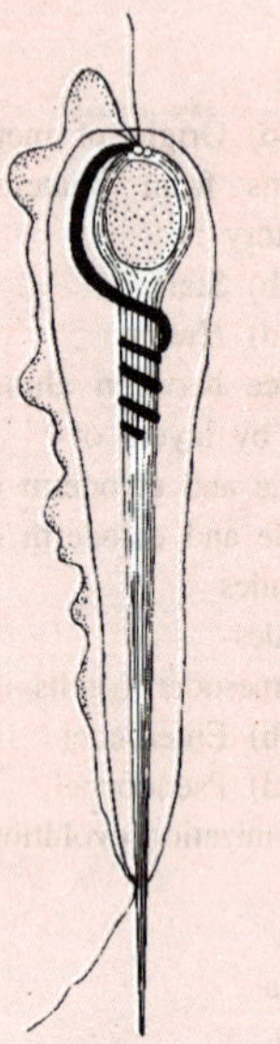

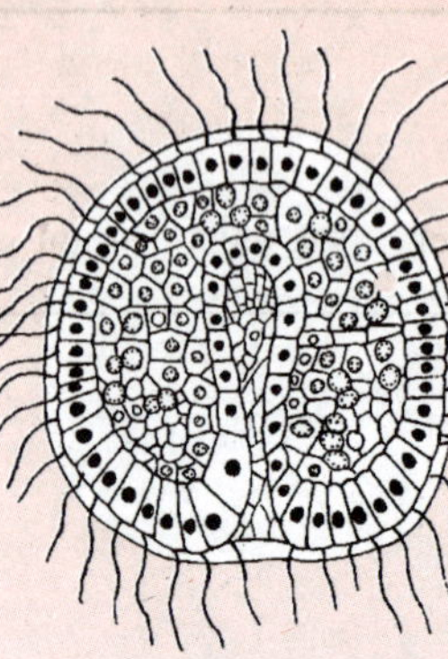

INTRODUCTION

On a rough estimate, life began on Earth about 3.7 billion years ago. First cell to appear was a prokaryote cell without a definite nucleus. Later, the eukaryote cell developed having a well organized nucleus. The earlier unicellular or non-multicellular organisms were neither plant **(Protophyta)** nor animal (**Protozoa**) in nature. **Ernst Haeckel** (1866) called them **Protista** which may be defined as organisms, usually of small size, consisting of a nuclear apparatus (no definite nucleus, one nucleus or many nuclei) in a cytoplasmic body which is not divided into cells.

With the increase in complexity of life, multicellularity developed in plants and animals. As mentioned earlier, the term **Metaphyta** is used for multicellular plants and **Metazoa** for multicellular animals.

PREMETAZOAN ANCESTORS

There is no direct proof about the ancestors of Metazoa. But all zoologists agree that the metazoans have evolved from some unicellular or acellular organisms, such as Protozoa. This is strongly evident by the facts of embryonic development in which each metazoan passes from an acellular (zygote) to a cellular condition (morula, blastula, gastrula, etc.).

A unicellular protistan individual performs all of the functions of the living. From such a functionally complex single-celled protistan arose the multicellular animals. But, how single-celled Protozoa, which evolved possibly 2,000 million years ago, gave rise to the multicelled Metazoa, remains one of the fundamental mysteries of evolution shrouded in complete obscurity. It is difficult to solve it because the earliest metazoans, like most invertebrates, were soft-bodied and left no fossils.

THEORIES OF ORIGIN OF METAZOA

Different views have been expressed by many prominent scientists from time to time to explain the evolution of the Metazoa from different groups of Protozoa. All these views can be generalized into three principal viewpoints as below :

(1) The ancestral metazoan originated from a colonial flagellate through increasing cellular specialization and interdependence **(Colonial Theory)**

(2) The ancestral metazoan arose from a multinucleate syncytial ciliate which became compartmented or cellularized by the appearance of cell boundaries **(Syncytial Theory)**

(3) The metazoans had a polyphyletic origin from different unicellular groups **(Polyphyletic Theory)**

A. Colonial Theory

The **Colonial** or **Flagellate Theory** is the classic and most frequently repeated theory of the origin of multicellular animals. It maintains that hollow colonial flagellates, like *Volvox,* were probable ancestors of the metazoans. This idea was first conceived by **Butschli**, **Lankester** and **Haeckel** (1874), later modified by **Metschnikoff** (1886), and recently revived by **Hyman** (1940).

1. Proposition. Colonial Theory holds that multicellular animals came about through association of many one-celled flagellate individuals forming a colony. With increase in the number of individual cells, they became more and more specialized in structure and functions.

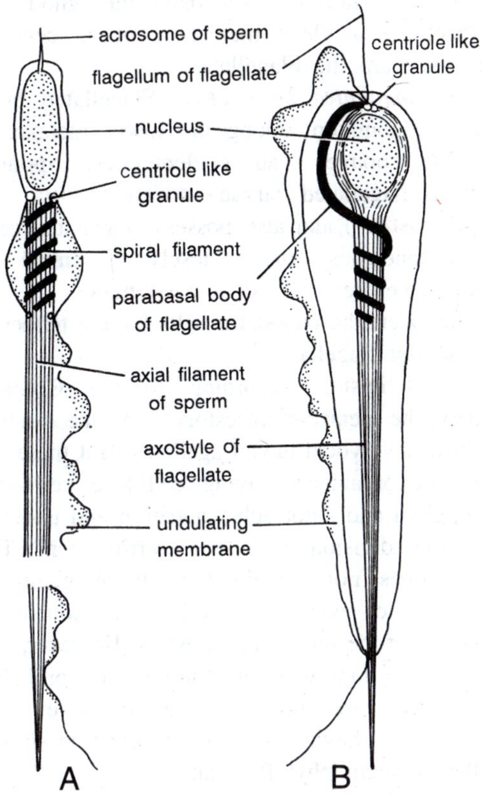

Fig. 1. Comparison between **A.** spermatozoan and a **B.** polymastigote flagellate.

Soon, individuality in the cells was lost and the whole colony itself became a single multicellular individual or a metazoan.

2. Support. The Colonial Theory has been greatly accepted, since it conforms best with the facts of embryonic development. The belief in the flagellate ancestry of Metazoa is also supported by the following facts.

(a) Plastic group. The flagellates are a highly plastic group, that is, capable of great changes. In all probability, the entire plant kingdom, and all other protozoan classes have originated from them.

(b) Compact colonies. The flagellates show a great tendency of forming compact colonies, and these clearly resemble various stages in the embryonic development of Metazoa.

(c) Flagellated spermatozoa. The metazoan tailed sperm cell very closely resembles a modified flagellate (Fig. 1). Therefore, the occurrence of flagellate or tailed spermatozoa throughout the Metazoa is a clear indication of descent of Metazoa from the Flagellata.

(d) Flagellated body cells. Flagellated body cells are common among the lower metazoans, that is, sponges and coelenterates. Sponges produce flagellated larvae (**parenchymula** and **amphiblastula**) and also possess flagellated cells or choanocytes that closely resemble a choanoflagellate (**Proterospongia**). In coelenterates, the endoderm cells have a tendency to give out flagella.

3. Criticism. According to the Colonial Theory, the metazoan ancestors were like existing freshwater volvocid phytoflagellates. But these are plant-like organisms having cellulose cellwalls, chlorophyll and autotrophic nutrition and undergo reduction division following fertilization. The theory does not explain how these plant-like characteristics were lost during the course of evolution of the metazoan ancestors. However, this objection is met with by holding that probably metazoans arose from some group of existing zooflagellates having colonial organization similar to the volvocid phytoflagellates.

The flagellate protozoan colonies are of several types such as linear, tree-like, plate-like, spherical and solid as well as hollow. Of these, which type gave rise to the ancestral Metazoa, has been a subject of great speculation among scientists. Some of the noteworthy views are listed below.

[I] Butschli's Theory

The plate-like form such as *Gonium* (Fig. 2) was favoured by **Butschli** as the ancestral metazoan type. He postulated that first this became two-layered by cutting off a lower plate of cells and later curved into a sphere. But this idea gets no support from the embryonic development of metazoans.

[II] Lankester's Theory

According to **Lankester**, the ancestral metazoan was a morula-like solid colony, such as *Synura* or *Pandorina* (Fig. 3). Food taken by outer surface of the external flagellated cells (**ectoderm**) was passed into their inner ends, a process still seen

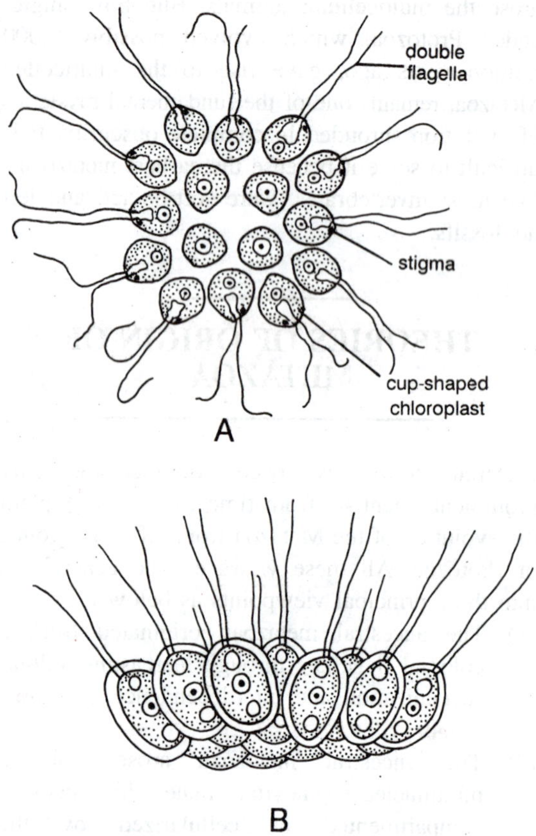

Fig. 2. *Gonium pectorale,* a Colonial phytomonad. **A.** From above, **B.** Inside view.

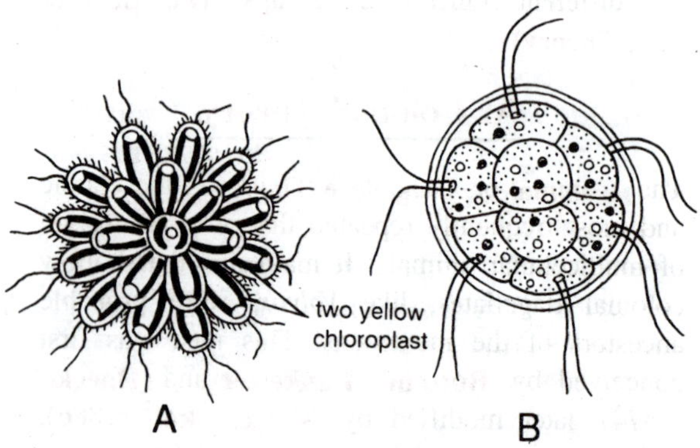

Fig. 3. A. *Synura vella,* a chrysomonad. **B.** *Pandorina,* a phytomonad.

in living sponges and coelenterates. Later, the inner ends of cells got separated forming an internal digestive layer or **endoderm** (Fig. 4). Lankester

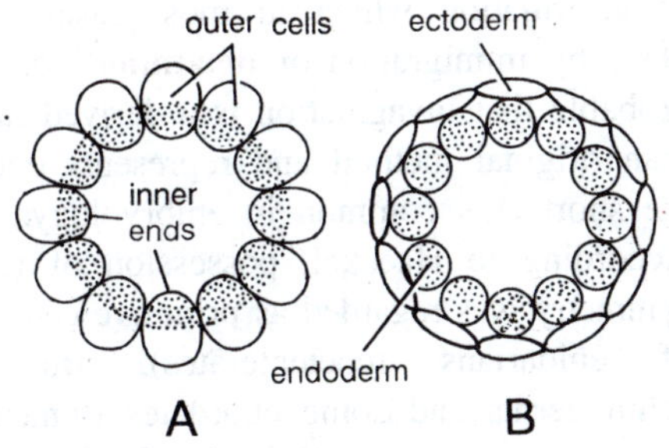

Fig. 4. A and **B.** Endoderm formation by primary delamination, according to Lankester's theory.

named this hypothetical ancestor as **planula** (Fig. 5), which was a solid organism. Later on, it developed mouth and digestive cavity.

Criticism. Lankester's theory found no acceptance, because :

(1) Endoderm formation by primary delamination is very rare in the Metazoa.

(2) Formation of internal digestive tube or endoderm will be of no use without a mouth for taking in food.

[III] Haeckel's Theory of Blastaea and Gastraea

The Colonial Theory is based mainly on **E.H. Haeckel's Gastraea Theory of metazoan ancestry**, presented in 1874.

1. Proposition. According to **Haeckel**, the metazoans evolved by aggregation of single protistan cells into a little hollow, spherical, flagellate colony, similar to *Volvox*. It possessed a distinct antero posterior axis and swam about with one pole directed forward. This hypothetical organism was named **blastaea** by Haeckel and was considered similar to the **blastula** or **coeloblastula stage** in the embryogeny of living metazoans. Blastaea showed an early differentiation between somatic and reproductive cells. Somatic cells were meant for perception, food capture and locomotion by the action of flagella. But, certain cells were specialized to become sex cells, a process already seen in *Volvox*. The embryonic archaeocytes of sponges and interstitial cells of coelenterates, which later become reproductive cells, are reminiscent of the same phenomenon.

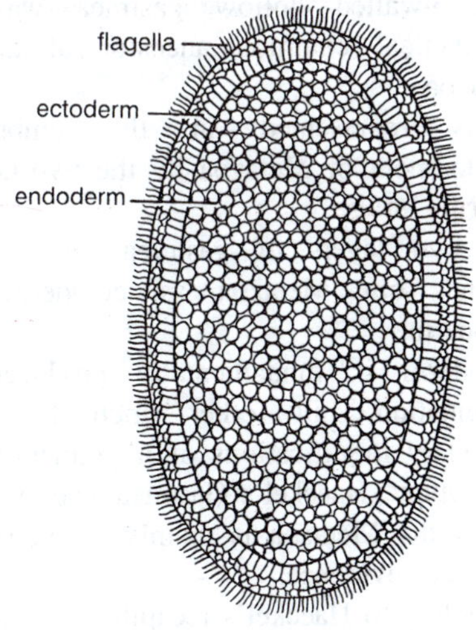

Fig. 5. Planula.

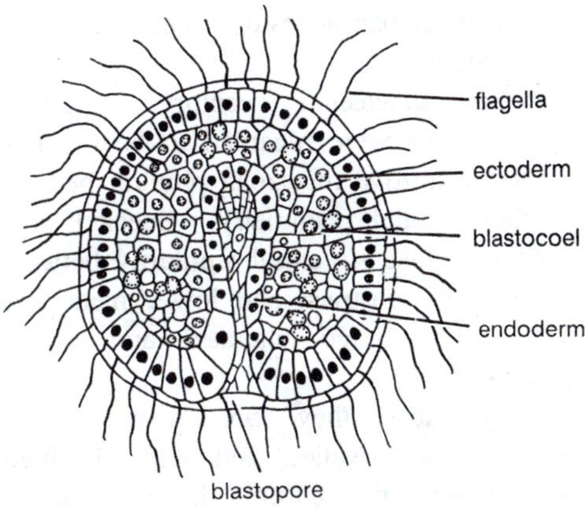

Fig. 6. A typical gastrula showing endoderm formation by embolic invagination (according to Haeckel).

The blastaea invaginated at the posterior pole to become a double-walled sac termed **gastraea** by Haeckel (Fig. 6). The inner endodermal sac thus formed was termed as primitive intestine or **archenteron** and its single opening to the exterior as primitive mouth, **protostoma** or **blastopore** (Fig. 6). The endodermal cells lining the digestive cavity were relieved from perceptive and locomotory functions and became specialized for engulfing and digesting food for the whole colony. According to the famous **Gastraea Theory** of **Haeckel** :

(1) A two-walled hollow gastraea was the hypothetical common ancestor of all the Metazoa.
(2) It is represented in the embryonic development of Metazoa by the two-layered **gastrula stage**
(3) The two layers of gastraea are retained through out Metazoa as ectoderm and endoderm.
(4) Endoderm is fundamentally produced by invagination, all other methods being secondary modifications of invagination.

2. Support. Haeckel's Gastraea Theory about ancestry of the Metazoa is mainly based on the embryological evidence.

(1) According to Haeckel's recapitulation theory, the blastula and gastrula stages in the embryonic development of higher metazoans represent repetition of the adult stages of the original metazoan ancestors, that is, blastaea and gastraea.
(2) In the nineteenth century, Cnidaria (Coelenterate) were regarded to be the first or most primitive metazoan phylum because of their radial symmetry. Haeckel also compared close structural similarity between the ancestral gastraea and lower metazoans like hydrozoan coelenterates and certain sponges.

Like gastraea, they are also radially symmetrical, double-walled and with a single opening (mouth or osculum) leading into a sac-like digestive cavity.

3. Criticism. Gastraea theory by **Haeckel** was widely accepted as it simplifies the embryologic and phylogenetic history of animals and gives a convincing explanation of the stages by which complex metazoan structure might have been achieved. However, it had to face some objections.

(1) Chief objection is that in metazoan embryology, endoderm is not always formed by **invagination**. Even in the Coelenterata, which are structurally nearest to the hypothetical gastraea, endoderm is formed by the inwandering of ectodermal cells **(multipolar ingression)**, and the larva is a parenchymula or planula rather than a gastrula. Invagination is an unusual method of gastrulation which in most cases takes place by immigration or **delamination**. It is probable that invagination is a derived rather than original method and represents one of the short cuts common in embryology.
(2) According to Haeckel, possession of radial symmetry was regarded as primitive character of cnidarians (coelenterates). But, in echinoderms and some chordates (tunicates) it is a secondary condition derived through metamorphosis of a bilaterally symmetrical larva. Thus, radial symmetry may be an adaptation due to a sedentary or sessile mode of life rather than a phylogenetically primitive condition, so that there is no compulsion to consider the ancestral metazoan to be a radially symmetrical, two-layered and hollow sac-like organism.
(3) Haeckelian recapitulation theory has been largely discarded now due to its many discrepancies.

[IV] Metschnikoff's Theory

Haeckel's Gastraea theory was later modified by **Metschnikoff** (1866).

1. Proposition. Metschnikoff noted that the primitive mode of gastrulation in coelenterates is by **ingression**, and not by invagination, producing a solid gastrula. He also discovered that digestion is intracellular and phagocytic in lower Metazoa. According to him, certain ectodermal cells of blastaea became phagocytic and wandered (multipolar ingression) into the inner blastocoel

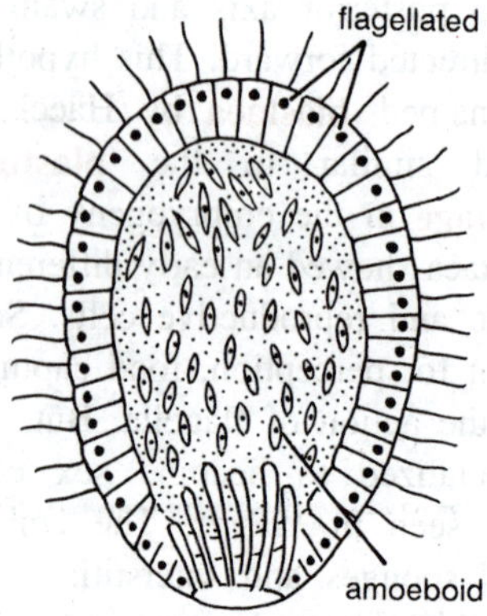

Fig. 7. Parenchymula larva.

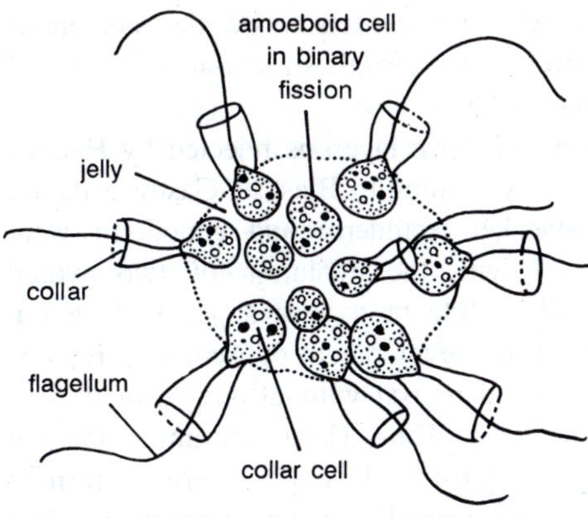

Fig. 8. *Proterospongia haeckeli.*

which became filled up with amoeboid cells embedded in a gelatinous material. Therefore, he argued that the diploblastic ancestral gastraea was a solid rather than a hollow organism, and he named it **parenchymula** (Fig. 7). It would not have required a digestive sac or mouth because of phagocytic intracellular digestion. Formation of archenteron and blastopore were regarded as later, secondary processes.

2. Support. According to Metschnikoff, his theory of solid gastraea was strongly supported by the discovery of **Proterospongia** (Fig. 8) by **Saville Kent** in 1880. It is a choanoflagellate colony embedded in a gelatinous matrix. Some outer collared cells frequently lose their collars and flagella and move into the jelly as amoeboid cells, indicative of ingression.

Facts of embryonic development also support Metschnikoffs theory rather than Haeckel's because gastrulation by inwandering of cells or multipolar ingression is far more common than invagination.

[V] Modern View by Hyman, etc.

Modern supporters of Colonial Theory, such as **Hyman** (1940), also derive Metazoa from an axiate hollow spherical flagellated colony, similar to **blastaea** of Haeckel. It showed an early differentiation into somatic and reproductive or germ cells. The somatic cells also became differentiated into locomotor-perceptive and nutritive types. Through the migration or wandering of nutritive cells inside (multipolar ingression), the originally hollow sphere became converted into a solid, ovoid and radially symmetrical **parenchymula** or **stereogastrula** (after Metschnikoff). It had no mouth and no digestive cavity. Food was caught, as in Protozoa, by the outer flagellated locomotor-perceptive cells and passed to the inner amoeboid nutritive and reproductive cells, for intra-cellular digestion. The anterior pole probably bore special sensory cells or a tuft of sensory cilia. This hypothetical organism has been called the **planuloid ancestor** (Fig. 9) because of its close similarity with the planula larva of coelenterates. According to this

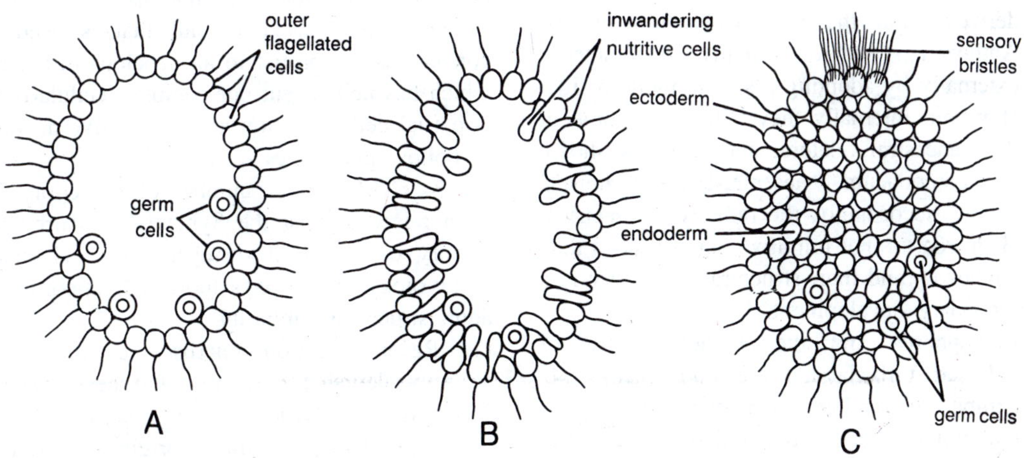

Fig. 9. Formation of endoderm according to Metschnikoff and Hyman. **A.** Hypothetical ancestral blastaea. **B.** Multipolar ingression. **C.** Hypothetical stereogastrula or a planuloid ancestor of Eumetazoa.

view, the radial symmetry of coelenterates is primitive and derived directly from this hypothetical planuloid ancestor.

B. Syncytial Theory by Hadzi

Resemblance between multinucleate ciliates and acoelous flatworms led **Hanson** (1958) and **Hadzi** (1963) to propose a ciliate origin for metazoans. Their theory may be called the **Syncytial Theory**. Syncytial refers to the multinucleate condition in which cell membranes are absent between adjacent nuclei. The theory was first proposed by **Sedgwick** many years ago but was ignored. Later, **Jovan Hadzi** (1963), a Yugoslavian, studied it for about 50 years and gave many logical reasons in its support.

1. Proposition. Hadzi's Theory states that Metazoa did not evolve by aggregation of uninucleate protistans, but by cellularization of a primitive multinucleate syncytial ciliate, thus producing a typical multicellular condition.

The main tenets of Hadzi's theory are as follows :

(1) The ancestral metazoan was at first syncytial in structure and bilaterally symmetrical. It gave rise to acoel, flatworms (Turbellaria) by process of cellularization. Acoel turbellaria are, therefore, regarded to be the most primitive living metazoans and their bilateral symmetry is considered to be primitive.

(2) Cnidaria, especially the Anthozoa, were derived from the rhabdocoel turbellarians. Anthozoa became radially symmetrical externally by adopting a sessile mode of life. However, internally they still retain some of their ancestral bilateral symmetry as shown by stomodaeum, mesenteries and muscle bands. Accordingly, the radially symmetrical Hydrozoa are not primitive but advanced than Anthozoa, and their structural simplicity is a secondary development.

(3) Ctenophores and the Muller's larva of polyclad turbellarians are both planktonic, exhibit octoradial symmetry, and their embryology is also similar in several important details (Lang and MacBride). Therefore, Hadzi held that Ctenophores evolved from polyclads by the neotenous retention of several structural features of Muller's larva.

(4) Theory of germ layers is rejected by Hadzi's theory. According to Blastaea-Gastraea theory of Haeckel, ectoderm and endoderm were formed with the evolution of two-layered gastrula, while mesoderm appeared later on. But, in the development of Acoela, 3 regions are formed *in situ* without any formation of germ layers. Thus, Hadzi presumes that the three germ layers were formed contemporaneously in the ancestral ciliate following cellularization.

(5) It is assumed that **conjugation** in ciliates is a form of sexual reproduction equivalent to copulation in Metazoa.

2. Support. Hadzi compared an acoelous turbellarian, such as *Convoluta,* with certain multinucleate ciliates. Both are in the same size range, have an antero-posterior axis, are ciliated and distinctly bilaterally symmetrical. Both tend to be syncytial and lack a digestive cavity. Multinucleate ciliates practice asexual conjugation, show a differentiation into inner, middle and outer material and organelles, and possess trichocysts which are similar to turbellarian sagittocysts and cnidarian nematocysts. Central parenchyma of Acoela *(Convoluta)* looks syncytial due to imperfect cellularization. Hadzi considered these similarities as evidence in support of the primitive position of acoelous turbellarians.

According to Hadzi, all that is needed to convert such a protist to an acoelous turbellarian, which has neither gut nor complete cellularization, is partial cellularization and a primitive nerve net. A hollow gut comes later.

3. Criticism. A number of objections have been raised against the Syncytial or Turbellarian Theory of Hadzi. It disregards the evidence of embryology, while pays too much importance to adult structural similarity.

(1) Recent electron microscopic studies have revealed that epidermis of Acoela is **cellular** and not **syncytial** as claimed by the supporters of ciliate origin of Metazoa. Parenchyma of Acoela may also prove to be cellular on closer examination.

(2) Syncytial theory considers biradial Anthozoa to be primitive and radial Hydrozoa to be most specialized. This is based on the presumption that bilaterally symmetrical acoelous turbellarians are the most primitive living metazoans. But evidence shows radial symmetry of coelenterates to be primary and not secondarily evolved from a bilateral ancestor.

(3) Body plan of primitive groups (or phyla) is generally more variable or less fixed. Naturally, Hydrozoa being more variable should be considered more primitive and low in evolutionary line than Anthozoa.

(4) Anthozoa, with few exceptions, are not hermaphroditic, although most sessile animals are. Thus, Anthozoa seem improbable descendants of hermaphroditic turbellarians, as claimed by Hadzi's theory.

(5) Flagellated sperms occur in Metazoa but not in ciliates. Therefore, it is difficult to explain a *de novo* origin of motile sperm in the ciliate metazoan ancestor. Hadzi's theory holds that in some ciliates, the male nucleus is already cellularized, that is, flagellated like a metazoan sperm. Thus, during conversion from Ciliata to Metazoa, cellularization of female gamete must not involve great difficulty.

(6) A **macronucleus** is present in ciliates but absent in acoels. It is assumed that macronucleus was absent in the multinucleate protociliate ancestor which gave rise to the metazoan ancestor. It probably developed later in the evolutionary line leading to higher ciliates.

(7) Turbellarians nemerteans, annelids and molluscs, etc., undergo **spiral cleavage** during embryonic development. On the other hand, coelenterates show a variety of cleavage patterns but not the spiral cleavage. Thus, derivation of coelenterates from turbellarians would require overnight abondonment of spiral cleavage which seems improbable.

Hadzi's Syncytial or Turbellarian theory has generated much discussion among zoologists, but it has failed to receive universal acceptance.

C. Polyphyletic Theory

More recently, **Greenberg** (1959) and **Preston** (1967) held that probably all Metazoa have not originated from any single ancestor. They have suggested a polyphyletic origin for the metazoans. It is fairly certain that sponges developed by way of colonial flagellates *(e.g., Proterospongia)*, whereas other multicellular groups originated from the cellularization of syncytial protociliates or perhaps the mesozoans.

Important Questions

➤➤*Long answer type questions*

1. Write an essay on the origin and evolution of Metazoa.
2. Describe in detail the Colonial or Flagellate Theory of origin of Metazoa.
3. Give a detailed account of Hadzi's Theory regarding origin of Metazoa.
4. Discuss that flagellate protozoans are a highly plastic group from which probably all other protozoan classes have originated.
5. Write short notes on : (i) Blastaea, (ii) Gastraea, (iii) Metschnikoff's theory of origin of Metazoa, (iv) Planuloid ancestor of Metazoa.

➤➤*Short answer type questions*

1. Who proposed the colonial theory of the origin of metazoan ?
2. Which theory of origin of metazoa supported by Hyman ?
3. Give the two failure of Lankestir's theory of origin of metazoa.
4. E. H. Haeckel presented a theory about evolution and origin of metazoa. What is the name of that theory?
5. What is blastoea ?
6. What is gastraea ?
7. What is the basic idea of the Haeckel's theory of origin of metazoan. ?

8. Who modified the Haeckel's theory ?
9. Who proposed the syncytial theory ?
10. What is meant by the word syncytial ?
11. What is the name of the theory of Hanson and Hadzi.
12. Polyphyletic theory proposed by whom.

➤➤*Multiple choice questions*

1. Syncytial theory explains :
   (a) formation of coelom
   (b) symmetry of animal
   (c) origin of life
   (d) origin of metazoa
2. On the basis of body organisation evolution of animal are grouped as :
   (a) Parazoa and metazoa
   (b) Protozoa and metazoa
   (c) Protozoa and parazoa
   (d) Metazoa and eumetazoa
3. Polyphyletic theory given by :
   (a) Greenberg and Perston
   (b) Lang and McBride
   (c) Hanson and Hadzi
   (d) Lankester
4. Colonial theory proposed by :
   (a) Butschli, Lankester and Haeckel
   (b) Hanson and Hadzi
   (c) Hymen L.H. (d) Haeckel E.H.
5. Proterospongia discovery by :
   (a) Hymen (b) Haeckel
   (c) Kent (d) Metschnikoff
6. Gastraea theory of metazoan ancestary presented by :
   (a) E. H. Haeckel (b) L.H. Hymen
   (c) J. Hadzi (d) Lankester
7. Syncytial theory proposed by :
   (a) Hanson and Hadzi (b) L.H. Hymen
   (c) Lankester (d) E. H. Haeckel
8. Which group of Protozoa support and give best explanation of colonial theory of origin of metazoan :
   (a) Ciliata (b) Sarcodina
   (c) Ciliophora (d) Flagellate

Answers

➤➤*Multiple choice questions*

1. (d) 2. (b) 3. (a) 4. (a) 5. (c) 6. (a) 7. (a) 8. (d)

18

Organization of Bilateria

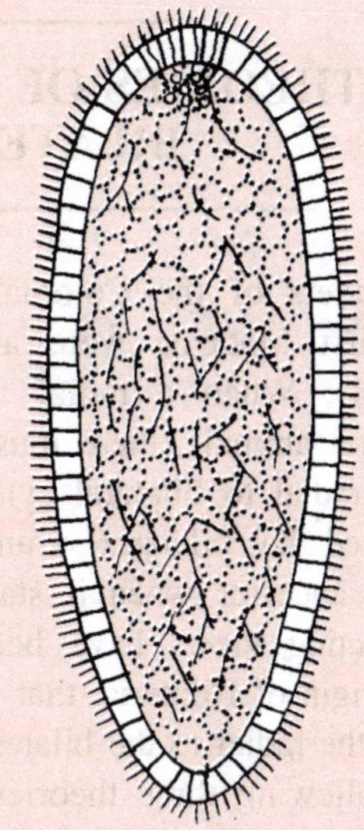

BILATERAL PHYLA

Earlier workers, especially **Hatschek** (1881-91), divided the Eumetazoa into two super-groups, **Radiata** and **Bilateria**. This division is based on the overall symmetry of the animal bodies. **Radiata** includes only two metazoan phyla, Coelenterata and Ctenophora, which display a fundamental radial symmetry. The remaining phyla, from Platyhelminthes upto Chordata, are bilaterally symmetrical and grouped together as the **Bilateria**. A section only in **median sagittal plane** (M.L.S.) divides their body into right and left halves which are mirror images of each other. Many zoologists agree that free-living acoeloid flatworms of phylum Platyhelminthes are the most primitive of all bilateral animals.

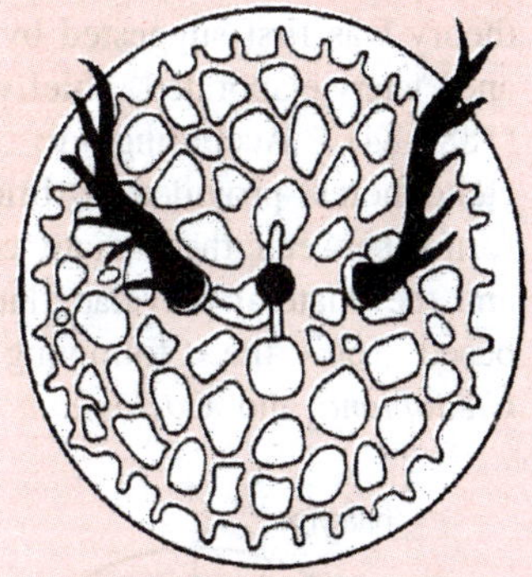

The Bilateria are defined as **bilaterally symmetrical** Eumetazoa with a main **antero-posterior body axis**, with **organ-systems** and a well defined or true **endodermal mesoderm**. Some characteristics, not common to all but present in many groups of Bilateria, are the presence of **coelom**, **segmentation** and a **definitive anus** or the posterior opening to the digestive tract.

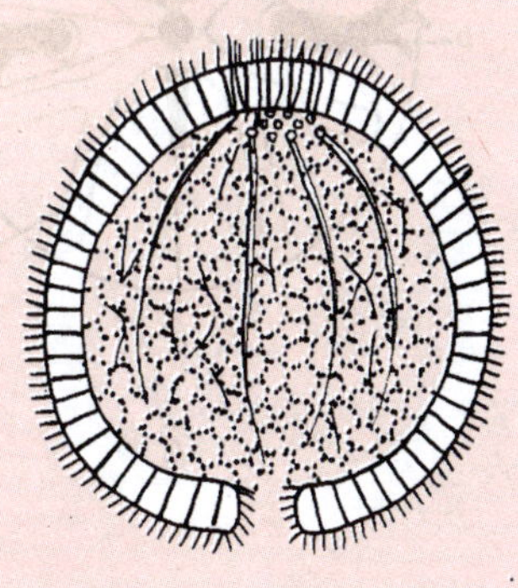

In larger bilateral animals we see the gradual appearance of **excretory**, **respiratory** and **circulatory** systems which are absent in Radiata. **Fertilization** in Bilateria is usually **internal** in contrast to **external** fertilization seen in Radiata.

THEORIES OF ORIGIN OF BILATERIA

Supporters of the Colonial Theory believe that the first Metazoa were almost certainly radial animals. When a radial ancestor gave rise to bilateral animals, there must have been a change from radial to bilateral symmetry. It is generally believed that bilateral symmetry developed in the radial ancestor when it started creeping. Several conflicting views have been put forward about the origin of Bilateria, that is, to explain the steps from the radial to the bilateral condition. Of these, the following three theories are noteworthy.

Ctenophore-polyclad Theory

With the discovery of aberrant creeping ctenophores, *Ctenoplana* and *Coeloplana,* this theory was first suggested by **Kowalevsky** (1880) and later elaborated chiefly by **Arnold Lang** (1881-84). According to this theory, these ctenophores provided a bridge or intermediate forms between the radiate coelenterate ancestors and the bilateral polyclad flatworms. This view is based on the following resemblances of *Ctenoplana* and *Coeloplana* with polyclads :

(1) Oval flattened shape with dorso-ventral differentiation.
(2) Creeping upon the entire ventral surface.
(3) Presence of 2 dorsal tentacles.
(4) Centrally located ventral mouth.
(5) Branched, blindly ending digestive canals.
(6) Radiating, anastomosing nervous system.
(7) Determinate cleavage and a large stomodaeal invagination contributing to the ventral surface.
(8) The swimming larval stage of many polyclads has 8 ciliated lobes comparable to 8 comb rows of ctenophores.

Figure 1 represents the stages showing a shift of nerve centre from middle to anterior side, conversion of branching elongated tentacles into simple small tentacles, and digestive canals branching extensively, during change from a biradially symmetrical Ctenophore to a bilaterally symmetrical polyclad.

Objections. Lang's ctenophore-polyclad theory is rejected on the following grounds :

(1) *Ctenoplana* and *Coeloplana* are highly aberrant ctenophores without phylogenetic significance. Their resemblances to polyclad flatworms are undoubtedly superficial.
(2) Cleavage is biradial in ctenophores but spiral in polyclads.

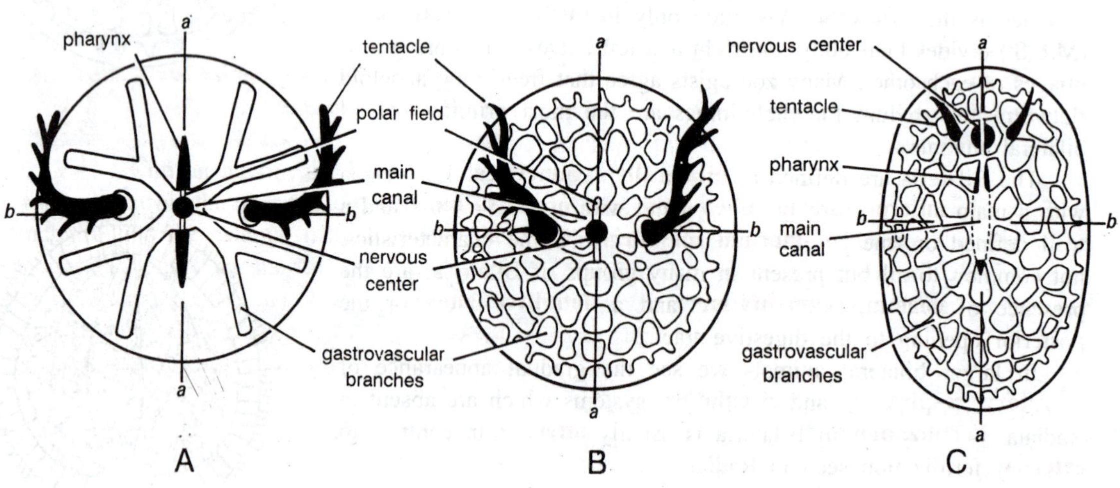

Fig. 1. Gradual change from a Ctenophore to a polyclad, according to Ctenophore-polyclad. **A.** Ctenophore. **B.** Hypothetical intermediate stage. **C.** Polyclad.

(3) According to Lang's theory, the polyclads were regarded as the most primitive Bilateria. Now a days, the acoel flatworms are regarded to be most primitive.

Ctenophore-trochophore Theory

This theory is based on the so-called resemblances between ctenophores and trochophore larva, and suffers from many lacunae.

(1) The theory derives annelids and higher Bilateria from ctenophores through trochophore, but fails to explain the origin of lower Bilateria.

(2) To bridge over this difficulty, some regard the larvae of acoelomate Bilateria (flatworms and nemertines) as earlier stages of trochophore larva. But then, such stages must logically resemble a Ctenophore more than a trochophore, which is not the case. Only a fully developed typical trochophore shows resemblances to a Ctenophore.

(3) Others try to justify that acoelomate Bilateria have degenerated from higher Bilateria. But free-living flatworms, especially the acoels, have been firmly established to be the most primitive Bilateria, and are not degenerated forms.

Planuloid-acoeloid Theory

This theory is based upon the remarkable features of acoelous flatworms and their phylogenetic significance. Some zoologists think that acoels are secondarily degenerate and polyclads are the most primitive group of turbellarians (Lang's ctenophore-polyclad theory). While others consider acoels to be the most primitive of all the turbellarian orders. This later view forms the basis of the planuloid-acoeloid theory which was first conceived by **Ludwig von Graff** (1882) and most recently elaborated by **Hyman** (1951). The theory places acoel flatworms at the base of the Bilateria and suggests the origin of Bilateria from a planuloid ancestor through an acoeloid form.

The hypothetical planuloid ancestor was very similar to the planula larva of coelenterates (Fig. 2-A). It was elongated, radially symmetrical and without mouth or archenteron. The exterior cells were ciliated or flagellated epithelio-muscular cells, while the solid mass of interior cells were digestive as well as reproductive. There was a nerve-net under epidermis and the anterior pole had a nervous centre or an accumulation of neurosensory cells. Many of these features are already present in the acoel flatworms.

According to **Hyman** (1951), some of the planuloid ancestors may have adopted a creeping mode of life. This could have resulted in their flattening out in the oral-aboral axis, in differentiation between dorsal and ventral surfaces, and .to the development of a ventral mouth.

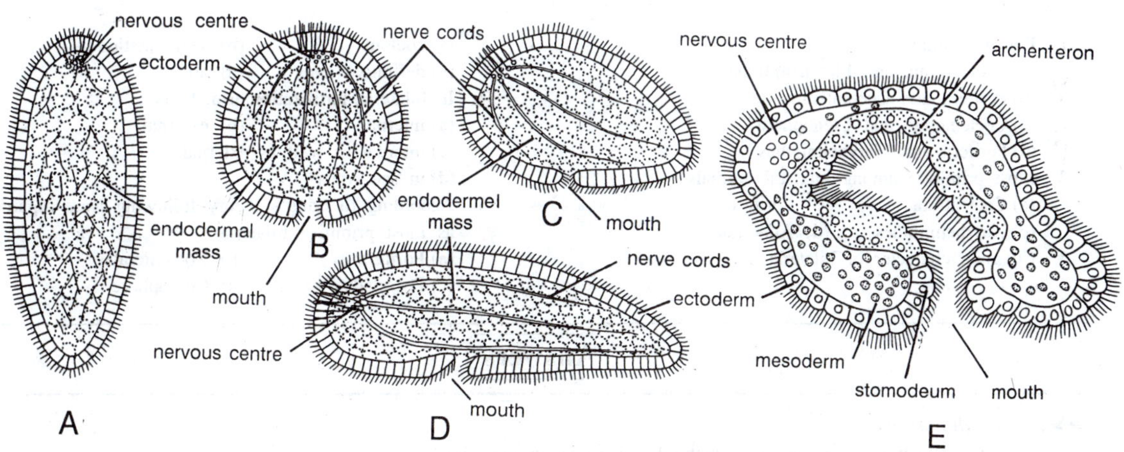

Fig. 2. Stages showing evolution of a planuloid ancestor into an acoel flatworm. **A.** Planula larva. **B.** Mouth formed. **C.** Body elongates and nervous centre shifts forward. **D.** Acoeloid stage. **E.** Young flatworm.

The forward shifting of the nervous centre would have led to the evolution of the acoeloid ancestral stage. Later on, the hollowing out of interior mass formed an archenteron, and the greater differentiation of anterior region formed a head which also sprouted a pair of tentacles. The result is the conversion of the radial planuloid ancestor into a simple flatworm with bilateral symmetry, ventral mouth, anteriorly located brain and sense organs (Fig. 2-E).

Important Questions

➤➤Long answer type questions

1. What are bilateral phyla ? Discuss the theories regarding their origin.
2. What is the significance of discovery of *Ctenoplana* and *Coeloplana* in the origin of Bilateria ?
3. Write short notes on : (i) Ctenophore-trochophore theory of origin if Bilateria. (ii) Planuloid-acoeloid theory.

➤➤Short answer type questions

1. What do you mean by bilateria ?
2. Explain ctenophora.
3. Describe the coeloplana ?
4. What is M.L.S. ?
5. Name the systems which are absent in radiata ?
6. Explain polyclad theory in brief ?
7. Name the theory which conceived by Graff ?

➤➤Fill in the blanks

1. Eumetazoa divided into ...... .and ....
2. Fertilization in bilateria is usually.......
3. Ctenophore polyclad theory was first suggested by......
4. Planuloid acoeloid theory was first conceived by.......

➤➤True and false statements

1. According to Flymen (1951) some of the planuloid ancestors may have adopted a creeping mode of life.
2. The hypothetical planuloid ancestor was very similar to planula larva of coelenterates.
3. Cleavage is biradial in ctenophores but spiral in polyclads.
4. Ctenophore polyclad theory put forward by L.H. Hymen.
5. Eumatozoa divided into radiata and biradiata.

➤➤Multiple choice questions

1. Radiata include :
   (a) Coelenterata and Platyhelminthes
   (b) Porifera and Coelenterata
   (c) Coelenterata and Ctenophora
   (d) Porifera and Platyhelminthes
2. Most primitive among bilateral animals :
   (a) Coelentrata (b) Porifera
   (c) Platyhelminthes (d) Annelida
3. Common character of bilateria :
   (a) coelome (b) segmentation
   (c) definitive anus (d) all
4. In bilateria fertilization usually is :
   (a) internal (b) external
   (c) internal as well as external
   (d) in sea water
5. According to Lang's theory following were regarded as most primitive bilateria :
   (a) Porifera (b) Coelenterates
   (c) Polyclads (d) Ctenophore

Answers

➤➤Fill in the blanks

1. radiata, Bilateria, 2. internal, 3. Kowalevsky, 4. Ludwing Von Graff.

➤➤True and false statements

1. true, 2. true, 3. true, 4. false, 5. false

➤➤Multiple choice questions

1. (c) 2. (c) 3. (d) 4. (a) 5. (c)

19

Leucosolenia : An Asconoid Sponge

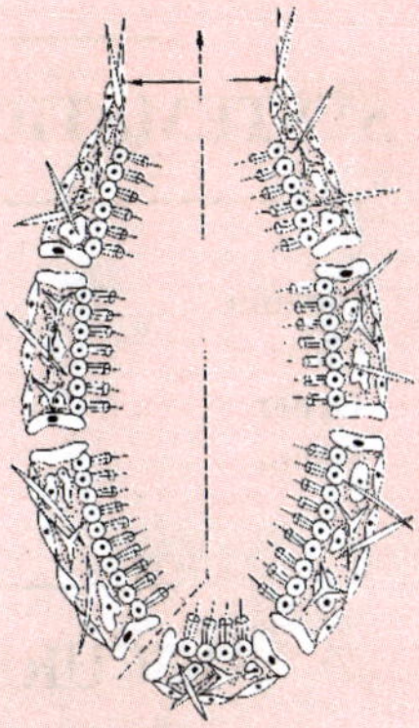

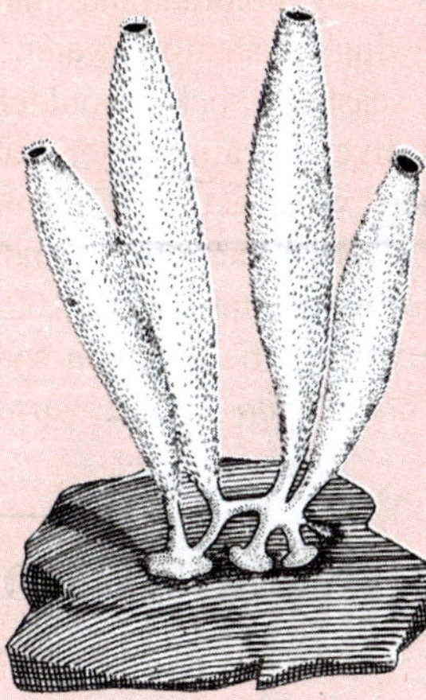

Phylum **Porifera** contains the most primitive of multicellular animals, called **sponges**. They are usually referred to as 'pore bearers', as their body walls contain tiny pores that are basic structures in their functional activity. The organization of sponges shows certain features to a degree not apparent in any of the other metazoan animal groups. Especially, the cell types and gross morphology are not at all comparable with other Metazoa. This led to place the sponges in an isolated branch of Metazoa, the **Parazoa**, whereas all the remaining phyla of multicellular animals are combined in the branch **Eumetazoa**.

Biology of sponges can best be understood by the study of a simple asconoid type. *Leucosolenia* is one of the simplest asconoid sponges.

Leucosolenia

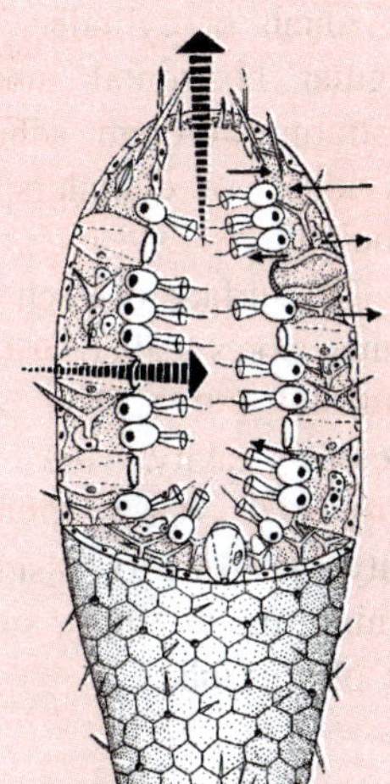

Leucosolenia (Gr., **leukos**, white + **solen**, pipe), as the name implies, has a pipe-like body. It is an **asconoid** sponge which represents the simplest and most primitive type of structural pattern. It is usually said to be the **olynthus type** as well, as its body organization resembles the olynthus stage in the **ontogeny** (developmental history) of certain other sponges, like *Scypha*.

SYSTEMATIC POSITION

Phylum	Porifera
Class	Calcarea
Order	Homocoela
Genus	*Leucosolenia*

OCCURRENCE

Leucosolenia is a small, delicate, sessile, branching, colonial and marine sponge. It is found growing in shallow water, below low-tide mark, on seashore rocks, boulders and jetties. It does not live in calm water but is found abundantly where wave action is intense. It is very sensitive to external conditions and will die if removed from its habitat. About 100 species occur all over the world. The common species are *L. botryoides, L. complicata* and *L. variabilis.*

EXTERNAL MORPHOLOGY

The colony of *Leucosolenia* is whitish or yellowish in colour. It consists of radially symmetrical, vaseshaped, vertical, hollow and tube-like individuals or cylinders, united at their bases by irregular horizontal tubes and attached to the substratum through adhesive discs. The upright individuals or cylinders may reach upto 25 mm in height.

The surface of each tube-like individual bears minute pores called **ostia** (singular, **ostium**) or **incurrent pores**. The pores lead into a central spacious cavity, the **paragastric cavity** or **spongocoel** (Gr., **spongos**, sponge.+ **koilos**, cavity). It opens to outside through a large circular opening, the **osculum**, situated at the free terminal end of the tube.

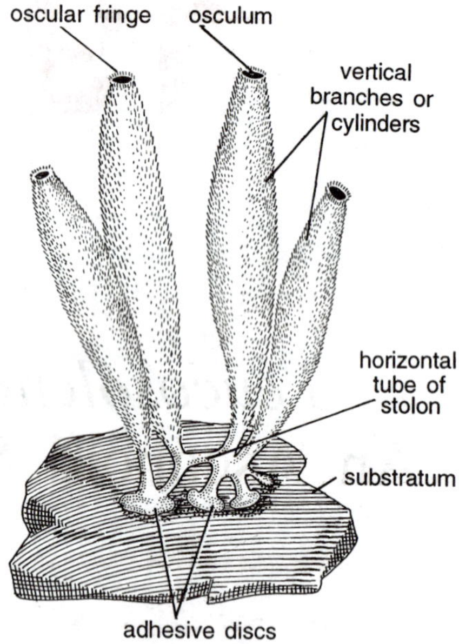

Fig. 1. *Leucosolenia.* A colony growing on a rock.

BODY WALL (HISTOLOGY)

The thin body wall, which encloses the spongocoel, is relatively simple. It consists of two cellular layers, the outer **pinacoderm** and inner **choanoderm**, with a non-cellular **mesenchyme** in between.

1. Pinacoderm. The body is covered externally by **epidermis** or **pinacoderm**. This is a single layer of thin and flat polygonal cells, the **pinacocytes** (Gr., **pinako**, plank + **kytos**, cell). It ensures protection to the internal organization of body. Pinacocytes are hexagonal in surface view with thin margins and a bulging central part containing a nucleus. The margins show contractility, so that the sponge can increase or decrease slightly in size.

Within the body wall are special, large and tubular cells, called **pore cells** or **porocytes**. They are supposed to be modified pinacocytes. Each

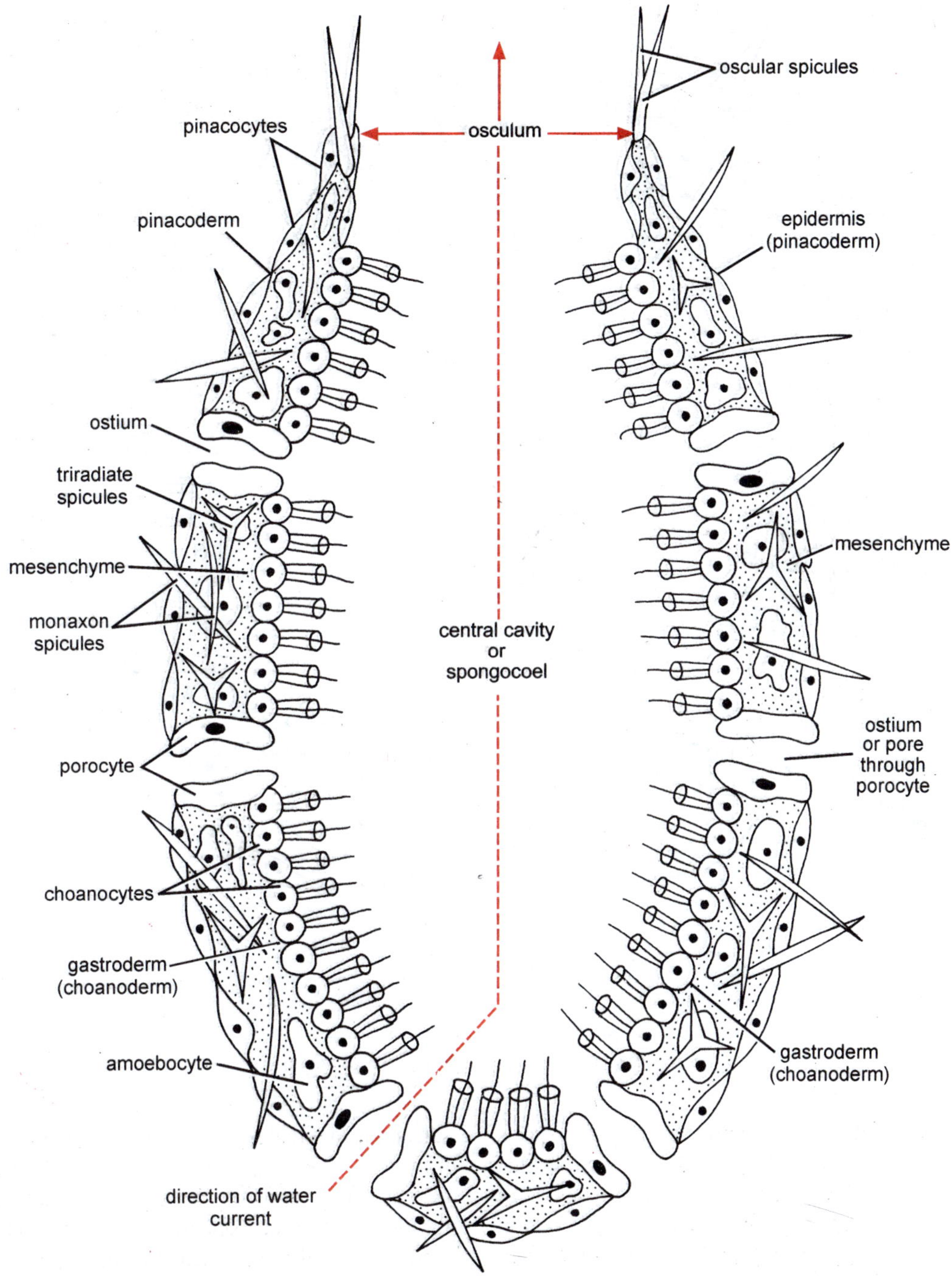

Fig. 2. *Leucosolenia.* Longitudinal section.

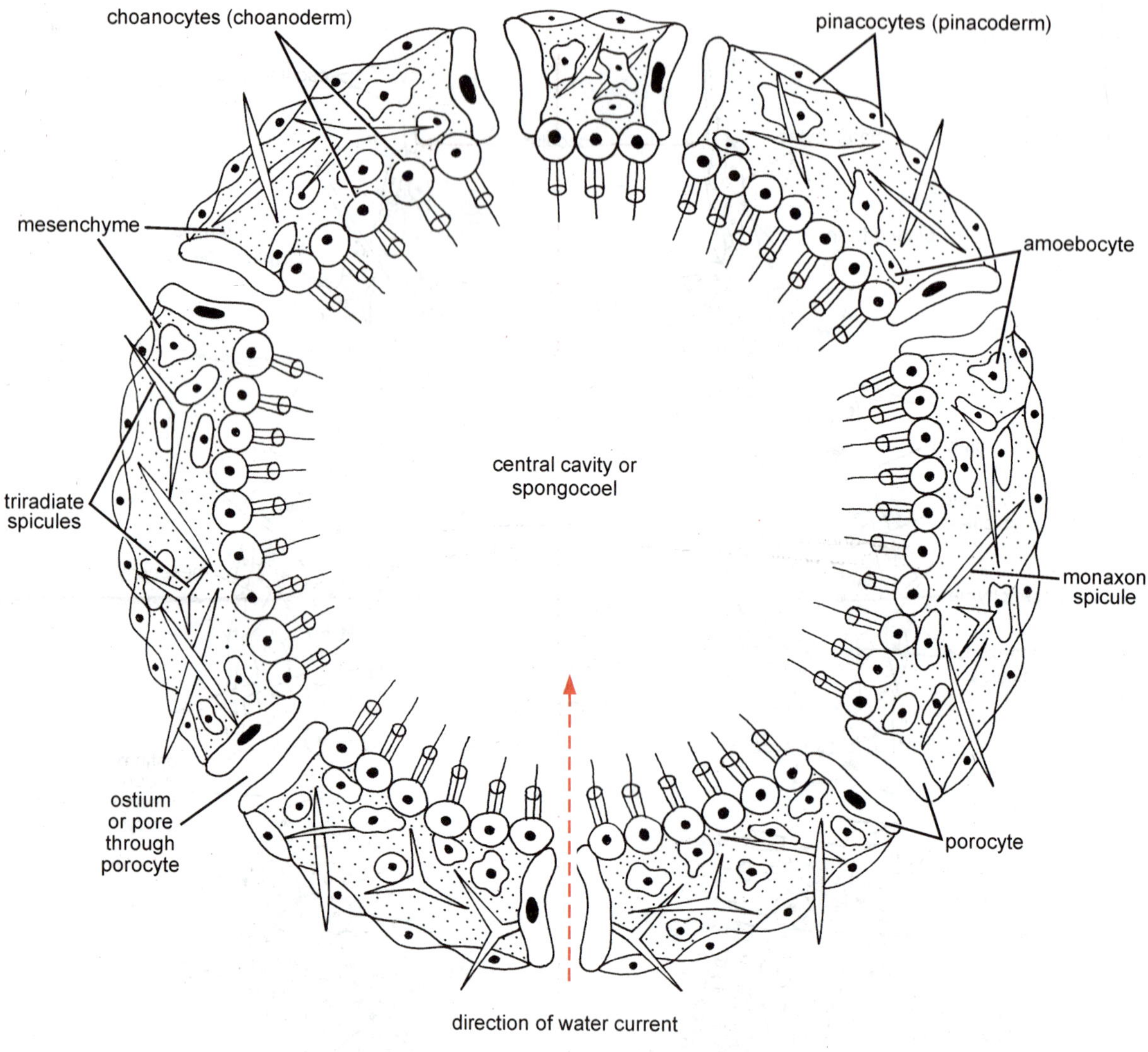

Fig. 3. *Leucosolenia.* Transverse section.

porocyte contains a central canal-like space, communicating with the outside as well as the spongocoel. These spaces are called **ostia** or **dermal pores** or **incurrent pores**. These permit water to flow from outside into the spongocoel.

2. Choanoderm. The spongocoel is lined internally by **gastrodermis** or **choanoderm**. It consists of a single layer of flagellated collar cells, the **choanocytes** (G., **choane**, funnel + **kytos**, cell). A choanocyte is an ovoid cell with its free end bearing a transparent contractile collar. It surrounds a single long **flagellum** which arises from a **basal granule** or **kinetosome**. The nucleus lies at the base or apex of its cell body. Protoplasmic processes of the cell body are embedded in the mesoglea. Choanocytes are used in feeding and for ensuring the flow of water within the animal's body by beating of their flagella.

3. Mesenchyme. In between pinacoderm and choanoderm is present non-cellular mesenchyme.

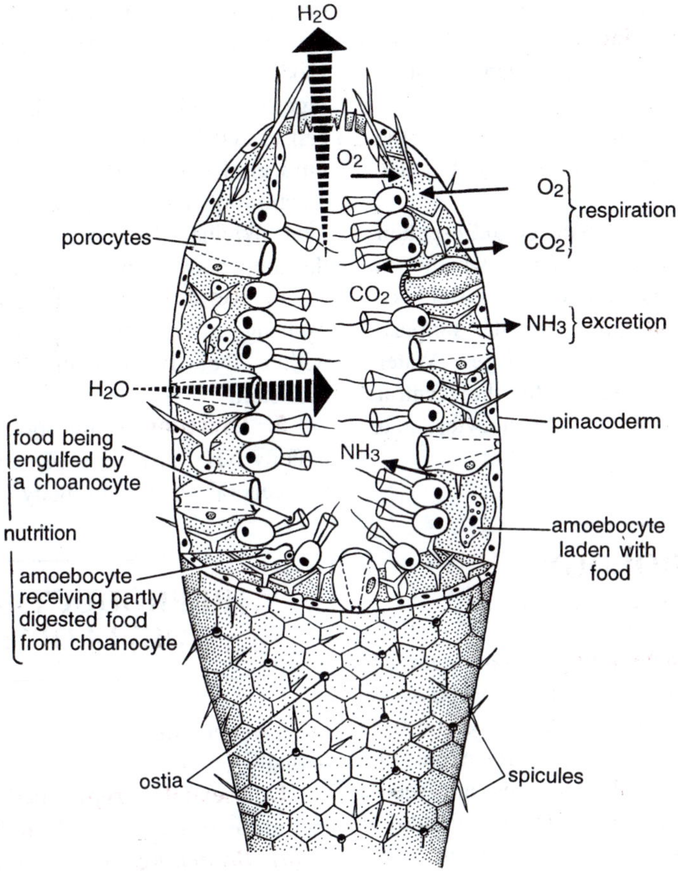

Fig. 4. *Leucosolenia.* An individual cylinder with a portion cut to show circulation of water, nutrition, respiration and excretion.

It is secreted by both body layers. It consists of a gelatinous matrix of protein, also called **mesogloea**. It contains several types of freely wandering cells like *Amoeba* or **amoebocytes**, and minute skeletal elements of $CaCO_3$, called **spicules** or **sclerites**.

(a) Amoebocytes. Amoebocytes are of many types depending upon the shape of their pseudopodia and function. Large-sized primary amoebocytes with blunt pseudopodium and large nucleus are the **archaeocytes**. These contain much RNA and they carry on all the functions essential for the life of sponge. They are self-replicating and also capable of giving rise to all other types of amoebocytes (totipotent). **Scleroblasts** are skeleton forming amoebocytes. They are called **calcoblasts** as they form calacareous spicules. **Thesocytes** are with rounded pseudopodia and have reserve food material. **Chromocytes** carry colour pigments and are responsible for colour of sponge. **Gland cells** secrete slime. Cells with thin branching pseudopodia are **collencytes** forming a sort of connective tissue and probably nerve cells. **Myocytes** are contractile thin cells, present around osculum and work as a sphincter.

(b) Spicules. These skeletal elements, small or large, are formed of crystalline calcium carbonate. Needle-like spicules are **monaxons** while some

are **tetraxons** with four rays. Some tetraxons secondarily become **triradiate** due to loss of one ray. Spicules remain embedded within mesenchyme, though many of them protrude through pinacoderm and impart roughness to the sponge's body surface. All spicules orient in the same direction.

Canal system. Ostia, spongocoel and osculum together form a canal system which is characteristic of all sponges. Canal system of *Leucosolenia* is of **ascon type**. It is the simplest type of canal system found in sponges. Water enters directly through ostia into the central spongocoel, which is lined by choanocytes, and leaves through osculum.

PHYSIOLOGY

1. Locomotion and behaviour. *Leucosolenia* is incapable of locomotion. It is supposed to possess local contractile powers that appear to be mostly restricted to the region of osculum. Reactions to stimuli are very slow and responses are seen several minutes after the application of stimulus.

2. Water current. Vital life processes of a sponge are dependent upon a continuous uninterrupted flow of water through its porous body. This is essential because they are sessile and have no other means of getting food and oxygen or getting rid of wastes. Water current is caused by constant beating of flagella of millions of choanocytes. Water enters spongocoel through ostia and exits through osculum. Flow of water is controlled by the closing and opening of ostia. Flow of water through a sponge can be demonstrated by adding carmine particles to water containing a living sponge. These particles are seen to enter body of sponge through ostia and pass out through osculum. The water current brings food and oxygen and removes excretory as well as reproductive elements.

3. Digestion. Food consists chiefly of plankton, *i.e.*, microscopic animals and plants and organic particles. Choanocytes capture and digest the food. Amoebocytes receive the partly digested food particles from choanocytes, complete digestion and distribute the digested food from cell to cell by diffusion. Amoebocytes digest food intracellularly within their food vacuoles. The reaction of food vacuoles is at first acidic and later alkaline. Undigested residue is eliminated by the amoebocytes into spongocoel.

4. Respiration. Respiration, involving exchange of O_2 from sea water and CO_2 produced within the living cells, is accomplished by diffusion.

5. Excretion. Elimination of metabolic wastes (chiefly NH_3) also takes place by diffusion through the general surface of body.

REPRODUCTION AND DEVELOPMENT

Leucosolenia reproduces asexually as well as sexually.

1. Asexual reproduction. *Leucosolenia* reproduces asexually by **branching** and **budding**.

(a) Branching. New horizontal branches arise from stolon, grow over rocks, and give rise to new erect vase-shaped cylinders. When an upright branch or cylinder attains sufficient size, its top breaks through as an osculum.

Regeneration is also common and a complete individual or colony will grow from almost any broken piece of sponge.

(b) Budding. This is accomplished by evagination of body wall near the base of a vertical tube in the form of a **bud**. The bud grows in size and breaks off an osculum at its free end, thus becoming an additional individual of colony. This may again bud off new individuals.

2. Sexual reproduction. *Leucosolenia* is monoecious, *i.e.*, male and female reproductive cells or gametes are formed in the same individual. No special gonads are formed. Sperm and ova are derived from **archaeocytes** by gametogenesis.

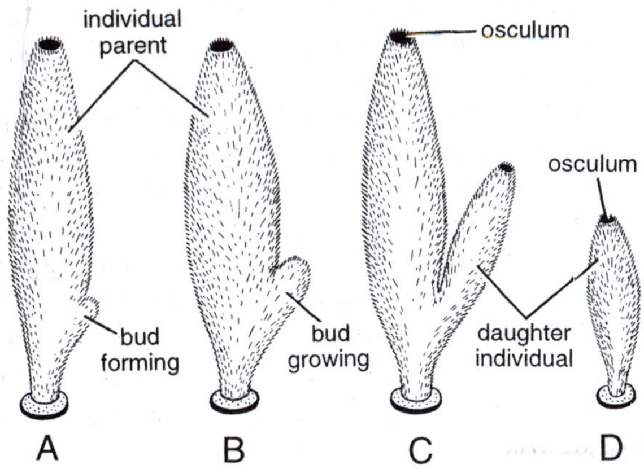

Fig. 5. *Leucosolenia.* Stages showing budding.

Sperms are released into sea water and they make their way into another sponge to a region of a mature ovum. Some workers believe that sperms are transported to mature ova by amoebocytes.

3. Development. Ovum is fertilized by a sperm within maternal body wall forming a diploid **zygote**. The zygote undergoes equal and holoblastic cleavage developing into a hollow blastula, called **coeloblastula**. Its wall consists of a single layer of narrow, elongated, columnar and flagellated cells. At the posterior pole of larva occurs a group of large, rounded, granular, non-flagellated cells. These are believed to be **archaeocytes** which form all future archaeocytes and reproductive cells of sponge. These granular cells later wander into the cavity of embryo. Some of the adjacent flagellate cells also lose their flagella, become amoeboid and pass into the internal cavity. The larva now reaches a stage corresponding to the **planula** larva of coelenterates. It is termed **stereogastrula** or

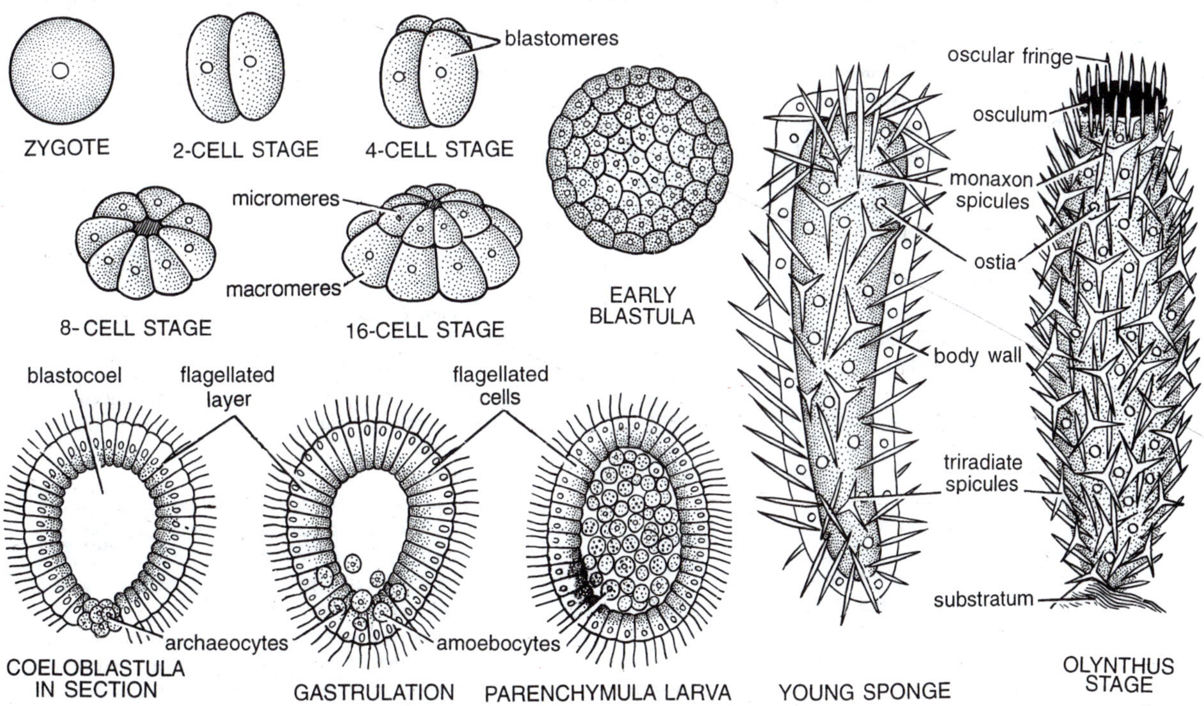

Fig. 6. *Leucosolenia.* Stages in development. **A.** Zygote. **B.** 2-cell stage. **C.** 4-cell stage. **D.** 8-cell stage. **E.** 16-cell stage. **F.** Early blastula. **G.** and **H.** Coeloblastula. **I.** Parenchymula. **J.** Young sponge. **K.** Adult sponge.

parenchymula and consists of an external layer of flagellated cells and an inner mass of amoeboid cells. It has no mouth opening.

Parenchymula swims freely for some hours. Then it becomes fixed by its anterior pole and develops into a flat plate with an irregular outline. Most of the amoeboid cells migrate to external surface, passing between flagellated cells, and form the pinacoderm and mesenchyme. Flagellated cells, thus enclosed, become the choanocytes. A central cavity or spongocoel appears which increases in size, becomes lined by choanocytes and opens to outside by an osculum. Certain non-flagellated cells in the wall of sponge, or porocytes, become perforated to form incurrent pores or ostia. Monaxon and triradiate spicules are secreted by the scleroblasts or modified amoeboid cells. Within a few days of its attachment, the larva is converted into the adult asconoid sponge.

Important Questions

➤➤*Long answer type questions*

1. Give an account of the structure, physiology and reproduction of an asconoid sponge studied by you.
2. Describe, with suitable diagrams, the development of *Leucosolenia.*
3. Draw a labelled diagram of the L.S. of *Leucosolenia.*
4. What are sponges ? Describe the cellular organization of *Leucosolenia.*
5. Briefly describe the structure and life history of *Leucosolenia.*
6. Write short notes on : (i) Choanoderm, (ii) Coeloblastula, (iii) Osculum, (iv) Ostia, (v) Parenchymula, (vi) Pinacoderm.

➤➤*Short answer type questions*

1. What is amoebocyte found in mesenchyme of *Leucosolenia* ? Write its various forms and their functions.
2. Describe the ascon type of sponge.
3. What is a choanocyte ? Relate its structure to its function.
4. Describe the various histological elements found in *Leucosolenia.*
5. Give the structure of pinacoderm in *Leucosolenia.*
6. Give the structure of choanoderm in *Leucosolenia.*
7. Describe the development and structure of parenchymula larva in *Leucosolenia.*

➤➤*Multiple choice questions*

1. *Leucosolenia* is :
   (a) fresh water colonial sponge
   (b) marine and solitary
   (c) solitary
   (d) colonial and firmly attached to the substratum
2. *Leucosolenia* is :
   (a) sessile (b) marine
   (c) colonial (d) all of these
3. Which is not known to occur in *Leucosolenia* :
   (a) respiration (b) egestion
   (c) asexual reproduction (d) locomotion
4. The free swimming larva of *Leucosolenia* is :
   (a) stomoblastula (b) amphiblastula
   (c) stereoparenchymula (d) olynthus stage
5. Which type of canal system is found in *Leucosolenia*?
   (a) ascon (b) rhagon
   (c) leucon (d) syconoid
6. In sponges food is ingested by :
   (a) scleroblasts (b) choanocytes
   (c) porocytes (d) pinacocytes
7. Digestion of food in *Leucosolenia* takes place :
   (a) in the spongocoel (b) in the amoebocytes
   (c) in the choanocytes
   (d) first in choanocytes and then in the amoebocytes
8. The internal skeleton of *Leucosolenia* is formed of calcareous :
   (a) monaxon (single-rayed) spicules
   (b) monaxon and tri-radiate (three-rayed) spicules
   (c) tri-radiate and quadri-radiate (four-rayed) spicules
   (d) monaxon, tri-radiate and quadri-radiate spicules
9. The principal cell types present in the body wall of *Leucosolenia* are the :
   (a) pinacocytes, porocytes, choanocytes, amoebocytes
   (b) pinacocytes, choanocytes, amoebocytes, nephrocytes
   (c) choanocytes, nerve cells, amoebocytes, nephrocytes
   (d) choanocytes, porocytes, nephrocytes, amoebocytes
10. The common bath sponge is :
    (a) *Leucosolenia* (b) *Euspongia*
    (c) *Sycon*
    (d) *Spongilla*

11. Gametes in *Leucosolenia* are derived from :
    (a) archaeocytes (b) choanocytes
    (c) porocytes (d) amoebocytes
12. Spicules of *Leucosolenia* originate from :
    (a) calcoblast (b) thesocyte
    (c) trophocyte (d) archeocyte
13. Nutrition in *Leucosolenia* is :
    (a) intracellular (b) extracellular
    (c) both (d) none
14. The chief excretory product of *Leucosolenia* :
    (a) ammonia (b) urea
    (c) uric acid (d) none
15. In *Leucosolenia* the sperm and ova develop from :
    (a) monocyte (b) amoebocyte
    (c) archaeocyte (d) gland cells
16. Young stage of *Leucosolenia* is :
    (a) coeloblastula (b) parenchymula
    (c) olynthus (d) stomoblastula
17. To which of the following classes would you place *Leucosolenia* ?
    (a) Calcarea (b) Hexactinellida
    (c) Desmospongia (d) Heterocoela
18. Which is the correct sequence of path of water current flowing through *Leucosolenia* ?
    (a) ostium-spongocoel-osculum
    (b) osculum-spongocoel-ostium
    (c) osculum-ostium-spongocoel
    (d) spongocoel-osculum-ostium
19. Exit for water current in *Leucosolenia* and other sponge is :
    (a) arboreal pore (b) ostium
    (c) osculum (d) cloaca
20. Symmetry in *Leucosolenia* is :
    (a) spherical (b) radial
    (c) bilateral (d) asymmetrical

Answers

➤➤*Multiple choice questions*

1. (d) 2. (d) 3. (d) 4. (c) 5. (a) 6. (b) 7. (c) 8. (d) 9. (a) 10. (b)
11. (a) 12. (a) 13. (a) 14. (a) 15. (c) 16. (c) 17. (a) 18. (a) 19. (c) 20. (b).

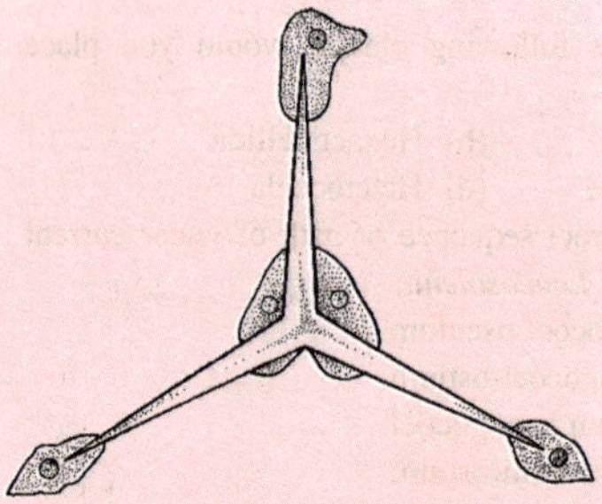

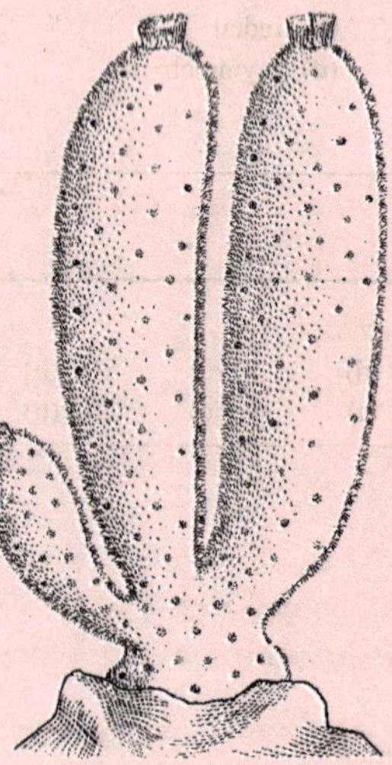

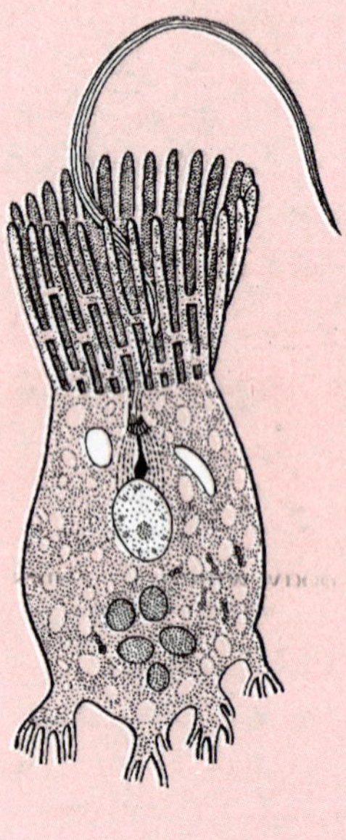

20

Scypha (=*Sycon*) : A Syconoid Sponge

The structure of *Leucosolenia,* described in preceding chapter, is of primitive asconoid type in which the body wall is not folded. In other types, the body organization is based upon the folding of the body wall and reduction in size of the spongocoel. The first stage of the folding of the body wall is displayed by sponges with syconoid type of structure in which the body wall has become folded horizontally, forming finger-like processes. Syconoid sponges include the well-known genera *Scypha* (=*Sycon*) and *Grantia.* The biology of *Scypha* has been treated in detail in this chapter.

Scypha or *Sycon*

Scypha is commonly known as **urn sponge**, as the individuals are shaped like an urn (a water vessel). It is also called the **crown sponge**, because the fringe of long and straight spicules at the top looks like a little crown. *Scypha* differs from *Grantia,* a European genus, in not having an outer covering, or the **cortex**, a characteristic feature of the latter. The best studied species of *Scypha* are *S. coronata, S. ciliatum, S. elegans, S. lingua* and *S. raphanus.*

SYSTEMATIC POSITION

Phylum	Porifera
Class	Calcarea
Order	Heterocoela
Family	Sycettidae
Genus	*Scypha* or *Sycon*

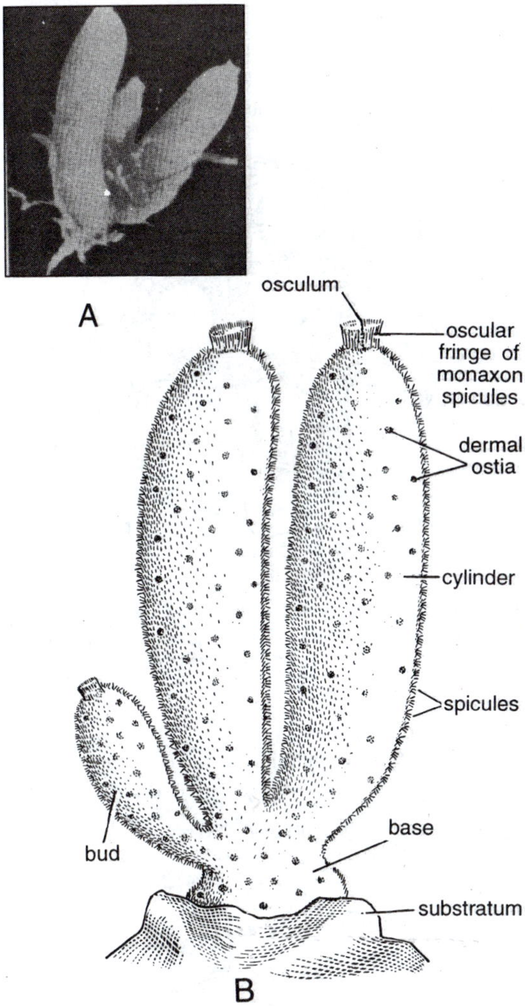

Fig. 1. *Scypha.* **A.** Colony in natural size. **B.** One individual magnified.

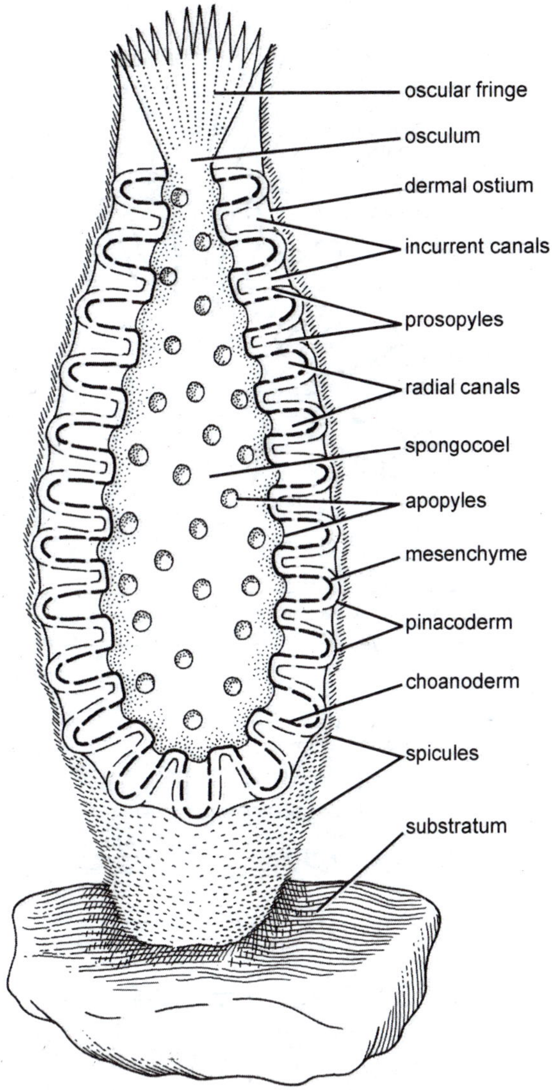

Fig. 2. *Scypha.* A diagrammatic L.S. of a cylinder showing gross internal structure and canal system.

HABITAT

Scypha (Gr. **skyphos**, cup) is a marine sponge which is widely distributed, but is best known from North Atlantic shores. It may be solitary or forms a colony by budding. Sessile colonies of cylindrical individuals are found permanently attached to submerged rocks or other solid substrata in shallow sea water along the coasts. They thrive well where wave action is not too strong and at low tide mark.

EXTERNAL MORPHOLOGY

Scypha or *Sycon* possesses a vase-shaped and radially symmetrical body which measures from 1 to 3 cm in height and 5 to 6 mm in diameter. The colour is not specific but varies from grey to light brown. Near the attached end are found a few small tubular projections, or **buds**, which are

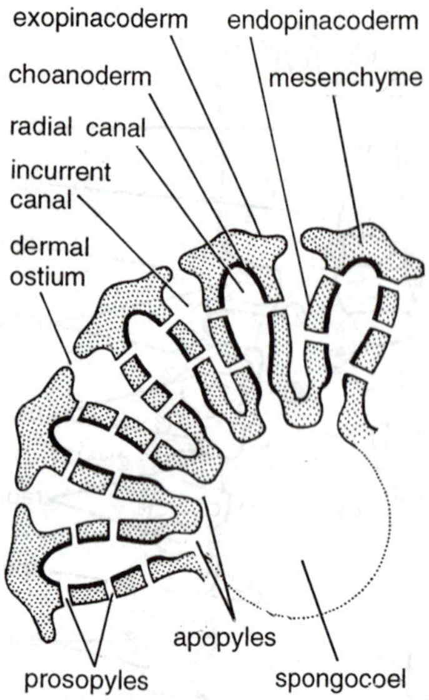

Fig. 3. *Scypha.* A portion of a T.S. (diagrammatic).

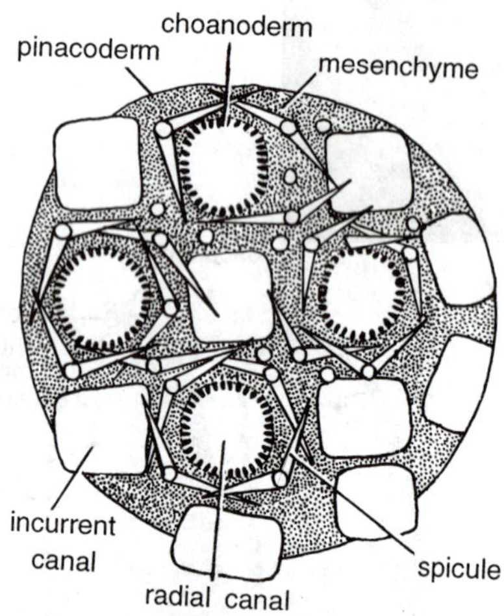

Fig. 4. *Scypha.* A tangential section of a portion of body wall showing arrangement of incurrent and radial canals.

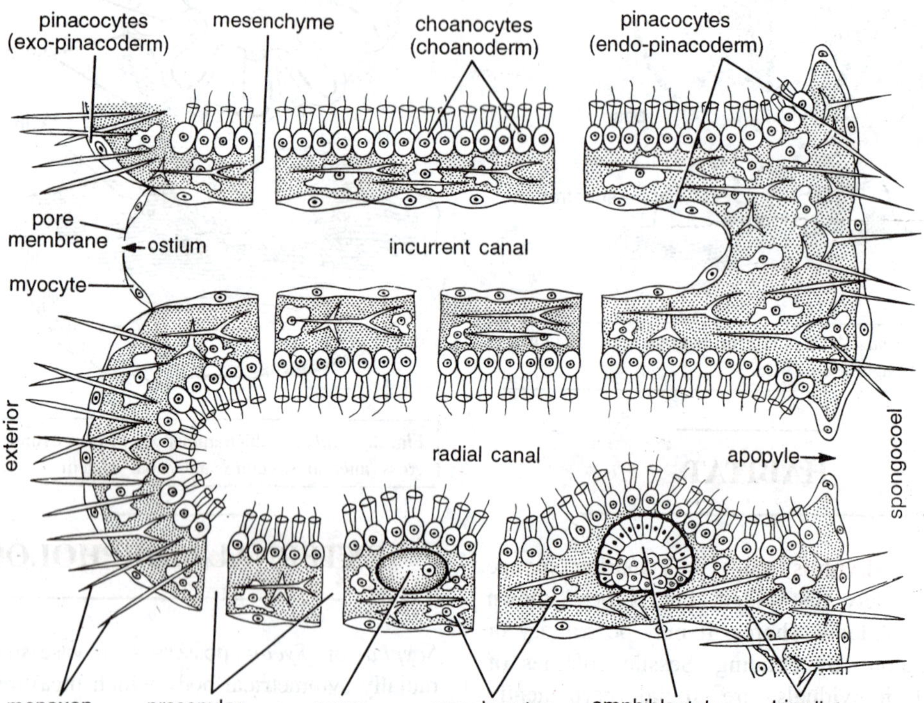

Fig. 5. *Scypha.* A diagrammatic sectional view of body wall showing one incurrent canal and one radial canal.

formed not so extensively as in *Leucosolenia* or other asconoid sponges. Free end of the vase-shaped individual bears a pore, the **osculum** (**exhalent** or **excurrent** pore), which is fringed with long, straight, needle-like calcareous monaxon spicules. This **oscular fringe** checks small animals from entering into the body. Just below osculum, the body becomes narrower forming a **collar region**. The body surface is thrown into regularly arranged polygonal elevations from which project spear-like spicules, called **oxeotes** (monaxon spicules), that impart a bristly appearance to the body. The polygonal elevations are separated by deep grooves, bearing minute pores, *i.e.*, **ostia**, which lead into the central body cavity, the **spongocoel** or **paragastric cavity**, through a system of canals.

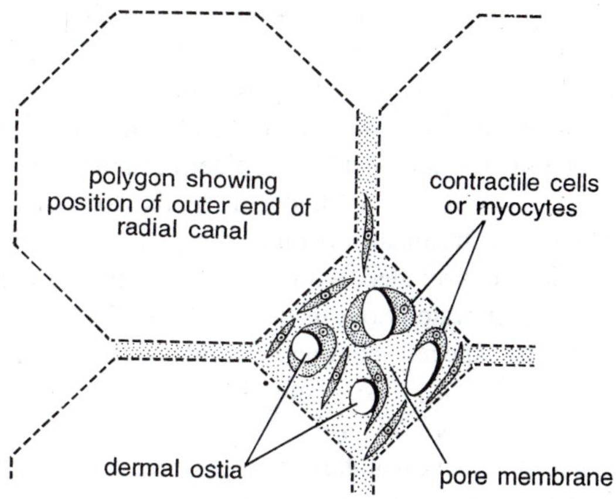

Fig. 6. *Scypha.* A surface view of pore membrane showing ostia.

Canal System

Body of *Scypha* is organized in such a manner as to form a complex system of pores and canals. This system is generally referred to as **canal system** or **aquiferous system**. Body wall has essentially the same cellular layers, **pinacoderm** and **choanoderm**, with a non-cellular gelatinous **mesenchyme** in between. But the body wall is so folded as to form regularly arranged alternating invaginations and evaginations, establishing the sycon type of canal system. Various components of canal system are as follows :

1. Ostia or dermal pores. The external grooves of body surface are stretched over by a thin **pore membrane.** It bears two or more openings for the ingress of outside water into the body of sponge. These pores are known as **ostia** (L., **ostium**, door) or **dermal pores** (Fig. 6). Because of the presence of contractile cells or **myocytes** around them, the ostia can reduce in diameter and thus regulate the amount of ingressing water.

2. Incurrent canals. These canals are the invaginated folds of body wall and are also called **inhalent canals**. These communicate with outside through ostia but end blindly at their inner ends. Pinacocytes line these canals throughout.

3. Prosopyles. Incurrent canals communicate with radial canals through intercellular spaces, called **prosopyles** (Gr., **pros**, near + **pyle**, gate).

4. Radial canals. Evaginations of body wall form thimble-shaped chambers lined by flagellated **choanocytes**. These chambers are called **flagellated** or **radial canals**. Incurrent and radial canals are parallel and alternate with each other, both vertically and radially. The arrangement is such that, in a vertical or tangential section through the wall of a cylinder, each radial canal is surrounded on four sides by incurrent canals, and each incurrent canal is surrounded likewise by four radial canals (Fig. 4).

Radial canals end blindly at their outer ends but lead at their inner ends into spongocoel.

5. Apopyles. Openings of radial canals into spongocoel are called **apopyles** (Gr., **apo**, away from + **pyle**, gate) or **internal ostia**. These are surrounded by contractile myocytes serving as a sphincter.

6. Spongocoel. It is the large central cavity of body forming the vertical axis of the cylinder

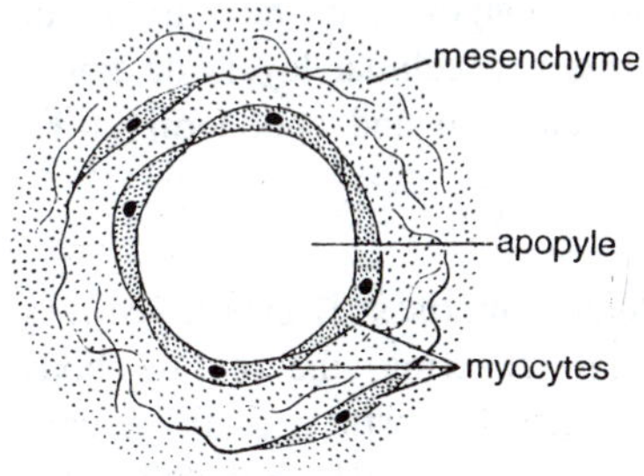

Fig. 7. *Scypha.* Apopyle lined by myocytes.

(Gr., **spongos**, sponge + **koilos**, hollow). In *Leucosolenia,* spongocoel is lined by flagellated collar cells or choanocytes. In *Scypha,* the choanocytes line the radial canals, whereas the spongocoel is lined with the epidermal pinacocytes.

7. Osculum. Spongocoel leads to outside through a terminal opening, the **osculum**. The oscula are provided with **sphincters** to regulate the rate of water flow in the body. Sphincters are lined by special contractile pinacocytes, called **myocytes**. (Gr., **myos**, muscle + **kytos**, cell).

8. Current of water. Flow of water inside canal system is maintained by continuous beating of flagella of collar cells lining the radial canals. Every beat of a flagellum consists of a normal active stroke and a recovery stroke. Electron microscopy has revealed that there is no coordination between the beating of flagella of adjacent cells. The course taken by water current into the canal system is as under :

Water from outside → (through dermal ostia) → Incurrent canals → (through prosopyles) → Radial canals → (through apopyles) → Spongocoel → (through osculum) → to outslide

The rate of flow of water in sponges body is about 0.01 mm/sec.

Cellular Organization (Histology)

Studies on structure and morphogenesis of *Scypha* have clearly revealed the presence of two types of cellular layers, the **pinacoderm** and **choanoderm**, with an intermediate **mesenchyme**. The former controls the interrelations between the mesenchyme and the external medium, while the latter controls mainly the nutrition of the animal. Asconoid sponges are not truly diploblastic because the two cellular layers do not correspond with the ectoderm and endoderm of Eumetazoa.

[I] Pinacoderm

Pinacoderm comprises (i) **exopinacoderm** (dermal epithelium) covering the entire body surface except dermal ostia and osculum, and (ii) **endopinacoderm** which includes the epithelial lining of incurrent canals and spongocoel. Pinacoderm is composed of large, flattened, polygonal cells, the **pinacocytes** (Gr., **pinako**, plank + **kytos**, cell). In profile, each cell presents a central bulging containing the nucleus. Margins of adjacent cells are closely cemented together. Pinacocytes are highly contractile and they can greatly increase or reduce the surface area of sponge body.

In the lining of incurrent canals, some pinacocytes are modified to form tubular cells, the **porocytes**. These connect the incurrent canals with radial canals through their intracellular channels, the **prosopyles**. Porocytes are thin-walled cells, open at both ends and with the nucleus present in the peripheral cytoplasm. According to some workers, porocytes are present only in young sponges. In the adult sponges, porocytes degenerate leaving empty spaces or intercellular prosopyles surrounded by contractile myocytes. Pinacocytes surrounding the osculum, outer (dermal) ostia and inner ostia (apopyles) are elongated and contractile and act as muscle cells, called **myocytes**. These cells form sphincters around them to regulate there openings.

[II] Choanoderm

Choanoderm, constituting the gastral epithelium, is formed of flagellated collar cells or **choanocytes** (Gr., **choane**, funnel + **kytos**, cell). The cells are oval or rounded and arranged in a loose layer upon the mesenchyme. Each cell contains a single nucleus, one or two contractile vacuoles, food vacuoles, reserve food, blepharoplast, rhizoplast and a single basal granule or kinetosome from which originates a long, whip-like flagellum. The flagellum is somewhat stiff towards the base and soft towards the tip. It is surrounded at its base by a thin cytoplasmic collar.

Electron microscopy has revealed the presence of all the intracellular organelles, such as mitochondria, Golgi bodies, endoplasmic reticulum, ribosomes, etc., in a choanocyte. The collar is shown to be formed by 20 to 30 cytoplasmic processes, the **microvilli** or **tentacles**, which are capable of contraction and often jointed together by side connections. The flagellum consists of contractile fibres arranged in the usual 9 + 2 pattern.

[III] Mesenchyme

Between pinacoderm and choanoderm lies the interconnecting gelatinous matrix, the **mesenchyme** (Gr., **mesos**, middle + **enchyme**, infusion) or **mesohyl**. It is supposed to be secreted by pinacoderm. It contains a variety of *Amoeba*-like cells, the **amoebocytes**. These are modified archaeocytes that migrate from outer cellular layer and carry on a variety of functions essential to the life of sponge. A few types of amoebocytes found in sponges are as follows :

1. Archaeocytes. These are undifferentiated embryonic amoebocytes, having blunt pseudopodia and large nucleus with conspicuous nucleolus. They can produce all other types of cells, needed by the sponge. Such cells are termed **totipotent** as they give rise to any other kinds of cells within an animal. They also give rise to sex cells, *i.e.*, **ova** and **sperms**, and play an important role in regeneration.

2. Collencytes. Most of the amoebocytes have branching pseudopodia often united into a syncytial network. These cells are **collencytes**.

3. Chromocytes. These are pigmented amoebocytes with lobose pseudopodia.

4. Thesocytes. These have lobose pseudopodia and are filled with food reserves, thus acting as storage cells.

5. Myocytes. These are fusiform contractile muscle cells present around ostia, oscula and other openings. These form a sphincter which regulates the size of these openings.

6. Scleroblasts. These manufacture the spicules and according to the nature of the product are known as **calcoblasts**, **silicoblasts**, or **spongioblasts**.

7. Gland cells. These are attached to the body surface by long strands and secrete slime.

8. Germ cells. Ova and **sperms** of sponges differentiate from amoebocytes (archaeocytes), though in some cases they have been stated to be modified choanocytes.

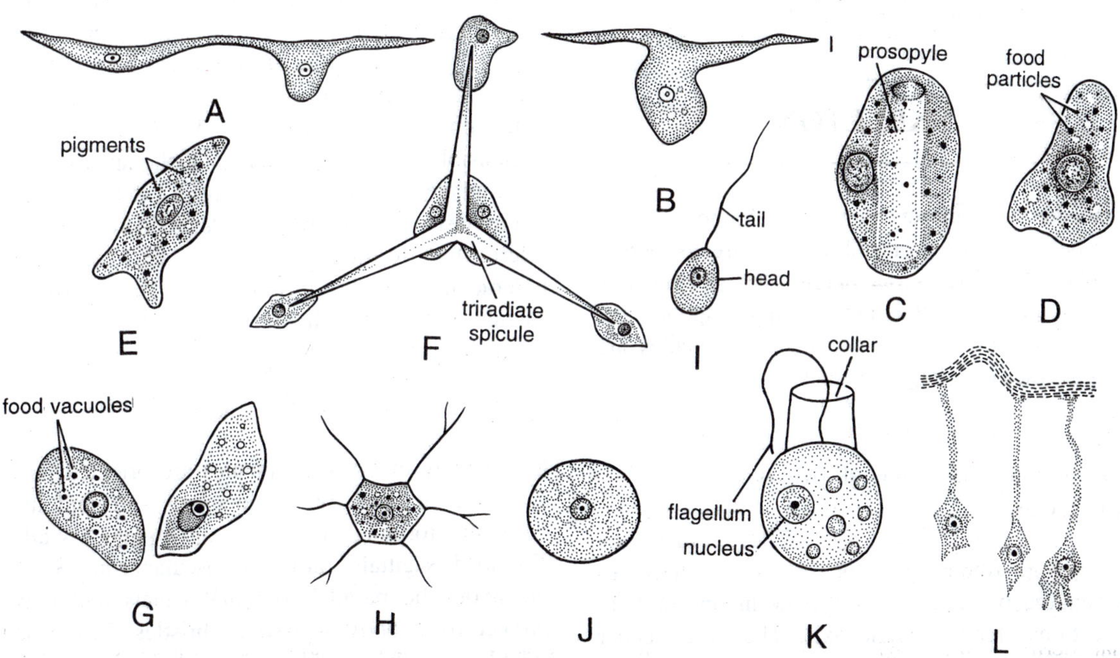

Fig. 8. *Scypha.* Kinds of cells. **A.** Pinacocytes. **B.** Myocyte, **C.** Porocyte. **D.** Thesocyte. **E.** Chromocyte, **F.** Sclerocytes, **G.** Archaeocytes, **H.** Collencyte, **I.** Sperm, **J.** Ovum, **K.** Choanocyte, **L.** Gland cells.

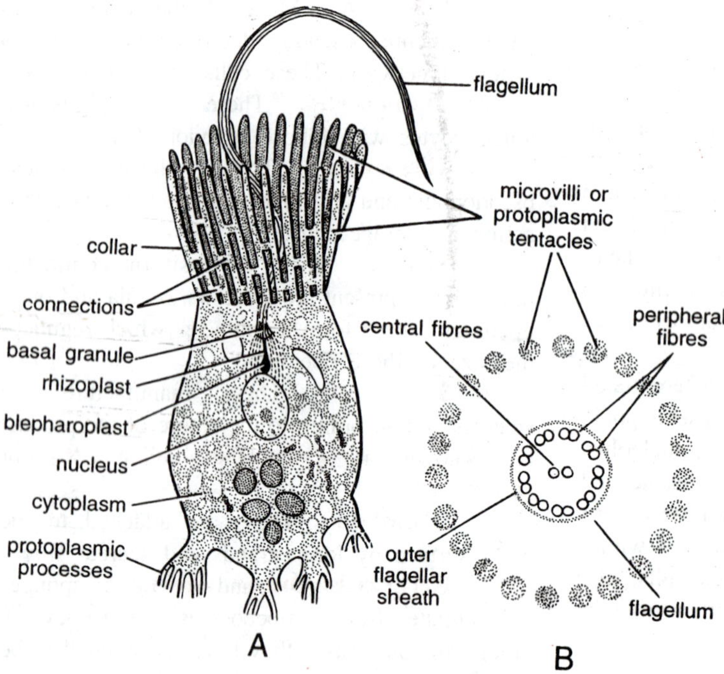

Fig. 9. A. Structure of a choanocyte based on electron micrography, **B.** T.S. through collar region of a choanocyte.

SKELETON

Embedded in mesenchyme are found a large number of minute, crystalline, calcareous bodies called **spicules** (L., **spica**, point). These constitute the **endoskeleton** that protects and supports the softer parts of the body. Calcareous spicules are all megascleres (large spicules).

[I] Types and Structure of Spicules

Spicules in *Scypha* are of two types, **monaxon** and **tetraxon**.

1. Monaxon spicules. These may be **uniradiate** (monactinal) or **biradiate** (diactinal) depending on whether they grow in one or both directions along a single axis. They are slender and straight like needles, or curved.

2. Tetraxon spicules. These are also known as **tetractines** and **tetraradiates**. Each consists of four rays not in the same plane. By loss of one ray, they become **triradiate**, which are very common. Three rays and angles may be equal (**regular**), or unequal (**sagittal Y-shaped**, or **T-shaped**).

Spicules consist very largely of crystalline calcium carbonate, but **Sollas** (1885) showed that they were a form of **calcite**. Qualitative analysis of spicules has shown the presence of magnesium, sulphates, sodium and water in addition to calcium carbonate.

[II] Development of Spicules

Spicules are secreted by specialized amoebocytes known as **scleroblasts**. According to **Minchin** (1896, 1898) the scleroblasts are derived from exopinacoderm. A monaxon spicule is formed by two **sclerocytes** produced by the division of one scleroblast. Outer sclerocyte is referred to as **thickener cell** and the inner one as **founder cell**. Thickener cell is responsible for the lengthening of the spicule ray. A triradiate spicule is formed by a group of three scleroblasts, each of which gets divided into two sclerocytes forming a **sextet** (six cells). A tetraradiate spicule develops like the triradiate spicule and the fourth ray is developed from the junction of the three.

[III] Arrangement of Spicules

In the body of *Scypha*, spicules are oriented in a characteristic way. Long monaxons surround the osculum forming the **oscular fringe**, while T-shaped sagittals form the oscular rim. Short monaxons lie parallel to radial canals and may project from body surface as bristles. Triradiate spicules mostly lie along the radial canals with one ray facing to outside. Tetra-radiate spicules occur along with the triradiates.

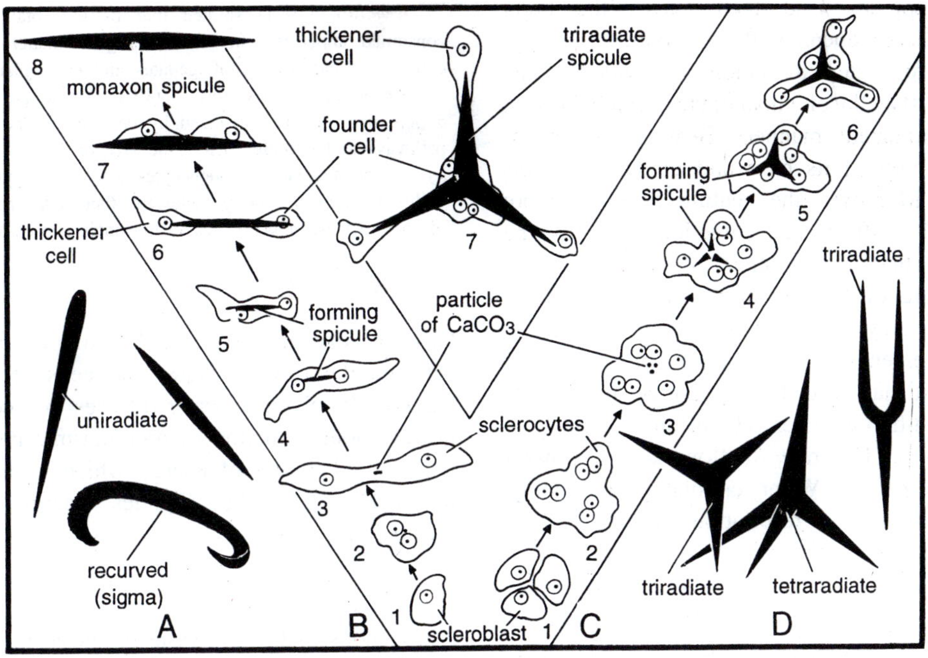

Fig. 10. *Scypha.* Structure and development of spicules. **A.** Kinds of monaxon spicules. **B.** Development of a monaxon spicule. **C.** Kinds of triradiate spicules. **D.** Development of a triradiate spicule.

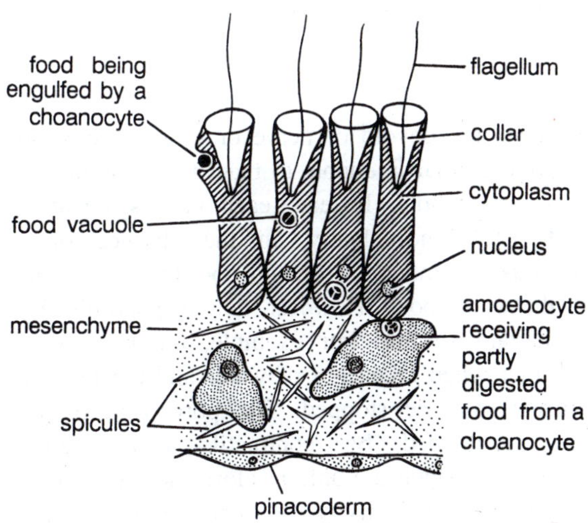

Fig. 11. *Scypha.* Ingestion of food by choanocytes and its digestion.

PHYSIOLOGY

[I] Movement

Scypha, like other sponges, possesses no locomotor organs for moving from one place to another. It is sessile being attached to the substratum. The whole body may show slow contractile movements as its outermost layer of cells, the pinacoderm, is highly contractile. Movements of cells are amoeboid in character, rather than muscular.

[II] Water Circulation

Water is drawn into the canal system through small ostia, present in the pore membranes. Passing through incurrent canals, the water current enters the flagellated radial canals through prosopyles, and the central spongocoel through apopyles, and then leaves the body through a large opening, the osculum.

Force for water circulation is provided by the beating of flagella of choanocytes which beat in an uncoordinated manner. Beating action of flagellum starts at its base and gradually travels to its tip. Moreover, the beating action is more strong and swift in one direction (towards the outer side) than in the other.

The rate of flow of water within the body of sponge may be increased or decreased by the enlargement or constriction of ostia and osculum. It has also been reported that water flows at different rates at different points of canal system. The rate of flow is the slowest into radial canals. Water circulation in sponge helps in nutrition, respiration, excretion and reproduction.

[III] Nutrition

Scypha is a filter feeder, subsisting on minute organisms (**planktons**) and organic particles. These enter the body with water through ostia which allow the entry of only small particles. Inside flagellated chambers, the beating of flagella of collar cells causes water to circulate through their collars, allowing the food particles to adhere to them. Microvilli of collars act as a filter for trapping food particles which move towards their bases. Consequently, the food particles are engulfed by pseudopodial action of choanocytes at the bases of their collars, and then taken up into food vacuoles.

The phase in food vacuoles is first acidic and then alkaline. Here, the food undergoes partial digestion and partly digested food is passed on to wandering amoebocytes in mesenchyme. Within amoebocytes, digestion of food is completed and indigestible residue is eliminated with the outgoing current of water. Amoebocytes also distribute the digested food to all other cells of body, while some of them (thesocytes) store some food for future use.

A number of enzymes have been isolated from sponges. They include protein, starch and fat-digesting enzymes.

Reisning (1971) studied that of the total particles consumed, about 80% consisted of organic matter and 20% consists of bacteria, dinoflagellates and other plankton.

According to **Weissenfel** (1976), food particles of 5–50μ size are phagocyotized by cells lining the inhalant pathways and particles below the size of 5μ are engulfed by the choanocytes. Amoebocytes and choanocytes have ability to transfer food particles to other cells and instead of choanocytes, amoebocytes are the main site of digestion.

[IV] Respiration

Gaseous exchange occurs by simple diffusion between the flowing water and cells of sponge. Oxygen dissolved in water diffuses into cells and brings about oxidation of protoplasmic molecules with the liberation of energy which is entrapped in ATP. The ATP supplies energy to the metabolizing cells.

[V] Excretion

Nitrogenous metabolic waste produced in sponges is largely ammonia. No special excretory tissue is present for excreting this to the outer medium. It leaves the body in the outgoing water current by diffusion. Some investigators claim that metabolic wastes are taken up by amoebocytes which discharge them into spongocoel.

[VI] Nervous System and Behaviour

Sponges are devoid of nerve or sensory cells, so that the animal is unable to react to a stimulus as a unified whole. Instead, each cell is sensitive and reacts individually. In the absence of a nervous system, there are no coordinated actions of the whole body. However, they respond directly to certain stimuli. For example, the body may contract when taken out of water and the feeding current may be stopped. The pores and oscula are surrounded by contractile cells, called **myocytes**, which are able to close these openings. Power of conductivity is very slight so that reactions to light, touch and chemical, etc., are very slow. Conductivity is best developed at the osculum.

However, **O. Tuzet** (1953) and **M. Pavans de Ceccatty** (1955) have attributed a nervous

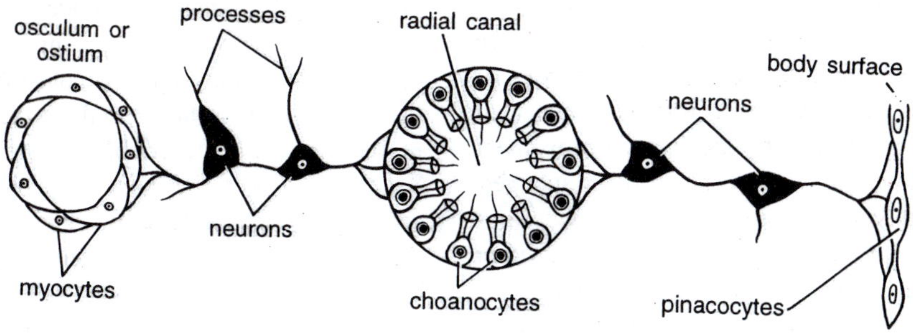

Fig. 12. Probable mechanism of reception and conduction of nerve impulses in a sponge.

function to **collencytes**. Some of them, acting as **neurons**, form a diffuse network connecting the choanocytes with pinacocytes and myocytes. The neurons are supposed to receive and conduct stimuli (Fig. 12).

REPRODUCTION

Scypha reproduces by asexual as well as sexual methods.

[I] Asexual Reproduction

Asexual reproduction is accomplished by **budding**. An outgrowth from the cylinder of sponge arises near its base or attached end to form a **bud**. An osculum is broken off at its free end. Fully grown bud may remain attached with the parent individual or is detatched to become free and forms a new sponge by fixing itself to the substratum.

[II] Sexual Reproduction

Scypha is a **monoecious** sponge but due to **protogyny**, only cross-fertilization occurs. Special sex organs are lacking. Male sex cells, or **sperms**, and female sex cells, or **ova**, are found in mesenchyme. They develop from undifferentiated amoebocytes, called **archaeocytes**, or from choanocytes.

1. Spermatogenesis. Sperm mother cell or **spermatogonium** is said to be an enlarged archaeocyte. However, it is supposed to be a modified choanocyte by **Gatenby**, who described the transformation of an entire flagellated chamber into spermatozoa. Soon the spermatogonium is surrounded by one or more flattened **cover cells** to form the **spermatocyst**. Cover cells are derived either by division of mother cell or from other

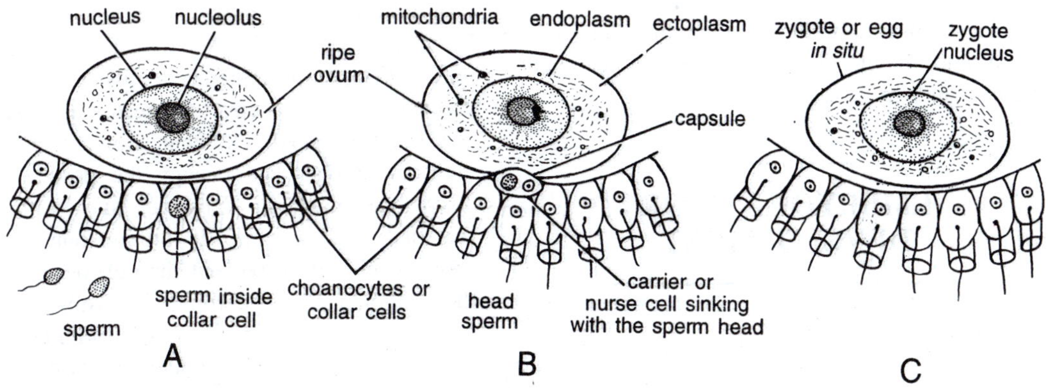

Fig. 13. Stages of fertilization *in situ* in a calcareous sponge.

amoebocytes. Spermatogonium undergoes two or three divisions to form **spermatocytes** which give rise to **spermatozoa**. A mature sperm or **spermatozoon** consists of a rounded nucleated head and a vibratile tail, by the lashing movement of which it moves in water to reach other sponges.

2. Oogenesis. Egg mother cell or **oocyte** is derived from a large archaeocyte with a distinct nucleus. This may sometimes arise by transformation of a choanocyte which stores some food, loses its flagellum and sinks into mesenchyme. Oocyte moves like an amoebocyte and grows by engulfing other cells which may be amoebocytes or special nurse cells (**trophocytes**). When full grown, the oocyte undergoes usual two maturation divisions to form the **ovum** which lies in the wall of a radial canal, ready to be fertilized by a sperm from another sponge.

3. Fertilization. Fertilization is internal and **cross**. Sperms from one individual enter other sponges with water current, and the **ova** are fertilized *in situ*. Process of fertilization is very remarkable and probably occurs in all the sponges. The spermatozoon enters first a choanocyte or collar cell which lies adjacent to a ripe ovum. It loses its tail and its swollen head becomes surrounded by a **capsule**. The choanocyte also loses its collar and flagellum and becomes amoeboid. It is now known as the **carrier cell** or **nurse cell**. Outer surface of ovum, at the point of contact, invaginates so that the carrier cell is received in a conical depression. The capsule containing sperm head now penetrates into the ovum. According to **Gatenby** and other, carriers cell fuses with ovum. But, according to **Duboscq** and **Tuzet**, it simply departs after the transfer of sperm into ovum. Fusion of sperm's head and ovum results in a **zygote** or **egg**

EMBRYOGENY OR DEVELOPMENT

[I] Early Embryonic Period

1. Cleavage. Fertilized egg begins its development *in situ*, that is, in the mesenchyme of maternal parent. It undergoes equal and holoblastic cleavage. First three divisions are vertical, resulting into 8 blastomeres. Fourth division, which is horizontal and unequal, results in the formation of 16 blastomeres in two tiers. 8 cells of lower tier next to parent choanoderm are larger and called **macromeres**. They are destined to form the pinacoderm. Other 8 cells of upper tier are smaller and called **micromeres** They form the choanoderm. The macromeres remain undivided for the time being but become somewhat rounded. The micromeres undergo rapid mitosis, forming several micromeres that acquire flagella at their inner ends facing the blastocoel. The embryo is now called **blastula** with a **blastocoel** developed between both the tiers. **Duboscq** and **Tuzet** have called this stage as **stomoblastula**

2. Stomoblastula. One side of stomoblastula is composed of many small, elongated and flagellated **micromeres**, and the other side is composed of eight rounded and non-flagellated **macromeres**. The inner cavity or blastocoel communicates to outside through an opening, the **mouth**, in the centre of macromeres. This mouth is used for engulfing the surrounding amoebocytes for nutrition.

[II] Larval Period

1. Amphiblastula. The stomoblastula undergoes a process called **inversion**, in which it turns itself out through the mouth, as in *Volvox*, so that the flagella of micromeres become directed towards outside. The embryo is now called the **amphiblastula** (Gr., **amphi**, both + **blastos**, germ), because its one half bears flagella and the other half does not. The flagellated cells or micromeres are narrow and occupy the greater anterior part. The non-flagellated cells or macromeres are large, eight in number and occupy the posterior part. Amphiblastula larva does not stay within the maternal mesenchyme. It breaks out into the spongocoel and passes out through osculum with the outgoing water current. It swims freely in water for some time. While swimming, the flagellated pole is directed anteriorly and the force for swimming is supplied by the beating of flagella.

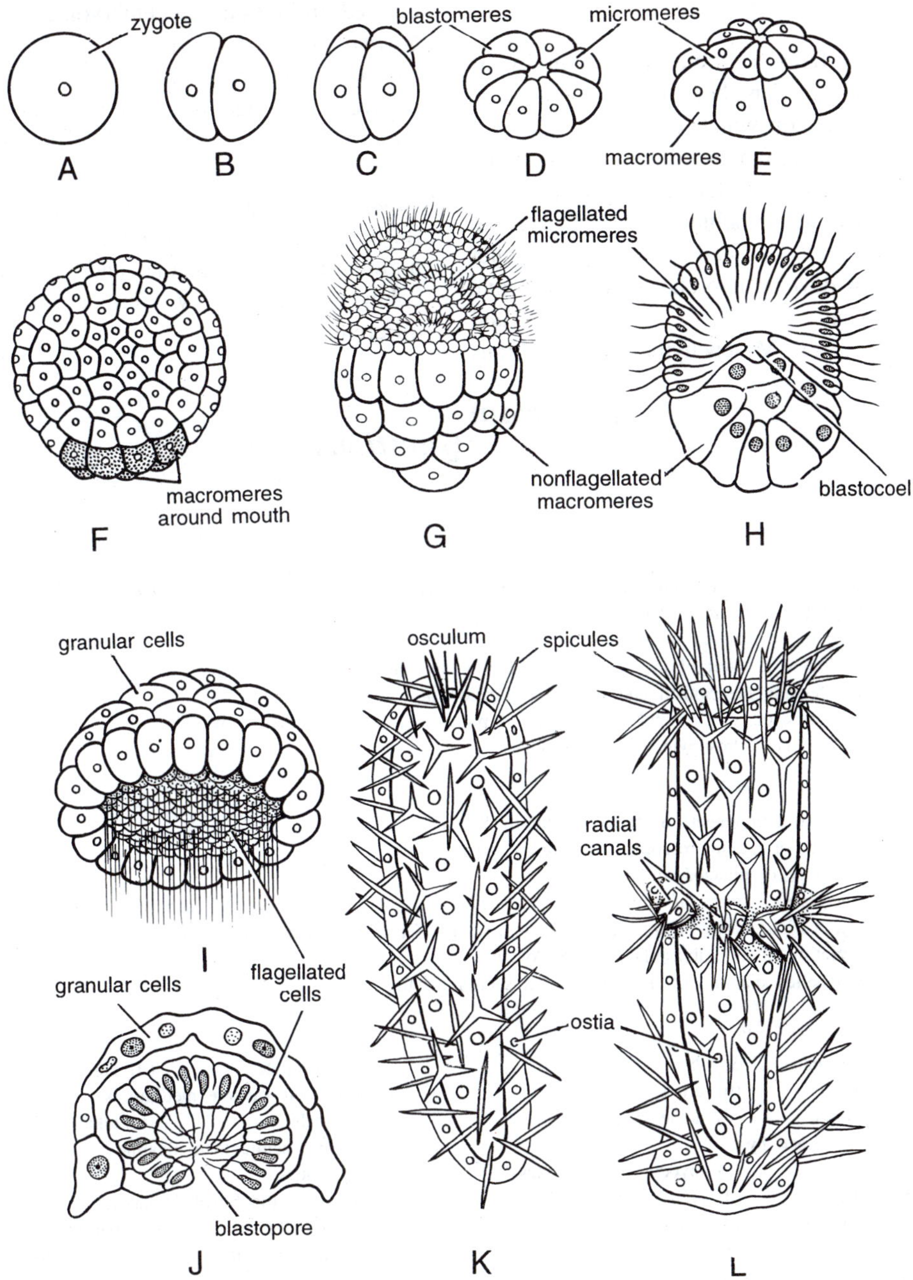

Fig. 14. *Scypha.* Stages of development. **A.** Zygote. **B.** 2-cell stage. **C.** 4-cell stage. **D.** 8-cell stage. **E.** 16-cell stage. **F.** Stomoblastula in surface view **G.** Amphiblastula in surface view. **H.** Amphiblastula in section. **I.** Invagination. **J.** Gastrula in section. **K.** Olynthus stage. **L.** Young sponge.

2. Gastrula. Ultimately the amphiblastula settles down and undergoes gastrulation. Now the macromeres multiply more rapidly than micromeres, so that the flagellated half of larva is invaginated into and overgrown by the granular non-flagellated half. The larva now resembles a typical double-walled **gastrula** with a **blastopore** at the invaginated side.

[III] Post Larval Period or Metamorphosis

Soon the larva fixes to some substratum by its blastoporal end and lengthens into a cylinder, at the free distal end on which opens the **osculum**. Several small perforations on the wall of cylinder become **ostia**. Outer non-flagellated granular cells give rise to dermal epithelium or exopinacoderm and to **scleroblasts** and **porocytes**. Inner flagellated cells develop into choanoderm and give rise to functional **choanocytes**, **archaeocytes** and other **amoebocytes**. Mesenchyme cells are thus derived from both the embryonic layers. Young *Scypha* now reaches the **olynthus stage**, which closely resembles a simple ascon sponge. Adult or syconoid stage is derived by the budding of radial canals first at the middle of young sponge which further grows vertically. Flagellated choanocytes shift into radial canals and body wall further, increases in thickness by the growth of middle layer with spicules and canal system. The colony develops by further branching and differentiation.

Important Questions

➤➤*Long answer type questions*

1. Describe the canal system of *Sycon* and explain the mechanism and importance of the circulation of water through this system.
2. Give an account of the structure and functions of various cells found in *Sycon* (=*Scypha*).
3. Give in detail the structure of the body wall of *Sycon.*
4. Give an account of skeleton in *Sycon.*
5. Give an illustrated account of the sexual reproduction and development of *Sycon.*
6. Describe the feeding mechanism and mode of nutrition in *Sycon.*
7. Write short notes on : (i) Amphiblastula, (ii) Apopyle, (iii) Archaeocytes, (iv) Choanocytes, (v) Olynthus stage, (vi) Pinacocytes, (vii) Prosopyle, (viii) Scleroblasts.

➤➤*Short answer type questions*

1. Write the various components of canal system in *Sycon.*
2. Give an account of the microscopic structure of *Sycon.*
3. Give the structure of pinacoderm in *Scypha.*
4. What are the various forms of amoebocytes in *Scypha* and write the function of each one of them.
5. What are the various types of spicules found in *Sycon* ? Give the structure of each.
6. Give the structure of amphiblastula larva of *Sycon.*
7. Draw fully labelled diagram of the longitudinal section of the *Sycon.*

➤➤*Multiple choice questions*

1. The common name of *Scypha* :
   (a) bath sponge (b) urn sponge
   (c) bowl sponge (d) horse sponge
2. sphincters are lined by :
   (a) theocyte (c) myocyte (pinacocytes)
   (b) thesocyte (d) archaeocyte
3. Among them which is food storage :
   (a) theocyte (b) myocyte
   (c) thesocyte (d) chromocyte
4. The food vacuole medium is :
   (a) acidic (b) alkali
   (c) first acidic then alkaline
   (e) neutral
5. Name of the Scypha's larva :
   (a) amphiblastula (b) parenchymula
   (c) stomablastula (d) all of them

Answers

➤➤*Multiple choice questions*

1. (b) 2. (c) 3. (c) 4. (c) 5. (a).

21

Porifera : Characters, Classification and Types

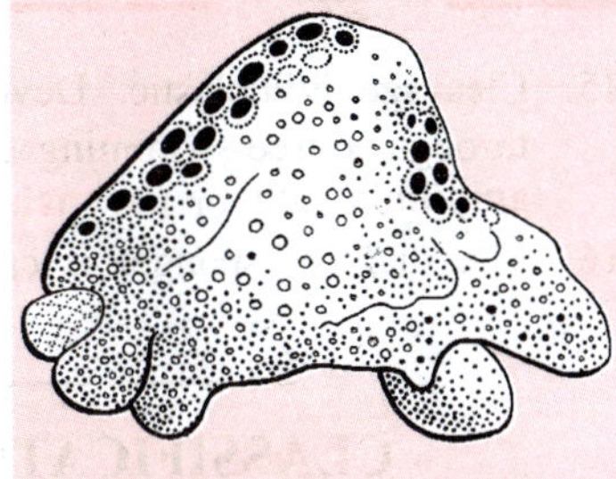

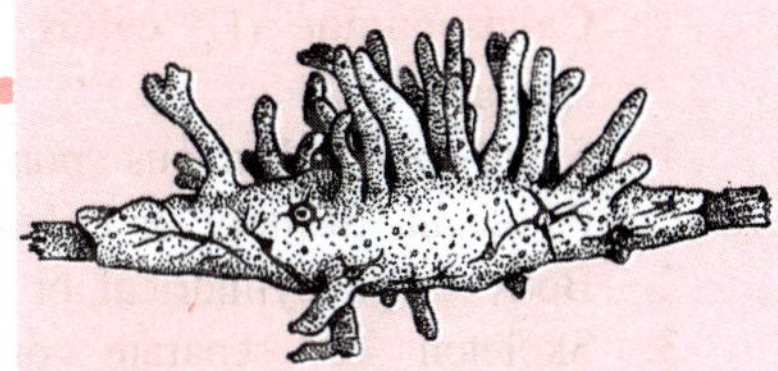

GENERAL CHARACTERS

1. Multicellular organisms with **cellular level** of body organization. No distinct tissues or organs.
2. Mostly marine, a few freshwater, all aquatic.
3. Solitary or colonial, all sessile.
4. Body form vase-like, cylindrical, tubular, cushion-shaped, many-branched, etc.
5. Symmetry radial or no symmetry.
6. Body wall with outer **pinacoderm** (dermal epithelium), inner **choanoderm** (gastral epithelium) and gelatinous non-cellular **mesenchyme** in between. Mesenchyme consists of skeletal elements and free amoeboid cells.
7. Cells loosely arranged and do not form definite layers, hence regarded not truly diploblastic.
8. Body with many pores (**ostia**), canals and chambers that serve for the flow of water. One or more water exits or **oscula** present.
9. Choanocytes or flagellated collar cells usually line special chambers. Sponges are the only metazoans having choanocytes.
10. Skeleton of calcareous or siliceous spicules or of protein spongin fibres, or of both, or absent.
11. Digestion intracellular. No respiratory or excretory organs. Contractile vacuoles in some freshwater forms.
12. Primitive nervous system of neurons arranged in a definite network of bipolar or multipolar cells in some, but is of doubtful status.
13. All sponges are hermaphrodite but cross-fertilization is the rule.
14. Asexual reproduction by **buds** or **gemmules**. Sexual reproduction by ova and sperms. All show regeneration power.

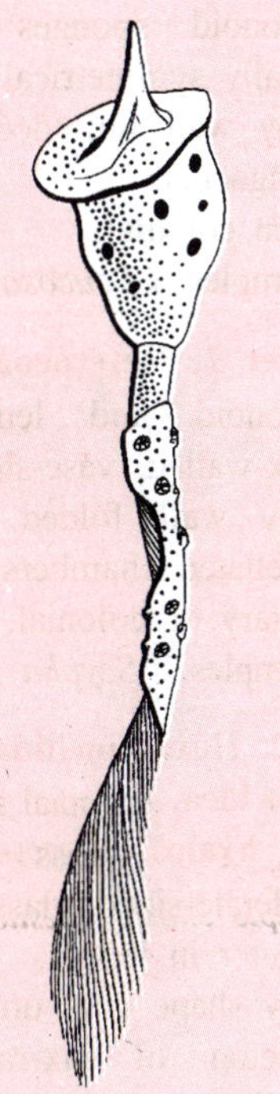

15. Cleavage holoblastic. Development indirect through a free-swimming ciliated larva, the **amphiblastula** or **parenchymula**.
16. The study of sponges is called spongiology.

CLASSIFICATION

The phylum includes about 11,055 species of sponges, grouped in 3 classes depending mainly upon the nature of skeleton they possess.

CLASS 1. Calcarea (L., calcarius, limy) or Calcispongiae (L., calcis, lime + spongia, sponge).

1. Small-sized calcareous sponges, below 10 cm in height. Solitary or colonial.
2. Body shape cylindrical or vase-like.
3. Skeleton of separate one or three or four-rayed calcareous spicules.
4. Body organization of asconoid, syconoid or leuconoid type.
5. Exclusively marine.

Order 1. Homocoela (= Asconosa)

1. Asconoid sponges with cylindrical and radially symmetrical body.
2. Body wall unfolded. Choanocytes line the spongocoel.
3. Often colonial.

Examples : *Leucosolenia, Clathrina.*

Order 2. Heterocoela (= Syconosa)

1. Syconoid and leuconoid sponges with thick-walled, vase-shaped body.
2. Body wall folded. Choanocytes line the flagellated chambers (radial canals) only.
3. Solitary or colonial.

Examples : *Scypha (= Sycon), Grantia.*

CLASS 2. Hexactinellida (Gr., hex, six + actin, ray + idea, terminal suffix) or Hyalospongiae (Gr., hyalos, glass+spongos, sponge).

1. Moderate-sized glass sponges. Some reach 1 meter in length.
2. Body shape cup, urn or vase-like.
3. Skeleton of six-rayed triaxon siliceous spicules.
4. No dermal epithelium or exopinacoderm.
5. Choanocytes restricted to finger-shaped chambers.
6. Exclusively marine, many in deep sea.

Order 1. Hexasterophora

1. Spicules are hexasters, *i.e.*, star-like in shape with axes branching into rays at their ends.
2. Flagellated chambers regularly and radially arranged.
3. Usually attached to substratum directly.

Examples : *Euplectella* (Venus' flower basket), *Farnera, Staurocalyptus.*

Order 2. Amphidiscophora

1. Spicules are amphidiscs, *i.e.,* with a convex disc, bearing backwardly directed marginal teeth at both ends.
2. Flagellated chambers slightly different from typical shape.
3. Usually attached to substratum by root tufts.

Examples : *Hyalonema* (glassrope sponge), *Pheronema* (bowl sponge).

CLASS 3. Demospongiae (Gr., demos, frame + spongos, sponge).

1. Small to large-sized, solitary or colonial.
2. Body shape like a vase, cup or cushion.
3. Skeleton of siliceous spicules or spongin fibres, or both, or absent.
4. Spicules monaxon or tetraxon, never triaxon.
5. Body organization leuconoid. Choanocytes restricted to small rounded chambers.
6. All marine except freshwater family Spongillidae.

Subclass 1. Tetractinellida

1. Spicules tetraxon or absent. No spongin.
2. Body flattened or rounded. Dull to brightly coloured.
3. Mostly in shallow water.

Order 1. Myxospongida

Structure simple. Spicules absent.

Examples : *Oscarella, Halisarca.*

Order 2. Carnosa

Equal-sized spicules present.

Examples : *Plankina, Chondrilla.*

Order 3. Choristidia

Both large and small spicules present.

Examples : *Thenea, Geodia.*

Subclass 2. Monaxonida

1. Spicules monaxon. Spongin present or absent.
2. Body form variable.
3. Mostly in shallow waters, some in deep sea, some in freshwater.

Order 1. Hadromerina

Large spicules knobbed at ends. Small spicules star-like when present. Spongin absent.

Examples : *Cliona* (boring sponge), *Suberites.*

Order 2. Halichondrina

Large spicules of many kinds, usually 2-rayed. Spongin present and scanty.

Example : *Halichondria* (crumb-of-bread sponge).

Order 3. Poecilosclerina

Large spicules united by spongin into a regular network. Small spicules C —shaped, curved or bow-shaped.

Example : *Microciona.*

Order 4. Haplosclerina

Large spicules 2-rayed. Usually no small spicules. Spongin usually present.

Examples : *Chalina* (mermaid's gloves), *Haliclona* (finger sponge), *Spongilla* and *Ephydatia* (freshwater sponges).

Subclass 3. Keratosa

1. Horny sponges with skeleton of spongin fibres. No spicules.
2. Body form usually rounded and massive, with a leathery surface and dark colour.

   Examples : *Euspongia* or *Spongia* (bath sponge), *Phyllospongia* (leaf sponge), *Hippospongia* (horse sponge).

In India, sponge fauna is dominated by Desmospongia, Hexactinellida and Calcarea (Gulf of Mannar, Palk bay, Andaman-Nicobar Islands and Lakshadweep).

OTHER TYPES OF PORIFERA

1. Olynthus stage. It is a stage present during the development of all syconoid sponges. The larva of syconoid sponges is an amphiblastula. After a brief period of free-swimming, the larva fixes to some substratum by its blastoporal end, and lengthens into a cylinder. It perforates at the free distal end forming an opening known as **osculum**. Several small perforations in the side wall of cylinder become the incurrent pores or **ostia**. This simple stage in the development of syconoid sponges resembles a simple ascon sponge and is known as the **olynthus stage**. Its inner cavity forms the future **spongocoel** lined by **choanoderm** made of a single layer of flagellated collar cells or choanocytes. The body surface is covered by **pinacoderm**, made of a single layer of flat cells or pinacocytes. The complex canal system of adult syconoid sponges is derived by budding, folding and further increase in thickness of body wall. The basic architecture of sponges and their archetypical functioning can be best understood by the study of olynthus stage.

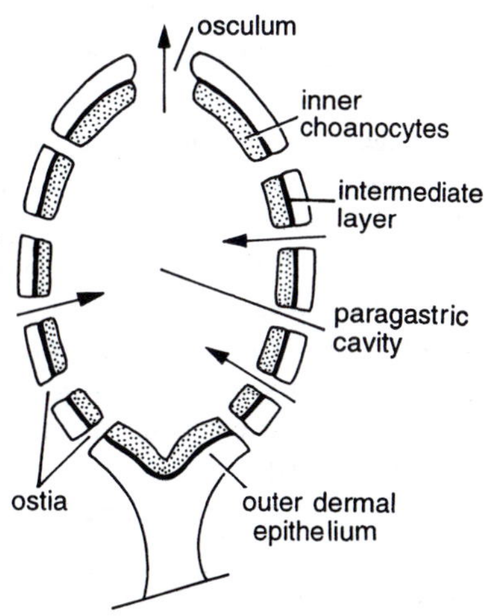

Fig. 1. Olynthus stage. A portion of bodywall cut away to show the spongocoel.

2. *Euplectella.* Perhaps the most beautiful of glass sponges belong to the genus *Euplectella,* popularly known as the 'Venus's flower baskets'. Their dried specimens or intricately-latticed skeletons are often seen in museums. They grow in deep water. *Euplectella aspergillum* is especially abundant near the Philipine islands, while *E. suberea* occurs off the West Indies. These are long, curved, thin-walled and cylindrical tubes, 15 to 30 cm long and 2.5 to 5 cm in diameter. Curved and rigid structure is an adaptation to slow constant water current found at depths from 500 to 5000 meters. Body is held up by skeleton made of four and six-rayed siliceous spicules fused at their tips and much united by cross-bars forming the intricate lacework. Beautiful to look at, the Venus's flower baskets are unpleasant to handle as the spicules projecting through the surrounding protoplasmic material prick the fingers. Gleaming white skeletons look fragile and delicate but they are surprisingly strong, and nearly perfect ones are often washed up on the beaches. Terminal opening at the upper end is closed by an **oscular sieve plate** formed of fused spicules. Openings or **parietal gaps**, seen in the meshwork of spicules, although connected with spongocoel, do not form part of canal system which is of simple sycon type. Flagellated chambers lie radially in the sponge wall.

Sponge is fastened in mud of sea bottom by a tuft of root-like, long and siliceous threads arising from lower end of body. Root spicules are adapted to anchor sponge to its unstable substratum of slimy oceanic ooze.

Species of *Euplectella* exhibit a commensal association with certain species of shrimps. Young male and female shrimps enter the spongocoel of sponge and after growth are unable to escape through the minute sieve plates of osculum. They lead their entire life within the sponge, feeding on plankton brought in with the ingressing water current. The sponge with its imprisoned shrimps makes a good wedding gift in Japan, symbolizing the idea 'till death us do part'.

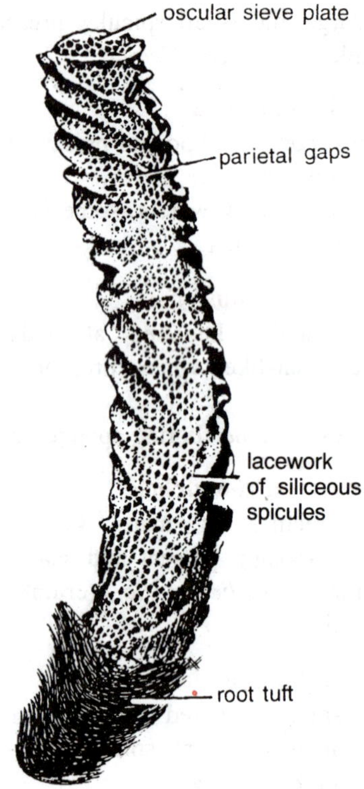

Fig. 2. *Euplectella.* Dried skeleton.

3. *Hyalonema.* One of the commonest hexactinellid sponges is the glassrope sponge *Hyalonema* with several widely distributed species. It lives only 10 to 15 meters deep in sea and is about 38 cm long. *Hyalonema longissimum,* an american species, lives at depths from 100 to 150 meters, off the New England coast. Body is variable in form but usually rounded or oval and radially symmetrical. Superficially it looks like a ball of glass wool with projecting tufts of glassy spicules. Small branching, and five-rayed spicules extend from all over the body. Besides, six-rayed spicules and large and small amphidiscs, like those of freshwater sponges, are also found.

Body is elevated from substratum by a stalk-like **root-tuft** often spirally twisted like rope. It is made of a bundle of very long, monactine anchoring spicules. Root-tuft also traverses the body as an axis or **columella** and often projects

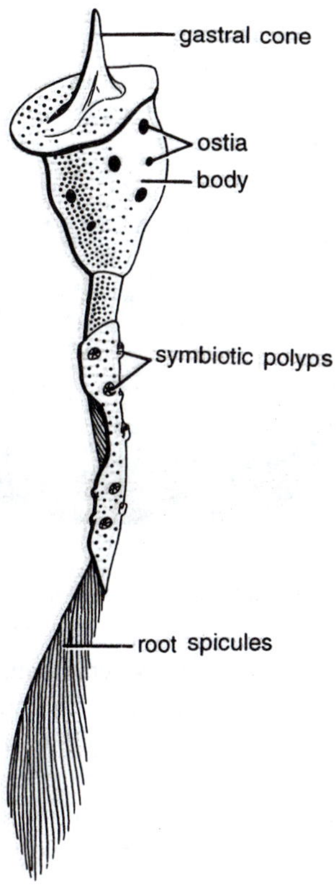

Fig. 3. *Hyalonema.* Dry specimen.

above as **gastral cone**. Middle part of root-tuft commonly bears epizoic or symbiotic anemones of the genus *Epizoanthus* or *Polythoa.*

Excurrent canals open on the gastral surface which may be flush with body surface or depressed to form a cavity or spongocoel. If gastral cone is present no spongocoel exists.

4. *Cliona.* Boring sponges belong to the family Clionidae. *Cliona* is a cosmopolitan genus, forming low encrustation on rocks, coral skeletons, shells of Mollusca and other calcareous objects. Often clam shells, particularly of *Venus,* are completely riddled by *Cliona.* Species may be green or purple but more often of a sulphur-yellow colour. Their larvae bore into dead or living shells or limestone in some unknown manner, forming extensive burrows and tunnels and growing out into compact masses, often 15 to 70 cm in diameter. The surface shows several small elevations, each opening by an osculum. Compact

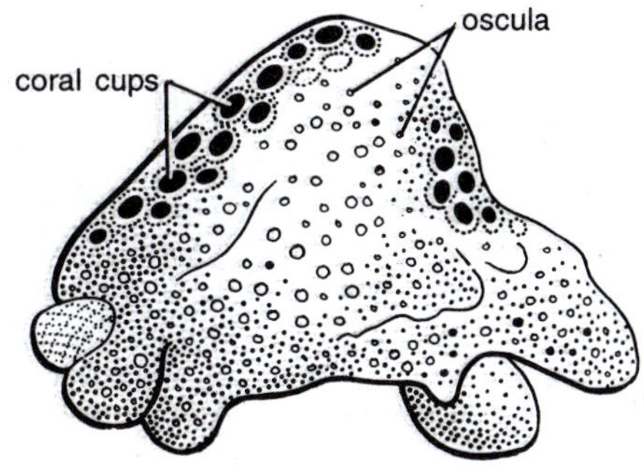

Fig. 4. *Cliona* encrusting a piece of coral.

substance of sponge contains spongin fibres and monaxial siliceous spicules of various forms. It has a sulfurous odour.

5. *Chalina.* *Chalina* is a deep-water form. It is orange, yellowish brown or red in colour. Except when broken from its stalk and washed ashore, it does not occur in shallow waters. The sponge is popularly known as 'the mermaid's gloves' or 'dead-man's fingers' because it is shaped like a hand with several fingers perforated with oscula. Skeleton consists of spongin with siliceous spicules (oxea) embedded in it.

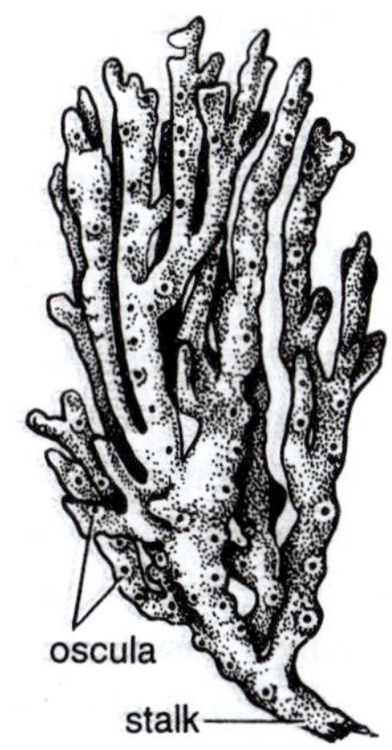

Fig. 5. A colony of *Chalina.*

6. *Spongilla.* *Spongilla* is a common, widely distributed freshwater sponge belonging to the family Spongillidae in the subclass Monaxonida. It is found growing on submerged plants and twigs in ponds and lakes and shows various shades of

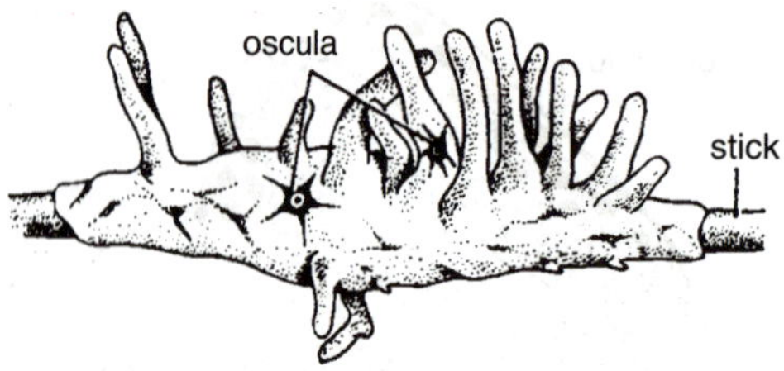

Fig. 6. *Spongilla lacustris.* A colony growing on a submerged twig.

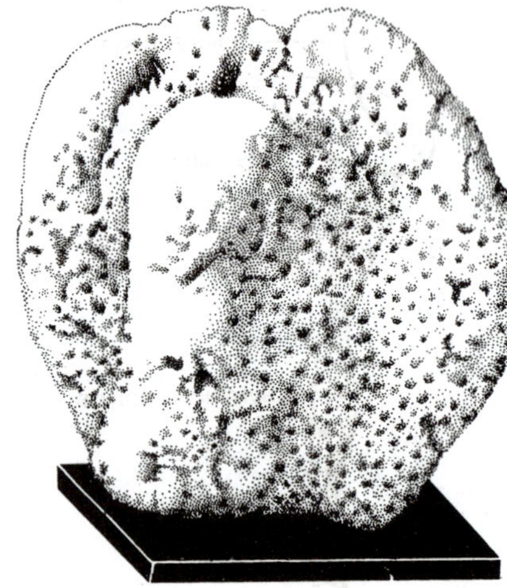

Fig. 7. *Euspongia officinalis,* the toilet sponge.

green because of symbiotic algae present inside. Its colony is more profusely branched than any other freshwater sponge and has a characteristic odour. *Spongilla lacustris* is the most abundant species. It prefers running water and is usually green. *S. fragilis* is also a common species, but prefers standing water and avoids light.

Canal system in *Spongilla* is essentially of rhagon type, with choanocytes restricted to small rounded chambers. Siliceous spicules are of monaxon type. Large spicules, called **megascleres**, are held together by spongin fibres and provide support for the colony. Small spicules, the **microscleres**, are embedded in cells of sponge. Asexual reproduction occurs by regeneration of fragments or by specialized structures, called **gemmules**. Protective covering of gemmules is without amphidiscs or peculiar dumbbell-shaped spicules, although typical needle or rod-like spicules are present. Interior of gemmules is filled with archaeocytes, which develop into a new sponge. Gemmules are produced in late summer and early fall and remain quiescent throughout winter. Characteristics of gemmules form the chief means of indentification of freshwater sponges.

Sexual reproduction also takes place. An unusual free-swimming larva is peculiar to *Spongilla*. Gastrulation does not take place by invagination as in other sponges.

7. *Euspongia.* Commercial bath sponges belong to the family Spongiidae of subclass Keratosa. They occur in warm shallow waters with a rocky bottom. *Euspongia* (= *Spongia),* which furnishes the best quality, is abundant in the Mediterranean. It also occurs near Bahamas, West Indies and Australia in waters from 2 to 200 meters deep. Body is massive with variable shape, often globose, lobed, cup-shaped or lamelliform. Colour in life is dark brown, almost black. Surface of sponge shows several small projections, the **conuli**, which are made of strands of spongin fibres. Body is traversed by several inhalent, exhalent and flagellated chambers and other spaces. Bath sponges are collected by hooking or diving and allowed to decay. Then they are squeezed and washed and their skeletons bleached or dried. Dried skeleton is yellowish or brown. Bath sponges are called horny sponges due to horny consistency of their skeleton made of spongin fibres, forming a close-meshed reticulum. Thickest spongin fibres run radially and each terminates in a conulus. The familiar bath sponge of household use consists only of spongin skeleton and its capacity for holding water is due to capillary forces of meshes of spongin reticulum. Most valuable commercial bath sponge is the light yellow and cup-shaped *Euspongia mollissima* from Asia Minor, while least valuable is the so-called glove sponge of America. All sponges have great powers of regeneration. Cut pieces of a bath sponge become large enough to be marketable in about three years. Sexes are separate but male specimens are rare.

Important Questions

➤➤*Long answer type questions*

1. Classify Porifera giving diagnostic characteristics and at least two examples of each group.
2. Write short notes on : (i) Bath sponge, (ii) *Cliona,* (iii) *Euplectella,* (iv) *Hyalonema,* (v) Olynthus stage, (vi) *Spongilla.*

➤➤*Short answer type questions*

1. What is the name given to the excurrent pore of sponges ?
2. What is the common name of *Didinium* ?
3. What are the common name of poriferans ?
4. Write the scientific name of 'venus flower basket'.
5. Which are the special flagellated cells of sponges that are not met within any other phylum ?
6. Mention the two types of larval forms of sponges.
7. Give the distinguishing characters and one example of the group keratosa.
8. Name the distinguishing features of *Euplectella.*
9. Describe the skeleton of sponges and its significance.
10. Describe the parenchymula larva.
11. Define parazoology ?
12. What is the meaning of hexactinellida ?
13. Describe olynthus stage ?
14. What is the common name of *Hylonema* ?
15. Write short notes on bath sponge.

➤➤*Fill in the blanks*

1. *Chalina* is commonly known as.....
2. Canal system in *Spongilla* is.....
3. ......is leaf sponge.
4. .....is a fresh water sponge.
5. Spicules absent in .....sponge.

➤➤*True and false statements*

1. All the sponges are exclusively marine.
2. Fresh water sponge belong to family spongillidae.
3. *Sycon* is heterocoela.
4. Spicules are triaxon in demospongia.
5. *Euspongia* is known as bath sponge.

➤➤*Multiple choice questions*

1. Which of the following is true about sponges :
   (a) innumerable mouth and one exit
   (b) one mouth and innumerable exit
   (c) spicules are made of chitin
   (d) a large spaceous stomach
2. Flagellated chambers are found in :
   (a) *Leucosolenia* (b) leucon type
   (c) ascon type (d) *Sycon*
3. Parazoology is the study of :
   (a) Protozoa (b) Porifera
   (c) Coelenterata (d) Parasite
4. Internal budding in some sponges is for :
   (a) reproduction
   (b) survival during adverse conditions
   (c) asexual reproduction
   (d) growth
5. Common bath sponge is :
   (a) *Spongia* (b) *Spongilla*
   (c) *Cliona* (d) *Euplectella*
6. Body organisation in sponges is :
   (a) protoplasmic grade (b) tissue grade
   (c) cellular grade (d) organ grade
7. The costliest Turkish Bath sponge is :
   (a) *Spongia officinalis* (b) *Hippospongia*
   (c) *Spongilla* (d) *Grantia*
8. Paragastric cavity is found in :
   (a) *Leucosolenia* only (b) all sponges
   (c) marine sponges
   (d) fresh water sponges
9. A sponge differs from other metazoans in having :
   (a) numerous mouthlets (b) one exit
   (c) both (d) none

10. A sponge can be distinguished from other animals in having :
(a) spicules (b) water circulating system
(c) choanocytes (d) all of these

11. Sponges are characterised by :
(a) choanocytes (b) canal system
(c) innumerable mouthlets, one exit
(d) all of these

12. Type of canal system exhibited by Demospongiae is :
(a) ascon (b) sycon
(c) rhagon (d) leucon

13. Ascon type of canal system is found in :
(a) *Grantia* (b) *Leucosolenia*
(c) *Hyalonema* (d) *Euspongia*

14. Which one of the following is the most distinctive characters of sponges :
(a) that they are acellular
(b) that they possess special cells called choanocytes
(c) that they reproduce asexually
(d) that they are all marine

15. The sponge Hexactinellida lives in :
(a) deep water (b) fresh water
(c) shallow water (d) brackish water

16. Check the appropriate answer. Indicate the Fresh water sponge from the following assemblage :
(a) *Euplectella* (b) *Hyalonema*
(c) *Spongilla* (d) *Euspongia*

17. The cells surrounding the osculum in sponges is :
(a) archaeocytes (b) myocytes
(c) choanocytes (d) gland cells

18. An example of the class Hexactinellida is :
(a) *Spongilla* (b) *Oscarella*
(c) *Euplectella* (d) *Grantia*
(e) *Leucosolenia*

19. Glass sponges belong to the class :
(a) demospongia (b) tetractinellida
(c) hexactinellida
(d) calcarea

20. Middle layer of the body wall of porifera is :
(a) mesenchyme (b) mesoderm
(c) mesogloea (d) mesentery

21. The type of canal system in *Clathrina* is :
(a) *Sycon* (b) ascon
(c) leucon (d) rhagon

22. Skeleton in Demospongia :
(a) silicious (b) calcareous
(c) spongin fibers (d) none

23. Who describe the animal nature of Porifera :
(a) Ellias (b) Ross
(c) Robert Grant (d) None

24. Classification of sponge is based on :
(a) locomotory organs (b) skeleton
(c) size
(d) numbers of ostia

25. Common 'bath sponge' is :
(a) *Scypha* (b) *Euplectella*
(c) *Cliona* (d) *Euspongia*

26. Fertilization in sponge is :
(a) internal (b) external
(c) both (d) none

Answers

➤➤*Fill in the blanks*

1. Mermaid's gloves, 2. Rhagon type. 3. *Phylospongia,* 4. *Spongilla,* 5. Bath sponge.

➤➤*True and false*

1. true, 2. true, 3. false, 4. false, 5. true.

➤➤*Multiple choice questions*

1. (a)	2. (a)	3. (b)	4. (c)	5. (a)	6. (c)	7. (a)	8. (b)	9. (c)	10. (d)
11. (d)	12. (b)	13. (b)	14. (b)	15. (d)	16. (d)	17. (b)	18. (c)	19. (c)	20. (a)
21. (b)	22. (c)	23. (a)	24. (b)	25. (d)	26. (b).				

22

Porifera : General Account

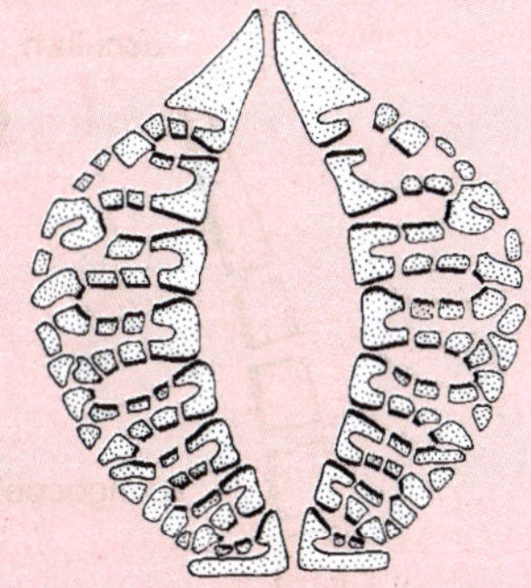

CANAL SYSTEM IN SPONGES

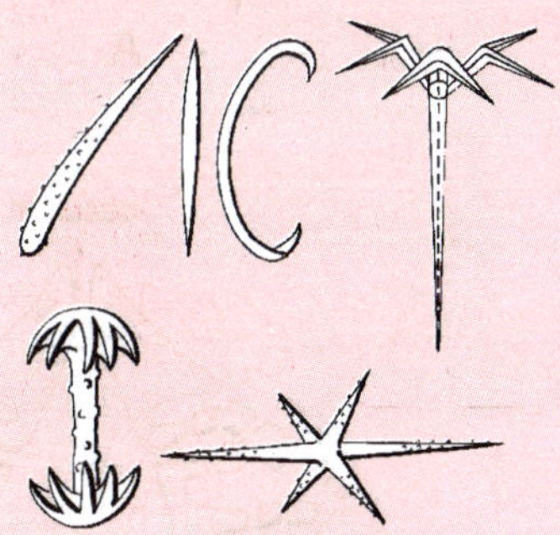

What is Canal System

A distinguishing feature of all sponges is the perforation of body surface by numerous apertures for the ingress and egress of water current. Inside body, the water current flows through a certain system of spaces collectively forming the **canal system**

Function of Water Current

The most vital role in the physiology of sponges is played by water current on which their life depends. All exchanges between sponge body and external medium are maintained by means of this current. Food and oxygen are brought into body and excreta and reproductive bodies carried out. This current is caused by beating of flagella of collar cells.

Types of Canal System

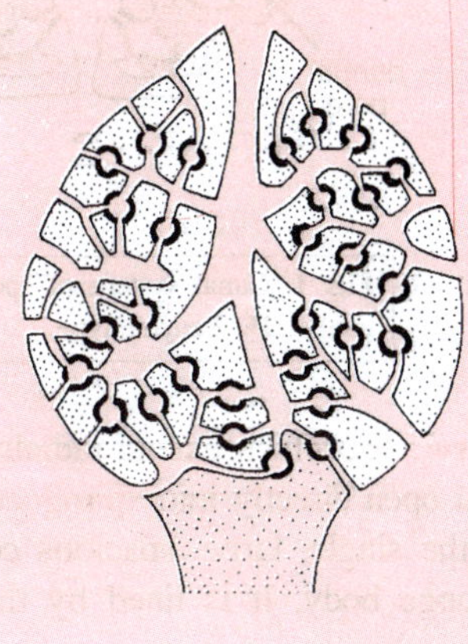

The arrangement and complexity of internal channels vary considerably in different sponges. Accordingly, the canal system has been divided into three types : **ascon**, **sycon** and **leucon**.

1. Ascon type. It is the simplest type of canal system which is found in asconoid sponges, like *Leucosolenia,* and in olynthus stage in the development of all syconoid sponges. Its body surface is pierced by a large number of minute openings called **incurrent pores** or **ostia**. These pores are intracellular spaces within tube-like cells, the

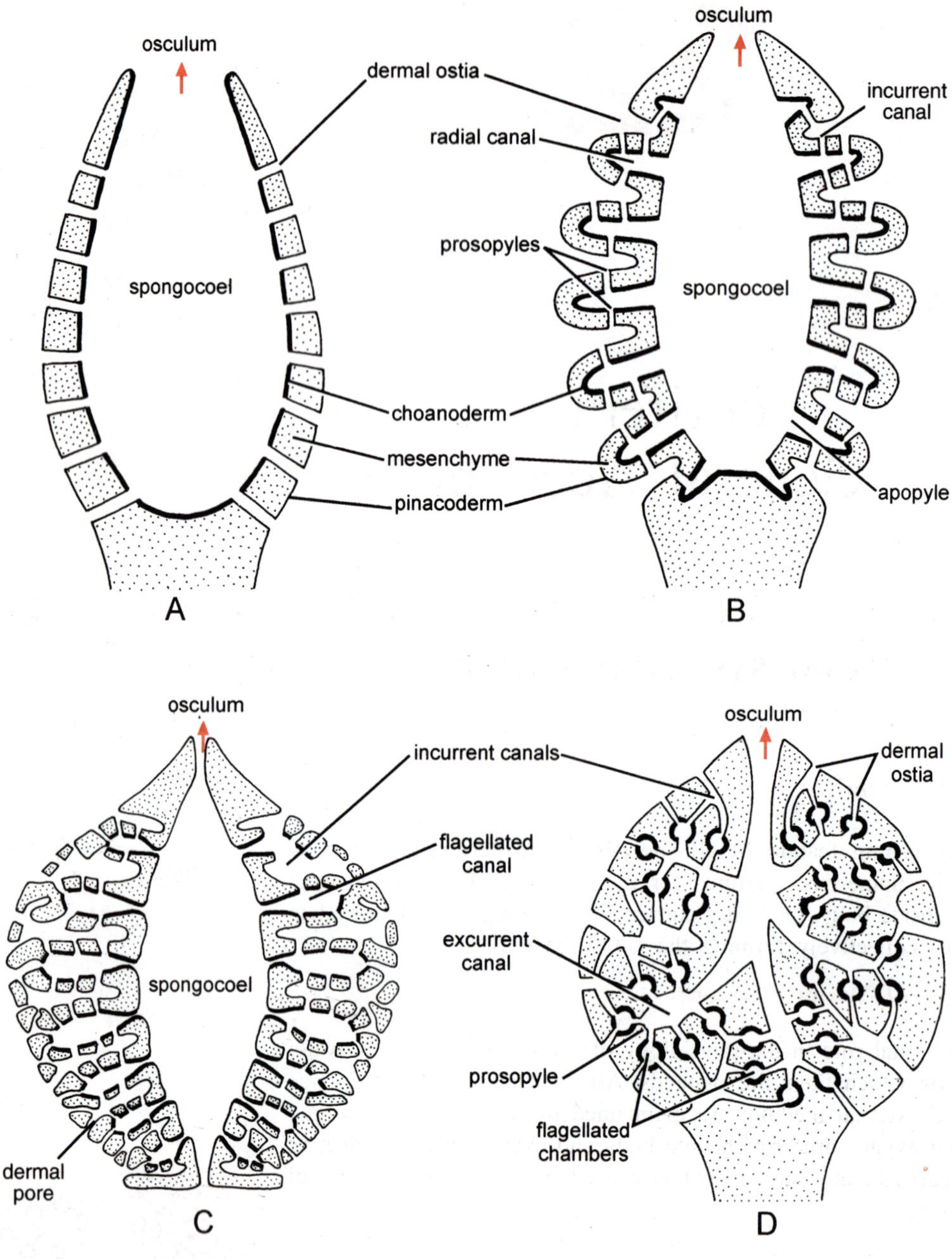

Fig. 1. Canal system of sponges. **A.** Ascon type. **B.** Simple sycon type. **C.** Complex syconoid type with cortex, **D.** Leucon type.

porocytes, which extend radially into mesenchyme, and open directly into **spongocoel**. The spongocoel is the single, large, spacious central cavity in the sponge body. It is lined by the flagellated collar cells or **choanocytes**. Spongocoel opens to outside through a narrow circular opening, the **osculum**, located at the distal free end, and often fringed with large monaxon spicules.

Surrounding sea water enters the canal system through ostia. Flow of water is maintained by the beating of flagella of collar cells. Rate of water flow is slow, because the large spongocoel contains much water which cannot be pushed out readily through a single osculum. Course taken by water current in the body of sponge may be shown as under :

Ingressing water → (through Ostia) → Spongocoel → (through Osculum) → To outside

2. Sycon type. Sycon type of canal system is a more complex system of pores and canals and is characteristic of syconoid sponges, like *Scypha* (= *Sycon*) and *Grantia*. It can be theoretically derived from the asconoid type by horizontal folding of its wall. Embryonic development of *Scypha* clearly shows the asconoid pattern converting into syconoid pattern. Body wall of syconoid sponges includes two types of canals, **incurrent** and **radial**, paralleling and alternating with each other. Both types of canals end blindly in body wall but are interconnected by minute pores. Incurrent pores or **dermal ostia**, found on the outer surface of body, open into the incurrent canals. These canals are non-flagellated, as they are lined by pinacocytes, and lead into adjacent radial canals through minute openings, called **prosopyles**. It is not clear whether prosopyles are channels through porocytes but it is definite that, in the adult, they are simple intercellular spaces. Radial canals are flagellated chambers, as only they are lined by choanocytes. These canals open into the central spongocoel by **internal ostia** or **apopyles**. Spongocoel is a narrow, non-flagellated cavity lined by pinacocytes. It opens to exterior through an excurrent pore, the **osculum**, similar to that of ascon type.

Course of water current may be represented as given below :

Ingressing water → (Dermal ostia) → Incurrent canals → (Prosopyles) → Radial canals → (Apopyles) → Spongocoel → (Osculum) → To outside

In more complex sycon type, as illustrated by *Grantia*, the incurrent canals are irregular, branching and anastomosing, forming large **sub-dermal spaces**. This is due to development of **cortex**, involving pinacoderm and mesenchyme, spreading over the entire outer surface of sponge.

3. Leucon type. As a result of further folding of body wall, the sycon type gives rise to a still more complex canal system, the **leucon type**. This is characteristic of leuconoid sponges, such as *Spongilla*. Here radial symmetry is lost and canal system has become very irregular. Flagellated chambers are small, spherical and lined by choanocytes. All other spaces are lined by pinacocytes. Incurrent canals open into flagellated chambers through prosopyles. Flagellated chambers, in their turn, communicate with **excurrent canals** through apopyles. Excurrent canals are developed as a result of shrinkage and division of **spongocoel** which has disappeared. Thus excurrent canals communicate with the outside through an **osculum**.

Course taken by water current is as follows :

Ingressing water → (Dermal ostia) → Incurrent canals → (Prosopyles) → Flagellated chambers → Apopyle → Excurrent canals → (Osculum) → To outside

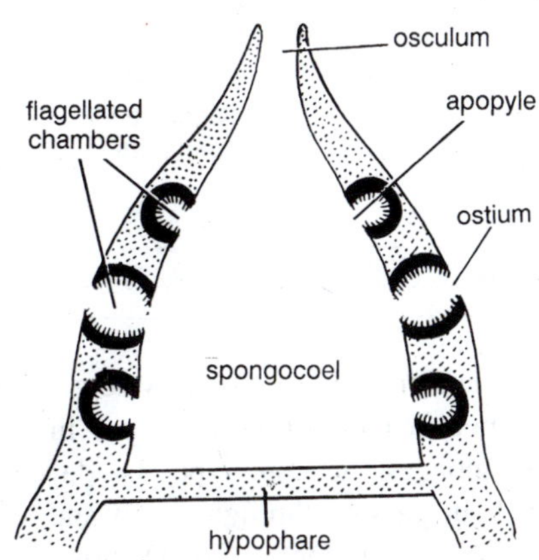

Fig. 2. *Rhagon* larva in V.S.

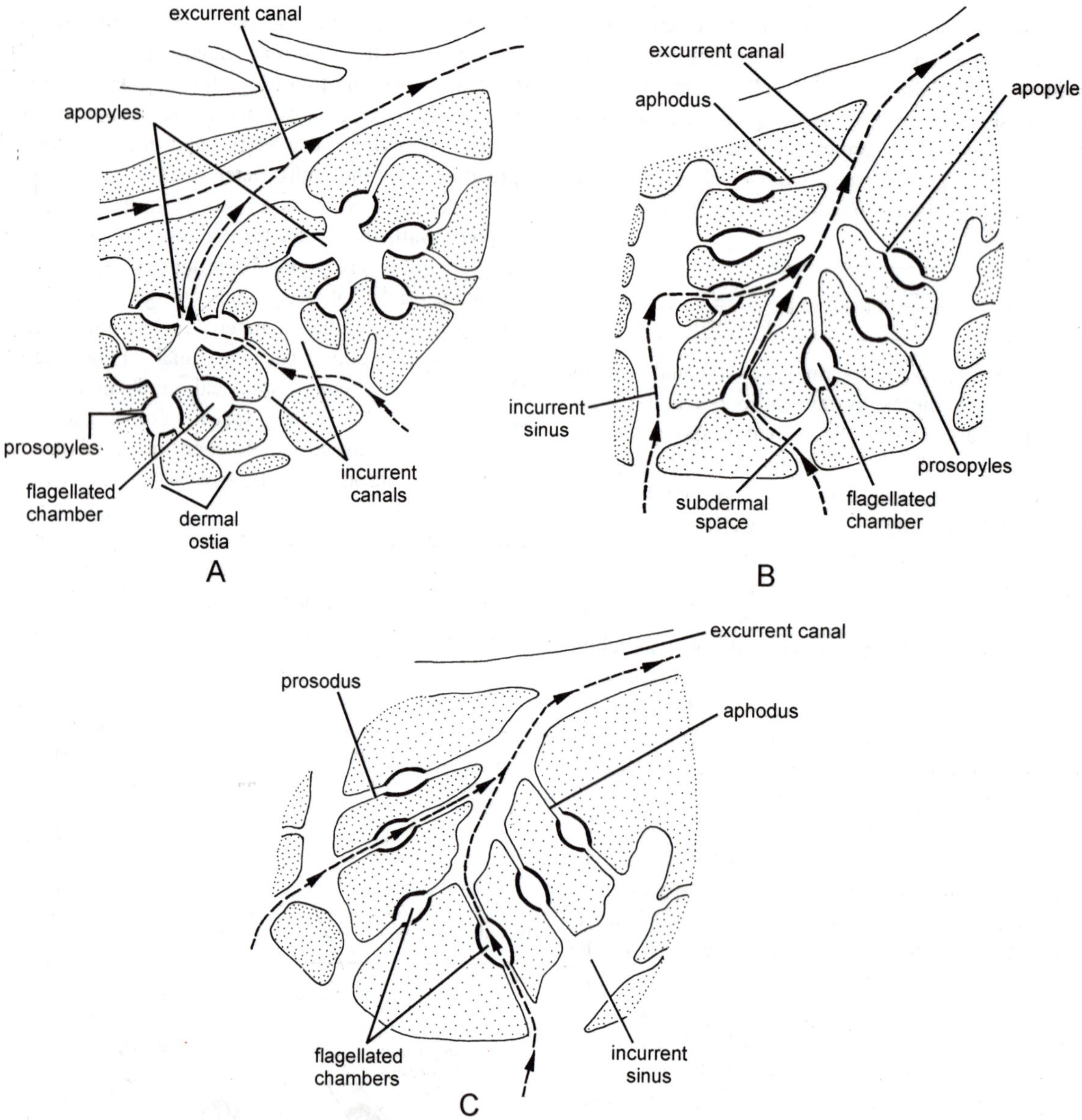

Fig. 3. Grades in leucon type of canal system. **A.** Eurypylous. **B.** Aphodal type. **C.** Diplodal type.

Though leucon type of canal system appears to be the modification of sycon type, in many calcareous sponges, leucon type is developed directly without passing through ascon and sycon types in their embryogeny. In Demospongiae, leuconoid condition is derived from a larval stage, called **rhagon**. Sponogocoel of rhagon is surrounded by flagellated chambers opening into it through very wide apopyles. A single osculum opens at the top of spongocoel. Canal system of rhagon larva does not occur in any adult sponge. In Demospongiae, lecuon type of canal system is also termed the **rhagon** type because of its derivation from rhagon stage (Fig. 2).

Leucon type of canal system presents three successive grades in its evolutionary pattern :

(a) Eurypylous type. It is the simplest and most primitive leucon type of canal system. In this type, the flagellated chambers communicate directly by broad apertures, the **apopyles**, with excurrent canals. Ex. *Plakina.*

(b) Aphodal type. In this type, the apopyle is drawn out as a narrow canal, called **aphodus**. This connects the flagellated chamber with excurrent canal. Ex. *Geodia.*

(c) Diplodal type. In some sponges, besides aphodus, another narrow tube, called **prosodus**, is present between incurrent canal and flagellated chamber. The pattern is called the diplodal type, Ex. *Spongilla, Oscarella.*

SKELETON IN SPONGES

Almost all sponges are provided with a skeleton, found embedded in mesenchyme. This may consist of separate **spicules**, or of interlacing **spongin fibres**, or of both. Skeleton supports and protects the soft body parts and also serves as the basis of classification of sponges.

Spicules

1. Structure and Types

Spicules are crystalline structures consisting of spines or rays that radiate from a point. These are secreted by special mesenchymal amoebocytes, called **scleroblasts**. All kinds of spicules have a core of organic material around which is deposited either calcium carbonate (calcite) or colloidal silica (silicon). Accordingly, the spicules are of two types : (i) calcareous, which are characteristic of the class Calcarea, and (ii) siliceous, characteristic of class Hexactinellida.

According to size, the larger spicules, constituting the main skeleton, are called **megascleres**; and the smaller spicules, which occur interstitially, are called **microscleres**. Further, spicules may occur in several forms; they may be simple rod-like or may take the form of forks, anchors, shovels, stars, plumes, etc. These forms depend upon the number of axes and rays present. Accordingly, they can be divided into the following types :

1. Monaxon. Monaxon spicules are formed by growth along one axis. These may be straight needle-like or rod-like or may be curved. Their

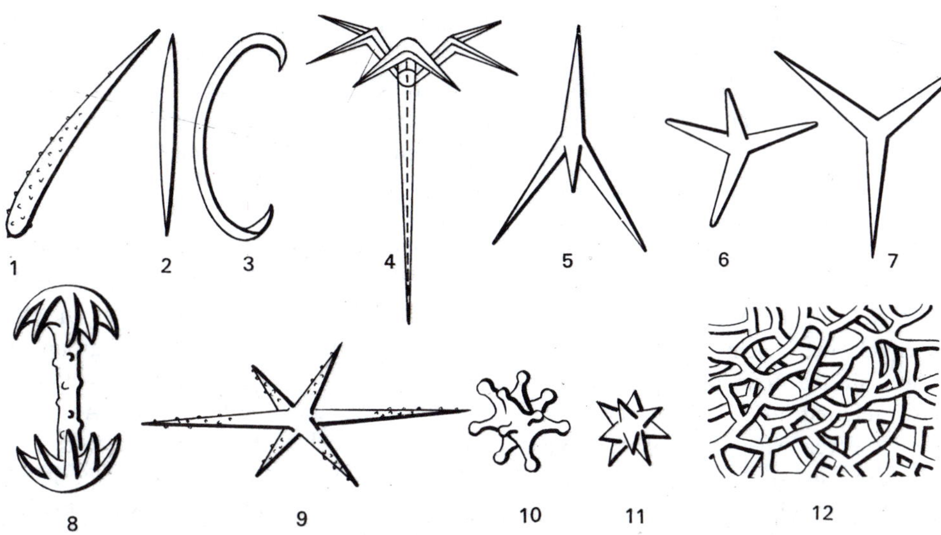

Fig. 4. Spicules and spongin 1–9. Megascleres. 10–11. Microscleres. 1. Monactinal monaxon. 2. Diactinal monaxon. 3. Curved monaxon. 4. Triaenes. 5–6. Tetraxon calthrops. 7. Triradiate. 8. Monaxon with ends hooked (amphidisc). 9. Hexactinal triaxon. 10–11. Polyaxons. 12. Spongin fibres.

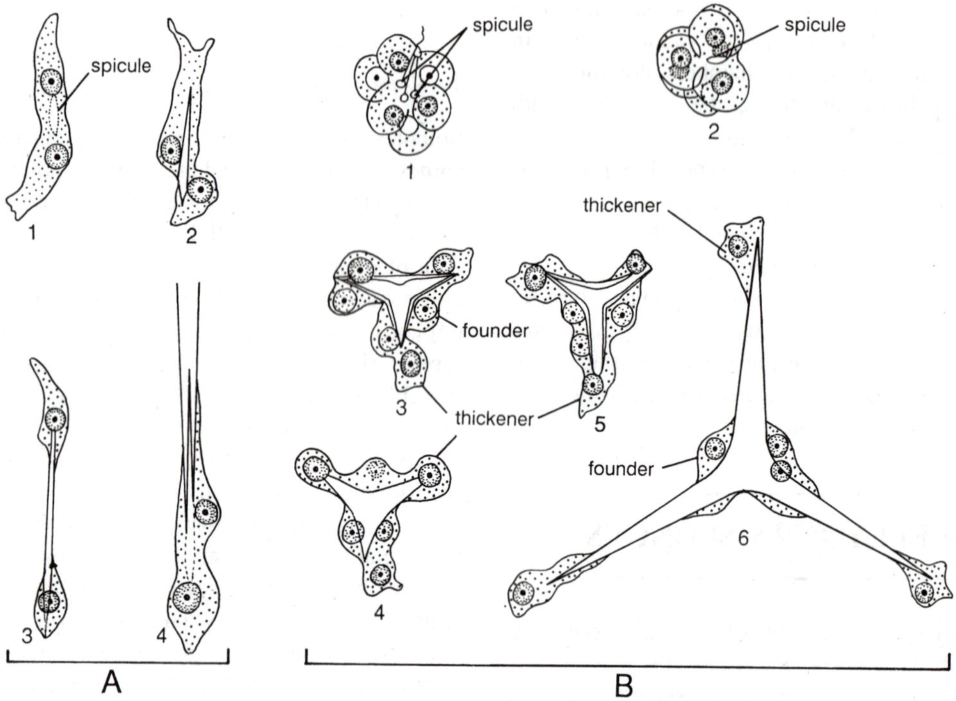

Fig. 5. Secretion of a A. monaxon and a B. triradiate spicule.

ends may be pointed, knobbed or hooked. They are of two kinds : (i) **monactinal**, in which growth takes place in one direction only, and (ii) **diactinal**, in which growth occurs in both directions. Monaxons are both calcareous and siliceous types.

2. Tetraxon. Tetraxon spicules are with typically four rays, each pointing in a different direction. Usually one of the four rays is elongated giving the appearance of a crown of 3 rays. Such spicules are termed **triaenes**. However, when all the rays are equal, the spicule is termed the **calthrops**. When all the four rays persist, the tetraxon is also referred to as **tetraradiate** or **quadriradiate**. However when one of the four rays (usually the elongated one) is lost, the spicule becomes **triradiate**, which is characteristic of calcareous sponges. If the elongated ray bears a disc at both ends, it is called an **amphidisc**.

3. Triaxon. Triaxon spicule has three axes that cross one another at right angles to produce six rays. It is thus **hexactinal**. Triaxons are characteristic of glass sponges (class Hexactinellida) only.

4. Polyaxon. These are spicules with several equal rays, radiating from a central point. They may be grouped to give bur or star-like appearance. Polyaxons are common among microscleres.

2. Development of Spicules

Calcareous spicules are secreted by special cells, called **sclerocytes**, derived from the mesenchymal **scleroblasts**. A monaxon spicule, or each ray of triradiate spicule, is secreted by a group of two sclerocytes, one acting as a **thickener cell** and the other as the **founder cell**. A binucleate scleroblast is believed to give rise to these cells. Formation of spicule begins as a deposition of a particle of calcium carbonate between two nuclei. The particle grows drawing apart first the two nuclei and then the two sclerocytes. Thickener cell lays down additional layers of $CaCO_3$ adding to the thickness of spicule. When spicule is fully formed, both the cells wander into mesenchyme.

Scleroblast secreting a calcareous spicule is called **calcoblast**, while that producing a siliceous spicule is called **silicoblast**. A siliceous monaxon is believed to be secreted by a single silicoblast, while the six-rayed triaxon is secreted within a multinucleate mass formed by repeated nuclear division of a single silicoblast.

Spongin

1. Structure

Spongin is an organic, horny, elastic substance, resembling with silk in chemical composition. Like nails, hair and feathers, "it is a scleroprotein containing sulphur and chemically related to collagen, a horny protein. It is insoluble, chemically inert and resistant to protein digesting enzymes. Spongin contains a large amount of iodine, reaching 8 to 14% in certain tropical species of the Spongiidae and Aplysinidae. It is interesting to know that old herb-doctors for centuries used bath-sponge as a cure for croup, a throat condition of children, resulting from inflammation and partial obstruction of larynx.

Spongin occurs in various forms in class **Demospongiae**. It may occur as a cement connecting together siliceous spicules (many **Monaxonida**). More usually it is found in the form of branching or anastomosing fibres, in which siliceous spicules are embedded (**Monaxonida**). In *Keratosa,* spicules are completely absent and spongin alone is formed. Spongin fibres are fine threads consisting of a soft granular axial core or **medulla**, surrounded externally by concentric layers of spongin.

2. Development

Spongin fibres are secreted by flask-shaped mesenchyme cells, called **spongioblasts**, which are seen coating the fibres. Spongioblasts become arranged in rows and the spongin rod secreted by each fuses with those of neighbouring cells to form a long fibre (Fig. 6). Spongioblasts later become vacuolated and degenerate after having secreted a certain quantity of spongin.

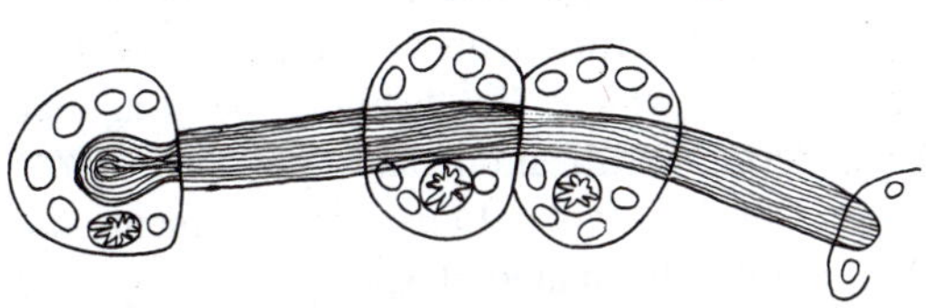

Fig. 6. Spongioblasts in series secreting a spongin fibre.

REPRODUCTION IN SPONGES

Sponges reproduce both by asexual as well as sexual methods.

Asexual Reproduction

Asexual reproduction in sponges is by **budding, fission** or asexually formed embryos like **reduction bodies** and **gemmules**.

1. By budding. By this method, number of individuals is increased in sponge colonies or new colonies are developed. In budding, numerous archaeocytes collect at the surface and pinacoderm bulges out to receive them. The bud, so formed, grows in an adult individual. It either remains attached with the parent individual or gets detached and attached nearby to develop into an independent colony.

2. By fission. In some sponges, multiplication takes place by fission, throwing off parts of the body. The sponge is hypertrophied over a limited area, developing a line of weakness. Along this weak line, splitting occurs and a part is separated from the main parental body which develops into a new individual. By budding this new individual takes the form of a colony.

3. Formation of reduction bodies. Another very unusual method of asexual reproduction the formation of **reduction bodies**. Most freshwater and marine sponges disintegrate in adverse circumstances. Ailing sponge will usually collapse leaving small rounded balls, called **reduction bodies**. Each body consists of an internal mass of amoebocytes, covered externally by a pinacoderm. When favourable conditions return, these reduction bodies grow into complete new sponges.

4. Formation of gemmules. All freshwater and some marine sponges (*Ficulina, Tethya, Suberites,* etc.) have a regular and peculiar mode

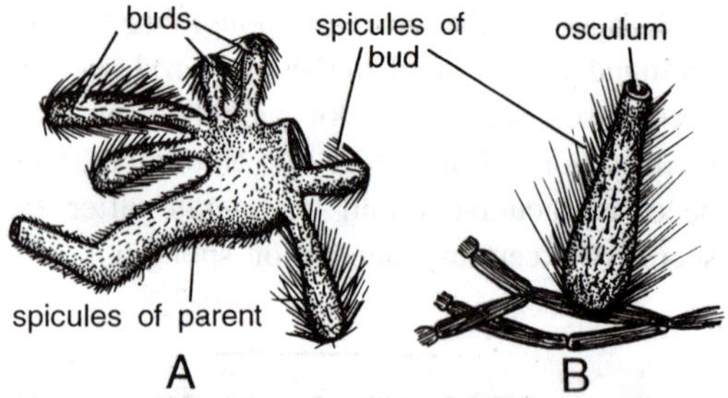

Fig. 7. Budding in *Leucosolenia.* **A.** A piece of sponge bearing buds. **B.** A liberated bud fixed by its free end.

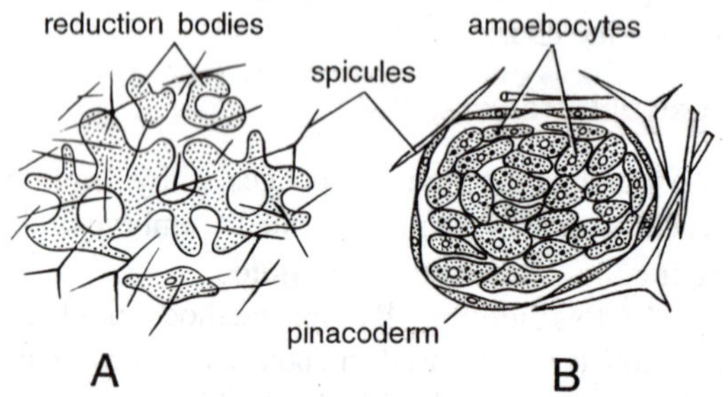

Fig. 8. Formation of reduction bodies. **A.** Reduction bodies of a calcareous sponge. **B.** Reduction body in section.

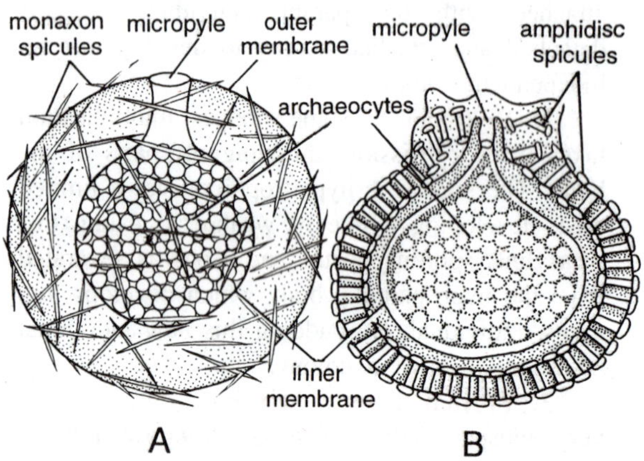

Fig. 9. A. Complete gemmule of *Spongilla.* **B.** Gemmule of *Ephydatia* in section.

of asexual reproduction by internal buds, called **gemmules** (Gr., **gemma**, bud). These eventually detach and develop into new individuals. Gemmules enable the sponges to tide over unfavourable conditions, such as excessive cold or draught, because they can withstand freezing and considerably greater degrees of desiccation than do adult sponges.

A full-grown gemmule is a small hard ball, consisting of an internal mass of food-laden archaeocytes, surrounded by resistant chitinous covering, envelope or capsule, made of two membranes (outer and inner) and usually pierced at one end by an outlet, the **micropyle**. Protective envelope may be strengthened with siliceous amphidisc spicules (*Ephydatia*) or by monaxon spicules (*Spongilla*).

In autumn, freshwater sponges suffering from cold and food scarcity, die and disintegrate, leaving behind a large number of gemmules which remain inert throughout winter. As the body of parent sponge decays, the gemmules are set free and these either sink to the bottom or simply float away. In April or May (spring), when conditions again become suitable, the gemmules begin to hatch. Their living contents (**histoblasts**) escape through micropyles and develop into new sponges. These produce ova and spermatozoa and give rise to a summer generation of sponges, which, in turn, die away in autumn, leaving behind the gemmules. The life history of such sponges, thus, illustrates an alternation of generations.

Sexual Reproduction

Most sponges are **monoecious** (hermaphrodite) but **dioecious** (unisexual) sponge species are also known. Sperms and ova are derived from archaeocytes or choanocytes which undergo gametogenesis. Although most sponges are hermaphrodite but cross-fertilization is the rule, because eggs and sperms are produced at different times.

1. Fertilization. Sperms are released to outside through outflowing water and make their way into another sponge along with the ingressing water current. Choanocytes, acting as nurse cells, transport the sperms, with tails left behind, to the mature ova which lie behind the flagellated choanoderm. Fertilization takes place *in situ.*

2. Development. Early development takes place within the maternal sponge leading to the formation of a larval stage. Larvae of two types have been described in sponges.

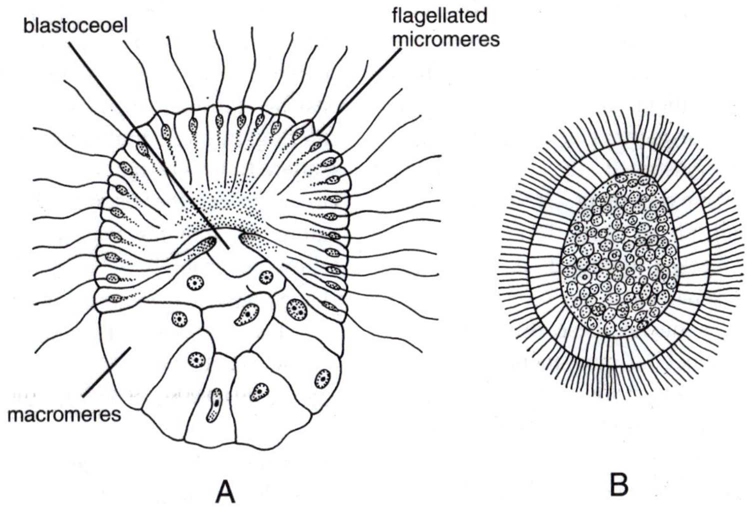

Fig. 10. Larval forms of sponges. A. Amphiblastula. B. Parenchymula.

(a) **Amphiblastula.** It is a hollow, oval larval stage, characteristic of many calcareous sponges (*e.g.*, *Scypha*). Anterior one-half of amphiblastula bears flagella, whereas the posterior half does not.

(b) **Parenchymula.** It is a solid, oval or flattened larval stage. It is characteristic of some Calcarea, Hexactinellida and most Demospongiae. Its entire outer surface bears flagella.

With the help of external flagella, the motile larvae escape from the parental body and swim for a few hours to many days. Finally, they settle down, become attached to some solid object, metamorphose and grow into an adult.

REGENERATION IN SPONGES

All animals, particularly the less specialized ones, can replace their lost or injured parts. The process is known as **regeneration** It is greater in simple animals and in simple tissues. Thus, epithelial tissues regenerate readily, whereas highly differentiated tissues, such as muscle or nerve tissue, have very limited powers of regeneration. The sponges, correlated with their low grade of organization, have a high capacity to regenerate.

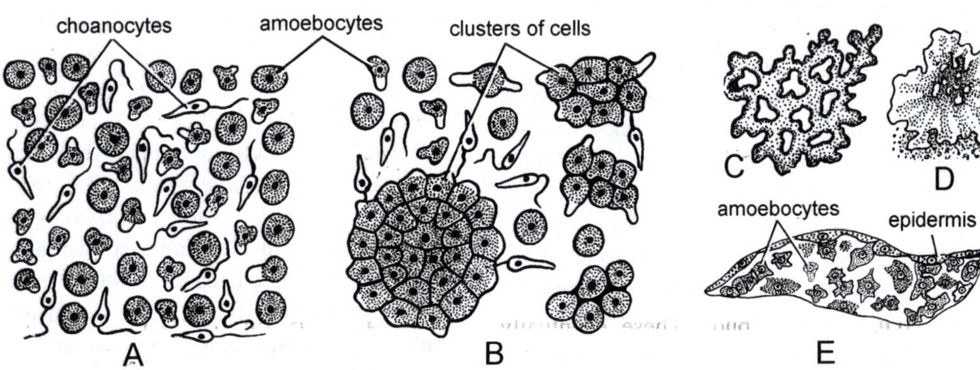

Fig. 11. Wilson's experiment on regeneration in sponges. **A.** cells of *Microciona* separated by squeezing living sponge through bolting cloth. **B.** Cells aggregating into small masses. **C.** A reticulate reunion mass. **D.** Later stage forming a young sponge or spongelet. **E.** Section through a stage like D.

If an individual is cut into pieces, each piece, however small but carrying amoebocytes and choanocytes, will grow into a complete or full-size sponge, if conditions are favourable. The regenerative power of certain sponges is quite noteworthy as demonstrated by **H.V. Wilson** (1907) by a simple experiment. If a sponge is chopped into small pieces, run through a meat grinder and then squeezed through fine silk bloting cloth, its cells are separated from one another. In a suitable medium, some of these disunited cells come together by amoeboid movements and unite to form small aggregates or **spongelets**. In course of several days these reunited masses acquire canals, flagellated chambers and skeleton, thus growing up into new sponges. Cells from different species of sponges may adhere temporarily but later separate, without re-forming a sponge.

According to the experiments conducted by **Bergquist** (1978), if a tissue is grafted in a sponge from another sponge of the same species, the host and graft will grow together. If the graft is from a different species, then the host will reject the graft (Fig. 11).

According to Humphrey, calcium and magnesium ions are necessary for regeneration. Some unknown aggregation factors from cell surface are also supposed to be necessary for the process of regeneration.

ANIMAL NATURE OF SPONGES

Sponges have managed to conceal their true animal nature for several centuries. They are sessile, variously branched, have no apparent way of capturing food or eliminating wastes, and show little response to stimuli. Some forms are even green due to the presence of unicellular symbiotic algae. Thus, earlier workers, like Aristotle, Pliny, Gerade, Grew, etc., were misled to consider sponges as plants and some like Peyssonnel even looked upon them as houses of non-living material secreted by worms and other marine animals that seek shelter in them.

Robert Grant (1857) was the first to recognize and prove the true animal nature of sponges. He demonstrated that water current passing through them takes up a fixed route, entering through numerous fine pores, the **ostia**, scattered all over the surface, and leaving through larger holes; the **oscula**. By 1857, the animal nature of sponges was well established on the following grounds :

(1) Sponges feed on inwafted solid particles. Their mode of nutrition, thus, is truly holozoic.

(2) Sponge cells are devoid of cellulose cell walls.

(3) Life cycle of sponges includes free-swimming ciliated larval stages resembling those of other marine animals. Such larval stages do not occur in plants.

INDIVIDUALITY OF SPONGES

Whether sponges are individuals or colonies, has been much debated and divergent views have been expressed regarding what makes the individual in a sponge. Majority of older writers regarded cell as the unit of individuality in a sponge. Discovery of collar cells by **James-Clark** (1867) led him and others to regard sponge as a colonial choanoflagellate. **Carter** and **Haeckel** considered the flagellated chamber, lined with choanocytes to be the individual. Individuality in sponges is hinted by numerous species, each with a specific construction and specific types of spicules arranged in a particular way. No doubt, absence of a nervous system and slight conductivity render this individuality to the lowest possible grade. The most natural concept, based on morphological and physiological considerations, is that an osculum, with its associated canal system, represents a sponge individual. Based on this concept, sponge with a single osculum is a solitary individual, while that with several oscula may represent a colony by vaguely indicated individuals.

ORIGIN OF SPONGES

Origin of sponges is a subject of much controversy. According to one view, they probably represent the non-progressive descendant of some primitive, sluggish and gastrula-like ancestor. According to another much advocated view, based on the presence of choanocytes, sponges evolved from choanoflagellate Protozoa. Among choanoflagellates, the genus *Proterospongia* closely resembles a sponge. It shows considerable morphological distinctiveness. Some of these remain typical choanocytes while some, which remain embedded in a jelly-like secretion, become amoeboid, and also give rise to reproductive cells. *Proterospongia* is considered to represent the probable first stage of evolution of choanoflagellate Protozoa in the direction of Porifera. But choanocytes are external in *Proterospongia,* while they are internal in Porifera. This objection is met with by arguing that choanocytes become internal in Porifera only at a later stage of life history. Migration of choanocytes in the interior of body, in the absence of a mouth and a digestive tract, resulted in the evolution of canal system, which is a unique anatomical feature of sponges. Evolutionary changes in sponges concern mainly with further elaboration of their canal system and development of skeleton.

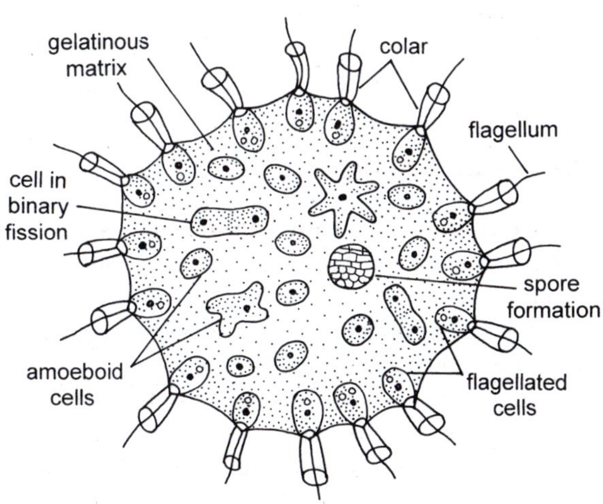

Fig. 12. *Proterospongia.* A choanoflagellate protozoan.

Adoption of sedentary life and retention of protozoan habits by their cells have probably checked the evolution of sponges beyond the stage of colonial Protozoa.

According to another view, like Metazoa, sponges, might have evolved from an ordinary flagellated protozoan (other than a choanoflagellate), through a process of colony formation. This view is founded on the absence of choanocytes and presence of ordinary and non-collared flagellated cells in sponge larvae. But workers, who consider **Protozoa** as acellular or non-cellular (instead of unicellular), strongly oppose this view.

It appears quite resonable to suppose that sponges have either evolved from a different group of flagellates than did the rest of **Metazoa**, or they have separated from the general metazoan stalk at a very early period. In either case, divergence took place very early, when metazoan organization was still in a very primitive stage. Thus, embryology of sponges cannot be expected to resemble that of other Metazoa.

Although origin of sponges is uncertain, but all scientists agree to the view that sponges have not given rise to any other members of animal kingdom. Several facts, such as absence of mouth and digestive cavity and water vascular system, indicate that sponges are a dead-end phylum.

AFFINITIES AND SYSTEMATIC POSITION OF SPONGES

Their wide distribution, large variety and vast numbers (approximately 5,000 species), have established sponges as a successful group of animals. They show affinities with Metazoa as well as Protozoa.

Metazoan Affinities

A sponge develops by the cleavage of a fertilized egg and its body is composed of numerous cells. It is, therefore, definitely a multicellular animal, though the simplest, illustrating the beginning of tissue grade of organization.

1. Resemblances with Metazoa. Among Metazoa, coelenterates show some superficial resemblances with Porifera as follows :

(1) Both are sedentary.

(2) Both are diploblastic and acoelomate. Instead of a cellular mesoderm, a gelatinous non-cellular mesogloea is present between two body layers.

(3) Spongocoel of sponges, leading to exterior through an osculum, can be compared with gastrovascular cavity of coelenterates, that opens to outside through mouth.

(4) Both reproduce asexually and form colonies by budding.

(5) Their gastrulae retain radial symmetry, and parenchymula larva of sponges resembles the planula of coelenterates.

(6) Biochemical studies have shown the presence of almost same nucleic acids and amino acids as in Metazoa. Both possess phosphoaginine (an invertebrate phosphagen) and phosphocreatine (a chordate phosphagen).

2. Differences with Metazoa. Sponges differ from coelenterates, as well as other Metazoa, in several important respects :

(1) In sponges, cells are far less specialized and less interdependent than in Metazoa.

(2) Tissue-formation in sponges is restricted to the production of epithelia on free surfaces. There are no organs as found in most higher animals.

(3) Sponges are said to be diploblastic, but the term is not strictly applicable because of difference in the manner of development of both body layers.

(4) Middle layer of sponges is not differentiated as in triploblastic animals. Moreover, epidermis (pinacoderm) and mesenchyme arise from the same embryonic cells, so that mesenchyme may be considered theoretically to be an ectomesoderm.

(5) Sponges do not have an anterior end or head, although each osculum physiologically serves as controlling region.

(6) Principal opening is not exhalent in Metazoa, like osculum of sponge.

(7) Sponges lack a true mouth. Osculum does not correspond with the mouth of coelenterates. Blastoporal end is aboral in Porifera while oral in other radiate phyla.

(8) Body surface, perforated by inhalent apertures (ostia) and body permeated with water channels, are some of the unique features of sponges.

(9) Stinging cells of Coelenterata are lacking.

(10) Sponges lack a proper digestive cavity, and digestion is intracellular, as in Protozoa.

(11) Sponges have a greater number of fats and fatty acids which are of a higher molecular weight than those found in Metazoa.

(12) Choanocytes, lining flagellated chambers, are peculiar to sponges.

(13) There is no nervous system to coordinate the various activities of sponges.

(14) Development of sponges is unique in many ways. During gastrulation, animal half invaginates and not the vegetal half as in other Metazoa. Anterior end of larva becomes the attached end of adult, as in other radiate phyla. But, in sponges, it also bears the blastopore, while in Radiata the blastoporal end is oral. A reversal of germ layers occurs at the time of metamorphosis of larva. Surface epithelium transforms into choanocyte layer, which lines the inner cavities, and serves for digestion and for circulating water through body. While inner cell mass comes to lie on surface of body and forms epidermis and mesenchyme. Thus, body layers of sponges do not correspond exactly to those in other Metazoa, in which ectoderm, mesoderm and endoderm retain from the beginning the same relative positions in body.

Protozoan Affinities

1. Resemblances with Protozoa. Sponges resemble colonial Protozoa in the following characters :

(1) Lack of organs and a digestive cavity. Digestion is intracellular.

(2) Simplicity of tissues and relative independence of cells.

(3) Production of skeleton concerning single cell or group of cells.

(4) Possession of collared and amoeboid cells, similar to those of colonial Choanoflagellata.

(5) Amphiblastula larva of syconoid Calcarea undergoes an inversion process similar to that occurring in the development of *Volvox*. The flagellated sponge larva is considered to be reminiscent of some flagellated colonial protozoan ancestor.

2. Differences with Protozoa. Sponges differ from Protozoa in possessing a canal system and a characterisitc skeleton, in developing from a fertilized egg by cleavage, and relatively more advanced condition of specialization of somatic cells. In cellular differentiation or division of labour among somatic cells, sponges become more complex than a colony of protozoans and resemble multicellular animals or Metazoa. Development of specialized cells such as pinacocytes, choanocytes and porocytes increases efficiency of these animals by division of labour. These specialized cells are, therefore, the forerunners of true tissues in higher animals.

Systematic Position

Porifera certainly have not given rise to any of the other metazoan phyla. They are thus regarded as on a by-road of evolution instead of being on the direct line of evolution. They constitute a blind branch which diverged early from the main evolutionary series near the foot of family tree and which gave rise to nothing but sponges. In spite of certain superficial resemblances, sponges bear no real similarity to, and probably no genetic affinity with Metazoa. To emphasize this factor, animal kingdom is often divided into three subkingdoms, and sponges are placed in the subkingdom **Parazoa** (Gr., **para**, beside + **zoios**, animal), of the same rank or at par with the other two subkingdoms, **Protozoa** and **Eumetazoa**. Diagnostic characters of Parazoa, as revealed by sponges, are as follows :

(1) Body a loose aggregation of cells without definite tissue layers and organs.
(2) No definite form or symmetry.
(3) No true gut. Only internal space in the form of cavities, pores and canals.
(4) Presence of flagellated collar cells or choanocytes.
(5) Water current flowing into the canal system subserves the functions of nutrition, respiration, excretion and reproduction.
(6) Skeletal elements isolated spicules.
(7) Power of absolute regeneration.
(8) Larval stage is a parenchymula or amphiblastula.

ECONOMIC IMPORTANCE

Beneficial Sponges

Since times immemorial, sponges have been put to a variety of uses, not only by man but also by other creatures dwelling in water.

1. As food. Sponges are generally unpalatable due to their unpleasant taste and odour and prickly nature of their spicules. However, some crustaceans parasitize them while nudibranchs molluscs depend upon them for their regular diet.

2. As commensals. Being unpalatable, sponges serve as secure houses for several crustaceans worms, molluscs, small fishes, etc., which seek protection in them against predaceous fishes and other enemies. In addition, these animals get a rich food supply from water circulating into sponge. Some animals even spend their entire life in these 'charitable hotels', providing free protection and food.

A large loggerhead sponge, the size of a wash-tub, may serve as a living hotel for as many as 17, 128 animals of 19 different species mostly of one kind of shrimp, as counted by **A.S. Pearse** in one case. Since these small creatures do not cause apparent harm to the sponge, they are considered **commensals**.

Sponges themselves may be commensals. Certain species of crabs deliberately pick up and attach bits of sponges to their back and legs. As the sponges grow, they provide a protective covering to crabs which become well camouflaged and eventually concealed completely. The sponge **Suberites**, frequently grows on snail shells occupied by hermit-crabs. After some time, sponge absorbs the shell so that crab comes to inhabit a cavity in sponge itself. Sponge suffers no harm but may even derive benefit by being transported from one place to another.

3. Other uses. Ancient Greeks used dried, fibrous skeletons of sponges for the purpose of bathing, washing, scrubbing floor and padding shields and armour. Romans used them for painting and mopping. Some even used them to absorb fluids to be drunk later on. Today, synthetic sponges seem to be gradually replacing the natural ones; but the latter, due to their durability, elasticity and absorbing capacity are still preferred for washing and polishing. Sponges of inferior quality are used for stuffing the sound absorbing walls, garments, saddles, cushions, furnitures, etc., and also as a packing material. In glass manufacture, sponges are used in wiping and cleaning hot glass.

Some sponges, such as Venus's flower baskets, are of ornamental interest, because they build up a beautiful latticed skeleton of 'spun-glass' in the form of a cylinder. Some species are of economic importance as their siliceous spicules form large flint deposits.

Harmful Sponges

Only a few sponges are really harmful. They may kill other sessile animals by growing over them and cutting off their food and oxygen supply. Boring sponges, such as *Cliona,* attach themselves to shells of oysters, clams, barnacles, etc., boring them so full of holes that their contained animals are destroyed, and in time, the shells are entirely broken up. Lime of these shells gets dissolved into sea water to be used over again by other animals. Sometimes, these boring sponges may prove great nuisance to oyster beds. In Europe, oysters are grown in large wire frames, which are raised at times and sprayed with fresh water in order to kill sponges.

Some sponges produce poisonous secretions or give off strong unpleasant odours, or sting with sharp spicules. Certain species may prove as dangerous as the poisonous ivy, when touched by man.

SPONGE INDUSTRY

Sponge Fishing

Although there are now many synthetic substitutes, yet natural sponges have an extensive market so that sponge-fishing is still quite a fascinating and important industry. Millions of sponges are sold every year, making sponge-fishing a big industry in many places on seacoasts. Valuable commercial sponges occur on rocky bottoms in warm and shallow waters of Mediterranean, Gulf of Mexico off Florida coast and waters around West Indies and Australia. Tarpon Springs in Florida is the best commercial sponge-fishing centre of United States, where more than a million dollars worth of sponges have been collected in a single year.

Sponges are collected in several ways, by drag hooks, by trawlers or by divers. In deeper waters, they are brought up by dredging or trawling, or more frequently, by divers, either with or without a diving suit, the latter being known as skin divers. Diving in deeper regions is dangerous because the divers often suffer from a pressure disease called **bends**. Sharks are another danger to divers. In shallow waters, sponges may be gathered by long-handled hooks. It is carried on usually by two men. While one man rows, the other watches the bottom of ocean through a glass-bottomed bucket. Any marketable sponge is pulled up with a two-pronged hook, fastened to a long pole.

A living sponge is a black, smooth and slimy mass with a leathery covering. Internally it looks like a piece of raw beef-liver in both consistency and colour. After a preliminary cleaning to remove rough dirt, sponges are hung from the boat's rigging and their living protoplasm allowed to decay. Then they are thoroughly washed, beaten or pounded with a wooden mallet to break up any solid material, such as shells of various invertebrates, and finally dried in sun. Cleaned sponges are cut or trimmed with shears. These are then bleached, graded, sorted and sold by auction. On account of their high elasticity, sponges can be easily pressed into compact bales for economy in shipping.

From commercial point of view 'bath sponges' are important. They belong to the family Spongiidae and the genera *Spongia* (= *Euspongia)* and *Hippospongia.* Their skeletons of spongin fibres are used for removing dirt from body surface while bathing and for other uses as mentioned earlier. Bath sponges reach marketable size in about 7 years. Sponges up to 20 or more years old have been reported to be caught for marketing purpose.

Sponge Cultivation

Sponges have great power of regeneration. Therefore, sponge culture is carried on in some areas to re-stock grounds depleted by excessive sponge-fishing. Small pieces of a living sponge are planted on cement blocks and submerged in favourable waters having abundant food and oxygen. This is a slow process, of course, but in several years these grow to marketable size. Sponge culture is little practised today, partly due to difficulties in preventing theft of crop and partly due to manufacture of many synthetic substitutes.

Important Questions

➤➤*Long answer type questions*

1. Give an account of the canal system in sponges.
2. Describe, in brief, the various methods of reproduction in sponges.
3. Give an account of the origin and systematic position of Porifera.
4. Write an essay on 'Sponge Industry'.
5. Distinguish between (i) Monaxon and triaxon spicules. (ii) Osculum and ostium. (iii) Radial and incurrent canals.
6. Write notes on : (i) Affinities of Porifera, (ii) Animal nature of sponges, (iii) Gemmules, (iv) Individuality in sponges, (v) Reduction bodies, (vi) Regeneration in sponges, (vii) Spicules.

➤➤*Short answer type questions*

1. Give the name of the cell type in sponge which is very important physiologically for metabolism.
2. What is the function of choanocytes in porifera ?
3. Name the cell that are filled with reserve food in sponges.
4. Which are the special flagellated cells of sponges that are not met within any other phylum ?
5. Mention the two types of larval forms of sponges.
6. Give the distinguishing character and one example of the group keratosa.
7. Name the distinguishing features of *Euplectella.*
8. Describe the skeleton of sponges and its significance.
9. Describe the parenchymula larva.
10. Write five points in which the leucon grade of canal system differs from the ascon grade.
11. Mention five different types of spicules that occur in sponges. Write two points : on each one of them.
12. Mention five different types of cells that occur in sponges body and give one-feature about each one of them.
13. Give the different types of spicules met within sponges and give at least one example for each group of the subkingdom.
14. How do the fresh water sponges escape from desication in summer ? Does the drying up of the pond effect their propagation.
15. How do the sponges obtain food and water ?
16. What is an Amphiblastula larva and a gemmule ? What functional purpose do they serve in the life of a sponge ? What part do they play in dispersal and in desication.
17. Can Proterospongia be the ancestor of sponge ? Give three reasons for your statement.
18. How does a monaxon spicule develop in sponges ?
19. Describe the leucon type of sponge.
20. Define parazoology.

➤➤Multiple choice questions

1. Many sponges may be bigger than the tiny fish yet the sponges have no .special respiratory organs. This is so because :
   (a) the sponges have a highly porous body through which the water flows bathing each cell
   (b) the sponges lead a stationary life and thus their oxygen requirement is too little
   (c) the respiratory organs would not be functional under water
   (d) sponges respire anaerobically
2. The strong circular sphincter around the osculum of sponge is formed by :
   (a) porocytes
   (b) trophocytes
   (c) spongin fibres
   (d) myocytes
3. The mouthlets or ostia in sponge, are the :
   (a) intracellular pores
   (b) pores between the cells of pinacoderm
   (c) pores surrounded by several cells
   (d) pores surrounded by several spicules
4. The 'totipotent' cells of sponges are :
   (a) thesocytes (b) trophocytes
   (c) archaeocytes (d) choanocytes
5. In sponge, gemmules are produced as a result of :
   (a) external budding (b) internal budding
   (c) regeneration (d) sexual reproduction
6. Simplest type of canal system is :
   (a) sycon type (b) ascon type
   (c) diplodal type (d) euryphylous type
7. Which of the following cells in sponges catch food ?
   (a) pinacocytes (b) choanocytes
   (c) porocytes (d) archaeocytes
8. Various canal systems of sponges have :
   (a) excurrent canal (b) paragastric cavity
   (c) flagellated chamber (d) none of these
9. Skeleton of sponge is made of spicules that can never be :
   (a) triaxon (b) monaxon
   (c) diaxon (d) amphidisc
10. Sponges transport their food by :
    (a) pinacocytes (b) trophocytes
    (c) choanocytes
    (d) porocytes
11. In the leucon type of canal system when the flagellated chambers open directly by wide apopyles into the excurrent canal, it is called :
    (a) euryphylous (b) apodal
    (c) diplodal (d) rhagon
12. Monaxon spicule with a knob at one end is called :
    (a) acanthostyle (b) monactyle
    (c) telostyle (d) style
13. In sponges skeletal secreting scleroblasts are modified :
    (a) choanocytes (b) pinacocytes
    (c) amoebocytes (d) porocytes
    (e) myocytes
14. Which cells in sponges have food reserves ?
    (a) archaeocytes (b) myocytes
    (c) thesocytes (d) choanocytes
15. In sponges first food is ingested by choanocytes and digestion is completed in :
    (a) amoebocytes (b) chromocytes
    (c) thesocytes (d) archaeocytes
16. Calcareous spicules are formed by :
    (a) calcoblasts (b) spinoblasts
    (c) silicoblasts (d) spongiblasts
17. The middle layer of the body wall of porifera is an :
    (a) mesenchyme (b) mesoderm
    (c) mesogloea (d) mesentery
18. One of the following cells takes part in reproduction of sponges :
    (a) archeocytes (b) thesocytes
    (c) myocytes (d) choanocytes
19. Which one of the following cells maintain a current of water in sponge ?
    (a) porocyte (b) myocyte
    (c) chromocyte (d) choanocyte
20. Skeleton in Demospongia :
    (a) silicious (b) calcareous
    (c) sponging (d) fibers
21. 'Parazoology' is the study of :
    (a) Protozoa (b) Porifera
    (c) Cnidaria (d) parasite
22. Who describe the animal nature of Porifera :
    (a) Ellias (b) Ross
    (c) Robert Grant (d) none
23. Amphiblastula is the larva of :
    (a) *Leucosolenia* (b) *Clathrina*
    (c) *Sycon* (d) none
24. Rhagon type canal system present in :
    (a) Echinodermeta (b) all sponge
    (c) adult sponge
    (d) larval stage of some sponges
25. Skeleton of sponge is secreted by :
    (a) chromocyte (b) sclerocyte
    (c) myocyte (d) thesocyte

Answers

➤➤Multiple choice questions

1. (a)	2. (d)	3. (b)	4. (c)	5. (b)	6. (b)	7. (c)	8. (b)	9. (c)	10. (c)
11. (a)	12. (b)	13. (c)	14. (c)	15. (a)	16. (a)	17. (a)	18. (a)	19. (d)	20. (c)
21. (b)	22. (a)	23. (c)	24. (d)	25. (b).					

23

Hydra

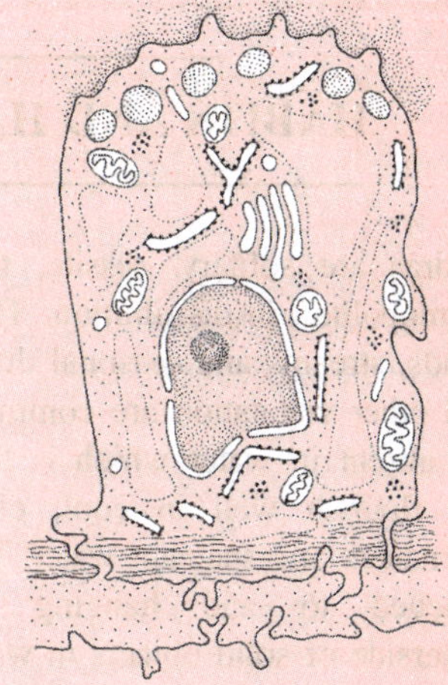

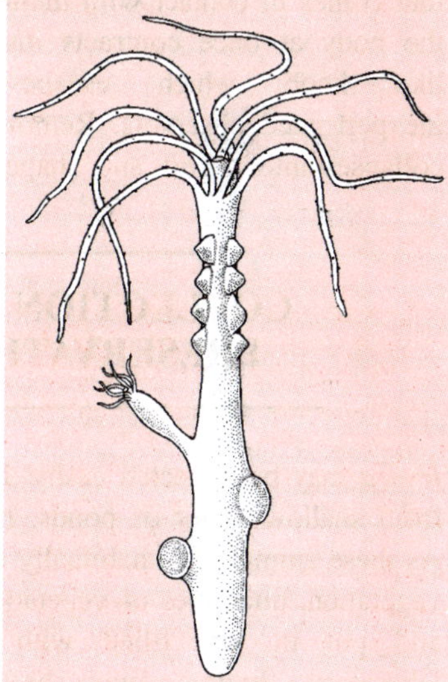

Coelenterata (or **Cnidaria**) is the phylum of acoelomate and radially symmetrical lower invertebrates (**Radiata**). Coelenterate animals are represented by two morphologically different types of individuals, polyps and medusae, (i) **Polyps** are sessile with a tubular body (*e.g., Hydra*), whereas (ii) **medusae** are free swimming with an umbrella or bell-shaped body (*e.g., Aurelia, Metridium*). Some coelenterates pass through both stages in their life-cycle with an alternations of generations (*e.g., Obelia*).

Hydra belongs to the most primitive class Hydrozoa of phylum Coelenterata. It is simple in form and structure and serves as a good example for the study of coelenterate organization by beginners.

Hydra vulgaris

There are several species of hydras. Following description applies in particular to *H. vulgaris,* the common Indian species, and in general to all the species of hydras.

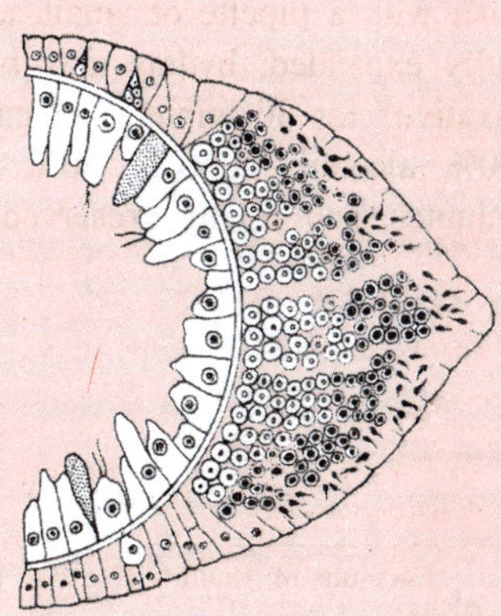

SYSTEMATIC POSITION

Phylum	Coelenterata
Class	Hydrozoa
Order	Hydroida
Suborder	Anthomedusae
Genus	*Hydra*
Species	*vulgaris*

HABITS AND HABITAT

Hydras are solitary, sessile, freshwater animals, cosmopolitan in distribution. They occur in lakes, ponds, streams and seasonal ditches, where weeds and other vegetation are commonly present. They are absent in water which is foul and too warm, but flourish well in cool, clear and relatively permanent and stagnant water. They may be found attached to and hanging downwards from underside of solid objects in water such as leaves, sticks, stones, weeds, etc. When hungry, their body and tentacles are stretched to the maximum limit and swayed in water in order to capture any prey that comes in contact with them. When disturbed, the body at once contracts into a minute jelly-like knob which escapes detection by inexperienced observers. Removed from water, it collapses into a soft and shapeless mass.

COLLECTION AND PRESERVATION

Hydra may be collected easily during early winter from shallow water of ponds, lakes and streams. As these animals live habitually clinging to aquatic vegetation, quantities of vegetation may be picked and put in jars filled with pond water. In laboratory, hydras may be dislodged from vegetation by squirting them with a jet of water from a pipette. They are then transferred to a dish with a pipette or small aquarium net. When fully expanded, hydras may be fixed in Bouin's fixative* for 30 minutes, given several washes in 30% alcohol, kept in 50% alcohol for 10-15 minutes, and finally preserved in 70% alcohol.

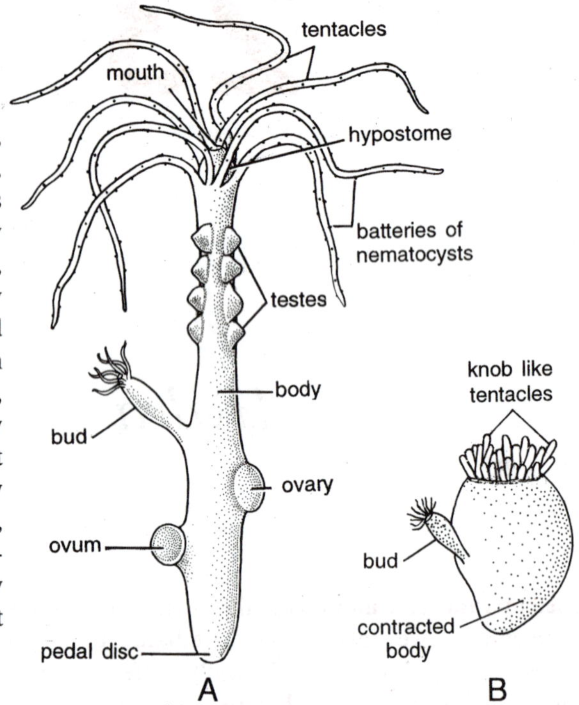

Fig. 1. *Hydra.* External morphology. **A.** *Expanded* body with bud and gonads. **B.** Contracted body bearing bud.

EXTERNAL MORPHOLOGY

[I] Shape and Size

Hydra is a polpy-like or polypoid coelenterate with a tubular or cylindrical body. When fully extended, it becomes elongated and slender and never measures more than 1 cm in length. When retracted, body becomes shortened and somewhat globular and measures only a few millimeters. Symmetry of body is typically radial, comprising an oral-aboral axis, with different parts arranged concentrically around it.

* **Formula of Bouin's fixative.** Picric acid (saturated aqueous sol.) 75 parts, Formalin (40%) 25 parts, and Acetic acid (glacial) 5 parts.

[II] Colouration

Most commonly occurring Indian species *H. vulgaris* is colourless. *H. gangetica,* found in ponds along river Ganges is white or pink in colour. *Pelmatohydra oligactis,* formerly known as *H. fusca,* is brown in colour. *Chlorohydra viridissima,* formerly called *H. viridis,* is the green *Hydra.* Its bright green colour is not because of chlorophyll-containing chloroplasts, but due to presence of symbiotic zoochlorellae *Chlorella vulgaris,* a unicellular green alga, that lives in its gastrodermal cells. The *Hydra* derives oxygen from the photosynthetic activity of this alga while the latter receives its carbon dioxide supply from the metabolic activity of the former.

[III] Pedal Disc

Proximal or aboral end of its body is closed, flattened and termed **pedal disc** or **basal disc**. This is used for temporary attachment to a substratum. It consists of a glandular zone which secretes adhesive substances for attachment and also a gas bubble for floating.

[IV] Hypostome, Mouth and Tentacles

Distal or free opposite end of body is produced into a conical elevation, the **hypostome**. It contains a circular aperture or **mouth** at its apex. Hypostome bears. at its base, a circlet of 6 to 10 slender, contractile and tubular thread-like processes, called **tentacles** (L. **tentare** to feel). These can be stretched to several millimeters, when the animal is hungry. Tentacles assist the animal in feeding and locomotion.

[V] Buds

Body of *Hydra* in some individuals may bear proximally lateral **buds** in various stages of development. A well developed bud possesses its own mouth, hypostome and tentacles. On detachment, it gives rise to a new individual.

[VI] Gonads

Other structures, occasionally seen on the external surface temporarily are the **gonads**. Conical **testes** occur near the oral end, while rounded **ovaries** are situated near the aboral end of animal.

INTERNAL STRUCTURE

Internal or histological structure of *Hydra* is best seen in its longitudinal and transverse sections.

Gastrovascular Cavity

Sections of *Hydra* show a central cavity or **coelenteron** (Or., **koilos**, hollow + **enteron**, gut), functionally referred to as the **gastrovascular cavity**. It is surrounded by the body wall. Mouth leads into this cavity which also extends into tentacles as their lumen. There is no anus and no excretory pore.

Body Wall (Histology)

Hydra is a **diploblastic** animal, *i.e.*, it is derived from two germ layers, the **ectoderm** and **endoderm**. These germ layers form two distinct cellular layers, outer **epidermis** and inner **gastrodermis**, respectively, of body wall and tentacles. Between these two layers is a thin, delicate, transparent and non-cellular **mesogloea**.

[A] Epidermis

Outer cellular layer of body wall, or **epidermis**, is composed of small, more or less cuboidal cells. It is a protective and sensory layer and is enveloped by a thin coating of **cuticle**. Various epidermal cells are of the following types :

1. Epithelio-muscle cells. Most of epidermis is formed by roughly conical or pear-shaped epithelio-muscle cells that serve both for epithelial covering and for muscular contraction. Thus each cell has two functional parts : epithelial and muscular. Outer **epithelial part** extends up to body surface, while inner or basal **muscular part** is drawn out into two **muscle processes** along the longitudinal axis of body. Muscle processes contain a contractile fibril, the **myoneme**. Muscle processes of all cells are embedded in surface folds of mesogloea. Contraction of these processes shortens the body and tentacles. A pair of non-contractile **supporting fibrils** extend into epithelial part as well as muscle processes of cell.

Electron microscopy has revealed the detailed structure of epithelio-muscle cells. In addition to the presence of usual intracellular organelles, like

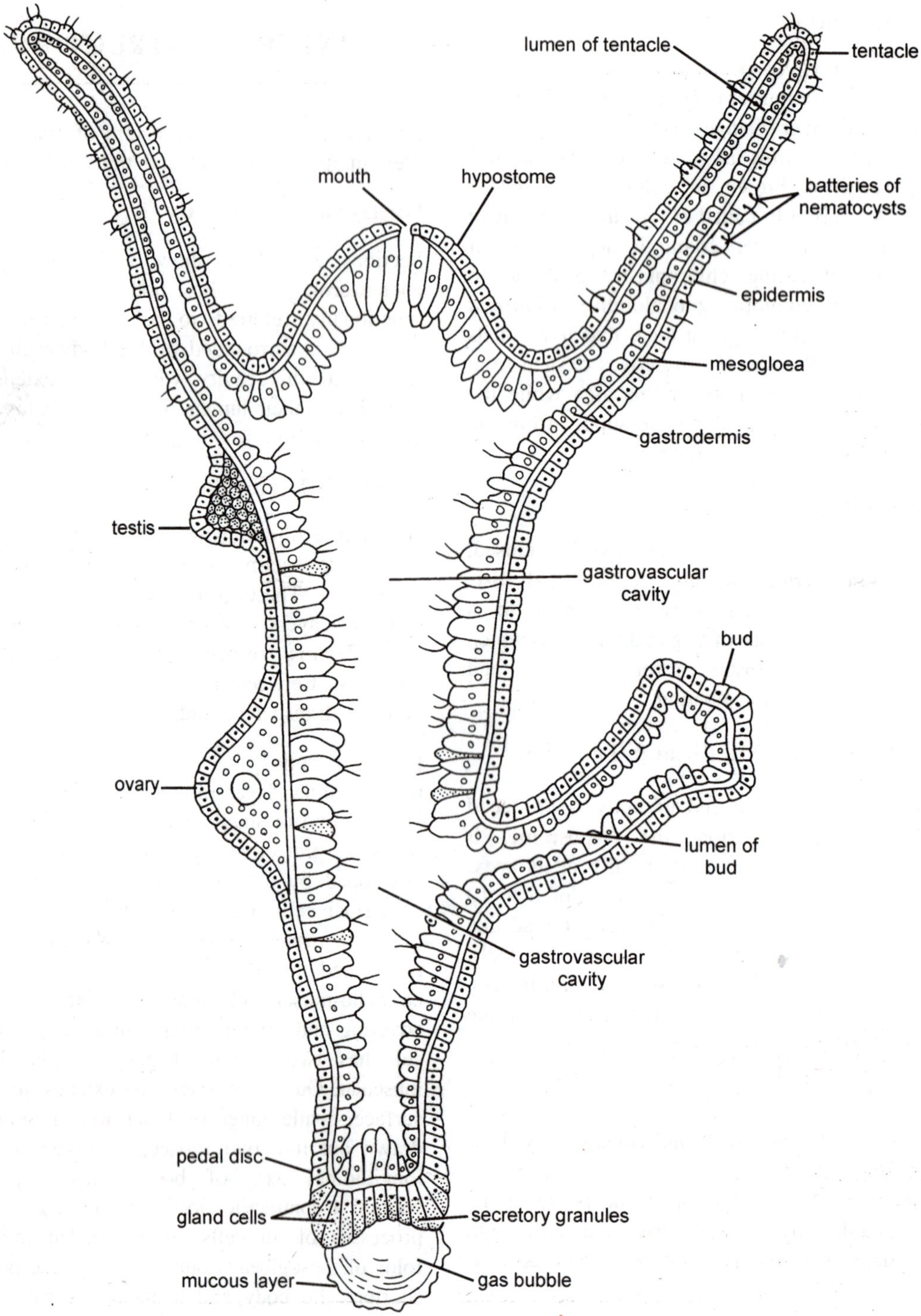

Fig. 2. *Hydra.* Longitudinal section of entire animal.

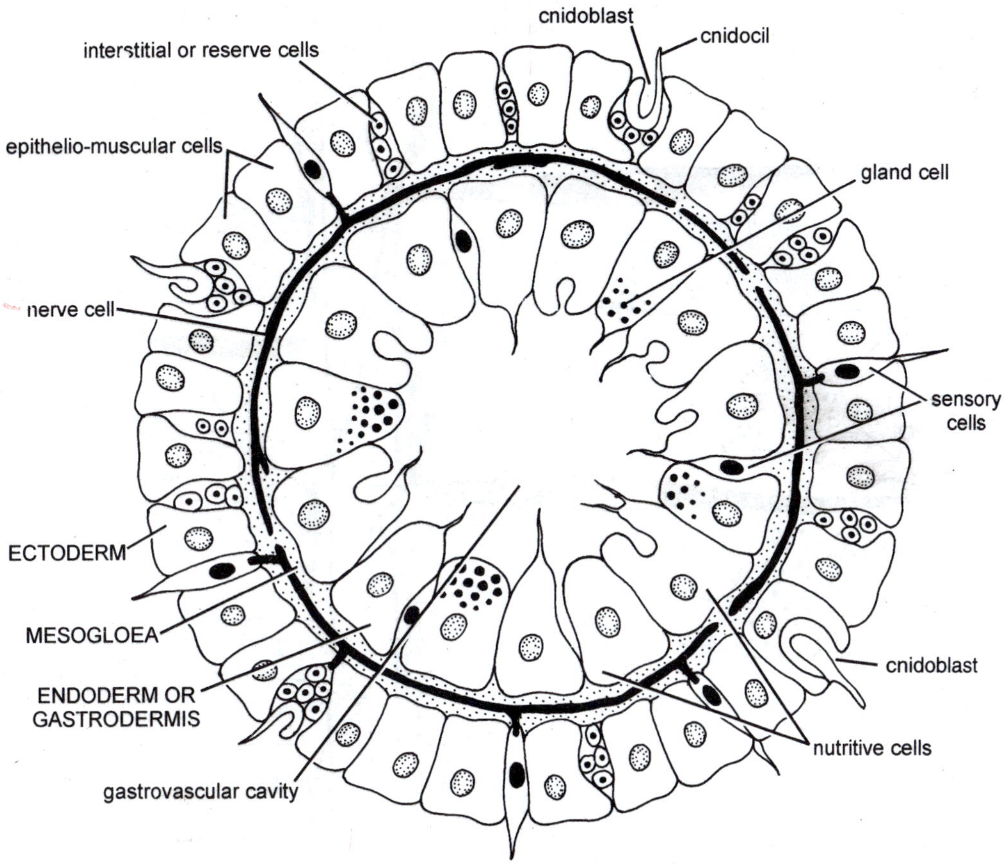

Fig. 3. *Hydra.* Transverse section.

nucleus, Golgi bodies, mitochondria, endoplasmic reticulum, ribosomes, lysosomes and vacuoles, there are present some other organelles as well. The cell membrane, at the outer free surface, has a few outward projections or **microvilli**. At the periphery below this membrane, are present a few **mucous bodies**. These secrete a finely granular material to form a thick, protective mucous layer on the cell surface. Basal part of the cell is produced into **muscle processes**, which run parallel to the long axis of animal body. These contain contractile **myofilaments** which constitute the **myoneme**

2. Gland cells. These are tall cells found chiefly on pedal disc and around mouth region. These secrete a mucus-like sticky material which serves for attachment, protection and entanglement of prey. Sometimes they secrete a gas bubble by which *Hydra* can rise to water surface or float.

3. Interstitial cells. These occupy the interstices or spaces between narrow basal ends of epithelio-muscle cells, hence their name. They are small, rounded, **undifferentiated cells**, measuring about 5μ in diameter. Each cell contains a clear cytoplasm and a relatively large nucleus with one nucleolus. Electron microscope also reveals smooth endoplasmic reticulum, scattered free ribosomes and a few small mitochondria. Interstitial cells are capable of developing into any other kind of cells such as reproductive, glandular, stinging, and buds etc., as required. They are thus **totipotent**, or reserve cells. According to **Brien** (1955), over a period of 45 days, all the cells of *Hydra* are replaced by

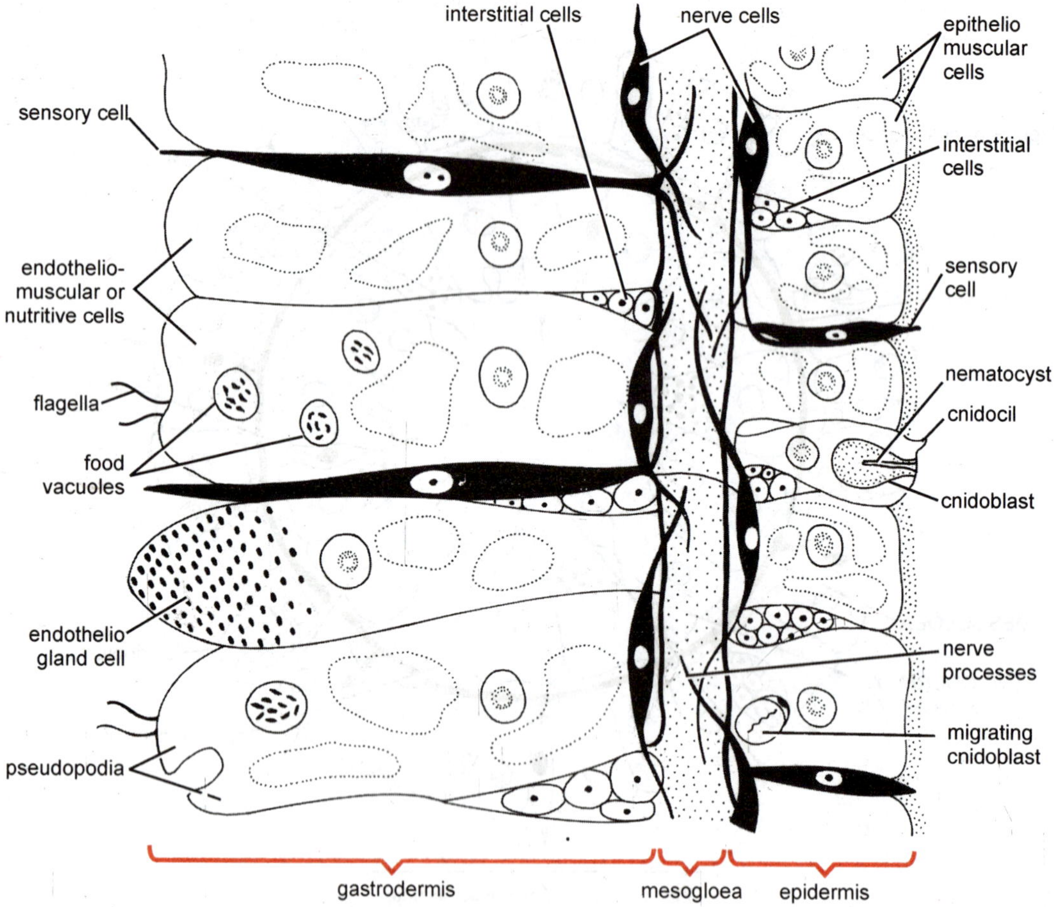

Fig. 4. *Hydra.* A portion of body wall in longitudinal section (magnified).

interstitial cells. Thus interstitial cells play an important role in regeneration, growth, budding and sexual reproduction, etc.

4. Cnidoblasts. Many of the interstitial cells of epidermis become specialized to form **stinging cells**, called **cnidoblasts** (Gr., **knide**, nettle + **blastos**, germ). These are especially abundant on tentacles arranged in clusters or batteries. A cnidoblast is somewhat oval with a basal nucleus and contains the sac-like organoid, the **nematocyst** or **stinging cell**. It is in the form of a capsule enclosing a coiled tube or thread. Nematocysts are characteristic of Cnidaria. These form organs of offence and defence of *Hydra* and also help in food-capture, locomotion and anchorage.

Nematocysts have been described in detail elsewhere in this chapter.

5. Sensory cells. Sensory cells occur scattered throughout epidermis among epithelio-muscle cells, especially on tentacles, hypostome and pedal disc. These are tall, narrow and columnar, thread-like cells, usually bearing a delicate hair-like process (apical cilium) at their outer free tips. Their basal or inner ends are connected by fine nodulated processes with nerve

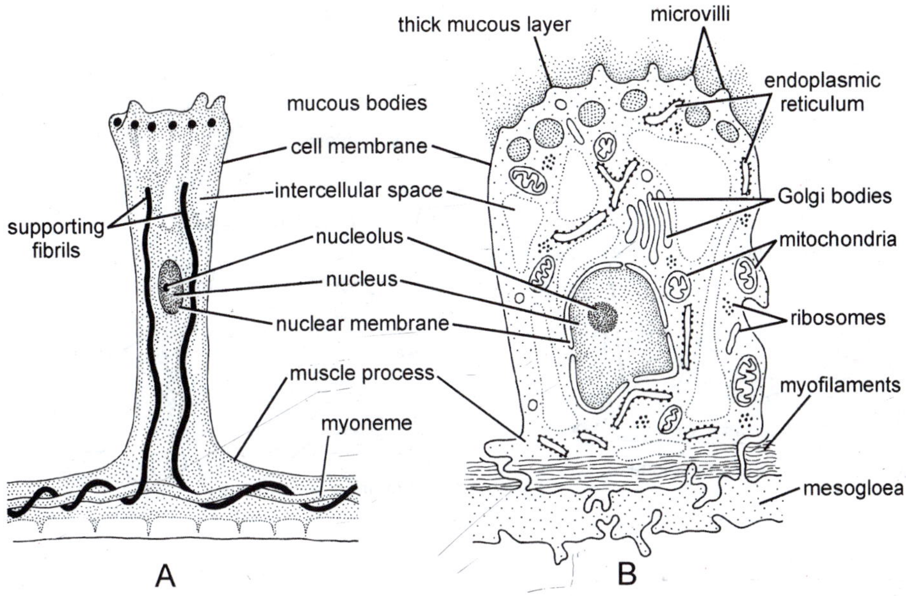

Fig. 5. *Hydra.* Epithelio-muscle cell. **A.** Under light microscope. **B.** Under electron microscope.

cells. Sensory cells serve as undifferentiated receptors for sensitivity to touch, temperature, chemical stimuli and light, etc. They receive and transmit impulses.

Electron microscope reveals that the apical hair-like process is in fact a cilium arising from a notch at the apex of sensory cell. It contains the usual 9 peripheral and 2 central microfibres arising from a basal granule from which small rootlets extend into cytoplasm.

6. Nerve cells. True nerve cells or **ganglion cells** occur for the first time in coelenterates. They

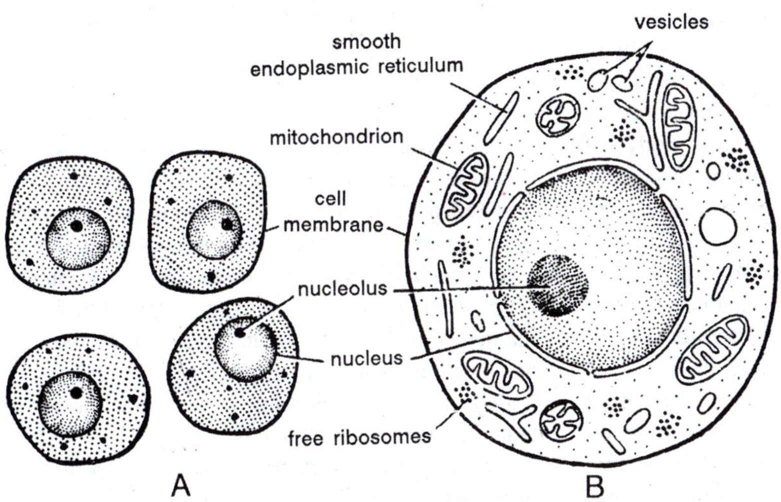

Fig. 6. *Hydra.* Interstitial cells. **A.** Cells under light microscope. **B.** A cell under electron microscope.

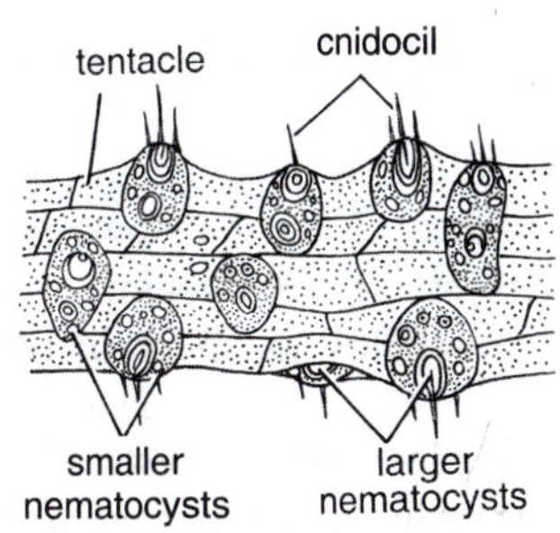

Fig. 7. *Hydra.* A portion of tentacle showing epidermal cells with batteries of nematocysts.

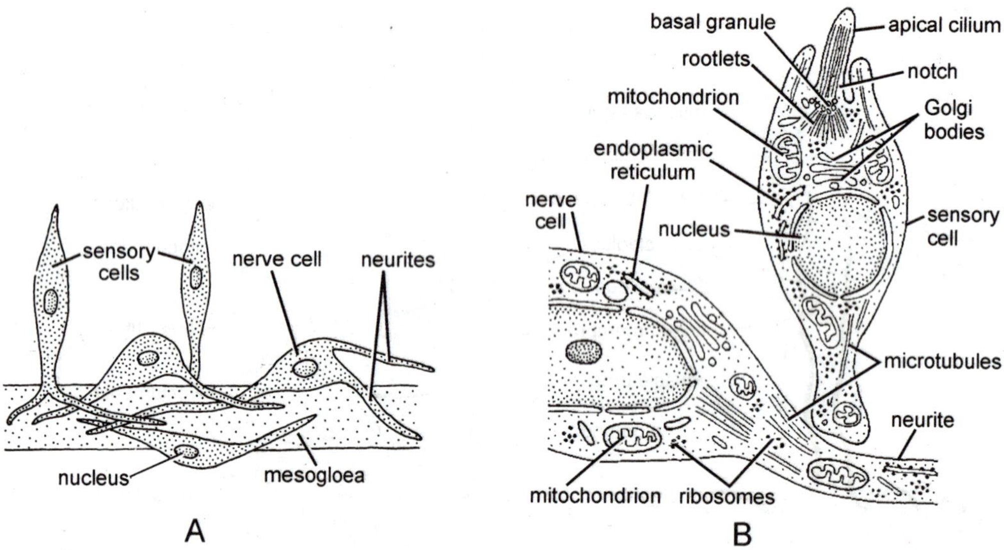

Fig. 8. *Hydra.* Sensory and nerve cells. **A.** Under light microscope **B.** Under electron microscope.

are derived from interstitial cells of epidermis. Nerve cells occur at the base of epithelio-muscle cells just above their muscle processes, forming a **nerve net** or **nerve plexus**. Each nerve cell consists of a small cell-body containing nucleus, and gives off two to several nerve processes or **neurites**, which are not differentiated into dendrites and axons as found in higher animals. The neurites of adjacent nerve cells form **synaptic contacts** *i.e.*, they lie very close together with microscopic gaps between them. Nerve cells, linked up with synaptic contacts, thus form an epidermal **nerve net** throughout the body of *Hydra.* The nerve net conducts impulses equally well in all directions due to restricted or no polarization.

Electron microscopy of nerve cells shows presence of usual cytoplasmic organelles. Nerve cells of the basal region of *Hydra* are devoid of microtubules and poor in ribosomes.

7. Germ cells. During summer, interstitial cells in certain restricted regions of body repeatedly divide and proliferate like reproductive cells forming **gonads**, which later differentiate into either **testes** or **ovaries**.

[B] Gastrodermis

Inner cell layer of body wall, called **gastrodermis**, lines the hollow and bag-like gastrovascular cavity. It constitutes nearly two-third of entire thickness of body wall. It is formed chiefly of large, typical columnar cells. This layer is mainly nutritive in function. Following types of cells are included in it :

1. Endothelio-muscle or nutritive-muscle cells. These are most numerous and conspicuous cells forming the bulk of gastrodermis. These resemble the epithelio-muscle cells of epidermis in all respects except that their basal contractile processes are single, much more delicate and oriented at right angles to the long axis of body next to mesogloea, thus forming a circular muscle layer. Each of these processes contain a muscle fibre or **myoneme**. Their contraction reduces the diameter of body and tentacles, which become narrower and longer. Contractile processes are highly developed around mouth and bases of tentacles to form sphincters.

Free end of a nutritive-muscle cell, projecting into gastrovascular cavity, may bear long, whip-like **flagella**, usually 2 in number, by means of

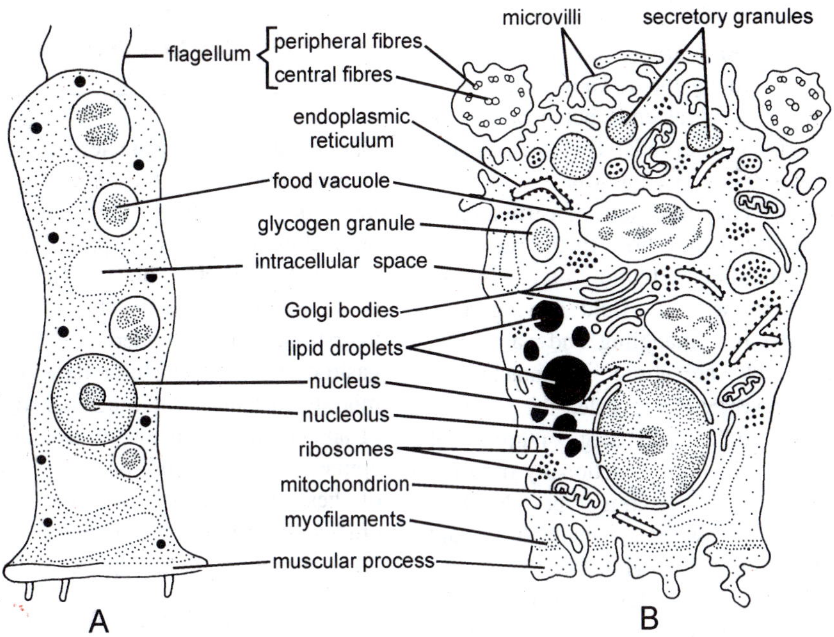

Fig. 9. *Hydra.* Nutritive-muscle cell. **A.** Under light microscope. **B.** Under electron microscope.

which the liquid food inside body-cavity is kept in motion. Besides flagella, blunt pseudopodia, like those of *Amoeba,* may also be put out from the free end to engulf particles of food. In a starving *Hydra,* protoplasm of nutritive cell remains much vacuolated. After a meal, however, the cells become gorged with nutritive particles.

Electron microscopic studies have shown that the free end of nutritive-muscle cell produces microvilli and two or more flagella that contain the usual 9 + 2 pattern of fibres. Apical cytoplasm includes, in large number, the mitochondria, glycogen granules, secretory granules and food vacuoles. Both smooth and rough endoplasmic reticulum, free ribosomes and lipid droplets are in abundance in cytoplasm. Nucleus lies centrally or basally and includes a single nucleolus. Small Golgi apparatus lies close to the nucleus. At the base of cell occur circularly oriented muscle processes containing myofilaments. Pinocytotic channels are also reported to run into cytoplasm from apical surface, which suggests occurrence of cell drinking phenomenon like that of *Amoeba.*

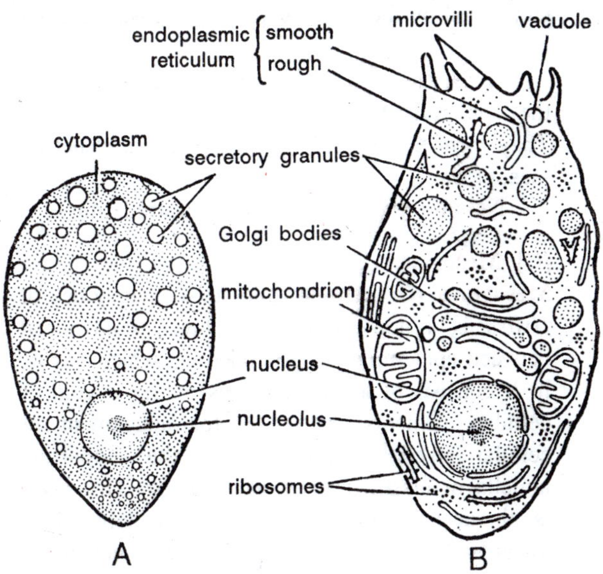

Fig. 10. *Hydra.* Endothelio-gland cell. **A.** Under light microscope. **B.** Under electron microscope.

2. Endothelio-gland cells. Endothelio-gland cells are smaller than nutritive-muscle cells and occur interspersed among them. They lack muscle tails at their tapering basal ends but bear one or two flagella at their free ends. Endothelio-gland cells are of two types. **Enzymatic gland cells** secrete digestive enzymes into gastrovascular cavity for extracellular digestion. In the region of hypostome and mouth are found **mucous gland cells**, which secrete a slimy fluid serving as a lubricant and also for entangling and paralyzing the prey. Gland cells are absent in tentacles.

Electron microscopy has revealed the presence of many Golgi bodies and large number of secretory granules, which are functionally divisible into mucus and enzyme-secreting types. Nucleus is basal, with or without nucleolus, and endoplasmic reticulum is rough-surfaced.

3. Interstitial cells. A few interstitial cells occur between the bases of endothelio-nutritive-muscle cells. These are **totipotent** as they may transform into other types of cells whenever the need arises.

4. Sensory cells. Large sensory cells are also found in gastrodermis. They are supposed to be stimulated by the entry of prey into gastrovascular cavity.

5. Nerve cells. These are similar to those of the epidermis but occur in far fewer numbers. They form a separate (gastrodermal) nerve net on mesogloea.

Nematocysts are absent in gastrodermis.

[C] Mesogloea

Mesogloea (Gr., **mesos**, middle + **glea**, glue) of *Hydra* is a non-cellular, thin layer, sandwiched between epidermis and gastrodermis, which secrete it. It consists of a proteinaceous matrix devoid of cellular elements. It serves for attachment of cellular layers, thus serving as the supporting lamella or elastic framework of body. This layer is thickest in pedal disc and gradually thins towards the tentacular ends. This arrangement is meant to enable the pedal region to withstand mechanical strain and to give flexibility to tentacles.

NEMATOCYSTS

One of the most characteristic structures of all coelenterates are the **stinging cells**, called **nematocysts**. These are not cells but cell organelles found in specialized cells called **cnidocysts** or **cnidoblasts** (Gr., **knide**, nettle + **blast**, germ). These develop only from modified interstitial cells of epidermis and are not found in gastrodermis. When fully developed, cnidoblasts migrate to tentacles through mesogloea by means of amoeboid movement. Projecting in between the epithelio-muscle cells, or lying inside the bodies of these cells (**host cells**), the cnidoblasts act as organs for offence and defence. They also serve for locomotion, food capture and anchorage.

[I] Structure of a Cnidoblast

A cnidoblast is an oval or rounded cell with a conspicuous basal **nucleus** lying on one side. The interior of cnidoblast is filled by a peculiar oval or pyriform sac or bladder, the **stinging capsule** or **nematocyst** (Gr., **nema**, thread + **kystis**, bladder). Nematocyst consists of a tiny bulb or

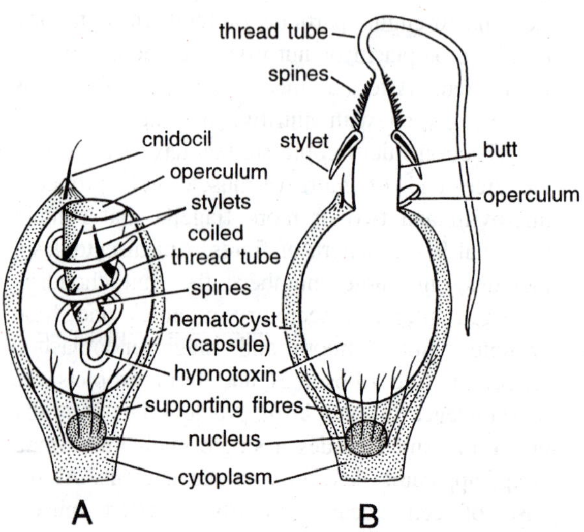

Fig. 11. *Hydra.* A cnidoblast. **A.** Undischarged. **B.** Discharged.

capsule, 5μ to 50μ in length, and composed of a material similar to chitin. It is filled with a poisonous fluid, or **hypnotoxin**, which is chemically a mixture of proteins and phenols. The narrowed outer end of capsule is invaginated into a long, hollow and tubular **filament** or **thread tube**, coiled like a watch-spring inside the sac itself. The base of thread tube is swollen to form the **butt** or **shaft**. Inside the butt are three large spines, called **barbs** or **stylets**, and three spiral rows of minute spines, called **barbules** or **spines**. The butt is covered externally by a little lid-like structure, the **operculum**. Outer end of cnidoblast projects freely beyond the epidermal surface, as a tiny, pointed and hair-like process, the **cnidocil**. (Gr., **knide**, nettle + **cilium**, hair) or **trigger**. Groups of supporting rods surround the cnidocil. Cytoplasm of cnidoblast may contain contractile muscle fibrils. In some, a restraining thread, called **lasso**, is attached to the base of cnidoblast. It prevents the nematocyst from being thrown out of it. Nematocyst described above is of **penetrant** type.

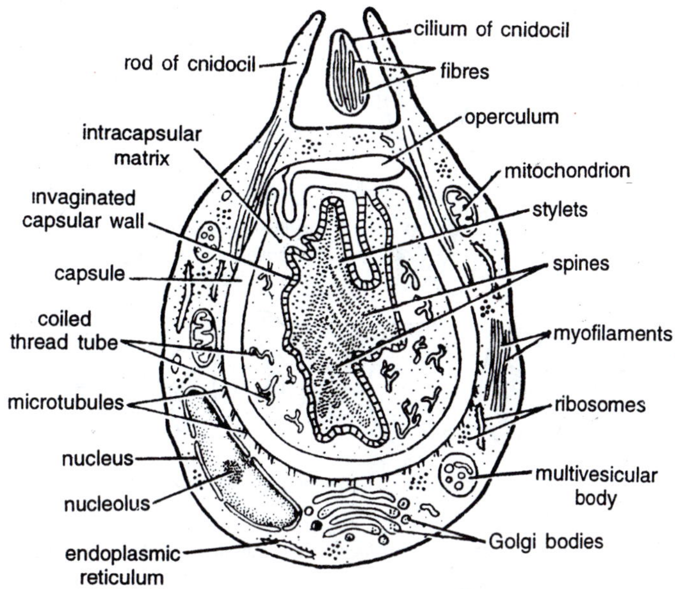

Fig. 12. *Hydra.* A cnidoblast as seen under electron microscope.

Electron microscopic studies have shown the presence of endoplasmic reticulum, free ribosomes, basally located Golgi bodies, mitochondria and multivesicular bodies in cytoplasm. A bundle of small myofilaments (probably **lasso**) extends from nematocyst capsule up to the basal part of cell. **Cnidocil** is composed of a central core surrounded by large rods. The core is structurally like a cilium as it contains fibres in 9 + 2 pattern. Fine microtubules are attached to, the base of capsule which is double-walled.

[II] Occurrence of Nematocysts

Nematocysts occur scattered, mostly singly, throughout the epidermis of body but remain absent on basal disc. They are especially abundant in the oral region and on tentacles where they form 'nematocyst batteries'. A battery consists of one or two large central nematocysts, surrounded by 10-12 smaller ones, all enclosed within a single large epithelio-muscle cell. Cnidoblasts are never formed in tentacles where they literally migrate from their seat of origin in the epidermis of body, or they may enter body cavity, whence they are passively transported in large quantities to the tentacles where they encyst in clusters to become batteries. None seems to encyst in gastrovascular cavity.

[III] Mechanism of Discharge

Explosion or discharge of nematocyst takes place when cnidocil is stimulated by food, prey or enemy. Neither touch alone, nor presence of food, causes discharge, but touch and presence of food together cause it. Thus, both mechanical stimulation, such as contact with prey, as well as chemical stimulation, emanating from an approaching prey, are involved in the mechanism of discharge.

Exact nature of discharge remains unknown. Response is wholly local and not under the control of nervous system. Enzymes involved also remain unknown. Further, it is not known why a well fed

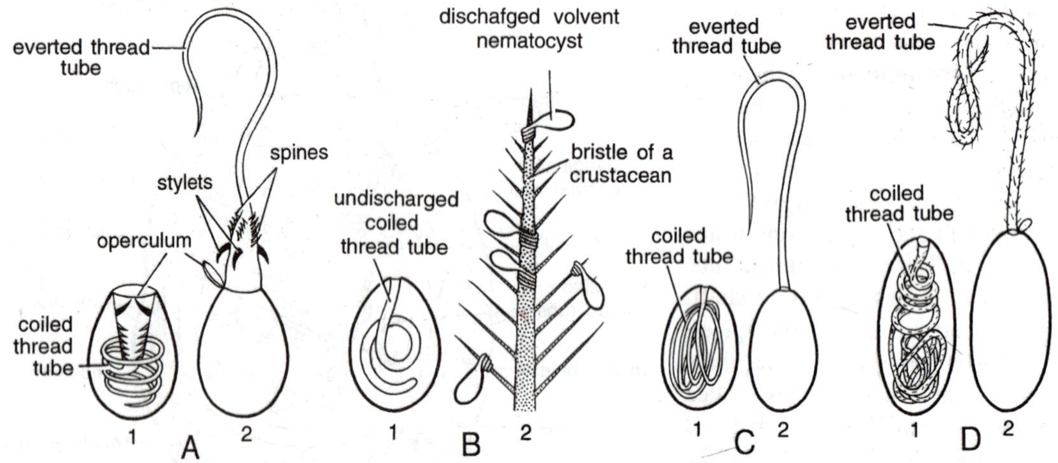

Fig. 13. *Hydra.* Types of nematocysts 1. Undischarged. 2. Discharged. **A.** Penetrant. **B.** Volvent. **C.** Stereoline glutinant. **D.** Streptoline glutinant.

Hydra fails to discharge nematocysts in the usual manner. Wall of nematocyst remains impermeable to water excepι at discharge. On stimulation the capsule wall suddenly increases its permeability. This causes a rapid intake of water and greatly increased osmotic pressure inside the capsule. As a result, the lid or operculum is forced open, the coiled thread tube turns inside out, and the entire nematocyst explodes to the outside. Contraction of contractile microtubules, surrounding the capsule, is also attributed by some workers to help in discharge. Probably the neuronal connections bring about coordinated discharge of nematocysts. As the thread tube everts, the barbs and spines present inside butt unfold to the outside.

Thread tube once discharged cannot be withdrawn, so that exploded nematocysts cannot be used again. Their cnidoblasts migrate to the gastrovascular cavity and are digested. Discharged nematocysts are replaced within 48 hours.

[IV] Types of Nematocysts

About 30 different kinds of nematocysts are found among different cnidarians. There kind is constant for the species and is of taxonomic value. *Hydra* has four basic types of nematocysts serving different functions, as follows :

1. Penetrant. Penetrant or **stenotele** is the largest (16μ in diameter) and most complex nematocyst. It is pear-shaped occupying almost the entire inner space of cnidoblast in which it lies. Its thread is a long and hollow tube, coiled transversely and bearing three large stylets and three rows of small spines at its stout base. Thread tube is open at the far end like a hypodermic needle. When discharged, it shoots out with such explosive force that it pierces the body of prey and injects the poisonous fluid (**hypnotoxin**) that paralyzes the victim or kills it outright. *Hydra* then seizes its prey with tentacles and draws it into mouth.

2. Volvent. Volvent or **desmoneme** is a small (9μ long) and pear-shaped nematocyst resembling a miniature bola. It contains a short, thick, spineless, smooth and elastic thread tube forming a single loop and closed at the far end. When discharged, it tightly coils around small projections such as hairs or bristles of prey, thus impeding its movements. Volvents are also useful in capturing the prey.

3. Stereoline glutinant. Small glutinants or **atrichous isorhizas** (9μ) are oval or elongate nematocysts. Butt is absent. They discharge a straight and unarmed thread tube open at the tip and used for attachment.

4. Streptoline glutinant. Large glutinants or **holotrichous isorhizas** (9μ) are oval or cylindrical. Thread tube is long with a narrow butt and forms three or four transverse coils. It is pointed and open at the tip. It bears a spiral row of small spines and may coil upon discharge.

Glutinants secrete a sticky substance possibly used in locomotion by fastening the tentacles of *Hydra* to solid objects, when somersaulting as well as in food-capture. They are also used to impede the progress of small animals when the projectiles are stuck to their appendages.

LOCOMOTION

Hydras are essentially sessile animals. They live attached by their pedal discs to objects in water. Brown and white hydras can remain fixed at one spot for a considerable time, but green hydras often move about from place to place by several ingenious devices. They twist about or perform movements to change their location either in response to light or some chemical stimulus or to obtain food. As gastrodermal muscle fibres are less developed, movements are largely due to contractions of the epidermal muscle fibres. *Hydra* shows movements of the following type :

[I] Expansion and Contraction

Hydra, attached to a substratum in water, frequently expands and contracts itself at intervals. This behaviour of *Hydra* is called **contraction burst**. It is initiated by a **pacemaker** located near the base of hypostome. These movements help to bring food organisms in contact with tentacles which are waved all around in water. Also, contraction of one side and elongation of other side of body or tentacles result in the bending and swaying movements which assist in the caputre of prey.

[II] Looping

Hydra can also move from place to place in search of food. Usually the body first extends and then bends over, so that the tentacles attach to the substratum with the help of adhesive glutinant nematocysts. Then the pedal disc is released and brought up closer to circlet of tentacles and than attached. Now tentacles loosen their hold and body becomes erect again. The whole process, which is repeated again and again, appears like a series of looping movements of a caterpillar or leech.

[III] Somersaulting

In another common and rapid method of progression, *Hydra* somersaults like an acrobat. The tentacular end is brought forward and attached to the substratum. The pedal disc is freed and moved upwards, thus causing *Hydra* to assume an inverted posture. Now, pedal disc is moved forward and attached to a new position. By freeing the tentacular end the animal again assumes its upright position. The animal performs a series of somersaults by repeating the process.

[IV] Gliding

For moving a short distance along a smooth surface, *Hydra* simply slides along on its basal disc like a snail, aided by secretions from mucous glands. It can slide or glide slowly over substratum due to creeping amoeboid movement of cells of pedal disc. Considerable distances are sometimes covered in this manner.

[V] Walking

Occasionally, *Hydra* becomes inverted, stands on its tentacles and moves in an inverted condition, using its tentacles as legs. This type of movement takes place on some object such as leaf and in leisurely hours.

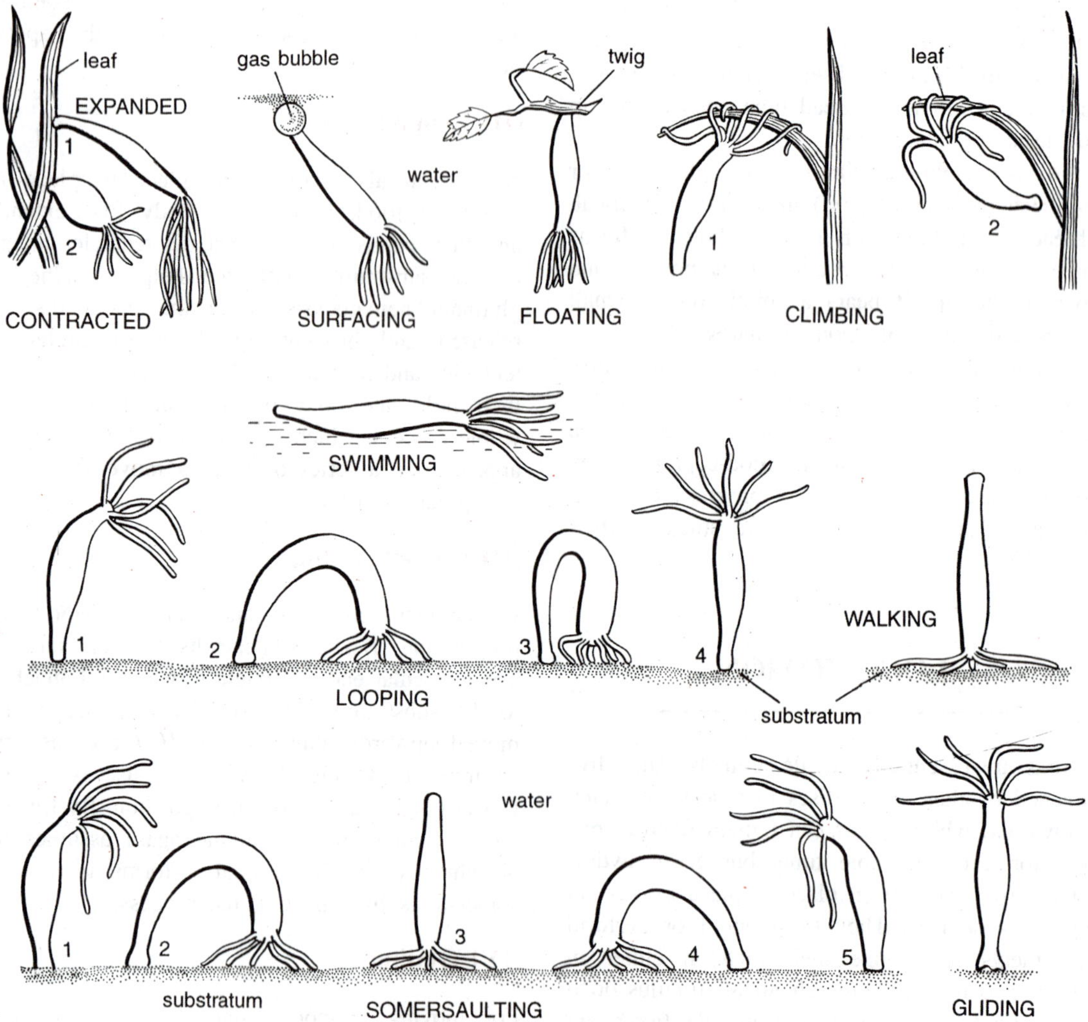

Fig. 14. Types of locomotion in *Hydra.*

[VI] Climbing

While changing location in a limited area, *Pelmatohydra oligactis* can even climb by attaching its long tentacles to some object, releasing the foot, and then contracting the tentacles, so that the body is lifted up.

[VII] Floating

Occasionally, *Hydra* throws its body free and floats on water surface with waves. Sometimes *Hydra* attaches to a floating leaf or twig by its pedal disc.

[VIII] Surfacing

Sometimes *Hydra* uses a gas bubble secreted in mucus by the cells of pedal disc, to rise in water and float at the surface. If gas bubble bursts, the mucous threads sustain the body on water surface due to surface tension.

[IX] Swimming

It is said that sometimes *Hydra* frees itself from substratum and swims in water by undulating, wave-like movements of tentacles and body.

NUTRITION

[I] Food

Hydra is almost exclusively carnivorous. Small aquatic animals such as insect larvae, crustaceans (*e.g.*, *Cyclops, Daphnia*) and annelid worms form the bulk of its food. It may swallow preys larger than itself, such as young fish and tadpoles. *Hydra* can damage trout nurseries by preying upon the newly hatched young. It can also feed on small bits of meat in an aquarium.

[II] Ingestion

Coelenterates are the first animals to use projectiles, called **nematocysts**, for capturing animals. A hungry *Hydra* waits for its food to come to it. It normally rests with its basal disc attached to some object, while its body and tentacles extend out in water, controlling a considerable amount of hunting territory. When a small passing organism, touches a tentacle, dozens of nematocysts are discharged into it at once. The volvents coil around bristles and other appendages on the prey, while glutinants fastens to its surface, thus holding it fast. The penetrants puncture the victim and inject the paralyzing **hypnotoxin**. The tentacles, holding the prey, now contract and bend inwards drawing the paralyzed prey towards the mucus-lined mouth, which opens widely to swallow it. Mucous secretions help in swallowing and the mouth can be greatly distended. Contractions of hypostome and body wall (peristaltic movements) force the food down into gastrovascular cavity where digestion takes place.

The food reaction varies greatly according to the state of the animal. A very well-fed *Hydra* will not react to food while a hungry *Hydra* will respond even when a chemical stimulus, such as beef juice, is added to the water. A chemical, called **glutathione**, usually found in the tissue fluids of most animals, is necessary to evoke feeding reaction. Thus, *Hydra* engulfs only those animals which have glutathione in their body.

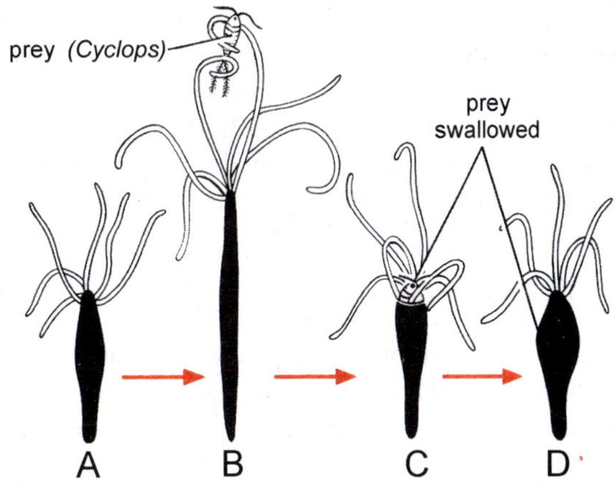

Fig. 15. *Hydra* capturing and ingesting a *Cyclops.*

[III] Digestion

Digestion is both **extra-cellular** and **intracellular**, and occurs in two stages.

1. Extracellular digestion. First, the prey is killed by the action of digestive juices secreted by the gland cells of gastrodermis. Churning movements caused by the expansion and contraction of body wall and lashing movements of flagella of nutritive-muscle cells thoroughly mix up the digestive juices with food which is broken into smaller particles suspended in a meaty broth. Digestive enzymes now act upon the disintegrating food. A proteolytic enzyme similar to trypsin partly digests proteins into polypeptides. This type of digestion, occurring in the cavity, outside gastrodermal cells, is called **extracellular digestion** and is purely proteolytic. It also takes place in the stomach and intestine of most multicellular animals like frog, earthworm, etc. It is met with in *Hydra* for the first time.

2. Intracellular digestion. Smaller fragments of food are engulfed by nutritive-muscle cells by means of pseudopodia and digested within food vacuoles. This is **intracellular digestion** as it occurs regularly in Protozoa and Porifera. Studies have revealed that food vacuoles undergo both acidic as well as alkaline phases, and digestion of protein is completed by other proteolytic enzymes.

Digestion in *Hydra,* therefore, is quite interesting, since it combines the digestive procedures of forms lower (Protozoa) as well as higher (Vertebrates) than itself. Retention of intracellular digestion is probably due to its aquatic mode of life, as the digestive juices get diluted in the gastrovascular cavity.

[IV] Absorption

Soluble products of digestion are distributed by diffusion from cell to cell. Some gastrodermal cells containing food vacuoles may become detached and move freely to distribute the digested food to all parts of body. Gastrovascular cavity thus serves for both digestion and circulation.

Hydra can digest proteins, fats and some carbohydrates but not starch. Reserve food materials, chiefly glycogen and fats, are stored in some of the gastrodermal cells.

[V] Egestion

Indigestible residues, like exoskeleton of Crustacea, are egested through mouth, for there is no anus. Egestion occurs by a sudden squirt due to muscular contraction of body, so that the debris is thrown at a distance.

RESPIRATION, EXCRETION AND OSMOREGULATION

Hydra has neither blood and blood vessels, nor organs of excretion and respiration, which are carried on individually by each cell. Due to thinness of body wall and circulation of water in gastrovascular cavity, most cells of body remain freely exposed to the surrounding water. Therefore, exchange of oxygen and carbon dioxide, and excretion of waste nitrogenous matter (chiefly ammonia) occurs directly by diffusion through cell membranes in the outside world.

Condition in *Hydra* favours an influx of surrounding water into its body cells. There is no evidence of any organs of osmoregulation. Therefore, it is not understood how water balance is maintained in the body cells of *Hydra.*

NERVOUS SYSTEM

Hydra possesses a very primitive type of nervous system. It includes bipolar and multipolar nerve cells or neurones lying immediately above the muscle processes and forming an irregular and discontinuous **nerve net** or **nerve plexus**. Neighbouring nerve cells are not fused together, but their processes or neurites form synaptic junctions. Such a nerve net is called a **synaptic nerve net**. Nerve cells are numerous around mouth and on pedal disc but show no groupings in the form of a nerve controlling centre like brain or nerve ring. A difference from higher animals is that the nerve net of *Hydra* is **unpolarized** so that impulses can pass in all directions

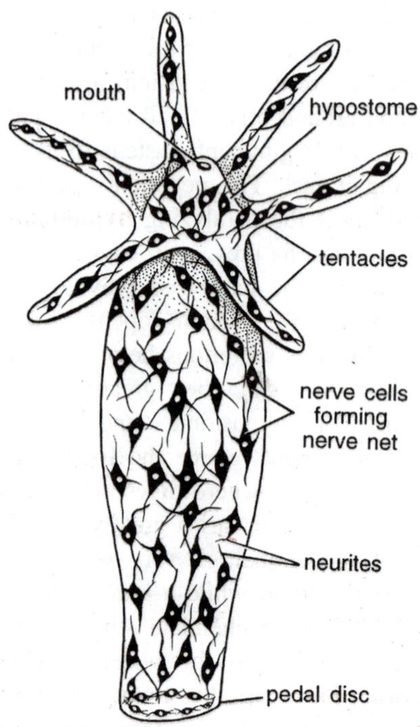

Fig. 16. Nerve net in *Hydra.*

(**diffuse transmission**). In brief, nerve net shows diffuse unpolarized transmission, autonomy of parts and paucity of reflexes.

It has been reported that nerve cells of epidermis and gastrodermis form two separate nerve nets which are interconnected. Their processes are connected to the sensory cells, which act as receptors for external stimuli, and to epithelio and endothelio-muscle cells which act as effectors by contracting their muscle processes. Such a combination of muscle processes of epithelio-muscle cells, sensory cells and nerve nets is referred to as **neuro-muscular system**. This is just the beginning in the evolution of nervous system.

BEHAVIOUR

Hydra reacts to various kinds of internal as well as external stimuli.

[I] Reactions to Internal Stimuli

In response to internal stimuli, *Hydra* shows **spontaneous movements** of body and tentacles. At regular intervals body, or tentacles, or both contract suddenly and rapidly and then slowly expand in a new direction. In a well-fed *Hydra,* such movements are slow. These movements are produced by muscle processes when they are stimulated through nerve net.

[II] Reactions to External Stimuli

Hydra responds positively as well as negatively to an external stimulus, depending upon its type and intensity. If the stimulus is strong, the animal usually responds negatively. Movements are produced by the contractile muscle processes in the wall, when they are stimulated through nerve net.

1. Contact. When a floating or swimming *Hydra* comes in contact with a substratum, it gets attached to it with its pedal disc. Application of a localized stimulus, such as touching of a tentacle with a pointed needle, causes the contraction of one or all the tentacles together with or without body. A mild stimulus affects only the part of body touched, while a strong stimulus affects the whole body.

2. Light. *Hydra* responds negatively to both strong as well as weak light. It prefers to accumulate in regions of moderate light intensity. Thus, animal has an optimum with regard to light. By trial and error, hydras seek areas of suitable light intensity.

3. Temperature. *Hydra* prefers cool or cold waters and disappears from surface waters when temperature reaches 20°C to 25°C.

4. Electric current. When *Hydra* is subjected to a weak and constant electric current, its tentacular end bends towards anode and pedal disc towards cathode. Whole of body and tentacles then contract. When the animal is inverted with tentacles attached to substratum, the pedal disc bends towards anode and tentacular end towards cathode.

5. Water current. *Hydra* shows no reactions to water currents.

6. Chemicals. *Hydra* avoids strong and injurious chemicals.

REPRODUCTION

Hydra reproduces asexually by budding and sexually by formation of gametes.

[I] Budding

During summer months, when the animal is well-fed and healthy, asexual budding is the usual method of reproduction. Near the middle or in basal part of body, a bulging appears as a result of repeated multiplication of epidermal interstitial cells. This grows as a bud with its wall consisting of epidermis and gastrodermis and the interior lumen in continuation with the parent's gastrovascular cavity. The bud enlarges, develops a mouth and a circlet of tentacles at its free end. When full grown, the bud constricts at the base and finally separates from the parent body. It feeds and grows into an adult *Hydra.*

Occasionally, several buds occur at the same time on a single parent, and these in turn may

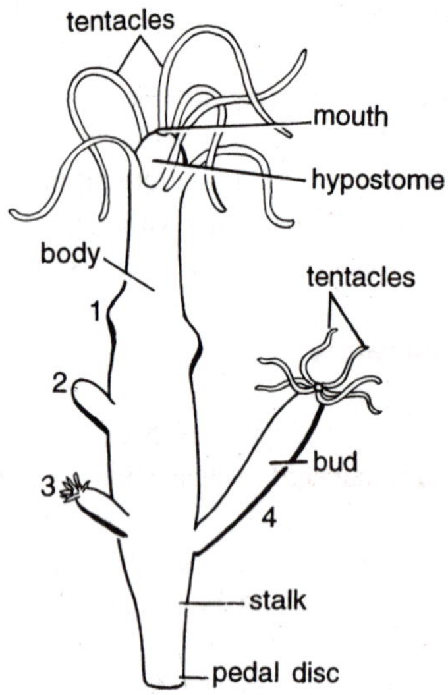

Fig. 17. *Hydra* showing stages (1–4) of budding.

develop secondary buds, so that a group is formed which temporarily resembles a colonial hydroid.

[II] Sexual Reproduction

Hydra reproduces sexually by the fusion of gametes, generally in autum. Gonads develop temporarily from interstitial cells of epidermis which accumulate to form bulges in the body wall. In some species, male and female gonads occur in the same individual which is then known as **hermaphrodite** or **monoecious** (Fig. 1). Usually testes develop towards the distal end and ovaries towards the basal end of body. Most species (*Pelmatohydra oligactis*), however, are **unisexual** or **dioecious**, *i.e.*, the individuals are either male or female and can be distinguished. The male is smaller in size and bears one to eight testes, while the larger female has only one or two ovaries. Even in hermaphrodite species, self-fertilization is avoided, because spermatozoa and ova mature at different times. As a rule, testes mature earlier than ovaries (protandrous condition).

1. Testes. Testes are conical elevations of body wall, varying in number from a few to many. They are usually located near the distal or oral end of body, but when numerous, they may cover the greater length of body. Each testis is formed by local proliferation of interstitial cells of epidermis, which get covered from outside by a capsule formed of large epidermal cells. Interstitial cells at the base are **spermatogonia** They undergo typical spermatogenesis, and passing through **primary spermatocyte**, **secondary spermatocyte** and **spermatid** stages become **spermatozoa** or **sperms** Each sperm has a head containing nuclear

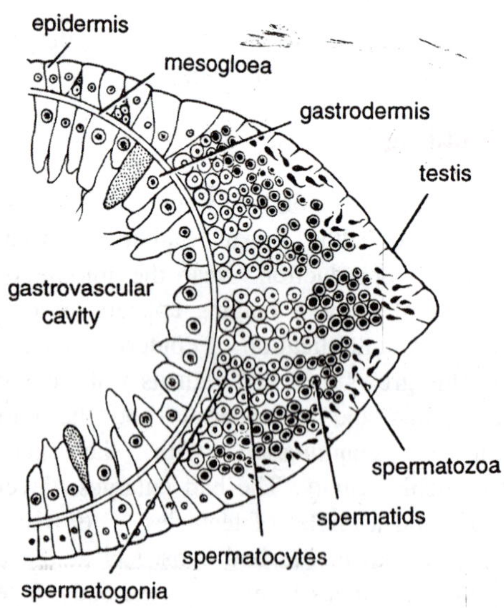

Fig. 18. *Hydra.* A testis in T.S. showing stages of spermatogenesis.

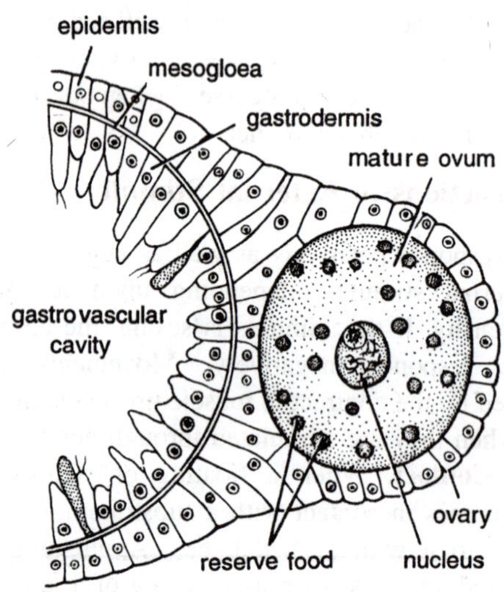

Fig. 19. *Hydra.* An orary in T.S. showing mature ovum.

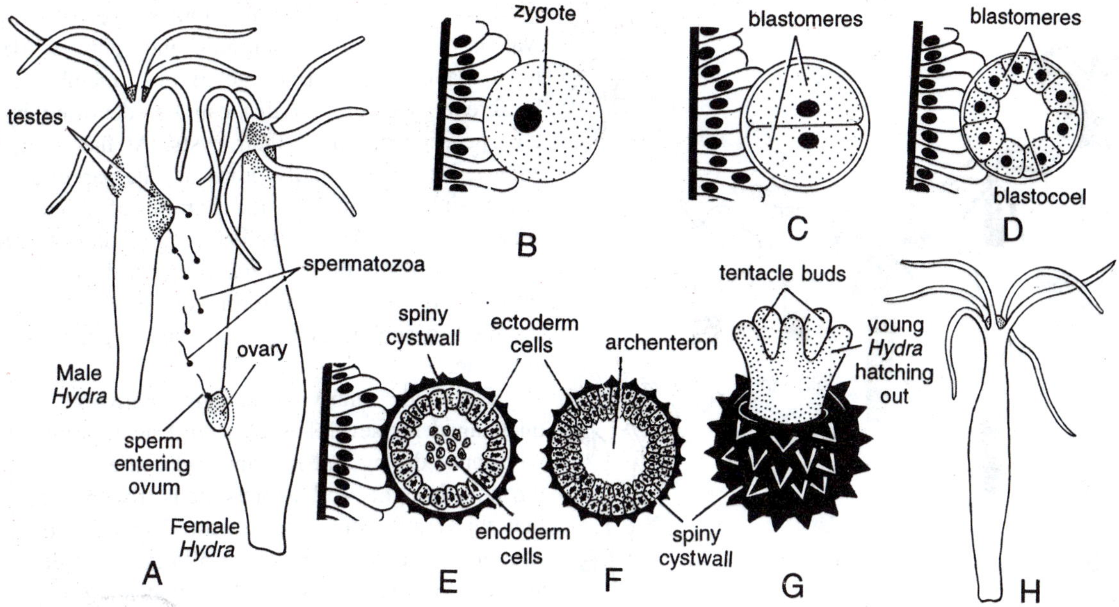

Fig. 20. *Hydra.* Fertilization and stages of development and hatching. **A.** Fertilization of an ovum. **B.** Zygote. **C.** 2-cell stage. **D.** Blastula. **E.** Early gastrula and encystation. **F.** Encysted gastrula. **G.** Young *Hydra* hatching out of cyst. **H.** Young *Hydra.*

material and a long vibratile tail. When mature, sperms are discharged by the rupture of testis wall at its apical nipple-like protuberance.

2. Ovaries. Ovaries develop a little later than testes. In green *Hydra,* usually one ovary is present, while in brown *Hydra* as many as eight ovaries are formed. These are ovoid structures, located near the basal end of body.

Ovary too is formed by the multiplication of interstitial cells which constitute the primary **oogonia**. But after sometime one centrally located cell, called **oocyte**, becomes larger and amoeboid, with a big nucleus. It feeds upon its smaller neighbouring interstitial cells, which become **yolk** or reserve food, to be used up later while young *Hydra* is still without a mouth to feed. As a result, oocyte increases greatly in size. It undergoes two maturation divisions resulting in the production of two polar bodies and reduction of chromosomes to haploid number which is 15 in case of *Pelmatohydra.* The sperm also has 15 chromosomes.

Mature egg or **ovum** is a large spherical mass, laden with yolk granules and occupying most of the space inside ovary. One ovary usually contains a single ovum, but rarely there are found two (*Chlorohydra viridissima*), or even more (*Hydra dioecia*) ova. When an egg is ripe, the epidermis over it ruptures and withdraws to form a cup-like receptacle containing the exposed egg. Ovum is not set free at once but remains attached to the parent by a broad base. A gelatinous protective sheath is secreted by the ovum around itself.

3. Fertilization. Ripe sperms, discharged from testes, swim about in water until they approach an extruded ovum and surround it. Several sperms may penetrate the gelatinous covering, but only one enters the ovum and fuses with it completely. This process is called **fertilization** *in situ,* and the fertilized egg is called **zygote**. Male and female nuclei unite to form the **fusion** or **zygote nucleus**. The ovum will die and degenerate, unless fertilization occurs within a short time after it is exposed.

4. Development. Development begins soon after fertilization, while the egg or zygote is still attached to the parent, by undergoing **cleavage** or **segmentation**.

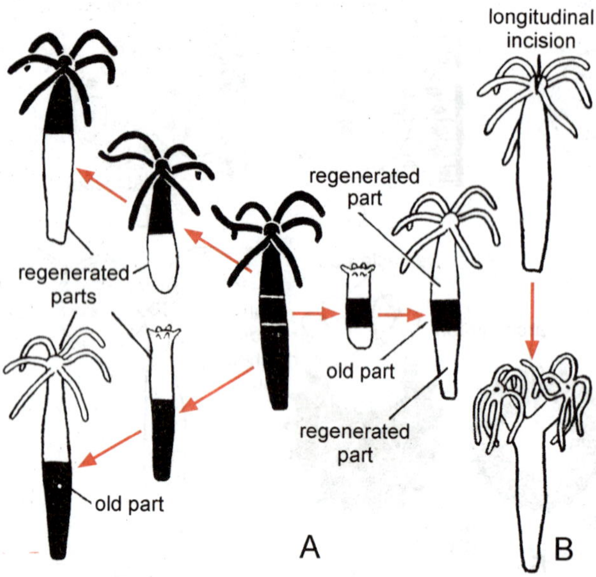

Fig. 21. *Hydra.* Regeneration. **A.** Regeneration of three individuals from one *Hydra.* **B.** Formation of two-headed *hydra* by regeneration.

(a) Cleavage. As egg has little yolk, cleavage is indeterminate, total and equal (**holoblastic**), resulting in equal-sized cells called **blastomeres**.

(b) Blastulation. Cleavage results in the formation of a hollow spherical ball, called **blastula** or **coeloblastula**. Its central narrow cavity is termed **blastocoel**. Equal-sized cells or blastomeres are arranged in a single surface layer.

(c) Gastrulation. Some cells of blastular wall detach and migrate inwards (**multipolar ingression**). While other cells form outer and inner cells by tangential division (**primary delamination**). As a result blastocoel is completely filled by new cells and the hollow blastula converts into a **solid gastrula**. It is made of a single layer of outer cells forming **ectoderm**, and an inner solid mass of cells forming **endoderm**. As described above, gastrulation or formation of endoderm in *Hydra* is partly by multipolar ingression and partly by primary delamination.

(d) Encystation. Soon a new cavity, called **gastrocoel** or **archenteron**, appears in the central solid mass of endodermal cells. Meanwhile, the ectodermal cells secrete a two-layered protective **shell** or **cyst** around gastrula. Outer layer of cyst wall is thick, horny or chitinoid and spiny, while inner layer is a thin gelatinous membrane. Different species of *Hydra* can be identified by the specific pattern of their cysts. At this stage, encysted gastrula usually drops off the parent and either sinks in mud at the bottom of pond or adheres by means of its spikes to any solid object it contacts with. In *P. oligactis,* gastrula is shed from ovary before the formation of shell, and fastened to aquatic plants by its sticky gelatinous covering.

Encysted embryo remains dormant and unchanged for several weeks, until next spring. It withstands drying and freezing and carries the race through droughts and winters. It is also probable that this resting stage serves for dispersal, for it can be carried by wind, or in mud on feet of animals to other ponds in which water is present.

(e) Hatching. With the advent of favourable conditions of water and temperature, development is resumed. **Interstitial cells** arise in ectoderm and **mesogloea** is secreted between two cellular layers. The embryo elongates, and a circlet of **tentacle buds** develops at one end with a **mouth** appearing in their midst. As embryo increases in size, the shell or cyst ruptures and a young *Hydra* with tentacles hatches out. Soon it grows into an adult. There is no free larval stage in the development of *Hydra.*

REGENERATION

Regeneration is the ability of an organism to replace its lost or damaged body parts. It usually occurs after an accident, but can be induced artificially by mutilation. *Hydra* has great power of regeneration as discovered long ago by **Abraham Trembley** in 1745. Since then, *Hydra* has been a favourite material for experimentations on regeneration.

If a living *Hydra* animal is cut into two, three or more pieces, each missing part grows and becomes a complete animal. A fragment of *Hydra,* measuring 1/6 mm, or more in diameter, is capable of regenerating an entire individual called morphallaxis. Regeneration is not reproduction because it is not a normal method of multiplication. Regeneration of hydras is made possible by the amazing generative powers of totipotent interstitial cells.

One characteristic feature of regenerating piece in *Hydra* is that it retains **polarity**. End nearer to mouth develops mouth and tentacles, while the end nearer to base forms a new pedal disc.

Parts of one *Hydra* may be easily grafted upon another provided they are of the same species. Grafting can be done in various arrangements producing bizarre effects.

Trembley observed that if head end of a *Hydra* is split in two and the parts are separated slightly, it results into a Y-shaped specimen, or 'two-headed' individual. By further splitting heads, Trembley succeeded in producing a 'seven-headed' *Hydra.* It was the great regenerative power of these animals which won for them the name '*Hydra*', after a Greek mythical monster which was finally destroyed by Hercules. According to the legend, '*Hydra*' had nine heads and no sooner did Hercules cut one off, instead two grew in its place.

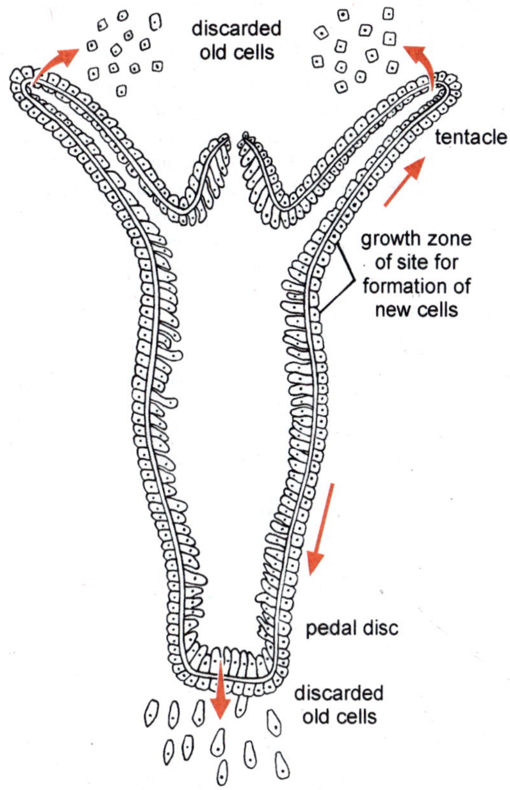

Fig. 22. Cell replacement in *Hydra.*

IMMORTALITY IN *HYDRA*

P. Brien (1955) and others have observed that a *Hydra* is at least potentially immortal. Just below the tentacles there is a growth zone, where interstitial cells give rise to all other cells of body. With the formation of new cells, old cells are pushed towards the end of tentacles and pedal disc, from where they are shed outside. In about 45 days, the older body cells are replaced by new cells. This process of cell replacement is an endless process. It has also been shown that, if interstitial cells of growth zone are destroyed, then *Hydra* lives only for a few days.

Important Questions

➤➤*Long answer type questions*

1. Give an account of histological structure of the body wall of *Hydra.*
2. Describe the various modes of locomotion in *Hydra.*
3. Give the structure of a nematoblast cell of *Hydra,* as revealed by electron microscope. Add a note on the various types of nematocysts and their functions.
4. Give a detailed account of reproduction in *Hydra.*
5. Write short notes on : (i) Behaviour of *Hydra,* (ii) Immortality in *Hydra,* (iii) Nutrition in *Hydra,* (iv) Regeneration in *Hydra.*

➤➤*Short answer type questions*

1. What is mesogloea ?
2. What are the five types of cells found in gastrodermis of *Hydra* and mention their functions.
3. Illustrate the structure of a cnidoblast.
4. Compare the life cycle of *Hydra* with that of *Obelia.*
5. List down 6 types of ectodermal cells present in the body wall of *Hydra* and mention their functions.
6. Draw a neatly labelled diagram of the T.S. of *Hydra.*
7. *Hydra* usually hermaphrodite, but cross fertilization occur. How can *Hydra* prevent the self fertilization ?
8. What are the food for a *Hydra* ?
9. Give the names of the movement of a *Hydra.*
10. Who discovered the regeneration power of *Hydra.*

➤➤*Multiple choice questions*

1. The body of *Hydra* is :
   (a) asymmetrical (b) bilaterally symmetrical
   (c) radially symmetrical
   (d) diploblastic and radially symmetrical
2. A chemical substance hypnotoxin is found in :
   (a) *Entamoeba histolytica*
   (b) ectoderm of olynthus
   (c) nematocyst of *Hydra*
   (d) venum of snake
3. Number of tentacles in *Hydra* is :
   (a) 2-4 (b) 6-10
   (c) 10-40 (d) 43-50
4. In *Hydra,* sperms and ova are formed from :
   (a) cnidoblasts (b) interstitial cells
   (c) myotomes (d) tentacles
5. The most primitive nervous system is found in :
   (a) *Hydra* (b) *Amoeba*
   (c) sponge (d) earthworm
6. Which of the following animals has a nervous system but no brain :
   (a) *Amoeba* (b) cockroach
   (c) earthworm (d) *Hydra*
7. The nematocysts of *Hydra* are important for :
   (a) catching prey (b) paralysing prey
   (c) testing the quality of food
   (d) testing the quality of water
8. The cavity in the body of *Hydra* is called :
   (a) hydrocoel (b) haemocoel
   (c) coelom (d) coelenteron
9. *Hydra* is an example of the type :
   (a) monoblastic (b) diploblastic
   (c) triploblastic (d) none of the above
10. Distinguishing feature of order Hydroida to which *Hydra* belongs is :
    (a) usually 4-12 tantacles
    (b) both polyp and medusa are included in life cycle
    (c) polyp phase dominant and medusa phase is usually absent
    (d) medusa phase is dominant
11. *Hydra* was first described by :
    (a) Trembley (b) Linnaeus
    (c) Leeuwenhock (d) Lamarck
12. Internal cavity in body of *Hydra* is coelenteron :
    (a) not communicated with outside
    (b) not extended into tentacles
    (c) communicated with outside by mouth and aboral aperture
    (d) communicated with outside by mouth only
13. *Hydra* can suddenly contract its body due to contraction in :
    (a) nutritive muscular cells
    (b) epithelio-muscular cells
    (c) gland cells (d) gastrodermis
14. Battery located in tentacles of *Hydra* are group of :
    (a) non-functional cnidoblasts
    (b) interstitial cells
    (c) cnidoblasts
    (d) functional cnidoblasts
15. Most complicated nematocysts of *Hydra* are :
    (a) penetrants (b) volvents
    (c) holotrichous (d) atrichous
16. Leech like locomotion in *Hydra* is :
    (a) walking (b) looping
    (c) gliding (d) somersaulting
17. Glutathione is a chemical released from :
    (a) prey tissues (b) *Hydra* to start ingestion
    (c) tissues of injured prey
    (d) tentacles of *Hydra*
18. Digestion in *Hydra* is initially extracellular and then intracellular, the undigested food is egested through :
    (a) individual cells after intracellular digestion
    (b) opercula of nematocyst
    (c) aperture of the pedal disc
    (d) mouth
19. In embryonic development of *Hydra* the stereogastrula is :
    (a) three layered solid mass
    (b) two layered with coelenteron
    (c) solid embryo
    (d) single layered with blastocoel
20. Pseudopodia of *Amoeba* corresponds to which structure in *Hydra* regarding feeding :
    (a) mouth (b) nematocysts
    (c) tentacles (d) epithelio-muscular cells

21. A piece of *Hydra* will regenerate if :
(a) there is part of basal disc and mouth
(b) the part has epidermis, gastrodermis and tentacles
(c) epidermis and gastrodermis is present in that
(d) only epidermis is present

22. Mouth serves for both ingestion and egestion in :
(a) *Leucosolenia* (b) *Fasciola*
(c) *Hydra* (d) *Ascaris*

23. The nitrogenous wastes in *Hydra* are :
(a) urea and uric acids, removed from whole body surface
(b) ammonia, removed from general body surface
(c) ammonia and urea, removed through mouth
(d) only urea, excreted through mouth

24. Discharge of nematocysts in *Hydra* depends upon :
(a) nervous control
(b) entrance of water in capsule
(c) enzymes and mechanical stimulation
(d) all these

25. Nerve cells of *Hydra* are :
(a) unipolar (b) bipolar
(c) nonpolar (d) none of these

26. The cell of *Hydra* that has poisonous fluid Hypnotoxin is :
(a) cnidoblast (b) interstitial cell
(c) glandulo-muscular cells
(d) none of these

27. Digestion in *Hydra* is :
(a) extracellular (b) intracellular
(c) both (d) none

28. *Hydra* has no special structures for :
(a) locomotion (b) respiration
(c) nutrition (d) reproduction

29. In *Hydra* larval stage is :
(a) planula (b) amphiblastula
(c) scyphistoma (d) none

30. In development of *Hydra* gastrulation is the result of :
(a) emboly (b) epiboly
(c) delamination (d) unipolar immigration

31. *Hydra* is :
(a) fresh water polyp (b) solitary and diploblastic
(c) radially symmetrical (d) all of these

32. Tentacles of *Hydra* are :
(a) solid (b) eight
(c) not for protection (d) all of these

33. Tentacles of *Hydra* are contractile and does not help in :
(a) locomotion (b) reproduction
(c) food capture (d) protection

34. Body wall of *Hydra* is made of :
(a) ectoderm (b) mesogloea
(c) endoderm (d) all of these

35. The correct statement for *Hydra* is :
(a) spongocoel as body cavity
(b) pedal disc with tentacles
(c) coelenteron is body cavity
(d) nonfibrous mesogloea

36. Epidermis of *Hydra* is made of :
(a) cnidoblast (b) epithelio-muscular cells
(c) interstitial cells (d) all of these

37. The incorrect statement for *Hydra* is :
(a) cnidocil receives mechanical stimulus
(b) functional nematocysts are found in both cellular layers
(c) largest nematocyst is penetrant
(d) action of nematocyst depends upon enzyme

38. *Hydra* is a :
(a) triploblastic, radially symmetrical, coelomate organism
(b) diploblastic, radially symmetrical, pseudocoelmate organism
(c) triploblastic, bilaterally symmetrical, coelomate organism
(d) diploblastic, radially symmetrical, acoelomate organism

39. Excretion of nitrogenous wastes in *Hydra* takes place through :
(a) flame cells
(b) nephridia
(c) namatocysts
(d) diffusion across the general body surface

40. The body wall of *Hydra* has :
(a) neither sensory nor nerve cells
(b) sensory cells but no nerve cells
(c) nerve cells but no sensory cells
(d) both sensory and nerve cells

41. *Hydra* has the ability to assume an inverted position and to move on its tentacles. Such a method of locomotion is called :
(a) somersaulting (b) gliding
(c) walking (d) looping

42. The developing embryo in *Hydra* drops down from the body of the parent :
(a) soon after formation of the zygote
(b) after formation of the blastula
(c) after formation of the stereogastrula
(d) only after it has developed into a young *Hydra*

43. In the nutritive muscular cells of *Hydra* algae such as *Zoochlorellae* and *Sooanthallae* are found. They live as :
(a) parasites (b) symbionts
(c) saprophytes
(d) commensals

44. Which one of the following types of cells in *Hydra* bears both flagella and pseudopodia :
(a) secretory cells (b) gland cells
(c) epithelial cells (d) nutritive cells

45. *Hydra* can be called a :
(a) coelomate (b) acoelomate
(c) pseudocoelomate (d) none of these

46. *Hydra* has no special structure for :
(a) respiration
(b) food capture
(c) attachment
(d) offence and defence

47. *Zoochlorella* in musculo-nutritive cells of *Hydra* live as :
(a) symbionts
(b) commensals
(c) parasites (d) predators

48. The rapid method of locomotion in *Hydra* is :
(a) gliding (b) looping
(c) somersaulting (d) walking

49. The largest nematocyst of *Hydra* is :
(a) penetrants (b) holotrichous isorhizas
(c) volvents (d) atrichous isorhizas

50. In *Hydra,* the mucous for attachment is secreted by :
(a) cnidoblasts (b) glandulo-muscular cells
(c) interstitial cells (d) epithelio-muscular cells

51. In *Hydra,* both flagella and pseudopodia are found in :
(a) gland cells (b) epithelio-muscular cells
(c) interstitial cells (d) nutritive-muscular cells

52. *Hydra* was discovered by :
(a) Leeuwenhock (b) Linnaeus
(c) Grant (d) Trembley
(e) Aristotle

53. Which of the following is not found in *Hydra* ?
(a) regeneration
(b) fertilization
(c) gastrovasculation
(d) segmentation

54. In *Hydra* the coelenteron serves for :
(a) locomotion and excretion
(b) excretion and circulation
(c) digestion and circulation
(d) digestion and storage

55. The mesoglea of *Hydra* contains :
(a) nerve cells (b) sensory cells
(c) muscle cells
(d) no cells

56. Regeneration in *Hydra* was discovered by :
(a) Lavern (b) Hymen
(c) Huxley (d) Tembley

57. The gastrula of *Hydra* is known as stereo-gastrula because it is :
(a) solid (b) hollow
(c) two-layered (d) encapsulated

58. In *Hydra* between the epidermis and gastrodermis an intermediate structureless layer is found. This is known as :
(a) mesogloea (b) musculo-endothelial
(c) myo-epithelial layer (d) endoderm

59. Spermatogonia in *Hydra* develop from :
(a) interstitial cells (b) gastrodermis
(c) musculo-endothelial cells
(d) gland cells

60. The digestion of food within the gastrovascular cavity of *Hydra* is called :
(a) intercellular digestion
(b) intracellular digestion
(c) extracellular digestion
(d) extracorporeal digestion

61. The proximal end of *Hydra* has been an adhesive disc which secretes a sticky substance for :
(a) protection
(b) defence from enemies
(c) sexual attraction
(d) attachment to the substratum of *Hydra*

Answers

►►*Multiple choice questions*

1. (d)	2. (c)	3. (b)	4. (b)	5. (a)	6. (d)	7. (b)	8. (d)	9. (b)	10. (c)
11. (a)	12. (d)	13. (b)	14. (d)	15. (a)	16. (b)	17. (a)	18. (d)	19. (c)	20. (c)
21. (c)	22. (c)	23. (b)	24. (b)	25. (b)	26. (a)	27. (c)	28. (b)	29. (d)	30. (c)
31. (d)	32. (d)	33. (b)	34. (d)	35. (c)	36. (d)	37. (b)	38. (d)	39. (d)	40. (d)
41. (c)	42. (c)	43. (b)	44. (d)	45. (b)	46. (a)	47. (a)	48. (c)	49. (a)	50. (b)
51. (d)	52. (d)	53. (d)	54. (c)	55. (d)	56. (d)	57. (a)	58. (a)	59. (a)	60. (c)
61. (d).									

24

Obelia : A Sea-Fur

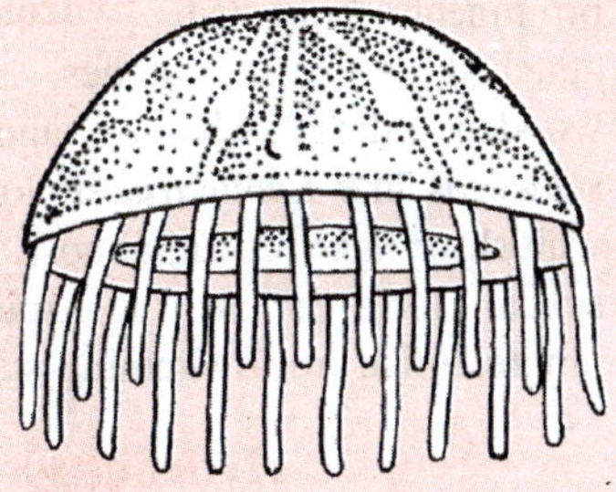

Unlike *Hydra,* which is a solitary freshwater polyp, most hydrozoans are colonial, strictly marine and having both polyp and medusa stages in their life histories. A typical example is *Obelia* which is a marine colonial hydrozoan of very small size. Its most common example is *Obelia geniculata.*

Obelia geniculata

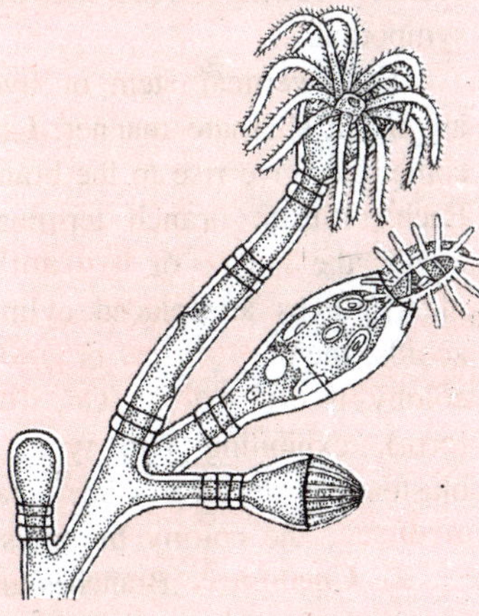

SYSTEMATIC POSITION

Phylum	Coelenterata
Class	Hydrozoa
Order	Hydroida
Suborder	Leptomedusae
Family	Eucopidae
Genus	*Obelia*
Species	*geniculata*

HABITS AND HABITAT

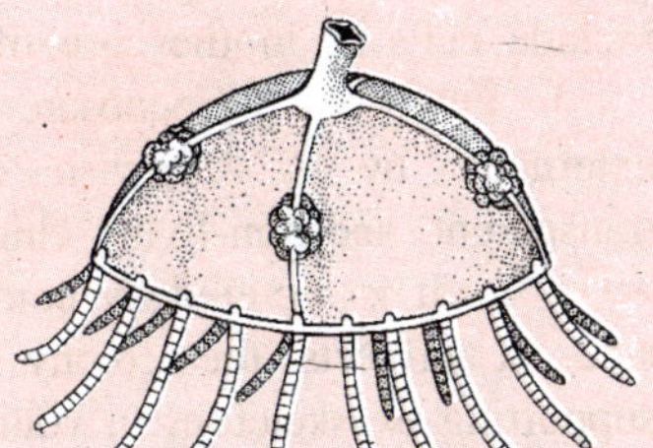

Obelia is a typical sedentary, marine and colonial hydroid having cosmopolitan distribution. It is abundant in both Atlantic and Pacific coastal waters and found up to a depth of 80 meters. It occurs as asexual and sexual forms. The asexual form is a prominent branched **hydroid colony** found attached to rocks, stones, shells of animals, wooden piling, wharves and fronds of large seaweeds like *Laminaria.* It looks like a delicate, whitish or light brown, almost fur-like growth. The sexual form is an inconspicuous bell or an umbrella-like free-swimming stage, called **medusa**.

1. Hydroid Colony

[I] Morphology of Colony

1. External features. Hydroid colony of *Obelia* is delicate, semitransparent, and whitish or light brown in colour. It consists of vertical branching stems, called **hydrocauli** (singular, **hydrocaulus**), rising 2 to 3 cm. above a root-like **stolon** or **hydrorhiza**. Both are of the thickness of an ordinary sewing thread. Growth of the colony is sympodial.

Each vertical stem or hydrocaulus branches are in an alternate manner. Lateral branches may sometimes give rise to the branches of third order. Each ultimate branch terminates in a nutritive zooid, the **polyp** or **hydranth**. In the axils of older polyps are placed cylindrical reproductive zooids, the **blastostyles** or **gonangia**. Thus *Obelia* colony is **dimorphic** (Gr., **dis**, two + **morphe**, form), exhibiting two types of zooids. When blastostyles develop saucer-shaped bodies, called **medusae**, the colony becomes **trimorphic**.

2. Coenosarc. Branches and zooids of colony consists of an inner, tubular and living portion, the **coenosarc**. It consists of a cellular wall enclosing a canal, the **coenosarcal** or **gastrovascular cavity**, which is continuous with those of zooids. The cellular wall consists of two layers (diploblastic), an outer **epidermis** and an inner **gastrodermis**, with a gelatinous mesogloea in between. Both epidermis and gastrodermis include cells as in the case of *Hydra.*

3. Perisarc. Coenosarc is surrounded externally by a yellowish or brown, tough, transparent, and non-living chitinous layer, called **perisarc**. It is secreted by epidermis on its outer side. It protects the colony and serves as a supporting exoskeleton. In young colony, perisarc is in close contact with coenosarc; but in older colony, it becomes separated by a space, except at occasional spots where the epidermal cells extend outwards to meet it.

At the base of each zooid, perisarc bears annular constrictions, called **perisarcal annuli**. Usually a single annulus is also present on the main stem, just below the base of each lateral branch. Perisarc provides rigidity to the colony, but annular constrictions permit limited swaying movements under the influence of water currents.

4. Zooids. The term 'zooid' (Gr., **zoon**, animal + **oid**, form) is used for an individual form of a coelenterate colony. As already noted, hydroid colony of *Obelia* is **dimorphic**, exhibiting two types of individuals or zooids : (i) **polyps** or **hydranths** and (ii) **gonangia**. These two types of zooids differ morphologically as well as physiologically.

(a) Polyp or Hydranth. Polyp (Gr., **polypus**, many-footed) or hydranth (Gr., ***hydra***, water serpent + **anthos**, flower) is the nutritive zooid of colony. A nutritive zooid is also called a **gastrozooid** or **trophozooid**. It is yellowish in colour and has the form of a radially symmetrical and cylindrical or conical hollow sac resembling a miniature *Hydra.* Its narrow proximal end is continuous with coenosarc of hydrocaulus. Its distal end is produced into a conical elevation, the **manubrium** or **hypostome**, which measures about one-third of the total length of hydranth. Hypostome bears a terminal aperture, the **mouth**, which is capable of dilation and contraction. Rising from the base of hypostome is a circlet of upto 30 filiform tentacles which are much longer than hypostome. Tentacles of *Obelia* are solid, unlike those of *Hydra,* which remain hollow.

Perisarc, surrounding the hydranth, dilates to form a loose, cup-like, transparent protective sheath, the **hydrotheca**. It remains open at the distal end. At the base, it is produced internally into a ring-like horizontal **shelf** on which rests the base of hydranth. As hydranth is capable of contraction through the distal open end of hydrotheca, the horizontal shelf checks it from being withdrawn into the perisarc of hydrocaulus from where it may not be protruded again.

(b) Blastostyle or gonangium. When hydrocaulus has reached full development, it

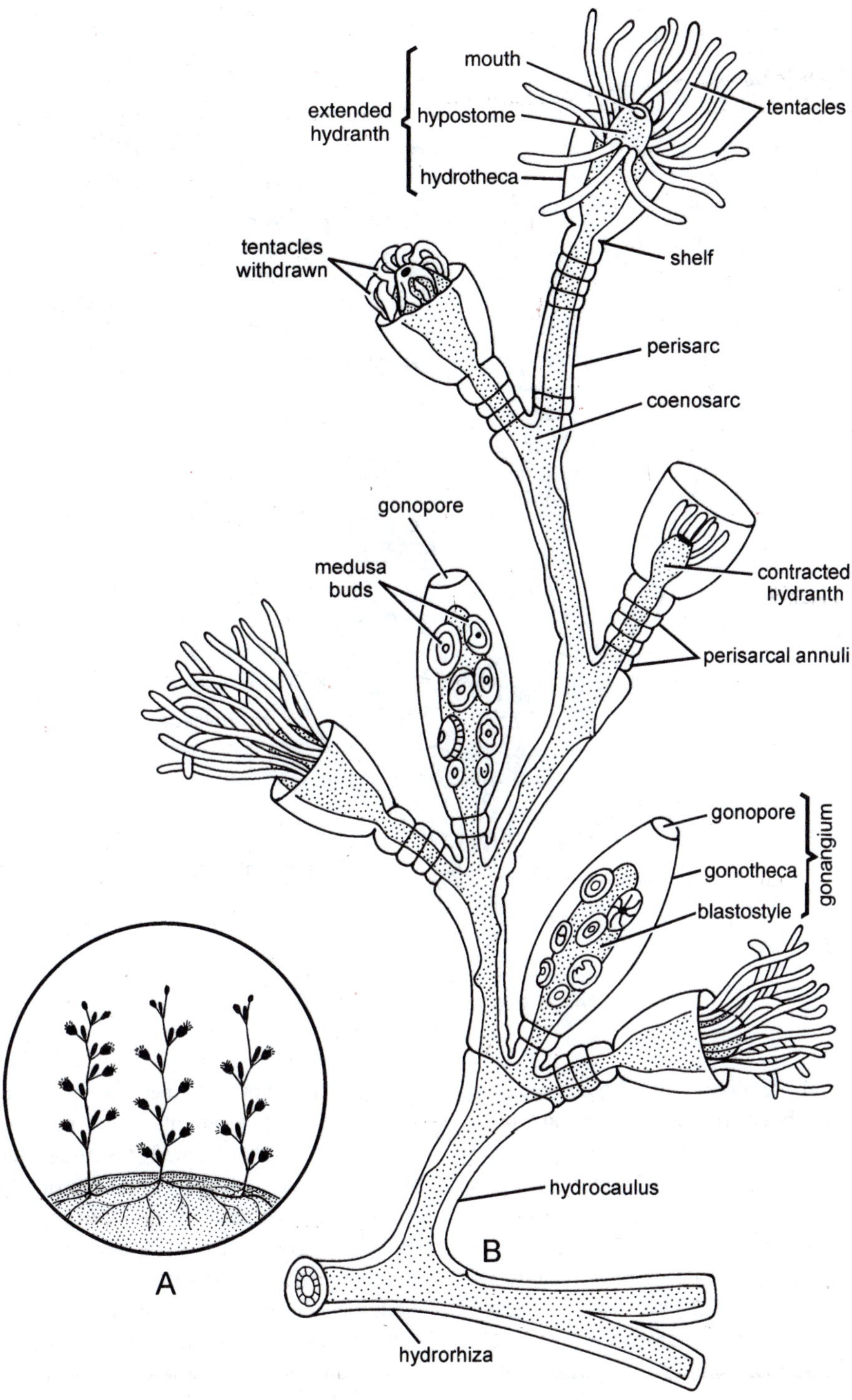

Fig. 1. *Obelia.* **A.** Natural size of colony. **B.** A part of colony under microscope.

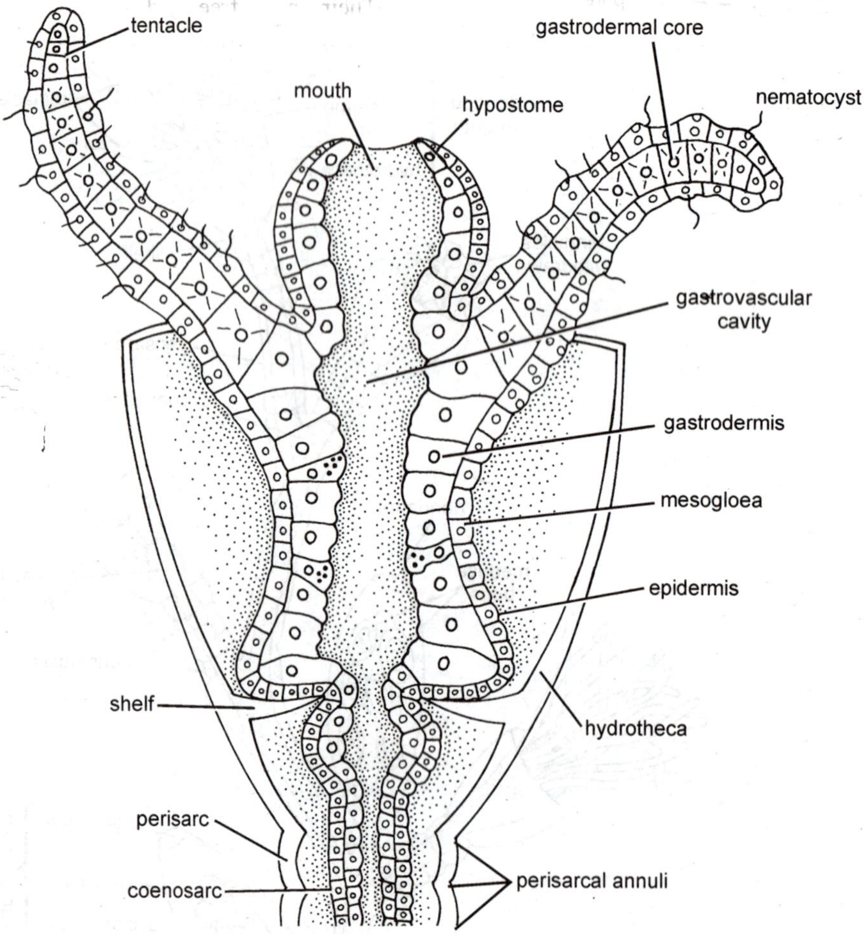

Fig. 2. *Obelia.* A hydranth in V.S.

produces special club-shaped bodies called **blastostyles**, or **blastozooids**, or **gonozooids**. These are less numerous than hydranths and occur in the axils of older hydranths. A blastostyle has the form of a narrow elongated tube. Its distal end ends blindly. It is devoid of mouth and tentacles and has a much reduced gastro-vascular cavity. The perisarc, covering blastostyle, forms a loose, transparent, vase-like capsule, the **gonotheca**, having a terminal collar-like constriction.

The blastostyle, by lateral asexual budding, produces sexual individuals, called **medusae** or **gonophores**. Several medusa buds, in various stages of development, can be seen on a blastostyle. Fully formed medusae detach from blastostyle to escape into surrounding water through an aperture, the **gonopore**, formed by the rupture of gonotheca at its distal end. Gonotheca together with blastostyle and gonophores or medusae is referred to as the **gonangium** (Gr., **gonos**, seed + **angeion**, vessel) (Fig. 3).

Hydroid *Obelia* colony with medusae becomes **trimorphic**. *i.e.*, it bears three types of zooids, namely **hydranths** (gastrozooids), **blastostyles** (gonozooids) and **medusae** (gonophores).

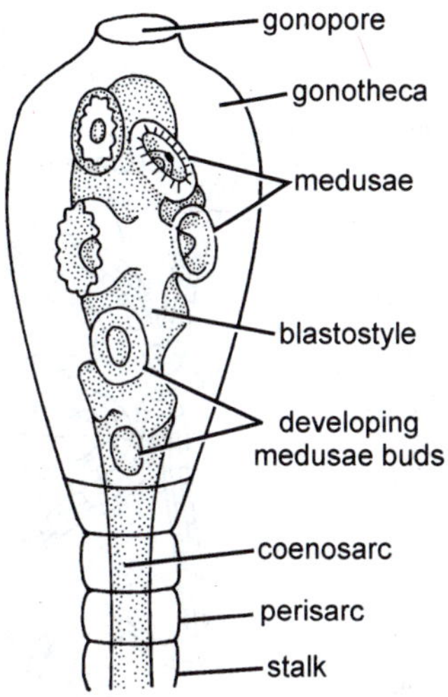

Fig. 3. *Obelia.* A gonangium.

[II] Histology of Colony

Cellular structure of hydranths, blastostyles and coenosarc is similar to that of *Hydra.* Their body wall is composed of two layers of cells, outer **epidermis** and inner **gastrodermis**. Between them, and secreted by them, is a thin delicate, transparent, non-cellular gelatinous layer called **mesogloea** or **supporting lamella**.

1. Epidermis. Epidermis is thin and chiefly made of large, conical and columnar **epithelio-muscle** cells with their broad bases outwards. Their narrow inner ends are prolonged into unstriped **muscle fibres**, arranged longitudinally between epidermis and mesogloea. They serve for rapid shortening of body and tentacles. Small and round **interstitial cells** are practically absent in spaces between inner ends of epithelial cells. Epidermis also contains stinging cells or **nematocysts**, which are specially abundant on tentacles forming annular batteries. Nematocysts of *Obelia* are of **penetrant** type. A nerve-net, composed of large and branched **nerve cells**, is present on each side of the mesogloea. Unlike *Hydra,* germ cells are not formed in the epidermis of polyps.

2. Gastrodermis. Gastrodermis lines the gastrovascular cavity throughout colony. It consists chiefly of large columnar **nutritive-muscle cells.** Their inner free ends can form pseudopodia to engulf and digest food particles, as in *Amoeba.* Sometimes, pseudopodia are replaced by long flagella, which bring about a constant movement of food-particles in gastrovascular cavity and circulation of digested food in coenosarcal canal. In hypostome, outer ends of these cells are produced into unstriped muscle fibres, arranged circularly, and serving to close mouth or reduce gastrovascular cavity. Among nutritive-muscle cells are present smaller and narrower **gland cells** with granular protoplasm. They secrete digestive juices in gastrovascular cavity. As already mentioned, tentacles are solid, having a gastrodermal core made of a single row of small, cylindrical and greatly vacuolated cells with thick cell-walls.

[III] Physiology of Colony

1. Movement. Hydroid *Obelia* colony, being sessile, does not exhibit bodily movements from one place to another. Annular constrictions of perisarc, however, permit slight swaying movements under the influence of water currents. The zooids, specially hydranths, can contract and expand their body and bend their tentacles considerably.

2. Nutrition. *Polyps* or *Hydranths* are the **gastrozooids**, *i.e.*, the nutritive or feeding zooids of colony. They are mostly carnivorous and prey upon small aquatic crustaceans, nematodes and other worms. Their tentacles, armed with nematocysts, capture the prey and transfer it to mouth. Digestive juices, secreted by gland cells of gastrodermis, bring about partial **extracellular digestion** of food in gastrovascular transfer of polyps. Flagella of gastrodermis beat and polyps contract rhythmically to circulate partly-digested food through gastrovascular cavity of the entire colony. Gastrodermal cells with their pseudopodial processes engulf small pieces of the partly digested food and digest them **intracellularly**. Digested food diffuses into the cells of entire colony. Undigested food is **egested** through mouth as there is no anus. Zooids of colony exhibit a great deal of **physiological coordination** as food captured by polyps is shared by the entire colony, including blastostyles.

3. Respiration and excretion. There are no special organs of respiration and excretion. Oxygen from surrounding water diffuses directly into cells and carbon dioxide and nitrogenous excretory products (chiefly ammonia) diffuse out. Water regularly enters the gastrovascular cavity of colony through mouths of polyps, thus establishing direct contact with the gastrodermal cells to facilitate the exchange of materials.

4. Asexual reproduction. *Obelia* colony does not reproduce sexually, but it propagates by the asexual method of **budding**. Horizontal roots or hydrorhizae sprout new vertical stems or hydrocauli which increase the number of individuals by budding. Blastostyles are specialized reproductive zooids forming medusae by budding. A special mode of asexual reproduction may occur, when water temperature exceeds 20°C. In this case, buds destined to form gonangia break off, settle on substratum and each gives rise to a new colony by stolon-like growth.

2. Medusa

Medusae are modified zooids meant for sexual reproduction. They arise from blastostyles by a process of asexual budding.

[I] Development of Medusa

In spring and summer, a large number of medusa buds, in various stages of development, can be seen on a blastostyle (Fig. 4). Medusa-formation begins as a small **outpushing** or hollow **protuberance** on the wall of blastostyle. It soon enlarges into a **vesicle**, connected with blastostyle by a narrow stalk. Its cavity is continuous with that of blastostyle and its wall has same cellular layers as blastostyle, *i.e.*, outer epidermis, inner gastrodermis and an intermediate mesogloea. Apical epidermis of vesicle now splits into two layers. Inner layer again splits and acquires a small cavity called **bell-rudiment**. Innermost epidermis and gastrodermis now evaginate to form the prospective **manubrium**, projecting into bell rudiment which subsequently enlarges to become **sub-umbrellar cavity**. It is lined throughout by epidermis and closed externally by two layers of outer epidermis, running from one margin of umbrella to other. As manubrium elongates the outer epidermis is broken through, leaving a narrow circular shelf, called **velum**, which projects inwards from the margin of umbrella. Velum is permanent and conspicuous in most hydroid medusae, but it diminishes to become vestigial in *Obelia*. Later, the **mouth** opens at the apex of manubrium, **tentacles** arise as finger-like hollow outgrowths from circumference of umbrella and finally the stalk connecting the medusa bud with blastostyle breaks. The young medusa, now fully formed, is free and it escapes through **gonopore** of **gonangium** to lead a free-swimming existence. In a few months it grows to full size.

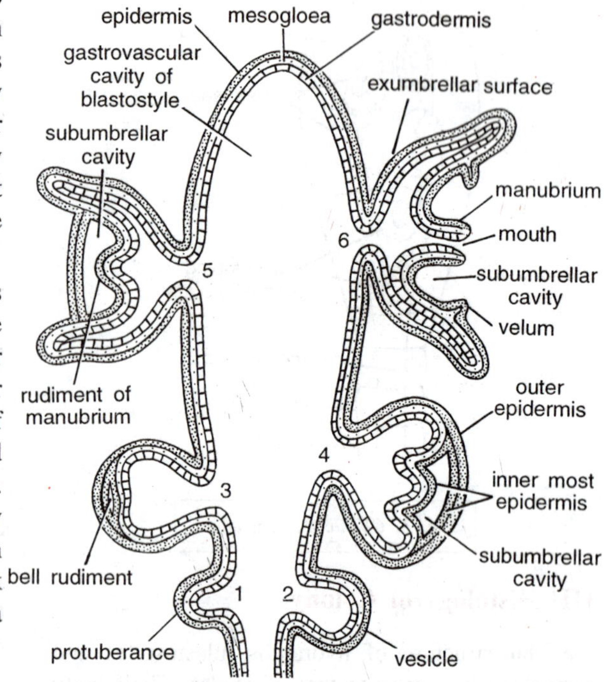

Fig. 4. *Obelia,* Blastostyle in V.S. showing various stages (1 to 6) of development of a medusa.

[II] Morphology of Medusa

1. Shape and size. A full grown medusa of *Obelia* is like a radially symmetrical tiny umbrella, bell or shallow saucer, measuring 1 or 2 mm in diameter. Outer convex surface of the umbrella is known as **ex-umbrella**, while inner concave surface as **sub-umbrella**. Sub-umbrellar surface shows four **radial canals** and a **circular canal**

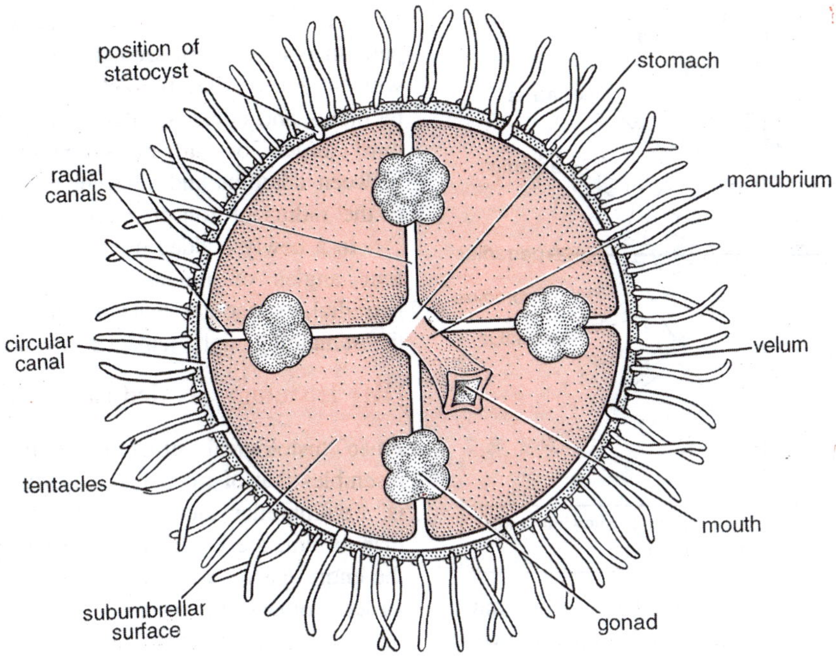

Fig. 5. *Obelia.* A medusa in oral view.

of gastrovascular system. A mature medusa bears four **gonads**, one in the middle of each radial canal.

2. Manubrium. From the centre of concave sub-umbrellar surface hangs down a short, hollow, handle-shaped process, the **manubrium** (L., **manus**, handle), bearing at its free distal end a four-sided **mouth** surrounded by four **oral lobes**. Handle-like manubrium and circular bell together make the medusa appear like a complete umbrella.

3. Velum. Circular edge of umbrella is produced inwards into a very narrow, rudimentary fold or shelf, called **velum** (L., **veil**). It is quite prominent in other hydrozoan medusae, but is insignificant in *Obelia* medusae.

4. Tentacles. Rim or margin of umbrella also bears numerous short **tentacles**. These are highly contractile and beset with nematocysts. Their bases are somewhat swollen to form **tentacular bulbs** that may lodge the sense organs **statocysts** and serve as sites for nematocyst-formation. A freshly budded medusa bears only 16 tentacles, the number of which gradually increases with age. Medusa derives its name due to presence of tentacles which remind of snake-entwined hair of 'Medusa', the female monster of Greek mythology.

5. Gastrovascular cavity. The rectangular **mouth** leads into a narrow passage running through the manubrium, called **gullet**. It is followed by a dilated **stomach** lying at the base of manubrium and occupying the central part of umbrella. Four narrow **radial canals** extend from stomach to the margin of umbrella. Radial canals run equidistant and at right angles to each other. They open into a **circular canal** or **ring canal**, running close and parallel to the free margin of umbrella. Manubrium, stomach and canals are lined by gastrodermis.

6. Nervous system. Nervous system is essentially like that of *Hydra.* On each side of mesogloea, nerve cells belonging to epidermis as well as gastrodermis, form nerve nets. Nerve cells are especially concentrated along the margin of bell forming two **circular nerve rings**, one just above and other just below the base of velum. Formation of nerve rings along the bell margin is correlated with the concentration of muscle-tails (muscle ring) and presence of sense organs

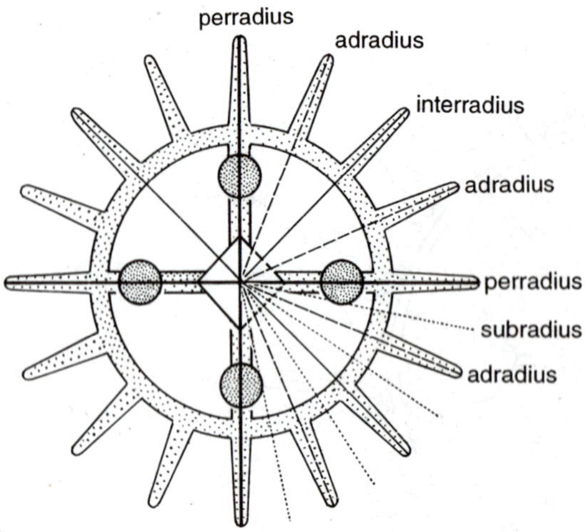

Fig. 6. *Obelia.* A young medusa showing radial symmetry and various orders of radii.

(statocysts) in this region. Upper or inner ex-umbrellar nerve ring supplies the tentacles while lower or outer nerve ring supplies the sub-umbrellar musculature and statocysts.

7. Radial symmetry and radii. Like polyp, medusa is radially symmetrical so that the position of tentacles and other body parts can be shown with reference to particular radii of the bell.

Four radial canals mark out the four principal radii, called **perradii**. Tentacles placed against the four radial canals are thus termed **perradial tentacles**. Bisecting the angle between any two perradii is a radius of second order or **interradius**, with an **interradial tentacle**. Midway between a perradius and adjacent interradius is the radius of third order or **adradius**. Similarly, midway between adradius and adjacent per or inter-radius is the radius of fourth order or **subradius**.

In a newly budded medusa, four radial canals, four angles of mouth and four tentacles are perradial, four other tentacles are interradial and the remaining eight tentacles are adradial.

[III] Histology of Medusa

Basic histological structure of medusa closely resembles that of hydranth. All exposed parts, that is, ex-umbrellar and sub-umbrellar surfaces and manubrium, are covered by **epidermis**. Similarly, the entire gastrovascular canal system, comprising gullet, stomach, radial canals and circular canal is lined by **gastrodermis**. Gastrodermal cells lack contractile extensions, so muscular system is restricted to epidermal layer only. It is better developed than in *Hydra*. Both epidermis and gastrodermis are continuous along the margin of mouth. Gelatinous **mesogloea**, lying between epidermis and gastrodermis is much thickened, particularly towards ex-umbrella, forming the main bulk of body. It is demarcated from the epidermis and gastrodermis by a thin layer of jelly.

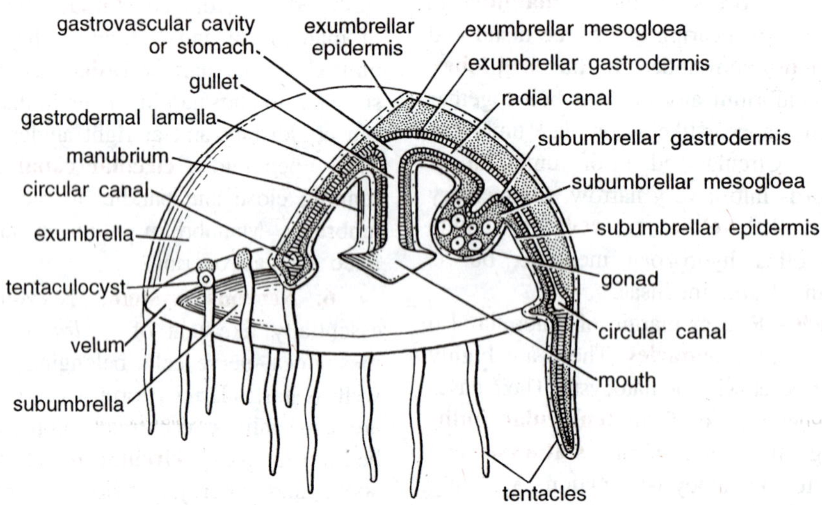

Fig. 7. *Obelia.* A medusa partly cut away to show internal structures and relations of parts.

Mesogloea consists of about 95% water. It is devoid of cells but contains fibre that are secreted by both epidermis and gastrodermis.

Interstitial cells are mainly accumulated at the swollen bases of tentacles and give rise to **cnidoblasts** which are particularly abundant on margin of umbrella, on tentacles and around mouth. **Sensory cells** are most abundant around mouth and tentacles.

Inside the bell, between radial canals and between ex-umbrellar and sub-umbrellar layers of epidermis, there is a thin sheet of gastrodermis, called **gastrodermal lamella**. It is presumbly formed by the fusion of upper and lower layers of gastrodermis. Velum is composed of a double layer of epidermal cells enclosing a middle narrow layer of mesogloea, there being no gastrodermis. This type of velum is called **true velum**. Tentacles are solid, each containing a core of vacuolated gastrodermal cells covered by epidermis.

[IV] Physiology of Medusa

1. Movement. Unlike sedentary hydroid colony, medusa usually floats passively in water. It is simply drifted here and there by water currents with manubrium hanging downward and tentacles swaying freely. It also swims actively by muscular contractions started by impulses which originates in the nerve at the umbrellar margin. By rhythmic contraction and expansion of bell, the water in sub-umbrellar cavity is propelled behind (**hydropropulsion**), and medusa moves forward in a series of jerks. Closure of bell is effected by contraction of epidermal muscle-tails that are best developed on sub-umbrellar surface, specially along the margin of bell, where they form a muscle ring, but remain undeveloped on exumbrellar surface. Opening of bell is brought about mainly by the elastic mesogloea regaining its original shape and to some extent due to contraction of muscle tails in the middle of upper surface. During swimming, body may be tilted, thrown aside, or more commonly turned inside out, so that the sub-umbrellar surface becomes outer and convex, with the manubrium springing up from its apex.

2. Nutrition. Medusa is strictly carnivorous. Food includes minute worms, nematodes, insects, crustaceans, etc. These are captured by nematocyst-bearing tentacles and ingested by the highly contractile mouth. Prey is digested exclusively in stomach. Digestion is both extracellular and intracellular like that of hydranth. Digested food is distributed to entire medusa through the system of radial and circular canals.

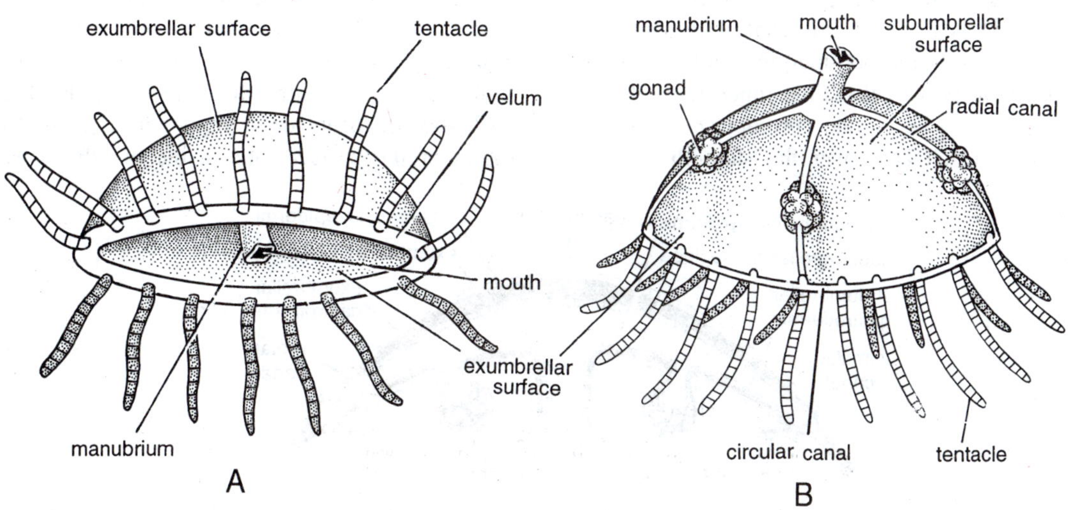

Fig. 8. *Obelia.* A swimming medusa. **A.** Swimming in normal way. **B.** Swimming with everted umbrella.

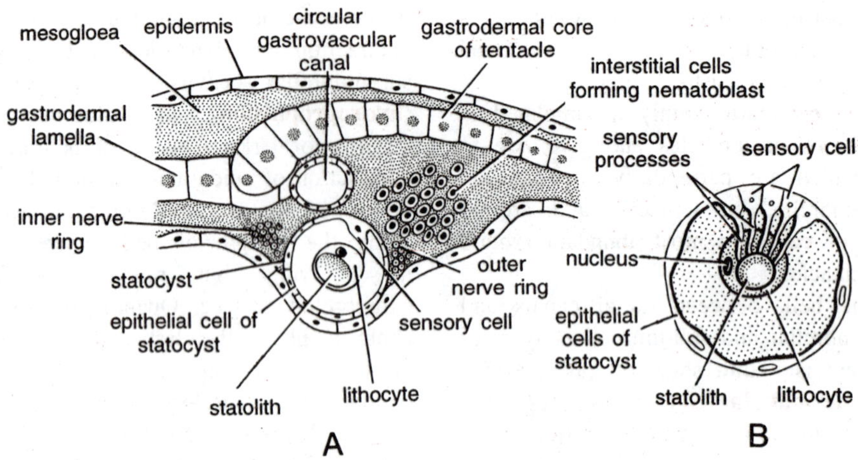

Fig. 9. *Obelia.* **A.** Base of an adradial tentacle in L.S. showing statocyst. **B.** Statocyst magnified.

3. Respiration and excretion. Respiration and excretion are carried on individually by each cell by diffusion, as in hydranth, and special organs are absent.

4. Statocysts and equilibrium. Medusa has eight marginal receptor organs or **statocysts**, situated at the bases of eight adradial tentacles on sub-umbrellar surface, just inside the bell-margin. Each consists of a minute fluid-filled ectodermal sac. Its cavity contains a movable round particle of calcium carbonate, called **statolith** or **otolith**, secreted by a large cell, termed **lithocyte**. Wall of statocyst is made of epithelial cells, which are sensory towards the bell-margin, being connected basally with nerve cells. Free inner ends of these sensory cells bear fine protoplasmic sensory processes which arch over the statolith.

Statocysts are considered to be the organs of **equilibrium** and muscular **coordination**. Their presence in medusae is associated with their active free-swimming habit. During swimming if body becomes tilted, the statolith falls over the tilted side against the processes of sensory cells which become stimulated. In this way, nerve impulse is created and transmitted to the nerve ring. As a result, the muscle tails of the stimulated side contract more rapidly and medusa is brought back to its normal horizontal position.

5. Sexual reproduction. Medusae are the sexual reproductive zooids, or **gonozooids**, possessing **gonads**. They are **dioecious**, *i.e.*, testes and ovaries are borne by separate individuals. Each medusa bears only four gonads situated on sub-umbrellar surface, one in the middle of each

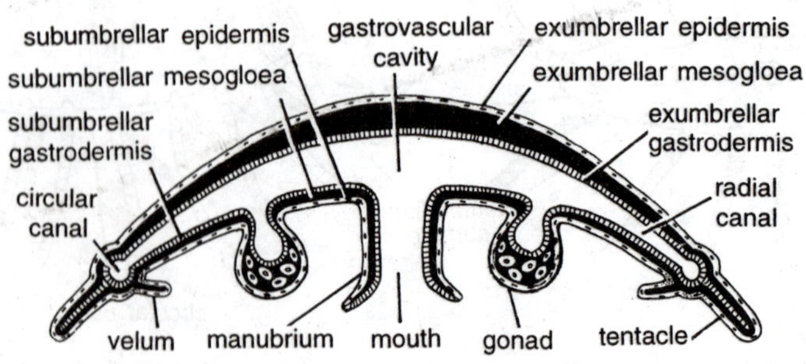

Fig. 10. *Obelia.* Vertical section of medusa showing gonads.

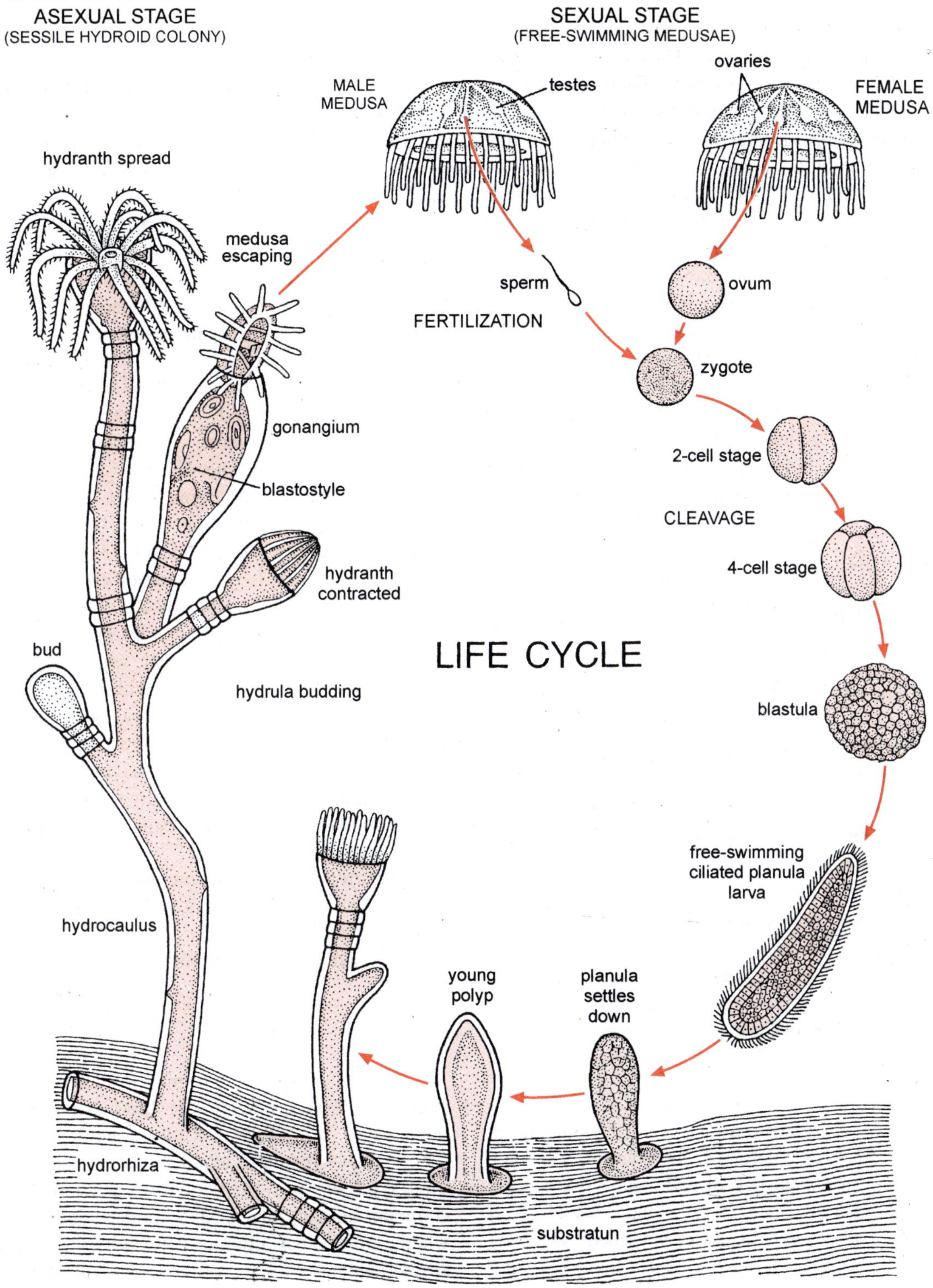

Fig. 11. Stages in the life cycle of *Obelia.*

radial canal. Gonads are formed as ventral diverticula of radial canals. They mature after the medusae escape from gonangia. Each gonad (testis or ovary) is an ovoid, knob-like body, having an outer covering of epidermis, continuous with that of sub-umbrella, and an inner lining of gastrodermis, continuous with that of radial canal. The space between these two layers is filled with a mass of **sex cells**, which become differentiated into sperms or ova, as the case may be.

Sex cells originate in the epidermis of manubrium, while medusa is still attached to the blastostyle. They soon pass into gullet through gastrodermis and finally make their way into gonads. Here they undergo maturation divisions and become gametes. These gonads are not true gonads but only aggregations of developing gametes. Outer wall (epidermis) of mature gonads ruptures to release the gametes in water.

LIFE HISTORY OF *OBELIA*

As already described, life-history of *Obelia* includes both asexual (hydroid colony) and sexual (medusa) generations, which regularly alternate with each other to complete the life-cycle. There are separate male and female medusae producing sperms and ova, respectively.

[I] Fertilization

Fertilization usually occurs externally in sea-water where the gametes are set free, or sperms may be carried by water currents to female medusae, where they fertilize eggs *in situ*. Parent medusae die soon after liberating the gametes.

[II] Development

1. Cleavage. Fertilized egg or **zygote** undergoes equal and complete or holoblastic cleavage resulting in a solid ball of cells, the **morula**. This is followed by the **blastula** stage which is a hollow ball of cells. Its cavity is termed **blastocoel** and the single layer of cells lining it, the **blastomeres**. Gastrulation occurs by two processes. First, the inner surface of blastomeres cut off new cells into blastocoel. This is called **delamination**. Second, cells detatch from one pole *(***unipolar ingression***)*, migrate into blastocoel and finally fill it. Thus, the embryo becomes a **solid gastrula** or **stereogastrula**. Its outermost layer is known as **ectoderm** and the inner mass of cells as **endoderm**.

2. Planula larva. The gastrula elongates, the ectodermal cells acquire cilia and an elongated free-swimming ciliated **planula larva** results. Anterior end of planula is broader than its posterior end. Soon the solid endoderm splits and develops a cavity, the **enteron**. Now planula becomes a truly **two-layered** larva with an outer ciliated ectoderm and an inner endoderm. Histologically, the larva possesses columnar ectodermal, sensory, nerve and gland cells, muscular processes and nematocysts.

3. Hydrula. After a brief and active free-swimming existence, planula larva settles down, attaches itself by its anterior end to stone, weed, wood or some other solid object in sea and undergoes metamorphosis. Its proximal end forms a basal disc for attachment, while the distal free end develops a manubrium with a mouth and a circlet of tentacles. The larva now closely resembles a simple Polyp or *Hydra* and is called **hydrula**. By repeating an extensive process of asexual budding, hydrula gradually changes into a new complex of branching *Obelia* colony similar to the parent.

Occurrence of free-swimming medusa and larva in the life-history of a fixed organism, like *Obelia,* is of distinct advantage, as it helps in dispersal and prevents overcrowding of the species.

[III] Alternation of Generations and Metagenesis

Alternation of generations may be defined as a phenomenon whereby, in the life-history of an organism, a **diploid asexual phase** and a **haploid sexual phase** regularly alternates with each other. This type of true alternation of generations is common among plants, like mosses and ferns, where an asexual diploid (**saprophytic**) and a sexual haploid (**gametophytic**) generation alternate regularly in life-cycle. In fern, the plant *(***diploid saprophyte**) produces **haploid spores**, which develop into flat, green, small heart-shaped **haploid gametophytes**. These produce **haploid ova** and **sperms**. After fertilization, they give rise to a new **diploid sporophyte**. Thus completing one life-cycle.

In *Obelia,* life-cycle includes two clearly defined phases : a fixed polypoid phase (hydroid colony) and a pelagic medusoid phase. Hydroid colony has no gonads and reproduces by asexual budding to give rise to medusae. On the other hand, medusae reproduce exclusively by sexual method (ova and sperms) to give rise to new hydroid colonies. This fact apparently seems to have given rise to the idea of **alternation of generations**, also called **metagenesis**, in coelenterates, in which an asexual polypoid generation appears to alternate regularly with a sexual medusoid generation.

But, in *Obelia,* medusoid phase does not represent a true haploid sexual generation, because: (i) Medusa arises from blastostyle (diploid) by a process of asexual budding. It implies that medusa too is a diploid zooid. (ii) Sex cells do not originate in medusa, but in the epidermis of blastostyle, from where they migrate into gonads of medusa. These facts show that medusa does not represent a sexual generation. It is simply a free-swimming diploid zooid specialized for dispersal of gametes of the sedentary hydroid colony. In fact, the so-called sexual generation in *Obelia* is indistinct and represented by haploid gametes only.

Thus, it is clearly impossible to differentiate between sexual and asexual generations in *Obelia.* Asexual hydroid colony and sexual medusa merely represent different phases or zooids, an example of polymorphism, and belong to a single diploid generation, so that a true alternation of generations can not be said to occur in *Obelia.*

In coelenterates (*e.g.*, *Obelia),* a regular alternation between fixed asexual hydroid and free-swimming medusoid phases, both of which are diploid, has been termed **metagenesis** by some workers. But, according to Hyman, concept of metagenesis should be discarded as there are no haploid and diploid generations in coelenterates. According to this view, medusa is regarded to be a completely evolved coelenterate while polyp is probably a persistent larval stage.

COMPARISON OF POLYP (HYDRANTH) AND MEDUSA

1. Dissimilarities between polyp and medusa. The main differences between a polyp and a medusa can be summarised as in *Table 1*.

2. Similarities or homology between polyp and medusa. Striking as is the difference between a polyp and a medusa, they are strictly homologous or fundamentally similar structures. The notable features of similarity between polyp and medusa are as given below :

(1) Body is radially symmetrical.

(2) Both are diploblastic, derived from two germ layers, ectoderm and endoderm.

(3) Exumbrellar surface of medusa corresponds with the base of polyp, providing attachment with the parental stem.

(4) Mouth is homologous in both cases, being situated on a similar process, called manubrium. Anus is absent in both.

(5) Stomach, radial canals and circular canal of medusa correspond with the gastrovascular cavity of polyp, lined by gastrodermis in both cases and serving for digestion of food.

(6) Both are carnivorous, capturing and ingesting food with the help of tentacles. Digestion is extracellular as well as intracellular and digested food diffuses throughout body without a circulatory system.

3. Advancement of medusa over polyp. Free-swimming habit is mainly responsible for the complexity of medusa accompanied by a differentiation of structures lying along certain radii. Thus, medusa shows several morphological advancements over polyp, some of which are as follows :

(1) Epidermis resembles the epithelium of higher Metazoa, forming a thin, protective and sensitive layer, formed by smaller cells with reduced muscle-tails.

(2) Great power of contractility is due to muscle-cells abundantly present in sub-umbrella and velum.
(3) Enormous development of mesogloea adds to the lightness and buoyancy of floating body and also reduces the gastrovascular cavity to a system of canals.
(4) Nervous system shows the beginning of differentiation into a central nervous system in the form of two distinct nerve rings, and a peripheral nervous system, formed by a diffused network of highly branched nerve cells.
(5) Presence of marginal sense organs at the bases of adradial tentacles is of distinct advantage to medusa as a free-swimming organism.
(6) Development of gonads, accompanied by a free-swimming habit, brings about reproduction with greater chances of dispersal.
(7) Physiologically, medusa shows a distinct improvement in swimming freely on the surface of sea by muscular action and is not sessile like polyp.

DERIVATION OF MEDUSA FROM POLYP

Homology and structural similarity of polyp and medusa can be best explained by the fact that one can be readily derived from the other. Suppose a polyp is inverted so that its manubrium is directed downwards. If its tentacular region is pulled out, a disc-like form results. With further differentiation of basal region, the disc will assume a cup or saucer-like form similar to that of a medusa. It will have an outer convex or

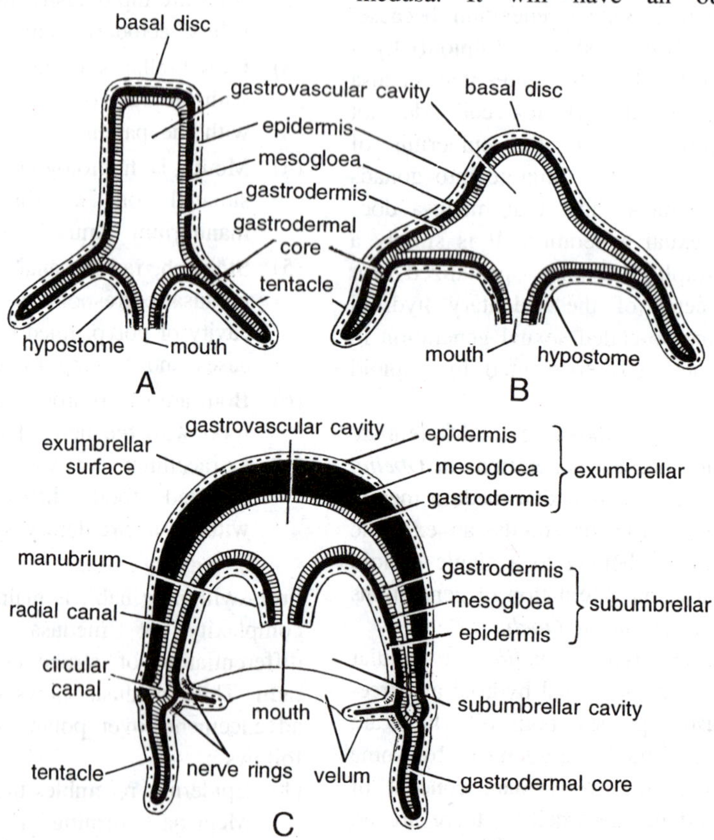

Fig. 12. *Obelia.* Derivation of medusa from polyp. **A.** Polyp in L.S. **B.** Polyp oral-aborally compressed. **C.** Medusa in V.S.

exumbrellar surface and a lower concave or sub-umbrellar surface with manubrium in its centre and tentacles hanging down from the periphery of the saucer, as in medusa. The saucer has a double wall of epidermis, enclosing a narrow space in continuation of the gastrovascular cavity of the manubrium and lined by gastrodermis. Mesogloea between epidermis and gastrodermis develops enormously to form the jelly. Actual medusa-like condition would be produced by the coalescence of upper and lower layers of gastrodermis so as to form the gastrodermal lamella, except in the region of a central gastric cavity, four radial canals and a marginal circular canal. In this way, a medusa is completely derived from a polyp.

Similarly, if exumbrellar surface of a medusa is pulled out and elongated, the two layers of gastrodermal lamella split apart, and the mesogloea reduced, the cylindrical polyp-like form will result.

Table 1. **Differences of Polyp (Hydranth) and Medusa of *Obelia.***

Polyp	Medusa
1. Fixed, rarely free.	1. Free-swimming.
2. Body cylindrically elongated.	2. Body saucer-shaped or umbrella-like.
3. Base attached below so that manubrium is directed upwards.	3. Base above so that manubrium hangs downwards.
4. Tentacles usually 24.	4. 16 tentacles in young medusa, numerous in adult.
5. Mesogloea poorly developed.	5. Mesogloea enormously developed.
6. Body structure simple. Muscles and nervous system simple.	6. Body structure complicated. Muscles and nervous system more developed.
7. Velum absent.	7. Velum present around the margin of umbrella.
8. Mouth circular, without oral lobes.	8. Mouth rectangular, with oral lobes.
9. Gastrovascular cavity simple, without radial and circular canals.	9. Gastrovascular cavity represented by stomach, four radial canals and one circular canal.
10. Sense organs absent.	10. Bases of eight adradial tentacles possess marginal sense organs, called statocysts.
11. Without gonads.	11. With four gonads on radial canals.
12. Reproduces asexually by budding.	12. Reproduces sexually by gametes.

Important Questions

➤➤*Long answer type questions*

1. Give a detailed account of the structure and life-history of *Obelia.*
2. Describe the various types of zooids found in the *Obelia* colony.
3. What is meant by 'metagenesis' ? Explain it with reference to the life-cycle of *Obelia.*
4. Describe the development, structure, physiology and reproduction of a hydrozoan medusa studied by you.
5. How can a medusa be derived from a polyp ?
6. Justify the statement, "striking as is the difference between a polyp and a medusa, they are strictly homologous structures."
7. Draw diagrams of the following :
   (i) Hydranth of *Obelia* in vertical section, (ii) Medusa of *Obelia.* (iii) Life cycle of *Obelia.*
8. Write short notes on : (i) Gonangium, (ii) Metagenesis, (iii) Medusa.

➤➤*Short answer type questions*

1. What is mesogloea ?
2. Is the pneumatophore a modified medusa of polyp ?
3. Write a note on planula.
4. Give the life cycle of *Obelia* and compare it with that of *Hydra.*
5. Illustrate the structure of cnidoblast.
6. Justify the difference seen between a polyp and a medusa of *Obelia* in relation to their different modes of life.

7. Define polymorphism. Explain it with reference to the structure of *Obelia* colony.
8. Illustrate the life cycle of *Aurelia* and *Obelia.* Point out 5 major differences between the two.
9. What are homologous organs ? Explain that a polyp of *Obelia* is homologous with its medusa.
10. List out the differences between polyp and a medusa.
11. Describe type of zooid in *Obelia.*
12. Justify, giving atleast 5 reasons that metagenesis is a significant phenomenon seen in the life cycle of *Obelia.*
13. Describe the homology between polyp and medusa of *Obelia.*
14. Explain the formation of medusa in *Obelia* within 10 sentences.
15. Explain how the medusa of *Obelia* moves in the water ?
16. Which of the following is the most acceptable statement regarding similarities of medusa of *Obelia* and *Aurelia.*
    (a) Mesogloea is noncellular in both.
    (b) Tentaculocyst are present in both.
    (c) Both are ciliated free swimming.
    (d) In both reproductive cells are derived from ectoderm.

➤➤*Multiple choice questions*

1. Based on the feeding habit of Obelian medusa it is called as :
   (a) herbivorous (b) carnivorous
   (c) omnivorous (d) detritous feeder
2. Alternation of generation as seen in *Obelia,* is termed as :
   (a) metamerism (b) dimorphism
   (c) metagenesis (d) metamorphosis
3. *Obelia* is :
   (a) polymorphic (b) dimorphic
   (c) trimorphic (d) monomorphic
4. Medusa of *Obelia* differs from *Aurelia* in having :
   (a) gonads (b) a tentacle
   (c) a rhopalium (d) tentaculocysts
5. The number of tentacles of *Obelia* medusa is :
   (a) 8 (b) 16
   (c) 32 (d) 64
6. Larva of *Obelia* is :
   (a) amphiblastula (b) scyphistoma
   (c) planula (d) parenchymula
7. *Obelia* is a hydroid of :
   (a) fresh water (b) marine water
   (c) brakish water (d) all the water
8. The stolen of *Obelia* is known as :
   (a) hydranth (b) hydrorhiza
   (c) hyporhiza (d) root
9. Coenosarc is a :
   (a) living (b) non-living
   (c) inner, tubular, living
   (d) none-cellular
10. Nutritive zooid of *Obelia :*
    (a) hydranth (b) gastrozooid
    (c) trophozooid (d) all
11. Blastozooid is :
    (a) gonozooid (b) trophozooid
    (c) gastrozooid (d) polyp
12. Manubrium and velum are the parts of :
    (a) hydranth (b) medusa
    (c) both of them (d) none of them
13. Haploid stage in life history of *Obelia :*
    (a) hydranth (b) medusa
    (c) gametes (d) planula
14. How many gonads present in a medusa :
    (a) two (b) four
    (c) eight (d) numerous
15. Statocyst is a sense organ present in :
    (a) polyp (b) medusa
    (c) both (d) none
16. Gonads of *Obelia* occur :
    (a) in hydrula stage (b) on blastocyst
    (c) bases of tentacles of medusa
    (d) radial canals of medusa
17. Most appropriate term to describe the life cycle of *Obelia* is :
    (a) alteration of generation
    (b) neoteny
    (c) metagenesis (d) metamorphosis
18. *Obelia* belongs to the class :
    (a) Siphonophora (b) Hydrozoa
    (c) Coelonterata (d) Acoelomate

Answers

➤➤*Multiple choice questions*

1. (b) 2. (c) 3. (b) 4. (c) 5. (b) 6. (c) 7. (b) 8. (b) 9. (c) 10. (d)
11. (a) 12. (b) 13. (c) 14. (b) 15. (b) 16. (d) 17. (c) 18. (b).

25

Aurelia : A Jelly-Fish

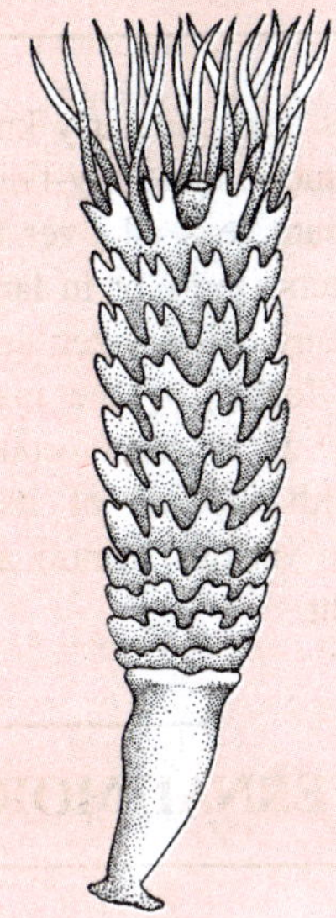

Jelly-fishes are animals which belong to the class **Scyphozoa** of phylum **Coelenterata**. In this class, medusa is the dominant and conspicuous zooid in life cycle while polypoid form is restricted to a short larval stage. The most common scyphozoan jelly-fish is *Aurelia aurita,* whose biology has been treated in detail in the following text. There are 2,000 different species of jellyfish found in world.

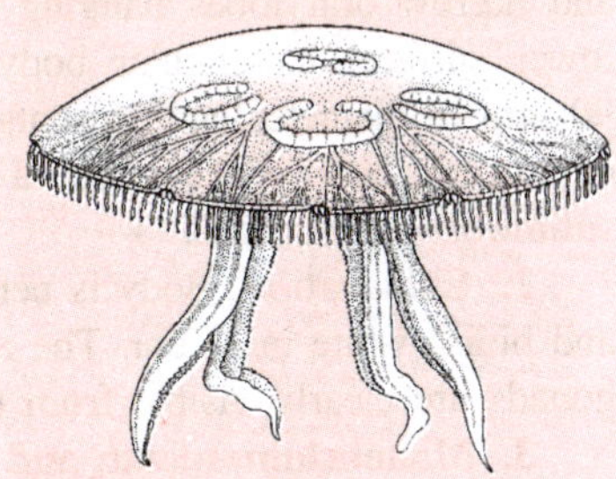

Aurelia aurita

A jelly-fish is not a true fish which is a vertebrate animal with a backbone. Whereas the name jelly-fish is given to this invertebrate coelenterate animal because it is made of a jelly-like substance.

SYSTEMATIC POSITION

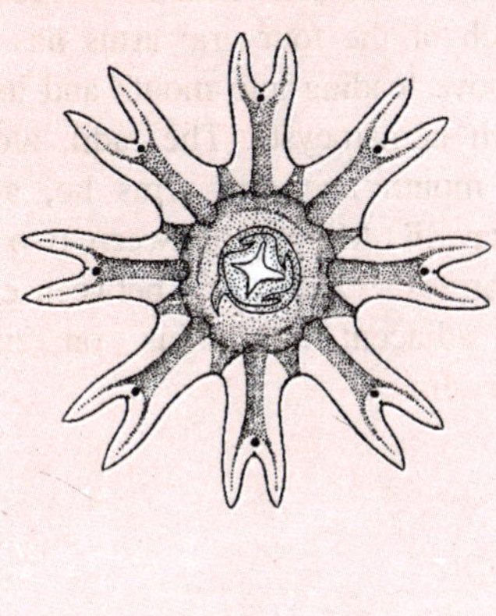

Phylum	Coelenterata
Class	Scyphozoa
Order	Semaeostomae
Family	Ulmaridae
Genus	*Aurelia*
Species	*aurita*

HABITS AND HABITAT

A. aurita is also popularly known as 'moon-jelly'. It is a cosmopolitan jelly-fish, occurring in warm and temperate seas all over the world. It lives in coastal waters singly or in large shoals. It is found either floating with water currents and waves or swimming feebly by the contraction movements of its bell. It is carnivorous, feeding on small animals with the aid of its long oral arms. It responds to various stimuli and is most active in diffuse light.

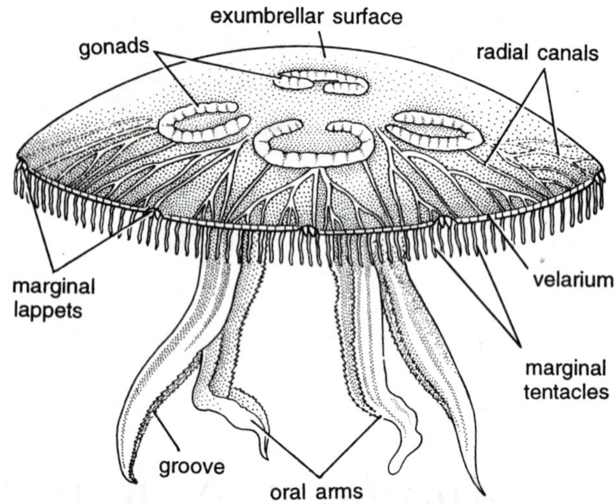

Fig. 1. *Aurelia aurita.* External morphology in dorso-lateral view.

EXTERNAL MORPHOLOGY

1. Shape and size. In general structure, *Aurelia* is like a large version of the medusa of *Obelia.* It is easily recognized by its soft bell or umbrella-shaped body with four red or purple horseshoe-shaped gonads on its upper surface and four long and narrow oral lobes hanging downwards from lower surface. Its circular body measures about 90 mm in diameter and presents a convex aboral or **exumbrellar surface** and a concave oral or **subumbrellar surface**.

2. Colouration. Body is perfectly transparent and bluish-white in colour. The reddish or pinkish gonads are clearly visible from the body surface.

3. Manubrium, mouth and oral arms. From the centre of subumbrellar surface hangs down a very short and inconspicuous **manubrium**. At its free distal end is a square **mouth**, from each corner of which hangs down a long, tapering much-frilled and delicate process, the **oral arm**. Each of the four oral arms has a ventral ciliated groove leading into mouth and its edges are armed with nematocysts. The radii, along which angles of mouth and oral arms lie, are referred to as **perradii**. Midway between two adjacent perradii is an **interradius**, and between each perradius and its adjacent interradius, on either side, is an **adradius**.

4. Subgenital pits. On each interradius, a little distance from mouth, the sub-umbrellar surface bears a circular aperture. It leads into a small shallow cavity, the **subgenital pit**, lying immediately beneath a gonad and of uncertain function.

5. Gonads. Just above each subgenital pit, within umbrella, is a horseshoe-shaped and frilled **gonad**, red or purple in colour. Free arms of all the four gonads are directed towards the centre of umbrella. There is no connection between gonads and subgenital pits.

6. Lappets and tentaculocysts. The circular margin of umbrella or bell is broken into 8 lobes by 8 indentations or notches, 4 of which are perradial and the other 4 interradial. In each notch, there are two delicate leaf-like processes, called the **marginal lappets**. Between lappets lies a small sensory organ, the **tentaculocyst** or **rhopalium**.

7. Marginal tentacles. Between notches or rhopalia, the free edge of umbrella is beset closely with a row of numerous small, delicate and hollow threads or **marginal tentacles**. The tentacles bear batteries of stinging cells or nematocysts.

8. Velarium. Margin of subumbrellar surface, bearing lappets and tentacles, forms a thin and flexible flap, called **velarium** or **pseudovelum**. It

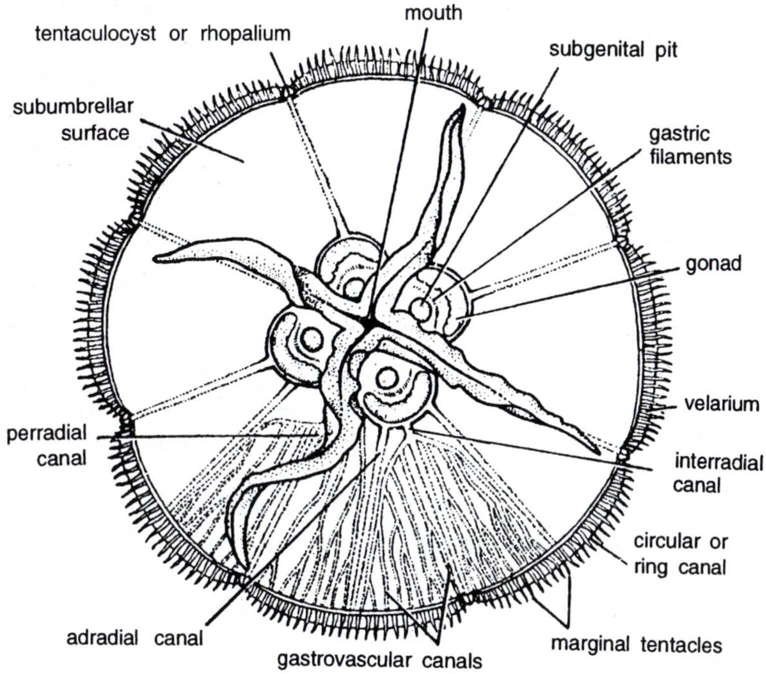

Fig. 2. *Aurelia aurita.* Ventral view.

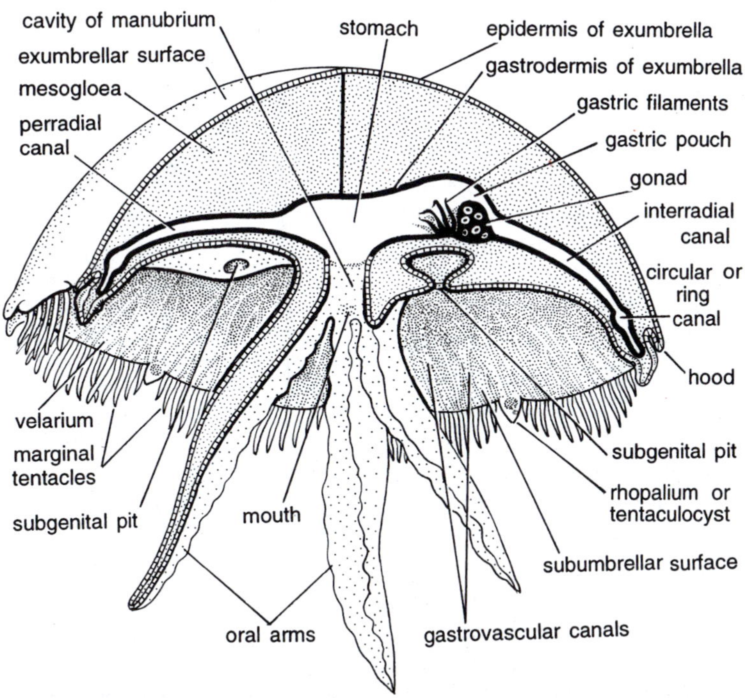

Fig. 3. *Aurelia aurita.* Lateral view with one-third of umbrella cut away to show internal structures and relations of parts.

differs from true velum of *Obelia* in having gastrodermal canals running into it. Such a medusa with a pseudovelum (*e.g.*, *Aurelia)* is called **acraspedote medusa**, while a medusa with a true velum (*e.g.*, *Obelia)* is called **craspedote medusa**.

HISTOLOGY

Basic histological plan of *Aurelia* medusa is more or less the same as that described for *Obelia* medusa. It is **diploblastic** and derived from two embryonic layers, ectoderm and endoderm.

1. Epidermis. All the exposed parts of umbrella, that is, exumbrellar surface, velarium, tentacles, subumbrellar surface including subgenital pits, oral arms and manubrium, are covered by **epidermis**. Gullet is formed by the invagination of epidermis. It consists of epithelial cells (on exumbrellar surface), epithelio-muscle cells (confined to subumbrellar surface) besides sensory cells, nerve cells, gland cells and cnidoblasts. Sensory cells form a sensory epithelium between epithelial cells.

2. Gastrodermis. All the parts of gastrovascular canal system, except gullet, are lined by **gastrodermis**. Gastric filaments are formed by a thin core of mesogloea and a double layer of gastrodermis. Gonads are also gastrodermal structures. Cavities of tentaculocysts and marginal tentacles, being extensions of gastrovascular canal system, are also lined by gastrodermis.

Gastrodermis mainly consists of flagellated columnar endothelial cells. Gland cells are present but nerve cells and muscle processes are wanting. Cnidoblasts are confined to gastric filaments.

In the interspaces between gastrovascular canals, gastrodermis of ex-and sub-umbrellar surfaces fuse to form a thin sheet, of **gastrodermal lamella**.

3. Mesogloea. It constitutes the main bulk of body, forming a thick gelatinous layer between epidermis and gastrodermis. Unlike *Hydra* and *Obelia,* the mesogloea of *Aurelia* is not structureless, but contains numerous branching **elastic fibres** and wandering **amoeboid cells** derived from epidermis. This type of mesogloea is more or less like connective tissue and is known as **collenchyma**. It is **ectomesodermal** in origin and not **endo-mesodermal** as in higher metazoans. The mesogloeal fluid resembles sea water except that it has more potassium and less sulphate.

NEMATOCYSTS

Nematocysts occur on oral arms, ex- and subumbrellar surfaces, marginal tentacles, as well as gastric filaments. They are of three types.

1. Atrichous isorhizas. Capsule is elongated. Thread tube is open at the tip and is without a butt and spines.

2. Holotrichous isorhizas. Capsule is oval and butt is narrow. Thread tube is long, open at the tip and armed with a spiral row of small spines.

3. Heterotrichous microbasic euryteles. Capsule is small. Thread tube is open at the tip and covered by minute spines. Butt is short and its distal dilated portion bears unequal spines.

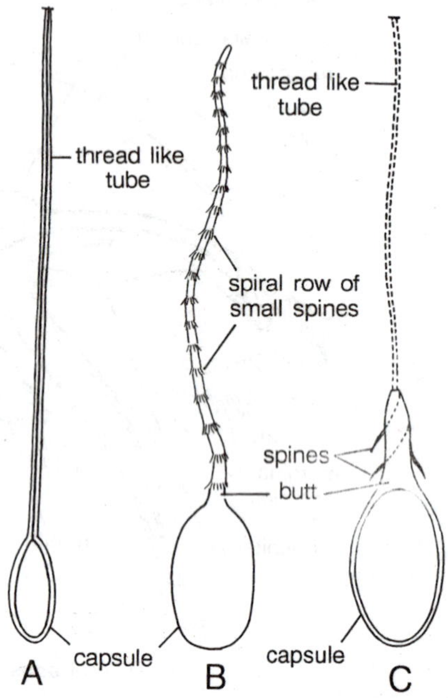

Fig. 4. *Aurelia.* Types of nematocysts. **A.** Atrichous isorhiza. **B.** Holotrichous isorhiza. **C.** Microbasic eurytele.

MUSCULATURE AND LOCOMOTION

1. Musculature. *Aurelia* possesses a well-developed musculature mainly confined to subumbrellar surface. It is formed by muscle processes of epithelio-muscle cells of epidermis. A broad, circular and peripheral muscle band, known as **coronal muscle**, extends along the periphery of sub-umbrella. Conspicuous **longitudinal muscles** are present in the tentacles, manubrium and oral arms. From manubrium to coronal muscle extend **radial muscles** along the main radii of umbrella. Coronal and longitudinal muscles of tentacles are striated, while others are unstriated.

2. Locomotion. The highly developed musculature brings about swimming movements of *Aurelia,* during which the ex-umbrellar surface is kept upwards. Rhythmic contractions of circular muscle force water out from the sub-umbrellar cavity, like a jet. As a result, body is propelled forward or upward. This type of jet-propulsion is known as **hydropropulsion**. When contractions stop, the body gradually sinks to bottom. Horizontal movements depend on wave-action and currents. Gelatinous mesogloea also helps in buoyance because of its low density. If the body is tilted, equilibrium is maintained with the help of 8 tentaculocysts.

DIGESTIVE SYSTEM

[I] Gastrovascular Canal System

The rectangular **mouth** leads into a short **gullet**, within the manubrium, which opens into a spacious, four-lobed **stomach** occupying the centre of umbrella. Extending laterally from stomach are four wide interradial **gastric pouches**. Within each gastric pouch is a C-shaped **gonad** and a row of small **gastric filaments** or **phacellae**, bearing nematocysts. Gullet communicates with each gastric pouch through a groove-like **gastro-genital canal** that runs between the two free ends of a gonad. Around each gonad runs an **exhalent channel** that also communicates with gullet through **gastro-oral canals**. **Radial canals** branch out from pouches to join the **circular** or **ring canal** in bell margin.

From each gastric pouch lead three radial canals through three small apertures. Middle or **interradial canal** divides at once into three branches out of which the outer two produce sub-branches before meeting the ring canal. Outer two **adradial canals** do not branch but meet the ring canal directly. Thus 8 unbranched adradial canals and 4 branched interradial canals emerge from 4 gastric pouches. Besides, 4 **perradial canals**, each emerging from an aperture between two adjacent gastric pouches, run to meet the ring canal, after showing branching pattern like that of interradial canals. All the canals are lined by ciliated gastrodermis.

[II] Nutrition

1. Food. *Aurelia* is exclusively carnivorous. It is mainly a suspension or ciliary feeder. It feeds upon planktonic organisms and small marine invertebrates, such as crustaceans, worms, their eggs and larvae.

2. Ingestion. Small planktonic organisms are carried directly into stomach with the entering water current. Some small organisms that may get entangled in mucus of sub-umbrellar surface are collected by oral arms. Sometimes, as medusa sinks slowly or swims gently downwards, prey is captured on contact with tentacles and oral arms. Nematocysts of oral arms paralyse and entangle these organisms, which are then swept up along the lateral tracts of oral arms and passed into mouth. Undesirable particles are rejected and dropped on the way.

If the prey, still is in living condition, reaches the gastric pouches, it is paralysed and killed by the nematocysts of gastric filaments.

3. Digestion. Digestion takes place in stomach and gastric pouches. Gland cells of these regions secrete most of the digestive enzymes for **extracellular digestion** of proteins, fats, carbohydrates and even chitin. Partly digested food particles circulate through canal system and are ingested by gastrodermal cells for **intracellular digestion** in food vacuoles. Undigested food leaves the body with the outgoing current of water.

4. Distribution. Water circulating through canal system transports digested food to the gastrodermis of all parts. Further, wandering amoeboid cells in mesogloea transports food from gastrodermis to epidermis.

5. Reserve food. Reserve food, in the form of glycogen and fat droplets, is stored in the gastrodermal cells of gastric filaments.

CIRCULATION OF WATER

Beating of cilia of gastrodermal cells lining the gastrovascular canals, sets up a current of water. It circulates through the gastrovascular canal system along a fixed route. Inhalent water current enters the mouth and passes through narrow gastrogenital canals into gastric pouches, and finally through unbranched adradial canals enters the circular canal. From circular canal, exhalent water current returns through perradial and interadial canals. Perradial canals convey it directly into stomach from where it passes into gullet and finally exits along the **basal grooves** of oral arms. Interradial canals convey it to the basal grooves of the oral arms via exhalent and gastro-oral canals. This arrangement greatly prevents the mixing of water currents entering and leaving the gastrovascular canal system. Bell contractions during locomotion and movements of oral arms help in circulation. One complete circulation takes about 20 minutes.

Circulation of water helps in nutrition, respiration, excretion and reproduction.

RESPIRATION AND EXCRETION

There are no special organs for respiration and excretion. Oxygen dissolved in water diffuses directly into epidermis as well as gastrodermis, both of which are constantly bathed by water.

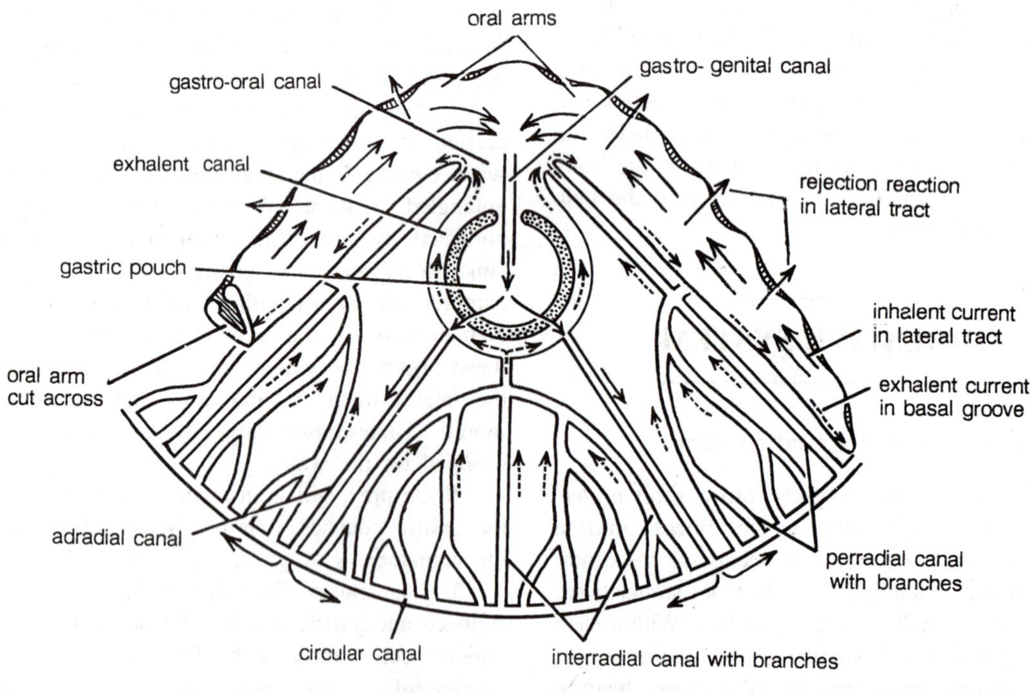

Fig. 5. *Aurelia.* Diagrammatic representation of gastrovascular canal system. Only one quadrant and two arms shown. Circulation of water shown by arrows.

Carbon dioxide and nitrogenous wastes diffuse out into surrounding water. Some workers are of the opinion that the subgenital pits facilitate gaseous exchange. This is based on the observation that, at the time of swimming, foul water constantly leaves the pits and fresh water (with dissolved oxygen) enters into them.

Water forms up to 99% of the weight of jelly-fish. Thus, compared to its weight, the oxygen requirement of *Aurelia* is quite insignificant. Almost nothing is known about the nature of excretory products.

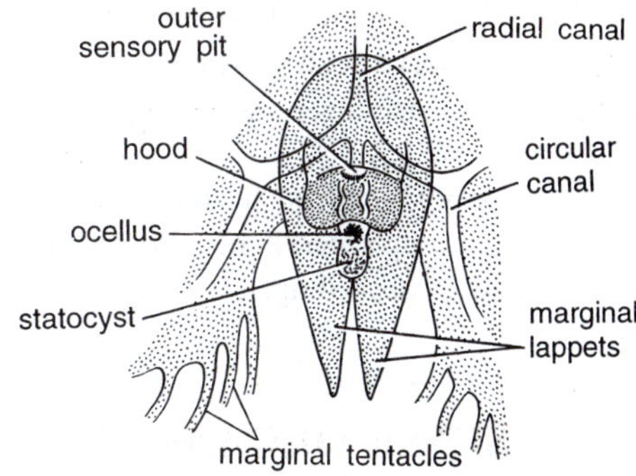

Fig. 6. *Aurelia. A.* portion of umbrella edge showing one rhopalium with its related parts.

NERVOUS SYSTEM

It consists of (i) a **main nerve net**, (ii) a **diffuse nerve net**, and (iii) eight **rhopalial ganglia**.

1. Main nerve net. Each nerve net or plexus consists of nerve cells and fibres. The main nerve net is more developed. It lies on the subumbrellar surface and extends into tentacles, rhopalia, manubrium and oral arms. Presence of main nerve net on subumbrellar side is correlated with the presence of well-developed musculature on that side; the former co-ordinates muscular movements during locomotion. Its nerve elements form a sort of **nerve ring** along the margin of umbrella near circular canal. Along per- and interradii, main nerve net is somewhat thickened due to the concentration of its nerve elements along these radii. Each radial thickening, near the margin of umbrella, is connected with the **rhopalial ganglion**, situated near the rhopalium on that radius.

2. Diffuse nerve net. Diffuse nerve net lies in the epidermis of subumbrella as well as exumbrella. Its nerve elements consist of smaller cell bodies. It is also connected with rhopalial ganglia. It controls local responses, like feeding, and can inhibit contractions of the umbrella.

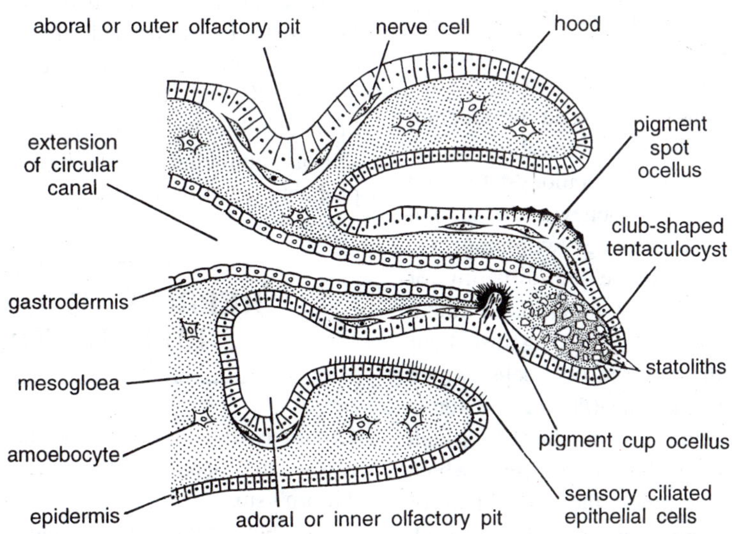

Fig. 7. *Aurelia.* Vertical section through a rhopalium showing the hood and various sensory areas.

3. Rhopalial ganglia. These are formed by aggregation of nerve cells. There are 8 such rhopalial ganglia, one near each sense organ or rhopalium.

Nerve impulses received by the sense organs, are conducted through nerve nets to the muscle fibres which react accordingly.

SENSE ORGANS

Sense organs of *Aurelia* are eight **rhopalia**, situated one in each of the per- and interadial marginal notches. Each rhopalium comprises **tentaculocyst** or **statocyst**, two **ocelli** and two **olfactory pits**.

1. Tentaculocyst or Statocyst. It is a hollow club-shaped structure, situated in the marginal notch between two **marginal lappets**. It is covered on the outer side by a process of bell margin, termed **hood**, which also connects the bases of two marginal lappets. Just below the club is a pad of tall ciliated **sensory epithelial cells** which are connected with the subumbrellar nerve net lying below epidermis. Tentaculocyst is a specialized hollow tentacle. Projecting into tentaculocyst is an extension of circular canal lined by gastrodermis. Lying in the distal part of tentaculocyst is a mass of polygonal **statolith cells** of gastrodermal origin. Each statolith cell contains a self-secreted particle, the **statolith** (Gr., **statos**, standing + **lithos**, stone), composed of calcium sulphate and calcium phosphate. Statoliths act as weight, causing the club of tentaculocyst to bend up and down at its base, whenever the animal tilts to one side or other, during swimming.

Tentaculocysts control the **equilibrium** of umbrella during swimming. If umbrella is tilted, the clubs of tentaculocysts press against their sensory pads beneath, the sensory cells which become stimulated. Higher the tilt, the greater is the stimulation. The impulse is conducted through subumbrellar nerve net to the muscle fibres which react accordingly. In response the upper half of umbrella drives less water than lower half at each beat, so that the umbrella automatically rights itself (Fig. 8).

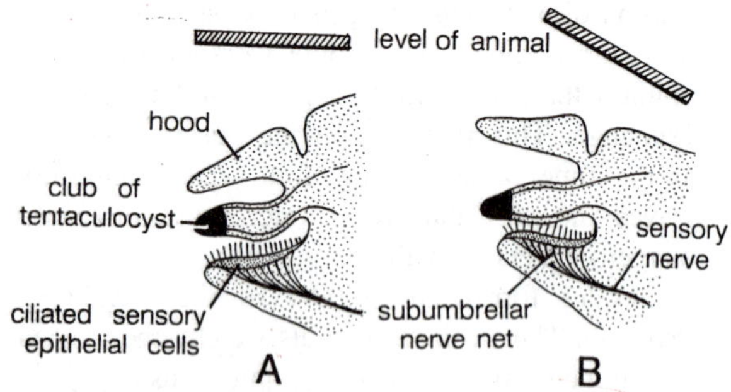

Fig. 8. *Aurelia.* Diagrams showing working of a tentaculocyst. **A.** Bell or umbrella in horizontal position. **B.** Bell tilted so that club of tentaculocyst presses down on pad of sensory epithelial cells.

2. Ocelli. There are two ocelli, one of ectodermal and other of endodermal origin. The former, known as **pigment spot ocellus**, consists of a patch of pigmented and sensory epidermal cells on the outer side of the club of tentaculocyst. The latter, known as **pigment cup ocellus**, consists of a cup-shaped cavity lined by pigmented and sensory gastrodermal cells and is situated on the inner side of tentaculocyst, in association with statoliths. Sensory cells of both the ocelli are connected with their respective underlying nerve nets. Ocelli are **photoreceptors**.

3. Olfactory pits. These are in the form of depressions of thickened epidermis containing sensory cells. One such depression lies at the base of hood. It is termed the **outer** or **aboral olfactory pit**. The other, known as **inner** or **adoral olfactory pit**, is situated on the inner side of tentaculocyst at the base of pad of ciliated sensory epithelial cells. These olfactory pits are probably **chemoreceptors**.

REPRODUCTION AND LIFE HISTORY

Aurelia is **dioecious**, that is, the male and female sexes are separate but there is no sexual dimorphism.

1. Sex organs. Testes and ovaries are similar in appearance. A medusa has 4 horseshoe-shaped gonads lying on the floor of stomach periphery, that is, one in each gastric pouch. They are reddish violet in colour. They are visible through the semitransparent jelly of umbrella as frilled organs with their concavities facing inwards. On maturity, ova and sperms break into the gastrovascular cavity and pass out of mouth with the outgoing water current. The ova or eggs are lodged in the frills of oral arms.

2. Fertilization. Spermatozoa, swimming about in water, reach the ova and fertilize them either in stomach of female or in the frills of oral arms. Thus, fertilization is either **internal** or **external**.

3. Formation of planula larva. Frills of oral arms serve as temporary brooding members. Here each fertilised egg or zygote undergoes early development into a ciliated larval stage, called **planula**. The zygote undergoes holoblastic and equal segmentation to produce a solid ball-like **morula**. Soon it is transformed into a single-layered **blastula**, enclosing a central fluid-filled cavity or **blastocoel**. Two-layered **gastrula** develops by invagination, having an outer ectoderm and an inner endoderm lining an enteron cavity, with its **blastopore** or **gastral mouth** not completely closed. Thus, it differs from the gastrula of Hydrozoa (*e.g.*, *Obelia)* which develops by a process of delamination and multipolar ingression of cells into blastocoel having no blastopore. The embryo now elongates, its outer cells become ciliated, blastopore closes and the typical **planula** larva is formed. At this stage, masses of planulae are visible as minute opaque patches on the oral arms of female individual.

4. Formation of scyphistoma. The ciliated planula eventually escapes and after a short free-swimming existence, attaches itself to a stone or seaweed by its broad anterior, aboral end. Cilia are lost and a **mouth** opens at its free distal end where blastopore had closed. The larva now becomes elongated and metamorphosed into a small trumpet-shaped or *Hydra*-like polyp, about 5 mm high. Its proximal part is narrowed into a stalk-like organ, attached to the substratum by an adhesive **basal disc**. **Tentacles** bud out around the mouth. First 4 tentacles are perradial, subsequent 4 interradial and then 8 adradial in position. Thus, 16 long and slender tentacles are formed. Mouth becomes square in outline and its edges become elongated to form a short **manubrium**. The larva now looks like a trumpet-shaped polyp or *Hydra,* and is called **hydratuba**

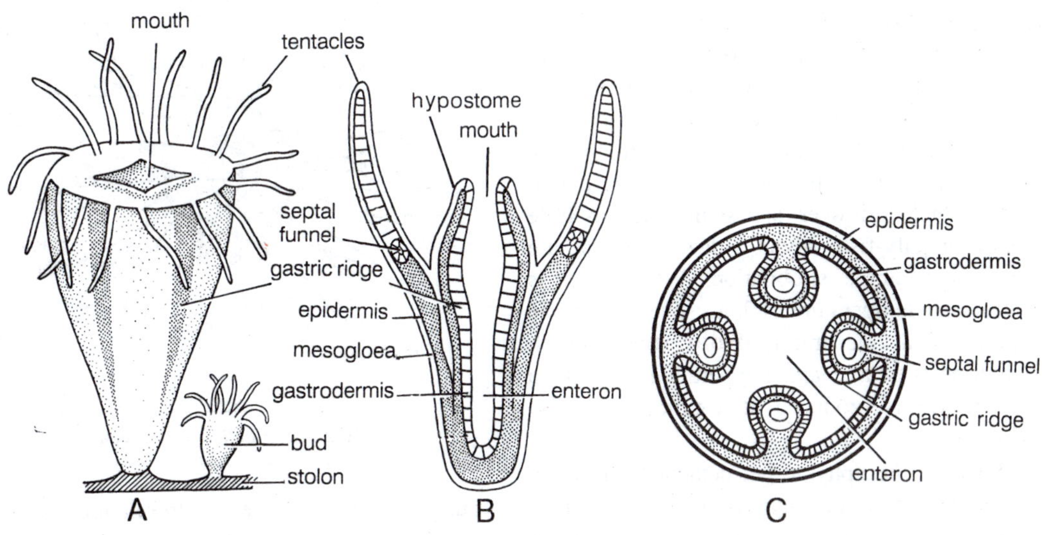

Fig. 9. *Aurelia.* Young scyphistoma or hydratuba. **A.** Outer view. **B.** Vertical section passing through an interradius. **C.** T.S. through septal funnels.

or **young scyphistoma** (Gr., **skyphos, cup** + **stoma**, mouth).

Endoderm of its enteric cavity is raised into four interradial longitudinal **gastric ridges** or **taenioles**, characteristic of **Scyphozoa**, which divides the enteric cavity into four perradial diverticula or pouches. Simultaneously, the ectoderm between mouth and tentacles also becomes invaginated as four interradial funnel-like depressions, known as **septal funnels** or **infundibula**, which sink into the four gastric ridges.

Scyphistoma feeds and grows up to 12 mm in height and may survive in this stage for several months. Sometimes, it multiplies either by lateral budding or by growing horizontal creeping stolons, which bud off fresh hydratubae. These buds eventually separate from the parent, as in *Hydra*.

5. Formation of ephyrae (strobilation). In autumn and winter, scyphistoma undergoes a remarkable process of budding or transverse fission of oral end, called **strobilation**. Distally, body develops a series of ring-like transverse constrictions or furrows which gradually deepens so that the organism resembles a pile of minute saucers or discs, placed one above the other. At this stage, scyphistoma with a segmented body is called a **strobila** and each of the segments is called an **ephyra** larva. The **ephyrae** or **ephyrulae** are connected together by muscular strands. As they grow older, their muscular strands contract violently and break at intervals. So one by one the distal ephyrae are pinched off from the parent strobila, which turn over, and swim away as little medusae or jelly-fish. About a dozen ephyrae are formed in a single strobilation. When food is plentiful and temperature is low, several ephyrae are produced at one time (**polydisc strobilation**). When food is scarce and temperature is high, a single ephyra is produced at a time (**monodisc strobilation**).

When all the ephyrae get detached, the basal unsegmented part of scyphistoma grows new tentacles and continues to live as a polyp or hydratuba. It may live for several years, feeding, growing and multiplying by budding in summers, but producing ephyrae by strobilations in winters. Under exceptional circumstances the whole scyphistoma may be metamorphosed into a single adult *Aurelia*, without forming ephyrae.

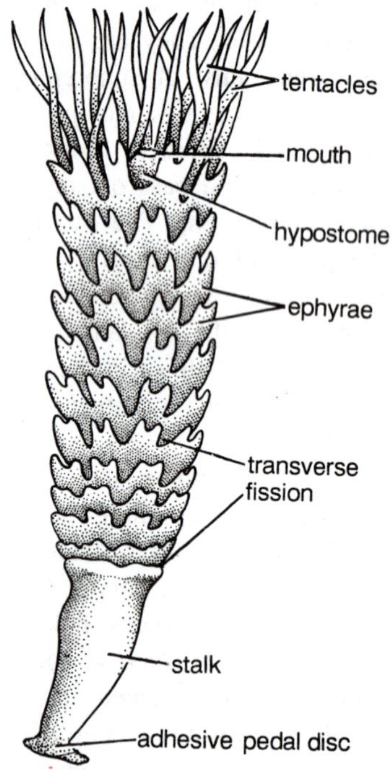

Fig. 10. *Aurelia.* A strobila.

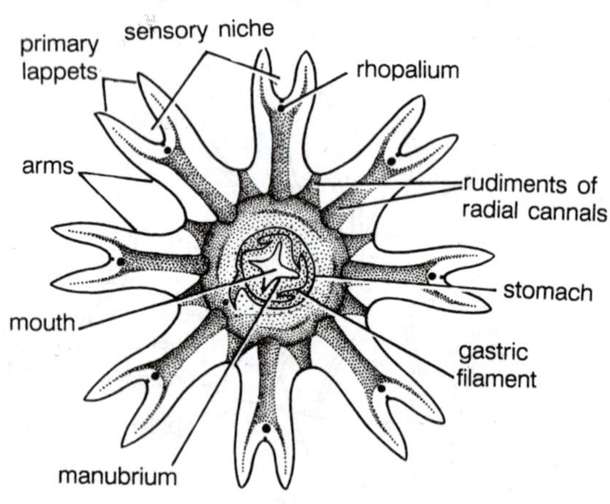

Fig. 11. *Aurelia.* A free ephyra.

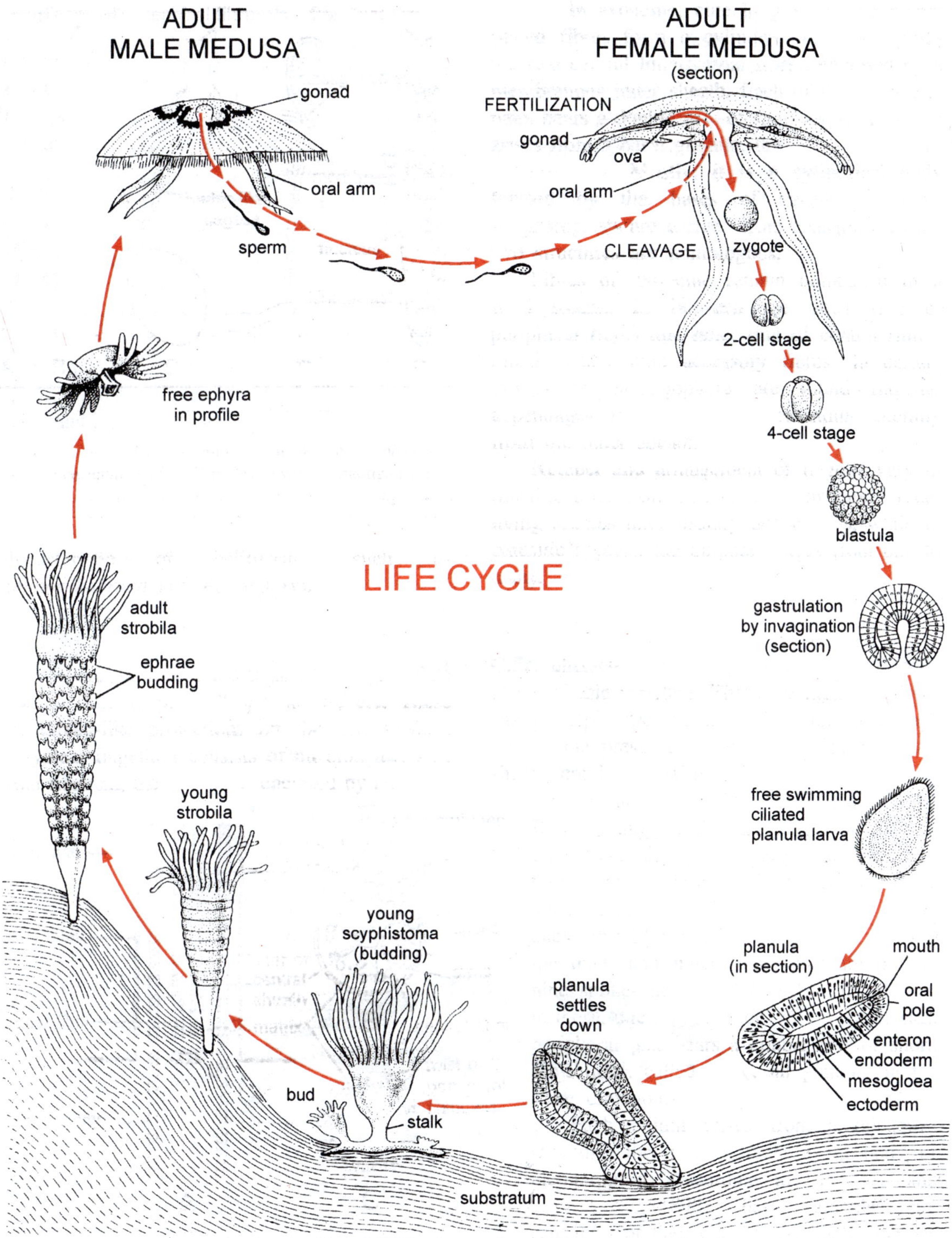

Fig. 12. *Aurelia.* Stages in development and life history.

6. Ephyra. A newly released ephyra is a microscopic gelatinous creature, about 1 mm in diameter and with a well-developed tetramerous symmetry. The edge of its umbrella is greatly fluted, being produced into eight **bifid lobes** or **arms** (4 per and 4 interradial), separated by 8 deep adradial indentations or **clefts**. Distal extremity of each lobe is deeply notched to form a pair of primary **lappets**. Notch or groove between lappets is a sensory recess or niche bearing a short tentacle, which becomes **tentaculocyst** or **rhopalium**. Ephyra contains a small segment of stomach of scyphistoma with gastric ridges. On the exumbrellar side, stomach becomes closed, while on subumbrellar side, it remains open and its edges grow out to form a short **manubrium**, bearing a four-sided **mouth** at its apex. Spacious enteric cavity projects into 8 arms as branched perradial and interradial canals. Adradial canals appear later on. Gastric ridges or mesenteries are replaced by **gastric filaments**. Four interradial septal invaginations probably persist as 4 **subgenital pits**. First ephyra differs from the rest in bearing the original tentacles of scyphistoma.

Ephyrae swim actively in sea-water feeding on minute organisms, such as protozoans, which are caught by lappets and transferred to mouth.

7. Metamorphosis. As growth proceeds, mesogloea increases enormously, so that two layers of endoderm (gastrodermis) fuse to form a solid **gastrodermal lamella**, except in regions of gastrovascular canals. Adradial regions grow more rapidly, gradually filling up their wide clefts, so that the umbrella of eight-rayed ephyra becomes circular and saucer-shaped, as in an adult medusa. With the appearance of 4 **oral arms** and numerous **marginal tentacles**, ephyra is finally transformed into adult *Aurelia*. An ephyra formed in winter becomes a sexually reproducing adult medusa by spring or summer.

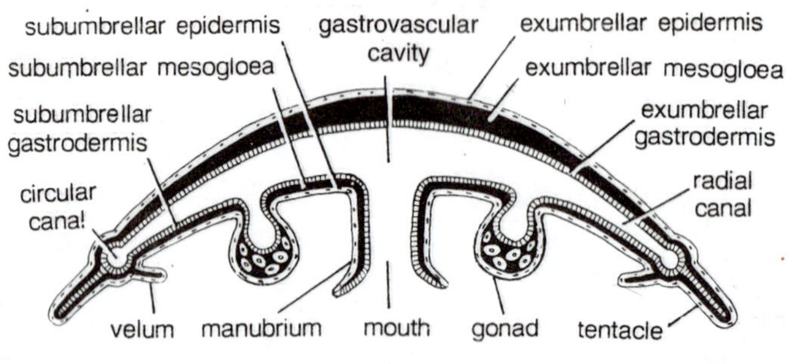

Fig. 13. *Obelia.* Medusa in diagrammatic vertical section.

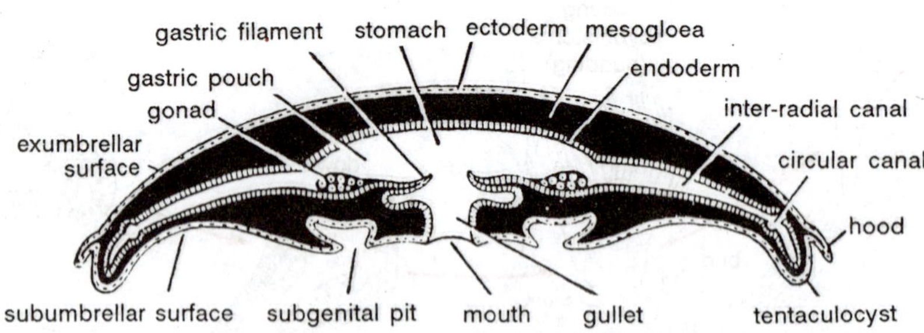

Fig. 14. *Aurelia.* Diagrammatic vertical section.

Table 1. **Contrast between *Obelia* and *Aurelia*.**

Obelia	*Aurelia*
A. POLYPOID STAGE	
1. Polypoid stage (hydroid colony) is dominant phase in life history. It is adult, colonial, polymorphic and covered with perisarc.	1. Polypoid stage or scyphistoma is small, reduced, larval, solitary and without perisarc.
2. Blastostyles produce medusae directly by asexual budding, there being no intermediate ephyra larval stage.	2. Scyphistoma produces, by transverse fission called strobilation, a series of disc like ephyrae, each of which metamorphoses into an adult *Aurelia* medusa.
B. MEDUSA STAGE	
3. Medusa phase is reduced, small-sized, called swimming bell.	3. Medusa phase is dominant, large-sized, called jelly-fish.
4. Medusae arise by asexual budding from blastostyle of polypoid colony.	4. Medusae arise by metamorphosis of ephyrae which are produced by strobilation of scyphistoma.
5. Bell margin not scalloped for tentaculocysts.	5. Bell margin having eight notches for tentaculocysts.
6. Bell margin with a few long, solid tentacles.	6. Bell margin with many small, hollow tentacles.
7. True velum present, projects inwards from bell margin and lacks endodermal canals (Craspedote medusa).	7. True velum absent. Inconspicuous velarium present containing endodermal canals (Acraspedote medusa).
8. Oral arms absent. Oral lips form small oral lobes.	8. Corners of mouth form 4 long oral arms.
9. Manubrium comparatively elongated, handle-like.	9. Manubrium comparatively short, inconspicuous.
10. Canal system simple, without gastric pouches. Canals unbranched. No gastric ridges and filaments.	10. Canal system complicated, with gastric pouches.Canals branched Gastric ridges bearing filaments present.
11. Mesogloea structureless, non-cellular.	11. Mesogloea with amoeboid cells and fibres.
12. Muscle strands derived from both ectoderm and endoderm.	12. Muscle strands derived only from ectoderm.
13. A double nerve ring around bell margin.	13. No marginal nerve ring. A sub-umbrellar nerve-plexus and radial nerve centres present.
14. Statocysts ectodermal, without hood and marginal lappets.	14. Tentaculocysts endodermal, protected by a hood and marginal lappets.
15. No ocelli and olfactory pits.	15. Ocelli and olfactory pits present.
16. Nematocysts confined to manubrium and tentacles only.	16. Nematocysts occur all over the bell, tentacles, oral arms and gastric filaments.
17. Gonads lie on radial canals bulging externally.	17. Gonads lie on floor of gastric pouches internally.
18. Reproductive cells derived from ectoderm.	18. Reproductive cells derived from endoderm.
19. Sub-genital pits absent.	19. Sub-genital pits present.
C. DEVELOPMENT	
20. Fertilization external, occurring in water.	20. Fertilization internal or external, effected inside gastro-vascular cavity or frills of oral arms of female.
21. Gastrula formed by delamination and multipolar immigration of ectodermal cells into blastocoel.	21. Gastrula formed by invagination of outer ectoderm.
22. Planula larva solid, without a blastopore and coelenteron.	22. Planula larva hollow, with a blastopore and a coelenterons.
23. Planula grows into a simple *Hydra*-like hydrula stage which produces directly, by budding, a new complex branching *Obelia* colony.	23. Planula grows into a trumpet-like scyphistoma representing the asexual polypoid generation.

ALTERNATION OF GENERATIONS (METAGENESIS)

There are two contrasting views regarding the life-history of *Aurelia.*

According to the older view, development of *Aurelia* illustrates, in a sense, the phenomenon of alternation of generations. Free-swimming adult *Aurelia* represents the **sexual generation**, producing ova or spermatozoa, like medusoid phase of *Obelia.* Fertilized egg gives rise to a small fixed polypoid form, or scyphistoma, after passing through a free-swimming ciliated planula larval stage. Scyphistoma represents the **asexual generation**, equivalent to the polypoid *Obelia* colony. It reproduces asexually by strobilation, budding off ephyrae, each of which metamorphoses into an adult *Aurelia.*

According to the modern view, development of *Aurelia* (or coelenterates) does not show true alternation of generations or metagenesis, as discussed earlier in *Obelia.* However, the development of *Aurelia* does not truely correspond with that of *Obelia.* In *Obelia,* medusoid stage develops as a bud on a branched polypoid colony, whereas the *Aurelia* medusa is formed by the metamorphosis of an ephyra larva, produced as one of the several transverse segments of the polypoid scyphistoma. The difference is mainly because *Aurelia* has a dominant medusoid phase and a suppressed polypoid phase in its life history, while the reverse is true in *Obelia.* Thus, life cycle of *Aurelia* may be considered as a simple continuous process involving a prolonged metamorphosis further complicated by the multiplication of scyphistoma larva into several ephyrae (Figs. 13 and 14).

CONTRAST BETWEEN *OBELIA* AND *AURELIA*

Genus *Obelia* belongs to the class Hydrozoa, and its life cycle includes both hydroid (polypoid) and medusoid stages. On the other hand genus *Aurelia* belongs to the class **Scyphozoa**, and medusa is the most conspicuous phase in its life history. ***Table 1*** includes some of the outstanding differences between these two genera.

Important Questions

➤➤*Long answer type questions*

1. Give an account of the structure and life history of *Aurelia.*
2. Describe the gastrovascular system and mechanism of feeding in *Aurelia.*
3. Describe the process of reproduction and strobilation in *Aurelia.*
4. Give a comparative account of structure of *Aurelia* and medusa of *Obelia.*
5. Draw neat and labelled diagrams of : (i) *Aurelia* in ventral view, (ii) Ephyra.
6. Write short notes on : (i) Ephyra, (ii) Mesenteries, (iii) Rhopalium, (iv) Scyphistoma, (v) Strobilation, (vi) Rhopalial ganglia, (vii) Hydratuba.

➤➤*Short answer type questions*

1. Name an animal where marginal lappet is present.
2. What is ephyrula ?
3. What are sense organs of *Aurelia* ?
4. What is strobilization ?
5. Describe the ephyra larva of *Aurelia.*
6. Describe the strobilation in *Aurelia.*
7. Give an account of the structure of sense organs of *Aurelia.*
8. Describe the life history of *Aurelia.*
9. Describe the structure and life history of *Aurelia.*
10. Give a diagrammatic representation of the various stages in the life cycle of *Aurelia.* No description is required.
11. Draw a labelled sketch of a vertical section through adult *Aurelia.*
12. Draw a labelled sketch of Ephyra.

13. Give a diagrammatic representation of the various stages in the life cycle of *Aurelia* (No description).
14. Draw and label the structure of ephyra.
15. What is the function of tentaculocyst ?
16. Give a diagrammatic representation of working of tentaculocyst.

➤➤*Multiple choice questions*

1. The sense organ in *Aurelia* is called as :
   (a) nematocyst (b) tentaculocyst
   (c) taster (d) tentilla
2. Process of gastrulation in the life history of *Aurelia* is by :
   (a) delamination (b) epiboly
   (c) emboly (d) invagination
3. Larva of *Aurelia* is :
   (a) planula (b) stereogastrula
   (c) scyphistoma (d) parenchymula
4. *Aurelia* belongs to order :
   (a) coronatae (b) semaeostomae
   (c) rhizostomae (d) none
5. *Aurelia* generally known as :
   (a) jelly fish (b) jolly fish
   (c) marine fish (d) true fish
6. Which zooid is dominant in *Aurelia* :
   (a) polyp (b) medusa
   (c) both (d) none
7. *Aurelia* maintain the equilibrium with the help of :
   (a) elastic fibre (b) nematocyst
   (c) tentaculocysts (d) radial muscles
8. *Aurelia* is :
   (a) herbivorous (b) carnivorous
   (c) omnivorous (d) all the above
9. Nerve net present in :
   (a) *Sycon* (b) *Hydra*
   (c) *Aurelia* (d) all
10. Ephyra is :
    (a) a Porifera (b) a Cnidaria
    (c) young *Hydra* (d) young *Aurelia*
11. Collencyhma is found in :
    (a) *Paramecium* (b) *Aurelia*
    (c) *Obelia* (d) *Euglena*
12. In origin colenchyma is :
    (a) mesodermal (b) endodermal
    (c) ectomesodermal
    (d) endo-mesodermal

Answers

➤➤*Multiple choice questions*

1. (b) 2. (d) 3. (a) 4. (b) 5. (a) 6. (b) 7. (c) 8. (b) 9. (c) 10. (d)
11. (b) 12. (c).

26

Metridium : A Sea Anemone

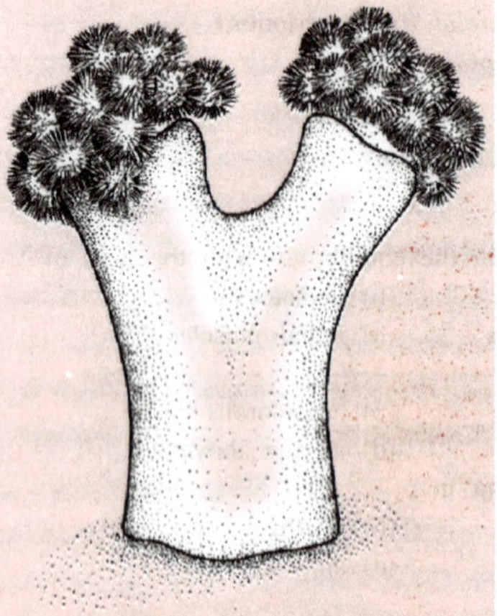

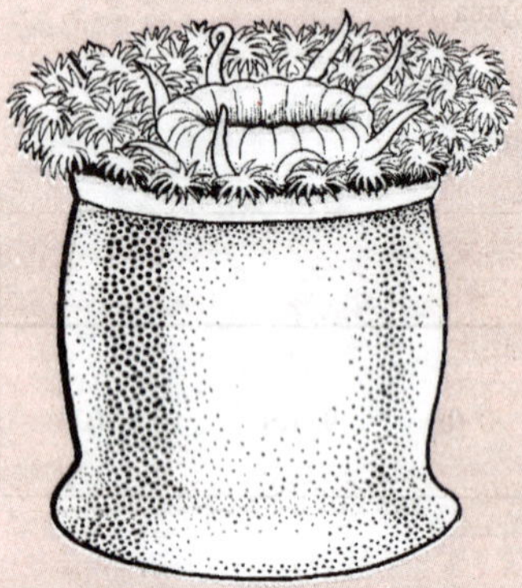

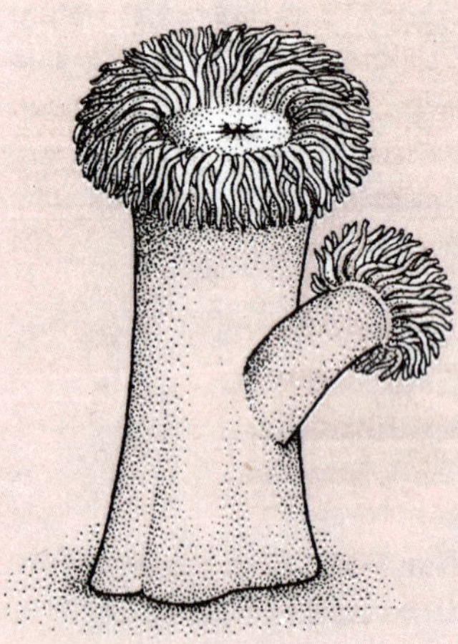

The largest class of phylum Coelenterata is **Anthozoa** or **Actinozoa**. They include exclusively marine and polypoid coelenterates, solitary as well as colonial, in which the medusoid stage is totally absent. An anthozoan polyp differs considerably from hydrozoan polyps (*e.g., Hydra, Obelia)* in several structural peculiarities which have been enumerated in ***Table 1***, in the end of this chapter.

Colonial Anthozoa are mostly corals of different kinds. Solitary Anthozoa are sea anemones, belonging to the order **Actiniaria**. They are quite in abundance and best known animals inhabiting warm coastal waters throughout the world. Common genera are *Adamsia, Edwardsia, Metridium, Urticina,* etc. Most studied example is *Metridium* (L., *metricus,* rhythm), and its most common species is *M. marginatum.* There are more than 1000 species of metridium known in the world.

Metridium marginatum

SYSTEMATIC POSITION

Phylum	Coelenterata
Class	Anthozoa (Actinozoa)
Subclass	Hexacorallia
Order	Actiniaria
Genus	*Metridium*
Species	*marginatum*

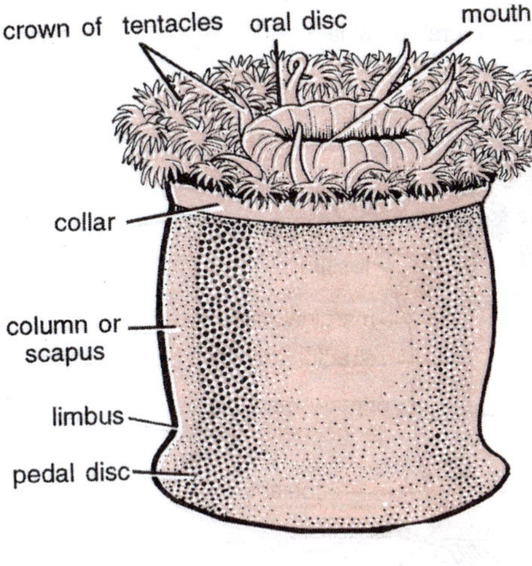

Fig. 1. *Metridium.* External structure.

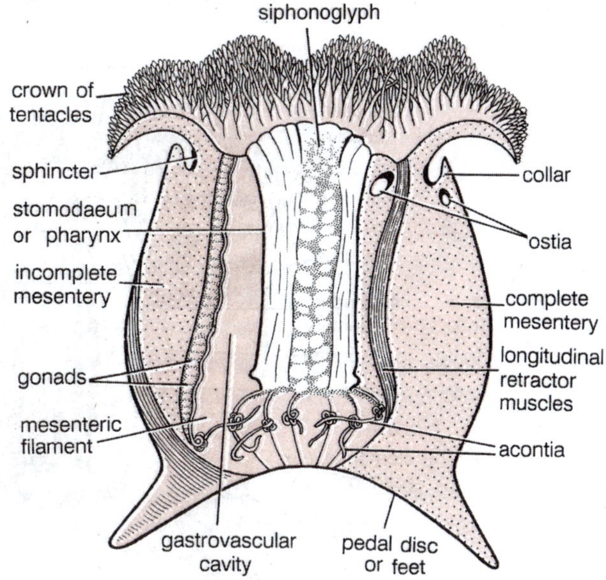

Fig. 2. *Metridium.* Vertical section to show internal structures.

HABITS AND HABITAT

Metridium is a solitary marine sea anemone, inhabiting warm coastal waters along the North Atlantic and Pacific coasts. It lives in shallow water or littoral zone, mostly attached to rocky bottom and other hard substrata, like wooden wharves and piers. It can creep slowly on the substratum. It is carnivorous feeding on minute organisms, crustaceans, worms, etc. It is very sensitive and highly contractile.

EXTERNAL MORPHOLOGY

1. Shape, size and colour. *Metridium* is a cylindrical and radially symmetrical animal which measures 5 to 7 cm in length. Its colour is variable, but usually brownish or yellowish. Body is clearly divisible into three regions : oral disc, pedal disc and column.

2. Oral disc. Oral end of body or **oral disc** is slightly convex and bears a slit-like central **mouth**. Encircling the mouth are a large number of hollow **tentacles** forming a sort of **crown**.

3. Pedal disc. Aboral end of body is broad, flat and called the **basal** or **pedal disc** or **foot**. With pedal disc the animal gets attached to the substratum.

4. Column. Middle part of body, which extends between oral disc and pedal disc, is referred to as the **scapus** or **column**. It can be greatly expanded or contracted. Column is demarcated from pedal disc by a groove called **limbus**. Column bears a prominent circular fold, the **collar** or **parapet**, at its junction with oral disc. Skin is soft but tough and is without skeletal structures.

INTERNAL MORPHOLOGY

Internal structure of *Metridium* is known from its longitudinal and transverse sections.

[I] Gastrovascular System

1. Pharynx or stomodaeum. Mouth leads into a tubular **gullet** or **pharynx** or **stomodaeum**. On either side of the pharynx is a ciliated groove called **siphonoglyph** (Gr., **siphon**, a tube + **glyphe**, carving). Individuals with 2 siphonoglyphs are called **diglyphic**. In some species there is only one siphonoglyph (**monoglyphic**). Sea anemone

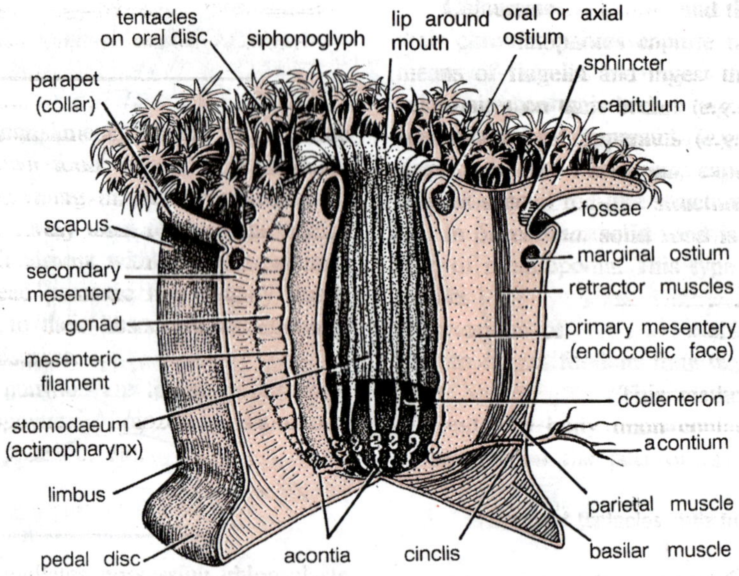

Fig. 3. *Metridiun.* Diagrammatic V.S.

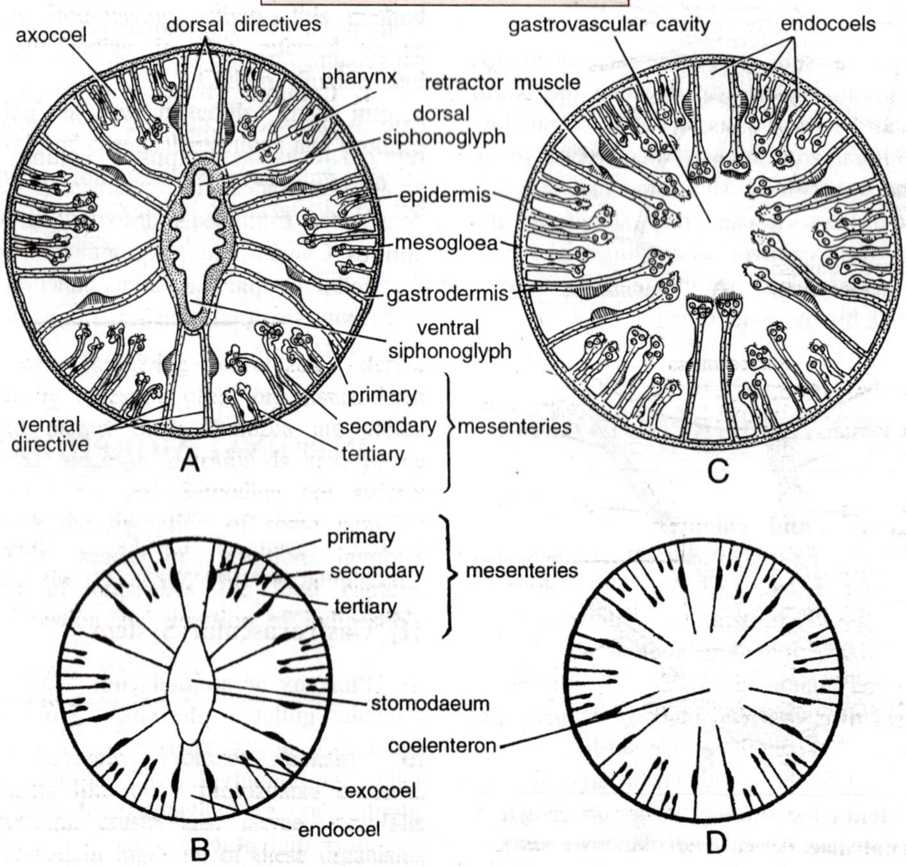

Fig. 4. *Metridium.* **A.** T.S. through pharyngeal region. **B.** Diagrammatic T. S. through pharyngeal region showing arrangement of mesenteries. **C.** T.S. below pharynx. **D.** Diagrammatic T. S. below pharynx showing arrangement of mesenteries.

with one siphonoglyph is considered bilateral, and with 2–siphonoglyphs, biradial. Besides siphonoglyphs, cilia are also present on other parts of pharynx. Cilia of grooves beat to create water currents, which flows through mouth into body cavity and ensure a constant supply of oxygen for respiration. Cilia on other parts of pharynx carry food in and waste products out.

2. Gastrovascular cavity or coelenteron. Pharynx extends downwards about 2/3 of the way into the inner body cavity called **gastrovascular cavity**. It is partitioned into radial chambers by vertical partitions of body wall, called **septa** or **mesenteries**.

3. Mesenteries. Mesentries of *Metridium* must not be confused with the mesenteries of coelomate animals (*e.g.*, frog, rabbit). There are two types of mesenteries, **complete** and **incomplete**. Complete or **primary mesenteries** are 6 pairs and extend vertically from body wall to the wall of pharynx. Two pairs of complete mesenteries on the sides of siphonoglyphs are called **directives**. In the region of pharynx, complete mesenteries bear openings or **ostia** for circulation of water from one chamber to another. Below pharynx, their edges are free which recurve towards the body wall. Six pairs of complete mesenteries constitute the **primary mesenteries**. Spaces between these pairs are called **exocoels**. Lying in exocoels are pairs of **secondary** and **tertiary mesenteries**, which are **incomplete**. They are connected only to the body wall and do not reach the pharynx. Space between two mesenteries of each pair is called the **endocoel**.

Below pharynx, free edges of mesenteries are called **mesenteric filaments** which are trilobed in appearance.

Attached to the bases of mesenteries are thread-like **acontia** (Gr., **akontion**, dart) which bear **nematocysts** and **gland cells**. Acontia can be protruded through mouth or through pores (**cinclides**) in body wall, to overcome the prey. Otherwise, in ordinary life, acontia serve to kill any food remaining alive into the gastrovascular cavity.

[II] Histological Structure

Histology of *Metridium* is more or less typical and similar to that of other coelenterates.

1. Body wall. Body wall consists of the usual two layers, outer **epidermis** and inner **gastrodermis**, with a gelatinous **mesogloea** in between.

(a) Epidermis. It covers the outer body surface and consists of tall columnar supporting cells along with sensory, glandular and stinging cells. Subepithelial muscle cells and nerve cells are also present.

(b) Gastrodermis. It lines whole of the gastrovascular cavity and includes epithelio-muscle, sensory, nerve, glandular and stinging cells.

(c) Mesogloea. It is much thicker than that of *Hydra* and *Obelia*, and contains a large number of **fibres** and wandering **amoebocytes**.

2. Pharynx or stomodaeum. It is derived from invaginated ectoderm, so that it is lined by epidermis on its inner side and covered by

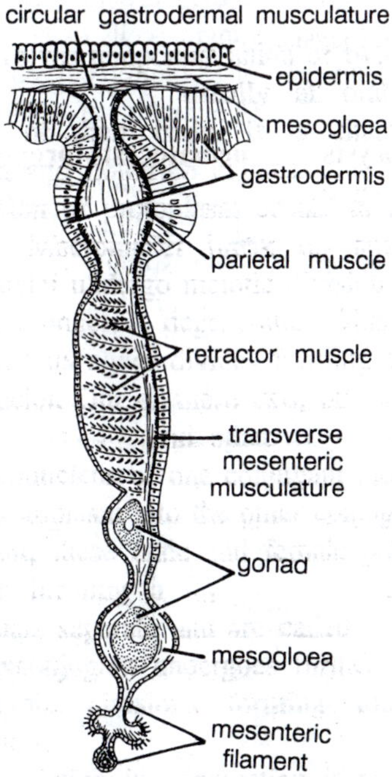

Fig. 5. *Metridium.* T.S. of mesentery.

gastrodermis on outer side. Epidermis is ciliated and rich in mucous gland cells.

3. Mesenteries. Each mesentery is composed of two layers of gastrodermis, enclosing a central layer of mesogloea. Gastrodermal layers are continuous with the gastrodermis of body wall. Upper portion of a mesenteric filament appears trilobed in cross section, consisting of a central ridge or **cnidoglandular band** or **tract**, bearing nematocysts and gland cells secreting digestive enzymes, and two lateral ridges (**ciliated bands** or **tracts**) bearing tall ciliated cells. This upper trilobed part of mesenteric filament serves for digestion and water circulation.

Lower part of each mesenteric filament is without ciliated bands and represented only by the cnidoglandular band having nematocysts and gland cells. It is exclusively digestive. Some of its cells are phagocytes, which engulf food particles to carry on intracellular digestion. Cnidoglandular band, forming the lower part of mesenteric filament, is produced into a thread-like acontium at the base of mesentery.

4. Nematocysts. Stinging cells or nematocysts of sea anemone are long, faintly curved and do not bear cnidocils. Two types of nematocysts occur, **spirocysts** and **nematocysts proper**.

(a) Spirocysts. These are limited to tentacles and oral disc. Each has a thin single-walled capsule and a long, smooth, spirally coiled thread tube of even diameter, devoid of spines. It stains with acid dyes and is permeable to water.

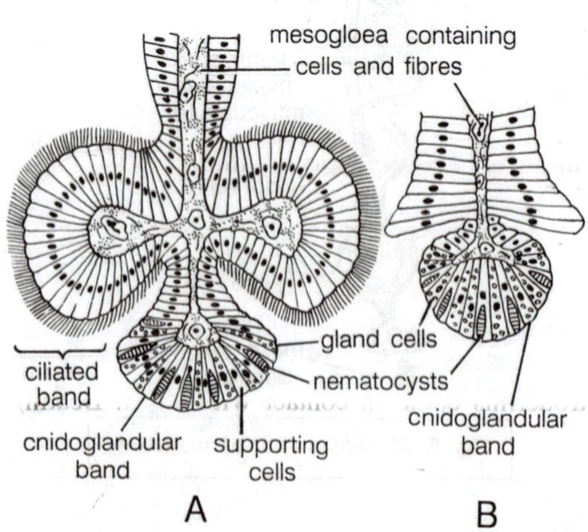

Fig. 6. *Metridium.* T.S. mesenteric filament. **A.** Upper region. **B.** Lower region.

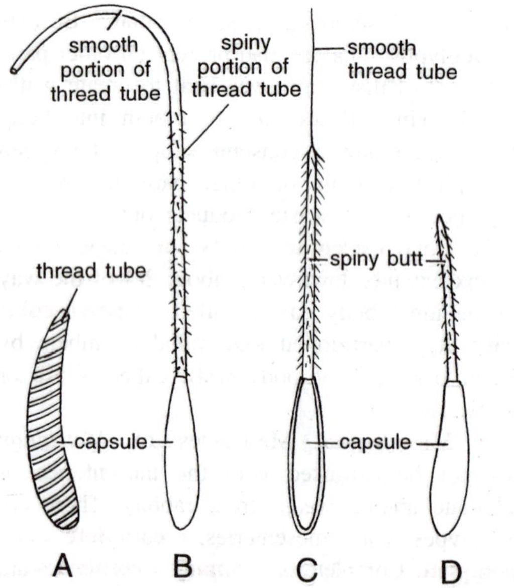

Fig. 7. *Metridium.* Types of nematocysts. **A.** Spirocyst. **B.** Basitrichous isorhiza. **C.** Microbasic mastigophore. **D.** Microbasic amastigophore.

(b) Nematocysts proper. These occur every where and each has a thick double-walled capsule, stains with basic dyes, is impermeable to water except at discharge and contains a tube of variable nature but usually armed by spiral rows of spines. Nematocysts proper are of 3 types : (i) **basitrichous isorhizas** with thread tube spiny at the base only, without butt and opening at the tip, (ii) **microbasic mastigophores** with a rounded capsule, long butt with spiral spines and a long thread tube closed at the tip, and (iii) **microbasic amastigophores** with an oval capsule, short butt with spiral spines and without a thread.

There is no skeleton in anemones.

[III] Musculature

Sea anemones perform various characteristic movements, so that their muscular system is better developed than in *Obelia* and *Aurelia,* particularly the gastrodermal musculature. Myoepithelial cells are organized into well developed muscles. In *Metridium,* **epidermal musculature** is limited to longitudinal fibres in tentacles and radial fibres in oral disc, but absent in column and pharynx. **Gastrodermal musculature** consists of circular muscles in tentacles, oral and pedal discs, gullet and column wall. At the junction of the column and oral disc, circular fibres form a distinct **sphincter**. It covers the oral disc and tentacles when retracted.

In a mesentery runs a thick **longitudinal** or **retractor muscle**, which appears as a prominent projection or bulging in a transverse section. Retractors of directive pair of mesenteries are exocoelic in position, while those of remaining pairs of mesenteries are endocoelic. Contraction of retractors shortens the mesenteries and the body column.

[IV] Nervous System

Specific sense organs do not occur in anemones. Nervous system is very simple, basically similar to that of other coelenterates, and is synaptic. It is represented by a typical diffuse **nerve net** with no indication of a centralized nervous control. It consists of an **epidermal plexus**, between epithelial and muscular layers, and a **gastrodermal plexus**, at least in septa; the two are connected through mesogloea. Each plexus consists of delicate **nerve fibres** and large **ganglion cells** occurring chiefly in tentacles, oral disc and pharynx. Reflex behaviour is poor due to lack of a centralized nervous system.

[V] Gonads

Metridium, like majority of sea anemones, is **dioecious**, *i.e.*, sexes are separate. Gonads (testes and ovaries) are simple and occur as thickened bands, generally on the incomplete mesenteries, lying parallel to and behind the mesenteric filaments.

PHYSIOLOGY

[I] Movements and Locomotion

Sea-anemones are sluggish creatures. Well developed musculature enables *Metridium* to perform many characteristic movements. It can extend and contract its column and tentacles and partly evert its gullet and mesenteries. It frequently retracts its oral disc, along with tentacles in gullet and closes the margin over it by means of the sphincter. In the process, gullet is transversely folded, water is discharged through mouth and acontia are protruded through cinclides.

Sea anemone is capable of slight locomotion. It can slide about slowly and change its position by slow-creeping movements performed by muscular undulations of its pedal disc. This movement, however, is too slow *i.e.*, about 8 cm per hour.

[II] Nutrition

Nutrition has been extensively studied in *Metridium* because of its large size and availability in abundance.

1. Food. It is quite voracious and feeds upon molluscs, crustaceans, worms, sea-urchins, fish, etc.

2. Ingestion. The prey is captured and partly paralyzed, before ingestion, by the nematocysts of tentacles, oral disc and acontia, then carried through the greedy mouth, down the gullet into gastrovascular cavity or coelenteron. Some prey is gripped directly by mouth and gullet, both of which can gape widely. Cilia of gullet, other than those on siphonoglyphs, reverse their beat, so that the captured food is swept down the gullet, into coelenteron, which is hardly more than a digestive sac. It has been known that reversal of ciliary beat is due to direct stimulation by chemical agents (food). Food is held by mesenteries and broken into smaller particles.

3. Digestion. Digestion is both **extracellular** and **intracellular**. Food is digested by hydrolysing enzymes secreted by acontia and gland cells of mesenteric filaments, and absorbed by gastrodermis. Undigested wastes are ejected through mouth.

[III] Respiration and Excretion

Oxygen dissolved in water directly diffuses in and carbon dioxide and nitrogenous wastes diffuse out. This is facilitated as all parts of epidermis and gastrodermis come in contact with water. Beating of cilia of gullet and siphonoglyphs set up a constant current of water entering through siphonoglyphs and leaving through the sides of gullet. Gastrodermal cilia also helps in circulation inside the gastrovascular cavity.

[IV] Reproduction

Reproduction in *Metridium* is both **asexual** and **sexual**.

1. Asexual reproduction. Pedal laceration or **fragmentation** is the predominant method. In this process, as the animal moves about, pieces of pedal disc are torn off and left behind, each regenerating into a small anemone at the old site. This accounts for the common occurrence of a large mother *Metridium,* surrounded by a whole brood of little anemones produced by this asexual method. Anemones regenerated from such pedal fragments naturally present many irregularities in the number and arrangement of mesenteries and siphonoglyphs.

A few instances of **budding** from column or pedal disc have been reported but these may be misinterpretations. **Longitudinal fission** has also been reported. **Transverse fission** is known only in the young individuals of primitive sea anemone, *Gonactinia prolifera.*

2. Sexual reproduction. *Metridium* is **dioecious** having separate male and female individuals. Sex cells originate from the interstitial cells of gastrodermis and mature in mesogloea of mesenteries. **Gonads** (testes or ovaries) form longitudinal bands just behind mesenteric

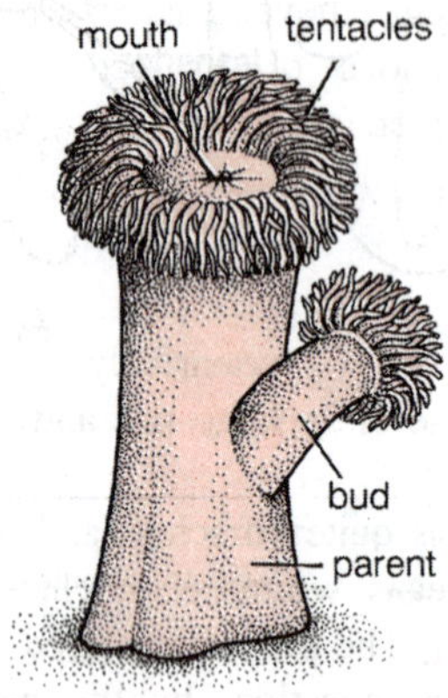

Fig. 8. *Metridium.* Budding.

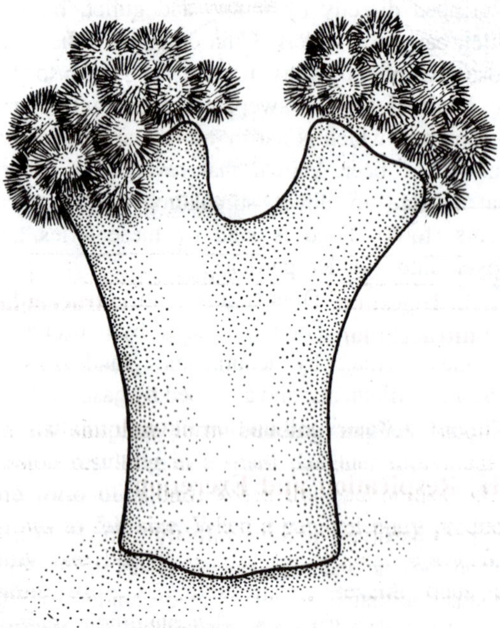

Fig. 9. *Metridium.* A stage in longitudinal binary fission.

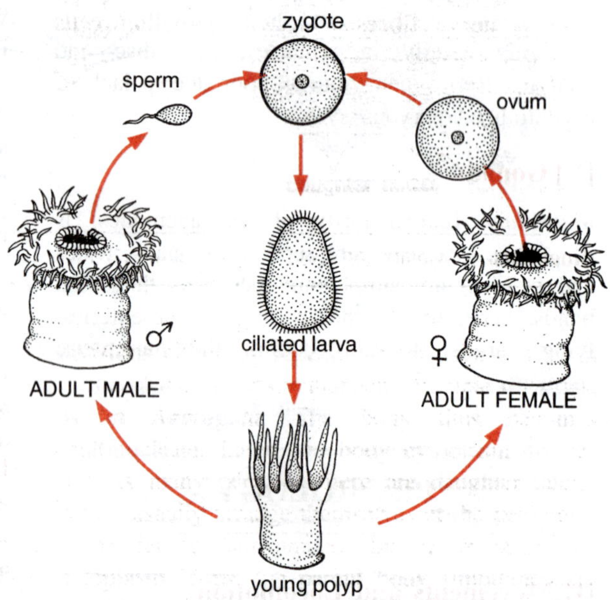

Fig. 10. *Metridium.* Stages in sexual reproduction and life cycle.

filaments. Mature eggs and sperms, released from gonads, leave through mouth and **external fertilization** occurs in sea-water. Fertilized egg or **zygote** develops into a slender, bilaterally symmetrical, free-swimming and ciliated **planula** larva. Its blastopore becomes the **mouth** and it feeds upon minute organisms. It creeps on the bottom and finally settles down in some rocky crevice attaching with its aboral end. With the growth of tentacles, septa and mesogloea, it metamorphoses into a miniature anemone. Pharynx is formed as a stomodeal invagination and mesenteries develop from the column wall. A medusa stage is totally absent. Thus, life cycle of *Metridium* is simple and does not show an alternation of generations.

[V] Regeneration

Regeneration has been considerably studied in *Metridium* and others. If the column is cut across, the aboral part regenerates a new oral disc, but the oral part fails to regenerate a pedal disc. In some cases, the oral portion may regenerate tentacles, thus, exhibiting a case of **heteromorphosis** or **revised polarity**.

COMPARISON OF HYDROZOAN POLYP AND SEA ANEMONE

Class Anthozoa differs from other classes (Hydrozoa and Scyphozoa) of coelenterates in several features. Some of these may be considered advancements. These are as follows :

1. Exclusively polypoid form, without any medusa stage.
2. Symmetry biradial or bilateral.
3. Presence of a disc-shaped oral end.
4. Formation of a true stomodaeum lined with invaginated ectoderm.
5. Presence of mesenteries dividing gastrovascular cavity.
6. Mesogloea made of amoeboid cells.

The main differences between hydrozoan and anthozoan polyps have been shown in ***Table 1***.

Table 1. **Differences between a Hydrozoan Polyp and a Sea Anemone.**

Hydrozoan Polyp *(Hydra, Obelia)*	Anthozoan Polyp or Sea Anemone *(Metridium)*
1. Marine or freshwater. Solitary or colonial.	1. Exclusively marine and solitary.
2. Polyp small, tall, slender and radially symmetrical.	2. Polyp large, broad, squat and bilaterally symmetrical.
3. Mouth circular, opens at the apex of an elevated oral cone or manubrium.	3. Mouth slit-like, opens in the centre of a flat disc-like oral end.
4. Tentacles less in number.	4. Tentacles numerous, usually arranged in multiples of six.
5. Mouth leads directly into gastrovascular cavity. Stomodaeum or pharynx absent.	5. Mouth leads into a tubular stomodaeum formed by invaginated ectoderm, which opens into gastrovascular cavity.
6. Gastrovascular cavity simple, remains undivided due to lack of septa or mesenteries.	6. Gastrovascular cavity divided into radial compartments by longitudinal septa or mesenteries.
7. Mesogloea thin, acellular, without endoskeleton.	7. Mesogloea thick, with amoeboid cells, and may contain calcareous spicules.
8. Musculature less developed.	8. Musculature more developed.
9. Nematocysts possess operculum and cnidocil.	9. Nematocysts do not have operculum and cnidocil.
10. Gonads ectodermal in origin and develop externally on body wall (*Hydra*).	10. Gonads gastrodermal and develop internally on mesenteries.

Important Questions

➤➤Long answer type questions

1. Describe the external and internal morphology of a sea anemone studied by you.
2. Illustrate the histology of a sea anemone with the help of transverse sections passing through its pharynx and the region below pharynx.
3. Draw a full page and well-labelled diagram of the longitudinal section of *Metridium.*
4. Give an account of the physiology and life history of *Metridium.*

➤➤Short answer type questions

1. To which order sea anemones belongs ?
2. What are the two types of nematocysts in *Metridium* ?
3. What are the mesenteric filaments in *Metridium ?*
4. Describe the structure and function of primary mesenteries in *Metridium* (Anthozoa).
5. Draw a labelled sketch of the L.S. of sea-anemone.

➤➤Multiple choice questions

1. *Metridium* belongs to order :
   (a) Actiniaria (b) Zoanthidia
   (c) Madreporaria (d) Antipatharia
2. Cinclides are :
   (a) nematocysts (b) mesenteric filaments
   (c) pores in body wall (d) none
3. Mesenteries are :
   (a) vertical partitions of body wall
   (b) chambers of gastrovascular cavity
   (c) partitions of pharynx
   (d) none
4. Groove in pharynx is called :
   (a) stomodaeum (b) limbus
   (c) siphonoglyph (d) cnidoglandular groove
5. Lower thread like filaments of mesenteries are :
   (a) cinclides (b) acontia
   (c) cnidoglandular bands
   (d) tentacles
6. Complete absence of medusa stage is found in :
   (a) *Obelia* (b) *Aurelia*
   (c) *Metridium*
   (d) all above
7. Pedal disc present in :
   (a) *Obelia* (b) *Aurelia*
   (c) *Metridium*
   (d) *Hydra*
8. Much thicker mesogloea with amoebocyte present in :
   (a) *Metridium* (b) *Aurelia*
   (c) *Obelia* (d) *Hydra*
9. Spirocyte is a :
   (a) nematocyst (b) tentaculocyst
   (c) muscle fibre (d) ciliated band
10. *Metridium* generally known as :
    (a) sea pen (b) sea fan
    (c) sea anemone (d) sea fur

Answers

➤➤Multiple choice questions

1. (a) 2. (c) 3. (a) 4. (c) 5. (b) 6. (c) 7. (c) 8. (a) 9. (a) 10. (c)

27

Coelenterata : Characters, Classification and Types

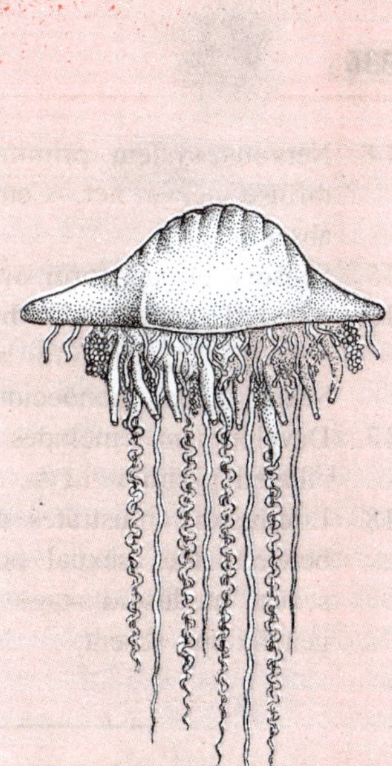

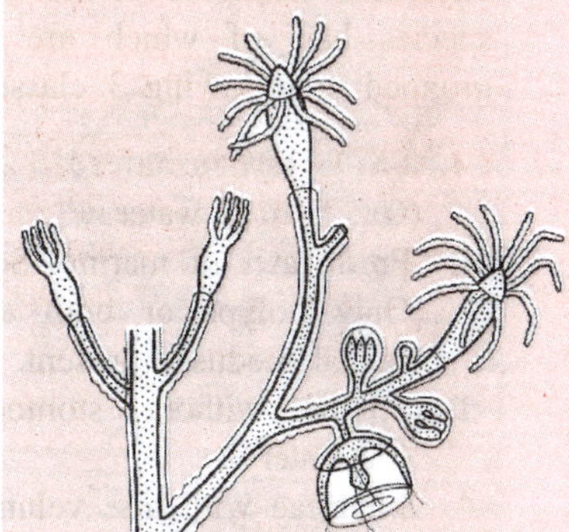

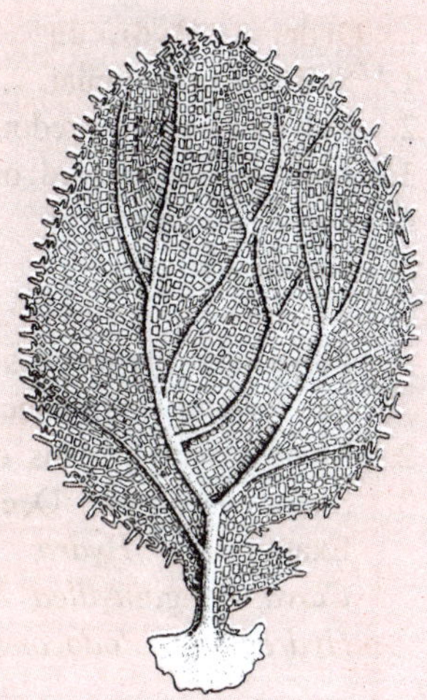

GENERAL CHARACTERS

1. All aquatic, some freshwater, mostly **marine**.
2. Solitary or colonial. Sedentary or free-swimming.
3. Symmetry **radial** or **biradial** about a longitudinal oral-aboral axis.
4. Body organization of **cell-tissue grade**. Cells mostly scattered and specialized for different functions. Some cells form tissues like **nerve net** or **nervous tissue**.
5. Exoskeleton chitinous (perisarc) or calcareous (corals).
6. Body wall **diploblastic** with two cellular layers —outer epidermis and inner gastrodermis — with a gelatinous acellular mesogloea in between. In advanced types mesogloea with cells and connective tissue, hence **triploblastic**.
7. Two types of individuals occur, attached **polyps** and free-swimming **medusae**. Some species are notable for **polymorphism** or variety of forms.
8. Mouth of polyps and bell margin of medusae often encircled by short and slender **tentacles**.
9. Coelom and respiratory, circulatory and excretory systems wanting.
10. Muscular system includes longitudinal and circular fibres formed by epithelio-muscle and endothelio-muscle cells.
11. A single internal cavity, lined with gastrodermis, called **gastrovascular cavity** or **coelenteron**, into which mouth opens. Anus is absent.
12. Digestion intracellular as well as extracellular.
13. One or both body layers with peculiar stinging cell organelles or **nematocysts**, which serve for adhesion, food capture, offence and defense.

14. Nervous system primitive, consisting of a diffuse **nerve net**. Central nervous system absent.
15. Sensory organs form **ocelli** and **statocysts**.
16. **Asexual reproduction** by budding or fission. **Sexual reproduction** by ova and sperms. Sexual forms monoecious or dioecious.
17. Development includes a free-swimming ciliated **planula larva**.
18. Life-history illustrates a regular alternation between the asexual polypoid stage and a sexual medusoid stage. True alternation of generations absent.

CLASSIFICATION

Phylum Coelenterata includes nearly 17,702 known species half of which are extinct. These are grouped in following 3 classes :

CLASS 1. Hydrozoa

(Gr., **hydra**, water + **zoon**, animal)

1. Freshwater or marine. Solitary or colonial.
2. Only polyps or both asexual polyps and sexual medusae present.
3. Polyps without stomodaeum and septa (mesentery).
4. Medusae with true velum (**craspedote**).
5. Mesogloea non-cellular.
6. Gonads epidermal. Sex cells shed directly on outside.

Order 1. Hydroida

1. Solitary or colonial.
2. Polyploid stage predominant.
3. Medusae short-lived or absent.
4. Sense organs of medusae exclusively ectodermal.

Suborder I. Anthomedusae (Gymnoblastea)

1. Polyps and blastostyles **athecate**, *i.e.*, perisarc not forming hydrothecae and gonothecae.
2. Medusa with gonads on manubrium.
3. Statocysts absent. Ocelli present in medusa. Examples : *Hydra, Ceratella, Tubularia, Clava, Bougainvillea, Pennaria, Eudendrium, Hydractinia, Podocoryne,* etc.

Suborder II. Leptomedusae (Calyptoblastea)

1. Polyps and blastostyles **thecate**, *i.e.*, with hydrothecae and gonothecae, respectively.
2. Medusa with gonads on radial canals.
3. Both statocysts and ocelli present in medusa. Examples : *Obelia, Sertularia, Campanularia, Plumularia, Aglaophenia.*

Order 2. Trachylina

1. Medusoid stage large, dominant, free-swimming and may develop directly from fertilized egg.
2. Polypoid stage reduced or absent.
3. Marginal sense organs or statocysts with endodermal statoliths.

Suborder I. Trachymedusae

1. Tentacles inserted above bell margin.
2. Gonads develop on radial canals. Examples : *Gonionemus, Craspedacusta.*

Suborder II. Narcomedusae

1. Tentacles arise between bell margin and vertex of exumbrella.
2. Gonads present on manubrium or on stomach floor. Examples : *Cunina, Cunarcha, Polycolpa.*

Order 3. Hydrocorallina

1. Fixed, colonial polypoid Hydrozoa in which coenosarc secretes a massive exoskeleton of calcium carbonate.
2. Polyps **dimorphic**, with slender **dactylozooids** and short plump **gastrozooids**.

Suborder I. Milleporina

1. Dactylozooids hollow, with capitate tentacles.
2. Medusae free and devoid of mouth, digestive canals and tentacles. Example : *Millepora.*

Suborder II. Stylasterina

1. Dactylozooids solid, without tentacles.
2. Gonophores reduced to sporosacs. Medusae not free. Example : *Stylaster.*

Order 4. Siphonophora

1. Pelagic colonial Hydrozoa showing extreme polymorphism of zooids.
2. Polyps without oral tentacles.
3. Medusae incomplete and rarely freed.

Suborder I. Calycophora

1. Pneumatophore absent.
2. Upper end of colony has one or more swimming bells (nectophores).
Examples : *Diphyes, Praya, Abyla.*

Suborder II. Physophorida

Upper end of colony forms a large gas-filled float (pneumatophore).
Examples : *Physalia, Halistemma, Stephalia.*

Order 5. Chondrophora

1. Pelagic, polymorphic polypoid colony.
2. Upper end forms a chitinous, gas-filled, oval float (pneumatophore).
3. Gonozooids produce free medusae.
Examples : *Porpita, Velella.*

CLASS 2. Scyphozoa

1. Exclusively marine and solitary forms.
2. Medusa stage dominant. Polyp stage reduced or absent.
3. Gastrovascular cavity with gastric pouches and endodermal gastric filaments. No stomodaeum.
4. Medusa **acraspedote**, without distinct velum.
5. Mesogloea extensive, gelatinous, with fibres and cells.
6. Gonads gastrodermal. Sex cells released in digestive cavity.

Order 1. Stauromedusae (Lucernariida)

1. Bell goblet or trumpet-shaped.
2. Sessile, attached by an aboral stalk.
3. No marginal sense organs or tentaculocysts.
Examples : *Lucemaria, Halidystus.*

Order 2. Cubomedusae (Carybdeida)

1. Bell cubical, with 4 flattened sides.
2. Four perradial tentaculocysts present.
Examples : *Charybdea, Chiropsalmus.*

Order 3. Coronatae

1. Bell conical, divided by a deep circular coronary groove.
2. Tentacles long, borne on pedalia.
3. Four to sixteen tentaculocysts present.
Examples : *Periphylla, Pericolpa.*

Order 4. Discomedusae (Semaeostomae)

1. Flat saucer or disc-like umbrella.
2. Eight tentaculocysts present.
3. Square shaped mouth extending into 4 long oral arms.
Examples : *Aurelia, Pelagia, Cyanea, Chrysaora.*

Order 5. Rhizostomae

1. Bell usually hemispherical, without marginal tentacles.
2. Typically 8 or more tentaculocysts.
3. No central mouth. Oral arms fused with several small mouths.
Examples : *Rhizostoma, Cassiopeia, Stomolophus.*

CLASS 3. Anthozoa (Actinozoa)

(Gr., **anthos**, **flower** + **zoios**, animal)

1. Exclusively marine. Solitary or colonial.
2. All polyps, no medusa.
3. Mouth leads into a tubular stomodaeum.
4. Gastrovascular cavity subdivided by 8 or more septa or mesenteries.
5. Mesogloea stout and cellular.
6. Mesenteries with nematocysts and gastrodermal gonads.

Subclass I. Octocorallia (Alcyonaria)

1. Exclusively colonial.
2. Polyps with 8 pinnate tentacles and 8 septa.
3. Gullet with one ventral siphonoglyph.

Order 1. Stolonifera

1. Polyps arising independently from a creeping mat or stolon.
2. Skeleton of calcareous tubes or separate calcareous spicules or absent.
Examples : *Tubipora* (organ pipe coral), *Clavularia.*

Order 2. Telestacea

1. Lateral polyps on simple or branched stems arising from a creeping base.
2. Skeleton of calcareous spicules.
Example : *Telesto.*

Order 3. Alcyonacea

1. Polyps proximally embedded in a fleshy mass or coenenchyme.
2. Skeleton of separate calcareous spicules.
Examples : Soft corals. *Alcyonium.*

Order 4. Coenothecalia

1. Polyps embedded and connected by solenial tubes.
2. Skeleton massive, calcareous and blue-green from iron salts.
   Example : *Heliopora* (blue coral).

Order 5. Gorgonacea

1. Colony usually of plant-like branching form bearing short polyps.
2. Axial skeleton composed of horn-like **gorgonin**, separate or fused calcareous spicules, or both.
   Examples : *Gorgonia* and *Corallium*.

Order 6. Pennatulacea

1. Colony elongated, sessile. Lower part embedded in mud. Upper part consists of a very long axial polyp with lateral branches bearing dimorphic polyps.
2. Axial skeleton of separate calcareous spicules or horny substance.
   Examples : *Pennatula*, and *Pteroeides*, (Sea pens), *Renilla* (sea pansy).

Subclass II. Hexacorallia (Zoantharia)

1. Solitary or colonial.
2. Tentacles usually unbranched, numerous but never 8.
3. Gullet commonly with 2 siphonoglyphs.

Order 1. Zoanthidea

1. Solitary or colonial. No skeleton. Mostly epizoic.
2. Polyps small and usually united by basal stolons.
3. Only single ventral siphonoglyph present.
   Examples : *Zoanthus, Epizoanthus*.

Order 2. Actiniaria

1. Simple, often large-sized, solitary anemones.
2. Body muscular, often with an aboral pedal disc.
3. Skeleton absent.
4. Tentacles and mesenteries are numerous.
5. Siphonoglyps usually one or two.
   Examples : Sea-anemones. *Metridium, Edwardsia, Adamsia, Actinia, Urticina*.

Order 3. Ceriantharia

1. Long, solitary, anemone-like forms, without pedal discs and skeleton.
2. Tentacles simple, numerous, arranged in two whorls —oral and marginal.
3. Siphonoglyphs single and dorsal.
   Example : *Cerianthus*.

Order 4. Antipatharia

1. Colonial and tree-like.
2. Tentacles and mesenteries comparatively few (6-24) in number.
3. Skeleton as branched, chitinoid axis.
4. Siphonoglyphs two.
   Examples : Black corals. *Antipathes*.

Order 5. Madreporaria

1. Solitary or colonial.
2. Exoskeleton hard, compact, often massive, calcareous.
3. Polyps small, living in cup-like cavities on exoskeleton.
4. Siphonoglyph absent and muscles feeble.
   Examples : True or stony corals. *Flabellum, Fungia* (mushroom coral), *Astrangia, Astraea* (star coral), *Favia, Oculina, Acropora* or *Madrepora* (stag horn coral), *Meandrina* or *Meandra* (brain coral).

Subclass III. Tabulata

Extinct colonial anthozoans with heavy calcareous skeletal tubules containing horizontal platforms or tubulae.
Examples : *Favosites, Halysites*.

OTHER TYPES OF COELENTERATA

1. *Bougainvillea.* *Bougainvillea* is a colonial hydrozoan which forms a tree-like branching colony. Creeping hydrorhiza gives off vertical stems or hydrocaulii, bearing polyps or hydranths and gonophores or medusae. Hydranth possesses a single whorl of filiform tentacles and is not enclosed within hydrotheca. Blastostyles do not

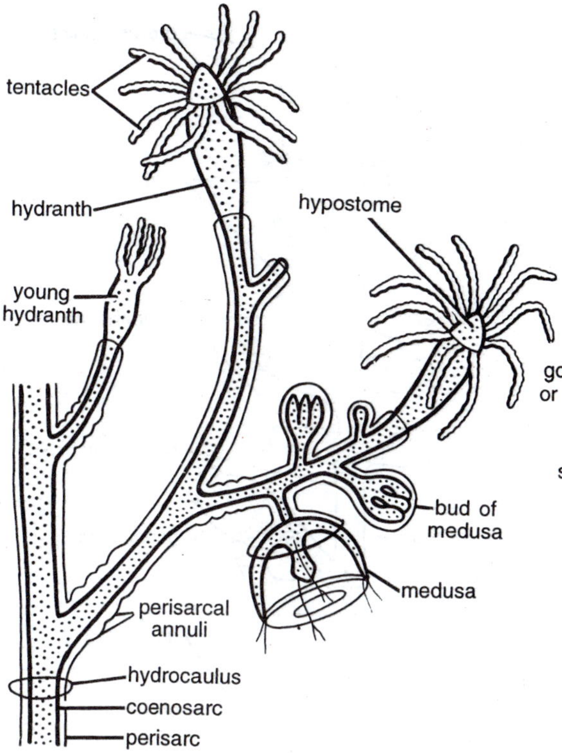

Fig. 1. *Bougainvillea.* A portion of colony.

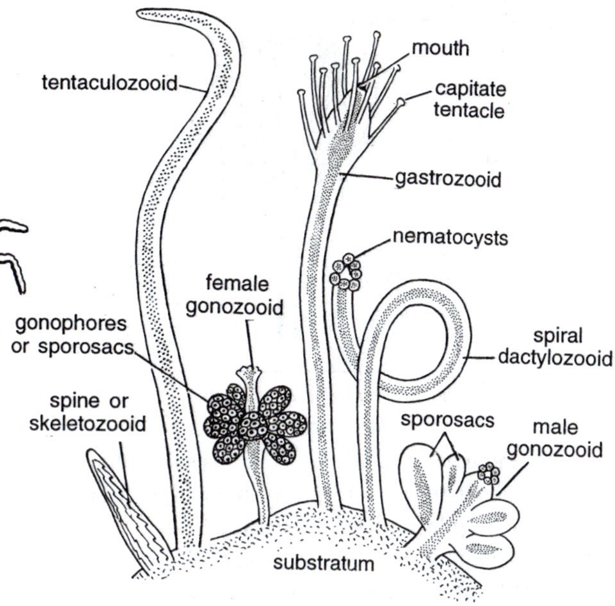

Fig. 2. *Hydractinia.* A polymorphic colony.

occur and medusa buds arise directly from coenosarc of hydrocaulus. Fully developed free-swimming medusa is shaped like a deep bell, with a thick gelatinous wall, has four radial canals, a wide velum and four pairs of marginal tentacles. Gonads develop on its manubrium and statocysts are absent.

2. *Hydractinia.* *Hydractinia,* commonly called hedgehog hydroid, is a small, marine, colonial hydrozoan. It commonly grows on gastropodal shells occupied by hermit-crabs. Hydrorhiza becomes fused to form firm brown encrusting mat covered by perisarc. Colony is polymorphic and remarkable in that polyps, of five functional types, arise directly from hydrorhiza and are devoid of perisarc. These are : (i) large feeding **gastrozooids** with a mouth, conical hypostome and filiform tentacles, (ii) defensive spiral **dactylozooids**, (iii) long sensory dactylozooids or **tentaculozooids**, (iv) long and slender **skeletozooids** and (v) reproductive male and female **gonozooids** with clusters of gonophores or sporosacs, containing eggs or spermatozoa, depending on their sex. Medusae are never formed.

3. *Tubularia.* *Tubularia* is a marine hydroid, occurring solitary or as large pink colonies, attached to wharf pilings. Hydranths are large, brilliantly coloured, flower-like and each bearing two rows of tentacles. Smaller distal filiform **oral tentacles** arise from the tip of oral cone or hypostome, surrounding the terminal mouth. Large proximal **aboral tentacles** surround the base of hydranth. Perisarc does not form hydrotheca around polyp. It does not grow beyond a swelling of stalk at the base of hydranth. Colony is dioecious. Male or female gonophores or medusae remain attached to hypostome in clusters and never set free. They are deep bell-shaped with manubrial gonads and four knob-like tentacles. Fertilized eggs remain in female gonophore. Ciliated, mouthless planula never becomes free-swimming but metamorphoses into a polypoid, tentaculate **actinula** larva which is then set free. It is of unusual interest, since it represents the structure of the ancestral coelenterate.

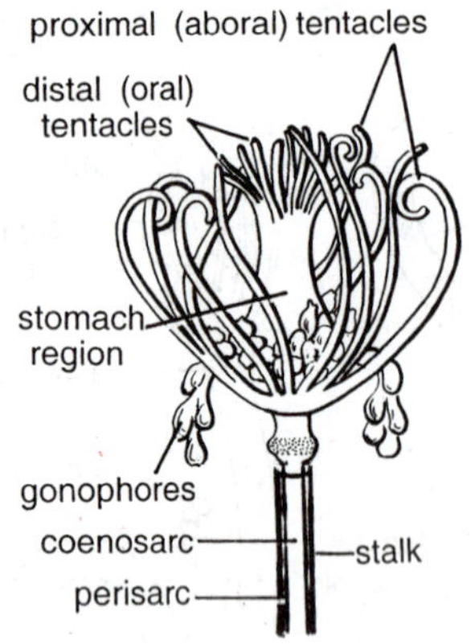

Fig. 3. *Tubularia.* A hydranth.

4. *Pennaria.* *Pennaria* is a marine colonial hydrozoan. It forms a regularly branching colony attached to piles, rocks or seaweeds in shallow water. Hydranths are without hydrothecae and bear two kinds of tentacles. Oral tentacles, scattered irregularly, are short, with a terminal knob packed with nematocysts. Longer filiform aboral tentacles form a regular proximal circlet around the base of hydranth. Medusae (gonophores) develop on the side of hydranth. When gonads mature, medusae may remain sessile or become free, each having four radial canals and rudimentary tentacles.

5. *Sertularia.* *Sertularia* is a colonial hydrozoan commonly found on floating seaweeds, or attached to rocks along sea coasts.

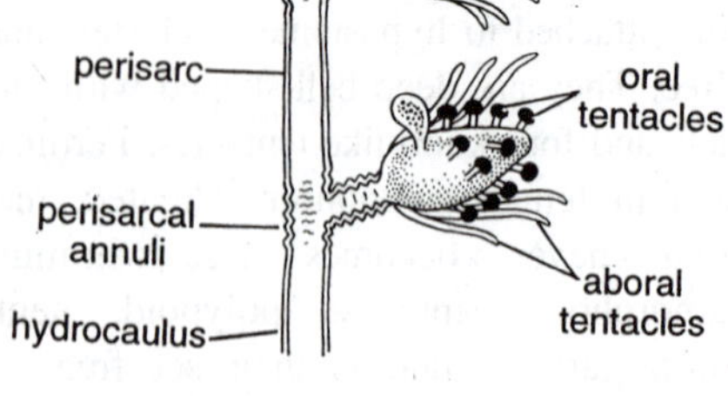

Fig. 4. *Pennaria.* A portion of colony.

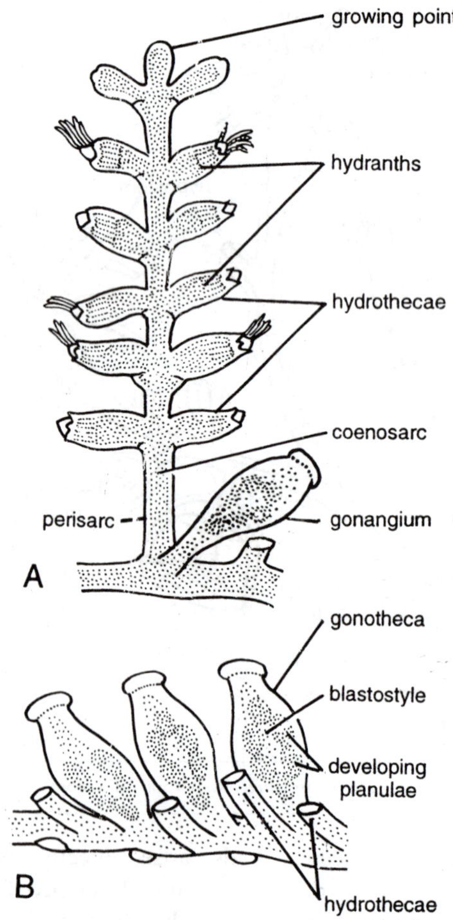

Fig. 5. *Sertularia.* **A.** Branch with hydranths. **B.** Branch with gonangia.

Hydrothecae containing polyps are arranged on opposite sides of branching stems (hydrocaulii) that come out from the creeping **hydrorhiza**. On each internode of stem there is one pair of hydrothecae. **Gonothecae** are large and vase-like. Each gonotheca encloses the blastostyle with developing planulae.

6. *Plumularia.* Marine and sessile colony of *Plumularia* consists of a creeping hydrorhiza giving off plume-like stems, each bearing alternate or pinnately arranged side branches, called **hydrocladia** (singular, hydrocladium). Sessile hydrothecae grow in a series only on one side of a hydrocladium. They are small so that polyps

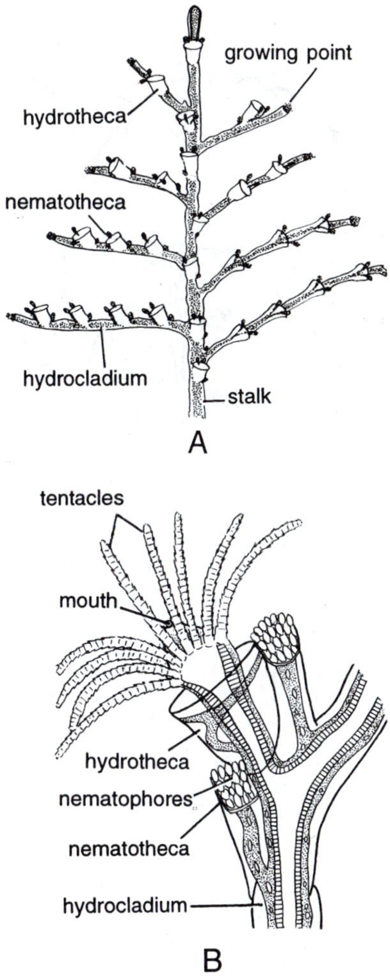

Fig. 6. *Plumularia.* **A.** A portion of colony. **B.** A polyp (magnified).

cannot be completely retracted within them. Besides nutritive polyps or gastrozooids, also occur smaller movable polyps or dactylozooids, called **nematophores**, contained into smaller thecae, called **nematothecae**. Nematophores are without mouth but with long amoeboid processes that engulf decaying polyps, diatoms, protozoans, minute larvae and other epizoic organisms. Usually three nematophores occur on the hydrotheca of each gastrozooid. Gonangia are much elongated and protected by gonothecae.

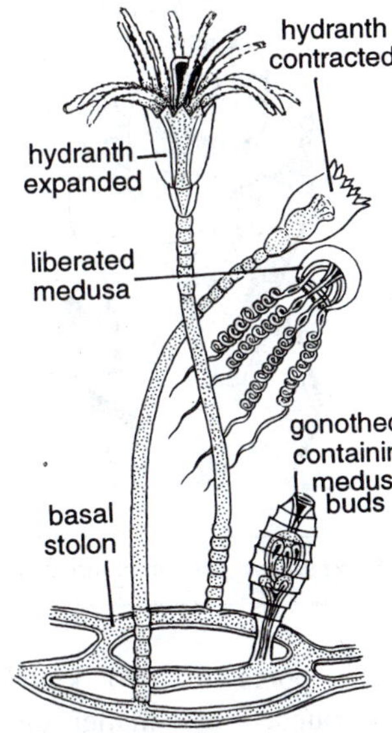

Fig. 7. *Campanularia.* A portion of colony.

7. *Campanularia.* *Campanularia* is a marine and sessile colonial hydrozoan, which occurs attached to piles, etc. Colony is either branched or simple, bearing usually stalked and bell-shaped open hydrothecae, which are with or without marginal teeth. Stalks bearing hydrothecae arise separately from basal stolon or hydrorhiza, and may be wholly or partially annulated. Hypostome of hydranth is trumpet-shaped. Gonangia, placed at the ends of short stalks, are large, oval or truncate and may be annulated. Blastostyles generally do not liberate free medusae. Sexual reproduction and early part of development takes place within gonotheca which retain medusae and later planulae are liberated.

8. *Millepora.* *Millepora* is a hydroid coral, mostly associated with coral reefs in tropical seas. It is also called fire coral or stinging coral because of its powerful nematocysts which are painful to man. Colony consists of much branched, anastomosing hydrorhiza, forming a broad basal mass encrusted on rocks. Colour of colony is yellow-brown due to presence of symbiotic zooxanthellae. Perisarc forms a massive solid calcareous exoskeleton (coenosteum). Its surface

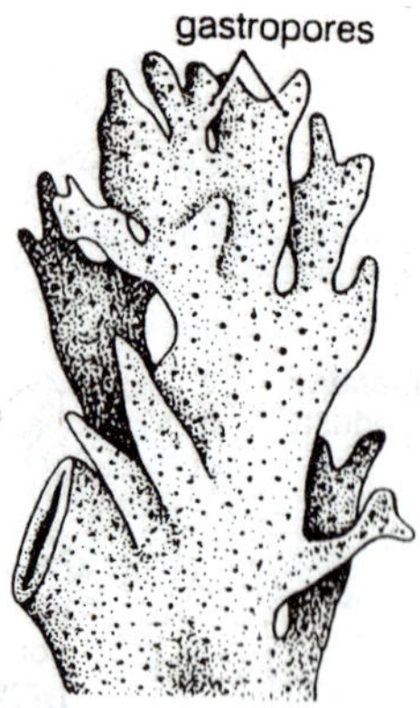

Fig. 8. *Millepora. A* portion of dried colony.

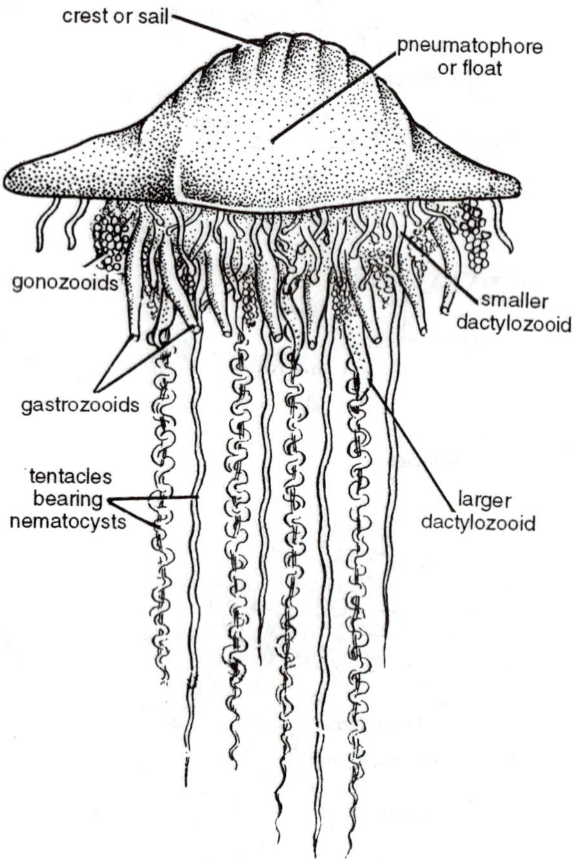

Fig. 9. *Physalia.* A colony.

bears numerous pores of two kinds —larger gastropores surrounded by smaller dactylopores. These pores lead into tubes, crossed at intervals by horizontal calcareous partitions, the **tabulae**. In living conditions two kinds of polyps protrude from these pores into water. From each gastropore protrudes a short, plump nutritive polyp or **gastrozooid**, with hypostome and mouth which is surrounded by 4 to 6 knobbed tentacles bearing nematocysts. Gastrozooids nourish the colony. From each dactylopore projects a long, slender and mouthless defensive polyp or **dactylozooid**. It is also provided with alternating short hollow and capitate tentacles, with batteries of nematocysts. Dactylozooids catch prey and pass it on to the central gastrozooids for digestion. Medusae are simple, free and with four or five rudimentary tentacles. They are budded off directly from coenosarc and lodged in pits of the stony mass, called **ampullae**. Their free existence is brief.

9. *Physalia.* *Physalia* (Gr., **physallis**, bladder), commonly called the 'Portuguese man-of-war', forms a massive, free-floating pelagic colony. It is characterised by the presence of an enormous bladder-like and brightly coloured **float** or **pneumatophore**, which supports the colony on water surface. Bright blue floats of *Physalia* are blown along by wind on warmer seas throughout the world. Float is produced dorsally into a **crest** or **sail**. A gas in the float chamber contains up to 90% nitrogen and 1.5% argon along with other gases. An oral disc on the underside of float contains the gas secreting tissue and a wide gastrovascular cavity. Below the disc hang groups of zooids or **cormidia**. Each cormidium consists of : (i) a **gastrozooid** with mouth but no tentacle, (ii) a **small dactylozooid** with a long slender tentacle, (iii) a **large dactylozooid** with an enormous nematocyst-bearing fishing tentacle, and (iv) a branched **gonozooid**, which bears both male and female **gonophores**. Tentacles have a great capacity for contraction and expansion. They may extend as long as 9 to 12 meters and contract to about 10 cm. They are useful in capturing prey, chiefly fish. They can also inflict serious and sometimes fatal injuries on man.

10. *Velella.* *Velella* is a close relative of *Physalia.* It is also common in warm seas, whence

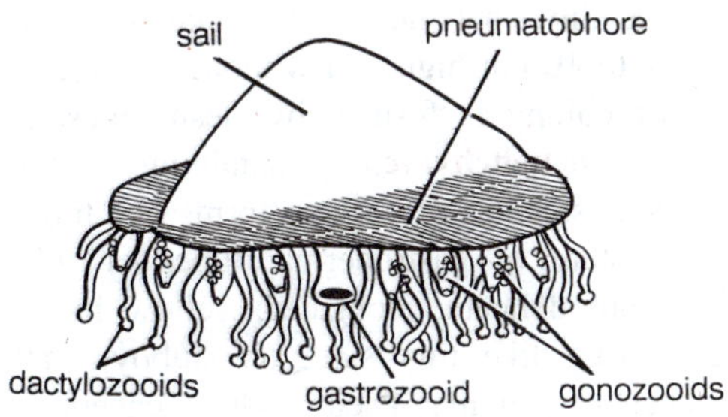

Fig. 10. *Velella.* A colony, pneumatophore or disc gonozooids.

whole fleet of them are often driven into a bay by a strong wind. Sometimes referred to as "sail-by-the-wind", or "the little sail", it is among the most beautiful and conspicuous animal of the open sea. Colony of *Velella* is flat, oval or rhomboidal and disc-like with a bright blue colour. Penumatophore is a flat, chambered, rigid and chitinous disc, 5 to 8 cm in diameter. Chambers of pneumatophore open on the upper side by pores and on the lower side by canals called **tracheae**. From the ventral surface of pneumatophore hangs a single, large, central gastrozooid, surrounded by numerous gonozooids bearing gonophores, and a fringe of dactylozooids at the margin. Colony has a clear, transparent and vertical ridge passing obliquely across the dorsal surface of pneumatophore, which acts as a **sail** in driving the colony along with the wind. Both the disc and sail contain gas-filled chambers which keep the colony permanently afloat. Gonozooids produce buds which actually escape as free medusae. *Velella* catches prey considerably smaller than that of *Physalia* and fishes only immediately below the surface. Whole colony looks like a single, highly organized, individual animal.

11. *Porpita.* The free-swimming colony of *Porpita,* like *Velella,* is also found in warm seas. Disc-like float or pneumatophore is large, flat, circular and containing a chitinoid shell with many concentric gas chambers. On ventral side of disc is a single large central **gastrozooid**. It is surrounded by clusters of small **gonozooids** or **blastostyles**. These bear sexual medusae which lead a short free-swimming life. On the edge of disc are present numerous long, tentacle-like **dactylozooids** armed with nematocysts.

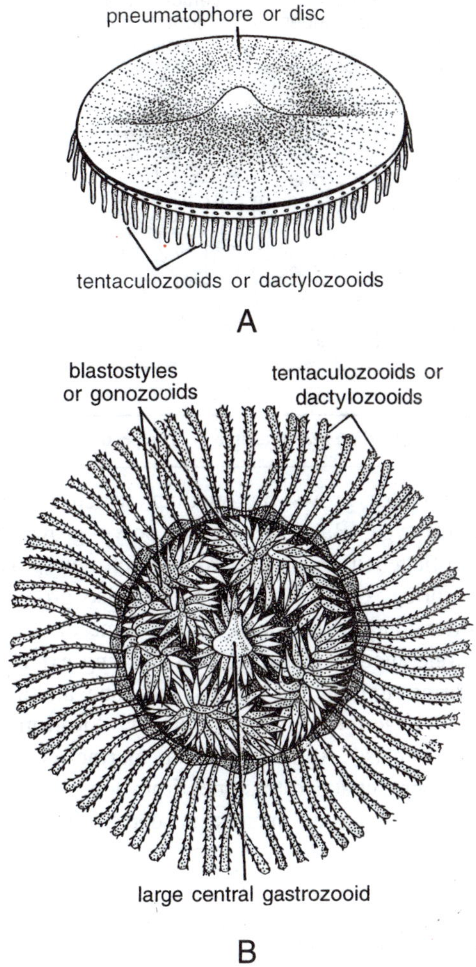

Fig. 11. *Porpita.* **A.** Colony in dorsal view. **B.** Colony in oral or ventral view.

12. *Tubipora.* *Tubipora* or 'organ-pipe coral' (class Anthozoa) is widely distributed on coral reefs in warm waters of seas. Skeleton consists of mesogloeal spicules of calcium carbonate fitted together very closely to form vertical tubes, one tube for each polyp, which are united by transverse platforms. Polyps are bright green in colour but skeleton is dull red due to presence of iron salts. Skeleton is internal in the living stage, as it is covered externally by epidermis. Gastrovascular canals, lined by gastrodermis, run through the calcareous tubes and horizontal platforms.

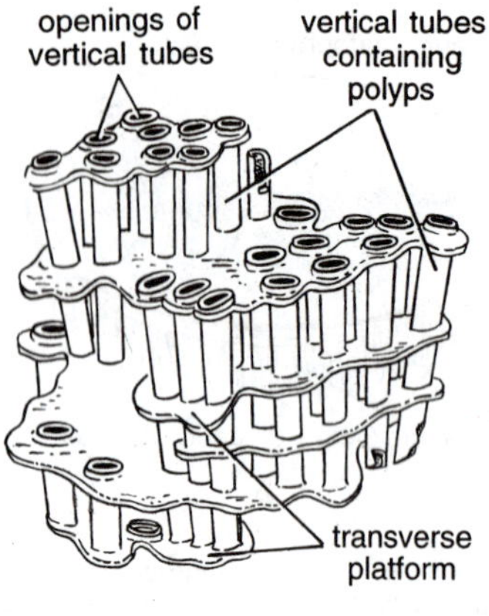

Fig. 12. *Tubipora.* Skeleton in perspective view.

Tubipora grows by budding which is of a peculiar type. Base of original polyp grows out into a flattened expansion from which new polyps may spring. Upper ends of polyps also grow out into horizontal expansions, which come in contact with one another and fuse, thus forming extensive platforms, uniting the polyps with one another.

13. *Alcyonium.* *Alcyonium* or 'dead-man's fingers' (class Anthozoa) is a marine colonial form, living attached to stones from low tide-mark to several hundred meters in depth. It usually forms a tiny, but heavy, tree-like broad-lobed mass, 4 to 10 cm high and without a central axis. Bulk of colony is formed by fleshy mesogloea, called **coenenchyme**, containing isolated calcareous spicules of varying forms and traversed by a system of gastrodermal tubes or **solenia**. Polyps are delicate, elongated, yellow in colour and looking like tiny stars on stubby, leathery branches of colony. Their bodies, except distal ends, remain completely buried in coenenchyme, connected together by gastrodermal tubes. Distal end of each polyp or **anthocodium**, includes a circular oral disc with a central mouth surrounded by a circlet of 8 pinnate tentacles. Mesenteries are eight in number. On slightest stimulus, anthocodium is completely pulled inside the gastrovascular cavity. Whole column of polyp becomes introverted, that is, turned outside in, as with the fingers in a glove. Gonads develop in breeding season on mesenteries of polyps. Fertilization is external. Life history includes free-swimming planula larva.

14. *Gorgonia.* *Gorgonia* or 'sea fan' forms a branching tree-like colony, occurring in shallow tropical and subtropical waters. In the common sea fan, *Gorgonia flabellum,* all the branches are in one plane and united by numerous cross

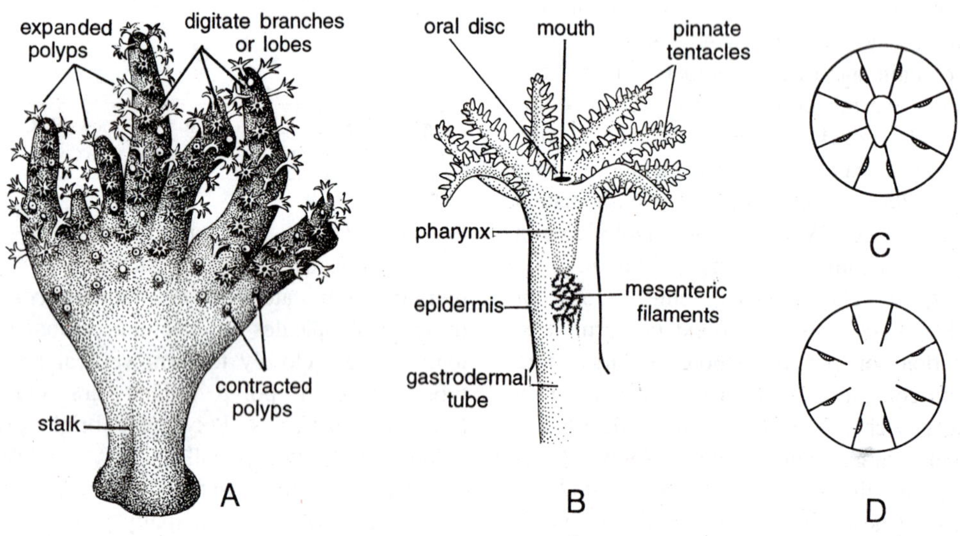

Fig. 13. *Alcyonium.* **A.** Colony. **B.** Isolated polyp. **C.** T.S. of polyp through pharyngeal region. **D.** T.S. of polyp below pharynx.

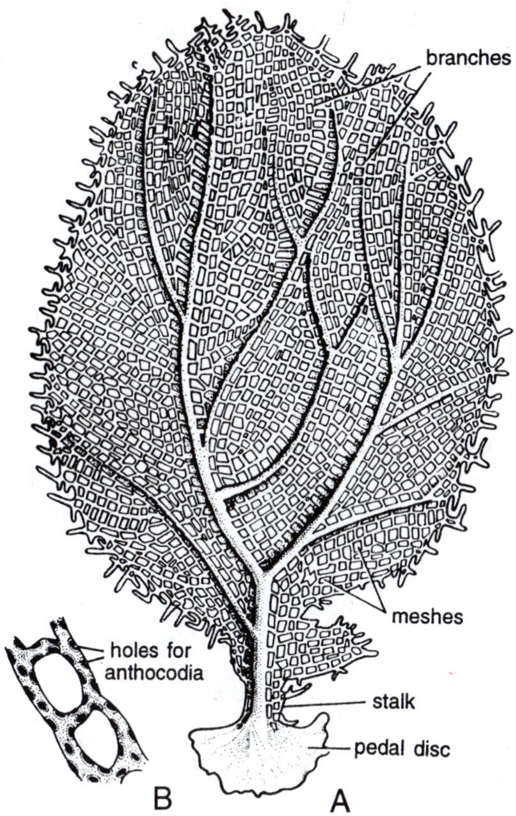

Fig. 14. *Gorgonia.* **A.** Entire colony. **B.** A portion (magnified).

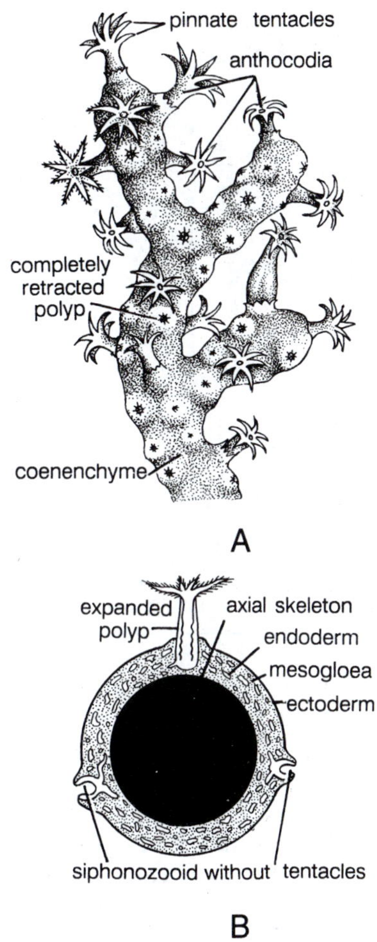

Fig. 15. *Corallium.* **A.** Portion of colony. **B.** T.S. of a branch.

branches, thus forming a network with meshes 2 to 5 mm wide. Numerous minute and retractile polyps (anthocodia) occur in rows on two sides of the stem branches. They come out at night and gather food with their tentacles. Skeleton consists of a central axial rod covered by a shallow layer of coenenchyme containing gastrovascular cavities of the polyps, numerous solenia and loose calcareous spicules of various shapes. The supporting axial rod is not calcareous but made of a flexible horny proteinous substance, called **gorgonin**, which is ectodermal in origin. With soft colouration and flexible skeleton, gorgonians sway gracefully with currents and form the most conspicuous and attractive features of coral reefs, adding much to their beauty. Dried skeletons of sea fans are often displayed as ornaments or curiosities.

15. *Corallium.* The red coral, *Corallium nobile* (**rubrum**), inhabits rocky bottoms chiefly in the central and western Mediterranean Sea. It forms a profusely branching colony reaching a height of about 30 cm. Polyps are white in colour, retractile and dimorphic. Nutritive polyps or **autozooids** bear 8 pinnate tentacles, mesenteries and gonads. Small **siphonozooids** are without tentacles and with reduced mesenteries. They pump water into canals of coenosarc. Axis of branches is made of a solid core of red coloured, hard, stony skeleton. It is formed as a result of cementing together of spicules by a red coloured deposit of calcium carbonate. It is the axial skeleton which is the

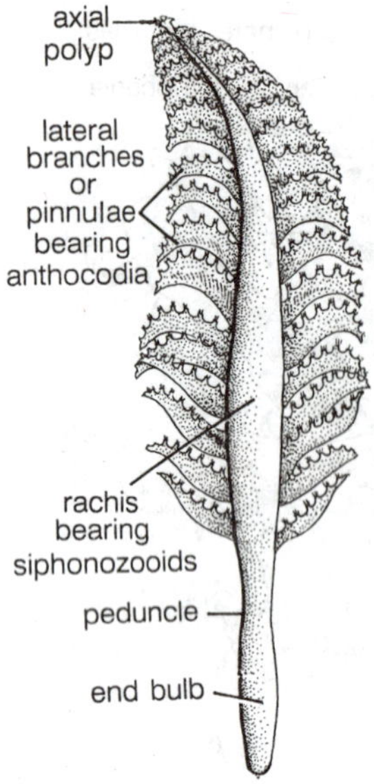

Fig. 16. *Pennatula.* Entire colony.

precious red coral of commerce used so extensively in jewellery.

16. *Pennatula.* *Pennatula* is commonly known as 'sea pen' or 'sea feather', because of its appearance like a pen or feather. Colony is not fixed but capable of independent, though limited, movement. Elongated stem of colony is an **axial polyp**. It is differentiated into two parts. Basal or proximal part is called **stalk** or **peduncle**. It remains embedded in sand or mud by an enlarged **end bulb**. Colony burrows in sea-mud or changes position by peristaltic movements of the peduncle. Upper or distal part, called **rachis**, bears two rows of long flattened, lateral branches, called **pinnulae**, thus providing the feather-like shape. Pinnulae may be 20 to 50 in number on each side. Colony is dimorphic, bearing two types of individuals. Normal polyps, called **autozooids** or **anthocodia**, feed the colony. They occur in a single row on the upper margins of pinnulae. In each row, outermost polyp is the oldest and nearest the dorsal side, while others are budded successively toward the ventral side of stem. Autozooids bear tentacles, mesenteries and gonads. Smaller polyps, called **siphonozooids**, lack tentacles or gonads, have reduced mesenteries but enlarged siphonoglyphs. They mainly serve to cause circulation of water in gastrodermal canal of colony. They are confined on the dorsal side of rachis as well as on the sides between pinnule bases.

Common species is *Pennatula aculeata*, inhabiting deeper waters, 180-900 meters off the eastern coast of North America.

17. *Fungia.* *Fungia*, commonly called 'Mushroom coral', is a large and solitary coral. Skeleton or **corallite** is discoid, convex on the upper and concave on the lower side. Theca or cup is confined to the lower surface and numerous septa are connected together by small calcareous rods, the **synapticulae**. Young polyp, called **anthocyathus**, is attached by the stalk, called **anthocaulus**. But the adult is free, large and lies at the bottom. It bears short tentacles and is without siphonoglyph.

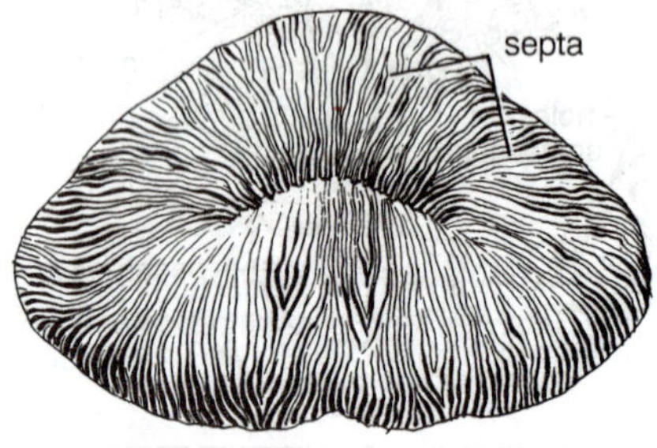

Fig. 17. *Fungia.* Skeleton.

18. *Madrepora (=Acropora).* *Madrepora* or *Acropora* is commonly called the 'Stag-horn coral'. Corallium is branched, extremely porous and of loose construction. It bears numerous small and crowded polyps in elevated cylindrical cups. Terminal polyps have six tentacles, while lateral polyps bear 12 tentacles. Mesenteries are arranged in bilateral pairs. Cups are small and deep, without columella and with 6 or 12 septa. Coenosarc is represented by a network of interlacing canals connecting the polyps.

19. *Astraea.* *Astraea* or star-coral is a colonial coral with closely-placed, polygonal cups bearing polyps. Thecae or cups are so close together as to have common walls. Colony is compact and massive and produced by budding, the buds growing more or less parallel with one another. Coenenchyme is formed by calcification of

Fig. 18. *Madrepora.* A portion of colony.

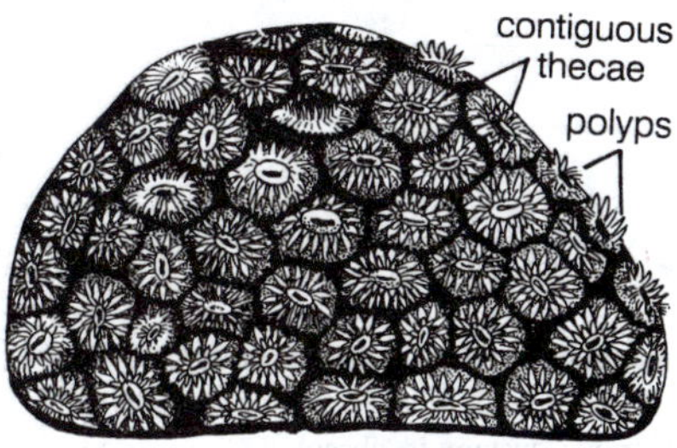

Fig. 19. *Astraea.* Entire colony.

coenosarc and gives origin to individual corallites. Polyps in cups are elevated, rounded, widely separated and spirally arranged.

20. *Meandrina (=Meandra).* *Meandrina* or 'brain-coral' forms encrusting gigantic masses of limestone, measuring 2 to 3 meters in diameter and weighing several tons. Each mass is an accumulated deposit of several generations of polyps. Surface of colony is marked by curious, long, curved or sinuous depressions or grooves, running more or less parallel to each other and recalling convolutions (fissures) of human brain. They become quite distinct in a cleaned skeleton, appearing like winding valleys and present the confluent corallites or thecae, which were formerly

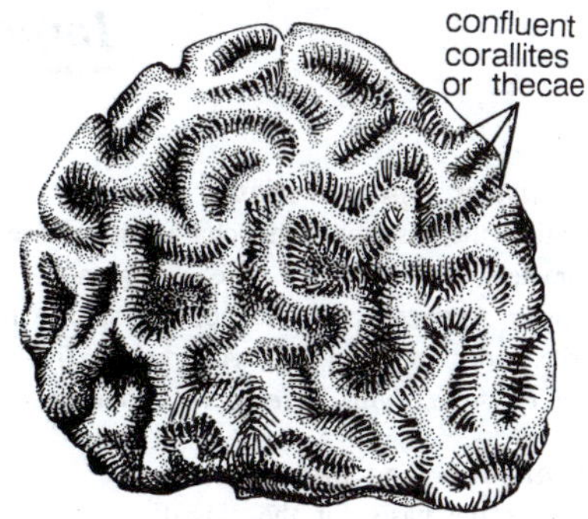

Fig. 20. *Meandrina.* Dried skeleton.

occupied by live polyps. Thus, in a living brain coral, polyps do not occupy separate cups, but several of them become confluent, having one common fringe of tentacles and rows of septa and mesenteries. But, mouths of these compound and enormously elongated polyps remain separate and lie at regular intervals along the bottom of grooves.

21. *Favea*. *Favea* is a stony coral, included among brain corals. It is found in colonies among the coral reefs of Australia. Skeleton is composed of calcium carbonate crystals secreted by column as well as basal disc. Polyps are arranged in rows. The rows are well separated, but polyps comprising a row are so close together that their cups are confluent. Mesenterial filaments contain only one glandular lobe with nematocysts. Polyps of *Favea* feed at night and remain contracted during day.

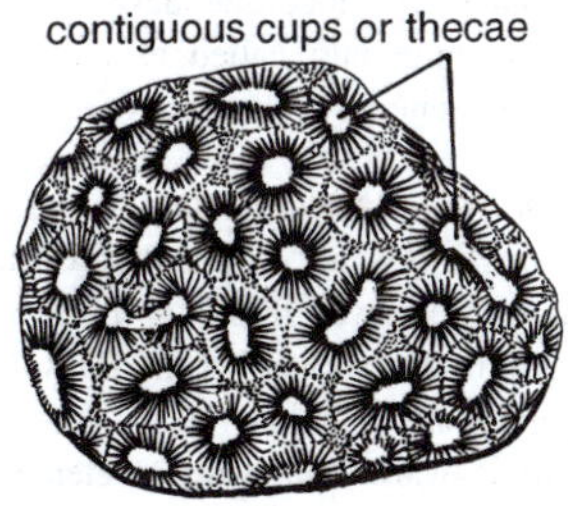

Fig. 21. *Favea.* Dried skeleton showing contiguous cups.

Important Questions

➤➤Long answer type questions

1. Mention the distinguishing characters of the phylum Coelenterata. Classify the phylum up to orders giving diagnostic characters and examples of each order.
2. Classify Anthozoa (= Actinozoa), stating the distinctive features of the various subdivisions with examples of each.
3. Write short notes on : (i) *Alcyonium,* (ii) *Gorgonia,* (iii) *Hydractinia,* (iv) *Madrepora,* (v) *Millepora,* (vi) *Pennatula,* (vii) *Physalia,* (viii) *Velella,* (ix) Planula, (x) Portuguese of war.

➤➤Short answer type questions

1. What is the basis for the classification of phylum Coelenterata in Cnidaria and Acnidaria ?
2. What is the other name given to tentaculocysts.
3. Name the animal to which the following larvae belong planula.
4. What do you understand by the term zooid ?
5. What is the symmetry found in a majority of coelenterates ?
6. Name some South Indian examples of coelenterates.
7. What is the scientific name of brain coral ?
8. Name the larva of zoanthus.
9. Mention the zoological name of the salt coral.
10. Give an example of a viviparous coral.
11. What is the scientific name of mushroom coral ?
12. List at least 5 major characteristics of phylum Coelenterata.
13. Name the distinguishing feature of *Fungia.*
14. List down 5 differences between Hexacorallia and Octacorallia.
15. What are the two types of zooids seen in *Pennatula* ? How are they different from each other ? Give two distinguishing characters.
16. Explain the three types of arrangement of zooids in the order Pennatulacia giving suitable examples.
17. Describe the various types of zooids found in *Physalia.*
18. In which group of animal nerve net present ? While the central nervous system is absent ?
19. Give the name of gelatineous layer between epidermis and gastrodermis in coelenterates ?
20. Which systems are absent in coelenterata ?
21. What are the names of asexual and sexual stages of a coelenterate ?
22. Noncellular mesogloea present in which class of coelenterata ?
23. Which is the main feature of octocorallia ?

➤➤Fill in the blanks

1. *Hydroctinia* commonly known as......
2. *Turbellaria* is a ......animal.
3. In *Pennaria* the medusae develop on the side of......
4. *Physalia* commonly called as ....
5. *Gorgonia* generally known as ......

➤➤True and false statements

1. *Bougainvillea* is a colonial hydrozoan.
2. Medusae are never found in *Hydractinia.*
3. The gas in float chamber in *Physalia* is oxygen.
4. *Velella* found in cold water.
5. *Porpita* is warm sea water coelenterate.
8. *Gorgonia* commonly known as sea fan.

➤➤Multiple choice questions

1. Nematocysts are found in :
   (a) Porifera (b) Coelenterata
   (c) Platyhelminthes (d) Annelida
2. Mesenteries attached to the stomodeum are known as :
   (a) primary (b) secondary
   (c) tertiary (d) directives
3. Function of a dactylozooid in polymorphic colony is :
   (a) nutrition (b) reproduction
   (c) locomotion (d) offence and defence
4. Polymorphism is found in :
   (a) *Halistemma* (b) *Leucernaria*
   (c) *Zoanthus* (d) *Alcyonium*

5. Pneumatophore is absent in :
(a) *Diphyes* (b) *Halistemma*
(c) *Physalia* (d) *Porpita*
6. Gonozooid with a mouth is found in :
(a) *Physalia* (b) *Velella*
(c) *Porpita* (d) all the above
7. Hypnotoxin is a poisonous fluid produced by :
(a) parasitic protozoa (b) nematocysts
(c) *Ascaris* (d) sponges
8. Gastric filaments occur in :
(a) *Hormiphora*
(b) *Aurelia*
(c) *Obelia*
(d) sea anemone
9. Which of the following inhabit fresh water bodies ?
(a) Scyphozoans (b) Anthozoans
(c) Hydrozoans (d) Ctenophores
10. The animals without sexual medusae :
(a) *Hydra* (b) *Millepora*
(c) *Velella* (d) *Halistemma*
11. The function of a tentaculocyst in *Aurelia* is :
(a) photoreception (b) chemoreception
(c) balancing (d) reproduction
12. Float is not found in :
(a) *Physalia* (b) *Velella*
(c) *Porpita* (d) *Obelia*
13. An example of a soft coral :
(a) *Tubipora* (b) *Heliopora.*
(c) *Alcyonium* (d) *Gorgonia*
14. Which of the following is connected with Coral formation ?
(a) *Halistemma* (b) *Millepora*
(c) *Adamsia* (d) *Rhizostoma*
15. The scientific name of precious red coral :
(a) *Tubipora* (b) *Fungia*
(c) *Heliopora* (d) *Corallium*
16. A sea anemone was found growing on a Gastropodan shell occupied by a hermit crab :
(a) symbiosis (b) commensalism
(c) competition (d) neutralism
17. Which of the following is the probable reason or such an association ?
(a) Sea anemone cannot live without the hermit crab
(b) Hermit crab cannot live without the sea anemone
(c) Hermit crab benefits by being protected by the sea anemone
(d) Both sea anemone and the hermit crab are competing to occupy the empty shell
18. Fringing reef is usually located :
(a) very near the shore (b) away from the shore
(c) in the deep sea (d) in the Island
19. The larva of *Physalia* is :
(a) a planula (b) an ephyra
(c) a cydippid larva (d) a scyphistoma
20. Example for a permanently sessile Scyphozoan is :
(a) *Aurelia* (b) *Lucernaria*
(c) *Pelagia* (d) *Pilema*
21. 'Hedgehog hydroid' commonly known to :
(a) *Tubularia* (b) *Hydractinia*
(c) *Pennaria* (d) *Physalia*
22. 'Fire coral' is known as :
(a) *Tubularia* (b) *Pennaria*
(c) *Millepora* (d) *Physalia*
23. The nature of *Hydra* is :
(a) herbivorous (b) carnivorous
(c) omnivorous (d) none
24. Mesogloea secreted by :
(a) ectoderm (b) endoderm
(c) both of them (d) none of them
25. Most sensitive region of the body of *Hydra :*
(a) mouth (b) tentacle
(c) basal disc (d) bud
26. Pennatula generally known as :
(a) sea fan (b) sea fish
(d) sea disc (d) sea pen
27. Which is known as mushroom coral :
(a) *Fungia* (b) *Tubipora*
(c) *Adamsia* (d) *Aurelia*
28. Dead man's finger is :
(a) *Fungia* (b) *Alcyonium*
(c) *Corallium* (d) *Heliopora*
29. Coelenterates differ from other animals in having :
(a) Skeleton (b) Polymorphism
(c) Nematocysts (d) Bilateral Symmetry
30. Phylum Coelenterata has remained at the level of organisation :
(a) Organ System (b) Organ
(c) Tissue (d) Cellular
31. Precious red coral used in jewellry is :
(a) *Corallium* (b) *Meandrina*
(c) *Astraea* (d) *Favia*

Answers

➤➤*Fill in the blanks*

1. hedgehog hydroid 2. marine, 3. hydranth, 4. Portuguese man of war, 5. sea fan.

➤➤*True and false statements*

1. true, 2. true, 3. false, 4. false, 5. true, 6. true.

➤➤*Multiple choice questions*

1. (b) 2. (a) 3. (d) 4. (a) 5. (a) 6. (d) 7. (b) 8. (b) 9. (c) 10. (a)
11. (c) 12. (d) 13. (c) 14. (b) 15. (d) 16. (b) 17. (c) 18. (a) 19. (a) 20. (b)
21. (b) 22. (c) 23. (b) 24. (c) 25. (a) 26. (a) 27. (a) 28. (b) 29. (c) 30. (c)
31. (a).

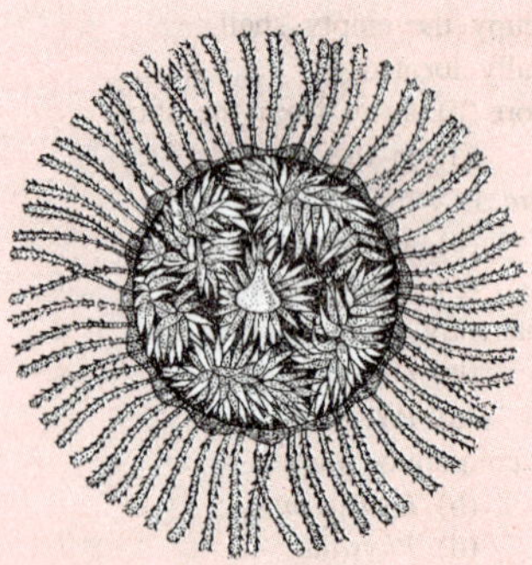

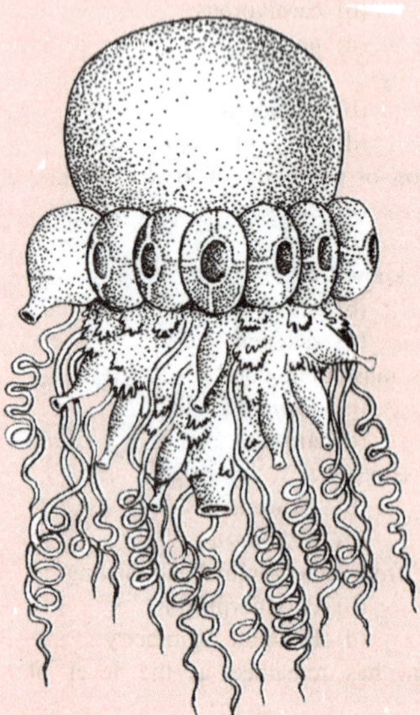

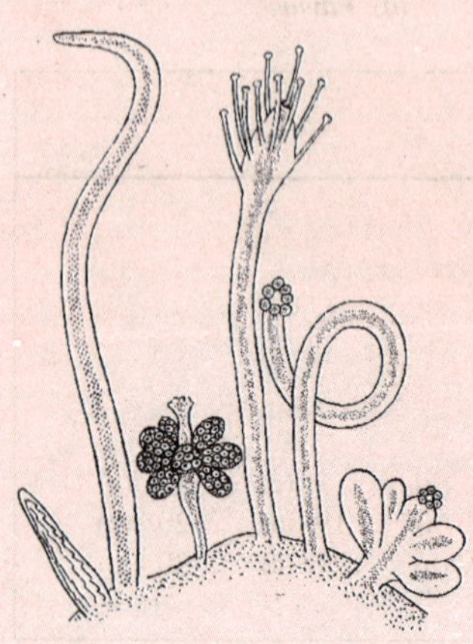

28

Coelenterata : General Account

POLYMORPHISM

Meaning of Polymorphism

Occurrence in the same species of more than one type of individuals, which differ in form and function, is known as **polymorphism** (Gr., **polys**, many + **morphe**, form). This ensures an efficient division of labour between the several individuals.

Different individuals of a species may remain separate, as represented by various castes in termites, certain ants and cuban snail (*Polymita*). This is also known as genetic **polymorphism**. However, in coelenterates the different individuals or zooids often get united in the form of a colony. Thus, polymorphism is an important feature of hydrozoan colonies which provide some of the best examples.

Two Basic Forms

In Hydrozoa (or coelenterates), which may be single or colonial, there occur two main types of individuals or zooids —**polyps** and **medusae**.

1. Polyps. A polyp has a tubular body with a mouth surrounded by tentacles at one end. Other end is blind and usually attached by a pedal disc to the substratum.

2. Medusa. A medusae has a bowl or umbrella-shaped body with marginal tentacles and mouth centrally located on a projection (manubrium) of the lower concave surface.

Although, polyps are typically sessile, and medusae are generally motile, there exists a homology between the two in their basic features.

Importance of Polymorphism

Polymorphism is essentially a phenomenon of division of labour. Different functions are assigned to different individuals, rather than to parts or organs of one individual. Thus, polyps are concerned with feeding, protection and asexual reproduction, while medusae are concerned with sexual reproduction.

Patterns of Polymorphism

Degree of polymorphism varies greatly in different groups of Hydrozoa.

1. Dimorphic. Simplest and commonest pattern of polymorphism is exhibited by many hydrozoan colonies like *Obelia, Tubularia, Campanularia,* etc. They have only two types of zooids (individuals). **Gastrozooids** or **hydranths** are concerned with feeding, while **gonazooids** or **blastostyles** with asexual budding forming sexual **medusae** or **gonophores**. Such colonies, bearing only two types of individuals are called **dimorphic**, and the phenomenon is termed **dimorphism**.

2. Trimorphic. Some forms, like *Plumularia,* are **trimorphic**. Besides gastrozooids and gonozooids, they also possess a third type of individuals, the **dactylozooids**. These are functionally non-feeding and defensive polyps bearing batteries of nematocysts.

3. Polymorphic. Coelenterates having more than three types of individuals are called **polymorphic**. A somewhat greater degree of polymorphism is found in the encrusting colony of *Hydractinia* with five types of polyps, each performing a specialized function. These are : (i) **gastrozooids** for feeding, (ii) **spiral dactylozooids** for protection, (iii) long sensory **tentaculozooids** with sensory cells, (iv) **skeletozooids** as spiny projections of chitin, and (v) **gonozooids** or reproductive individuals, bearing male or female gonophores (sporosacs) or medusae for sexual reproduction.

Extreme examples of polymorphism are seen in the pelagic or swimming colonies of the orders Siphonophora (*Diphyes, Halistemmia, Stephalia, Physalia*) and chondrophora (*Porpita, Velella*). As in *Hydractinia,* both polypoid and medusoid individuals, specialized for various vital functions, occur in the same colony. Polymorphism reaches its peak in siphonophora.

(a) Modifications of polyps. **Polypoid** individuals include :

(1) **Gastrozooid** or feeding polyp with a mouth and a long tentacle.
(2) **Dactylozooid** or protective polyp without mouth and usually with a long basal tentacle.
(3) **Gonozooid** or reproductive polyp which produces sexual medusae or gonophores.

(b) Modifications of medusae. The **medusoid** individuals are of the following types :

(1) **Nectophore** or **nectocalyx** or swimming zooid with a muscular bell without manubrium or tentacles.
(2) **Pneumatophore** or **float** as a bladder-like medusa filled with secreted gas.
(3) **Phyllozooid** or **bract**, usually leaf-like and studded with nematocysts, serving for protection of other zooids.
(4) **Gonophore** bearing gonads, which may be either male, producing sperms, or female producing ova.

Notable Polymorphic Colonies

1. Order Siphonophora. In calycophoran siphonophores, like *Diphyes,* colonies are linear with one or more nectophores located at the apical end. Polypoid and medusoid individuals are grouped as units, called **cormidia**, which are repeated in a linear succession. A typical cormidium consists of a gastrozooid with a tentacle bearing nematocysts, a phyllozooid or bract, and medusoid gonophores of one sex which are never freed. Dactylozooids are lacking.

In physophoran siphonophores, there is a pneumatophore or float at the apex of colony above water level. This is filled with gas, secreted by the gas-secreting tissue, enclosed within an oval disc. In *Physalia,* underneath this disc bears groups of cormidia, each including a gastrozooid, a small and a large dactylozooid, both with long tentacles, and a branched gonozooid with both male and female gonophores. Nectocalyces or swimming bells and bracts are altogether absent.

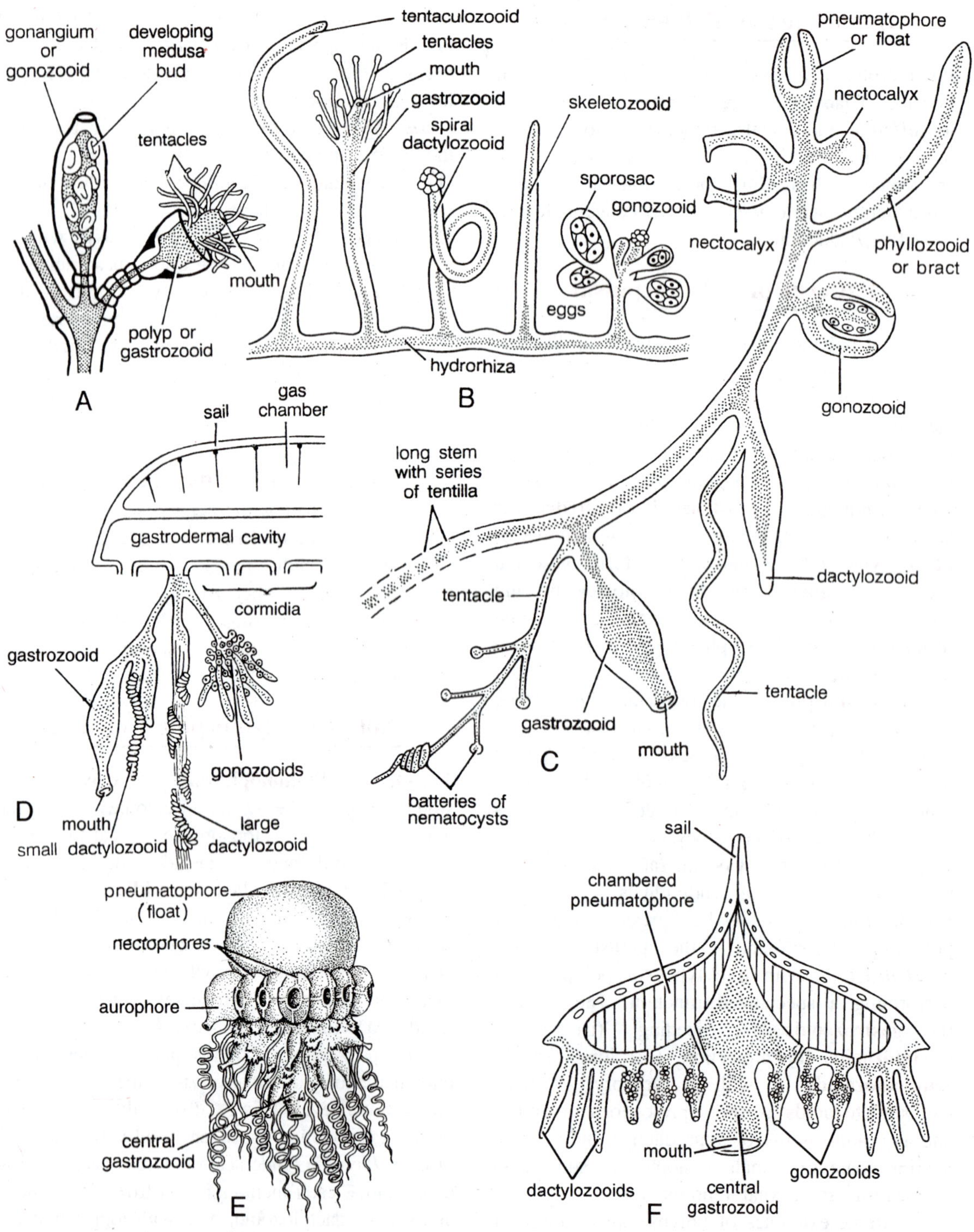

Fig. 1. Polymorphic colonies of Hydrozoa. **A.** *Obelia.* **B.** *Hydractinia.* **C.** Generalized calycophoran Siphonophora showing a single cormidium. **D.** *Physalia* showing a single cormidium. **E.** *Stephalia* showing swimming bells and aurophore. **F.** *Velella.* in V.S.

In *Nectalia* and *Stephalia,* swimming bells are highly developed. *Stephalia* is peculiar in that a portion of float or pneumatophore is constricted off as a bell-like body, called **aurophore**. Its function and homology remains uncertain.

2. Order Chondrophora. In *Velella* and *Porpita,* colony seems to be highly organized. There is a single central gastrozooid with a mouth. Around it are concentric rows of gonozooids surrounded by a few rows of dactylozooids. Entire colony looks like a single individual animal.

Origin of Polymorphism

As we have seen, colonies of Siphonophora represent the most specialized of Hydrozoa, attaining the highest degree of polymorphism and presenting the greatest number of medusoid and polypoid types.

There are two views regarding which came first, polyp or medusa, during the evolution of polymorphism in Coelenterata.

According to one view, the ancestral coelenterate was a *Hydra*-like **polyp** (archhydra of Haeckel) which arose from gastraea. It gave rise to hydroid colony by asexual budding. In the sessile colony some polyps became modified into medusae for sexual reproduction and pelagic life. Thus, through division of labour, the hydroid colony became polymorphic.

According to second view (**Brooks**, 1886), which seems to be more acceptable, the ancestral Coelenterata was a primitive **medusa**. It arose from metagastraea by developing tentacles and becoming free-swimming. According to **Huxley**, **Eschscholtz** and **Metschnikoff**, the manubrium, tentacles and umbrella of this primitive medusoid individual were multiplied and shifted from their original positions to become various zooids of the polymorphic colony. According to this view, polypoid stage is considered the persistent larval stage and medusoid the completely evolved coelenterate.

According to **Moser**, various zooids of Siphonophorae are merely organs that have not attained the grade of polymorphic individuals (**poly-organs**). She regards siphonophora to be ancestral to Hydrozoa which has fully differentiated zooids (**poly-persons**). Moser believes that **poly-organs** of Siphonophora by further differentiation became the **poly-persons** of Hydrozoa. However, Moser's views err too much to deny the full colonial nature to Siphonophora.

CORALS

Meaning of Coral

Coral animals or **corals** are marine, mostly colonial, polypoid coelenterates, looking like miniature sea anemones and living in a secreted skeleton of their own. Their calcareous or horny skeleton is also commonly known as **coral**. Some corals grow into massive, solid structures; others form large, branched colonies. Most of the corals

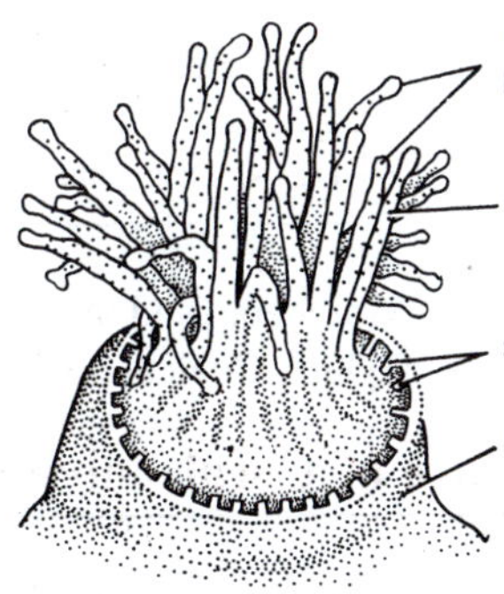

Fig. 2. A coral polyp *(Astrangia)* extended from theca.

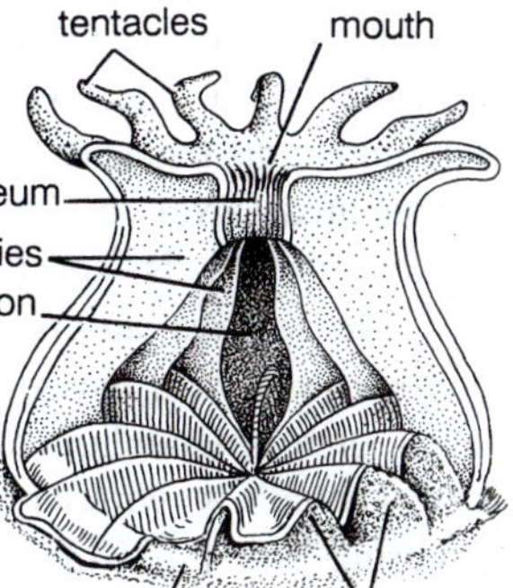

Fig. 3. Internal structure of a coral polyp in semi-diagrammatic V.S.

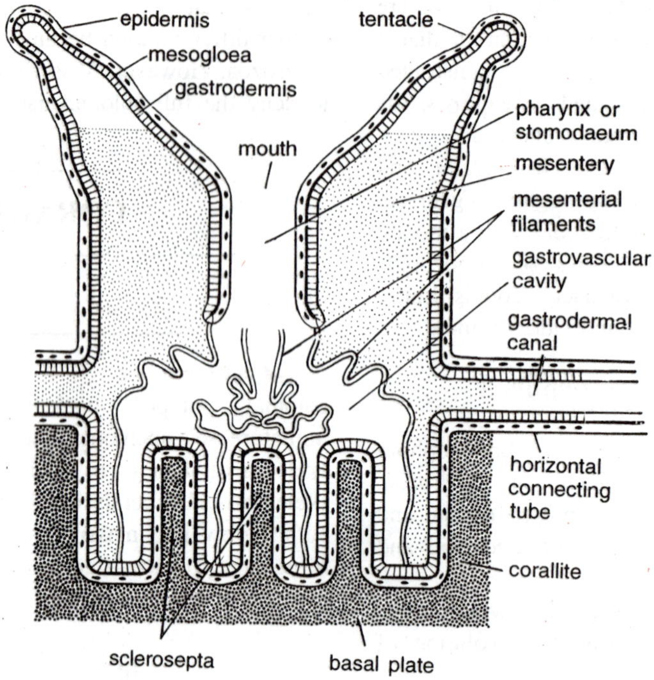

Fig. 4. Diagrammatic V.S. of a coral polyp with its corallite.

belong to the class Anthozoa and a few to the class Hydrozoa of phylum Coelenterata.

Structure of Coral Polyp

Soft structure. A typical coral polyp from a colony is a small organism about 10 mm long and 1 to 3 mm in diameter. Solitary coral polyps are much larger reaching up to 25 cm in diameter. A basal disc is absent because the basal region of polyp is surrounded by a calcareous exoskeleton. Oral disc bears numerous tentacles, in several rows around an elongated, oval or circular mouth. Pharynx or stomodaeum is short and without siphonoglyphs. Mesenteries are restricted to the upper part of coelenteron and mesenterial filaments contain only one glandular lobe bearing nematocysts. Bodywall is without cinclides and nematocyst bearing structures (acontia). Muscles are poorly developed while little is known about nervous system.

Living polyps are found only on surface layers of coral masses. They feed at night both by raptorial and suspension feeding. When not feeding, they withdraw into cup-like cavities of skeleton.

Coral Skeleton

1. Structure of coral skeleton. Skeleton of a solitary coral is known as **corallite**. It is a calcareous exoskeleton secreted by epidermis. In a colonial coral, corallites of individual polyps fuse together to form a skeletal mass, called **corallum**. Each corallite is like a stony cup with a basal part or **basal plate**, and a cup wall or **theca**, enclosing the aboral portion of polyp. Cavity of cup contains a number of vertical radiating ridges called **sclerosepta**, proceeding from theca towards the centre of cup. Inner ends of sclerosepta are fused to form an irregular central skeletal mass or **columella**.

2. Formation of coral skeleton. In coral polyps, sexual reproduction takes place by fusion of gametes. Zygote develops into a free-swimming ciliated **planula** larva which settles down and metamorphoses into a young coral **polyp**. There is no medusa stage. By asexual budding, single polyp becomes the parent of all other members of the colony. The coral polyp begins to secrete a skeletal rudiment or **prototheca**. It is secreted by ectoderm, first as a basal plate. Following it, radial folds develop which secrete sclerosepta. At the same time, a rim is built up as a thecal wall around the polyp, lying at the top. Meanwhile further skeletal material is added into the gaps between sclerosepta of skeleton which usually alternates with mesenteries of the polyp.

Coral colony grows in size continuously by budding of new polyps, particularly along the margins and on surface layers of coral masses. Variety in form of compound corals results due to various patterns of budding.

Types of Corals in Different Groups

1. Hydrozoan corals. Order Hydrocorallina includes few genera, like *Millepora, Stylaster* and *Distichopora,* which are colonial and secrete massive branched calcareous exoskeletons. These are found in coral reefs with other corals. Skeleton is secreted by a modified epidermis, called **calicoblastic layer**. Living within the skeleton occur two types of polyps, large feeding gastrozooids and defensive dactylozooids.

2. Octocorallian corals. Order Alcyonacea includes marine, colonial and soft corals. A well-known genus is *Alcyonium,* popular as 'dead man's fingers' because of its resemblance to a human hand. It has an endoskeleton of separate calcareous spicules embedded in a massive mesogloea or **coenenchyme**.

Order Stolonifera includes the organ pipe coral, *Tubipora musica,* widely distributed on coral reefs in warm waters. Skeleton is made of mesogloeal calcareous spicules forming parallel and vertical tubes, each occupied by one polyp, and connected together by lateral platforms. Skeleton is dull red in colour due to presence of iron salts.

Order Coenothecalia includes a single genus *Heliopora,* commonly known as blue coral. Its massive calcareous, skeleton or **corallium** is secreted by polyps living in large erect, cylindrical solenial tubes on the surface of skeleton.

Order Gorgonacea includes plant-like colonies of sea fans or horny corals. In *Gorgonia,* colony branches in one plane only. Its axial skeleton is made by horny proteinaceous material intermixed with calcareous spicules arranged around the polyps. In precious red coral, *Corallium nobile,* the branching colony has canaliferous coenenchyme of coenosarc containing dimorphic polyps. Axial skeleton consists of spicules embedded in $CaCO_3$ forming precious hard red coral which is used in jewellery.

3. Hexacorallian corals. Order Madrepora includes stony corals or true corals, which are the principal builders of coral reefs. While some of them are solitary, most are colonial, assuming a great variety of forms.

(a) Solitary corals. *Fungia, Flabellum, Caryophylla,* etc., are the solitary corals or cup corals. The corallite is disc-like, cup-like or mushroom-shaped in form and measures 5 mm to 25 cm across. It is often without a theca.

(b) Colonial corals. Most of stony corals are colonial with plate-like, cup-like, spherical, or vase-shaped skeleton (**coralium**). Polyps live at the surface of the calcareous skeleton. Typical examples of colonial madreporarian corals are *Acropora, Oculina, Favia, Madrepora, Meandrina,* etc. Some of the colonies are branched. In stag-horn coral, *Acropora,* there is always a primary

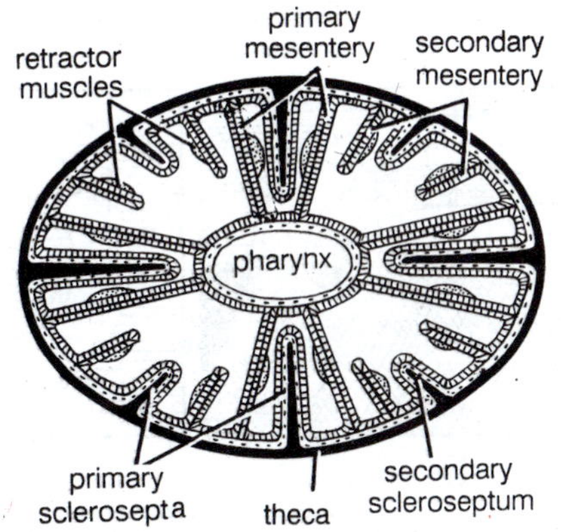

Fig. 5. T.S. of a simple coral polyp with corallite.

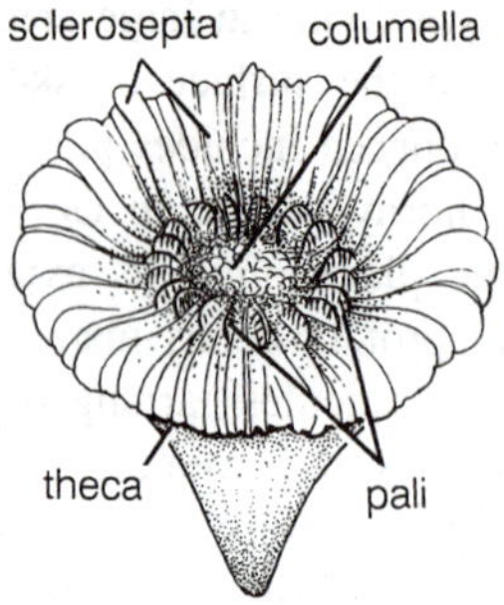

Fig. 6. Corallite of a solitary coral (*Flabellum, Caryophyllia*).

polyp at the top of colony with lateral branches on either side. In some corals, like *Oculina,* polyps remain widely separated, each occupying a separate theca. In others, like *Favia* and *Astraea,* thecae are so close together as to have common walls. In the brain-coral, *Meandrina,* polyps as well as thecae become confluent, occupying valleys separated by ridges, on the surface of coralium.

CORAL REEFS

Meaning of Coral Reef

Coral colonies grow continuously in size by budding of polyps and often form extensive masses, known as **coral reefs**. According to **T. Wayland Vaughan** (1917), a coral reef is a ridge or mound of limestone, the upper surface of which is near the surface of sea and which is formed chiefly of $CaCO_3$ secreted by coral polyps. Principal builders of coral reefs are stony corals (Madreporaria), but other important contributors are the hydrocorallines and alcyonarians. Coralline algae and Foraminiferan Protozoa also take part in the formation of coral reefs.

Reef building corals require warm shallow waters (normally above 20°C). They are therefore limited to the Indo-Pacific, the Central-Western Pacific, and the Caribbean regions north of

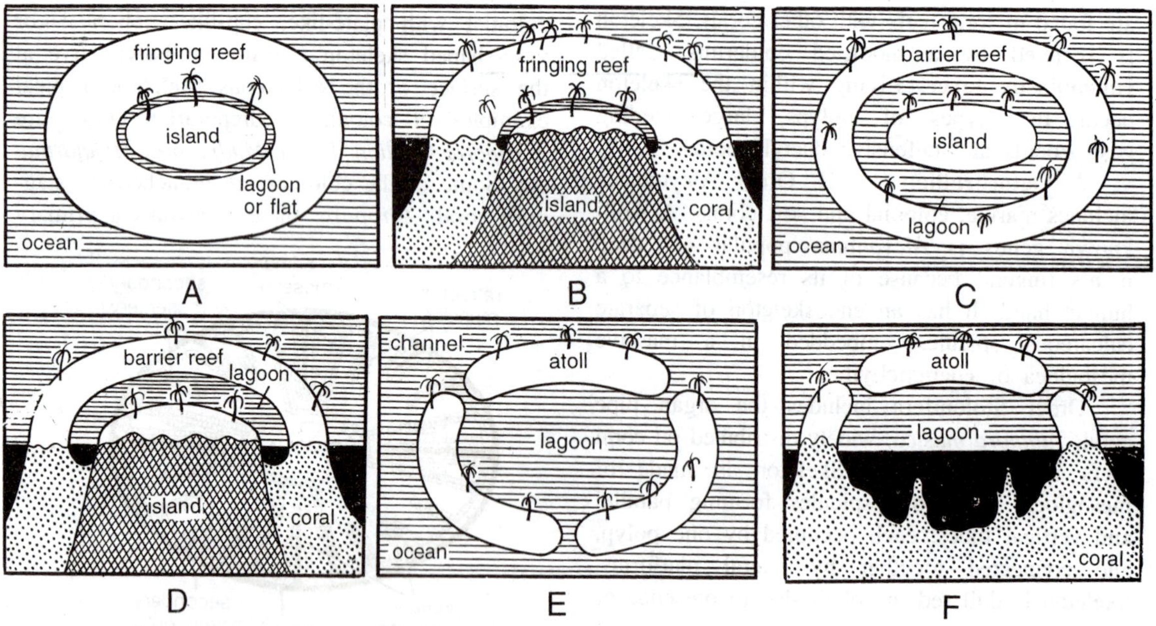

Fig. 7. Coral reefs. **A.** Fringing reef. **B.** Fringing reef in section. **C.** Barrier reef. **D.** Barrier reef in section. **E.** Atoll. **F.** Atoll in section.

Bermuda. About 50 species of corals contribute in the formation of reefs along the Florida Keys and in the West-Indies.

The Great Barrier Reef of Australia is largest runs in 2,600 km. India has them in Andaman-Nicobar, Gulf of Kutch, Gulf of Mannar and Lakshadweep.

Kinds of Coral Reefs

The coral reefs are of three kinds, depending on how they are formed.

1. Fringing reefs. Coral reefs lying close to the shores of some volcanic island or part of some continent are termed **fringing reefs**. A fringing reef may extend out to a distance of a quarter mile from the shore with the most active zone of the coral growth facing the sea. This seaward zone is commonly called the **edge** or **front**. A shallow water channel, 50 to 100 meters broad, lies between the reef-edge and shore. At low tide, water of channel recedes at quickly exposing a flat bottom surface, called **reef flat**. It is largely composed of coral sand, mud, dead and living coral colonies and other animals.

2. Barrier reef. Barrier reefs are like fringing reefs but they are located some distance away from the shore. The stretch of water, separating the barrier reef from land, may be half a 16 km or more in width. It is called a **lagoon**. It is 20 meters to 100 meters deep and suitable for navigation.

Most notable example of barrier reef is the Great Barrier Reef along the North-eastern coast of Australia. It is about 2,000 km long and up to 150 km from shore.

3. Atoll. An **atoll** is also termed a **coral island** or **lagoon island**. It is a ring-like or horse-shoe-shaped reef that encircles a lagoon but not an island. The lagoon varies from a few to about 90 km across. It may be complete or broken by a number of channels, of which only a few are navigable. Outer side of the reef slopes off rather steeply into the depth of ocean.

The atoll of Bikine, famous for atomic and hydrogen bomb tests, lies in the Pacific Ocean.

Formation of Coral Reefs

Many theories have been advanced to explain coral-reef formation, but none are entirely satisfactory. Two theories seem to be of some convincing importance.

1. Subsidence theory by Darwin. According to this theory, as put forth by **Darwin** (1831), fringing reef was first formed on the sloping shore of an island. Subsidence of sea-floor then commenced in the regions of reef followed by upward and outward growth of coral. Thus, the fringing reef became the barrier reef. By gradual sinking the island ultimately vanished and the barrier reef became a coral atoll with a central lagoon. In time it acquired a growth of vegetation.

2. Glacial-control theory by Daly. Another theory, as propounded by Daly, accounts for the lowering of the ocean level by the withdrawal of water for glacial formation. This resulted in the exposing of several flat platforms cut out by the action of waves. When the glaciers melted and the temperature became favourable, corals began to grow on these platforms, building higher as the ocean level rose.

Most reefs grew at the rate of 10 to 200 mm each year. Most of the existing reefs could have formed with in a period of 15,000 to 30,000 years.

Economic Importance of Coral Reefs

Corals of the remote geological past formed reef structures that were highly favourable sites for the accumulation of petroleum deposits. Thus coral reefs are of much importance to oil industry. Large quantities of corals are shipped every year for the curio trade. The coral reefs serve as habitats for many plants and animals like sponges, molluscs, echinoderms, fishes, etc. Some coral reefs are used as habitations by man as well. Some corals are highly priced for their decorative value. *Corallum rubrum* is considered to be a precious stone in India and China and treated as auspicious. The red coral and organ pipe coral are used in some indigenous system of medicine in S. India. Chunks of coral skeleton belonging to species *Porites* are used as building material. Coral skeletons serve as raw material for the preparation of lime, mortar and cement because of their calcium carbonate and magnesium carbonate content. Coral skeletons are also helpful in making ridges that may act as natural barriers against sea erosion and cyclonic storms. Coral reefs serve as good nursery grounds for commercially important fishes. Reef fish varieties are more colourful than others.

Important Questions

➤➤*Long answer type questions*

1. Write an essay on Polymorphism in Hydrozoa.
2. Describe a corallite. How are corals formed ? Give an account of the principal corals studied by you.
3. What are coral reefs ? Give an account of the various forms of coral reefs met with all over the world.
4. Write short notes on : (i) Atoll, (ii) Barrier reef, (iii) Fringing reef, (iv) Coral polyp, (v) Coral reefs.

➤➤*Short answer type questions*

1. How many times can a nemate cell be used ?
2. What is cnidoblast ?
3. Which are the original cells of nematocysts of coelenterates ?
4. Mention an example for a solitary coral.
5. Give an example of commensalism from Anthozoa.
6. Mention the use of coral reefs to other animals.
7. Mention four coral building coelenterates.
8. What are the three types of coral reefs ?
9. Define polymorphism. Illustrate it with a suitable example from phylum Coelenterata.
10. How will you differentiate a hydrozoan polyp from an anthozoan polyp. List at least 5 differences.
11. Describe the sense organs in coelenterates.
12. Explain the concept of polymorphism with reference to Coelenterata.
13. Trace the evolutionary sequence of the gastrovascular system in Coelenterata.
14. One of the important steps in the evolution of metazoans is the differentiation of cells to perform different functions. Justify this statement in relation to the cellular organization of *Hydra.*
15. What are coral reefs ? How does a barrier reef differ from a fringing reef ?
16. Give the structure of a corallite.
17. How are coral reefs formed ?
18. Name the animal living in close association with the stinging Portuguese man of war. What is the type of association exhibited here ? Explain the association.
19. Explain the polyps.
20. Define the medusa.
21. What is polymorphic polymorphism ?
22. Describe polymorphism in order siphonophora.
23. What is corallite ?
24. Describe an atoll.
25. Who proposed glacial control theory to formation of coral reefs ?

➤➤*Fill in the blanks*

1. Most reefs grew at the rate of.......to......mm each year.
2. ....is considered to be a precious stone in India.
3. Coral reefs lying close to the stone of some volcanic island are termed as........
4. *Alcyonium* popular as.......
5. *Heliopora* commonly known as.......

➤➤*True and false statements*

1. Marine coelenterates known as coral.
2. Most of the corals belongs to class schyphozoa.
3. Skeleton of a solitary coral is known as corallite.
4. Principal builders of coral reefs are stony corals.
5. Reef building corals require warm shallow water.

➤➤*Multiple choice questions*

1. Division of labour between the several individuals is known as :
   (a) vital activity (b) labour division
   (c) polymorphism (d) poly-functions
2. Polymorphism occurs in :
   (a) termite (b) certain ants
   (c) certain snail (d) coelenterates
   (e) all
3. Which zooid is generally motile :
   (a) polyp (b) medusae
   (c) both (d) none
4. Gastrozooid, gonozooid and dactylozooid are the characteristics of :
   (a) dimorphic (b) trimorphic
   (c) polymorphic (d) all the above
5. Modification of polyps are :
   (a) gastrozooid (b) dactylozooid
   (c) gonozooid (d) all
6. Modification of medusa :
   (a) phyllozooids
   (b) nectophore
   (c) pneumatophore
   (d) all
7. Corallite is a/an :
   (a) animal (b) individual
   (c) a coelenterata
   (d) skeleton of a solitary coral
8. Prototheca is a :
   (a) nutritive secretion (b) skeleton rudiments
   (c) larvae (d) exoskeleton
9. Reef building coral require water warm above :
   (a) 10°C (b) 20°C
   (c) 40°C (d) 100°C
10. Most notable example of barrier reef is :
    (a) Florida keys (b) West Indies
    (c) Great Barrier Reef (d) None

Answers

➤➤*Fill in the blanks*

1. 10 to 100, 2. *Corallum rubustum,* 3. fringing reefs, 4. dead man's finger, 5. blue coral.

➤➤*True and false statements*

1. false, 2. false, 3. true, 4. true, 5. true.

➤➤*Multiple choice questions*

1. (c) 2. (c) 3. (b) 4. (b) 5. (d) 6. (d) 7. (d) 8. (b) 9. (b) 10. (c)

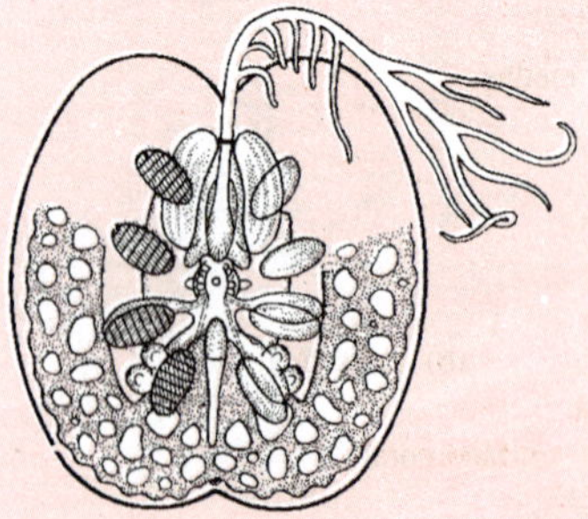

29

Ctenophora

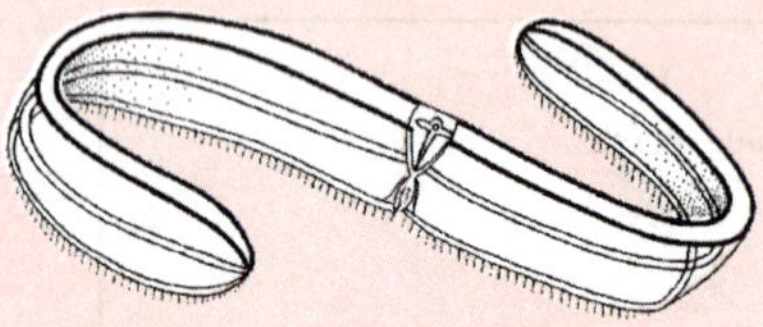

Ctenophora is a small phylum of marine animals, which are commonly known as **sea walnuts** or **comb jellies**. The phylum takes its name from two Greek words, **ktenos** = comb and **phoros** = bearing, as they possess 8 comb-like plates for locomotion. In previous classifications, ctenophores have been placed in subphylum *Cnidaria* under the phylum Coelenterata. But, the present tendency is to consider them as a separate phylum. Ctenophores were recognized as a distinct group by Eschscholtz and placed under a distinct phylum by Hatschek.

A new species ***Tamoya ohboya***, Bonaire Banded Box Jelly has been described in 2011. It was caught in the sea around Netherlands.

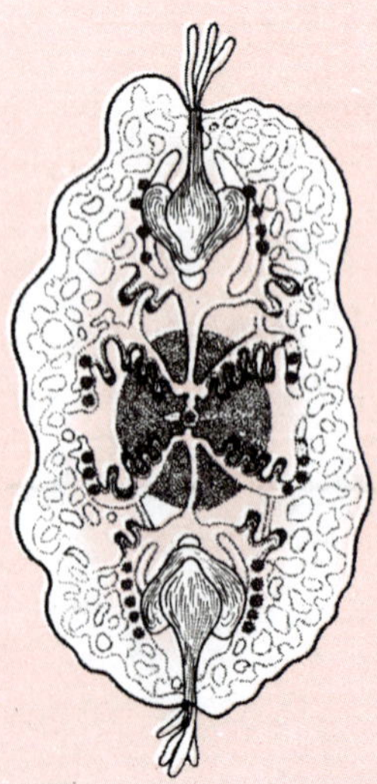

GENERAL CHARACTERS

1. Marine, solitary, free-swimming or pelagic. No polymorphism and no attached stages.
2. Body transparent. Symmetry **biradial** along an oral-aboral axis.
3. External surface with 8 vertical rows of **comb plates** of fused cilia, for locomotion. Hence the name **comb jellies**.
4. A pair of long, solid, retractile **tentacles** present.
5. Cell-tissue grade of body organization.
6. Body **acoelomate** and **triploblastic**, with an outer epidermis, inner gastrodermis, and middle jelly-like mesogloea with scattered cells and muscle fibres.
7. Digestive system with mouth, stomodaeum, complex gastrovascular canals and 2 aboral **anal pores**.
8. **Nematocysts** absent. Instead, special adhesive and sensory cells, called **colloblasts**, present on tentacles, help in food capture.

9. Skeletal, circulatory, respiratory and excretory organs absent.
10. Nervous system **diffuse**. Aboral end bears a sensory organ, the **statocyst**.
11. All **monoecious** (hermaphrodite). Gonads develop side by side on digestive canals and develop from endoderm.
12. Development usually includes a characteristic **cydippid larva**.
13. **Asexual reproduction** and **alternation of generations** absent.
14. **Regeneration** and **paedogenesis** common.

CLASSIFICATION

Phylum Ctenophora contains about 199 known species grouped in 2 classes, as follows :

CLASS 1. Tentaculata

Ctenophores with 2 long aboral tentacles.

Order 1. Cydippida

1. Body rounded or oval.
2. Tentacles branched, retractile into pouches.
   Examples : *Pleurobrachia, Hormiphora, Mertensia.*

Order 2. Lobata

1. Body oval, laterally compressed.
2. Two large oral lobes and 4 slender flap-like auricles around mouth.
3. Pouched tentacles in larva, reduced and without pouch in adult.
   Examples : *Mnemiopsis, Bolinopsis.*

Order 3. Cestida

1. Body elongated, flat, ribbon-like.
2. Two main tentacles in sheaths but reduced. Many small lateral tentacles along oral edge.
3. Four comb plates prominent, four rudimentary.
   Examples : *Velamen, Cestum.*

Order 4. Platyctenea

1. Body much flat, oral-aborally compressed.
2. Tentacles well-developed, with sheath.
3. Comb plates reduced or absent in adult.
4. Adapted for creeping.
   Examples : *Ctenoplana, Coeloplana.*

Order 5. Thalassocalycida

Found from the surface waters down up to 2,765 meters in the Atlantic Ocean and Mediterranean sea.

(1) The body is shaped like the bell of a medusa and may be up to 15 cm diameter.
(2) A central cone-shaped peducle holds the mouth slit.
(3) A pair of small tentacles hang from the sides of the peduncle. With its transparent, colorless body, this comb jelly is usually very difficult to see.
(4) *Thalassocalyce* holds the "bell" wide open to capture zooplankton prey.
(5) Presumbly hermaphroditic.
(6) Compared to other comb jellies, this species has limited swimming ability.
   Example : *Thalassocalyce inconstans.*

CLASS 2. Nuda

Ctenophores without tentacles.

Order 5. Beroida

1. No tentacles and oral lobes.
2. Body conical and laterally compressed.
3. Mouth large. Stomach voluminous.
   Example : *Beroe.*

AFFINITIES

Different workers have dissimilar views regarding origin and relationships of Ctenophores. No fossil record is available due to their soft bodies. In the absence of fossils, their origin remains obscure.

1. Affinities with coelenterata. In the beginning, **Eschscholtz** (1829-1833) regarded Ctenophora as a class under phylum Coelenterata.

(a) Resemblances with Coelenterata. There are certainly many morphological similarities between Ctenophora and Coelenterata.

(i) Biradial symmetry.

(ii) Body parts arranged along an oral-aboral axis.
(iii) Presence of gelatinous mesogloea.
(iv) Lack of organ-systems (tissue grade).
(v) No coelom. Single gastrovascular cavity.
(vi) Diffuse nerve net or plexus.
(vii) Presence of statocysts.
(viii) Endodermal origin of gonads.

(b) Differences from Coelenterata.

(i) Tentacles oppositely placed. Symmetry bilateral.
(ii) No polymorphism. No colony formation.
(iii) Presence of 8 comb plates for locomotion.
(iv) Mesenchymal muscles present. No epithelio-muscular fibres.
(v) Nematocysts absent. Instead, special sensory cells or colloblasts present on tentacles.
(vi) Statocyst present aborally, not marginally.
(vii) Gastrovascular system with anal pores more organized.
(viii) Development of determinate type.

2. Affinities with Platyhelminthes. Platyctenea Ctenophores (*Ctenoplana* and *Coeloplana*) show certain resemblances with polyclad turbellarians *(Cestoplana).*

(i) Dorsoventrally flattened body.
(ii) Crawling mode of life.
(iii) Ectoderm ciliated.
(iv) Lobed gastrovascular cavity, especially in embryos.
(v) Gelatinous mesenchyme with muscle fibres and cells.
(vi) Similar earlier stages of segmentation and gastrulation.

On account of these similarities, *Ctenoplana* and *Coeloplana* have been considered the missing links between Coelenterata and Platyhelminthes. This view is no longer supported because *Ctenoplana* and *Coeloplana* are now considered typical Ctenophores adapted for a creeping mode of life.

3. Position in animal kingdom. Ctenophores have certain characteristics in common with coelenterates. Ctenophora are believed to have diverged very early from the ancestral medusoid coelenterate which was a spherical animal with concentration of cilia along 8 meridional rows which later developed into comb plates. However, they represent a blind offshoot which gave rise to no higher forms. Ctenophores also present certain advancements over Radiata, such as prominence of apical region, musculature derived from mesoderm, presence of gonoducts and determinate type of cleavage. This implies that Ctenophora are intermediate between Radiata and Bilateria and have undergone considerable specialization with many striking characteristics of their own. Therefore, it seems quite logical to treat Ctenophora as a separate phylum rather than a class or subphylum of phylum Coelenterata.

TYPES OF CTENOPHORA

1. *Pleurobrachia* and *Hormiphora*. *Pleurobrachia* (Gr., **pleuron**, side + **brachia**, arms) and the closely allied genus *Hormiphora*, are common marine pelagic forms, found all over the world. Body is glassy transparent and pear-shaped or walnut-shaped, hence commonly known as **sea walnuts**. *Pleurobrachia pileus* is about 20 mm in diameter, while *Hormiphora plumosa* is slightly smaller and 5 to 20 mm in diameter. External surface bears 8 equally spaced **paddle plates** or

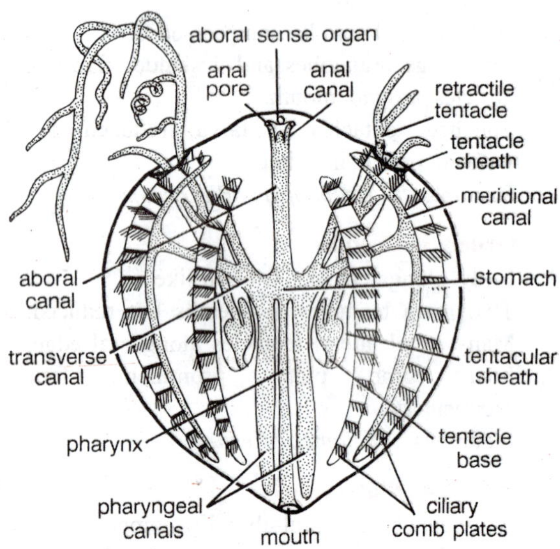

Fig. 1. *Pleurobrachia.*

comb plates, each made of transverse rows of long fused cilia. For this reason they are also called **comb jellies**. Near aboral end, on opposite sides, are attached two very long (15 cm), solid and highly extensile **tentacles** provided with lateral branches. Each tentacle can be completely withdrawn into a deep cavity or **tentacle sheath**. At the aboral end is a shallow depression with a sense organ or **statocyst**. A slit-like **mouth** is present at the oral end. It leads into a long, narrow pharynx or stomodaeum, opening into the stomach, which is connected to a system of gastrovascular canals. Comb jelly is strictly carnivorous. Food is captured by tentacles with the help of **colloblasts**. Undigested material comes out of the mouth. The animal is hermaphrodite. Gonads develop in the meridional gastrovascular canals. Development includes a free-swimming **cydippid** larva.

2. *Velamen.* *Velamen* is commonly known as 'Venus girdle'. It occurs in warmer seas, especially the Mediterranean. Body is about 45 cm long and 3 cm wide. It is laterally compressed and looks like a transparent celluloid belt or ribbon. Of 8 comb plates, 4 are very long and the other 4 are very short. Locomotion takes place by means of long comb rows and muscular undulations. At the middle of the oral surface of ribbon is a mouth and the whole oral margin bears lateral tentacles. Sense organ or statocyst is present in the middle of aboral surface. Along the middle of ribbon run two sub-tentacular meridional canals and along the oral edge, the pharyngeal canals.

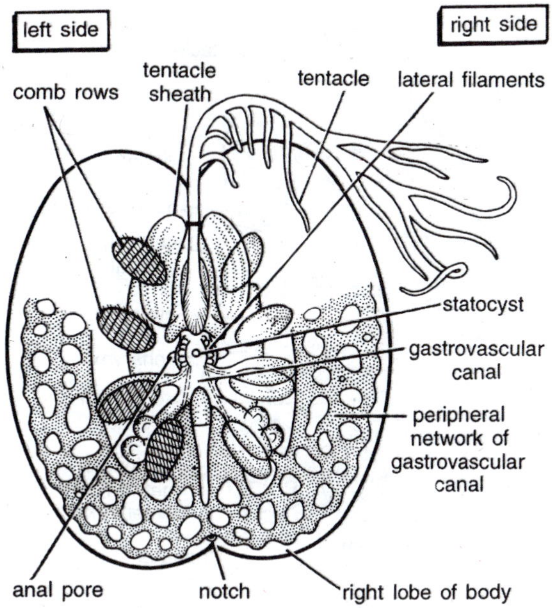

Fig. 3. *Ctenoplana.* Other tentacle not shown.

3. *Ctenoplana.* *Ctenoplana* is a small marine ctenophore living in the surface waters of sea. Body is flattened, oval or disc-shaped and measures about 6 mm in diameter. Dorsal surface is mostly olive green, brown or reddish. Body has a central thick portion and two rounded lobes in sagittal plane. At each end, in the notch between two lobes, there is a long, pennate, retractile tentacle. In the centre of aboral or dorsal surface is a typical polar body

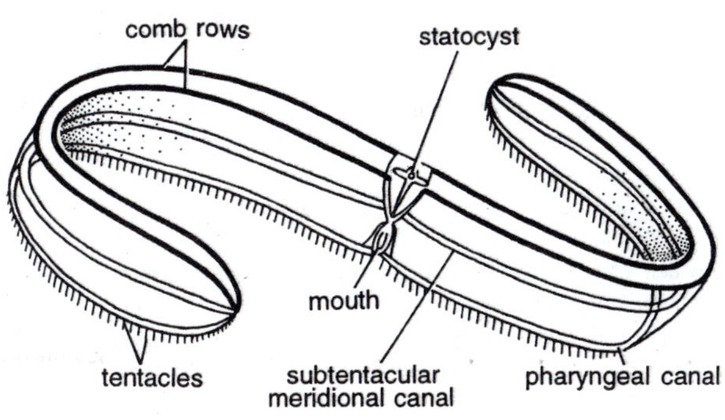

Fig. 2. *Velamen.*

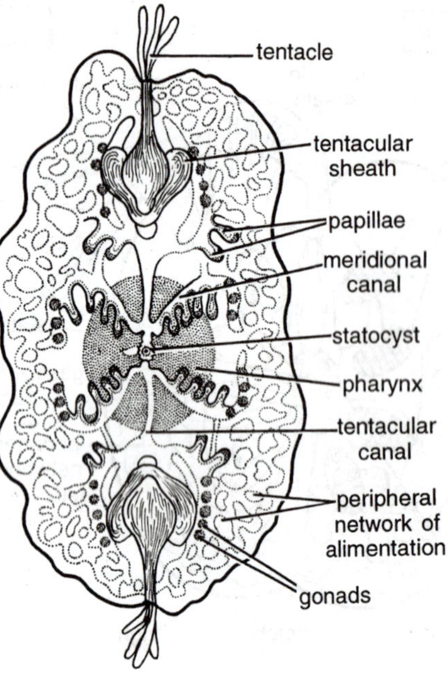

Fig. 4. *Coeloplana.*

or **statocyst** surrounded by a ring of sensory papillae or filaments arranged bilaterally into two groups of eight each. There are eight short comb rows of several plates, connected to the statocyst by ciliated furrows. *Ctenoplana* occurs along the coasts of Indo-China and Japan. It is a creeping animal and was first discovered by **Korotneff** in 1886.

4. *Coeloplana.* *Coeloplana* resemble *Ctenoplana* in most details. It is a flat oval animal, elongated in tentacular plane and about 60 mm in length. Comb rows do not occur, although traces of ciliated furrows exist. Tentacular sheaths project only slightly. Movement takes place only by creeping and as ectocommensal on specific kinds of alcyonarians. Presence of 12 to 60 erectile papillae is the diagnostic character of animal. Gonads are found in the walls of eight meridional canals. Development includes a typical cydippid larva with comb plates. *Coeloplana* is found on the coasts of Japan and was first discovered by **Kowalevsky** in 1880.

Important Questions

➤➤*Long answer type questions*

1. Mention the general characters of Ctenophora. What is its relationship with Coelenterata ?
2. Classify Ctenophora giving characters and examples of each group and discuss its affinities.
3. Write short notes on : (i) *Coeloplana,* (ii) *Ctenoplana,* (iii) *Pleurobrachia,* (iv) *Velamen.*

➤➤*Short answer type questions*

1. Name the larva of ctenophores.
2. Which are the animals that possess an aboral sense organ ?
3. To which class does cydippids (*e.g., Pleurobrachia, Hormiphora, Mertensia)* belong.
4. Give the common name for ctenophores.
5. Give the name of a parasitic ctenophore.
6. What are the type of animals that class Nuda includes ?
7. What is the basis on which Ctenophora is divided into two subclasses ?
8. What type of diet is seen in Ctenophora ?
9. What is the function of the aboral sense organ of a ctenophore ?
10. What is the function of comb plates ?
11. Mention the features in which ctenophores resemble and differ from coelenterates.
12. Give an account of the *Coeloplana* and *Ctenoplana.*
13. Give an example of a flat worm-like ctenophore and describe the same. Give 5 points.
14. Ctenophores when observed under dark condition are found to emit light through the walls of the meridonial canals. What is this phenomenon known as ?

15. Write notes on the affinity of ctenophores to other coelenterates.
16. Describe the gastrovascular system in *Hormiphora.*
17. Describe the gastrovascular system and sense organs in *Pleurobrachia.*
18. Describe the structure and mode of life of ctenophora and add a note on their systematic position.
19. Differentiate ctenophora from the other coelenterates.
20. Draw the aboral sense organ of a ctenophore and label the parts fully.

➤➤Multiple choice questions

1. The free swimming larval form of *Pleurobrachia* is called as :
   (a) Planula (b) an Ephyrula
   (c) a Semper's larva (d) a Cydippid
2. Select the creeping ctenophore from the following :
   (a) *Ctenoplana* (b) *Coeloplana*
   (c) *Deiopea* (d) *Hormiphora*
3. Ctenophores are :
   (a) unisexual creatures (b) hermaphrodites
   (c) asexual creatures (d) neotenic forms
4. The ctenophore *Cestum* is commonly known as :
   (a) venu's flower basket
   (b) dead man's fingers
   (c) brain coral (d) venus's gridle
5. Development is indirect in ctenophores and the larva is known as :
   (a) Planula (b) Miracidium
   (c) Cydippid (d) Ephyra
6. One of the following Ctenophore is without tentacles :
   (a) *Coeloplana* (b) *Ctenoplana*
   (c) *Beroe* (d) *Pleurobrachia*
7. Entoprocts are :
   (a) acoelomate (b) pseudocoelomate
   (c) coelomate (d) none of the above
8. Ectoprocta is also known as :
   (a) Polyzoa (b) Bryozoa
   (c) both of these (d) none
9. The technical name of lamp shell is :
   (a) *Phoronis* (b) *Lingula*
   (c) *Bugula* (d) *Ectinorhynchus*
10. Male lives inside female body. This condition is seen in :
    (a) *Sipunculus* (b) *Bonellia*
    (c) *Priapulus* (d) *Echiurus*
11. One of these is also known as 'glass worms' :
    (a) *Urechis* (b) *Glossiponia*
    (c) *Sagitta* (d) *Echiurus*
12. Chaetognaths are also called as :
    (a) glow worms (b) proboscis worms
    (c) arrow worms (d) thread worms
13. Lophophore is found in the phylum :
    (a) Ectoprocta (b) Phoronida
    (c) Branchiopoda (d) all of these
14. Bugula belongs to :
    (a) Brachiopoda (b) Gastrotricha
    (c) Ectoprocta (d) Nemertinea
15. Members of which phylum are called wheel animal cule ?
    (a) Rotifera (b) Gastrotricha
    (c) Kinorthyncha (d) Nematoda
16. Body cavity is absent in the phylum :
    (a) Platyhelminthes (b) Gnathostomulida
    (c) Nemertinea (d) all of these
17. Members of which phylum are called bear animalcules?
    (a) Pentastomida (b) Tardigrada
    (c) Chordata (d) Protozoa
18. Actinotroch larva is a modified :
    (a) veliger (b) tadpole
    (c) planula (d) trochophore
19. Which is commonly known as 'Sea walnuts' :
    (a) Porifera (b) Coelentrata
    (c) Ctenophora (d) Echinodermata
20. 8 comb like plates for locomotion in :
    (a) Arthropoda (d) Platyhelminthes
    (c) Mollusca (d) Ctenophora
21. Calloblast is adhesive and sensory cells in :
    (a) Porifera (b) Ctenophora
    (c) Coelenterata (d) Mollusca
22. Body acoelomate and triploblastic in :
    (a) Coelenterata (b) Ctenophora
    (c) Annelida (d) Porifera
23. Which is commonly known as 'venus girdle':
    (a) *Physalia* (b) *Porpita*
    (c) *Hyalonema* (d) *Velamen*
24. *Cestum* and *Beroe* belong to :
    (a) Protozoa (b) Porifera
    (c) Coelenterata (d) Ctenophora
25. Statocyst present in :
    (a) Ctenophora (b) Coelenterata
    (c) both of them (d) none

Answers

➤➤Multiple choice questions

1. (d)	2. (a)	3. (b)	4. (d)	5. (c)	6. (c)	7. (b)	8. (b)	9. (b)	10. (b)
11. (c)	12. (c)	13. (d)	14. (c)	15. (a)	16. (d)	17. (b)	18. (d)	19. (c)	20. (d)
21. (b)	22. (b)	23. (d)	24. (d)	25. (c).					

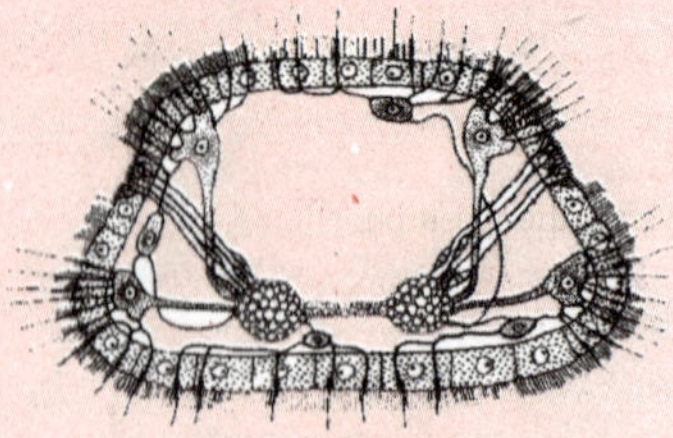

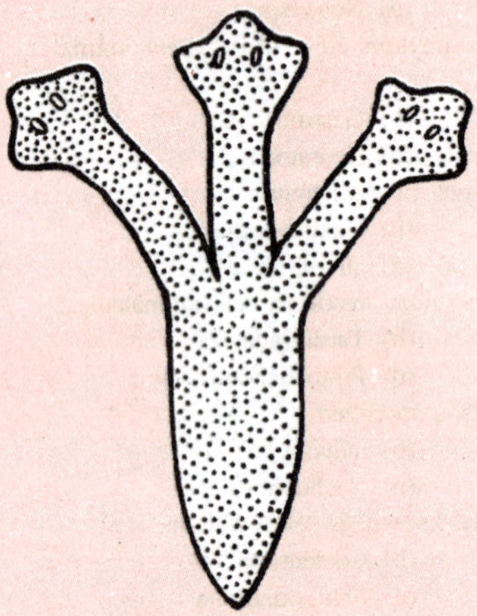

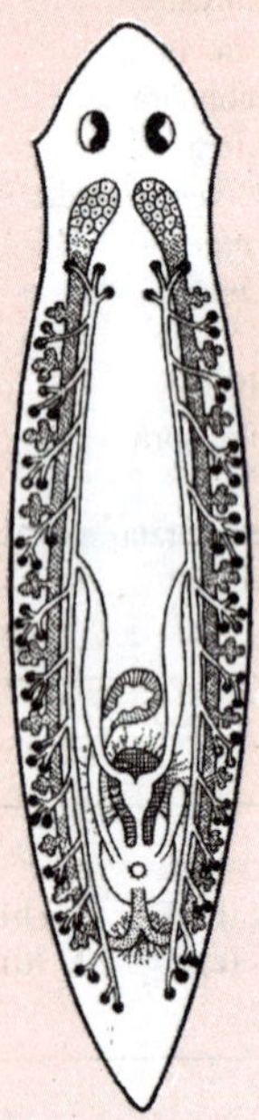

30

Dugesia (*Euplanaria*) : A Planarian

Phylum **Platyhelminthes** includes primitive, bilaterally symmetrical and acoelomate worm-like animals, commonly known as 'filaments' (Gr., **platys**, flat + **helmins**, worm). They show many phylogenetic advances over Coelenterata and Ctenophora, such as definitely triploblastic, definite head and tissue-organ level of organization. Class **Turbellaria** includes free-living planarians, whereas the other two classes are exclusively parasitic. Class **Trematoda** includes 'flukes' and class **Cestoda**, the 'tapeworms'.

Trematodes and cestodes having undergone parasitic adaptations, the basic plan of platyhelminth organization is exhibited by turbellarians. Commonly studied turbellarians by undergraduate students, belong to the genus *Dugesia.* Following account mostly relates to *Dugesia tigrina,* formerly called *Planaria maculata.* But the account is sufficiently general to apply to species of other allied genera, such as *Planaria, Euplanaria, Polycelis, Dendrocoelum.*

Dugesia tigrina

SYSTEMATIC POSITION

Phylum	Platyhelminthes
Class	Turbellaria
Order	Tricladida
Suborder	Paludicola
Family	Planaridae
Genus	*Dugesia*
Species	*tigrina*

HABITS AND HABITAT

Dugesia is found commonly in freshwater ponds, lakes, streams and shallow rivers of cold running water. It prefers to keep to the bottom, where it is found resting or gliding on the undersurface of stones and on aquatic leaves. Planarians are gregarious, *i.e.*, they live in groups. They are carnivorous and extremely cannibalistic.

COLLECTION OF PLANARIA

Planarians may be collected in quantity for laboratory by baiting shallow streams with pieces of raw beef or liver. When pieces of meat are kept overnight between rocks, submerged logs or any other debris, the planarians adhere to them. Planarians thus collected are transferred to clean, wide-mouthed jar filled with pond water. Small stones for hiding place and algae and meat or small crustaceans (as *Daphnia)* for feeding are introduced into the jar. It is advisable to change water after every three or four days.

EXTERNAL MORPHOLOGY

1. Shape, size and colour. *Dugesia* has a distinct bilateral symmetry. Body is small, thin, highly flattened, elongated, leaf-like, with a broader anterior end and pointed posterior end. It measures about 1.5 cm in length. Body wall is slimy and somewhat transparent, through which some of the internal organs are visible. Colour is dull, being greyish, brownish or blackish. Dorsal surface is darker, bearing spots or streak, which accounts for the species name *Dugesia tigrina.*

2. Head end. Anterior end of body is differentiated into a broad, blunt and triangular **head**. Dorsally, the head bears a pair of dark

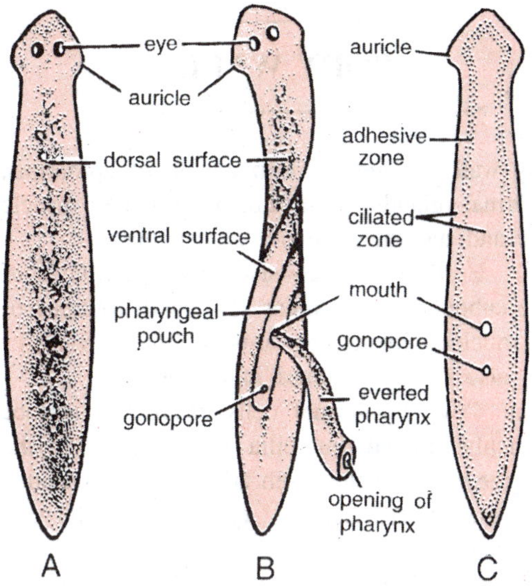

Fig. 1. *Dugesia.* External features. **A.** Dorsal view. **B.** Lateral view (partly). **C.** Ventral view.

spots, the **eyes**, and laterally on either side a small projection, the **auricle**. Behind auricles, body is somewhat constricted to form the so-called **neck**, beyond which it gradually tapers towards the posterior end.

3. Cilia. These are fine, hair-like, locomotory structures covering the ventral surface of body. However, a narrow strip all along the margin of ventral surface is non-ciliated and adhesive in function. It is known as the **adhesive zone**.

4. External apertures. Following apertures are visible externally.

(a) Mouth. It is an oval aperture situated on the ventral surface, in median line, a little behind the middle of body. Pharynx protrudes frequently through the mouth for capturing prey. The everted pharynx is also called **proboscis**. **Anus** is absent.

(b) Genital aperture. Common genital aperture or **gonopore**, present only in sexually mature individuals, lies a little behind mouth on the ventral surface.

(c) Nephridiopores. These are numerous minute excretory apertures on the dorsal surface along the two lateral sides of body.

BODY WALL

Body wall of *Dugesia* is made up of epidermis, epidermal glands, basement membrane, muscle layer and mesenchyme or parenchyma.

1. Epidermis. It consists of a single layer of large cuboidal or columnar cells, with rounded or oval nuclei, situated towards the base. **Sensory cells** also occur scattered in epidermis.

(a) Cilia. Epidermal cells of ventral surface are richly ciliated, but cilia are absent from the adhesive zone. Each cilium arises from a rounded body, the **basal granule**, lying below the cell membrane.

(b) Rhabdites. Several rod-shaped, hyaline bodies, the **rhabdites**, occur in epidermal cells. They are more abundant on dorsal side and are altogether absent in the adhesive zone and in sensory cells. Presence of rhabdites is a distinctive feature of turbellarian epidermis. Rhabdites are produced by special **rhabdites forming gland cells**, situated in epidermis or in mesenchyme. The exact function of rhabdites is still doubtful. It is, however, believed that, when epidermis is chemically irritated, say by acetic acid, the rhabdites discharge in water to form a thick, adhesive layer around the body. It protects the animal and also helps in capturing the prey.

(c) Subepidermal glands. Numerous **mucus gland cells** and **adhesive gland cells** of epidermal origin lie deep in the mesenchyme. They are provided with narrow ducts or necks, the openings of which lie between epidermal cells. Mucus gland cells are more abundant on the ventral surface in anterior part of body. They secrete a slime trail for locomotion. Adhesive gland cells produce a sticky secretion which helps in attachment of body and eggs to substratum, and capturing prey.

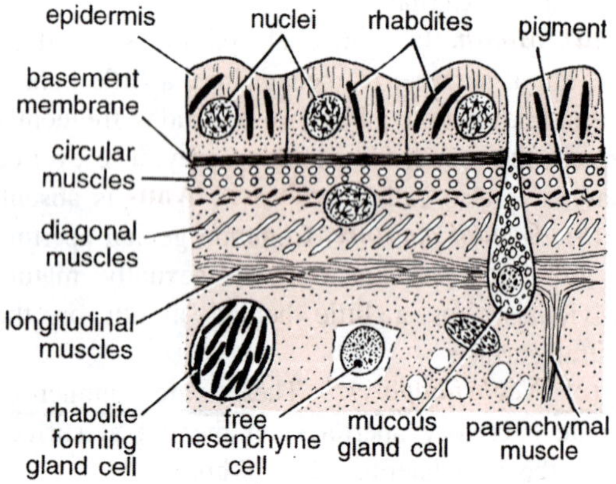

Fig. 2. *Dugesia.* V.L.S. dorsal body wall.

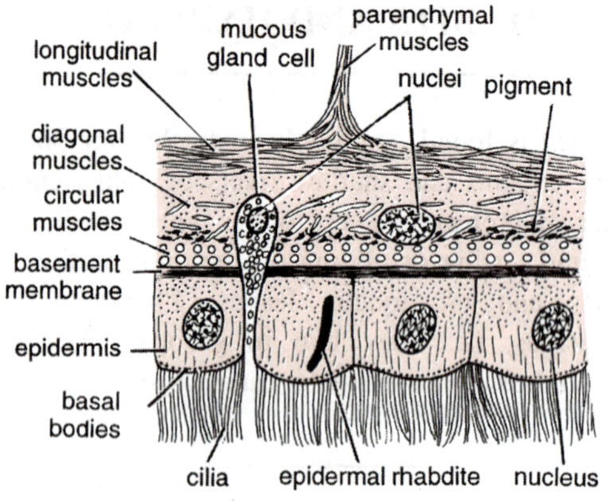

Fig. 3. *Dugesia.* V.L.S. ventral body wall.

2. Basement membrane. It is a thin, structureless membrane on which rests the epidermal cells. It serves as a surface for attachment of the underlying muscles and for maintaining the general form of body.

3. Muscle layer. This layer lies below the basement membrane. It consists of elongated contractile **muscle cells** produced by special cells, called **myoblasts**, which are of mesodermal origin. Immediately below the basement membrane, the muscle cells form an outer layer of **circular muscles**, a middle layer of **diagonal muscles** and an inner layer of **longitudinal muscles**. There are present two series of diagonal muscle fibres which run at right angles to each other. Longitudinal muscle fibres are more developed on the ventral side. Extending across the body between dorsal and ventral surfaces are also present **dorso-ventral muscles**.

Associated with the muscle layer are also present **pigment cells** that impart colouration to the body. These are more abundant on dorsal side.

4. Mesenchyme or parenchyma. This is a mesodermal connective tissue characteristic of Platyhelminthes. It fills up the interior of body and serves as a sort of **packing tissue** between its various organs. Mesenchymal cells are vacuolated, syncytial and have irregular processes, which form a loose network with fluid-filled intercellular spaces. These cells serve to transport digested food and excretory materials. Mesenchyme thus performs the role of circulatory system in which the animal lacks. In mesenchyme lie numerous free, amoeboid **formative cells**, or **neoblasts**, which rush to injured areas and bring about regeneration. Various gland cells are seated in mesenchyme.

LOCOMOTION

Dugesia, though aquatic, does not swim in water. It always moves in contact with some substratum or on the underside of water surface film. Movements are of two types : **gliding**, and **muscular**.

1. Gliding or ciliary movements. These are performed with the help of **cilia**. A thin layer of mucus is first poured on substratum. Cilia then beat backwards in the mucous layer resulting in forward movement of body. Head is kept somewhat raised during gliding.

2. Muscular movements. These are responsible for the bending and twisting of the body in different forms and for crawling.

During crawling, the body first elongates by contraction of circular muscles and the head adheres to some substratum with the help of secretion of adhesive glands. This is followed by waves of contraction of longitudinal muscles, passing from anterior to posterior end, alternately along the right and left sides, results in the forward movement of body in a wavy manner. Contraction of diagonal muscles stiffens the worm by increasing its internal turgor.

DIGESTIVE SYSTEM

Alimentary Canal

Alimentary canal of *Dugesia* is distinct, extensive but **incomplete**, as there is no anal opening. Since an anus is absent, it is comparable to the gastrovascular cavity of Coelenterata. It includes mouth, pharynx and intestine.

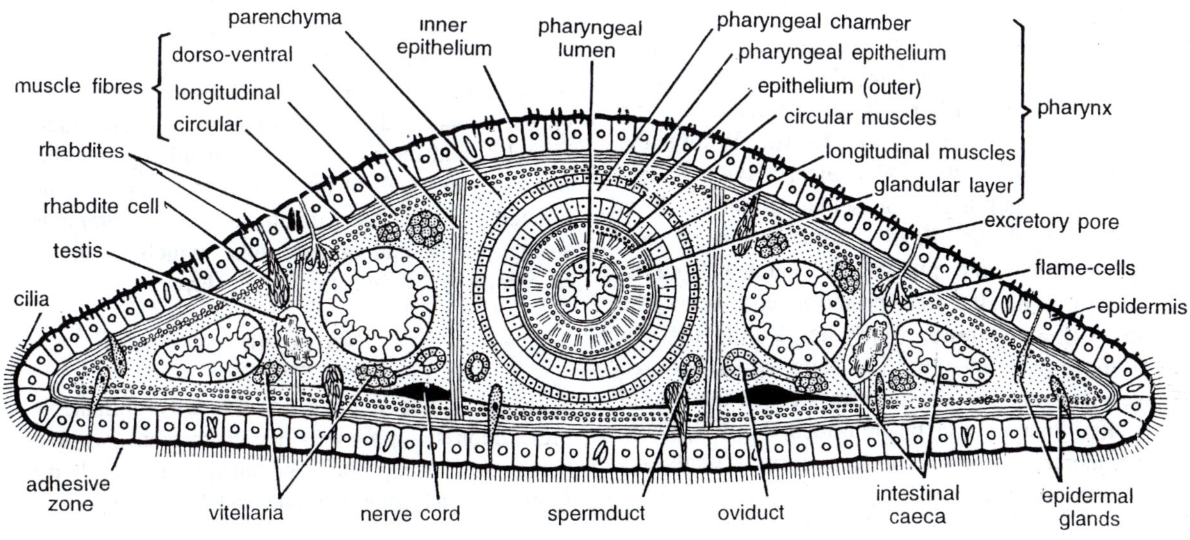

Fig. 4. *Dugesia.* T.S. through pharyngeal region (semi-diagrammatic).

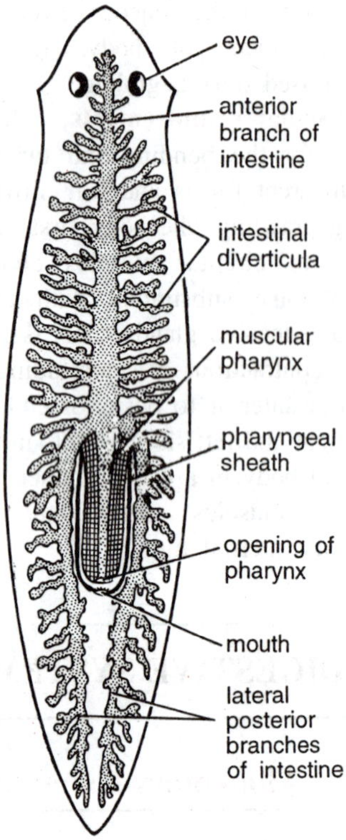

Fig. 5. *Dugesia.* Showing the alimentary canal.

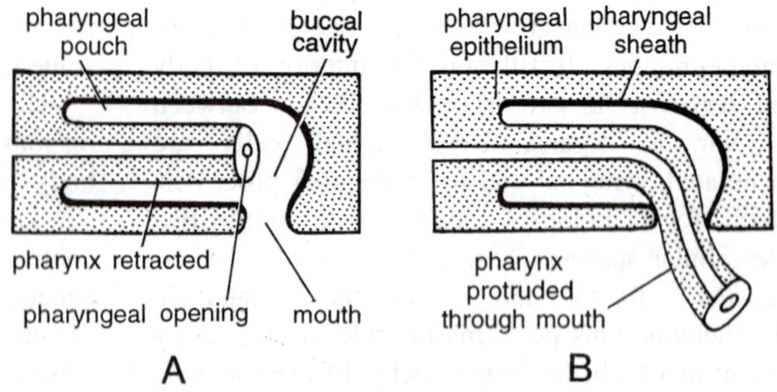

Fig. 6. *Dugesia.* Diagrammatic representation of pharynx. **A.** Retracted. **B.** Protruded.

1. Mouth. Mouth is an oval or rounded aperture located mid-ventrally a little behind the centre of body.

2. Pharynx. Mouth opens into a large, elongated and non-muscular cavity, called **pharyngeal chamber**. It is bounded by an epithelium or **pharyngeal sheath**. Within the chamber lies the elongated, thick-walled, cylindrical and highly muscular **pharynx**. It is called **folded** or **plicate pharynx** and projects behind freely into pharyngeal chamber from its anterior wall. Its free posterior end bears the **pharyngeal opening**, which lies a little in front of the mouth opening. The short space between these two openings is sometimes called the **buccal cavity**. Pharynx can be protruded through mouth considerably, when feeding, to form a **proboscis** for enveloping prey (Fig. 6).

The pharyngeal wall consists of outer and inner epithelial layers, between which lie muscle layers, gland cells and nerve plexus (Fig. 4). Inner epithelial layer is ectodermal. Nuclei of some of these ectodermal cells lie sunk into mesenchyme and are connected with their respective cells by long branching cytoplasmic processes.

3. Intestine. Anteriorly, pharynx leads through an inconspicuous **oesophagus** into the **enteron** or **intestine** which immediately forms three main branches, one median and two lateral. Median branch extends to the anterior end, and each lateral branch immediately bends backwards and extends to the posterior end. This type of **three-forked** intestine is a characteristic feature of the turbellarian order Tricladida.

Each main branch of intestine gives out numerous lateral branches, the **diverticula** or **caeca**, which further branch and ramify. The main branches and their diverticula end blindly in solid mesenchyme. Wall of intestine is made up of non-ciliated, vacuolated columnar cells, granular storage cells and gland cells, which is lined externally by a delicate muscle layer.

Nutrition

1. Food. *Dugesia* is a **carnivorous** animal. It preys upon small animals like crustaceans, worms, nematodes, insects, etc., and subsists also on body fragments of larger animals, dead or living.

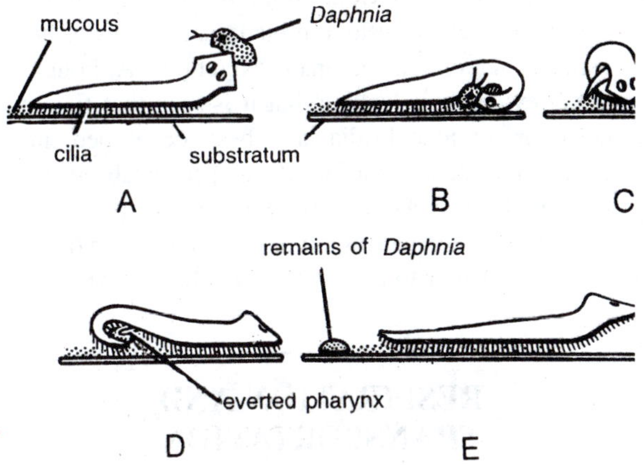

Fig. 7. A planarian capturing the prey (*Daphnia*) and feeding.

2. Ingestion. Food is detected, from some distance, by special sensory cells, located at the sides of head. When it approaches the food, it begins waving its head from side to side. Then it moves over the prey to trap it by pressing its body between its ventral surface and substratum, and by pouring over it the sticky secretion of adhesive glands and rhabdites. Now pharynx is everted to **ingest** the food by peristaltic movements. Smaller prey is ingested as such, while larger ones are first reduced to smaller bits by the pumping movements of pharynx aided by a digestive juice which is secreted by the pharyngeal glands.

3. Digestion. Digestion is both **intracellular** and **extracellular**. Fats are probably digested extracellularly in the lumen of intestine. Intracellular digestion is brought about by amoeboid cells, lining the intestine, which engulf microscopic particles and incorporate them in food vacuoles. Digested food diffuses through the walls of intestine into surrounding mesenchyme. Highly branched intestine and mesenchyme make possible the distribution of digested food to all the parts in the absence of a circulatory system. **Reserve food**, mostly in the form of fat and sometimes as protein globules, is stored in the epithelial cells of intestine.

4. Egestion. There is no anus, so that undigested food is egested through the mouth.

A planarian prefers consuming its own body parts to starving. The order of preference is eggs, vitelline glands, ovaries and testes, mesenchyme, and in extreme conditions, even gut and muscles, sparing only the nervous tissues and sense organs. These substances pass on into the lumen of the gut where they are digested. A planarian under these conditions shrinks greatly in size, and body may be reduced to as little as 1/300 of the original. When it feeds again, it regenerates the lost parts and becomes normal in size.

EXCRETORY SYSTEM

Excretory system of *Dugesia* consists of a large number of excretory cells, the **protonephridia** or **flame cells** and a system of **excretory tubules**.

1. Flame cells. Flame cells occur in large numbers along the length of body on each side. Each flame cell is large and gives out numerous branched protoplasmic processes (resembling pseudopodia) in the surrounding mesenchyme. In the centre of cell is a conspicuous bulbous cavity or **cell lumen**. The cytoplasm thus occurs in the periphery. It contains a rounded or oval nucleus and some excretory globules and vacuoles. Lying suspended in cell lumen is a tuft of numerous, long, hair-like **flagella** or **cilia**. Each cilium arises from a rounded body or basal granule, situated in cytoplasm. In a living animal cilia vibrates in the manner in which a candle flame flickers. This is why these excretory cells are called **flame cells**.

2. Excretory tubules. The bulbous cavity of flame cell narrows down into a fine vessel or **capillary duct**, which in turn opens into a lateral **longitudinal excretory duct** or **canal**. There are two such longitudinal excretory canals on each side. They extend along the entire length of body, forming a network. Longitudinal excretory canals of right and left sides are united anteriorly, in front of eyes, by a **transverse vessel**. Each longitudinal canal opens to the exterior through a number of minute **excretory pores** or **nephridiopores**, situated on the dorsal surface.

Physiology of excretion. Excretory materials diffuse from the surrounding mesenchyme into flame cells and eventually pass into their bulbous

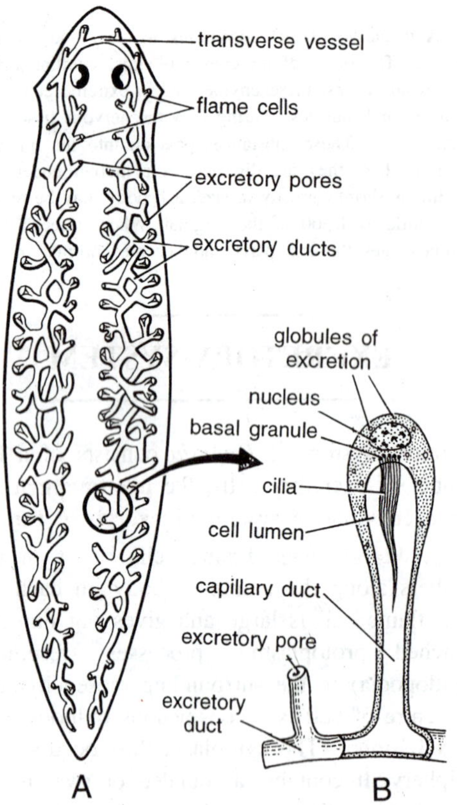

Fig. 8. *Dugesia.* **A.** Excretory system, **B.** A single flame cell in section, connected with an excretory duct.

cavities. Frequently, loss of some useful substances like salts and sugars also occur along with the excretory materials. Vibrations of flagella or cilia cause these materials, dissolved in water, to move into longitudinal excretory canals. Walls of canals reabsorb useful substances, while excretory materials, along with excess of water, are expelled out through the nephridiopores. A considerable amount of excretory waste is also expelled through mouth. Walls of lateral excretory canals are also provided with bundles of long flagella, here known as **lateral flames**, which keep the fluids moving through them.

In some turbellarians, special vacuolated cells with prominent nuclei, known as **athrocytes** or **paranephrocytes**, occur around the excretory tubules and help in transferring excretory materials into them.

3. Osmoregulation. Main function of flame cells or excretory system in planarians seems to be osmoregulation rather than excretion. Although direct evidence is lacking, but it is seen that flame cells or protonephridia are best developed in freshwater species but less developed or absent in marine species. As the excretory system plays an important role in getting rid of excess of water from body, it is better to call it **osmoregulatory system**

RESPIRATION AND TRANSPORTATION

There are no special organs for respiration. Gaseous exchange takes place through general body surface by diffusion. A circulatory system is also lacking. The function of transportation of gases within body, of digested food from intestine to different parts and of excretory wastes from different parts to flame cells, is performed by mesenchyme described earlier.

The intestine, being highly branched, is not too far from any part of body. Moreover, the body being highly flat, the various organs are not far from body surface. These features limit the scope for transportation and an elaborate system for the same would have been superfluous.

NERVOUS SYSTEM

Nervous system of planarians marks the beginning of a centralized nervous system encountered in higher animals.

At the anterior end, below epidermis and a little behind the eyes, is a prominent bilobed mass of nervous tissue, called **brain** or **cerebral ganglia**. Two lobes of brain are connected by several transverse fibres. From the brain arise two **lateral longitudinal nerve cords**, which extend to the posterior end along the ventral side of body. Each nerve cord is in fact a strand of concentrated nerve cells.

From each lobe of brain arise numerous **peripheral nerves** which supply the various sense organs located at the anterior end. Each nerve cord also gives out numerous branched nerves on both the sides all along its length. Those arising from the inner side frequently anastomose with similar ones from the other nerve cord to form **transverse commissures** or **connectives**. This is sometimes called the **ladder type** of nervous system.

In addition to the centralized nervous system, *Planaria* also possesses a **sub-epidermal nerve net** like that of coelenterates. Brain receives stimuli from the sense organs and conveys them to different parts of body. It, however, does not appear to exercise any control over muscular movements concerned with locomotion.

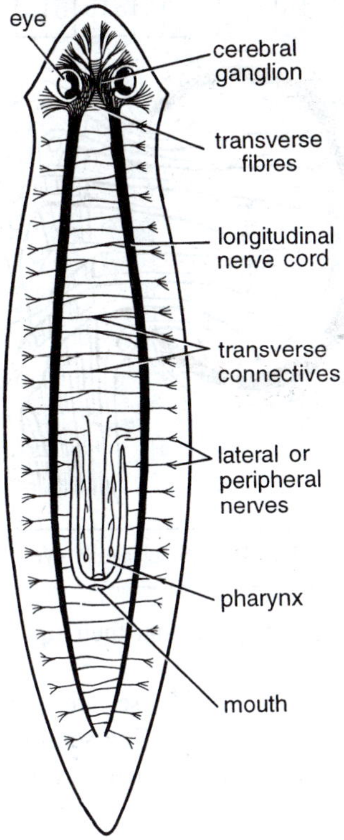

Fig. 9. *Dugesia.* Nervous system.

SENSE ORGANS

Sense organs or **receptors** occurring in *Planaria* are : (i) a pair of eyes (photoreceptors), (ii) ciliated pits, grooves and auricular organs (chemoreceptors), (iii) some scattered tangoreceptors, and (iv) rheoreceptors. Most of these receptors lie concentrated at the anterior end of the body (head), which remains foremost during locomotion. Thus, planarians show the beginning of **cephalization** which is highly developed in vertebrates.

1. Eyes. Eyes or **ocelli** are a pair of conspicuous, black, rounded spots situated dorsally on head. Each eye has a cup-like form with its opening directed antero-laterally. The cup is lined by a layer of **pigment cells**. Projecting into its hollow are numerous bipolar neuro-sensory light-receptive **photosensitive** or **retinal cells**. In this type of eye, known as **inverse eye**, the light has first to pass through photosensitive cells before reaching the pigmented cells. Expanded distal end of each photosensitive cell, projecting into cup longitudinally striated forming a **striated border** of uncertain function. Proximal ends of these cells run side by side forming the so-called nerve fibres which communicate with the brain. Eye of *Planaria* lacks a lens. It cannot form an image but can perceive the difference between light and dark. Planarians with eyes removed can react to light but slower and less accurately.

2. Ciliated pits, grooves and auricular organs. Anterior end of body bears several grooves and pit-like depressions. A pair of whitish grooves, situated on auricles, are known as **auricular organs**. Ciliated pits, grooves and auricular organs are devoid of gland cells and rhabdites and the sensory cells lining them are **chemoreceptors**, *i.e.*, sensitive to chemicals. Their sensory processes project just a little above the epidermis. These receptors help in detecting food and orienting the body, towards it, hence are called **taste receptors**. If auricles are removed, the animal can not locate food.

3. Tangoreceptors. These are **touch-sensitive** or tactile cells found all over the body. They are distributed abundantly on ventral surface, especially around mouth, on lateral margins and at the anterior end. Their sensory processes project

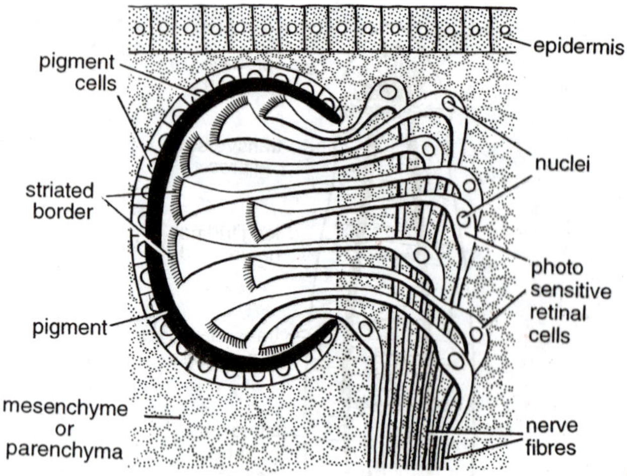

Fig. 10. V.S. planarian eye.

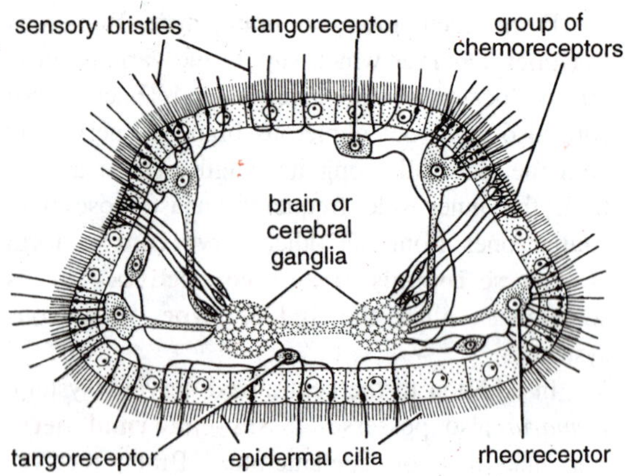

Fig. 11. A diagrammatic T.S. of the anterior end of a planarian showing the various receptors.

beyond the epidermis so that their tips lie slightly beyond the level of cilia.

4. Rheoreceptors. These are sensory cells sensitive to water currents. The sensory processes of rheoreceptors project much beyond the level of cilia.

REPRODUCTIVE SYSTEM

Dugesia reproduces **sexually** as well as **asexually**.

[I] Asexual Reproduction

Asexual reproduction occurs by the appearance of a constriction behind the pharynx. It gradually deepens and finally divides the animal into two parts, each of which regenerates the wanting structures. This mode of asexual reproduction is termed **transverse binary fission**.

[II] Sexual Reproduction

In *Dugesia,* both male and female reproductive organs are present in the same individual. It is thus **hermaphrodite** or **monoecious** or **bisexual**. Both male and female gonoducts open into a common chamber, the **genital atrium**, which in turn opens to the exterior by a common **gonopore**, situated ventrally in the median position, just behind mouth.

1. Male reproductive system. It consists of testes, vasa efferentia, vasa deferentia and a cirrus or penis. **Testes** occur as numerous, small and

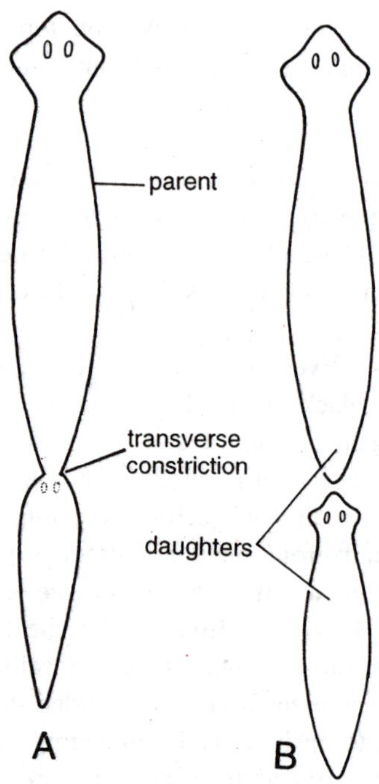

Fig. 12. *Dugesia.* Transverse binary fission.

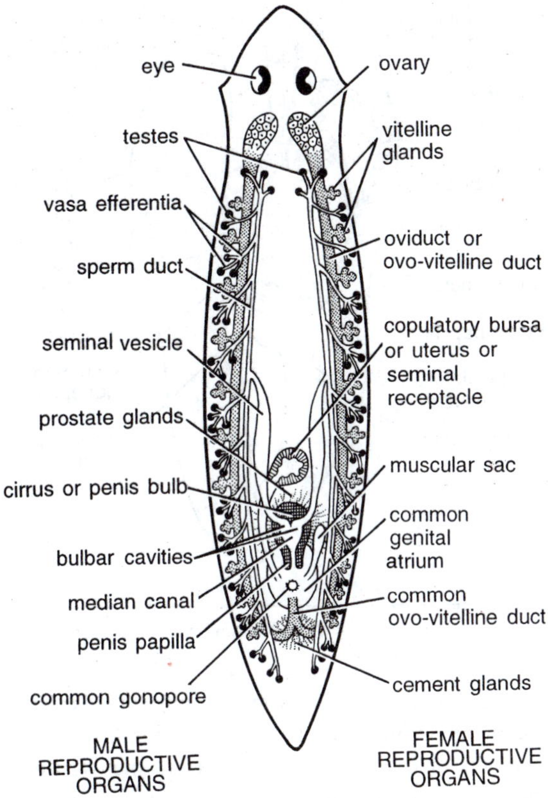

Fig. 13. *Dugesia.* Reproductive system.

rounded bodies lying along either side of body. Gonads of this type are called **follicular**. From each testis arises a minute **sperm ductule** or **vas efferens**. It usually unites with a few similar ductules from the neighbouring testes before opening into the **vas deferens** or **sperm duct** of that side. **Vasa deferentia**, two in number, are narrow longitudinal tubes, one on either side. Each vas deferens, all along its length, receives vasa efferentia from the testes of its own side. From each vas deferens arises a dilated **seminal vesicle** for storage of sperms. Posteriorly, the two seminal vesicles unite together and open into a thick-walled, muscular copulatory sac, called **cirrus** or **penis**, which consists of two parts. Proximal musculo-glandular part is known as **penis bulb**. It is surrounded by numerous unicellular **prostate glands**, opening into its lumen formed by two **bulbar cavities**. Distal conical part is called **penis papilla**. Its lumen or **median canal** opens into the **common genital atrium** which opens outside through **common gonopore**. During copulation, penis papilla extends through the gonopore. **Sperms** are long, filamentous, and each provided with 2 long flagella. Arrangement of axoneme is 9-0 in the flagella of sperms.

2. Female reproductive system. It consists of ovaries, oviducts, vagina, uterus or bursa copulatrix and vitelline or yolk glands. **Ovaries** are a pair of rounded bodies situated at the anterior end, one behind each eye. Posteriorly, each ovary is connected with a narrow elongated tubule, the **oviduct** or **ovo-vitelline duct**, running parallel and dorsal to the nerve cord of its side. Posterior ends of two oviducts unite to form a

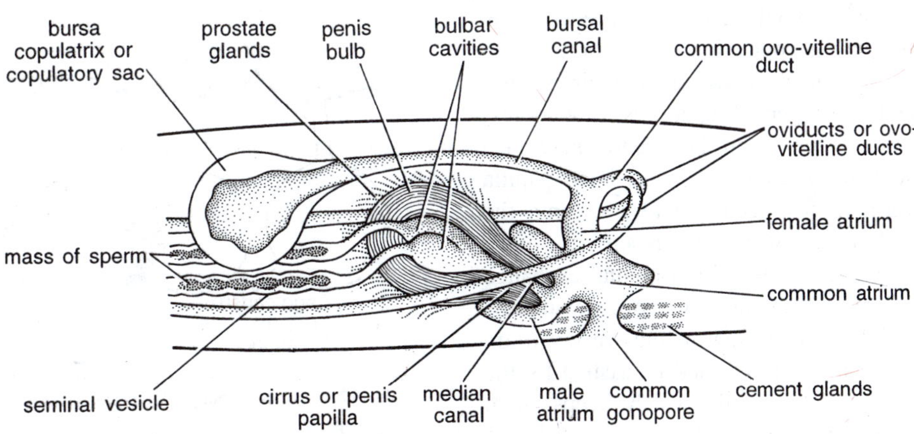

Fig. 14. *Dugesia.* A part of the reproductive system in lateral view.

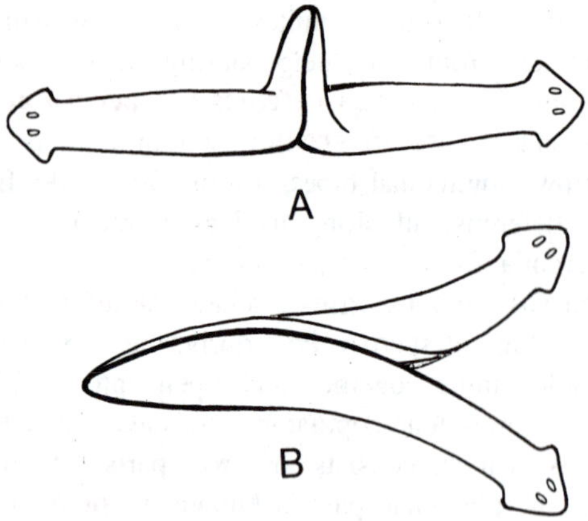

Fig. 15. *Dugesia.* Copulation. **A.** Head ends facing in opposite directions. **B.** Head ends facing in same direction.

median vagina or **common ovo-vitelline duct**. It opens into the genital atrium lying in front of it. Each oviduct, all along its length, is surrounded by numerous follicular **vitelline** or **yolk glands** which open into it. It is for this reason that the oviduct is also referred to as **ovo-vitelline duct**. Numerous minute **cement glands** surround as well as open into vagina and genital atrium.

An elongated and club-shaped sac, termed variously as **uterus**, **bursa copulatrix** or **seminal receptacle**, opens into the genital atrium. It receives sperms from the other individual during copulation. Into the genital atrium also opens a much smaller **muscular sac** of uncertain function.

3. Copulation. Planarians are hermaphrodite, but self-fertilization does not occur. Instead, they practice **cross fertilization** preceded by **copulation**. During copulation, two worms come in temporary contact along with their ventral surfaces and their genital apertures (gonopores) opposing each other. Their head ends may face in the same or in opposite directions. Penis papilla of each worm is protruded and inserted into the gonopore of other worm to reach its bursa copulatrix. This is followed by a mutual exchange of sperms, after which the two worms withdraw their penis and separate. Sperms thus received are stored in seminal receptacle and nourished by the secretion of prostate glands which keeps them active for fertilizing ova.

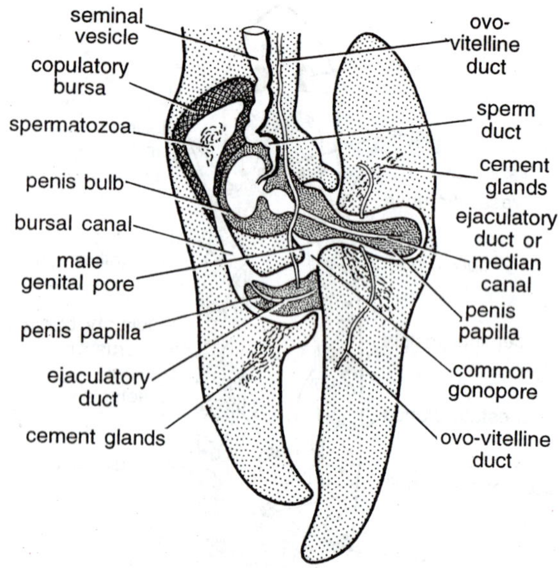

Fig. 16. *Dugesia.* A copulating pair in section.

4. Fertilization. After copulation sperms leave the seminal receptacle and migrate into the oviducts or ovo-vitelline ducts. Here they fertilize the eggs, as they are released from the ovaries. Thus fertilization is **internal** in *Dugesia.*

5. Cocoon-formation. Fertilized eggs or **zygotes** move backwards through ovo-vitelline ducts along with yolk-laden **yolk cells** released by the yolk glands. In genital atrium, about ten zygotes and thousands of yolk cells get enclosed in a protective, proteinaceous shell called **cocoon**

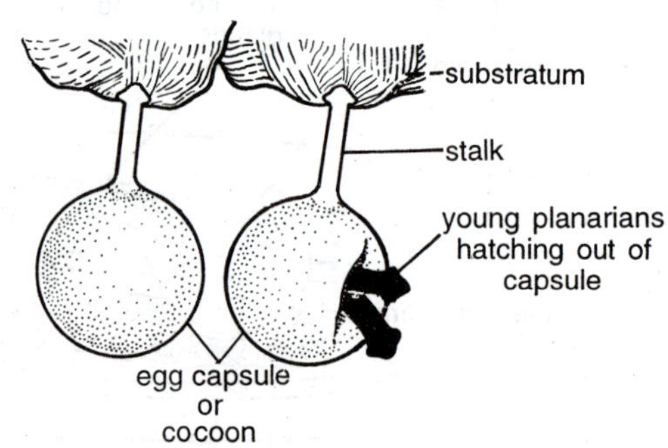

Fig. 17. *Dugesia.* Young planarians hatching from an egg capsule.

or **egg capsule**. The capsule is spherical and is provided with a little adhesive stalk secreted by cement glands. One such egg capsule is formed after every few days in a planarian copulating is frequently during the breeding season. Capsules are laid on the undersurface of some submerged object, like aquatic weed, stone, etc., to which they adhere by means of adhesive stalks.

6. Development. Development starts soon after the egg capsule is laid and is completed in two to three weeks. It is **direct**, *i.e.*, without the occurrence of any larval stage. Cleavage is **spiral** and **determinate**, *i.e.*, future of each blastomere is pre-determined. Each zygote develops into a young one. Yolk cells in egg capsule serve as food for the developing young ones. On completion of development, the egg capsule ruptures, probably along a definite line, and the young planarians or **juveniles** glide out of it. A newly-hatched planarian lacks reproductive organs and is smaller in size than the parents.

Reproductive organs degenerate and disappear after each mating, after which the worms reproduce asexually by fission. However they develop again during the next breeding season.

REGENERATION

Planarians possess a tremendous power of regeneration. If cut across into two, three or more parts, each part regenerates into a complete and normal individual. Regeneration, thus, involves two complementary processes, *viz.*, **epimorphosis**, in which the missing parts are formed and **morphollaxis**, in which the original parts are fit to function with regenerated parts are in the new individual.

When the anterior end of *Planaria* is cut along the length into two or more parts, each part develops into a new head, resulting in a many-headed monster or **heteromorph**.

A note-worthy observation is that a piece from the middle always regenerates a head towards its anterior side and tail towards its posterio side. In other words, each piece maintains its original **linear polarity**. This can be explained by the theory of **metabolic** or **axial gradient** by **C.M. Child**. The theory holds that metabolic activity is highest in head and gradually decreases towards the tail end.

Correspondingly, anterior end of each piece, having greater metabolic activity, regenerates the anterior part of body and posterior end piece, having lesser metabolic activity, regenerates the posterior part of body. Gradual difference in intensity of metabolic activity, and hence the capacity of regeneration, from anterior to posterior end of body (along the antero-posterior axis of body) is spoken of as the **axial** or **metabolic gradient**

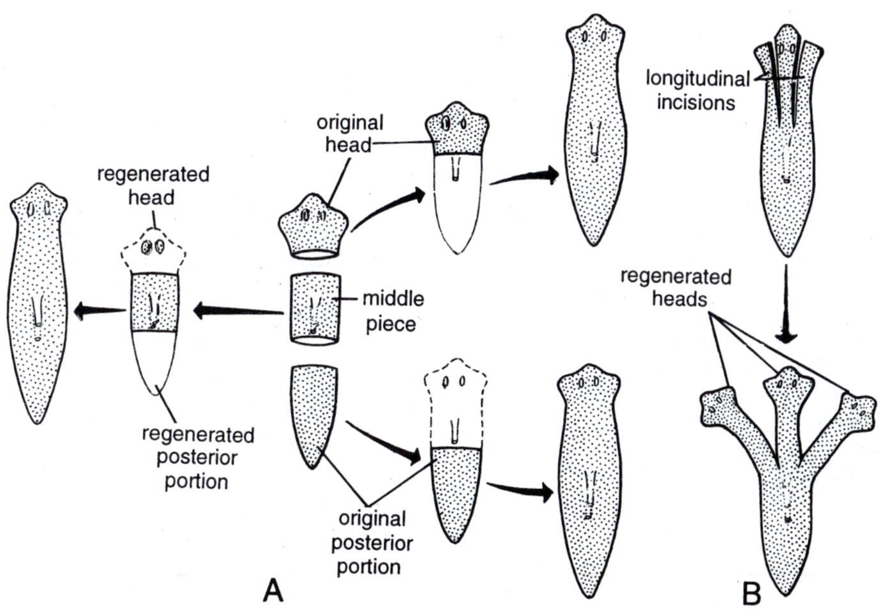

Fig. 18. Regeneration in planarians. **A.** Three individuals regenerate from an individual cut into three parts. **B.** Formation of a heteromorph with three heads.

Regeneration, however, is not always perfect. A sexually mature *Planaria,* cut across between its pharynx and copulatory apparatus, loses its genitalia by degeneration. The two halves fail to regenerate the sex organs and each grows into an **asexual** individual.

Recent investigations have revealed that, when body of *Planaria* is cut across, free cells from mesenchyme, the **neoblasts** or **formative cells**, migrate to the cut surface to form a **blastema** that develops into the new parts. Radiations destroy neoblasts and thus prevents regeneration.

Important Questions

➤➤*Long answer type questions*

1. Give an account of the structure, physiology and reproduction of a planarian.
2. Draw a labelled diagram of transverse section of the body passing through the pharynx of a planarian.
3. Describe the histological structure of the body wall of *Planaria.*
4. Give an illustrated account of the reproductive system of any planarian and add a note on regeneration.
5. Write notes on : (i) Eye of *Planaria,* (ii) Feeding in *Planaria, (in)* Flame cell, (iv) Locomotion in *Planaria,* (v) Regeneration in *Planaria,* (vi) Rhabdites.

➤➤*Short answer type questions*

1. What are the structures that produce mucus in *Planaria* ?
2. Describe regeneration in *Dugesia.*
3. Explain the female reproductive system of *Planaria* with a sketch.
4. Discuss regeneration with reference to *Planaria.*
5. Give an account of the structure and life history of a planarian.
6. Draw and label T.S. of *Dugesia.*

➤➤*Multiple choice questions*

1. Planarians have extra power of regeneration due to the presence of :
   (a) parenchyma (b) rhabdites
   (c) neoblasts or formative cells
   (d) interstitial cells
2. Ciliated epidermis is found in :
   (a) *Dugesia* (b) *Fasciola*
   (c) Taenia (d) *Schistosoma*
3. Rhabdites are found in :
   (a) *Fasciola* (b) *Taenia*
   (c) *Echinoocus* (d) *Planaria*
4. Platyhelminth which live in fresh water :
   (a) *Fasciola* (b) *Taenia*
   (c) *Schistosoma* (d) *Dugesia*
5. Rhabdite occurs in epidermis of :
   (a) Turbillarian (b) Trematoda
   (c) Cestoda (d) all above
6. Oesophagus lead into intestine :
   (a) Rotifera (b) Ctenophora
   (c) Planarian (d) all
7. Cephalization found in :
   (a) Porifera (b) Coelenterata
   (c) Ctenophora (d) Planarian
8. Cocoon formation occurs in :
   (a) *Taenia* (b) *Fasciola*
   (c) *Schistosoma* (d) *Dugesia*
9. Regeneration power can be studied in :
   (a) *Dugesia / Planaria* (b) *Amphioxus*
   (c) Earthworm (d) *Ascaris*
10. Eyes are present in :
    (a) *Dugesia* (b) *Pheretima*
    (c) *Taenia* (d) *Fasciola*
11. Planarians are :
    (a) Solitary (b) Gregarious
    (c) Parasites (d) Colonial
12. Regeneration can best be described as :
    (a) Removal of unwanted parts
    (b) Deformation of lost parts
    (c) Rehealing of a wound
    (d) Replacement of last parts
13. Pharynx can be everted in :
    (a) Planaria (b) Liver fluke
    (c) Tapeworm (d) None

Answers

➤➤*Multiple choice questions*

1. (c) 2. (a) 3. (d) 4. (d) 5. (a) 6. (c) 7. (d) 8. (d) 9. (a) 10. (a)
11. (b) 12. (a) 13. (a)

31

Fasciola hepatica : The Sheep Liver Fluke

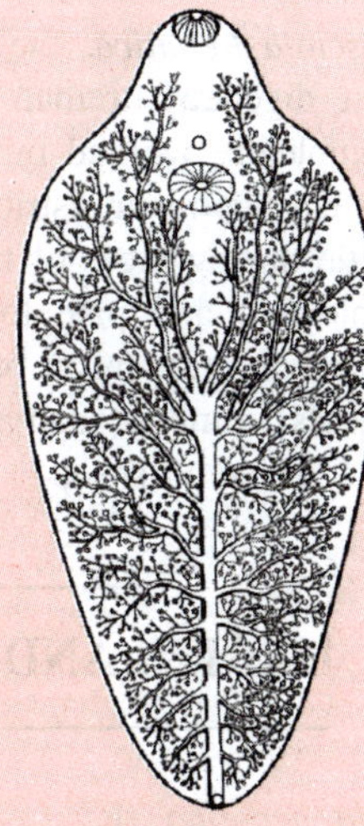

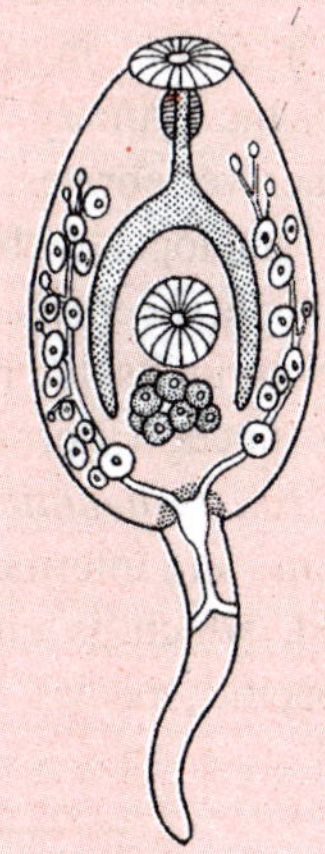

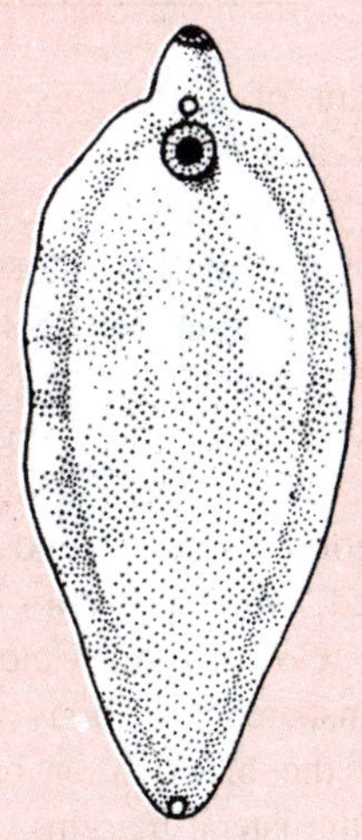

The trematodes, belonging to the class Trematoda, are all parasitic. They are commonly called 'flatworms' or 'flukes' (Anglo Saxon, **floc** = flat), on account of their flat, leaf-like form. Flukes inhabit liver and bile duct of vertebrates like cattle, sheep, goat, rabbit, pig, dog and man. *Fasciola hepatica* is the common liverfluke of sheep. *F. gigantica* (=*F. indica*), the liver fluke of cattle, is however a more familiar species in India. *Fasciolopsis buski* occurs as an intestinal parasite in man and pig, while *Clonorchis* (=*Opisthorchis*) *sinensis,* the Chinese liver fluke is the causative organism of **clonorchiasis** in human beings in Oriental countries like Japan, Vietnam, Korea and South-eastern China. The present account mostly relates to *Fasciola hepatica,* the sheep liver fluke which is one of the most widely studied trematodes. Its structure and life history are well-known, but physiology is not much understood, because it is difficult to keep the parasites outside their hosts in a situation where they may be observed and used in experiments. There are 14 species of Liver fluke in world.

Fasciola hepatica

SYSTEMATIC POSITION

Phylum	Platyhelminthes
Class	Trematoda
Order	Digenea
Family	Fascioliodae
Genus	*Fasciola*
Species	*hepatica*

Fasciola hepatica, the sheep liver fluke was the first digenetic trematode whose life history was completely worked out by **Thomas** in 1883. It has a cosmopolitan distribution throughout sheep and cattle raising areas of the world. It is of great parasitological and pathological importance as it is the causative organism of **fascioliasis**, a disease that causes immense damage to the liver-tissues and bile ducts of sheep.

HABITS AND HABITAT

Fasciola hepatica (L., **fasciola**, small bandage+ Gr., **hepar**, liver) is an endoparasite which completes its life history in two hosts. Adult *F. hepatica* is known as the sheep liver fluke because it occurs in the liver and bile passages of sheep, the primary host. It may also occur in some other vertebrates, like goat, horse, dog, ass, ox, deer, antelope, rabbit, elephant, man, monkey, etc. A single sheep may harbour as many as 200 adult flukes in its liver, which may consequently cease to function normally. This effect is known as **liver rot**. *F. hepatica* spends a part of its life history in an intermediate host, a freshwater gastropod, which is either *Limnea truncatula* or some specific species of *Planorbis* or *Bulinus.*

EXTERNAL MORPHOLOGY

Structure of a fluke is more or less similar to that of a planarian.

1. Shape and size. Body of *F. hepatica* is soft, oval in outline, dorsoventrally flattened and leaf-like. It is about 1.8 to 3 cm long. Maximum width is about 0.4 to 1.5 cm, which is a little infront of the middle region of body. From this region, body tapers away anteriorly as well as posteriorly. Anterior end is somewhat broad and rounded, while posterior end is bluntly pointed.

2. Colouration. Colour is usually pinkish, but the transparency of body wall enables the observer to see the blackish or brownish vitelline glands along the lateral margins, and the alimentary canal,

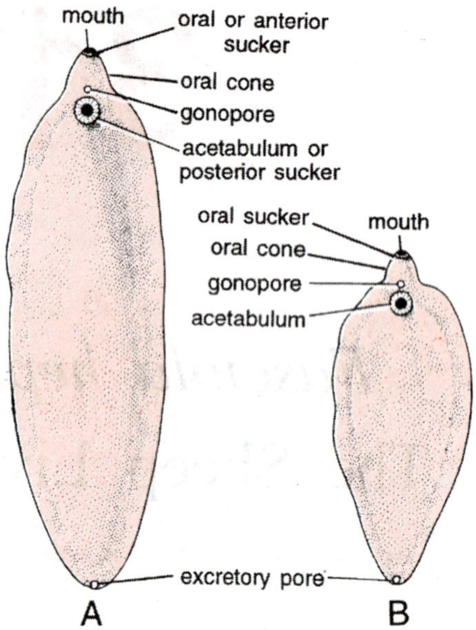

Fig. 1. Liver flukes. **A.** *Fasciola gigantica.* **B.** *Fasciola hepatica.* External features in ventral view.

which appears brownish due to ingested bile of the host.

3. Oral cone. Anterior end of body is drawn out into a prominent conical projection, termed the **oral cone** or **head lobe**, bearing at its tip a somewhat triangular aperture, the **mouth**.

4. Suckers. There are two small suckers, anterior and ventral. Both are devoid of hooks and spines.

(a) Anterior sucker. Mouth is situated at the centre and bottom of a cup-shaped muscular organ, known as the **anterior** or **oral sucker**. It has a diameter of about 1 mm. Muscles of oral sucker radiate from margin of mouth to the periphery of sucker. Oral sucker acts as an ideal suctorial organ serving for adhesion as well as ingestion.

(b) Ventral sucker. About 3 to 4 mm behind the oral sucker, situated mid-ventrally, is another bowl-like adhesive sucker, the **ventral** or **posterior sucker**, also known as **acetabulum**. It is without an aperture and has a diameter of about 1.6 mm.

5. Apertures. In addition to **mouth**, there are two more permanent apertures on body. A small common **genital aperture** or **gonopore**, situated mid-ventrally a little in front of the acetabulum. A single **excretory pore** lies at the posterior end, slightly towards the ventral surface. During breeding season, a temporary **opening of Laurer's canal** appears on dorsal surface, a little anterior to the middle of body. The alimentary canal being incomplete as there is no **anus**.

6. Scales. Body of liver fluke is covered with **tegument** (cuticle), from which project numerous minute backwardly directed **spinules** or **scales**. These anchor the body to the walls of host's bile duct, protect the body and help in locomotion.

BODY WALL

Unlike turbellarians, a ciliated epidermis is absent in trematodes. Body wall of *Fasciola* consists of the following succession of layers : (i) tegument, (ii) basement membrane, (iii) musculature and (iv) mesenchyme.

[I] Tegument

Body is covered externally by a thick, non-ciliated cytoplasmic syncytium, the **tegument**. When observed earlier with the ordinary light microscope, it appeared non-cellular and homogeneous and was regarded as a non-living **cuticle**. However, recent studies made with electron microscope, by **Threadgold** (1963) and others, have revealed that it contains mitochondria, endoplasmic reticulum, vacuoles and pinocytotic vesicles. Thus, the old term 'cuticle' is rejected in favour of 'epidermis' or 'tegument' for this outer cytoplasmic layer. It is continuous with the cytoplasmic processes of certain **tegument secreting cells** (earlier cuticle secreting cells) lying in mesenchyme.

Tegument is thick, tough and capable of withstanding the action of host's digestive juices. It thus serves as an ideal body covering of the endoparasite. Further, it bears all over numerous microscopic and broad backwardly directed **spinules** or **spines** or **scales**, which anchor the fluke in host's bile passage, render protection to body and facilitate locomotion.

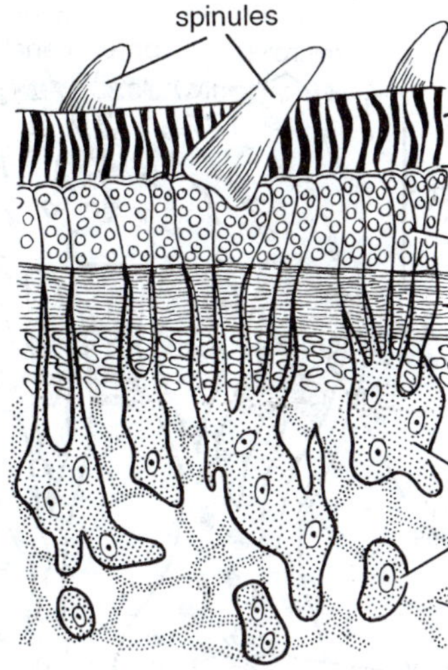

Fig. 2. *Fasciola.* V. L. S. body wall under light microscope.

[II] Basement Membrane

Immediately beneath tegument, the light microscope shows a thin but well-defined acidophilic basement membrane. However, with electron microscope only its outer edge is visible, while inner edge appears to merge imperceptively with underlying tissues. Inside the basement membrane are the integumentary muscles.

[III] Musculature

Body wall musculature or integumentary muscles include an outer **circular layer** and an inner **longitudinal layer**. In Digenea (*Fasciola*), a third layer of **oblique** or **diagonal muscle fibres** is also present inside the longitudinal layer. In suckers, muscles form stout bundles of **radial fibres**, imparting radial striations to them. All muscle fibres are smooth.

[IV] Mesenchyme or Parenchyma

This peculiar tissue of mesodermal origin surrounds the various internal organs serving as a sort of packing material between them. It consists of numerous loosely arranged, irregular, uninucleate and binucleate cells with large fluid-filled intercellular spaces. Some of these cells are large and have processes extending to the base of the tegument which they are believed to secrete.

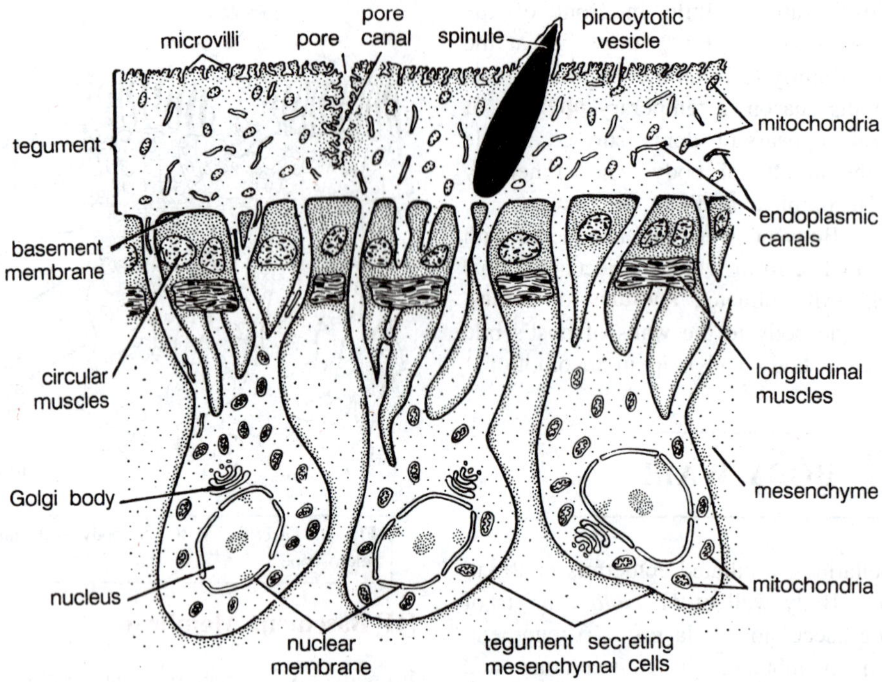

Fig. 3. *Fasciola.* V. L. S. Body wall under electron microscope.

In addition to its skeletal function, mesenchyme also serves as an important transport medium since flatworms are devoid of a blood vascular system.

DIGESTIVE SYSTEM

[I] Alimentary Canal

Fasciola hepatica has an incomplete alimentary canal as it lacks the anus. **Mouth** is situated at the anterior end surrounded by oral sucker. It leads into an ovoid **pharynx** having a small narrow lumen and thick walls provided with radial muscles and **pharyngeal glands**. It is followed by a short, narrow **oesophagus**, which is lined by a single layer of epithelial cells and opens into the large **intestine**. Intestine at once forks into right and left main branches or limbs. Each branch terminates blindly near the rear end of body and

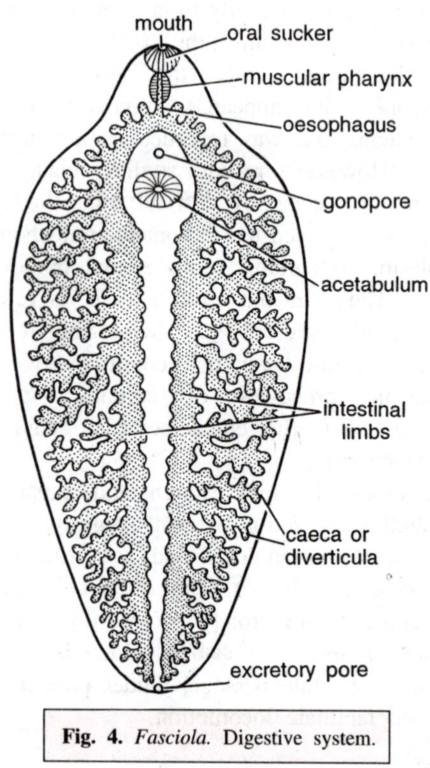

Fig. 4. *Fasciola.* Digestive system.

gives out numerous irregular side branches called **caeca** or **diverticula**, all along its length. Caeca of outer side are large and further branched, while those of inner side are short and simple.

Pharynx and oesophagus are coated internally by cuticle, while intestine is lined by endodermal columnar epithelial cells, around which is a thin muscle layer consisting of circular and longitudinal fibres. Caecal epithelium has secretory gland cells.

[II] Food and Feeding

Migration occurs during feeding. It has been found that hungry flukes migrate into smaller bile ducts and capillaries for feeding. They suck up food, lymph, bile and tissue pieces rasped off by oral sucker from the walls of bile passages. Oral sucker and muscular pharynx serve as an efficient suctorial apparatus.

[III] Digestion

Digestion, which is **extracellular**, takes place in the intestine. Distribution of digested food is accomplished by the ramifying diverticula of intestine aided by mesenchyme, there being no circulatory system. Waste materials diffuse into the surrounding mesenchyme or egested through mouth. Monosaccharide sugars, like glucose and fructose, can diffuse directly into the body of fluke through its general surface. Absorption of some digested food and amino acids also takes place through tegument. This is facilitated by the numerous fine folds formed by the tegument on its outer surface. Food is reserved in the form of glycogen and fats in mesenchyme and muscles.

RESPIRATION

Oxygen content in bile being extremely low, respiration in *Fasciola hepatica* is usually **anaerobic** or **anoxybiotic**.

In this process glycogen (present as reserve food), breaks up by anaerobic glycolysis into carbon dioxide and fatty acids. Lactic acid is the end product of glycolysis. CO_2 diffuses out through general body surface, whereas fatty acids (lactic acid) get eliminated through excretory system. This is an exothermic reaction involving release of energy (heat). If free oxygen is available, aerobic respiration takes place.

PROTONEPHRIDIAL OR EXCRETORY SYSTEM

This system concerned with **excretion** and **osmo-regulation** is similar to that of planarians. It comprises of a large number of **flame cells** or **flame bulbs** or **protonephridia**, connected with an intricate system of **excretory ducts** of various orders.

[I] Flame Cells

Flame cells or protonephridia, which themselves are modified mesenchymal cells, are distributed throughout mesenchyme in a specific pattern, called the **flame cell pattern** by **Faust** (1919). They are usually of irregular shape and send out **pseudopodial processes** into the surrounding tissue. Each flame cell has an **intracellular lumen**

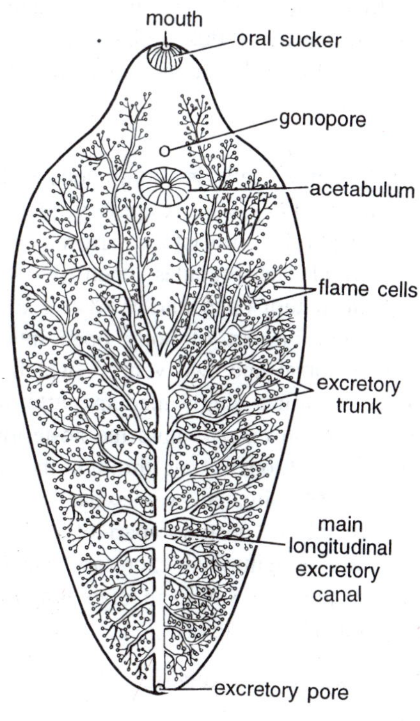

Fig. 5. *Fasciola.* Protonephridial system.

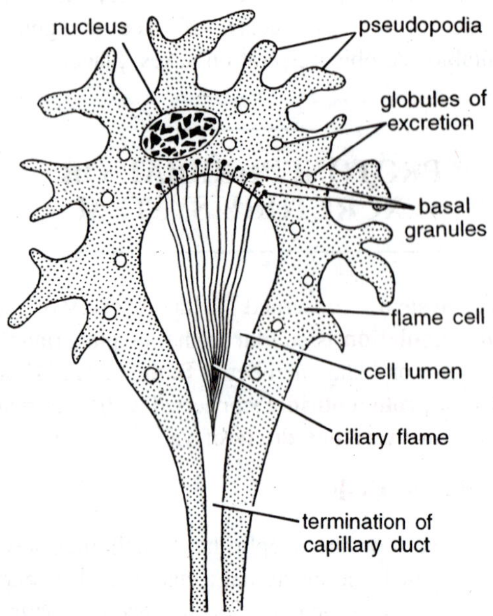

Fig. 6. *Fasciola,* A flame cell.

or **cavity**, in which hang a few long **cilia**, each arising from a **basal granule** situated in cytoplasm. In the living fluke, cilia vibrate like the flickering of a flame; hence the name 'flame cell'.

[II] Excretory Ducts

Lumen of flame cell is continuous with a microscopic **capillary duct**. Capillary ducts from a few protonephridia open into a narrow **collecting tubule**. Several such tubules open into larger **twigs**, which in turn open into **vessels**. Excretory vessels of anterior part of body open into four **trunks**, two dorsal and two ventral, which unite posteriorly to form a single median **longitudinal excretory canal**. It extends upto the posterior end of body where it opens out through the single **excretory pore** situated somewhat ventrally. Excretory vessels of posterior part of body open directly in the longitudinal excretory canal. All the ducts, except the single median longitudinal canal, are lined with cilia.

[III] Physiology of Excretion

Main excretory products of fluke are fatty acids, carbon dioxide and ammonia. These substances

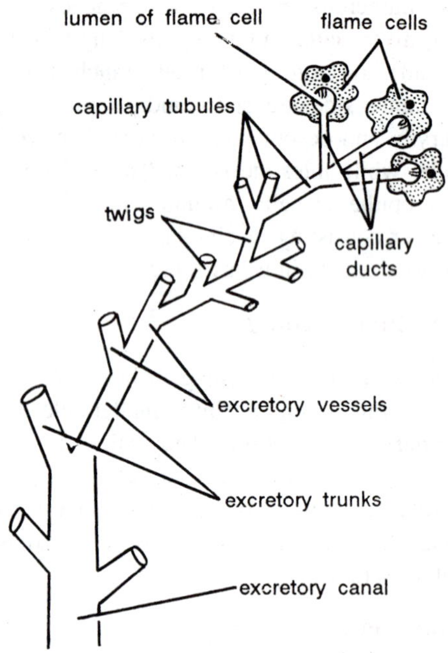

Fig. 7. *Fasciola.* Arrangement of flame cells and excretory ducts.

diffuse into protonephridia (flame cells) from the surrounding mesenchyme. Excretory fluids are kept moving through tubules by the action of cilia, and finally squeezed out through the excretory pore by contractions of the body. The enzyme alkaline phosphatase, found in flame cells and collecting tubules serves in the selective transfer of chemical substances. Some nitrogenous wastes are also passed to the exterior through the surface tegument.

Protonephridial system is excretory as well as osmoregulatory, since it controls the amount of fluid (water) in the body.

NERVOUS SYSTEM

Being parasitic, adult fluke lacks any **sense organs**. Nervous system however, is considerably developed and resembles the turbellarian plan.

Brain forms a **collar** or **ring (cerebral ring)** around pharynx and bears a pair of lateral

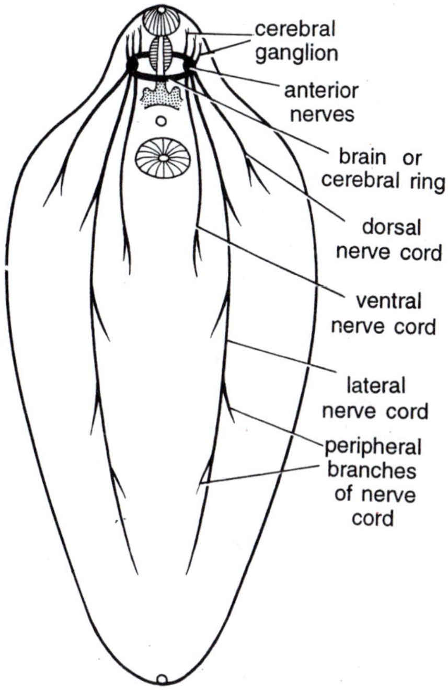

Fig. 8. *Fasciola.* Nervous system.

cerebral ganglia and a **ventral ganglion**. Fine nerves arise from brain and supply the anterior and posterior regions. Suckers receive a more liberal supply. Three pairs of longitudinal **nerve cords** extend posteriorly, giving out numerous fine peripheral branches to various organs. Of these, one pair is **dorsal**, one **ventral** and one **lateral**. Unlike dorsal and ventral cords, lateral cords are well-developed and connected by a few **transverse commissures**.

REPRODUCTIVE SYSTEM

Fasciola hepatica is **hermaphrodite**. Gonads are well developed and male and female genital ducts open into a common chamber, the **genital atrium**. It is situated anteriorly in body and opens to the exterior through the common **genital aperture** or **gonopore**, located ventrally just in front of acetabulum.

[I] Male Reproductive System

1. Testes. A pair of highly lobed tubular **testes** lie in tandem arrangement (one behind the other), occupying a large space in the middle and posterior regions of body. Anterior and posterior testes are morphologically the left and right testes, respectively. Epithelium lining the testis gives rise to sperms.

2. Vasa deferentia and seminal vesicle. Leading from each testis is a narrow, slender **sperm duct** or **vas deferens**. The two vasa deferentia run side by side anteriorly upto the level of acetabulum. Where they become united and form a large pear-shaped **seminal vesicle** for storage of sperms.

3. Cirrus sac. Anteriorly, from seminal vesicle, arises a narrow and somewhat twisted tube, the **ejaculatory duct**, which extends forward and enters a stout muscular and eversible copulatory organ, the **penis** or **cirrus**. It extends upto the common **genital atrium** into which it opens by the **male genital aperture**. Genital atrium in turn opens outside through common **gonopore**.

Numerous unicellular **prostate glands** surround and open into the ejaculatory duct. Their alkaline secretion helps in the free movement of sperms during copulation. Cirrus, prostate glands and seminal vesicle are enclosed in a **cirrus pouch** or **cirrus sac**.

[II] Female Reproductive System

1. Ovary. A solitary, highly branched and tubular **ovary** or **germarium** lies anterior to the testes on right side in the anterior one-third of body.

2. Oviduct and uterus. All the branches of ovary gather posteriorly to form a short narrow tube, the **oviduct**. It extends backwards towards the middle line of body, and joins the **median vitelline duct** (described ahead) to form the **ovo-vitelline duct** or **uterus**. In some species, before joining median vitelline duct, the oviduct swells up to form a small rounded central chamber, the **ootype**. However, ootype is absent in *F. hepatica,* **Uterus** is a long, wide and highly convoluted tube that extends up to the genital atrium, opening into it through the female genital aperture, close to

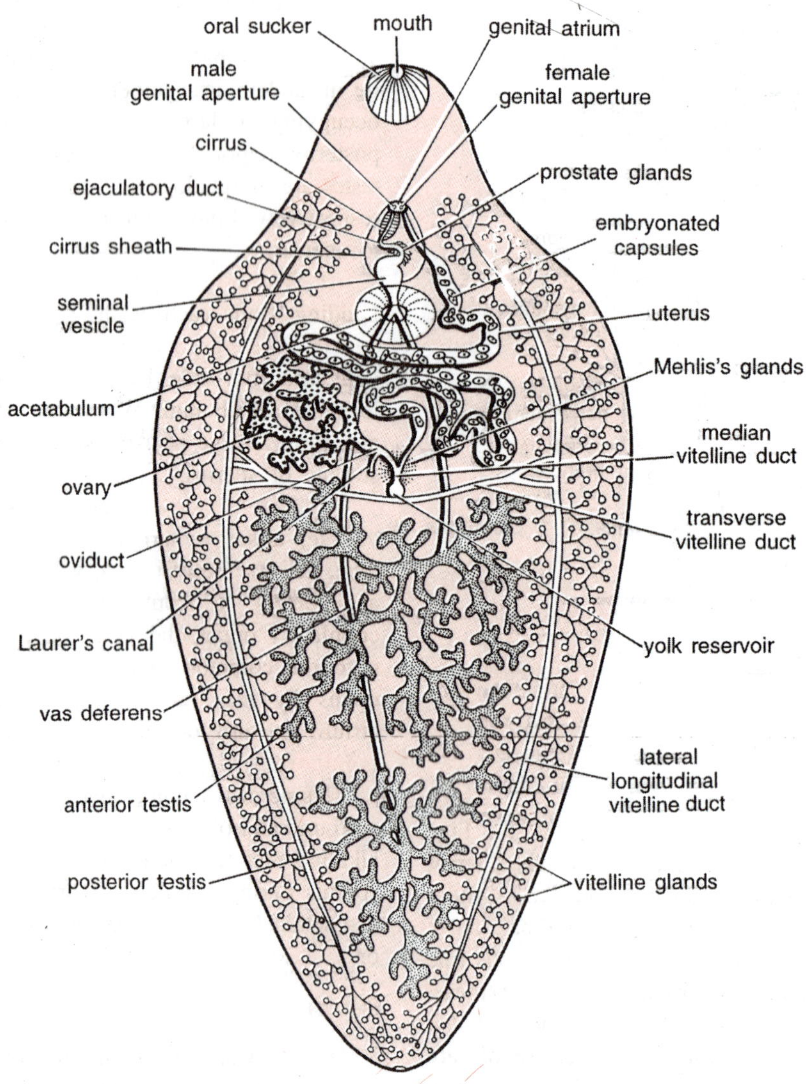

Fig. 9. *F. hepatica.* Reproductive system in ventral view.

the male genital aperture on the left side. Uterus contains a large number of **capsules** containing fertilized eggs. Enormous size of uterus is in fact corelated with the high rate of egg production, a feature characteristic of endoparasitic mode of life. A short muscular **copulatory tube**, also known as the **Laurer's canal** or **Laurer-Stiedia canal**, arises from oviduct. During breeding period it opens through a temporary median dorsal pore, a little in front of the middle of the body. It serves as a **vagina** during copulation.

3. Vitellaria. Associated with female reproductive system are two important glands : (i) **vitelline glands** or **vitellaria**, and (ii) **Mehlis's glands**. Vitellaria occur as clusters of follicles on the right and left sides, almost all along the entire length of body. Follicles on each side are connected, by fine interconnected **ductules**, to the **longitudinal vitelline duct**. A little behind the origin of uterus, longitudinal vitelline ducts of both sides are connected by a **transverse duct**, the three together form the letter 'H'. Parts of each

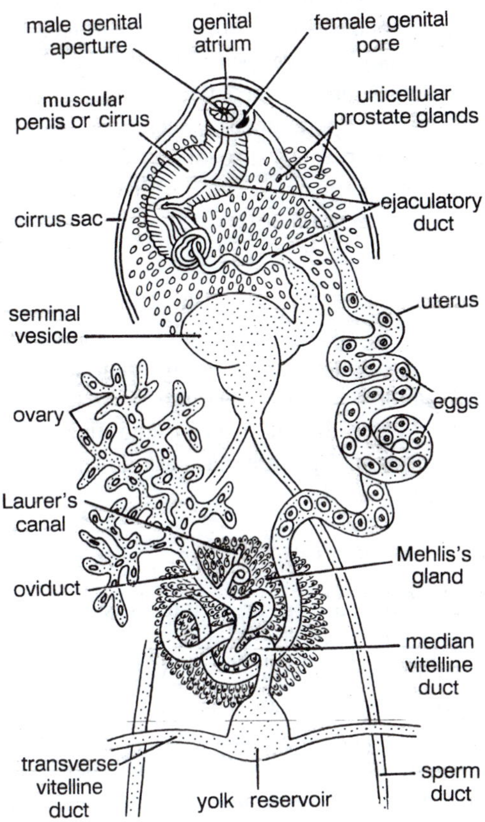

Fig. 10. *F. hepatica.* Details of male and female reproductive organs in the anterior region of body (ventral view).

longitudinal duct in front and behind the transverse duct are called anterior and posterior limbs, respectively. Transverse vitelline duct swells up in the centre to form a **yolk reservoir**. From this arises anteriorly a **median vitelline duct**, which joins oviduct to form the uterus. Vitellaria produce

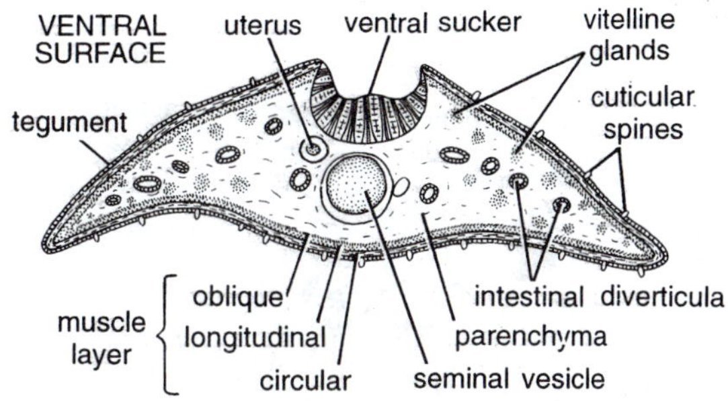

Fig. 11. *F. hepatica.* T.S. body through seminal vesicle.

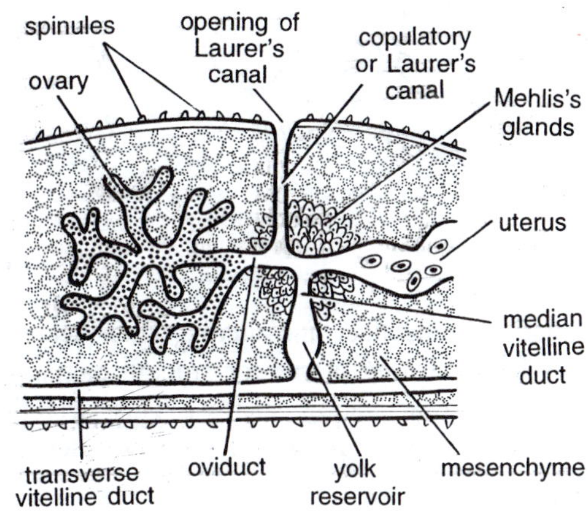

Fig. 12. *F. hepatica.* T.S. showing relations of ovary and genital ducts.

special **yolk cells** or **vitelline gland cells**. These cells contain abundant yolk for nourishing the embryo and numerous large **shell globules** which form egg shells.

Mehlis's glands. Mehlis's glands also known as **shell glands** and does not play any role in shell-formation. They are a cluster of unicellular glands and surround the junction of oviduct,

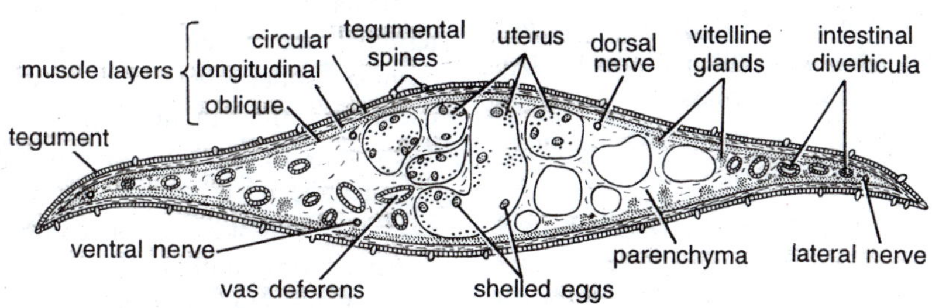

Fig. 13. *F. hepatica.* T. S. body through uterus.

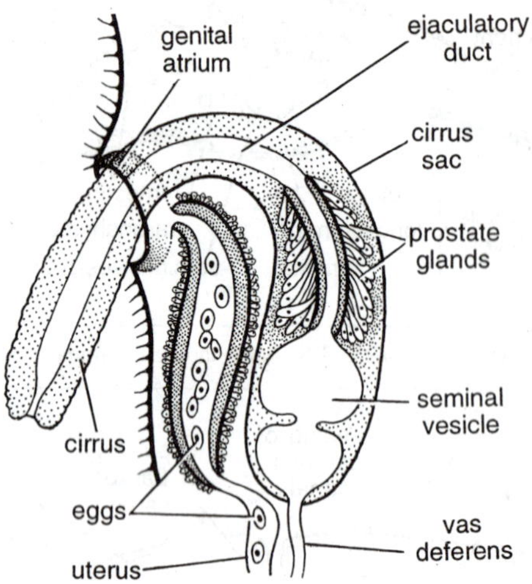

Fig. 14. *F. hepatica.* Cirrus protruding through gonopore.

median vitelline duct and uterus. Their secretion helps in lubricating uterus for smooth passage of eggs and in activating sperms. This secretion, perhaps, also causes release of shell globules from vitelline gland cells.

LIFE CYCLE AND DEVELOPMENT

[I] Digenetic Life Cycle

Life cycle of *F. hepatica* is complex and completed in two hosts. **Primary host**, in which the adult fluke lives, is sheep. While the **intermediate host**, in which numerous larval stages are passed, is a snail *(Lymnaea, Planorbis,* etc.). This type of life cycle, involving two different kinds of hosts, is termed **digenetic**.

[II] Copulation

Self-fertilization is of rare occurrence in liver flukes though they are hermaphrodite. **Cross-fertilization** preceded by **copulation** is of normal occurrence. In *F. hepatica,* copulation takes place in bile ducts of the host.

Two flukes in copulation bring their genital pores in opposition. Cirrus of one fluke, everted through its gonopore, penetrates the Laurer's canal of the other through the latter's temporary opening, and injects spermatozoa. Secretion of prostate glands, and perhaps also of the Mehlis's glands, keep the sperm active for fertilization.

[III] Fertilization

Fertilization is **internal**. In **cross-fertilization**, sperms received in Laurer's canal during copulation, enter the distal end of oviduct where fertilization occurs. During **self-fertilization**, sperms enter the uterus of same fluke through female genital aperture and pass down to fertilize the eggs.

[IV] Capsule Formation

Each fertilized egg or zygote is surrounded by yolk cells, which provide yolk and shell material. Shell-globules of yolk cells contain proteins and a phenol. According to **Stephenson** (1947), phenol is oxidized to a quinone in the proximal part of uterus. Quinone then tans the protein, producing a hard, resistant and leathery **sclerotin** like that of insects. This sclerotin forms the shell around fertilized eggs. Above finding of Stephenson in liver flukes is perhaps true for all platyhelminths.

[V] Capsules

Shelled eggs are termed **capsules** or simply **eggs**. A shell or capsule is yellow or brown, in colour and oval in shape. It is about 130 to 150μ long and 60 to 90μ wide. It is operculate, *i.e.*, provided with a lid or **operculum**. Situated immediately beneath the operculum, at the terminal end of egg is a viscous and granular cushion. About 3,000 or more such capsules may occur at a time in the uterus of a single fluke. There may be as many as 200 flukes in the liver of one sheep. If each fluke produces 500,000 eggs (in 10 years), a single infected sheep may disperse 100 million fertile eggs. This vast capacity for egg production is necessary in view of the complicated life cycle and slim chances of survival.

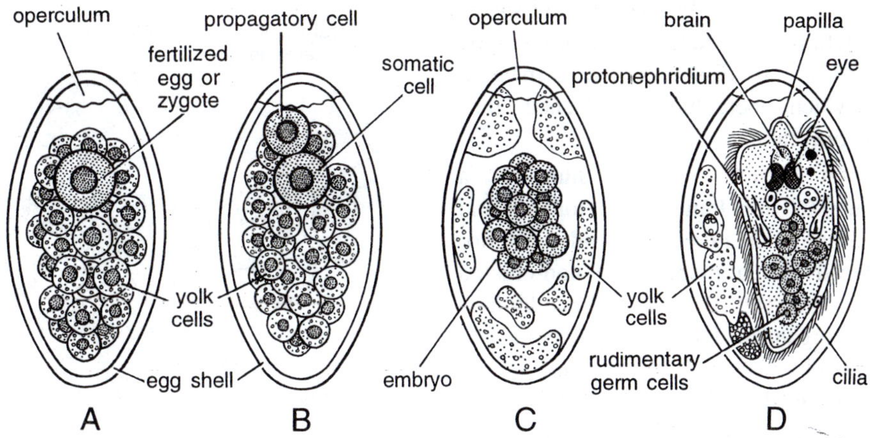

Fig. 15. *F. hepatica.* Early or embryonic development. **A.** Zygote in capsule. **B.** 2-Cell stage. **C.** Many-cell stage. **D.** Miracidium in capsule.

[VI] Cleavage and Embryonic Development

Cleavage starts while eggs are still in uterus. Cleavage is **holoblastic** and **unequal**. First division of zygote results in two **unequal** cells, a larger **somatic cell** and a smaller **propagatory cell**. Subsequent divisions of somatic cell form larval ectoderm and tissues. Propagatory cell divides further into two daughter cells. One daughter cell by its divisions finally produces the larval body. Other daughter cell divides several times to form a mass of smaller **germ cells** which cluster in the posterior part of larval body.

Encapsulated embryos or **capsules** or simply **eggs** do not develop further in fluke's uterus. A very large number of capsules leave fluke's body through its gonopore into host's intestine, and finally ejected out with its faeces. Further development takes place when capsules come in contact with water (or damp areas with at least 60% moisture content) which is slightly acidic (pH 6.5). Optimum temperature for development ranges from 22°C to 25°C.

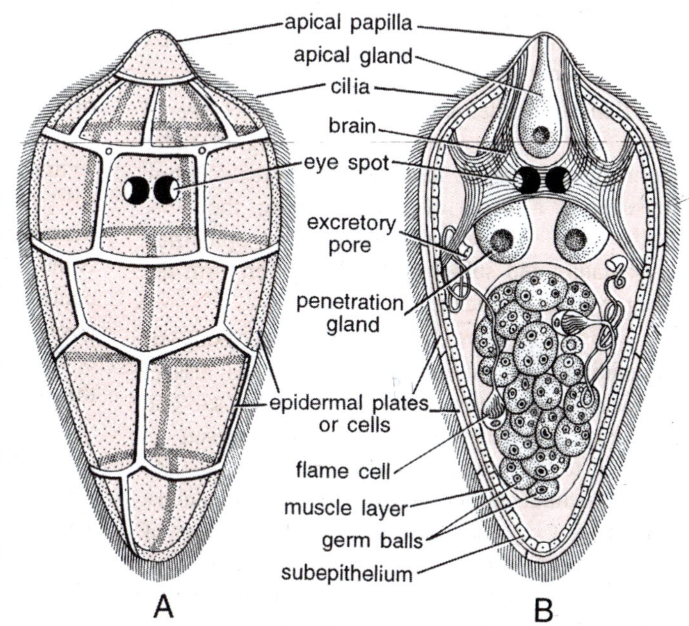

Fig. 16. *F. hepatica.* Miracidium larva. **A.** External structure. **B.** Internal structure.

[VII] Miracidium Larva

It is the first larval stage involved in life cycle. When suitable conditions become available, the encapsulated embryo, in 4-15 days, differentiates into a **miracidium larva**. It hatches out and swims in water. Hatching is initiated by a proteolytic **hatching enzyme**. It dissolves the cementing material by which operculum is attached, thus releasing the operculum.

1. External structure. Miracidium is a minute, about 0.07 mm long, oval, elongate and richly ciliated active creature. Its broader anterior end is produced into a mobile and non-ciliated **apical papilla** or **terebratorium**. Miracidium is a multicellular organism. Its body is covered with flattened ciliated **epidermal plates** or cells, 21 in number and arranged in five rows or **tiers**.

Number and arrangement of cells in each tier is fixed. **First tier** (anterior most) has 6 plates, two dorsal, two lateral and two ventral. **Second tier** also has 6 plates, three dorsal and three ventral. **Third tier** has 3 plates one dorsal and two ventro-lateral. **Fourth tier** has 4 plates, two right and two left. **Fifth tier** (posterior most) has 2 plates, one left and one right.

Beneath epidermal plates is a fine layer of **sub-epidermal musculature**, consisting of outer circular and inner longitudinal fibres. Below muscles is a layer of cells forming the **sub-epithelium**. Epidermal plates, sub-epidermal musculature and sub-epithelium together form the body wall of miracidium.

2. Internal structure. Within the body of miracidium are present glands, nervous tissue, protonephridia and germ cells. A sac-like multinucleate mass of granular protoplasm is attached to the centre of apical papilla by a stalk. This structure, earlier thought to be rudimentary gut, is now regarded an **apical gland**. A pair of large, unicellular **cephalic** or **penetration glands** open by their narrow ends near the apical papilla. A large **brain** with several associated nerve fibres lies dorsally below epidermal cells of second tier. Situated above the brain is an '*x*' shaped larval **eye**, consisting of two crescentic pigmented cells or **eye spots**, with their concavities facing away from each other. The concavities contain a clear refractile material serving as lens. (It should be noted here that a photoreceptor of any sort is absent in adult due to its or the parasitic mode of life). A pair of long tubular **protonephridia** or **flame cells** open to the exterior through two

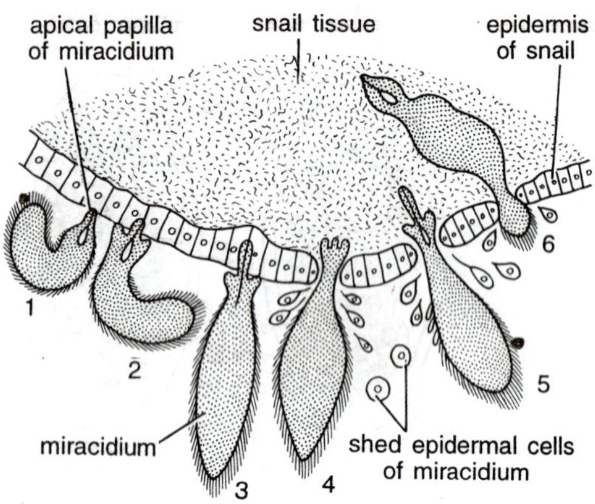

Fig. 17. Miracidia of *Fasciolopsis buski.* Stage of penetration through snail epidermis.

nephridiopores or **excretory pores**, situated laterally in the posterior half of body. **Germ cells** lie in groups, called **germ balls**, in the rear part of body.

3. Physiology. Miracidium does not feed. It swims about desperately as if "**seeking something with feverish haste**" (**Barlow**, 1925). Miracidia tries to penetrate any object or organism they may come across, but only those succeed that come in contact with a specific intermediate snail host (*Lymnaea, Succinea, Planorbis, Bulinus, Possaria,* or *Practicolella).* Those, which do not come across the suitable host, in about 24 hours after hatching, die invariably. For penetration, miracidium attaches its apical papilla and performs boring movements together with contractions and expansions of body. This process, aided by the action of flesh-dissolving larval secretions, results in a minute opening in the host's tissue. The larva then squeezes through this opening, casting off its ciliated epidermis while doing so. It soon makes its way into the digestive gland of snail, where it undergoes various changes and, in about 14 days, develops into the second larval stage, the **sporocyst larva**

[VIII] Sporocyst Larva

It looks like an elongated sac, about 0.7 mm long. Its body wall retains all the layers of body wall of miracidium except the ciliated epidermis, which

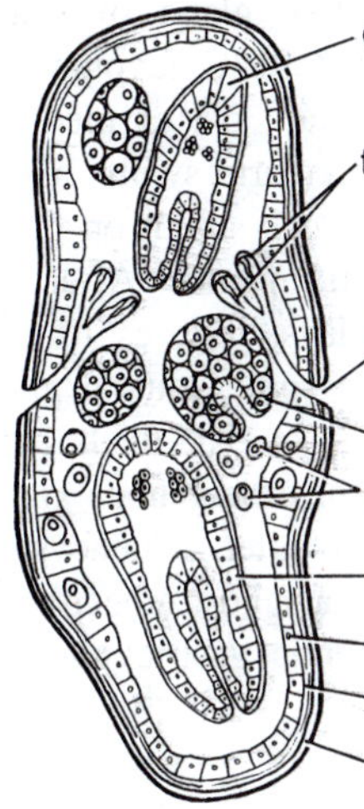

Fig. 18. *E hepatica.* Sporocyst larva.

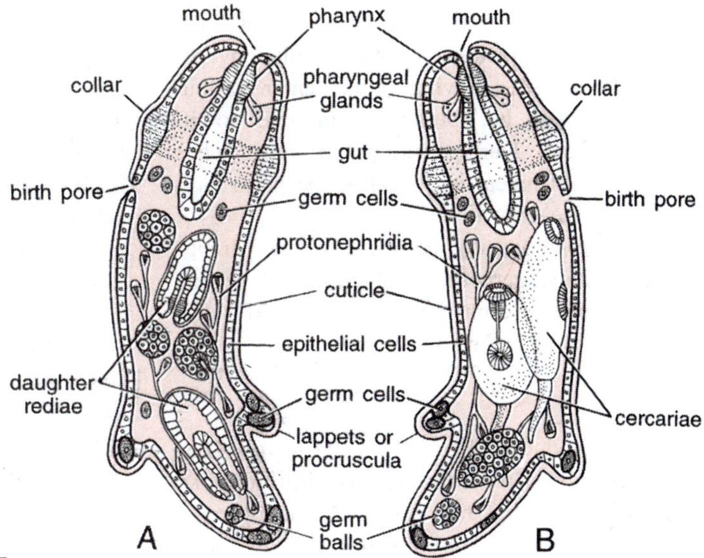

Fig. 19. *F. hepatica.* **A.** Redia with daughter rediae. **B.** Redia with cercariae.

is lost in the process of penetration and soon replaced by a thin **cuticle**. Glands, brain, eye spots and apical papilla of miracidium degenerate and disappear in sporocyst. **Protonephridium** of each side divides into two **flame cells** which open outside by a common excretory pore. In addition, the sporocyst contains **germ balls**. Sporocyst moves about in the tissue of host, absorbing nutrition from it. Its germ balls develop into the next larval generation, the **rediae**. Each sporocyst produces 5 to 8 rediae.

[IX] Redia Larva

Rediae emerges from the sporocyst by rupture of its body wall. Each redia is an elongated and cylindrical creature, about 1.3 mm to 1.6 mm long. It bears a **mouth** at the anterior end, a ring of muscular swelling or **collar** near anterior end, a permanent **birth pore** a little behind collar, and a pair of projections called **lappets** or **procruscula** ventrolaterally near the posterior end. Body wall consists of the usual layers, *viz.*, cuticle, musculature of outer circular and inner longitudinal fibres, and subepithelium. Mouth leads into a short muscular *pharynx,* followed by an elongated sac-like **intestine**, **enteron** or **gut** lined by a single layer of cells. Numerous unicellular **pharyngeal glands** open into pharynx.

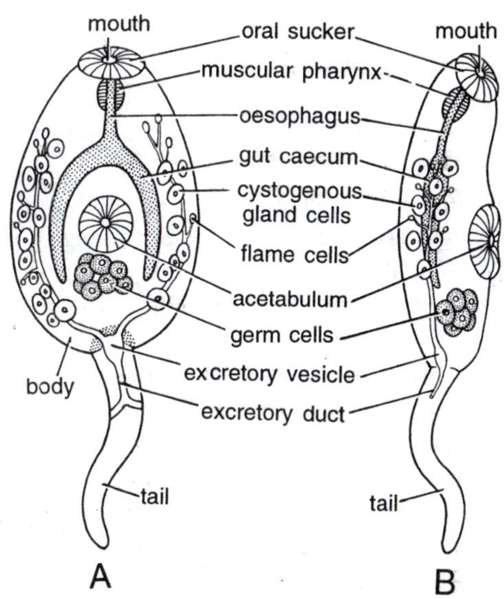

Fig. 20. *F. hepatica.* **A.** Cercaria in ventral view. **B.** Cercaria in lateral view.

Protonephridia divide further and form a much branched system. However, all the flame cells of each side open out through a common excretory duct. Body of larva is packed with **germ balls** and mesenchyme cells.

Redia moves through the host's tissues on which it also feeds. Movements are brought about by muscular contractions of body, aided by the collar and lappets. Moving rediae enter various organs of snail but prefer to migrate to its digestive gland. During summer months, when sufficient nourishment is available, the germ balls of rediae give rise to a **second generation of rediae**, morphologically identical to the parents. During winters, germ balls of rediae of second generation develop into larvae of the next stage, known as **cercaria larvae**.

[X] Cercaria Larva

Each redia produces 14 to 20 cercaria larvae. They leave the body of redia through its birth pore and enters the snail's digestive gland. Morphologically, cercaria bears a close resemblance with the adult fluke. It has an oval body, 0.25 mm to 0.35 mm long, with a long simple tail for swimming. Layers of its body wall are the same as of sporocyst and redia. Cuticle bears backwardly directed spines. Below body wall lie numerous **cystogenous gland cells**, which secrete cyst for the next larva (metacercaria). Well formed **anterior (oral)** and **ventral suckers** like those of adult are present. Rudiments of adult's digestive, excretory and genital systems can be seen in cercaria. Mouth leads into a muscular **pharynx**, followed by oesophagus and intestine, the latter forking in front of the ventral sucker to form two tubular limbs. **Flame cells** occur in large numbers along the lateral zones, opening into a pair of **excretory tubules**, which unite in front of the tail to form an **excretory vesicle** or **bladder**. An **excretory duct** arises from bladder and extends into tail, where it bifurcates and opens out through a pair of **nephridiopores**. Lying in the body are groups of **germ cells**, which are direct descendants of propagatory cell of the capsule. These cells represent rudiment of adult's genital system.

Mature cercaria makes its way through the host's tissues, often migrates to its pulmonary sac, and from there escapes to the surrounding water. It takes 35 to 65 days, after entry of miracidium into the body of snail to the exit of cercariae from the host's body. Rate and extent of larval development in snail's body depends upon the availability of nutrients, primarily those stored in the hepatopancreas or digestive gland of snail. Temperature between 9°C and 26°C favours the emergence of cercaria from snail; pH of medium has no effect in the matter. Cercaria swims about in the surrounding water or crawls about on grass blades or some other aquatic plants. After an active life of two or three days, it loses its tail and undergoes encystment on some aquatic plant to become the **metacercaria larva**. The thick, hard and whitish cyst is a product of cystogenous gland cells which degenerate soon after. Unencysted carcariae ingested by the primary host (sheep) are destroyed by action of its acidic gastric juice.

[XI] Metacercaria Larva

As many as a thousand metacercariae may be found attached to a single grass blade. They have a rounded form with a diameter of about 0.2 mm. Metacercaria is in fact the **juvenile fluke**, also called **marita**. It differs from cercaria in that it has a rounded form, a thick hard cyst and large number of flame cells. It lacks a tail and cystogenous gland cells and its excretory bladder opens out directly through a single pore. Germ cells or the genital rudiments are present as such. Cyst provides protection against short periods of desiccation.

[XII] Infection of Primary Host

Metacercaria develops into adult fluke only inside its definitive host or sheep. The latter gets infection by grazing on leaves and grass

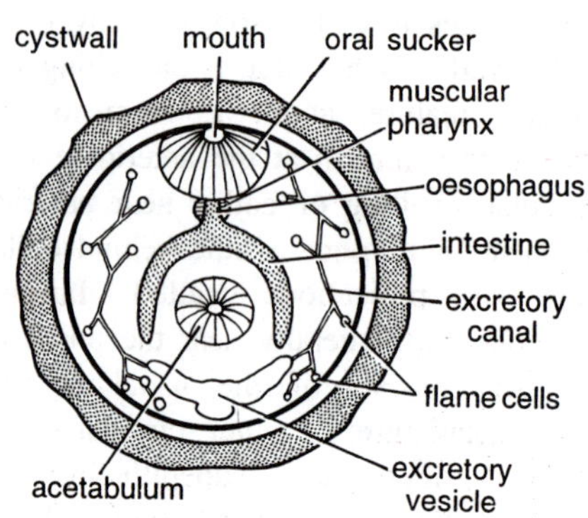

Fig. 21. *F. hepatica.* Metacercaria.

blades to which the cysts are attached. Metacercaria survives action of host's gastric juice as its cyst is insoluble in it. Cyst wall finally dissolves in proximal part of intestine and liberates the larva. It penetrates the intestinal wall and gets into coelomic cavity. Now it infects the liver, feeds on its tissue, and grows in size in five to six weeks. It then takes up its position in the bile duct, where it finally attains sexual maturity. In 11 to 13 weeks, after entering the body of host, it starts laying eggs (capsules.)

FEATURES OF SIGNIFICANCE IN LIFE CYCLE OF *F. HEPATICA*

1. Complex life cycle

Involvement of several larval stages complicates the life cycle of liver fluke. Further, liver fluke is digenetic, that is life cycle is completed in two alternating hosts. Primary or definitive host is sheep and secondary or intermediate host is a freshwater snail.

2. Chances of death

Due to complicated and prolonged life cycle, developmental stages have to encounter several hazards. First hazard is faced by **encapsulated embryos** as soon as they come out with faeces of primary host. They cannot develop further if water (or moisture), suitable pH and temperature are not available. Second hazard is encountered by **miracidia** which emerge from the capsules. If they are unable to make contact with a suitable soft part of the specific intermediate host within 24 hours, they perish. **Cercariae** which leave the intermediate host and swim into surrounding water are the next to face an uncertain future. If ingested by the vertebrate host, before they have encysted, they are destroyed by the action of host's gastric juice. Finally, **metacercariae** are destroyed if they have to face a long period of desiccation before they are ingested by a suitable vertebrate host.

3. High rate of reproduction

Rate of reproduction of an animal, as a general rule, is directly proportional to the **chances of death** it has to face at various levels. The ratio, in fact, is the deciding factor for perpetuation of a species. *Fasciola hepatica,* exposed to several threats to its survival, is highly prolific. Each fluke in its lifetime produces more than 200,000 eggs. Further, each sporocyst produces 5-8 rediae, each of which in turn produces 8-12 rediae in the second generation, each of which further produces 14 to 20 cercariae. Thus, each egg is **capable** of producing 1000 to 2000 cercariae but the actual number produced is far less due to high mortality. Out of those produced, a very small number successfully encysts and infects the final host.

4. Heterogamy

Grobben (1882) and some others believed that germ cells in the sporocysts and rediae were eggs which developed **parthenogenetically** into subsequent larval forms. This kind of asexual parthenogenetic reproduction by larval forms is known as **heterogamy**. Reproduction in immature or larval stages is called **paedogenesis** (**Gr.**, **pais**, child; **genesis**, origin). This view, however, is now considered erroneous and has been given up in favour of the **polyembryony**, concept is described further.

5. Polyembryony

Germ cells in sporocysts and rediae are direct descendants of **preparatory cells** produced by first division of zygotes. These germ cells multiply mitotically and produce subsequent larval stages within sporocysts and rediae. In doing so, they (germ cells or propagatory cells) behave as several embryos. Thus, this process of reproduction in sporocysts and rediae has been looked upon as **polyembryony** by **Ishii** (1934), **Chen** (1937), **Rees** (1940) and **Cart** (1944).

6. Metagenesis

Life cycle of *Fasciola* involves a period of asexual reproduction during immature stages (sporocysts and rediae) followed by a period of sexual reproduction in the adult stage. **Steenstrup** (1942),

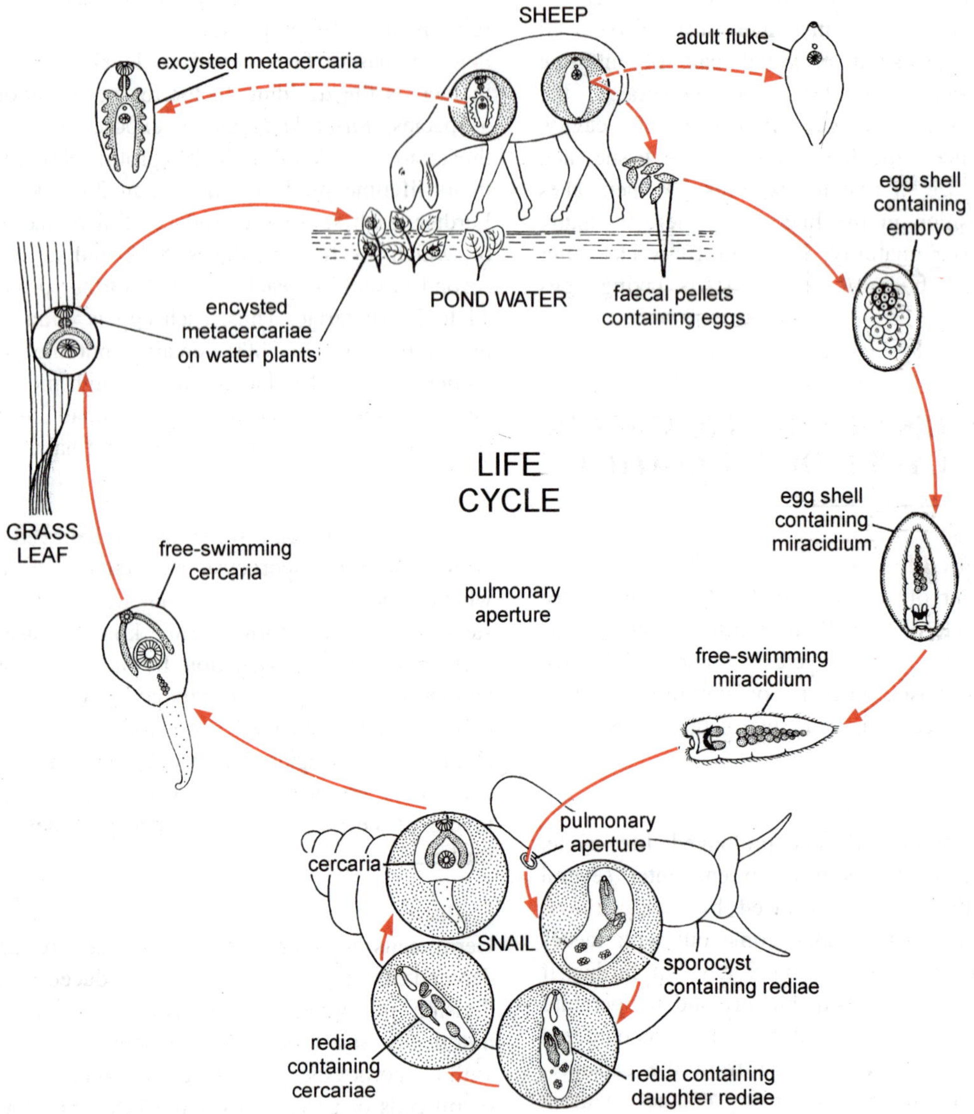

Fig. 22. *F. hepatica.* Life cycle.

and some others interpret this as an alternation of asexual and sexual generations in the life cycle. But **Hyman** is of the view that it is a continuous ontogeny (life history) involving asexual multiplication in larval stages.

7. Advanced larval stages

Miracidium and cercaria, being free-living larvae, exhibit more advanced features than the adult which has undergone degeneration in many respects to suit its parasitic mode of life. Body cavity, locomotory organs, sense organs, and a cellular epidermis are lacking in adult but are present in larvae.

PARASITIC ADAPTATIONS OF *FASCIOLA*

Liver fluke has undergone great modifications, morphological as well as physiological, to suit its existence as an endoparasite in the bile ducts of sheep.

(1) Outer **tegument** is thick, permeable to water, but enzyme-resistant, so that parasite is not digested by digestive juices of the host.

(2) Locomotory organs are absent as not required by adult. However, free-swimming larvae, such as miracidium has cilia and cercaria has a locomotory tail.

(3) Oral sucker, acetabulum and spines of body wall of adult worm serve as organs of attachment in the host's body.

(4) Alimentary canal is without anus as there is no undigested food for egestion. Suctorial pharynx helps in sucking bile etc. and much-branched intestine serves to distribute digested food to all parts of the body.

(5) Adult lacks circulatory, respiratory and sensory organs as they are not needed. Nervous system is poorly developed for the same reason. However, free-swimming miracidia has sensory eye spots.

(6) Respiration is **anaerobic** as free O_2 is not available.

(7) Reproductive system is highly developed. Number of eggs produced is enormous (about 200,000 eggs per fluke). This is necessary to offset several hazards resulting in great mortality.

(8) Resistant egg shells around zygotes provide further safety from unfavourable environmental conditions.

(9) Hermaphroditism ensures self-fertilization even in the absence of another companion for copulation. It is necessary for survival of the species.

LIVER ROT

When sheep are infected by the liver fluke *Fasciola hepatica,* the liver of sheep is seriously affected in structure and function. This disease is known as 'Liver rot', or 'Fascioliasis'.

Infection

The vertebrate host (sheep, goat, etc.) gets the infection by grazing on grass, leaves and other vegetation to which metacercarial cysts are attached. The invertebrate host (snail) acquires infection when a miracidium, at random, establishes contact with a suitable part of its body.

Pathogenesis or symptoms

Infection of invertebrate host (snail) results in a partial or complete destruction of the affected site, which is preferably the digestive gland (liver) or gonad. In case of heavy infections, snail considerably increases in size.

Of significant economic importance is the effect of *F. hepatica* on its vertebrate host, whose bile ducts as well as liver may be damaged. In bile ducts, it causes **inflammation** and **hepatitis**, resulting in loss of its epithelium and thickening of wall, followed by calcification and formation of **gall stones**. Heavy infections upset the normal metabolism of liver. This is due to **haemorrhage** caused and irritation inflicted by cuticular spines. The disease thus caused is called **liver-rot** or **fascioliasis**.

Symptoms of **liver-rot**, are more acute in lambs than in sheep, appear about a month after infection. Frequently, death may soon result due to **cerebral apoplexy**. However, if the host survives few weeks of infection, it falls a victim to acute **anaemia** and falls even at mild contact. Its appetite declines, rumination (chewing the cud) becomes irregular and at times there is fever and increase in respiratory activity. Conjunctiva becomes whitish-yellow, and dry and brittle wool falls off. After three months of infection comes the fatal period. Large **oedemas** or swellings ('watery poke') appear on jaws. Lactation and breeding are greatly reduced. Rarely does the host survive this period. In case it does, the fluke may migrate to the duodenum and finally escape to the outside world with faeces. When this happens, or when fluke somehow dies *in situ,* the host recovers considerably.

Infection by *F. hepatica* takes a huge toll of sheep annually. In England it caused the death of about one and a half million sheep in 1830 and about double in number in 1879-80. Ireland lost 60% of its flocks in 1882.

Therapy or treatment

Drug for **human fascioliasis** is **Triclabendazole** (Egaten tablets). Other drugs are **Nitazoxanide** (Nitax, Zox, Nitazox, Niazid) **Bithion Mirazid**.

Foranimals Triclabendazole (Fasinex liquid) is considered as the most common drug due to its high efficacy against adult as well as juvenile flukes. A new drug called **'Compound Alpha'** similar to triclabendazole is being tested.

Prophylaxis or prevention. It is better to prevent infection by control of the vector or intermediate host. Preventive measures include : (i) killing heavily infected sheep, (ii) destroying eggs and manure of infected sheep, (iii) feeding infected sheep with salt and little dry food, and (iv) killing or checking snail population. Snails are killed by adding copper sulphate solution in ponds and ditches or by draining their pastures as they are unable to survive long dry periods. Ducks feed on snails and can be usually employed in removing their population. Breeding of snails can be checked by removing vegetation from ponds and streams they inhabit. Man can avoid infection by consuming thoroughly washed and adequately cooked vegetables.

Important Questions

➤➤*Long answer type questions*

1. What is a digenetic life cycle ? Explain it with reference to the life history of *Fasciola hepatica.*
2. Give an account of the reproductive organs of *Fasciola hepatica.*
3. Give a detailed account of the parasitic adaptation in *Fasciola.*
4. Give a detailed account of the life cycle of *Fasciola hepatica* and discuss its economic importance.
5. Describe the anatomy of *Fasciola hepatica* with special reference to its parasitic mode of life.
6. Draw neat and fully labelled diagrams of the following : (i) T.S. of *Fasciola* passing through the ovary, (ii) V.S. of Body wall of *Fasciola.* (iii) Full page, well-labelled diagrams of life history of *Fasciola.* (iv) Reproductive system of *Fasciola hepatica.*
7. Write an essay on the economic importance of *Fasciola hepatica.*
8. Write notes on : (i) Cercaria, (ii) Control measures against *Fasciola* infection, (iii) Distinction between digenetic and monogenetic parasites, (iv) Effects of infection of *Fasciola* on primary host, (v) Flame cell, (vi) Metacercaria, (vii) Miracidium, (viii) Ootype, (ix) Polyembryony, (x) Redia, (xi) Sporocyst.

➤➤*Short answer type questions*

1. What is miracidium ?
2. What are flame cells ?
3. Where can you locate flame cells ?
4. In *Fasciola* write the name of the canal that leads from the junction of the oviduct and combined vitelline duct and open to the exterior on the mid-dorsal surface.
5. Explain the structure of the body wall in *Fasciola.* Draw and mention three adaptive features.
6. Define polyembryony and mention its example.
7. Define parthenogenesis. Explain it with reference to development in *Fasciola.*
8. Draw a labelled sketch of Miracidium of *Fasciola hepatica.*
9. Draw a neat labelled diagram T.S. of liver fluke through middle region.
10. In *Fasciola,* draw neatly stages from Redia to Metacercaria and label all the parts.
11. Draw a sketch showing the reproductive system of *Fasciola hepatica* and label it fully.
12. Give an account of the reproductive system of *Fasciola.*
13. Give an account of the life history of the common liver fluke.
14. Give a detailed account of the reproductive organs of *Fasciola.* How is it adapted to the parasitic mode of life ?

➤➤*Multiple choice questions*

1. In *Fasciola* miracidium develops into the next stage inside :
   (a) *Bulimes* (b) *Limnea truncutata*
   (c) *Pila globosa* (d) *Planorbis*
2. *Planorbis* and *Lymnaea* are the intermediate host of :
   (a) *Fasciola* (b) *Schistosoma*
   (c) *Trichinella* (d) *Echinococcus*
3. Which of the infective stage to primary host of *Fasciola* ?
   (a) Miracidium (b) Sporocyst
   (c) Metacercaria (d) Cercaria
4. One of the following larval stage of *Fasciola* does not produce several larvae ?
   (a) sporocyst (b) secondary Redia
   (c) primary Redia (d) Cercaria
5. Which layer of *Dugesia* contains rhabdites :
   (a) epidermis (b) circular muscle
   (c) parenchyma (d) cuticle
6. Miracidium is one of the stages in the development of :
   (a) *Fasciola hepatica* (b) *Taenia solium*
   (c) *Ascaris lumbricoides*
   (d) *Planaria*
7. In *Fasciola* the ragion where the shell gland opens into is the :
   (a) oviduct (b) ovovitelline duct
   (c) uterus (d) oocyte
8. *Fasciola hepatica* is a parasite that lives in the :
   (a) intestine of sheep (b) liver of sheep
   (c) spleen of sheep (d) pancreas of sheep
9. In *Fasciola* the germ cells of the redia gives rise to :
   (a) daughter Cercaria (b) daughter Sporocyst
   (c) daughter Redia
   (d) daughter Metacercaria
10. Liver fluke is not affected by enzymes of host because of :
    (a) cuticle (b) suckers
    (c) epidermis (d) hooks
11. The stage of life history of the liver fluke when it infects the primary host is :
    (a) Redia (b) Cercaria
    (c) Sporocyst (d) Cysticercus
12. Which stage of Liver fluke infects the intermediate host :
    (a) Redia (b) Cercaria
    (c) Miracidium (d) Sporocyst
13. Primary host of *Fasciola* is :
    (a) man (b) pig
    (c) sheep (d) dog
14. For attachment *Fasciola* has :
    (a) four suckers and hooks
    (b) two suckers .
    (c) two suckers and hooks
    (d) hooks only
15. Correct sequence of larval stages during development of *Fasciola* is :
    (a) Miracidium, Redia, Sporocyst and Metacercaria
    (b) Cercaria, Miracidium, Sporocyst and Metacercaria
    (c) Miracidium, Sporocyst, Redia, Cercaria and Metacercaria
    (d) Redia, Cercaria, Metacercaria and Sporocyst
16. In *Fasciola* germ balls of Redia give rise to :
    (a) daughter Sporocyst (b) daughter Cercaria
    (c) daughter Redia (d) Metacercaria
17. Which of the infectious stage to secondary host of *Fasciola :*
    (a) Miracidium (b) Sporocyst
    (c) Redia (d) Cercaria
18. Which is the first stage of larva of *Fasciola* :
    (a) Miracidium (b) Sporocyst
    (c) Redia (d) Cercaria
19. Protonephridia perform the function in *Fasciola* :
    (a) excretion (b) osmoregulation
    (c) both
    (d) none the above
20. *Fasciola* has :
    (a) salivery gland (b) pharyngial gland
    (c) gastric gland
    (d) intestinal gland
21. The cause of 'liver rot' in sheep :
    (a) *Dugesia* (b) *Fasciola*
    (c) *Taenia* (d) *Schistosoma*
22. Liver Fluke belongs to the class :
    (a) Cestoda (b) Nematoda
    (c) Turbellaria (d) Trematoda
23. Which one is not the larval stage of *Fasciola*?
    (a) Cystecercus (b) Cercaria
    (c) Miracidium (d) Redia

Answers

➤➤*Multiple choice questions*

1. (b)	2. (a)	3. (c)	4. (d)	5. (a)	6. (a)	7. (b)	8. (b)	9. (c)	10. (a)
11. (b)	12. (b)	13. (c)	14. (b)	15. (c)	16. (c)	17. (a)	18. (a)	19. (c)	20. (b)
21. (b)	22. (d)	23. (a).							

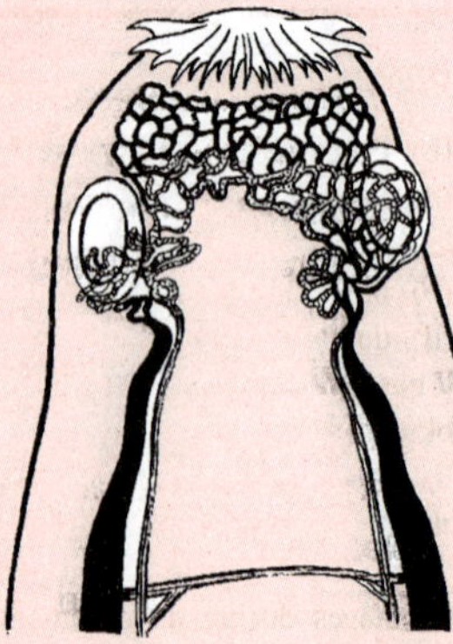

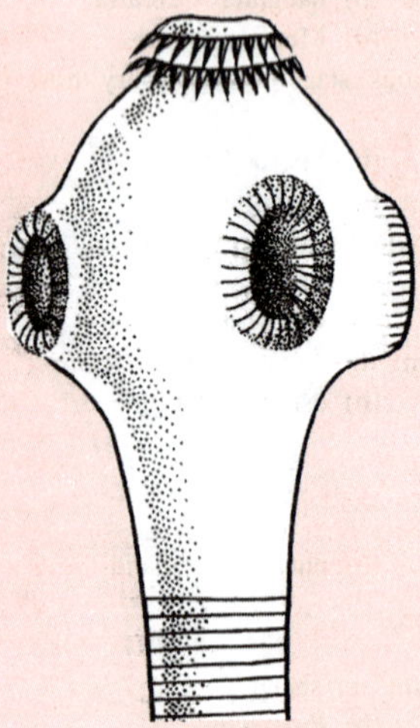

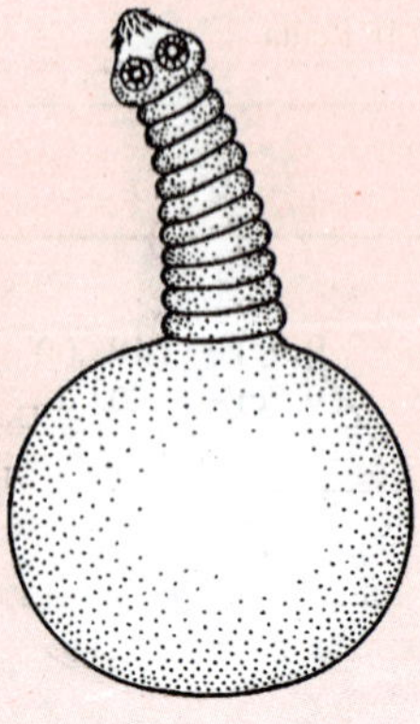

32

Taenia solium : The Pork Tapeworm

Cestodes belong to the class cestoda of phylum Platyhelminthes and are exclusively endoparasites. They differ from turbellarians and trematodes in the total absence of a digestive tract. They are commonly known as **tapeworms** for their long, flat and ribbon-like form. Genus *Taenia* includes several species parasitizing man and domestic animals. Common examples are *T. solium* (the pork tapeworm) and *T. saginata* (the beef tapeworm).

The present text mostly relates to *T. solium,* the pork tapeworm of man.

Taenia solium

SYSTEMATIC POSITION

Phylum	Platyhelminthes
Class	Cestoda
Subclass	Eucestoda
Order	Taenioidea (= Cyclophyllidea)
Family	Taenidae
Genus	*Taenia*
Species	*solium*

History of Discovery

In history tapeworm infections have been recorded from 1500 B.C. and it has been recognized as one of the earliest human parasite. Life cycle of the parasite has been identified by the 1850s. Two species *T.solium* and *T.saginata* are very small. *T.solium* (extra intestinal infection) is far more serious and dangerous.

HABITS AND HABITAT

Taenia solium, like *Fasciola hepatica,* completes its life cycle in two hosts. The adult dwells as internal parasite in the small intestine of man (primary or final host) where it is anchored to the intestinal mucosa by its **scolex**. It has no mouth or digestive cavity but absorbs the host's digested food through its body wall. The larval stage occurs in the tissues of a secondary or intermediate host which is usually pig and sometimes dog and sheep. A number of other animals such as goat, cattle, horse, bear and monkeys have also been mentioned as intermediate hosts.

Taenia solium is found in all those parts of the globe, where pig is domesticated and consumed as food. Its distribution, thus, is cosmopolitan.

EXTERNAL MORPHOLOGY

[I] Shape, Size and Colouration

Taenia solium is usually opaque white in colour but creamish, yellowish or greyish colouration is also common. Body is 1 to 5 meters long, and flattened like a ribbon or tape. The two flat surfaces represent the dorsal and ventral surfaces respectively, but from external examination it is

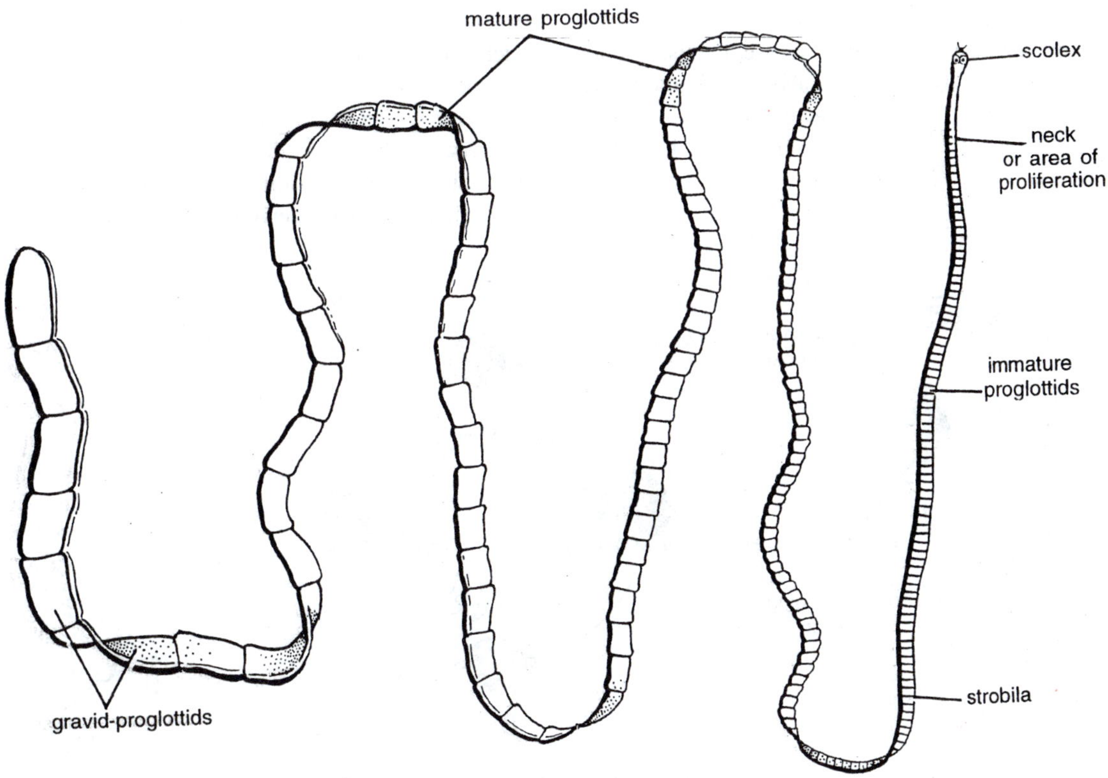

Fig. 1. *Taenia solium.* Entire animal.

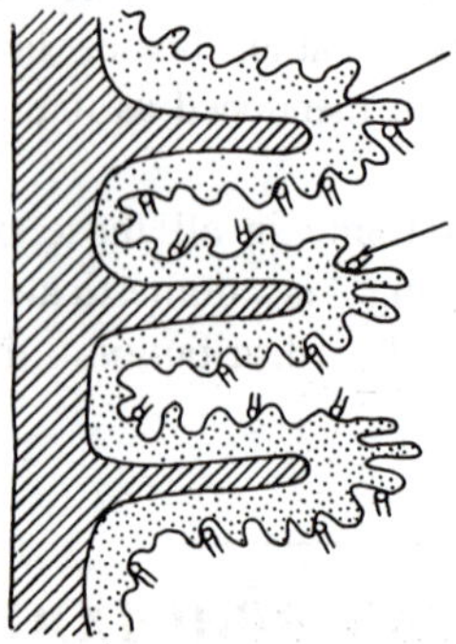

Fig. 2. Scolex of *Taenia* attached to the host's intestinal wall.

not possible to identify them. Internal view reveals that surface closer to testes is dorsal and nearer the female reproductive organs, the ventral surface. Elongated body is extremely narrow anteriorly and gradually broadens towards the posterior end.

[II] Segmentation

Elongated body of tapeworm is divided into a great number of parts or segments, called **proglottids**, possibly upto about 850 altogether. Segmentation of tapeworms is called **pseudometamerism** in contrast to the **true metamerism** of annelids and arthropods (chapter 15). Therefore, the term 'segment' in tapeworm is being used solely for convenience and in no way implies a true metameric segment.

Entire body is divisible into three distinct parts : (i) an anterior **scolex** or head, (ii) a short unsegmented **neck**, and (iii) a segmented **strobila**.

[III] Scolex

It represents the anterior end of body. It is knob-like, biradially symmetrical and 0.6 to 1 mm wide. In en-face view it appears roughly quadrangular. At the apex is a prominent rounded mobile cone, the **rostellum**. It is armed with 22 to 32 curved chitinous **hooks**, arranged in two rows or circles around its base. Hooks of anterior circle are larger, each measuring 0.14 to 0.18 mm and those of posterior circle are smaller each measuring 0.11 to 0.14 mm. Each hook consists of a **base** by which it is fixed, a blunt projection or **handle** directed towards the apex, and a conical outwardly directed **blade**. The broadest part of scolex bears four hemispherical highly muscular suctorial organs, the **true suckers** or **acetabula**, one at each angle of the quadrangle. Suckers are devoid of hooks or spines.

The scolex with the help of its hooks and suckers, lies buried in the host's intestinal mucosa, providing firm adhesion to the body against the loosening action of peristaltic movements of the

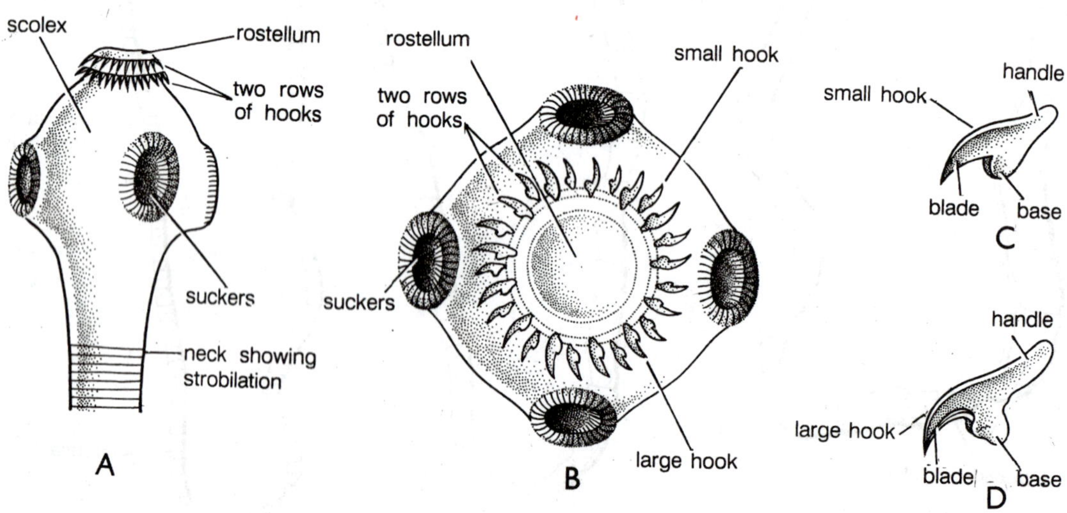

Fig. 3. *Taenia solium.* **A.** Scolex and neck in side view. **B.** Scolex in en-face view. **C.** Small hook. **D.** Large hook.

host's intestine. It is thus an organ of attachment or **holdfast**. It plays no role in perceiving or catching food. The term 'head' frequently used for the scolex is, thus, inappropriate.

[IV] Neck

It is a well-defined, short, narrow and unsegmented region behind scolex. Unlike scolex, it is dorsoventrally flattened. It has been variously termed the **budding zone**, **growth zone**, **area of proliferation** and **area of segmentation**, because it is this region where the segments or proglottids, constituting **strobila**, are budded off and pushed backwards.

[V] Strobila

It forms the main bulk of body. It consists of 800 to 1,000 segments or **proglottids** arranged in a linear series in a **chain-like** fashion. (The species *solium*, according to **Leuckart**, has got its name from **schuschel**, a **Syrian** word referring to a **chain**). A proglottid is a unit part of the body enclosing a complete set of genitalia. Linear repetition of proglottids, and hence of genital organs, is termed **proglottisation**. Proglottids are budded off in the neck region and pushed backwards due to addition of more proglottids in front. Thus, in a strobila, anterior proglottids are the youngest, while those at the posterior end are the oldest. Adjacent proglottids remain attached together by longitudinal muscles, excretory ducts and nerve cords, which extend along the entire body length. According to the degree of development, the strobila includes the three kinds of proglottids : **immature**, **mature** and **gravid**.

1. Immature proglottids. These comprise of about 200 anterior proglottids just behind the neck. They are the youngest, sexually immature and devoid of reproductive organs. They are short, broader than long and rectangular in outline.

2. Mature proglottids. There are about 450 mature proglottids forming the middle part of strobila. These are large and squarish in outline. The anterior 100 to 150 proglottids contain only male reproductive organs, while the posterior 250 mature proglottids develop both male and female reproductive organs making them hermaphrodite.

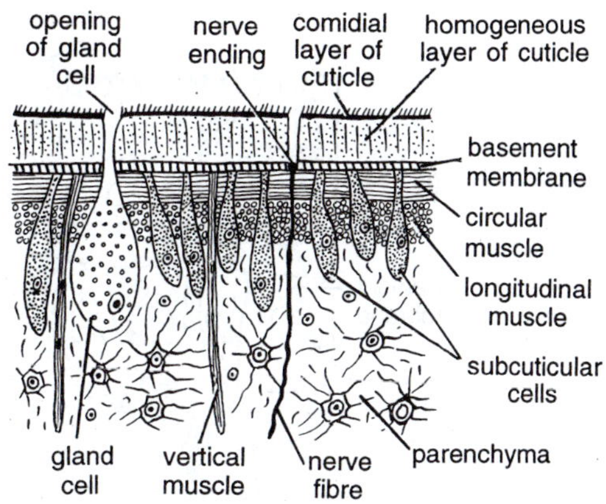

Fig. 4. *T. solium.* V.S. body wall under light microscope.

Each mature proglottid, on one side bears a tiny protuberance, the **genital papilla**, at the tip of which is situated the **common genital pore**. These pores, in the successive proglottids, are situated alternately on the right and left sides. A mature proglottid is a complete reproductive unit and produces eggs which are fertilized by its own sperms *(***self fertilization***)* or by those of other mature proglottids *(***cross fertilization***)*.

3. Ripe or gravid proglottids. The oldest and the last 150 to 350 proglottids, upto the posterior emd of body, are termed **gravid** or **ripe**. They are longer than broad in outline. All the male and female reproductive organs have degenerated except the highly branched **uterus** full of fertilized eggs (Fig. 12).

[VI] Apolysis

Small groups of gravid proglottids regularly detach from the posterior end of strobila and pass out with the host's faeces. Shedding of gravid proglottids is termed **apolysis** and the tapeworm exhibiting this phenomenon is called **apolytic** in contrast to the **anapolytic tapeworms** (most pseudophyllids), which retain all their proglottids throughout life. Apolysis serves a twofold purpose. (i) It serves to transfer the developing embryos to the exterior, where they can be ingested by the secondary host, and (ii) it limits the size of body which may otherwise attain enormous length due to continued proliferation in the neck region.

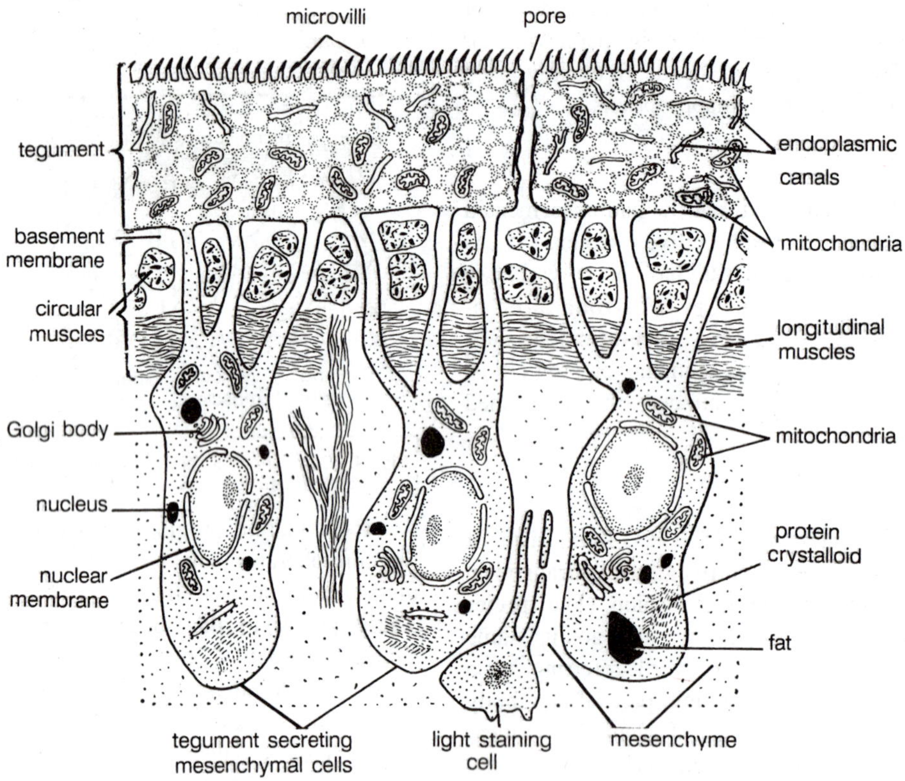

Fig. 5. *T. solium.* V.L.S. body wall based on electron microscope.

BODY WALL

Like flukes, tapeworms also lack a cellular or ciliated epidermis. Layers of body wall from surface are : (i) **tegument**, (ii) **basement membrane**, (iii) **integumentary muscles**, and (iv) **parenchyma**.

[I] Tegument

It is a thick resistant layer clothing the body in the absence of a cellular epidermis. It is composed of protein impregnated with calcium carbonate and is perforated by numerous fine canals. Under light microscope, three layers can be distinguished in this layer : (i) outermost hair-like or fringe-like **comidial layer**, (ii) middle thick **homogeneous layer**, and (iii) innermost **basement membrane**.

26. (Z–1)

As no living structures are visible, this layer used to be regarded as non-living **cuticle**. Studies under electron microscope by **Threadgold** and others have shown that outermost cuticle is in fact a thick, living and syncytial layer called **tegument**, continuous with **tegument secreting cells**, of mesenchyme. Tegument contains mitochondria and lysosomes and gives out **microvilli-like** processes, called **microtriches**, on its outer surface. The Microvilli serve two purposes : (i) they facilitate absorption of host's food by increasing the surface area of body, and (ii) they partially act as holdfast organs (**A. H. Rothman**, 1963) by interlocking with microvilli of cells lining the host's intestine. Tegument is also perforated by numerous fine **pore canals** through which substances (in solution) are absorbed from the host's intestine.

Tegument is constantly renewed by the underlying **tegument secreting cells** with which it is continuous.

[II] Basement Membrane

With a light microscope, region beneath tegument (then cuticle) appears as a thin, well-defined acidophilic **basement membrane**. However, with electron microscope, its outer edge is identified, while its inner edge merges imperceptively with the underlying mesenchyma.

[III] Integumentary Musculature

Beneath the basement membrane musculature consists of well-developed outer **circular** and inner **longitudinal** fibres. Lying in the mesenchyme also is the **mesenchymal musculature** consisting of **longitudinal**, **transverse** or **circular**, and **vertical** or **dorsoventral muscle** fibres.

[IV] Mesenchyme or Parenchyma

It consists of loosely-packed cells with fluid-filled interspaces, forming a **packing** around various internal organs. In young proglottids and in neck region, however, mesenchyme is more compact. Numerous round or oval calcareous bodies, composed of concentric layers of calcium carbonate, are present in mesenchyme. These are secreted by special mesenchymal **lime cells**, which later atrophy, setting the bodies free in mesenchyme. Besides, it also contains a large number of **tegument secreting cells**. Circular muscle fibres of the mesenchymal musculature, except at the margins, divide the mesenchyme into an outer **cortex** or **cortical zone** and inner **medulla** or **medullary zone**.

In addition to its skeletal function mesenchyme acts as an important transport medium in the absence of a blood vascular system.

NUTRITION

As already mentioned, tapeworm completely lacks alimentation in all stages of life-history. The predigested food in host's small intestine (especially ileum) is the chief source of nourishment for tapeworm. Soluble nutrients, like glucose, amino acids, glycerol, etc., diffuse in directly through general body surface (tegument). As noted above, the absorptive surface of parasite is greatly increased by the microvilli of tegument. The swift efficiency with which absorption takes place has been compared with the soaking action of blotting paper. Some tissue fluids from host are probably absorbed by scolex of tapeworm insinuated deeply into intestinal mucosa. Stored food consists mainly of glycogen and of some lipoid substances. Glycogen content of *T. solium*, by net weight, is 2.17%.

For its mode of obtaining nutrients, tapeworm does not require an alimentary canal, which is thus totally absent in all stages of its life history. Moreover, presence of an alimentary canal would not have suited the mode of life of tapeworm as it would have rendered impossible the process of apolysis is indispensable for it. The alternative mode of nutrition described above is thus a parasitic adaptation of tapeworms, unique in the animal kingdom.

RESPIRATION

Respiration of the tapeworm is mainly **anaerobic** or **anoxybiotic**. Glycogen, the principal reserve food and the chief source of energy, undergoes glycolysis, producing carbon dioxide and fatty acids. Steps involved are same as in liver fluke. In addition to fatty acids, other organic acids, like lactic acid, are also produced.

Free oxygen, whenever available, is also consumed by tapeworms. Rate of consumption is maximum in the anterior proglottids and declines gradually towards the posterior end.

EXCRETORY AND OSMO-REGULATORY SYSTEM

It consists of : (i) lateral longitudinal canals, (ii) secondary canals, (iii) capillaries, and (iv) flame cells.

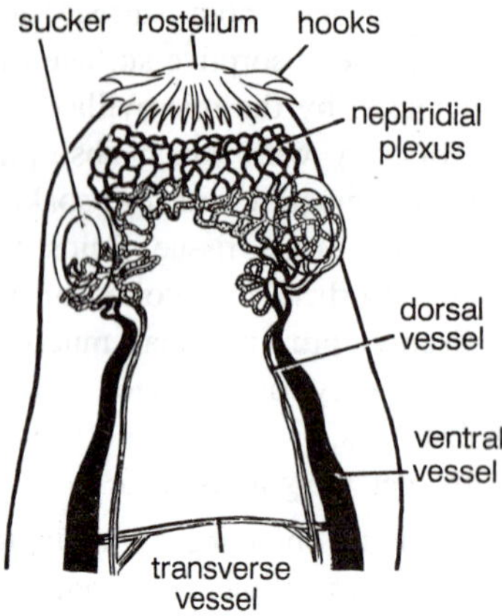

Fig. 6. *Taenia solium.* Excretory system at the anterior end.

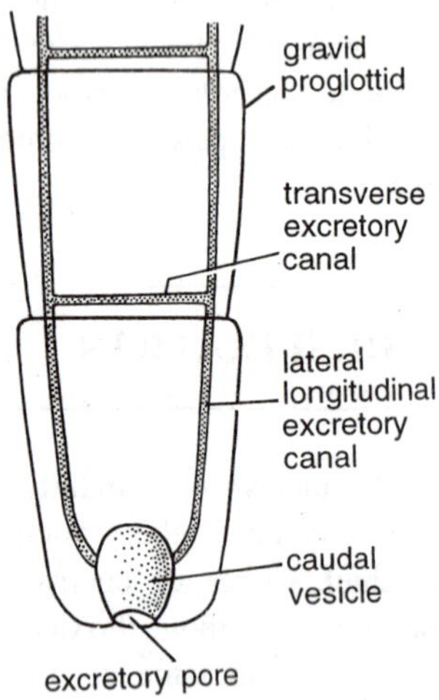

Fig. 7. *Taenia solium.* Caudal vesicle in last proglottid.

[I] Excretory Canals

There are, on each side, two **lateral longitudinal excretory canals** or the **collecting tubules** of which one is dorsal and the other ventral. In scolex they get connected with one another by a network of tubules or **nephridial plexus**. The canals run through proglottids, just inner to the mesenchymal musculature. **Dorsal canals** are produced and confined to the anterior part of strobila. **Ventral canals** are large and extend along its entire length. Two ventral canals are connected by a **transverse canal** at the posterior part of each proglottid (except the last). In the last

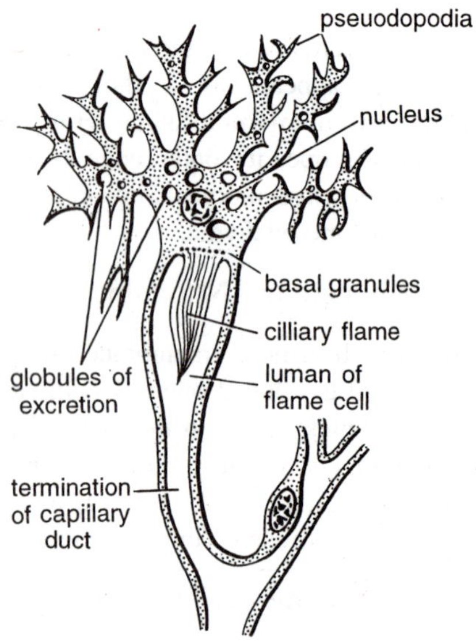

Fig. 8. *Taenia solium.* A flame cell.

proglottid, they join to form a pulsatile **caudal vesicle**, opening to the exterior by a single **excretory pore**. When the last proglottid is shed off (apolysis), the terminals of the two ventral canals behave as independent excretory pores.

Each longitudinal excretory canal receives numerous **secondary canals** all along its length.

[II] Flame Cells

A secondary canal is formed by the union of several fine **capillaries**, each connected to a **flame cell**. These are scattered throughout parenchyma from which they remove metabolic wastes. A flame cell is of irregular shape, with granular cytoplasm and a nucleus. Bundle of cilia, or flame, arises from basal granules near nucleus. Cilia are enclosed into a funnel-shaped lumen formed by the terminal blind end of a capillary.

[III] Physiology

Longitudinal canals are lined internally by cuticle, while secondary canals and capillaries have a ciliated lining. Cilia set up a hydrostatic pressure which drives out the excretory products through excretory canals and out of excretory pores. Exact nature of excretory products is, however, not known. Fluid contents of body are also regulated

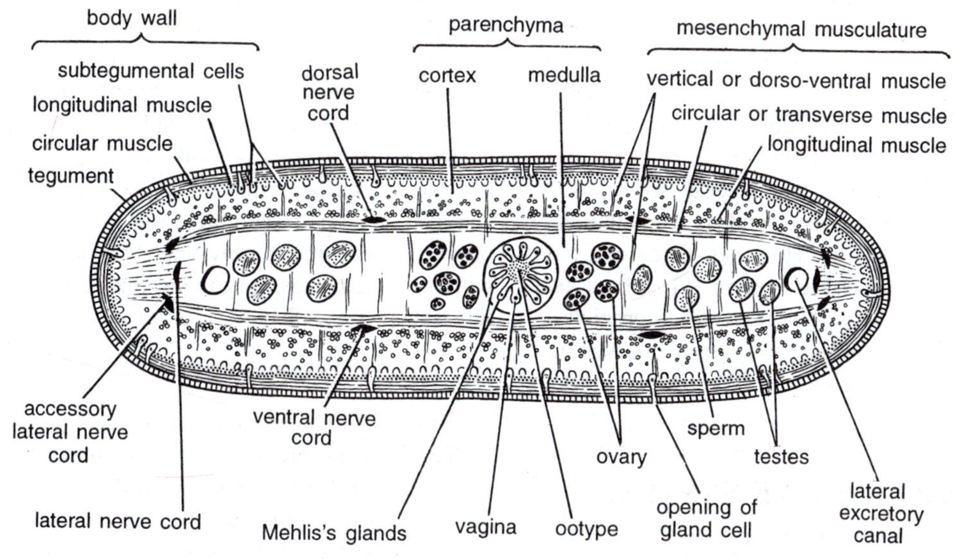

Fig. 9. *Taenia solium.* T.S. mature proglottid through ootype showing positions of 10 longitudinal nerves.

by this system, which is, therefore, also **osmoregulatory**.

NERVOUS SYSTEM

Nervous system of *Taenia solium* has not been thoroughly examined so far. Among the taenoid tapeworms, *Moniezia* (**Tower**, 1900) and *Anoplocephala* (**Becker**, 1921) have been best studied for this system. Their findings in these tapeworms are probably true for other taenoid tapeworms also. Nervous system of *Moniezia* is, however, described below.

In the scolex lies a pair of stout **cerebral ganglia** connected by (i) a ring consisting of a **dorsal** and a **ventral commissure**, and (ii) a thick, ganglionated **cross commissure** or the **transverse commissure**. From the **brain complex** so formed, 8 nerves are given out anteriorly (4 from ring commissure and 4 from transverse commissure). These terminate into another smaller ganglionated **rostellar nerve ring**. Fibres from ganglia in scolex supply the suckers and rostellum. From brain complex, 3 pairs of longitudinal nerve cords arise and extend posteriorly into strobila. These are : (i) **lateral nerve cords** arising from cerebral ganglia and extending through medullary region of mesenchyme, outer to the excretory canals, (ii) **dorsal nerve cords** arising from dorsal commissure, and (iii) **ventral nerve cords** arising from ventral commissure. In each proglottid, all the 6 nerve cords are connected by a **ring commissure** or **transverse connective** behind the transverse excretory canal.

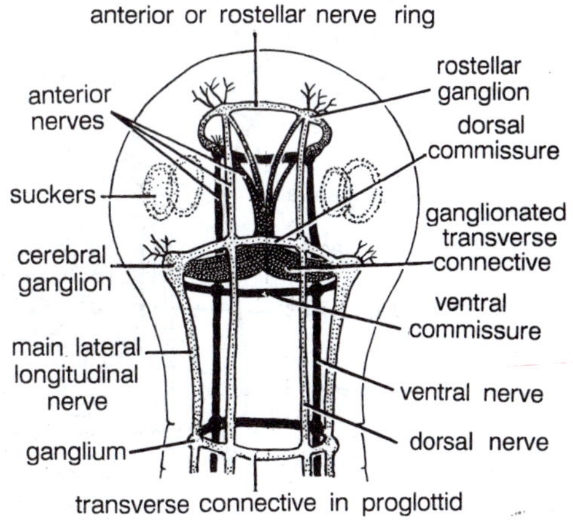

Fig. 10. *Moniezia.* Nervous system in scolex.

One feature, in which nervous system of *Taenia solium* and most other tapeworms surely differs from that of *Moniezia,* is that 10 **nerve cords** instead of 6 (three pairs) arise from brain complex and run through strobila. The 4 additional nerve cords are the **accessory lateral nerve cords**, 2 of which accompany each lateral nerve cord.

Special receptors, as found in turbellarians, are lacking in tapeworm. However, numerous free sensory nerve-endings are present throughout the body specially in the scolex. A detached proglottid, passing out with faeces, shows some movement and sensitivity to stimuli.

REPRODUCTIVE SYSTEM

All tapeworms, except *Dioecocestus* (a genus allied to *Taenia),* are **hermaphrodite**. Each proglottid, at maturity, contains a complete set of male and female reproductive organs. These organs develop from mesenchyme and lie embedded in it. Male organs differentiate before the female organs (**protandrous** condition), so that in the mature zone of strobila, consisting of about 450 proglottids, the anterior 100-150 proglottids possess only male system while the remaining posterior ones possess both male and female systems. After a period of intense reproductive activity, mature proglottid loses all its genital organs except the highly-branched uterus filled with fertilized eggs. Segments that have reached this stage, are called **ripe** or **gravid** segments.

[I] Male Reproductive System

1. Testes. Testes are numerous small, spherical bodies scattered throughout the mesenchyme, close to the dorsal surface. (Some workers are of the opinion that the numerous rounded bodies, referred to as testes here, constitute a single and highly subdivided testis).

2. Vasa efferentia. From each testis arises a fine ductule, the **vas efferens**, which frequently gets interconnected with similar ductules from the surrounding testes. All the vasa efferentia finally

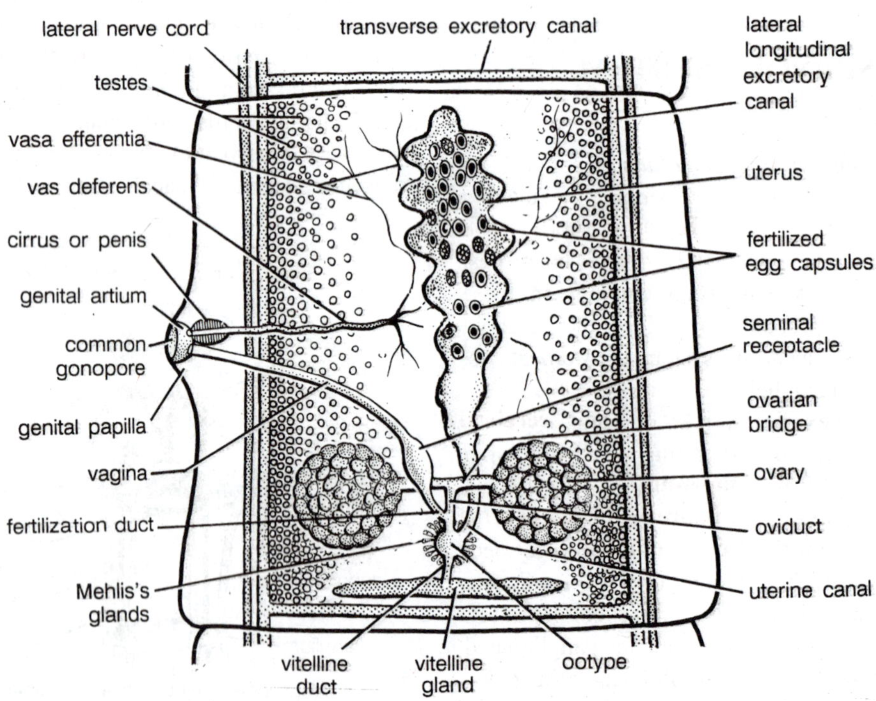

Fig. 11. *Taenia solium.* Reproductive system in a mature proglottid.

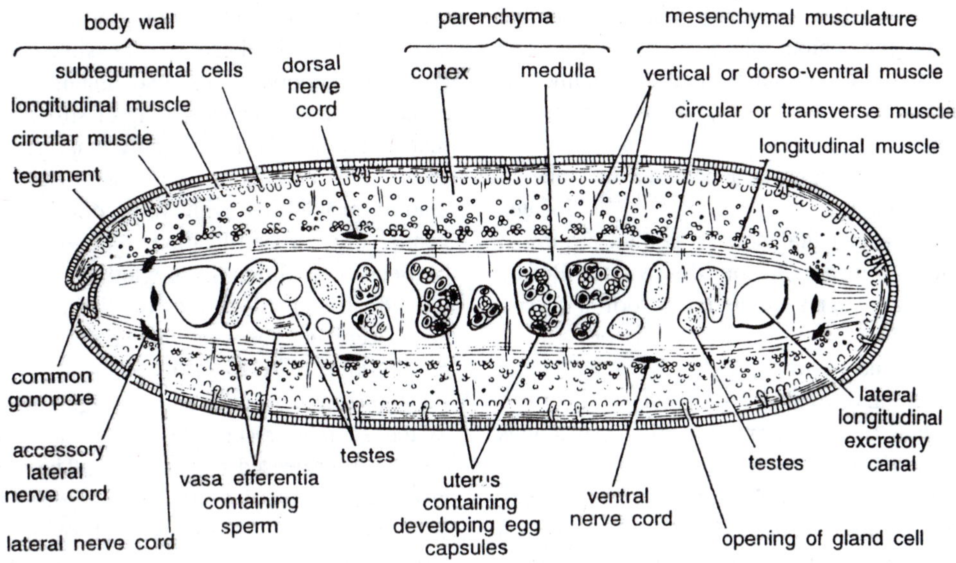

Fig. 12. *Taenia solium.* T.S. mature proglottid through uterus and genital atrium.

unite, approximately in the middle of proglottid, to form a common **sperm duct** or **vas deferens**.

3. Vas deferens. It is a thick and convoluted tube, extending upto the lateral margin of proglottid, at right angles to it.

4. Cirrus. Outer end of **vas deferens** forms the lumen of a thick, muscular, eversible, copulatory organ, the **cirrus**, enclosed in a firm **cirrus sheath**. There is no seminal vesicle.

5. Genital atrium. Cirrus opens into a cup-shaped **genital atrium** through the **male genital pore**. Genital atrium, in turn, opens to the exterior through the **common gonopore** situated at the peak of a tiny protuberance, the **genital papilla**, in the middle of the lateral margin of proglottid. Gonopore is provided with a sphincter muscle. Common gonopores of successive proglottids lie alternately on both the sides.

[II] Female Reproductive System

It resembles with the corresponding system of liver flukes.

1. Ovary. A single bilobed **ovary** or **germarium** lies ventrally in the posterior part of proglottid. Each lobe of the ovary (regarded by

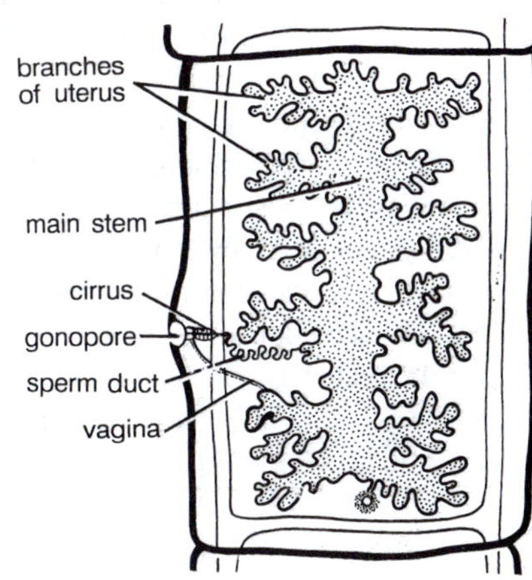

Fig. 13. *Taenia solium.* A gravid proglottid showing branched uterus.

some as one complete ovary) is dorso-ventrally flattened and consists of a number of radially-arranged **germinal cords** or **follicles**. The right and left lobes are connected medially by a transverse, tubular bridge, the **ovarium isthmus**.

2. Oviduct. A short oviduct arises from about the middle of ovarian isthmus. It runs backwards, joins another slender tube, the **vagina** and then forms a swollen chamber, the **ootype**.

3. Ootype. It is a small rounded chamber formed by the union of oviduct, uterus and vitelline duct. It is surrounded by numerous unicellular **Mehlis's glands**, known after its discoverer.

4. Vagina. Vagina is a bent narrow tube. It originates from the **female genital pore** located behind the male genital pore in genital atrium. It runs obliquely inwards to join the oviduct. Before joining oviduct, vagina swells up to form a sperm-storing sac, the **receptaculum seminis** or **seminal receptacle**. The narrow part of vaginal duct between sac and oviduct is called **fertilization duct**. Vagina of tapeworm probably corresponds to the Laurer's canal of liver fluke.

5. Vitelline gland. A large lobulated gland, the **vitelline gland** or **vitellarium**, forms a compact elliptical mass behind the ovary. It is connected with the ootype by a short median **vitelline duct**. Vitelline gland consists of numerous follicles secreting yolk cells.

6. Uterus. From ootype arises a blind and cylindrical tube, the **uterus**, extending upto the anterior part of proglottid. It consists of a proximal short and narrow tubular portion, the **uterine duct**, and a distal broad part forming the **uterine expansion**. Uterus is the most characteristic feature of reproductive system of tapeworms. It is meant to accommodate thousands of fertilized eggs and in the process of accomplishing this, it forms 7 to 13 lateral .branches on each side. The branched uterus is the only genital structure presisting in a gravid proglottid.

LIFE HISTORY AND DEVELOPMENT

[I] Copulation and Fertilization

Life cycle of *T. solium* is **digenetic**, involving two hosts as in case of a fluke. But life cycle of tapeworm is much simpler and without a free larval stage. Presence of a single tapeworm in a host diminishes possibility of cross-fertilization. Fertilization is preceded by **copulation** which is accomplished by insertion of cirrus into vagina of the same or other proglottid to release spermatozoa. It becomes possible when the common gonopores of two mature proglottids come in contact due to folding of strobila. Anterior mature proglottids, having only male genitalia, can enter into copulation only with the posterior mature proglottids with fully developed female genital organs. Fertilization, following copulation between two different proglottids, is sometimes termed **cross-fertilization** to distinguish it from that occurring between gametes of the same proglottid *(***self-fertilization***)*.

Spermatozoa injected into vagina, swim down to the seminal receptacle where they are stored till **ova** are released by the ovary. The two finally meet in the **fertilization duct** (part of vagina between seminal receptacle and oviduct) where fertilization takes place and **zygotes** are formed. Thus, fertilization is **internal**.

[II] Capsule Formation

Zygotes or **egg cells** pass into the ootype, where each becomes associated with a large **yolk cell** or **vitelline cell** provided by the vitelline gland. The two become enclosed in a thin **shell** or **chorionic membrane**, formed by material exuded by the yolk cell. The **capsule** so formed passes into uterus, where further development takes place. Passage of capsules into uterus is lubricated by the secretion from Mehlis glands. As more and more capsules pass into uterus, it develops lateral branches to accommodate them.

[III] Onchosphere Formation

1. Cleavage. Zygote or egg cell undergoes **cleavage** when the capsule is in uterus. Cleavage is holoblastic and unequal. First **unequal** division results in a larger **megamere** and a smaller **embryonic cell**.

2. Morula. Megamere divides further and forms several similar **megameres**, while embryonic cell divides repeatedly producing two types of embryonic cells, larger **mesomeres** and smaller **micromeres**. Thus, three types of cells result from the zygote; small micromeres, medium mesomeres and large megameres. Micromeres form a rounded mass, the **morula**, surrounded by an inner envelope of mesomeres and an outer envelope of

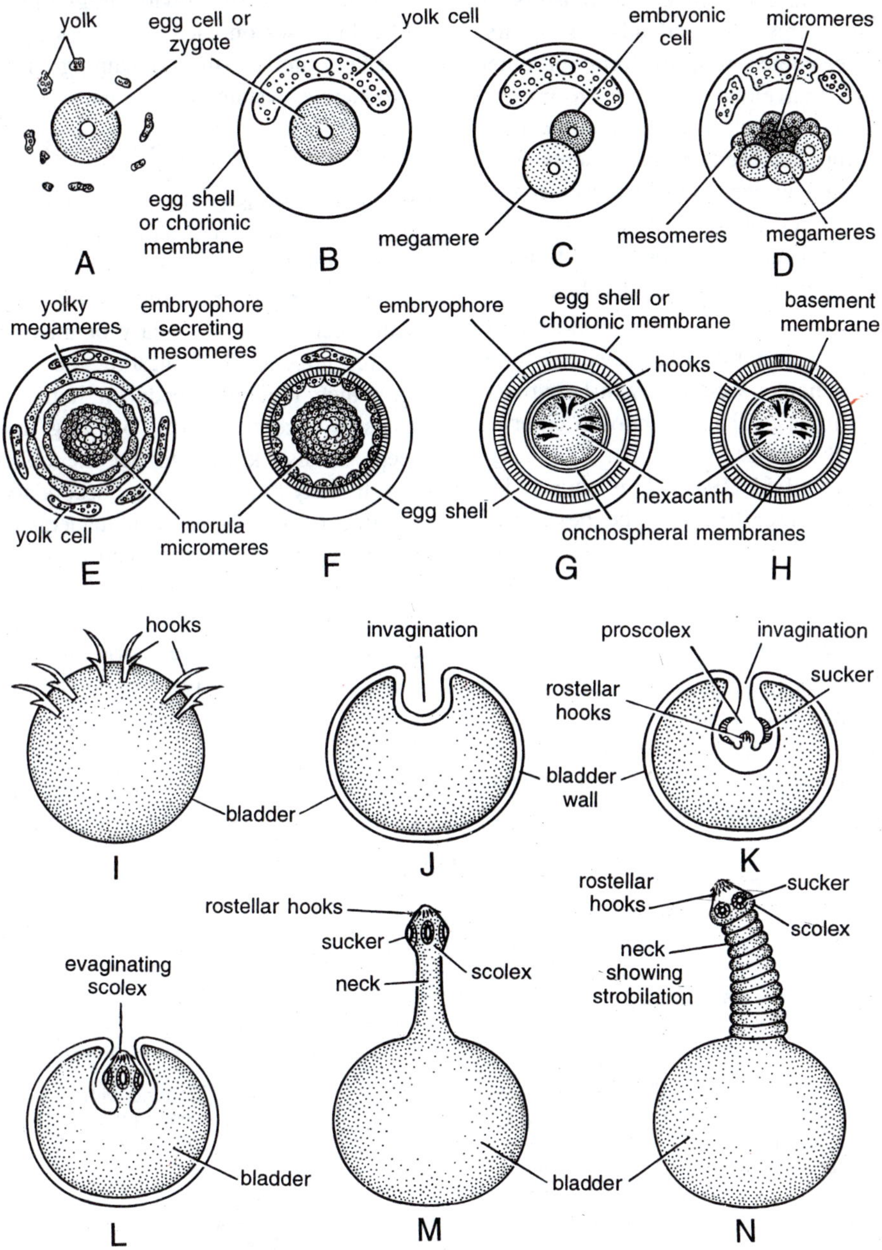

Fig. 14. *Taenia solium.* Stages in development. **A.** Zygote. **B.** Zygote in egg shell. **C.** Two-cell stage. **D.** Early morula **E.** Late morula. **F.** Formation of embryophore. **G.** Young onchosphere. **H.** Onchosphere without egg shell. **I.** Free hexacanth. **J.** Bladderworm with invagination of scolex. **K.** Bladderworm with proscolex. **L.** Evagination. **M.** Young cysticercus. **N.** Neck budding off proglottids.

megameres. The yolk or vitelline cell transfers its yolk to the megameres and gradually disappears. Large yolky megameres fuse to form an outer syncytial nutritive envelope or **outer embryonic membrane**, which nourishes the embryonic cells and finally disappears. Medium mesomeres form a thick, hard, cuticularized and radially striated shell, known as **embryophore** or **inner embryonic membrane** surrounding morula. Beneath embryophore is a thin **basement membrane**.

3. Hexacanth and Onchosphere. Morula, at its morphologically posterior end, develops three pairs of chitinous **hooks** secreted by differentiated cells, called **onchoblasts**. This six-hooked embryo, called **hexacanth**, possesses a pair of large penetration glands (**Reid**, 1947). It is surrounded by two **hexacanth membranes**. The hexacanth, together with all the membranes surrounding it, is known as **Onchosphere**. It loses the original thin shell or chorionic membrane so that embryophore forms its outermost covering.

By the time onchospheres are formed, the proglottid becomes gravid and increases in size. Its uterus forms 7-13 lateral branches on each side and contains 30,000 to 40,000 onchospheres. In *Taenia saginata,* another intestinal tapeworm of man, uterus of gravid proglottid has 16-20 lateral branches on each side. Leaving the highly-branched uterus, the remaining structures of reproductive system degenerate.

[IV] Infection to Secondary Host (Pig)

Gravid proglottids at the posterior end of strobila detach (apolysis) in groups of 4 or 5 and pass out with the host's faeces. On ground, proglottids eventually disintegrate, setting free thousands of onchospheres. The secondary or intermediate host acquires infection by ingesting the onchospheres. Pig, which regularly feeds on human excreta is the usual secondary host, but dog, monkey and sheep are also known to get the infection. Man himself may serve as the secondary host by ingesting onchospheres with inadequately cooked or raw vegetables. Auto-infection may take place in a person already serving as a primary host. This happens when due to **reverse peristalsis** the detached proglottids are carried to the stomach, where the onchospheres are liberated.

[V] Migration within Secondary Host

In the stomach of secondary host (pig) **Onchosphere** loses its embryophore and basement

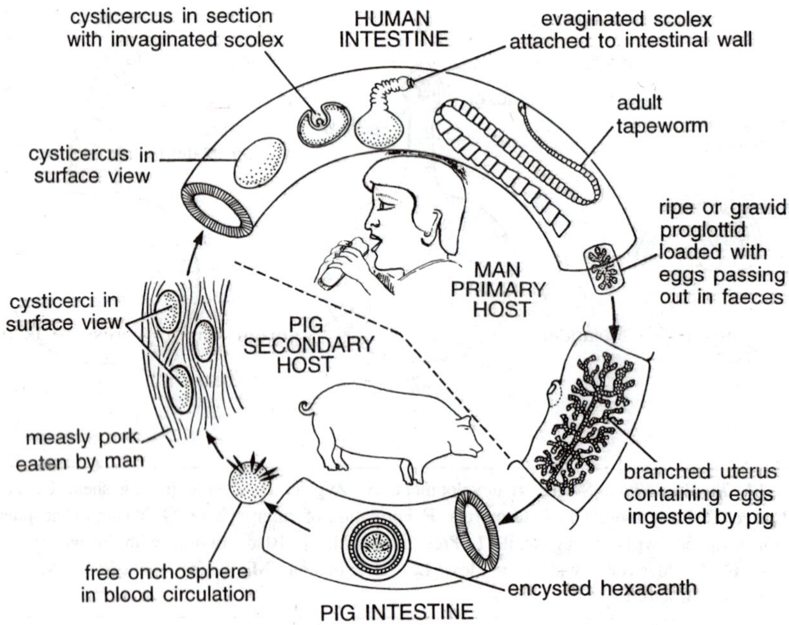

Fig. 15. *Taenia solium.* Life cycle.

membrane by the action of acidic juices (acid pepsin). The free hexacanth embryo then passes into the small intestine, where the two persisting hexacanth membranes are also lost by the action of alkaline juices. Hexacanth, now activated by the presence of bile salts, bores its way through the intestinal epithelium to reach a submucosal blood or lymph vessel. This is accomplished, perhaps jointly by the six hooks and penetration glands. Hooks merely anchor the hexacanth to the intestinal wall, while secretion of penetration glands dissolves the intestinal tissues. Entire process takes about 10 minutes, after which the hooks, are of no further use, and are shed off. Submucosal blood vessel carries hexacanth to liver via hepatic portal vein. From liver it reaches heart and enters the arterial circulation. It finally reaches the striped (voluntary) muscles usually of the tongue, shoulder, neck, thigh, heart, etc., where it settles to develop into a **bladder-worm** or **cysticercus**. However, these may also develop in other organs such as lungs, liver, kidney or brain. Non-muscular vital organs like eyes, brain or liver may frequently become the sites of cysticercus formation.

[VI] Cysticercus or Bladderworm Formation

Hexacanth, now devoid of hooks, absorbs nourishment from host's tissue and grows in size attaining a diameter of about 18 mm. A central cavity appears as cells in that region break down. It enlarges and becomes filled with a fluid consisting mainly of blood plasma of the host. The fluid-filled vesicle or **bladder**, as it is now called, has a thin wall consisting of an outer layer of thick syncytial protoplasmic mass (the so-called cuticle) and an inner mesenchymal or germinal layer. At a point, morphologically the anterior end (*i.e.*, opposite the side where hooks were present), the wall thickens and invaginates. The invagination, which looks like a hollow knob, differentiates into an **inverted scolex** possessing suckers, hooks and rostellum. It is called **proscolex**, the embryo at this stage is called a **bladderworm**. In *T. solium,* bladderworm is of **cysticercus** type which is characterised by a large vesicle and one scolex. That is why the bladderworm is also referred to as **cysticercus**. It appears to have a wall of cellulose and for this reason, sometimes called **cysticercus cellulosae**. Formation of cysticerci is completed in about 10 weeks in the pig.

Cysticercus develops in adult tapeworm only when ingested by the human host. In pig's body it leads quite an inactive life and remains viable for several years, after which it dies and becomes calcified. Pork (pig's flesh) containing viable cysticerci is called **measly pork** for its spotted appearance. One kg gm of measly pork may contain 500 or more cysticerci.

[VII] Infection to Primary Host (Man)

Man gets infection by eating undercooked measly pork. Cysticercus becomes active on reaching the small intestine. Proscolex everts or evaginates and anchors to the intestinal wall. Neck begins to proliferate proglottids and the bladder, sooner or later, gets detached and digested. In 10 to 12 weeks the parasite attains adulthood and possesses gravid proglottids ready for apolysis.

PARASITIC ADAPTATIONS OF *TAENIA*

Tapeworm shows several adjustments to its internal parasitic life, in comparison with a free-living animal.

(1) External body covering or **tegument**, is freely permeable to water and nutrients, but protects against digestion by host's alkaline digestive juices.
(2) Internal osmotic pressure is higher than that of the surrounding host's fluid or tissue, and pH tolerance is high, 4 to 11.
(3) Adult as well as larva lack cilia and other organs of locomotion, which are not needed.
(4) Scolex, with suckers and spines, serves for attachment with the epithelial lining of the host's intestine, so that parasite may not be ejected from intestine due to its peristaltic contractions.
(5) Alimentary canal is totally absent as the parasite absorbs readily available digested

(Contd. on page 414)

Table 1. **Comparative study of *Fasciola* and *Taenia*.**

Characters	*Fasciola hepatica* (The Sheep Liver Fluke)	*Taenia solium* (The Pork Tapeworm)
Habits and habitat	1. Endoparasitic in two hosts (digenetic). Several adults live in bile passages of a single sheep or goat (primary host). Larval stages are passed in a freshwater snail (secondary host).	1. Endoparasite in two hosts (digenetic). A single adult lives in intestine of a single primary host (man). Larval stages occur in pig (secondary host).
External morphology	2. Body bilaterally symmetrical, dorso-ventrally flattened, oval and leaf-like. Length 1.2 to 5 cm and width about 1.2 cm. Colour pinkish.	2. Body bilaterally symmetrical (except scolex which is radially symmetrical), dorso-ventrally flattened, tape or ribbon-like, up to several meters in length. Colour opaque white.
	3. Body unsegmented. Anterior end produced into a conical projection, the head lobe.	3. Body segmented divisible into (i) scolex or holdfast, (ii) neck, and (iii) strobila of about 800 segments or proglottids.
	4. Adhesive organs are : (i) an anterior or oral sucker and (ii) a posterior or ventral sucker or acetabulum.	4. Adhesive organs are : (i) four suckers and (ii) two circular rows of hooks on scolex.
	5. Body apertures are a mouth, a common gonopore and an excretory pore. A temporary opening of Laurer's canal appears at the time of copulation.	5. Body apertures are a common gonopore on each proglottid and one (two after apolysis starts) excretory pore on the last proglottid.
Tegument cuticle	6. With minute spinules and without prominent cuticle microvilli.	6. Without spinules and with prominent microvilli.
Musculature	7. Sub-epidermal musculature of circular, longitudinal and diagonal fibres. Mesenchymal musculature lacking.	7. Sub-epidermal musculature of circular and longitudinal fibres and a mesenchymal musculature of longitudinal, circular (transverse) and dorso-ventral fibres.
Nutrition	8. Holozoic. Alimentary canal comprises of a mouth surrounded by oral sucker, suctorial pharynx, short oesophagus and a forked, much-branched intestine.	8. Saprozoic. Gut altogether absent.
Excretory system	9. It consists of flame cells and a system of ramifying excretory ducts. Excretory pore is single situated at the posterior and apolysis does not occur.	9. Includes flame cells and four main excretory ducts, only two of which extend up to last proglottid and open through a common excretory pore. When last proglottid is lost by apolysis, two excretory ducts open through separate apertures.
Nervous system	10. There is a nerve ring bearing a pair of dorso-lateral ganglia and a ventral ganglion. The ganglia give out numerous fine nerves anteriorly and three pairs (dorsal, ventral and lateral) posteriorly.	10. Scolex has a pair of cerebral ganglia connected by a transverse connective and a ganglionated ring commissure. Five pairs of longitudinal nerves are given out into strobila.
Reproductive system	11. Hermaphrodite. A single set of reproductive organs persists throughout life.	11. Hermaphrodite. One complete set of reproductive organs in each mature proglottid. After capsule-formation, only uterus persists, while remaining genitalia degenerate.
Male reproductive system	12. Only one pair of large lobed testes, lying one behind the other in middle region of the body.	12. Testes numerous, rounded (follicular), scattered throughout the proglottid.

	13. Two **vasa deferentia** arise from two testes, and unite into a **seminal vesicle**, from which a convoluted **ejaculatory duct** arises and enters a muscular copulatory organ, the **cirrus** or **penis**.	13. Numerous fine **vasa efferentia**, one from each testis, unite into a single **vas deferens** which enters the **cirrus** or **penis**. **Seminal vesicle** is lacking.
	14. **Prostate glands** and **cirrus sheath** present.	14. Absent.
Female reproductive organs	15. A single, many branched **ovary** is situated before testes on the right side.	15. A single **ovary** lies in the posterior part of proglottid. It has two lobes united by a transverse ovarian bridge or isthmus.
	16. **Oviduct** starts from inner side of ovary and joins the **median vitelline duct**. **Ootype** is absent.	16. **Oviduct** starts from the middle of isthmus and opens into the **ootype**, which also receives the **vitelline duct**.
	17. A **Laurer's canal** extends from oviduct to the dorsal side of the body.	17. Absent.
	18. **Uterus** is a long, much-convoluted tube. It starts from the point of union of median vitelline duct and oviduct. It opens into genital atrium as the female genital aperture. It enlarges considerably to store the capsule.	18. **Uterus** is an unconvoluted, blind tube arising from the ootype and extending towards the anterior side. It not only enlarges, but also becomes much-branched to store the capsules.
	19. **Seminal** receptacle is lacking.	19. **Vagina** is a slender tube extending from genital atriurn to oviduct. Before opening in oviduct, it dilates to form the **seminal receptacle**.
	20. Numerous rounded **vitelline glands** occupy lateral zones of body. They give two **lateral vitelline** ducts connected by a **transverse vitelline duct**. The latter swells up in the middle to form a yolk reservoir, from which a **median vitelline duct** arises to meet the oviduct.	20. **Vitelline gland** is a single, large, flat mass in the posterior part of proglottid. A single **vitelline duct** arises from it and opens into the ootype.
	21. **Mehlis glands** are numerous minute bodies around the junction of oviduct, median vitelline duct and uterus.	21. **Mehlis glands** occur around the ootype and open into it.
	22. Capsules are released into host's bile duct, and finally excreted out side with host's faeces.	22. Gravid proglottids reach the exterior with host's faeces and degenerate to liberate capsules.
Life history	23. Capsules with **operculum**.	23. Capsules without **operculum**.
	24. **Development** begins outside parent's body.	24. **Development** begins within parent's body,
	25. A conical, ciliated **miracidium** hatches out from capsule, swims about for some time and finally penetrates into the body of snail the host.	25. Zygote develops into a six-hooked **hexacanth** which, together with its protective covering, is called **onchosphere**. It is ingested by the pig host.
	26. In snail, miracidium develops into a **sporocyst**: which produces the first generation of **rediae**. The latter produce a second generation of rediae which in turn produce **cercariae**. Cercaria leaves snail and encyst on some aquatic plant as **metacercaria**.	26. Hexacanth loses its protective envelope, bores into blood vessels of gut and migrates to voluntary muscles, where it encysts as a **cysticercus**. Thus, a complicated series of larval stages do not occur.
	27. Sheep acquires infection by ingesting **metacercariae** alongwith aquatic vegetation.	27. Man acquires infection by eating inadequately cooked measly pork containing **cysticercus**.
	28. Disease caused in sheep is called '**Liver rot**' or '**Fascioliasis**'.	28. Disease caused in man is called '**Taenniasis**'.

food of the host through its body surface. Microvilli of tegument also increase the absorptive surface.

(6) Circulatory, respiratory and sensory organs are absent. Nervous system poorly developed.

(7) Respiration is anaerobic as free O_2 is not available.

(8) Reproductive system is immensely developed. Production of very large number of eggs (40,000 per gravid proglottid) suitably faces many challanges to survival of species.

(9) Resistent covering, shell or capsule around zygotes and embryos provides protection from unfavourable conditions.

(10) Hermaphroditism and proglottization ensures self-fertilization within the same proglottid or cross-fertilization with another proglottid, in the same worm.

CESTODIASIS

Cestodiasis is the disease in man caused by infection due to tapeworms. The scientific study of the occurrence (distribution), transmission, frequency and control of infectious diseases in man is called **epidemiology**.

1. Infection. Human beings get infected by eating raw and inadequately cooked measly pork containing cysticerci of tapeworm.

2. Pathogenesis (symptoms). Man may fall a victim to adult tapeworm as well as its cysticerci; effects of the former are referred to as **taeniasis** and those of the latter as **cysticercosis**.

(a) Taeniasis. It is indicated by a variety of symptoms including pain in the abdomen, nausea, anaemia, increased appetite, indigestion, increase of eosinophil cells in blood, and above all, nervous disorders of the type occurring in epilepsy. These serious disorders are caused by toxins produced by the parasite and not, as believed earlier, by its continued drain on the host's digested food. Hooks and suckers may cause mechanical irritation in intestine, which may initiate reverse peristalsis leading to auto-infection. Usually a single tapeworm is found to parasitize a host. This is because the presence of one tapeworm provides a kind of **immunity** or **premunition** to the host against fresh infection.

(b) Cysticercosis. It is far more dangerous than taeniasis. Encystment of bladder may take place in the host's voluntary muscles, cardiac muscles and even in some more delicate vital organs like liver, eyes and brain. Removal of cysticerci from these delicate tissues is extremely difficult. Cysticercosis of the brain results in several degenerative changes and necrosis in the brain and the patient shows epileptic behaviour.

3. Therapy (treatment). Infection of tapeworm can be tackled by several anti-helminth drugs. Drug for human tapeworms is **Praziquantel** (Biltricide). Drug for neurocysticercosis is **Albendazole** (Zentel). Under the action of drugs, strobila is removed but the embedded scolex persists which again buds off a new strobila. Removal of scolex can be brought about by surgery. Removal of cysticerci, especially from delicate organs like brain, eyes and liver is extremely difficult.

4. Prophylaxis (prevention). Consumption of undercooked measly pork should be avoided. Faeces of infected persons should be properly disposed off and destroyed, preventing, pigs having access to them and ingesting hexacanth embryos.

5. Diagnostic tests. Taeniasis (Tapeworm disease) can be diagnosed by recovering eggs or gravid proglottids from the faeces of the infected human host. Since *T. solium* and *T. saginata* are morphologically very similar, identification at the species level can be made by the number of lateral uterine branches in the gravid proglottids and differing scolexes.

Diagnosis of Cystecercosis can be done by confirming the presence of hooks of the scolex in case *T. solium* under the microscope. Calcified larvae in the muscle tissues can be identified by X-ray examination. Brain lessions can be detected by CT scan and MRI. However diagnosis of neurocystecercosis is difficult, often times doctors rely on the presence of brain tumors.

Important Questions

➤➤*Long answer type questions*

1. Describe the anatomy of *Taenia solium* with special reference to its parasitic mode of life.
2. Give an account of the reproductive organs of *Taenia.* What is meant by a gravid proglottid ?
3. Give an account of the life cycle of *Taenia solium* highlighting its characteristic features.
4. What is parasitism ? Give an account of the parasitic adaptations of *Taenia.* Also write a paragraph on its economic importance.
5. Compare and contrast the structure and life history of *Fasciola hepatica* and *Taenia solium.*
6. Make full-page, well-labelled diagrams of the following. No description is required.
   (a) Section of body wall of *Taenia* based on electron microscope.
   (b) T.S. mature proglottid of *Taenia* through ootype.
   (c) A mature proglottid of *Taenia* showing reproductive organs.
   (d) Stages in development of *Taenia.*
7. Write short notes on : (i) Apolysis, (ii) Bladder worm, (iii) Cercaria, (iv) Cysticercus, (v) Gravid proglottid. (vi) Hexacanth, (vii) Measly pork, (viii) Redia, (ix) Scolex of *Taenia solium,* (x) Sporocyst.

➤➤*Short answer type questions*

1. In *Taenia solium,* write the name of the stage which pierce the mucous membrane of the intestine and enters the blood stream ?
2. Where can you locate flame cell ?
3. What are flame cells ?
4. What is the name of the organ of attachment in tape worm ?
5. Describe with diagram the internal organization of a proglottid of *Taenia.*
6. Describe strobilation.
7. Mention five possible methods of eradicating tapeworm at different stages.
8. Write descriptive note on hydatid cyst.
9. Give a description of the head of *Taenia.*
10. Describe briefly the shell gland.
11. Discuss whether *Taenia solium* is metamerically segmented.
12. Give an account of the life cycle of *Taenia.*
13. Describe briefly *Taenia.*
14. Draw a labelled diagram of the mature proglottid of *Taenia solium.*
15. Rewrite the following in order :
    (a) Cysticercus (b) Proglottid (c) Onchosphere (d) *Taenia*

➤➤*Multiple choice questions*

1. The smallest tapeworm is :
   (a) *Taenia* (b) *Moniezia*
   (c) *Echinococcus* (d) *Schistosoma*
2. In *Taenia,* cross fertilization between different proglottids of the same individual occurs and this is called :
   (a) self-fertilization (b) cross-fertilization
   (c) incestuosus (d) apolysis
3. Tapeworms with undivided body, without hooked scolex, and larva are included in subclass :
   (a) Eucestoda (b) Monogenea
   (c) Digenea (d) Cestodaria
4. *Taenia* belongs to order :
   (a) Cyclophyllidea (b) Pseudophyllidea
   (c) Amphilinidea (d) Echinostomida
5. A proglottid is called gravid proglottid when it has :
   (a) both male and female reproductive units well developed
   (b) only female reproductive unit well developed
   (c) only male reproductive unit well developed
   (d) branched uterus filled with fertilized eggs
6. In *Taenia,* scolex bears in the middle a prominent :
   (a) cup shaped sucker (b) head
   (c) rostellum (d) recurved hooks
7. In *Taenia solium* the proximal portion of the oviduct that leads into the vagina is called :
   (a) receptaculum semines
   (b) uterus
   (c) fertilization canal
   (d) vitellarium
8. Who are more likely to get tape-worm ?
   (a) vegetarians (b) non-vegetarians
   (c) both (d) none
9. Bladder worm is found in :
   (a) human muscles (b) muscles of cow or pig
   (c) human faeces (d) soil
10. Tape worm infesting intestine of animals obtain their nutrition :

(a) by ingesting food particles through their suckers
(b) by scraping the intestinal walls with their hooks
(c) by preparing food in their own bodies
(d) by absorption of liquid food through their integument

11. A gravid proglottid of *Taenia* has :
(a) male and female reproductive system
(b) well developed female reproductive system and degenerate male reproductive system
(c) well developed male reproductive system and degenerate female reproductive system
(d) only branched uterus

12. Cysticercosis is a disease caused by accidental infection of :
(a) hexacanth of *Taenia*
(b) bladder worm of *Taenia*
(c) encysted cercaria of *Schistosoma*
(d) eggs of *Enterobius*

13. Area for proliferation in body of *Taenia* is :
(a) scolex (b) neck
(c) strobila (d) gravid proglottid

14. Apolysis is a process in *Taenia* for :
(a) regular detachment of proglottid from strobila
(b) embryonic development
(c) ingestion (d) infection to man

15. Suckers in tape worm are meant for :
(a) attachment to intestinal wall
(b) ingestion of food
(c) attachment during locomotion
(d) all these

16. Secondary host of *Taenia* is :
(a) pig (b) man
(c) snail (d) sheep

17. *Taenia* lack of :
(a) digestive system (b) reproductive system
(c) excretory system (d) all of these

18. The excretory product of *Taenia :*
(a) ammonia (b) fatty acid
(c) urea (d) ammonia and fatty acid

19. Larval forms of *Taenia :*
(a) oncosphere (b) hexacanth
(c) cysticercus (d) all

20. In which forms *Taenia* pass through feacal matter of man :
(a) oncosphere (b) hexacanth
(c) cysticercus (d) young tape worm

21. Pig acquired infection of *Taenia* by ingestion of :
(a) oncosphere (b) hexacanth
(c) cysticercus (d) none of above

22. Cysticercus in pig muscle can remain for :
(a) two year (b) four year
(c) seven year (d) ten year

23. *Taenia saginata* is commonly known as :
(a) pork tapeworm (b) beef tapeworm
(c) human tapeworm (d) none of them

24. Unarmed tapeworm is :
(a) *T. saginata* (b) *T. solium*
(c) none

25. The encystment of bladder worm occurs in man is :
(a) caretiac muscle (b) liver
(c) eye (d) brain
(e) all

26. *Taenia* is found in :
(a) large intestine (b) small intestine
(c) rectum (d) duodenum

27. *Taenia* does not show locomotion because it has
(a) suckers for attachment
(b) no digestive system
(c) suckers and hooks for attachment
(d) complete set of male and female reproductive organ in each segment

28. Shedding of proglottid in tapeworm is called :
(a) *histolytica* (b) topolysis
(c) apolysis (d) detachment

29. Alimentary canal in tapeworm is :
(a) straight (b) coiled
(d) branched (d) absent

30. Tapeworm has no alimentary canal because :
(a) it does not require nutrients
(b) it feeds only when young
(c) it absorbs nutrients from body surfaces
(d) none of these

31. Copulation in *Taenia* occur between :
(a) two *Taenia*
(b) two proglottids
(c) male and female *Taenia*
(d) two mature proglottid of different species

32. Indicate the correct statement pertaining to *Taenia* :
(a) only anterior proglottids contain male organs, whereas ripe proglottids contain only uterus filled with eggs
(b) male organs are confirmed to posterior proglottids
(c) all proglottids contain both male and female organs
(d) female organs are confined to anterior proglottids

Answers

➤➤*Multiple choice questions*

1. (c)	2. (a)	3. (d)	4. (a)	5. (a)	6. (c)	7. (a)	8. (b)	9. (b)	10. (d)
11. (c)	12. (b)	13. (b)	14. (a)	15. (a)	16. (a)	17. (a)	18. (d)	19. (d)	20. (a)
21. (a)	22. (c)	23. (b)	24. (a)	25. (d)	26. (b)	27. (c)	28. (c)	29. (d)	30. (c)
31. (b)	32. (c).								

33

Platyhelminthes : Characters, Classification and Types

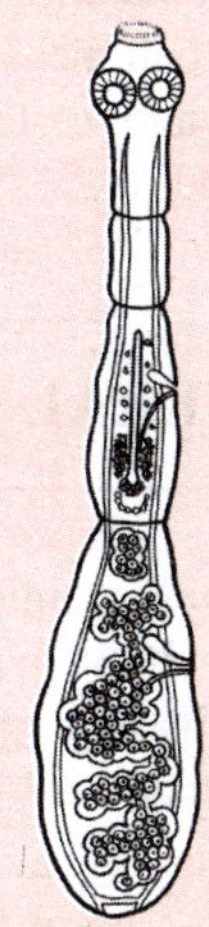

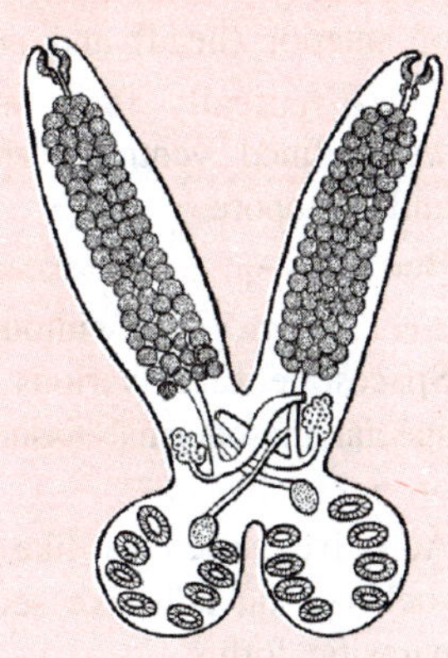

DEFINITION

The term **Platyhelminthes** (Gr., **platys, flat+helmins**, worm) was first proposed by **Gegenbaur** (1859) for the flatworms. These may be defined as acoelomate, triploblastic, bilaterally symmetrical, vermiform, dorso-ventrally flattened organisms; devoid of a definite anus; without skeletal, respiratory and circulatory systems; with flame cells or protonephridia and with mesenchymal cells representing mesoderm, filling up spaces between various organs. Its 29,488 species are known.

MORPHOLOGICAL ADVANCES

Platyhelminthes shows an advancement over the previous metazoan phyla (coelenterata and ctenophora) in the following respects :

1. Tissue-organ level of organization.
2. Triploblastic body. Third germ layer mesoderm instead of mesogloea.
3. Symmetry definitely bilateral, at the lowest level.
4. A definite head and tail.
5. Mesodermal muscles in bundles and layers.
6. More complex and efficient gastrovascular cavity.
7. Nervous system with large, anteriorly situated ganglia and nerve cords running along the body.
8. Better developed sensory organs.

9. Internally situated gonads with permanent gonoducts and copulatory structures.
10. Behaviour organization through learning.

PHYLOGENETIC SIGNIFICANCE

Platyhelminthes or flatworms are supposed to have evolved from a coelenterate-like ancestor, which underwent a change from radial to bilateral symmetry. Coelenterate ancestor was like a **planula larva** as primitive acoel worms share many characteristics with planula, such as : (i) absence of epidermal basement membrane, (ii) absence of gut, (iii) absence of excretory organs, (iv) absence of gonads, (v) a centro-ventrally located mouth, and (vi) a nerve plexus under epidermis. Trematodes are thought to have evolved from some commensal or parasitic rhabdocoel turbellarians and cestodes evolved independently from trematodes.

GENERAL CHARACTERS

1. Free-living, commensal or parasitic forms.
2. **Tissue-organ grade of organization**, *i.e.*, body cells aggregate into definite tissues and tissues make up organs.
3. **Triploblastic**, *i.e.*, body derived from three embryonic germ layers; ectoderm, mesoderm and endoderm.
4. **Bilaterally symmetrical** with definite polarity of anterior (head) and posterior (tail) ends.
5. **Dorso-ventrally flattened.** Usually with a well-defined ventral surface bearing mouth and gonopore.
6. Body **unsegmented** (except in class Cestoda).
7. **Acoelomate**, *i.e.*, without any body cavity. Spaces between various organs filled with special mesodermal tissue, the **mesenchyme** or **parenchyma**.
8. Adhesive structures like hooks, spines and suckers, and adhesive secretions common in parasitic forms.

9. **Epidermis** cellular or syncytial, frequently ciliated. Absent in some.
10. **Muscular system** of mesodermal origin. Longitudinal, circular and oblique muscle layers beneath epidermis.
11. **Digestive system** branched and incomplete without anus. Altogether absent in Acoela and Cestoda.
12. Skeletal, respiratory and circulatory systems are wanting.
13. **Excretory system** includes lateral canals and protonephridia (flame cells). Absent in some primitive forms.
14. **Nervous system** primitive, ladder-like. Comprises a pair of anterior ganglia with longitudinal nerve cords connected by transverse nerves.
15. **Sense organs** simple. Eye-spots or photo-receptors in free living forms.
16. Mostly **monoecious** (hermaphrodite) with complex reproductive system. Well-developed gonads, gonoducts and accessory organs. Eggs mostly devoid of yolk. Yolk produced separately in **yolk** or **vitelline glands**.
17. **Fertilization** internal, may be cross or self.
18. **Development** direct or indirect. Usually indirect in endoparasites with a complicated life cycle involving many larvae and hosts.

CLASSIFICATION

CLASS 1. Turbellaria

(L., *turbella,* a stirring)

1. Usually non-parasitic, free-living worms are called **planarians**.
2. Terrestrial marine or freshwater.
3. Body unsegmented, flattened and covered with ciliated cellular or syncytial epidermis, containing mucus secreting cells and rod-shaped bodies called **rhabdites**.
4. Mouth ventral. Intestine preceded by muscular pharynx.
5. Suckers absent. Tango , chemo and photo-receptors common in free-living forms.
6. Mostly hermaphroditic. Some reproduce asexually. Development usually direct.

Order 1. Acoela

1. Minute, exclusively marine, less than 2 mm.
2. Ventral mouth; no muscular pharynx and without intestine.
3. Flame cells, definite gonads, gonoducts and yolk glands wanting.
4. Mostly free-living, found under stones, algae or on bottom mud. Some dwell in intestine of sea-urchins and sea-cucumbers.

Examples : *Convoluta, Amphiscolops, Ectocotyl, Afronta.*

Order 2. Rhabdocoela

1. Small, usually less than 3 mm.
2. Simple pharynx and sac-like intestine.
3. Protonephridial excretory system.
4. One or two gonads. Yolk glands present or absent.
5. Marine, freshwater or terrestrial. Free-living, commensal or parasitic.

Examples : *Stenostomum, Microstomum, Actinodactyletta, Catenula, Macrostomum, Mesostoma.*

Order 3. Alloeocoela

1. Moderate-sized, between 1 and 10 mm.
2. Pharynx simple, bulbous or plicate. Intestine straight or branched.
3. Protonephridia paired, usually branched.
4. Testes numerous. Penis papilla mostly present.
5. Mostly marine, common in littoral sand and mud. Some freshwater.

Examples : *Prorhynchus, Plagiostomum, Geocentrophora.*

Order 4. Tricladida

1. Large 2 to 60 cm in length.
2. Mouth mid-ventral, pharynx plicate and intestine with three branches, each with many diverticula.
3. Protonephridia as lateral networks with many nephridiopores.
4. Testes numerous, ovaries two. Yolk glands present.
5. Marine, freshwater or terrestrial.

Examples : *Dugesia, Gunda, Bdelloura, Geoplana, Bipalium, Euplanaria.*

Order 5. Polycladida

1. Moderate-sized, 2 to 20 mm.
2. Pharynx plicate. Intestine highly branched.
3. Gonads many, scattered. Yolk glands absent.
4. Male and female gonopores separate.
5. Marine, many bottom dwellers of littoral zone.

Examples : *Leptoplana, Notoplana, Cestoplana, Planocera, Thysanozoon.*

CLASS 2. Trematoda

(Gr., **tremta, hole+eidos**, form)

1. Ecto-or endoparasitic flatworms, called **flukes**.
2. Body unsegmented, dorso-ventrally flattened, leaf-like. Tegument thick but without cilia and rhabdites.
3. Suckers and sometimes hooks present.
4. Alimentary canal with anterior mouth, simple pharynx and two main branches.
5. Three pairs of longitudinal nerve cords.
6. Mostly monoecious. Development direct (in ectoparasites) or indirect (in endoparasites) with alternation of hosts.

Order 1. Monogenea

1. Mostly ectoparasites in cold blooded aquatic vertebrates.
2. Posterior adhesive organ or opisthaptor with suckers armed with hooks or spines.
3. Excretory pores two, situated anteriorly on dorsal side.
4. Vagina one or two. Uterus small with a few shelled eggs.
5. Only a single host in life cycle.
6. Free-swimming ciliated larva called **onchomiracidium**.

Examples : *Gyrodactylus, Dactylogyrus, Polystoma, Diplozoon.*

Order 2. Digenea

1. Mostly endoparasites in vertebrates and invertebrates.
2. Two suckers, oral and acetabulum, both devoid of hooks.
3. Single posterior excretory pore.
4. No vagina. Uterus long with numerous shelled eggs.
5. Life cycle complex with numerous larval stages in two to three intermediate hosts.

6. Larval forms reproduce asexually before metamorphosis.
   Examples : *Bucephalus, Fasciola, Paramphistomum, Paragonimus, Schistosoma, Opisthorchis* (= *Clonorchis*).

Order 3. Aspidocotylea (= Aspidogastraea)

1. Oral sucker absent.
2. Large ventral sucker subdivided into several suckers without hooks.
3. Only one testis in male system.
4. Endoparasites in gut of fishes and reptiles.
   Examples : *Aspidogaster, Cotylapsis, Stichocotyle.*

CLASS 3. Cestoda (= Cestoidea)

(Gr., **kestos**, girdle + **eidos**, form)

1. Endoparasitic flatworms, called tapeworms.
2. Body segmented, elongate, flat, ribbon-like. Tegument with microvilli.
3. Scolex (head) with suckers, or hooks, or both.
4. No alimentary canal. No sense organs.
5. Each mature segment or proglottid monoecious, with male and female organs.
6. Life cycle complicated involving one or more intermediate hosts. Embryos with hooks.

Subclass A. Cestodaria

1. Body unsegmented, leaf-like, without scolex and strobila (**monozoic**).
2. Only one set of monoecious reproductive system.
3. Larva **lycophore**, with 10 hooks.

Order 1. Amphilinidea

1. No suckers. Pharynx protrusible.
2. Male genital pore and vagina situated posteriorly. Uterus coiled.
3. Endoparasitic in coelom of primitive fishes.
   Example : *Amphilina.*

Order 2. Gyrocotylidea

1. An anterior sucker and a posterior rosette-shaped adhesive organ present.
2. Eversible proboscis at the anterior end.
3. Endoparasites in chimaeroid fishes.
   Example : *Gyrocotyle.*

Subclass B. Eucestoda

1. Body long, ribbon-like. Divided into scolex, neck and strobila with many proglottids (**polyzoic**).
2. Mostly with several sets of monoecious reproductive organs.
3. Larvae with six hooks.

Order 1. Proteocephalidea

1. Scolex with 4 cup-shaped suckers.
2. Ovary bilobed. Uterus branched. Vitellaria scattered.
3. Parasitic in freshwater fishes, amphibians and reptiles.
   Examples : *Proteocephalus, Ophiotaenia.*

Order 2. Tetraphyllidea

1. Scolex with 4 leaf-like **bothria**.
2. Testes anterior to ovaries. Vitelline glands scattered.
3. Parasitic in intestine of elasmobranch fishes.
   Example : *Phyllobothrium.*

Order 3. Disculicepitidea

1. Scolex with large cushion-like pad at anterior end.
2. Female gonopore, anterior to male gonopore. Testes numerous. Uterus lobed.
3. Endoparasites of Selachii.
   Example : *Disculiceps.*

Order 4. Lecanicephaloidea

1. Scolex divided by a transverse groove. Upper disc-like lower with 4 suckers.
2. Vitellaria as two lateral bands.
3. Intestinal parasites in elasmobranch fishes.
   Examples : *Lecanicephalum, Tetragonocephalum.*

Order 5. Pseudophyllidea

1. Scolex with 2 to 6 bothria.
2. Testes numerous. Ovary bilobed. Vitellaria follicular.

3. Parasitic in freshwater fishes (teleosts).
   Examples : *Dibothriocephalus, Haplobothrium.*

Order 6. Trypanorhyncha

1. Scolex with 2 to 4 bothria and 4 spiny tentacles.
2. Vitellaria in continuous layer in cortical parenchyma.
3. Parasitic in elasmobranch fishes.
   Example : *Tetrahynchus.*

Order 7. Cyclophyllidea (= Taenioidea)

1. Scolex with 4 large deep suckers and hooks.
2. Ovary lobed. Uterus blind. Vitellaria follicular.
3. Parasitic in amphibians, reptiles, birds and mammals.
   Examples : *Taenia, Moniezia, Echinococcus, Dipylidium, Hymenolepsis.*

Order 8. Aporidea

1. Scolex with 4 suckers. Rostellum armed.
2. No external segmentation.
3. Ootype absent. Vitallaria absent or present. Sex ducts and genital apertures absent.
4. Parasites in birds.
   Examples : *Nematoparataenia, Gastrotaenia.*

Order 9. Nippotaeniidea

1. No scolex but well-developed terminal sucker.
2. Proglottids few. Vitellaria few.
3. Parasites in freshwater fishes of Japan.
   Examples : *Nippotaenia, Amurotaenia.*

Order 10. Caryophyllidea

1. Scolex without true suckers or bothria.
2. Eggs non-embryonated when laid.
3. Parasites in fishes.
   Examples : *Caryophyllaeus, Archigetes, Glaridacris.*

Order 11. Spathebothridea

1. Scolex without suckers or bothria.
2. Testes are medullary. Ovary median.
3. Parasites in primitive fishes.
   Example : *Spathebothrium.*

OTHER TYPES OF PLATYHELMINTHES

1. *Convoluta.* *Convoluta* is the most common acoel turbellarian. It is a minute and marine creature of gregarious habit, found mostly among sea-weeds between tidemarks. Groups of pigment cells impart characteristic patches to the body surface. Some species are cylindrical. *Convoluta* is so-called because during locomotion, its anterior end, with out-turned margins, looks more or less like a horn or **cornucopia** after the horn of Amalthea, the goat that suckled Zeus, the supreme God of Greek mythology. For feeding, it protudes its pharynx through **mouth**, situated ventrally near head. Five pairs of nerve cords, interconnected by cross strands, are present. It is protandrously hermaphroditic. Male and female genital apertures are situated close to each other about the middle of body on ventral side. Gonads are not differentiated. Some species are associated

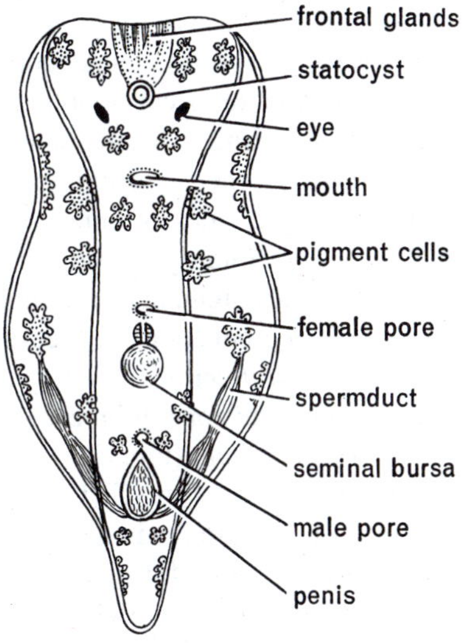

Fig. 1. *Convoluta.*

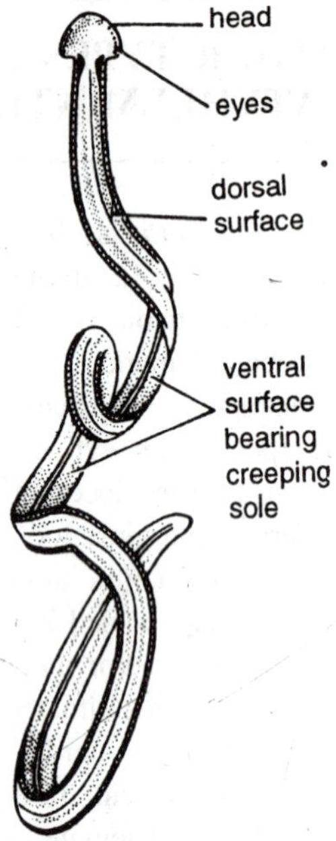

Fig. 2. *Bipalium.*

symbiotically with green or brown algae dwelling within them.

2. *Bipalium.* This genus is a triclad turbellarian and includes terrestrial species, found commonly in greenhouses. Distribution is cosmopolitan. Body is much elongated attaining a length of up to 50 cm. Head is expanded. There are numerous eyes along head's margin and sides of anterior part of body. Dorsal surface is variously striped and ventral surface bears a median longitudinal creeping sole. General mode of reproduction is asexual by fragmentation. *B. adventitium* is a North American species living in temperate zone. It breeds sexually. *B. kewense* is cosmopolitan (Fig. 2).

3. *Thysanozoon.* It is a polyclad turbellarian found in cool waters. Body is elongate and somewhat oval with a pair of anteriorly-situated marginal tentacles bearing eyes. Behind tentacles are numerous cerebral eyes, behind which is the ventral mouth leading into a ruffled pharynx. Intestine is highly branched, each branch extending into a projection of the body wall, known as papilla or tubercle. This is a characteristic feature of this animal. Two male genital apertures are situated near about the middle of body behind pharynx. A single female pore is situated behind

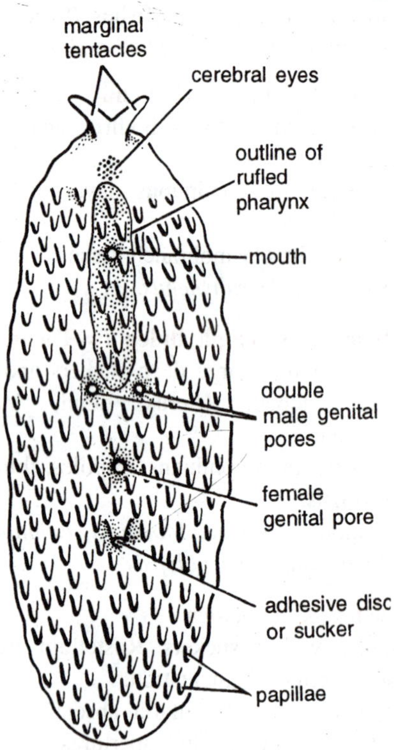

Fig. 3. *Thysanozoon.*

the male pores and in front of an adhesive disc or 'sucker', which is helpful in holding the two individuals together during copulation.

4. *Diplozoon.* It is a trematode flatworm, ectoparasitic on the gills of some freshwater fishes. Most peculiar feature of *Diplozoon* is that, during life history, two larvae (**diporpa larvae**) fuse, both morphologically and physiologically, and remain as such throughout life. Fusion takes place in the region of genital atria of the two, situated a little behind the middle of body. The pair, which appears like an X, thus seems to be engaged in a permanent copulation.

Each individual bears an **oral sucker** at the anterior extremity and a pair of **lateral suckers** just behind. There are 4 pairs of suckers, arranged in two rows, on **opisthaptor** or posterior part of each individual. Mouth is surrounded by the oral sucker. **Pharynx** is muscular. **Intestine** is not forked but has numerous diverticula. **Testis** is

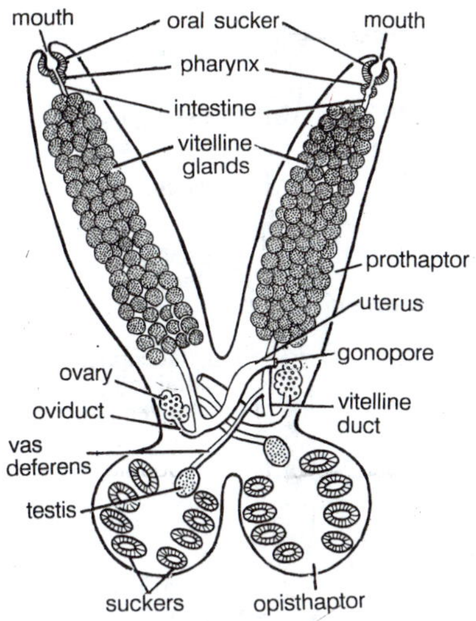

Fig. 4. *Diplozoon.* Two individuals.

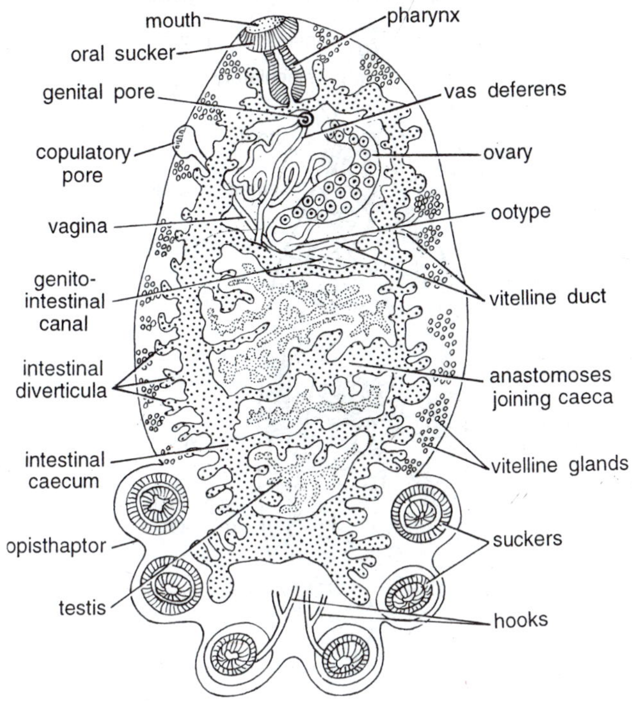

Fig. 5. *Polystomum.*

single, compact and situated posteriorly. A **vas deferens** arises from testis and crosses over to join the vitelline duct of other individual of the pair. **Ovary** is single, loop-like and situated in front of testis. **Oviduct** joins the vitelline duct of the same individual. **Uterus** is surrounded by Mehlis glands and contains a single egg. Genital ducts of one individual cross with those of the other of the pair. Vitelline glands are in the form of numerous follicles filling up the middle of body in front of the point of fusion.

5. Polystomum. It is a monogenean trematode endoparasite in the urinary bladder of frogs, toads and turtles. Body is flat and leaf like. Posterior end of body or opisthaptor contains 3 pairs of suckers, arranged in a circle and 2 to 4 large chitinous hooks. Mouth surrounded by an oral sucker, leads into pharynx. Intestine is divided into two main limbs with diverticula, and connected by transverse connectives. A nephridiopore opens on either side of mouth. *Polystomum* is bisexual. Gonopore opens ventrally a little behind mouth. Several testes are located in the posterior region of body. Ovary is single, having a short oviduct, which communicates with the right intestinal caecum by a short **genito-intestinal canal** of unknown function. There are two vaginae, which open internally into a transverse vitelline duct and externally or lateral funnel-like prominences, one on either side near the anterior end of body.

Life cycle of *Polystomum* shows a remarkable synchronization with that of the amphibian host (frog). Eggs produced in the bladder of frog are passed into water along with the faeces of host. Embryo hatches in the form of a larva called **onchomiracidium**. It has four eyes, cilia in transverse bands, two protonephridia and a disc-shaped opisthaptor with 16 hooklets. Larva attaches to the gills of a tadpole. When tadpole metamorphoses, larva leaves the gill chamber and reaches the bladder and develops to adult stage. Eyes and cilia are lost. Neoteny has been reported in some cases of *Polystomum* larva. If larva attaches to a young tadpole, then it may become sexually mature and lays eggs. But this ectoparasitic generation dies when metamorphosis occurs.

6. Opisthorchis (=Clonorchis) sinensis. *Opisthorchis* (= *Clonorchis*) *sinensis* is the common liver fluke of man. It is also known as the 'Chinese Liver Fluke' for its common

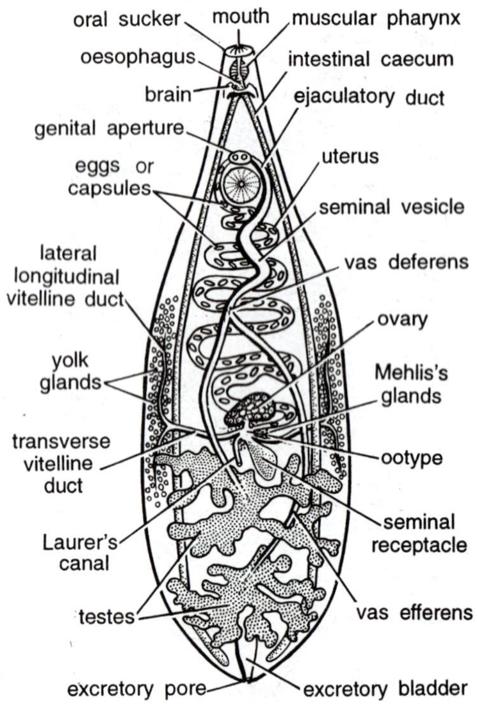

Fig. 6. *Opisthorchis (=Clonorchis) sinensis.*

occurrence in China and the neighbouring countries including India. It is smaller and narrower than *Fasciola* which it resembles morphologically. Adults, in thousands, inhabit bile ducts and sometimes pancreatic ducts and duodenum of man. They have been reported from pig and dog also. Life cycle involves two intermediate hosts, a freshwater snail and a freshwater fish.

Eggs hatch when ingested by a snail of the genus *Bythinia*. Earlier larval stages **(Miracidia, sporocysts** and **Rediae)** are passed in snail. **Cercariae** leave the snail host and bore into a freshwater fish-carp or minnow, in which they encyst as **metacercariae**. Final host (man) acquires infection by consuming infected and uncooked fish.

Mild infections often produce no symptoms but heavy infections may cause diarrhoea, enlarged liver, abdominal discomforts, jaundice and reduced immunity to bacterial infection. The disease is termed **clonorchiasis**. Drugs commonly employed for treatment are coated tablets of gentian violet, chloroquine and antimony compounds.

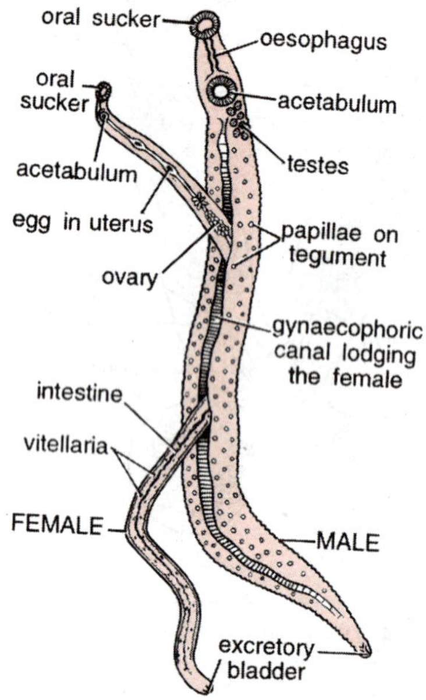

Fig. 7. *Schistosoma.* Male and female.

7. *Schistosoma.* Blood-flukes of man and other animals are placed in the genus *Schistosoma* (Gr., **schistos**, divided + **soma**, body).

Schistosoma or *Bilharzia* was discovered by **Bilharz** in 1851. Three species are known to infect man : (i) *S. haematobium* in blood vessels of bladder and urinary tract, (ii) *S. mansoni* in small branches of mesenteric and portal veins, and (iii) *S. japonicum* in small veins of portal and mesenteric vessels.

Unlike most other flukes, *Schistosoma* is **dioecious** with well-defined **sexual dimorphism** and the two intestinal limbs become united into one at the posterior end of body. Female is more slender and cylindrical and lies permanently lodged in the **gynaecophoric canal** of male formed by the rolling of sides of body behind the ventral sucker.

Life cycle involves a single intermediate snail host and in general resembles that of other flukes, except that the **cercariae** infact the final host (man) by directly penetrating his skin.

Infection of *Schistosoma* causes various symptoms of allergy like itch, rash, aches, fever, eosinophilia, etc. When infection is heavy, eggs may be deposited in arterioles of lungs causing **cardio-pulmonary schistosomiasis** which may

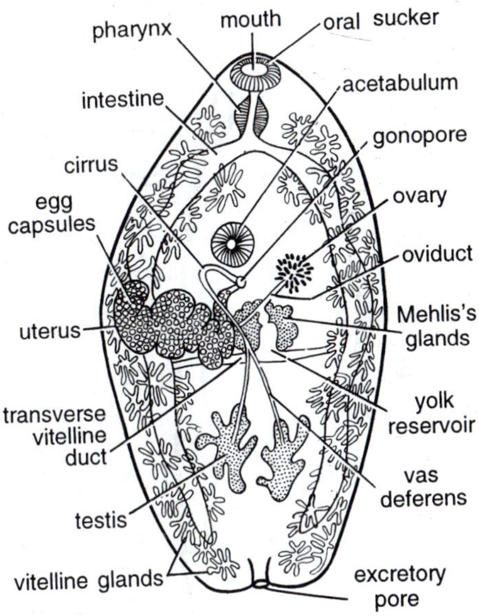

Fig. 8. *Paragonimus.*

lead to congestive heart failure. **Schistosomiasis** is treated with antimony drugs, such as tartar emetic, fuadin and anthiomaline.

8. *Paragonimus.* Flukes of this genus *Paragonimus* parasitize lungs of carnivorous mammals including man, tiger, dog, pig, rat, goat, etc. They are commonly called 'lung flukes'. They have been reported from India, China, Japan, Philippines, New Guinea, Africa, etc. First species to be described was *P. westermanni* from Bengal tigers. First human species obtained was named *P. ringeri.* It is still uncertain whether the two species named above are distinct or the same species.

Adult lung-fluke is oval in shape, reddish brown in colour, about 8-12 mm long and 4-6 mm in diameter. General plan organization is more or less the same as that of *Fasciola.*

Adult flukes live in cyst-like pockets near the surface of lung. When these pockets rupture, eggs are liberated which reach the exterior with the host's sputum or excreta. Under suitable conditions, eggs hatch into miracidia into surrounding water. Life cycle involves two intermediate hosts. First intermediate host is a suitable amphibious snail into which miracidia enter. Earlier larval stages are passed in it. Cercariae leaving snail encyst in the second intermediate host, usually a freshwater crab or crayfish. Man (primary host) gets infection by consuming the undercooked second intermediate host.

Effects of *Paragonimus* infection on human host are not very serious but can sometimes lead to tuberculosis. Common symptoms of infection are intermittent cough, streaks of blood in sputum, anemia, slight temperature and fatigue.

9. *Moniezia.* The sheep tapeworm includes about 13 species of intestinal parasites of ruminants (sheep, goat, cattle, etc.) and other mammals. *Moniezia expansa,* perhaps the most common species attains a length of 3 to 6 meters. Small scolex has well-developed suckers but is without rostellum and hooks. Proglottids are much wider than their length and each has a double set of reproductive organs. Testes are distributed throughout the central region of proglottid. Ovaries are fan-like. Uteri of two sides form a meshwork. Vagina lies ventral to the cirrus sac on one side

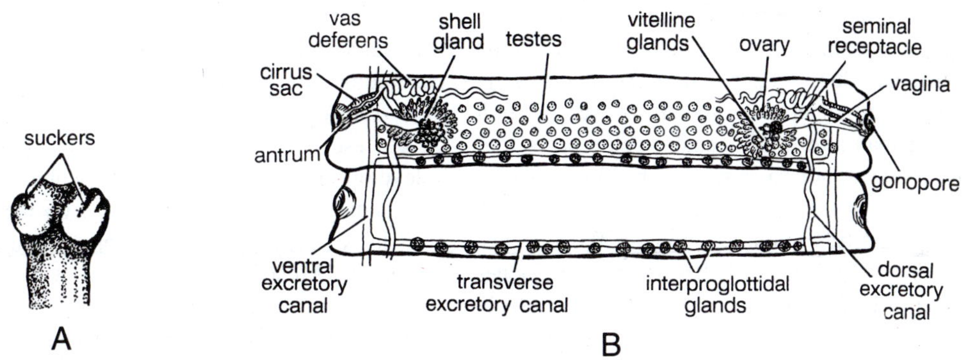

Fig. 9. *Moniezia.* **A.** Scolex. **B.** Two proglottids.

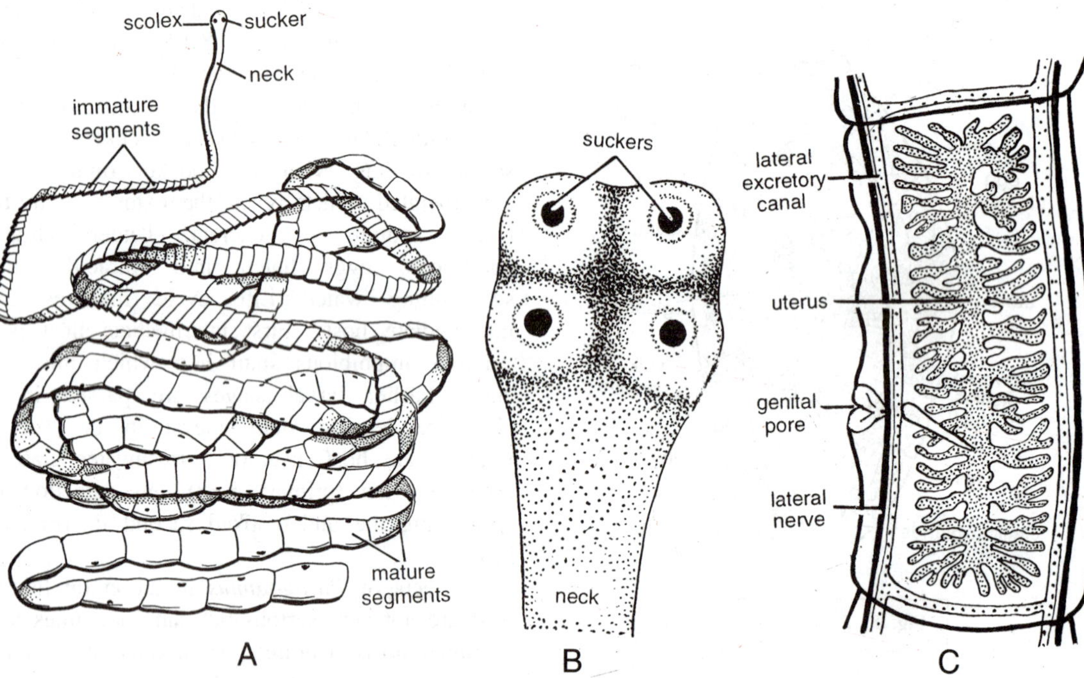

Fig. 10. *Taenia saginata.* **A.** Entire. **B.** Scolex. **C.** Gravid proglottid.

and dorsal to it on the other side. Intermediate host is a non-parasitic mite *(Galumna).* Ingested eggs develop into cysticercoids. These mites live about the roots of grass and are often ingested by the final host (sheep or goat) when the latter comes to graze.

10. *Taenia saginata.* The beef tapeworm *Taenia saginata* is similar to the pork tapeworm *Taenia solium,* in structure and life history. It is the commonest tapeworm of man with a much greater incidence than that of *T. solium.* It enjoys a cosmopolitan distribution infecting the beef-eating population. Its intermediate hosts are cattle and buffaloes. It is longer than *T. solium,* usually attaining a length upto 12 meters or more. Adult lives in human alimentary canal. Scolex bears four strong, rounded, adhesive suckers but lacks hooks. Strobila comprises up to 2,000 proglottids. A gravid proglottid contains about 100,000 eggs. Uterus of gravid proglottids has 15 to 35 branches on either side. Proglottids are usually shed singly with faeces and disintegrate. Scattered eggs on soil, grass, etc., are swallowed by grazing cattle. Eggs hatch in host's intestine freeing 6-hooked onchospheres which encyst in muscles forming cysticerci or bladder worms. Man is infected by eating such measly meat.

The dog tapeworm *T. pisiformis* is widely used as a type for study in the laboratory.

11. *Echinococcus granulosus.* The hydatid worm *Echinococcus granulosus* may be dangerous to man. Adult worms, *(Taenia echinococcus),* in hundreds or even thousands, parasitize the intestine of dog, wolf, fox, cat, etc. Man sometimes serves as an accidental final host. Adult measures 3-6 mm in length. It is made of scolex and 3 to 4 proglottids. Scolex has 4 suckers, rostellum and a double row of hooks.

Life cycle involves a single intermediate host. Eggs escape with the faeces of primary host and develop into **onchospheres**, which infect the secondary host (man and cattle) through his food or drink or careless association with dogs, sheep, goat, etc. Most harm is done by **cysticercus** larva. In liver or lung and in a few cases in brain, heart, bone, kidney, spleen, muscle, etc., the larva develops into a watery bladder or **hydatid vesicle** or **hydatid cyst**, having a two-layered cyst wall.

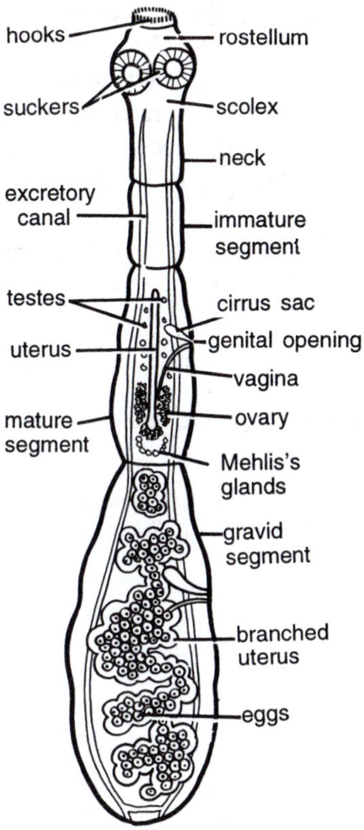

Fig. 11. *Echinococcus granulosus.*

The **cyst** produces by asexual budding, **exogenous** and **endogenous** buds containing brood chambers and many internal scolices. A hydatid cyst may grow upto football size in 12 to 20 years in man, necessitating surgery. Dogs acquire infection when they feed upon slaughtered bodies of infected secondary hosts.

Large hydatid cysts cause inflammation, diarrhoea, abdominal pains, vomiting, eosinophilia, etc. Presence of cysts in brain, eye or-kidney may prove fatal. Brood capsules and scolices can be killed by injecting formalin solution into the cysts. Removal of adult from the dog's intestine can be affected by Areclin hydrobromide.

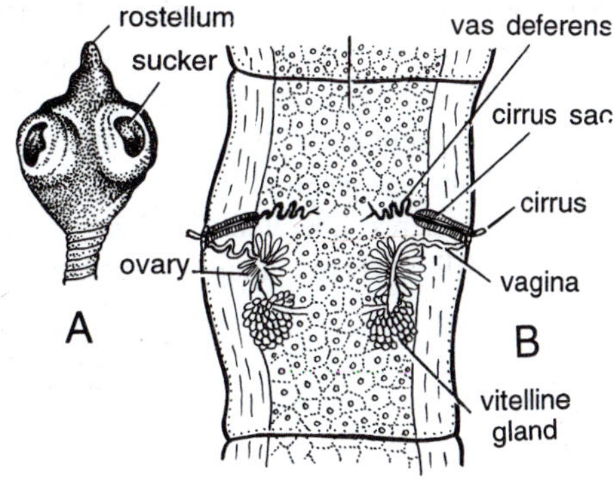

Fig. 12. *Dipylidium caninum.* **A.** Scolex, **B.** Mature proglottid.

12. *Dipylidium.* The dog tapeworm *Dipylidium caninum* found universally in pet dogs, cats and sometimes children. Adult worm is nearly 25 cm long with about 200 proglottids, each with 2 sets of reproductive organs. Scolex contains 4 suckers, while rostellum is retractile bearing 4 rows of hooks. Embryonated eggs are passed out and become dried on skin. These are swallowed by the intermediate hosts which are common dog louse *Ctenocephalus canis* and dog-flea *Pulex irritans,* in which larval forms occur. Dog or cat become infected by licking or chewing these intermediate hosts. Human beings, especially children, become infected by accidentally swallowing infected lice or fleas. It takes about 2 weeks for the worms to mature. It is not very pathogenic.

13. *Hymenolepis.* The dwarf tapeworm *Hymenolepis nana* is the smallest of human tapeworms, especially among children, throughout the world. Adult is 2 to 4.5 cm long with 100

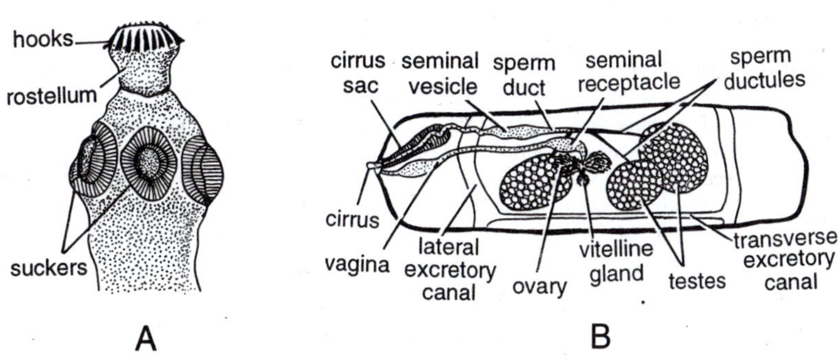

Fig. 13. *Hymenolepis.* **A.** Scolex. **B.** Mature proglottid.

to 200 proglottids. Scolex contains 4 suckers. Rostellum is short with a single row of 20 to 30 hooks. There are 3 testes in each proglottid. Life cycle is simple without any intermediate host, but flour beetles and fleas may serve as such. Onchospheres liberated from eggs live in intestinal villi where they transform into cysticercoid larvae. These reenter lumen of intestine to become attached by evaginated scolices and mature into adults (auto-infection). Unsanitary toilet habits spread infection from person to person. Dwarf tapeworm is pathogenic causing abdominal pain, diarrhoea and eosinophilia.

Important Questions

➤➤Long answer type questions

1. Give general characters of phylum Platyhelminthes and classify it giving characteristic features and at least two examples of each group.
2. Name and characterize the classes of phylum Platyhelminthes.
3. Write short notes on : (i) *Bipalium,* (ii) *Convoluta,* (iii) *Diplozoon,* (iv) *Dipylidium,* (v) *Echinococcus,* (vi) *Hymenolepis,* (vii) *Moniezia,* (viii) *Opisthorchis (=Chlonorchis),* (ix) *Paragonimus,* (x) *Polystomum,* (xi) *Schistosoma,* (xii) *Taenia saginata,* (xiii) *Thysanozoon.*

➤➤Short answer type questions

1. Give an example for a polyclad flatworm.
2. What is the name of the cestodarian embryo.
3. How many number of proglottids are present in *Echinococcus* ?
4. Name the intermediate host of *Echinococcus granulosus.*
5. Write the name and the characteristics of excretory organ seen in Platyhelminthes.
6. What is apolysis ?
7. Which subclass among cestodes does not possess a scolex ?
8. What is the characteristic of cestodarian embryos ?
9. Mention ten characteristic features of phylum Platyhelminthes.
10. Classify *Polystomum* upto its order.
11. Mention three main classes of phylum Platyhelminthes and write three features of each class with an example.
12. Comment on the importance of cestodes to man.
13. Mention a parasitic Platyhelminthes in man, where both the male and female live in close association.
14. Give the distinguishing features of the Turbellaria, Trematoda and Cestoda.
15. Differentiate between parthenogenesis and polyembryony.
16. Who proposed the term Platyhelminthes ?
17. Which system is absent in the animal of Platyhelminthes ?
18. What is the name of the class of Platyhelminthes in which segmented animals are placed ?
19. What is rhabdites ?
20. How many orders in the class Turbularia ?
21. Give the name of orders in the class Trematoda.
22. What is the generic name of Chinese liver fluke ?
23. Describe the taxonomic condition of *Echinococcus.*

➤➤Fill in the blanks

1. In Platyhelminthes the body organisation is ....
2. Skeletal, respiratory and circulatory system are .... in Platyhelminthes.
3. In Platyhelminthes, the fertilization is ....
4. In order Rhabdocoela the animal are ....
5. Endoparasitic flatworm known as ....
6. The meaning of Cestoidea is ....
7. *Paragonimus* is commonly known as ....

➤➤True and false statements

1. *Thysanozoon* is a polyclad turbellarian found in cold water.
2. Coelome is present in Platyhelminthes.
3. Fertilization in Platyhelminthes is internal may be cross or self.
4. The total Platyhelminthes are segmented body's animals.
5. *Schistosoma*, blood fluke of man discovered by Bilharz in 1851.

➤➤Multiple choice questions

1. The disease in man caused by infection of bladder-worm is known as :
(a) schistosmiasis (b) cysticercosis
(c) teaniasis (d) enterobiasis
2. Trematodes are usually described as :
(a) blood flukes (b) flukes
(c) worms (d) flatworms
3. Regeneration can be best described as :
(a) removal of unwanted parts
(b) replacement of lost parts
(c) deformation of lost parts
(d) rehealing of a wound
4. Spines are seen in the cuticle of :
(a) *Planaria* (b) *Taenia*
(c) *Echinococcus* (d) *Fasciola*
5. In platyhelminthes, vitellarium is generally found in association with :
(a) ovary (b) testis
(c) both ovary and testis
(d) shell glands
6. Flat worms are characterised by presence of :
(a) protonephridia or flame cells
(b) renette cells
(c) canals (d) nephridia
7. A digestive system is absent in :
(a) Turbellaria (b) *Ascaris*
(c) *Taenia* (d) *Schistosoma*
8. Sporocyst is found in :
(a) sponge (b) *Fasciola*
(c) *Hydra* (d) all these
9. Anaerobic respiration is likely to occur in :
(a) earthworm (b) tape worms
(c) flat worms (d) echinoderms
10. Liver fluke belongs to class :
(a) Trematoda (b) Nematoda
(c) Turbellaria (d) Cestoda
11. Cercaria is found in the life cycle of :
(a) *Taenia* (b) *Fasciola*
(c) *Ascaris* (d) *Planaria*
12. Parasitic animals have well developed :
(a) respiratory system (b) reproductive system
(c) digestive system (d) nervous system
13. Common feature among tapeworm, liver fluke and planarian is that all are :
(a) segmented (b) found in night
(c) coelomates (d) flat
14. *Echinococcus* generally known as :
(a) pig tapeworm (b) sheep tapeworm
(c) man tapeworm (d) dog tapeworm
15. Intestinal fluke is :
(a) *Fasciola hepatica* (b) *Fasciola gigantica*
(c) *Fasciola buski* (d) all
16. *Paragonimus* infect the :
(a) liver (b) heart
(c) lungs (d) brain
17. Who was suggested the name 'platyhelminthes' :
(a) Robert Grant (b) Leucart
(c) Gegenbaur (d) Von Sciblod
18. Classification of flatworm is based on :
(a) mode of life (b) presense of gut
(c) body coverage (d) all above
19. Pseudometamerism is found in :
(a) *Fasciola* (b) *Taenia*
(c) *Planaria* (d) *Ascaris*
20. Eyes are present in :
(a) *Dugesia* (b) *Taenia*
(c) *Fasciola* (d) all above
21. Smallest of human tapeworm :
(a) *Taenia* (b) *Echinococcus*
(c) *Hymenolepis* (d) *Polyforma*
22. Sucker absent in :
(a) Turbullaria (b) Trematoda
(c) Cestoidia (d) all
23. Lycophore is :
(a) endoparasite
(b) larvae of cestodarian
(c) parasite of fish
(d) parasite of man
24. Opisthapter is a :
(a) anterior sucker (b) posterior sucker
(c) hook (d) spine
25. *Polystomum* is endoparasite of :
(a) frogs (b) fish
(c) toad (d) all

Answers

➤➤Fill in the blanks

1. tissue organs level, 2. absent, 3. internal, 4. free living, 5. tapeworms, 6. girdle worm, 7. lung fluke.

➤➤True and false statements

1. true, 2. false, 3. true, 4. false, 5. true.

➤➤Multiple choice questions

1. (b) 2. (b) 3. (d) 4. (d) 5. (a) 6. (a) 7. (c) 8. (b) 9. (c) 10. (a)
11. (b) 12. (b) 13. (d) 14. (d) 15. (c) 16. (c) 17. (c) 18. (d) 19. (b) 20. (a)
21. (c) 22. (a) 23. (b) 24. (b) 25. (d).

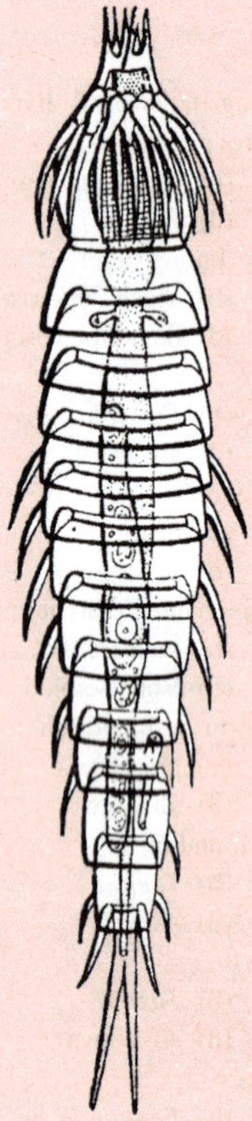

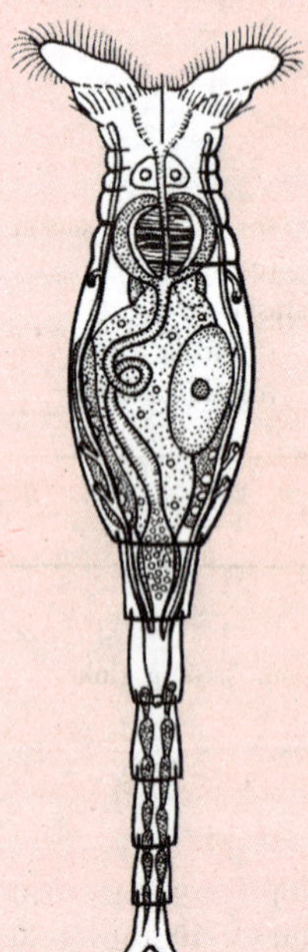

34

Superphylum Aschelminthes : The Pseudocoelomate Phyla

There are several groups of animals in which the space inside the body is a **pseudocoel**. It is not a true coelom because it is not lined with mesodermal epithelium, and represents a persistent embryonic blastocoel. These groups of animals are known as **pseudocoelomates**, and they are often placed within a single superphylum **Aschelminthes**.

Very recently they are considered most hardy creatures by NASA, (USA) because a group of small (1mm adult) *Caenorhabditis elegans* worms, living in petridishes enclosed in aluminium canisters, survived re-entry in the atmosphere and impact with the ground and were recovered weeks after the space shuttle blew up into small pieces, killing all the seven astronauts. These nematodes were part of a Biological Research in Canisters (BRIC) experiment by NASA Ames Research Center, designed to study the effect of weightlessness on physiology.

DERIVATION OF NAME

The name **Aschelminthes** was proposed by **Grobben** (1910), in place of the older name Nemathelminthes. It has been derived from two Greek words : **askos**, cavity + **helmins**, worm.

GENERAL CHARACTERS

1. Mostly aquatic, free-living or parasitic.
2. Usually small, even microscopic. Some reach a meter or more in length.
3. Body slender, vermiform, unsegmented, flat or cylindrical, bilaterally symmetrical and triploblastic.
4. **Organ system grade** of body organization.
5. Head not distinctly formed with well-defined sense organs.
6. Body wall with a **syncytial** or cellular epidermis, externally covered with thick **cuticle** of scleroprotein.
7. **Cilia** absent except anterior cilia of rotifers.
8. Musculature includes mostly **longitudinal fibres**.
9. Body cavity a **pseudocoel** not lined by mesoderm.
10. Digestive canal complete with mouth, specialized pharynx, straight non-muscular intestine and posterior anus.
11. No circulatory and respiratory systems.
12. Excretory system of protonephridia (in some) and canals. Cloaca present in some.
13. Nervous system of cerebral ganglia, or of circumenteric nerve ring with anterior and posterior nerves.
14. Mostly **dioecious**. Male usually smaller than female. Gonads and ducts single or double. Eggs with chitinous shell. Cleavage determinate and spiral.
15. Development usually direct with no larval stages, or indirect with a complicated life history.

CLASSIFICATION OR MAJOR GROUPS

Aschelminthes includes one very large group (Nematoda) and four smaller groups (Rotifera, Gastrotricha, Kinorhyncha and Nematomorpha). According to some zoologists such as **Hyman** (1951), Aschelminthes is regarded as a distinct **phylum**, and the various groups included in it as classes. However, others treat different groups as separate phyla and the name of Aschelminthes as a **superphylum** or without any taxonomic rank. Such an approach has been followed in the following scheme of classification.

1. Phylum Rotifera

(G., **rota**, wheel + **ferre**, to bear)

1. Microscopic animals found in ponds, lakes and streams. Rarely in oceans.
2. Body wall thickened into stiff plates or **lorica** into which head may retreat.
3. Anterior end with ciliary disc or **corona** (wheel organ), used for feeding and locomotion.
4. Post-anal tail or with two toes and adhesive glands for attachment.
5. Body musculature includes longitudinal and transverse muscle bands and strands.
6. Digestive system with a grinding organ, **mastax**, lined internally by strong cuticle.

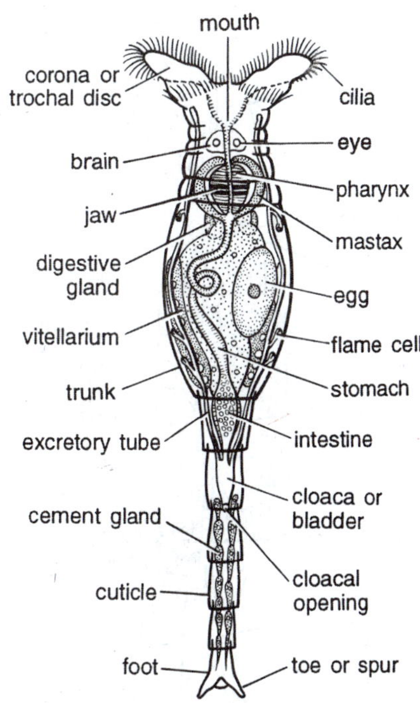

Fig. 1. *Philodina.*

7. Excretory system of protonephridia and two protonephridial tubes which empty into a bladder.
8. Nervous system of three major ganglia and nerves.
9. Sensory organs antennae and eye spots.
10. Males smaller than females. Parthenogenesis common. No larval stage.
Examples : *Philodina, Asplanchna, Rotaria, Epiphanes (= Hydatina).*

2. Phylum Gastrotricha

(L., **gaster**, stomach + **trichos**, hair)

1. Microscopic, marine or freshwater.
2. Body wall with cuticle bearing short, curved, dorsal spines.
3. Corona absent cilia on ventral surface for locomotion.
4. Posterior end forked and with adhesive tubes and glands for attachment.
5. Body musculature includes six pairs of longitudinal muscles.
6. Mouth surrounded by bristles. Pharynx triradiate and muscular.
7. Excretory system with two protonephridia.
8. Nervous system with a saddle-shaped ganglion and two lateral nerves.
9. Dioecious or monoecious. Freshwater females parthenogenetic.
10. Development direct. Young and adult alike.
Examples : *Chaetonotus, Macrodasys.*

3. Phylum Kinorhyncha (= Echinodera)

(Dr., **kineo**, to move + **rhynchos**, beak)

1. Marine, microscopic animals.

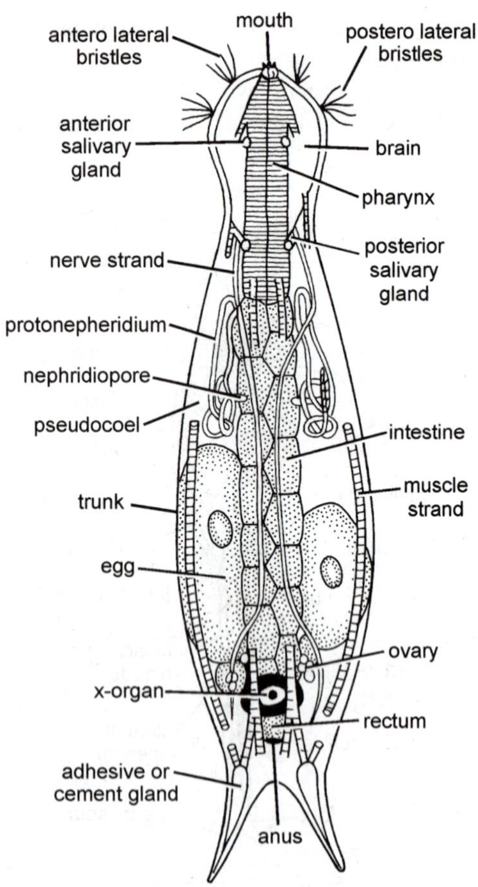

Fig. 2. *Chaetonotus.*

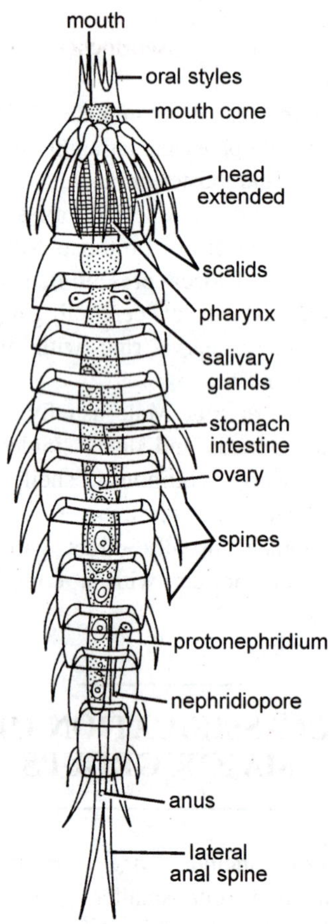

Fig. 3. *Echinoderes.*

2. Superficial segmentation of body into 13 or 14 overlapping rings (zonites).
3. Body surface with spiny cuticle but no cilia.
4. **Mouth cone** or head protrusible and covered with **scalids**.
5. A pair of adhesive tubes in front part of the ventral surface.
6. Pseudocoel with fluid containing amoebocytes.
7. A nerve ring with ventral cord with a ganglion in each zonite. Eyespots in some.
8. Digestive system complete, with salivary glands.
9. Dioecious. Gonads as a pair of tubular sacs. Penial spicules in males.
10. Fertilization internal. Metamorphosis with several larval stages.

Examples : *Echinoderes, Pycnophyes.*

4. Phylum Nematomorpha (= Gordiacea)

(Gr., **nema**, thread + **morphe**, shape)

1. Hair worms, found in fresh water. One genus *(Nectonema)* marine.
2. Body long, slender and cylindrical.
3. Cuticle thick bearing small papillae. Epidermis cellular, single layered.

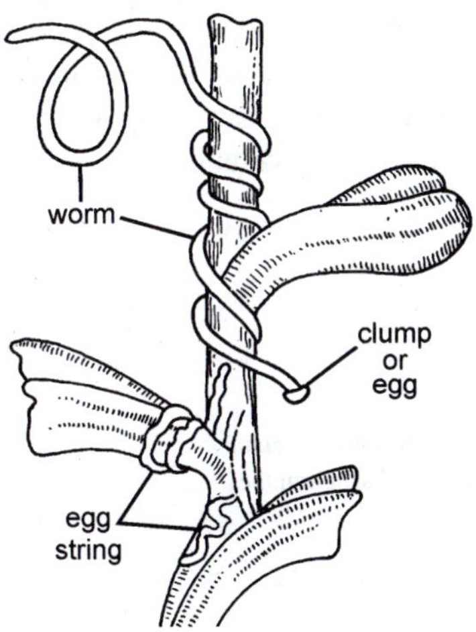

Fig. 4. *Paragordius* (female).

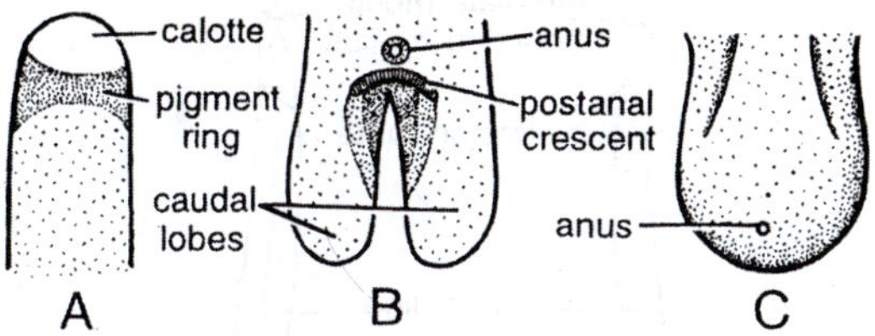

Fig. 5. *Paragordius.* **A.** Anterior end. **B.** Posterior end of male. **C.** Posterior end of female.

4. Digestive system complete in larva but degenerates in non-feeding adults. Cloaca present.
5. Pseudocoel mostly filled with parenchyma.
6. No circulatory, respiratory and excretory systems.
7. Nervous system with a circumenteric nerve ring and a mid ventral nerve cord.
8. Dioecious. Gonads and gonoducts paired. Oviducts also open into cloaca.
9. Juveniles parasitic in grasshoppers, crickets and other insects.

Examples : *Gordius, Paragordius, Nectonema.*

5. Phylum Nematoda

(Gr., **nema**, thread + **eidos**, form)

1. Aquatic, terrestrial or parasitic roundworms.
2. Body elongated, cylindrical and unsegmented.
3. Body wall with thick cuticle, cellular or syncytial epidermis and longitudinal muscle cells in four bands.
4. No cilia, circulatory and respiratory systems.
5. Digestive system complete with muscular pharynx and glands.
6. Excretory system of glandular organs, canals or both.
7. Nervous system with circumenteric ring and anterior and posterior nerves. Sense organs simple.
8. Dioecious. Male with penial spicules and smaller than female. Gonads one or two. Male gential ducts lead into cloaca, female genital duct with a separate opening.

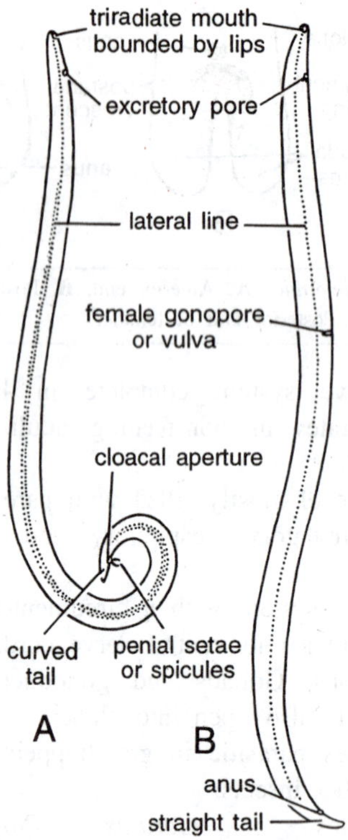

Fig. 6. *Ascaris.* **A.** Male. **B.** female.

9. Fertilization internal. Development usually direct. No asexual reproduction or regeneration.
Examples : *Ascaris, Necator, Wuchereria, Trichinella, Enterobius,* etc.

INTER-RELATIONSHIPS

Phylogenetic position of Aschelminthes is obscure. Its relations are not clear with other major phyla with which it has few affinities. Rotifera and Gastrotricha show some resemblances to Turbellaria and may have arisen from a primitive flatworm.

Within Aschelminthes itself, various groups show some relationships. They are usually small to microscopic, worm-like, bilaterally symmetrical, with a thick and tough resistant cuticle, with only longitudinal muscle fibres, and with a pseudocoel not derived from mesoderm. For this reason, various groups are treated by a number of zoologists as classes under a single phylum Aschelminthes.

However, despite these similarities, the pseudocoelomate groups are extremely diverse in other respects, and lack any well-defined distinctive unifying characters to justify their grouping together into one phylum. They are not as clearly related to one another as are the classes in other well-established phyla. For this reason, many zoologists prefer to use the name Aschelminthes as a convenient term without any taxonomic rank. They may call it a superphylum, and treat its different groups as independent phyla. In that case, the numerous structural similarities existing between them are supposed to be the result of convergent evolution.

Important Questions

➤➤*Long answer type questions*

1. Describe the general characters of phylum Aschelminthes and discuss its taxonomic status.
2. Give salient characters and examples of major animal groups belonging to Aschelminthes.

➤➤*Short answer type questions*

1. Define the term pseudocoelome.
2. What is the meaning of Aschelminthes ?
3. Which grade of body organisation found in Aschelminthes ?
4. Name the groups in which Aschelminthes are classified ?
5. Define Rotifera.
6. Give taxonomic condition of Nematoda ?
7. Name the larval form in Rotifera ? Is this group of animal have any larva ?
8. Systematised to Gastrotricha.
9. Define interrelationship among Aschelminthes ?
10. No asexual reproduction among Aschelminthes ?

11. Describe the systemic position of *Ascaris.*
12. What is syncytial epidermis ?

➤➤*Fill in the blanks*

1. Digestive system of Rotifera contain a grinding organ, it is known as ......
2. Parthenogenetic female present in ......
3. In phylum Kinorhyncha head protrusible and covered with ......
4. Pseudocoel mostly filled with ...... in Nematomorpha.
5. No asexual reproduction or regeneration in phylum ......
6. The name aschelminthes was proposed by ....
7. Male is smaller than female and no larval stage in ......
8. *Gordius* belong to phylum ......

➤➤*True and false statements*

1. *Philodina* belong to the phylum Rotifera.
2. *Chaetonatus* is an animal of phylum Gastrotricha.
3. *Enterobius* belong to Nematoda.
4. *Wuchereria* is belong to Nematoda.
5. Aschelminthes is a term which proposed by **Grobben** (1910).

➤➤*Multiple choice questions*

1. Pseudocoelome present in :
   (a) Aschelminthes (b) Platyhelminthes
   (c) Coelenterata (d) Annelida
2. Organ system grade of body organization present in :
   (a) Porifera (b) Platyhelminthes
   (c) Aschelminthes (d) Coelenterata
3. In Rotifera the sense organs are :
   (a) eye spot (b) antinnae
   (c) statocyst and eye spot
   (d) eye spot and antennae
4. Development in Gastrotricha is :
   (a) direct (b) indirect
   (c) parthenogenetic (d) regenerative
5. One of these known as hair worms :
   (a) Rotifera (b) Kinorhyncha
   (c) Nematomorpha (d) Nematoda
6. Juvenile parasite in grasshopper and other insects :
   (a) Rotifera (b) Kinorhyncha
   (c) Nematoda (d) Nematomorpha
7. Digestive system present complete with muscular pharynx and glands :
   (a) Nematoda (b) Nematomorpha
   (c) Kinorhyncha (d) Gastrotricha
8. Digestive system with a grinding mastax in animal of phylum :
   (a) Rotifera (b) Gastrotricha
   (c) Kinorhyncha (d) Nematomorpha
9. The circumentric nerve ring of *Ascaris* is located around :
   (a) buccal chamber (b) pharynx
   (c) oesophagus (d) stomach
10. How many ganglia are attached to the nerve ring in *Ascaris* :
    (a) One (b) Two
    (c) Nine (d) Numerous
11. Median unpaired structure in female *Ascaris* is :
    (a) vagina (b) uterus
    (c) oviduct (d) ovary
12. Eggs of *Ascaris* that pass out of the host body are :
    (a) unfertilized
    (b) unfertilized and unsegmented
    (c) fertilized
    (d) fertilized and segmented
13. During its migration in host body, *Ascaris* pass through
    (a) spleen (b) kidney
    (c) skeletal muscle (d) lung
14. *Waucheria bancrofti* is transmitted by :
    (a) *Culex* (b) *Anopheles*
    (c) Tsetse fly (d) Sandfly

Answers

➤➤*Fill in the blanks*

1. mastax, 2. gastrotricha, 3. scalids, 4. parenchyma, 5. Nematoda, 6. Grobben, 7.Rotifera, 8. Nematomorpha.

➤➤*True and false statements*

1. true, 2. true, 3. false, 4. false, 5. true.

➤➤*Multiple choice questions*

1. (a) 2. (c) 3. (d) 4. (a) 5. (c) 6. (d) 7. (a) 8. (a) 9. (b) 10. (c)
11. (a) 12. (c) 13. (d) 14. (a).

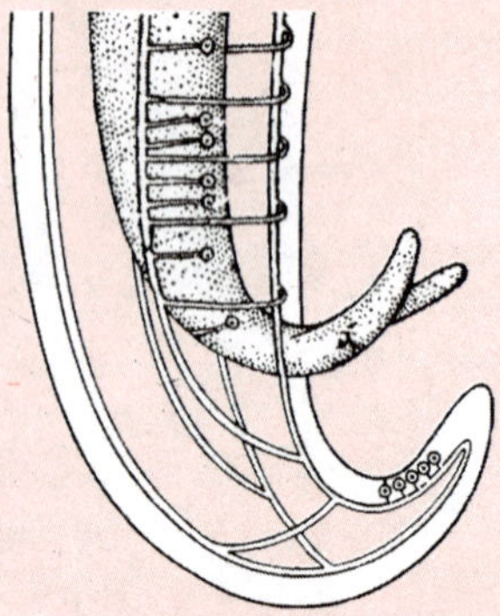

35

Ascaris lumbricoides : The Common Roundworm

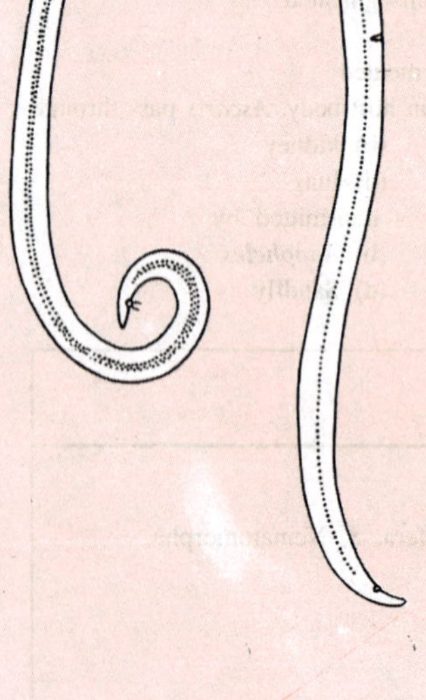

Nematodes, commonly called roundworms, constitute the largest phylum **Nematoda** of super phylum **Aschelminthes**. They are distinct from flatworms and tapeworms (platyhelminthes) in having a **cylindrical body**, a **pseudocoel**, and a **complete digestive tract** lined by endodermal epithelium. Majority of them are free-living in soil and water, while a good number of species are parasitic on plants and animals. Species of genus *Ascaris* are large-sized and inhabit the intestines of various mammals (vertebrates). The most common and best known member is *Ascaris lumbricoides,* a gastro-intestinal parasite of human.

Ascaris lumbricoides

Present chapter deals with the morphology, physiology, life history and pathogenicity of *Ascaris lumbricoides.* It is selected for study of nematodes because of its large size, abundance and wide distribution. It is used in laboratory to illustrate the processes of fertilization, early cleavage and mitosis.

SYSTEMATIC POSITION

Phylum	Nematoda
Class	Phasmidia
Order	Ascaroidea
Suborder	Ascarinae
Family	Ascaridae
Genus	*Ascaris*
Species	*lumbricoides*

DISTRIBUTION, HABITS AND HABITAT

Ascaris lumbricoides is one of the most familiar endoparasites of man. It has also been reported from pigs, cattle, sheep and squirrels (**Goody**, 1936). It inhabits the small intestine, more frequently of children than of adults. It has a cosmopolitan distribution but chiefly found in India, China, Korea, Philippines and Pacific Islands. Its incidence varies in different parts of the globe. **Stoll** (1947) estimated that there are 644 million human infections in the world, of which 3 million are found in North America, 42 million in tropical America, 59 million in Africa, 488 million in Asia, 32 million in Europe, 19.9 million in Russia and its neighbouring countries and 500,000 in the Pacific Islands.

EXTERNAL MORPHOLOGY

[I] Shape and Size

Body is elongate, cylindrical and gradually tapering at both ends, the anterior end being more slender than the posterior. Sexes are separate (**dioecious**) with a distinct sexual dimorphism. Female measures 20 to 40 cm in length and 4 to 6 mm in diameter. It has a straight tail. While male is smaller, upto 20 cm in length and 2 to 4 mm in diameter. Its tail is curved ventrally. The cuticle covering the body surface bears minute transverse striations, giving a pseudosegmented appearance to the worm.

[II] Colouration

Fresh specimens are light yellow to light pink in colour. Semitransparency of body wall enables some of the internal organs to be seen through it.

[III] Longitudinal Streaks or Lines

Running along the entire length of body are four longitudinal **streaks** or **lines** — one **mid-dorsal**, one **mid-ventral** and two **lateral**. Dorsal and ventral lines appear pure white, while lateral lines, which are more conspicuous, appear brown. These lines are simply the impressions of syncytial epidermis which is much thickened all along the length at the median dorsal, median ventral and two lateral positions.

The **excretory pore** is situated mid-ventrally, a little behind the mouth. In female, the **genital pore** and **anus** open separately; the **genital pore** is situated mid-ventrally at about one-third distance from the anterior end.

[IV] Anterior End

Anterior end is rounded bearing a small terminal triradiate **mouth** aperture guarded by three broad **lips** or **labia**. These are best seen **en face**, that is, in frontal view. One lip is **mid-dorsal** and broadly elliptical and two are **sub-ventral** or **latero-ventral** and oval. Inner or oral margin of each lip has a forked fleshy core made of two

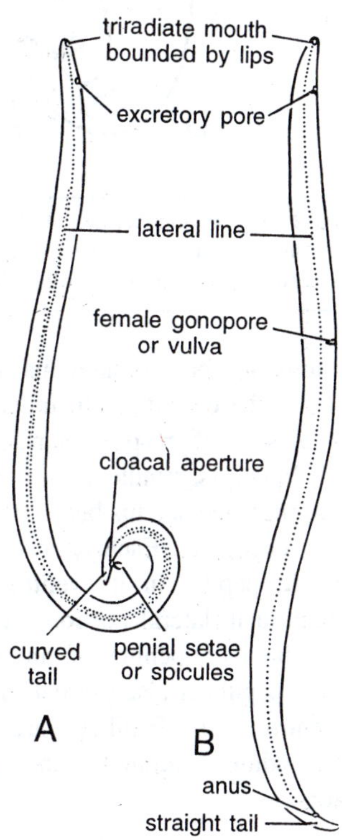

Fig. 1. *Ascaris lumbricoides.* Adult worms in lateral view. **A.** Male. **B.** Female.

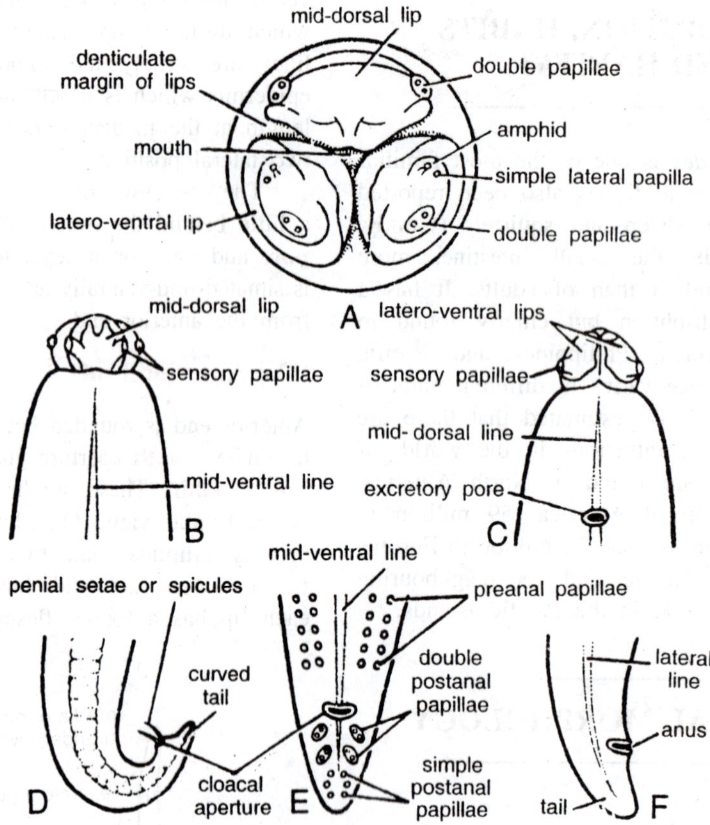

Fig. 2. *Ascaris* lumbricoides. **A.** En face view of mouth and lips. **B.** Anterior end in dorsal view. **C.** Anterior end in ventral view. **D.** Posterior end of male in side view. **E.** Posterior end of male in ventral view showing papillae. **F.** Posterior end of female in side view.

anterior extensions of labial parenchyma, and bears minute **denticles**. These denticulate inner margins of lips form a sort of rasping surface. Outer surface of each lip bears minute **sensory papillae** (L., **papilla**, nipple). Dorsal lip bears two double papillae, one on each side. Each latero-ventral lip bears one double papilla in its ventral portion, and a single papilla in lateral portion, near which is an **amphid**, that is, opening of an amphidial gland. A cervical papilla is also present just behind each latero-ventral lip. Papillae are perhaps tangoreceptors while amphids are probably chaemoreceptors.

[V] Posterior End

As already stated, posterior or tail end of body shows clear sexual dimorphism. It is straight in female, but curved ventrally in male. In female, a little in front of the tail end lies a mid-ventral transverse aperture, or **anus**, guarded by thick lips. Only digestive tube opens to outside through anus.

In male, anus is replaced by a **cloaca**. It is the common aperture for digestive and genital tubes. Sometimes, two chitinous spiculate processes of equal size are seen protruding out of the cloacal aperture. These are called **peneal setae** or **spicules** which serve to transfer- sperms into female vagina during copulation. Part of caudal end behind anus or cloacal aperture may be called the **tail**.

Tail end of male *Ascaris* is characterized by the presence of numerous **genital papillae** on ventral surface. There are 50 pairs of **preanal papillae** in front of cloaca, and 5 pairs of **postanal papillae** behind it. Genital papillae of male help in copulation.

Two anterior pairs of postanal papillae are double, while the rest are single.

[VI] Excretory Pore

A single excretory pore lies mid-ventrally at a distance of about 2 mm from the anterior end.

[VII] Female Gonopore

The genital pore or vulva of female lies mid-ventrally at about one-third distance from the anterior end. But the genital pore of male opens into cloaca.

BODY WALL

Body wall of *Ascaris* is made up of 3 layers : (i) an outer **cuticle**, (ii) a middle **epidermis**, and (iii) an inner layer of **longitudinal muscles** lining the body cavity.

[I] Cuticle

It is a thick, tough, transparent and glossy layer secreted by the underlying epidermis and continuous with the cuticular lining of pharynx and rectum. In the young worm it is shed off (moulted) to permit growth. Under light microscope, four distinct layers of different chemical composition and different structural arrangements can be identified in the cuticle. A fifth lipid layer has also been revealed under electron microscope.

The fi e layers of cuticle, along with their sub-divisions as revealed by electron microscope, are as follows :

1. Lipoid layer. It is in the form of a thin osmophilic membrane, about 1000 A° thick.

2. Cortical or cortex layer. It consists of a dense material, or **heratin**, resistant to the digestive juices of host. It includes : (i) an **outer cortical layer** which lies as discontinuous strips or rings around body and (ii) an **inner cortical layer**.

3. Matrix layer. It contains sulphur-rich **mat ricin** and consists of (i) an outer **fibrillar layer** traversed by branching canals, (ii) a **homogeneous layer** which shows several radial striations, and (iii) a **boundary layer** resembling the outer cortical layer.

4. Fibre layer. It consists of collagen fibres crossing each other and arranged in three strata.

5. Basement membrane. It is a thin layer forming the inner limit of cuticle.

Cuticle is permeable to salts, water products of metabolism. It is believed that enzymes are present in cuticle of *Ascaris* which control the passage of different metabolites. Some anti-enzymes are also secreted by cuticle of *Ascaris*, which neutralize the effect of digestive juices of host upon the parasite.

[II] Epidermis or Hypodermis

It forms a syncytial layer below cuticle. Along the entire length of body, at median dorsal, median ventral and two lateral positions, it is much thickened to form the **dorsal**, **ventral** and **lateral**

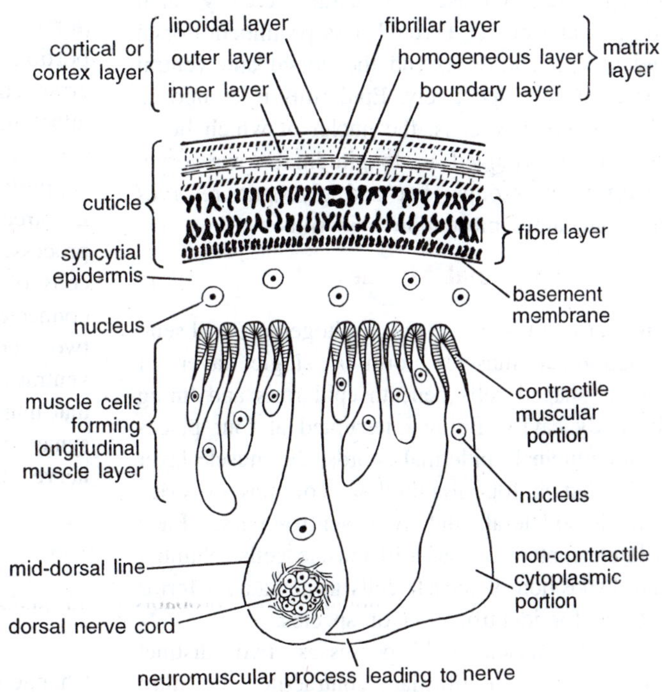

Fig. 3. *Ascaris.* T. S. body wall (a portion).

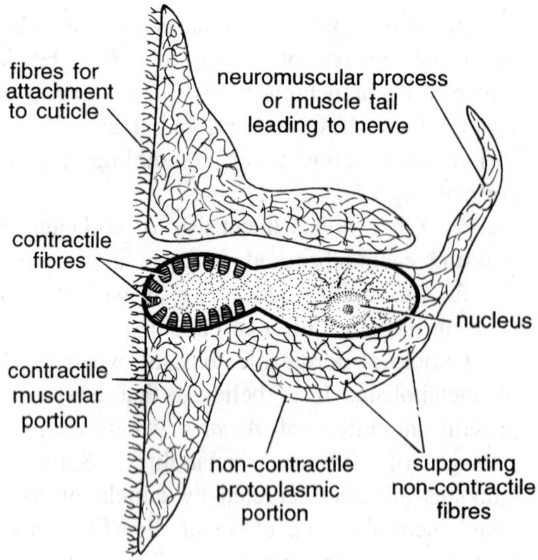

Fig. 4. *Ascaris.* A single muscle cell. The cut part shows structure in T.S.

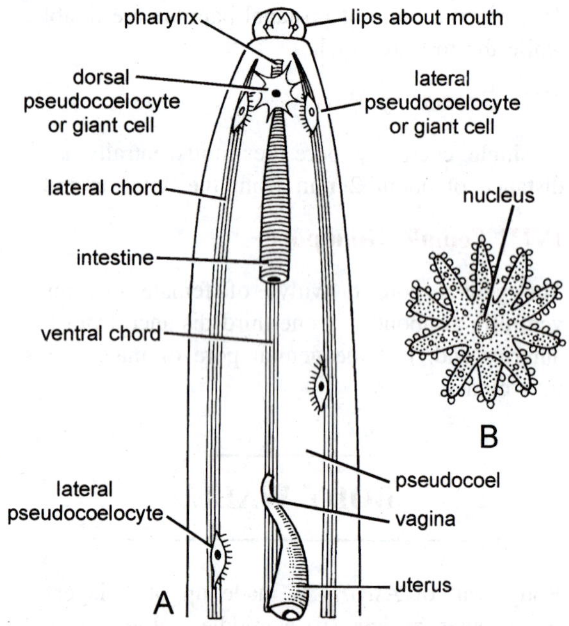

Fig. 5. *Ascaris* **A.** Position of giant cells in anterior region of pseudocoel. **B.** Single giant cell or pseudocoelocyte.

lines or **chords**, respectively, bulging into body cavity. Lateral chords are more conspicuous and seen on the surface as yellow lines. Running through each of these is a lateral excretory canal and a lateral nerve. Through less prominent dorsal and ventral chords extend the dorsal and ventral nerve cords, respectively. Epidermis is composed of relatively few cells, the nuclei of which lie in the chords arranged in longitudinal rows, often occurring in groups. Fat and glycogen reserves are abundantly present in epidermis.

[III] Longitudinal Muscles

Circular muscles are altogether absent. Longitudinal muscles form a single layer of spindle-shaped cells beneath epidermis and lining the body cavity. Being interrupted at four places by longitudinal epidermal chords, the muscle layer is divided in four longitudinal columns or strips two dorso-lateral and two ventro-lateral. Each column contains about 150 muscle cells. Number and disposition of muscle cells in nematodes forms a basis for identification of species.

Each muscle cell possesses two distinct portions : (i) a fibrillar, contractile **muscular portion**, lying lengthwise against the epidermis, and (ii) a granular non-contractile **protoplasmic portion** projecting into body cavity. Muscular portion contains longitudinal **contractile fibres** arranged at intervals. It also has fibres for attachment to cuticle. **Non-contractile protoplasmic portion** contains non-contractile supporting fibres and a nucleus and gives out at its free surface one or more neuromuscular processes or **muscle tails**. Muscle tails of all the cells of two dorso-lateral columns are ultimately connected with dorsal nerve cord, while those of two ventro-lateral columns are connected with ventral nerve cord. Electron microscopy has shown that muscle tails are cellular extensions which form synapsis on the motor nerves of dorsal and ventral nerve cords (**Debell**, 1965).

MOVEMENTS

Change in body length is restricted due to absence of circular muscles in body wall. However, changes in length to about 15% are made possible by the fibres of cuticle. These fibres, though themselves inelastic, are disposed in a spiral and mesh-like arrangement which permits limited change of length.

Undulating movements, in dorso-ventral waves, can be performed by alternate contractions of dorso-lateral and ventro-lateral muscles in the anterior end of body. These movements help the parasite to counteract the peristaltic activity of the host's intestine.

BODY CAVITY OR PSEUDOCOEL

Between body wall and visceral organs is a spacious fluid-filled cavity. This cavity is not true coelom as (i) it is not lined by coelomic epithelium, (ii) it has no relation with reproductive and excretory organs, and (iii) it develops from blastocoel, *i.e.*, between mesoderm and endoderm of embryo. This body cavity is referred to as **pseudocoel** or false coelom.

Pseudocoel of *Ascaris* contains five giant stellate mesenchymal, cells, known as **pseudo-coelomocytes**, occupying fixed positions. Out of these, the largest cell lies on the dorsal side of pharynx, and from it extend numerous thin **cytoplasmic strands** in the form of **fenestrated membranes**. These membranes cross the cavity at various places and form delicate layers over the visceral organs and muscles of body wall. Remaining stellate giant cells occur in the anterior third of the body, two in relation with each lateral epidermal chord. In fact, pseudocoelom is a large intracellular space formed by the union of these coelomocytes.

Pseudocoel is filled with an odorous protein rich fluid, the **perienteric** or **pseudocoelomic fluid**, which serves for transportation of metabolites and keeps the body distended. It is composed of about 93% water and the rest of it is in the form of solids including protein glucose, non-protein nitrogenous substances, sodium chloride and phosphate.

DIGESTIVE SYSTEM

[I] Alimentary Canal

Alimentary canal is a straight and complete tube extending from mouth at one end to anus on other end of body. It comprises a short **pharynx** or oesophagus representing the **foregut**, a long intestine or **mid-gut**, and a short **rectum** or **hind gut**.

1. Mouth. It is a triradiate aperture, situated at the anterior tip guarded by three lips or **labia** (already described).

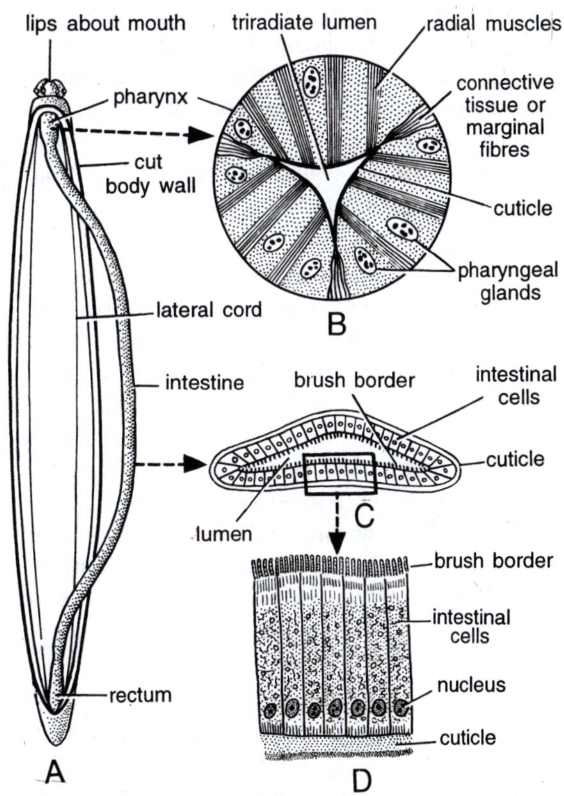

Fig. 6. *Ascaris.* **A.** Alimentary canal. **B.** Pharynx in T.S. **C.** Intestine in T.S. **D.** A portion of intestinal wall showing brush border.

2. Pharynx. Mouth opens into a short, characteristic, cylindrical, thick-walled and muscular **pharynx**. Wall of pharynx, consists of a syncytial epithelium traversed by radial muscle fibres and containing glands. It is bounded externally by a membrane. Internally it is lined by cuticle which, at the margin of mouth, is continuous with the cuticle of body wall. Lumen of pharynx is tri-radiate. Rays are in the form of deep grooves which demarcate the pharyngeal wall into three sectors —one dorsal and two sub-ventral. Dorsal sector contains a pinnately branched **pharyngeal** or **oesophageal gland**, while each sub-ventral sector contains a similar palmately branched gland. Numerous groups of radial muscle fibres extend between inner cuticular lining and outer bounding membrane. Contraction of these muscles dilates the lumen. Between outer end of each ray of lumen and outer bounding membrane extend the connective tissue fibres, known as **marginal fibres**, which help in preserving the tri-radiate form of lumen.

3. Intestine. Pharynx is followed by **intestine** which extends almost along the entire body length. A small valve is present at the junction of pharynx and intestine which stops the food from going back into pharynx. Intestine is dorsoventrally flattened and its wall consists of a single layer of tall columnar cells, lined externally by a basement membrane and a thin layer of cuticle. Inner free margin of each cell is produced into several hair-like projections, the **microvilli**, forming a sort of **brush border**, to increase the absorptive surface area. The microvilli, in fact, are formed by a bacillary layer. Muscle fibres are wanting in intestine.

4. Rectum. Intestine is followed by a short **rectum** which too is dorso-ventrally flattened. Wall of rectum consists of tall columnar cells and is lined internally by cuticle and externally by muscle tissue. In male, rectum opens into **cloaca** which also receives the ejaculatory duct. While in female, rectum opens out through anus. Which is a transverse slit guarded by anterior and posterior lips and is provided with a special muscle, the **depressor ani**. Anus or cloaca lies at a distance of about 2 mm from tail end.

[II] Food, Feeding and Digestion

Food of *Ascaris lumbricoides* consists of blood and fully or partially digested food occurring in fluid form in the host's gut. It is sucked by the rhythmic pumping action of pharynx. Digestion is completely extracellular in intestine and is aided by the enzymes **proteases**, **amylase**, and **lipase**. Digested food is absorbed by the intestinal cells and distributed by the pseudocoelomic fluid. Excess of food is stored mainly as glycogen and a little fat in syncytial epidermis. Some intestinal cells also engulf small solid particles by phagocytosis, and digest them **intracellularly**. **Defaecation** of undigested food whenever it occurs, is facilitated by the **depressor ani** muscle which raises the dorsal wall of rectum and posterior lip of anus or cloaca.

RESPIRATION

Ascaris lumbricoides, like most other endo-parasites, respire **anaerobically** or **anoxybiotically** because the oxygen content in host's intestine is usually poor. In this process, glycogen undergoes glycolysis and finally yields carbon dioxide, fatty acids and **energy**. Details of the process are the same as described for *Fasciola hepatica* and *Taenia solium.* Main fatty acids produced are valerianic butyric and caproic acids. These are excreted through cuticle and impart a characteristic smell like that of canned pineapple.

Aerobic respiration probably occurs whenever free oxygen is available in host's intestine. This is indicated by the presence of a small amount of **cytochrome** in parasite. According to some workers (**Hyman**), haemoglobin present in small amounts in the pseudocoelomic fluid and body wall, serves to transport oxygen.

EXCRETORY SYSTEM

[I] Excretory Organs

Excretory system is quite simple due to absence of flame cells or protonephridia. There is a H-shaped tubular excretory system. It is supposed to be formed by a huge excretory **renette cell** at the juvenile stage. Into its cytoplasm tunnel-like structures form canals of the excretory system. It consists of two lateral **longitudinal excretory canals**, right and left, connected anteriorly, below pharynx, by a **transverse canalicular network**. Each longitudinal canal extends posteriorly, along the entire body length through a lateral epidermal chord and is closed at both ends. Externally, their location is marked by the two lateral lines. Left canal is slightly wider than the right. Anterior limbs of H are reduced. Lumen of canals is devoid of cilia. A short **terminal excretory duct** extends from left side of transverse canalicular network to the **excretory pore** situated mid-ventrally, a little behind the anterior tip. The nucleus of the excretory cell lies anteriorly on the left longitudinal canal, two more smaller nuclei have also been located, one on the terminal duct and one on the transverse canalicular network, indicating that the canal system is evolved probably by more than one cell.

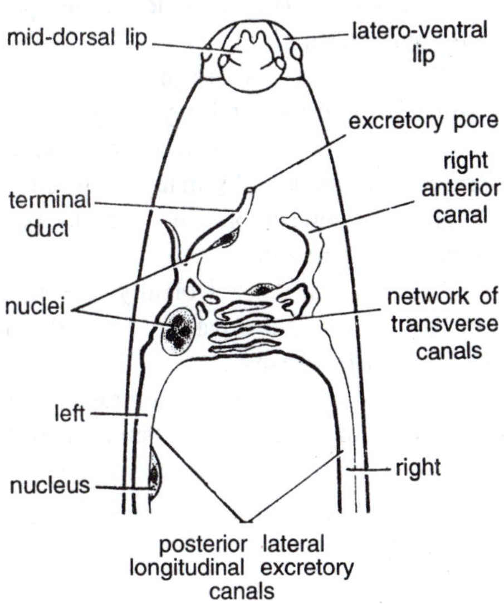

Fig. 7. *Ascaris.* 'H' type excretory system in anterior part of the body (ventral view).

[II] Physiology

Available data suggests that excretory product of *Ascaris* is mainly urea, thus it is a **ureotelic animal**. Excretory canals collect the excretory products from different parts of body. Pressure of pseudocoelomic fluid helps in ultrafiltration. Excretory products are eliminated through the excretory pore. Some ammonia and urea are also passed out with faeces through anus.

NERVOUS SYSTEM

Nervous system of *Ascaris* was exhaustively studied by **Goldschmidt** (1908-1910). According to his findings, confirmed by other workers, nerve cells forming the system are constant in number, position, form and course of fibres.

[I] Central Nervous System

It comprises of a richly ganglionated **nerve ring** or **circumenteric ring** encircling pharynx. The **ganglia** located on nerve ring are (i) a single **dorsal ganglion**, (ii) a pair of **sub-dorsal ganglia**, one on each side of dorsal ganglion, (iii) a pair of **lateral ganglia**, each subdivided to form six **lateral ganglia**, and (iv) a pair of large **ventral ganglia**.

[II] Peripheral Nerves

From the nerve ring 8 nerves run anteriorly. Out of these, 6 supply the 6 labial papillae and each bears a **papillary ganglion**, near its base. Remaining 2 nerves, known as amphidial nerves, supply the amphids. Each amphidial nerve arises from one of the 6 lateral ganglia, called **amphidial ganglion**, of its side.

Posterior part of body is supplied by nerves. A **dorsal nerve cord** (motor) runs through the dorsal epidermal chord. A **ventral ganglionated nerve cord** runs through the ventral epidermal chord. It terminates at the posterior end of body,

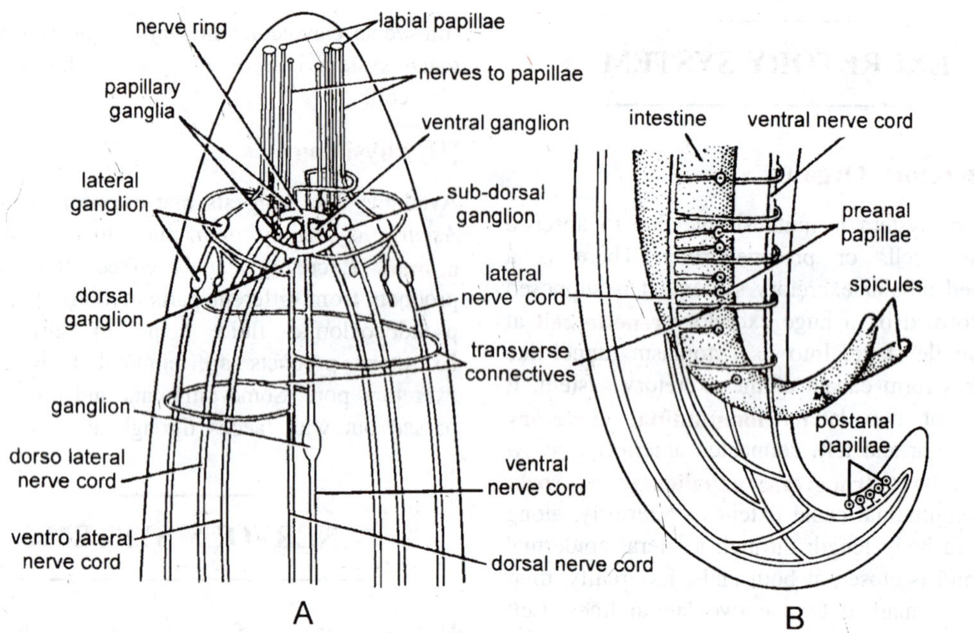

Fig. 8. *Ascaris.* Nervous system. **A.** In anterior end (dorsal view). **B.** In posterior end of male (lateral view).

after forming an **anal ganglion**. There are 2 thin, delicate **lateral nerve** chords (sensory), each extending through a lateral epidermal chord. Besides 2 **dorso-lateral** and 2 **ventro-lateral nerve cords** are present, one above and one below the lateral nerve cord, on either side. Several **commissures** or **connectives**, arranged asymmetrically along the entire length connect ventral nerve cord with lateral and dorsal nerve cords.

[III] Rectal Nervous System

In male, at the posterior end of body, (i) lateral nerve cords supply the pre-anal papillae, (ii) ventral nerve cord supplies the post-anal papillae, and (iii) dorsal, ventral and lateral nerve cords become interconnected. No such complications occur at the posterior end of the female.

SENSE ORGANS

The sense organs of *Ascaris* are simple elevations supplied by nerves. They include various papillae, amphids and phasmids.

1. Papillae. Papillae are in the form of small villi situated on different body parts.

(a) Labial papillae. These are formed by sensory cells surrounded by many supporting cells. These are gustatoreceptors and present on three lips surrounding mouth.

(b) Cervical papillae. Dorsally, about 2 mm behind lips, are present a pair of cervical papillae. These are tactile organs.

(c) Anal papillae. Present ventrally below the posterior end of male are 5 pairs of post-anal papillae and 50 pairs of pre-anal papillae. They help in copulation.

2. Amphids. On each latero-ventral lip, a single amphid is present near a single papilla. These are gustatory sensory or chemo-receptors.

3. Phasmids. These are chemo-receptor unicellular glands, opening one on either side of tail.

REPRODUCTIVE SYSTEM

Sexes in *Ascaris* are separate and **sexual dimorphism** is well defined. Males are smaller than females. They possess a **recurved tail** with pre- and post anal **papillae**, a **cloaca**, and a pair of spicules or **penial setae**. Gonads are typically long, coiled and tubular. They are attached at the genital pore in female and at cloaca in male.

[I] Male Reproductive Organs

These are confined to the posterior half of pseudocoel.

1. Testis. There is a single **testis**, a condition termed **monorchic** in contrast to the **diorchic** or two testis condition found in some nematodes. It extends to the middle of body and has the form of a long, thread-like, highly twisted tube. Wall of testis is made up of a single layer of cuboidal cells covered by basement membrane. Central axis of testis is in the form of a solid **cytoplasmic rachis**, around which are clusters of amoeboid **sperms** in various stages of development. There is no lumen.

2. Vas deferens. Distal part of testis continues into a short and thick twisted tube, the **vas deferens**. It is distinguished from testis in having a lumen in place of the central cytoplasmic rachis.

3. Seminal vesicle. Vas deferens is followed by a much thicker, wider, somewhat muscular straight tube, the **seminal vesicle**. It lies in the posterior third of pseudocoel below intestine.

4. Ejaculatory duct. Terminal part of seminal vesicle narrows to form a highly muscular **ejaculatory duct**. It joins rectum to from the cloaca. Ejaculatory duct contains a number of

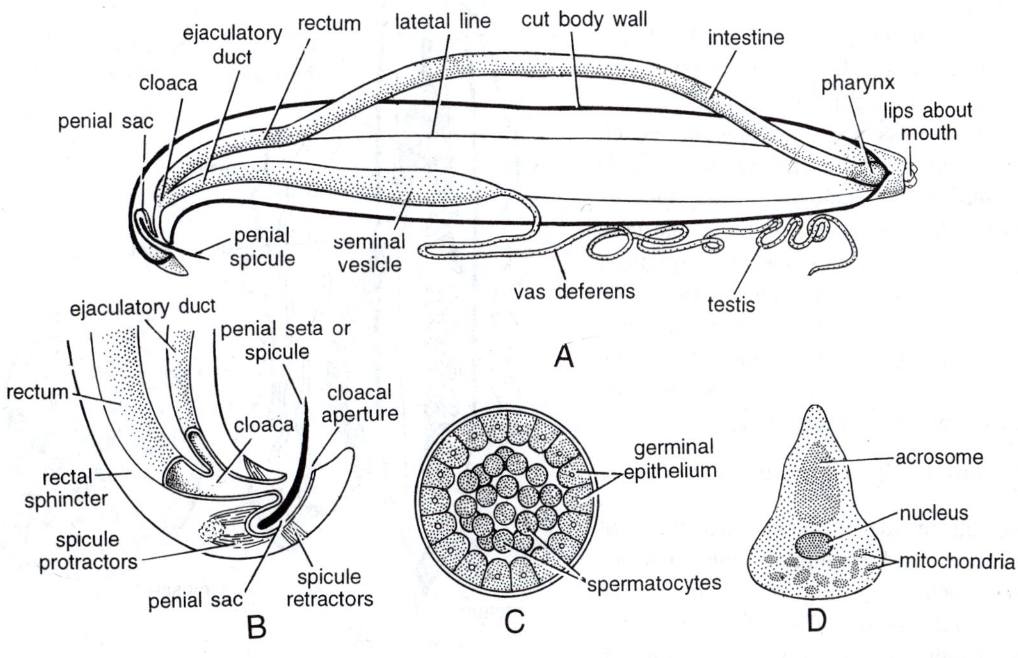

Fig. 9. *Ascaris.* **A.** Male reproductive system. **B.** Posterior end of male *Ascaris* in lateral view showing cloaca and spicules. **C.** T.S. vas deferens. **D.** A sperm.

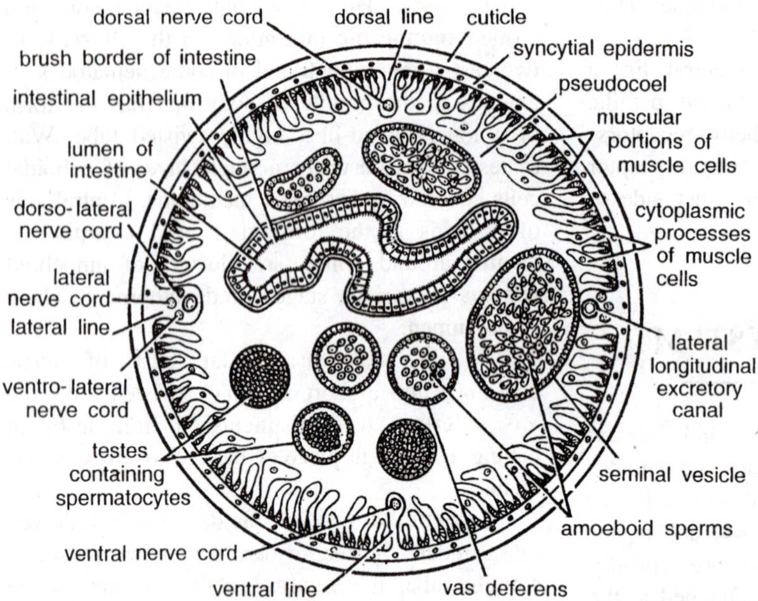

Fig. 10. *Ascaris.* T.S. of mature male.

prostatic glands whose secretions help in copulation.

5. Penial setae. Situated on the dorsal side of cloaca, and formed by evaginations of the latter, is a pair of small penial sacs or **spicular pouches**. Each spicular pouch secretes and houses a club-shaped **spicule** or **penial seta**, enclosed in a **spicular sheath**, and consists of a cytoplasmic core surrounded by thick cuticle. The spicules, each 2 to 3.5 mm long, can be protruded and retracted through the **cloacal aperture** by special **protractor** and **retractor muscles**. Spicules help in opening the female gonopore for copulation.

[II] Female Reproductive Organs

These lie in the posterior two third of pseudocoel and comprise two parallel **tracts**, each consisting of an ovary, an oviduct and a uterus. The **didelphic** or two tract condition is most common in nematodes, though **monodelphic** (one tract) and **polydelphic** (many tracts, upto 10 or 11) conditions are also met with.

1. Ovaries. These are two in number. Each **ovary** of *Ascaris* is a long, thread-like, much twisted and blind tubule. Its wall consists of a single layer of cuboidal epithelial cells lined externally by a basement membrane. Its central axis is in the form of a **cytoplasmic rachis**, around which are groups of ova. There is no lumen.

2. Oviducts. From the distal end of each ovary arises a thicker, wider and twisted **oviduct**, with a similar wall as that of the former and a lumen instead of solid cytoplasmic rachis.

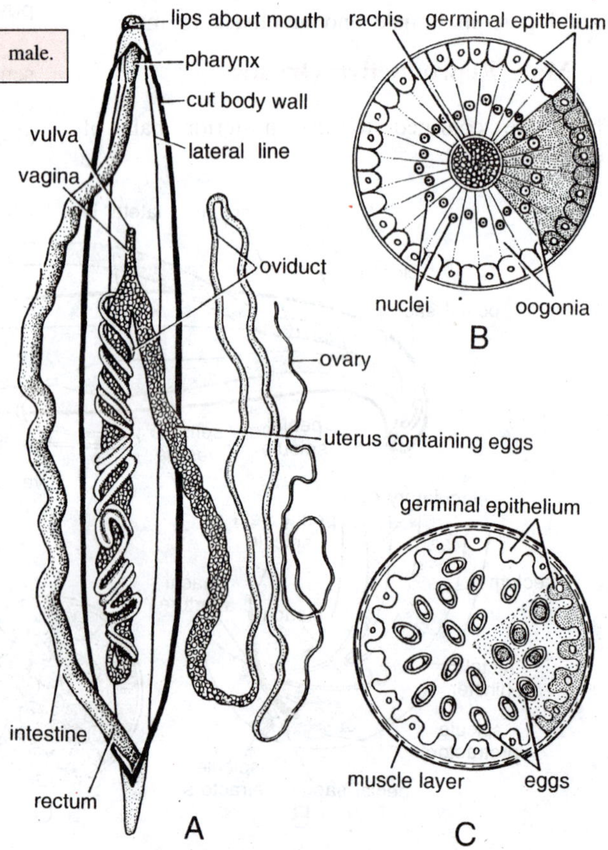

Fig. 11. *Ascaris.* **A.** Female reproductive system. **B.** T.S. ovary. **C.** T.S. uterus.

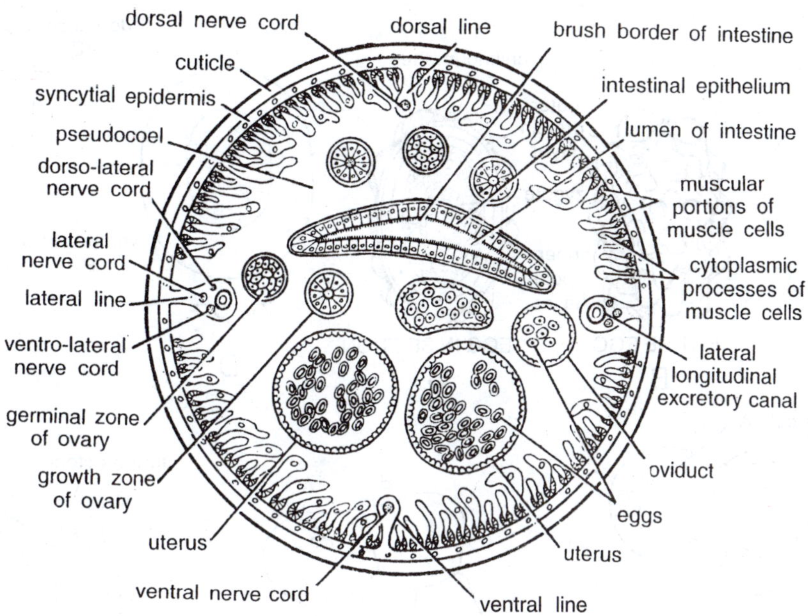

Fig. 12. *Ascaris.* T. S. of a mature female.

3. Uteri. Each oviduct continues into a much wider, thicker and almost untwisted tube, the **uterus**. Their wall is composed of tufted cells, surrounded by a muscle layer with inner **circular** and outer **oblique** fibres. The uteri serve to store fertilized eggs enclosed in shells.

4. Vagina. Near the anterior one third of body, the two uteri unite to from a median, short and highly muscular tube, the **vagina**, having an inner lining of cuticle. It opens to exterior through female gonopore or **vulva**, situated mid-ventrally at about one third distance from the anterior end.

[III] Gamete Formation

In *Ascaris,* **gametogonia** are budded off from the blind proximal end of gonad and undergo gametogenesis as they move towards the distal end. This mode of germ cell formation, termed **telogonic**, occurs in most nematodes. The alternative mode termed **hologonic**, is confined to only two orders, Trichuroidea and Dioctophy moidea. In this latter mode, the germ cells proliferate along the entire length of gonad. In a **telogonic gonad**, three zones can be distinguished :

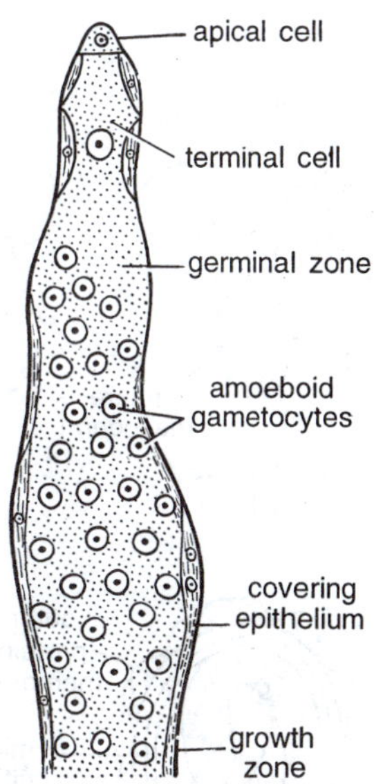

Fig. 13. *Ascaris.* Telogonic origin of sex cells in the proximal end of gonad. Maturation zone of gonad not shown.

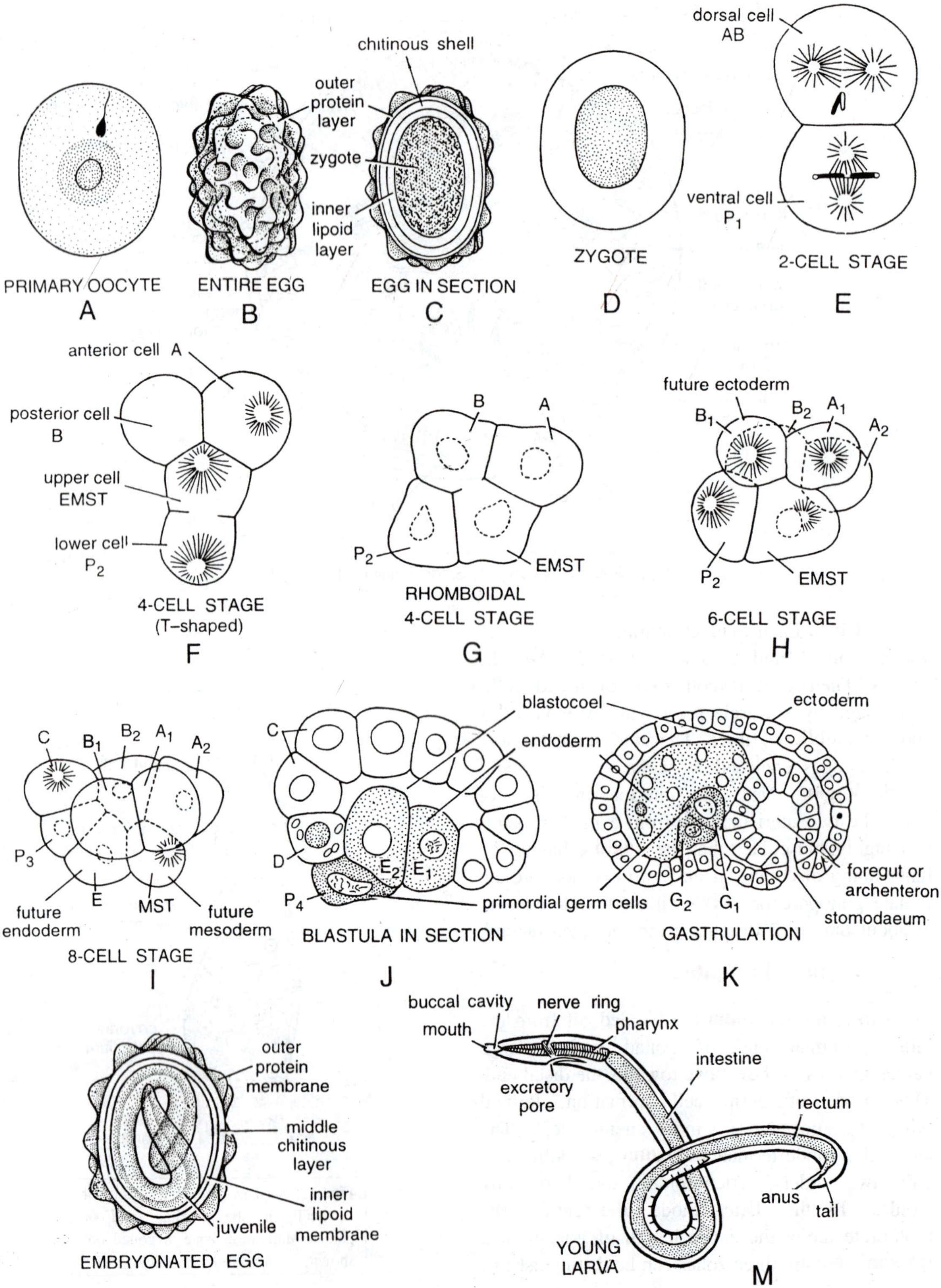
chitinous shell
outer protein layer
zygote
inner lipoid layer
PRIMARY OOCYTE
A
ENTIRE EGG
B
EGG IN SECTION
C
ZYGOTE
D
dorsal cell AB
ventral cell P_1
2-CELL STAGE
E
anterior cell A
posterior cell B
upper cell EMST
lower cell P_2
4-CELL STAGE (T–shaped)
F
B
A
P_2
EMST
RHOMBOIDAL 4-CELL STAGE
G
future ectoderm
B_1
B_2
A_1
A_2
P_2
EMST
6-CELL STAGE
H
C
B_1
B_2
A_1
A_2
P_3
E
MST
future endoderm
future mesoderm
8-CELL STAGE
I
C
D
P_4
E_2
E_1
blastocoel
endoderm
primordial germ cells
BLASTULA IN SECTION
J
ectoderm
G_2
G_1
foregut or archenteron
stomodaeum
GASTRULATION
K
outer protein membrane
middle chitinous layer
inner lipoid membrane
juvenile
EMBRYONATED EGG
L
buccal cavity
nerve ring
mouth
pharynx
excretory pore
intestine
rectum
anus
tail
YOUNG LARVA
M

(*Fig. 14. Contd...*→)

1. Germinal zone. Also known as **proliferation zone**, it lies at the proximal blind end of gonad. Usually a single large **terminal cell** constitutes this zone. A large **apical cell** which is in fact an epithelial cell of the gonad wall, lies in front of the terminal cell, from which gametogonia arise by budding.

2. Growth zone. It follows the germinal zone. Here **gametogonia**, which lie attached to the cytoplasmic rachis, grow and differentiate into **amoeboid gametocytes**.

3. Maturation zone. It is the distal most zone, where **gametocytes** separate from rachis and undergo **maturation** to form **gametes**. It is followed by the **gonoduct**.

Spermatozoa (sperms) are amoeboid, while the **ova** are elliptical. Ova at this stage are actually **secondary oocytes**, having resulted from oogonia by a single maturation division. Second maturation division occurs after **fertilization**.

LIFE HISTORY

[II] Copulation and Fertilization

Copulation takes place in the small intestine of host (man) where the adult worms live. The male orients its body at right angles to that of female in such a way that its cloacal aperture apposes the vulva of female. Penial setae of male help to open the vulva and sperms are transferred into vagina, from where thay pass on to the proximal ends of uteri. **Fertilization**, which is effected by the entry of a sperm into ova, takes place in the proximal ends of uteri or distal ends of oviducts. After fertilization, ova undergo second maturation division.

[II] Zygote

Unfertilized egg contains globules of glycogen and fat. Immediately after fertilization, glycogen globules migrate to the surface and form a **fertilization membrane**, which hardens into a thick, clear and **chitinous shell**. Next, the fat globules form a thin **lipoid layer** below shell. As the fertilized egg or **zygote** moves down the uterus, the uterine wall secretes around it a thick hard, yellow or brown **albuminous** (proteinous) **coat** having a typical wavy surface *(rippling)*. Fertilized egg at this stage is elliptical, measuring 60-70μ by 40-50μ.

Uteri of a single mature female may contain as many as 27 million eggs (**Cram**, 1925) with an average daily production of some 2,00,000 eggs. These leave mother's body through gonopore into host's intestine and finally pass out with the host's faeces. Shelled eggs are remarkably resistant and remain alive in moist soil for several years under adverse conditions.

Under suitable conditions of temperature, moisture and oxygen, eggs undergo cleavage and develop into the infective stage. Optimum temperature for development is 85°F. Below 60°F development stops and above 100°F the eggs gradually degenerate.

[III] Cleavage and Early Development

Cleavage is of **spiral** and **determinate** type. First division results in a dorsal (**AB**) and a ventral cell (**P_1**). Dorsal cell divides into an anterior (**A**) and a posterior cell (**B**), while ventral cell divides into an upper (**EMST**) and a lower cell (**P_2**). The four-celled embryo, thus formed, is first **T-shaped**, but soon becomes **rhomboidal** as **P_2** comes to lie posterior to **EMST**.

In the next cleavage, **A** and **B** divide into right and left cells (**A_1, A_2** and **B_1, B_2**), **EMST** into **MST** and **E**, and **P_2** into **P_3** and **C**. Further, **E** divides into **E_1** and **E_2** and **P_3** divides into **P_4** and **D**. **P_4** further divides into **G_1** and **G_2**. Fate of various cells at this stage is fixed. (i) Descendants of A and B give rise to the **ectoderm**, (ii) **MST** gives rise to **mesoderm** and to **ectoderm** of foregut. (iii) **E_1** and **E_2** give rise

Fig. 14. *Ascaris*. Stages of development. **A.** Primary oocyte. **B.** Entire egg. **C.** Egg in section. **D.** Zygote without shell. **E.** 2 cell stage. **F.** T shaped 4-cell stage. **G.** Rhomboidal 4- cell stage. **H.** 6-cell stage. **I.** 8-cell stage. **J.** Blastula in median sagittal section. **K.** Early embryo in median sagittal section. **L.** Embryonated egg with first stage larva. **M.** Second free stage larva.

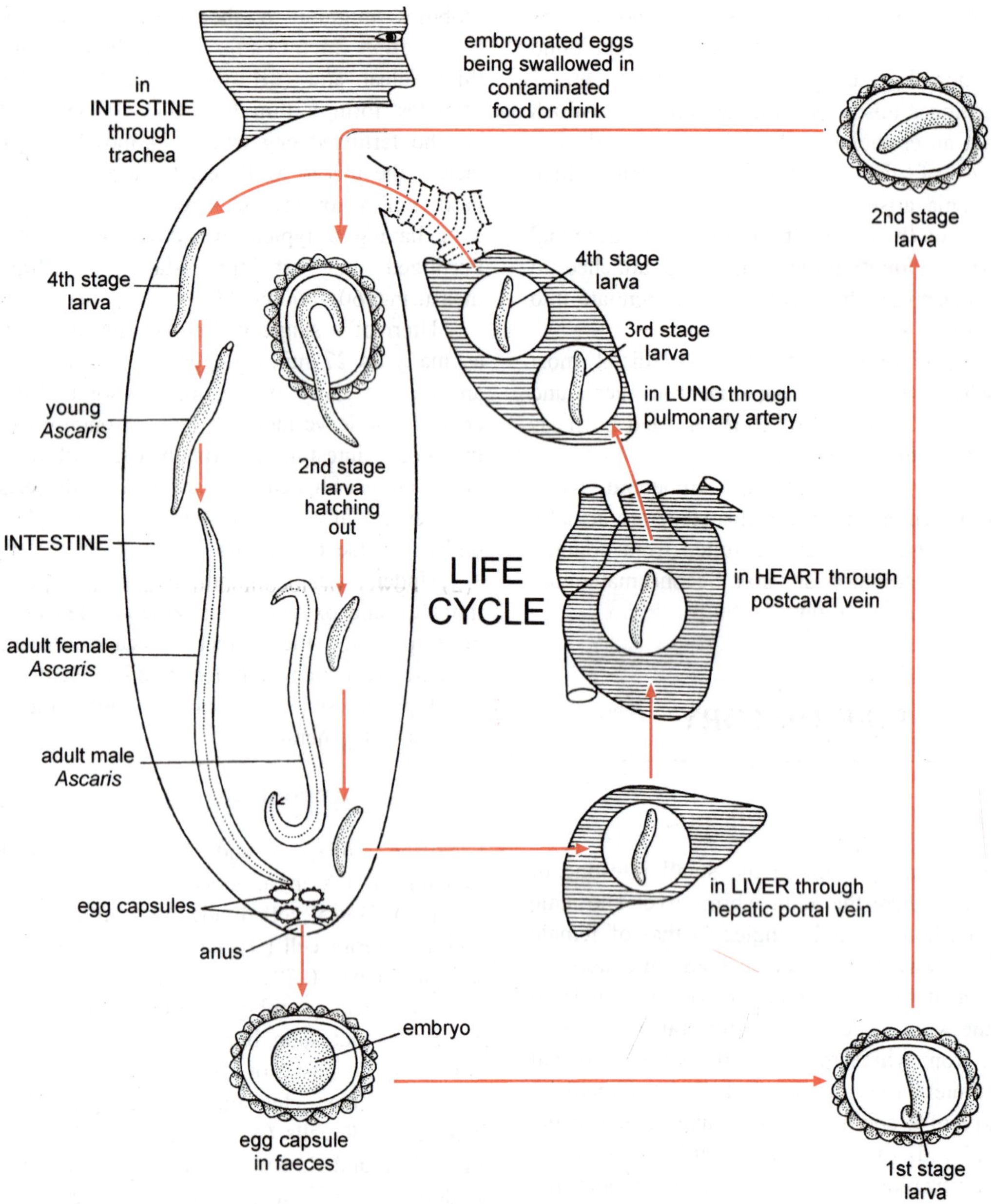

Fig. 15. *Ascaris.* Life cycle.

to **endoderm**. (iv) G_1 and G_2 form the germ cell primordium and (v) **C** and **D** together give rise to the **ectoderm** and **mesoderm**.

16-celled embryo attains the form of a hollow ball. It is termed **blastula**, and its cavity, the **blastocoel**. Blastula undergoes a process of invagination and becomes **gastrula**. It grows in length to become an active **juvenile** in 10-14 days from the start of cleavage. Juvenile has an alimentary canal, a nerve ring and a larval excretory system. For its close resemblance with the nematode genus *Rhabditis,* found in soil and human faeces, the juvenile is also termed **rhabditoid** or **rhabditiform larva** of **first stage**. It is not infective. In another week's time, it moults within the egg shell and becomes the **second stage rhabditoid**, which is capable of infecting the host. Under suitable conditions of moisture, oxygen and temperature, infective eggs of *Ascaris lumbricoides* are known to remain viable for about six years.

[IV] Infection of New Host

There being no intermediate host, man acquires infection by directly ingesting *Ascaris* eggs, containing the infective second stage rhabditoid larva, with contaminated food or water. In the small intestine, by the action of host's digestive juices, the egg shells dissolve and the **juveniles** (second stage larvae) hatch out.

[V] Later Development and Migration

At hatching, a juvenile measures 0.25 to 0.3 mm in length and 13μ to 15μ in diameter. It performs active thrashing movements and bores through the epithelium of host's intestine and starts its migration in host's body.

1. Primary migration. Larva enters the hepatic portal circulation which carries it to the liver. From liver it finally reaches the heart through post-caval vein. From heart it is transported to the lung via pulmonary artery. Larva generally remains in lung for a few days and increases in size. Then it ruptures out of blood capillary and finally bores its way into **alveoli**. After about 6 day's stay there, the second stage larva moults to become the **third stage larva**. Another moult after 4 days results in the **fourth stage larva**, which has by this time grown to a length of 2 to 3 mm.

2. Secondary migration. Fourth stage larva leaves its temporary home in lung alveoli and through trachea reaches the pharynx, from where it is coughed up and then swallowed for the second time into gut. In the intestine, it moults for the fourth and the last time to become an **adult**.

Adult attains sexual maturity within 8-10 weeks. The average life-span of *Ascaris lumbricoides* in the host is 9-12 months.

3. Aberrant migration. Sometimes larva of *Ascaris* does not follow its usual migration path but reaches the brain or spinal cord or any such organ. Larva is not able to survive in these organs and a calcareous cyst is formed around it.

PARASITIC ADAPTATIONS OF *ASCARIS*

Ascaris shows several adaptations to its parasitic mode of life in human intestine.

(1) Tough, thick and resistant cuticle, covering the body, shields against the action of hosts digestive enzymes and antitoxins. *Ascaris* also secretes anti-enzymes that protect it from hosts digestive enzymes.

(2) Power of locomotion counteracts peristalsis of host's intestine and compensates for the absence of adhesive suckers and spines. This helps the worm from getting disloged and remain in host's intestine.

(3) Muscular pharynx facilitates ingestion of food by sucking action.

(4) There is a continuous supply of food, as a result the alimentary tract is simple without provision for storage.

(5) Ingested food is pre-digested, so that there are no elaborate digestive glands.

(6) Pseudocoelomic fluid serves for absorption, transport and distribution of food, oxygen and wastes, so that there is no circulatory system.

(7) Extremely low metabolic rate and anaerobic respiration enables the worm to live inside the host's intestine, where free oxygen is negligible.

(8) Parasite remains well protected inside the host's intestine, from predators, so that there is no need for complicated sensory organs which are quite simple.

(9) Reproduction is wasteful in the extreme as transfer to a new host is full of many hazards. Firstly, the transfer to a new host is an entirely passive process depending on an accidental ingestion of embryonated eggs by some unsuspecting host. Chances become all the more difficult because the host must be

a specific one. Eggs laid by a pig *Ascaris* must be ingested by another pig, and similarly, the eggs laid by a human *Ascaris* must be picked up by another human. Moreover, the eggs must have undergone some embryological development before ingestion by another host. If an egg is not embryonated and ingested prematurely, it will not develop into a new worm. Further, only a warm, shady and moist environment is suitable for proper development of the eggs. Dehydration and very high (above 100°F) or low temperatures (below 60°F) are usually fatal to the developing embryos. These and many other hazards are overcome by these parasites by producing enormous quantity of eggs to ensure greater chances for survival. Thus, a single mature female may contain as many as 27 millions of eggs with a daily production of 234,000 eggs or about 162 eggs per minute, which also compensates for the lack of asexual multiplication and hermaphroditism.

(10) Resistant covering or shell provides safety to the zygotes and embryonated eggs from unfavourable environmental factors. They remain viable for years.

(11) Minute size and resistant nature of eggs affords far and wide dispersal of the parasite.

(12) Infection of a new human is direct, without an intermediate host, which makes transfer from one host to another easier and safer.

ASCARIASIS

[I] Infection

Disease caused by *Ascaris* is commonly called **Ascariasis**. Man gets infection by consuming food and water contaminated with infective eggs. Incidence of Ascariasis is greater in children than in adults.

[II] Pathogenesis

Larvae often prove more injurious than adult worms by causing haemorrhages. They sometimes bore through intestinal epithelium and enter general circulation, which may land them in some vital organ structures like kidneys, spinal cord, brain or muscles, where they may cause serious injuries. Severe haemorrhagic conditions develop in lungs where they cause petechial haemorrhages and inflammation of alveolar tissue followed by oedema. In severe infection, they may cause acute pneumonia with fatal consequences. Frequently, infection is immediately followed by temperature, anaemia, leucocytosis and eosinophilia.

Adults usually cause **enteritis** and through their migration into vermiform appendix, gall bladder and common bile duct, may cause inflammation of these structures.

Adults draw their nourishment from contents of intestine and may suck blood from its walls. Worm produces **toxins** which may cause irritation of mucous membrane, nervous symptoms like convulsions (involuntary contraction of voluntary muscles), delirium (light headedness), coma (deep sleep), and nervousness. A substance produced by parasite combines with trypsin, thus interfering with protein digestion which leads to protein deficiency and hence stunted growth especially among childern. Presence of even a few parasites in intestine may result in intense colic pains, abdominal discomforts, diarrhoea, vomiting and mild temperature. Administration of drugs often irritates the parasites, which consequently get entangled in masses to block the lumen. About a hundred to a thousand or even more (up to 5,000 have been recorded) worms may be involved and may prove fatal, if not promptly removed by surgery.

[III] Therapy

Infection may be diagnosed by examining the stool for *Ascaris* eggs. Infection can be treated with a dose of **hexylresorcinol crystals** in a gelatin capsule after about 12 hours fasting. The dose, followed by a fast of another four hours, kills the worms, which can be finally expelled by a purgative like sodium sulphate. Some anti-helminth drugs, like oil of chenopodium, are very effective by being highly toxic, while some like tetrachlorethylene simply irritate the worms, which thus entangle and block the lumen. But a mixture of tetrachlorethylene and oil of chenopodium is very effective. Some modern successful antihelminth drugs for Ascariasis are **Mebendazole** (Mebex), **Piprazine, Pyrantel Pamoate** (Nemocid), and **Albendazole** (Zentel).

[IV] Prophylaxis

(i) Soil pollution, being the chief source of infection, should be prevented. People, specially children, should be made to observe sanitary habits. (ii) Vegetables grown in polluted soil should be thoroughly washed and boiled before consumption. (iii) Finger nails should be regularly cut to avoid accumulation of egg below them and hands should be properly soap-washed before eating. If reinfection is completely checked, the worms can be got rid of in 9-12 months even if treatment is not taken.

Important Questions

➤➤ *Long answer type questions*

1. Give an account of the morphology and physiology of *Ascaris lumbricoides.*
2. Describe the reproductive organs of male and female *Ascaris.* Mention differences between the two.
3. Give an illustrated account of the life history of *Ascaris.*
4. What disease is caused *by Ascaris*'? Write an account of its pathogenicity, therapy and prophylaxis.
5. Write a detailed note on parasitic adaptations in *Ascaris.*
6. Make full-page labelled diagrams of T.S. of male and female *Ascaris* passing through the middle of body. No description is needed.

➤➤ *Short answer type questions*

1. What is the shape of sperm of *Ascaris* ?
2. What are the accessory structures that are present in the penial sac of *Ascaris* ?
3. Why is the living *Ascaris* not digested in the human intestine ?
4. Explain ten features of the Rhabditi form larva of *Ascaris.*
5. Trace the journey of *Ascaris* larva through the human host.
6. Define extra-intestinal migration ? Explain with reference to *Ascaris.*
7. With the aid of sketches, mention the sexual dimorphic features of *Ascaris lumbricoides.*
8. Give an illustrated account of male and female reproductive organs of *Ascaris.*
9. Draw labelled sketch of the T.S. through the middle of the body of female *Ascaris.*
10. Draw the T.S. of male *Ascaris* and mark all the parts.
11. Explain the structure of muscle fibre in nematode with a suitable sketch.
12. Draw T.S. of female *Ascaris,* label it and mention five points in which it differs from that of *Fasciola hepatica.*

➤➤ *Fill in the blanks*

1. The innermost layer of mammilated eggs of *Ascaris* is formed of.....
2. Gonads of *Ascaris* are .....
3. Body cavity in *Ascaris* is .....
4. The sperm of *Ascaris* is .... in shape.
5. The lumen of *Ascaris* pharynx is ....
6. The hydrostatic skeleton of *Ascaris* consists of.....

➤➤ *True and false statements*

1. Female *Ascaris* is didelphic. True / False / Do not know
2. Sperm of *Ascaris* is amoeboid in shape. True / False / Do not know
3. Male *Ascaris* is longer than the female and has anus as in female.
4. Caudal end of all male round worms is curved. True / False / Do not know
5. Rhabditiform larva performs extra intestinal migration. True / False / Do not know

➤➤ *Multiple choice questions*

1. In *Ascaris,* the reserve food is stored as glycogen and fats in :
   (a) intestine (b) muscles
   (c) hypodermis or epidermis
   (d) all of these
2. The mammilated eggs of *Ascaris* are enclosed in :
   (a) one layered membrane
   (b) two layered membrane
   (c) three layered membrane
   (d) none of the above

3. Sheathed microfilaria normally circulate at night :
   (a) in the arteries
   (b) in the caval veins
   (c) in the peripheral blood vessels
   (d) deep in the lymph vessels
4. The mouth of *Ascaris* is guarded by :
   (a) two lateral lips
   (b) one dorsal and two ventro-lateral lips
   (c) one ventral.and two dorso-lateral lips
   (d) one ventral, one dorsal and two lateral lips
5. When egg of *Ascaris* is swallowed by man it settles in intestine after :
   (a) 3 days (b) one week
   (c) two weeks
   (d) three weeks
6. While migrating the larva of *Ascaris* takes its 4th moult in :
   (a) intestine (b) liver
   (c) lungs (d) heart
7. In which of these animals, the syncytial epidermis follows the cuticle :
   (a) *Fasciola* (b) *Ascaris*
   (c) *Taenia* (d) Earthworm
8. Pineal satae are present in :
   (a) earthworm (b) *Ascaris*
   (c) cockroach (d) none of these
9. The pseudocoelom of *Ascaris* is :
   (a) an intracellular space in mesodermal cells
   (b) a space between few mesodermal and endodermal cells
   (c) a space in between botryoidal tissue
   (d) intercellular space between false mesodermal cells
10. *Ascaris* has a body cavity which is :
    (a) intracellular (b) pseudocoel
    (c) intercellular and pseudocoel
    (d) highly obliterated by cells
11. *Ascaris* has no :
    (a) cephalisation (b) alimentary canal
    (c) receptors (d) larva
12. Maximum number of eggs released by single female *Ascaris* in **any** day is :
    (a) 2,000 (b) 20,000 (c) 5,000 (d) 2,00,000
13. Final moult in life cycle of *Ascaris* takes place in :
    (a) soil (b) lung
    (c) intestine (d) intestine before migration
14. *Ascaris* is :
    (a) host (b) bisexual
    (c) dioecious (d) digenetic
15. Infective stage of *Ascaris* for new host :
    (a) I stage juvenile (b) embryonated egg
    (c) II stage juvenile (d) larva after I moult
16. Embryonated egg is :
    (a) shelled egg (b) shelled zygote
    (c) shelled II stage juvenile
    (d) II stage juvenile
17. The stage hatched from the ingested egg of *Ascaris* is called :
    (a) bladder worm (b) hexacanth
    (c) maggot (d) rhabditiform larva
18. Ovaries of *Ascaris* are :
    (a) monodelphic (b) diadelphic
    (c) opisthodelphic (d) none of these
19. The lips of *Ascaris* are :
    (a) denticulate (b) smooth
    (c) cartilaginous (d) bony
20. Muscle cells are found in :
    (a) body wall of *Ascaris*
    (b) circular muscle layer of *Taenia*
    (c) medullary parenchyma of *Fasciola*
    (d) mesogloea of *Hydra*
21. The correct statement for *Ascaris* is that it has :
    (a) diploblastic body (b) segmented body
    (c) radial symmetry (d) denticulate lips
22. Epidermis in body wall of *Ascaris* is :
    (a) cellular (b) syncytial
    (c) with antienzymes (d) none of these
23. Excretory organs in *Ascaris* are :
    (a) nephridia (b) kidneys
    (c) flame cells (d) none of these
24. Two sexes of *Ascaris* can be identified as :
    (a) post, end of female is straight
    (b) anal papillae are found in male
    (c) female has vulva (d) all of these
25. Penial setae arise from :
    (a) cloaca of male (b) anus of female
    (c) vulva of female (d) none of these
26. In *Ascaris* eggs are fertilized in :
    (a) oviduct (b) uterus
    (c) vagina (d) outside human body
27. Eggs of *Ascaris* passed out of hosts body are :
    (a) unfertilized
    (b) fertilized
    (c) fertilized and shelled
    (d) with a larva
28. During development *Ascaris* needs :
    (a) one intermediate host(b)two definitive hosts
    (c) no intermediate host(d) no definitive host
29. In life cycle of *Ascaris* the juvenile hatches out of egg in :
    (a) lung (b) liver
    (c) intestine (d) none of these
30. The final moult in Juvenile of *Ascaris* occurs in :
    (a) lung (b) intestine
    (c) liver (d) none of these
31. *Ascaris* is not :
    (a) endoparasite (b) monogenetic
    (c) free living (d) dioecious
32. *Ascaris* has well developed :
    (a) nervous system (b) reproductive system
    (c) receptors (d) digestive system

33. In the development of *Ascaris,* the second stage juvenile is formed :
    (a) within the eggs itself
    (b) in the intestine of man
    (c) in the lung of man
    (d) in the liver of man
34. The musculature in the body wall of *Ascaris* comprises :
    (a) outer longitudinal and inner circular muscles
    (b) outer circular and inner longitudinal muscles
    (c) longitudinal muscles only
    (d) circular muscles only
35. The reproductive organs of *Ascaris* comprise :
    (a) one pair of testes in male and one pair of ovaries in female
    (b) a single testis in male and one pair ovaries in female
    (c) one pair of testes in male and a single ovary in female
    (d) a single testis in male and a single ovary in female
36. The kind of body cavity found in *Ascaris* is known as :
    (a) haemocoel (b) enterocoel
    (c) coelenteron (d) pseudocoel
37. Syncytial epidermis is found in :
    (a) *Hydra*
    (b) *Ascaris*
    (c) earthworm
    (d) star fish
38. The adult male of *Ascaris* can be identified externally by the presence of :
    (a) pineal setae
    (b) fifty pairs of pre-anal papillae
    (c) five pairs of post anal papillae
    (d) all of the above
39. The circumentric nerve ring of *Ascaris* is located around:
    (a) buccal chamber (b) pharynx
    (c) oesophagus (d) stomach
40. How many ganglia are attached to the nerve ring in *Ascaris*?
    (a) one (b) two
    (c) nine
    (d) numerous
41. Median unpaired structure in female *Ascaris* is:
    (a) vagina (b) uterus
    (c) oviduct (d) ovary

Answers

➤➤*Fill in the blanks*

1. lipid, 2. telogonic coiled and tubular, 3. pseudocoel, 4. amoeboid, 5. triradiate, 6. Pseudocoel filled with proteinous fluid.

➤➤*True and false statements*

1. True, 2. True, 3. False, 4. True, 5. True.

➤➤*Multiple choice questions*

1. (c)	2. (c)	3. (b)	4. (b)	5. (c)	6. (a)	7. (b)	8. (b)	9. (b)	10. (b)
11. (a)	12. (d)	13. (c)	14. (c)	15. (b)	16. (c)	17. (d)	18. (b)	19. (a)	20. (a)
21. (d)	22. (b)	23. (d)	24. (d)	25. (a)	26. (a)	27. (d)	28. (c)	29. (c)	30. (b)
31. (c)	32. (b)	33. (a)	34. (c)	35. (a)	36. (d)	37. (b)	38. (d)	39. (b)	40. (c)
41. (a).									

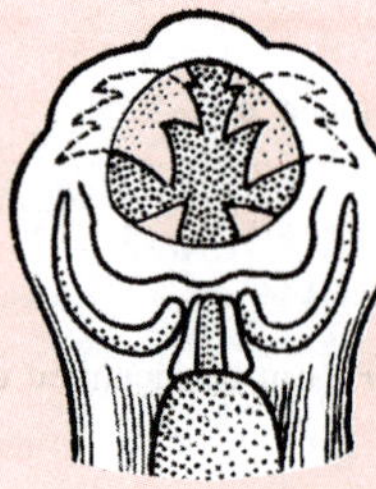

36

Ancylostoma duodenale : The Common Hookworm

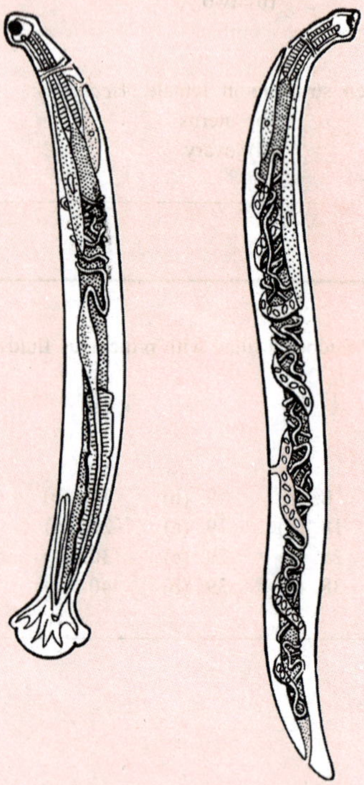

Hookworms are the most dangerous parasitic roundworms, causing serious infections in humans. They belong to the order Strongyloidea of phylum Nematoda. They differ from ascaroid nematodes in the absence of **lips** but in the presence of a cuticulate buccal capsule of taxonomic importance. An outstanding morphological feature is a **copulatory bursa** in male, supported by rays, disposition of which is also of taxonomic significance. Following description applies mainly to *Ancylostoma duodenale,* the common hookworm. It was discovered in 1838 in an autopsy of an Italian peasant woman by **Dubini**.

Ancylostoma duodenale

SYSTEMATIC POSITION

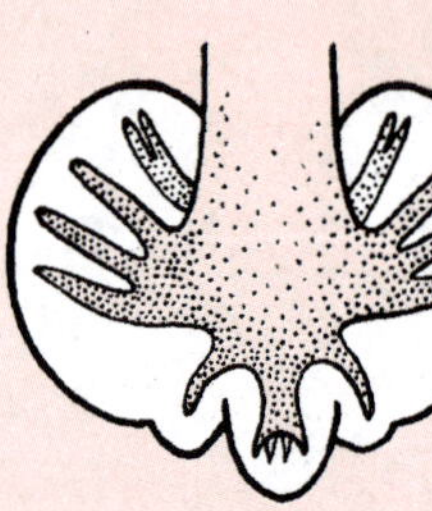

Phylum	Nematoda
Class	Phasmidia
Order	Strongyloidea
Family	Ancylostomidae
Genus	*Ancylostoma*
Species	*duodenale*

GEOGRAPHICAL DISTRIBUTION

About one-half billion people, or nearly 25% of world population, are infected by hookworms. *Ancylostoma duodenale,* is commonly known as the 'Old World Hookworm'. It is common in tropical, subtropical and temperate regions of Asia, Africa, Europe, Pacific Islands and two Americas. However, the incidence is greater in Europe and Asia. Some other species, namely, *A. caninum* and *A. braziliense,* both in dogs and cats, and *A. malayanum* in bears, are also common.

HABITS AND HABITAT

Hookworms flourish under primitive conditions where people go barefoot, modern sanitary conditions do not exist and human faeces are deposited on ground. The adult worms live parasitically in the intestine of man where they suck blood, lymph, bits of mucous membrane and tissue fluids from the lining of the intestinal wall. The infective juveniles enter the human host percutaneously from the soil contaminated by faeces in which they abound.

MORPHOLOGY

[I] External Features

They are called 'hook-worms' because the male has a hook-shaped body. Besides, prominent hooks are present in the genital bursa of male, while smaller hooks occur within the mouth of both sexes.

1. Shape and size. Sexes are separate. Mature worms are cylindrical in shape and white or grey in colour. Adult male is 8 to 11 mm long and 0.4 to 0.5 mm in diameter; while adult female is generally larger, 10 to 13 mm long and 0.6 mm in diameter.

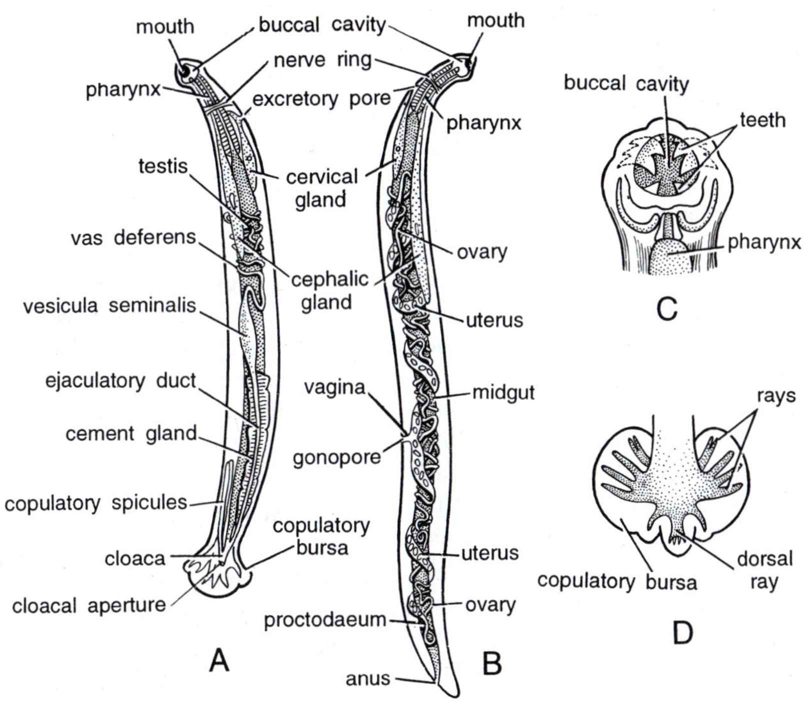

Fig. 1. *Ancylostoma duodenale* : **A.** Adult male. **B.** Adult female. **C.** Anterior end. **D.** Posterior end of male.

2. Buccal capsule. Anterior end of male and female is similar, somewhat narrower and curved dorsally, probably the reason for their being called the hookworms. Mouth leads into a large cup-shaped cavity or **buccal region** or capsule, lined internally by cuticle. Entrance to mouth is armed with a pair of chitinous ventro-lateral cutting plates bearing teeth. Two sharp teeth or lancets are also present at the base of buccal capsule which serves for attachment with the intestinal wall of host. Dorsal curving of anterior end of body presumably permits a better grip of chitinous plates and lancets on the hosts intestinal epithelium.

3. Copulatory bursa. Posterior end of female tapers, while that of male is in the form of a broad, umbrella-like structure, the **copulatory bursa**, surrounding the cloaca. Bursa is supported by fleshy bursal rays having a definite arrangement. Bursal rays and teeth in buccal capsule are of taxonomic value.

[II] Body Wall

In cross section, bodywall includes from outside a cuticle, an epidermis and a longitudinally directed musculature. Body cavity is a pseudocoel surrounding the following organ systems.

[III] Digestive System

Digestive system comprises a mouth, a buccal capsule, a muscular pharynx with a tri-radiate lumen lined with cuticle, an oesophageal bulb, an intestine, a rectum and an anus in female but a cloaca in male. Food consists of mucous membrane of host's intestine, blood cells and serum. Blood and other fluids are sucked by the sucking action of pharynx from holes cut by tiny teeth or lancets, in the intestinal wall of host, that has been pinched into buccal capsule by the cutting plates. Worms secrete an anti-coagulant in wound that prevents blood from clotting. Wounds keep oozing blood for some time even though the worms have moved to another location. Each adult worm sucks about 0.8 cc of blood per day from the host, resulting in severe anaemia. Digestion occurs in the intestine.

[IV] Excretory System

Excretory system is similar to that of *Ascaris*. It includes two longitudinal canals in the lateral cords and short anterior excretory canal which opens to the exterior through the excretory pore.

[V] Reproductive System

Male has a single tubular thread-like testis twisted around the intestine in the middle of body. It leads posteriorly into a vas deferens, followed by a seminal vesicle. Posterior end of seminal vesicle tapers into an ejaculatory duct which enters the cloaca.

Female has two much convoluted ovarian tubules, one anterior and the other posterior, to the level of gonopore. Each ovary leads into a short oviduct that opens into a dilated seminal receptacle. From the latter arises a muscular uterus that terminates into a vagina. Two vaginae become confluent and open through the common gonopore or vulva.

LIFE CYCLE

Copulation and Fertilization

Hookworms mate in the host's intestine. During copulation, bursa of male is applied on the vulva of female and sperms transferred. Fertilization occurs in the seminal receptacles.

Eggs

Fertilized eggs pass out with the host's faeces. Each female produces about 9,000 fertile eggs per day. Eggs are oval in shape and protected by hyaline chitinous shells. Zygote within has already cleaved into a four or eight-celled embryo.

Larvae

Under favourable environmental conditions of moisture, oxygen supply and temperature (68-85°F), embryo develops into the **first-stage juvenile** or **rhabditiform larva** which hatches out within 24 to 48 hours. Newly hatched larva has a mouth, a buccal capsule, an elongated pharynx with an oesophageal bulb, and an intestine. It feeds on bacteria of faeces or other organic debris of soil for 4 to 5 days and moults twice to form a **third-stage juvenile** or **filariform larva**. This juvenile is about 0.5 mm long and infective for

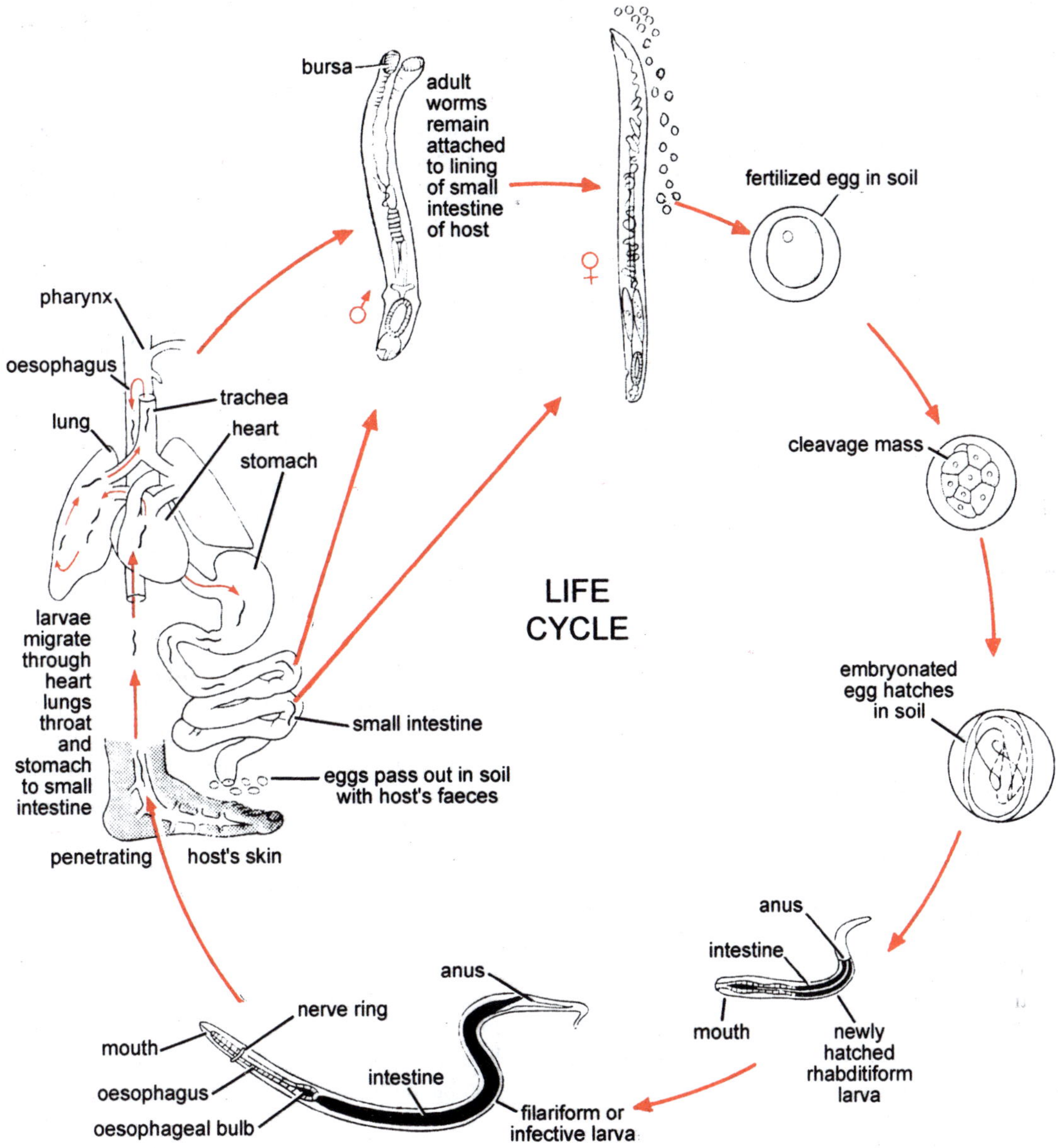

Fig. 2. *Ancylostoma duodenale.* Life cycle.

man. It has a non-patent mouth and therefore does not feed, But it remains alive and infective for several weeks under favourable conditions.

Infection of a New Host

Filariform larva **infects** a new host (man) by chance contact with his skin. Its anterior end is equipped with oral spears which enables larva to penetrate skin of a potential human host. Larvae may bore through skin in any part of body. Generally they penetrate the soft skin on the sides of feet and hands, through hair follicles. Their penetration is generally accompanied by a severe dermatitis called 'ground itch', characterized by ulceration of skin about wounds. Besides percutaneous entrance, infection is also possible through ingestion of contaminated food and water.

Larval Migration

Within 24 hours of infection, larvae reach blood vessels and follow the same migratory course described for the larvae of *Ascaris.* In blood stream they are carried first to the right ventricle of heart and then to lungs by way of pulmonary arteries. In lungs they break out of pulmonary capillaries into alveolar spaces. From lungs larvae ascend trachea to reach pharynx, become swallowed through oesophagus and finally reach the small intestine (ileum), where they become attached to the mucous lining and feed on blood of host. In 5 to 6 weeks, they moult twice to become adult. Male and female copulate and female starts laying eggs which pass out in the faecal material of human host.

Normal life-span of hookworms is 5 to 7 years that can extend to a maximum of 16 years (**Taylor**, 1933).

ANCYLOSTOMIASIS

Ancylostomiasis is the name given to infection caused by the Old World hookworm *Ancylostoma duodenale.* Larvae which occur in contaminated soil, invade the host (man) and burrow through skin of feet into blood.

Diagnosis

Hookworm infection is diagnosed by the observation of hyaline 'eggs' in a faecal smear from infected person.

Pathogenesis (Symptoms)

Pathogenicity is fairly extensive involving skin, lungs and small intestine. Penetrating **larvae** cause characteristic tiny and irritating sores, called **dew sores** or **ground itching** and inflammation of skin. In lungs, migratory **larvae** cause petechial haemorrhages and bronchial pneumonitis. But greatest damage occurs in small intestine by the adult worms. Here, the punctured wounds continue to bleed for some time so that more blood is lost than worms can consume. This results in severe anaemia, decrease in general immunity and bloody stools. Intestinal wounds may lead to various forms of detrimental infections. Deleterious toxins secreted by glands in head region of worms cause stomach pain, food fermentation, diarrhoea, constipation, dyspnea, palpitation of heart, eosinophilia, loss of health and collapse. Mental and physical growth is retarded in children and growing youth. Patient tries to ease the epigastric pain by eating, even dirt, hence the term 'dirt eaters'. Unchecked infection may lead .to fatty degeneration of heart, liver and kidneys, ending in death.

Therapy (Treatment)

Treatment involves administration of drugs, such as carbon tetrachloride, thymol, oil of chenopodium, hexylresorcinol, etc. Most commonly recommended drug is tetrachloroethylene or blephenium because of its high efficiency and low toxicity. These days the drugs for human hookworm are **Albendazole** (Zentel), **Mebendazole** (Mebex) **Pyrantel Pamoate** (Nemocid). Treatment is followed by a purge to flush out the dead worms. Iron is usually prescribed to overcome the haemoglobin deficiency due to anaemia.

Prophylaxis (Control)

The Hookworm disease is responsible for considerable economic loss as well as poor health. Important prophylactic or preventive measures include wearing shoes and sanitary disposal of human faeces in affected areas so that soil may not be polluted.

Necator americanus

Another important human hookworm is the 'New World hookworm' or *Necator americanus,* which literally means the 'American killer.' It is the causative agent of the hookworm disease or **Necatoriasis** in man in Africa, Asia, Sri Lanka and tropical America. It is believed to be introduced into America, before their civil war, with the infected slaves from central and southern Africa. It was first discovered in 1902 by **Stiles**.

Necator americanus is smaller than *Ancylostoma duodenale.* Adult female attains a length of 10 mm, whereas male is still shorter. Buccal capsule has a pair of ventral plates and a pair of relatively smaller dorsal plates which lack teeth, unlike *Ancylostoma.* Each pair of plates presents a halfmoon appearance, hence the name semilunar plates. Besides plates, there is a dorsal cone and two pairs of lancets in the base of buccal capsule. Bursa of *Necator* is longer and narrower than that of *Ancylostoma.* Female *Necator* is capable of delivering about 15,000 eggs per day. Rest of Morphology, life cycle, diagnosis, pathogenicity, therapy and prophylaxis are similar to those of *Ancylostoma.*

Important Questions

➤➤Long answer type questions

1. Describe the structure and life-history of *Ancylostoma duodenale.*
2. What disease is caused *by Ancylostoma* ? Describe its pathogenesis, therapy and prophylaxis.
3. Enumerate the main differences between *Ancylostoma* and *Necator.*

➤➤Short answer type questions

1. Why *Ancylostoma* is known as hookworm ?
2. Where adult *Ancylostoma duodenale* lives ?
3. Describe the structure of copulatory bursa of male *Ancylostoma.*
4. Draw a well labelled diagram of male *Ancylostoma.*

➤➤Fill in the blanks

1. Infective juveniles enter the human host .... from the contaminated soil.
2. At the posterior end copulatory bursa is found in .... hook worm.

➤➤Multiple choice questions

1. Rhabditiform larvae of *Ancylostoma duodenale* grows by feeding is parasite of :
   (a) man (b) horse
   (c) cow (d) pig
2. Infective stage of larva is :
   (a) filariform larva (3rd stage juvenile)
   (b) rhabditiform larva (1st stage juvenile)
   (c) 4th stage juvenile
   (d) zygote
3. Rhabditiform larva hatches out in :
   (a) 24 to 28 hrs. (b) 36 to 48 hrs.
   (c) 48 to 60 hrs. (d) 3rd day
4. *Ancylostoma* live in intestine of man and feed upon :
   (a) digestive juice (b) blood
   (c) food (d) all of these
5. No secondary host in life history of :
   (a) *Fasciola* (b) *Taenia*
   (c) *Ancylostoma* (d) *Plasmodium*
6. Who discovered *Ancylostoma* :
   (a) Dubini (b) Manson
   (c) Ross (d) Huxley
7. The cause of Ancylostomiasis :
   (a) *Wuchereria* (b) *Trichuris*
   (c) *Ancylostoma* (d) *Enterobius*
8. The 'Old World Hookworm' :
   (a) *Ancylostoma* (b) *Wuchereria*
   (c) *Enterobius* (d) none
9. *Ancylostoma* generally known as :
   (a) pinworm (b) whip worm
   (c) eye worm (d) hook worm
10. Maximum people infected by :
    (a) *Taenia* (b) *Ascaris*
    (c) *Ancylostoma* (d) *Trypanosoma*

Answers

➤➤Fill in the blanks

1. skin, 2. mail.

➤➤Multiple choice questions

1. (a) 2. (a) 3. (a) 4. (b) 5. (c) 6. (a) 7. (c) 8. (a) 9. (d) 10. (c)

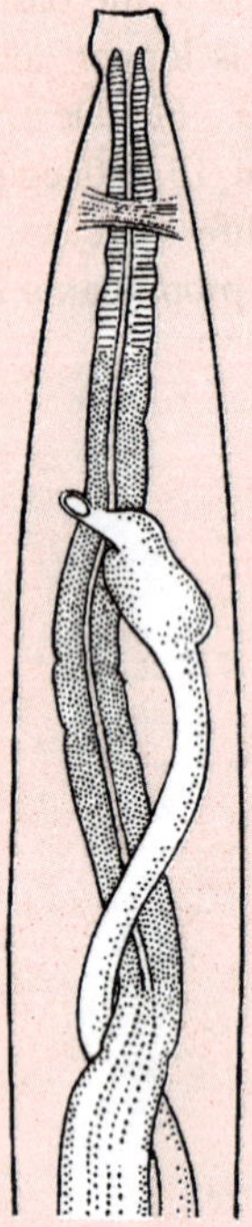

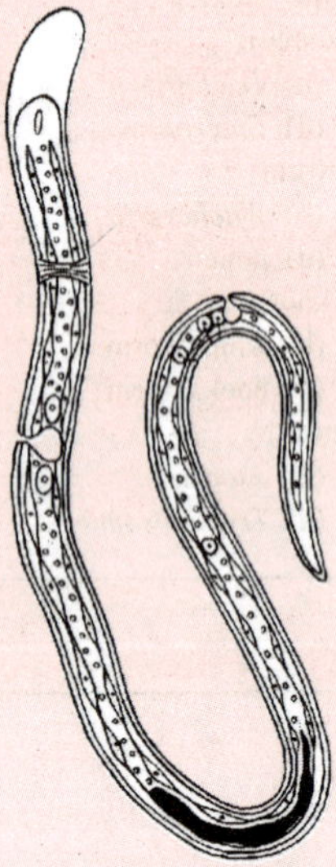

37

Wuchereria bancrofti : The Filarial Worm

The **filarial nematodes** belong to the order Filaroidea. They are lipless and filiform worms in which pharynx is not bulbous, but muscular anteriorly and glandular posteriorly. Genital aperture in female is present in the pharyngeal region, and males are non-bursate. Following description belongs to *Wuchereria* (= *Filaria*) *bancrofti*, the human filarial worm causing **filariasis** or **elephantiasis**.

Wuchereria bancrofti

SYSTEMATIC POSITION

Phylum	Nematoda
Class	Phasmidia
Order	Filaroidea
Family	Filariidae
Genus	*Wuchereria*
Species	*bancrofti*

GEOGRAPHICAL DISTRIBUTION

Wuchereria bancrofti has a widespread distribution throughout world except Polar regions. It is found in Arabia, India, Malaysia, China,

Korea, Japan, East Indies, Brazil and South Pacific Islands. It is relatively absent from Europe, North America and Africa.

HABITS AND HABITAT

Filarial worm is a dreaded human parasite of human blood and lymph. It is a digenetic parasite completing its life cycle in two hosts. Final host is man harbouring the adult worms, while intermediate host is a blood-sucking insect, usually a mosquito. Adult worms live coiled up in the lymph glands and lymph passages of man, where they often obstruct the flow of lymph.

MORPHOLOGY

Adult worms are filiform and cylindrical in shape and both body ends terminate bluntly. They are

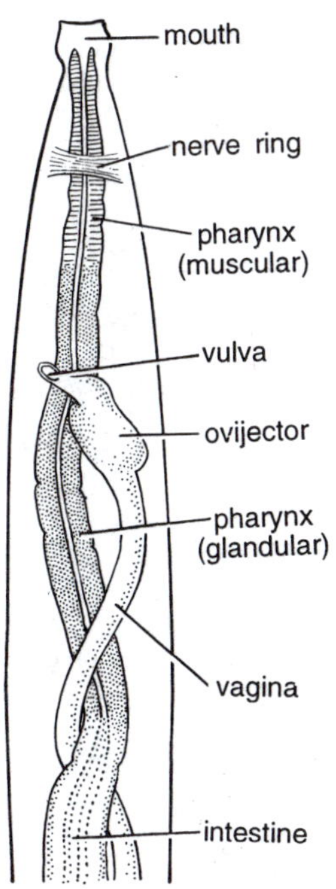

Fig. 1. *Wuchereria bancrofti.* Anterior part of female.

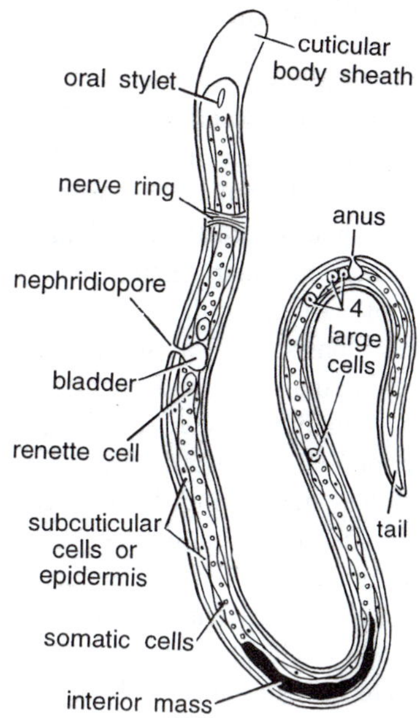

Fig. 2. *Wuchereria bancrofti.* Juvenile or microfilaria.

coloured creamy white. Sexes are separate and there is a distinct **sexual dimorphism**. Female measures 65 to 100 mm in length and 0.25 mm in diameter, while male 40 mm and 0.1 mm, respectively. Mouth aperture is simple, without lips. Pharynx or oesophagus is divisible into an anterior muscular portion and a posterior glandular portion. An oesophageal bulb is lacking. Intestine is simple, as in other nematodes. Posterior end of male is sharply curved ventrally, containing a number of genital papillae, caudal alae and two unequal copulatory spicules. Vulva or genital pore of female is located ventrally in the pharyngeal region, and provided with pyriform ejector mechanism or ovijector.

LIFE CYCLE

[I] Copulation

Copulation takes place when individuals of both sexes are present in the same lymph gland.

[II] Larval Development in Man

Female is viviparous (probably ovoviviparous) releasing numerous Juveniles called **microfilariae**. They are born in a very immature state, being in fact embryos rather than juveniles. They are microscopic, about 0.2 to 0.3 mm long, surrounded by a delicate cuticular sheath and containing rudiments of various adult structures. Body of a microfilaria consists of a surface covering of flattened epidermal cells and an inner column of cytoplasm containing nuclei. Important structures from anterior end backwards are : future mouth or oral stylet, nerve ring band, nephridiopore, renette cell, darkly staining inner mass, 4 large cells and future anus. Microfilariae, discharged into lymph vessels, soon enter blood vessels and circulate with blood showing active movements. They migrate to reside ultimately in deeper blood vessels of thorax. But they do not undergo further development until sucked by the intermediate host, *i.e.*, mosquito*.

In blood of man, microfilariae show day and night periodicity, called diurnal rhythm. By day they live in large deep-seated blood vessels, but at night or during sleep they come into superficial or peripheral vessels in skin, to be sucked by nocturnal mosquitoes (*Culex* or *Aedes*) which serve as intermediate host. In places where there are diurnal mosquitoes, they reverse this periodicity. Microfilariae in human blood eventually die unless ingested by the intermediate host sucking blood from infected humans.

[III] Development in Mosquito

In the stomach of mosquito microfilariae lose their sheaths, penetrate the stomach wall and migrate to thoracic muscles or wing musculature where they undergo metamorphosis and grow. They change first to a plump sausage-shaped organism, later to a more elongated form, and finally to a long slender juvenile of the third infective stage. Microfilariae undergo two moults in about 10 days

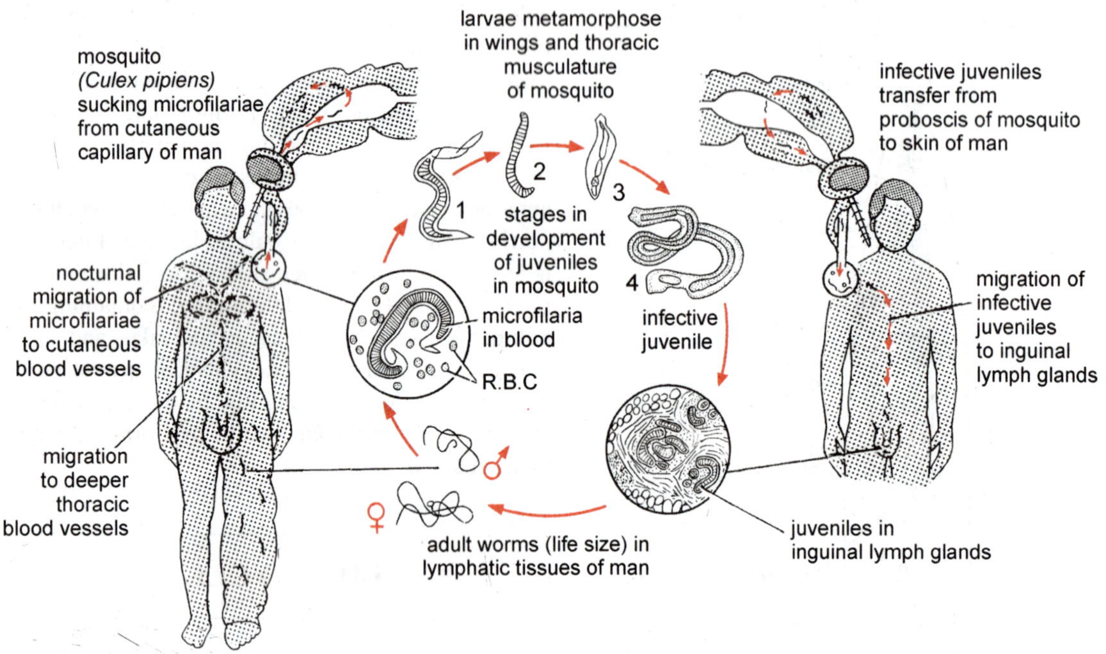

Fig. 3. *Wuchereria bancrofti.* Life cycle.

* Filarioidea differ from all other nematodes in that they require transmission through skin by a blood sucking intermediate host. This was discovered by Manson, in 1878, for *Wuchereria bancrofti.* Incidently, this discovery was the first proof of transmission of a human blood parasite by an insect.

to reach the third stage larvae which are about 1.5 mm long. Infective juveniles now migrate into mosquito's labium (proboscis).

[IV] Infection of New Human Host

When this mosquito pierces its proboscis into another potential human host, the infective juveniles come out of labium on the skin of victim and enter it. They probably penetrate through the wound made by mosquito and enter blood of human host. In new human host, juveniles pass into lymph glands and lymph passages, where they coil up and develop into adult forms. Adults copulate and the females deliver microfilariae.

FILARIASIS OR ELEPHANTIASIS

Filarial worms (*Wuchereria bancrofti*) live in the lymphatic system of man, where they obstruct the flow of lymph, causing a severe condition known as **elephantiasis**, in which the limbs or other body parts grow to enormous size.

[I] Diagnosis

Diagnosis of filarioids involves the study of microfilariae after staining. Microfilariae of different species are identified after their specific shape and morphological characters.

[II] Pathogenesis

Light infection produces no serious symptoms. It causes filarial fever, mental depression, headache, etc., In heavy infection, accumulation of living or dead worms eventually blocks the lymphatic vessels and glands, resulting in various pathological conditions. Most specta cular is the immense swelling of the affected body parts, termed **elephantiasis** or **filariasis**. Due to lymphatic obstruction lymph cannot get back into circulatory system, accumulates into organs and causes them to swell or enlarge to fantastic proportions (lymphedema). Generally lower limbs, scrotum in male, legs and mammary glands are affected. Concurrently there occurs inflammations of lymphatic vessels (lymphangitis) and lymphatic glands (lymphadentis). In severe cases, abnormal connective tissue form in the affected areas to further complicate the condition.

[III] Theraphy and Control

Infection may be reduced or eliminated by eradication of microfilariae from circulation by administering heterazan and compounds of antimony and arsenic. Modern drugs for filariasis are **Albendazole** (Zental), **Diethylcarbamazine** (Hetrazan). An important preventive measure is protection from mosquito bites.

Important Questions

➤➤*Long answer type questions*

1. Describe the habitat, structure and life history of any nematode studied by you.
2. Describe the habit, mode of infection, effect on the human host and other important features in the life history of *Wuchereria bancrofti.* How would you account for its periodicity ?

➤➤*Short answer type questions*

1. Name the intermediate host of *Wuchereria bancrofti.*
2. Write the name of the insect vector that is involved in the transmission of elephantiasis.
3. Explain the externals of *Wuchereria bancrofti* and write the points of differentiation between male and female.
4. Describe the life history of *Wuchereria bancrofti.*

➤➤*True and false statements*

1. Mode of transmission in filaria is inoculative.
   True / False / Do not know
2. Eradication of mosquitoes will eradicate spreading of parasitic disease elephantiasis.
   True / False / Do not know

3. Microfilariae occur in peripheral blood vessels of the man throughout the day.
   True / False / Do not know
4. Females of *Wuchereria bancrofti* give rise to microfilaria by parthenogenesis.
   True / False / Do not know

➤➤Multiple choice questions

1. Elephantiasis is a common disease in :
   (a) subtropical countries
   (b) subtropical and tropical countries
   (c) tropical countries
   (d) polar regions only
2. The vector of elephantiasis is :
   (a) *Culex* mosquito (b) female mosquito
   (c) female *Anopheles* mosquito
   (d) infected *Culex* mosquito
3. *Wuchereria bancrofti* is transmitted by :
   (a) Tse tse fly (b) *Anopheles*
   (c) *Culex* (d) Sandfly
4. Microfilarie during night resides in :
   (a) deep seated blood vessels
   (b) peripheral vessels
   (c) caval veins (d) mesenteric veins
5. Viviparity generally occurs in :
   (a) *Wuchereria* (b) *Ancylostoma*
   (c) *Enterobius* (d) *Trichuris*
6. Microfilaria is larva of :
   (a) tape worm (b) pin worm
   (c) round worm (d) filarial worm
7. Filariasis result as the *Wuchereria* due to :
   (a) block lymp gland
   (b) block blood capillary
   (c) block the veins
   (d) block the artery
8. The place where adult *Wuchereria bancrofti* :
   (a) human blood vessels
   (b) human lymph gland
   (c) muscles of *Culex*
   (d) salivary gland of *Culex*
9. *Wuchereria* was discovered by :
   (a) Huxley
   (b) Manson
   (c) Ross
   (d) Leibe
10. Transmission of *Wuchereria* to new host :
    (a) larval stage
    (b) encyst stage
    (c) juveniles stage
    (d) none of the above
11. A viviparous nematode is :
    (a) *Filaria* (b) *Enterobius*
    (c) *Ascaris*
    (d) *Trichinella*
12. Elephantiasis is a common disease of :
    (a) tropical countries
    (b) subtropical countries
    (c) temperate countries
    (d) both (a) and (b)
13. Microfilaria refers to :
    (a) female *Wuchereria*
    (b) male *Wuchereria*
    (c) juvenile *Wuchereria*
    (d) none
14. A digenetic nematode parasite is :
    (a) *Fasciola* (b) *Filaria*
    (c) *Ancylostoma* (d) *Enterobius*
15. The place where adult Wuchereria bancrofti lives in :
    (a) human lymph glands
    (b) human subdermal spaces
    (c) muscles of *Culex*
    (d) salivary glands of *Culex*

Answers

➤➤True and false statements

1. true, 2. true, 3. false, 4. false.

➤➤Multiple choice questions

1. (b) 2. (d) 3. (c) 4. (b) 5. (a) 6. (d) 7. (a) 8. (b) 9. (b) 10. (c)
11. (a) 12. (d) 13. (c) 14. (b) 15. (a).

38

Phylum Nematoda : Characters, Classification and Types

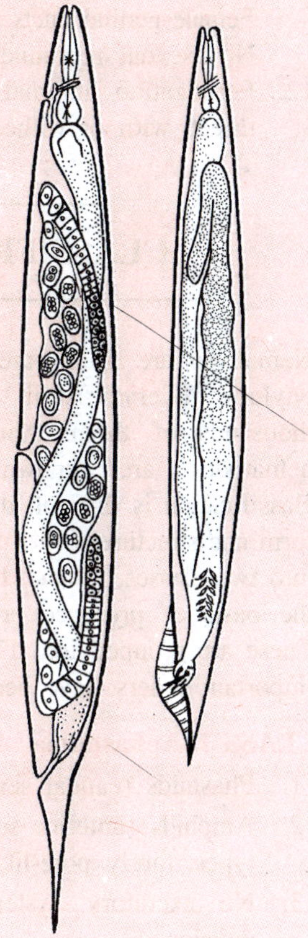

Nematodes (Gr., **nema** thread + **eidos**, form) are commonly referred to as non-segmented roundworms, threadworms or pinworms, as distinct from lower flatworms and higher segmented annelids. They constitute the largest phylum of pseudocoelomate groups combined under the super phylum Aschelminthes. It has total 25,043 species.

GENERAL CHARACTERS

Main distinguishing features of phylum Nematoda are as follows :

1. Widely distributed, aquatic or terrestrial, and parasitic or free living.
2. Body elongated, cylindrical, unsegmented, worm-like, bilaterally symmetrical, and tapering towards both ends.
3. Body triploblastic with organ-system grade of organization.
4. Body wall with thick resistant cuticle, cellular or syncytial epidermis, and only longitudinal muscle fibres in four bands.
5. True coelom absent. Persistent blastocoel or pseudocoel present not lined by mesoderm.
6. No cilia, no circulatory and no respiratory system.
7. Digestive system complete with anus, with muscular pharynx and non-muscular intestine.
8. Excretory system of glandular organs or canals or both. Flame cells absent.
9. Nervous system with circumenteric ring and anterior and posterior nerves.
10. Sense organs poorly developed, in the form of small papillae and amphids near on two body ends.

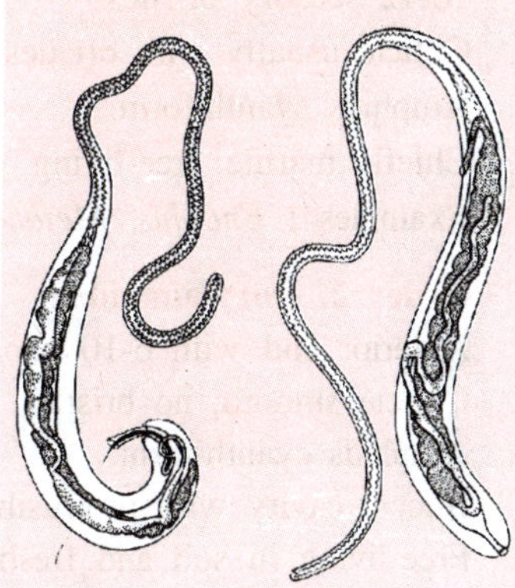

11. Dioecious with sexual dimorphism. Male smaller than female. Gonads simple and coiled. Male genital ducts lead into cloaca. Female genital ducts with a separate opening. No asexual reproduction.
12. Fertilization internal. Development usually direct, with or without an intermediate host.

CLASSIFICATION

Nematodes are among the most numerous of any phylum. Every fistful of soil may contain thousands of them. About 15,000 species of nematodes are known at present. Their classification is difficult due to much diversity in form and structure. **Chitwood** (1933) divided them into two classes, **Phasmidia** and **Aphasmidia**, on the basis of presence or absence of phasmids. These are grouped into 17 orders, but only some important orders have been described here.

CLASS 1. Aphasmidia (Adenophorea)

1. Phasmids (caudal sensory organs) absent.
2. Amphids (anterior sense organs) of various types, rarely pore-like.
3. No excretory system. If present, poorly developed.
4. Mesenterial tissue well developed.
5. Caudal adhesive glands present.

Order 1. Enoploidea

1. Anterior end with six labial papillae and 10-12 sensory bristles.
2. Cuticle usually with bristles.
3. Amphids cyanthiform.
4. Chiefly marine, free-living.

Examples : *Enoplus, Metonchdiamus.*

Order 2. Dorylaimoidea

1. Anterior end with 6-10 papillae.
2. Cuticle smooth, no bristles.
3. Amphids cyanthiform.
4. Buccal cavity with protrusible spear.
5. Free-living in soil and fresh-water.

Examples : *Dorylaimus, Tylencholaimus.*

Order 3. Mermithoidea

1. Large-sized.
2. Anterior end with 16 labial papillae and no sensory bristles.
3. Cuticle smooth, no bristles.
4. Amphids cyanthiform or reduced.
5. Oesophagus long leading into blind intestine.
6. Larvae parasitic in invertebrates, adults free-living.

Examples : *Mermis, Agamermis, Paramermis.*

Order 4. Chromadoroidea

1. Small-sized.
2. Cuticle smooth or ringed; with heavy bristles.
3. Amphids spiral.
4. Buccal cavity with teeth.
5. Pharynx with posterior bulb.
6. Mostly marine; free-living.

Example : *Halichoanolaimus.*

Order 5. Monohysteroidea

1. Small-sized.
2. Amphids circular.
3. Cuticle smooth or slightly ringed, often with bristles.
4. Anterior end with 4, 6, 8, or many sensory bristles.
5. Free-living; mostly marine, some fresh-water, some terrestrial.

Examples : *Monohystera, Plectus.*

Order 6. Desmoscolecoidea

1. Small-sized.
2. Amphids cresent-shaped or pump-shaped.
3. Cuticle heavily ringed.
4. Anterior end with four sensory bristles, head armoured.
5. Marine; free-living.

Example : *Desmoscolex, Epsilonema.*

CLASS 2. Phasmidia (Secernentea)

1. Phasmids present.
2. Amphids pore-like.
3. Excretory system developed.
4. Mesenterial tissue weakly developed.
5. No caudal adhesive glands.

Order 1. Trichuroidea (Trichinelloidea)

1. Body filiform anteriorly.
2. Mouth without lips; pharynx slender.

3. Female with one ovary and male with one or none spicule.
4. Parasites of vertebrates.
Examples : *Trichuris, Trichinella.*

Order 2. Dioctophymoidea

1. Large worms.
2. Mouth without lips; with 6, 12 or 18 papillae.
3. Pharynx elongated and with no bulb.
4. Female with one ovary and male with one spicule and muscular bursa.
5. Parasites of mammals and birds.
Examples : *Dioctophyma, Hystrichis.*

Order 3. Rhabditoidea

1. Small to moderate sized.
2. Cuticle smooth or ringed.
3. Sensory bristles as papillae in two rings, an inner ring of 6 and outer ring of 4, 6 or 10.
4. Copulatory spicules in male accompanied by gubernaculum.
5. Free-living or parasitic in animals and plants.
Examples : *Rhabditis, Heterodera.*

Order 4. Rhabdiasoidea

1. Medium-sized.
2. Cuticle smooth
3. No pharyngeal bulb.
4. Parasite stage in vertebrates is either hermaphroditic or parthenogenetic.
Examples : *Rhabdias, Strongyloides.*

Order 5. Oxyuroidea

1. Pin-shaped small worms.
2. Mouth surrounded by 3-6 simple lips.
3. Pharynx with valvular posterior bulbs.
4. Female with long pointed tail.
5. Parasitic in invertebrates and vertebrates.
Examples : *Oxyuris, Enterobius.*

Order 6. Ascaroidea

1. Large-sized worms.
2. Mouth surrounded by 3 lips.
3. Pharynx without bulb.
4. Male with ventrally coiled tail.
5. Parasitic in vertebrates.
Examples : *Ascaris, Ascaridia.*

Order 7. Strongyloidea

1. Mouth without lips but with leaf crowns.
2. Buccal capsule well developed.
3. No pharyngeal bulb.
4. Male with expanded copulatory bursa; female usually with ovijector.
5. Parasites of vertebrates.
Examples : *Necator, Ancylostoma, Strongylus.*

Order 8. Spiruroidea

1. Mouth with two lateral lips.
2. Pharynx with bulb, muscular anteriorly and glandular posteriorly.
3. No copulatory bursa in male but with spirally coiled tail.
4. Parasites in animals.
Examples : *Thelazia, Gnathostoma, Spiroxys.*

Order 9. Dracunculoidea

1. No lips or buccal capsule.
2. Mouth surrounded by ring of papillae.
3. No pharyngeal bulb; pharynx muscular anteriorly and glandular posteriorly.
4. Male without copulatory bursa.
5. Parasites of vertebrates.
Examples : *Dracunculus, Philometra.*

Order 10. Filarioidea

1. Filiform slender worms.
2. No lips and buccal capsule.
3. Six labial papillae present.
4. No pharyngeal bulb.
5. Male small with coiled tail.
6. Microfilariae in blood or skin and develop in blood-sucking insects.
7. Parasites of vertebrates.
Examples : *Wuchereria, Microfilaria, Loa.*

COMMON NEMATODE PARASITES

1. *Trichuris trichiura* (The Whip Worm). This parasitic nematode dwells in the caecum and large intestine of man, especially children, in tropical regions. Adult female measures 30 to 50 mm in length, male being slightly shorter. Anterior two-third of body is narrow and presents a thread-like or **whip-like** appearance, hence the name 'whip worm'. This whip-like portion contains the long oesophagus and anchors worm to wall of host's

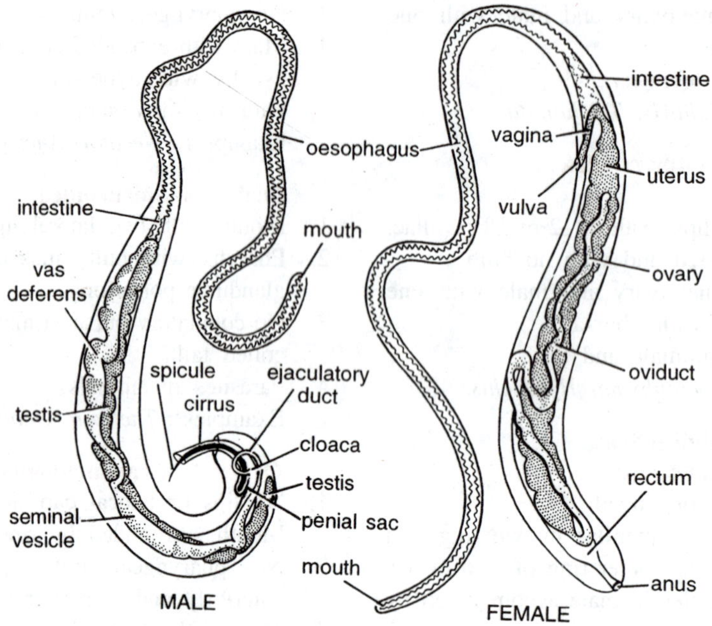

Fig. 1. *Trichuris trichiura.* Male and female individuals.

intestine. Life cycle is direct. Eggs pass out with faeces and embryonate in about three weeks time in moist soil. Host acquires infection by ingesting embryonated eggs with contaminated food or drink. Larvae hatch out near caecum and develop into adults in about two months. Infection usually causes gastro-intestinal troubles. Heavy infections cause severe diarrhoea and even anaemia due to blood sucking habit of adults.

2. *Trichinella spiralis* (The Trichina Worm). *Trichinella spiralis* or Trichina worm is an important parasitic nematode of man, and other mammals like pig, dog, cat, rat, black bear and polar bear. It causes a disease known as **trichinosis**, common in U.S.A., Europe and the Arctic regions. Parasite occurs as a coiled, encysted larva in voluntary muscles of host, usually in limbs, chest, diaphragm, tongue, eye and neck. Larva measures about 1.0 mm in length, while the cyst has a diameter of 0.25 to 0.5 mm. One ounce of flesh may contain thousands of such larvae. A new host acquires infection by eating raw or improperly cooked flesh containing encysted larvae. Cysts dissolve by the action of digestive juices and larvae are liberated in the intestine of new host. They soon move to safer places between villi of intestine. Within about two day's time they attain sexual maturity and become adults. Adults are short-lived, surviving only for a week or two but encysted larvae retain viability for several years.

Males are smaller than females. Male is 1.4-1.6 mm long and female 3-6 mm. In both sexes, anterior end is pointed and posterior end fleshy. Cuticle is transversely striated. Mouth leads into pharynx followed by the capillary oesophagus embedded in large granular cells, called **stichosomes**. Male has a pair of conical appendages or **copulatory lobes** at the posterior end.

Copulation takes place in intestine, after which males usually die and pass out. Females burrow into intestinal wall and live there for some time, each producing about 1,500 young larvae, about 0.1 mm long. These enter lymphatics or blood circulation and reach voluntary muscles, where they grow to a length of about 1.0 mm, coil up and encyst. Formation of cyst is completed in about 9 weeks. Larvae, within cysts, may remain viable for 10 to 20 years or even more, but finally

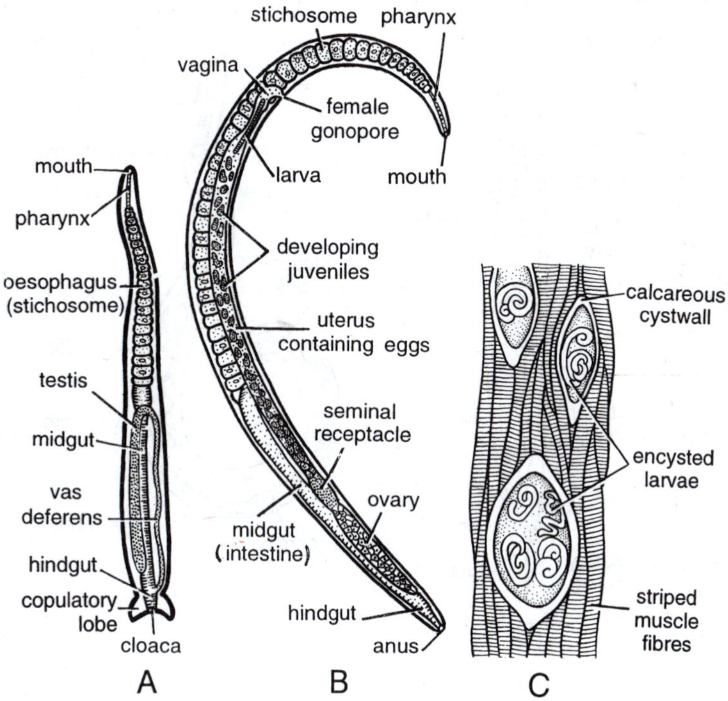

Fig. 2. *Trichinella spiralis.* **A.** Adult male. **B.** Adult female. **C.** Encysted larvae in striped (voluntary) muscle of pig.

undergo calcification and die. Calcified cysts persist in host's muscles throughout life. Viable encysted larvae attain sexual maturity only when ingested by a new host. Human infection is practically a 'dead end' for the worm, because consumption of human flesh is rare.

Mild infection usually produces no symptoms. Heavy infection causes the disease **trichinosis**, producing a variety of symptoms like diarrhoea, nausea, abdominal pain, hypereosinophilia, thrombosis, muscular pains etc. As the larvae spread, there is fever and muscles swell and harden. Difficulty in breathing and swallowing may also be felt. Heavy infections may be fatal. There is no sure treatment for the disease. However, piperazine citrate has been found effective in the removal of adult worms from intestine.

3. *Rhabditis.* This genus includes several free-living and semi parasitic forms. *R. hominis* is found in human stool. *R. pellio* lives in human vagina and its larvae escape in urine. *R. maupasi* lives in soil and its juveniles are found in earthworms. Adult male and female of this species measure 1.7 and 2.0 mm in length, respectively. Male has a single tubular testis and the female has two ovaries. Life cycle is simple and direct. Eggs pass into earthworms in which they become juveniles. These attain maturity in decaying flesh after death of earthworm.

4. *Enterobius vermicularis* (The Pin Worm of Man) *Enterobius vermicularis* is the human 'pin worm' or 'seat worm' and is perhaps the most common parasitic nematode of man throughout world. It has been reported to occur in chimpanzee also. Its incidence, however, is greater in women and children. Adult worms live in caecum, appendix and at the junction of small and large intestines. Sexes are separate and there is distinct sexual dimorphism. Worms are slender and cream-coloured. Male is 2 to 5 mm long, with a diameter of 0.1 to 0.2 mm. Female is 8 to 13 mm long with a diameter of 0.3 to 0.5 mm. At the anterior end, in both sexes, are 3 lips and a pair of cephalic expansions. Posterior or tail

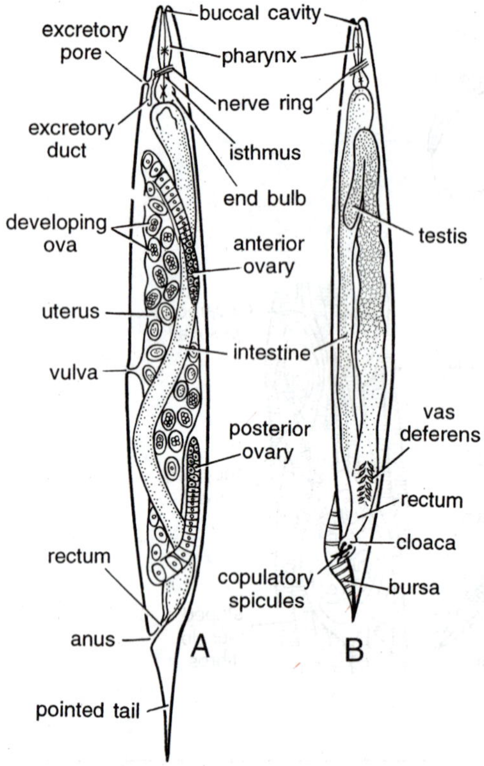

Fig. 3. *Rhabditis.* **A.** Adult female. **B.** Adult male.

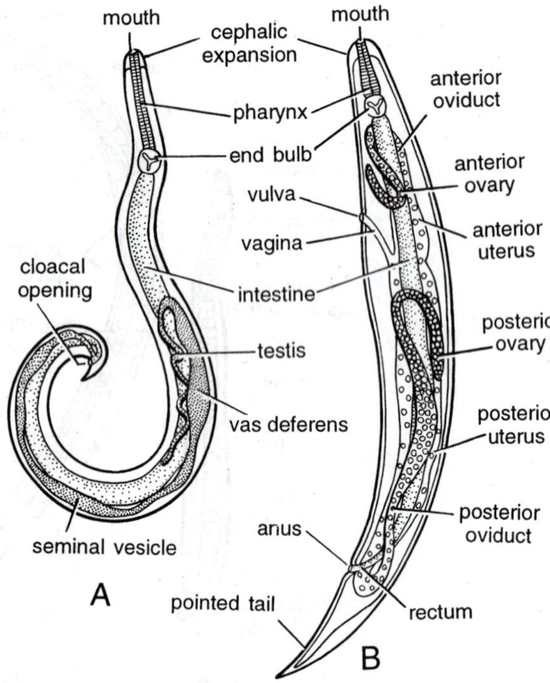

Fig. 4. *Enterobius vermicularis.* **A.** Adult male. **B.** Adult female.

end of female is long and pointed, while that of male is curved, blunt, and with a bursa-like expansion and a single spicule. There is no gubernaculum. Papillae support caudal algae. Male is monorchic and female didelphic.

Life cycle is simple and direct. No intermediate host is involved. At night, females with eggs migrate to anus and deposit eggs. This causes severe itching at the anal and perineal regions and the host is tempted to scratch. Tiny eggs, about 55μ by 30μ and each containing a much developed larva, thus associate with fingers and finger nails and finally enter its gut with food to cause a reinfection. Eggs hatch in duodenum. Larvae reach small intestine, where they moult twice and become adults.

Adults live in host for 20 to 30 days. Infection thus automatically dies out within a month provided there is no reinfection. Good sanitary habits are the best weapon against infection. Symptoms of infection are loss of appetite, insomnia, hysteria, restlessness and inflammation of mucous membrane of infected regions.

5. *Dracunculus medinensis* (The Guinea Worm). It is a common human parasite, known to man since Biblical times, when it was known as the 'Fiery Serpent'. It is of common occurrence in dry regions of Africa, India, Arabia, China, etc. It also occurs in some other mammals. In United States, it has been found in dogs, raccoons, foxes and minks. Adult worms occur in subcutaneous tissue, especially of arms, shoulders and legs. Females attain a length of 100 to 400 cm. Males, are rare and 12 to 29 mm in length. At the anterior end is a cuticular ring bearing a mouth, an inner circlet of six labial papillae and an external circlet of four double papillae. Posterior end of male bears 10 pairs of papillae (4 pairs pre-anal and 6 pairs post anal). Oesophagus has two regions, an anterior narrow muscular region and a posterior broad glandular

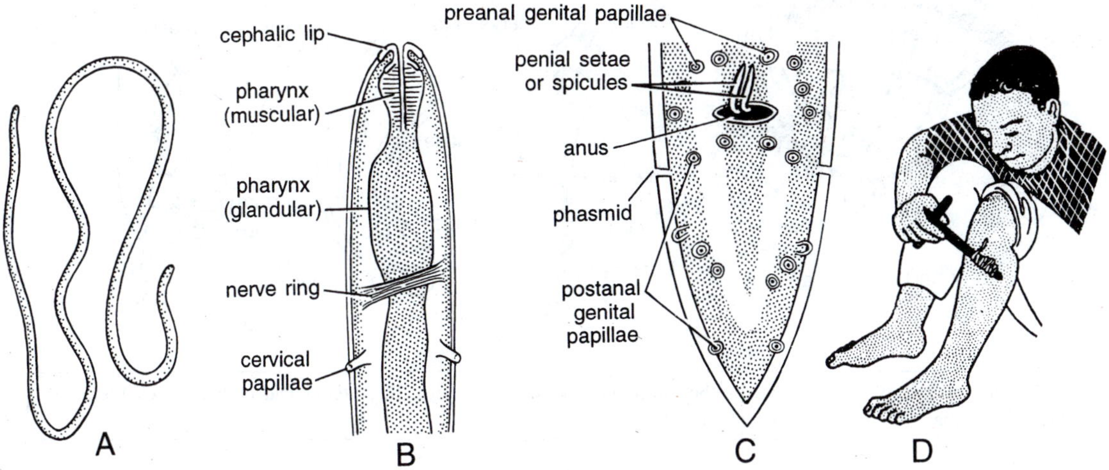

Fig. 5. *Dracunculus medinensis* **A.** Entire worm. **B.** Anterior end of male. **C.** Posterior end of male in ventral view. **D.** Guinea worm being wound on a stick.

region. Female is didelphic with apposed uteri and vulva near the middle of body. Male has equal-sized filiform spicules and gubernaculum.

Mature female contains numerous embryos in its much enlarged uterus which fills up the body cavity. Egg-laying female worm takes up its position just below skin in such a part of body which regularly comes in contact with cold water, like hands, feet, legs, etc. Here the parasite produces toxic secretions which produce a blister on skin. Soon an ulcer is formed. On coming in contact with water, the ulcer breaks and worm freely projects its uterus and releases tiny coiled embryos in water. Larvae, after swimming freely for a short time, penetrate the body of intermediate host *Cyclops,* a copepod crustacean. In *Cyclops,* larva moults twice and becomes infective in three weeks time. When infected *Cyclops* is taken in by man with water, the larvae escape and bore their way into subcutaneous tissue where they attain adulthood.

Primitive method of removing the worm from body is by winding it on a thin stick, a few turns a day. An attempt to remove the worm more rapidly results in its rupture leading to serious bacterial infection. Guinea-worm infection causes itchiness, nausea, diarrhoea, vomiting, asthma, eosinophilia, etc. These effects are due to toxic substance produced by the parasite. Infection can be treated with injections of phenothial zinc emulsified in olive green.

6. *Loa loa* (The Eyeworm). The other important human filaria is the African eye-worm *Loa loa,* chiefly found in Africa. They commonly invade subcutaneous tissue and during their migration may pass across the eye-ball, hence the name eye-worm. Male and female measures 20 to 35 mm and 20 to 70 mm in length, respectively. Body is covered by numerous warts. Gravid females produce ensheathed juveniles or microfilariae. They show diurnal periodicity, the microfilariae appearing only in day times in peripheral circulation but disappearing at night. Intermediate hosts are mango flies, *Chrysops dimidia* or *C. silacea.* Larvae moult and develop up to infective stage in fly and are again infected in the blood stream of man through proboscis.

Loa microfilariae cause very injurious and fatal cases when they penetrate brain and spinal cord and perhaps carry neurotropic viruses. During their migration they cause intense itching and swelling. They also cause swelling and pain in eyes, known as 'calabar swellings'.

7. *Tococara canus* (The Dog Ascarid). The dog ascarid, *Toxocara canis,* is quite interesting. Larvae hatching from eggs ingested by a pregnant

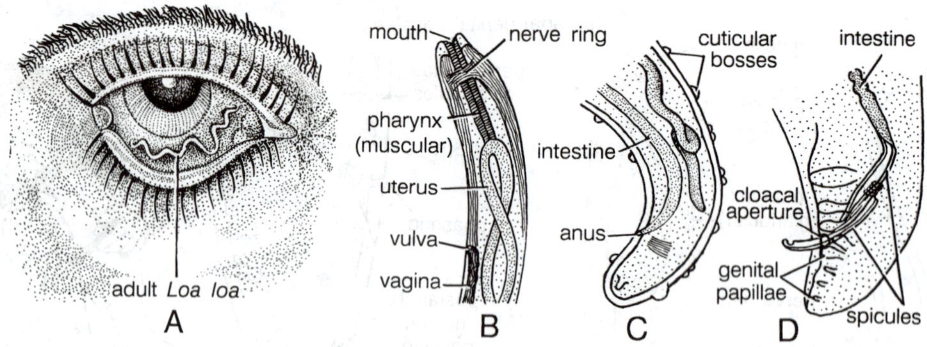

Fig. 6. *Loa loa.* **A.** Adult across the bulbar conjunctiva of human eye. **B.** Anterior end of female. **C.** Posterior end of female. **D.** Posterior end of male.

female dog, migrate through placenta to enter the developing embryos or foeti. Puppies born may show a heavy infection of these worms. Adult individuals generally have fewer worms. Adult male and female worms reach up to 10 and 20 cm respectively. Eggs start development in host's faeces and become infective in 4 or 5 days. When ingested, they hatch in the new host's intestine to liberate juveniles. They penetrate the intestinal wall, reach lung through hepatic portal circulation, break into alveoli, migrate up the air passage, become swallowed in pharynx and re-enter intestine. Dog ascarid is also important medically. Embryonated eggs in contaminated soil may be ingested by children, especially during the dirt eating stage, when children are one to three year old. Larvae, known as **larva migrans**, continue to live in liver of infected children indefinitely. Damage depends on the number of larvae present. Usually, they are not enough to produce symptoms, but may prove fatal when present in larger numbers.

Important Questions

➤➤*Long answer type questions*

1. What are the general characteristics that describe the roundworms as a group ?
2. Classify Nematoda giving diagnostic characters and examples of each group.
3. Name common human nematode parasites. Give a brief account of any one of them.
4. In what ways are the Nematoda more advanced than the Platyhelminthes ?
5. Write an essay on medical importance of nematodes.
6. Write notes on : (i) *Ancylostoma duodenale,* (ii) *Enterobius vermicularis*, (iii) *Trichinella spiralis,* and (iv) *Wuchereria bancrofti.*

➤➤*Short answer type questions*

1. Name the nematode parasite whose juvenile enters the human host by active penetration.
2. Name a helminth parasite which is unisexual.
3. What is aberrant visceral migration?
4. Name a helminth parasite whose primary host is man and the intermediate host is *Cyclops*.
5. Name five human nematode parasites causing diseases.
6. Classify the animal *Trichinella spiralis* upto order.
7. Classify the following, giving two peculiar features in their structure : *Dracunculus, Trichinella spiralis.*
8. Write the common name of *Dracunculus medinensis* ? Mention its mode of infection, and explain its development.
9. Give an account of nematode parasites in man, and discuss the parasitic adaptations of any one of them.
10. Describe the structure, habit, and disease caused by any four important nematode parasites in man.

➤➤Fill in the blanks

1. *Schistosoma haematobium*, causes ....
2. *Schistosoma haematobium* is parasitic in ....

➤➤Multiple choice questions

1. Which larval stage is not found in *Schistosoma* ?
   (a) Miracidium (b) Sporocyst
   (c) Redia (d) Cercaria
2. In the development of *Schistosoma* (blood flukes) one of the following stage is skipped :
   (a) Cercaria (b) Sporocyst
   (c) Miracidium (d) Redia
3. *Dracunculus medinensis* (Guinea worm of man) is a parasitic worm which invades :
   (a) subcutaneous connective tissue
   (b) lymphatic vessel
   (c) small intestine
   (d) colon and rectum
4. Microfilariae refers to :
   (a) female *Wuchereria bancrofti*
   (b) male *Wuchereria bancrofti*
   (c) living embryo of *Wuchereria bancrofti*
   (d) embryo set free in water by female mosquito
5. All roundworms differ from all flatworms in having :
   (a) longitudinal nerve cord
   (b) segmented body
   (c) pseudocoel
   (d) metamorphosis in their life history
   (e) blood circulatory system
6. There is only single host in the life history of :
   (a) flatworm (b) tapeworm
   (c) *Schistosoma* (d) *Fasciola*
7. One of these is different from others :
   (a) *Ancylostoma* (b) *Trichinella*
   (c) *Enterobius* (d) *Ascaris*
8. One of these is different from others :
   (a) *Ascaris* (b) *Ancylostoma*
   (c) *Enoplus* (d) *Enterobius*
9. *Dracunculus medinensis* is transmitted to man by :
   (a) mosquito (b) fly
   (c) crustacean (d) infected *Cyclops*
10. Which of the following parasites have no intermediate host ?
    (a) tape-worm (b) *Ascaris* (round worm)
    (c) liver-fluke (d) malarial parasite
11. Oral infection occurs in :
    (a) malaria (b) filaria
    (c) ascariasis (d) none of these
12. The animal that never performs locomotion :
    (a) *Ascaris*
    (b) *Leucosolenia*
    (c) both
    (d) *Hydra*
13. The term Nematoda was coined by :
    (a) Gegenbaur (b) Grobben
    (c) Rudolphi (d) Lamarck
14. Which is parasite of man :
    (a) hook worm (b) pin worm
    (c) whip worm (d) eye worm
    (e) all
15. The word 'Aschelminthes' was proposed by :
    (a) Gegenbaur (b) Grobben
    (c) Rudalphi (d) Lamark
16. Sexual dimorphism in :
    (a) Porifera
    (b) Coelenterata
    (c) Platyhelminthes
    (d) Aschelminth
17. Free living nematodes :
    (a) *Trichuris*
    (b) *Trichinella*
    (c) *Rhabdite*
    (d) *Wuchereria*
18. The eye worm of man is :
    (a) *Wuchereria* (b) *Trichuria*
    (c) *Loa loa* (d) none
19. Commonly known as hook worm :
    (a) *Ancyclostoma* (b) Necator
    (c) both (d) none
20. Match the following :
    (a) *Ascaris* (a) Whip worm
    (b) *Ancylostoma* (b) Pin worm
    (c) *Wuchereria* (c) Filarial worm
    (d) *Enterobius* (d) Hook worm
    (e) *Trichuris* (e) Round worm

Answers

➤➤Fill in the blanks

1. Schistosomiasis, 2. snail.

➤➤Multiple choice questions

1. (c) 2. (d) 3. (a) 4. (c) 5. (c) 6. (a) 7. (d) 8. (c) 9. (d) 10. (b)
11. (c) 12. (a) 13. (a) 14. (e) 15. (b) 16. (d) 17. (c) 18. (c) 19. (c)
20. [(a-e), (b-d), (c-c), (d-a), (e-a)].

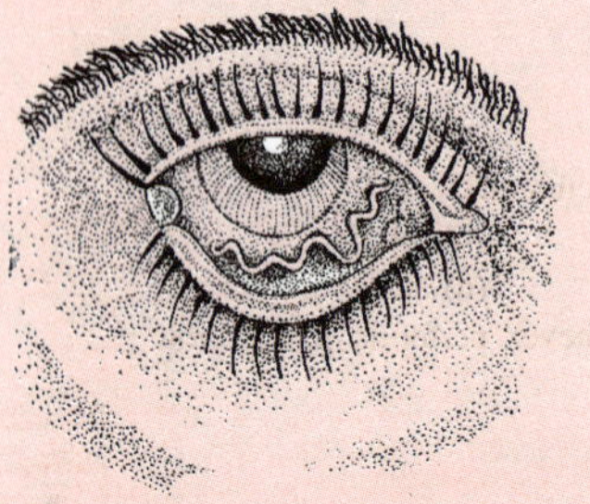

39

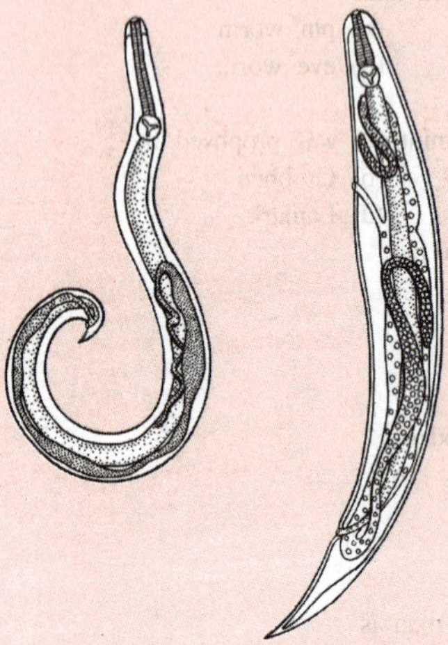

Helminthes : General Account

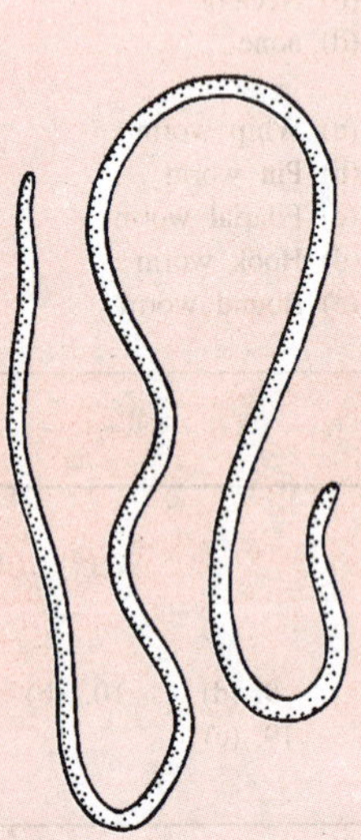

EVOLUTION OF PARASITISM IN PLATYHELMINTHES

There are several conflicting views about the origin of Platyhelminthes. According to **Ciliate Theory**, ciliate-like multinucleate ancestors became bilaterally symmetrical syncytial metazoans and led to the modern Acoela. According to **Planula Theory**, a planuloid ancestor became flattened and acquired a ventral mouth leading either into a semisolid interior (Acoela) or into a central cavity (Polyclada). Acoela are regarded by some as primitive platyhelminths most nearly related to the ancestral platyhelminth. Others think Acoela to have been secondarily simplified polyclad ancestors.

Parasitic platyhelminths (trematodes and cestodes) have clearly been derived from free-living turbellarian progenitors which were ectoparasites on molluscan, crustacean and echinoderm hosts. It is quite possible that monogenetic and digenetic trematode parasites had separate origins.

According to **Hyman**, molluscs could have been the original hosts of digenetic trematodes. Some ancestral rhabdocoels with a suctorial pharynx to feed on soft tissues of relatively large molluscs could have entered the mollusc body through feeding wound and became endoparasites. Thus, endoparastic forms resulted from those free-living forms that were accidentally introduced into the invertebrate host with the evolution of vertebrates, they invaded the new hosts but connection with the original hosts were retained which became intermediate hosts. During the course of evolution the endoparasites acquired certain adaptations to endoparasitism for survival in the intestine of a predatory vertebrate host. These included changes in structure and physiology as a result of continuous mutations. Well-marked changes include : (i) degeneration of alimentary canal in flukes and absent in tapeworms, (ii) nervous degeneration, (iii) development of suckers and hooks for attachment, (iv) greatly developed reproductive system and (v) complex life-cycles. Thus, a second or vertebrate host added to life-cycle became obligatory. Second and even third intermediate invertebrate hosts, which formed organisms in the food chain of vertebrate host, were eventually incorporated in life cycle of parasite. Hazards of a two or multiple host cycle may have been compensated for by the development of polyembryony and larval multiplication in molluscan host. In this manner, modern digenean endoparasites, with their complicated life-cycles, may have emerged.

Flukes and tapeworms which involve many intermediate hosts, are more advanced, since the parasites became adapted, in time, to several species of hosts. With the elimination of some intermediate hosts, still more advanced parasitic forms now restrict to one or two intermediate hosts.

It seems likely that monogenean parasites of vertebrates arose from free-living rhabdocoels long after the digenean parasites.

PARASITIC ADAPTATION OF FLATWORMS

Adaptation. Adaptation is the fitness of an organism to its environment. It is the characteristic which results in suitable and convenient morphological and functional correlation between an organism and its environment.

Parasitic adaptations. Parasitic flatworms (trematodes and cestodes) have undergone profound adaptations to suit their parasitic mode of life. These adaptations termed parasitic adaptations in such cases, are of morphological as well as physiological nature.

[I] Morphological Adaptations

1. Body covering. Thick **tegument**, frequently provided with scales, affords suitable protection to the parasite. It is probable that this thick protoplasmic layer is continually renewed by the mesenchymal cells forming it.

2. Organs of adhesion. For a firm grip on or in the host's body, some special organs of adhesion are necessary. Flatworms, for this purpose, are variously armed with suckers, hooks and spines. Suckers themselves may be with or without hooks and spines.

3. Organs of locomotion. Locomotion is actually an effort for procuring food. But parasites habitually inhabit such places in host's body, where sufficient food is available without effort. Thus, organs of locomotion, such as cilia of free-living turbellarians, are absent in parasitic forms. It is interesting to note that locomotory organs are duly present in free-living larvae of parasitic forms. Miracidium possesses cilia and cercaria possesses a tail for locomotion.

4. Organs of nutrition (trophic organs). Food of parasite comprises readily available digested and semi-digested food of the host. Elaborate organs for nutrition are thus not needed. Trematodes have an incomplete gut and, in most cases, a suctorial pharynx for sucking food. An eversible pharynx present in free-living turbellarians is absent in this case, as the parasite has not to capture a large prey. In cestodes, parasite freely bathes in digested food of host which is absorbed directly. There is thus total absence of alimentation in tapeworms.

5. Neuro-sensory system. Need for quick and efficient 'response to stimuli' is associated with free active life and not with a quiet parasitic life in a safe environment. In parasites, therefore there is profound reduction of nervous system and total absence of sense organs. But, the free-living miracidium possesses eye spots.

6. Reproductive system. It is the best developed system in helminth parasites, designed and perfected to meet the need for tremendous egg production. Parasitic flatworms, with a few exceptions like *Schistosoma,* are monoecious (hermaphrodite). Hermaphroditism is of distinct advantage to the parasite, because (i) it ensures copulation even when a few individuals are present, (ii) after copulation both individuals lay eggs, thus doubling the rate of reproduction and (iii) in the absence of a companion parasite can reproduce offspring. In cestodes reproductive system is much more elaborate and each mature proglottid possesses one (*e.g., Taenia solium*) or two (*e.g., Cotugnia, Moniezia, Dipylidium*) complete sets of male and female genitalia. In a gravid proglottid all other organs of the system degenerate to make room for the uterus which becomes highly enlarged and branched to accommodate large number of eggs.

[II] Physiological Adaptations

1. Protective mechanism. Alimentary canal of parasites have to protect themselves from the action of digestive juices of host. Tapeworms accomplish this : (i) by stimulating walls of gut to secrete mucus, which then forms a protective covering around the parasite (ii) by secreting antienzymes to neutralize the digestive enzymes of host, and (iii) by probably continually renewing their protective body covering *i.e.,* tegument.

2. Anaerobic respiration. Environment in gut and bile ducts is devoid of free oxygen. Flatworms inhabiting these places, therefore, respire anaerobically by breaking down glycogen.

3. Osmoregulation. Osmotic pressure of endoparasite's body fluids, especially in case of trematodes is almost the same as that of host. This renders osmoregulation unnecessary. But in intestinal tapeworms, osmotic pressure is a little higher. This permits ready absorption of host's digested food by tapeworms.

4. High fertility. Eggs produced by a parasitic flatworm face a very uncertain future. While passing through the complex life cycle, these potential offsprings face several hazards as a result of which a very small percentage of total eggs produced reaches adulthood. This threat to the very existence of species is suitably met by the parasite which in its life time may produce eggs in millions. Reproductive organs of flatworms, as already noted, are accordingly developed.

Additional multiplicative phases in life-cycle of some flatworms further increase the output of potential offspring. Several cercariae develop from a single miracidium of liver fluke and a single hexacanth of *Echinococcus* produces several scolices, each of which is a potential tapeworm.

HELMINTHS AND HUMAN DISEASES

Helminths are animals that belong to the phyla Platyhelminthes and Nematoda (Nemathelminthes.) Many parasitic forms of this group, popularly known as parasitic worms, are endoparasites of gut and blood in human body which cause diseases, collectively called **helminthiasis**. Following three forms of helminthiasis are the most widespread : (i) nematodiasis, (ii) trematodiasis and (iii) cestodiasis.

1. Nematodiasis

Diseases caused by nematode helminths constitute **nematodiasis**. Most widespread human diseases due to infestation with namatodes are as follows :

1. Ascariasis. Ascariasis is a highly prevalent disease caused by the giant intestinal roundworms, *Ascaris lumbricoides.* It is off white in colour, resembles an earthworm and is most frequently seen in children. Female is longer than male and lays about 200,000 eggs daily that pass out with human faeces and may remain alive in the soil for several days. In man infection follows the ingestion of embryonated eggs with contaminated food (raw vegetables, fruits) and drinking water. Among childern playing in contaminated soil, there is also hand to mouth transfer of eggs. In intestine, juveniles hatch out of eggs and cycle of larval development begins. Juveniles penetrate the intestinal wall and migrate to liver and from there to heart and lungs. Finally they return back into intestine and develop into adult parasites.

With adult *Ascaris* residing in intestine, the patient complains of abdominal pains, vomiting, headache, irritability, dizziness and night terrors. Sometimes there is diarrhoea and salivation. Often the patient grits his teeth in sleep. When adult worms migrate through intestinal wall, thay cause severe **peritonitis**. They may find their way into other areas causing appendicitis, gall bladder trouble and liver disease. In lungs, juveniles may cause bronchitis.

Treatment of human ascariasis has been fairly successful through oral administration of **Piperazine citrate syrup** (two teaspoonful twice a day for one week followed by another course after a gap of one week) and **Hexylresorcinol tablets** (100 mg taken at one time with water). **Santonin** and **Oil of Chenopodium** are also useful.

2. Ancylostomiasis. Ancylostomiasis is caused by two hook worms *Ancylostoma duodenale* and *Necator americanus.* Both nematodes are parasites within the intestine. They are most frequent in rural areas. Female hookworms produce 5000 to 10,000 eggs per day which pass out with stool. Man acquires infection when eggs hatch and juveniles penetrate through soft skin of hands and feet. They enter the blood vessels and are carried to heart and lungs. Now they make their way to one of the bronchial tubes, pass into small intestine and finally develop into the adult worms.

Characteristic symptoms of ancylostomiasis are gastro-intestinal disturbances, anaemia and nervous disorders. Patients appear pale and are often weak. They complain of dizziness, ringing in ears and headache. Nausea and vomiting are frequent. In more severe cases, men may become impotent and women may cease to menstruate.

In the treatment of hook-worm, safe and effective drugs are **tetrachloroethane** and **carbon tetrachloride**.

3. Enterobiasis. This disease in caused by *Enterobius vermicularis,* commonly called pin-worms or seatworms. These are small white worms about 6 or 7 mm long and inhabit the upper part of colon. Females migrate out through colon and rectum and deposit enormous number of eggs in the skin folds about anus, where they cause intense itching. When skin about the anus is scratched, eggs are easily picked upon fingers and under nails from where they find their way to food and are swallowed. They hatch in stomach and the juveniles migrate to colon and develop into adult worms.

Pin-worm infection is more frequent in children, than adults. Symptoms of disease include severe itching around anus, loss of appetite, sleeplessness, bed-wetting, grinding of teeth, nausea and vomiting. In girls irritation of genitalia by pin-worms leads to an early sexual desire.

Piperazine is the most effective drug in treatment of pin-worms. The patient must wash his perianal region with warm water and soap before going to and on rising from bed.

4. Trichuriasis. This disease is caused by *Trichuris trichiura,* commonly called whip-worms. These worms inhabit the large intestine, mainly caecum or vermiform appendix. Females lay enormous number of eggs daily that pass in stools. Eggs gain entry to human body with contaminated drinking water and raw fruits and vegetables. They may also be spread by flies. In the intestine eggs hatch and larvae develop into adult whip-worms.

Patient suffering from whip-worm disease complains of nausea, vomiting, constipation, headache, slight fever, and even paroxylmal pains resembling appendicitis. In more severe cases, anaemia and eosinophilia develop.

Drugs most commonly used for expulsion of whip-worms are **osarsol** (acetarsone) and **dithiazanine**.

5. Trichinosis. This disease is caused by *Trichinella spiralis.* The trichina worms are widespread in America. These are transmitted by raw meat, especially pork. Man becomes infected following ingestion of encysted larvae with raw or under-cooked pork. In the intestine, larvae hatch out of cysts and develop into sexually mature adult worms. Female lays eggs that hatch and the resulting young larvae enter blood stream and carried to muscles of chest and legs. Here they form cysts.

Early symptoms of trichinosis are nausea, vomitting, oedema of face and eyelids and fever. After this, muscular pains are felt. Patient complains of pains in chewing, swallowing, breathing and in moving his arms and legs. There is no specific treatment for this disease.

6. Strongyloidosis. This disease is caused by *Strongyloides stercoralis,* commonly called thread-worms. These invade the lining of alimentary canal. Female lays large number of eggs that pass out in faeces and develop into **rhadbitiform larvae**. These transform into **filariform larvae** that penetrate the skin of bare feet, and thus enter human body.

Patient with thread-worm disease complains of nausea, dizziness and bloody diarrhoea resulting from ulcerations due to parasites lodged within the intestinal mucosa and glands. There may also be vomiting, cough and fever.

Most effective drug for treatment of thread worms is **gentian violet**. **Dithiazamine** also works well.

7. Filariasis or Elephantiasis. Causative organism for this disease is a nematode, *Wuchereria bancrofti,* commonly called the filaria worm. These tiny worms live in lymphatic system and connective tissues of body and are also found in circulating blood at night. Infection is spread through *Culex* mosquito. Following copulation, female worm delivers juveniles called **microfilariae**. These, at night, get into blood capillaries of skin to be sucked up by mosquito with blood meal. Mosquito transmits them into another human body where they enter the lymph nodes and develop into adult parasites.

Infection of filaria worms causes enlargement of limbs, scrotum and mammae. Swelling takes place due to blockage of lymph circulation by parasitic worms, resulting into inflammation of lymph vessels and lymph glands.

There is no effective drug for eradication of filaria worms. *Cyanine dyes* and *Diethylcarbamazine* may be used for some success. Very large swellings can sometimes be removed by surgery.

2. Trematodiasis

The diseases caused by trematode helminths constitute **trematodiasis**. Some common such human diseases are as follows :

1. Opisthorchiasis (= Clonorchiasis). This disease is caused by *Opisthorchis* (*Clonorchis*) *sinensis* which inhabits the bile ducts. Disease is widespread in China, Japan, Korea, Vietnam and India. Human infections are acquired through eating raw or under-cooked fish which harbour **metacercariae**. Thousands of adult flukes are found in biliary ducts, causing the thickening of duct walls. Severe cases usually lead to cirrhosis and ultimate death.

Gentian violet and **chloroquine** proves helpful in curing infections. Most important is prevention. All freshwater fish should be thoroughly cooked before eating.

2. Fasciolopsiasis. This disease is caused by the intestinal fluke *Fasciolopsis fuelleborni* in India. It utilizes snail as intermediate host in which it passes through an elaborate developmental cycle, producing metacercariae that leave the snail to be located on water plants. Infections are acquired by eating these water plants, particularly the water nuts. In intestine these metacercariae develop into adult worms within three months.

Worms cause erosion of intestine lining, resulting in bleeding and pain. This is followed by diarrhoea, nausea, and vomiting, *Hexylresorcinol* and 'Crystaloids' anthelmintic are helpful in the eradication of the intestinal flukes.

3. Schistosomiasis. This type of disease is caused by three species of blood flukes-*Schistosoma mansoni, S. japonicum* and *S. haematobium*. These live in the blood stream. Disease is widespread throughout the world. Females lay enormous number of eggs in blood stream from where they find their way either into intestine (*S. mansoni, S. japonicum*) or urinary bladder (*S. haematobium*) and are ejected in faeces or urine. When eggs come in contact with water, miracidia hatch out and get into snail. Here these develop into cercariae which leave the snail, swim freely in water and penetrate the skin of man. Through blood circulation cercariae reach lungs to become adult worms. Adults then get into blood stream.

Blood flukes cause asthmatic attacks and hepatitis. This may be followed by fever, sweating, diarrhoea, weight loss and lack of appetite.

For treatment of disease, **antimony compounds** are recommended. Sanitary disposal of all human faeces and urine is essential for the control of Schistosomiasis.

4. Paragonimiasis. This disease is caused by the lung fluke, *Paragonimus westermani.* Infection is wide-spread in Asia, Africa and South and Central America. Flukes are found in lungs in encapsulated form. Eggs are expelled out in sputum and when they come in contact with water these develop into miracidia. Miracidia find their way into snail host where they develop into cercaria. Cercariae emerge and get into crabs and crayfish and encyst into metacercarial cysts. Man acquires infection from eating raw or under-cooked crabs and crayfish. Excystment occurs in the man's intestine and juveniles migrate into lungs where they attain maturity and become encapsulated.

Lung flukes cause chronic cough with emission of bloody, sputum. Heavy infections cause chest pain with pleurisy, shortness of breath, fever and anaemia. **Emetine hydrochloride** and **sulpha drugs** are recommended for the treatment of lung flukes.

3. Cestodiasis

Cestodiasis is caused by tapeworms. The two common forms of this disease are as under :

1. Taeniasis. This disease is caused by the species belonging to the genus *Taenia,* which include mainly *Taenia solium* (pork tapeworm) and *Taenia saginata* (beef tapeworm). Man acquires infection by eating raw or under cooked pork or beef that contains the cysticerci. In intestine cysticerci develop into adult tapeworms.

Presence of tapeworms in intestine causes gastro-intestinal disorders. Some patients complain of hunger pain. Anaemic conditions may also develop. **Atabrin** or **quinacrine hydrochloride** is the drug of choice. Prevention requires the eating of thoroughly cooked pork or beef.

2. Hydatid disease. The disease is caused by the hydatid worm, *Echinococcus granulosus.* Primary host of this worm is dog in whose intestine the eggs are set free. These pass out in faeces and develop into **onchospheres**. Man acquires infection on eating food or drinking water contaminated with onchosphere-containing eggs. In man, hydatid cysts develop in liver, lungs and other tissues and represent the end of parasitic life-line. Cysts cause inflammation of tissues. Presence of cysts in brain and kidney may prove fatal. Recommended drug is the same as for Taeniasis.

NEMATODE PARASITES OF MAN AND ANIMALS

Nematodes are elongated, unsegmented, triploblastic, pseudocoelomate animals, commonly known as 'roundworms.' They are widely distributed free-living as well as parasitic animals. Free-living nematodes are found in all sorts of environments in sea, in fresh water and in soil on land. Parasitic nematodes parasitize all kinds of plants and animals and display all degrees of parasitism. About 50 species of nematodes are known to occur in man, but only a dozen of these are of pathogenic importance. They are of utmost economic and medical importance to man. Some of these nematode parasites of man and animals have already been described in the previous chapter. These are *Trichuris trichiura, Trichinella spiralis, Rhabditis, Enterobius vermicularis, Dracunculus medinensis, Loa loa, Toxocara canis, Ancylostoma duodenale, Necator americanus, Wuchereria bancrofti* and *Ascaris lumbricoides.* Readers may refer to previous pages for their ecology, morphology, pathology and prophyllaxis, etc. To describe them here again will be merely repetition of what has already been written.

Important Questions

➤➤*Long answer type questions*

1. Write an essay on "Helminths and Human Diseases".
2. Give an account of some of the human diseases caused by nematodes.
3. Describe some nematode parasites of man and domestic animals.
4. Name common human nematode parasites. Give a brief account of any one of them.

➤➤*Short answer type questions*

1. What is nematodiasis ?
2. Clarify the trematodiasis.
3. What are the flukes and how they infect the human ?
4. How you understand the term cestodiasis ?
5. How you define the parasite ?
6. Distinguish between exo and endo-parasite.
7. Give a definition of adaptation ?
8. What is helminthiasis ?
9. What is the treatment of ascariasis ?
10. What is the way of infection of pin worms ?
11. What is the cause for the disease enterobiasis ?
12. Give any five point of parasitic adaptation of helminthes.
13. Explain the causes of hydatid disease ?
14. What are round worms ?
15. Describe polyclada.

➤➤*Fill in the blanks*

1. Parasitic platyhelminthes are trematodes and......
2. Molluscs could have been the original hosts of..... trematodes.
3. The oil of chenopodium is useful in treatment of......
4. Female larva worm produce 5000 to......eggs per day.
5. Blood fluke cause asthamatic attack and......

➤➤*True and false statements*

1. Hydatid disease is caused by *Echinococcus.*
2. *Peragonimus* is known as lung fluke.
3. Fasciolopsiasis is caused by *Fasciola hepatica.*
4. Elephantiasis is caused by *Trichinella.*
5. *Enterobius* is commonly known as pin worms.

➤➤*Multiple choice questions*

1. Treatment of human ascariasis fairly successful through :
   (a) oral administration of drugs
   (b) vaccine
   (c) inject the medicine into muscles
   (d) none the above
2. Characteristics symptoms of ancylostomiasis :
   (a) gastro-intestinal disturbance
   (b) anaemia
   (c) nervous disorders (d) all the above
3. Pinworm is a common name of :
   (a) *Ancylostoma* (b) *Enterobius*
   (c) *Trichuris* (d) *Trichinella*
4. *Trichuris trichiura* is commonly known as :
   (a) *Ascaris* (b) pin worm
   (c) whip worm (d) round worm
5. Elephantiasis spread through the vector :
   (a) *Anophelese* (b) *Culex*
   (c) *Musca-domestica* (d) contaminated water
6. *Opisthorchis* inhabit into :
   (a) blood (b) liver
   (c) bile duct (d) intestine
7. Antimony compounds recommended to the treatment of :
   (a) elephantiasis (b) fasciolopsiasis
   (c) schistosomiasis (d) none to any

8. Beef tapeworm is :
   (a) *T. solium* (*b*) *T. saginata*
   (c) both of them (d) none of them
9. In male *Ascaris* :
   (a) unpaired vas deferens is present
   (b) paired vas deferens are present
   (c) paired testes are present
   (d) none of these
10. Sexual dimorphism is found in :
    (a) *Pheretima* (b) *Amoeba*
    (c) *Ascaris* (d) all of these
11. *Cyclops* is the intermediate host of :
    (a) *Ancylostoma* (b) *Echinococcus*
    (c) *Planaria* (d) *Dracunculus*
12. Blisters are produced on the body by the infection of
    (a) *Trichinella* (b) *Dracunculus*
    (c) *Wuchereria* (d) *Echinococcus*
13. The roundworm can be called specialized and not degenerate with reference to parasitism because :
    (a) it has no respiratory organ
    (b) it is dioceious
    (c) it has a straight, uncoiled alimentary canal
    (d) the cuticle over its body wall is resistant to digestive juices of the host
14. *Ascaris* exhibits :
    (a) polymorphism
    (b) anisogamy
    (c) regeneration
    (d) sexual dimorphism
15. Syncytial (=coenocytic) epidermis is present in:
    a) Housefly
    (b) *Ascaris*
    (c) *Pheretima*
    (d) *Periplaneta*

Answers

➤➤*Fill in the blanks*

1. Cestodes, 2. digenetic, 3. ascariasis, 4. 10000, 5. hepatitis.

➤➤*True and false statements*

1. true, 2. true, 3. false, 4. false, 5. true.

➤➤*Multiple choice questions*

1. (a) 2. (d) 3. (b) 4. (c) 5. (b) 6. (c) 7. (c) 8. (b) 9. (a) 10. (c)
11. (d) 12. (b) 13. (d) 14. (d) 15. (b).

40

Neanthes or *Nereis* : A Clamworm

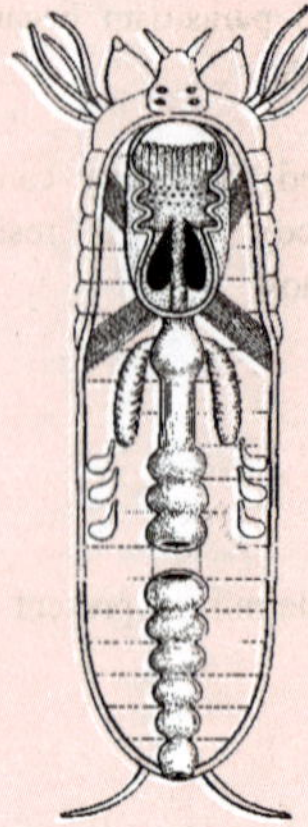

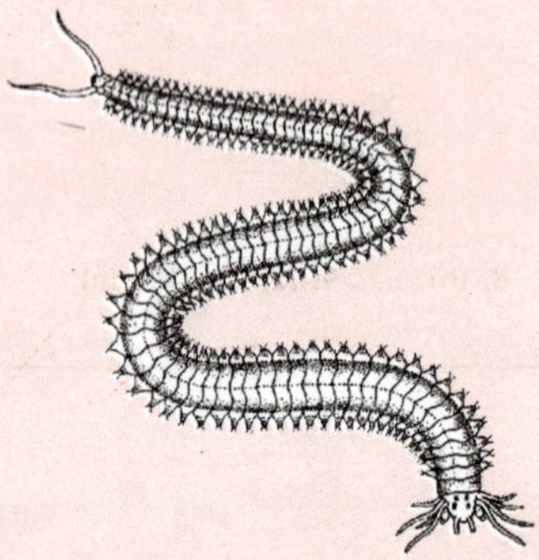

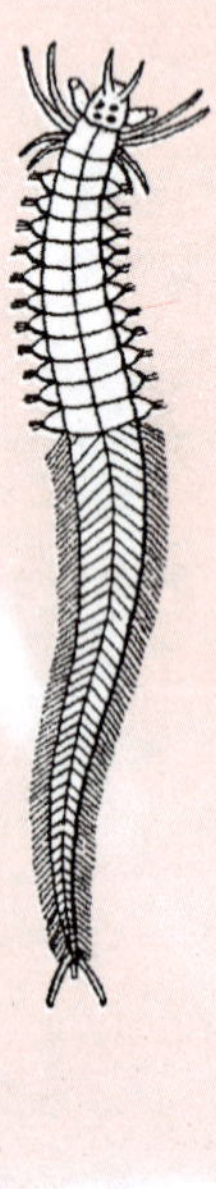

Phylum Annelida (L., **annulus**, a little ring), sometimes called Annulata, comprises of 3 main classes of segmented worms - like invertebrate animals. All are characterized by **metameric segmentation**, which means their body is typically divided into a linear series of similar segments, called **somites** or **metameres**. Previously described flatworms and roundworms do not have this type of body segmentation. Classes Oligochaeta and Hirudinea include the familiar earthworms and leeches, respectively. Third and the largest Class **Polychaeta** includes marine worms of which most persons are not aware. As their name implies, each body segment bears numerous setae on a pair of lateral appendges the **parapodia**. *Nereis* is a typical polychaete genus, living in burrows in sand or mud, often with clams, a reason for which its species are commonly known as **clamworms** or **sandworms.** Several authors have placed these worms under the genus *Neanthes* but both (*Nereis* and *Neanthes*) are distinct and well established genera with many species belonging to each one.

Nereis includes several species, of which more common are *N. virens, N. cultrifera, N. dumerilli, N. diversicolor, N. pelagica* and *N. succinea.* The following account largely pertains to *Nereis virens.*

Neanthes : Nereis virens

SYSTEMATIC POSITION

Phylum	Annelida
Class	Polychaeta
Subclass	Errantia
Family	Nereidae
Genus	*Neanthes* or *Nereis*
Species	*virens*

HABITS AND HABITAT (ECOLOGY)

Nereis is a cosmopolitan marine polychaete, usually found on sandy shores between tidemarks. It remains overlooked by casual observers because of its secretive habits. Most of the time it lives in burrows, which it forms with the help of its jaws. Burrow is U-shaped, up to 60 cm deep and coated with mucus, secreted from body, which binds together fine particles of sand. Worm maintains a constant current of water through the burrow by dorso-ventral undulations of its body. Water current brings oxygen for respiration. Worm is nocturnal and carnivorous. At night it keeps its head protruded out of the burrow in search of prey, which is usually a small crustacean, mollusc or annelid. Prey is captured by protruding the pharynx provided with chitinous teeth and jaws, dragged into burrow and then devoured. Occasionally, the worm leaves its burrow and creeps about beneath stones and rocks or among seaweeds, in search of food. During breeding season, the worm leaves its burrow permanently and swims about actively in surface layers of water. At this stage, it is known as *Heteronereis.*

Clamworms are frequently used for fish bait. In Greek mythology, nereids were sea-nymphs or beautiful human figures who sometimes lured sailors into captivity.

EXTERNAL MORPHOLOGY

1. Shape, Size and Colour

Body is elongated, slender, bilaterally symmetrical, somewhat broad anteriorly and tapering posteriorly. It is slightly flattened dorso-ventrally, dorsal surface being convex, while ventral surface flat or even somewhat concave. Length of adult worm varies from a few to about 40 cm or even more. Different species are differently coloured. *N. pelagica* is brownish, *N. cultrifera* greenish and *N. virens* steel blue. But colour may vary with age and sexual maturity. Cuticle imparts an iridescent sheen to the surface.

2. Body Divisions

Body of *Nereis,* as in other annelids, is metamerically divided into a number of **metameres** or **segments** arranged in a linear series. Adjacent metameres are demarcated externally by circular grooves. Number of segments is fairly constant for a species; about 80 in *N. cultrifera* and *N. dumerilli* and about 200 in *N. virens.* Three distinct regions can be identified in the body : (i) head, (ii) trunk, and (iii) pygidium.

[I] Head

Nereis possesses well-developed head because of its active life and predaceous habit. It lies at the anterior end and consists of two parts : (i) **peristomium**, and (ii) **prostomium**.

1. Peristomium. Peristomium is the first segment of body, distinct from others. It is large, ring-like and surrounds the ventral, transversely elongated slit-like **mouth**, hence its name (Gr., **peri**, around + **stoma**, mouth). It is formed by the fusion of first two embryonic segments during cephalization. It differs from a trunk segment in being longer, in lacking **parapodia** and in the presence of two pairs of thread-like **peristomial cirri** on each side. The dorso-lateral pair of cirri is longer than the ventro-lateral pair. Peristomial cirri are homologous with the

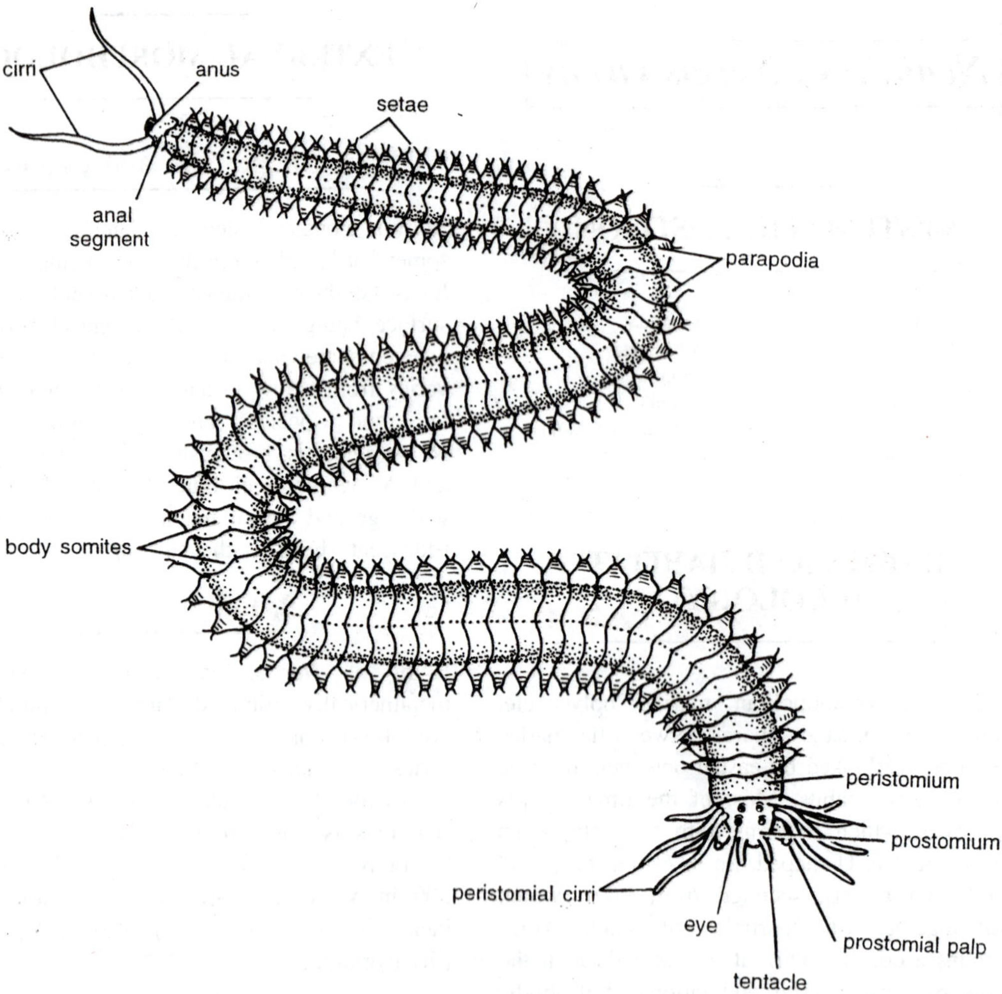

Fig. 1. *Nereis.* External features in dorsal view.

notopodial and neuropodial cirri of parapodia of trunk segments. Each cirrus is a long, slender, tactile structure, having a short proximal joint and a long distal joint.

2. Prostomium. It is not a true segment of body but rather dorsal, anterior projection of peristomium, presumably of its first fused segment. It is a roughly triangular, dorso-ventrally flattened, fleshy lobe lying above and in front of **mouth** (Gr., **pro**, anterior + **stoma**, mouth). It bears two pairs of simple pigmented **eyes** on dorsal surface, a pair of short cylindrical, sensory **prostomial tentacles** anteriorly, and a pair of short, stout, fleshy, and two-jointed **palps** ventro-laterally. Palps are somewhat contractile and their small distal joints can be retracted into their large proximal joints.

Tentacles, palps and cirri, all serve as sensory organs.

[II] Trunk

It comprises practically the entire body excluding head and last body segment or **pygidium.** It consists of 80-200 similar segments, each broader than being long and characterized by the presence of a **parapodium** on each lateral side.

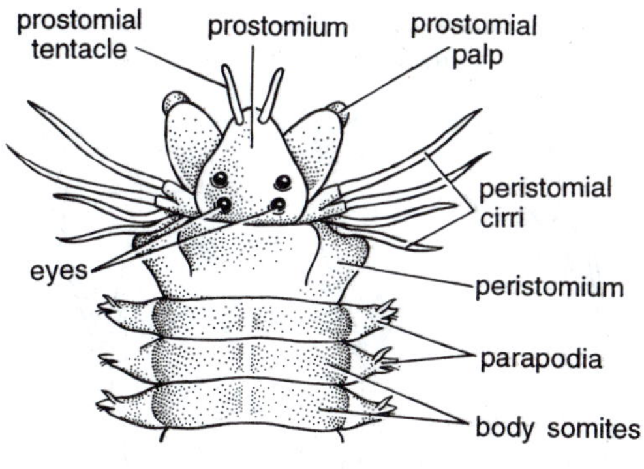

Fig. 2. *Nereis.* Head in dorsal view.

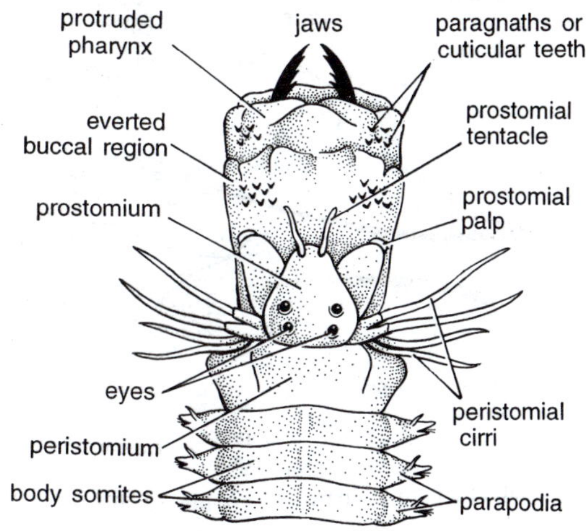

Fig. 3. *Nereis.* Head in dorsal view with everted proboscis.

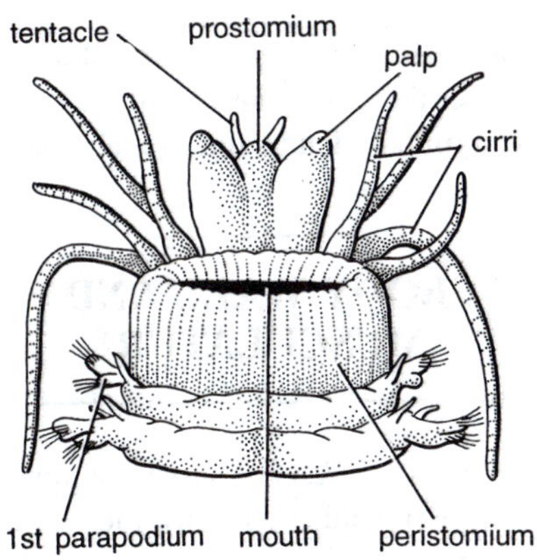

Fig. 4. *Nereis.* Head in ventral view.

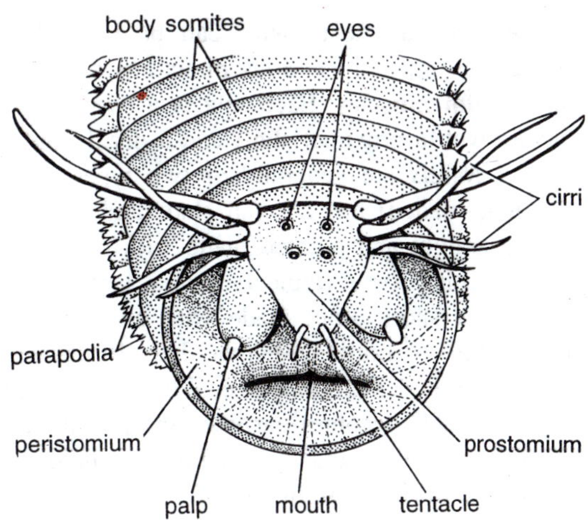

Fig. 5. *Nereis.* Head in frontal view.

1. Parapodia. Parapodia (Gr., **para**, beside + **podos**, foot) are flattened, fleshy, vertical flap-like outgrowth of body wall on lateral sides of trunk segments. They are hollow structures enclosing coelom which is continuous with that of trunk segments. Structurally, a parapodium is typically **biramous**, consisting of a proximal **basal region** bearing distally two rami or parts, a dorsal **notopodium** and a ventral **neuropodium**. Each part or ramus is further subdivided into two leaf-like lobes or **ligulae**, a dorsal **superior ligula** and a ventral **inferior ligula**.

Each part bears at its base a slender, tentacular process known as **cirrus**. Notopodial or **dorsal cirrus** is slightly larger than neuropodial or **ventral cirrus**. Each part is supported internally by a deeply-embedded, long, stout and black chitinous rod, known as **aciculum**. It also serves for attachment of **setal muscles**, thus serving as a sort of endoskeleton. Each part also bears a bundle of long, fine, stiff, chitinous bristles, the **setae** or **chaetae,** which project beyond its margin. Each seta is lodged in a **setal** or **setigerous sac**, formed by an inpushing of epidermis and arises from a single **formative cell,** lying at the base of sac. New setae are continuously produced by setal sac

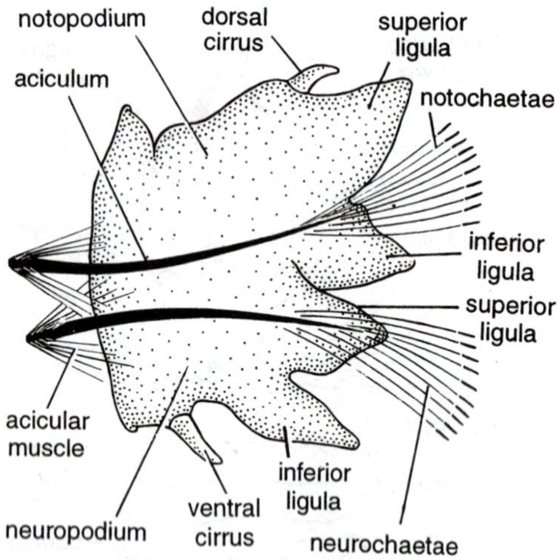

Fig. 6. *Nereis.* Parapodium.

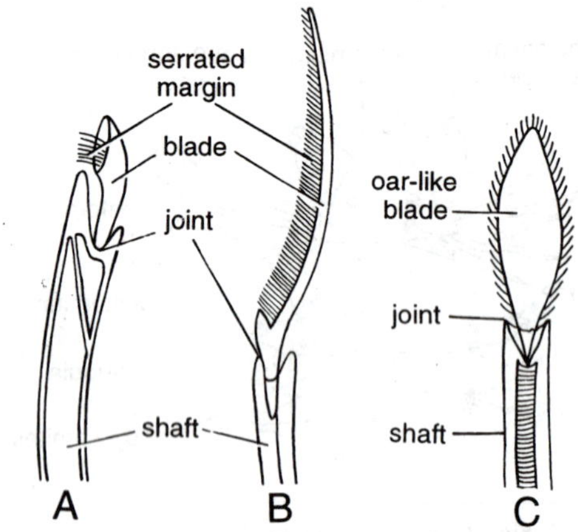

Fig. 7. *Nereis.* Kinds of setae. **A.** Typical. **B.** Long-bladed. **C.** Oar–shaped of heteroneries.

as older ones are lost. Setae can be protruded, retracted or turned in various directions with the help of **setal muscles**. Each seta is a two-jointed structure with a proximal **shaft** and a distal **blade**. Setae of *Nereis* are generally of two types. One type (long-bladed) has a small shaft and a long, slender, straight and pointed blade with one edge serrated. Other type (typical) has a large stout shaft and a short, stout blade with an incurved notched tip. A third type (oar-shaped) of seta occurs in the sexual phase or *Heteronereis* in which the blade becomes oar-like. Sharp setae are used for protection and for obtaining a hold on the smooth inner walls of burrow.

Parapodia near the middle region of trunk are the largest and gradually decreases in size towards the two ends. All parapodia are essentially similar except that the first two pairs lack notopodial setae. Parapodia serve the dual purpose of locomotion and respiration.

2. Nephridiopores. These are minute excretory pores through which the nephridia communicate with the exterior. Each parapodium of a nephridia-bearing segment bears a nephridiopore near the base of its ventral cirrus.

[III] Pygidium

This is represented by the last body segment, also known as **tail** or **anal segment**. It bears a terminal **anus**, a pair of long filamentous, ventral appendages, the **anal cirri**, and several minute **sensory papillae**. Parapodia are absent.

BODY WALL AND MUSCULATURE

The body wall of *Nereis* consists of four layers : (i) cuticle, (ii) epidermis, (iii) musculature, and (iv) peritoneum.

[I] Cuticle

It is the outermost, thin, tough and chitinous layer, with a system of intersecting striations which impart iridescence to it. It is secreted by the underlying epidermis and is perforated by numerous minute openings of epidermal gland cells.

[II] Epidermis

It lies beneath cuticle and rests on a thin **basement membrane**. It consists of a single layer of tall columnar **supporting** cells and some scattered **glandular** and **sensory cells**. Epidermis on ventral surface, especially near bases and on lobes of parapodia, is somewhat thicker and more glandular. Epidermal gland cells secrete **mucus**,

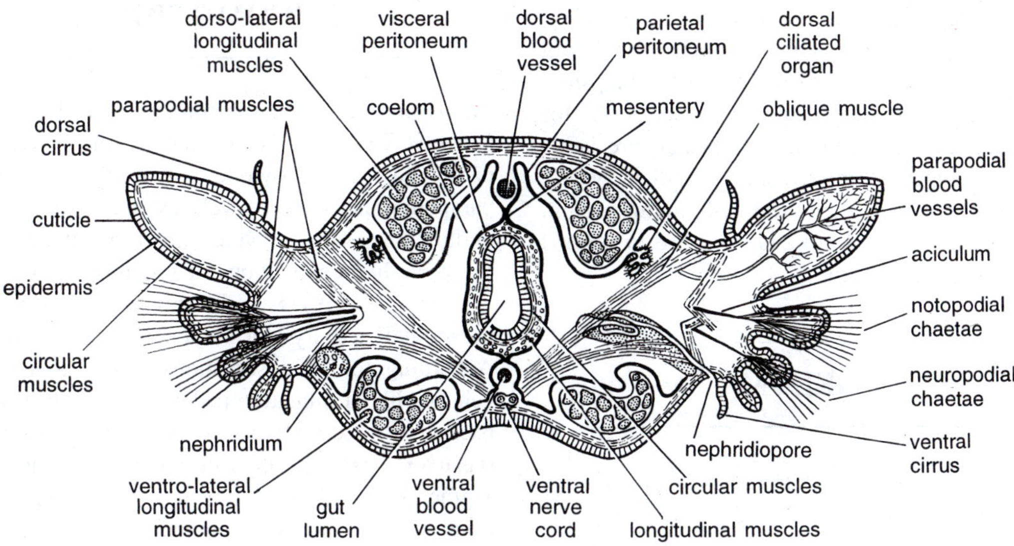

Fig. 8. *Nereis.* T.S. of a segment.

which lines the U-shaped tube in which the worm lives.

[III] Musculature

Nereis has a well-developed musculature consisting of (i) **circular**, (ii) **longitudinal**, and (iii) **oblique** muscles. These are composed of unstriped muscle fibres.

1. Circular muscles. These form a continuous layer beneath epidermis, which is slightly thicker on ventral side. In parapodia, circular muscles are modified to form a complicated system of parapodial muscles comprising the **protractor** and **retractor** muscles. Protractors radiate from bases of setigerous sacs to the layer of circular muscles on all sides. Retractors extend from outer part of setigerous sacs to the dorso-lateral body wall.

Contraction of circular muscles make the body longer and thinner, while that of protractors and retractors causes respectively the protrusion and retraction of parapodial acicula and lobes.

2. Longitudinal muscles. These do not form a continuous layer beneath circular muscles, but occur in four powerful longitudinal bundles, two **dorso-lateral**, one on either side of dorsal blood vessel, and two **ventro-lateral**, one on either side of ventral nerve cord. Contraction of longitudinal muscles causes shortening and thickening of body.

3. Oblique muscles. These lie in each segment in two pairs. First pair is situated at the level of anterior limit, and second at the level of posterior limit of parapodia. Each oblique muscle arises between ventral nerve cord and a bundle of ventro-lateral longitudinal muscles. It soon bifurcates into a dorsal and a ventral branch, each extending to the base of corresponding parapodium.

Oblique muscles are responsible for bending movements of parapodia and their complete retraction.

[IV] Peritoneum

Muscles are lined internally by a thin, delicate layer of **coelomic epithelium** or peritoneum. This layer also lines the coelom externally and is thus also described as the somatic or parietal layer of coelomic epithelium. It secretes the coelomic fluid.

[V] Functions of Body Wall

Body wall of *Nereis* performs a variety of functions, (i) Cuticle protects the body against desiccation and mechanical injuries. (ii) Epidermis, being highly vascular, serves for respiration, (iii) Epidermal glands secrete mucus for lining the burrow so as to prevent it from collapsing, (iv) Epidermal sensory cells are responsible for perceiving external stimuli, (v) Musculature helps in various types of movements, (vi) Setae serve for locomotion, (vii) Peritoneal cells lining coelom secrete the coelomic fluid and give rise to gonads during breeding season.

COELOM

Body cavity of *Nereis* is a true coelom of schizocoelic origin, *i.e.*, it is formed by the splitting of embryonic mesoderm. It forms a spacious perivisceral cavity between the body wall and alimentary canal. Numerous membranous transverse **septa** or **coelosepta** divide the entire coelom into a linear series of compartments, each corresponding to a segment of body. Septa correspond in position to the external intersegmental grooves. They are perforated and each is composed of connective tissue and muscles sandwiched between two layers of coelomic epithelium. Coelom of each segment is further divided into right and left compartments by dorsal and ventral mesenteries. The latter, however, disappears in older segments.

Coelom is filled with a colourless coelomic fluid containing amoeboid corpuscles (leucocytes) and, during breeding season, numerous reproductive cells in various stages of development. It is internally lined by thin, delicate **coelomic epithelium** divided into two layers : a **parietal layer** beneath body wall and a **visceral layer** covering the alimentary canal. Coelom communicates with the exterior through nephridia. Coelomic fluid acts as an **hydrostatic skeleton** and for transportation of digested food, gases and waste materials.

LOCOMOTION

Nereis can crawl on substratum as well as swim actively. **Crawling** is affected by the activity of parapodia alone, while **swimming** involves parapodial activity as well as snake-like lateral undulations of body which are brought about by the wave-like contractions of longitudinal muscles.

Gray (1939) has described the pattern of parapodial activity during locomotion in *Nereis.* According to his observations, each parapodium behaves like a miniature paddle, alternately beating backwards (**effective stroke**) and forwards (**recovery stroke**). Opposite parapodia of adjacent segments always perform the same stroke at a time, while opposite parapodia of the same segment perform reverse strokes at the same time. Movements of parapodia are controlled by oblique and parapodial muscles and coelomic fluid which may be forced into, or withdrawn from them.

DIGESTIVE SYSTEM

[I] Alimentary Canal

It is a straight tube extending from anterior to posterior end of the body suspended in the body cavity by the dorsal mesentery and the intersegmental septa. It is open at both ends, the anterior opening is the **mouth** and posterior opening, the **anus**. It can be distinguished into three distinct regions : (i) **stomodaeum** or **foregut** comprising buccal cavity and pharynx, (ii) **mesenteron** or **mid-gut** comprising oesophagus and stomach-intestine, and (iii) **proctodaeum** or **hind gut** comprising rectum. Foregut and hind gut are lined internally by ectodermal epithelium and cuticle which is continuous with those of body wall, while the mid-gut is lined by endodermal epithelium.

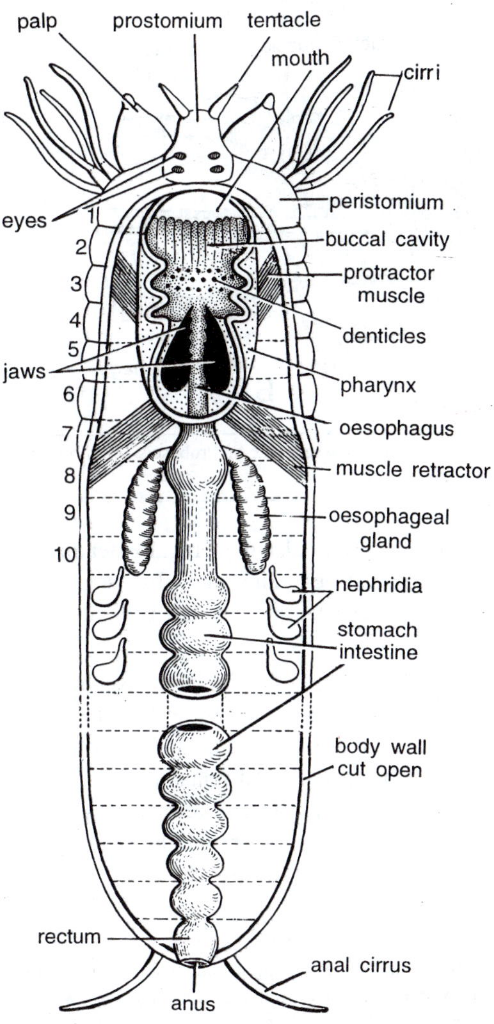

Fig. 9. *Nereis.* Diagrammatic view of the anterior and posterior ends dissected dorsally to show the alimentary canal.

1. Mouth. It is a transversely elongated slit, opening ventrally below prostomium and bordered by peristomium. It leads into buccal cavity.

2. Buccal cavity and pharynx. Buccal cavity is a wide chamber succeeded by the highly muscular **pharynx**. Cuticle-lining their lumen is thickened at places to form **denticles** or **paragnaths** or **teeth**, which have fixed numbers and specific arrangement in each species. Posterior part of pharynx has a thick muscular wall and a narrow lumen. Embedded laterally in the wall of this region (sometimes called **dentary region**), is a pair of stout, mobile and dark, chitinous **jaws**. Each jaw has a hollow **base** and an incurved, pointed, somewhat notched **apex**. Inner margins of jaws are serrated. Buccal cavity and pharynx are wrapped in a common muscular coat and together they extend up to 4th or 5th trunk segment. They can be fully everted to form a **proboscis** or **introvert**. This operation exposes tips of jaws for capturing prey.

3. Oesophagus. Pharynx narrows posteriorly to lead into oesophagus. It is a narrow tube extending through 5 segments behind pharynx. A pair of long, unbranched, sacculated glandular pouches, the **oesophageal glands** or **caeca**, open into it laterally at the anterior end. They probably secrete proteolytic enzymes. Oesophagus is followed by the stomach-intestine. Opening between the two is guarded by a sphincter muscle. A distinct stomach is absent.

4. Stomach-intestine. It is a segmentally constricted, straight, thin-walled tube extending up to the last trunk segment. Epithelial lining of mid-gut contains scattered **gland cells** which secrete digestive enzymes.

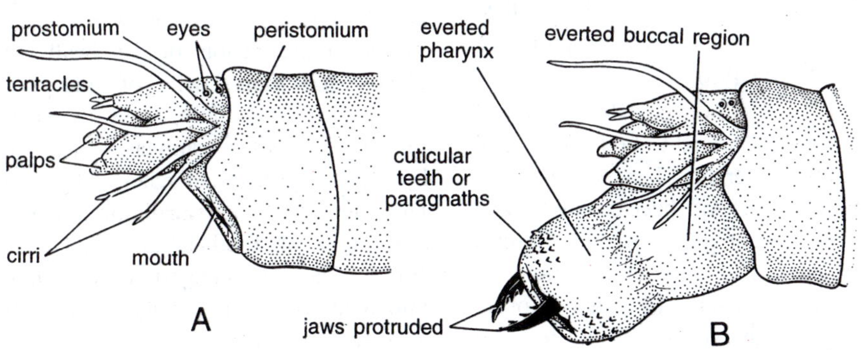

Fig. 10. *Nereis.* Anterior end in lateral view. **A.** Pharynx retracted. **B.** Pharynx everted.

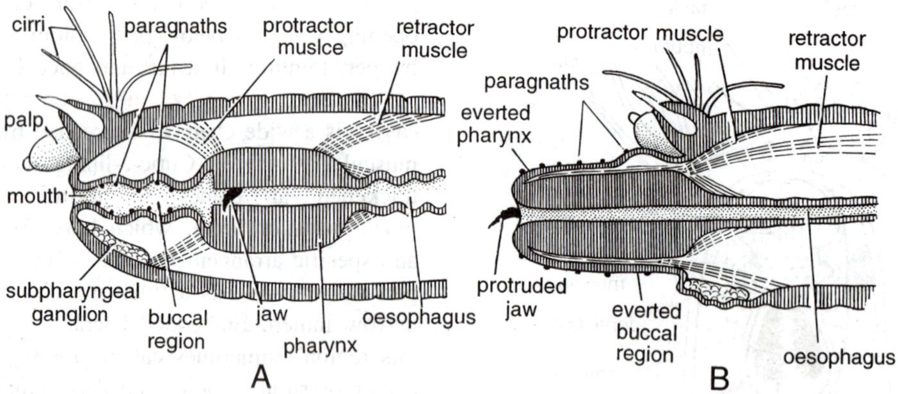

Fig. 11. *Nereis.* Diagrammatic representation of the action of introvert in sagittal section. **A.** Proboscis or introvert at rest. **B.** Introvert protruded.

5. Rectum. Rectum or hindgut occupies the last body segment or pygidium and opens to the exterior through a terminal **anus**.

[II] Histology of Alimentary Canal

Gut wall consists of : (i) an outermost layer of **visceral peritoneum**, (ii) a layer of **longitudinal muscles**, (iii) a layer of **circular muscles**, (iv) an **enteric epithelium** which is endodermal in case of mid-gut but ectodermal in fore- and hind guts, and (v) a layer of **cuticle** in fore- and hind guts.

[III] Food and Feeding Mechanism

Nereis is **carnivorous** and a raptorial feeder. It preys upon small crustaceans, worms, larvae, etc., which are actively captured by the exposed jaws of proboscis. Worm lives in the burrow with its head protruded and suddenly everts its proboscis to capture any minute creature passing by. Jaws seize the prey like a pair of forceps. Worm then withdraws into burrow and retracts its proboscis to swallow the helpless prey.

1. Eversion of pharynx. Eversion of proboscis or introvert is brought about by special **protractor muscles**, extending between the wall of peristomium and that of buccal cavity and pharynx. Combined action of protractor muscles and the pressure generated on coelomic fluid by the contraction of body wall musculature forces the buccal cavity and pharynx inside out to form a proboscis. A muscular diaphragm extending from body wall between 2nd and 3rd segments to buccal cavity checks the eversion beyond a certain limit.

2. Retraction of pharynx. Retraction of introvert is brought about by **retractor muscles** extending backwards from posterior end of pharynx to body wall, and relaxation of body wall musculature. As proboscis retracts, jaws come closer crossing each other and holding the prey which is also withdrawn inside.

Nereis diversicolor frequently exhibits the mechanism of **filter feeding**. It secretes a **mucous cone** at one end of its burrow and by dorso-ventral undulations of its body sets up a constant current of water entering the burrow through that end and leaving through the other. Mucous cone, behaving like a strainer, holds up the food particles. Food-laden mucous cone is finally ingested by the worm.

[IV] Digestion, Absorption and Egestion

Engulfed food passes backwards through midgut by peristaltic action of gut wall. Oesophageal glands and gland cells of mid-gut epithelium secrete juices containing digestive enzymes which include proteases, amylases and lipases for hydrolysing proteins, starches and fats, respectively. Digestion is **extracellular**. Digestion and absorption of food takes place mainly in the stomach-intestine. Indigestible food substances pass into rectum and are finally egested through anus as faecal matter.

RESPIRATION

Gills or any other special organs of respiration are lacking in *Nereis.* However, respiration is carried on by the whole body surface, but more specially by the thin, flattened lobes of parapodia, which possess extensive capillary-networks, lying very close to the surface. Blood running through them gives up carbon dioxide collected from tissues and receives oxygen dissolved in surrounding water. Water is constantly renewed by gentle dorso-ventral undulations of body which cause a steady current to flow through the burrow.

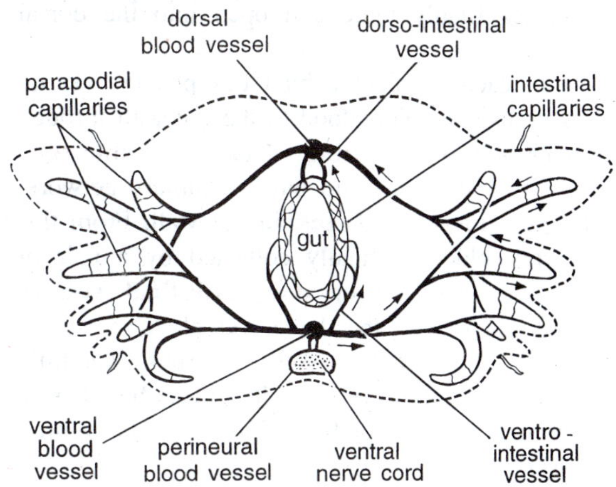

Fig. 12. *Nereis.* Blood-vascular system, representation in T.S. of a segment.

BLOOD-VASCULAR SYSTEM

Transportation of materials in annelids and higher animals is brought about by a specialized system, the **circulatory** or **blood vascular system**. It consists of a fluid tissue, called **blood**, which circulates throughout the body through a system of closed tubes or **blood vessels**. This type of blood-vascular system is called **closed type**.

[I] Blood

It consists of a fluid medium called **plasma** containing numerous nucleated, colourless amoeboid cells or **corpuscles** and dissolved **haemoglobin**, which serves as the respiratory pigment and imparts bright red colouration to blood. Blood circulates in body throughout life of the animal (hence the name **circulatory system**) and brings about transportation of gases, food, excretory materials, etc.

[II] Blood Vessels

There are three main longitudinal blood vessels, *viz*., **dorsal, ventral** and a **peri-neural**.

1. Dorsal vessel. It runs above the alimentary canal, through dorsal mesentery, from hind end forward to the fifth segment. Its walls are highly contractile and surrounded at several places by rings of muscles. Strong peristaltic waves pass from its posterior to anterior end, driving blood anteriorly. Behind oesophagus, it acts as a **collecting vessel** receiving blood from body wall, parapodia, nephridia and alimentary canal. In the region of oesophagus, it bifurcates and each branch forms a **capillary meshwork** to supply the oesophageal wall. Here, it acts as a **distributing vessel**. Blood from capillary meshwork in oesophageal wall is collected by a single median vessel which opens into the ventral vessel.

2. Ventral vessel. It runs mid-ventrally below alimentary canal almost the entire length of body. Its walls have little contractility and hence peristalic waves do not pass through it. Blood flows in it from anterior to posterior end. It is a **distributing vessel** except at the anterior end, where it collects blood from the oesophageal wall. In the last segment, it communicates with dorsal vessel by a simple **circum-rectal ring**.

In each segment, behind the oesophagus, the ventral vessel is connected with the dorsal vessel by a pair of loop-like **lateral commissural vessels**. Each of these vessels shortly divides into **afferent** branches (afferent cutaneous, afferent nephridial and afferent parapodial) which carry blood to the body wall, nephridium and parapodium of its side. In these organs, the afferent branches break up into a network of capillaries which reunite to form the corresponding **efferent** branches. The efferent branches finally unite and open into the dorsal vessel.

In each segment behind oesophagus, ventral vessel also supplies blood to the gut wall through a pair of **afferent intestinal** or **ventro-intestinal vessels**. These break up into a capillary network, the **peri-intestinal plexus**, in gut wall. From this network blood is finally collected by a pair of **efferent intestinal** or **dorso-intestinal** vessels, which carry it to the dorsal vessel.

3. Peri-neural vessel. It runs ventrally surrounding the nerve cord. It collects blood from ventral body wall and conveys it to the ventral vessel.

EXCRETORY SYSTEM

[I] Excretory Organs

Excretion is brought about by **nephridia**, one pair of which is found in each segment except a few anterior and a few posterior ones.

Each nephridium is distinguished into a somewhat oval, curved **body** and a narrow **neck**. Body lies somewhat transversely in the ventro-lateral aspect of segment, while neck penetrates the intersegmental septum in front and extends a short distance into adjacent anterior segment. Body of nephridium consists of a syncytial mass of connective tissue, containing a coiled **excretory** or **nephridial tubule** open at both ends. One end of tubule runs through neck and opens into coelom of adjacent anterior segment by a funnel-like **nephrostome**. Margin of nephrostome is produced into a number of long, delicate processes provided with **cilia**. A nephridium possessing such a ciliated nephrostome is termed **metanephridium**. Behind nephrostome the excretory tubule becomes much coiled in the body region and finally forms the **terminal duct**, which opens to the exterior through **nephridiopore**, situated near the base of ventral cirrus of parapodium. Nephridiopore is guarded by a sphincter. Excretory tubule is ciliated at most places. Terminal duct is without cilia.

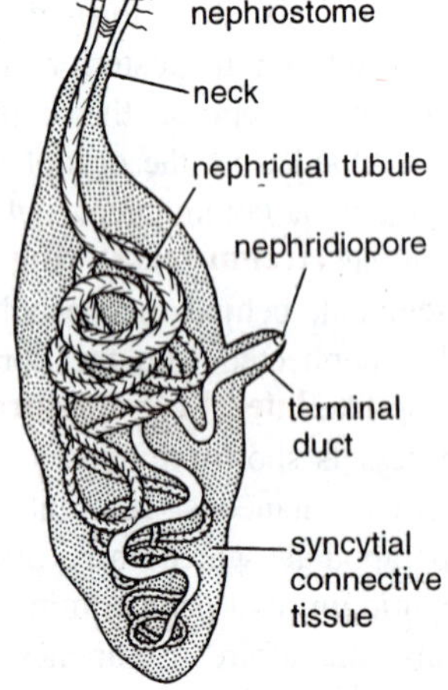

Fig. 13. *Nereis.* A single nephridium.

[II] Physiology of Excretion

Excretion is carried out in two ways. Outer surface of nephridium is densely covered by blood capillaries. Gland cells, forming the wall of ciliated excretory tubule, remove waste material from blood and finally discharge it through nephridiopores. Nitrogenous waste is mainly ammonia (**ammonotelic**). Ciliated funnels of nephridia, opening into coelom, also remove dead or dying coelomic corpuscles which have eaten up, or have been destroyed by foreign bodies such as bacteria. Any useful substance entering the excretory tubule is reabsorbed by its cells and returned to the blood capillaries. This is known as **selective resorption**.

NERVOUS SYSTEM

Correlated with its active predaceous life, *Nereis* possesses a well-developed nervous system. It is bilaterally symmetrical, metamerically segmented and distinguished into **central**, **peripheral** and **visceral** nervous systems.

[I] Central Nervous System

Cerebral ganglia or **brain** is a large bilobed mass located dorsally in prostomium. It consists of nerve cells surrounding a centrally-placed mass of nerve

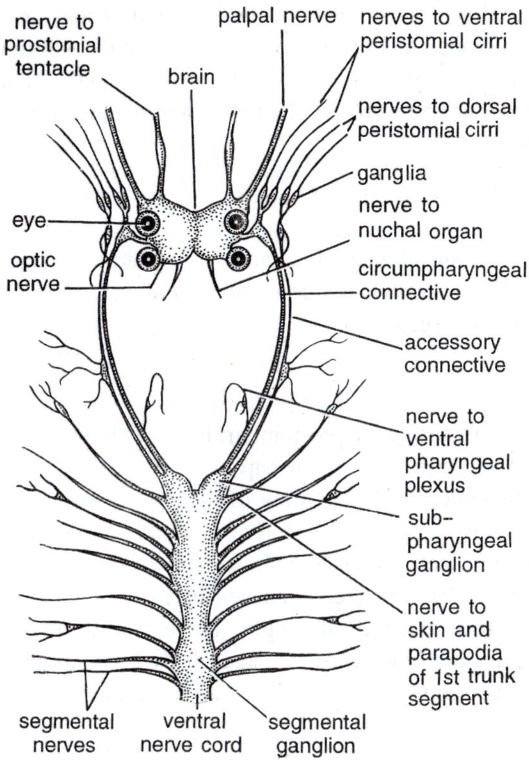

Fig. 14. *Nereis.* Nervous system.

fibres. In brain, three **centres** are mainly recognised. These are **anterior**, **middle**, and **posterior centres**. Middle centre bears a pair of small lobes called **corpora pedunculata**. These are **association centres** that coordinate all impulses entering the brain.

A **sub-pharyngeal** ganglion, formed by fusion of two pairs of ganglia of ventral nerve cord, lies ventrally below the pharynx in first trunk segment. Brain is connected with sub-pharyngeal ganglion by a pair of stout nerve strands or **circum-pharyngeal connectives**, running one on either side of pharynx, forming a nerve collar or nerve ring. Connectives bear small ganglia near their points of attachment with brain. These ganglia probably represent nerve centres that were originally located in ventral nerve cord but have shifted to their present position due to cephalization.

Ventral nerve cord runs posteriorly from sub-pharyngeal ganglion throughout body along the mid-ventral line, immediately beneath the ventral blood vessel. It is actually formed by the fusion of two very closely apposed cords enclosed in a common sheath of connective tissue, and joined together by segmental cross connections. Its dual nature is revealed by a careful examination of its transverse section. In each segment, nerve cord is dilated into a **segmental ganglion** which actually represents a pair of ganglia very closely fused together. A segmental ganglion occupies about two-third of the length of a segment. Cell bodies of nerve cord are confined to the segmental ganglia.

Throughout the length of ventral nerve cord run 5 **giant fibres** or **axons**, 3 central and 2 lateral in position. These fibres are very thick and long, some extending the entire length of nerve cord without synapses. Through such fibres impulses travel very fast, about 25 cm in a hundredth of a second, as against 25 cm in 10 seconds in case of the usual ventral nervous system that has several ganglion and synapses. Such giant fibres, found in most annelids (including earthworm), enable the worm to respond to an alarm by sudden violent shortening of the entire body.

[II] Peripheral Nervous System

It consists of nerves connected to brain, circum-pharyngeal connectives and segmental ganglia of ventral nerve cord. Brain gives out : (i) from its anterior centre, a pair of short nerves to prostomial palps, (ii) from its middle centre, two pairs of stout nerves to eyes and a pair of nerves to prostomial tentacles, and (iii) from its posterior centre, a pair of slender nerves to nuchal organs. Ventral pair of peristomial cirri receives nerves from ganglia on circum-pharyngeal connectives. Sub-pharyngeal ganglion gives out anteriorly a pair of long nerves, the **accessory connectives**, which run parallel to the circum-pharyngeal connectives through much of their course and supply the dorsal pair of peristomial cirri. Each accessory connective, before giving out a pair of nerves to peristomial cirri, swells into a ganglion. It lies close to the corresponding ganglion on circum-pharyngeal connectives, and like the latter, represents the nerve centre of ventral nerve cord. Posteriorly, the sub-pharyngeal ganglion gives out a pair of nerves to body wall and parapodia of third segment (first trunk segment).

Each segmental ganglion gives out four pairs of peripheral nerves. I and IV pairs supply the longitudinal muscles and body wall. II pair supplies the parapodia, and III pair is composed of fibres from proprio-receptors in muscles. Each peripheral nerve contains afferent and efferent fibres.

[III] Visceral Nervous System

It consists of a network of fine nerves **(stomogastric nerves)** and a few ganglia supplying the dorsal and ventral walls of pharynx (proboscis). These are involved in the motor control of proboscis. Dorsal and ventral networks are connected with the ganglia and ventral ends of circum-pharyngeal connectives, respectively.

Nereis possesses a slight power of learning including finding its way and avoiding unpleasant stimuli. However, when brain is removed, the ability to retain learnt behaviour is greatly reduced.

SENSE ORGANS

[I] Prostomial Tentacles

These are a pair of small, cylindrical projections from anterior border of prostomium. They are probably **tactile**. Surface of tentacles bears numerous sensory spiral organs, each consisting of about 100 photoreceptive cells, spirally arranged within a cuticular pit.

[II] Prostomial Palps

These are a pair of stout highly muscular, two-jointed structures arising one from each ventro-lateral aspect of prostomium. Proxima or basal joint is large, while distal one is small and can be retracted into the former. In addition to serving as lateral lips, the palps probably serve as organs of **touch**, **taste** and **smell**.

[III] Nuchal Organs

These are a pair of small pits on the posterior dorsal surface of prostomium behind eyes. Pits are lined by ciliated epithelium with gland cells and connected by nerves to the hind part of brain. They serve as organs of **smell** and **chemoreception** and help the worm in detecting prey.

[IV] Peristomial Cirri

These are elongated, slender, unjointed structures, two pairs on each side of peristomium, serving as **tactile** organs.

[V] Eyes

Dorsal surface of prostomium bears 4 simple black eyes. Each eye is a cup-like structure with a pigmented wall consisting of radially arranged tall, narrow and light-receptive **retinal cells**. Each retinal cell is differentiated into three distinct parts : (i) An outer nucleated part drawn out into a nerve fibre of the optic nerve, (ii) A highly pigmented middle part or main body, (iii) An inner part forming a clear hyaline rod projecting towards lens.

Cup is filled with a gelatinous substance forming the **lens** which, according to Andrews, is secreted by the retinal cells. Outer exposed surface of eye is covered by a layer of flattened epidermal cells and a transparent protective cuticular layer, together forming the **cornea**. Retinal cells are modified ectodermal cells and become continuous with epidermis at the edges, so that the small opening of cup towards cornea functions as a **pupil**.

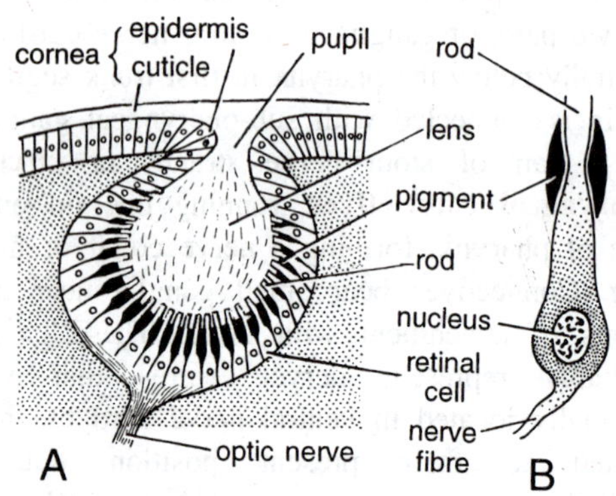

Fig. 15. *Nereis.* Eye. **A.** Entire eye in V.S. **B.** An isolated retinal cell.

Eyes are not image-forming but enable the worm to detect changes in light intensity. Clamworms are photonegative.

REPRODUCTIVE SYSTEM

[I] Gonads

Some freshwater nereids are known to be hermaphrodite. However, most species of *Nereis* are dioecious (unisexual). Gonads (testes and ovaries) are neither distinct nor permanent organs. They are masses of developing gametes formed only during breeding season. These masses develop as projections of swellings by proliferation of ventral septal peritoneum in all trunk segments of body, leaving a few anterior segments.

Gametes are shed as **spermatogonia** in male and as **oogonia** in female into coelomic fluid where they undergo maturation to develop into spermatozoa and ova, respectively. Spermatozoa or sperms are small cells with a minute rod-shaped head and a long vibratile tail. Ova are somewhat large and rounded and packed with yolky globules. In a mature worm, coelom remains packed with gametes.

[II] Gonoducts

In *Nereis* there are no gonoducts. Ripe sperms and ova are discharged to outside sea water mostly through metanephridia by the action of cilia borne by nephrostomes and nephridial tubules. **Goodrich** (1945) proposed the name **nephromixia** or **mixonephridia** for such nephridia, acting both as excretory and genital ducts. In some species, gametes are liberated by the breakdown of body wall. After the rupture of body wall the adults die.

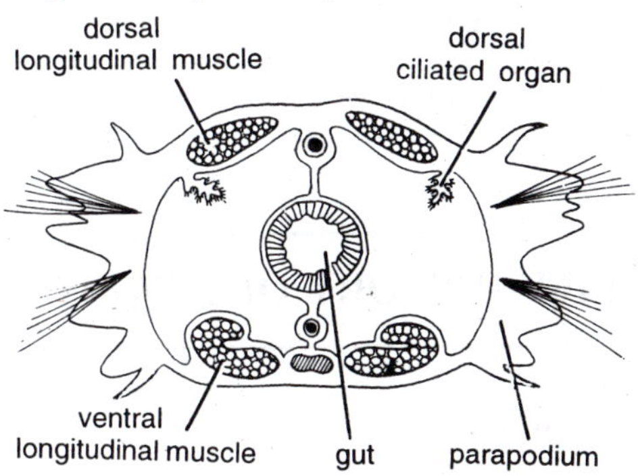

Fig. 16. *Nereis.* T.S. of a segment to show the ciliated organs.

[III] Dorsal Ciliated Organs

A pair of dorsal ciliated organs occurs in each segment in close relation to the dorsal longitudinal muscles. Each ciliated organ is a ciliated tract of coelomic epithelium appearing as a short and much folded funnel with a wide opening into body cavity but lacks an opening to the exterior. They resemble the **coelomoducts** of other polychaete annelids. Formerly regarded as excretory organs, they are now looked upon as gonoducts. It has been said that during sexual maturity, funnels open to the exterior by temporarily acquired minute apertures opposite to them in body wall. But due to lack of adequate evidence, the real function of ciliated funnels remains controversial.

HETERONEREIS

[I] Epitoky

At sexual maturity, most of the posterior segments, filled with gametes, exhibit morphological and anatomical differentiation. These constitute the sexual region or **epitoke** of worm. Few anterior segments which do not take part in gamete-formation, constitute the asexual region or **atoke**. Sexually mature worm with these two regions is known as *Heteronereis* and the phenomenon involving transformation of non-sexual individual into sexual individual is referred to as **epitoky**.

In *N. virens* and some other species, the sexually mature phase differs so widely from the immature phase that it has long been considered to be a member of a separate genus, which was given the name *Heteronereis*. Its correct identity was known only after **Malamgren** recognized it as merely the sexual individual of *Nereis*.

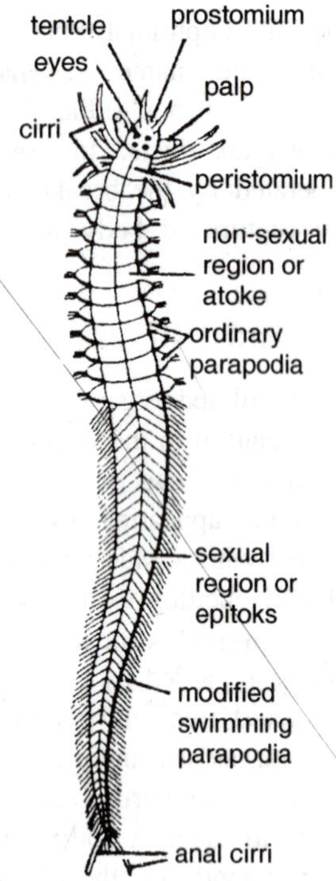

Fig. 17. *Heteronereis.*

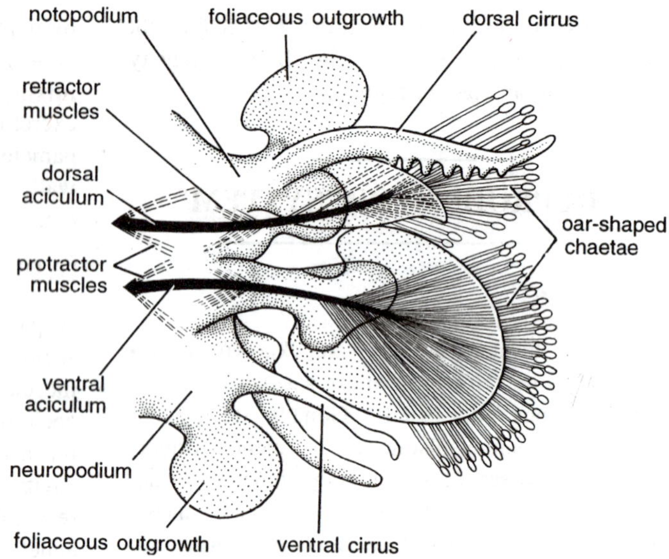

Fig. 18. *Heteronereis.* Parapodium.

[II] Characteristic Features of *Heteronereis*

(1) Instead of creeping about on sea bottom or living in burrows like *Nereis,* the *Heteronereis* swims actively in surface waters.
(2) Body of *Heteronereis* is divisible into two distinct regions : an anterior asexual **atoke**, and a posterior sexual **epitoke**.
(3) Parapodia of posterior sexual region becomes larger, more vascularized and develop flattened leaf-life outgrowths for more rapid respiration. Their normal setae are replaced by flattened oar-shaped setae arranged in a fan-like manner, to offer a larger surface for swimming. Dorsal cirri are altered. Leucocytes break down and digest the original parapodial muscles and new muscles are formed.
(4) The eyes become greatly enlarged and conspicuous.
(5) Prostomial palps and tentacles become reduced, but peristomial cirri become longer.
(6) Intestine becomes compressed and functionless due to much developed gonads.
(7) Sensory projections become shrunken while pygidium or anal segment develops special sensory papillae.
(8) In some species, such as *Nereis virens, Heteronereis* shows sexual dimorphism, male having less unaltered anterior segments than female.

It has been shown by **Durchon** that immature worm is prevented from metamorphosing into *Heteronereis* by inhibitory action of neurosecretions produced by cerebral ganglia. Sexual maturity is attained only when secretory activity in cerebral ganglia is declined. Some other hormones from cerebral ganglia are supposed to be responsible for sexual maturity of the worm.

LIFE-HISTORY OF *NEREIS*

[I] Swarming

Sexually mature individuals or *Heteronereis* swim to the surface of sea water in order to shed sperms or ova. This behaviour is called **swarming**. It generally occurs at night and some species, while

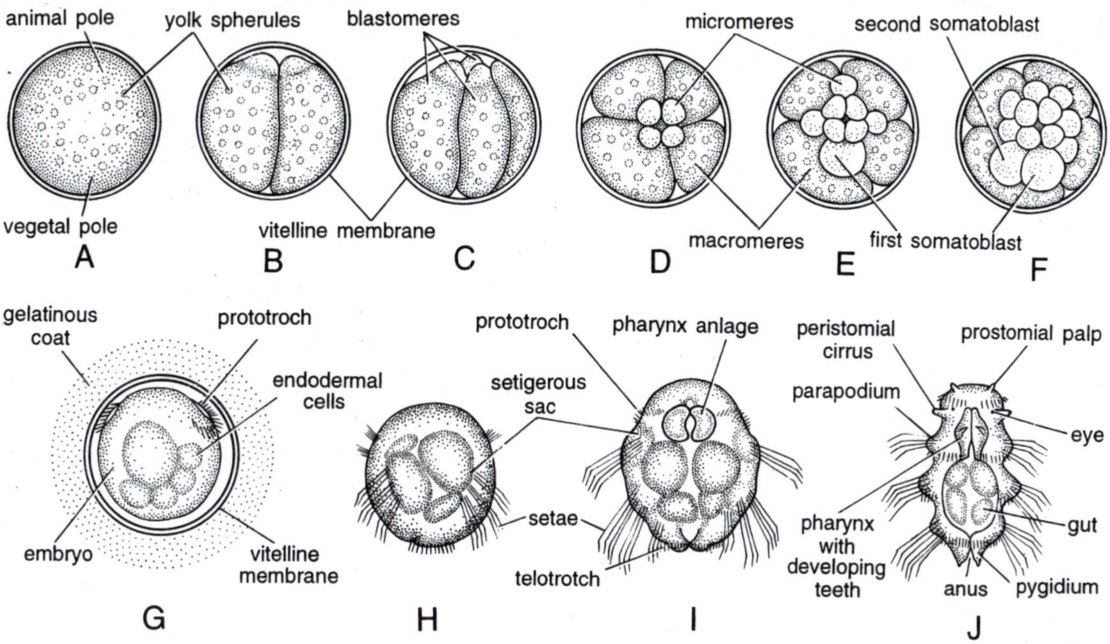

Fig. 19. *Nereis.* Stages in development.
A. Fertilized egg. **B.** Two-cell stage. **C.** Four-cell stage. **D.** Eight-cell stage. **E.** Ist somatoblast formed. **F.** 2nd somatoblast formed. **G.** Young trochophore before hatching. **H.** Post-trochophore or nectochaete larva with three setigerous segments. **I.** Older larva. **J.** Later larva after three weeks.

swarming, perform a nuptial dance in which both males and females swim rapidly in a circle. Females produce a substance, called **fertilium**, which attracts the males and stimulates shedding of sperms, which in turn excites the females and stimulates shedding of eggs. In case of *N. succinea,* males first swim to the surface and wait for females. When the latter appear, males swim around them shedding sperms. In response, the females get excited and shed ova.

[II] Fertilization

Fertilization is mostly external in *Nereis virens* and takes place in sea water. *Platynereis megalops* has an altogether different mechanism. The male wraps tightly around female, inserts his anus into her mouth and injects sperms. Gut of both having been eroded by phagocytes, sperms pass directly into coelom of female where fertilization of eggs occurs. Fertilized eggs are shed at once from the posterior end of female's body. This mechanism of internal fertilization has probably evolved in *Platynereis megalops* because its eggs become unfertilizable just after 30 seconds contact with sea water. In case of *N. diversicolor,* which does not attain epitokous form, fertilization occurs in burrow of female into which male enters, or on the surface of mud.

[III] Development

Development of *Nereis* comprises three distinct periods :

1. Pre-larval period. Unfertilized Egg of *Nereis* contains numerous yolk spherules and oil droplets. It is covered by a thick, radially striated membrane, called **zona radiata**, which in turn has another thin delicate membrane around it. Outside these membranes is a thick gelatinous coating. Soon after fertilization, zona radiata disappears, yolk spherules from animal pole move into vegetal pole (**telolecithal** condition), egg extrudes two polar bodies, and undergoes cleavage.

First two cleavages of zygote are **equal** and **vertical** and result in four cells or **blastomeres**, lying in the same plane. Cleavage is **determinate**, *i.e.*, fate of blastomere is fixed and after four-cell stage, each blastomere gives rise to only one quadrant (quarter) of embryo.

Third cleavage is unequal and horizontal, *i.e.,* at right angles to the first two cleavages. It produces four small, yolk-free **micromeres** towards the animal pole and four large, yolky **macromeres** towards the vegetal pole. Remaining cleavages are also unequal. Fourth, fifth and sixth cleavages are also horizontal and cut off three more quartettes of micromeres from megameres. Cells of adjacent quartettes lie alternately. This type of cleavage is termed **spiral cleavage**. One micromere of second quartette and one of fourth are larger than the others. These are respectively termed the **first** and **second somatoblasts**.

Micromeres (except the second somatoblast) are the forerunners of ectoderm, megameres of endoderm, and second somatoblast of mesoderm. Cleavage results into a stereoblastula. Gastrulation takes place by invagination of the gut pouch resulting in a ciliated gastrula.

2. Larval period (Trochophore). After gastrulation, ciliated embryo rapidly develops into a larval stage, called **trochophore** or **trochosphere**. Only after 24 hours of development, the trochophore structure begins to appear. Embryo takes up a top-like form. Ringing the larva just above its equator develops a girdle of ciliated cells, called **prototroch** or **preoral ciliated band**.

A typical trochophore larva does not occur in *Nereis.* Instead, trochophore stage is embryonic and passed inside the egg membrane. It differs from a typical trochophore in the absence of a blastocoel. Newly hatched larva of *Nereis* is termed **post-trochophore** larva or **nectochaete**. It possesses three segments with bristles and is without greatly developed larval structures.

3. Post-larval period or metamorphosis. Post-trochophore larva swims actively for a few days, feeding on microorganisms. While swimming it undergoes metamorphosis to change into the adult. Its preoral apical portion develops into the **prostomium** of adult. First segment becomes the **peristomium** and the last segment the **pygidium**. Ciliary bands disappear and larva grows in size with the addition of new segments just in front of pygidium. Segments soon develop parapodia with long provisional or larval setae. The resulting young worm settles to the bottom at the low tide line and starts forming its tubular burrow. In the burrow it grows further adding new segments and becomes the burrowing adult worm.

TYPICAL TROCHOPHORE LARVA

In some polychaetes, the young hatches from egg in a characteristic larval stage called **trochophore**. It is not only characteristic of polychaetes, but a similar larva also occurs in molluscs and other phyla.

[I] Structure of Trochophore

It is a minute, ciliated, unsegmented, and almost pear-shaped pelagic creature, with oral and aboral surfaces recognizable. In a full grown typical trochophore larva, there is a sensory **apical organ** or **plate** bearing a **tuft of cilia**. Brain rudiments, as a **ganglion**, are usually evident beneath the apical organ. A characteristic feature of trochophore is the presence of a preoral ciliated girdle of cells, just above equator, and called **prototroch**. Digestive tract is complete. **Mouth** lies ventrally just beneath the prototroch and **anus** is near the lower apex. A **pastoral ciliated band** or **metatroch** lies behind the mouth, and in some there is also a, **telotroch** just infront of anus. These ciliated rings help in feeding and locomotion. Gut is regionated into oesophagus, stomach and intestine. Mesoderm is a pair of undifferentiated masses of cells located in the lower cone. Lying beside these is a pair of protonephridia that develop from ectoderm. Trochophore, at its early stage of development, lacks a coelom. Its body is composed primarily of an outer ectoderm with ectodermal derivatives (nervous tissue and scattered ectodermal elements), and inner endoderm forming the gut. Space enclosed between gut endoderm and ectoderm is called **blastocoel**.

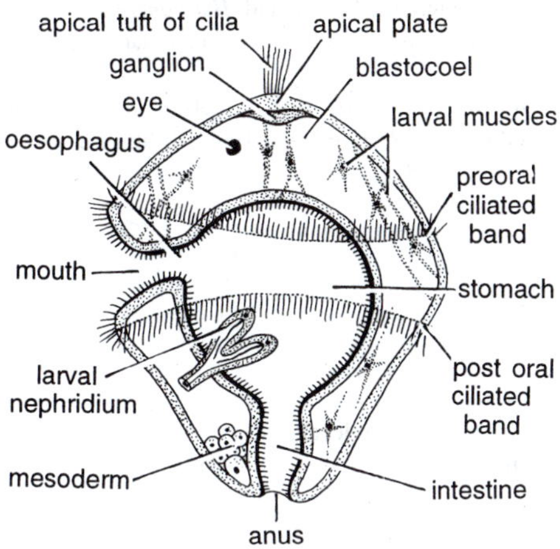

Fig. 20. A typical trochophore larva.

[II] Metamorphosis of Trochophore Larva

Trochophore is a pelagic creature drifting about in sea. Metamorphosis results in immediate termination of planktonic existence, and loss of many larval structures, such as protonephridia, ecto- mesodermal muscle bands and ciliated girdles. Cells of apical plate form the prostomium and brain. Larva gradually grows in length with the formation and development of trunk segments in front of the terminal pygidium. Formation of body segments is marked externally by the development of setal sacs and setae. Mouth region fuses with the first trunk segment to form the peristomium without setae. Larva eventually sinks to the bottom to complete post larval development and assume the habits of the adult.

[III] Significance of Trochophore

Trochophore larva serves for dispersal of the species. Besides, it has great phylogenetic significance. A trochophore larva appears in the development of animals belonging to several phyla, such as Annelida, Mollusca, Bryozoa, etc. This led **Hatschek** (1878) and other embryologists to conclude that these groups of animals are all descended from a common hypothetical ancestor called a **trochozoon**. This is known as **Trochophore theory**. However, strong arguments have been given both for and against this theory.

ADAPTATIONS OF *NEREIS*

Structural and behavioural adaptations of *Nereis* or *Neanthes,* for its burrowing and carnivorous habits are as follows :

(1) Body elongated and flattened for burrowing in mud.

(2) Secretion of mucus for cementing together sand grains in the formation of burrow and prevent its collapse.

(3) Burrowing and nocturnal habits protect from predators.

(4) Eversible pharynx and chitinous sickle-shaped jaws help in capture of fast-moving prey.

(5) Needle-like setae on flat oar-like parapodia help in swimming and crawling.

(6) Thin and highly vascular integument of parapodia serves for gaseous exchange, in the absence of respiratory organs.

(7) Head with sense organs well-developed corresponding to its active mode of life.

(8) Nervous system well-developed correlated with active predaceous life.

(9) Swarming of sexually mature individuals (*Heteronereis*) ensures fertilization of ova.

(10) Occurrence of free-swimming trochophore larva in development serves for far and wide dispersal of species.

Important Questions

➤➤ *Long answer type questions*

1. Describe the external morphology of *Nereis.* Compare it with that of *Heteronereis.*
2. Give an account of the feeding mechanism of *Nereis* and explain clearly the physiology of digestion in this animal.
3. Give an account of the process of reproduction and development in *Nereis.*

4. What do you know of the term *Heteronereis* ? Give a comparative account of *Nereis* and *Heteronereis.*
5. Draw labelled diagrams of : (i) T.S. of *Nereis* passing through parapodia, (ii) Typical trochophore larva, (iii) Dorsal view of anterior end of *Nereis* with everted proboscis.
6. Write short notes on : (i) Adaptations of *Nereis,* (ii) Ciliated organs of *Nereis,* (iii) Epitoky, (iv) *Heteronereis,* (v) Locomotion in *Nereis,* (vi) Parapodium of *Nereis,* (vii) Parapodium of *Heteronereis,* (viii) Swarming in *Nereis,* (ix) Trochophore.

➤➤*Short answer type questions*

1. Mention two functions performed by parapodia in *Nereis.*
2. Describe the parapodium of *Neanthes.*
3. Mention the locomotory organ of *Nereis.*
4. Describe how '*Nereis* feeds'.
5. Give an account of the alimentary canal and mode of feeding in *Nereis* and leech.
6. Give an account of excretory organs and the process of excretion in *Nereis* and leech.
7. Classify the *Nereis* giving two peculiar features in structure and life history.
8. Draw and mark all the parts in a T.S. of *Nereis.*

➤➤*True and false statements*

1. One of the examples of unisexual polychaetes is *Nereis.*
2. Neuropodium is dorsal lobe of parapodium.

➤➤*Multiple choice questions*

1. What function does the coelomoduct perform in *Nereis*?
(a) respiratory and excretory
(b) secretory and excretory
(c) nutritive and locomotory
(d) reproductive and excretory
2. What is the origin of the coelomoduct in a *Nereis :*
(a) germinal (b) ectodermal
(c) mesodermal (d) endodermal
3. Peristomium in *Nereis* is composed of :
(a) one segment (b) two segments
(c) three segments (d) six segments
4. Excretory organs of *Nereis* are :
(a) nephridia (b) metanephridia
(c) dorsal ciliated organ
(d) coelomoduct
5. Sexually mature worm is called :
(a) *Nereis* (b) *Heteronereis*
(c) both (d) none
6. Worm having epitoke and atoke regions is called :
(a) *Heteronereis* (b) Sexually mature *Nereis*
(c) Heteronephron (d) none
7. Fertilium is secreted by :
(a) *Nereis* (b) male *Heteronereis*
(c) female *Heteronereis* (d) both
8. *Nereis* is commonly called :
(a) rag worm (b) clam worm
(c) sand worm (d) all of them
9. Locomotory organs in *Nereis* is :
(a) podia (b) setae
(c) parapodia (d) all the above
10. Who made the study about locomotion of *Nereis :*
(a) Peter (b) Robert Hook
(c) Gregenbaur (d) Gray
11. Respiration in *Nereis* perform by :
(a) gill (b) trachia
(c) body surface (d) lungs
12. The main excretory product in *Nereis* is :
(a) ammonia (b) urea
(c) uric acid (d) all these
13. The chemoreceptor organs in *Nereis* is :
(a) prostomial palp (b) prostomial tentacle
(c) nuchal organ (d) peristomial cirri
14. The name of the larva of *Nereis :*
(a) Trochophore (b) Bipinaria
(c) Zoea (d) Cercaria
15. *Nereis* is :
(a) herbivorous (b) carnivorous
(c) omnivorous (d) parasite

Answers

➤➤*True and false statements*

1. true, 2. false.

➤➤*Multiple choice questions*

1. (d) 2. (b) 3. (a) 4. (b) 5. (b) 6. (a) 7. (c) 8. (d) 9. (c) 10. (d)
11. (c) 12. (a) 13. (c) 14. (a) 15. (b)

41

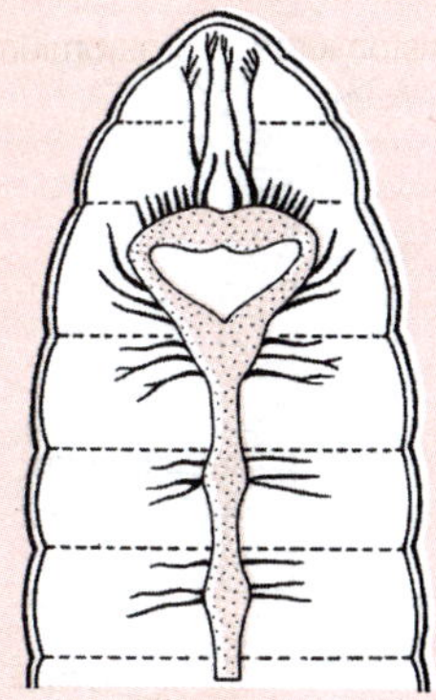

Pheretima posthuma : The Indian Earthworm

Class **Oligochaeta** includes terrestrial earthworms and some other species that live in fresh water. As compared to many setae of polychaetes, they possess few locomotor setae borne directly by body segments which are devoid of parapodia (L., **oligos**, few + **chaete**, bristles). There are several genera of earthworms. Common genus of Europe and North America is *Lumbricus*. In India, *Drawida* and *Megascolex* occur in South India and *Eutyphaeus* is found in the Gangetic plain of North India. *Pheretima,* which is commonly found in South East Asia, Japan, Sri Lanka and Australia, is represented by 13 species in the Indian soil. Following description applies to *Pheretima posthuma,* whose anatomy has been worked out by late **Professor K.N. Bahl** (1926) and others. An earthworm is usually studied as a type of Annelida becuase it is easily available almost every where.

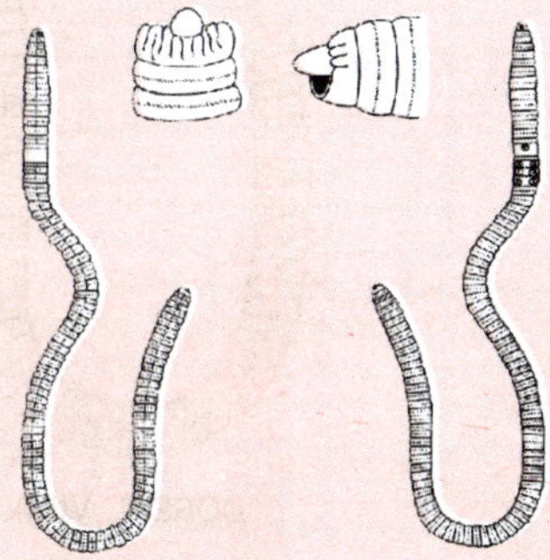

Pheretima posthuma

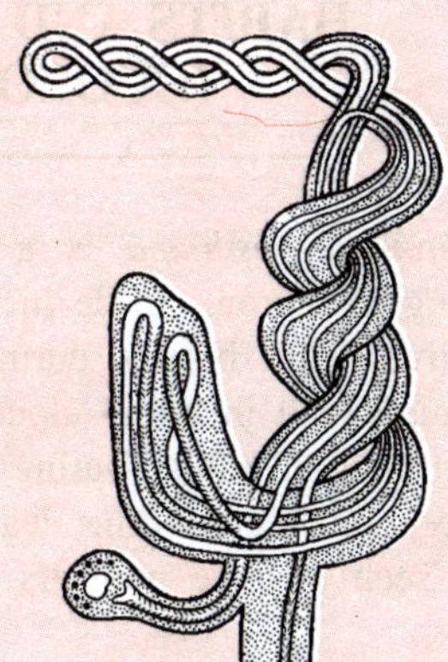

SYSTEMATIC POSITION

Phylum	Annelida
Class	Oligochaeta
Order	Opisthopora
Family	Megascolecidae
Genus	*Pheretima*
Type	*posthuma*

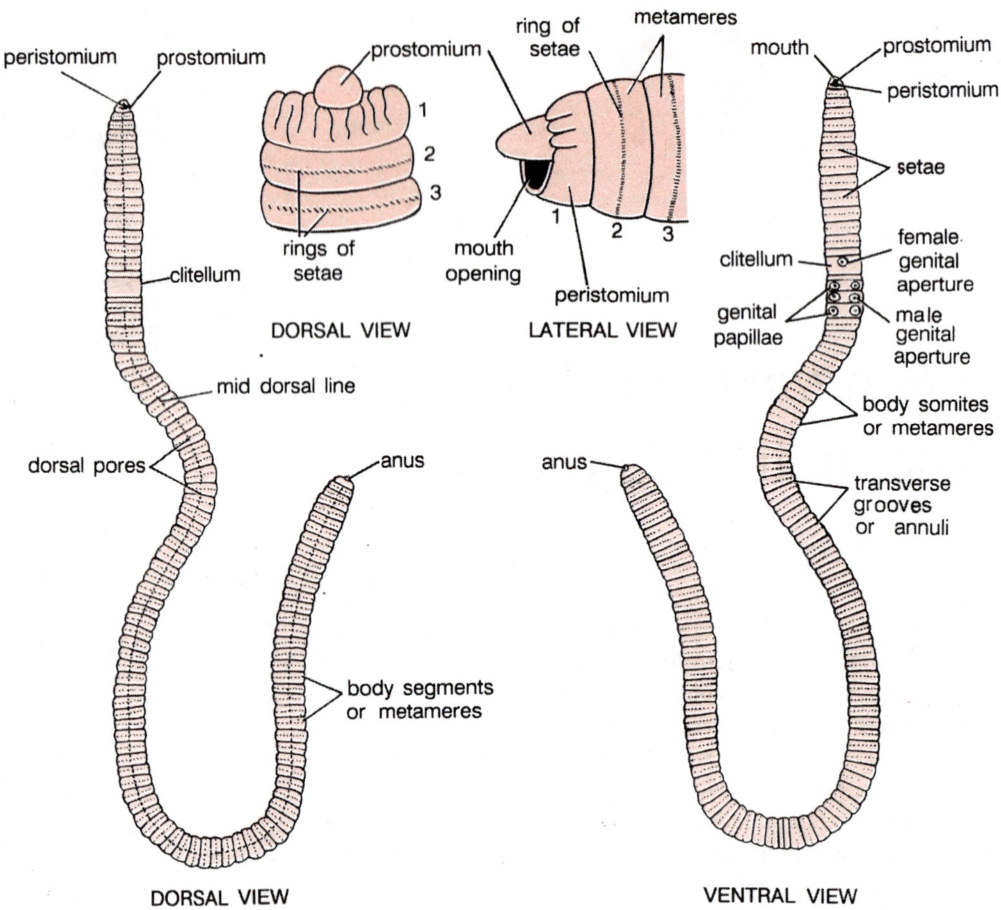

Fig. 1. *Pheretima posthuma.* **A.** Entire worm in dorsal view. **B.** Entire worm in ventral view **C.** Anterior end in dorsal view. **D.** Anterior end in lateral view.

HABITS AND HABITAT (ECOLOGY)

Pheretima posthuma is a terrestrial earthworm living in burrows made in moist earth. It prefers to live in the burrow during day and comes out at night and in damp cloudy weather. It is thus nocturnal in habit. During rainy season, after a heavy fall, earthworms leave their burrows and are seen in large numbers crawling on ground.

Earthworm makes its burrow partly by boring with its pointed anterior end and partly by sucking and swallowing the earth. It feeds on dead organic matter present in soil. Food and soil are ingested together and the latter, along with undigested food is finally egested in the form of **worm castings**. Earthworm are hermaphrodite, but they undergo copulation for exchange of their spermatozoa. Fertilization and development occur inside a cocoon. Trochophore larva does not occur as young worm hatches out of cocoon. Earthworm possess great power of regeneration.

EXTERNAL MORPHOLOGY

1. Shape and size. Earthworm is a bisymmetrical animal. Its body is cylindrically elongated, pointed in front, blunt behind and thickest a little behind the anterior end. It is well-adapted for burrowing. A mature worm measures about 150 mm in length and 3 to 5 mm in width.

2. Colour. Earthworm is of a glistening deep brown or clay colour. Dorsal surface is darker than the ventral surface and carries a dark coloured median line due to dorsal blood vessel which is seen through the integument. Brown colour of worm is due to the pigment **porphyrin** present in bodywall and it protects the body against bright and strong light.

3. Segmentation. Soft and naked body of earthworm is .divided into 100 to 120 similar segments, called **metameres** or **somites**. These are without parapodia. Segments are separated from each other by distinct ring-like grooves. External segmentation corresponds with the internal segmentation of body.

4. Head. Earthworm lacks a distinct head and sense organs like eyes, cirri and tentacles. First segment at the anterior end of the body is called **buccal segment** or **peristomium** bearing the terminal, crescentic mouth. It is prolonged anteriorly into a fleshy lobe, the **prostomium**, which overhangs the mouth.

5. Clitellum. In mature worms, a conspicuous external feature is a girdle-like thick band of glandular tissue, the **clitellum**, which completely and permanently surrounds the segments 14 to 16. Due to its presence, the body is distinguished into **peri-clitellar, clitellar** and **post-clitellar** regions. Clitellum is a glandular organ which secretes mucus, albumen and an egg case or cocoon for eggs.

6. Setae. About the middle of each segment there is a ring of tiny curved bristles, called **setae** or **chaetae**, formed of a horny nitrogenous organic substance, known as **chitin**. About 80 to 120 setae are present on each segment. But they are absent on peristomium, pygidium and the clitellum. Each seta is embedded in a small pit in body wall, called **setigerous** or **setal sac**. It is formed by a single **formative cell** present in the basal part of sac. It has a faint yellow colour and is shaped like an elongated 'S' with a swollen middle part, called **nodulus**. About one-third of its length projects above the surface of skin in a contracted segment.

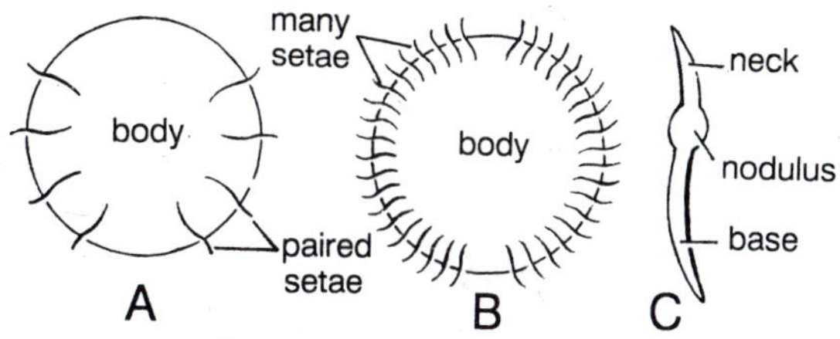

Fig. 2. Diagrammatic representation of arrangement of setae in earthworms. **A.** Lumbricine arrangement in *Lumbricus.* **B.** Perichaetine arrangement in *Pheretima.* **C.** Parts of a single seta.

> Arrangement of numerous setae in a ring in each segment is known as **perichaetine** arrangement, as found in *Pheretima.* In **lumbricine** or **octochaetine arrangement,** setae are arranged in two pairs on each side in each segment, as in *Eutyphaeus,* and the European earthworm, *Lumbricus.*

7. Apertures. These include : (i) **Mouth**, a crescentic anterior aperture, surrounded by peristomium and overhung by prostomium. (ii) **Anus**, a vertical slit-like aperture at the posterior terminus, (iii) **Female genital pore**, a single median aperture of oviducts on ventral surface of 14th segment in clitellar region, (iv) **Male genital pores**, a pair of crescentic openings of common prostatic and spermatic ducts on ventral surface of 18th segment, one on each side, (v) **Spermathecal** pores, four pairs of apertures of spermathecae in the grooves of 5/6, 6/7, 7/8 and 8/9 segments, one on each side in ventro-lateral position, (vi) **Nephridiopores**, a large number of very minute openings of integumentary nephridia scattered all over the body except the first two segments, (vii) **Dorsal pores**, minute apertures of coelomic chambers located mid-dorsally, one in each intersegmental groove, behind 12th segment, except the last groove.

8. Genital papillae. These are two pairs of conspicuous rounded elevations, one pair each in the 17th and 19th segments, on ventral surface. Each papilla bears a shallow cup-like depression at its top which acts as sucker during copulation.

BODY WALL

Body wall of earthworm comprises a thin cuticle, an epidermis, a well-developed musculature and a coelomic epithelium or parietal peritoneum.

1. Cuticle. Cuticle is an elastic, non-cellular and finely striated layer which is secreted by the underlying epidermis.

2. Epidermis. It is single-layered and just beneath cuticle. Cells of epidermis are of various types, performing different functions. **Supporting cells**, forming bulk of epidermis, are of columnar type. **Gland cells** include numerous **mucous cells** and a few **albumen cells** packed with secretory granules. **Basal cells**, which are small and rounded or conical, lie in spaces between inner ends of supporting cells and gland cells. **Receptor cells** occur in groups with their outer ends giving out fine hair-like processes. Epidermis rests on a thin basement membrane.

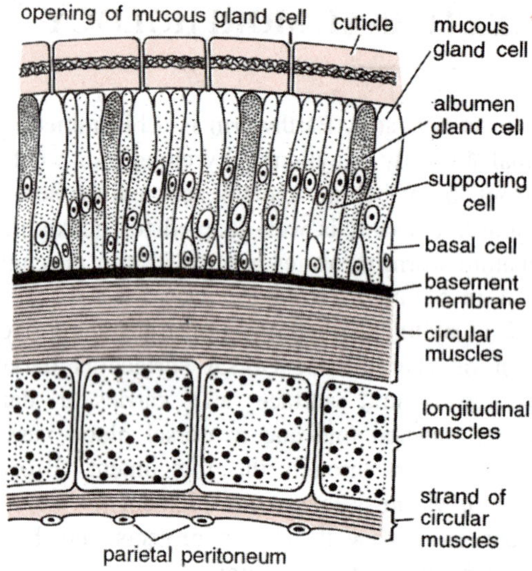

Fig. 3. *Pheretima.* T.S. of a portion of body wall.

3. Muscles. Musculature lies below epidermis. It consists of an outer thin layer of **circular muscle fibres** running around the body, and an inner thick layer of **longitudinal muscle fibres** running along the length of body.

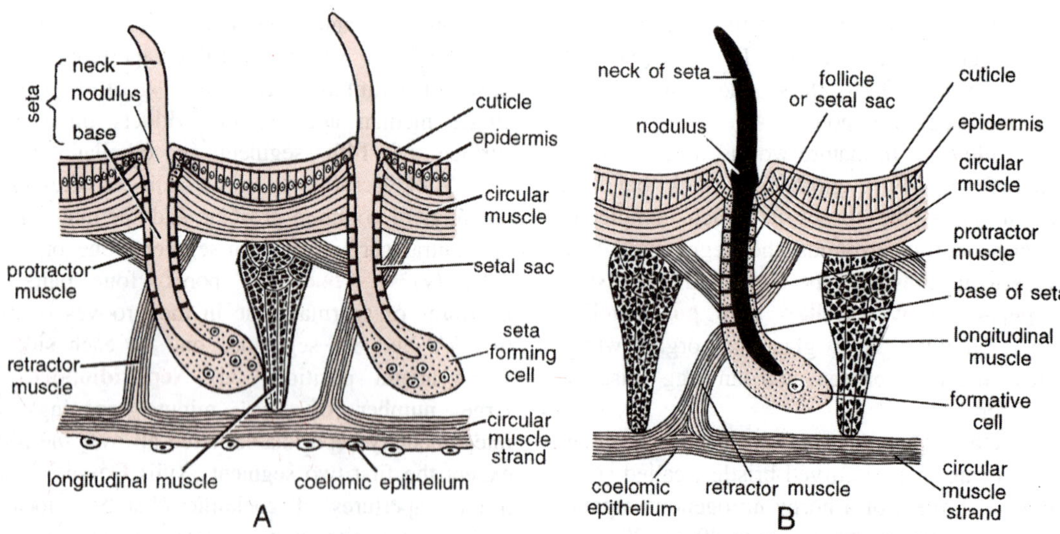

Fig. 4. *Pheretima posthuma.* V.S. body wall through setal sacs showing setal musculature. **A.** According to **Bahl**; **B.** According to **Vijay Kumar**.

Longitudinal muscle fibres lie in parallel bundles, separated by connective tissue and strengthened by collagen fibres. **Setal musculature.** Two additional types of muscle also occur inserted at the base of each setal sac bearing a seta. These are a pair of **protractor muscles** passing outwards to join the circular muscle layer, and a single **retractor muscle**, passing inwards to join another thin sheet of **circular muscles** forming a ring (parallel to the ring of setae) below the longitudinal muscles. All the muscle fibres are unstriped, long and spindle-shaped.

According to **K.N. Bahl** (1926), protractor muscles radiate outward to join the circular muscle layer. According to **Vijay Kumar** (1972), protractor muscle fibres pass independently through both longitudinal and circular muscle layers and distally become attached on the basement membrane lying beneath epidermis [(Fig. 4 B)].

4. Coelomic epithelium. Between the body wall musculature and coelom is a thin layer of **coelomic epithelium** or **parietal peritoneum**, consisting of flat cells, which are recognizable by their nuclei only.

Functions of Body Wall

(1) Maintains body form due to its elasticity.
(2) Protects against mechanical injuries.
(3) Secretes mucus which helps in plastering the internal walls of burrow, keeps body surface slimy and kills harmful bacteria.
(4) Sensory epidermal cells serve for reception of external stimuli.
(5) Lodges setae which help in locomotion.
(6) Body wall musculature helps in movements.
(7) Moist and vascular body wall helps in gaseous exchange.
(8) Albumen secreted by clitellar glands helps in nutrition of embryos developing inside cocoons.

COELOM

Body cavity of earthworm is a true coelom which lies between body wall and alimentary canal. It is lined by coelomic epithelium derived from mesoderm.

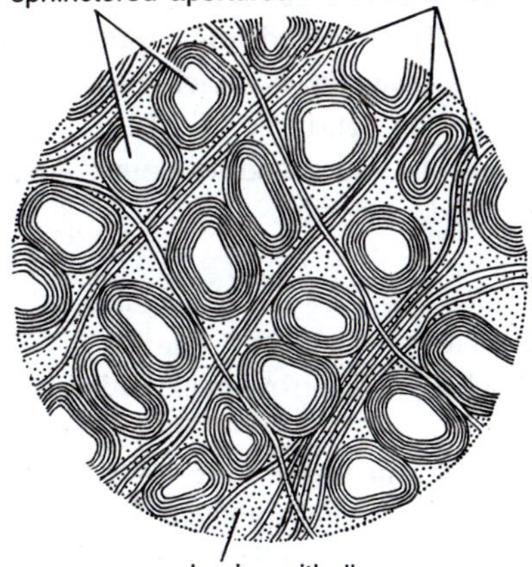

Fig. 5. *Pheretima.* A part of intersegmental septum.

[I] Septa

Coelom is partitioned into a series of coelomic chambers by transverse **intersegmental septa**. Each septum consists of a thin layer of interlacing muscle fibres, covered on both surfaces by coelomic epithelium. Septa are absent in first four segments. First septum, lying between segments 4 and 5, is thin and membranous. Next five septa are thick and muscular. All these six septa are cone-like and run obliquely backwards from body wall to gut wall. There is a single septum between segments 8 and 10, the other being absent. Contraction of these cone-like septa increases pressure on coelomic fluid, thus making the anterior body segments turgid and elongated during locomotion and burrowing. First nine septa, *i.e.*, up to septum 13/14, are without perforations. Remaining septa beginning from septum 14/15 are perforated by numerous sphinctered oval or circular apertures, through which adjacent coelomic chambers maintain continuity.

[II] Coelomic Fluid

Coelom is filled with an alkaline, colourless or milky **coelomic fluid** containing water, salts, some proteins and at least four types of coelomic corpuscles as follows :

1. Phagocytes. Largest and more numerous are the nucleated **amoeboid corpuscles**, or **phagocytes**. Each has several folds on surface, a deep concavity on one side, and contains many ingested granules such as bacteria.

2. Mucocytes. These are elongate cells, each having a broad, fan-like process, attached to a narrow nucleated body.

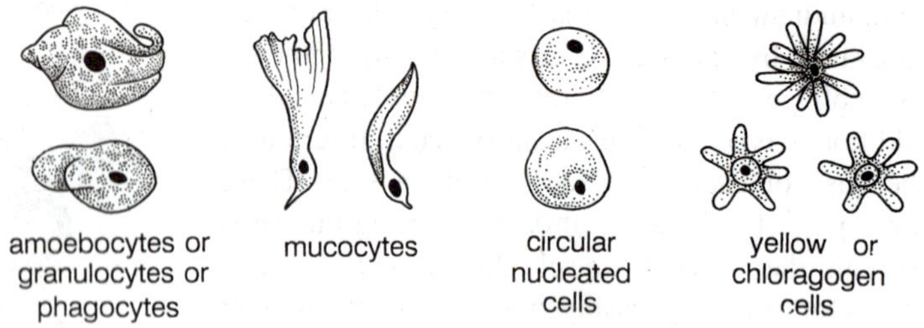

Fig. 6. *Pheretima.* Kinds of coelomic corpuscles.

3. Circular nucleated cells. About 10% of coelomic corpuscles are rounded, nucleated and blood corpuscle-like cells possessing clear protoplasm and characteristic markings on surface.

4. Chloragogen cells. Also known as yellow cells, these are star-shaped, small-sized cells. They become deep yellow when stained with iodine solution. They are supposed to be excretory in function removing excretory products from coelomic fluid.

Functions of coelomic fluid

(1) Helps in locomotion by turgescence.
(2) Its circulation from one chamber to another helps in distribution of digested food.
(3) Exceding through dorsal pores, it keeps body surface moist thus helping in respiration.
(4) It destroys harmful bacteria and other parasites in soil.
(5) Forms a protective, shock-proof covering around internal organs of body.
(6) Its chloragogen cells help in removing excretory products out of body.
(7) It causes luminescence in some earthworms.

LOCOMOTION

Movement in earthworm involves the musculature of body wall and setae. According to the studies of **Gray** and **Lissman** (1938), the worm's body undergoes extension, anchoring and contraction during the course of its progression. A wave of contraction, affecting circular muscles, begins at the anterior end and travels posteriorly. This causes the body to become thinner and longer. This is followed by another wave of contraction affecting longitudinal muscles causing thickening and shortening of body. This is again followed by the wave of thinning and the process is repeated alternately. Each wave of circular contraction causes the segments affected to move forward. But the segments in a state of longitudinal contraction do not move as they are anchored to the ground by the protruded setae. Setae always protrude during longitudinal contraction and retract during circular contraction. It has been calculated that, by this method, the earthworm travels a distance of about 25 cm in one minute.

When the direction of waves is reversed, the worm crawls backwards.

During locomotion, coelomic fluid serves as a kind of hydraulic skeleton. When compressed due to contraction of circular muscles, it provides stiffness to body and aids in relaxation of longitudinal muscles.

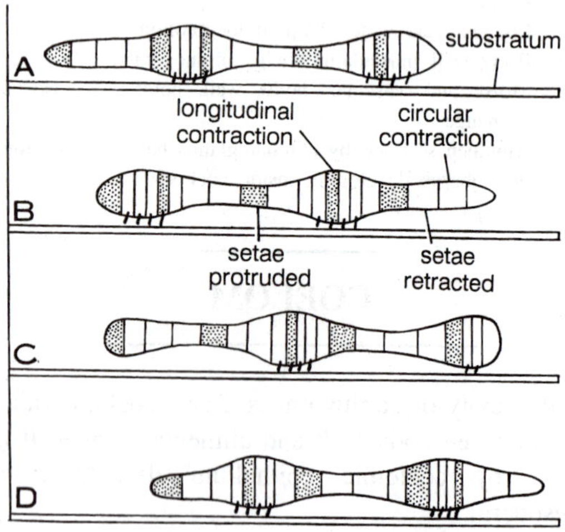

Fig. 7. *Pheretima.* Stages in locomotion.

DIGESTIVE SYSTEM

[I] Alimentary Canal

Alimentary canal of *Pheretima* is a complete and straight tube running along the entire length of body. Mouth and anus constitute its anterior and posterior openings, respectively. It is functionally regionated into various parts which are buccal chamber, pharynx, oesophagus, gizzard, stomach and intestine.

1. Buccal chamber. Crescentic mouth, situated ventral to prostomium at the anterior end of peristomium, leads into a short and narrow protrusible buccal chamber, extending up to middle of third segment. Its lining epithelium is thrown into longitudinal folds.

2. Pharynx. Buccal chamber leads into a spacious pear-shaped muscular **pharynx**, which extends up to the fourth segment. Its anterior end is marked by a nerve ring placed in a transverse groove between it and buccal chamber. Its cavity is somewhat dorso-ventrally compressed due to the presence on its roof of a large glandular **pharyngeal mass** producing a **salivary secretion**. Lateral walls of pharynx are pushed inside forming a narrow **horizontal shelf** on each side. The two shelves meet anteriorly and posteriorly, thus dividing the pharyngeal cavity into a dorsal or **salivary chamber** and a ventral or **conducting chamber**. **Salivary secretion** contains **mucus** and proteolytic enzymes which are poured into the salivary chamber.

3. Oesophagus. Behind pharynx lies the **oesophagus** or **gullet**. It is a short, narrow, thin-walled tube. It extends up to the seventh segment.

4. Gizzard. Oesophagus is modified into a prominent, oval, hard and thick-walled muscular organ, the **gizzard**, lying in eighth segment. Its muscular wall consists of circular muscle fibres. It is internally lined by a tough cuticle.

5. Stomach. Gizzard is followed by a short narrow tube, the **stomach**, which extends from segments 9 to 14, with a sphincter at each end. Its walls are highly vascular and glandular and thrown into internal transverse folds.

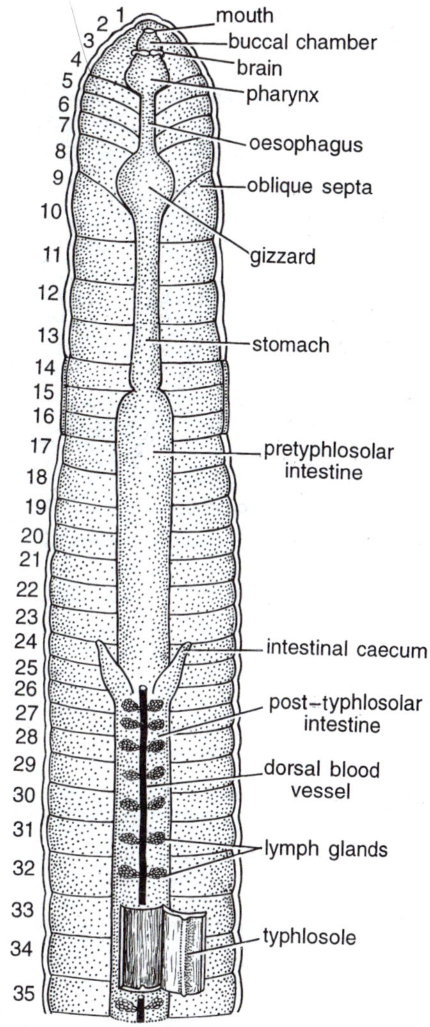

Fig. 8. *Pheretima.* Dissection of alimentary canal.

6. Intestine. Region next to stomach is the **intestine** which is a long, wide and thin-walled tube extending from 15th segment to the last. It has a beaded appearance due to circular constrictions corresponding to septa. Its internal lining is ciliated, folded, vascular and glandular. Intestine is divisible into three parts :

(a) Pre-typhlosolar region. First or anterior part lying between segments 15 to 26 is known as **pre-typhlosolar region**. Its wall is internally folded to form minute processes, the **villi**, and is highly vascular. From 26th segment are given out

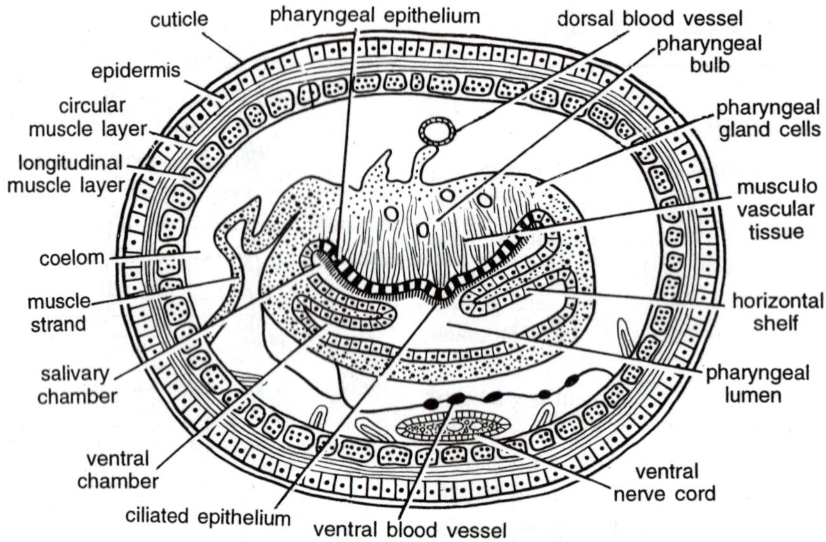

Fig. 9. *Pheretima.* T.S. body through pharynx.

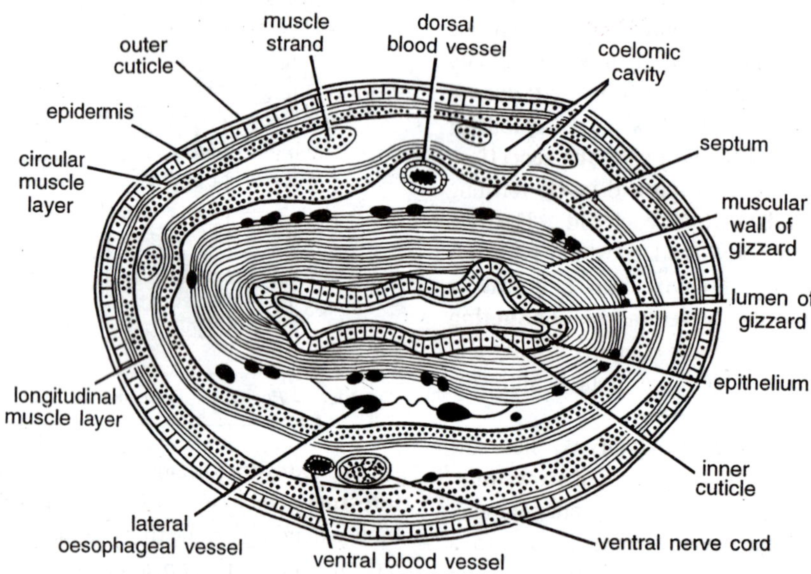

Fig. 10. *Pheretima.* T.S. body through gizzard.

externally a pair of forwardly-directed lateral conical outgrowths, the intestinal caeca, which run up to 22nd or 23rd segment. These are richly vascular and internally thrown into villi-like processes.

(b) Typhlosolar region. Second or middle part lies between 27th segment up to 23-25 segments in front of anus. This is characterized by the presence of a highly glandular and vascular

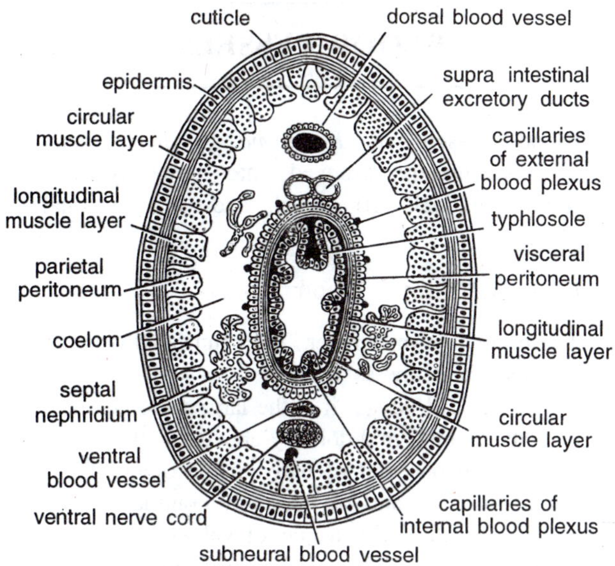

Fig. 11. *Pheretima.* T.S. body through intestine (typhlosolar region).

longitudinal ridge, arising as a median in growth of the dorsal aspect of the intestinal cavity. This is called the **typhlosole**.

(c) Post-typhlosolar region. Third or the last part, also known as **rectum**, is of about 23-25 segments. It is internally marked by the presence of longitudinal folds. It opens to outside through a terminal **anus**.

[II] Histology of Gut Wall

Wall of alimentary canal consists of four layers in succession.

1. Peritoneum. It is the outermost layer consisting of tall and narrow cells. On the intestine and sometimes on stomach, some of these cells contain yellow refractile granules, hence called **yellow** or **chloragogen** cells. Their exact function is controversial. They are said to serve for storage of reserve food, deamination of proteins, formation of urea from ammonia, excretion, etc.

2. Muscles. These lying below peritoneum include outer layer of **longitudinal** and inner layer of **circular muscle fibres**. They are well developed in pharynx and oesophagus but ill-developed in intestine. In the gizzard, circular muscles are much developed, while longitudinal muscles are absent. Muscles are unstriped or involuntary.

3. Enteric epithelium. It consists of a single row of columnar cells which become ciliated on the roof of pharynx, glandular in stomach and absorptive and glandular in intestine. It is internally thrown into folds in oesophagus, stomach and intestine.

4. Cuticle. It is present in buccal cavity as a thin lining and in gizzard as a thick lining.

[III] Food and Feeding Mechanism

Earthworm feeds on dead organic matter, particularly vegetation along with soil. It also feeds directly upon leaves, grasses, seeds, algae, etc.

Earthworm ingests food by the pumping action of its pharynx. It presses its mouth against soil and the contractile sucking action of pharyngeal wall draws fragments of soil into buccal chamber. Pharyngeal action is augmented by the action of strands of muscle fibres, extending from pharynx to body wall.

[IV] Physiology of Digestion

Ingested food is pressed to move posteriorly. No digestion takes place inside buccal chamber. While passing through the ventral conducting chamber of pharynx, it meets a salivary secretion produced by the glandular cells of pharyngeal mass and poured into the salivary chamber of pharynx. It contains **mucin**, which lubricates the food and an enzyme **protease** which digests the proteins. Food then passes back through oesophagus into gizzard. The gizzard, acting as a grinding machine, pulverises the food masses. This is facilitated by the contractile movements of its muscular wall which cause the food to roll about, and the internal cuticular lining, striking against which the food particles are ground up fully. In stomach, a chalky secretion of **calciferous glands** located in stomach wall, neutralizes the humic acids present in soil. Intestine is the principal site of digestion. Intestinal wall consists of glandular cells which secrete digestive juice containing **pepsin, trypsin, amylase, lipase** and **cellulase**. Pepsin hydrolyses proteins into proteases and peptones, and trypsin hydrolyses these products into amino acids. Amylase hydrolyses starch into maltose, lipase brings about hydrolysis of fats into glycerol and fatty acids, while cellulase causes the digestion of cellulose into cellulobiose.

Thus digestion is **extracellular** in earthworm, as in higher animals such as frog and rabbit.

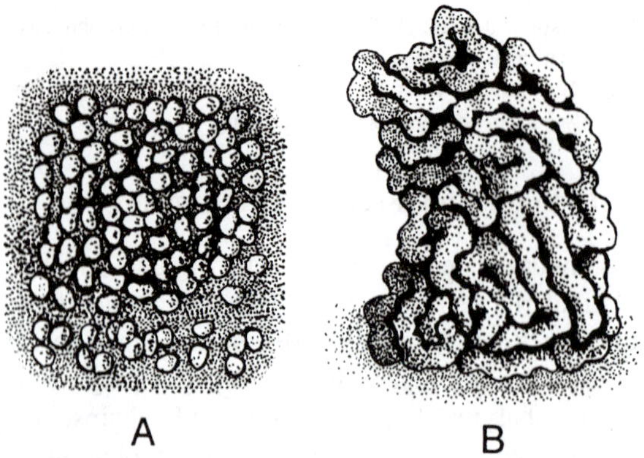

Fig. 12. Worm castings. **A.** Of *Pheretima.* **B.** Of *Eutyphaeus.*

Intestine also functions for absorbing the digested nutrients. After being absorbed by the absorptive cells of intestinal epithelium, nutrients are passed to blood capillaries in the intestinal wall. Presence of typhlosole in greater part of intestine increases the surface both for digestion and absorption.

Undigested food and the soil are eliminated through anus to outside in the form of **worm castings**. Castings of *Pheretima* consist of distinct, small and rounded pellets or balls, while those of *Eutyphaeus* make a large tower-like heap with an open passage in the middle.

CIRCULATORY SYSTEM

Circulatory or blood vascular system of earthworm is a closed system consisting of blood vessels and capillaries which ramify to all parts of the body. **Blood** is composed of a fluid plasma and colourless corpuscles, physiologically comparable to the leucocytes of the vertebrates. The red respiratory pigment, **haemoglobin** (or **erythrocruorin**) occurs dissolved in plasma. It gives a red colour to blood and aids in the transportation of oxygen for respiration.

BLOOD VESSELS

Blood vessels of *Pheretima posthuma* may be conveniently grouped into three types : (i) longitudinal, (ii) lateral, and (iii) intestinal plexuses.

[I] Longitudinal Blood Vessels

These are five in number. They run lengthwise in the body and are as follows :

1. Dorsal vessel. It is the largest blood vessel of body running mid-dorsally above the alimentary canal, from one end of body to the other. It has thick, muscular and rhythmically contractile wall and is provided with a pair of valves in front of the septum in each segment. Blood flows through it from backward to forward.

Behind 13th segment, dorsal vessel is a collecting vessel, receiving blood through two pairs of **dorso-intestinal vessels** from intestine and a pair of **commissural vessels** from **sub-neural vessel** in each segment. In front of 13th segment (anteriorly), it distributes blood to the anterior regions of alimentary canal and through the so-called **hearts** to ventral vessel. Extending up to cerebral ganglia, dorsal vessel trifurcates and the three branches are distributed over the pharyngeal bulb and roof of buccal chamber. In each of the 3rd, 4th, 5th, 6th and 8th segments, a pair of stout pulsating branches send blood to the pharyngeal nephridia, oesophagus and gizzard.

2. Ventral vessel. It is a large vessel that runs mid-ventrally below alimentary canal and above nerve cord from one end of body to another. Its walls are thin and non-contractile and valves are altogether absent. Blood flows through it posteriorly.

Ventral vessel is principally a distributing vessel. It supplies blood, in each segment, through a pair of **ventro-tegumentary vessels** to integumentary nephridia, body wall, septa and

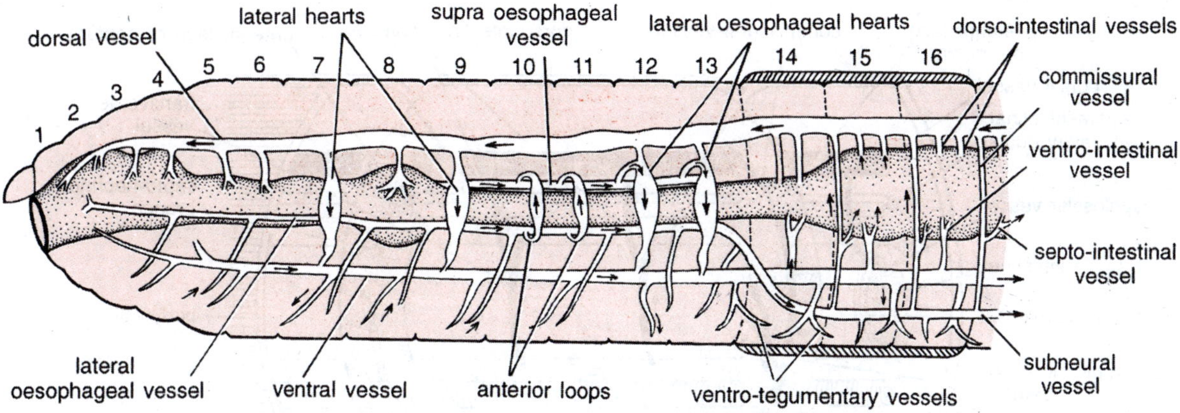

Fig. 13. *Pheretima.* Blood-vascular system in first 13 segments.

reproductive organs. Behind 13th segment, each ventro-tegumentary vessel sends a small branch, a **septo-nephridial branch**, supplying the septal nephridia. Besides in each segment behind 13th, ventral vessel gives off a median **ventro-intestinal vessel** to intestine.

3. Lateral oesophageal vessels. These are two vessels lying one on either ventro-lateral side of gut, running from the anterior end of body up to 13th segment. These receive a pair of **ventro-tegumentary vessels** in each segment, collecting blood from body wall, septa, nephridia and reproductive organs. Flowing posteriorly, some of its blood passes to the **supra-oesophageal vessel** through a pair of **anterior loops** in each of the segments 10 and 11 and through several ring vessels running through the wall of stomach. Rest of blood flows backward into **sub-neural vessel.**

4. Sub-neural vessel. It is a slender vessel which runs immediately beneath the nerve cord in mid-ventral position. It extends from 14th segment up to the posterior end and is formed by the union of two **lateral oesophageal vessels.** Flow of blood is from in front backwards. It is a collecting vessel and receives blood from ventral nerve cord and ventral body wall in each segment through a pair of small branches. It pours blood via a pair of **commissurals**, in each segment, into dorsal vessel.

5. Supra-oesophageal blood vessel. It is a short thin-walled collecting vessel lying mid-dorsally above stomach and confined to segments 9 to 13. It is connected to lateral oesophageal vessel through two pairs of **anterior loops** and to ventral vessel through two pairs of **latero-oesophageal hearts**. At places it divides into separate vessels which reunite to form a single vessel. It collects blood from stomach, gizzard and (through anterior loops) from lateral oesophageals and pumps it through lateral oesophageal hearts into ventral vessel.

[II] Lateral or Transverse Blood Vessels

All the longitudinal blood vessels are interconnected with one another, directly or indirectly, through numerous segmentally arranged **transverse** or **lateral blood vessels**. Lateral blood vessels of anterior region (first 13 segments) and those of posterior or intestinal region (behind 13th segment) of body are described separately as below.

[A] Lateral blood vessels of anterior region (first 13 segments)

1. Hearts. In each of the segments 7, 9, 12 and 13 is found a pair of large, thick, muscular and rhythmically contractile vertical vessels, called **hearts**. They pump blood from dorsal to ventral vessel, while flow in opposite direction is prevented by internal valves. Hearts of 7th and 9th segments connect dorsal and ventral vessels only and are called **lateral hearts**. Those of 12th

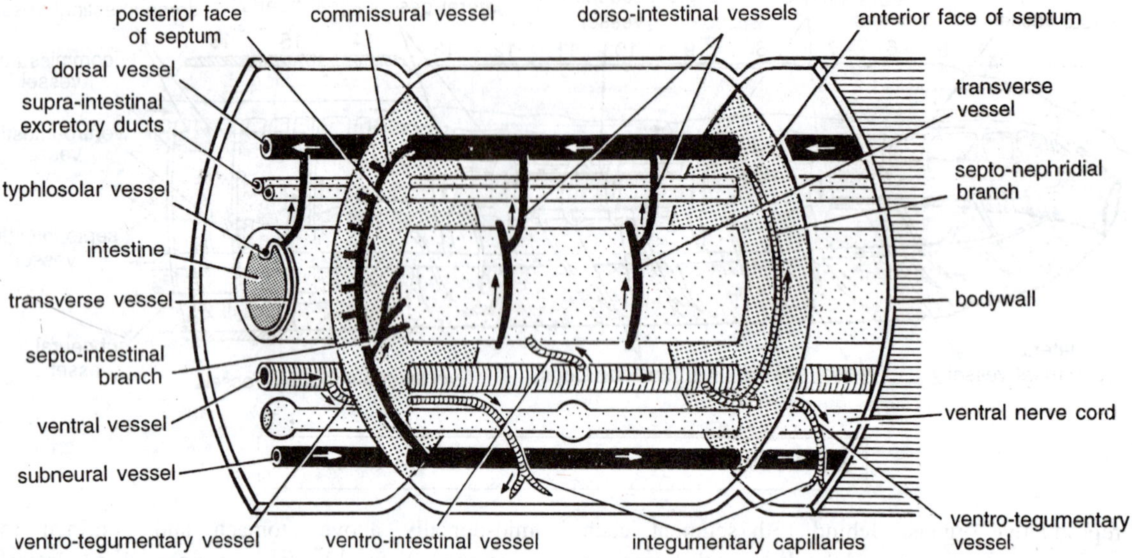

Fig. 14. *Pheretima.* Blood vascular system in a segment of intestinal region.

and 13th segments connect both dorsal and supra-oesophageal vessels with ventral vessel, and are designated as **latero-oesophageal hearts** (Fig. 15).

2. Anterior loops. There is a pair of thin-walled, non-pulsatile, non-muscular and loop-like broad vessels, without valves, in each of the 10th and llth segments. These vessels, known as **anterior loops**, convey blood from lateral-oesophageals into supra-oesophageal vessel.

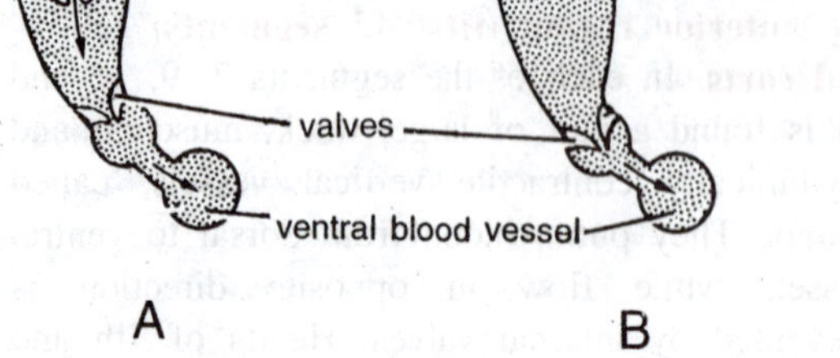

Fig. 15. *Pheretima.* Diagrammatic representation of hearts in section. **A.** Lateral-heart. **B.** Latero-oesophageal heart.

3. Ring vessels. These are characteristic circular vessels of stomach situated within its muscular coat, about 12 vessels per segment. Through these vessels, blood of lateral-oesophageals reaches the supraoesophageal.

4. Ventro-tegumentary vessels. Ventral vessel gives off a pair of **ventro-tegumentary vessels** in each segment to body wall, septa, nephridia and reproductive organs of the same segment.

[B] Lateral vessels of intestinal region (behind 13th segment)

1. Ventro-tegumentary vessels. Behind 13th segment, ventral vessel gives off a ventro-tegumentary on either side in the posterior region of segment. It sends a delicate **septo-nephridial branch** to septal nephridia, which runs upwards along the anterior face of septum. Then it pierces the septum to enter the segment behind to supply its body wall and integumentary nephridia.

2. Commissural vessels. Behind 13th segment, there is a pair of these vessels, in each segment. They run upwards along the posterior face of septum, one on either side, conveying blood from sub-neural to dorsal vessel. Each commissural gives off a small **septo-intestinal**

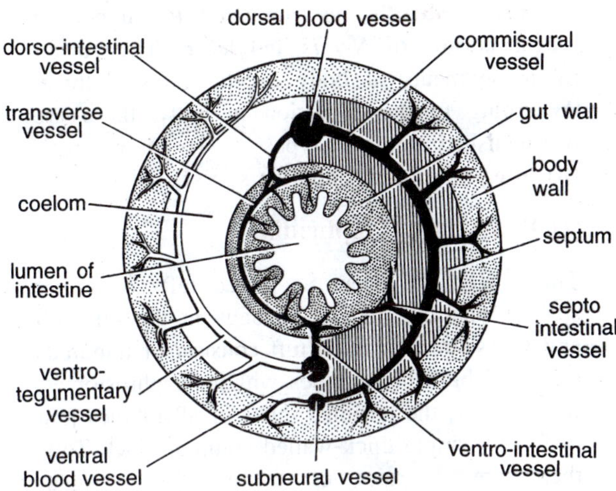

Fig. 16. *Pheretima.* T.S. body through intestinal region showing arrangement of blood vessels. Section of the right half passes through a septum.

branch to the intestinal plexus and receives several branches from septal nephridia and body wall.

3. Dorso-intestinals. In the region of intestine, two pairs of these vessels, in each segment convey blood from intestinal wall to dorsal vessel.

4. Ventro-intestinals. Similarly, a single median ventro-intestinal vessel in each segment of intestinal region carries blood from ventral vessel to intestinal wall.

[II] Intestinal Plexuses

Wall of intestine contains many blood capillaries arranged in two networks, plexuses or plexes. One, the **external plexus**, lies on the surface of gut. It receives blood from ventral vessel through ventro-intestinals and septo-intestinals, and passes it on to the **internal plexus**. The latter is situated between circular muscles and enteric epithelium. Internal plexus passes on blood, along with absorbed nutrients, to dorsal vessel through dorso-intestinals.

CIRCULATION OF BLOOD

Blood flows from behind to forward in dorsal vessel and from front to backwards in ventral, latero-oesophageal, supra-oesophageal and sub-neural vessels.

Ventral vessel is the main distributing vessel, supplying blood to all parts of body. In first 13 segments, it supplies blood to body wall, septa, nephridia and reproductive organs through ventro-tegumentaries. Behind 13th segment, it supplies blood to body wall and nephridia through ventro-tegumentaries and to gut wall through ventro-intestinals.

Sub-neural, **lateral oesophageals** and **supra-oesophageals** are the main collecting vessels. Lateral oesophageals collect blood in first 13 segments from alimentary canal, body wall, nephridia, septa and reproductive organs and discharge into supra-oesophageal through anterior loops and ring vessels. Supra-oesophageal also collects blood from gizzard and stomach and pours it into ventral vessel through latero-oesophageal

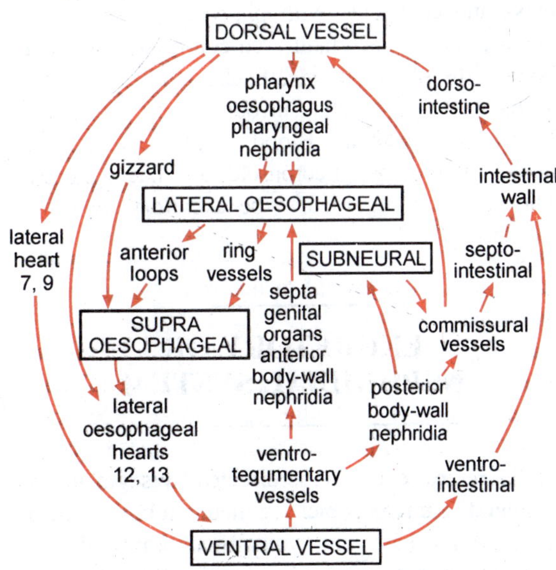

Fig. 17. *Pheretima.* Course of circulation of blood.

hearts. Sub-neural collects blood in the intestinal region from ventral body wall and nerve cord and sends into dorsal vessel through commissurals which also receive blood from body wall, septa and nephridia. Commissural also pours some blood into gut wall through septo-intestinals.

Dorsal vessel functions both as a collecting and a distributing vessel. In the intestinal region, it collects blood through dorso-intestinals from gut wall and through commissurals from sub-neural vessel, septa and nephridia. In first 13 segments, it distributes some blood through branches to alimentary canal and pours the remaining blood through hearts into ventral vessel.

Digested food absorbed through the intestinal wall is distributed to different parts of body by the circulatory system, whereas CO_2 and nitrogenous wastes are carried to nephridia, skin and coelomic fluid for elimination.

BLOOD GLANDS

In segments 4, 5 and 6 lying above pharyngeal mass and connected with pharyngeal or salivary glands, are found small, red-coloured, follicular bodies, the **blood glands**. Each gland consists of a mass of loose cells surrounded by a capsule with a syncytial wall. Blood glands serve for the manufacture of blood corpuscles and haemoglobin. They are also regarded to be excretory by some workers.

EXCRETORY OR NEPHRIDIAL SYSTEM

In *Pheretima,* excretion is effected by segmentally arranged, microscopic, coiled tubes called **nephridia**. These are typically unbranched and their inner ends open into coelom by a ciliated funnel, called **nephrostome**. Such a nephridium, opening by a ciliated funnel, is termed a **metanephridium**. But, in some cases, nephrostome is secondarily lost and nephridium becomes branched. In *Pheretima,* nephridia are small sized or **micronephridia**, as compared to large-sized meganephridia of *Nereis* and leech. They occur in all segments of body except the first three. According to their location in body, these are distinguished into 3 types : (i) **pharyngeal**, (ii) **integumentary**, and (iii) **septal**.

[I] Pharyngeal Nephridia

These nephridia occur as paired tufts on either side of pharynx and oesophagus in the 4th, 5th and 6th segments. Each tuft consists of hundreds of coiled branched tubules without nephrostomes. In each tuft, the terminal ducts of all tubules join to form a single thick-walled common duct. Thus there are 3 pairs of **common pharyngeal nephridial ducts**, which run anteriorly parallel to the ventral nerve cord. Ducts of 4th and 5th segments open into pharynx, while those of 6th segment open into buccal chamber. Pharyngeal nephridia are thus **enteronephric**.

[II] Integumentary Nephridia

These nephridia lie scattered on the entire inner or parietal surface of body wall in each segment, except the first two. There are 200-250 nephridia

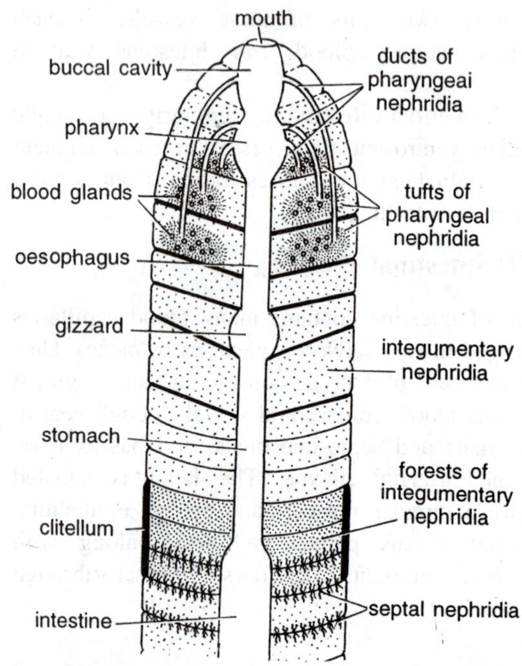

Fig. 18. *Pheretima.* Location of the three types of nephridia.

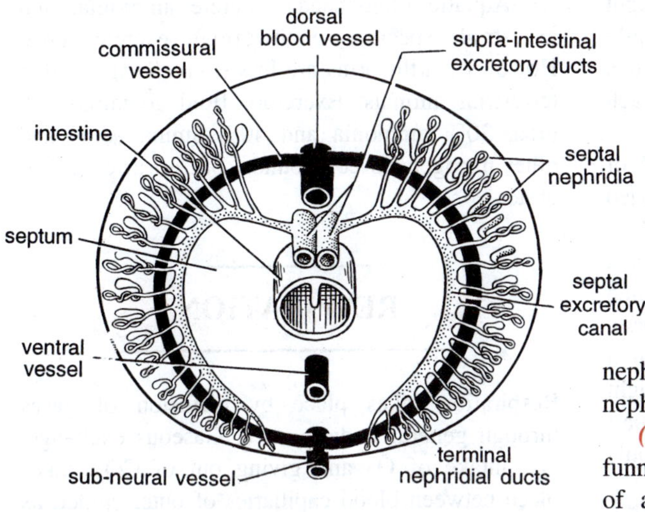

Fig. 19. *Pheretima.* A diagrammatic representation of arrangement of septal nephridia on a septum.

in each segment but in the segments of clitellum (segments 14 to 16), their number increases to more than 2000 constituting the 'forests of nephridia'. Integumentary nephridia are micro-scopic V-like in shape and lack nephrostomes. Their terminal ducts open on body surface independently through minute openings known as **nephridiopores**. Integumentary nephridia are thus **exonephric**.

[III] Septal Nephridia

These are the largest nephridia of *Pheretima.* They are attached to both the faces of each intersegmental septum behind 15th segment.

1. Structure. A typical septal nephridium consists of three main parts : (a) nephrostome, (b) body, and (c) terminal duct.

(a) Nephrostome. Nephrostome is a ciliated funnel communicating with the coelom. It consists of an elliptical pore bounded by the so-called upper and lower lips. Upper lip is formed of a large **central cell** and 8 or 9 **marginal cells**, whereas lower lip is formed of 4 or 5 **compact cells**. All the cells are ciliated.

(b) Body of nephridium. Nephrostome leads into the main body of nephridium through a short, narrow and ciliated tube-like **neck**. Body consists of two parts, a short **straight lobe** and a long

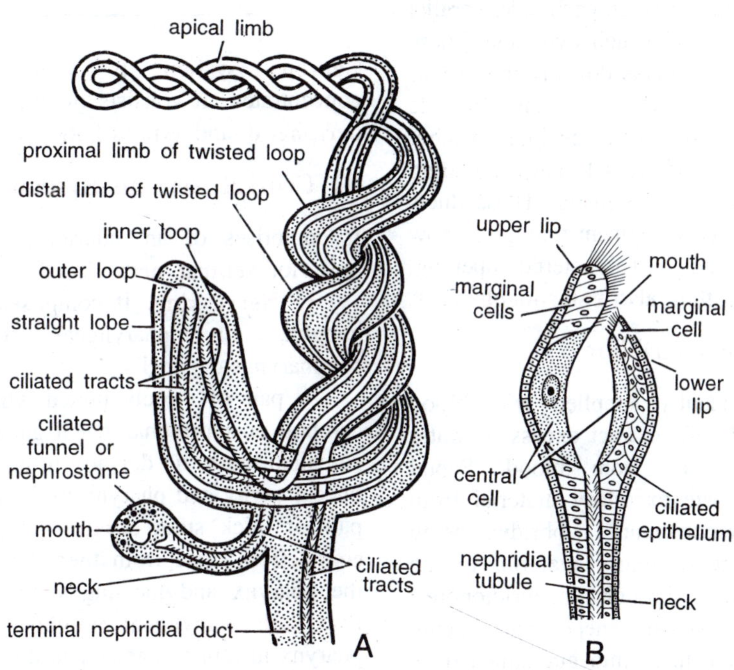

Fig. 20. *Pheretima.* septal nephridium. **A.** Entire nephridium. **B.** Nephrostome in L.S.

twisted lobe with a narrow apical part. Straight lobe is one half of the twisted lobe's length. Twisted lobe consists of a **proximal limb** and a **distal limb**, which are spirally twisted upon each other. Proximal limb is joined to the neck.

(c) Terminal duct. Distal limb of body of nephridium ends in a short and narrow duct, called **terminal duct**.

Nephridial tubule. Nephridium consists of a syncitial glandular mass traversed by a coiled tubule, having four ciliated tracts in its course, one in neck, two in body and one in terminal duct. There are four parallel tubules in the straight lobe, 3 in the basal part and 2 in the apical part of each limb of twisted loop, and a single tubule in each of the neck and terminal duct.

2. Arrangement. Each septum, behind 15th segment, bears 4 rows of septal nephridia, 2 on its anterior face and 2 on posterior face. Each row may contain 20 to 25 nephridia, so that there are 80 to 100 nephridia on each septum or in each coelomic compartment. Nephridia remain suspended freely in coelom of each segment, while their terminal ducts open into a pair of **septal excretory canals**, which run inwards along the posterior face of septum, one on each side, parallel to the commissural vessel of their own side. These canals discharge their contents dorsally into a pair of **supra-intestinal excretory ducts**, situated side by side mid-dorsally just above the intestine, but beneath the dorsal blood vessel, and extending from 15th to the last body segment. These ducts open into intestine in each segment through narrow **ductules**, each having a sphinctered opening. Septal nephridia are thus also **enteronephric**.

[IV] Physiology of Excretion

Nephridia are abundantly supplied with blood vessels. Their gland cells extract excess of water and nitrogenous wastes from blood. Septal nephridia also eliminate excretory material from coelomic fluid. Integumentary nephridia, being exonephric, discharge excretory material to the outer body surface through nephridiopores. Pharynegeal and septal nephridia, being endonephric, discharge them into gut lumen from where they are eliminated with faeces.

Aquatic oligochaetes excrete ammonia, and terrestrial species (earthworms) excrete urea. However, earthworms are less **ureotelic** than other terrestrial animals. Excretory fluid contains 40% urea, 20% ammonia and 40% amino acids and other nitrogenous compounds, but no uric acid or urates.

RESPIRATION

Respiration takes place by diffusion of gases through general body surface. Gaseous exchange, *i.e.*, intake of O_2 and giving out of CO_2, takes place between blood capillaries of outer epidermis and surface film of moisture contributed by secreted mucus, excreted wastes and coelomic fluid. Haemoglobin dissolved in plasma of blood acts as a respiratory pigment, transporting O_2 to the body tissues.

NERVOUS SYSTEM

Nervous system is well developed and concentrated. It consists of three parts : **central**, **peripheral** and **sympathetic** nervous systems.

[I] Central Nervous System

It comprises of an anterior **nerve ring** and a posterior **ventral nerve cord**.

1. Nerve ring. It comprises paired cerebral ganglia, circumpharyngeal connectives and subpharyngeal ganglia.

A pair of closely united white, pear-shaped, cerebral or supra-pharyngeal ganglia, forming the so-called brain, lie dorsally in depression between buccal cavity and pharynx in the third segment. A pair of thick stout **circum**- or **peri-pharyngeal connectives** arise from them laterally, embracing the pharynx and meeting ventrally in a pair of fused **sub-pharyngeal ganglia** lying beneath the pharynx in fourth segment. In this way, a complete **nerve ring** is formed around pharynx.

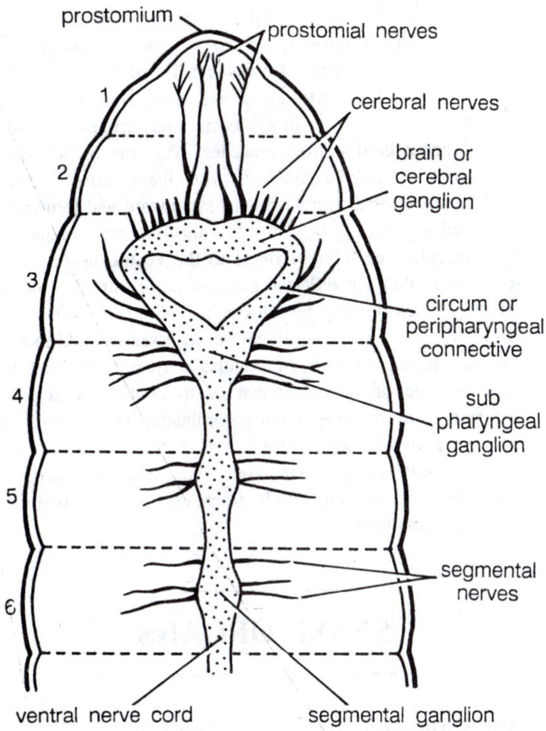

Fig. 21. *Pheretima.* Nervous system in the anterior end of body in dorsal view.

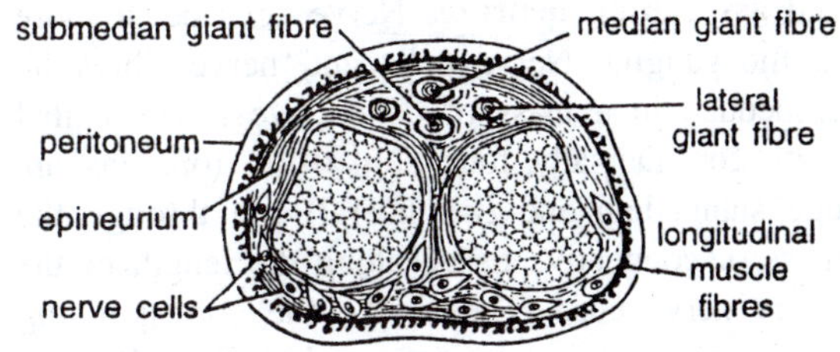

Fig. 23. *Pheretima.* Ventral nerve cord in T.S.

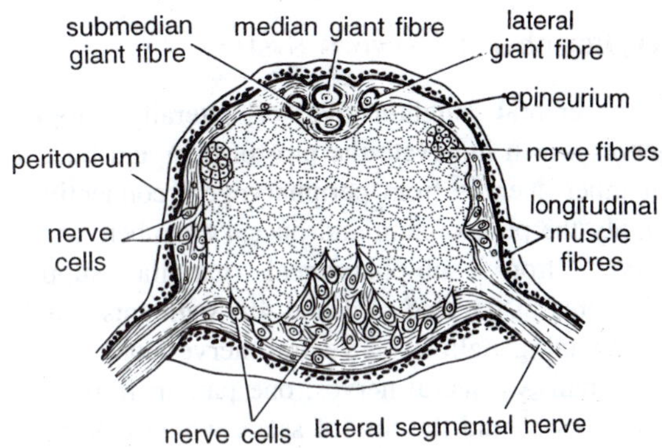

Fig. 24. *Pheretima.* A segmental nerve ganglion in T.S.

2. Ventral nerve cord. Ventral nerve cord arising from the sub-pharyngeal ganglia, runs backwards in mid-ventral line to the posterior end of body. In each segment, it presents a slight enlargement or a **ganglion**. Ventral nerve cord appears to be single but it is really double, consisting of two compactly united right and left cords, as seen in a transverse section. Each **segmental ganglion** also represents the fusion of a pair of ganglia, one belonging to each cord of the double ventral nerve cord.

Histologically, the nerve cord consists of **nerve fibres** and **nerve cells**. Externally, the nerve cord is covered by a layer of **visceral peritoneum**, beneath which lies a thin layer of **longitudinal muscle fibres**, surrounding a fibrous capsule of **epineurium**. Fibres form the core of the cord. Two such cores are visible in a section through the cord. This indicates the double nature of the cord. In the regions of segmental ganglia, two cores of nerve fibres are completely fused along the middle line. On the sides and below cores of nerve fibres lie the nerve cells. These are of two types, **motor neurons**

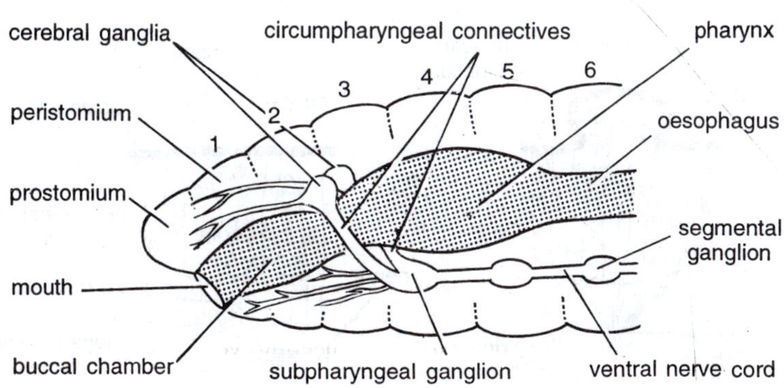

Fig. 22. *Pheretima.* Nerve ring and ventral nerve cord in the anterior end of body in lateral view.

and **association neurons**. Nerve cells occur more in the ganglia. Nerve cells and nerve fibres lie embedded in a mass of connective tissue, called **neuroglia**. Dorsally, four **giant fibres** (one median, one submedian and two lateral) run through the mass of connective tissue along the length of the entire nerve cord. These, like the corresponding fibres of *Nereis,* are responsible for effecting sudden violent contractions of body in response to alarms. They are filled with a homogeneous plasma-like fluid which helps in their contractions.

[II] Peripheral Nervous System

Each cerebral ganglion gives off laterally 8 to 10 nerves which innervate the prostomium and buccal chamber. Nerves from peripharyngeal connectives supply the peristomium and buccal chamber, while nerves from sub-pharyngeal ganglia supply structures in the 2nd, 3rd and 4th segments. Each segmental ganglion of ventral nerve cord gives off 3 pairs of lateral nerves, one pair in front and two pairs behind the row of setae, which innervate the gut wall, body wall and other internal organs of their segments.

Nerves are of **mixed** type, consisting of both **afferent** or **sensory fibres** and **efferent** or **motor fibres**.

[III] Sympathetic Nervous System

It consists of an extensive **nerve plexus** spread beneath epidermis, within muscles of body wall and on alimentary canal. These plexuses are connected with the peri-pharyngeal connectives.

Working mechanism. All the activities of earthworm are under the control of nervous system, though not necessarily of the brain. Nerves consist of both **afferent** and **efferent** fibres. Afferent fibres start from a **sensory cell** or **receptor organ** in epidermis and terminate in the ventral nerve cord in fine branches. Near the branches in cord, and forming a **synapse** with them, arise similar branches of efferent fibres which run outward and terminate in muscles. **Stimuli** or **sensory impulses** are conducted from receptors by afferent fibres to the ventral nerve cord, from where they are reflected back as motor **impulses** along efferent fibres to the muscles which contract. Often an **adjuster neuron** may be present between sensory and motor neurons, through which an impulse may be conveyed to opposite side of same segment or to another or several segments, thus bringing about co-ordination of activities of different parts of body. **Giant fibres** of nerve cord conduct impulses more rapidly than other fibres and cause sudden movements of the entire body when one point on body is strongly stimulated.

SENSE ORGANS

Earthworms have well-developed sense **organs** or **receptor organs** which are quite simple in structure, consisting of a single cell or a group of specialized ectodermal cells. *Pheretima* has three types of sense organs : (i) **epidermal receptors**, (ii) **buccal receptors** and (iii) **photo-receptors**.

1. Epidermal receptors. These are distributed all over epidermis but are more abundant on the lateral sides and ventral surface of body. Each receptor has an elevated cuticle covering a group of tall, slender and columnar

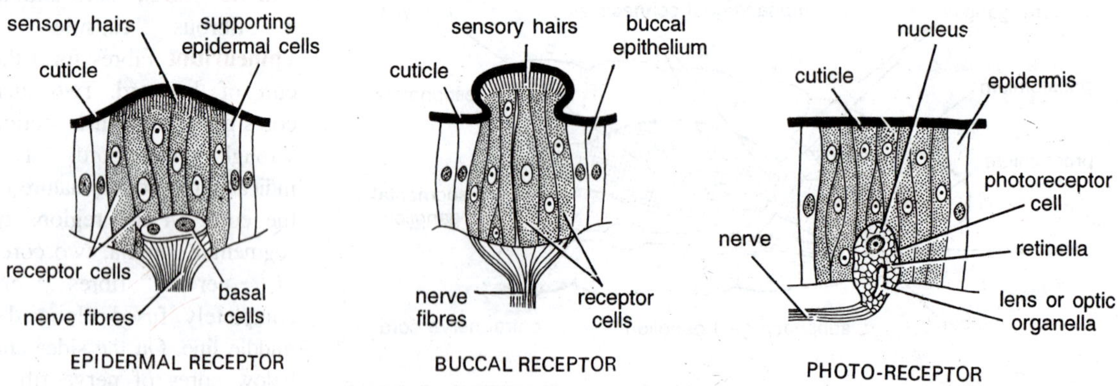

Fig. 25. *Pheretima.* Types of sensory organs in V.S. **A.** Epidermal receptor. **B.** Buccal receptor. **C.** Photo-receptor.

receptor cells, bearing small hair-like processes at their outer ends and connected with nerve fibres at their inner ends. They are surrounded on all sides by ordinary supporting epidermal cells, are separated from each other by spaces, have nuclei at different levels and possess internally a few basal cells. They are tactile (relating to touch) in function and according to some, they also respond to chemical stimuli and changes in temperature.

2. Buccal receptors. These are confined to the epithelium of buccal chamber. They are similar to epidermal receptors except that they possess broader outer ends, better developed sensory hairs, and more deeply situated nuclei. They are gustatory and olfactory (relating to taste and smell) and probably also respond to chemical stimuli.

3. Photo-receptors. Photo-sensitive organs, restricted only to dorsal surface, are more numerous on prostomium and peristomium and gradually reduce in number towards posterior end of body. They are totally absent in clitellum. Each photoreceptor consists of a single ovoid cell, with a nucleus and clear cytoplasm containing a network of neurofibrillae and a small transparent L-shaped lens or optic organelle or phaosome, made up of a hyaline substance. Lens focusses light rays from all directions on neurofibrils. Neurofibrils converge to an afferent nerve fibre which leaves the cell at its base to join the central nervous system. Photoreceptors enable worms to judge the intensity and duration of light.

GENERAL BEHAVIOUR

Earthworms do not have special sense organs, yet they show some sort of behaviour to all kinds of stimuli such as touch, jarring, light, and noxious chemicals. As already stated, their epidermal receptors are extremely sensitive to touch (tactile) and mechanical vibrations, such as heavy foot falls on ground, which cause them to retire quickly in their burrows. When touched, their body immediately rolls over.

With the help of buccal receptors, earthworms display chemical responses comparable to taste and smell, in the choice of food. Unpleasant and irritating chemical vapours cause them to withdraw immediately into the burrow. When irritated they eject coelomic fluid through the dorsal pores.

Photoreceptors, occurring in clusters enable worms to judge the intensity and duration of light. Worms are negatively phototrophic to strong light and positive to weak light. They avoid strong day light and will atonce recede into burrow if flashed with a torch.

Earthworms respond to very low and very high temperatures by burrowing deeper in soil. They love moisture and avoid dryness, but come out of their burrows when they get flooded during rainy season. They give no evidence of sense of hearing and evidently do not perceive mere sound vibrations in air.

Experiments prove that earthworms have some learning ability. They have been taught to make right and left turns at burrow intersections by giving a mild electric shock when they make a mistake. Darwin credited worms with some intelligence in pulling leaves into their burrows. However, others consider that the process is mainly one of trial and error.

REPRODUCTIVE SYSTEM

Earthworms do not reproduce asexually. Reproductive organs are somewhat complicated. The earthworms are monoecious (hermaphroditic) but they cannot fertilize their own eggs because they are protandrous. As a rule, cross-fertilization takes place. It is preceded by copulation and cocoon formation.

[I] Male Reproductive Organs

These include testes, testis sacs, seminal vesicles, vasa deferentia, prostate glands and accessory glands.

1. Testes. There are 2 pairs of very minute, white and lobed testes, one pair in 10th and the other in 11th segment. They lie ventro-laterally beneath the alimentary canal, close to mid-ventral line, on either side of nerve cord and attached to the anterior wall of their respective testis sacs. Each testis is made of a compact narrow base from which arise 4 to 8 small digitate processes containing rounded cells or spermatogonia. Testes are well formed only in young worms but become degenerated in adults.

2. Testis sacs. Both the testes of each segment are enclosed within a wide, thin-walled testis sac, which is a coelomic space cut off from general body cavity. Thus, there are two testis

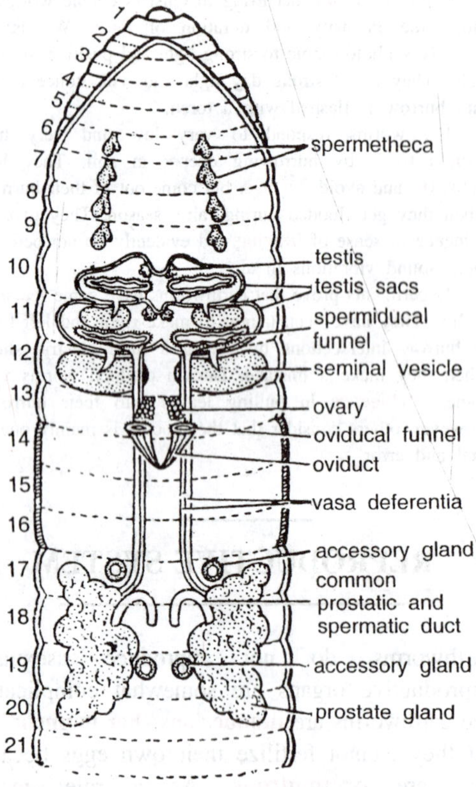

Fig. 26. *Pheretima.* Reproductive system.

sacs situated in segments 10 and 11, ventro-laterally one behind the other below alimentary canal. Each testis sac encloses a pair of testes and a pair of ciliated **spermiducal funnels** and also communicates behind, by a pair of tubular connections, with two seminal vesicles of succeeding segment. Testis sac of llth segment is large enough so as to enclose also the seminal vesicles of that segment.

3. Seminal vesicles. There are two pairs of large, white **seminal vesicles** lying in segments 11 and 12, respectively. They are also referred to as **septal pouches** since they grow as outgrowths of the septa. Testis sac of 10th segment communicates with seminal vesicles of llth segment, and testis sac of llth segment with seminal vesicles of 12th segment. Seminal vesicles of llth segment lie enclosed within testis sac of the same segment, while those of 12th segment lie free.

4. Spermiducal funnels. There are two pairs of ciliated **spermiducal funnels** (or spermrosettes), one of them lying behind each testis in the same segment and enclosed within the same testis sac.

5. Vasa deferentia. Each funnel leads behind into a slender, ciliated, thread-like **sperm duct** or **vas deferens**. Two vasa deferentia of same side run close together posteriorly along the ventral body wall up to 18th segment to join the **prostatic duct**.

6. Prostate glands. Prostate glands are a pair of dirty white, flat, solid, irregular and lobulate masses, lying one on either side of gut and extending from 16th or 17th segment up to 20th or 21st segment. Each gland consists of a big glandular part and a small non-glandular part. Glandular part is a racemose gland consisting of several lobes closely fixed together. Non-glandular portion consists of several small ductules which unite to form a short, thick, muscular and curved **prostatic duct** in 18th segment. Immediately on its emergence from the inner side of gland, it is enclosed in a common muscular sheath, along with the two vasa deferentia of its own side, forming a **common spermatic** and **prostatic duct**, within which the three tubes remain separate. Both common ducts curve inwards to open to the exterior independently by a pair of **male genital pores** ventrally on 18th segment. Prostate glands manufacture a fluid, the prostatic fluid, of unknown function.

7. Accessory glands. In each of the segments 17th and 19th is found a pair of rounded, white fluffy masses, the **accessary glands**, on ventro-lateral body wall, one on either side of nerve cord. They open to the exterior by a number of ducts on two pairs of **genital papillae**, situated externally upon the 17th and 19th segments, one on either side of mid-ventral line. Their secretion helps probably in uniting the two worms during copulation.

From testes, spermatogonia or sperm-mother cells are shed into testis sacs. From here they enter seminal vesicles to undergo maturation and develop into spermatozoa. Mature sperms move

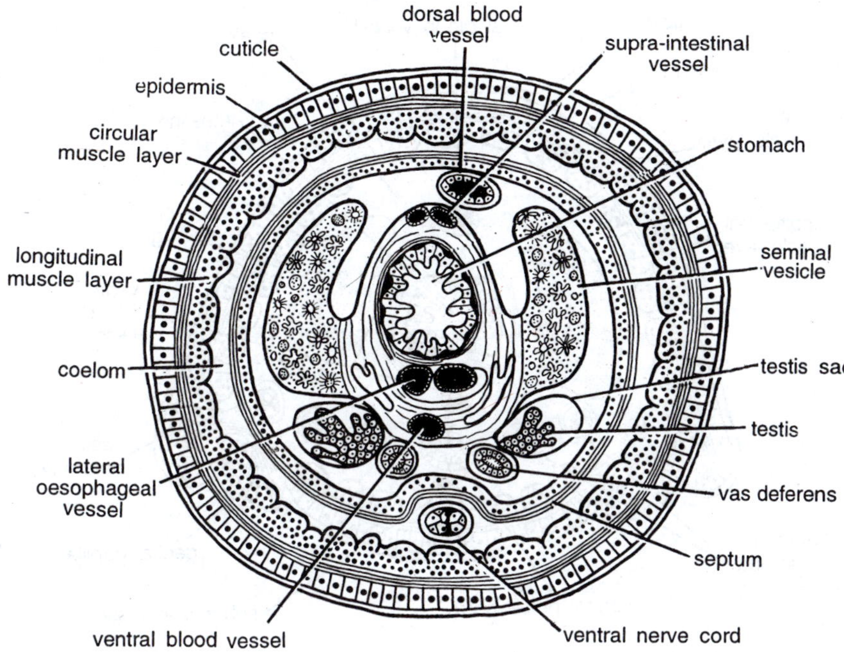

Fig. 27. *Pheretima.* T.S. 11th segment of body passing through the testis sacs, testes and seminal vesicles.

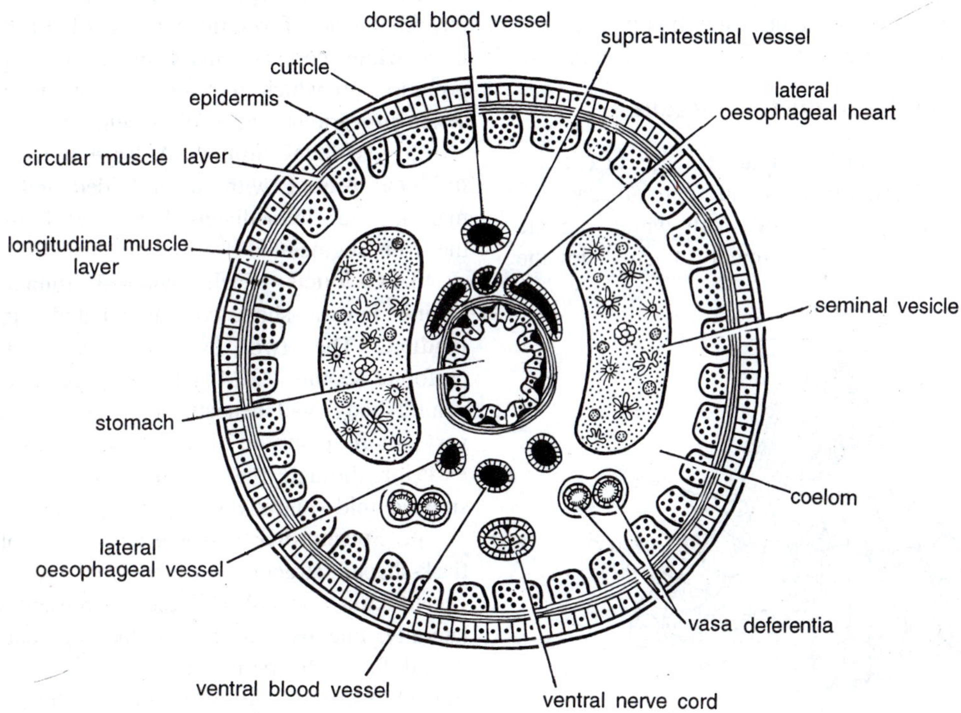

Fig. 28. *Pheretima.* T.S. 12th segment of body, passing through seminal vesicles and vasa deferentia.

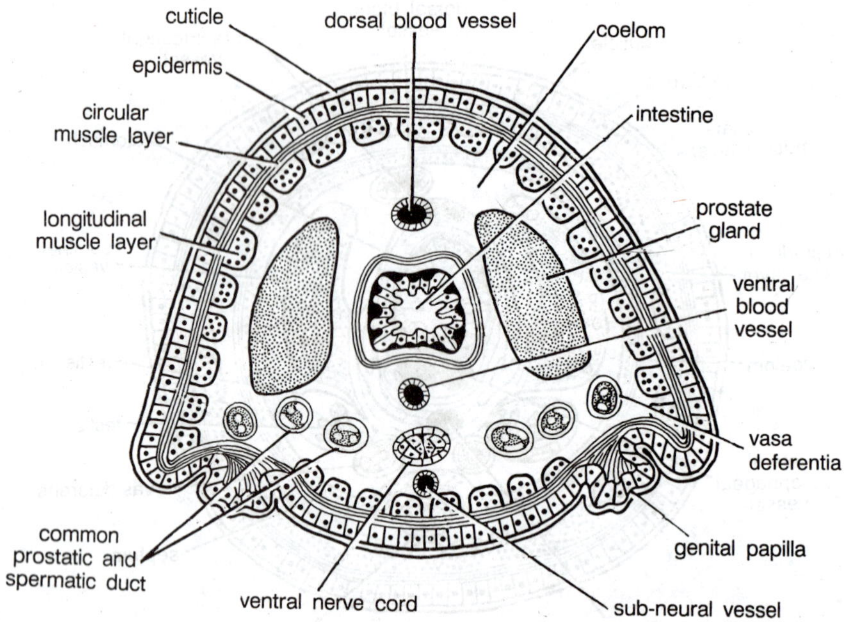

Fig. 29. *Pheretima.* T.S. 17th segment of body passing through prostate glands and genital papillae.

back into testis sacs, enter spermiducal funnels, travel along vasa deferentia and finally pass out through the male genital pores during copulation.

[II] Female Reproductive Organs

The female reproductive organs consists of ovaries, oviducal funnels, oviducts and spermathecae.

1. Ovaries. A pair of small white, digitate **ovaries** lies in 13th segment attached to the posterior face of septum 12/13 in front of it, one on either side of ventral nerve cord. Each ovary is a white compact mass made of finger-like processes in which **ova** are arranged in a linear series in various stages of development.

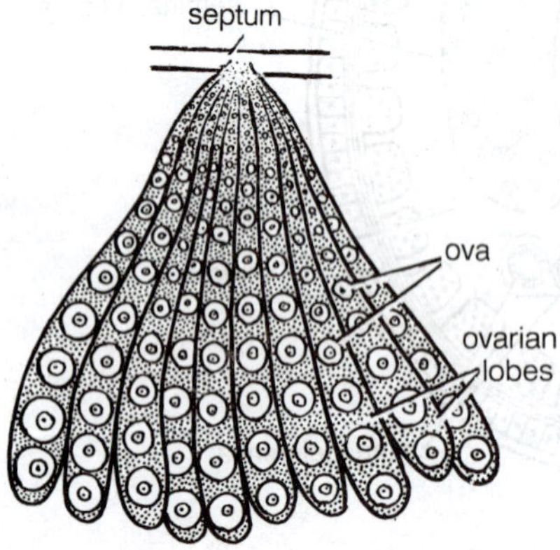

Fig. 30. *Pheretima.* Ovary.

2. Oviducal funnels. A large saucer-shaped **oviducal funnel**, with much folded and ciliated margins, lies immediately behind each ovary in the 13th segment.

3. Oviducts. Each oviducal funnel leads behind into a short conical ciliated tube, the **oviduct**. The two oviducts run backwards, perforate septum 13/14 and converge to meet in ventral body wall beneath nerve cord, forming a very short **common oviduct**. It opens to the **exterior** through the **female genital aperture**, situated mid-ventrally on 14th segment.

4. Spermathecae. There are 4 pairs of small flask-shaped structures, called **spermathecae** or **receptacula seminales**. These are present ventro-laterally, one pair in each of the segments 6, 7, 8 and 9. Each spermatheca has a broad, pear-shaped body, the **ampulla**, and a short narrow **neck**, which give off a narrow elongated blind **caecum** or **diverticulum** before opening to the

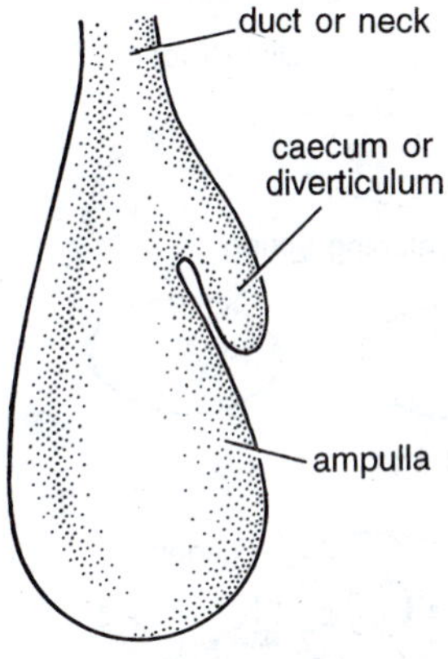

Fig. 31. *Pheretima.* A spermatheca.

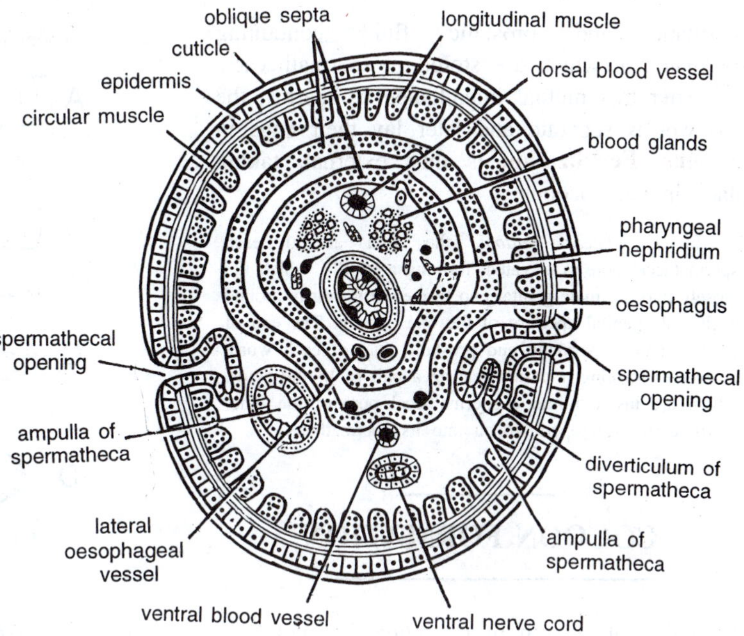

Fig. 32. *Pheretima.* T.S. body through spermathecae.

exterior. Thus 4 pairs of spermathecae open to outside by 4 pairs of separate **spermathecal pores** situated ventro-laterally in the grooves between 5/6, 6/7, 7/8 and 8/9 segments, respectively. Spermathecae receive sperms from another worm during copulation, and store them in their diverticula in *Pheretima* and in ampullae in other earthworms.

Mature **ova** shed from ovaries are entangled by oviducal funnels, travel along oviducts, and pass out to the exterior through the female genital aperture, to be laid inside the cocoon.

COPULATION AND FERTILIZATION

Earthworms are bisexual, still self-fertilization does not occur because they are **protandrous**. A reciprocal cross-fertilization occurs between two worms and spermatozoa of one worm are transferred to another during a process, termed **copulation**. It has not been studied so far in the Indian earthworm *Pheretima posthuma,* as it probably takes place underground, but has been studied in *Pheretima communissimma.* It generally takes place at night, during rainy season, and lasts for about one hour. There is no penis or vagina for the transfer of sperms. During copulation, two worms apply to each other by their ventral surfaces with head ends pointing in opposite directions, so that the male genital pores of each lie against a pair of spermathecal pores of other. Areas surrounding the male genital apertures are raised into papillae, which are inserted successively from behind to forward into the spermathecal pores of the other worm and discharge the

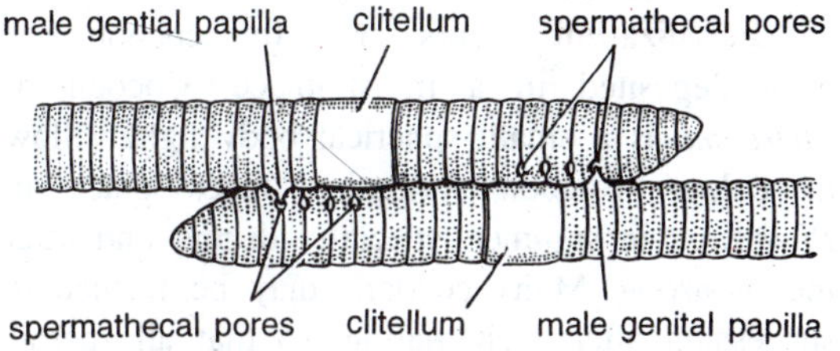

Fig. 33. *Pheretima.* Two worms in copulation.

spermatic and prostatic fluid containing spermatozoa, which are stored in spermathecae.

After this mutual interchange of sperms, the two worms separate and later lay their eggs in **cocoons**. **Fertilization** is thus external, taking place in the cocoons.

In *Eutyphaeus waltoni,* the area around each spermathecal pore is elevated, forming a papilla, which fits closely into the male genital cup formed temporarily around each male genital aperture of the other worm. The end of each common spermatic and prostatic duct of each worm is everted forming a **penis**, bearing a pair of **penial setae**, which are inserted into a spermathecal pore of the other worm to transfer spermatozoa into its spermatheca.

COCOON FORMATION

Formation of cocoon in *Pheretima* has not been studied as yet. In other worms, such as *Eisenia* and *Rhynchelmis,* etc., it is secreted as a viscid and gelatinous substance by clitellar glands, forming a broad membranous band or girdle around clitellum. It hardens gradually on exposure to air into a tough but elastic tube which becomes the **cocoon** or **egg capsule**. A **slime tube** is also secreted by epidermal mucous cells of clitellum over cocoon. As the worm wriggles behind, the slime tube and cocoon are slipped forward over the head. On its way the cocoon receives ova from female genital aperture and sperms of other worm, from spermathecae, so that cross-fertilization is ensured and zygotes are formed. An albuminous fluid is also deposited inside cocoon by the glands of anterior segments of body. Finally, when cocoon is thrown off the head, its elastic ends close up and a yellowish rounded cocoon is formed.

Fertilization occurs after the cocoon has been deposited in a moist place. Cocoon of *Pheretima* is a small, spherical body light yellow in colour. Cocoon formation takes place in *Pheretima* in summer, specially during and after the monsoon. Many cocoons may be formed in succession after each mating, so that all sperms stored in the spermathecae are not passed out at once.

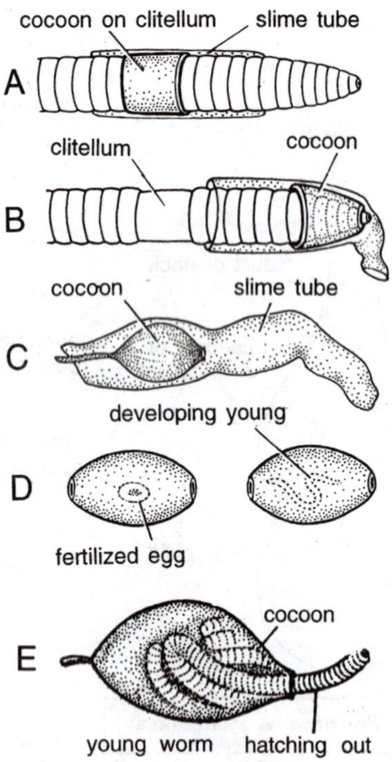

Fig. 34. *Eisenia foetida.* Stages of Cocoon-formation. **A.** Slime tube and cocoon formed. **B.** Earthworm withdrawing posteriorly. **C.** Cocoon left in its slime tube. **D.** Cocoons with fertilized egg and developing young. **E.** Young worm hatching out of the cocoon.

DEVELOPMENT

A cocoon may contain many fertilized eggs but only one embryo develops, growing at the expense of other eggs serving as **nurse cells** and albumen stored in cocoon. **Cleavage** is holoblastic and unequal, and **development** is direct without any free larval stage. A hollow blastula is formed and later a gastrula by invagination. Mesoderm develops from two large cells of blastula, called **mesoblasts**. They divide to form two mesoblastic bands which later give rise to the coelomic epithelial lining. Young worm, when fully grown, crawls out of cocoon in about two or three weeks. Newly hatched young worm receives no parental care and resembles the adult except for size and absence of clitellum.

REGENERATION AND GRAFTING

Earthworms do not reproduce asexually, but they have great capacity for **regeneration**. If ends of body are cut accidently or removed experimentally, a head or tail is regenerated to replace the lost part.

Earthworms can also be grafted like *Planaria.* Experimentally, some abnormal types have been produced such as worms with two tails, short worms by grafting together two terminal parts, or exceptionally long worms by joining together end to end pieces of several worms. However such freaks do not survive as they cannot feed.

ADAPTATIONS OF EARTHWORM

Earthworms are well-adapted for a subterranean or burrowing mode of life.

(1) Elongated, slender, cylindrical and streamlined body is well-suited for burrowing in soil.

(2) Setae and musculature serve for locomotion as well as for anchoring body firmly in burrow.

(3) Secretes mucus for plastering the internal walls of burrow.

(4) Coelomic fluid oozing through dorsal pores keeps skin moist for gaseous exchange in the absence of respiratory organs.

(5) Amoebocytes of coelomic fluid kill harmful bacteria and other parasites and protect body.

(6) Nocturnal and burrowing habits provide safety from predators.

(7) Special sensory organs, such as eyes and ears, are absent due to burrowing life.

(8) Hermaphroditism and regeneration ensure continuity of species against many hazards in life.

(9) Copulation followed by formation of cocoons for fertilization and development are adaptations for reproduction on dry land.

ECONOMIC IMPORTANCE

Earthworms are simple, common place creatures of great economic importance to man. Small they certainly are, but directly or indirectly they are useful to us. Our earth would be a far different place without them.

1. As bait and food. All over the world they are used as bait for fishing. Various methods are employed to drive them out of their burrows for making large collections. Some of these include jarring soil by beating a stick driven into it, pouring poisonous chemical solutions such as mercuric chloride on ground and even using an electric current. They form the best food of fish in aquaria. A small white earthworm *(Enchytraeus albidus)* is often grown in soil and used to feed aquarium fish and small laboratory animals. They are actually used as food by uncivilized people in many parts of the world. They are eagerly hunted as food by some birds, notably robins and chickens. Large numbers of them are eaten by frogs, moles, lizards, small snakes, centipedes and other predatory invertebrates.

2. In agriculture. Earthworms are in general beneficial to agriculture. Although they may sometimes do damage to young and tender plants, yet they are good friends to the gardener and farmer as they are continually ploughing and manuring the soil. Their habit of burrowing and swallowing earth increases fertility of soil in many ways. Their burrows permit penetration of air and moisture in porous soil, improve drainage, and make easier the downward growth of roots. The thorough grinding of soil in gizzard constitutes an effective kind of soil 'cultivation'. Earthworms are continuously dragging dead leaves into their burrows, partly to make the underground retreat more comfortable and partly to eat them. They are only partially digested and their remains are thoroughly mixed with the castings (adding humus to soil). Thus, in course of time they have made most of the vegetable moulds of the world. However, the claim, that fertility of an unproductive land will greatly increase by the addition of earthworms, is not correct. Excretory wastes and other secretions of worms also enrich soil by adding nitrogenous matters that form important plant food. Their effect on turning over of soil is considerable. One acre of ground may contain 50,000 earthworms and the quantity of earth brought up from below and deposited on the surface as worm castings has been estimated by Darwin to be as 18 tons per acre per year. In recent years, there is stress on the importance of culturing earthworms to build up soil to a high degree of fertility.

3. In medicines. Earthworms were used variously as medicines in the past. **Hamdullah Mustaufi** of **Qazwin** in ***'Naizat-ul-Qutub'*** written in A.D. 1340, and **Damari** in ***'Hayat-ul-Haiwan'*** (The Life of Animals), written in A.D. 1371, tell us of medicines prepared from earthworms to cure stones in bladder, yellowness of jaundice, pyorrhoea, piles, rheumatism or gout, diarrhoea, weakness after pregnancy and sexual impotency. Even to this day the Chinese, the Japanese and the Indians are said to use earthworms in various fancy medicines.

4. In laboratories. Earthworms are easily obtained and are of convenient size for dissections. They are, therefore, universally employed for class studies and for investigations in general and comparative physiology.

5. Harmful worms. In some cases, earthworms become harmful. Exceptionally, their burrows may cause loss of water by seepage from ditches in irrigated lands. Their castings on sloping lands tend to be washed away by rain and thus contribute to soil erosion, though to a lesser extent. Certain species live as external parasites of frogs and man. Sometimes, they bury in the dead bodies of buried animals and bring the disease-germs to surface, where they may infect other animals. Earthworms are said to serve as intermediate hosts in the transmission of some parasites, such as tapeworm (*Amoebotaenia sphenoides)* and gapeworm (*Syngamus*) of chicken and lung nematode (*Metastrongylus elongatus*) of pigs. The latter parasite carries a virus, which together with a bacterium, causes hog influenza or swine influenza. Some species become pests of plants. *Pheretima elongata* is suspected of damaging the roots of the Betel-Vine (Piper Betel) in Coimbatore. *Malabaria podudicola* and *Aphanascus oryzivorus* are said to injure the roots of paddy in Malabar. A species of *Perionyx* damages Cardamom stems grown on the Anamalai Hills.

Important Questions

➤➤*Long answer type questions*

1. Describe the digestive system of earthworm.
2. With the help of proper diagrams describe in brief the blood-vascular system of *Pheretima.*
3. Give an account of the reproductive organs of *Pheretima.*
4. Describe the nephridial system of *Pheretima.*
5. Describe the septal nephridium of *Pheretima.* How do the pharyngeal and integumentary nephridia differ from it ?
6. Draw full-page labelled diagrams of the following : (i) T.S. *Pheretima* through spermathecae. (ii) T.S. *Pheretima* through seminal vesicles, (iii) V.L.S. *Pheretima* through first 20 segments.
7. Write an essay on the Economic Importance of earthworms.
8. Write short notes on : (i) Adaptations of earthworm, (ii) Behaviour of earthworm, (iii) Coelom in earthworm, (iv) Cocoon-formation in earthworm, (v) Locomotion in earthworm, (vi) Sense organs of *Pheretima.*

➤➤*Short answer type questions*

1. Name the animal where the chloragogen cells are present.
2. Where is haemoglobin found in the earth worm blood ? How does it compare with the blood of vertebrates ?
3. What are lateral warts ? Mention their number, location, function. Write within four sentences.
4. How are setae embedded in the body wall of earth worm. Write in three sentences only.
5. With suitable illustration, show all the external apertures seen on the body of earthworm.
6. Give characteristic morphological differences between *Pheretima* and *Nereis.*

7. Give an account of the blood vascular system in the anterior thirteen segments of *Pheretima.*
8. Compare the excretory organs of *Pheretima* with that of *Nereis.*
9. Sketch the circulatory system in earthworm and name all the parts.

➤➤*True and false statements*

1. Excretion in earthworm is performed by means of *Nephredia.*
   True / False / Do not know
2. Ovary in *Pheretima* is present in the fourteenth segment.
   True / False / Do not know
3. Body cavity of earthworm is a haemocoel.
   True / False / Do not know
4. Syncitium is present in earthworm.
   True / False / Do not know
5. Typhlosole in earthworm supports the intestine.
   True / False / Do not know
6. In earthworm blood flows backwards in the dorsal blood vessel.
   True / False / Do not know
7. In earthworm haemoglobin is dissolved in plasma.
   True / False / Do not know

➤➤*Multiple choice questions*

1. In Earthworm, the spermathecae are used for :
   (a) development of ovum
   (b) development of sperm
   (c) storing spermatozoa
   (d) development of zygote
   (e) storing ova
2. In annelida formation of larva is usually absent. But when present the larva is called :
   (a) tadpole
   (b) Planula
   (c) trochophore
   (d) Ephyra
3. Metamerism is characteristic of :
   (a) Platyhelminthes (b) Nematoda
   (c) Annelida (d) Arthropoda
4. The body cavity found in Annelida is :
   (a) enterocoel (b) schizocoel
   (c) pseudocoel (d) none of these
5. Oxygen is circulated to various tissues of earthworms by :
   (a) blood corpuscles
   (b) plasma
   (c) blood corpuscles and plasma
   (d) none of the above
6. In *Pheretima,* spermathecae are found in segments :
   (a) 4th to 7th (b) 6th to 9th
   (c) 9th to 12th (d) 17th to 20th
7. Two pairs of hearts in earthworm occur in segments :
   (a) 6, 7 and 9, 10 (b) 7, 9 and 12, 13
   (c) 9, 10 and 14, 15 (d) none of these
8. In earthworm, blood from seminal vesicle is collected by :
   (a) ventral vessel (b) lateral oesophageal vessel
   (c) dorsal vessel (d) supra-oesophageal vessel
9. Earthworm is :
   (a) ammonotelic (b) ureotelic
   (c) uricotelic (d) aminotelic
10. Earthworm appears brown due to the presence of :
    (a) chloragogen cells (b) chloragosomes
    (c) porphyrin (d) all of these
11. The urine of earthworm contains :
    (a) urea (b) ammonia
    (c) creatinine (d) all of these
12. Exonephric nephridia of earthworm is :
    (a) septal (b) pharyngeal
    (c) integumentary (d) all of these
13. The earthworm body is devoid of :
    (a) cuticle (b) epidermis
    (c) seta (d) appendages
14. The body cavity of earthworm is :
    (a) true coelom (b) pseudocoelom
    (c) acoelom (d) haemocoel
15. In earthworm the clitellar region helps the process of :
    (a) copulation (b) cocoon formation
    (c) digestion (d) locomotion
16. In earthworm the testes are enclosed in the segments :
    (a) 9 & 10 (b) 8 & 9
    (c) 12 & 10 (d) 11 & 10
    (e) 6 & 9
17. In earthworm the ovary is situated in the :
    (a) 10th segment (b) 13th segment
    (c) 11th segment (d) 14th segment
18. Flow of blood in the dorsal blood vessel of *Pheretima* is :
    (a) from infront backwards
    (b) from behind forwards
    (c) in both direction (d) none of these
19. Lateral and latero-oesophageal hearts in *Pheretima* are situated in the segments :
    (a) 7, 9, 10, 12 (b) 7, 9, 13, 12
    (c) 7, 8, 13, 12 (d) 7, 9, 11, 12
20. The earthworm moves with the help of :
    (a) setae alone (b) setae and muscles
    (c) muscles alone (d) parapodia

21. In earthworm, the typhlosole is a part of the :
(a) reproductive tract (b) intestine
(c) circulatory system (d) nephridium
22. Locomotion in earthworm is directly facilitated by :
(a) mucus secreted by the epidermis
(b) segmentation of the body
(c) rhythmic contraction of individual segments
(d) setae
23. The body cavity of earthworm represents a true :
(a) coelenteron (b) haemocoel
(c) coelom (d) blastocoel
24. The nephridia found most in number are :
(a) septal (b) pharyngeal
(c) integumentary (d) oesophageal
25. The male genital pore of earthworm is found in :
(a) I4th segment (b) 18th segment
(c) 16th segment (d) 17th & 19th segment
26. Which of the following is called natural ploughman :
(a) tapeworm (b) liver fluke
(c) earthworm (d) leech
27. Septal nephridia are :
(a) respiratory in function
(b) excretory in function
(c) circulatory in function
(d) nervous in function
28. The blood of earthworm is red because :
(a) haemoglobin is present in blood cells
(b) haemoglobin is dissolved in plasma
(c) both (d) none
29. Chloragogen cells of earthworm are comparable to vertebrate :
(a) liver (b) spleen
(c) kidney (d) none
30. Nephrostome in earthworm is found in :
(a) septal nephridia (b) integumentary nephridia
(c) pharyngeal nephridia (d) in all nephridia
31 Earthworm moves with the help of :
(a) sphincter muscles (b) muscular contraction
(c) setae only
(d) setae and muscles, supported by hydrostatic pressure of coelomic fluid
32. Excretory product of earthworm is :
(a) ammonia (b) uric acid
(c) urea (d) ammonia and urea
33. The lateral hearts of earthworm are present in :
(a) 5th and 8th segment
(b) 7th and 9th segment
(c) 10th and 12th segment
(d) 13th and 14th segment
34. In earthworm, urea is mainly formed in :
(a) coelomic fluid (b) chloragogen cells
(c) chromophil cells (d) nephridia
35. Chief excretory waste of earthworm is :
(a) water (b) ammonia
(c) urea (d) creatinine
36. Earthworm has no special structures for :
(a) locomotion (b) nutrition
(c) respiration (d) circulation
37. Segments of earthworm having spermathecae are :
(a) five to eight (b) six to nine
(c) seven to ten (d) four to nine
38. Cocoon of earthworm is secreted :
(a) around clitellum
(b) before fertilization and development
(c) both (d) after oviposition
39. A larval stage does not occur in :
(a) sponge (b) *Hydra* and earthworm
(c) *Hydra* (d) cockroach
40. In earthworms the body surface is kept moist with the help of coelomic fluid which oozes out from :
(a) nephridiopores (b) genital pores
(c) dorsal pores (d) ventral pores
41. Testes sacs in earthworm are located in which segments :
(a) 13 & 14 (b) 9 & 10
(c) 11 & 12 (d) 10 & 11
42. Intestinal caeca secretes :
(a) proteolytic enzyme (b) amylase
(c) lipase (d) all of the above
43. Metamorphosis does not occur in :
(a) frog (b) butterfly
(c) house fly (d) earthworm
44. In earthworm spermathecae are used for :
(a) development of ovum
(b) storing spermatozoa
(c) development and nourishment of sperms
(d) none of these
45. Skeleton-like function during locomotion of *Pheretima* is performed by :
(a) setae
(b) alimentary canal full of sand
(c) coelomic fluid (d) none of these
46. In *Pheretima,* nephrostome is absent in :
(a) septal nephridia (b) pharyngeal nephridia
(c) integumentary nephridia and pharyngeal nephridia
(d) all these
47. Clitellum in *Pheretima* includes segments :
(a) 12, 13, 14 (b) 13, 14, 15
(c) 14, 15, 16 (d) 15, 16, 17
48. Segment of earthworm in which mouth is found :
(a) peristomium (b) prostomium
(c) II segment (d) none of these
49. While burrowing, earthworms do not tear apart the soil, and ingest :
(a) grass only (b) insects only
(c) soil (d) dead organic particles only
50. Castings of *Pheretima posthuma* are in the form of :
(a) miniature heaps of small pellets or balls of soil
(b) a long coiled string of soil
(c) a fine powder of soil
(d) none of these
51. Due to burrowing of earthworms the subsoil :
(a) is rendered more fertile
(b) brought to the surfare of top soil
(c) is constantly passed through the gut
(d) all of these

52. 'Natural Ploughman' is said for :
(a) snakes (b) earthworms and snakes
(c) rabbits (d) earthworm
53. Male genital aperture of earthworm in :
(a) dorsal surface (b) ventral surface and single
(c) ventral surface and paired
(d) 14th segment
54. Setae are found all over body in earthworm except :
(a) peristomium (b) clitellum
(c) anal segment (d) all of these
55. Locomotion in earthworm is helped by :
(a) setae (b) mucous
(c) muscles (d) all of these
56. In earthworm unicellular salivary glands are found in :
(a) oesophagus (b) pharynx
(c) stomach (d) none of these
57. Alimentary canal of *Pheretima* is internally lined by cuticle in :
(a) buccal chamber (b) gizzard
(c) both (d) none
58. Typhlosole of *Pheretima* is in :
(a) middle and posterior intestine
(b) middle intestine and for absorption
(c) anterior intestine and for absorption
(d) intestine
59. Chloragogen cells of earthworm are in :
(a) parietal peritoneum and perform excretion
(b) visceral peritoneum and perform digestion
(c) visceral peritoneum and perform excretion
(d) none of these
60. Nephridia of earthworm are :
(a) absent in first three segments
(b) for excretion
(c) for osmoregulation (d) all of these
61. Which is not enteronephric nephridia of earthworm :
(a) integumentary (b) pharyngeal
(c) septal (d) both 1 and 2
62. Pharyngeal nephridia of earthworm are located in segments :
(a) 6, 7, 8 (b) 7, 8, 9
(c) 3, 4, 5 (d) 4, 5, 6
63. Earthworm respires through its :
(a) moist skin (b) typhlosole
(c) gills (d) clitellum
64. Blood is red but there are no RBCs in :
(a) leeches (b) earthworm
(c) earthworm and leeches
(d) rabbit
65. In earthworm, exonephric excretion takes place by :
(a) septal nephridia (b) pharyngeal nephridia
(c) integumentary nephridia
(d) all of these
66. Segments of earthworm bearing accessory glands are :
(a) 19, 20 (b) 17, 20
(c) 17, 19 (d) 17, 18
67. Leech is :
(a) unisexual (b) dioecious
(c) hermaphrodite (d) vector
68. Blood from seminal vesicles of earthworm are collected by :
(a) dorsal vessel
(b) subneural vessel
(c) lateral oesophageal vessel
(d) ventral vessel
69. Enteronephric nephridia of earthworm open into :
(a) coelom (b) body surface
(c) intestine
(d) stomach
70. Typhlosole of earthworm starts from segment :
(a) 14 (b) 27
(c) 20 (d) 40
71. Organ of earthworm analogous to kidney of rabbit is:
(a) testis (b) nephridium
(c) lateral heart (d) salivary gland
72. Earthworm belongs to group :
(a) Oligochaeta (b) Polychaeta
(c) Hirudinea (d) Arthropoda
73. Earthworm has :
(a) many eyes (b) no eyes
(c) single eye (d) two eyes
74. Neurons in earthworm are :
(a) sensory
(b) adjustor
(c) sensory, motor & adjustor
(d) adjustor
75. The septal nephridia of *Pheretima* expell their wastes into :
(a) coelom (b) blood
(c) pharynx (d) intestine
76. The female genital aperture in *Pheretima* are found in segment :
(a) 18 (b) 14
(c) 13 (d) 16
77. Fertilization in *Pheretima* occurs in :
(a) soil (b) cocoon
(c) seminal vesicle (d) oviduct
78. The clitellum in a mature *Pheretima* is a thick girdle which is :
(a) glandular in nature and surrounds segments 16-18
(b) non-glandular in nature and surrounds segments 16-18
(c) glandular in nature and surrounds segments 14-16
(d) non-glandular in nature and surrounds segments 14-16
79. The intestinal caecae in *Pheretima* arise from the intestine in segment :
(a) 24 and extend forward
(b) 26 and extend backward
(c) 24 and extend backward
(d) 26 and extend forward
80. The flow of blood in the lateral hearts present in segments 7 and 9 of *Pheretima* is from the :
(a) dorsal to the ventral blood vessel
(b) ventral to the dorsal blood vessel
(c) dorsal to the lateral oesophageal blood vessel
(d) lateral oesophageal to the dorsal blood vessel

81. Which of the following statements is NOT correct with respect to the pharyngeal nephridia of *Pheretima*?
    (a) they are present in 3 paired groups
    (b) one paired group of these nephridia lies in each of the segments 4-6
    (c) they are exonephric
    (d) they are without nephrostomes
82. There are 4 pairs of spermathecae in *Pheretima.* One pair of these is present in each of the segments :
    (a) 4, 5, 6 and 7 (b) 5, 6, 7 and 8
    (c) 6, 7, 8 and 9 (d) 7, 8, 9 and 10
83. Lateral oesophageal hearts in earthworm connect :
    (a) dorsal and ventral vessels
    (b) dorsal and supra-oesophageal vessel with ventral vessel
    (c) dorsal and subneural vessel
    (d) lateral oesophageal and supra oesophageal vessels
84. Red colour of blood of earthworm is due to :
    (a) haemoglobin (b) chlorocruorin
    (c) erythrocruorin (d) haemocyanin
85. Typhlosole of *Pheretima* serves :
    (a) to increase the absorptive area of intestinal epithelium
    (b) to slow down the rate of passage of food
    (c) to secrete digestive juice
    (d) no purpose
86. The clitellum of earthworm is concerned with :
    (a) copulation (b) formation of cocoon
    (c) storage of sperms
    (d) release of sperms from male genital pore
87. Functioning of which one of the following is the nearest comparable with the chloragogen cells of *Pheretima* ?
    (a) kidney (b) liver
    (c) spleen (d) sweat glands
88. A ventral nerve cord is found in :
    (a) rabbit (b) snake
    (c) earthworm (d) *Amoeba*
89. Photoreceptors in *Pheretima* are found in :
    (a) dorsal surface (b) ventral surface
    (c) lateral sides (d) all over body
90. The generic name '*Pheretima*' was first used by :
    (a) Kingberg (b) Kimball
    (c) Lamark (d) K.N. Bahl
91. A detailed study about *Pheretima* made by :
    (a) Kingberg (b) Kimball
    (c) K.N. Bahl (d) Lamark
92. The pigment in skin of *Pheretima* is :
    (a) melanin (b) porphyrin
    (c) carofin (d) cutinin
93. Male genital papillae are present on segment :
    (a) 17 (b) 19
    (c) 17 and 19 (d) 18
94. 'Forests of nephridia' present in the region of :
    (a) precliteller (b) cliteller
    (c) post cliteller
95. Coelomic fluid contains :
    (a) granulocytes (b) mercocytes
    (c) leucocyte (d) all

Answers

➤➤True and false statements

1. true 2. false 3. false 4. false 5. true 6. false 7. true.

➤➤Multiple choice questions

1. (c)	2. (c)	3. (c)	4. (b)	5. (c)	6. (b).	7. (b)	8. (b)	9. (b)	10. (c)
11. (a)	12. (a)	13. (d)	14. (a)	15. (b)	16. (d)	17. (b)	18. (b)	19. (b)	20. (b)
21. (b)	22. (d)	23. (c)	24. (c)	25. (b)	26. (c)	27. (b)	28. (b)	29. (a)	30. (a)
31. (d)	32. (d)	33. (b)	34. (b)	35. (c)	36. (c)	37. (b)	38. (b)	39. (d)	40. (c)
41. (d)	42. (b)	43. (d)	44. (b)	45. (a)	46. (c)	47. (c)	48. (a)	49. (c)	50. (a)
51. (d)	52. (d)	53. (d)	54. (d)	55. (a)	56. (b)	57. (b)	58. (b)	59. (d)	60. (b)
61. (a)	62. (d)	63. (a)	64. (c)	65. (c)	66. (c)	67. (c)	68. (c)	69. (a)	70. (b)
71. (b)	72. (a)	73. (b)	74. (c)	75. (d)	76. (b)	77. (b)	78. (c)	79. (d)	80. (a)
81. (c)	82. (c)	83. (b)	84. (a)	85. (c)	86. (b)	87. (a)	88. (c)	89. (a)	90. (a)
91. (c)	92. (b)	93. (c)	94. (b)	95. (d)					

42

Hirudinaria granulosa : The Indian Cattle Leech

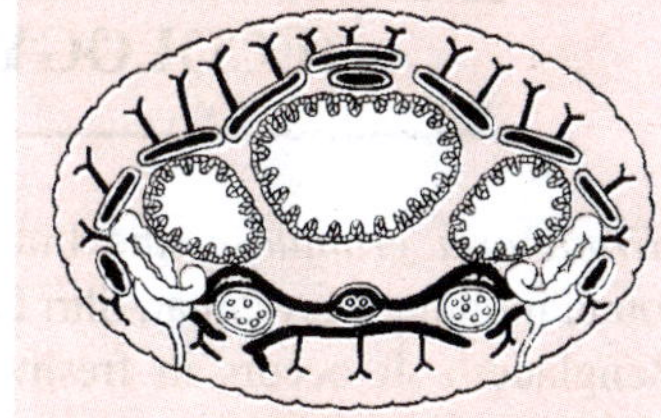

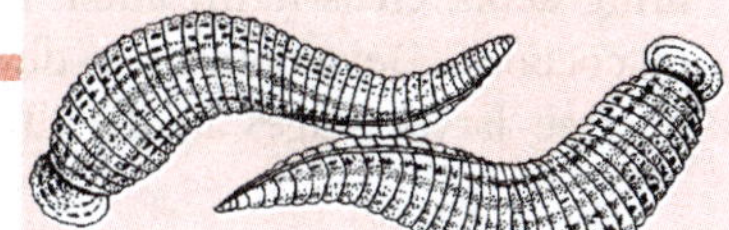

Class *Hirudinea* comprises of leeches which are perhaps the most specialized annelids without parapodia and setae but with suckers. About 300 species of leeches are known to occur in the tropical and temperate parts of the globe. Most of them are freshwater, while some are marine or terrestrial. Most of leeches are ectoparasitic, living on the blood of vertebrates, while some are predaceous, feeding on worms, snails, insect larvae, etc.

The genus *Hirudinaria* includes four species of Indian cattle leeches : *H. viridis, H. javanica, H. manilensis* and *H. granulosa*. **Late Prof. M.L. Bhatia** (1941) has given a detailed monographic morphology of *H. granulosa,* the common Indian cattle leech. The description given below however, applies broadly to other species as well.

Hirudinaria granulosa

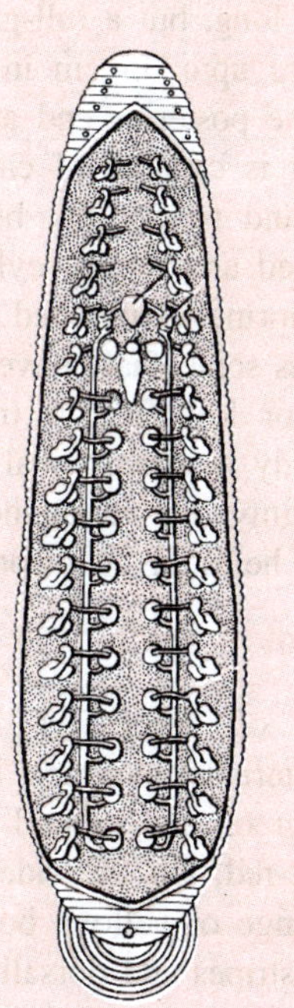

SYSTEMATIC POSITION

Phylum	Annelida
Class	Hirudinea
Order	Gnathobdellida
Family	Hirudinae
Genus	*Hirudinaria*
Species	*granulosa*

HABITS AND HABITAT (ECOLOGY)

Hirudinaria granulosa, the Indian cattle leech, is found in India, Mayanmar, Shri Lanka, Pakistan and Bangladesh. It occurs in freshwater ponds, lakes, tanks, swamps and slow streams, where it either swims by vertical undulations or grips objects with its suckers and moves by looping. Like most ectoparasitic leeches, it has a blood-sucking (**sanguivorous**) habit, feeding on the blood of fish and frogs and also of cattle and men, who happen to enter the water inhabited by it.

Though hermaphroditic, leeches copulate to bring about cross-fertilization. Eggs are deposited in cocoons. Development is direct, *i.e.*, there are no free larval stages in the life history.

EXTERNAL MORPHOLOGY

[I] Shape and Size

Hirudinaria has a soft, vermiform, elongated and dorso-ventrally flattened body, ordinarily 10 to 15 cm long, but a full-grown specimen may even measure upto 35 cm in length. Body is broader near the posterior end and narrow at the anterior end. It is capable of causing great alterations in form and proportion, being ribbon-shaped when extended and almost cylindrical when contracted. In a normally-stretched leech, the dorsal surface remains somewhat convex with the ventral surface more or less flat. A transverse section through the body is almost oval in outline. Skin is moist and slimy due to abundant secretion of mucus, which helps in cutaneous respiration.

[II] Colouration

Body of leech is brightly coloured with characteristic markings. Dorsal surface is coloured olive-green and ventral surface orange-yellow or orange-red. On the sides are conspicuous stripes of orange or yellow, bounded ventrally by broad black stripes and dorsally by black spots arranged

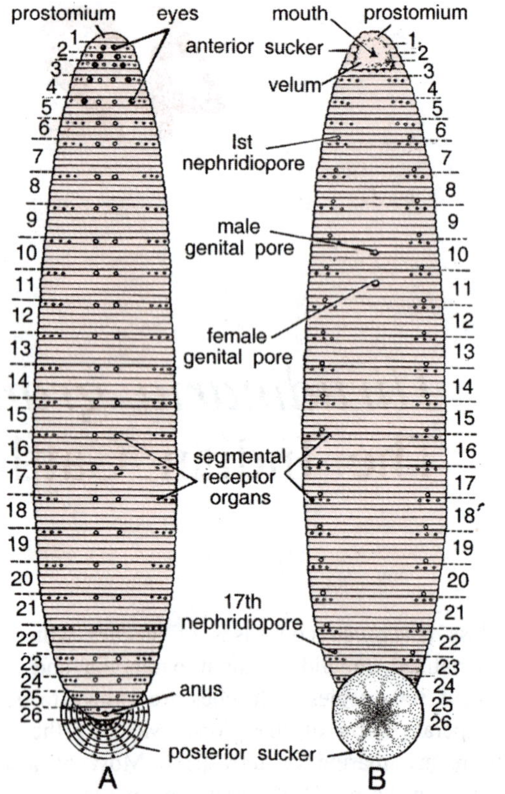

Fig. 1. *Hirudinaria.* External features. **A.** Dorsal view. **B.** Ventral view.

in a series on the second and fifth annuli of each segment.

On the dorsal side, a continuous or interrupted **median stripe** extends from first segment to anus. On either side of the median stripe run four interrupted wavy stripes. Along the course of the outer two wavy stripes are present supra-marginal spots on the first and fifth annuli of each segment. These spots are of taxonomic significance. Individual variations in colour also exist.

[III] Segmentation

Body of leech is metamerically divided into 33 segments or somites. Except the first two and the last seven, each segment is further superficially subdivided into **rings** or **annuli** by closely-set grooves or furrows. A typical segment possesses five annuli. Segments with less than five annuli are referred to as incomplete. A temporary

clitellum is formed by segments 9th, 10th and 11th, during breeding season.

[IV] Receptors

Surface of each annulus is divided by fine longitudinal furrows into rectangular areas, each bearing a minute and elevated sensory papilla, the **annular receptor**. There are about 18 such receptors on dorsal and 18 on ventral side of each annulus. Besides these, there are large sensory papillae, called **segmental receptors** or **sensillae**, 4 pairs on dorsal and 3 on ventral side in the first annulus of each segment. A pair of **eyes** is borne on dorsal surface by the first annulus of each of the five segments, which are termed the **ocular segments**. The 5 pairs of eyes appear as a semicircle of black dots at the anterior end.

[V] Suckers

Each end of the body bears a hollow muscular organ, the *sucker* :

1. Anterior sucker. Anterior or **cephalic sucker** comprises the prostomium and three anterior body-somites. It is oval in outline with a ventrally directed, cup-like hollow, called the **pre-oral chamber**, at the bottom of which lies the **mouth**.

2. Posterior sucker. Posterior end of body also bears a circular and highly muscular disc, the **posterior** or **anal sucker**, also directed downwards. It is formed by the complete fusion of the last seven body segments (27 to 33) arranged in concentric rings.

Both the suckers are primarily meant for adhesion and locomotion. The anterior sucker also helps in feeding.

[VI] External Apertures

(i) **Mouth** is a narrow, triradiate aperture opening centrally in the preoral chamber of anterior sucker. (ii) **Anus** is a small aperture opening mid-dorsally on the 26th segment at the base of posterior sucker. (iii) Nephridia open to the exterior by 17 pairs of **nephridiopores**, of which one pair lies ventrally on the last annulus of each of the segments from 6 to 22. (iv) **Male genital pore** is a mid-ventral opening, situated in a groove

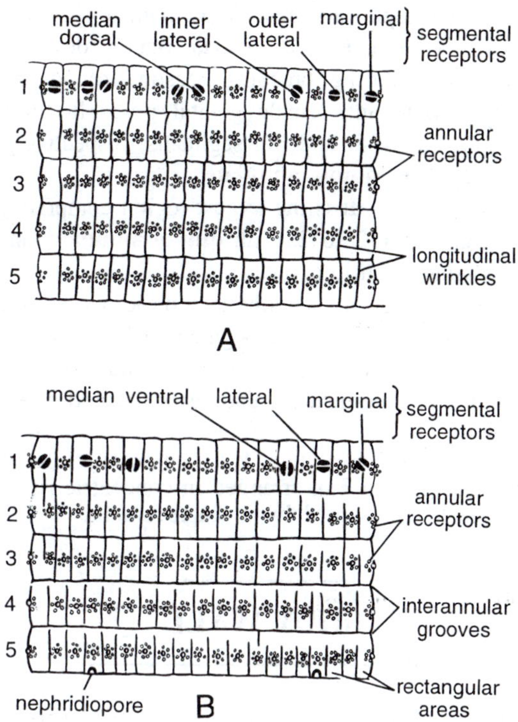

Fig. 2. *Hirudinaria..* A segment from middle region of the body showing the receptors. **A.** Dorsal view. **B.** Ventral view.

between second and third annuli of 10th somite. Sometimes, a thread like **penis** is seen protruding through this aperture. (v) **Female genital pore** lies midventrally in between the second and third annuli of llth somite.

[VII] Body Divisions

Body of leech can be distinguished into following six regions :

1. Cephalic. It is composed of first 5 or the ocular segments. It includes prostomium, anterior sucker, mouth and eyes. Nephridiopores are absent. The segments one and two are **uniannulate**, segment three is **biannulate** and segments four and five are **triannulate**. The first three segments with prostomium form the **upper lip**.

2. Pre-clitellar. It is formed by three segments (6,7 and 8), all bearing nephridiopores. Segment six is **triannulate** and the remaining two are **quinquannulate**, each having five annuli.

3. Clitellar. It comprises of three segments (9,10 and 11), which possess highly glandular walls and nephridiopores. A permanent **clitellum** does not exist in *Hirudinaria* but it develops temporarily during breeding season around this region. Male and female genital pores lie mid-ventrally on 10th and llth segments, respectively. Segments are **quinquannulate**.

4. Middle. It is the largest region comprising 11 segments (12 to 22), all **quinquannulate** and possessing nephridiopores.

5. Caudal. It is without nephridiopores and includes four incomplete segments (23 to 26), of which the 23rd is **triannulate** and the rest **biannulate**. Segment 26 also bears the mid-dorsal **anal aperture**.

6. Posterior sucker. It is composed of seven segments (27 to 33), arranged in concentric rings, and each represented by a single annulus.

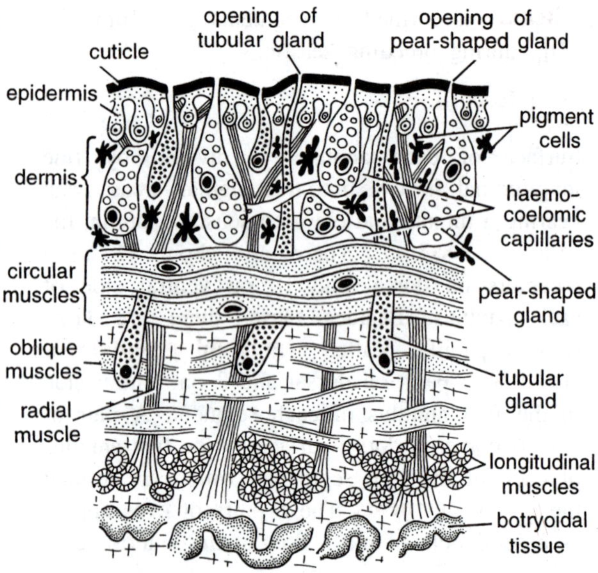

Fig. 3. *Hirudinaria.* T.S. of a portion of body wall.

BODY WALL

Body wall of leech includes five layers : (i) cuticle, (ii) epidermis, (iii) dermis, (iv) muscular layer and (v) botryoidal tissue.

[I] Cuticle

It is the outermost, thin, delicate, transparent, colourless, and moderately elastic protective covering, secreted by the underlying epidermis. It is slightly thicker on dorsal surface and is perforated all over by openings of epidermal glands. It is cast off from time to time in the form of thin transparent shreds especially after a heavy meal, and renewed. It is also shed if the leech is placed in dirty water.

[II] Epidermis

Lying below cuticle is the **epidermis**, which consists of a single layer of hammer-shaped cells. These cells are broad, flattened and pentagonal towards the outer, and narrower towards their inner ends. Outer ends fit closely while the nucleated inner ends have interspaces containing fibrous connective tissue, pigment cells and haemocoelomic capillaries, forming a vascular and respiratory membrane.

Some epidermal cells become modified to form unicellular **glands** and multicellular **receptor organs**.

1. Epidermal glands. Several kinds of glands are found in the epidermis.

(a) ***Slime glands*** are distributed all over the body. They secrete a slimy mucus which covers the entire body. Their **pear-shaped** or **tubular** body extends deep into dermis. It contains a nucleus and numerous secretory granules in its cytoplasm. From it, a narrow, neck-like and cylindrical prolongation leads outwards to open to the exterior by a minute aperture.

(b) **Sucker glands** occur in clusters in the two suckers. They are pear-shaped in the anterior sucker and rounded or spherical in the posterior-sucker. Their secretion clears and smoothens the surface of substratum for adhesion.

(c) ***Prostomial glands.*** These per shaped glands forms a series of distinct groups in the prostomium. They manufacture the two **plugs** of the cocoon or egg-case.

(d) ***Clitellar glands***, found in clitellar region (segments 9 to 11), become active during breeding season. They are of two special types : **Chitogenous glands** form the wall of cocoon, and **albumen glands** secret albumen that fills up the cocoon for nourishment of the developing embryos.

2. Epidermal receptors. These have been described latex with other sense organs.

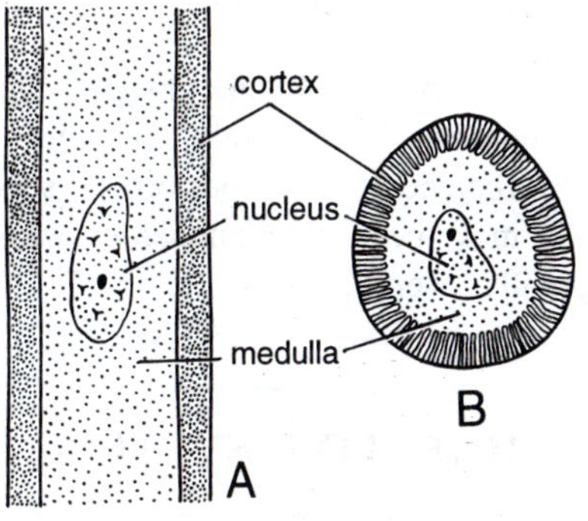

Fig. 4. *Hirudinaria.* A muscle fibre in L.S. (A) and T.S. (B).

[III] Dermis

It lies between epidermis and muscles and consists of a network of fibrous connective tissue. Besides, it contains connective tissue cells, pigment and fat cells, scattered muscle fibres, haemocoelomic capillaries and basal portions of epidermal slime glands.

[IV] Muscles

Musculature is well developed and forms the largest part of body wall. Muscles form either continuous layers or separate bundles, (i) **Circular muscles** form the outermost thin layer lying beneath the dermis. Next to them is a layer of (ii) **oblique muscles**. Beneath them are found many layers of (iii) **longitudinal muscles**, which form the greatest part of musculature. Besides, a number of isolated muscle bundles run either diagonally through body or in straight lines from dorsal to ventral surface. A pair of such (iv) **dorsal ventral muscles** run vertically in each segment by the sides of alimentary canal, while those lying near the flanks are termed the (v) **vertical muscles**. Their ends spread out and terminate just beneath the epidermis. Numerous (vi) **radial muscles** run from the interior of body to the surface.

A **muscle fibre** of leech is an elongated, fusiform and tubular cell consisting of an outer contractile **cortex** or **myoplasm** enclosing an inner, non-contractile **medulla** or **sacroplasm**, which is composed of finely granular protoplasm containing an oval nucleus.

[V] Botryoidal Tissue

Beneath the longitudinal muscles and immediately surrounding the alimentary canal is found the characteristic **botryoidal tissue**. This tissue fills the entire coelom except a few spaces termed the **haemocoelomic spaces**. It is composed of a network of large branching tubular cells arranged end to end. Walls of cells are loaded with a dark brown pigment while their intracellular canals contain a red fluid (haemocoelomic fluid) in life. It is probably excretory in function.

LOCOMOTION

Leeches performs two types of movements : (i) **looping** or **crawling movement** on a substratum, and (ii) **swimming** in water.

[I] Looping or Crawling Movements

These are performed with the help of muscles and suckers which serve for attachment (Fig. 5).

First, the leech fixes its posterior sucker firmly onto the substratum, aided by the slimy secretion of sucker glands (A). This initiates contraction of circular muscles and relaxation of longitudinal

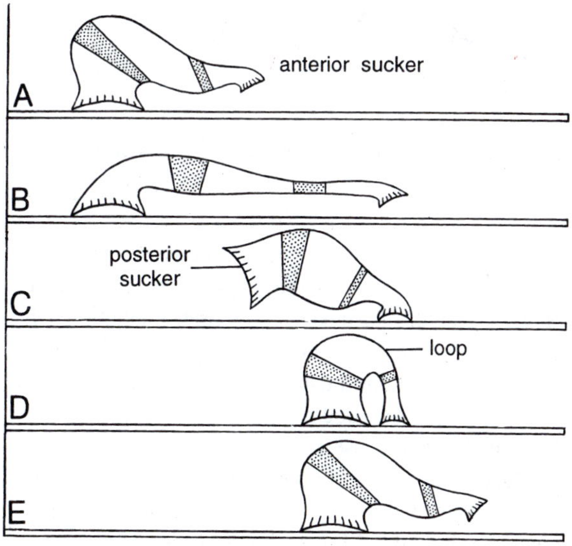

Fig. 5. *Hirudinaria* Stages of looping locomotion.

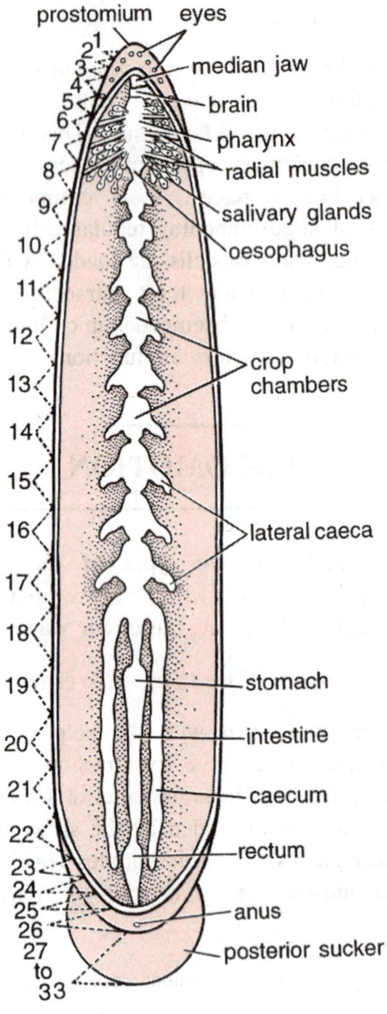

Fig. 6. *Hirudinaria.* Alimentary canal dissected in dorsal view.

muscles so that anterior part of body is extended forward as far as possible (B). Next, the leech fixes its anterior sucker. Now the relaxation of circular muscles and contraction of longitudinal muscles release the posterior sucker from its hold and shortens, the body (C). As posterior end is brought forward close to the anterior end, the body is raised up in the middle forming a loop (D). Once more the posterior sucker is fixed, anterior sucker released and the whole process repeated (E). This is a type of activity often called 'leech-like' locomotion.

[II] Swimming Movements

Leech swims very actively and gracefully in water. During swimming, the body becomes dorso-ventrally flattened and performs successive undulating movements. The undulating waves pass longitudinally over the body.

DIGESTIVE SYSTEM

[I] Alimentary Canal

Alimentary canal of leech is a complete straight tube of varying diameter, extending from **mouth** to **anus**. It is differentiated into **buccal cavity, pharynx, oesophagus, crop, stomach, intestine** and **rectum**. Buccal cavity, pharynx and rectum are lined internally by ectodermal epithelium covered by cuticle. They form **stomodaeum** and **proctodaeum**, respectively, while the rest of gut is lined internally by endodermal epithelium and forms the **mid-gut** or **mesenteron**.

Greater part of alimentary canal (crop) is concerned with storage of uncoagulated ingested blood, while only a small portion is concerned with digestion and absorption. This modification is strictly in accordance with the sanguivorous habit of leech.

1. Pre-oral chamber and mouth. The pre-oral chamber is a cup-like depression on the ventral side of oral sucker. Its roof is formed by a membrane-like **velum** bearing a triradiate opening, the **mouth**, in the centre. One ray or **chink** of mouth is medio-dorsal, while the other two are ventro-lateral. Velum forms three **lips** around the mouth, one between two adjacent rays, so that there are two dorso-lateral and one mid-ventral lip.

2. Buccal cavity. Mouth leads into buccal cavity, which is a short chamber behind velum. In its mucous lining are embedded three cresentic **jaws**, arranged in a triangle. There is one jaw behind each ray or chink of mouth, so that one jaw is mid-dorsal and the other two ventro-lateral in position. Each jaw is a laterally compressed muscular cushion, covered with fine cuticle which

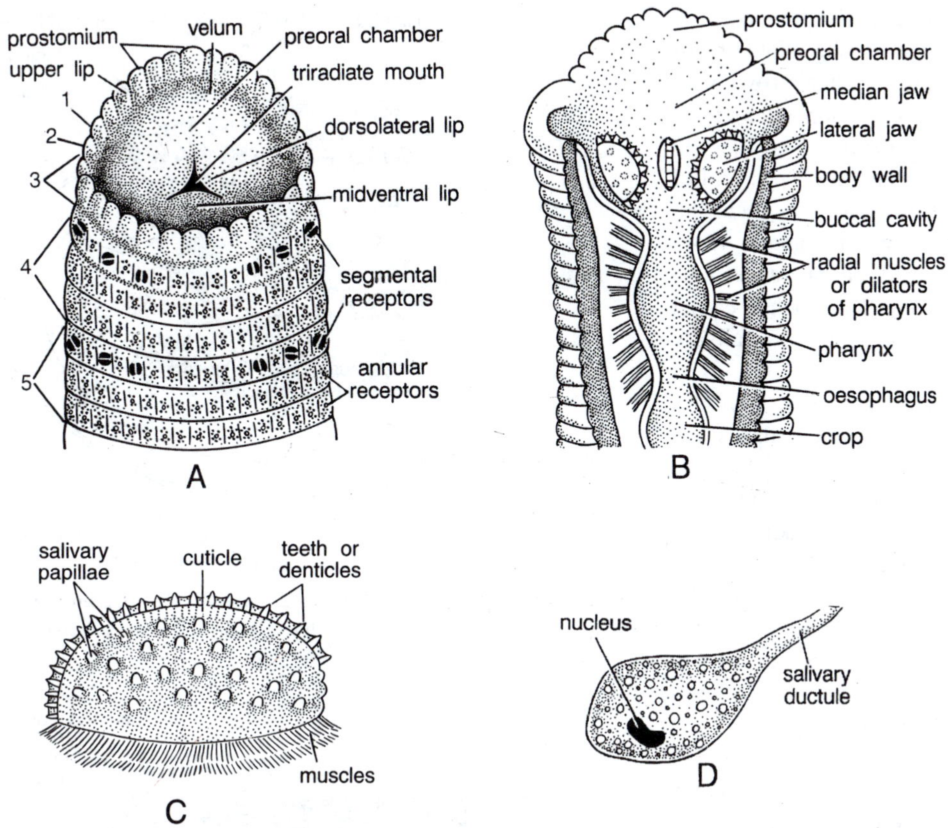

Fig. 7. *Hirudinaria.* **A.** Cephalic region (anterior end) in ventral view. **B.** Anterior end dissected ventrally to show the 3 jaws in buccal cavity and pharynx. **C.** A single jaw in lateral view. **D.** A single salivary gland with its duct.

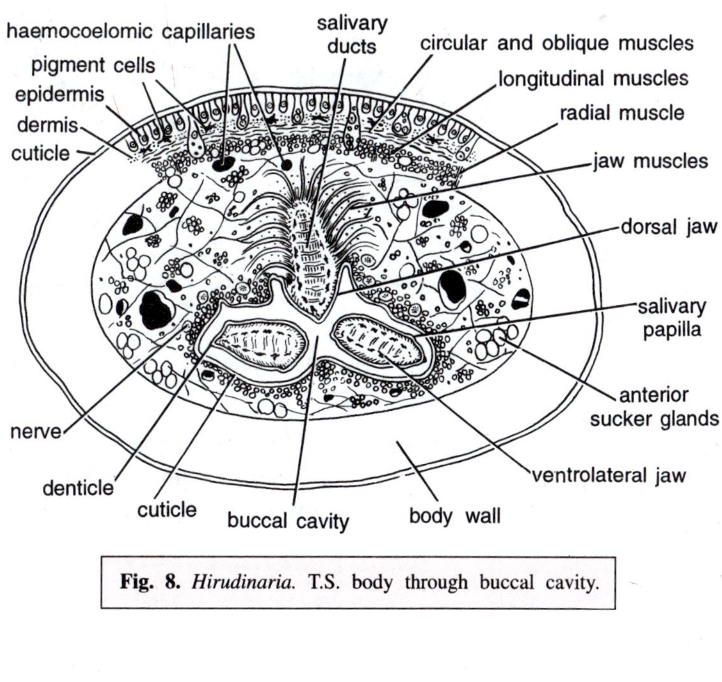

Fig. 8. *Hirudinaria.* T.S. body through buccal cavity.

is thickened at the free edge to form a ridge bearing minute **teeth** or **denticles** in a single row. Such jaws are termed **monostichodont**. Number of denticles is 103-128 on the median jaw and 85-115 on a lateral jaw. On each flat side of a jaw are about 42-45 button-shaped protuberances, the **salivary papillae**, each .bearing numerous openings of salivary glands.

Jaws are moved by muscles and work as semicircular saws. Acting together, they produce the characteristic triradiate bite or Y-shaped wound in the skin of host.

3. Pharynx. Buccal cavity leads into a highly muscular pharynx. It is an oval sac extending from the 5th to 8th segments. Its inner surface bears longitudinal folds. Externally, it is surrounded by large masses of unicellular pyriform **salivary glands**, whose ductules, in bundles, run anteriorly and enter the three jaws to open on their salivary papillae. Secretion of these glands contains **hirudin** or **anticoagulin** which prevents coagulation of blood while the leech is feeding. Numerous **radial muscles** connect the wall of pharynx with the body wall.

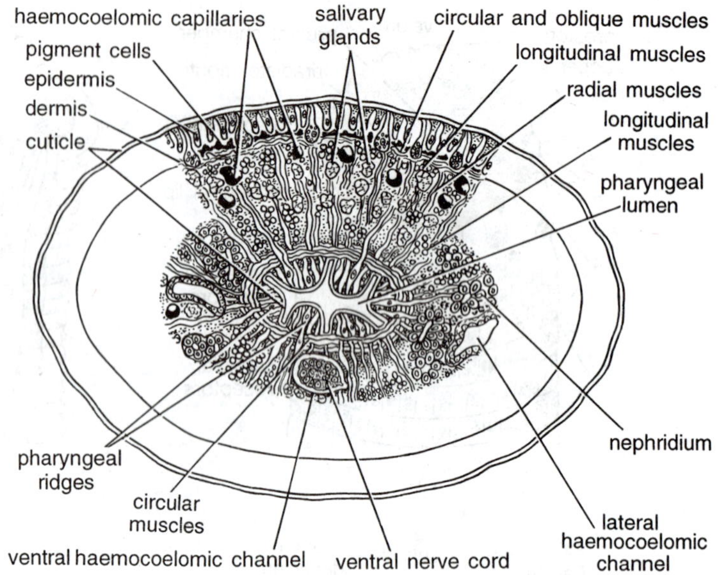

Fig. 9. *Hirudinaria.* T.S. body through pharyngeal region.

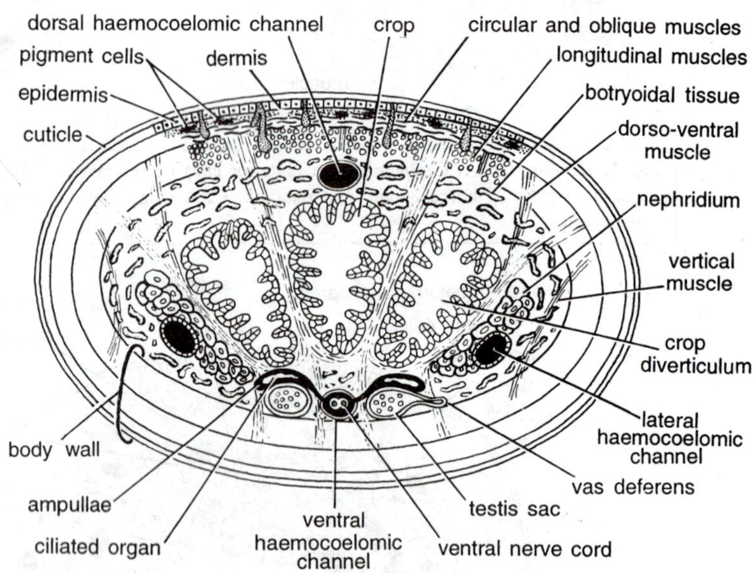

Fig. 10. *Hirudinaria.* T.S. body through crop.

4. Oesophagus. It is a remarkably short and narrow tube through which pharynx leads into **crop**. It has a very narrow lumen and a much folded epithelial lining.

5. Crop. It comprises the largest portion of alimentary canal. It is a thin-walled extensive tube, occupying about two-third of the visceral space and extending from 9th to 18th segment. It is divided by narrow constrictions into a series of 10 (some times 11) chambers, communicating by more or less circular apertures surrounded by sphincters. Each chamber

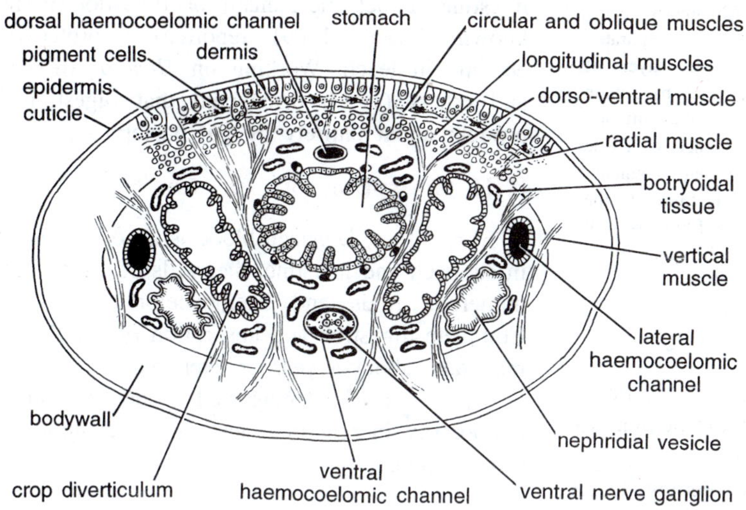

Fig. 11. *Hirudinaria.* T.S. body through stomach.

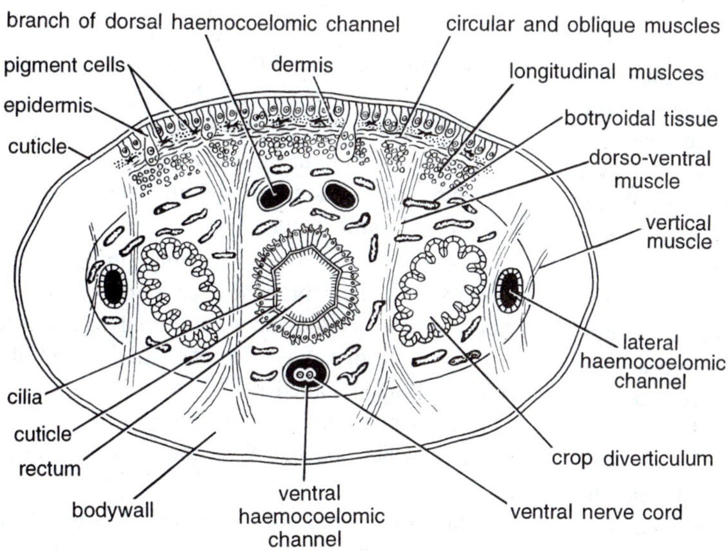

Fig. 12. *Hirudinaria.* T.S. body through rectum.

consists of a small anterior part and a broad posterior part, bearing a pair of lateral backwardly directed blind outgrowths, or storing pockets, called **caeca** or **diverticula**. Crop chambers and caeca gradually increase in size towards the posterior side. The last or 10th chamber is the biggest, and its caeca extend as two greatly elongated blind sacs along the intestine, as far as, or even beyond the 22nd segment.

Crop and its diverticula are capable of great dilation to store enormous quantity of blood which can be digested slowly.

6. Stomach. Posteriorly, the last chamber of crop ends into a funnel-shaped tube. It leads through a sphinctered aperture into a small heart-shaped stomach, lying in the 19th segment. Mucous lining of stomach is thrown into anastomosing transverse folds.

7. Intestine. Posteriorly, the stomach leads into intestine. It is a thin-walled, straight narrow tube running from 20th to 22nd segment. Its inner lining is thrown into numerous spiral folds and villi-like processes which increase its absorptive surface.

8. Rectum. Intestine is followed by a somewhat dilated and thin-walled rectum lying from 23rd to 26th segment. It runs slightly upwards and opens to the exterior through a small **anus** placed mid-dorsally on the 26th segment. Rectum has an internal ciliated cuticular lining and is without folds.

Histology of alimentary canal. Alimentary canal consists of a layer of **columnar epithelial cells** separated by a **basement membrane** from an outer layer of **connective tissue**. Columnar epithelium is lined by cuticle in fore and hind guts. It is also ciliated in hind gut and contains scattered goblet cells in crop. Connective tissue layer contains haemocoelomic capillaries (specially abundant in stomach and intestine) and circular muscle fibres. In the pre-oral chamber it contains circular muscle fibres as well as longitudinal muscle fibres, while in the pharynx it contains circular, longitudinal and radial fibres.

[II] Food, Feeding and Digestion

1. Food. *Hirudinaria* has a **sanguivorous** habit, sucking blood of cattle and other domestic animals which visit waters harbouring them.

2. Feeding or ingestion. A feeding leech firmly adheres to the victim by its posterior sucker and then applies its anterior sucker. The three jaws are protruded through mouth and moving like saws, make the characteristic triradiate or Y-shaped painless incision in the skin of host. Cutaneous vessels are laid open and the oozing blood is sucked in with the help of pre-oral chamber, buccal cavity and muscular pharynx. Contraction of radial muscles, around pharynx, dilates the pharyngeal cavity resulting in the suction of blood. The active substance, **hirudin** or **anticoagulin**, present in salivary secretion poured on wound, prevents the clotting of blood which continues to flow regularly for some time, even after the leech has detached itself.

A leech can suck, at a single meal, several times its own weight of blood, which is enough for the animal for several months or even a year. After feeding it drops off its victim, seeks a dark and concealed shelter and digests at leisure.

3. Haemolysis and digestion. The ingested blood is stored uncoagulated in crop and its diverticula. Blood gets **haemolysed** in crop, *i.e.*, the blood corpuscles break down and their haemoglobin goes into solution. Water from haemolysed blood is absorbed and blood becomes jelly-like and dark red in colour.

Blood from crop passes drop by drop through the sphinctered aperture into stomach, where it turns green. Digestion takes place in stomach and intestine. Exact mechanism of digestion is not known. **Heise** (1909) reported a proteolytic enzyme in leech. Working on *Hirudo*, **Bupsing** and others (1953) discovered that digestion is brought about entirely by gut bacteria.

[III] Absorption and Egestion

1. Absorption. Digested blood is absorbed slowly in stomach and intestine, the walls of which have numerous folds and haemocoelomic branches. Complete digestion and absorption of a full meal may take about a year or even more.

2. Egestion. Undigested food is stored temporarily in rectum. Egestion, is accomplished through anus.

COELOM

A true perivisceral **coelom** or a well-defined **body cavity** around alimentary canal, as found in earthworm and *Nereis* is absent in leeches. It is greatly obliterated and filled with the characteristic **botryoidal tissue**. However, the original coelom in *Hirudinaria* is represented by four **longitudinal channels**, their branches and capillaries, and numerous **spaces**, some of definite, but others of doubtful coelomic nature.

The **coelomic fluid**, running through the channels and their branches, contains colourless amoeboid **corpuscles** and is coloured red due to **haemoglobin** dissolved in it. This blood-like coelomic fluid is called the **haemocoelomic fluid** and the channels are termed the **haemocoelomic channels**, which together form the **haemocoelomic system.**

Thus, in *Hirudinaria,* there are no true blood vessels, their function being taken up by the haemocoelomic system.

Coelomic spaces enclosed within testis sacs and ovisacs, around vasa deferentia and lying in between viscera, contain a colourless fluid without respiratory pigment or haemoglobin. These spaces represent the true, coelom in *Hirudinaria.*

HAEMOCOELOMIC SYSTEM

[I] Haemocoelomic Channels

Haemocoelomic system of *Hirudinaria* mainly consists of four longitudinal sinuses or channels : one **dorsal**, one **ventral** and two **lateral**, and their **branches**. As already referred to, these channels are not true blood vessels but remnants of original coelom. Dorsal and ventral channels have non-contractile thin walls made of connective tissue and coelomic epithelium. Lateral channels secondarily acquire contractile and thick muscular walls and appear like true blood vessels.

Through the haemocoelomic channels circulates the red-coloured **haemocoelomic** fluid containing amoeboid corpuscles and dissolved haemoglobin. The amoeboid corpuscles or coelomocytes are phagocytic in nature.

1. Dorsal channels. It runs mid-dorsally, below body wall, attached to the alimentary canal. It takes up a sinuous course. Anteriorly it runs up to 6th segment, where it breaks up into smaller branches and capillaries, which extend into the first five segments. Posteriorly, it bifurcates in 22nd segment, the two branches passing ventrally around rectum to join the posterior dilatation of ventral channel.

Dorsal channel has a thin, non-contractile, wall and devoid of valves and its haemocoelomic fluid runs from posterior to anterior side. It is a **distributing** channel. It gives out two types of branches.

(a) Dorso-laterals. Two pairs of dorso-lateral branches arise ventro-laterally from dorsal channel in each, segment. Each branch runs outwards to form a capillary plexus in dorsal and dorso-lateral regions of the body wall.

(b) Dorso-intestinals. These are numerous small branches arising mid-ventrally from dorsal channel, all along its length. These supply the gut wall.

2. Ventral channel. It runs mid-ventrally beneath alimentary canal, from one end of body to other, along a straight course. It is wider than dorsal channel and encloses the entire central

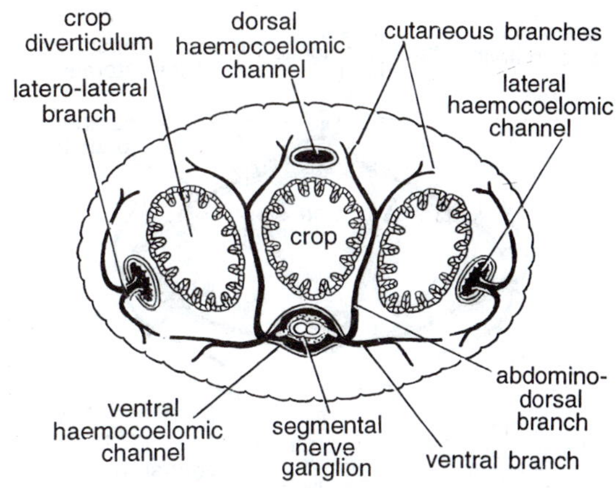

Fig. 13. *Hirudinaria.* T.S. body through abdomino-dorsal branches.

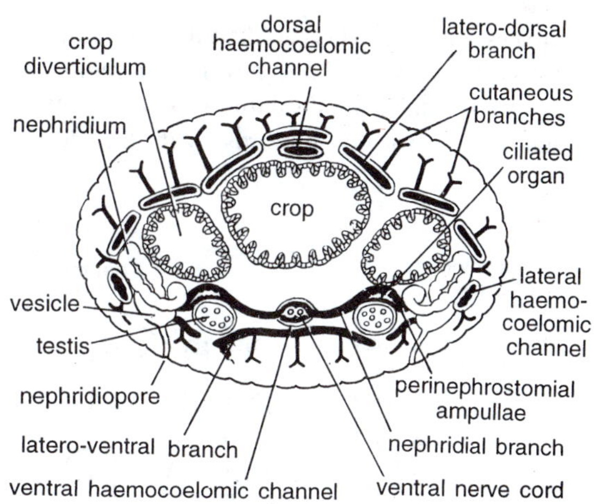

Fig. 14. *Hirudinaria* T.S. body through nephridial branches.

nervous system including nerve ring and ventral nerve cord. It is somewhat dilated around the segmental nerve ganglia and the terminal ganglionic mass.

Like dorsal channel, the ventral channel has a non-contractile and is also devoid of valves and acts as a distributing channel. Haemocoelomic fluid in it flows from anterior to posterior side. It gives out two pairs of branches in each segment.

(a) First pair. First pair or **cutaneous branches** are given out at the level of segmental nerve ganglion. Branch of each side at once

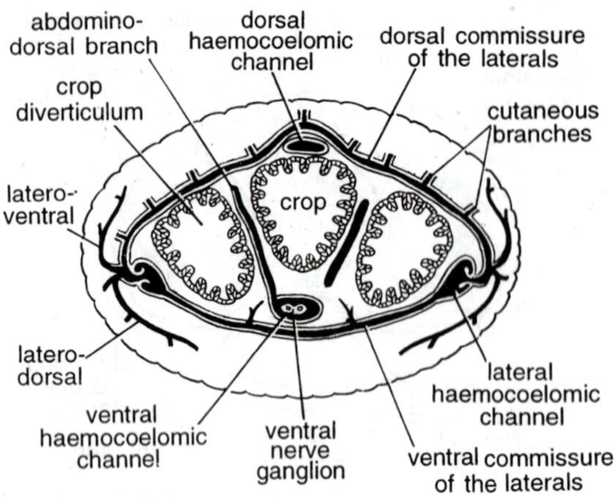

Fig. 15. *Hirudinaria.* T.S. body through dorsal and ventral commissures of lateral haemocoelomic channels.

bifurcates into a **ventral branch**, forming a capillary network in the ventro-lateral region of body wall, and an **abdomino-dorsal branch**, which runs vertically upwards with the dorso-ventral muscle and forms dorso-lateral cutaneous plexus.

(b) Second pair. Second pair or **nephridial branches** are given out just behind the segmental ganglion of ventral nerve cord. Nephridial branch of each side runs outwards, and reaching over the testis sac of its side, it widens into two or three closely set saccules, the **peri-nephrostomial ampullae**, containing a **ciliated organ**. The branch then continues outwards to supply the body wall, sending on its way, a small branch to the nephridium.

Ciliated organ manufactures **coelomic corpuscles** for the haemocoelomic system and its cilia are believed to help in the circulation of haemocoelomic fluid.

There are only 11 pairs of nephridial branches, one pair in each of the segments 12 to 22, which contain testis sacs.

3. Lateral channels. There are two lateral haemocoelomic channels placed one on either side of alimentary canal. They are wide and uniform in diameter, with thick muscular and contractile walls and possess valves like true blood vessels. Haemocoelomic fluid flows in them from behind forwards. They are **collecting** as well as **distributing** channels.

In each segment, each lateral channel receives on its outer side two branches, **latero-lateral** and **latero-dorsal**, and gives off on the inner side one branch, the **latero-ventral**.

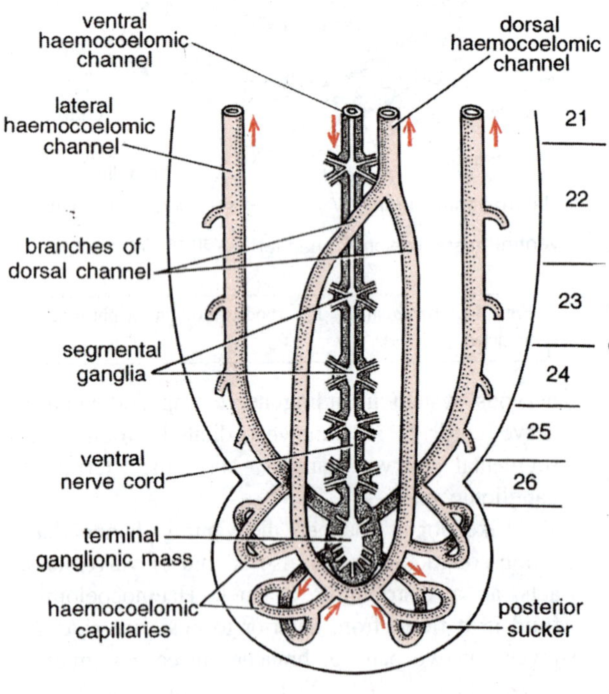

Fig. 16. *Hirudinaria.* Hind end of the body showing union of the four longitudinal haemocoelomic channels.

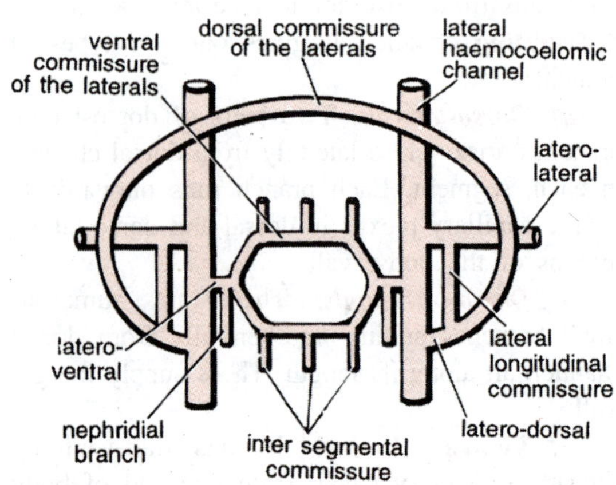

Fig 17. *Hirudinaria.* Diagrammatic representation of the lateral channels and their branches in a segment in dorsal view.

(a) Latero-lateral. A short latero-lateral is formed by branches from lateral region of body wall and nephridia. It joins the lateral channel at the level of nephridial vesicle.

(b) Latero-dorsal. A large **latero-dorsal** is formed by branches from dorsal and dorso-lateral regions of body wall, gut wall and nephridium. It joins the lateral channel at the level of main lobe of nephridium. Two latero-dorsals of opposite sides are connected by a **transverse loop** above the dorsal channel. There are seventeen such loops, called the **dorsal commissures of the lateral channels**, one in each segment from segments 6 to 22. Latero-dorsal is also connected with the latero-lateral of its own side by a longitudinal **lateral commissure**.

Latero-laterals and latero-dorsals are collecting branches and their openings into lateral channels are guarded by valves.

(c) Latero-ventral. A latero-ventral arises from the inner side of lateral channel and at once gives off a branch to supply nephridium and ventro-lateral regions of body wall. Then it bifurcates into two diverging branches, **anterior** and **posterior**. These unite with their corresponding fellows of opposite side, beneath ventral channel, by the **ventral commissures of the laterals**, forming a characteristic rhomboidal figure. There are 18 such rhomboids, one in each segment from 6th to 23rd segment. Rhomboids of adjacent segments are also connected together by three longitudinal **intersegmental commissures**, one median and two lateral. Latero-ventrals supply branches to nephridia, ventral side of alimentary canal and reproductive organs.

Anteriorly, both lateral channels break up in 6th segment into capillaries, while posteriorly they open into the dilatation of ventral channel where all the four channels are in direct communication.

[II] Circulation of Haemocoelomic Fluid

Haemocoelomic fluid flows in a definite direction, from behind to front in dorsal and two lateral channels, and from in front to backwards in ventral channel. All the four channels are directly

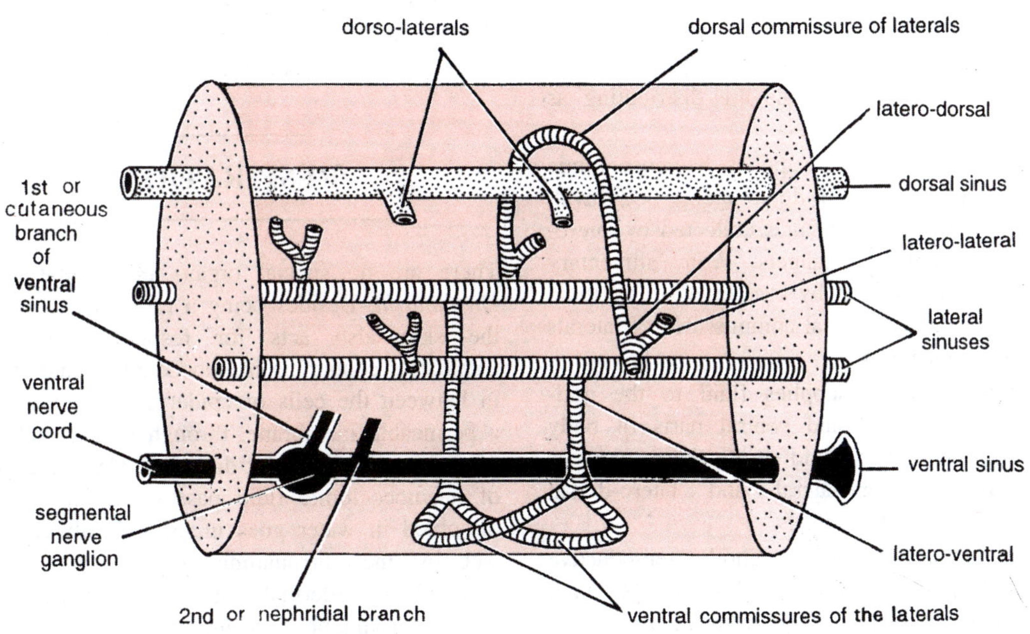

Fig. 18. *Hirudinaria.* Diagrammatic representation of the haemocoelomic system of a single segment in lateral view.

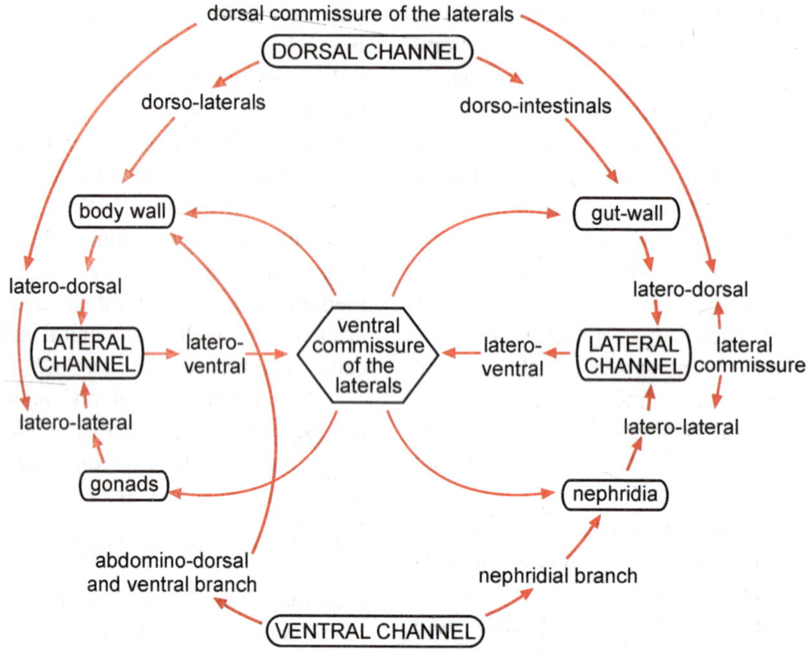

Fig. 19. *Hirudinaria.* Diagrammatic representation of the course of circulation of haemocoelomic fluid.

connected together posteriorly in 26th segment. Dorsal and ventral channels are only distributing channels. While two laterals are distributing as well as collecting channels.

The dorsal channel supplies haemocoelomic fluid to the dorsal and dorso-lateral body wall and gut wall from where it is collected by latero-dorsals into lateral channels. From alimentary canal, fluid is drained by two **latero-intestinals** that pour it through dorsal commissures of laterals into lateral channels.

Ventral channel supplies fluid to the mid-dorsal, ventro-lateral and ventral parts of body wall and nephridia. Fluid from these parts is collected by latero-lateral and latero-dorsal branches of lateral channels.

Lateral channels supply nephridia, reproductive organs, floor of gut and ventral body wall through latero-ventral branches. Fluid from all these parts is brought back to the lateral channels through latero-lateral and latero-dorsal branches.

RESPIRATION

There are no special organs of respiration in, *Hirudinaria.* Besides being a protective covering, the skin also acts for respiratory function. Capillaries containing haemocoelomic fluid extend in between the cells of epidermis, which acts as a permeable membrane, through which exchange of gases takes place by diffusion. Carbon dioxide of haemocoelomic fluid passes out and oxygen dissolved in water goes in. Skin is always kept wet by the surrounding water, while mucus secreted by epidermal slime glands also prevents it from drying on exposure.

EXCRETORY SYSTEM

Excretory system consists of 17 pairs of small coiled tubes, the **nephridia**, arranged segmentally, one pair in each segment from 6th to 22nd (Fig. 26). Nephridia are of two types : (i) **testicular**, and (ii) **pre-testicular**.

[I] Testicular Nephridia

Posterior 11 pairs of nephridia, lying one pair in each segment from 12th to 22nd, are termed testicular nephridia, due to the presence of a pair of testis sacs in each of these segments. A typical testicular nephridium is a horse shoe-shaped structure traversed by a complicated system of canals. It consists of 6 parts : (i) **main lobe**, (ii) **vesicle** and **vesicle duct**, (iii) **apical lobe**, (iv) **inner lobe**, (v) **initial lobe**, and (vi) **ciliated organ**.

1. Main lobe. Main lobe, forming the horse shoe proper, lies in a ventro-lateral position between two adjacent caeca of crop. It forms the major part of nephridium and consists of two unequal limbs. One limb is longer and anterior in position, and the other shorter and posterior. Cells of main lobe are big and polyhedral in shape.

2. Vesicle and the vesicle duct. A narrow **vesicle duct** arises from the inner ventral end of anterior limb of main lobe. It runs posteriorly to open into a large **bladder** or **terminal vesicle**, situated ventro-laterally behind the rest of nephridium. Vesicle is a large oval sac, with a non-contractile thin wall, internally lined by a ciliated epithelium. A short and narrow **excretory duct**, lined with a non-ciliated epithelium, leads from vesicle to open to the exterior through a rounded aperture, the **nephridiopore**. At its origin from vesicle, the excretory duct is provided with a sphincter muscle that controls the flow of excretory substances out of the vesicle. According to **M.L. Bhatia** (1940), bladder is lined by cilia. But, recent work by **B. Dev** has confirmed that the so-called cilia are in fact non-motile bacteria, the **nephridial microflora**, 2.8 to 7μ in length.

3. Apical lobe. Inner free end of posterior limb of main lobe is continued to form a stout **apical lobe**, lying antero-posteriorly beneath the crop. Its anterior end is slightly swollen and bent on itself like the handle of a walking stick. Its cells are big and traversed by regular intracellular canals.

4. Inner lobes. Extending between the anterior and posterior limbs of main lobe is a short **incurrent lobe** or **inner lobe**, which also runs forward along the outer side of apical lobe for about half of its length.

5. Initial lobe. It is a long, narrow, transparent and cord-like structure, formed of a single row of elongated tubular cells and closely coiled around the apical lobe. Its posterior end joins the main lobe, while its anterior end runs inwards and reaches over the testis sac of its own side, where it ends blindly close to the **perinephrostomial ampullae**. The intracellular canal of initial lobe gives off many diverticula in each cell.

6. Ciliated organ. Inside peri-nephrostomial ampullae lies a peculiar structure, the **ciliated organ**. It is suspended from the inner walls of ampullae by 4 to 5 **strands** or **trabeculae**. It corresponds to the **funnel** or **nephrostome** of a typical annelid nephridium, but is a greatly modified, and compound structure.

Ciliated organ consists of a spongy central **reservoir** and **ciliated funnels**. The reservoir contains the central mass of connective tissue cells which manufacture the coelomic corpuscles. Outer wall of central mass, made of a single layer of cells, bears numerous minute pores. A ciliated funnel fits into each pore on the outside. Each funnel is like an ear lobe, with about one-fourth of its margin incomplete. Funnel covered with outwardly directed cilia on its outer margin and inner surface. In the embryo, ciliated organ has a distinct cellular connection with the nephridium. But, in adult *Hirudinaria,* it loses the connection as well as excretory function and becomes a part of the haemocoelomic system for which it manufactures coelomic corpuscles.

Histology of nephridium. Lobes of nephridium consist of a mass of gland cells, traversed by an anastomosing system of **canals**. The initial lobe, formed by a single row of tubular cells arranged end to end, encloses a continuous

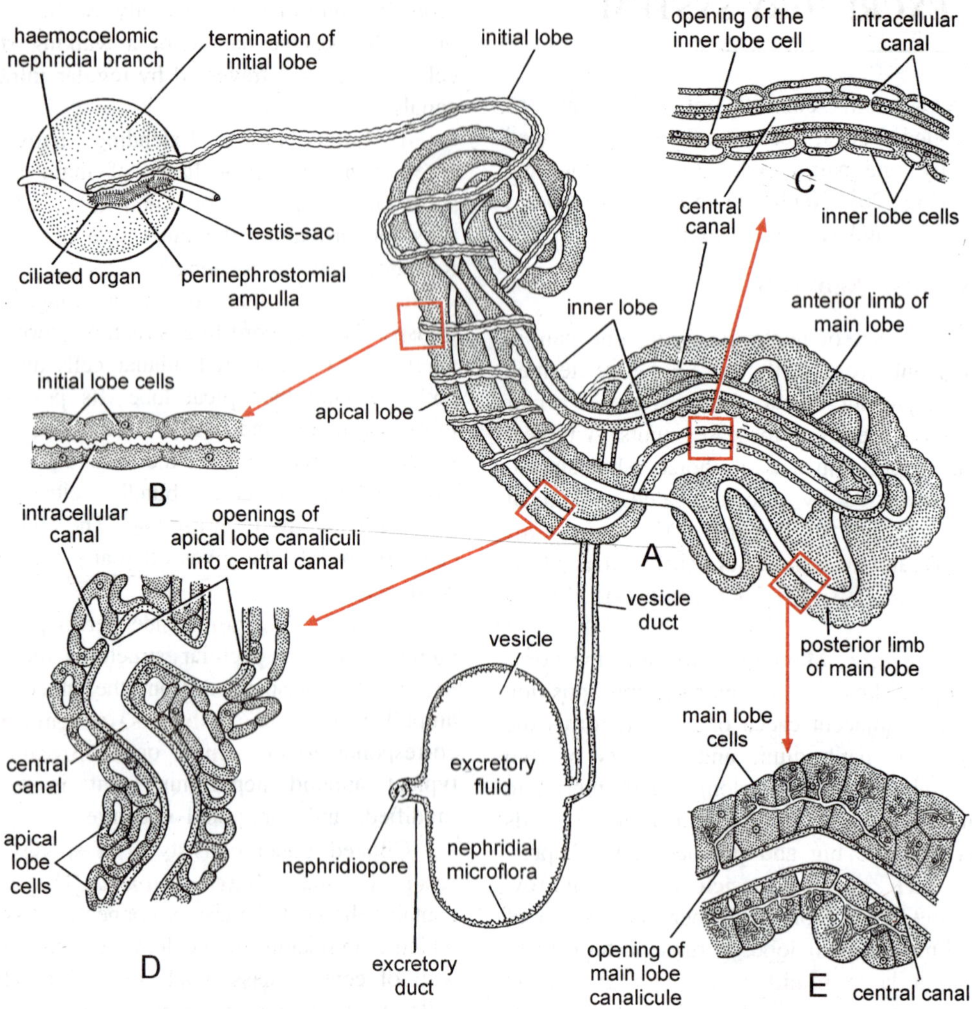

Fig. 20. *Hirudinaria.* **A.** Complete testicular nephridium. **B.** Portion of initial lobe in L. S. **C.** Portion of inner lobe in L.S. **D.** Portion of apical lobe in L.S. **E.** Portion of main lobe in L.S.

intracellular canal. It opens into the central canal of the main lobe just before its entry into the vesicle duct. All the other lobes are traversed by a conuinuous **central canal**, which starts at the anterior part of apical lobe and after taking up a complicated route through the various lobes, enters the vesicle duct near the anterior limb of main lobe. All along its length, the central canal receives numerous fine intracellular **canaliculi** from the surrounding cells. Canaliculi of adjacent cells anastomose to form an intricate meshwork throughout the nephridium.

[II] Pre-testicular Nephridia

First six pairs of nephridia are termed pre-testicular nephridia owing to their location in

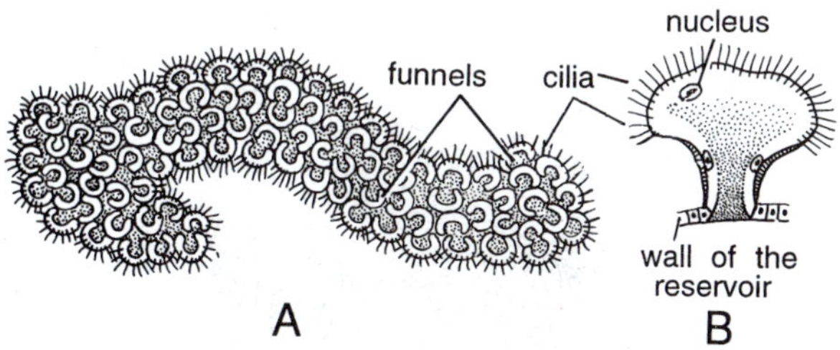

Fig. 21. *Hirudinaria.* **A.** Entire ciliated organ. **B.** A single funnel.

segments 6 to 11 without testis sacs themselves, but in front of those containing testis sacs. These nephridia resemble testicular nephridia in all respects except that their initial lobes end loosely in general connective tissue on their side of ventral nerve cord. There are no testis sacs, peri-nephrostomial ampullae and ciliated organs in their segments.

[III] Physiology of Excretion

The ciliated organ, as already noted, is separated from nephridium in the adult. It has no excretory function, but manufactures coelomic corpuscles or coelomocytes of the haemocoelomic system. Coelomocytes are phagocytic and engulf particulate matter, but their ultimate fate is uncertain.

Nephridium proper is truly excretory in function and serves to eliminate excess of water and nitrogenous wastes (mostly ammonia and small quantities of urea). Thus, leech is predominantly **ammonotelic**. Nephridium is profusely supplied with branches of haemocoelomic channels. Its gland cells separate waste products from haemocoelomic fluid. Excretory fluid is finally collected into terminal vesicle to be discharged to outside through the nephridiopore.

Several workers have also attributed an excretory function to the **botryoidal tissue**, the intracellular capillaries of which communicate with the haemocoelomic vessels.

NERVOUS SYSTEM

It is of the typical annelidan type, consisting of usual three parts : (i) **central**, (ii) **peripheral** and (iii) **sympathetic**.

[I] Central Nervous System

Entire central nervous system lies within the ventral haemocoelomic channel and consists of : (a) an anterior nerve ring, (b) a ventral nerve cord, and (c) a terminal ganglionic mass. Brain is rather smaller and segmental ganglia more distinct than those of earthworm.

1. Anterior nerve rin. It includes a small dorsal **brain**, formed of a pair of fused **cerebral** or **supra-pharyngeal ganglia**, lying dorsally above the anterior part of pharynx, just behind the median dorsal jaw in 5th segment. It is connected by a pair of short and stout lateral **peripharyngeal connectives**, one on either side of pharynx, with a **ventral sub-pharyngeal ganglionic mass** situated beneath pharynx in 5th segment. The more or less triangular sub-pharyngeal ganglionic mass is a composite structure made up of 4 pairs of embryonic ganglia fused together. Cerebral ganglia and sub-pharyngeal ganglionic mass represent the ganglia of first 5 segments.

2. Ventral nerve cord. Ventral nerve cord arises from the posterior end of sub-pharyngeal ganglionic mass and runs backwards along the mid-ventral line, from 6th to 26th segment. Although it appears single, it is really double as in earthworm. It carries, at wide intervals, 21 well-formed ganglia, each located in the first annulus of its own segment.

Nerve cord is made up of **nerve cells** and their processes or **nerve fibres**. Nerve cells are confined to ganglia, where they surround the nerve fibres. The entire nerve cord is covered with a protective sheath, the **neurilemma**.

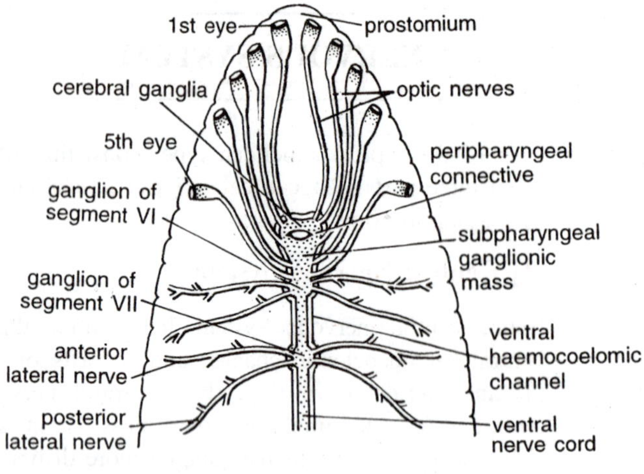

Fig. 22. *Hirudinaria.* Nervous system in the anterior end of body.

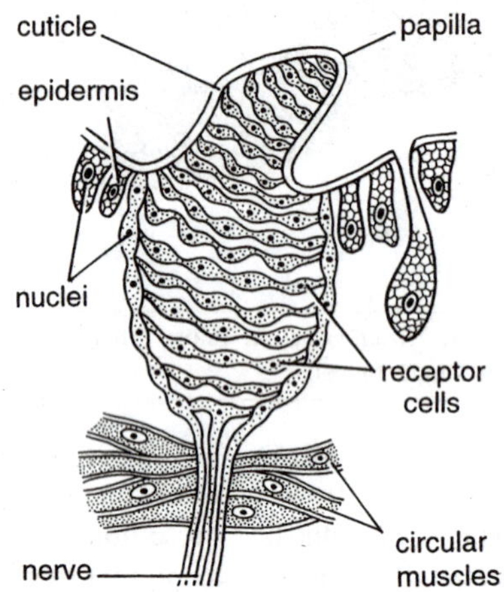

Fig. 23. *Hirudinaria.* An annular receptor in V.S.

3. Terminal ganglionic mass. Posteriorly, the ventral nerve cord ends in a large ovoid **terminal ganglionic mass**, situated within the posterior sucker. It is formed by the fusion of 7 pairs of embryonic ganglia of the last 7 segments that constitute the posterior sucker.

[II] Peripheral Nervous System

It consists of paired nerves arising from ganglia of central nervous system.

A pair of stout **optic nerves** arises anteriorly from brain or cerebral ganglia and runs forward to supply the 1st pair of eyes, prostomium and roof of buccal chamber. Four pairs of **optic nerves** arise laterally from sub-pharyngeal ganglionic mass and supply the 2nd, 3rd, 4th and 5th pairs of eyes, respectively. A few nerves arise ventrally from this ganglionic mass and supply the floor of buccal cavity, muscles of body wall, and segmental receptor organs of anterior 5 segments.

Each ganglion of ventral nerve cord gives off two pairs of nerves, the anterior laterals, and posterior laterals. **Anterior laterals**, arising anteriorly from ganglion, are stout nerves and branch to supply the nephridium vesicle, vas deferens, muscles of body wall, ventral receptors and two outer dorsal receptors of their own side. **Posterior laterals**, arising from the posterior part of ganglion, are also stout nerves which branch to supply the viscera, dorsal body wall, rest of dorsal receptors and testis sacs of their own side. Terminal ganglionic mass sends off several nerves supplying the receptor organs and other structures found in the posterior sucker.

3. Sympathetic or autonomic nervous system. It consists of an extensive **nerve plexus** lying beneath epidermis, within muscles and on gut wall. It joins, on one hand, with certain cells on both sides of peripharyngeal connectives, and on the other hand, with certain multipolar ganglion cells irregularly distributed over the entire plexus of gut wall.

SENSE ORGANS

Sense' organs or receptors of leech consist of modified epidermal cells. These are of four types (i) nerve endings, (ii) annular receptors, (iii) segmental receptors, and (iv) eyes.

[I] Free Nerve Endings

Free nerve endings occur all over the body, between epidermal cells, with their ganglion cells lying beneath the epidermis. These are probably **chemoreceptors**. detecting chemical changes in the surrounding water.

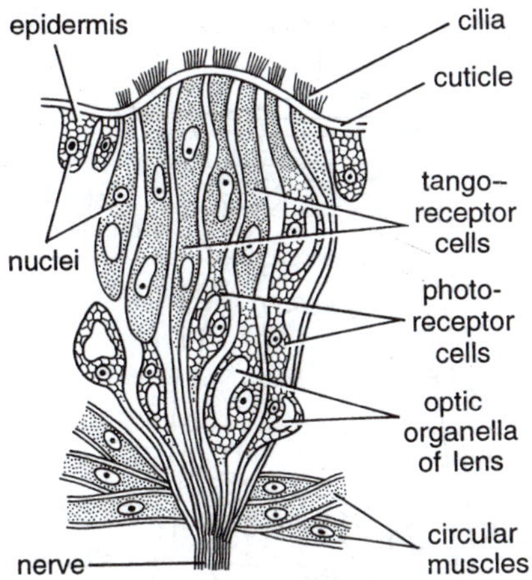

Fig. 24. *Hirudinaria.* A segment receptor in V.S.

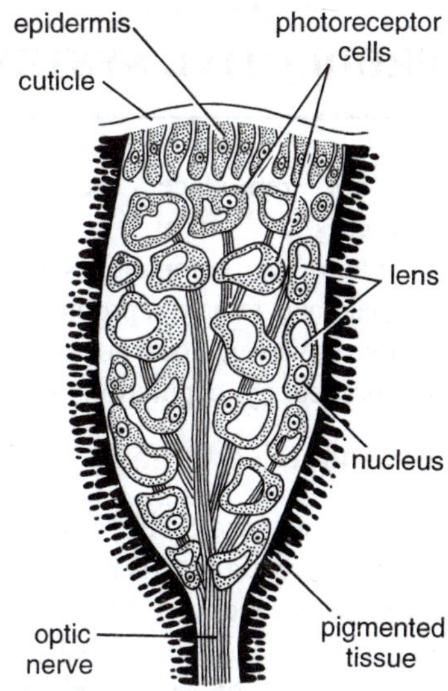

Fig. 25. *Hirudinaria.* Vertical section of an eye.

[II] Annular Receptors

Each annulus of body bears 36 very minute annular receptors, 18 on dorsal and 18 on ventral side, arranged in a transverse row. Each receptor projects as a minute papilla and consists of many flattened overlapping cells that receive their nerve supply from the lateral branches. They function as **tangoreceptors** or **tactile organs**.

[III] Segmental Receptors

These are small whitish elliptical papillae borne upon the first annulus of each body segment, 4 pairs dorsally and 3 pairs ventrally. Each receptor consists of two types of cells : (i) tactile cells or **tangoreceptors** and (ii) light-sensitive cells or **photoreceptors**. There are 5 to 10 long, slender tactile cells, separated from one another and provided with hair-like processes at their outer free ends. The light-perceiving or photoreceptor cells, found only in the dorsal receptors, contain a cresentic hyaline substance in their cytoplasm, the **optic organelle** or **lens**. Each receptor receives a nerve branch and functions both as **tangoreceptor** and **photoreceptor**.

[IV] Eyes

There are 5 pairs of eyes, arranged as a semicircle of black spots, along the dorsal margin of the anterior sucker, one pair in the first annulus of each of the first 5 segments. Each eye is in the form of a cylinder or cup with its long axis perpendicular to body surface. Wall of cylinder is formed of black pigmented tissue, enclosing a large number of clear, refractile, photoreceptor cells arranged in longitudinal rows. Each cell has a small rounded **nucleus** embedded in a thin peripheral layer of cytoplasm, surrounding a crescentic hyaline **lens** or **optic organelle**. Outer free convex surface of eye is covered by transparent epidermal cells and cuticle forming a sort of **cornea**. An **optic nerve** enters each eye basally and runs along its median axis distributing branches to all the photoreceptor cells.

All the eyes are not of equal size. They become smaller posteriorly so that the 5th pair is the smallest. Eyes are differently directed and each can receive light only from one direction. It is not known whether the eyes form any image. Probably they only enable the leech to distinguish light from darkness and to locate the direction of source of light.

On the basis of their metameric arrangement and histological structure, Whiteman regards eyes to be serially homologous with the segmental sense organs. Moreover, the segments with eyes lack the usual segmental receptors.

REPRODUCTIVE SYSTEM

Leeches do not reproduce asexually. Like earthworms, leeches are **hermaphroditic**, *i.e.*, each individual contains a complete set of well differentiated male and female reproductive organs. Self-fertilization, however, does not occur. Cross-fertilization preceded by copulation, is a rule.

[I] Male Reproductive Organs

These are : (i) testis sacs, (ii) vasa efferentia, (iii) vasa deferentia, (iv) epididymes, (v) ejaculatory ducts, and (vi) atrium.

1. Testis sacs. There are usually 11 pairs (rarely 12 or 13 pairs) of small, spherical **testis sacs**, one pair in each segment from 12 to 22, located ventrally, one on either side of the ventral nerve cord. Each testis sac is really a constricted off portion of coelom. From the walls of testis sacs, **spermatogonia** or **sperm-mother cells** bud off and give rise to **spermatozoa** that float in the enclosed coelomic fluid.

2. Vasa efferentia. Spermatozoa pass from each testis sac into a short sinuous duct, the vas efferens. It arises from the postero-lateral border of testis sac and runs outwards to join the common **vas deferens** of its side. All the vasa efferentia of one side open into the common vas deferens of that side.

3. Vasa deferentia. Each **vas deferens** is a slender, longitudinal, wavy duct, lying on ventral body wall parallel to the nerve cord, and extending forward from 22nd to llth segment. Each vas deferens is enclosed within a tubular coelomic space, containing amoeboid corpuscles similar to those of haemocoelomic fluid.

4. Epididymes. Each vas deferens, in 10th segment, swells to form a highly convoluted mass, the **epididymis** or **sperm vesicle**. The two epididymes serve to store spermatozoa brought by the vasa deferentia.

5. Ejaculatory ducts. From the anterior and inner end of each epididymis arises a short and narrow **ejaculatory duct** or **ductus ejaculatorious**.

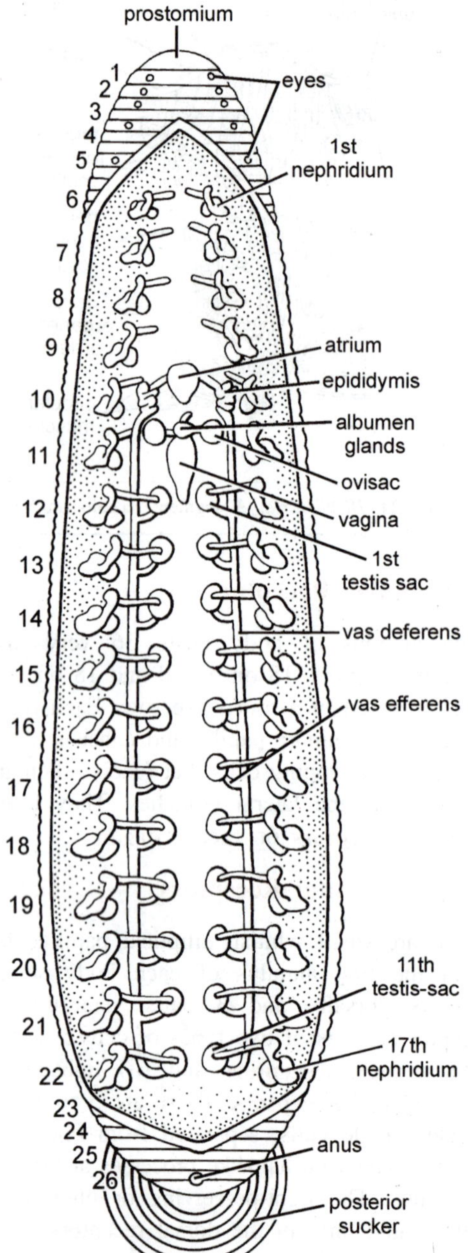

Fig. 26. *Hirudinaria.* Reproductive and nephridial systems.

6. Atrium. Ejaculatory ducts of both sides join a median pyriform sac, the **atrium**, extending into 9th and 10th segments, and opening outside by the male genital pore. Atrium consists of two

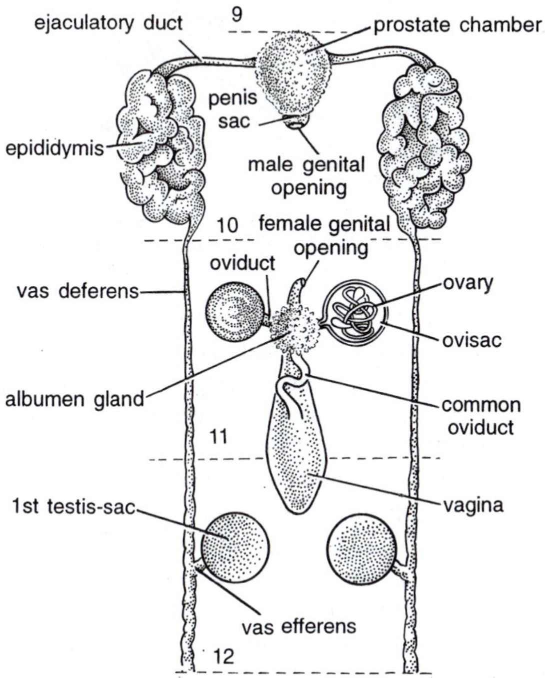

Fig. 27. *Hirudinaria.* Reproductive organs in the 10th, llth and 12th segments of body.

parts : a vase-like broad anterior **prostate chamber** and a neck-like narrow posteriorly directed **penis sac**. Prostate chamber has thick muscular walls which are covered externally by

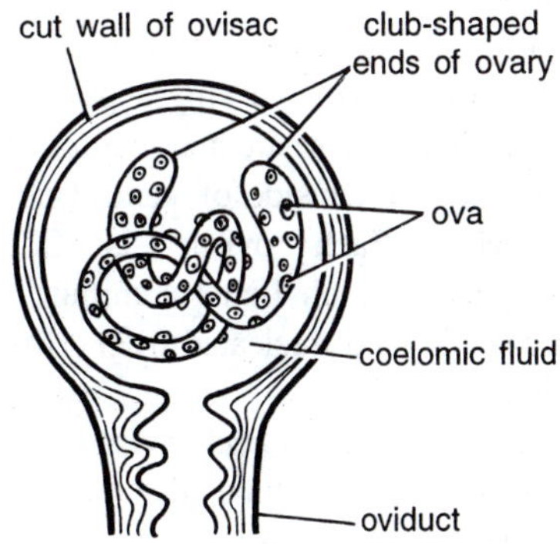

Fig. 28. *Hirudinaria.* Ovisac dissected to show an ovary.

numerous unicellular **prostate glands**. Penis sac is a highly muscular tube, containing a filamentous coiled tubular **penis**, which can be frequently seen protruding through the male genital pore.

From epididymes, spermatozoa pass into prostate chamber where they are glued together by the prostate secretion into bundles or packets, forming **spermatophores**, which are transferred through penis into vagina of the partner leech during copulation.

[II] Female Reproductive Organs

Female reproductive organs are comparatively compact and include : (i) ovisacs, (ii) ovaries,

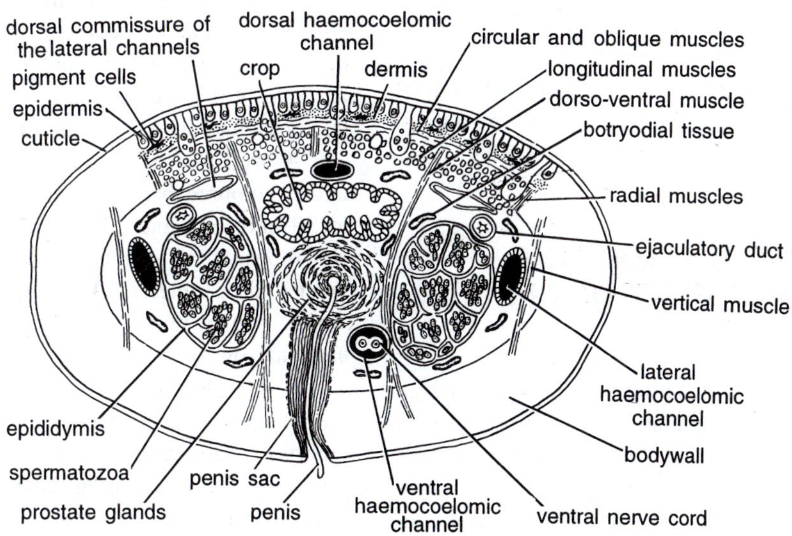

Fig. 29. *Hirudinaria.* T.S. body through epididymes, prostate glands and penis-sac (penis is enclosed with in the penis sac).

(iii) oviducts, (iv) common oviduct, and (v) vagina.

1. Ovisacs. There is a single pair of hollow, globular, coelomic sacs situated ventrally in 11th segment, one on either side of ventral nerve cord.

2. Ovaries. Each ovisac is filled with coelomic fluid in which floats a minute, delicate, coiled, thread-like and nucleated **ovary** from which ova are budded off.

3. Oviducts. Base of each ovisac is continued into a short and slender tube, the **oviduct**, which runs inwards and backwards.

4. Common oviduct. Right oviduct passes beneath the ventral nerve cord and unites with the left one in 11th segment to form a single and median **common oviduct**. Place of their union is covered by a thick layer of unicellular **albumen glands**, opening into the common oviduct. Posterior part of common oviduct, which is free from albumen glands, is folded and leads behind into vagina.

5. Vagina. Vagina is a large and pear-shaped muscular bag, lying mid-ventrally in the posterior part of 11th segment. It increases in size and also becomes internally folded during breeding season. Anteriorly it bends upon itself and narrows into a short duct, which opens to the exterior mid-ventrally through the female genital pore on 11th segment.

Ova are budded off from ovaries into their ovisacs. From here they pass down the oviducts to the common oviduct where each is coated with albumen secreted by the albumen glands. Ova then are passed into vagina, where they are stored till **fertilization** takes place during copulation.

LIFE HISTORY AND DEVELOPMENT OF LEECH

1. Copulation. Copulation in leeches, which possess a penis, is similar to direct transmission of sperms, as seen in earthworms. Copulation takes place during March and April. Two leeches come in contact by their ventral surfaces in head-to-tail position, so that the male genital pore of one lies

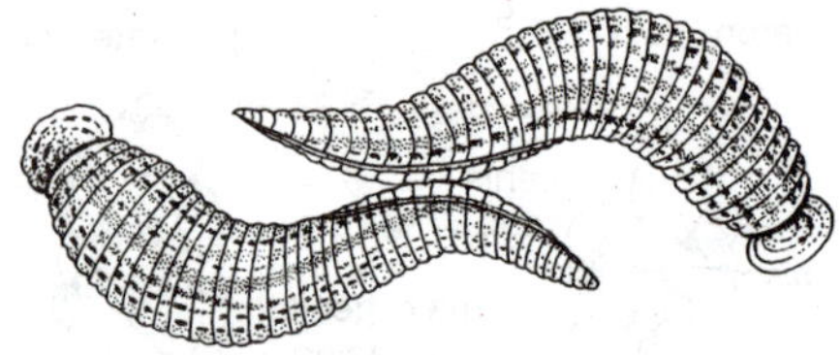

Fig. 30. *Hirudinaria.* Two leeches in copulation.

against the female genital pore of other. Penis of one leech enters the female genital pore of other and there is a mutual exchange of seminal fluid containing spermatophores, which enter the vaginae of both the leeches. Copulation lasts for an hour after which the mates separate.

2. Fertilization. Internal fertilization takes place in the vagina of each individual and the fertilized ova are discharged into a cocoon for further development.

3. Cocoon formation. Cocoons are formed following copulation, during April, May and June. The cocoon or ootheca or egg-case is secreted in the form of a snow-white frothy girdle by the clitellar glands around segments 9, 10 and 11. Clitellar glands also secrete an albuminous fluid which is deposited, with the fertilized ova, inside cocoon. The cocoon is then passed over the head of each individual. As the leech withdraws its anterior end by rhythmic movements, the prostomial glands secrete the two **polar plugs** of cocoon. Complete cocoon is formed in about six hours, after which it is laid in a moist place by the side of a pond or pool, but never in water. It becomes hardened on exposure to air.

A well-formed cocoon is a light yellow or amber-coloured barrel- shaped structure, about 25 to 30 mm long and 12 to 15 mm in diameter. Its wall consists of an inner thin tough membranous layer and an outer thick spongy layer. There is a distinct polar plug with a conical projection at each narrow end.

4. Development. Development proceeds within the cocoon, which may contain one to 24 embryos. Albumen contents of cocoon serve as food for the developing embryos. **Cleavage** is unequal and development is direct without any larval forms. At the time of hatching, polar plugs

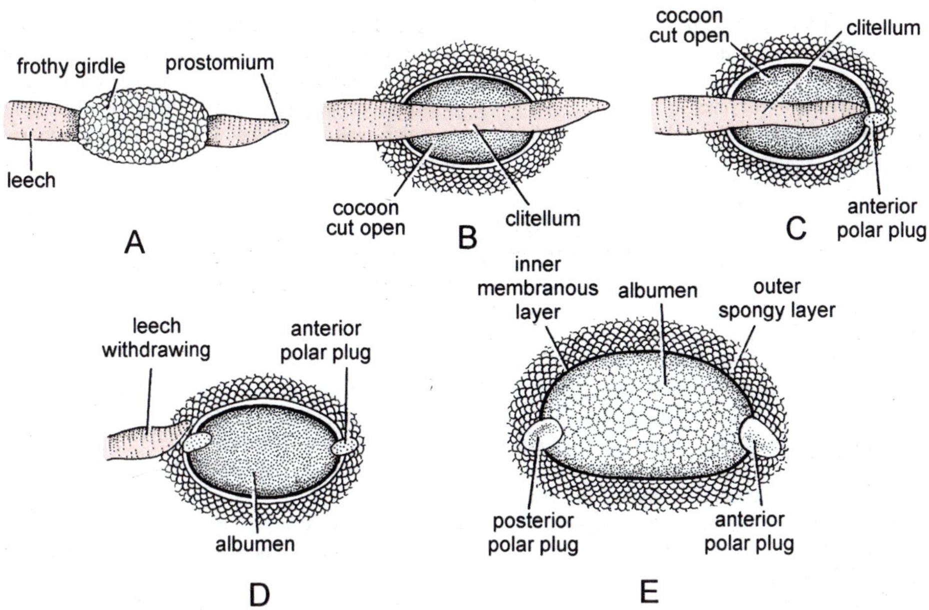

Fig. 31. *Hirudinaria.* Stages in cocoon-formation. **A.** Formation of frothy girdle around clitellum. **B.** Same in section. **C.** Formation of anterior plug. **D.** Formation of posterior plug. **E.** Cocoon in sectional view.

drop off and young leeches, resembling the adults, emerge. The whole process of development takes about a fortnight for completion.

PARASITIC ADAPTATIONS OF LEECHES

Most leeches lead a semi-parasitic life, sucking blood of vertebrates. Correspondingly, they show several parasitic adaptations in their habits, habitat and structure.

1. **Habitat.** It inhabits freshwater ponds and pools visited frequently by men and cattle. Its habitat is thus most favourable for sure food supply. Moreover, it prefers to live in shallow waters where it can easily hide under weeds, logs and stones.

2. **Habits.** Its active swimming habit is ideal for searching the host and escaping from predators.

3. **Shape.** Body is long, flattened and devoid of appendages, like parapodia, and hence suitable for aquatic life.

4. **Slime glands.** These are present in body wall. Their secretion keeps the body moist and slippery to prevent desiccation and reduces friction during locomotion.

5. **Suckers.** These serve as organs of locomotion and also provide firm adhesion to the host's body at the time of feeding.

6. **Jaws.** The three jaws are specialized weapons for inflicting a painless triradiate wound on the host's skin for sucking blood.

7. **Suctorial pharynx.** Mouth leads into a muscular pharynx connected with body wall by radiating muscles. Cavity of pharynx alternately expands and contracts to serve as a suction pump so that blood, oozing from the host's wound, is sucked in.

8. **Hirudin.** Salivary glands of leeches secrete an anticoagulating substance, called **hirudin** or **anticoagulin**. It pervents clotting of blood while the leech is taking its meal, thus ensuring a continuous supply of blood. (Vampire bats and hookworms also secrete a similar substance to facilitate their feeding).

9. **Spacious crop.** Chances of getting a host are remote, so that supply of nutrient is very irregular and a matter of chance. Digestive tract is accordingly modified. At the time of a single meal a leech obtains many times of its own weight of blood, which suffices for several months or even a year. To accommodate this blood, the crop is spacious, thin-walled, elastic and capable of great dilatation. It is also produced into lateral pouches or storing pockets. It can, therefore, store enough blood to last for months. However, some species that parasitize aquatic vertebrates remain attached to their hosts.

10. **Slow digestion.** After a meal, leech drops off, remains torpid and digests blood at a very slow pace. Last chamber of crop opens behind into stomach through a sphinctered aperture, so that blood passes from crop into

stomach with extreme slowness. Digestion of a "crop full of blood" takes many months and thus a single meal lasts for several months. For this reason also leeches lack elaborate digestive juices and enzymes.

11. Sense organs. These are well-developed and provide the animal with greater opportunities of life.

12. Hermaphroditism. This doubles the rate of reproduction as, after copulation, both individuals lay eggs.

13. Development It takes place within the cocoon which serves as a protective covering. Further, development is quick, and completed within a fortnight to ensure the maintenance of a regular population.

Important Questions

➤➤*Long answer type questions*

1. Describe the food, feeding mechanism and physiology of digestion in leech.
2. Describe the haemocoelomic system of *Hirudinaria,* and discuss the nature of the sinuses in the animal.
3. Give an account of the structure and function of a typical nephridium *of Hirudinaria.*
4. Give an account of the receptor organs of the leech.
5. Describe the reproductive organs of the leech *Hirudinaria.*
6. Write an essay on the parasitic adaptations in leeches.
7. Draw labelled full page diagrams of the following : (i) T.S. through the middle region *of Hirudinaria.* (ii) T.S. through the buccal cavity *of Hirudinaria.* (iii) T.S. through the epididymes and penis sac of *Hirudiaria.* (iv) Reproductive organs of *Hirudinaria.*
8. Write short notes on : (i) Botryoidal tissue, (ii) Ciliated organ of leech, (iii) Cocoon-formation in leech, (iv) Locomotion in leech, (v) Medicinal importance of leech.

➤➤*Short answer type questions*

1. Name the animal where the botryoidal tissue is present.
2. What is hirudin ?
3. What is the number of nephridia in *Hirudo* ?
4. How many segments are present in the cephalic region of leech ?
5. How many pairs of testes are there in *Hirudo ?*
6. Enumerate the significance of crop in Hirudinea, within 5 sentences.
7. With the help of a diagram, describe an annulus in leech in 7 lines.
8. Compare the meganephridium of *Megascolex* with that of *Hirudinaria.*
9. Describe the features indicating the adaptations of leech to lead an ectoparasitic life.
10. Give an account of excretory system of leech.
11. Describe how 'leech moves'.
12. Give an account of the *Hirudinaria.*
13. Describe the haemocoelomic system of *Hirudinaria.*
14. Make a labelled diagram of the T.S. of leech through crop.
15. Make fully labelled sketches of general internal anatomy of *Hirudinaria.*
16. Draw and label nephridium of *Hirudo.*

➤➤*Fill up the blanks*

1. The connective tissue in the body cavity of *Hirudinaria* is known as.....
2. Salivary secretion of leech contains an active substance called ....
3. Epidermis that lies below the cuticle of leech consists ....
4. In leech, the space between the body wall and the alimentary canal is filled with a kind of tissue called .....
5. The connective tissue within the body cavity of *Hirudinaria* is known as .....
6. The number of nephridia in *Hirudo* are .....
7. (a) The male reproductive opening in leech occurs in the annulus of the .... segment.
   (b) The female reproductive opening in leech occurs in the annulus of the .... segments.

➤➤*True and false statements*

1. Hirudin helps in digestion.
   True / False / Do not know
2. The rumber of segment is fixed in *Hirudo.*
   True / False / Do not know

➤➤*Multiple choice questions*

1. Saliva of leeches contain an anticoagulant called :
   (a) haemoglobin (b) hirudin
   (c) heparin (d) histamine
2. Common cattle leech belongs to the genus :
   (a) *Megascolex* (b) *Neanthes*
   (c) *Arenicola* (d) *Hirudo*
3. The saliva of leech has an anticoagulant called :
   (a) heparin (b) hirudin
   (c) chloragosomes (d) none of these
4. How many chambers are found in the crop of leech?
   (a) six (b) eight
   (c) nine (d) ten
5. Leech is a blood sucking animal, nutritionally, therefore leech is :
   (a) herbivorous (b) carnivorous
   (c) sanguivorous (d) omnivorous
6. In which segment do you find the ovisac in leech ?
   (a) 8th (b) llth
   (c) 15th (d) 13th
7. Which segment possess nephridia in *Hirudinaria* ?
   (a) 7-10th segments (b) 6-22 segments
   (c) 10-20 segments (d) 7-18 segments
8. Female genital opening of *Hirudinaria* is present in :
   (a) 1st annule of 10th segment
   (b) 1st annule of llth segment
   (c) 2nd annule of 10th segment
   (d) 2nd annule of llth segment
9. The posterior sucker of *Hirudinaria* is formed by the union of :
   (a) 1 segment (b) 7 segments
   (c) 25 segments (d) 33 segments
10. The number of segments in *Hirudinaria* is :
    (a) 109 (b) 33
    (c) 26 (d) numerous
11. Anterior sucker of *Hirudinaria* is formed by the union of :
    (a) 1 segment (b) 3 segments
    (c) 7 segments
    (d) prostomium and 3 segments
12. In *Hirudinaria* space between body wall and alimentary canal is filled not the :
    (a) botryoidal tissue
    (b) connective tissue
    (c) haemocoelomic fluid
    (d) none
13. Chitogenous glands are found in :
    (a) clitellar region forming wall of cocoon
    (b) prostomium
    (c) suckers
    (d) epidermis
14. The detailed study about *Hirudinaria* made by :
    (a) K. N. Behl (b) S. N. Dash
    (c) M. L. Bhatia (d) none
15. *Hirudinaria* is :
    (a) herbivorous (b) carnivorous
    (c) omnivorous (d) sangivorous
16. A temporary clitellum is form in *Hirudinaria* on segments :
    (a) 7th 8th 9th (b) 9th to 11th
    (c) 11th to 13th (d) none
17. Number of annular recepter in each annulus :
    (a) 18
    (b) 18 on dorsal and 18 on ventral
    (c) 9
18. *Hirudinaria* contains :
    (a) slime gland
    (b) sucker gland
    (c) cliteller gland
    (d) all above
19. Velum present in :
    (a) *Pheritema* (b) *Hirudinaria*
    (c) *Ascaris* (d) all
20. In *Hirudinaria* there is no special organs far :
    (a) digestion (b) excretion
    (c) respiration (d) reproduction
21. Number of testes in *Hirudinaria* are :
    (a) 10 pairs (b) 11 pairs
    (c) 12 pairs (d) 13 pairs

Answers

➤➤*Fill in the blanks*

1. botryoidal tissue, 2. hirudin, 3. of a layer of hammer-shaped cells, 4. botryoidal tissue, 5. botryodial tissue, 6. 17 pairs, 7. (a) second and third segments, (b) 11th segment.

➤➤*True and false questions*

1. false. 2. true.

➤➤*Multiple choice questions*

1. (b) 2. (d) 3. (b) 4. (d) 5. (c) 6. (b) 7. (b) 8. (d) 9. (b) 10. (b)
11. (d) 12. (a) 13. (a) 14. (c) 15. (d) 16. (b) 17. (b) 18. (d) 19. (b) 20. (c)
21. (b).

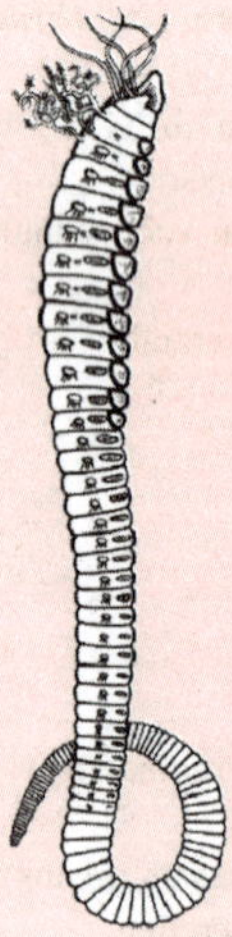

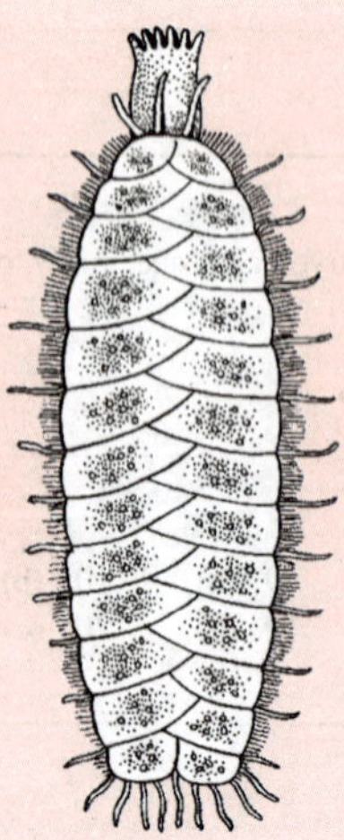

43

Annelida : Characters, Classification and Types

Name of phylum Annelida was first coined by **Lamarck** for the higher segmented worms (L., **annelus**, little ring+ Gr., **eidos**, form). In general, annelids are elongated, bilaterally symmetrical and highly organized animals, in which the organs have grouped into definite systems. Appearance of **metamerism** represents their greatest advancement, so that they are called **segmented worms** in order to distinguish them from flatworms (Platyhelminthes) and roundworms (Nematoda) which are not segmented. Their paired appendages, when present, are never jointed. Their coelom, nephridia and cephalization are better developed than those of the unsegmented worms. They are the first animals to have a closed vascular system. Nervous system is fundamentally similar to that of Arthropoda and embryology is not much different from that of Mollusca.

GENERAL CHARACTERS

1. Mostly aquatic, some terrestrial. Burrowing or tubicolous. Some commensal and parasitic.
2. Body elongated, bilaterally symmetrical, triploblastic, truely coelomate and **metamerically segmented** into similar metameres.
3. **Epidermis** of a single layer of columnar epithelial cells, covered externally by a thin **cuticle** not made of chitin.
4. Body wall **dermo-muscular**. Outer muscle fibres circular, inner longitudinal.
5. Locomotory organs are segmentally repeated chitinous bristles, called **setae** or **chaetae**, embedded in skin. May be borne by lateral fleshy appendages or **parapodia**.

6. **Coelom**, true, schizocoelous. Mostly well-developed except in leeches. Usually divided into compartments by transverse septa. Coelomic fluid with cells or corpuscles.
7. Digestive system straight and complete. Digestion entirely extracellular.
8. Blood vascular system closed. Respiratory pigments either haemoglobin or erythrocruorin dissolved in blood plasma.
9. Respiration by moist skin or gills of parapodia and head.
10. Excretory system consisting of metamerically disposed coiled tubes, called **nephridia**.
11. Nervous system with a pair of cerebral ganglia (brain) and a double ventral nerve cord bearing ganglia and lateral nerves in each segment.
12. Sensory organs include tactile organs, taste buds, statocysts, photoreceptor cells and sometimes eyes with lenses in some.
13. **Hermaphroditic** or sexes separate cleavage spiral and determinate. Larva, when present, is a **trochophore**. Regeneration common.

CLASSIFICATION

About 17,426 known species of Annelida are divided into four main classes, primarily on the basis of the presence or absence of parapodia, setae, metameres, and other morphological features.

CLASS 1. Polychaeta

(Gr., **polys**, many + **chaite**, hair)

1. Chiefly marine, some in fresh water.
2. Segmentation internal and external.
3. Head distinct with eyes, palps and tentacles.
4. Setae numerous, on lateral parapodia.
5. Clitellum absent.
6. Sexes separate. Gonads temporary and in many segments.
7. Trochophore larva present.

Attempts to arrange families into orders has not proved satisfactory. It is usual, therefore, to divide polychaetes into two subclasses, *Errantia* and *Sedentaria,* after **Fauvel** (1959). However according to **Dab** (1963), this subdivision is artificial and not a natural one.

Subclass I. Errantia

1. Free-swimming, crawling, burrowing or tube-dwelling and predatory polychaetes.
2. Segments numerous and similar, except for head and anal region.
3. Prostomium distinct with sensory structures.
4. Parapodia with acicula and compound setae.
5. Pharynx protrusible, enlarged and usually with jaws and teeth.
   Examples : *Aphrodite* (sea mouse), *Polynoe, Phyllodoce, Tomopteris, Syllis, Nereis, Glycera, Eunice, Diopatra, Histriobdella.*

Subclass II. Sedentaria

1. Sedentary polychaetes living in burrows or tubes.
2. Body made of two or more regions, with dissimilar segments and parapodia.
3. Prostomium small.
4. No acicula and compound setae.
5. Pharynx without jaws and teeth.
   Examples : *Chaetopterus, Arenicola, Owenia, Sabella, Sabellaria, Terebella, Amphitrite, Pomatoceros, Spirorbis, Serpula.*

CLASS 2. Oligochaeta

(Gr., ***oligos***, few + ***chaite***, hair)

1. Mostly terrestrial, some in fresh water.
2. Segmentation external and internal.
3. Head indistinct, without sensory organs.
4. Setae few, embedded in skin. Parapodia absent.
5. Glandular clitellum present for cocoon-formation.
6. Hermaphroditic. Testes anterior to ovaries.
7. Fertilization external (in cocoon); development direct, no larval stages.

Order 1. Plesiopora plesiothecata

1. Mostly aquatic.
2. Male gonopores on segment immediately following that which contains testes.
3. Spermathecae in the testes-containing segments, or nearby.
   Examples : *Aelosoma, Nais, Dero, Chaetogaster, Tubifex.*

Order 2. Plesiopora prosothecata

Spermathecae far anteriorly to the segment containing testes.

Example : *Enchytraeus.*

Order 3. Prosopora

1. Mostly aquatic.
2. Male gonopores on the same segment containing testes, or on segment containing the second pair of testes.
   Example : *Branchiobdella* (parasitic).

Order 4. Opisthopora

1. Mostly terrestrial earthworms.
2. Male gonopores some distance behind the testes-containing segments.
   Examples : *Lumbricus, Eisenia, Pheretima, Megascolex, Allolobophora, Dendrobaena.*

Chaetopoda (Gr., **chaite**, hair + **pous**, foot). Chaetopoda is a super group which includes both Polychaeta and Oligochaeta. This creation is because of the fact that both the classes are provided with setae.

CLASS 3. Hirudinea

(L., **hirudo,** leech)

1. Freshwater, marine or terrestrial. Generally ectoparasitic, blood-sucking or carnivorous.
2. Body with fixed number of segments (33). Each segment subdivided externally into annuli.
3. Segmentation external without internal septa. Parapodia and setae absent.
4. Both anterior and posterior ends of body with suckers.
5. Coelom much reduced due to its filling by botryoidal tissue, and forms haemocoelomic sinuses.
6. Hermaphroditic with one male and one female gonopore. Fertilization internal. Development in cocoons, direct without larval stages.

Order 1. Acanthobdellida

1. Primitive, without anterior sucker, proboscis and jaws.
2. Setae present in 5 anterior segments.
3. Coelom with compartments.
   Example : A single Russian genus and species (*Acanthobdella*) parasitic on salmon.

Order 2. Rhynchobdellida

1. Only aquatic leeches, ectoparasitic.
2. A protrusible proboscis with no jaws.
3. Coelom without compartments.
4. Blood vascular system separated from coelomic sinuses. Blood colourless.
   Examples : *Glossiphonia, Placobdella, Helob della, Piscicola, Pontobdella, Branchellion, Ozobranchus.*

Order 3. Gnathobdellida

1. Aquatic or terrestrial. Ectoparasitic blood-sucking leeches.
2. Pharynx non-eversible with 3 pairs of jaws.
   Examples : *Hirudo, Hirudinaria, Haemadipsa.*

Order 4. Pharyngobdellida

1. Terrestrial and aquatic. Some predaceous.
2. Pharynx non-protrusible. No teeth but one or two styles may be present.
   Examples : *Erpobdella, Dina.*

CLASS 4. Archiannellida

(Gr., **arch**, first)

1. About one dozen genera of small, marine worms of unknown affinities.
2. Segmentation chiefly internal. No parapodia and setae.
3. Sexes usually separate.
4. Usually trochophore larva.
   Examples : *Polygordius, Dinophilus, Protodrilus.*

OTHER TYPES OF ANNELIDA

1. *Aphrodite.* *Aphrodite,* commonly called the 'sea mouse', is a marine polychaete. It is found crawling or burrowing in mud or sand of sea bottom. Its body is oval, broad, dorso-ventrally flattened, about 12 cm long, and consists of 30-35 segments. Dorsal surface is covered by 15 pairs of partly overlapping scales or **elytra**, which are supposed to be modified dorsal cirri of certain parapodia. Elytra are partly covered by a felt of fine, long and hair-like setae that arise from notopodia. In addition, there are brown protective, **stiff spine-like setae** and lateral **iridescent**

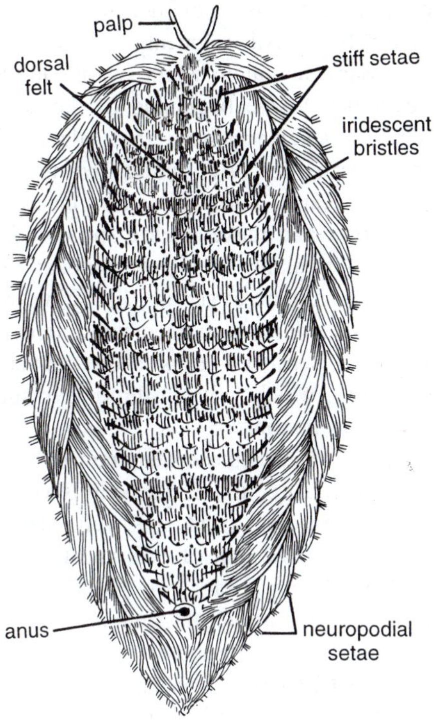

Fig. 1. *Aphrodite* in dorsal view.

bristles, variously coloured in different species. As the animal moves, the body colour changes from gold to peacock blue.

Head is small and remains hidden beneath the dorsal felt. It consists of a peristomium and a prostomium bearing a pair of eyes, a single median tentacle and a pair of long lateral palps. Proboscis is an eversible buccal tube and leads into a muscular pharynx, provided with two pairs of jaws. Eighteen pairs of lateral caeca open into the intestine along its entire length. Their openings are guarded by epithelial cells, which act as sieves allowing only small particles of food to enter into caeca. Digestion is completed within caeca. Respiration takes place through the dorsal body surface. Water current is produced by the movement of elytra and pumping action of dorsal body wall. The current passes through the channel between body wall and elytra and escapes near the posterior end of body.

The most common species is *Aphrodite aculeata,* inhabiting the Atlantic and the Mediterranean.

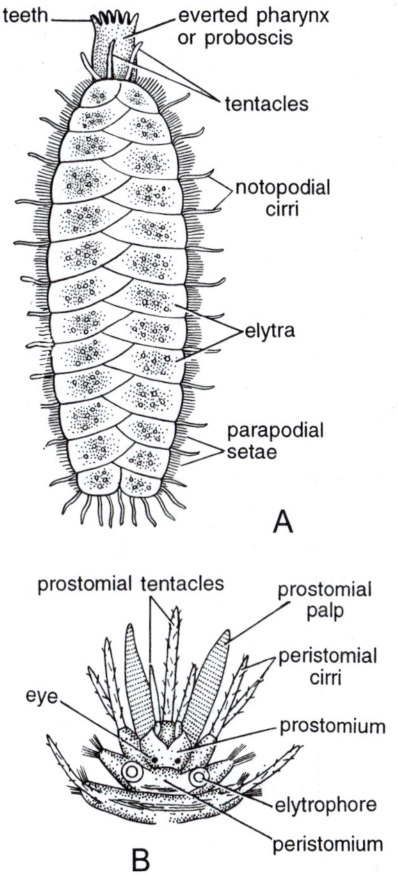

Fig. 2. *Polynoe.* **A.** Entire worm. **B.** Anterior end.

2. *Polynoe.* *Polynoe,* commonly called the 'scale worm', is a close ally of *Aphrodite.* It is a marine crawling polychaete living under stones on seashore. Body is short, oval and dorso-ventrally flattened. As in *Aphrodite,* the back of body is covered by plate-like scales or **elytra**, which exist in pairs and overlap one another. When the animal is irritated, elytra may be shed and new ones may be regenerated later on. The overlaying felt of long setae is wanting.

Head is small and includes a peristomium bearing two pairs of cirri, and a prostomium with three tentacles (one median long and two lateral small), a pair of long palps and two pairs of eyes. Pharynx is protrusible and bears teeth and jaws, suggesting that the worm is carnivorous. Gas exchange is restricted to the dorsal surface. Cilia on dorsal surface create a water current that flows posteriorly beneath the elytra. Several species of *Polynoe* are bioluminescent, the elytra being brilliantly illuminated. Luminescent material is secreted by the gland cells located on surface of elytra.

3. *Eunice.* *Eunice* is a marine polychaete, living in coral rocks and crevices below low tide mark. Some of its species are commonly known as 'palolo worms'. Body is elongated and with numerous segments. In *E. viridis,* the Pacific or Salmoan palolo worm, anterior region or **atoke** is non-sexual, while posterior region or **epitoke** is sexual, with each segment bearing an eyespot on the ventral side. Small head consists of a peristomium and a prostomium. Peristomium bears a pair of cirri and prostomium bears two pairs of tentacles, one pair of palps and one pair of eyes. Parapodia are biramous and the ventral cirri are wanting. Proboscis is provided with a single pair of lateral chitinous jaws. Gills are pectinate and arise from the bases of dorsal cirri, usually beginning from sixth segment.

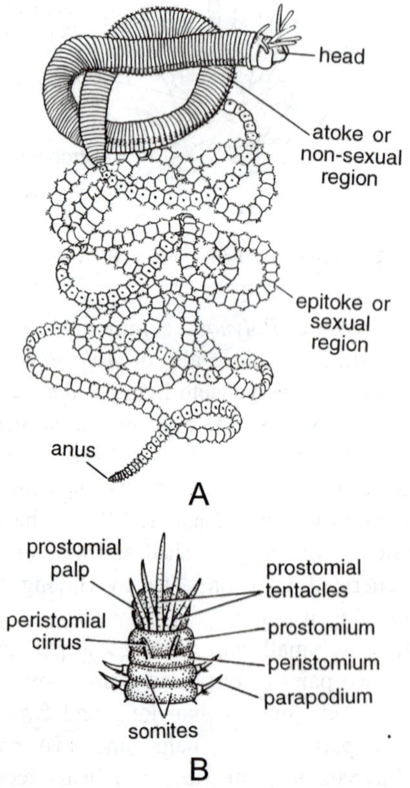

Fig. 3. A. Entire Pacific palolo worm *Eunice viridis.* **B.** Anterior end.

The Salmoan palolo worm swarms in October or November, at the beginning of the last lunar quarter. *E. schemacephala,* the Atlantic palolo worm, swarms in July during the third lunar quarters. In early morning, worms emerge from their burrows and make intense spiral swimming movements breaking free their epitokal regions. The epitokes rupture at sunrise, liberating ova and spermatozoa, which is immediately followed by fertilization. Fertilized eggs soon convert into young larvae, which, with in three days sink and settles at the bottom.

4. *Phyllodoce.* Like other errant polychaetes, it is marine and free-living. Its body is very much elongated and its head is well developed. Prostomium bears a pair of eyes and two pairs of palps. Peristomium bears two pairs of cirri on each side. It is carnivorous and ingests its prey by everting its great proboscis.

5. *Chaetopterus.* *Chaetopterus,* commonly called the 'paddle worm', is a tubicolous, marine polychaete 15 to 30 cm long. It lives in a U-shaped parchment tube, formed in sand or mud in the intertidal zone. Two openings of the tube project slightly above mud into sea water. Its parapodia show extensive specialization for its tube-dwelling existence. Body may be divided into three regions : (i) Flat **anterior region**, consists of 10-12 segments. Prostomium is much reduced. Mouth is surrounded by a wide funnel-shaped

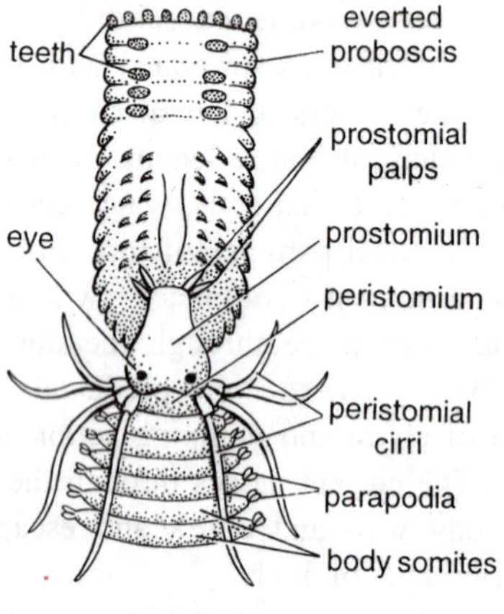

Fig. 4. Anterior end of *Phyllodoce.*

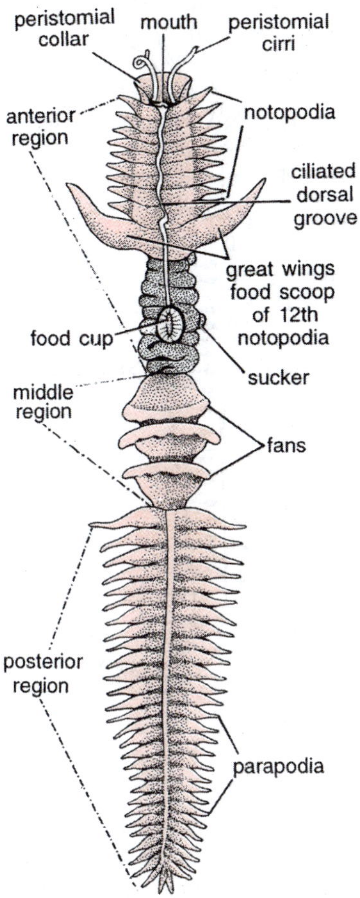

Fig. 5. *Chaetopterus* in dorsal view.

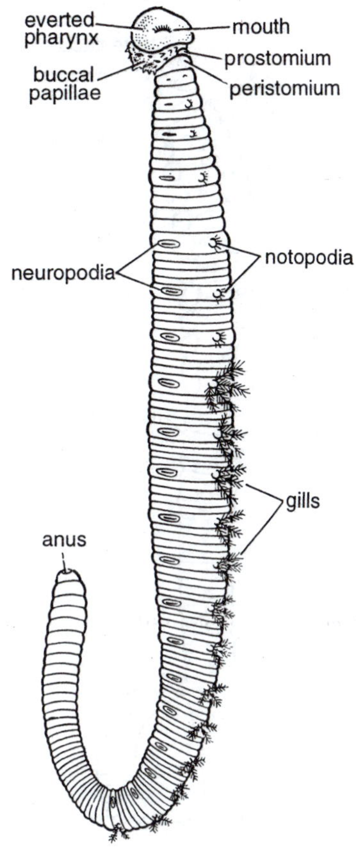

Fig. 6. *Arenicola.*

peristomial collar with a pair of peristomial cirri. The following nine segments bear simple parapodia represented by conical and lobe-like notopodia with delicate setae. Tenth segment bears a pair of large **aliform** or wing-like notopodia. Succeeding segments have greatly reduced notopodia. But their neuropodia become fused to form ventral **suckers** that grip the tube, keeping the worm in position. A **ciliated groove** runs backwards from mouth along the dorsal surface of anterior region. It ends in a dorsal ciliated **food cup**. (ii) **Middle region** consists of three segments have broad, flat neuropodia which are fused in the mid-line forming 3 sets of parapodial **fans**, (iii) **Posterior region** consists of several segments, each with paired biramous parapodia.

Chaetopterus is a filter feeder. It sets up a feeding current of water through the tube, propelled by the fan-like notopodia. Minute organisms, contained in water current, are collected by two great winged notopodia, protruding like scoops. Food particles, entangled in mucus, collect in the **food cup** forming a ball which is moved forward along the dorsal ciliated groove to mouth. Worm is **strongly** luminescent. When removed from the tube, its whole body glows brilliantly. Worm possesses extraordinary power of regeneration. It can regenerate the whole body from a single segment.

6. *Arenicola.* *Arenicola,* commonly known as tentacles 'lugworm', lives just below the tide mark in sea prostomium or estuaries having less water salinity. It lives in mud or sand in U or J-shaped burrows lined by mucus. Body is elongated, cylindrical, brownish-green in colour and measures

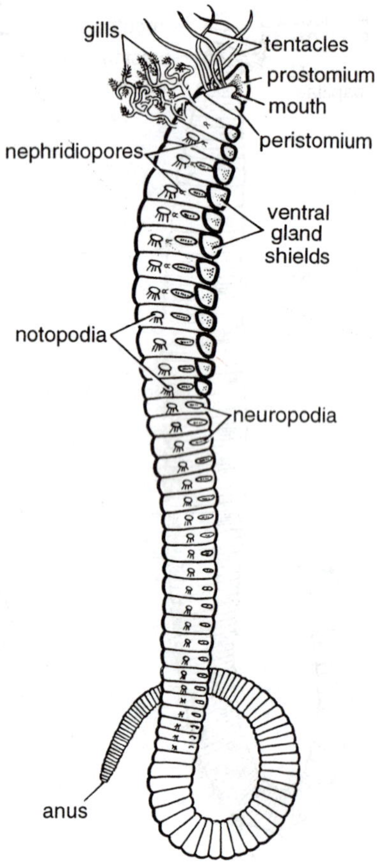

Fig. 7. *Amphitrite.*

about 20 cm in length. External segmentation is indistinct due to further division of segments into annuli. Whole body is divisible into three regions. Thicker **anterior region** consists of a much reduced trilobed prostomium without eyes or tentacles, a biannulate peristomium with no appendages, an achaetous segment and six segments bearing biramous parapodia with setae. Each parapodium bears dorso-laterally a conical notopodium with needle-like setae and ventro-laterally a long neuropodium with hook-like **uncini**. **Middle region** consists of 13 segments, each bearing parapodia and branched gills which are modified dorsal cirri. **Posterior region** comprises a variable number of segments which are devoid of gills and parapodia. Last segment bears a terminal anus.

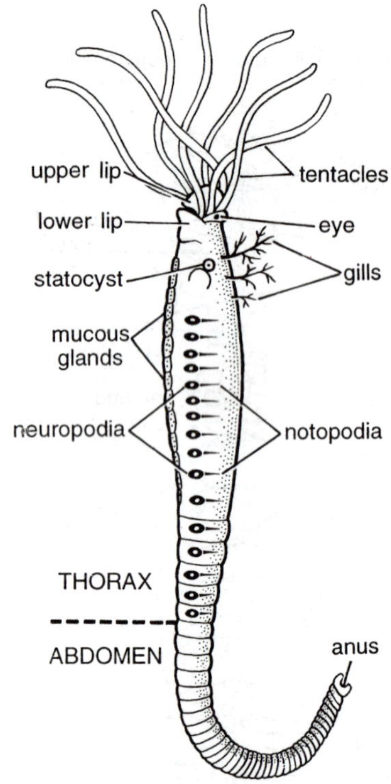

Fig. 8. *Terebella.*

Mouth is transverse and ventral to the prostomium. Through it can be everted the buccal cavity and pharynx, forming a **proboscis** which is covered with minute chitinous papillae. The worm ingests mud or sand to nourish upon its organic matter and egests it through anus as castings. There are only six reproductive segments. Eggs are laid as mucous masses anchored to the surface of sand bottom. Gills act as respiratory surfaces and a water current is maintained through the burrow. *Arenicola* is used as fish bait.

7. *Amphitrite.* *Amphitrite* is a marine tubicolous polychaete, living permanently in a tube of mud or sand between tide marks. It has a long cylindrical body, measuring upto 30 cm in length. Body has three distinct regions. **Anterior region** constitutes the head formed by prostomium and peristomium. Peristomium bears a number of long filiform tentacles which protroude from the tube. **Middle region** consists of many segments bearing notopodia with setae and neuropodia with **uncini**. Its 12 anterior segments are thickened ventrally

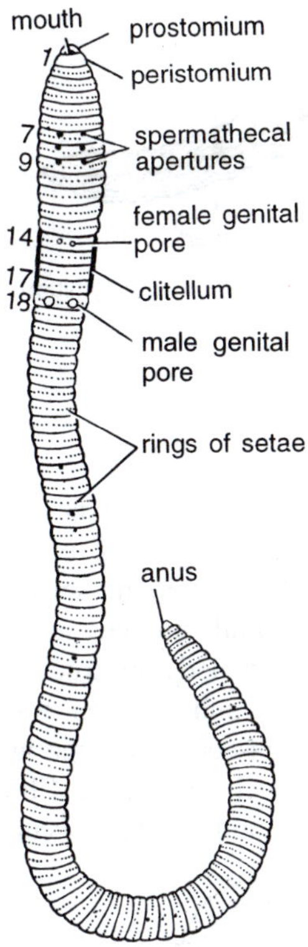

Fig. 9. *Megascolex.*

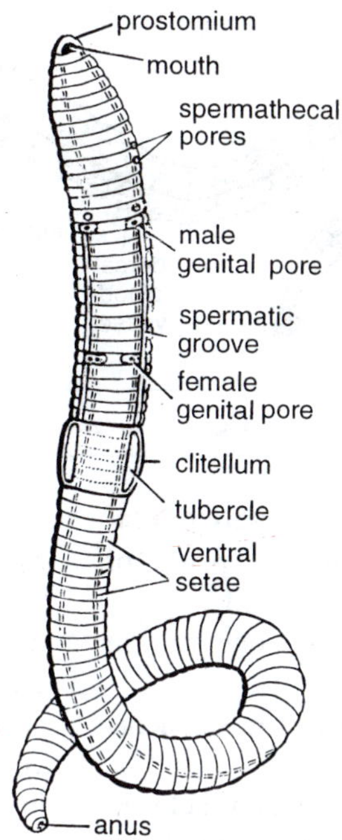

Fig. 10. *Lumbricus.*

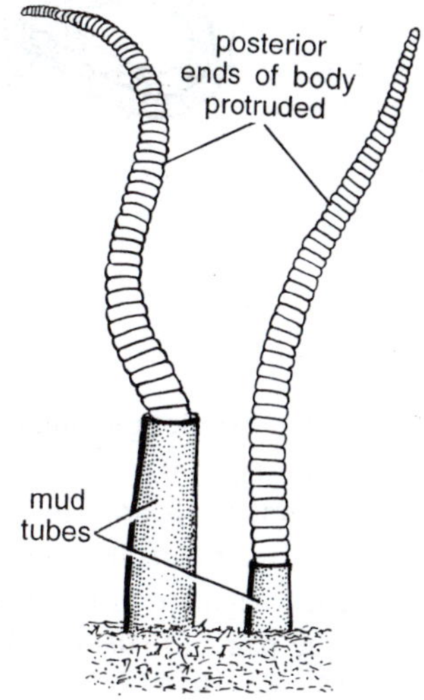

Fig. 11. *Tubifex.* Posterior parts of body protruding from tube and waving to facilitate gaseous exchange.

by **gland shields**, which act as mucus-secreting structures for tube formation, and the first 3 segments bear paired dorso-lateral branched **gills**. **Posterior region** comprises many segments without notopodia, neuropodia and setae. Anus lies at its posterior.

Worm lacks a proboscis. Feeding organs are the contractile tentacles. Food particles are adhered to mucous secretion on the surface of tentacles and then move towards mouth down their ciliated grooves. Coiling of intestine increases the surface for digestion and absorption. Gills act as respiratory surfaces. There are a few gamete-bearing segments.

8. *Terebella.* *Terebella* closely resembles *Amphitrite* but can be distinguished from the latter in having its third pair of gills smaller than the first two pairs.

9. *Megascolex.* *Megascolex* is the earthworm commonly found in South India. Its two common species are *M. mauritti* and *M. konkonensis.* Worm measures 8 to 12 cm in length and bears upto 100 or more segments.

Dorsal surface is dark purplish brown and ventral surface is somewhat pale in colour. A large glandular swelling, or **clitellum**, covers segments 14 to 17. External apertures include : (i) an anterior terminal **mouth**, overhung by prostomium, (ii) a posterior terminal **anus**, (iii) **dorsal pores** in mid-dorsal line, one in each intersegmental groove behind the 10th segment, (iv) ventral **spermathecal openings**, one pair in each intersegmental groove between segments 6/7, 7/8 and 8/9, (v) ventral **female genital pores** on segment 14, (vi) ventral **male genital pores** on segment 18, and (vii) numerous scattered **nephridiopores** on each segment.

Anatomy of *Megascolex* closely resembles that of *Pheretima* but for the following differences :

(1) **Pharyngeal nephridia** present as paired tufts in segments 5 to 9 **integumentary nephridia** in all segments behind 14th and **meganephridia** in place of septal nephridia one pair in each of the segments behind 18th.

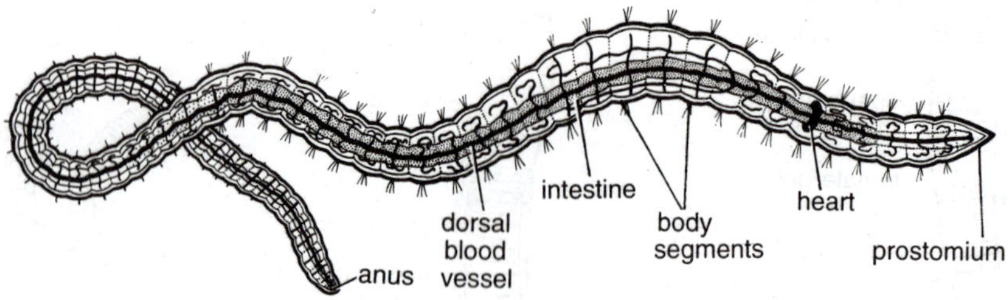

Fig. 12. *Tubifex*. Entire worm.

(2) Only two longitudinal blood vessels **dorsal** and **ventral**, connected by 8 pairs of contractile hearts, in segments 6-13.

(3) Two pairs of **seminal vesicles**, one pair in each segment from 9 to 12.

(4) One pair of **prostate glands** lies in segments 18 and 19.

(5) Two pairs of **penial setae** which help in copulation, are present in the neighbourhood of male genital pores.

10. *Lumbricus.* *Lumbricus terrestris* is an earthworm commonly found in Europe and America. Body is cylindrical, upto 30 cm in length and made of 100 to 180 segments. Prostomium is **tanylobus**, dividing the peristomium into two parts. A permanent clitellum develops in segments 33 to 37. Each segment, except the first (peristomium) and the last, bears S-shaped chitinous setae arranged in 4 pairs, of which 2 pairs are ventral and two ventro-lateral (octochaetine arrangement). Tail end is flattened. Post-terminal anus is ventral. Prostrate glands are absent. Oesophagus is with 3 pairs of calciferous glands. Five pairs of contractile hearts lie in segments 7 to 9. Nephridia are **meganephridia**, one pair in each segment except the first three and the last. Seminal vesicles are three pairs in segments 9, 11 and 12. Spermathecae are two pairs in segments 9 and 10. Two male genital pores lie ventrally on segment 15, and two female genital pores on segment 14.

11. *Tubifex.* It is a freshwater oligochaete living in tubes at the muddy bottoms of lakes, ponds and streams. Tubes are made of mud and minerals glued together by mucus. Body is cylindrical, red-coloured, 3 to 4 cm long, and resembles an earthworm. It is commonly known as 'blood worm' because of its bright red colour. Numerous short and blunt setae are located in dorsal and ventral bundles on each segment. Posterior part of body protrudes from the tube and sways back and forth in water to facilitate ventilation for gaseous exchange in stagnant water in which O_2 level falls down. Clitellum develops in segments 11 and 12. Male genital pore lies on

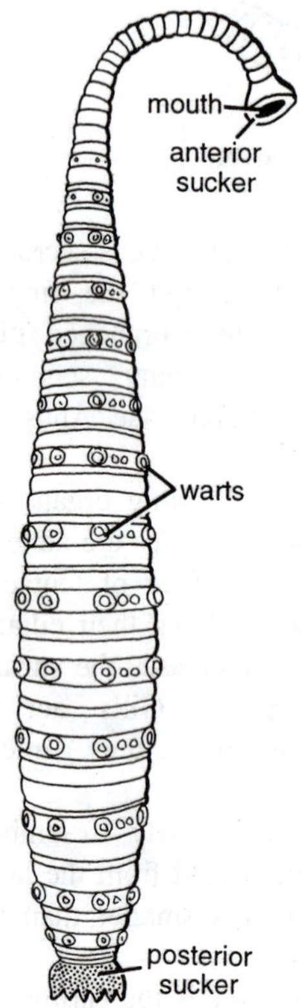

Fig. 13. *Pontobdella.*

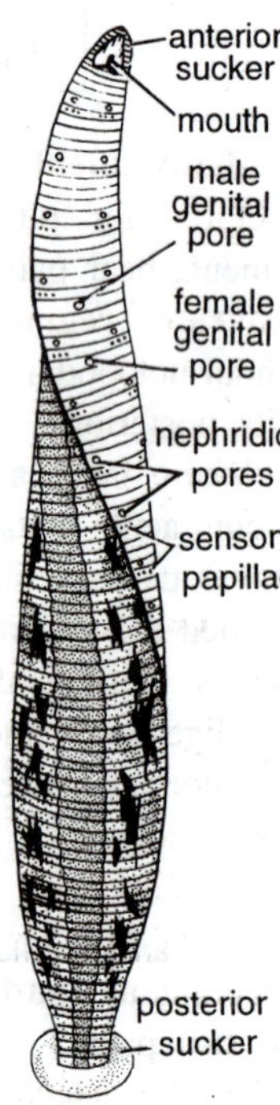

Fig. 14. *Hirudo medicinalis.*

segment 11, while female genital pore on segment 12. There is a single pair of circumintestinal hearts in segment 8. *Tubifex* reproduces only sexually. Cocoons secreted by clitellum measure 0.85 mm to 1.06 mm. *Tubifex* helps in purification of polluted fresh water.

12. *Pontobdella.* *Pontobdella* or **skate-sucker** is a marine leech attacking chiefly the skates, rays and sharks. Body is about 20 cm long, cylindrically elongated, without eyes and gills, but having strong adhesive suckers. Body surface is rough and leathery being covered by conspicuous green tubercles or warts. Anterior sucker is saucer-shaped and its action is so strong that it leaves a deep circular scar upon the body of host. Posterior sucker is deep and cup-like. Pharynx is protrusible but jaws are absent. Nephridia are absent but a complex ventral network of tube bears a pair of ciliated nephrostomes and a pair of nephridiopores in each body segment. *P. muricata* lays velvety eggs in empty shells of molluscs, and mounts to guard over them for more than 100 days till they hatch.

13. *Hirudo medicinalis* (The Medicinal leech). The most famous of all the Hirudinea is the medicinal leech *Hirudo medicinalis* of Europe. It is so called because it was widely used in medicine during the 17th and 18th centuries. Now almost extinct, at one time, it was a common British species. Habits and structure are similar to *Hirudinaria granulosa* the common Indian cattle leech.

1. Habits and habitat (Ecology). The leech lives in freshwater ponds and marshes and sucks the blood of fish, frogs, men, buffaloes and other domestic and wild animals visiting its haunts.

2. External characteristics. Body is about 10 cm long, dorso-ventrally flattened and divided into 33 or 34 true segments with 95 annuli. Colour is yellowish brown. It has great power of contraction and elongation and can stretch upto 20 or 30 cm. Setae, parapodia and tentacles are lacking. There are two suckers. Small **anterior** or **oral sucker** surrounds the mouth. Large **posterior sucker** serves for attachment with the host.

3. Internal characteristics. Alimentary canal is well adapted for ingestion and digestion of vertebrate blood on which it feeds. Mouth contains three **jaws** armed with chitinous teeth for biting. Blood is sucked up by the dilatation of strong muscular **pharynx** Its salivary glands secrete a histamine-like substance which dilates the blood capillaries of the host. Saliva also contains an anticoagulant, called **hirudin**, that makes the blood flow freely. The large **crop** with lateral diverticula is capable of ingesting 3 to 10 times its own weight in blood, and requires months

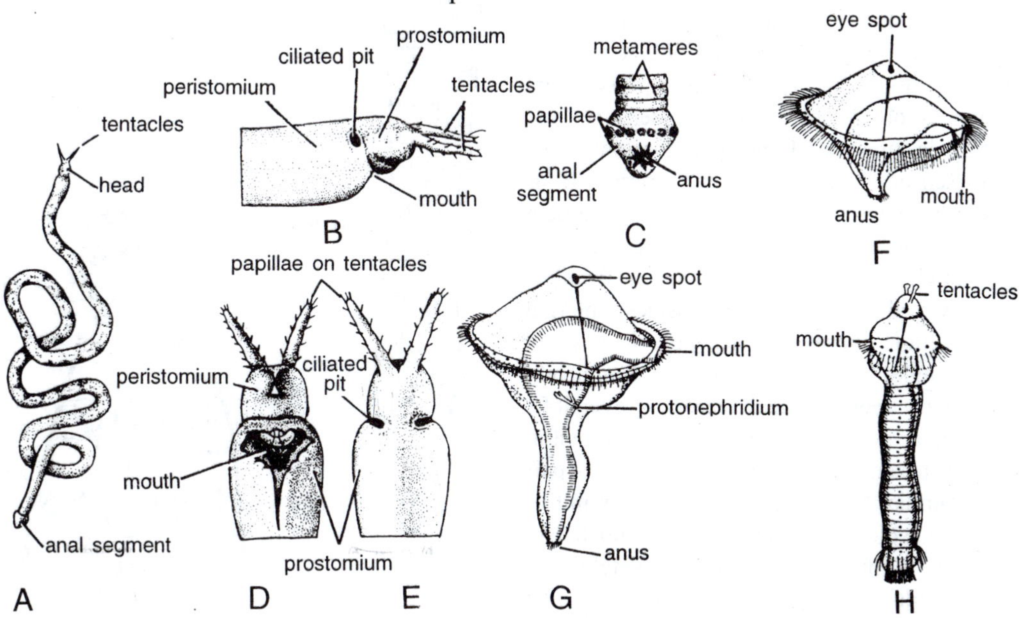

Fig. 15. *Polygordius.* **A.** Entire worm. **B.** Head in lateral view. **C.** Posterior end in ventral view. **D.** Head in ventral view. **E.** Head in dorsal view. **F.** Trochophore larva. **G, – H.** Stages of transformation into adult worm.

for digestion, short rectum opens dorsally on the posterior sucker by **anus**. Blood is digested presumably by the action of the bacterium *Pseudomonas hirudinis*.

Coelom is greatly reduced due to the development of **botryoidal tissue**. The spaces left form haemocoelomic sinuses containing blood with dissolved haemoglobin. **Respiration** occurs through the moist skin. **Excretion** of waste products from blood and coelomic fluid takes place by 17 pairs of **nephridia**, one pair in each segment from 7 to 23. **Nervous system** is typically annelidan and there are **sensory organs** of taste, touch, smell and sight.

4. Reproduction. Leech is **hermaphroditic,** with several pairs of testes and one pair of ovaries, but cross fertilization occurs by mutual transfer of sperms during copulation. Development occurs inside a cocoon, as in earthworm, and there is no larval stage.

5. Phlebotomy. Phlebotomy or 'blood-letting', painlessly achieved by the application of leeches, was a common, though erroneous method of medical treatment in Europe in the early nineteenth century. By blood-letting, the balance could be restored and health regained. When a person got a 'black eye' or conspicuous black and blue spots on the body, the doctor used to apply a lean hungry leech to the skin for sucking the impure blood.

14. *Polygordius.* The best known genus of class Archiannelida is *Polygordius* (Gr., **polys**, many + **gordius**, knot). It is a marine worm with a narrow, elongated and cylindrical body which is 3 to 10 cm long. It lives in bottom sand of European seas. External segmentation is indistinct, faintly marked by grooves. Parapodia, setae, cirri and gills are altogether absent. Small prostomium bears two short fleshy tentacles, sensory and respiratory in function. Large peristomium bears a pair of sensory ciliated pits often referred to as eyes. Mouth opens ventrally on the first segment and anus on the last segment. Anal segment also bears a circle of adhesive papillae in front of anus. Body wall lacks circular muscles. Internal organs resemble those of other polychaetes but are of simple form. Coelom is completely subdivided by septa. Buccal region is eversible. Blood vascular system includes dorsal and ventral vessels. Each segment of coelom contains a pair of metanephridia. Nervous system is most primitive and found in epidermis. Nerve ring is situated in prostomium. Ventral nerve cord is single and without ganglia. *Polygordius* is unisexual, the ovaries or testes occurring in the posterior segments. There are no special gonoducts. Development includes the so-called **Loven's larva**. It is a typical trochophore. During metamorphosis, its anal region elongates posteriorly. This elongation becomes segmented and by continued growth, the larva transforms into adult worm.

Relationships. Class Archiannelida comprises a heterogeneous grouping of largely unrelated genera. Class name was created with the presumption that these worms are primitive. But some authorities believe that they are not primitive but specialized due to simplification and reduction, and represent modifications from several groups of polychaetes. Others think that they represent remnants of the ancestral annelid stock. In either case, archiannelid genera may be arranged in a series starting from *Polygordius* and ending with forms, such as *Nerilla,* close to Polychaeta.

15. *Sipunculus.* *Sipunculus,* commonly known as 'pea-nut worm', was once placed in Annelida but now treated in a separate phylum,

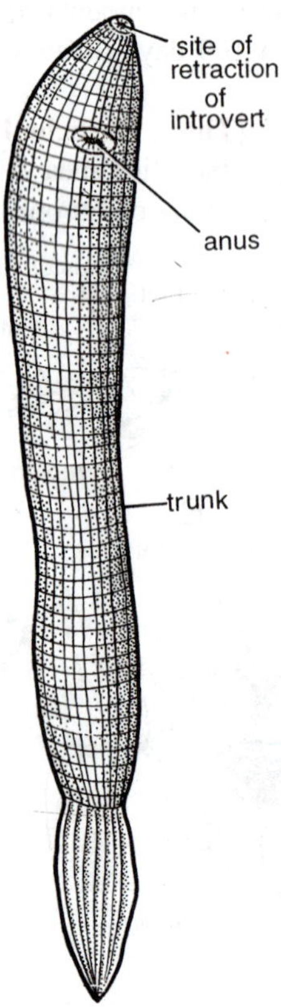

Fig. 16. *Sipunculus.* Entire.

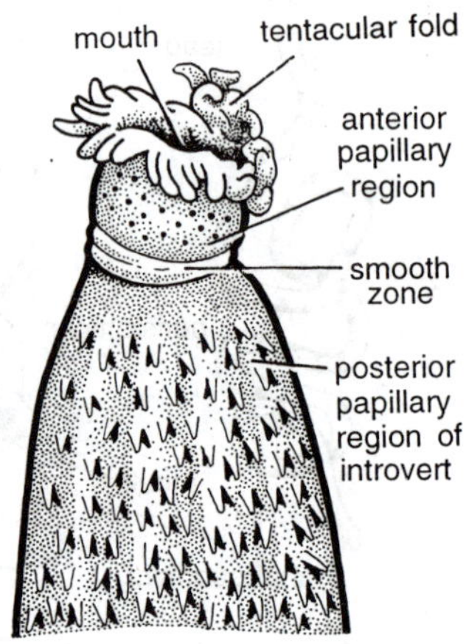

Fig. 17. *Sipunculus.* Anterior end with everted proboscis.

Sipunculida. The unsegmented body measures around 25 cm in length and is characterised by an anterior narrow retractile **introvert**, bearing mouth, papillae and a circlet of short fringe-like tentacles. The digestive system is complete and U-shaped. The circulatory system is absent. The nervous system is well developed; there is a brain, circumo-esophageal connectives and a mid-ventral nerve cord. The coelom is large and undivided. One pair of sac-like metanephridia are the excretory organs. Sexes are separate but alike. Sex cells develop in breeding season on retractor muscles of introvert. Fertilization is external and the zygote develops into a typical trochophore larva.

Sipunculus is a marine, bottom-dwelling burrowing worm. It is generally sedentary and feeds upon detritus.

Important Questions

➤➤Long answer type questions

1. Classify Annelida, giving diagnostic characters and common examples of each group.
2. Write short notes on : (i) *Amphitrite,* (ii) *Aphrodite,* (iii) *Arenicola,* (iv) *Chaetopterus,* (v) *Hirudo,* (vi) *Megascolex,* (vii) *Polygordius,* (viii) *Sipunculus,* (ix) *Tubifex.*

➤➤Short answer type questions

1. List down five diagnostic features of tubicolous polychaetes.
2. What are the special features of an Archiannelida ?
3. List down the different regions of the tube dwelling polychaete *Chaetopterus* and the functions performed by them.
4. Explain why earthworm and leech are included under one phylum.
5. Differentiate between Errantia and Sedentaria.
6. Who coined the term Annelida first time ?
7. What is metamerically segmentation ?
8. Define parapodia ?
9. Give an account of sense organs in annelids ?
10. Give the name of larvae of Annelida.
11. Establish the taxonomic position of *Aphrodite.*
12. Give the classification of leech ?

➤➤Fill in the blanks

1. The eyes become large, parapodia become enlarged and oar-shaped. The animal at this stage is known as .....
2. The segmentation of Annelida is known as .....
3. In *Chaetopterus,* the fans are modified .....
4. Gills in terebellid worms are the modified .....
5. Respiratory pigment.......in blood of Annelidans.
6. The known species of Annelida.
7. Name of the larva of Annelida is.......
8. *Protodrilus* belong to the class .... of Anmnelida.

➤➤True and false statements

1. *Eunice* is commonly known as 'polalo worm'.
   True / False /Do not know
2. The name of 'scale worm' is *Polynoe.*
   True / False /Do not know
3. *Chaetopterus* is a cilliary feeder.
   True / False /Do not know
4. *Amphitrite* is a fresh water animal.
   True / False /Do not know
5. *Megascolex* is earthworm found in Himalaya.
   True / False /Do not know
6. *Lumbricus* is earthworm found in Europe.
   True / False /Do not know

7. *Hirudinaria* is commonly known as medicinal leech.
True / False /Do not know
8. The 'pea nut worm' is commonly known as *Sipunculus*..
True / False /Do not know
9. 'Lug worm' is common name of *Arenicola.*
True / False /Do not know

➤➤Multiple choice questions

1. *Aphrodite* is commonly known as :
(a) lugworm (b) sea fan
(c) sea mouse (d) polalo worm
2. Tube dwelling polychaetes are known as :
(a) burrowing (b) planktonic
(c) pelagic (d) tubicolous
3. In Polychaeta, the nature of seta is :
(a) numerous (b) occur singly
(c) occur in bundles (d) fused
4. Presence of coelom and metamerism are the most important characters in :
(a) Helminthes (b) Arthropods
(c) Annelids (d) Coelenterates
5. The origin of the nephridium is :
(a) germinal (b) ectodermal
(c) mesodermal (d) endodermal
6. Cuticle of annelids is :
(a) non-chitinous and albuminoid
(b) chitinous
(c) chitinous and albuminoid
(d) non-chitinous
7. The body cavity of *Hirudo* is filled with :
(a) connective tissue (b) parenchyma tissue
(c) botryoidal tissue (d) coelomic fluid
8. Tube within a tube body plan is found in :
(a) *Pheretima posthuma*
(b) *Hydra*
(c) *Rana tigrina*
(d) *Lepus cuniculus*
9. *Chaetopterus* is commonly known as :
(a) lugworm (b) paddle worm
(c) sea pen (d) polalo worm
10. *Nereis* is commonly known as :
(a) clam worm (b) earthworm
(c) skate sucker (d) sea fan
11. Sucking of human blood by the leech is called :
(a) blood sucking (b) phlebotomy
(c) autotomy (d) phrynotomy
12. The coelom in Annelids is :
(a) pseudocoelic (b) enterocoelic
(c) schizocoelic
13. Lug worm is commonly known to :
(a) *Nereis* (b) *Arenicola*
(c) *Aphrodite* (d) *Polynoe*
14. *Sabela* generally known as :
(a) rag worm (b) scale worm
(c) peacock worm (d) none
15. Which is known as blood worm :
(a) *Terebella* (b) *Tubifex*
(c) *Lumbricus* (d) *Nereis*
16. The '*Tubifex*' is :.
(a) blood parasite (b) sand dweller
(c) fresh water (d) parasite
17. Medicinal leech is :
(a) *Hirudinaria* (b) *Hirudo*
(c) *Hirudo medicinalis*
(d) none of the above
18. *Sipunculus* commonly known as :
(a) nut worm (b) pea nut worm
(c) peacock worm
19. The annelidan larva is :
(a) Bipinnaria (b) Trochophore
(c) Cercaria (d) none

Answers

➤➤Fill in the blanks

1. *Heteronereis*, 2. metamerism, 3. neuropodia, 4. prostomial tentacles, 5. haemoglobin or erythrocruorin, 6. 8, 7000, 7. trochophore, 8. Archiannellida.

➤➤True and false statements

1. true, 2. true, 3. true, 4. false, 5. false, 6. true, 7. false, 8. true, 9. true.

➤➤Multiple choice questions

1. (c) 2. (c) 3. (d) 4. (c) 5. (c) 6. (b) 7. (d) 8. (c) 9. (a) 10. (b)
11. (a) 12. (b) 13. (c) 14. (b) 15. (c) 16. (b) 17. (c) 18. (c) 19. (b)

44

Annelida : General Account

COMPARATIVE ACCOUNTS OF VARIOUS SYSTEM OF ANNELIDA

Table 1. **Comparative Account of Digestive System** *of Nereis, Pheretima* **and** *Hirudinaria.*

Characters	*Nereis* (Clamworm)	*Pheretima* (Earthworm)	*Hirudinaria* (Leech)
1. Food and feeding habits	(1) **Carnivorous** and **predaceous**. Diet consists of small crustaceans, molluscs, sponges etc., which it captures actively.	(1) **Scavengers** or **omnivorous**. Food comprises dead organic matter (humus), bits of leaves, algae, etc., present in the soil which it ingests.	(1) Ectoparasitic **sanguivorous** Food consists of vertebrate blood.
2. Alimentary canal	(2) Alimentary canal is suspended in coelom by dorsal mesentery and held in position by septa.	(2) Alimentary canal is held in position in coelom by septa.	(2) Alimentary canal is held in position by the botryoidal tissue and mesenchyme.
3. Parts of alimentary canal	(3) Alimentary canal is a straight tube extending from one end of the body to other. It is differentiated into following parts : (i) Mouth, (ii) Buccal cavity, (iii) Pharynx, (iv) Intestine, (v) Rectum, (vi) Anus.	(3) Alimentary canal is a straight tube extending end to end in body. It is differentiated into following parts : (i) Mouth, (ii) Buccal cavity, (iii) Pharynx, (iv) Oesophagus, (v) Gizzard, (vi) Stomach, (vii) Intestine, (viii) Rectum, (ix) Anus.	(3) Alimentary canal is a straight tube extending up to 26th segment. It is differentiated into following parts : (i) Mouth and pre-oral chamber, (ii) Buccal, cavity, (iii) Pharynx, (iv) Oesophagus, (v) Crop, (vi) Stomach, (vii) Intestine, (viii) Rectum, (ix) Anus.
4. Mouth	(4) Mouth is a **transverse** slit, situated antero-ventrally in peristomium. Suckers and lips absent.	(4) Mouth is a **crescentic** slit being anteriorly in peristomium. Suckers and lips absent.	(4) Mouth is a **triradiate** slit, situated ventrally at the bottom of oral sucker. It is bounded by three **lips** formed by velum.

Characters	*Nereis* (Clamworm)	*Pheretima* (Earthworm)	*Hirudinaria* (Leech)
5. Buccal cavity	(5) Short and wide muscular chamber with cuticular denticles.	(5) Short, narrow, thin-walled chamber without denticles or jaws.	(5) Buccal cavity is short chamber containing three jaws, one behind each ray of mouth.
6. Pharynx	(6) Wide muscular chamber with denticles and a pair of stout, curved, chitinous jaws. Buccal cavity and pharynx enclosed in a common muscular sheath.	(6) Pear-shaped muscular chamber differentiated into a dorsal salivary and a ventral conducting chamber. Buccal cavity and pharynx are distinct.	(6) Oval muscular chamber connected by radial muscles with body wall. Buccal cavity and pharynx are distinct.
7. Proboscis and ingestion	(7) Buccal cavity and pharynx are everted as proboscis to capture and ingest the prey.	(7) Proboscis absent. Ingestion of soil is facilitated by the pumping activity of muscular pharyngeal wall.	(7) Proboscis absent. Jaws puncture the host's skin and blood is sucked in by pumping action of pharynx.
8. Oesophagus	(8) Long and narrow tube extending through five segments (6-10).	(8) Long and narrow extending through segments (5-8).	(8) Short and narrow extending not even through one full segment. Its inner lining is much folded.
9. Gizzard	(9) Since crushing of food is accomplished in buccal cavity and pharynx provided with denticles, a separate gizzard is not required and hence absent.	(9) Oesophagus is followed by a thick-walled, muscular, oval gizzard lined internally by tough cuticle. It serves as the grinding organ.	(9) Foood being blood, an organ for grinding or crushing is not at all required. Thus, gizzard is absent.
10. Crop	(10) Food being readily available, and elaborate storage organ for pre-digested food is not required. Thus, crop is absent.	(10) Food is readily available. Hence, crop is absent.	(10) Availability of vertebrate hosts being uncertain. leech, takes a heavy blood meal that lasts for a longer period. It has an elaborate storage organ, the crop, forming the largest part of alimentary canal.
11. Stomach and intestine	(11) Stomach and intestine combined into a stomach-intestine is a segmentally constricted thin-walled tube extending from oesophagus to the last but one segment.	(11) Stomach extends as a narrow tube from segments 9-14. Its both ends are sphinctered is followed by the intestine extending upto the last but 25-26 segments. A part of it is with typhlosole. A pair of intestinal caeca is present.	(11) Stomach is a small tubes, much folded internally. It is situated in 19th segment. It is followed by intestine extending upto 22nd segment. Its inner wall is thrown into villi.
12. Typhlosole and intestinal caeca	(12) Both are absent.	(12) 26th segment gives out a pair of lateral intestinal caeca. Middle part of intestine bears a mid-dorsal ridge or typhlosole.	(12) Both are absent.
13. Rectum	(13) Rectum is a short tube in the last segment.	(13) Rectum extends through the last 25 or 26th segments	(13) Rectum extends from 22 to 26th segment.
14. Anus	(14) Anus is a terminal aperture in the last segment.	(14) Anus is a terminal aperture in the last segment.	(14) Anus is situated middorsally on the 26th segment.
15. Digestive glands	(15) Digestive glands are : (i) A pair of oesophageal glands or caeca opening in the anterior part of oesophagus,	(15) Digestive glands are : (i) Pharyngeal or salivary gland cells (chromophil cells) on pharynx which secrete saliva. (ii) Gland cells in the	(15) Include : (i) Unicellular salivary glands on pharynx. (ii) Mucous gland cells in the lining of crop, stomach and intestine.

Characters	*Nereis* (Clamworm)	*Pheretima* (Earthworm)	*Hirudinaria* (Leech)
	(ii) **Gland cells** in the epithelial lining of oesophagus stomach and intestine.	epithelial lining of stomach and intestine.	
16. Hirudin	(16) Absent.	(16) Saliva does not contain hirudin.	(16) Saliva contains an anticoagulant of blood, called **hirudin**.

Table 2. **Comparative Account of Excretory System of *Nereis, Pheretima* and *Hirudinaria*.**

Characters	*Nereis* (Clamworm)	*Pheretima* (Earthworm)	*Hirudinaria* (Leech)
1. Excretory organs	(1) Excretory organs are coiled tubes of considerably large size, called **meganephridia**.	(1) Excretory organs are minute coiled tubes, called **micronephridia**.	(1) Nephridia are large coiled tubes or **meganephridia**.
2. Occurrence	(2) A pair of nephridia occurs in each segment except a few anterior and posterior ones.	(2) Numerous nephridia occur in each segment except the first two which totally lack them.	(2) There are 17 pairs of nephridia, one pair in each segment from 6 to 22.
3. Types of nephridia	(3) All the nephridia are similar.	(3) Three types of nephridia occurs : (i) **Integumentary** in all segments except the first two. (ii) **Pharyngeal** in segments 4, 5, and 6. (iii) **Septal** on both faces of septa behind 15th segment.	(3) Two types of nephridia are present : (i) **Pre-testicular nephridia** in segments 6 to 11. (ii) **Testicular nephridia** in segments 12 to 22.
4. Nephro-stome	(4) One end of nephridium opens in coelom of segment in front by a ciliated funnel or **nephrostome**.	(4) Each septal nephridium opens in the coelom of its own segment by a nephrostome. Integumentary and pharyngeal nephridia lack nephrostome.	(4) Nephridia of the adult lack nephrostomes which are probably represented by the ciliated organs.
5. Nephri-diopore	(5) Other end of nephridium opens to outside through an opening called **nephridiopore**. Such a nephridium is called **exonephric**.	(5) Integumentary nephridia possess nephridiopores opening to the exterior (**exonephric**). Pharyngeal and septal nephridia lack nephridiopores and oepn into alimentary canal (**enteronephric**).	(5) All nephridia possess nephridiopores opening to the exterior (**exonephric**).
6. Parts of nephridium	(6) Each nephridium can be distinguished into : (i) a small narrow neck, and (ii) a broad oval body. Ciliated organ does not form.	(6) A typical septal nephridium has : (i) Nephrostome or ciliated funnel. (ii) Main body comprising a straight lobe and a spirally twisted lobe, and (iii) A short terminal duct. Ciliated organ does not form.	(6) Each testicular nephridium has : (i) Main lobe, (ii) Vesicle with vesicle duct, (iii) Apical lobe, (iv) Inner lobe, (v) Initial lobe, and (iv) Ciliated organ. A pre-testicular nephridium has no ciliated organ.
7. Function of Nephridia	(7) Nephridia remove excretory wastes from blood as well as from coelomic fluid.	(7) All nephridia remove exretory wastes from blood. Only septal nephridia remove the same from coelomic fluid.	(7) All the nephridia remove excretory wastes from the haemocoelomic fluid.

Characters	*Nereis* (Clamworm)	*Pheretima* (Earthworm)	*Hirudinaria* (Leech)
8. Nature of excretion	(8) Mainly excrete ammonia, hence ammonotelic.	(8) Excrete 40% urea. Hence more or less ureotelic.	(8) Mostly excrete ammonia. Hence predominantly ammonotelic.
9. Accessory excretory organs	(9) There are no accessory excretory organs.	(9) Chloragogen cells also bring about excretion.	(9) Botryoidal tissue is also considered to have an excretory function.

Table 3. **Comparative Account of Reproductive System of *Nereis, Pheretima* and *Hirudinaria.***

Characters	*Nereis* (Clamworm)	*Pheretima* (Earthworm)	*Hirudinaria* (Leech)
1. Sexual nature	(1) Unisexual, exhibits sexual dimorphism.	(1) Bisexual, hence no dimorphism.	(1) Bisexual, hence no dimorphism.
2. Nature of gonads	(2) Gonads are temporary, appearing only during breeding season.	(2) Reproductive organs are permanent structures present in all sexually mature individuals.	(2) Permanent reproductive organs are present in all sexually mature individuals.
3. Sexual phase	(3) Sexually mature individual or sexual phase distinct from asexual phase and termed *Heteronereis*.	(3) Distinct sexual phase or *Heteronereis* absent.	(3) No distinct sexual phase.
4. Male reprodu-ctive organs	(4) These are represented by testes only, as loose masses of cells during breeding season.	(4) These comprise : (i) Testes, (ii) Testis-sacs, (iii) Seminal vesicles, (iv) Vasa deferentia, (v) Prostate glands, (vi) Accessory glands.	(4) These comprise : (i) Testes, (ii) Testis sacs, (iii) Vasa efferentia and vasa deferentia, (iv) Epididymes, (v) Ejaculatory ducts, (vi) Atrium, (viii) Prostate glands.
5. Testes	(5) Occur in all posterior segments as temporary masses of cells formed by cells of coelomic epithelium.	(5) There are two pairs of testes, one pair each in 10th and 11th segments.	(5) There are 11 pairs of testes, one pair each in segments 12th to 22nd.
6. Testis sacs	(6) Testis sacs are absent.	(6) Each pair of testis is enclosed in a testis sac.	(6) Each testis is actually the inner lining of a rounded testis sac.
7. Seminal vesicles	(7) Seminal vesicles are absent.	(7) Two pairs of seminal vesicles lie one pair each in segments 11 to 12. Serve for maturation and storage of sperms.	(7) Seminal vesicles are absent or epididymes lies in the 10th segment. They serve for storage of sperms.
8. Male gonoducts (Vasa-deferentia)	(8) Male gonoducts are absent. Sperms are shed through temporary ruptures in body wall.	(8) From each testis starts a narrow sperm duct or vas deferens. It extends behind upto 18th segment.	(8) From each testis sac arises a fine duct or vas deferens. 11 vasa deferentia of same size form a longitudinal vas deferens.
9. Dorsal ciliated organs	(9) One pair of ciliated funnels attached to dorso-lateral body wall of a reproductive segment are said to serve as gonoducts by acquiring temporary external opening.	(9) Ciliated organs are absent.	(9) Ciliated organs are absent.
10. Male genital openings	(10) No permanent opening.	(10) Two vasa deferentia along with a prostatic duct on either side unite to open through a pair of male genital apertures ventrally on 18th segment.	(10) Penis sac opens out through a male genital pore mid ventrally on 10th segment.

Characters	*Nereis* (Clamworm)	*Pheretima* (Earthworm)	*Hirudinaria* (Leech)
11. Spermi-ducal funnels	(11) Anterior end of each vas deferens forms a ciliated funnel like openinig called spermiducal funnel.	(11) Spermiducal funnels absent.	(11) Spermiducal funnels absent.
12. Epidi-dymes	(12) Absent.	(12) Absent	(12) Each vas deferens forms a convoluted mass or epididymis in 10th segment. Both epididymes serve to store sperms.
13. Ejaculatory duct	(13) Absent.	(13) Absent.	(13) Each epididymis gives out a narrow ejaculatory duct opening into an atrium.
14. Atrium and penis	(14) Atrium and penis are absent.	(14) Atrium absent. Terminal parts of vasa deferentia and prostatic duct of each side act as penis, during copulation.	(14) Atrium a pyriform sac in 10th segment, is distinguished into an anterior prostatic chamber and a posterior penis sac containing a filamentous penis, often seen protruding through the male gentital pore.
15. Prostate and accessory glands	(15) Prostate and accessory glands are absent.	(15) A pair each of multicellular prostate and accessory glands present. Function of prostatic secretion is not certain.	(15) Unicellular prostate glands present on prostatic chamber. Accessory glands absent. Prostatic secretion binds sperms into bundles called spermatophores.
16. Female reprodu-ctive organs	(16) These are represented by ovaries alone, as loose masses of cells during breeding season.	(16) These comprise (i) Ovaries, (ii) Oviducts, (iii) Spermat-hecae or seminal receptacles.	(16) These comprise : (i) Ovaries, (ii) Ovisacs, (iii) Oviducts, common oviduct and vagina, (iv) Albumen glands.
17. Ovaries	(17) Occur in all segments except the first few. Formed by cells of coelomic epithelium.	(17) There is a single pair of ovaries in 13th segment. Each consists of numerous finger-like lobes.	(17) There is a pair of delicate, thread-like coiled ovaries in 11th segment.
18. Ovisacs	(18) Ovisacs are absent.	(18) Ovisacs are absent.	(18) Each ovary is enclosed in a spherical ovisac.
19. Female gonoducts	(19) Female gonoducts absent. Ova escape through temporary openings of body wall or dorsal ciliated organs.	(19) Behind each ovary runs an oviduct. Both oviducts unite to form a very short common oviduct in 14th segment.	(19) Two oviducts, arising from two ovisacs, unite into a common oviduct which opens into a large, pear shaped vagina.
20. Female genital openings	(20) No permanent openings.	(20) Common oviduct opens through the female genital pores mid-ventrally on 14th segment.	(20) Vagina opens out through the female genital aperture mid-ventrally on 11th segment.
21. Oviducal funnels	(21) Absent.	(21) The anterior end of each oviduct forms an oviducal funnel.	(21) Oviducal funnels absent.
22. Sperma-thecae	(22) Spermathecae are absent.	(22) There are four pairs of spermathecae, one pair each in segments 6 to 9. These store sperms after copulation till fertilization occurs.	(22) Spermathecae are absent. After copulation sperms are stored in vagina till fertilization occurs.

Characters	Nereis (Clamworm)	Pheretima (Earthworm)	Hirudinaria (Leech)
23. Albumen glands	(23) Albumen glands absent.	(23) Albumen glands absent.	(23) Albumen glands present at the junction of two oviducts.
24. Copulation	(24) Copulation does not occur.	(24) Copulation takes place.	(24) Copulation takes place.
25. Fertilization	(25) Fertilization occurs in sea water, externally.	(25) Fertilizaton occurs in cocoon, externally.	(25) Fertilization occurs in vagina, internally.
26. Development	(26) Development indirect involving free swimming trochophore larva.	(26) Development direct. No free larval stage.	(26) Development direct. No free larval stage.

ORIGIN OF COELOM AND METAMERISM

Coelom and metamerism, and the various theories explaining their origin and evolution, have already been discussed in chapter 16 on the 'Organization of Metazoa'. Phylum Annelida represents the first group of metazoan animals developing a **true coelom** with **true metameric segmentation**.

Segmentation. When body is marked by transverse constrictions on A/P axis it is called segmentation.

(a) Superficial. A series of ring like creases develop in the cuticle and the body wall which facilitate the bending and telescoping of body. It is strictly ectodermal.

(b) Metameric. When the body is divided into smaller compartments on its A/P axis. Each compartment is separated by septum; when most of the mesodermal derivatives are also segmentally arranged.

Advantages. Enough number of structures and once the repeat pattern is available the units of the pattern can become elaborate and specialized.

ADAPTIVE RADIATION IN ANNELIDA

The evolution and spread from a single ancestral species to a variety of forms, which occupy different habitats, is called **adaptive radiation** or

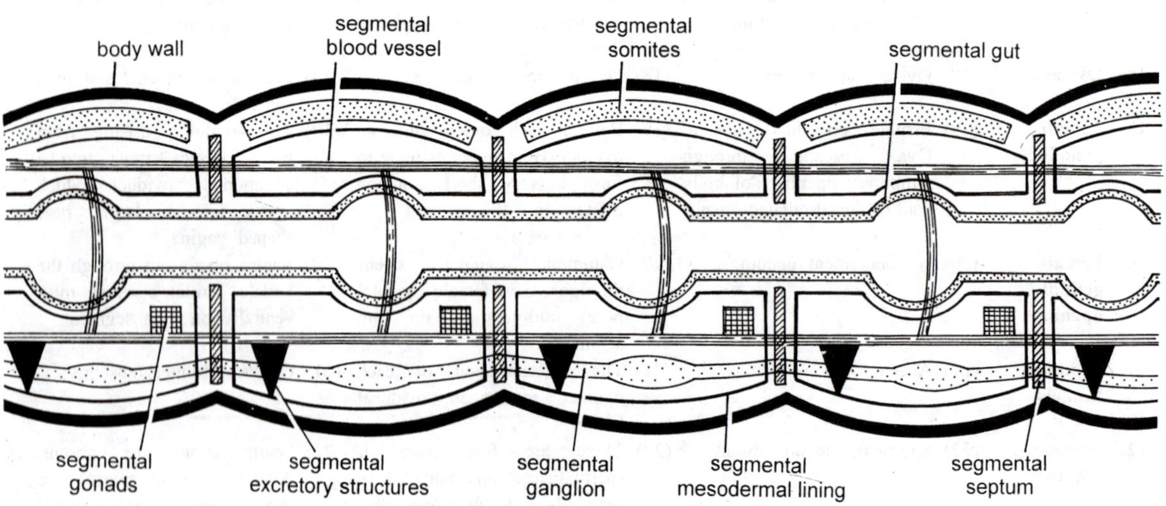

Fig. Metameric segmentation in Annelids.

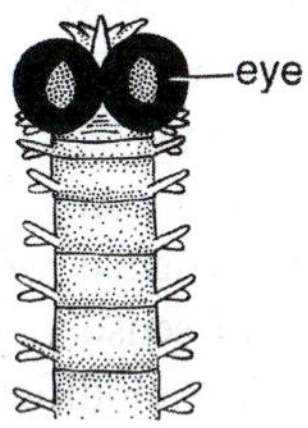

Fig. 1. *Vanadis grandis,* showing highly developed eyes.

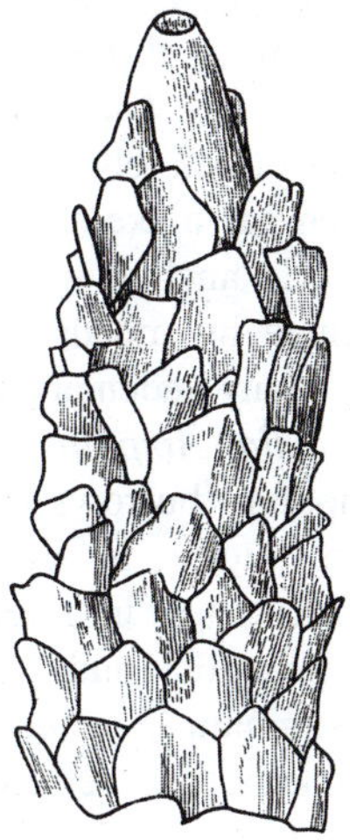

Fig. 2. *Owenia.* Anterior part of the tube with overlapping sand grains.

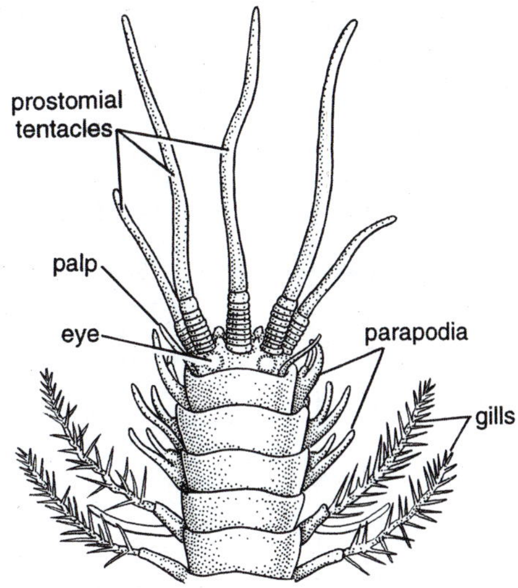

Fig. 3. *Diopatra.* Anterior end showing head and two .gill-bearing segments.

divergent evolution. This concept was developed by **H.F. Osborn** (1898). It states that starting from a common ancestral type, several different forms of evolutionary adaptations may occur, leading to evolutionary divergence, so that occupation of many ecological niches becomes possible depending on the adaptive nature of the invading species. Examples often cited as evidence include Darwin's finches of the Galapagos Islands, varied limb structure of mammals and Australian mammals, etc.

Phylum Annelida also displays a notable adaptive radiation or diversity in form and structure. The following account discusses adaptive radiation separately for each of the 3 major classes of Annelida, that is, Polychaeta, Oligochaeta and Hirudinea.

Adaptive Radiation in Polychaeta

Polychaeta is the largest class of phylum Annelida. It comprises some 64 families, 1600 genera and more than 5,000 species, most of which are marine. General form of body is extremely variable, depending on whether the worm is sedentary, free-living or pelagic. Ecologically, polychaetes are usually divided into 2 groups : *Errantia* and *Sedentaria.* Errant polychaetes include free-swimming, pelagic, crawling, actively burrowing and partially tubicolous forms. Sedentary polychaetes are permanent burrowers or tube dwellers. But this division is not natural. Adaptive diversity in polychaetes may be discussed dealing with two main aspects : (i) **habitat**, and (ii) **nutrition** or **feeding**.

[I] Diversity According to Habitat

1. Crawling polychaetes. Crawling polychaetes or bottom dwellers live under stones, submerged vegetation and sessile organisms. They most nearly approach the 'typical' or 'generalized' body form as represented by *Nereis, Eunice, Syllis, Colycera, Phyllodoce, Polynoe, Aphrodite,* etc. They possess well developed head bearing several kinds of sensory organs such as eyes, palps, antennae and cirri. Parapodia are large and used in crawling or swimming. Body segments are generally similar. In scale worms (*e.g., Polynoe, Lepidonotus*), dorsal or notopodial cirri become broad scale-like **elytra** forming a protective covering on entire dorsal surface. In sea-mice (*e.g., Aphrodite, Laetomonice),* elytra are covered by a 'felt' formed by fine hair-like notopodial setae.

2. Planktonic or pelagic polychaetes. These are adapted to live in open sea and tend to be transparent. *Tomopteris* has large, biramous

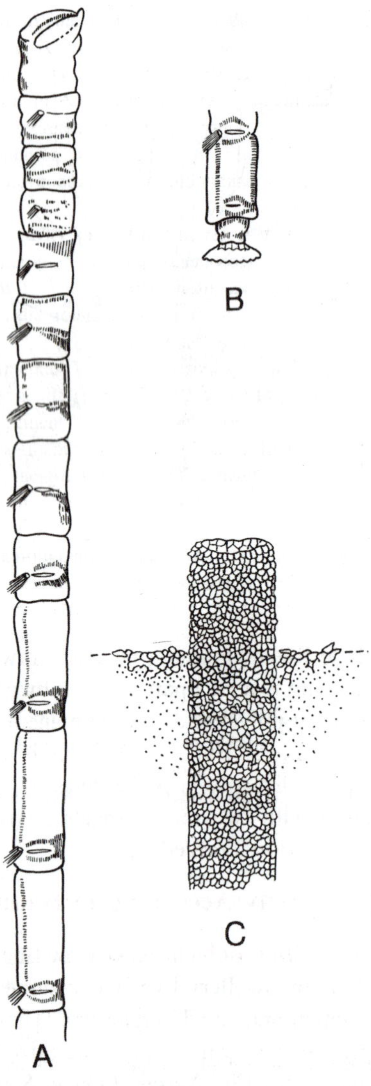

Fig. 4. *Clymenella* (Bamboo worm). **A.** Anterior part of worm. **B.** Posterior end. **C.** Sand grain tube with attached egg mass.

parapodia with membranous pinnules, but without setae. The Alciopidae (*e.g. Vanadis)* have very large eyes. Longitudinal muscles better developed than circular muscles, and septa reduced in crawling and swimming polychaetes.

3. Burrowing polychaetes. Many errant polychaetes such as *Glycera, Ophelia, Capetella,* etc., are adapted for burrowing in sand, like earthworms. Body is elongated. Prostomium is reduced or absent. Eyes, palps and tentacles are usually absent. Parapodia tend to become reduced. They move through the substratum by peristaltic contractions. Circular muscles are well-developed. Septa effectively compartment the coelomic fluid which has a skeletal function in locomotion. Setae serve to anchor against the wall of burrow which is lined by mucus to prevent its collapse.

Many sedentary burrowers such as *Arenicola, Terebella, Amphitrite,* etc., occupy more or less temporary burrows or tubes and move about very little. Body is cylindrical, plump and often differentiated into regions based on size, nature of parapodia and presence or absence of gills. Prostomium is generally devoid of sensory organs. Parapodia are reduced and setae modified into hook-like **uncini** that help in gripping the wall of burrow.

4. Tubicolous or tube-dwelling polychaetes. A tube-dwelling habit has developed in many families of polychaetes, errant as well as sedentary. They live in **temporary** or **permanent** tubes which are secreted by them or built from gathered materials. The tubes serve as a protective retreat or as a lair from which the worm may pounce upon a passing prey. The tubes vary greatly in form and construction, so that, the tubicolous worms fall into various adaptive groups, as briefly described below.

(a) Mucus-lined burrows. Some errant polychaetes such as *Eunice* and *Perinereis,* excavate mucus-lined burrow or tubes in sand and mud. They are similar to crawling polychaetes having well-developed prostomial sense organs and parapodia. Worms are carnivorous and extend from the openings of tubes to seize passing prey.

(b) Shell and sand grain tubes. Tubes are usually straight, planted vertically in sand or mud, and composed of sand grains or shell pieces cemented together with mucus. Conical tube of *Pectinaria* has a smooth masonary like surface. Tube of *Owenia* is made of flat pieces of shell or sand which appear like overlapping tiles. *Owenia* carries its tube about and uses its chimney-like end like a screw. Tube of *Diopatra* projects like a funnel-shaped chimney camouflaged with bits of shell, seaweed and other debris. Sandgrain tubes of bamboo worms, or *Clymenella,* occur in intertidal zones. These worms have a truncate head and parapodia reduced to ridges, looking like cane joints.

(c) Parchment tubes. Membranous or parchment tube of *Chaetopterus* is U-shaped, upto 70 cm long and 2.5 cm in diameter. Some terebellids also have membranous tubes. *Platynereis australis* covers the surface of tube with shell fragments and sand grains.

(d) Calcareous tubes. In fanworms (*e.g., Sabella, Serpula,* etc.), peristomial collar acts as a mould in forming the tube. Serpulid tubes are calcareous. Two large glands beneath the collar fold secrete the calcium carbonate which is added as a ring to the end of the tube while sabellids secrete a noncalcareous tube of sand grains embedded in mucus.

Some other tubicolous polychaetes have been described in connection with feeding.

[II] Diversity According to Feeding

Polychaetes have exploited a wide range of feeding niches, and several modifications of structure are directly correlated with the evolution of different methods of feeding.

1. Carnivores or raptorial feeders. Most crawling, some burrowing and tubicolous and all pelagic polychaetes are carnivores. They feed on small invertebrates including other polychaetes. These are captured by means of an eversible pharynx or proboscis. In *Nereis* and *Nephthys,* pharynx is armed with a single pair of chitinous jaws, while *Syllis* has a single tooth. Some have a long, tubular and retractile proboscis bearing 4 teeth at the tip in *Glycera.* Several teeth all over in *Phyllodoce,* and a circlet of teeth at the tip in *Autolytus.*

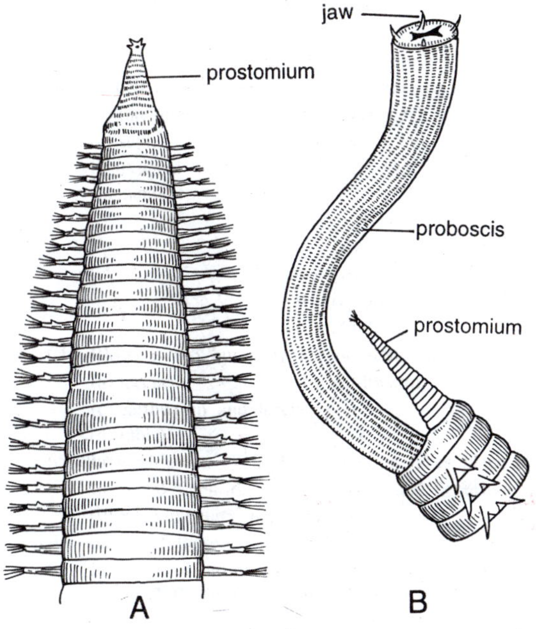

Fig. 5. *Glycera.* **A.** Anterior end. **B.** Head with protruded pharynx bearing four teeth.

Some jaw-bearing errant polychaetes, especially some nereids, are **herbivores**. They feed on algae, using their jaws to tear off small pieces.

2. Sand and mud detritus feeders. The bottom sand or mud of shallow seas is a source of food of great nutritional value. It contains bacteria, diatoms, other micro-organisms, as well as dead organic matter. Many polychaetes, especially sedentary species, rely on this organic deposit or detritus of sand.

(a) Direct deposit feeders. Some polychaetes obtain their food by directly-swallowing sand or mud as they burrow through it. Organic matter is digested while sand egested as castings. Ingestion is carried by a simple, non-muscular pharynx, which is everted as a proboscis, by elevated coelomic fluid pressure. Direct deposit feeders include burrowing and tube-dwelling species of ophellids, capitellids, maldanids, etc.

(b) Indirect deposit feeders. Indirect deposit feeders lack a proboscis. Instead, they use highly extensile ciliated grooved tentacles, secreting mucus. Small food particles moving down the

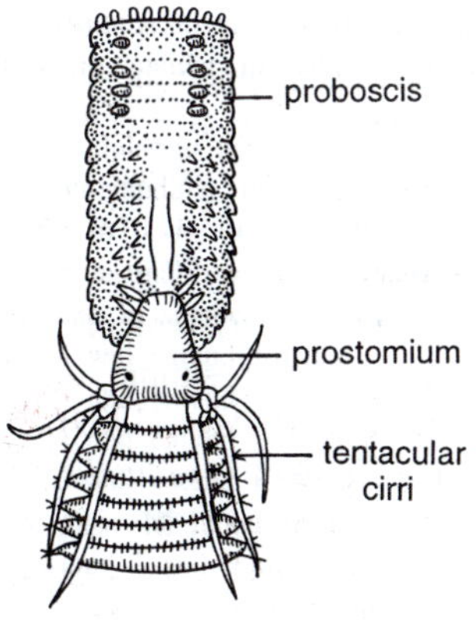

Fig. 6. *Phyllodoce.* Proboscis bearing several teeth.

groove by ciliary action, accumulate at the base of tentacles and conveyed to mouth by wiping the tentacles across the lips. Tentacle-feeding has evolved in *Amphitrite, Terebella, Pectinaria, Owenia,* etc. Some spionids and owenids are transition between indirect deposit feeders and filter feeders. They not only feed on bottom deposits, but also use palps or crown to collect suspended detritus.

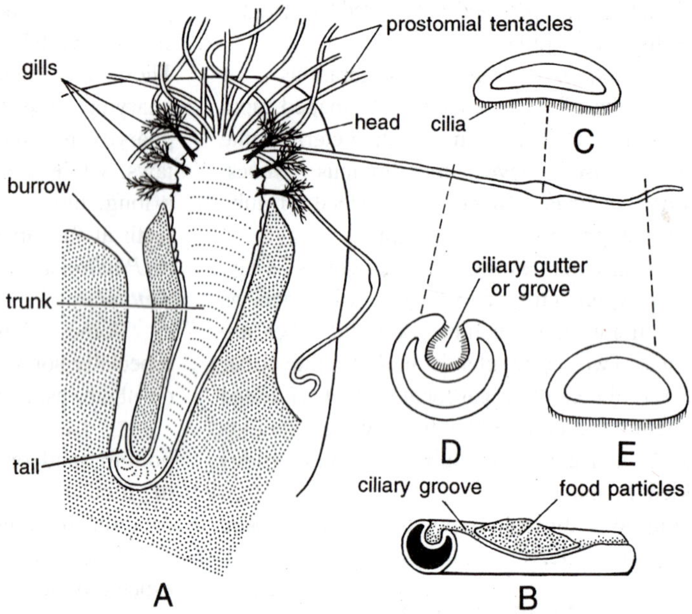

Fig. 7. *Terebella.* **A.** Worm at the aperture of burrow in feeding posture. **B.** Cross section of tentacle creeping over substratum. **C.** Section of tentacle showing ciliary groove. **D.** Section of tentacle transporting deposit material.

3. Filter feeders. Many sedentary and tubicolous polychaetes (*e.g. Sabella, Serpula, Owenia,* etc.) are filter feeders. Like indirect deposit feeders, a proboscis is lacking. Head is provided with long bipinnate filaments or tentacles, called **radides**, with a ciliated groove running along their oral surface. These are used for collecting food particles suspended in water.

Chaetopterus, which lives in a U-shaped parchment tube, has evolved a unique method of straining food particles from water. Notopodia of segments 14-16 are modified to form the fans. Beating of fans produces a water current entering the tube at the anterior end and flowing out of the other. Notopodia of 10th segment are extremely long and wing-like or aliform. Their ciliated glandular epithelium secretes a sheet of mucus forming a bag in which the suspended food particles are caught. Mucus bag ends in a ciliated food cup where the food is rolled up into a ball. It is passed forward along a mid-dorsal ciliary groove to the mouth.

The lungworms (Arenicolidael) were previously thought to be simple direct deposit feeders. But studies of **Kriiger (1959)** have shown them to be filter feeders. *Arenicola* excavates a L-shaped burrow and periodically ingests sand by means of a simple proboscis. This causes sand to cave in forming a funnel-shaped depression at the surface. Sand filters suspended food particles from water percolating down the funnel. This organically rich sand is then ingested by the worm. Periodically, the worm stops feeding, backs up the vertical limb of tube, and defaecates at the

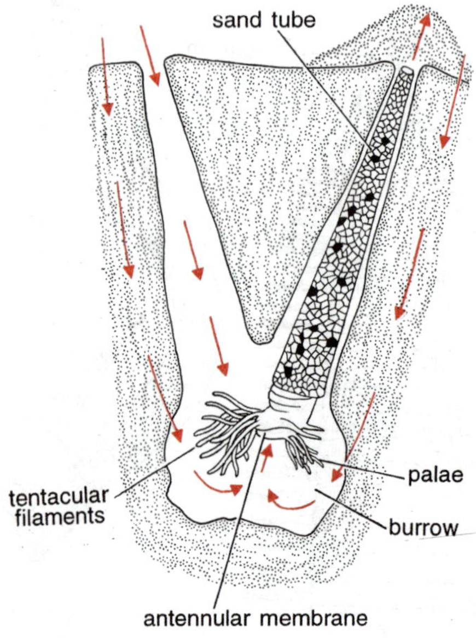

Fig. 8. *Pectinaria.* Sand grain tube in normal buried position. Arrows indicate path of water current.

surface, leaving its characteristic castings near the entrance of the burrow.

[III] Adaptive Radiation in Oligochaeta

1. Habitats. Aquatic oligochaetes are small and live in all types of freshwater habitats, but are most abundant in shallow waters. The only exception is the family Tubificidae (*e.g. Tubifex)* which inhabit the bottoms of deep lakes. In general, aquatic forms possess **longer setae** (*e.g., Ripistes).* Free-living species **crawl** about submerged plants and other objects feeding on surface debris. Most species **burrow** in mud and debris on substratum. Species of *Aulophorus* construct tubes, and the tropical species *Aulophorus carteri* builds its tube with spores of aquatic ferns. *Dero* and *Tubifex* live in mud tubes. Few species can encyst during unfavourable environmental conditions.

Terrestrial oligochaetes include earthworms which belong to four families. They are usually large and all **burrowers** found in soil, in leaf mould, and under stores, logs and tree barks. Members of Enchytraeidae are **amphibious** and live both in moist soil and in water.

2. Nutrition. Small aquatic species are **raptorial**. *Chaetogaster diaphanus* preys upon cladocerans *(Daphnia)* and other small aquatic invertebrates by the sucking action of pharynx. **Ripistis** is an **indirect deposit feeder**. While feeding it extends its anterior end from the tube. Its long setae moving about in water collect detritus which is periodically wiped off into mouth. Most oligochaetes, aquatic as well as terrestrial, are **scavengers**. They feed on dead organic matter which may be ingested directly or with mud or sand in the process of burrowing. For all soil-dwellers, the primary source of food is the plant litter or leaf mould. **Parasitism** is confined to the family Branchiobdellidae, all members of which are parasitic or commensal on freshwater crayfish.

[IV] Adaptive Radiation in Hirudinea

Some leeches are marine or terrestrial, but most of the aquatic species occur in shallow fresh water. Leeches are either **predaceous** or **blood-sucking ectoparasites**. Both types occur in all the three principal orders of leeches.

1. Predaceous leeches. Predaceous or predatory leeches outnumber the blood-sucking parasites in temperate climate. Family Erpobdellidae (order Pharyngobdellida) contain the greatest number of predaceous leeches. Their pharynx is non-protrusible and jaws or teeth are absent. They always feed on small invertebrates such as worms, snails and insect larvae. Feeding is relatively frequent and the prey is usually swallowed whole. Predaceous species remain hiding during day under submerged plants and rocky bottoms and become active at night.

Predaceous leeches are found in other families also. *Amphibious leeches,* such as *'Haemopis,* belonging to the blood-sucking family Hirudidae, are **carnivorous scavengers**. They live in soil or beneath stones, up-to a considerable distance from water. Their teeth are too small to pierce the skin of the prey, so that they can not suck blood. They can eat all kinds of protein food such as worms, insect larvae, tadpoles, small fish, carion and even

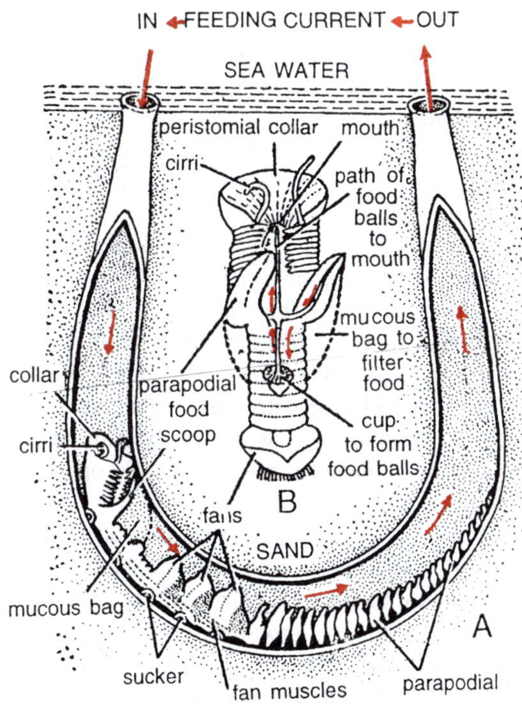

Fig. 9. *Chaetopterus.* **A.** Worm inside tube. **B.** Anterior end of the body in dorsal view. Direction of water current indicated by arrows.

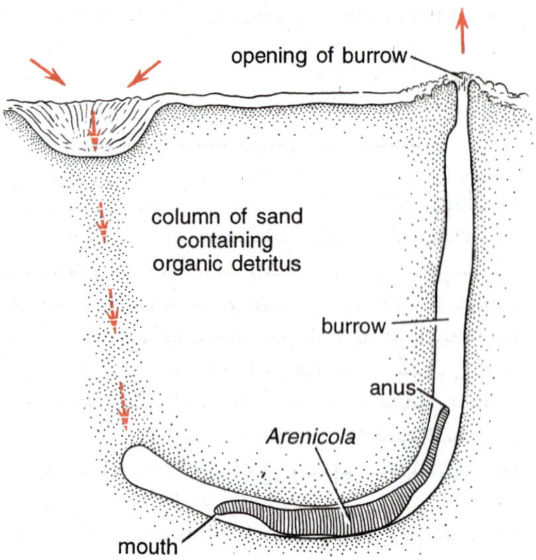

Fig. 10. *Arenicola.* Worm in its burrow, showing feeding mechanism. Arrows indicate direction of water flow.

young of their own species. Further, adults of some leeches are blood-sucking, but their juveniles are predaceous. Many glossiphoniids suck soft tissues from their hosts and are regarded as specialized predators.

2. Blood-sucking leeches. Blood-sucking or **sanguivorous leeches** can be arranged in two groups, **proboscidate** and **non-proboscidate**.

(a) Proboscidate leeches. These leeches possess a strong eversible muscular **proboscis** and lack jaws or teeth. They belong to the orders Acanthobdellida and Rhynchobdellida. Proboscis, when retracted, lies within a proboscis cavity. It is an unattached tube with a triangular lumen and lined internally as well as externally with cuticle. Ducts from unicellular salivary glands open into the proboscis.

When feeding, the proboscis is protruded out of their mouth and forced into the body of the host to suck blood and other tissues. Penetration of host's skin by proboscis is not properly understood. Probably it is aided by enzymatic action. Proboscidate leeches attack a variety of invertebrate and vertebrate hosts. Piscicolidae (*e.g., Piscicola, Pterobdella)* parasitise both freshwater and marine fish, sharks and rays. Members of other families (*e.g., Placobdella, Hemiclepsis)* suck blood from amphibians, turtles, snakes, crocodiles, birds and mammals.

(b) Non-proboscidate or jawed leeches. Non-proboscidate blood-sucking leeches belong to the order Gnathobdellida. Instead of an eversible proboscis, they possess a muscular sucking pharynx with three oval, blade-like, toothed **jaws**. When feeding, their anterior sucker containing mouth is attached to the host's body surface. Edges of jaws then make an incision in the hosts skin. Incision is anaesthetized by a substance of unknown origin. Secretion of **hirudin** in saliva prevents coagulation of oozing blood which is continually sucked by the muscular activity of pharynx.

Blood suckers feed infrequently depending on chance availability of their host. They can consume an enormous quantity of blood which is stored in their large crop. **Hirudo** normally ingests 2 to 5 times its own weight of blood, while *Haemodipsa,* a truely terrestrial leech from tropics, can take in ten times its own weight. Symbiotic bacteria found in their digestive tract provide the digestive enzymes, a unique arrangement with few parallels in the Animal Kingdom.

SEGMENTAL ORGANS IN ANNELIDA (COELOMODUCTS AND NEPHRIDIA)

In annelids are found certain tubes called **segmental organs**, as they are repeated in successive segments. These tubes serve to convey the excretory and reproductive products from coelom to the exterior. Segmental organs are primarily divided into two types :

(1) **coelomoducts** derived from mesoderm, and
(2) **nephridia** derived from ectoderm.

[I[Coelomoducts

Coelomoducts are normally wide tubes of mesodermal origin, developed as evaginations from coelom to the exterior. Typically, a coelomoduct opens, to the exterior by a **genital pore** and into coelom by a relatively large ciliated funnel, the

coelomostome. It is easily distinguished from the small ciliated funnel or **nephrostome** of the nephridium.

Coelomoducts primarily function as **gonoducts** and are confined to only a few reproductive segments. In the *Oligochaeta* (earthworms), the reproductive funnels and ducts, both male and female, are coelomoducts. However, in some forms, the coelomoducts may secondarily function as excretory organs. The uriniferous tubules of vertebrate kidney are coelomoducts.

[II] Nephridia

Nephridia are also segmentally arranged coiled tubes of ectodermal origin developed as invaginations from ectoderm into coelom. They communicate with the exterior through laterally placed small apertures called **nephridiopores**. Internally, they may end blindly **(protonephridia)** or may open by small ciliated funnels, or **nephrostomes**, into coelom **(metanephridia)**. Nephrostomes may open into coelom of the same segment in which the nephridia lie, or of the segment just in front. Nephridia are primarily excretory in function but may secondarily serve to convery the genital products to the exterior. Polychaete excretory organs are either protonephridia or metanephridia.

1. Protonephridia. The 'closed' or protonephridium seems to be more of the primitive type. It terminates in the coelom as a blind tube. Protonephridia, made of a few syncytial cells with an intracellular blind ending tubule, always develop in the larval polychaetes, irrespective of the kind of adult nephridia. The closed end or other parts of the tube are provided with peculiar specialized excretory **tube cells** or **solenocytes**. These are similar to the flame cells of Platyhelminthes and Rotifera. Solenocytes may occur singly or in groups. A solenocyte is a rather rounded ciliated cell connected to the protonephridium by a thin tube, the lumen of which encloses a long, vibratile flagellum. Excretory fluid enters through the walls of nephridial tubules which are internally ciliated. This fluid is driven into the lumen of nephridium by flagellum and forced to the exterior through nephridiopore. Protonephridia are found in some

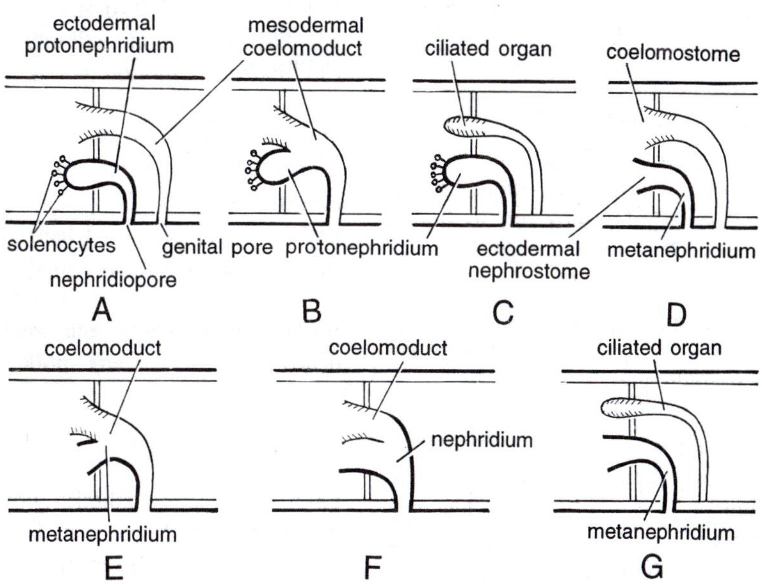

Fig. 11. Diagrammatic representation of types of nephridia, coelomoducts and nephromixia in various adult polychaetes (After **Goodrich**). **A.** Protonephridium and coelomoduct of *Vanadis*. **B.** Protonephromixium of *Phyllodoce*. **C.** Protonephridium and ciliated organ of *Glycera* and *Nephthys*. **D.** Metanephridium and coelomoduct of *Capitellidae* and *Notomastus*. **E.** Metanephromixium of Hesione. **F.** Mixonephridium of *Arenicola*. **G.** Metanephridium and ciliated organ of *Nereis*.

adult polychaetes such as *Vanadis, Phyllodoce Tomopteris, Glycera, Nephthys,* etc.

2. Metanephridia. The 'opened' or metanephridia are far advanced and found in the majority of polychaetes (*Neanthes*), all the oligochaetes (*Lumbricus*) and leeches. Instead of solenocytes, the inner end of metanephridium opens into coelom by a ciliated funnel or **nephrostome**. The other end opens to the exterior through the **nephridiopore**. A metanephridium is thus open at both ends. A typical metanephridium occurs in *Nereis,* those of other polychaetes differ only in minor details. Archiannelida usually possess one pair of nephridia which may be protonephridia (*Dinophilus*) or metanephridia (*Polygordius, Protodrilus*). Principal nitrogenous waste in polychaetes is ammonia. Excretory wastes diffuse from coelomic fluid or blood into the lumen of nephridial tubule and discharged to outside through nephridiopore.

(a) Micro and Meganephridia. Nephridia may be micronephridia or meganephridia on the basis of their size and number. **Micronephridia** or **meronephridia** are smaller in size, sometimes microscopic, and are numerous in each segment. They are networks of fine tubes lying on the body-wall and septa in each segment. All the nephridia of *Pheretima* are micronephridia. **Meganephridia** or **holonephridia** are larger in size and generally one pair per segment. They usually extend over two segments and their nephrostomes open into the segments next in front. They are represented in Polychaeta and Hirudinea. In *Neanthes, Nereis* and *Hirudo,* they are typical metanephridia with internal ciliated funnels. But in *Hirudinaria,* a ciliated funnel or nephrostome is lacking due to its modification into **ciliated organ**. The two kinds of nephridia may exist in the same worm and even in the same segment, as in **Megascolex**. In *Serpula* and some other tubicolous worms, a division of labour exists. Nephridia in the anterior region of body are large and excretory in function, while those in the posterior region are small and serve as gonoducts.

(b) Exo and enteronephridia. Nephridia are termed **exonephric** or **ectonephric** when they directly open to the exterior through **nephridiopores**, such as the meganephridia of *Nereis, Hirudinaria* and *Lumbricus,* and integumentary micronephridia of *Pheretima.* They are termed **enteronephric** when they lack nephridiopores and open into the excretory canals or alimentary canal, as septal and pharyngeal nephridia of *Pheretima.*

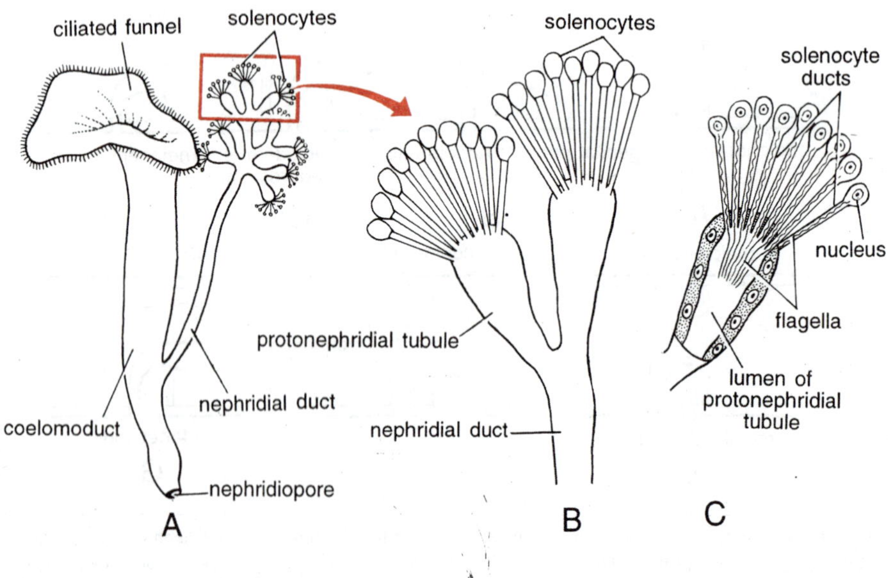

Fig. 12. Protonephromixium *of Phyllodoce paretti.* **A.** Relation of protonephridium and coelomoduct. **B.** Two branches of protonephridium. **C.** One branch with solenocytes in section.

[III] Nephromixia

In Oligochaeta, Hirudinea and the more primitive Polychaeta, nephridia and coelomoducts are separate. In some Polychaeta, coelomoducts do not remain independent but become fused, partially or wholly with the nephridia forming compound segmental organs or **nephromixia**. They consist both of ectoderm and mesoderm and used both as genital as well as excretory duct.

Nephridia and the coelomoducts show various degrees of combination. Either they share only the same external opening, or their fusion may be more intimate so that they share most of the same duct.

1. Protonephromixium. Coelomoduct is united with a protonephridium. It conveys both reproductive and excretory products to the exterior. Protonephromixia occur in *Phyllodoce.*

2. Metanephromixium. Coelomoduct is grafted on to a metanephridium. A good example is seen in *Hesione.*

3. Mixonephridium. Coelomoduct and nephridium are intimately fused to form a simple composite organ. Its funnel is formed by coelomoduct and its duct by nephridium. Mixonephridia occur in *Arenicola* where they are restricted to the 6 segments of the second tegma of the divided body. Each organ has a frilly funnel as an internal opening, a rich supply of blood vessels, and the gonad tissue in close proximity.

4. Ciliated organs. In some forms, coelomoducts are reduced to **ciliated organs**. In *Nereis,* they are attached to the dorso-lateral longitudinal muscles and are known to open externally.

REGENERATION

[I] Meaning of Regeneration

Regeneration (L., **re**, again + **generate**, to beget) is the normal response of organisms to replace the lost, injured, worn out or used up parts of body, by cellular multiplication and differentiation. Many lower forms (*e.g.*, *Hydra, Planaria)* have this capacity to a marked degree. But higher the animal in the scale of life, the more restricted is its power of regeneration.

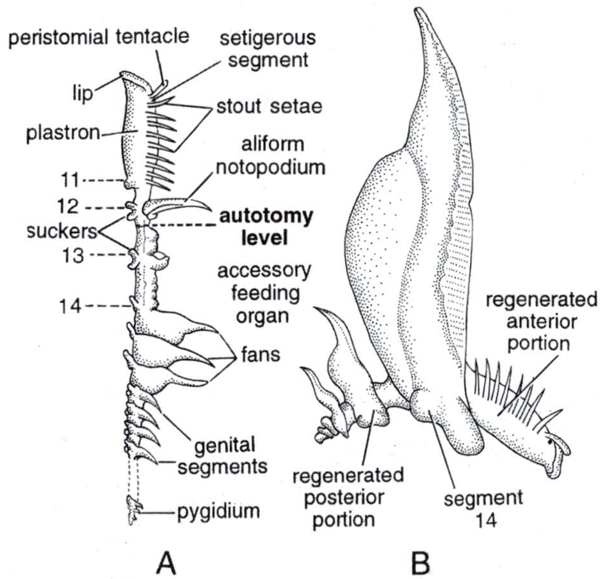

Fig. 13. Regeneration in *Chaetopterus.* **A.** Anterior region showing autotomy level. **B.** Anterior and posterior regeneration from 14th segment.

The capacity to regenerate lost parts occurs in all the annelids more so in the polychaetes and oligochaetes, but not in leeches.

[II] Process of Regeneration

At the cut surface or end of stump, the wound is first sealed with the formation of a dense mass of undifferentiated cells, called a **blastema**- (Gr., **blastema**, bud). From blastema then proliferates an appropriate number of segments to restore the adult pattern. In fact, the power of regeneration of lost parts, apparently absent in some animals, is actually present but is prevented by normal healing without formation of a blastema.

[III] Types of Regeneration

Capacity for regeneration is generally well developed in the Polychaeta. Tentacles, palps, cirri, elytra and other attenuated parts that are lost are soon replaced. In general, new trunk segments are more readily regenerated in worms without differentiation of body into regions.

In worms with a thorax and abdomen (or tail), caudal regeneration may occur but a new head does not form so readily. Thus nereids and phyllodocids seldom regenerate their lost head. However, syllids, sabellids and serpulids can regenerate the lost head but leading to some imperfections in the restorative process. For example, in *Autolytus* (a syllid polychaete), a cut made between segments 5 and 13 will cause restoration of head plus 5 segments, a posterior cut upto 42 segment will result in head only, and a more posterior cut without any regeneration of head. In **syllis spongicola**, head plus 2 segments only regenerate, no matter where the amputation is done. Similarly, in *Sabella,* prostomium, peristomium and a single additional segment are regenerated, even if the cut is made towards the posterior end of the body.

Some species the can regenerate the entire body from a few segments only, but surprisingly enough, worms with a high degree or complexity or segmental specialization such as *Chaetopterus* and *Myxicola,* can do so from a single segment. A single isolated segment of *Chaetopterus* upto first 14 segments will regenerate a complete worm, whereas if 15 anterior segments are cut off, there is no regeneration.

Some polychaete worms even display self-amputation or **autotomy**. For example, when disturbed, the elytra of certain scale worms (*Polynoe*) and the posterior trunk segments of some eunicids (*Diopatra, Marphysa*) become detached readily. The lost parts are later regenerated.

[IV] Control of Regeneration

Experimental studies, especially with oligochaetes, have indicated that the **nervous system** plays an important inducive role in regeneration. **Okada** (1934) found that severing the nerve cord alone can induce formation of a new head where the cut is made. A lateral secondary head plus few anterior segments will form if the severed end of nerve cord, cut just behind the supraoesophageal ganglion, is pulled through a hole in the lateral body wall. This indicates that regeneration is a **neurosecretory phenomenon**.

[V] Advantage of Regeneration

Regeneration along with autotomy is of great advantage to the animal. For the worms having an elongated body, injury inflicted by a predator is a common hazard. Thus, the ability to shed the damaged parts (autotomy) and then to replace them by developing new organs (regeneration) is of great **survival value** to the worm.

Same worms, such as *Autolytus,* undergo fragmentation. The fragments subsequently regenerate the missing parts to become complete animals. Thus regneration along with fragmentation results in **asexual reproduction**.

ASEXUAL REPRODUCTION IN POLYCHAETA

Replication is a fundamental property of all living organisms. It is expressed in two forms in the life cycles of animals : **asexual reproduction** and **sexual reproduction**. Here it is restricted to asexual reproduction only. Asexual reproduction in polychaetes takes the form of **fragmentation** or **budding** processes that involve both growth and regeneration of missing parts.

[I] Fragmentation

When an individual breaks into two or more parts, each capable of growing into a complete new individual, such a division is called **fragmentation**. Fragmentation is used in natural production of new individuals in two ways by polychaetes.

1. Spontaneous fragmentation. In *Dodecaceria, Ctenodrilus,Phyllochaetopterus, Autolytus,* etc., division of body into single segments or short sections occurs spontaneously. This is followed by regeneration of a complete new individual from each fragment. But instances of spontaneous fragmentation are very rare.

2. Orderly fragmentation. Orderly fragmen-tation occurs in some spionids (*Pygospio*), sabellids (*Sabella spallanzanii*), serpulids and syllids. Posterior part of body breaks off by a

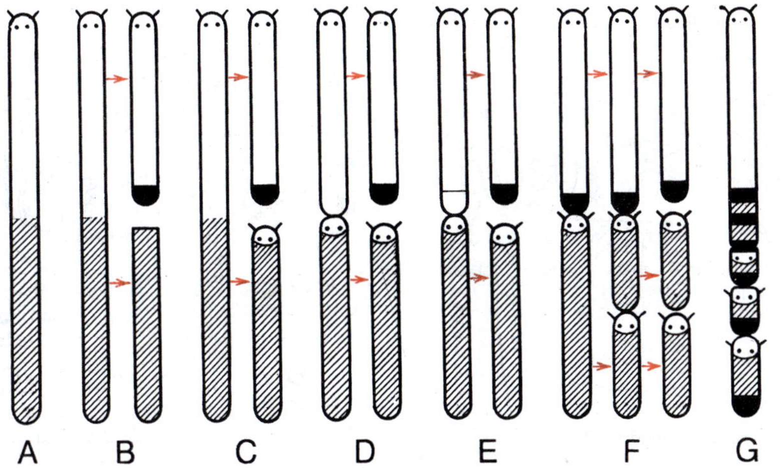

Fig. 14. Budding and stolon-formation in polychaetes. Modified after **Herlant-Meewis**. Gamete-bearing part striped, pygidial regeneration black.
A. Heteronereid or heterosyllis *(Odonotosyllis)*.
B. Epitokal region breaking off (Palolo worm, *Eunice)*.
C. Epitoke regenerating a head after liberation *(Syllis gracilis)*.
D. Formation of head before liberation of epitoke *(Syllis amica)*.
E. Formation of new head and pygidium of stock before liberation of epitoke *(Syllis vittata)*.
F. Formation of a second stolon before liberation of the first.
G. Formation of multiple stolons *(Autolytus)*.

transverse fission to form a new individual. In some instances, position of break is often precisely determined. It is recognized by the formation of a **macroseptum** internally and a **white line** externally.

New individuals formed may be similar to the parent or but quite dissimilar meant for sexual reproduction. They are often better adapted for swimming up to the surface for spawning. Formation of sexual individuals by fragmentation of an existing body is known as **stolonization**.

Simplest type of fission is seen in *Salmacina* and *Filograna* (Serpulidae), in which a constriction is formed near the posterior end of body, dividing the worm into two. Anterior part regenerates a new anal region, while posterior part forms a new cephalic region. Both parent and the resulting offspring are similar and sexual (hermaphrodite), so that there is no alternation of generations.

The process is somewhat complicated in Syllidae and occurs in a great variety of ways. In *Odontosyllis* and *Eusyllis monilicornis,* the heteronereid or heterosyllid individual undergoes

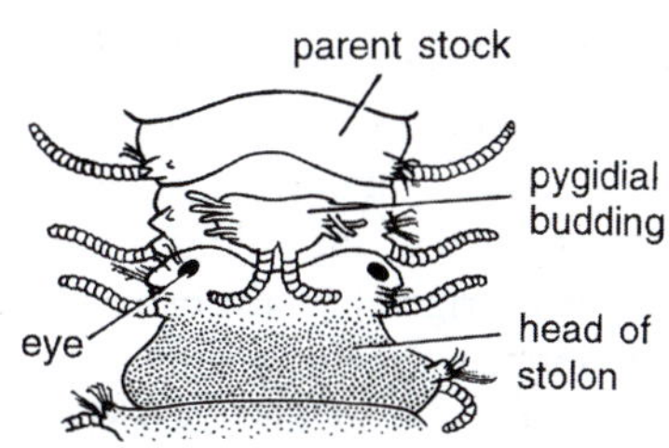

Fig. 15. *Syllis vittata.* Pygidial regeneration in stock in front of the stolon head.

metamorphosis in the posterior gamete-producing segments which do not break off [Fig. 14 (A)]. In *Syllis hyalina, Syllis spongicola* and palolo worms (*Eunice*), the sexual hind region, called **stolon** or **epitoke**, breaks off but does not develop a head because its independent life is brief, concerned only with swarming (B). In others, such as *Syllis gracilis,* the liberated epitoke may form an incomplete head lacking pharynx and jaws (C). The anterior non-sexual portion *(**stock** or **atok**)* regenerates the lost hind portion which again acquires gonads and becomes detached as the

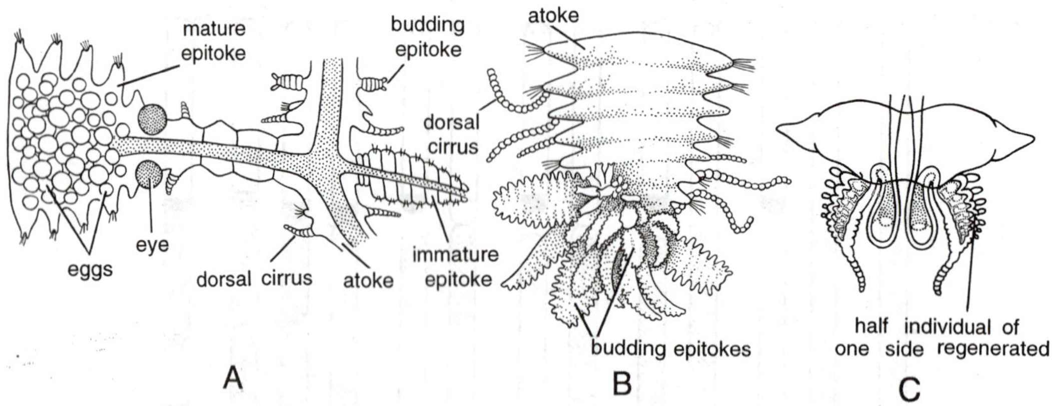

Fig. 16. Budding or gemmiparity. **A.** Portion of colony of *Syllis ramosa* showing development of sexual zooids (epitokes) from the sides of atoke. **B.** Posterior end of *Trypanosyllis crosslandi* showing a cluster of budding epitokes. **C.** Posterior regeneration in *Trypanosyllis prolifera* showing independence of (two sides).

sexual eiptoke in the next breeding season. In *Syllis* **amica**, regeneration of head actually precedes fragmentation, *i.e.*, the epitoke develops its head while still attached to the anterior non-sexual parent stock (D). This may also be accompanied by pygidial budding in the parent stock in *Syllis* (Fig. E and F). Further, in *Autolytus* and *Myrianida,* multiple stolons formation results in chains of individuals (G). All stolons in a chain are of the same sex. They remain connected together for some time in a linear series, the posterior-most being the oldest and the most developed. The original worm or parent swims at the water surface dragging after it the stolons which break-off one by one as sexually mature individuals. According to **Malaquin** (1893), a total of 29 individuals or stolons, complete with heads, have been observed attached to the parent stock in the syllid, *Myrianida.* They do not produce gametes before attaining some size, so that there is a true **alternation of generations**, the parent (stock) individual of the chain functioning as the asexual phase.

[II] Budding or Gemmiparity

Gemmiparity or **budding** is the process of proliferating new individuals, without division of the parent body. It is particularly well-developed in syllids, and there is a bewildering variety of methods in which this may occur. Genus *Trypanosyllis,* in particular, shows great variation from species to species. *Trypanosyllis prolifera* can produce heads from almost any segment and chains of new heads occur on consecutive segments. *Trypanosyllis asterobia* forms clusters of stolons each growing out from a single segment. *Trypanosyllis crosslandi* has a proliferating prepygidial zone producing bunches of stolons [Fig. 16 (B)]. In *Trypanosyllis prolifera,* in posterior regeneration of a new individual, the two sides of the body proliferate as two separate parts which later fuse together in the mid-line. *Syllis ramosa,* which is commensal within a certain deep sea hexactinellid sponge, produces lateral sterile buds from various segments. New individuals grow and remain attached to the parent sufficiently long forming a much branched colony. Some of these may themselves produce secondary lateral sexual individuals before separation.

ECONOMIC IMPORTANCE OF ANNELIDA

In terms of influence on. human welfare or economic importance, the earthworms (Oligochaeta) and leeches (Hirudinea) are more important than the polychaetes.

[I] Economic Importance of Polychaeta

1. As food. The sexual parts or epitokes of palolo worms are used as food by the native people. These are collected during swarming which takes place once a year. Epitokes are highly nutritive consisting of almost pure yolk-laden eggs. The natives attending these feasts follow certain traditional rites. These are considered a great delicacy and even given as presents by native chiefs. Palolo worm of Pacific is *Eunice viridis,* that of Atlantic is *Eunice schemacepha,* and of Japan is *Ceratocephala orawi.*

2. Transporter of soil. Lugworm *(Arenicola)* is a transporter of soil and considered even more effective than the common earthworm.

3. Fish bait. *Arenicola* is one of the commonest baits for certain fish. Other polychaetes used are species of *Neanthes* and *Glycera.*

4. Reef-building agents. Some sand and lime-concreting tubicolous polychaetes are important reef-building agents in some parts of the world. As a result of selective action in construction of tubes, these reefs may be pure sand or lime particles of homogeneous size.

5. Harmful polychaetes. Species of *Polydora* are **oyster-pets**, causing mud blisters in the nacreous layers of shells and making the oysters unfit to be sold. Oyster growers call it 'worm disease'. Some sedentary polychaetes cause **fouling** on the bottoms of ships, dikes and other harbour installations. They cause destruction to the building materials and add to the submerged weight resulting in lessening of speed of vessels. Species of **Fireworms** *Odontosyllis* (Fam. Amphinomidae) are large polychaete worms, about 30 cm long, found along tropical shores, are **injurious** to people. They cause severe burning from contact if picked up by the unwary collector because of their beautiful colours.

[II] Economic Importance of Oligochaeta

Economic importance of earthworms has already been dealt with in detail in Chapter 40, so that it would be unwise to repeat here the same information. Readers may refer to the relevant pages.

[III] Economic Importance of Hirudinea

Similarly, detailed imformation on the economic importance of leeches is available in Chapter 41.

Important Questions

►►*Long answer type questions*

1. Give a comparative account of the digestive system of *Nereis*, earthworm and leech.
2. Give a comparative account of the excretory organs of the annelidan types studied by you.
3. Tabulate the differences of reproductive organs in *Nereis, Pheretima* and *Hirudinaria.*
4. Discuss adaptive radiation based on nutrition, in the three main classes of Annelida.
5. Describe adaptive radiation based on habitats in Polychaeta.
6. What are segmental organs ? Give their account in Annelida.
7. Give an account of asexual reproduction in Polychaeta.
8. Write short notes on : (i) Adaptive radiation in leeches, (ii) Coelomoducts, (iii) Crawling polychaetes, (iv) Filter feeder polychaetes, (v) Regeneration in Annelida, (vi) Tubicolous polychaetes.

►►*Short answer type questions*

1. Mention the types of the nephridia in *Megascolex.*
2. Define 'Forests of nephridia' in three sentences.
3. Describe a typical nephridium in *Megascolex.*
4. Write explanatory notes on the coelom and haemocoel.
5. Differentiate micronephridium from meganephridium.
6. Compare the nephridia and coelomoduct in Annelida by mentioning three important features for each.
7. Describe the tubes based on their structure and formation in Annelida.
8. Write explanatory notes on the following : (i) Tube-dwelling polychaetes. (ii) Receptor organs in Annelida.
9. Discuss the relation of nephridia and coelomoducts in Annelida.

10. Give a detailed description of alimentary canal of any annelid studied by you.
11. Describe the structure of a typical parapodium of Polychaeta. What are its principal modifications ?
12. Give an account of locomotion in polychaetes.
13. Define true coelom ?
14. What is the difference in food and feeding habit of *Nereis* and *Pheretima* ?
15. Compare the digestive gland of *Pheretima* and *Hirudinaria* ?
16. What are chloragogen cells ?
17. Give a brief account about fertilization in different annelids ?
18. Comments on trochophora larval.
19. How many family consist the phylum Annelida?

➤➤Fill in the blanks

1. *Sabella* and *Serpula* are........
2. In *Hirudinaria* fertilization is........
3. *Pheretima* is........on excretory point of views.
4. Typhlosole and intestinal caeca are absent in........
5. Seminal vesicles are absent in.........

➤➤Multiple choice questions

1. The feeding habit of *Nereis* :
   (a) carnivorous (b) predaceous
   (c) both (d) none
2. Typlosole present in intestine of :
   (a) *Nereis*
   (b) *Hirudinaria*
   (c) *Pheretima*
   (d) All
3. Which is the more or less ureotelic :
   (a) *Pheretima* (b) *Nereis*
   (c) *Hirudinaria* (d) all
4. Trochophore larva present in the life history of :
   (a) *Hirudinaria* (b) *Nereis*
   (c) *Pheretima* (d) all
5. *Sabella* is :
   (a) cilliary feeder (b) bottom dweller
   (c) raptarial feeder (d) filter feeder
6. Asexual reproduction occurs in :
   (a) Protozoa (b) Porifera
   (c) Coelenterata (d) Polychaete
   (e) all
7. Which is use as fish bait :
   (a) *Hirudinaria* (b) *Nereis*
   (c) *Pheretima* (d) *Arenicola*
8. Harmful oyster pets is :
   (a) *Polydora* (b) *Arenicola*
   (c) *Syllis* (d) *Lumbricus*
9. *Hirudo* normally ingested the blood.......times its own weight of blood :
   (a) 1 to 2 (b) 2 to 3
   (c) 2 to 5 (d) 3 to 6
10. Generally known as fire worm :
    (a) *Odontosyllis* (b) *Syllis*
    (c) *Sabella* (d) *Autolytus*

Answers

➤➤Fill in the blanks

1. free swimming, 2. internal, 3. ureotelic, 4. *Nereis,* 5. *Nereis*

➤➤Multiple choice questions

1. (c) 2. (c) 3. (a) 4. (b) 5. (d) 6. (e) 7. (d) 8. (a) 9. (c) 10. (a)

45

Palaemon or *Macrobrachium malcolmsonii*: The Indian Freshwater Prawn

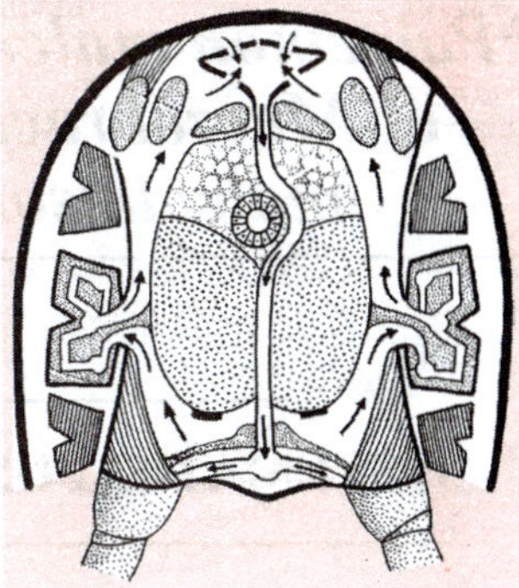

Phylum **Arthropoda** is the largest phylum of the Animal Kingdom, comprising of more than 999059 species. **Crustacea** is the largest and important class of the phylum Arthropoda. It includes prawns, crabs, lobsters, shrimps, barnacles, water-fleas and related forms. Also comprising of 35536 brown species. They get their name from the hard shells or exoskeleton they bear (L., **crusta**, hard shell). Tiny crustaceans called copepods (less than one centimenter long) are the most abundantly found animals in the ocean. They are so many, if their population in ocean could be weighed they may outweigh all the whales of the sea. There are some more crustaceans which shows remarkable power of distribution and adaptation e.g., *Gammarus wilkitzkii* makes a living on the underside of the Arctic ice pack, feeding on other crustaceans. There are crabs that live more than a mile beneath the ocean surface in the harsh environment around hydrothermal vents. They make their livings in total darkness and can withstand intense pressure, servere temperatures, and the caustic chemicals that are characteristics of these vents. Cleaner shrimp run underwater 'service stations', where the parasites and old scales of fishes are removed by the attending shrimps. The parasitic crustacean, *Cymothoa exigua,* lives in fish's mouth and attached with its tongue feasting on its blood.

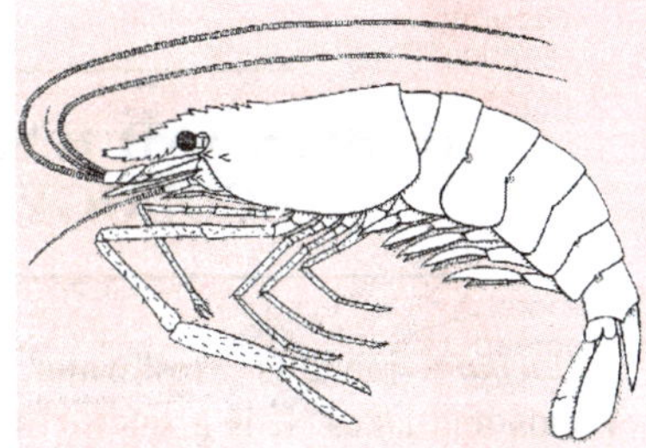

Palaemon (The Freshwater Prawn) is studied in the Indian Universities as a typical representative of the class Crustacea. Common Indian species of *Palaemon* are *P. carcinus, P. idae, P. rudis, P. lamarrei,* and *P. malcolmsonii.* The following account mostly relates to *P. malcolmsonii.* It is quite large in size, easily available and quite suitable for study. A monograph on *P. malcolmsonii* was written by **S. S. Patwardhan** and published in 1937, under the series of 'The Indian Zoological Memoirs.'

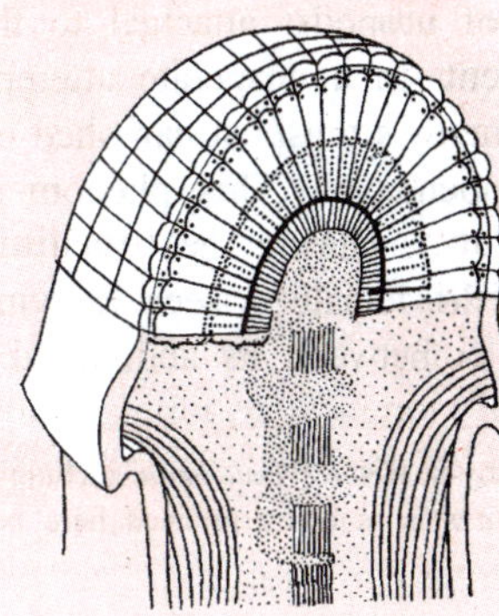

**Palaemon malcolmsonii* or *Macrobrachium malcolmsonii*

SYSTEMATIC POSITION

Phylum	Arthropoda
Subphylum	Mandibulata
Class	Crustacea
Subclass	Malacostraca
Order	Decapoda
Suborder	Natantia
Family	Palaemonidae
Genus	*Palaemon*
Species	*malcolmsonii*

HABITS AND HABITAT (ECOLOGY)

Palaemon inhabits freshwater streams, rivers, ponds and lakes. It is a **nocturnal** creature, hiding at the bottom during the day and coming to the surface at night in search of food. It is **omnivorous**, feeding on small organisms, like algae, mosses, minute insects, debris, etc. It walks slowly at the bottom with the help of its 10 walking legs and swims actively to the surface with the help of its 10 pleopods. When disturbed, it suddenly springs backwards with the help of a pair of uropods, attached to the last abdominal segment. In a desperate attempt to escape from the enemy's grasp, it can shed off one or more of its appendages. This phenomenon is known as **autotomy**. During the **breeding period** (May to July) the female is seen carrying a large number of eggs between its abdominal appendages.

EXTERNAL MORPHOLOGY

[I] Shape and Size

Body is elongated, more or less spindle-shaped and bilaterally symmetrical. It offers least resistance in swimming. Size of adult varies from species to species. *P. malcolmsonii,* now *Macrobrachium malcolmsonii,* found in Central India and Tamil Nadu, measures 25 to 40 cm in length. The giant prawn *P. carcinus* from Kerala is upto 90 cm long. While the dwarf prawn *P. lamarrei,* found almost throughout India, is 2.5 to 5 cm long.

[II] Colouration

Young stages are translucent and white, but the adults are differently tinted according to the species. Usual colour is dull pale-blue or greenish with brown orange-red patches. Preserved specimens become deep orange-red.

[III] Segmentation and Body Divisions

Body of adult prawn is distinctly divided into 19 segments or somites, all bearing jointed appendages. The segments are arranged into two main regions : an anterior **cephalothorax** (fused head-thorax) and a posterior abdomen.

1. Cephalothorax. Cephalothorax is large, rigid, unjointed and more or less cylindrical in shape. It consists of 13 segments. The joints between segments are obliterated. Cephalothorax is formed by the union of two regions : (i) **head** and (ii) **thorax**. Head consists of 5 segments, while thorax includes 8 segments, all bearing jointed appendages.

2 Abdomen. Well-developed abdomen is jointed, unlike cephalothorax. It is composed of 6 distinct movable segments, and a terminal conical piece, the **tail-plate** or **telson**, which is not considered a segment because of post-segmental

* Many species of *Palaemon,* including *P. malcolmsonii,* are now ascribed to the Genus *Macrobrachium.* But the generic name *Palaemon* is being retained here because of its familiarity and to avoid confusion at this stage.

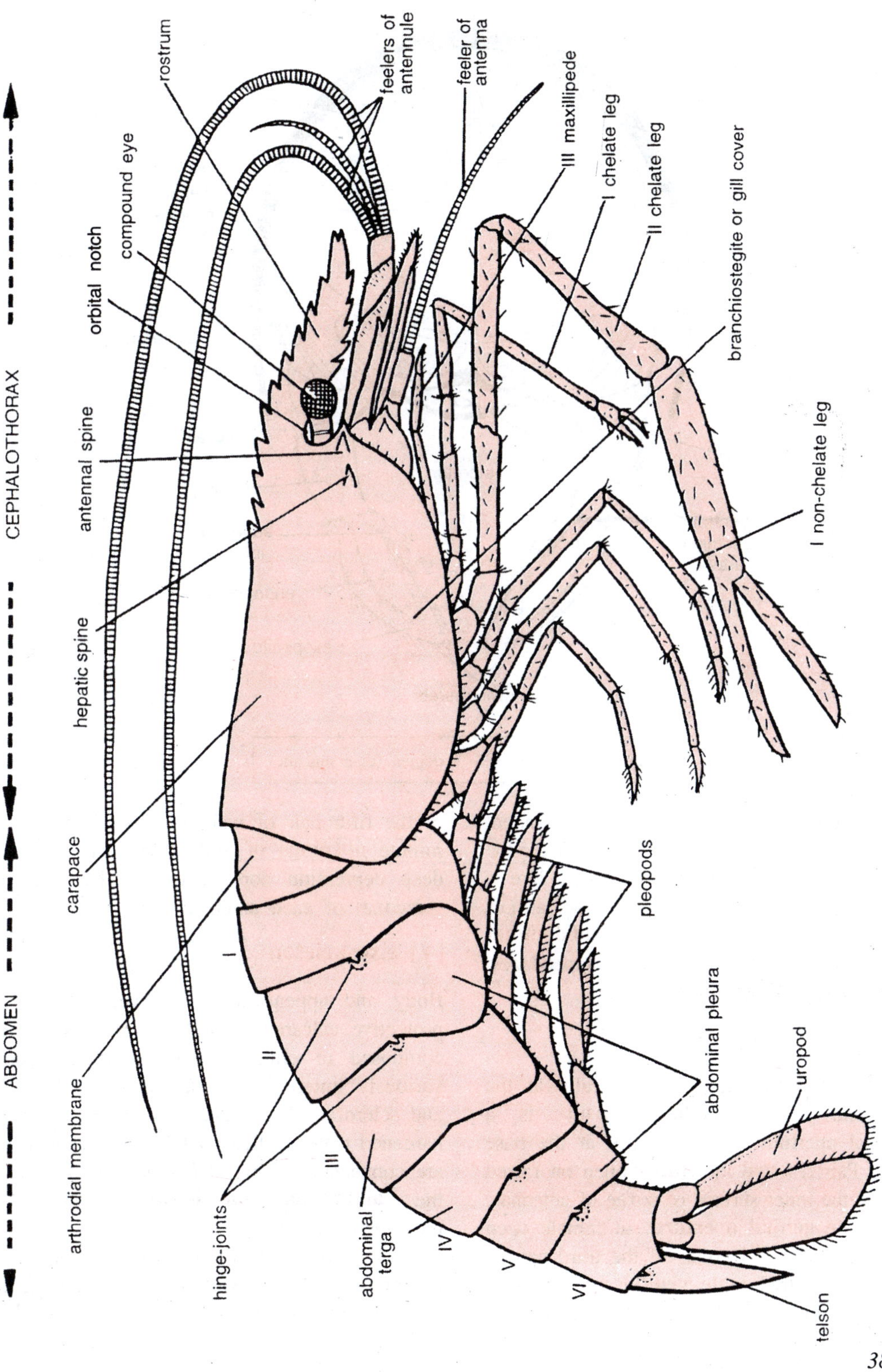

Fig. 1. *Palaemon.* External features of male in lateral view.

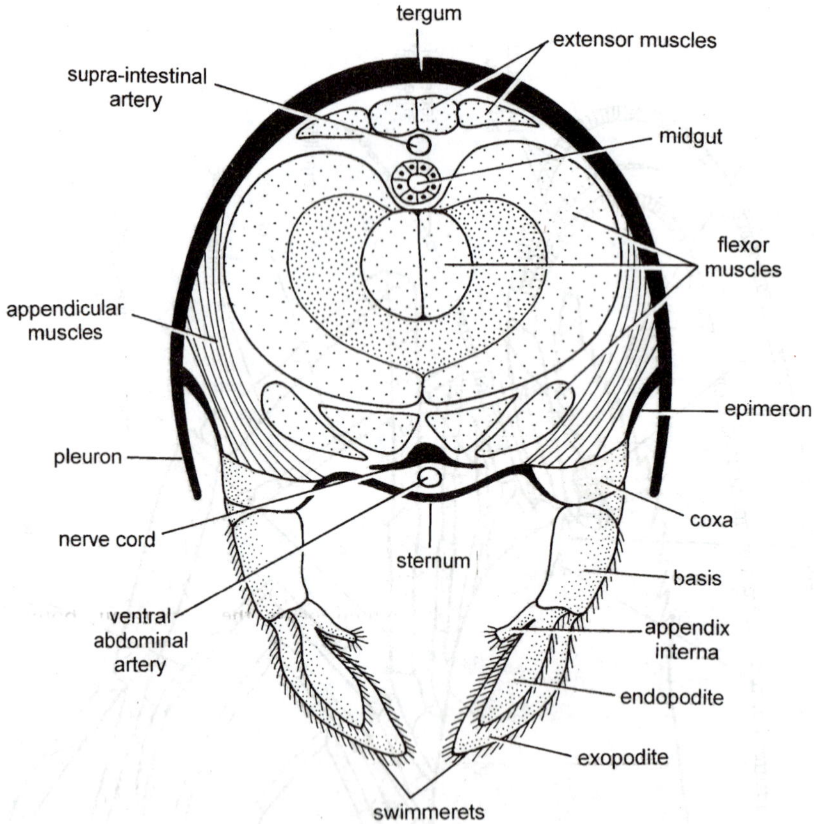

Fig. 2. *Palaemon.* T.S. abdomen (diagrammatic).

origin. Abdominal segments are dorsally rounded, laterally compressed and normally bent under the cephalothorax, so that the animal looks like a comma (,) in shape. The abdomen looks almost circular in a cross section. Each **abdominal** segment carries a pair of jointed appendages, called **pleopods** or **swimmeret**.

[IV] External Apertures

The slit-like **mouth** opens mid-ventrally at the anterior end of cephalothorax. **Anus** is a longitudinal aperture lying ventrally at the base of telson. Paired **renal apertures** open on raised papillae on the inner surface of coxae of antennae. Paired **female genital apertures** in female open on the inner surface of coxae of the third pair of walking legs. Paired **male genital apertures** in the male are situated on the inner surface of coxae of the fifth pair of walking legs. There are two minute **openings of statocysts**, one lying in a deep depression dorsally on the basal segment (precoxa) of each antennule.

[V] Exoskeleton

Body and appendages are covered by a hard protective calcareous **shell** or **exoskeleton**. It is composed of chitinous cuticle which becomes variously tinted by the deposition of lime salts and sclerotin. The exoskeleton comprises several hardened plates, called **sclerites**. Adjacent sclerites are connected by thin, soft, uncalcified cuticle or the **arthrodial membranes**, making the movements feasible.

1. Cephalothoracic sclerites. All the sclerites of dorsal and lateral sides of cephalothorax unite to form a single, large and coinuous **dorsal shield**.

The anterior and somewhat triangular region of dorsal shield is termed **dorsal plate**. It extends forward over the head as a laterally compressed and serrated vertical process, called **rostrum**. At the base of rostrum, on either side, is an **orbital notch**, which accommodates a stalked, jointed and movable **compound** eye. Just behind and below each orbital notch are two spine-like outgrowths, the anterior **antennal spine** and the posterior **hepatic spine**. The posterior region of dorsal shield is termed **carapace**. On either side of thorax, it hangs down freely as **branchiostegite** or **gill-cover** which encloses a gill-chamber housing the gills.

2. Abdominal sclerites. The sclerite of each abdominal segment is separate, ring-like and articulates with the adjacent sclerites by thin, flexible, uncalcified **arthrodial membranes** providing movable joints. In each abdominal sclerite, its dorsal broad plate is called as **tergum**, the ventral narrow transverse bar-like plate as **sternum**, and the two lateral flap-like plates as **pleura**. An appendage is connected with the pleuron of its side by a small plate the **epimeron**.

Tergum and pleura of an abdominal segment slightly cover the corresponding parts of the succeeding segment. This overlapping is known as the **imbricate arrangement** of terga and pleura. However, the pleura of second abdominal segment are much developed and overlap the pleura of both the first and third segments, thus disturbing the imbricate arrangement. Pleura of sixth abdominal segment are greatly reduced.

Two adjacent abdominal segments articulate with each other by means of a pair of **hinge joints**, one on either side. A hinge joint consists of a small round peg, fitting into a socket on the succeeding segment. However, the hinge joints are lacking between the third and fourth segments. Abdominal segments can move upon each other only in a vertical plane due to presence of arthrodial membranes and hinge joints between them.

APPENDAGES

Each segment of body bears a pair of jointed appendages. Thus, there are 19 pairs of appendages in *Palaemon*. They show considerable variations, depending on the functions they perform. However, they all are of a **biramous type**, as they are built on the same fundamental biramous plan.

Each appendage consists of a common base or **protopodite**, bearing two **ramii** or branches, an inner or median **endopodite** and an outer or lateral **exopodite**. Any appendage composed of two branches is called **biramous** (L. **bi**, two + **ramus**, branch). Typically, the basal **protopodite** is composed of two segments, a proximal **coxa** for attachment with the body and a distal **basis** which bears the two ramii, both comprising of several segments or **podomeres**.

With the exception of antennules which are uniramous, all the appendages of *Palaemon*, are **homologous structures**, regardless of their functions, because they are all biramous and have similar embryonic origin. As they occur in a serial

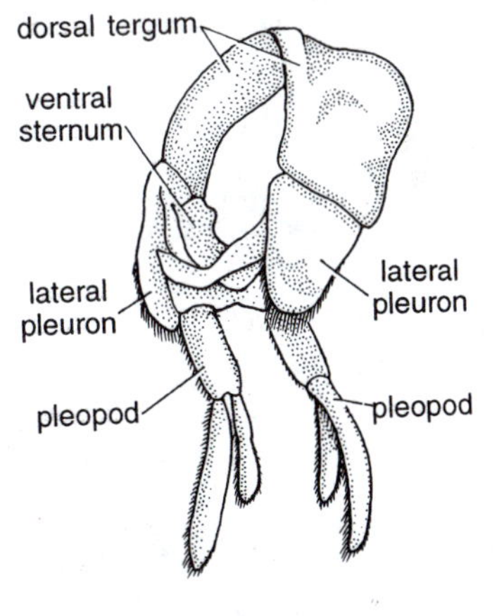

Fig. 3. *Palaemon.* Exoskeletal ring or sclerite of an abdominal segment, with appendages.

sequence on body, they also illustrate an example of **serial homology**.

In prawn, there are 19 pairs of appendages, 13 in cephalothorax and 6 in abdomen. Cephalothoracic appendages further include 5 pairs of anterior **cephalic appendages** and 8 pairs of posterior **thoracic appendages**.

1. Cephalic Appendages

There are 5 pairs of cephalic or head appendages. Beginning from the anterior end of head they are the **antennules**, **antennae**, **mandibles**, **maxillulae**, and **maxillae**. Antennules and antennae are **pre-oral**, while mandibles, maxillulae and maxillae are **post-oral**.

1. Antennules. The antennules are attached, one on either side, below the bases of eyes-stalks. The **protopodite** consists of three segments-a large proximal **precoxa**, middle **coxa** and distal **basis**. Precoxa bears a depression, containing the opening of **statocyst**, on its dorsal side. It also bears a basal spiny lobe called **stylocerite** and a distal spine on its outer margin. Coxa is short and cylindrical. Basis is elongated and without setae. It carries two long and many jointed, whip-like **feelers**, which are probably not homologous with the exopodite and endopodite. Outer feeler is further divided into an inner smaller branch and an outer larger branch. The feelers of antennules bear sensory setae and are **tactile** in function.

2. Antennae. The antennae lie, one on either side, just below the antennules. The **protopodite** is greatly swollen due to presence of excretory organ within, which opens by a minute **renal aperture** on the inner margin of **coxa**. Basis bears a **spine**. **Endopodite** is represented by a many-jointed sensory feeler, while **exopodite** is in the form of a broad and leaf-like plate, the **squama** or **scale**. It bears setae along its inner and distal

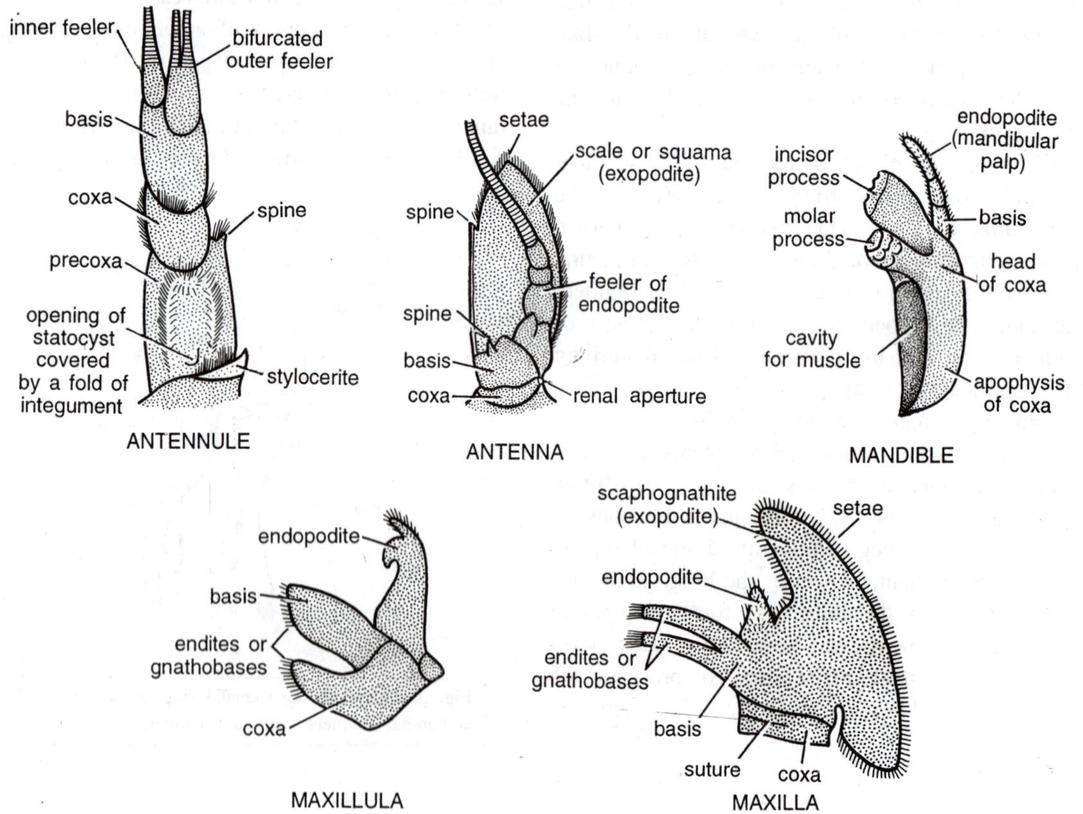

Fig. 4. *Palaemon.* Cephalic appendages.

margins, while the outer smooth margin bears a small spine. Squama probably serves as a balancer during swimming. Thus, the antennae are **sensory**, **excretory** and **balancing** in function.

3. Mandibles. The two mandibles are strong calcified bodies, lying one on either side of the mouth. Almost the entire mandible consists of **coxa**, which is differentiated into a proximal, triangular and hollow **apophysis**, and a distal solid **head**. The head forms two processes, a stout **molar process** bearing 5 to 6 dental plates, and a plate-like **incisor process** ending in 3 teeth. Outer margin of head carries a **mandibular palp** made of 3 segments. The proximal segment represents the **basis**, while two distal segments represent the **endopodite**. The **exopodite** is absent. Mandibles constitute the biting jaws and are **masticatory** in function.

4. Maxillulae. These are small, thin and leaf-like appendages. Free borders of **coxa** and **basis** are covered with pointed spines and project inwards as jaws or gnathobases (Gr., **gnathos**, jaw). **Endopodite** forms a curved process bifurcated at the apex. The **exopodite** is absent. Maxillulae help in the **manipulation** of food.

5. Maxillae. These are also thin and leaf-like mouth appendages. The small **coxa** is partially divided, while the large **basis** forms a bifurcated **gnathobase** internally. **Endopodite** is quite small, while **exopodite** forms a large expanded, fan-shaped **scaphognathite** or **baler**, the movements of which create a water current passing over the gills. The whole free margin of scaphognathite is beset with setae. Maxillae help in **respiration** and in the **manipulation** of food.

2. Thoracic Appendages

There are 8 pairs of thoracic appendages. These are differentiated into anterior 3 pairs of **maxillipedes** (Gr., **maxilla**, Jaw+**podos**, foot) or **foot-jaws**, and posterior 5 pairs of **paraeopods** or walking legs.

1. First maxillipedes. These are thin and leaf-like. Inner borders of coxa and basis form endites or **gnathobases**. Outer side of coxa bears a bilobed respiratory primitive gill or **epipodite**. **Endopodite** is smaller than **exopodite**, which gives out a plate-like process from its base. Margins of expodite and endopodite are fringed with setae.

2. Second maxillipedes. Coxa bears an **epipodite** and a **gill** (podobranch) on its outer margin. **Basis** carries a long, slender and unjointed **exopodite**, covered with setae along its distal half, and a 5- segmented **endopodite**. The segments or podomeres of endopodite are named from the base as **ischium**, **merus**, **carpus**, **propodus** and **dactylus**. The last two podomeres are bent backwards and inwards and possess cutting margins.

3. Third maxillipedes. These look leg-like in appearance and have the same parts as second maxillipedes. Outer border of **coxa** bears an **epipodite**. **Basis** supports a long, slender and

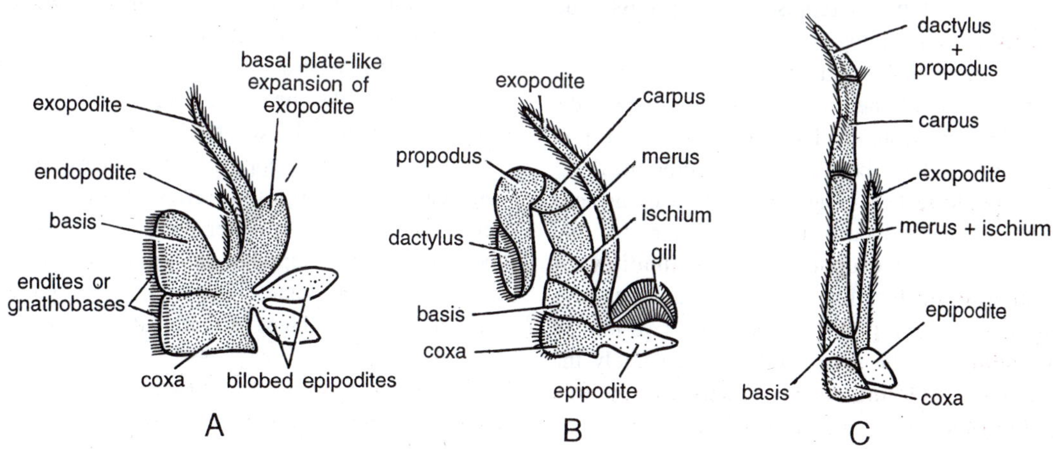

Fig. 5. *Palaemon.* Maxillipedes. **A.** First maxillipede. **B.** Second maxillipede. **C.** Third maxillipede.

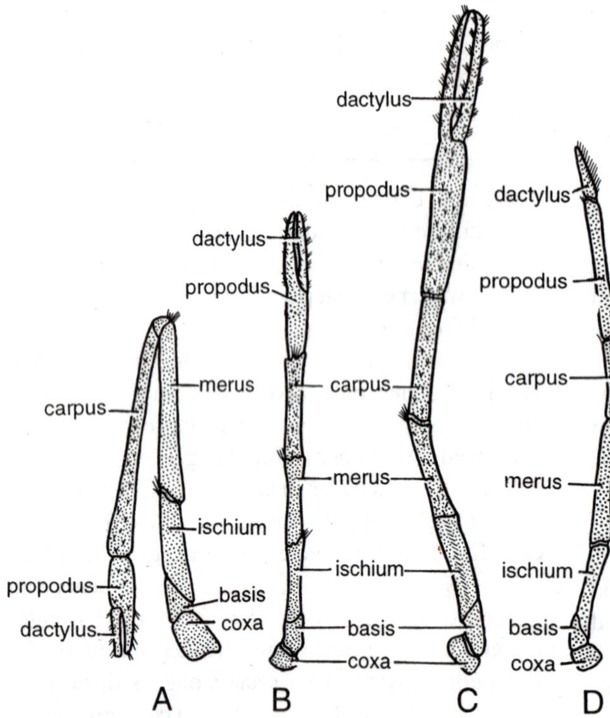

Fig. 6. *Palaemon.* Thoracic legs. **A.** First chelate leg. **B.** Second chelate leg of female. **C.** Second chelate leg of male. **D.** Typical or fourth non-chelate leg.

unsegmented **exopodite** covered with setae and a three-jointed **endopodite**. Proximal podomere of endopodite represents **ischium** and **merus** fused together, the middle podomere is the **carpus** and the distal podomere represents **propodus** and **dactylus** combined together.

The three pairs of maxillipedes take part in feeding and hold the food in position while the mandibles masticate it. They are also helpful in respiration as they bear gills and epipodites.

4. Walking legs. The 5 pairs of walking legs differ from maxillipedes in their greater size and in the absence of **exopodites** and **epipodites**. A **typical walking leg,** like the fourth, consists of a two-jointed **protopodite** and a five-jointed **endopodite**. All the seven podomeres, namely the **coxa, basis**, **ischium, merus, carpus, propodus** and **dactylus** are arranged in a linear series and are movably hinged together.

In the **first** and **second** pairs of legs, propodus is prolonged beyond its articulation with dactylus, so that the two podomeres work one against the other like the blades of a pair of forceps and form a **chela** or **pincer**. Such legs are termed **chelipeds** or **chelate legs**. They are used to grasp food and pass it on to the mouth. They also serve as organs of offence and defence. The second chelate legs in male are larger and more powerful than in female.

The **third**, **fourth** and **fifth** pairs of legs are **non-chelate** and typical. In female, each third leg bears a female reproductive aperture on the inner side of the coxa. While in male, each fifth leg bears a male genital aperture on the arthrodial membrane between the leg and thorax.

3. Abdominal Appendages

Abdomen bears 6 pairs of abdominal appendages, one pair in each of its segments. First 5 pairs, are the swimming **pleopods** or **swimmerets**, used as paddles, while the 6th pair are the **uropods** which, along with the post-segmental **telson**, form the tail fin. All these appendages are of simple biramous type.

1. Typical abdominal appendages. In a typical appendage, like the 3rd, 4th or 5th, the **protopodite** consists of a ring-like **coxa** and a cylindrical **basis**. The basis bears flattened leaf-like smaller **endopodite,** and larger **exopodite**. From the inner basal margin of endopodite arises a small rod-like structure, the **appendix interna,** with a knob-like head bearing many hook-like processes. In female, during breeding season, the appendix internae of opposite appendages articulate with each other forming a series of bridges which serve to carry eggs. Outer surface of basis and the margins of endopodite and exopodite are beset with numerous setae. The remaining pairs of abdominal appendages slightly differ from this typical structure.

2. First abdominal appendages. Appendix interna is absent and endopodite is greatly reduced in size. Rest of the structure is typical.

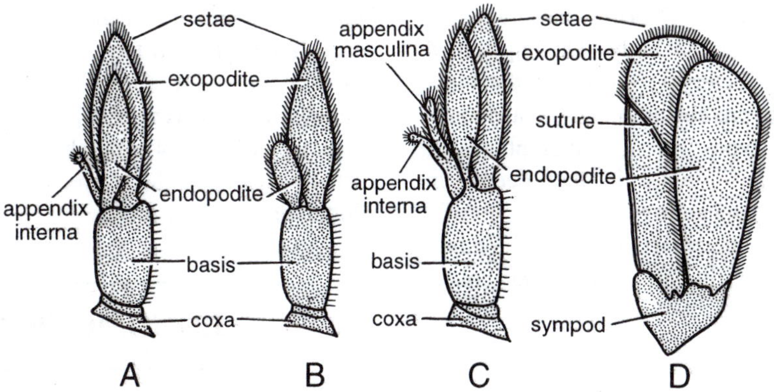

Fig. 7. *Palaemon.* Abdominal appendages. **A.** Typical. **B.** First. **C.** 2nd of male. **D.** Uropod.

3. Second abdomial appendages of male. Second pleopod of female is typical. But, in the second pleopod of male, there is an additional rod-like and setae-bearing process, the **appendix masculina,** lying in between the appendix interna and endopodite. The rest of the structure is typical.

4. Uropods. The 6th pair of abdominal appendages are called **uropods**. These are large and lie one on either side of the telson. Together with telson, they form **tail-fin** which enables the prawn to take backward spring in water. In each uropod, coxa and basis fuse together to form a triangular **sympod,** bearing the oar-shaped endopodite and exopodite. Exopodite is bigger than the endopodite and incompletely divided in the middle by a transverse-suture. Their margins, except the outer border of exopodite, are fringed with numerous setae.

BODY WALL

Body wall consists of an outer **cuticle**, a middle **epidermis** and an inner **dermis**.

[I] Cuticle

The outer layer of cuticle, forming the **exoskeleton,** is thick and non-cellular. It is further divisible into a thin, non-chitinous, outer **epicuticle,** and a **thick, chitinous, inner endocuticle**.

1. Epicuticle. It is made of an outer **lipoid layer** and an inner **protein layer**. Lipoid layer is permeable to gases but impermeable to water. Protein layer is relatively thick and hard, and is pigmented. Epicuticle is produced into spines of varying forms and bears, at places, fringes of setae. It is secreted by the tegumental glands lying in dermis.

2. Endocuticle. Endocuticle is elastic and permeable to gases and some solutes. It is secreted by epidermal cells. It is differentiated into three successive layers-a **pigmented layer,** a **calcified layer** and an **uncalcified layer**. All these three layers contain chitin.

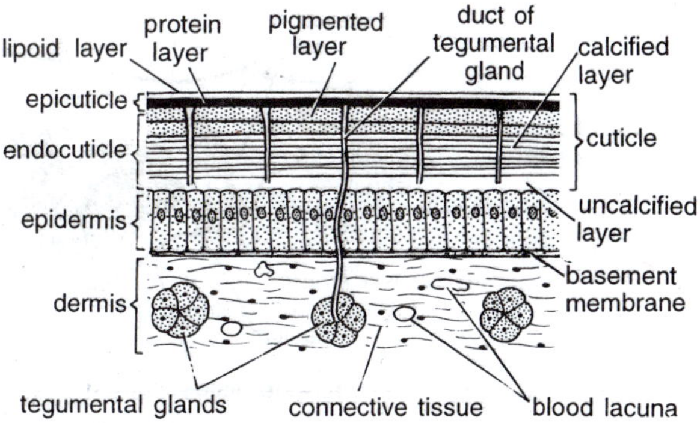

Fig. 8. *Palaemon.* V.S. of a portion of body wall.

The whole cuticle forms an external supporting structure of the body. The pigmented layer imparts a characteristic colour to the body due to the pigments present in the chromatophores. There are two types of chromatophores : (i) **primary chromatophores** located deeper in the body, and (ii) **secondary chromatophores** lying in chitinous layer. It is the secondary chromatophores which imparts a particular colour to the animal.

[I] Epidermis

It comprises a single layer of glandular columnar epithelium with centrally placed nuclei. It is lined by a thin basement membrane. This layer secretes the overlying endocuticle.

[II] Dermis

Dermis is made up of loose connective tissue beset with **blood lacunae**. It contains three types of **tegumental glands,** each of which opens to the outside through a fine duct.

BODY CAVITY

In contrast to annelids, the arthropods have a much reduced coelomic cavity. The space between gut and body wall is mostly occupied by muscles and organs with blood containing interspaces. These spaces together form the **haemocoel**, which is not a true coelom as it is not lined by the mesodermal epithelium. However, greatly reduced true coelom exists in the form of a number of separate spaces, such as enclosing the excretory (nephrocoel) and genital organs (gonocoel).

LOCOMOTION

The prawn crawls at the bottom of the river or pond by means of its walking legs. It can swim forward in a leisurely manner by beating its swimmerets or the abdominal appendages. It may take a quick backward spring by sudden contraction of the muscles which pulls the uropods and telson ventrally with a powerful stroke.

DIGESTIVE SYSTEM

1. Alimentary Canal

Alimentary canal consists of three distinct regions : (i) **Foregut,** comprising mouth, buccal cavity, oesophagus and stomach, (ii) **midgut** including intestine, and (iii) **hindgut** or rectum. Foregut and

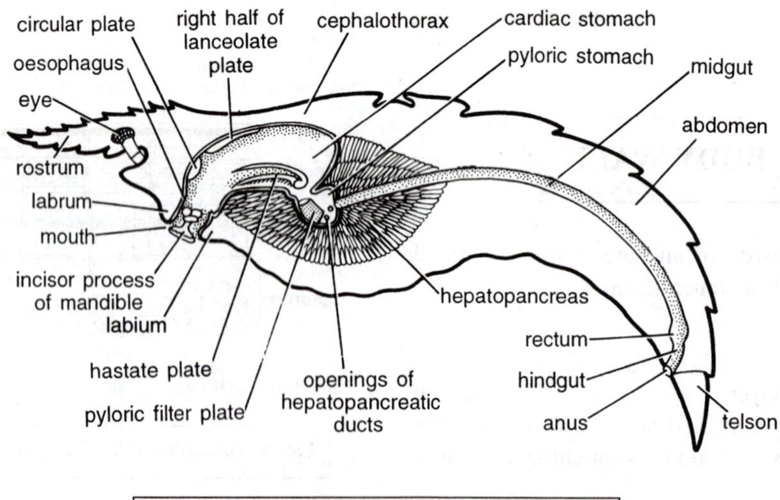

Fig. 9. *Palaemon.* Alimentary canal in lateral view.

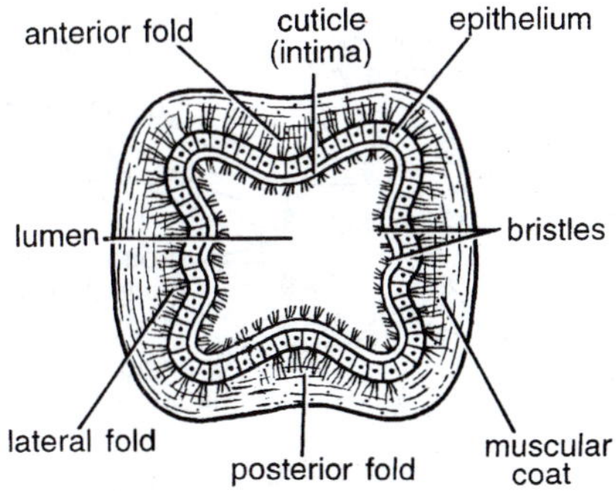

Fig. 10. *Palaemon.* T.S. oesophagus.

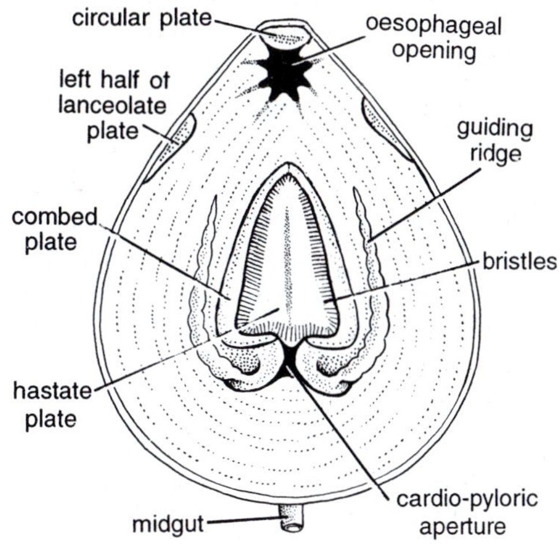

Fig. 11. *Palaemon.* Floor of cardiac stomach (dorsal view).

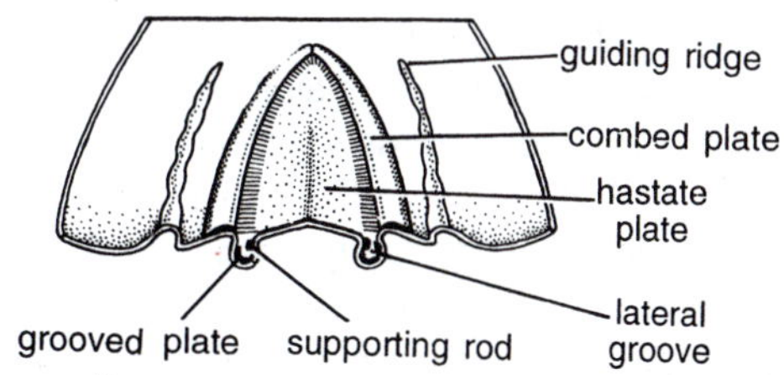

Fig. 12. *Palaemon.* Floor of cardiac stomach cut across, the hastate plate.

hindgut are lined internally by cuticle, called **intima**, which is shed with the exoskeleton when the animal moults. Midgut is lined internally by **endoderm**.

[I] Mouth

It is a large, slit-like aperture lying mid-ventrally below the anterior end of head. It is bounded in front by the shield-like fleshy **labrum**, laterally by the plate-like **incisor processes of mandibles** and behind by the bilobed **labium**.

[I] Buccal Cavity

Mouth leads into a short buccal cavity. It is antero-posteriorly compressed and has a thick cuticular lining which is irregularly folded. The **molar processes** of mandibles lie opposite each other in the buccal cavity to crush the food between them.

[III] Oesophagus

The short tubular oesophagus runs vertically upwards from the buccal cavity to the floor of cardiac stomach. Internally the thick muscular wall of oesophagus is thrown into four prominent longitudinal folds, one anterior, one posterior and two lateral.

[IV] Stomach

Stomach occupies most of the cephalothrac cavity. It remains buried laterally, ventrally and posteriorly in the hepatopancreas. Stomach of prawn is thin-walled and double-chambered, consisting of two parts : (i) a large anterior bag-like **cardiac stomach**, and (ii) a much smaller posterior **pyloric stomach**.

1. Cardiac stomach. The inner cuticular lining, or **intima**, of cardiac stomach presents numerous, inconspicuous, longitudinal folds covered by minute bristles. The wall of stomach is supported by some cuticular plates which remain embedded in it. Forming the anterior wall of the oesophageal opening is a **circular plate**. Behind it, on the roof of the stomach, is a **lanceolate plate**. A large triangular plate is embedded in the mid-ventral floor of cardiac stomach. It is called the **hastate plate**, because it looks like the head of a spear. Upper surface of hastate plate has a thick growth of delicate setae and carries a distinct **median ridge** with gradually sloping sides. The posterior triangular part of hastate plate is depressed and fringed with setae along its edge.

It forms the anterior border of the **cardio-pyloric aperture**. Each lateral side of hastate plate is supported beneath by a longitudinal cuticular **supporting rod**. A narrow **lateral groove** runs along either lateral border of the hastate plate. Floor of each lateral groove is covered by a cuticular plate which resembles an open drain pipe and is called the **grooved plate**. Each lateral groove is bounded on its inner side by the supporting rod and, on outer side, by a long cuticular **ridged plate**. Inner border of each ridged plate is fringed all along with a row of delicate bristles, forming a comb-like structure, so that it is also named as a **combed plate**. The bristles bridge over the lateral groove and partially overlap the lateral margin of the hastate plate, where they constantly keep moving in a living prawn. The two combed plates are united anteriorly, thus completely enclosing the hastate plate except that their incurved posterior ends remain separated by the cardio-pyloric aperture. Outside the combed plates, on either side, the lateral wall of cardiac stomach is folded inwards to form a prominent **lateral longitudinal fold**. The two folds are very low anteriorly but gradually increase in height posteriorly and also bend inwards to form the sides of the cardio-pyloric aperture. These folds are also known as the **guiding ridges** because they guide the food towards the cardio-pyloric aperture.

Cardio-pyloric aperture is narrow, X-shaped and leads into the pyloric stomach. It is guarded by four valves. **Anterior valve** is formed by the depressed posterior part of hastate plate, posterior **valve** by a semilunar fold of stomach wall, and two **lateral valves** by the large flap-like posterior ends of guiding ridges.

2. Pyloric stomach. Pyloric stomach is a small and narrow chamber lying below the posterior end of cardiac stomach. Its lateral walls are thick, muscular and prominently folded inwards, so that its cavity is imperfectly divided into a big **ventral chamber** and a small **dorsal chamber**, which are connected by a narrow vertical passage. Floor of ventral chamber is raised into a median longitudinal ridge, dividing it into two lateral compartments. Floor is covered by a **filter plate**. It is made of two rectangular surfaces

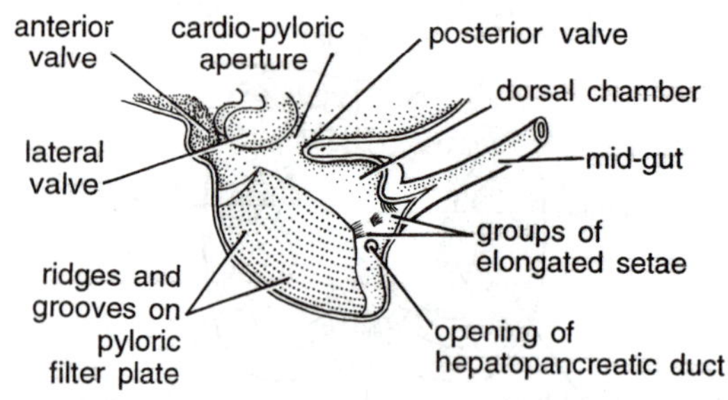

Fig. 13. *Palaemon.* Structure of the pyloric stomach (left wall removed).

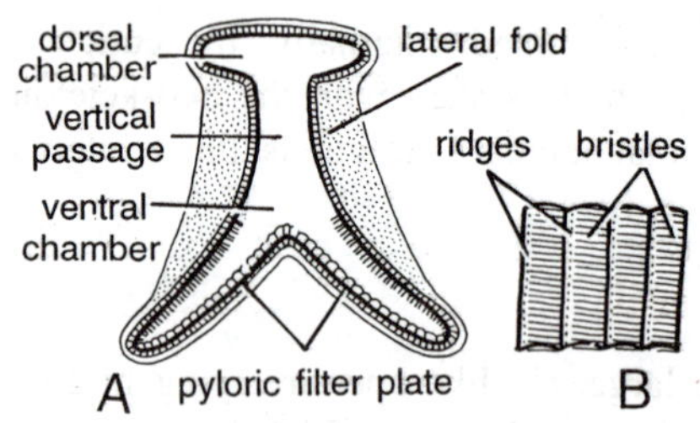

Fig. 14. *Palaemon.* **A.** T.S. of the pyloric stomach. **B.** Part of the pyloric filter.

and appears V-shaped in cross section. Each rectangular surface bears a series of alternating longitudinal ridges and grooves. The ridges bear rows of bristles forming a felt-like covering over the grooves. The side walls of ventral chamber are also covered with closely-set bristles which, together with the filter plate, form an efficient **strainer** or **filter**. This **pyloric filtering apparatus** allows only liquid food to pass through it. The paired **openings of the hepatopancreatic ducts** lie behind the filtering apparatus, just below the junction of the dorsal chamber of pyloric stomach and midgut. These openings are guarded by a group of **elongated setae** arising from the posterior end of the median ridge of filter plate.

Dorsal chamber gives out a small blind **caecum** dorsally and then leads behind into midgut. The junction of the two is guarded by one median dorsal and two lateral groups of **elongated setae** that project backwards into the midgut. These groups of setae strain the food entering the midgut and prevent its regurgitation into the dorsal chamber.

[V] Midgut

Midgut or intestine is a long, narrow and straight tube running back along the median line, between the extensor and flexor muscles, upto the 6th abdominal segment. Its lumen is wide at the anterior end but reduced posteriorly due to the presence of longitudinal folds.

[VI] Hindgut

It is the shortest portion of the alimentary canal, leading from midgut to anus. Its anterior swollen muscular part, called the **intestinal bulb** or **rectum**, bears many internal longitudinal folds. The terminal narrow, tubular part opens to the exterior through **anus**, which is a sphinctered mid-ventral longitudinal slit-like opening, situated on a raised papilla at the base of telson.

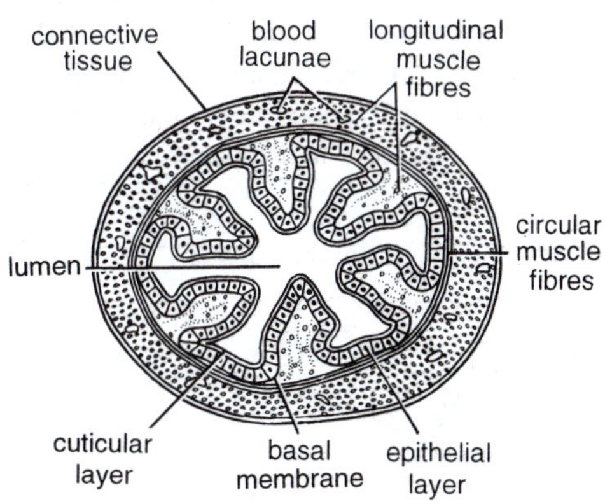

Fig. 15. *Palaemon.* T.S. of midgut.

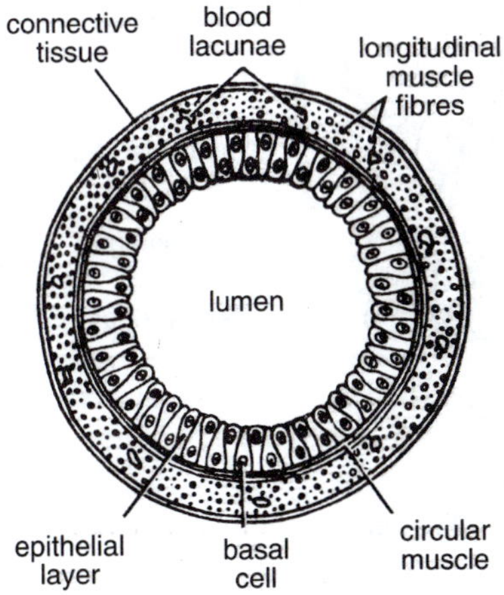

Fig. 16. *Palaemon.* T.S. of hindgut (rectum).

Hepatopancreas

It is a large, bilobed, dense and orange glandular mass, which lies below gonads and nearly fills up the cephalothoracic cavity. It surrounds stomach on its lateral, ventral and posterior sides. Hepatopancreas consists of numerous branching tubules completely held together by connective tissue. Wall of tubules consists of a single layer of columnar epithelium which is made up of : (i) granular cells, (ii) ferment cells, (iii) hepatic cells with globules of fat, and (iv) replacing or basal cells. The epithelium rests on a basement membrane. The tubules rejoin to form larger and larger canals, finally forming two large **hepatopancreatic** ducts, which open into the ventral chamber of pyloric stomach just behind the pyloric filter plate.

Hepatopancreas combines in itself the functions of pancreas, small intestine and liver of higher animals. Functioning as pancreas, it secretes digestive enzymes which can digest carbohydrates, proteins and fats. As midgut, it absorbs the digested food material, and as liver it serves as an important storage organ for glycogen, fat and calcium. Some intracellular digestion also seems to take place in hepatopancreas.

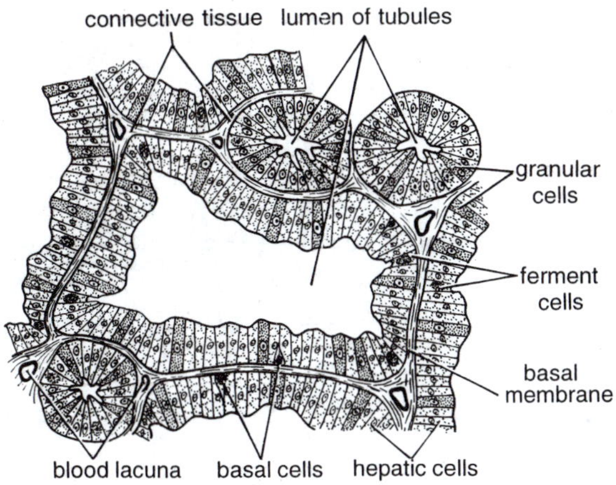

Fig. 17. *Palaemon.* A part of hepatopancreas in section.

Food and Feeding

Prawn feeds mainly on algae, moss and other aquatic weeds. It occasionally feeds on small aquatic animals such as insects, snails, tadpoles, fish, and debris of the bottom. It feeds at night, being more active at dawn and dusk than at any other time. Chelate legs, aided by the third maxillipedes, capture and convey food to the mouth. Coxae of second maxillipedes hold the food, while incisor processes of mandibles cut it into smaller pieces, which are swallowed with the help of maxillipedes, maxillulae and maxillae. Inside the buccal cavity, molar processes of mandibles masticate the food, which is then conveyed to the cardiac stomach through oesophagus. Passage of food through oesophagus is facilitated by the peristaltic activity of oesophagus and the sucking action of cardiac stomach.

Digestion and Absorption

The enzymatic digestive secretion of hepatopancreas flows through the two hepatopancreatic ducts into the ventral chamber of pyloric stomach, from where it reaches the cardiac stomach and mixes with food. Cardiac stomach expands and contracts to effect the churning of food and to facilitate its digestion by the action of digestive enzymes. As food passes over the hastate plate, the moving bristles of combed plates cut it into smaller particles. The semi-liquid and semi-digested food is filtered through the bristles of combed plates, into lateral grooves below, whence it is carried into the ventral chamber of pyloric stomach through the cardiopyloric aperture. Here the digested and liquefied food is filtered again through the pyloric filtering apparatus. Thus, only the finest food particles enter through hepatopancreatic ducts into

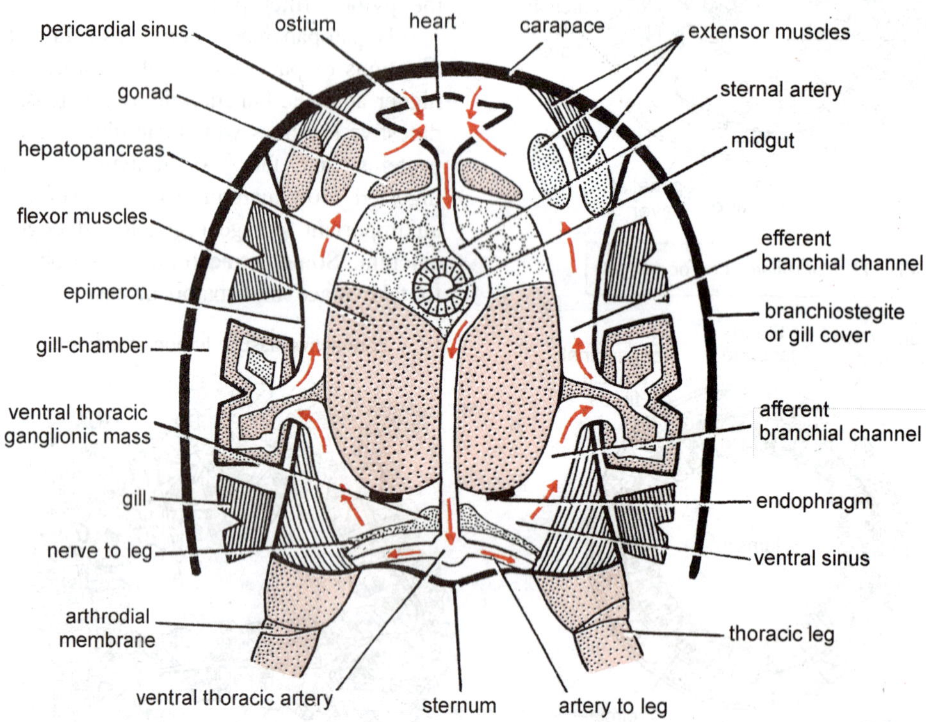

Fig. 18. *Palaemon.* A transverse hand section through the cephalothorax to show the two gill chambers.

the large digestive gland where they are hydrolysed and absorbed. The residual food, consisting of undigested and coarser particles, ascends up the dorsal pyloric chamber and from there enters the midgut for digestion and absorption. Undigested residual matter passes on to the hindgut. Here water is absorbed from it and the dry **faeces** thus formed is egested through the sphinctered anus.

RESPIRATORY SYSTEM

1. Respiratory Organs

Respiratory system is well developed and consists of : (i) **lining of branchiostegites** or gill covers, (ii) three pairs of **epipodites,** and (iii) eight pairs of **gills** or **branchiae**. These are sheltered in two large and compressed **gill-chambers,** one on either side of thorax. Each gill-chamber is bounded internally by **epimeron** or lateral wall of thorax, and externally by the curving pleural side of carapace or **branchiostegite**. The gill-chambers open on the anterior, ventral and posterior sides.

[I] Lining of Branchiostegites

Inner lining of branchiostegites or gill-covers is thin, membranous and highly vascular containing minute blood lacunae. These form large respiratory surfaces which absorb oxygen (O_2) dissolved in water and give out carbon dioxide (CO_2).

[II] Epipodites

These are 3 pairs of simple, foliaceous and highly vascular outgrowths of integument, given out from the coxal segments of 3 pairs of maxillipedes. They occupy the anterior part of gill-chambers beneath the scaphognathites of maxillae. Epipodites of 1st pair are bilobed and larger than others. Epipodites also serve as respiratory organs like primitive gills.

[III] Gills

There are 8 gills inside each gill-chamber. Only 7 of them are exposed on removing the gill-cover as the 8th gill lies concealed beneath the dorsal part of the 2nd gill.

1. Types of gills. Gills are of three kinds according to their place of origin and attachment.

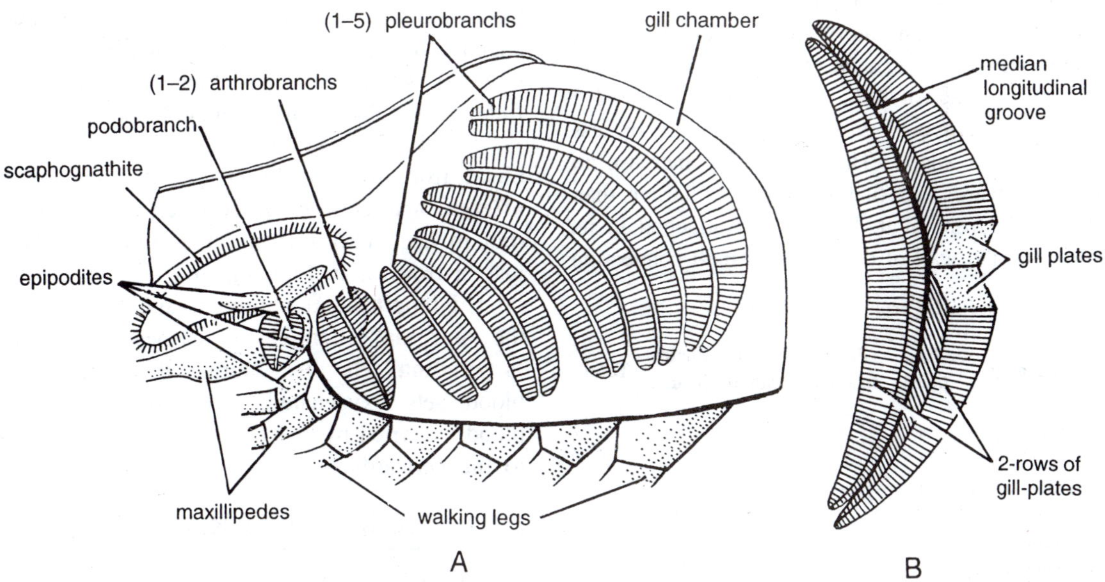

Fig. 19. *Palaemon.* **A.** Left gill-chamber exposed to show the gills. **B.** A phyllobranch.

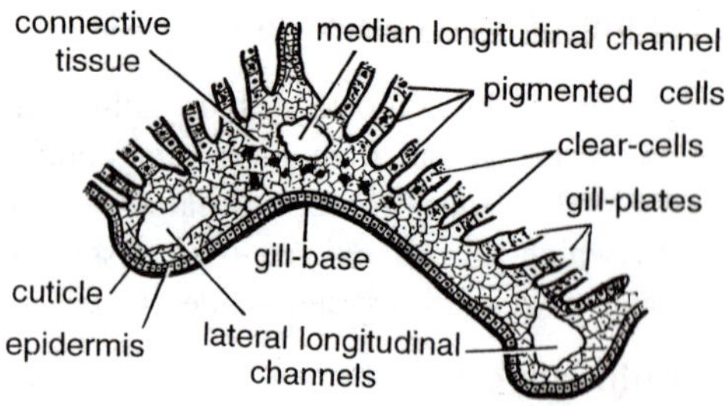

Fig. 20. *Palaemon.* Gill in oblique T.S.

(a) Podobranch or foot-gill. It is attached to the coxa of an appendage. In *Palaemon,* one podobranch is carried by the coxa of each second maxillipede.

(b) Arthrobranch or joint-gill. It is attached to the arthrodial membrane joining a limb with the body. Each third maxillipede bears two arthrobranchs. Second arthrobranch is the smallest and remains concealed beneath the first arthrobranch.

(c) Pleurobranch or side gill. It is attached to the lateral wall of segment bearing the limb. Last 5 gills on each side are pleurobranchs, attached to the lateral wall of thoracic segments bearing the 5 walking legs.

2. Branchial formula. Number and disposition of respiratory organs of each gill chamber in prawn can be represented in the form of a **branchial formula** as shown in *Table 1.*

3. Structure of gills. Gills are more or less crescentic in shape. They gradually increase in size backwards, so that each gill is larger than the one in front of it. Each gill is attached in its middle to the wall of thorax by a small connection

Table 1. **Branchial Formula of *Palaemon.***

Appendage	Epi-podite	Podo branch	Arthro-branch	Pleuro-branch	Total
I Maxillipede	1	–	–	–	1
II Maxillipede	1	1	–	–	2
III Maxillipede	1	–	2	–	3
I Walking leg	–	–	–	1	1
II Walking leg	–	–	–	1	1
III Walking leg	–	–	–	1	1
IV Walking leg	–	–	–	1	1
V Walking leg	–	–	–	1	1
Total	3	1	2	5	11

called the **gill-root,** through which nerves and blood channels enter and leave the gill. All the gills of *Palaemon* are **phyllobranchs,** *i.e.*, each of them consists of two rows of leaf-like rhomboidal **gill-plates** arranged like leaves of a book, at right angles to the long narrow **axis** or **base** of gill. Gill-plates are largest in the middle but become gradually smaller towards the two ends. A deep median longitudinal **groove** runs between the two rows of diverging gill-plates.

Histologically, a gill-plate is seen to be made of a single layer of cells with thin cuticle on both sides. The cells are of two types, **pigmented** and **transparent,** alternating with each other. Gill-base appears roughly triangular in a cross section, consisting of connective tissue bounded by an epidermis which is externally protected by a thin cuticle.

4. Blood supply in a gill. Three longitudinal **blood channels** run through the gill-base from one end of gill to the other. Two are **lateral longitudinal channels** running along the lateral margins, one on each side. The third is **median longitudinal channel** running through the apex of gill-base, beneath the outer median groove of gill. Lateral channels are connected together by a series of **transverse connectives,** forming a ladder-like structure. In each gill-plate the lateral channel of that side gives off a slender **marginal channel** which runs all along its margin and finally joins the median longitudinal channel.

5. Blood circulation in a gill. Deoxygenated blood from body is brought to the gill by an **afferent branchial channel** which enters the gill-root to open into a **transverse connective** lying just in front of it. Flowing first through the two **lateral longitudinal channels,** and then through the **marginal channels** it reaches the **median longitudinal channel.** During this journey the blood gets oxygenated. From median channel, blood is carried by an **efferent branchial channel** to the pericardium.

Mechanism of Respiration

The **scaphognathite** of each maxilla lies anteriorly inside the gill-chamber. By its constant vibrating movements, it bales out water from the anterior open end of gill chamber. Action of

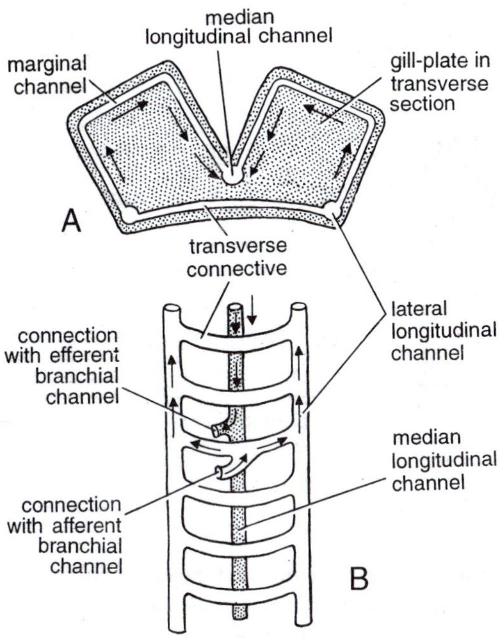

Fig. 21. *Palaemon.* **A.** Blood supply in gill plates. **B.** Diagrammatic representation of the ladder of blood channels in a gill.

scaphognathites is supplemented by the exopodites of maxillipedes. Fresh water enters the gill chamber from behind in the form of a current. This current of water flows over the lining of branchiostegites, gills and epipodites which are richly supplied with blood, so that exchange of gases takes place. The extremely delicate and thin gill-plates act as excellent permeable membranes for the passage of gases to and fro by diffusion. O_2, dissolved in water, is taken in by blood and CO_2 from blood diffuses out in the water.

BLOOD-VASCULAR SYSTEM

Blood Vascular Organs

Unlike annelids which have a 'closed type' of blood vascular system, prawn has an 'open type' or **lacunar type** of blood vascular system. Strictly speaking, it is partly closed and partly open. This type of blood vascular system is characterised by the absence of capillaries so that blood flows through open spaces, the **lacunae** or **sinuses,** in body. Blood vascular system of prawn includes : (i) **pericardium,** (ii) **heart,**. (iii) **arteries,** (iv) **blood lacunae** or **sinuses,** (v) **blood channels** and (vi) **blood.** There are no veins and capillaries as in vertebrates.

[I] Pericardium

Heart lies dorsally in the posterior part of thorax, enclosed in a spacious haemocoelic chamber, the **dorsal sinus** or **pericardium**. Floor of pericardium is in the form of a thin **horizontal septum,** lying just above hepatopancreas and gonad. This septum is attached in front and behind to the dorsal body wall and laterally to the thoracic wall.

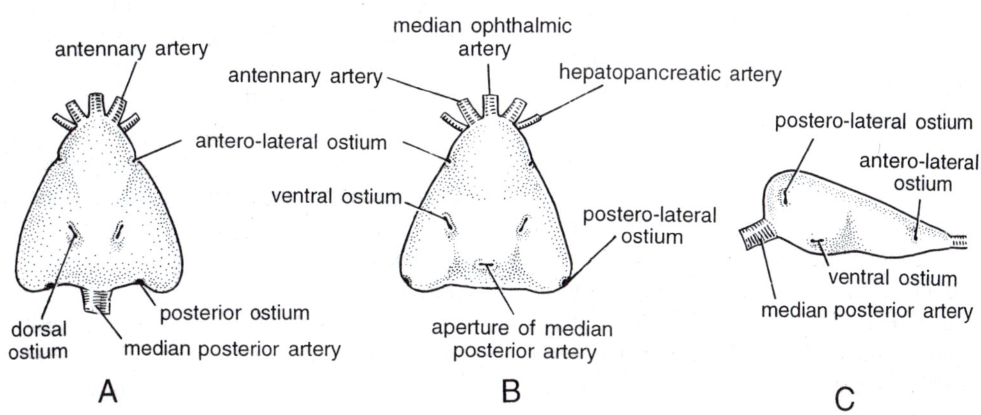

Fig. 22. *Palaemon.* Heart showing ostia. **A.** Dorsal view. **B.** Ventral view. **C.** Lateral view.

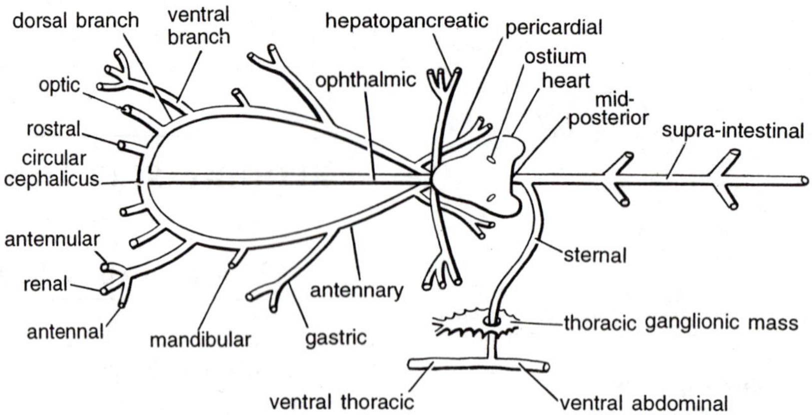

Fig. 23. *Palaemon.* Heart and principal arteries.

[II] Heart

Heart is a muscular and somewhat triangular organ with its **apex** directed anteriorly and the broad **base** posteriorly. A median longitudinal **cardio-pyloric strand** of fibrous tissue runs from its apex to the pyloric stomach. Two **lateral strands** extend from postero-lateral angles of heart to the body wall. The three strands keep the heart in position inside the pericardium. Thick and muscular wall of heart is perforated by five pairs of valvular, slit-like apertures, called **ostia** (L., **ostium**, a door). Blood from pericardial sinus enters the heart through ostia. Ostia are so distributed that the 1st pair lies dorsally, 2nd pair ventrally, 3rd pair posteriorly, 4th pair anterio-laterally and the 5th postero-laterally.

In a section, cavity of heart appears sponge-like, filled with numerous interlacing muscle fibres with blood in the interspaces.

[III] Arteries

Heart pumps blood to the body through narrow tube-like arteries which are provided with thick, strong and muscular walls. The principal arteries are as follows. Five of them arise from the anterior end and one from the posterior end of the heart.

1. Median ophthalmic. A single, slender **median cephalic** or **ophthalmic artery** arises from the apex of heart. It runs forward mid-dorsally along the renal sac to supply the cardiac stomach, oesophagus and head. It joins the two antennary arteries above oesophagus.

2. Antennary. A pair of **lateral cephalic** or **antennary arteries** also spring from heart's apex, one on either outer side of ophthalmic. Each antennary runs forward somewhat obliquely, passing along the outer border of mandibular muscle. It sends a **pericardial branch** to pericardium, a **gastric branch** to cardiac stomach and a **mandibular branch** to mandibular muscle. Then it bifurcates into a **dorsal** and a **ventral branch**. The **ventral branch** further divides to supply the antennule, antenna and renal organ. The **dorsal branch** sends an **optic artery** to the eye. Then it bends inwards to meet its fellow of the opposite side as well as the median opthalmic, thus forming a circular loop-like artery, called **circulus cephalicus**, which gives off a pair of **rostral arteries** to the rostrum.

3. Hepatic. A pair of **hepatic** or **hepatopan-creatic arteries** arise from heart ventro-laterally, one on each side just behind the antennary. They plunge downwards into the hepatopancreas within which they divide and subdivide.

4. Mid-posterior. A short but stout mid-posterior artery arises from the postero-ventral surface of heart and at once bifurcates into a **supra-intestinal** and a **sternal artery.**

Supra-intestinal or **dorsal abdominal artery** runs backwards along the dorsal surface of midgut

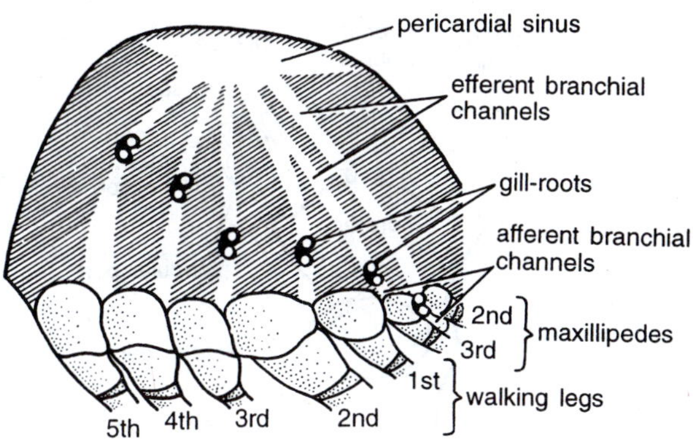

Fig. 24. *Palaemon.* Blood channels in cephalothorax.

up to hindgut. It supplies blood to midgut and dorsal abdominal muscles.

The large **sternal artery** is the stoutest of all. It runs straight downwards through the hepatopancreas. It passes through an aperture in the middle of the ventral thoracic ganglionic mass to reach the ventral side. Then it divides into two branches, (i) The **ventral thoracic** runs anteriorly up to mouth supplying the sternal region, first three pairs of walking legs, maxillae, maxillulae, oesophagus, gonads, etc. (ii) The **ventral abdominal** runs posteriorly upto anus and supplies blood to the ventral abdominal region, last two pairs of legs, pleopods, uropods, hindgut, etc.

[IV] Blood Sinuses

The heart and arteries comprise the closed portion of circulatory system. Arteries repeatedly branch in various organs of body. True **capillaries** and **veins** are absent. Minute arterial branches open freely into **blood sinuses** or **lacunae** of the haemocoel. All the sinuses of body eventually meet into a pair of elongated and ill-defined **ventral sinuses** lying below hepatopancreas on the floor of thorax. The two sinuses communicate with each other at various places.

[V] Blood Channels

The channels are lacunar tubes without proper walls. The sinuses and channels comprise the open portion of circulatory system. **Six afferent branchial channels** carry venous blood from each ventral sinus to the gills of that side, where it is aerated. As blood flows through gills, it gives off CO_2 and receives a fresh supply of O_2 from water

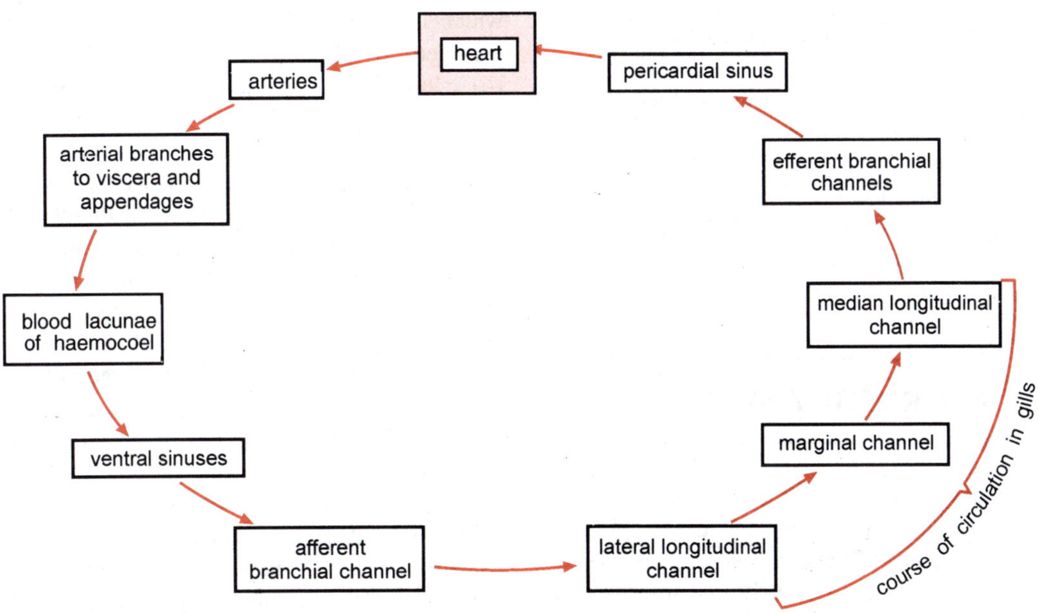

Fig. 25. *Palaemon.* A diagrammatic representation of the course of circulation of blood.

in the gill-chamber. Afferent channels run upwards along the inner side of the lateral thoracic wall and enter gills through their gill roots. First channel feeds the podobranch and two arthrobranchs. Of the remaining five channels, each supplies blood to a pleurobranch. Aerated blood from gills of each side is conveyed to the pericardium through another series of six **efferent branchial channels,** which also leave gills through their gill- roots.

[VI] Blood

Blood is colourless, thin and watery fluid, containing floating amoeboid white corpuscles or **leucocytes** which are phagocytic. There are no red blood cells. The respiratory pigment is **haemocyanin** which is dissolved in plasma. It has the same function as **haemoglobin** of other animals, but its metallic base is copper instead of iron. Haemocyanin becomes blue when combined with oxygen.

Blood of prawn has remarkable clotting properties. If an appendage is removed forcibly, there is hardly any noticeable loss of blood. The clot forms almost at once and fills the large wound opening.

Course of Blood Circulation

The heart, by means of its rhythmic contractions, forces blood through the arteries to all the parts of body. Before being returned to the heart, the blood is distributed to the gills and blood sinuses. The course of circulation of blood in the body of prawn can be diagrammatically represented as shown in Figure. 25.

EXCRETORY SYSTEM

The excretory system of adult *Palaemon* consists of (i) a pair of **antennary** or **green glands,** (ii) a pair of **lateral ducts,** (iii) an unpaired **renal** or **nephroperitoneal** sac, and (iv) the **integument.** True nephridia do not occur.

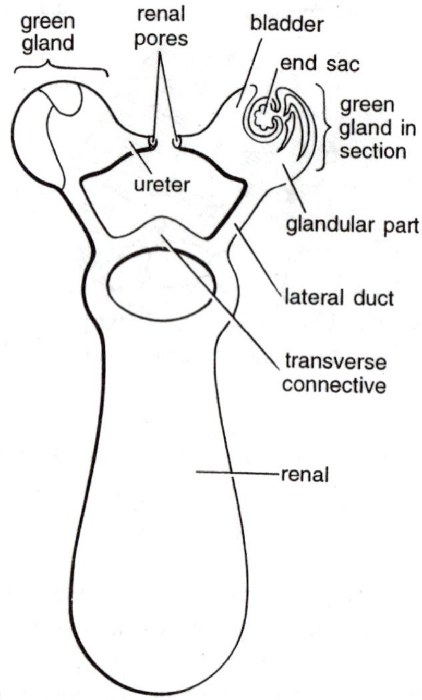

Fig. 26. *Palaemon.* Excretory organs in dorsal view.

[I] Antennary Glands

Coxa of each antenna encloses an antennary gland which is opaque white in colour and as big as a pea-seed. It includes three parts : (i) **end-sac,** (ii) **labyrinth** or **glandular plexus,** and (iii) **bladder**.

1. End sac. The bean-shaped end-sac is the smallest part lying between bladder and labyrinth. Internally, it contains a large central **blood-lacuna**. Its wall, made of two layers projects into central cavity in the form of radial **septa**. Outer thick layer of wall consists of connective tissue containing numerous small blood-lacunae, while inner thin layer consists of large excretory epithelial cells.

2. Labyrinth. Labyrinth or glandular plexus is relatively larger than the end-sac and lies on its outer side. It consists of numerous narrow, branching and greatly coiled **excretory tubules,** embedded in a mass of connective tissue containing blood lacunae. Tubules are lined by a single layer of large excretory epithelial cells. They

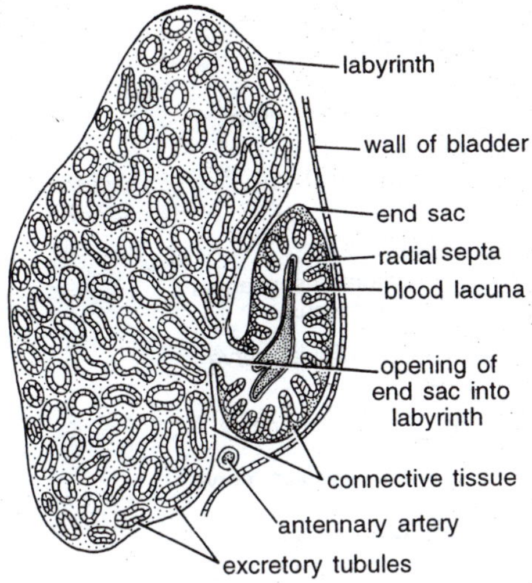

Fig. 27. *Palaemon.* Antennary gland in section.

open by a single aperture into end-sac and by many apertures into bladder.

3. Bladder. Bladder is the largest of all, lying on the inner side of end-sac. It is a thin-walled sac made of a single layer of excretory epithelial cells. Its inner wall is prolonged as a short excretory duct or **ureter,** which opens to outside through a small rounded **renal pore,** situated on a papilla on the inner surface of coxa of antenna.

[II] Lateral Ducts

A narrow lateral duct runs posteriorly from the bladder of each antennary gland. Lateral ducts of both sides are connected by a **transverse connective** just in front of the brain. The two ducts run backwards along the oesophagus to open into the **renal sac.**

[III] Renal-Sac

It is large thin-walled sac lying above the cardiac stomach, just beneath the carapace and extending posteriorly up to the gonads. Its wall is made of a single layer of flattened excretory epithelial cells.

Physiology of excretion. The complex, nephridia-like antennary glands extract nitrogenous wastes and excess water (osmoregulation) from blood in the same manner as the vertebrate kidneys. The end-sacs excrete mainly compounds of ammonia, but uric acid and other nitrogenous compounds are excreted by other parts. The excretory fluid from end sacs passes into labyrinths in which the useful materials are taken back by blood (selective resorption). The remaining fluid (urine) passes into bladders and finally expelled out through the renal apertures.

[IV] Integument

When the non-living chitinous covering or integument is cast off at each moult, the nitrogenous products secreted by body and deposited on the integument are also expelled. Thus, integument is believed to be an important excretory organ.

NERVOUS SYSTEM

The nervous system of prawn is of the **annelidan type**. However, it is somewhat larger and has more fusion of ganglia. It consists of : (i) the **central nervous system** including brain connected with a ventral ganglionated nerve cord through a pair of circum-oesophageal commissures, (ii) the **peripheral nervous system** including nerves, and (iii) the **sympathetic nervous system**.

[I] Brain or Supra-Oesophageal Ganglia

Brain lies at the base of rostrum, anterior to oesophagus and surrounded by a thick mass of fat. It is a bilobed structure derived from the fusion of several ganglia. On each side the brain gives off : (i) an **antennulary nerve** to antennule, into which it sends a **statocystic branch** to the statocyst, (ii) a stout **optic nerve** to compound eye, (iii) an ophthalmic nerve to muscles of eye-stalk, (iv) an **antennary nerve** to antenna and (v) a slender **tegumental nerve** to labrum.

[II] Circum-Oesophageal Commissures

Posteriorly, the Brain gives off a pair of stout nerves or the **circum-oesophageal commissures**.

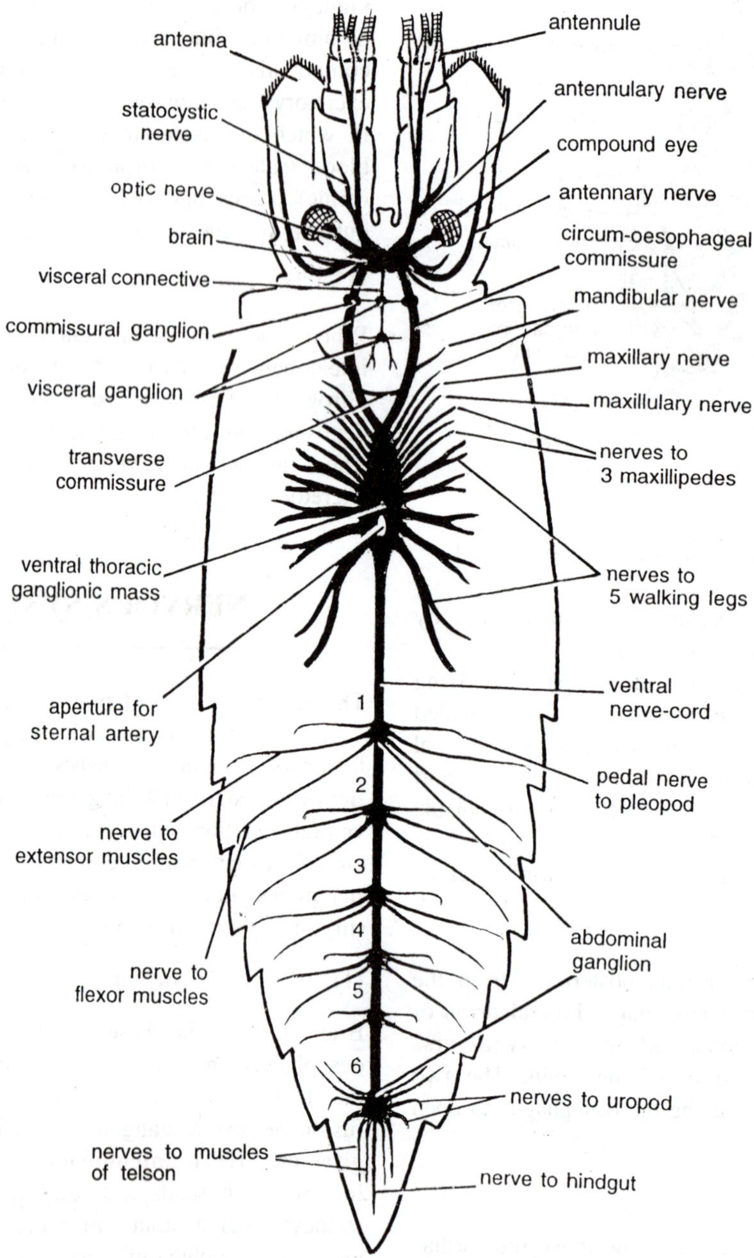

Fig. 28. *Palaemon.* Nervous system in dorsal view.

These run backwards and downwards, encircle the oesophagus and unite ventrally with the **sub-oesophageal ganglia**. The latter form an indistinguishable anterior part of the **ventral thoracic ganglionic mass**. The two commissures are crossed over, just behind oesophagus, by a double bridge of tough connective tissue, called **endosternite**. Each commissure bears a small **commissural ganglion** near its anterior end, sends a small **mandibular** nerve to mandible of its side.

Both oesophageal commissures are connected together by a slender transverse commissure near their posterior ends.

[III] Ventral Thoracic Ganglionic Mass

Like segments, the segmental nerve ganglia of cephalothorax also become fused to form an elongated **ventral thoracic ganglionic mass,** lying mid-ventrally on the floor of cephalothorax. It represents fusion of 11 pairs of ganglia and gives off laterally 11 pairs of nerves. First three pairs are the **cephalic nerves,** supplying the mandibles, maxillulae and maxillae, respectively. Last eight pairs are the **thoracic nerves,** of which first three pairs supply the three pairs of maxillipedes, and the remaining five pairs supply the five pairs of walking legs. Each nerve to a leg becomes bifurcated before entering the legs.

[IV] Ventral Nerve Cord

Ventral thoracic ganglionic mass gives off from its hind end a stout **ventral** or **abdominal nerve cord**. It runs along the mid-ventral line of abdomen. In each abdominal segment, it enlarges to form an **abdominal ganglion**. Each of the first five abdominal ganglia gives off three pairs of nerves : (i) one pair of **pedal nerves** to pleopods, (ii) one pair of nerves to extensor muscles, and (iii) one pair of nerves to flexor muscles of succeeding segment. The last, **stellate** or **sixth abdominal ganglion** is the largest composed of several fused ganglia. It supplies two pairs of nerves to flexor muscles, two pairs to uropods, two pairs to telson and a single median nerve to hindgut.

[V] Sympathetic Nervous System

Sympathetic, visceral or **autonomic nervous system** comprises a few ganglia and nerves. A small nerve, arising mid-posteriorly from brain, bears two **visceral ganglia** lying one behind the other. First ganglion is joined with the two **commissural ganglia** by connectives. Second ganglion gives off two pairs of nerves to the walls of oesophagus and cardiac stomach.

SENSE ORGANS

The most conspicuous sense organs are the eyes, antennules and antennae.

Compound Eyes

1. **Structure.** Prawn has one pair of black and hemispherical eyes. Each eye is mounted on a short, movable and two-jointed **stalk,** which is lodged in an **orbital notch** at the base of rostrum. Each eye is made of a large number of independent visual elements or units, called **ommatidia** (Gr., **ommation,** little eye). Such eyes are called the **compound eyes**. These are characteristic of Arthropoda and do not occur

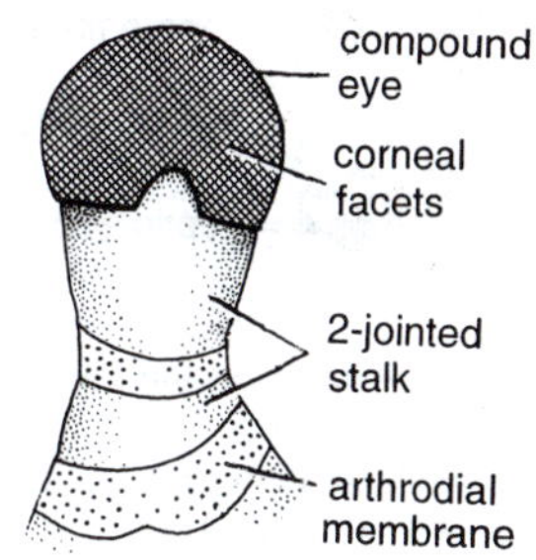

Fig. 29. *Palaemon.* Compound eye.

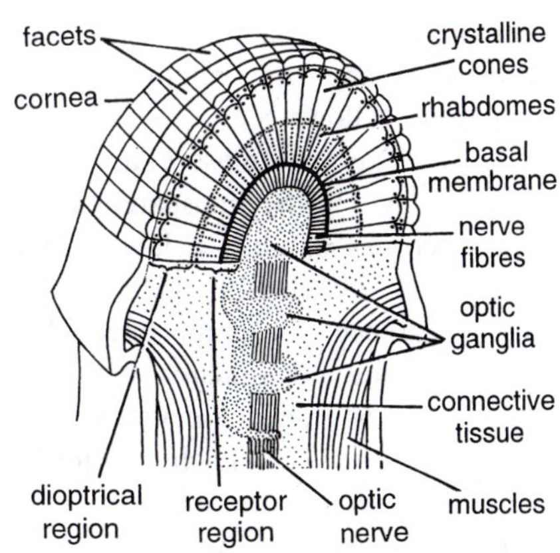

Fig. 30. *Palaemon.* L.s. of compound eye showing arrangement of ommatidia.

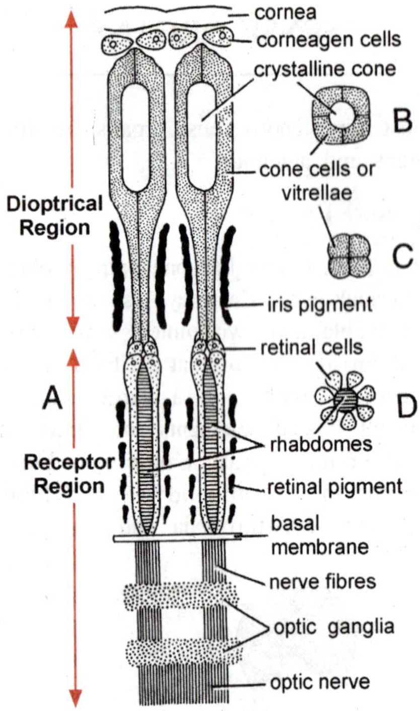

Fig. 31. *Palaemon.* Histological structure of compound eye. **A.** Two ommatidia in L.S. (semi-diagrammatic). **B.** T.S. of an ommatidium through cone cells. **C.** T.S. through basal ends of cone cells. **D.** T.S. through retinal cells.

elsewhere in the animal kingdom. All the ommatidia (about 2,500) are arranged radially and are similar in structure, each consisting of many cells arranged along its central axis. Their description is as follows :

(a) Cornea. The outermost convex layer of eye forming **cornea** is the transparent cuticle. In surface view, cornea exhibits a large number of squares or **facets** by clearly visible lines, thus giving the appearance of a graph paper. In insect eyes, the facets are not squares but hexagons. Below each facet lies one ommatidium.

(b) Corneagen cells. Each corneal facet thickens in the centre to form a **biconvex corneal lens**. Beneath the lens lie two **corneagen** cells which are modified epidermal cells and secrete a new cornea as soon as the old one is cast off in moulting.

(c) Cone cells. Beneath the corneagen cells lie four elongated **cone cells** or **vitrellae** which constitute a transparent, homogeneous **crystalline cone**. Inner ends of cone cells are long and tapering.

The part of eye, from cornea up to extreme ends of cone cells, is known as the **dioptrical region,** which focusses light upon the inner sensitive part or **receptor region** of eye.

(d) Rhabdome and retinal cells. Inner ends of cone cells lie upon an elongated, spindle-shaped rod, the **rhabdome**. It has a transversely striated appearance. Rhabdome is secreted and surrounded by a group of seven elongated **retinal cells.** Rhabdome and retinal cells together form the **receptor region** of eye. Inner ends of retinal cells rest upon a **basal membrane** beyond which they are continuous with sensory nerve fibres of **optic ganglia** which are connected with brain by the **optic nerve**.

(e) Chromatophores. Each ommatidium is cut off from its neighbours by a sheath of movable, amoeboid, dark **pigment cells** or **chromatophores** which are arranged in two series. Outer series lying along the cone cells is called **iris pigment,** and inner series separating the rhabdomes is called **retinal pigment**. Amoeboid pigment cells take up different positions according to the variations in the intensity of light.

2. Mosaic vision. Working of compound eye is very complex. It is deficient in focussing ability and clarity of image. But, such an eye is efficient for picking up motion and for **peripheral vision**. It functions as a very efficient organ for photo-reception. Mounted on a movable stalk, it can move on the head in much the same manner as the antenna of radar, and gives the animal almost 360-degree vision. Each ommatidium is capable of producing a separate image of a small part of the object seen. Therefore, in prawns and other arthropods possessing compound eyes, the image of the object viewed consists of several dark and light tiny pieces or spots, so that the total image of an object formed is a sort of a flat **mosaic.** Moving objects can thus be detected. The vision effected is said to be **mosasic vision** because of its similarity to mosaic art work.

The nature of composite image formed varies according to different intensities of light. Thus two types of images are formed. This is made possible by the movement of pigment cells.

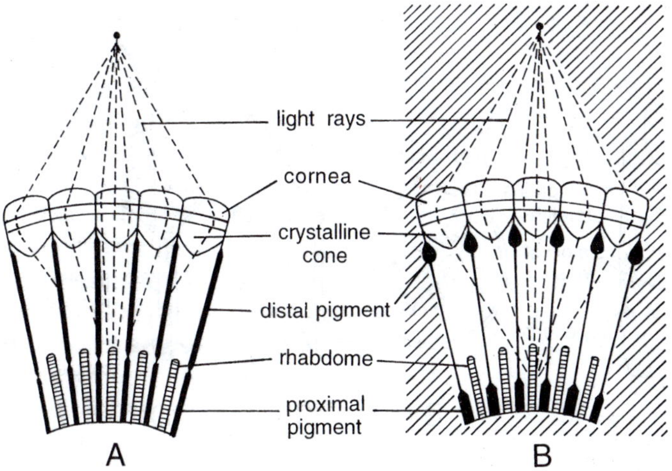

Fig. 32. *Palaemon.* Diagrammatic representation of image formation by a compound eye. **A.** Apposition image in bright light (day vision). **B.** Superposition image in dim light (night vision).

(a) Apposition image. In **bright light** (during daytime), the pigment cells spread in such a way that they completely isolate optically the adjacent ommatidia. No light can pass through from one visual unit to the other. In this condition the rays of light, which strike the cornea obliquely, are absorbed by the pigment cells without producing a visual effect. Only those rays of light which fall perpendicularly upon the cornea, can travel through the ommatidium and reach the rhabdome to form a point of image. As a result, the complete image formed is a mosaic of several components placed in juxtaposition in which the slightest movement is readily detected. In other words, each ommatidium responds to a fragment of the total field of vision and then these fragmentary images are fitted together into a single general picture. It is known as a mosaic or **apposition image**. Its sharpness depends upon the number of ommatidia involved and the degree of their isolation from one another. In butterflies, which are night-blind, the eyes are permanently set in this condition and are suited to see only in bright light. The image formed by this type of eye is never very good. It functions best at short distances only. Thus, most arthropods are always short-sighted.

(b) Superposition image. In **dim light** (during night), the pigment cells migrate and become separated into distal and proximal pigments, so that the neighbouring ommatidia no longer remain optically isolated but work in unison. In this condition even oblique rays of light are capable of forming a point of image after passing through a number of ommatidia in their way. As a result, an overlapping of the adjacent points of image occurs so that a continuous or **superposition image** is obtained. It is not sharp but the animal gets some sort of idea of the objects moving about in the surrounding. In some insects, like moths and fireflies, the eyes are permanently set like this, so that they are well adapted to see at night but are day-blind.

The prawns, like most arthropods, seem to adjust their eyes to form both types of images according to the prevailing intensity of light.

The **optic nerve** carries impulses (electro-chemical waves of energy) to the brain, where they are interpreted and registered as an upright **mental** image.

[II] Statocysts

1. Structure of statocyst. Statocysts are a pair of small, white, bead-like cuticular and hollow

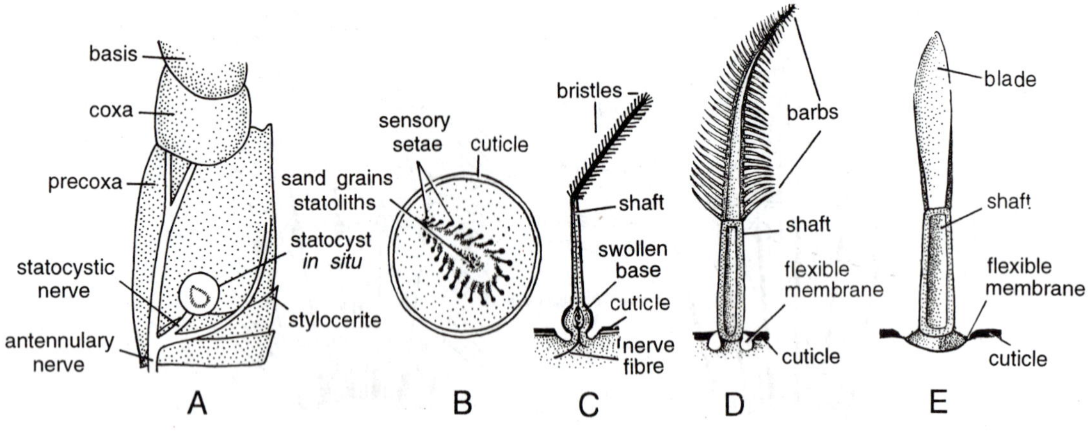

Fig. 33. *Palaemon.* **A.** Precoxa of antennule showing statocyst *in situ*. **B.** T.S. of statocyst. **C.** A single receptor seta (highly magnified).' **D.** A tactile seta (tangoreceptor). **E.** An olfactory seta (chemoreceptor).

spherical sacs. A statocyst lies inside the basal segment or precoxa of each antennule, attached to its dorsal wall. It opens dorsally on the concave surface or depression of precoxa through a minute **statocystic aperture**, which remains covered by a small fold of integument. A small **statocystic branch** of antennular nerve supplies the statocyst. Cavity of statocyst is oval, filled with minute sand particles, and lined by a number of elongated delicate sensory hairs or **receptor setae**. Each receptor seta is innervated by a small branch of the statocystic nerve. It consists of a swollen **base** and a long tapering **shaft,** which points towards the centre. The shaft is bent in the middle and bears fine bristles beyond the bend.

2. Function of statocysts. Statocysts perceive the direction of the force of gravity and function as the organs of **orientation** and **equilibrium**. The sand particles function as **statoliths**. Any change in the position of the swimming prawn causes a corresponding displacement of sand particles, which press against some of the sensory setae and stimulate them. Stimulated setae convey the information to brain through nerves, so that the animal corrects its position. At each moulting (ecdysis), statoliths are also shed along with the chitinous lining of statocysts, so that freshly-moulted individuals lose much of the power of orientation. However, the animal acquires new sand particles through the dorsal pore when the statocyst lining is renewed.

[III] Other Sense Organs

1. Tangoreceptors. The prawn is without a sense of hearing. However, the animal is sensitive to touch by means of tactile organs or tangoreceptors. These are in the form of **plumose setae** fringing the flattened portions of appendages, like the ramii of pleopods. Each tactile seta is a hollow cuticular outgrowth supplied with a nerve fibre. It consists of two segments. Basal segment or **shaft** is slightly swollen and attached to the integument by a membrane. Distal segment or **blade** gradually tapers and bears two rows of small barbs.

The elongated feelers of both **antennae** are also said to be tactile in function.

2. Chemoreceptors. Chemoreceptors or olfactory organs respond to chemical stimuli. They occur on **mouthparts, flagella of antennules** and inner wall of **gill chambers**. Inner smaller branch of outer feeler of each antennule bears a longitudinal groove containing numerous **olfactory setae**. Each seta consists of a basal segment or **shaft** which is attached to the integument by a flexible membrane, and a distal segment or **blade** which is bluntly rounded. A small nerve fibre from the olfactory branch of antennulary nerve innervates each seta.

3. Proprioceptors. These occur internally throughout the body. They perceive internal stimuli such as related to posture and muscular function.

ENDOCRINE SYSTEM

Palaemon, like other crustaceans, produces a large number of hormones. It is believed that the **sinus gland,** located at the base of eyestalk, secretes many hormones. They are believed to regulate : (i) the spread of pigment in chromotophores of epidermis and in compound eyes, (ii) deposition of lime salts in the exoskeleton, and (iii) moulting. Recent investigations have shown that the hormones that regulate moulting are of two types. The moulting-inhibiting hormones are secreted by **X organ** in the eyestalk and moulting- accelerating hormone by the **Y organ** beneath the adductor muscle of mandible. The latter hormone also induces metamorphosis.

Secretion of male sex hormones **(androgens)** has been reported by **H. Chariaux-Cotton** (1954) from androgenic glands located between, muscles of coxal segments of the last pair of walking legs. These hormones control the male sex characters.

REPRODUCTIVE SYSTEM

[I] Sexual Dimorphism

The **sexes** are separate (dioecious) and **sexual dimorphism** is well marked :

(1) Male is bigger in size than female.

(2) The male possesses a narrower abdomen than female.

(3) In male, bases of thoracic legs are more closely approximated than in female.

(4) In male, second chelate legs are longer, stronger and more spiny than in female.

(5) In male, each second pleopod bears an additional process, the **appendix masculina,** in between endopodite and appendix interna.

(6) In male, epimera of abdominal segments are smaller than in female.

(7) In male, paired genital openings lie on the coxae of 5th pair of legs, while they lie on the coxae of 3rd pair of legs in female.

A pair of **gonads** are similar in position, shape, size and general disposition in both the sexes. They lie in the posterior region of thorax, dorsally above the hepatopancreas and below pericardium. They extend anteriorly up to the renal sac and posteriorly up to the first abdominal segment.

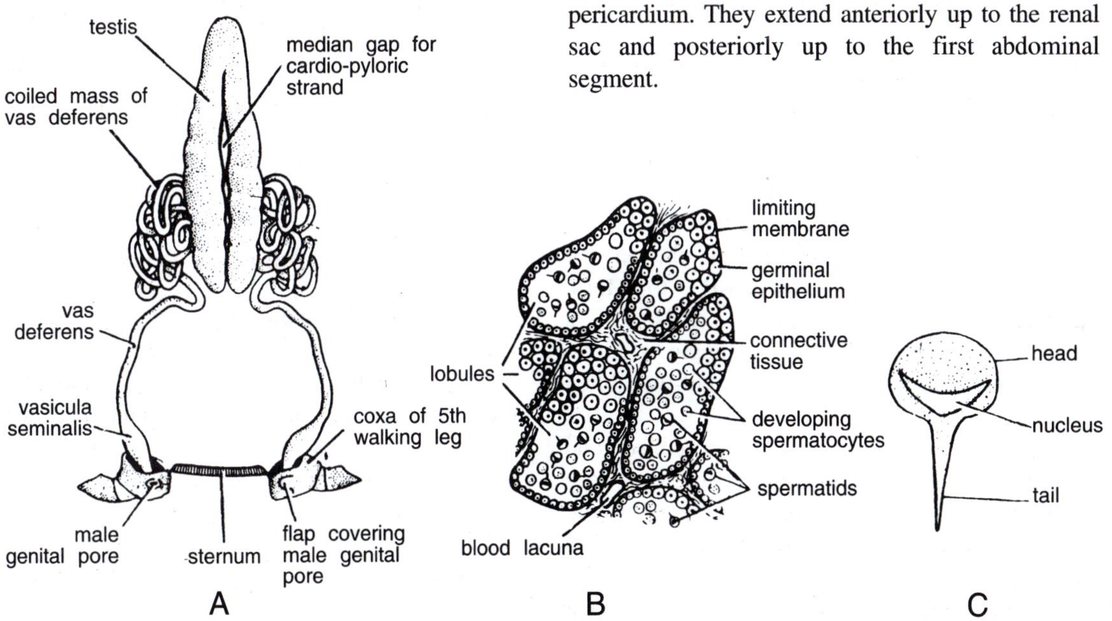

Fig. 34. *Palaemon.* **A.** Male reproductive organs. **B.** A portion of testis in section showing seminiferous tubules. **C.** A sperm.

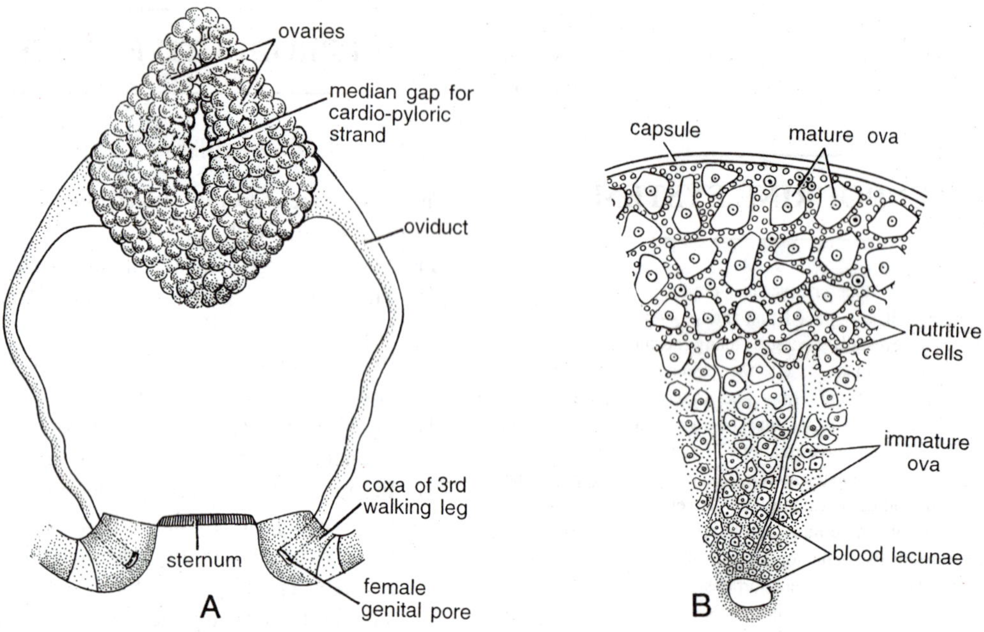

Fig. 35. *Palaemon.* **A.** Female reproductive organs. **B.** A portion of ovary in section (magnified).

[II] Male Reproductive System

1. Testes. The two testes are soft, white and elongated bodies which fuse at their anterior ends to form a common lobe. They enclose between them a gap for the passage of the **cardio - pyloric strand** connecting heart to pyloric stomach. Histologically, each testis consists of a large number of coiled, narrow and thin-walled **seminiferous tubules** embedded in connective tissue. The cavity of each tubule is lined by a single layer of **germinal epithelium,** the cells of which undergo **spermatogenesis** to form spermatozoa. A mature **sperm** consists of a rounded cytoplasmic body, containing a large, dark, crescentic nucleus, and a tail-like blunt process.

2. Vasa deferentia. A long, coiled and narrow tube, the **vas deferens,** arises from each testis near its posterior end. On emerging out the vas deferens of each side at once forms a much coiled mass and then runs vertically downwards between the abdominal flexor muscles on the inner side and thoracic wall on the outer side.

3. Vesicula seminalis. Each vas deferens reaching ventrally near the base of fifth leg, swells to form a club-shaped **vesicula seminalis**. These store spermatozoa in the form of white compact, bodies, called **spermatophores**. Each vesicula seminalis or seminal vesicle opens to the exterior through a **male genital aperture** situated on the inner side of coxa of fifth walking leg of its side. Each male genital aperture is covered by a small flap of integument.

[III] Female Reproductive System

1. Ovaries. The two ovaries are white, compact and sickle-shaped bodies touching each other at both the ends but leaving a gap in the middle for the passage of the cardiopyloric strand. The shape and size of ovaries vary with age and the season of year. Each ovary is enclosed within a membranous capsule and is made of numerous radial rows of **ova** in various stages of development. Immature ova lie towards the centre while mature ova towards the surface of ovary. Mature ova or **eggs** are large nucleated cells with plenty of yolk material (centrolecithal).

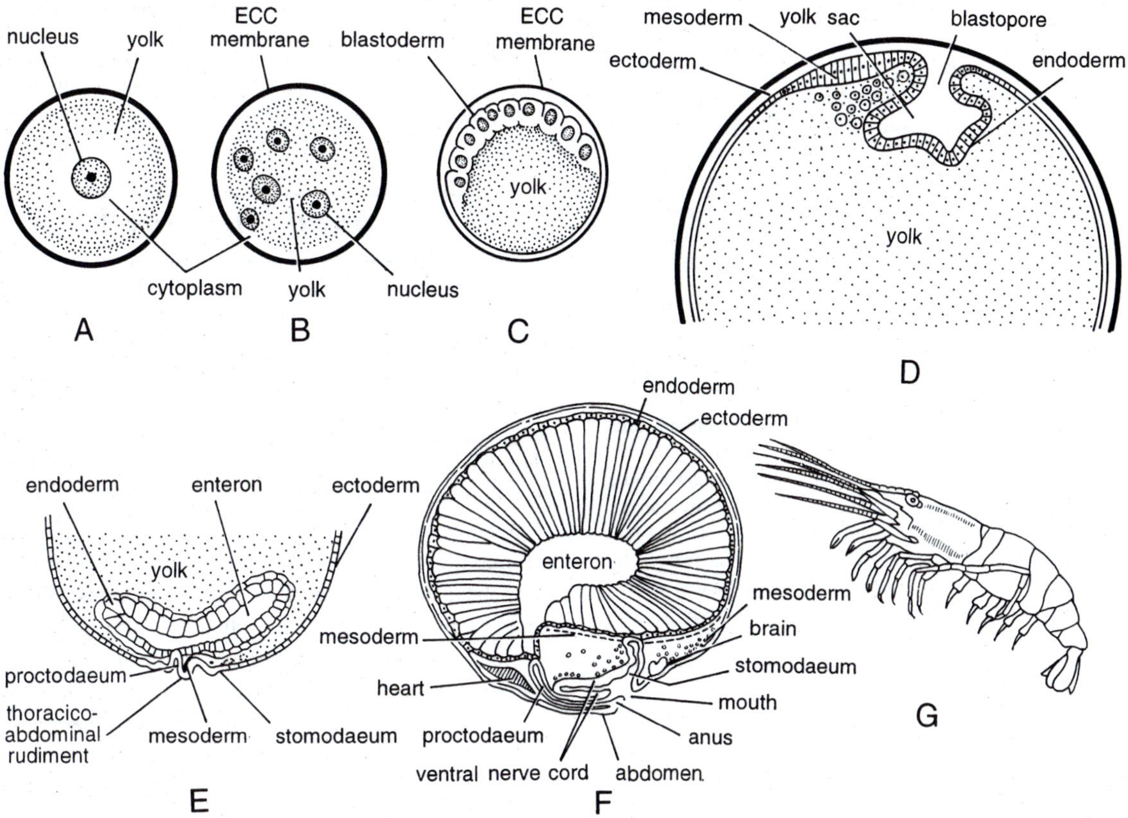

Fig. 36. *Palaemon.* Stages of development (diagrammatic) in sections. **A.** Zygote. **B.** Early cleavage. **C.** Early blastula. **D.** Early gastrula. **E.** Early embryo. **F.** Late embryo. **G**.Young prawn.

2. Oviducts. A short, wide and thin walled tube, the **oviduct**, originates from the outer middle border of each ovary. It runs vertically downwards to open through a **female genital aperture** on the inner side of the coxa of third walking leg of its side.

LIFE HISTORY AND DEVELOPMENT

1. Fertilization. *P. malcolmsonii* breeds during May, June and July. About two to three hundred mature eggs are laid by the female at one time in slimy strings. The male deposits sperms (spermatophores) near the genital openings of the female and the eggs are fertilized as they come out. Thus, **fertilization** is external, or *in situ*. After fertilization, the eggs are fastened to the pleopods by the sticky secretion of certain tegumental glands. The eggs hanging from pleopods look like berries or bunches of grapes. During breeding season, a female carries hundreds of eggs in this way, until they hatch. She carries them wherever she goes and the eggs are kept aerated by the slow back and forth movements of pleopods. The female is now said to be 'in berry'.

2. Development. Development is direct as there is no free larval form involved. The offspring or juvenile hatching out of the egg resembles the adult except in size. The female bends down her abdomen to protect first the eggs, and later the young, which hatch in 5 to 6 weeks and cling to the pleopods for some time. Growth occurs in short periods between moulting and the adult form is reached after a series of moults. Prawns usually live for 3 to 5 years.

Development of *Palaemon malcolmsonii* has not been worked out. The brief description that follows is based on the development of the cray fish *Astacus* or *Cambarus.*

(a) Cleavage. Segmentation or cleavage is superficial which is characteristic of the centrolecithal eggs of arthropods in general. The centrally located nucleus of fertilized egg divides repeatedly by mitosis forming many nuclei, resulting in a syncytial mass. Cleavage is incomplete as there is no corresponding protoplasmic division into cells.

(b) Blastulation. The nuclei then migrate to the peripheral cytoplasm where cell membranes are laid forming a peripheral cellular layer, called **blastoderm,** corresponding to the blastula stage. The central part of egg consists only of undivided yolk material. Blastoderm appears first on the ventral side of egg representing the future ventral surface of the animal. It gradually spreads and completely encloses the undivided central yolk.

(c) Gastrulation. While blastoderm is still under formation, an invagination takes place at the most posterior end of its ventral surface. This marks the beginning of gastrulation. The invagination forms a small sac, the **archenteron,** which opens to outisde through an aperture, the **blastopore.** As the blastopore closes, the archenteron is converted into a blind sac, called **endodermal vesicle.** Cells arising from the floor of endodermal vesicle pass into yolk where they multiply. Later, they become arranged on the surface of yolk forming the **midgut epithelium.** The yolk contained within the endodermal sac (yolk sac) becomes arranged in a number of radiating **pyramids.** By this process the embryo passes into the **gastrula** stage. Cells of endodermal sac form the **endoderm,** while the rest of blastoderm forms the **ectoderm.** Cells budding off from the lateral walls of endodermal vesicle, in the neighbourhood of blastopore, accumulate between ectoderm and endoderm to form the mesenchyme or **mesoderm.** The embryo now becomes triploblastic or three layered.

(d) Organogenesis. Organogeny or the formation of organs starts with the formation of midgut or mesenteron. Two ectodermal invaginations in the region of closed blastopore, the **stomodaeum** or **foregut** with its aperture as mouth and **proctodaeum** or **hindgut** with its aperture as anus, eventually, open into midgut thus completing the enteric canal. Ectoderm also gives rise to the epidermis of integument, the nervous system, part of eyes, external skeleton, and epithelia of gills and statocysts. Endoderm forms the epithelium of midgut and hepatopancreas. While mesoderm or mesenchyme gives rise to the connective tissue, muscles, vascular system and gonads.

Gradual consumption of yolk leads to growth in size. The young hatching out in 4 to 6 weeks looks like a miniature adult prawn.

Penaeus indicus

1. Structure. *Penaeus indicus* is a small common marine prawn found in Indian waters. It differs from the freshwater Indian Prawn *(Macrobrachium* or *Palaeomon malcolmsonii)* in minor details. It is smaller, about 15-20 cm in length. It has thin integument and the pigmentation is lighter. Antennae are longer than antennules. Endopodite of maxilla is elongated and segmented. Thoracic legs bear small exopodites and epipodites. First three pairs of legs are chelate. Abdomen is laterally compressed and straight unlike *Palaeomon* where it is bent sharply. Pleura of first abdominal segment overlap those of the second. Pleopods are biramous but lack the finger-like process, called **appendix internae**, on their endopodites. In male, the endopodites of first pair of pleopods are membranous and interlocked by hooks forming an apparatus, called **petasma**. It is meant for transference of sperms during copulation. In female, the sternum of the last thoracic segment gives out a hollow outgrowth, called **thelycum**. Spermatophores are deposited by the male into the cavity of thelycum. Gills are **dendrobranchiate**, each composed of an axis with numerous branched lateral filaments on either side.

2. Life history. Development is indirect. The egg hatches into a **nauplius** larva. It metamorphoses into adult only after passing through several larval stages such as metanauplius, protozoaea, zoaea and mysis. These larvae have been described in Chapter 53 on 'General account of Arthropoda', under the Heading 'Larvae of Crustacea', along with other forms of crustacean larvae which do not occur in the life-history of *Penaeus*.

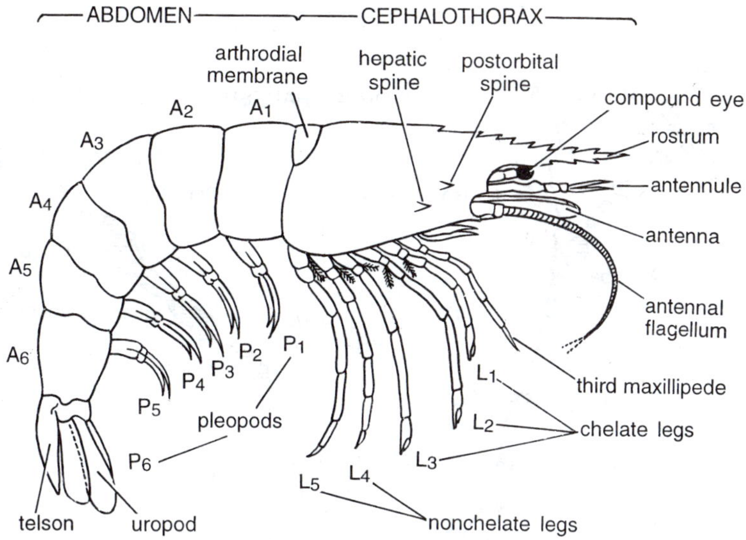

Fig. 37. *Penaeus* in lateral view.

CRUSTACEANS AS SOURCE OF FOOD

Protein is the stuff which builds body tissues; we need it daily. Yet millions of people in many countries suffer from protein malnutrition, for protein, especially of animal origin, is costly. Protein deficiency in human food can be best .solved by obtaining marine food which is rich in its protein content. The chief items of marine food consumed by man are fish, bivalve molluscs and the decapod crustaceans.

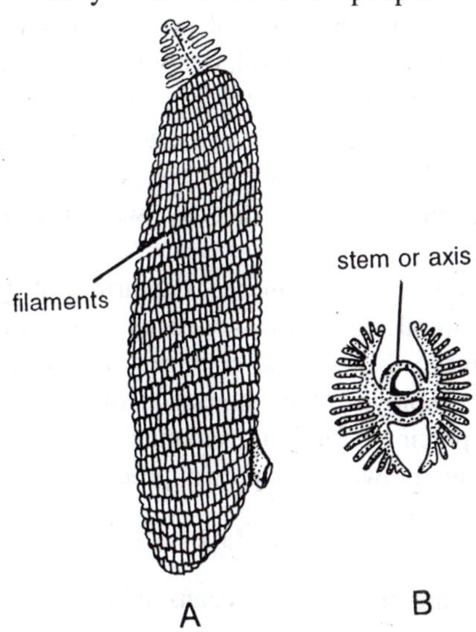

Fig. 38. *Penaeus.* **A.** Complete gill or dendrobranch. **B.** Same in T.S

A large number of crustaceans are consumed by man, especially the lobsters, shrimps, prawns, squiilae, crabs, crayfishes, etc. They form an important diet of man with great nutritive value. The most edible portions are their tails (abdomens) almost all of which are composed of muscles. The flexor muscles are particularly bulky. There is also some good 'meat' in their chelipeds. In mud crabs, the claws are the best part of the animal to eat. The muscles are either freshly cooked or canned. The blue crab *(Callinectes)* is held captive until it molts, then sold in the soft-shelled condition. After removal of viscera, the whole animal is cooked and eaten. Shrimps and prawns are captured with seines, but crabs, lobsters and crayfishes are lured into baited traps of wire, wood or net. Prawn fishery, which means capturing as well as culturing them, has advanced in many countries, including India. Commercial farming of

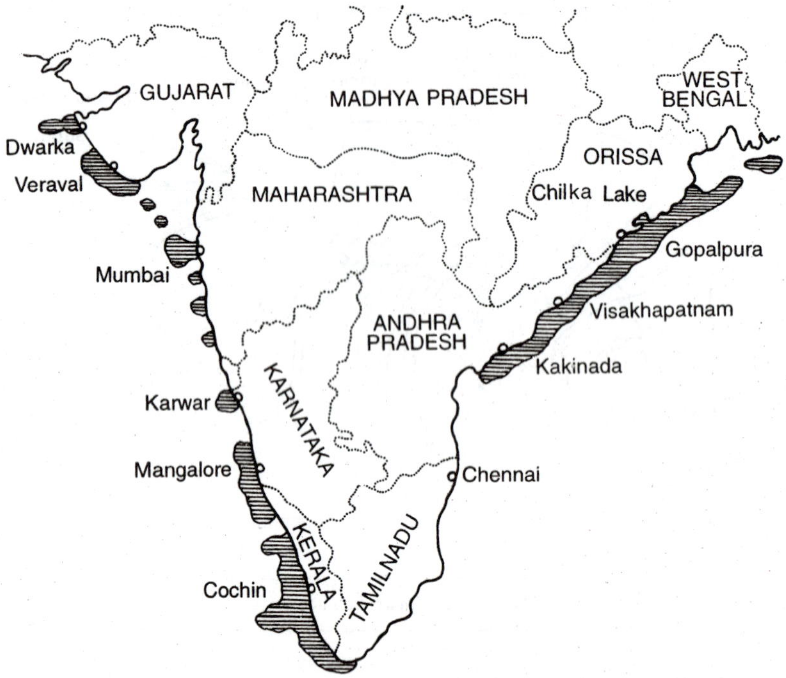

Fig. 39. Indian coastline showing prominent prawn productive areas.

freshwater prawns is being tried in Australia to meet the rising demand of this delicacy.

The smaller species of Crustacea form the bulk of the zooplankton which plays a vital role in the food chains of both salt and freshwater fishes and other aquatic animals that eventually come to our table. Man eats fish, and there is a stage in the development of all fishes, when they must feed on some forms of small Crustacea, such as a larval stage of some larger form like crayfish, or a minute adult such as *Cyclops* and *Daphnia.* The whales which are hunted by man for their various economic products depend on crustaceans for their food. Amphipods and euphausids form a great part of diet of other animals like the seals, sea gulls and penguins. Two tons of *Calanus,* a marine copepod about 5 mm long, have been found in the stomach of a blue whale. The aquarium dealers collect the adults and eggs of *Artemia* and *Daphnia* and sell them as fish food.

PRAWN FISHERY IN INDIA

Prawns are a rich source of proteins and vitamins A and D. Their muscles contain considerable amounts of glycogen and free amino acids, making their flesh sweet and tasty. Their fat content is also very low. Thus they are considered a delicacy and favourite protein diet by weight conscious aristocracy.

Prawn fishery means capturing of prawns from natural resources as well as their culture. It has advanced in several countries including India, giving employment to thousands of people. India earns precious foreign exchange every year by exporting prawns and prawn products.

[I] Habitat and Habits of Prawns

Prawns are fragile crustaceans inhabiting marine, fresh and estuarine waters. They are bottom

Table 1. **Some important species of prawns in the order of their abundance in landings.**

A. Penaeid prawns		B. Non-penaeid prawns	
Metapenaeus dobsoni	33.8%	*Acetes indicus*	39.5%
Parapenaeopsis stylifera	17.1%	*Palaemon tennipes*	17.8%
Penaeus indicus	11.6%	*Palaemon styliferrus*	11.5%
Metapenaeus affinis	9.6%	*Hippolysmata ensirostris*	7%
Metapenaeus brevicornis	6.8%	*Macrobrachium malcolmsonii*	...
Metapenaeus monoceros	4.4%	*Macrobrachium rude*	...
Parapenaeopsis hardwickii	3.6%		

dwellers feeding mostly on plant and animal detritus setting over the bottom. Among the marine prawns, older specimens prefer deeper waters, whereas the young and immature ones live in shallow waters, less than 20 meters in depth, and contribute to the bulk of catch along the coasts. All penaeid prawns migrate into estuaries and back waters where they feed and grow for some months. Then they move back into the sea for further growth and sexual maturity. Prawns pass most of their time crawling over the muddy bottom, occasionally swimming with the aid of swimmerets. They grow by moulting, periodically shedding their exoskeletons. During the soft-shell stage they retire to safer spots to avoid enemies. Sexes are separate and male and female prawns are easily distinguishable. Number of eggs produced by a single female penaeid prawn varies from 30 to 40 thousand per prawn. Some species (*e.g., Metapenaeus dobsoni* and *Parapenaeopsis stylifera)* liberate eggs in shallow coastal waters, whereas others (*e.g., P. indicus)* prefer deeper water. Spawning season also varies considerably according to species. The prawn fishery depends greatly on the particular habitats and habits of the different species of prawn.

[II] Kinds of Prawn Fishery

Depending on the areas and nature of water, from where prawns are captured in India, there are three kinds of prawn fishery : **marine, estuarine** and **freshwater**. About two-thirds of the total catch is made by marine prawn fishery. It is confined to shallow coastal areas upto 40 metres deep. Important prawn productive areas along the Indian coastline include the backwaters of Cochin-Kanayanpur, Parur, Crangannore and Vaikam along the west coast, and Ennore, Pulicat, Collair and Chilka lakes and low-lying areas close to the major estuaries along the east coast. The marine, prawns migrate into them for feeding and breeding. Rich offshore and deep-fishing grounds are located off Veraval, Mumbai, Karwar, Mangalore, Cochin, Visakhapatnam, and Chilka lake. Freshwater prawns are caught from rivers, lakes and ponds throughout the country.

[III] Seasonal Variations in Prawn Fishery

Freshwater and marine prawns are caught throughout the year, but there are one or two peak seasons. The peak season along the west coast is confined to the months of November to May. Offshore fishing is more profitable from July to October along the coasts of Mumbai and Gujarat. Along the east coast, active fishing begins in December and extends upto August. In Chilka, Collair and Pulicat lakes, the palaemonids dominate the catch in rainy season, while penaeids dominate in summer.

[IV] Indian Prawns of Commercial Value

The commercial prawns of India belong to about 73 species, 20 genera and 5 families which are Penaeidae, Pandalidae, Hippolytidae, Sergestidae and Palaemonidae. Some important genera of commercial Indian prawns are *Palaemon,*

Table 2. **Statewise landings of prawns in India.**

Maharashtra	48%	Gujarat	5%
Kerala	27%	West Bengal	4%
Tamil Nadu	7%	Orissa	—
Andhra	6%	Karnatak	2%

Macrobrachium, Penaeus, Metapenaeus, Parapenaeopsis, Acetes, etc. Penaeids form more than 50% of the total catch and harvested mostly along the Southern coasts especially from Kerala state. In non-penaeids, about 90% are hauled from Maharashtra and Gujarat coasts.

Freshwater prawns are mostly species of palaemonids such as *Palaemon* and *Macrobrachium.* They are found all over the country in inland river regions and hill streams. They are mostly fished off Bombay coasts and in the Ganges estuary. Some of them, such as *P. carcinus, P. idae, M. malcolmsonii,* etc., depend on brackish water for breeding.

All marine prawns belong to penaeids. *Penaeus carinatus* is the largest species (25 cm) but seldom fished in large numbers. *P. indicus,* which is slightly smaller (23 cm) is the most important commercial species. *Metapenaeus monoceros* (18 cm) is common in estuaries and brackish water lakes like Chilka, Pulicat and Collair. *M. brevicornis* (12.5 cm) is abundant in low lying paddy fields of Bengal. *M. dobsoni* (12.5 cm) is caught in backwaters of Kerala. *Parapenaeopsis stylifera* (14 cm) is caught off the Malabar coast. Species of *Acetes,* 2-4 cm long, are caught in countless numbers in certain river mouths and coastal waters and sun-dried. Larvae of marine prawns migrate into estuaries and backwaters for growth and development. The rich prawn fisheries owe greatly to this habit.

Besides prawns, there is great demand for lobsters. These inhabit rocky sea beds. They are massive in size and weight, compared to prawns. The 'lobster tail' sold in market is the abdominal region of spiny lobsters. Important commercial species of Indian lobsters belong to the genera *Panulirus, Palinustus* and *Puerulus.* Lobster fishery is greatly contributed by the species *Panulirus polyphagus, P. ornatus* and *P. homarus,* caught off the coasts of Kerala, Mumbai, Tamil Nadu and Bengal.

[I] Fishing Methods or Collection of Prawns

Fishing normally requires **crafts** and **gears**.

1. Fishing crafts. Locally made traditional crafts such as primitive wooden **catamarans** and **boats** and **canoes** have been in use since long in India. They are known by different names according to the regions. These are most efficient for shallow waters and account for a major part of the total catch.

But mechanised fishing on modern lines has been introduced recently. Mechanization of craft and gear ensures larger hauls and greater safety to the fisherman. **Motorised boats** were first of all introduced in the state of Maharashtra. Motorisation has enabled fishing in a shorter time and up to greater depths. Steel **trawlers** have come under use recently. They have large size, great speed, **mechanical trawls**, refrigeration and cold-storage facilities, air-conditioned rooms, laboratory, acoustic instruments, besides electronic, radar and special navigational equipment. But, most of the Indian shrimp trawlers are much less sophisticated.

2. Fishing gears. Several types of gears or devices are in use to catch prawns and lobsters, like **nets**, **traps, hooks** and even hands. They are made and employed depending upon the local conditions and the behavioural traits of prawns to be captured. Different types of gears have their own local names in different linguistic parts of India. Some gears are operated from vessels, while others from shore. In India, **cast-nets, stake-nets,** and a variety of **traps** are used in estuaries, shallow inlets of sea and backwaters. Movable nets which are kept hanging by attaching floats on the upper margin and sinkers on the lower margin are termed **seines**. **Boat-seines, shore-seines** and **drift nets** are conventionally used to catch prawns in the coastal waters. Smaller types of nets in use are the **scoop nets** and **hand nets**. **Traps** made of bamboo mats are the earliest devices to catch prawns, lobsters, crabs and fishes. They are left overnight in shallow waters against the water current to collect prawns. Several types of them have been in use in different parts of India. Lobsters are also caught by **metal hooks**. Along the canals of Malabar, an age-old method of collecting prawns is by **chain-dragging**. A heavy chain of iron is dragged over the bottom by two canoes connected by short bamboo poles in front and rear. The moving chain disturbs the bottom-feeding prawns which jump in agony and fall into the canoes.

[VI] Prawn Culture

In India, prawns are reared by trapping juveniles in paddy fields and allowing them to grow for a few months till they reach possible maximum size. Popularly this is known as 'paddy-field culture'. This is carried out on a small scale in certain regions of the backwaters of Kerala, north of Cochin harbour. The yield ranges from 600 kg to 1600 kg per hectare which is quite profitable. In India, there are more than 2 million hectares of brackish water areas which can be utilized for rearing prawns.

In paddy-field culture, bunds are raised around the fields and sluices fixed. The sluice are kept open during high tide to let in saline water containing huge populations of prawn larvae. During low tide the water is let out, but escape of prawn larvae is prevented by placing bamboo mats inside the sluice gates. The process is repeated several times from November till April. When prawns have grown to a marketable size, the water in the field is filtered with bag-nets tied to the outer edges of the sluices to collect the prawns. In paddy-fields, juveniles of 5 species of marine prawns are trapped in varying proportions. A prawn culturing factory was started by the Central Marine Fisheries Research Institute in September 1975 at Narakkal, Cochin. Now there are so many culturing and processing units in coastal region of India.

An improved method of culturing prawn is employed in Taiwan, Philippines and Japan. It includes keeping male and female prawns together in artificially made tanks and inducing them to spawn. The eggs and very young juveniles are collected and reared to bigger size by transferring them to bigger culture ponds. They are provided ideal conditions for nourishment and growth till they attain adult size. Commercial farming of freshwater prawns is being tried in Australia.

[VII] Chikara and Prawn Fishery

Along the western coastline of Kerala semicircular mud banks, called **Chikara** in Malyalam, extend into the sea. They enclose calm and quiet water undisturbed by turbulent sea waves. They are so rich in fishes and prawns that fishermen reach there for fishing from several miles away. Prawns worth crores and crores of rupees are caught every year from these mud banks.

[VIII] Pollution and Prawn Fishery

Prawn fishery is adversely affected by pesticides, effluents, metals and oil. Industrialization leads to pollution which is a great problem in many places. The waste products of some industries, such as tannery, textile and paper mills, wood distillation and iron works, and sugar and distillery mills are extremely poisonous. Prawns were found dead or dying where such effluents were drained into rivers. Dilution of wood distillate effluent to 200 times has been found safer but dilution of paper mill waste has not been effective. Concentration of 1 ppb of DDT in water can kill or paralyse commercial prawns. Many metals (*e.g.*, $HgCl_2$) are toxic to marine organisms. Oil and its components can also affect prawns in many ways. At least 50% of the ship cargo of the world is in the form of oil. Burning is an effective method of dealing with oil spill.

[IX] Preservation and Processing of Prawns

Prawn catching and processing has become leucerative business now in India. A part of the catch is marketed fresh for consumption, but a major portion is processed and preserved to meet the demand of the market at home and abroad.

1. Spoilage. Freshly caught prawns are subject to putrefaction or spoilage of mainly two types. In **chemical spoilage,** tissue contents, such as free amino-acids, are leached out leading to loss of taste, nutritive value and weight. **Bacterial spoilage** is due to bacterial contamination from outside, and affects the quality of flesh by causing rancidity.

Prawns deteriorate faster than fish. They keep up their freshness at 28°C up to 4 hours only, but after that, they may become unfit for eating in 6 to 8 hours only. Prevention of spoilage is known as **preservation**.

2. Chilling and freezing. Chilling and freezing are the most practical and effective means of controlling spoilage. The freshly caught prawns should be stored in ice till they are processed for packing and transport. The storage time in which prawns remain in good condition is known as **shelf time**. The quantity of ice used should be in the proportion of 1:1, ice to prawn by weight. The ice used must be free from bacteria and prepared from chlorinated water.

Refrigerated sea water at 0°C to 1°C serves as a good preservative medium. Faster freezing prevents spoilage quickly and the prawns retain their natural taste. This is done by **jet-freezing** in which cryogenic nitrogen (-320°F) freezes the product to the core within no time, in some cases in less than a minute. Peeled prawns store in a better condition than the whole prawns, whereas beheaded prawns show intermediate characters.

In storage, the prawns begin to blacken within a few days **(melanosis)** by enzymatic action which is an oxidative change. This is prevented by **glazing** which means covering iced prawns with water. It prevents evaporation and loss of taste and brightens the colour of prawns.

3. Beach-drying. The simplest and the most widely practised method of curing prawns along our coasts is beach-drying. The prawns are simply spread over the sand for drying by the solar heat and marketed as such or after shelling.

4. Pulp-making. Prawns are boiled in weak brine, dried in sun over coir mats, packed into jute bags and beaten with clubs. The shells separate and later removed by winnowing. The prawns are reduced to a paste or pulp like mass which has a good market in the Far East.

5. Pickling. Along the coasts of Malabar and Chennai, the prawn pulp obtained after boiling and shelling is pickled with vinegar, oil and spices, and marketed as such.

6. Blanching. Before freezing and packing, fresh prawns are cleaned and dipped in boiling brine for 1.5 to 3 minutes, a process called **blanching.** It is followed by immersing in water cooled around 2°C with ice, called **glazing.** Cooling time is greatly reduced by cooling in a chamber with a blast of air from a fan.

7. Smoking. Smoking of prawns is practised to some extent in Orissa and Andhra Pradesh. Prawns are first boiled in salt water followed by drying in kilns by burning moist bark. The purpose of smoking is to impart the particular desirable smoked flavour and colour to the prawns.

8. Semi-drying. This process has been developed by the Chennai Fisheries Department. Cleaned prawns are blanched in brine, shelled by hand, treated by 25% brine, and then half-dried in solar heat or by artificial drives. They can be stored for 2 months and, if sealed in tins with CO_2, for 8 to 10 months.

9. Freezing and canning. Abdomens of larger prawns are frozen directly without being boiled and shelled. First they are chilled down to about 40°-50°F and then placed in a 'freeze storage' at 10°F until transportation.

Canning is a costly and sophisticated method by which prawns can be kept indefinitely without the use of any preservative. Several processes are involved in canning. Prawns are cleaned, cooked at high temperature under pressure, and packed in cans or boxes with the addition of oil to add to their taste. The cans are then properly sealed airtight and sterilized by heat to kill bacteria so that prawns are no longer liable to decompose.

[X] Prawn Meal and Manure

Prawn-manure includes semi-dried prawns and heads, tails and chitinous body shells of prawns which are separated during the preparation of prawn pulp. It contains 5-6% of nitrogen and 3-4% of phosphates and some lime.

The chitinous shells, if dried and powdered, provide the **prawn-meal** which is a valuable protein-rich poultry feed. Mixed with shark-meal the prawn-meal can be used as a food for pigs.

The Central Institute of Fisheries Technology (GIFT) at Veraval in Gujarat has developed a crystal-like material called 'Dhatosin' from prawn shells. It can supplement an imported material used by the textile industry for sizing and would save crores of rupees in foreign exchange. About 37,000 tons of prawn shells are available from the Saurashthra coast alone.

[XI] Indian Prawn Fishery and Its Future

Total catch of prawns in India, during the last 15 years, has increased from 68,000 to 3,00,000 tonnes. In terms of value it accounts for 60% of total Indian fisheries. Since 1973, India occupies first position among prawn-fishing countries.

Prawns are also good foreign exchange earners of India because of their export to several countries. The export products are prawn powder, prawn bits, prawn curry, prawn meal and prawn pulp. Prawn pulp is exported to Mayanmar and Malasia from earliest times, whereas Individual Quick Frozen (IQF) and canned prawns have been exported to U.S.A., Japan and other countries in recent years.

A very large number of people are engaged in Indian prawn fishery. Their economic condition has improved greatly. A number of freezing and canning units for processing prawns have been established all along sea coast. But prawn landing in India has not been steady throughout the past years. Besides, there is occasional depletion of prawn fishery. Apart from natural agencies, the human agency has also contributed a lot to the depletion of prawn production. Overfishing with innumerable small mechanised boats strongly disturbs the prawn beds. This results in migration of prawn to deeper waters where small boats are unfit for fishing. Due to lack of restriction of mesh size, juveniles are also caught and destroyed in large numbers.

The following remedial measures have been suggested to avoid destructive prawn fishing :

(1) Restricting the number of small boats engaged in inshore prawn beds.

(2) Stoppage of more reclamation of backwater areas for paddy crops to save the prawn nursery grounds.

(3) Legislation to fix the mesh size of stake-nets and paddy field filtration nets.

(4) Prevention of rivers, estuaries and seas by industrial effluents, pesticides and oil.

Important Questions

➤➤*Long answer type questions*

1. Describe the structure of a typical biramous appendage and derive the structure of abdominal appendages *of Palaemon* from this basic plan.
2. Describe the cephalic and thoracic appendages of prawn.
3. Give an account of the digestive organs of prawn and its mechanism of feeding. What is the role of hepatopancreas in prawn ?
4. Describe the respiratory organs of prawn. How do they differ from those of cockroach ?
5. Describe the circulatory system and the course of circulation in *Palaemon.*
6. Describe the excretory organs and the process of excretion in *Palaemon.*
7. What is a compound eye ? Give the microscopic structure of the eye of prawn and explain the theory of mosaic vision.
8. Give a brief account of life history and development of prawn.
9. Distinguish between : (i) Epicuticle and endocuticle. (ii) Simple eye and compound eye. (iii) Coelom and haemocoel. (iv) Chelate and non-chelate leg. (v) 2nd pleoped of male and female prawns, (vi) Podobranch and arthrobranch. (vii) Apposition and superposition image. (viii) *Palaemon* and *Penaeus.* (ix) Phyllobranch and dendrobranch.
10. Draw a neat and well labelled diagram of T.S. of *Palaemon* through branchial region.
11. Write short notes on : (i) Pyloric filter, (ii) Statocyst, (iii) Hastate plate, (iv) Pancreas, (v) Antennary gland, (vi) Compound eye, (vii) Scaphognathite, and (viii) Sexual dimorphism in *Palaemon.*

➤➤*Short answer type questions*

1. What are green glands ?
2. Describe the biramous appendage in prawn.
3. Write a note on the ommatidium of prawn.
4. Describe the appendages in cephalothoracic region of prawn with suitable diagrams.
5. Describe any four differences between marine and fresh water prawns.
6. Describe the digestive system and feeding mechanism of *Palaemon.*
7. Explain the thoracic appendages found in prawn.
8. Describe the sense organs of prawn.
9. Make a labelled sketch of the median longitudinal section of prawn.

10. Make labelled diagrams of the T.S. of prawn through cephalothorax.
11. Draw neat labelled diagrams of the reproductive organs of male *Palaemon.*
12. Draw and label the following parts accurately in prawn : (i) Antennule (ii) II Maxilla, (iii) II Maxillepede (iv) Chelate leg. Mention the function of each appendage.
13. Discuss the respiratory system of *Palaemon.*
14. Describe the mouth parts of prawn.

➤➤*Multiple choice questions*

1. In the second maxillipede of prawn which part is not present :
   (a) protopodite (b) scaphognathite
   (c) gnathobases (d) endopodite
2. The crustacean biramous appendages have a basal part known as :
   (a) exopodite (b) epipodite
   (c) protopodite
   (d) endopodite
3. The sex of prawn can be identified externally by observing these appendages :
   (a) II Chelate legs
   (b) I Maxillipedes
   (c) I Abdominal appendages
   (d) II Non-chelate legs
4. The total number of Appendages present in the prawn is :
   (a) sixteen
   (b) nineteen
   (c) seven
   (d) thirteen
5. Cephalic appendages of prawn are :
   (a) antennules, antennae, mandibles, maxillipedes, and maxillae
   (b) antennules, antennae, mandibles, maxillulae, maxillae
   (c) antennules, antennae, pleopods, uropods, telson
6. Male prawn can be distinguished from female by presence of :
   (a) appendix masculina in 2nd pleopod
   (b) appendix masculina and appendix interna in 2nd pleopod
   (c) appendix interna in typical appendage
   (d) absence of appendix masculina in 2nd pleopod
7. Excretory organ of prawn is :
   (a) green glands (b) coxal glands
   (c) malpighian tubules (d) tracheae
8. A monograph on *Palaemon malcolmsonii* was written by :
   (a) S.N. Das (b) Ronald Ross
   (c) S.P. Patwardhan (d) P. Maheswari
9. How many segments in cephalothorax :
   (a) 11 (b) 13 (c) 15 (d) 19
10. Variation among appendages based on :
    (a) structure (b) function
    (c) adaptation (d) modification
11. Statocyst present with in :
    (a) antennule (b) antenna
    (c) mandible (d) maxilla
12. The gill attached on outer margin of :
    (a) I maxillipede (b) II maxillipede
    (c) III maxillipede (d) all maxillipede
13. Walking legs in Prawn :
    (a) 5 pairs (b) 8 pairs
    (c) 13 pairs (d) 19 pairs
14. Tegumentary gland present in :
    (a) epidermis (b) dermis
    (c) cuticle (d) legs

Answers

➤➤*Multiple choice questions*

1. (c) 2. (c) 3. (a) 4. (b) 5. (b) 6. (b) 7. (a) 8. (c) 9. (b) 10. (b)
11. (a) 12. (d) 13. (a) 14. (b)

46

Palamnaeus : The Indian Scorpion

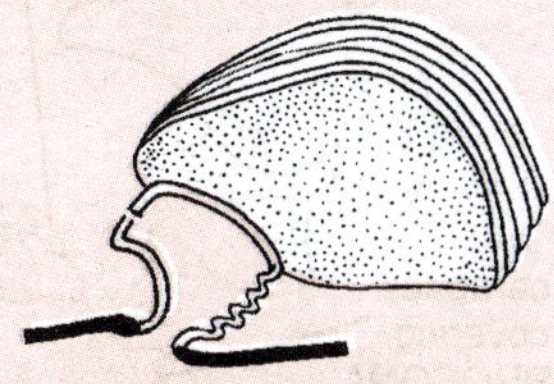

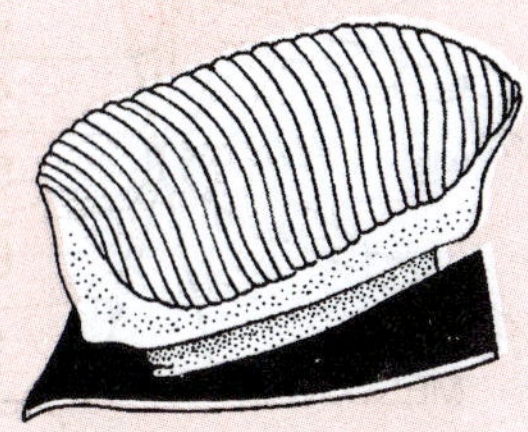

Class **Arachnida** of phylum Arthropoda includes such well-known and common forms as **scorpions**, **spiders**, **mites** and **ticks**. These are well-adapted for terrestrial life.

The scorpions are readily recognized by their elongated body with a flexible tail terminating in a sharp **poison sting**. They are the oldest known terrestrial arthropods. They belong to the order **Scorpionida** which includes 7 families, nearly 73 genera. The common living genera of scorpions are *Buthus, Parabuthus, Palamnaeus, Scorpio, Hormurus, Androctonus,* etc. The following description of scorpion is generalized but mainly relates to the Indian genus *Palamnaeus*. There are about 2,000 species of scorpions are found in the world.

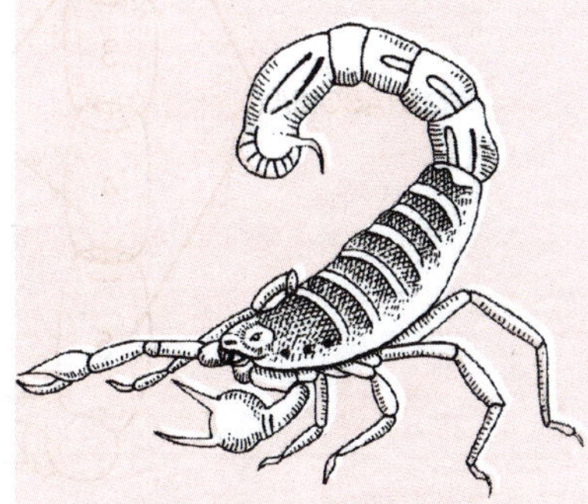

Palamnaeus

SYSTEMATIC POSITION

Phylum	Arthropoda
Subphylum	Chelicerata
Order	Scorpionida
Genus	*Palamnaeus*

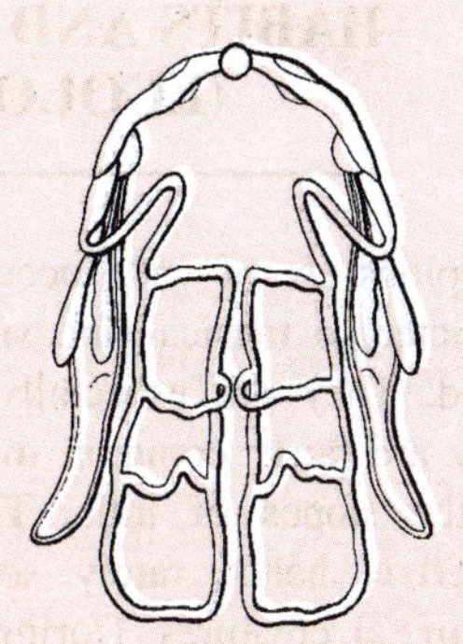

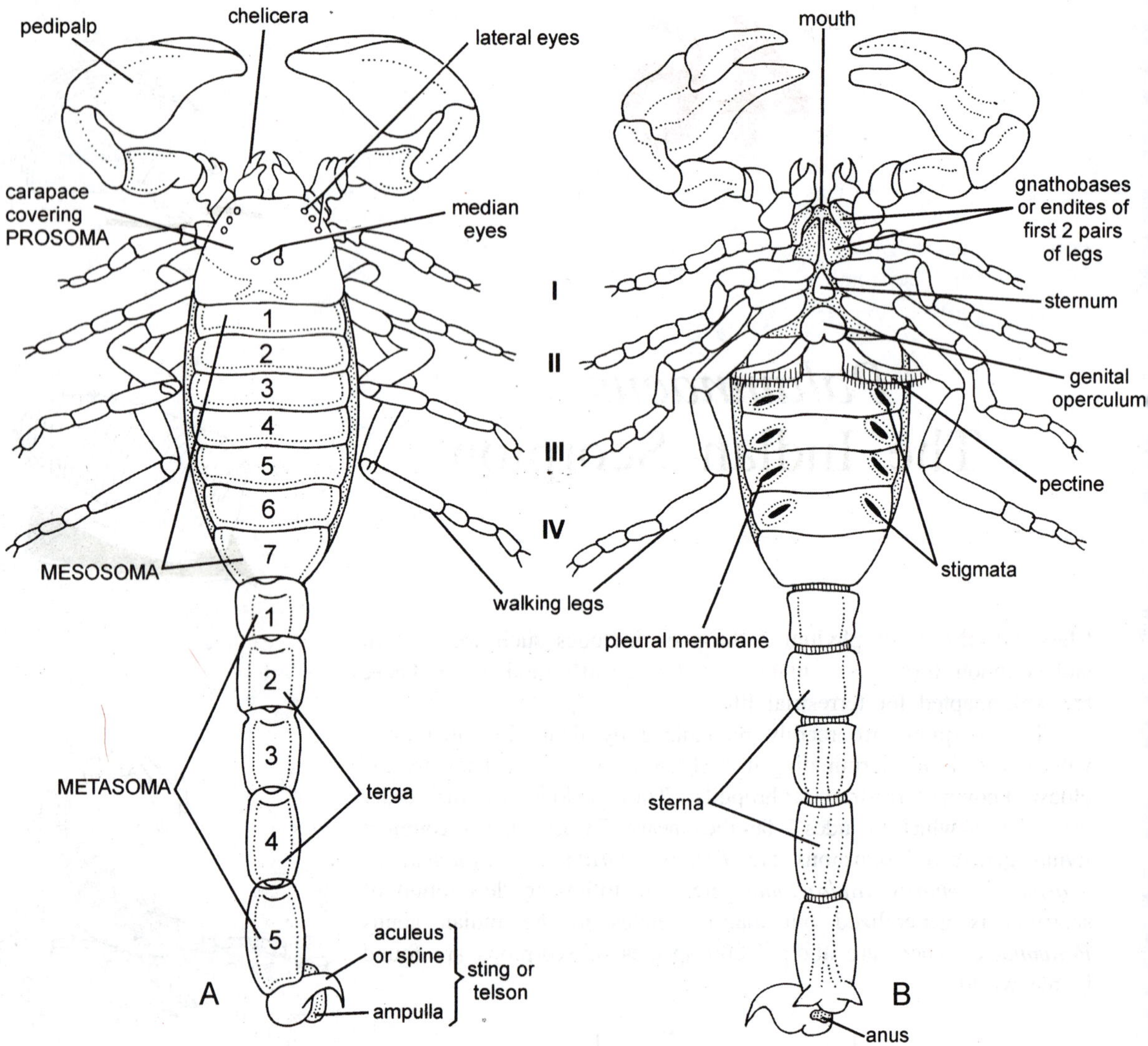

Fig. 1. Scorpion. External features. **A.** Dorsal view. **B.** Ventral view.

HABITS AND HABITAT (ECOLOGY)

Scorpions are of wide occurrence but are largely restricted to tropical and subtropical areas of the world. They are especially abundant in deserts. They are fairly common in India, more abundant on the slopes of hills. They are **solitary** and **secretive**, hence rarely seen. They are largely **nocturnal** creatures. During day, they hide under stones, logs, debris, old mattings, etc., in crevices or holes, and underground in burrows. They become active at night when they come out of their dens to hunt and seek their prey. They are **carnivorous** and **predaceous** feeding chiefly upon spiders, nocturnal insects and other smaller animals. They also resort to **cannibalism**. They kill the struggling prey with their **poisonous sting**. Sexes are separate and during courtship, the male and female individuals seize each other's claws and perform a **mating dance (promenade a deux)**.

EXTERNAL MORPHOLOGY

[I] Shape and Size

Body of scorpion is long, narrow, segmented and dorso-ventrally flattened. Size varies from 3 to 8 cm in length. The smallest species is the Middle Eastern *Microbuthus pusillus* which is about 13 mm long. The largest one is African *Pandinus imperator* which measures about 18 cm in length. The largest Indian species, *Palamnaeus swammerdami,* grows to a length of about 15 cm, while *Buthus tamulus,* a common species, reaches a length of nearly 8 cm.

[II] Colouration

Colour varies from shining black (in species infesting the tropical forests) to pale yellow (in forms living in sand). Dorsal surface is usually darker than the ventral surface.

[III] Body Divisions

Body is covered with chitinous **exoskeleton**. It is made of 18 segments which are divided into two main regions : (i) a short anterior **prosoma** or **cephalothorax**, and (ii) a long posterior **opisthosoma** or **abdomen**, ending in a stinging apparatus.

[IV] Prosoma

Anterior prosoma or cephalo-thorax is broad and formed by the fusion of **head** with **thorax**. Its segmented character is indicated by the presence of 6 pairs of appendages on the ventral side. These include small chelate chelicerae, large pedipalpi and 4 pairs of walking legs. Dorsally prosoma remains covered by a single, large, four-sided chitinous plate or **carapace**. Its anterior margin is divided into two **frontal lobes** by a median notch. Carapace bears a pair of conspicuous simple **median eyes**, each raised on a small tubercle and closely placed together in the middle. In addition, 2 to 5 pairs of smaller lateral eyes are present along the antero-lateral margins of carapace. Besides appendages, the ventral surface of prosoma also bears a single, small, narrow and usually triangular plate, the **sternum**, between the coxae of third and fourth pairs of legs.

[V] Opisthosoma

Posterior opisthosoma or abdomen is long and narrow. It is made of 12 distinct segments and is further differentiated into two parts : (i) an anterior broad **preabdomen** or **mesosoma**, and (ii) a posterior narrow **postabdomen** or **metasoma**.

1. Mesosoma. Pre-abdomen or mesosoma is as broad as prosoma anteriorly but slightly narrower posteriorly. It is made of 7 segments, each covered dorsally by a chitinous plate **tergite** or tergum, and ventrally by a **sclerite** or **sternum**. Both tergum and sternum are joined together laterally by a soft arthrodial membrane.

(a) Genital operculum. Sternum of first mesosomal segment bears a median, plate-like, rounded, bifid and movable lid, called genital **operculum**, under which lies the **genital orifice**.

(b) Pectines. Sternum of second segment bears a pair of remarkable comb-like sensory appendages, the pectines. These are used for exploring the ground and for recognition of sex.

(c) Stigmata. Sternum of each of the 3rd, 4th, 5th and 6th preabdominal segments bears a pair of lateral, oblique and slit-like respiratory apertures, the **stigmata** or **spiracles**, which lead into book-lungs. Sternum of 7th segment is without any appendage.

2. Metasoma. The slender metasoma or postabdomen is often wrongly called a **tail**. It is held upraised, arching over the back in a moving scorpion. It consists of 5 distinct cylindrical segments, each enclosed within a complete chitinous ring. In a transverse section, each section appears octagonal in shape. The last metasomal segment bears the **anus** on its post-ventral side and a **stinging apparatus** or **telson**. The latter consists of a swollen base, the **vesicle** or **ampulla**, and a curved and pointed **spine** or **aculeus**. It remains attached to the last segment by means of a flexible membrane. Inside the vesicle lies a pair of **poison glands**, the ducts of which open by a pair of minute apertures at the tip of the spine. The poison or venom of most scorpions is not

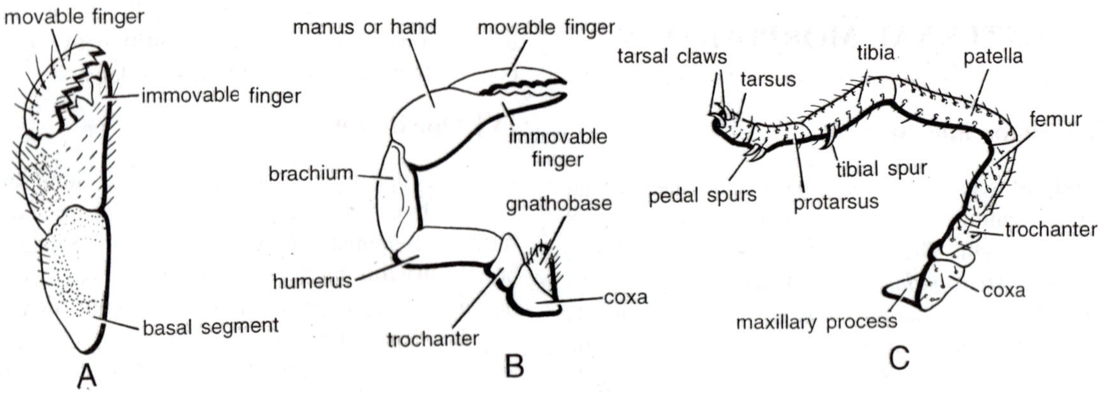

Fig. 2. Scorpion. Prosomatic appendages. **A.** Chelicera. **B.** Pedipalp. **C.** First walking leg.

harmful to man. But incertain species (*Androctomus australis* of Sahara Desert), it is highly toxic and can be fatal to man.

[VI] External Apertures

(i) **Mouth** is a small transverse aperture, opening mid-ventrally at the anterior end of body, (ii) **Anus** is a small median ventral aperture, lying at the base of telson. (iii) **Genital aperture** is situated mid-ventrally on the first segment of mesosoma, covered by the genital operculum. (iv) Openings of book-lungs, *i.e.*, **stigmata**, are 4 pairs of oblique slits lying ventro-laterally, one pair on each segment of mesosoma from 3rd to 6th. (v) Two small **openings of poison glands** lie at the tip of the sting. (vi) Coxae of 5th pair of legs bear two minute **openings of coxal glands**.

APPENDAGES

6 pairs of appendages are borne ventrally by the prosoma or cephalothorax. These are one pair of **chelicerae**, one pair of **pedipalpi** or **pedipalps**, and 4 pairs of **walking legs**.

[I] Chelicerae

Chelicerae are anterior-most, small, pincer-like and secondarily preoral appendages, lying close together on either side of mouth. They are said to be homologous with the second pair of antennae of Crustacea. Each chelicera is small, 3-segmented and chelate. Basal segment is small, ring-like and remains concealed beneath the carapace. Two distal segments form the **chela** or **small pincer**. 2nd segment is large, swollen and produced on the inner side into a pointed and toothed process, which forms the **immovable finger** of chela. 3rd segment, arising from the outer side of 2nd, is also curved and toothed and forms the movable finger. Chelicerae are prehensile appendages used for holding or tearing the prey.

[II] Pedipalps

Behind the chelicerae are the 2nd pair of appendages, called **large pincer claws, pedipalpi** or **pedipalps**. They are large, post-oral, powerful and efficient clawed weapons, characteristic of scorpions. Each pedipalp is 6-segmented consisting from the base of **coxa**, **trochanter**, **humerus**, **brachium**, **manus** (hand) and **movable finger**. Basal joint or coxa bears, towards mouth, a blunt blade-like process, the **gnathobase**, which works as a jaw into the preoral cavity. It bites against the corresponding **gnathobase** of the opposite pedipalp and helps in squeezing the body of prey. Last two segments are opposed to one another, like a finger and thumb, to form the **chela**, as in a chelicera, and serves to seize the prey. In a walking scorpion, the two pedipalps are held horizontally in front and used as tactile and raptorial organs for seizing prey, for fighting and for joining hands during the remarkable mating dance or '**Promenade a deux**'.

[III] Walking Legs

Pedipalps are followed by 4 pairs of walking legs, which are closely alike and used for walking. Each leg is composed of seven podomeres : the **coxa, trochanter, femur, patella, tibia, protarsus** and **tarsus**. Tarsus **terminates** into 2 or 3 curved and **pointed horny claws**. Coxae of 1st and 2nd pairs of legs are movable and each provided with a forwardly directed triangular **maxillary process** or **gnathobase**. Thus, 4 maxillary processes from pedal coxae lie below mouth and triturate the food as it is held by chelicerae. While coxae of 1st pair remian separated, those of 2nd pair meet each other in mid-line. Coxae of, 3rd and 4th pairs of legs are immovable, without maxillary processes and separated from those of opposite side by sternum. Besides walking, legs are also used in feeding.

BODY WALL

Body-wall of scorpion consists from outside of three layers : **cuticle**, **hypodermis** and **basement membrane**.

1. Cuticle. It is the outer, non-living, thick protective layer of chitin covering the entire body and forming the **exoskeleton**. It is secreted by the underlying hypodermis. It is composed of three layers : outer **tetostracum**, middle **epiostracum**, and inner **hypostracum**. Cuticle is traversed by numerous **canals**, which open to outside on its surface.

2. Hypodermis. It consists of a single layer of cubical epithelial cells, just beneath the cuticle. Some hypodermal cells become modified to form hairs, which project above the cuticle and are extremely sensitive to touch.

3. Basement membrane. The basement membrane, lying beneath the hypodermis, is a thin and structureless layer.

4. Projections. Projections of integument are of two types, (i) **Dermatidia** are solid processes

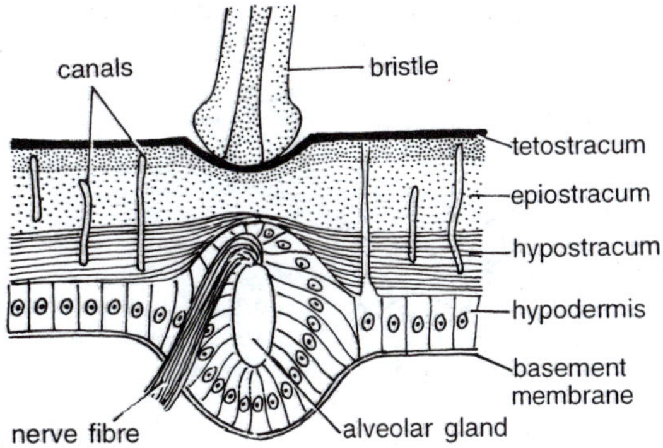

Fig. 3. Scorpion. V.S. through body wall.

of chitin alone, such as denticles on chelae, **(ii) Coelodermatidia** are hollow processes of integument such as spines, spurs, tarsal claws, etc.

BODY CAVITY, COELOM AND MUSCULATURE

As in cockroach, body cavity is an extensive **haemocoel** filled with blood and surrounding the heart, alimentary canal, hepatopancreas, gonads, and other internal parts. **Coelom** is reduced to vestiges associated with reproductive system and certain glands. **Musculature** includes one pair of muscles in cephalothorax and eight pairs of dorso-ventral muscles in pre-abdomen, running from dorsal to ventral surface of body, on either lateral side of body-wall.

ENDOSKELETON

Internal skeleton consists of a cartilaginous plate called **endosternite**. It lies inside the body somewhat obliquely between prosoma and mesosoma, dorsally to nerve cord and ventrally to alimentary canal. It is roughly a triangular structure shaped somewhat like a bird with outstretched wings. It is produced into several paired processes (anterior, posterior and horizontal)

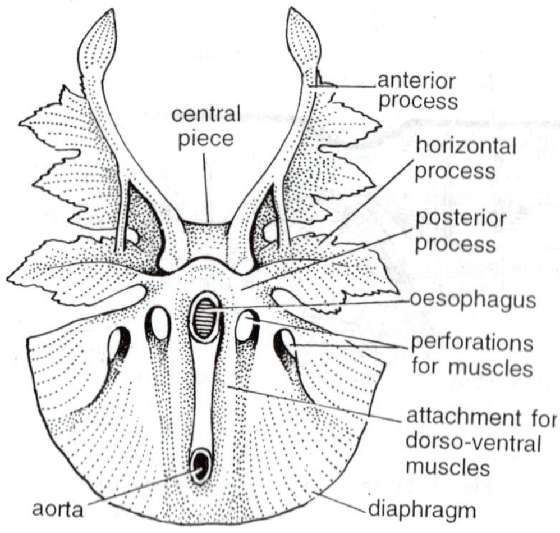

Fig. 4. Scorpion. Endosternite and diaphragm.

which provide attachment to the muscles. Histologically, the cartilage of endosternite differs from the vertebrate cartilage. Posteriorly, the body of endosternite becomes fused with a chitinous plate, called **diaphragm**, which is pierced by nerve cord, alimentary canal, muscles and dorsal aorta.

DIGESTIVE SYSTEM

[I] Alimentary Canal

Alimentary canal is a fairly uniform and straight tube extending from mouth to anus. It may be divided into four regions : (i) preoral cavity, (ii) foregut, (iii) midgut, and (iv) hindgut.

1. Preoral cavity. A large open preoral cavity is formed in front of mouth surrounded by coxae of first four pairs of prosomatic appendages. Dorsally, it is bounded by two **chelicerae** and a median **rostrum** or **labrum**. Laterally, it is guarded by the coxae of **pedipalps**. Ventrally, the floor is formed by two pairs of **maxillary processes** of first two pairs of legs.

2. Foregut or stomodaeum. Foregut is restricted to prosoma. It includes mouth, pharynx and oesophagus. These are internally lined by cuticle which is shed at moulting.

(a) Mouth. Mouth is a small, narrow and transverse aperture situated behind the gutter-like preoral cavity below the base of labrum, and leads into pharynx. The narrow mouth can only admit juices and pulps.

(b) Pharynx. Pharynx is a large, pear-shaped, muscular and suctorial chamber with elastic walls. It is capable of great dilatation due to many radiating bundles of muscles running outwards from it to the wall of prosoma. These muscles enable pharynx to work as a sucking organ so that the liquid food is sucked through mouth.

(c) Oesophagus. Pharynx is followed by a small narrow tube, the **oesophagus**. It passes through the cephalic nerve ring, and then pierces the endosternite. It extends into the stomach to form a **sleeve-valve** which prevents regurgitation of food.

3. Midgut or mesenteron. It includes stomach, intestine and two digestive glands. It is internally lined by epithelium.

(a) Stomach. Oesophagus leads into a short and dilated stomach, extending up to the diaphragm.

(b) Intestine. Intestine is the longest part of alimentary canal, running up to the last segment of abdomen, where it passes into the hindgut. Intestine is a wide tube with glandular walls and differentiated into two distinct regions : (i) pre-abdominal **pars-tecta-intestini**, and (ii) post-abdominal **pars-nuda-intestini**. The junction of

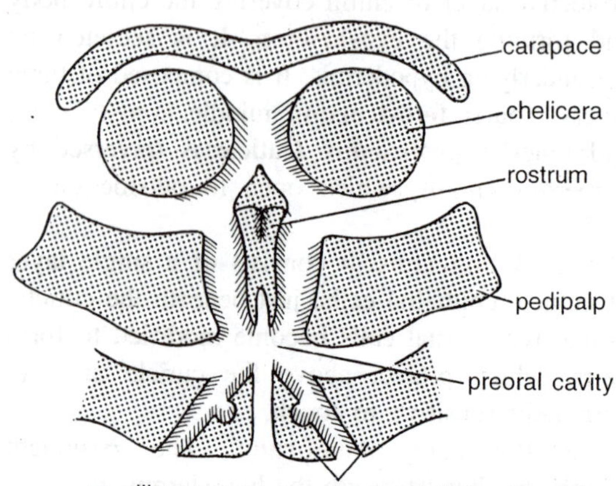

Fig. 5. Scorpion. T.S. Through preoral cavity.

these two regions is marked by a narrow constriction from where arise one or two pairs of narrow, elongated and blind **Malpighian tubules**, which are excretory in function.

4. Hindgut or proctodaeum. It is the smallest part of alimentary canal, restricted to the last metasomatic segment. It is internally lined by chitin and opens to the exterior through the **anal aperture** situated ventrally at the base of the stinging apparatus.

[II] Digestive Glands

1. Salivary gland. A trilobed brownish **gland**, lying in prosoma, opens into the stomach. **Huxley** regarded it as **salivary gland**, but **Blanchard** and **Pavlovsky** did not accept this view and called it the **stomach gland**.

2. Hepatopancreas. A large, brownish and lobulated glandular mass, called **chylenteron, liver** or **hepatopancreas**, occupies nearly the whole of pre-abdominal cavity. The heart lies in a mid-dorsal longitudinal groove of hepatopancreas, while intestine and other internal organs are found embedded in it. Five pairs of lateral narrow **hepatic ducts** lead from hepatopancreas to open into the intestine. Hepatopancreas is a racemose gland, consisting of a mass of small tubules, which are in communication with the hepatic ducts in the same way as the twigs of a tree are with the main stem. Its hepatic function is doubtful, but due to its large bulk it seems to perform other important functions, besides secreting digestive juices.

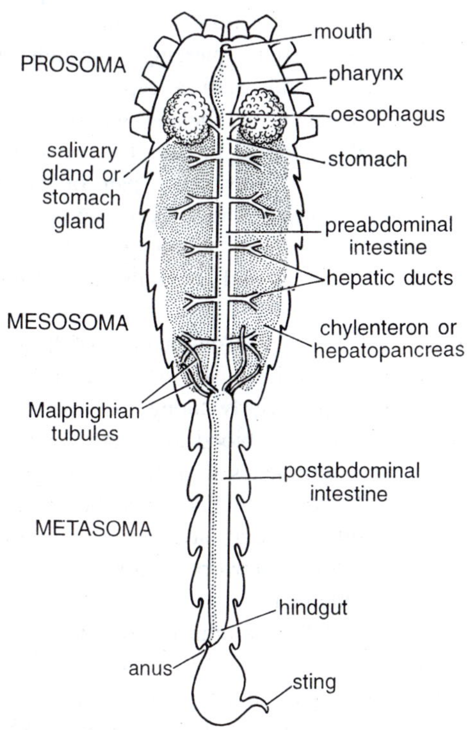

Fig. 6. Scorpion. Digestive system in dorsal view.

[III] Food and Feeding Mechanism

Scorpions are carnivorous and predaceous and they feed upon small invertebrates like insects, spiders, etc. They can remain alive without food for nearly six months. They are also cannibals, often feeding upon other smaller scorpions. Mouth of scorpion is very small, so that it cannot feed on solid particles. It only sucks liquids from the body of preys.

The prey is seized and held by chelate pedipalps while being paralysed by the sting. It is then transferred to the preoral cavity and torn to pieces by chelicerae. Basal segments of appendages, forming the preoral cavity, are slowly pressed upon the prey which are gradually squeezed and the oozing liquids are sucked in by the muscular pharynx. Process of feeding is very slow requiring at least two hours to devour a cockroach.

[IV] Digestion, Absorption and Egestion

Inside preoral cavity, food is reduced to a pulp. Its proteins are partially digested by secretions of certain **alveolar glands** found in maxillary processes of first two pairs of legs. Inside stomach, this partially digested food is mixed with the secretion of stomach gland containing the enzymes amylase, trypsin, lipase, etc. The digested food is absorbed into intestine while undigested portion reaches the hindgut to be excreted through the anus.

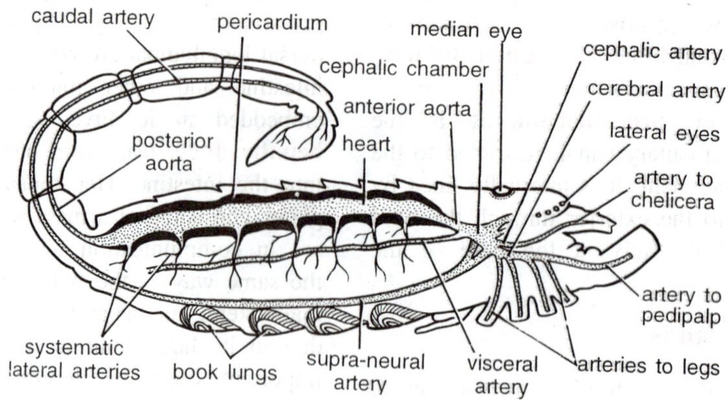

Fig. 7. Scorpion. Circulatory system in lateral view.

CIRCULATORY SYSTEM

[I] Organs of Circulation

Circulatory or blood-vascular system is open, as is typical for arthropods. It consists of : (i) **heart** and **pericardium**, (ii) **arteries**, (iii) **sinuses** and (iv) **veins**.

1. Heart and pericardium. Heart of scorpion is an elongated and muscular tube of a faint greenish colour. It lies in the pre-abdominal region, just beneath the chitinous terga, in a deep median dorsal groove of hepatopancreas. The heart is indistinctly divided into seven chambers by shallow transverse constrictions. The heart has seven pairs of dorso-lateral valvular apertures or **ostia**, each chamber possessing one pair.

The heart lies within a membranous sheath, the **pericardium**. It is closely attached to the heart by means of many ligaments which divide the pericardial cavity into a dorsal, a ventral and two lateral compartments.

2. Arteries. Heart is continued anteriorly as the **anterior aorta** and posteriorly as the **posterior aorta**. A series of paired lateral **systemic arteries** are also given off from the heart.

(a) Anterior aorta. It arises from the anterior end of heart and runs anteriorly along the dorsal surface of alimentary canal. It sends a pair of small **visceral arteries** to intestine and hepatopancreas. After piercing through the diaphragm, it expands into a cephalic chamber over the oesophagus and behind the brain. The chamber sends arteries to all the prosomatic appendages. It also gives off a pair of **ventro-lateral** arteries which embrace the oesophagus and unite below it to form a single **supraneural artery**, running behind and above the ventral nerve cord.

(b) Posterior aorta. Continuation of heart on the posterior side is the **posterior aorta** or **caudal artery**. It runs posteriorly along the dorsal side of intestine. It supplies blood to intestine, muscles and telson and terminates into the poison glands.

(c) Systemic arteries. Each chamber of heart gives off on either lateral side a systemic artery. These form extensive networks and supply various parts of pre-abdomen.

3. Sinuses. Ultimate branches of arteries open into blood spaces or **lacunae** among the visceral organs. From these small perivisceral lacunae, blood is gathered up into bigger sinuses, which are 5 in number. **Pericardial sinus** completely surrounds the heart. **Dorsal sinus** lies just above

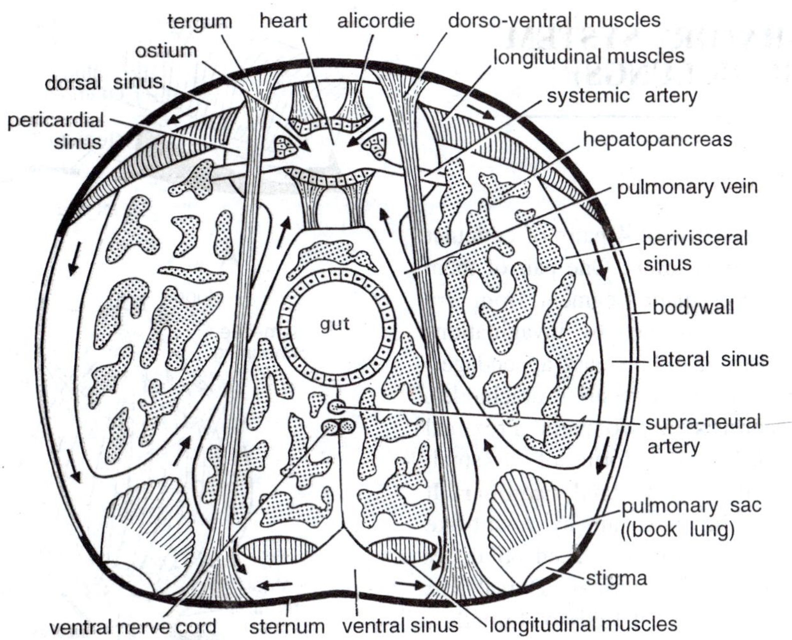

Fig. 8. Scorpion. T.S. of mesosoma showing pulmonary veins and the main sinuses of the body.

the pericardial sinus. Two **lateral sinuses** are found one on either lateral side of body. A large **ventral sinus** occurs on the ventral side of body.

4. Pulmonary veins. From each book-lung, oxygenated blood is sent to the pericardial sinus by a thin-walled pulmonary vein. Thus there are 4 pairs of pulmonary veins.

[II] Blood

Blood or **haemolymph** of scorpion is colourless, with a slight bluish tinge and containing numerous nucleated corpuscles or leucocytes. A blue-coloured respiratory pigment, the **haemocyanin**, is found dissolved in the blood plasma. Its molecular structure is similar to haemoglobin except that it is a copper-containing protein.

[III] Course of Circulation

With the contraction of ligaments, cavity of heart enlarges so that blood of pericardial sinus rushes into heart through ostia. With the contraction of heart, ostia are closed and blood is pumped to various parts of body through arteries which communicate with blood lacunae among the viscera. Impure blood of body is collected finally into the ventral sinus which sends diverticula to surround each book-lung. Oxygenated blood from book-lungs is carried by pulmonary veins to pericardial sinus to enter the heart through ostia. Course of circulation of blood may be represented diagrammatically as follows :

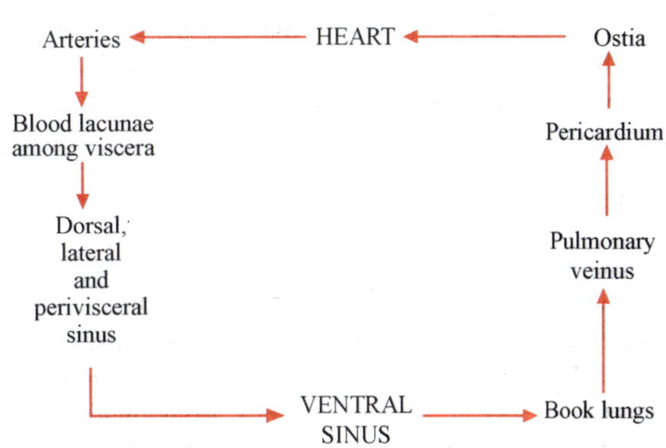

RESPIRATORY SYSTEM (BOOK-LUNGS)

1. Structure of Book-Lungs

Respiratory system consists of 4 pairs of cuticular **pulmonary sacs** or **book-lungs**. One pair of them lies inside each mesosomatic segment from 3rd to 6th. Book lungs are formed by invaginations of cuticle at the bases of rudimentary appendages. They are peculiar to the terrestrial arachnides. Each book-lung consists of two parts. Proximal or ventral part is in the form of a small compressed air cavity, called **atrial chamber**. It communicates with the outer air by a slit-like opening, the **spiracle** or **stigma**, placed obliquely on the ventro-lateral side of sternum. Dorsal part is made of nearly 150 vertical folds or **lamellae**, running parallel and arranged like leaves of a book. Each lamella is a hollow structure, made of two thin layers of cuticle united at their edges. A thin **air-space** is bounded in between two adjacent lamellae. Roof of atrial chamber is perforated by many linear, slit-like openings. The atrial chamber communicates with the inter lamellar air-spaces through these openings. Internal spaces of lamellae are continuous with body cavity and thus filled with blood.

2. Blood Supply of Book-Lungs

Venous blood from ventral sinus is sent to each book-lung by a **diverticulum** from which it enters the lamellae at their base. Aerated blood from lamellae of each book-lung is collected by a **pulmonary vein** which runs dorsally to open into the pericardium.

3. Mechanism of Respiration

Inflow and outflow of air in book-lungs seems to be controlled by the action of **dorso-ventral** and **atrial muscles**. On contraction of these muscles, book-lungs are compressed and air of inter-lamellar spaces is forced out into atrial chamber and then to exterior through **stigmata**. When the muscles relax the book-lungs resume their normal shape so that fresh air enters through stigmata, first into the artrial chamber and then into the inter-lamellar spaces.

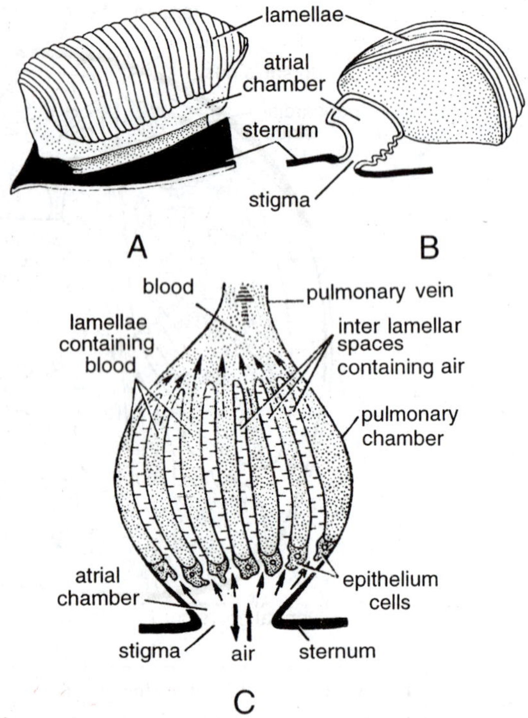

Fig. 9. Scorpion. **A.** Book-lung. **B.** Book-lung in V.S. **C.** Book-lung in a diagrammatic V.S. showing the mechanism of respiration.

Exchange of gases takes place between air of interlamellar spaces and venous blood through the thin membranous walls of lamellae. Blood becomes oxygenated and its CO_2 is passed out into air.

EXCRETORY SYSTEM

[I] Excretory Organs

Excretory organs of scorpion are : (i) Malpighian tubules, (ii) coxal glands, (iii) hepatopancreas, and (iv) nephrocytes and lymph tissue organs.

1. Malpighian tubules. One or two pairs of Malpighian tubules are attached at the junction of pre-abdominal and post-abdominal parts of intestine. It is remarkable that a scorpion has only one or two pairs of these tubes while an insect of equal size, such as cockroach, requires a large number of them. They are ectodermal in cockroach but endodermal in origin in scorpion. They are restricted to pre-abdomen but may enter the prosoma. Their diameter is irregular indicating a differential absorptive and diffusive capacity. The excretory wastes collected from blood by these tubules are poured into gut from where they are passed out alongwith faeces. Chief waste is **guanine**, but adenine and uric acid have also been found in the rectal contents of scorpions.

2. Coxal glands. A pair of shining white coxal glands are situated one on each side in prosoma, close in front of the endosternite. Coxal glands are derived from coelomoducts, five pairs of which are represented in segments 2, 4, 5, 6 and 8 in the embryo. But in adult scorpion only those of the 5th segment persist as coxal glands. Each gland consists of three main parts : (i) a large excretory **saccule**, **vesicle** or **end-sac**, (ii) a coiled tube or **labyrinth**, and (iii) a swollen terminal reservoir or **bladder**. A small exit duct leads from bladder to open to the exterior by a small orifice on the posterior face of the 5th walking leg.

Coxal glands of scorpion are homologous with the green glands of Crustacea and probably function as excretory organs. Nitrogenous wastes from blood are collected by saccule and labyrinth of each coxal gland and discharged through the excretory pore. Urate crystals have been found in saccule. **Buxton** noticed that the glands soon pick up carmine injected in the body cavity.

3. Hepatopancreas. Hepatopancreas of scor-pion is also understood to serve as an excretory organ. According to Pavlovsky, ammonia carmine injected into the body cavity of **scorpion** is deposited in the form of minute, bright red granules inside the cells of hepatopancreas.

4. Nephrocytes. Large nephrocytes and lymph tissue organs are specialized structures found underneath the body wall in the region of mesosoma. They are both excretory and phagocytic in function.

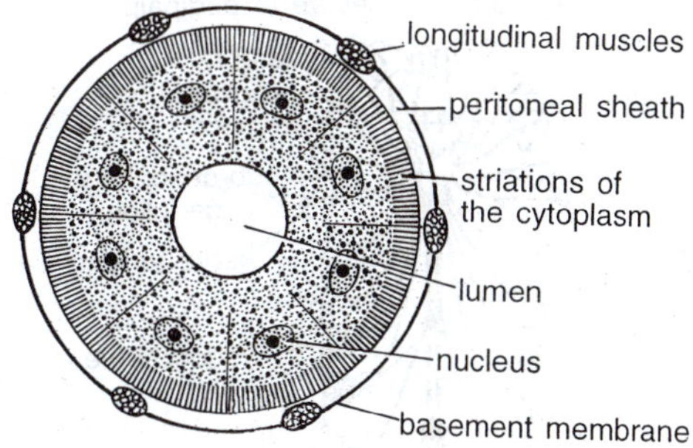

Fig. 10. Scorpion. A Malpighian tubule in T.S.

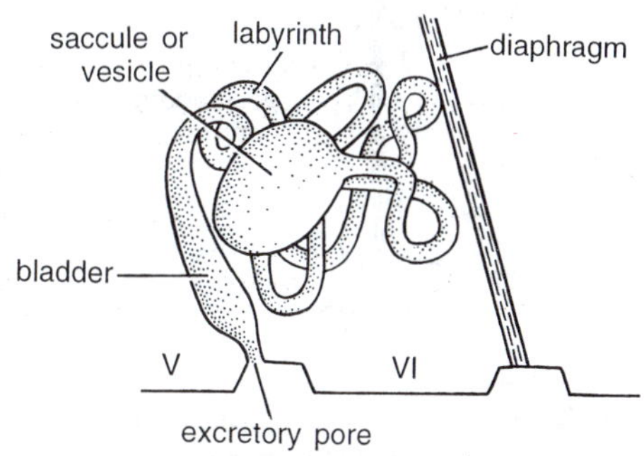

Fig. 11. Scorpion. A coxal gland.

NERVOUS SYSTEM

The nervous system is of annelidan type.

1. Brain. Brain or **cerebral ganglion** is a small bilobed mass, situated in prosoma just beneath the median eyes. It gives off a pair of **optic nerves** to the median and lateral eyes. Besides, the brain gives off numerous delicate nerves to rostrum, pharynx and oesophagus.

Brain also gives off a pair of thick, short and stout **circum-oesophageal connectives**. These form a collar around oesophagus and unite

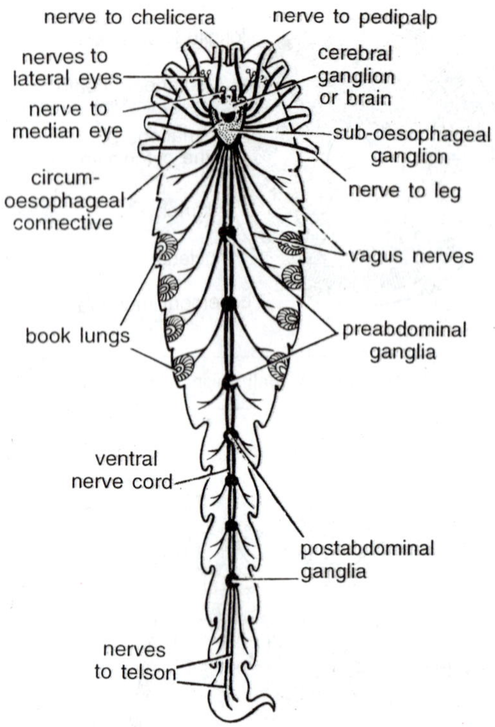

Fig. 12. Scorpion. Nervous system in dorsal view.

ventrally into a **sub-oesophageal ganglion**. From connectives and the ganglion arise 6 pairs of **lateral nerves**, which innervate the 6 pairs of **prosomatic appendages**. Besides, the sub-oesophageal ganglion gives off 2 to 4 pairs of **vagus nerves** which run posteriorly into pre-abdomen to supply the genital operculum, pectines, and first two pairs of book-lungs.

2. Ventral nerve cord. The sub-oesophageal ganglion is followed by a double ventral nerve cord which runs up to the 4th segment of post-abdomen. The nerve cord is slender and rounded in pre-abdomen but flattened and ribbon-shaped in post-abdomen. It bears 3 ganglia in pre-abdomen and 4 in post-abdomen. The nerves arising from these segmental ganglia supply the neighbouring parts of body.

SENSE ORGANS

Sensory organs include : (i) eyes, (ii) pectines, (iii) stridulating organs and (iv) sensory hairs.

[I] Eyes

Eyes are of two types **median** and **lateral**. One pair of conspicous **median eyes** are present dorsally on carapace, closely placed together in the middle. Each eye is raised on a small tubercle. In addition, 2 to 5 pairs of smaller **lateral eyes** are placed on the antero-lateral margines of carapace. Their number varies according to the species. Both kinds of eyes are simple in structure. Lateral eyes are similar in character to the simple eyes or ocelli of insects. Median eyes are more complicated in structure. They resemble the lateral eyes in the presence of a single cuticular lens but differ from them in having photoreceptor **retinal cells**. These are cylindrical in shape, have a nucleus proximally and a light sensory **rod** or

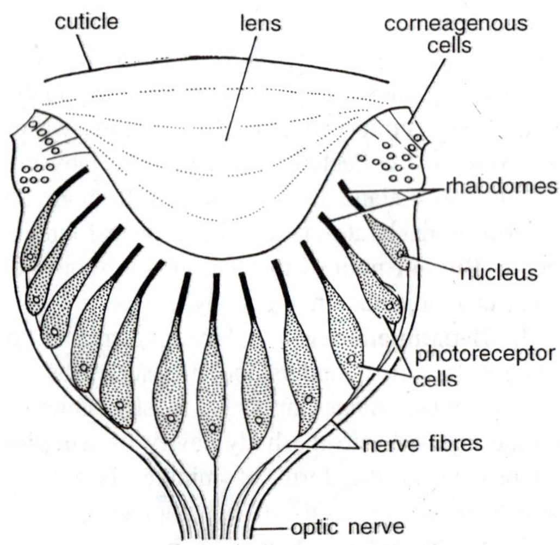

Fig. 13. Scorpion. Median eye in V.S.

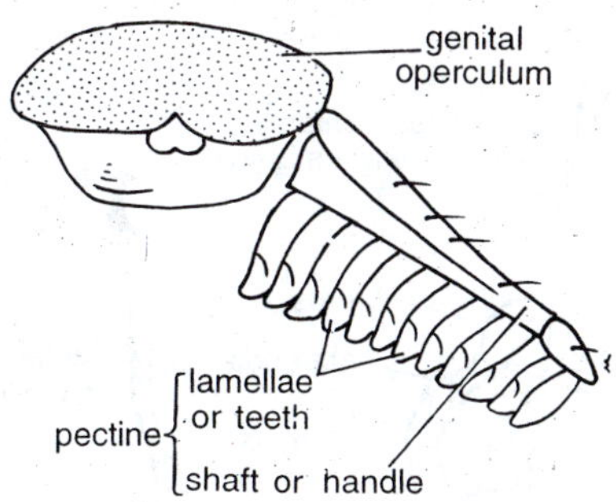

Fig. 14. Scorpion. Genital operculum and the left pectine in ventral view.

bacillus at the distal end. Such eyes with distal receptive rods are termed **direct eyes**. Retinal cells are grouped in retinule which secrete a rhabdome as in a compound eye. Nerve fibres from individual cells join to form the optic nerve which connects the optic ganglion of brain. However, a few scorpions are blind.

[II] Pectines

A pair of remarkable comb-like sensory appendages, called **pectines**, are found on the sternum of 2nd segment of pre-abdomen or mesosoma. Each pectine consists of a three-segmental **stem**, **shaft** or **handle**. Along the posterior margin of shaft is a row of 4 to 36 narrow movable processes, just like teeth of a comb. Pectines are tactile or perhaps olfactory organs which enable the animal in finding the nature of the substratum. **Pocock** noticed a scorpion walking over a cockroach until the pectines came in contact with it, when it at once moved back and ate it. Pectines are generally larger in male and sometimes curiously modified in female, so that sexual and other roles have also been suggested for them. According to **Alexander**, the function of pectines is the selection and exploration of a suitable mating site by the male.

[III] Stridulating Organs

In some species of scorpions efficient **Stridulating organs** are found on coxae of pedipalps or first pair of legs in the form of ridges, across which file-like surface can be drawn to produce sound. It indicates that scorpions probably have some perception of sound.

[IV] Sensory Hair

Long and straight, delicate sensory hairs, called **trichobothria**, arise from saucer-like depressions richly supplied with nerves. They are usually present on appendages in arachnids, but also occur on the 'tail' of scorpions. These are processes of modified sensory cells of hypodermis. Trichobothria are able to detect air borne vibrations.

REPRODUCTIVE SYSTEM

Sexes are separate. **Sexual dimorphism** is not distinct so that two sexes cannot be easily distinguished externally. However, the male is usually smaller, with longer pedipalps and pectines, and a narrower abdomen than those of female. Each opercular plate of male also carries a copulatory hook. **Gonads** form a network of tubules and are essentially similar in both sexes.

[I] Male Reproductive System

1. Testes. Male scorpion has one pair of mesh-like **testes** lying embedded in hepatopancreas and extending from 3rd to 6th segments of pre-abdomen. Each testis consists of a pair of narrow, longitudinal tubules or trunks, connected by transverse ducts so as to form three squares. **Spermatozoa** develop from cells of **germinal epithelium** lining the tubules of testes. Mature spermatozoa are filiform and motile, each having an oval body and a long tail. They are contained within a **spermatophore**.

2. Vasa deferentia. Anteriorly, a narrow duct, called **vas deferens**, arises from the outer angle of each testis.

3. Genital chamber. Each vas deferens runs forward to open into the **genital chamber** of its side. Before entering the genital chamber each vas deferens is somewhat dilated in some cases to form the terminal ampulla, into which open

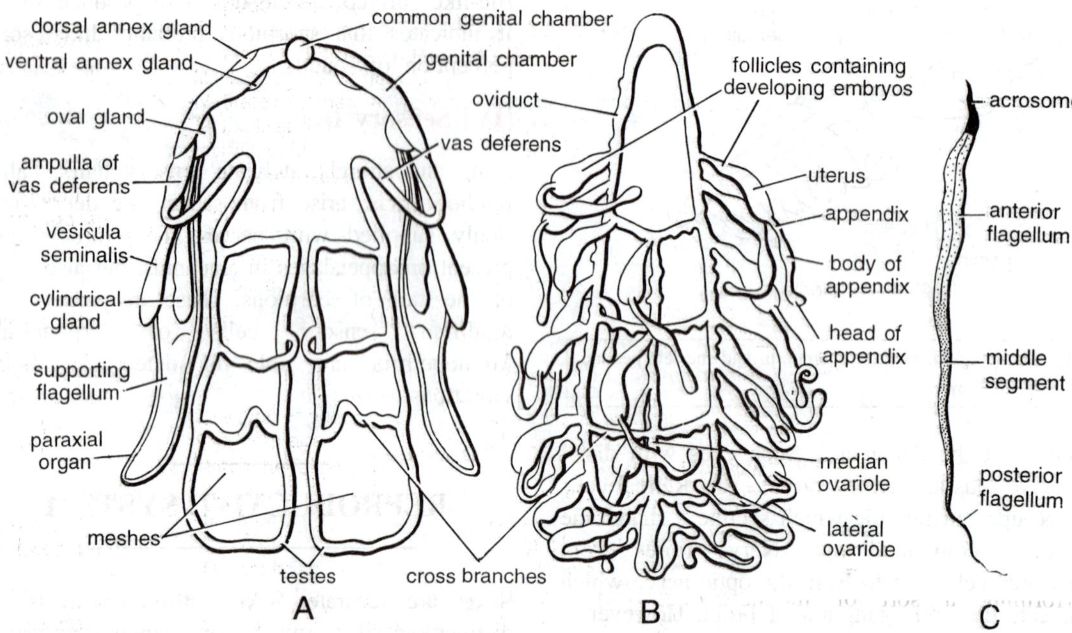

Fig. 15. Scorpion (*Palamnaeus*). **A.** Male reproductive system. **B.** Female reproductive system. **C.** A mature sperm of *Buthus occitanus*.

the **accessory glands** and a club-shaped **vesicula seminalis**. Accessory glands secrete a fluid which helps in reproduction, while the vesiculae seminales serve to store the mature spermatozoa.

4. Paraxial organs. Each genital chamber is produced behind into a **paraxial organ** which contains a tightly-fitting chitinous rod, the **flagellum**. Structure of flagellum varies in different scorpions. It is provided with many spines and a longitudinal groove on the inner side. The two flagella together form the so-called 'double penis' of scorpions and they are said to evert out through the male reproductive aperture to serve as claspers during copulation.

5. Common genital chamber. Two genital chambers open anteriorly into a small median **common genital chamber**. It opens to outside through a **male genital aperture** lying beneath the genital operculum on the sternum of first pre-abdominal segment.

[II] Female Reproductive System

1. Ovary. Female scorpion has a single ovary occupying the same position inside pre-abdomen as the testes in male. It consists of three narrow, longitudinal tubules or **ovarioles**, one median and two lateral. The shorter median ovariole is connected with the longer lateral ovarioles by 3-4 cross branches so as to enclose three wide meshes on each side. Ovarioles and their cross branches are lined by germinal epithelium, cells of which form the ova. Developing ova project into hollow, bud-like processes on the surface, called **diverticula** or **follicles**. Form and structure of follicles vary in different species and their size depends upon the size of the developing embryo within. Follicles remain spherical throughout in *Buthus*. Typically they are fusiform (*Palamnaeus*), with a proximal swollen base or **'uterus'** containing the embryo and a distal narrow **appendix** with a body and an enlarged tip called **head**.

2. Oviducts. The two lateral ovarioles are continued anteriorly as two **oviducts** which converge to open into a small median **genital chamber**.

3. Genital chamber. Median genital chamber opens to outside by the **female reproductive**

aperture lying beneath the bilobed operculum on the sternum of first pre-abdominal segment.

LIFE HISTORY

In India scorpions breed during the monsoon months.

[I] Fertilization and Copulation

Fertilization is internal and preceded by **courtship** and **copulation**. **Fabre** has given a picturesque account of their mating habit. According to him, the two scorpions stand face to face with their tails upraised and intertwined. Male grasps the pedipalps of female with its own and the two go on moving round and round for an hour or more performing a sort of mating dance termed '**promenade a deux**' by **Fabre**. It became famous due to its picturization by **Walt Disney** in one of his old nature films, '**The Living Desert**'.

Eventually the male deposits a **spermatophore** containing spermatozoa, on ground. A wing-like lever projects above it. Male now drags the female so that her genital aperture is over the spermatophore. Downward pressure on its lever ejects the sperm mass into the female opening. This indirect sperm transmission by spermatophore is characteristic of most arachnids and represents an adaptation for terrestrial existence.

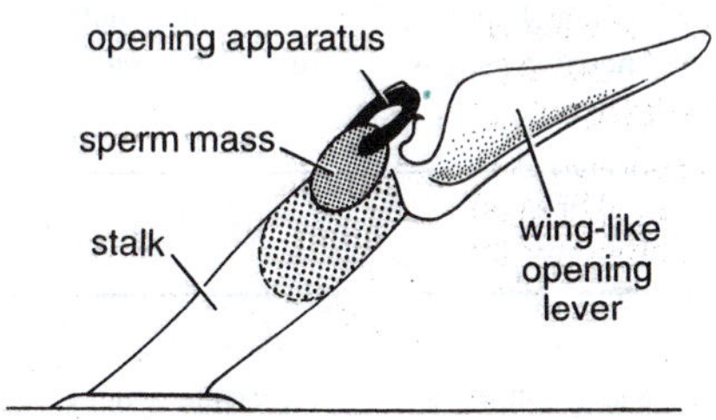

Fig. 16. Scorpion. A spermatophore attached to ground.

At last, the male burrows a hole under a flat stone and both disappear into it. After mating, the female often devours her mate.

[II] Development

Development is always internal. Early development takes place in ovarian follicles but later stages are passed through the ovarioles which function as uteri. Cleavage is discoidal. Development is very slow and some species require several years to reach maturity. Scorpions with large telolecithal eggs are **ovoviviparous**, while those with little eggs poor in yolk are **viviparous**.

In viviparous species, the mother scorpion gives birth from 6 to 90 young ones at a time. The newly born, delicate, miniature scorpions are only a few millimetres long. They climb up the body of the mother scorpion for getting shelter and food. She carries them on her back for some days till they can take care of themselves. After a week, they undergo first moult. Then they gradually leave the mother and become independent. They undergo several moults to grow into adults. The popular belief that the young ones devour the mother is wrong. The fact is that babies' jaws are very weak, while the mother's back is hard and without perforations.

Important Questions

➤➤*Long answer type questions*

1. Make full page, properly labelled diagram, illustrating the external features of scorpion.
2. Describe the digestive system of scorpion and compare it with that of prawn.
3. Give an account of the organs that subserve respiration in scorpion.
4. Describe the male and female reproductive organs of scorpion.
5. Write short notes on : (i) Book-lung, (ii) Coxal gland, (iii) Endosternite, (iv) Pectines, (v) Pedipalp, (vi) Promenade a deux, (vii) Trichobothrium.

➤➤*Short answer type questions*

1. What is the function of the coxal glands ?
2. What are the 3 body divisions of scorpion ?
3. Describe the mechanism of stinging and injecting poison by scorpion.
4. Briefly explain food and feeding apparatus in the scorpion.
5. Discuss the respiratory system of *Palaemon* and *Palamnaeus.*
6. Show how the scorpion is adapted for its terrestrial mode of life.
7. Describe with a neat labelled diagram of a chelicerae.
8. Draw a well labelled diagram of a pedipalp and give its function.

➤➤*Fill in the blanks*

1. The respiratory organ in scorpion is known as .....
2. Excretory organs in scorpion are .....
3. Mating dance of male and female scorpion is also called as .....
4. Coxae of 5th pair of legs of scorpion bear openings of ....

➤➤*Multiple choice questions*

1. Body divisions of scorpion are :
   (a) prosoma and opisthosoma
   (b) prosoma, mesosoma and metasoma
   (c) head, thorax and abdomen
2. Exploring of ground and recognition of sex is done by :
   (a) spiracles (b) pectines
   (c) chelicera (d) pedipalp
3. Solid processes of chitin in scorpion are :
   (a) dermatidia (b) coelodermatidia
   (c) epiostracum (d) telostracum
4. Hollow processes of chitin in scorpion are :
   (a) hypostracum (b) coelodermatidia
   (c) coelostomata (d) tracheinida
5. Respiratory organs of scorpion :
   (a) coxal organ (b) book lung
   (c) antennary organ (d) tracheae
6. Scorpions are :
   (a) oviparous (b) ovoviviparous
   (c) viviparous (d) both
7. How man podomeres in walking legs of *Palamnaeus* :
   (a) 5 (b) 7
   (c) 9 (d) 11
8. The heart chambers in *Palamnaeus :*
   (a) 1 (b) 3
   (c) 5 (d) 7
9. Blood pigment in scorpion :
   (a) haemocyanin (b) erythrocyanin
   (c) haemoglobin (d) none
10. The excretory system of scorpion includes :
   (a) Malpighian tubules (b) coxal gland
   (c) hepatopancreas (d) nephrocyte
   (e) all

Answers

➤➤*Fill in the blanks*

1. Book-lungs, 2. Malpighian tubules, Coxal glands, Hepatopancreas and Nephrocytes, 3. promenade a deux, 4. Coxal glands.

➤➤*Multiple choice questions*

1. (b) 2. (b) 3. (a) 4. (b) 5. (b) 6. (d) 7. (b) 8. (d) 9. (a) 10. (c)

47

Periplaneta americana : The Common Cockroach

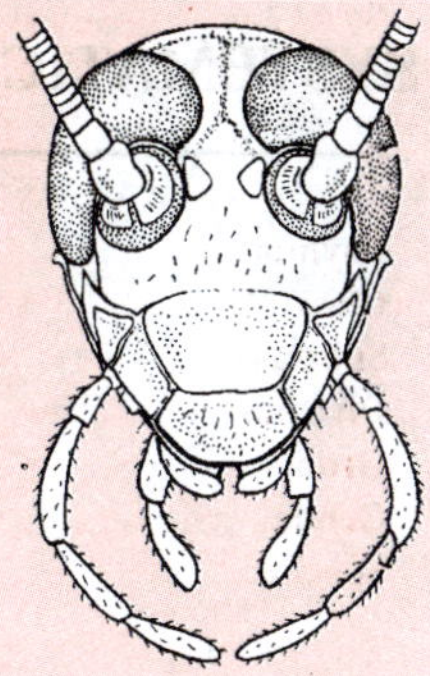

Arthropods are incredibly diverse. They account for more than 83% of all animal species and out of that 80% are insects. Class **Insecta** of phylum Arthropoda is the largest in the Animal Kingdom, as there are more species of insects (about 80,000) than all other species of animals combined. Insects are air-breathing and usually flying arthropods. They are distinguished from other arthropods by the possession of 3 pairs of legs and usually 2 pairs of wings. They include familiar examples such as cockroaches, grasshoppers, house flies, honey bees, silkworms, etc. Cockroaches belong to the order **Orthoptera**. They are common pests of food industries, hotels, kitchens, warehouses, etc.

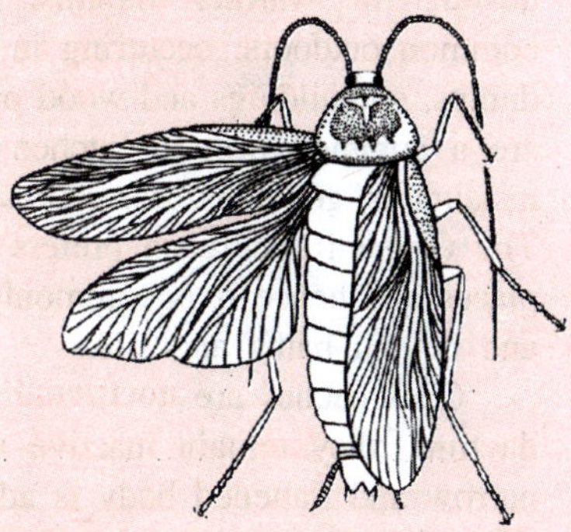

The word cockroach is derived from the Spanish 'Cucaracha'. Its four common species found in India are : (i) *Blattella germanica,* the German cockroach, (ii) *Blatta orientalis,* the Oriental cockroach, (iii) *Periplaneta americana,* the American cockroach, and (iv) *Periplaneta australasiae,* the Australian cockroach. The following text deals mainly with the biology of *P. americana.* There are about 30 species of cockroaches found in world.

Periplaneta americana

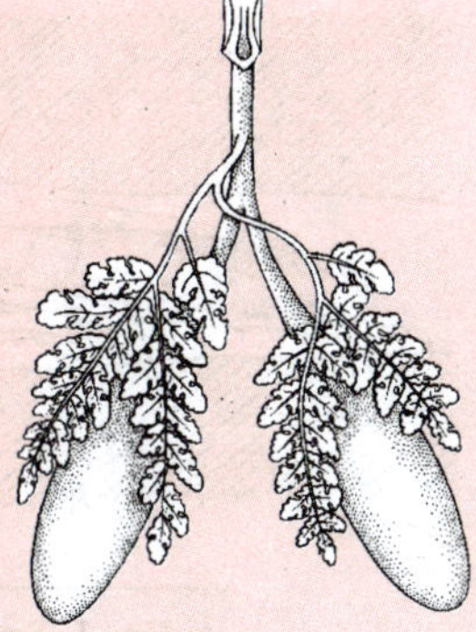

Periplaneta americana was originally named *Blatta americana* by **Linnaeus** (1758). **De Geer** (1773) called it *Blatta kakeralac.* The generic name *Periplaneta* was assigned to it by **Burmeister** (1838). Its original home is supposed to be tropical Africa and not America, as previously believed. Today, it exists in all parts of the world. It has become well established almost throughout India.

SYSTEMATIC POSITION

Phylum	Arthropoda
Class	Insecta
Subclass	Pterygota
Division	Exopterygota
Order	Orthoptera
Genus	*Periplaneta*
Species	*americana*

HABITS AND HABITAT (ECOLOGY)

Cockroaches are found in places where there is warmth, dampness and plenty of organic food to devour. In warmer months, cockroaches are common outdoors, occurring in sewer, manholes, dumps, outbuildings and wood piles. Indoors, they are a common pest in kitchens, latrines, hotels, restaurants, godowns, storerooms, board ships, etc. The Oriental cockroach prefers somewhat cooler places and hence it is commonly found in toilets and behind baths and sinks.

Cockroaches are **nocturnal** creatures. During daytime, they remain inactive and hiding. Their narrow and flattened body is adapted to slip into narrow creeks and crevices. During night, they show much activity and run here and there in search of food. They are **cursorial** insects, *i.e.*, fast runners, and rarely resort to flight. It has been reported that, at 25°C, *Periplaneta* runs from 70-130 cm per second. Being **omnivorous** and **scavengerous** in diet, they devour any animal or vegetable substance and even non-living materials like leather, paper, cloth, etc., causing great loss.

EXTERNAL MORPHOLOGY

[I] Shape, Size and Colour

Body of cockroach is narrow elongated, bilaterally symmetrical and dorso-ventrally flattened. The adult measures from 28 to 44 mm in length and 8 to 10 mm in width. Colour is shining reddish-brown with a paler yellow area around the edge of tergum of pronotum of prothorax and two dark patches over it.

[II] Exoskeleton or Cuticle

The entire body is covered externally by a non-living brown-coloured, hard, jointed and chitinous **exoskeleton** composed of several plates or **sclerites**. It is formed by the surface layer or cuticle of body wall and is secreted by the underlying hypodermis.

[III] Divisions of Body

Body of cockroach is distinctly divided into segments. These are grouped into 3 well-defined regions or tegmata : (i) **Head**, (ii) **thorax** and (iii) **abdomen**.

[IV] Head

Anterior region of body or head is small and roughly triangular or pear-shaped. It is held at right angles (perpendicularly) to the long axis of

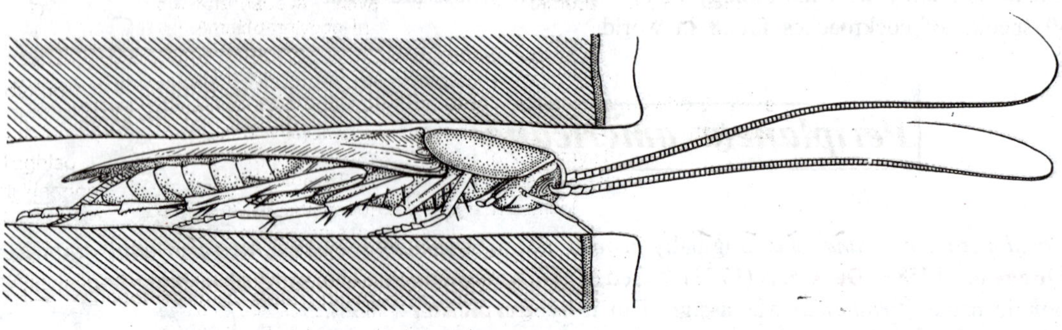

Fig. 1. Sectional view of a crevice showing a cockroach in resting posture.

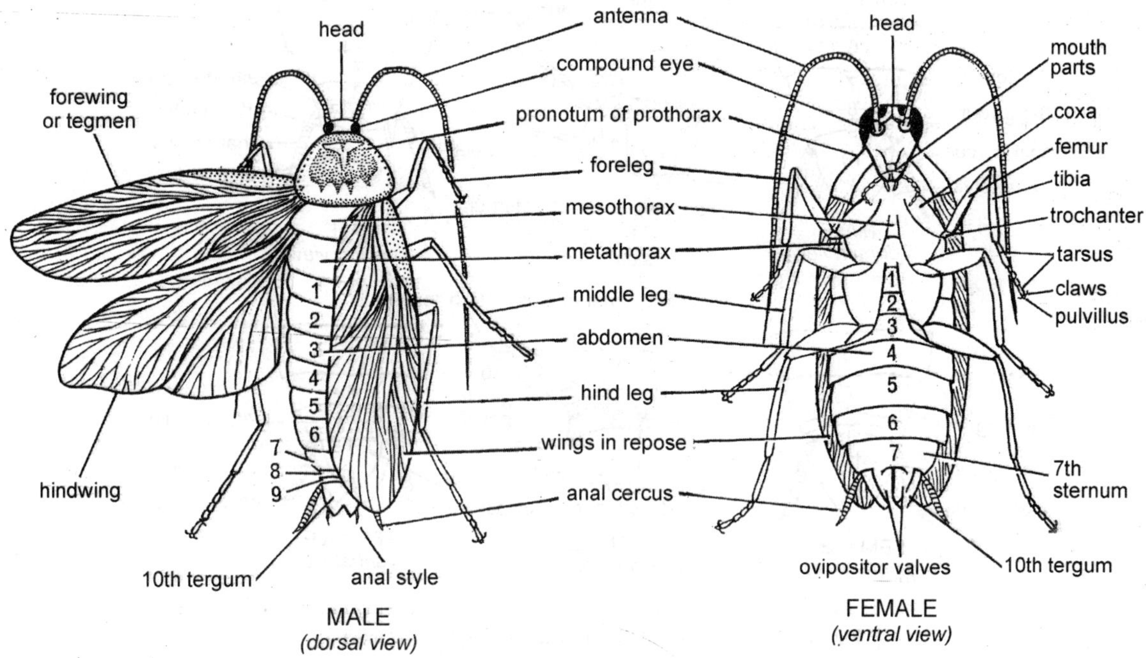

Fig. 2. *Periplaneta.* External features. **A.** Male in dorsal view. **B.** Female in ventral view.

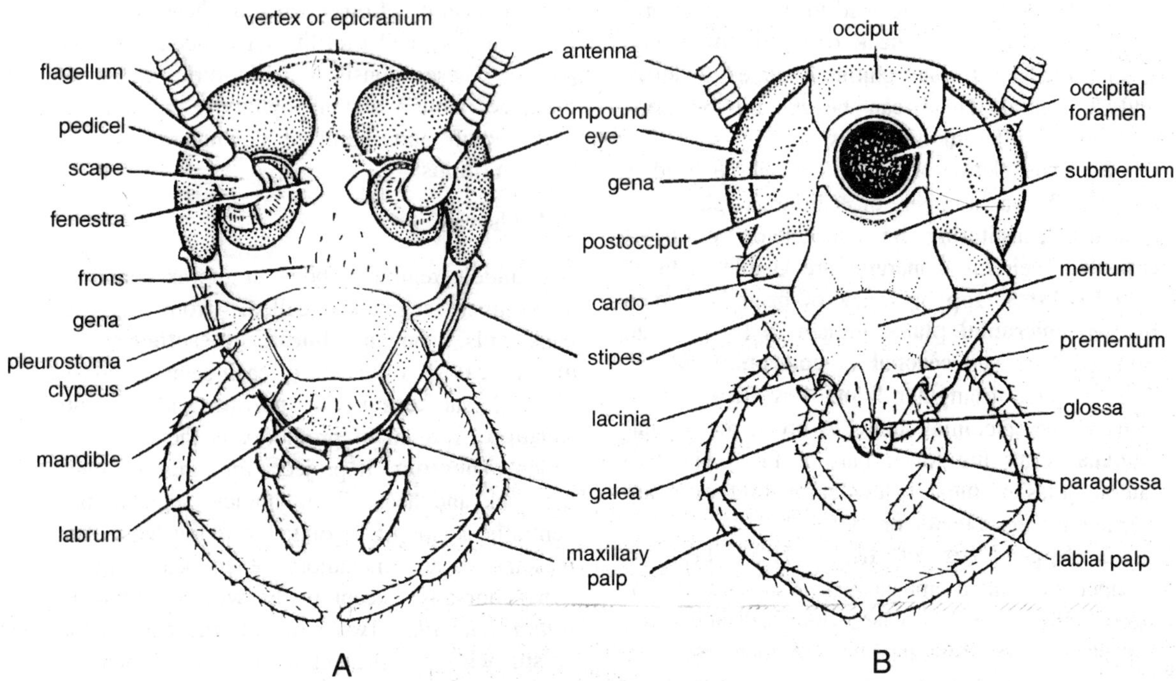

Fig. 3. *Periplaneta.* Head. **A.** Anterior view. **B.** Posterior view.

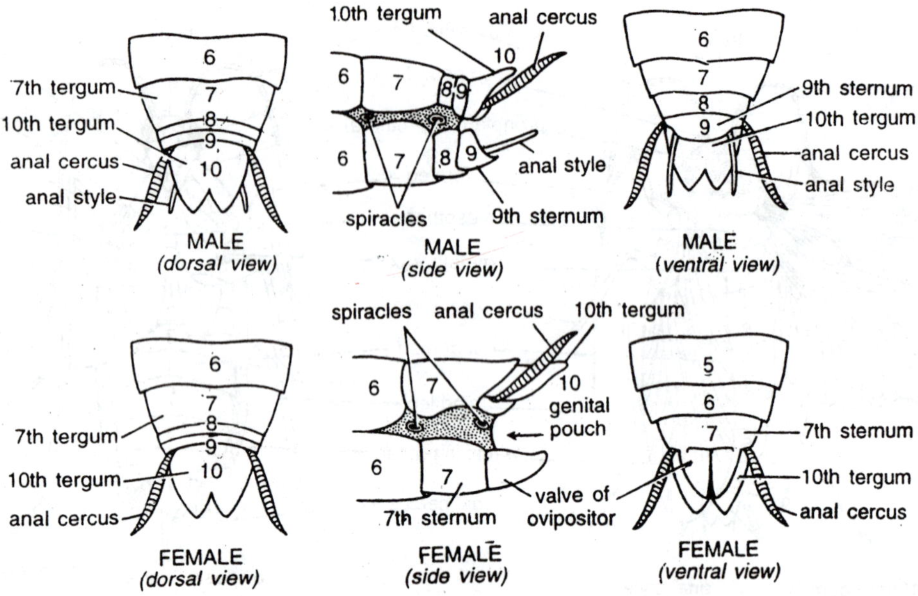

Fig. 4. *P. americana.* Posterior segments of the abdomen.

body and scarcely visible in dorsal view. It is formed by the fusion of 6 embryonic segments.

1. Neck. Head is attached to thorax anteriorly by a short and narrow **neck** or **cervicum**, which is supported by 4 small chitinous plates, 2 dorsal and 2 ventral. Head can be moved by neck muscles in different directions.

2. Exoskeleton of head. The whole head is covered by a number of chitinous plates or sclerites, constituting its exoskeleton or **head capsule**. Their fused margins are visible as lines, called **sutures**. Top or vertex of head is formed by two **epicranial plates** jointed in front by an inverted Y-shaped **coronal** or **epicranial suture**. The unpaired triangular **frons** lies between the arms of the epicranial suture. A broad rectangular **clypeus** forms the lower part of face. The two lateral sides of head below compound eyes are formed by the **cheeks** or **genae**.

3. Appendages of head. Head bears a number of jointed appendages. On each dorso-lateral side of head is a large reniform and black **compound eye**. Just in front of each eye is a long, slender and multi-segmented **antenna**. Inner to the base of each antenna is a small, rounded and whitish area or **fenestra**, representing a simple eye or ocellus. Lower end of head bears the preoral cavity and **mouth**, surrounded by **mouth parts**. These consist of labrum, mandibles, maxillae, labium and hypopharynx. Such a head with mouth parts directed downwards is called **hypognathous**.

[V] Thorax

The middle region of body or thorax consists of 3 distinct segments called, from in front backwards, the **prothorax**, **mesothorax** and **metathorax**. Exoskeleton of each segment consists of a dorsal **tergum** or **notum** and a ventral **sternum**. Tergum of prothorax is the largest and called **pronotum**. It projects forwards to conceal the neck and head. Each thoracic segment bears ventrally a pair of 9 jointed walking **legs**. While mesothorax and metathorax each bear a pair of **wings** dorsally, except in the wingless female of *Blatta orientalis*. Two pairs of tracheal openings or **spiracles** are present on the lateral pleura of thorax.

[VI] Abdomen

The posterior region of body is called abdomen. It is broader than thorax and dorso-ventrally flattened. Abdomen of adult consists of 10 segments while embryo has 11 segments. Not all segments are visible externally, the hinder segments remain somewhat concealed. 8 pairs of tracheal openings or **spiracles** are present on lateral pleura of first 8 abdominal segments.

1. Exoskeleton of abdomen. Exoskeleton of each abdominal segment consists of a dorsal **tergum** and a ventral **sternum**, united laterally by soft cuticle. Tergum of 7th segment overlapps those of 8th and 9th segments. 10th tergum projects behind freely as a deeply notched bifid plate. **Anus** lies below 10th tergum between 4 **podical plates**, a dorsal **epiproct**, a ventral **hypoproct** and two lateral **paraprocts**, which may represent the vestiges of llth embryonic segment.

2. Abdominal appendages and sexual dimorphism. 10th segment in both sexes bears a pair of small, filamentous and sensory **anal cerci**. They project laterally, one on either side, at the posterior end. They probably represent the llth segment. Each anal cercus is made of 15 segments. Terminal part of abdomen differs in two sexes.

In male, in addition to anal cerci, 9th sternum bears a pair of short, unjointed thread-like **anal styles** which are absent in female. The **genital aperture** of male is placed between 9th and 10th sterna, surrounded by complicated processes known as **gonapophyses**.

In female cockroach, abdomen is broader than in male. The **genital aperture** is located on 8th sternum surrounded by **gonapophyses** forming an **ovipositor**. 7th sternum of female forms a boat-shaped **genital pouch** and is split behind forming two **gynovalvular plates**. Frequently, an egg cocoon may be seen projecting from this.

APPENDAGES OF COCKROACH

Main appendages include : **antennae** and **mouthparts** on head, **legs** and **wings** on thorax, and **external genitalia** on abdomen.

[I] Antennae

Antennae are a pair of long, slender, whip-like, and many-jointed movable appendages, found on head. They are embryologically preoral and hence homologous with the antennules of prawn (Crustacea). Male cockroach has longer antennae than female.

An antenna is attached close to the inner side of compound eye in a circular socket enclosed by a ring-like **antennal sclerite**, and freely articulating by a thin membrane. It consists of a large basal segment, the **scape**, followed by a short and cylindrical **pedicel**, and a long many-jointed **flagellum**. The antennae bear tactile and olfactory receptors and are sensitive to touch and smell.

[II] Mouth Parts

Mouth parts are head appendages surrounding the mouth. These consist of the **labrum** or upper lip, a pair of **mandibles**, a pair of **maxillae**, the **labium** or lower lip and the **hypopharynx**. Mouth parts of cockroach have remained unspecialised due to its omnivorous habit. They are of the **biting** and **chewing** type, also known as of **mandibulate** type. They enable cockroach to bite and chew hard stuffs, consume soft stuffs and lap up liquids.

1. Labrum. Labrum or upper lip is the appendage of 3rd head segment. It is a broad and oblong plate, movably attached to the lower margin of clypeus and forms the anterior margin of mouth cavity. At its junction with clypeus, it bears a small sclerite, called **torma**, on each inner side. Its inner surface also bears two rows of sensory (gustatory) setae, one on each side. Labrum overlaps the toothed portions of mandibles.

2. Mandibles. These are appendages of 4th head segment. These are small, triangular, and strongly sclerotised structures. They are attached with the sides of head capsule and articulate by means of a **condyle** with the gena of same side.

Inner margin of each mandible is made of two small cutting or serrated lobes, distal incisor and proximal molar, each bearing 3 tooth-like denticles of thickened cuticle. Proximal to molar lobe is a membranous lobe, the **prostheca**, that bears sensory hairs. Mandibles work like jaws.

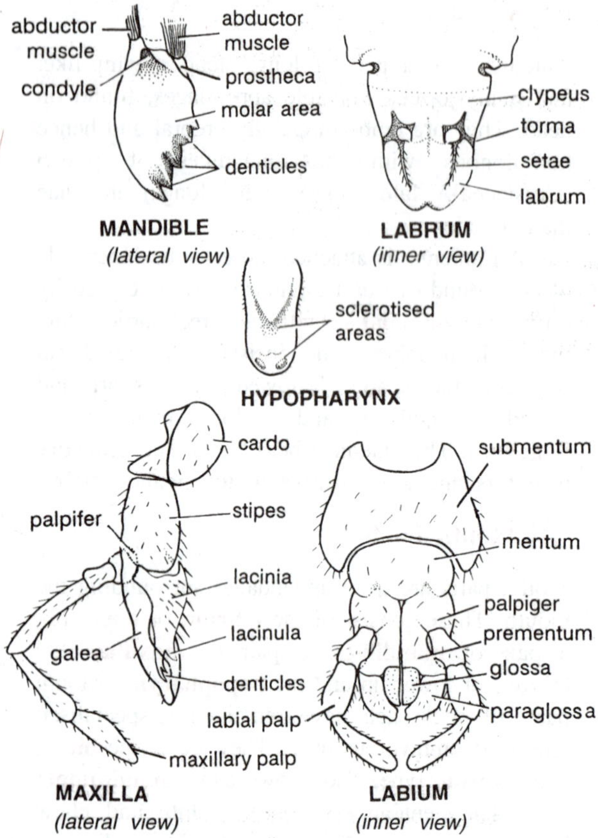

Fig. 5. *P. americana.* Mouth parts.

Denticles act as the interlocking structures whereas smooth molar areas as masticatory surfaces. Both the mandibles show horizontal sideways movements brought about by the action of **adductor** and **abductor muscles**.

3. Maxillae. These are appendages of 5th head segment and known as first pair of maxillae. They lie beneath mandibles and articulate with the posterior surface of head capsule, one on either side. They are biramous in structure. The basal part or protopodite consists of a distal cylindrical **stipes** hinged to a proximal basal **cardo** at an obtuse angle. Stipes bears three processes : (i) inner **lacinia** which is sclerotised with a pair of sharp **denticles** and a blunt **lacinula**, (ii) middle **galea** which acts as a hood for lacinia and (iii) outer long **maxillary palp**, which is 5-segmented and sensory in nature. It is borne on a small basal sclerite, the **palpifer**. Maxillary palp represents exopodite while lacinia and galea represent endopodite. Both maxillae act as accessory jaws. They work sideways to masticate food and to convey food morsels into oral cavity. They also serve to clear antennae and forelegs.

4. Labium. Labium or lower lip is the appendage of 6th head segment. It is the posterior-most of all mouth parts. It is formed by the fusion of two maxilla-like embryonic appendages or second pair of maxillae. Fused protopodite is 3-segmented consisting of a large proximal **submentum**, a small middle **mentum** and a distal **prementum**. The partially fused endopodites form a tongue-like **ligula**. Each half of ligula consists of an inner **glossa** and an outer **paraglossa**, corresponding with lacinia and galaea respectively. Prementum bears, on each side, a 3-jointed sensory **labial palp** borne on a basal projection, the **palpiger**. Terminal segment of palp is covered thickly with sensory setae.

5. Hypopharynx. It lies between the maxillae and above the labium, inside the mouth cavity. It is a median some what cylindrical structure also known as **lingua**. Its distal end bears two comb-like plates, the **superlingua**. The common salivary duct opens at its front end.

[III] Walking Legs

There are three pairs of walking legs, one pair attached ventrally to sternum of each of prothorax, mesothorax and metathorax and called for, mid- and hind legs, respectively. All the legs are similar, and each consisting of a linear series of 9-segments or **podomeres**. Leg is articulated with its thoracic segment by a stout flattened, basal **coxa**, followed by a smaller triangular **trochanter**. The latter is fused to a stout spiny **femur**, followed by a slender spiny **tibia**, which is the longest segment of leg. Finally, the **tarsus** is made of 5 very small movable podomeres or **tarsomeres** bearing fine hairs. Basal tarsomere is longer and called **metatarsus** while terminal tarsomere is named **pretarsus**. Inner edges of first four

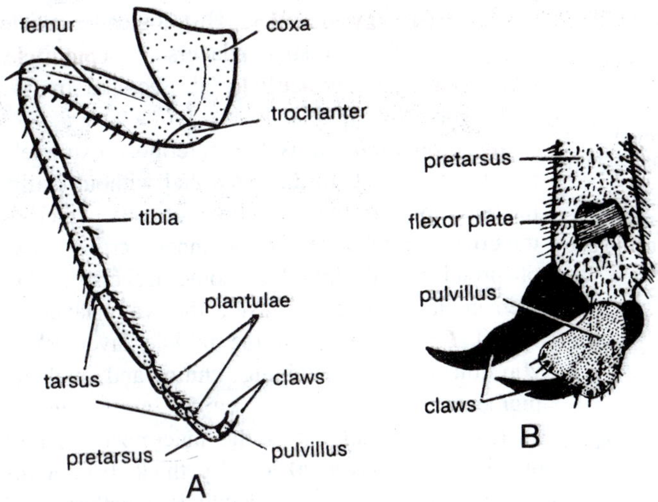

Fig. 6. *Periplaneta.* **A.** Complete leg. **B.** Magnified view of pretarsus.

tarsomeres bear soft adhesive pads, the **plantulae**. Pretarsus ends in two lateral **claws** and between them a median, delicate hair-covered porous pad, the **arolium**, which prevents from slipping. All the legs are used for walking, running and climbing.

[IV] Wings

Cockroach has two pairs of wings. One pair of wings are attached dorso-laterally between tergum and pleuron of both mesothorax and metathorax. In *P. americana* wings reach up to the tip of abdomen in female, and a little beyond in male cockroach.

Each wing consists of two membranous layers of cuticle enclosing tubular tracheae. Chitin thickens around tracheae to form **nurvures** or **veins** that strengthen the wings. Wing veins are of such constant pattern in each insect as to form a basis of their classification.

1. Forewings. Mesothoracic or forewings are heavily sclerotized. They are narrow, dark opaque and leathery in texture. They are not used in flight but serve to protect the hindwings at rest. Hence they are known as **elytra, wing covers** or **tegmina**.

2. Hindwings. Metathoracic or hindwings are thin, membranous, transparent, broad, delicate and used in flight. They are kept folded like a fan under the tegmina, when at rest. Movements are due to special muscles attached to wing bases and working at a very high speed.

[V] External Genitalia

Reproductive organs (such as testes and ovaries) concerned with production and transmission of gametes (ova and sperms) are collectively termed **internal genitalia**. While organs involved in sexual dimorphism, mating and deposition of eggs are known as **external genitalia**. In cockroach, these are appendages of the terminal segments of abdomen. **Anal cerci** and **anal styles** have already been described earlier. Whereas **gonapophyses** shall be dealt with the reproductive system of cockroach.

BODY WALL AND EXOSKELETON

Body wall or integument consists of : (i) **cuticle**, (ii) **hypodermis** and (iii) **basement membrane**.

[I] Cuticle (Exoskeleton)

Entire body including appendages is covered externally by a thick, brown-coloured, non-living, hard and chitinous cuticle which forms the **exoskeleton** of insect. It is secreted by the hypodermis underneath.

1. Structure of cuticle. Cuticle or exoskeleton is principally made of **chitin**, a horny proteinous substance, chemically an amino-polysaccharride ($C_{32}H_{54}N_4O_{21}$), insoluble in water and resistant to most solvents. Cuticle of insects becomes hard not due to calcium salts as in some other classes, but due to certain proteins and

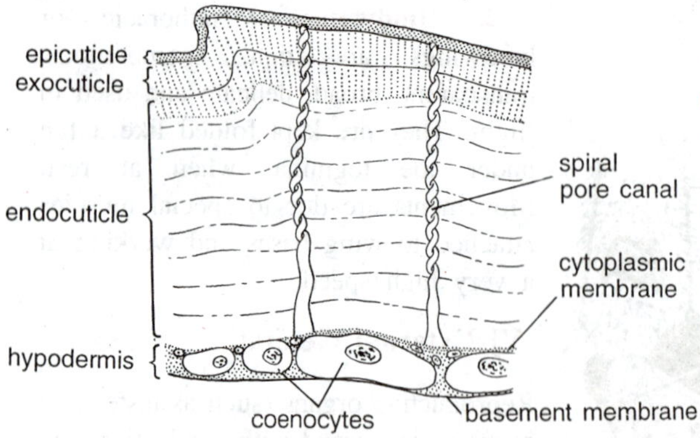

Fig. 7. *Periplaneta.* V.S. of body wall (semi-diagrammatic).

chitin. Hardening of cuticle due to chitin is called **sclerotization** and its hardened plates are called **sclerites**. At the joints (**sutures**), sclerites are connected by soft and flexible **arthrodial membranes** which allow movements of body and appendages. Head armour is made of several plates as already described. In each thoracic and abdominal segment, there are 4 sclerites : a dorsal **tergum**, a ventral **sternum** and two very small lateral **pleura**.

Cuticle also lines the foregut, hindgut, tracheae and genital ducts. Numerous invaginations of cuticle form a sort of endoskeletons. That of head is called as **tentorium** while those of thorax and abdomen as **apodemes**. Further, the stiff, immovable **bristles** or **spines** covering the body and its appendages are in fact the outgrowths of cuticle.

2. Histology of cuticle. Histologically, cuticle is made of 3 distinct layers : **epicuticle**, **exocuticle**, and **endocuticle**, all secreted by the underlying hypodermis.

(a) Epicuticle. It is the outermost extremely thin layer 1 to 2μ in thickness and without chitin. Its outer part is made of lipid or wax which is impermeable to water, while inner part is made of proteins and lipids. In some insects, a thin 'cement layer' occurs outside the waxy layer.

(b) Exocuticle. It is the middle layer 10 to 20μ thick and containing chitin and melanin pigments.

(c) Endocuticle. It is the innermost, thickest and laminated layer 20 to 30 μ thick. It contains chitin. Exocuticle and endocuticle together form the **procuticle**.

(d) Pore canals. Electron microscopic studies by **Richards** and **Anderson** have revealed presence of **pore canals** in exocuticle and endocuticle. These are hollow, spirally coiled tubes extending from hypodermis up to, but not into, the epicuticle. Their number is about 12,00,000 per sq. cm. Pore canals are probably filled with salt solution in ionic equilibrium with the hypodermal cells. Exact function of these canals is not known but they are supposed to influence the flexibility and permeability of cuticle.

3. Functions of cuticle (exoskeleton). Thick cuticle of arthropods is like steel to modern man, and is primarily responsible for their great success. It is useful in many ways as follows :

(1) Provides support, rigidity and protection.

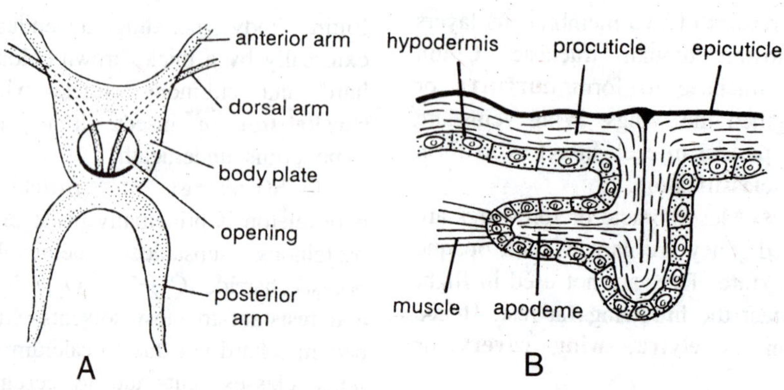

Fig. 8. Cockroach. **A.** Tentorium. **B.** An apodeme.

(2) Waxy layer prevents loss of water due to evaporation, thus enabling insects to live in relatively dry environments.
(3) Cuticular invaginations offer points of attachment for muscles.
(4) Jointedness of cuticle of body segments and appendages makes accurate complicated movements possible.
(5) Cuticular lining of fore and hind guts protects their epidermis from abrasions by passage of food.
(6) Melanin pigments within cuticle provide protective colouration.
(7) Many of the cuticular outgrowths form sensory receptors.

[II] Hypodermis

Hypodermis (or epidermis) lies beneath cuticle which it secretes. It is a highly organised epithelium made of a single layer of columnar cells resting on a basement membrane and anchored to it by hemidesmosomes. Besides secreting cuticle, hypodermis resorbs endocuticle before each moult or ecdysis.

1. Glands. Dermal glands found in hypodermis are of two kinds. They are concerned with the production of cement layer and different kinds of cuticle. They probably also secrete pheromones and some hormones. Swollen body of glands lie in hypodermis while their ducts run through cuticle to open on its surface.

2. Oenocytes. In adult cockroach large irregular cells, more than 100 μ in diameter, appear between hypodermal cells and basement membrane. They probably secrete wax or lipids (**Kramer** and **Wigglesworth**) that is transmitted to the surface of cuticle.

3. Trichogen cells. Specialized, movable and hair-like sensory setae, projecting above the surface of cuticle, are secreted by special trichogen cells of hypodermis.

[III] Basement Membrane

A typical stout and continuous basement membrane bounds the inner surface of hypodermis. It consists of an amorphous granular material, probably a mucopolysaccharide.

COELOM AND BODY CAVITY

The large perivisceral body cavity of cockroach, enclosing alimentary canal and other viscera, is not a coelom as found in earthworm. It is a **haemocoel** containing blood, and shall be described with the circulatory system. True coelom in cockroach is much reduced and represented only by the cavities of gonads.

[I] Digestive System

It is the most conspicuous organ system in the body. It includes the **mouth parts** already described, a long **alimentary canal** and a pair of **salivary glands**.

[II] Alimentary Canal

It is a long and somewhat coiled tube of uneven diameter and divisible into three regions : **foregut**, **midgut** and **hindgut**. Foregut and hindgut are ectodermal and lined with a thin cuticle secreted by ectoderm. Whereas midgut is endodermal, devoid of cuticular lining and capable of absorbing digested food.

A. Foregut or stomodaeum

It includes mouth cavity, pharynx, oesophagus, crop and gizzard.

1. Mouth cavity. Mouth cavity or **preoral chamber** is a small, ill-defined space outside mouth, surrounded by mouth parts. Hypopharynx divides preoral cavity into a posterior part, called **salivarium**, into which common salivary duct opens, and an anterior part or **cibarium**, which leads towards mouth as a narrow food passage. Food is crushed and acted upon by the salivary secretion or **saliva** in mouth cavity.

2. Mouth. True mouth is a small opening at the base of preoral cavity and leads into pharynx.

3. Pharynx. It is short and tubular and its cuticular lining is more folded posteriorly.

4. Oesophagus. From pharynx arises a long, straight narrow and laterally compressed tube, the **oesophagus**. It passes through the nerve collar,

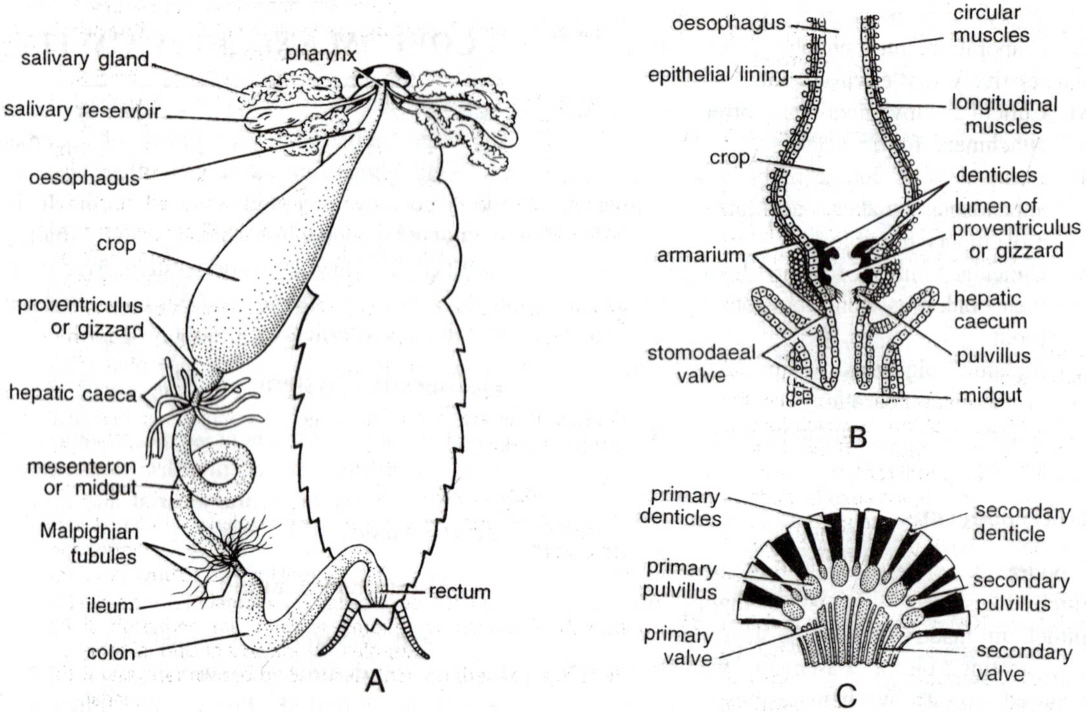

Fig. 9. *P. americana.* **A.** Alimentary canal and salivary apparatus. **B.** Crop and gizzard in L.S. **C.** Proventriculus split. longitudinally and laid open.

runs through the neck and enters thorax to merge with crop.

5. Crop. It is a large, thin-walled, pear-shaped sac, which extends well up to the third or fourth abdominal segment. It is the largest part of foregut. Its internal epithelial and cuticular lining is very much folded. Its outer surface is covered by a network of tracheae. Crop serves as a reservoir for storing food.

6. Gizzard. Crop leads behind into a small, cone-shaped, muscular and thick-walled chamber, the **gizzard** or **proventriculus**, which marks the end of foregut. It consists of two parts, an anterior **armarium** and a posterior **stomodaeal valve**.

(a) Armarium. The armarium possesses internally six longitudinal folds that greatly reduce its lumen. Longitudinal folds alternate with six longitudinal grooves which also bear small secondary folds. Cuticular lining of each longitudinal fold forms, (i) in the anterior part, a **thick plate** produced centrally into strong, sharp **teeth** or **denticles**, (ii) in the posterior part, a

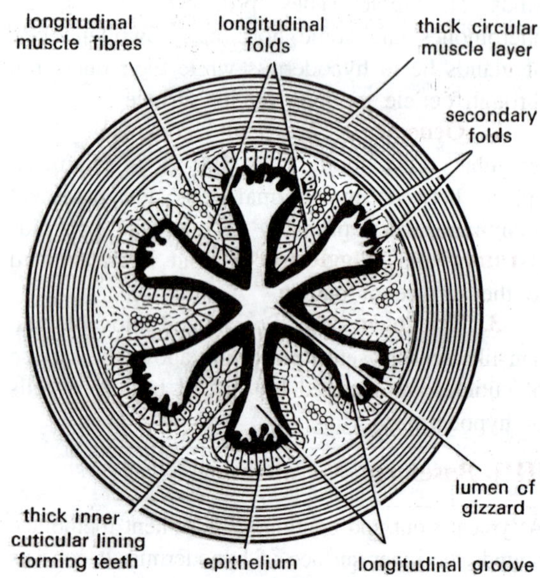

Fig. 10. *P. americana.* Gizzard in T.S.

thin plate. Behind each longitudinal fold the cuticular lining of the gizzard forms a soft cushion-like lobe, the **pad** or **pulvillus**, with long, backwardly directed hairs which act as strainers and allow only finer food particles to pass into midgut.

(b) Stomodaeal valve. Behind pads, the posterior end of **gizzard** extends into the lumen of midgut as a spout-like narrow tube, the **stomodaeal valve**. The latter folds back on itself, and is thus double-walled. Stomodaeal valve prevents regurgitation of food from midgut into gizzard.

B. Midgut or mesenteron

Midgut is the short and narrow tube-like middle part of alimentary canal also known as **ventriculus** or **mesenteron**. It is internally lined by glandular epithelium and forms the true **stomach** serving mainly for digestion and absorption.

1. Hepatic caeca. Opening into the anterior end of midgut are 7 or 8 short, narrow, blindly ending hollow tubes, called **enteric** or **hepatic caeca**. These are internally lined by epithelium and secrete digestive enzymes.

2. Malpighian tubules. From the junction of midgut and hindgut arise 80 to 90 very narrow, thread-like, yellow-coloured blind tubules projecting freely into haemocoel. These are called **Malpighian tubules** after the great Italian anatomist Malpighi. In fact these tubules arise from hindgut and are excretory in function. In *Periplaneta,* these have been shown to contain certain intracellular enzymes also (**E. Schlottke**, 1937).

C. Hind gut or proctodaeum

The posterior 1/3 part of alimentary canal forms the hindgut. It is divided into three regions : **ileum**, **colon** and **rectum**.

1. Ileum. Ileum is a narrow and short tube and its posterior end is characterised by the possession of six tiny triangular lobes internally, bearing spicules and acting as a sort of sphincter.

2. Colon. Colon is longer and wider and with an irregular shape.

3. Rectum. Rectum is an oval or spindle-shaped sac with external ridges alternating with

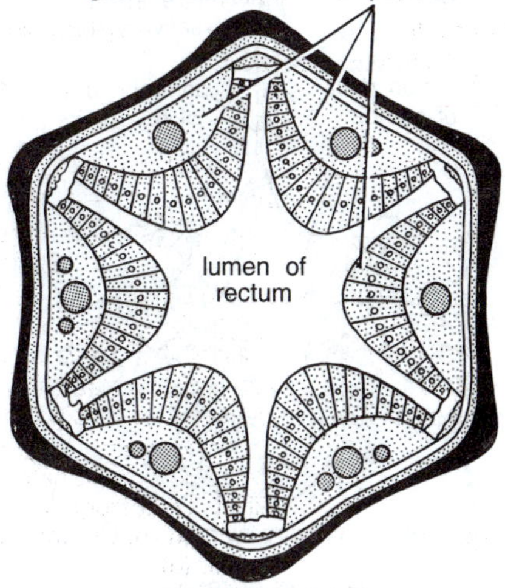

Fig. 11. *P. americana.* Rectum in T.S.

internal longitudinal thickenings called **rectal pads** or **rectal papillae**. These are 6 in number and also known as **rectal glands**. Rectum opens to outside by **anus**. Lining of hindgut is also cuticular, but is more pervious to water than that of foregut.

[III] Salivary Apparatus

Lying in thorax, on either dorso-lateral side of oesophagus, is a pair of bipartite, diffuse and whitish **salivary glands**. Each gland consists of several secreting **lobules** or **acini** in grape-like clusters and connected together by fine tubules. Acini contain two types of cells, described by Day as **zymogenic cells** and **ductule-containing cells**. Both types of cells secrete **saliva**, that consists of an enzyme **zymase** and mucoid substance. Ductules from ductule-containing cells open into an **intercalary duct**, that leaves acinus to join similar ducts of other acini, to form a larger duct. Associated with each gland is an elongated, sac-like, thin-walled **reservoir** or **receptacle** for the storage of saliva.

Ducts from the two salivary glands unite to form a single **common glandular duct**. Similarly, ducts from two reservoirs unite with each other forming a common **reservoir duct**. Both common

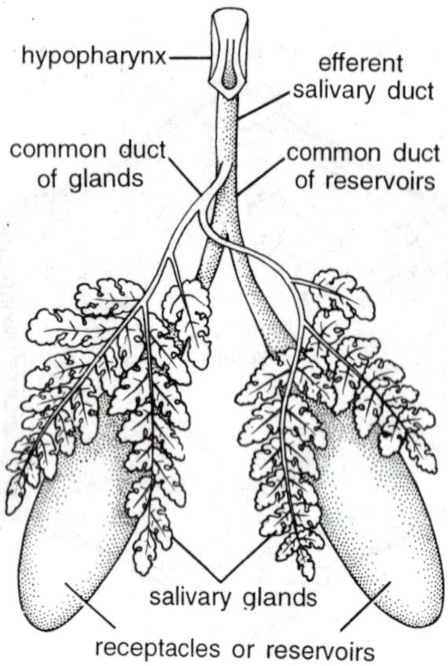

Fig. 12. *P. americana.* The salivary apparatus.

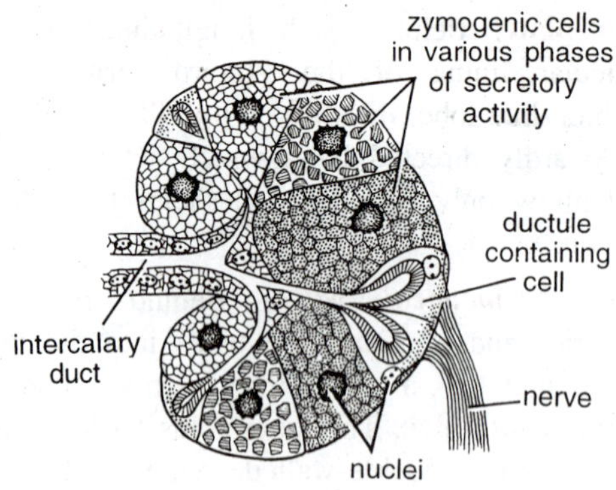

Fig. 13. *P. americana.* A salivary acinus in section.

ducts then join to form a common **efferent salivary duct** that opens at the base of hypopharynx in the salivarium of preoral cavity.

[IV] Nutrition

1. Food. Cockroach is **omnivorous**, feeding on any kind of animal or vegetable matter, including wood, book-bindings, cloth, leather, paper, pastes, glues, hair, and even its own cast cuticle. It usually feeds at night.

2. Feeding. Food is searched by the sweeping antennae tasted by maxillary and labial palps, and seized by the forelegs, labrum and labium. During eating, mandibles (and maxillae) undergo a sideways movement brought about by the action of the adductor and abductor muscles, and masticate the food into finer particles.

3. Digestion. In the **salivarium** of mouth cavity, crushed food mixes with salivary secretion. **Mucus** of saliva lubricates food while its digestive enzyme **zymase** hydrolyses the starchy matter of food. Lubricated food is then pushed into pharynx through mouth with the help of labium. Food now starts moving through alimentary canal by **peristalsis**. Most of digestion occurs in crop by saliva and digestive secretions from hepatic caeca and midgut. Upon entering proventriculus, food is subjected to further mastication by internal denticles worked upon by longitudinal and annular muscles. Hair on pulvilli act as a filter, allowing only finer particles to move ahead. In midgut, partly digested food is mixed with enzymatic secretions of epithelial secretory cells of enteric caeca and midgut itself. These secretions contain : (i) a **trypsin-like** enzyme and **peptidases** that break down proteins, (ii) **amylase** that completes the breaking down of starches, and (iii) **lipase** that breaks down fats. A thin transparent chitinous **peritrophic membrane** lines the midgut internally in the form of a tube. This membrane is secreted by the stomodaeal valve of **gizzard**. It protects the delicate lining of midgut from abrasion by the hard indigestible components of food. It is, however, permeable to enzymes as well as digested food and, therefore, does not hinder their normal course of digestion and absorption.

4. Absorption. Epithelial cells of enteric caeca and midgut are also absorptive. They absorb digested food which is stored in the diffuse fat bodies. According to **Abbott** (1926), large quantities of fat are absorbed by the epithelial cells of crop.

5. Egestion. Undigested food is passed first into ileum and then into colon. In rectum, the rectal pads absorb water and the faeces is eliminated to outside through anus as dry pellets.

CIRCULATORY SYSTEM

Cockroach has an **open** or **lacunar** circulatory system as blood, also called **haemolymph**, flows freely within the body cavity or **haemocoel**. There is a heart and aorta but no capillaries and veins.

[I] Haemocoel

Body cavity of cockroach is called **haemocoel** (Gr., **haima**, blood + **koila**, cavity), as it is filled with blood or haemolymph.

1. Diaphragms. Haemocoel is divided by a dorsal and a ventral diaphragm into three sinuses : (i) **dorsal pericardial**, (ii) middle **perivisceral** and (iii) ventral **perineural**. Diaphragms are provided with pores or **fenestrae** to permit flow of haemolymph from one sinus to another. Ventral diaphragm also extends into each leg as a septum dividing its cavity into two sinuses, one for the outward and the other for inward flow of haemolymph.

2. Dorsal pericardial sinus. It encloses heart and aorta and paired fan-shaped (triangular) **alary muscles**, one pair in each segment, one on either side of heart. Apices of these muscles are attached to the terga and their broad bases to the dorsal diaphragm.

3. Perivisceral sinus. Middle or perivisceral sinus contains the alimentary canal. It is mostly occupied by a whitish mass of tissue or **fat body**. It consists of several types of cells. Trophocytes store reserve food in the form of fat globules, protein globules and glycogen. The latter is used during starvation. **Urate cells** are excretory in nature. **Oenocytes** produce wax while **myocytes** contain intracellular symbiotic bacteria.

4. Perineural or sternal sinus. It encloses the ventral nerve cord.

[II] Heart and Aorta

Enclosed by dorsal pericardial sinus is the heart. It lies mid-dorsally beneath the terga of thorax and abdomen. It is a long narrow tube with anterior end open and posterior end closed. It

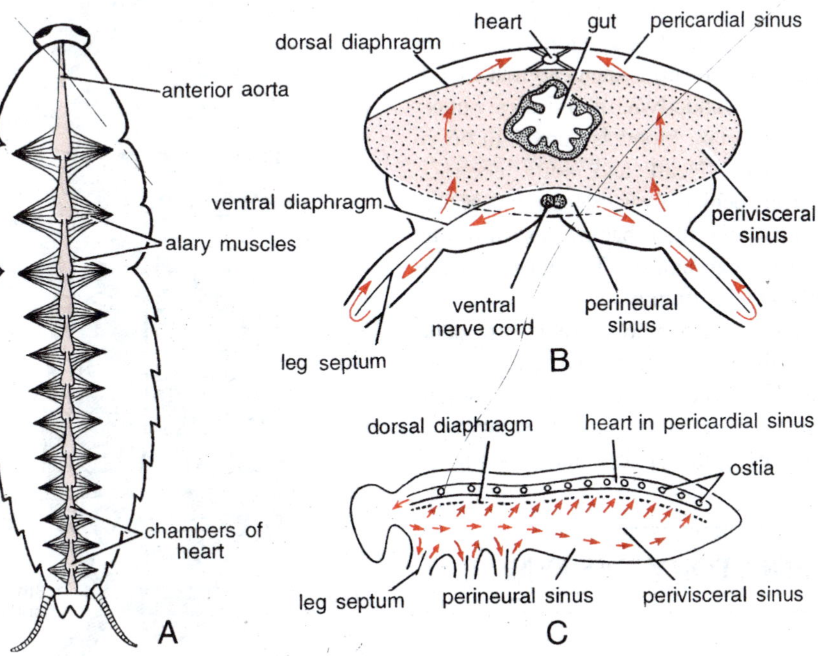

Fig. 14. *P. americana.* **A.** Heart in dorsal view. **B.** Course of circulation of blood in T.S. of a thoracic segment. **C.** Course of circulation of blood in L.S. of body.

consists of 13 funnel-shaped chambers or segments, each communicating by a valvular opening with the one in front of it. Hinder end of each chamber has a pair of minute lateral openings, the **ostia**. These allow flow of haemolymph from pericardial sinus into heart only and not in a back gear. The anterior narrow and tubular part of heart is called **anterior aorta**. It is without ostia and forwards into the **head sinus**.

[III] Haemolymph

The haemolymph or blood of cockroach consists of a clear colourless plasma rich in amino acids, uric acid, and numerous different types of cells, called **haemocytes**. According to **Wigglesworth** (1965), there are seven types of such cells. Some cells transfer, by ingestion, the food material from blood to the tissues, while others act as phagocytes in removing the metabolic wastes from tissues. Haemolymph is devoid of a respiratory pigment and hence does not assist in respiration.

[IV] Blood Circulation

Blood or haemolymph circulates by the contraction and relaxation of heart assisted by the paired fan-shaped **alary muscles**. Contraction of these muscles enlarges the pericardial sinus so that blood flows into it from the underlying perivisceral sinus. When the muscles relax, blood is forced through ostia into the heart. Heart and aorta contract peristaltically from behind forwards driving blood into head sinuses and then backwards into the perivisceral and sternal sinuses. The rate of heart beat in *Periplaneta* is 49/min.

From head sinuses, haemolymph is also sent into antennae by the pumping activity of two small ampullae, located in head. In case of wings, small pulsatile organs at their bases cause haemolymph to flow through the veins of fore and hindwings.

RESPIRATORY SYSTEM

[I] Respiratory Organs

Respiratory system of cockroach is well-developed to compensate the poorly developed circulatory system. It consists, as in other insects, of **tracheae, tracheoles** and **spiracles**.

1. Tracheae. Haemocoel contains a network of elastic, closed and branching air tubes or **tracheae**. There are three pairs of large, parallel, longitudinal **tracheal trunks**, one dorsal, one ventral and one lateral in position, which are connected together by transverse commissures. Tracheae are formed as invaginations of outer integument, hence they are made of an outer epithelial wall lined by an inner chitinous cuticle. The cuticular lining is spirally thickened forming **intima** or **taenidia** which prevents the tracheal tubes from collapsing. When cockroach is dissected under water, the tracheae, filled with air, presents a glistening appearance.

2. Tracheoles. The profusely branching trachae anastomose and penetrate to all parts of body. The ultimate finer branches of tracheae are called **tracheoles** which come in contact with the individual body cells. They have a diameter of

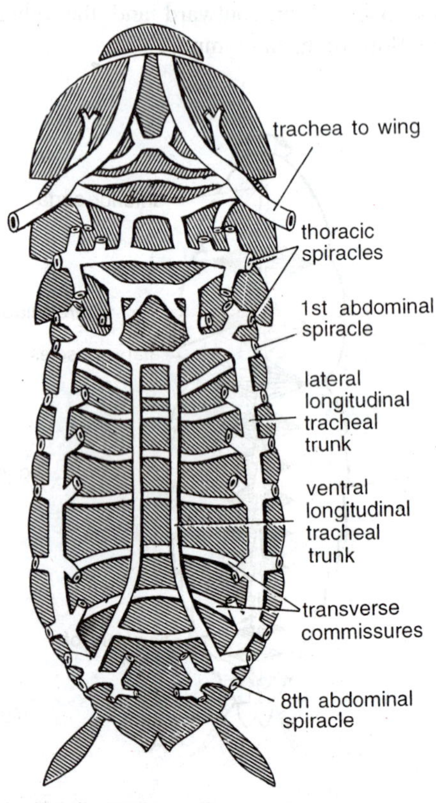

Fig. 15. *P. americana.* Tracheal system in dorsal view.

only 1 micron. Their cavities are intracellular, that is, each tracheole is made of a single cell. Their walls are very thin and devoid of cuticular spiral thickening, instead they are lined by a protein called **trachein**. They are permeable to water. Their tips are usually filled with a fluid in which oxygen dissolves and diffuses to the tissues. However, some zoologists think that tracheoles end blindly and remain without fluid. Thus, the elaborate tracheal system carries oxygen directly to all the body cells. This very well compensates for the inability of blood to transport oxygen due to absence of a respiratory pigment.

3. Spiracles. The main tracheal trunks open to the exterior on body surface through 10 pairs of segmentally arranged apertures termed **spiracles** or **stigmata**. Two pairs of spiracles are thoracic, one between pro- and mesothorax and the other between meso- and metathorax. Eight pairs of spiracles are abdominal, one pair in each of the first eight abdominal segments. They are present laterally in the soft cuticle between terga and sterna. A spiracle is guarded by bristles or hair to keep out dirt. It is surrounded by an annular sclerite, the **peritreme**. It has a closing device in the form of a simple *valve* which prevents undue loss of water and can be closed or opened to regulate the flow of air. Each spiracle internally leads into a short tracheal chamber or **atrium** from which arises a main tracheal trunk.

[II] Respiratory Mechanism and Gaseous Exchange

Alternate contraction and relaxation of the abdominal muscles (tergo-sternal muscles) cause rhythmic contraction and expansions of abdomen. Such movements cause change in diameter of tracheae and force air in and out of tracheal tubes through spiracles. 1st and 3rd pairs of spiracles always remain open while the remaining eight pairs open only during inspiration. Respiratory movements depend on the activity of insect and temperature. Greater the muscular activity, the more vigorous is the pumping in-and-out of air. Respiratory movements are coordinated and regulated by nerve centres in thoracic ganglia which are stimulated by low O_2 and higher CO_2 concentrations in tissue fluids.

Gaseous exchange occurs by simple diffusion between air in tracheae and dissolved gases in blood or tracheolar fluid which has been shown to rise and fall. When insect is resting, the tips of tracheoles remain filled with fluid so that O_2 diffuses slowly through fluid into body cells and tissues. When the metabolic activity is greater, fluid from tracheoles is withdrawn into surrounding tissue due to a rise in osmotic pressure of cells. This exposes more surface walls of tracheoles to oxygen, so that more oxygen is supplied to the surrounding tissues. O_2 entering the tissues brings

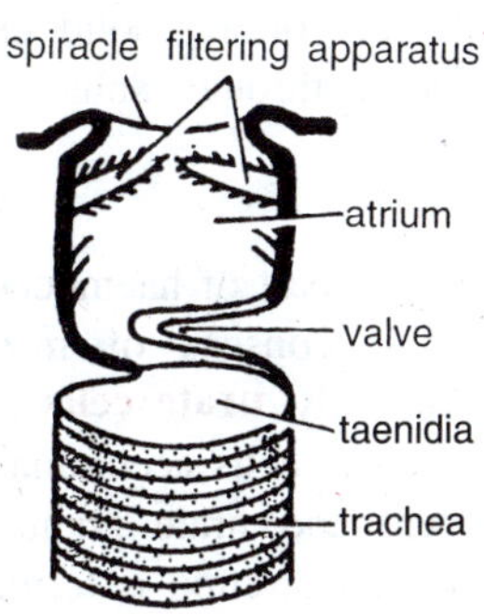

Fig. 16. A spiracle with atrium, filtering apparatus and valve.

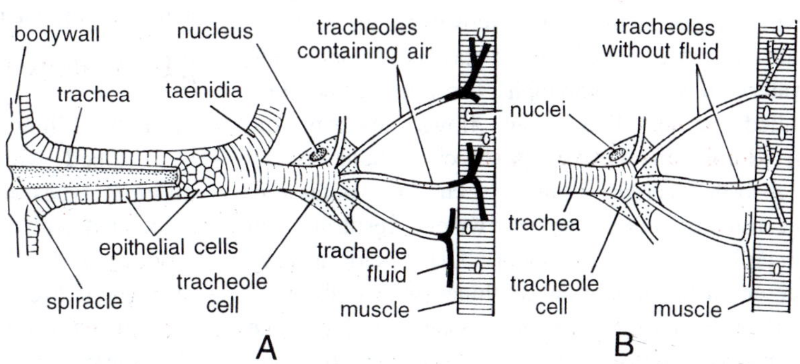

Fig. 17. *Periplaneta.* Role of tracheoles in gaseous exchange. **A.** Tracheoles with fluid at rest. **B.** Tracheoles without fluid after work.

about oxidation of energy-rich food molecules with the release of energy and production of CO_2 and water. Some of the CO_2 leaves the body through tracheae and spiracles. But major part of CO_2 leaves through the cuticular covering of body because it can diffuse more readily through chitin than O_2.

EXCRETORY SYSTEM

Excretory system regulates the amounts of nitrogenous material, inorganic salts and water in blood or haemolymph. As a result of the protein metabolism, nitrogen is produced in excess which is excreted as uric acid. The main structures that play the role of excretion are : (i) **Malpighian tubules**, (ii) **fat body cells**, (iii) **uricose glands**, and (iv) **cuticle**.

[I] Malpighian Tubules

These are attached to the alimentary canal at the extreme anterior end of hindgut. These are fine, long, unbranched, yellowish and blind tubules lying freely in the haemolymph. These are between 60 to 150 in number and are arranged in 6-8 bundles. Each tubule is about 16 mm long and 0.5 mm in diameter and is lined by glandular epithelium with a characteristic **brush border**.

A Malpighian tubule has two functional parts. Glandular cells of **distal secretory part** extract nitrogenous wastes (mostly in the form of salts of uric acid, *e.g.*, potassium urate) and water from haemolymph forming a solution called **urine**. The urine flows towards the **proximal absorptive part** of tubule which reabsorbs certain salts, such as potassium bicarbonate, and some water resulting in precipitation of uric acid. Uric acid already present in haemolymph combines with reabsorbed potassium bicarbonate and water to form the relatively soluble potassium urate which again becomes available to be actively transported from haemolymph into the lumen of distal portion of Malpighian tubule. From Malpighian tubule uric acid moves into ileum by gentle

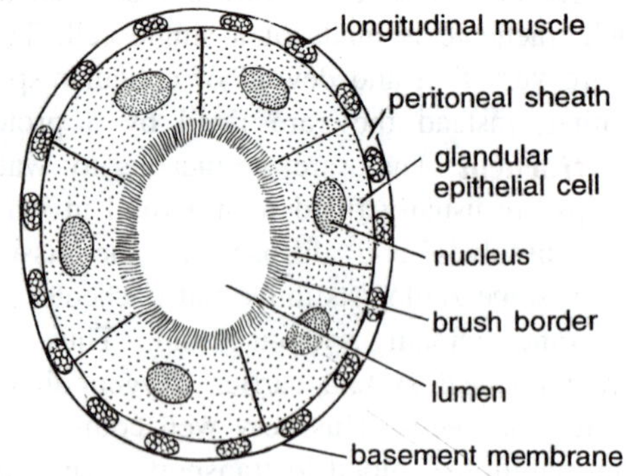

Fig. 18. *P. americana.* T.S. of Malpighian tubule.

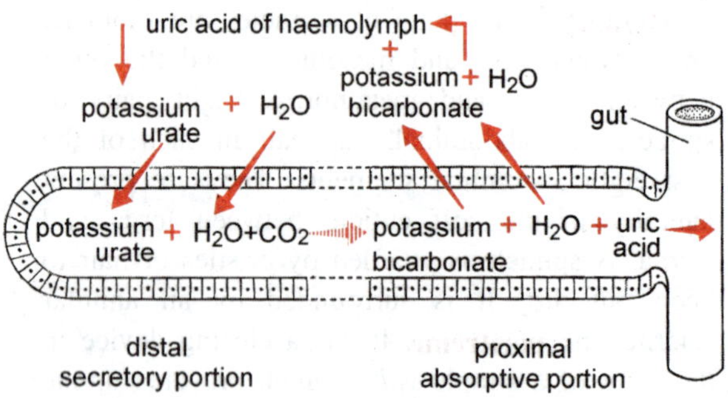

Fig. 19. *P. americana.* Diagrammatic representation of Physiology of Malpighian tubule.

peristaltic waves. More water is reabsorbed in colon and rectum so that more or less solid uric acid is eliminated with faeces through anus.

[II] Fat Body Cells

Fat body, filling up the greater part of haemocoel, is a lobed white tissue. It consists of many different types of cells, but only **urate cells** are associated with excretion. These cells accumulate, produce and store uric acid and urate granules throughout life. This mode of excretion is termed **storage excretion**.

Some other important types of cells in the fat body are : (i) **Trophocytes** that store reserves of fat, glycogen and protein, (ii) **Oenocytes** that are believed to produce lipoprotein (wax) for formation of new epicuticle at moulting, and

(iii) **Mycetocytes** that are filled with **bacteroids** (symbiotic intracellular bacteria). **Henry** and **Block** have shown that these bacteroids contribute in the synthesis of amino acids. Further, **Keller** is of the opinion that these bacteroids can breakdown uric acid to remobilise its nitrogen.

[III] Uricose Glands

The mushroom gland of male cockroach possesses long, blind tubules at its periphery, called **uricose glands** or **utriculi majores**. These tubules store uric acid (storage excretion) and discharge it over the spermatophore during copulation. Thus they serve as 'storage excretory organs' between matings and as 'active excretory organs' during copulation (**Roth** and **Dateo**).

[IV] Cuticle

Cuticle also acts as a site where nitrogenous waste material is deposited and then eliminated with its shedding at each moult.

NERVOUS SYSTEM

Nervous system comprises the **central**, **peripheral** and **sympathetic** or **stomatogastric** nervous systems.

[I] Central Nervous System

It consists of the **brain** and **ventral nerve cord** with its ganglia.

1. Brain. Brain or **supra-oesophageal ganglion** is a large and bilobed mass located in head above oesophagus. It represents 3 pairs of ganglia of head region fused together. It is divided into three parts : **protocerebrum**, **deutero-cerebrum** and **tritocerebrum**. Remaining three pairs of ganglia of head fuse to form the **sub-oesophageal ganglion** which lies below oesophagous. Brain and sub-oesophageal ganglion are connected together, on either side of oesophagus, by a **circumoesophageal commissure**.

2. Ventral nerve cord. From sub-oesophageal ganglion runs posteriorly a double **ventral nerve cord** along the mid-ventral line of thorax and abdomen. It bears 9 ganglia, 3 in thorax and 6 in abdomen. The last abdominal ganglion is the largest, representing many fused ganglia of the posterior abdominal segments.

[II] Peripheral Nervous System

The nerves, given off from ganglia to all parts of body, constitute the **peripheral nervous system**. Protocerebrum of brain gives off paired **optic nerves** to eyes, deutero-cerebrum gives off paired **antennary nerves** to antennae, and tritocerebrum supplies nerves to frons and labrum. Sub-oesophageal ganglion supplies **mandibular**, **maxillary** and **labial nerves** to the mandibles, maxillae and labium, respectively. Nerves from

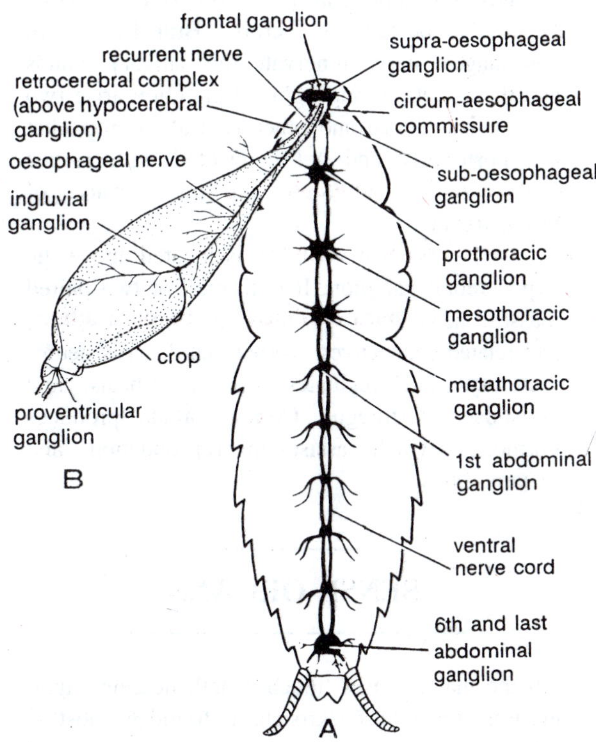

Fig. 20. *P. americana.* **A.** Central and peripheral nervous system. **B.** Stomatogastric nervous system.

thoracic ganglia innervate thoracic muscles, mainly those of wings and legs. First 5 abdominal ganglia send nerves to the dorsal and ventral muscles of body wall, spiracles and heart, whereas the last abdominal ganglion supplies nerves to muscles of last three abdominal segments, reproductive organs, copulatory appendages, and anal cerci.

[III] Sympathetic or Stomatogastric Nervous System

It comprises four ganglia and a retro-cerebral complex, (i) A **frontal ganglion** lies above pharynx, in front of brain. It sends nerves to pharynx, clypeus and labrum. It is connected to protocerebrum of brain by a median **nervous connectives** and to tritocerebrum by a pair of **frontal connectives**. Frontal ganglion is also connected, by a median unpaired **recurrent nerve**, with (ii) a **hypocerebral ganglion** on oesophagus. It sends an oesophageal nerve, to (iii) an **ingluvial ganglion** located on crop. Branches from oesophageal nerve innervate the salivary glands and their ducts. From ingluvial ganglion arise two nerves, one dorsal and other ventral to crop, that run posteriorly and get connected with (iv) a **proventricular ganglion** on the surface of proventriculus.

Retro-cerebral complex is located above the hypocerebral ganglion. It is formed of two paired masses, the **corpora cardiaca** and **corpora allata**, and related connectives. Corpora cardiaca is neuro-secretory and regulates the heart-beats and peristalsis of foregut. Corpora allata produces hormones which assist in reproduction and metamorphosis.

SENSE ORGANS

All the major senses (touch, smell, hearing, sight, etc.) are found in cockroach, as found in most of other.

[I] Photoreceptors or Eyes

Visual receptors or eyes are of two types, **compound eyes** and **ocelli**.

1. Compound eyes. Of all the organs of special senses, the most important are a pair of large, sessile, **compound eyes** in the form of black, kidney-shaped organs on lateral sides of head. Each consists of about 2,000 visual elements or units, called **ommatidia**, similar in structure to those already described for the prawn. The same may be referred to for details.

Compound eye of cockroach and other insects is a very elaborate structure. In complexity, it challenges the eyes of vertebrates, but falls far behind in efficiency. The pigment does not seem to be retractable in the eyes of cockroach. Owing to its form and permanently extended pigment sheath, the only light rays which reach the retinular cells are those nearly parallel to its long axis. Thus, the image seen by the whole eye is made up of many spots of light, each contributed by one ommatidium. This is known as a **mosaic** or **apposition image**. Compound eyes are specially adapted to perceive movements of objects.

2. Ocelli. At the base of each antenna is a fenestra which represents a **simple eye** or **ocellus**. Each comprises of a single corneal facet. It is mainly concerned with light-collecting rather than image-forming. Probably it enhances the sensivity of compound eyes.

[II] Sensillae

Sense organs, other than eyes and ocelli, are scattered all over the body, but are especially numerous on antennae, mouth parts, legs and wings. These sense organs or receptors are called **sensillae** (**Sing, sensillum**). Which are isolated or collective modifications of epidermal cells. A simple sensillum comprises a sensory cell connected with the fibre of a sensory nerve. Sensillae take different forms and perform various functions.

1. Thigmoreceptors. These are tactile hair sensitive to touch which are present on body surface, antennae, bristles of legs and maxillary palps. They help in searching food.

2. Chemoreceptors. These are organs of taste and smell, mainly confined to the tips of maxillary palps, labial palps, labium and hypopharynx.

3. Auditory receptors. Hair sensillae present on anal cerci respond to the air or earth-borne sound vibrations.

4. Proprioreceptors. Campaniform sensillae, present on the joints of maxillary palps and legs, perceive strains set up in the cuticle during feeding and movement.

REPRODUCTIVE SYSTEM

Cockroach is **dioecious**, *i.e.,* the sexes are separate. **Sexual dimorphism** is evident in the internal reproductive organs as well as external morphology.

[I] Male Reproductive System

Male reproductive organs include testes, vasa deferentia, ejaculatory duct, mushroom gland, seminal vesicles, phallic gland, genital pouch and external genitalia, all located in abdomen.

1. Testes. Male gonads are a pair of testes, lying embedded in fat body, one on either dorso-lateral side of 4th to 6th abdominal segments. Each testis is three-lobed and consists of numerous (30-40) small whitish transparent **follicles** arranged in a longitudinal series around a narrow, delicate **vas deferens**, into which they discharge their products (sperms). In adult cockroach testes become non-functional and greatly reduced in size.

2. Vasa deferentia. The paired **vasa deferentia**, one from each testis, run posteriorly and downwards to open into the **ejaculatory duct**.

3. Ejaculatory duct. It is a single, median and muscular duct that extends posteriorly and opens into the genital pouch through the **male genital pore**, lying immediately below anus and between 9th and 10th sterna. The glandular wall of ejaculatory duct secretes the middle layer of spermatophore.

4. Mushroom gland. The junction of two vasa deferentia and ejaculatory duct is surrounded by an elaborate **mushroom gland** consisting of numerous compact, finger like blind tubules arranged in two distinct groups, and opening into the anterior part of ejaculatory duct.

(a) Utriculi majores. These are long, slender and peripheral tubules. Their secretion forms the inner layer of spermatophore.

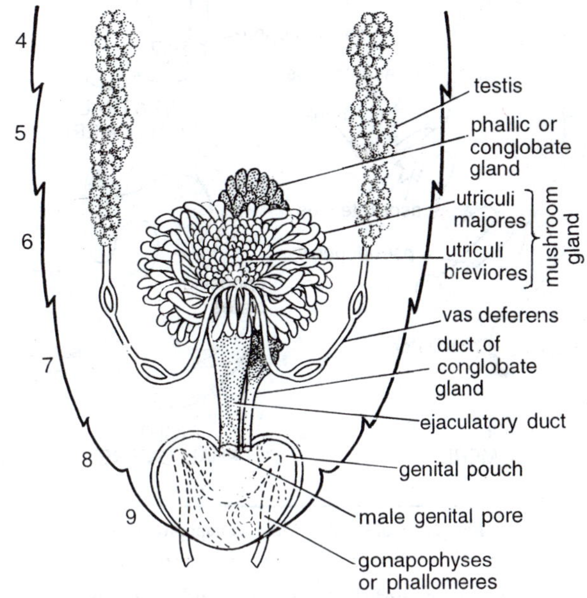

Fig. 21. *P. americana.* Male reproductive organs in dorsal view.

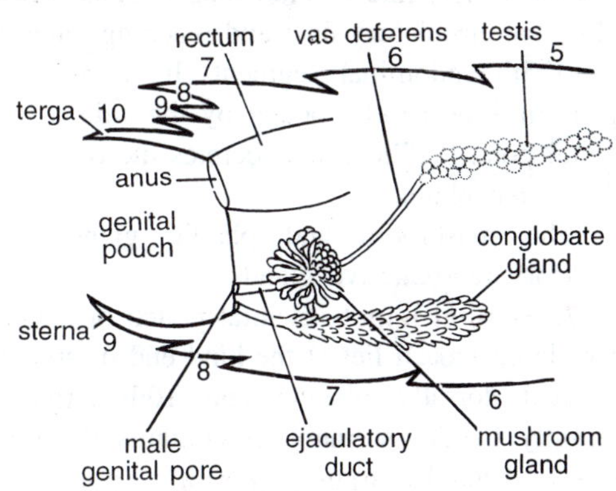

Fig. 22. *P. americana.* Male reproductive organs in lateral view.

(b) Utriculi breviores. These are short tubules forming the bulk of mushroom gland. Their secretion nourishes the sperms.

5. Seminal vesicles. Present on the ventral surface of anterior part of ejaculatory duct are two groups of numerous small glistening white sacs constituting the **seminal vesicles** meant for storage of sperms. Seminal vesicles can be distinguished from **utriculi breviores** of mushroom

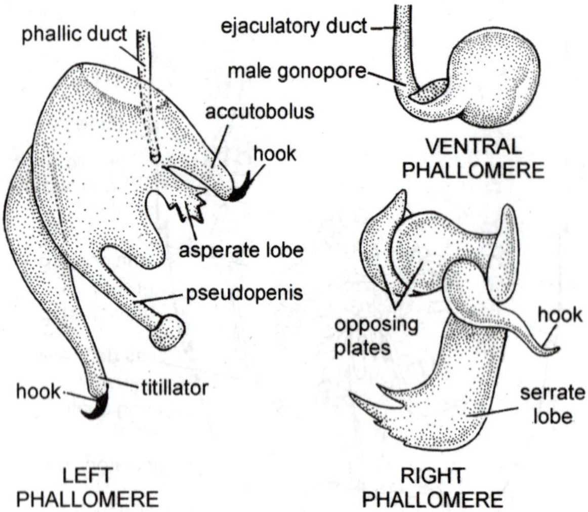

Fig. 23. Cockroach. Male gonapophyses.

gland by their slightly larger size and more opaque whiteness.

6. Phallic or conglobate gland. It is a large, multilobed, leaf-like or club-shaped gland present below the ejaculatory duct and reaching anteriorly up to 5th abdominal segment. Its narrow duct opens into the genital pouch by the side of male genital pore. Phallic gland secretes the outer layer of spermatophore.

The mushroom and phallic glands form accessory reproductive glands.

7. Genital pouch. Genital pouch or chamber of male cockroach lies at the hind end of abdomen bounded dorsally by 9th and 10th terga and ventrally by 9th sternum. It contains the dorsal **anus**, ventral **male genital pore** and **gonapophyses**.

8. External genitalia or gonapophyses. In the genital pouch, surrounding the male genital pore, are present the **phallic organs** or **gonapophyses** which help in copulation. These consist of 3 small irregular chitinous plates (right, left and ventral) known as **phallomeres**.

(a) Right phallomere. It consists of two horizontal opposing plates, a sickle-shaped hook and a serrated lobe terminating in two prongs.

(b) Left phallomere. It is made of a broad base from which arise 4 lobes of sclerites. Outer most is the short **pseudopenis** with a broad hammer-like head. To its. left is a long, slender **titillator** with a curved hook. Inner most is the **accutobols** with a large curved hook. Outer to this is the serrated **asperate.** Opening of phallic gland lies between acutobolus and asperate lobes.

(c) Ventral phallomere. It is simple, broad and brown plate without lobes and lying beneath the right phallomere. Opening of ejaculatory duct lies at its base in an intromittent muscular region, called **phallus** or **edeugus**.

[II] Female Reproductive System

Female reproductive organs include ovaries, oviducts and vagina, spermatheca, collaterial gland, genital pouch and external genitalia, all confined to abdomen.

1. Ovaries. A pair of yellow-coloured ovaries lie in 2nd to 6th abdominal segments, one on either side of hindgut, and embedded in fat bodies. Each ovary consists of 8 elongated, tapering and beaded blind tubes called **ovarian tubules** or **ovarioles**. Each ovariole contains a linear series of ova in various stages of development. Each ovariole consists of anterior' region or **germarium**, containing immature ova in early stages of egg-formation, and a posterior region of **vitellarium** containing maturing eggs. The anterior tapering apical ends or **terminal filaments** of the ovarioles of each ovary unite to form a single thread or **suspensory ligament** which is lost in the fat body.

Parts of an ovariole. Each ovariole includes the following parts :

(a) Terminal filament. It is the anterior-most tapering thread-like terminal end of ovariole. Filaments of ovarioles of each ovary are united in a single suspensory ligament attached to the fat body.

(b) Germarium. The anterior 1/3 part is the germarium containing only the primordial germ cells and oogonia only.

(c) Vitellarium. It is the posterior part containing ova in various stages of development and maturation. As the ova grow in size they pass down the ovariole. The wall of ovariole encloses each ovum in a definite sac or **follicle**. These follicles give a beaded appearance to ovariole. Cells of follicle secrete the egg shell or **chorion** of egg.

(d) Egg chamber. The last follicle of vitellarium is the largest and contains at a time a single fully mature ovum which is the oldest.

(e) Plug. Beyond egg chamber or last follicle the rear end of vitellarium closed by a moss of follicle cells forming a plug. At the time of egg-laying, the cells of plug dissolve to create a passage through pedicel into lateral oviduct.

(f) Pedicel. Beyond plug, the ovariole forms a short narrow stem or pedicel which opens into a lateral oviduct.

This type of ovariole having only eggs and no nutritive or nurse cells is known as **paniostic type**. The developing ova are nourished by the secretions of the epithelial cells of follicle.

2. Oviducts and vagina. Posteriorly, all the ovarioles of an ovary unite to form a short and wide **lateral oviduct**. The two lateral oviducts in their turn unite in 7th segment to form a very short median **common oviduct**. Its posterior **wider** part is called **vagina** which is formed as an invagination of body wall. It opens into a large **genital pouch** by a vertical slit-like **vulva** or **female genital pore** on the 8th sternum.

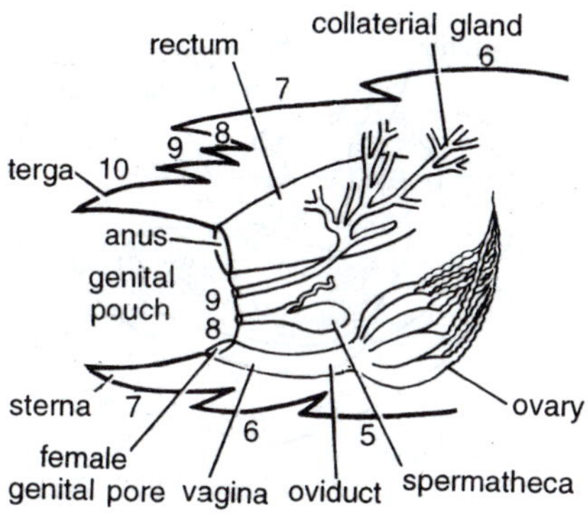

Fig. 24. *P. americana.* Female reproductive organs in lateral view.

3. Spermatheca. A spermatheca or **receptaculum seminis**, consisting of a left sac-like and a right filamentous caecum, opens by a median aperture in the dorsal wall of genital pouch on 9th sternum, at the tip of a small **spermathecal papilla**. In a fertile female, spermatheca is found filled with spermatozoa, received during copulation, from the male.

4. Collaterial gland. A pair of much branched accessory or **collaterial glands** lies behind and above the ovaries. Left gland is opaque and more developed, while the right one is transparent and less developed. The two open into the genital pouch through two separate openings, lying close together, a little behind and above the spermathecal opening. Secretion of two collaterial glands forms the hard egg-case or **ootheca** around groups of eggs.

5. Genital pouch. It is located at the hind end of abdomen. 7th sternum is large, boat-shaped and split behind into two **apical lobes** or

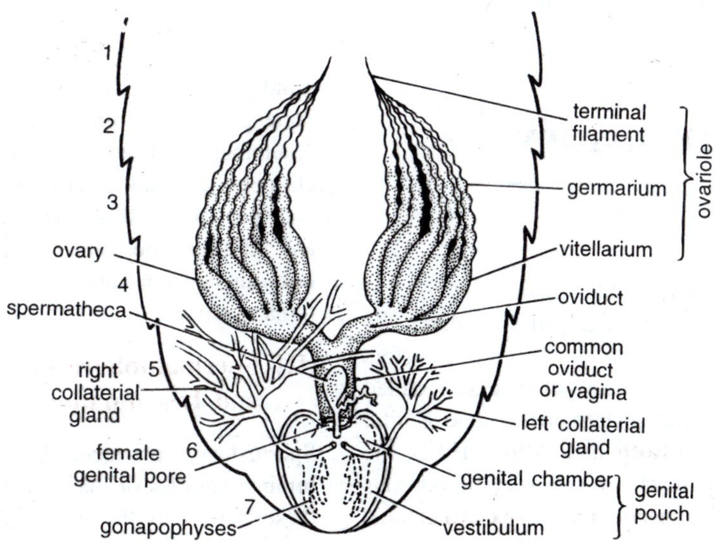

Fig. 25. *P americana.* Female reproductive organs in dorsal view.

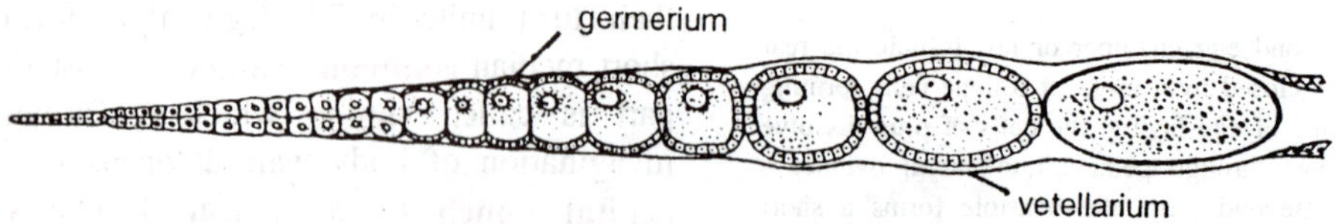

Fig. 26. Parts of a paniostic ovariole.

gynovalvular plates. These plates form the lateral, ventral and posterior limits of genital pouch, whose anterior and dorsal limits are formed respectively by the 8th and 9th sterna. Genital pouch can be divided into two parts. Smaller anterior part containing female gonopore and pores of spermatheca and collaterial glands is termed **genital atrium**. While its larger posterior part in which ootheca is formed is called **vestibulum**.

6. External genitalia. A sclerite around female genital pore is called **valvular plate**. Besides, three pairs of plate-like chitinous **gonapophyses** are present between the female genital pore and anus. Of these one pair arises laterally from 8th sternum (**I basivalvulae**), while the other two pairs (**valvifers** and **II basivalvulae**) from 9th sternum. These form the three pairs of **ovipositor valves**. These assist in copulation, in laying eggs and in the formation of ootheca. A median sclerite of 8th sternum forms a **spermathecal papilla** on which opens the spermathecal pore.

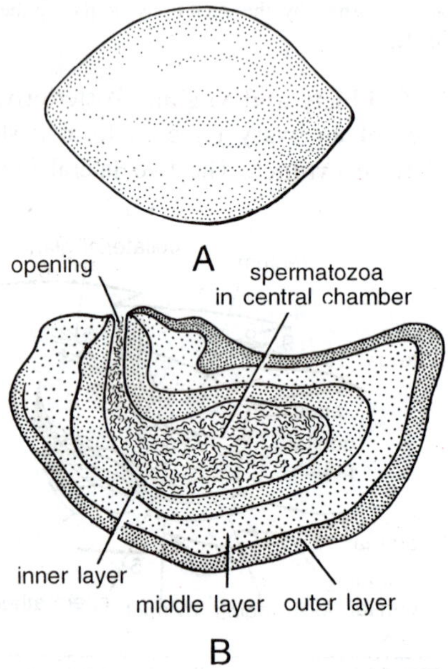

Fig. 27. *P. americana.* A spermatophore. **A.** Complete. **B.** In section.

LIFE HISTORY

[I] Copulation

Sexually mature male and female cockroaches generally mate during night from March to September. During copulation, the male transfers its spermatozoa in the form of packets, called **spermatophores**, directly into the female genital pouch. According to **Roth** and **Willis** (1952), the mature female cockroach emits a specific odorous volatile chemical, a **pheromone** or **sex attractant**, which is detected by the antennal chaemoreceptors of male. In response, the sexually excited male raises its tegmina (fore wings) by 45° to 90° and rapidly flutters its hindwings. In mating, first the male gets under female, extends his left hooked phallomere to insert into female genital chamber, clasping a small sclerite in front of ovipositor and then moves from beneath her. Both now gain an end-to-end position. The process of copulation lasts an hour or more after which the male and female separate.

[II] Spermatophore Formation and Insemination

Spermatozoa produced by testes get stored in seminal vesicles of male. At the time of copulation these pass into the ejaculatory duct where they immediately get mixed with a nourishing fluid secreted by utriculi breviores. At the same time utriculi majores pours its secretion that hardens

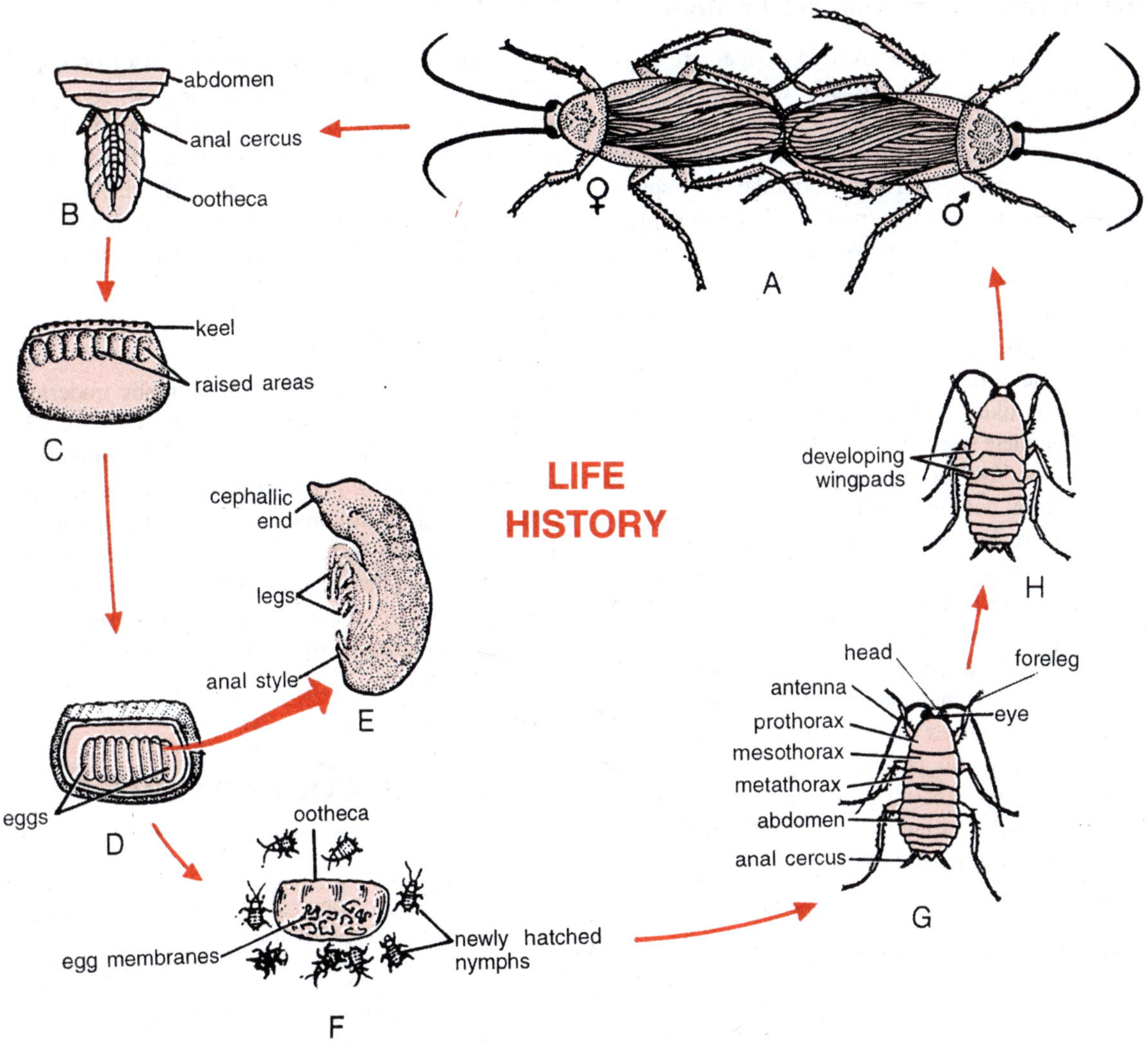

Fig. 28. *P. americana.* Life history. **A.** Copulation. **B.** Laying of ootheca. **C.** A single ootheca. **D.** Ootheca in section showing eggs. **E.** Early embryo. **F.** Hatching. **G.** Early nymph. **H.** Late nymph with wing pads.

to form the inner layer of spermatophore. As this one-layered spermatophore passes backwards, glandular wall of ejaculatory duct secretes around it another (middle) layer. During copulation, the two-layered spermatophore is deposited into female genital pouch on the spermathecal papilla. Here the secretion of phallic or conglobate gland forms its third (outer) layer.

Fully formed spermatophore is a three-layered, tough and pear-shaped capsule about 1.5 mm long. Its central chamber contains spermatozoa in a spermatic fluid. Its capsule has a simple opening that comes in close contact with the female spermathecal opening. Within 24 hours, sperms enter the spermatheca where they remain stored until required, while empty spermatophore drops down.

[III] Ootheca Formation and Fetilization

Ripe eggs are elongate oval in shape. 16 eggs, one from each ovariole usually descend together through vulva into vestibulum of genital pouch. As they are laid they become fertilized by sperms ejected simultaneously from spermatheca. A sperm enters through a minute pore, called **micropyle**, of egg shell and fertilizes the egg inside. Fertilized eggs become surrounded by the secretion of collaterial glands, which hardens to form an **egg case** or **ootheca**. 16 eggs are arranged in two rows in one ootheca, assisted by ovipositor valves, very much like cigarettes packed in a cigarettes case. Complete formation of an ootheca requires about two days during which it is protruding behind from her abdomen. A total of about 15 oothecae are laid by the female cockroach.

An ootheca is about 8 mm long. It is shaped like an oblong bag fringed with a dorsal serrated keel and each lateral wall raised into 8 elongated areas due to 16 enclosed eggs. An ootheca is moulded into its characteristic shape by the action of ovipositor valves and the inner configuration of walls of genital pouch. Colour of ootheca is first white, but later turns to pink or chocholate brown. Ootheca contains no chitin, but protein hardened by quinone, both secreted by collaterial glands.

[IV] Laying of Ootheca

The female cockroach carries the ootheca, protruding from tip of her abdomen, for several days, till it is deposited in a warm, sheltered and dark place.

[V] Embryonic Development

Fertilized eggs are slightly curved, concave on one side and convex on the other. Each contains a diploid nucleus and sufficient **yolk**, providing enough food for embryonic development. Each egg undergoes superficial **cleavage** characteristic of insects, resulting in a **blastula** followed by a **gastrula** stage. Gastrula is differentiated into embryo which gets separated from blastoderm by an amniotic cavity. Its three germ layers (ectoderm, mesoderm and endoderm) form the various parts of the **nymph**.

[VI] Hatching

When **hatching** occurs, the dorsal keel of ootheca splits and the young cockroaches or **nymphs** emerge out leaving their egg membranes behind. Freshly-hatched nymphs are delicate, transparent and almost colourless creatures with black eyes. They possess nearly all adult characters but differ in size and colouration, in being sexually immature and lacking wings.

[VII] Metamorphosis

Immediately after hatching, the nymphs undergo first moulting or **ecdysis** and then 6 or 7 successive moults follow. As nymphal development proceeds, the wing pads arise, body increases in size, colouration becomes darker and ultimately the adult takes its form with fully developed wings and genitalia. Thus development of cockroach is simple and **direct** and includes **incomplete** or **gradual** metamorphosis which is known as **paurometaboly**.

OTHER COCKROACHES

1. *Blatta orientalis.* It is commonly known as the 'Oriental cockroach' or 'Black beetle'. It was originally a native of North Africa, but today it is found all over the world. It commonly occupies indoor cooler places and prefers temperature between 20° and 29°C.

Body measures 20-24 mm in length, Colour varies from reddish brown to black. Colouration on pronotum is uniform. In male, fore and hind wings reach almost up to the posterior end of abdomen. While in female, tegmina are very short and hind wings are absent. Neither sex has power of flight.

2. *Blatella germanica.* It is commonly known as the 'German cockroach' or 'Steamfly'. It was originally a native of Africa, but today it is very common in almost all the countries including India. It prefers to live in warm and moist conditions and hence inhabits indoor places like kitchens, restaurants and dining halls.

Cockroach is small is size, measuring 10 to 15 mm in length. Body colouration is pale ochreous buff (straw-coloured) with two dark longitudinal bands on . pronotum. Body of male is thin and slender and its wings do not reach up to the posterior end of abdomen. While female body is stout and robust and wings reach up to the end of abdomen. Female carries ootheca for 10 to 15 days or more until the eggs hatch. The cockroach does not fly but can run swiftly by gliding flight.

3. *Periplaneta australasiae.* It is commonly called the 'Australian cockroach'. It was also originally native of Africa but, like other cockroaches, it has now spread to whole of the world. In India it is found mostly in South. It prefers to live in warm conditions and thus is a common pest of buildings. It has not been reported from sewers, unlike *P. americana*.

Cockroach is large in size and measures 30-35 mm in length. Body colour is reddish brown with two distinct and separate dark patches on pronotum. In both sexes, wings extend a little beyond the apex of abdomen. A diagnostic feature of this cockroach is the pale basal margins of tegmina.

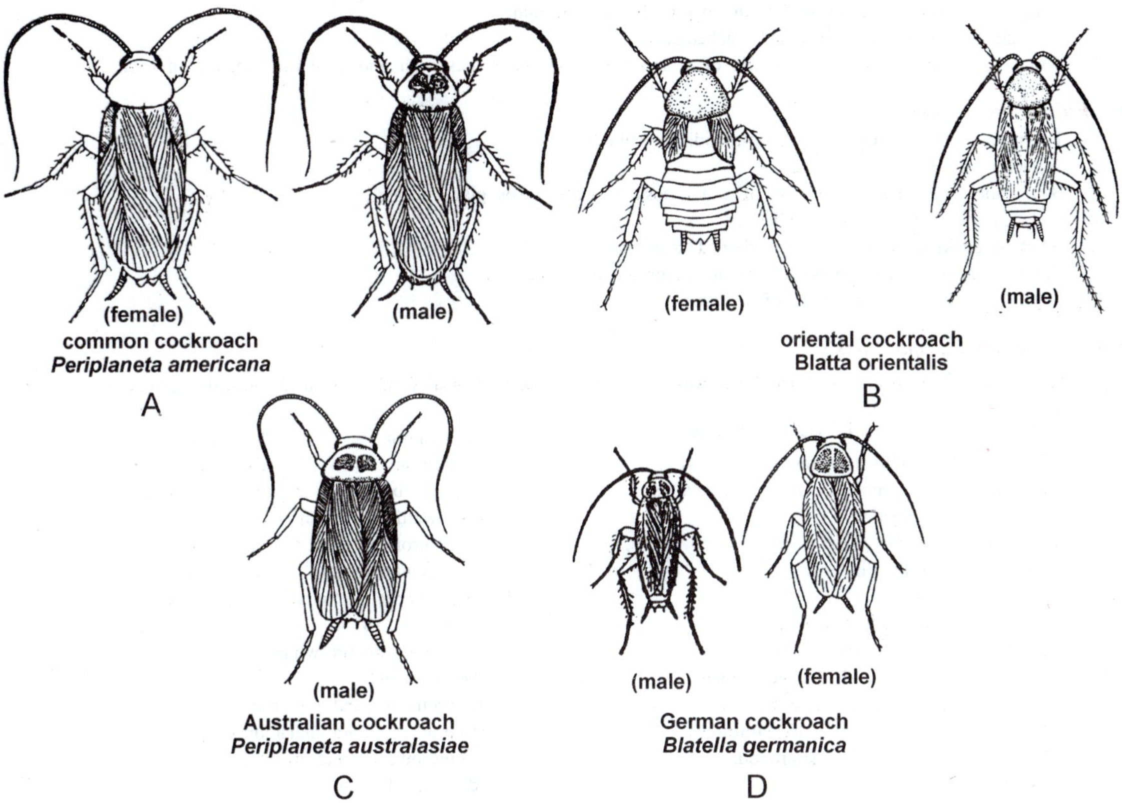

Fig. 29. Common types of cockroaches found in India. **A.** *Periplaneta americana.* **B.** *Blatta orientalis* (female and male), **C.** *Periplaneta australasiae* (male), **D.** *Blatella germanica* (male and female).

Important Questions

➤➤*Long answer type questions*

1. Describe the mouth parts of cockroach and add a note on its feeding mechanism.
2. Give an account of the digestive system *of Periplaneta americana.*
3. Describe the respiratory organs and mechanism of respiration of the cockroach.
4. Describe the reproductive organs of any cockroach studied by you.
5. Give an account of the life history of common cockroach.
6. Write short notes on : (i) Malpighian tubules, (ii) Nymph, (iii) Ootheca, (iv) Salivary apparatus, (v) Spermatophore.

➤➤*Short answer type questions*

1. What is the name of the rings that strengthen the trachea ?
2. How do the cockroach feed ?
3. Describe the compound eye of *Periplaneta* and clearly mention its adaptation for bright and dim-light.
4. Draw neatly labelled diagrams of the T.S. through the abdomen of cockroach.
5. Draw male and female reproductive organs of cockroach. Label them. How is ootheca formed in it ?
6. Cockroach and scorpion are both terrestrial forms. Why should then cockroach have a tracheal system which supplies oxygen directly to the tissues for respiration and scorpion have book lungs functioning differently. Answer within 30 words.
7. Compare the organs of respiration in cockroach and scorpion.
8. Distinguish between a trachea and tracheole.
9. Write the names of external genitalia or gonapophyses of male cockroach and give the function of each.

➤➤*Fill in the blanks*

1. The type of image that falls on ommatidia is called ...
2. Spiracles are also called as .....
3. Hard cuticle prevents evaporation of water, loss of water takes place through .....
4. Body cavity of cockroach is called as .....
5. Heart of cockroach has .... funnel-shaped chambers.
6. Uricose glands of male cockroach is an .... organ and stores .....
7. Metamorphosis in cockroach is of .... type.

➤➤*Multiple choice questions*

1. Generic name of cockroach *Periplaneta* was assigned by :
   (a) Linnaeus (1758) (b) De Beer (1973)
   (c) Burmerister (1838) (d) Bentham (1664)
2. The function of spermatheca in *Periplaneta* is :
   (a) to help in copulation
   (b) to store eggs
   (c) to secrete musky fluid
   (d) to store sperms
3. Egg case of cockroach is known as :
   (a) embryophore (b) ootheca
   (c) gonophore (d) female gamete
4. *Periplaneta* belongs to the phylum :
   (a) Annelida (b) Platyhelminthes
   (c) Molluscs (d) Arthropoda
   (e) Oncychophora
5. Male and female cockroach can be identified by :
   (a) anal cerci (b) anal styles
   (c) antennae (d) compound eyes
6. Anal styles are found in :
   (a) male cockroach (b) housefly
   (c) male mosquito (d) female cockroach
7. The skeleton of cockroach is formed of :
   (a) chitin (b) cuticle
   (c) cartilage (d) amino acid
8. Vision of cockroach is :
   (a) monocular (b) binocular
   (c) mosaic (d) all
9. The part of head of cockroach between and behind the eyes is :
   (a) gena (b) ocellus
   (c) clypeus (d) epicranium
10. Dorsal surface of an abdominal segment of cockroach is :
    (a) sternum (b) pleuron
    (c) tergum (d) none of these
11. The out border of tergum in cockroach bends downwards to attach with :
    (a) pleuron (b) sternum
    (c) mesenteries (d) muscles
12. The number of thoracic segments in cockroach are :
    (a) 2 (b) 4 (c) 3 (d) 5
13. The two pairs of wings in cockroach are situated in the segments :
    (a) prothorax and mesothorax
    (b) prothorax and metathorax
    (c) mesothorax and metathorax
    (d) none of these
14. The chambers in the heart of cockroach are :
    (a) 13 (b) 9 (c) 12 (d) 15
15. The egg of cockroach is :
    (a) Isolecithal (b) Telolecithal
    (c) Microlecithal (d) Centrolecithal
16. Conglobate glands arefound in the reproductive organs of :
    (a) male cockroach (b) female cockroach
    (c) frog (d) earthworm
17. The respiratory passage in cockroach during inspiration:
    (a) spiracle trachea
    (b) longitudinal respiratory tube
    (c) air chamber (d) trachea
18. Palpiger is a portion of :
    (a) labrum (b) labium
    (c) maxilla (d) mandible

19. Lacunar type of circulatory system is present in :
(a) *Periplaneta americana*
(b) *Blatta orientalis*
(c) *Blatella germanica*
(d) all of the above
20. Corpora cardiaca are :
(a) an endocrine organs in insects
(b) another name of salivary gand of bugs
(c) a part of nervous system of insects
(d) excretory organ of a insect
21. Function of the colleterial gland in cockroach is to :
(a) store sperms (b) secrete the egg-case
(c) keep vagina moist (d) store eggs
22. The respiratory pigments in cockroach is absent. Hence in cockroach :
(a) O_2 diffuses directly in the tissues
(b) colleterial respiration takes place
(c) respiration occurs by book lungs
(d) none of the above
23. Ootheca of cockroach is secreted by :
(a) mushroom glands (b) conglobate glands
(c) collaterial glands
(d) genital pouch of female
24. The excretory waste of cockroach and other insect is:
(a) ammonia (b) uric acid
(c) urea (d) guanine
25. Malpighian tubules in cockroach are used for :
(a) respiration (b) excretion
(c) reproduction (d) digestion
26. The body cavity of cockroach is :
(a) coelom (b) pseudocoel
(c) haemocoel (d) coelenteron
27. The Nymph grows into a cockroach by :
(a) incomplete metamorphosis
(b) complete metamorphosis
(c) no metamorphosis (d) two of the foregoing
28. The male cockroach can be distinguished from the female in having :
(a) anal cerci (b) anal styles
(c) large antenna (d) large wings
29. The chitinous structures found around the genital aperture of cockroach are :
(a) gonapophyses (b) odontoid processes
(c) anal cerci (d) sternum
30. The salivary gland of cockroach secretes :
(a) ptyalin (b) amylase
(c) trypsin (d) pepsin
31. How many teeth are present in the gizzard of cockroach ?
(a) 6 (b) 4 (c) 3 (d) none
32. Number of segments in leg of cockroach is :
(a) 3 (b) 5 (c) 6 (d) 9
33. In cockroach, the outer border of a tergum turns downwards to become attached with :
(a) sternum (b) pleuron
(c) muscles (d) mesenteries
34. Class of cockroach is :
(a) Insecta (b) Arachnida
(c) Gastropoda (d) Arthropoda
35. One common feature in the trachea of cockroach and the trachea of rabbit is that :
(a) their inner lining is ciliated
(b) their walls are strengthened by hard structures
(c) both start in the head region
(d) both are paired
36. Mouth parts of cockroach are :
(a) cutting and chewing (b) sucking
(c) piercing (d) all these
37. Wings are vestigeal in :
(a) female *Periplaneta* (b) male *Periplaneta*
(c) *Blatta* (d) Housefly
38. Plantulae are found in :
(a) parapodia of *Nereis* (b) maxilla of cockroach
(c) leg of cockroach (d) proboscis of housefly
39. Alary muscles are present in connection with :
(a) wing of cockroach (b) ovary of cockroach
(c) heart of cockroach (d) tracheae of cockroach
40. Phallic glands in cockroach :
(a) secrete ootheca
(b) secrete outer covering of spermatophore
(c) store excretory products
(d) secrete ecdysone
41. Common salivary duct of cockroach opens at base of :
(a) hypopharynx (b) mandibles
(c) second maxillae (d) labrum
42. Egg of cockroach is :
(a) centrolecithal (b) telolecithal
(c) microlecithal (d) megalecithal
43. The two sexes are externally demarcated in :
(a) earthworm (b) cockroach
(c) *Amoeba* (d) *Hydra*
44. Fixed number of body segments are found in :
(a) cockroach (b) earthworm
(c) *Taenia* (d) none of these
45. Number of segments that constitute thorax of cockroach is :
(a) 6 (b) 3 (c) 8 (d) 2
46. The most broad segment in the leg of cockroach is :
(a) femur (b) coxa
(c) trochanter (d) tibia
47. Each ovary of cockroach is made of ovarioles, number is :
(a) 4 (b) 30 (c) 8 (d) 16
48. Mandibles of cockroach are :
(a) long
(b) short with teeth
(c) short without teeth
(d) syringe like
49. Colleterial glands of cockroach are to help :
(a) secretion of ootheca
(b) secretion of genital chamber
(c) oviposition (d) fertilization

50. The common feature among cockroach, scorpion and centipede :
(a) compound eyes (b) throrax
(c) tracheae (d) jointed legs
51. The common feature between a cockroach and earthworm :
(a) nephridia (b) moulting
(c) larva (d) ventral nerve
52. Number of paired spiracles in cockroach are :
(a) 3 (b) 10 (c) 6 (d) 2
53. Dorsal plate in an abdominal segment of cockroach is termed :
(a) tergum (b) pleurite
(c) sternite (d) none of these
54. Outer part of a tergite bends downwards in cockroach and it is attached with :
(a) sternite (b) pleurite
(c) both of these (d) none of these
55. Animals active during day time are called :
(a) nocturnal (b) diurnal
(c) carnivorous (d) herbivorous
56. Rhabdome is found in eyes of :
(a) mollusc (b) frog
(c) man (d) insects
57. Image chiefly formed by compound eyes of insects is :
(a) superposition (b) apposition
(c) both of these (d) monocular
58. Chief excretory wastes of insects is :
(a) urea (b) ammonia
(c) uric acid (d) amino acids
59. The portion of head in between and behind the eyes in cockroach is :
(a) epicranium (b) gene
(c) clypeus (d) frons
60. Vision in cockroach is called :
(a) binocular (b) monocular
(c) mosaic (d) stereoscopic
61. Wings of cockroach usually help the insect in :
(a) escaping from enemies
(b) oviposition
(c) copulation
(d) feeding
62. Both cockroach and housefly have :
(a) two pairs of wings (b) compound eyes
(c) liquid diet (d) pupa stage
63. Both male and female cockroach do not have :
(a) anal styles (b) anal cerci
(c) three segmented thorax
(d) bitting mouth parts
64. Skeleton of cockroach and other insects is made of :
(a) spicule (b) fluid
(c) coral (d) chitin
65. Cockroach has chitinous teeth for grinding in its :
(a) gizzard (b) mandible
(c) both (d) pharynx
66. Salivary glands of cockroach secrete :
(a) amylase (b) chitinase
(c) cellulase (d) all of these
67. Alary muscles of cockroach are related with :
(a) nutrition (b) circulation
(c) respiration (d) all of these
68. Heart of cockroach is :
(a) multichambered (b) single
(c) longitudinally beaded(d) all of these
69. Blood of cockroach has :
(a) haemoglobin (b) no pigment
(c) R.B.C. (d) oxygen
70. Malpighian tubules of cockroach are for :
(a) excretion (b) osmoregulation
(c) both (d) digestion
71. Ommatidium is :
(a) structural unit of compound eye
(b) visual unit
(c) both (d) false
72. Ootheca of cockroach contains :
(a) one zygote
(b) 16 zygotes and is secreted by colleterial gland
(c) 16 zygotes (d) a larva
73. Nymph grows into an adult cockroach by :
(a) complete metamorphosis
(b) incomplete metamorphosis
(c) less than 7 moults
(d) less than 12 moults
74. The tracheal system in *Periplaneta* communicates to the outside by :
(a) 10 pairs of spiracles (b) 12 pairs of spiracles
(c) 8 pairs of spiracles
(d) 6 pairs of spiracles
75. The function of the phallic gland in *Periplaneta* is to secrete :
(a) a secretion which helps the sperms in the spermatophore to stick together
(b) the innermost layer of the wall of the spermatophore
(c) the middle layer of the wall of the spermatophore
(d) the outermost layer of the wall of the spermatophore
76. Which one of the following regions of the alimentary canal in *Periplaneta* is characterised by the presence of a peritrophic membrane ?
(a) Ileum (b) Mesenteron
(c) Crop (d) Gizzard
77. The oxygen carrying pigment in the blood of cockroach is :
(a) haemocyanin (b) haemoglobin
(c) xanthophyll (d) not present at all
78. Number of ganglia located over ventral nerve cord of cockroach is :
(a) 6 thoracic and 3 abdominal
(b) 3 thoracic and 6 abdominal
(c) 1 thoracic and 1 abdominal
(d) 1 thoracic and 8 abdominal
79. Metamorphosis in cockroach is regulated by :
(a) corpora allata (b) temperature
(c) food (d) brain
80. Body is covered by :
(a) cuticle (b) chitin
(c) sclerites (d) suberin

81. Gloss and paraglossa are collectively called :
(a) tergum (b) natum
(c) lingula (d) pleura

82. Respiratory pigment in *Periplanata :*
(a) haemoglobin (b) haemocynin
(c) both (d) none

83. Excretory system of cockroach :
(a) trophocytes (b) mycetocytes
(c) oenocytes (d) urate cell
(e) all of the above

84. Juvenile hormone secreted by :
(a) corpora allata (b) intercerebral gland
(c) prothoracic gland (d) none the above

85. The period between two mould is :
(a) stadium (b) incubation
(c) latent (d) zero

86. Each ovary contain how many ovarioles :
(a) 4 (b) 6 (c) 8 (d) 10

87. Number of chromosomes in male :
(a) 33 (b) 34 (c) 32 (d) 30

88. The blood of cockroach contains no respiratory pigment. It means that :
(a) cockroach does not respire
(b) respiration is anaerobic
(c) oxygen goes into tissues by intracellular capillary system
(d) oxygen goes directly into tissues through tracheal system

89. Biggest phylum with reference to number of species is :
(a) Insecta (b) Arthropoda
(c) Chordata
(d) Platyhelminthes

90. The type of antenna in cockroach is:
(a) Aristotle (b) Filiform
(c) Clubbed (d) Plumose

Answers

➤➤*Fill in the blanks*

1.Mosaic or apposition, 2. Stigmata, 3. Spiracle valve, 4. Haemocael, 5. 13, 6. Excretory organ / uric acid, 7. Incomplete type.

➤➤*Multiple choice questions*

1. (c) 2. (d) 3. (b) 4. (d) 5. (b) 6. (a) 7. (a) 8. (c) 9. (a) 10. (c)
11. (b) 12. (c) 13. (c) 14. (a) 15. (c) 16. (a) 17. (a) 18. (b) 19. (d) 20. (a)
21. (b) 22. (a) 23. (c) 24. (b) 25. (b) 26. (c) 27. (a) 28. (b) 29. (a) 30. (b)
31. (a) 32. (d) 33. (a) 34. (a) 35. (b) 36. (a) 37. (a) 38. (c) 39. (c) 40. (b)
41. (a) 42. (a) 43. (b) 44. (a) 45. (b) 46. (b) 47. (c) 48. (b) 49. (a) 50. (d)
51. (d) 52. (b) 53. (a) 54. (a) 55. (b) 56. (d) 57. (b) 58. (c) 59. (d) 60. (c)
61. (a) 62. (b) 63. (a) 64. (d) 65. (a) 66. (a) 67. (b) 68. (a) 69. (b) 70. (a)
71. (b) 72. (b) 73. (b) 74. (a) 75. (d) 76. (b) 77. (d) 78. (b) 79. (a) 80. (c)
81. (c) 82. (d) 83. (e) 84. (a) 85. (a) 86. (c) 87. (a) 88. (d) 89. (b) 90. (b).

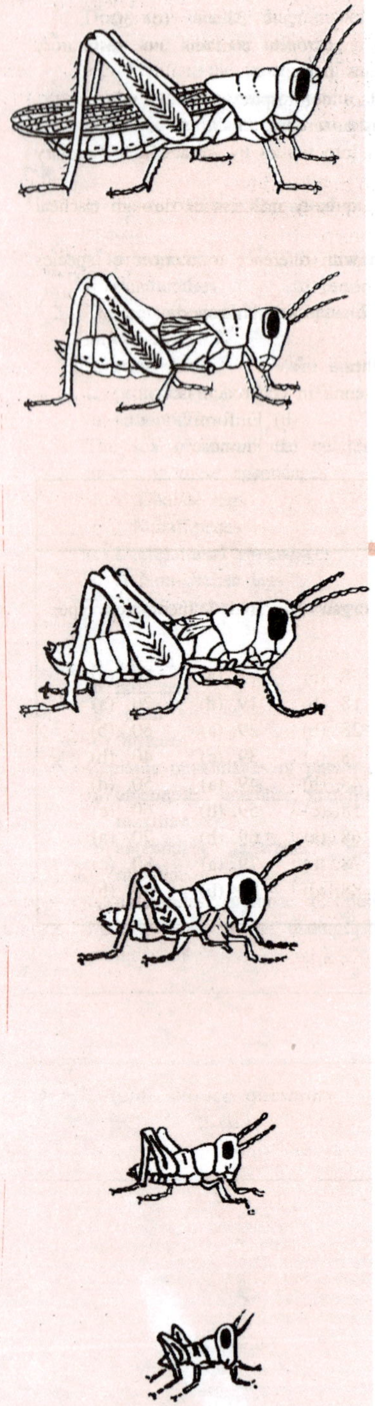

48

The Grasshopper or Locust

Like cockroaches, the well-known grasshoppers and locusts also belong to the order **Orthoptera** of class Insecta. Long-horned grasshoppers or katydids belong to the family **Tettigonidae**, whereas short-horned grasshoppers and locusts belong to the family **Acrididae**. They are comparatively large, with large prothorax and jumping hind legs and commonly studied as an example of a generalized insect. The term '**locust**' is applied to the gregarious and migratory form of grasshoppers. The common species of grasshoppers and locusts are *Schistocerca americana* (American grasshopper), *Schistocerca gregaria* (migratory desert locust), *Romalea microptera* (lubber grasshopper), *Cyrtacanthacris succinata* (Bombay locust) and *Poecilocerus pictus* (Ak grasshopper). Following text applies to grasshoppers in general. There are 11,000 known species of grasshopper on the earth.

SYSTEMATIC POSITION

Phylum	Arthropoda
Class	Insecta
Subclass	Pterygota
Division	Exopterygota
Order	Orthoptera
Family	Acrididae
Type	Grasshopper or Locust

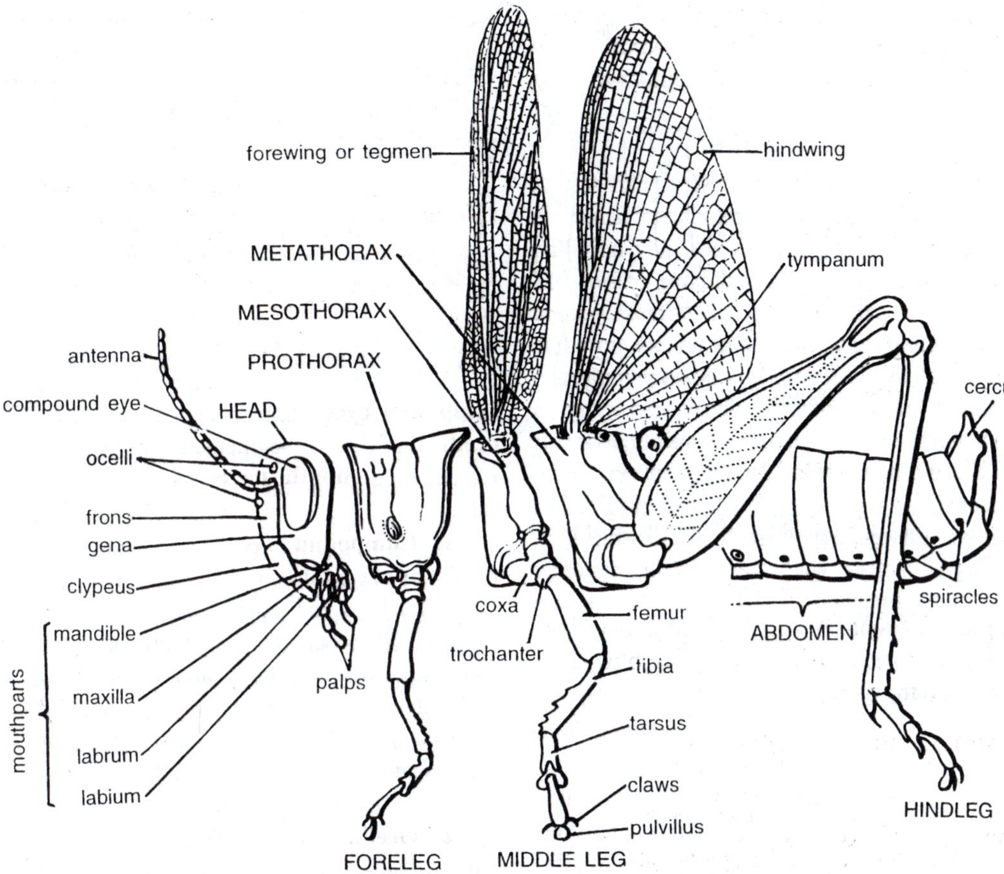

Fig. 1. Grasshopper. External features of male in lateral view.

DISTRIBUTION, HABITS AND HABITAT

Grasshoppers are **voracious herbivores** having a worldwide distribution. They are found in greatest abundance in places with open grass lands and abundant leafy vegetation, where there is plenty of food and place to breed.

Most African and Asian species are **gregareous** and **migratory**, the most common being *Schistocerca gregaria* and *S. migratoria* (desert locusts) extending from North Africa to North India. They have long wings and move in swarms from one region to another, feeding upon leafy vegetation, grasslands, trees and crops on the way. Whereas short-winged (or wingless) grasshoppers are **solitary** and confined to their respective ranges.

EXTERNAL MORPHOLOGY

[I] Shape and Size

Grasshopper is relatively a large insect reaching up to 8 cm in length. Body is narrow, elongated, cylindrical and bilaterally symmetrical.

[II] Colouration

Usual body colouration is yellowish or brownish with different colour spots and markings. This provides a variegated pattern that enables them to

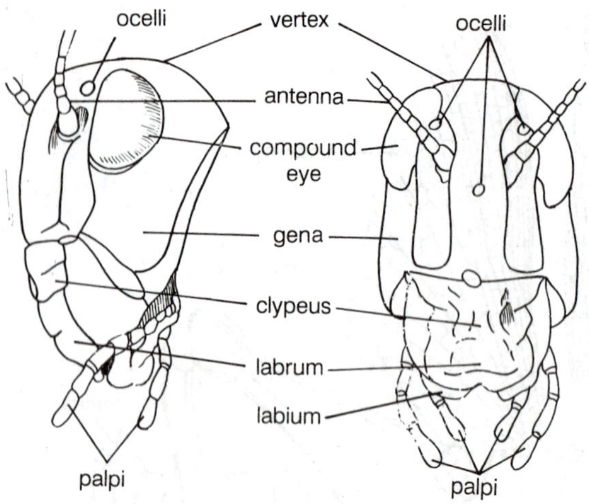

Fig. 2. Grasshopper. **A.** Head in lateral view, **B.** Head in front view.

resemble the environment in which they live. The Ak plant grasshopper (*Poecilocerus*) is blue and yellow in colour.

[III] Exoskeleton

The whole body is covered by a cuticular exoskeleton having chitin. It is secreted by the underlying layer of cells, the **hypodermis**. While the adult grasshopper does not moult, the exoskeleton of the young stages or nymphs is periodically moulted to permit the growth and increase in size. Each body segment is made up of separate hard exoskeletal plates, called **sclerites**. Cuticle is soft between segments; these softer regions are known as **sutures**, which permit movements of body segments and appendages.

Pigments in and under cuticle provide a **protective colouration** by which grasshoppers resemble their surroundings.

[IV] Segmentation

Body of grasshopper is distinctly segmented and externally divided into three typical regions : (i) **head** (6-fused segments), (ii) **thorax** (3 segments) and **abdomen** (11 segments).

[V] Head

The anterior movable region or head appears as a single structure, but it is made up of 6-embryonic fused segments, as indicated by the presence of paired appendages on adult head. Head is usually pear-shaped and of **hypognathous type**, *i.e.,* lying vertically and its broader mouth-bearing side directed downwards.

Exoskeleton of head (**head capsule** or **epicranium**) is made of firmly-jointed chitinous plates or sclerites. Its dorsal region is known as **vertex**, lateral sides as cheeks or **genae** and the front portion as **frons**. A broad rectangular plate below frons is the **clypeus**.

Head bears one pair of dorso-lateral **compound eyes**, three frontal simple eyes called **ocelli**, one pair of segmented **antennae** and a set of chewing **mouth parts** ventrally surrounding mouth.

1. Compound eyes. Two large compound eyes, are placed dorso-laterally one on either side of head. They are appendages of the first head segment. They occupy a relatively large area, curve round the sides of head and command a broad field of vision. Structurally, they are similar to those of a prawn but they are sessile as the stalks are absent. Some insects, possibly the grasshoppers are able to distinguish colours.

2. Ocelli. Three simple eyes or ocelli are situated on top of head between compound eyes. An ocellus consists of a thick, transparent cuticle forming **lens**, and a group of deeper light-sensitive cells forming **retina. Pigment** cells are also present. A simple eye has a single light-condensing apparatus for all the sensory cells. True function of ocelli is not properly understood.

3. Antennae. The second head segment bears a pair of antennae lodged in **antennal sockets** in front of the compound eyes. They are homologous to the antennules of prawn. Antennae are slender, thread-like, unbranched and several jointed. Each consists of a large basal joint, the **scape**, a short **pedicel** and a many-jointed **flagellum**. They bear tactile bristles and olfactory pits and make efficient sensory organs. Length of antennae varies in different species of grasshoppers.

4. Mouth parts. Mouth parts are attached ventrally to the head surrounding the preoral cavity. These are of **chewing** or **mandibulate type** and enables the insect to bite and chew vegetable matter. These consists of :

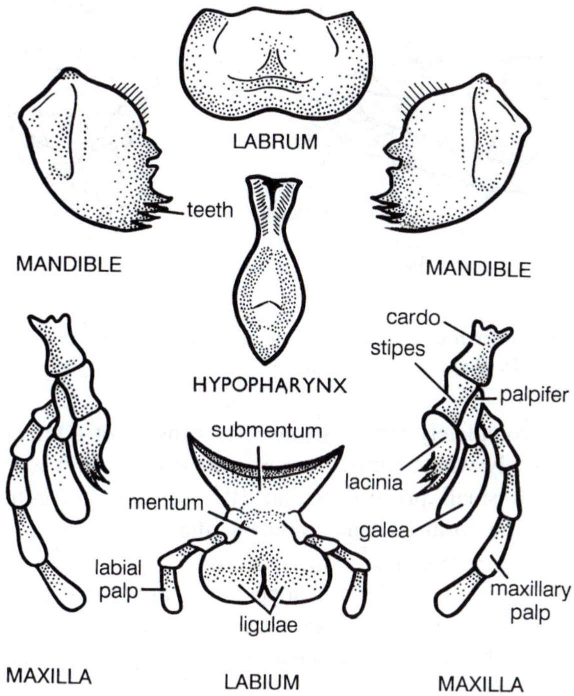

Fig. 3. Grasshopper. Mouth parts.

(a) Labrum. Upper lip or labrum is a broad, somewhat rectangular plate attached to the ventral edge of clypeus.

(b) Hypopharynx. It is a median, membranous, tongue-like structure lying beneath labrum.

(c) Mandibles. On either side below cheek is a large, heavy, somewhat triangular and dark-coloured jaw or mandible. It is hard, horny and with toothed inner margin for chewing food.

(d) Maxillae. Lying behind mandibles are a pair of maxillae. Each maxilla consists of several parts (basal **cardo**, middle **stipes**, distal elongated curved **lacinia** and rounded **galea**) and a slender five-segmented **sensory maxillary palp** which arises from a **palpifer**. Maxillae are used in manipulating food as it enters mouth.

(e) Labium. The broad, median lower lip or labium is considered as a fused second pair of maxillae. It consists of a basal **submentum**, a central **mentum**, two movable flaps, the **ligulae**, and a three-segmented sensory **labial palp** on either side. Labrum and labium serve to hold food between the mandibles and maxillae which move laterally to grind it.

[VI] Thorax

It is the middle region of body attached to head and abdomen by flexible joints. The three segments of thorax are : (i) large anterior **prothorax**, (ii) middle **mesothorax** and (iii) posterior **metathorax**. Exoskeleton of each segment consists of several sclerites. Dorsal region or **tergum** of prothorax, called **pronotum**, is very large saddle-like and consists of 4 fused sclerites in a row (prescutum, scutum, scutellum and postscutellum) as indicated by 3-transverse grooves. Lateral region or **pleuron** on each side consists of three sclerites, the episternum, epimeron and parapteron. Ventral part or **sternum** includes a single sclerite.

Two pairs of small slit-like **spiracles** are located on the sides of thorax near the ventral margin. First pair lies between pro- and mesothorax below pronotum, and second pair between meso- and metathorax.

1. Legs. Each thoracic segment has a pair of jointed legs attached ventro-laterally. Each leg consists in a linear series of 5-segments : (i) a short basal **coxa** which articulates with thorax, (ii) a smaller trochanter which is fused with, (iii) a large and stout **femur**, (iv) **an elongated, slender, spiny tibia**, and (v) **tarsus** made of 3-short segments or **tarsomeres** bearing tiny pads (**plantulae**) on ventral side. Terminal tarsomere also bears a median fleshy pad (**pulvillus** or **arolium**) between two lateral **claws**.

All the legs are used in walking and climbing, but the powerful metathoracic legs, having large muscular femur and elongated tibia, are specialized for jumping or leaping (**saltatorial legs**). Tarsal pads and pulvilli serve to hold on to smooth surfaces whereas the tarsal claws and tibial spines are used in clinging to rough surfaces.

2. Wings. In grasshopper, each of the mesothorax and metathorax bears a pair of wings. An insect wing arises dorso-laterally, between tergum and pleuron, as a double-layered sac-like projection of epidermis. But, in a fully-developed wing, the epidermal cells disappear so that the hard, dry and lifeless wing consists of a double

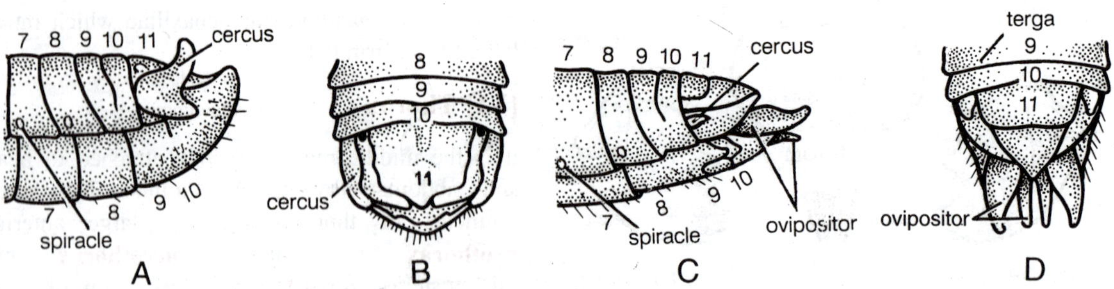

Fig. 4. Abdomen of male and female grasshoppers showing differences and sexual dimorphism. **A.** Lateral view (Male). **B.** Dorsal view (Male). **C.** Lateral view (Female). **D.** Dorsal view (Female).

layer of cuticle secreted by these cells. Each wing is supported by numerous longitudinal cuticular thickenings, called **nervures** or **veins**, which are narrow tubes originally occupied by tracheae, nerves and blood. Number and arrangement of these veins (**wing venation**) differ in different species of insects, but remain constant and characteristic in the individuals of a single species. Wing venation, therefore, is useful in classification.

In grasshopper, both pairs of wings are different from each other. Mesothoracic or forewings, called **tegmina**, are narrow, hardened, leathery or parchment-like, not used in flight but simply serving as covers for the hindwings. Metathoracic or hindwings are large, broad, membranous and used in flight. At rest they become folded like a fan under the forewings.

[VII] Abdomen

Abdomen is slender, cylindrical and consisting of 11 segments. Each segment typically has a dorsal tergum and a ventral sternum, there being no pleura. 1st segment is incomplete as its sternum is fused with that of metathorax. It consists of a tergum only.

1. Tympanum. On each lateral side of 1st abdominal segment is an oval **tympanic membrane** that covers an **auditory sac** or organ of hearing.

2. Spiracles. 8 pairs of small **spiracles** are present. Those of the first pair are larger than the rest and lie in first segment in front of tympanum, remaining 7 pairs of spiracles are situated antero-laterally on segments 2 to 8. In segments 9 and 10, terga are partially fused, while sterna are completely fused. Segment 11 consists of a tergum only which forms the **supra-anal plate** over anus. A small process behind 10th segment on either side forms a small process, the **anal cercus**.

3. Sexual dimorphism and external genitalia. Male grasshopper is smaller than female in size. Terminal abdominal segments are modified in both sexes for mating and egg-laying, and indicating **sexual dimorphism**.

In male grasshopper, tip of abdomen is rounded and anal styles like those of cockroach are absent. 9th sternum enclosing copulatory organs is elongated and curved upwards. It bears a **sub-genital plate** which covers the male genital pore and terminates dorsally in two short projections. **Anal cercus** is somewhat bigger and bifid.

In female grasshopper, abdomen is more tapering or pointed than in male. 9th sternum is elongated on which opens the female genital pore. Abdomen terminates in 3 pairs of chitinous plates, two pairs larger and one pair smaller. These form 3 pairs of valves of the egg-laying apparatus or **ovipositor**.

BODY CAVITY AND MUSCLES

Most of the internal space or body cavity is not a true coelom but a **haemocoel** filled with **haemolymph** or blood. Organ systems lie within the haemocoel. Vestiges of true coelom can be identified only in the early stages of development. In adult, coelom is represented by cavities of gonads.

Muscles are of Striated Type

They are soft, delicate but strong. In abdomen they are arranged segmentally for respiratory and reproductive movements. They are grouped especially for the movements of mouth parts, wings, legs and ovipositor.

DIGESTIVE SYSTEM

[I] Alimentary Canal

The alimentary canal consists of three principal parts : **foregut, midgut** and **hindgut**. Foregut and hindgut are lined with cuticle.

1. Foregut. It starts at **mouth** situated at the dorsal limit of a **preoral cavity**, which is a narrow space bounded anteriorly by labrum and clypeus, laterally by mandibles and maxillae and posteriorly by labium. Lying suspended into it is the tongue or **hypopharynx**, bearing the opening of salivary glands. Mouth leads into a very short **pharynx**. It is followed by a short, narrow, thin-walled **oesophagus** which enlarges into a thin-walled sac or **crop**, extending up to the middle of thorax. Below crop is present a pair of small, branched **salivary glands**, the ducts of which open into preoral cavity at labium. It produces an enzyme containing secretion. Behind crop is a short **gizzard** or **proventriculus** extending up to the hind end of thorax. It has thick muscular walls with a cuticular lining. In its anterior part cuticle forms a ring of grinding **teeth**. Behind teeth lie two rows of **pads** or **pulvilli**, beset with **bristles** or **hair**, and act as strainers, which permit only the finely ground food to enter midgut.

2. Midgut. Next to gizzard is the midgut. It includes the **ventriculus** or **stomach** which reaches up to the middle of abdomen. It is thin-walled, devoid of a cuticular lining and sphinctered at both ends. Opening into the anterior end of stomach is a ring of six pairs of elongated, cone-shaped pouches called hepatic or gastric caeca. One pouch of each pair extends anteriorly over proventriculus and the other posteriorly over ventriculus. Gastric caeca produce digestive enzymes and facilitate absorption of digestive foods.

3. Hindgut or intestine. It is internally lined by cuticle. Its junction with stomach is marked by the attachment of numerous long thread-like yellowish **Malpighian tubules** which open into it. The uncoiled intestine consists of an anterior broad tapered **ileum**, a slender middle **colon** and a posterior enlarged, thin-walled **rectum** which opens to outside through **anus** lying beneath a supra-anal plate on the last or llth abdominal segment.

[II] Nutrition

1. Food and feeding. Grasshoppers devour green leafy vegetation. They move from plant to plant by walking, hopping and flying in order to feed upon their leaves. Food is held by forelegs, labrum and labium, lubricated by salivary secretion and chewed by mandibles and maxillae, as in cockroach before being swallowed by mouth into pharynx.

2. Digestion. Saliva not only facilitates chewing and swallowing but also contains an amylolytic enzyme that helps to digest starch. Masticated food passes from pharynx down oesophagus into crop which is a temporary storage chamber. Now food is channelled into the muscular **gizzard** which acts as a grinding chamber. Gizzard brings about physical digestion of food which is broken into smaller pieces to increase their surface area for the action of enzymes. Hairy pads of gizzard allow only finer food particles to strain into midgut or stomach. Anterior sphincter of stomach checks regurgitation of food. Food undergoes chemical digestion inside stomach. It is acted upon by enzymes such as tryptase, amylase, maltase, invertase, lipase, etc., present in juices secreted by epithelia of hepatic caeca and midgut. Anterior end of midgut also secretes a chitoproteinous substance. It forms a thin, tubular **peritrophic membrane** which protects midgut wall from abrasions by hard food particles. It, however, does not hinder digestion as it is permeable to digestive enzymes as well as to products of digestion. Both digestion and absorption take place in midgut or stomach while undigestible matter is passed on to hindgut. Some

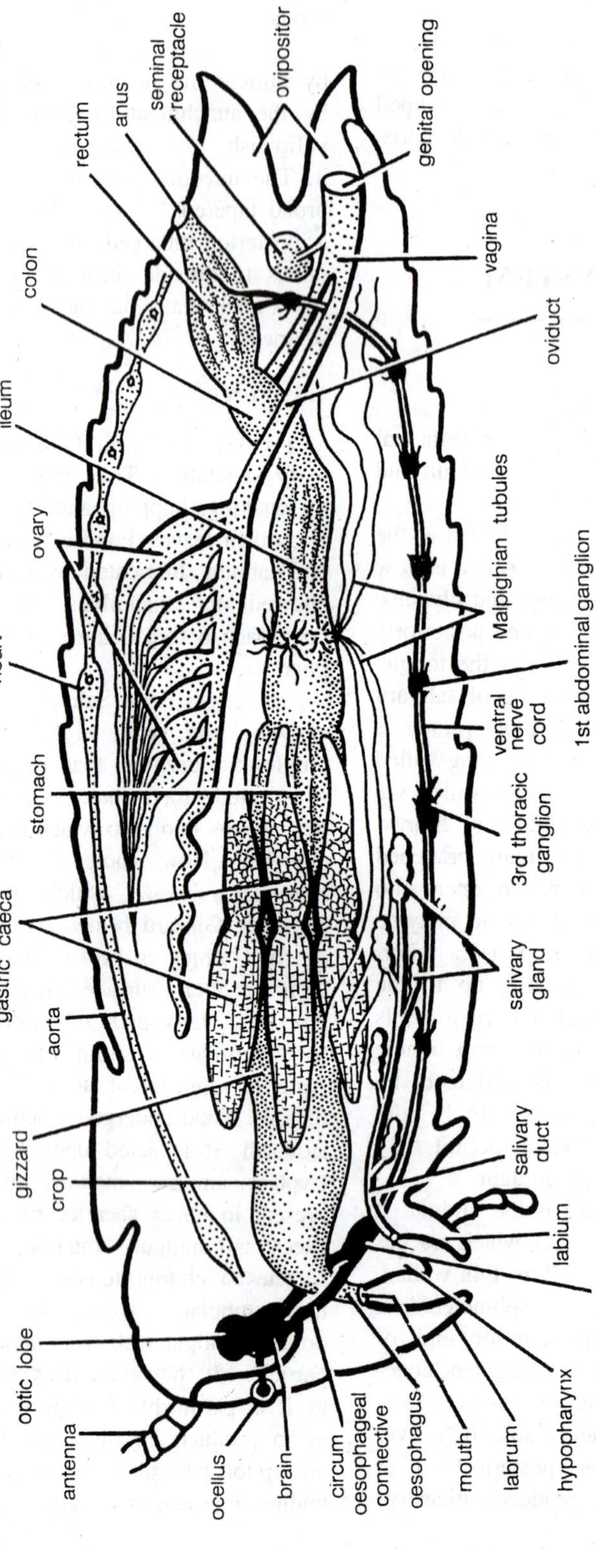

Fig. 5. Internal anatomy of female grasshopper seen from left side.

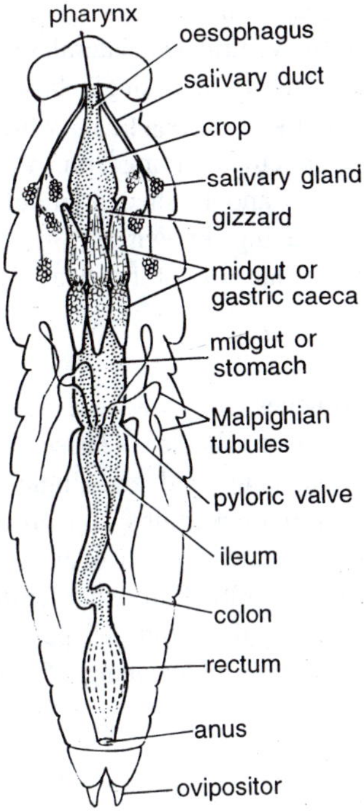

Fig. 6. Grasshopper. Alimentary canal in dorsal view.

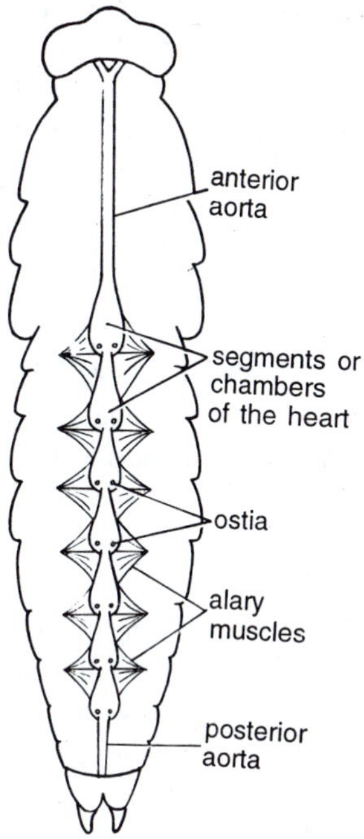

Fig. 7. Grasshopper. Circulatory system.

digestion is also brought about in the crop by salivary enzymes and by digestive juices that are passed forward from the midgut. Some fats may be absorbed in the crop. Regurgitation of undigestible matter is checked by the posterior sphincter of stomach. Water from undigested food is absorbed in rectum and resulting dry and elongated **faecal pellets** are egested through anus.

CIRCULATORY SYSTEM

Circulatory or blood-vascular system is of open or lacunar type. It bears a close resemblance to that of cockroach. Blood is confined to vessel (heart and aorta) during circulation. Remainder of its journey takes place within body cavity or **haemocoel**. In view of its minor role in respiration, circulatory system is much less extensive than in prawn.

[I] Heart and Aorta

There is a single, slender, tubular and pulsating **heart** situated mid-dorsally in abdomen. It is suspended in a shallow **pericardial sinus** which is separated from the main body cavity (**perivisceral sinus**) by a thin, delicate and perforated transverse partition known as **dorsal septum** or **diaphragm**. A similar **ventral diaphragm** separates a narrow **ventral sinus** or **perineural sinus** (surrounding nerve cord) from perivisceral sinus. Heart is divided into a linear row of chambers, one in each segment, and each perforated by a pair of minute, slit-like lateral openings, the **ostia**, which are closed by valves when the heart contracts. Heart is supported by a series of paired, triangular **alary muscles** that extend between terga and dorsal diaphragm.

[II] Course of Circulation

By rhythmic contrac-tions of alary muscles, blood enters the heart through ostia and pumped forward into its anterior prolongation, the **aorta**. From anterior open end of aorta, blood drains into head, perineural sinus and perivisceral sinus. From perineural sinus, it enters the legs and returns into perivisceral sinus. Wings also receive their blood supply from perivisceral sinus. From perivisceral sinus blood enters pericardial sinus through perforations in dorsal diaphragm, and finally enters the heart for next cycle. Circulation of blood through body is also facilitated by various body movements such as abdominal contractions during gaseous exchange and muscular movements during flight.

[III] Haemolymph

Blood, or haemolymph consists of a clear liquid **plasma** and white blood cells or **haemocytes**, which act as phagocytes to remove foreign organisms. It has no pigment and is not concerned with respiration but serves mainly to transport food, hormones and wastes and protects body against disease germs. Blood also contains the **fat body** as a loose network of tissue which acts as a reservoir for food. Besides, when under pressure, blood helps in hatching from egg, in moulting, in expansion of wings and in breathing movements.

RESPIRATORY SYSTEM

[I] Spiracles

Air enters and leaves the body through tiny breathing pores, the **spiracles**, located at the sides of thorax (2-pairs) and abdomen (8-pairs) in the thin membrane between segments. Spiracles are guarded by hairs to keep out dirt and foreign objects and by minute valves which can be opened or closed to regulate the flow of air at will. Closure of spiracles also helps in reducing evaporation of water.

[II] Tracheae

Each spiracle opens into a small chamber, the **atrium**, which opens into a network of fine, elastic, ectodermal air tubes, the **tracheae**, which branch and ramify to all parts of body. Main tracheal tubes form **longitudinal tracheal trunks** in the dorsal, ventral and lateral regions of body. A trachea consists of a single layer of squamous epithelial cells (**ectotrachea**) which secretes a chitinous cuticular lining on the inner side (**endotrachea**) and is shed at moulting. It forms a spiral thickening, known as **intima** or **taenidia**, to prevent collapsing of tracheal walls.

[III] Tracheoles

Finest branches of tracheae are called air capillaries or tracheoles. These are only one micron in diameter and are single celled. These have no spiral thickenings and are usually filled with tissue fluid which serves in internal respiration, like blood in other animals. Oxygen gets dissolved in this fluid and passes by diffusion to the tissue cells in exchange for the unwanted carbon dioxide.

[IV] Air Sacs

In grasshopper and certain other insects (Honey bee), some of the tracheae become expanded into large thin-walled **air sacs**, which serve for storage

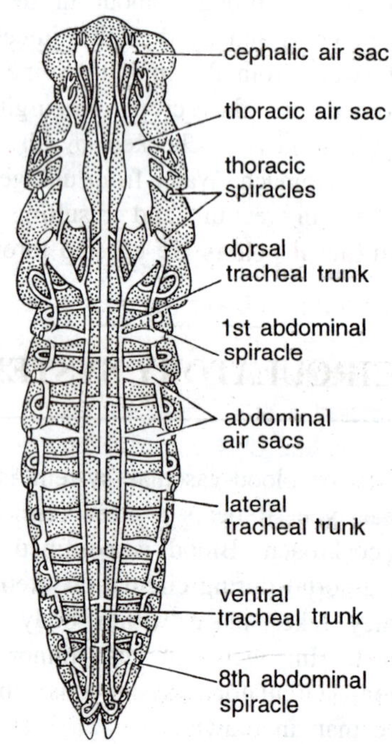

Fig. 8. Grasshopper. Respiratory system.

and better circulation of air. Air sacs are devoid of taenidia.

[V] Gaseous Exchange

The elaborate tracheal system permits diffusion of oxygen directly into the innermost tissues of body. It shows a remarkable adaptation to life in air, as oxygen has been shown to diffuse 3,00,000 times faster in air than through water, or 10,00,000 times faster than through living tissue. Rhythmic contraction and expansion of abdomen, due to muscular activity, pump air in and out of the tracheal system. In grasshopper, a definite circulation of air is maintained. Air is inhaled through the first 4 pairs of spiracles which remain open only during inspiration and are known as **inspiratory spiracles**. Air is exhaled through the remaining 6-pairs of spiracles which open only during expiration and are known as **expiratory spiracles**.

EXCRETORY SYSTEM

The principal excretory organs are a number of minute, slender, blind and thread-like **Malpighian tubules**, lying coiled about in haemocoel. They open into alimentary canal at the juncture of stomach and intestine, while remain closed at their free ends. Wall of tubules is made of a single layer of cells covered externally by a basement membrane and internally with numerous **microvilli**, forming a characteristic **brush border**. Cells extract metabolic wastes such as urates from haemolymph, convert them into uric acid and pass on to hindgut for elimination with faeces. Water is reabsorbed by the proximal absorptive part of the Malpighian tubule. Removal of dry excretory wastes is characteristic of small land animals that have only a limited water supply.

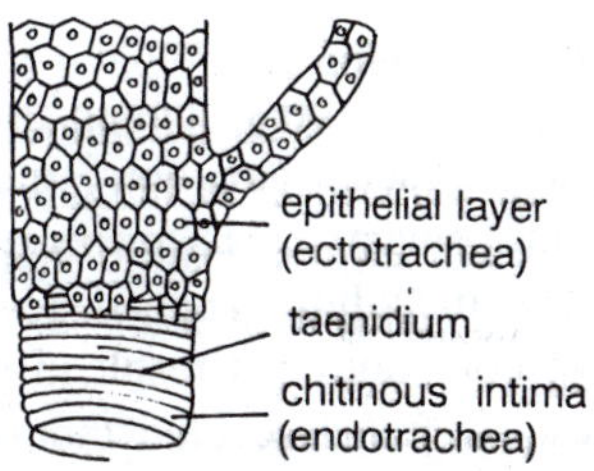

Fig. 9. A part of trachea showing spiral intima (endotrachea).

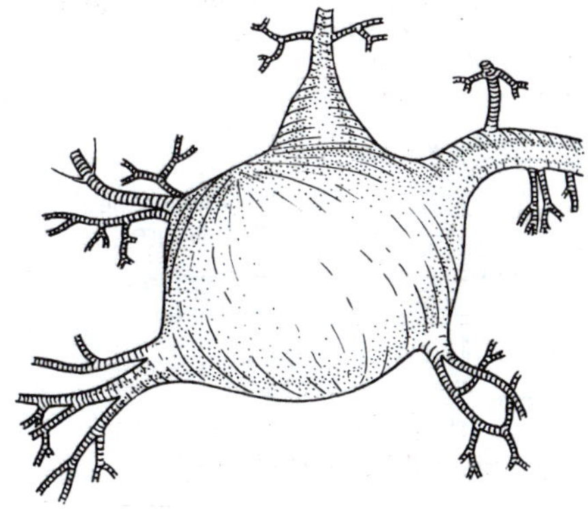

Fig. 10. An air sac.

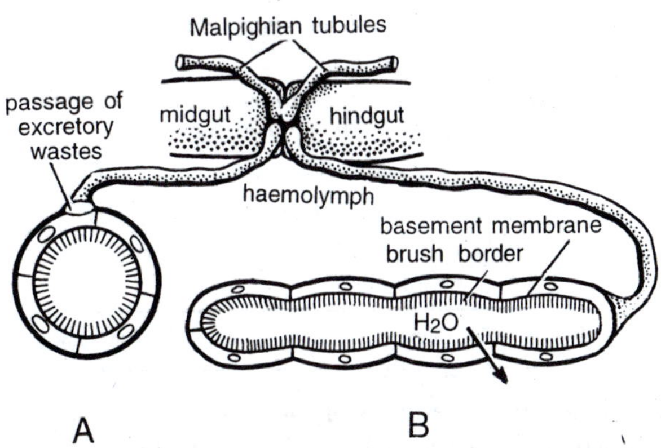

Fig. 11. Grasshopper. Malpighian tubule. **A.** T.S. **B.** L.S.

NERVOUS SYSTEM

[I] Central Nervous System

It includes a **brain** two **circum-oesophageal connectives**, a **sub-oesophageal ganglion** and a **ventral nerve cord**.

1. Brain. Brain lies dorsally above oesophagus and consists of 3-pairs of fused supra-oesophageal ganglia. Brain is distinguished into :

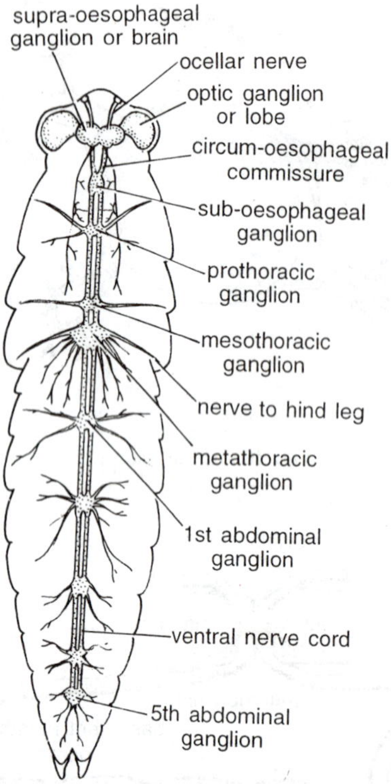

Fig. 12. Grasshopper. Nervous system in dorsal view.

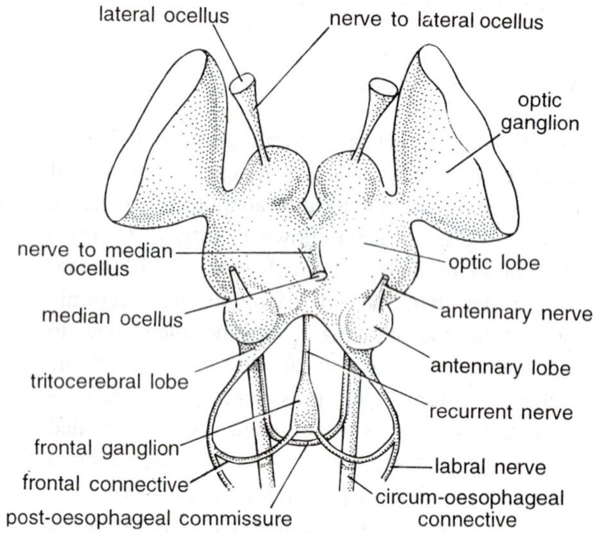

Fig. 13. Grasshopper. Brain in dorsal view.

(i) a large anterior **protocerebrum** consisting of two large optic lobes, (ii) a **deutocerebrum** comprising of two small lobes situated ventrally on the posterior side of protocerebrum, and (iii) a small **tritocerebrum** of two lobes behind deutocerebrum. A loop-like **tritocerebral commissure** runs along the inner sides of circum-oesophageal connectives and connects the two sides (lobes) of tritocerebrum. Brain chiefly serves as a relay centre which receives stimuli from sense organs and thus directs the movements of body. It also exerts an inhibiting influence but does not seem to coordinate muscular activity for the animal can jump, walk and fly even after removal of the head or brain.

2. Circum-oesophageal connectives. Brain is connected by a pair of circum-oesophageal connectives around oesophagus to a ventral sub-oesophageal ganglion.

3. Sub-oesophageal ganglion. It also represents fusion of 3-pairs of ganglia (mandibular, maxillary and labial) and innervates the mouth parts. It also plays some role in maintaining balance.

Both the brain and sub-oesophageal ganglia contain **neurosecretory cells** that secrete a hormone which activates other hormones to control moulting and some other aspects of metamorphosis.

4. Ventral nerve cord. A double ventral nerve cord arises from the hind end of sub-oesophageal ganglion and extends down the mid-ventral region of body. It bears paired or bilobed ganglia 3-thoracic and 5-abdominal. Third or metathoracic ganglion is the largest as the first three abdominal ganglia are also fused with it. Similarly, 5th abdominal ganglion is the largest representing last 4-pairs of abdominal ganglia fused together. All segmental ganglia are coordinated by nerve fibres running in the ventral cord. Besides, each ganglion independently controls the movements of its respective segment (or segments) including appendages. Thus, a severed nerve cord does not result in paralysis as in vertebrates, so that a severed thorax can walk off by itself and an isolated abdominal segment can perform breathing movements.

[II] Peripheral Nervous System

It comprises nerves arising from ganglia of central nervous system. From protocerebrum arises a pair of stout **optic nerves** and three fine **ocellary nerves**, which supply the compound eyes and ocelli, respectively. Deutocerebrum sends off a pair of **antennary nerves** to antennae. From tritocerebrum arises a pair of **labro-frontal nerves**, each of which divides into **a labral** branch to labrum and a **frontal ganglion connective** to **frontal ganglion**. From sub-oesophageal ganglion arise a pair each of **hypopharyngeal, mandibular, maxillary** and **labial nerves** and a few pairs of **cervical nerves** to neck.

From each ganglion of ventral nerve cord arise several paired nerves and supply their respective segments. Metathorax and first three abdominal segments are supplied by nerves from metathoracic ganglion, while last four abdominal segments receive nerves from the last (fifth) abdominal ganglion.

[III] Sympathetic Nervous System

Sympathetic, autonomic, visceral or stomato-gastric nervous system includes certain ganglia and nerves connected to brain. Ganglia include : (i) a small **frontal ganglion** lying above pharynx in front of brain, (ii) an **occipital ganglion** above oesophagus behind brain, and (iii) a pair of **ingluvial ganglia** in front of gastric caeca. Sympathetic nervous system controls the involuntary activities of alimentary canal, heart, aorta and reproductive organs.

SENSE ORGANS

1. Tactile organs. Tactile hairs sensitive to touch are located on various body parts, particularly the antennae, palps, cerci and distal leg segments.

2. Olfactory organs. Antennae are supplied with the organs of smell.

3. Gustatory organs. Organs of taste are present on mouth parts, particularly the palps.

4. Visual organs. Three **ocelli** are sensitive to light but may form crude images at close range.

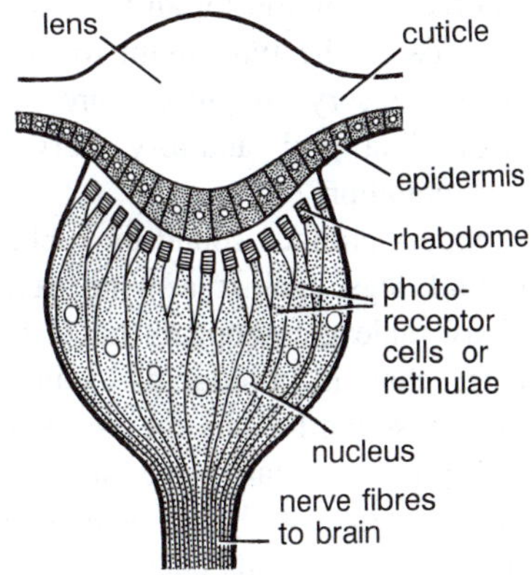

Fig. 14. Grasshopper. Ocellus in V.S.

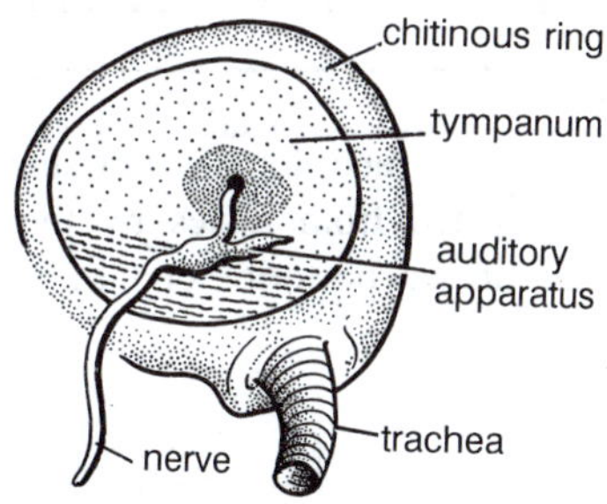

Fig. 15. Grasshopper. Interior view of tympanum and adjacent trachea.

An ocellus consists of a group of **photoreceptor cells** or **retinulae**, each ending in nerve fibre which leads to the brain. Outer end of each photoreceptor cell forms a **rhabdome**. Cuticle covering the group of photoreceptor cells forms a thick, biconvex, transparent **lens**. Real function of ocelli is not clearly known. **Compound eyes** are similar to those of prawn or cockroach in structure as well as in function.

5. Auditory and stridulatory organs. An organ of hearing is present on either lateral side in the tergum of first abdominal segment. It consists essentially of a tympanic membrane stretched within an almost circular chitinous ring. Attached to inner surface of tympanum is a whitish, sensory **auditory apparatus** or **Muller's organ** consisting of numerous three-celled columns, called **scolopidia**, which are connected with metathoracic ganglion by an **auditory nerve**. Sound waves cause the tympanum to vibrate and the resultant sensory impulses are conveyed through scolopidia and auditory nerve to the metathoracic ganglion.

Grasshoppers produce sound by rubbing tibial spines of hind legs against a wing vein which vibrates. Only male grasshoppers have the ability to stridulate, that is, producing sound. Some insects can hear sound beyond the range of human ears. Sound produced has some significance in mating as experimental study has demonstrated that insects can recognize sounds characteristic of their own species.

REPRODUCTIVE SYSTEM

Sexes are separate and female grasshoppers are easily distinguished from males by the presence of ovipositor at the posterior end of abdomen.

[I] Male Reproductive System

In male grasshopper, spermatozoa develop in a pair of **testes** situated above intestine in 3rd, 4th and 5th abdominal segments. Each testis consists of a series of slender tubules or **follicles**. These open by fine ducts, the **vasa efferentia**, in a long and coiled **vas deferens** arising ventral from testis. Two vasa deferentia meet below hindgut in 9th abdominal segment forming a common median muscular tubule or **ejaculatory duct**, which opens to outside through a large ventral copulatory organ, the **penis** or **aedeagus**. In front of ejaculatory duct, and opening into its anterior end, is a pair of **accessory glands**, each consisting of a bundle of elongated tubules. These glands discharge into ejaculatory duct a fluid in which sperms are conveyed to the female during mating. Associated with each accessory gland is a much longer tubule, the **seminal vesicle**, which stores spermatozoa to finally discharge them into the ejaculatory duct at the time of mating.

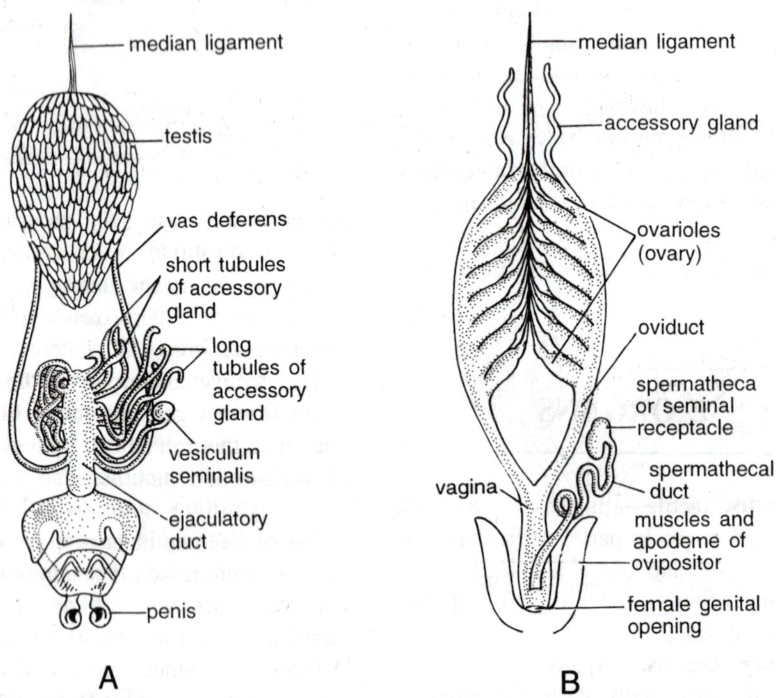

Fig. 16. Grasshopper. **A.** Male reproductive system. **B.** Female reproductive system.

[II] Female Reproductive System

Female grasshopper has two **ovaries** located over the intestine and connected to the dorsal body wall by a median ligament. Each ovary consists of 6 to 8-tapering oblique egg tubes or **ovarioles**, each containing a linear series of developing ova. As an oogonium descends down the ovariole, it enlarges and accumulates yolk and becomes a primary oocyte. It becomes surrounded by a **chorion** or **egg shell** secreted by epithelium of posterior end of ovariole. Shell has a minute pore, the **micropyle**, at its upper pole. After mating, sperm enters the ovum through micropyle to bring about fertilization.

Ovarioles of each ovary open into a wide **oviduct**. It runs along the outer margin of its ovary and then extends backwards and downwards to join its fellow of the opposite side below rectum in 7th abdominal segment. Their union forms a wide muscular tube, the **vagina**, which leads to the female genital opening between the plates of ovipositor. Leading dorsally from vaginal pore is a small tube that connects to a storage chamber known as **seminal receptacle** or **spermatheca**. Lying in front of each oviduct and opening into it is an elongated, curved **accessory** or **colleterial gland**. These glands secrete a cement-like substance that holds the eggs together when they are laid.

LIFE CYCLE OF GRASSHOPPER

1. Copulation. Mating occurs during late summer. Male clings to the back of female and inserts his penis into her vagina to discharge the seminal fluid containing sperms. The sperms pass into seminal receptacle or spermatheca where they are temporarily stored, waiting to fertilize eggs as they pass along the vagina. Mating may occur several times before the female begins to lay eggs.

2. Fertilization. Mature egg passing down the oviduct is 3 to 5 mm long. It is supplied with yolk and covered with a thin **vitelline membrane** and a tough brownish outer shell or **chorion**. A sperm coming from seminal receptacle enters the egg through a small pore in chorion, called **micropyle**. Sperm nucleus fuses with egg nucleus and a blastoderm is formed around the periphery of egg from which an embryo develops.

3. Ovipostion. Egg-laying begins at short intervals after copulation and continues into the autumn. Female uses her ovipositor to dig a short tunnel in ground, in which about 20-eggs are deposited in a cluster, connected together by a sticky secretion as an egg-pod. A single female may lay about 200-eggs in 10-clusters. Both parents die some days after mating and egg-laying.

4. Development. Embryonic development continues for about three weeks. It is then arrested and embryo enters into a rest period or **diapause** to tide over the adverse conditions of cold and food shortage in winter. With the onset of next spring, growth is resumed and the young grasshopper, hatching from egg, is called **nymph**. It is like a miniature adult and has a relatively large head and lacks wings and reproductive organs. It feeds upon vegetation and grows rapidly. But the inflexible chitinous exoskeleton cannot stretch very much, so that it is shed periodically. This complex process is termed **moulting** or **ecdysis**. Wings are developed gradually from external wing pads, and after 5 to 6 moults the young grasshopper becomes a full-fledged adult. This type of development is called **gradual metamorphosis**, as already seen in cockroach.

ECONOMIC IMPORTANCE OF GRASSHOPPER

1. As crop pests. Both nymphs and adults devour many kinds of vegetation, particularly the succulent types. Locusts migrate into new feeding grounds and may damage farm and garden plantings. The migratory locust (*Locusta migratoria*) has caused famines since Biblical times. During scarcity of food, the grasshoppers will eat cotton, woollen fabrics, wood and even their disabled fellows. Grasshoppers feeding on grass also cause heavy damage to crops and pasture lands. The annual crop loss due to grasshoppers and locusts amounts to several million rupees.

2. As food. Eggs, nymphs and adults of grasshoppers provide food for several predatory insects, spiders, frogs, reptiles, birds and mammals. They make good fish bait, either living or dead. They are sometimes used even for human food. They continue to be used as food in Mexico, Japan and Phillippines. They are commonly eaten by North American Indians and by primitive tribes in other parts of the world. In 1875, when the United States was over run by their devastatings hordes, the people were advised to make food of them, either boiled, fried or roasted.

3. As Intermediate host. Certain flukes and roundworms attack grasshoppers which serve as the intermediate hosts for these parasites. When these grasshoppers are devoured by certain birds, the parasites reach into their vertebrate hosts.

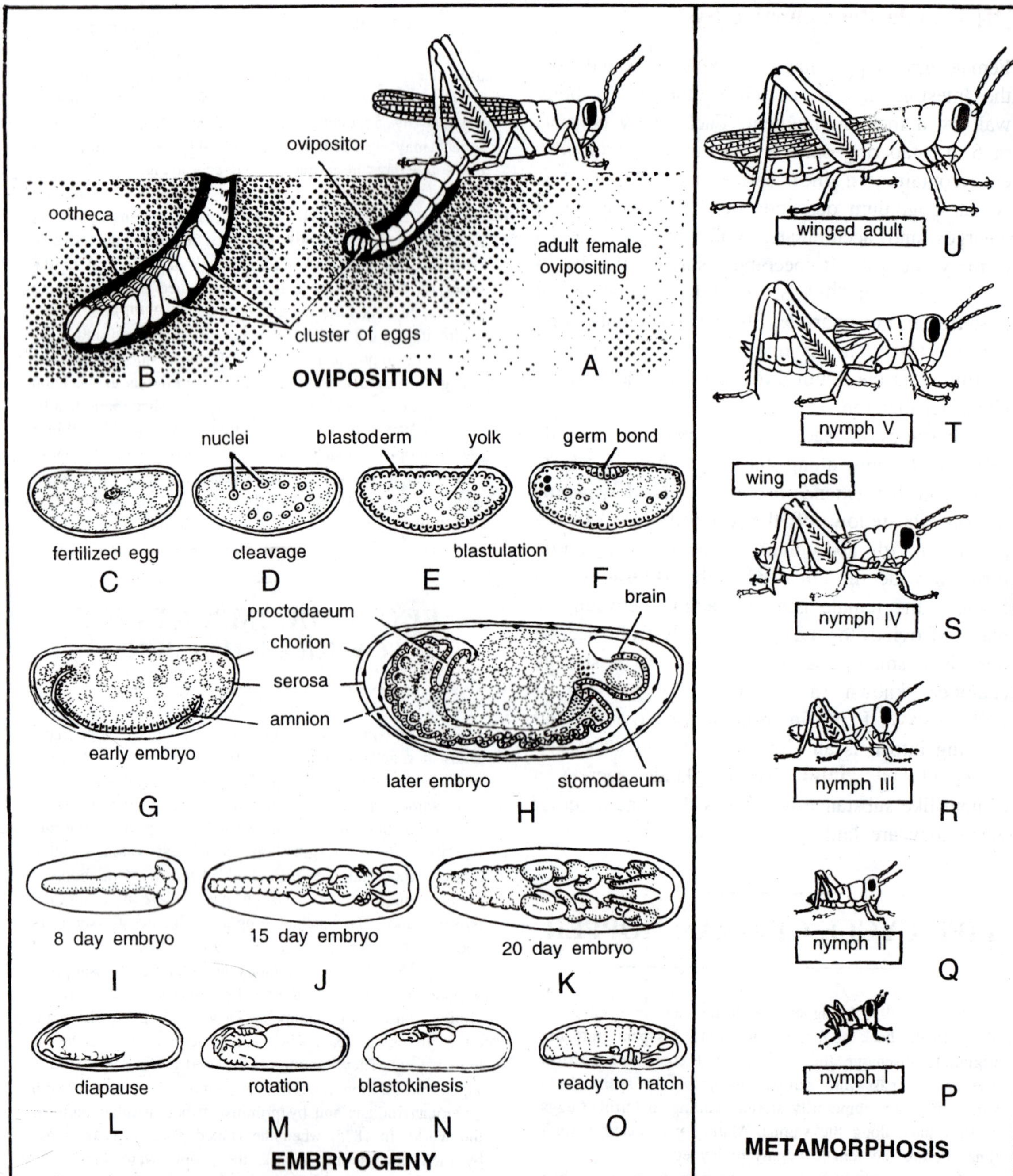

Fig. 17. Grasshopper. Stages in oviposition **(A-B)**, embryogeny **(C-O)**, and metamorphosis **(P-U)**.

GRASSHOPPER CONTROL

(i) The old method of controlling grasshoppers was to give them food mixed with arsenic or some other insect poison. Various insecticides are used either in the form of chemical sprays or dusts or poisoned baits that kill by contact or when eaten. Some recently employed insecticides are B.H.C., aldrin, chlordane, dieldrin, heptachlor and toxaphene, Methoxychlor leaves no residue harmful to man or domestic animals and hence is good for protecting fruits and vegetables and pasture lands, (ii) Ploughing of weed and stubbled fields exposes the egg masses and nymphs of grasshoppers to sun. (iii) Parasitic insects laying eggs or larvae on the grasshopper eggs, nymphs and adults also constitute a factor in their control, (iv) Both fungus and bacterial disease also take a heavy toll of grasshoppers.

LOCUST MIGRATION

Under favourable conditions in their natural breeding places, the locusts occassionally increase enormously in number and the consequent crowding induces them sometimes to form long migrations. Once, a swarm of migratory locusts flying over the Red Sea was found to be 2,000 square miles in extent. Individual locusts may travel great distance in air during migration.

LOCUST PHASE THEORY

The term 'locust' is given to a few species of short-horned grasshoppers (fam. Locustidae), which are capable, under certain conditions not fully under stood, of forming large swarms that migrate over considerable areas causing great devastations of natural and cultivated vegetations where they feed.

It was **Uvarov** who proposed the *Phase Theory of Locusts,* in 1921, as a working hypothesis to explain this irregular periodicity of locust outbreaks.

1. Two phases of locusts. According to **Uvarov** and **Zolotarevsky** (1929), species of true locusts are polymorphic, existing in a series of forms. Of these, two extreme forms or phases, which differ from each other in behaviour, appearance (colour) and morphological features, are : (i) **Gregarious** or **migratory phase** (or **phasis gregaria**), and (ii) **solitary phase** (or **phasis solitaria**). They are often so strikingly distinct that they were regarded earlier by taxonomists as separate species, thereby resulting in confusion. Their main characteristics as well as differences have been listed briefly in *Table 1.*

Phasis transiens. A third phase also occurs during the transition of population from one extreme phase to another. It is called **phasis transiens**. It has no definite form but is represented by a continuous series of transitional or **intermediate** forms. When transformation is from solitary to gregarious phase, these forms are termed **phasis**

Table 1. **Differences between Solitary and Gregarious phases of Locusts.**

Characteristics	Solitary phase	Gregarious phase
(1) Diapause	(1) Eggs develop without a diapause (suspended growth).	(1) Eggs require a diapause at low temperature for their development.
(2) Habit	(2) Do not betray gregarious habit	(2) Nymphs and adults are definitely gregarious.
(3) Colouration	(3) Nymphs extremely variable in colour (green, grey, brown, etc). Usually similar to the of colouration of their normal environment.	(3) Nymphs have a bold colour pattern, mainly black, yellow or orange irrespective of environment.
(4) Pronotum	(4) Pronotum of adult is large and with a mid-longitudinal crest or carina. There is no constriction.	(4) Pronotum of adult is shorter, and saddle-shaped. It is laterally constricted and dorsally concave with no carina.
(5) Wings	(5) Wings are normal.	(5) Wings are conspicuous and proportionately large.
(6) Femur	(6) Hind femur is relatively longer.	(6) Hind femur is relatively shorter.
(7) Maturity	(7) Attain sexual maturity without a diapause or migratory flight.	(7) Do not attain sexual maturity without a migratory flight, oweing to an imaginal diapause.
(8) Behaviour	(8) Solitary locust behaves like a large grasshopper and does little damage.	(8) Gregarious locust is very active and swarming but very destructive to green foliage.

congregans. If tendency is in opposite direction, it is termed **phasis dissocians**.

2. Experimental observations. Laboratory studies have shown that by giving suitable temperature, humidity and food, it is possible to induce one phase to develop from the other. Nymphs reared in isolation develop evident characters of solitary phase, while those reared collectively in a large number or crowd, develop into gregarious phase.

3. Factors governing phase production. Environmental factors which result in phase production in field are not adequately understood. Under specific conditions in their natural breeding places, the solitary locusts increase enormously in numbers leading to their gregarization. The conditions or factors may be meteorological (temperature) physical (water) or physiological (food, fat, hormone), etc.

In tropical regions increase in number of locusts may be provided by **dryness** or **lack of water**, and in temperate areas by a **high temperature**. As food dwindles, the solitary locusts become congregated in restricted areas (out break areas) where there is enough water to keep the vegetation green. The. consequent over crowding induces them sometimes to enter upon gregarious phase. In such overcrowded conditions, a **juvenile hormone** is secreted in large quantities by a gland in the head of solitary locusts, which speeds up reproduction and effects changes in progeny leading to characteristics of gregarious phase resulting in their swarming. Gregarious locusts migrate sometimes to very long distances in vast swarms what is known as a **locust invasion** or **plague**, under the influence of winds. Ultimately environmental conditions (*e.g.*, fall of temperature) cause the migrating swarms to settle down and breed. Distribution over an enormous area and different conditions reduce number of their generations. The area affected by swarms becomes smaller, the locust plague ends, and the locust enter upon the solitary phase.

4. Conclusion. Phase theory of locusts, although an established fact, still largely remains hypothetical. However, (i) It lends an ecological interpretation of locust outburst, (ii) On its basis, it is possible to foretel the transformation of locust into gregarious phase, (iii) Identification of favoured breeding grounds of migratory locusts can lead to their destruction by aeroplane dusting of their vast numbers, (iv) Ecological study opens up possibility of altering the character of such breeding grounds by cultivation or other means.

Important Questions

➤➤*Long answer type questions*

1. Describe the external features of the grasshopper.
2. Give an account of the circulatory, respiratory and excretory systems of grasshopper.
3. Describe the mode of fertilization and the development of the grasshopper studied by you.
4. Give an account of the digestive system of a grasshopper or locust.
5. What is locust migration ? Discuss the phase theory of locusts.

➤➤*Short answer type questions*

1. What is the zoological name of grasshopper and locust ?
2. How many ocelli are present in grasshopper. Give its structure.
3. What type of mouth parts are present in grasshopper ? Give the names of various parts.
4. How many pairs of spiracles present in grasshopper ?
5. What is the function of scolopidia in the auditory apparatus ?
6. What is the function of Malpighian tubules?
7. Explain the Muller's organ work.

➤➤*Multiple choice questions*

1. Locusts form big migratory swarms when :
   (a) they are hungry and come out in search of food
   (b) environmental conditions are very favourable and there is plenty of food in their breeding places
   (c) crops are ready to eat
   (d) large number of migratory forms accumulate at one place
2. Grasshopper belongs to the order :
   (a) Orthoptera (b) Hemiptera
   (c) Odonata (d) Dermaptera
3. Mouth parts of locusts are of :
   (a) bitting and chewing type
   (b) sponging type
   (c) siphoning type
   (d) piercing and sucking type
4. Auditory organ of locust is :
   (a) Muller's organ
   (b) tympanum attached laterally on the first abdominal segment

(c) Scolopidia numerous 3-celled columns
(d) none

5. How many segments in the body of locust :
(a) 19 (b) 20 (c) 21 (d) 13
6. The spiracles present in locust's abdomen :
(a) one pair (b) two pairs
(c) 4-pairs (d) 8-pairs
7. Haemolymph is consist of :
(a) plasma (b) white blood cells
(c) fats body (d) all
8. Pigment in blood of locust :
(a) Haemoglobin
(b) Haemocyanin
(c) Haemoerythonin
(d) None of the all
9. Malpighian tubules perform the function of :
(a) digestion
(b) respiration
(c) excretion
(d) reproduction

Answers

➤➤*Multiple choice questions*

1. (b) 2. (a) 3. (a) 4. (d) 5. (b) 6. (d) 7. (d) 8. (d) 9. (c)

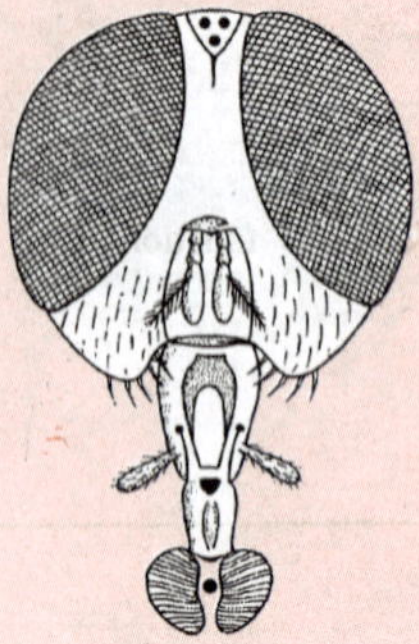

49

Musca domestica : The Housefly

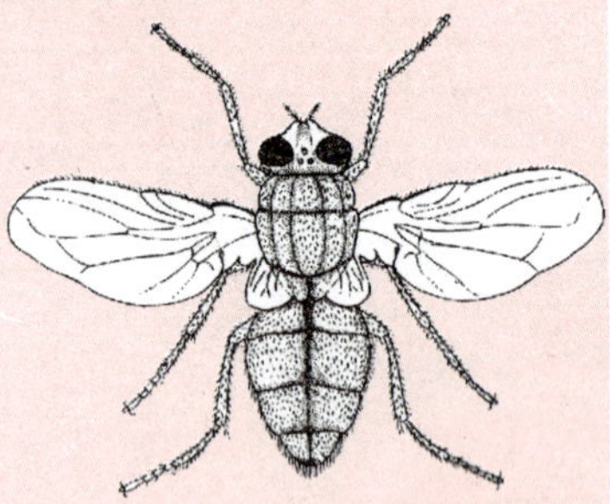

Flies are insects that belong to the order **Diptera** (Gr., **dis**, two + **pteron**,wing). These are characterised by the presence of one pair of wings, sponging or lapping type of mouthparts and short antennae. Common flies which constitute the pests of mankind throughout the world, include the flesh flies, screw-worm flies, botflies, filth flies, vinegar flies, fruit flies, etc. Wingless fly is amongs one of the very few land insects of Antarctica. It measures about 10–15 mm and is one of the largest flies found on the earth. The most familiar houseflies are belonging to the genus *Musca.* Its three species (*Musca domestica, M. nebula* and *M. vicinia*) are well represented in India.

Musca domestica

SYSTEMATIC POSITION

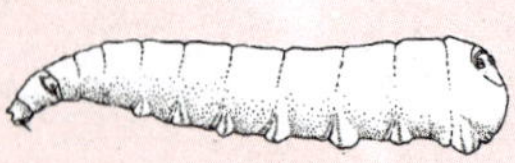

Phylum	Arthropoda
Subphylum	Mandibulata
Class	Insecta
Subclass	Pterygota
Division	Endopterygota
Order	Diptera
Family	Muscidae
Genus	*Musca*
Species	*domestica*

HABITS AND HABITAT

The common housefly, *Musca domestica* is of world-wide distribution, abundantly found around human habitation and filthy and dirty places. It is an active insect flying freely from one place to another and feeding upon human debris and other decaying organic matter. It is a dangerous house pest as it serves as a carrier of many disease-producing organisms on its body surface or hairs. The fly is **diurnal** as it shows much activity during daytime and in sunlight. It neither bites nor stings but its mere presence becomes intolerable to a conscious person, fearful of invisible enemies adhering to it.

With the help of its sponging type of mouthparts, the fly licks up liquid food. It feeds very frequently, approximately every 10 or 15 minutes. It is very sensitive and leaves the resting or feeding spot on a mere sight or sound of any object approaching it. It is also in the habit of passing out faecal matter frequently. Its faecal matter contains several microorganisms. Its development undergoes complete metamorphosis involving the egg, larva, pupa and adult stages.

EXTERNAL MORPHOLOGY

[I] Shape, Size and Colouration

Housefly is a heavily built somewhat oval insect of about 6-8 mm in length. Its dark grey body, with yellowish tinge on ventral side, has four longitudinal lines on thorax and one black streak on abdomen.

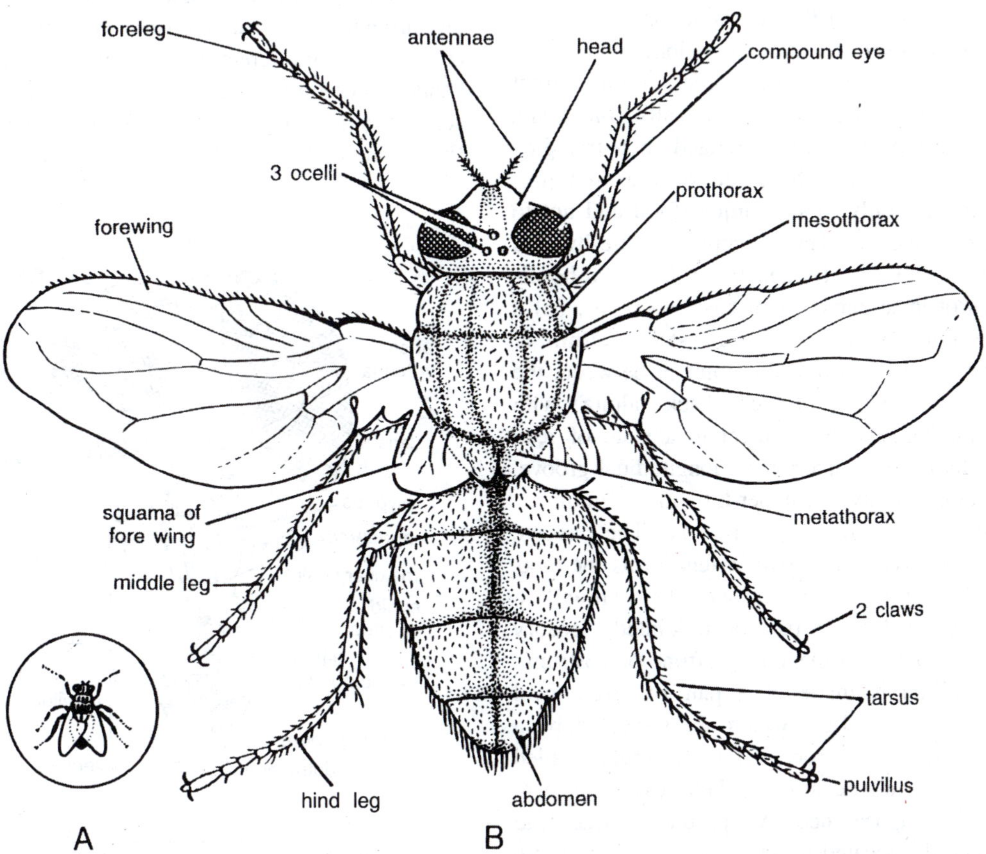

Fig. 1. *M. domestica.* External features. **A.** Fly in natural size. **B.** Fly magnified (dorsal view).

[II] Body Divisions

Body is distinctly segmented and divisible into three regions : **head**, **thorax** and **abdomen**.

1. Head. It is small, semicircular in outline and bears :

(a) Antennae. These are two short flexible and highly sensitive structures arising from the anterior end of head, close together from a median depression of the frontal exoskeletal plate or frons. Each antenna is 3-jointed, made of a small basal **scape**, the middle **pedicel** and a large distal **flagellum** which bears a three-jointed process called **arista**. Two basal segments of arista are extremely small, while the large distal segment bears two series of long hairs or **spinulae**. Antennae serve as sense organs (chemoreceptors and rheoreceptors).

(b) Compound eyes. These are two large, prominent, spherical and black structures present dorso-laterally on head. Each compound eye consists of about 4,000 visual units or **ommatidia**. They are adapted for diurnal vision.

(c) Ocelli. These are three simple, light sensitive spots located mid-dorsally on head, between compound eyes, arranged in a triangle.

(d) Mouthparts. These are of the **sponging type**, adapted only to suck liquid food and not to pierce the tissues. They comprise a fleshy and retractile **proboscis** formed by **labium**. Proboscis consists of three parts : (i) a basal **rostrum**, (ii) a middle **haustellum**, and (iii) a distal **labellum** or **oral disc**. Due to a movable joint in between, the haustellum can be folded back along with labellum, beneath the rostrum in a state of rest. Haustellum carries anteriorly a longitudinal groove. It contains two stylets, upper **labrum-epipharynx** and lower blade-like **hypopharynx**. The passage bounded between these two stylets forms the **food channel**. **Mandibles** are absent. **First maxillae** are represented by a pair of unjointed, sensory **maxillary palps** borne on the front of rostrum.

Labellum is made of a pair of large, oval and fleshy **oral lobes**, which are traversed by a network of fine grooves or channels, called **pseudotracheae**, because of their resemblance to tracheae in appearance. All pseudotracheae lead to a centrally located **mouth aperture**. Lower tip of food channel can be applied to mouth aperture so that liquid food collected by pseudotracheae can be sucked up by the pumping action of pharynx. Oral lobes are provided on the undersurface with minute, teeth-like structures for breaking up solid food and with sensory organs to examine its taste and smell.

Feeding. Fly feeds very frequently. During feeding the proboscis is extended and two expanded oral lobes of labella are placed on the exposed liquid food. If food material is solid, such as a sugar crystal, the fly first pours a little saliva or regurgitates droplets of liquid from its crop to liquefy it and then sucks the liquid which fills the tubular pseudotrachaeae by capillary action. Pseudotracheae then converge into mouth, whence the food rises up the food channel due to suction force set up in muscular pharynx. Fly is attracted by odours of ammonia and fermenting substances.

2. Thorax. Thorax is indistinctly segmented and bears a pair of strong, membranous **wings** dorsally and three pairs of jointed **legs** ventrally.

(a) Wings. Broad and transparent **forewings** are covered with minute hairs and have veins with an arrangement characteristic for the species. **Hindwings** are greatly reduced to drumstick-shaped structures, called **halteres**. These carry sense organs and serve as balancing organs during flight. These also emit buzzing sound during flight.

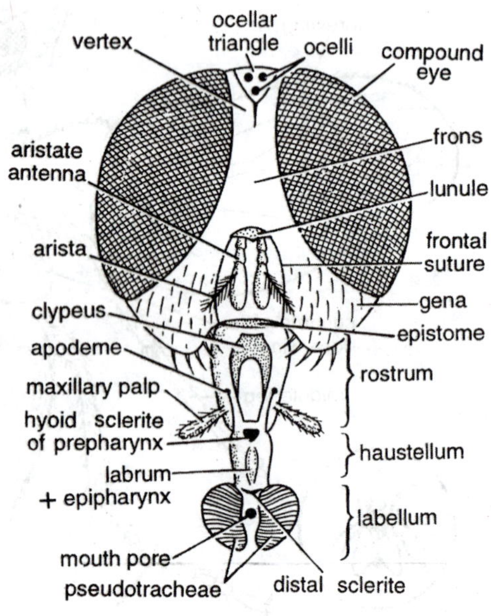

Fig. 2. Housefly. Head and mouthparts in frontal view.

Flies can fly very fast with a high frequency of wing beat (200 beats per second).

(b) Legs. Legs are thickly covered with hairs. Each leg consists of **coxa, trochanter, femur, tibia** and five-jointed **tarsus**, terminating in a pair of curved **claws** with a pair of glandular pads, called **pulvilli** between them. Each **pulvillus** is thickly beset with minute, hollow hair secreting a sticky fluid which enables the fly to climb smooth surfaces or to walk easily on walls and ceilings.

3. Abdomen. Abdomen is also covered with fine hair and comprises 8 segments in male and 9 in female. 6th to 9th segments are telescoped into 5th segment. Last 4 segments form **genital pouch** and other accessory reproductive organs in male. 9th segment in male represents a pair of claspers with an **aedeagus** (or penis) between them. 10th segment in both sexes remain fused and represented by a pair of **anal cerci**. In female, the last 4 segments form a peculiar and tubular retractile **ovipositor**. During copulation, ovipositor is inserted into genital pouch of male to receive sperms. Later, the ovipositor helps in laying eggs.

INTERNAL ANATOMY

Body cavity is a haemocoel and is very much reduced. The well developed flight muscles in body have perfected the flying mechanism. Alimentary canal is long and coiled with pharynx, oesophagus, crop, gizzard, stomach or midgut and rectum. Two small salivary glands pour saliva on food to hydrolyse starch. Digestion and absorption mainly take place in midgut. Organs of respiration consist of longitudinal tracheal trunks with transverse connective and 10 pairs of spiracles. Nitrogenous waste, which is mainly uric acid, is removed from haemolymph by four long Malpighian tubules that pour it into rectum to be expelled with excreta. Ventral nerve cord has a single large ganglion representing fusion of all the thoracic and abdominal ganglia.

LIFE HISTORY OF *MUSCA*

Sexes are separate and female is slightly bigger than male.

[I] Copulation

Flies breed during spring and summer (June to October) when they become very active. Greatest breeding activity occurs in the warmest months of August and September. Pairing or mating of male and female takes place on ground when at rest, and lasts for a few minutes. Male alights on the back of female, which inserts its ovipositor into the genital atrium of male to receive spermatozoa.

[II] Oviposition

6 to 8 days after copulation the female fly starts laying fertilized eggs on any kind of decaying organic matter such as horse manure, cow dung, chicken manure, decomposing fruits and vegetables, seeds and grasses, human faeces, etc. Eggs are deposited in a cluster by ovipositor about 12 mm below the surface of organic debris, where they get protection, darkness, warmth and moisture to develop. Each female lays about 500-600 eggs in five to six batches, each of 100-125 eggs in one season.

[III] Development and Metamorphosis

Development is indirect with complete metamorphosis (holometaboly) including 4 stages as follows :

1. Eggs. Eggs are small, nearly white and elongate with one end slightly broader than the other. Each egg is about 1 mm in length and bears two dorsal, rib-like, curved and longitudinal thickenings. Eggs hatch in 12 to 24 hours, depending upon temperature, and the larvae come out from a longitudinal split on the dorsal side of eggs.

2. Larvae. The legless and headless larvae of flies are popularly known as **maggots** or **gentles**. While growing up, the larva moults twice, thus including three stages or **instars**.

(a) First instar larva. The newly hatched or first instar larva is a soft, white, cylindrically elongated and worm-like limbless creature, tapering anteriorly. Body is covered externally by chitinous cuticle. It measures about 2 mm in length and consists of 13 segments. First or cephalic segment,

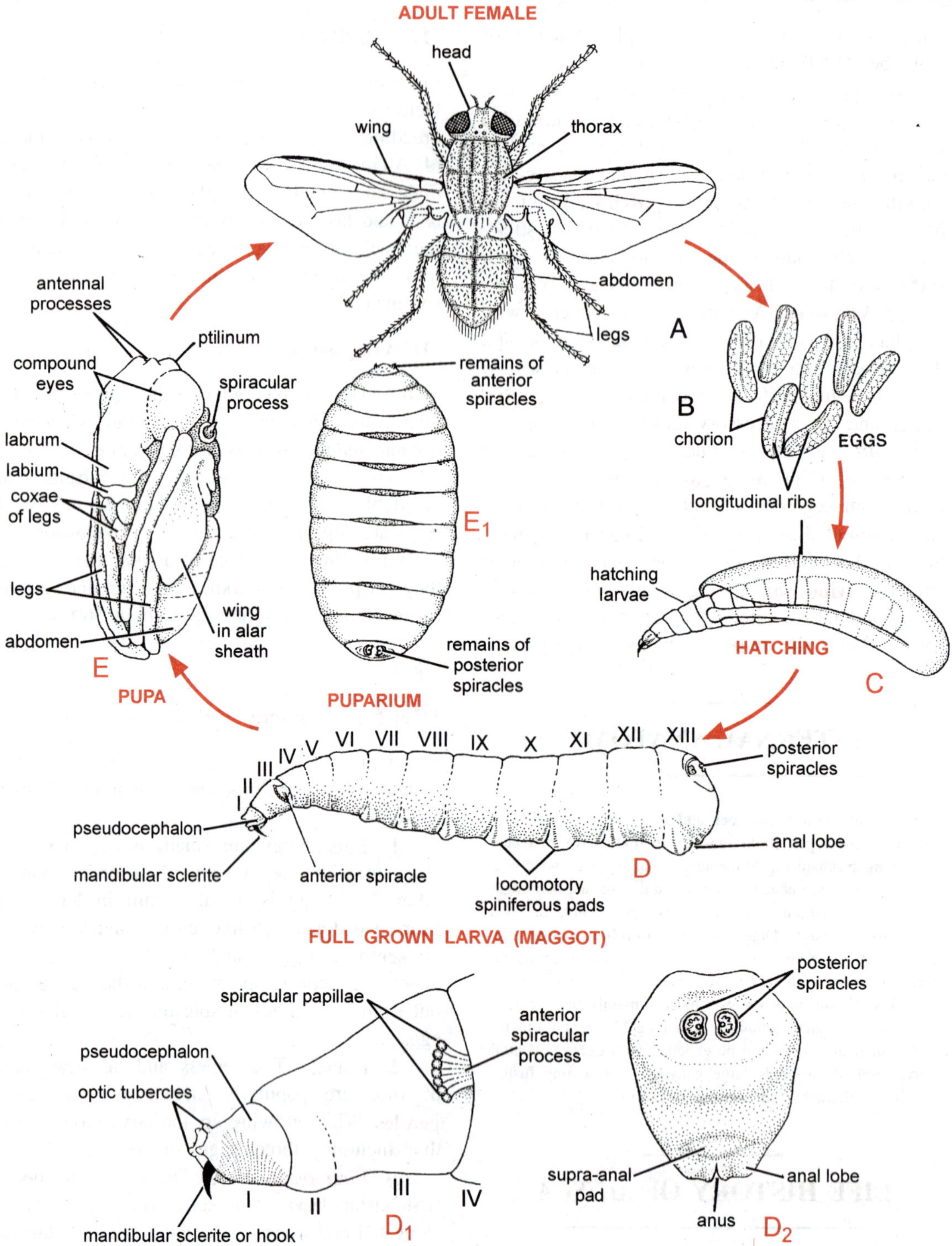

Fig. 3. *M. domestica.* Life history. **A.** Adult (female), **B.** Eggs, **C.** Hatching larva, **D.** Full grown larva, $\mathbf{D_1}$. Anterior end of larva, $\mathbf{D_2}$. Posterior end of larva, **E.** Full grown pupa, $\mathbf{E_1}$. Puparium.

which can be drawn into second, is called **pseudocephalon**. It carries the ventral **mouth**. It is surrounded dorsally by two **oral lobes** and anteriorly by a chitinous, black and hook-like **mandibular sclerite**, which helps in locomotion and in tearing up food. Ventral surface of oral lobes is traversed by numerous food channels, while each bears dorsally a pair of conical, sensory processes termed **optic tubercles**. Last trunk segment (13th) bears mid-dorsally a pair of D-shaped **posterior spiracles**, which are the external openings of body tracheae. **Anus** is located ventrally on the last segment between two prominent and locomotory **anal lobes**. On the anterior ventral border of each of 6th to 12th trunk segments, there is a semilunar pad beset with minute, stiff spines. **Spiniferous pads** are locomotory organs and help the larva in moving backwards and forwards. First instar larva is very active feeding voraciously on decomposing liquid organic matter, so that it rapidly grows in size.

(b) Second instar larva. After a day or so, the larva undergoes first moulting and becomes the second instar larva. It also develops a pair of fan-shaped **anterior spiracles** dorsally upon the third segment.

(c) Third instar larva. After a day or two, the larva undergoes second moulting and becomes the third instar larva, which lasts for 3-5 days. The full grown larva measures 8 to 12 mm in length and looks creamy-white in colour. It avoids light and is greatly affected by humidity, temperature and food. It now transforms itself into a differently shaped creature, the **pupa**,

3. Pupa. Before transforming to the pupal stage, the larva seeks a drier area of breeding material. Body of larva shrivels and the anterior segments are withdrawn, so that it becomes cylindrically ovoid with both the ends evenly rounded. Larval skin becomes dark brown and forms a protective covering, called **pupal case** or **puparium**. Pupa has no mouth and anus and it respires through one pair of anterior and one pair of posterior spiracles.

Externally the motionless pupa looks inactive but internally a rapid breaking-down process, called **histolysis**, takes place as the **phagocytes** break down most of the larval tissues into a creamy mass. Certain groups of cells, called **imaginal discs** or **buds**, now feed upon the elemental cream and produce organs of the adult. This rebuilding process is known as **histogenesis**, which is completed in 4-5 days in favourable conditions but in several months during cold weather.

4. Adult. By the end of histogenesis, pupa transforms into the adult fly or **imago**. It hatches by knocking out the anterior end of puparium in the form of a lid with the help of a bladder-like structure, called **ptilinum**, situated on the front part of its head. The newly-emerged imago is colourless with small delicate wings. Very soon the ptilinum is withdrawn into head, the normal colour is attained, the wings become fully expanded and hard on exposure to air, and the imago starts on its flight as a full-fledged fly.

ECONOMIC IMPORTANCE OF HOUSEFLY

Houseflies are associated with filth of all kinds. At one time they were allowed to walk freely over the human food, as they were considered to be very useful scavengers, feeding on all kinds of debris and domestic refuge. But in the light of modern scientific knowledge, people have been so thoroughly aroused to the menace of these dangerous insects that they are no longer tolerated.

1. Conveyance of diseases. Housefly is the most important single agent in spreading disease. It is, therefore, one of the deadliest enemies of humanity. It contaminates our food owing to its insanitary habits. On one hand it feeds on human excrement, sputum, exudates of sores, diseased bodies, manures etc., and on human food on the other hand. Rough and hairy surface of its body and legs are well adapted for carrying disease-causing bacteria and protozoans from refuge to articles of food and drink. It is believed to disseminate the agents of a number of fatal diseases such as typhoid fever, dysentery, cholera, yaws, tuberculosis, leprosy, trachoma, gonorrhoea, anthrax and others. Eggs and larvae of several parasitic worms, like *Taenia solium* and *Ascaris lumbricoides,* may also be transmitted by housefly. It also serves as an intermediate host of **poultry tapeworm** and may transfer **roundworm** from one horse to the other from their sores, lips and eyes. Conveyance of disease germs may be brought about in the following two ways :

(a) External transference. When the fly sits on debris, refuge or excreta, the disease germs become entagled with the numerous hairs covering its body, wings, legs and mouthparts. These germs are then dropped in or rubbed with the human food, or are scrapped or washed off as they walk over sugar or drown in tea or milk.

(b) Internal transference. The disease germs are ingested with food and they live and multiply in the crop of fly. They are transferred to human food either with faecal spots or with regurgitated food.

2. Myiasis in man. Invasion of a part of body of man or other animals by the eggs or larvae of flies is called **myiasis**. *M. domestica* is known to oviposit or larviposit on and in human tissues. The flies sometimes lay eggs on open wounds where the maggots hatch out. Maggots have been isolated from inflamed leg of a man of 80 suffering from varicose veins. Sometimes eggs are laid in the nasal passages, mouth, anus, vagina and orbit of eye, causing serious disorders. Eggs and larvae are very often ingested with contaminated food, causing intestinal myiasis. This causes intestinal disorders.

3. Diseases spread by housefly. The following diseases are spread by housefly : Anthrax, Leprosy, Typhoid, Gangrene, Trachoma, Cholera, Diarrhoea, Plague, Dysentry, Tuberculosis and Gonorrhoea etc.

FLY CONTROL

We have seen that housefly is a constant menace to public health. It has been said that by spreading diseases of man, cattle and other animals, they cause more deaths than the reptiles and the wild beasts put together. Therefore, it is very essential that the growth and increase of houseflies and the infection caused by them must be stopped. It can be done in the following way :

[I] Prevention of Breeding

Elimination of breeding media is the first and most important step in fly control. Its significance can be well understood in view of their great power of reproduction. According to **Prof. C.F. Hodge**, a single pair of houseflies may produce at the normal rate such a huge number of flies within only five month that they would cover the earth 14 meters deep. To eliminate breeding grounds and to kill immature forms is the easiest, cheapest, and therefore, the best method of fly control. Measures for prevention of fly breeding may be either physical or chemical.

1. Physical methods. The garbage, stable manure and other organic materials, in which flies may lay eggs, should be kept covered and systematically removed at least once a week. The kitchens, compounds, streets, markets and other crowdy places must be kept free of filth. In villages and on farms, the manure should be either kept covered in manure pits or spread thinly over fields and pastures, every two or three days, to dry. Exposure to sunlight kills the eggs and larvae. Outdoor toilets should be eliminated or made fly-proof as far as possible.

2. Chemical methods. If rubbish or manure is accumulating near human dwellings and cannot be disposed off immediately, it should be treated chemically with lime, crude oil, copper sulphate, formaldehyde or any other poisonous substance to prevent breeding of flies. Manure to be used as a fertiliser should be treated with a solution of borax (454 gm in 72 litres of water) or creosote (454 gm in 6 litres of water) or sodium fluo-silicate, so that useful bacteria may not be killed. However, the practice of treating manure with borax or other chemicals is not too satisfactory. Regardless of measures employed there will be some breeding and supplementary control measures needed.

[II] Destruction of Adult Flies

1. Mechanical methods. Flies can be attracted by **baits** or **traps** and then killed by other means.

They are easily attracted by banana and milk or cheese or brown sugar, in equal parts kept wet and fermenting. **Electrified traps** and **electrified screens** also aid in reduction of fly population.

2. Chemical methods. Use of **fly-papers** and **fly-poisons** is also very helpful. Blotting papers and cotton wicks soaked in a mixture of formaline, sugar, milk and water, attract flies which lick them and die. A hot mixture of castor oil and resin painted on papers, wires and bars is quite useful in killing flies inside houses. **Repellants** like creosote oil and borax solution, sprinkled in and around animal sheds, keep away flies. A mixture of coal-tar and carbon disulphide applied to the wounds of animals, proves useful. **Fumigation** with dry neem leaves or carbolic acid repels flies. **Anti-fly candles** can be burnt in places from where flies are to be removed. Household **sprays** containing poisons like pyrethrums, derris and lethane give a quick knock-down to the flies that get into houses, and provide temporary relief. DDT residual sprays seem the best of all insecticides for indoor use, which kill not only the houseflies but other common house pests like moths, bed-bugs, mosquitoes, etc., and are also harmless to man. However, DDT and other chlorinated hydrocarbon insecticides, like chlordane, lindane, toxaphene and methoxychlor are not very satisfactory in the control of DDT-resistant flies. Certain organic phosphorus compounds of sufficiently low mammalian toxicity have been found to be efficient in housefly control. Malathion is currently recommended for use under some conditions.

[III] Protection from Infection by Flies

Flies and filth go together so that all filth must be kept well isolated and covered. Cleanliness must be observed strictly, especially with small children and the flies must not be allowed to sit on their bodies. Houses must be properly screened and food, fruits and vegetables must not be exposed to flies in houses or in markets. Sweets, fruits, milk, vegetables and other eatables, on which flies are allowed to sit freely, must not be used. Excreta containing disease germs must be burnt or treated chemically so that flies may not pick up germs from it.

Important Questions

➤➤*Long answer type questions*

1. Give an account of the morphology of the common housefly.
2. Describe the life history *of Musca domestica* and measures for its control.
3. Write short notes on : (i) Histolysis, (ii) Histogenesis, (iii) Maggots, (iv) Mouth parts of housefly, (v) Myiasis.

➤➤*Short answer type questions*

1. What is the type of mouth parts in housefly ?
2. What is the function of halteres in housefly ?
3. Draw and label mouth parts of *Musca.*
4. Name the disease spread by housefly.
5. Which larva is amphipneustic ?

➤➤*Multiple choice questions*

1. The adult housefly which emerges from the puparium has :
   (a) no wings (b) two pairs of wings
   (c) one pair of wings
   (d) one pair of wings and pair of halters
2. The larva of *Musca domestica* is known as :
   (a) Maggot or gentles (b) Caterpillar
   (c) Wrigglers (d) Tumbler
3. The housefly (*Musca domestica*) undergoes metamorphosis in the following order :
   (a) egg-pupa-larva-adult (b) egg-larva-nymph-adult
   (c) egg-larva-pupa-adult (d) egg-nymph-adult
4. The type of metamorphosis exhibited by *Musca domestica* (Housefly) is :
   (a) complete (b) incomplete
   (c) parthenogenetic (d) retrogressive
5. Housefly differs from mosquitoes in having :
   (a) hind legs resting on surface while sitting
   (b) a proboscis which is quite broad
   (c) all the three pairs of legs present on mesothorax

6. Larva of housefly is called :
(a) nymph (b) grub
(c) tumbler (d) pupa
7. Male mosquito can be identified from female mosquito by :
(a) body size (b) size of head
(c) size and number of bristles over antennae
(d) longer legs
8. Larvae of housefly respires by :
(a) tracheae (b) skin
(c) gills (d) anaerobically
9. Larva of housefly is suitable to live in :
(a) water (b) dung
(c) vegetable (d) soil
10. *Musca domestica* is the zoological name of :
(a) mosquito (b) housefly
(c) honey bee (d) silkworm
11. Housefly can take its food only when it is :
(a) semisolid (b) semiliquid
(c) solid (d) liquid
12. Housefly lays eggs in :
(a) water
(b) stagnant water
(c) decomposing organic matters
(d) comb
13. In housefly pseudotracheae are found in :
(a) trachea (b) thorax
(c) larva
(d) mouth part
14. The first instar larva of the housefly :
(a) is limbless and has one pair of posterior abdominal spiracles
(b) is limbless and has one pair of anterior prothoracic spiracles
(c) is limbless and has one pair of anterior prothoracic and one pair of posterior abdominal spiracles
(d) has three pairs of legs and one pair of posterior abdominal spiracles
15. Disease spread by housefly is :
(a) elephantiasis (b) dengue
(c) gangrene
(d) bubonic plague
16. The commonest Indian housefly :
(a) *M. domestica* (b) *M. nebula*
(c) *M. vicinia* (d) all
17. How much ommatidia in a compound eye :
(a) 2000 (b) 3000
(c) 4000 (d) 5000
18. Housefly is :
(a) herbivorous (b) carnivorous
(c) saprophagous (d) none
19. The total larval period in housefly is :
(a) 5-8 days (b) 6-8 days
(c) 7-9 days (d) 4-10 days
20. The disease caused by maggot :
(a) anthrax (b) cholera
(c) myiasis (d) plague
21. Larvae of housefly respire by mean of :
(a) spiracle (b) tracheal
(c) gill (d) book lungs
22. The antennae present in housefly are :
(a) plumose (b) pilose
(c) filiform (d) aristate
23. Housefly possesses :
(a) two pairs of wings (b) one pair of wings
(c) three pair of wings (d) none of these
24. The reduced hindwings of the housefly are called :
(a) Elytra (b) Tarsi
(c) Halteres (d) Cerci
25. Mouthparts of housefly are adapted for :
(a) absorbing fluid like a sponge
(b) pierching and sucking
(c) cutting and biting
(d) sucking flower nector
26. The wings of housefly are located on :
(a) cephalothorax (b) prothorax
(c) mesothorax (d) metathorax
27. Maxillary palps in housefly is :
(a) unsegmented (b) 3-segmented
c) 5-segmented (d) absent
28. In housefly, pseudotracheae are found in:
(a) rostrum (b) basiproboscis
(c) haustellum
(d) labellum
29. Housefly feeds on sugar by :
(a) sucking sugar crystals through pseudotracheae
(b) dissolving sugar crystals by saliva and sucking the solution
(c) crushing sugar crystals with mandibles before sucking it
(d) crushing sugar crystals with its legs and then transferring it to pseudotracheae
30. Which of the following is totally absent in the mouthparts of housefly ?
(a) labrum (b) mandibles
(c) labium (d) hypopharynx

Answers

➤➤*Multiple choice questions*

1. (d)	2. (a)	3. (c)	4. (a)	5. (b)	6. (b)	7. (c)	8. (a)	9. (b)	10. (b)
11. (d)	12. (d)	13. (d)	14. (a)	15. (d)	16. (b)	17. (c)	18. (c)	19. (b)	20. (c)
21. (b)	22. (d)	23. (b)	24. (c)	25. (a)	26. (c)	27. (a)	28. (d)	29. (b)	30. (b).

50

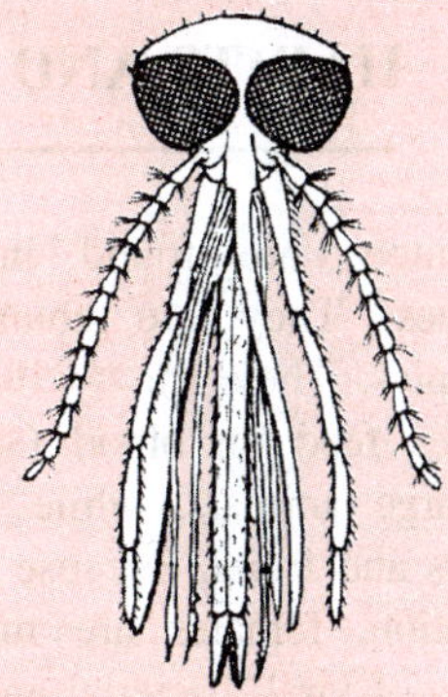

Culex and *Anopheles* : The Mosquitoes

About a billion (10^{18}) (1,000,000,000,000,000,000) insects are alive at any time. That's about 160 million insects for each person on the earth! Everyone is familiar with mosquitoes because of their painful and irritating bites, especially at night. Like houseflies, mosquitoes also belong to the order **Diptera** of class Insecta. However, they differ from flies in having a shorter body, longer legs, piercing and sucking mouth parts and many-segmented antennae. Important genera are *Anopheles* (fam. Anophelidae) and *Culex* and *Aedes* (fam. Culicidae). Following description pertains mainly to *Culex* and *Anopheles*. There are approximately, 3,500 species of mosquitoes are known to us.

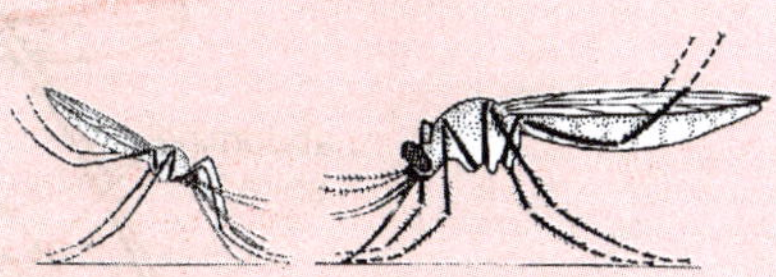

Culex and *Anopheles*

SYSTEMATIC POSITION

Phylum	Arthropoda
Class	Insecta
Subclass	Pterygota
Division	Endopterygota
Order	Diptera
Genus	*Culex* or *Anopheles*

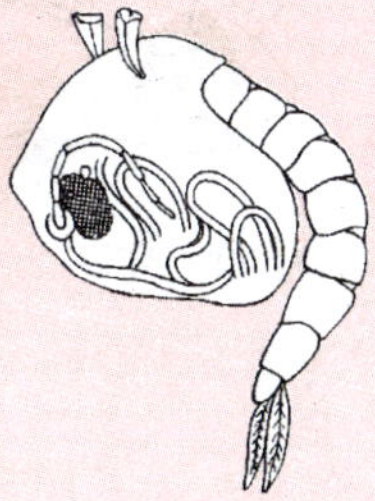

HABITS AND HABITAT

Mosquitoes are found in damp and marshy localities. They are abundant in tropics and subtropics. They are **nocturnal** in feeding habit. Females feed on blood (**sanguivorous**) of man and large animals, while males suck juices of flowers and fruits. Because of their blood-sucking adaptation, females are medically important as carriers of viral, bacterial and protozoan infections. *Anopheles* carries malaria, *Culex* carries dengue and elephantiasis and *Aedes* carries yellow fever. Adult mosquitoes fly well and can cover great distances. They are able to fly without rest for 2 or 3 hours and easily cover 20 to 30 kilometers in a night. Females become aggressive after copulation. They lay eggs in water where development of larvae and pupae takes place.

Two common mosquito genera, (*Anopheles* and *Culex*) can be easily identified by their sitting postures. When sitting, the abdomen of *Anopheles* is always held at an angle to the surface while that of *Culex* is held parallel to the surface.

EXTERNAL MORPHOLOGY

[I] Shape, Size and Colouration

Body of mosquito is small, soft, slender and covered with scales. It measures 3 to 6 mm in length. Body colour is greyish black. Wings of

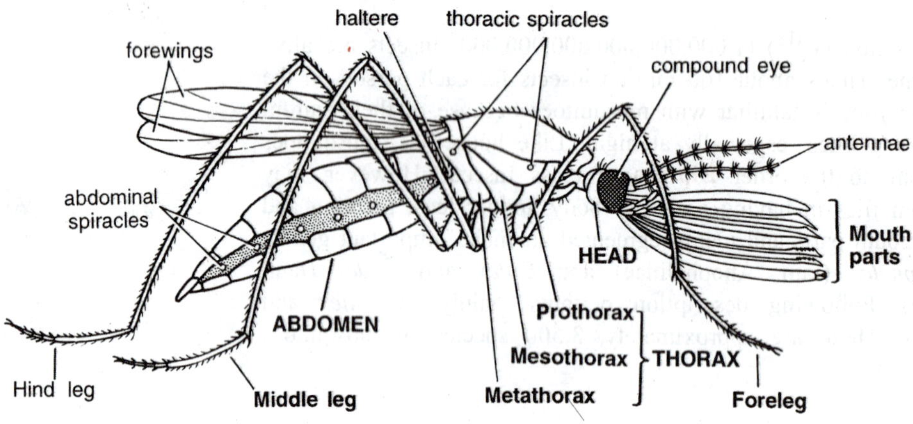

Fig. 1. *Culex.* Female in lateral view.

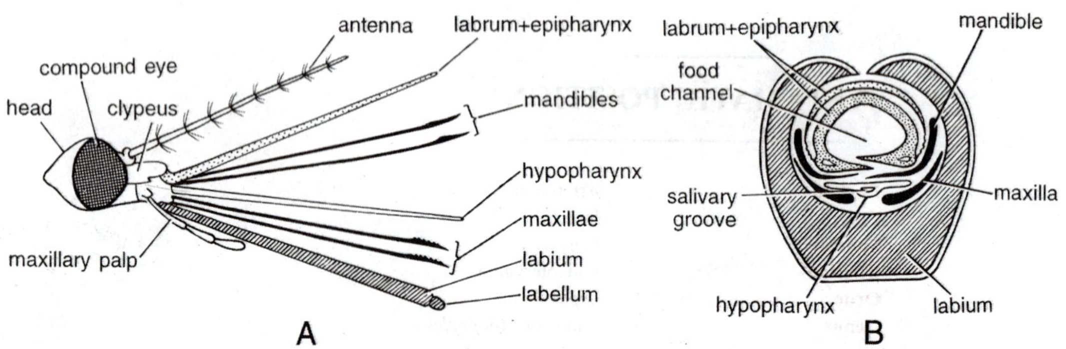

Fig. 2. *Culex.* Head and mouth parts of female. **A.** In lateral view, **B.** In T.S.

Anopheles carry dark spots. *Aedes* has a black and white striped body.

[II] Divisions of Body

Body is segmented and like other insects, divided into 3 distinct parts : **head**, **thorax** and **abdomen**.

1. Head. Small and almost spherical head is connected with thorax by a short, narrow **neck**. It bears the following parts :

(a) Compound eyes. These are two enormous, kidney-shaped structures almost touching each other.

(b) Antennae. Below eyes, in front of head, a triangular plate, called **clypeus**, projects much forward and bears a pair of long, many-jointed and feathery **antennae**, bearing delicate hair arranged in several whorls. Antennae of male are densely clothed with long hair (**plumose antennae**) while those of female are with only a few short hair (**pilose antennae**).

(c) Mouth parts. Mouth parts form a long piercing and sucking tubular **proboscis**.

(*i*) *Labium.* In female mosquito, **labium** forms the **proboscis-sheath**. It ends in a pair of white, pointed lobes, the **labellae**, bearing tactile hair. It bears a dorsal groove, which lodges all the other mouth parts modified into six needle-shaped **piercing stylets**, meant for puncturing the host skin.

(*ii*) *Labrum-epipharynx.* A long, pointed and stiff rod is called **labrum-epipharynx**. It is formed by the fusion of labrum or upper lip and epipharynx, a projection from the roof of mouth. Ventral surface of labrum-epipharynx bears a longitudinal groove or **food channel**.

(*iii*) *Hypopharynx.* Food channel is closed below by a long pointed and flattened plate like a double edged sword, called **hypopharynx**. It is traversed by a minute median **salivary channel**, down which flows the secretion of salivary glands.

(*iv*) *Mandibles and maxillae.* Within proboscis also lie paired **mandibles** and **maxillae (galeae)**, forming long and needle-shaped stylets. Mandibles end in sharp tiny blades while maxillae into saw-like blades bearing teeth. A pair of long and segmented **maxillary palps**, bearing tactile hair, project from sides at the base of proboscis.

In male mosquito, the piercing organs are reduced. Mandibles and maxillae are absent and hypopharynx fused with labium.

Male and female sexes of mosquitoes can easily be determined by the form of their antennae and maxillary palps. In *Anopheles,* palps are nearly as long as proboscis in both sexes, but in male, their ends are club-shaped. In *Culex,* palps are much shorter than proboscis in female, but slightly longer in male.

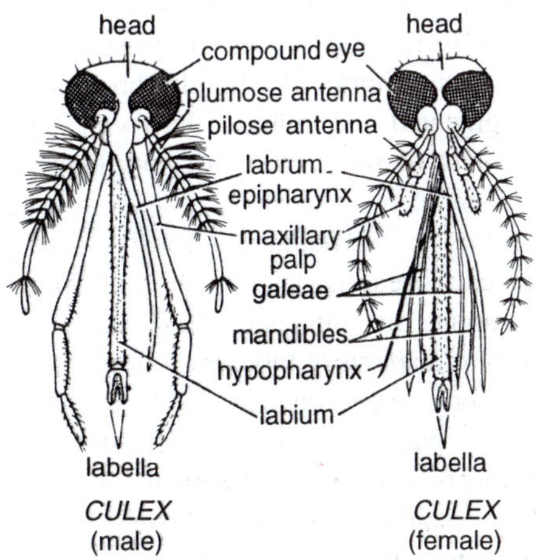

Fig. 3. *Culex.* Head and mouthparts.

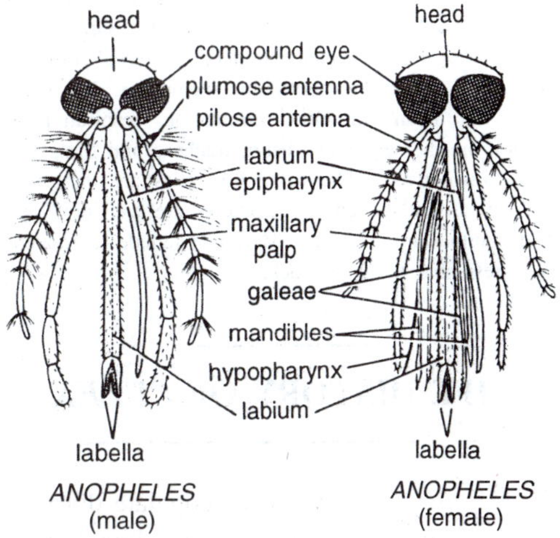

Fig. 4. *Anopheles.* Head and mouthparts.

Feeding. To feed upon blood meal, the female presses the labellae of proboscis against the skin of man. Maxillae and mandibles pierce deep into skin in order to puncture the blood capillaries. Saliva, acting as an anticoagulant, is injected down the hypopharynx into the wound. Labrum-epipharynx and hypopharynx form a feeding tube to suck up blood. Pharyngeal suction force is believed to be responsible for sucking of blood.

2. Thorax. Thorax is made of a very small **prothorax**, a very large **mesothorax** and a small **metathorax**, each bearing ventro-laterally a single pair of long, slender and jointed legs. A single pair of wings are attached dorso-laterally to mesothorax. These are narrow, membranous and covered with scales in a characteristic manner that differs from species to species. Metathorax bears a pair of small drumstick-shaped or club-shaped processes, called **halteres** or **balancers**. These represent the rudimentary hindwings, useless for flight and probably sensory in function. These are supposed to produce characteristic sound.

3. Abdomen. The abdomen is slender and 9 segmented. It bears **anus** on 8th segment and a **genital pore** on 9th segment.

INTERNAL ANATOMY

The alimentary canal is constructed on the typical plan as described for cockroach. Crop has the capacity of absorbing water and thus concentrating blood. Gizzard or proventriculus has no cuticular teeth. Internal lining of midgut does not secrete peritrophic membrane. As in other insects, the organ for blood circulation, respiration and excretion are the segmented heart, tracheae and Malpighian tubules, respectively. The compound eyes are very much prominent and other sensory structures are well developed. All mosquitoes are dioecious. Male and female reproductive organs resemble those of cockroach.

LIFE HISTORY OF *CULEX*

1. Copulation. The two sexes copulate during the flight. Female takes a meal of blood before mating. Male gets attached to the female by her high pitched whine. Female mates only once in her life time which is enough for laying eggs 4-5 times of her remaining life for a month or two.

2. Oviposition. Female takes a meal of blood before laying eggs. Before laying, she fertilizes her eggs with the spermatozoa stored in her spermatheca. She lays nearly 200 to 400 eggs in a cluster, in standing polluted water of ponds, ditches, marshes, street gutters or water barrels. By her hindlegs, she arranges the eggs, which are cemented together to form a small **floating raft**.

3. Eggs. Eggs are white, elongated and cigar-shaped. They are pointed at their upper ends and broad and rounded at the base. Air bubbles, entangled among the upper narrowed ends, assist in floating. After some time the eggs become dark grey.

4. Larva. Eggs hatch within 2 to 3 days, and a small transparent larva, measuring about 1 mm, emerges from the lower end of each egg. Aquatic larva of mosquito is popularly known as the **wriggler**, as it swims actively in water by wriggling movements. Its body is distinctly divided into : head, thorax and abdomen.

(a) Head. The large head of larva bears a pair of **compound eyes**, a pair of **ocelli**, a pair of two- jointed **antennae** and chewing **mouthparts**. Larva feeds upon minute aquatic plants and animals. In front of mouth, a pair of special branches of hair, the **feeding brushes**, move rapidly to set up a current of water, carrying small particles of food into mouth. Mouth has a pair of **mandibles** and a pair of **maxillae**.

(b) Thorax. Thorax is slightly broader than head, unsegmented and limbless, but bears 3 pairs of lateral **tufts of hair**, each tuft springing from a small tubercle.

(c) Abdomen. Abdomen consists of 9 segments, each bearing lateral **tufts of bristles**. Dorsally, upon the 8th abdominal segment is located a long respiratory tube or **siphon**. Tracheal system of body opens on outside through a pair of **spiracles**, placed at the tip of this siphon. Spiracles can be closed by 5 terminal **flaps**. Larva thrusts its respiratory siphon above the water surface at intervals to breathe. Besides, 4 small leaf-like **tracheal gills** are attached to the last abdominal segment, surrounding the **anus**. These

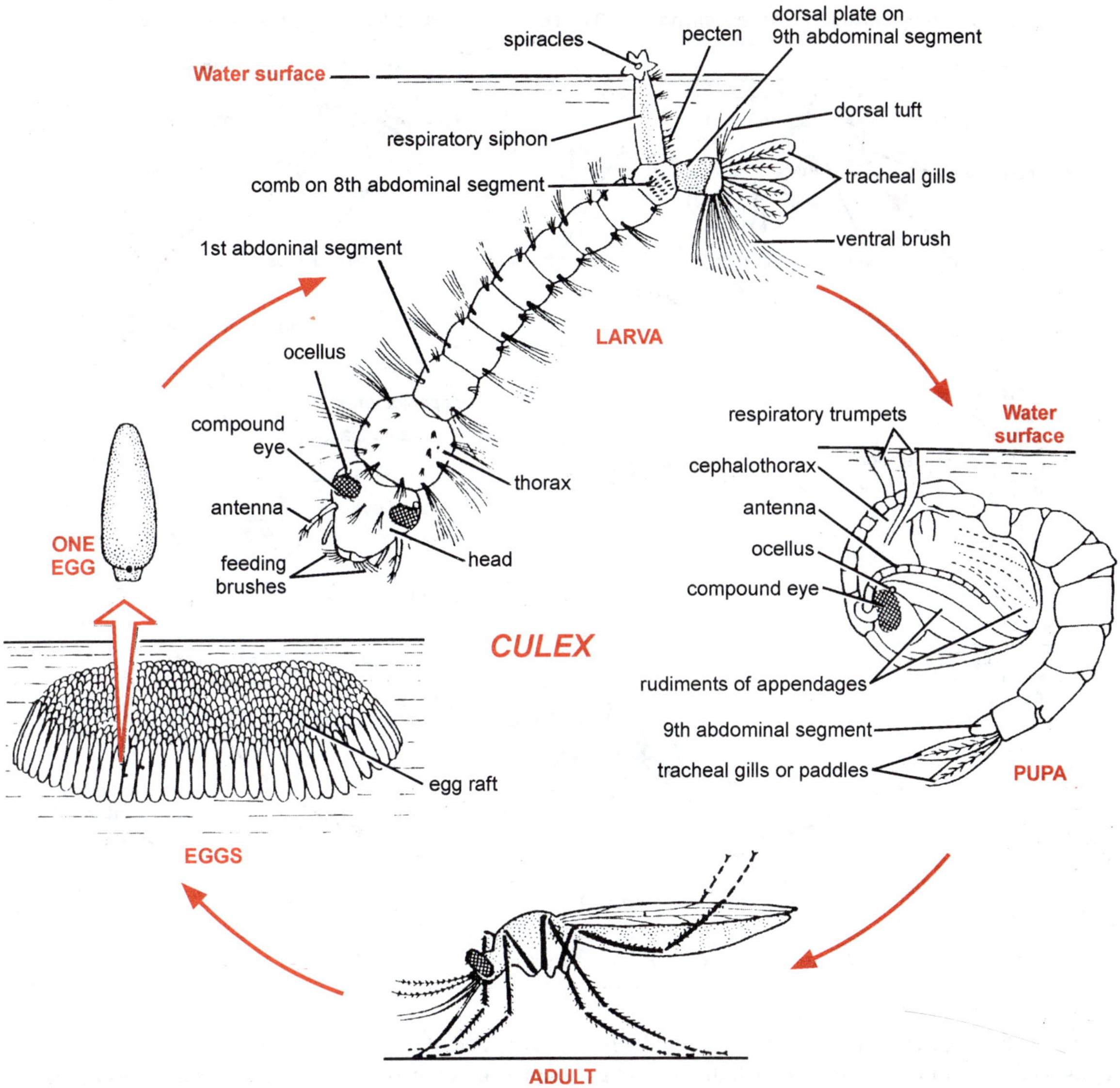

Fig. 5. *Culex.* Stages of life cycle.

tracheal gills contain tracheae and probably take oxygen dissolved in water. 9th abdominal segment also bears two tufts of bristles, **ventral brush** and a **dorsal brush**, which help in swimming. When at rest, the larva hangs with its head downwards at an angle in water with the top of its respiratory siphon projecting above the surface of water.

In a fortnight or so, the larva moults for the 3rd or 4th time and measures 1.25 mm in length. At the last moult, it changes into **pupa**.

5. Pupa. Aquatic pupa of mosquito is commonly known as **tumbler**. Unlike most insect pupae, it is quite active. But like them, it does not feed. It has a comma-shaped body, consisting of a swollen unsegmented **cephalothorax**

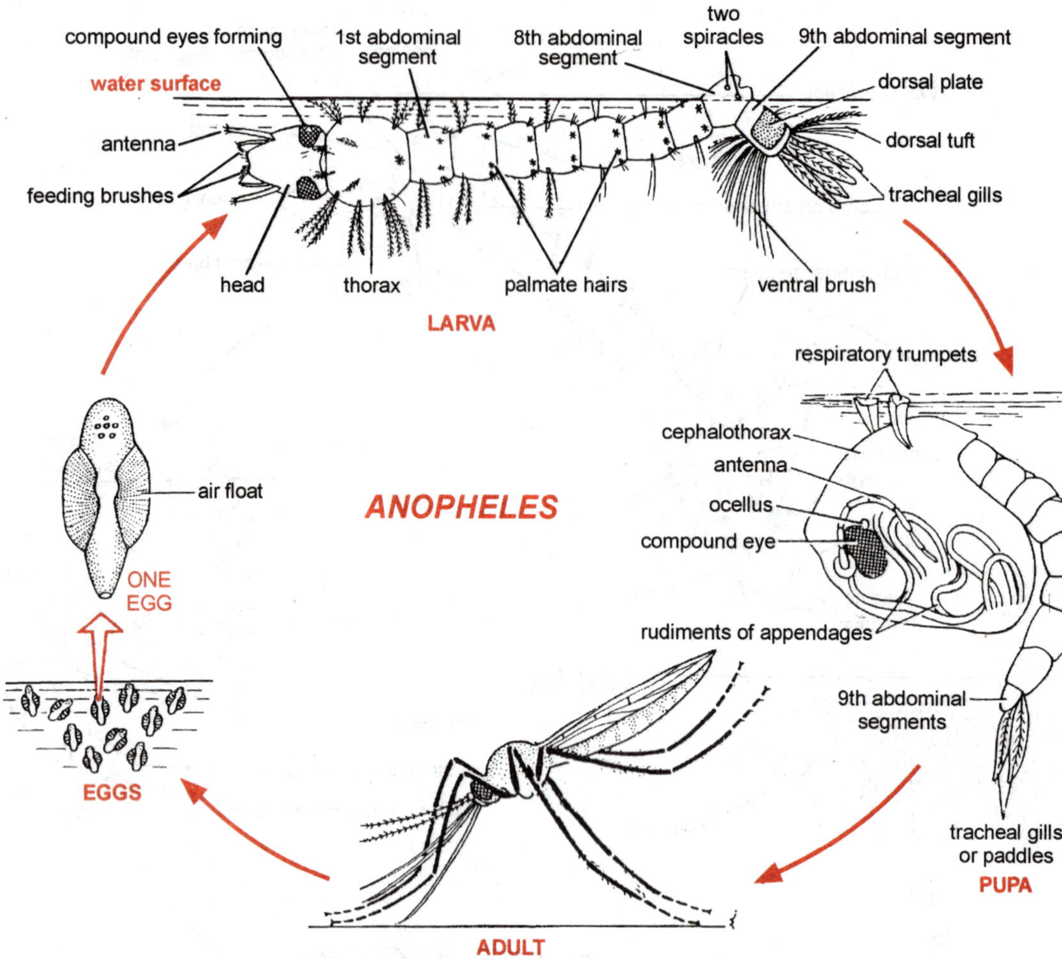

Fig. 6. *Anopheles,* Stages of life cycle.

(head + thorax) and a slender, depressed 9-segmented **abdomen** under cephalothorax. 8th abdominal segment bears a pair of large **paddles**, used in swimming. Pupa clings to the surface film with its head upwards. If disturbed, it at once darts downwards with a tumbling motion and rises to the surface again after some time. Pupa respires with air through a pair of small, trumpet-like tubes or **respiratory trumpets**, located dorsally on cephalothorax. These trumpets are connected with the tracheae of body.

Pupa does not feed as there is no **mouth** or **anus**, but active changes of metamorphosis are going on internally. Certain phagocytotic cells cause **histolysis** of larval organs, which break down and degenerate into a milky pulp. At the same time, certain other cells, called **imaginal discs**, carry out **histogenesis** by producing organs of the adult from the pulpy mass. Transformation of internal organs can be observed through **pulpal skin**, which encloses the cephalothorax.

6. Imago. Pupal stage lasts for a short duration, usually of 2 or 3 days. After completion of metamorphosis, pupa is transformed into the perfect adult insect, called **imago**. Pupal skin now splits along the back between the two respiratory trumpets. The imago emerges out with well-developed wings and after a few moments flies away.

Complete life cycle, from egg to adult, takes 10 to 15 days. The female lives for one month or more and the male for about a week. Differences in the stages of life cycles of *Culex* and *Anopheles* have been given in *Table 1*.

MOSQUITO-BORNE DISEASES

1. Malaria. Caused by species of *Plasmodium* carried by female *Anopheles*. Symptoms include recurrence of fever every third or fourth day.

2. Elephantiasis. Caused by a parasitic roundworm, *Wuchereria bancrofti,* carried by *Culex fatigans*. Symptoms include enlargement of limbs, scrotum or mammae.

3. Dengue fever. Caused by a virus and carried by *Culex fatigans* and *Aedes*. Symptoms include aching limbs, high temperature and minute rash.

4. Yellow fever. Caused by a virus and carried by *Aedes*. Symptoms include high fever, internal bleeding, anaemia and destruction of liver cells.

5. Encephalitis. Caused by a virus and spread by various species of *Culex* and *Aedes*. It can kill its victim or cause serious brain damage.

6. Leprosy. Thought to be spread only by contact, it can be spread by Indian mosquitoes also, according to the scientists at the **Jawaharlal Nehru Institute for Postgraduate Education and Research** (JIPMER) at Pondichery.

MOSQUITO CONTROL

In order to prevent and control malaria and other diseases caused by mosquitoes, the following methods for the destruction of mosquitoes must be adopted.

Table 1. **Comparison of Life Histories of *Culex* and *Anopheles*.**

Stages	*Culex*	*Anopheles*
A. Eggs	(1) Usually 200-400 eggs are laid in dirty water near human habitation.	(1) Usually 40-100 eggs are laid in fresh and clean water.
	(2) Eggs lie vertically on the surface of water in a cluster forming a raft.	(2) Eggs lie horizontally and singly on the surface of water. Raft is not formed.
	(3) Cigar-shaped and light brown.	(3) Boat-shaped and dark brown or black.
	(4) Without lateral air-floats.	(4) Each egg has two lateral air-floats which help in floatation.
B. Larva	(5) Bottom feeder. Relatively larger and lighter in colour.	(5) Surface feeder. Relatively smaller and darker in colour.
	(6) During intake of air the head hangs downwards at an angle of 45° to the surface of water.	(6) During intake of air, the head lies horizontally, parallel to the surface of water.
	(7) With a long conical respiratory siphon on 8th abdominal segment, bearing 2 spiracles.	(7) No respiratory siphon. Instead, 8th abdominal segment bears a raised chitinous plate with two spiracles.
	(8) A comb plate present on 8th abdominal segment.	(8) Comb plate absent.
	(9) Without palmate hairs on abdomen.	(9) With palmate hairs on abdomen.
C. Pupa	(10) Larger and colourless. Abdomen more curved.	(10) Smaller and green. Abdomen more curved.
	(11) Respiratory trumpets longer and narrower.	(11) Respiratory trumpets shorter and broader.
	(12) No palmate hairs on abdomen.	(12) With palmate hairs on abdomen.
	(13) Each paddle bears a long bristle.	(13) Each paddle bears a long and a short bristle.
D. Adult	(14) Body well built. Legs stouter without hairs.	(14) Body slender. Legs delicate with hairs.
	(15) At rest, body lies parallel to the surface and both ends of body deflexed.	(15) When at rest, the body is inclined at an angle of 45° to the surface.
	(16) Wings of uniform colour. Can fly for long.	(16) Wings spotted. Cannot fly for long.
	(17) Maxillary palps shorter in female (3-segmented), but longer (5-segmented) than proboscis and pointed in male.	(17) Maxillary palps 5-segmented and equal with proboscis in both sexes. Clubbate palps in male.
	(18) Scutellum trilobed.	(18) Scutellum evenly rounded.
	(19) Female *Culex* transmits filarial parasite.	(19) Female *Anopheles* transmits malarial parasite.

Table 2. **Different diseases spread by various mosquitoes with their pathogens.**

Disease	Pathogen	Mosquito
Malaria	*Plasmodium*	*Anopheles* female
Filariasis	*Wuchereria*	*Culex* and *Aedes*
Encephalitis	Virus	*Culex* and *Aedes*
Dengue fever	Virus	*Aedes* female
Yellow fever	Virus	*Aedes* female

1. Elimination of breeding places. Mosquito larvae and pupae develop in water; therefore, elimination of breeding places is of primary importance. Swampy areas and stagnant water should be drained out. Water should also be prevented from standing in gutters, drains and depressions. Domestic species can be largely controlled by eliminating receptacles that hold water such as tin-cans, buckets, cisterns, barrels, etc. If possible the breeding grounds may be filled up. The bushes and shrubs should be cleared off.

2. Destruction of larvae and pupae. Wherever it is not possible to fill up or drain out swamps or lakes and ponds, the larvae and pupae may be killed by covering or spraying the surface of water with petroleum, paraffin oil, crude oil or kerosine oil. These substances produce a thin film of oil on the surface, so that the larvae and pupae die due to suffocation or lack of respiration. Oil also enters the spiracles, blocks the tracheae, and poisons the body. Oil solutions or emulsions of DDT, DDD and benzene hexachloride are effective larvicides. Dusts containing paris green, DDT or BHC are effective in the control of surface feeding *Anopheles* larvae. Larvicidal fishes like sticklebacks, minnows, trouts etc., should be introduced in the ornamental fountains, ditches, ponds, lakes, canals, tanks, etc. Aquatic nymphs and adult insects (dragon-flies), which are natural enemies of mosquitoes, should be encouraged. Keeping the lakes, reservoirs and streams free of aquatic vegetation and other floating material provides the top-feeding minnows (*Gambusia*) better opportunities to search out and feed upon the larvae and pupae.

3. Destruction of adult mosquitoes. Adult mosquitoes are most effectively combated in dwellings. Mosquitoes should be driven out of the houses or killed by fumigation with pyrethrum, sulphur, etc. Occasionally spraying of houses and adjoining lands with DDT, flit, pyrethrum and other insecticides proves useful. Residual sprays of DDT may remain effective for several months.

4. Defence against mosquito bite. Houses in mosquito-infested areas should be made mosquito-proof by screening all the doors and windows. While sleeping, mosquito nets should be used to keep away the mosquitoes. Exposed parts of body may be protected by the use of veils, gloves and boots or by the application of repellents, such as antimosquito creams, mustard oil, mixture of kerosine oil and eucalyptus oil or citrinella and cocoanut oil, and dimethyl phthalate or dimethyl carbate, etc.

Important Questions

➤➤*Long answer type questions*

1. Describe, with the help of a well-labelled diagram, the external features of *Culex* mosquito.
2. Describe the mouth parts of mosquitoes and show how they are adapted for the feeding habits of these insects.
3. Give an account of the life cycle of *Culex* mosquito and compare it with that *of Anopheles*.
4. Write an essay on mosquito control.

➤➤*Short answer type questions*

1. Describe the mouth parts of blood sucking insects and their adaptation to their mode of life.
2. What are the important genera of mosquitoes ?
3. Name the mosquitoes which carry the parasites of malaria, dengu, elephantiasis, and yellcw fever.
4. What is the difference in the mouth parts of male and female *Culex* ?

➤➤*Multiple choice questions*

1. Piercing and sucking type of mouth parts are found in :
   (a) housefly
   (b) mosquito
   (c) mosquitoes and bugs
   (d) wasps
2. Female *Anopheles* lays eggs :
   (a) in dark (b) in dung
   (c) in decaying organic matter
   (d) in water
3. The *Anopheles* larva remains at the water surface :
   (a) at right angle to the surface
   (b) at 45° angle to the surface
   (c) horizontal to the surface
   (d) all the positions
4. *Anopheles* eggs float in water due to the presence of :
   (a) nucleus (b) air floats
   (c) air bubbles (d) yolk
5. The eggs of *Anopheles* are found :
   (a) singly with air floats
   (b) in rafts without air floats
   (c) singly without air floats
   (d) in rafts with air floats
6. The larva of *Culex* mosquito floats in water :
   (a) vertically with head downwards and tail upwards
   (b) vertically with head upwards and tail downwards
   (c) horizontally with both ends submerged in water
   (d) horizontally with tail end exposed to air
7. Respiratory siphons are found in :
   (a) cockroach nymph (b) tadpole of frog
   (c) housefly larva (d) mosquito larva
8. Which mouth part is better developed in female *Anopheles* than in the male ?
   (a) maxillae (b) mandibles
   (c) proboscis (d) hypopharynx
9. *Anopheles* and *Culex* mosquitoes can be distinguished by their :
   (a) sitting posture (b) external genitalia
   (c) size (d) colour
10. *Anopheles* larva differs from that of *Culex* in having :
    (a) no respiratory siphon
    (b) a developed respiratory siphon
    (c) two respiratory siphons
    (d) none of these
11. Mosquitoes possess :
    (a) two pairs of long membrane wings, one mesothoracic and other metathoracic
    (b) one pair of long membranous wings of mesothorax and those of metathorax modified into halters
    (c) mesothoracic wings modified to form halters and metathoracic pair is long and membranous
    (d) mesothoracic pair forming calypteron and metathoracic well developed
12. The puncturing elements in female *Anopheles* are :
    (a) maxillae and mandibles
    (b) labium and hypopharynx
    (c) labrum epipharynx
    (d) labium and proboscis
13. An important distinction between *Anopheles* larva and *Culex* larva (while floating on the surface of water) is that :
    (a) the head of *Anopheles* larva hangs vertically downwards
    (b) the head of *Culex* larva hangs vertically downwards
    (c) the respiratory siphon at the posterior end is long in *Anopheles*
    (d) the respiratory trumpets at the cephalothorax in *Anopheles* larva are very prominent
14. Female *Anopheles* lays eggs in :
    (a) water (b) dung
    (c) dark (d) soil
15. Female *Anopheles* can be identified with female *Culex* as it sits :
    (a) parallel to surface of substratum
    (b) at an angle with substratum
    (c) none of these
    (d) at 90° angle to surface of substratum
16. Eggs of *Anopheles* are found :
    (a) singly with airfloats (b) in raft with airfloats
    (c) in raft without airfloats
    (d) singly without airfloats
17. Eggs of *Anopheles* are laid :
    (a) isolately (b) in group
    (c) isolately and float vertically
    (d) none of these
18. Respiratory siphon is found in larva of :
    (a) housefly (b) *Anopheles*
    (c) *Culex* (d) mosquito
19. Which is well developed in a male mosquito :
    (a) mandible (b) labium
    (c) maxilla (d) none of these
20. The larva of *Anopheles* mosquito :
    (a) feeds below the water surface and has a tubular respiratory siphon
    (b) feeds below the water surface and respiratory siphon is developed
    (c) feeds on the water surface and has a tubular respiratory siphon
    (d) feeds on the water surface and the respiratory siphon is undeveloped
21. A pair of wings in *Culex* mosquito is reduced to form a pair of halteres. These halteres arise from the :
    (a) prothorax (b) mesothorax
    (c) metathorax (d) abdomen
22. Young one of a mosquito is known as :
    (a) maggot (b) imago
    (c) nymph (d) caterpillar

23. Male mosquito cannot suck blood due to absence of :
(a) labrum (b) mandibles
(c) hypopharynx (d) labium
24. The male mosquito feed on :
(a) blood (b) nector
(c) jelly (d) fruit juice
25. The puncturing elements in female mosquito :
(a) maxillae (b) mandibles
(c) both (d) none
26. The larva of mosquito is :
(a) maggot (b) instar
(c) wriggler (d) imago
27. The pupa of mosquito is known as :
(a) maggot (b) imago
(c) tumbler (d) none
28. Insectivorous fish use for mosquito control is :
(a) *Hilsa* (b) *Labio*
(c) *Gambusia* (d) *Spherna*
29. Dengue fever spread by :
(a) *Anopheles* female (b) *Culex* female
(c) *Adis* female
30. The disease transmitted by mosquitoes :
(a) filarial (b) encephalitis
(c) yellow fever (d) all
31. Which of the following insect is much more close to mosquito in taxonomy?
(a) Cockroach (b) Housefly
(c) Butterfly (d) Dragonfly
32. Common characteristics of cockroach, housefly and mosquito are :
(a) Compound and simple eyes
(b) Two pairs of legs and two compound eyes
(c) Three pairs of legs and one pair
(d) One pair each of wings and of developed wings
33. There is fixed number of segments in :
(a) *Culex* (b) *Taenia*
(c) *Pheretima* (d) *Ascaris*
34. Johnson's organ is found in :
(a) spiders (b) antennae of mosquitoes
(c) head of cockroach (d) abdomen of houseflies
35. Halteres in mosquitoes and houseflies develop from:
(a) prothorax (b) mesothorax
(c) metathorax (d) head
36. Ovulation in female mosquito is induced by :
(a) Temperature (b) Humidity
(c) Photoperiodism (d) A meal of blood
37. The eggs of *Anopheles* are :
(a) boat-shaped (b) club-shaped
(c) cigar-shaped (d) spherical
38. Mosquito larva moults :
(a) 3 times (b) 4 times
(c) 5 times (d) many times

Answers

➤➤*Multiple choice questions*

1. (c)	2. (d)	3. (c)	4. (b)	5. (a)	6. (a)	7. (d)	8. (c)	9. (a)	10. (a)
11. (b)	12. (a)	13. (b)	14. (a)	15. (b)	16. (a)	17. (c)	18. (c)	19. (d)	20. (a)
21. (c)	22. (c)	23. (c)	24. (b)	25. (c)	26. (c)	27. (c)	28. (c)	29. (c)	30. (d)
31. (b)	32. (c)	33. (a)	34. (b)	35. (c)	36. (d)	37. (a)	38. (b).		

51

Apis : The Honey Bee

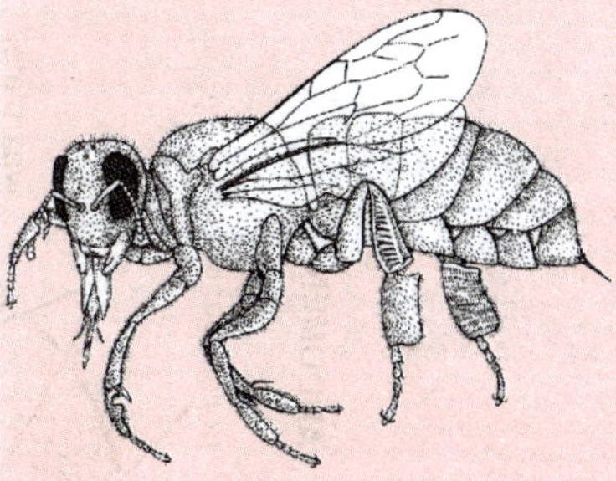

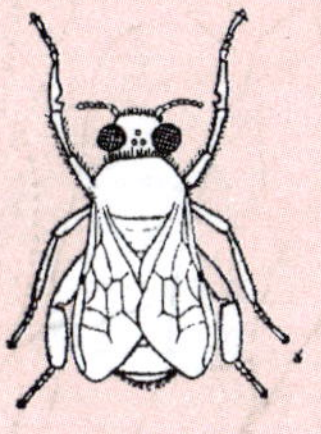

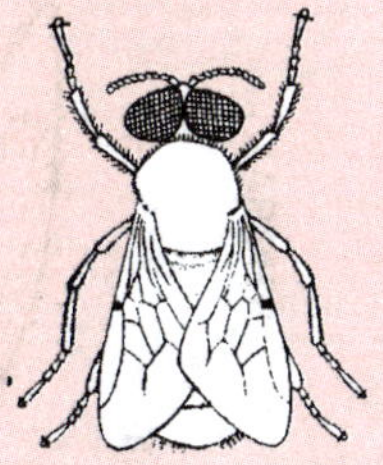

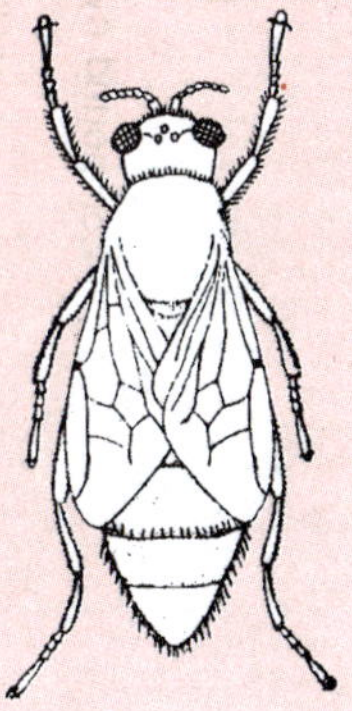

The insect order **Hymenoptera** includes bees, ants, wasps and some other insects. These are characterized by the possession of two pairs of membranous wings, chewing or chewing and lapping type of mouthparts and, in many, a sting in the female.

Honey bees are the most familiar members of the order Hymenoptera. The honey bee usually described in textbooks is *Apis mellifera* (=*A. mellifica*), the common European honey bee. The small-sized honey bee, employed for commercial bee-keeping (apiculture) in India, is *Apis indica.* Other honey bees commonly found in India are *A. dorsata,* the giant honey bee and *A. florae,* the little honey bee. Approximately 20,000 species of honey bees are known to human.

SYSTEMATIC POSITION

Phylum	Arthropoda
Class	Insecta
Subclass	Pterygota
Division	Endopterygota
Order	Hymenoptera
Family	Apidae
Genus	*Apis*

HABITS AND HABITAT

Honey bees are found all over the world and are known for their art of manufacturing honey and bee-wax. They are highly specialized insects, both in structure and habits. Sense organs, mouthparts, wings,

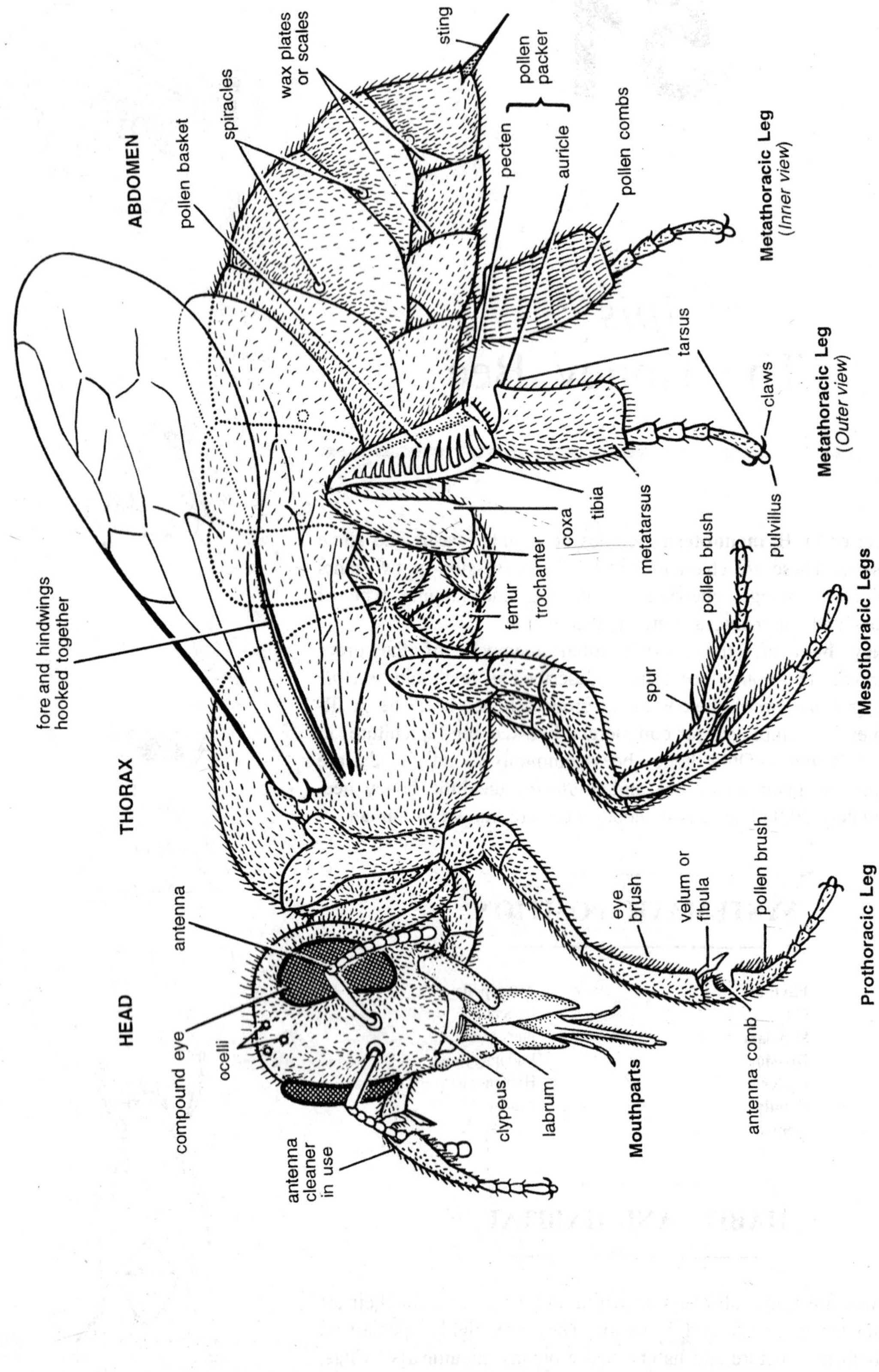

Fig. 1. Worker honey bee in lateral view.

legs and many internal organs are more diversified and specialized than in cockroach or grasshopper. They are social insects living in colonies and exhibiting polymorphism and division of labour. The nests or **beehives** of honey bees, harbouring thousands of individuals, are seen hanging down the tree branches or ceilings of houses and old buildings. These are built by their cooperative efforts and manifest a spectacular engineering feat. Honey bees feed on pollen and nectar of flowers. They communicate with each other through a sign language. Mating occurs in a nuptial flight and development includes metamorphosis.

BEE COLONY (CASTES)

A colony of honey bees consists of three kinds of individuals or **castes** : (i) **workers** which are sterile females, (ii) **drones** which are fertile males, and (iii) **queen** which is a fertile female. Population of an average-sized colony consists of one adult queen, about 100 drones and 60,000 workers. Drones and queen are concerned solely with reproduction. Hence, the workers perform all other duties such as producing royal jelly for feeding the community, producing wax and building the beehives, rearing the larvae, cleaning and ventillating the hive, disposing of the debris and dead bees, etc.

EXTERNAL MORPHOLOGY

[I] Worker Bee

Worker bee is the smallest member of colony and makes up the largest number of colony individuals. It is black or brownish in colour with body densely covered with hairs.

Like all other insects, the body of worker bee is divisible into three regions : **head, thorax** and **abdomen**.

1. Head. It is a wide triangular structure with the apex pointed below. It bears dorso-laterally a pair of large **compound eyes** and three **ocelli** on the middle of its top. A pair of short and 13-jointed **antennae** (12-jointed in male) are borne on the middle of face and probably serve as tactile and gustatory organs. From the bottom of head project the specialized mouthparts.

(a) Mouthparts. Mouthparts of honey bee are of **chewing** and **lapping type**, adapted for sucking nectar from flowers and moulding the wax. A broad and short **labrum** lies below clypeus and a fleshy **epipharynx** projects beneath it. The spoon-shaped and smooth-edged **mandibles** are situated below and one on either side of labrum. They work sideways and are used for moulding wax and manipulating pollen. In each **first maxilla**, lacinia is absent, maxillary palp is reduced and galea forms an elongated blade. Glossae of **second maxillae** or **labium** form a tongue-like process or **ligula** with a spoon-like depression at its tip, called **labellum** or **honey spoon**. It has a mid-ventral groove (food channel) and constitutes a tube or proboscis for sucking up the nectar. **Labial palps** are well developed and elongated. Paraglossae are very much reduced.

(b) Feeding. When the bee sucks nectar, it extends its ligula or tongue into the nectary of flower. The nectar rises up through its ventral groove or food channel by capillary action.

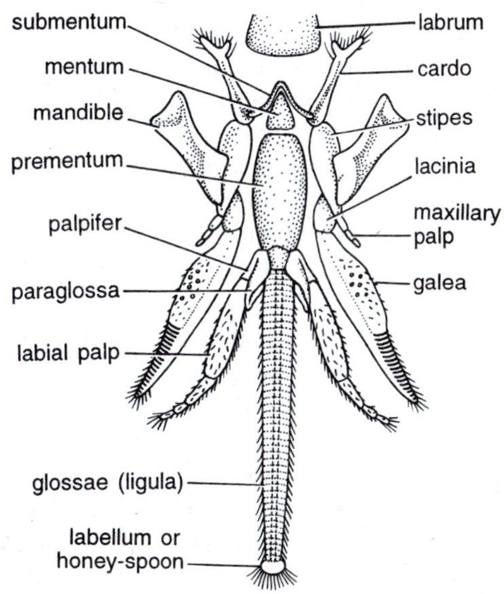

Fig. 2. Honey bee. Mouthparts.

Moreover, the shortening of ligula and the sucking action of pharynx helps in sucking up the nectar.

2. Thorax. Thorax is divided into the usual 3 segments : an anterior **prothorax,** a middle **mesothorax** and a posterior **metathorax**. Each of these segments bears a pair of **legs** and a pair of **wings** is borne by each of the mesothorax as well as metathorax.

(a) Legs. Three pairs of legs are densely covered with hair and are variously adapted. Because of their complexity and differences, each leg will be considered separately.

(i) Prothoracic legs. The segments of a prothoracic or foreleg are : (1) an oblong **coxa**, (2) a short **trochanter,** (3) a long **femur** provided with pollen-carrying hairs, (4) a **tibia** with a fringe of stiff hairs, or **eye brush,** along its medial edge for cleaning the compound eye, and a movable spine-like **velum** and a **pollen brush** on opposite margin at the distal end, and (5) a 5-segmented **tarsus** terminating in a pad-like **pulvillus**, which secretes a sticky substance for adherence, and a pair of **claws**. Its proximal segment or **metatarsus** bears a semicircular notch, called **antennal comb.** This comb, along with tibial velum, forms the **antenna cleaner** that serves to clean the antenna drawn in between them.

(ii) Mesothoracic legs. A mesothoracic or mid-leg has all the segments as seen in the foreleg. It has a spike-like **pollen spur** at the distal end of tibia and a **pollen brush** on the inner surface of metatarsus. Spurs of both the mid-legs are used to remove pollen from **pollen baskets** of hindlegs and to dislodge wax from wax pockets on the ventral surface of abdomen.

(iii) Metathoracic legs. Segments of a metathoracic or hind-leg too are the same. The expanded tibia bears a **pollen basket** on its outer concave surface which is partially covered by rows of long curved bristles along both its margins. Lower end of tibia bears a row of stout bristles forming the **pecten**. Proximal end of metatarsus bears a smooth, concave, lip-like plate, the **auricle**. Pecten and auricle together form a **pollen packer** to convey and pack pollen into the pollen basket. Inner surface of metatarsus bears many transverse rows of stiff bristles, forming the **pollen combs**. These brush off pollen from the body parts and handle wax.

(b) Wings. Two pairs of wings are double-layered, small, narrow, membranous and transparent. They have a much modified and reduced venation. Front margin of each hindwing bears a row of minute hooks, called **hamuli**, that fasten to the grooves along rear margin of forewing forming a single flight blade. Wings may vibrate over 400 times per second during flight. Workers may travel as far as 15 km to gather nectar and pollen.

3. Abdomen. It consists of 6 visible segments and bears the **wax glands** and the **sting**.

(a) Wax glands. The glandular area secreting wax lies on the ventral surface of the last 4 visible segments of abdomen. Wax is secreted through minute pores in the form of flat scales. It is masticated by the mandibles before its use for building the 'cells' of the honeycomb.

(b) Sting. Sting is the modified ovipositor of the insect and is used for injecting poison for protection. It is composed of two straight grooved **stylets** or **lancets**. Their distal free ends are provided with many anteriorly-pointed spines or **barbs** which prevent the removal of sting. The muscles associated with the sting help in the operation of lancets. A set of 3 chitinous plates on either side act as levers to move the barbs. A pair of filiform **poison glands** secrete the acidic

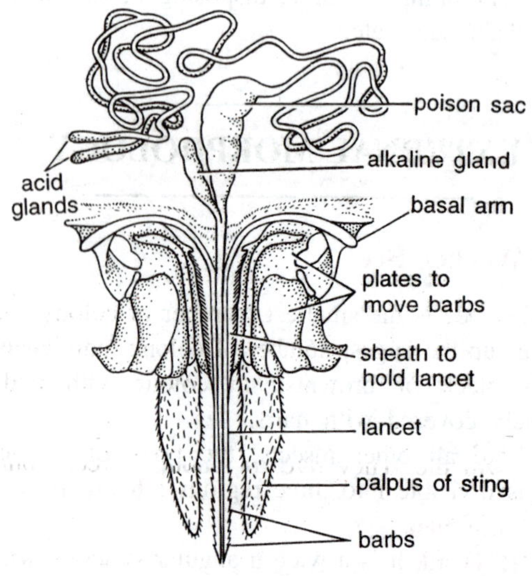

Fig. 3. Honey bee. Sting.

material that is stored in a sac-like storage **poison sac** located at the base of the sting. Associated with it is an elongated **alkaline gland** that secretes an alkaline material. The two materials mix to form the poison or **bee venom** that flows down the sting into the wound of the victim. After stinging, the worker bee leaves the sting apparatus and dies. Sting of queen bee is longer but less barbed and can be withdrawn without injury to her own body. Males do not have a sting because they have a copulatory apparatus instead of an ovipositor.

[II] Queen

Queen is the only fertile female in a beehive, having immensely developed ovaries. She is elongated, 15-20 mm long and is easily distinguished by her long tapering abdomen, short legs and wings. She has pointed mandibles and shorter mouthparts. She can sting repeatedly as there are no barbs on her sting. She is unable to produce wax and honey or gather pollen or nectar. She is a specialized and degenerate individual with a small brain and without salivary glands.

The queen arises from a fertilized egg and larva especially fed on **royal jelly**. She alone lays eggs and is the mother of almost all the members of the hive. She lives for several successive years laying about 2,000 or more eggs a day and up to about 1,500,000 eggs during her lifetime.

[III] Drones

Flying idly near the hive in the sunshine can be seen the male bees or drones. There are usually 100 drones in a typical colony, depending upon the season of the year. They are intermediate in size 15-17 mm long, but considerably stouter and broader. They possess very large eyes, small pointed mandibles and lack wax-producing glands, pollen-collecting apparatus and a sting.

Drones do not work and may be seen begging for honey from workers. If not fed by workers, they will die. They live for about 5 weeks only. They develop parthenogenetically from unfertilized eggs laid by the queen and exist only to mate with the queen of their own or some other colony.

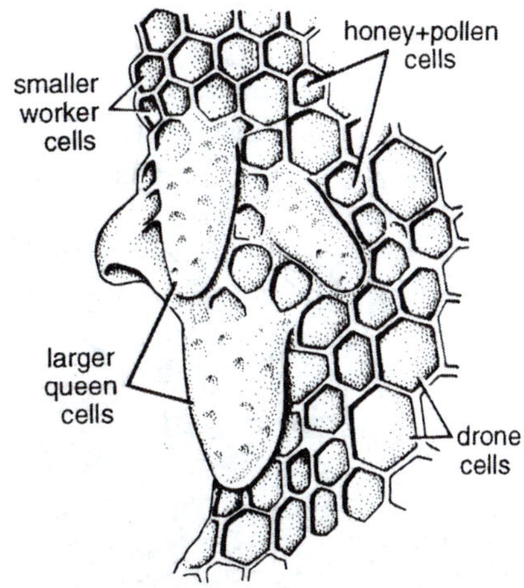

Fig. 4. A portion of the honeycomb showing various types of cells.

BEEHIVE OR HONEYCOMB

The honeycomb or nest is commonly built hanging down vertically from a rock, building or branch of a tree. It consists of two layers of hexagonal chambers of cells made by the beeswax secreted from the abdominal glands of worker bees. Their walls are extremely thin and fragile. **Storage cells,** containing honey and pollen, are usually built near the top and margins of the comb. **Brood cells,** generally occupying the lower and central positions, contain the young stages. In *A. dorsata,* brood cells are similar in shape and size but in other species they may be of three types. **Worker cells**, in which workers are reared, are small like honey cells. **Drone cells** are slightly larger, while **queen cells** are enormous, irregular, cylindrical or vase-shaped structures hanging down from the bottom. Queen cells cannot be used again like other cells.

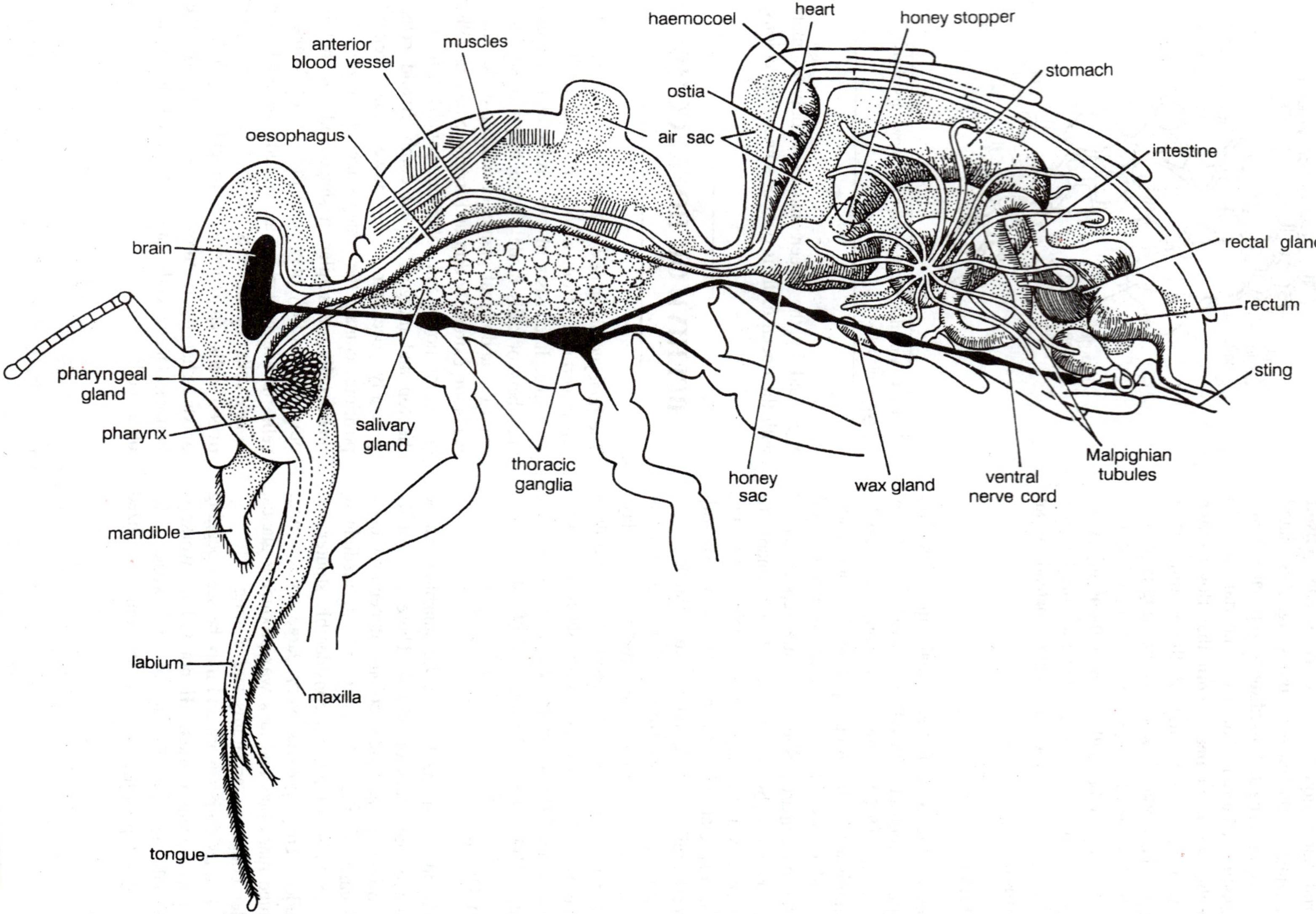

Fig. 5. Internal anatomy of the worker honey be showing : **A.** Digestive system, **B.** Circulatory system, **C.** Respiratory system and **D.** Nervous system.

INTERNAL ANATOMY

1. Body cavity. Body cavity is a haemocoel. True coelom is greatly reduced.

2. Digestive system. Mouth parts of honey bee are fitted for both chewing and sucking pollen and fluids which are sucked through **mouth** into a small **pharynx**. An elongated thin **oesophagus** joins a globular **crop** or **honey sac** in the anterior part of abdomen, where nectar of flowers is chemically changed into honey. Honey sac empties into a large cylindrical **stomach,** followed by a narrow **intestine** and an expanded **rectum**, which opens at the **anus**.

3. Circulatory system. A long, tubular, muscular and 5-chambered **heart** lies mid-dorsally in a **pericardial sinus**, receiving blood through 5 pairs of **ostia**, with valves to prevent backflow. Blood is pumped into head region through an anterior **aorta,** and it diffuses through haemocoel back into pericardial sinus. Blood is colourless, with **white blood corpuscles** in its plasma.

4. Respiratory system. 7 pairs of very small lateral **spiracles** (2 thoracic and 5 abdominal) admit air into a branched system of **tracheae** which convey O_2 to all body parts. Certain tracheae enlarge to become **air sacs** which are larger than those of grasshopper.

5. Excretory system. Numerous hollow, glandular and thread-like **Malpighian tubules** excrete wastes from haemocoel into the anterior end of intestine, much in the same manner as in grasshopper.

6. Nervous system. It is similar but **brain** is proportionately larger than that of grasshopper. **Ventral** nerve cord is double containing 2 **thoracic** and 5 **abdominal ganglia**.

7. Sensory system. Sense organs are more highly developed than in grasshopper. The **antennae** bear end organs of smell **(olfactory),** hearing **(auditory)** and touch *(***tactile***)*. Bristle-like taste setae **(gustatory)** are located near mouth and on the so-called 'tongue'. There are two large **compound eyes (visual)** and 3 **ocelli** on top of head.

8. Reproductive system. The sterile worker honey bee contains only vestigial reproductive organs. The **female reproductive system** of queen bee includes a pair of large **ovaries** almost filling the long abdomen, a pair of **oviducts** that unite to form the **vagina**, and a large **spermatheca** for storing spermatozoa and arising from the dorsal wall of vagina. The **sting** is a modified **ovipositor** formed by the external genitalia.

The **male reproductive system** of male or drone consists of one pair each of **testes, vasa deferentia, seminal vesicles** and **accessory glands,** and an **ejaculatory duct** that leads to the **copulatory apparatus**.

During **spermatogenesis,** there is no reduction division or **meiosis** because drones develop parthenogenetically from unfertilized eggs so that their cells are haploid containing 16 chromosomes. Whereas in **oogenesis**, meiosis occurs because the cells of female are diploid containing 32 chromosomes.

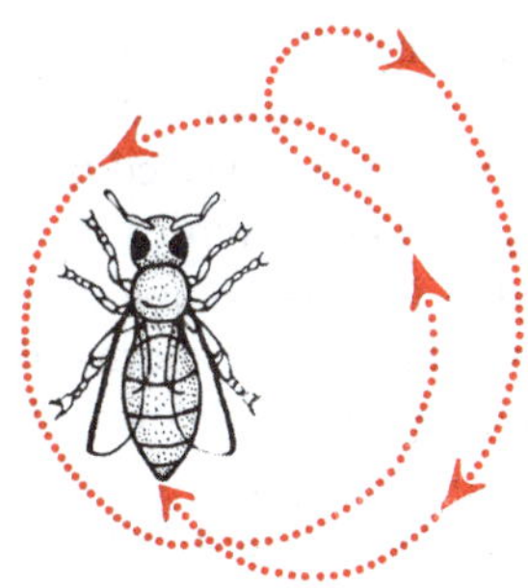

A

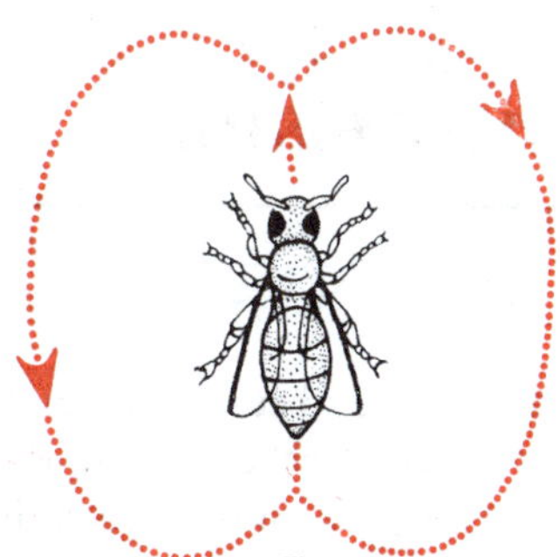

B

Fig. 6. Bee dances. **A.** Round dance. **B.** Tail-wagging dance.

BEE LANGUAGE AND COMMUNICATION

Bees find flowers both by scent and sight. According to the brilliant Austrian Scientist, **Prof. Karl Ernst Von Frisch** (1948), the foraging bees returning to the hive indicate the new source of nectar and pollen by performing certain rhythmic body movements, called **bee dances** or **dance language**, which are readily understood by other bees of the colony. The two extreme kinds of bee dances are as follows :

1. Round dance. The bee turns in a circle, once to left and then once to right. The round or circling dance indicates that the source of food is near the hive.

2. Wag-tail dance. If the source of food is more than 100 metres away, the informant bee performs a wag-tail dance on the vertical comb. It runs for a short distance in a straight line swiftly wagging or vibrating its tail, and then moves in two semicircles one on either side of this straight line. Direction of food is indicated by the direction of tail-wagging. Running vertically up the comb means the food is in the direction of sun. Running vertically down the comb means that food is present opposite the sun. However, if the straight run is oblique, then source of food is at the corresponding angle. If it is dark, the worker bees touch the runner with their antennae to determine the direction of its run.

LIFE HISTORY

1. Swarming

The behaviour of honey bees to come out of the hive in large numbers is called **swarming**. It takes place during the spring or early summer. It relieves the over crowding and provides a means of colony reproduction, *i.e.,* founding of new colonies. On a fine forenoon, the **old queen** leaves the hive accompanied by a large number of **old workers** and **drones** in a swarm. The swarm soon settles down, often on the branch of a tree, and builds a new nest. Left behind in the old hive are the young workers and several new queens still in their cells but approaching the time of emergence. Only one queen survives by stinging to death the other newly-hatched queens.

2. Nuptial or Marriage Flight

The **prime swarm** is led by the old queen while the **second swarm** is accompanied by the newly-emerged virgin queen. About a week after emergence, the new queen takes her first aerial flight followed by a swarm of drones. The queen flies very high and the drones gradually drop out of the race. The last drone left in the race, mates with her. **Mating** occurs in mid-air, during which the queen receives spermatophores from one of the drones. The queen mates only once in a life time. The sperms stored in her spermatheca fertilize her eggs as long as she lives. The genital parts of male are forced out with such a great pressure that he dies after mating. The pair then

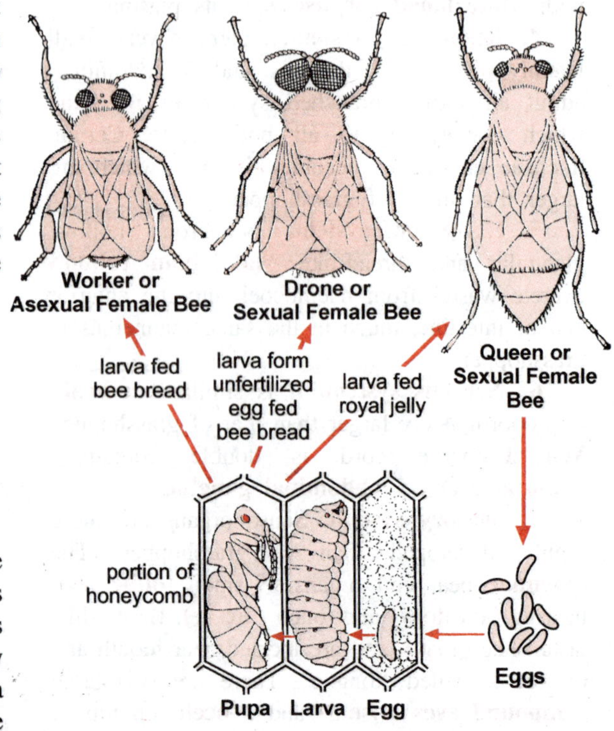

Fig. 7. Honey bee. Stages of Life history and castes.

falls to the ground and the queen after pulling herself away returns to the hive not to leave it again, until she grows old and leads a prime swarm.

3. Development and Caste Determination

Development takes place in the cells of hive and undergoes complete metamorphosis through larval and pupal stages. 3 or 4 days after mating, the young queen begins to oviposit.

1. Eggs. As an egg passes down the ovary through oviduct, it receives a **shell** and gets fertilized by a sperm. The queen has a remarkable capacity of controlling the fertilization of its eggs. A fertilized egg is laid in a worker or queen cell of honeycomb, while an unfertilized egg in a drone cell, the latter developing parthenogenetically. Eggs are small, oblong and bluish-white. A queen may lay one million eggs in her life time at the rate of 1500 eggs per day.

2. Larvae. After 3 or 4 days, small larvae or **grubs** hatch out from the eggs. They have no legs or eyes. The formation of a queen or worker depends on the diet which the larvae receive. All larvae are fed by the workers for the first few days on **royal jelly**, which is composed of a predigested mixture of honey and pollen. The larvae destined to become queens receive this food throughout their life. While, after 3rd day, the worker and drone larvae are fed upon a mixture of honey and pollen, called **bee bread**. If a worker larva is transferred to a queen cell before 3rd day, it develops into a queen, and vice versa.

3. Pupae. The grubs grow and moult several times and after 5 or 6 days develop into pupae. A pupa is enclosed in a silken **cocoon** secreted by the larva. It takes the queen on an average 13 days, the worker 18 days and the drone 21 days to complete metamorphosis and emerge out as adults.

At the approach of winter, drones are driven out of the hive by workers. They are either stung by workers or die of cold or hunger. The queen and the workers bridge over the winter season by feeding upon the stored honey and pollen.

ECONOMIC IMPORTANCE OF HONEY BEE

(1) The **honey** makes an important food for man and other animals. It is the crop or honey-sac where the nectar is changed into honey by losing a certain amount of water and becoming chemically altered. The salivary enzyme converts the complex sugar of nectar into the simple sugar of honey.

(2) **Bee wax** is used for medical purposes.

(3) The honey bee is probably of greatest importance to agriculture in the **pollination** of plants. It is the only pollinating insect which can be controlled by man and, therefore, is extremely valuable to him. Certain orchard trees yield very little without the aid of the bees.

(4) Lastly, to some extent, they can **annoy** man by their stings.

APICULTURE

[I] Apiculture in India

Before man learned to manufacture sugar, he depended for sweets largely upon honey, which formed an important element in his diet. Honey is mentioned in ancient Indian books like Vedas, Purans, Ramayan and Mahabharat, etc. Foreign travellers, like Fahiyan, have described its medicinal use. Honey is produced by honey bees which has been most intimately associated with mankind since very ancient times, and has reached the highest degree of domestication. Bee-keeping therefore, is one of the oldest agricultural pursuits of man. The modern scientific and commercial method of breeding and care of bees for the production of honey and bees wax is known as **apiculture**.

Bee-keeping or apiculture is widely practiced in U.S.A., Canada, Australia and New Zealand. In India also it is a thriving cottage industry widespread in South India as well as in some northern states where it is providing gainful employment to several lakhs of people. The Khadi and Village Industries Commission (KVIC) and the Indian Council of Agricultural Research (ICAR) are making efforts to raise the industrial status of apiculture in India.

[II] Choice of Flora

A place where bees are kept is known as **apiary**. For setting up an apiary, the local flora is an important part of apiculture. Bees can collect nector and pollen from far distant places but it will be better to choose more nector-yielding plants such as neem, jamun, soapnut, etc. Plants like maize, rose and sorghum are rich in pollen whereas others like cherry, apple, sheesham, coconut, guava and musturd serve both for nector and pollen.

[III] Choice of Bees

For starting an apiary, it is important to select a good variety of bees for domestication. (i) They should be gentle in temperament and have a colonial habit. (ii) They should be able to protect from enemies. (iii) Their workers should be smart and energetic and be able to suck juice from a variety of plants. (iv) They should be high honey yielding. (v) They should be such that can form a hive anywhere.

Four species of honey bees (*Apis*) commonly found in India are :

1. *Apis dorsata.* Popularly known as Giant Honey Bee, it is the largest in size, about 20 mm in length. It yields maximum honey of all, about 25 kg from each comb. But they could not be domesticated because of their ferocious and irritable nature and migratory habits.

2. *A. indica.* The small Indian Bee (about 15 mm long) inhabits forests and plain regions throughout India. It can be easily domesticated because of gentle nature, but the yield of honey is much less, only 3 to 4 kg per comb.

3. *A. florea.* The Little Bee is still smaller in size (8 mm) and its honey-yielding capacity is 250 gm per comb. It is docile in nature and stings rarely, so that removal of combs for honey extraction is easier.

4. *A. mellifera.* The European Bee, is also docile in nature and easy to domesticate.

Out of the four species of honey bee mentioned above, *A. indica* is the best, used in India for apiculture industries.

[IV] Apiculture Methods

The main aim of apiculture is to obtain more and more honey in pure form. There are two methods of apiculture in practice : (i) **old** or **indigenous** and (ii) **modern**.

1. Old or indigenous method. The primitive old method is very crude and unplanned. The hives used are : (i) natural **fixed combs** prepared by bees on walls or the branches of trees, and (ii) artificial **movable hives** prepared of hollow wooden logs, empty boxes and earthen pots, etc. For collecting honey by this primitive method, the bees are first killed or made to escape by smoking or by bringing fire flames near the bee hives especially during night when the bees are at rest.

The main drawbacks of this old crude method of bee-keeping are :

(i) Honey thus extracted is not pure because the brood cells containing larvae and pupae and pollen cells are also extracted.

(ii) It destroys the old colony affecting future yield of honey and the bees have to spend a lot of energy afresh in the construction of new hives.

(iii) Bees may not construct new hive at the same old place again.

(iv) Natural hives may be attacked by enemies (rats, ants, wasps and monkeys) or damaged by climatic factors.

(v) Chances of improving the race are not possible.

The crude indigenous method of apiculture despite many drawbacks still persists in many parts even today. But it is being gradually replaced by a better and scientific modern method.

2. Modern method. In modern method, honey bees are reared in movable artificial hives invented by **Longstroth** in 1951. This has converted apiculture into a cottage industry providing employment to lakhs of village and urban people.

A brief description of the various **appliances** used in modern method is as follows :

(a) Artificial hive. Artificial hives are made of wooden boxes which can be shifted easily. A typical movable hive has many parts arranged in 2 or 3 tiers.

(i) Brood chamber. Base of hive is formed by a **bottom board** with two gates for entry and exit of worker bees. The bottom board carries a wooden box made of two chambers. The larger lower chamber is called the **brood chamber**, which contains the queen bee, and the upper smaller one the **super chamber**. The two are separated by a perforated zinc sheet called **queen excluder**. The size of perforations is 3.75 mm so that workers and drones can pass easily through them. But the queen bee with a 4.3 mm to 4.5 mm thorax size can never pass through them. This prevents her from laying eggs in the super chamber which would create problem in obtaining pure honey.

The super chamber is covered by a wooden **inner cover** bearing several holes for proper ventilation. Finally a **top cover** fitted with a plain and slopping zinc sheet, protects the bee hive from rains. The entire behive is mounted on a **stand** which is adjustible for creating the slope needed.

(ii) Comb foundations. The brood chamber contains 5 to 10 vertical **frames** in parallel position with their upper ends touching the queen excluder partition and lower ends in contact with the bottom board. In each frame is kept a bees wax sheet in vertical position, having hexagonal chambers on both sides. These wax sheets are called **comb foundations** as they provide the basis for construction of comb by the worker bees. The queen bee lays eggs in the hexagonal chambers of these comb plates which can be used repeatedly for getting a regular brood of worker bees.

The super chamber also contains additional frames with comb foundations in which the bees collect only honey.

(iii) Models of hives. Various models of artificial beehives have been developed for use in different parts of India. The **Longstroth model** having 10 frames (44.8 cm by 23 cm) is used in Jammu and Kashmir, Punjab and Himachal Pradesh. The **Newton Model** with 7 to 10 frames (21 cm by 14.5 cm) is popular in south, east and central India. The **Jeolikote model** contains 8 frames (30 cm by 18 cm). The Indian Standard Institute (ISI) has even standardized two models, one with smaller frames (21 cm by 14.5 cm) and the other with larger frames (31 cm by 24.4 cm). In India, comb foundations of different sizes are manufactured at the Central Bee Research Station, Poona.

(b) Honey extractor. It is large drum of tin containing revolving rods bearing pockets of netted cloth. Extraction of honey is done on the principle of centrifugal force. Comb foundations full of honey are kept in pockets of netted cloth and the rods rotated manually at a high speed, as in a centrifuge. The pure honey thus extracted is collected through a basal outlet. The comb foundations remain undamaged and can be used again and again.

(c) Uncapping knife. In the beehive, chambers full of honey are sealed with a wax seal by worker bees. Before placing these combs in honey extractor, their wax seals are uncapped or scraped with a knife heated over steam.

(d) Miscellaneous equipment. A few other things needed are : (i) a **smoking apparatus** which is a simple tin box having an outlet at its top for the smoke to come out. (ii) **rubber gloves** to protect hands from the sting of bees, (iii) **bee net** made of synthetic or cotton cloth to protect from the attacks of bees.

[V] Products of Apiculture

Two main products are **honey** and **bee wax.**

1. Honey. This is the most important product of apiculture. It is used as food as well as in medicines.

Honey is believed to be highly nutritious. 200 gm of honey provides the same nourishment as 330 gm meat or 11.5 litre of milk or 1.6 kg cream. It has been estimated that 2.1 gm of honey contains 67 K. calories of energy. Our body systems readily absorb the sugar, mineral, vitamin and other elements of honey. Honey is useful to a healthy as well as sick person of all ages, even the newly born, and can be taken without the consideration of time. It is also used in the preparation of cakes and biscuits, etc., and also taken with or without milk.

Honey has its great medicinal value. It is generally used in Ayurvedic and Unani systems of medical treatment. It is a mild luxative, antiseptic and sedative, and helps in the formation of haemoglobin in anaemic patients. It prevents cough, cold and fever. It is used as a blood purifier, and to cure the ulcers of tongue and alimentary tract. It is also recommended in severe heart attack and to cure diabeties. It is helpful in indigestion. It is a quick killer of germs of typhoid, dysentry and branchiopneumonia. It also give quick energy.

Besides food and medicinal value, honey is used in distillaries for making alcoholic drinks, and in poultry and fishing industries. It is of great value in laboratories where it is used as growth stimulant of plants, for the bacterial culture, for inoculating the seeds of clove, etc. It forms the diet of many insects and is used in making poison baits to save many fruits from fruitflies.

Honey is a sugar rich compound having the following constituents : (i) Levulose – 38.9%; (ii) Dextrose – 21.28%; (iii) Maltose and other sugars – 8.81%; (iv) Enzymes and pigments – 2.21%; (v) Ash – 1.0%; (vi) Water – 17.20%.

2. Bees wax. It is useful in the manufacture of cosmetics, face creams, paints, polishes, plastic works, ointments, carbon papers and many lubricants. It is of common use in the microtomy work in laboratories to prepare blocks of tissue where it is mixed with common wax. Bees consume 10-20 kg of honey to produce 1 kg of wax.

Chibnall (1934) has reported that all insect waxes are complex mixture of varying proprortion of : (i) Even numbered alcohols ranging from C_{24} to C_{36}. (ii) Even numbered normal fatty acids from C_{24} to C_{34} and (iii) Odd numbered normal paraffins ranging from C_{23} to C_{37}.

The various bees waxes differ only due to change in the proportions of these constituents.

[VI] Bee Enemies

Colony of honey bees is harmed by enemies of bees in different ways. The wax moths (*Galleria melonella* and *Achoria grisella*), the bee eaters (*Merops orientalis*) and Kingcrow (*Dicrurus macrocercus*) are common enemies of bee's comb and honey.

Important Questions

➤➤*Long answer type questions*

1. Give an account of the external features of a worker honey bee.
2. Describe the life history and economic importance of the honey bee.
3. Honey bee is a social insect. Discuss.
4. Write short notes on : (i) Bees wax, (ii) Honey, (iii) Mouth parts of honey bee, (iv) Polymorphism in honey bee, (v) Sting apparatus of honey bee.

➤➤*Short answer type questions*

1. Write note on pollination by honey bees.
2. Describe the economic importance and life history of honey bee.
3. Briefly explain food and feeding apparatus of honey bee.
4. What are colonial insects ? Give two examples. What are the different type of bees that are found in a colony ? Mention their 'duty'. Do you consider the colonial forms a highly evolved unit ? Reason out within 40 words.
5. Describe waggel dance ?
6. Describe the polymorphic condition in honey bee.
7. What difference between drone, worker and queen on genetical basis ?

➤➤*Fill in the blanks*

1. In honey bee sex differentiation is by .....
2. The role of *Apis* queen is exclusively .....
3. If the source of food is too far the informant bee shows a .....

➤➤*Multiple choice questions*

1. The 'royal jelly' of the bees is the food given to the :
   (a) larva that become king and queen
   (b) queen
   (c) drone (king)
   (d) larva that becomes the workers
2. An example for best domesticated honey bee is in Europe and America :
   (a) *Apis dorsata* (b) *Apis florea*
   (c) *Apis mellifera* (d) *Apis indica*
3. Which of the following products are obtained from bees ?
   (a) cochineal dye (b) wax
   (c) honey (d) honey and wax
4. The scientist, who decoded the language of bees was :
   (a) Charles Darwin (b) William Harvey
   (c) Karl von Frisch (d) Carolus Linnaeus
5. A worker bee from another nest of same species be introduced in a hive :
   (a) it will be welcome to add to the population
   (b) it will be shut in a special cell for becoming familiar with smell of this hive
   (c) it will be killed and kicked away
   (d) it will be killed and eaten up
6. A bee has flown about 3 kilometers from its hive, her direction of returning to the hive will depend on the :
   (a) memory of the position of sun when she had set out
   (b) memory of the roads, buildings, trees etc.
   (c) visibility of the hive
   (d) attraction by the smell of the hive
7. What guide the subsequent bees to visit the same food source :
   (a) smell of earlier bees left on the flowers
   (b) smell of the flower
   (c) smell of the flower and of the earlier bees
   (d) shape of the flower
8. The members of a bee colony recognise each other by :
   (a) smell (b) vision
   (c) dance (d) touch
9. In honey bee the drones (males) are produced from :
   (a) unfertilized eggs (b) fertilized eggs
   (c) larvae from fertilized eggs, which are fed on royal jelly
   (d) larvae from unfertilized eggs, which are not cared by the workers at all
10. The sterile individuals in honey bee are :
   (a) drones (b) queens
   (c) workers (d) none of these
11. Patterns of 'bee dances' used for foraging and communication are :
   (a) round and tail wagging dance
   (b) spiral and round dance
   (c) elliptical and spiral dance
   (d) tail wagging and wing flippering dance
12. The venom of honey bee is used for treatment of :
   (a) ulcers (b) arthritis and snake bite
   (c) whooping cough (d) inflammations
13. Bee-keeping is commonly known as :
   (a) aquaculture (b) apiculture
   (c) sericulture (d) none
14. Mouth parts of honey bee is :
   (a) chewing type (b) cutting type
   (c) pickring type (d) chewing and lapping type
15. Arrhenotoky is :
   (a) haploid parthenogenesis
   (b) diploid parthenogenesis
   (c) none (d) both of these
16. Royal jelly is secreted by :
   (a) worker (b) drone
   (c) queen (d) all
17. Wax secreted by :
   (a) worker (b) drone
   (c) queen (d) all
18. Dance of honey bee massage include :
   (a) source of food (b) direction of food
   (c) richness of food (d) all the above
19. Queen honey bee is adapted for :
   (a) controlling other bees
   (b) punishing worker bees
   (c) laying eggs and rearing the youngs
   (d) preparing honey
20. The stimulation through which a dancing scout bee communicates the location of a food source to foraging bees in a hive is :
   (a) visual (b) acoustic
   (c) contact (d) contact and sound
21. Honey bees store honey in :
   (a) stomach (b) cell of comb
   (c) crop (d) salivary gland
22. Which one is the real product of honey bee ?
   (a) bee wax (b) honey
   (c) propalis (d) pollen
23. A queen honey bee lays eggs of :
   (a) one type from which all castes develop
   (b) two types, one forming queen and workers and second forming drones
   (c) three types forming queen, drone and workers
   (d) unfertilized eggs die while fertilized ones form all castes

24. Pollen basket occurs in honey bee on :
(a) prothoracic leg (b) mesothoracic leg
(c) metathoracic leg
(d) union of thorax and abdomen

25. In honey bee barbless sting occurs in :
(a) worker (b) drone
(c) queen bee
(d) both queen and worker bee

26. Which of the following are produced by parthenogenesis in honey bee ?
(a) workers
(b) drones
(c) workers and drones
(d) only queen

27. Which castes of bees visit flowers ?
(a) drones (b) queens
(c) workers (d) both drones and workers

28. In honey, which of the following sugars is present ?
(a) levulose (b) glycogen
(c) sucrose (d) cellulose

Answers

➤➤*Fill in the blanks*

1. sting, 2. laying eggs, 3. wag tail dance.

➤➤*Multiple choice questions*

1. (b)	2. (c)	3. (d)	4. (c)	5. (c)	6. (d)	7. (a)	8. (a)	9. (a)	10. (a)
11. (a)	12. (b)	13. (b)	14. (d)	15. (a)	16. (a)	17. (a)	18. (d)	19. (c)	20. (d)
21. (b)	22. (a)	23. (a)	24. (c)	25. (b)	26. (b)	27. (c)	28. (a)		

52

Bombyx mori : The Silkworm

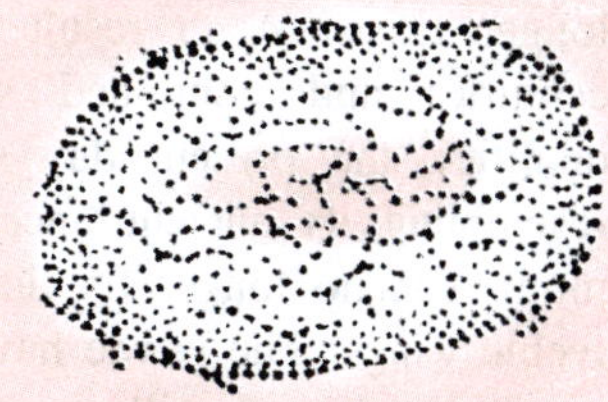

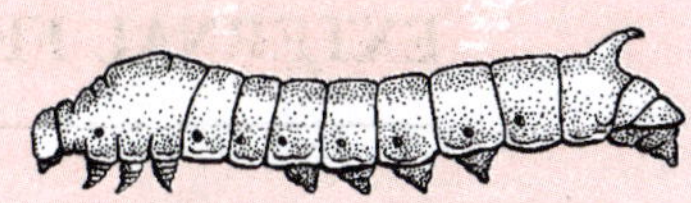

Pure or raw silk is obtained from cocoons of insects belonging to the order *Lepidoptera.* They are popularly known as **silkworms** or **silkmoths**. The most common silkworm is *Bombyx mori.* From the commercial point of view, it is the most important insect in the world, and huge industries have centred around the silkworm for many centuries.

Bombyx mori

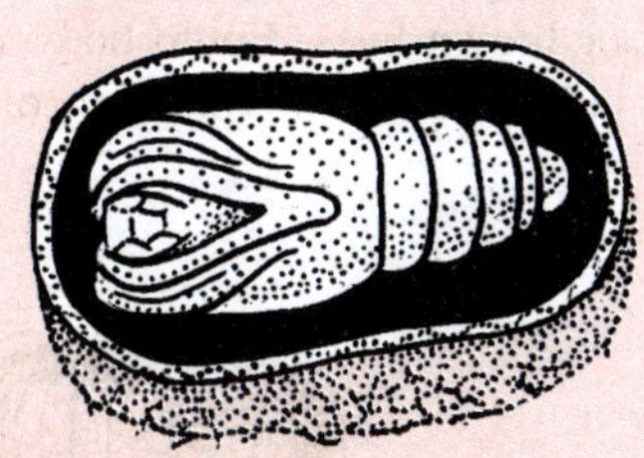

SYSTEMATIC POSITION

Phylum	Arthropoda
Class	Insecta
Subclass	Pterygota
Division	Endopterygota
Order	Lepidoptera
Family	Bombycidae
Genus	*Bombyx*
Species	*mori*

HABITS AND HABITAT

Silkworm has been domesticated for so long that it can no longer survive in nature without human care. The original stock probably no longer exists. As a result of domestication for centuries and continued selection for certain qualities, the present day silkmoth is far from resembling its ancient ancestors. The adult silk moths can no longer fly due to heavy body and feeble wings. The larvae have lost the capacity to search food, and will starve unless the food is placed right near them. They are reared in domestication on a diet of mulberry leaves. But, the adult moth does not feed during its short span of 4 to 5 days only. Commercial silk is secreted by the silk glands of larvae.

EXTERNAL FEATURES

Adult silkmoth is a medium-sized insect, about 25 mm long and with a wing span of 40-50 mm. Body is robust, creamy-white or yellow in colour, and divided into usual 3 regions : head, thorax and abdomen. **Head** is small and bears a pair of compound eyes, a pair of branched or plumed antennae, and degenerate siphoning mouth parts. **Thorax** bears 3 pairs of small legs and 2 pairs of large but feeble wings marked by several faint or brown lines. Entire body is covered by minute coloured scales. **Abdomen** of female is larger and much distended due to great number of eggs it contains.

LIFE HISTORY

Silkworm is **dioecious,** *i.e.,* the sexes are separate. **Fertilization** is internal, preceded by **copulation**. Development includes a complicated **metamorphosis**.

1. Eggs

Soon after fertilization, each female lays about 300-500 eggs in clusters upon the leaves of mulberry tree. Female covers the eggs by a gelatinous secretion which sticks them to the leaves. Small, smooth and spherical eggs are first yellowish white and become darker later on. After laying, the female does not take food and dies within 4-5 days. In India, and other tropical countries, the silkmoth lays **non-diapause** eggs with continuous growth and development, which enables to raise 2 to 7 generations within a year. In temperate (cold) countries, **diapause** eggs are laid which rest until spring when growth is resumed, so that a single generation is produced per year.

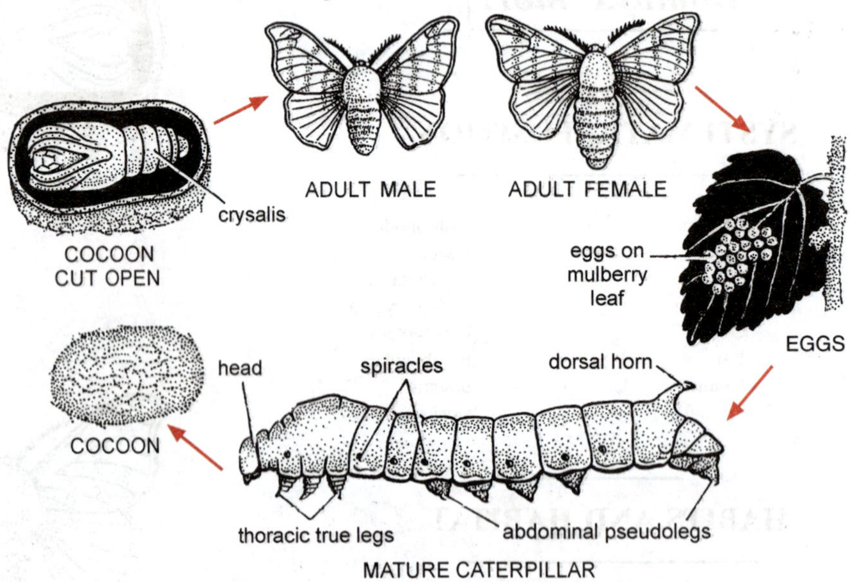

Fig. 1. *Bombyx mori.* Stages of life history.

2. Caterpillar Larva

The larva which hatches from egg in about 10 days, is known as the **caterpillar**. The 1st instar larva is a tiny creature, about 6 mm long, and moves about in a characteristic looping manner. It has a rough and wrinkled whitish or greyish body, which is made of 12 segments. Its head bears mandibulate mouth parts with which it at-once starts feeding on mulberry leaves and grows very quickly. It can also be fed on lettuce and osage orange (*Maclura*) leaves. After 4 or 5 days, it stops feeding and becomes inactive. **Moulting** or **ecdysis** then takes place. Its older skin bursts and a new caterpillar emerges out with a new skin and slightly bigger in size than the older caterpillar. It resumes eating voraciously and growing until the skin is again cast off after a week. The larva repeats this process four times.

A full grown 5th instar caterpillar is now 21-25 days old, about 8 cm long, and weighs nearly 4 to 6 gm. Its elongated body is 12-segmented. Small **head** bears biting and chewing mouth parts. Unsegmented **thorax** bears 3 pairs of jointed **true legs.** The 10-segmented **abdomen** is provided with five pairs of unjointed, stumpy **prolegs** or **pseudolegs,** a short dorsal **anal horn** on the 8th segment and a series of respiratory **spiracles** or **ostia** on either lateral side. A pair of **salivary glands** develop in the lateral sides of its body.

3. Pupa or Chrysalis

Mature caterpillar stops feeding and returns to a corner among the leaves. Its salivary glands now secrete a sticky fluid through a narrow pore of its spinning apparatus, called **spinneret**, situated on **hypopharynx**. The sticky substance turns into a fine, long and solid thread of **silk** in contact with air. Silk thread is made of 5 filaments stuck together by a gummy substance or **sericin**, which is secreted by two other glands. It becomes wrapped around the body of caterpillar forming a **pupal case** or covering known as **cocoon**. This process goes on for about 3-4 days, at the end of which the silkworm is enclosed within a thick, oval, white or yellow silken cocoon. A single caterpillar is said to produce nearly 1000-1500 metres of silk thread in this manner. Within a fortnight the silkworm transforms inside, cocoon into a tubular brownish organism, the **pupa** or **chrysalis**.

4. Imago

Active metamorphic changes take place during pupation. Unsegmented abdominal prolegs disappear, while thorax develops two pairs of wings. The pupa finally metamorphoses into the **baby insect** or **imago**. It secretes an alkaline fluid to moisten one end of cocoon and then escapes by forcing its way out of the softened silk. Soon after emergence the male and female moths mate, lay eggs and die within 3 to 4 days.

SERICULTURE

Commercial **silk** is secreted by the silk glands (salivary glands) of larval silkworms. The rearing of silkworms for the production of raw silk is known as **sericulture**. It is an industry in a few countries including India.

[I] Historical

Silk was manufactured for the first time probably in China, about 3500 B.C., and was kept a closely guarded secret. Somehow the secret got smuggled out of China through some monks. It was introduced in Europe in 552 A.D. Today, sericulture is an important industry of several Asian and European countries, but China and Japan are the only great producers of raw silk. Silk-growing has never been profitable in America due to high cost of hand labour.

Indian sericulture has a long history and silk seems to have been used by the Indians several centuries prior to the Christian era. Historical records testify to the existence of a flourishing export trade in silk goods between India and

European countries during the second century B.C.

[II] Types of Silkworms

A large number of wild and semi-domesticated silk-moths belong to the family **Saturniidae**. Many of them have been partly cultivated for centuries to produce usable silk in Oriental Asia.

1. Muga silkworm (*Antheraea assamensis*). It is a wild and semi-domesticated species in Assam, W. Bengal, Bihar and Orissa. Cocoons are amber or white and the caterpillars feed on *Machilus,* cinnamon, etc.

2. Tussore or Tassah silkworm (*Antheraea mylitta* or *A. paphia*). It is found in China, India, and Shrilanka, etc. Its caterpillars feed on leaves of fig, oak, ber, sal, etc. Cocoons are brown, red or yellow and collected from forests, as the moths do not breed easily in captivity.

3. Oak silkworm. *A. pernyi* of China and Japan, *A. roylei* of the Himalaya and *A. yamamai* of Japan have been collected and reared for centuries. They produce silk of a fine quality.

4. Giant silkworm. *Attacus atlas* of India and Malaysia, is nearly the largest of all living insects reaching up to 28 cm in wing span.

5. Eri silkworm. *Attacus rechinii* or *Phlosamia ricinii* of S.E. Asia, feeds on castor and produces a rough and strong silk locally known as 'Arandi silk'.

6. Chinese or Mulberry silkworm. The commonest of all, which produces most of the raw silk we use, is *Bombyx mori.* It belongs to the family *Bombycidae.* It is found on mulberry leaves in India. Although a native of China, it is almost universally domesticated. It has been reared for so many centuries that it no longer exists in a wild state. By careful selection and hybridization many races have been developed to meet the various needs of climate, quality and quantity of the silk obtained. The silk obtained from this silkworm is white or yellow in colour.

[III] Chemistry and Uses of Silk : Economic Importance

The diameter of raw silk fibres is from 4.5 to 8.2 μ, and elasticity is 20%. Three-fourth part of fibres is made of a tough, elastic protein insoluble in water. One-fourth part consists of a gelatinous protein readily soluble in warm water. Wax and carotenoid pigments are also found in traces.

Raw silk threads woven into cloth has several uses. Sarees, kurtas, jackets, blouses, shawls, curtains and other garments are prepared from it. It is knitted into goods such as gloves, stocking, socks, vests, etc. Articles manufactured include tyres of racing cars, insulation coils for telephones and wireless receivers, sieves for flour mills, fishing lines, parachute cords, parachutes, etc.

[IV] Sericulture Industry (Management)

The major steps and requirements of sericulture industry are briefly as follows :

1. Requirements and appliances. The prime requirement for commercial production of good silk, are genetically improved races of *Bombyx mori,* keeping in view the climatic conditions of the region. Another need of the industry is the plantation of good nutritive varieties of mulberry for a continuous supply of their leaves for feeding the larvae.

The following appliances are required :

(i) **Machana.** The proper place for rearing the silkworms.

(ii) **Rearing trays.** For keeping the laid eggs along with the mulberry leaves.

(iii) **Spinning or Chandrakis tray.** For keeping the caterpillars at the time of full grown stage ready for pupation.

(iv) **Dalas.** For fetching the mulberry leaves.

(v) **Chopping knife.** For cutting the mulberry leaves into smaller pieces.

(vi) **Baskets.** For the distribution of mulberry leaves.

(vii) **Hygrometer.** To know the % humidity in the atmosphere.

(viii) **Thermometer.** To take the reading for the room temperature.

(ix) **Oven.** To regulate the different stages of life cycle at different temperature.

(x) **Refrigerator.** For the storage of eggs (seed) for next generation.

2. Grainage management. Maintenance of good quality races of silkworms and the supply of good quality seeds to rearers is known as **grainage management**. Soon after emergence, the males and females are segregated before they can mate. Females of one lot are later kept with males of another lot for mating and they die soon after laying eggs. The eggs, now known as 'seeds', are kept in sterilized trays at 4°C and moved periodically with feather to ensure 100% hatching. The seeds are kept in diapause and their hatching time is artificially coincided with the best season of mulberry. The rearers are provided with either seeds or the 2nd instar larvae, depending on their knowledge and managing capacity for obtaining good quality cocoons.

3. Stiffling. For obtaining the commercial silk, the cocoons are treated with hot water or placed in a hot oven to kill the pupae inside. If allowed to hatch, they would cut the silk threads in emerging. This process of killing the cocoons is called **stiffling** in the industry.

4. Reeling and spinning. Removal of silk thread from cocoons is termed **reeling**. After boiling, the cocoons become soft and their fibre become loose so that they can be easily unwoven or reeled off. Each cocoon is made of a single fibre. A single cocoon yields 1,000 to 1,500 metres of unbroken silk fibre. 454 gm of silk is obtained from about 25,000 cocoons. A few alive cocoons are kept as 'seeds' for the next crop.

4 or 5 silk fibres are passed through eyelets of wooden guides and twisted into one thread. It is wound round a large wheel and later transferred to spools. This is called **raw silk** or **reeled silk**. For achieving its characteristic lustre, raw silk is again boiled, purified by acid or by fermentation, and thoroughly washed. Several such silk threads are twisted to get **fibre silk**, through the process known as **spinning**. The damaged cocoons and waste threads are also teased and spun into threads, called **spun silk**.

5. Enemies (diseases) of silkworm. The caterpillars of silkworm are subject to a severe hereditary disease, known as **Pebrine**. It is caused by a protozoan parasite *(Nosema bombycis)* which is transmitted through the eggs. The disease can be eliminated by a microscopic examination of the blood of females and avoiding the eggs containing the parasite. Caterpillars are also subject to parasitism by certain tachina flies, which lay their eggs upon mulberry leaves and thus get inside the body of caterpillars through their alimentary canal.

SERICULTURE IN INDIA

[I] Beginning

In India, it was **Lefroy** who first began investigations on silkworm and sericulture, in 1904-05, at Pusa Institute, New Delhi. But the planned development of this industry was taken up only after the establishment of the Silk Board in April, 1949. Today, sericulture is a well-nit cottage industry achieving continued progress through improved technology.

[II] Economics

Sericulture is unique and profitable in many ways, in India. According to experts the annual yield per hectare of land in India is roughly Rs. 3,840 under jute, Rs. 4,060 under paddy, Rs. 1,425 under wheat, but Rs. 15,750 under sericulture. It provides larger opportunities to unskilled labour at the minimum cost. Even the old and handicapped members of a family can be employed gainfully. It requires a small capital investment of a few hundred rupees and the returns are quick and adequate. After considering the increased returns many farmers have switched over to sericulture, replacing jute in West Bengal, cotton in Tamil Nadu, grapes in Andhra Pradesh, and sugarcane in Karnataka.

Besides India, other natural silk producing countries are Japan, China, South Korea and U.S.S.R. India is the only country producing all the four commercial varieties of natural silk. Production of **mulberry silk** is confined to Karnataka, West Bengal, Jammu and Kashmir, Andhra Pradesh and Tamil Nadu. India is fifth in the world production of mulberry silk. The major **tussore silk** producing centres are in Bihar, Madhya Pradesh, Orissa and West Bengal. India is second in the world production of tassar silk. The bulk of **eri silk** is produced in Assam which also holds the virtual monopoly over the world **muga silk** production. The states of Manipur, Assam, Nagaland, Uttar Pradesh, Arunachal Pradesh, Meghalaya and Mizoram are producing a new variety of tassar known as **oak tassar**. Presently sericulture is practised in over 27,500 villages, in 207 out of 398 districts, providing employment to nearly 4 million people, mostly tribal.

At present India produces about 6,200 tonnes of raw silk which amounts to a little over 5-6% of the total world production. The whole production is consumed in the country itself, whereas 25 to 30% of it is exported in the form of fabrics, dress material and garments. Indian hand-spun and hand-woven silk fabrics have a good market both at home and abroad. Scarves, sarees, dress materials, carpets, ready-made garments, raw silk, silk yarn and silk waste are exported to Japan, U.S., U.K., Germany, France, Italy and Switzerland. The foreign exchange earnings which were just Rs. 8.30 crore in 1972-73 rose to Rs. 33.06 crores during 1977-78. Now the foreign exchange earnings have reached to Rupees 2800/= crores in the year 2012-2013. At present sericulture industry is employing more than 75 Lac people all over India. Silk production in Japan, which contributed 45% of the global output, has declined considerably in recent years due to high labour costs and a drift to electronic and major industries. This has created good chances for larger exports of silk goods from India.

The important Silk-Research Institutes are : Central Sericultural Research and Training Institute (CSR and TI) Mysore (Karnataka); Central Sericultural Research and Training Institute (CSR and TI) Berhampur (West Bengal); Central Tasar Research and Training Institute (CTR and TI) Ranchi (Bihar); Central Silk Technological Research Institute (CSTRI) Banglore (Karnataka).

[III] Problems

At present, the Indian silk industry faces many problems of productivity, marketing and finance. Per hectare yield and quality of raw silk are poor in comparison to other major silk producing countries. The industry is introducing several measures to boost silk production, such as supply of quality silkworm eggs to sericulture centres, replacement of local mulberry by other varieties and employing improved and latest reeling technology by using semi-automatic reeling units and multi-end basins.

[IV] Research and Development

For carrying out Research and Development activities, the Central Silk Board has established 3 major Research Institutes for Mulberry at Mysore (Karnataka), Berhampore (West Bengal) and Pampore (Jammu and Kashmir). The Institute of Ranchi (Jharkhand) deals with Tasar, whereas, the Institute at Lahdoigarh (Assam) is looking after research and development related to Muga and Eri sericulture. Post cocoon R and D activities are carried out by Central Silk Technological Research Institute (CSTRI), Bangalore.

The Silkworm Seed Technology Laboratory (SSTL), Banglore (Karnataka), is engaged in research on issues relating to silkworm seed. While the Central Sericulatural Germplasm Resource Centre (CSGRC), Hosur (Tamil Nadu) has the

Unit : Metric Tonnes

Year	Mulberry Hecterage	Mulberry Coccon	Non-mulberry Raw silk	Total raw Raw silk	Per Ha. Silk	Productivity (in kg) Mulberry Raw silk
2007-08	184928	132038	16245	2075	18320	87.84
2008-09	177943	124838	15610	2760	18370	87.73
2009-10	188773	131661	16332	3368	19690	88.82
2010-11	170314	130714	16360	4050	20410	96.06
2011-12 (P)	171697	43395	5384	783	6167	

P : Provisional

responsibility of maintaining sericulture Germplasm, the Seri-Biotech Research Laboratory (SBRL), Bangalore, attends to R and D in frontier areas like molecular characterization of silkworm races and mulberry varieties. In order to increase productivity and quality of silk, the chain of Research Extension Centres are engaged for transfer of technologies from Research Institutes of the field. The following table indicates the increase in productivity of silk during 4 years of 11th Plan.

Important Questions

Long answer type questions

1. Give an account of the structure and life history of silk worm.
2. Describe the requirements and major steps of sericulture industry.
3. Write an essay on sericulture and its position in our country.
4. Write short notes on : (i) *Bombyx mori,* (ii) Pebrine, (iii) Tussore silk, (iv) Uses of silk.

Short answer type questions

1. What are the two types of eggs laid by silkmoths ?
2. Where the spinneret is located ?
3. What are the various types of silkworms ?
4. Define stiffling.
5. What is the name of the gummy substance which stick the silk filaments.

Multiple choice questions

1. Silk is a secretion of silkworm from its specialized :
   (a) Malpighian tubules (b) salivary glands
   (c) fat bodies (d) poison glands
2. Commercial silk is secreted by silk glands or salivary gland of :
   (a) larvae (b) pupae
   (c) adults (d) none
3. Pore of spinning apparatus situated on hypopharynx is called :
   (a) spinneret (b) spiracle
   (c) gustapore (d) none
4. Silk thread is made of five filaments which are stuck together by :
   (a) glue (b) waxy substance
   (c) sericin (d) none
5. Rearing of silkworm is known as :
   (a) apiculture (b) sericulture
   (c) aquaculture (d) floriculture
6. Silk thread contain :
   (a) fibronin (b) sericin
   (c) both (d) none of these
7. The biggest centre of silk production in India :
   (a) Banglore (b) Mysore
   (c) Hyderabad (d) Bhubneswer
8. Central sericulture institute is situated in :
   (a) Raipur (b) Rajpur
   (c) Bharmpore (d) Shimla
9. From which stage of silkmoth, the silk is obtained ?
   (a) adult (b) egg
   (c) caterpiller (d) cocoon
10. Natural silk contains :
    (a) nitrogen (b) potassium
    (c) magnesium (d) phosphorus
11. Which one is the best silk ?
    (a) Erisilk (b) Mulberry silk
    (c) Tassar silk (d) none of the above
12. Chemical nature of silk is :
    (a) protein (b) lipid
    (c) carbohydrate (d) chitin
13. Silkworm is a :
    (a) helminth (b) fly
    (c) beetle (d) moth

Answers

Multiple choice questions

1. (b) 2. (a) 3. (a) 4. (c) 5. (b) 6. (c) 7. (a) 8. (c) 9. (d) 10. (a)
11. (b) 12. (a) 13. (d)

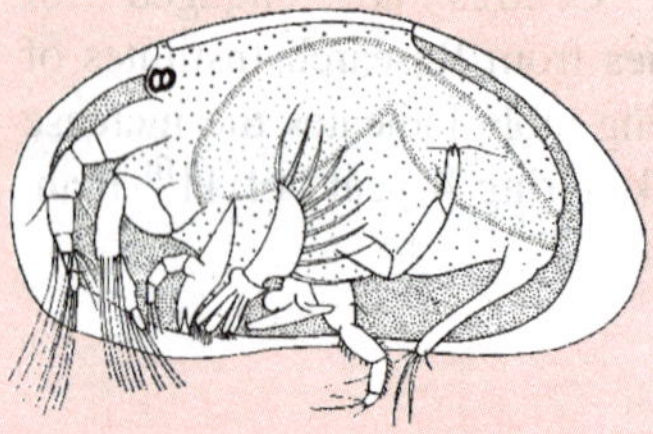

53

Arthropoda : Characters, Classification and Types

Phylum Arthropoda (Gr., **arthros**, joint + **podos**, foot) is the largest phylum of Animal Kingdom including about 12,57,040 species in all habitats, which constitute about 83% of all the known species of animals.

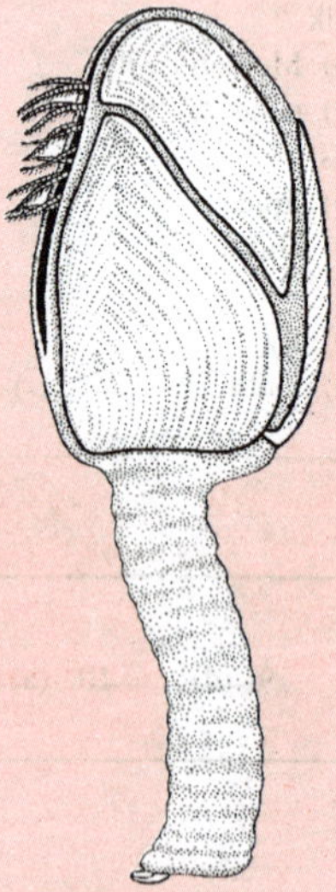

GENERAL CHARACTERS

1. Organ-system level of body organization.
2. Body bilaterally symmetrical, triploblastic and metamerically segmented.
3. Appendages jointed, usually one pair to a somite, and with varied functions as jaws, gills, legs, etc.
4. Exoskeleton of dead chitinous cuticle that is shed at intervals, called ecdysis or moulting, for growth and development.
5. Body divisible into head, thorax and abdomen. Head and thorax often fused to form a cephalothorax.
6. True coelom reduced and largely replaced by a blood-filled haemocoel.
7. Muscles mostly striated, usually capable of rapid contraction.
8. Digestive system complete with mouth and anus. Mouth parts adapted for various modes of feeding.
9. Circulatory system open with a dorsal often many-chambered heart, arteries and blood sinuses or haemocoel.
10. Respiration by general body surface, gills, tracheae or book-lungs.
11. Excretory organs are green glands or Malpighian tubules.

12. Nervous system typically annelidan, with a dorsal brain connected with a nerve ring to a double ventral nerve cord.
13. Sensory organs comprises of eyes (simple and compound), chemo- and tactile receptors, balancing and auditory organs.
14. Sexes usually separate (dioecious). Reproductive organs and ducts paired. Fertilization usually internal. Oviparous or ovoviviparous.
15. Development direct or indirect with one to many larval stages. Parthenogenesis in some.
16. Cilia and flagella absent except in Onychophora.
17. Parental care often well-marked.

CLASSIFICATION

Arthropoda is a much heterogeneous group including a variety of animals with divergent views concerning their phylogeny. Because of this reason, no difinitive system of classifying this phylum exists. The classification adopted here is a synthesis of several views so that such a large and diverse phylum may be conveniently grouped. In fact, arthropod classification is still in a state of flux, and may always remain so.

Seven subphyla are recognized in the following classification. Of these, only Trilobitomorpha, Chelicerata and Mandibulata are definitely arthropods. Trilobitomorpha includes a number of extinct classes only. Onychophora, Tardigrada and Pentastomida show only doubtful or superficial relationships with other arthropods, so that some books treat them as independent minor phyla. We have also described the type *Peripatus* under a separate phylum Onychophora, to accommodate this view. Pycnogonida is sometimes included as a class within the subphylum chelicerata. The old class Myriapoda is retained here within the subphylum Mandibulata, otherwise, every order of Myriapoda is equivalent in status to other classes of Mandibulata.

Subphylum I. Trilobitomorpha

(Gr., tri, three + lobos, lobe + morphe, form)

1. Fossil trilobites. Mostly marine and bottom-dwellers. Cambrian to Permian.
2. Body 3-lobed, due to 2 longitudinal furrows.
3. Head distinct. Probably one pair of antennae.
4. Biramous appendages on all segments except the last one.

   Examples : *Triarthrus, Dalmanites.*

Subphylum II. Chelicerata

(Gr., chele, claw + keros, horn + ata, group)

1. Body divided into an anterior cephalothorax (prosoma) and a posterior abdomen (opisthosoma).
2. Prosomatic appendages 6 pairs. First pair of preoral chelicerae with claws, followed by postoral pedipalps and 4 pairs of walking legs.
3. Antennae and true jaws absent.

CLASS 1. Merostomata

(Gr., **meros**, thigh + **stoma**, mouth)

1. Marine with median simple and lateral compound eyes.
2. 5 to 6 pairs of abdominal appendages with book-gills.
3. Abdomen ending in a sharp telson or spine.
4. Excretion by coxal glands. No Malpighian tubules.

Subclass 1. Xiphosura

(Gr., **xiphos**, sword + **aura**, tail)

1. Cephalothorax with large extended, convex, horseshoe-shaped carapace.
2. Abdomen unsegmented with a long terminal telson.

   Example : *Limulus* (horseshoe or king crab).

Subclass 2. Eurypterida

(Gr., **eurys**, broad + **pteryx**, wing)

1. Extinct marine, giant water scorpions.
2. Cephalothorax small. Carapace plain, not extended.
3. Abdomen 12-segmented and narrowed behind.

   Examples : *Eurypterus, Pterygotus.*

CLASS 2. Arachnida

(Gr., **arachne**, spider)

1. Terrestrial or aquatic. Eyes simple. No compound eyes.
2. Cephalothorax (prosoma) with 2 chelicerae, 2 pedipalps and 4 pairs of walking legs.
3. Abdomen generally without appendages.
4. Respiration by tracheae, book-lungs or both.
5. Excretion by coxal glands and Malpighian tubules.
6. Dioecious. Mostly oviparous courtship before mating.

Order 1. Scorpionida (= Scorpiones)

1. Elongated, fair-sized true scorpions.
2. Small prosoma broadly joined to large opisthosoma, which is made of a broad anterior 7-segmented mesosoma, and a narrow posterior 5-segmented metasoma.
3. Metasoma ending in a telson and poison sting.
4. Two ventral comb-like sensory pectines on 2nd abdominal segment.
5. Respiration by 4 pairs of book lungs.
   Examples : *Buthus, Palamnaeus, Androctonus, Centruroides* (= *Centruus*).

Order 2. Pseudoscorpionida (= Chelonethida)

1. Tiny false scorpions.
2. Abdomen 11-segmented, without sting and telson.
3. Chelicerae 2-jointed, with comb-like serrations.
   Examples : *Chelifer, Microcreagris.*

Order 3. Araneae

1. True spiders, prosoma and opisthosoma without visible segments and joined by a narrow pedicel.
2. Chelicerae 2-jointed, with a poison duct in terminal claw.
3. Pedipalps leg-like, used for transfer of sperms in male.
4. Opisthosoma with 3 pairs of spinnerets. No telson.
   Examples : *Lycosa* (wolf spider), *Agelena* (funnel-web spider), *Latrodectus* (black widow), *Achaearanea* (house spider), *Argiope* (writing spider).

Order 4. Solifugida (=Solifuge)

1. False spiders, sun spiders or wind spiders.
2. Prosoma divided into a large anterior and a small posterior part.
3. Opisthosoma of 10 or 11 segments. No spinnerets.
4. Chelicerae very large forming heavy pincers. No poison glands.
5. A flagellum on each chelicera of male, for sperm transfer.
   Example : *Galeodes.*

Order 5. Palpigradi

1. Small-sized microwhip scorpions. Without eyes.
2. Prosomal carapace made of large anterior and small posterior portions.
3. Opisthosoma 11-segmented, ending in a large 15-jointed telson or flagellum.
   Example : *Koenenia.*

Order 6. Pedipelpi (= Uropygi)

1. Elongate whip scorpions with 1 pair of eyes.
2. Prosomal carapace entire.
3. Opisthosoma 12-segmented. Last segment with a long flagellum or telson.
4. Pedipalps large, heavy, with terminal pincer.
   Examples : *Mastigoproctus, Thelyphonus.*

Order 7. Amblypygi (= Phrynichida)

1. Flattened scorpion-spiders or taillers whip scorpions.
2. Carapace undivided. Pedipalps large and rhaptorial.
3. Abdomen 12-segmented, without flagellum.
4. First pair of legs, long, whip-like, sensory.
   Example : *Charinus.*

Order 8. Ricinulei (= Podogna)

1. Rare, small, tick-like, heavy-bodied arachnids.
2. Carapace with an anterior hood-like movable plate.
3. Opisthosoma 6-segmented. Narrow anteriorly with a posterior tubercle bearing anus.
4. Third pair of legs in male form copulatory organs.
   Examples : *Ricinoides, Cryptocellus.*

Order 9. Opioliones (= Phalangida)

1. Spider-like Harvest-men, Harvest-spiders or daddy longlegs.

2. Body small, oval. Legs extremely long, slender.
3. Prosoma broadly joined to Opisthosoma.
4. Scent glands under carapace. Two eyes.
Examples : *Phalangium, Leiobunum.*

Order 10. Acarina

1. Ticks and mites. Free-living or parasitic.
2. Body small, oval, unsegmented, with no distinction between prosoma and opisthosoma.
3. Largest arachnid order with 20,000 species.
Examples : *Sarcoptes* (*Itchomite*), *Ixodes* (Sheep tick), *Dermacentor* (Dog tick), *Argas* (Bird tick).

Subphylum III. Mandibulata

(L., mandibula, mandible + ata, group)

1. Body divisible into head, thorax and abdomen.
2. Head appendages are 1 or 2 pairs of antennae, 1 pair of jaws or mandibles and 1 or 2 pairs of maxillae.
3. Compound eyes common.

CLASS 1. Crustacea

(L. crusta, shell)

1. Head often joined with thorax to form cephalothorax.
2. Exoskeleton chitinous, hard, limy (calcareous).
3. Head 5-segmented, bearing 2 pairs of antennae, 1 pair of mandibles and 2 pairs of maxillae. Appendages typically biramous.
4. Respiration by gills or body surface.
5. Excretion by antennal glands.
6. Sexes usually separate. Development with nauplius stage.

Subclass 1. Cephalocarida

1. Body made of a horseshoe-shaped head and 19 trunk segments. Only anterior 9 trunk segments bear appendages that appear triramous.
2. Antennae short. Eyes absent.
3. Hermaphrodite. Larva a metanauplius.
Example : *Hutchinsoniella.*

Subclass 2. Branchiopoda

1. Primitive, small-sized, mostly freshwater.
2. Trunk appendages leaf-like, serving for respiration (gills), locomotion and filter-feeding.
3. Antennules and 2nd maxillae reduced or absent.
4. Abdomen ends in a pair of jointed or unjointed caudal styles or cercopods.

Order 1. Anostraca

1. Fairy shrimps with 19 or more trunk segments. Only anterior 11 to 19 segments bear appendages.
2. Carapace absent. Eyes stalked. Styles unjointed.
Examples : *Artemia, Eubranchipus.*

Order 2. Notostraca

1. Tadpole shrimps with 25-45 trunk segments. Anterior half with 35 to 71 pairs of appendages.
2. Carapace shield-like. Eyes sessile. Styles jointed.
Examples : *Apus, Lepidurus.*

Order 3. Diplostraca

1. Clam shrimps and water fleas with a bivalved carapace enclosing body with or without head.
2. Eyes fused, sessile. Styles unjointed, claw-like.
Examples : *Daphnia, Cyzicus* (= *Estheria*).

Subclass 3. Ostracoda

1. Minute mussel or seed shrimps with poorly segmented body entirely enclosed in a bivalved carapace.
2. Trunk appendages 2 pairs, leg-like.
3. Antennules and antennae large, used in swimming.

Order 1. Myodocopa

1. Carapace with antennal notches.
2. Antennae biramous, enlarged at base.
Example : *Cypridina.*

Order 2. Podocopa

1. Carapace unnotched. Trunk appendages 2 pairs.
2. Antennae uniramous, clawed at tips.
Examples : *Cypris, Darwinula.*

Order 3. Platycopa

1. Carapace unnotched. Trunk appendages 1 pair.
2. Antennae uniramous.
   Example : *Cytherella.*

Order 4. Cladocopa

Carapace unnotched. Antennae biramous.
Example : *Polycope.*

Subclass 4. Mystacocarida

1. Primitive. Body microscopic. Antennules and Antennae prominent.
2. A single median eye. No compound eyes.
3. Abdomen limbless. A pair of caudal styles.
   Example : *Derocheilocaris.*

Subclass 5. Copepoda

1. Body small, made of head, thorax and abdomen.
2. No carapace. No compound eyes but a median eye.
3. Antennules long. Antennae smaller.
4. Abdomen limbless. Telson with two caudal styles.
   Examples : *Cyclops, Ergasilus, Caligus.*

Subclass 6. Branchiura

1. Fish lice. Temporarily ectoparasites of skin and gill chambers of fishes and some amphibians.
2. Body dorso-ventrally flattened.
3. Shield-like carapace covers head and thorax.
4. A pair of sessile compound eyes. Mouth suctorial.
5. Antennules and antennae reduced.
6. First maxillae modified into suckers.
7. Abdomen unsegmented, bilobed. Caudal claws minute.
   Examples : *Argulus, Dolops.*

Subclass 7. Cirripedia

1. Barnacles. Adults sessile, attached or parasitic.
2. Carapace forms two folds of mantle surrounding body and covered externally by calcareous plates.
3. Thoracic limbs 6 pairs, biramous and cirriform.
4. Antennules become cement glands for attachment.
5. Antennae and compound eyes are lost in adult.
6. Abdomen rudimentary with caudal styles.
7. Nauplius larva passes through a cypris stage.

Order 1. Thoracica

1. Non-parasitic. With or without stalk, Hermaphrodite.
2. Mantle present with calcareous plates.
3. Thoracic appendages 6 pairs, cirriform.
   Examples : *Lepas, Balanus.*

Order 2. Acrothoracica

1. Sessile. Bore into mollusc shells or corals. Unisexual.
2. Mantle reduced to a chitinous attachment disc.
3. Trunk appendges usually 4 pairs, cirriform.
   Examples : *Alcippe, Cryptophialus.*

Order 3. Ascothoracica

1. Parasitic in echinoderms and corals.
2. Mantle bivalved or saccular.
3. Often appendages are lost but antennules remain present.
   Examples : *Synagoga, Dendrogaster.*

Order 4. Apoda

Parasitic without mantle and appendages.
Example : *Proteolepis.*

Order 5. Rhizocephala

1. Adult parasitic, degenerate, sac-like.
2. Peduncle forms root-like absorptive branches ramifying throughout host's tissues.
   Example : *Sacculina.*

Subclass 8. Malacostraca

1. Body large-sized. Typically made of 19 segments.
2. Head and one or more thoracic segments form cephalothorax.
3. Carapace well-formed or vestigial or absent.
4. Paired compound eyes stalked or sessile.
5. Abdomen ends in a telson. No caudal styles.
   Important orders are as follows :

Order 1. Nebaliacea

1. Carapace bivalved with an adductor muscle.
2. Abdominal segment 7 instead of 6.
   Example : *Nebalia.*

Order 2. Mysidacea

1. Body elongated. Uropods form fan tail.
2. Carapace thin, covering mostly of thorax.
   Example : *Mysis* (opossum shrimp).

Order 3. Cumacea

1. Head and thorax greatly enlarged.
2. Carapace fused to 3-4 thoracic segments.
3. Abdomen narrows. Uropods slender.
   Examples : *Diastylis, Cumopis.*

Order 4. Isopoda

1. Wood lice. Body dorso-ventrally flattened.
2. Head and 1 or 2 thoracic segments form cephalothorax.
3. Carapace absent. Gills and heart abdominal.
   Examples : *Oniscus, Asellus, Limnoria.*

Order 5. Amphipoda

1. Sand hoppers. Body laterally compressed.
2. Carapace absent. Gills thoracic. Eyes sessile, lateral.
   Examples : *Gammarus, Caprella, Cyamus.*

Order 6. Stomatopoda

1. Mantis shrimps. Body flattened. Carapace small.
2. Abdomen large, broader than cephalothorax.
3. Second maxillipedes raptorial. Gills abdominal.
   Example : *Squilla.*

Order 7. Decapoda

1. Shrimps, crayfishes, Lobsters, prawns, crabs, etc.
2. Carapace well-developed. Usually enclosing gill chambers on sides of cephalothorax.
3. First 3 pairs of thoracic limbs form maxillipedes.
4. Gills usually in 3 series present on thorax.
5. Statocyst present. Larva typically a zoaea.

Suborder (a) Natantia

1. Body laterally compressed. Rostrum prominent.
2. Pleopods well-developed. Modified for swimming.
   Examples : Prawns (*Palaemon, Penaeus, Macrobrachium*), Shrimp (*Leucifer, Crangon*).

Suborder (b) Reptantia

1. Body dorso-ventrally flattened. Rostrum short or absent.
2. Pleopods reduced. Not modified for swimming.
   Examples : Lobsters (*Palinurus, Scyllarus, Homarus*), Crayfish (*Astacus, Cambarus*), True crabs (*Cancer, Carcinus*), Hermit crab (*Eupagurus*), coconut crab (*Hippa*), Spider crab (*Inachus*).

CLASS 2. Myriapoda

(G., myrios, ten thousand + **podos**, foot)

1. Exclusively terrestrial, air-breathing mandibulate arthropods.
2. Body worm-like, made of head and elongated trunk with many similar leg-bearing segments.
3. Antennae 1 pair, jaws 3 pairs, legs more than 11 pairs.
4. Respiration by tracheae. Spiracles arranged segmentally.
5. Excretion by 1 or 2 pairs of Malpighian tubules.
6. Sexes separate, Gonad single. Gonoducts paired.

Note : The following 4 orders of group Myriapodia are not related to one another and exhibit marked dfferences. Therefore, each one of them is nowadays treated as a separate class of mandibulate arthropods. However, the old class Myriapoda has no systematic status and is retained here just for convenience.

Order (or class) 1. Diplopoda

(Gr., **diplos**, double + **pous**, foot)

1. Millipedes. Body elongate, sub-cylindrical and divisible into 5-segmented head, 4-segmented thorax and 11 to 100-segmented trunk.
2. Legs 2 pairs on each trunk segment (Diplopoda).
3. Mandibles and maxillae 1 pair each (Dignatha).
   Examples : *Spirobolus, Julus.*

Order (or class) 2. Chilopoda

(Gr., **cheilos**, lip + **pous** (foot)

1. Centipedes. Body dorso-ventrally flattened and divisible into head and 15 to 173 trunk segments.

2. Legs 1 pair on each trunk segment.
3. Mandibles 1 pair. Maxillae 2 pairs (Trignatha).
4. First pair of legs form poison claws.
Examples : *Scutigera, Lithobius, Scolopendra.*

Order (or class) 3. Pauropoda
(Gr., **pauros**, small + **pous**, foot)
1. Minute grub-like body divisible into head and 11-12 trunk segments with 9-10 pairs of legs.
2. No eyes.
Example : *Pauropus.*

Order (or class) 4. Symphyla
(Gr., **Syn**, together + **phylon**, tribe)
1. Body slender made of head and 15-22 trunk segments with 10-12 pairs of legs.
2. No eyes.
Example : *Scutigerella.*

CLASS 3. Insecta
(L. **insectus**, cut or divided)
1. Body made of head (6 fused segments), thorax (3 segments) and abdomen (up to 11 segments).
2. Head with compound eyes (1 pair), antennae (1 pair) mandibles (1 pair) and maxillae (2 pairs).
3. Mouth parts modified for different feeding habits.
4. Thorax with 3 pairs of jointed legs and 1 or 2 pairs of wings which may be absent.
5. Respiration by tracheae. Spiracles lateral.
6. Excretion by Malpighian tubules.
7. Unisexual. Fertilization internal. Development usually with metamorphosis.

Subclass 1. Apterygota (Ametabola)
1. Primitively wingless insects.
2. Abdomen with cerci and style-like appendages.
3. Little or no metamorphosis.

Order 1. Protura
1. No antennae, true eyes and metamorphosis.
2. Abdomen of 11 segments plus a telson.
Example : *Acerentulus.*

Order 2. Collembola
1. No eyes, tracheae, Malpighian tubules and metamorphosis. Mouth parts chewing or sucking.
2. Abdomen 6-segmented, with a springing organ.
Examples : Springtails, *Achorutes, Sminthurus.*

Order 3. Thysanura
1. Body covered by minute silvery scales.
2. Antennae long. Mouth parts chewing.
3. Abdomen 11-segmented. Cerci and telson long.
Example : *Lepisma* (silver fish).

Subclass 2. Pterygota (Metabola)
1. Wings present. Secondarily lost in some.
2. No abdominal appendages except cerci.
3. Metamorphosis complete or incomplete.

Division (a). Exopterygota (Heterometabola)
1. Wings develop externally as buds.
2. Metamorphosis gradual. Young stages are nymphs.

Order 1. Orthoptera
1. Wings 2 pairs. Forewings straight and leathery. Hindwings membranous and folded at rest.
2. Mouth parts chewing. Prothorax large. Hindlegs jumping.
Examples : *Romalia* and *Poecilocercus* (Grasshoppers), *Schistocerca* (Locust), *Periplaneta* (Cockroach), *Gryllus* or *Acheta* (Cricket), *Mantis* (Praying mantis), *Phyllium* (Leaf insect), *Carausius* (Stick insect).

Order 2. Isoptera
1. Wings 2 pairs; Held flat on back. Or wingless.
2. Mouth parts chewing. Social insects with many castes.
Examples : Termites or white ants.

Order 3. Dermaptera
1. Forewings small, leathery. Hindwings large, semicircular.
2. Mouth parts chewing. Forcep-like cerci at the tip of abdomen for offense and defense.
Example : *Forficula* (Earwig).

Order 4. Ephemeroptera
1. Wings 2 pairs, membranous. Forewings longer and triangular. Hindwings smaller and rounded.

2. Adult mouth parts vestigial. Mandibulate in nymphs.
3. Abdomen carries long cerci and caudal filament.
   Example : *Ephemera* (Mayfly).

Order 5. Odonata

1. Wings 2 pairs, membranous. Eyes very large.
2. Mouth parts chewing. Predaceous.
   Examples : Dragon flies, Damsel flies.

Order 6. Plecoptera

1. Wings 2 pairs, membranous, longer than body.
2. Antennae long. Mouth parts chewing.
3. Naiads usually with tracheal gills.
   Example : *Isoperla* (Stonefly).

Order 7. Psocoptera (= Corrodentia)

1. Wingless or forewings larger than hindwings.
2. Antennae long. Mouthparts chewing. Cerci absent.
   Examples : Book lice (wingless), Bark lice (winged).

Order 8. Mallophaga

1. Wings absent. Body small. Head large. Mouth parts chewing. Eyes degenerate. Legs clasping.
2. Ectoparasitic on skin, hairs and feathers of mammals and birds.
   Examples : Bird lice, Biting lice (on mammals).

Order 9. Anoplura (= Siphunculata)

1. No wings. Body broad, flat. Head small. Mouth parts piercing and sucking.
2. Claws clinging to hairs. Ectoparasitic on mammals.
   Example : *Pediculus* (Human louse).

Order 10. Thysanoptera

1. Wings 2 pairs, similar, fringed with long hairs.
2. Mouth parts rasping and sucking.
   Example : Thrips.

Order 11. Hemiptera

1. Wings 2 pairs or wingless. Forewings thickened at base, membranous at tip (**hemelytra**).
2. Mouth parts piercing-sucking, forming jointed beak.
   Examples : Bedbug (*Cimex*), Giant water bug (**Belostoma**), Water scorpion (*Ranatra*).

Order 12. Homoptera

1. Wingless or 2 pairs of uniform membranous wings.
2. Mouth parts form a piercing and sucking beak.
   Examples : Cicadas, Aphids, Scale insects.

Subclass 3. Endopterygota (Holometabola)

1. Wings develop internally in pupal case.
2. Metamorphosis complete with larval and pupal stages.

Order 1. Neuroptera

1. Wings large, membranous, many-veined.
2. Antennae long. Mouth parts chewing. Cerci absent.
3. Larvae carnivorous. Abdominal gills in aquatic larvae.
   Examples : *Crysopa* (Lacewing), *Myrmeleon* (Antlion).

Order 2. Coleoptera

1. Forewings leathery (elytra). Hindwings membranous, folding.
2. Antennae variously modified. Mouth parts chewing.
   Example : Beetles.

Order 3. Mecoptera

1. Wings long, similar, narrow, membranous.
2. Mouth parts chewing, on a prolonged beak.
3. In male, tip of abdomen curved sting-like.
   Example : *Panorpa* (Scorpion fly).

Order 4. Trichoptera

1. Wings long, hairy, folded roof-like over abdomen.
2. Antennae long. Mouth parts rudimentary.
3. Larva pupates within a tube of foreign particles.
   Example : Caddis flies.

Order 5. Lepidoptera

1. Wings membranous, covered with over-lapping scales.
2. Mouth parts sucking, coiled under head.
3. Larva a caterpillar with chewing mouth parts.

Examples : Butterflies (antennae filamentous), Moths (antennae feathery).

Order 6. Diptera

1. Wings 1 pair. Hindwings as knob-like halteres.
2. Mouth parts piercing-sucking or sponging.
3. Larva limbless, wormlike, called maggot.
   Examples : *Musca* (House fly), *Culex* (Mosquito), *Drosophila* (Fruit fly).

Order 7. Hymenoptera

1. Wings 2 pairs, similar, membranous. On each side hooked together during flight.
2. Mouth parts sucking or chewing. Ovipositor of female usually forms a piercing sting.
3. Highly specialized. Some social in behaviour.
   Examples : *Apis*, wasp / *Vespa* and Ants etc.

Order 8. Siphonaptera

1. Small. Laterally flattened. Secondarily wingless.
2. Mouth parts piercing-sucking. Legs long, leaping.
3. Ectoparasites on birds and mammals.
   Examples : *Pulex* and *Xenopsylla* (Fleas).

Minor Doubtful Arthropoda
Subphylum IV. Onychophora
(G., onychos = claw + phoros, bearing)

1. Terrestrial, primitive, worm-like, unsegmented.
2. Single pairs of antennae, eyes and jaws.
3. Numerous stumpy, unjointed clawed legs.
   Example : *Peripatus*

Subphylum V. Tardigrada

1. Minute, aquatic. Segmentation indistinct. No antennae.
2. Mouth retractile, with a pair of horny stylets.
3. Four pairs of stumpy, unjointed, clawed legs.
4. No respiratory, circulatory and excretory organs.
   Example : *Macrobiolus* (Water bear).

Subphylum VI. Pentastomida (= Linguatulida)

1. Vermiform, unsegmented, parasitic worms. No antennae.
2. Two pairs of ventral retractile hooks near mouth.
3. No respiratory, circulatory and excretory organs.
   Example : *Linguatula* (Tongue worm).

Subphylum VII. Pycnogonida (= Pentopoda)

1. Small, marine, spider-like. Abdomen vestigial.
2. Mouth on a long proboscis. 4 simple eyes.
3. Appendages include chelicerae, pedipalps, ovigerous legs (1 pair) and long walking legs (4 to 12 pairs).
4. No respiratory and excretory systems.
   Example : *Pycnogonum, Nymphon* (Sea spiders).

OTHER TYPES OF ARTHROPODA

1. *Limulus.* *Limulus* or King crab belongs to the subclass Xiphosure and class Merostomata of subphylum chelicerata. It is a large-sized, marine animal, up to 60 cm long. It lives in shallow water, burrowing in sand or mud, along the coast. It is a living member of very ancient (200 million years) primitive chelicerates and hence called a 'living fossil'. Dorso-ventrally flattened body is composed of an anterior **prosoma** (6 fused segments) and a posterior **opisthosoma** (9 segments). Prosoma is covered by a large horseshoe-shaped **carapace** marked by two lateral longitudinal grooves and bearing median simple eyes (ocelli) and 2 lateral compound eyes. Ventrally, prosoma bears a pair of 3-jointed preoral **chelicerae** followed by 5 pairs of 6-jointed chelate legs. In male, 1st pair of legs form clasping

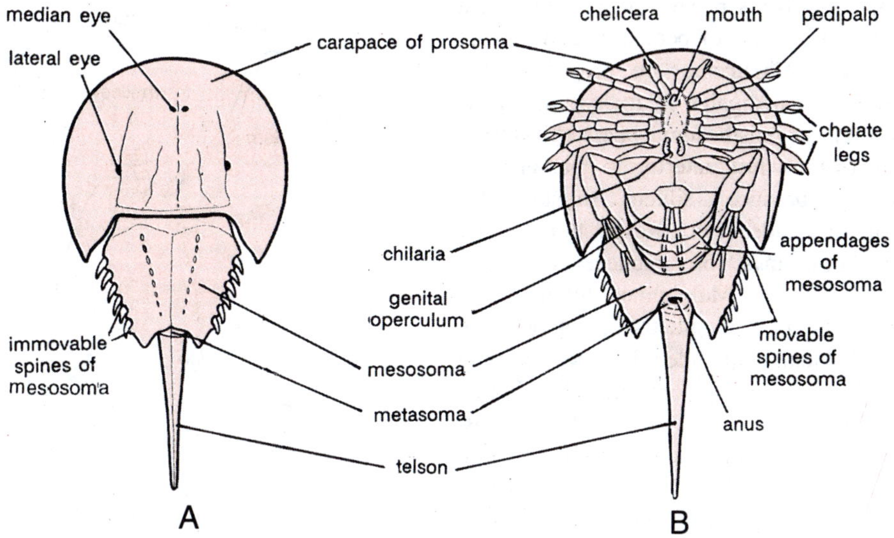

Fig. 1. *Limulus.* **A.** Dorsal view **B.** Ventral view.

organs. A small 7th pair of **chilaria** of doubtful function are also present.

The broad hexagonal opisthosoma is divisible into an anterior **mesosoma** (6-segments) covered by a dorsal shield, and a posterior **metasoma** (3-segments) ending in a long pointed **telson**. Mesosoma bears 6 pairs of thin, broad appendages fused in the median line. Last five pairs carry 5 pairs of **book-gills**.

The animal feeds upon molluscs, worms and bottom algae. Excretion takes place through 4 pairs of **coxal glands**. Sexes are separate. In breeding season males and females copulate in shallow water. Eggs, laid by females in holes in sand, are fertilized by sperms. *Limulus* (=*Xiphosura*) *Polyphemus* is found on the Atlantic American Coast while *Tachypleus* and *Carcinoscorpus* along coasts of south-east Asia. *Paralithodes comtschatica* is an edible species found off the North Pacific coasts.

2. Spiders. Spiders belong to the order Araneae of class Arachnida. Body is divisible into prosoma and opisthosoma, without visible segments, and connected together by a short and narrow pedicel or 'waist'. Prosoma is covered by carapace that bears 8 simple eyes anteriorly.

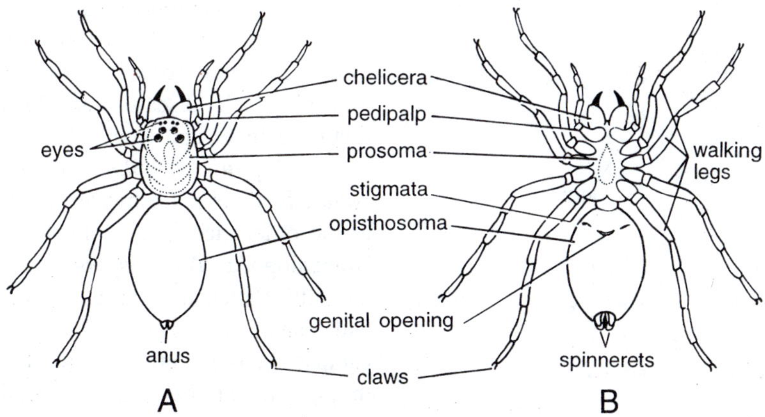

Fig. 2. Spider. **A.** Dorsal view. **B.** Ventral view.

Ventrally prosoma bears 6 pairs of appendages, a pair of chelicerae, a pair of pedipalps and 4 pairs of walking legs. Chelicerae are provided with poison glands and pedipalps in male function as copulatory organs. Opisthosoma bears **spinnerets** or **spinning organs** just anterior to the terminal **anus**. These produce silken threads for construction of spider-web. Respiration is by means of book-lungs or tracheae or both. Excretory organs are Malpighian tubules. All spiders are predators. The soft tissues of prey are liquefied and sucked up aided by the pumping stomach. Most male spiders have elaborate courtship displays. The silk secreted by most spiders is used for constructing webs to trap insects, as guide lines for males in finding sexual partners and for the construction of egg sacs.

Common spiders are trap-door spiders (*Pachlomerus*), funnel-web spiders (*Agelena*), wolf-spiders (*Lycosa*), oral web spiders (*Argiope*), black-widow spider (*Latrodectus*), etc.

3. Mites. Mites belong to the order Acarina of class Arachnida. These are minute (1 mm) sac-like animals without body divisions and abdominal segmentation. Mouthparts produce a head-like structure, the **gnathosoma**, that bears a pair of chelicerae and a pair of pedipalps. Legs are 4 pairs. The whole body bears long pointed spines which are sensory in function.

Mites are free-living as well as parasitic. The majority of parasitic mites are ectoparasites of both vertebrates and invertebrates. Many are plant parasites as well. The most common parasitic species are the human itch-mites (*Sarcoptes*). Other

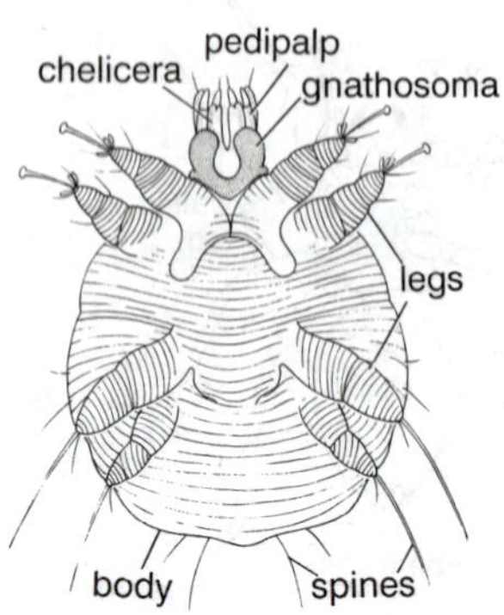

Fig. 3. *Sarcoptes scabiei.* Human itch-mite (ventral view).

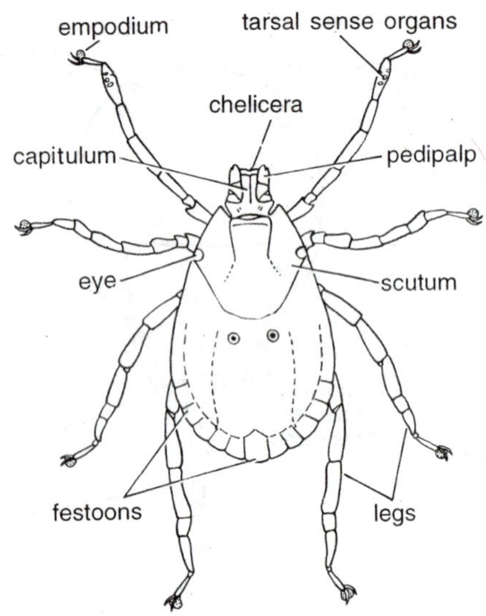

Fig. 4. Tick. Dorsal view.

species include spider-mites, chigger-mites, feather-mites, etc.

4. Ticks. Like mites, ticks are also members of the class arachnida, order Acarina. These are so minute creatures but larger than mites. There is no external visible segmentation and prosoma (cephalothorax) is fused with opisthosoma (abdomen). Most ticks are ectoparasites of vertebrates and provided with blood-sucking mouthparts with recurved teeth adapted for piercing. Larvae are present in bushes and attach themselves to a host as it passes. Respiration takes place by tracheae and a tracheal spiracle is located behind third or fourth pair of coxae. Common genera of ticks are *Argas* (bird tick). *Dermacentor* (dog tick) and *Ixodes* (sheep-tick). Ticks may transmit such diseases as Texas cattle fever, relapsing fever and Rocy Mountain spotted fever.

5. *Daphnia.* *Daphnia,* commonly known as 'water flea', is a freshwater crustacean belonging to the order cladocera (= Diplostraca) of subclass Branchiopoda. Body is about 2 mm long, oval, laterally flattened with a ventral beak on head and a sharp, posterior caudal spine. Thin bivalved carapace cover entire body but not head, and through it can be seen the beating heart. Both sessile compound eyes are united into one large structure. The brood pouch in female lodges the

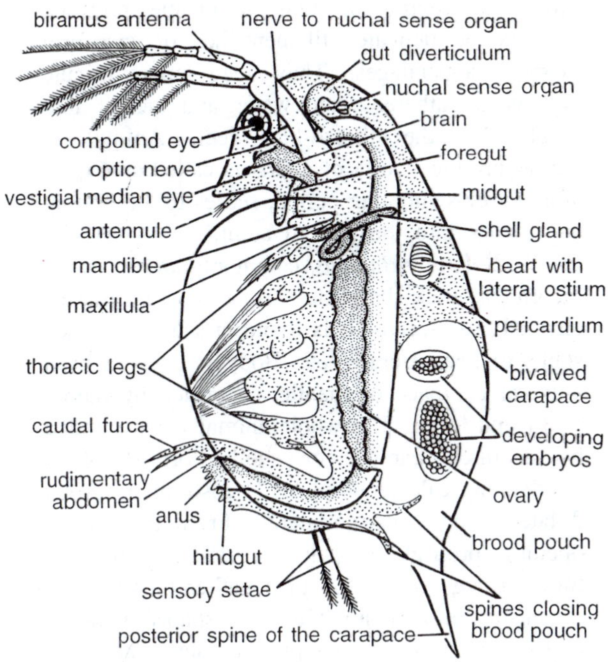

Fig. 5. *Daphnia.* Female in lateral view.

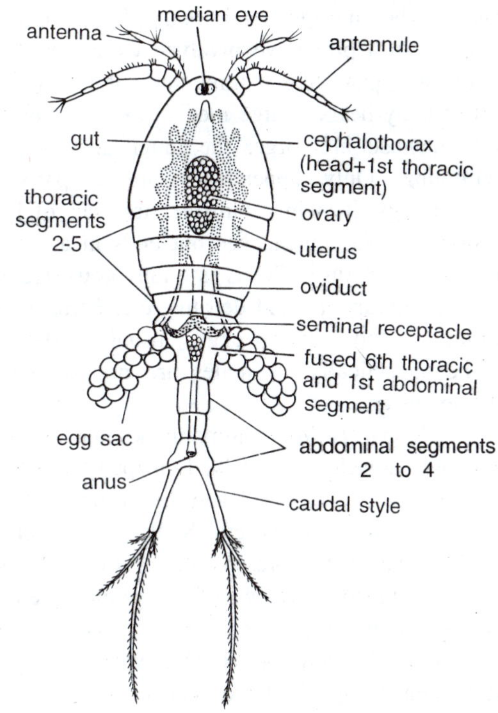

Fig. 6. *Cyclops.* Female in dorsal view.

eggs where the embryos develop. Five pairs of thoracic appendages form an efficient food gathering mechanism. *Daphnia* swims by rapid jerks of the two very large biramous antennae. Antennules are much reduced and uniramous.

6. *Cyclops.* *Cyclops,* because of the possession of a simple median eye, is named after the one-eyed giant of Greek mythology called 'cyclops'. It is the most familiar copepod crustacean found in freshwater ponds. Elongated or pear-shaped body measures 1.5 to 5 mm in length. Head and first thoracic segment become fused to form cephalothorax, which is covered dorsally by a carapace. 4 appendages-bearing thoracic segments are free, whereas the 6th one becomes fused with the 1st abdominal segment. 4th abdominal segment bears a pair of caudal styles or forked tail. The uniramous antennules are long and conspicuous, whereas the antennae are very small. Mature female is easily identified as it carries a pair of posterior **egg sacs**, filled with eggs. *Cyclops* serve as intermediate hosts for the guineaworm (*Dracunculus*) of man.

7. *Cypris.* *Cypris,* commonly known as 'mussel shrimp', is a small ostracodan crustacean. It occurs in large numbers in stagnant freshwater

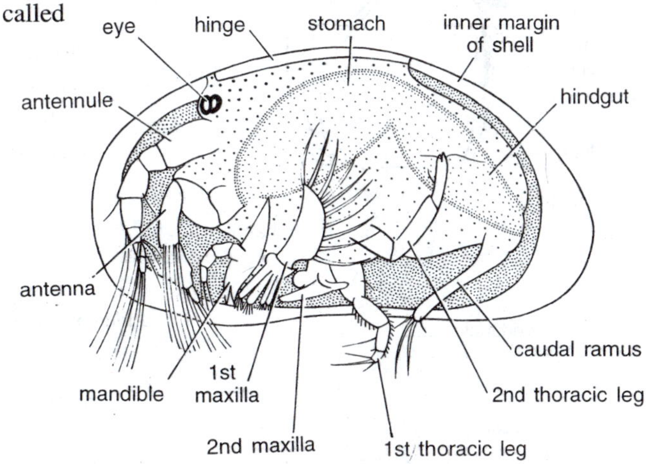

Fig. 7. *Cypris.*

ponds. The unsegmented body is completely enclosed in a protective bivalved carapace which makes it appear like a miniature clam. Anterior end of body bears a large median eye. Antennules and antennae are large, the latter also help in swimming. Only appendages are 2 pairs of thoracic legs. It swims with the first pair of legs. Abdomen lacks appendages but ends in a pair of small **caudal ramii**. *Cypris* is omnivorous, feeding on small organisms and detritus, etc. Females are common and their eggs usually develop parthenogenetically, as males are not known in certain genera.

8. *Lepas*. *Lepas*, commonly known as 'ship or goose barnacle', is a stalked cirriped crustacean. It is found in seas attached by a peduncle to wooden ships, wharves, etc. Stalk or peduncle is highly movable. It represents the oral end of body and contains the vestiges of antennules (cement glands). Body proper or **capitulum** is protected by a bivalved mantle or carapace, which is further strengthened by 5 large calcareous plates : 2 proximal **scuta**, 2 distal **terga** and single dorsal **carina**. Because of these calcareous plates early zoologists considered *Lepas* to be a mollusc. From the ventral opening of mantle protrude 6 pairs of delicate, filament-like or cirriform thoracic appendages. They feed upon minute organisms gathered from water and kicked into mouth by the thread-like feet. Animal is hermaphrodite. Larva of *Lepas* is called as *Cypris* after a free-swimming period, settles to the bottom and grows into the stalked adult.

9. *Balanus*. *Balanus,* commonly known as the 'acorn or rock barnacle', is a sessile, marine cirriped crustaceans. It is found attached to rocks, shells of hermit crabs, etc. Body is covered by a valvular carapace or mantle, supported by paired **scuta** and **terga**. The whole animal is surrounded by a thick parapet or **shell**, formed of 6 overlapping calcareous plates (1 rostrum, 1 carina, 2 lateral and 2 carino-lateral). From the mantle opening protrude 6 pairs of delicate, fringed, **thoracic legs** or **cirri**, which gather minute food organisms into mouth. Gut is U-shaped with an expansion forming a simple stomach. There are no blood vessels or heart. Nervous system is simple and concentrated in trunk region. Adult excretory organs are maxillary glands. The animal is hermaphrodite. When the tide is out, the mantle

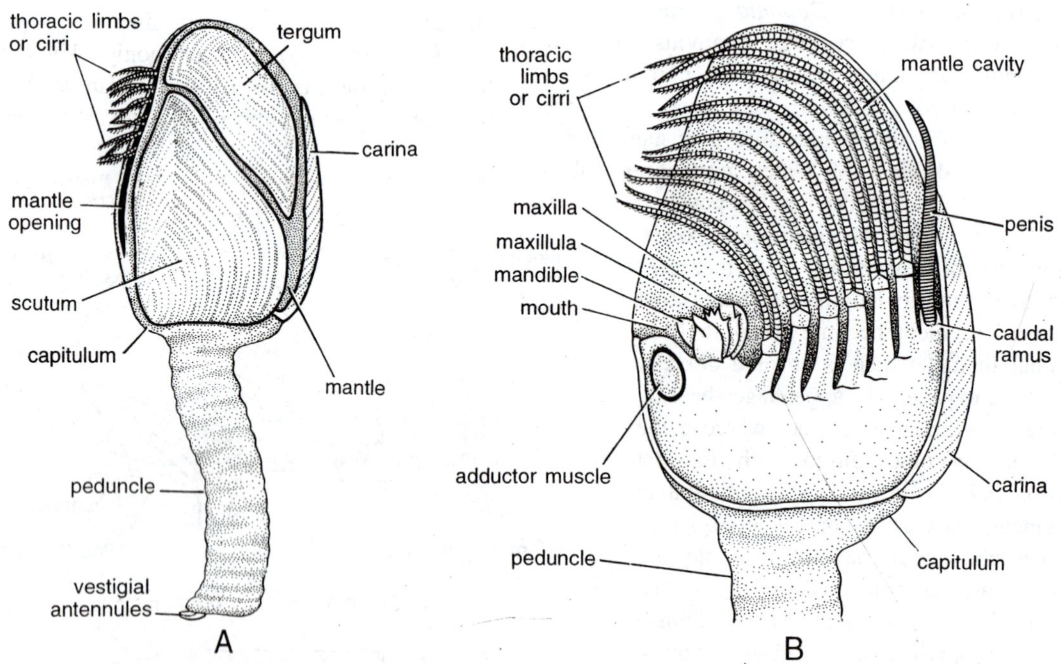

Fig. 8. *Lepas.* **A.** Lateral view. **B.** Lateral view of capitulum after removing a scutum and a tergum.

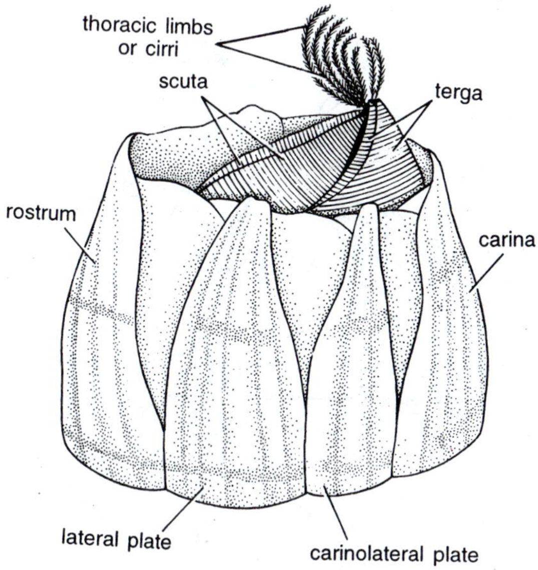

Fig. 9. *Balanus.* Lateral view.

plates close for protection. Rock barnacles are serious pests on the bottom of ships.

10. *Sacculina.* (a) *Structure.* The best known of parasitic barnacles are the rhizocephalans (subclass Cirripedia), and the famous example is *Sacculina,* parasitic on crab. It shows extreme degeneration due to parasitic mode of life. The adult looks like a large tumour or ovoid sac fastened to the abdomen of crab by a short stalk. It has no segmentation, appendages, sense organs and alimentary canal. It comprises only gonads enclosed in a thin bag-shaped mantle. A root-like system of numerous delicate filaments or rhizoids, arising from its peduncle or stalk, extends into all parts of the body of host crab and absorbs nutrients.

(b) Life-history. Life history begins with a free swimming **nauplius** larva having characteristic frontal horns, a median eye and 3 pairs of unjointed appendages. It develops into a **cypris** larva, which attaches by one of its antennules to a bristle on the limb of crab. The cypris loses its trunk and appendages and becomes a rounded mass of cells, called **kentrogen** larva. It enters the host's body through a hollow bristle and attaches to its intestine. The endoparasite now sends a system of branching rootlets to all parts of crab's body for getting nourishment. When the host crab moults, the endoparasite becomes the adult ectoparasite.

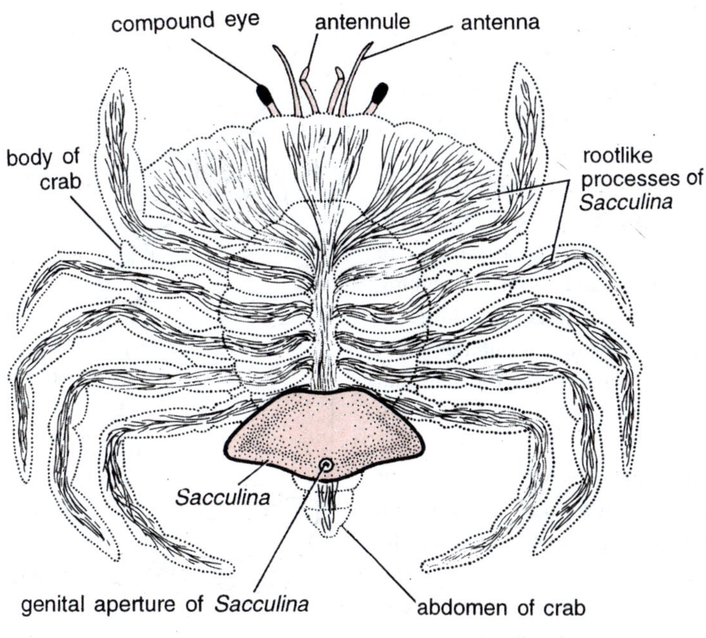

Fig. 10. *Sacculina,* parasitizing a crab.

(c) *Effects of parasitism on host.* Crabs infested by *Sacculina* exhibit striking effects.

(*i*) *Moulting ceases.* When the parasite becomes external, further moulting or ecdysis stops.

(*ii*) *Parasitic castration.* Reproductive organs of infested crab become reduced or completely degenerated. This phenomenon is known as **parasitic castration**. The infested males acquire female characteristics. Their abdomen becomes longer and broader, chelae become smaller, copulatory styles reduce or disappear, and they develop ovigerous appendages as in female crab. Parasitized female crabs revert to a juvenile stage. Their gonads atrophy and ovigerous appendages reduce, but they do not develop male secondary sexual characters.

(*iii*) *Hermaphroditism due to parasitism.* If the *Sacculina* drops off, the male host crab recovers and develops a hermaphrodite gland. However, the female crab does not develop a hermaphrodite gland after recovery from the disease.

11. *Squilla.* *Squilla* belongs to the order stomatopoda of subclass Malacostraca. It is a large marine crustacean living in burrows. Body is dorso-ventrally flattened and about 25 cm long. Small shield-like carapace leaves the last 4 thoracic segments uncovered. Anteriorly, the head has 2 movable segments, the first bearing the large, stalked and bilobed compound eyes and the second one bearing the antennules. First 5 pairs of thoracic limb are turned forward to serve as maxillipedes of which the second pair is exceptionally large, sub-chelate and raptorial. Remaining 3 pairs of thoracic appendages are biramous, having exopodites, and form the walking legs. The 6 segmented abdomen is very large in proportion to cephalothorax and terminates in a broad telson armed with three pairs of marginal spines. Abdominal appendages or pleopods are large and biramous of which the first 5 pairs have gills attached to their exopodites while the last or 6th pair forms the large uropods. Larvae are pelagic and in their general form resemble the zoaea larva of crab. *Squilla* or the 'mantis shrimp' feeds on small fishes, crustaceans, molluscs, etc.

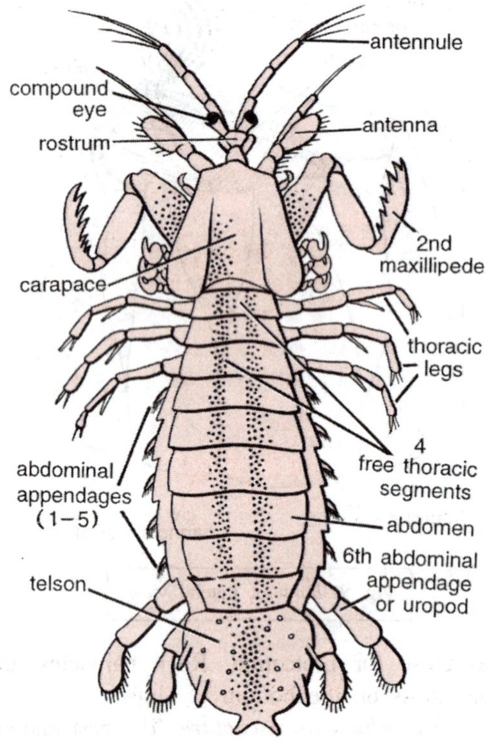

Fig. 11. *Squilla.* Dorsal view.

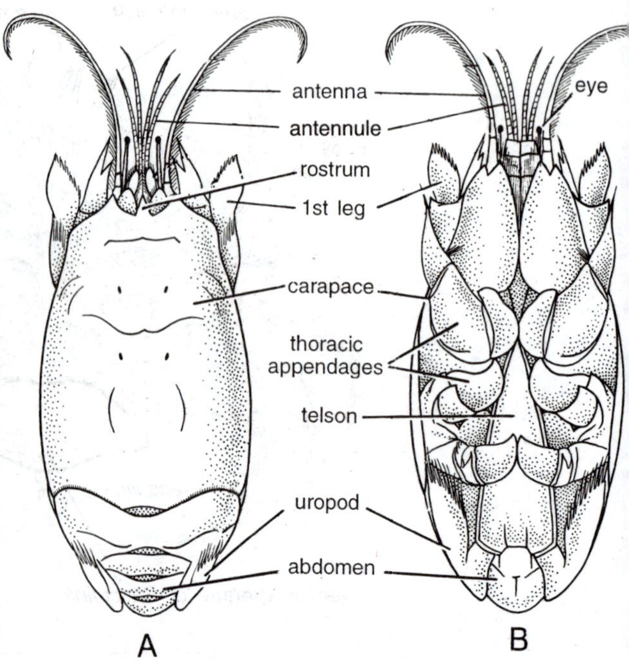

Fig. 12. *Hippa* or *Emerita.* **A.** Dorsal view. **B.** Ventral view.

12. *Hippa* or *Emerita.* *Hippa* or *Emerita* are commonly known as 'mole crab', is found in bottom sand near sea coasts. It is also known as *Emerita*. It belongs to the order Decapoda of

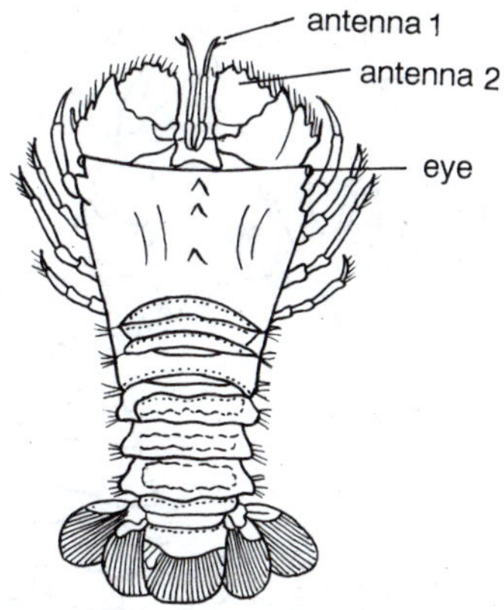

Fig. 13. *Palinurus* (spiny lobster), in dorsal view.

subclass Malacostraca. Body is oval with a large, more or less cylindrical cephalothorax with similar carapace. Head bears a pair of stalked compound eyes, short antennules and long hair antennae. Rostrum is very much reduced. The 5 pairs of thoracic walking legs are adapted for digging, being curved and flattened. The first legs are chelate. A gill is attached with each leg. Abdomen is 6-segmented, symmetrical, flexed underneath thorax and terminating in a long, tapering telson. Anterior pleopods produce water currents to bathe the gills. Uropods are directed posteriorly and used for darting.

13. Lobster (*Palinurus*). Crayfishes (*Cambarus, Astacus*) and lobsters (*Homarus, Palinurus*) are large-bodied, marine, bottom-dwelling crustaceans belonging to the order Decapoda of subclass Malacostraca. They are greatly priced for their meat. Their description is similar since, they differ from each other only in minor structural details.

Spiny or rock lobsters (*Palinurus*) have their bodies covered with spines. Size ranges from 15 to 45 cm. They differ from other lobsters and cryfishes in having no antennal scale, all legs similar and non-chelate, and pleopods broad and fin-like. 1st abdominal segment is without pleopods. They inhabit the sea bed in rocky areas. They are secretive, nocturnal and omnivorous feeding on worms, molluscs corals, algae, etc.

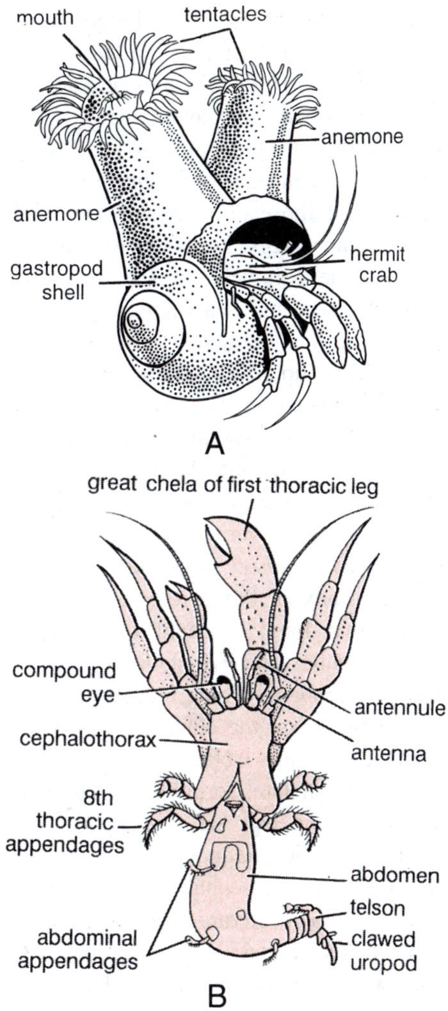

Fig. 14. *Eupagurus.* **A.** Commensalism between hermit-crab and sea-anemone. **B.** Hermit crab taken out of mollusc shell.

They emit a peculiar sound like filing a saw or cracking of leather, by rubbing antennal pad against a sternal keel. The larva, called **phyllosoma**. Broad and flattened schizopod, species of *Palinurus* also occur in Indian waters. Their meat is excellent and they are also exported.

14. *Eupagurus.* *Eupagurus* is commonly called the 'hermit crab'. It lives in empty shells of marine snails, for protection, and moving to larger shells as it grows. Cephalothorax is broad, flat and with a hard carapace. Antennules are short but antennae are long. Eye stalks are also elongated. Thoracic legs are 5 pairs, of 1st pair chelate and the last 2 pairs reduced. Right chela is greater than the left. Abdomen becomes asymmetrical and spirally twisted to fit into the coils of molluscan shell. Abdominal appendages of short side become atrophied, and those of long side are retained in females to carry eggs. Uropods become hook-like to hold on the columella of shell.

Hermit crab and sea-anemone often form a close association, termed **commensalism**. Sea-anemone lives upon the molluscan shell, while hermit crab inside it. Sea-anemone protects the hermit crab by keeping its enemies away by its offensive odour and unpalatable taste. In return, the hermit crab carries the sea-anemone from place to place, providing varieties of food.

15. *Cancer*. *Cancer* is a true crab known as rock crab. Body is flat and oval. Cephalothorax is broader than long. Antennules and antennae are small. Antennules and eye stalks are contained in sockets of carapace. Third maxillipedes are flat, plate–like and cover the other mouth parts. 5 pairs of thoracic legs are well developed and clawed. First legs are chelate, forming the large pinching claws. Remaining legs are non-chelate. Abdomen is segmented but reduced and remains permanently bent under cephalothorax fitting into the groove in thoracic sterna, thus remaining almost invisible in the dorsal view of animal. Abdomen is narrower in male but somewhat broader in female. Pleopods are greatly reduced. Male retains only 2 pairs of them, which serve as copulatory organs. Female has 4 pairs of pleopods for carrying eggs. Uropods are absent. The young hatches in the zoaea stage which passes through

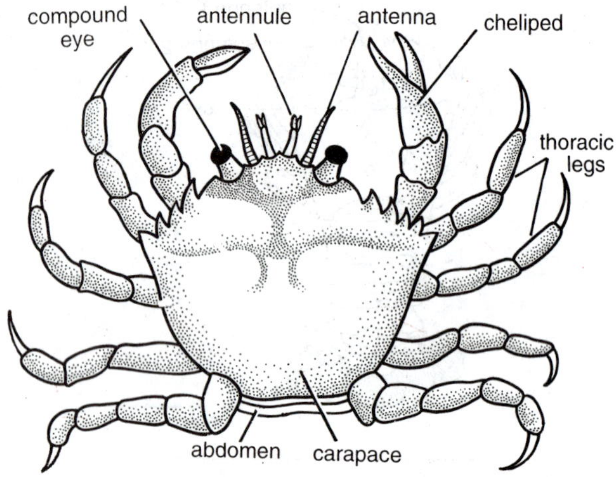

Fig. 15. *Cancer* in dorsal view.

a megalopa stage before reaching maturity, *Carcinus maenas* (dark-green shore crab) and *Callinectes hastatus,* (Atlantic blue crab) are other examples of true crabs.

16. Millipedes. Millipedes belong to the order (or class) Diplopoda of Myriapoda. Millipede are called 'thousand leggers' because of possession of numerous walking legs. Most common millipedes are *Julus* and *Spirobolus,* that live underneath leaves, barks, logs and stores, especially in tropical countries. Body is elongated, cylindrical, about 15 cm long, made of small head and about 40 trunk segments which are **diplosegments**, that is, each with 2 pairs of jointed legs. Anterior 4 segments constitute the **thorax**. Its 1st segment is legless (**collum**) while other are with only a single pair of legs. In some

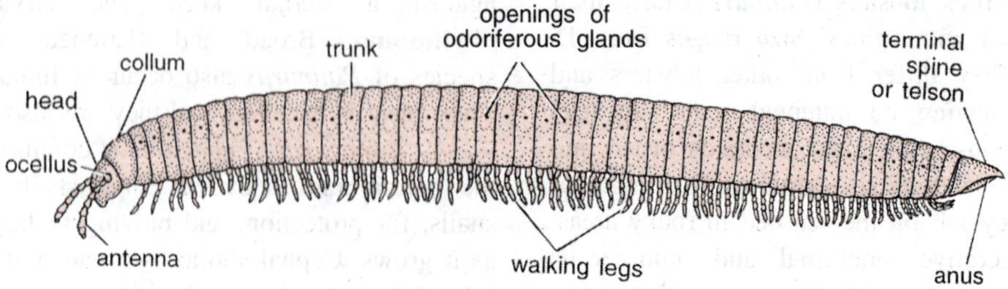

Fig. 16. *Spirobolus.* A millipede.

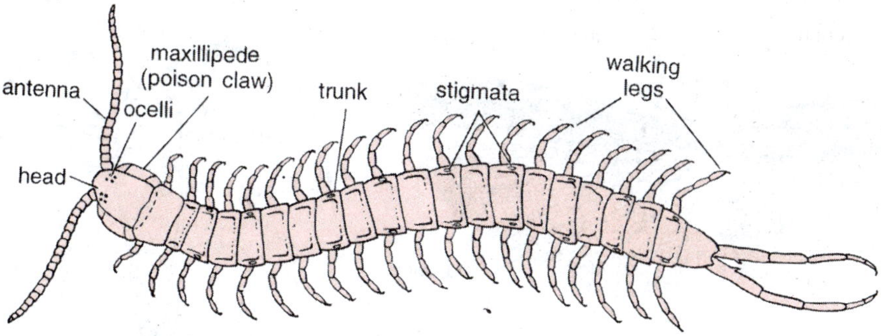

Fig. 17. *Scolopendra.* A centipede.

millipedes last 1-5 segments are also legless. Head bears 2 clumps of simple eyes (**ocelli**) and one pair each of antennae, mandibles and maxillae. There are no poison claws.

On most of the trunk segments are located **stink glands** or **odoriferous glands**, opening to outside through minute openings. Their secretions are toxic to other animals.

Millipedes are sluggish, timid and secretive. Inspite of their numerous legs, they move very slowly. When touched or lifted they curl up in a flat spiral like a watch spring. Some may roll themselves into a ball. They are mostly herbivorous (vegetarian) or saprophytic (scavenger). Some attack living roots causing serious damage to plants in gardens and greenhouses.

17. Centipedes. Centipedes or 'hundred leggers' are most familiar myriapod arthropods. They are cosmopolitan in occurrence and live in protected damp places such as under bark, stones or logs. The familiar centipedes belong to the order Scolopendromorpha. *Scolopendra* is a common genus. Its body is elongated, dorsoventrally flattened and about 20 cm long. It is made of a small head and 21-segmented trunk. Head bears two long many-jointed antennae, two groups of ocelli and mouth appendages. Each trunk segment bears a single pair of legs, except the one posterior to head that bears a pair of **poison claws** or **maxillipedes**. Legs are attached ventro-laterally to the segments.

Centipedes are agile, fast-moving and carnivorous in diet. They kill their prey with their poison claws. The common house centipede is *Scutigera,* which bears 15 pairs of long legs. It is not harmful to man. But some larger tropical species inflict a painful bite causing fever, **dizziness** and headache.

18. *Lepisma.* *Lepisma* is commonly known as 'silver fish' because of its glistening silvery white, fish-like body. It belongs to the insect order Thysanura. It is a familiar small-sized (13 mm) household pest, residing in damp cool places and feeding on starch of starchy clothes and paste or glue in bookbindings. Abdomen is 11-segmented that bears two pairs of abdominal appendages, two long anal cerci and a long thread-like telson. Antennae are long and wings are wanting. Eggs

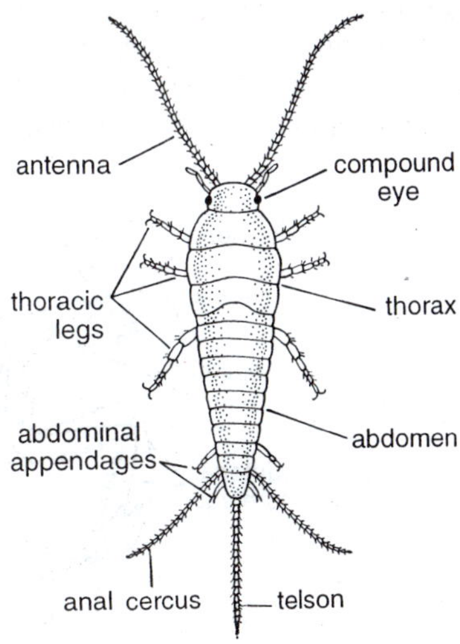

Fig. 18. *Lepisma* (Silver fish).

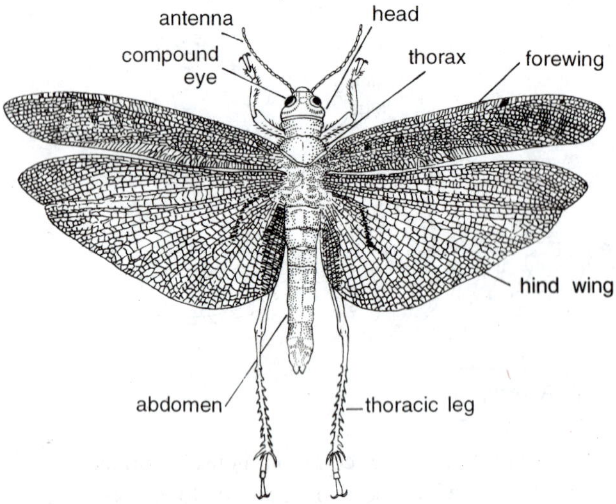

Fig. 19. *Schistocerca* (Locust).

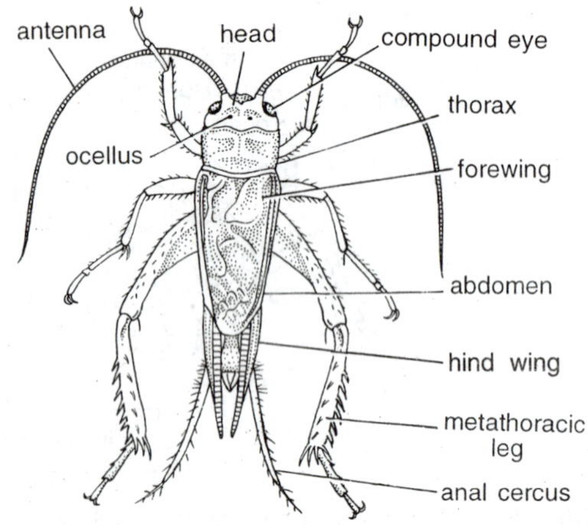

Fig. 20. House crickets *Acheta* (= *Gryllus*) *domestica.*

are laid in crevices and the larva resembles the adult.

19. Locust. The locust are migratory allies of grasshoppers, belonging to the order Orthoptera. The most common locust is *Schistocera gregaria,* the desert locust, which occurs in tropical Africa, Arabia, Iran, Palestine, Afghanistan, Pakistan and North-west India. Body is robust with head broadly articulated to thorax. Forewings are hardened to form wing covers (**tegmina**). Hindwings are large and membranous. Hind legs are long and meant for jumping. Locust is herbivorous in diet and gregarious in nature, migrating or swarming in great numbers. It is the worst destroyer of standing crops and orchards.

20. *Gryllus* (= *Acheta*). Crickets belong to the order Orthoptera. *Gryllus* is the common house-cricket which lives in damp warm places like under logs, boxes and stones, behind books, in holes and crevices and in kitchens. The male produces the familiar sound with its stridulating organs. Its antennae are very long, hind legs are long and stout and anal cerci are unsegmented. Wings do not help in flying. Female possesses a well-developed ovipositor that serves for depositing eggs in holes or crevices. It is a nocturnal insect and, being omnivorous, devours all sorts of animal and vegetable matter such as clothing, paper, fruits, vegetables, skin, etc. Nymph hatching out of egg

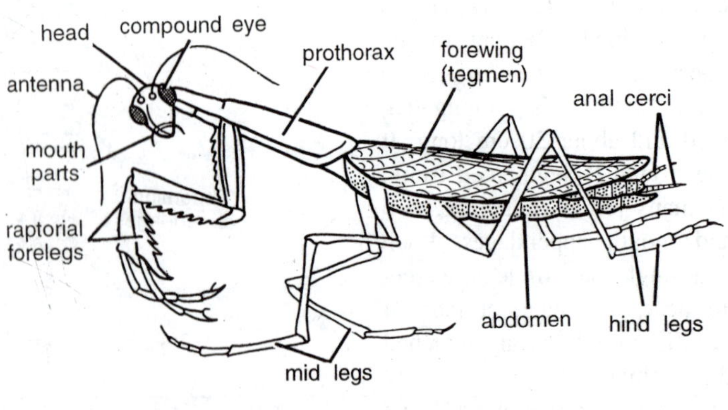

Fig. 21. *Mantis religiosa* (Praying mantis).

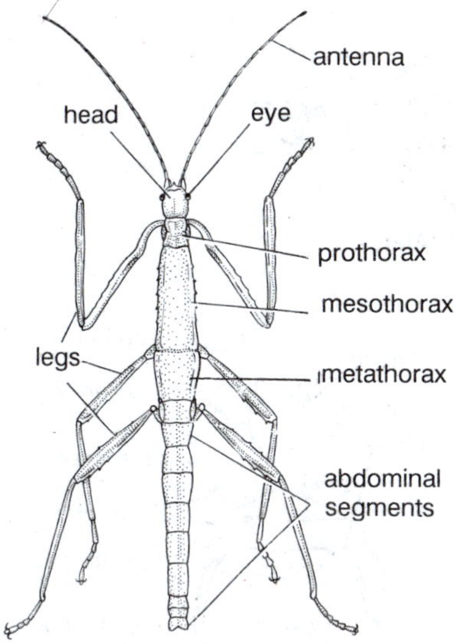

Fig. 22. *Carausius* (Stick insect).

is similar to adult but smaller in size and with out wings. Another common cricket in houses is *Gryllodes sigillatus*.

21. *Mantis religiosa* or Praying mantis. Praying mantis is a large insect (5-10 cm), with a small triangular head, a long prothorax and abdomen consisting of 10 segments. Wings are well developed and pincer-like, forelegs are raptorial and modified for grasping prey. It is a voracious carnivore, devouring living insects, fruits and vegetables. It usually inhabits plantation area. Female eats up the male after copulation, she lays about 200 eggs in a frothy mass which soon hardens into a waterproof *egg case*.

22. *Carausius* or Stick insect. The stick insect or walking stick (order Orthoptera) is an elongated insect that commonly mimics a twig to escape detection by enemies. Mesothorax is much enlarged. There being no wings, it does not fly but it does a little walking. It is herbivorous in diet and possesses chewing mouthparts. *Carausius* is a stick insect of tropical forests.

23. Termites. Termites (order Isoptera) are social insects and exhibit the most complex social life of all insects. Although termed white ants, they are neither white nor ants. They are brown or pale-coloured soft-bodied insects and are readily distinguished from ants by the absence of a pedicel between thorax and abdomen. They live in colonies with well-organised **castes** or **polymorphism**.

(a) Castes. There are two main castes of termites : **fertile** males and females and **infertile sterile** males and females. These are differentiated into winged (alate) and non-winged (dealate) reproductive individuals and workers, soldiers and nasutes. (i) **Workers** are wingless forms with no or under-developed eyes. They collect food, look after the young, attend to the reproductive forms and construct nests and passage ways etc. (ii) **Soldiers** are large-headed, bigger-bodied individuals possessing large and strong mandibles. They protect the colony from the enemies. In some termites (*e.g., Eutermes*) the soldiers are replaced by (iii) **Nasutes**, which are also wingless and possess a rostrum that pours a sticky secretion for destroying enemies and dissolving hard substances, (iv) Reproductive male and female termites are called **king** and **queen**, respectively. They are wingless but develop from sexual winged forms and live in royal chambers. The queen has a much enlarged abdomen to house the reproductive organs and eggs.

(b) Nests. Most termites build underground nests. A good number of termites build high and large mounds raised high up the ground. Nests are generally made of chewed soil and contain small chambers with entry galleries and tiny holes serving as exits. There is also a large or royal chamber for the reproductive pair. Certain termites build their nests inside wood or on branches of trees.

(c) Food. Cellulose is the principal foodstuff of termites. They feed on dead or living woods, cardboards and papers. Workers and young nymphs leave the nest for feeding. After returning into nest they vomit the ingested food which is then used by other members of colony. Royal pair is served with a semi-digested food. Cellulose is supposed to be digested either by bacterial or protozoan decomposition or by the enzyme **cellulase**.

(d) Odorants. Termites are known to secrete special odour-substances which serve as guide lines

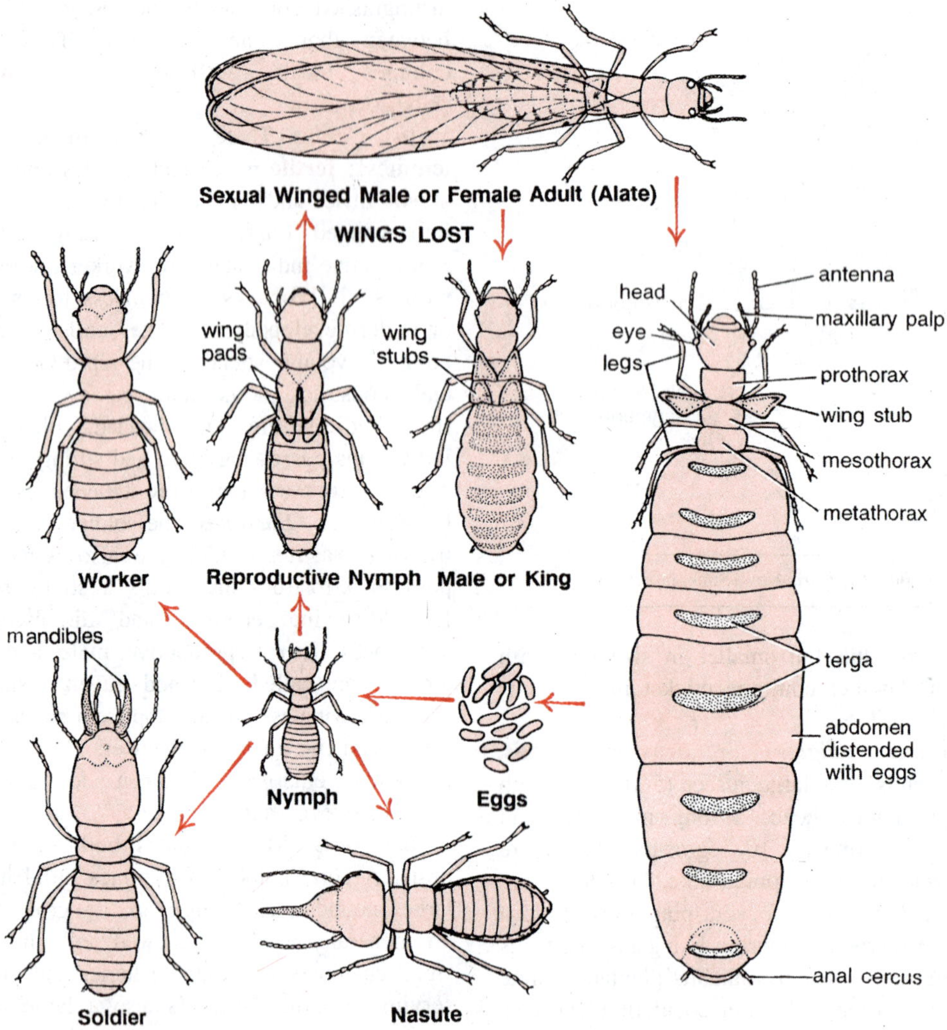

Fig. 23. Termites. Castes and life cycle.

for their back journeys into the nest. Odorants are complex organic compounds secreted by sternal glands lying in the 5th abdominal segment.

(e) Life-history. At the beginning of rainy season, the winged sexual forms (kings and queens) come out of their nests in huge numbers, called a **swarm**. After descending to ground, they mate in pairs. The royal couple (king and queen) then excavate a small **chamber** in soil which marks the beginning of a new colony. Both of them now shed their wings. Mating is repeated at regular intervals as the male (king) cohabits with the female (queen) for life. The first cluster of eggs laid by the queen is looked after by both. The nymphs, hatched out of these eggs, develop into many workers, and a few soldiers. In early years only workers and soldiers are formed, but later alates (winged forms) are also produced.

Abdomen of queen grows enormously so that she becomes 20 to 30 thousand times larger than a worker. On an average, the queen lays 4,000 eggs per day and about one million eggs per year. She lives for about 6 to 9 years.

In termites, determination of castes depends upon the complex interactions between pheromones, hormones, variable food supply, and social behaviour, etc.

24. Dragon fly. The dragon flies (order Odonata) are commonly found in the vicinity of water. They have long brightly coloured body, large head and huge compound eyes that often contain as many as 30,000 lenses (ommatidia). Their gauze-like 2 pairs of wings are large, membranous and each with a joint-like nodus or notch near the middle, and a coloured spot or **pterostigma** near the apex of front margin. Antennae are small and mouth parts are chewing. Abdomen is greatly elongated. They are strong fliers and skilful hunters. They are also known as the 'mosquito-hawks' as mosquitoes form their main diet. Both the aquatic nymphs and adults are predaceous. Common genera are *Agrian, Petalura,* etc.

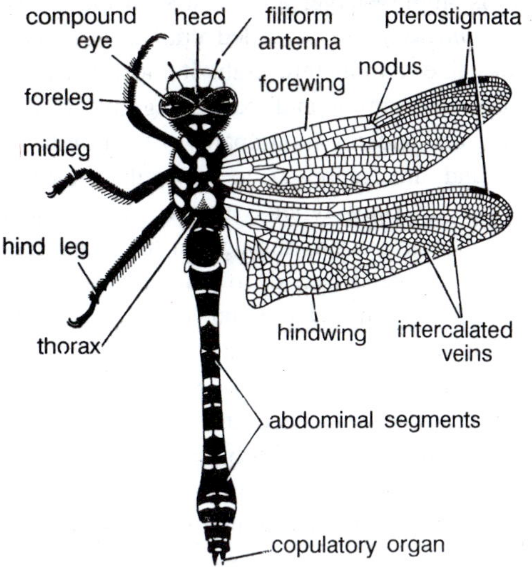

Fig. 24. Dragon fly.

25. *Pediculus* (Human louse). *Pediculus* (order Anoplura) is a true louse with minute, dorsoventrally flat and wingless body, free horizontal head, claws for clinging to hair, and degenerated eyes. It is an ectoparasite of mammals. *Pediculus humanus capitis* is the human head-louse that occurs clung to the head hairs, whereas, *Pediculus humanus corporis,* the human body-

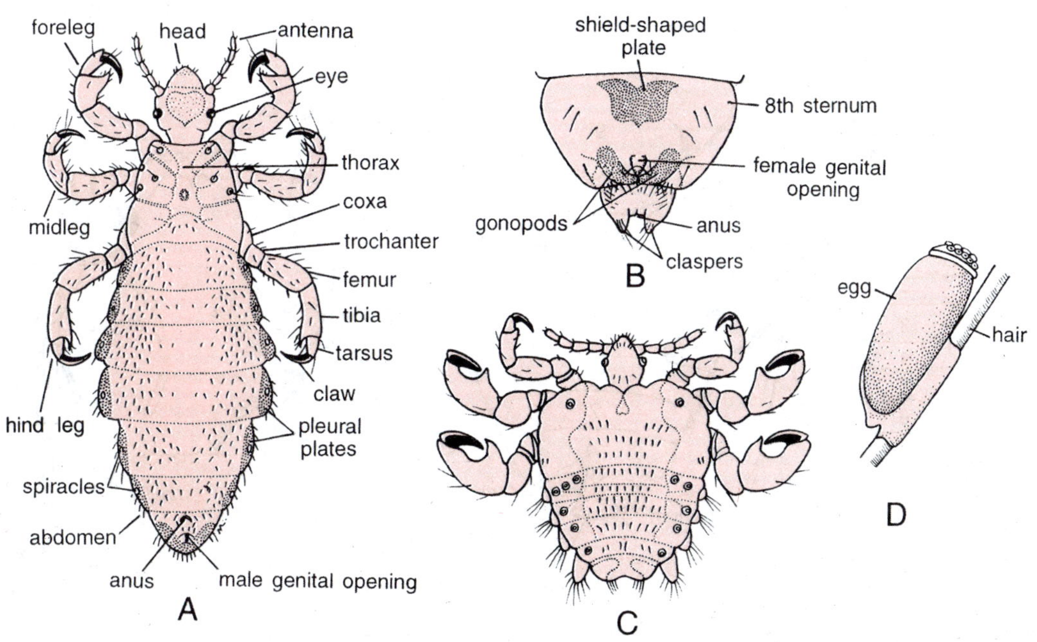

Fig. 25. Human lice. **A.** Human head-louse, male (*Pediculus humanus capitis*). **B.** Posterior end of the female head-louse. **C.** Human pubic louse (*Phthirius pubis*). **D.** An egg or nit.

louse, is slightly larger and occurs on the hairs of body. *Pthirius pubis* is called crab louse because it looks like a miniature crab. It mostly harbours the hairs of arm-pit and pelvic region. All have piercing and sucking mouthparts to feed on human blood and are supposed to inocculate bacterial pathogens causing typhus, trench and relapsing fevers. Breeding occurs throughout the years. Eggs, called **nits**, are attached to hair by glue from ovipositor. Nymphs hatching out in 8 days resemble the adult and mature in about 11 days.

26. *Cimex* (Bed bug). *Cimex rotundatus* (order Hemiptera), the bedbug, is an ectoparasitic insect feeding upon human blood. When in sound sleep on bug-infested cot, mat or durry, everyone of us has experienced this uninvited guest of the night.

(a) Habits and habitat. Bedbug inhabits human dwellings such as hotel rooms, old buildings and unclean houses, hiding in crevices. It is a *nocturnal* insect but often comes out during day. It is strongly attracted by the warmth and odour of human body.

(b) Structure. Body is oval, dorsoventrally flattened and chestnut brown in colour, becoming deep purple or red after blood meal. Head is small and broad. It bears two lateral compound eyes and two short 4-jointed antennae. Mouthparts are piercing and sucking and form a hollow **beak** or proboscis beneath head. The beak is formed by 3-jointed labium, which bears a mid-dorsal groove for enclosing the needle-shaped 4 stylets of 2 mandibles and 2 maxillae. Blade-like mandibular stylets are used for making puncture in the skin, whereas maxillae enclose 2 channels for the inward flow of blood and downward flow of saliva. Maxillary and labial palps are absent.

Thorax is 3-segmented with a somewhat large prothorax. Mesothorax bears two small pads covering the metathorax and representing the vestigeal forewings. Three pairs of legs are strongly clawed and adapted for swift running. Abdomen of male bug is narrower and more pointed than that of female.

(c) Life history. Breeding occurs almost throughout the year. Before mating, the female takes a meal of human blood. While copulating, the male takes up a position diagonally across the body of the female and introduces its penis into the slit of her **organ of Berlese** (on 4th abdominal segment ventrally) to transfer the spermatophores. Fertilization is internal.

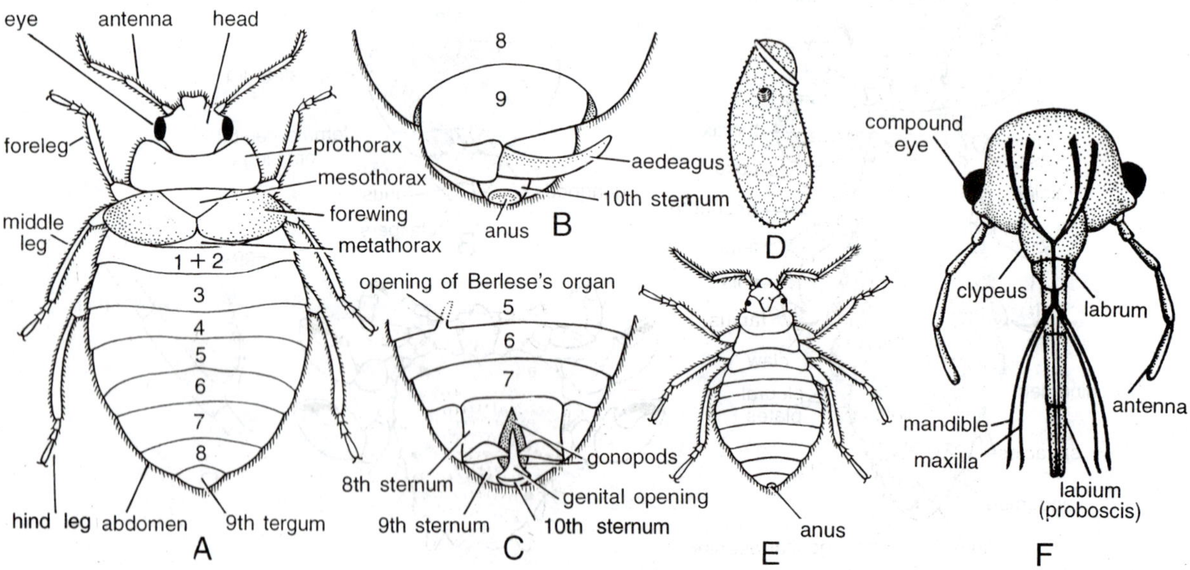

Fig. 26. *Cimex.* **A.** Adult male (Dorsal view). **B.** Genitalia of male. **C.** Posterior end of abdomen of adult female (Ventral view). **D.** Egg or nit. **E.** Young nymph. **F.** Head and mouth parts.

A female lays about 200 or more **eggs**, at the rate of 3 or 4 eggs per day, in a crevice or crack of wall, furniture, cot, etc. Eggs are pearly white, cylindrical and provided with a cap-like lid. In about 6 to 10 days, young **nymphs** hatch out of the eggs in warm weather but take longer time in winter. The nymph is like the adult but is small, paler in colour and with thick antennae and stout legs. It moults 5 times to become an adult.

Bedbugs are thought to transmit organisms of kala-azar, relapsing fever, plague, typhoid and tuberculosis. Common insecticides, such as 5% DDT (dichloro-diphenyl-trichlorethane) kill the bugs.

27. Cicada. Cicadas are large-sized, homopterous insects well-known for their sound-production and interesting breeding habits. Male has at the base of abdomen a drum-like apparatus capable of producing a loud, shrill, vibrating sound supposed to be a sexual call. Life cycle undergoes gradual metamorphosis and requires from 2 to 17 years for its completion. Eggs are embedded in tree barks by the sharp ovipositor. They hatch in a few weeks and the young nymphs fall to the ground, burrow in soil, and feed on plant roots. After 2-17 years, having moulted about 25 times, the nymph comes out of the ground, crawls up a plant, sheds its skin, and emerges as the adult. Cicadas are often called 'seventeen-year locusts' on this account. However, they are not true locusts or they would be placed in the order Orthoptera. The sound or song is produced by two sets of special vibrating membranes, or drums, present ventrally on the basal abdominal segments, with the help of special muscular arrangement.

28. Earwig. The earwig (order Dermaptera) has earned its name because of the superstitious belief that it would crawl into the ear of a sleeping person. Body is elongated with slender antennae and biting and chewing mouth parts. Forewings are smaller and leathery, whereas the hindwings are large, membranous, semicircular with radial veins and folded lengthwise. The characteristic feature is the presence of two pincer-like appendages (anal cerci) at the tip of the abdomen. Common genus of earwigs is *Forficula.*

29. Butterfly. Butterflies (order Lepidoptera) are familiar insects seen in gardens. Their body is covered by scales with many colour patterns. Body is slender. Antennae are long and filamentous. Wings are 2 pairs, membranous and covered with overlapping scales. At rest, these are held vertically. Mouthparts form a coiled tube or proboscis for sucking up the nectar of flowers. Metamorphosis is complete and the larva is called the **caterpillar**. Common examples are dead leaf butterflies (*Kallima*), monarch (*Danaus*), swallow tail (*Papilio*), painted lady (*Vanessa*), cabbage butterfly (*Pieris*), etc.

30. Moth. Moths (order Lepidoptera) are close allies of butterflies. Their body is stout and

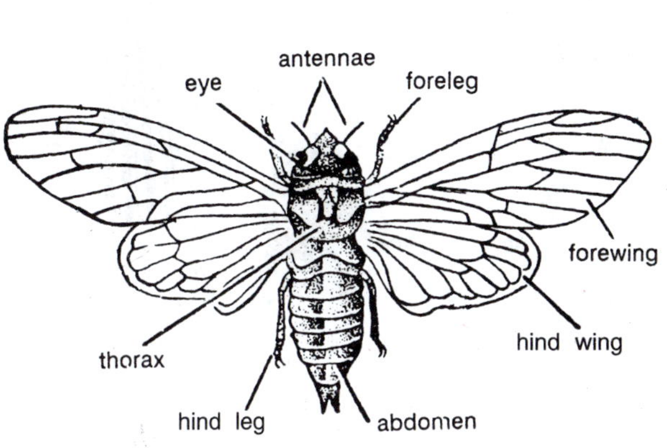

Fig. 27. Cicada.

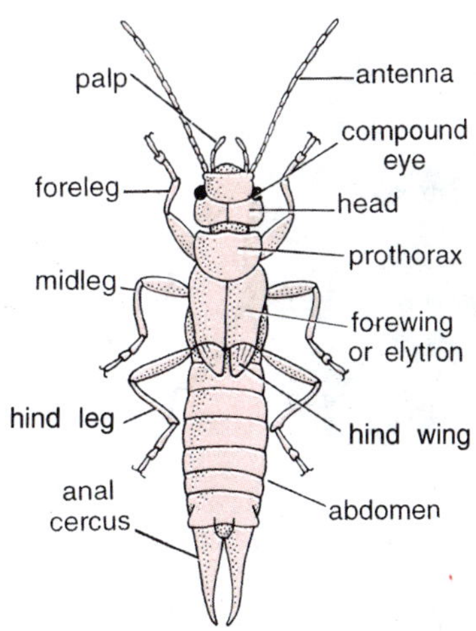

Fig. 28. Earwig (*Forficula*).

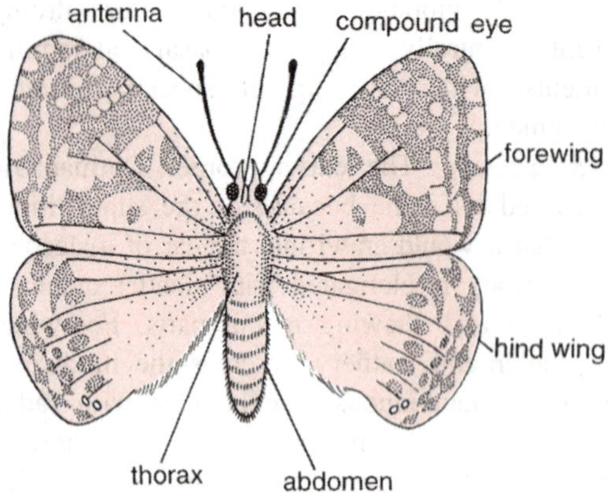

Fig. 29. Butterfly.

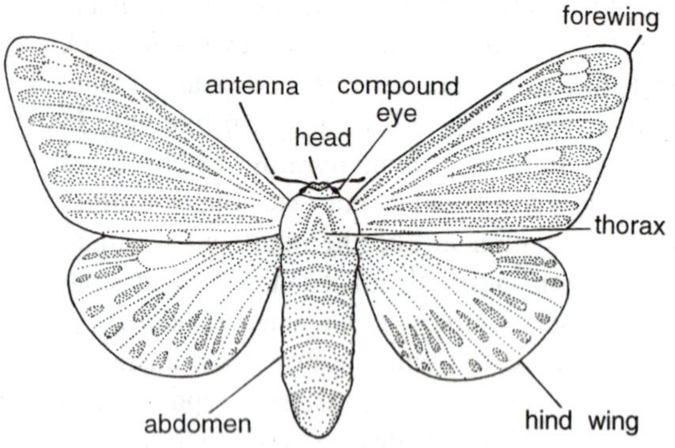

Fig. 30. Moth.

antennae are short and feather-like. Mouthparts are long, coiled of siphoning type. At rest, their wings are held horizontally. Examples of moths are cercopia moth (*Hyalophora*), sphinx moth (*Celerio*), cloth moth (*Tinea*), rice moth (*Corcyra*), Indian meal moth (*Plodia*), silk moth (*Bombyx*), etc.

31. Beetle. Beetles (order Coleoptera) are hard-bodied insects with heavy cuticle and mandibulate mouthparts. Forewings form hard, opaque, horny and veinless elytra, usually covering the entire abdomen. Hind wings are membranous with a few veins and folded under elytra at rest. Metamorphosis is complete. Larvae called **grubs**, may be legless or with 3 pairs of legs. About 2,80,000 species or 40% of all insects are beetles. Among the most familiar examples are carpet beetle (*Anthrenus*), tiger beetle (*Cicindela*) diving beetle (*Dytiscus*), blister beetle (*Epicauta*), ladybird beetle (*Coccinella*), scrab or dung beetle (*Scarabaeus*), rhinoceros beetle (*Dynastes*), meal worm beetle (*Tenebris*), rice weevil (*Sitophilus*), etc.

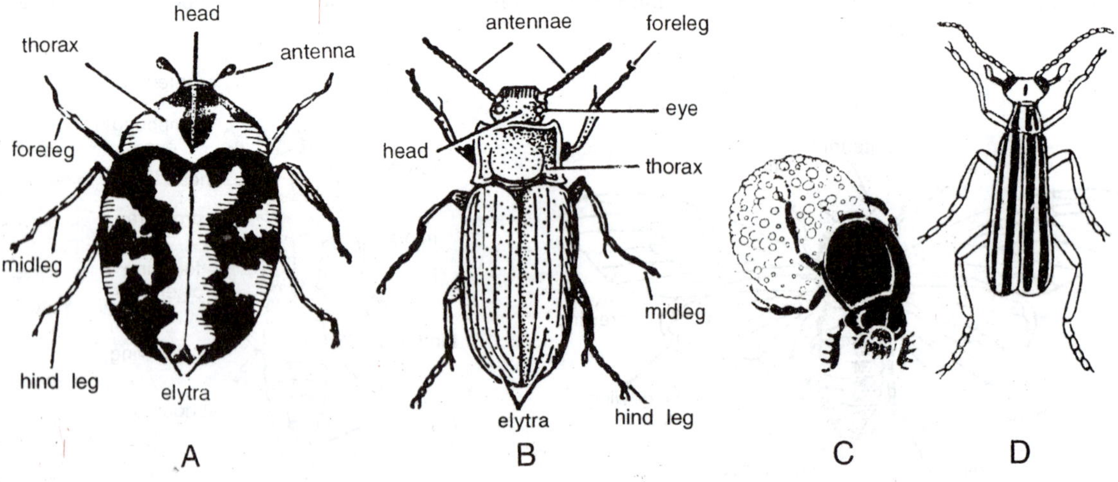

Fig. 31. Beetles. **A.** Carpet beetle (*Anthrenus* sp.). **B.** Meal-worm beetle (*Tenebris molitor*). **C.** Dung beetle (*Scarabaeus* sp.). **D.** Blister beetle (*Epicuta* sp.).

Carpet beetles attack hair, silk, feathers, leather, cotton, wool and the articles making holes in them. They can be controlled by the use of naphthalene, DDT, etc. Buffalo carpet beetle (*Anthrenus scrophulariae*), sometimes called buffalo moth, is about 3 mm long, motted with red, yellow, brown and black scales. When disturbed it feigns death, folding legs close to the body. Adults are often found on window sill.

The meal-worm beetle (*Tenebris molitor*) is a brownish black beetle, 20 mm long, commonly seen around granaries, flourmills, stores and barns. The larva, known as the meal-worm, is raised in large numbers for feeding birds, etc. on a commercial basis.

32. *Drosophila.* *Drosophila,* commonly known as **fruit fly** or **vinegar fly** belongs to the insect order **Diptera** and family **Drosophillidae**. It is a small insect about 1/4 the size of a house fly. Body is divided distinctly into head, thorax and abdomen. Head carries 2 plumose antennae, 3 ocelli, 2 large compound eyes, and sucking mouthparts. Broad thorax bears bristles, and abdomen has dark posterior bands. There is a single pair of large membranous forewings with few veins. It feeds upon fermenting fruits and often seen buzzing around bunches of bananas, vegetable bins and garbage cans in large numbers. Metamorphosis is complete.

Drosophila was first described in 1823 by **Fallen**. It is an excellent organism for genetic research because of its short life cycle (10 days), large number of offsprings produced (about 200) and only 4 pairs of chromosomes in its somatic cells.

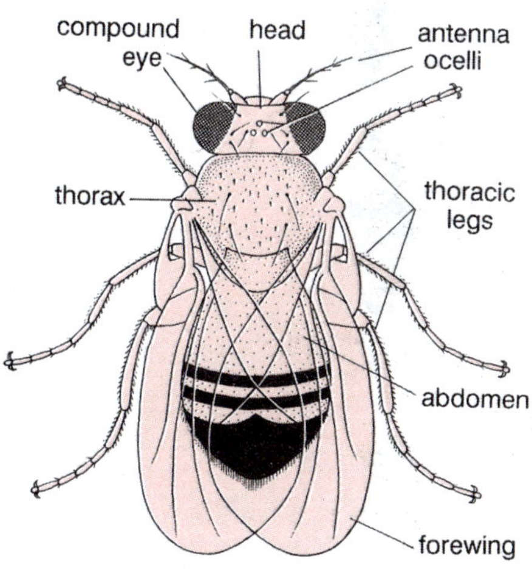

Fig. 32. *Drosophila melanogaster.*

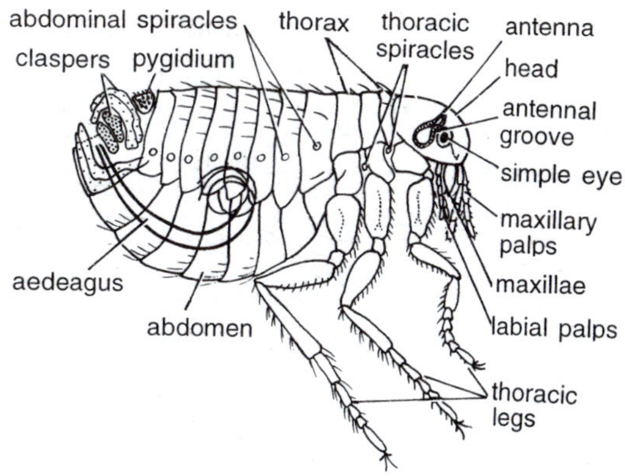

Fig. 33. Human flea (*Pulex irritans*).

33. Fleas (*Pulex*). Fleas (order Siphonaptera) are ectoparasites of birds and mammals, feeding on their blood. Body is small (3-5 mm), oval, laterally compressed, wingless, and covered with bristles. Head and thorax are small but abdomen is large. Antennae are short and thick, eyes are simple, mouth parts piercing and sucking and legs stout and large. Metamorphosis is complete. Larvae spin cocoons to undergo pupal stage. These blood sucking pests act as serious vectors of many diseases. The human flea, *Pulex irritans,* attacks men, cats, dogs, rats and horses. The Oriental rat flea, *Xenopsylla cheopsis,* resembles the human flea and is the chief transmitter of bubonic plague.

34. Wasp. Wasps (order Hymenoptera) are those familiar social insects that build nests of mud, paper, wood pulp, etc., hanging over walls, ceilings, trees, and in porch or verandah by means of a stalk. The common Indian wasp is *Vespa orientalis,* a deep brown or chestnut-red coloured insect. A smaller variety, called *Polistis* or yellow wasp, resembles the red wasp (*Vespa*) but differs by having a plain yellow or golden colour. Body is slender with long legs. Head bears a pair of antennae, 3 ocelli, 2 compound eyes, and chewing mouthparts. Abdomen is pedicellate (unlike honey bee), that is, thorax and abdomen are joined by a small narrow waist, pedicel or petiole, which is formed by the 1st narrowed down abdominal

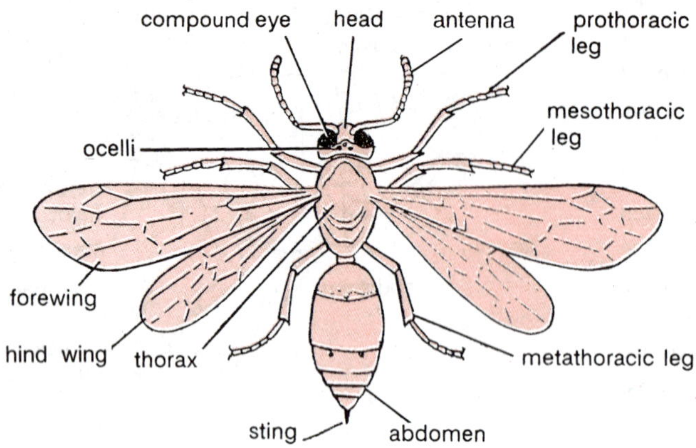

Fig. 34. Wasp (*Vespa*). Female.

segment. Wings are membranous. Forewings are larger than hind wings with which they remain hooked during flight. Adults feed on small insect larvae and plant juices. Female possesses a sting which is very painful and sometimes dangerous to human beings.

At any moment, there are about 10^{15} (1,000,000,000,000,000) ants alive on the planet. If each ant were half a centimeter long, they would form a chain long enough to circle the Earth at least 125,000 times!

35. Ants. Like honey bees and termites, all the species of ants are highly **evolved** social, colonial and polymorphic insects (order Hymenoptera). They are distinguished from other insects by the possession of elbowed antennae, small, vestigial or rarely absent eyes, deeply petiolated abdomen (gaster) near its junction with thorax (propodeum) and absence of wings except in sexually mature forms.

Main polymorphic forms or castes of ants are workers, soldiers, queen and aners. Workers are sterile females, possessing reduced thorax, small gaster and small eyes. **Soldiers** are sterile

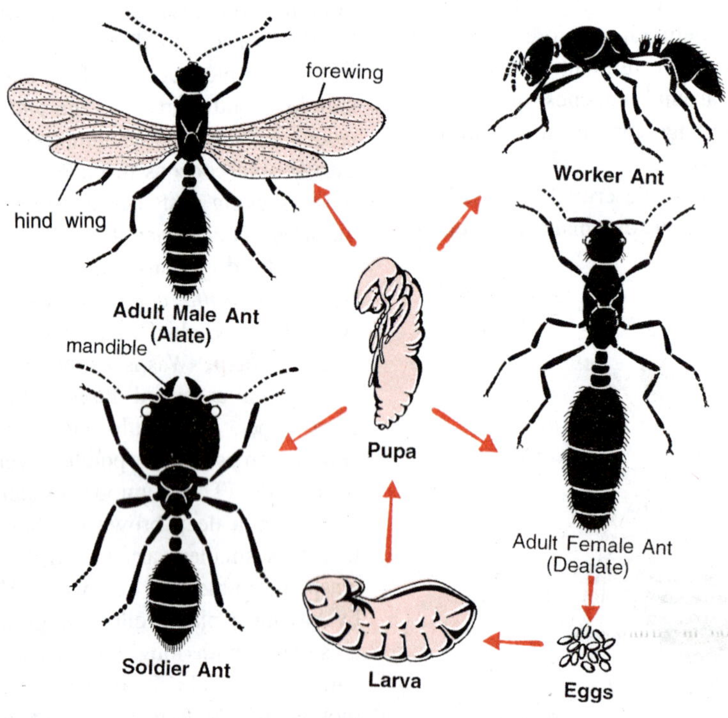

Fig. 35. Ants. Castes and life cycle.

females with large heads and powerful mandibles. **Queens** are fertile females with a larger gaster than other members. Unlike bees, a colony of ants contain several queens. **Aners** are slender fertile males provided with two pairs of wings.

Ants, mostly construct underground **nests**. Females lay eggs, followng mating in a nuptial flight (swarming). After mating, the males usually die. Eggs hatch into larvae called **grubs**. The pupae are naked, *i.e.*, not enclosed within puparia. Female ants secrete pheromones in air to attract their mating partners. Common Indian ants are :

1. Monomorium. Large black ants found in crevices of walls, tree trunks, etc.

2. Camponotus. Common black house ants.

3. Solenopsis. Small red ants of house.

4. Dorylus. Wasp-like winged ants which appear around light after rains.

5. Aenictus. Common army ants that are gregarious.

Important Questions

➤➤ *Long answer type questions*

1. Give distinctive features of the phylum Arthropoda and its main subdivisions upto classes.
2. Classify Arthropoda up to classes giving their diagnostic characters and familiar examples.
3. Describe briefly the caste system in social insects studied by you.
4. Mention salient features giving important examples of :
   (i) Xiphosura, (ii) Arachnida, (iii) Chilopoda, (iv) Diplopoda.
5. Write short notes on : (i) *Cyclops,* (ii) *Daphnia,* (iii) *Drosophila,* (iv) Earwig, (v) *Limulus,* (vi) Millipede, (vii) *Sacculina,* (viii) *Scolopendra,* (ix) Silver-fish, (x) Spiny crab.

➤➤ *Short answer type questions*

1. What is the study of insects called as ?
2. Name any three freshwater crustaceans found in Delhi.
3. Name any four diseases spread by dipterous insects.
4. Name the animal group where the crystalline cone is present.
5. Give an example of parasitic copepod.
6. What is the respiratory organs of Crustacea ?
7. How many classes does phylum Arthropoda have ?
8. What is the respiratory organ in spider ?
9. Give an example for sexual dimorphism in Decapoda.
10. Give one point of external identification between an arachnid and an insect.
11. What is the main economic importance of order Decapoda ? Outline with atleast two examples.
12. Give the distinguishing character and one example of the diplopoda.
13. Describe briefly the *Sacculina.*
14. Show why the parasitic *Sacculina* should be a hermaphrodite. Do not exceed 50 words.
15. Name the distinguishing feature of the following organism. *Sacculina* and *Trichonympha.*
16. Mention four distinctive characters of Arachnida.
17. Explain the effect of the sedentary habit in the structure of *Lepas.*
18. Distinguish between moths and butterflies.
19. *Lepas* is a sessile crustacean. How does it obtain its food? Explain in not more than 50 words.
20. How will you identify a male crab from a female crab externally. Outline the importance of gastric mill within 30 words.
21. Describe a typical appendage and its various modifications in *Apus.*
22. Describe the mouth parts of scorpion, centipede and bed bug.
23. Explain how aquatic insects have adapted themselves to their environment ?
24. Explain exoskeleton in Arthropoda.
25. Describe *Trilobitomorpha.*
26. Comments on *Xiphosura.*
27. Give important characteristics of Arachnida.
28. What are the characters of the animals which belongs to order Acarina ?
29. What do you mean by mandibulata ?

30. Give an example of order Apoda.
31. Differentiate the Isoptera and Dermaptera.
32. Distinguish between Myriapoda and Insecta.
33. Give a classification of Insecta.
34. Compare the Hemiptera and Homoptera.
35. Comment on Oxychophora.
36. Give a brief account of Tardigrada.
37. What are Linguatulida ?
38. Define specific features of Pychnogonida.
39. Comment on *Cyclops.*
40. Comments upon given :
    (a) *Daphnia,* (b) *Balanus,* (c) *Sacculina,* (d) *Hippa,* (e) *Eupagurus,* (f) *Cimex,* (g) *Pediculus,* (h) *Drosophila.*
41. Write the taxonomic condition of the following :
    (a) ant, (b) wasp, (c) moth, (d) dragon fly, (e) termite, (f) mantis, (g) locust, (h) millipedes.

➤➤Fill in the blanks

1. A crustacean larva characterized by anterior rostral spine, median dorsal spine is .....
2. *Sacculina* causes .... in crabs.
3. A crab that lives in the gastropod shell is called as .....
4. The mouth parts of butterfly is of .... type.
5. Cypris larva has a .... shell.
6. Extinct trilobites are a primitive group of.....
7. In aquatic insects, respiration occurs through .....

➤➤Multiple choice questions

1. Statolith is a sense organ that helps in :
   (a) tactile stimulus (b) vision
   (c) equilibrium
   (d) chemical stimulus
2. The larva of *Sacculina* is called as :
   (a) cypris larva
   (b) megalopa larva and nauplius
   (c) nauplius, cypris and kentrogen
   (d) zoea
3. *Daphnia* is also called as :
   (a) clam shrimp (b) fairy shrimp
   (c) water fleas (d) tadpole shrimp
4. *Lepus* is also known as :
   (a) acorn barnacle (b) goose or ship barnacle
   (c) sea barnacle (d) Rock barnacle
5. Kentrogon larval stage is exhibited by :
   (a) *Squilla* (b) *Palaemon*
   (c) *Eupagurus* (d) none of these
6. Most of the insects excrete in the form of uric acid because :
   (a) most of them are aquatic
   (b) most of them are terrestrial
   (c) the circulatory system is open
   (d) the blood is colourless
7. Which of the following species of mosquitoes is responsible for filariasis :
   (a) *Culex* (b) *Anopheles*
   (c) *Aedes* (d) none of the above
8. The scientific name of blood red ants is :
   (a) *Formica sanguinea* (b) *Polyergus rufesseens*
   (c) *Formica reufa* (d) *Formica fusca*
   (e) *Solenopsis*
9. What is the scientific name of tse tse fly ?
   (a) *Aedes aegypti* (b) *Glossina palplis*
   (c) *Culex palplis*
   (d) *Glossina gambiense*
   (e) *Periplaneta americana*
10. The term Arthropoda was coined by :
    (a) Linnaeus (b) De Monet
    (c) Cuvier (d) Von Siechold
11. Rhabdome is found in the eyes of :
    (a) earthworm (b) cockroach
    (c) frog (d) rabbit
12. The respiratory pigment of insects is :
    (a) haemoglobin (b) haemoerythrin
    (c) haemocyanin (d) none
13. Complete metamorphosis occurs in :
    (a) silver fish (b) locust
    (c) bedbug (d) mosquitoes
14. Histolysis and histogenesis occurs during :
    (a) larval stage (b) pupal stage
    (c) adult stage (d) none of these
15. The moulting hormone or ecdysone is secreted from :
    (a) corpora allata (b) prothoracic glands
    (c) corpora adiposa (d) intercerebral glands
16. The juvenile hormone is secreted from :
    (a) corpora allata (b) corpora cardiaca
    (c) corpora adipose (d) none of these
17. Crustacea are distinguished from insects & myriapods by the presence of :
    (a) 1st pair of antennae
    (b) 2nd pair of antennae
    (c) number of legs
    (d) jointed appendages

18. Horse shoe or king crabs are included under the order :
(a) Enlypterida (b) Arachnida
(c) Pychnida (d) Pycnogonida
(e) Xiphosura
19. *Lepas* and *Balanus* are included under the order :
(a) Decapoda (b) Thoracica
(c) Rhizocephala (d) Amphipoda
20. Which one of the following groups come under social insects ?
(a) cockroaches (b) houseflies
(c) termites (d) mosquitoes
21. Arthropoda include animals with jointed legs, they have a body cavity called :
(a) coelom (b) haemocoel
(c) gastrovascular cavity(d) pseudocoel
22. Antennae are absent in :
(a) dragon fly (b) spider
(c) prawn (d) *Peripatus*
23. Halters refer to :
(a) reduced wings of house fly
(b) forewings of beetles
(c) balancing organs
(d) wings of hemiptera
24. The type of mouth parts in Bed bug is :
(a) piercing & sucking (b) biting & chewing
(c) biting (d) sponging
25. For an insect feeding on body fluids/blood of the prey the mouth parts should be of :
(a) sucking type (b) sponging type
(c) piercing and sucking type
(d) biting type
26. Adult butterflies can ingest :
(a) only liquid food
(b) only solid food
(c) both solid and liquid food
(d) only gases
27. The integument of an arthropod is covered with :
(a) smooth muscles (b) mucus
(c) chitinous cuticle (d) calcareous shell
28. Which of the following animals has a ventral nerve cord :
(a) frog (b) cockroach
(c) *Hydra* (d) *Amoeba*
29. Which of the following scientists was awarded the noble prize for deciphering the language of bees :
(a) Charles Darwin (b) William Harvey
(c) Karl von Frisch (d) Har Gobind Khorana
30. The visual unit of the compound eye of an insect is :
(a) ommatidium (b) rods and cones
(c) retina (d) cornea
31. Respiratory organs of crustaceans are :
(a) gills (b) lungs
(c) general surface
(d) both gills and general surface
32. Excretory product in an insect is :
(a) uric acid (b) urea
(c) ammonia (d) guanine
33. Which one of these is good example of metamorphosis :
(a) regeneration of broken tail of lizard
(b) growth and development of young one of Kangaroo in its marsupium
(c) development of adult from pupa of insects
(d) hatching of maggot from the egg of housefly
34. In insects, oxygen is carried to various tissues by :
(a) blood plasma
(b) blood pigment
(c) tracheal tubes
(d) diffusion through body surface
35. Metamorphosis occurs when :
(a) larva changes into adult
(b) growth occurs .
(c) adults are developed parthenogenetically
(d) adult reproduces sexually
36. *Aedes* female transmits :
(a) yellow fever
(b) dengue fever
(c) both (d) none
37. One of the following is a vector for the germ causing sleeping sickness :
(a) tse-tse fly (b) sand-fly
(c) *Wuchereria* (d) *Culex*
38. Paurometabolous development occurs in :
(a) housefly (b) cockroach
(c) mosquito (d) moth
39. What is common between the septal nephridia of earthworm and Malpighian tubules of cockroach ?
(a) both excrete the nitrogenous wastes in the form of urea
(b) both discharge their contents to the outside of the body through the alimentary canal
(c) both have their free ends opening into the body cavity
(d) both are segmentally arranged in the body
40. Compound eye in arthropods form apposition images :
(a) in dim light (b) in bright light
(c) in dim and birght light
(d) such images are never formed
41. Chitinous cuticle is an identification of :
(a) Annelida (b) Arthropoda
(c) Mollusca (d) all
42. Respiratory organs in Arthropoda :
(a) trachea (b) gills
(c) book lungs (d) all
43. Which is viviparous :
(a) sacculina
(b) scorpion
(c) spider
(d) peripatus

44. Termitarium is :
(a) bunch of termite (b) house of termite
(c) colony of termite (d) larvae of termite
45. Which is the pest of vegetable :
(a) aphids (b) pyrilla
(c) grasshopper (d) beetle
46. Pest of domestic animals :
(a) fleas (b) lice
(c) bugs (d) mosquitoes
(e) all
47. Termite casts have :
(a) king
(b) queen
(c) worker
(d) soldiers
(e) all of these
48. The study of insect is called :
(a) Entomology (b) Enthrology
(c) Epidomology
(d) none the above

Answers

➤➤*Fill in the blanks*

1. Zoaea larva, 2. Parasitism, 3. Hermit crab, 4. Siphoning, 5. Bivalved, 6. Subphylum – Tribitomorpha, 7. Tracheal gills.

➤➤*Multiple choice questions*

1. (c)	2. (c)	3. (c)	4. (b)	5. (d)	6. (b)	7. (a)	8. (e)	9. (b)	10. (d)
11. (b)	12. (d)	13. (d)	14. (b)	15. (b)	16. (a)	17. (d)	18. (e)	19. (b)	20. (c)
21. (b)	22. (b)	23. (a)	24. (a)	25. (c)	26. (a)	27. (c)	28. (b)	29. (c)	30. (a)
31. (a)	32. (a)	33. (c)	34. (c)	35. (d)	36. (a)	37. (a)	38. (b)	39. (b)	40. (b)
41. (b)	42. (d)	43. (b)	44. (b)	45. (a)	46. (e)	47. (e)	48. (b).		

54

Arthropoda : General Account

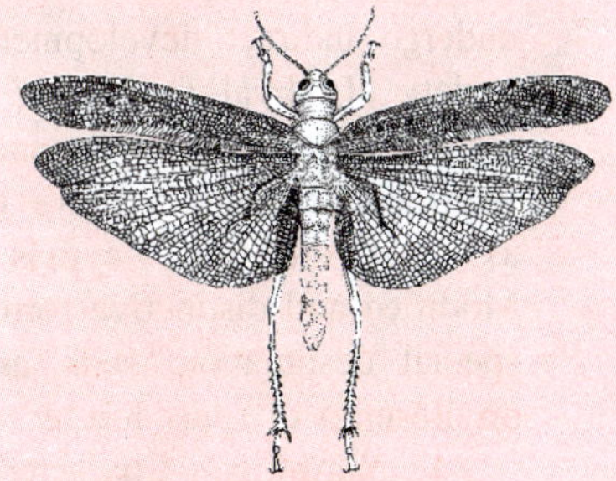

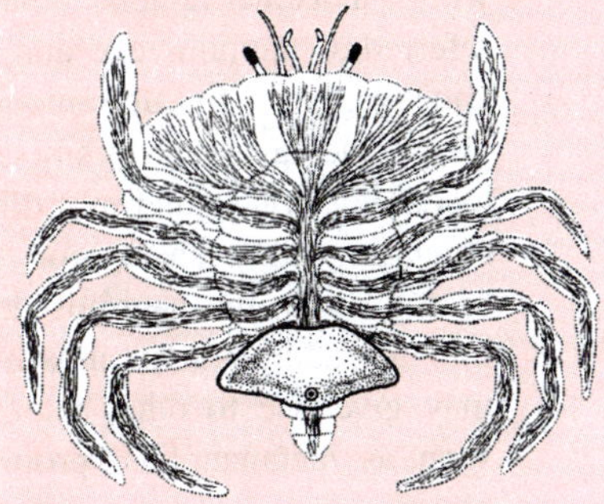

BIOLOGICAL SUCCESS

Without question, phylum Arthropoda is the most successful of all the invertebrate phyla. Some of the criteria for their success are : (i) largest number and variety of species (12,57,040), (ii) variety of habitats occupied, (iii) widespread distribution, (iv) variety of food habits, (v) capability of defending against enemies, (vi) power to adapt themselves to changing conditions, etc.

Some structural and physiological patterns (advances) which have contributed to their big biological success are : (i) thick chitinous exoskeleton, (ii) jointedness of body and limbs, (iii) locomotion by small, striated, extrinsic muscles, (iv) various specializations of somites, alimentary canal and respiratory organs, (v) cephalization with concentration of ganglia and sense organs in head region, (vi) behaviour patterns with primitive intelligence and social instincts in some group.

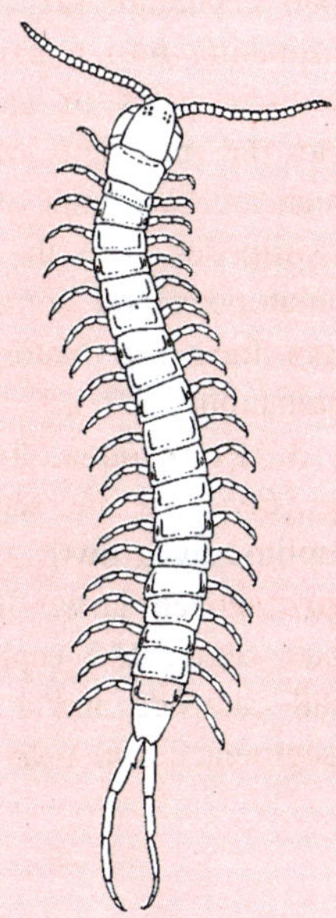

LARVAE OF CRUSTACEA

Crustaceans show both direct and indirect development. In **direct development** (*e.g., Palaemon,* crayfish), the adult is attained by progressive growth and differentiation of the embryo, so that the newly hatched young resembles the parents in general structure. In **indirect development**, there is a larval stage which differs from the adult in many features and acquires adulthood through metamorphosis. Many of the crustaceans undergo indirect development, involving a wide variety of larval forms. Of these 3 main larval forms are **nauplius, zoaea** and **megalopa**. Intermediate stages receive different names such as **metanauplius, cypris** and **protozoaea**. Modified and distinctive forms of zoaea are given special designation, such as **mysis** of lobsters, **phyllosoma** of spiny lobster, and **allima** of squilla.

1. Nauplius. Characteristic of the class, **nauplius** is the simplest and commonest type of larva, found in most marine crustaceans and a few malacostracans. When development proceeds through many larval forms, the **nauplius** is the earliest and the basic larva. The body is minute with 3 indistinct regions, a single median eye often referred as nauplius eye and three pairs of jointed appendages — the uniramous **antennules**, mainly the balancing organ; biramous **antennae**, principal locomotor organs and **mandibles,** which alongwith antennae may share for food collection. In branchiopods the nauplius develops straight away into the adult, but in mostly other crustaceans it may give rise to other intermediate larval forms, such as metanauplius, protozoaea, zoaea, mysis, etc.

2. Metanauplius. Metanauplius is the later nauplius instar and results by the process of moulting and growth. Its body is divisible into a broad cephalothorax and an elongated abdomen, terminating into a pair of caudal forks. Besides the three pairs of nauplius appendages, it also bears the rudiments of four pairs of appendages, which are two pairs of maxillae and two pairs of maxillipedes of the adult. Some decapods, stomatopods and some notostracans (*e.g., Apus*) begin their life history with the free-swimming metanauplius larva.

3. Protozoaea. In case of marine prawns (*e.g., Penaeus*), and sergestid decapods, the earliest nauplius, by growth and moulting, develops into a **protozoaea** larva. Its body is divisible into a broad segmented cephalothorax covered with a small carapace and a slender abdomen which is unsegmented and bear no appendages terminating in a forked telson. There is a single median nauplius eye and the appendages comprise of the antennules, antennae, mouthparts and first and second maxillipedes. The protozoaea later modifies into the **zoaea**.

4. Zoaea. In almost all marine decapods, except peneids and sergestids, hatching takes place at the **zoaea** stage (as in true crabs). Zoaea has a broad cephalothorax and a curved abdomen, which assists in swimming, is provided with a forked telson. Helmet-like carapace bears two long spines, a **median dorsal** and a **median rostrum**, two **lateral spines** are often met with. A pair of large stalked movable compound eyes are present. In addition to protozoaeal appendages, there appear rudiments of thoracic appendages. Biramous maxillipedes are used for swimming.

5. Cypris. In Cirripedia (*e.g. Lepas, Sacculina*), the nauplius larva passes into the **Cypris** stage. In this form the body and appendages are enclosed within a bivalved shell provided with adductor muscle as is seen in an ostracod adult, **Cypris**. Its modified antennules have cement glands at their bases. All other cephalic appendages with a compound eye only, except antennae, are present. Six pairs of biramous thoracic limbs are formed. It has abdomen with 4–5 segments.

6. Mysis or Schizopod. In peneid decapods (*e.g. Penaeus*) and lobsters zoaea is modified into **Mysis** or **Schizopod** larva. It bears 13 pairs of appendages and resembles adult **Mysis**. It has 5 pairs of posterior biramous thoracic appendages. Abdomen is posterior similar to that of adult with 5 pairs of biramous pleopods and a pair of uropods and telson. In some lobsters mysis marks the begining of the life history as the naupleus and zoaea are passed within the egg but at the same time it marks the end of the life history of a prawn.

7. Megalopa. In brachyuran decapods (true crabs), zoaea metamorphoses into the **Megalopa** larva. It resembles, to some extent, the adult crab and possesses all 13 pairs of appendages. Abdomen bears 6 pairs of pleopods and is placed straight in line with cephalothorax. In crabs

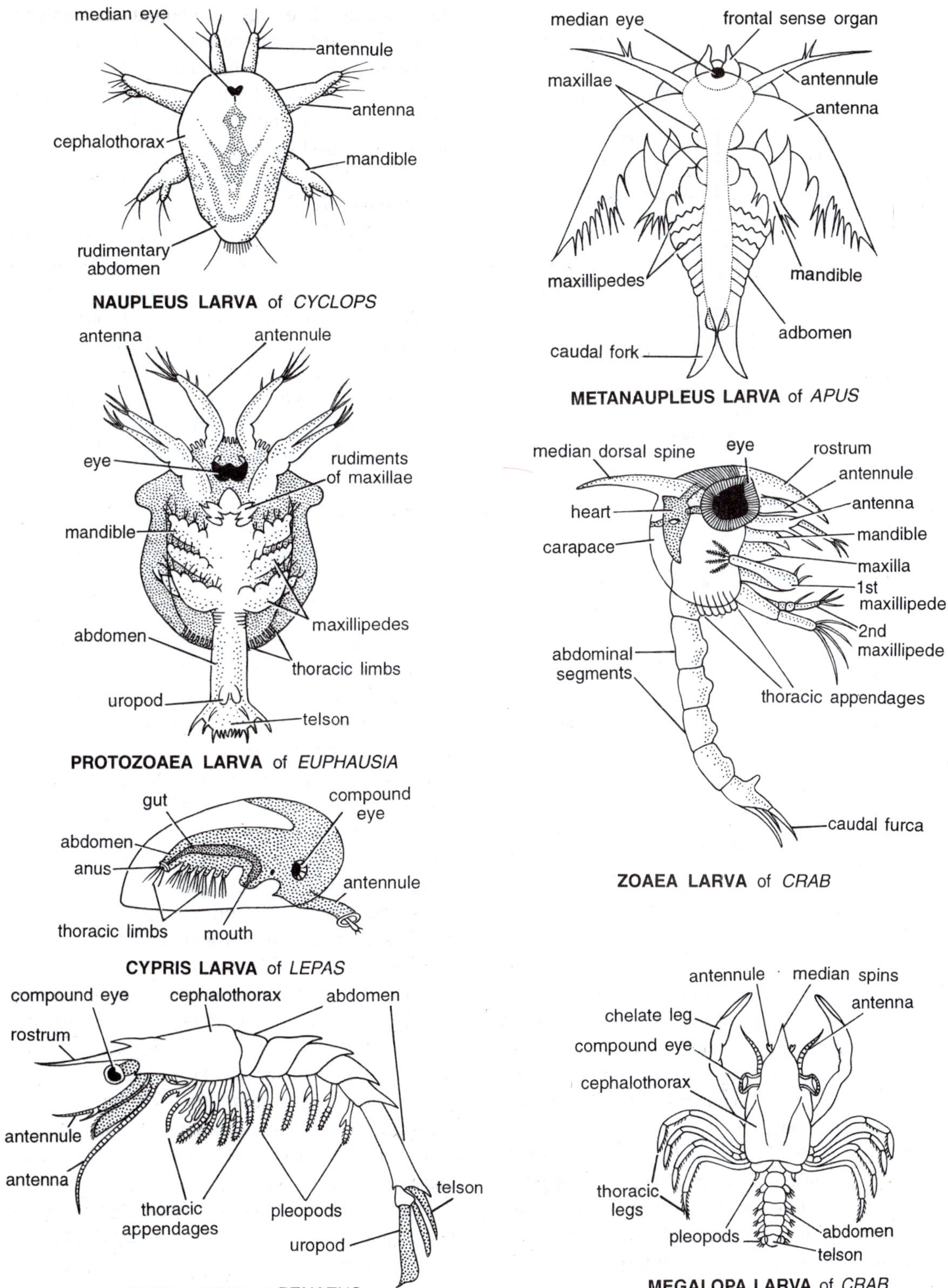

Fig. 1. Larval forms of Crustacea.

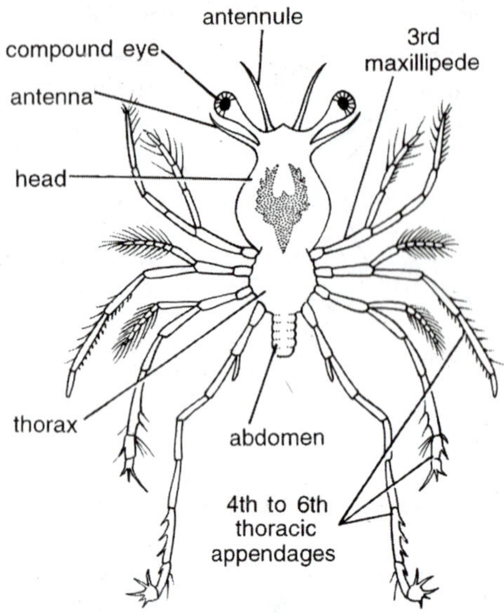

PHYLLOSOMA LARVA of *PALINURUS*

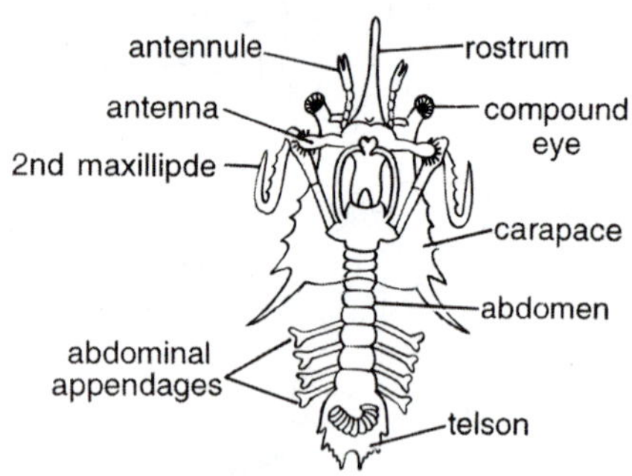

ALIMA LARVA of *SQUILLA*

Fig. 1. Contd. (Phyllosoma and Alima larva of Crustacea).

nauplius stage is passed within egg which hatches as zoaea. It then by moulting forms megalopa to be metamorphosed into adult.

In hermit crabs, the **glaucothoe** corresponds to a megalopa with symmetrical abdomen and swimming pleopods.

8. Phyllosoma. Larva of *Palinurus,* the spiny crab or rock lobster, is called **Phyllosoma** or **glass crab**. It is a modified mysis stage. It is remarkably large, flattened, leaf like, delicate and glassy. Body is distinguished into head, a transparent thorax and abdomen. Eyes are compound and stalked. Out of six pairs of thoracic appendages, the first or maxillipedes are rudimentary, second are uniramous, third well formed biramous succeeded by rest 3 (4th, 5th and 6th) pair of long biramous legs. A segmented but limbless abdomen is present. Before reaching an adult stage, it undergoes several moultings.

9. Alima. It is modified form of zoaea found in some malacostracan (*e.g., Squilla*) which hatches from egg. It is a pelagic form with glassy transparency having a slender body. It bears short and broad carapace. It has all the cephalic appendages but only first two thoracic ones. A six segmented abdomen with 4 or 5 pairs of pleopods, is present. It differs from zoaea in having well formed second maxillipedes and the armature of the telson.

Significance of Larval Forms

According to the **biogenetic law** or **recapitulation theory** of **Haeckel**, every organism, during its development **(ontogeny),** repeats to some extent its evolutionary history (**phylogeny**). In other words, successive stages of individual development correspond with successive adult ancestors in the line of evolutionary descent. Due to its occurrence in the development of all Crustacea, the nauplius was previously regarded to be representing the ancestral form of Crustacea. It was presumed that from this ancestral form the present day crustaceans evolved phylogenetically. In other words, the other larval forms (**zoaea, megalopa,** etc.) show stages of evolution of the higher crustaceans from nauplius-like ancestors. But the old idea of recapitulation stands greatly modified nowadays and the crustacean larval forms are now regarded to be the larval reversions to the types much simpler than the crustacean ancestors. The larval stages are useful for finding out the homologies and the affinities among various groups. The animals which pass through similar stages are closely related.

Larvae are helpful in the wide distribution of species and in keeping the food reserves of eggs to a minimum.

MOUTHPARTS OF INSECTS

Insects feed on animals and plants in a diversity of ways and their mouthparts have become modified for these purposes. Mouthparts are essentially the paired appendages of mandibles and maxillary and labial segments of the head. These have evolved into a variety of forms, which have been perfected to meet different kinds of highly specialized feeding habits. Important types of insect mouthparts may be described as follows :

(1) Biting and Chewing Type

Biting-chewing mouthparts represent the most primitive and unspecialised type which appeared in early insects in course of evolution. These consist of **labrum** or upper lip, a pair of **mandibles,** paired **maxillae** (first maxillae), **labium** (second maxillae) or lower lip, **epipharynx** and **hypopharynx**.

Maxillary palps act as sensory feelers to locate the food. Lacinia are often employed for grasping the food and cutting or chewing it. Mandibles, worked by two sets of muscles, masticate food with their teeth like processes. Ligula, formed by paired and glossae paraglossae, help in pushing the food into pharynx.

Biting and chewing type of mouthparts are common among orthopteran insects such as grasshoppers, cockroaches and crickets. They also occur in silver fish (*Lepisma*), earwigs, termites, book-lice, bird lice, beetles, some Hymenoptera and many larval forms, specially the caterpillars of Lepidoptera.

(2) Piercing and Sucking Type

Mouthparts of insects which feed on fluids are modified in various ways to form a tube through which liquid can be drawn and saliva can be injected. This results in the elongation of some parts and in the loss of some of the typical structures. Piercing-sucking mouthparts are found in blood-sucking insects like the mosquitoes, the bugs such as the bedbug and kissing bug, and the herbivorous insects such as aphids, which feed on plant juices. In this type the **mandibles** and **maxillae** resemble to fine needles, meant for piercing the skin or plant tissues. **Labium** forms the hollow grooved channel which encloses these needles. Open part of the groove is covered by the labrum. Hypopharynx, when present is hollow and needle-like, through which flows the saliva.

In **mosquito**, mouthparts consist of a long **proboscis** or beak, which is composed of the **labium**, forming an elongated, fleshy and mid-dorsally grooved tube. It encloses the needle-like stylets formed by the modifications of the **mandibles, maxillae** and **hypopharynx.** Needle-like **labrum** is fused with the **epipharynx** and forms the long covering of the open groove of proboscis. Proboscis bears, at its tip, two small labellae, which are used as feelers and enables the mosquito to select the appropriate part of its victim to attack. Mouthparts are well developed in female mosquitoes as they feed on blood.

In **bedbug**, labium forms a three-jointed proboscis. Stylets are four in number, consisting of two **mandibles** and two **maxillae,** former with blade-like and the latter with saw-like tips. Labrum is a flap-like structure, covering the groove of proboscis at the base only. Of the four stylets, the maxillae are doubly grooved on their inner faces, one acting as a food canal for the flow of blood and the other as a salivary canal for the flow of saliva.

(3) Chewing and Lapping Type

Chewing-lapping mouthparts occur in the honey-bees and bumble-bees. These consist of a long tongue which is formed from the **glossae** of the **labium,** ending in a spoon-shaped **labellum** or **flabellum**. **Galeae** of the maxillae form blade-like structures and the maxillary palps are very small. A temporary food channel is formed by the proboscis, galeae and labial palps fitting together. Through this food channel the liquid food is sucked up, assisted by the pumping action of the pharynx. Labrum and mandibles act for chewing.

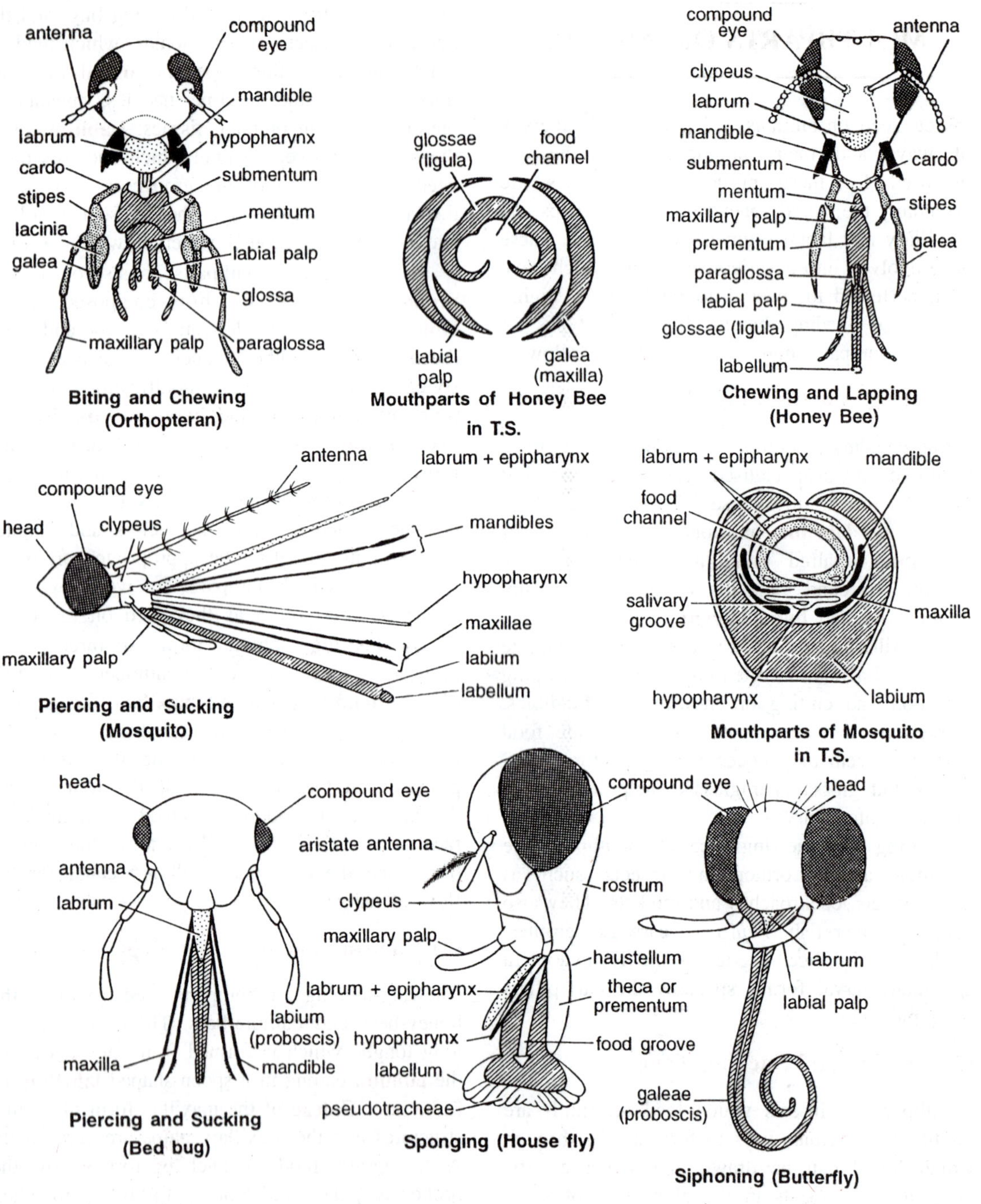

Fig. 2. Mouthparts of insects.

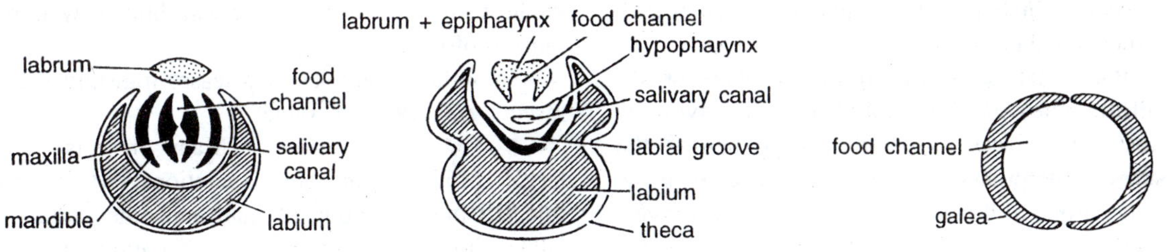

Mouthparts of Beg bug in T.S. **Mouthparts of House fly in T.S.** **Proboscis of Butterfly in T.S.**

Fig. 2. Contd. Mouthparts of insects.

(4) Sponging Type

Sponging mouthparts are found in the housefly and some other flies to suck up the liquified food. These flies lack the cutting weapons of the insects that chew the food. **Mandibles** are altogether absent, while the **maxillae** are represented only by two maxillary palps, each made of a single piece. **Labium** is greatly modified to form the so-called **proboscis,** which is divisible into three parts : (i) a proximal cone-like **rostrum** bearing the maxillary palps, (ii) middle **haustellum** with a mid-dorsal groove, serving as food passage and a ventral heart-shaped plate called **theca,** and (iii) distal **labellum** (oral disc) consisting of two expanded lobes or **labellae,** underside of which is made of numerous incomplete cylindrical tubes, the **pseudotrachea**, all converging in the centre into the mouth aperture, which leads into the food groove. Labellum is used to lick the food by its contractile activity. It is first collected into the pseudotracheae and then passed on to the food canal, formed by the **labium, epipharynx** and **hypopharynx**, lying in mid-dorsal groove of the haustellum.

(5) Siphoning Type

Butterflies and moths are adapted for feeding on nectar, like the bees, but in their mouthparts the maxillae form the main proboscis and not the labium. Mandibles and labium are much reduced, the maxillary palps are rudimentary and labium forms a triangular plate bearing labial palps. Galeae are much elongated and coiled, each forming a half tube, which makes complete tube when both are locked together. When not in use, proboscis is coiled into position beneath the head and when insect wants to feed, it becomes uncoiled to reach the nectary. It is the rise in blood pressure which uncoils the proboscis.

ECONOMIC IMPORTANCE OF INSECTS

Most people think that insects are pestiferous creatures and should be destroyed at all cost. But few people know the exact nature of damage they do. Even fewer people know the many acts of insects that are indispensible to man. Unfortunately, beneficial acts performed by insects are far outweighed by the tremendous amount of damage done by them.

[I] Injurious Insects

1. Pests of plants, fruits and stored grains. Insects attack leaves, stems, buds, flowers, seeds, fruits, barks, woods, roots, vegetables, stored products, wool, feathers, cigars, tools and even minerals. **Chewing insects**, like the cabbage-worm, hoppers and potato beetles, chew and swallow the external parts of the plants. Grasshoppers and locusts have invaded the green crops since times immemorial. Locusts, on the war-path, sometimes move on in swarms extending for many kilometers. Grass and leaves are devoured and even branches are broken by weight of insects settling on them. Plant-bugs, aphids, scale-insects and others possess an extremely sharp-pointed, beak-like process or proboscis, which is thrusted into the plant tissues to suck up their juices.

Bark beetles destroy timber in the forests, while termites damage timber after it leaves the forests. Boll weevils spoil cotton before the harvest. Caterpillars and Japanese beetles strip the foliage from millions of shady trees every year.

2. Household pests. Several insects are unwanted guests in the house. Mostly annoying, sometimes they become destructive. Bedbugs, mosquitoes and stable-flies are much annoying. Ants, crickets, cockroaches, weevils, fruit-flies and silver-fish, etc., spoil the food. Clothing, carpets, furs and feathers may be damaged by cloth moths and carpet beetles, etc.

3. Injurious to domestic animals. Domestic animals are often seriously injured by insects. Many of them live more or less as parasites either externally, such as fleas, lice, bugs, mosquitoes, etc., or internally, such as larvae of boltfly in sheep. The bird-lice *(Mallophaga),* feeding upon feathers of chicken, cause irritation and loss of flesh. The blood-sucking horn-fly is a serious pest of cattle. The grubs of ox warble-fly cut holes in the skin of cattle, thus causing damage of hide and flesh. The larvae of horse boltfly sometimes cause serious disturbances in stomach.

4. Disease carriers. Many insects play a great role in spreading serious diseases of man and domestic animals. They often act as vectors for transmitting various disease producing organisms either by infecting them in blood stream *e.g.,* malaria, or by contaminating the food *e.g.,* bacteria. The various species of the genus *Anopheles* of mosquitoes have been found to convey one or other parasitic Protozoa causing **human malaria**. Certain **Culicine** mosquitoes spread the nematode worm, *Filaria bancrofti,* which causes **filariasis** in man. Similarly, **yellow fever** is spread by a mosquito **Stegomyia**; **surra** disease among horses, camels, etc. of tropical countries by **Tabanus fly**; **African sleeping sickness** by **tsetse fly**; **typhoid**, **diarrhoea**, **cholera**, etc. by common housefly *Musca domestica;* **bubonic plague** by fleas and **relapsing fever** by bedbug and body louse, and so on. Undoubtedly, the insects carrying diseases are greatest enemies of man, affecting human welfare most profoundly.

Important examples of diseases transmitted by insects are given in Table 1.

5. Poisonous insects. Many insects and larvae produce poisonous secretions, which are injected into the body of man and other animals either through a bite or sting. Irritation of skin pain and swelling may result. The common examples of poisonous insects are honey-bees, wasps, hornets, fire-ants, bedbugs, mosquitoes and a few lepidopterous and other larvae.

[II] Productive Insects

Human beings are greatly indebted to certain insects, which supply them with useful products. Many commercial products produced by insects are indispensable to modern man.

1. Honey. Honey is produced by honey bees (*Apis*). In India approximately 70,000 tonnes of honey is produced annually. Out of this approximately 27,000 tonnes is being exported outside India.

2. Beeswax. Wax produced by honey bees is used in polishes, churches, modelling and to wax the thread. A few scale insects also produce wax.

3. Raw silk. Silk is produced by silkworms (*Bombyx mori* and others). A cocoon yields about 300 meter silk fibre, and about 25,000 cocoons are unwound to spin one pound of silk thread. The number of silkworms employed in service of man can be imagined by the fact that nearly 40–50 million kg silk are used in the world every year.

4. Lac. The shellac of commerce is obtained from waxy secretions of lac insects (*Tacchardia lacca*) or scale insects (fam. Coccidae) of India, the females of which secrete lac. It is useful as a foundation of lacquers and shellacs.

5. Dyes. The dyes known as **tannin, cochineal** and **crimson lake** are derived from the dried bodies of certain scale insects living on cacti. Cochineal is no longer of value since synthetic aniline dyes have largely taken its place.

6. Medical products. Certain medical products like **Cantheridine** are also derived from blister beetles.

Table 1. **Diseases Transmitted by Insects.**

Insect vector	Disease	Host	Causative organism
1. Mosquito *Anopheles*	Malaria	Man	*Plasmodium,* (protozoan)
2. Mosquito, *Aedes aegypti*	Yellow fever	Man	Virus
3. Mosquitoes, *Culex* and others	Filariasis	Man	*Wuchereria* bancrofti (nematode)
4. Tsetse fly, *Glossina palpalis*	African sleeping sickness	Man	*Trypanosoma gambiense* (protozoan)
5. Bug, *Triatoma*	Chagas	Children	*Trypanosoma cruzi* (protozoan)
6. Tsetse fly, *Glossina morsitans*	Nagana	Domestic animals	*Trypanosoma brucei* (protozoan)
7. Tabanus fly, *Tabanus*	Surra	Horse, camel	*Trypanosoma evansi* (protozoan)
8. Sandfly, Phlebotomus	Kala-azar	Man	*Leishmania donovani* (protozoan)
9. Flea, *Xenopsylla*	Bubonic plague	Rats and man	*Pasteurella pestis*
10. Bodylouse, *Pediculus*	Typhus fever	Man	*Rickettsia*
11. Housefly, *Musca*	Cholera	Man	*Vibrio cholera* (bacteria)
12. Housefly, *Musca*	Diarrhoea	Man	*Giardia* intestinalis (protozoan)
13. Bedbug, body louse	Relapsing fever	Man	

[III] Helpful Insects

1. Pollinators of flowers. 'Insect friends' of man renders him the greatest service in pollination of flowers. Plants depend upon certain insects for cross-pollination or cross-fertilization which is very necessary for their fertility and the vigour. Various insects and flowers are mutually dependent since many insects feed upon their nector and pollen grains. Chief pollinating insects are bees, wasps, beetles, ants, flies, etc.

2. Scavengers. Insects feed upon waste material such as dead bodies and debris of plants and animals, thus preventing decay and obnoxious odours. Common examples of insect-scavengers are silver-fish *(Lepisma),* termites, housefly, blowfly, maggots, dung-beetles, carrion beetles, fleas, cockroaches and many larvae.

3. Insects as food. Insects provide an abundant food supply for animals like frogs, lizards, snakes, fishes, etc. Blue-birds, meadow-larks and house sparrows depend chiefly on insects. Moles, shrews, armadillos and ant-eaters live wholly upon small insects. Man consumes many insects and their larvae, eggs, etc., only accidently with fruits and other foods. Natives of Amazon valley eat suab-ants. Termites form favourite food in tropics. Eggs of *Coriza femorata* are eaten in Mexico, while larvae of Goliath beetle of Africa are a fine food morsel. Greeks ground locusts in mortars and made flour of them. American Indians used to dry or smoke larger caterpillars and preserve them for later use.

4. Insects in medicine. Cochineal insects contain carminic acid, coccerin, myrestin, fat and fatty-acids and are used in the treatment of neuralgia, and whooping cough. Blowfly larvae are used in treating decay of tissues. Cantheridin oil, made out of blister-beetles, serves as hair restorer. The body extract of the cocoons of silk moth, *Bombyx mori* is used for checking profuse menstruation and in treating leucorrhoea and chronic diarrhoea. Bee-venom has been used with some degree of success in treatment of some forms of arthritis. Bee-venom has also been used in the preparation of an anti-venom to counteract snake bite. Honey is a natural antiseptic which prevents infection if applied to a wound. It is also applied to cure ulcers. Beeswax is used as a base for ointment.

5. Biological control. An important service that insects render is to exercise **biological control** over other harmful species. Insects which attack and eat up other insects are called **predators**. Many harmful plant-eating insects are devoured by a host of predaceous insects, such as ground beetles, syrphid flies and wasps. Aphids and scale insects, which are pests of citrus and other trees, are eaten by larvae of lady bird beetles (*Coccinella,* etc.). Predaceous insects are even reared or imported and liberated in orchards to control scale insects. Predatory tiger beetles and nocturnal carnivorous ground beetles are also considered beneficial as they destroy serious insect pests (*e.g.,* cutworms, armyworms) and large numbers of detrimental slugs and snails.

Larvae of parasitic insects which attack the larvae of plant-feeding insects, are known as **parasitoids**. They are in turn subject to hyper-parasites, which reduce their effectiveness.

6. Insects in fine arts. Insects produce noises in various ways. Whether their sounds can be called musical is disputable. In Japan, **cicadas** and **crickets** are placed in small cages, like birds, in the houses. In Tokyo, about 50 markets were dealing with them prior to the war. Japanese celebrate an annual function, the **'festival of the singing insects'**.

Beautifully coloured elytra and wings of some Coleoptera and Lepidoptera are used in jewellery and pictures in Central America, in crafts by Indians of America, in embroidery, pottery, baskets, metals, alloys, etc. by Red Indians of America, and in ear-rings by **Jivaros** of Equador.

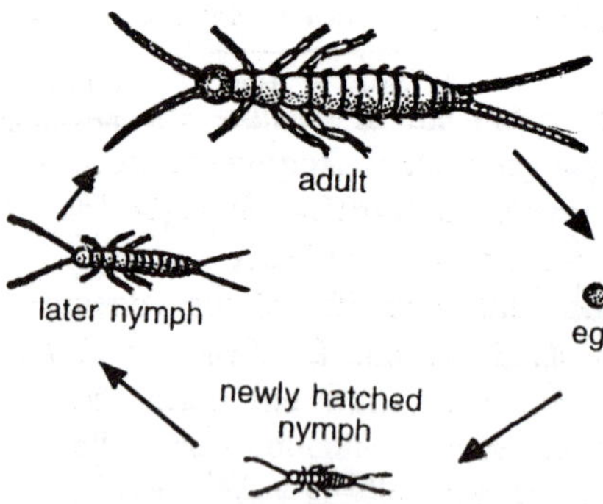

Fig. 3. No metamorphosis *(Lepisma).*

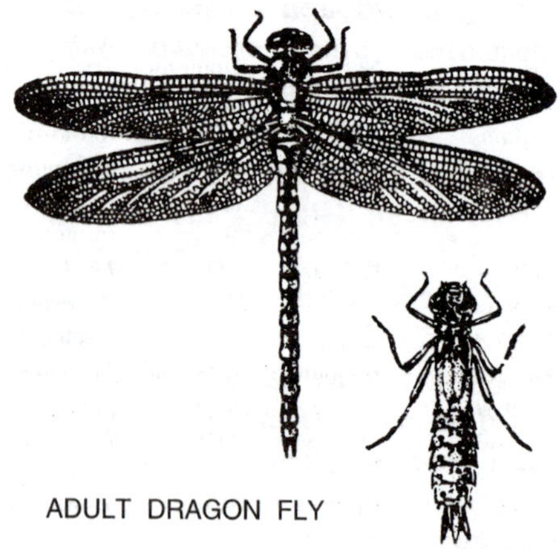

Fig. 4. Incomplete metamorphosis (dragonfly).

METAMORPHOSIS IN INSECTS

Transformation of an immature larval individual into a sexually mature reproducing adult of very different form, structure and habit, is called metamorphosis.

Types of Metamorphosis

Insects display 4 types of metamorphosis.

1. No-metamorphosis or ametabolous development. In case of no-metamorphosis, newly hatched creature looks like an adult except in size and differences in armature of spines and setae. Examples : Silver-fish, spring-tails.

2. Incomplete metamorphosis or hemi-metabolous development. In case of incomplete metamorphosis, immature stages are the **nymphs** or **naiads,** which are aquatic and respire by tracheal gills, whereas the adults are terrestrial or aerial and respire by tracheae. Examples : Mayflies, dragonflies, stone-flies.

3. Gradual metamorphosis or pauro-metabolous development. In case of gradual metamorphosis, the newly hatched creature resembles an adult in general body form, but lacks wings and external genital appendages. Young or the **nymph** undergoes several nymphal stages through successive moultings to become an adult, Examples : Grasshoppers, aphids, stink bug, etc.

4. Complete metamorphosis or holometabolous development. This type of metamorphosis includes four developmental stages : **egg, larva, pupa** and **adult**. Larva, after hatching, moults several times to become a fully grown one. It later becomes a pupa within a secreted case, called the **puparium**. Pupa differentiates into the young adult that breaks the puparium open and emerges outside. It grows to a mature form. Examples : Housefly, mosquito, butterfly.

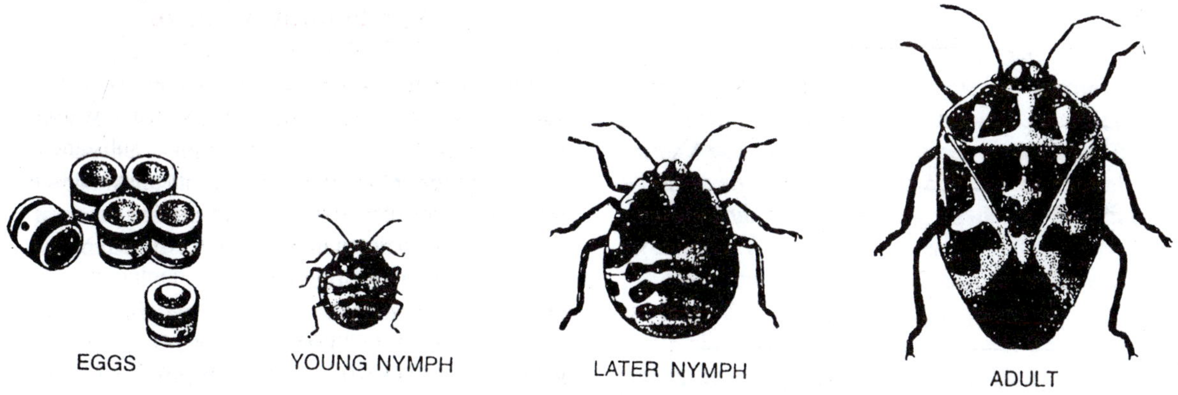

Fig. 5. Gradual metamorphosis (stink bug).

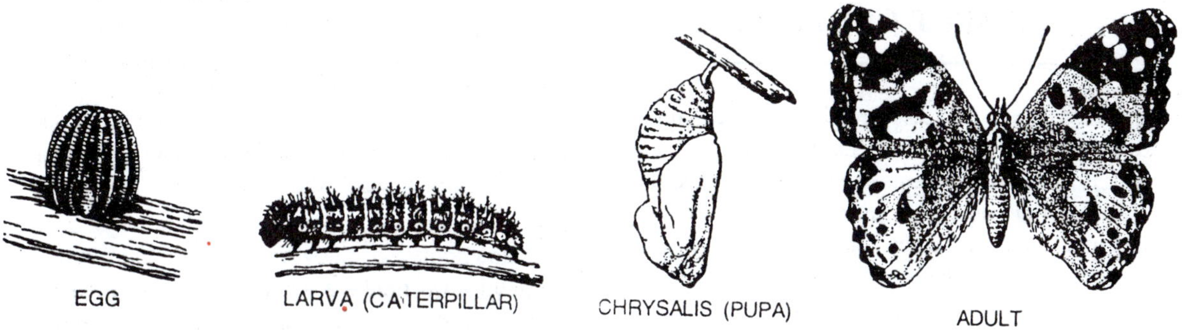

Fig. 6. Complete metamorphosis (butterfly).

HORMONAL CONTROL OF METAMORPHOSIS

Metamorphosis or the post-embryonic growth of insects, as in most higher organisms, is under hormonal control. The various hormones secreted are as follows :

1. Brain hormone (BH). Brain hormone is secreted by the neurosecretory cells of the brain. Chemically it is a lipid. This hormone serves to activate the **corpora cardiaca**, a component of the retro-cerebral complex of the stomatogastric nervous system.

2. Prothoracicotropic hormone (PTTH). This hormone is secreted by the corpora cardiaca, which in turn stimulates the **prothoracic glands.**

3. Prothoracic gland hormone (PGH). This hormone is secreted by the paired, bilateral sheet of cells in the thorax, constituting the prothoracic glands. Chemically it is **ecdysone**. This hormone is known to trigger moulting as it acts on the tissues to promote all of the changes characterizing a moult.

4. Juvenile hormone (JH). This hormone is secreted by another component of the retrocerebral complex, the **corpora allata.** Chemically it is an unsaponifiable, non-sterolic lipid. This hormone regulates morphogenesis and so promotes

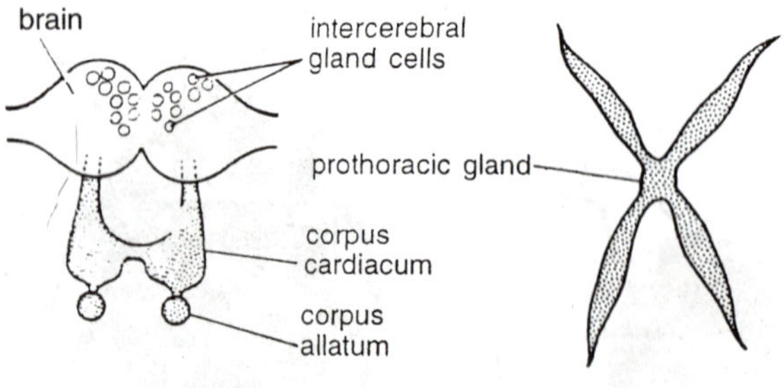

Fig. 7. Endocrine glands of insects.

metamorphosis, that is, development of the larva into adult through pupal stage.

INSECT CONTROL

Insect control in its broadest sense implies everything that makes life hard for insects, tends to kill them and prevents their increase or spread. Value of Entomology to mankind is largely based on insect control. Insect control is at least as old as agriculture, as there has always been a need to keep crops from being destroyed by the agency of Insects (pests). In order to maximize food production, it is important to protect crops from insects. This is where the role of Entomology comes in. Crops worth Rs. 50,000 crore are lost every year due to insects and plant diseases. Not only this, insects are competing with man in all the spheres. Therefore, to prevent the damage to agricultural, horticultural and forest produces, and to protect the health of man and his domestic animals, insect control is imperative.

METHODS OF INSECT CONTROL

Several methods are employed for insect control. **Physical control** involves use of hands, traps, fire and water, etc., to kill insects, as in case of locusts, mosquitoes, houseflies and ants. But these are not very effective. **Chemical control** and **biological control** are better developed methods.

[A] Chemical Control

Killing insects by using insecticides is called chemical control. It is the most extensively used method at present. All those chemical substances (toxicants) that kill insects by their chemical action are known as **insecticides** or **pesticides**.

Chemical insecticides date back, 4,500 years, when the Sumerians used sulfur compounds as insecticides. The Rig Veda, which is about 4,000 years old, also mentions the use of poisonous plants for insect control. It was only with the Industrialization in the 18th and 19th century, and the introduction of the insecticides **pyrethrum** and **derris** (plant origin) that chemical insect control became widespread. In the 20th century, the discovery of several synthetic insecticides, such as **DDT,** boosted this development chemical insect control is still the predominant type of insect control today, although its long-term effects led to a renewed interest in traditional and biological insect control towards the end of the 20th century.

[I] Types of Insecticides

Spraying insecticides by handheld units, or trucks that carry the spraying equipment, is a common method of pest control. Spray insecticides (pesticides) to kill off insects that would threaten the crops. However, some insecticides may cause cancer and other health problems, to the persons handling them.

1. Stomach poisons. Insects with chewing, sponging, lapping or siphoning mouthparts are readily killed when they ingest treated plants. Insects ingest the insecticide while feeding on the plants. Common stomach poisons are as follows :

(a) Lead hydrogen arsenate. Lead arsenate was the most extensively used arsenical insecticide. Two formulations were marketed : **basic lead arsenate and acid lead arsenate.** As an insecticide, it was first used against the gypsy moth and later mostly against the potato beetle. It was used mainly on apples, but also on other fruit trees and garden ornamental plants.

(b) Sodium arsenate. It is a inorganic colourless solid salt that is highly toxic soil poison against termites.

(c) Sodium fluoride. It is also an inorganic colourless salt which is effective against cockroaches and poultry lice.

(d) Sodium fluosilicate. This chemical is employed in the preparation of poison baits for the control of moths that destroy garments.

(e) Cryolite. It is an uncommon mineral insecticide was first discovered on the shores of greenland.

2. Contact poisons. These constitute those toxicants which kill the insects on coming in direct contact with them. These are employed primarily in the control of insects with piercing and sucking mouthparts although they can kill insects with other types of mouthparts as well. The common contact poisons are as follows :

(a) DDT (Dichlorodiphenyltrichloroethane). It is a colorless, crystalline, tasteless and almost odorless organochloride. Its agricultural insecticidal property was discovered by the Swiss chemist **Paul Hermann Muller** in 1939. DDT is nearly insoluble in water but makes good solutions in xylene/petrol, emulsions/lotions in fats/oils, powders and vaporizers. DDT have been linked to diabetes, Parkinson's disease. Alzheimer's disease and asthma. DDT causes a number of cancers in humans. Once the mainstay of anti-malaria campaigns, Mosquito resistance to DDT's effectiveness has greatly reduced its use.

(b) BHC (Hexachlorobenzene, perchlorobenzene or gammexene). It is a white crystalline solid that has negligible solubility in water and variable solubility in different organic solvents. It is used mainly to kill aphids on fruit and nut trees.

(c) Malathion. It is an organophosphate insecticide of relatively low human toxicity that is widely used in agriculture, residental landscaping, public recreation areas, and in mosquito and fruit fly eradication control programs.

(d) Pyrethrum. Pyrethrum is the name of a natural insecticide made from the dried flower heads of *Chrysanthemum cinerariifolium* and *Chrysanthemum coccineum.* It is a contact poison effective against many household, livestock and vegetable pests.

(e) Aldrin. It is colourless solid organochloride insecticide heavily used to treat seed and soil against grasshoppers and cotton pests.

(f) Toxaphene. It is a mixture of approximately 200 organic compounds, Toxaphene is usually seen as a yellow to amber waxy solid. Toxaphene was used as an insecticide in the cotton and soybean growing areas.

3. Fumigants. Fumigants are gaseous pesticides/fumigants that completely fills an area to suffocate or poison the pests within. It is utilized for control of pests in buildings (structural fumigation), soil, grain, and produce.

4. Repellents. Repellents are substances applied to skin, clothing, or other surfaces which discourages insects from landing or climbing on that surface. Trichlorobenzene is used to protect buildings from termites. Bordeux mixture and lime is used to ward off leaf hoppers. Oil of Cider keeps carpet beetles and clothes moths at bay.

[II] Hazards of Insecticides

Insecticides are proved to be a mixed blessing. In late 1940, DDT was discovered in the tissues of fish, wildlife and humans. Later studies showed it could cause blood cancer (leukaemia) in laboratory animals. In 1972, U.S.A. banned its use except in cases of serious epidemics or infestations. Similar discoveries were made with regards to several other pesticides. BHC was found to cause nervous disorders and fumigants led to cell mutations. Other diseases, such as cirrohosis, carcinoma, hypertension, tumour formation and sterility are also alleged to the insecticides polluting human food. Introduction of pesticides has resulted in the abandonment of such traditional and sound agricultural practices as rotating and diversifying crops. Moreover, because of over-exposure, insects were becoming immune to chemical pesticides. 156 out of the 1000 or so species of crop pests have built up resistance. As a result, farmers have been forced to use even greater amounts of increasingly expensive insecticides. But, they have been receiving an even smaller return from their heavy investments. The pesticides frequently kill the beneficial insects also that help keep pests under control.

To meet the hazards of insecticides, the entomologists advocate a different approach, called **integrated pest control**. They no longer talk of eradicating species, but advocate that insect depredations can be kept small enough to be acceptable economically. They realize that some pesticides will always be needed to cope with insect problems that suddenly get out of hand.

[B] Biological Control

Scientists now believe that the most promising weapons for the battle are **biological controls**. These can be aimed at specific pests without adversely affecting either human beings or the environment. The more important biological controls are the following :

1. Hormones. During the larval stages, **juvenile hormone** is secreted by the corpora allata, paired glands associated with the brain. It permits larva-to-larva moult but inhibits metamorphosis. Some trees secrete substances with juvenile hormone-like activity. When ingested, these prevent the harmless larvae from maturing into annoying adults and reproducing, thus protecting the trees against certain insects. Zoecon Corporation of Palo Alto, California, is marketing **Altosid Insect Growth Regulator**, a compound chemically similar to juvenile hormone.

Another insect hormone, **ecdysone**, secreted by the prothoracic glands, induces moulting accompanied by metamorphosis. When the juvenile hormone declines in the blood, the ecdysone is free to bring about first larva-to-pupa moult and then pupa-to-adult moult. Leaves of certain few trees and weeds produce substances with ecdysone-like activity. Moth larvae eating the leaves and ingesting these substances undergo premature moulting and die. **Professor William Bowers** has isolated two substances from ageratum, a flowering plant; these interfere with an insect's production of juvenile hormone. Cotton stainers treated with these compounds moult precociously into tiny sterile adults and die, Mexican bean beetles also become sterilized, and Colorado potato beetles enter hibernation and die.

2. Pheromones. Pheromones are substances secreted by the exocrine glands and released into the external environment. They cause specific reactions in other animals of the same species. Pheromones include **sex attractants,** exuded by females of same species that can draw males of the same species from a distance as great as 4.5 km. Synthetic forms of such chemicals are used in the detection and control of insect pests which are collected by baiting traps with attractants and killed with potent insecticides. If these attractants are spread in large quantities over an insect-infested field, the confused males might get killed. In the absence of males, mating cannot occur.

3. Predation and parasitism. Idea of using insects to combat insects has achieved some striking successes. Many harmful plant pests are devoured by a host of **predaceous insects** such as ground beetles, syrphid flies and wasps. Larvae of ladybird beetles (Coccinellidae) eat up aphids and scale insects that feed on and damage citrus and other trees. They have been imported, reared and liberated in orchards to control scale insects.

Parasitic insects oviposit in the eggs, larvae, or adults of plant feeding insects, so that the hatching larva uses up the host for food. At Cornell University, entomologists Maurice Tauber and Robert Helgesen are using a tiny wasp to control the white fly which is a serious pest of poinsettia plants. Over the past 20 years, a total of 42 insect species have been controlled or their damage reduced, by introducing parasites of various types.

4. Sterilization. Insect birth rate can be reduced by tricking females into mating with males that have been sterilized by exposure to radiation.

5. Genetic control. According to the theory of 'incompatibility' by **Professor Hannes Laven** of West Germany, certain strains of mosquitoes when crossed cannot breed. He found in 1947 that if a Paris female *(Culex fatigans)* was crossed with a Hamburg male, the eggs would never hatch. Experiments are being conducted in India also by the WHO-ICMR unit for genetic control of mosquitoes in consulation with Prof. Laven.

Uneasy Detente

Achieving effective and environmentally accepted methods of biological control of insects will be quite expensive. The cost of production of even a tiny amount of pheromone runs into thousands of rupees. The expenses involved in sterilizing insects and in identifying and isolating their hormones, are exorbitant. The ever adaptable insects also adjust to man's latest weapons against them. Laboratory tests show that after 15 generations, houseflies and mosquitoes develop resistance to juvenile hormone insecticides.

So the battle between human beings and insects goes on, with some hope that man will continue to maintain an useage detente with the insect world for centuries to come.

RESPIRATION IN INSECTS

Gas exchanges with environment with uptake of oxygen and release of carbon dioxide while minimising the concurrent water losses, constitute respiration. It includes chemical and physical processes. Chemical phase of respiration is oxidation of tissues with the formation of water and carbon dioxide. Physical phase comprises of the transportation of the oxygen to different tissues and removal of carbon dioxide.

Various types of respiratory organs are known among the insects according to their habits of life.

Tracheal Respiration

[I] Tracheal System

Great majority of insects breathe by means of an elaborate and most efficient gas exchange system made of branching, elastic air tubes or **tracheae,** called the **tracheal system**. In its general plan, tracheal system resembles circulatory system of vertebrate animals, as it is distributed throughout the body, so that air is brought in direct contact with tissues and cells of the body. In insects tracheal system serves for transport of oxygen and carbon dioxide and blood has lost its respiratory function. With the exception of **chironomus** larvae (midgeflies) which have haemoglobin in blood plasma, insect blood has a very poor capacity for carrying oxygen.

1. Spiracles. Air enters the tracheae through a number of elongated or circular apertures, called **stigmata** or **spiracles,** opening on the lateral sides of the body. Their usual number is ten pairs of which two are thoracic and eight abdominal. Each spiracle is supported by a small annular sclerite or **peritreme** and capable of being closed by a **sphincter** or **spiracular muscle**. Spiracle may be with or without a chamber **(atrium),** a filtering apparatus and a closing apparatus. The filtering apparatus includes fine hair or bristles, to keep out dirt. The closing apparatus is in the form of a valve to prevent undue loss of water and to regulate the in and out flow of air.

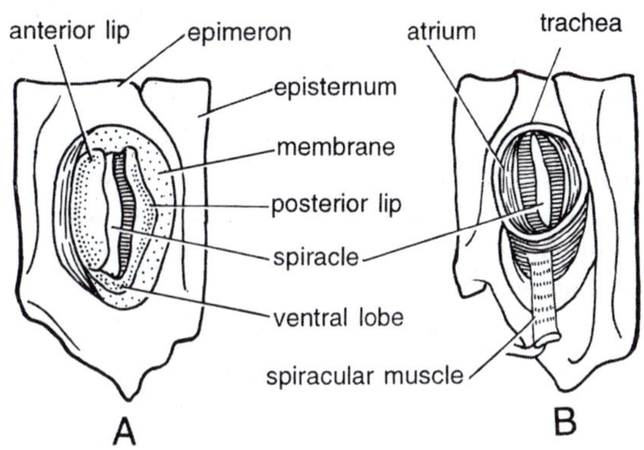

Fig. 8. Spiracles of an insect. **A.** Outer view. **B.** Inner view.

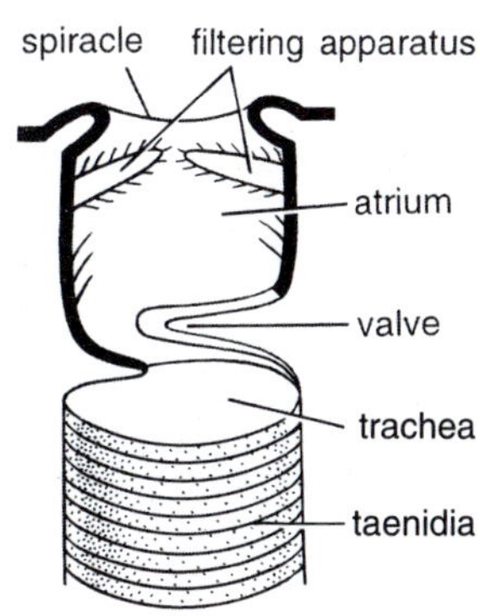

Fig. 9. A spiracle with atrium, filtering apparatus and valve.

The spiracles normally open for the shortest time for efficient respiration in order to minimise the water loss from tracheal system. Their closer results from sustained contraction of the closer muscles, while their opening results from the elasticity of the surrounding cuticle when the closer muscle is relaxed. The muscle is controlled by the central nervous system. They may also respond to local chemical stimuli which interact with the central control.

2. Tracheae. Tracheal system is an invagination of the cuticle. It is represented by an extensive network of branching and anastomosing tubes. The general plan varies in different species and orders of insects, but there is usually a pair or more of longitudinal trunks with cross-connections. The wall of each trachea is formed of three layers, an inner chitinous layer or **intima** with a spiral thread-like thickening or **taenidia**. Taenidia prevents the collapse of the trachea if the pressure within gets reduced. Intima consists of outer epicuticle with a protein-chitin layer below it. In the middle is a cellular layer of **epithelium** and an outer delicate supporting layer or **basement membrane**.

3. Tracheoles. At various points along their length, and especially distally, the tracheae gives rise to finer tubes called tracheoles. There is no sharp distinction between tracheae and tracheoles but the latter are always intracellular and retain their cuticular lining at moulting, which is not usually true with tracheae. Tracheoles are broader proximally and taper down distally. They are formed in cells called tracheoblasts derived from the epidermal cells lining the tracheae. They have thick intima but are devoid of taenidia. They are lined by a protein called trachein. The tracheoles are very intimately associated with the tissues and in fibrillar muscles, for instance, they may indent the muscle plasma membrane and penetrate deep into the fibre, but it is probable that they never become truely intracellular. Distally the tracheoles end blindly or they may anastomose. Ultimate tracheoles enter the individual cells of the body, and are partly filled with a fluid called **tracheole fluid** at their distal end. The amount of fluid present is believed to be reduced, by osmotic movement of water to the surrounding tissues, when metabolism is most active. By reduction in the amount of fluid, more of the surface walls of the tubules is exposed to oxygen, and more oxygen is immediately supplied to the tissues.

4. Air sacs. In grasshoppers, butterflies, bees, cicadas, etc., the tracheae dilates into **air sacs** of different sizes. These may be found in places where muscle movements will fill and empty them. Scarabaeid beetles also have small air sacs along the tracheae of the elytra. These chambers are innervated and invested with the tracheal coverings but in the reverse order, with the cuticle outside. The air sacs lack the spiral thickening, allow an increased supply of oxygen and afford a greater breathing capacity to the insect.

[II] Types of Tracheal Systems

Tracheal system can be classified on the basis of number and distribution of functional spiracles out of total number of ten pairs.

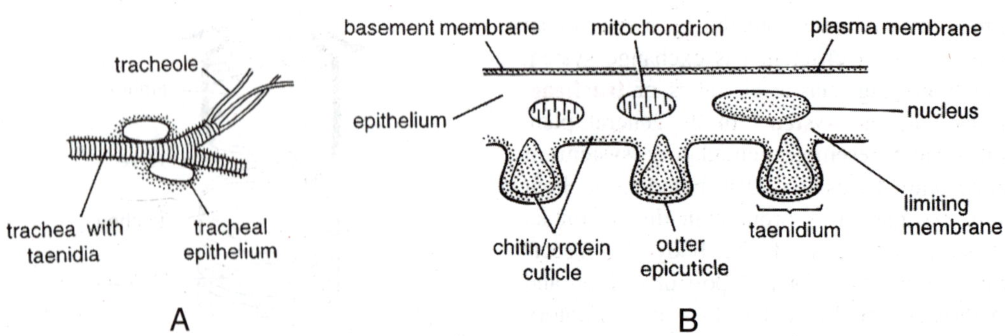

Fig. 10. A. Tracheal system showing tracheae. **B.** L.s. of tracheal wall with body cavity above and lumen of trachea below.

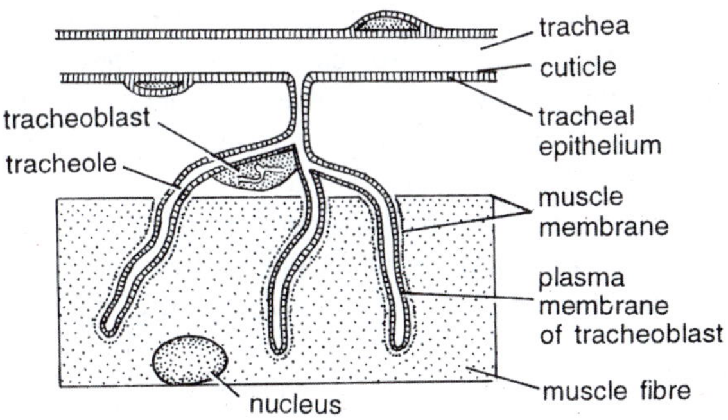

Fig. 11. An intracellular tracheole (diagrammatic).

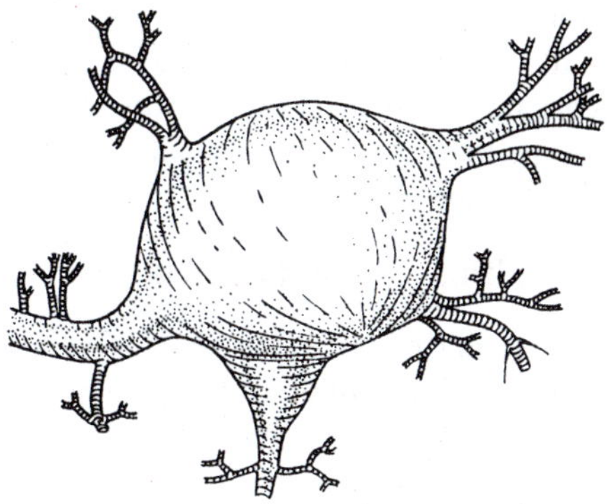

Fig. 12. An air sac.

1. Holopneustic. When all the spiracles remain open *e.g.,* most adults and many larval insects.

2. Hemipneustic. When one or pair of spiracles remain closed *e.g.,* some flies, beetles, butterflies, etc.

3. Apneustic. When the spiracles are altogether lacking *e.g.,* endoparasitic and aquatic larval insects. In such cases the gaseous exchange takes place either through the body surface or through its out growths, called gills.

Wigglesworth (1954) has described six types of tracheal systems as follows :

(1) Two thoracic and eight abdominal spiracles open, *e.g.,* cockroach.
(2) All spiracles open. Air sacs present, *e.g.,* grasshopper, honeybee.
(3) All spiracles closed except the most posterior ones, *e.g.,* larval insects.
(4) All spiracles closed (apneustic). Gaseous exchange through integument, *e.g.,* Collembola, parasitic larval Hymenoptera.
(5) Spiracles closed. Tracheal gills on abdomen, *e.g.,* larvae of mayflies, dragonflies, etc.
(6) Spiracles closed. Gills present inside hindgut.

[III] Mechanism of Gaseous Exchange

Respiratory exchange in the tracheal system is partly by **diffusion** and partly by **ventilation,** as found in a vertebrate lung. Inward diffusion of oxygen from the spiracles depends on partial pressure within the tracheoles being lower than in the outside air and this will arise from the passage of O_2 into the tissue when it is utilized. Only small insects such as *Drosophila* may get sufficient oxygen by diffusion alone, for flight muscles, but in larger insects diffusion by itself cannot meet the demands of highly active tissues. The high demand of large, active insects for oxygen, is supplemented by convection produced by changes in the volume of the tracheal system. This is known as ventilation.

Most tracheae are circular in cross-section and resist any change in form, but some, such as longitudinal trunks of *Dytiscus* (Coleoptera) larvae, are oval in cross-section and are subjected to collapse. Collapse of trachea forces air out of the tracheal system, while its subsequent expansion sucks in air again. But by changes in shape of the trachea, only small volume changes. Much larger changes, hence better gaseous exchange, are produced by the alternating collapse and expansion of air sacs. The air enters the tracheal system through the spiracles by the movements of the abdomen. The spiracles show controlled movements to allow the desired amount of air to enter in. The control is supposed to be by the neurosecretions as well as the respiratory centres of the central nervous system.

As the air enters the tracheal system **(inspiration),** tracheole fluid takes some oxygen which diffuses inwards to the tissues or active cells. Greater the need for oxygen, more is the withdrawal of tracheal fluid into the tissue and vice versa. When the air is expelled to outside **(expiration),** tracheal fluid again fills into tracheole tips. CO_2 produced mostly diffuses out through the tracheae and the integument, and partly through the spiracles.

[IV] Moulting of the Tracheal System

Cuticular lining of trachea is cast at each moult, and a new larger **intima** is formed in its place. The longitudinal trunk breaks at predetermined points, called the nodes, between adjacent spiracles and shed with rest of the exuviae. In early *Sciara* larvae having relatively simple tracheal system the whole of the cuticular lining is shed, but in older larvae having more complex system, the lining of finest branches are persistent (**Keister,** 1948). In *Rhodnius,* the lining of the tracheoles are never shed (**Wigglesworth,** 1945), instead, when a new tracheal lining is formed it pinches in on the old lining at the margin of the tracheoles. At the point of constriction the old cuticle breaks and new tracheal intima becomes continuous with the original tracheole lining, leaving behind a marked discontinuity.

[V] Tracheal Respiration in Semi-aquatic Insects

Semi-aquatic insects have tracheae and dwell for the most part on or near the surface of water.

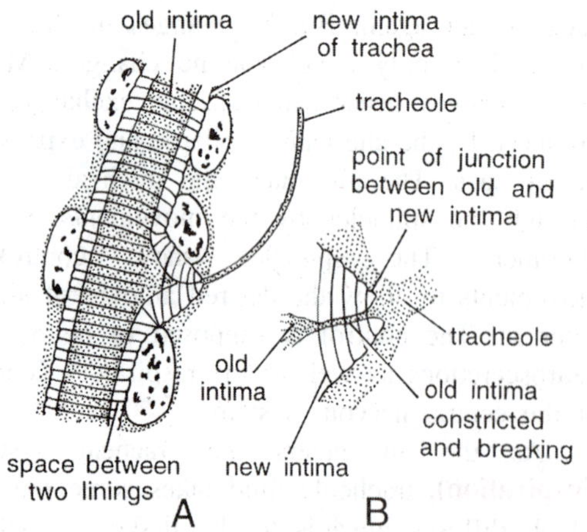

Fig. 13. A. Development of new intima and its junction with original lining of a tracheola. **B.** Magnified view of junction between the two.

They are dependent on atmospheric air and cannot breathe oxygen dissolved in water. They have developed various structural adaptations for carrying down a supply of air from the surface of water. Thus an air bubble or film of air is entangled either in closely set hairs, or at the tip of the abdomen, or below the hard elytra. Larvae and pupae of mosquitoes are provided with respiratory siphons protruding above water surface. Water scorpions have a breathing tube at the tip of abdomen, which can be extended to the surface of water. Rat-tailed maggot *(Eristalis)* carries a telescoped breathing tube that can be extended at will to the surface for considerable distance.

[VI] Plastron Respiration

Some insects have specialized structures which hold a permanent thin films of air on the outside of the body in such a way that an extensive air water interface is present for gaseous exchange. Such a film of gas is called **plastron** (**Thorpe,** 1950). The tracheae opens into the plastron so that oxygen can pass directly to the tissue. The volume of plastron is constant and usually small since it does not provide a store of air, but acts as a gill. The constant volume is maintained by various hydrofuge devices spaced very close together so that water only penetrates between them when under considerable pressure. Excess external pressure which may develop through the normal insect life, are resisted. Such pressures may develop through the utilization of the oxygen from the plastron so that the internal pressure is reduced, or through the insect being in deep water and therefore, subjected to high hydrostatic pressure.

In adult insects the plastron is held by a very close hair pile in which the hairs resist wetting as a result of their hydrofuge properties and their orientation. Hair lying parallel with the surface of the body, as in *Aphelocheirus* provide most efficient resistance to wetting. Hair in this insect are bent at the tips and the plastron covers the ventral and part of the dorsal surface of the body. Hair are packed very close together and can withstand a high pressure before they colapse. Such a plastron is said to be stable one and would only be displaced by water of great depths. The

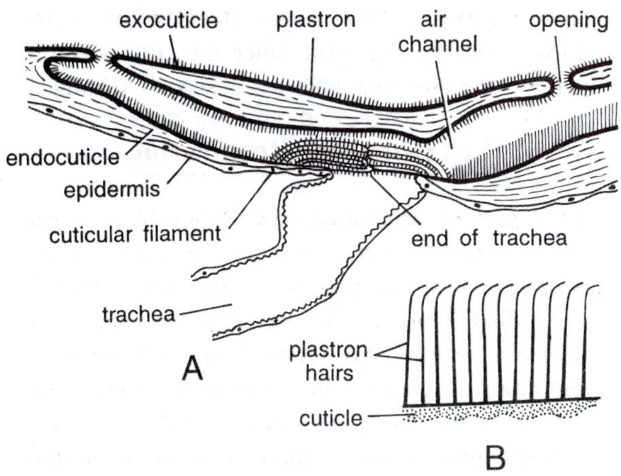

Fig. 14. *Aphelocheirus.* **A.** Junction of trachea with system of channels in the cuticle. **B.** Part of plastron magnified to show hairs.

spiracles open into the plastron by small pores along a series of radiating canals in the cuticle. These canals are lined with hair to prevent the entry of water into the tracheal system. Basal rate of oxygen utilization at 20°C is about 6 μl/h individual and this is readily provided by plastron, so that, except in water poor in oxygen, insect never needs to come to the surface.

Plastron forms an essential part of respiratory apparatus of many insect eggs and also of the pupae of many aquatic insects. In the later the plastron is held by respiratory gills.

[VII] Tracheal Gills

To obtain oxygen dissolved in water, nymphs and larvae of many aquatic insects possess special adaptations. Trachial gill are one such adaptation which develop as leaf-like or filiform outgrowths covered by very thin cuticle with a network of tracheoles. They are located on various parts of body, frequently abdomen, less frequently thorax and rarely head. For instance, most zygopteron larvae have 3 caudal gills, trichopteran larvae have filamentous abdominal gills, while larval Plecoptera have tracheal gills. Larval Anisoptera have gills in the anterior part of the rectum, which is known as the branchial chamber. Water is drawn in and out over the gills by muscular pumping resulting largely from the activity of muscles unconnected with the gut. Contraction of the dorsoventral muscles in the abdomen reduce the volume primarily by drawing up the sterna of segments 5 to 7. Haemocoel in the posterior part of abdomen is isolated from the rest of the body

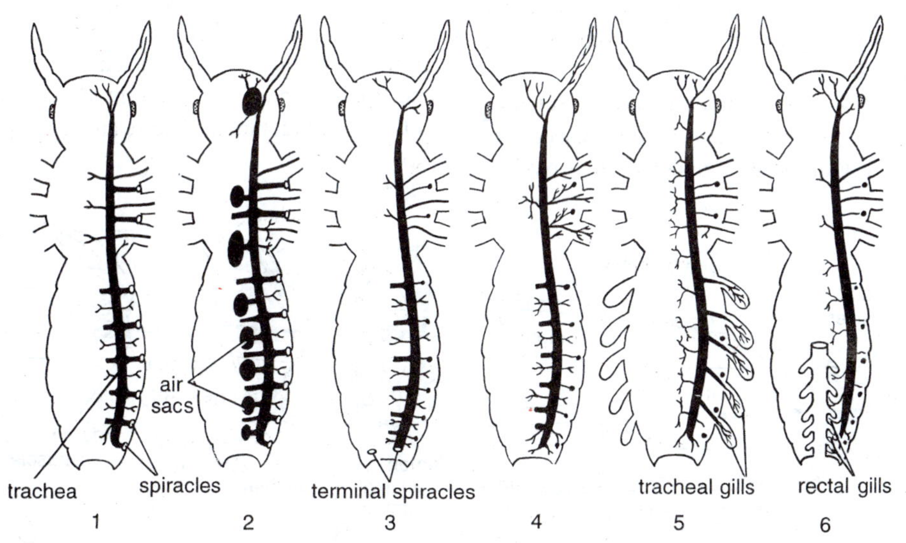

Fig. 15. Types of tracheal system in insects according to Wigglesworth.

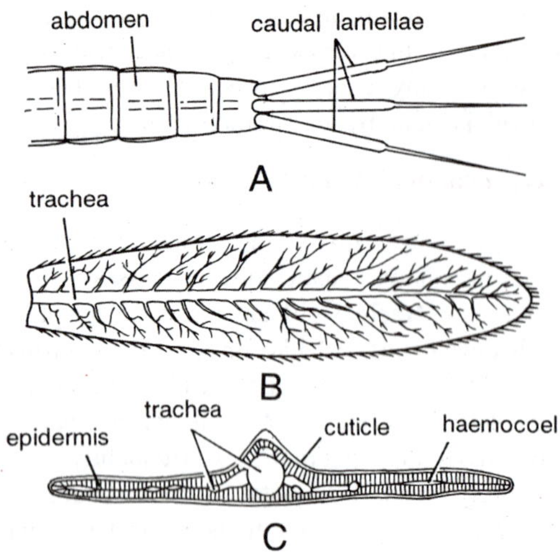

Fig. 16. Tracheal gills of larval Zygoptera. **A.** Posterior end of larva of *Coenagrion* in dorsal view. **B.** One lamella in lateral view. **C.** Caudal lamella of *Synlestes* in T.S.

by a muscular diaphragm across the fifth abdominal segment. As a result pressure is exerted on the branchial chamber. Water is forced out through the partially open anus and redrawn when the volume of the abdomen is restored partly by the elasticity of the cuticle and partly by contraction of muscles in the diaphragm and in a transverse subintestinal muscle. During inspiration the anal valves are wide open. A muscular prebanchial valve prevents water from moving forward into the midgut.

In general a good deal of gaseous exchange may take place through tracheal gills, insects are able to survive without gills under normal oxygen tensions. When the oxygen tension of water is low, they are important, since they considerably increase the area available for gaseous exchange.

Integumental Respiration

Some gaseous exchange takes place through the cuticle of most insects, but this does not amount to more than a few percent of total movement of gas. Most Collembola and Protura, have no tracheal system and depend on integumental respiration together with transport from body surface to the tissue by haemolymph. This type of respiration is also important in eggs, aquatic insects coupled with an apneustic tracheal system, and endoparasitic insects. Integumental respiration without an associated apneustic tracheal system can only suffice for very small insects with a large surface/volume ratio.

Impermeability of most insect cuticles to oxygen arises from the epicuticle, but not from the wax layer which renders the cuticle impermeable to water. Permeability to carbon dioxide may be greater and the loss of this gas through the intersegmental membranes may be appreciable.

Blood Gills

Some aquatic insects such as trichopterus and tipulid larvae, possess gills devoid of tracheae but containing blood. These are termed as **blood gills**.

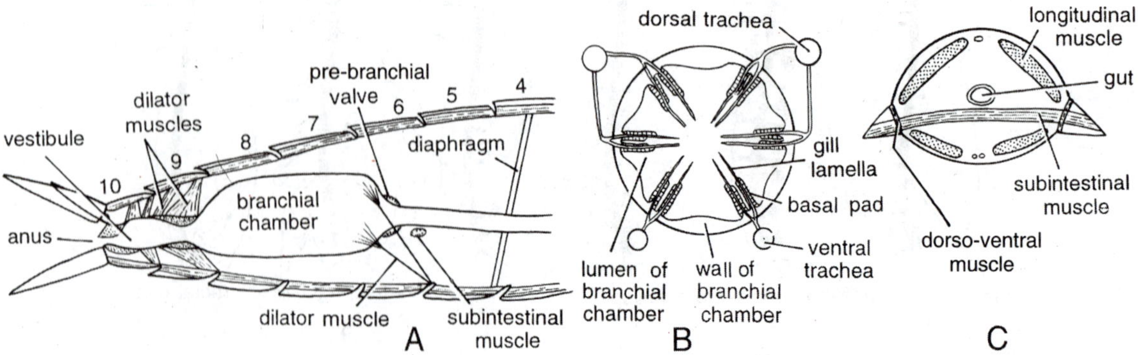

Fig. 17. A. Abdomen of dragonfly larva in L.S. **B.** Branchial chamber in cross section. **C.** T.S. of sixth abdominal segment.

Spiracular Gills

A spiracular gill consists of an extension of the cuticle surrounding a spiracle and bearing a plastron connected to the spiracle by aeropyles. In water the plastron provides a large gas/water interface for diffusion, while in air, the interstices of the gill provides a direct route for the entry of oxygen, and water loss is limited because the gill opens into the atrium of spiracle. Thus, in air, water loss through the spiracle is scarcely greater than in terrestrial insects (**Hinton,** 1968).

Spiracular gills occur in the pupal stages of many flies and beetles living intertidally or at the edges of streams. They also occur in larvae of a few coleoptera and in *Canace* (Diptera). When they occur in pupae, the basal structure of the gill is modified to permit respiration by the pharate adult.

Spiracular gills of most dipteran pupae are prothoracic and contact with the prothoracic spiracles as in *Taphrophila.* In this insect, there is a single gill on each side with eight branches. In the pupae of *Psephenoids volatilis,* the gills are associated with abdominal spiracles on abdominal segments 2 to 7. Abdominal spiracular gills, also occur in larvae of *Torridincola, Sphaerius* and *Hydroscapha.* The spiracular gills can resist high pressure.

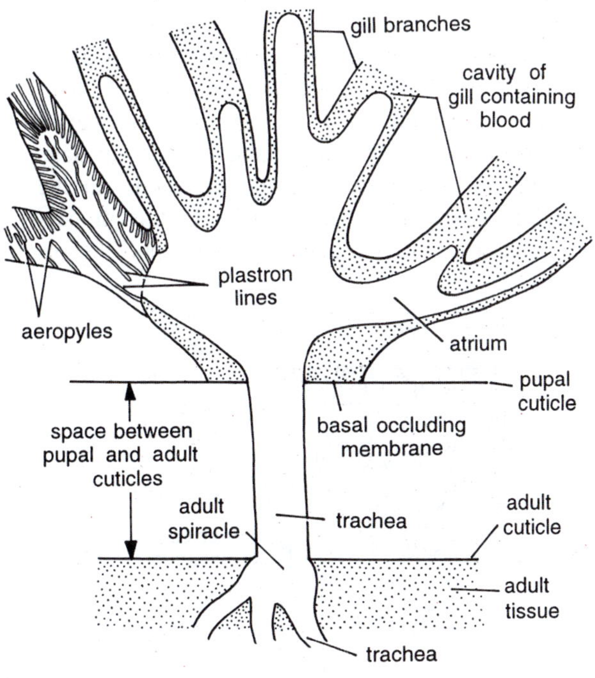

Fig. 18. *Taphrophila.* Spiracular gills of pharate adult.

BODY WALL OF ARTHROPODS

Body wall, body covering, external skeleton, exoskeleton or the integument are the various names given to the outer layer of body which is a unique feature of arthropods. It comprises of a single layer of cells, the epidermis **(hypodermis),** an internal **basement membrane** and an external **cuticle** secreted by the hypodermis. The cuticle, a characteristic feature of arthropods, is thick, tough, non-living and horny or chitinous and forms the exoskeleton.

Epidermis

Epidermis also called **hypodermis** is the outer cell layer of the arthropodan body wall. It is one cell thick layer. The epidermal cells are held together near their apices by zonulae adhaerens lower down by gap junctions. During and just after a moult the cells may have lone cytoplasmic processes on the outside.

Hypodermis has a typical stout **basement membrane**. The epidermal cells stand on the basement membrane. It forms a continuous sheet to which muscles can be attached. At the points of attachment of muscles, the basement membrane, incase of insects, is continuous with the sarcolemma. Basement membrane is said to be produced by epidermal cells.

Insects possess **oenocytes** which are often large cells, more than 100 μm in diameter, found between the bases of the epidermal cells and basement membrane. Oenocytes probably produce the lipids in cuticle and synthesize ecdysone, an insect hormone.

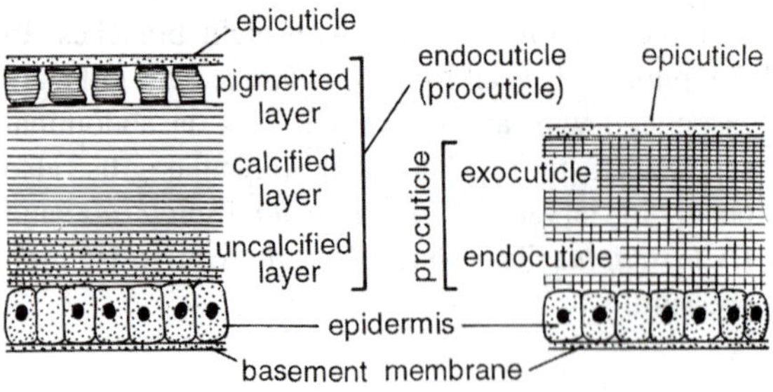

Fig. 19. Integumentary patterns in **Arthropoda**. **A.** Decapod crustacean. **B.** Insects.

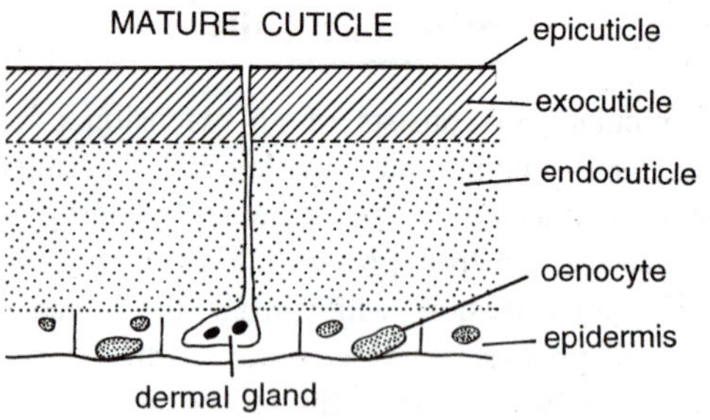

Fig. 20. A mature insect cuticle showing oenocytes.

Cuticle

Cuticle is a secretion of the epidermis and covers the whole of the outside of the body as well as lining ectodermal invaginations such as the stomodaeum and proctodaeum. It is only the midgut region which lacks cuticle. The tracheae of insects, chilopods, diplopods and some arachnids, the book lungs of scorpions, spiders and also the parts of reproductive system of some groups of Arthropoda are also lined with cuticle. It takes various forms in different groups of the phylum but basically consists of two layers, **epicuticle** and **procuticle**.

1. Epicuticle. Extremely thin outer layer, composed of proteins and lipids, without chitin, is called **epicuticle**. It is 1 to 4 µm thick. It is usually made of two layers, the outer lipid (waxy) layer and the inner layer of proteins and lipids. Epicuticle of insects is very complex when it is made up of several layers. Here, the thickest inner layer is called **inner epicuticle** lying immediately outside the procuticle. Outside the inner epicuticle lies the very thin **outer epicuticle**, which is followed by a waxy layer of variable thickness. In some insect a thin 'cement' layer occurs outside the wax. The waxy layer makes the cuticle impermeable to water. It is related to the osmoregulation problems of life on land and as well as in water.

2. Procuticle. The much thicker layer of cuticle is termed as **procuticle**. It is upto 200 µm thick. Procuticle is further differentiated into an outer **exocuticle** and an inner **endocuticle**. Procuticle consists primarily of a chitin-protein complex that varies greatly in different arthropods.

Chitin is the principal constituent of procuticle and forms the bulk of cuticle. It is insoluble in water and resistant to most solvents. It protects the body from loss of water, provides rigidity and offers points of attachment to body muscles. In many arthropods procuticle is impregnated with meniral salts such as in Crustacea in which calcium carbonate and calcium phosphate deposition takes place in it. It is an impermeable covering except in certain areas where it is thin and allows the passage of gases or the absorption of water.

Generally, over every segment of the body or a limb, cuticle (exoskeleton) becomes thick, hard and non-flexible due to the process called **sclerotization** or by deposition of calcium salts as in decapod crustacean and diplopods, in the middle layer of the endocuticle. In this process of sclerotization, addition of proteins, polyphenols

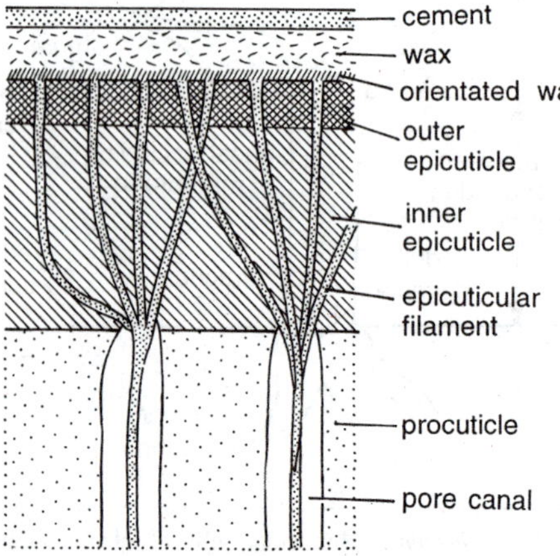

Fig. 21. Section through epicuticle (Diagrammatic).

and polyphenol-oxidases, takes place. In insects, it is the exocuticular layers and in crustaceans, the pigmented layers of the endocuticle, which are most highly sclerotized. However, at joints, it remains thin, soft and flexible, so that movements are possible.

Running through cuticle at right angles to the surface are very fine **pore canals**. They extend from the hypodermis to the inner epicuticle and, at least early in the development of the cuticle, they contain cytoplasmic extensions of the epidermal cells. In insects like adult beetles and bugs the pore canals are more abundant in the outer procuticle. They are one micron or less in diameter. These epicuticular filaments run from the wax layer through the epicuticle and into the pore canal. In many insects several filaments converge on each pore canal unlike *Rhodnius* where there is only one to each canal.

Effects of Thick Cuticle

Utility of thick cuticular exoskeleton to arthropods can be the same as that of steel to civilized man. It is partly to the possession of this hard cuticle that arthropods owe their great success.

1. Growth and moulting. A thick and generally inflexible exoskeleton created problems of growth and locomotion, to the arthropods. Problem of growth has been eliminated by the phenomenon of **moulting,** a characteristic of the arthropods.

Process both of ridding the body of old exoskeleton or cuticie and of secreting a new cuticle is called **moulting** or **ecdysis**. This process is under hormonal control.

2. Metamorphosis. Moulting is a drastic process, and it is often related to other drastic changes. Often the mode of life of the animal changes, accompanied by a change in body shape and feeding habit. Hence the life cycles of arthropods often involve transformations or metamorphosis from one shape to another.

3. Joints and movements. Thickness of cuticle brings with it the necessity for joints in the body and limbs, needed for locomotion. Joints divide the cuticle into segmental plates connected by articular membranes in which the cuticle is thin, soft and flexible. Two advantages of jointed limbs may be mentioned. Firstly, they have become variously modified as walking legs, sensory structures, gills, genitalia, etc. Secondly, limbed locomotion conserves energy since the entire body is no longer used through this energy-wasting process.

4. Discontinuous muscles. Annelids have somatic musculature in the form of continuous muscular layers. In arthropods, on the other hand, due to much jointedness of body and limbs, the musculature consists of separate striated muscles, attached to the inner surface of the exoskeleton. The jointed exoskeleton on body and appendages, with its associated discontinuous muscles provides for a great variety of specialized movements. Each joint has flexor muscles which bend it, and extensor muscles which straighten it out.

5. Coelom. Turgescence of the body wall can no longer be a factor in locomotion with a thick cuticle. Therefore, in arthropods, a closely compartmented perivisceral coelom, serving as a hydrostatic skeleton, becomes useless and is much reduced. It is restricted to the gonocoel and the nephrocoel.

6. Haemocoel. Coelom in arthropods has been replaced by an extensive haemocoel, that is, merely sinuses of spaces in the tissues filled with blood.

7. Gills and tracheae. Thick cuticle is impermeable to water. It can not serve for respiration by osmosis. This fact probably led to the evolution of tracheae and gills in the arthropods.

8. Nephridia. Thick cuticle lead to the reduction of coelom which probably resulted in the loss of nephridia.

9. Chaetae. With a thick and rigid cuticle, chaetae would be useless and hence are absent in the arthropods.

10. Invasion of land. Ability of arthropods to live on dry land is due to their thick, chitinous and impermeable or waterproof cuticle. It cuts loss of water by evaporation and prevents drying of the watery tissues within. Thus there is no danger of death due to desication. The possession of legs with an exoskeleton was a pre-adaptation of the arthropods to a terrestrial mode of life. The legs support the animal above the ground, reduce friction, and permit more rapid movement. In the insects, development of wings opened further opportunities of rapid movement. This has enabled the arthropods very successfully to invade and exploit the dry land. They face practically no serious competition on land from other invertebrate groups, most of which are largely aquatic.

Functions of Cuticle

Cuticle is one of the features of arthropods which is primarily responsible for their success.

(1) It provides support, an essential requirement of terrestrial animals.
(2) The presence of hard, jointed appendages makes accurate movements possible with a minimum of muscle, by lifting the body off the ground, facilitating rapid movement.
(3) Cuticle provides rigidity. In insects, the flight depends on the possession of rigid wings, which is provided by cuticle.
(4) Cuticle provides protection. Arthropods have hardened heavily selerotized cuticle which makes them difficult for predators to catch or parasites to parasitise.
(5) Cuticle also provides protection from environment as in case of beetles, the upper cuticle of the abdomen protected by the elytra, is very thin, but the cuticle of ventral surface, exposed and subjected to abrasion by the substratum, is very thick.
(6) Cuticular lining of fore- and hindguts also protects the epidermis from abrasion by the food.
(7) Wax layer of cuticle restricts the water loss.
(8) Cuticle is also modified to form sense organs like sensilla on the antennae of various insects in the form of hair, (*Bombyx*), pegs (locusts), flat plates (bees and aphids).
(9) Cuticle has an importance in production of colour of arthropods. Colouration results from the deposition of brown, yellow, orange, and red melanin pigment within the cuticle. The iridescent greens, purples, and other colours result from fine striations of epicuticle which cause light refraction and give the aspect of colour.

RESPIRATORY ORGANS IN ARTHROPODA

Structures adapted to facilitate the passage of oxygen or air from the environment across the body surface are included in respiratory system. In many arthropods these organs are protected in the pouches, and in aquatic animals, irrigated by water currents produced by the associated appendages.

There are arthropods which lack specialized respiratory structures and the diffusion of gases in such cases, across their body surfaces become adequate for their requirements.

Different types of respiratory organs found in different groups of Arthropoda are as follows :

Chelicerata

1. Merostomata. In xiphosuran, *Limulus,* the respiratory organs are **book-gills** which occur on the posterior wall of the plate like appendages of five posterior segments of mesosoma. They become modified as gills. On each appendage are found, some 1,500 thin-walled lamellae formed by folding of posterior integument. The lamellae project from the surface and since they lie parallel to each other resembling the pages of a book, they are characteristically known as **book-gills**. The beating of the appendages causes a current of water to pass over the book gills. The coxa of the last pair of legs bears a short spatulate flabellum which cleans the gills and also serves to sense the oxygen content of the water current. The respiratory pigment is haemocyanin.

In eurypteridan genus *Slimonia,* the gills are not borne on the genital operculum or mesosomal plates, but are represented by a highly vascularised area on the ventral body (**Moore,** 1941).

2. Arachnida. Organs of respiration are either book-lungs or tracheae or both. **Book-lungs** are modified book-gills due to migration of arachnids to a terrestrial environment. Unlike book gills the book-lungs are internal. There are four pairs of book-lungs each pair being located inside each mesosomatic segments from 3 to 6.

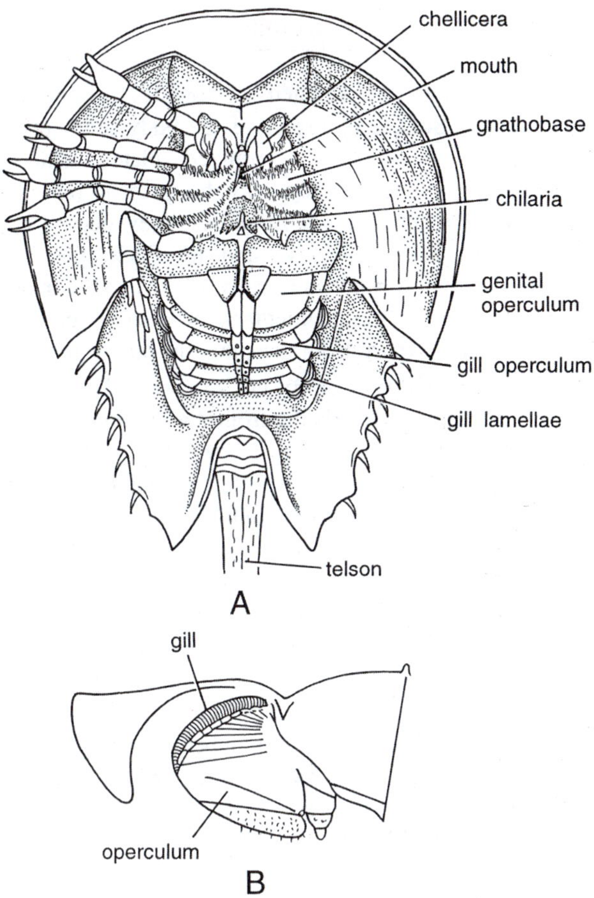

Fig. 22. *Limulus.* **A.** Gills on appendages of posterior segments. **B.** Posterior surface of gill operculum.

In Amblypygida and Uropygida book-lungs are found on the third and fourth segments of the opisthosoma (mesosoma). In Araneae book-lungs occur on 3rd and 4th segments also in the listomorph and mygalomorph spiders. **Tracheae** also forms the respiratory organ of some arachnids. It is very much like the tracheae of insects except their independent evolution. In some cases the tracheal system seems to be derived from book-lungs. This type of tracheal system is found in spiders. In two small groups of spiders book-lungs are replaced by tracheae.

In Araneae the spiracles lead to a small atrial cavity from which four primary tracheal tubes extend anteriorly. The spiracles may migrate posteriorly and fuse across the mid-ventral line to form simple opening lying just infront of the spinnerets.

In pseudoscorpions only two pairs of spiracles are present leading to tracheae. In opiliones there is a single pair of spiracles leading to a well developed tracheal system. In solifugae the tracheal system opens to the exterior by three pairs of spiracles. In Ricinulei there is only a single pair of spiracles leading to an atrial cavity from which a large number of minute sclerotized tubes conduct air to the organs of the prosoma. In Acari, a well developed tracheal system is present in many forms where the tracheae open to the exterior by 1 to 4 pairs of spiracles.

Crustacea

Gills are the organs of respiration in most crustaceans but in primitive crustaceans the gaseous exchange takes place over the general body surface. In many crustaceans **carapace** is present as their characteristic feature, the inner thin lining acts as gill as in Malacostraca, Myscidacea, Tanaidacea, and Decapoda. It is highly vascularised enabling the exchange of gases from the water current passing over it. Cephalocarida and Mystacocarida, the primitive crustaceans groups however either lack carapce or it is poorly developed. In Cirripedia, the **mantle** is thought to be the main site of gaseous exchange.

In Malacostraca **gills** are found as well defined set of respiratory organs. Primitive malacostracan *Nebalia* and anapsides, gills are **epipodites** found on thoracic limbs. In stomatopod, *Squilla,* besides epipodites on thoracic limbs, large branching **gills** develop on the exopodites of the pleopods. In Cumacea the epipodites are filamentous. In Isopoda, the epipodites are lacking and the abdominal pleopods become modified for respiration. In Amphipoda simple plate like epipodites on thoracic limbs act as gills. It is in Decapoda that the fully developed gills are found. For instance in Palaemon, there are 8 gills present inside each gill chamber. They may be of three types based on their place of origin and attachment, (i) **Podobranch** or foot gills, (ii) **Arthrobranch** or joint gills,

(iii) **Pleurobranch** or side gills. A fourth type of gill called **mastigobranch** is also reported in other decapods which does not function as gill but serves to clean them. In addition to their number which may reach upto 32, gills themselves show adaptations to respiratory exchange. In peneid shrimps, *Palaemon* etc., gill are **phylobranchs** *i.e.,* they have a central axis bearing biserially arranged main branches which are themselves sub-branched. Such gills occur in Caridea, Anomura, and Branchyura. Another type of gills, the **trichobrancheate** gill made up of many unbranched filamentous projections, is found in many Macura, some anomurans, and primitive Brachyura.

Gills may be reduced or altogether absent, and the respiration is entirely through the inner surface of the carapace in some terrestrial brachyurans.

Pauropoda

Majority of pauropods lack special respiratory organs. A few have a very short tracheae, the spiracles are situated on the coxae of the walking legs.

Symphyla

There is only a single pair of spiracle situated on the head just above the base of the mandibles. A few branches of tracheae from head pass to the anterior trunk segments.

Diplopoda

Each trunk segment has a pair of spiracles and two pairs per diplo-segment. They are situated on sternites anterior and lateral to each pair of walking legs. They open to an atrial cavity which is thin walled in *Polyzonium* and in some Colobgnatha and heavily sclerotized in other Diplopoda. Bundles of tracheae arise from atrium, usually unbranched, do not anastomose and do not taper. Nematomorphs show the anastamosis of tracheae in the vicinity of spinning glands. Their branching occurs in Pselaphognatha. They are very long in *Glomeris.*

Chilopoda

The spiracles are situated in the pleural region, except in Scutigeromorpha, of the body segments, usually on separate sclerotized plates called **stigmatopleurites**. They may be present on all trunk segments, but in *Lithobius,* they occur on segments 3, 5, 8, 10, 12 and 14. Each spiracle leads to an atrium the walls of which bear a number of fine cuticular processes called **trichomes**. Apart from keeping the atrial lumen open, the trichomes may act as physical gills.

The tracheae arising from the atrium go to the tissues and organs near by. In **scolopendra** longitudinal tracheae have developed; in Geophilomorpha, the longitudinal tracheae extend the whole length of the body.

In *Scutigera,* the respiratory arrangement is much different. The spiracles are slit openings in the mid-dorsal line on the posterior tergites on all the trunk segments that have long tergal plates. Each spiracle leads to an atrium from each side of which arises a large number (600) of fine tubules. These may branch once or twice near their bases and end blindly to form a compact kidney-shaped organ. Air is drawn into this organ. This type of tracheal organ occurs sporadically in other arthropods and may be referred to as a **tracheal lung**.

Insecta

1. Apterygota. Tracheal system constitutes the main organs of respiration found in various degrees of development in different orders.

In *Protura* no tracheae occur in Acerentomidae (*Acerentulus*) but they are present in Eosentomidae. In *Eosentomon* two pairs of spiracles are present, one pair each on mesothorax and metathorax. Tracheae arising from each do not interconnect. Abdominal spiracles are absent.

In *Collembola,* no tracheae are indicated in majority of them. Cuticle forms the respiratory source. Tracheae are found only in the species of the family Sminthuridae where there is only a single pair of spiracle situated between head and thorax.

In *Thysanura,* nine pairs of spiracles in the Machilidae of which one pair is on mesothorax, one pair on metathorax and seven pairs on the second to eighth abdominal segments. 1st abdominal segment is without spiracle. In Lepismidae, there are ten pairs of spiracles. The tracheae arising from spiracle do not anastomose but those in Lepismidae anastomoses occurs.

2. Pterygota. Tracheal system varies in various orders due to adaptation for various purposes. Primitive orders have 10 pairs of spiracles with 2 pairs on thorax and eight on abdomen. Tracheae arising from spiracles have numerous longitudinal and transverse anastomoses welding the whole into a complex efficient respiratory system.

The pterygotes tracheal system consists of two distinct organs, the **tracheoles** and the **tracheae**. Former are small intracellular tubules filled with fluid at rest, later are intercellular tubes of relatively larger diameter as compared with tracheoles. They are lined with cuticle, have a wax layer in longer branches. Accept in aquatic insects and during embryonic development, they do not contain fluid. They are kept open by spiral lining, the taenidea, and conduct air to the tissues.

3. Other groups. Special respiratory organs are absent in the *Pycnogonida, Tardigrada* and *Pentastoma.*

ECONOMIC IMPORTANCE OF ARTHROPODA

Merostomata

The American species *(Limulus polyphemus)* is sometimes fed to chickens and pigs. The female specimens are preferred on account of their eggs of which half-a-pint may be crowded into a cephalic shield. There is a belief that this diet makes the poultry lay more eggs. Undoubtedly, it fattens both fowls and pigs, but also imparts a shocking flavour to their flesh. Asian species of horseshoe crabs are also consumed by human beings.

Arachnida

Scorpion are studied as a type of arachnid in most of the universities. In tropical countries, it entres house and often becomes a nuisance. It is a poisonous animal and its bite can be extremely painful, sometimes resulting in death, especially in children.

Scorpions are beneficial also to some extent, as they feed largely on unwanted insects, as cockroaches and beetles, etc. Besides, the scorpion venom is used for pharmacological, biochemical and immunological researches. Chief centres for the study of above in India are Haffkine Institute, Bombay; School of Tropical Medicine, Calcutta and Institute of Preventive Medicine, Hyderabad. The scientists at these centres 'milk' the scorpions of their venom by giving them weak electric shocks, while holding their stings in glass pipettes.

Certain mites cause damage to the crop fruits like apple, pear and grapes *e.g.,* blister mite (*Eriophyee*). Ticks suck the blood of man and domestic animals. They also act as vector of protozoa causing Taxas fever in cattle and tick fever in man.

Crustacea

Crustacea are of considerable economic significance to man. Group is of great value directly or indirectly for his health and economic progress.

1. As food. A large number of crustaceans are consumed by man, especially the lobsters, shrimps, prawns, squillae, crabs and crayfishes, etc. They form an important diet of man with great nutritive value. Most edible portions are their tails (abdomens) almost all of which are composed of muscles. There is also some good 'meat' in their chelipedes. In mud crabs the claws are the best part of the animal to eat. Muscles are either freshly cooked or canned. Blue crab (*Callinectes*) is held captive until it molts, then sold in the soft-shelled condition; after removal of the viscera the whole animal is cooked and eaten.

Decapods and the bivalve molluscs together comprise the so-called shell-fish industry. Animals worth more than 500 million dollar are marketed annually in U.S.A. alone. Prawn fishery is very much advanced in several countries including India, giving employment to thousands of people. India is exporting prawns / shrimps etc. approximately 59000 tonnes annually.

Smaller species of Crustacea form the bulk of zooplankton which plays a vital role in food chains of both salt and freshwater fishes and other aquatic animals that eventually come to our table. Man eats fish, and there is a stage, when they must feed on some forms of small Crustacea, such as a larval stage of some larger form like crayfish, or a minute adult such as *Cyclops* and *Daphnia.* Whales which are hunted by man for their various economic products depend on crustaceans for their food. Amphipods and euphausids form a great part of diet of other animals like the seals, sea gulls and penguins. Two tons of *Calanus,* a marine copepod about 5 mm long, have been found in the stomach of a blue whale. Aquarium dealers collect the adults and eggs of *Artemia* and *Daphnia* and sell them as fish food.

2. As fish bait. In most parts of America, the caryfishes, especially the soft-shelled individuals, are quite popular among fisherman as a fish bait. Soft-shelled individuals are kept soft for a week or so on ice since refrigeration slows metabolism so that the shell develops slowly.

3. As scavengers. Some Crustacea, such as crayfish, are beneficial, as they serve as an agency in the destruction of decaying vegetables and animal bodies in water.

4. As intermediate host. Although some crustaceans are parasites of aquatic animals, none is a parasite of man or other land animals. However, a handful of them serve as intermediate hosts to certain dangerous worm parasites of man and other vertebrates. Dreaded, human lung-fluke, *Paragonimus westermani,* uses a crayfish for this purpose. Species of *Cyclops* which live in freshwater, serve as intermediate hosts for the human guineaworm *(Dracunculus medinensis)* and for the broad tapeworm *(Diphyllobothrium latum).* Guineaworm disease was once very common in India, Egypt and central Africa. Thus, the crustacean which has done the greatest harm to man is the *Cyclops.*

5. As pests. Certain crustaceans may become serious pests when present in large numbers. Thus crayfishes damage cultivated crops by eating young corn and cotton plants. Sow-bugs and pill bugs, which also feed on vegetation, may turn pests in green houses and fields when sufficiently numerous.

Crayfishes occasionally burrow, making holes in dams and levees and weaken them, thus causing serious damage.

Some crustaceans bore into marine timber structures (jetties, piles, poles, props, etc.) and destroy them causing loss of several crores of rupees. Some crustacean wood borers are *Chelura terebrans* (Amphipoda) and species of *Spheroma* and *Limnoria* (Isopoda). They burrow into and damage wharves in salt water. Branacles form one of the fouling animals by attaching to the hulls of ships.

Diplopoda

Millipedes may be useful as they are scavengers and can dispose of the dead organic matter. They cause demage to plants as they feed also on plant roots and destroy the green houses and gardens.

Chilopoda

They are beneficial to man as they feed on insects some of which may be injurious also. They are also reported to feed on snakes too. Some species are harmless to man, but some longer tropical ones inflect a painful bite and cause fever, dizziness and headache. *Scolopendra gigantea* cause human death, especially if the victim is a child.

Insecta

Insects are very well known for their economic value. A detailed study of their economic importance has been dealt separately.

Onycophora. It forms a taxonomically important group as it forms a link between Annelida and Arthropoda. They are otherwise of no economic value.

SOCIAL LIFE IN INSECTS

Social Behaviour

In a broader sense, any interaction between two or more individuals constitutes **social behaviour**. Usually, social relationship implies interactions among members of the same species. The mere presence of more than one individual does not mean that the behaviour is social. Various types of associations occur among insects.

1. Solitary insects. When each individual is more or less independent, insects are called **solitary**. They forage independently and the two sexes come in contact only to mate. The female deserts her eggs or dies after laying, and does not look after the offsprings.

2. Gregarious insects. Many solitary insects are **gregarious,** that is, they form dense but temporary populations or aggregations in response to factors of physical environment or to share certain common needs or tracts. Thus, light at night stimulates large numbers of moths and other insects to collect around it. High humidity under a log causes aggregations of wood lice. Caterpillars of the same or different species may live together and act in mutual cooperation. Ladybird beetles assemble together for hibernation. Locust and may-flies come out in huge swarms. None of these groups is strictly speaking, social. Gregariousness being a temporary habit it does not involve any association of the parents and the offsprings and has nothing to do with the evolution of the social behaviour among insects.

3. Social insects. On the other hand, insects of a given species that live together in organised groups or colonies are known as **social insects**. In a social organisation many individuals of species live together in an integrated manner so that each contributes in some specialized way to the welfare of all.

Evolution of Social Habit

Among the oldest and most highly developed societies in the animal kingdom are those of insects. Three hundred million years ago these societies were already in existence. Social habit has arisen independently in several orders of insects. Transition from solitary to social life in these usually short-lived animals was made possible by the prolongation of adult or parental life and increasing parental care. First the progeny depended on the parents, then the parents on the progeny.

Orders of Social Insects

Social species of insects belong to seven orders, namely Orthoptera (cockroaches, gryllotalpa), Dermaptera (earwigs), Isoptera (termites), Embioptera (web spinners), Psocoptera (book lice), Coleoptera (beetles) and Hymenoptera (bees, wasps and ants). About 6,000 species of insects in all exhibit social instincts, including nearly 500 species of bees, 800 species of wasps, 1,000 species of termites and 3,500 species of ants.

Gradations of Social Behaviour

Various gradations between the two extremes of solitary and social insects are evident in existing species.

Solitary insects, such as mosquitoes and dragonflies, drop their eggs anywhere and go about their business. Butterflies and fleshflies lay their eggs on food suitable for the young. Solitary digger wasp digs a hole, provisions it with food such as paralysed caterpillars and spiders, lays an egg, seals the entrance and departs, never to see its offspring which later hatches out and grows independently.

Sub-social insects represent a further step towards social habit. They provide a mass of food for each egg but remain to guard the nest or young. In some species of dung beetles female collects and rolls a ball of dung, excavetes a burrow, drops the ball in, lays eggs and departs. In other species, the male assists by guarding the dung balls while the female excavates. In still another species, both sexes dig chambers, stock them with dung on which the female lays eggs, then guard them until the eggs hatch. At this time all disperse. A female earwig guards her eggs and later the young. Cockroaches, crickets, some bugs, web spinners and book lice do likewise.

True social insects, on the other hand, forage for food for the colony continuously, the two parents live longer so as to come in association with many generations of their progeny, and the youngs cooperate in caring for the next generation. Only a few insect species have been able to develop social habit completely. Most highly developed and complex of insect societies are found in the termites (Isoptera), and the ants, bees and wasps (Hymenoptera).

Characteristics of Social Insects

All the social insects possess certain characteristics in common, which are as follows :

1. Large populations (colonies). Many individuals of a species of social insects live together in an integrated manner in a comparatively large group or population which is not gregarious. Term 'colony' is commonly applied to the complex society they form. Number of individuals forming a colony ranges from 35,000 to 50,000 in honey bees, upto 600,000 in ants and several millions in termites. Colonies are **matriarch**, *i.e.*, all members of a colony are the offspring of a single female and hence all have very similar genotypes. Ordinarily, a society of social insects does not accept members from other colonies of the same species.

2. Elaborate nests. Social insects construct more or less elaborate nests for protection, storage of food and maintenance of broods. Some interesting habits of bees, ants and, termites regarding their nest and broods are summarized in the adjacent table.

3. Extra population of nests. Some small crustaceans, mites, beetles and other insects are attracted by the high temperature and surplus food of the nests of the ants and termites and get protection from their enemies. They live in close **symbiotic** relationship with the hosts. **Guests** or the outsiders feed upon the debris or waste of the nests and the dead bodies of the hosts. Slave-making ants bring eggs, pupae and adults of other ants and take from them functions of **slaves.**

Habits	Honey bees	Ants	Termites
(1) Position of nest	Trees, etc.	Leaves, wood, ground	Wood, ground
(2) Material of nest	Wax	Leaves, wood, earth	Wood soil
(3) Shape of nest	Hexagonal cells	Chambers & galleries	Chambers & galleries
(4) Nest started by	Female & workers	Female or male and workers	Female, male & workers
(5) Number of population	35-50 thousand	600 thousand	Several millions
(6) Nature of brood	Perennial	Perennial	Perennial
(7) Food of brood	Pollen and nectar	Vegetables, wood & insects	Wood & insects
(8) Type of feeding	Progressive	Progressive	Progressive
(9) Swarming	Yes	Only in some species	Yes

Intruders or **thieves** rob the social insects of their food and brood. Some beetles (*e.g., Atemeles*) live in brood chambers of the ants, where the beetle larvae eat the ant larvae with impunity. Both larvae and adults of the beetle secrete pheromone-like substances which serve to appease the aggressive tendencies of the ants.

Predator insects prey upon the social insects, stylopids attack wasps and bees, while bee moth attacks the wax of honey bee hives.

4. Polymorphism (caste system). Typically social insects have a division of labour. Members are differentiated into distinct **castes**, which are specialized in structure, function (reproduction, feeding, guarding, etc.) and behaviour. Principal castes are the **reproductives** (king and queen) and the **sterile** members (workers and soldiers). Workers are the smallest in size. Queen is the largest with a long abdomen and lays eggs. It lives for several years. Males are intermediate in size and develop parthenogenetically from unfertilized eggs. In social wasps, bees and ants, workers are only sterile females. In termites and some higher ants, workers and soldiers belong to both sexes. In termites, a special soldier-caste exists, called the **nasute,** with an elongated projection on the head. The greatest diversity of castes (polymorphism) is found among ants, and all forms may have large and dwarf individuals.

Caste determination depends upon a number of factors. In Hymenoptera (bees, wasps and ants), genetics and nutrition form the basis of differentiation. Males are haploid and develop parthenogenetically from unfertilized eggs. Queens, workers and soldiers are diploid females which develop from fertilized eggs. Differences between queen and worker are mainly due to differences in the quantity and quality of their food. Bee larvae destined to become queens are fed on royal jelly for a few days and then are fed bee bread (honey and pollen). Bee larvae destined to become workers and drones are fed entirely on beebread.

In Isoptera (termites), determination of castes is due to extrinsic factors rather than to genetical ones. Reproductives and soldiers secrete ectohormones containing inhibitory substances. When fed to nymphs, these substances prevent them from developing into like forms (soldiers and reproductives).

5. Cohesiveness of colony. Division of labour or separation of function requires great coordination, for the group must perform as a biological unit. All the members of an insect society live in an integrated or cohesive manner, subordinated to the life of the community. As a result, various castes, which differ in structure and physiology, can not live independently. They work in cooperation and with mutual benefit. For instance, many females mutually cooperate so that one worker looks after an egg (larva) laid by another female (queen) and so on. Success with them is measured in terms of the colony and not of the individual. Different castes are bound together by chemical and physiological mechanisms rather than structure.

6. Parental care. Basis for the family relationship is the provision of shelter, food and defence for the young. Thus social life in insects is correlated with the lengthening of the adult or parental life and increasing parental care. It provides, greater association of parents and young. Parental care includes various activities such as provisioning of food, cleaning and feeding of young and queen, removal of debris and bodies, taking away and putting eggs in proper chambers, cooling of chambers in summer and protecting queen from winter by clinging and clustering about her, by workers. Parental care is instinctive behaviour.

7. Progressive provisioning of food. Stingless bees and potter wasps lay their eggs singly or in small groups and provide sufficient mass of food at the same time for the complete development of the larvae which hatch out from the eggs. This is known as **mass provisioning of food**. But true social insects (bees, ants and termites) feed their young extensively and continuously from day to day until they metamorphose into the adults. This is known as **progressive provisioning of food**.

Ants show a progression of food habits such as probably occurred in man's history. Lowest kinds (army ants) hunt insects or flesh. Pastoral ants feed on the honey dew produced by aphids. They carry these 'ant cows' (aphids) into overwintering quarters and protect them from predators. Harvester ants gather and store seeds in summer to tide them through the winter. Finally, the leaf-cutter or fungus growing ants (*Atta*) grow their own pure crops of fungi in underground gardens fertilized with organic debris. These ants cut leaves and carry them underground to serve as a substrate for growing pure strains of fungi on which they feed. A young queen, upon setting out to find a new colony, carries a seed stalk of fungal hyphae in a pouch below the mouth.

8. Trophallaxis. Exchange of food between one insect and other is known **trophallaxis**. Ants and termites feed one another from mouth to mouth. Young exchange food with the adult. Some ants feed some beetles, coccids and aphids and in return imbibe a fluid secreted by them.

In termites, trophallaxis plays an important role in the regulation and determination of castes. Ectohormones containing inhibitory substances are secreted by the reproductives and the soldiers. During mutual feeding (trophallaxis), these are passed on to the nymphs, and prevent them from developing into individuals of the same sex or

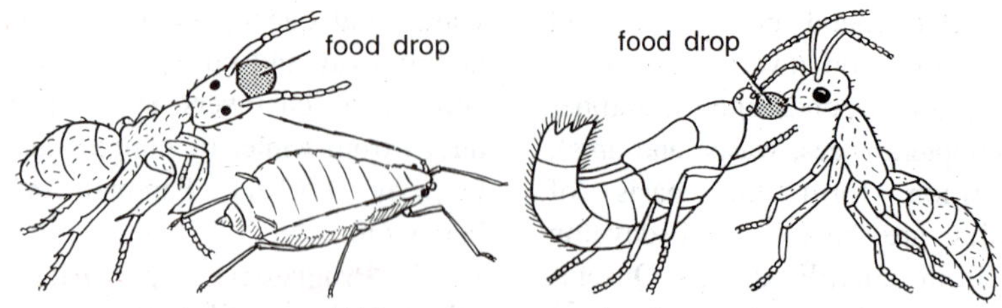

Fig. 23. Trophallaxis.

caste. This tends to keep caste numbers within bounds. Some undifferentiated nymphs may not come under the influence of trophallaxis and may become additional members of the same caste.

9. Swarming. Swarming occurs as a means of alleviating congestion in the overcrowded colony, or as a means of distribution. In many, swarming occurs for feeding, migration and mating. Mostly mating takes place between the queen and the males during swarming, called the **nuptial** or **marriage flight.** Honeybees propagate colonies by swarming. Each swarm consists of an old queen and many workers and produces a new colony. **C.G . Butler** (1961) found that the queen has over her body a so-called queen substance, secreted by her mandibular glands. It inhibits workers from becoming queens when they share this substance (trophallaxis). Swarming may come about in an overcrowded colony because this substance may not be distributed properly to all workers.

10. Protective devices. Social insects develop several devices. **Stings** are developed in most bees and certain ants. **Jaws** are highly developed in stingless bees and soldiers of ants and termites to ward off enemies. Sometimes, a few **guards** are posted at certain convenient places near the nests. Guards protect the nest and attack the intruders in a few cases. Nests are also made in **protective localities** such as ground, hollow trees, mud, paper etc., and have numerous side exits through which rapid escape is possible at the time of danger.

11. Communication. Both social and non-social insects utilize chemical, tactile, visual and auditory signals as methods of communicating with each other. Chemical communication occurs with the help of body secretions called **pheromones**. These are secretions of exocrine glands that pass to the outside of the body and play an important role in regulating and coordinating the activities of a colony of the same species (**E.O. Wilson**, 1965). Pheromones include sex attractants, the queen substance and other types whose influences are useful to the colony. Substances deposited on the ground by ants returning from a foraging trip serve as a trail marker for other ants. Substances released by the dead body of an ant within the colony stimulate other workers to remove the body.

In honey bees, a **queen substance** produced by the queen's mandibulate glands, controls nursing behaviour of the workers, caste determination and swarming. **Language** of the bees (**Von Frisch,** 1950) represents a most revealing method of communication known among insects. Some aspects, such as the tail wagging dance, set the honeybees apart from all other social insects.

Comparison of Human and Insect Societies

Highly evolved organization and cohesiveness of insect societies often prompts comparison with human societies.

1. Similarities. Some notable similarities are as follows :

(1) Human societies originally began as discrete families (one female and her offspring), but this is not repeated phylogenetically. Insect societies also arise from discrete families and by swarming frequently found new colonies.

(2) Human societies comprise integrated groups of like individuals (castes) that specialize in different trades or professions. They benefit both the individuals and the group by their cooperative effort. Insect societies also have distinct castes performing various functions and living together in an integrated manner.
(3) Man has tradition and social heredity. Ants bequeath fungi and real estate (hunting grounds).
(4) Man constructs and uses tools. The Indian red tree ants *(Oecophylla smaragdina)* use their larvae as spinning shuttles to fasten leaves together with larval silk.
(5) Man controls and modifies the environment to his own advantage, including the production of food. Ants cultivate their own specific strains of fungi, and bees and termites also control the temperature of their nests.
(6) Man domesticates animals. Ants domesticate aphids (ant-cows).
(7) Men has a language. Bees and ants communicate by dances and pheromones.

2. Differences. Basic difference between human and insect societies is in the manner in which the two adapt to different requirements.

(1) Man has evolved learning and abstract intelligence whereby he meets different situations. Adaptation provides him with a **flexible society**. Insects evolved different castes each of which meets the specific need. Their society, therefore, is a **rigid** one.
(2) Success in human society is measured both in terms of the individuals as well as the community. In an insect society, success is measured in terms of the colony and not the individuals.

CRUSTACEAN PARASITES

An association between two organisms of the same or different species in which the smaller one, called the **parasite**, lives inside or upon the body of the larger one, called the **host,** and obtains its nourishment from the latter, without actually destroying it, is called **parasitism.** Parasitism differs from **predation** in which the prey (host) is killed by the predator.

Parasitic Groups of Crustacea

Parasitism prevails in several groups of Crustacea, such as Copepoda, Branchiura, Cirripedia, Isopoda and Amphipoda. Parasites, however, are not known among the Branchiopoda and Ostracoda.

Peculiarities of Crustacean Parasites

Peculiarities of the crustacean parasites in relation to parasitism are as follows :

1. Degeneration. Parasites have lost organs no longer used in the adult form. Consequently, parasitic Cirripedia and Copepoda look similar in shape (sac like) due to parasitic degeneration. According to one view, parasites must not be considered degenerate, rather they have become specialized both in morphology and function.

2. Adhesive organs. Adhesive organs in the form of suckers and hooks, have specially developed for attachment to the host, externally or inside cavities.

3. Feeding. There are various modifications in nutrition. Feeding occurs by suction or absorption through organs which do not always represent true organs.

4. Reproduction. Reproduction is vast and wasteful to ensure chances of fertilization.

5. Sexual dimorphism. Usually a remarkable parallelism occurs between the degree of parasitism and the degree of sexual dimorphism.

6. Hermaphroditism. Hermaphroditism is induced in some cases due to parasitism.

7. Life-cycle. A striking feature of parasitism in general is the complexity of life cycle to overcome various hazards in life. But this is rare in crustacean parasites. There is no complicated wandering from host to host, no alternation of hosts and no alternation of generations.

8. Effect on host. General effects of parasitism have many variables. In some cases, the parasites seem to have no noticeable effect on the host. In other cases, hosts may show marked effects caused by some interference with their normal metabolism.

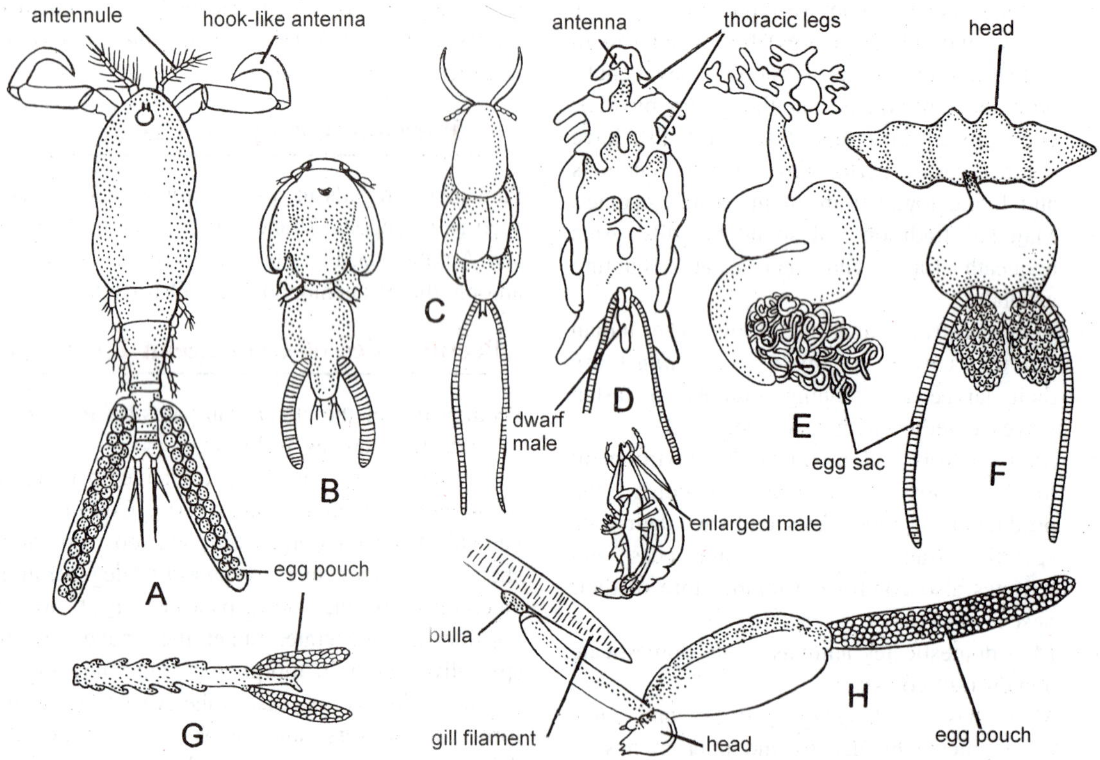

Fig. 24. Parasitic Copepoda. **A.** *Ergasilus,* **B.** *Caligus,* **C.** *Anthosoma,* **D.** *Chondracanthus,* male and female, **E.** *Lernaea,* **F.** *Leisteria,* **G.** *Mytilicola.* **H.** *Salmincola.*

Parasitic Copepoda

Majority of crustacean parasites belong to the Copepoda. At least four orders (Notodelphyoidea, Monstrilloida, Coligoida and Lernaeopouoida) are exclusively parasitic. They are called **fish lice**. They live upon the gills, fins or skin or burrow into the flesh of fishes or live within or upon other animals including other crustaceans. They remain immovably attached to their hosts, although some are unattached at times and swim in the plankton.

Parasitic Copepoda present every grade of morphological modification from normal free-living forms to the most degenerated parasites. The commensals undergo very little modification, but some endoparasites are so greatly modified as to become worm-like. This is illustrated by the general survey of some parasitic copepods.

1. ***Ergasilus.*** Adult looks like a normal copepod. It lives attached to the gills of several species of fishes. Trunk appendages are greatly reduced. Antennae terminate into hooks for holding on to the host. Eyes are absent.

2. ***Notodelphys.*** *Notodelphys, Ascidiocola* and *Doropygus,* belonging to the order Notodelphyoidea, live commensally in the branchial cavities or pharynx of tunicates, whose food they share. They are only very slightly modified from free-living copepods. They possess true prehensile antennae and normal mouthparts, but the appendages are usually small. The exopodite of the thoracic limbs bears a claw for attachment. The body of female is clumsy with a large egg pouch on the 5th and 6th thoracic somites. Both the sexes swim actively and leave the host at will.

3. ***Monstrilla.*** The adults lack mouthparts and antennae. Life history is remarkable. Nauplius is free swimming but other immature stages are parasitic in a sedentary polychaete worm, absorbing nourishment through antennae. Adult, after emerging from the host, throws off antennae to assume a brief pelagic existence as its digestive tract is vestigial. Adult lives on stored food. Two dorso-lateral and one median eyes are present.

4. *Caligus.* Sexual dimorphism is not well marked. Both sexes live ectoparasitically on the skins or inside gill chambers of fishes, but can swim rapidly. Body is clumsily built. Mouthparts are adapted for piercing and sucking. Labrum and labium together form a sucking proboscis containing the sickle-like or style-like mandibles for piercing. In the fish louse, *Lepeophtheirus pectoralis,* as well as in others, the antennae, maxillae and maxillipedes are also used to lacerate the surface of the host.

5. *Anthosoma.* It is parasitic inside the mouth cavity of fishes. Appendages are recognisable but body is much modified due to curious overlapping lobes.

6. *Chondracanthus.* It is parasitic on the gills of various marine fishes and exhibits sexual dimorphism. Female is large, depressed, unsegmented and bearing paired irregular lobes. It remains permanently attached to the host and can not swim. Appendages are degenerate. Proboscis is absent but mouth is provided with 3 pairs of sickle-shaped jaws. Male is much smaller, typically copepodan in structure and attached permanently to the female near her genital aperture, by means of prehensile antennae.

7. *Lernaea.* It also exhibits marked sexual dimorphism. Male is unmistakenly copepodan. Female is vermiform with a curiously lobed anterior end. Maxillae are piercing. Thoracic appendages are greatly reduced. Two egg-sacs are attached to the hind end. In *Mytilicola,* living in gut of mussels, body is maggot-like with poorly developed appendages. In *Salmincola,* the frontal gland on head produces a special attachment button and thread for temporary attachment with gill filaments of the host. Later another button of bull is produced into which second maxillae are inserted for permanent attachment.

8. *Lesteria.* Morphological degradation is even more marked. Female is about 70 mm long with a swollen head embedded in skin of fish, while remainder of body hangs freely in water.

Parasitic Branchiura

Subclass Branchiura includes fish lice or carp lice, temporarily ectoparasitic on the skin or in gill chambers of freshwater and marine fishes and on some amphibians. More than 80 species are known, all included in the order Arguloida. The most common genus of carp-lice is *Argulus.*

Body is made of a large flattened oval or disc-like cephalothorax and a small unsegmented limbless and bilobed abdomen ending in a minute caudal furca. A large shield-like carapace covers the cephalothorax. Eyes are paired, lateral, sessile and compound. Mouth is suctorial. Labrum and labium form a sucking tube, the mouth cone, proboscis or siphon. It encloses the piercing mandibles. They feed on blood and other tissue fluids by making rasping wounds with their mandibles. There is a median tube located in front of the proboscis and ending in a large hollow poison-spine used for puncturing the skin of the host. Both pairs of antennae are reduced. Each antennule is modified into a claw for attachment. First maxillae are modified into a pair of sucking discs. Second maxillae are heavy, uniramous and clawed. There are no maxillipedes. Four pairs of thoracic appendages are large, biramous, with swimming setae, and with a proximal extension of the exopodite. Backwardly directed spines present on the under surface prevent the parasite from being brushed off by the passage of the host through water. Both sexes crawl freely over the surface of the host and leave regularly at the breeding season in spring and autumn. Fertilization is internal. Eggs are fastened in rows to stones and other objects. Development into the adult is direct, there being no nauplius or larval stages. Sexual maturity is attained after seven moults.

Branchiura were formerly regarded parasitic copepods due to superficial resemblance. But they differ from copepods in several important features, such as dorso-ventrally flattened body, large shield-like carapace, paired sessile compound eyes, lateral head lobes, opening of genital apertures at the bases of 4th pair of thoracic legs, a phyllopod or flagellar process arising from thoracic exopodities, no egg-sacs and moulting continuing in the adult stage.

Parasitic Cirripedia

Many cirripedes are parasitic and exhibit various degrees of degeneration according to the nature of their parasitism.

1. *Syngoga.* *Syngoga* (Order Ascothoracica), parasitic in Anthozoa, is essentially like the non-parasitic forms.

2. *Laura.* *Laura* (Order Ascothoracica) is parasitic on corals. Mouthparts are modified for piercing and sucking. Thoracic limbs are less than 6 pairs and reduced. The intestine and ovaries have extensive diverticula into the mantle. Males are much smaller than females.

3. *Trypetesa.* *Trypetesa* (Order Acrothoracica), which bores into certain mollusc shells, is a little barnacle with only 4 pairs of thoracic limbs and no abdominal segments.

4. *Proteolepas.* It is the only genus of the order Apoda and was found by Darwin in the mantle cavity of the stalked barnacle, *Alepas.* It is small, maggot-like, without mantle, trunk appendages or anus, and body divided into rings, as in an annelid.

5. *Sacculina.* The order Rhizocephala is exclusively parasitic, nearly always on decapod crustaceans, and of worldwide distribution. *Peltogasterella* is parasitic on hermit crabs *(Eupagurus),* whereas *Sacculina* is parasitic on other crabs. The adult is a tumour-like or sac-like structure attached to the abdomen of its host by a short peduncle. Adult body has no appendages, segmentation, sense organs or digestive system. A thin bag-like mantle encloses the visceral mass which is made chiefly of hermaphrodite gonad. A rootlike system of numerous delicate filaments or rhizoids arises from the peduncle to penetrate to all parts of the host's body and absorb nutrients. Thus, Rhizocephala shows an extreme case of morphological degeneration due to parasitism.

Sacculina, definitely a cirriped crustacean is shown by its very remarkable larval history. The fertilized eggs hatch inside the brood pouch into typical *Nauplius* larvae with characteristic frontal horns and without mouth or alimentation. They escape into the sea by an opening. The free-swimming nauplius develops into a **cypris** larva. It attaches to the bristle of a crab by means of its antenna and becomes a **kentrogon** larva which penetrates the host's body to become an endoparasitic cell mass. In a few weeks the endoparasite becomes external as a small sac having female gonad. A cypris larva, destined to become male attaches itself to the female and injects into its brood pouch a cell mass that later differentiates into a testis. Female is now converted into a hermaphrodite.

Decapods parasitised by *Sacculina* exhibit two striking effects. Firstly, **moulting** or ecdysis ceases from the moult at which the parasite becomes external. Secondly, the generative organs of the host becomes reduced or completely degenerated, a phenomenon known as **parasitic castration**. Infested males assume female secondary sexual characters. Thus, their abdomen becomes longer and broader, chelae become smaller, copulatory styles reduce or disappear and ovigerous appendages or swimmerets develop, as in the female crab. If *Sacculina* drops off and the male host recovers, it develops a hermaphrodite gonad. Thus, the male host also displays hermaphroditism due to parasitism.

Parasitised females, on the other hand, show merely parasitic castration and revert to a juvenile type. Their gonads atrophy, the ovigerous appendages may become reduced, but they develop neither male secondary sexual characters, nor a hermaphrodite gland after recovering from the disease.

Such effects on the hosts may be due to metabolic disturbances or hormonal imbalance. Parasite causes effects by digesting away the androgenic gland, which regulates sex in crustaceans.

Parasitic Isopoda

Most isopods are free-living but several of them are parasitic and display every grade from well-organized temporary ectoparasites to mere sacs of eggs.

1. *Aega.* It is a fish louse with ordinary isopod form, heavy built, piercing mouthparts, some legs hooked, and broad uropods forming a tail fan.

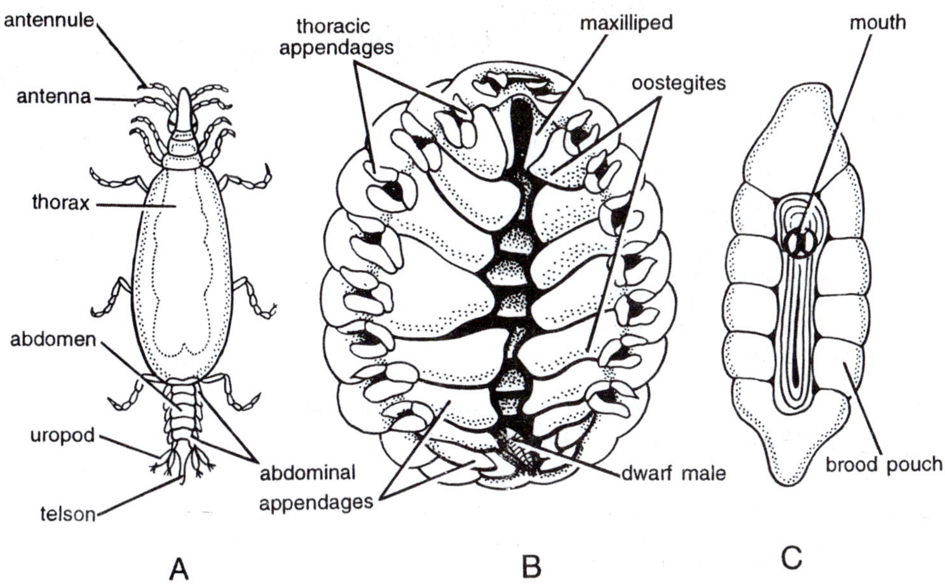

Fig. 25. Parasitic Isopoda. **A.** Pranzia larva of *Gnathio*. **B.** Female *Bopyrus*. **C.** *Cryptoniscus*.

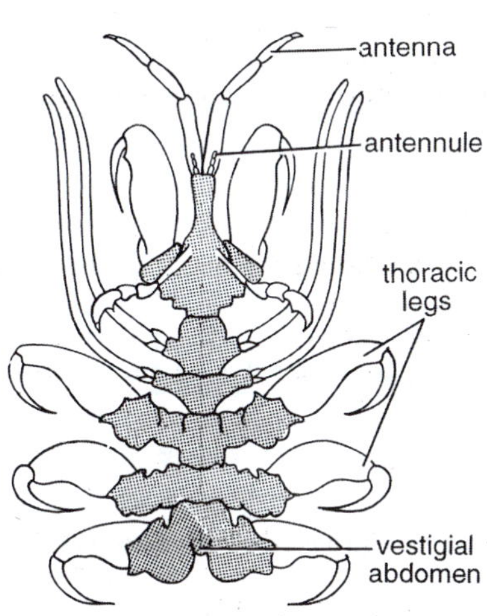

Fig. 26. Parasitic Amphipoda *Cyamus*.

2. *Cymothoa.* It is an ectoparasite on fishes, holding the mucous membrane of the host by their clawed legs. Mouthparts are sucking. Form is typically isopodan. Perhaps each, individual is hermaphrodite, acting as male when young and free swimming and becomes a female after settling down.

3. *Gnathia.* Adults are clearly isopods. However, certain stages in life-history, called **pranzia larvae**, are intermittent ectoparasites of marine fishes. Their mouthparts are modified for piercing and sucking.

4. *Bopyrus.* It is parasitic inside the gill chambers of prawns. Mouthparts are suctorial and mandibles piercing. It shows a notable degree of sexual dimorphism. Female is large, shows greater degeneration and asymmetry, though still recognisable as an isopod. Maxillae are vestigial or absent. Male is very small, typical and permanently attached to female.

5. *Cryptoniscus.* It is a hyperparasite on rhizocephalan cirripedes and a protandrous hermaphrodite. Adult female degenerate with not leaving any trace of crustacean organization. Only its developmental history reveals its true crustacean affinities.

Parasitic Amphipoda

Parasitism is less prevalent in amphipods than in isopods. There are a few ectoparasites of fish, such as *Lafystius* and *Trischinostoma.* Whale louse or *Cyamus* lives ectoparasitically on whales. Body is depressed and broad and abdomen is reduced. Mouthparts are suctorial and the legs are clawed, adapted for clinging to the skin of the host.

Important Questions

➤➤Long answer type questions

1. Describe the various types of larval forms found in Crustacea.
2. Describe the various kinds of mouthparts of insects.
3. Discuss 'respiration in insects'.
4. Write an essay on 'metamorphosis in insects'.
5. Give an account of the role played by insects in human welfare.
6. Write a brief essay on insecticides.
7. Discuss biological control of insects.
8. Write an essay on 'Parasitism in Crustacea'.
9. Discuss social life in insects.
10. Discuss the bodywall in Arthropoda.
11. Give an account of various types of respiratory organs found in Arthropoda.
12. In what ways the Arthropoda are important ? Discuss.

➤➤Short answer type questions

1. What is the typical larval form of crustaceans ?
2. What is the larval stage of *Squilla* known as ?
3. Name the animal to which the **Kentrogon** larvae belong.
4. Name the two insects that could bring typhus fever to man.
5. Which fly transmits sleeping sickness ?
6. What is the adult insect that lives for one day ?
7. Write a note on the T.S. of ommatidium in Arthropoda.
8. Describe how lobster feeds.
9. What is parthenogenesis ? Give an example from class Insecta. Explain the phenomenon in 30 words.
10. What are beneficial insects ? Outline the appropriateness of the term with reference to one example in 50 words.
11. Write notes on beneficial insects and harmful insects.
12. Name the insects useful to man in South India.
13. Write note on destruction of wood by termites.
14. Write note on the gastric mill in lobster.
15. When ecdysis occurs, it takes sometime for the newly formed skin to harden. During this time, the arthropod is said to 'hide'. Give atleast three reasons, why in your opinion, it should hide ?
16. What is migration ? In which insect do you find it ? Briefly explain it in 50 words.
17. Outline the earlier stages of attack by the *Sacculina* larval form. What are the changes that are brought about in the adult.
18. Describe the segmentation of the arthropodan head.
19. Describe the cephalic structure in centipede.
20. Describe the morphology, ecology and significance of crustacean larvae.
21. With the help of neat diagrams describe the sponging and piercing sucking mouth parts found in insects. Give examples.
22. What are the internal factor regulating metamorphosis in insects ?
23. Compare the habit, habitat, adaptations and mode of living of the centipede and the millipede.
24. It is usually felt that in the battle between man and insects, man has so far not succeeded in completely eliminating them. Comment on it.
25. Explain how parasitic arachnids could be a menace to man.
26. Describe social insects with special reference to termites.
27. Give a brief account of the structure of a typical trachea of an insect. In what respect does the tracheal system of an insect differ from that of an onychophoran. What is a tracheal gill. Give example.
28. Give an account of the respiratory organs found in Arthropoda.
29. Mention atleast three poisonous arthropods, preferably from various classes. What are their different poison apparatus ? Explain one such poison apparatus with a diagram in not more than 20 words.
30. Describe the various system of *Apus.* Add a note on its same organs.
31. Give an account of the different types of mouth parts in insects and explain how they are adapted for their feeding habits.
32. Define biological succession.

33. What is cephalization.
34. Describe Nauplius.
35. List out the arthropods which show direct development ?
36. Compare Zoaea and Protozoaea.
37. Characterised the Megalopa larva.
38. What is an Alima ?
39. How many types of mouthparts are found in insects ?
40. Describe the spongin type of mouthparts.
41. List any five which act as vectors?
42. List any five which are known as harmful insects.
43. Some insects are used as medicine name any two.
44. Give brief account on metamorphosis in insects.
45. What is the hormonic control of metamorphosis in insects ?
46. Name any five chemical which are used as pest or insect control.
47. Define biological control.
48. Comment on cuticle of arthropods.
49. Define the trachial respiration in insects.
50. Give check list of respiratory organs in Arthropoda.
51. Comments on intermediatory host.
52. Which Crustacean use as food by man.
53. Given the name of Crustacean which function as pest ?
54. Give the name of vector of following disease :
    (a) malaria (f) kala azar
    (b) yellow fever (g) plague
    (c) filaria (h) cholera
    (d) sleeping sickeness (i) diarrhoea
    (e) chagas (j) typhus fever.

➤➤Fill in the blanks

1. Silk produced by........
2. Blow flower's larva is used in the treatment of........
3. Butterflies and moths have mouth parts of........types.
4. Moths are close allies of........
5. Organs of Berlese present........bed bug.
6. The dragon fly related to order ....
7. Termites belong to order........
8. Cantheridine is derived from........
9. Two thoracic and 8 abdominal spiracle in........
10. Epidermis also called........in arthropods.

➤➤Multiple choice questions

1. Alary muscles in insects refers to :
   (a) heart muscles (b) leg muscles
   (c) wing muscles (d) gut muscles
2. In insects each compound eye consists of small units known as :
   (a) ommatidia (b) corneal lens
   (c) lenticular cells (d) pterostigma
3. Respiratory organs of crustaceans are :
   (a) gills (b) lungs
   (c) general surface
   (d) both gills and general surface
4. Excretory product in an insect is :
   (a) uric acid (b) urea
   (c) ammonia (d) guanine
5. The unit of compound eye is :
   (a) eye-let (b) ommatidium
   (c) retinulae (d) eye spot
6. Which one of these is good example of metamorphosis:
   (a) Regeneration of broken tail of lizard
   (b) Growth and development of young one of Kangaroo in its marsupium
   (c) Development of adult from pupa of insects
   (d) Hatching of maggot from the egg of housefly
7. Ecdysone regulates :
   (a) periodical shedding of epidermis of reptiles
   (b) ecdysis of cuticle in adult animals
   (c) metamorphosis in insects
   (d) germination of sponges by gemmule

8. Complete metamorphosis occurs in :
(a) termite (b) silver fish
(c) housefly (d) bed bug
9. Most simple and commonest type of larva is :
(a) Nauplius (b) Zoaea
(c) Alima (d) Megalopa
10. It is also known as modified form of zoaea :
(a) Mysis (b) Megalopa
(c) Cypris (d) Alima
11. Biting and chewing type of mouth parts occurs in :
(a) grasshopper (b) cockroach
(c) cricket (d) all
12. One cocoon give the fibre of :
(a) 10 feet (b) 100 feet
(c) 1000 feet (d) 10000 feet
13. Bee-venom is used in the treatment of :
(a) arthritis (b) malaria
(c) myopia (d) cholera
14. Brain hormone in insect secreted from :
(a) pituitary gland
(b) neurosecretory cells of the brain
(c) thoracic gland (b) nephridia
15. Ecdysone is :
(a) BH (b) PTTH
(c) PGH (d) JH
16. Genetic control of insect proposed by :
(a) C. Laven (b) H. Laven
(c) R. Ross (d) Helgesen
17. Tracheal system consist of :
(a) spiracle (b) trachae
(c) tracheoles (d) air sac
(e) all the above
18. Means of respiration in insects :
(a) integuments (b) blood gills
(c) spiracular gills (d) tracheal gills
(e) all
19. Scorpions are beneficial because these are used in :
(a) biological control of insect
(b) pheumacological research
(c) immunological research
(d) all

Answers

➤➤*Fill in the blanks*

1. silk worms, 2. decaying tissue, 3. siphoning, 4. butterflies, 5. in female, 6. Odonata, 7. Isoptera, 8. blister beets, 9. cockroach, 10. hypodermis.

➤➤*Multiple choice questions*

1. (a) 2. (a) 3. (a) 4. (a) 5. (b) 6. (c) 7. (c) 8. (c) 9. (a) 10. (d)
11. (d) 12. (c) 13. (a) 14. (b) 15. (c) 16. (b) 17. (e) 18. (e) 19. (d)

55

Peripatus

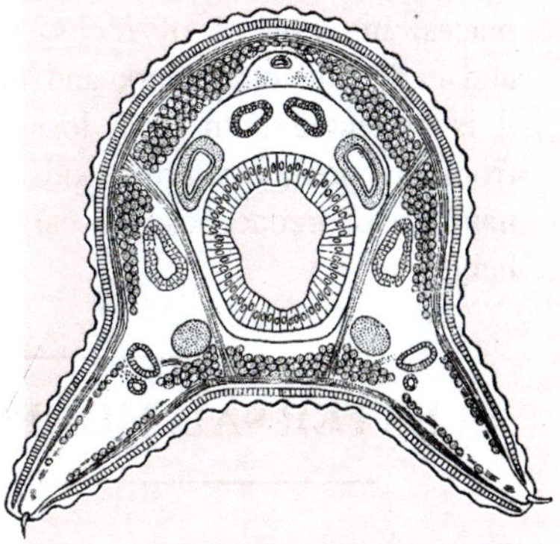

From the standpoint of phylogenetic relationship between two taxonomic groups, few animals are more interesting than the small caterpillar-like form, called *Peripatus*. This animal displays some of the characteristics of both the annelid worms as well as arthropods, besides its own typical features. Owing to its resemblance with two different phyla, *Peripatus* is often referred to as the connecting link between them.

Peripatus and related animals belong to the phylum *Onychophora* (Gr., **onyx**, **claw** + **pherein**, to bear), which is conventionally classified with Arthropoda as one of its classes. But the modern system of animal classification has given to it the status of an independent phylum. There are about 70 species in this phylum belonging to a few genera or possibly to one genus, *Peripatus*.

GEOGRAPHICAL DISTRIBUTION

Peripatus exhibits discontinuous distribution with its species scattered in most of the warmer parts of the world : Africa, Australia, New Zealand, Central America, Mexico, West Indies, Malaya Archipelago, India and other localities. Some common forms of *Peripatus* are :

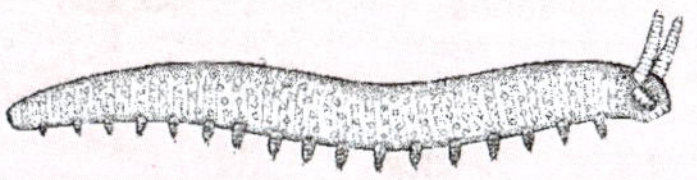

(1) *Opisthopatus* — Chile and South Africa.
(2) *Mesoperipatus* — Congo and Caribbean Islands.
(3) *Peripatopsis* — South Africa.
(4) *Ooperipatus*—Australia.

HABITS AND HABITAT

Peripatus is a terrestrial animal, living in moist places, in crevices of rocks, under stones, logs and bark, and other dark and damp places where it is protected both from loss of water and also from the predatory arthropods. It is nocturnal in nature and predaceous and carnivorous in feeding habit.

EXTERNAL MORPHOLOGY

Body of *Peripatus* is elongated, cylindrical and bilaterally symmetrical, measuring between 5-7 cm in length. The external segmentation is unclear and the legs, which number from 14 to 43 pairs, are unjointed. Each leg is a stumpy protuberance ending in two claws and provided with 3 to 6 pads on ventral side which serve as walking soles which contact the substratum. That is why they are also called walking worms. Body is actually lifted off the ground by the legs which move in a series of steps. Skin surfaced with a thin cuticle is soft and velvety and thrown into transverse ridges bearing numerous conical **papillae** or **tubercles** armed with chitinous spines. Below the cuticle are the typical layers of muscles and connective tissue of a worm like body. Surprisingly the muscle plays no role in the process of locomotion which is carried out by the movement of appendages alone. The muscle of body wall takes part in changes of length and shape of the body (Sidnie and Manton). Anterior end which is not differentiated as a distinct head, bears a pair

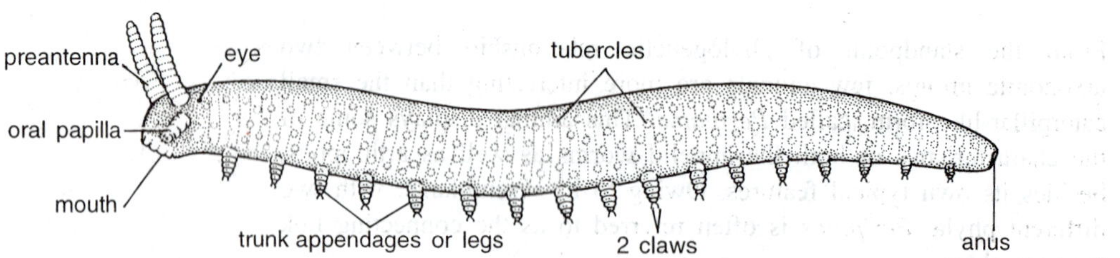

Fig. 1. *Peripatus.* External features in lateral view.

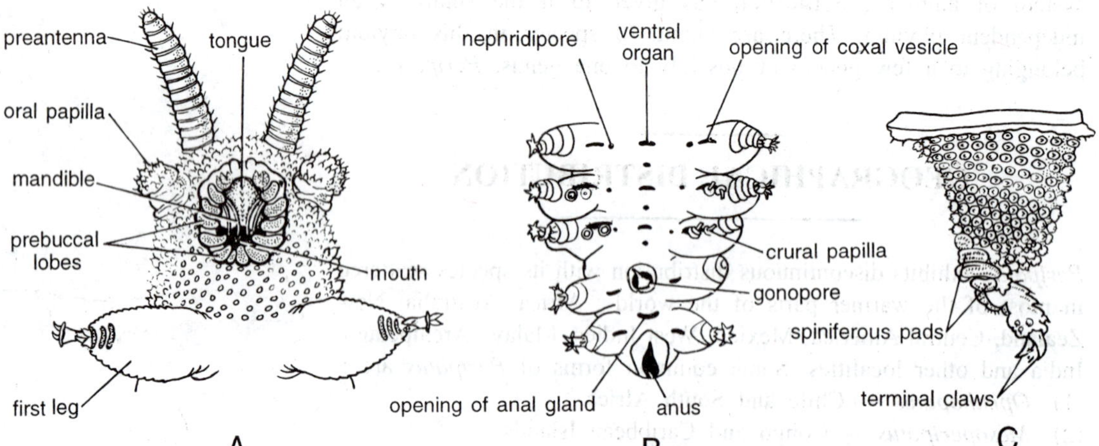

Fig. 2. *Peripatus.* **A.** Anterior end in ventral view. **B.** Posterior end in ventral view. **C.** A leg.

of dorsal eyes, a pair of **pre-antennae** or **preoral antennae**, a pair of slime-secreting **oral papillae**, and a mid-ventral mouth with a pair of chitinous jaws. Both the jaws and the oral papillae are modified legs. The unjointed legs are short and stumpy and each ends in two claws. The anus lies at the posterior end, slightly towards the ventral side, and the genital aperture is located in front of the anus. *Peripatus* shows sexual dimorphism; males are smaller than females.

ANATOMICAL FEATURES

(1) The body wall is dermo-muscular, consisting of cuticle, epidermis, dermis, and striped circular and longitudinal muscles.

(2) The coelom is in the form of small cavities around gonads and metanephridia.

(3) The body cavity is a haemocoel, lined with epithelium.

(4) A pair of **slime glands** are located, one on either side of the body cavity. These open on the oral papillae and secrete an adhesive slime for entangling the prey.

(5) The mouth leads into the alimentary canal, which comprises of a tongue with rows of sensory spines, muscular pharynx into which opens a pair of large salivary glands, short foregut or oesophagus, long midgut or **stomach-intestine** and short hind gut or rectum.

(6) The respiratory organs are unbranched tracheae, communicating to outside through minute spiracles.

(7) The excretory organs are coxal glands, one pair in each segment, opening at the base of the legs.

(8) A dorsal tubular contractile heart lies within the pericardial cavity.

(9) The nervous system consists of a pair of suprapharyngeal ganglia or brain with two circumpharyngeal connectives and a pair of widely separated lateral longitudinal nerve cords connected together by transverse commissures. Ganglia on the cords are indistinct.

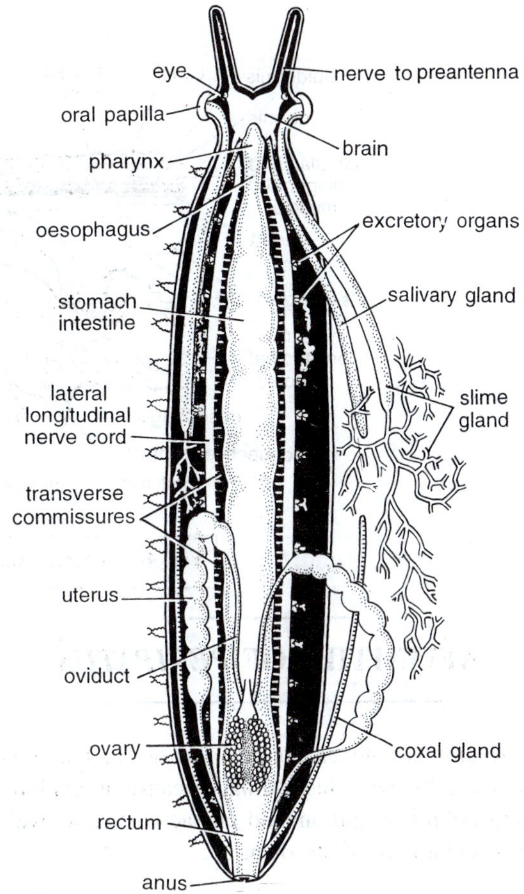

Fig. 3. *Peripatus.* Internal structures.

(10) Sensory organs are a pair of eyes near the base of the antennae, taste spines on the lips and preoral cavity and tactile spines on the surface tubercles.

(11) Female reproductive organs include a pair of ovaries, a pair of oviducts and a pair of beaded uteri. The uteri join to open to outside through a vaginal opening. The male reproductive system has paired testes, seminal vesicles, vasa deferentia and genital openings.

(12) Fertilization is internal. The female produces about 30 or more young in a year. The young resembles the adult.

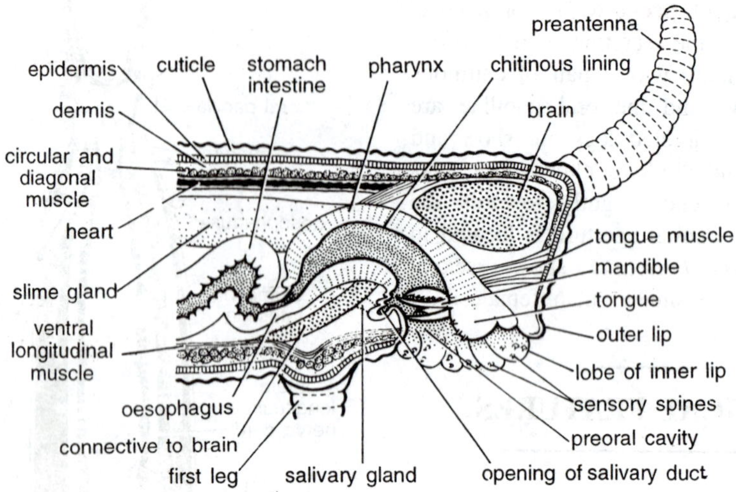

Fig. 4. *Peripatus.* Anterior end in sectional view.

AFFINITIES OF *PERIPATUS*

Peripatus has no economic importance; but it is zoologically very interesting, because it exhibits both arthropod and annelid characteristics as well as peculiarities of its own.

[I] Annelidan Characteristics

(1) Vermiform body with truncated extremities.
(2) Absence of a true head.
(3) Dermo-muscular body-wall, consisting of a thin flexible cuticle and underlying circular and longitudinal muscles.
(4) Locomotion slow and by peristalsis as in an earthworm.
(5) Structure of the simple eyes as in polychaetes.
(6) Unjointed, hollow, stumpy appendages of the nature of extensions of the body-wall, like parapodia of Polychaeta.
(7) Simple, straight alimentary canal with terminal mouth and anus.
(8) Segmentally arranged paired nephridia.
(9) Slime and coxal glands correspond with similar glands of Chaetopoda.
(10) Presence of cilia in the excretory and reproductive ducts.

[II] Arthropodan Characteristics

(1) Presence of antennae.
(2) Jaws are modified appendages provided with striped muscles.
(3) Locomotion by definite legs, having definite musculature and provided with claws.
(4) Cuticle has a thin deposit of chitin, like that of arthropods.
(5) Body-cavity is a haemocoel.
(6) Coelom reduced to small cavities that surround the excretory and reproductive ducts.
(7) Peculiar salivary glands, supposed to be modified nephridia.
(8) Dorsal tubular heart with lateral ostia.
(9) Presence of a tracheal respiratory system.
(10) Brain is large and typically arthropodan.
(11) General structure of the reproductive organs and development mainly arthropodan.

[III] Onychophoran Characteristics

Following features are peculiar to *Peripatus,* in which they differ from other phyla :

(1) Body shows no or indistinct external segmentation.

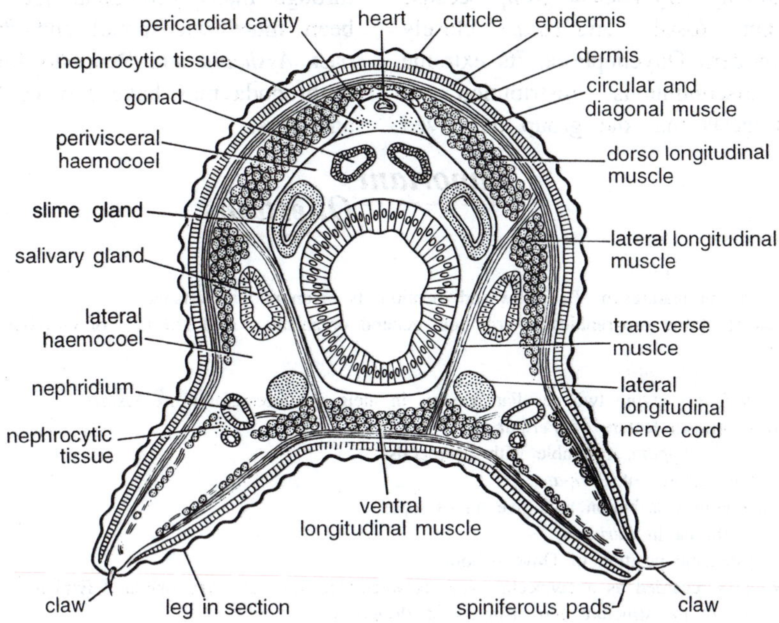

Fig. 5. *Peripatus.* Body in T.S.

(2) Texture of skin. A rough cuticle covered with numerous velvety processes not known in other phyla.
(3) Antennae not homologous to the antennae of other arthropods.
(4) Three-segmented head of *Peripatus* shows a condition mid-way between that of Annelida and Arthropoda.
(5) Restriction of jaws to a single pair. Movement of jaws is antero-posterior.
(6) Presence of non-jointed legs with claws.
(7) Irregular distribution of spiracles or tracheal openings.
(8) Two ventral nerve cords widely separated and without true ganglia.
(9) Structure of eyes is less complicated.
(10) Distribution of reproductive organs.

[IV] Molluscan Characteristics

Peripatus was previously included with Mollusca due to slug-like appearance of its body and ladder-like nervous system, as found in Amphineura and Prosobranchiata. But these are only superficial resemblances.

TAXONOMIC POSITION

Onychophora have both annelidan and arthropodan characteristics. Therefore, they are regarded to be an intermediate stage or connecting link between Annelida and Arthropoda. However, they appear to be more closely allied to arthropods than to annelids and perhaps arose as an offshoot from near the base of the arthropod line. Based on such phylogenetic considerations, **Manton** (1958) and other contemporary zoologists have included onychophorans within the Arthropoda as a class or a subphylum. But, lack of an exoskeleton and jointed limbs and the presence of primary annelidan characters in Onychophora create serious problems. Others claim that *Peripatus* is definitely an annelid. In fact, onychophorans are neither worms nor arthropods but have distinctive characteristics of their own. Therefore, Onychophora are now-a-days treated as a separate phylum in their own right.

It is undoubtedly very ancient group because a mid-Cambrian fossil, *Aysheaia,* closely resembles the modern Onychophora. Its extreme isolation or discontinuous distribution at present also suggests that the group has lived through many geological ages and that it had been more widespread and diversified in the past. *Aysheaia* and *Peripatus* further suggest that Arthropoda may have evolved from annelid-like ancestors.

Important Questions

➤➤Long answer type questions

1. Describe the salient features of *Peripatus* and mention its taxonomic importance.
2. Give an account of the geographical distribution, general organization and affinities of Onychophora.

➤➤Short answer type questions

1. Classify the *Peripatus* giving two peculiar features in their structure and/or life history.
2. Mention the annelidian feature of *Peripatus.*
3. Which features of *Peripatus* resemble with arthropods.
4. Mention peculiar features of *Peripatus.*
5. Define discontinuous distribution and give an example.
6. Describe the nephridia in *Peripatus.*
7. Discuss the systematic position of Onychophora.
8. Why *Peripatus* is regarded as a connecting link between Annelida and Arthropoda ? Explain with specific facts.
9. Give an account of the structure and affinities of *Peripatus.*

➤➤True and false statements

1. *Peripatus* is a connecting link between Annelida and Arthropoda.
2. *Peripatus capensis* is the best known species of Onycophora

➤➤Multiple choice questions

1. *Peripatus* breaths by :
   (a) trachea (b) gills
   (c) book lungs (d) branchiostegite
2. The number of legs found in *Peripatus* is :
   (a) 10-20 pairs (b) 15 to 40 pairs
   (c) 14 to 43 pairs (d) 10-15 pairs
3. The excretory structures in *Peripatus* are :
   (a) Malpighian tubules (b) coxal glands
   (c) nephridia (d) solenocytes
4. *Peripatus* shows :
   (a) metameric segmentation
   (b) scant metamerism
   (c) pseudometamerism (d) none of these
5. The nephridia of *Peripatus* open :
   (a) into the gut (b) at the base of each leg
   (c) near the genital opening
6. *Peripatus* is :
   (a) oviparous (b) viviparous
   (c) ovo-viviparous (d) none
7. One of the following shows discontinuous distribution:
   (a) *Schistocerca* (b) *Peripatus*
   (c) *Periplaneta* (d) *Musca*
8. In Australia the form of *Peripatus* is :
   (a) Ophisthopaters (b) Mesoperipatus
   (c) Peripatopsis (d) Ooperipatus
9. *Peripatus* is :
   (a) aquatic (b) terrestrial
   (c) parasitic (d) none
10. An animal without economic importance is :
    (a) *Peripatus* (b) hony bee
    (c) *Pheritima* (d) none
11. *Peripatus* is :
    (a) herbivorous (b) carnivorous
    (c) omnivorous (d) parasite
12. *Peripatus* is connecting link between :
    (a) Annelida and Arthropoda
    (b) Annelida and Platyhelminthes
    (c) Annelida and Mollusca
    (d) Mollusca and Arthropoda

Answers

➤➤True and false statements

1. true, 2. true.

➤➤Multiple choice questions

1. (a) 2. (c) 3. (b) 4. (b) 5. (b) 6. (b) 7. (b) 8. (d) 9. (b) 10. (a)
11. (b) 12. (a)

56

Pila : An Apple Snail

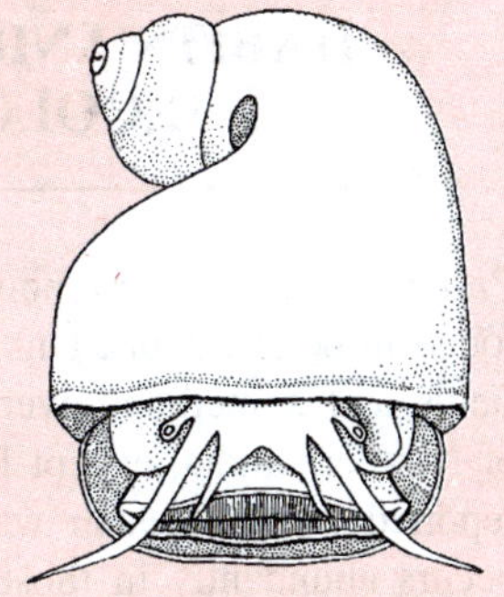

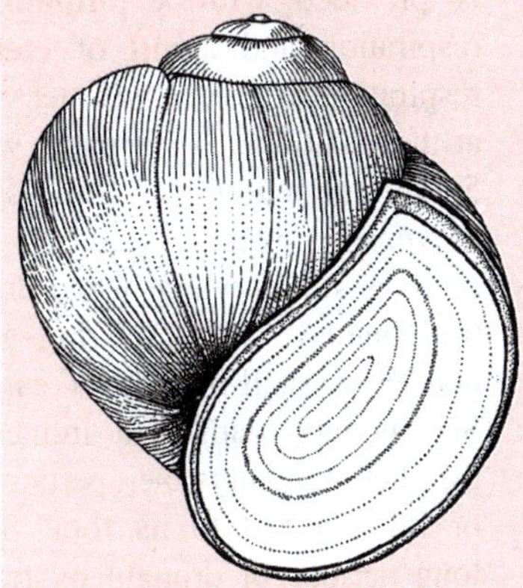

Phylum **Mollusca** (L., **molluscus**, soft) includes soft-bodied invertebrate animals, such as snails, slugs, mussels, clams, oysteres, tusk-shells, squids, octopods, etc. It is a very successful, diverse and widespread group, with about 112,000 species. Largest of the molluscan classes is **Gastropoda**, which is represented by about 35,000 living and some 15,000 fossil species. It includes the most familiar forms, such as snails and slugs. Apple snail, *Pila,* is a freshwater snail and is represented in India by eight species. Most common species found in Northen India is *Pila globosa* which forms the subject matter of this chapter.

Pila globosa

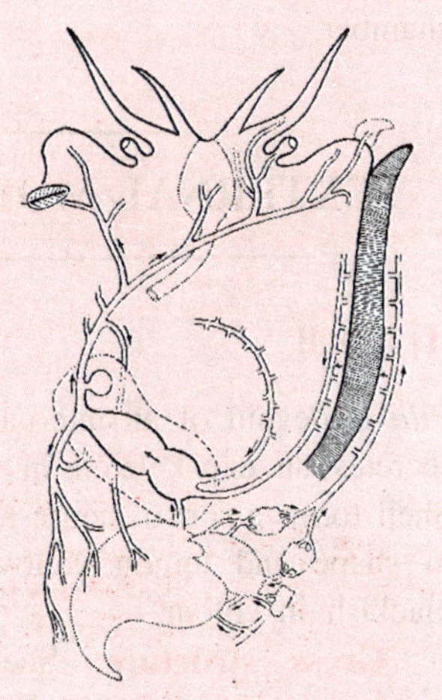

SYSTEMATIC POSITION

Phylum	Mollusca
Class	Gastropoda
Subclass	Prosobranchia
Order	Mesogastropoda
Family	Pilidae
Genus	*Pila*
Species	*globosa*

HABITS AND HABITAT (ECOLOGY)

Pila globosa is a common freshwater apple snail found in ponds, pools, tanks, lakes, marshes, paddy fields and sometimes even in streams and rivers of Northern India, except Punjab. It has also been reported from brackish waters of low salinity. It occurs abundantly in those waters which harbour succulent aquatic plants like *Vallisnaria* and *Pistia,* forming the food of this snail. *Pila* is a voracious eater and feeds upon plant scrappings, cut with the help of chitinous jaws and radular teeth.

Pila is adapted to lead an amphibious life; it is provided with a **pulmonary sac** for aerial respiration and a **gill** or **ctenidium** for aquatic respiration. It creeps with its ventral muscular foot at the traditional 'snail pace' which averages about 5 cm a minute at full speed. To facilitate movement over a hard dry surface, the snail secretes slime, which when dried, leaves a silvery trail behind. While moving and feeding, it does not leave shell-house but carries it on its back. Sensing any danger, it immediately retreats into the shell and shuts the aperture with the operculum or lid attached to its foot. Snail may tide over long periods of drought by remaining torpid with the shell aperture tightly closed. It is then said to be in **summer sleep** or **aestivation**. During rains, it leaves the ponds and makes long excursions on land, breathing air directly through a pulmonary chamber.

EXTERNAL MORPHOLOGY

[I] Shell

Pila is devoid of an internal skeleton; instead, it secretes an exoskeleton in the form of a **shell**. Shell forms a sort of house to live in. It is globose in shape and lemon yellow, brownish or even blackish in colour.

Gross structure. Shell is an elongated structure consisting of tubular whorls coiled around

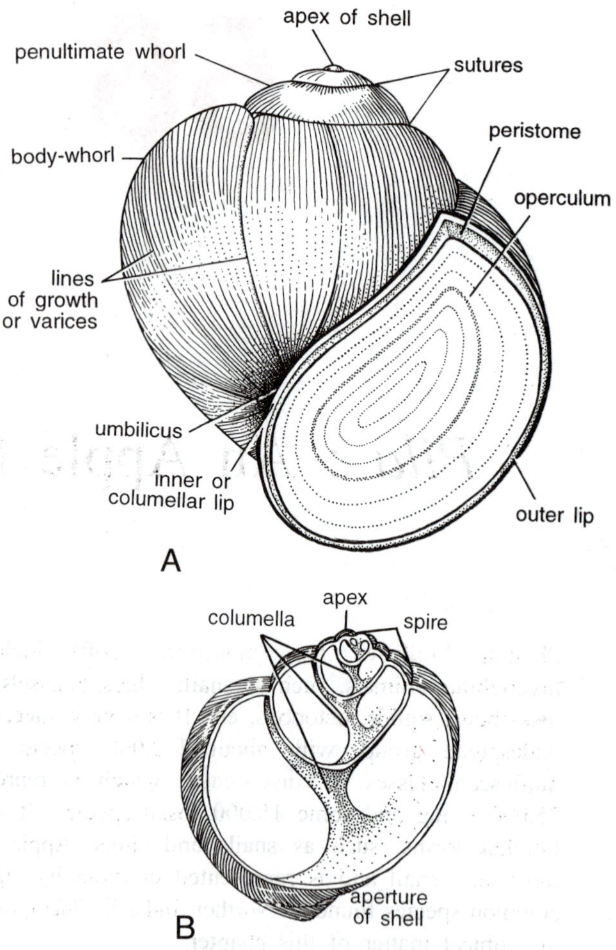

Fig. 1. *Pila.* **A.** The shell with operculum (ventral view). **B.** Shell in a section (dorsal view).

a central axis called the **columella**. Various whorls lie in different planes so that the shell looks like a **conical spire**. This type of shell is called **conispiral** in contrast to the **planospiral** shell found in protozoan, *Elphidium* and the mollusc, *Nautilus*. There are usually 6.5 whorls in the shell of *P. globosa*. Smallest and the oldest whorl lies at the apex of the shell, which represents the **first shell** or the **protoconch** laid down by the larva. Starting from it, the whorls successively increase in size and end in the largest whorl or the **body whorl**. Immediately preceding the body whorl is a **penultimate whorl**, which is also large. These two whorls enclose the greater part of the animal's body. Whole series of whorls, exclusive of the

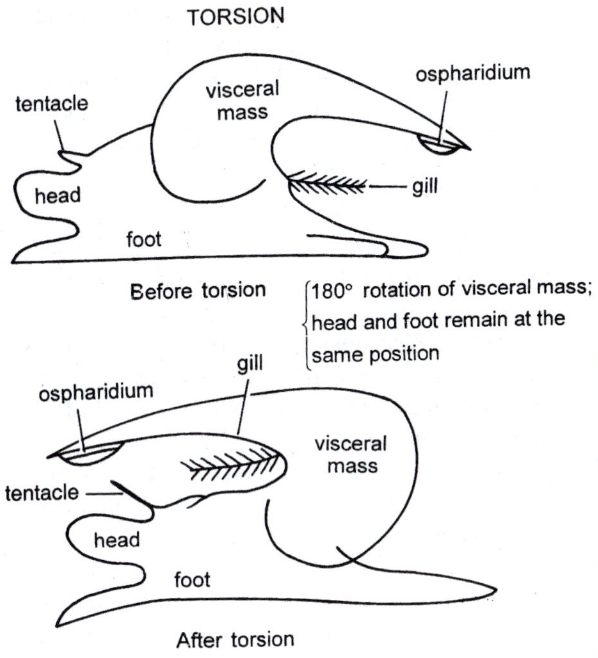

Fig. 2. *Pila.* The position of *Pila* before and after the torsion.

body whorl, is known as the **spire**. Internally, all the whorls communicate freely with one another, there being no separating partitions between them; such a shell is called **unilocular.** Externally, successive whorls are demarcated by lines, called **sutures**.

Body whorl opens to the exterior by a wide opening, the **aperture** or **mouth,** which is situated on the ventral surface of the shell. Smooth and continuous margin of the aperture is called the **peristome**. Its outer margin is called the **outer lip**, while the inner one next to the columella is the **inner** or **columellar lip**. Columella is a hollow, twisted and rod-like structure, which opens to the exterior through a narrow aperture, the **umbilicus,** situated near the collumellar lip at the end opposite the apex of the shell. Shell with an umbilicus is known as **umbilicated** or **perforate**, as in *Pila,* whereas one without umbilicus is called **unumbilicated** or **imperforate,** as in *Triton.*

Surface of the shell is marked by numerous, somewhat raised **lines of growth** running parallel to the margin of the aperture; some of these are in the form of well marked ridges and are called

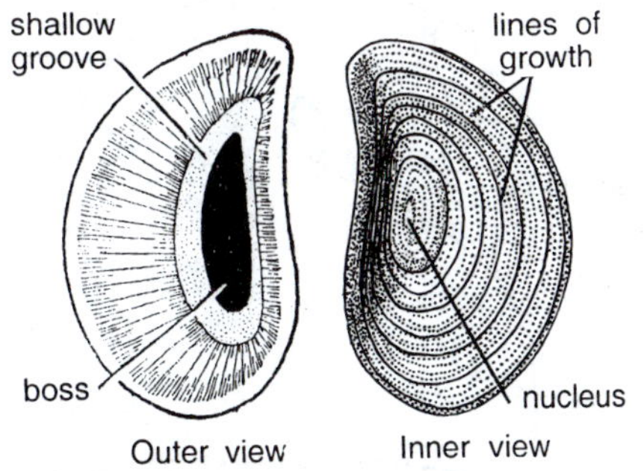

Fig. 3. *Pila.* Operculum. **A.** Inner view. **B.** Outer view.

varices (singular, **varix**); these indicate periods of temporary cessation of shell-secretion.

When the shell of *Pila* is held with the apex upward and the aperture facing the observer, it displays a clockwise coiling from the apex to the aperture so that the latter comes to lie to the right-hand side. Such a shell is called **dextral, clockwise** or **right-handed** and is of normal occurrence in *Pila* and a big majority of gastropods. But abnormally in these gastropods and normally in a few species, the shell displays anti-clockwise coiling and is called **sinistral, anti-clockwise** or **left-handed.**

Torsion. Torsion is a 180° counter clockwise rotation, or twisting, of the visceral mass, shell, mantle, and mantle cavity with respect to the head and foot. It is a developmental event that occurs in the larva, not in the adult. The head and foot remain unaltered by this rotation, and except for the change in orientation. Structure such as nerves and gut that pass between the head, foot and the visceral mass are twisted.

Torsion is a defining characteristic of the gastropods and it is a developmental event that continues to occur in the life cycle of each individual gastropod.

Prior to torsion, the mantle cavity and its organs are posterior. After torsion, the mantle cavity and organs are anterior, immediately behind and above the head. Internally, the digestive tract is looped into another U superimposed on that resulting from flexure. The stomach is now posterior and dorsal whereas the mouth and anus are anterior and ventral. The nervous system has

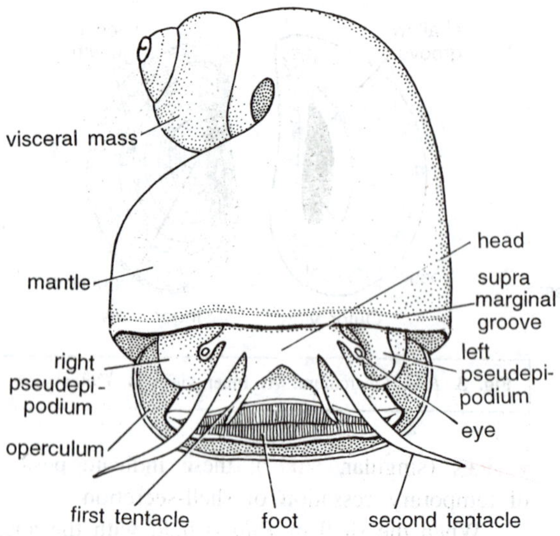

Fig. 4. *Pila.* Body after removing the shell.

been twisted into a figure of eight. The head and foot retain the original untorted bilateral symmetry. The chemical analyser and gill comes in front.

Microscopic structure. Chemically, a molluscan shell is composed of : (i) **conchiolin,** an albuminoid horny substance which serves as the **matrix**, and (ii) **calcium carbonate**, which occurs as crystals of calcite or aragonite. Depending upon the disposition of the crystals in the matrix, the shell exhibits three distinct layers in section :

(a) Periostracum. It is the thin outermost layer made up of conchiolin.

(b) Ostracum. It is the middle layer made up of calcium carbonate, disposed as **plates** running in the direction of shell coiling; the plates are further made up of fibrils.

(c) Hypostracum or nacreous layer or mother of pearl. It is the innermost layer made up of plates (consisting of fibrils) of calcium carbonate that run parallel to the margin of aperture. Shell is secreted by the underlying mantle.

Operculum. When the snail is inside the shell, aperture gets tightly closed by a thick oblong plate, the **operculum,** attached dorsally to the hinder part of the foot. It is a calcareous plate formed by cuticular secretion of the glandular cells of the animal's foot. Its outer chitinous covering shows many concentric rings of growth around a small sub-central **nucleus**. Inner surface exhibits a distinct elliptical area of a creamy colour, the **boss**, to which the opercular muscle is attached, boss is surrounded by a shallow groove.

[II] Body

Body of *Pila* is divisible into four distinct regions: **head, foot, visceral mass** and **mantle** or **pallium**. Head and foot together form the **head-foot complex** which retains the bilateral symmetry. Visceral mass and pallium together form the **visceropallium** which has secondarily lost the bilateral symmetry.

1. Head. Head is the anterior fleshy part of the body overhanging the foot. It is prolonged into a contractile **snout** and bears a **mouth,** two pairs of **tentacles** and a pair of **eyes**. First pair of tentacles or **labial palps** are small, tapering and highly contractile. They arise from the sides of the head. Second pair of tentacles are long fleshy and cylindrical and arise from behind the snout from the dorsal surface of the head. They are also contractile and can be withdrawn by invagination. From near the base of each tentacle projects a small, stumpy eye-stalk or **ommatophore** bearing a small but prominent eye at its tip. Mouth is a vertical slit-like aperture, lying ventrally between the bases of the first pair of tentacles.

2. Foot. Large, strongly muscular, ventral part of the body forms the **foot**. When fully expanded, it is roughly triangular in shape with the apex directed backwards. Foot has a broad, flat, smooth and grey ventral surface, or the **sole**, for creeping. Foot bears the operculum dorsally on its posterior end, called the **operculiferous lobe.** When the foot is withdrawn, operculum completely fits into the mouth of the shell and closes it.

Foot is the locomotory organ of snail. It is highly contractile with the muscle fibres arranged crosswise and lengthwise. Inside the foot are present numerous pedal gland-cells which secrete a slime trail during locomotion. Head-foot complex is attached to the visceral mass by a short inconspicuous **neck.**

3. Visceral mass. It constitutes a sort of hump on the dorsal side, containing all the visceral organs. It is soft and grey to dark brown in colour. It is spirally coiled like the shell in which it lies, occupying the body whorl as well as the spire. Coiling results from faster growth on the side of the visceral mass away from the columella. Visceral mass also exhibits the phenomenon of **torsion**.

4. Mantle. Skin of the visceral mass forms a thin and delicate covering, called the **mantle** or **pallium,** which is a characteristic molluscan organ. Anteriorly, the mantle becomes thickened and pigmented and serves as a protective cloak or hood over the head and its appendages when the animal is retracted. A long, **narrow supra-marginal groove** runs along the thickened free edge of the mantle. Behind the groove is a thick band, containing shell-secreting glands. Mantle secretes and lines the shell. Periostracum and the ostracum layers of the shell are secreted by the groove and shell glands respectively, while the hypostracum is secreted by the general epithelial covering of the mantle. At the sides of the head and over the foot, the mantle is prolonged into highly contractile and fleshy processes, the **nuchal lobes** or the **pseudepipodia**. Left one is much longer than the right and forms a respiratory siphon during aerial breathing.

MANTLE CAVITY AND PALLIAL COMPLEX

Mantle encloses a large cavity which is dorso-lateral in position and called the **mantle** or **pallial** cavity. Animal retracts its head and foot into this cavity for protection and rest; the head is retracted first and is followed by the foot. On the floor of this cavity, arising near the anterior edge of the right nuchal lobe, runs a prominent ridge, the **epitaenia,** which extends up to the extreme posterior end of the cavity. It divides the mantle cavity into two chambers — the right

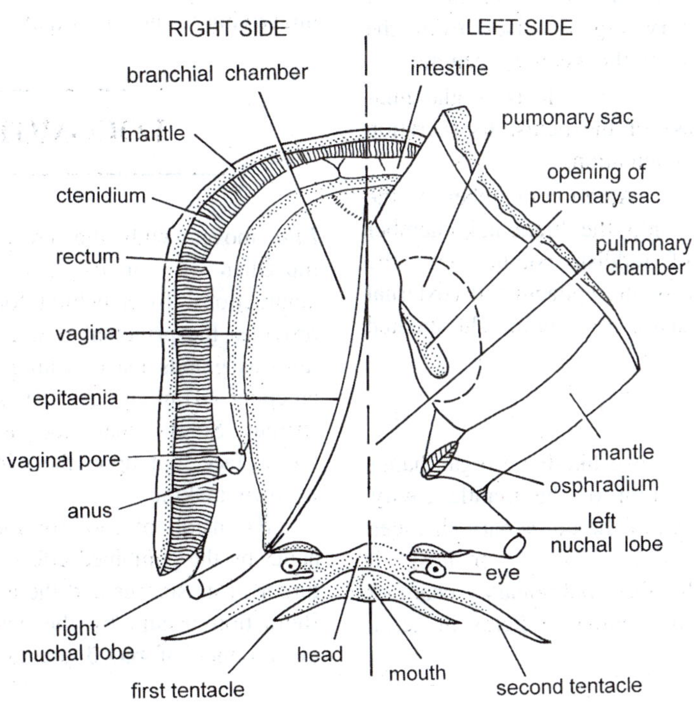

Fig. 5. *Pila.* Mantle cavity and pallial complex.

branchial chamber and the left **pulmonary chamber**. These two chambers play an important role in respiration.

Mantle cavity lodges many important structures, together known as the organs of **pallial complex**. These are described here :

[I] Organs of Branchial Chamber

1. Ctenidium or gill. It lies at the extreme right side of the mantle cavity hanging vertically downwards from its dorso-lateral wall. Being **monopectinate** (comb-like), its gill filaments or **lamellae,** which are somewhat triangular in shape, hang freely in the branchial chamber.

2. Rectum. Rectum lies to the left of the ctenidium on the floor of the branchial chamber. It is a raised tube-like organ, extending from the extreme posterior end of the mantle cavity and terminating a little behind the right nuchal lobe. Its external opening is the anus which is surrounded by a rosette of minute papillae.

3. Genital opening. Male or female genital duct lies close to the rectum. In male, the **penis**, which is a copulatory organ, arises from the mantle edge in front of the genital opening.

4. Hypobranchial gland. It is a glandular thickening at the base of the penis. Its secretion helps in the act of copulation.

5. Renal opening. Anterior chamber of the renal organ projects into the branchial chamber as a small area of reddish colour, near the posterior termination of the epitaenia. Its external opening is like a transversely oblique slit situated in a shallow depression.

[II] Organs of Pulmonary Chamber

1. Pulmonary sac. A bag like-large organ hangs downwards from the roof of the mantle cavity, occupying a larger area of the pulmonary chamber. This organ is the **pulmonary sac** or **lung.** It communicates with the pulmonary chamber through an elongated opening. It helps in aerial respiration.

2. Osphradium. Arising from the mantle, adjacent to the left nuchal lobe and situated on the left side of the pulmonary chamber is a gill-like **osphradium**. It is **bipectinate** (feather-like) and consists of 22 to 28 fleshy and somewhat triangular leaflets. It helps in testing the physical and chemical qualities of the entering water and aids in the selection of food.

INTEGUMENT

Skin varies greatly in thickness. It is made of an **epidermis** and a more complex **dermis**. Epidermis, covering the whole body, consists of a single layer of **epithelium** which is ciliated in parts not protected by the shell. Numerous epithelial cells become modified into unicellular glands from which mucus, pigments and lime are secreted. Gland cells are specially abundant on the mantle edge. Dermis or the **corium** includes muscle fibres and connective tissue and is not clearly marked off from the underlying tissues. It is quite rich in pigment cells.

LOCOMOTION

Pila moves with the creeping activity of the muscular **sole** of its foot. Waves of muscular contractions, from behind forward or vice versa, provides the principal power of movement. Pedal mucous glands leave behind a **slime trail** during creeping. Trail is at first sticky and later on stiffens. So the snail does not move directly on the surface, but upon the slime path, a carpet of its own making.

Extension of foot for locomotion is brought about by the combined effect of the blood pressure turgor in its sinuses and the activity of its muscles. Retraction occurs by the action of muscles and the absence of blood pressure turgor.

DIGESTIVE SYSTEM

[I] Alimentary Canal

Alimentary canal of *Pila* is a coiled tube extending from the mouth and terminating at the anus. Its anterior part is especially modified. Entire canal may be divided into three regions : the **foregut** comprising of buccal cavity and oesophagus, the **midgut** including stomach and intestine, and the **hindgut** consisting of rectum.

1. Buccal cavity. Chamber into which the mouth leads is the buccal cavity. It is lined by cuticle and surrounded by a large, thick-walled, highly muscular and pear-shaped structure, the **buccal mass**. Its wall is provided with several sets of muscles for its movement and the movement of **radula.**

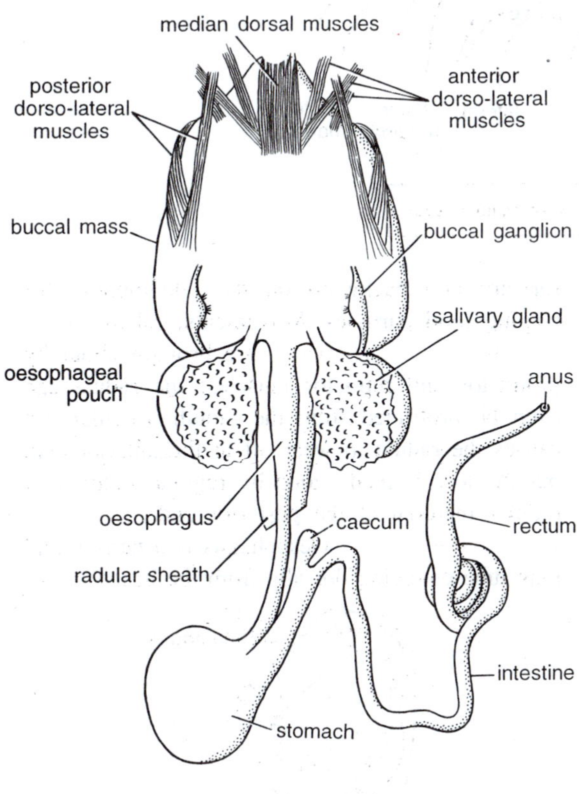

Fig. 6. *Pila.* Alimentary canal (Radular sac shifted backwards below the oesophagus).

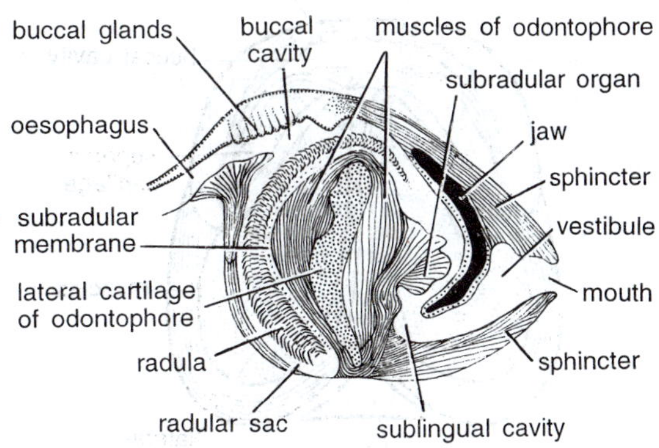

Fig. 7. *Pila.* Buccal mass in V.L.S.

(a) Buccal musculature. Out of several sets of muscles, the **protractors** are well developed. They include : (i) a **median dorsal**, three pairs of **anterior dorso-laterals** and two pairs of **posterior dorso-laterals** on the dorsal surface, and (ii) three **anterior muscles** and a pair of long and strong **latero-ventral** forward muscles on the ventral surface. These muscles are mainly concerned with the protrusion of lateral cartilages and also assist in the protrusion and depression of the buccal mass.

(b) Vestibule and jaws. Buccal cavity is regionated into an anterior tubular part, called **vestibule,** and a posterior part. The posterior limit of the short vestibule is marked by a pair of thickened **jaws,** placed dorso-laterally one on each side and connected together by a thin cuticular membrane. Anterior cutting edge of each jaw is truncated and serrated, bearing numerous small and two or three large tooth-like processes. Wall of the vestibule is beset with longitudinal muscle fibres that form the **mouth sphincter**. Sphincter regulates the opening of the mouth and operates the jaws at the time of feeding.

(c) Odontophore. In the posterior part of the buccal cavity the floor is raised into a thick muscular structure called **tongue mass** or **odontophore.** Structure is supported by two sets of cartilages : (i) a pair of more or less triangular **superior cartilages** lying below the epithelium at the top of the odontophore, and (ii) a pair of

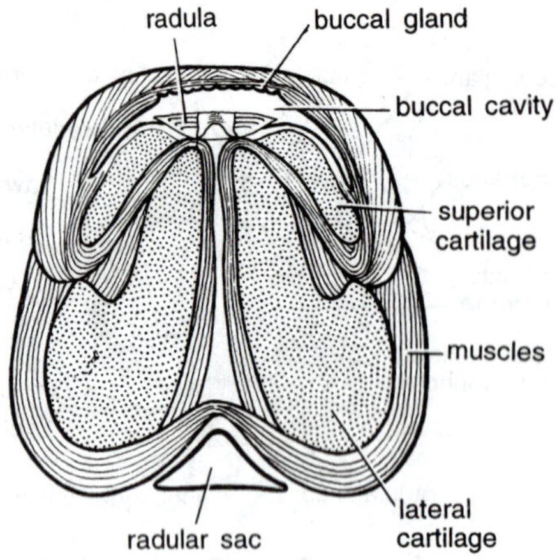

Fig. 8. *Pila.* Buccal mass in T.S.

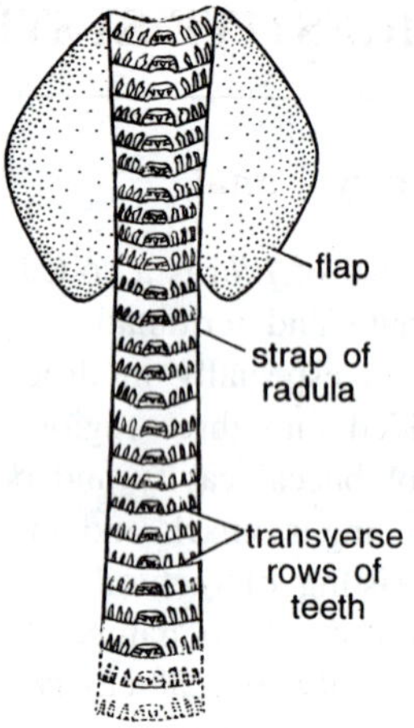

Fig. 9. *Pila.* The radula.

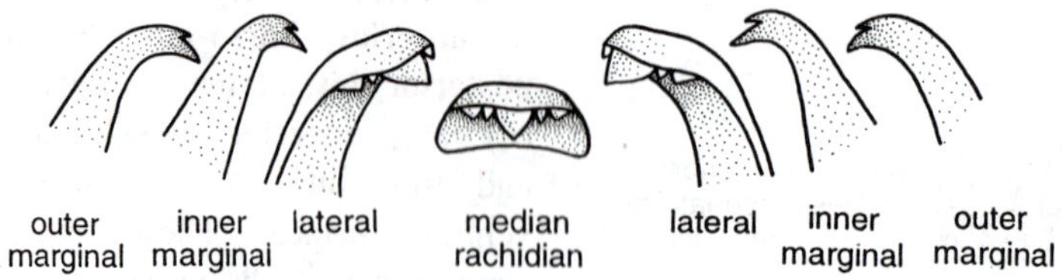

Fig. 10. *Pila.* A single row of radular teeth.

S-shaped **lateral cartilages**, with thick ventral edges and thin dorsal edges, lying on the sides. Anteriorly the odontophore forms a small process, the **sub-radular organ,** roofing a narrow space called the **sub-lingual cavity**.

(d) Radula. Buccal cavity contains a brownish, chitinous, curved, ribbon-like structure, called the **radula** or **lingual ribbon**. Its anterior end, bearing a pair of wing-like flaps, runs longitudinally over the summit of the odontophore. Its posterior end is lodged in a band-like, 2 mm wide **radular sac** flexed behind and below the buccal mass. Radula itself is formed by secretion of the epithelial lining of the radular sac. Below the radula lies a delicate and elastic, **sub-radular membrane**. Dorsal surface of the radula bears **teeth** arranged in numerous transverse rows. Each row contains seven teeth, one **central** (rachidian), and one **lateral** and two **marginals** on its either side, giving the formula 2, 1, 1, 1, 2. Radula is moved forward and backward on the odontophore for rasping food particles. Movements, called **chain-saw movements** (**Huxley**), are brought about by protractor and retractor muscles; the radula can even be protruded from the mouth. Regular use causes the radula to wear off at the anterior end, but the loss is made good by regular addition of radular material at the posterior end.

2. Oesophagus. Oesophagus is a narrow and long tube emerging dorsally from the buccal mass.

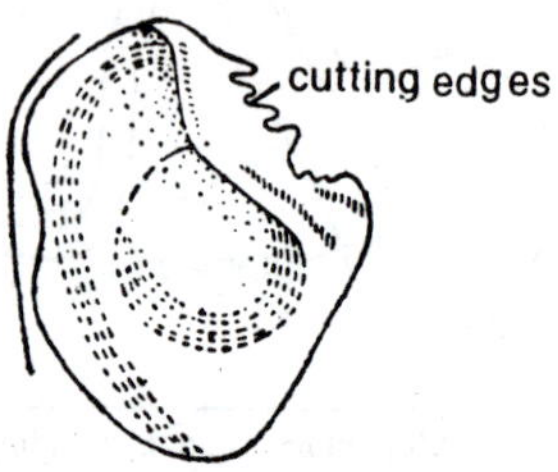

Fig. 11. *Pila.* A jaw.

Running posteriorly for a short distance, it turns to left and enters the visceral mass to open into the stomach.

3. Stomach. Stomach lies on the left side of the visceral mass, below the pericardium. Its cavity is U-shaped which is regionated into a broad posterior **cardiac chamber** that receives the oesophagus and a narrow anterior **pyloric chamber** from which the intestine takes its origin. Lining of the stomach is folded; folds of the cardiac stomach are low and run from right to left, while those of the pyloric stomach are somewhat prominent and run transversely. A short rounded and blind pouch, the **caecum**, arises from the lower outer wall of the pyloric chamber. At the junction of two chambers of the stomach opens a duct from the digestive gland.

4. Intestine. Pyloric stomach is followed by a long and coiled intestine. It runs backward into the visceral mass where it makes 2.5 or 3 coils, between the gonad in front and the digestive gland behind, before joining the rectum.

5. Rectum. It comprises of a thick-walled tube which extends into the branchial chamber of the mantle cavity between the ctenidium and genital duct. Its external opening, the **anus**, lies about 6 mm away from the edge of the right nuchal lobe.

[II] Digestive Glands

1. Salivary glands. These are two in number and lie on either side of the posterior part of the buccal mass. Each gland looks like a branching white mass. A duct from each gland enters the muscles of the buccal mass and then opens into the buccal cavity in the area of the dorsal buccal glands. Salivary secretion contains mucin-like substance and a carbohydrase enzyme.

2. Digestive gland. A somewhat triangular plate or cone with a convex outer and more or less flattened inner surface occupies the greater part of the coiled visceral mass. This structure is a **digestive gland** (often referred to as the **liver** or **hepatopancreas**), which is also coiled and is brownish to dirty green in colour. It has two main

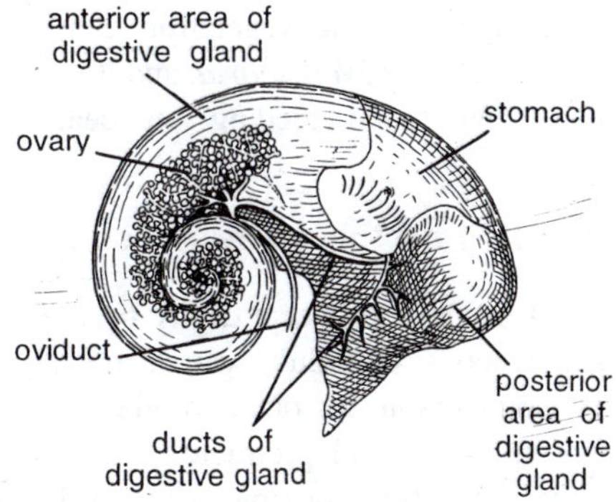

Fig. 12. *Pila.* Digestive gland.

lobes, smaller in contact with the stomach and larger extending to the apex of the spiral. Two separate ducts arise from two lobes which unite together to form a common duct before opening into the stomach. These ducts, open into the digestive gland, branch repeatedly and end blindly in a very large number of small tubes, the **alveoli**. Alveoli are lined with a digestive epithelium made up of three types of cells : (i) **secretory cells** which secrete a cellulose digesting enzyme, (ii) **resorptive cells** which digest proteins intracellularly, and (iii) **lime cells** which store calcium phosphate. Semi-digested food enters into these alveoli, where digestion of cellulose and proteins takes place.

3. Oesophageal pouches. A pair of simple, rounded, cream-coloured oesophageal pouches lies below the salivary glands. Each pouch opens by a narrow duct at the junction of the buccal cavity and oesophagus. These pouches probably secrete digestive enzymes.

4. Buccal glands. These are a pair of glandular areas in the roof of the buccal cavity, a little in front of its junction with the oesophagus. Each glandular area consists of two pads, demarcated by an oblique longitudinal furrow; each pad bears a row of transverse grooves. Exact function of these glands is not known; they are probably of the nature of accessory digestive glands.

[III] Food

Pila feeds upon aquatic vegetation consisting of succulent plants like *Vallisnaria* and *Pistia.* Snail has also been found feeding on dead animal tissues.

[IV] Feeding or Ingestion of Food

Food is taken into the buccal cavity by the **'chain-saw' movements**, of radula, which are limited in *Pila* in comparison to other molluscs. By the action of sphincter and protractor muscles of the buccal mass, the two jaws move up to the mouth opening and cut up leaves of aquatic plants. Meanwhile the radula is also brought forward, the pieces of leaves caught by its teeth and thrown backward into the buccal cavity. Thus, the food is cut and masticated inside the buccal cavity.

[V] Digestion

Salivary glands pour their secretion, by means of their ducts, into the buccal cavity where it mixes with the food. It contains a carbohydrase enzyme, which converts starch into sugar. In the stomach the food is digested by the secretion of the digestive gland, containing enzymes comparable to those of the pancreas in vertebrate animals. Thus, extracellular digestion takes place in the stomach. Digestion also occurs intracellularly inside the digestive gland. Absorption of digested food takes place chiefly in the digestive gland and the intestine. Undigested food from the rectum passes out through the anus into the branchial chamber and finally to the exterior along with the outgoing current of water through the right nuchal lobe or pseudepipodium.

RESPIRATORY SYSTEM

Pila shows both aquatic as well as aerial modes of respiration. To carry on with these two modes of respiration efficiently and successfully, the animal possesses a gill or **ctenidium** for aquatic respiration and a pulmonary sac or lung for aerial respiration. Left and right nuchal lobes act as accessory structures.

[I] Respiratory Organs

1. Ctenidium. It is a **monopectinate** gill (comb-like) and is situated on the right side of the branchial chamber hanging from its dorsolateral wall. It consists of a long axis, the ctenidial axis, which remains attached to the mantle wall. Ctenidial axis is traversed by an afferent blood vessel and an efferent blood vessel. Afferent blood vessel carries deoxygenated blood to the gill and the efferent blood vessel carries oxygenated blood from the gill to the heart. Axis bears a long series of flat triangular leaflets known as **lamellae.** Each lamella is attached by a broad base to the ctenidial axis from which it hangs down into the branchial chamber. Two free sides of the lamella are unequal; the shorter side receives blood from the afferent vessel and is thus called the **afferent side** while the other is called the **efferent side** as from it the blood flows into the efferent vessel. Further, all the lamellae are not of the same size. In the middle the lamellae are larger and decrease in size towards the two ends. Anterior and posterior faces of each lamella are provided with transverse ridges or **pleats,** containing branches of blood vessels, through which the blood flows from the afferent blood vessel to the efferent blood vessel.

Ctenidium is innervated by nerves from the left pleural and supra-intestinal ganglia. This nerve supply indicates that it is actually an organ of the left side but has shifted to the right side during development (torsion).

Histologically each lamella is formed of a double layer of epithelium which encloses a narrow cavity. Each epithelial layer consists of : (i) **non-ciliated columnar cells,** (ii) **ciliated columnar cells,** and (iii) **glandular cells.** At the base of the layers are connective tissue cells and oblique muscle fibres. Next to these lies the pallial epithelium.

2. Pulmonary sac. It is a bag-like structure and developed from the mantle wall. It hangs into the pulmonary chamber of the mantle cavity into which it opens through a large, oblique **pulmonary aperture**. Dorsal wall of the pulmonary chamber is pigmented, while the ventral wall is creamy white. These walls are muscular and highly vascular, consisting of blood spaces, throughout (Fig. 13).

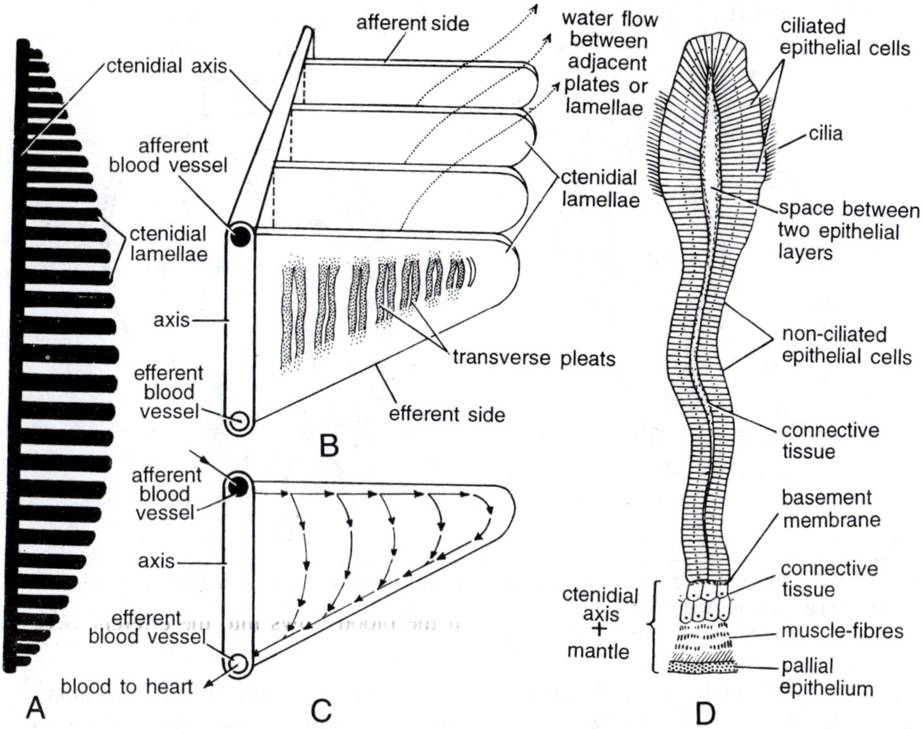

Fig. 13. *Pila.* Respiratory organs. **A.** A monopectinate ctenidium. **B.** Stereogram to show water current through gill lamellae. **C.** A single lamella to show flow of blood wintin it. **D.** A lamella in T.S.

3. Nuchal lobes. These are highly contractile structures of the mantle; situated one on either side of the head. Two nuchal lobes form a sort of drain for the entry and exit of water in and out of the mantle cavity.

[II] Mechanism of Respiration

1. Aquatic respiration. Aquatic respiration takes place when the snail is at the bottom or is floating or lying suspended in midwater or attached to plants and weeds in water. At this time the animal is found to be fully expanded with the head and foot fully extended. Two nuchal lobes form channel-like structures. Left lobe forms a sort of drain through which water flows inwards. Water current first flows beneath the osphradium to the posterior part of the pulmonary chamber, crossing over the epitaenia and then enters the branchial chamber. After washing the entire length of the ctenidium, it flows out through the right nuchal lobe. Flow of water current is maintained by (i) the alternate lowering and raising of the floor of the mantle cavity; this is brought about by alternate protrusion and retraction of the head, and (ii) by the beating of cilia on the lamellae of the ctenidium.

2. Aerial or pulmonary respiration. This mode of respiration is observed when the animal is in water or out on land. When it is in water, it comes nearer the water surface and then extends its left nuchal lobe, which rolls up to form a tube like structure, called **siphon**. Opening of the siphon projects above the water surface and through this air enters the pulmonary sac through the pulmonary aperture. Air always backs through this path. To maintain the in-and-out flow of air, the alternate expansion and contraction of the pulmonary sac takes place. Epitaenia is raised to push the mantle wall and hence the air is not allowed to enter the branchial chamber.

Sometimes, when the water surface is clear, the siphon's margin broadens and forms a sort of

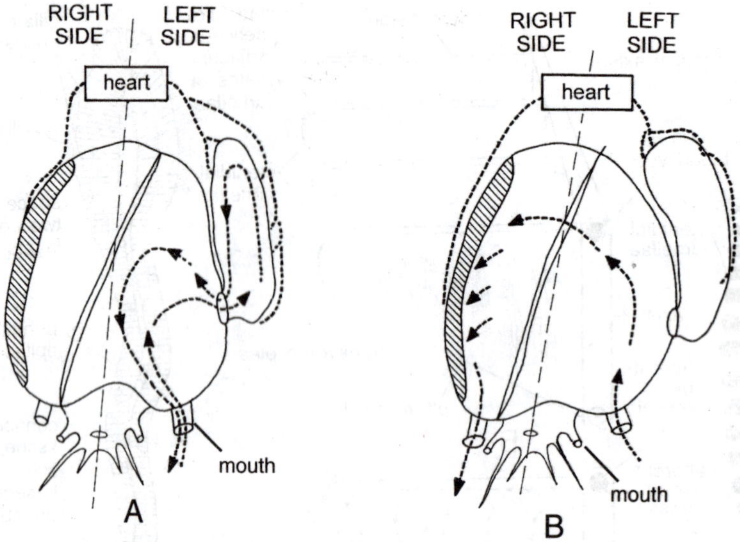

Fig. 14. *Pila.* Mechanism of respiration. **A.** Stereogram showing air current in aerial respiration. **B.** Stereogram showing water current in aquatic respiration.

funnel, which helps to maintain the position of the snail comfortable. Wall of the pulmonary sac contains blood vessels and sinuses whose blood takes up oxygen of the air and gives out carbon dioxide.

Generally, the two modes of respiration go on alternately but one mode of respiration may be prefered to the other depending upon the prevailing conditions.

When the snail travels on land from one source of water to the other, or for the purpose of laying eggs, it may breathe air by the pulmonary method, without forming a siphon of the left nuchal lobe which is simply expanded.

During aestivation it shows extremely decreased activities. Body gets completely within the shell and the aperture is tightly closed with the operculum. Respiration takes place with that much of air which is stored in the pulmonary sac.

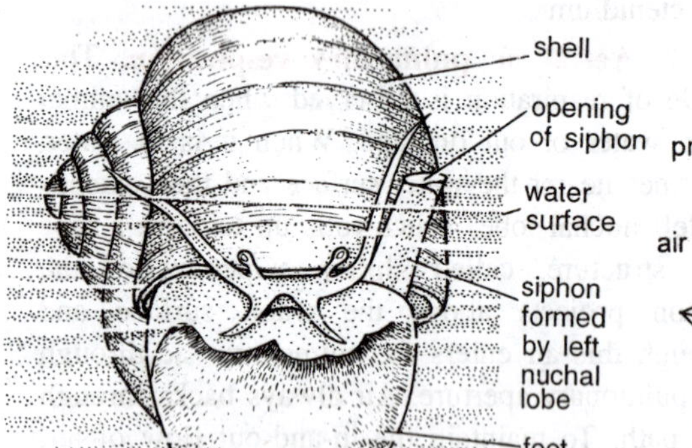

Fig. 15. *Pila* with its siphon up for breathing air.

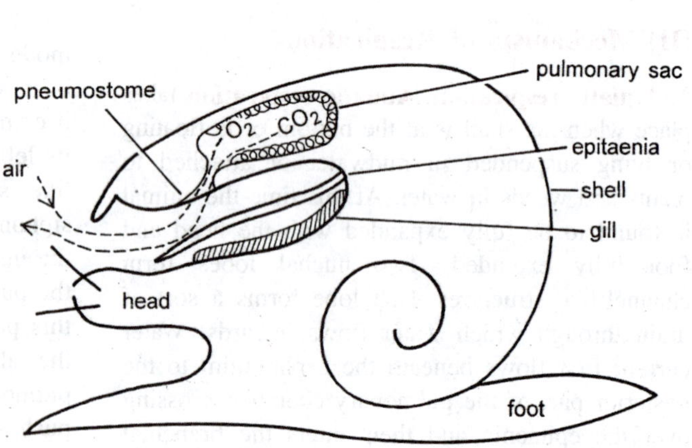

Fig. 16. *Pila.* Pulmonary respiration.

BLOOD-VASCULAR SYSTEM

Circulatory system of *Pila* has attained a great complexity due to its double mode of respiration, involving the gill as well as lung. It consists of : (i) **pericardium,** (ii) **heart,** (iii) **arteries,** (iv) **sinuses,** and (v) **veins.**

1. Pericardium. Pericardium is a thin-walled, roughly ovoid chamber lying dorsally on the left side of the body whorl behind the mantle cavity. It extends anteriorly up to the stomach and the digestive gland. Its fairly deep cavity communicates with the posterior renal chamber through a **reno-pericardial aperture.** Pericardial cavity represents the true coelom as it communicates with the kidneys and in the embryo, reproductive cells develop from its wall.

2. Heart. Heart, lying enclosed within the pericardium, consists of two chambers, an **auricle** and **ventricle.**

(a) Auricle. It is a thin-walled, highly contractile and roughly triangular sac, situated in the dorsal part of the pericardium. Dorsal apex of the auricle receives blood from three main veins, the **efferent ctenidial vein** from the gill, **efferent renal vein** from the posterior renal chamber, and the **pulmonary vein** from the pulmonary sac. Ventrally, the auricle opens into the ventricle through an **auriculo-ventricular opening,** provided with two semilunar valves, so arranged as to permit the flow of blood into the ventricle but not vice versa.

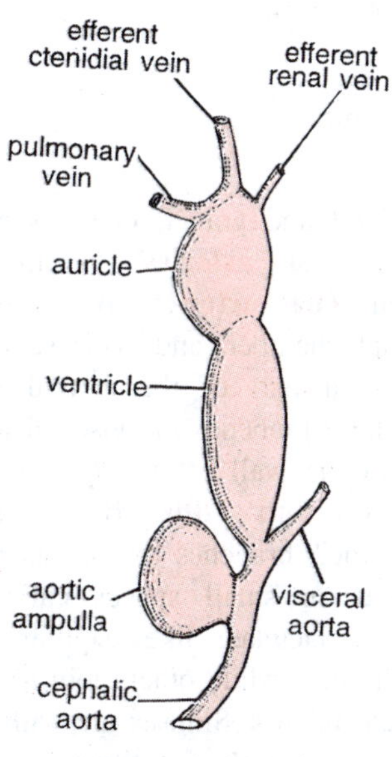

Fig. 17. *Pila.* Heart.

(b) Ventricle. It is an ovoid sac lying below the auricle. It has thick, spongy and muscular walls and a reduced cavity due to a coarse meshwork of muscular strands. From the lower end of the ventricle arises a large artery, the **aortic trunk.** Opening between the ventricle and aorta is guarded by two semilunar valves which prevent the flow of blood back into the ventricle.

3. Arteries. Aortic trunk divides immediately into two branches, an anterior **cephalic aorta** and a posterior **visceral aorta,** each breaking up into numerous arteries.

(a) Cephalic aorta. Base of the cephalic aorta bears a bulb-like, thick-walled and highly contractile structure, the **aortic ampulla,** the presence of which is characteristic of all the members of the family *Pilidae.* Rhythmic contractions of the ampulla help in the circulation of blood. Opening of the ampulla into the aorta is without valves.

Immediately beyond the ampulla, the cephalic aorta gives off, on its outer side, three arteries : (i) a fine **cutaneous artery** to the skin, (ii) a thick **oesophageal artery,** to the oesophagus, and (iii) a stout **left pallial artery** to the left side of mantle, the left nuchal lobe and the osphradium. On the inner side, it gives off a **pericardial artery,** which supplies blood to pericardium and then enters the posterior renal chamber and supplies blood to both the renal chambers (through **renal branches**) and to a part of the genital organs.

Main trunk of cephalic aorta now traverses the perivisceral sinus, runs for a while along the left side of the oesophagus and then crosses over to its right. It now sends numerous small branches to the oesophageal wall and the floor of the mantle cavity. On its right, it gives off a large branch which further divides into three arteries : (i) a **right pallial artery** to the right part of mantle, (ii) a **right siphonal artery** to the right nuchal

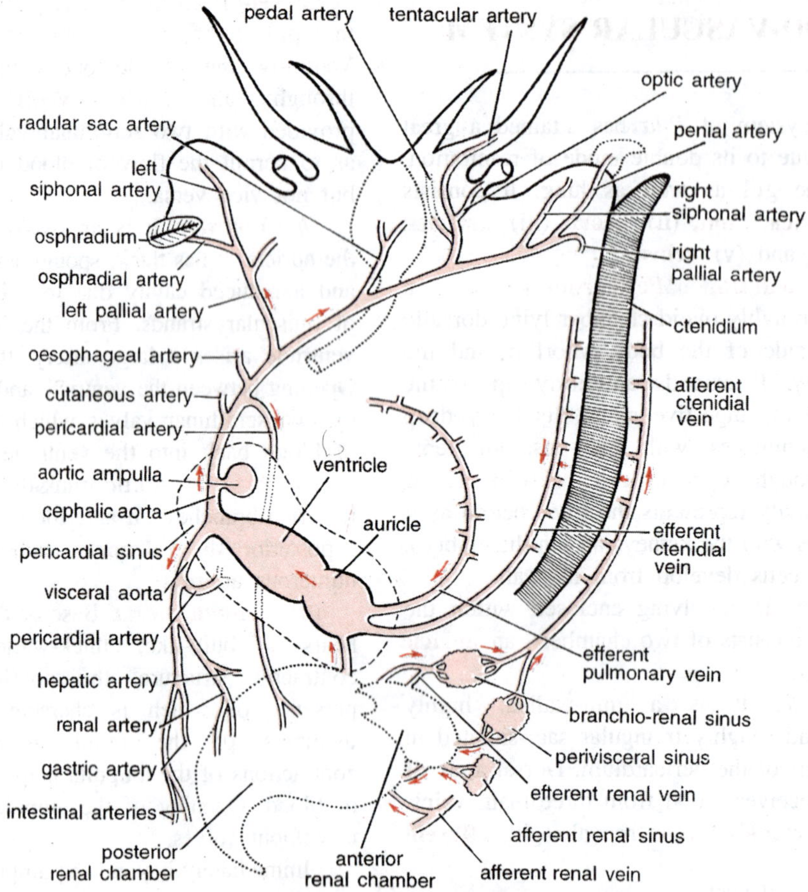

Fig. 18. *Pila.* Heart and blood-vascular system.

lobe, and (iii) a **penial artery** to the copulatory organ.

Main trunk further sends. (i) a **radular sac artery** to the radular sac, (ii) **optic arteries** to the eyes and eye-stalks, (iii) **tentacular arteries** to the tentacles, and finally (iv) **pedal arteries** to the foot into which they form an irregular network.

(b) Visceral aorta. Visceral aorta runs posteriorly into the visceral mass and supplies the visceral organs through a series of numerous branches. These are : (i) a small **pericardial artery** to the pericardium, skin and digestive gland, (ii) a large **gastric artery** to stomach, (iii) several **intestinal arteries** to the intestine, (iv) several **renal arteries** to the roof of the posterior renal chamber, (v) a large artery to the digestive gland and gonad, (vi) a second set of several **intestinal arteries** to the intestine, (vii) several **renal arteries** to the roof of the anterior renal chamber, and (viii) several arteries to the terminal part of the genital duct. After giving out these branches the visceral aorta finally terminates in the wall of the rectum.

4. Sinuses and veins. Blood supplied by arteries and their branches to various parts of the body collects in small spaces called **lacunae**. Some of the lacunae, like capillaries, connect arteries with veins while others join to form larger spaces called **sinuses.** Sinuses are without definite walls. They collectively constitute the **haemocoel**. Blood from the sinuses is drained by **veins**. Veins carry it to the auricle of the heart, either directly or through ctenidium, mantle or kidney. Chief sinuses and the veins draining them are as follows :

(a) Peri-visceral sinus. It lies below the floor of the mantle cavity, surrounding the anterior part of the alimentary canal. It collects blood supplied by the cephalic aorta (except pericardial artery) and some branches of the visceral aorta. Blood from it is mainly drained into the anterior renal chamber and the pulmonary sinus through sinuses. However, some blood from the rectal wall and terminal part of the genital duct passes directly into the afferent ctenidial vein through a number of branches.

(b) Peri-intestinal sinus. It lies along the coils of the intestine. It collects blood from the viscera, digestive gland and greater part of the reproductive system. It extends anteriorly in a tubular form and supplies blood to renal chambers through two branches. Of these, the right branch forms the **afferent renal sinus** in the floor of the anterior renal chamber and the left branch forms the **afferent renal vein** in the roof of the posterior renal chamber.

Blood from the posterior renal chamber is carried by an **efferent renal vein** to the auricle of the heart. Some blood, however, also flows from the anterior renal chamber to the posterior renal chamber.

(c) Branchio-renal sinus. It lies to the right of anterior renal chamber. It collects blood from the renal chambers through efferent branches from their roofs. It extends forward as the **afferent ctenidial vein** and supplies the blood to the ctenidium through numerous branches (one branch to each gill lamella) for aeration. Afferent ctenidial vein receives numerous branches from the rectal wall and terminal part of the genital duct (perivisceral sinus).

Aerated blood from the gill lamella is collected by an **efferent ctenidial vein** and conveyed to the auricle of the heart. Efferent ctenidial vein also receives branches from the mantle and the copulatory organ.

(d) Pulmonary sinus. It lies in the wall of the pulmonary sac. It collects blood from the perivisceral sinus through numerous large veins **(afferent pulmonary veins).** After aeration the blood passes into the main **pulmonary vein** (efferent **pulmonary vein**) that conveys it to the auricle of the heart.

Blood

The blood serves to transport nutrients, oxygen and waste products from one part of the body to the other. It contains some colourless stellate amoeboid cells and a respiratory pigment, called **haemocyanin,** which is dissolved in plasma and

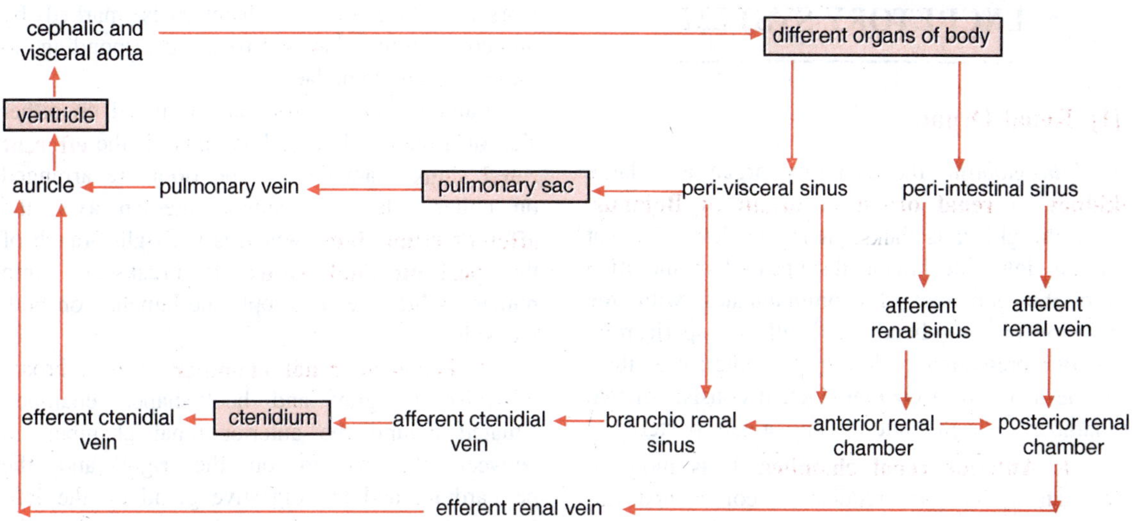

Fig. 19. Diagrammatic representation of the course of circulation in *P. globosa.*

is very common among molluscs and provides a blue tint to the oxidised blood. It is similar to haemoglobin but contains copper rather than iron, and is less efficient. Amoeboid cells are phagocytic in nature. They aid in the removal of waste substances as well as in the intracellular digestion.

Course of Circulation

Ventricle of the heart pumps blood through branches of the **cephalic aorta** to the head, mantle, buccal mass, oesophagus, copulatory organ, foot and columellar muscle and through the **visceral aorta** to the visceral mass. From all parts of the body the venous blood is collected into **perivisceral** and **peri-intestinal sinuses**. From these sinuses the blood is conducted back to the auricle, passing on its way either through the kidney for removal of its nitrogenous waste products or through the ctenidium or pulmonary sac. Major part of the blood passes into the ctenidium, during **aquatic respiration**, and into the pulmonary sac, during **aerial respiration**. Auricle thus gets **aerated** blood from the ctenidium and pulmonary sac, and **non-aerated** blood from the posterior renal chamber. It is this **mixed** blood that passes into the ventricle to be pumped again to all parts of the body.

EXCRETORY SYSTEM

[I] Renal Organ

In *Pila globosa,* the excretory organ is a large **kidney** or **renal organ** or **organ of Bojanus**. Like the gill, it is the organ of the left side, that of the right side having disappeared or modified into the gonoduct. It communicates with the exterior on one hand and with the pericardial cavity representing coelom on the other; it is, thus, of the nature of a coelomoduct. It consists of two chambers, a right anterior and a left posterior.

1. Anterior renal chamber. It is more or less an oval organ, reddish in colour and lies anterior to the pericardium. It opens into the branchial chamber of the mantle cavity through a slit-like opening near the epitaenia. At the other

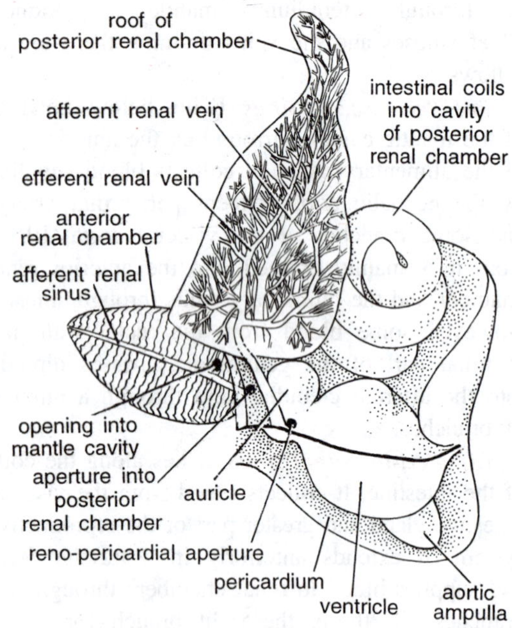

Fig. 20. *Pila.* Excretory organs.

end, it communicates with the posterior renal chamber through an internal opening. Internal cavity of the anterior chamber is very much reduced due to the presence of many triangular leaf like processes or **lamellae,** those arising from the roof alternating with those from the floor. Dorsal surface of the chamber is marked by numerous transverse grooves, corresponding to these internal lamellae.

Lamellae on the roof are arranged on either side of a median longitudinal axis, or the **efferent renal sinus**. Lamellae on the floor are arranged on either side of a similar median axis, the **afferent renal sinus**, which is the right branch of the **peri-intestinal sinus**. It breaks up into numerous branches to supply the lamellae on both the sides.

2. Posterior renal chamber. It is a broad, brownish to grey and hook-shaped chamber, situated behind the anterior renal chamber, in between the rectum on the right and the pericardium and the digestive gland on the left. Its large internal cavity encloses a part of the genital duct and a few coils of the intestine. At one end, it communicates with the anterior renal chamber through an aperture, and at the other with the pericardium through an elongated slit-like **reno-pericardial aperture,** perforating a thin vertical **reno-pericardial septum**, separating the two **afferent** and **efferent renal vessels** profusely branch in the roof of this chamber.

[II] Physiology of Excretion

Two renal chambers are richly supplied with blood from which the nitrogenous waste products are separated. Excretory fluid from the posterior chamber is also transferred to the anterior chamber, from where it is discharged through the external renal aperture into the mantle cavity and finally passed out of the body through the right nuchal lobe along with the outflowing water. Excretory fluid contains mostly ammonia and some ammonium compounds, urea and uric acid. *Pila* shows an adaptation for water conservation during terrestrial phase by converting ammonia into the insoluble uric acid. During aquatic phase, *Pila* excretes ammonia, but during terrestrial phase, it excretes uric acid. So it is both **ammonotelic** as well as **uricotelic.**

Digestive gland of most gastropods also forms an organ of excretion. It contains some excretory cells which engulf and store the excretory wastes and later pass out through the stomach and intestine.

NERVOUS SYSTEM

Nervous system of *Pila* consists of paired ganglia, commissures and connectives uniting them, and nerves running from these central organs to all parts of the body.

1. Ganglia

Paired ganglia, which are aggregations of nerve cells, are as follows :

1. Cerebral ganglia. A pair of roughly triangular ganglia, situated anteriorly on the dorso-lateral sides of the buccal mass.

2. Buccal ganglia. A pair of small, triangular ganglia, lying dorso-laterally one on either side at the junction of the buccal mass and oesophagus, partly embedded in the muscles.

3. Pleuro-pedal ganglia. A pair of large, somewhat triangular ganglionic masses present one on either ventro-lateral side of the buccal mass. Each one is formed by the fusion of an outer **pleural** and an inner **pedal** ganglion, separated by a faint notch. Right pleuro-pedal ganglionic mass also consists of the **infra-intestinal** ganglion fused with it.

4. Supra-intestinal ganglion. An unpaired fusiform ganglion, lying in a sinus behind the left pleuro-pedal ganglionic mass.

5. Visceral ganglia. A single ganglionic mass representing two fused ganglia, situated at the lower end of the visceral mass.

2. Commissures

Commissures are those nerves which establish connections between two similar ganglia. In *Pila* these constitute :

1. Cerebral commissure. A thick band of nerve connecting two cerebral ganglia and lying dorsally to the buccal mass.

2. Buccal commissure. A fine nerve which connects the two buccal ganglia and runs transversely on the ventral side of the oesophagus.

3. Pedal commissures. Two thick nervous bands that lie one above the other underneath the buccal mass and connect the two pedal ganglia together.

3. Connectives

Connectives are those nerves which connect two different ganglia. In the nervous system of *Pila,* these are :

1. Two cerebro-buccal connectives. These connect, on either side, the cerebral ganglion and buccal ganglion together.

2. Two cerebro-pleural connectives. These connect, on either side, the cerebral and outer pleural ganglion of the pleuro-pedal ganglionic mass.

3. Two cerebro-pedal connectives. These connect, on either side, the cerebral and inner pedal ganglion of the pleuro-pedal ganglionic mass.

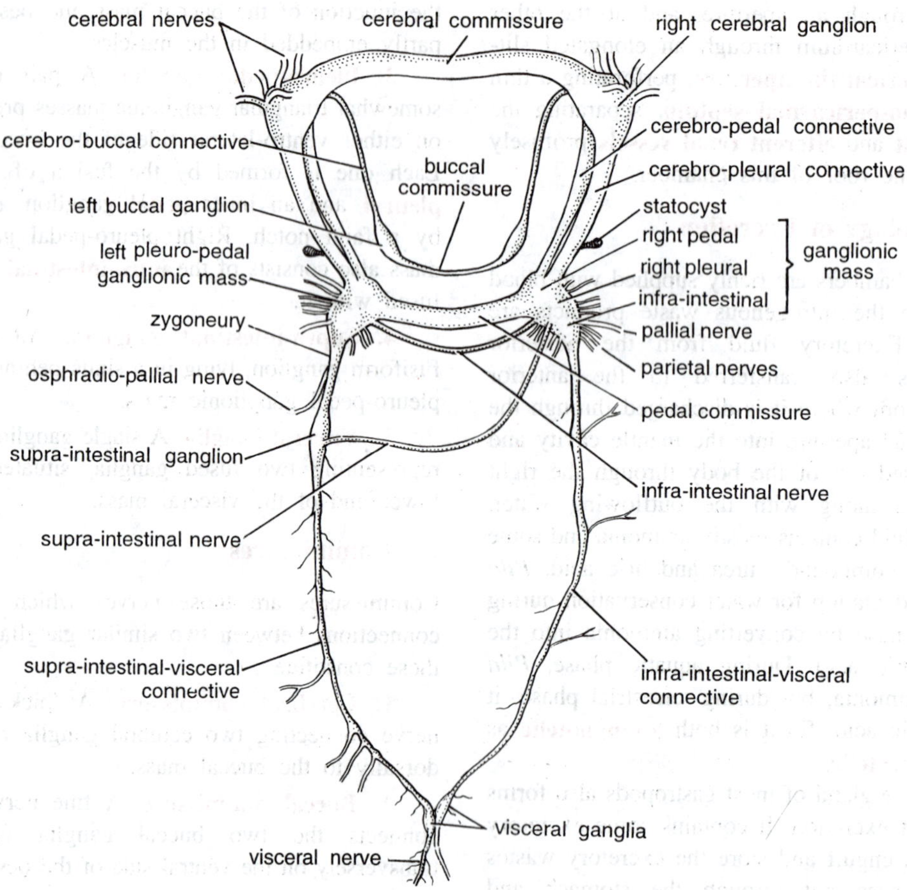

Fig. 21. *Pila.* Nervous system.

4. Pleuro-infra intestinal connective. Also called **infra-intestinal nerve**, it is a nerve connection between the pleural ganglion of the left pleuro-pedal mass and the infra-intestinal ganglion which is fused with the right pleuro-pedal mass.

5. Infra-intestinal visceral connective. Running below the intestine, it is a long nerve that connects the visceral ganglion with the infra-intestinal part of the right pleuro-pedal-infra-intestinal ganglionic mass.

6. Supra-intestinal visceral connective. Running above the intestine, it is a slender nerve that connects the visceral ganglion with the supra-intestinal ganglion.

7. Supra-intestinal-pleural connective. Also called the **supra-intestinal nerve,** it connects the supra-intestinal ganglion with the pleural part of the right pleuro-pedal-infra-intestinal ganglionic mass.

8. Zygoneury. It is a nerve connection between the pleural part of the left pleuropedal ganglionic mass and supra-intestinal ganglion.

4. Nerves to Different Parts of the Body

Various ganglia send nerves to innervate different parts of the body : (i) Each cerebral ganglion gives off nerves, supplying the snout, skin, tentacle and buccal mass anteriorly and the tentacle, eye and statocyst, posteriorly, (ii) Buccal ganglion, of each side, sends nerves to innervate the buccal mass, radular sac, salivary glands, oesophagus and oesophageal pouches, (iii) Pedal ganglia give off nerves, anteriorly as well as posteriorly, to innervate the foot. *Statocyst,* on each side, is also connected, by a band of, connective tissue, to each pedal ganglion. (iv) Left pleural ganglion innervates the parietal wall, mantle, osphradium, left nuchal lobe, columellar muscle and anterior part of the gill. (v) Right pleural ganglion innervates the parietal wall, epitaenia, right nuchal lobe, copulatory organ, columellar muscle and rectum. (vi) Supra-intestinal ganglion gives off a stout nerve to innervate the mantle and the anterior part of ctenidium, while its connective with the left pleural ganglion sends a few nerves to the parietal wall, (vii) Nerves from the visceral ganglion supply the renal organ, genital organs, pericardium, stomach, intestine, digestive gland, etc.

SENSE ORGANS

A snail is diffusely sensitive for the sensory cells are distributed all over the head, foot and various other parts of the body. Special organs of sense include a single **osphradium** and paired **eyes, statocysts, labial palps** and **tentacles**.

[I] Osphradium

Single **osphradium** is situated on the left side of the animal suspended from the roof of the mantle cavity close to the entrance through the left nuchal lobe. It is a small, elongated, oval structure, about 6-7 mm long. While broadest in the middle, its inner or right end is bluntly rounded and the outer or left end is somewhat pointed. It is bipectinate consisting of 22-28 thick, fleshy, and roughly, triangular **leaflets,** arranged in two rows along a slightly raised **median** or **central axis**. Leaflets are largest in the middle of the osphradium. Each leaflet is attached to the mantle wall by its broad base, to the central axis by its smaller inner side, while its outer longer side remains free. Osphradium is supplied by a nerve from the left pleural ganglion.

In a transverse section, the osphradium consists of an outermost covering of a single layered **epithelium,** internally lined by a thin **basement membrane**, the interior filled up with **nerves, connective tissue** and **blood spaces**. Epithelial cells are elongated possessing basal nuclei, and they are of three types : (i) **sensory,** (ii)**ciliated** and (iii) **glandular.** The ciliated cells line the attached margin, while the sensory cells devoid of cilia cover the osphradium. Flask-shaped glandular cells are found scattered among the sensory cells.

Osphradium hangs like a curtain in the path of the respiratory water current, and probably serves as an **olfactory organ**. Its name has been derived from a Greek word, meaning to **smell.** It serves to test the chemical nature of the inspiratory water current. In case the water is foul, its entry into the mantle cavity is stopped by the closure of the left nuchal lobe. It may also help in the selection of the food material.

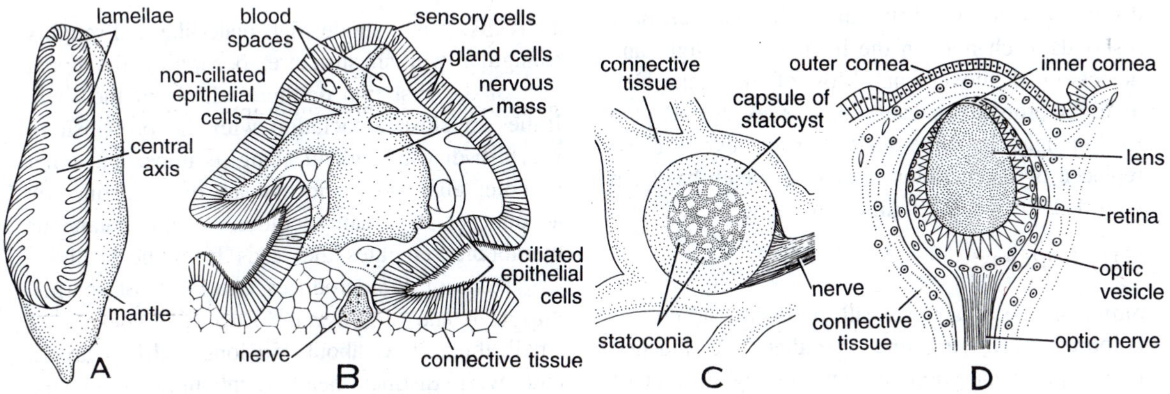

Fig. 22. *Pila.* **A.** Osphradium. **B.** Osphradium in T.S. **C.** Statocyst. **D.** H.L.S. of eye.

[II] Eyes

Snail's head carries a pair of short fleshy and stump-like **stalks** or **ommatophores**, one on either side, behind the second pair of tentacles. Each ommatophore bears a small, black and circular **eye**, slightly below its tip on the outer side.

In a section, an eye looks a pyriform cup-like invagination of epidermis, called the **optic vesicle,** surrounded by a firm **sheath of connective tissue. Optic nerve** enters the vesicle somewhat obliquely at its posterior end. Cavity of the vesicle is filled up by an oval, structureless hyaline body, the **lens**. Wall of the vesicle, formed by a single layer of epithelium, is modified posteriorly to form a sensory and pigmented **retina** containing two kinds of large cells : (i) broad **visual cells** having a brush of hair like processes on their outer ends; (ii) narrow **supporting** or **packing cells** lying in between the visual cells. Anterior continuation of retina, in front of the lens, becomes transparent to form the **inner cornea** or **pellucida interna**, consisting of low cubical and non-pigmented cells having small spherical nuclei. General epithelial covering of the body and the eye-stalks becomes thin and transparent in front of the eye, forming its outer wall termed the **outer cornea** or **pellucida extema,** consisting of similar small, squarish, non-sensory, non-pigmented and non-glandular cells with homogeneous protoplasm.

In spite of their elaborate structure, eyes of *Pila* are probably not true organs of sight. Sense of sight is greatly limited in range and the snail does not seem to distinguish objects, but only responds to changes in the intensities of light and detects quick movements. Most of the snails feed at night probably because their eyes are adapted to dim light. In some snails the lost eyes can be replaced by regeneration, a process which has been recorded to occur 20 times in succession.

[III] Statocysts

Statocysts are a pair of small, pyriform and cream-coloured sacs, lying one on either side attached to the pedal ganglion of that side by a band of connective tissue. Each statocyst lies in a depression, posterior and outer to the ganglion. Each statocyst is a hollow capsule surrounded by an outer thick, tough, leathery **covering** of connective tissue. Wall of the capsule is made of a single layer of **ectodermal cells**, and supplied by a **nerve** from the cerebral ganglion. Cavity of the capsule is filled with a **fluid** and a variable number of minute, oval and calcareous particles, the **statoconia.** Statocysts are organs of **equilibrium.**

[IV] Tentacles

As already referred to the snout of *Pila* is anteriorly prolonged into a pair of short, contractile and conical processes bordering the mouth. These are the **labial palps** or the **anterior** or **first pair of tentacles.** Behind them arise, one on either side, a pair of long, tapering, filamentous and highly contractile whip-like processes, the **true** or **second** pair of **tentacles.**

Tentacles are of the same colour as the snout and **tactile** in function. A sense of **taste** is doubtfully attributed to the labial palps.

REPRODUCTIVE SYSTEM

Pila is **dioecious**, *i.e.*, sexes are separate. **Sexual dimorphism** is slight but distinct. Male has a smaller shell, a less swollen body whorl, and a more developed copulatory organ than the female.

[I] Male Reproductive System

1. Testis. It is a single, flat, plate-like and roughly triangular whitish structure, occupying the upper part of the first two or three whorls of the shell. It lies in close association with the brownish or greenish digestive gland along its upper and inner or columellar sides and appears quite distinct on account of its creamy colour. A thin cutaneous membrane separates the testis from the shell.

Testis of *Pila* produces two kinds of sperms, **eupyrene** and **oligopyrene**. **Eupyrene sperms** are small, thread-like, about 25μ long and 1.2μ broad, and with distinct head (containing a twisted nucleus), middle piece and tail with a single cilium. They are motile and can fertilize the ova.

The **oligopyrine sperms** are large, spindle-like, about 32.5μ long 3μ broad with a distinct head (containing a broad curved nucleus), middle piece, and tail of 4 or 5 cilia. They are non-motile, incapable of fertilizing ova and probably have some other function.

2. Vasa efferentia. Several fine ducts, the **vasa efferentia**, arise from the different regions of the testis and unite to form a large common duct, the **vas deferens.**

3. Vas deferens. It emerges from the posterior end of the testis and is differentiated into three morphologically distinct regions : (i) proximal tubular part, (ii) vesicula seminalis, and (iii) terminal glandular part. **Proximal** part is narrow and thin-walled. It runs anteriorly along the inner or columellar border of the digestive gland upto the posterior renal chamber, and then it runs to the left. After reaching the pericardium it rises upwards and turning to the right opens into the ventral side of the **vesicula seminalis**, situated to the right of the pericardium below the junction of the two renal chambers. Vesicula seminalis is a somewhat curved, swollen and flask-shaped structure with a blind rounded posterior prolongation. It serves as a storehouse for the sperms. Its narrow tubular end on the left side leads into the terminal, thick-walled, glandular part of the vas deferens, which runs forward into the left side of the rectum. It finally opens by the **male genital aperture,** a little behind the anus, at the tip of a claw shaped **genital papilla** which is a miniature second penis, according to Bahl.

4. Copulatory organ. Copulatory organ or **penis** arises separately from the mantle edge in front of the anus. It is a long, stout, slightly curved and flagellar structure with a swollen base and a tapering free end. It is deeply grooved throughout its length on the inner side. It is highly extensile, being about 1.5 cm long but attains a length of 4 cm during copulation.

Penis remains enclosed in a **penis sheath** formed by a thick, glandular flap of a yellow colour attached to the mantle along its side, while its free left edge is rolled up to form a spout-like covering for the penis.

5. Hypobranchial gland. At the base of the penis sheath is an oval glandular thickening with pleated surface, the **hypobranchial gland**. It consists of tall cells containing small basal nuclei. Gland is not provided with a duct so that its secretions are released directly upon its surface.

[II] Female Reproductive System

1. Ovary. Ovary occupies the same position as the testis in male, but is less extensive and lies mostly towards the inner side of the digestive gland. It is much branched and coloured light-orange in the young, becoming darker in mature individuals. Branches of the ovary, or **acini,** are single layered, and flask-shaped structures, with their closed rounded ends projecting outwards and the slender tubular necks uniting into a single oviduct.

2. Oviduct. Narrow and transparent **oviduct** arises from about the middle of the ovary. It runs anteriorly just below the skin along the inner margin of the digestive gland. Near the renal organ

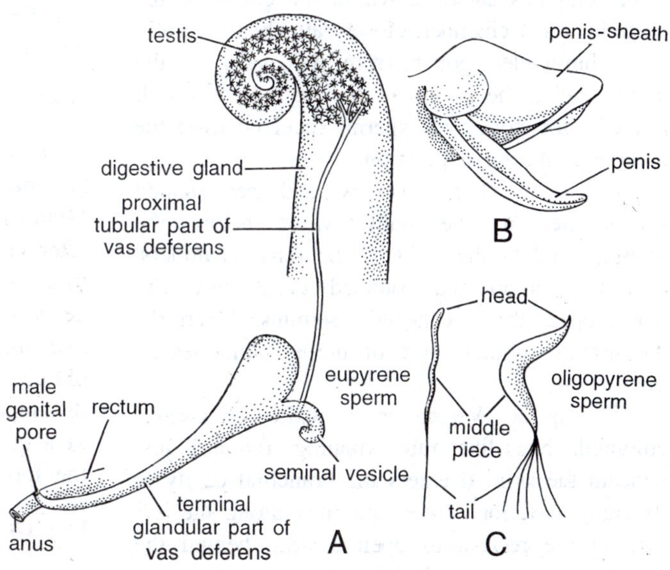

Fig. 23. *Pila.* **A.** Male reproductive organs. **B.** Male copulatory organs. **C.** Sperms.

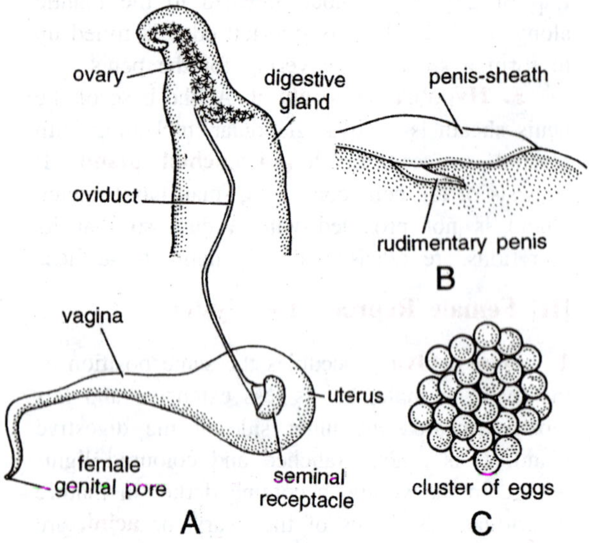

Fig. 24. *Pila.* **A.** Female reproductive organs. **B.** Rudimentary penis. **C.** Eggs.

it turns downwards and then upwards to open into the **receptaculum seminis**.

3. Receptaculum seminis. Small, bean-shaped **receptaculum seminis** (or seminal receptacle) lies enclosed within the cavity of the posterior renal chamber, closely attached ventrally to a thin-walled pouch from the walls of the uterus, called the **pouch of the receptaculum**. It is meant for storing the sperms received from the male *Pila* during copulation.

4. Uterus. Large yellow and pear-shaped **uterus** lies in the body whorl below the intestine and to the right of the renal chambers. It has a broad and rounded basal part into which opens the receptaculum seminis. Anteriorly, the narrow tubular apex of uterus continues as the **vagina**.

5. Vagina. Vagina is a white or cream-coloured, band-like tube running forward just beneath the skin. It enters the branchial cavity at its right posterior corner and runs along the left side of the rectum to open a little behind the anus by a narrow, slit-like **female genital aperture** situated on a small **papilla**.

6. Copulatory apparatus. Rudimentary **penis** is not enclosed within a well-developed **penis-sheath**, but lies beneath a glandular fold of the mantle. It is a thin, flagellar structure nearly 6 mm long and uniform in thickness but pointed at the tip. It also bears a rudimentary groove along its inner surface, like the penis of the male.

7. Hypobranchial gland. Hypobranchial gland of the female is poorly developed with a rudimentary glandular thickening.

[III] Copulation

Pila breeds during the rainy season, when the long period of aestivation comes to an end. **Copulation** occurs under water or on the moist ground of the banks and may continue for three or four hours. During copulation, the male and female snails approach each other in such a manner that the right nuchal lobe of one comes to lie opposite to that of the other. In the male the genital papilla, bearing the male genital aperture, becomes connected to the base of the penis which, together with its sheath, extends into the mantle cavity of the female. Tip of the penis enters the female genital aperture and the sperms from the vesicula seminalis of the male are transferred to the receptaculum seminis of the female.

[IV] Ovulation

Fertilization is internal but the **development** of the embryo takes place outside the body of female. **Ovulation** or **oviposition** commences two days after copulation. 200 to 800 eggs are laid at a time in a sheltered cavity on the moist bank but never under water. Eggs come out one by one from the female genital aperture, travel down an oblique tube formed by two folds temporarily developed on the right side of the foot, and laid in a cavity in the earth, over which the sole of the foot forms a dome.

[V] Development

Parents do not incubate or look after the eggs. These are left to develop by themselves. They are round and as big as the pea-seeds. Each egg consists of an outer whitish shell, enclosing a double shell membrane, a thick mass of solid white albumen and a little central mass of liquid albumen containing the embryo. During development the visceral mass and the shell of embryo become spirally coiled and a characteristic phenomenon of **torsion** takes place resulting in the asymmetry of the body. Young snails emerging from the eggs are similar to the adults.

Important Questions

➤➤*Long answer type questions*

1. Describe the morphology and histology of the shell of *Pila.*
2. Give an account of the various structures found in the mantle cavity of *Pila.*
3. Describe the radular apparatus and digestive organs of *Pila.*
4. Describe the digestive system and feeding mechanism of *Pila.*
5. Describe the respiratory organs and mechanism of respiration in the common apple snail.
6. Give an account of the blood vascular system of *Pila.*
7. Describe the nervous system of *Pila* and explain how it has been affected by torsion.
8. Give an account of the reproductive organs of *Pila.*
9. Write short notes on : (i) Radula, (ii) Osphradium, (iii) Organ of *Bojanus,* (iv) Organs of pallial complex, (v) Odontrophore.

➤➤*Short answer type questions*

1. Classify *Pila* upto order.
2. What term has been given to the crossed condition of the pleuro-visceral connectives.
3. Mention the physiological role of osphradium.
4. How do *Pila* survive even in paddy fields ?
5. Describe the heart of *Pila.*
6. Describe the adaptations in *Pila* to an amphibious mode of life.
7. Describe the nervous system in *Pila.*
8. Draw a labelled diagram of the organs seen in the mantle cavity of *Pila.*
9. What are the three distinct layers of a molluscan shell ?

➤➤*Fill in the blanks*

1. In *Pila* mantle secretes .....
2. In *Pila* aerial respiration occurs through .....
3. Chemically molluscan shell is formed of .... and ....

➤➤*Multiple choice questions*

1. Fresh water apple snail is :
   (a) *Unio* (b) *Pila* (c) *Doris* (d) *Aeolis*
2. Aerial respiration in *Pila* occurs from :
   (a) mantle (b) ctenidium
   (c) pulmonary sac (d) branchiae
3. Excretory organ of *Pila* is :
   (a) Keber's organ (b) kidney
   (c) renal organ (d) none
4. Organs of *Bojanus* is the excretory organ of :
   (a) *Unio* (b) *Pila* (c) both (d) none
5. Respiratory pigment in Mollusca is :
   (a) haemoglobin (b) erythrocin
   (c) haemocyanin (d) none
6. How many teeth are found in a row of radula :
   (a) 7 (b) 5 (c) 9 (d) 11
7. Torsion is shown in :
   (a) *Unio* (b) *Pila* (c) *Loligo* (d) *Mytilus*
8. The shell coiling right handed is known as :
   (a) dextral (b) sinistral
   (c) anticlockwise
9. Ctenidium perform respiration is :
   (a) air (b) water
   (c) both the place (d) terristrial
10. Blood of *Pila* contains :
    (a) RBC (b) WBC
    (c) amoebocyte (d) monocyte
11. *Pila* is :
    (a) monoecious (b) dioecious
    (c) hemaphrodite
12. In *Pila* fertilization is internal and development :
    (a) internal (b) external
    (c) partially internal
13. Shell is dextral, if the coiling is :
    (a) right handed (b) left handed
    (c) anticlockwise (d) none of these
14. *Pila* is :
    (a) carnivorous (b) herbivorous
    (c) scangivorous (d) omnivorous

15. Osphradium is the associated with :
    (a) digestion (b) respiration
    (c) excretion
    (d) olfactory sensation
16. Statocyst organs of :
    (a) excretion (b) olfactation
    (c) secretion (d) equilibrium
17. Gill is monopectinate in :
    (a) *Pila* (b) *Unio*
    (c) *Chiton* (d) *Mytilus*
18. Mode of respiration in *Pila* is :
    (a) aquatic (b) aerial
    (c) none (d) both
19. Aerial respiration in *Pila* performed is performed by :
    (a) book lungs (b) gills
    (c) pulmonary (d) pulmonary chamber
20. Radula is present in :
    (a) *Pila* (b) *Unio*
    (c) *Loligo* (d) *Mytilus*

Answers

➤➤*Fill in the blanks*

1. shell, 2. siphon, 3. conchiolin and calcium carbonate.

➤➤*Multiple choice questions*

1. (b)	2. (c)	3. (c)	4. (c)	5. (c)	6. (a)	7. (b)	8. (a)	9. (b)	10. (c)
11. (b)	12. (b)	13. (a)	14. (b)	15. (d)	16. (d)	17. (a)	18. (d)	19. (d)	20. (a)

57

Unio (*Lamellidens*) : Freshwater Mussel

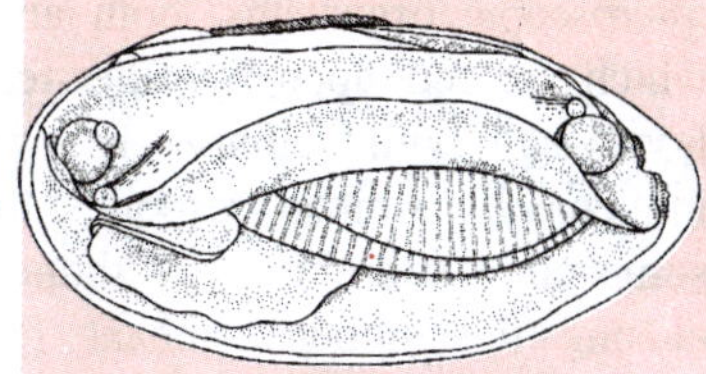

Class **Bivalvia** or **Pelecypoda** (Gr., **pelekus**, hatchet + **podos**, foot) of phylum **Mollusca**, includes forms which are enclosed within a bivalve shell and possess hatchet-like foot. Most familiar members of this class are mussels, clams, oysters and scallops. *Unio* and *Lamellidens* are the common freshwater mussels, found in the Indian lakes and rivers. The following account relates to *Unio, Lamellidens* and *Anodonta.*

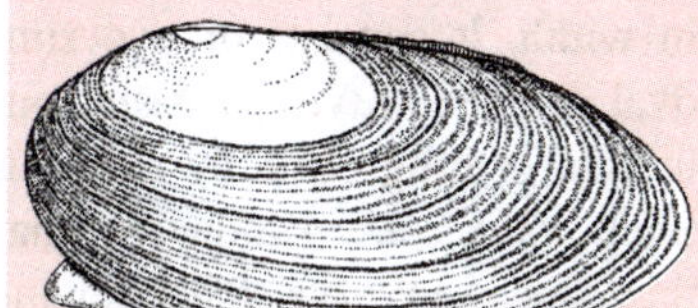

SYSTEMATIC POSITION

Phylum	Mollusca
Class	Pelecypoda (Bivalvia)
Order	Eulamellibranchia
Family	Unionidae
Genus	*Unio* or *Lamellidens* or *Anodonta*

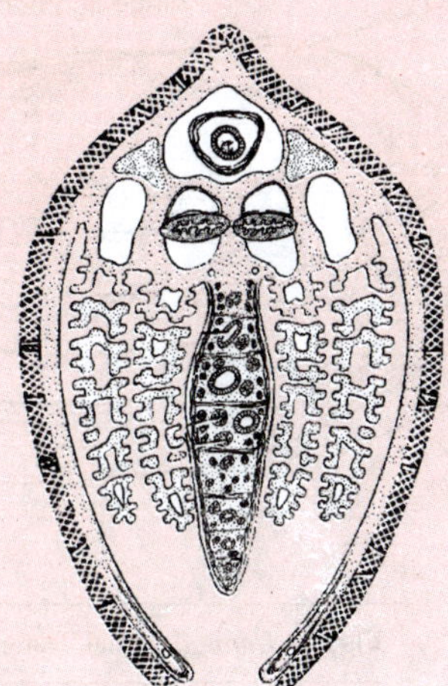

HABITS AND HABITAT

Freshwater mussels are found in freshwater lakes, rivers and streams, inhabiting the surface layers of the muddy beds of rivers and lakes. In a buried position the posterior tip of its shell remains exposed in water to facilitate entry and exit of the water current, which is helpful both in breathing and feeding. It crawls slowly with the help of its plough-like, wedge-shaped muscular foot that leaves a deep trail all along its journey. It usually moves to shallow places by night and retires to deeper places by day. Food consists of microscopic organisms, both animals and plants, which are fed upon by filter-feeding mechanism involving both ciliary as well as mucoid movements. Animal responds to light, touch or some other stimulus by withdrawing its foot and closing the siphons, meant for incoming and outgoing water currents.

MORPHOLOGY

[I] Shell

Soft-bodied animal is completely enclosed within a calcareous **shell** which represents its exoskeleton. Shell measures about 10 cm in length and 5 cm in width. It consists of two similar, more or less oval, convex **valves,** that are joined at their dorsal margin or **hinge line** by a strong **hinge ligament**. Shell valves are similar in shape and size, *i.e.,* they are **equivalve**. This ligament is elastic and causes the shell to gape ventrally. Dorsally and somewhat anteriorly, each shell valve has a slightly raised part, called the **umbo**. It represents the oldest part of the shell and concentric lines around it are the **lines of shell growth,** representing intervals between successive growth stages. Anterior end of the shell is somewhat rounded and through its antero-ventral margin may be protruded the muscular foot for ploughing into the mud or sand. Posterior end is tapering and projecting, behind it can be seen two short tubes or **siphons,** one for the entry and other for exit of water current.

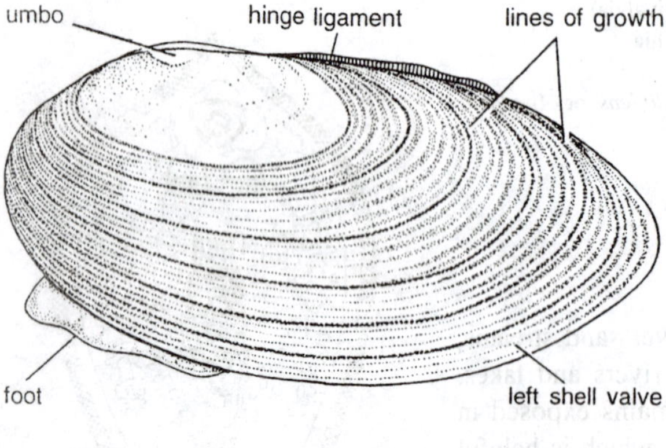

Fig. 1. *Unio.* External features (lateral view).

Internally, the shell valves are attached to the mantle of the body with three anterior and two posterior sets of muscles, constituting the anterior **protractor, retractor** and **adductor** muscles, and posterior **retractor** and **adductor** muscles. On the inner surface of each valve are seen characteristic markings or **scars** of the attachment of these muscles. Near each end, anteriorly and posteriorly are two large and oval scars of the **anterior** and **posterior adductor muscles**. Near the anterior one are two additional small scars, the dorso-posterior scar being of the **anterior retractor muscle** and postero-ventral scar of the **anterior protractor muscle.** A small scar of the **posterior retractor muscle** also lies dorsal to the scar of the posterior adductor muscle. Running inside and parallel to the free ventral margin of the valve, from one adductor's scar to the other, is a fine line, the **pallial line**. This line marks the area of the valve, along which the muscles in the free border of the mantle lobe were attached. At the dorsal margin of the inner surface occur two blade-like processes, the **hinge teeth,** one near the umbo and the other along the hinge line, which hold the valves from slipping out of position. This becomes possible by locking together of the two valves by hinge teeth and opposing sockets.

Histology of the shell. The shell of *Unio* is composed of three layers : (i) The outer **periostracum,** a thin horny layer of an organic substance, **conchiolin,** which imparts brownish colour to the shell. It is subjected to erosion by the action of carbonic and humic acids present in water and is thus often absent on the umbo which represents the oldest part of the shell. (ii) Middle **prismatic layer**, composed of columns of crystalline calcium carbonate separated by conchiolin layers. The columns lie perpendicular to the surface of the shell, (iii) The inner **nacreous layer** of iridescent **nacre** or 'mother-of-pearl' which is formed of many thin and alternating layers of calcium carbonate and conchiolin. These layers lie parallel to the surface of the shell. The

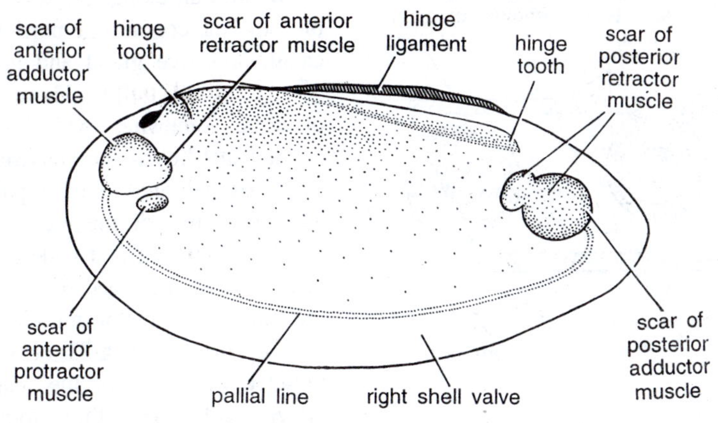

Fig. 2. *Unio.* Internal surface of right shell valve.

first two layers are secreted by the edge of the mantle whereas the innermost layer by its whole surface.

Pearl formation. When a foreign object, such as a sand grain or an egg or a parasite, gets in between the shell and mantle, the mantle is stimulated to secrete concentric layers of nacre around it. This results in the formation of a pearl. The pearl of *Unio* is not commercially precious. The bivalves which produce commercial pearls are called pearl oysters *e.g., Pinctada vulgaris, Meleagrina,* etc. The pearls are of variable shapes and sizes. Their colour may be white or cream red or pink red. Spherical pearls of rainbow colour are rarely found. The best quality of Pearl is known as 'LINGHA PEARL' and they are obtained from marine oysters. Pearl comprises of water, organic matter, calcium carbonate and the residue. (i) water : 2-4%; (ii) organic matter : 3.5-5.9%; (iii) calcium carbonate : 90%; and (iv) residue : 0.1-0.8%.

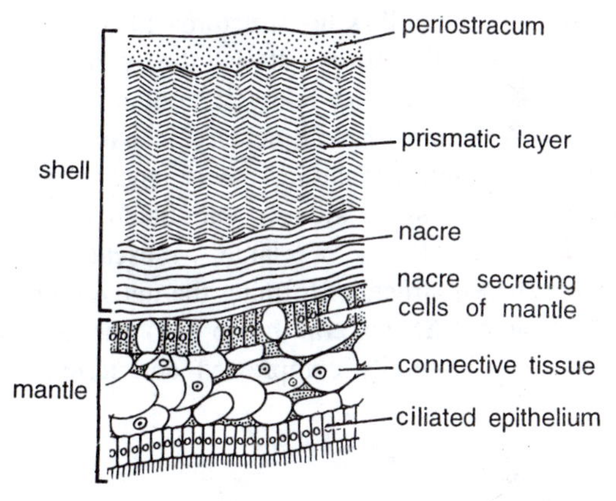

Fig. 3. *Unio.* A part of shell and mantle in T.S.

[II] Mantle and Mantle Cavity

The **mantle** or **pallium** consists of two fleshy lobes, which are attached dorsally to the body and ventrally to the shell valves, along the pallial line. At the posterior end, the edges of two mantle lobes touch each other at four points and extend slightly forming two tubes. The dorsal narrow tube is **excurrent** or **exhalent siphon,** through which water leaves the body and the ventral somewhat broader and longer tube is the **incurrent** or **inhalent siphon**, through which water enters the body. The latter's inner margin is papillated; the papillae test the incoming water. On the dorso-posterior side the two mantle folds also form a **dorsal pore**.

Histologically, the mantle consists of : (i) an outer **columnar epithelium** beset with nacre-secreting cells, (ii) a middle fibrous **connective tissue,** and (iii) an inner **ciliated epithelium** containing mucus-secreting cells.

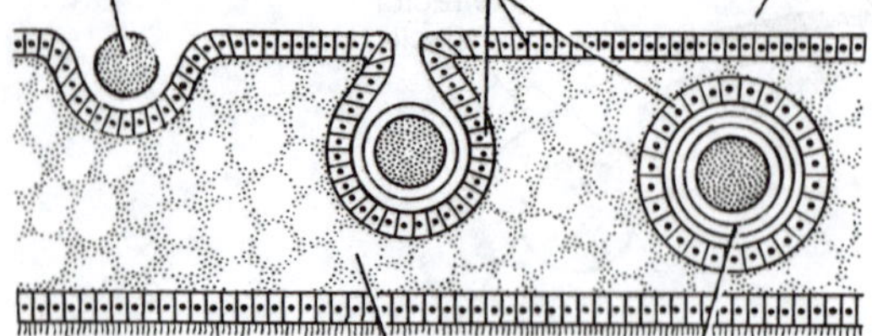

Fig. 4. *Unio.* Pearl formation.

The two mantle lobes enclose a bilaterally compressed cavity, the **mantle** or **pallial cavity**. It contains the following structures :

1. Visceral hump. It is the compressed mid-dorsal part of the body. It contains various organs including the alimentary canal, gonad, digestive gland, pericardial cavity enclosing the heart, blood vessels, kidneys and bladders.

2. Foot. It is a large, muscular and wedge-shaped structure, forming the antero-ventral extension of the visceral hump into the mantle cavity. It is laterally compressed and terminates below into an elongated **keel.** The thick basal part of the foot contains a portion of the alimentary canal, digestive gland and gonad.

3. Labial palps. These comprise two pairs of large somewhat oval, flattened fleshy lobes, located just below the anterior protractor muscles. The external and internal palps form the upper and lower lips of the mouth, respectively.

4. Mouth. It is a wide transverse slit, located in a depression formed by the labial palps between the anterior adductor muscle and the foot.

5. Gills. The gills or **ctenidia** are two in number and hang in the mantle cavity, one on either side of the foot. Each gill is characteristically folded to form two laminae each comprising two lamellae. The gills are plate-like structures, formed by the fusion of successive branchial or gill filaments. These are perforated by minute pores and are surfaced with cilia. Of each gill, the outer gill lamina is somewhat smaller than the inner.

The mantle cavity is divided into two chambers, the large ventral, **infra-branchial chamber** and the smaller dorsal **supra-branchial chamber**. The bases (dorsal margins) of the gills form a horizontal partition between these two chambers.

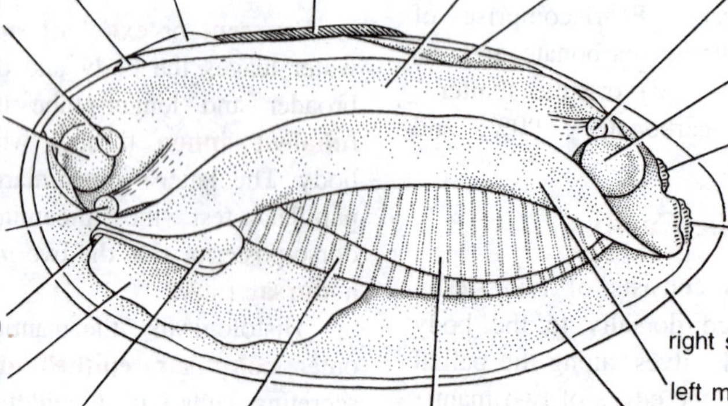

Fig. 5. *Unio.* Internal features after removal of left shell valve.

MUSCULATURE

The muscles are mainly of the slow contracting **unstriped** type and arranged in distinct bands or sheets. The two shell valves are closed by the contraction of two large, strong cylindrical, transverse muscles, situated one close to either end dorsally and passing across the body from one valve to another. They are called **anterior** and **posterior adductor** muscles. When these muscles relax, the elastic hinge ligament opens the valves. Near these muscles are two smaller muscles, the **anterior** and **posterior retractor muscles,** which run from the foot to the shell and serve to withdraw the foot, during locomotion. A small **protractor muscle,** closely behind the anterior adductor, serves to compress the visceral mass, thus causing the protrusion of the foot. The complex **intrinsic muscle** of the foot also serves as a protractor of that organ.

The mantle contains delicate **pallial muscles** inserted upon the shell along the pallial line. These contain three types of fibres : (i) **Dorso-ventral fibres** which extend dorso-ventrally through the mantle and form circular muscles around the siphons, (ii) **Longitudinal fibres** which extend through the mantle along the longitudinal axis and form longitudinal muscles around the siphons, (iii) **Transverse fibres** which extend across the mantle, perpendicular to its surface and form radial muscles around the siphons. The pallial muscles serve to retract the edge of the mantle.

LOCOMOTION

The **foot** is chiefly a locomotory organ and is adapted for burrowing. It protrudes between the gaping shell valves and ploughs through the mud or sand, as the case may be. The movement of foot is effected by : (i) The anterior and posterior retractor muscles and the anterior protractor muscles, which are also called **pedal muscles**, and (ii) blood pressure. The contraction of protractor muscles causes the extensions of the foot into the mud. This causes an influx of blood into the foot cavity (pedal sinuses), resulting in the dilation of the foot tip and anchoring of foot into the mud. Now, the anterior and posterior pairs of retractor muscles contract. This action tends to withdraw the foot, but since its tip is firmly anchored, the body is pulled forward. In this action the blood is forced out of the foot which now becomes narrower. The repeated movements of the foot results in a slow progression of the animal.

BODY CAVITY

The general body cavity is a **haemocoel** filled with blood or **haemolymph**. The true coelom is represented only by three small cavities : (i) A single ovoidal chamber, the **pericardium,** which lies dorsally, containing the heart and a part of the rectum and lined by the coelomic epithelium, (ii) The **gonocoels** or the cavities of the gonads, (iii) The **urocoels** or the cavities of the excretory organs.

RESPIRATORY SYSTEM

[I] Respiratory Organs

For its aeration, the animal depends upon oxygen dissolved in the surrounding water. The process is accomplished by two structures : the **mantle** and a pair of **gills** or **ctenidia**. The gills are infact the highly specialized derivatives of the mantle itself.

1. Gills. These lie in the mantle cavity, one on either side of the body and are suspended from the visceral mass and the mantle lobe of their sides.

(a) Structure. The gills are of typical eulamellibranch type. Each gill consists of two 'half-gills' or **demibranchs** or **laminae,** lying side by side. The two laminae of each gill are similar in structure except that the inner one is a little broader than the outer one. Each lamina is made up of two similar, elongated, somewhat rectangular

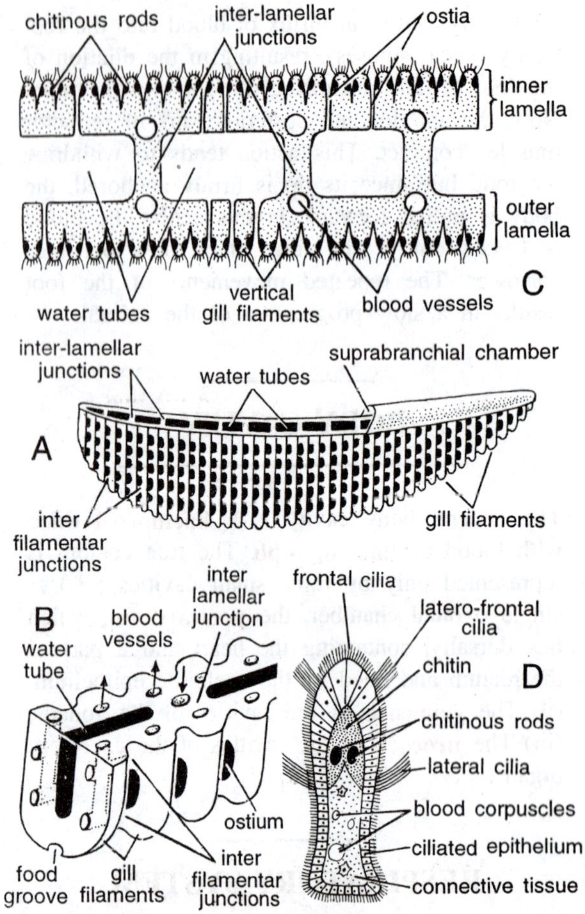

Fig. 6. *Unio.* Structure of gill. **A.** A demibranch seen from outside. **B.** A portion of gill (enlarged). **C.** A portion of gill lamina in T. S. **D.** Cross section of one gill filament.

plates, called the **lamellae**. These lie side by side attached to each other along their anterior, ventral and posterior edges. It resembles a man's wallet with a single pocket. The narrow cavity between the two lamellae is partitioned by several thin vertical septa, the **interlamellar junctions**, into a number of vertical compartments called **water tubes**. All the water tubes of a lamina open dorsally into a common **supra-branchial chamber** extending above the lamina.

A lamina is in fact made up of a series of vertical V-shaped rods, called **gill filaments.** The corresponding limbs of successive V-shaped gill filaments are connected by numerous longitudinal bars of connective tissue, called **inter-filamentar junctions**. The inter-filamentar junctions between adjacent gill filaments bound microscopic openings, the **ostia,** through which the mantle cavity (in fact the infra-branchial chamber of the mantle cavity) communicates with the water tubes of the lamina. Each V-shaped gill filament bears a notch at its angle; the notches of the successive gill filaments line up to form a longitudinal **food groove** which runs along the ventral edge of the lamina.

All the gill filaments of the lamina are alike. This condition, known as **homorhabdic**, is met within most pelecypods. In some pseudo-lamellibranchs, the condition is **heterorhabdic**, *i.e.*, two types of gill filaments occur. Each gill filament of *Unio* is composed mainly of connective tissue and is supported by a skeleton of chitinous rods; the entire structure is bounded by an epithelium which is richly ciliated. The cilia are of three main types : (i) **frontal cilia** lining the external, ridge-like face of the filament, (ii) **lateral cilia** lining the lateral sides of the filament which bounds an ostium, and (iii) **latero-frontal cilia** which lie on either side, between the frontal and lateral cilia.

The inner lamella of the outer lamina and the outer lamella of the inner lamina of each gill are attached with each other dorsally, along their entire length. The line of attachment is called **ctenidial** or **gill axis**. Thus each gill presents a typical bipectinate structure consisting of a ctenidial axis from which arise two series of V-shaped gill filaments; each gill filament consists of a **descending limb** arising from the ctenidial axis, and an **ascending limb**. In transverse section, each gill appears W-shaped.

(b) Blood supply. The gills receive deoxygenated blood from the kidneys through the **afferent branchial vessel** which runs through the ctenidial axis and gives out branches into the inter-lamellar junctions. These branches open into an **efferent branchial vessel** which also runs through the ctenidial axis. During its flow from the afferent to the efferent blood vessel, the blood is oxygenated. The oxygenated blood is finally conveyed to the heart by the efferent blood vessel.

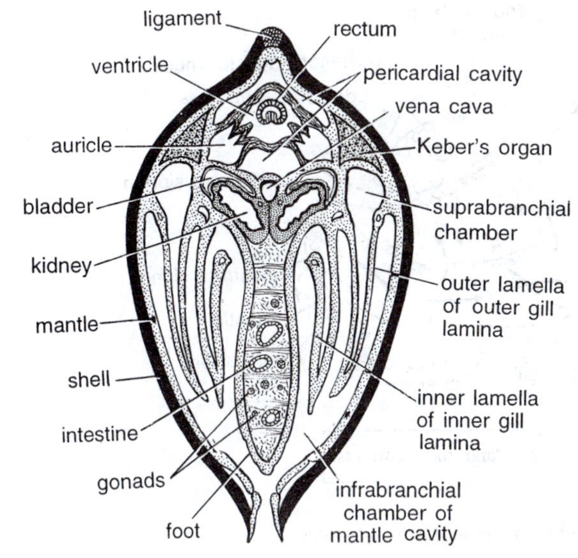

Fig. 7. *Unio.* T.S. Body through anterior region of gills.

(c) Attachment. The ctenidial axis (formed by the attachment of the inner and outer lamellae of the outer and inner laminae) of each gill is attached, along its entire length, to the side of the visceral mass. The outer lamella of the outer lamina is attached to the mantle; the inner lamella of the inner lamina is attached to the side of the visceral mass anteriorly, is free for a short distance near the middle, and is attached to its corresponding lamella of the other gill in the posterior region.

Each gill divides the mantle cavity of its side into : (i) a spacious **infra-branchial chamber** that lies below the attached dorsal margins of the lamellae and communicates with the exterior through the **inhalent siphon**, and (ii) two narrow longitudinal **supra-branchial chambers** (one above each lamina), partitioned from each other by the ctenidial axis. In the posterior region, the inner supra-branchial chambers of the two gills become continuous due to the attachment of their inner lamellae with each other. Posteriorly the supra-branchial chambers of both the gills open into a so-called **cloaca** into which also opens the rectum. The cloaca communicates with the exterior through the exhalent siphon.

2. Mantle. The mantle lobes, in addition to secrete the shell, also carry on respiration. Being thin and highly vascular, they are well suited for the purpose.

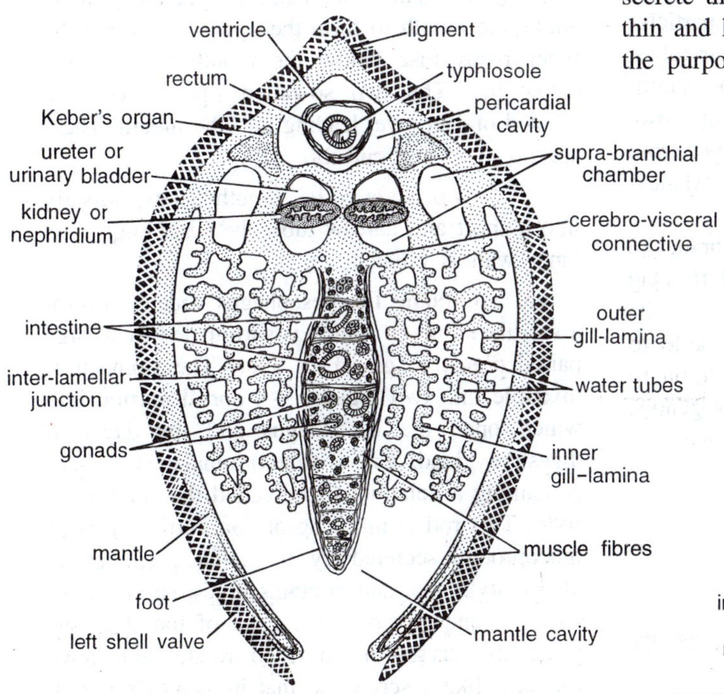

Fig. 8. *Unio.* T.S. Body through middle region of gills.

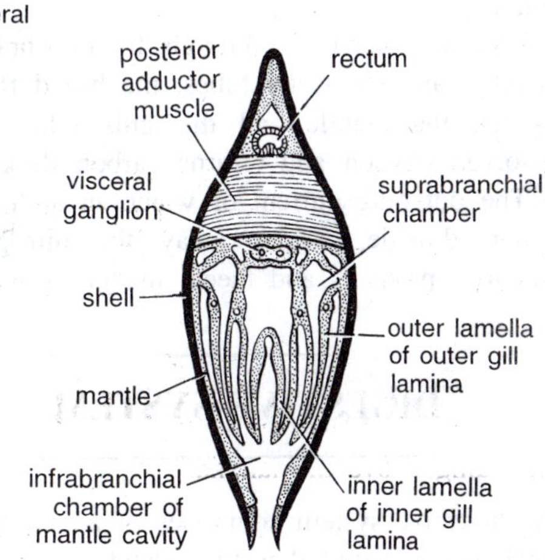

Fig. 9. *Unio.* T.S. Body through posterior region of gills.

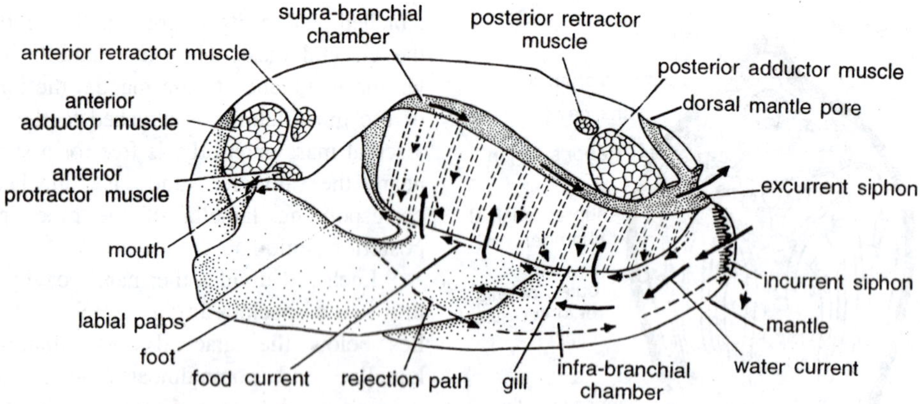

Fig. 10. *Unio.* Diagram showing water and food currents.

[II] Physiology of Respiration

Lateral cilia of the gill filaments and those lining the mantle, set up a current of water which enters the infra-branchial chamber through the inhalent siphon. The incoming water contains dissolved oxygen and microorganisms which constitute the food of the animal. From the infra-branchial chamber the water enters the water tubes of the gill laminae through the ostia; the food particles do not enter the water tubes. In the water tubes, the water flows upwards and enters the supra-branchial chambers. In these chambers it flows backwards and reaches the cloacal chamber from where it leaves the body through the exhalent siphon.

As the water flows through the infra-branchial chamber and the water tubes, the blood flowing through the mantle and the gills take up its dissolved oxygen and returns carbon dioxide to it. The outgoing current of water, in addition to carbon dioxide, carries away the nitrogenous excretory products and faecal matter as well.

DIGESTIVE SYSTEM

The digestive system comprises of an alimentary canal and a paired digestive gland.

[I] Alimentary Canal

The alimentary canal is a long coiled tube and consists of the following parts :

1. Mouth. The **mouth** is a transverse slit, lying immediately postero-ventrally to the anterior adductor muscle. On each side of the mouth is a pair of oval and flattened fleshy flaps, called the **labial palps**. The outer palps on each side unite above the mouth to form the **upper lip**, while the inner palps fuse below the mouth to form the **lower lip**. The two so-called lips enclose the ciliated oral groove, leading into the mouth. There are no **jaws** and **radula**.

2. Oesophagus. The mouth leads dorsally into a short and narrow tube, the **oesophagus**. Its inner wall is ciliated.

3. Stomach. The oesophagus widens to form a sac-like **stomach** which is surrounded by a large paired **digestive gland** or **liver.** The stomach is divisible into two regions —a dorsal portion into which open the oesophagus and the ducts of digestive gland, and a ventral tubular **style sac**, containing a gelatinous rod, called the **crystalline style**. This rod is made up of concentric layers of mucoprotein secreted by the ciliated epithelium of the style sac, and contains digestive enzymes, **amylase** and **glycogenase**. Cilia of the style sac cause the crystalline style to rotate and move forward (like a screw) so that its head (free end)

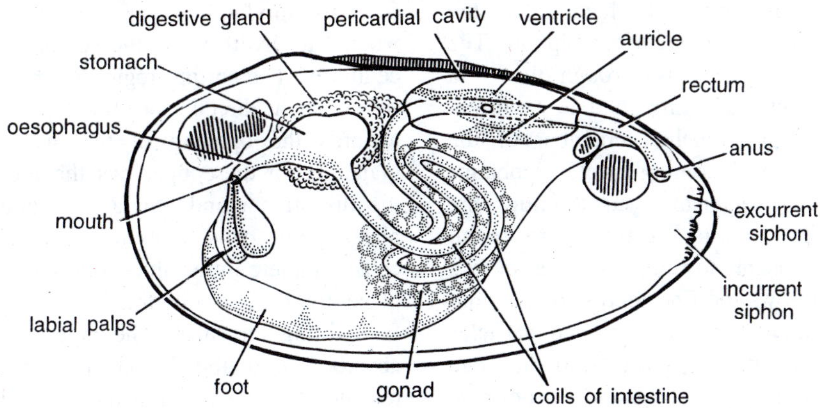

Fig. 11. *Unio.* Alimentary canal and digestive gland.

strikes against the opposite wall of the stomach; the cuticle lining of this region of the stomach is thickened to form a **gastric shield**. As the head of the crystalline style strikes against the gastric shield, small fragments containing enzymes break off from it and get mixed with the food particles. When feeding stops or when the animal is removed from water, the entire crystalline style gradually dissolves.

The dorsal portion of the stomach is folded around the openings of the ducts of digestive gland. These folds help in the sorting mechanism. They conduct the useless substances to the intestine, fine and partially digested food particles to the ducts of the digestive gland, and waste materials (returning from the digestive gland) to the intestine without letting them remix with the food particles present in the stomach.

4. Intestine. The **intestine** arises from the posterior end of the stomach. It runs ventrally into the foot coiling its way through the visceral mass, where it is closely surrounded by the gonad. The coiling is of a characteristic pattern — the ascending limb runs parallel to the descending limb and continues into the rectum.

5. Rectum. The post-terminal part of the intestine is called the **rectum**. Its wall is produced internally into a longitudinal, mid-ventral fold called **typhlosole**. It runs posteriorly through the pericardial cavity, embraced by the ventricle of the heart and then passes over the posterior adductor muscle and finally opens through the **anus** into the posterior part of the supra-branchial chamber or the **cloaca**.

[II] Digestive Gland

It is also called the **digestive diverticulum** or **liver**. It is a large and paired structure of dark brown or green colour surrounding the stomach. It opens into the dorsal portion of the stomach by many ducts. **Owen** (1955) has reported that in lamellibranch molluscs, such as *Unio,* the ducts of the digestive gland displays two tracts, the non-ciliated inhalent tract for carrying food particles into the gland and the ciliated exhalent tract for sending the wastes back into the stomach. This gland not only secretes digestive enzymes but its cells readily ingest fine food particles, bringing about their intracellular digestion.

[III] Food and Feeding

Unio is a planktonic feeder. Its food consists of diatoms, Protozoa, other microorganisms and organic detritus, brought in by the circulating water current.

Unio displays **filter-feeding** that involves straining food from large quantities of water. The beating of lateral cilia of the gill filaments draws water into the infra-branchial chamber of the mantle cavity through the incurrent siphon. The water current enters the gills through the ostia and egresses to outside through the excurrent siphon via the supra-branchial chamber of the mantle cavity. The incoming water contains millions of microorganisms, constituting the animal's food. The beating of the latero-frontal cilia of the gill filaments towards the outer surface of the lamellae throws the fine food particles onto the lamellar surfaces. The heavier sand particles are conveyed from the gill surfaces to the edge of the laminae, which touch the mantle. From here these are dropped to the bottom of the mantle cavity. The lamellar surfaces secrete a mucous sheet to trap the food particles. The frontal cilia of the gill filaments beat downwards towards the food groove, moving the mucous sheet downwards. The mucus from both sides of each gill lamina moves into the food groove. Here, ciliary beating moves the food towards the mouth. While reaching in between the labial palps, further sorting of food particles takes place. This is achieved by a series of overlapping folds on the inner surface of labial palps. In the gutters between them, a set of ciliar tracts beats outwards, causing the heavier particles to drop out of the mucus into the gutters. Another set of ciliary tracts allows the finer particles to reach the mouth. Through the mouth the food passes into the oesophagus and stomach.

When a good amount of heavier particles are accumulated on the labial palps, these fling to sweep them to the bottom of the mantle cavity. A sudden thrust of water outward through the incurrent siphon carries the rejected food particles and the sediment out of the mantle cavity.

[IV] Digestion, Absorption and Egestion

Within the stomach the food particles are subjected to sorting, maceration, digestion and partial absorption. The crystalline style performs the function of a stirring rod and a windlass. Its rotation by the style sac cilia causes the detachment of the food particles from the string-like mucous sheet entering the stomach. The detachment of food particles from the mucous sheet is facilitated by the lower pH of the stomach which decreases the viscosity of mucous. The style frees its **amylase enzyme** into the stomach fluid where it hydrolyses the carbohydrates extra-cellularly. The sorting region of the dorsal portion of stomach directs the finer particles of food towards the opening of the ducts of digestive gland. From these openings the food moves into the digestive gland where it is engulfed by the phagocytotic cells of its epithelium. Within these cells complete intracellular digestion of food takes place by the carbohydrate, protein and possibly fat-splitting enzymes. The digested nutrients are absorbed into the blood and the undigestible residue is sent back into the stomach. From here the coarse food particles and undigestible residue pass on to the lumen of intestine. The amoeboid cells, present throughout the digestive tract, are said to engulf and digest food particles, a method that recalls the part played by the amoebocytes of sponges.

In the intestine, the food contents are moulded as faecal pellets that are conveyed into the rectum and finally voided through the anus into the cloaca, to be carried away with the outgoing water current.

BLOOD-VASCULAR SYSTEM

[I] Blood-vascular Organs

The blood-vascular system is well developed and is of the open type. It consists of : (i) the **heart** and **pericardium**, (ii) **arteries**, (iii) **sinuses**, and (iv) **veins**.

1. Heart and pericardium. Near the dorsal mid-line and just in front of the posterior adductor muscle, there is a thin-walled triangular chamber, the **pericardium**, which encloses the heart. The pericardium, lined by an epithelium and filled with **pericardial** or **coelomic fluid,** represents a portion of coelom and communicates with the supra-branchial chambers through the kidneys.

The **heart** is highly contractile and three-chambered, consisting of a single **ventricle** and lateral paired **auricles**. The two **auricles,** lying one on either side of the ventricle, are transparent, thin-walled, roughly triangular and distensible to

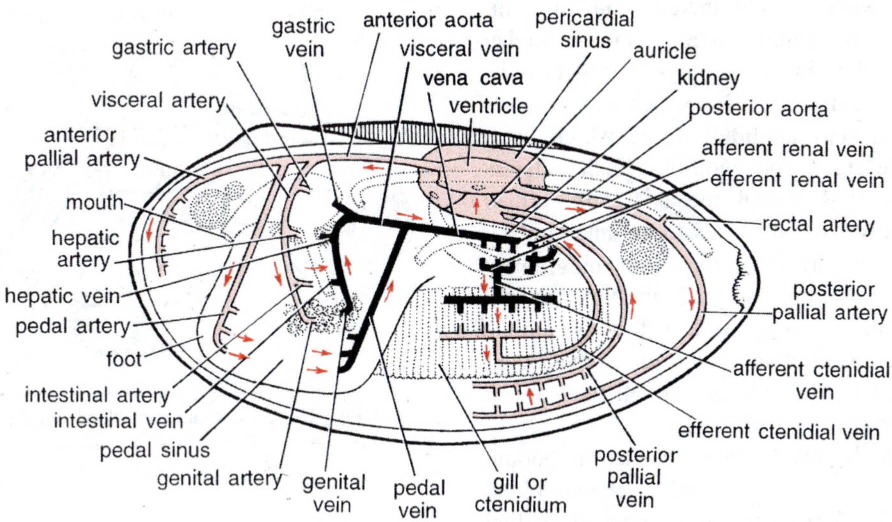

Fig. 12. *Unio*. Blood vascular system.

a great extent. Each auricle is attached to the pericardial wall by broad base and opens into the ventricle by an **auriculo-ventricular aperture** having valved opening towards the ventricle. The auricles receive blood from the gills, kidneys and mantle, and pour it into the ventricle. The single median **ventricle** is a large, thick-walled and muscular chamber, wrapped around the rectum. It pumps blood to the body. The muscular heart beats 20 to 100 times per minute. The circulation is rather slow but it seems to be adequate for such a sedentary animal.

2. Arteries. The ventricle pumps blood, both forward and backward, through two main arteries, known as the **anterior** and **posterior aortae,** respectively. The anterior aorta runs anteriorly, dorsal to the intestine and divides into three main branches : (i) an **anterior pallial artery** which supplies branches to the mantle lobes, (ii) a **pedal artery** which supplies the foot and (iii) a **visceral artery** which supplies the various visceral organs through numerous branches, of which the main are a **gastric branch** to the stomach, a **hepatic branch** to the digestive gland, an **intestinal branch** to the intestine, and a **gonadial** or **genital branch** to the gonad. The posterior aorta runs posteriorly ventral to the rectum. It sends out fine branches to the pericardium and kidneys, a prominent **rectal artery** to the rectum and a **posterior pallial artery** to the mantle.

3. Sinuses. The arteries break up into a network of vessels all over the body. These vessels, instead of forming small capillaries, lead into irregular cavities, the blood **sinuses.** The sinuses lack the epithelial lining of true blood vessels and pour their blood directly into veins. The freshwater mussel, therefore, has an 'open type' of circulatory system. Among molluscs, the capillaries occur only among the advanced cephalopods.

4. Veins. The venous blood from various parts of the body is collected by several veins; the veins themselves are in fact large sinuses. Blood from the various visceral organs is collected by a **visceral vein** through **genital, intestinal, hepatic** and **gastric** branches. Blood from the **pedal sinus** in the foot is collected by a **pedal vein** which joins the visceral vein to form a large longitudinal vein, the **vena cava,** which lies beneath the pericardium in between the kidneys. The **vena cava** supplies all its blood through **afferent renal veins** to the kidneys where the nitrogenous waste is eliminated from it. This reminds us of the **renal portal system** of higher animals. Blood from the kidneys is collected by numerous **efferent renal veins** which unite to form a pair of **afferent ctenidial veins** which carry blood to the gills for aeration. Each afferent ctenidial vein runs through the axis of the gill of its side and gives out branches to the interlamellar junctions. These branches open into an **efferent ctenidial vein** which also runs through the gill axis and finally conveys the blood to the auricle of its side. Some efferent renal veins convey their blood directly to the auricles through the efferent ctenidial veins.

The mantle also serves as an important respiratory organ and it sends aerated blood directly to the auricles through posterior pallial veins.

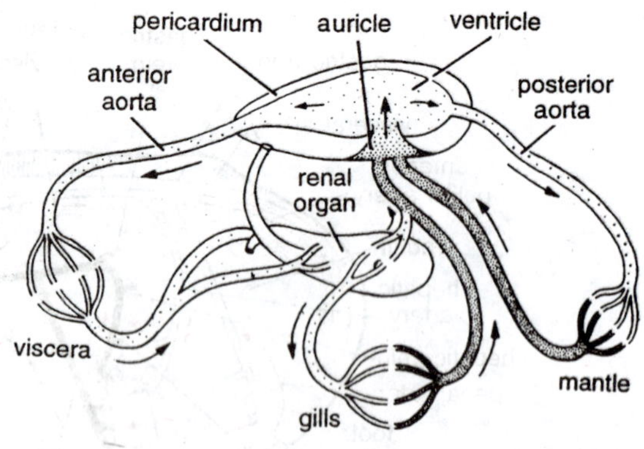

Fig. 13. *Unio.* Course of blood circulation.

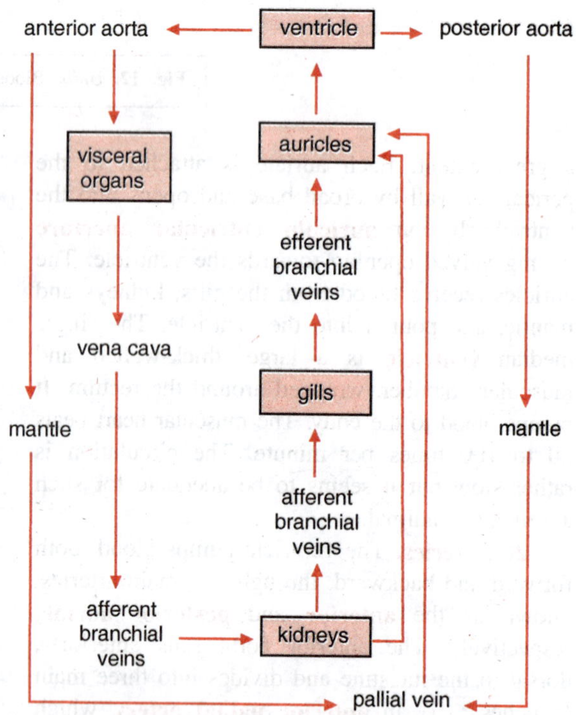

Fig. 14. *Unio.* Flow chart for the course of blood circulation.

[II] Blood

The blood consists of colourless **plasma** containing a respiratory pigment, **haemocyanin,** and numerous white amoeboid cells or **leucocytes,** floating in it. Haemocyanin has a 'copper base' in contrast to haemoglobin which has an 'iron base', and imparts faint blue colouration to the blood. Leucocytes are of granular and agranular types. The blood distributes oxygen and nutrients to all parts of the body and transfers CO_2 and other waste products of metabolism to the gills, mantle and kidneys.

[III] Course of Circulation

The ventricle drives the blood forward through the anterior aorta and backward through the posterior aorta. A part of the blood goes into the mantle, where it is oxygenated and then returns directly to the auricles. The rest of the blood circulates through numerous spaces or sinuses in the body carrying food, oxygen, carbon dioxide and nitrogenous metabolic wastes. Finally collected by the vena cava, the deoxygenated blood first passes into the kidneys which extracts the nitrogenous wastes and then into the gills, where it picks up O_2 and releases CO_2 and finally through the auricles into the ventricle. Some blood from the kidneys goes directly to the heart without going around the gills. The blood is kept travelling in the proper direction by means of valves which are present in the heart and the veins.

EXCRETORY SYSTEM

Excretion is taken care of by : (i) a pair of kidneys or **organs of Bojanus**, and (ii) the **Keber's organs**.

[I] Organs of Bojanus

The two **kidneys** or **nephridia** are often termed the **organs of Bojanus** after the name of their discoverer. They are situated beneath the floor of the pericardial cavity, one on each side of the vena cava. They are derived from the true coelom **(urocoels)**.

Each kidney is a long, dark and glandular tube open at both ends. It is bent upon itself like a broad U-shaped tube, with the loop posterior, the two ends anterior and the two limbs lying parallel and one above the other. The lower arm is brown, spongy, glandular and thick-walled, forming the **kidney proper**, which opens anteriorly into the fluid-filled pericardial cavity by a small ciliated **renopericardial aperture**. The dorsal arm is small, non-glandular, lined by ciliated epithelium and thin-walled known as the **ureter** or **urinary bladder,** which opens anteriorly into the supra-branchial chamber of the inner gill-lamina by a small **renal aperture**. The bladders of both the kidneys intercommunicate by an oval aperture.

Physiology of excretion. The ventral glandular portion of the kidney extracts guanin and other nitrogenous waste products of metabolism from the coelomic pericardial fluid as well as the blood supplied to the kidneys. The walls of the pericardial sinus are also glandular, and suppossed to secrete waste materials from the blood into the coelomic cavity.

The ciliated epithelial lining of the bladder produces an outgoing current, thus conveying excretory fluid from the glandular part of the kidney to the supra-branchial chamber which leads to the excurrent siphon. There is reabsorption of salts in the two kidneys which also serves for maintaining the blood concentration level by removing excessive water from it.

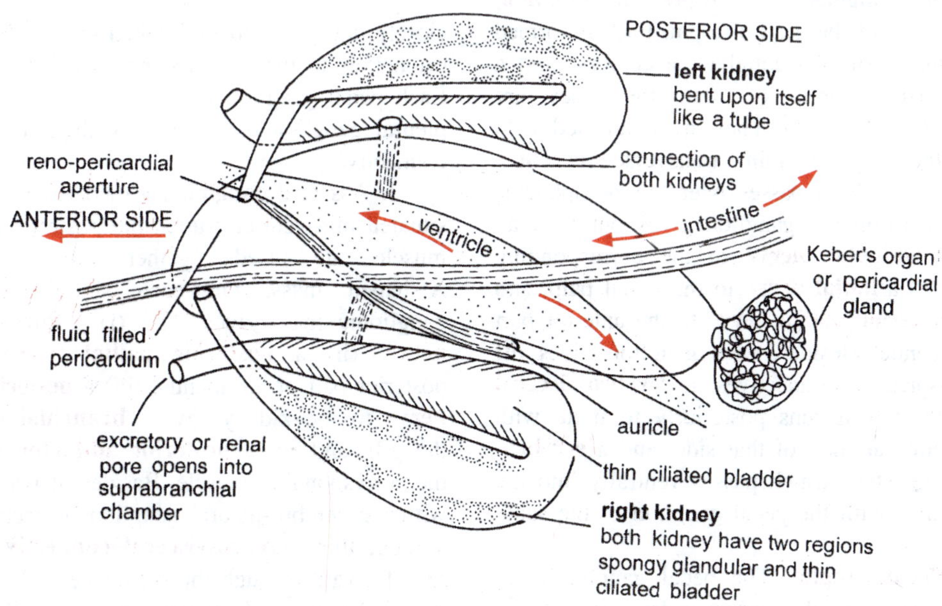

Fig. 15. *Unio.* Excretory system. Coelomic fluid as well as the blood supplied through the reno-pericardial aperture, glandular part extracts NH_3, Guanin, bladder stores and reabsorbs inorganic salts.

[II] Keber's organ

The **Keber's organ** or the **pericardial gland** is a large reddish-brown, glandular mass situated in front of the pericardium. It probably helps in excretion, discharging waste products into the pericardial cavity to be finally eliminated through the kidneys.

NERVOUS SYSTEM

Freshwater mussel has a muscular body performing coordinated movements, but the nervous system is greatly reduced due to sluggish and sedentary mode of life, and there is little evidence of the brain. It consists of : (i) a **central nervous system** including three pairs of **ganglia** for three main regions of the body : head, foot and viscera, (ii) their **connectives,** and (iii) small **nerves**.

1. Cerebro-pleural ganglia. At the base of the labial palps, just outside the corners of the mouth on either side, is placed a small yellowish, somewhat triangular **cerebro-pleural ganglion**, about the size of the head of a pin. Each is formed by the fusion of two ganglia : a cerebral and a pleural. The ganglia of both the sides are equivalent of the brain. They are connected with each other by a thin transverse **cerebral commissure**, which passes over the oesophagus. Each cerebro-pleural ganglion gives out : (i) an **anterior adductor nerve** to the corresponding muscle, (ii) a **labial nerve** to the labial palp, and (iii) an **anterior pallial nerve** to the anterior part of the mantle. Besides, each ganglion gives off two conspicuous connectives, a **cerebro-visceral connective** which runs posteriorly to unite with the visceral ganglion of that side, and a **cerebro-pedal connective** which passes ventrally into the foot to unite with the pedal ganglion of the same side.

2. Pedal ganglia. The pedal ganglia lie in the foot, about one-third distance from its anterior end, dorsal to its muscular portion and just below the visceral mass. The two pedal ganglia are closely united to form a bilobed mass, which gives off nerves to the foot, its muscles and statocysts. Each pedal ganglion is connected to the cerebro-pleural ganglion of its side by the cerebro-pedal connective.

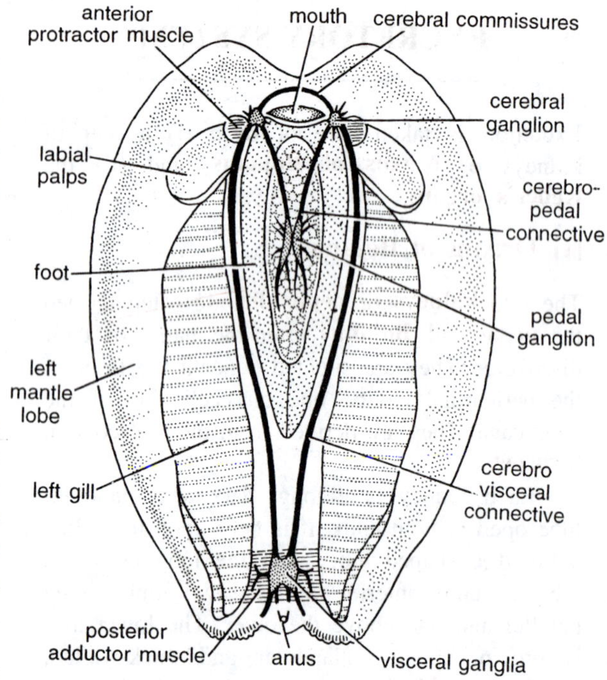

Fig. 16. *Unio.* Nervous system in ventral view.

3. Visceral ganglia. The two visceral ganglia, situated mid-ventrally upon the posterior adductor muscle, are fused together into a flattened, X-shaped mass. On each side the visceral ganglionic mass gives off : (i) a **dorsal pallial nerve**, (ii) a **posterior pallial nerve** to the posterior part of the mantle, (iii) a **posterior renal nerve** to the kidney, (iv) a **branchial nerve** to the gill, and (v) a **posterior adductor nerve** to the corresponding muscle. Besides, it is connected with the cerebro-pleural ganglion of each side by a long, thin **cerebro-visceral connective**, which runs forward through the substance of kidney just below the place of attachment of the gill lamina, giving off several small nerves to the visceral mass on its way.

SENSE ORGANS

The sense organs are poorly developed due to slow, sluggish and sedentary habits. The eyes and tentacles are altogether absent. The main sensory structures are : (i) **statocysts,** (ii) an **osphradium,** and (iii) scattered epithelial **sensory cells**.

1. Statocysts. The foot contains a pair of minute hollow vesicles, the **statocysts**, one close to each pedal ganglion. They are innervated by the cerebro-pedal connectives. The statocyst is lined by sensory cells and contains a mass of lime, called **statolith**, the movement of which stimulates the sensory cells. The statocysts are thought to be organs of equilibrium.

2. Osphradium. At the base of the gills, on the ventral surface of the posterior adductor muscle and covering the visceral ganglia is a pair of dark yellow patches of sensory epithelial cells, forming the **osphradium**. It is probably used for testing the chemical nature of the water entering the mantle cavity through the inhalent siphon.

3. Sensory cells. The edges of the mantle lobes are provided with sensory cells, especially abundant on the inhalent siphon. They probably respond to touch **(thigmoreceptors)** and also seem to be sensitive to light **(photoreceptors)**.

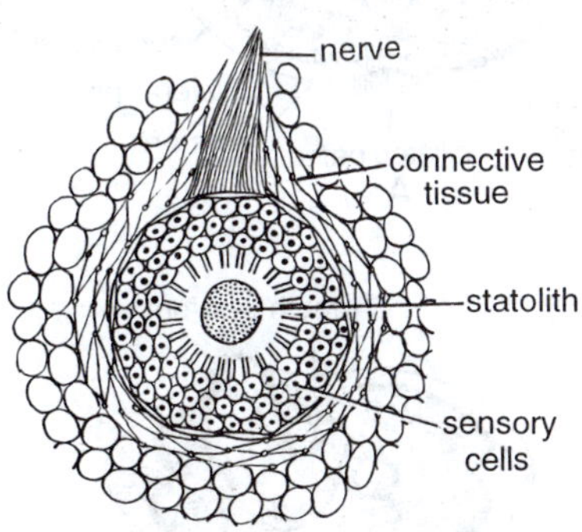

Fig. 17. *Unio.* T.S. Statocyst.

REPRODUCTIVE SYSTEM

Unio is **dioecious,** *i.e.,* the sexes are separate, but there is no **external dimorphism**.

Gonads

The gonads, **testes** in male and **ovaries** in female, are a pair of large, simple, branching tubes lying among the intestinal coils, in the visceral mass, just above the foot. During breeding season, the gonads become greatly enlarged, conspicuous and brightly coloured. The testes become whitish and the ovaries reddish. The coelomic epithelium lining their tubules proliferates spermatozoa and ova, respectively. The gonad of each side has a short duct, the **vas deferens** in male and the **oviduct** in female. Each duct opens into the supra-branchial chamber of the inner gill-lamina of its own side. This opening, called **genital aperture,** is situated just in front of the **renal aperture** of the ureter. There are no accessory reproductive organs like glands and copulatory structures.

Fertilization

In male, *sperms* are shed through the genital apertures into the supra-branchial chambers and expelled from the body along with the outgoing water current through the exhalent siphon. Some of these sperms enter through the inhalent siphon of a female purely by chance. The **eggs** shed through the genital apertures into the supra-branchial chambers do not leave the body, but are carried into the water tubes of the outer gill laminae. Here these are held by mucus and fertilized by sperms which enter the water tubes through ostia. The zygote is surrounded by a **vitelline membrane** bearing a small perforation, the **micropyle**.

Development

Development of the fertilized eggs or **zygotes** proceeds within the water tubes of the outer gill laminae which greatly enlarge .due to accumulation of a large number of embryos. Thus, the water tubes of the outer gill laminae serve as the **brood pouches** or **marsupia**. The females are easily recognized at this stage by the swollen appearance of their outer gill laminae. In *Anodonta,* the eggs are usually fertilized in August and the embryos, which develop from them, remain in the gills of the mother, throughout winter.

The development of *Unio* is not properly known, but the following account of *Anodonta* will apply almost equally well to it.

Early Development

The zygote undergoes complete but unequal **cleavage**, resulting in a hollow ball of cells, the **blastula.** It consists of two types of cells : (i) small, clear, yolkless and rapidly multiplying **micromeres** on the upper side, and (ii) large, yolk-filled and slowly dividing **macromeres** on the lower side. The invagination of the macromeres into the micromeres results in a **gastrula** having a small **archenteron**. The gastrula bears a tuft of cilia posteriorly. Some large cells are budded off into the **blastocoel** to form the mesoderm. Simultaneously, a deep invagination on the posterior dorsal, side of the embryo forms the characteristic molluscan organ, the **shell gland**, secreting first an **unpaired** embryonic shell which is replaced later by a bivalved shell. The valves of the shell are triangular and the ventral angle of each bears a serrated hook. An **adductor muscle** is formed by the mesoderm cell. The mouth of the gastrula closes and a definite mouth is subsequently formed by an invagination of the ectoderm. The body of the embryo splits from ventral to dorsal side and gradually changes into a larva, called **glochidium**, characteristic of fresh water mussels.

Glochidium Larva

Glochidium means the 'point of an arrow". It is a minute larva, 0.1 to 0.4 mm wide. The **shell** consists of two triangular and porous **valves**, united dorsally and free ventrally. The ventral free end of each valve is produced into a conspicuous **hook**, beset with many **spines**. The shell encloses the body with the right and left **mantle lobes**. The mantle lobes are very small and their margins bear on each side three or four groups of peculiar brush-like **sensory bristles**. The valves clasp together by the action of a single

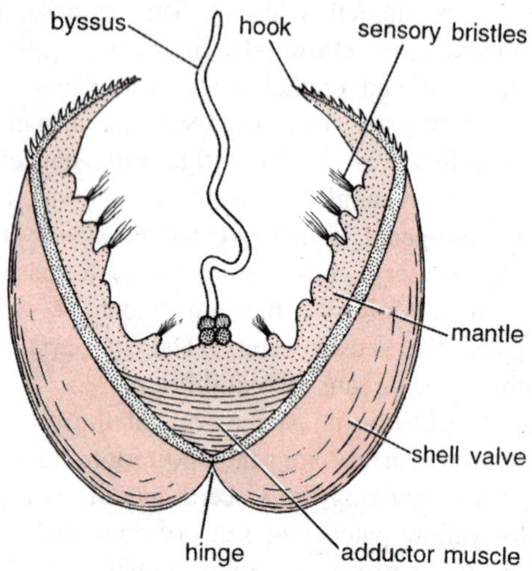

Fig. 18. *Unio.* A glochidium.

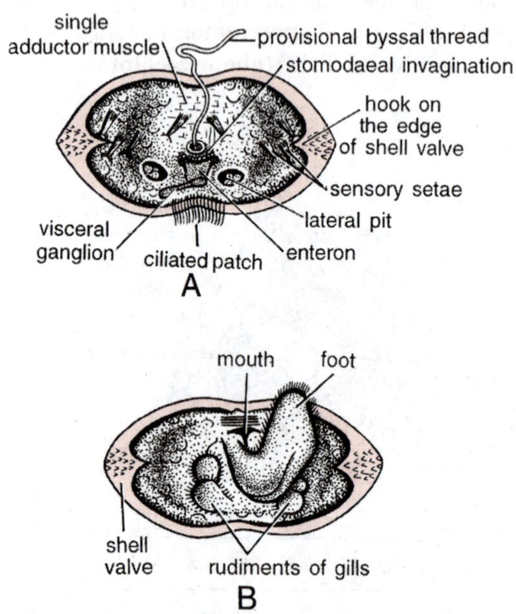

Fig. 19. *Unio.* Stages of metamorphosis of glochidium.

massive **adductor muscle**, extending transversely between the two valves. The **foot** is not yet developed, but the mid-ventral surface of the body has a **glandular pouch**, which secretes a long sticky thread, the larval **byssus**.

Tremendous number of glochidia are produced which fill the outer gill laminae. Inside them the glochidia remain entangled by their byssuses, and nourish on mucus secreted by the gills. When sufficiently mature, they are expelled through the exhalent siphon into the surrounding water, where they slowly sink to the bottom or are scattered by water currents. The glochidia cannot move about or feed. They can open or close their shells but cannot move independently. They grow further and to continue their existence they must now come in contact with a freshwater fish as ectoparasites. Many of them die for want of proper host. However, some happen to come in contact with the proper species of fish and anchor its skin, fins or gills, by closing their hooked valves and penetrating the host-tissues by larval byssuses. The closure of valves seems to be due to chemical stimulation, by salts escaping from the tissue of the fish. In this position they soon become encysted by a pathological overgrowth of the skin or mucous membrane of the host. The skin of fish, growing around the glochidia, forms the 'black-heads'. In this condition, they lead a truly **ectoparasitic life** receiving nourishment from the host by absorption for about 10 weeks during which they undergo a peculiar **metamorphosis** to develop into young adults.

Metamorphosis of glochidia. The parasitic larvae feed and grow by absorbing nutrients from the host's body fluids. They are also nourished by tissue severed from the host at the time of attachment and by the disintegrated larval organs.

During metamorphosis, the larval structures, such as sensory bristles and byssus disappear; the single adductor muscle is replaced by two; the mantle lobes are replaced; a stomodaeum is

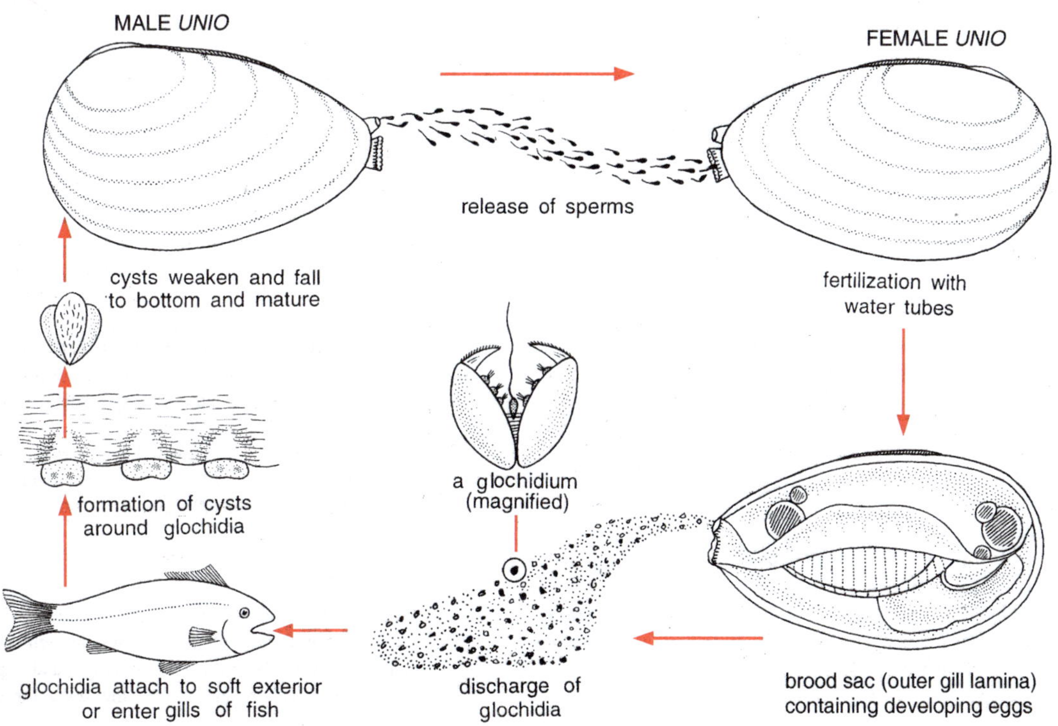

Fig. 20. *Unio.* Life cycle.

formed by invagination behind the position of byssus and opens into the enteron; the ectoderm lying close to the posterior end of the enteron ruptures to form the anus : thus the proctodaeum is not formed; the foot makes its appearance mid-ventrally behind the mouth and gills arise from its sides; and rudiments of visceral organs make their appearance.

After metamorphosis the cyst wall weakens and breaks, liberating the immature mussels that fall to the bottom. Here they feed in the normal adult manner and gradually acquire adult size and shape.

Significance of glochidium. The life cycle, that includes a parasitic glochidium larva on a fish host has many advantages. Besides affording **protection** and a means of **nourishment,** it ensures a far wide and more rapid **dispersal** of the species. A fish may carry these tiny parasites to great distances before they drop off. Considering the sluggish habits and poor locomotory ability of the mussels, this is probably the only way to ensure their proper distribution.

***Table 1.* Comparative Study of Respiratory Systems of *Pila* and *Unio*.**

***Pila* (Apple snail)**	***Unio* (Freshwater mussel)**
1. Being **amphibious**, exhibits **aquatic** as well as **aerial** modes of respiration.	1. Being **aquatic**, exhibits only **aquatic** mode of respiration.
2. Depends upon incoming water only for oxygen.	2. Depends for food and male gametes (in female only) on water current.
3. Organs of respiration are : single **gill** or **ctenidium, a pulmonary** sac or **lung** and a pair of **nuchal lobes**.	3. Respiration is brought about by : a pair of **ctenidia** or **gills** and **mantle**.
4. Gaseous exchange by **gill** or **ctenidium** in case of aquatic respiration and by the **pulmonary sac** in case of aerial respiration.	4. Gaseous exchange by a pair of **gills** or **ctenidia** and the **mantle**.
5. Accessory respiratory organs are the right and left nuchal lobes which are modifications of mantle.	5. Accessory respiratory organs are absent.
6. (a) The single gill lies asymmetrically in the branchial chamber of the mantle cavity.	6. (a) The two gills lie symmetrically in the body, one on each side, in the mantle cavity.
(b) The gill is simple and monopectinate. It has a **gill axis** attached to the roof of the branchial chamber. From the gill axis hang down numerous triangular **gill lamellae**, one behind the other, in a single series. Each lamella has two layers of epithelial cells enclosing a narrow cavity; the cavity does not communicate with the exterior. The adjacent lamellae are not connected with one another. Water tubes, **ostia** and **supra-branchial chambers** are absent.	(b) The gills are complex and bipectinate. It has a **gill axis** attached to the side of visceral mass. From the gill axis hang down two series of V-shaped **gill filaments**. Corresponding limbs of the adjacent gill filaments of each series are connected by **inter-filamental junctions** with minute pores, called **ostia**, between them, to form a **gill lamella**. Two lamellae of each series are attached with each other along their anterior, ventral and posterior margins to form a **lamina**. The cavity of lamina is divided into vertical water tubes by **inter-lamellar junctions**. Water tubes of each lamina open dorsally into a **supra-branchial chamber**.
(c) Afferent and efferent blood vessels run through the gill axis.	(c) Same.
(d) Cilia of the gill lamellae set up a water current which bathes the gill lamellae for gaseous exchange.	(d) Cilia of gill filaments set up a water current which enters the gills through ostia, flows through water tubes (to bring about gaseous exchange) and supra-branchial chambers and leaves the body through exhalent siphon.
7. The pulmonary sac lies in the roof of the pulmonary chamber. It is formed by the mantle. Its roof is highly vascular and brings about gaseous exchange during aerial respiration.	7. The mantle does not form a pulmonary sac due to absence of aerial respiration. It is, however, highly vascular and brings about gaseous exchange during aquatic respiration.

Table 2. **Comparative Account of Nervous Systems of *Pila* and *Unio*.**

***Pila* (Apple snail)**	***Unio* (Freshwater mussel)**
1. Nervous system well developed.	1. Nervous system less developed due to sedentary habit.
2. Nervous system asymmetrical due to twisting of visceral nerves into a figure of 8. This condition known as **chiastoneury,** is the result of torsion.	2. Nervous system bilaterally symmetrical as there is no torsion.
3. Some nerves and ganglia concentrate at the anterior end to form a sort of **nerve ring** around the buccal mass.	3. No nerve ring is formed at the anterior end which usually remains buried.
4. The ganglia, especially at the anterior end, lie close together due to shortening of nerves between them.	4. The ganglia are wide apart with long connectives between them.
5. Paired ganglia are : **cerebral, buccal, pleural, pedal** and **visceral.** Unpaired are **supra-intestinal** and **infra-intestinal.**	5. Paired ganglia are : **cerebro-pleural, pedal** and **visceral.** Unpaired ones are absent.
6. **Cerebral ganglia** are broad and triangular, situated one on either dorso-lateral side of the buccal mass. A broad **cerebral commissure** connects the two. **Pleural ganglia** are not fused with them.	6. **Cerebral ganglia** are small and triangular situated one at the base of each labial palp. A thin **cerebral commissure** connects the two. Each is fused with a **pleural ganglion** to form the **cerebro-pleural ganglion.**
7. There are two **pedal ganglia,** each fused with a **pleural ganglion** to form a **pleuro-pedal ganglionic mass,** situated on the ventro-lateral side of the buccal mass. The two are connected by two thick **pedal commissures.**	7. The two **pedal ganglia** are fused to form the bilobed ganglionic mass located at the junction of the foot and visceral mass; a pedal commissure is not necessary and hence absent.
8. Each cerebral ganglion is connected with the pleuro-pedal mass of its side by a thick **cerebro-pleural** and a thick **cerebro-pedal connective.**	8. Each cerebro-pleural ganglion is connected with the pedal ganglion of its side by a fine **cerebro-pedal connective.**
9. A pair of **buccal ganglia** lies dorsally on the posterior side of the buccal mass. They are connected with each other as well as with the cerebral ganglia of their side.	9. **Buccal ganglia** are absent.
10. Two **visceral ganglia** are fused into a bilobed mass situated in the visceral mass, close to the heart. They are connected with the right pleuropedal mass and the supra-intestinal ganglion by an **infra** and a **supra-intestinal visceral connective,** respectively.	10. Two **visceral ganglia** are fused to form a flattened X-shaped mass below the posterior adductor muscle. They are connected with the cerebro-pleural ganglia of their sides by **cerebro-visceral connectives.** There is no direct connection between visceral and pedal ganglia.
11. Two **intestinal ganglia** are present : (i) A **supra-intestinal ganglion** lies on the left side and is connected with the two pleuro-pedal ganglionic masses and the visceral ganglion of the left side, (ii) An **infra-intestinal ganglion** lies fused with the right pleuro-pedal ganglionic mass and is connected with the pleural ganglion of the left pleuro-pedal mass by an **infra-intestinal** nerve.	11. **Intestinal ganglia** are absent.
12. Total commissures are only three : cerebral, buccal and pedal.	12. The only commissure found is **cerebral.**
13. Total number of connectives found are : **cerebro-pleural, cerebro-pedal, cerebro-buccal** (all paired), **supra intestinal, infra-intestinal, zygoneury, supra-intestinal visceral** and **infra-intestinal visceral** (all unpaired).	13. Only two paired connectives are found : **cerebro-pedal** and **cerebro-visceral.**

Important Questions

➤➤Long answer type questions

1. Give an account of the morphology and histology of shell of *Unio*.
2. Describe the respiratory system of *Unio* and compare it with that *of Pila.*
3. Describe the digestive organs and feeding mechanism of *Unio.*
4. With the help of a well-labelled diagram, describe the circulatory system of *Unio.*
5. Give an account of the reproduction and development of *Unio.*
6. Write short notes on : (i) Organ of **Bojanus**, (ii) **Glochidium**, (iii) Gill lamina, (iv) **Keber's** organ, (v) Mantle.
7. Draw a labelled diagram of a T.S. of *Unio* passing through the gill region.

➤➤Short answer type questions

1. What are Keber's organ ? Where do you find them ?
2. Define the terms : (i) Monopectinate, (ii) Bipectinate.
3. State the position and function of the organs of Bojanus in *Lamellidens.*
4. Describe with suitable diagrams the organs found in the mantle cavity of freshwater mussel and *Pila.*
5. Describe the microscopical structure of the shell of a fresh water mussel and add a note on the formation of pearl.
6. Describe the development and life history of the fresh water mussel and mention how dispersal is brought about.
7. Draw labelled diagrams of the T.S. of shell and mantle of *Unio* (No description).

➤➤Fill in the blanks

1. Organs of Bojanus is another name for the .... of *Lamellidens.*
2. Organ of Bojanus in freshwater mussel is .... in function.

➤➤Multiple choice questions

1. *Unio* belongs to which class :
   (a) Placophora (b) Scaphopoda
   (c) Pelecypoda (d) Gastropoda
2. Keber's organ is the excretory organ of :
   (a) *Pila* (b) *Unio*
   (c) *Neopilina* (d) *Doris*
3. Larva of *Unio* is :
   (a) trochophore (b) glochidium
   (c) veliger (d) planula
4. The outer layer of shell of *Unio* contains :
   (a) Chitin (b) Conchiolin
   (c) $CaCO_3$ (c) Si_2O_3
5. The muscles which responsible to open and close the value in *Unio :*
   (a) pallial muscles
   (b) protracter muscles
   (c) adductor muscles
6. Gills in *Unio* are :
   (a) a part of visceral mass
   (b) derivatives of mental
   (c) derivatives of foot
   (d) modification of tantacle
7. Typlosole present in *Unio* in :
   (a) stomach (b) intestine
   (c) rectum (d) oesophagous
8. Heart beat of *Unio* per minute :
   (a) 10-20 (b) 20-40
   (c) 40-80 (d) 20-100
9. 'Organs of Bojanus' are derivatives of :
   (a) spleen (b) kidney
   (c) gall bladder (d) liver
10. Statolith is a part of :
    (a) statocyst (b) organs of Bojanus
    (c) liver (d) intestine

Answers

➤➤Fill in the blanks

1. kidneys, 2. excretory.

➤➤Multiple choice questions

1. (c) 2. (b) 3. (b) 4. (b) 5. (c) 6. (b) 7. (c) 8. (c) 9. (b) 10. (a)

58

Neopilina

Neopilina belongs to a group of molluscs known as **Monoplacophora**. Before the discovery of living specimens of *Neopilina,* the Monoplacophora had been known only as fossils from the Lower Cambrian to the Devonian and classified into 2 orders, 6 families and 21 genera and 11 species. All of them carried a single symmetrical shell, bearing 6 pairs of muscle scars on the undersurface. Hence, they were earlier classified with either the gastropods or the amphineurans.

DISCOVERY

Four living species of *Neopilina* have been discovered so far. *Neopilina galatheae* was discovered in 1952. Ten living specimens were dredged from a depth of 3,500 meters off the Pacific Coast of Costa Rica (Mexico) by the Danish research ship **Galathea**. They were first reported in February 1957 by **Dr. Henning Lemche** as *Neopilina galatheae.* Their detailed anatomy was worked out by **Lemche** and **Wingstrand** (1959). *N. ewingi* was dredged in the Peru-Chile Trench at 6,000 meters, off Peru in 1959. 14 more specimens were collected off the southern tip of Lower California in 1959 and 1960. A minute species, *N. veleronis,* was brought up from the Cedros Trench, Mexico, in 1961. *N. adenensis* was discovered from the Gulf of Aden in 1967.

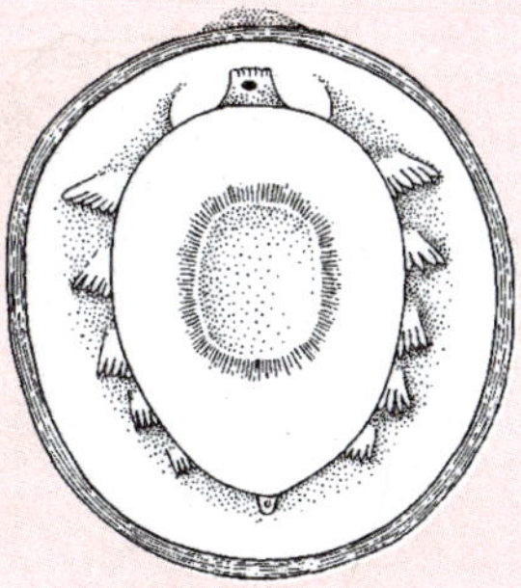

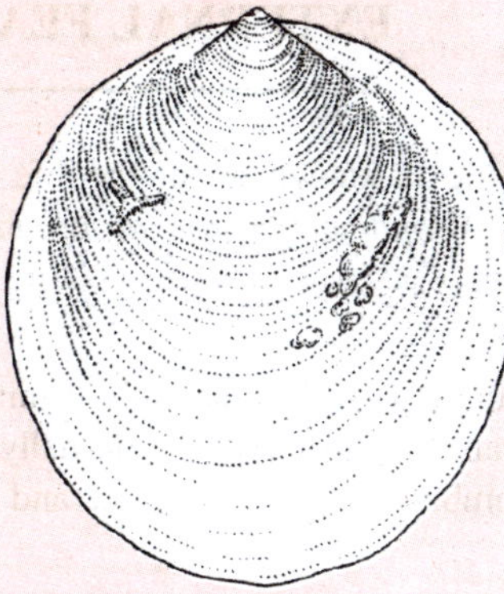

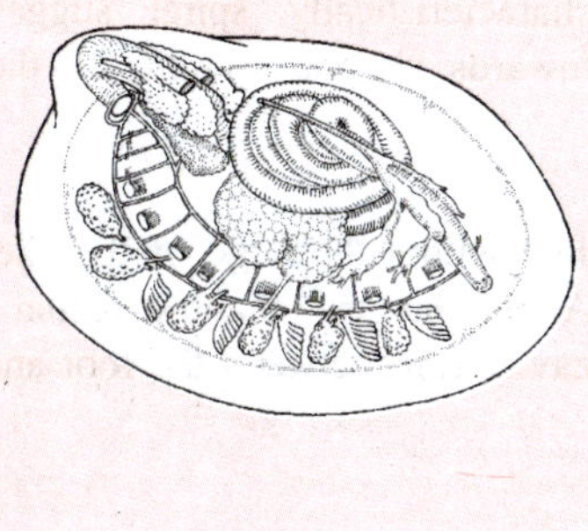

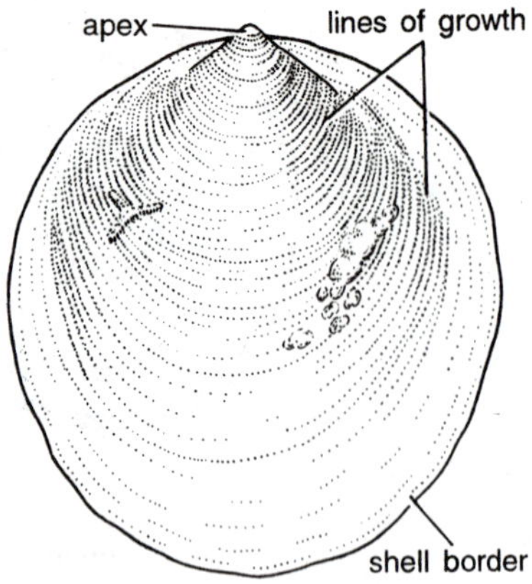

Fig. 1. *Neopilina galatheae.* Dorsal view of shell.

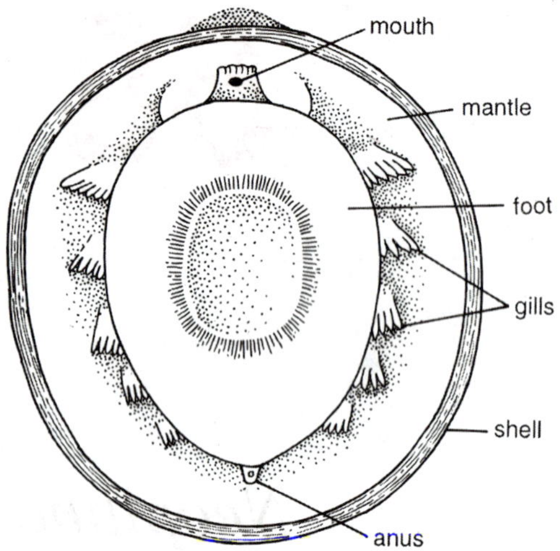

Fig. 2. *Neopilina galatheae.* Ventral view.

EXTERNAL FEATURES

Little is known about the life history and habits of *Neopilina* because of its abyssal habitat.

Shape and Size

Body is oval, bilaterally symmetrical and more than 3 cm long. Externally, it resembles a combination of gastropod and *Chiton.*

Shell

Dorsally, mantle secretes a single, large, cup-shaped, bilateral shell much like that of a limpet. It is thin, subcircular, about 35 mm long and its apex is raised and curved, both anteriorly and ventrally. Shell is 3-layered, made of an outer periostracum, a middle prismatic layer and an inner nacreous layer. Mode of shell formation is typically molluscan. Larval shell is characteristically spiral suggesting a tendency towards visceral spirality in the larva.

Mantle and Mantle Cavity

Just within the edge of the shell lies the mantle. A circular and shallow pallial groove or mantle cavity runs between the foot and the mantle edge.

Foot

Ventral surface carries a broad, flat, muscular and almost circular foot, evidentally meant for clinging and creeping. It is ciliated and has many gland cells.

Organs of Pallial Cavity

Ventrally, the **head** region, in front of the mouth is greatly reduced. Mid-ventral **mouth** opens anterior to foot and the median **anus** behind, in the pallial groove. In front of the mouth is a preoral ciliated flap, or **velum**, extending laterally on each side as a large palp-like appendage. Behind the mouth lies a postoral fold, forming a pair of tentacular tufts, which are considered homologous with the cephalopod arms and the scaphopod captacula, by **Lemche** and **Wingstrand**. Velum and the tentacles may be associated with feeding. In *N. galatheae* the pallial groove contains 5 pairs of **gills** (6 pairs in *N. ewingi),* placed symmetrically and lateral to the foot. Gills are unipectinate, rather than bipectinate, each composed of 5 to 8 finger-like branches or lamellae on one side of the gill-stem; vestigial lamellae may be present on the other side. It is probable that the ventilating current is similar to

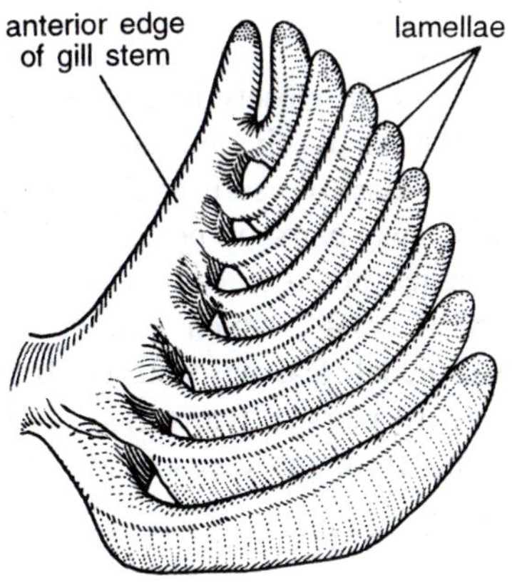

Fig. 3. *N. galatheae.* A gill in ventral view.

that of chitons. 5 pairs of minute **nephridiopores** open at the bases of the gills, while the 6th pair opens anteriorly. Gills are probably not true ctenidia but represent prectenidial structures or pallial gills.

INTERNAL STRUCTURE

Digestive System

Digestive system has both amphineuran and gastropodan characters. Digestive tract begins at the antero-ventral mouth. Inside the buccal cavity is a long and coiled subradular organ or radular sac, containing a well-developed food-rasping organ or **radula**, bearing 16 rows of comb-like teeth. Pharynx has two large diverticula and a single salivary gland. Straight oesophagus joins a somewhat triangular **stomach** which has a large digestive gland or **liver** lobe on each side and gives out a short caecum containing a **crystalline style**. **Intestine** is greatly coiled and the **rectum** opens into the mantle cavity through the posterior and mid-ventral **anus**. Mollusc feeds on radiolarians possibly by filter mechanism.

Musculature

8 pairs of pedal retractor muscles are present laterally. Their dorsal ends are inserted into the interior of the shell. Buccal musculature is complicated.

Vascular System

Blood vascular system resembles that of a molluscan including a dorsal **heart** and associated vessels and a series of sinuses. Heart, surrounded by a paired pericardial coelom, consists of two pairs of **atria** or **auricles**, opening into a pair of **ventricles**. Auricles receive arterial blood through **efferent gill vessels**, the anterior pair draining the first 4 pairs of gills and the posterior pairs draining the posterior pair of gills. Two ventricles

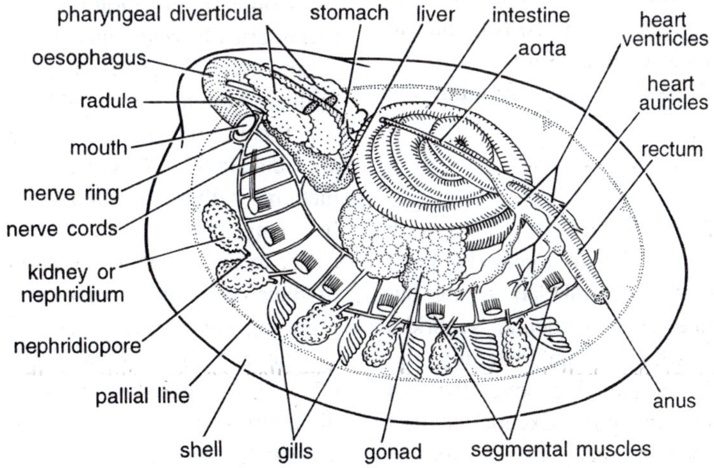

Fig. 4. *Neopilina.* Internal anatomy in side view.

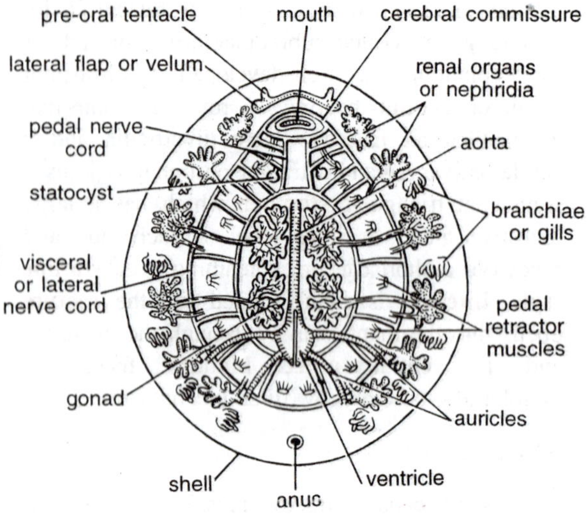

Fig. 5. *N. galatheae.* Internal anatomy in ventral view.

are divided by the rectum, which they surround but do not communicate with it. Each ventricle gives rise to an anterior **aorta**. Two aortae soon unite to form a typical **dorsal aorta** which extends forward to open into the head sinuses.

Coelom

Coelom is relatively extensive and made of three components : pericardial, dorsal and ventral. As already stated, the heart is surrounded by a paired **pericardial coelom**. In front of it, but not communicating with the pericardium, lies the paired **dorsal** or **body coelom**. It corresponds to the gonadial sacs in the chitons. **Ventral coelom** is represented by 2 pairs of sacs forming the gonads.

Nephridial System

There are 6 pairs of renal organs or **nephridia**, 6 of them located on each side of the body. Each nephridium consists of a central sac giving out several secretory lobules and diverticula. Nephridia have short ducts, which open laterally by **nephridiopores** into the pallial groove, the last five pairs located near the gills. Except the first pair, all nephridia open internally into coelom, 3 pairs into the body coelom and 2 pairs into the pericardial coelom.

Nervous System

The nervous system closely resembles that of the chitons. There is a circumoral nerve ring or **cerebral commissure** surrounding the mouth. From it arise two pairs of posterior longitudinal cords, the inner **pedal cords** and the outer or lateral **visceral cords**, which join posteriorly to form loops. Pedal and visceral cords are also connected by 10 pairs of **transverse connectives**. Nerves extend from the cords to the gills.

Reproductive System

Sexes are separate. Gonads are two pairs lying in the middle of the body on either side of the intestine and open in between the middle 2 pairs of nephridia which also serve as gonoducts. Fertilization is external. Nothing is known about the development.

RELATIONSHIPS OF MONOPLACOPHORA AND ORIGIN OF MOLLUSCA

Neopilina is undoubtedly an ancient form and the only living representative of the class Monoplacophora, which existed about 450,000,000 years ago. Its nearest relatives are fossils from Cambrian to Devonian. It is radically different from other living molluscs in that it is internally metamerically segmented and not quite bilaterally symmetrical. There is no common name of this mollusc which is truly a living fossil.

Precise relationships of the Monoplacophora are debatable. Whether the group represents metamerically segmented molluscs is not decided. In many features, the monoplacophorans come closer to the Amphineura, particularly in the nature of their circumpedal mantle groove, flat creeping foot, digestive system and nervous system.

Several theories are prevalent to explain the origin and phylogeny of the Mollusca. Until the discovery of the living monoplacophoran *Neopilina* in 1952, the Mollusca have been firmly and unanimously regarded as unsegmented animals. For many years it was thought that they have been derived from some marine flatworms, because of the following similarities.

(1) Soft, flattened and unsegmented body with primitive ciliation.

(2) Foot is ventral, muscular and creeping.

(3) Embryogeny of two groups is similar.

On the other hand, the affinity between Mollusca and Annelida, particularly the polychaete annelids, has long been recognized. They exhibit spiral cleavage and virtually identical trochophore larvae. The discovery of *Neopilina* lends further support to a molluscan-annelid relationship. Most striking feature of *Neopilina* is the apparent metameric plan of structure, having 8 pairs of retractor muscles, 5 or 6 pairs of gills, 2 pairs of auricles and 6 pairs of nephridia. Many features of *Neopilina* are archaic and it is regarded by many biologists to represent the ancestral molluscan structure with particular emphasis on its apparent metamerism. Several zoologists, such as **Lemche** and **Wingstrand** (1959) and **Fretter** and **Graham** (1962), believe that *Neopilina* is truly metameric, like annelids and arthropods, and therefore, consider Mollusca to be a fundamentally metameric phylum. This view holds that molluscs evolved from the annelids. There are, on the other hand, several objections to this view :

(1) *Neopilina* does not show external metamerism for there is a single shell and a single foot, without a vestige of segmentation.

(2) External signs of metamerism, shown by annelids, such as parapodia and intersegmental grooves, are unknown in Mollusca.

(3) Apparent internal metamerism of *Neopilina* is quite irregular. Duplicated structures differ in numbers and sequence. Therefore, it is not possible to determine structurally what would represent a monoplacophoran segment. Thus, the apparent metamerism of *Neopilina* cannot be compared with the classical metamerism of the Annelida.

(4) In *Neopilina,* the multiplication of structures seems to resemble that of *Chiton* and *Nautilus* and can be explained on functional grounds rather than implying inherent metamerism.

(5) Not all the metameric structures in *Neopilina* are primitive. For instance, the gills are unipectinate, not bipectinate. Gill-lamellae differ from those of the other molluscs.

(6) If Mollusca are basically metameric, it is difficult to understand why no other group of Mollusca, except Monoplacophora, retains any embryonic or anatomical vestige of metamerism. Even the 8 shell plates of chitons develop from a single embryonic shell gland.

Conclusions. Taking together all these facts, we come to several conclusions :

(1) Metameric nature of *Neopilina* is debatable. Its apparent metamerism probably shows a secondary replication of structures, such as has occurred independently in chitons and *Nautilus* (**Morton**, 1963; **Vagvolgyi**, 1967). Hence it is difficult to support the view that the Mollusca is a basically metameric phylum.

(2) Only the embryological studies of *Neopilina* can reveal whether the origin of its adult structure is basically metameric or not.

(3) Affinity between Mollusca and Annelida is apparent even without *Neopilina,* as shown by spiral cleavages and trochophore larva. But this does not prove that the Mollusca evolved from the Annelida. On the other hand, it may reflect only a common early ancestry. It is likely that the two groups diverged from the common ancestor before either the molluscan characters or annelidan characters, particularly metamerism, had evolved.

(4) **Vagvolgyi** (1967) holds that the coeloms of the Mollusca and Annelida have evolved independently. It may be argued that the ancestral coelom became reduced in the molluscan line.

(5) *Neopilina* certainly displays a number of archaic features, but its ancestral nature is also debatable. It does not resemble, nor it can be derived from the basic plan of the hypothetical ancestral mollusc as provided by **Yonge**, that is recognisable throughout the phylum. If *Neopilina* represents the ancestral form, much secondary simplification must

have occurred before the other molluscan classes evolved from it.

Present debate on the origin and evolution of Mollusca is nicely summed up by two important observations recently made by Morton (1963). (i) If *Neopilina* is the survivor of molluscs that were primitively metameric, it would be difficult to suggest from it what was the composition of a single generalised segment, or indeed to balance its segmental arithmetic at all. (ii) And *Neopilina* is a compellingly interesting animal, for which the malacologist will never cease to be grateful, but we fail to do it justice if, in our phylogenetic enthusiasm, we base upon it conclusions it will not easily carry.

Important Questions

➤➤*Long answer type questions*

1. Discuss the relationships of Monoplacophora emphasising the origin of Mollusca.

➤➤*Short answer type questions*

1. State the position and functions of the crystalline style of the Bivalves.
2. Why is the discovery of *Neopilina* of interest to the zoologist ?
3. Give an account of *Neopilina*.

➤➤*Multiple choice questions*

1. *Neopilina* belong to :
   (a) monoplacophora (b) bivalvia
   (c) snail (d) none
2. *Neopilina* discovered by :
   (a) Rich (b) Huxley
   (c) Hanning (d) Miller
3. Which show abyssal habitat :
   (a) *Unio* (b) *Pila*
   (c) *Mytilus* (d) *Neopilina*
4. Eight pairs of pedal retracter muscles are present in :
   (a) *Chiton* (b) *Neopilina*
   (c) *Sepia* (d) *Octopus*
5. Six pairs of renal organs is identification of :
   (a) *Loligo* (b) *Unio*
   (c) *Helix* (d) *Neopilina*
6. *Neopilina* is a connecting link between :
   (a) Arthropoda and Mollusca
   (b) Annelida and Mollusca
   (c) Mollusca and Echinodermata
   (d) Mollusca and Helminthes
7. The most remarkable characteristic of *Neopilina* is :
   (a) presence of shell
   (b) presence of gill
   (c) metameric segmentation
   (d) nephridia
8. Which of the following mollusc is a living fossil ?
   (a) *Limulus* (b) *Latimeria*
   (c) *Neopilina* (d) *Chaetoderma*
9. The primitive living mollusca :
   (a) *Architeuthis* (b) *Neopilina*
   (c) *Chiton* (d) *Dentalium*

Answers

➤➤*Multiple choice questions*

1. (a) 2. (c) 3. (d) 4. (b) 5. (d) 6. (b) 7. (c) 8. (c) 9. (b)

59

Sepia : The Cuttle-fish

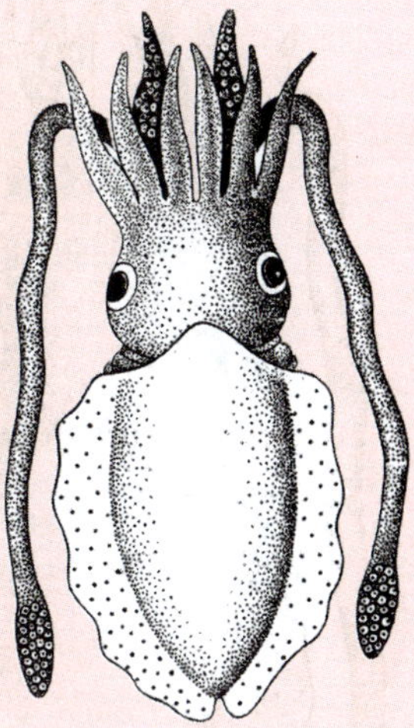

Cephalopods, whose living representatives include cuttle-fish, squid, *Octopus, Nautilus,* etc., are marine molluscs very different from the other molluscs in general. Adapted for a swimming existence, they are the most highly organised of all molluscs and include the largest species of the invertebrate animals. Head region, as the name implies, is large and well developed. A very common cephalopod is cuttle-fish.

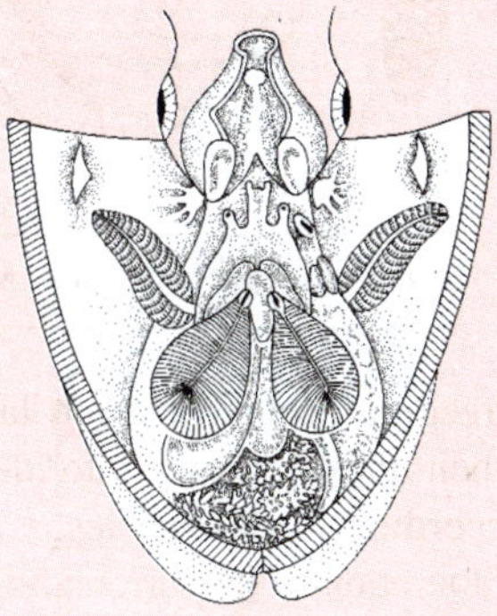

SYSTEMATIC POSITION

Phylum	Mollusca
Class	Cephalopoda
Order	Decapoda
Family	Sepiidae
Genus	*Sepia*

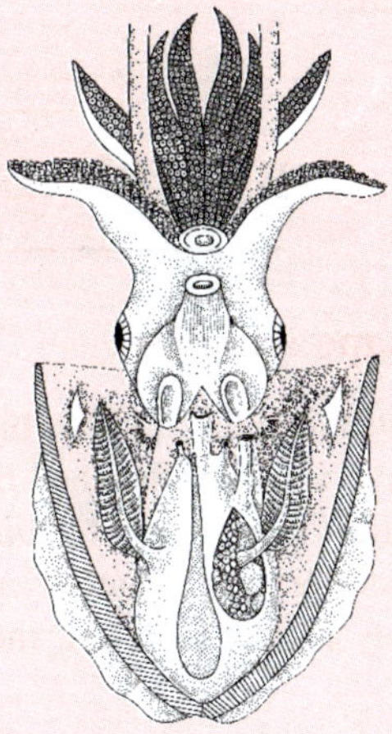

HABITS AND HABITAT

Cuttle-fish, like squid (*Loligo*), is a marine mollusc living usually in shallow coastal waters. It is widely distributed especially in warmer seas like Mediterranean. It is not a bottom-dweller. It is quite a good swimmer and can swim either forward or backward by its fins and funnel. They are found in groups, either swimming freely or resting on the sea bottom, where it can bury itself. Various species live in different depths of the sea, some as far down as 3,000 meters.

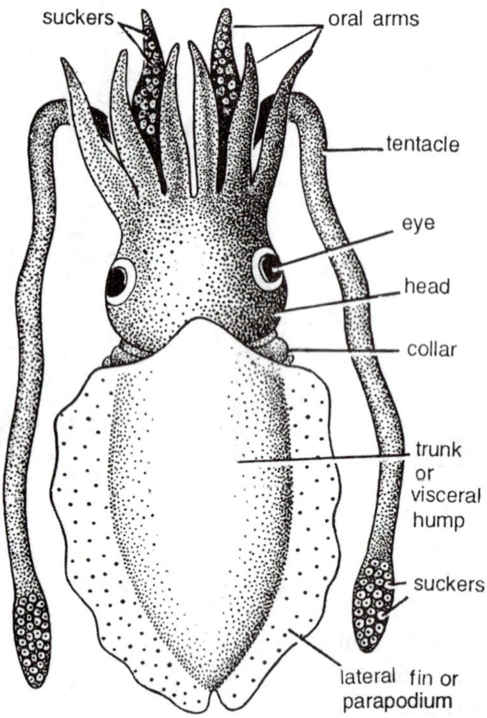

Fig. 1. *Sepia officinalis.* Male in dorsal view.

Breeding occurs between late winter and summer when they migrate into deep or shallow water, according to the species. Cuttle-fish can adapt its colouration to its surroundings and, like most other cephalopods, may exhibit **luminescence**. It feeds on fish, crabs, shrimps and prawns, etc., and can eject a jet of ink to distract the attention of the enemy.

EXTERNAL FEATURES

Shape and Size

Cuttle-fish has a fishy, bilaterally symmetrical and dorso-ventrally flattened body, which is shorter and flatter than that of *Loligo,* and tapers towards posterior end. Anterior and posterior ends of the body in fact represent the dorsal and ventral ends due to much elongation of the dorso-ventral axis of the body. Average size is about 30 cm. Smallest cuttle-fish, belonging to genus *Idiosepius,* measures about 15 mm in length.

Body is divisible into an anterior prominent **head** and a posterior **trunk**, united by a constricted **neck**.

Head

Head bears a pair of large, highly developed **eyes** at the sides, and a **mouth** at the free extremity, surrounded by 5 pairs of tapering, muscular **circumoral appendages**. These are differentiated into 4 pairs of short and stout **arms** and one pair of long **tentacles**, retractile into large pits at their base. Tentacles are used in the capturing of prey and in copulation. Bases of all but ventral pair of arms are united by an interbranchial **web** of integument. Inner flat surface of each arm bears four longitudinal rows of **suckers**. Each sucker is a muscular, shallow cup with a narrow, horny rim and supported on a short, thick stalk. Suckers can be firmly applied to body of the prey or to any other object by creating a partial vacuum inside.

In male cuttle-fish, left fourth or ventral arm is modified or **hectocotylised** to serve as an intromittent organ, by suppression of suckers in its basal part. In tentacles, suckers are found only on their terminal expanded portions. Apex of each tentacle bears a curious small terminal **pad**.

Trunk

Rest of the body or trunk is elongated and shield-shaped, with its base directed anteriorly and aboral end or apex posteriorly. It is slightly convex above and flat below, and bordered by a narrow frill-like fin on either lateral side. Fins are separated by a cleft at the aboral end and are used in leisurely locomotion.

Mantle

Trunk is covered by a thick, muscular **mantle,** enclosing on the ventral side a large **mantle cavity,** which contains viscera. Towards oral end, free mantle edge, encircling the narrow neck, forms a rounded lobe above and a **collar** below.

Funnel

Below the head lies a large conical muscular tube, **siphon** or the **funnel**, projecting beyond neck. It opens externally by a narrow aperture, but internally by a wide aperture, into mantle cavity. A pair of cartilaginous knobs on mantle fits into corresponding sockets on posterior ventral surface of the funnel. A typical molluscan foot is not present. It is represented, in part at least, by the siphonal apparatus. Old concept, that circumoral arms and tentacles correspond to the forefoot or epipodia of Gastropoda, seems to be erroneous; they are probably true cephalic appendages. Therefore, the name **Siphonopoda** has been suggested, instead of **Cephalopoda**, for the class.

Skin

Living cuttle-fish undergoes frequent **changes of colour**. These are brought about by numerous, large, pigment cells, the **chromatophores** and **iridocytes**, which lie in the dermis of skin. Chromatophores are of three kinds : reddish, yellowish-brown and orange. They can be detailed by the contraction of radiating muscle fibres attached to cell walls at one end, and to skin at the other. By alternate contraction and expansion of the chromatophores, blushes of different hues pass rapidly over the surface of the body. The chromatophores, are less on the ventral side, which is much paler than dorsal. Iridocytes, lying beneath the chromatophores, are transparent cells with a reticulate structure. They refract light, producing characteristic iridescence of the skin. The colour changes are controlled by central nervous system. Preserved specimens usually loose their natural colour.

Shell

The shell of cuttle-fish is internal, lying embedded in upper side, completely enclosed in a sac of the mantle and secreted by its epithelial lining. It is flat, broad and oval in shape, represented by **phragmocone** with a broader and rounded oral end, called **pro-ostracum**, and a narrow, pointed aboral end, called **rostrum**, projecting into a spine. Shell is entirely dead and composed of calcareous

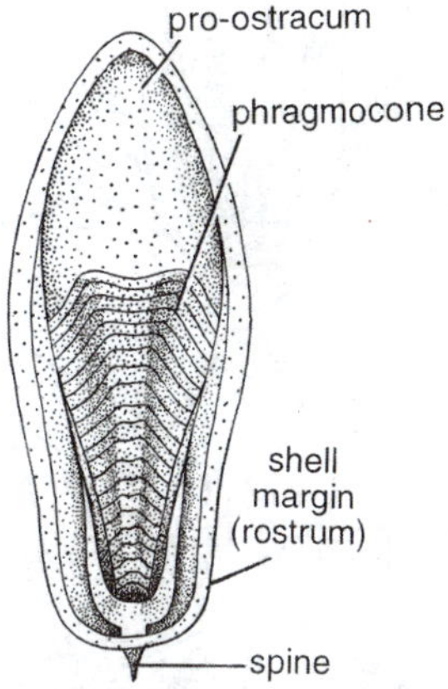

Fig. 2. *Sepia.* Shell.

rather than horny matter. Hard and resistant shell provides rigidity to trunk, like an endoskeleton. Calcareous matter is arranged in fine parallel layers, septa or laminae, enclosing spaces containing fluid and gas, so that light shell serves as a hydrostatic organ or float and, owing to its dorsal position, helps in maintaining the equilibrium of body.

Shell of cuttle fish, or **cuttle-bone**, is a familiar object of the seashore. It is rather soft and spongy. Being light, it floats in water, and during monsoon, these bones are drifted ashore in such great numbers that they have been named as the **sea-foam**. It is given as a bill sharpener as well as a source of calcium to caged birds.

MANTLE CAVITY

When mantle is cut mid-ventrally by a longitudinal incision and reflected, voluminous mantle cavity is exposed. It extends posteriorly up to apex of the trunk. Funnel is seen towards oral end with its narrow external and wide internal openings. Basally, on its ventral surface, funnel bears a pair of cartilaginous depressions or sockets, called **funnel cartilages**. Opposite to them, on internal surface of the reflected mantle folds, are seen a pair of oval cartilaginous knobs, the **mantle**

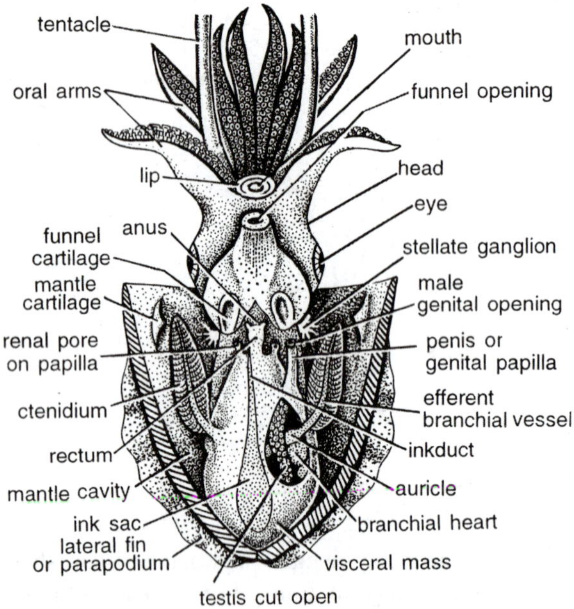

Fig. 3. *Sepia.* A male specimen with mantle cavity laid open.

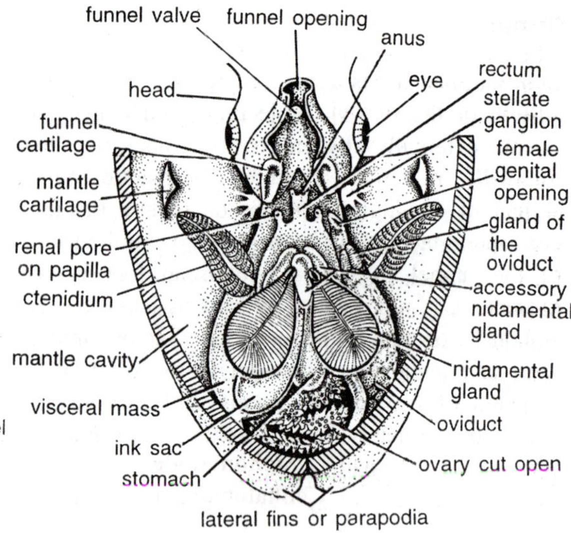

Fig. 4. *Sepia.* Mantle cavity of a female specimen. Visceral sac partly dissected.

cartilages. Similarly, dorsal surface of neck carries **nuchal cartilage**, which fits against **dorsal cartilage** on mantle.

Mantle cavity communicates with outside in two ways, through the wide opening around the neck and by means of the funnel. When the mantle cavity enlarges, the **inhalent water** is drawn into mantle cavity by the wide opening between mantle-collar and neck. When mantle cavity contracts, an interlocking occurs between mantle and funnel cartilages ventrally and nuchal and dorsal cartilages dorsally, so that collar space is tightly closed. Thus, **exhalent** water has to pass out through funnel in the interior of which a flap-like **valve**, opening outwards, allows water to run out from mantle cavity but not in the reverse direction.

Bulk of mantle cavity is occupied by **visceral dome**, consisting of various internal organs. The **digestive gland** lies anteriorly, bordered on either side by the **retractor muscles** of head and funnel. Median **rectum** opens by central **anus** at the base of the funnel. On either side of the rectum extends a **renal sac**, opening into mantle cavity by an external **renal aperture** upon a **renal papilla**. On left side only is the **genital aperture**, also at the end of a papilla. A little posterior lie the two large plume-shaped **ctenidia**, one on each side. Two large **stellate ganglia** lie one on each side on the mantle wall, where the neck meets the trunk. Posteriorly, the round **ink-sac** can be recognized by its metallic colour. Its duct runs forward ventral to rectum and opens into it dorsally close to anus. In **male**, testis lies partly covered by the ink-sac. In **female** renal sac is hidden from view by a pair of **nidamental glands** and a pair of **accessory nidamental glands**. **Genital duct** (**penis** in male, **oviduct** in female) opens on left side near left renal aperture.

LOCOMOTION

Cuttle-fish swims about gently by undulating movements of fins which are also used in directing the course of animal. But, most important movements are the swift, darting movements with the help of the funnel, caused by the rhythmical contractions of mantle. When mantle relaxes, mantle cavity is enlarged and water is taken in through collar space around neck. When mantle contracts the mantle-collar tightly fits on the neck closing this opening, so that water in the mantle cavity is shot violently out through the siphon like a jet, and animal is propelled rapidly through water in opposite direction. The resulting backward movement is so sudden and rapid that cuttle-fish seems to vanish instantly. The siphon may also determine the course of direction; when it is pointed forward, the cuttle-fish darts backward, and when it is pointed backward, animal is driven forward. Quick reaction is made possible by giant fibres in nerves, over which impulses travel very rapidly as in giant fibres of earthworm. Man has employed jet propulsion in rockets and jet aircraft.

Coelom

The coelom is represented by **viscero-pericardial coelom** and cavities of **renal sacs**. Former is a larger bag-like cavity extending backwards and divided by a constriction into two parts. Anterior part or **pericardium** encloses the hearts and communicates by two reno-pericardial apertures with cavities of renal sacs. Posterior part or **gonocoel** encloses the gonad.

DIGESTIVE SYSTEM

Alimentary Canal

Mouth, lying in the midst of the oral arms, is surrounded by a fleshy, circular **lip**, beset with numerous papillae. Just within lip is a pair of sharp, powerful, horny **jaws**, looking like the inverted beak of a parrot. Mouth leads into a large thick-walled, muscular **pharynx** or **buccal cavity**, containing **tongue** or **odontophore** and **radula**. **Oesophagus** is a long narrow tube running straight backwards between two lobes of liver to open into a rounded thick-walled muscular bag, the **stomach**. Large thin walled and slightly coiled pouch, **caecum**, is connected to stomach close to starting point of **intestine**. Short intestine runs anteriorly nearly parallel with oesophagus, and merges into **rectum** which opens into mantle cavity by **anus**. A pair of leaf-like **anal valves**, of uncertain function, project at the sides of anus.

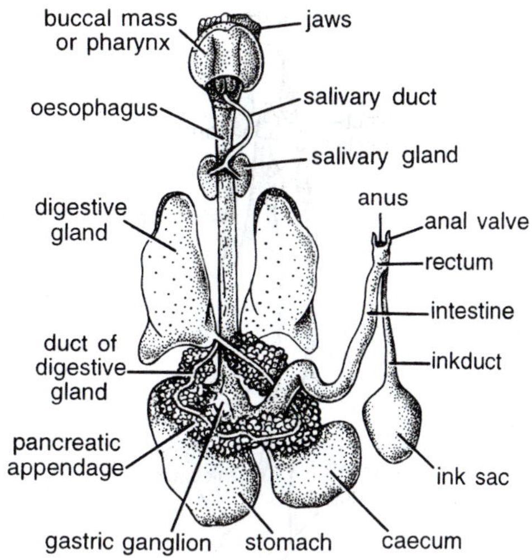

Fig. 5. *Sepia.* Alimentary canal.

Digestive Glands

Sublingual glandular tissue of unknown function lies on ventral side of tongue. A pair of **anterior salivary glands** lie within buccal mass opening on either side of radula. A pair of posterior **salivary glands**, lying in front of liver, one on either side of oesophagus, open by a common duct at the tip of tongue in buccal cavity. Salivary glands have been misnamed because they are really **poison glands** and their secretion is used to paralyse the prey. The large, brown **digestive gland** or **liver** extends from near posterior salivary glands up to the posterior end of body. It is a solid, bilobed gland, giving off one duct from each lobe. Two ducts pass through dorsal renal chamber and unite to open into vestibule or chamber where stomach, caecum and intestine meet. Two hepatopancreatic ducts bear minute vesicles which constitute the **pancreas**.

Ink Gland

As already stated, a pear-shaped **ink sac** lies over the posterior ventral surface of visceral dome and opens by a duct dorsally into the rectum close to the anus. Terminal part of the duct forms an

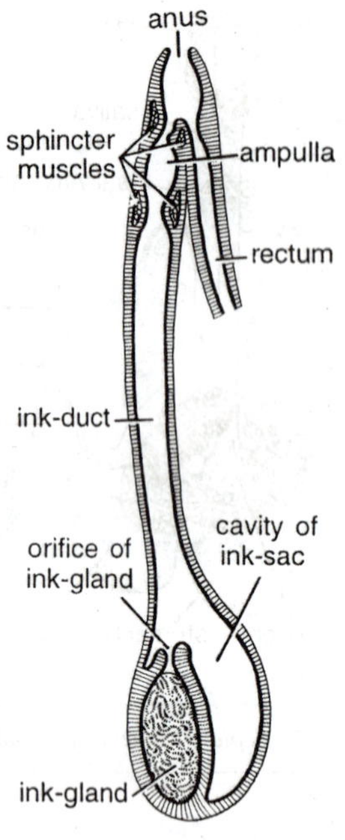

Fig. 6. *Sepia.* Ink-sac in M.L.S.

ejecting ampulla. **Ink gland**, lying inside the wall of the large reservoir or ink sac, secretes a brown or black fluid or **ink**. It contains a high concentration of melanin pigment and is stored in the ink sac. When the cuttle-fish is startled, it discharges the ink through the funnel as a black cloud, which forms a sort of smoke-screen or a dummy, under the cover of which the animal escapes from an enemy or approaches a prey. Ink of *Sepia* provides sepia pigment used by artists for hundreds of years.

Food and Feeding Mechanism

All cephalopods are carnivorus. **Food** consists chiefly of Crustacea, Mollusca and small fishes. Tentacles are rapidly extended and attached to the living prey by the suckers. Then the tentacles retract, so that the food is brought within the reach of the arms which hold it. Prey is paralysed by the poisonous secretion of the salivary glands, broken into pieces by the jaws and swallowed probably by the aid of the radula.

Digestion

Inside the stomach, the food mixes with the fluids from the liver and pancreas. Semi-digested food passes to the spiral caecum, where digestion is completed. Liquid products of digestion are absorbed in the caecum, while undigested food is passed on to the intestine where some absorption may take place. Residual food is expelled out of the anus.

Hepatopancreas not only secretes digestive enzymes but also serves for the absorption and storage of food and the excretion of waste products. Cephalopods differ from the majority of invertebrates in that probably the food stored in the liver is not absorbed directly but received through the blood by the liver.

RESPIRATORY SYSTEM

Respiratory Organs

These include a pair of large, plume-shaped **ctenidia,** lying in the mantle cavity, one on either lateral side. Each ctenidium or gill is bipinnate, with numerous delicate **lamellae** on either side of a central axis. Surface of the lamellae is much folded to increase the respiratory surface. Cilia are absent as removal of sediment is not a problem in pelagic animals and water current is created by mantle contractions. Each gill receives venous blood through an **afferent branchial vessel** from the branchial heart of its side. Inside the gill it passes through a system of minute branches through the lamellae and is collected finally into an **efferent branchial vessel** leading to the auricle.

Respiratory Mechanism

Muscular mantle rhythmically expands and contracts, so that the mantle cavity alternately increases and diminishes in size. Consequently, the oxygen bearing inhalent current of water enters the mantle cavity through the wide aperture around the neck and the exhalent current escapes through the funnel. Exchange of gases occurs when the water passes over the ctenidia which are richly vascular.

CIRCULATORY SYSTEM

Blood vascular system is well developed with a complete separation of venous and arterial blood. It consists of the heart, arteries, veins, and a system of capillaries.

There are three hearts in *Sepia*, as in all the dibranchiate cephalopods. **Systemic** or **arterial heart** lies in the middle of the visceral mass enclosed in the **pericardium**. It consists of a thick-walled median **ventricle**, and two thin walled lateral **auricles**, all spindle-shaped. Ventricle is slightly constricted into two lobes. It supplies arterial blood through a large **anterior** or **cephalic** or **oral aorta** and a small **posterior** or **aboral aorta,** to the anterior and posterior regions of the body, respectively. Aorta branch into **arteries** which lead into a system of **capillaries** and then into **veins**.

Venous blood of the head returns by a large **vena cava,** which bifurcates in front of the rectum into right and left **branchial veins,** leading into the so-called **branchial hearts**, lying one at the base of each ctenidium. Each branchial or gill heart also receives directly a small **pallial vein** from the mantle and an **abdominal vein** from the posterior body region. Unpaired **genital** and **ink sac veins** join the right branchial vein. Each branchial heart pumps blood to the ctenidium of its side through an **afferent branchial vein,** running through the axis of the gill and giving off branches as it goes. Oxygenated blood of a ctenidium is returned by an **efferent branchial vein** first to the auricle of that side, and then to the ventricle.

Blood containing **haemocyanin** and **amoebocytes,** is colourless when venous and pale-blue when oxygenated.

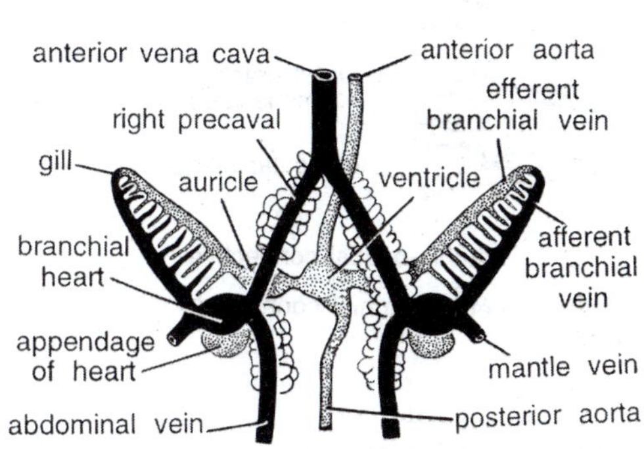

Fig. 7. *Sepia.* Circulatory system.

EXCRETORY SYSTEM

It includes a **kidney** or **renal sac** consisting of three thin-walled chambers, two ventral and one mid-dorsal, which communicate with one another. Two ventral chambers open, at one end, to the exterior by **renal apertures** placed on **renal papillae,** lying one on either side of the rectum, and at the other, communicate with the pericardium by **reno-pericardial apertures**. Through each ventral chamber passes the corresponding branchial vein, formed by the bifurcation of the vena cava. Vein is covered by excretory glandular epithelium which extracts the nitrogenous waste products from the blood. Dorsal chamber encloses the **pancreatic follicles** covering and opening into the ducts of the digestive gland. They are richly vascular and are said to serve an excretory function. Nitrogenous

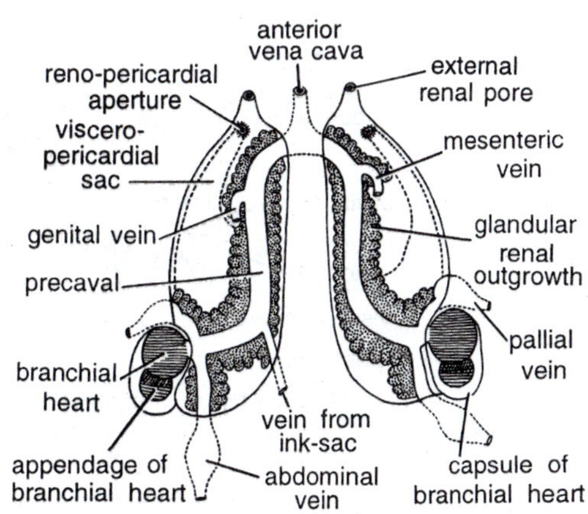

Fig. 8. *Sepia.* Excretory system.

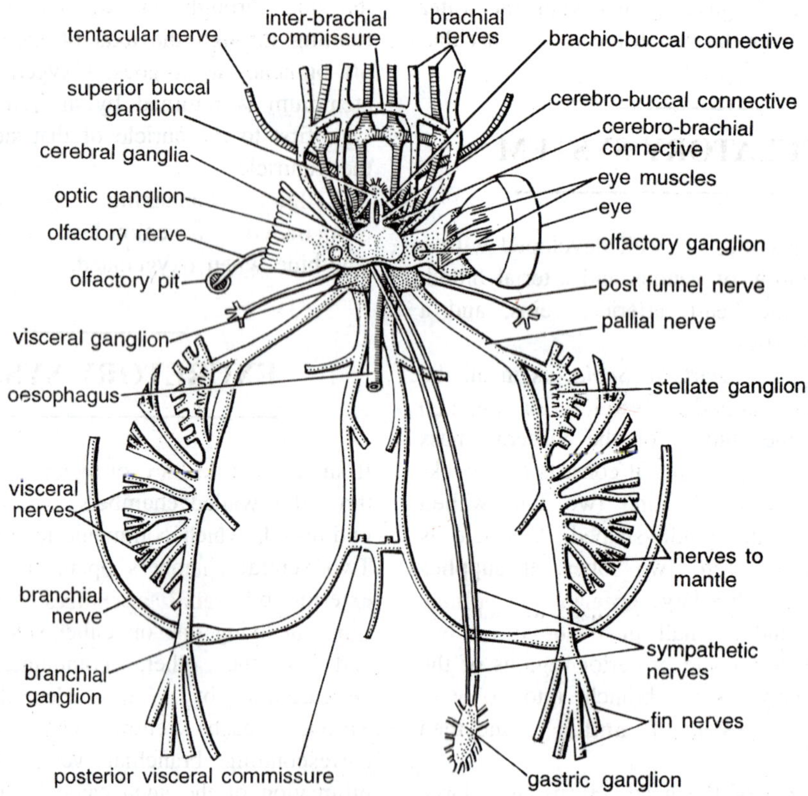

Fig. 9. *Sepia officinalis.* Nervous system in dorsal view.

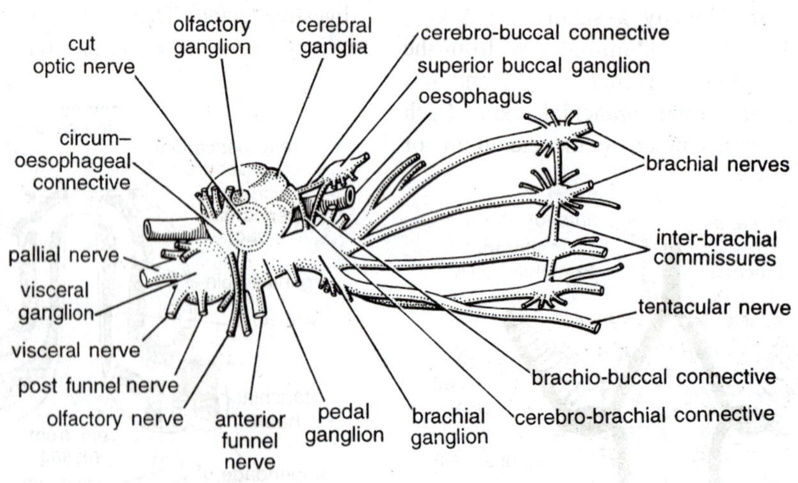

Fig. 10. *Sepia officinalis.* Nervous system in lateral view.

excretory substance has been detected in the cavities of the renal sac in the form of **guanin** which is discharged into the mantle cavity.

NERVOUS SYSTEM

Nervous system of *Sepia,* as of all the cephalopods, shows a high grade of organization, attained only by some insects and arachnids among the other invertebrates.

Brain

Brain consists of four typical molluscan ganglionic masses, all concentrated in the head, round the oesophagus, behind the buccal mass, and protected by a cartilaginous 'skull'. A pair of **cerebral** or **supra-oesophageal ganglia** are fused together into a rounded mass, lying dorsal to the oesophagus. Laterally, they give off a pair of extremely stout **optic nerves** which at once expand into large kidney-shaped **optic ganglia** of the eyes. A small **olfactory ganglion** lies on the dorsal side of each optic nerve. Anteriorly, a pair of slender **cerebro-buccal connectives** connects the cerebral ganglia to a pair of **superior buccal ganglia,** which are situated dorsal to the buccal mass, and connected by circum-oesophageal connectives to a pair of **inferior buccal ganglia** lying below the buccal mass. A pair of stout **circum-oesophageal connectives** connect the cerebral ganglia to rest of the brain lying ventral to the oesophagus.

Sub-oesophageal ganglionic mass, lying beneath the oesophagus, is partly divided into an anterior lobe, the **brachial ganglion**, and a posterior lobe, the **pedal ganglion**. From the brachial ganglion run forward ten **brachial nerves** to the arms; this is one of the reasons to regard these arms as portions of the foot. It is also connected to the superior buccal ganglia by paired **brachio-buccal connectives** and to the cerebral ganglia by **cerebro-brachial connectives**. Pedal ganglion supplies the funnel.

A pair of **pleuro-visceral ganglia** is also united to form a single mass lying in contact with the pedal, behind the oesophagus. They give off two pairs of main nerves, directed posteriorly. The **visceral nerves,** supplying the various internal organs, form a visceral loop from which springs a pair of **brachial nerves**, innervating the gills and bearing each a **brachial ganglion** at the base of the gill. Stout **pallial nerves** innervate the mantle. To bring about rapid and synchronous muscular movements of the arms, siphon and mantle, there is present a system of giant motor neurons, with its centre lying in a median ventral lobe of the fused visceral ganglia.

Stellate Ganglia

Pallial nerve on either side runs backwards through the neck to the inner surface of the mantle cavity, were it divides into two branches. **Outer branch** immediately terminates into a large, roughly triangular, **stellate ganglion** which can be seen without dissection, in front of the ctenidium when the mantle cavity is opened. Several nerves, arising from its outer border, innervate the mantle. **Inner branch** is connected to the stellate ganglion by two commissures, and innervates the fin.

Sympathetic System

A pair of **sympathetic nerves**, arising from the inferior buccal ganglion, runs posteriorly along the oesophagus to join **gastric ganglion** lying between the stomach and the caecum. Gastric ganglion sends nerves to the liver, stomach, caecum and intestine.

SENSE ORGANS

Special sense organs are very well developed and comprise of a paired **eyes**, **statocysts**, ciliated **olfactory pits** and an unpaired **gustatory organ**.

Eyes

Paired eyes are large, efficient and bulge from the dorso-lateral sides of the head. They bear striking resemblance to those of a vertebrate in that a cornea, iris, lens and retina are present. Lens projects an inverted image on the retina, as in the vertebrate eye. External muscle attachments enable limited movements of the eye. But the embryological development of the cephalopod eye is entirely different from that of the vertebrate eye, so that homologically they are different, for the vertebrate eye is formed as an outgrowth of the brain, while the cephalopod eye is formed by an ectodermal invagination. Similarity between the two is due to convergent evolution, that is, similarity which is not due to phylogeny.

Each eye lies within an **orbit** formed by cartilages. Outer wall of the eyeball, or the **sclerotic coat**, is strengthened by cartilage, and covered by a silvery membrane. It extends in front as the contractile **iris** presenting a large central opening, the **pupil**, which can be increased or diminished by muscles. Just internal to the iris lies a large, almost spherical **lens**, consisting of two plano-convex halves, and held in place by a **ciliary body**. A **choroid** is absent. Inner sensitive layer, or **retina**, is somewhat complicated in structure and is composed of a layer of parallel rods, there being no cones. Close behind the eyeball, the **optic nerve** swells up into the **optic ganglion**, from which several bundles of nerve fibres are distributed on the posterior surface of the retina. A small **optic gland** or **white body** of unknown function lies near the optic ganglion. A **true cornea** is also lacking. Transparent horny portion of the skin, covering the exposed surface of the eye, is termed as the **false cornea**. Skin also forms protecting **lids**. Cavity of the eye is divided by the lens into a small anterior chamber filled with a water-like **aqueous humour**, and a large posterior chamber containing a jelly-like **vitreous humour**. Cephalopod eye can accommodate itself to light changes both by modification in the pupil's size and by the migration of pigment in the retina. It can probably detect colour.

Statocysts

A pair of **statocyst**, which are organs of equilibrium or balance, lies ventral to the pleuro-visceral ganglion, enclosed in the cranial cartilage. Each statocyst is a small spherical body containing a large statolith and a fluid. Inner surface, lined with a flattened epithelium, is raised into numerous

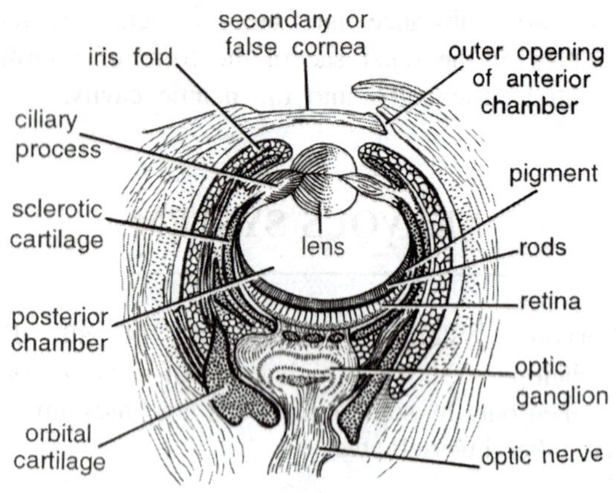

Fig. 11. *Sepia.* Eye in section.

processes, so that the cavity of the statocyst has a very irregular shape.

Olfactory Pits

Osphradia of the usual type are lacking. Instead, a small ciliated olfactory pit is situated posterior to each eye. Sensory cells of the pit are innervated from the small olfactory ganglion lying close to the optic ganglion.

Gustatory Organ

On the floor of the buccal cavity, just in front of the odontophore, is a small elevation covered with papillae. It is said to be the organ of taste.

Finally, **tactile** or otherwise sensitive cells are also found on the arms, tentacles and elsewhere.

REPRODUCTIVE SYSTEM

Sexes are separate. Males are usually smaller, less rounded dorsally, and possess slightly longer arms.

Male Reproductive System

Large, oval, yellowish and saccular **testis** lies near the apex of the visceral mass. Sperms produced in the testis are passed into its lumen which opens into a coelomic sac by a small ciliated aperture on its ventral side. Coelomic sac leads on the left side into a long narrow coiled tube, the **spermiduct** or **vas deferens**, which opens into a long **seminal vesicle**. Here, the sperms are rolled up into long and narrow bundles enclosed in elaborate chitinous capsules, called the **spermatophores**. A spermatophore is like an automatically explosive bomb; at one end it has a complex spring-like arrangement which ruptures its wall and discharges the sperms after copulation. Terminal part of seminal vesicle gives off two blind folds, one of them being the **prostate** or **accessory gland**. Seminal vesicle terminates into a wide reservoir, the sperm sac or **Needham's sac**, which opens into the mantle cavity by the **genital aperture** lying on a papilla or the **penis** to the left of the anus.

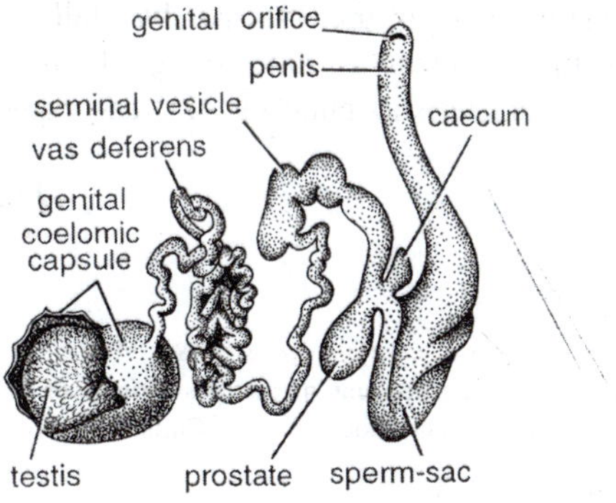

Fig. 12. *Sepia.* Male reproductive system.

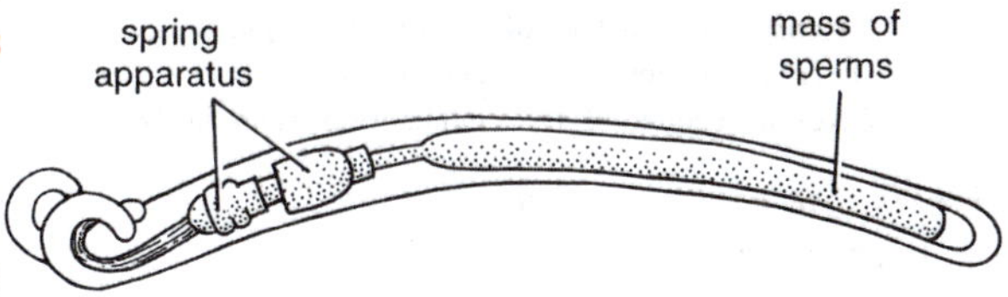

Fig. 13. *Sepia.* A spermatophore.

Female Reproductive System

Large rounded white and saccular **ovary** is also situated, like the testis in the male, in a coelomic sac near the apex of the visceral mass. **Oviduct**, leading from the coelomic sac, is a short, thin-walled, wide tube which opens into the mantle cavity to the left of the anus. Narrower distal end of the oviduct has thick glandular walls forming the oviducal gland, the function of which is to secrete the outer coat of the ova. A pair of large oval and flattened **nidamental glands** lies one on either side of the ink duct; each opens by its duct anteriorly into the mantle cavity. A pair of orange coloured **accessory nidamental glands** is situated in front of the nidamental glands, opening into the mantle cavity by numerous minute pores. All these glands serve to secrete the elastic egg capsules.

Fig. 14. *Sepia.* A cluster of eggs.

Copulation

In *Sepia,* the fourth left arm of the male is specially modified as an intromittent organ for copulation by the suppression of some basal rows of suckers, and is called the **hectocotylus**. Male inserts it in its own mantle cavity to extract the spermatophores, or the elastic capsules filled with sperms. Next, this arm is thrust into the mantle cavity of the female with its load of spermatophores which are deposited on the **bursa copulatrix**, or a modified part of the funnel. In some octopods, as in *Argonauta* and *Philonexis,* the hectocotylus breaks off and remains in the mantle cavity of the female. This detached arm was originally considered to be a parasitic worm and described as **hectocotylus**. Spermatophores burst by an elaborate arrangement and liberate the sperms after copulation. When ripe, the ova escape one by one from the ovary and, as they pass through the funnel, **fertilization** is effected. Viscid secretion of the nidamental glands and ink gland provide the tough gelatinous covering of eggs.

Masses of eggs are attached by stalks to a twig of marine plant or any foreign body and form the characteristic bunches of 'sea-grapes'. Eggs are larger and the developing embryos feed on large amount of stored food yolk. Young ones that hatch out are like the adults.

Important Questions

➤➤Long answer type questions

1. Give an account of nervous system of *Sepia.*
2. Write short note on : (i) Cuttle bone (ii) Ink sac.

➤➤Short answer type questions

1. State the position and functions of the statocyst in *Sepia.*
2. Write an account of the structure of the eye and the chromatophores of *Sepia.*
3. Describe how *Sepia* moves.
4. Describe the structure of the eye of *Sepia.*
5. Describe the nervous system in *Sepia.*
6. Give an account of the circulatory system in *Sepia.*

➤➤Multiple choice questions

1. Protective device of *Sepia* is :
   (a) ink secreted from ink gland
   (b) melanin pigment
   (c) smoke screen (d) brown fluid
2. Secretion of 2 pairs of salivary glands of *Sepia* are :
   (a) digestive (b) lubricant
   (c) poisonous
   (d) poisonous used to paralyse the prey
3. *Sepia* belongs to class :
   (a) Monoplacophora (b) Cephalopoda
   (c) Scaphopoda (d) Gastropoda
4. *Sepia* generally known as :
   (a) devil fish (b) mussel
   (c) cuttle-fish (d) snail
5. Luminescence exhibit in :
   (a) *Pila* (b) *Snail*
   (c) *Sepia* (d) *Neopilina*
6. Ink gland present in :
   (a) *Unio* (b) *Pila*
   (c) *Mytilus* (d) *Sepia*
7. Which belong to cephalopoda :
   (a) *Sepia* (b) *Unio*
   (c) *Mytilus* (d) *Pila*
8. *Sepia* belong to the order :
   (a) Octopoda (b) Decapoda
   (c) Pteropoda (d) Parasita
9. Siphonopoda is the term synonyms of :
   (a) Cephalopoda (b) Palecepoda
   (c) Gastropoda (d) none
10. *Sepia* show the type of development :
    (a) direct (b) indirect
    (c) none (d) both
11. Which of the following eats its prey ?
    (a) Leech (b) Starfish
    (c) *Sepia* (d) both (b) and (c)
12. Foot is displaced to the neighbourhood of the mouth and divided into arms is:
    (a) *Sepia* (b) *Chiton*
    (c) *Pila* (d) *Neopilina*
13. Squid, cuttle-fish and *Octopus* belong to the class :
    (a) Decapoda (b) Scaphopoda
    (c) Cephalopoda (d) Apoda
14. Cuttle-fish is a :
    (a) Mollusca (b) Annelid
    (c) Cnidarian (d) fish
15. Jet propulsion is the mode of locomotion in :
    a) Starfish (b) Dogfish
    (c) Cuttle-fish (d) Jelly-fish

Answers

➤➤Multiple choice questions

1. (a) 2. (d) 3. (b) 4. (c) 5. (c) 6. (d) 7. (a) 8. (b) 9. (a) 10. (a)
11. (d) 12. (a) 13. (c) 14. (a) 15. (c)

60

Mollusca : Characters, Classification and Types

'Mollusca' is derived from Latin word **mollis** or **molluscs** which mean soft bodied. This term was first applied by **Aristotle** to the cuttle-fish of the Aegean Sea.

DEFINITION

Mollusca which include clams, snails, slugs, squids, octopods and nautili are triploblastic, bilaterally symmetrical animals with anus and coelom and without segmentation. They usually have shell and a characteristic ventral muscular foot. There are 84,978 known living species in this phylum.

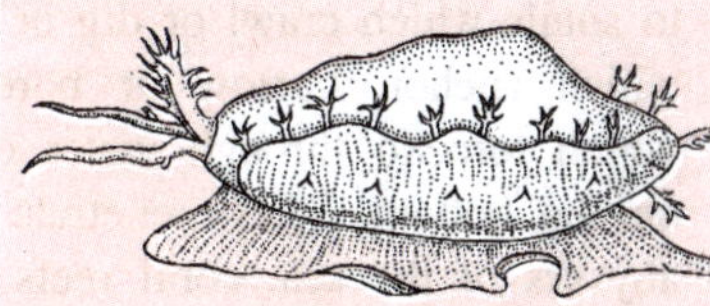

DERIVATION OF NAME

In Latin, **mollusca** is the name of a soft nut with a thin shell, referring to the bivalve shell and the soft bodied animal within the shell. The term Molluscs are among the most **abundant** of all animals. In number of species, the Mollusca is the second largest phylum after Arthopoda. It is not possible to assess accurately the total of the known species. But probably 100,000 living and a good number of fossil species (35,000) exist, as compared with ten times as many insects but only half as many vertebrates. Three quarters (80,000 species) of the Mollusca are the gastropodes with about 1,700 genera. Molluscs have no uniform plan as well as no specific shape. In an

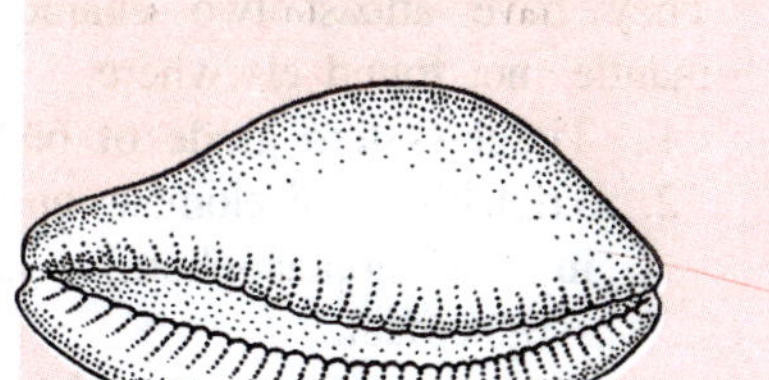

evolutionary sense they are plastic material as the out lines of the body are freely altered as new habitats are acquired and new structures are needed. Most of them are slow moving and confined to rather special habitats.

EVOLUTIONARY CHARACTERS

Molluscans exhibit few evolutionary characteristics which are :

1. Some molluscs are herbivorous while others are carnivorous. The digestive system is largely extracellular.
2. They mostly have ganglionated nervous system and the ganglia have a tendency to become concentrated at the anterior end.
3. The nephridial wall tends to become evaginated and folded to effect an increase in the surface area for tubular secretion of waste picked up from circulatory blood.
4. The gonads have lost their primitive association with the pericardial cavity and have mounted on special axis to the outside.

GENERAL CHARACTERS

Molluscs range from limpets clinging to the rocks, to snails which crawl or dig or swim, to bivalves which anchor, burrow or bore, to cephalopods which torpedo through water or lurk watchfully on the bottom. They penetrate all habitats : the abysses of the sea, coral reefs, mudflats, deserts and forests, rivers, lakes and underground. They may be hidden as parasites in the interior of other animals. They feed on every possible food and vary in size from giant squids and clams to little snails, a millimeter long. They form one of the most definitely characterized group of animals. They have atleast two characters 'radula and mantle' not found elsewhere.

1. Tissue-system grade of body organization.
2. Triploblastic, coelomate, unsegmented (except in Monoplacophora) and bilaterally symmetrical.
3. Body divisible into head, mantle, foot and visceral mass.
4. Shell, when present, usually univalve or bivalve, constituting an exoskeleton, internal in some.
5. Coelom reduced and represented mainly by pericardial cavity, gonadial cavity and kidney.
6. Digestive system complete with a digestive gland or liver (hepatopancreas); a rasping organ, the radula, usually present.
7. Circulatory system mainly of closed type, but some emptying into sinuses; heart with one or two auricles and one ventricle; blood with amoebocytes and haemocyanin.
8. Respiration direct or by gills or lungs or both.
9. Excretion by paired metanephridia (kidneys).
10. Nervous system of paired ganglia, connectives and nerves. Ganglia usually form a circumenteric ring.
11. Sense organs include eyes, statocysts and receptors for touch, smell and taste.
12. Dioecious or monoecious; one or two gonads with gonoducts, opening into renal ducts or to exterior.
13. Fertilization external or internal; development direct or through free larval forms.
14. Terrestrial or aquatic (freshwater or marine).

CLASSIFICATION

Molluscs are classified into six classes according to their symmetry and the characters of food, shell, mantle, gills, nervous system, muscles and radula.

CLASS 1. Monoplacophora

(Gr., **monos**, one + **plax**, plate + **pherein**, bearing)

1. Body bilaterally symmetrical, with a dome-shaped mantle.
2. Flattened limpet-shaped shell with spirally coiled protoconch.
3. Foot broad and flat, with 8 pairs of pedal retractor muscles.
4. Five pairs of gills in pallial grooves.
5. Six pairs of nephridia, two of which are gonoducts.

6. Radula in a raddular sac; intestine much coiled.
7. Heart of two pairs of auricles and a single ventricle.
8. Nervous system with longitudinal pallial and pedal cords.
9. Internal segmentation.
10. Marine.

Example : *Neopilina galatheae.*

CLASS 2. Amphineura

(Gr., **amphi**, both + **neuron**, nerve)

1. Elongated body with reduced head.
2. Radula present.
3. Shell as 8 dorsal plates or as spicules.
4. Foot ventral, large, flat and muscular.
5. Non-ganglionated nerve ring around mouth with 2 pairs of interconnected nerve cord.
6. Fertilization external; larva trochophore.
7. Marine.

Subclass I. Aplacophora

1. Body worm-like with a mantle but no shell and foot.
2. Calcareous spicules buried in cuticle.
3. Radula simple; mantle cavity posterior, some with a pair of bipectinate ctenidia.

Examples : *Neomenia, Nematomeina, Chaetoderma.*

Subclass II. Polyplacophora

1. Dorso-ventrally flattened body; small head (no eyes and tentacles); radula, mantle, foot and external gills present.
2. Mantle cavity posterior.
3. Shell as 8 calcareous dorsal plates.

Order 1. Lepidopleurina

1. Valves of shell without insertion plates.
2. Ctenidia a few and posterior.

Example : *Lepidopleurus.*

Order 2. Chitonida

1. Valves of shell with insertion plates.
2. Gills along whole length of mantle groove.

Examples : *Chaetopleura, Chiton, Ischnochiton.*

CLASS 3. Scaphopoda

(Gr., **scapha**, boat+**podos**, foot)

1. Tusk-shells.
2. Body within a tubular shell, open at both ends.
3. No head; mouth with tentacles; no eyes.
4. Foot conical, radula present; no gills.
5. Kidneys paired; gonad single.
6. Dioecious; larva trochophore.
7. Marine.

Examples : *Dentalium, Cadulus, Pulsellum.*

CLASS 4. Gastropoda

(Gr., **gaster**, belly + **podos**, foot)

1. Snails and slugs.
2. Torsion (coiling) of body mass at sometime in development.
3. Well developed head with eyes and tentacles; radula present.
4. Foot large and flat.
5. Shell present or absent; univalve and usually coiled.
6. Mostly marine; some freshwater or terrestrial.

Subclass I. Prosobranchia (=Streptoneura)

1. Body mass torted.
2. Head with a single pair of tentacles.
3. Shell closed by an operculum borne on foot.
4. Two ctenidia in mantle cavity situated anterior to the heart.
5. Nervous system streptoneurous with pleuro-visceral connectives twisted in a figure of 8.
6. Sexes separate; gonad single; larvae trochophore or veliger.
7. Marine, freshwater or terrestrial.

Order 1. Archaeogastropoda (= Aspidobranchia)

1. One or two bipectinate ctenidia.
2. Two kidneys and heart with two auricles.
3. Shell usually coiled.
4. Nervous system not concentrated, with pedal cord.
5. Genital products conveyed to outside through right kidney. Fertilization external.

Examples : *Fissurella* (key-hole limpet), *Patella* (limpet), *Trochus* (top shell), *Acmaea* (limpet).

Order 2. Mesogastropoda (= Pectinibranchia)

1. One auricle, one kidney and one monopectinate ctenidium.
2. Siphon, operculum and penis present ; osphradium developed.
3. Nervous system without pedal cords.
4. Fertilization internal; larva usually a free-swimming veliger.
5. Mostly marine, some freshwater.

Examples : *Pila* (apple snail), *Littorina* (periwinkle), *Crepidula* (slipper shell), *Cypraea* (cowrie), *Natica* (star shell).

Order 3. Neogastropoda (= Stenoglossa)

1. Shell with a short to very long siphonal canal.
2. Nervous system concentrated.
3. Osphradium large.
4. Free-swimming veliger suppressed.

Examples : *Buccinum* (whelk), *Murex, Nassarius, Oliva, Magilus, Malongena.*

Subclass II. Opisthobranchia

1. Shell small without operculum or no shell.
2. Body mass torted or detorted.
3. Gills posterior to heart.
4. One auricle, one kidney, and one gonad.
5. Nervous system euthyneurous, *i.e.,* without twisted pleuro-visceral loop.
6. Monoecious; larva veliger.
7. Exclusively marine.

Order 1. Cephalaspidea

1. Shell moderately developed.
2. Head with tentacular shield.
3. Lateral parapodial lobes prominent.

Examples : *Acteon, Bulla.*

Order 2. Anaspidea

1. Shell usually reduced or internal.
2. Well developed parapodial lobes.
3. Head with a pair of rhinophores.

Examples : *Aplysia* (sea-hare).

Order 3. Pteropoda

1. Shell present or absent.
2. Parapodial fins for swimming.
3. With or without mantle cavity.
4. Head with a pair of rhinophores.

Examples : *Cavolina, Clione, Corolla.*

Order 4. Sacoglossa

1. Shell present or absent.
2. Pharynx suctorial. Spermduct closed.

Examples : *Oxynoe, Elysia.*

Order 5. Acochlidiaceae

1. Minute to small-sized; no shell; no gill.
2. Visceral mass separated from foot and covered with spicules.
3. Inhabit coarse sand.

Examples : *Acochlidium, Unela.*

Order 6. Notaspidea

1. Shell external or reduced and internal.
2. Mantle, but no mantle cavity.
3. Bipectinate osphradium on the right side.

Example : *Pleurobranchus.*

Order 7. Nudibranchia

1. No shell, gill, mantle cavity and osphradium.
2. Various dorsal outgrowths.

Examples : *Doris, Aeolis.*

Order 8. Pyramidellacea

1. Shell typically spirally coiled.
2. An operculum but no gill and radula; proboscis long.
3. Semi-parasitic.

Example : *Pyramidella.*

Order 9. Philinoglossacea

1. Small naked snails without gill and head appendages.
2. Visceral mass separated from foot only by a groove.

Examples : *Philinoglossa, Sapha.*

Order 10. Rhodopacea

1. Vermiform snails, without external appendages.
2. Anus on right side of body.

Example : *Rhodope* (Single genus).

Order 11. Onchidiacea

1. Slug-like without shell and with pulmonary sac.
2. Posteriorly located anus.
   Examples : *Onchidium, Onchidella.*

Order 12. Parasita

1. Endoparasitic in holothurians.
2. Shelled embryos.
   Examples : *Entoconcha, Thyonicola.*

Subclass III. Pulmonata

1. Detorted body mass.
2. With or without shell.
3. Mantle cavity or pulmonary sac with a pore on right side anteriorly.
4. Hermaphrodites; mostly freshwater or terrestrial, a few marine.

Order 1. Basommatophora

1. One pair of non-invaginable tentacles.
2. Eyes at tentacular bases.
3. Male or female gonopores generally separate.
   Examples : *Siphonaria, Lymnaea, Planorbis.*

Order 2. Stylommatophora

1. Two pairs of invaginable tentacles.
2. Second pair of tentacles with eyes at their tips.
3. Male and female gonopores usually united.
   Examples : *Partula, Helix* (land snail), *Limax* (slug).

CLASS 5. Pelecypoda

(= **Bivalvia**, **Lamellibranchia**)

(Gr., **pelekus**, batchet + **podos**, foot)

1. Body enclosed in a bivalve shell and laterally compressed.
2. No head, tentacles, eyes, jaws and radula.
3. Foot often hatchet-shaped and extending between mantle lobes.
4. Mostly filter-feeding.
5. Usually dioecious, veliger or glochidium larva.
6. Mostly marine, a few freshwater.

Order 1. Protobranchia

1. Gill filaments not folded.
2. Mouth placed at the base of muscular proboscides.
3. Stomach with style sac.
   Examples : *Nucula, Nuculina, Yoldia, Malletia.*

Order 2. Filibranchia

1. Gill filaments reflexed but incompletely fused.
2. Chitinous gastric shield in stomach developed.
3. Style sac with crystalline style.
   Examples : *Ostrea, Mytilus* (mussel), *Spondylus* (edible oyster), *Pecten* (scallop).

Order 3. Eulamellibranchia

1. Gill filaments reflexed and fused completely to form tissue sheets.
2. Gills function for food gathering.
3. Style sac short.
   Examples : *Teredo* (ship-worm), *Ensis* (razor clam), *Unio, Anodonta, Lamellidens.*

Order 4. Septibranchia

1. Gills absent.
2. Stomach lined by chitin; style sac reduced.
3. Marine.
   Examples : *Poromya, Cuspidaria.*

CLASS 6. Cephalopoda = (Siphonopoda)

(Gr., **kephale**, head + **podos**, foot)

1. Body elongated dorsoventrally.
2. Shell external, internal or absent.
3. Head distinct and large with well-developed eyes; foot as tentacles and siphon; radula present.
4. Dioecious; development direct.
5. Marine and free-swimming.

Subclass I. Nautiloidea (=Tetrabranchia)

1. Shell external, coiled or straight; without complex sutures.
2. Recent species with many suckerless tentacles.
3. Two pairs of gills; two pairs of nephridia.
   Example : *Nautilus.*

Subclass II. Smmonoidea

1. Extinct.
2. Shell external and coiled with complex septa and sutures.
   Example : *Pachydiscus.*

Subclass III. Coeloidea (=Dibranchia)

1. Shell internal or absent.
2. Tentacles a few with suckers.
3. One pair of gills, one pair of nephridia.

Order 1. Decapoda

1. Ten arms – two elongated and called tentacles, and eight short arms. Examples : *Loligo* (squid), *Sepia* (cuttle-fish), *Spirula* (spiral shell).

Order 2. Octopoda

1. Body globular and without fins.
2. Eight equal arms. Examples : *Octopus, Argonauta.*

A FEW OTHER MOLLUSCS

1. *Chiton.* *Chiton*, or '**coat of mail shells**' is a marine polyplacophoran mollusc, well adapted for life on hard and uneven surface in the littoral zone. The characteristic feature is the presence of eight articulated **shell plates** borne on the mantle dorsally. The mantle cavity consists of two mantle or **pallial grooves**, bearing numerous gill lamellae in lateral series. The small head bears a long **radula** for grazing of attached algae and diatoms from rock surfaces. The ventral muscular locomotory foot is flat and elongated. The nervous system consists of a double ladder of pallial and pedal cords, and the only ganglia are sub-radular. The multiple sense organs are : (i) a pair of **osphradia**, one on either side of anus, (ii) **tactile receptors** in the mantle girdle and on the snout, (iii) **taste receptors** in the buccal region, (iv) **otocysts** near the pedal cords, and (v) **pit organs** in the shell plates. Sexes are separate, fertilization is external and the zygote develops into a **trochophore larva**.

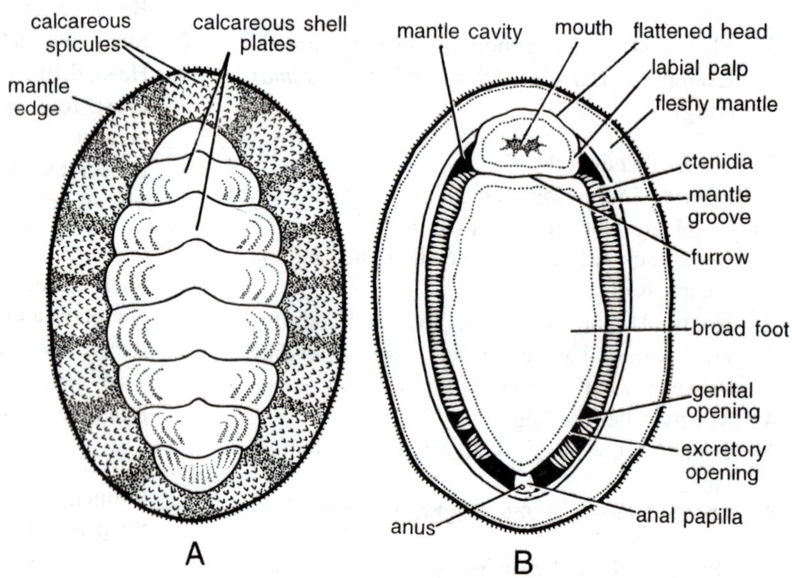

Fig. 1. *Chiton.* **A.** Dorsal view. **B.** Ventral view.

2. *Dentalium.* *Dentalium* is commonly known as '**elephant's tusk-shell**'. It is a marine mollusc, living at moderate depths in sublittoral zone. *D. entalis* is found burrowing in sand or shell gravel. The shell is a slightly curved and tapered tube, opening at both ends and resembling an elephant's tusk. The mantle cavity is long and the gills are entirely wanting. The foot is pointed and the mouth is located at its base in a projection of the pharynx. The radula is simple and its feeding function is assisted by groups of filamentous tentacles, called the **captacula**. The heart is without auriclcs, the excretory organs are paired nephridia and the nervous system is typically molluscan. The sexes are separate and the sex cells are discharged through right nephridium.

3. *Patella.* *Patella* is a small and sluggish marine gastropod, inhabiting the rocky beaches and feeding on algae. The dorsally placed shell is rounded and conical and the foot is a broad creeping sole, with no operculum. The head bears a pair of eyes and a pair of large sensory tentacles. The respiratory organs are leg-like secondary gills found in a circle between the mantle and foot. There are two unequal kidneys, opening into the

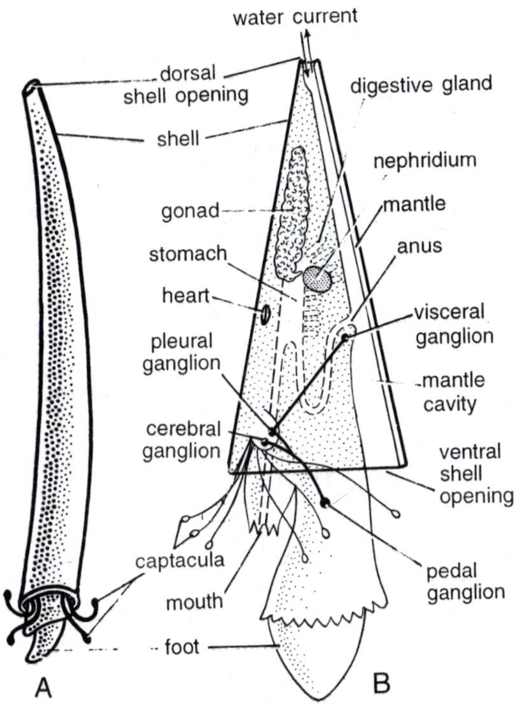

Fig. 2. *Dentalium.* **A.** External features. **B.** Internal (anatomical) features (diagrammatic).

anteriorly located mantle cavity by separate openings. A pair of osphradia projects into the manlte cavity, which serve to test the quality of incurrent water. The common species of *Patella* are *P. tarentina* and *P. vulgata.*

4. *Cypraea.* *Cypraea* is commonly known as 'cowrie.' It is a marine gastropod, found in Indian and Pacific oceans among coral reefs. The shell is univalve with a large single smooth and polished whorl. The shell aperture is narrow longitudinal and frilled slit on its ventral flat surface. Because of torsion, the body includes a single osphradium and single kidney. Cowrie shells are used in India in an indoor game, called 'Chowpad'.

5. *Aplysia.* *Aplysia* is commonly known as 'sea-hare'. It is a tectibranch gastropod, found crawling over the surface of rocks and boulders. The body is without an external shell and is large-sized. The small internal shell is covered externally by the mantle. The head bears two pairs of tentacles which are grooved on their outer sides; the posterior pair becomes ear-like and is called the **rhinophores**. At the base of each of these tentacles is located a simple eye. The foot is broad and bears a pair of lateral, folds, the **parapodia**, used in swimming, and a posterior small tail. The mantle cavity opens to the right side of body through a longitudinal slit. The anus is posteriorly located. *Aplysia,* when disturbed, secretes a purple fluid which makes the animal invisible to its enemy. The animal is bisexual, with a single gonoduct and a single genital aperture.

6. *Helix.* It is the common 'sand-snail' dwelling in moist and shady places like lawns and gardens. During the colder months it retires below the soil to hibernate. It remains more active at night (nocturnal) when it creeps about at its famous 'snail speed'; during creeping it secretes a film of mucus over the substratum to reduce friction. It is herbivorous, feeding mainly on leaves and fruits. *Helix pomatia* is the 'Frenchedible-snail' consumed in western countries.

The body is about 4 cm long, with a thin shell having a low conical spire and distinct lines of growth. The head bears two pairs of tentacles; the posterior pair is large and bears eyes. The

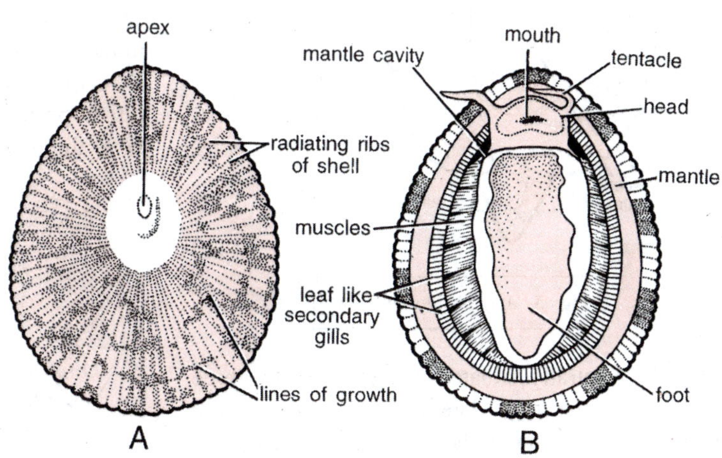

Fig. 3. *Patella.* **A.** Dorsal view. **B.** Ventral view.

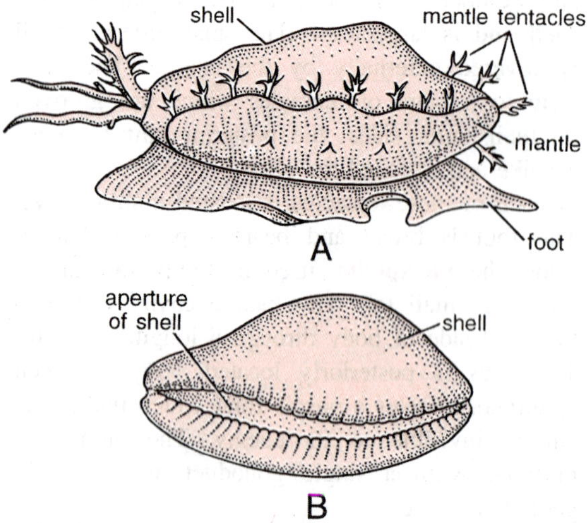

Fig. 4. *Cypraea* (cowrie). A. Living cowrie. B. Shell of cowrie.

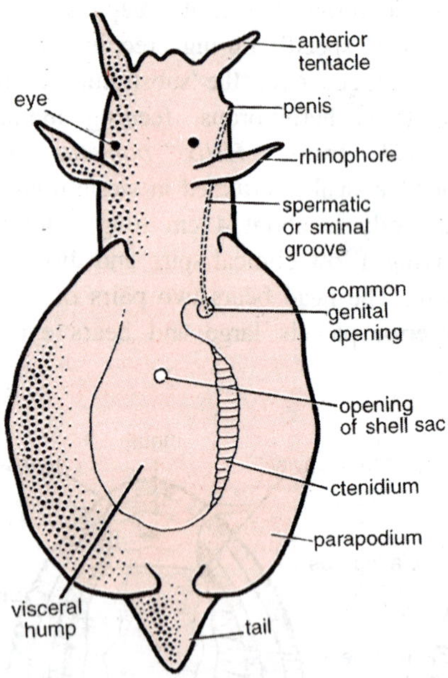

Fig. 5. *Aplysia.* Dorsal view.

mouth is ventrally situated and guarded by three lips; above the right lateral lip is situated the genital aperture. Below the shell, on the right side, is the pulmonary aperture or **pneumostome** which opens into the mantle cavity; the latter lies within the shell and functions like a lung during aerial

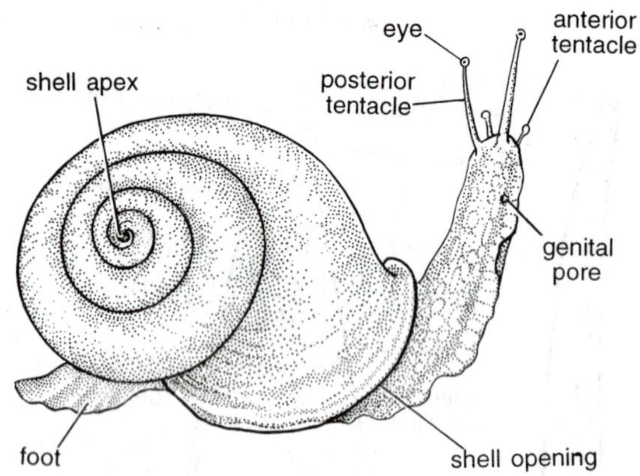

Fig. 6. *Helix.*

respiration. Anus lies behind the pneumostome. It is hermaphrodite. Oviposition takes place in July and August. The eggs laid in the ground, under cover, hatch into tiny snails within a month.

7. *Doris.* *Doris* also known as 'sea-lemon', is a marine, nudibranch gastropod. It creeps sluggishly over the substratum and feeds on encrusting organism like sponges. It has a short (5-10 cm long) oval body with a convex dorsal surface and a flat ventral surface; the latter is marked by the presence of a broad foot. The body is covered by a tough, pigmented mantle beset with calcareous spicules and bearing **papillae** or **tubercles** on the dorsal side. Due to **detorsion** (complete reversion of torsion), the shell, mantle cavity and primary gill are absent. The head is indistinct but bears a pair of prominent, short, retractile **tentacles** or **rhinophores** which are olfactory in nature. The **anus** lies mid-dorsally near the posterior end and is surrounded by a

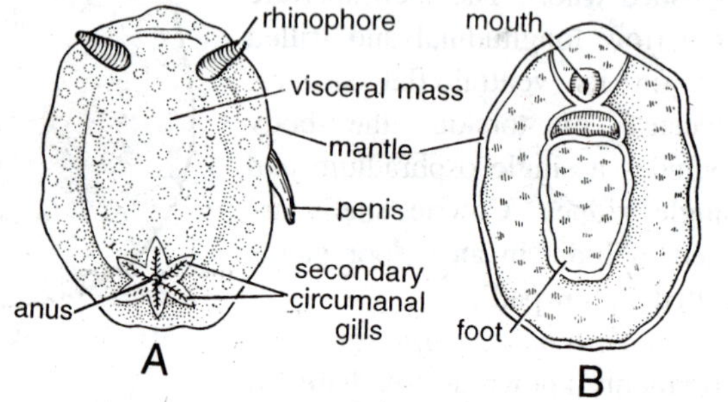

Fig. 7. *Doris.* A. Dorsal view. B. Ventral view.

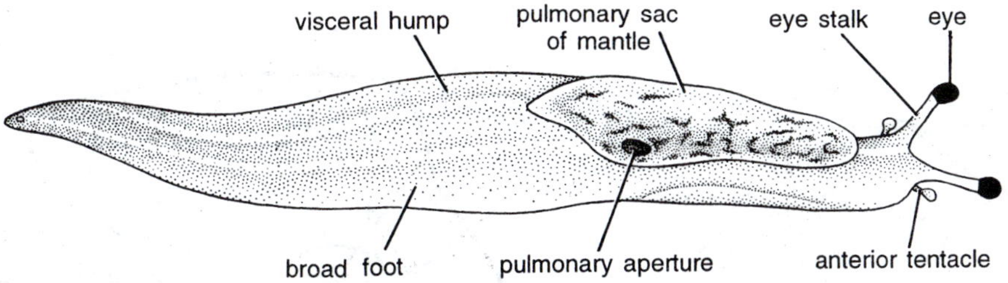

Fig. 8. *Limax.*

circlet of feathery, retractile **secondary gills**, called **cerata**. A median **renal aperture** lies near the anus. The animal is hermaphrodite; the penis and the genital aperture lie asymmetrically on the right side.

8. *Limax.* *Limax* is commonly known as 'grey-slug'. It is a terrestrial gastropod, living in gardens and moist woody region. It feeds voraciously on succulent plant material. The body is elongated without an external shell. The head bears a pair of retractile tentacles, the posterior pair bearing eyes at their tips. The mantle is a small antero-dorsally placed shield with a pulmonary opening, leading into the pulmonary sac, meant for aerial respiration. A small and thin shell lies beneath the mantle. The ventral muscular foot is broad and flat and serves for creeping. The slug is nocturnal in habit and lays eggs in damp sheltered places.

9. *Mytilus.* *Mytilus* is commonly known as 'sea-mussel'. It is a cosmopolitan, marine and sedentary animal, found attached to rocks between tide-marks. It is a filter feeder, filtering planktons from the incurrent water. The shell is bivalve with the anteriorly placed umbo. The ventral foot is tongue-like with **byssus** threads, serving as organs for attachment. A pair of simple eyes is found anterior to the inner gill lamella. The sexes are separate.

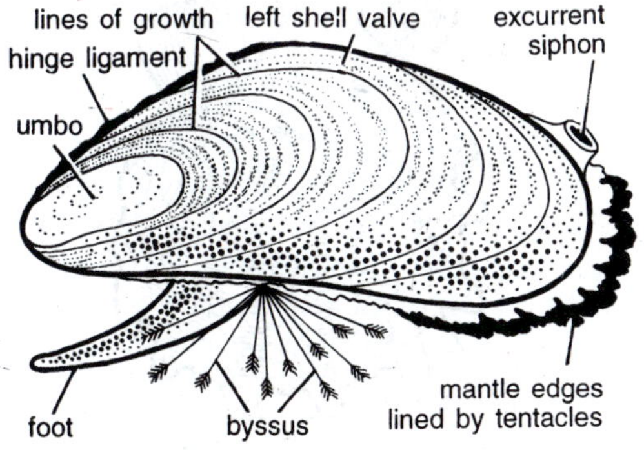

Fig. 9. *Mytilus.*

10. *Ostrea.* They are bivalve sedentary marine oysters, found attached to rocks and stones. The two valves are unequal. The left valve being large, thick and convex and remains permanently attached to a rock, piling on another shell. The right, smaller, thinner and flattened valve form a sort of lid. The shell surface is coarse, irregular and ruffled. There is a single adductor muscle. The foot is absent. There is no byssus. Heart with two fused auricles lies below the rectum. The same individual can alter the sex as male and female. A single female lays about half a billion eggs in a season. This high rate of egg laying is due to high rate of mortality in eggs. Veliger larvae after setting at the bottom are called 'spat'. *Ostrea,* commonly known as edible oysters

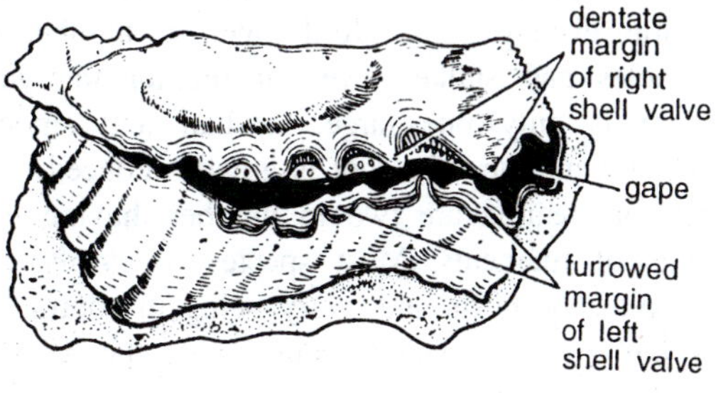

Fig. 10. *Ostrea*

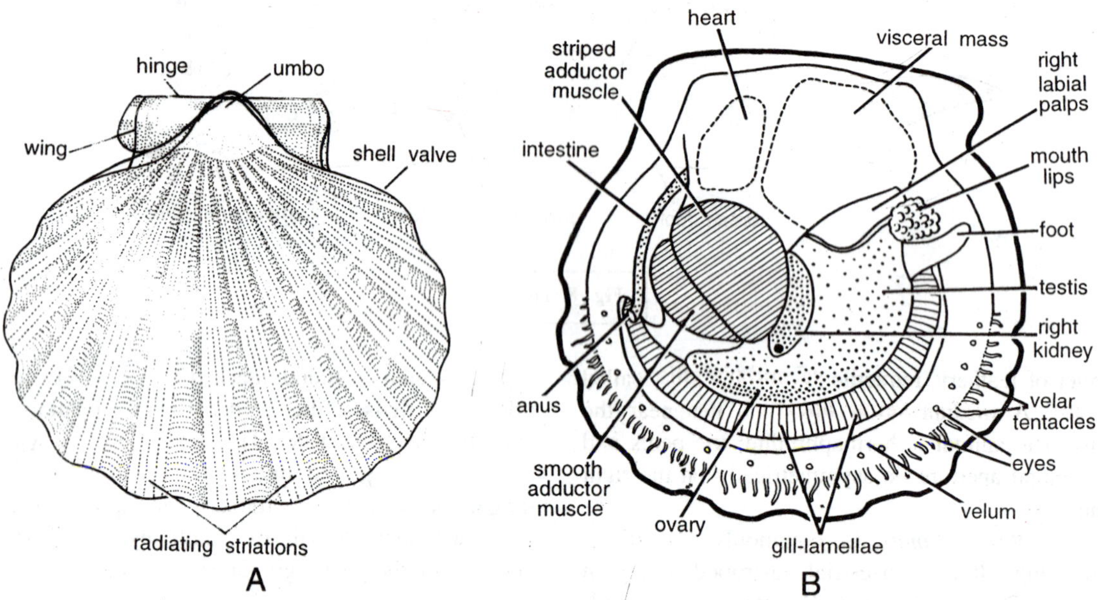

Fig. 11. *Pecten.* **A.** External features. **B.** Internal organs after removal of right shell valve.

are nicely flavoured, rich in vitamins and minerals. The Indian edible oyster is *Ostrea virginiana.* The shells of *Ostrea* are grinded up which form a source of calcium for poultry.

11. *Pecten.* *Pecten* is commonly known as 'scallop'. It is a worldwidely distributed marine bivalve. The unequal shell valves are beautifully sculptured and bear radiating ribose striations. The hinge ligament is straight and toothless. The animal rests on the larger right valve which is having an anterior notch. Locomotion takes place in spurts, the swimming is affected by the rapid opening and closing of shell valves. The margin of the mantle displays velar folds with a number of tentacles and stalked eyes at regular intervals. There is only one large, median and smooth adductor muscle. The foot is greatly reduced. The gills or ctenidia are two in number, having two kinds of filaments. Most species of *Pecten* are bisexual, the testis and ovary being in continuity. *Pecten irridians,* which occurs along the Atlantic Coast, makes a delicious food article.

12. Pearl oyster. Oysters are sedentary marine bivalves found in all seas except in colder ones. The Indian pearl oyster is *Pinctada vulgaris* that is known for yielding precious pearls. The two shells are unequal, with the left one comparatively larger than the right one and remains attached to a rock. The adductor muscle is single and large. The foot is lacking in the adult. Pearl is formed as a result of nacreous secretion from the mantle around a sand particle.

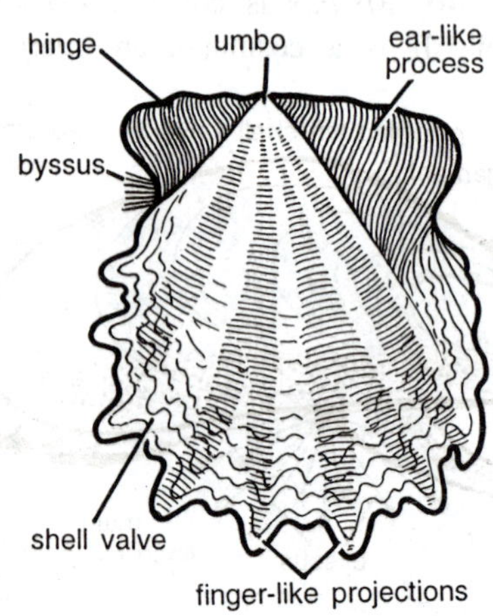

Fig. 12. Pearl oyster *(Pinctada vulgaris).*

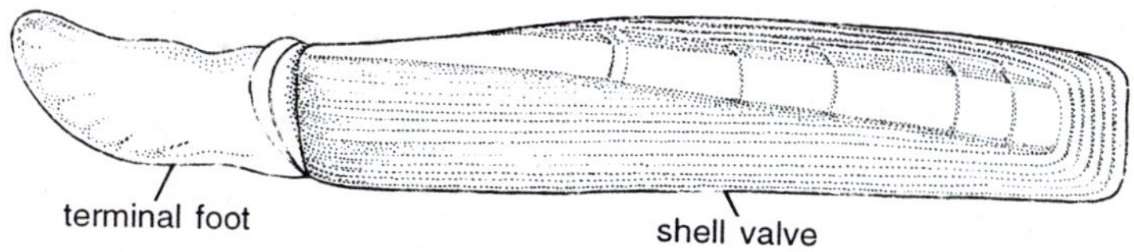

Fig. 13. *Solen.* Razor mollusc.

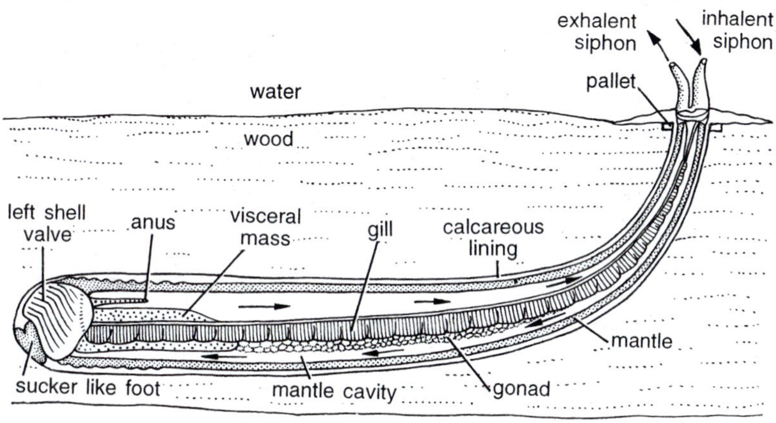

Fig. 14. *Teredo.* Anatomical features (diagrammatic).

13. *Solen.* *Solen ensis* is commonly known as 'razor mollusc' because of its superficial resemblance to a barber's razor. It is a marine mollusc that burrows in sand. The body is elongated and narrow, enclosed in a thin bivalve shell. The umbo is located at its anterior end. Each valve has its own hinge tooth. The mantle lobes are united ventrally to form an elongated tubular mantle cavity, bearing a pair of long and narrow ctenidia. The foot is muscular and cylindrical and constitutes the burrowing apparatus.

14. *Teredo*. *Teredo* is commonly known as 'shipworm'. It is a highly specialized marine bivalve which is very destructive to wood in sea water. The body is long and slender with a small anterior shell. The shell is used for burrowing in the wood of ships or wharves. The foot is extremely reduced and acts as an adhesive organ for attachment to the wall of its hole-habitat. The mantle cavity bears paired and elongated gills and gonads. The inhalent (incurrent) and exhalent siphons are also elongated.

15. *Loligo.* *Loligo* is commonly known as 'squid' or 'sea-arrow'. It is also a decapod cephalopod, like *Sepia,* and is found in warmer seas. Its body is long, like a *Torpedo,* and resembles very much that of *Sepia* in form except that it is narrower than that of *Sepia.* The head bears a circle of 10 oral arms provided with suckers devoid of horny rims. Two of these arms are long and called the tentacles which are more retractile than others and bear suckers only at their distal ends. The lateral fins (parapodia) cover about half of the trunk surface and meet at the posterior end to form a sort of 'arrow-head'. *Loligo* is a fast swimmer in the open waters of the sea. The shell is a narrow internal pen-like strip. The ink sac secretes the ink. Sexes are separate. The copulation is followed by spawning and death.

16. *Octopus.* *Octopus* or 'devil fish' is an octopod cephalopod, possessing eight oral arms. The body is globular and bag-like and there are no lateral fins and internal shell as are seen in

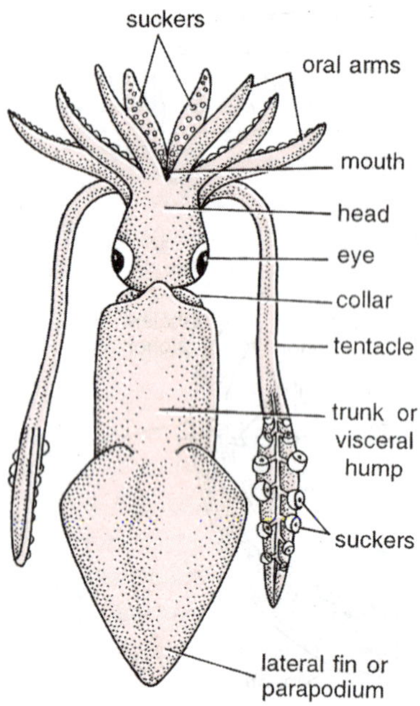

Fig. 15. *Loligo.* Dorsal view.

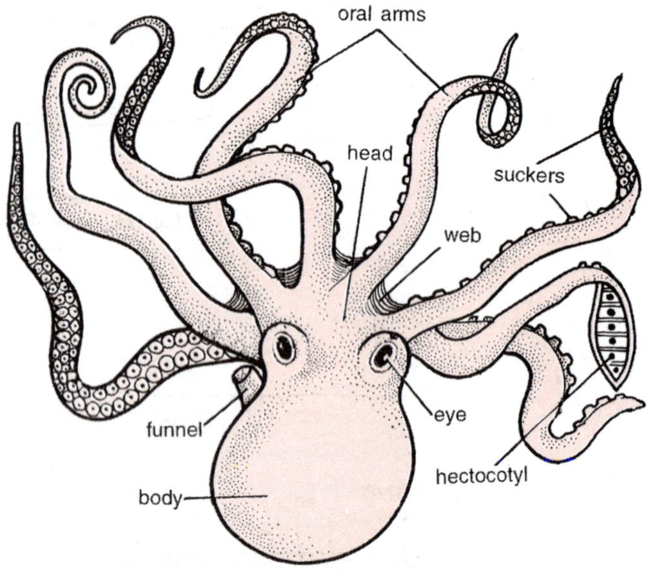

Fig. 16. *Octopus.* Dorsal view.

Sepia or *Loligo.* The head bears a mouth and two large prominent eyes. The oral arms are much elongated and bear suckers in two rows on their inner surfaces. These arms are joined together at their bases by a **web**. *Octopus* occurs in dark crevices and among corals in sea water. It is a benthic hunter.

17. *Nautilus.* *Nautilus* is a tetrabranch cephalopod, found in sea waters in Indo-Pacific region. The shell is external and is coiled over the head in a bilaterally symmetrical planospiral. It is divided into internal chambers by the presence of transverse septa. The living *Nautilus* lives only in the last chamber. The septa are perforated in the middle and through the perforations emerges a cord mantle, called **siphuncle**, through which the animal secretes gas into the empty chambers. This makes the shell buoyant. The inner layer of the shell is pearly. The body of *Nautilus* is

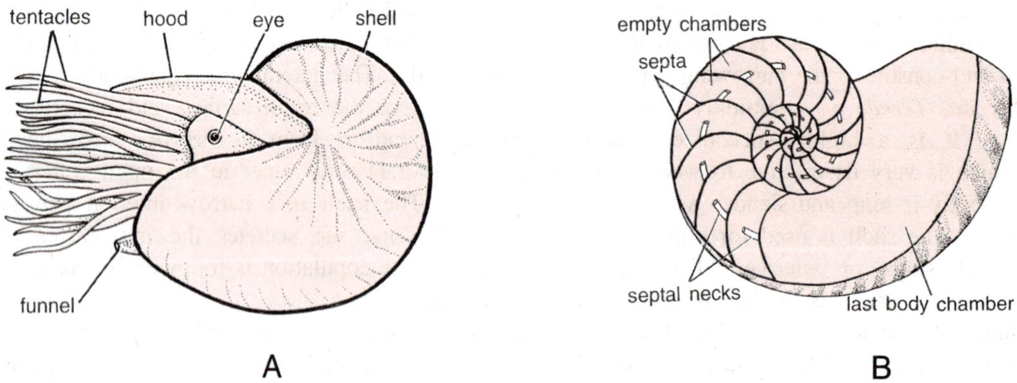

Fig. 17. *Nautilus.* **A.** Animal within shell. **B.** Shell in section.

distinguished into a large conical head, bearing a mouth, a pair of eyes and about 90 filiform tentacles without suckers and a round bag-like trunk which has no ink sac. Osphradia are present and the respiratory current is generated by muscular pulsations of the funnel. *Nautilus* is always active at night and rests on bottom during the day.

Important Questions

➤➤Long answer type questions

1. Classify Mollusca up to orders, giving diagnostic characters and representative examples.
2. Write short notes on : (i) *Patella,* (ii) *Chiton, (iii) Dentalium,* (iv) Sea-hare, (v) Razor mollusc, (vi) *Loligo,* (vii) *Teredo,* (viii) *Nautilus.*

➤➤Short answer type questions

1. Name the animal to which the glochidium larva belongs.
2. Name the animal group where the hectocotylyzed arm is present.
3. Mention two wood boring molluscs.
4. Name the distinguishing features of the *Nautilus.*
5. What is the 'sea devil' ? Give reason for such a name; should limit your answer in five lines.
6. Define (i) Monotocardia, (ii) Diotocardia.
7. Mention the name of a mollusc that bores through the wood and has a microfauna related to its mode of nutrition. What would be your evaluation from the above statement. Write in five sentences.
8. Give reason for the popular name Paper *Nautilus.*
9. Present a clear illustration to describe the shell of cephalopoda. Add a note on its evolutionary significance.
10. What are Keber's organ ? Where do you find them ?
11. Justify the inclusion of *Pila* and *Lamellidens* in the same phylum.
12. Give a short account of archimollusc.
13. Give an account of various kinds of foot adaptations found in Mollusca.
14. Write an essay of 'Foot in Mollusca'.
15. Briefly explain the peculiarities of the nervous system in Gastropoda.
16. Define term Mollusca.
17. Describe the habit of Mollusca.
18. Describe the pigmentation in blood of Mollusca.
19. Give an account about larval stage of Mollusca.
20. Define Monoplacophora.
21. Describe Scaphopoda.
22. What do you mean Gastropoda ?
23. Define the Bivalvia.
24. Give the major characteristic of Cephalopoda.
25. What are the identical characteristic of Mollusca.

➤➤Fill in the blanks

1. *Chiton* commonly known as........
2. *Aplysia* is commonly known as..,.....
3. Sea lemon is the common name of........
4. *Mytilus* generally called as........
5. Shipworm is the common name of........

➤➤Matching type questions

(i) *Teredo*	(a) scallop
(ii) *Loligo*	(b) pearl oyaster
(iii) *Octopus*	(c) shipworm
(iv) *Solen*	(d) devil fish
(v) *Pinctada*	(e) squid
(vi) *Pecten*	(f) razor worm

➤➤*Multiple choice questions*

1. Identify the mollusc having the monopectinate ctenidium, in the list given below :
(a) *Pila* (b) *Lamellidens*
(c) *Nautilus* (d) *Chiton*
2. Choose the tetrabranchiate mollusc from the given list below :
(a) sea squid (b) *Area*
(c) *Anodonta* (d) *Nautilus*
3. Check the list and choose the most suitable item that resembles the hare :
(a) *Acolis*
(b) *Phyllobranchus*
(c) *Doris* (d) *Aplysia*
4. Identify the larva which is characteristic of Gastropoda and Scaphopoda :
(a) trochophore (b) veliger
(c) Muller's larva (d) bipinnaria
5. An example of a wood boring mollusca is :
(a) *Pirolas* (b) *Avicula*
(c) *Tridachna* (d) *Trigonia*
6. The word 'Mollusca' was coined by :
(a) Lamarck (b) Johnston
(c) Palseneer (d) Aristotle
7. One of the following molluscs is worm like and without shell :
(a) *Neopilina* (b) *Chaetoderma*
(c) *Mytilus* (d) *Cypraea*
8. The bivalve burrowing in hard clay or soft rocks is :
(a) *Teredo* (b) *Pholas*
(c) *Pecten* (d) *Ensis*
9. The freshwater clam or mussel is :
(a) *Mytilus* (b) *Anodonta*
(c) *Ensis* (d) *Pholas*
10. Radula is found in all molluscs except :
(a) bivalves (b) cephalopods
(c) scaphopods (d) aplacophorans
11. One of the following is a living fossil :
(a) *Unio* (b) *Dentalium*
(c) *Nautilus* (d) *Pecten*
12. Which one is a connecting link between Annelida and Mollusca ?
(a) *Nautilus* (b) *Chaetoderma*
(c) *Neopilina* (d) *Lymnaea*
13. The technical name of razor shell is :
(a) *Chiton* (b) *Loligo*
(c) *Dentalium* (d) *Solen*
14. Which one of the following groups of animals has the maximum economic importance ?
(a) Palecypoda (b) Cephalopoda
(c) Gastropoda
(d) Myriapoda
15. *Pinctada* (pearl oyster) belongs to the class :
(a) Cephalopoda (b) Scaphopoda
(c) Gastropoda (d) Bivalvia
16. Worm like and without shell mollusc is included in class :
(a) Monoplacophora (b) Polyplacophora
(c) Aplacophora (d) Scaphopoda
17. Locomotion with jet propulsion mechanism is found in class :
(a) Scaphopoda (b) Cephalopoda
(c) Gastropoda (d) Bivalvia
18. In which of the animal is the ctenidium absent and pallial branchiae act as respiratory structures ?
(a) *Haliotis* (b) *Eolis*
(c) *Ariophanta* (d) *Doris*
19. The study of Mollusca is generally known as :
(a) Malacology (b) Concology
(c) Helminthology (d) Herpatology
20. Concology is the study of :
(a) Mollusca (b) foot of Mollusca
(c) shell of Mollusca (d) behaviour of Mollusca
21. Digestive gland in Mollusca :
(a) salivary gland (b) pancreas
(c) liver (d) hepatopancreas
22. Shell of Mollusca secreated by :
(a) ectodermis (b) endodermis
(c) mantle (d) all the above
23. The larvae of Mollusca :
(a) veliger (b) glochidium
(c) both (d) none
24. Commonly known as tusk shell :
(a) Gastropoda (b) Scaphopoda
(c) Palecypoda (d) Aplacophora
25. Torsion is the characteristics of :
(a) Gastropoda (b) Scaphopoda
(c) Palecypoda (d) Aplacophora

Answers

➤➤*Fill in the blanks*

1. Coat of nail shell, 2. Sea-hare, 3. *Doris*, 4. Sea-mussel, 5. *Teredo*.

➤➤*Matching type questions*

(i–c), (ii–e), (iii–d), (iv–f), (v–b), (vi–a)

➤➤*Multiple choice questions*

1. (a)	2. (d)	3. (d)	4. (a)	5. (a)	6. (b)	7. (b)	8. (d)	9. (b)	10. (a)
11. (c)	12. (c)	13. (d)	14. (a)	15. (d)	16. (c)	17. (b)	18. (b)	19. (a)	20. (c)
21. (d)	22. (c)	23. (c)	24. (b)	25. (a)					

61

Mollusca : General Account

PODIUM OR FOOT IN MOLLUSCA

Ventral muscular **podium** or **foot**, distinct from the rest of the body, is one of the most characteristic organs of Mollusca. Foot and its appendages get their nerve-supply from the pedal ganglia or pedal nerve cords. Foot presents various modifications in adaptation to different modes of life and locomotion in different groups of Mollusca.

[I] In Monoplacophora

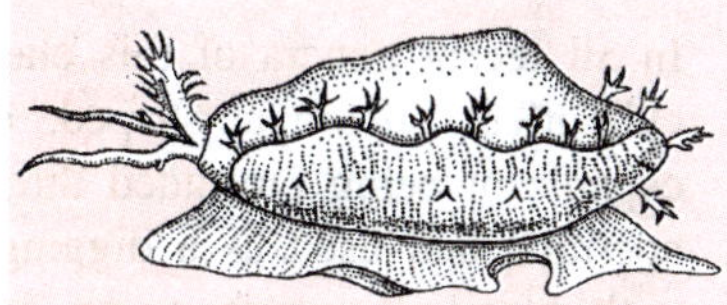

Ventral surface of *Neopilina* very much resembles that of a chiton. Foot is present in centre and shows a primitive condition. It is broad, almost circular, with a flat sole on which the animal creeps by muscular waves.

[II] In Amphineura

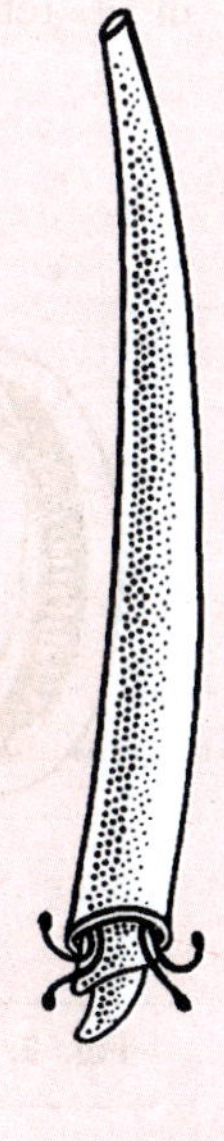

In subclass **Polyplacophora** (chitons) the foot is large and muscular, occupying greater part of ventral surface. Flat creeping sole of foot is usually broad but narrow in *Chitononellus* and *Cryptoplax* etc. It also serves as a sucker to adhere firmly to rocks, when at rest. A relatively small amount of mucus is secreted by the foot. Propulsion is accomplished by mucus and waves of muscular activity.

In subclass **Aplacophora**, in the order **Neomeniomorpha** (*e.g., Neomenia,* etc.), a ciliated, longitudinal mid-ventral **groove** runs from near the mouth to cloacal aperture, containing a small longitudinal, ciliated **ridge**, representing the vestigial foot. In order **Chaetodermomorpha** (*e.g., Chaetoderma,* etc.) ventral foot groove is absent.

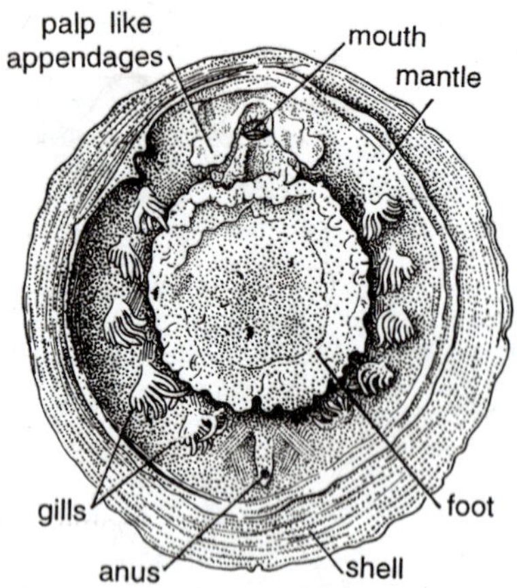

Fig. 1. Foot in Monoplacophora (*Neopilina*).

[III] In Scaphopoda

In all recent genera of this class, foot is narrow, cylindrical or tongue-shaped, directed forwards, capable of being protruded through the oral shell aperture, and used as a digging organ. Engorged with blood, it serves as an anchor. Lower free end of the foot is conical and trilobed in *Dentalium,* carrying on each lateral side a wing-like fold or **pleat** which is comparable to an **epipodium**. In *Antalis entalis,* erectile side lobes become distended to help anchor the foot in sand. Body is then pulled down to the foot and the process repeated. In *Siphonodentalium,* foot terminates in a retractile **disc** with papillated margins, while in *Pulsellum,* a filiform **tentacle** springs from the centre of the disc.

[IV] In Gastropoda

1. Typical creeping condition. Typically, gastropod foot has a simple, elongated and undivided flat ventral creeping sole, similar to that of ancestral mollusc. It acts passively as a holdfast organ to secure attachment, or actively to produce movement. This typical condition is found in limpets (*e.g. Patella*) which are adapted for clinging and moving on rocky surfaces, like chitons. Many marine and freshwater gastropods, moving on soft sand or mud bottoms or seaweeds, and pulmonates living on terrestrial vegetation, rotten leaves and logs, retain the primitive foot.

(*a*) ***Division of foot.*** In typical gastropods, the foot is usually differentiated into three regions or parts : a small anterior **propodium,** a large middle **mesopodium,** and a posterior **metapodium**. These are variously modified in different gastropods. Anterior margin of foot may form small tactile papillae (*e.g., Trochus*), or may project as a small fleshy process (*e.g., Turbonilla*), called the **mentum**, or may give out a pair of **pedal tentacles** (*e.g., Valvata, Vermetus, Aeolis,* etc.).

(*b*) ***Appendages of foot.*** Gastropodan foot bears several kinds of appendages, of which the most important are the **epipodia** and **parapodia**. Epipodia are prominent folds, arising from sides or base of the foot along its entire length. They may be beset with papillae, eyespots and tentalces

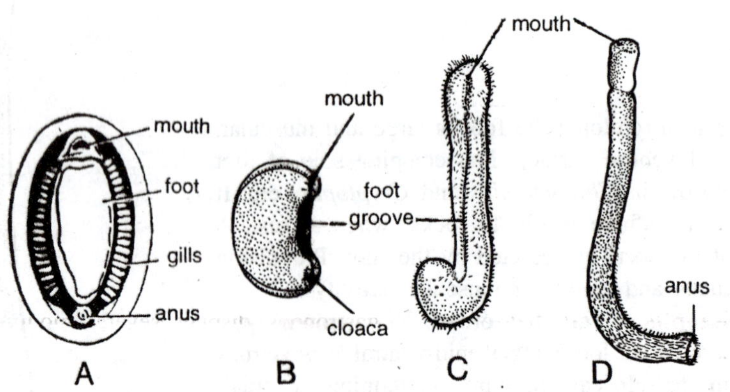

Fig. 2. Foot in Amphineura. **A.** *Chiton,* **B.** *Neomenia.* **C.** *Proneomenia.* **D.** *Chaetoderma.*

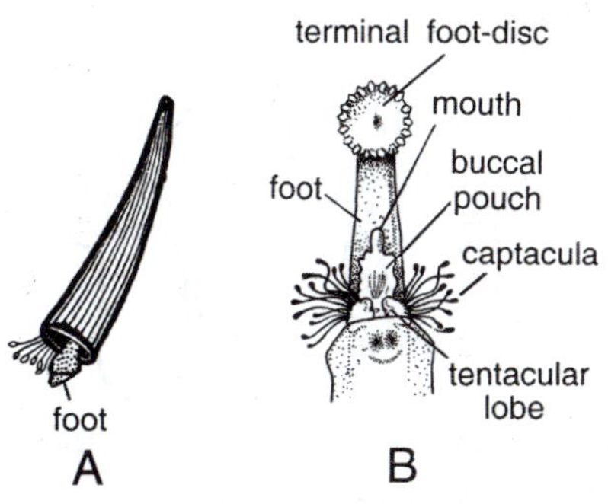

Fig. 3. Foot in Scaphopoda. **A.** *Dentalium.* **B.** *Siphonodentalium.*

(*e.g.*, *Fissurella, Monodonta, Trochus,* etc.). They are most strongly developed in *Haliotis,* bearing ringed lobes and sensory tentacles.

(c) Pedal glands. Foot contains scattered unicellular as well as aggregated multicellular mucus-secreting pedal glands. The later include : (i) **labial glands** in a groove at its anterior end, (ii) **supra-pedal gland** between head and anterior end of foot, (iii) **sole gland** opening mid-ventrally in the middle of foot, and (iv) **posterior glands** at the hind end of foot.

(d) Mechanism of movement. Gastropods, with typical primitive condition of foot, generally creeps by muscular wave like contractions of longitudinal muscles passing over the sole, leaving a mucous-slime trail behind. Muscular contractions may be **direct,** *i.e.,* passing forwards along the foot from behind or may be **retrograde**, *i.e.*, passing backwards.

Waves may be **monotaxic,** *i.e.*, a single series traversing the foot. In marine snails, *Lacuna* and *Littorina,* the foot is divided into right and left halves by a mid-longitudinal groove and muscular waves are **ditaxic**, *i.e.,* the snail moves by advancing alternately the right and left sides of foot in a co-ordinated manner. A similar gait is exhibited by the land snail, *Pomatias.* In marsh-dwelling pulmonate, *Melampus,* first, the anterior end of foot is extended and then the rest of the body pulled behind it. *Helminthoglypta dupetihouarsi* shows galloping locomotion. Anterior end of body is raised, extended forward and anchored, while the foot is raised as an arch and then moved forward. Some smaller snails, like *Caecum,* glide along merely by cilia working through the mucous layer, as in some flatworms. *Bulla* and *Pyrula* can crawl rapidly over wet sand as their foot is widely extended on all sides. *Harpa* cuts off the posterior lobe of its foot (autotomy), when hard pressed, with the sharp edge of the shell.

Gastropods show great variations in form and size of foot, in relation to different conditions of life. Important modifications are met in those gastropods which burrow or swim and are sedentary or parasitic.

2. Burrowing gastropods. Some burrowing gastropods (*e.g. Terebridae*), simply extend the anterior extremity of foot with flow of blood, thus forming an anchor, as seen in scaphopods and many bivalves. In fossorial gastropods, which crawl about in wet sand, anterior region, or propodium, is distinctly separated from the rest of the foot by a constriction (*e.g.*, *Harpidae*) or by a transverse furrow (*e.g.*, *Olividae*). It is circular and disc-like in *Oliva.* In *Polinices, Sigaretus,* and *Natica,* etc., propodium acts like a powerful digging organ or plough, while a dorsal flap-like fold of foot covers the head as a protective, fleshy cephalic shield. In *Natica,* propodium also forms a sort of siphon, on the left side to conduct water to the respiratory chamber.

3. Leaping gastropods. As a rule, the posterior region or metapodium is not sharply marked off from rest of the foot. In most shell-bearing prosobranchs, it forms an operculigenous lobe, bearing dorsally a solid, horny and calcareous disc, called the **operculum**, which serves to close the shell aperture, as a protective door or lid, when the animal is retracted. In Strombidae (*e.g.*, *Strombus, Rostellaria, Xenophorous,* etc.), the operculum becomes sharp like a dagger or claw. As the claw digs into the sand, the somewhat reduced and hook-like foot becomes flexed. As a result, the animal pulls forward, thus moving in a series of strong leaps.

4. Swimming gastropods. Swimming gastropods display several modifications of the foot, which forms an efficient swimming organ. Several families of mesogastropod prosobranchs, known collectively as **heteropods**, are permanently pelagic and swim upside down. Foot is carinate,

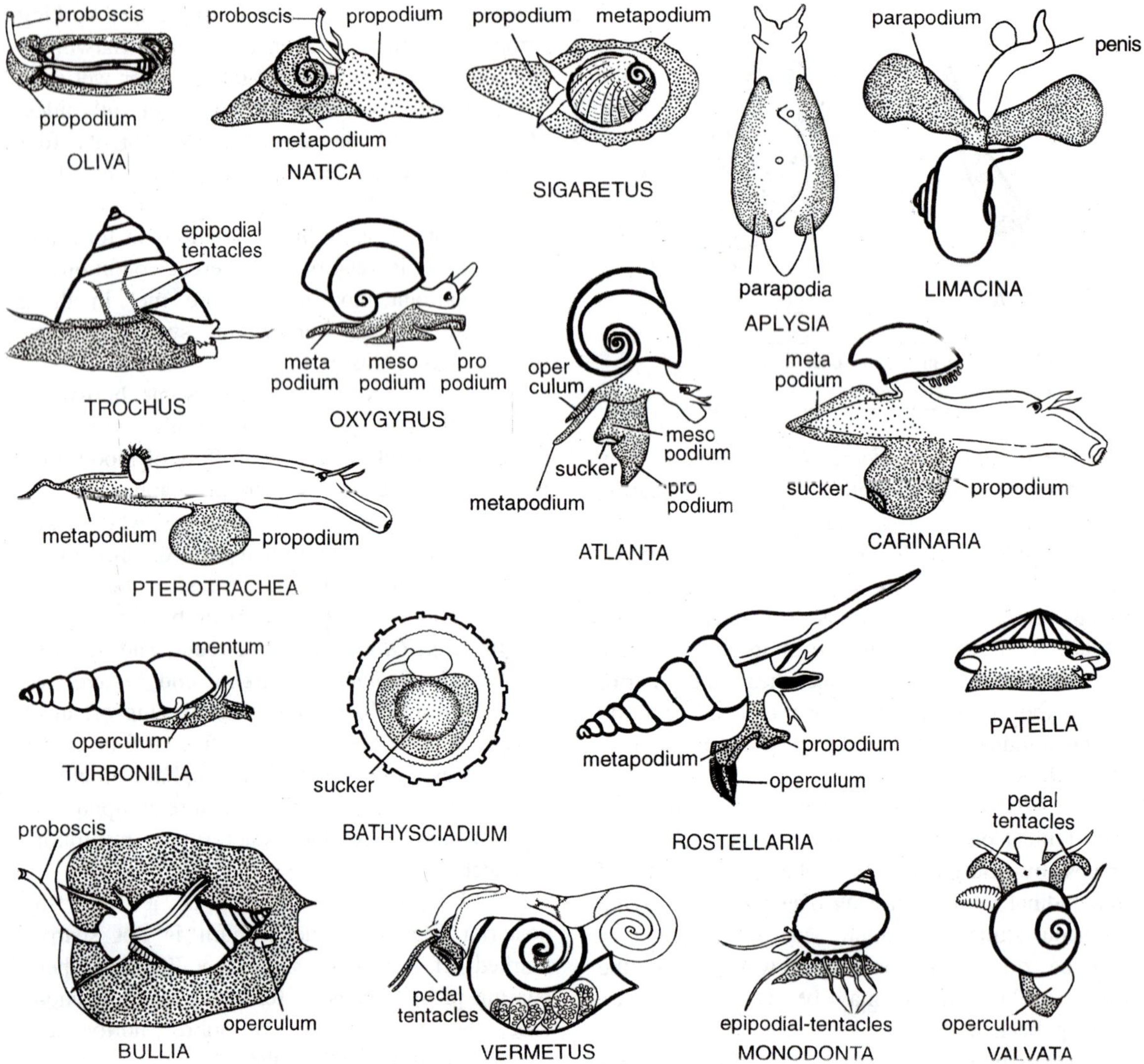

Fig. 4. Various types of foot (stippled) met within Gastropoda.

i.e., the foot is transformed into a vertical natatory lobe or fin which helps in swimming by its undulating movements. Its development can be traced almost step by step in the group. In *Oxygyus* and *Atlanta,* shell is still prominent laterally compressed and produced into a sharp keel. Propodium is fin-like, the mesopodium bears a sucker and the metapodium bears an operculum. In *Carinaria,* the fin further enlarges, while the sucker, metapodium and shell are reduced. In *Pterotrachea,* sucker persists only in male, while the metapodium loses the operculum and becomes reduced to a short, filament-like tail.

Swimming is far more widespread in opisthobranchs with different modifications of foot. In the common sea-hare *Aplysia* and *Notarchus,* two lateral projections of the foot, called **parapodia**, form broad fins. These can be folded and united over the head forming a sac, through which water can be expelled forcibly like a jet. Swimming is achieved in short bursts by rhythmic waves which pass along the parapodia. Broad wing-like or oar-like parapodia, confined to anterior part of the body, form the largest part of foot in the shelled pteropods, or sea-butterflies (order Thecosomata). *Limacina* swims upwards spirally using them as oars, and drops down by holding them motionless over the head. Naked pteropods (order Gymnostomata) are the fastest gastropod swimmers, in which ventrally located parapodia perform a rapid sculling motion, each describing a figure of eight. In nudibranchs (order Acoela), forward locomotion is achieved by the passage of undulating waves down the thin lateral margins of foot and mantle; and an upward motion by flexure of the body in segital plane.

5. Sessile gastropods. In sedentary and clinging forms (*e.g. Patella, Bathysiadium, Crepidula,* etc.), foot is reduced and the ventral sole becomes an effective sucker of attachment. In truly sessile tube-dwellers (*e.g., Vermetus*), foot is further reduced. In worm shells (*e.g., Vermicularia*), shells are immobile; they are attached to other shells or entangled in sponges.

6. Parasitic gastropods. Parasitic gastropods, like *Stylifer, Thyca* and *Entoconcha,* endoparasitic in echinorderms, show reduction or loss not only of foot but of their other body organs.

[V] In Pelecypoda

Pelecypods are mostly sedentary and principal modifications of foot are due to its employment as a burrowing organ. Foot often contains a part of viscera (intestine, liver, gonad, etc.).

1. Typical or hatchet-shaped. Typically the foot is laterally compressed and wedge-like or tongue-like or blade-like and directed anteriorly. It is also called **ploughshare-like** or **hatchet-shaped** and is well-adapted for slowly making its way through sand or mud, by alternate contraction and expansion. Ventrally, foot terminates into an elongated **keel** with either a single anterior point (*e.g., Unio, Tellina* and *Cardium*) or with two points, an anterior and a posterior (*e.g. Trigonia*). Anterior end of foot forms a long, cylindrical tentacle in **poromya**, with a swollen free extremity in *Axinus.*

Burrowing mechanism is affected by a coordinated action of blood pressure and muscular cavity. Blade of foot is extended, with the aid of protractor muscles into sand or mud. Blood acts as a hydrostatic endoskeleton. Distal part of foot is dilated as an anchor and contraction of the retractor muscles pulls the animal downwards.

2. Primitive or gastropodan. In primitive, Protobranchia (*e.g., Nucula, Solenomya, Yoldia*) and some Toxodonta (*e.g. Arca*), the foot has a flat ventral surface but not for creeping. Two sides of foot can be folded together producing a wedge-like or blade-like edge, which is thrusted into the mud or sand. Sole then opens to serve as an anchor and the body drawn down into substratum.

3. Springing. Not all bivalves burrow; some have become bottom-dwellers, moving by means of leaping motion. In forms, such as *Trigonia* and *Cardium,* the large and narrow foot is bent

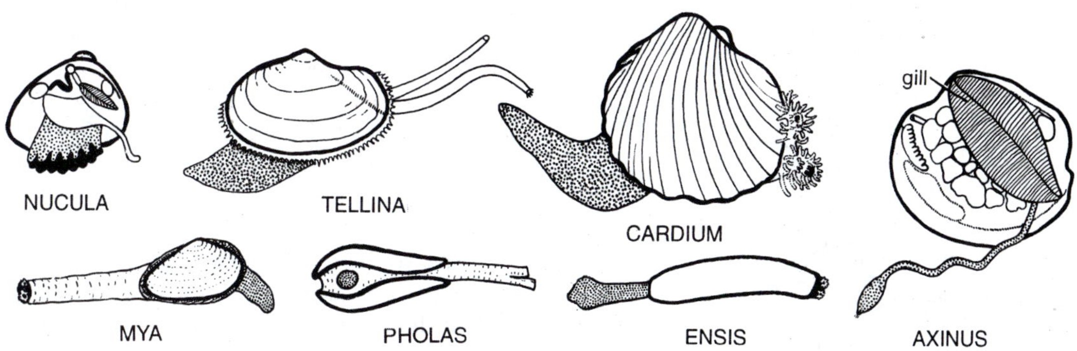

Fig. 5. Modifications of foot (stippled) in the class Pelecypoda.

upon itself. When suddenly straightened it works like a leaping organ, lifting the animal up in a surprisingly strong leap.

4. Boring. In gapers and clams (*Mya*), which burrow but slightly, the foot is feebly developed and pedal opening is narrow. In fast burrowers, like *Tellina,* foot is large and it can be outstretched into a wide sheet of muscle, as thin as blade of a knife, as it plunges into the sand. In deep burrowers, like *Solen* and *Ensis,* foot is proportionately large and anterior in position, accompanied by a corresponding narrowing of the shell. It is thick and liguiform in *Solecurtus,* club-shaped and truncated at the tip in *Ensis* and cylindrical with an egg-shaped tip in *Solen.* Piston-like foot, occupying nearly half the mantle cavity can be pointed and plunged into sand, where it is made turgid with blood swells into a thick tube, and pulls the animal after it.

5. Vestigial. In sedentary or attached genera (*e.g., Pecten, Chama, Anomia,* etc.), having limited or no locomotory powers, foot becomes rudimentary. A jerky kind of swimming has evolved in scallops (*e.g. Pecten*) by use of water jets. In piddocks or rock-borers (*Pholas*) and shipworms (*Teredo*), mantle is closed except for a small pedal aperture, through which the small foot can be protruded to grip the end of burrow,

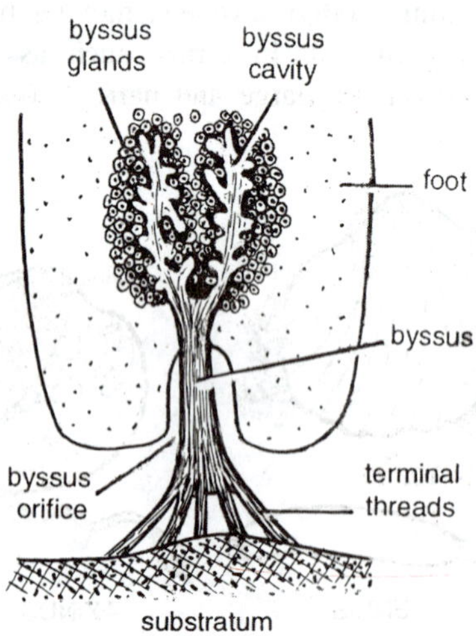

Fig. 6. Diagrammatic T.S. of a Lamellibranch through byssus orifice.

like a sucker, while boring. Foot is altogether absent in Ostreidae (*Ostrea*).

6. Byssus apparatus. A byssus gland is characteristic of many pelecypods. It is situated immediately behind or at the base of the foot. It is associated with a drastic reduction of foot and its muscles. Secretion of the gland, as it passes out through its aperture, hardens on contact with the water forming a bundle of tough, horny, silk-like threads, which serve to anchor the animal permanently or temporarily. In Ostreidae (*e.g., Ostrea*), secretion of the byssus gland permanently cements the animal by its left shell valve, after which gland atrophies. In Mytilidae (*e.g., Mytilus*), byssus threads are formed in posterior groove of the much reduced foot and attach the animal firmly to wharf pilings, rocks and oyster beds. In *Anomia,* byssus thread becomes calcified and extends through a hole in the right shell valve, which lies against the substratum. Byssus described above is not homologous with provisional byssus of larval *Anodonta,* but the byssus gland seems to be homologous with pedal gland of Gastropoda. In *Arca* and *Mytilus,* old byssus may be shed and a new one formed, while in *Pinna* the threads are fine enough to be woven in a fabric. In some species of *Lima,* the byssal threads form a sort of protective nest round the animal, consolidated by fragments of shells and stones.

[VI] In Cephalopoda

Cephalopodan foot is no longer recognisable as a typical molluscan foot. It becomes modified partly into 8 to 10 cephalic or circumoral **arms** for seizing prey, and partly into ventral **siphon** or **funnel** which helps in swimming activity.

1. Cephalic arms. Decapoda (*e.g., Sepia, Loligo,* etc.) have 5 pairs of arms, bearing stalked suckers provided with horny rims. The 4th pair of arms, called **tentacles,** are long, prehensile into basal pits and bearing suckers only at their free club-shaped ends. Octopoda (*e.g., Octopus*) have only 4 pairs of similar arms bearing sessile suckers without horny rims. Tetrabranchia (*e.g., Nautilus*) do not have arms; instead, the pedal crown, round the head, is divided into lobes, each bearing numerous, short, prehensile tentacles which are strongly adhesive but without suckers.

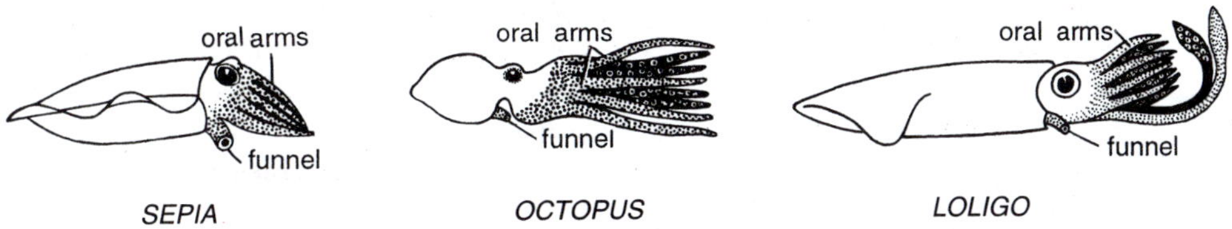

Fig. 7. Modification of foot (stippled) into arms and funnel, in *Cephalopoda.*

In mantle, one of the arms (left 5th in *Sepia*) becomes **hectocotylized**, *i.e.,* it is slightly modified by suppression of suckers and serves as copulatory organ. In some species (*e.g., Argonauta, Tremoctopus*), it is entirely detached from male and left in the manlte cavity of female; in others it is not detached.

Old concept of pedal nature of arms, is proved by their innervation from the brachial ganglion which is a division of the pedal ganglion; and by their development in embryo, ventrally in between the mouth and anus. Later on they gradually migrate forward to encircle the mouth. However, the concept may be erroneous and the arms and tentacles probably represent true cephalic structures. The name 'Cephalopoda' may thus be strictly speaking, a misnomer.

2. Siphonal apparatus. Siphon or funnel is a complete tube in all the modern dibranchiates, but in *Nautilus,* it consists of two separate, lateral and muscular lobes, overlapping but not fused together. When unrolled and flattened out, they greatly resemble the foot of a Gastropoda.

RESPIRATION IN MOLLUSCA

Types of respiration in Mollusca are **cutaneous, branchial** and **pulmonary.** Several modifications of each type are met with in different classes.

A. Cutaneous or Pallial Respiration

Respiration by moist integument or mantle is the simplest method in those Mollusca which have no differentiated respiratory organs. It occurs in Scaphopoda, small Aplacophora, some Opisthobranchia, terrestrial forms (Vaginulidae) and parasitic forms (*e.g., Entoconcha*). Respiration occurs through whole body surface of the soft-skinned parts of the body. It must be very restricted when body surface is covered with a shell; or when the body is thick and bulky. Cutaneous respiration may also take place through the skin of those molluscs having other respiratory mechanisms (*e.g.,* pulmonate gastropods). In all bivalves, the inner mantle surface also contributes the oxygenation, although respiration is mainly branchial, through gills.

B. Branchial or Ctenidial Respiration

[I] True Gills or Ctenidia

Most of the molluscs are aquatic and respire by means of gills or ctenidia. Gills are the most important of all the pallial organs in Mollusca. They serve a vital role in the economy of these animals and remain well-protected inside the mantle cavity. Throughout the division of Mollusca, the gill is a homologous organ, derived from the gill of a common ancestral type. It is termed as a **ctenidium**.

1. Structure of a ctenidium. Ctenidia are the projections of body surface or mantle. They are fundamentally of same structure in all the primitive members, such as *Polyplacophora* (Amphineura), *Rhipidoglossa* (Gastropoda), *Protobranchiata* (Lamellibranchia) and Cephalopoda.

A ctenidium consists of a horizontal **main stem** or **axis** attached to the body and bearing on either side a row of alternately arranged, delicate, flexible, respiratory **lamellae** or **filaments**. Surface

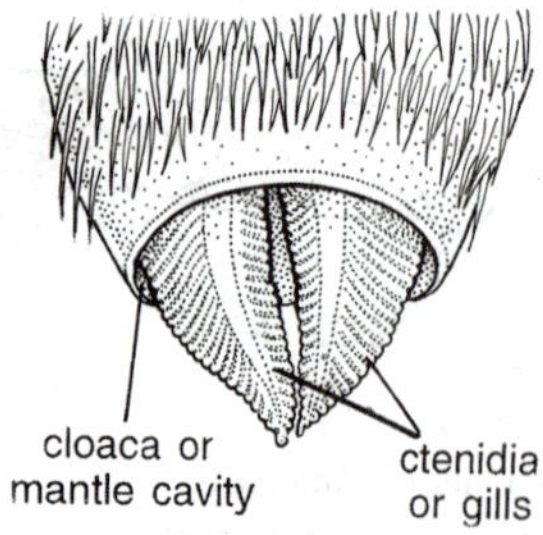

Fig. 8. *Chaetoderma.* Cloacal region with ctenidia.

epithelium of gills is usually covered by **cilia,** the movements of which are responsible for continuous renewal of water over the gills. Ctenidium receives venous blood from the body through an **afferent branchial vein,** and after becoming oxygenated in the filaments, it is sent back to heart through an **efferent branchial vein**.

Water flows over the gills in a direction opposite to that taken by blood within the filaments. This counterflow, involving the countercurrent principle, ensures a more efficient respiratory exchange of gases. Oxygen of water passes through the wall of gill filaments into blood; in return, carbon dioxide (waste of blood) passes into the water. Amount of gaseous exchange is proportional to respiratory surface area and to concentration gradient. Gaseous exchange in molluscs is a matter of simple diffusion without involving active transport.

2. Number, form and position of ctenidia. Most of molluscans possess one pair of lamellar ctenidia. Near the ctenidia are found **osphradia** which test the nature of incoming water current.

Classification of Mollusca is based mainly on character of the gills. **Cephalopoda** is divided into **Dibranchia** and **Tetrabanchia** according to number of gills they possess. **Gastropoda** has three subdivisions : **Prosobranchia**, **Opisthobranchia** and **Pulmonata** : according to the position and character of the gills. Plate-like form of gills has earned the name **Lamellibranchia** for the bivalves.

In class **Monoplacophora** (*e.g., Neopilina*), 5 pairs of unipectinate gills, composed of finger-like lamellae, are present in pallial groove. Vestigial lamellae are supposed to be present on the opposite side of the axis. Water current circulates probably as in chitons. Gill of Monoplacophora are probably not true gills; more likely, they represent prectenidial structures or pallial gills.

In class **Amphineura**, mantle cavity is primarily posterior and the gills are true ctenidia. In **Placophora** (chitions), ctenidia are arranged in two rows, one in each pallial groove, the number varying from 6 to 80 according to species. They are bipectinate and structure of gill filaments is essentially like that of the ancestral mollusc, but their lateral margins are more rounded in chitons. In **Aplacophora** the mantle cavity is represented by a mid-ventral longitudinal groove widening posteriorly into a small cloacal cavity. Gills are either absent (*Neomenia*) or reduced to two large plume-like ctenidia (*Chaetoderma*), one on either side of the anus.

In class **Gastropoda,** gills are shifted to the front of the body with mantle cavity, as a result of torsion. Most **Prosobranch Archaeogastropoda** (*e.g., Haliotis, Fissurella*) retain a single pair of gills, that of the right side usually smaller in size. These gills have primitive bipectinate condition of the ancestral mollusc. In one-gilled **Archaeogastropoda** (*e.g., Trochus, Turbinus, Acmaea,* etc.) the gill is bipectinate. In remaining prosobranchs, **Mesogastropoda** and **Neogastropoda** (*e.g. Triton, Pila*), right gill has completely disappeared. A single gill has developed on the left side which is monopectinate, carrying a single row of leaflets on its right side. Respiratory water current enters the mantle cavity on the left, sweeps over the ctenidium and leaves on the right. This arrangement has been affected due to torsion in these gastropods.

In subclass **Opisthobranchia**, mantle cavity tends to become shorter and moves back along the right side to its posterior position, owing to shell reduction and detorsion. This involves the decreased importance of ctenidium. In *Aplysia,* single ctenidium is still retained on the right side. It probably represents the original single left ctenidium of higher prosobranchs. In opisthobranchs, there has been a tendency towards loss of original gill, and a greater demand on the mantle for respiratory needs, resulting in development of secondary gills. **Nudibranchs** or

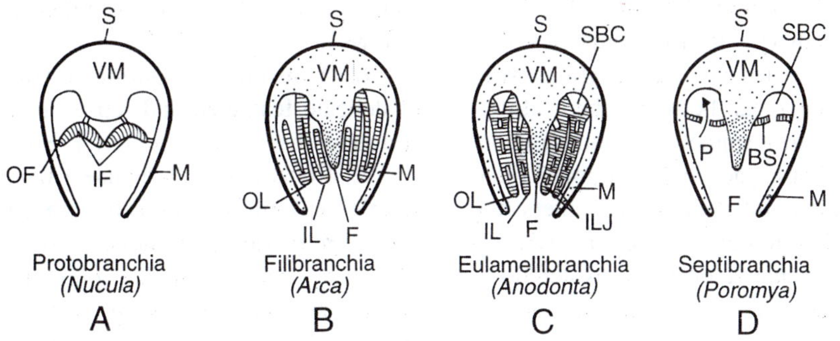

Fig. 9. Diagrammatic T.S. to show chief kinds of gills and their arrangement in different groups of Bivalvia. BS-branchial septum. F-foot. IF-inner-gill filaments. IL-inner gill-lamina. ILJ-inner lamellar junctions. M-mantle, OF-outer gill-filaments, OL-outer gill- lamina, P-perforation, S-shell, SBC-supra-branchial chamber, VM-visceral mass.

Acoela (*Aeolis, Doris*) have altogether lost the true ctenidium with mantle cavity. In terrestrial snails (subclass **Pulmonata),** gills disappear and the mantle forms a lung for aerial respiration.

Class **Pelecypoda** have the most complex, molluscan gills. Besides respiration, gills make important organs for nutrition and serve as brood pouches. Bivalves have retained the single pair of gills of the ancestral mollusc. Gills are bipectinate and equal on both sides. In subclass **Protobranchia** (*Nucula*), gills are relatively smaller, situated at the back of manlte cavity, and their filaments, relatively few, form small, compressed and horizontally disposed, triangular leaflets, free from one another. Leaflets are arranged in two opposite rows in a gill. There are tufts of loosely interlocking cilia at the ends of leaflets of each gill, connecting them to the sides of foot and to the inner surface of mantle. Arrangement divides the mantle cavity into a large inhalent chamber below each gill and a small exhalent chamber above. Such gills are essentially respiratory in function. Food particles are collected by long labial palps and not by the cilia on the gills. In other lamellibranchs, gills have also taken over the function of obtaining food. As a result, the gills become more specialized, the degree of specialization corresponding with general evolutionary level. Many filaments have been added so that the ctenidia extend further anterior, than in *Nucula*.

In **Filibranchia**, gill filaments become elongated and thread-like (*Amusium*). In *Arca*, each gill filament is bent upon itself to form an elongated V. Thus, each gill forms a W in section and consists of two V-shaped demibranchs. Each demibranch is two-lamellae thick, made of a descending limb and an ascending limb. A food groove appears all along the underside of each gill at the angle of flexure of its filaments. In *Mytilus,* the adjacent filaments of each demibranch are loosely united to each other, side to side, by groups of large, stiff and interlocking cilia. Thus, single ctenidium of *Nucula* is replaced, in *Mytilus,* by two plate-like laminae, each made of an outer and an inner lamella. In addition, non-vascular inter-lamellar junctions run between the descending and ascending limbs and hold together the two lamellae of each demibranch.

In **Pseudolamellibranchia** (*Ostrea, Pecten, Pteria,* etc.), gills have a greater cohesion than in filibranchs and reflected distal tips of filaments become coalesced laterally with the mantle.

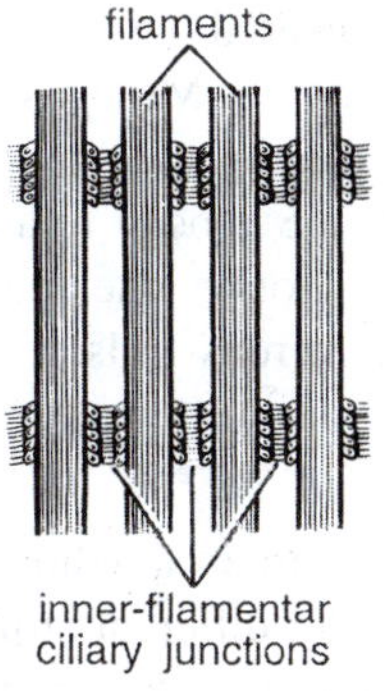

Fig. 10. *Mytillus.* Four gill-filaments.

In **Eulamellibranchia** (*Unio, Anodonta, Lamellidens,* etc.), ciliary inter-filamentar junctions are replaced by vascular cross-connections with narrow openings or ostia between them. Inter-lamellar junctions also become large and vascular, so that interlamellar space is partitioned into vertical water tubes. Inhalent water enters the water tubes through minute pores, or **ostia,** in lamellae. Gaseous exchange occurs as water flows into dorsal suprabranchial chamber and out of exhalent opening.

Finally in subclass **Septibranchia** (*Poromya, Cuspidaria,* etc.), gills degenerate, being replaced by a horizontal, perforated, muscular septum, extending from the base of foot to the mantle. Septa move up and down so that a water current enters through inhalent siphon. Respiratory function is taken up entirely by the mantle.

In class **Cephalopoda**, gills are simple and bipectinate or plume-like, suspended on either side of the anus. Lamellae are delicate, leaf-like and arranged in a linear row on the axis. Cilia are absent so that water is pumped in and out by the coordinated activities of the muscular mantle, funnel and the inlet valves. There are two ctenidia in Dibranchia and four in Tetrabranchia. *Nautilus* has an osphradium which is absent in other cephalopods.

3. Relation between branchiae and auricles. Blood purified in gills reaches the heart through auricles, so that a close structural relation is found between the two. Archaeo-gastropoda, Pelecypoda and dibranchiate Cephalopoda have one pair each of the gills and auricles. *Nautilus* has 4 gills and, to correspond, two right and two left auricles. Unpaired gill of Docoglossa, Monotocardia or Mesogastropoda and Tectibranchia, is accompanied by a single auricle on the side of the body retaining the gill. Chitonidae possess only one pair of auricles, inspite of their numerous gills.

[II] Secondary or Adaptive Gills

In certain molluscs, true ctenidium is absent and other morphologically different structures develop, termed as **secondary gills** or **adaptive gills**. Accessory aquatic respiratory organs, or adaptive gills, are met with in Docoglossa, Nudibranchia and a few others. Their main forms are as follows :

1. Anal gills. In Dorididae (*Doris*), a rosette of delicate feathered and retractile secondary gills surround the anus at posterior end of the body.

2. Cerata. In Aeolididae (*Aeolis*), numerous simple or pinnately branched secondary gills or cerata are carried upon dorsal surface of the body. They are richly vascular, present a variety of beauty and form, serve to protect by concealment, are capable of breaking off easily and regenerated quickly. Diverticula of digestive gland open to exterior at their tips, which also store stinging cells with stinging capsules. Cerata may occur with anal gills in Polyceridae.

3. Pallial gills. In *Patella,* a series of adaptive gills occur in a row on each lateral side in the pallial groove. They may be found together with true ctenidia in *Pneumoderma* and *Lottia.*

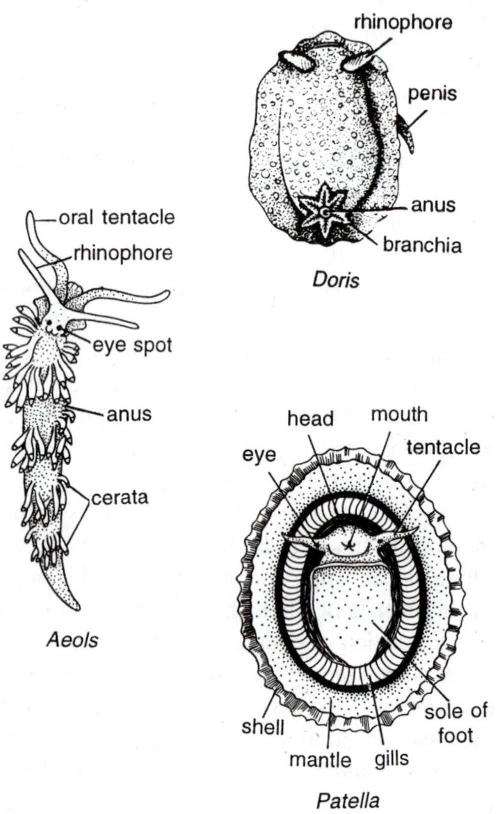

Fig. 11. Gastropods having adaptive or secondary gills.

C. Pulmonary Respiration

In terrestrial pulmonates (*Limax, Arion*), a true ctenidium disappears and the mantle cavity is transformed into a pulmonary sac or lung for aerial respiration. Roof of pulmonary sac is richly supplied with vessels. Pulmonate lungs are an adaptation for terrestrial existence. Alternate muscular contraction and relaxation of mantle floor, lower or raise it, causing the air to rush in and out of mantle cavity. Air can enter or leave the sac through a small rounded pulmonary aperture on right side provided with a valve. Compression of manlte cavity increases the partial pressure of O_2 and facilitates its absorption. Some pulmonates derive O_2 from water in the mantle cavity and do not come to the surface to breathe. Majority of lower pulmonates (*Lymnaea, Planorbis,* etc.) is amphibious and their mantle cavity serves both for aerial and aquatic respiration. They are said to drown if prevented from coming periodically to the surface to fill the lung with air. In some amphibious Prosobranchia (*Pila, Ampullaria, Siphonaria,* etc.), mantle cavity is divided by an incomplete septum (epitaenia) into a left pulmonary chamber and a right ctenidial chamber containing a gill. Animal can breathe in water by its gill and by its lung in the air.

TORSION IN GASTROPODA

All the living molluscs, except the Gastropoda, retain ancestral bilateral symmetry of the body, with mantle cavity lying posteriorly or laterally. Gastropods, on the other hand, possess an asymmetrical body with mantle cavity lying anteriorly, and the shell and the visceral mass coiled spirally and directed posteriorly. In spite of their asymmetry, Gastropoda are generally believed to have descended from an unsegmented and bilaterally symmetrical ancestor with a low conical shell, a straight alimentary canal ending in a posterior anus, and mantle cavity posterior, that is, lying behind the visceral dome. Anterior situation of the mantle cavity in gastropods is due to **torsion** or **twisting** of the visceral mass during development.

What is Torsion ?

Torsion or **twisting** is a process, during larval development of gastropods, which rotates the viscero-pallium anti-clockwise through 180° from its initial position, so that mantle cavity, with its pallial complex, is brought in front of the body, in adult.

Site of Torsion

In larval gastropods, only visceral mass undergoes rotation through 180°, whereas head and foot remain, fixed. Actual site of torsion is neck, behind the head-foot, through which oesophagus, rectum, aorta, visceral nerve loop and shell muscles pass. Thus, actual twisting involves the neck tissue and structures within it.

Torsion Versus Coiling

Torsion is often confused with spiral coiling of visceral mass and shell, but the two are entirely distinct and quite independent. Torsion is not coiling of the shell which starts even much before coiling.

Coiling is achieved by a more rapid growth of one side of the visceral mass than the other. Torsion and coiling are, therefore, separate evolutionary events. Torsion was a much more drastic event than the spiralling of shell.

How Torsion Occurs ?

Torsion is not merely an evolutionary hypothesis. Its occurrence can be seen in the embryogeny of living gastropods. Before torsion, larva is quite symmetrical, the mantle cavity faces backwards and downwards, alimentary canal is straight and anus opens posteriorly in middle line. A ventral **flexure** of the body results in looping of alimentary canal and approximation of mouth and anus. Shell and visceral mass, originally saucer-shaped, become first cone-shaped and later spirally coiled. Shell lies dorsally and forms a coil on the anterior side; such a shell is called **exogastric**.

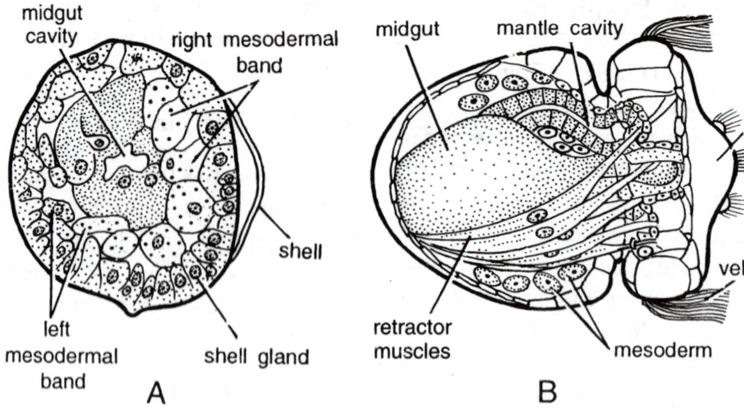

Fig. 12. Mechanism of torsion. **A.** T.S. early veliger *of Haliotis* showing disproportionate growth of right mesodermal cells. **B.** 48 hour larva of *Patella vulgata* showing a symmetrical retractor muscle.

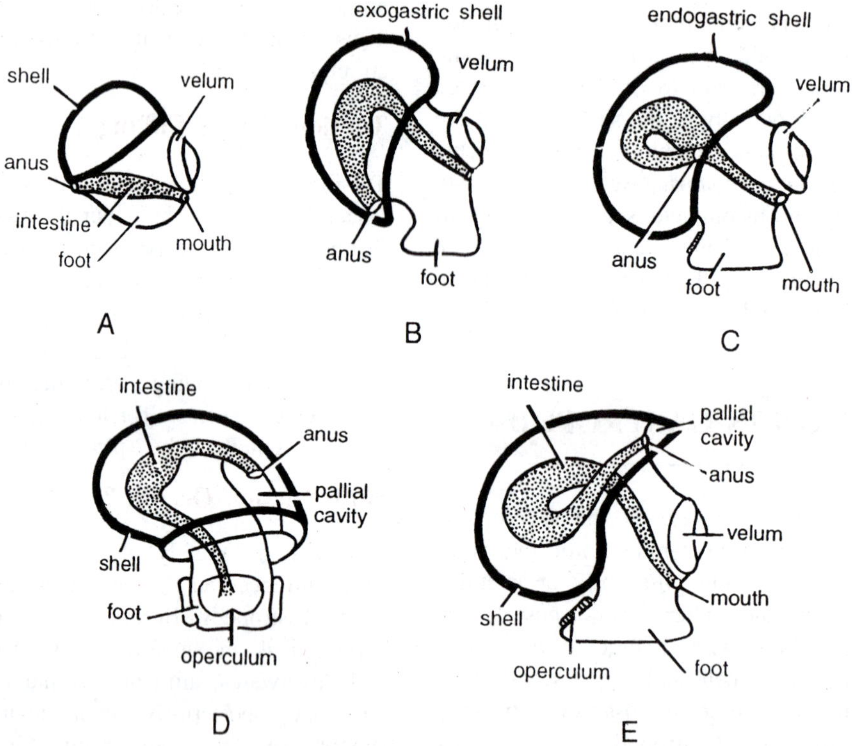

Fig. 13. Five successive stages in the development of a gastropod to show occurrence of torsion. **A.** Early veliger larva or pretorsional stage in lateral view. **B.** Larva with ventral flexure and an exogastric shell in lateral view. **C.** Stage showing 90° of lateral anticlockwise torsion. Shell becomes endogastric. Mantle cavity and anus move on to right side. **D.** 90° torsion stage in posterior view. **E.** Adult stage with complete or 180° torsion in lateral view.

Ventral flexure is followed by a lateral **torsion,** so that dorsal or exogastric shell becomes ventral or **endogastric**. Lateral torsion is probably due to arrest of growth on one side and active extension on the other. Generally, growth of the right side becomes retarded so that mantle cavity and pallial complex gradually pass round to right side, and so to the anterior side, on account of greater growth of the visceral sac towards the left. But the whole process completes in 2 or 3 minutes in *Acmaea,* so that it cannot be regarded due to differential growth. On the contrary, it is due to muscular contractions. Actual mechanism of torsion is supposed to be the asymmetrical position and pull of the larval retractor muscles running from the velar lobes to the shell. They are present only on the right side, there being no related muscles on the left side. Contraction of larval retractor muscles brings about the rotation or torsion. Only narrow neck of the larva is actually twisted. Consequently, everything between the head and anus undergoes an anticlockwise rotation or torsion through an angle of 180° around a vertical axis passing in a dorso-ventral direction.

Thomson (1958) after careful study recognises five ways in which torsion can be brought about :

(1) Complete or 180° rotation, achieved by muscles contraction alone, is known only for *Acmaea* (Archaeogastropoda).

(2) 180° rotation achieved in two stages, first 90° movement by contraction of laryal retractor muscles and later a slower 90° rotation by differential growth. It is the commonest mechanism which is known today, *e.g., Haliotes, Patella.*

(3) 180° rotation by differential growth processes alone, *e.g., Vivapara.*

(4) Rotation by differential growth processes, with anus coming to a position appropriate to adult state, *e.g., Aplysia.*

(5) Torsion no longer recognisable as a movement of viscero-pallium, the organs in post-torsional position from their first appearance, *e.g., Adalaria.*

Effects of Torsion

Torsion is a fundamental feature of gastropods and represents their greatest departure from the ancestral molluscan plan. Peculiarities of organization of gastropods due to torsion were first realized by **Spengel** (1881).

1. Displacement of mantle cavity. Mantle cavity was primitively posterior in position. Increase in length of ventral foot, which primitively was very short, tends to remove the mantle cavity and pallial complex away from the head. After torsion the mantle cavity opens just behind the head and its associated parts are shifted forwards.

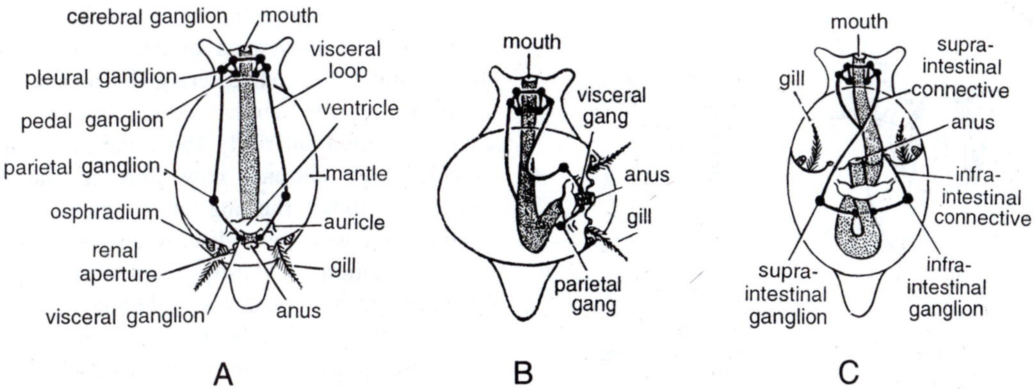

Fig. 14. Effects of torsion upon position of gills, digestive tract and nervous system, **A.** Hypothetical primitive stage before torsion. **B.** Intermediate stage showing 90° torsion with mantle cavity and pallial complex displaced to the right side of body. **C.** Final stage showing 180° torsion.

2. Changes in relative positions. Before torsion, anus, ctenidia and renal orifices point backwards, and the auricles lie behind ventricle. After torsion, anus, ctenidia and renal orifices project forward, and the auricles lie in front of ventricle. Original posterior face of visceral sac becomes the actual anterior face, so that visceral organs morphologically of the original right side become placed topographically on the left side, and vice versa.

3. Looping of alimentary canal. Digestive tract, which was originally straight from mouth to anus, is thrown into a loop.

4. Chiastoneury. Long, uncoiled pleuro-visceral nerve connectives become twisted into a figure of 8. Right connective with its ganglion passes over the intestine to become supra-intestinal, while left connective passing underneath the intestine to become infra-intestinal.

5. Endogastric coil. Coil of visceral sac and the shell, which primitively was dorsal or **exogastric**, becomes ventral or **endogastric**, after torsion.

6. Loss of symmetry and atrophy. Anus is displaced towards right side of the pallial cavity so that original symmetry of organisation disappears. Another characteristic feature involving asymmetry is reduction or atrophy of the paired parts of primitively left or topographically right side. In primitive Archaeogastropoda or Diotocardia (*Patella, Haliotis, Fissurella,* etc.) maximum symmetry is preserved by retaining two gills, two auricles and two kidneys, but the right kidney serves as a gonoduct. In more specialized Mesogastropoda or Monotocardia (*Pila, Buccinum, Littornia,* etc.), ctenidium, auricle, osphradium, hypobranchial gland and kidney of the topographically right side disappear in dextral forms, but reverse process occurs in the sinistral forms. Remaining gill may bear one row of filaments (monopectinate gill).

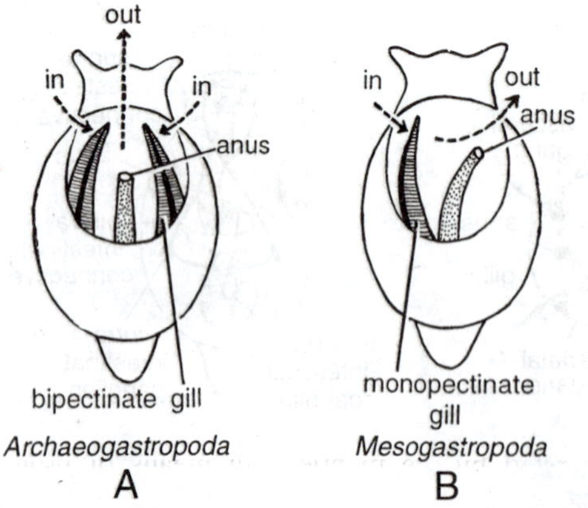

Fig. 15. Diagrams to illustrate the loss of symmetry and atrophy due to torsion. The arrows indicate the course of respiratory water currents.

Significance of Torsion

According to **Garstang** (1928), torsion first occurred as a larval mutation of advantage to the larva adapted to pelagic life but of little direct use to the adult. Before torsion, the untwisted swimming larva fell an easy victim to its predators because the posterior mantle cavity could receive the delicate head and velum only after the foot was already inside. After torsion, the mantle cavity became anterior, so that the sensitive parts *i.e.,* head and velum could withdraw first followed by the foot. Operculum sealed the aperture, the cilia of velum stopped beating, so that larva could fell to the sea bottom and avoid its enemies swimming in the water.

According to **Morton** (1958), main advantage of torsion must be to the adult. Firstly, torsion promotes stability in the adult by placing bulky mass of animal nearer the substratum. Secondly, in primitive Molluscs, the mantle cavity containing gill was situated posteriorly, so that when the animal moved upstream, the water-flow and the current due to movement of the animal opposed the respiratory current entering the mantle cavity from behind the animal. After torsion, mantle cavity is curled anteriorly above the head, so that all the three currents now flow in the same direction, thus flushing the mantle cavity continuously with fresh clean water and increasing its ventilation. Thirdly, anteriorly placed chaemoreceptive organs (osphradia) can also continuously sample the sediment and incurrent water stream and the animal orients itself properly with the help of the sense organs on the head. Once the shell is lost, gills become exposed to the external currents and their anterior position remains of no advantage, so that the mantle cavity

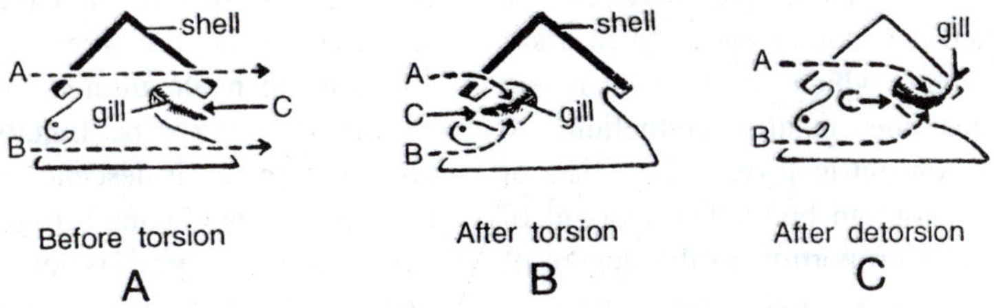

Fig. 16. Diagrams illustrating the possible advantage of torsion. **A.** Current due to flow of water. **B.** Current due to movement of gastropod. **C.** Respiratory current.

and the pallial complex shift back to their original posterior position (**detorsion**).

Detorsion

Changes occurring in torsion are to a certain extent reversible. This reversion is known as **detorsion** and it is very characteristic of the whole group of the **Euthyneura**. As a result, pallial complex travels back towards the posterior end along the right side, ctenidia point backwards, auricles move behind the ventricle, and the visceral loop becomes untwisted and symmetrical. In this way, a secondary external symmetry is re-established. Torsion must be disadvantageous to adult snails, as many of them have undergone detorsion processes. Various degrees of detorsion are met within the Euthyneura. In the least specialized Opisthobranchia and Pulmonata (*Acteon, Bulla,* etc.), detorsion is not complete, so that the visceral loop remains partly twisted and the anus and ctenidium are directed laterally, instead of anteriorly. Formerly, this condition was looked upon as an arrested stage in the torsion, but there is the same reduction of the paired parts of the pallial complex as in the specialized **Streptoneura**. Total detorsion, as shown by the typical **Opisthobranchia** (*Aplysia*), is accompanied by a reduction of disappearance of the shell. In extreme cases, as in *Pterotrachea,* the mantle and the visceral sac also disappear and the body elongates to become worm-like. The mantle cavity, visceral hump, external shell and even ctenidia may be lost, as in **Nudibranchia** (*Eolis, Doris,* etc.).

The phenomenon of detorsion can thus be elaborated as follows :

(1) In some cases the right ctenidium (originally left) and the osphradium are absent.

(2) In *Eolis,* there is veliger larva with a coiled visceral hump that undergoes torsion but adults do not show any sign and the pallial complex is posteriorly placed in adult. Naturally, the detorsion must have occurred during the course of further development.

(3) In **pulmonata**, the pallial complex is shifted but there is no chiastoneury as a result of shortening of visceral commissures. The pleurovisceral mass and so the chiastoneury is secondarily lost.

PEARL CULTURE

Pearl is also called 'Moti'. It is white, highly shining globular in shape and made by the clam, a mollusc called *Oyster* within its shell. Pearls are prized as gems from ancient times. Pearls are among the most beautiful and valuable of our jewels. **Kokichi Mikimoto** of Toba (Japan) is known to be the father of Pearl Industry. He discovered a method to induce foreign particle between the mantle and the shell of the pearl *Oyster* and thus stimulated pearl formation.

Pearl Formation or Pearl Fisheries

A pearl is a result of an injury to molluscs. It is secreted by the mantle as a means of protection against some foreign body. Whenever a foreign body such as a particle of sand or a small parasite (a trematode or cestode larva), a small animalcule or alga or even bit of shell, gets between the mantle and the shell it becomes enclosed in a sac of mantle epithelium which is thus irritated. Irritation stimulates the mantle epithelium to secrete nacre thin concentric layers of mother of pearl all around the foreign body. The amount of deposition is in direct proportion to the degree of irritation. At the end of several years, a pearl is formed.

Pearl Molluscs

Pearls are often found in clams and edible oysters but these are not nacreous and therefore, they are of little value. Most precious pearls are found in pearl oysters of the genus *Pinctada.* Important species for pearls are *P. vulgaris, P. chemnitzi, P. margaritifera, P. anomioides* and *P. atropurpurea* found in Indian waters. *P. vulgaris* which is closely allied to freshwater mussel is common species distributed in gulf of Kutch, gulf of Mannar and the Palk Bay and Baroda.

Artificial Pearl

Japanese have developed a technique of producing pearls artificially by inserting foreign bodies, such as glass beads, into the mantle of oysters. The oysters are retained in wire cages or crates until pearls are produced, which can be later removed and sold in markets. This was Mikimoto's discovery which had made him a wealthy person. It takes about 3 to 4 years to produce a pearl of considerable size but a large one takes 7 years. Cultured pearls are genuine pearls but are less valuable than uncultured pearls which can be identified by experts. Imitation pearls are beads coated with an iridescent substance called pearl essence that is obtained from scales of fishes. The best quality of pearl is known as 'Lingha pearl' obtained from marine oysters.

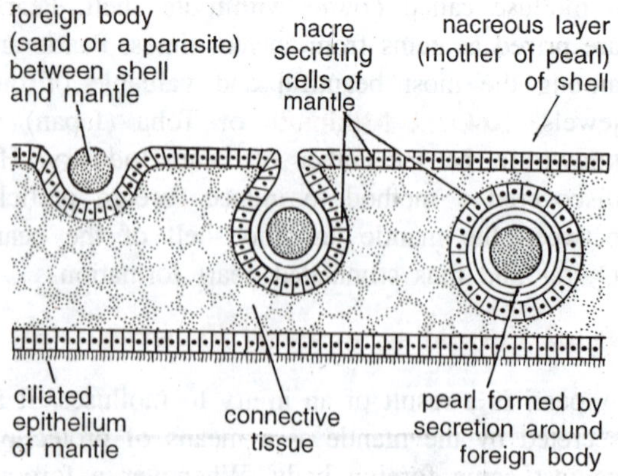

Fig. 17. Stages in pearl formation.

Culture of Oysters

The woman divers in Japan are called 'The Girls of the Sea'. From early morning till noon, they collect the oysters from the bottom of the sea with the help of nets. Collection of oysters is best done in the two months of the summer season when the sea is Calm and the water is nearer the shore. Oysters are collected, stocked and reared in rearing cages. Each cage is divided into 4-6 smaller chambers and is covered with metal mesh provided with cotton netting. The oysters are cleaned before being placed in the culture cages for about 10 to 20 days for acclimatizing them to shallow water conditions. After which they are processed for artificial pearl formation.

CHANK FISHERIES

Chank Fishery, a great source of revenue is based on a turbinellid gastropod, *Xanchus pyrum,* the shell of which is called the chank. Chank is used as trumpet in temples and is considered sacred in our country. It is a raw material for the manufacturing of ladies bangles and thus supports a number of industries especially in Nadia, Bangkura and Burdwan. Raw material for industries in West Bengal is obtained from South. Tuticorin shell is the best for bangles. Chanks or *Xanchus pyrum,* feed on polychaete worms and hence inhabit sandy bottoms. Breeding months are January to March. Fishing is done from October to May, by divers picking by hand about 100 shells per day. Chank beds on southeast coast are confined to Ramnad, Tanjore, Tirunelveli, Chingleput and Nellore districts. Chanks are abundantly found in Palk Bay and Gulf of Mannar.

Chank beds are also found on West Coast of Travancore and Kathiawar.

LIME FISHERIES

Molluscan shells yielding lime for industries constitute lime fisheries. Dead shells drifted ashore at estauries by currents, sub-fossil deposits in lakes and back waters and broken shells of surf beaten beaches, form the raw material for the lime fisheries. Large quantities of shells of bivales and Cerethidea are collected from estuaries during monsoon. Pulicat lake (Tamil Nadu), Surla lake (Ichapur) and Vembanad lake (Kerala) and Tuticorine supply the main bulk as subfossil deposits. The molluscs, the shell of which form the massive collection several feet thick are *Xancus, Arca, Vellorita, Meretrix, Murex, Rapana, Natica, Thais, Umborium* etc.

ECONOMIC IMPORTANCE OF MOLLUSCA

Molluscs are of major interest to man as about 10,000 species are of economic importance. Mostly they are beneficial to man although there are some molluscs which are indirectly harmful.

Beneficial Molluscs

1. As food. Chitons formed the main food of Red Indians. The gastropods are consumed by numerous predators chiefly fish, birds and mammals. The large land snail *Helix pomatia,* large foot of *Haliotis* and apple snail (*Pila*) form common food in New York, California and South India, respectively. Oysters, scallops, marine mussels and clams have been often used for food. Romans cultivate oysters to meet their great demand for countries like USA. Oysters are served fried, the hard shell marine clam, *Venus mercenaria* as whole on half shelled or cooked in chowder, *Mya arenaria,* a soft shelled clam, is steamed in shells and served with butter. *Mytilus edulis* is used in chowder, and adductor muscle of *Pecten* are served in flour and fried. Pelecypods also furnish food for star fish, boring sponges, drilling snails, some marine leeches, fish and shore birds.

Squids, cuttle fish and devil fish are popular as food articles in Oriental and Mediterranian countries. *Sepia* is used as food either cooked or dried in open air in European countries. *Loligo* are split, sundried and preserved for later use. *Octopus,* the devil fish is used in Canada and Alaska on many occasions. *Nautilus pompillus* is much prized as food by Pacific islanders. Cephalopods also form food for other animals like marine mammals and large fish.

2. As bait. Many gastropods are very useful to man, as bait for catching fish. Squids make an excellent bait for marine fishes especially cod in United States. Small *Octopus* are used as bait by the line fishermen of Palk Bay.

3. Money. Red Indian tribes of America used the common *Dentallium indianorum* as sawampum or money. Value of shells varied lengthwise. Gastropodan shells were source of money for various native races, including Vampum of American Indians. Shell of gastropods made media of barter in Africa and other countries. American *Oyster, Crassostria viginica,* is commercially cultivated and harvested and provides millions of dollars to the industry. Squids, cuttle fish and devil fish earn money as they are sold in market for food in China, Japan, India and Italy.

4. Ornamentation. Scaphopod, *Dentallium indianorum,* tooth shells are valued as ornaments. Tools, utensils and objects of delight have been formed from gastropodan shells. Some marine snails of South Pacific (turban shells) have a calcarious operculum so rounded and coloured as to resemble a vertebrate eye. These opercula known as 'cat's eye', are sought as curios. *Nautilus* shell is much used for decoration, art and for many other useful purposes.

5. Useful dyes and ink. Some gastropods like *Nucella* (*Purpura*) and *Murex,* are sources of Tyrian purple from their juices. Dye for royal purple in Biblical literature originally came from a gland of the snail, *Murex truncutus.* Secretion is colourless but becomes a beautiful purple by exposure to the air. Contents of ink-sac of cuttle-fish provides a rich brown pigment called 'sepia', used by the artists. Originally Indian ink was obtained from the ink of a cuttle-fish, *Sepia cubrata.* Now-a-days a certain brown finish of photograph is termed as sepia finish.

6. Buttons and pearls. Gastropodan shells are used to manufacture buttons and other articles. Shells of certain bivalves have been used for mother-of-pearl layer, also for buttons, knife handles etc. Buttons are made by hand by cutting shells of freshwater bivalves and some marine clams. Pearls are made by clam and pearl oysters themselves and are among the most beautiful and valuable of our jewels.

7. In art and medicine. Shell cameos are made mostly from snails notably that of *Cypraea tigris* and *Cases tuberosa. Nautilus* shell is commonly used in art. It is a pretty object thrown ashore during monsoon storms on the Indian coasts.

A rather odd and unexpected use for fossil cephalopods is found among the Red Indians of Montana and Wyoming. Their medicine men collect specimens of beautifully preserved fossil ammonoids from Cretaceous Strata and keep them as 'medicine'. The internal calcarious shell of *Sepia,* the 'cuttle-bone' is used as medicine as well as for other purposes.

8. In literature. There are stories about giant squids and octopuses cited to play exaggerated role in popular literature. One such story, pictures a huge squid dragged a small ship beneath the waves and grabbed the helpless sailors in its cruel, snake-like arms and crushed them to death. Large squids and octopuses are feared more rightly for their dangerously powerful beaks and ability to sieze men from boats or grip persons under water by their deadly tentacles.

9. Animal inventions. Cephalopods get credit for two animal inventions. One of them is the principle of 'jet propulsion' only recently discovered by man but used by squids and octopuses for millions of years. The second novel invention is the use of a 'smoke screen', in both offence and defence, is another novel invention by cephalopods. A smoke screen is formed by ejecting a brownish ink into the water. This diffuses into a large area and allows cephalopod to stalk stealthily through the 'smoke' searching for its prey, or to escape in cloudy water if persued by an enemy. Man could use such a device in warfare not earlier than first World War.

Harmful Molluscs

Destructive Forms

The destructive activities of some molluscs are of great importance to global economy.

1. Herbivores. Some gastropods like land slugs and snails cause damage to gardens, orchards, green houses and mushroom beds by feeding upon the succulent parts of seedlings and mature plants. They do not spare the vegetable crops, flowering plants and other decorative plants. Certain pelecypods burrow or bore into wood and stone. The *Teredo,* the shipworm, is not a worm instead a bivalve mollusc and does more harm to wooden boats and ships. According to an estimate, the shipworm caused 2,500,000 dollar's damage to wooden structures in San Francisco Bay in four years time (1917-1921).

2. Carnivores. Some gastropods are ferocious predators (*Natica, Buccinum, Murex, Urosalpinx*), that bore into and fed on other molluscs. Marine snail *Urosalprinx,* causes serious losses to oyster industry. It is known as 'oyster drill'. Cephalopods are all predaceous and carnivorous molluscs, devouring great number of fish, crustaceans and other molluscs and are much destructive to fisheries. Crabs fall favourite food of *Octopus.* Giant squid (*Architeuthis princeps*) are known to engage in battles with whales and devour their tongue by their sharp beaks.

3. Parasites. Members of the families Pyramidellidae (*e.g,. Brachystomia*) and Eulimidae of gastropoda are ectoparasites and suck blood from bivalve molluscs, polychaetes and echinoderms. *Stylifer* (family Styliferidae) is an endoparasite in the wall of echinoderms. Among the pelecypod parasites, *Entovalva* lives within the gut of a sea-cucumber (*Synapta*) and absorbs predigested food of the host.

4. Intermediate hosts. Snails are of considerable importance from a medical point of view as many of them serve as intermediate hosts for parasitic flat worms, such as *Fasciola* and *Schistosoma.*

5. Commensals. There are commensal pelecypods which include : *Phylyctaenachlamys* living in burrows of a shrimp of Great Barrier Reef, *Lepton* in burrows of shrimps and Polychaetes, *Mediolaria* in the test of sea-squirts and *Vulsella* in sponges.

MOLLUSCAN LARVAE

Larvae may serve two purposes : finding new settling sites and gaining access to the rich food supply of phytoplankton. The earliest molluscan larva was no doubt a trochophore like that of an annelid. It was a top-shaped creature with a tuft of cilia above and a ciliated band around the middle. This is the only closest resemblance of Mollusca to the Annelida. Molluscan life histories do not perfectly correspond with taxonomy, but in general it is true that the archaeogastropods and the bivalves begin life as a trochophore and rapidly pass on to a veliger.

There are different types of molluscan larvae according to the importance of the pelegic phase and amount of planktonic food taken. **G. Thorson** (1950) recognised three ecological types of larvae in Mollusca.

1. Planktotrophic larvae with long larval life. Such larvae have larval life of two or three months *e.g.,* Lamillibranchs, Prosobranchs. They are capable of wide distribution. They are usually found in tropical, subtropical and a few in high Arctic seas. Such molluscan veligers are all ciliary feeders. The large vellar cilia collects particles which are thrown on to a tract at the base of the velum leading to a mouth. Coarse or unsuitable particles are removed by rejectory tracts upon the foot.

2. Planktotrophic larvae with short swimming life. Such larvae have larval life of not more than a week in the plankton *e.g.,* Nudibranch larvae, *Gibbula cineraria, Hydrobia ulva, Turitella communis* and *Bela trevelyana.* The velum never elaborate. Planktonic feeding is of secondary importance. Distribution is the main

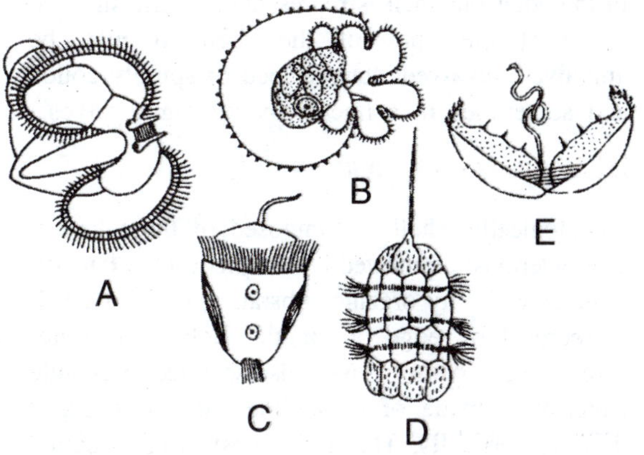

Fig. 18. Molluscan larvae. **A.** Veliger *of Aolis* (Mesogastropoda). **B.** Veliger of *Lamellaria* with *Echinospira* shell. **C.** Trochophore stage of a bivalve. **D.** Yolk larva of *Yoldia* (Protobranchia). **E.** Glochidium *of Anodonta.*

object of their life. There is little growth between hatching and settling. Being less dependent on food, they are surprisingly adaptable to unfavourable conditions, and serve mainly for dispersal.

3. Yolk larvae. Such larvae take no food in plankton. They are **lecithotrophic**, as they hatch from very yolky eggs and develop into large, 'yolky larvae'. They swim little and are passively carried about in the plankton. Gastropods show only few examples of yolk larvae. Yolk larvae are normally found in Amphineura, Scaphopoda and protobranchiate Lamillibranchia. In *Chiton,* the yolk larvae are modified egg-shaped trochophores with a broad ciliary ring. They spend six hours to few days in the plankton. *Neomenia* has three such rings, *Dentalium* four. In protobranchs such as *Yoldia* and *Nucula* they form a large barrel-shaped ciliated test, which is thrown off when the larva settles.

SHELL IN MOLLUSCA

The molluscs are characterised by having as a rule, a protective exoskeleton in the form of a **shell**. Shell is usually external, sometimes internal. It is derived from the mantle of the veliger larva.

In the adult the shell is represented by the smallest whorl at the apex of the shell. It may be univalved, bivalved, cone-shaped or spirally coiled and sometimes in a linear row of eight valves.

[I] Structure of Shell

Histologically, shell is composed of three layers. The outermost pigmented layer called **periostracum** is made of horny organic substance, **conchiolin**. It is secreted by edge of mantle. Below the outer layer is the **prismatic** layer, also secreted by mantle edge. It is made of crystelline calcarious plates running vertically. The inner most-layer is called **nacreous** or **pearly** layer. It is made of conchiolin and calcarious plates running alternately and longitudinally. It is secreted by whole outer surface of manlte and presents a lustrous surface. It is also called, '**mother of pearl**'.

[II] Variety of Shells

A large variety of shells are found in molluscs. Shell of one representative type from each class of Mollusca are discussed below.

1. *Chiton.* Shell forms the most characteristic feature of chitons. It is made of eight transverse, overlapping, calcareous **plates** or **valves**, arranged in a longitudinal row. It forms the solid armour covering the dorsal surface. Name of order Polyplacophora is derived from this nature of shell. Shell plates are moveable upon one another allow the animal to roll up like a wood-louse. First

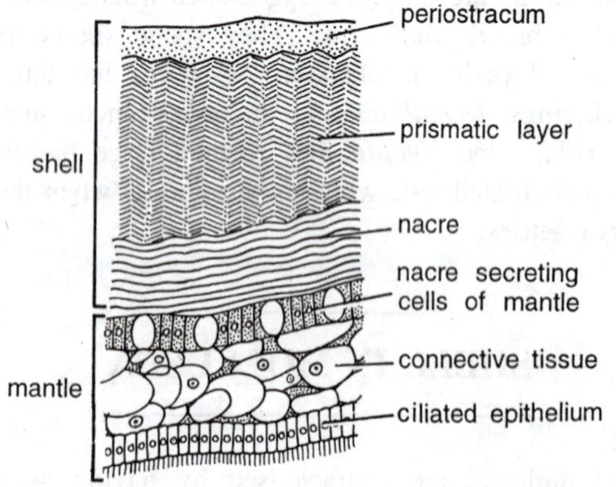

Fig. 19. *Unio.* A part of shell and mantle in T.S.

(**cephalic**) and eighth (**anal**) shell plates are hemispherical in shape, while others, called **intermediate** shell plates are somewhat rectangular and often keeled mid-dorsally. The posterior edge of each plate overlaps anterior edge (**wing**) of the next behind. Each shell plate is made up of two distinct layers. **Tegmentum** or upper layer, consists of an organic conchiolin matrix and lower thicker and denser layer called **articulamentum** consists of calcium carbonate only.

2. *Dentalium.* Shell is external in Dentalium. It is cylindrical tubular, slightly curved and tapering. Which is open at both ends. It is shaped like a trumpet or elephant's tusk, hence often called 'tusk shells'. During life, shell is buried obliquely in mud with wider anterior end lying deepest and narrow posterior end of apex projecting above the surface of mud. Shell of *D. octogonum* is snowy-white with 8 grooves running down the entire length, that of *D. formosum* is mottled with shades of dark red. There is no operculum and shell grows at larger anterior end. The scaphopod shell differs from that of other molluscs in being unchambered and open at both ends and that of tubiculous annelids in having 3 instead of 2 layers.

3. *Cypraea.* Shell in *Cypraea* or cowries, is oval, rounded on the top and flattened beneath and consists mostly of a large whorl. The shell opening is long and narrow, toothed on both sides and channelled at each end. Their extremely smooth, highly polished and variously coloured shells make them very conspicuous and are prized by the collectors.

4. *Mytilus.* Shell is equivalved and wedge-shaped being pointed in front and rounded behind. Two shell valves are united antero-dorsally by a **hinge-ligament**. Hinge-ligament is made of conchiolin and is brown, tough, elastic and non-calcareous. *Umbo,* a whitish knob-like swelling in each valve lies anteriorly. *Umbo* is the thickest and oldest portion of the shell. **Lines** or **rings of growth** run around umbo as centre and run parallel to the free margin of the shell.

5. *Sepia.* Shell of cuttle fish or *Sepia* is internal and lies embedded in upper side, completely enclosed in a sac of mantle. It is secreted by its epithelial lining. It is flat, broad

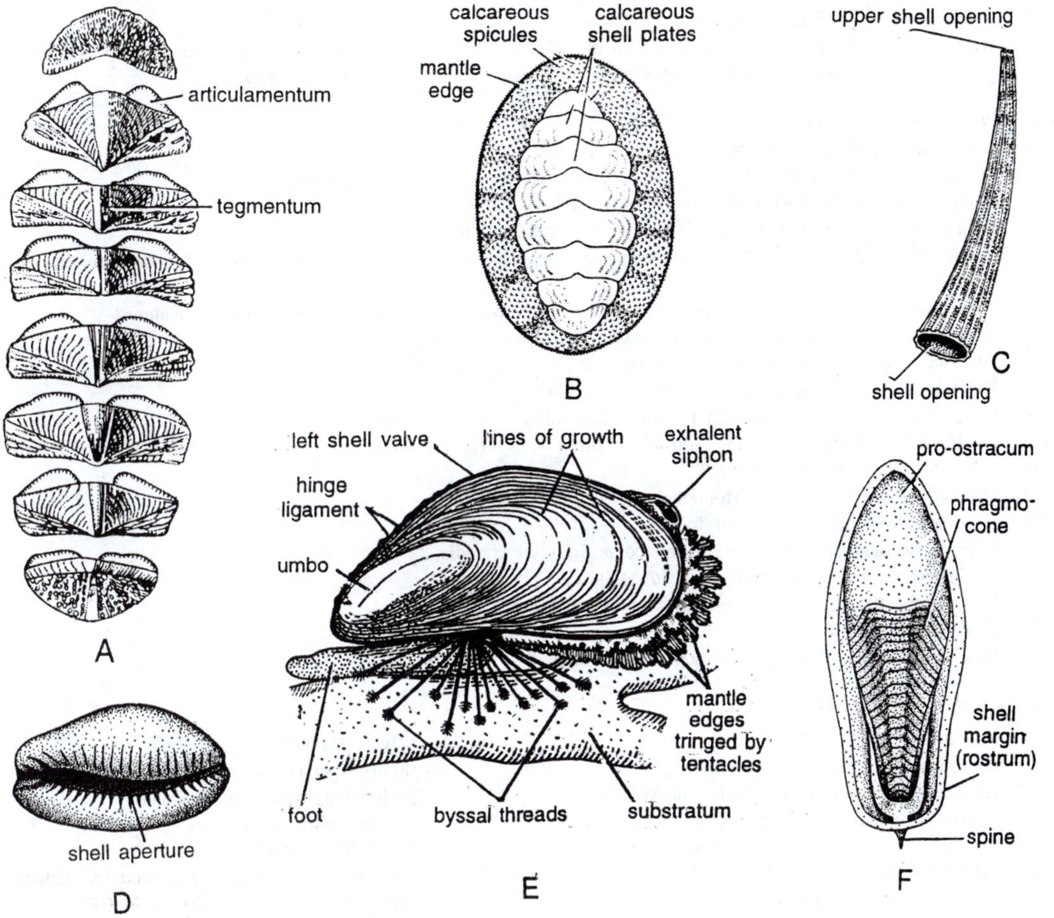

Fig. 20. Molluscan shells. **A.** *Chiton* with shell. **B.** *Chiton* shell plates separated. **C.** Shell of *Dentalium.* **D.** *Cypraea* shell in ventral view. **E.** *Mytilus* shell. **F.** *Sepia* shell.

and oval in shape, represented by **phragmocone** with a broader and rounded oral end, called **pro-ostracum** and a narrow, pointed aboral end called **rostrum**, projecting into a spine. Shell is dead and composed of calcareous rather than horny matter. Hard and resistant shell provides rigidity to trunk, like an endoskeleton. The calcareous matter is arranged in fine parallel layers called septa or laminae enclosing space containing fluid and gas, so that light shell serves as a hydrostatic organ or float. It is commonly-spoken as **cuttle bone** and is a familiar object of seashore. Being light it floats in water, and during monsoon these bones are drifted ashore in such a great number that they have been named as **sea-foam**.

Important Questions

➤➤*Long answer type questions*

1. Give an account of various types of podium found in Mollusca.
2. Discuss in detail different kinds of respiration met within different Molluscs.
3. What is torsion ? How does it affect the gastropods ?
4. What is pearl ? How the pearl formed ?

5. How are molluscs important to mankind ? Discuss.
6. How many kinds of larvae occur in molluscs ? Discuss.
7. Write short notes on : (i) Detorsion, (ii) Chank fisheries, (iii) Lime fisheries, (iv) Ctenidium, (v) Glochidium, (vi) Pedal glands, (vii) Torsion, (viii) Byssus apparatus, (ix) Pearl culture, (x) Secondary gills.

➤➤Short answer type questions

1. Name the group/groups where the radula is present.
2. Mention a mollusc where both the mantle and shell are open at both the ends.
3. Name the two respiratory pigments in the molluscan blood.
4. Define sinistral and dextral shells. Give an example for each.
5. Enumerate the most significant feature of cephalopodan eye and mention the reasons for its similarity to that of vertebrate.
6. Give a general account of the form of shells of Mollusca with a note on their adaptive features.
7. Explain the adaptations to pelagic life met with amongst molluscs ?
8. Give an account of the adaptive radiation met within Gastropoda.
9. Explain the phenomenon of torsion in Gastropoda.
10. Locate the two specimens which resemble each other (form the same line).
    (a) *Hermaca, Doris* (b) *Notrachus, Bulla* (c) *Dolabella, Aplysia* (d) *Maginulus, Notobranchus*
11. What do you mean by podium ?
12. Give an account of foot in Monoplacophora.
13. How you define the division of foot ?
14. Give an account about number, form and position of ctenidia ?
15. Comments on pulmonary respiration in Mollusca.
16. List out the molluscs which is used as food of man ?
17. Describe ornamental mollusca.
18. How many kinds of larvae are found in Mollusca ?

➤➤Multiple choice questions

1. Malacology is the study of :
   (a) mollusca (b) mantis
   (c) shell of mollusca (d) marine animal
2. Digestive gland in Mollusca is :
   (a) liver (b) pancreas
   (c) hepatopancreas (d) all
3. Conchology is the study of :
   (a) Coelenterata (b) Mollusca
   (c) shell of Mollusca (d) Mentle of Mollusca
4. Pedal glans of mollusca consist of :
   (a) labial gland (b) supra-pedal gland
   (c) sole gland (d) all the above
5. Mode of respiration in Mollusca :
   (a) cutaneous
   (b) branchial
   (c) pulmonary
   (d) all the above
6. Effect of torsion :
   (a) displacement of mentle cavity
   (b) looping in alimentary canal
   (c) endogastric cool
   (d) all of the above
7. Food items of Mediterranian country include :
   (a) squid (b) cuttle fish
   (c) devil fish (d) all
8. Blood of Mollusca contains :
   (a) RBC (b) WBC
   (c) amoebocytes (d) none
9. Chank is :
   (a) *Xanchus* (b) *Penctada*
   (c) *Aplysia* (d) none
10. The best quality of pearls is known as :
    (a) real moti (b) sweta moti
    (c) lingha moti (d) none

Answers

➤➤Multiple choice questions

1. (a) 2. (c) 3. (c) 4. (d) 5. (d) 6. (d) 7. (d) 8. (c) 9. (a) 10. (c)

62

Asterias : A Sea Star

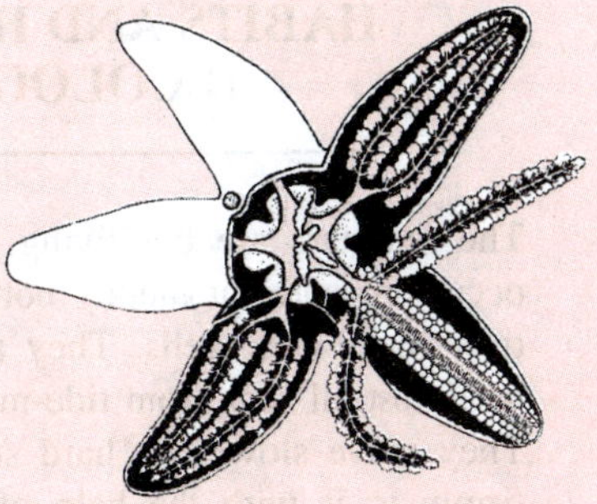

Phylum Echinodermata (Gr., **echinos**, hedge-hog + **derma**, skin + **ata**, characterized by) includes exclusively marine invertebrates displaying pentamerous radial symmetry and an endoskeleton of calcareous plates and spines. They constitute some of the most beautiful members of sea fauna, such as starfishes or sea stars, brittle stars, sea cucumbers, sea urchins, sand dollars and sea lillies, etc. Familiar starfishes belong to class Asteroidea, characterized by the presence of five or more arms not sharply set off from a central disc. The name 'starfish' is, however, misleading as these animals are not true fishes. A more suitable name suggested for them is 'sea star'.

Common genera of sea stars are *Asterias, Luidia, Pentaceros, Astropecten, Solaster, Heliaster, Asterina,* etc. Genus *Asterias* includes about 1,500 species of which most common is *A. rubens.* Following description is generalized, though based mainly on *A. rubens.*

SYSTEMATIC POSITION

Phylum	Echinodermata
Subphylum	Eleutherozoa
Class	Asteroidea
Subclass	Euasteroidea
Order	Forcipulata
Genus	*Asterias*
Species	*rubens*

HABITS AND HABITAT (ECOLOGY)

The sea stars are free-living marine animals that occur on sandy or muddy bottoms or crawl about over rocks and shells. They are well represented in almost all seas from tide-marks to deep waters. They move slowly on hard substratum or adhere firmly to it with the help of their characteristic locomotory **podia** or **tube feet**. All sea stars are carnivorous, feeding mainly on crustaceans, polychaetes and molluscs. They also feed on detritus and plankton entrapped in mucus, secreted by body and carried to the mouth by cilia. Sea stars, in general, exhibit remarkable powers of autotomy and regeneration.

EXTERNAL FEATURES

Shape, Size and Colour

Most sea stars possess a pentamerous radial symmetry, secondarily derived from a larval bilateral symmetry. In *Asterias,* body consists of an indistinct **central disc**, from which radiate out five elongated and tapering **rays** or **arms**. In some genera, number of arms may be more than five; for example, there are 7-14 arms in **Solaster** and more than 40 arms in **Heliaster**.

Size averages from 10 to 20 cm in diameter, though some forms may be much smaller or larger. Colouration is variable including shades of brown, yellow, orange, pink and purple. There may also be a combination of colours. Body is strongly flattened with distinct oral and aboral surfaces.

Oral Surface

Flat lower surface of the body which is directed to the substratum is the **oral** or **ventral surface**. Central disc on this surface bears a centrally located aperture, the **actinostome** or **mouth**. It is a pentagonal aperture with five angles, each directed towards an arm. The mouth is surrounded by a soft and delicate membrane, the **perioral membrane** or **peristome**, and is guarded by five groups of **oral spines** or **mouth papillae**. From each angle of mouth extends radially a narrow groove, called the **ambulacral groove**, which runs all along ventral surface of the corresponding arm. Each groove is guarded on each side by two or

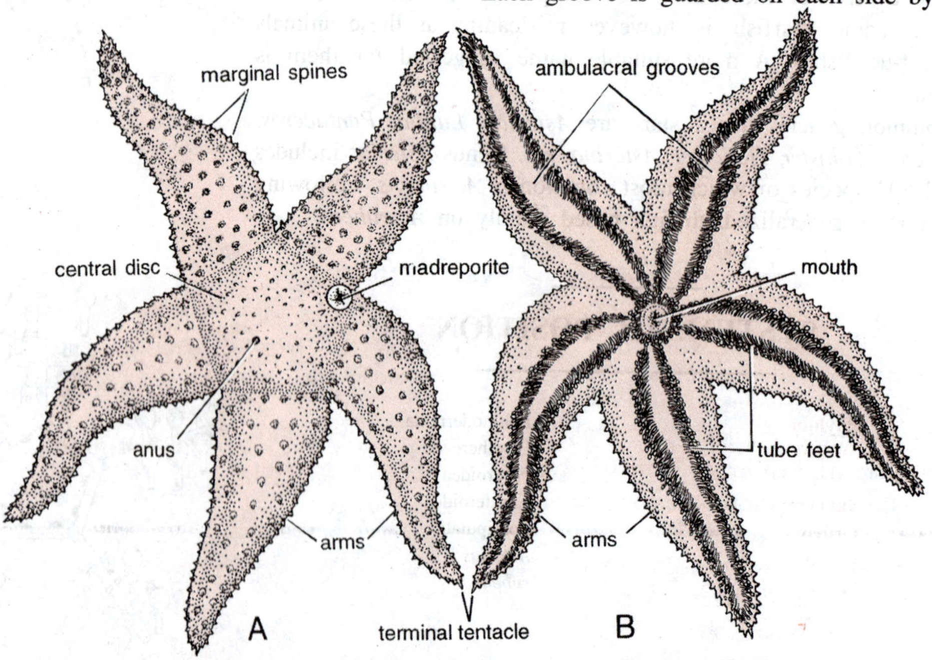

Fig. 1. *Asterias.* External features. **A.** Aboral view. **B.** Oral view.

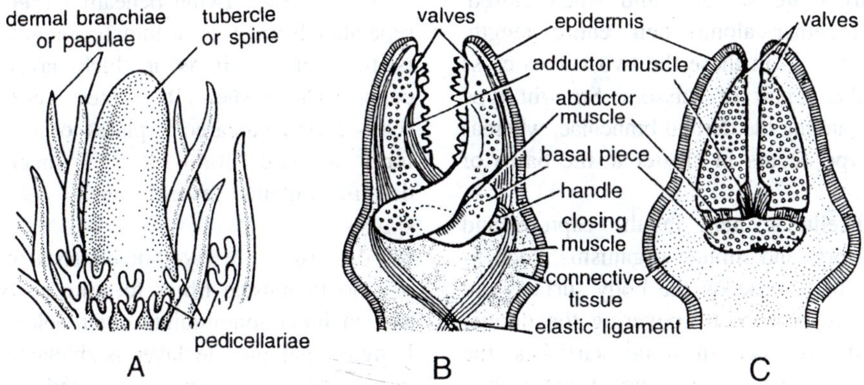

Fig. 2. *Asterias.* **A.** A cluster of pedicellariae, papulae and tubercle. **B.** Crossed type pedicellaria. **C.** Straight type pedicellaria.

three rows of movable, calcareous, **adambulacral spines**, which are capable of closing over the groove. External to these spines are three rows of stout immovable spines, beyond which occurs another series of marginal spines demarcating oral and aboral surfaces. Each ambulacral groove contains two double rows of short, tubular retractile projections, called as **podia** or **tube feet**, that end in suckers. Tube feet are characteristic organs of echinoderms, serving variously for locomotion, capturing of food and respiration, etc.

Tip of each arm bears a small median, non-retractile and hollow projection, the **terminal tentacle**. It acts as a tactile and olfactory organ. At its base occurs a bright red photosensitive **eye spot** made up of several ocelli.

Aboral Surface

It is the upper and slightly convex dorsal surface of the body. It bears a large number of short and blunt, immovable, calcareous spines or **tubercles** arranged in irregular rows parallel with long axes of arms. Around and between these spines occur tiny pincer-like structures, the **pedicellariae**. These are grasping organs used for cleaning or protecting the body surface. They also occur on oral surface attached to the bases of spines or in spaces between them. There also occur, on both oral and aboral surfaces, minute, finger-like, hollow and retractile processes, the **dermal branchiae**, or **gills**, or **papulae**. These project through minute **dermal pores** of integument and serve for respiration as well as excretion. A minute circular aperture, the **anus**, is situated close to the centre on the aboral surface of central disc. A distinct, flat, circular area, the **madreporite**, is present in an interradius between two arms. These two arms are called the **bivium** and the remaining three, the **trivium**. Madreporite is a sieve-like plate, with numerous narrow and radiating grooves bearing pores which lead into the water-vascular system inside the body.

PEDICELLARIAE

Pedicellariae of sea stars are minute, whitish jaw-like structures, found on both the body surfaces, in association with spines. Pedicellariae may be stalked or sessile. But only stalked type occurs in *Asterias.* Each consists of a short, fleshy and a movable stalk, bearing two articulating calcareous **blades** or **valves**, resting upon and articulating with a third calcareous plate, the **basal** or **basilar piece**. Such pedicellariae (consisting of three calcareous plates) are termed **forcipulate**. Opposite surfaces of two valves are serrated. Valves open and close with the help of one pair of **abductor** and two pairs of **adductor muscles**, respectively. According to disposition of valves, two types of pedicellariae occur in *Asterias* : (i) **Straight type** — in these the valves are straight, and when closed, these meet together along their entire length, (ii) **Crossed type** — in these the two valves cross each other like a pair of scissors. Straight type occurs largely among the dermal branchiae, whereas the crossed type occurs in cluster at the bases of the spines.

The pedicellariae help in the capture and removal of debris and minute organisms, such as larvae, which may settle on the body surface and interfere with respiration by covering the dermal branchiae and tube feet. In some starfishes, the pedicellariae may help in capturing of small prey.

BODY WALL

The body wall of *Asterias* consists of epidermis, dermis, muscular layer and parietal peritoneum.

1. Epidermis. It extends all over the body surface but wears off over the spines. It consists of ciliated columnar **epithelial cells** and a few scattered **gland cells** and **neurosensory cells**. Gland cells secrete a protective mucous sheet around the body. The spindle-shaped neurosensory cells are connected to a **nerve plexus** or nerve net beneath the epidermis. Epidermis is covered externally by a two-layered thin **cuticle** secreted by it.

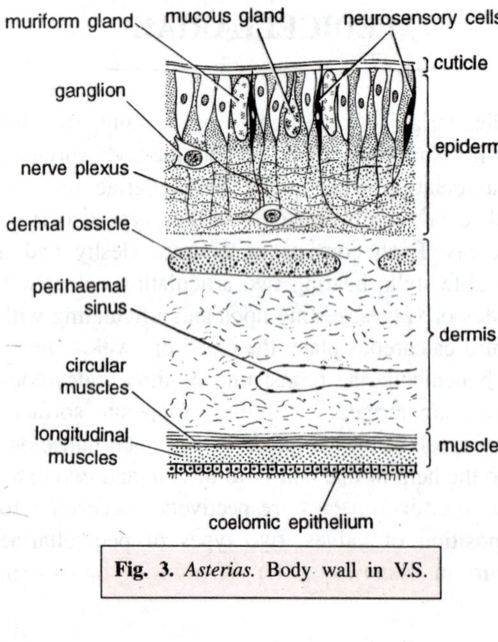

Fig. 3. *Asterias.* Body wall in V.S.

2. Dermis. Lying beneath the epidermis, and separated from it by a thin **basement membrane,** is the dermis. It is a thick layer of fibrous connective tissue. Its outer region contains self-secreted calcareous plates or **ossicles,** which form a hard **endoskeleton.** Inner region of dermis contains several **perihaemal** or **dermal spaces.**

3. Muscular layer. It consists of involuntary or smooth muscle fibres forming an outer circular and an inner longitudinal layer below the dermis. Longitudinal muscle layer is thicker than circular muscle layer and more developed on aboral than on the oral side. Musculature is responsible for bringing about bending of the arms and opening and closing of the ambulacral grooves. Pedicellariae are worked by special muscles, comprising of the adductor and abductor muscles.

4. Parietal peritoneum. It is the inner most layer of the body wall, lining the coelom and consisting of ciliated cuboidal epithelium.

ENDOSKELETON

Endoskeleton of *Asterias* (or echinoderms in general) is unique in being a mesodermal structure instead of ectodermal, as in other invertebrates. It

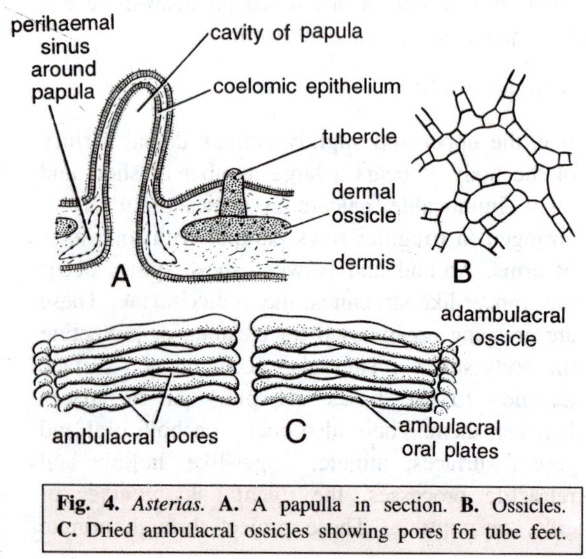

Fig. 4. *Asterias.* **A.** A papulla in section. **B.** Ossicles. **C.** Dried ambulacral ossicles showing pores for tube feet.

consists of numerous calcareous **ossicles** secreted by dermis and lodged in its outer region. Ossicles are of varied but definite forms, such as spines, rods, cones or plates. These are not united into a continuous structure but lie at intervals, connected together by muscles and connective tissue. This arrangement provides rigidity, and at the same time, permits movement. Arrangement of ossicles is irregular on the aboral side but more or less regular on the oral side. Surrounding the mouth are five **oral ossicles.** Above each ambulacral groove are two rows of large rod-shaped **ambulacral ossicles**. Ossicles of the opposite rows are arranged like an inverted 'V'. Apex of the 'V' forms a prominent **ambulacral ridge** projecting into the coelom. Ambulacral ossicles are devoid of spines or tubercles. Each bears a notch or **half-pore** on its inner as well as outer margin resulting in a double row of **ambulacral pores,** on each row of the ambulacral ossicles. From each ambulacral pore protrudes a **tube foot**. Outer to each row of ambulacral ossicles is a row of **adambulacral ossicles** bearing movable spines and pedicellariae. Outer to this is a row each of **supra** and **infra-marginal ossicles**.

COELOM

Sea star possesses a true and spacious coelom which is an **enterocoel**. A greater part of it extends into the central disc and the arms as a continuous **perivisceral coelom** between the body wall and visceral organs. **Coelomic epithelium**, consisting of ciliated cuboidal cells, lines the body wall as well as viscera. In addition to perivisceral coelom there are other restricted coelomic spaces or compartments like — **water-vascular system**, **axial sinus**, **perihaemal canals** and **sinuses**, **genital sinuses**, etc.

Coelomic epithelium secretes a colourless **coelomic fluid**, which fills the coelomic spaces. Coelomic fluid resembles sea water in composition but is, however, less alkaline. It contains phagocytic amoeboid cells, termed as **amoebocytes** or **coelomocytes,** which are formed by budding from coelomic epithelial cells.

In absence of a true blood-vascular system, the coelomic fluid, which is kept in continuous circulation by the action of cilia of coelomic epithelium, brings about transportation of materials between various parts of the body. Food materials are carried from pyloric caeca of alimentary canal to other parts. Oxygen is carried from dermal branchiae to various parts, while carbon dioxide and nitrogenous wastes are carried reversely. Some coelomocytes even possess a respiratory pigment. Phagocytic coelomocytes ingest wastes and foreign particles and exit through the dermal branchiae.

DIGESTIVE SYSTEM

Alimentary Canal

Sea star possesses a complete digestive tract situated in the central disc. It lies along the oral-aboral axis and is quite short due to flattening of the body along that axis. Between the mouth and anus it can be differentiated into oesophagus, stomach and intestine. Histologically, the alimentary canal resembles the body wall.

Mouth

It is a pentagonal aperture surrounded by a delicate perioral membrane called **peristome,** situated on the oral side in the centre of central disc. It is guarded by five groups of tubercles called **mouth papillae** or **oral spines**, lying in five inter-radii. Muscles of the body wall surrounding mouth, act as a **sphincter** for closing and opening the aperture.

Oesophagus

Mouth opens into a short and wide oesophagus. It extends vertically to open into the stomach.

Stomach

It is a spacious sac occupying most of the interior of the central disc. It is differentiated into a larger oral part or **cardiac stomach** and a smaller aboral part or **pyloric stomach**, the two separated by a constriction.

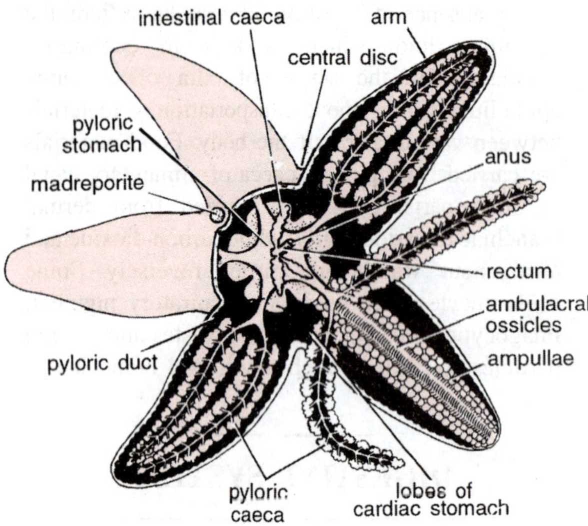

Fig. 5. *Asterias.* Dissection to show alimentary canal.

1. Cardiac stomach. It is a large sac having five lobes along the five radii. Its wall is thin muscular and highly folded. Entire cardiac stomach can be everted through mouth at the time of feeding. This is brought about by contraction of the body muscles and consequent pressure of coelomic fluid. Retraction is brought about by five pairs of **retractor muscles**, which connect the cardiac stomach with ambulacral ridges of five arms. These retractor muscles are composed mostly of connective tissue and a few muscle fibres and originate like mesenteries. They are thus also called **mesenteries** or **gastric ligaments**. Cardiac stomach is glandular and secretes mucus.

2. Pyloric stomach. It is a small, flat pentagonal sac opening aborally into intestine. Angles of the pentagon lie along the radii and each receives a duct, called the **pyloric duct**, from the corresponding **pyloric caeca** or **digestive glands**.

3. Pyloric caeca or digestive glands. Each arm of sea star contains a pair of digestive glands or pyloric caeca extending up to its tip, and each is suspended from the aboral body wall by a pair of longitudinal mesenteries. Each pyloric caecum has a hollow longitudinal axis, from which arise two series of short, lateral, hollow branches, each terminating into several small bladder-like products or lobules of greenish or brownish colour. The hollow axes of the two pyloric caeca of an arm open into a common pyloric duct that leads into pyloric stomach.

Epithelial lining of pyloric caeca consists of four types of flagellated columnar cells : (i) **secretory** or **granular cells** which secrete proteolytic, amylolytic and lipolytic enzymes, a reason for which the pyloric caeca are compared with the pancreas of vertebrates, (ii) **mucous cells** which secrete mucus, (iii) **storage cells** which store and reserve food as lipids, glycogen and protein-polysaccharide complex, and (iv) **current producing cells** which bear longer flagella that maintain a steady flow of enzymes towards the stomach along the aboral side and of digested food into pyloric caeca along the oral side.

Intestine

It is a short, five-sided tube extending vertically to aboral surface, where it opens out by an anus. It gives out two or three small, branched and brownish appendages, the **intestinal** or **rectal caeca**, at interradial positions. Intestinal caeca secrete a brownish fluid, probably excretory in nature. Part of the intestine after caeca is sometimes called the **rectum**.

Anus

It is a small, rounded aperture, situated somewhat eccentrically on aboral surface of the central disc.

Food and Feeding Mechanism

Sea star is a voracious carnivore. It prefers to prey on sedentary marine animals like clams, oysters, mussels, snails, crabs, barnacles, worms, and even sea urchins and other smaller starfish. At times it also feeds on small fishes and injured and dead animals.

Sea star ingests and digests food in a fascinating manner. Prey is captured and held in position by the arms and tube feet. It may be interesting to note how a sea star devours clams and other bivalves. It conveniently creeps over the clam, arches its body over it and firmly attaches its tube feet to two shell valves in such a way that the ventral margin of the clam comes to lie in front of its mouth. It then tries to pull apart the two valves held tightly together by the powerful adductor muscles. A few tube feet at the tips of arms are also attached to the substratum to aid in the process. Pull is steadily maintained till the adductor muscles of clam are exhausted and give way. Adductor muscles cannot, as a rule, remain in a continuous state of contraction for a long time. When the valves finally gape, cardiac stomach of sea star is everted into the mantle cavity of the clam to devour it.

Digestion, Absorption and Egestion

When cardiac stomach is everted over the captured prey, secretions of stomach and pyloric caeca are poured over it. Enzymes proteases, amylases and lipases, digest the proteins, starches and fats, respectively. Digestion, thus, takes place outside the body. Digested materials are then carried into alimentary canal by retraction of cardiac stomach. This is brought about by contraction of gastric ligaments. Digestion of semi-digested materials, if any, is completed in stomach and pyloric caeca. In the latter, some intracellular digestion is also believed to occur. Some small animals are, however, ingested as such and digested only in stomach and pyloric caeca.

Digested food is absorbed mainly by pyloric caeca and distributed to various parts of the body by coelomic fluid. Excess food is stored in the storage cells of the pyloric caeca.

As the absorption of food is more or less completed in pyloric caeca, it does not require a long intestine. Hence, the intestine is very short.

As animal ingests partially digested food, it has little undigested material, which is mainly eliminated through the mouth itself. Little, if any, egestion takes place through the anus.

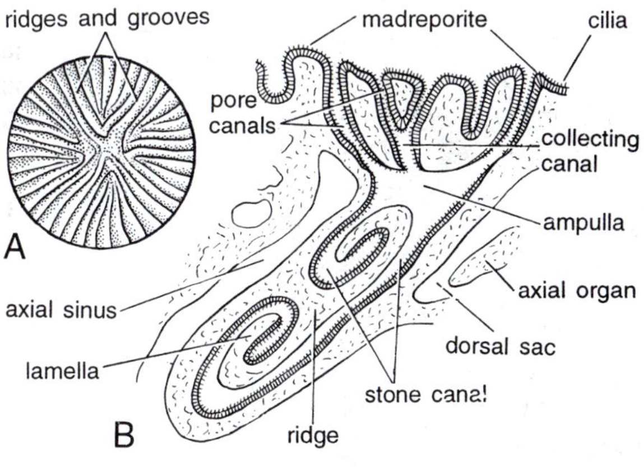

Fig. 6. *Asterias.* **A.** Madreporite seen from outside. **B.** V.S. through madreporite.

WATER VASCULAR OR AMBULACRAL SYSTEM

It is a unique system of echinoderms which helps mainly in locomotion. It is in fact a modified part of coelom consisting of a system of canals containing sea water and amoeboid corpuscles. All the canals have muscular walls and are lined internally by ciliated epithelium. Water-vascular system of sea star consists of the following parts :

1. Madreporite. It is a thick, rounded, sieve-like calcareous plate, situated on the aboral surface of the central disc. It lies on an interradius near the bases of two adjacent arms forming the **bivium**. Its surface bears numerous, fine radiating furrows permeated by as many as 250 minute **pores**. Each pore leads into a **pore canal**. Pore canals unite to form larger **collecting canals** within the substance of madreporite. Collecting canals finally lead into a sac-like **ampulla** lying below the madreporite. Ampulla continues into stone canal.

2. Stone canal. Also known as **madreporic canal**, is a S-shaped tube opening on the oral side into a **ring canal** around the mouth. Its walls are supported by a series of calcareous rings, hence the name stone canal. Inner lining of the walls is of tall cells bearing cilia or flagella which draw water into canal. In a young sea star, stone canal is like a simple tube but as it grows older a prominent **ridge** and two spirally coiled **lamellae** develop within its lumen.

Stone canal, along with an **axial organ**, is enclosed in a coelomic sac, the **axial sinus**. Three together form the **axial complex**.

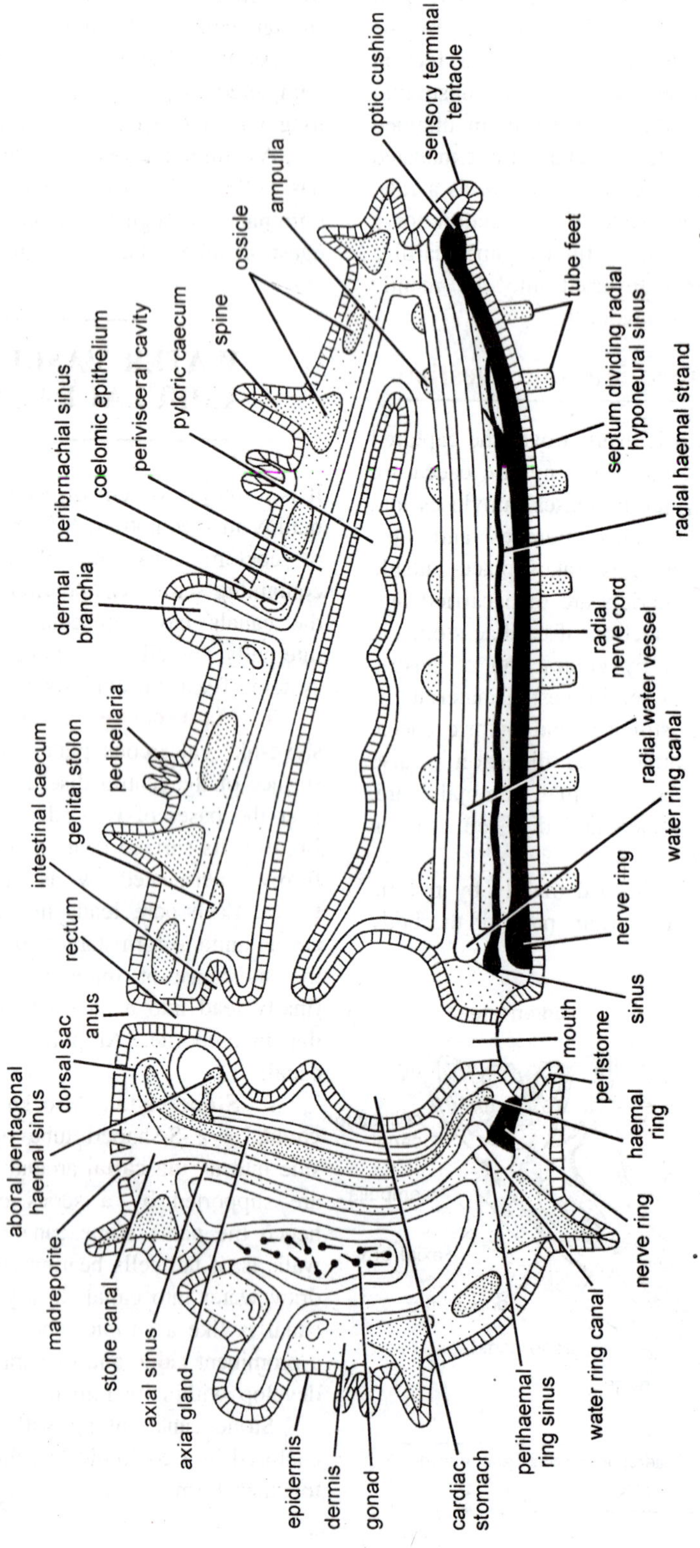

Fig. 7. *Asterias.* Diagrammatic V.S. of oral disc and arm.

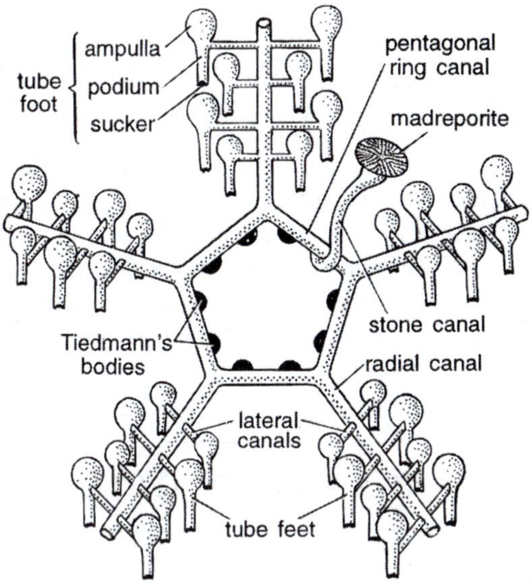

Fig. 8. *Asterias.* Water-vascular system.

3. Ring canal. It is a wide, somewhat pentagonal canal forming a ring around the oesophagus. Angles of pentagon lie in the radial positions.

4. Tiedemann's bodies. These are also known as **racemose glands**. These are small, rounded, yellowish, glandular sacs opening into ring canal on its inner side. There is one Tiedemann's body between each radius and innerradius except in one such position where stone canal enters the ring canal. There are thus, nine Tiedemann's bodies. Each Tiedemann's body consists of a bounding peritoneum enclosing a stroma of connective tissue and muscle fibres containing numerous radiating tubules. Electron microscope has revealed that each cell of the radiating tubules bears a flagellum in midst of microvilli like processes. Exact function of Tiedemann's bodies is still unresolved. However, some workers consider them as filtering devices, some look upon them as enzyme-forming bodies and others as lymphatic glands, probably manufacturing phagocytic amoebocytes which are released into water-vascular system.

5. Polian vesicles. These are pear-shaped, thin-walled contractile bladders, situated along the interradii and opening into the ring canal on its outer side. They occur in most sea stars. Their number varies from one to four in each interradius, in different species. They probably store water and thus help in regulating pressure in the water-vascular system. Polian vesicles are, however, wanting in families Asteridae (which includes *Asterias*) and Echinasteridae.

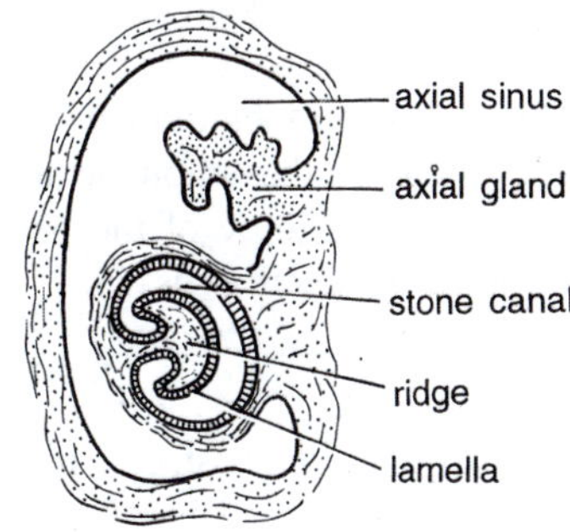

Fig. **9**. *Asterias.* T.S. of axial complex.

6. Radial canals. Along each radius, the ring canal gives out a radial canal which extends up to the tip of the corresponding arm. Radial canal lies below the ambulacral ossicles and terminates as lumen of the terminal tentacle.

7. Lateral canals. Each radial canal, in its corresponding arm, gives out two series of narrow **lateral** or **podial canals**, along its entire length. The lateral canals of two series are alternately long and short in such a way that each short canal has a long canal on its outer and inner side but a short canal on its opposite side, and vice versa. Each lateral canal opens into a **tube foot**, the opening being provided by a valve to prevent back flow of fluid into radial canal.

8. Tube feet. There are two double rows of tube feet in each arm, one double row in relation to each series of alternately placed long and short lateral canals. Each tube foot has the form of a closed thin-walled tube. It extends through a gap, called the **ambulacral pore**, which lies between two adjacent ambulacral ossicles. Each tube foot can be distinguished into three regions : (i) a rounded sac-like **ampulla**, situated above the ambulacral ossicles and projecting into the coelom, (ii) a middle tubular **podium** extending through the ambulacral groove, and (iii) a cup-like **sucker** at the lower end of the podium. Wall of the tube feet possesses strong longitudinal muscles. In addition, the walls of the ampullae also possess circular muscles, while those of the podia possess rings of inelastic connective tissue.

Function

Most peculiar and interesting role of the water-vascular system is in bringing about locomotion by providing a hydraulic pressure mechanism. Thin walls of tube feet may serve for respiratory exchange of gases. Tube feet also help in anchoring the body to the substratum and in capturing and handling the food.

LOCOMOTION

Locomotion is performed with the help of water-vascular system which sets up a hydraulic pressure. By the action of cilia lining the ambulacral canals, sea water enters through madreporite and fills up all canals of the system including the tube feet. Body is moved by the stepping action of tube feet which are alternately adhered to and released from the substratum. One or two arms, in the desired direction of movement, are raised from the substratum. Simultaneously the ampullae of tube feet to these arms contract by the action of their circular muscles. This increases the hydraulic pressure within the tube feet, which consequently elongate, extend forward and adhere firmly to the substratum by vacuum action of their suckers. Adhesion is further strengthened by mucus secreted by the tips of the tube feet. Then, by muscular activity, tube feet assume a vertical posture, dragging the body forward. Tube feet then shorten by contracting their longitudinal muscles and forcing some water back into their ampullae. Consequently, the suckers release their hold on the substratum.

During locomotion, one or two arms serve as leading arms, and all the tube feet extend in the same direction in a coordinated manner. However, the tube feet may not work in unison. As a result, the sea star moves forward steadily but slowly, at a speed of about 15 cm per minute. Sea stars can also climb up the rock by the combined action of their tube feet.

If a sea star is accidentally turned upside down, it can correct its posture by folding or arching its arms. In folding action, tips of one or two arms twist to bring their tube feet in contact with the substratum, thus permitting the whole body to fold over and right itself. In arching the upturned body is first raised on its arm tips and then rolled over.

CIRCULATORY SYSTEM

A true blood vascular system is wanting in the echinoderms. In a sea star, the circulatory system is represented by a **haemal system** which is enclosed within a **perihaemal system**.

[I] Haemal System

It is greatly reduced in a sea star and consists of intercommunicating sinuses or channels filled with a coelomic fluid containing coelomocytes. Sinuses are devoid of an epithelial lining and thus resemble the haemocoel of arthropods and molluscs. Main haemal sinuses are as follows :

1. Oral haemal ring. It is the circular haemal sinus, located around the mouth just below the ring canal of the water vascular system.

2. Radial haemal sinuses or strands. These arise radially from the oral haemal ring and one extends into each arm, along the floor of the ambulacral groove just below the radial canal of the water vascular system. They send out short branches to the tube feet.

3. Axial gland. This is the principal part of the haemal system, often referred to as the 'heart' or 'brown gland'. It remains enclosed within the axial sinus of perihaemal system. It is composed of connective tissue containing many small intercommunicating spaces filled with a fluid with amoeboid coelomocytes having a brown pigment. Axial gland is connected with the oral and aboral haemal sinuses at its oral and aboral ends

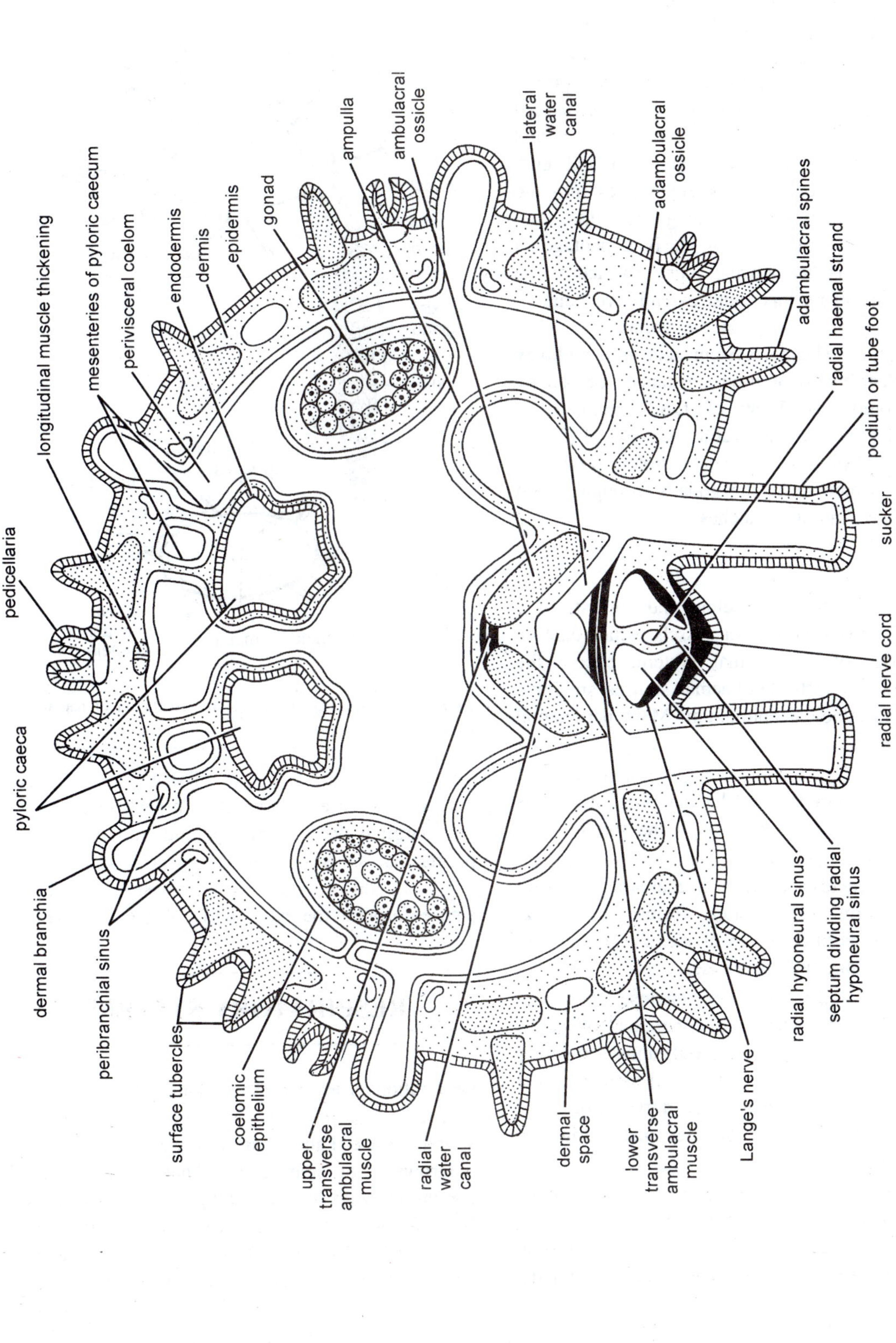

Fig. 10. *Asterias.* Diagrammatic T.S. of an arm.

respectively. From its aboral end arises a small terminal **head process**, which runs along the stone canal and is enclosed within a small coelomic sac, the **dorsal sac** located below the madreporite. A pair of **gastric tufts** also arise from the haemal sinuses in the wall of the cardiac stomach and open into the axial gland near its aboral end. Digested food from the stomach passes into the haemal circulation through the gastric tufts.

4. Aboral haemal ring. It is a pentagonal ring canal lying beneath the aboral surface of the central disc. From this canal extend five pairs of **genital haemal strands** to the gonads.

Function. Haemal system acts as a pathway for the distribution of food substances carried by the coelomocytes. Flow of fluid within it is maintained by the contractile activity of dorsal sac. Axial gland acts as a genital stolon, producing sex cells, which reach the gonads through aboral haemal ring and its branches.

[II] Perihaemal System

It is derived from the coelom and consists of a system of tubular sinuses or vessels, which enclose the sinuses of the haemal system, except the gastric haemal tufts. Perihaemal sinuses are lined by ciliated cuboidal epithelium. Main perihaemal sinuses are as follows :

1. Oral perihaemal or hyponeural ring sinus. It encloses the oral haemal ring sinus. The latter runs as a septum through the former.

2. Radial perihaemal sinuses or canals. Each canal encloses a radial haemal strand running into an arm. The latter forms a vertical longitudinal partition which divides the former into two lateral longitudinal sinuses.

3. Axial sinuses. They enclose the axial gland of haemal system and stone canal of water-vascular system. The three axial sinuses together constitute the **axial complex**. Axial sinus forms a small **dorsal sac** to enclose the dorsal process of the axial gland. Dorsal sac lies just below the madreporite.

4. Aboral perihaemal ring sinus. It encloses the aboral haemal ring sinus.

5. Genital perihaemal sinuses. Which enclose the gonads and the genital haemal strands.

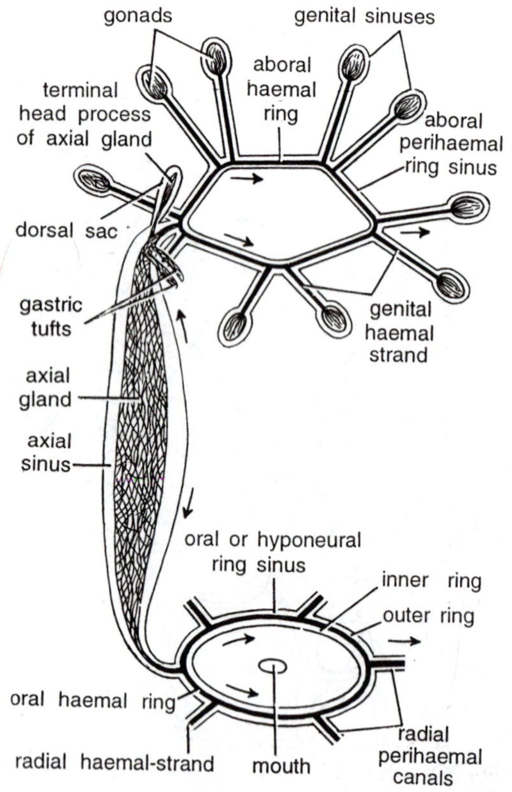

Fig. 11. *Asterias.* Diagrammatic representation of haemal and perihaemal systems.

6. Marginal sinuses. These are 10 in number, two in each arm, one on each side below the marginal nerve cord, opening into the radial perihaemal sinuses of their respective arms.

7. Peribranchial sinuses. These are circular sinuses around the basal parts of the papulae.

RESPIRATORY SYSTEM

Respiration is accomplished mainly by numerous **dermal branchiae** or **papulae** present on the aboral surface. Papulae are thin-walled, contractile outgrowths of the skin. Their cavities are continuous with coelom. Oxygen dissolved in the surrounding sea water diffuses into the papullae and carbon dioxide diffuses out. A respiratory water current is set up to pass over the papulae by cilia, lining their outer surface. Cilia, lining their cavities, cause the coelomic fluid to flow into them.

Some gaseous exchange also takes place through the thin-walled **tube feet**. According to **Mayer** (1935), in *Asterias rubens,* with covered ambulacral grooves, oxygen consumption decreases by nearly 60%.

EXCRETION

Organs specialized for excretion are lacking. Excretory products are mainly the ammonium compounds, urea and creatine. These diffuse from the body tissue into the coelomic fluid where they are engulfed by wandering amoeboid coelomocytes. Latter finally pass out of the body through the papulae, the tips of which pinch off to liberate the coelomocytes. Some excretion also takes place through the tube feet (by diffusion). Some wastes may be excreted by the cells of the pyloric caeca and leave the body through the anus.

NERVOUS SYSTEM

It is of simple and primitive type. It has the form of a **nerve net**, consisting of **nerve fibres** and a few **ganglion cells**, all confined to the body wall except the **visceral nerve plexus** situated in the gut wall. At certain places the nervous tissue is concentrated to form distinct **nerve cords**. Nervous system of *Asterias* can be distinguished into four units placed at different levels in the disc and arms.

1. Superficial or ectoneural nervous system. It is situated just beneath the epidermis and consists of : (i) a **circumoral nerve ring** (ii) five **radial nerves** and (iii) **subepidermal plexus**.

The **circumoral nerve ring** is pentagonal and lies in the periphery of the peristome. It supplies fibres to the peristome and oesophagus. From each angle of the pentagonal circumoral nerve ring arises a **radial nerve cord**, which runs in the ambulacral groove of its corresponding arm and terminates in a **sensory cushion** at the base of the terminal tentacle. In cross section the radial nerve cord appears V-shaped. **Radial nerves** supply fibers to the tube feet and ampullae. The **subepidermal plexus** is in the form of an elaborate nerve net throughout the body wall. In each form : (i) it is connected to the radial nerve cord by fine nerve fibres, (ii) it forms a pair of **adradial** or **marginal nerves**, one along each margin of the ambulacral groove, and (iii) it forms a **nerve ring** in the sucker of each tube foot.

The superficial nervous system contains sensory as well as motor fibres. Thus it serves to receive stimuli and coordinate responses like the central nervous system of other animals.

2. Deep or hyponeural nervous system. It consists of : (i) a double circumoral ring situated on the oral side just above the main or ectoneural nerve ring, and (ii) five pairs of **Lange's nerves** which arise from the former and extend into the arms, one pair in each. Lange's nerves are named after their discoverer. Each extends as a plate of nervous tissue in the outer oral wall of the radial hyponeural sinus and sends out branches to the muscles of the arm.

The deep nervous system arises from the mesoderm and is primarily motor in function.

3. Aboral or coelomic nervous system. It is represented by a thin nerve plexus situated in the aboral body wall, just above the parietal layer of coelomic epithelium. It is somewhat thickened to form an **anal nerve ring** in the central disc and a nerve in each arm. Aboral nervous system is connected with the marginal nerves by several **lateral nerves** in each arm. This system is also mesodermal in origin and motor in nature.

4. Visceral nervous system. It consists of a well-defined nerve plexus situated in the gut-wall. It innervates the muscles of the gut-wall and is connected with visceral receptors.

SENSE ORGANS

1. Neurosensory cells. These are distributed throughout the epidermis, especially concentrated in the suckers of the podia, at the bases of pedicellariae and spines and in the terminal tentacles. Each neurosensory cell is fusiform, bears a thread-like process at its free end, and is connected by fine fibres to the underlying sub-epidermal nerve plexus. Neurosensory cells are of two types : **tactile** and **olfactory**. Former are more abundant in the tube feet and the latter, around the mouth.

2. Eyes. Sea star possesses five bright red **eye spots**, one at the end of each arm at the base of the terminal tentacle, on the oral side. Each eye is formed by the special thickening (sensory cushion) of the radial nerve cord and thickening of the subepidermal nerve plexus (nervous layer) at the base of the terminal tentacle. It is, therefore, also referred to as the **optic cushion**. Bright red eye spot infact consists of numerous **photoreceptors** or **eye-pits** or **ocelli**. Each eye-pit is cup-shaped pocket of ectoderm filled with a transparent gelatinous tissue. It is covered externally by **cuticle**, below which is a transparent thickening or **lens**, formed by epidermis. Its wall is made up of **pigment cells** containing red pigment granules and **visual** or **retinal cells**. Inner end of each retinal cell is highly refractile and bulb-like and projects into the cavity of the eye-pit, while the outer end gives out a fine nerve fibre which joins the underlying radial nerve. Eye spots detect changes in light intensity.

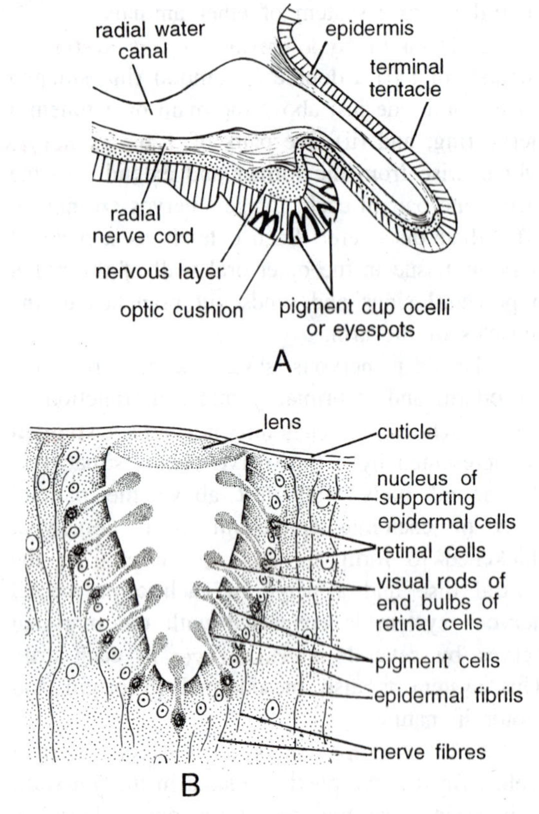

Fig. 12. *Asterias.* **A.** Section through a terminal tentacle and eye spot. **B.** V.S. of a single eye-pit or ocellus.

REPRODUCTIVE SYSTEM

Asterias is **unisexual**, but there is no sexual dimorphism. Some sea stars are hermaphrodite. *Asterina gibbosa* is protandric, the individuals being male when young but becoming female when adult. Reproductive organs lack copulatory organs, accessory glands, and reservoirs or receptacles for storing sperms and ova.

Gonads. There are five pairs of gonads, one pair in the base of each arm, lying freely between pyloric caeca and ampullae of the tube feet. Testes and ovaries are morphologically similar and occupy similar positions in the body. Colour variation however exists, the testes being greyish and the ovaries pinkish to orange.

Gonads develop periodically and their size varies with their degree of development. When mature, they occupy a considerable portion of perivisceral space, frequently extending up to tips of the arms. Each gonad is in fact the expanded terminal part of the genital haemal strand. It is an elongated, branched mass consisting of membranous and rounded follicles, thus resembling a bunch of minute grapes. It is enclosed in a genital sinus of perihaemal system. From its

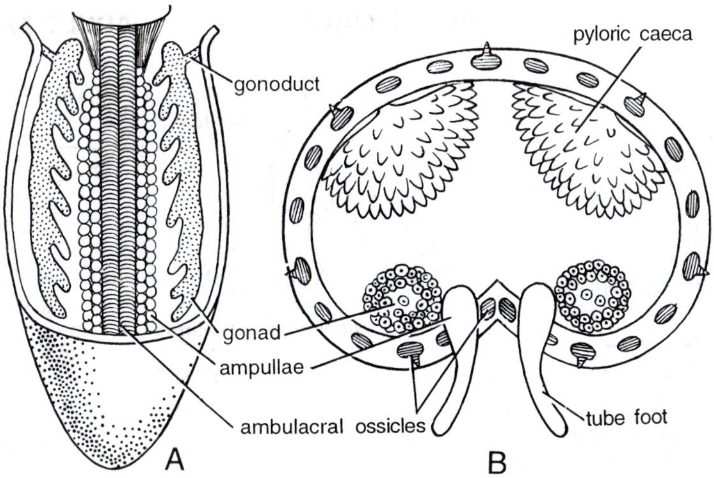

Fig. 13. *Asterias.* Gonads. **A.** An arm dissected to show gonads. **B.** T.S. of an arm showing gonads.

proximal end arises a short, ciliated **gonoduct** which opens out by a minute **gonopore**, situated laterally on the aboral surface.

Gonads, it is claimed, do not produce sex cells. These are believed to originate in the aboral end of axial gland (genital stolon) from where they migrate to gonads.

LIFE HISTORY AND DEVELOPMENT

Fertilization

Breeding takes place in spring season. Mature gametes (ova and sperms) are shed freely in surrounding sea water where **external fertilization** takes place. Sea star is highly prolific. Female lays about 1.25 million eggs per hour, producing in all, about 200 million eggs in a season. Male produces many times that number of sperms. First sperm, to come in contact with an egg, rapidly penetrates it, leaving its flagellum outside. This is followed by the fusion of their nuclei resulting in formation of a **zygote**. Zygote soon forms a delicate **fertilization membrane** around it.

Embryogeny

Development is indirect. Fertilized egg is **microlecithal**, *i.e.*, it contains a small quantity of yolk. **Cleavage** is rapid, holoblastic (total), practically equal, radial and indeterminate *i.e.,* the blastomeres do not have a fixed fate and if separated at a very early stage, develop into separate larvae. On the second day of cleavage the **blastula** or **coeloblastula** stage is reached. It is typically a spherical, hollow, one-layered and ciliated embryo which swims about freely. Its large cavity, called **blastocoel**, is filled with a **blastocoelic fluid**. Blastula undergoes **embolic invagination** and becomes a two-layered, cup-like **gastrula**. Its outer and inner layers are called **ectoderm** and **endoderm**, respectively. New cavity of the gastrula formed by invagination and lined by endoderm is called the **archenteron** or primitive gut. It opens to the exterior by a wide opening, the **blastopore**.

During gastrulation, the advancing end of archenteron (endoderm) buds off cells, into the

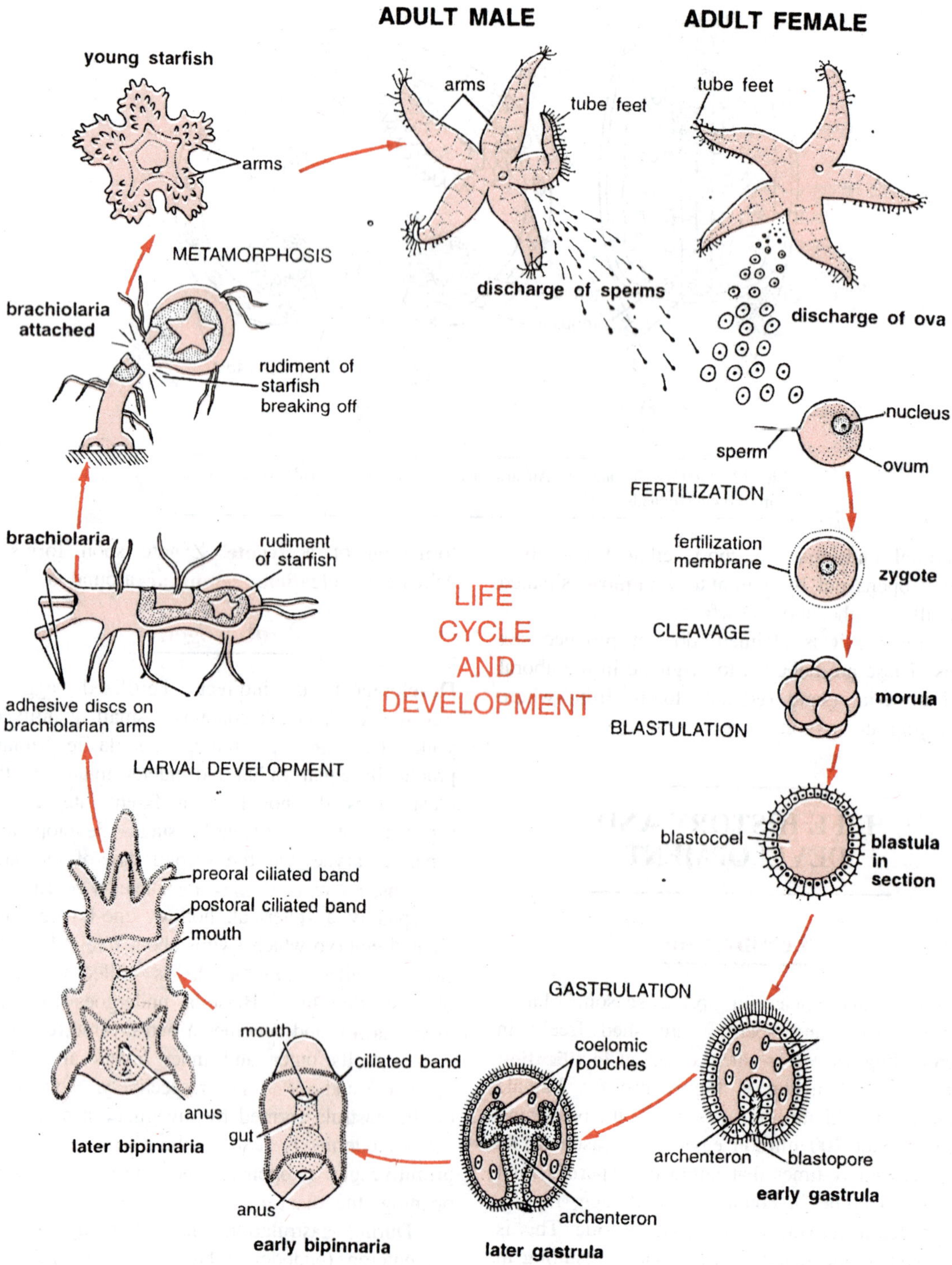

Fig. 14. Sea star. Life history and development.

blastocoel, which form the **mesenchyme** or **mesoderm**. Advancing archenteron also forms the lateral pockets, which soon pinch off as the right and left **coelomic pouches**. These later give rise to coelom, its mesodermal lining and the water-vascular system.

Embryo, at this stage, becomes a free-swimming larva.

Larval Development

Development of a sea star includes three larval stages : dipleurula, bipinnaria and brachiolaria.

1. Dipleurula larva or early bipinnaria. First larval stage in all echinoderms is called **early bipinnaria**. This larva closely resembles a hypothetical **dipleurula larva**. It is believed that all modern echinoderms have been derived from a dipleurula-like ancestor. Early bipinnaria is an egg-shaped and bilaterally symmetrical organism. An anterior mid-ventral ectodermal invagination, called **stomodaeum**, becomes continuous with the archenteron to form the larval **mouth**. Blastopore becomes the larval **anus**. Archenteron differentiates into a digestive tract made of oesophagus, stomach and intestine. Uniform ciliation of gastrula is replaced in dipleurula by two ciliary bands : a **perioral band** surrounding the mouth, and an **adoral band** lying inside the mouth. Larva feeds actively on unicellular algae, particularly diatoms. Food particles are collected from the water current produced by the adoral or stomodaeal cilia. As larva swims forward, with the help of perioral band of cilia, it rotates clockwise.

2. Bipinnaria larva. Dipleurula soon forms on its front side a large **preoral lobe**, which becomes bordered by a **preoral loop** of cilia. Simultaneously, on each lateral side, it forms three **lateral lobes** which become bordered by a **post-oral loop** of cilia. Pre- and postoral loops arise by the splitting of the perioral band of cilia. Larva thus formed is called **bipinnaria**. It is bilaterally symmetrical. It swims and feeds freely and after some weeks transforms into the next larval stage, the **brachiolaria larva**.

3. Brachiolaria larva. The lobes of the bipinnaria develop into long, slender, ciliated, contractile structures, called **larval arms**. From the preoral lobe arise three, short and non-ciliated appendages, each ending in a **sucker** or **adhesive disc**. These appendages are called **brachiolarian arms** or **fixing processes**. Larva is now called **brachiolaria**. It is also bilaterally symmetrical and swims and feeds actively. It gradually metamorphoses into a small sea star.

Metamorphosis

In about six to seven weeks time, the brachiolaria settles on some solid object with the help of adhesive discs on its fixing processes. Preoral lobe (representing the anterior end of larva) forms a sort of stalk for adhesion. Stalk later degenerates and becomes completely absorbed. Adult body develops from the rounded posterior end of the larva.

Larval mouth, anus and ciliated bands disappear. New mouth and anus are formed on the left and right sides, respectively. Five arm rudiments appear around the oral-aboral axis. Skeletal elements appear on the arm rudiments and radial canals grow into them. General body cavity (coelom) of adult develops from right and left coelomic sacs of the larva. In each arm,

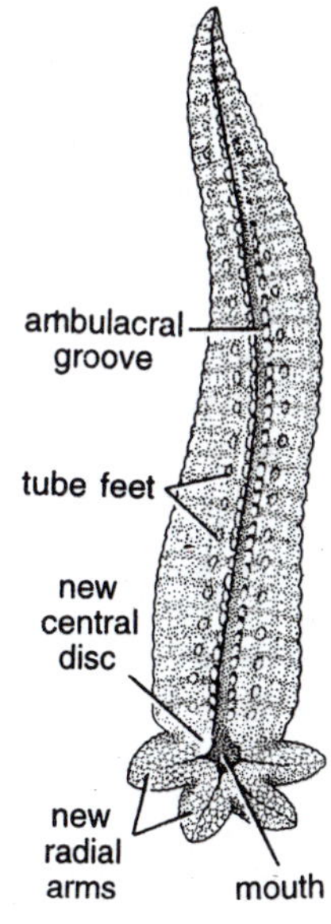

Fig. 15. Comet state *of Linckia.*

two pairs of outgrowths form the first tube feet which serve for attachment. Further complex reorganisational changes result in the formation of adult sea star. Newly detached rudiment of the body of sea star is less than 1 mm with short stubby arms.

REGENERATION AND AUTOTOMY

Sea stars have great power of **regeneration**. Whenever an arm is held, injured or unduly stimulated, the sea star readily breaks off its own arm. This process is termed **autotomy**. This ability is found in many other invertebrates. The arm is usually cast off near the base at the fourth or fifth ossicle. The dropped arm may then be regenerated. Autotomy serves as a means of protection to the animal. Whenever caught, it may throw off one or more arms and make good its escape. It is also a method of getting rid of injured body parts and replacing them with perfect ones.

Lost parts regenerate slowly, resulting in strange forms. A disc deprived of all its arms regenerates. In *Asterina vulgaris,* a single arm with a portion of disc regenerates an entire animal. But in *Linckia,* an arm totally devoid of disc can also regenerate complete animal. Specimens with small regenerating arms at the base of the large original arm are popularly called **comets**.

Important Questions

➤➤ *Long answer type questions*

1. Give an illustrated account of the external features of sea star.
2. Describe the water-vascular system of a sea star and comment upon its utility.
3. Describe the reproductive system, life history and development of a sea star.
4. Give an account of circulatory system of a sea star.
5. Describe the digestive system and mode of feeding in star fish.
6. Draw a neat and labelled diagram of T.S. of arm of a sea star.
7. Write short notes on : (i) Locomotion in a sea star, (ii) Pedicellariae, (iii) Tube feet, (iv) Bipinnaria, (v) Regeneration in sea star, (vi) Stone canal, (vii) Polian vesicles, (viii) Madreporite.

➤➤ *Short answer type questions*

1. Name the animal group where polian vesicles are present.
2. List down the coelomic spaces occurring in starfish.
3. Describe the functioning of tube feet in starfishes.
4. Write a short note on 'stone canal'.
5. Describe the following : (i) Tiedman's body and (ii) Dermal branchiae in starfish.
6. Give a diagrammatic representation to show the haemal system in starfish.
7. Justify how the water vascular system helps in the locomotion of starfish.
8. Describe the external features of starfish.
9. Write what you know about the nervous system or the sense organ of starfish.
10. Give an account of the development of starfish.
11. Give the alternative name of axial complex and mention its functions.
12. Draw and label the parts of bipinnaria larva.
13. Draw well labelled sketch of T.S. of arm of starfish.
14. Give a fully labelled sketch to show the course of the water current in the water vascular system in starfish.

➤➤ *Multiple choice questions*

1. *Asterias* belongs to :
   (a) Asteroidea (b) Echinoidea
   (c) Ophiurodea (d) Holothuroidea
2. Common name of *Asterias* :
   (a) sea pentagon (b) brittle star
   (c) starfish (d) basket star
3. Respiratory organs of starfish :
   (a) pedicellariae (b) papulae
   (c) body wall (d) podia
4. Function of pedicellariae :
   (a) capturing prey
   (b) capture and removal of debris

(c) capture, removal of debris and minute organisms
(d) help in respiration

5. Locomotory organs of starfish :
   (a) podia
   (b) polian vesicles
   (c) none
6. Characteristic feature of echinoderms :
   (a) water vascular or ambulacral system
   (b) haemal system
   (c) both (d) none
7. Epidermis of *Asterias* contains :
   (a) ephithelial cells (b) gland cells
   (c) neurosensory cells (d) all the above
8. Which is enterocoelic animal :
   (a) *Taenia* (b) *Pheretima*
   (c) *Palaemon* (d) *Asterias*
9. Larval stage in life history of starfish :
   (a) bipinnaria (b) brachiolaria
   (c) both (d) none
10. The eggs of *Asterias* are :
    (a) microlecithal
    (b) macrolecithal
    (c) megalecithal
    (d) oligolecithal
11. The females of *Asterias* lay eggs per hours approximately :
    (a) 100 (b) 10000
    (c) 100000 (d) 1000000
12. Stone cell connect the :
    (a) ring canal and polian vesicle
    (b) ring canal and radial canal
    (c) madreporite and ring canal

Answers

➤➤*Multiple choice questions*

1. (a) 2. (c) 3. (b) 4. (c) 5. (a) 6. (c) 7. (d) 8. (d) 9. (b) 10. (a)
11. (d) 12. (c)

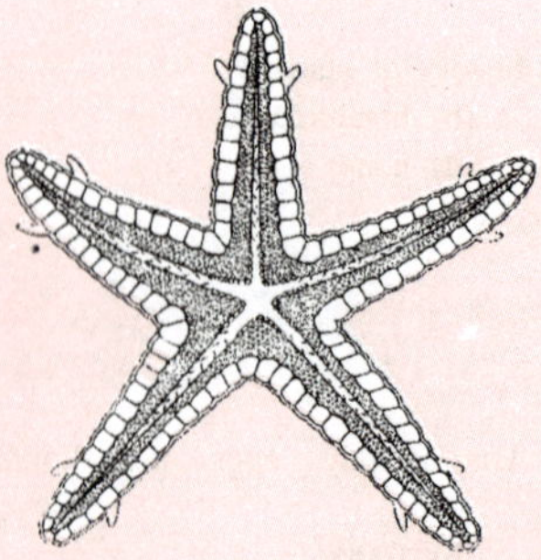

63

Echinodermata : Characters, Classification and Types

Echinoderms are one of the most beautiful and most familiar sea creatures. Forms such as the sea stars have become a symbol of sea life. Other forms such as brittle stars, sea urchins, sea cucumbers and sea lilies are also quite well known to the visitors on the sea-shore. There are 7,550 sps known in Echinodermata.

HISTORICAL

Echinoderms are known since very ancient times. Name of this phylum was introduced by **Klein** in 1734 for sea urchins. For many years echinoderms and coelenterates were included as a class among Radiata, largely because of the radial symmetry of the adults. **Echinodermata** were first recognized as a group distinct from the Radiata by **Leukart** in 1847.

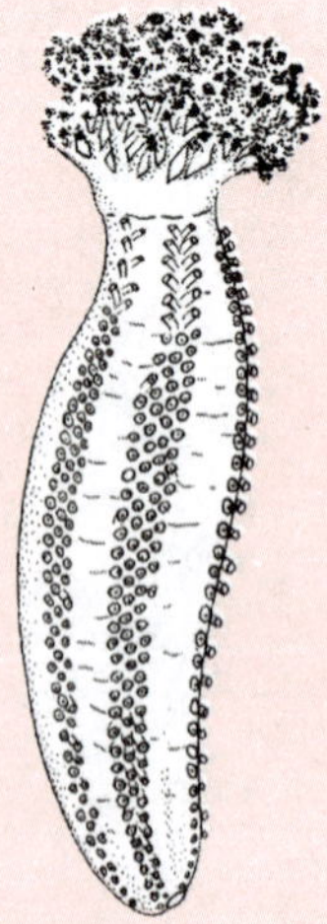

DERIVATION OF NAME

Echinodermata literally means 'spiny or prickly skinned' (Gr., **echinos**, hedgehog; **derma**, skin) and refers to the conspicuous spines possessed by their test or skin. **Jacob Klein** (1734) first used this name for echinoids. The Greeks applied the name **echinos** to the hedgehog as well as the sea urchin, both having a prickly appearance. Term **echinus** has been used for a certain sea urchin. Possession of spines is not diagnostic of the phylum because only better known members, such as sea urchin, brittle stars and starfishes, have spines.

DEFINITION

Echinoderms are exclusively marine and largely bottom dwellers enterocoelous coelomate, triploblastic animals. They have a pentamerous radial symmetry derived from an original bilateral symmetry. They possess an endoskeleton of calcareous plates or spicules embedded in the skin; a peculiar water-vascular system of coelomic origin; numerous podia or tube feet; an ectodermal nervous system; no definite head or brain; no nephridia; gonads open directly to the exterior by special ducts.

GENERAL CHARACTERS

Phylum Echinodermata contains some 5300 known species and constitutes the only major group of deuterostome invertebrates. **Bather** (1900) stated the phylum as "one of the best characterised and most distinct phyla of the animal kingdom". Echinoderms are distinguished from all animals by a number of characteristics.

1. Organ-system grade of body organization.
2. Triploblastic, coelomate and radially symmetrical; often pentamerous.
3. Body unsegmented with globular, star-like, spherical, discoidal or elongated shape.
4. Head absent; body surface is marked by five symmetrically radiating areas **(ambulacra)** and five alternating interradii **(inter-ambulacra)**.
5. Endoskeleton of dermal calcareous ossicles with spines, covered by the epidermis.
6. Water-vascular system of coelomic origin, including **podia** or **tube feet** for locomotion and usually with a **madreporite**.
7. Coelom of enterocoelous type constitute the perivisceral cavity and cavity of the water-vascular system; coelomic fluid with coelomocytes.
8. Alimentary canal straight or coiled.
9. Vascular system and haemal system, enclosed in coelomic perihaemal channels.
10. Respiratory organs include dermal **branchiae** **tube feet**, **respiratory tree** and **bursae**
11. Nervous system without a brain and with a circumoral ring and radial nerves.
12. Poorly developed sense organs include tactile organs, chemoreceptors, terminal tentacles, photoreceptors and statocysts.
13. No excretory organs.
14. Usually dioecious, gonads large and single or multiple; fertilization external; development indirect through free-swimming larval forms.
15. Regeneration of lost parts, a peculiarity.
16. Exclusively marine.

CLASSIFICATION

Subphylum I. Eleutherozoa
(Gr., eleutheros, free + zoios = zoon, animal)

Free-living echinoderms.

CLASS 1. Asteroidea

(Gr., **aster**, star + **eidos**, form)

1. Starfishes or sea stars.
2. Arms 5 or more and not sharply marked off from the central disc.
3. Tube feet in orally placed ambulacral grooves; with suckers.
4. Anus and madreporite aboral.
5. Pedicellariae present.
6. Free-living, slow-creeping, predaceous and scavengerous.

Subclass 1. Somasteroidea

Fossil Palaeozoic sea stars. *Platasterias latiradiata* is the only living species.

Subclass 2. Euasteroidea

Living sea stars.

Order 1. Phanerozonia

1. Body with marginal plates and usually with papulae, on aboral surface.
2. Pedicellariae sessile, not crossed.
3. Tube feet without suckers.

4. Mostly burrowers in soft bottom.
Examples : *Astropecten, Luidia, Goniaster, Oreaster* (= *Pentaceros*).

Order 2. Spinulosa

1. Usually without conspicuous marginal plates and with papulae on both surfaces.
2. Pedicellariae rare.
3. Tube feet with suckers.
4. Aboral surface with low spines.
Examples : *Asterina, Solaster, Pteraster, Echinaster.*

Order 3. Forcipulata

1. No conspicuous marginal plates.
2. Pedicellariae pedunculate and straight or crossed type.
3. Four rows of tube feet.
Examples : *Asterias, Heliaster.*

CLASS 2. Ophiuroidea

(Gr., **ophis**, snake + **oura**, tail + **eidos**, form)

1. Brittle-stars and allies.
2. Body star-like with arms sharply marked off from the central disc.
3. Pedicellariae absent.
4. Stomach sac-like; no anus.
5. Ambulacral grooves absent or covered by ossicles; tube feet without suckers.
6. Madreporite oral.

Order 1. Ophiurae

1. Brittle and serpent stars.
2. Small and five-armed.
3. Arms move transversely.
4. Disc and arms usually covered with plates.
Examples : *Ophiura, Ophiothrix, Ophioderma, Ophiopholis.*

Order 2. Euryalae

1. Arms simple or branched.
2. Arms move vertically.
3. Disc and arms covered by soft skin.
Examples : *Gorgonocephalus* (basket star), *Asteronyx.*

CLASS 3. Echinoidea

(Gr., **echinos**, hedgehog + **eidos**, form)

1. Sea urchins and dollars.
2. Body discoid, oval or semi-spherical and without arms.
3. Skeleton or test compact bearing movable spines and three-jawed pedicellariae.
4. Chewing apparatus or **Aristotle's lantern** with teeth.
5. Ambulacral grooves covered by ossicles; tube feet with suckers.
6. Gonads usually five or less.

Subclass 1. Bothriocidaroida

1. A single row of plates in each inter-ambulacral area.
2. Without typical lantern.
3. Madreporite radial.
Example : Single extinct Ordovician genus *Bothriocidaris.*

Subclass 2. Regularia

1. Body globular, pentamerous, with two rows of inter-ambulacral plates in existing members.
2. Mouth central.
3. Aristotle's lantern well developed.
4. Anus central on aboral surface with well-developed apical plates.
5. Madreporite oral.

Order 1. Lepidocentroida

1. Test flexible with overlapping plates.
2. Ambulacral plates extend up to mouth lip.
3. Inter-ambulacral plates in more than two rows in extinct forms.
Example : *Palaeodiscus.*

Order 2. Melonechinoida

1. Test spherical and rigid.
2. Ambulacral plates continue to mouth lip.
3. Inter-ambulacral plates in four or more rows.
4. Wholly extinct, carboniferous.
Example : *Melonechinus.*

Order 3. Cidaroida

1. Test globular and rigid.
2. Two rows of long narrow ambulacral plates and two rows of inter-ambulacral plates.
3. No peristomial gills.
4. Anus aboral and central.
Examples : *Histocidaris, Goniocidaris.*

Order 4. Diadematoida

1. Test globular usually with compound ambulacral plates.

2. Peristomial gills present.
3. Anus aboral and central.
   Examples : *Diadema, Echinus, Arbacia.*

Subclass 3. Irregularia

1. Body oval or circular, flattened oral-aborally.
2. Mouth central or displaced anteriorly on oral surface.
3. Anus marginal, outside the apical system of plates.
4. Tube feet generally not locomotor.

Order 1. Holectypoida

1. Test regular with simple ambulacral and centrally located peristome and apical system.
2. Lantern present.
3. Mostly extinct.
   Examples : *Holectypus, Echinoneus.*

Order 2. Cassiduloida

1. Aboral ambulacral areas petaloid, forming a five-armed figure like petals of a flower.
2. Lantern absent.
3. Mostly extinct.
   Example : *Cassidulus.*

Order 3. Clypeastroida

1. Test flattened with oval or rounded shape.
2. Mouth central, anus excentric.
3. Aboral ambulacral areas petaloid.
4. Aristotle's lantern present.
5. Gills absent.
6. Bottom dwellers.
   Examples : Sand dollars : *Clypeaster, Echinarachinus, Echinocyamus.*

Order 4. Spatangoida

1. Test oval or heart-shaped with excentric mouth and anus.
2. Four aboral ambulacral areas pataloid.
3. Lantern absent.
4. Gills absent.
5. Burrowing.
   Examples : Heart urchins; *Spatangus, Echinocardium lovenia, Hemipneustes.*

CLASS 4. Holothuroidea

(Gr., **holothurion**, sea cucumber + **eidos**, form)

1. Sea cucumbers.
2. No arms, no spines.
3. Body elongate on oral-aboral axis; body wall leathery.
4. Mouth anterior, surrounded by tentacles.
5. Ambulacral grooves concealed; tube feet with suckers.
6. Usually with respiratory tree for respiration.

Order 1. Dendrochirota

1. Tentacles irregularly branched.
2. Tube feet numerous, on the sole or all ambulacral or entire surface.
3. Respiratory tree present.
   Examples : *Cucumaria, Thyone.*

Order 2. Aspidochirota

1. Tentacles peltate or leaf-like.
2. Tube feet numerous, sometimes forming a well-developed sole.
3. Respiratory tree present.
   Examples : *Holothuria, Actinopyga.*

Order 3. Elasipoda

1. Tentacles leaf-like.
2. No respiratory tree.
3. Tube feet webbed together to form fins.
4. Deep-sea dwellers.
   Example : *Pelagothuria.*

Order 4. Molpadonia

1. 15 digitate tentacles.
2. No tube feet.
3. Posterior end tail-like.
4. Respiratory tree present.
   Examples : *Molpadia, Caudina.*

Order 5. Apoda

1. Worm-like sea cucumbers.
2. No tube feet and respiratory tree.
3. Burrowing.
   Examples : *Leptosynapta, Synapta.*

Subphylum II. Pelmatozoa

(Gr., pelmatos, stalk + zooios, animal)

Stalked, sedentary echinoderms.

CLASS 5. Crinoidea

(Gr., **crinon**, lily + **eidos**, form)

1. Sea lillies.
2. Body attached during part or whole of life by an aboral stalk.

3. Mouth and anus on oral surface.
4. Arms with pinnules.
5. Tube feet without suckers; no madreporite, spines and pedicellariae.
6. Ciliated ambulacral grooves on oral surface.

Order : Articulata

1. Living sea lillies and feather stars.
2. Feather stars non-sessile and free swimming. Examples : *Antedon* (sea lily), *Neometra* (feather star).

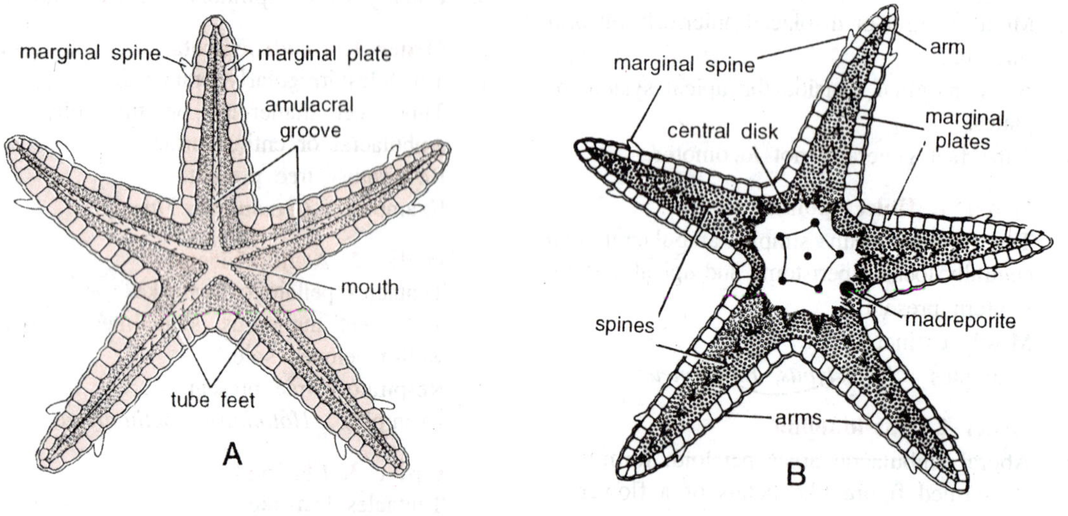

Fig. 1. *Pentaceros.* **A.** Oral view. **B.** Aboral view.

A FEW OTHER ECHINODERMS

1. *Pentaceros* (= Oreaster). *Pentaceros* is often referred to as 'sea pentagon' of Indian seas because of its resemblance to a pentagon. Central disc is large and the five arms are short and tapering. Two are not clearly demarcated. Aboral surface is convex and bears rows of definitely arranged spines. Aboral skeleton is reticulate, enclosing popular areas in its meshes. Upper marginal plates are smaller than the ventral ones and often concealed from view. Aboral surface bears immovable tubercles, a madreporite and anus. Oral surface is concave, having a central mouth, communicating with five ambulacral grooves, each bearing two double rows of tube feet. Pedicellariae are small and valvate type. *Pentaceros* or *Oreaster* is very harmful to pearl industry as it feeds on pearl oysters.

2. *Astropecten*. *Astropecten* is a common starfish, occurring in all seas. It inhabits the sandy bottom to lead a burrowing life. It comes out once in the morning and once in the evening in

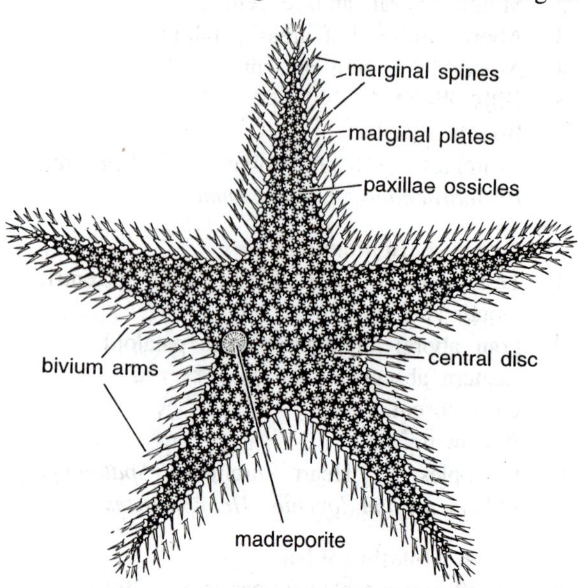

Fig. 2. *Astropecten.* Aboral view.

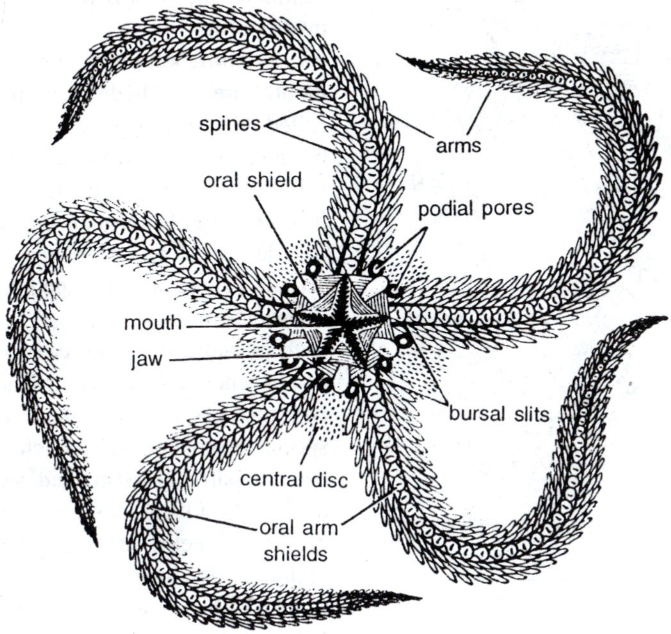

Fig. 3. *Ophiothrix.* Oral view.

search of food. Central disc is large and pentagonal, while the arms are short and tapering. Arms are bordered by the marginal spines and the marginal plates are large and elongated. Pedicellariae are sessile and the tube feet are without suckers. Water-vascular system includes two to four polian vesicles in each interradius and the stone canal is structurally complicated. There occurs no **brachiolaria** larva. *A. auranciacus* of the Mediterranean is often used for embryological studies.

3. ***Ophiothrix.*** *Ophiothrix* is a common spiny brittle star, with a small rounded central disc and five slender jointed arms, arising from the lower surface of the disc. Arms are covered on all sides by the plates or shields and fringed with spines. Pedicellariae, ambulacral grooves and dermal branchiae are altogether absent. Oral surface of the disc bears a madreporite and a mouth, with five movable, plates, serving as jaws. Tube feet are without suckers and present on the lower plates of arms. Base of each arm bears a pair of deep grooves, called the **bursal slits**, through which pass to outside the mature sex cells. Oral surface of the disc also bears five oral shields and podial pores. It has a great power of regenerating its lost arm. *Ophiothrix* is common along the Atlantic coast.

4. ***Gorgonocephalus.*** *Gorgonocephalus* is commonly called the basket star. Body consists of a large pentagonal disc and five elongated and much branched arms. Oral surface of each arm is spiny, whereas the aboral surface is annulated. Oral and tooth papillae are present and the teeth bear a continuous series of minute spines. Aboral surface of the disc shows long and spoke-like radial shields from the centre to the periphery. Disc margin between the radial shields bears a series of plates. One madreporite is usually present in each interradius. Basket star can walk on the tips of its arms.

5. ***Echinus.*** *Echinus* is commonly known as 'sea urchin'. It is a benthonic animal, occurring from intertidal zone to a depth of 5000 m. It has a globe-shaped body with very long movable spines. Visceral organs are contained within a shell or **corona** of fused endoskeletal plates, with rounded tubercles. Surface of corona is divided into alternating five ambulacral and five inter-ambulacral areas. Ten rows of podia or tube feet

Fig. 4. *Gorgonocephalus.* Oral view.

run from one pole of the animal to another. Oral pole has a mouth with five teeth surrounded by the peristome. Among spines are found stalked pedicellariae, with three jaws. An anus lies in the centre of the aboral pole. Close to anus lies the madreporite.

Aristotle's lantern. Five teeth surrounding the mouth are attached to a masticatory apparatus, called **Aristotle's lantern**, after its discoverer and because of its resemblance to an ancient Greek ship-lantern. It is situated within the test and projects slightly through the mouth. It consists of five large calcareous plates, called **pyramids** or **alveoli**. These are arranged radially by means of transverse muscle fibres. Along the inner side of each pyramid is a long calcareous band. Its upper end is enclosed within a dental sac and the oral end projects out as a hard tooth. By means of special protractor and retractor muscles the lantern can be partially protracted and retracted through the mouth. Other muscles control the opening and closing of teeth. Aristotle's lantern is used by the urchin in feeding.

6. *Clypeaster.* *Clypeaster* is commonly called 'cake urchin' because of its flattened body. It is found in tropical and subtropical seas. It creeps on the bottom or remains partly buried in sand. Oral surface is covered by short and delicate spines, between which are found stalked pedicellariae. Mouth is central in position and from it radiate five ambulacral grooves. These possess locomotor suckered tube feet. Anus also lies on the oral surface. Slightly convex aboral surface is covered by minute delicate spines. It bears a central madreporite, from which radiate five **petaloid ambulacral areas**, each of which is

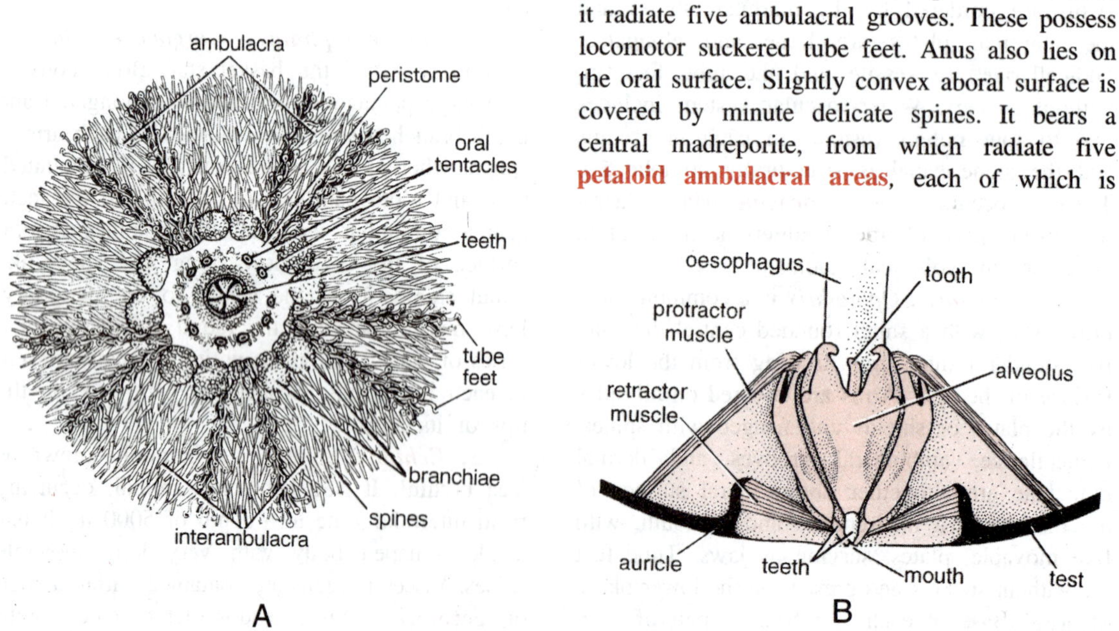

Fig. 5. *Echinus.* **A.** Oral view. **B.** Aristotle's lantern *in situ.*

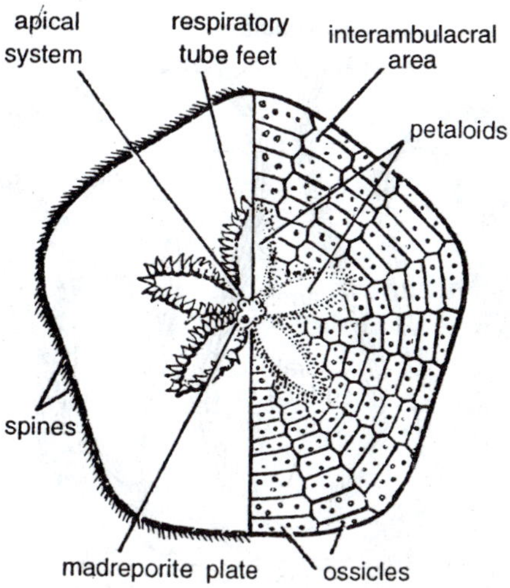

Fig. 6. *Clypeaster.* Aboral view.

bordered by two rows of flattened **respiratory tube feet**. Aristotle's lantern is present whose teeth project out of mouth.

7. ***Cucumaria.*** *Cucumaria* is very common among sea cucumbers, found along the sea bottom in shallow tropical waters. Body is long and cylindrical and is covered by a flexible and leathery skin. Terminal mouth, at the oral end, is surrounded by a circular row of 10 dendritic or tree-like tentacles, used for feeding. Anus lies at the posterior aboral end. Animal lies with its ventral side against the substratum. This side consists of three ambulacral areas (trivium) constituting the sole and possessing locomotory podia. Dorsal surface is composed of two ambulacral areas (bivium) and three inter-ambulacral areas. Tube feet of this surface are without suckers. Tube feet, on both the surfaces, are restricted to the five ambulacral areas. *Cucumaria* displays a creeping movement on the bottom surface and feeds upon detritus and plankton. Gas exchange is accomplished by a system of tubules, called **respiratory tree**. These are two in number, lying on either side of the gut and emerge from the upper end of cloaca. Water flows through its tubules to bring about gaseous exchange. Polian vesicle is only one in number. Sexes are separate and development is indirect with an auricularia larva. On being frightened the

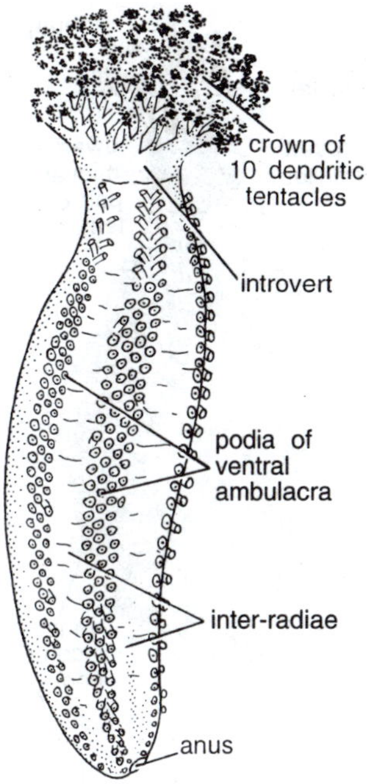

Fig. 7. *Cucumaria.* Ventral view.

animal sheds almost all or part of its viscera which regenerate again.

8. ***Thyone.*** *Thyone* like *Cucumaria,* is also a common sea cucumber. It is free-swimming but generally found buried in muddy and sandy sea bottom. Body is elongate with 10 dendritic tentacles surrounding the terminal mouth. Podia are scattered over the entire body surface. Paired respiratory trees are well represented. These arise from the cloaca by way of a common trunk. The number of polian vesicles is four. Sexes are separate.

9. ***Antedon.*** *Antedon,* commonly called 'feather star', is somewhat plant-like in superficial resemblance. It occurs in sea waters of the Atlantic, Western Africa, the Mediterranean and West of Tropical America. It is an unstalked but attached crinoid, possessing a central cup-shaped disc which is covered by leathery skin. Oral surface is directed upwards, whereas the aboral surface, downwards. From the central disc arise five long, slender and movable arms that branch dichotomously to form 10 arms and each of them bears a row of small and slender branches, called **pinnules**, on each side. From the mouth

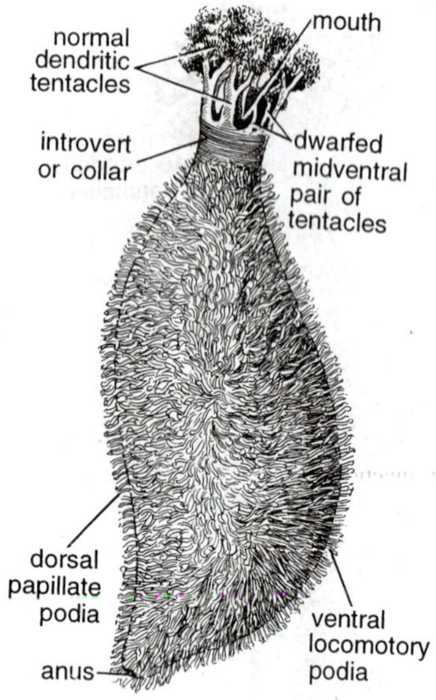

Fig. 8. *Thyone.*

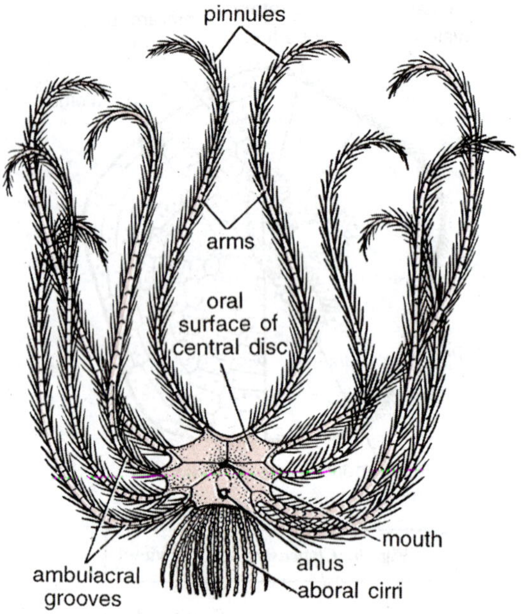

Fig. 9. *Antedon.*

radiate five ambulacral grooves which divide to form 10 grooves, one for each arm. Along the sides of each ambulacral groove are two rows of finger-like podia, without suckers. On the oral surface is a central mouth and an anus on a prominent papilla. Aboral surface bears numerous long, cylindrical appendages, the **cirri**, meant for gripping the substratum.

Important Questions

➤➤*Long answer type questions*

1. Classify Echinodermata giving the main characters and familiar examples.
2. Write short notes on : (i) *Astropecten,* (ii) Brittle-star, (iii) *Echinus,* (iv) Aristotle's lantern, (v) *Cucumaria,* (vi) *Antedon.*

➤➤*Short answer type questions*

1. Mention an Echinoderm with no spines, and no pedicellariae.
2. Define the term trivium and bivium.
3. Define enterocoelic type of coelom formation.
4. What specific association do holothurians exhibit. Describe in 3 sentences.
5. Give the distinguishing characters and one example of the class Crinoidea.
6. Classify the *Ophiothrix,* giving two peculiar features in their structure and/or life history.
7. Distinguish between Asteroidea and Ophiuroidea.
8. List the basic features of Echinoderms that have on affinity to early chordates. Mention atleast 4 points.
9. Classify Phylum Echinodermata, giving distinguishing characters and example of each class.
10. Give characteristic morphological differences between starfish and sea-urchin.
11. Discuss the problem of symmetry in echinoderms.
12. What are sea urchin and sea lilies ?
13. Define the term 'Echinodermata' ?
14. Classify *Asterias.*
15. Which basic factors classified the Echinodermata.

➤➤*Fill in the blanks*

1. *Pentaceros* generally known as........
2. *Ophiothrix* commonly known as.......
3. Zoological name of basket star is........
4. Sea urchin is common name of........
5. *Antedone* generally known as........
6. In Echinodermata........organs absent.
7. Sea lillies put in the class...........
8. The class of *Holothuria* is.........
9. Sea stars belongs to the class........
10. Clypeaster is commonly known as........

➤➤*Multiple choice questions*

1. Identify a Holothurian with no tube feet, no respiratory trees, cuvierian glands and no radial canals :
   (a) *Thyone* (b) *Synapta*
   (c) *Holothuria* (d) *Cucumaria*
2. The larva that occurs in the development of Holothurians is :
   (a) Auricularia (b) Brachiolaria
   (c) Doliolari (d) Bipinnaria
3. Aristotle's lantern is found in :
   (a) *Cucumaria* (b) *Antedon*
   (c) *Echinus* (d) *Opiothrix*
4. Tiedmann bodies produce free :
   (a) amino acids (b) amoebocytes
   (c) haemocytes (d) thesocytes
5. Excretory organs are absent in :
   (a) Nematoda (b) Echinodermata
   (c) Crustacea (d) Onychophora
6. The word Echinodermata was first used by :
   (a) Linnaeus (b) Aristotle
   (c) Jacob Klein (d) Grant
7. Brachiolaria larva is characteristic of the class :
   (a) Crinoidea (b) Asteroidea
   (c) Ophiuroidea (d) Echinoidea
   (e) Holothuroidea
8. Brittle star belongs to the class :
   (a) Asteroidea (b) Ophiuroidea
   (c) Crinoidea (d) Echinoidea
9. *Thyone* is commonly known as :
   (a) brittle star (b) sea cucumber
   (c) sea lily (d) cake urchin
   (e) none of these
10. The heart urchins are included in order :
    (a) Holosteroidea (b) Spatangoida
    (c) Nucleolitoida (d) Pedinoidea
11. The locomotor organs of Echinoderms are called :
    (a) parapodia (b) pseudopodia
    (c) tube feet (d) setae
12. The stone canal does not open to the exterior but ends in a madreporite hanging down into the body cavity in the case of :
    (a) Ophiuroidea (b) Echinoidea
    (c) Crinoidea (d) Holothuroidea
13. The Aristotle's lantern of Echinoid is an apparatus that functions as :
    (a) respiratory (b) excretory
    (c) masticatory (d) accessory
14. Coelom in starfish is :
    (a) schizocoelous (b) enterocoelous
    (c) both (d) none
15. Excretory products :
    (a) urea (b) urea and creatine
    (c) ammonia (d) uric acid
16. Autotomy is found in :
    (a) molluscs (b) *Aurelia*
    (c) leech (d) sea stars
    In *Linckia* arm devoid of disc regenerates an entire animal Comets are the small regenerating arms at the base of original arm.
17. Respiratory organ in Holothuroidea :
    (a) papulae
    (b) dermal branchiae
    (c) respiratory tree (d) bursae
18. Sea lily and feather star belong :
    (a) Echinoidea (b) Crinoidea
    (c) Holothuroidea (d) Asteroidea
19. Which phylum of animal is exclusively :
    (a) Mollusca (b) Coelentrata
    (c) Echinodermata (d) Porifera
20. Water vascular system is identification of :
    (a) Porifera (b) Coelentrata
    (c) Arthropoda
    (d) Echinodermata
21. Respiratory organs in echinodermata :
    (a) branchiate (b) tube feet
    (c) respiratory organs (d) bursae
    (e) all above
22. Free moving echinodermata :
    (a) Subphylum-Eleutherozoa
    (b) Subphylum-Pelmetozoa
    (c) both of them (d) none of them
23. Larval forms of echinodermata :
    (a) Bipinaria (b) Branchiolaria
    (c) Auricularia (d) Doliolaria
    (e) all above

24. Ophiopluteus is :
   (a) an echinoderm (b) a poriferian
   (c) an annelida
   (d) a larva of echinodermata

25. The animal which commonly known as 'sand dollar' :
   (a) Antidon (b) Clypeaster
   (c) Ophiothrix (d) Echinus

Answers

➤➤*Fill in the blanks*

1. sea pentagon, 2. spiny brittle star, 3. Gorgonocephalus, 4. echinus, 5. feather star, 6. excretory, 7. crinoidea, 8. holothuroidea, 9. asteroidea, 10. cake urchin.

➤➤*Multiple choice questions*

1. (b)	2. (a)	3. (c)	4. (c)	5. (b)	6. (c)	7. (b)	8. (b)	9. (b)	10. (b)
11. (c)	12. .(d)	13. (c)	14. (b)	15. (b)	16. (d)	17. (c)	18. (a)	19. (c)	20. (d)
21. (c)	22. (a)	23. (e)	24. (d)	25. (b)					

64

Echinodermata : General Account

LARVAL FORMS IN ECHINODERMATA

No other group of animals has such complicated metamorphosis in the course of development. Development may be direct or indirect. In direct one, the larval stages are missing while in indirect one, various types of free-swimming larvae are formed. In each class, a few members, are viviparous, that is, they brood their young in a sort of brood pouch on the surface of their body. The development of larva takes place in a typical deuterostomous fashion. In most cases the characteristic free swimming larvae develop externally which are of great phylogenetic significance.

Echinoderm larva is strikingly bilaterally symmetrical in marked contrast to radially symmetrical adult. It swims about by means of a ciliated band, which may be complicated by a number of short or long slender projection or arms from the body wall. Based upon the nature and position of the arms or their absence, larvae of different classes of Echinodermata may be distinguished. After a free-swimming planktonic existence, the bilateral larva undergoes a metamorphosis, in which the radial symmetry of the adult is developed. In different classes of echinoderms, different types of larvae complete the development.

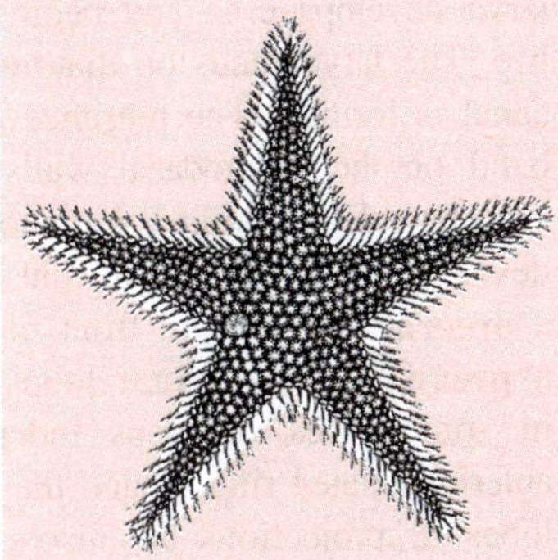

(1) Class I. Asteroidea

Bipinnaria Larva

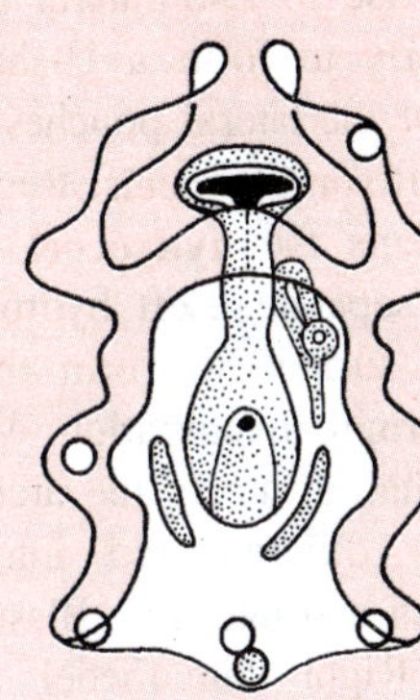

Two types of development occurs in asteroids. The **direct** type has large, yolky eggs and no free swimming larval stage. The **indirect**

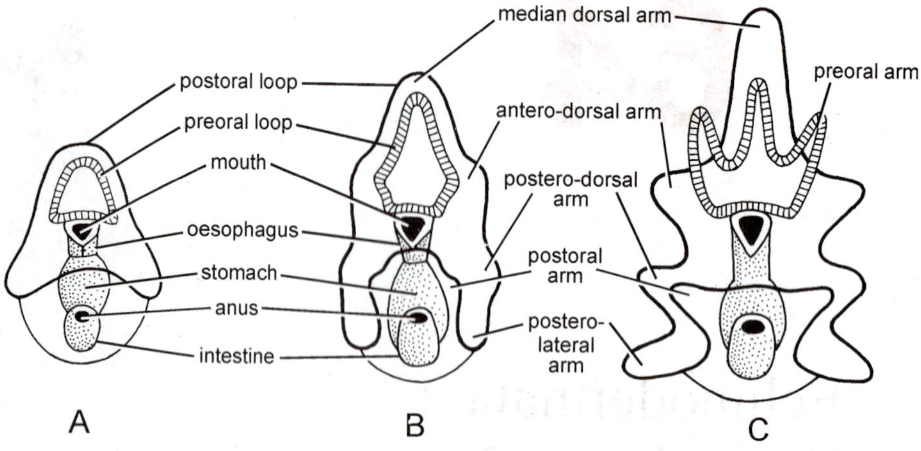

Fig. 1. Stages in development of a bipinnaria larva.

type has homolecithal eggs with little yolk and a free swimming larval stage. After hatching the larva develops cilia and begins a free-swimming life. The larva feeds on diatoms as an alimentary canal is formed. The presence of powerful ciliary band on the stomodaeal walls helps in feeding. Two lateral longitudinal locomotory ciliated bands develop which connect infront of mouth, forming a **preoral loop** and in front of the anus, to form a **preanal loop**. Preoral loop later, separates or in some cases develops independently into an anterior ciliated ring around the body. Three lateral lobes or projections are also developed on each side of the body bordered by ciliary bands. This larva is known as **bipinnaria** and develops in 2 to 7 days.

Internal development of bipinnaria. Tip of larval archenteron forms the mesenchyme and later gives rise to two lateral pouches which connect arteriorly to form a U-shaped coelom. Posterior ends of the lateral pouches pinch off to form right and left **somateocoels**. Remaining anterior portion represents the **hydrocoel** and **axocoel**, but they never separate. Left hydrocoel connects with the dorsal surface to form the **hydropore**, without ectodermal invagination. Ventrally an ectodermal invagination meets the archenteron and the larval gut is differentiated into mouth, oesophagus, stomach and intestine. Blastopore remains as larval anus. Right somatocoel and axohydrocoel get reduced in metamorphosis, while left axohydrocoel gives rise to water ring and radial canals. Axocoel separates from hydrocoel and contributes to stonecanal. Madreporite or dorsal sac originates either from rearrangement of mesenchyme cells or from ectodermal invagination or from right axohydrocoel. Bipinnaria larva, after free-swimming existence for a few weeks, changes to next larval stage, called **brachiolaria larva**.

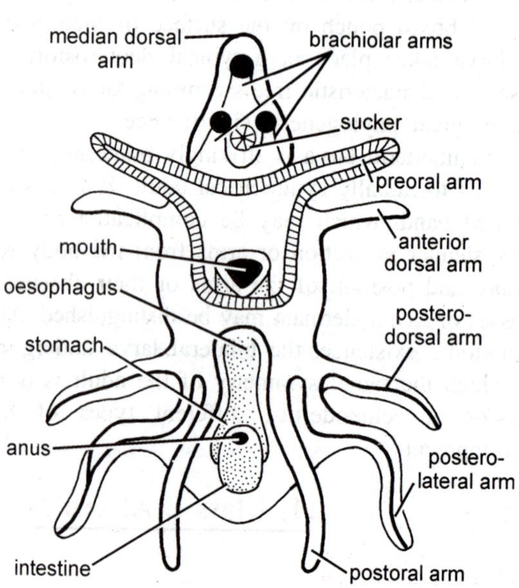

Fig. 2. Brachiolaria larva.

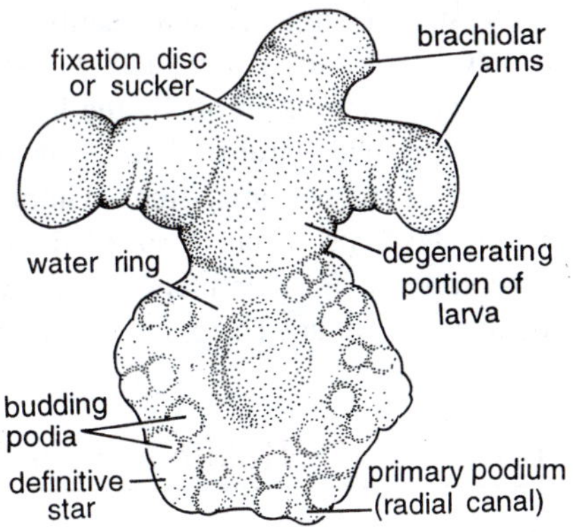

Fig. 3. Metamorphosis of brachiolaria in sea star, *Leptasterias hexactis.*

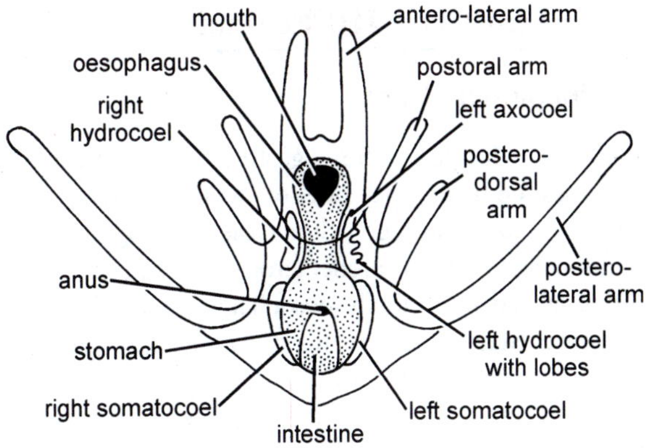

Fig. 4. Ophiopluteus larva of *Ophiocomina.*

Brachiolaria Larva

Bipinnaria transforms into brachiolaria larva which develops three short arms at preoral lobe, known as **brachiolar arms** (one median and two lateral arms). They contain coelomic extensions and adhesive cells at their tips. An adhesive glandular area at their base acts as a sucker. Appearance of the sucker marks the beginning of metamorphosis.

Metamorphosis of brachiolaria. With the help of adhesive structures it attaches to some object. Anterior region acts as stalk for sometime, while posterior part, having gut and coelomic chambers, converts into a young star. This star detaches itself and starts leading a free life.

Some species cut short the development as a result of deletion of some larval stages. In *Astropecten,* the brachiolaria stage is missed with the result bipinnaria directly metamorphoses into adult with in 2-3 months. In *Asterina gibbosa,* bipinnaria stage is omitted, larva develops an adhesive apparatus, as brachiolar arms and sucker, and undergoes metamorphosis. Still in *Luidia,* a giant and peculiar is formed which is called as **bipinnaria asterigeara**.

(2) Class II. Ophiuroidea

Ophiopluteus Larva

Pluteus is the free swimming larva in brittle stars which is known as **ophiopluteus**. It is similar to echinopluteus of echinoids with the only difference that the former has fewer arms than the later. The posterolateral arms are the longest and directed forward. After gastrulation the arms develop gradually. Posterolateral arms are formed first. After 4, 10 and 18 days, anterolateral, postoral and posterodorsal arms develop, respectively. Ciliated bands accompany the arms edges. Internally the larva contains coelomic chambers and archenteron. Internal development proceeds in the same way as in other classes. While free swimming metamorphosis of the larva starts, there being no attachment stage. Tiny serpent star sinks to the bottom to begin its adult existence.

Amphiura vivipara, a viviparous form, omits pluteus stage. In *Ophionotus hexactis,* development takes place in ovary and the aborted pluteus larva is devoid of arms and anus.

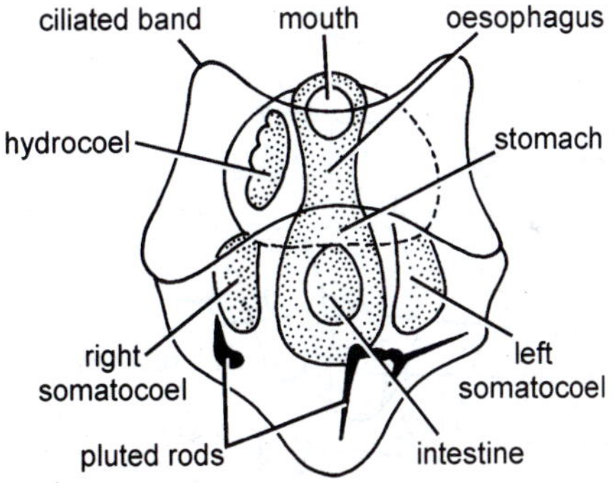

Fig 5. Aborted pluteus of *Ophionotus hexactis.*

(3) Class III. Echinoidea

Echinopluteus Larva

Larva is formed after gastrulation. Gastrula becomes conical, one side of which flattens to form the oral surface. Stomodaeal invagination communicates with archenteron and the gut is differentiated into mouth, oesophagus, stomach and intestine. Blastopore remains as larval anus. Larva begins to form projections which develop into arms. There are six arms namely, preoral, anterolateral, anterodorsal, postoral, postero-dorsal and posterolateral. Posterolateral arms are very short and directed outwards or backwards. In some cases, anterodorsal arms may also not develop. Thus a fully developed echinopluteus may have 5 or even 4 pairs of arms instead of usual six. Tips of the arms are pigmented and are supported by calcareous skeletal rods. Locomotion is by ciliated bands, which in some case become thickened and known as **epaulettes**. In *Arbacia* and *Cidaris,* larva develops special ciliated lobes, between the arm bases known as vibratile lobes, auricular lobes or auricles.

Internal development. Archenteron gives off hydro-enterocoels which contribute to axocoels, hydrocoels and somatocoels. A vestibule is formed by the enlargement of an ectodermal invagination on the left side. Hydrocoel and vestibule form the oral side of the adult. Five radial arms and five primary podia are given off by the hydrocoel. Lantern is formed from left somatocoel. Echinopluteus is microscopic, free swimming in water and it develops within 7 to 30 days.

Metamorphosis is extremely rapid, taking place in about an hour. There is no attachment stage in echinoids.

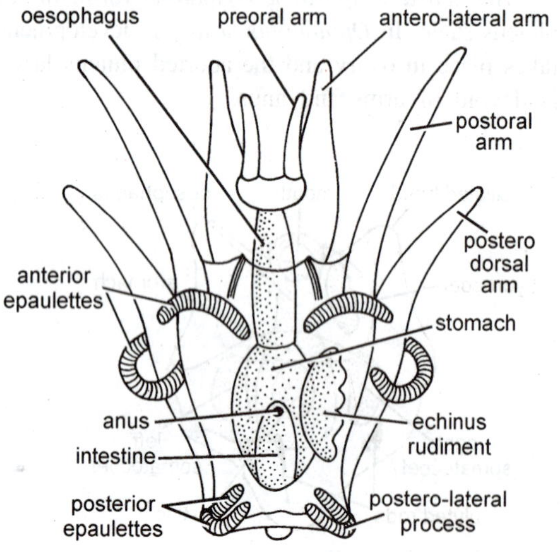

Fig. 6. Echinopluteus larva of *Strongylocentrotus franciscanus.*

(4) Class IV. Holothuroidea

Auricularia Larva

After gastrulation and formation of coelomic sacs and gut, the embryo becomes a free-swimming larva called **auricularia larva**, within 3 days. It is transparent, pelagic about 0.5 to 1 mm in length. It swims about by a ciliated band which forms **preoral loop** and an **anal loop**.

Internally, larva has a curved gut with sacciform stomach, hydrocoel and right and left somatocoels.

Some giant auricularians of unknown adults reported from Bermuda, Japan and Canary islands measure about 15 mm in length and possess a frilly flagellated band.

Doliolaria Larva

It is a transitional stage from auricularia and appears barrel-shaped and alike doliolaria of crinoids.

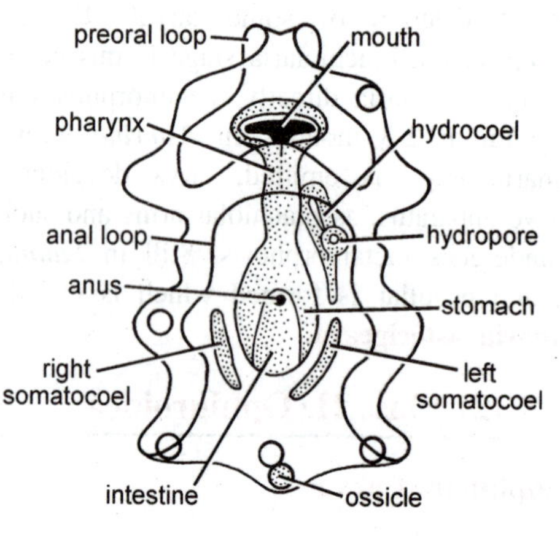

Fig. 7. Auricularia larva.

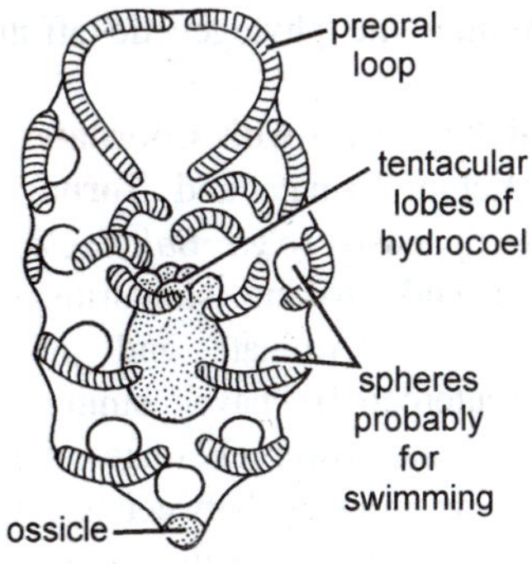

Fig. 8. Transitional stage from auricularia to doliolaria.

Continuous ciliated band breaks in 3 to 5 flagellated rings. Mouth is shifted to anterior and anus to posterior pole.

Metamorphosis is gradual during which it acquires 5 tentacles and 1 to 2 functional podia. As such it is sometimes known as 'pentactula'. After appearence of more podia and tentacles, sea cucumber settles to the sea bottom and leads an adult mode of life.

Other forms of this class show marked peculiarities in larval development. In *Cucumaria planci* and *C. quinquesemita,* etc., there is no auricularia stage and embryo directly develops into **doliolaria** larva. In others like *C. saxiola, C. frondosa,* both of these, larval stages are omitted and the larva only swims about having an oval ciliated shape. In *Holothuria floridana,* embryo hatches directly into a young.

(5) Class V. Crinoidea

Doliolaria Larva

It hatches as a free-swimming larva. Body has 4 to 5 ciliated bands with an apical sensory plate at the anterior end provided with a bunch of cilia. There is an adhesive pit over the first ciliated band, near the apical plate in the mid ventral line. Between second and third ciliated band lies the stomodaeum or vestibule. Skeleton also develops at this larval stage. After the differentiation into prospective organs, larva attaches itself and internal organs rotate at an angle of 90 degrees from ventral to posterior position. Larva forms a stalk and is now referred as

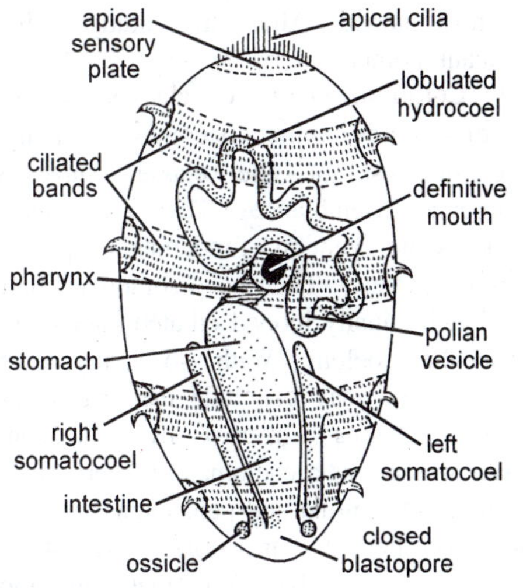

Fig. 9. Mature doliolaria larva of *Leptosynapta inhaerens.*

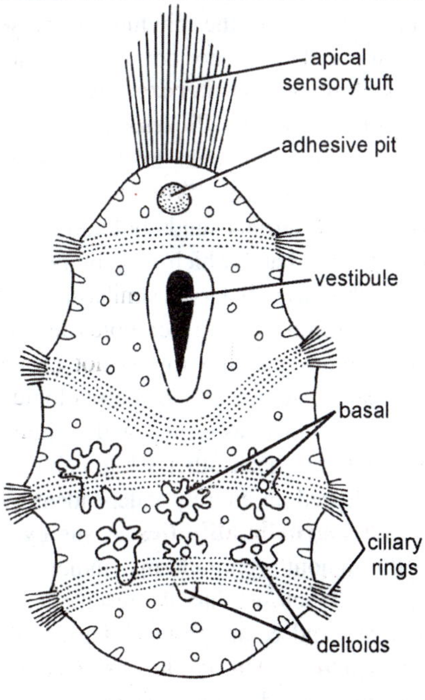

Fig. 10. Doliolaria of *Antedon bifida.*

cystidean or **pentacrinoid** larva which after sometime metamorphoses into adult.

(6) Significance of Echinoderm Larvae

It is seen that different classes of echinoderms have somewhat different larvae which are differently named. Alter their study, following significant points can be drawn.

1. Common origin of classes. Except the larva of crinoidea which becomes sedentary, the larvae of rest of the classes have some fundamental resemblances. They are constructed on the same general fundamental plan with bilateral symmetry. They have somewhat flattened body, longitudinally looped ciliated bands, gut and enterocoelic coelom. With so many common characters, one may conclude the origin of their respective classes (groups) from a common ancestor which was a coelomate, bilateral and free-swimming. Dipleurula and pentactula larva are two such hypothetical ancestors suggested by zoologists. It is believed that all modern echinoderms have originated from them.

2. Taxonomic affinities. Closely looking at the classification of the phylum, it is seen that the larval similarities do not indicate taxonomic affinities. Among **Eleutherozoa**, two well marked larval forms occur : (i) **Pluteus group** is common to ophiuroids and echinoids, bilaterally symmetrical with long arms, (ii) **Auricularia group**, is common to asteroids and holothurians, has a winding ciliated band which may be produced into lobes. On the basis of larval similarities ophiuroids should be placed near to echinoids and asteroids near to holothurians. But this is not in agreement with the palaeontological and morphological result, according to which asteroids and ophiuroids are closely related to each other while echinoids seem to have followed an entirely independent evolution.

3. Phylogenetic affinities. A survey of larval types throughout echinoderms indicates several examples of close larval resemblances *e.g.*, ophiopluteus and echinopluteus. This must be due to **convergent larval evolution**. Occurrence of convergence in development is seen among unrelated groups such as Asteroidea, Holothuroidea and Crinoidea. Similarly, larva of closely related forms such as asteroids and ophiuroid, exhibit major differences, which must be due to **divergent larval evolution**. Occurrence of divergent type of development is seen within related groups (ophiuroidea). Therefore, the larval structures in echinoderms, cannot serve the purpose of determining the phylogenetic affinities in the phylum.

4. Relationship with Chordates. Auricularia larva of Echinodermata and **Tornaria larva** of some enteropneusts (*e.g., Balanoglossus)* shows very close and striking similarities. Moreover, cleavage is indeterminate and mesoderm and coelom (enterocoel) have similar origin in echinoderms and lower chordates. Serology also indicates a relationship between the two groups. In view of all this and other evidences, echinoderms and chordates have been regarded as phylogenetically related groups.

5. Aid in dispersal and feeding. Since the adult echinoderms are somewhat sluggish, their larvae are the main dispersive phase for them. They remain in plankton for sufficient time to be swept from the place of their birth to new areas, or to restock the original areas. In addition to their dispersive function, larvae will aid the species in feeding from a different source from their adults, and thus when food is short larvae and adult will not compete.

AUTOTOMY AND REGENERATION

Breaking off the injured or unduly stimulated body parts, is termed as **autotomy** or **self amputation**. It serves as a means of protection to the animal. It is also a method of getting rid of injured body parts and replacing them with perfect ones. Replacement of the lost parts is brought about by another process known as **regeneration**. Echinoderms possess remarkable powers of autotomy and regeneration.

[I] Asteroidea

Sea star immediately shed off or break off its arm whenever it is injured or attached to some base. The arm is deteched from its base at the IV or V ambulacral ossicle. This process is known as **autotomy**. Whenever caught, a starfish may throw off one or more of its arms.

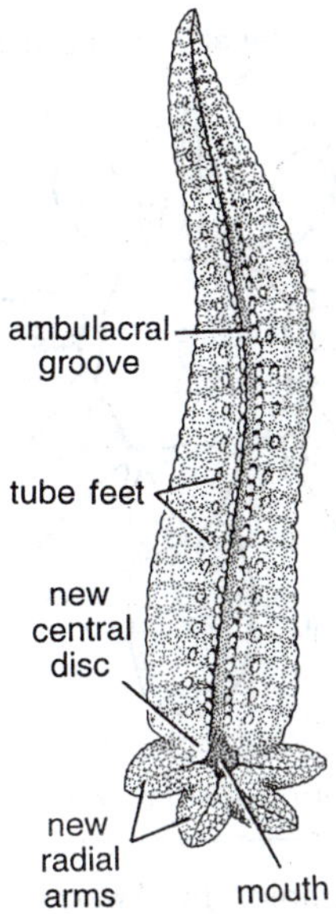

Fig. 11. *Linckia.* Comet stage.

The starfish has remarkable power of **regeneration**. A single arm with small part of disc or a disc alone can regenerate the whole body. Process of regeneration is slow and may take even one year. In *Asterina vulgaris* at least one fifth of disc must be left attached to the arm, for regeneration. But, in *Linckia,* regeneration of complete animal from a cast off arm without disc retention has been observed. Such regenerating specimens with small regenerating arms at the base of original arm are called **comet** forms. A number of asteroids show, the breaking-apart, a normal rather than accidental phenomenon and regenerating the missing parts leading to a regular aspect of asexual reproduction.

[II] Ophiuroidea

Term **brittle star** is derived from the fact that these animals break off their arms if they are disturbed, seized or injured and they speedily regenerate. Autotomy often allows the animal to escape from its enemies. In a number of species the aboral covering of the disc is normally cast off, probably for reproductive purposes. In most ophiuroids, disc have regenerative powers of the arms. In some species the entire disc and at least one arm must be present for regeneration of the other arms.

[III] Holothuroidea

Sea cucumbers possess remarkable powers of autotomy and regeneration. When the animal is irritated it contracts the muscles of the body wall with such a great force that the viscera is extruded through the anus or cloaca or sometimes the mouth. This is called **Evisceration** which is more common and more drastic phenomenon. Respiratory tree, one or both is extruded first. It is of protective importance because some branches (cuvierian organ) of it may swell up into an entangling mass. Parts lost in this process are eventually regenerated.

[IV] Crinoidea

Crinoids exhibit considerable powers of autotomy and regeneration like asteroids and ophiuroids. Part or all of an arm can be cast off if seized or subjected to unfavourable environmental conditions. Lost arms, pinnules and cirri are easily regenerated. In *Antedon,* one fifth of the disc with corresponding arm can be regenerated. It can sustain the simultaneous loss of four or five pairs of arms without death. In this animal all of the visceral mass can be regenerated if aboral nervous system remains intact. Coelomocytes supply food to the wound and phagocytize tissue debris.

WATER-VASCULAR SYSTEM

Water-vascular system is a division of the coelom and most distinctive feature of echinoderms. It is a system of canals and appendages of body wall. It is also termed as **ambulacral system**. Since the entire system is derived from coelom, the canals are lined with a ciliated epithelium and filled with watery fluid. Water-vascular system functions as a means of locomotion. It consists of **madreporite** or **sieve plate**, **stone canal, water ring** or **ring canal**, **radial canals** and **podia** or **tube feet**. In different classes of Echinodermata, water-vascular system is variously modified.

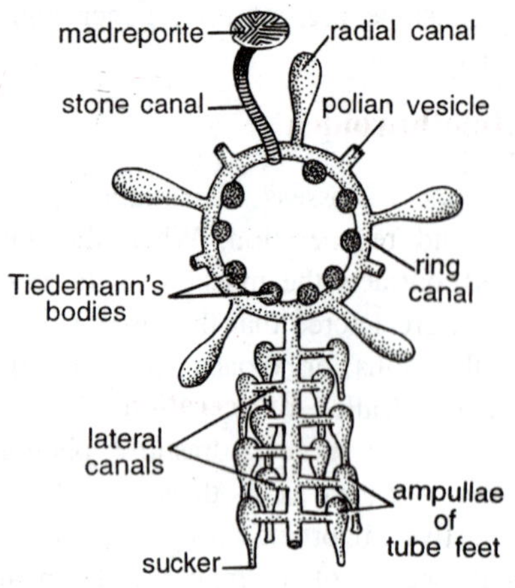

Fig. 12. Water-vascular system of a sea star.

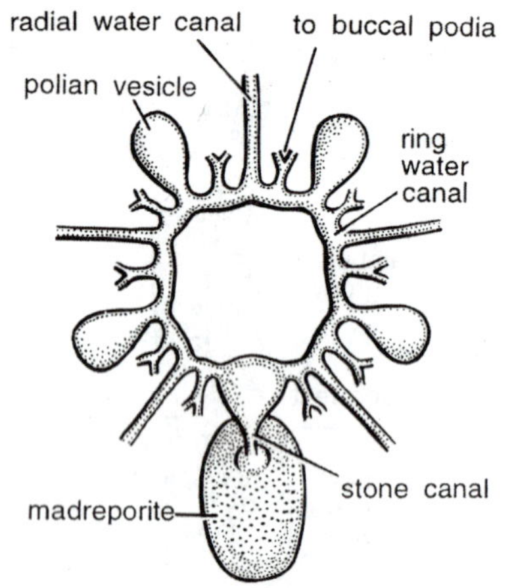

Fig. 14. Water-vascular system of an ophiuroid (Schematic).

[I] Asteroidea

In asteroids, water-vascular system is based on the general echinoderm plan, with certain modifications.

1. Madreporite. Internal canals of the system connect with sea water outside through button-shaped **madreporite** or **seive plate** located on the aboral surface in interradial position. Seive plate is calcareous having numerous ridges and furrows. Furrow have about 200 pores, each of which leads into a **pore canal**. Pore canals join to form a common canal to open into an ampulla beneath the madreporite. Although a great majority of asteroids have but one madreporite, multiplication of madreporites is not rare.

2. Stone canal. It is so named because of calcareous deposites located in its walls. A vertical S-shaped **stone canal** extends towards oral side. A longitudinal ridge divides the lumen of the stone canal into two passages. In *Henricia*, the internal

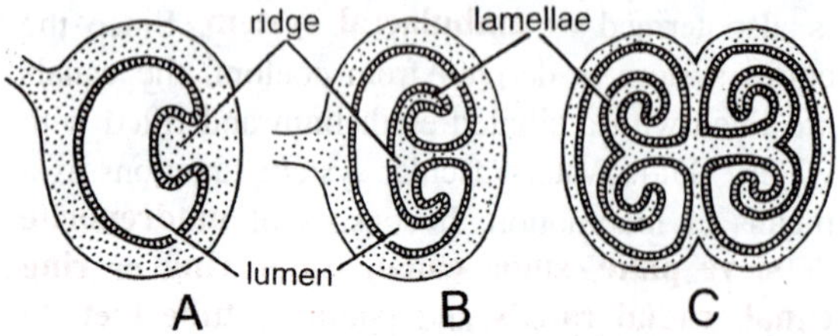

Fig. 13. T.S. stone canals of some asteroids : **A.** *Henricia.* **B.** *Asterias,* **C.** *Astropecten.*

ridge is very simple. In *Asterias* and *Asterina*, ridge is bifurcated into two rolled vertical lamellae. In *Asteropecten*, ridge meets the opposite wall dividing the interior into tubes provided with a pair of scrolls each. The lumen of stone canal is ciliated.

3. Ring canal. Stone canal opens into a circular **ring canal** or **water ring** situated just internal to the peristomial ring of ossicles around the mouth margin. Often the walls of the ring canal are folded to divide its lumen into a more or less separate channels.

4. Radial canals. From the outer margin of ring canal are given out five **radial canals**. Each radial canal runs into each radial arm and terminates into the terminal tentacle at the tip of the arm. The radial canal runs on the oral side of ambulacral ossicles covering the ambulacral groove.

5. Lateral canals. From each side of the radial canal, **lateral canals** are given off alternately which pass between the ossicles on each side to enter the coelom. Each lateral canal has a valve and terminates into a bulb or ampulla connected to tube foot.

6. Polian vesicles. The ring canal, forms five polian vesicles on the outerside. They store water which is to be utilized when sea star comes out side the water.

7. Tiedemann's bodies. The ring canal on its inner side also gives off interradially five pairs

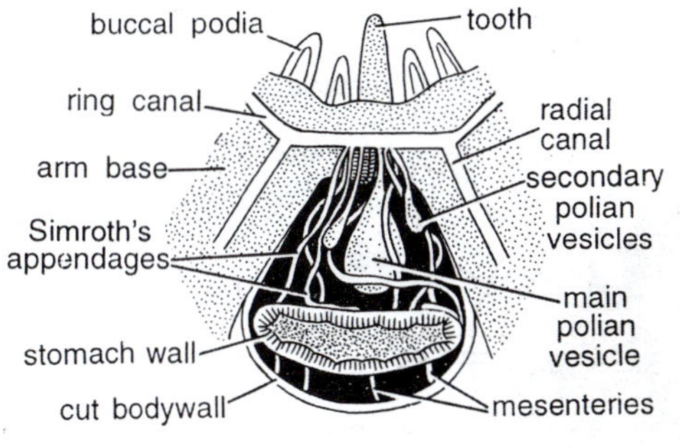

Fig. 15. *Ophiactis virens.* Sector of disc showing water-vascular system.

of small, irregularly shaped bodies, known as Tiedemann's bodies. Some forms have only 9 such bodies. They produce coelomic corpuscles which are passed into ring canal as such giving the system, the name **water-vascular system.**

8. Tube feet. A **podium** or **tube foot**, is a short hollow, elastic, thin walled closed cylinder located in the ambulacral groove. Tip of the podium is flattened forming a sucker for attachment. Inner or basal end of each podium pierces the ambulacral ossicle, through a tiny ambulacral pore and expands to form a little rounded bulb or bladder called **ampulla**, lying into the aboral side of the coelom. The walls of ampulla contain longitudinal and circular muscle fibres where as the tube foot has longitudinal fibres only, there being no circular muscles in the tube foot. Podia are arranged in four rows along the length of ambulacral groove. As lateral canal on each side are alternately long and short, the podia look like in four rows instead of two. Species having only two rows of podia have lateral canals of equal length.

Water-vascular system helps in locomotion. Animal moves by tube feet. Water after entering the madreporite passes into stone canal, ring canal, radial canal, podial canal and finally into ampullae of podia. As ampulla contracts the water is forced into the tube foot which gets elongated. The sucker-like tip of tube foot touches the substratum and contracts again. Fluid goes back into ampulla and the body is drawn forwards bringing about the locomotion of sea star.

[II] Ophiuroidea

Madreporite lies on the oral surface. It bears a single pore and a **pore canal**. Some species have more than one madreporite. **Stone canal** ascends to the **water ring**, which is located in a groove on the aboral surface of the jaws. Stone canal gives off a madreporic **ampulla** just beneath the madreporic plate. Water ring or **ring canal** gives off four **polian vesicles** in each inter radius except one which contains stone canal. Sometimes polian vesicle are accompanied by long slender, tubular appendages, known as **Simroth's appendages**. Ophiuroids lack **Tiedeman's bodies**. Water ring

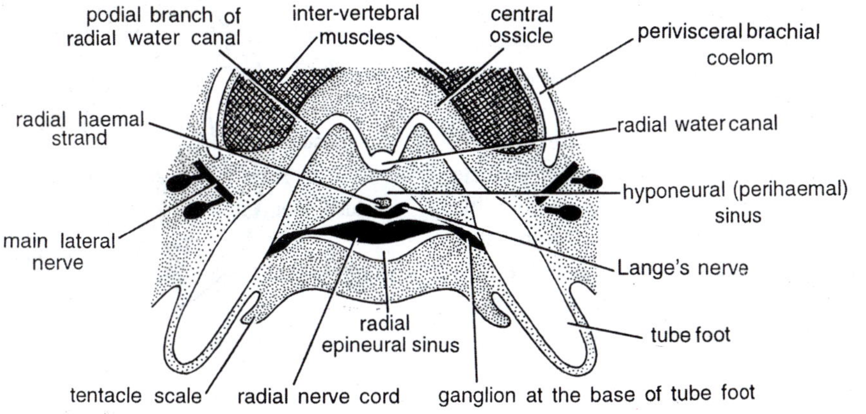

Fig. 16. *Ophiothrix.* Partial section of arm showing podial water canals.

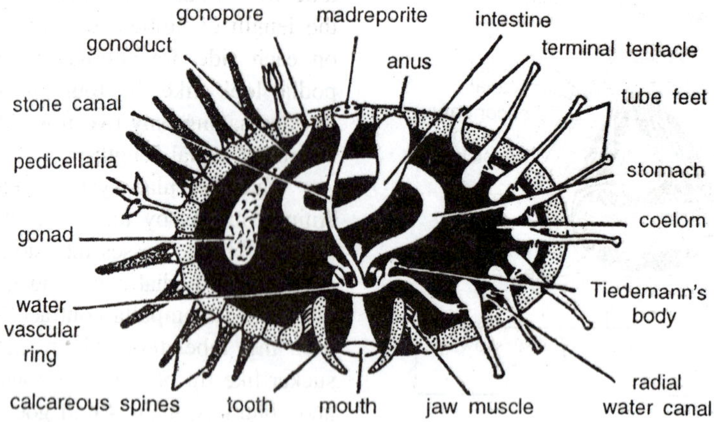

Fig. 17. Echinoidea. A section through a sea urchin showing water-vascular system.

gives rise to a **radial canal** in each radius which descend, towards oral side, runs along the whole arm to terminate into terminal tentacle. Radial canals penetrate through the lower side of vertebral ossicles of the arms. In each ossicle, radial canal gives a V-shaped **podial canal** or **lateral canal** that enters into a pair of podia. **Podia** do not bear **ampulla**. There is a valve between podium and the lateral canal. Podia are reduced. Entire water-vascular system is lined with ciliated peritonium.

[III] Echinoidea

Water-vascular system of echinoids is like that of sea stars. One of the genital plates around the periproct bear pores and **pore canal** and also serves as **madreporite**. **Stone canal** descends orally to a **water ring** surrounding the oesophagus aboral to the lantern. Stone canal is accompanied by axial gland. Five **radial canals** arise from the water ring and terminate into the terminal tentacle. Water ring also gives rise to five sponge like bodies called **Tiedemann's bodies**. From either side of radial canal alternately come out **lateral canals** to the bases of **ampullae**. Suckers of podia in sea urchins are highly developed.

[IV] Holothuroidea

They have water-vascular system like that of other echinoderms but **madreporite** is peculiar. Neither it is connected with the body surface nor it is attached in the coelom. It hangs just below the base of the pharynx and is connected to the **water ring** by a short **stone canal**. Pores and **pore canals** still persist in madreporite but in place of sea water, coelomic fluid circulates in it.

Water ring encircling the base of the pharynx gives rise to elongated **polian vesicles** which hang into the coelom. Their number may be one in **cucumaria**, 3 or 4 in **thyone**, or 10 or 11 in

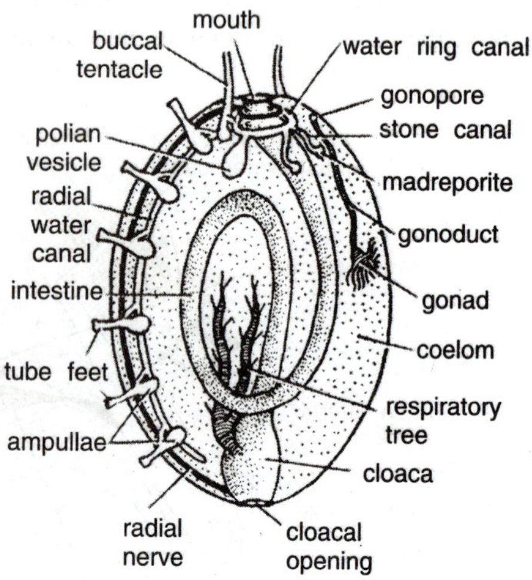

Fig. 18. L.S. of a holothurian, showing water-vascular system (diagrammatic).

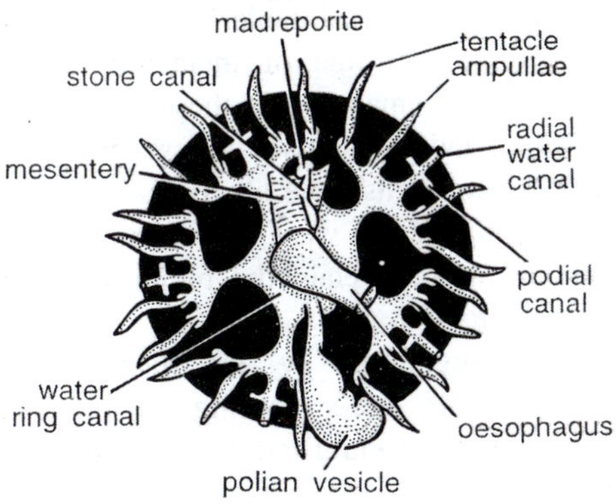

Fig. 19. *Holothuria.* Water-vascular system.

certain Apoda. Five **radial canals** arise from water ring and pass upward to the inner side of the calcareous ring and then outward through a notch at the end of each radial plate. Before leaving the calcareous ring, each radial canal gives off smaller branches to the buccal tentacles. After leaving the ring, radial canals pass posteriorly within the body wall all along the length of ambulacral groove where they supply the **podia**. **Ampullae** are found on both podia and tentacles.

In Apoda, which lack podia, water-vascular system is confined to oral water ring, polian vesicles and buccal tentacles.

[V] Crinoidea

Crinoids lack **madreporite**. Water-vascular system is restricted to a **ring canal**, 5 **radial canals** and the **podia**. Pentagonal ring canal encircles the mouth and gives off at each inter radius, a large number (30-50) of short **stone canals**, that open into coelom. Each radial canal extends into each arm, underneath the ambulacral groove and bifurcates with the arm and gives off alternate branches into pinnules and **podia**. **Ampullae** absent. Peculiar for crinoids is the presence of 500 to 1500 minute ciliated funnel like canals perforating the wall of tegmen to open into coelomic cavity, called as **ciliated funnels**. These perforations or openings compensate for the

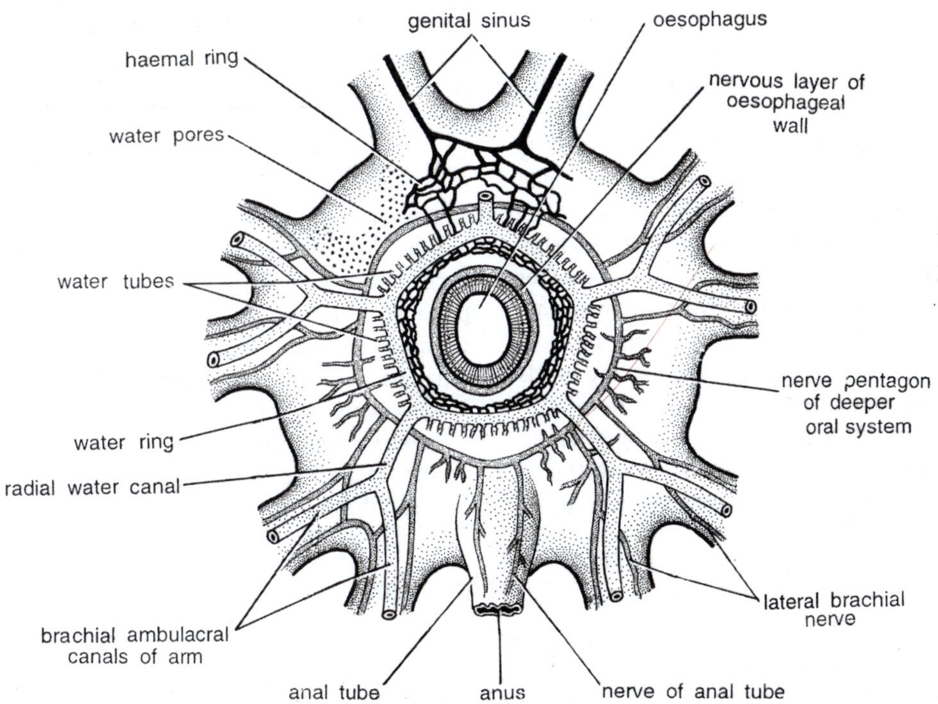

Fig. 20. *Antedon.* Water-vascular and nervous system (diagrammatic).

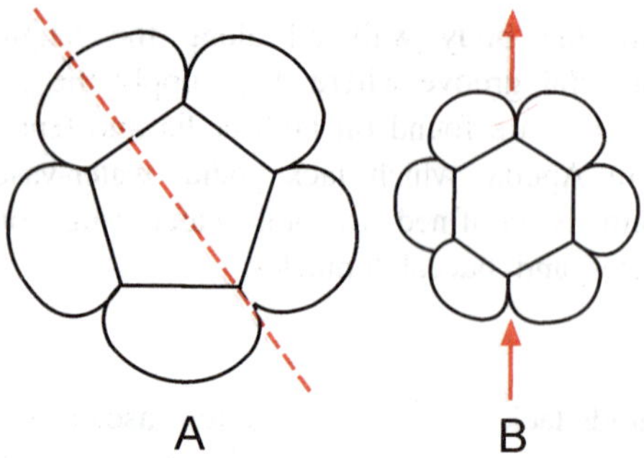

Fig. 21. Radial symmetry. A. Involving five ossicles. B. Involving six ossicles.

absence of a madreporite by permitting sea water to enter the coelom and maintain a proper fluid pressure in the body and the water-vascular system.

SYMMETRY IN ECHINODERMATA

Phylum has almost universally adopted and retained pentamerous radial symmetry. Their radial symmetry is secondarily evolved since echinoderms are supposed to have been derived from bilaterally symmetrical ancestors. To a great extent, radial symmetry with mouth up is functionally useful to sessile animals which have the same relationship to their environment on all sides. On the other hand bilateral symmetry with a front and rear end, upper and lower surfaces, left and right pairs of effector organs, is functionally suited to the needs of a travelling animal.

Apart from the echinoderms, only other group of metazoan animals having radial symmetry, is coelenterates, many of which are fixed. Now, in more archaic living echinoderms an attachment is developed at the time of metamorphosis. It is probable that many of these were sessile organisms, and it is also almost probable that earliest echinoderms were all fixed. Thus it seems that all echinoderms were at one time fixed and those now mobile retain the radial symmetry of these ancestors.

It is suggested by the fact the larval echinoderms are free-swimming and bilateral and possess a more centralized nervous system than adults. In a sense, the adults are more premitive than the larvae. Radial symmetry and lack of definite head or brain are characteristics of lower phyla such as sponges, coelenterates and ctenophores. But echinoderms are in no way related to these three radiate phyla. It is note worthy that echinoderms are triploblastic and true coelomate and with a much higher structural level than that of other radiate groups. Evolution involving reversion to primitive characteristics is called **retrogression**.

Echinoderms skeleton is in sessile forms, probably must have evolved as a supportive and protective structure. According to **Nichols** (1966), pentamerous form of radial symmetry arose in conjunction with the skeleton. Future planes, *i.e.*, junction point between two skeletal plates, represent a weak point in the body wall from a structural stand point, and it would have been advantageous to the animal not to have had two such suture planes opposite each other. This is possible only by an odd number of ossicles forming the circumference of the body wall. Smallest number, therefore, would be five if the animal were to be truly radial. This stage of echinoderm symmetry is illustrated by some extinct groups, as well as by certain of the living crinoids, or sea lillies.

Important Questions

►► *Long answer type questions*

1. Give an account of larval forms found in Echinodermata. What is their significance ?
2. What do you understand by autotomy and regeneration ? Discuss in reference to echinoderms.
3. Discuss the water-vascular system and its importance in Echinodermata.
4. Discuss the pentamerous radial symmetry in echinoderms.
5. Write short notes on : (i) Bipinnaria larva, (ii) Echinopleuteus larva, (iii) Doliolaria larva, (iv) Madriporite, (v) Tube feet, (vi) Radial symmetry of echinoderm.

➤➤ *Short answer type questions*

1. What are the different types of Echinoderm larvae studied by you ?
2. What is Aristotle lantern ?
3. What are tube feet ?
4. Write two functions for each item given below :
   (i) Tube feet (ii) Polian vesicles
   (iii) Axial sinus (iv) Spines in sea urchin
5. Name the echinoderm larvae and the classes to which they belong.
   Or Form a list of the echinoderm larvae and their classes to which they belong.
6. Compare the water-vascular system of a starfish with that of a holothurian.
7. Give an account of water-vascular system in Echinodermata.
8. Draw and label longitudinal section of an arm of starfish.
9. Describe how a holothurian respires ?
10. Describe the water-vascular and respiratory systems in a holothurian.
11. Comments on bipinnaria larva.
12. Discuss the brachiolarea larva.
13. Echinopleuteus larva belongs to which class of phylum Echinodermata.
14. Match A and B

(A)	(B)
(i) Brachiolaria	(i) crinoidia
(ii) Bipinnaria	(ii) holothuroidea
(iii) Ophiopluteus	(iii) echinoidea
(iv) Echinopluteus	(iv) ophiuroidea
(v) Auricularia	(v) asteroidea
(vi) Doliolaria	

➤➤ *Fill in the blanks*

1. .... is the larva of starfish.
2. In starfish the water-vascular system is derived from ....
3. The name 'pen' is given to the internal shell of ....
4. A stalked but single polian vesicle occurs in .... in contrast to other echinoderms in general.

➤➤ *Multiple choice questions*

1. Madriporite absent in :
   (a) holothuroidea (b) crinoidea
   (c) echinoidea (d) asteroidea
2. Asteroidea contain the larval forms :
   (a) bipinnaria (b) branchiolaria
   (c) both (d) none
3. Bipinnaria have :
   (a) somateocoel (b) hydrocoel
   (c) axocoel (d) all
4. Viviparous Echinodermata is :
   (a) *Amphiura* (b) *Antidoni*
   (c) *Holothuria* (d) none
5. Doliolaria larva belongs to the class :
   (a) Echinoidea (b) Holothuroidea
   (c) Crinoidea (d) none the above
6. The power of regeneration occurs in :
   (a) Coelenterata (b) Porifera
   (c) Echinodermata (b) all
7. Water-vascular system contains :
   (a) madreporite
   (b) stone canal
   (c) ring canal
   (d) radial canal
   (e) all these
8. Radial symmetry occur in metazoans :
   (a) Poifera Coelenterata
   (b) Coelenterata Platyhelminthes
   (c) Annelida Echinodermata
   (d) Coelenterata Echinodermata
9. Sea pantagon is a :
   (a) Porifera (b) Coelenterata
   (c) Mollusca (d) Echinodermata
10. Aristotle's lantern present in :
    (a) sea lilies (b) sea coccumber
    (c) sea urchin (d) sea pen

Answers

➤➤ *Fill in the blanks*

1. Bipinnaria, 2. Coelom, 3. Loligo, 4. Cucumaria.

➤➤ *Multiple choice questions*

1. (b) 2. (c) 3. (d) 4. (a) 5. (c) 6. (d) 7. (e) 8. (d) 9. (d) 10. (c).

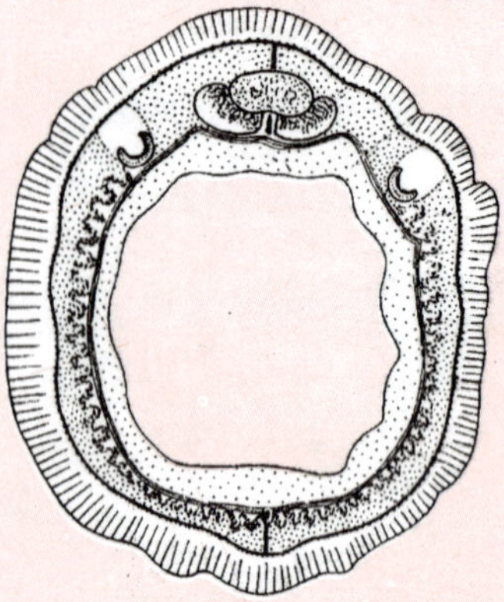

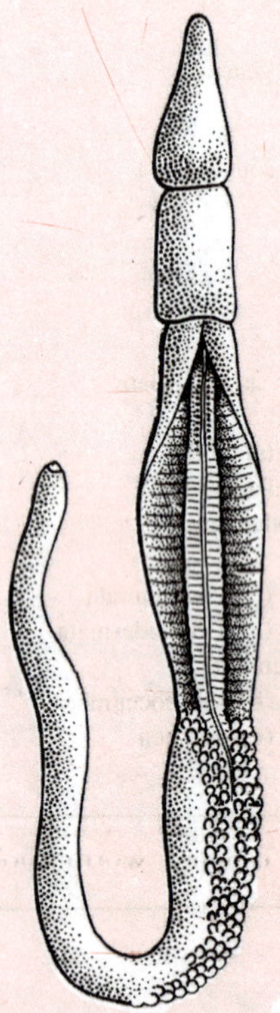

65

Balanoglossus : A Tongue Worm

Hemichordata (Gr. **hemi** half; **chorde**, cord), till recently treated as a subphylum of the phylum Chordata (or Protochordata), are now regarded to be an independent phylum of invertebrates close to Echinodermata. It includes deuterostome animals with enterocoelous coelom, pharyngeal gill-slits, a buccal diverticulum (considered earlier as 'notochord' and a vermiform body divisible into three regions : proboscis, collar and trunk. Hemichordates includes a small group of soft, marine and primitive chordates, commonly called the 'acorn worms'or 'tongue worms'. Most familiar hemichordate genus is *Balanoglossus* which belongs to the class Enteropneusta. Other closely related genera are *Saccoglossus (=Dolichoglossus), Glossobalanus, Ptychodera, Spengelia,* etc.

SYSTEMATIC POSITION

Phylum	Chordata
Subphylum	Hemichordata
Class	Enteropneusta
Family	Ptychoderidae
Genus	*Balanoglossus*

DERIVATION OF NAME

Balanoglossus clavigerus was recorded and named by **Delle Chiaje** in 1829. Its generic name was derived from two Greek words, **balanos** and **glossus**. Term **balanos** means an 'acorn' (fruit of oak) and refers to the proboscis projecting from collar looking like an acorn-nut, hence the common name 'acorn worm'. Term *glossus* means 'tongue' and refers to the shape of its proboscis, collar and genital wings bearing a close resemblance to an ox tongue, hence the common name 'tongue worm'. Local fisher-men call *Balanoglossus* by the name 'ox tongue'.

GEOGRAPHICAL DISTRIBUTION

Balanoglossus, like all other hemichordates, is a marine animal having a world-wide distribution. About 20 species occur all over the world especially in the tropical and subtropical seas. Some species are *B. australiensis* (Australia), *B. carnosus* (Indo-Pacific), *B. misakiensis* (Japan), *B. jamaicensis* (West Indies), *B. gigas* (Brazil), *B. capensis* (South Africa) and *B. clavigerus* (Mediterranean and British isles).

HABITS AND HABITAT

Balanoglossus is a marine, tubicolous or burrowing hemichordate inhabiting shallow coastal waters of intertidal zone, but a few occur in deeper water.

Burrow. The animal may conceal under stones or sea weeds or excavate its own burrow in the bottom of sand or mud. *B. clavigerus* lives inside a U-shaped tube or burrow with the two vertical limbs, 50-75 cm deep and the two openings 10-30 cm apart. In some species *(Saccoglossus),* body of the animal and tube are twisted, whereas the anterior and posterior extremities remain straight. Anterior opening of the burrow is funnel-shaped and exposed. Anterior vertical limb may give out side branches each having its independent funnel-like opening. Posterior opening of the burrow is rounded and concealed below the spirally coiled faecal matter of the animal.

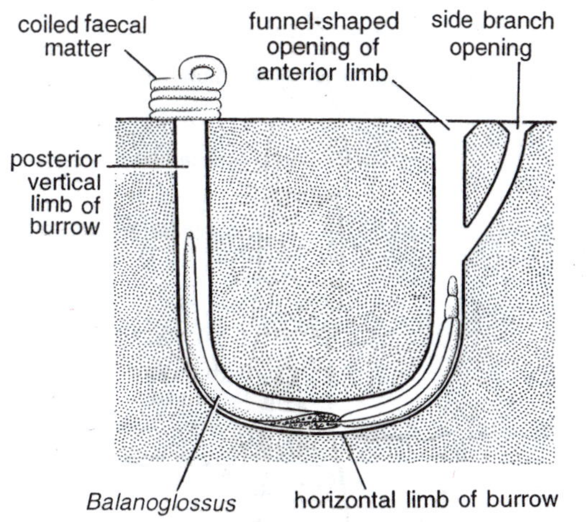

Fig. 1. *Balanoglossus clavigerous* in U-shaped burrow.

Protective device. Inner wall of the fragile tube is smoothly lined by sand particles cemented together into a tough cast with sticky mucus secreted by the skin mucous glands of the animal. This prevents the collapse of the burrow and protects the delicate animal from burial in loose sand. Another protective device is the secretion of a foul-smelling odour similar to iodoform. One species shows phosphorescence.

Movements. Worm is sluggish and little affected by external stimuli. It moves in its burrow by cilia covering its body surface. Most active part of the body is proboscis. It elongates and shortens by muscular activity and helps in burrowing. When the tide recedes, the tongue worm protrudes its anterior end out of the burrow to explore the surroundings, or its posterior end to cast out faecal matter.

Feeding and breeding. It swallows sand or mud to obtain diatoms, protozoans, other microorganisms and organic detritus on which it feeds. Sexes are separate. Males and females, living in separate tubes, shed their gametes in seawater where fertilization occurs. Life cycle includes a free-swimming planktonic ciliated larva, the **tornaria larva** Tongue worm does not reproduce asexually, but its fragile body has considerable power of regeneration.

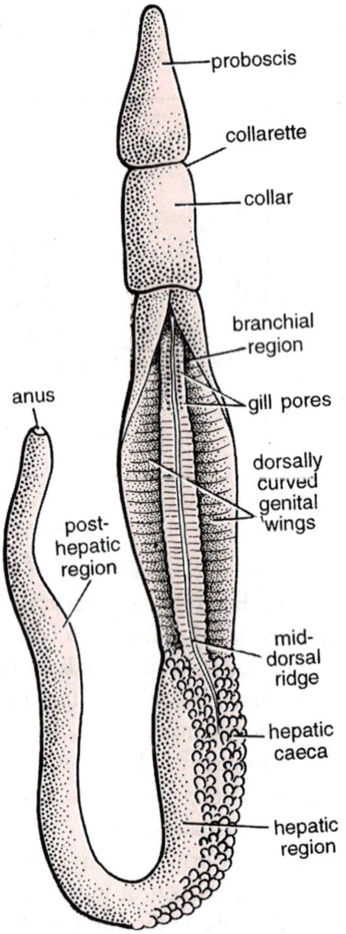

Fig. 2. *Balanoglossus.* External features in dorsal view.

EXTERNAL MORPHOLOGY

Shape, size and colouration. Body is soft, elongated, worm-like, cylindrical and bilaterally symmetrical. It measures 10 to 50 cm in length, according to species. *B. gigas* attains a length of 1.8 m (**Sawaya**, 1951) or 2.5 m (**Spengel**, 1893). Colour is bright or drab with reddish or orange tints. Body is uniformly ciliated and without any exoskeleton or external appendages.

Division of Body

Body is unsegmented but divisible into three distinct regions or parts : **proboscis, collar** and **trunk**.

[I] Proboscis

Proboscis or **protosome** is the anteriormost part of the body. It is short, club-shaped or conical and circular in cross-section. It has thick muscular wall and is hollow within. Its cavity or proboscis coelom communicates with the outside through a minute **proboscis pore** situated mid-dorsally near its base. In some species the proboscis pore ends blindly or there are two pores. Posteriorly, the proboscis narrows into a slender **neck** or **proboscis stalk** which is attached to the collar. Below the stalk, the base of proboscis bears a U-shaped ciliated epidermal depression, called the **preoral ciliary organ**, which tests the quality of food and water entering the mouth.

[II] Collar

Collar or **mesosome** is the middle, short and cylindrical part. Its flap-like or funnel-like anterior margin, termed **collarette**, completely surrounds and conceals the proboscis stalk and the posterior part of proboscis. Ventrally, below the proboscis stalk, the collarette or collar-rim encloses a permanently open wide aperture, the **mouth**. It opens into buccal cavity inside the collar. Posterior end of collar is well demarcated from the trunk by a circular constriction. Wall of collar is thick, highly muscular and encloses a cavity, the **collar coelom**. It opens to outside through a pair of **collar pores** into the first pair of gill pouches behind.

[III] Trunk

Trunk or **metasome** is the posterior and largest part of the body. It is rather flat and appears annulated due to circular constrictions on the surface. All along its length, trunk bears a **mid-dorsal** and a **mid-ventral ridge**, each accommodating its corresponding nerve and blood vessel. Trunk is further differentiated into three regions : an anterior **branchiogenital**, a middle **hepatic** and a posterior **post-hepatic**, **abdominal** or **caudal** region.

1. Branchiogenital region. Anterior or branchiogenital region of trunk is marked by a pair of lateral, thin, flat and longitudinal flaps, the **genital wings**, containing the **gonads**. **Gonopores** are microscopic and cannot be seen by the unaided eyes. Anterior half of branchiogenital region bears two longitudinal rows of small **branchial apertures** or **gill pores**. One row of gill pores is mounted on a prominent longitudinal ridge on each side of the mid-dorsal ridge. Number of gill pores increases with the age of the animal. Two genital wings can be curved to meet mid-dorsally so as to conceal the gill pores. In some species, a posterior prolongation of the collar, called the **operculum**, may cover the anterior-most gill pores.

2. Hepatic region. Middle or hepatic region of trunk is somewhat smaller than the genital region. It is greenish in colour and its dorsal surface marked by the presence of numerous irregular intestinal sacculations or **hepatic caeca**.

3. Post-hepatic region. It is the posterior-most and the longest part of the trunk also called **abdomen** or **caudal region**. It is more or less uniform in diameter but its posterior end slightly tapers and bears a terminal **anus**.

BODY WALL

Body wall is composed of epidermis, musculature and peritoneum.

1. Epidermis. Outermost layer or epidermis consists of a single layer of mostly tall, slender, columnar and ciliated cells. Three kinds of gland cells secreting mucus are present. **Goblet gland cells** are flask-shaped with fine granules. **Reticulate gland cells** have vacuolated cytoplasm. **Mulberry** or **granular gland cells** contain coarse granules and also secrete amylase. Gland cells are more abundant in the collar region. Besides, **neuro-sensory cells**, which stain darker, are present in the epidermis of proboscis and anterior

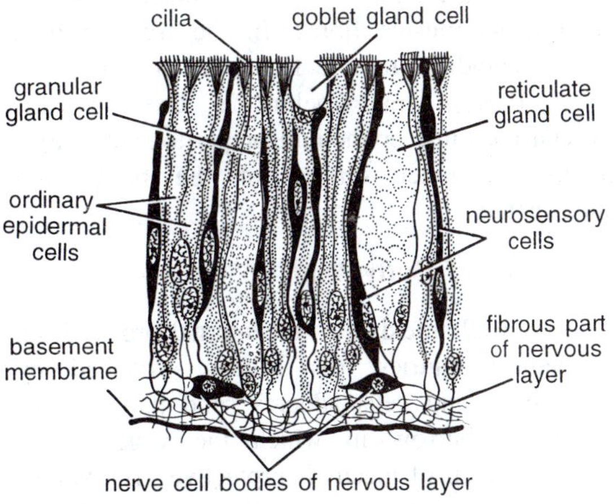

Fig. 3. *Balanoglossus.* V.S. of epidermis.

part of the collar. Dermis is absent. Below the epidermal cells is a thick **nervous layer** consisting of a network of nerve cells and nerve fibres. Below the nervous layer is a thick **basement membrane** which supports the epidermis and serves for attachment of the underlying musculature.

2. Musculature. Muscles are smooth, weak and mostly longitudinal. Proboscis and the anterior end of collar (collarette) have an outer layer of

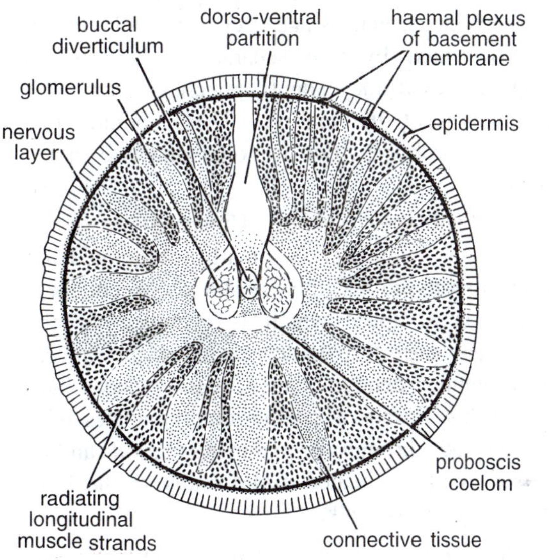

Fig. 4. *Balanoglossus.* T.S. through proboscis.

circular muscle fibres and an inner layer of longitudinal muscle fibres. In the trunk region, only longitudinal muscle fibres are present.

3. Peritoneum. Coelom is lined by the parietal coelomic epithelium or peritoneum which covers the inner surface of the body wall musculature.

Functions of Body Wall

(i) Body wall shields the soft internal organs from mechanical injuries, (ii) Mucus produced by epidermal gland cells adheres sand particles for lining the burrow in which the worm lives, (iii) Foul smell of mucus is protective, (iv) Neuro-sensory cells serve to receive the external stimuli, (v) Musculature helps in body movements.

COELOM

Balanoglossus has a spacious coelom lined by coelomic epithelium and enterocoelous in origin. Coelomic cavities of proboscis, collar and trunk are completely separated from one another. However, in the adult, the original coelom is greatly obliterated by connective tissue and muscle fibres derived from coelomic epithelium which becomes inconspicuous and irregular. Adult coelom is represented by five separate cavities which originate as independent pouches from the archenteron of the embryo. These include one cavity in proboscis, two in collar and two in the trunk.

1. Proboscis coelom. Unpaired **proboscis coelom** or **protocoel** is greatly obliterated by connective tissue and muscle strands except for a small central space which is occupied by the **proboscis complex**. Latter includes the buccal diverticulum, central sinus, heart vesicle and glomerulus. Proboscis coelom communicates with the outside through a **proboscis canal** and a **proboscis pore** situated mid-dorsally at the base of the proboscis stalk.

2. Collar coelom. Collar coelom or **mesocoel** is represented by two narrow lateral cavities, one on each side between the collar wall

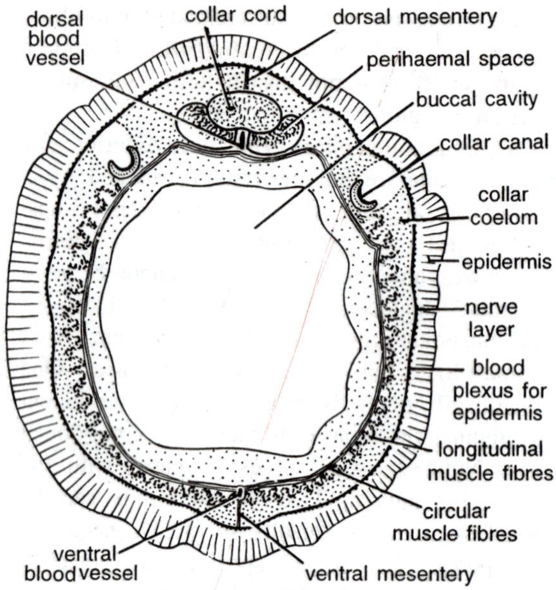

Fig. 5. *Balanoglossus.* T.S. through collar region.

and buccal tube. Two cavities are partitioned by incomplete mid-dorsal and mid-ventral mesenteries. Collar coelom does not communicate with the proboscis coelom, but posteriorly, its each cavity opens into the first gill sac of its side by a **collar canal** and a **collar pore**. Collar coelom is greatly obliterated by the collar musculature and connective tissue.

3. Trunk coelom. Trunk coelom or **metacoel** is represented by a pair of closed cavities between the body wall and gut wall. Two cavities are separated by an incomplete dorsal and a complete ventral mesentery. In the branchiogenital region each cavity is further divided by a **lateral septum** into a dorso-lateral and ventro-lateral compartment. Trunk coelom is partitioned from the collar coelom by a **collar-trunk septum**. Trunk coelom is obliterated by the trunk musculature.

Coelomic Fluid

Proboscis and collar coelom communicate with the exterior and are largely filled with sea water which keeps them turgid. Trunk coelom is filled with a watery coelomic fluid containing amoeboid **coelomocytes**, each with a single large vacuole. Coelomocytes originate from the coelomic epithelium. According to **Spengel**, they behave like leucocytes by secreting a membrane around any foreign body that may invade the animal.

ENDOSKELETON

Balanoglossus has no definite endoskeleton of bone or cartilage. However, the following four stiff structures are present : (i) **Buccal diverticulum**, (ii) **proboscis skeleton**, (iii) **branchial skeleton**, and (iv) **pygochord**.

1. Buccal diverticulum. From the roof of buccal cavity (lying inside collar region), a short stiff, thick-walled, hollow projection extends forward through the proboscis stalk into the proboscis coelom. Its wall is composed of a single layer of tall, slender, vacuolated endodermal cells. For a long time it was considered a **notochord** (**Bateson**, 1885), or **stomochord** (**Willey**, 1899 and **Dawydoff**, 1948). Histologically as well as developmentally, it is quite different from the true notochord of other chordates, but resembles the wall of the buccal cavity. To most modern workers it is only a preoral extension of buccal cavity, so that, Hyman preferred to use the non-committal term **buccal diverticulum** for this tubular outgrowth of buccal cavity.

2. Proboscis skeleton. Proboscis or **nuchal skeleton** is a Y-shaped chitinous structure formed by the thickening of the basement membrane. It consists of a broad, flat, roughly rectangular **median plate** produced ventrally into a **keel** and posteriorly into two diverging **horns**. Median plate lies below the buccal diverticulum in the proboscis stalk while the two horns extend backwards into the roof of the buccal cavity. Median plate remains embedded in a reticulum of stiff **chondroid tissue** which resembles vertebrate cartilage.

3. Branchial skeleton. It is also formed by the thickening of the basement membrane. It consists of numerous M-shaped chitinous skeletal rods that lie in the wall of the pharynx and support the U-shaped gill-slits that perforate it. Middle arm of the skeletal rod is thicker than the others and bifurcated at the free end which indicates that it is formed by the fusion of two arms of two adjacent 'inverted U-shaped' rods that join to form the M-shaped rod.

4. Pygochord. In the post-hepatic region of the trunk, mid-ventrally between the intestine and body wall, develops a rod-like thickening called **pygochord**. Its cells are vacuolated. It supports the post-hepatic region of the body but probably also performs some other function not yet understood.

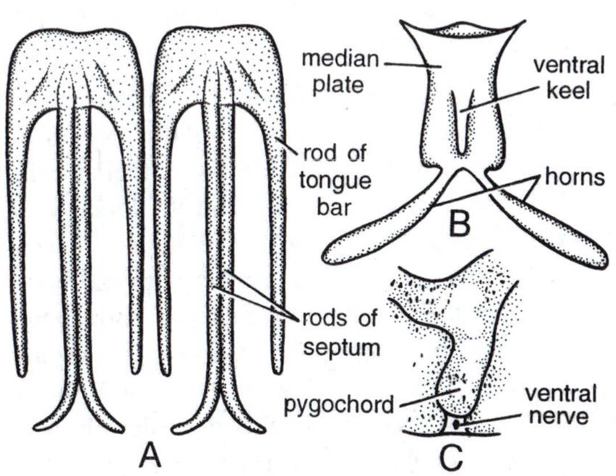

Fig. 6. *Balanoglossus.* Skeletal elements. **A.** Branchial skeleton. **B.** Proboscis skeleton. **C.** Pygochord.

DIGESTIVE SYSTEM

[I] Alimentary Canal

Alimentary canal is a complete and straight tube running between the mouth and anus. It is supported throughout its length by the dorsal and ventral mesenteries. Its wall is made up of ciliated epithelium covered externally by a basement membrane, but peculiarly, muscle layers are absent. Alimentary canal comprises : (i) Mouth, (ii) buccal cavity, (iii) pharynx, (iv) oesophagus, (v) intestine, and (vi) anus.

1. Mouth. It is wide and circular opening situated ventrally in a groove between the proboscis stalk and collarette. According to **Knight-Jones** (1952), it can be closed or opened and does not remain permanently open as previously supposed. It has two sets of muscle

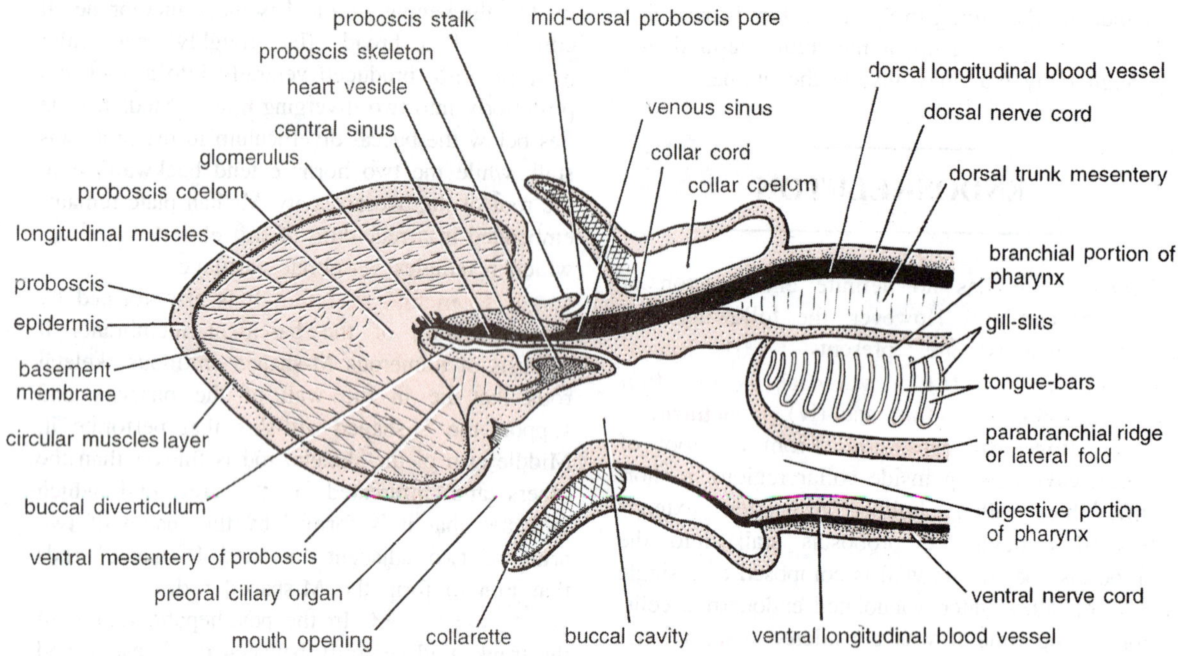

Fig. 7. *Balanoglossus.* V.L.S. Anterior region to show the alimentary canal.

fibres, the radial fibres to open it and the concentric fibres to close it. Mouth leads into buccal cavity.

2. Buccal cavity. Short buccal cavity occupies the collar region. Its epithelial wall contains glandular goblet cells. Anteriorly its dorsal wall forms a short, stiff and hollow **buccal diverticulum** that projects into the proboscis coelom. Posteriorly it extends up to the collar-trunk septum behind which it continues into the pharynx.

3. Pharynx. It lies in the branchial region of the trunk. Externally its wall bears a longitudinal constriction along each lateral side. These lateral constrictions project into its lumen as ridges, called **parabranchial ridges**, consisting of tall columnar cells. These ridges incompletely divide the pharynx into a dorsal **respiratory** or **branchial portion** and a ventral **digestive portion**. Dorsal branchial portion is perforated dorso-laterally by two rows of U-shaped gill-slits, and is concerned with respiration. Ventral digestive portion, lined with ciliated epithelium with gland cells, helps in food concentration.

4. Oesophagus. Behind the last pair of gill-slits the pharynx continues into the short **oesophagus**. Dorsal and ventral divisions of pharynx continue for some distance into oesophagus. In this region, the dorsal part is called **postbranchial canal** which possesses thick, folded and glandular epithelium. Posterior part of oesophagus reduces in diameter and has deeply furrowed epithelium.

5. Intestine. It occupies the hepatic and post-hepatic regions of trunk. **Hepatic region** of the intestine is highly vascular. Its epithelial cells are dark green or dark brown, and its dorsal wall forms numerous sacculations called **hepatic caeca**. Intestinal wall lies in close contact with the body wall, so that the intestinal sacculations correspond with those of the body wall. **Post-hepatic region** of intestine is connected with the ventral body wall by the pygochord described earlier. It is a simple and straight tube bearing a pair of dorso-lateral grooves lined by tall epithelial cells with long cilia.

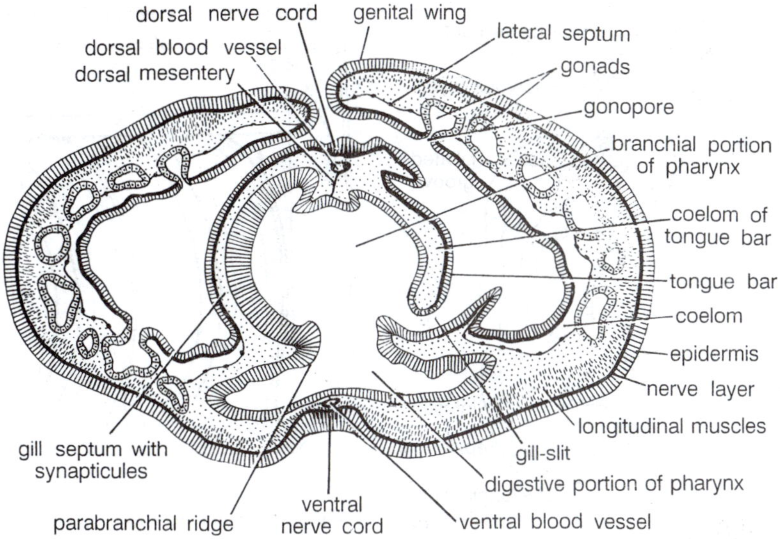

Fig. 8. *Balanoglossus.* T.S. through pharyngeal or branchio-genital region.

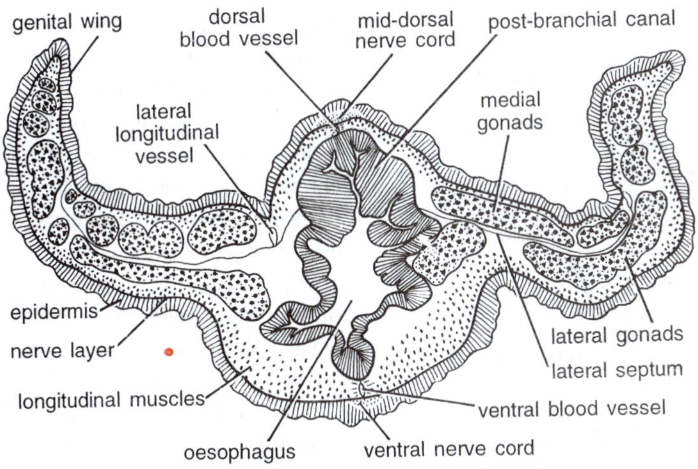

Fig. 9. *Balanoglossus.* T.S. through oesophageal region.

6. Anus. Posteriorly, the intestine opens to the exterior by a terminal circular aperture, the **anus**, at the tip of the trunk. It is often surrounded by a sphincter muscle.

[II] Food, Feeding and Digestion

Balanoglossus is a 'ciliary feeder'. Its food comprises of microscopic organisms and organic particles present in water and the bottom sand in which it makes its burrows. Lateral cilia lining the gill-slits set up a current of water which enters through the mouth, takes its course through the buccal cavity, pharynx, gill-slits and branchial sacs, and leaves through the gill pores. This is the **respiratory-cum-food current**. Some food particles directly enter the mouth with this current while some come in contact with the proboscis and get entangled in the mucus that covers it.

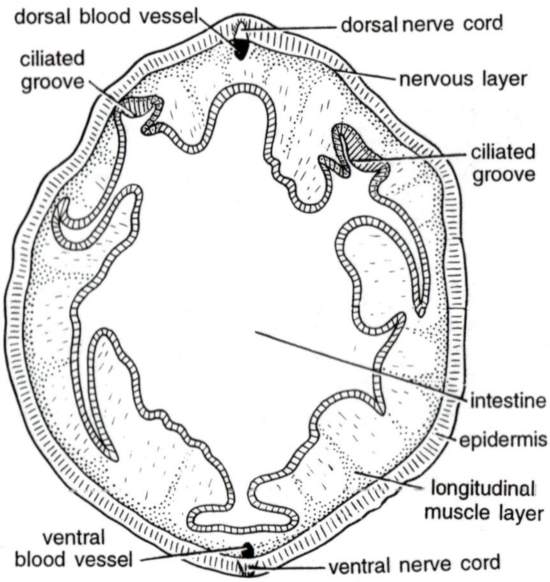

Fig. 10. *Balanoglossus.* T. S. through post-hepatic region of intestine.

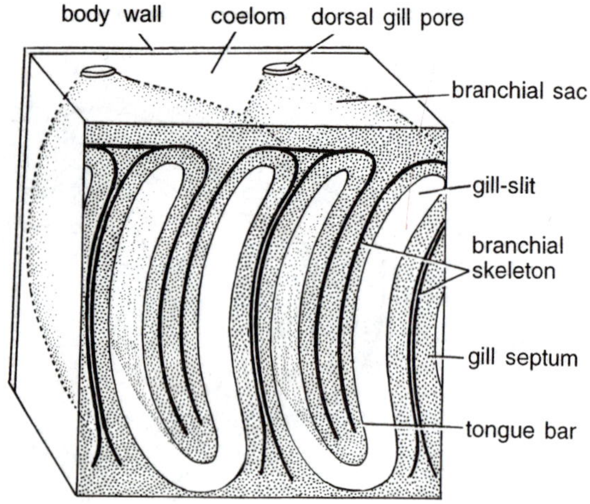

Fig. 13. *Balanoglossus.* 3-dimentional view of two gill-slits and two branchial sacs (Diagrammatic).

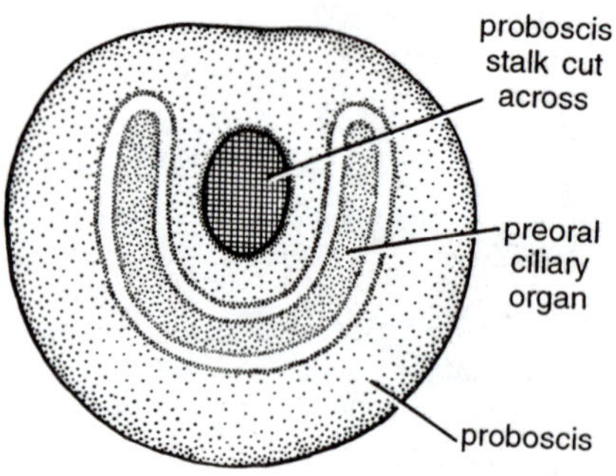

Fig. 11. *Balanoglossus.* Posterior view of proboscis.

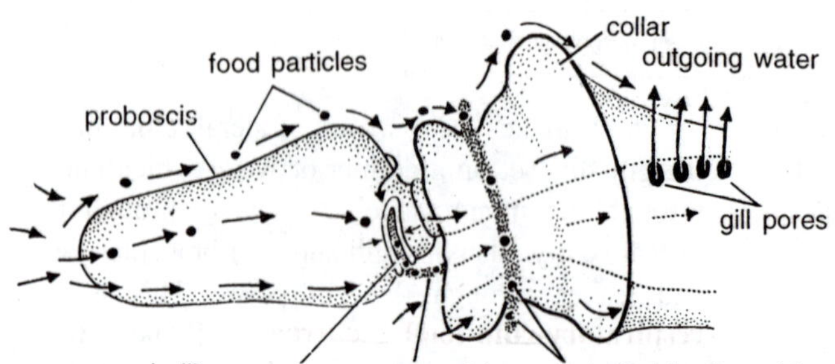

Fig. 12. *Balanoglossus.* Anterior end of body showing feeding current in lateral view. Arrows indicate the direction of the current.

Mucus is secreted by the gland cells of the proboscis epithelium. Cilia covering the proboscis direct the mucous string, containing food particles, towards the pre-oral ciliary organ at the base of the proboscis. From here the mucous string is passed back into the mouth by the action of the proboscis cilia, aided by the main water current entering the mouth. Organic particles present in the sand are ingested directly along with the latter at the time of burrowing.

U-shaped **pre-oral ciliary organ**, at the base of proboscis stalk, tests the quality of food and water entering the mouth. Undesirable substances are prevented from entering the mouth by the ventral part of the collarette which does so by covering the mouth. Thus, the rejected particles, instead of entering the mouth, pass back over the collar.

Backward movement of food through the alimentary canal is maintained by the cilia lining its walls. In the pharynx, the food moves through the ventral digestive portion. Digestion is brought about by enzymes secreted by gland cells of the pharynx, oesophagus and hepatic region of the intestine. Exact process of digestion in *Balanoglossus* is not known. Undigested substances, along with sand and silt, pass out through the anus as 'castings'.

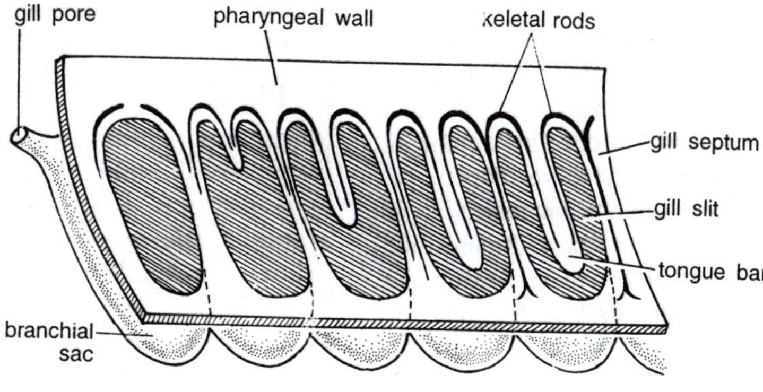

Fig. 14. *Balanoglossus.* Development of tongue bars.

RESPIRATORY SYSTEM

Respiratory apparatus of *Balanoglossus* comprises : (i) **branchial portion of pharynx** bearing **gill-slits**, and (ii) **branchial sacs** that open out through **gill-pores**.

[I] Branchial Pharynx

As already described, two lateral longitudinal **parabranchial ridges** divide the pharyngeal cavity into a ventral digestive portion and a dorsal respiratory or **branchial portion**. Dorsolaterally, on each side, the branchial portion is perforated by a longitudinal series of numerous U-shaped openings, the **gill-slits**. Their number varies and increases as the animal grows older. To start development a gill slit is a broad oval slit. Later, a hollow projection of dorsal pharyngeal wall, called **tongue bar**, grows into the slit making it U-shaped. Hollow tongue bars enclose coelomic cavity and do not touch the ventral side of gill-slits. Portions of the pharyngeal wall between two adjacent **gill-slits** are termed **gill septa**. They are solid, without enclosing coelom. A tongue bar is connected with adjacent gill septa by short transverse or horizontal connections, the **synapticula**. Development and arrangement of gill-slits is identical with that found in *Branchiostoma.*

As described earlier, the tongue bars and septa are supported internally by a chitinous skeleton forming M-shaped rods. A septum contains the middle arm of a M-rod which is bifurcated at its lower free end. Lateral arms of M-rod lie in adjacent tongue bars so that each bar contains two arms of two adjacent skeletal rods. Gill-slits are richly lined by cilia, called lateral **cilia**.

[II] Branchial Sacs

Gill-slits do not open directly outside. Each gill-slit opens into a gill pouch called **branchial**

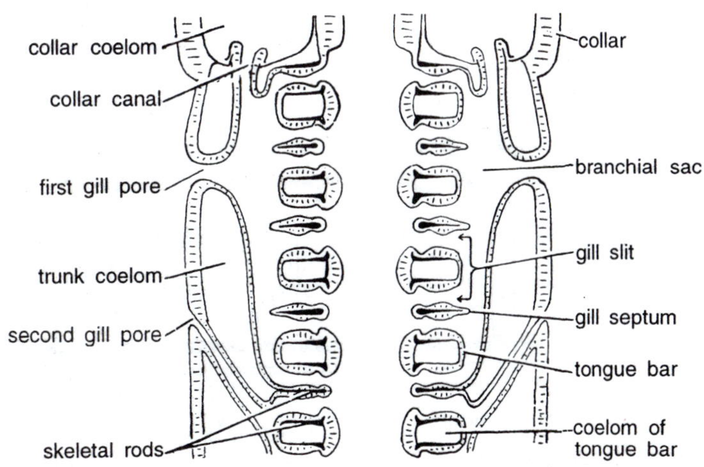

Fig. 15. *Balanoglossus misakiensis.* H.L.S. of branchial portion of pharynx to show first four gill-slits opening by a common gill pore.

sac, which lies between the body wall and the pharynx. Each branchial sac in turn opens to the exterior by a small, independent **gill pore**. However, in one species *(B. misakiensis)* the first four pouches become united to open by a common gill pore to outside. Collar coelom also communicates with the common branchial sac of its side through a **collar canal**. Gill-pores are visible externally in two longitudinal rows one on each side of the mid-dorsal ridge in the branchiogenital region of the trunk.

Mechanism of Respiration

Lateral cilia lining the gill-slits set up a food-cum respiratory current of water. It enters the pharynx through mouth, then passes through gill-slits into the branchial sacs and finally leaves through the gill pores. Tongue bars are richly vascular and participate in gaseous exchange. Blood of their capillary network takes up the oxygen dissolved in water and returns carbon dioxide to it.

BLOOD VASCULAR SYSTEM

Blood vascular system of *Balanoglossus* is of the **open** or **lacunar type**. It consists of : (i) a colourless **blood**, (ii) a **central sinus** and a **heart vesicle**, (iii) distributing vessels or **arteries** and **sinuses**, and (iv) collecting vessels or **veins**.

[I] Blood

Blood is colourless fluid containing few white corpuscles which are possibly detached endothelial cells. A respiratory pigment is probably absent. Functioning of the circulatory system is not properly understood.

[II] Central Sinus and Heart Vesicle

Central sinus is a small elongated non-contractile sinus situated in the proboscis just above the buccal diverticulum. Just above it is a closed triangular **cardiac sac** or **heart vesicle**. Its ventral wall is muscular and contracts rythmically thereby producing pulsations in the central sinus that help in the circulation of blood. Central sinus receives blood from collecting vessels that open into its posterior end. Anteriorly, it pumps its blood into several **afferent vessels** which form a plexus in the **proboscis gland** or **glomerulus**, lying in front of it. In glomerulus the blood gets rid of its excretory wastes.

[III] Distributing Vessels or Arteries

Blood from the glomerulus is carried away by four arteries. Of these, two arteries, a **mid-dorsal proboscis artery** and a **mid-ventral proboscis**

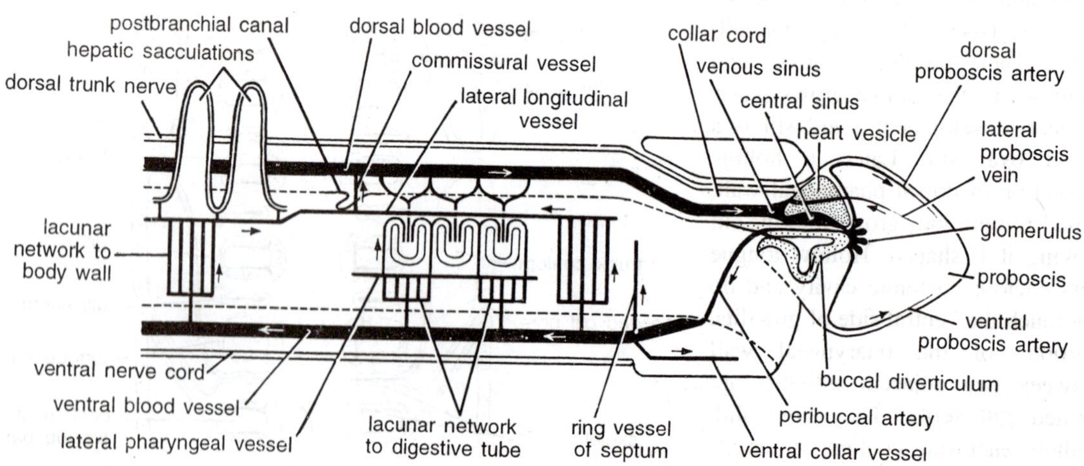

Fig. 16. *Balanoglossus.* Blood-vascular system in anterior end in lateral view.

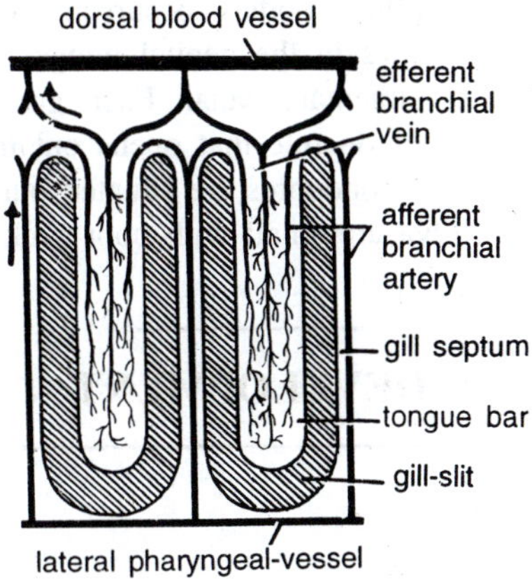

Fig. 17. *Balanoglossus.* Lymph sinusoids in a portion of the branchial wall.

artery, supply the proboscis. Other two, called **efferent glomerular arteries**, run backward along the two sides of buccal diverticulum, encircle the buccal tube as **peribuccal vessels** (which are actually of the nature of blood plexuses) and unite in a single longitudinal **ventral vessel** that runs up to the posterior end of the body through the ventral mesentery. On its way, the ventral vessel gives out a **ventral collar vessel** to the collar, a **ring vessel** to the collar-trunk septum and an **afferent branchial artery** to each gill septum in which it bifurcates to supply two adjacent tongue bars. All these branches break up into a system of sinuses in their respective structures. All along its length, the ventral vessel also supplies the body wall and gut wall by an elaborate network of sinuses. Ventral vessel has muscular contractile walls and the blood in it flows backwards.

[IV] Collecting Vessels or Veins

Blood from body wall, gut wall and branchial apparatus **(efferent branchial vessel)** is collected by a single median **dorsal vessel** which runs through the dorsal mesentery, from posterior end upto the collar. It has muscular and contractile wall and the blood flows in it forwards. At the anterior end of collar, the dorsal vessel dilates a little to form a **venous sinus**. Latter receives a lateral **proboscis vein** from each side of proboscis and then opens into the central sinus.

EXCRETORY SYSTEM

Excretory organ is **glomerulus** or **proboscis gland** lying in front of the central sinus and projecting into the proboscis coelom. It is made up of several blind tubular projections formed by the peritoneum covering the buccal diverticulum, central sinus and heart vesicle. Tubular projections contain blood confluent with that of central sinus. Excretory peritoneal cells of glomerulus contain yellow or brown granules, probably of excretory substances. From glomerulus the excretory substances pass on into the proboscis coelom and finally to the exterior through the proboscis pore.

NERVOUS SYSTEM

Nervous system is of primitive type resembling that of coelenterates and echinoderms. Throughout the body a plexus or layer of nerve cells and nerve fibres lie just below the epidermis. Nerve fibres are traversed by the filamentous basal portions of epidermal cells, and form synapses with the processes of nerve cells. Nervous layer

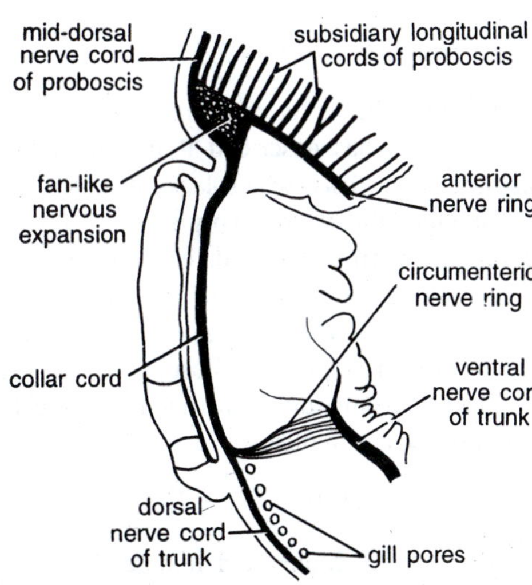

Fig. 18. *Balanoglossus.* Nerve cords in the anterior region of the body.

is thickened along definite strands to form two main **nerve cords**, one **mid-dorsal** and other **mid-ventral** which run along the entire length of the trunk. **Ventral cord** extends upto collar-trunk septum where it is connected with the dorsal cord by a circular strand, called **circumenteric nerve ring**. **Dorsal cord** extends anteriorly upto the base of proboscis where it is connected with another circular strand called **anterior nerve ring**. In collar region, dorsal cord leaves the epidermis and traverses the collar coelom as **collar cord**. It is supposed to be the nervous centre of the animal. But unlike brain, it has no concentration of nerve cells and also does not give out nerves. However, it contains some giant nerve cells which help in transportation of impulses over the body and in reflexes. Collar cord contains a cavity called **neurocoel**.

Sense organs of *Balanoglossus* are simple and comprise : (i) **neuro-sensory cells** in the epidermis of proboscis and anterior part of collar, sensitive to touch and light; and (ii) **preoral ciliary organ** situated ventrally at the base of proboscis, which is a chaemoreceptor.

REPRODUCTIVE SYSTEM

Asexual Reproduction

Asexual reproduction is rare in enteropneusts. **Gilchrist** (1923) has described it in one species, *Balanoglossus capensis,* which lacks hepatic caeca. During summer the young worms (juveniles) cut off small pieces from tail end, each regenerating into a complete sexual adult in winter.

Regeneration. *Balanoglossus* shows great power of regeneration. Proboscis, collar and isolated pieces from trunk can regenerate the lost part of the body completely.

Sexual Reproduction

Sexes are separate. Males and females cannot be identified externally except for the difference in colour of the ripe gonads in the living specimens. **Testes** of male and **ovaries** of female are similar. They are sac-like bodies occurring in several longitudinal rows in the genital wings, on either side of the alimentary canal. Each gonad has a narrow neck or ductule that opens out through a **gonopore**. All gonopores are situated outer to the gill pores. Ova are small with poor yolk content.

DEVELOPMENT

[I] Fertilization

During the breeding season (May to June) mature sperms and ova escape into the surrounding water where fertilization takes place. First the ova, as egg mass, are released by the female from the burrow and then the sperms are emitted by the male from its burrow. Sperms fertilize the eggs in open water.

[II] Pre-larval Development

Development is indirect with a larval stage, called **tornaria**. Zygote, produced as a result of fertilization, undergoes **cleavage** which is **holoblastic**, almost equal and of the radial type. It results into a sphere of blastomeres, the **morula**. Morula undergoes reorganization of its blastomeres and takes the form of a single-layered, hollow and spherical embryo, the **coeloblastula**. Its central fluid-filled cavity is called the **blastocoel**. Blastula results in about 6-15 hours after fertilization. Within 12 to 24 hours, an invagination starts in the blastula which deepens to form the **archenteron** that opens outside through a **blastopore**. Soon the blastopore closes and the embryo, now called **gastrula**, lengthens along the antero-posterior axis. Now the tip of the archenteron is pinched off as a coelomic vesicle called the **protocoel**. Remaining portion marks the future gut. The protocoel becomes triangular in shape. Its one end gets attached to the underside of the apical thickening and another end opens to outside through an aperture, the **hydropore**, towards the dorsal side of the embryo. Protocoel and hydropore are the future **proboscis coelom** and **proboscis pore**, respectively. Collar and trunk coeloms arise as solid evaginations of the hindgut, independent of the formation of protocoel.

[III] Larval Development

With the formation of the protocoel the inner end of the early gut moves towards the ventral surface and opens outside through a mouth. Gut is now regionated into the oesophagus, stomach and intestine; intestine opens to outside through an anus, formed at the side of the closed blastopore. By this time the embryo becomes uniformly ciliated and escapes from the egg membrane to

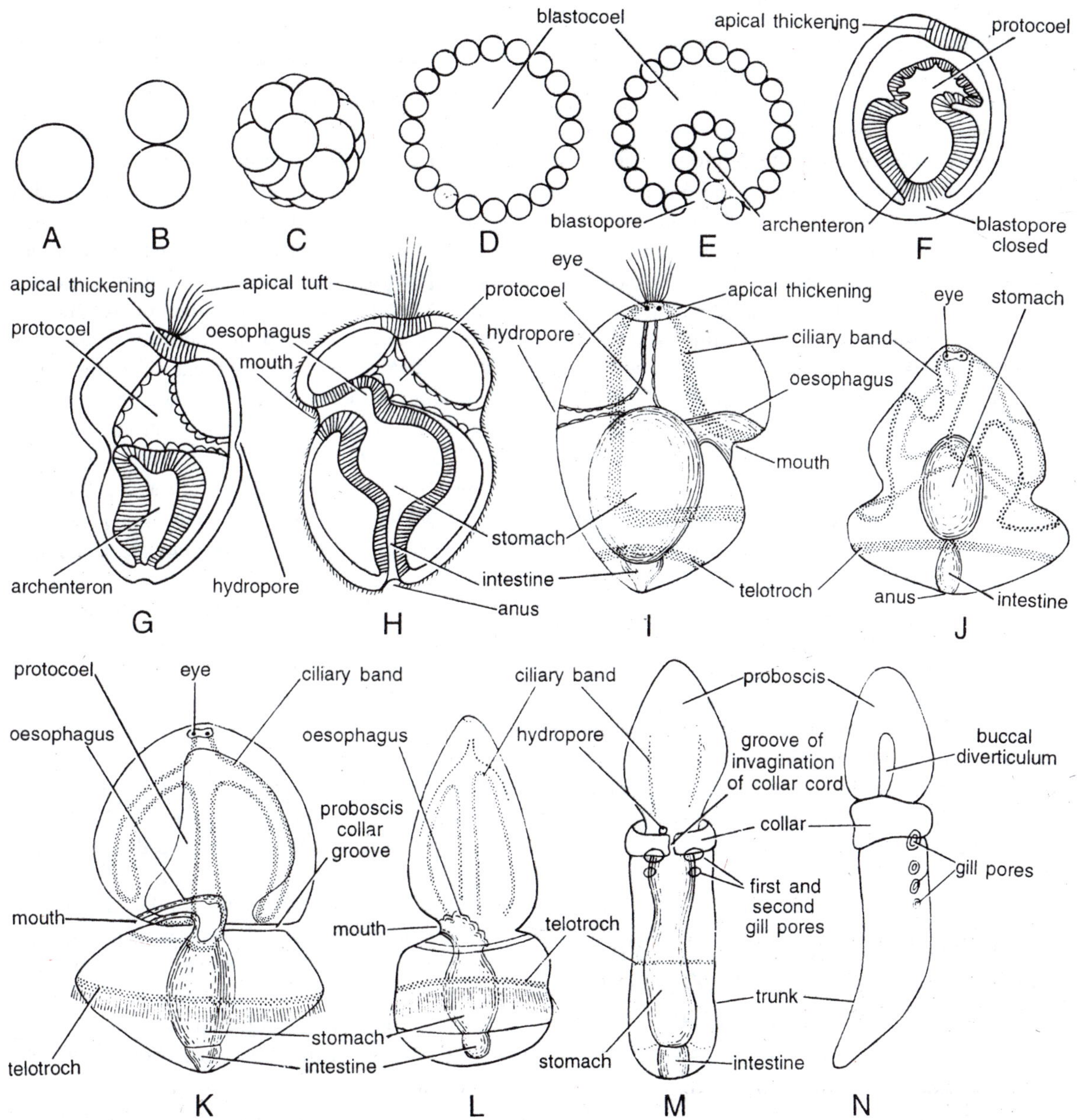

Fig. 19. *Balanoglossus.* Stages of development. **A.** Zygote. **B & C.** Cleavage. **D.** Coeloblastula. **E.** Early gastrula. **F** to **H.** Gastrula development. **I.** Young tornaria. **J.** Adult tornaria. **K** to **N.** Development of tornaria into a young worm.

lead a free swimming larval life. It is called **tornaria**.

Tornaria larva. Tornaria larva was first described by **J. Muller** in 1850 who suspected it to be the larva of some starfish. Later on it was known to belong to *Balanoglossus clavigerus.* It is so called because of its habit of rotating in circles. It is clear, glossy in appearance with an oval body ranging upto 3 mm in size. It has a ventral mouth, a posterior anus and the larval gut differentiated into an oesophagus, stomach and intestine. Cilia form two bands on the body surface. Anterior **ciliary band** or **circumoral band** takes up a winding course over the preoral surface and forms a postoral loop; its clilia are short and serve to collect food. Posterior ciliated band or **telotroch** occurs as a ring in front of the anus; its cilia are long and serve as locomotor organs. At the anterior end is an **apical plate** of thickened epidermal cells, which bears a pair of **eye spot** or **ocelli** and a tuft of sensory cilia called **apical tuft** or **ciliary organ**. **Protocoel** (proboscis coelom) in the form of a thin-walled sac, is present and opens to the exterior through a **hydropore** (proboscis pore). To the right of the hydropore lies a pulsating **heart**. Collar and trunk coeloms appear in the older larva.

Metamorphosis. Larva swims freely, leads a planktonic life feeding on minute organisms, and metamorphoses into an adult worm. During metamorphosis, the size is reduced, the body becomes differentiated into proboscis, collar and trunk by the appearance of two constrictions, the ciliary bands are lost and the trunk region is elongated. Animal sinks to the bottom to lead a benthonic life as an adult.

AFFINITIES & SYSTEMATIC POSITION OF *BALANOGLOSSUS* (HEMICHORDATA)

Group *Enteropneusta,* to which *Balanoglossus* belongs, was established by **Gegenbaur** in 1870. **Bateson** (1885) proposed the name *Hemichordata* in place of Enteropneusta. Since then, due to their peculiar anatomical organisation and embryology, Hemichordata (or *Balanoglossus*) have been considered closer to the chordata as well as most non-chordate phyla by different workers from time to time. Some of these views regarding the phylogenetic relationship (affinities) and taxonomic position of the Hemichordata are summarized below.

(1) Affinities with Chordata

Earlier zoologist such as **William Bateson** (1885) proposed closer affinities between Hemichordata and Chordata. Their resemblance was based on the presence of the three fundamental chordate characteristics in Hemichordata, that is, (i) a notochord, (ii) a dorsal hollow nerve cord and (iii) pharyngeal gill-slits (pharyngôtremy). Moreover, the structure and function of pharynx and branchial apparatus are similar to those of Urochordata and Cephalochordata. Also, the origin of coelom is enterocoelic in the form of five pouches from larval archenteron, as in *Branchiostoma.* Due to these similarities Hemichordata had been considered as a subphylum of the phylum Chordata till recently, representing its lowest group, and probably having a common ancestry.

Objections. However, the hemichordates are no longer included under chordates because they do not possess chordate characters in a typical condition. The main objections are :

(1) A true notochord does not occur in hemichordates. Unlike that of the chordates, the so-called 'notochord' is very short, confined to proboscis and without any supporting function. It is ventral to the main (dorsal) blood vessel and not covered by sheaths. Instead of being solid and made of vacuolated cells, it is hollow and lined by epithelial cells. It does not originate from the roof of larval archenteron but as a forward hollow projection of the foregut. Instead of being called notochord (**Bateson**, 1885) it is now termed the **stomochord** (**Willy**, 1899). Hyman prefers to name it **buccal diverticulum**.

(2) Nervous system is distinctly of the invertebrate type being intra-epidermal in position and having a ventral nerve cord and

a circumenteric nerve ring which are absent in chordates. In *Balanoglossus,* the dorsal tubular nerve cord is confined to the collar region only.

(3) Gill-slits of *Balanoglossus* are numerous and dorsal in position, whereas they are 5 to 7 and lateral in higher chordates.

Other differences. Hemichordates further differ from the chordates in :

(1) Lacking metameric segmentation, cephalization, paired appendages, postanal tail, exoskeleton, living endoskeleton, dermis, liver, haemoglobin, red blood corpuscles, etc.

(2) Having peculiar division of body and coelom (into proboscis, collar and trunk), single-layered ciliated epidermis, hepatic caeca, dorsal heart, open neurocoel, colourless blood, numerous gonads, etc.

(2) Affinities with Rhynchocephalia (Nemertinea)

Feeding and burrowing habits are similar in *Balanoglossus* and Nemertinea. Body in both is elongated, vermiform, without external metamerism, with terminal anus, with smooth skin containing unicellular glands and ectodermal nerve plexes, and having metamerically arranged simple gonads. But Nemertinea differ in lacking a dorsal nerve cord and in having lateral nerve cords and a protrusible proboscis.

(3) Affinities with Phoronida

Some zoologists like **A.T. Masterman** (1897) advocated relationship of *Balanoglossus* with *Phoronis* on the following grounds :

(1) Similar nature of epidermal nervous system.

(2) Paired gastric diverticula of *Phoronis,* like the **buccal diverticulum** of *Balanoglossus,* forming so-called notochord.

(3) Actinotroch larva of *Phoronis* has several enteropneust features of tornaria such as similar disposition of coelom, anus surrounded by a ciliary ring, presence of a proboscis pore and a sensory apical plate with cilia and eye spots.

(4) Both have great power of regeneration.

Objections. But, the chordate features of *Balanoglossus* like pharyngeal gills, are absent in *Phoronis* which also differs in having paired metanephridia. Moreover, **Selys-Long Champ's** (1940) account of development of *Phoronis* does not coroborate Masterman's observations, so that relationships of these two groups are rejected.

(4) Affinities with Pogonophora

Marcus (1958) tried to relate Hemichordata with Pogonophora due to following similarities :

(1) Enterocoelous formation of coelom.

(2) Body and coelom divided into three regions.

(3) Mesosome and metasome separated by a septum.

(4) Nervous system intra-epidermal.

(5) Pericardial sac in some pogonophores.

(6) Gonads found in trunk.

Objections. But pogonophores differ in having protocoelic nephridial coelomoducts and lacking an alimentary canal. Moreover, nervous system is concentrated in protosome in Pogonophora, but in mesosome in Hemichordata.

(5) Affinities with Annelida

Spengel (1893) first suggested affinities of Annelida and Hemichordata as follows :

(1) Body vermiform and coelomate.

(2) Burrowing habit, tubicolous life and ingesting mud which is passed out as castings through anus.

(3) Collar of *Balanoglossus* similar to **clitellum** of earthworm.

(4) Proboscis and prostomium similar and preoral.

(5) Similar arrangement of blood vessels with blood flowing anteriorly in dorsal vessel and posteriorly in ventral vessel.

(6) Dorsal position of heart.

(7) Tornaria larva of *Balanoglossus* shows several structural resemblances with the trochophore larva of Annelida in being pelagic, ciliated, with apical plate, eye spots, sensory cilia and well developed alimentary canal with similar parts.

Objections. However, the two groups show striking differences as follows :

(1) Annelids do not have pharyngeal gill-slits, stomochord or buccal diverticulum and dorsal tubular nerve cord found in *Balanoglossus.*

(2) *Balanoglossus* does not have double and solid ventral nerve cords and nephridia found in annelids.

(3) In tornaria larva of *Balanoglossus,* preoral or proboscis coelom is present, nephridia are absent and blastopore becomes anus of the adult (Deuterostomia). In trochophore larva of annelids, preoral coelom is absent, nephridia present and the blastopore becomes the mouth (Proterostomia).

Thus, compared to their great fundamental differences, the similarities of the two groups are only superficial and quite insignificant indicating probably a convergent evolution due to similar habits and habitat.

(6) Affinities with Echinodermata

Adult resemblances. Adult hemichordates and echinoderms are structurally quite different and it is difficult to suspect any phylogenetic relationship between them. They show few resemblances such as :

(1) Enterocoelic origin of coelom and its division into three successive parts filled with sea water to serve a hydraulic mechanism.

(2) Heart vesicle and glomerulus of enteropneusts are considered homologous to the dorsal sac and axial gland of echinoderms. Both the structures are related and combine vascular and excretory functions.

(3) Nervous system is poorly developed and forms epidermal nerve plexus.

(4) Proteins and phosphagens present in hemichordates closely resemble those of echinoderms.

(5) Common habits and ecological niches and remarkable power of regeneration.

Larval resemblances. Two groups show a strong affinity on embryological ground as the tornaria larva of *Balanoglossus* has a striking structural similarity with an echinoderm larva, in particular the bipinnaria larva of asteroids. In fact, the tornaria was regarded an echinoderm larva for a long time by **Johannes Muller** (1850), **Krohn** (1854), **Agassiz** (1864), etc., till **Metschnikoff** (1870) proved it to be an enteropneust larva. Larvae of the two groups possess the following common features :

(1) Small, pelagic, transparent and oval.

(2) Identical ciliated bands taking up a similar twisted course.

(3) Enterocoelic origin and similar development of coelom.

(4) Proboscis coelom opening to outside by proboscis pore of tornaria comparable to hydrocoel of echinoderm dipleurula.

(5) Blastopore becomes the anus (Deuterostomia) and digestive tract is complete with mouth, anus and same parts.

Objections. However, the tornaria larva shows presence of apical plate with sensory hairs and eye spots and telotroch which are absent in echinoderm larvae. Protocoel is single in tornaria but paired in echinoderm larva. This raises doubts about the echinoderm affinities of hemichordates. **Fell** (1963) and others believe that their larval similarities are only because of convergent evolution due to same mode of habits and habitat.

(7) Systematic Position and Phylogeny

Peculiar anatomical organisation of *Balanoglossus* or hemichordates makes their systematic position uncertain and controversial. Earlier workers (**Bateson**, 1885) placed them as a **subphylum** under the phylum Chordata representing its lowest group. But the only chordate feature shown by them is the presence of pharyngeal gill-slits. Therefore, some recent workers like **Van der Host** (1939), **Dawydoff** (1948), **Marcus** (1958) and **Hyman** (1959) have chosen to remove hemichordates from the phylum Chordata and treat them as an **independent invertebrate phylum**. Since the group comprises only about 80 species, it is included in the category of a **minor phylum**. Name 'Hemichordata' is however, retained for the group as it suggests that its members are related to chordates, *i.e.,* they are 'half' or 'part' chordates, a fact that is undisputed.

Regarding phylogeny, the close affinities of Echinodermata, Phoronida, Pogonophora, Hemichordata and Chordata have led to the conclusion that they have arisen from a common ancestral stock, probably the dipleurula larva (**Bather**, 1900). But **Berril** (1955), **Whitear** (1957), **Carter** (1958), **Marcus** (1958), **Hyman** (1959, **Bone** (1960) and many others do not contribute to this view. **Barrington** (1965) has interpreted their views based on the deuterostome line of chordate evolution. According to his explanation, the Echinodermata deviated greatly from the ancestral stock and formed a blind branch. Hemichordata also did not stand on the direct line of ancestry but formed a divergent offshoot from the main line of chordate evolution. Since the hemichordates arose from the ancestral line after the divergence of the ancient Echinodermata but before the rise of the true chordates, they are often called the **prechordates**

Important Questions

➤➤ *Long answer type questions*

1. Give a general account of the internal anatomy *Balanoglossus.*
2. Describe the alimentary canal and mode of feeding in *Balanoglossus.*
3. Describe the circulatory system *of Balanoglossus.*
4. Discuss the affinities and taxonomic position of *Balanoglossus.*
5. Draw neat and well-labelled following diagrams of *Balanoglossus :*
   (i) M.L.S. of anterior end, (ii) T.S. proboscis, (iii) T.S. collar region, (iv) T.S. branchio-genital region.
6. Write short notes on : (i) Buccal diverticulum, (ii) Pygochord, (iii) Proboscis skeleton, (iv) Tornaria larva.

➤➤ *Short answer type questions*

1. State the position and functions of the gonads in *Balanoglossus.*
2. What is the name of larva of *Balanoglossus* ?
3. What is proboscis complex ?
4. Discuss the systematic position of *Balanoglossus.*
5. Classify *Balanoglossus* with full justification.
6. Give an account of the structure of the pharynx and the process of respiration in *Balanoglossus.*
7. Discuss the affinities of *Balanoglossus.*
8. Describe the external characters of *Balanoglossus* and give your reasons for including it in the phylum chordata.
9. Draw a labelled diagram of the vertical longitudinal section through the anterior half of *Balanoglossus.*
10. Give the labelled diagram of tornaria larva of *Balanoglossus.*
11. Draw a labelled diagram of the T.S. through proboscis of *Balanoglossus.*

➤➤ *Multiple choice questions*

1. Who coined the name *Balanoglossus clavigerus :*
   (a) Della chiaja (b) Linnaeus
   (c) Aristotle (d) Darwin
2. Tongue worm is the name of :
   (a) *Saccoglossus* (b) *Balanoglossus*
   (c) *Polychodera* (d) *Spengelia*
3. Largest species of *Balanoglossus :*
   (a) *B. australiensis* (b) *B. misakiensis*
   (c) *B. capensis* (d) *B. gigas*
4. Integument of Hemichordates :
   (a) single layered epidermis and dermis
   (b) single layered epidermis, dermis absent
   (c) single layered epidermis, nervous layer and basement membrane
   (d) single layered epidermis, dermis, nervous layer and basement membrane
5. Coelom is :
   (a) enterocoelous (b) schizocoelous
   (c) both
   (d) none
6. Proboscis complex includes :
   (a) buccal diverticulum and heart vesicle
   (b) buccal diverticulum and central sinus
   (c) buccal diverticulum, central sinus, heart vesicle and glomerulus
   (d) central sinus, heart vesicle, glomerulus and protocoel
7. Nerve centre of the *Balanoglossus :*
   (a) ventral nerve cord
   (b) dorsal nerve cord
   (c) circumenteric nerve ring
   (d) collar cord

8. Larva of *Balanoglossus* :
   (a) Doliolaria (b) Trochophore
   (c) Tornaria (d) Parenchymula
9. *Balanoglossus* belongs to the class :
   (a) Pterobranchia (b) Planctospheroidea
   (c) Enteropneusta (d) Tetrabranchiata
10. Connecting link between invertebrate and chordata :
    (a) *Amphioxus* (b) *Balanoglossus*
    (c) *Echinodermata* (d) *Herdmania*
11. The endoskeleton of *Balanoglossus* contain :
    (a) buccal diverticulum (b) proboscis skeleton
    (c) branchial skeleton (d) pygochord
    (e) all the above
12. *Balanoglossus* is a :
    (a) predator (b) bottom dweller
    (c) ciliary feeder (d) passive feeder
13. Gill slit in *Balanoglossus* open into :
    (a) outside the body (b) in operculum
    (c) branchial sac
    (d) none of the above
14. Excretory organs in *Balanoglossus* :
    (a) glomerulus (b) kidney
    (c) nephridia (d) solenocytes
15. Regeneration power occurs in :
    (a) *Hydra* (b) *Asterias*
    (c) *Balanoglossus* (d) all

Answers

➤➤*Multiple choice questions*

1. (a)	2. (b)	3. (d)	4. (c)	5. (a)	6. (c)	7. (d)	8. (c)	9. (c)	10. (c)
11. (e)	12. (c)	13. (c)	14. (a)	15. (d)					

66

Hemichordata : Characters, Classification and Types

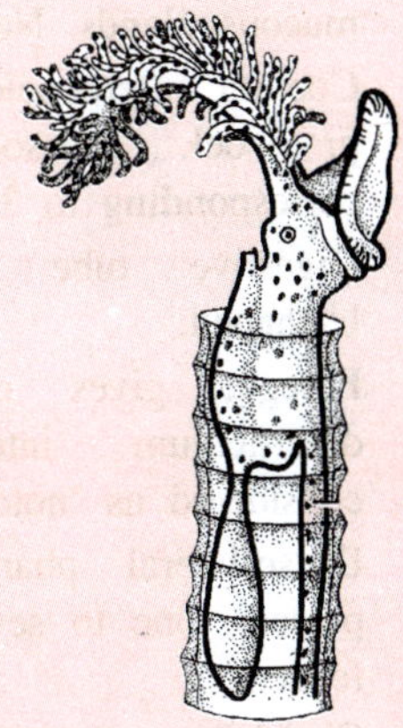

Hemichordates are usually described as, 'primitive chordates' or 'invertebrate chordates'. This phylum of wormlike animals of shallow ocean bottom was considered closely related to the chordates. Alliance with the chordates was based on the presence of gill slits and the so called notochord. It is now generally agreed that the hemichordate 'notochord' is neither analogous nor homologous with the chordate notochord and that except the common possession of pharyngeal clefts the two groups are dissimilar. Hemichordates have thus been removed from chordates and been given the rank of a separate phylum.

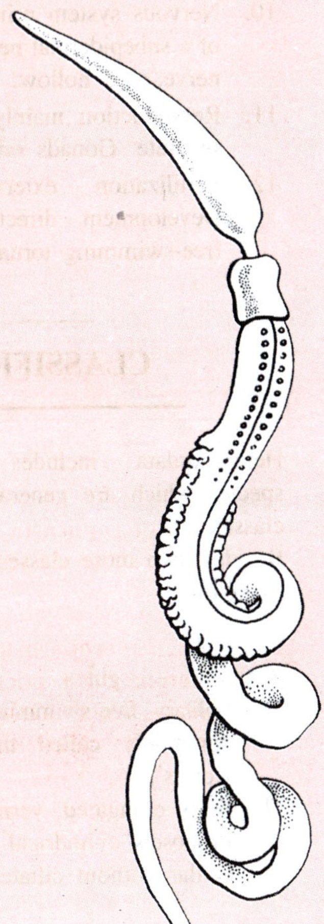

DEFINITION

Name **Hemichordata** (Gr., **hemi**, half; **Chorde**, Cord) means they are 'half' or 'part' chordates, a fact that is undisputed. They are vermiform, solitary or colonial enterocoelous coelomate animals having intra-epidermal nervous system and a pre-oral gut with or without gill-slits and without typical nephridia.

GENERAL CHARACTERS

1. Exclusively marine, solitary or colonial, mostly tubicolous.
2. Body soft, fragile, vermiform, unsegmented, bilaterally symmetrical and triploblastic.

3. Body wall of a single-layered epidermis with mucous glands. No dermis.
4. Coelom enterocoelous, usually divided into protocoel, mesocoel and metacoel, corresponding to 3 body regions.
5. Digestive tube complete, straight or U-shaped.
6. Foregut gives out a hollow buccal diverticulum into proboscis, earlier considered as 'notochord'.
7. Dorso-lateral pharyngeal gill-slits, when present, one to several pairs. Ciliary filter feeders.
8. Circulatory system simple and open, including a dorsal heart and two longitudinal vessels, one dorsal and one ventral.
9. Excretion by a single proboscis gland or glomerulus connected to blood vessels.
10. Nervous system primitive consisting mainly of a subepidermal nerve plexus. Dorsal collar nerve cord hollow.
11. Reproduction mainly sexual. Sexes usually separate. Gonads one to several pairs.
12. Fertilization external in sea water. Development direct or indirect with a free-swimming tornaria larva.

CLASSIFICATION

Hemichordata includes about 80 known species which are generally grouped under two classes, **Enteropneusta** and **Pterobranchia**. Besides, two more classes are included by some, as below :

CLASS 1. Enteropneusta

(Gr., **enteron**, gut + **pneustos**, breathed)

1. Solitary, free-swimming or burrowing animals, commonly called the 'acorn' or 'tongue worms'.
2. Body elongated, vermiform, with no stalk.
3. Proboscis cylindrical and tapering.
4. Collar without ciliated arms (lophophore).
5. Alimentary canal straight. Mouth and anus at opposite ends. Filter feeding.
6. Several pairs of U-shaped gill-slits.
7. Sexes separate. Gonads numerous, sac-like.
8. Development includes tornaria larva in some. Asexul reproduction lacking.

Examples : *Balanoglossus, Saccoglossus (= Dolichoglossus), Protoglossus, Ptychodera, Spengelia.*

CLASS 2. Pterobranchia

(Gr. **pteron**, feather + **branchion**, gill)

1. Solitary or colonial, sessile and tubicolous animals living inside secreted chitinous tubes.
2. Body short, compact, with stalk for attachment.
3. Proboscis shield-like.
4. Collar bearing ciliated arms (lophophore).
5. Alimentary canal U-shaped. Anus dorsal lying near mouth. Ciliary feeding.
6. Gill-slits one pair or absent, never U-shaped.
7. Sexes separate or united. Gonads 1 or 1 pair.
8. Development direct or with a larval stage. Asexual reproduction by budding in some.

Order 1. Rhabdopleurida

1. Colonial, zooids connected by a stolon.
2. Collar with two tentaculated arms.
3. Gill-slits absent.
4. Gonad single.

Example : Single genus *Rhabdopleura.*

Order 2. Cephalodiscida

1. Solitary or several zooids living unconnected in a common gelatinous case.
2. Collar with several tentaculated arms.
3. Gill-slits single pair present.
4. Gonads single pair present.

Examples : *Cephalodiscus, Atubaria.*

CLASS 3. Planctosphaeroidea

This class is represented by a few small, rounded, transparent and pelagic larvae, supposed to be specialized tornaria of some unknown hemichordate termed *Planctosphaera pelagica.* The larval body is covered by extensively branched ciliary bands and its alimentary canal is L-shaped.

CLASS 4. Graptolita

The fossil graptolites (*e.g., Dendrograptus*) were abundant in Ordovician and Silurian periods and often placed as an extinct class under Hemichordata. Their tubular chitinous skeleton and colonial habits show an affinity with *Rhabdopleura*.

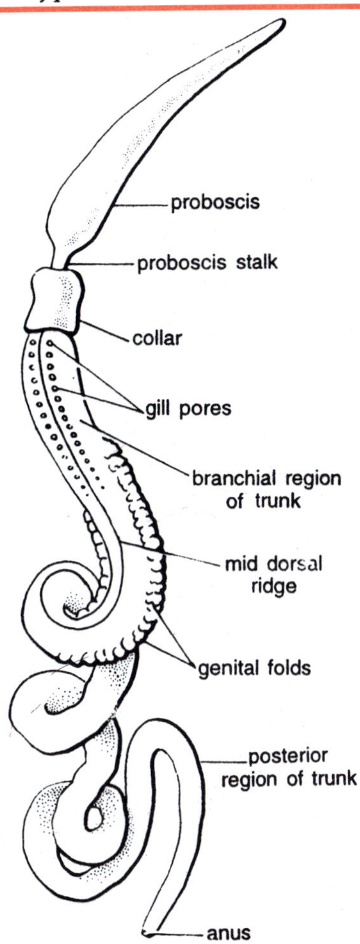

Fig. 1. *Saccoglossus.*

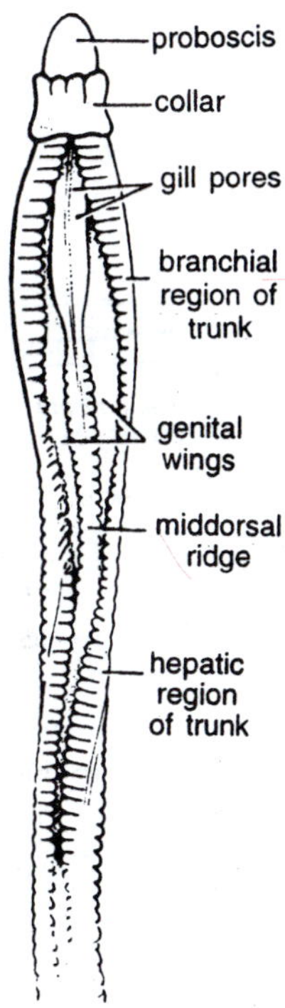

Fig. 2. *Ptychodera.*

OTHER HEMICHORDATES

1. *Saccoglossus* (=Dolichoglossus). It is a typical enteropneust genus very much similar to *Balanoglossus* in habitat, habits and structure. It is a marine, slender, soft-bodied tubicolous tongue worm living in spirally twisted burrows. Body has the usual three divisions : proboscis, collar and trunk. Proboscis is exceptionally longer and pointed than in other tongue worms. The posterior rim of collar hangs like operculum over the anterior end of trunk covering first 3 or 4 pairs of gill-pores. Genital wings and hepatic caeca, so well-developed in *Balanoglossus,* are absent. Mature gonads are yellow in male and grey in female and their position marked externally by dorso-lateral genital folds in the middle part of the trunk. Synapticula are not present so that tongue bars hang freely in their gill-slits. Development is direct without a free-swimming tornaria larva. It occurs almost universally. *Saccoglossus pygmaeus,* measuring 2 to 3 cm in length, represents the smallest known species of Enteropneusta.

2. *Ptychodera.* This genus also bears close resemblance to *Balanoglossus,* ecologically as well as morphologically and embryologically. Its proboscis and collar are somewhat shorter, but the trunk possesses conspicuous genital wings and hepatic sacculations. Development is indirect involving a free-swimming tornaria larva.

3. *Rhabdopleura.* It is a marine, sedentary and colonial pterobranch mainly found in the North Atlantic. The colony consists of horizontal branching gelatinous tubes, forming the **coenoecium**, which remain attached on hard substratum such as stones, corals, mollusc shells, sponges, etc. Erect tubes, about 6-7 mm in height, arise at short intervals, each housing an individual or zooid of the colony. The tubes are ringed, membranous and secreted by the zooids. A zooid is minute, hardly 1 mm long and occupies the distal part of the tube. Proboscis is disc-shaped. Collar bears a pair of hollow elongated arms beset with numerous fine ciliated tentacles for food collection. Alimentary canal is U-shaped so that anus lies near mouth. Gill-clefts and glomerulus are absent. Sexes are separate but a colony has both male and female individuals. A single gonad is present on the right side of the trunk. Basally the trunk of each zooid is attached by a long contractile muscular **stalk** to a common cord of

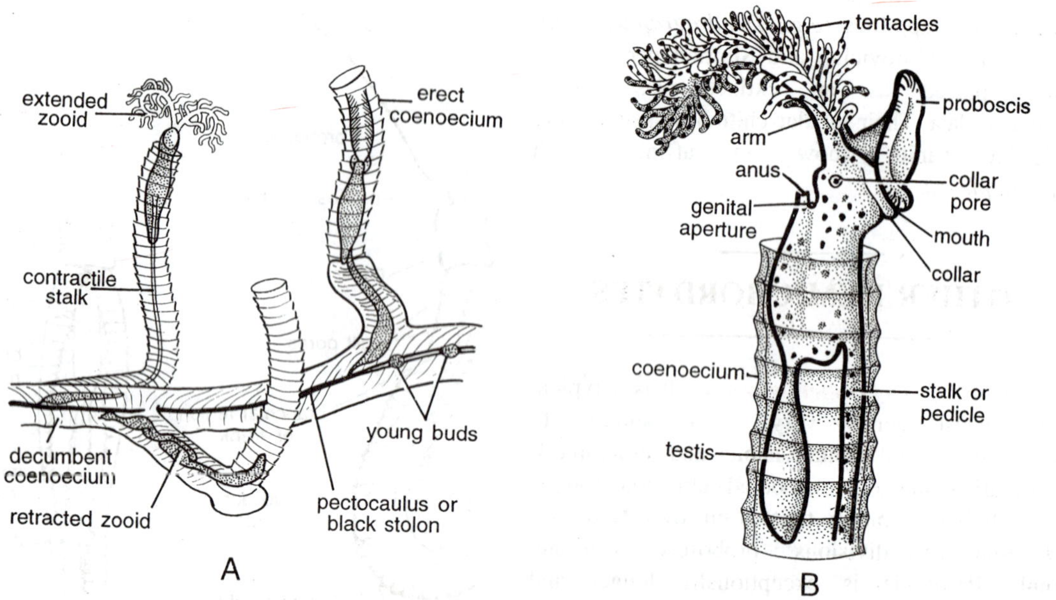

Fig. 3. *Rhabdopleura.* **A.** A portion of colony. **B.** An individual zooid.

living substance, called **black stolon** or **pectocaulus**, running inside the horizontal tubes. The stalks can quickly contract spirally so as to withdraw the zooids into coenoecium for protection. The black stolon also forms new individuals asexually by budding.

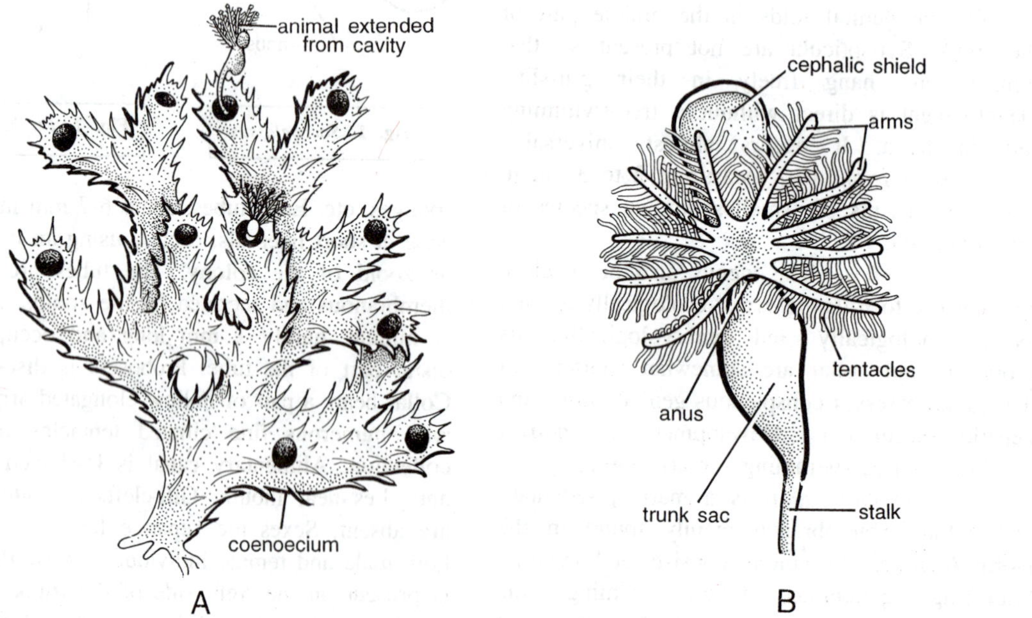

Fig. 4. *Cephalodiscus.* **A.** A part of colony. **B.** An individual zoioid.

4. *Cephalodiscus*. It is a sedentary and gregarious pterobranch found mainly in the seas of Southern Hemisphere at depths of 50 to 650 m. Several zooids live in separate upright gelatinous tubes secreted by them and embedded in a common matrix called **coenoecium**, fixed permanently to substratum. Foreign materials such as sand grains, sponge spicules, molluscan shells, etc. also adhere to the coenoecium. The zooids remain unconnected organically and thus do not constitute a true colony. Each zooid is 2 to 3 mm long and has the usual three body divisions : proboscis, collar and trunk. Proboscis is sheild-shaped overhanging the mouth. Its cavity, the proboscis coelom, opens out through two proboscis pores. The collar bears 8-16 hollow arms (lophophore), which in female are beset with numerous fine, pinnately arranged and heavily ciliated tentacles used for food capture. The tips of tentacles bear glandular knobs. The trunk is short and plump bearing a single pair of gill-slits without skeletal support. Alimentary canal is U-shaped with ventral mouth and dorsal anus located at the distal end. A narrow elongated contractile stalk arising from the aboral end attaches the zooid to its tube. Sexes are separate, gonads a single pair and development direct. Asexual reproduction also takes place by buds arising from stalk and soon becoming free.

5. *Atubaria*. Sato first described *Atubaria* in 1936. It is a sedentary and solitary pterobranch genus clinging to hydroid colonies by its long stalk and closely resembling *Cephalodiscus.* A coenoecium is lacking. The zooid measures 1-5 mm in length with usual three divisions of the body. The collar carries four pairs of tentaculated arms of which second pair distally has rod-like terminations devoid of tentacles. One pair of pharyngeal gill-slits are present.

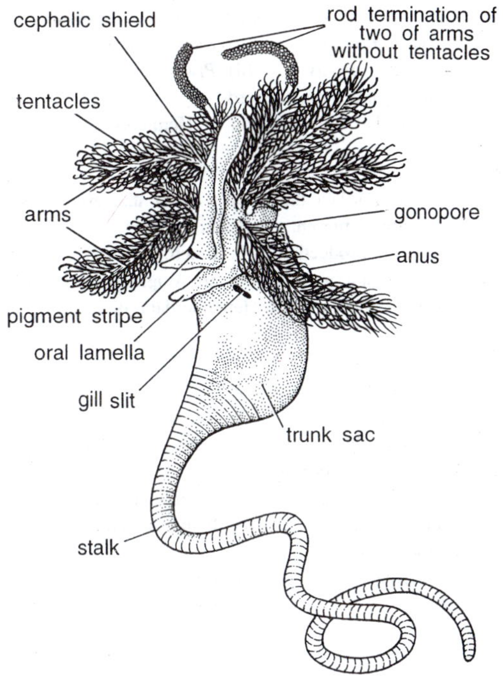

Fig. 5. *Atubaria.*

Important Questions

➤➤*Long answer type questions*

1. Give characters and classification of phylum Hemichordata up to orders.
2. Write short notes on : (i) *Balanoglossus,* (ii) *Cephalodiscus,* (iii) *Rhabdopleura.*

➤➤*Short answer type questions*

1. Define hemichorda.
2. What is the coelomic condition of Hemichordata ?
3. Give an account of classification of Hemichordata.
4. What do you mean by Enteropneusta ?
5. Give a short note on *Saccoglossus.*
6. Write down the taxonomic condition of *Rhabdopleura.*
7. Distinguish between Enteropneusta and Pterobranchia.
8. What is the excretory organs in Hemichordata ?
9. Where is the fertilization occurs in Hemichordata ?
10. Give classification of Cephalodiscus.

11. Give the taxonomic condition of the followings :
    (a) *Saccoglossus* (b) *Rhabdopleura*
    (c) *Cephalodiscus* (d) *Ptychodera*
12. Describe the stomatochord.
13. How is the body division in Hemichordata.

➤➤Fill in the blanks

1. In Hemichordata the body wall of a......epidermis.
2. ......coelome present in Hemichordata.
3. Asexual reproduction......in Enteropneusta.
4. Gill slits and gonads both are single pairs in order......
5. Name of the larva of Hemichordata is ....

➤➤Multiple choice questions

1. Larva of *Balanoglossus* is :
   (a) bipinnaria larva (b) tornaria larva
   (c) tadpole larva (d) trochophore larva
2. *Enteropneusta* belong to :
   (a) Hemichordata (b) Cephalochordata
   (c) Urochordata (d) Echinodermata
3. Protochordata are :
   (a) marine (b) fresh water
   (c) both (d) none
4. Stomochord present in :
   (a) *Amphioxus* (b) *Herdmania*
   (c) *Balanoglossus* (d) none the above
5. Proboscis gland is the excretory organ in :
   (a) *Balanoglossus* (b) *Herdmania*
   (c) *Amphioxus*

Answers

➤➤Fill in the blanks

1. single layered, 2. Enterocoelous, 3. lacking, 4. cephalodiscida, 5. tornaria.

➤➤Multiple choice questions

1. (b) 2. (a) 3. (a) 4. (c) 5. (a)

Glossary of Technical Terms

Abductor. A muscle that draws an appendage away from the mid-line of the body.
Aboral. Away from the mouth.
Absorption. Taking in of substances by cells or vessels.
Acanthocephala. A phylum of parasitic spiny-headed worms.
Acclimatize. To become accustomed to a new environment.
Acellular. Without cellular organization.
Acinus. A small terminal sac in a multicellular gland.
Acoelomate. Without coelom.
Adambulacral. Pertaining to the side towards or upon the ambulacral groove.
Adaptation. Fitness of an organism for its environment.
Adductor. A muscle that draws an appendage towards the mid-line of the body.
Adenosine triphosphate. Energy reservoir of the cell. Abbreviated as ATP.
Adipose. Referring to a tissue in which fat is stored.
Aerial. Inhabiting or travelling in the air.
Aerobe. Organism that requires free air (aerobic respiration).
Afferent. A vessel or structure leading to or towards a given position.
Agamete. An asexual structure in the embryogeny that develops into a complete individual.
Agamogony. A method of asexual reproduction.
Agamont. A gametes producing parent body.
Alternation of generations. Alternation of sexual (haploid) and asexual (diploid) generation in life cycle.
Ambulacral groove. A groove in the arm of an echinoderm lined by tube feet.
Amitosis. A type of cell division that does not involve nuclear changes.
Ammonotelic. An animal whose nitrogenous excreta is mainly ammonia.
Amoeboid. Putting forth pseudopodia like an *Amoeba.*
Amphibious. Capable of living on land as well as in water.
Amphiblastula. A larval form in the development of *Sycon* sponge.
Amphimixis. Union of male and female gametic nuclei to form a zygote nucleus.
Ampulla. A small bladder-like sac.
Anaerobic. Oxygen independent process or organisms.
Analogous. Features similar in function but different in origin.
Anal pore. Weak point in cell membrane for elimination of solid wastes.

Anastomosis. Union of two or more vessels.
Anatomy. Study of the gross structure of an organism.
Angstrom. One thousandth of a micron. Abbreviated as Å.
Anisogamy. A method of sexual reproduction involving fusion of dissimilar gametes.
Annelida. A phylum of segmented worms.
Anoxybiotic. Same as anaerobic.
Antenna. A sensitive feeler from the animal's head.
Anterior. Forward-moving end of animal.
Anticoagulant. A substance that prevents coagulation of blood.
Apical. At the apex or terminus.
Appendage. A movable projecting part of the body.
Arboreal. Tree dwelling.
Archenteron. The central cavity of gastrula, lined with endoderm, which forms the rudiment of the digestive system.
Areolar. Containing minute air-filled spaces.
Arthropoda. A phylum of segmented invertebrates with jointed legs.
Articulation. A joint between two structures.
Aschelminthes. A phylum of pseudocoelomate worms.
Asexual. Without sex, not related to gametes.
Assimilation. Process by which digested food is changed into protoplasm.
Asymmetrical. Without symmetry.
Atoll. Coral reef in the form of a ring.
Auditory. Relating to sense of hearing.
Autotrophy. Self manufacturing of organic nutrients from inorganic raw material.
Axon. Nerve fibre that conducts nerve impulses away from cell body.
Benthos. Flora or fauna of the bottom of oceans or lakes.
Bilateral symmetry. A symmetry with similarity on both sides.
Bilateria. Metazoans with bilateral symmetry.
Binary fission. Division of a unicellular organism, by mitosis, into two daughter individuals.
Binomial nomenclature. System of naming organisms by the combination of the names of genus and species.
Bioluminiscence. Emission of light by living organisms.
Biramous. Possessing two branches.
Bivalve. Possessing two shell-valves.
Bladder. A gas or fluid filled, thin-walled sac.
Blastocoel. Fluid-filled cavity of the blastula.
Blastopore. Mouth-like opening of the gastrula.
Blastula. An embryonic stage with blastocoel.
Blepharoplast. Basal granule (centriole) at the base of a flagellum or cilium.
Brownian movement. Motion of small particles in solution or suspension resulting from their being bumped by water molecules.
Bryozoa. A phylum of sea-weed like invertebrates : moss animals.
Bud. Extension of the body wall.
Budding. Form of asexual reproduction; growth from mother cell which remains attached to or separates from mother cell.
Calcareous. Made of calcium salt.
Carapace. Hard covering of the crustacean cephalothorax.

Carnivorous. Meat or flesh eating.
Castings. Undigested solids containing soil.
Caterpillar. Larval stage in the metamorphosis of moths and butterflies.
Caudal. Relating to the tail or posterior part of the body.
Cellular. Pertaining to or consisting of cells.
Cephalic. Relating to or towards the head.
Cephalothorax. The fused head and thorax.
Cercaria. A free swimming larval stage of trematode parasite, which encysts in a fish.
Cervix. Neck-like structure.
Chelate. Claw-possessing appendage.
Chelicera. A pair of pincer-like head appendages of arachinids.
Chemoreceptor. A sense organ or cell that responds to chemical stimulus.
Chemotaxis. A movement response to a chemical stimulus.
Chemotheraphy. Study of the use of drugs and other substances of chemical nature in the treatment of specific discases.
Chiasmata. A cross over; usually refers to opposite chiasma.
Chitin. A polysaccharide-protein horny substance, that forms the exoskeleton of arthropods.
Chloroplast. A chlorophyll bearing cytoplasmic organelle that acts as a site of photosynthesis.
Chromatid. Either of the two identical halves of a chromosome.
Chromatin. Deeply staining material of the cell nucleus, whether dispersed as a network or condensed as chromosomes.
Chromomere. An individual chromatin granule in a chromosome.
Chromonema. A slender thread of chromatin.
Chromosomes. Filamentous bodies in the cell nucleus which contain the hereditary units, the genes.
Chrysalis. Pupal stage of butterfly.
Cilia. Numerous small hair-like projections from the cell surface that beat in coordinated fashion.
Cirrus. Small and movable projections from a cell or body surface.
Cleavage. Early stages in division of an egg.
Clitellum. A thick glandular band around an annelid worm. It secretes the cocoon.
Cloaca. Common terminal portion of digestive, excretory and reproductive systems.
Cnidaria. A phylum of cnidoblast bearing invertebrates.
Cnidoblast. A coelenterate cell with nematocyst.
Cocoon. Protective covering around the cluster of eggs or pupa, etc.
Coelenterata. A phylum of invertebrates with persistent blastocoel (coelenteron) and cnidoblasts (stinging cells).
Coelom. Body cavity in between the body wall and gut lined by mesodermal epithelium.
Cold blooded. Having the same body temperature as that of the immediate environment.
Commensalism. Association of individuals of two or more different species to the benefit of one or more and harmful to none.

Community. A group of organisms living together by virtue of environmental needs.

Compound eye. An eye (in arthropods) with numerous ommatidia.

Conjugation. A method of sexual reproduction in which two individuals unite temporarily and exchange nuclear material: a method of reproduction found in *Paramecium*.

Copulation. A sexual act for the transference of sperm from one to another of mating partners.

Cortex. Outer layer of structure.

Cranial. Relating to the skull or to brain.

Crop. A thin walled part of the alimentary canal that serves to store the ingested food.

Cross-fertilization. Union of egg and sperm derived from different individuals of the same species.

Crossing-over. Process in which homologous chromosomes break and exchange corresponding parts.

Ctenophora. A phylum of marine invertebrates with eight rows of cilia, resembling combs.

Cuticle. Thin, non-cellular outermost covering secreted by underlying epidermis.

Cyclosis. Circular streaming of cytoplasm.

Cyst. A resistant, protective covering around the organism during unfavourable environmental condition.

Cytokinesis. Division of cytoplasm during mitosis or meiosis.

Degenerate. To lose generative ability.

Dendrite. Fibre of nerve cell that conducts impulses towards the cell body.

Dermal. Pertaining to the skin.

Diapause. Inactive stage of an insect during pupal stage.

Diffusion. Movement of molecules resulting from their kinetic energy which tends to distribute them throughout a medium.

Digestion. Conversion of complex substances of food into simple forms which can be absorbed.

Dimorphism. Occurrence of two forms within the same species.

Di-oecious. Species population in which male and female reproductive organs occur in separate individuals.

Diploblastic. Possessing two germ layers ectoderm and endoderm.

Diploid. Number of chromosomes in somatic cells.

Distal. Away from the point of origin.

Diurnal. Relating to day time.

Diverticulum. A sac-like extension of a tubular structure.

DNA. Deoxyribonucleic acid; contains hereditary information coded in specific sequence of nucleotides.

Dorsal. Pertaining to the back.

Duct. A tube for the flow of a fluid.

Ecdysis. Shedding of the outer body layer by an arthropod.

Echinodermata. A phylum of spiny-skinned marine invertebrates.

Ecology. Study of the interrelations between living organisms and their environment.

Ectoparasite. A parasite that lives on the outside of the host's body.

Efferent. A vessel or structure leading away from a given point.

Egestion. Discharge of undigested food from the alimentary canal.

Electron microscope. Microscope in which a beam of electrons is used to focus the highly magnified image of an object on a fluorescent screen or photographic plate.

Embryogeny. Development of an organism.

Embryology. Study of the development of an organism.

Encystment. Formation of a cyst wall around an organism or its any part.

Endoparasite. A parasite living within the host's body.

Endoskeleton. A supporting framework or structure within the body of an animal.

Enterocoel. A coelom formed by the outpouching of a mesodermal sac from the endoderm.

Entomology. The study of insects.

Enzyme. An organic substance (protein) that produces a specific chemical change.

Epiboly. A method of gastrulation by which the smaller blastomeres at the animal pole of the embryo grow over and enclose the cells of the vegetal hemisphere.

Epidermis. Outer layer of cells covering the surface of the body.

Epithelium. A sheet of cells, usually covering external or internal surfaces of the body.

Erythrocyte. A red blood cell.

Evagination. Outpocketing from any surface.

Excretion. Removal of metabolic wastes by an organism.

Expiration. To breathe out.

Extracellular. Outside the cell.

Fauna. Animal life of a particular area or period of time.

Fertilization. Union of a sperm with an ovum resulting in a zygote.

Filiform. Thread-like.

Fission. Process of asexual reproduction by division into two or more parts.

Flagellum. A long whip-like cytoplasmic organelle.

Food vacuole. An intracellular digestive organelle.

Fossil. Any remains of the past life.

Free-living. Capable of independent existence.

Fundus. Bottom or base of an organ.

Gamete. A reproductive cell, an egg or sperm.

Gametogenesis. Process of formation of gametes.

Ganglion. A mass of the cell bodies of nerve cells located outside the central nervous system.

Gastrodermis. Lining of coelenteron in coelenterates.

Gastrula. An embryonic stage with two germ layers, ectoderm and endoderm.

Gastrulation. Process by which a blastula becomes the gastrula.

Gemmule. An asexual reproductive body of some sponges.

Gene. A functional segment of DNA, containing hereditary information.

Genetics. Study of the heredity and variations.

Genital. Relating to reproductive organs.

Genus. A rank in taxonomic hierarchy in which closely related species are grouped together.

Glochidium. A larva of musssel and clam.

Glycolysis. Metabolic conversion of sugars into simpler compounds.

Golgi apparatus. A cytoplasmic organelle for the storage of cellular secretions.

Gonad. A gamete-producing gland, an ovary or testis.

Gregarious. Living in groups.

Habitat. Physical area of the occurrence of an animal species.

Haemocoel. Coelom filled with blood (haemolymph).

Haploid. With a single set of chromosomes, as normally present in a gamete.

Hepatic. Relating to the liver.

Herbivorous. Feeding on vegetation.

Hermaphrodite. With both male and female reproductive organs.

Heterogamy. Reproduction involving the union of two gametes that differ in size and structure.

Heterogenous. Of a different kind.

Hibernation. Dormant state of decreased metabolism.

Holoblastic. Cleavage in which an entire egg cell divides.

Holophytic. A type of nutrition involving photosynthesis.

Holozoic. A type of nutrition involving phagocytosis.

Homogeneous. Of a similar kind.

Homologous. Having the same embryonic and phylogenetic origin.

Hormone. Substance which is secreted by an endocrine gland directly into the blood stream, which controls actions in some part of the body or some body process.

Host. An organism upon which another organism (parasite) subsists for food, shelter or other benefits.

Hydranth. A coelenterate polyp with mouth and tentacles.

Hyperparasitism. Infection of a parasite by another parasitic species.

Imago. An adult insect.

Ingestion. Act of taking the food into the body.

Insecticide. Chemical agent that kills insects.

Insectivorous. Insect eating animal.

Integument. Covering of the body, the skin.

Intercellular. Between cells.

Intracellular. Within a cell.

Invagination. Infolding of any body part; infolding of a layer of tissue in gastrulation.

Invertebrate. An animal without vertebral column or backbone.

Irradiation. Exposure to ultraviolet or other type of rays.

Irritability. Ability to respond to a stimulus; a general property of protoplasm.

Isogamy. A sexual reproduction involving fusion of two similar gametes.

Jelly fish. Coelenterates belonging to the class Scyphozoa.

Joint. A place of union between two hardened structures.

Juvenile hormone. An insect hormone secreted by corpora allata of brain which is responsible for the maintenance of larval structures and inhibits metamorphosis.

Kinetodesmos. A fibril of a ciliate kinety.

Kinetosome. Basal body of a cilium or flagellum.

Kinorhyncha. A phylum of marine pseudocoelomates.

Larva. An embryonic stage which is free living and self feeding and unlike the parent.

Lamella. A thin leaf-like layer.

Lateral. Relating to the side.

Leucocytes. White blood cells.

Littoral: Region of shallow water near the shore between the high and low tide marks.

Lumen. Internal cavity within a body or structure.

Lymph. Part of the blood serum which is outside the blood vessels; bathes the cells.

Macronucleus. A large type of nucleus occurring in ciliated protozoans; controls all but sexual functions.

Madreporite. A sieve like opening on aboral surface of starfish for the entry of water into the water vascular system.

Maggot. Larval stage in the metamorphosis of the common house-fly.

Manubrium. An extension from the subumbrellar surface of jelly-fish.

Matrix. Non-living material secreted by the cells.

Medusa. A free-living umbrella like coelenterate (jellyfish); a jelly fish-like stage in the life cycle of certain coelenterates.

Meiosis. Reduction division; reduction of original number of chromosomes so that each gamete (egg or sperm) has half the original number of chromosomes, characteristic of the species.

Mesenchyme. Mesodermal connective tissue.

Mesoderm. Middle layer of cells formed during embryonic development.

Mesogloea. Gelatinous noncellular layer between the ectodermal and endodermal layers of coelenterate body.

Metamorphosis. A marked structural change from a larva into an adult.

Metazoa. Multicellular animals.

Micron (μ). One thousandth part of a millimeter.

Mitochondria. A cytoplasmic organelle, the main site of cellular respiration (Krebs cycle); the power house of a cell.

Mitosis. A method of cell division, characteristic of exact chromosome duplication.

Mollusca. A phylum of soft-bodied invertebrates.

Monera. A category of organisms without organised nucleus; bacteria and blue green algae.

Monoecious. With both male and female reproductive organs.

Morula. A solid ball of cells, the end product of cleavage.

Mutualism. Mutually helpful relationship between two organisms of different species.

Nacre. Mother of pearl; substance which forms a pearl in the irritated bivalve.

Nauplius. First in a series of larval stages in Crustacea.

Nematocyst. A structure in the nematoblast cell of a coelenterate; stinging cell.

Nephridium. A tubular excretory organ of the earthworm and other annelids which consists of a ciliated funnel, opening to the coelomic cavity.

Neuron. A nerve cell.

Nocturnal. Active at night.

Nucleic acid. A macromolecule composed of nucleotides; DNA or RNA.

Nucleous. A spherical body in the cell nucleus, rich in RNA and believed to be the site of synthesis of ribosomes.
Nucleus. An organelle within the cell which controls all cellular activity and governs heredity.
Nymph. An embryonic stage of insect that resembles the adult.
Ocellus. A small light perceiving eye, chiefly of arthropods.
Oesophagus. Gullet or food tube leading from the mouth to the stomach.
Olfactory. Pertaining to smell.
Ommatidium. A visual unit in the compound eye of arthropods.
Omnivorous. Feeding on both plants and animals.
Ontogeny. Development of the individual.
Onychophora. A phylum of caterpillar-like animals, structurally intermediate between annelids and arthropods.
Oogenesis. Process of the formation of ova.
Optic. Pertaining to light.
Oral. Relating to the mouth.
Organ. A group of different tissues working together to perform a specific function.
Organelle. Specialized structure within a cell.
Organism. A single complete living unit (animal or plant).
Osmosis. Diffusion through a differently permeable membrane.
Ostium. An opening to passage or to a canal system in sponges.
Otic. Relating to the ear.
Oviparous. Egg-laying.
Ovoviviparous. Production of eggs that are retained and hatched within mother's body.
Ovulation. Discharge of a mature ovum from the ovary.
Ovum. A female sex cell.
Oxidation. Chemical combination of oxygen with another element.
Palaeontology. Study of ancient life.
Palp. A feeler-like head appendage.
Papilla. A small nipple-like projection.
Paramylon (Paramylum). A carbohydrate, distinct from starch or glycogen, and a storage form of euglenoids.
Parapodia. Fleshy paddle-like segmented appendages of polychaete worms.
Posterior. Rear end of the body.
Parasitism. Relationship between a host and a parasite.
Parthenogenesis. Development of an egg without fertilization.
Pathogenic. Disease causing.
Pathology. Study of diseases in relation to their causative organisms.
Pedal. Relating to the foot.
Pedicillaria. Pincer-like structures on the echinoderm body for protecting skin gills.
Pelagic. Inhabitation in open water, as in mid-ocean.
Peristome. Region around an opening or mouth.
Phagocytosis. Engulfing of macromolecules by the invagination of cell membrane.
Pharynx. Part of the gut immediately behind the mouth cavity.
Pheromone. A substance secreted by an organisms to the external environment which influence the behaviour of other members of the same species.
Phototaxis. Movement response of an organism to light.
Phylogeny. Evolutionary history of an organism.
Physiology. Study of living functions.
Pinocytosis. 'Cell drinking'; engulfing of liquid by a cell.
Plankton. Minute free floating organism.
Planula. Ciliated free living larva of most coelenterates.
Plasmogamy. Mating union of cytoplasm of two individuals.
Platyhelminthes. A phylum of flat worms.
Polymorphism. Occurrence of several forms within a species.
Poikilothermic. Cold-blooded.
Polyp. A tubular coelenterate form.
Porifera. A phylum that includes sponges, pore bearing animals.
Proboscis. Any tubular prolongation of nose, lips or pharynx; the extended mouth parts of an insect.
Proglottid. A segment or body portion of tape worms.
Prostomium. Portion of the head situated before the mouth.
Protista. Unicellular organisms.
Proximal. Pertaining to nearest.
Pseudocoel. Body cavity not lined by the mesoderm.
Pseudocoelomate. With pseudocoel.
Pseudopodium. Projection of protoplasm; aids in locomotion of *Amoeba.*
Pulmonary. Relating to the lungs.
Pupa. Quiescent stage in the complete metamorphosis of an insect.
Puparium. Pupal case.
Quinine. Drug used in the treatment of malaria.
Radial symmetry. Pattern of an organism where similar parts are arranged about a common centre.
Radiata. A taxonomic group of radially symmetrical animals.
Radula. A horny rasping organ within the mouth of many molluscs.
Receptor. A sensory end organ.
Reflex. Simple, involuntary action in response to a stimulus.
Regeneration. Act of growing a new body part for one which has been injured or lost.
Renal. Pertaining to the kidney.
Reticulum. A network.
Rheotaxis. A movement response to air or water current.
Ribosome. A cytoplasmic organelle containing protein and RNA; a site of protein synthesis.
RNA. Ribonucleic acid, contains the sugar ribose and helps in protein synthesis.
Rudimentary. Pertaining to a very little development of any structure.
Saliva. Secretion of the salivary gland.
Saprophyte. Animal which lives on dead organic matter.
Schizocoel. Coelom formed by the splitting of embryonic mesoderm.
Secretion. Substances produced in the body by a cell or multicellular gland.
Segmentation. Division of a body into more or less similar parts.
Septum. An internal body partition.
Sinus. A hollow cavity.
Species. A population of interbreeding individuals.
Sperm. A male sex cell.

Spicule. Skeletal, often needle-like, element of sponges.

Spiracle. Surface opening of the tracheal system in insects.

Spore. An asexual reproductive element, usually a cell.

Statocyst. A sense organ of equilibrium.

Symbiosis. Association of two individuals of two different species for mutual benefits.

Syncytium. A multinucleate mass of cytoplasm formed by the merging of cells.

Syngamy. Union of gametes in sexual reproduction.

System. A group of organs working together to perform a particular function.

Taxis. A movement response.

Taxonomy. Study of classification and nomenclature.

Telson. Terminal body part of certain arthropods.

Tentacles. Flexible arm like projections of coelenterates and some other animals.

Testis. A male gonad.

Thermotaxis. A movement response to temperature.

Thigmotaxis. A movement response to touch.

Tissue. A group of similar cells that perform the same function.

Trachea. Respiratory tube of an insect.

Trichocyst. An adhesive cytoplasmic organelle of ciliate Protozoa.

Trochophore. A free swimming ciliated larva characteristic of Annelida and Mollusca.

Ultramicroscopic. So small that it is beyond the microscope.

Umbo. A protuberance of a shell valve in bivalves.

Univalve. Molluscs with a single shell.

Ureotelic. Animal whose nitrogenous excreta contains mainly urea.

Uricotelic. Animal whose nitrogenous excreta contains mainly uric acid.

Vacuole. A membrane bound space within the cell containing solid matter or fluid or both.

Valve. A structure that limits or closes an opening.

Veliger. A larval stage of many molluscs.

Velum. A membrane found on the subumbrellar surface of some jelly fish.

Ventral. Pertaining to the lower or belly side.

Vestibule. A passage.

Viviparous. An animal whose female gives birth to young.

Zoaea. A kind of crustacean larva.

Zoology. Science that deals with animal life.

Zygote. Resultant of the fusion of two gametes.

Selected References

Abbot, R.T. (1962) : Sea shells of the world, New York.

Al-Hussaini, A.H. and **Demian. E.S.** (1967) : Practical Animal Morphology, Macmillan & Co., London.

Allen, R.D. (1952) : Amoeboid movement, *Sci. Am.,* **206** : 112-122.

Andarson, J.M. (1965) : Studies on visceral regeneration in sea stars. III. Regeneration of the cardiac stomach in *Asterias forbesi* (Desor). *Biol Bull.* **129** : 454-470.

Bahl. K.N. (1925) : *Pheretima.* The Indian Zoological Memoirs, Lucknow Publishing House, Lucknow.

Baker, J.R. (1969) : Parasitic Protozoa, Hutchinson, London.

Barnes., R.D. (1968) : Invertebrate Zoology, W.B. Saunders Co., Philadelphia, London and Toronto.

Barrington, E.J.W. (1965) : The Biology of *Hemichordata* and *Protochordata* (Edinburgh : Oliver and Boyd).

Bayer, F.M. and Owre, H.B. (1968) : The Free Living Lower Invertebrates, The Macmillan Co., New York.

Beament, J.W.L. (1960) : Wetting properties of insect cuticle. *Nature, Lond.* **186** : 408-409.

Berril, N.J. (1957) : *Indestructible Hydra,* Sci. Am., December, 1957.

Bhatia, M.L. (1941) : *Hirudinaria,* The Indian Zoological Memoirs, Lucknow Publishing House, Lucknow.

Binyon, J. (1964) : On the mode of functioning of the Water-vascular system of *Asterias rubens. L.J. Marine Biol. Asso. U.K.* **44** : 577-88.

Borradaile, L.A., Potts, F.A. and **Eastham, L.E.S.** (1958) : *The Invertebrates,* Cambridge University Press, New York.

Brown, Jr., F.A. (1950) : *Selected Invertebrates Types,* John Wiley & Sons, New York.

Buchsaum, R. (1938) : *Animals without Backbone,* University of Chicago Press.

Bullough, W.S. (1958) : *Practical Invertebrate Anatomy,* The Macmillan Co., London.

Bury. H. (1895) : Pentaetula theory. *Q. J. microsc. Sci.,* **38** : 45-136

Cain, A.J. (1966) : *Animal Species and their Evolution,* Hutchinson University Library, London.

Carter, G.S. (1951) : *General Zoology of the Invertebrates,* Sidgwick and Jackson, London,

Chadwick, H.C. (1914) : L.M.B.C. *Memoir,* 22. (Echinoderm Larvae).

Chandler, A.C. and **Read, C.R.** (1961) : Introduction to Parasitology, John, Wiley & Sons, Inc.

Chen, S.H. (1946) : Evolution of the insect larva. *Trans. R. Ent. Soc.* Lond. **97** : 381-404.

Cheng, T.C. (1964) : *The Biology of Animal Parasites,* WB. Saunders Co., Philadelphia and London.

Clark, R.B. (1964) : *Dynamics in Metazoan Evolution.*

Oxford : Clarendon Press.

Cobb, N.A. (1915) : Nematodes and their relationships. *Year book, U.S. Dept. Agric.* **1914** : 457-490.

Cockbain, A.E. (1966) : Pentamerism in echinoderms and the calcite skeleton. *Nature,* **212** : 740-741.

Corfton, H.B. (1966) : *Nematodes,* Hutchinson University Library, London.

Cornwel, P.B. (1968) : *The Cockroach,* Vol. I, Hutchinson & Co. Ltd., London.

Dales, R., Philips (1963) : *Annelids,* Hutchinson University Library, London.

Darwin, C. (1962) : *The Structure and Distribution of Coral Reefs,* University of California Press, Berkely.

Dawes, B. (1956) : *The Trematoda,* Cambridge University Press.

De Bruyn, P.P.H. (1947) : *Theories of Amoeboid Movement,* Quart, Rev. Bio. **22** : 1-14.

De Robertis, E.D.P., Nowinski, W.W and **Saez, F.** (1965) : *Cell Biology,* W.B. Saunders Co. London.

Dellinger, O. (1906) : *Locomotion of Amoeba and Allied forms,* J. Exp. Zool., **3** : 337-358.

Diller, W.F. (1963) : *Nuclear reorganization process in Paramecium aurelia with descriptions of autogamy and hemixis,* J. Morphol., **59** : 11-67.

Dogiel, V.A. (1965) : *General Protozoology,* Oxford University Press, New York.

Edward E. Ruppert, Richard S. Fox, Robert D. Barnes : *Invertebrate Zoology : A Functional Evolutionary Approach.* ISBN : 13:978-0030259821

Edwards, G.A., Ruska, H. and **Harvcn, E.de** (1958) : The fine structure of insect tracheoblasts, tracheal and tracheoles. *Arch. Biol.* **69** : 351-369.

Ehret, C.F. and **Powers, E.L.** (1959) : *The cell surface of Paramecium,* Inc. Rev. Cytol, **8** : 97-133.

Franklin Watts and **Christine Petersen** (2005) : *Invertebrates.*

Fry, WG. (ed) (1970) : *The Biology of the Porifera,* Academic Press, Inc., New York.

Gray, J. and **Lissman H. W.,** (1938) : *Studies of Animal Locomotion,* VII., *Locomotory reflexes in the Earthworm,* J. Exp. Biol., 15 : 506-517.

Gray, J., Lissmann, H.W. and **Pumphrey, RJ.** (1938) : The method of locomotion in Leech (Hirudu medicinalis) *J. exp. Biol,* **15** : 408-30.

Green, J. (1961) : *A Biology of Crustacea,* Quadrangle Books, Chiacago.

Grell, K.G (1983) : *Protozoology,* Springer-Verlag, New York, Berlin.

Hadzi, J. (1963) : *The Evolution of Metazoa.* Macmillan, New York. 499.

Hall, R.P. (1953) : *Protozoology,* Prentice-Hall, Englewood Cliffs, N.J.

Hegner, R.W., and **Engemann, J.G** (1974) : *Invertebrate Zoology.* The Macmillan Co., New York.

Hickmann, C.P. (1973) : *Biology of the Invertebrates.* The C.V. Mosby Co., St. Louis.

Hinton, H.E. (1968) : Spiracular gills. *Adv. Insect Physiol.* **5**: 65-162.

Hirschfield, H.I. (ed). (1962) : *The Biology of the Amoeba,* Annals of the New York Academy of Sciences,. **78.** (2) : 401-704.

Honigberg, B.M. (1964) : *A revised Classification of Phylum Protozoa,* J. Protozool, **11** : 7-20.

Hyman, L.H. (1940) : *The Invertebrates,* Vol. I., *Protozoa through Ctenophora,* McGraw-Hill, N.Y.

Hyman, L.H. (1951) : *The Invertebrates,* Vol. II. *Platyhelminthes and Rhynchocoela,* McGraw-Hill, N.Y.

Hyman, L.H. (1951) : *The Invertebrates,* Vol. Ill, *Acanthocephala, Aschlminthes and Entoprocta,* McGraw-Hill, N.Y.

Hyman, L.H. (1955) : *The Invertebrates,* Vo. IV. *Echinodermata* McGraw-Hill, N.Y.

Hyman, L.H. (1959) : *The Invertebrates,* Vol. V. *Smaller Coelomate Groups,* McGraw-Hill, N.Y.

Hyman, L.H. (1967) : *The Invertebrates,* Vol. VI. *Mollusca* I, McGraw-Hill, N.Y.

Imms, A.D. (1975) : A General Textbook of Entomology, 10th ed. Methuen, London.

James L. Gould and **Carol Grant Gould** (2007) : *Animal Architects : Building and the Evolution of Intelligence.* ISBN : 13:978-0465028382.

Jan Pechenik (2009) : *Biology of the Invertebrates.* ISBN : 13:978-0073028262.

Jennings, H.S. (1906) : *Behaviour of Lower Organisms,* Columbia University Press, New York.

Julian Sprung : *Invertebrates : A Quick Reference Guide* : (Oceanographic Series) ISBN : 13:978-1983693008.

Kathleen Feeley (2005) : *Invertebrates,* Gareth Stevens Publishing.

Keister, M.L. (1948) : The Morphogenesis of the tracheal system of *Sciara. J. Morph.* **83** : 373-424.

Kennedy, D. (1965) : *The Living Cell,* Readings from Scientific American, Freeman, Sən Francisco.

Kimbell, R.F. (1943) : *Mating types in the ciliate Protozoa,* Quart. Rev. Biol. **18** : 30-45.

Kotpal, R.L. (1975) : *Minor Phyla* Zoology Phylum Books, Rastogi Publications, Meerut, India.

Kotpal, R.L. (1976) : Zoology Phylum Books, *Protozoa to Echinodermata,* Rastogi Publications, Meerut, India.

Kudo, R.R. *Protozoology,* Charles C. Thomas, Springfield, III.

Lanan, F. (1967) : *Euglenoid Flagellates,* Prentice Hall, Englewood Cliffs, New Jersey.

Larrington, E.J.W (1969) : Invertebrate Structure and Function, Thomas Nelson & Sons, Ltd., London.

Laverack, M.S. (1963) : *The Physiology of Earthworms,* The Macmillan Co., New York.

Leedale, G.F. (1964) : *Pellicle Structure in Euglena.* Brit, Phycol, Bull., **2** : 291–306.

Leedale, G.F. (1966) : *Euglena : A new look with the electron microscope,* Adv. of Science, 23.

Lemche, H. (1959) : *Molluscan Phylogeny in light of Neopilina,* Proc. 15th Int. Congr. Zool., London.

Lemche, H. and **K.G. Wingstrand.** (1959) : The anatomy of *Neopilina galatheae* Lemche, 1957. *Galathea Report,* **3** : 9-72, + 56P1.

Little, V.A. (1967) : *General and Applied Entomology,* Oxford and IBH Publishing Co., New Delhi.

Mann, K.H. (1962) : *Leeches (Hirudinea),* Pergamon Press, New York.

Manwell, R.D. (1961) : *Introduction to Protozoology,* St. Martin's Press, New York.

Mast, S.O. (1931) : *Locomotion in Amoeba proteus. Protoplasma,* **14** : 321-330.

Mayr, E. (1963) : *Animal Species and Evolution.* Harward University Press, Cambridge.

Meglitsch, P.A. (1967) : *Invertebrate Zoology,* Oxford University Press, New York.

Mercer, E.H. (1959) : *An Electron Microscopic Study of Amoeba proteus, Proc. R. Soc. Lond. B.* **150** : 216-232.

Michael F. Land and **Dan-Eric Nilsson** (2012) : *Animal Eyes* (Oxford Animal Biology) ISBN : 10:0199581142.

Mill, P.J. and **Pickard, R.S.** (1972) : Anal valve movement and normal ventilation in aeshnid dragonfly larvae. *J. exp. Biol.* **56** : 537-543.

Millot, N. (1967) : *Echinoderm Biology.* Academic Press. New York, London.

Moore, J.P. (1959) : Hirudinea. In W.T. Edmondson. *Fresh Water Biology,* 542-557 2nd ed. Wiley, New York.

Morton, J.E. (1967) : *Molluscs,* Hutchinson University Library, London.

Nichols, D. (1967) : *Echinoderms,* Hutchinson University, London.

Patton, R.L. (1963) : *Introductory Insect Physiology,* W.B. Saunders Co., Philadelphia.

Patwardhan, S.S. (1937) : *Palaemon,* The Indian Zoological Memoirs, Lucknow Publishing House, Lucknow.

Pelseneer, P. (1906) : *Mollusca,* Vol. 5. In *Atriatise on Zoology, ed.* E.R. Lankester. London.

Pfadt, R.E. (1962) : Fundamentals of Applied Entomology, 668 Macmillan, New York.

Pickard, R.S., and **Mill, P.J.** (1972) : Ventilatory muscle activity in insect preparations of aeshnid dragonfly larvae. J. exp. Biol. **56** : 527-536.

Pitelka, D.R. (1963) : *Electron Microscopic Structure of Protozoa.* Pergamon Press, New York.

Pon, R.D. (1968) : The Biology of the Mollusca, Pergamon Press, New York.

Prasad Beni. (1932) : *Pila.* The Indian Zoological memoirs, Lucknow Publishing House, Lucknow.

Richard C. Brusca, Gary, J. Brusca (2005) : *Invertebrates,* Sinauer Associates.

Rockstein, M. (ed). (1973-74) : The Physiology of Insecta. Academic Press, New York and London.

Russel-Hunter, W.D. (1968) : *A Biology of Lower Invertebrates,* The Macmillan Co., New York.

Russel-Hunter, W.D. (1969) : *A Biology of Higher Invertebrates,* The Macmillan Co., New York.

Saxena, B.B. (1955) : *Physiology excretion in the common Indian apple snail, Pila globosa, J. Anim, Morph. and Physiol,* **2** : 87- 95.

Simpson, G.G. (1961) : *Principles of Taxonomy,* Columbia University Press, New York.

Sleigh, M. (1973) : *The Biology of Protozoa,* Edward Arnold, London.

Smyth, J.D. (1966) : *The Physiology of Trematodes,* W.H. Freeman & Co., San Fransisco.

Snodgrass, R.E. (1952) : *A Text Book of Arthropod Anatomy,* Comstock Publishing Associates, New York.

Snodgrass, R.E. (1954) : Insect metamorphosis. *Smithson. misc. collns.* **122**, (9), 122 pp.

Stephenson, J. (1930) : *The Oligochaeta,* Oxford University Press.

Storer, T.L., and Usinger, R.L. (1965) : *General Zoology,* McGraw-Hill Book, Co., New York.

Tartar, V. : *The Biology of Stenor.* Pergamon Press, New York.

Taylor, J.W. (1894-1907) : Monograph of the land and freshwater Mollusca of the British Isles. 4 vols. Leeds.

Ted O'Hare (2005) : *Invertebrates,* Rourke Publishing.

Thorpe, W.H. (1950) : Plastron respiration in aquatic insects. *Biol Rev.* **25** : 344-390.

Thorpe, W.H. and **Crisp, D.J.** (1947a) : Studies on plastron respiration. I. The biology of *Aphelocheirus* and the mechanism of plastron retension. *J. exp. Biol,* **24** : 227-269.

Thorson, G. (1946) : *Reproduction and larval development of Denish marine bottom invertebrates.* Medd. Comm. Danm. Fiskeriog Havunders, Ser. Plankton 4 (1).

Tuzet, O. (1963) : *The Physiology of Sponges, In* E.C. Dougherty *(ed.). The Lower Metazoa,* University of California Press, Berkeley.

Vickerman, K. (1966) : *Genetic system in unicellular animals.* Science Progress, **54 :** pp. 13-26.

Vickerman, K., and **Cox, F.E.G.** (1967) : *The Protozoa,* John Murray, London.

Waterman, T.H., (ed). (1960-61) : *The Physiology of Crustacea,* (2 Vols.), Academic Press, New York.

Weis-Fogh, T. (1970) : Structure and formation of insect cuticle. *Symp. R. Ent. Soc. Loud.* **5** : 165-185.

Wells, H.S. (1931) : Observations on the blood-sucking activities of the hookworm, *Ancylostoma canium. J. Parasitol,* **17** : 167-182.

Whitten, J.M. (1972) : Comparative anatomy of the tracheal system. *A. Rev. Ent.* **17** : 373-402.

Wichterman, R. (1953) : *The Biology of Paramecium,* Blackiston, New York.

Wigglesworth, V.B. (1954) : Growth and regeneration in the tracheal system of an insect, *Rhodnius prolixus* (Hemiptera). *Q. J. Microsc. Sci.* **95** : 115-137.

Wigglesworth, V.B. (1966) : *Insect Physiology,* John Wiley & Sons, New York.